The Algebra of Organic Synthesis

Green Metrics, Design Strategy, Route Selection, and Optimization

The Algebra of Organic Synthesis

Green Metrics, Design Strategy, Route Selection, and Optimization

John Andraos

CRC Press
Taylor & Francis Group
Boca Raton   London   New York

CRC Press is an imprint of the
Taylor & Francis Group, an **informa** business

CRC Press
Taylor & Francis Group
6000 Broken Sound Parkway NW, Suite 300
Boca Raton, FL 33487-2742

First issued in paperback 2019

© 2012 by Taylor & Francis Group, LLC
CRC Press is an imprint of Taylor & Francis Group, an Informa business

No claim to original U.S. Government works

ISBN-13: 978-1-4200-9328-5 (hbk)
ISBN-13: 978-0-367-14964-2 (pbk)

Library of Congress Cataloging-in-Publication Data

Andraos, John.
    The algebra of organic synthesis : green metrics, design strategy, route selection, and optimization / John Andraos.
        p. cm.
    Includes bibliographical references and index.
    ISBN 978-1-4200-9328-5 (hardback)
    1. Organic compounds--Synthesis--Mathematical models. I. Title.

QD262.A529 2012
547.2--dc23

2011023884

Visit the Taylor & Francis Web site at
http://www.taylorandfrancis.com

and the CRC Press Web site at
http://www.crcpress.com

This book is dedicated to the students and process chemists whose hands make reliable and reproducible experimental procedures, and to their advisors who are not afraid to disclose them for all to use in their own work for the benefit of humankind.

Contents

Foreword

This book is very different from numerous chemistry texts published each year. It is not a traditional organic chemistry book, although it encompasses more than a thousand syntheses for industrially relevant target molecules. It is not a traditional green chemistry book either, although it actualizes and rejuvenates the principles of green chemistry. This book is a dramatic step forward in approaching organic chemistry by highlighting the greenness of a reaction through quantitative measurements. This is perfectly emphasized by the book title that fuses the idea of logic and meticulousness of algebra with the field of organic chemistry. This concept is in fact evident throughout all the chapters.

Since the birth of green chemistry, scientists have started to think greener, but the author poses an important question: how green are the green reactions really? It is obvious that this question, which is the focal point of this impressive work, represents an important passage from the green chemistry starting era to a more mature green chemistry, possibly to a more responsible and conscientious one.

The chapters of this book mirror perfectly this theme. The author carefully describes the organic and green chemistry state of the art, pointing out the challenges in modern synthetic methodologies but also underlining the problems inherent in the experimental procedures usually reported in the chemistry literature. He also gives a comprehensive overview of essential green metrics that provide information already in the design phase of a chemical process and indicating consumptions measured as material and energy flows, and waste or toxic release emissions. By their application, chemists can improve their awareness of environmental issues related to new products as well as to existing ones. In this perspective, the radial pentagon analysis that leads to calculations of green metrics is of invaluable help in assessing reaction "greenness."

However, the core of this work, covered by Chapters 5, 6, 9 and 10, is dedicated to green metrics, optimization, and the parameterization of synthesis strategy for route selection and reaction networks. This section is extremely important and educational for scientists as it is instructive in teaching chemists how to critique plans and in ultimately choosing the best possible (and possibly greener) synthetic methodology approach.

All chapters are illustrated with a rigorous and still easily comprehensible logic throughout the book, and arguments are always supported by numerous example organic reactions. More than a thousand syntheses are reported as references, each studied according to various metrics highlighting the "greenest" synthetic plans. The author is able to wisely combine green metrics with strategic syntheses providing a work that is a great tool for reaction optimization.

This book is therefore to be highly recommended to a wide readership, including green chemists, organic chemists, industrial chemists, and students. I personally believe that this book is a must for every chemist as a valuable tool to acquire a new vision of organic chemistry, in general, and a more realistic vision of green chemistry, in particular.

I think this book is the first stone that will pave the way to analyze organic chemistry and green chemistry together for a virtuous completion toward a sustainable development.

Piero Tundo
Venice, Italy

Acknowledgments

I would like to first acknowledge the following undergraduate students taking CHEM 3070 Industrial and Green Chemistry at York University from 2001 to 2008 who made the first green metrics analyses of some of the compounds contained in this database as part of their written assignment for the course. Their names are given along with their chosen compound in parentheses:

Barav Abbas (estrone), Samuel Akinwande (carvone), Fred Ampadu (cyclooctatetraene), Ghazaleh Arbabi (indigo), Abena Asenso (patchouli alcohol), Richard Bissessar (DDT), Kelvin Chan (isotretinoin), Gargi Dhora (lysergic acid), Jonathan Dudzik (phenobarbital), Marilyn Fernandes (thyroxine), Nikita Goussev (atrazine), Matthew Habil (tamoxifen), Phoevos Hughes (imatinib), Sameen Ijaz (oseltamivir phosphate), Ronen Itzkovitch (vanillin), Julia Izhakova (nicotine), Shadi Jahandideh (morphine), Kirti Johal (physostigmine), Thusyanthini Kandasamy (penicillin V), Andreas Katsiapsis (fluoxetine), Julia Krainova (saccharin), John Kriarakis (tropinone), Charlie Kuo (phenytoin), Truny Le (aspirin), Libby Lee (dimethoate), Hobart Leung (menthol), Amanda MacIntosh (quinine), Geoffrey Malana (alpha-thujone), Milos Markovic (sildenafil citrate), Joy McCourt (triclosan), Attia Momin (risperidone), Nathaniel Nowacki (vitamin C), Dyana Odisho (quinapril), Mario Orestano (warfarin), Rebecca Orth (ephedrine), Osawe Osayande (serotonin), Thushanthy Packiyanathan (linalool), Hyesoon Park (epibatidine), Mitesh Patel (cubane), Zainab Patel (alpha-irone), Nicolas Quadros (beta-carotene), Madlena Rabaev (acetaminophen), Delita Rafeson (drometrizole), Samira Rafiei (vitamin A), Tanya Rahman (tetracycline), Davin Ramsamujh (coniine), Rosanna Rita (permethrin), Trevor Russell (ibuprofen), Lukasz Sadowski (thienamycin), Jordan Schwartz (naproxen), Senthuri Senthivel (vitamin E), Hossain Seraj (pregabalin), Agnesa Shala (sertraline), Kulwinder Sidhu (longifolene), Christopher Steeves (metolachlor), Sujiththa Suntharalingam (flunitrazepam), Dunja Tisler (cocaine), Jiann-Yeou Wang (beta-ionone), Pooneh Vaziri (fluoxetine), Tara S. White (diethyl phthalate), and Yodita Zemicael (ferrocene). Particular mention should be made to the following students whose pointed questions led me to new ways of thinking about green chemistry: Anna Burdy, Neeshma Dave-Puzzo, Dragana Djokic, Julia Izhakova, Julia Krainova, Jelena Loncar, Maija Elina Lukkari, Rebecca Mallinson, Joy McCourt, Imir Metushi, Nathaniel Nowacki, Eran Obadia, Mario Orestano, Rosa Park, Matteo Perri, Murtuzaali Sayed, Lukasz Sadowski, Jordan Schwartz, Ethan Tumarkin, Yodita Zemicael, and Barbara Zimmerman.

I would also like to acknowledge the following people who have encouraged me and offered constructive criticism on several aspects of this work over its many incarnations, particularly Prof. Paul T. Anastas (Yale University), Dr. Neal G. Anderson (Practical Process R&D, Jacksonville, Oregon), Dr. Joseph D'Antuono (Row2 Technologies, Parsippany, New Jersey), Dr. Floyd H. Dean (Ontario Research Foundation, retired), Dr. Andrew P. Dicks (University of Toronto), Prof. C. J. Li (McGill University), Dr. Vivianne Massonneau (Merck & Co., Rahway, New Jersey), Dr. Christopher R. Schmid (deceased) (Eli Lilly and Company, Indianapolis, Indiana), Dr. Peter L. Spargo (Scientific Update, United Kingdom), Dr. Thuy van Pham, and Prof. John C. Warner (University of Massachusetts, Boston, Massachusetts).

From this list, I make special mention to the following people in the subsequent paragraphs. In 2009, I dedicated my tamiflu publication to the memory of Chris Schmid, the late editor of *Organic Process Research and Development*. Though I never had occasion to meet this man personally, I would like the world to know that he was the first person to recognize the value of quantitative analysis to total synthesis design and its consequential benefits to the fledgling field of green chemistry. He overcame the stereotypical thinking of his synthetic organic chemist colleagues and took a risk in giving me an opportunity to disclose my work on green metrics in the literature, thereby opening a new door in my career as a chemist. I thank him for his open-mindedness.

To Pete Spargo—I thank him for a question he asked me one early morning in June 2007 before we began our session to teach the green chemistry course offered by Scientific Update in Montreal. That was the day when he skipped breakfast to take an early morning call from a client in India! His question was: "Wouldn't it be nice if you could take a product structure and demarcate the source of all atoms back to their progenitor starting materials so you could see straight away how the target molecule was put together?" To Vivianne Massonneau who was a "student" taking part in the course—I thank her for asking me if there was some kind of diagram like the radial pentagon for individual reactions I had introduced that could be used to visually describe the efficiency of an entire synthesis plan. Their similar lines of thinking evidently bore some interesting fruit with respect to the development of tools for visualizing design strategy, especially when comparing multiple routes to a given target molecule. Pete's question led me to develop the target bond map. Vivianne's idea eventually led to the development of a radial hexagon that linked material and strategy efficiency metrics for any synthesis plan. This made me more convinced that visual tools are key to the understanding and implementation of green chemistry principles. More importantly, it is the best way to convince both chemists and non-chemists of the merits and weaknesses of synthesis strategies.

A special thank you to Thuy van Pham, whom I have known since we were graduate students at the University of Toronto. His generous gift of a laptop computer made the writing of this book possible and the navigation between software programs manageable. I thank him and his family, and Trong Than, for their kind hospitality and emotional support over the last decade. To Floyd H. Dean and his son Jason—I am deeply indebted to their constant offering of assistance in the way of tracking down references, answering my chemistry queries when I encountered exotic reactions in the literature, and for bouncing ideas off each other on reaction mechanisms especially when authors in the literature did not bother to offer explanations in their own published work! To Hilary Rowe of Taylor & Francis for her enthusiasm about green chemistry and her patience in waiting all this time until this work was finally completed.

I also would like to thank Profs. Adrian Brook and J. Bryan Jones at the University of Toronto for introducing me to organic synthesis when I was an undergraduate student. From Adrian, I learned about the theory of organic synthesis, how to pose problems, and how to present lectures clearly. I fondly remember him as the only professor I had who wore a lab coat while he lectured! From Bryan and his then graduate students Phil Hultin (now a professor at the University of Manitoba) and Eric Toone (now a professor at Duke University), after a three-month summer internship as a novice undergrad, I learned everything I needed to know about the practice of organic synthesis in the laboratory, how to keep a proper lab book, and how to write up an accurate record of experimental results. Bryan told me that writing a book is like a pregnancy—it is in you and you have to get it out. I hope my baby is worth it!

Finally, I thank my mother for providing me physical and emotional sustenance during this time by providing me space to carry out my work from a small corner table by the balcony window and a maple wooden chair from which all of the ideas contained in this work emerged.

Author

John Andraos received his PhD in 1992 from the University of Toronto in physical organic chemistry. He then did postdoctoral work at the University of Ottawa and at the University of Queensland studying kinetics of reactions in heterogeneous media and cumulene intermediates in low temperature matrices, respectively. Since his appointment as lecturer and course director at York University from 1999 to 2009, he has taught and developed courses in organic chemistry. In 2002, he launched the first industrial and "green" chemistry course in the history of the Department of Chemistry at York. His current research is broadly defined as reaction optimization and discovery, including application of reaction metrics for analysis of organic reactions and total syntheses of organic molecules; optimization of recycling and reagent retrieval protocols; discovery of new multicomponent reactions by structural combinatorial techniques; unified mathematical analysis of green metrics; synthesis tree analysis; molecular and topological complexity and connectivity; ring construction databases; and quantification and optimization of dynamic kinetic resolution relevant to stereoselective syntheses of pharmaceuticals. Currently, he is undertaking an ambitious project, which is the construction of a database of synthesis plans fully quantified by green metrics analysis for pharmaceuticals, important natural products, dyestuffs, agrichemicals, and molecules of theoretical interest.

Dr. Andraos has given invited addresses to Concordia University, the University of Western Ontario, and the University of Toronto. In 2000, he launched the CareerChem Web site (http://www.careerchem.com/MainFrame.html), which is an in-depth resource for tracking and cataloging all named things in chemistry and physics, chronicling the development of chemistry through scientific genealogies, and supplying career information to young researchers and students for placement in academic and industrial positions worldwide. Since 2000, he has given career workshops at the annual Canadian Society of Chemistry conference for students and postdoctoral fellows. He has given 15 invited addresses on green chemistry research and education at various locations, including the University of Western Ontario; the National Academy of Sciences in Washington, District of Columbia; the University of Toronto; the *15th Organic Process Research and Development Conference* in San Diego, California; Merck & Co. (Rahway, New Jersey); BioVerdant (San Diego, California); Apotex Pharmaceuticals (Brantford, Ontario); and the University of Ontario Institute of Technology. He is the author of 50 scientific papers in refereed journals and 6 book chapters on the subject of green chemistry metrics and education.

His awards of recognition include the Junior Research Award from the Australian Research Council in 1996, Distinguished Sigma Xi Lecturer at the University of Toronto (2004 and 2007), and President of the University of Toronto Sigma Xi Chapter (2004–2006).

Please find the downloadable resources available on
www.crcpress.com/9781420093285

1 Aims and Scope

1.1 WHAT DOES THIS BOOK OFFER?

Books on organic synthesis traditionally focus on reaction types and strategies in a qualitative sense citing vast numbers of example reactions and plans from the literature to substantiate points made by authors. No numerical analyses are performed on reaction or plan performance beyond using the elementary metrics of individual reaction and overall yield and number of steps. There is little quantitative analysis made using graph theoretical methods though the literature in this field is well developed. Books on the subject of synthesis plans for pharmaceuticals substances, in particular, give plans without reaction yield data and without balancing chemical equations. This does not allow quantitative comparisons between plans to a common target to be made particularly when the most material efficient one is to be sought. Interested readers need to go through the tedious task of looking up original papers and patents in the literature to gather the necessary data. Books published on the subject of green chemistry since the early 1990s have been of the "show and tell" genre where past environmentally bad processes and syntheses are compared to modern "greened up" versions that mainly utilize so-called "green solvents" such as ionic liquids, water, supercritical CO_2, etc., solvent-free reactions, and microwave irradiation. None of these books have discussions on quantitative measurements of just how "green" synthesis plans are, especially when comparing ones that utilize bio-feedstocks versus traditional petroleum-based feedstocks. Books and book chapters on graph theory applied to organic synthesis are few in number and emphasize the mathematical aspects with little practical applications that a synthetic organic chemist who is not receptive to the value of mathematical analysis can use in his or her work. The average synthetic and process chemist is largely unaware of this literature and if exposed to it would be immediately turned off by the mathematical jargon and esoteric language used. The bibliography of references at the end of this chapter highlights some of the hallmark works available. Though all of these references have great value for both the novice and seasoned chemist practitioner alike, there is room for something different.

The reader may balk at the phrase "algebra of organic synthesis" used in the title. It is often mentioned by the leading organic chemists that the subject is based on logic, yet the way the subject is presented in books does not reflect this. Rather than showing a "big picture" view of patterns of reactivity, structural motifs, and strategies of building up molecules, the subject is presented as an endless stream of example after example. It is up to you, the student or reader, to eventually "get it" by a process of osmosis. The word "algebra" originates from the Arabic *al-jabr*, which literally means "the reduction." Reduction implies logic. Associated with the arithmetic of algebra is the notion of writing and balancing equations. In chemistry, we are talking about chemical equations that have reactants written on the left-hand side and products written on the right-hand side. Antoine Lavoisier taught us that the law of conservation of mass guarantees that the sum of amounts of material on the left must equal that on the right. The underlying principle is that successful waste reduction in all its forms can only happen if we begin from the logic or algebra of balanced chemical equations. There is also another kind of algebra, more abstract, that is based on constructing pictures or graphs. In chemistry, the pictures refer to drawings of molecular structures. The construction refers to the additive operation of connecting atoms to make chemical bonds. Synthesis strategy is synonymous with bond making operations where nucleophilic minus nodes are joined with electrophilic plus nodes. The order, or sequence, of operations matters. Coming up with ways to construct a target structure is based on logical thinking. Coming up with ways to construct a target structure with the least number of additive operations in a direct manner with no residual bits left over is the basis of efficient and concise building up of molecules. This requires even harder thinking. Hence, this book is the first compilation to combine the aims, philosophies, and efforts of organic synthesis, reaction optimization, and green chemistry using the logic of algebra in both senses described above. It provides the first complete quantitative description of synthesis strategy analysis in the context of green chemistry and reaction optimization. Quantitative descriptions are powerful in describing reaction and synthesis plan strategic patterns. They also provide convincing numerical evidence of claims of synthesis efficiency. The present compilation aims to address directly each of the shortcomings cited above. The aims are therefore

1. To teach quantitative material efficiency to any practicing chemist who synthesizes molecules
2. To convince the chemistry community that metrics analysis is key to actualizing the principles of green chemistry
3. To explain in simple language the most useful ideas in graph theory that have direct application in organic synthesis
4. To put on firmer ground the value of simple mathematical analysis of organic synthesis so that any synthetic or process chemist can understand and apply these ideas to their own work
5. To illustrate how synthesis plans can be ranked in an unbiased way
6. To illustrate all of these ideas with example target plans to classical targets that are considered milestones in synthetic chemistry from pharmaceuticals, industrial commodity chemicals, dyestuffs, agrichemicals, flavorings, natural products, and molecules of theoretical interest
7. To create an updateable electronic database of the above

What will this book offer you, the reader? If you are a university professor or instructor teaching courses in organic chemistry and/or green chemistry, this book will provide the following:

- 1000+ worked out synthesis plans to important target molecules that can be used to enhance your lecture notes and give you ideas for problem sets highlighting synthesis efficiency and strategy
- A one-stop historical account of the milestone syntheses of classic target compounds
- A set of spreadsheet algorithms that you and your students can use in your undergraduate laboratories to test the "greenness" of your experiments

If you are a research professor engaged in original organic synthesis research, this book will provide the following:

- An up-to-date listing of synthesis plans for important molecules that are often revisited since they are the testing ground for new synthesis methodologies
- A quantitative methodology for ranking your new proposed plan to one of these target molecules with respect to novelty and material and strategy efficiencies
- A complete and systematic cataloging of ring construction strategies for monocyclic and fused bicyclic ring systems that you can use to propose new synthesis methodologies and as tool for reaction discovery for cases where there are no current examples in the literature
- A set of spreadsheet algorithms for you to use to test the efficiency and novelty of your proposed plan against prior published plans at the design stage before beginning laboratory work

If you are a student of organic synthesis this book will provide the following:

- An excellent reference for learning about the accomplishments of chemists worldwide in their endeavors to develop the field of total synthesis
- A resource that you can use to teach yourself the strategies of organic synthesis and green chemistry not found in other texts

If you are an industrial process chemist, this book will provide the following:

- An up-to-date and comprehensive database at your fingertips for finding out the latest syntheses of important compounds used in various aspects of the chemical industry
- A complete ranking of these synthesis plans according to waste production so that you can pick out those plans that have a "green" advantage and others that would be targets for improvement
- A set of spreadsheet algorithms for you to use to test the efficiency and novelty of your proposed plan against prior published plans at the design stage before beginning laboratory work

1.2 CHAPTER DESCRIPTIONS

Chapter 2 discusses the general state-of-affairs of organic and green chemistry from historical, educational, and research trend perspectives.

Chapter 3 is devoted to the problems in the reporting of experimental procedures in the chemistry literature.

Chapter 4 discusses some challenges faced in modern organic synthesis and green chemistry.

Chapter 5 gives a comprehensive overview of essential green metrics and their uses to evaluate reaction and synthesis plan performance. Full details are provided using illustrative examples where key spreadsheet algorithms are introduced to circumvent the drudgery of calculations. Step-by-step instructions on their use are also given.

Chapter 6 discusses the question of optimization, radial pentagon analysis to assess individual reaction "greenness," connectivity analysis as a complementary tool to retrosynthesis in the design of synthesis plans, and probability analysis to assess the likelihood that a given reaction can be called "green."

Chapter 7 gives a complete catalog of 500+ named organic reactions according to various metrics. All chemical equations are fully balanced. General trends among reaction classes are discussed in detail. Graphs give easy visual representations of trends.

Chapter 8 catalogs sacrificial reactions encountered in the extensive synthesis database.

Chapter 9 is a key section on parameterizing synthesis strategy. This is an important chapter of the book. Route selection and reaction networks are introduced. All sections are fully illustrated with worked out examples. The reader can glean the essential ideas so that they can then apply them to their own research problems.

Chapter 10 is a comprehensive discussion of ring construction strategies. This is the second important chapter of the book. All target compounds in the synthesis database that have rings are analyzed according to simple graphical methods to assess their construction strategies. For compounds with multiple plans, a general notation is introduced that allows patterns in synthesis plan strategies to be sorted and compared for novelty.

Chapter 11 gives vignettes on some key features of plans found in the database.

Chapter 12 gives a graphical summary of trends found in the 1000+ synthesis database according to various metrics. The "greenest" plans are highlighted.

Chapter 13 gives summary tables of the statistics of plan performances and rankings for those synthetic targets that had multiple plans.

The final part of the book is a series of appendices that the reader will find invaluable as references for various topics in organic and green chemistry. The most notable are as follows. Appendix A.1 lists densities of various solvents and aqueous solutions that are used in the radial pentagon analysis for assessing individual reaction material efficiency performance. Appendices A.2 and A.3 give lists of abbreviations for reagents and functional groups, respectively. Appendix A.6 is a summary of tables for trends in the named organic reaction database. Appendix A.7 is a complete monocyclic and bicyclic ring enumeration database that complements Chapter 10. Appendix A.10 is a glossary of terms used in organic synthesis. Appendix A.11 is a listing of the U.S. EPA President's Green Chemistry Awards since 1996. Appendix A.12 lists all 100% atom economical reactions cited in Trost's landmark publication in which he introduced the concept term "atom economy." Finally, in Appendix B, a list of challenging redox reactions is given from the synthesis database.

1.3   SYNTHESIS PLAN DATABASE

Tables 1.1 through 1.7 itemize the list of synthetic targets and synthesis plans covered in the present database in the following categories: (a) natural products, (b) pharmaceuticals, (c) agrichemicals and insecticides, (d) dyestuffs and colorants, (e) flavorings, fragrances, and sweeteners, (f) molecules of theoretical interest, and (g) miscellaneous industrial chemicals. Compounds are listed in alphabetical order and plans for each target molecule are listed in chronological order. Plans for individual target molecules that are designated as G1, G2, etc. refer to generation 1, generation 2, etc. versions from the same research group. Table 1.8 gives a numerical count of all plans and target structures in each category. The cataloging of synthesis plans according to various metrics was done in order to discover gems of synthesis and material efficiency as well as examples of strategy *faux pas*. This effort required literally combing thousands of plans over the entire range of published literature from Wöhler's 1828 synthesis of urea to the present day. All cited articles in this extensive compilation were actually read, particularly the experimental sections.

1.4   HOW TO USE SYNTHESIS DATABASE COMPILATION

The accompanying download includes a reproduction of the tables given in Section 1.3 with direct electronic links to files for all plans. Each file contains a list of references for the plan including sec-ondary references for the synthesis of starting materials if applicable, a complete synthesis scheme with fully balanced chemical equations, a synthesis tree, a summary of green metrics parameters, a target bond map and list of reagents that end up in the target molecule, and series of graphs depict-ing the plan performance in a visual sense. The graphs include (a) a radial hexagon, (b) a bar graph for the molecular weight first moment, (c) a percent kernel mass of waste distribution profile, (d) an atom economy profile, (e) a reaction yield profile, (f) a kernel reaction mass efficiency profile, (g) a hypsicity profile, and (h) a target bond making profile. These profiles are color-coded for easy ref-erence according to the description given in Table 1.9. For convergent plans, these profiles refer to the longest branch in the sequence. For pharmaceuticals, dyestuffs, agrichemicals, flavorings, and miscellaneous industrial chemicals, radial pentagons are also shown for each step. For these plans, global metrics calculations were performed in addition to the kernel metrics. These will be of high interest to process chemists in the chemical industry. All computations were greatly facilitated by automated spreadsheet algorithms that are included as template Excel files in the download: PENTAGON, LINEAR-KERNEL, LINEAR-COMPLETE, CONVERGENT-KERNEL, and CONVERGENT-COMPLETE. Chapter 13 gives summary tables of the statistics of plan perfor-mances and rankings for those synthetic targets that had multiple plans. For all plans analyzed at the kernel level of waste production, reaction yields were taken as reported in the experimental

TABLE 1.1
List of 529 Synthesis Plans for 82 Natural Products

Compound	Synthesis Plans
Absinthin	H. Zhai (2005)
Adociacetylene B	J. Garcia (1999), B.W. Gung (2001), B.M. Trost (2006)
Adrenaline (epinephrine)	F. Stolz (1904), H.D. Dakin (1905), E.M. Carreira (1997)
Allantoin	E. Grimaux (1877), W.W. Hartman (1933)
Amphidinolide P	B.M. Trost (2004)
Aspidophytine	E.J. Corey (1999)
Astaxanthin	Hoffmann-LaRoche G1 (1981), Hoffmann-LaRoche G2 (1981), M. Ito G1 (2001), M. Ito G2 (2001), S. Koo (2005)
Azadirachtin	S.V. Ley G1 (2008), S.V. Ley G2 (2009)
Bombykol	A. Butenandt G1 (1961), A. Butenandt G2 (1961), Bayer G1 (1962), Bayer G2 (1962), E. Negishi (1973), J.F. Normant (1975), H.J. Bestmann (1977), H. Suginome (1973), B.M. Trost (1980), A. Alexakis G1 (1984), A. Alexakis G2 (1989), F. Naso (1989), L. Dasaradhi (1991), J. Uenishi (2000)
Bupleurynol	M.G. Organ G1 (2003), M.G. Organ G2 (2004)
Caffeine	W. Traube (1900), H. Bredereck (1950)
Camphor	G. Komppa (1903)
β-Carotene	Hoffmann-LaRoche G1 (1947), H.H. Inhoffen (1950), Hoffmann-LaRoche G2a (1961), Hoffmann-LaRoche G2b (1961), Hoffmann-LaRoche G3 (1969), J.E. McMurry (1974), BASF (1977)
Caryophyllene	E.J. Corey (1964)
Cecropia juvenile hormone	B.M. Trost (1967), E.J. Corey (1968), Syntex (1968), W. Hoffmann (1969), H. Schulz (1969)
Cephalosporin C	R.B. Woodward (1966)
Cocaine	R. Willstätter (1923), J.J. Tufariello (1979), J.F. Casale (1987), H. Rapoport (1998), J.K. Cha (2000), W.H. Pearson (2004)
Codeine	M. Gates (1950), K.C. Rice (1980), L.E. Overman (1993), J.D. White (1997), J. Mulzer (1998), K. Ogasawara (2001), D.F. Taber (2002), B.M. Trost (2005), T. Fukuyama (2006), K.A. Parker (2006), T. Hudlicky (2007), N. Chida (2008), B. Iorga (2008), G. Stork (2009), P. Magnus (2009)
Coenzyme A	H.G. Khorana (1961)
Colchicine	A. Eschenmoser (1961), E.E. van Tamelen (1961), A.I. Scott (1963), R.B. Woodward (1963), Roussel–Uclaf (1965), D.A. Evans (1984), J. Cha (1998), M.G. Banwell (1996), H.G. Schmalz (2005)
Coniine	A. Ladenburg (1886), K. Aketa (1976), J.P. Husson (1983), T. Gallagher (1986), S.L. Buchwald (1988), T. Nagasaka (1988), C. Kibayashi (1992), A.I. Meyers (1995), G. Pandey (1997), J.C. Gramain (2000), M. Shipman (2001), P. Knochel (2004), J.P. Hurvois (2006), A. Couture (2007), A. Beauchemin (2009)
Cylindricine C	G.A. Molander (1999), B.M. Trost (2003)
Curcumin	V. Lampe (1918)
Dehydroabietic acid	G. Stork (1962)
Discodermolide	A.B. Smith III G1 (1995), S.L. Schreiber (1996), D.C. Myles (1997), J.A. Marshall (1998), I. Paterson G1 (2000), A.B. Smith III G2 (2000), A.B. Smith III G3 (2003), I. Paterson G2 (2003), Novartis (2004), I. Paterson G3 (2004), A.B. Smith III G4 (2005), J.S. Panek (2005), J. Ardisson (2007)
Eleutherobin	S. Danishefsky (1999), G. Gennari (2006)
Ephedrine	E. Späth (1920), R.H. Manske–T.B. Johnson G1 (1929), R.H. Manske–T.B. Johnson G2 (1929), F. Stolz G1 (1931), F. Stolz G2 (1932), F. Stolz G3 (1932), Hoffmann-LaRoche (1932), K. Freudenberg (1932)

(continued)

TABLE 1.1 (continued)
List of 529 Synthesis Plans for 82 Natural Products

Compound	Synthesis Plans
Epibatidine	C.A. Broka (1993), E.J. Corey (1993), A.C. Regan (1993), T.Y. Shen (1993), S.R. Fletcher (1994), F.I. Carroll (1995), D. Bai (1996), C. Szantay G1 (1996), C. Szantay G2 (1996), M.L. Trudell (1996), G.M.P. Giblin (1997), H. Kosugi (1997), C. Kibayashi (1998), G. Pandey (1998), H.F. Olivo (1999), D.A. Evans (2001), Y. Hoashi (2004), V.K. Aggarwal (2005), T.P. Loh (2005)
Estrone	W.E. Bachmann (1942), W.S. Johnson (1950), I.V. Torgov (1960), P.A. Bartlett (1973), S. Danishefsky (1976), T. Kametani (1977), P.A. Grieco (1980), G. Pattenden (2004), E.J. Corey (2007)
Fenchone	L. Ruzicka (1917), G. Komppa (1935), W. Cocker (1971), G. Buchbauer (1981)
Helminthosporal	E.J. Corey (1963)
α-Himachalene	E. Wenkert (1973), W. Oppolzer (1981), D.A. Evans (1997)
Hipposudoric acid	K. Hashimoto–M. Nakata (2006)
Hirsutene	S. Nozoe (1976), T. Hudlicky (1980), G. Mehta (1981), P.A. Wender (1982), R.D. Little (1983), R.L. Funk (1984), D.P. Curran (1985), A.T. Hewson (1985), S.V. Ley (1985), P. Magnus (1985), A.C. Weedon (1985), M. Iyoda (1986), J. Cossy (1987), D.H. Hua (1988), A.E. Greene (1990), L.A. Paquette (1990), T.K. Sarkar (1990), D.D. Sternbach (1990), T. Cohen (1993), M. Franck-Neumann (1992), K. Fukumoto (1993), K. Weinges (1993), W. Oppolzer (1994), L. Fitjer (1998), J. Leonard (1999), M.G. Banwell (2002), V. Singh (2002), M.J. Krische (2003), Z.X. Yu (2008), B. List (2008)
Hypoglycin A	Abbott (1958), D.K. Black (1963), J.E. Baldwin (1992)
Ibogamine	W. Nagata (1968), B.M. Trost (1978), M. Hanaoka (1981), D.M. Hodgson (2005)
Kainic acid	W. Oppolzer (1982), D.W. Knight (1987), S. Takano G1 (1988), S. Takano G2 (1988), J.E. Baldwin (1990), S. Benetti (1991), S. Takano G3 (1992), S. Takano G4 (1993), J.A. Monn (1994), S. Yoo (1994), S. Hanessian (1996), M.D. Bachi (1997), K. Ogasawara G1 (1997), J. Cossy (1998), A. Rubio (1998), J. Montgomery (1999), T. Naito (2000), K. Ogasawara G2 (2000), R.J.K. Taylor (2000), B. Ganem (2001), K. Ogasawara G3 (2001), J. Clayden (2002), B.M. Trost (2003), J.C. Anderson (2005), T. Fukuyama G1 (2005), D. Hoppe (2005), M. Lautens (2005), J.F. Poisson G1 (2005), J.F. Poisson G2 (2006), T. Fukuyama G2 (2007), K.W. Jung (2007), T. Fukuyama G3 (2008), S.G. Tilve (2009)
Laurallene	M.T. Crimmins (2000), T. Suzuki (2003), K. Takeda (2008)
Longifolene	E.J. Corey (1964), J.E. McMurry (1972), W.S. Johnson (1975), W. Oppolzer G1 (1978), W. Oppolzer G2 (1984), A.G. Schultz G1 (1985), A.G. Schultz G2 (1985), T. Money (1988), H.J. Liu (1992), A.G. Fallis (1993), S. Karimi (2003)
Lycopodine	G. Stork (1968), W.A. Ayer (1968)
Lysergic acid	R.B. Woodward (1956), W. Oppolzer (1981), R. Ramage (1981), J. Rebek Jr. (1984), I. Ninomiya (1985), T. Kurihara (1987), G. Ortar (1988), J.B. Hendrickson (2004), C. Szantay (2004), T. Fukuyama G1 (2009), T. Fukuyama G2 (2009)
Morphine	M. Gates (1950), K.C. Rice (1980), L.E. Overman (1993), J.D. White (1997), J. Mulzer (1998), K. Ogasawara (2001), D.F. Taber (2002), B.M. Trost (2005), T. Fukuyama (2006), K.A. Parker (2006), N. Chida (2008), G. Stork (2009)
Nakadomarin A	A. Nishida G1 (2003), A. Nishida G2 (2004), M.A. Kerr (2007), D.J. Dixon (2009)
Nepetalactone	T. Sakan (1960), G.W.K. Cavill (1963), K. Mori (1988), K. Sakai (1988), J.A. Pickett G1 (1996), J.A. Pickett G2 (1996)
Nicotinamide	A. Vogel (1978)
Nicotine	A. Pictet (1904), E. Späth (1928), L.C. Craig (1933), C.G. Chavdarian (1982), P. Jacob III (1982), T.P. Loh (1999), J. Lebreton (2000), A. Delgado (2001), S.V. Ley (2002), G. Helmchen (2005), K.R. Campos (2006)
Oleocanthal	A.B. Smith III G1 (2005), A.B. Smith III G2 (2007), R.M. Williams (2009)
Oxomaritidine	R.A. Holton (1970), S.V. Ley G1 (1999), S.V. Ley G2 (2006)

TABLE 1.1 (continued)
List of 529 Synthesis Plans for 82 Natural Products

Compound	Synthesis Plans
Paclitaxel	A.E. Greene G1 (1991), A. Commercon (1992), A.E. Greene G2 (1994), D.G.I. Kingston (1994), R.A. Holton (1994), K.C. Nicolaou (1995), S. Danishefsky (1996), C. Gennari (1997), P.A. Wender (1997), T. Mukaiyama (1999), I. Kuwajima (2000), T. Takahashi (2006)
Palau'amine	P.S. Baran (2010)
Papaverine	A. Pictet (1909), Decker–Wahl (1909, 1950), Kindler–Peschke–Pal (1934, 1958), Redel–Bouteville (1949), F.H. Dean (1974)
Patchouli alcohol	G. Büchi (1964), S. Danishefsky (1968), R.N. Mirrington (1972), M. Bertrand (1980), Firmenich (1981), T.V. Magee (1995), G.S.R.S. Rao (1997), G. Srikrishna (2005)
Physostigmine	P.L. Julian (1935), W.N. Speckamp (1978), S. Takano G1 (1982), K. Fukumoto (1986), K. Fuji (1991), S. Takano G2 (1991), J.P. Marino (1992), Q.S. Yu (1994), L.E. Overman (1998), M. Nakagawa (2000), F. Johnson (2003), C. Mukai (2006), B.M. Trost (2006), M. Nakada (2008), M.G. Kulkarni (2009)
Platensimycin	K.C. Nicolaou G1 (2006), E.J. Corey (2007), J. Mulzer (2007), K.C. Nicolaou G2 (2007), K.C. Nicolaou G3 (2007), B.B. Snider (2007), H. Yamamoto (2007), E. Lee (2008), K.C. Nicolaou G4 (2007), K.C. Nicolaou G5 (2008), A.K. Ghosh (2009), K.C. Nicolaou G6 (2009) J.T. Njardarson (2009)
Quinine	Woodward–Doering–Rabe (1945), M. Gates (1970), S.F. Martin (1972), Merck (1978), G. Stork (2001), E.N. Jacobsen (2004), Kobayashi (2004), M.J. Krische (2008), V.K. Aggarwal (2010)
Quisqualic acid	J.E. Baldwin (1985)
Reserpine	R.B. Woodward (1958), B.A. Pearlman (1979), P.A. Wender (1980), S.F. Martin (1985), B. Fraser-Reid (1995), C.C. Liao (1996), S. Hanessian (1997), G. Mehta (2000), K.J. Shea (2003), G. Stork (2005)
Resveratrol	E. Spath (1941), M. Moreno-Manas (1985), M. Cushman (1993), M. Yus G1 (1997), H. Meier (1997), M. Guiso (2002), Z.L. Shen (2002), C. Marra (2006), M. Yus G2 (2009), M. Yus G3 (2009)
Ricinine	E. Späth (1923), G. Schroeter (1932), E.C. Taylor Jr. (1956), T. Sugasawa (1974)
Salsolene oxide	L.A. Paquette (1997), P.A. Wender (2006)
α-Santonin	Takeda Pharmaceuticals (1956), J.A. Marshall (1978)
Serotonin	J. Harley-Mason (1954)
Silphinene	S. Ito (1983), L.A. Paquette (1983), D.D. Sternbach (1985), P.A. Wender (1985), M.T. Crimmins (1986), M. Nagarajan (1988), S. Yamamura (1989), M. Franck-Neumann (1991)
Sparteine	R. Raper (1949), E.F.L.J. Anet (1950), N.J. Leonard G1 (1950), N.J. Leonard G2 (1950), E.E. van Tamelen (1960), F. Bohlmann–T. Gallagher (1973, 2006), N. Takatsu (1987), G.J. Koomen (1996), J. Aube (2002), I. Fleming (2004), P.R. Blakemore (2008)
Stenine	D.J. Hart (1993), P. Wipf (1995), Y. Morimoto (1996), J. Aube (2005) A. Padwa (2005)
Strychnine	R.B. Woodward (1963), P. Magnus (1992), G. Stork (1992), V.H. Rawal (1994), M.E. Kuehne (1993), L.E. Overman (1993), J. Bonjoch (1999), K.P.C. Vollhardt (2000), S.F. Martin (2001), G.J. Bodwell (2002), M. Mori (2002), M. Shibasaki (2002), T. Fukuyama (2004), A. Padwa (2007), H.U. Reissig (2010)
Swainsonine	G.W.J. Fleet (1984), T. Takaya (1984), T. Suami G1 (1984), M. Hashimoto (1985), A.C. Richardson (1985), K.B. Sharpless (1985), T. Suami G2 (1985), J.K. Cha (1989), N. Ikota (1990), C. Kibayashi (1994), M. Hirama (1995), S.F. Kang (1995), W.S. Zhou (1995), W.R. Roush (1995), W.H. Pearson (1996), F. Ferreira (1997), F. Bermejo (1998), B.M. Trost (1999), J.C. Carretero (2000), D.R. Mootoo (2001), S. Blechert (2002), S.G. Pyne G1 (2004), O. Reiser (2005), A. Riera (2005), G.A. O'Doherty G1 (2006), G.A. O'Doherty G2 (2006), J.F. Poisson (2006), S.G. Pyne G2 (2006), J. Cossy G1 (2007), J. Cossy G2 (2007), GlycoDesign (2008), Y.D. Vankar (2009), W.H. Ham (2009)

(continued)

TABLE 1.1 (continued)
List of 529 Synthesis Plans for 82 Natural Products

Compound	Synthesis Plans
α-Terpineol	W.H. Perkin Jr. (1904)
Tetracyclines	M.M. Shemyakin (1963) (tetracycline), R.B. Woodward (1968) (sancycline), H. Muxfeldt (1979) (terramycin), G. Stork (1996) (12a-deoxytetracycline), K. Tatsuta (2000) (tetracycline), A.G. Myers G1 (2005) (tetracycline), A.G. Myers G2 (2007) (tetracycline)
Theobromine	H. Bredereck G1 (1950), H. Bredereck G2 (1950), Hoechst (1994)
Thienamycin	T.N. Salzmann (1980), S.M. Schmitt (1980), P.J. Reider (1982), D.G. Melillo, I. Shinkai (1986), S. Hanessian (1990), K. Tatsuta (2001)
α-Thujone	W. Oppolzer (1997)
Thyroxine	C.R. Harington (1927), J.R. Chalmers (1949)
Tropinone	R. Willstätter (1901) G1, R. Willstätter (1921) G2, R. Robinson (1917), P. Karrer (1947), G.I. Bazilevskaya (1958) G1, G.I. Bazilevskaya (1958) G2, G.I. Bazilevskaya (1958) G3, G.I. Bazilevskaya (1958) G4, W. Parker (1959)
Uric acid	R. Behrend (1889), E. Fischer (1895), W. Traube (1900), C.W. Bills (1962)
Vinblastine	J.P. Kutney (1988), M.E. Kuehne (1991), T. Fukuyama G1 (2002) T. Fukuyama G2 (2007), D.L. Boger (2008)
Vincristine	T. Fukuyama (2004)
Vindoline	T. Fukuyama G1 (2000), T. Fukuyama G2 (2002), D.L. Boger (2005)
Vitamin A (retinol)	Hoffmann-LaRoche G1 (1947), Hoffmann-LaRoche G2 (1949), Merck (1951), Hoffmann-LaRoche G3 (1976), Hoffmann-LaRoche G4 (1976), Hoffmann-LaRoche G5 (1976) (vitamin A ethyl ester), Hoffmann-LaRoche G6a (1976), Hoffmann-LaRoche G6b (1976), Hoffmann-LaRoche G6c (1976), Hoffmann-LaRoche G7 (1978)
Vitamin C (ascorbic acid)	W.N. Haworth (1933), T. Reichstein (1934), Biotechnical Resources (1991), Roche (1992), M.G. Banwell (1998)
Vitamin E (α-tocopherol)	P. Karrer (1939), Roche G1 (1963), Roche G2 (1979), L.F. Tietze (2006), B. Breit (2007), W.D. Woggon (2008), D.W. Knight (2009)
Welwitindolinone A	J.L. Wood (2004), P.S. Baran (2005)
Zingerone	H. Nomura (1917), T. Mukaiyama (1976)

sections of the original publications. For all plans analyzed at the global level of waste production, radial pentagon analyses trapped all errors in the original publications and these were corrected. These data should be taken as reliable. If authors did not cite how starting materials were made literature searches were conducted on those precise structures to obtain the requisite references. If literature searches failed to turn up any reliable results, then reaction yields were assumed (denoted by an asterisk after the value of yield). Such analyses are therefore lower limits on waste production are denoted with greater than (>) signs for *E*-factor values and less than signs (<) signs for kernel reaction mass efficiency (RME) values.

Sections 5.2, 5.3, and 5.5 give explanations on notation used in drawing out reaction schemes and in constructing synthesis tree diagrams. Sections 6.2 and 6.3 give explanations of notation used in constructing radial pentagons and hexagons, respectively. The introduction of Chapter 10 gives an explanation of notation used in enumerating ring construction strategies.

TABLE 1.2
List of 243 Synthesis Plans for 43 Pharmaceuticals

Compound	Synthesis Plans
Acetaminophen	D.E. Pearson (1953), Warner-Lambert (1961), A. Vogel (1978)
Actarit	Mitsubishi (1983)
Antipyrine	A. Vogel (1948)
Aprepitant	Merck G1 (1998), Merck G2 (2003)
Aspirin	A. Vogel (1948), W.L. Faith (1966)
Atiprimod	L.M. Rice (1973), SmithKline & French (1988), SmithKline Beecham (1995), A. Tanaka (2006)
Barbituric acid	J.B. Dickey–A.R. Gray (1943)
Dapoxetine	Lilly (1988), V. Gotor (2006), K.V. Srinivasan (2007), A.R.A.S. Deshmukh (2009)
L-DOPA	Hoffman-LaRoche (1973)
Drometrizole	S. Tanimoto (1986), A.G. Koutsimpelis (2005)
Fesoterodine fumarate	Schwarz-Pharma (2004), Pfizer (2009)
Flunitrazepam	Roche (1963)
Fluoxetine	Lilly G1 (1982), H.C. Brown (1988), K.B. Sharpless (1988), E.J. Corey (1989), A. Kumar (1991), Lilly G2 (1994), F. Bracher (1996), Sepracor (2001), W.H. Miles (2001), P. Kumar (2004), M. Panunzio (2004), M. Shibasaki (2004), Q. Wang (2005)
Ganciclovir	Ogilvie G1 (1982), Ogilvie G2 (1982), Ogilvie G3 (1982), Syntex G1 (1982), Syntex G2 (1983), Merck (1983), Syntex G3 (1988), Roche (1995), Reddy Laboratories (2009)
Ibuprofen	Boots (1968), DD113889 (1975), Upjohn (1977), J.T. Pinhey (1984), DuPont (1985), Hoechst-Celanese (1988), C. Ruchardt (1991), R. Furstoss (1999), T.V. RajanBabu (2009), D.T. McQuade (2009)
Imatinib	Novartis (2003), Natco Pharma (2004), B. Noszal (2005), C.L. Wang (2008), A.S. Ivanov (2009), S.V. Ley (2010)
Indomethacin	Merck G1 (1964), Merck G2 (1962), Merck G3 (1967), Sumitomo (1968)
Isotretinoin	C.F. Garbers (1968), B.C.L. Weedon (1968), Roche (1985), G. Solladie (1989), BASF (1995), Abbott (1999), M. Salman (2005)
Linezolid	Upjohn G1 (1996), Upjohn G2 (1998), Reddy (2003), A. Sudalai (2006), Pfizer (2007), V. Gotor (2008)
Methamphetamine	A. Ogata (1919), D.B. Repke (1978)
3,4-Methylene-dioxymethamphetamine	U. Braun (1980), N.N. Daeid G1 (2008), N.N. Daeid G2 (2008), N.N. Daeid G3 (2008)
Naproxen	Syntex G1 (1970), Syntex G2 (1972), DuPont (1985), Syntex G3 (1986), R. Noyori (1987), Zambon G1 (1987), Zambon G2 (1989), F. Effenberger (1997), C. Ruchardt (1991), A.S.C. Chan (1995), H. Brunner (2000)
Nelfinavir	Agouron-Lilly (1997), Japan-Tobacco-Agouron (1998)
Oseltamivir	Gilead (1998), Roche G1 (1999), Roche G2 (2001), Roche G3 (2004), Roche G4 (2000), Roche G5 (2000), M. Shibasaki G1 (2006)
	M. Shibasaki G2 (2007), M. Shibasaki G3 (2007), E.J. Corey (2006), J.M. Fang G1 (2007), T. Fukuyama (2007), N. Kann (2007), H. Okamura (2008), B.M. Trost (2008), J.M. Fang G2 (2008), J.M. Fang G3 (2008), M.G. Banwell (2008), Y. Hayashi (2009), T. Hudlicky G1 (2009), T. Fukuyama (2009), T. Mandai G1 (2009), T. Mandai G2 (2009), X.X. Shi (2009), Roche G6 (2009), M. Shibasaki G4 (2009), M. Shibasaki G5 (2009), T. Hudlicky G2 (2010), P. Chen–X.W. Liu (2010)
Paroxetine	Ferrosan (1975), M. Amat G1 (2000), M. Amat G2 (2000), M.S. Yu (2000), P. Beak (2001), V. Gotor (2001), T. Hayashi (2001), L.T. Liu (2001), J. Cossy (2002), S.L. Buchwald (2003), M.Y. Chang (2003), N.S. Simpkins (2003), C. Szantay (2004), K. Takasu–M. Ihara (2005), K.A. Jorgensen (2006), M.J. Krische (2006), T. Gallagher (2007), D.J. Dixon (2008), J. Vesely–A. Moyano–R. Rios (2009)

(continued)

TABLE 1.2 (continued)
List of 243 Synthesis Plans for 43 Pharmaceuticals

Compound	Synthesis Plans
Penicillins	J.C. Sheehan-penicillin V (1959), J.C. Sheehan-6-Aminopenicillanic acid (1962)
Phenacetin	A. Vogel (1948)
Phenobarbital	J. Stieglitz (1918), M.M. Rising G1 (1927), M.M. Rising G2 (1928)
	W.L. Nelson (1928), J.S. Chamberlain (1953), J.T. Pinhey (1984)
Phenytoin	H. Biltz (1908), Parke-Davis (1946), G. Olah (1998), D. Lambert G1 (2003), D. Lambert G2 (2003)
Pregabalin	R.B. Silverman G1 (1989), R.B. Silverman G2 (1990), Warner-Lambert G1 (1996), Parke-Davis G1 (1997), Parke-Davis G2 (1997), Parke-Davis G3 (1997), Parke-Davis G4 (1997), Warner-Lambert G2 (2001), Dowpharma-Pfizer (2003), E.N. Jacobsen (2003), W. Hu (2004), M. Shibasaki (2004), A. Armstrong (2006), Z. Hamersak G1 (2007), Z. Hamersak G2 (2007), Y. Hayashi (2007), H.S. Lee (2007), D.T. McQuade (2007), F. Sartillo-Piscil (2007), F. Felluga (2008), Pfizer G1 (2008), Pfizer G2 (2008), R.M. Ortuno (2008), A.M.P. Koskinen (2009)
Quinapril	Warner-Lambert-Parke-Davis (1986)
Rimonabant	Sanofi G1 (1997), Sanofi G2 (1997), Reddy (2007)
Risperidone	Janssen (1989), J.H. Jeong (2005)
Sertraline	Pfizer G1 (1984), Pfizer G2 (1999), M. Lautens G1(1999), S.L. Buchwald (2000), Gedeon-Richter (2002), Pfizer G3 (2004), M. Lautens G2 (2005), G. Zhao (2006), J.E. Bäckvall (2010)
Sildenafil	Pfizer (2000), Pfizer-Reddy (2004)
Sitagliptin	Merck G1 (2005), Merck G2 (2005), Merck G3 (2009)
Sulfapyridine	A. Vogel (1978)
Tamoxifen	ICI (1965), J.A. Katzenellenbogen (1982), Bristol-Myers (1984), R.B. Miller (1985), M.I. Al-Hassan (1987), NRDC (1987), R. McCague (1990), Dow Chemical (1995), R.W. Armstrong (1997), P. Knochel (1998), K. Itami (2003), A.G. Fallis (2003), M. Yus (2003), I. Shiina (2004), R.C. Larock (2005), M. Shindo (2005), M. Mori-Y. Sato (2006), D.F. O'Shea (2006), Y. Nishihara (2007), N. Taniguchi (2009)
Thalidomide	Chemie Grünenthal G1 (1957), Chemie Grünenthal G2 (1960), Chemie Grünenthal G3 (1970), H. Galons G1 (1999), H. Galons G2 (1999), Celgene (1999), J.A. Seijas (2001)
Veronal	A. Vogel (1978)
Warfarin	K.P. Link (1961), DuPont-Merck G1 (1996), DuPont-Merck G2 (1996), G. Cravotto (2001), K.A. Jorgenson (2003), M. Sodeoka (2006)
Zafirlukast	ICI (1990), MSN (2004), Reddy (2009)
Zamifenacin	J. Cossy (1997)

TABLE 1.3
List of 62 Synthesis Plans for 21 Dyestuffs and Colorants

Compound	Synthesis Plan
Alizarin (C.I. 58000)	C. Graebe–C. Liebermann (1869), A. Vogel (1948), H.E. Fierz-David G1 (1949), H.E. Fierz-David G2 (1949)
Aniline yellow (C.I. 11000)	A. Vogel (1948)
Congo red (C.I. 22120)	A. Vogel (1948), H.E. Fierz-David (1949)
Crystal violet (C.I. 42555)	A. Vogel (1948)
Eriochrome blue black R (C.I. 15705)	R.N. Shreve (1967)
Fluorescein (C.I. 45350)	O. Mühlhäuser (1887), A. Vogel (1948)
Indigo (C.I. 73000)	A. Baeyer G1 (1879), A. Baeyer G2 (1880), A. Baeyer G3 (1882), A. Baeyer G4 (1882), K. Heumann G1 (1890), K. Heumann G2 (1890), K. Heumann–H.E. Fierz-David (1891, 1949), T. Sandmeyer (1903), W. Madelung G1 (1914), W. Madelung G2 (1914), J.C. Cain (1920), A. Vogel (1948), J. Harley-Mason (1950), E. Ziegler G1 (1964) E. Ziegler G2 (1965), J. Gosteli G1 (1977), J. Gosteli G2 (1977), J. Gosteli G3 (1977)
Malachite green (C.I. 42000)	O. Mühlhäuser (1887), H.E. Fierz-David (1921)
Methyl red (C.I. 13020)	H.T. Clarke (1941), A. Vogel (1948)
Methylene blue (C.I. 52015)	H. Caro (1878), O. Mühlhäuser G1 (1887), O. Mühlhäuser G2 (1887), F. Kehrmann (1916), H.E. Fierz-David (1921), American Cyanamid (1980)
Orange I (C.I. 14600)	O. Mühlhäuser (1887)
Orange II (C.I. 15510)	O. Mühlhäuser (1887), A. Vogel (1948), H.E. Fierz-David (1949)
Orange III (methyl orange) (C.I. 13025)	A. Vogel (1948)
Orange IV (C.I. 13080)	H.E. Fierz-David (1949)
Phenolphthalein	A. Vogel (1948)
Phenolsulfonephthalein	A. Vogel (1948)
Phthalocyanine (Monastral fast blue B, C.I. 74160)	A. Vogel (1948), H.E. Fierz-David (1949)
Picric acid (C.I. 10305)	A. Vogel (1948), H.E. Fierz-David (1949)
Safranin O (C.I. 50240)	H.E. Fierz-David (1921)
Thioindigo (C.I. 73300)	H.E. Fierz-David (1949)
Tyrian purple (C.I. 75800)	F. Sachs (1903), P. Friedländer (1912), E. Grandmougin (1914), R. Majima (1930), H. Gerlach (1989), C.J. Cooksey G1 (1994), P. Imming (2001), Y. Tanoue (2001), C.J. Cooksey G2 (2005)

TABLE 1.4
List of 59 Synthesis Plans for 25 Agrichemicals and Insecticides

Compound	Synthesis Plan
2,4,5-T	A. Galat (1952)
2,4-D	Dolge (1941), ICI (1945), Haskelberg G1 (1947), Haskelberg G2 (1947), R.H.F. Manske (1949)
Atrazine	J.R. Geigy (1960)
Benfluralin	Lilly (1966)
Bifenazate	Uniroyal G1 (1993), Uniroyal G2 (2000), C.K. Witco (2001), Uniroyal G3 (2004), G.L. Chee (2006), H. Sajiki (2007), F.X. Felpin (2008)
Carbaryl	Union Carbide (1959)
Carbofuran	FMC G1 (1966), FMC G2 (1967), FMC G3 (1969), Union Carbide (1984)
Clothianidin	Novartis G1 (2000), Novartis G2 (2000)
DDT	A. Vogel (1948), W.L. Faith (1966)
DEET	U.S. Dept. Agriculture (1953), Dominion Rubber (1962), Virginia Chemicals (1971), B.J.S. Wang (1974), Showa Denko (1976), Pfizer (1979), V. Snieckus (1989), J.W. LeFevre (1990), E.G. Neeland (1998), A. Khalafi-Nezhad (2005), A. Bhattacharya (2006)
Dimethoate	American Cyanamid (1950), Montecatini (1958)
Fluridone	Lilly (1978), Hooker (1980)
Glyphosate	Monsanto (1972)
Imidacloprid	Nihon (1986)
Indoxacarb	DuPont G1 (2002), DuPont G2 (2002)
Isodrin	Shell (1955)
Malathion	American Cyanamid (1951)
Metolachlor	Ciba-Geigy G1 (1972), Ciba-Geigy G2 (1982), B.T. Cho (1992), Ciba-Syngenta (1999), S. Zhang (2006)
Parathion	American Cyanamid (1948)
Permethrin	Kuraray (1976)
Phorate	Bayer AG (1956)
Pirate	American Cyanamid (1992)
Rynaxypyr	DuPont G1 (2007), DuPont G2 (2007), DuPont G3 (2007)
Tebufenozide	Rohm & Haas (1991)
Trifluralin	Lilly (1966)

TABLE 1.5
List of 73 Synthesis Plans for 12 Flavorings, Fragrances, and Sweeteners

Compound	Synthesis Plan
Carvone	E.E. Royals (1951), A.M. Todd Co. (1958)
β-Ionone	Hoffmann-LaRoche (1946, 1958)
α-Irone	Givaudan G1 (1959), A. Yoshikoshi (1982), Firmenich G1 (1984), Givaudan G2 (1987), Givaudan G3 (1989), Y. Ohtsuka (1991), Firmenich G2 (1992), H. Kiyota (2000)
γ-Irone	J. Garnero–D. Joulain (1979), K. Mori (1983), O. Takazawa (1985), F. Leyendecker (1987), Givaudan (1988), H. Kiyota (2000), H. Monti G1 (1996), H. Monti G2 (2000), H. Monti G3 (2000), E. Brenna G1 (2001), E. Brenna G2 (2001), P. Gosselin (2001), G. Vidari G1 (2003), G. Vidari G2 (2003)
Limonene	W.H. Perkin Jr. (1904)
Linalool	L. Ruzicka (1919)
Menthol	W.R. Brode (1947), Tanabe–Seiyaku (1968), BASF (1973), Takasago G1 (1989), Takasago G2 (2002), N. Sayo–T. Matsumoto (2002), K. Mashima (2008)
Menthone	A. Kötz-A. Schwarz (1907), L.T. Sandborn (1929)
Rosefuran	G. Büchi (1968), O.P. Vig (1974), A.J. Birch (1976), G. Pattenden (1977), A. Takeda (1977), R. Okazaki (1984), S. Takano (1984), H.D. Scharf (1986), E. Wenkert (1988), H. Tsukasa (1989), R. Iriye G1 (1990), R. Iriye G2 (1990), J.A. Marshall (1993), B.M. Trost (1994), A. Barco (1995), Y. Butsugan (1995), G. Salerno (1997), H.N.C. Wong (1997)
Saccharin	C. Fahlberg–I. Remsen (1879), C. Fahlberg G1 (1884–1909), C. Fahlberg G2 (1895), A. Vogel (1948), Maumee Chemical Co. (1954), Maumee-Sherwin-Williams (1980s), Rhône-Poulenc (1973), BASF (1984)
Sorbitol	W.L. Faith (1966)
Vanillin	F. Tiemann–W. Haarmann (1874), J.R. Geigy (1899), A. Roesler (1907), C. Harries–R. Haarmann (1915), Givaudan (1921), J.D. Riedel (1927), H.O. Mottern (1934), H. Hibbert G1 (1936), H. Hibbert G2 (1937), N.I. Volynkin (1938), E. Delvaux (1947), E. Mayer (1949), V.A. Zasosov (1959), G.M. Lampman G1 (1977), G.M. Lampman G2 (1977), J.W. Frost (1998), Givaudan–Roure (1998)

TABLE 1.6
List of 42 Synthesis Plans for 13 Molecules of Theoretical Interest

Compound	Synthesis Plan
Adamantane	V. Prelog G1 (1941), V. Prelog G2 (1941), H. Stetter (1956), P.v.R. Schleyer (1957), DuPont (1960)
Barrelene	H.E. Zimmerman G1 (1969), H.E. Zimmerman G2 (1969)
Basketene	W.G. Dauben (1966), S. Masamune (1966)
Bullvalene	G. Schröder (1963), W. von Doering (1965)
Cubane	P.E. Eaton (1964), R. Pettit (1966)
Cyclooctatetraene	R. Willstätter–A.C. Cope (1913, 1948), W. Reppe (1948)
Dewar benzene	E.E. van Tamelen (1962)
Dodecahedrane	L.A. Paquette (1982)
Ferrocene	P.L. Pauson (1951), British Oxygen (1952), DuPont (1957), G. Wilkinson G1 (1963), G. Wilkinson G2 (1963), BASF G1 (1963) BASF G2 (1963)
[1.1.1]Propellane	K.B. Wiberg (1982)
Triquinacene	R.B. Woodward (1964)
Tropolone	J.W. Cook (1951), D.J. Cram (1951), W. von Doering (1951), E.E. van Tamelen (1956), J.J. Drysdale (1958), A.P. ter Borg (1962), O.L. Chapman (1963), H.C. Stevens (1965), M. Oda (1969), R.A. Minns (1988)
Twistane	H.W. Whitlock Jr. (1962, 1968), P. Deslongchamps (1967) G1, P. Deslongchamps (1969) G2, M. Tichy (1969), Hamon-Young (1976), J.B. Jones (1980)

TABLE 1.7

List of 58 Synthesis Plans for 25 Miscellaneous Industrial Chemicals

Compound	Synthesis Plan
Acetic anhydride	W.L. Faith G1 (1966), W.L. Faith G2 (1966)
Adipic acid	B.A. Ellis (1941), A. Vogel (1948), W.L. Faith G1 (1975), W.L. Faith G2 (1975)
Aniline	R.N. Shreve–W.L. Faith G1 (1966, 1967), R.N. Shreve–W.L. Faith G2 (1966, 1967), R.N. Shreve–W.L. Faith G3 (1966, 1967)
Anthraquinone	A. Vogel (1948), W.L. Faith G1 (1966), W.L. Faith G2 (1966)
Benzoic acid	W.L. Faith G1 (1966), W.L. Faith G2 (1966)
Bisphenol A	W.L. Faith G1 (1966), W.L. Faith G2 (1966), W.L. Faith G3 (1966), W.L. Faith G4 (1966), W.L. Faith G5 (1966), G.D. Yadav (2009)
ε-Caprolactam	C.S. Marvel (1943), W.L. Faith (1975), BP Chemicals (1976), P.A. Evans (2002), Y. Ishii G1 (2003), Y. Ishii G2 (2001), J.M. Thomas (2005)
Diethyl phthalate	W.L. Faith G1 (1966), W.L. Faith G2 (1966)
Epichlorohydrin	W.L. Faith (1966)
Ethylene oxide	W.L. Faith G1 (1966), W.L. Faith G2 (1966)
Glycerol	W.L. Faith G1 (1975), W.L. Faith G2 (1975), W.L. Faith G3 (1975)
Hexamethylenetetramine	A. Vogel (1948), W.L. Faith (1966)
Maleic anhydride	W.L. Faith (1966)
Melamine	W.L. Faith (1966)
Methyl methacrylate	W.L. Faith (1966)
Michler's hydrol	H.E. Fierz-David (1949)
Michler's ketone	H.E. Fierz-David (1949)
Pentaerythritol	W.L. Faith (1966)
Phenol	W.L. Faith G1 (1966), W.L. Faith G2 (1966), W.L. Faith G3 (1966), W.L. Faith G4 (1966)
Phthalic anhydride	W.L. Faith G1 (1966), W.L. Faith G2 (1966)
Resorcinol	O. Mühlhäuser (1887), W.L. Faith (1966)
Triclosan	Geigy (1970), G.J. Lourens (1997), Ciba G1 (1998), Ciba G2 (1998)
Urea	W.L. Faith (1966)
Vinyl acetate	W.L. Faith (1966)
Vinyl chloride	W.L. Faith (1966)

TABLE 1.8

Summary of Statistics for Synthesis Plan Database

Target Type	Number of Targets	Number of Plans
Natural products	82	529
Pharmaceuticals	43	247
Agrichemicals and insecticides	25	59
Dyestuffs and colorants	21	62
Flavorings, fragrances, and sweeteners	12	73
Molecules of theoretical interest	13	42
Miscellaneous industrial chemicals	25	58
Total	221	1070

TABLE 1.9

Color Code for Profiles Shown in Synthesis Plan Files in online download

Profile	Color Code
Molecular weight	Purple
Percent kernel mass of waste	Red
Percent total mass of waste	Red
Atom economy	Blue
Reaction yield	Red
Kernel reaction mass efficiency	Green
Hypsicity	Purple
Target bond making	Purple

BIBLIOGRAPHY

Books

Organic Synthesis

Anand, N.; Bindra, J.S.; Ranganathan, S. *Art in Organic Synthesis*, Holden-Day: San Francisco, CA, 1970.

Anand, N.; Bindra, J.S.; Ranganathan, S. *Art in Organic Synthesis*, 2nd edn., Wiley: New York, 1988.

Carlson, R. *Design and Optimization in Organic Synthesis*, Elsevier: Amsterdam, the Netherlands, 1992.

Carriera, E.M.; Kvaerno, L. *Classics in Stereoselective Synthesis*, Wiley-VCH: Weinheim, Germany, 2009.

Corey, E.J.; Cheng, X. *The Logic of Chemical Synthesis*, Wiley: New York, 1995.

Fleming, I. *Selected Organic Syntheses*, Wiley: New York, 1973.

Hudlicky, T.; Reed, J.W. *The Way of Synthesis: Evolution of Design and Methods for Natural Products*, Wiley-VCH: Weinheim, Germany, 2007.

Ireland, R.E. *Organic Synthesis*, Prentice-Hall: Englewood Cliffs, NJ, 1969.

Nicolaou, K.C.; Snyder, S.A. *Classics in Total Synthesis II*, Wiley-VCH: Weinheim, Germany, 2003.

Nicolaou, K.C.; Sorensen, E.J. *Classics in Total Synthesis*, Wiley-VCH: Weinheim, Germany, 1996.

Pirrung, M.C. *The Synthetic Organic Chemist's Companion*, Wiley: New York, 2007.

Serratosa, F. *Organic Chemistry in Action: The Design of Organic Synthesis*, Elsevier: Amsterdam, the Netherlands, 1990.

Smit, W.A.; Bochkov, A.F.; Caple, R. *Organic Synthesis: The Science Behind the Art*, Royal Society of Chemistry: Cambridge, U.K., 1998.

Taber, D.F. *Organic Synthesis: State of the Art 2003–2005*, Wiley: New York, 2006.

Taber, D.F. *Organic Synthesis: State of the Art 2005–2007*, Wiley: New York, 2008.

Tietze, L.F.; Brasche, G.; Gericke, K. *Domino Reactions in Organic Synthesis*, Wiley-VCH: Weinheim, Germany, 2006.

Tietze, L.F.; Eicher, T.; Diederischen, U.; Speicher, A. *Reactions and Syntheses: In the Organic Chemistry Laboratory*, Wiley-VCH: Weinheim, Germany, 2007.

Warren, S.; Wyatt, P. *Organic Synthesis: The Disconnection Approach*, 2nd edn., Wiley: New York, 2009.

Wender, P.; Miller, B.L., Toward the ideal synthesis: Connectivity analysis and multibond-forming processes. In *Organic Synthesis: Theory and Applications* (T. Hudlicky, ed.), Vol. 2, JAI Press: Greenwich, CT, 1993, p. 27.

Wyatt, P.; Warren, S. *Organic Synthesis: Strategy and Control*, Wiley: New York, 2007.

Synthesis of Pharmaceuticals

Budavari, S. (ed.), *The Merck Index: An Encyclopedia of Chemicals, Drugs and Biologicals*, 12th edn., Merck & Co.: Whitehouse Station, NJ, 1996.

Johnson, D.S.; Li, J.J. *The Art of Drug Synthesis*, Wiley: New York, 2007.

Kleeman, A.; Engel, J. *Pharmaceutical Substances*, Vols. 1, 2, Thieme: Stuttgart, Germany, 2004.

Li, J.J.; Johnson, D.S.; Sliskovic, D.R.; Roth, B.D. *Contemporary Drug Synthesis*, Wiley: New York, 2004.

O'Niel, M.J. (ed.), *The Merck Index: An Encyclopedia of Chemicals, Drugs, and Biologicals*, 14th edn., Merck & Co.: Whitehouse Station, NJ, 2006.

Tao, J.; Lin, G.Q. *Biocatalysis for the Pharmaceutical Industry*, Wiley: New York, 2009.

Process Chemistry

Abdel-Magid, F.A. *Chemical Process Research: The Art of Practical Organic Synthesis*, American Chemical Society: Washington, DC, 2004.

Anderson, N.G. *Practical Process Research and Development*, Academic Press: San Diego, CA, 2000.

Lee, S.; Robinson, G. *Process Development: Fine Chemicals from Grams to Kilograms*, Oxford University Press: Oxford, U.K., 1995.

Industrial Chemicals

Ashford, R.D. *Ashford's Dictionary of Industrial Chemicals*, Wavelength Publications, Ltd.: U.K., 1994.

Chenier, P.J. *Survey of Industrial Chemistry*, Wiley: New York, 2002.

Cook, G.A. *Survey of Modern Industrial Chemistry*, Ann Arbor Science Publishers: Ann Arbor, MI, 1975.

Faith, W.L.; Keyes, D.B.; Clark, R.L. *Industrial Chemicals*, 3rd. edn., Wiley: New York, 1966.

Green, M.; Wittcoff, H.A. *Organic Chemistry Principles and Industrial Practice*, Wiley-VCH: Weinheim, Germany, 2006.

Kent, J.A. (ed.), *Riegel's Handbook of Industrial Chemistry*, 10th edn., Kluwer Academic/Plenum: New York, 2003.

Kirk-Othmer Encyclopedia of Chemical Technology, 4th edn., Wiley: New York, 1991, http://www.interscience.wiley.com/kirkothmer

Lowenheim, F.A.; Moran, M.K. *Faith, Keyes, and Clark's Industrial Chemicals*, 4th edn., Wiley: New York, 1975.

Shreve, J.; Brink, A. Jr.; Shreve, R.N. *Chemical Process Industries*, McGraw-Hill: New York, 1977.

Ullmann's Encyclopedia of Industrial Chemistry, Wiley-VCH: Weinheim, Germany, 2000, http://www.interscience.wiley.com/ullmanns

Weissermel, K.; Arpe, H.J. *Industrial Organic Chemistry: Important Raw Materials and Intermediates*, Verlag Chemie: Stuttgart, Germany, 1978.

White, H.L. *Introduction to Industrial Chemistry*, Wiley: New York, 1986.

Wiseman, P. *An Introduction to Industrial Organic Chemistry*, Applied Science: London, U.K., 1979.

Wittcoff, H.A.; Reuben, B.G. *Industrial Organic Chemicals in Perspective: Parts 1 and 2*, Wiley: New York, 1980.

Green Chemistry Synthesis

Ahluwalia, V.K. *Green Chemistry: Environmentally Benign Reactions*, CRC Press: Boca Raton, FL, 2008.

Ahluwalia, V.K. *Green Solvents for Organic Synthesis*, Alpha Science International: Oxford, U.K., 2009.

Ahluwalia, V.K.; Kidwai, M. *New Trends in Green Chemistry*, Kluwer Academic: New York, 2004.

Allen, D.T. *Green Engineering: Environmentally Conscious Design of Chemical Processes*, Prentice-Hall: Upper Saddle River, NJ, 2002.

Anastas, P.T.; Warner, J.C. *Green Chemistry*, Oxford University Press: New York, 1998.

Anastas, P.T.; Williamson, T.C. (eds.), *Green Chemistry: Designing Chemistry for the Environment*, American Chemical Society: Washington, DC, 1996.

Anastas, P.T.; Williamson, T.C. *Green Chemistry: Frontiers in Benign Chemical Syntheses and Processes*, Oxford University Press: Oxford, U.K., 1998.

Ballini, R. *Eco-Friendly Synthesis of Fine Chemicals*, Royal Society of Chemistry: Cambridge, U.K., 2009.

Crabtree, R.H.; Anastas, P.T. (eds.), *Handbook of Green Chemistry: Green Catalysis*, Wiley-VCH: Weinheim, Germany, Vols. 1–3, 2009.

DeVito, S.C.; Garrett, R.L. (eds.), *Designing Safer Chemicals: Green Chemistry for Pollution Prevention*, American Chemical Society: Washington, DC, 1996.

Doble, M. *Green Chemistry and Processes*, Elsevier: Amsterdam, the Netherlands, 2007.

Dunn, P.; Wells, A.; Williams, M.T. (eds.), *Green Chemistry in the Pharmaceutical Industry*, Wiley-VCH: Weinheim, Germany, 2010.

Kaupp, G. *Solvent-free Organic Synthesis*, Wiley-VCH: Weinheim, Germany, 2009.

Kennepohl, D.K.; Roesky, H.W. *Experiments in Green and Sustainable Chemistry*, Wiley-VCH: Weinheim, Germany, 2009.

Lancaster, M. *Green Chemistry: An Introductory Text*, Royal Society of Chemistry: Cambridge, U.K., 2002.

Lapkin, A.; Constable, D.J.C. (eds.), *Green Chemistry Metrics: Measuring and Monitoring Sustainable Processes*, Wiley-Blackwell Scientific Publishers: Oxford, U.K., 2008.

Loupy, A. *Microwaves in Organic Synthesis*, Wiley-VCH: Weinheim, Germany, 2002.

Matlack, A.S. *Introduction to Green Chemistry*, Marcel Dekker, Inc.: New York, 2001.

Rothenberg, G. *Catalysis: Concepts and Green Applications*, Wiley-VCH: Weinheim, Germany, 2008.
Sheldon, R.A.; Arends, I. *Green Chemistry and Catalysis*, Wiley-VCH: Weinheim, Germany, 2007.
Tanaka, K. *Solvent Free Organic Synthesis*, Wiley-VCH: Weinheim, Germany, 2003.
Tundo, P.; Perosa, A.; Zecchini, F. *Methods and Reagents in Green Chemistry*, Wiley: New York, 2007.

JOURNALS

Graph Theory Applied to Organic Synthesis

Balaban, A.T. Reflections about mathematical chemistry. *Found. Chem.* 2005, *7*, 289.
Barone, R.; Petitjean, M.; Baralotto, C.; Piras, P.; Chanon, M. Information theory description of synthetic strategies: A new similarity index. *J. Phys. Org. Chem.* 2003, *16*, 9.
Bauer, J.; Fontain, E.; Ugi, I. Computer-assisted bilateral solution of chemical problems and generation of reaction networks. *Anal. Chim. Acta.* 1988, *210*, 123.
Bersohn, M.; Esack, A. Computers and organic synthesis. *Chem. Rev.* 1976, *76*, 269.
Bertz, S.H. The first general index of molecular complexity. *J. Am. Chem. Soc.* 1981, *103*, 3599.
Bertz, S.H. On the complexity of graphs and molecules. *Bull. Math. Biol.* 1983, *45*, 849.
Bertz, S.H. Branching in graphs and molecules. *Discrete Appl. Math.* 1988, *19*, 65.
Bertz, S.H. Complexity of synthetic reactions. The use of complexity indices to evaluate reactions, transforms and disconnections. *New J. Chem.* 2003a, *27*, 860.
Bertz, S.H. Complexity of synthetic routes: Linear, convergent and reflexive syntheses. *New J. Chem.* 2003b, *27*, 870.
Bertz, S.H.; Sommer, T.J. Applications of graph theory to synthesis planning: Complexity, reflexivity, and vulnerability, in *Organic Synthesis: Theory and Applications* (T. Hudlicky, ed.), Vol. 2, JAI Press: Greenwich, CT, 1993, p. 67.
Bonchev, D.; Rouvray, D.H. (eds.), *Chemical Graph Theory: Introduction and Fundamentals*, Taylor & Francis: London, U.K., 2003a.
Bonchev, D.; Rouvray, D.H. (eds.), *Complexity in Chemistry: Introduction and Fundamentals*, Taylor & Francis: London, U.K., 2003b.
Hall, L.H.; Kier, L.B. Issues in representation of molecular structure—The development of molecular connectivity. *J. Molec. Graphics Model.* 2001, *20*, 4.
Hendrickson, J.B. Synthesis design for substituted aromatics. *J. Am. Chem. Soc.* 1971, *93*, 6854.
Hendrickson, J.B. Systematic synthesis design. 3. The scope of the problem. *J. Am. Chem. Soc.* 1973, *97*, 5763.
Hendrickson, J.B. Systematic synthesis design. 4. Numerical codification of construction reactions. *J. Am. Chem. Soc.* 1975, *97*, 5784.
Hendrickson, J.B. Systematic synthesis design. 6. Yield analysis and convergency. *J. Am. Chem. Soc.* 1977, *99*, 5438.
Hendrickson, J.B.; Braun-Keller, E.; Toczko, G.A. A logic for synthesis design. *Tetrahedron* 1981, *37(Suppl. 1)*, 359.
Hendrickson, J.B.; Huang, P.; Toczko, G.A. Molecular complexity: A simplified formula adapted to individual atoms. *J. Chem. Inf. Comput. Sci.* 1987, *27*, 63.
Kier, L.B.; Hall, L.H. *Molecular Connectivity in Chemistry and Drug Research*, Academic Press: New York, 1976.
Randic, M. The nature of chemical structure. *J. Math. Chem.* 1990, *4*, 157.
Rouvray, D.H. Graph theory in chemistry. *RIC Reviews* 1971, *4*, 173.
Tanaka, A.; Kawai, T.; Matsumoto, T.; Fujii, M.; Takabatake, T.; Okamoto, H.; Funatsu, K. Construction of a statistical evaluation model based on molecular centrality to find retrosynthetically important bonds in organic compounds. *Eur. J. Org. Chem.* 2008, 5995.
Tanaka, A.; Kawai, T.; Fujii, M.; Matsumoto, T.; Takabatake, T.; Okamoto, H.; Funatsu, K. Molecular centrality for synthetic design of convergent reactions. *Tetrahedron* 2008, *64*, 4602.
Temkin, O.N.; Zeigarnik, A.V.; Bonchev, D. *Chemical Reaction Networks: A Graph Theoretical Approach*, CRC Press: Boca Raton, FL, 1996.
Trinajstic, N. *Chemical Graph Theory*, 2nd edn., CRC Press: Boca Raton, FL, 1992.
Ugi, I.; Bauer, J.; Bley, K.; Dengler, A.; Dietz, A.; Fontain, E.; Gruber, B.; Herges, R.; Knauer, M.; Reitsam, K.; Stein, N. Computer-assisted solution of chemical problems—The historical development and the present state of the art of a new discipline of chemistry. *Angew. Chem. Int. Ed.* 1993, *32*, 201.
Ugi, I.; Bauer, J.; Brandt, J.; Friedrich, J.; Gasteiger, J.; Jochum, C.; Schubert, W. New applications of computers in chemistry. *Angew. Chem. Int. Ed.* 1979, *18*, 111.

2 General Comments on Organic Chemistry and Green Chemistry

2.1 GETTING OUR HOUSE IN ORDER

2.1.1 EDUCATION

Chemistry is a science of compromise. A chemist needs to decide at any given time what are the advantages and what are the liabilities of each process for consideration. Moreover, that decision changes daily as economies governing the availability of starting materials change and as new advances and discoveries are made with respect to new synthetic methodologies and new reactions. Perfection can never be achieved though chemists strive to that ideal. The payoff of this drive to this virtual goal is the continuous and steady small steps of discovery that are made. From the Stone Age to modern times mankind's understanding of what constitutes matter and how to transform one substance into another has evolved and progressed over the millennia from magic to the modern science called chemistry as shown in Figure 2.1. Among the cast of players in this game are well-known, but not yet labeled, workers. These people hover somewhere between alchemists and chemists and are here referred to as "chefists." A "chefist" (combination of "chef" and "chemist") is a kind of chemist who synthesizes molecules by trial and error and is fixated on getting the correct structure of the target compound only, but puts mechanistic concerns and understanding at a low priority. His or her success is largely governed by drawing upon a vast memory bank of personal experiences of handling various reagents, using many kinds of laboratory equipment, and carrying out many kinds of reactions. They may have great chemical intuition and hands-on savvy but they suffer from inadequacies in basic numeracy, physical chemistry, and quantitative reasoning. In fact, they find the idea of incorporating any kind of calculations in synthesis beyond weighing out masses of materials repugnant, even contemptuous. The following reviewer comment from an industrial synthetic chemist puts it well: "I became a synthetic organic chemist precisely to get away from numbers."

The reality is that the only way to turn any kind of human activity into a science necessarily involves reducing and parameterizing that activity in terms of numbers that can be manipulated by some kind of calculation operations. The benefit of this approach is that the activity becomes reliable and predictable. One of the hallmarks of science, and chemistry in particular, is reproducibility of results. In synthesis, those results are the outcomes of following recipes for making substances. When a new kind of chemical reaction or sequence of reaction steps is discovered that connects one kind of molecular structure to another by a reliable methodology that is also efficient, then that is truly magic. It is the hope that this book will shift the chefistry–chemistry equilibrium toward the chemistry end when it comes to analyzing efficiencies of chemical reactions and synthesis plans. The way to cause this shift is education.

Though chemists have come a long way from fuzzy notions of what matter is, there are wider issues that need addressing. Green chemistry is forcing chemists to look in the mirror and question their progress. The prowess and arrogance that has come since chemists' triumphs in being able to synthesize any substance at will is now tempered with humility for the first time. The illusion of limitless resources is now being replaced with the reality of finiteness of resources. Modern discussions on sustainability address the sustainability of individual processes or operations, that is, how to balance

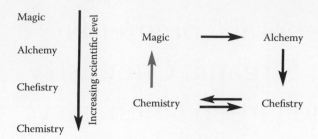

FIGURE 2.1   A cartoon showing the evolution of chemistry from magic.

material and energy flows of inputs and outputs, and how to reduce the component of outputs that is called waste. However, they fail to address the scale of those operations and the vast distribution networks to transport products. All models are resigned to the idea that the scales of all operations are on an ever-increasing upward trajectory. Steady growth at every level is considered a standard metric for economic and technological progress and it is here to stay. Other professionals are now challenging this flawed thinking [1], but not chemists. Chemists are still conscripted to use their ingenuity and technological powers to figure out ways to circumvent the environmental ills of pollution and waste production, but at the same time subscribing to the same flawed continuous growth models that disregard scale. This flawed thinking in particular ignores the continuous growth in world human population, which is the central parameter from which all others depend. Controlling the scale of human population is then paramount to achieve sustainability with respect to various parameters given the finite constraints imposed by living on planet Earth. The reality is that no matter how "green" an operation is there will be a scale beyond which it will not be sustainable. All technologies, old or new, become unsustainable when their scales exceed certain critical values. It is quite surprising then that a significant number of scientists of all stripes holding advanced degrees from prestigious universities, honoured with medals and prizes of high distinction, and holding high decision making positions in academia, industry, and government fail to recognize this obvious fact. For example, suppose we were to completely replace petrochemical based fuels with biofuels. Is there enough arable land available on the planet to grow the necessary crops to meet the insatiable global demand for fuel? This is a problem that subscribers to the new movement of green chemistry have not yet addressed. Its solution will necessarily require partnerships across many disciplines beyond the scientific community involving professional people who are humble, willing to learn from one another, share and exchange ideas to truly solve global problems, and who recognize that scale is the central issue in achieving any degree of sustainability for a given technology or operation.

With respect to education, an interesting article entitled "Ten things chemists should know about chemical engineering and ten things chemical engineers should know about chemists" addresses the polarization between chemists and chemical engineers who view each other as adversaries rather than as allies [2–4]. Also, pertinent is the comment [5] made by Cornforth in an interview after winning the 1975 Nobel Prize in Chemistry on his experiences in academic and industrial environments: "The most important thing for the cooperation of academic and industrial establishments, and individuals, is for each to understand the other's constraints, limitations and capacities. It is, for example, no good offering an elegant, difficult, and expensive process to an industrial chemist, whose ideal is something to be carried out in a disused bathtub by a one-armed man who cannot read, the product being collected continuously through the drain-hole in 100 percent purity and yield. Academe and Industry can collaborate if they have the will to understand each other. Nowadays it is becoming ever more necessary to retrain scientific and technical staff as technology advances and requirements change; and here I think is a very valuable area not only for cooperation but for the spreading of mutual understanding." As stated earlier, synthetic organic chemists need to change their attitude toward computation and to brush up their basic numeracy skills. The number of mistakes appearing in experimental sections of research papers on something as basic as reaction yield determinations is astounding. Many times authors report values that actually underestimate reaction performance contrary to popular belief that yields are only exaggerated in publications.

Before embarking on the green chemistry bandwagon it is imperative to realize that practicing green chemistry requires a chemist to be fluent in a number of areas of chemistry beyond synthesis. A green chemist is a complete chemist, certainly not a chefist, and is therefore the most prized of all chemists. Having said that, all is not well in how we prepare the future generation of chemists who will be entrusted with charting new territory in this broad discipline. The way organic chemistry has been taught for the last 100 years is wholly inadequate and is in need of an overhaul before even considering anything to do with green chemistry. In other words we as chemistry educators need to get our house in order before setting sail in this direction. A number of key issues on the fundamentals of organic chemistry will now be discussed. Firstly, there are problems with how the subject of organic chemistry is presented in introductory textbooks. A list of major concerns is the following:

- They are organized according to functional groups, not patterns of reactivity.
- Chemical equations are not balanced.
- Chapters on stereochemistry (based on 3D geometry) are far removed from earlier chapters on bonding and chemical structure (based on 2D geometry).
- There are no citations to the original literature, so students are forced to believe what the author says without checking for themselves the validity of the claims made.
- Examples are often not taken from the original literature especially when discussing fundamental concepts like bonding and molecular structure.
- There is a disconnect between concepts introduced and the originator(s) of the idea, and the circumstances around which an idea arose, that is, the mechanics of how research is done is not shown to the student.
- There is an overwhelming emphasis on individual examples (memorization) rather than on pattern recognition (understanding).

Table 2.1 summarizes the frequency of occurrence of certain key topics in a survey of 16 introductory organic textbooks [6–21].

Some facts are never stated upfront in chemistry textbooks. For example, the concept of the existence of "indivisible atoms" arose as a way of reconciling the paradox of having a finite entity that is composed of an infinite number of constituent elements (Zeno of Elea's paradox). Figure 2.2 shows a graphical representation of this paradox. If a frog is 1 m away from a tree and always hops half the distance between itself and the tree it will take theoretically an infinite number of hops to get there,

TABLE 2.1
Summary of Topic Frequencies for
16 Introductory Organic Textbooks

Topic	Percent
Balancing chemical equations	0.0
Chapter on synthesis strategy	25.0
Citations to original literature	31.3
Color coding as an instructional tool	50.0
Functional group class arrangement	93.8
Historical anecdotes	37.5
Illustrative experimental procedures	12.5
Oxidation number usage	18.8
Rules for writing reaction mechanisms	31.3
Summary/table of named rxns	18.8
Summary/table of reaction intermediate types	6.3
Summary/table of reaction mechanism types	12.5

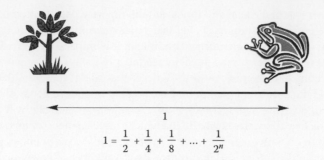

$$1 = \frac{1}{2} + \frac{1}{4} + \frac{1}{8} + ... + \frac{1}{2^n}$$

FIGURE 2.2   The infinite series with a finite sum that is the basis of Zeno's paradox.

yet the 1 m distance is finite. Thinking about this in the reverse sense, if we have a 1 m long loaf of bread and we slice it in half, and then slice that half in half, and continue on this way, what will we have left? Will we still have some piece that we can still call bread? This is what led Democritus and Leucippus to think about the indivisibility of matter up to the point of atom. What we clearly see here is that the basis of all of chemistry—its cornerstone, its pillar idea upon which all others are hinged—is pure mathematics, namely, the sum of a particular infinite series!!!

At this point, it is instructive to remind ourselves of the prediction made by the Scottish chemist Alexander Crum Brown in 1874 [22]:

> Chemistry will become a branch of applied mathematics; but it will not cease to be an experimental science. Mathematics may enable us retrospectively to justify results obtained by experiment, may point out useful lines of research, and even sometimes predict entirely novel discoveries. We do not know when the change will take place, or whether it will be gradual or sudden.

Following is a list of other understated facts:

- Chemistry is first and foremost an empirical science. Chemistry is a science of compromise.
- All scales and metrics used in chemistry are based on relative terms, not on absolute terms, with predefined (often arbitrary) reference states. It is the differences between values that counts, not so much the values themselves.
- Pauling's electronegativity scale is based on an arbitrary designation.
- All chemical structure drawings are graphs (mathematical term for dot and line diagrams).
- International Union of Pure and Applied Chemists (IUPAC) nomenclature rules and the rings and unsaturations formula are based on graph theory. The number of rings and unsaturations in any chemical structure is also known as the cyclomatic number.
- The sum of a set of elementary reactions constituting a reaction mechanism must equal the corresponding overall stoichiometrically balanced chemical equation.
- The Arrhenius equation (1889) has no theoretical basis—it is purely an empirically based relationship.
- An energy-reaction coordinate diagram is not a proper graph or plot. The "reaction coordinate" axis has quantitative meaning only when it refers to a single structural parameter undergoing change in a reaction such as a bond length, bond angle, or dihedral angle. The diagram is quantitative in both axes only for reactions involving diatomics. The multidimensional nature of structural changes taking place in reactions more complicated than this prevents us from drawing a quantitatively accurate multidimensional plot. For such reactions, the diagram is more properly called a cartoon.
- A rate law for a given reaction gives the stoichiometry of reacting species in the transition state (TS) for the rate-determining step.
- The concept of spin in quantum mechanics has NO classical analogue and therefore cannot be explained by ANY classical mechanical argument.

- The sum of exponents in hybridized orbital descriptors for a carbon atom in a given structure equals the number of neighboring atoms bonded to it. For example, an sp^3 carbon atom has 4 atoms bonded to it, an sp^2 carbon has 3 atoms bonded to it, and an sp carbon has 2 atoms bonded to it.
- Allowed resonance structures: open shell to open shell, or closed shell to closed shell. The following example for benzene illustrates this.

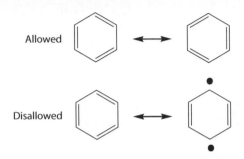

- All geometry-optimized calculations yield chemical structures under gas phase conditions at 0°K. The computed energy for such a structure is always a negative number and is referenced to the hypothetical state where its constituent atoms are infinitely spaced apart. Such a state is arbitrarily assigned an energy value of 0.
- Thermodynamic stability and control refer to energy differences between initial and final states.
- Kinetic stability and control refer to energy differences between initial and transition state (TS).
- Eyring TS theory is an "ad hoc" description that assigns thermodynamic state function variables to TS. This violates the fundamental tenet of thermodynamics that says that changes between initial and final states are path independent.

The problems cited earlier with respect to textbooks give rise to certain myths about organic chemistry ideas that prevail even in the brightest minds on the subject. Following is a list of the chief myths along with their refutations.

Myth 1

Inversion and retention of configuration are described by changes in R and S configurations at a given reaction center.

Figure 2.3 shows examples that refute this false notion. The correct definitions of both terms are as follows.

Inversion of configuration: the incoming group occupies a position *opposite* (i.e., 180°) to the leaving group with respect to other groups around an asymmetric center.

Retention of configuration: the incoming group occupies the *same* position spatially as the leaving group with respect to other groups around an asymmetric center.

Both of these definitions have nothing to do with the R and S configuration changes.

The apparent "conundrum" occurs when the priority of the incoming group differs from that of leaving group in the R/S context. Unfortunately, every textbook illustrates an example inversion reaction whose configuration at the asymmetric center changes from R to S, or vice versa. The ubiquitous example reactions are conversion of (*S*)-2-bromobutane to (*R*)-2-butanol, (*S*)-2-bromobutane to (*R*)-2-iodobutane, and (*R*)-2-bromooctane to (*S*)-2-octanol. In addition, every textbook illustrates an example of retention reaction whose configuration at the asymmetric center remains invariant. Can one be blamed for falling into this trap of faulty thinking if everywhere one looks there are always the same examples shown in every book over and over again?

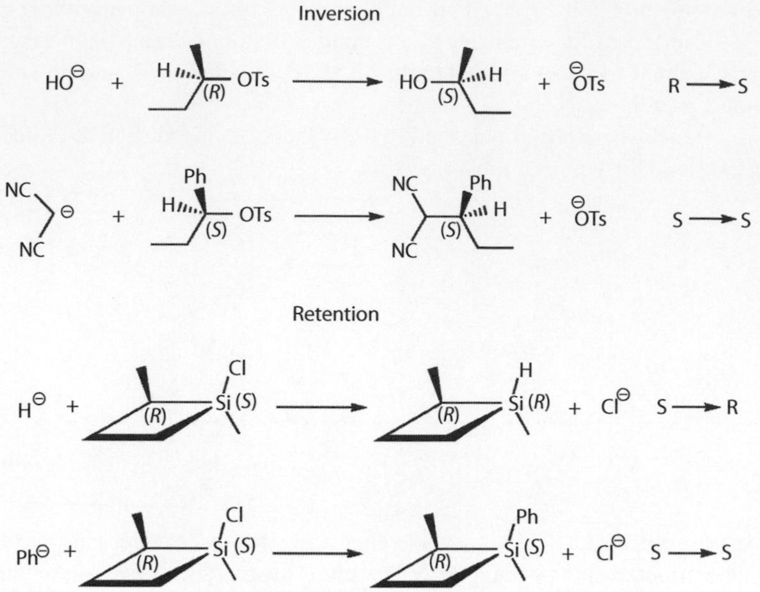

FIGURE 2.3 Example reactions showing inversion and retention of configuration.

Myth 2

> The L and *l* stereochemistry descriptors are equivalent.
> The D and *d* stereochemistry descriptors are equivalent.

Figure 2.4 shows example compounds where neither of these statements is true. The D/L nomenclature is based on the arbitrarily chosen template Fischer projection drawing of D-glyceraldehyde where the hydroxyl group appears on the right. The *d/l* nomenclature is linked to the direction of optical rotation: *d* = dextrorotatory (+ sign, clockwise rotation), *l* = levorotatory (− sign, anticlockwise rotation). The example of lactic acid is a fascinating demonstration of stereochemical nomenclatures.

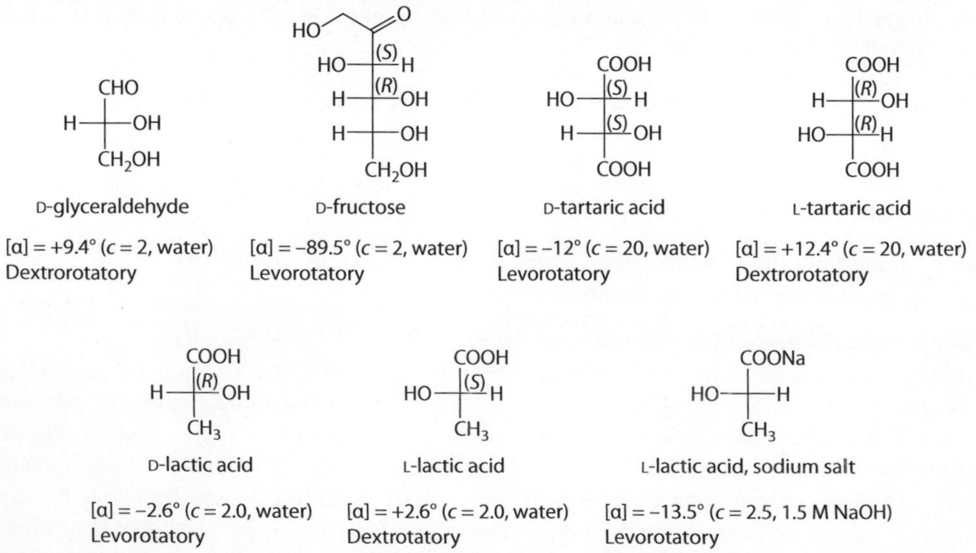

FIGURE 2.4 Example molecules showing D/L configurations according to Fischer projections along with their observed optical rotations.

Myth 3

The terms side product and by-product are synonymous.

A by-product arises as a mechanistic consequence of producing a target product as prescribed by a given target reaction.

A side product arises as a consequence of a competing side reaction that occurs by a different mechanism in parallel to the intended target reaction. Section 5.2 discusses this difference in the context of waste determination. Several example reactions are given. This example illustrates that we must be precise in our choice of words in describing what waste is.

Myth 4

Convergent synthesis plans are always more material efficient than linear plans.
Short synthesis plans are always more material efficient than long plans.

Section 9.8 gives a discussion showing that these statements are incorrect. There is no mathematical proof for either of these statements being true.

Myth 5

The amount of catalyst used in a reaction is much smaller than the substrate on which it acts.

The caveat phrase is "catalytic amount," which gives the mental picture that a "pinch" of some substance is added to a reaction mixture.

Following is a case where the mole amount of limiting reagent exceeds the mole amount of catalyst [23]. This is an example of low catalyst loading.

164	107	253
4.92 g (0.03 mol)	3.32 g (0.031 mol)	mol% catalyst = (0.00032/0.03) * 100 = 1.1%

Following is a case where the mole amount of catalyst exceeds the mole amount of limiting reagent [24]. This is an example of a high catalyst loading.

154	100	254
100 g (0.65 mol)	72 g (0.72 mol)	mol% catalyst = (1.46/0.65) * 100 = 225%

Myth 6

Oxidation numbers have little usefulness in organic reactions.

This myth is thoroughly dispelled in Section 5.4. The study of redox reactions in which protic acids act as redox reagents are a great training ground to apply the oxidation number concept and to learn how to balance chemical equations. For example, sulfuric acid and nitric acid act as oxidizing reagents in the following transformations. The first entry is an important reaction in the chemical industry to produce phthalic acid, which in turn is an important feedstock for dyestuffs, agrichemicals, and pharmaceuticals. The second entry is the traditional method of making adipic acid. For convenience, the balanced equation shows nitric acid being reduced to nitrogen dioxide. In reality, the byproduct will be a complex mixture of various nitrogen oxides.

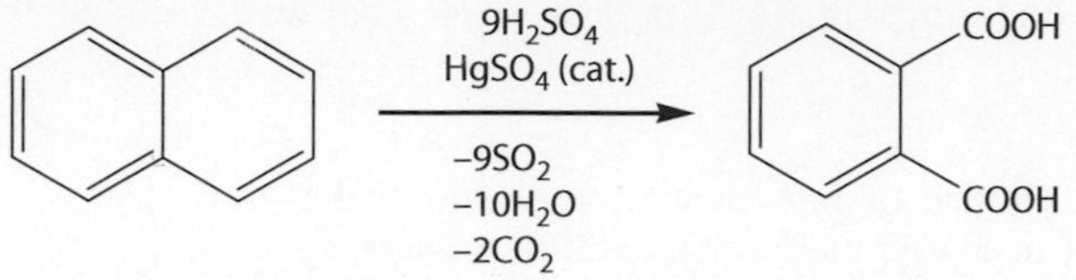

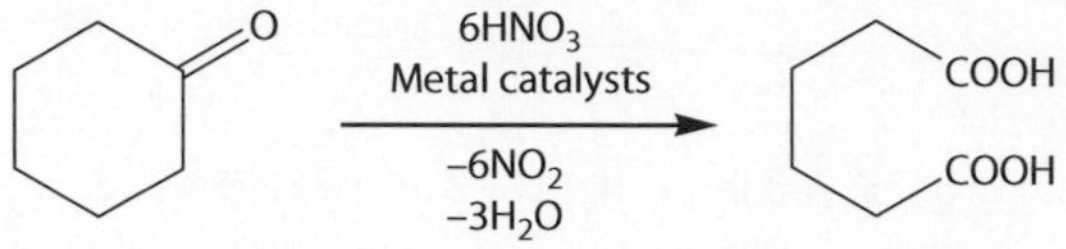

Formic acid or its ammonium salt acts as a reducing agent in the following transformation. Formate ion fragments to form carbon dioxide and hydride ion, which then acts as a nucleophile to displace the hydroxyl group formed in the intermediate, thereby reducing the hydroxymethyl group to a methyl group.

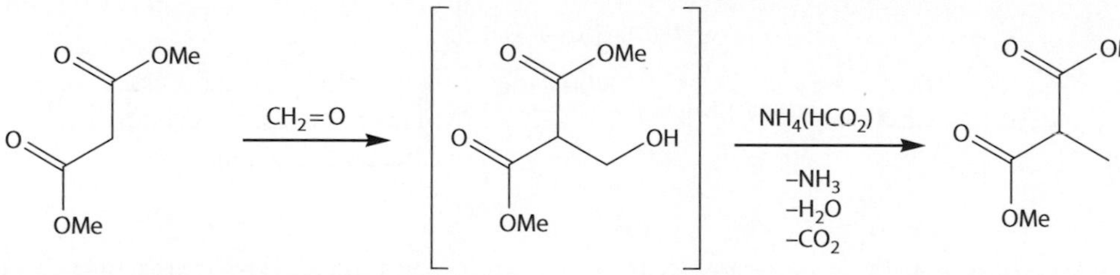

Using the ideas in Section 5.4 the student may work out the oxidation number changes of the key atoms involved in target bond forming steps, and link these with the redox couples.

If a chemist makes the claim that they know everything there is to know about a given reaction, then what they mean to say is that they know the structures of all reactants and products, they have determined the associated kinetic and thermodynamic parameters, they have direct and/or indirect experimental evidence for reaction intermediates existing between reactants and products, they are able to write out a mechanism that is consistent with all of these findings, they have optimized the reaction conditions to obtain the desired product in the highest yield, and they have considered alternative methods to make the same target molecule. These items are conveniently depicted in Figure 2.5. How many reactions can the reader name that are known to such a degree?

Figure 2.6 is a one-page representation of all the essential ideas in organic chemistry that are needed in elucidating a reaction mechanism for a given reaction. These ideas are applicable to all known reactions and to any future reaction yet to be discovered. This is convenient for students as they practice the logical thinking behind this important exercise in organic chemistry. By exercising these thought patterns with every new reaction studied, over time they will have committed these ideas to memory and the task will become second nature.

For the student who wishes to find the rules for writing out reaction mechanisms delineated in one place in compact form, the following section will be of interest. The book by Miller and Solomon is particularly useful [25]. The reader is also directed to the extensive compilation of

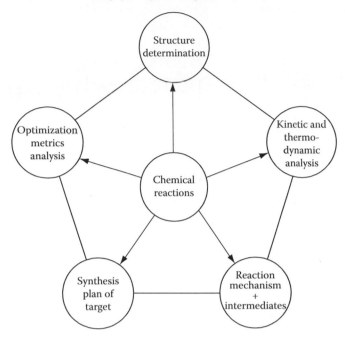

FIGURE 2.5 Pentad representation of concepts attributed to a given reaction.

reaction intermediates that catalogs their discovery [26]. The curly arrow notation chemists commonly use to track electron flow from one structure to another when writing out a reaction mechanism in order to follow which bonds are made and broken was invented by Sir Robert Robinson, a synthetic organic chemist, not a mechanistic chemist [27]. Arthur Lapworth, a mechanistic chemist who studied enol intermediates, also used a primitive form of this notation [28].

2.1.1.1 Rules for Writing Reaction Mechanisms

1. Graphs depicting chemical structures are composed of lines (bonds) and letters (chemical symbols of elements from periodic table). Lone pair electrons are drawn either as two dots or as a bar over an atomic nucleus. Two dots together mean that electrons are in the same molecular orbital; separated dots mean electrons occupy different orbitals.
2. Electron flow is depicted by an arrow such that the arrowtail corresponds to the electron rich source and the arrowhead corresponds to the electron deficient sink.
3. For two-electron movement, the arrowtail is drawn from (a) a lone pair of electrons (belonging to a nucleophilic center) occupying the same molecular orbital, or (b) from two electrons making up a bond.

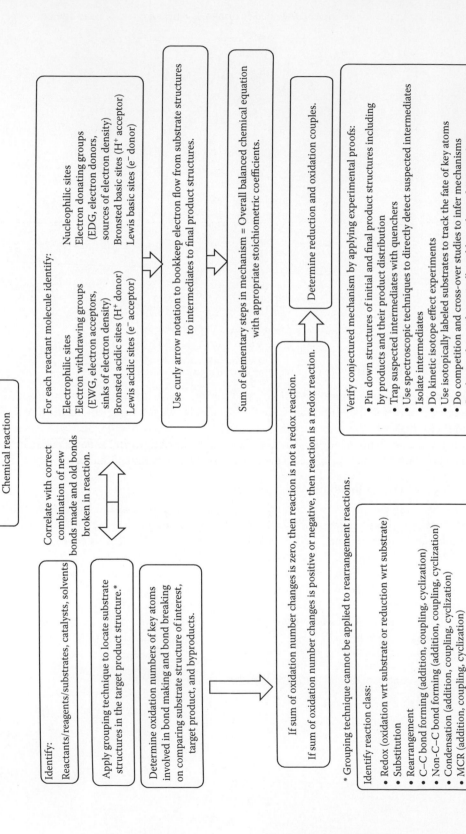

FIGURE 2.6   One-page concept sheet for conjecturing possible mechanisms for any given chemical reaction.

4. For two-electron movement, the arrowhead is drawn toward an electrophilic center.

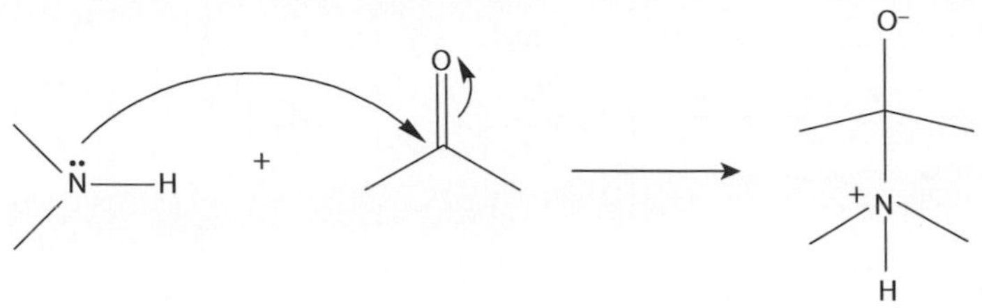

5. For single-electron movement, the arrowtail is drawn from (a) a bond made up of two electrons, or (b) from a center having a single valence electron occupying a molecular orbital (e.g., radical center, triplet carbene center).

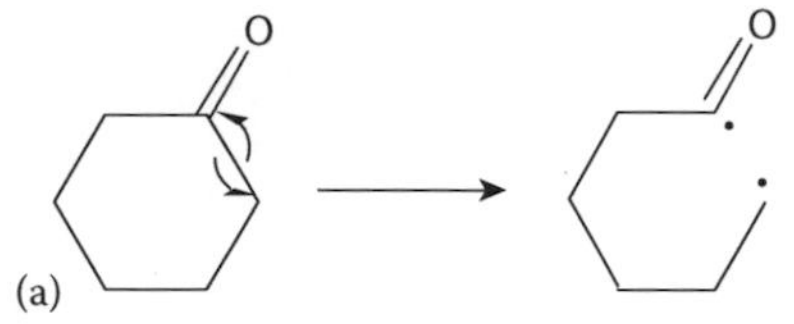

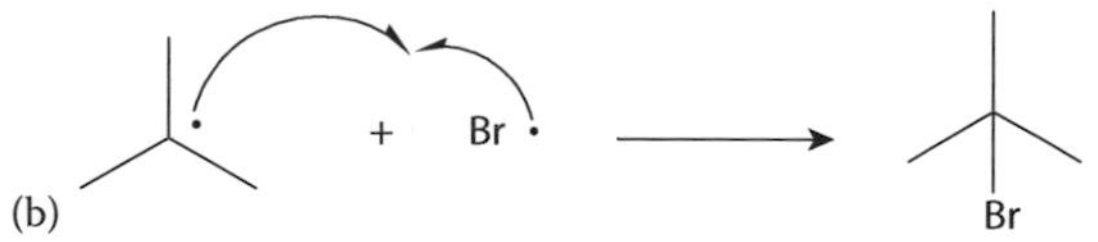

(a)

(b)

6. For single-electron movement, the arrowhead is drawn toward another center with the arrowhead being a half-arrowhead.

7. Two structures are *resonance* structures if the atomic connectivity in the two structures is identical (same sigma bonding) but the electron density distribution (bond order, lone pair electrons, and pi bonds) over the atomic nuclei is different.

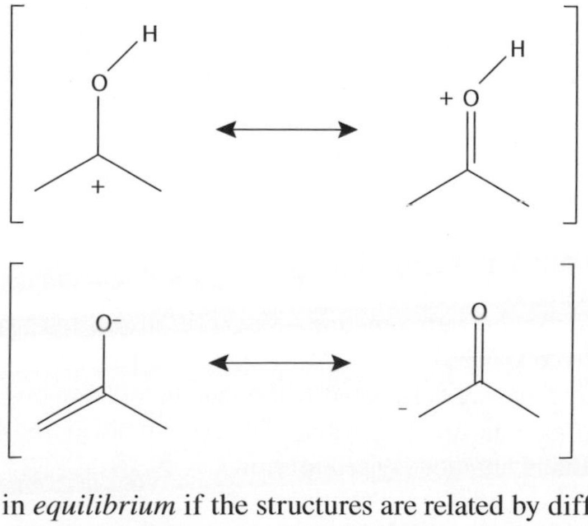

8. Two structures are in *equilibrium* if the structures are related by differing atomic connectivity (i.e., different sigma bonding pattern) and differing electron density distribution, as in the case of tautomers.

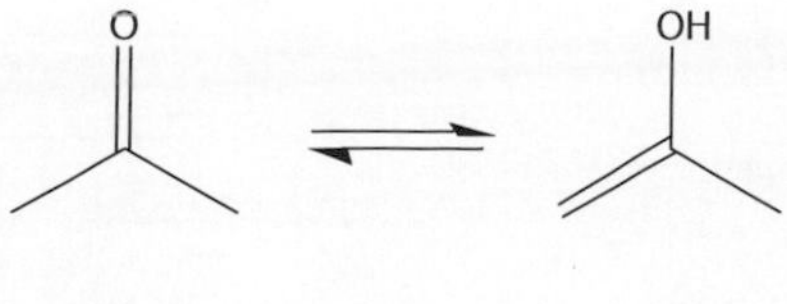

Pairs of stereoisomers and geometric isomers may also be in equilibrium.

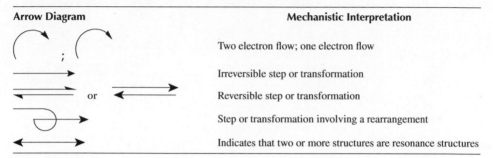

9. The following arrow diagrams have the stated meanings:

Arrow Diagram	Mechanistic Interpretation
	Two electron flow; one electron flow
	Irreversible step or transformation
or	Reversible step or transformation
	Step or transformation involving a rearrangement
	Indicates that two or more structures are resonance structures

10. TS structures are drawn between square brackets with the symbol ‡ shown on the outside top corner of the right bracket. Bond breaking and bond making are designated with broken lines. Localized charges on nuclei in TS structures are denoted as δ+ or δ−.
    Late TS: advanced toward products
    Early TS: advanced toward reactants

$$\left[\begin{array}{c} \overset{\delta+}{N} - - - \overset{\delta-}{C} - - - OTs \end{array} \right]^{\ddagger}$$

Long broken lines mean that either bond making is *not advanced* or bond breaking is *advanced* in TS; short broken lines mean that either bond making is *advanced* or bond breaking is *not advanced* in TS. Similar chemical structures mean they are similar in energy.

11. Reaction intermediates are drawn in the same way like reactants and products with solid lines representing bonds; they are sometimes also written within square brackets to signify that they are short-lived species.

12. Photolytic transformations are designated with the symbol hν above the arrow.
    Thermolytic transformations are designated with the symbol Δ above the arrow.

13. The following shorthand notation is used to show

Protonation	$+H^+$ →
Deprotonation	→ $-H^+$
Leaving group	→ $-X^-$
Elimination of small molecule	→ $-H_2O$

14. A reaction mechanism is a list of *elementary (one-step) reactions*. Each elementary reaction is mass balanced and charge balanced. The sum of all elementary reactions constituting a mechanism is the *stoichiometric equation*. The stoichiometric equation is also mass and charge balanced.

15. Best *structures* in a mechanism *minimize* charge and energy (in thermodynamic sense; e.g., molecular strain); and *maximize* conjugation, planarity, and *aromaticity*.

16. Best *mechanisms minimize* the number of steps and the number of intermediates involved. They also proceed with the least energy demanding pathway. The most favorable reaction is the one leading to products that arise via a pathway that overcomes small energy barriers between initial and transition states (*kinetic control*) and/or are products of lowest energy (*thermodynamic control*) relative to starting materials. The term *stability* is understood in relative terms and has two forms: *kinetic stability* and *thermodynamic stability*. Kinetically stable chemical structures are those that react slowly because the barrier for reaction is large; whereas, thermodynamically stable chemical structures are of lower energy because they exist in deeper potential energy wells.

17. Solvent separated species are drawn as

18. Two fragments in a solvent cage are drawn as

Triplet radical pair

19. Excited states are denoted with an asterisk (*) in the top right-hand corner of the structure. Structures not showing an asterisk are assumed to be in the ground state. Overall charge of a species and multiplicity of a species are also shown in the top right-hand corner of the structure as shown in the following. Localized charges on nuclei (+ or −) are drawn directly above specific nuclei.

Naphthalene triplet

Naphthyl radical cation

Naphthyl radical anion

Best mechanism must be consistent with experimental observations.

2.1.2   LITERATURE

We have seen problems with how we educate chemists. Now, we turn briefly to the research literature in synthetic organic chemistry. This topic will be more thoroughly discussed in Section 4.3. As a primer, the reader should take a look at Cornforth's famous paper entitled "The trouble with organic synthesis" [29]. The following list of pet peeve phrases seems to turn up everywhere. After each phrase a question is posed which should prompt the reader to realize the unscientific nature of these phrases.

- "In our hands" (How many fingers and hands do those workers have?)
- "Readily available" (Did the workers find the compound in a chemical catalogue?)
- "Obtained in good yield" (What kind of quantitative statement is this in a scientific field?)
- "Worked up in the usual way" (Is there such a thing as an unusual workup?)
- "Added to a dilute solution" (How concentrated is dilute?)
- "Excess reagent was used" (How much excess?)
- "Typical procedure" (Does this mean that it is a one-size-fits-all procedure?)

All of them are unacceptable in scientific papers. The "in our hands" phrase are code words for obfuscation in prior publications with respect to exaggerated yields or lack of disclosure of key experimental details in procedures. The question of how many fingers and hands these people have to come up with high reaction yields seems legitimate to ask—maybe they have an advantage! The second phrase encourages myths to develop—there is a detachment between the reality of how chemicals are made and what is "available" in a chemical catalog, akin to the illusion of the true origins of foodstuffs "readily" found in a supermarket. Sections 3.9, 9.4, and 11.8 discuss how the choice of starting materials influences the efficiencies of synthesis plans.

An excellent strategy for reading synthesis papers is to read them backwards, that is, to begin from the target molecule and work backwards to map out the thread of the synthesis back to the starting materials. This greatly facilitates laying out the plan so than an analysis of its efficiency can proceed. Authors normally give a detailed account of their work outlining all tangents and dead ends. For the purpose of quickly getting at the synthesis plan that worked, it can make for difficult reading of such papers in the forward sense.

When reading scientific papers in foreign languages, as was necessary in looking up syntheses for some classic dyestuff molecules for this synthesis compilation, the word for "yield" to look for in French is "rendement" abbreviated as "Rdt"; in German it is "Ausbeute" abbreviated as "Aus.," and in Italian it is "resa." In old French language publications one needs to be aware of the following symbols in chemical structures: Az = N (not azide) and Ph = P (not phenyl). Also, French publications used superscripting in writing out empirical formulas rather than subscripting numbers of elements. For example, the empirical formula for sucrose would be written as $C^{12}H^{22}O^{11}$. In both the old French and German publications a hexagonal ring in a chemical structure stands for an aromatic phenyl ring. Cyclohexane rings were drawn as a hexagon with "CH_2" labels written at each vertex.

Also, substituted benzene rings were very oddly drawn as shown in the following. The major reason for these odd depictions was the primitive typesetting technology available in the nineteenth century in depicting structures in print media.

Old graphic

Modern graphic

C_6H_3 — COOH (1)
CN (3)
OH (5)

NC — COOH
OH

3-cyano-5-hydroxy-benzoic acid

The Haworth projection for drawing carbohydrate structures was invented so that oligosaccharides could be written out in the same line and thus save space on the page. One needs to remember that the junction bonds are not methylene linkages.

* = Not methylene linkages

A positive point is all equations in the old German literature were properly balanced using empirical formulas on the left- and right-hand sides of the equation. An equal sign, not an arrow notation was used. This is more in line with the mathematical notation for equations. The examples shown in Figure 2.7a and b illustrate old and modern notations for a chemical reaction [30,31]. Also included is how a green chemist would write out chemical equations showing both structures and full chemical balancing. This is the proper way that any chemical equation should be written, which should be made standard in all textbooks and journal articles.

2.2 RESEARCH TRENDS IN ORGANIC SYNTHESIS

The only area of chemistry that has been elevated to an art form by its practitioners is organic synthesis. Though there are esthetic elements in common, one should remember that in art, each piece is unique—any replicas are considered either cheap imitations or forgeries. This may explain why a first synthesis to a target molecule is venerated more than later versions even though they may be

Old graphic:

$$[(CH_3)_2NC_6H_4]_3 \equiv C \text{-----} Cl + NaOC_2H_5 = [(CH_3)_2NC_6H_4]_3 \equiv C \text{-----} OC_2H_5 + NaCl$$

$$(H_2NC_6H_4)_3 \equiv C \text{-----} Cl + NaOC_2H_5 = (H_2NC_6H_4)_3 \equiv C \text{-----} OC_2H_5 + NaCl$$

Modern graphic:

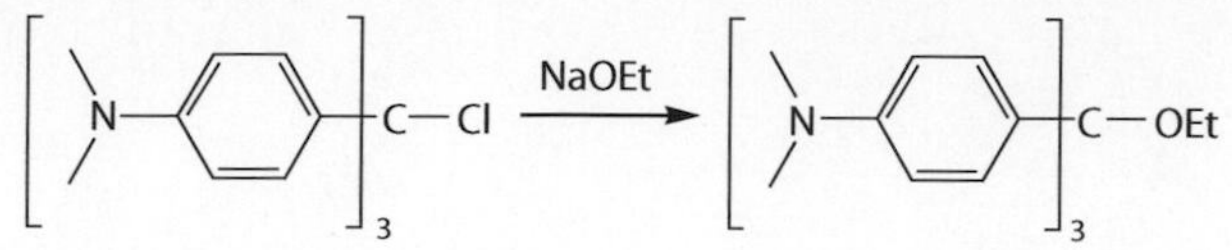

Green chemist's graphic:

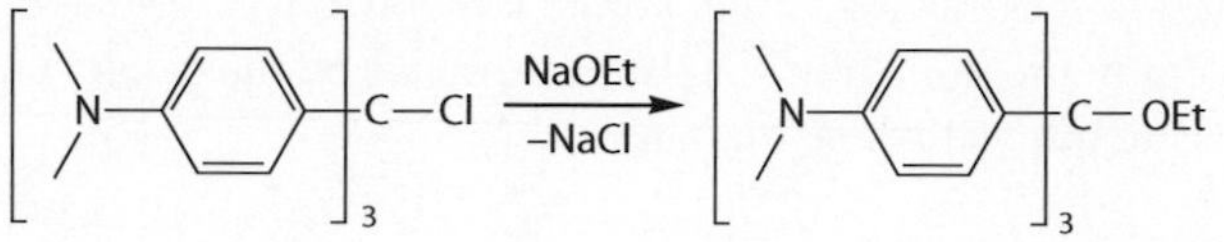

(a)

Old graphic:

$$C_{14}H_8(O_2)'' + 4Br = C_{14}H_6Br_2(O_2)'' + 2BrH$$

$$C_{14}H_6Br_2(O_2)'' + 4KHO = C_{14}H_6(OK)_2(O_2)'' + 2KBr + 2H_2O$$

Modern graphic:

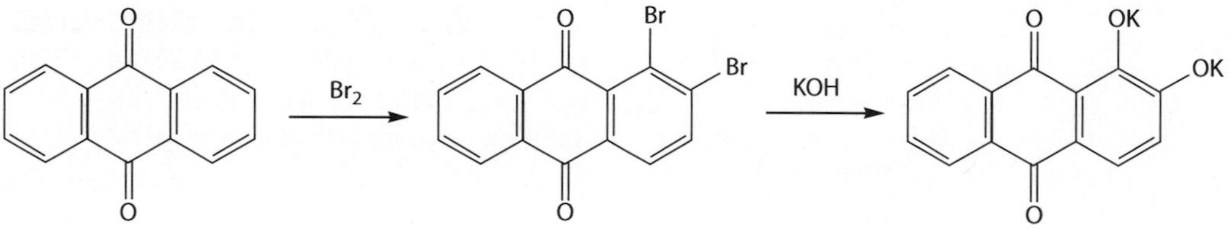

Green chemist's graphic:

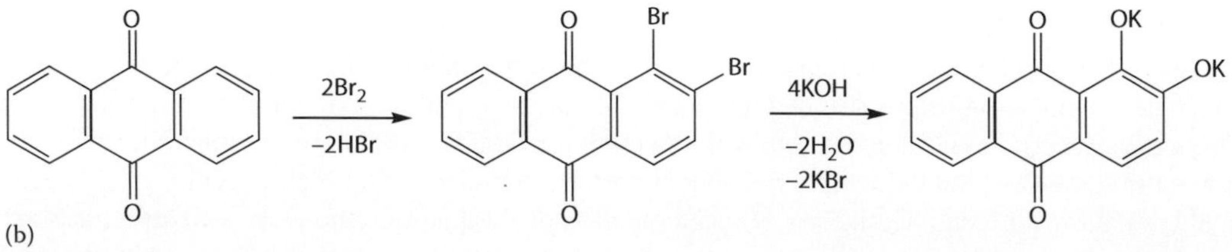

(b)

FIGURE 2.7   (a) Graphics for the synthesis of triphenylmethane dyes. (b) Graphics for the synthesis of alizarin.

more efficient than the first. Some of the great triumphs of organic chemistry include the following: (a) structure elucidation by functional group tests and spectroscopy; (b) proof of structure of complex natural products by degradation, total synthesis, and advanced spectroscopic techniques such as nuclear magnetic spectroscopy (NMR); (c) the development of a database of reliable reactions that chemists can draw upon to synthesize any target molecule; and (d) the concepts of nucleophicity and electrophilicity that help explain patterns of chemical reactivity. Structurally linear-type molecules are easy to synthesize and are easily amenable to automated combinatorial chemistry approach strategies. The only exception to this is linear chains having groups distributed along their backbone in regular stereo-oriented patterns, such as methyl or hydroxy groups as shown in the following.

Synthetic chemists are particularly attracted to complex ring systems, especially newly discovered ones that are exotic; for example, colchicine, strychnine, tetrodotoxin, and platensimycin. These are the sorts of structures that pose the greatest challenges and offer the most fertile ground to learn, discover, and develop new chemistry. Other attractive areas in the field of synthesis methodology involve the systematic evaluation of functional group transformation reactions for the purpose of increasing the size and repertoire of the synthetic chemist's toolbox. These include (a) the invention of methodologies for carrying out a given transformation under the three main reaction conditions: acidic conditions, basic conditions, and neutral conditions, (b) the invention of *umpolung* (reverse polarization) analogs for every functional group transformation reaction, and (c) the interconversion of any functional group into another. Often, advances are made by accident with a lot of luck. This is particularly true of unimolecular rearrangement reactions taking place under thermal, photochemical, or acidic or basic conditions.

Organic synthesis may be viewed as a giant combinatorial game where the deck size of building block reactions constantly increases thereby increasing the number of possible combinations of sequences that are possible and hence increasing the complexity of target structures produced. In exploring the full matrix of possibilities there is a bottom-up approach to create a database of multicomponent reactions (MCR), involving the coupling of three or more reagents, in order to find interesting targets made by an additive approach, and a top-down approach to find such target units in complex structures such as natural products via similarity matching analysis. MCR that involve in situ redox reactions after the C–C coupling reactions are complete are particularly interesting. The oxidizing agent is usually oxygen in air and the driving force may be achieving aromaticity in the final structure as shown by the following examples. Since reactions are written without balancing, authors omit writing oxygen as one of the reagents in such transformations.

Example 1 [32]

Example 2 [33]

Example 3 [34]

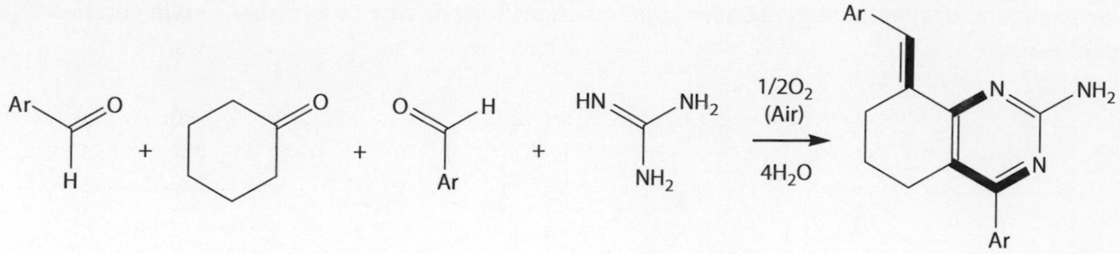

In other cases, the oxidizing agent is a metal salt as shown by the following example, which juxtaposes two different reaction outcomes.

Example 4 [35]

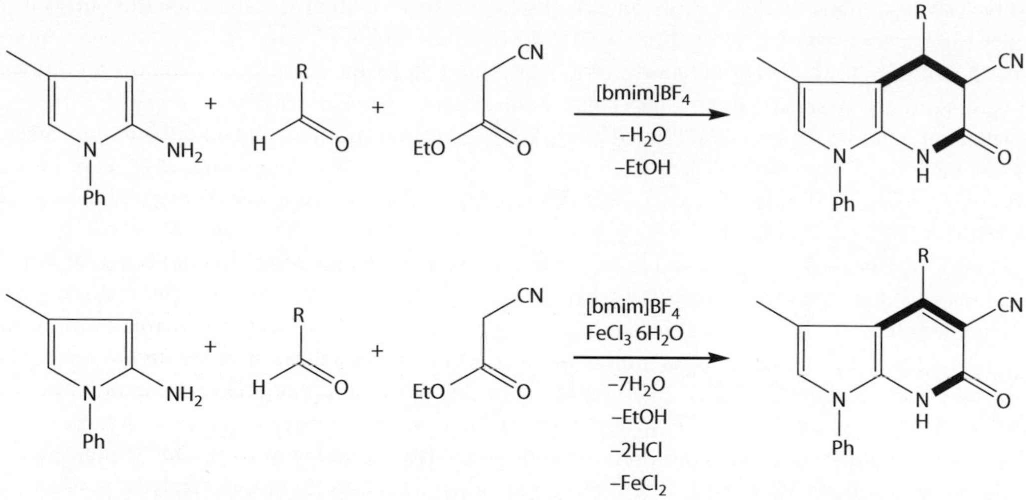

The modern challenge in organic synthesis is the orchestration of efficiency by deliberate action. The idea is to choreograph optimization of all metrics in the same direction, like making music and melodies from individual notes. Whatever the reaction is under consideration the following key questions are always asked in the subject of organic synthesis:

a. What does this structure remind me of?
b. What does this reaction remind me of?
c. Where do I find the starting materials in the target product structure?
d. Where are the nucleophilic and electrophilic sites in a given structure?
e. Where are the acidic and basic sites in a given structure?
f. How many ways can one construct a given ring system?
g. Are there any elements of symmetry in the target product structure that may be exploited with respect to its synthesis?

Any reaction combination is possible so long as a suitable set of reaction conditions is found to carry out the transformation. Usually this entails finding the magic catalyst that will to do the trick. Modern organic chemistry is peeling away any constraints on what is doable, so making such a statement that any reaction is possible may not be so far fetched as it sounds. One should always be open-minded and not prejudge reaction outcomes. Chemistry is an empirical science that has progressed by trial and error. One should consider himself/herself as a perpetual student as Prelog's autobiography suggests [36]. Thinking in this way ensures the flame of curiosity burns brightly throughout one's career and also introduces the humility factor to counteract the two enemies of a creative mind, complacency and arrogance. Favorite questions to ask are as follows:

a. Why not? (with respect to the possibility of carrying out a particular reaction).
b. What if? (with respect to the implications on the possibility of carrying out a particular reaction).

2.3 RESEARCH TRENDS IN GREEN CHEMISTRY

The Anastas-Warner working definition of "green chemistry" [37] is as follows:

> Green chemistry is the utilization of a set of principles that reduces or eliminates the use or generation of hazardous substances in the design, manufacture, and application of chemical products.

The subject's development from the early 1990s has already been documented [38,39]. Research in this area of chemistry may be conveniently summarized as shown by the "green matrix" in Table 2.2 where the left-hand column gives a selection of reaction media and the top row gives a selection of "green" technologies. Basically, publications involve some combination of these two categories. The scope of green chemistry possibilities in the literature may be pigeonholed into these slots. The bulk of green chemistry research has been in the area of replacing or reducing solvent demand for chemical processes. The word "benign" has been used to describe alternative "green" or environmentally friendly solvents. The use of ionic liquids has taken the lion share of research in this area. In terms of green methodologies, microwave-mediated reactions for organic synthesis, catalysis, and the invention of MCR are vigorous areas of research.

The novice reader may think that the subject is an emerging field; however, the "green chemistry" name is an umbrella for many well-established ideas and techniques that already exist in the literature. None of the techniques under the green technology label can therefore be considered "new." For example, the first room-temperature ionic liquid, ethylammonium nitrate $[EtNH_3]^+$ $[NO_3]^-$, which melts at 12°C was discovered by Paul Walden in 1914 [40]. Ionic liquids were also observed in Friedel–Crafts acylation and alkylation reactions [41,42]. The reader is directed to two reviews that chronicle the historical development of ionic liquids [43,44]. Microwave-assisted synthesis was developed by Gedeye and Westaway at Laurentian University in Sudbury, Canada [45–48].

TABLE 2.2
Green Matrix of Possibilities

Medium	Green Technology					
	Microwaves	Ultrasound	Grinding	Catalysis	MCRs	Biofeedstocks and Renewables
Ionic liquid	×	×	×	×	×	×
sCO₂ (supercritical carbon dioxide)	×	×	×	×	×	×
Solid supports	×	×	×	×	×	×
Solvent-free (neat reagents)	×	×	×	×	×	×
Water	×	×	×	×	×	×

Since these researchers where physical organic chemists with little experience in organic synthesis, their work was really a one-off curiosity. Their work focused on the physical properties of solvents when exposed to microwave irradiation using domestic ovens, and on working out appropriate hardware methodologies in carrying out the experiments. Moreover, they did not follow-up on their work as they clearly did not see any useful application of it at the time. The modern name "multicomponent reactions" can be traced back to reactions discovered in the nineteenth century. Ugi review has written a historical review [49] highlighting such reactions, which include the Strecker synthesis of α-cyanoamines (1850), the Hantzsch dihydropyridine synthesis (1882), the Radziszewski imidazole synthesis reaction (1882), the Riehm quinoline synthesis (1885), the Doebner reaction (1887), the Pinner triazine synthesis (1890), the Hantzsch synthesis of pyrroles (1890), the Biginelli reaction (1893), the Guareschi–Thorpe condensation (1896), and the Thiele reaction (1898). Section 7.4 gives a full listing of these and other modern MCR. Carrying out organic reactions in aqueous media is certainly not new. The discoveries of electrophilic aromatic substitution reactions by Kopp for the nitration of benzene (1856) [50] and Meyer for the sulfonation of benzene (1871) [51–53] in concentrated aqueous acidic media are probably the earliest examples. A pioneer in the use of enzymes such as pig liver esterase and horse liver alcohol dehydrogenase as catalysts for organic reactions is J. Bryan Jones from the University of Toronto [54–61]. In the 1980s, his work was considered cutting edge well before green chemistry got its name in the early 1990s. The idea of using enzymes in organic reactions goes back even further to the work of Fischer [62], Warburg [63], Pottevin [64], and Rosenthaler [65] who coined the term "asymmetric synthesis." Prelog [66–69] applied this technique in the resolution of racemic mixtures of intermediates used in the synthesis of steroids [70] in the late 1950s as shown in the example in the following.

Excellent accounts tracing the early beginnings of enzyme or microorganism usage in organic synthesis may be found in various books [71–73]. An early example of a solvent-free reaction was the synthesis of 2,4-dichlorophenoxyacetic acid, a constituent of Agent Orange, in 1949 [74]. A caveat with respect to the term "solvent-free" is that it refers specifically to no *reaction solvents* being used in carrying out the reaction. It does not automatically imply that no solvents were used in the work-up and purification phases as well. Often, authors will advertise their "green" methodology as solvent-free, but in fact it may have a significant auxiliary E-factor (E-aux) contribution to total E-factor (E-total) from nonreaction solvent auxiliary materials. The use of biofeedstocks and renewables goes back to the 1930s when this kind of chemistry was known as *chemurgy*. This will be discussed later in Section 9.6. Mechano-chemistry [75–84] is the broad title encompassing various techniques called ball milling, grindstone chemistry, and mortar and pestle chemistry. The earliest account of this technique goes back to Theophrastus of Ephesus (371–286 BC) who ground cinnabar with copper in a mortar and pestle, which resulted in the production of quicksilver according to the redox equation: HgS (cinnabar) + Cu → Hg (quicksilver) + CuS [80]. The father of modern mechanochemistry, Matthew Carey Lea, discovered in 1892 that silver halides could be decomposed by grinding them with mercury halides in a mortar and pestle according to AgCl + HgCl → Cl_2 + Ag(Hg) [75–77]. Finally, the technique of ultrasound or sonochemistry was pioneered by Loomis in 1927 as a possible way to activate molecules [85–91].

At its core, green chemistry is all about decision making regardless of the chemistry technique used. However, before any decision can be made among several options, some kind of standardized metrics and ranking analysis is required. No decision can be made without a thorough metrics analysis. Any decisions made without the benefit of metrics are meaningless. The current challenge is to convince chemists to use them routinely as part of their day-to-day work so that they are able to make credible claims of "greenness" that can be substantiated quantitatively. This is the chief aim of this book. The more false claims are made, the less credible will be the field of green chemistry and this backlash will play into the hands of skeptics who will readily take advantage of the situation. When assessing material efficiency there are three categories of metrics:

Category A: Metrics based on counting:
Number of reaction stages (N); number of reaction steps (M); number of input materials (I), degree of convergence (delta), degree of asymmetry (beta), hypsicity index (HI), oxidation length (|UD|), number of target bonds per reaction step (B/M), and number of sacrificial reactions per reaction stage (f(nb)).

Category B: Metrics based on molecular weights:
Atom economy (AE); fraction of sacrificial reagents not ending up in the target molecule (f(sac)); and molecular weight building up parameter (μl).

Category C: Metrics based on mass:
Kernel reaction mass efficiency (RME) and mass of kernel waste.

When assessing energy efficiency there are two metrics: the global energy input required for carrying out an entire synthesis plan, and the fraction of that global energy input that is directed to making the product of interest. All of these metrics are thoroughly discussed in Chapter 5. The word "economy" has been attached to key metrics and concepts: atom economy (Barry Trost), step economy (Paul Wender), and redox economy (Phil Baran). The common thread in all these is time efficiency, that is, how quickly can one reach the target molecule of interest at the same time making the maximum use of all material resources. In economics, time and economy are directly equated with cost. Roger Sheldon, a chemical engineer by training, invented the term global E-factor as a metric [92]. This examines the amount of waste produced relative to the amount of desired product. Both Trost and Sheldon independently promulgated their respective metrics from positive and negative points of view with respect to reaction performance for over a decade, as if they were separate issues. However, both concepts describe the same thing—the link is the balanced chemical equation and law of conservation of mass applied to a chemical reaction put forward by Antoine Lavoisier in 1775 [93]. The mathematical link between the Trost and Sheldon parameters was established in 2005 [94]. It is interesting to note that in Trost's 1991 publication [95], no calculations were performed and no formal mathematical expression was given for the concept. In fact, none of Trost's papers on the subject included atom economy calculations. This apparent paradox is explained by the fact that the atom economies of reactions Trost invented are 100% anyway, so no calculations are required. It is also interesting to note that the 1991 paper has been cited over 1200 times. It can be used as a model paper to teach students the importance of coining names and associating those coined phrases with a scientist's name by the force of repetition especially by other authors citing that scientist's work in the literature. Trost was awarded the Presidential Green Chemistry Challenge Award in 1998, which has the distinction of being the only such award given to a concept or idea. The citation for the award is as follows [96].

The general area of chemical synthesis covers virtually all segments of the chemical industry—oil refining, bulk or commodity chemicals, fine chemicals including agrochemicals, flavors, fragrances etc., and pharmaceuticals. Economics generally dictates the feasibility of processes that are 'practical.' A criterion that traditionally has not been explicitly recognized relates to the total quantity of raw materials required for the process compared to the quantity of product produced or, simply put, 'how much of what you put into your pot ends up in your product.' In considering the question of what constitutes synthetic efficiency, Professor Barry M. Trost has explicitly enunciated a new set of criteria by which chemical processes should be evaluated. They fall under two categories—selectivity and atom economy.

Selectivity and atom economy evolve from two basic considerations. First, the vast majority of the synthetic organic chemicals in production derive from non-renewable resources. It is self-evident that such resources should be used as sparingly as possible. Second, all waste streams should be minimized. This requires employment of reactions that produce minimal byproducts, either through the intrinsic stoichiometry of a reaction or as a result of minimizing competing undesirable reactions, i.e., making reactions more selective. The issues of selectivity can be categorized under four headings—chemoselectivity (differentiation among various functional groups in a polyfunctional molecule), regioselectivity (orientational control), diastereoselectivity (control of relative stereochemistry), and enantioselectivity (control of absolute stereochemistry). These considerations have been readily accepted by the chemical community at large. In approaching these goals, little attention traditionally has been paid to the question of what is required. In too many cases, efforts to achieve the goal of selectivity led to reactions requiring multiple components in stoichiometric quantities that are not incorporated in the product or reagents, thus intrinsically creating significant amounts of byproducts. Consideration of how much of the reactants end up in the product, i.e., atom economy, traditionally has been ignored. When Professor Trost's first paper on atom economy appeared in the literature, the idea generally was not adopted by both academia and industry. Many in industry, however, were practicing this concept without explicitly enunciating it. Others in industry did not consideration the concept since it did not appear to have any economic consequence. Today, all of the chemical industry explicitly acknowledges the importance of atom economy.

Achieving the objectives of selectivity and atom economy encompasses the entire spectrum of chemical activities—from basic research to commercial processes. In enunciating these principles, Professor Trost has set a challenge for those involved in basic research to create new chemical processes that meet the objectives. Professor Trost's efforts to meet this challenge involve the rational invention of new chemical reactions that are either simple additions or, at most, produce low molecular weight innocuous byproducts. A major application of these reactions is in the synthesis of fine chemicals and pharmaceuticals which in general utilize very atom uneconomical reactions. Professor Trost's research involves catalysis, largely focused on transition metal catalysis but also main group catalysis. The major purpose of his research is to increase the toolbox of available reactions to serve these industries for problems they encounter in the future. However, even today, there are applications for which such methodology may offer more efficient syntheses.

For completeness, Appendix A.12 contains the 100% atom economical reactions cited in Trost's landmark publication. All were ring-forming reactions such as cycloadditions and prototropic cycloisomerizations, which satisfy the research interests of leading synthetic organic chemists.

Water is the most often used reagent found in the present synthesis database of 1050+ synthesis plans. This is probably true for all synthesis plans in the literature. As a reagent, water is used either in hydration or protonation reactions where a hydroxyl group or a proton is added to a substrate, respectively. It also happens to be the most used substance in work-up procedures in the form of various aqueous acidic, basic, or salt solutions. There exists, however, a misguided notion that it should not be counted in calculations of E-factors. Perhaps this way of thinking stems from the belief that it is readily available and free for the taking. In turn, this has led to its widespread abuse. By the way, E-factor calculations that deliberately do not count aqueous waste are a good way to artificially lower the real E-factors for syntheses and chemical processes, so really it is a kind of cheating. In the present database of synthesis plans, no special exemption status was granted to water in either kernel or global calculations of waste production. A recent report on the cost of water for various municipalities around the world shows that water is not always a "readily available" commodity [97]. Figures 2.8 through 2.14 show bar graphs for the cost of water in various cities by continent. All values are in U.S. 2009 dollars per 100 U.S. gallons based on a monthly water consumption of 4000 U.S. gallons.

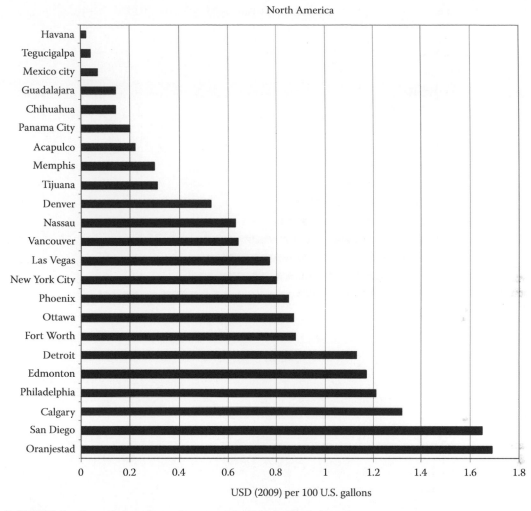

FIGURE 2.8 Cost of water for various municipalities in North America.

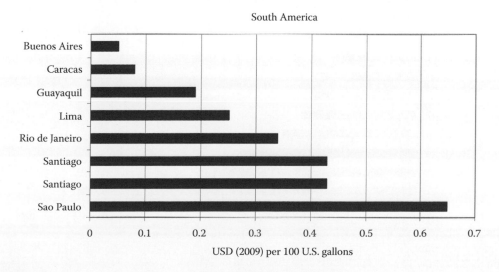

FIGURE 2.9 Cost of water for various municipalities in South America.

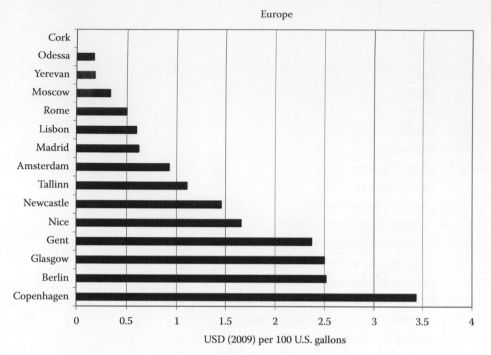

FIGURE 2.10   Cost of water for various municipalities in Europe.

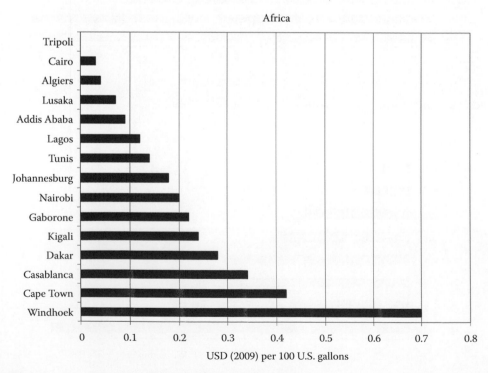

FIGURE 2.11   Cost of water for various municipalities in Africa.

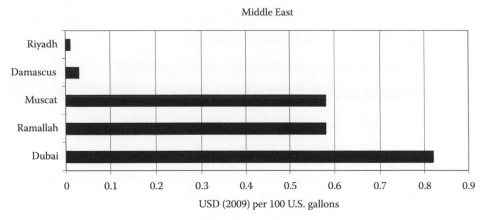

FIGURE 2.12 Cost of water for various municipalities in the Middle East.

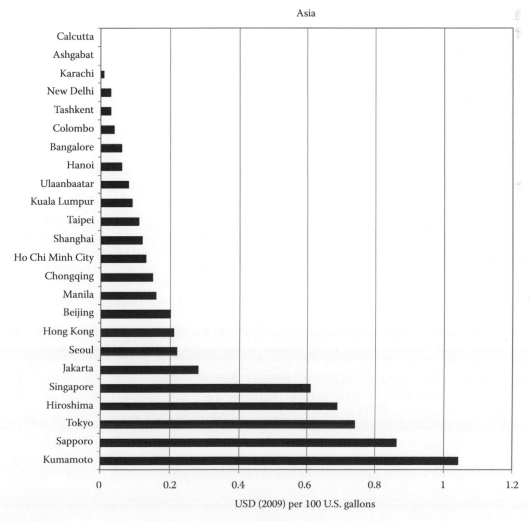

FIGURE 2.13 Cost of water for various municipalities in Asia.

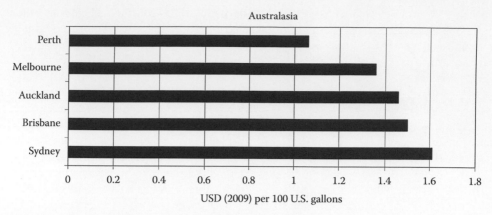

FIGURE 2.14   Cost of water for various municipalities in Australasia.

2.4 GREEN CHEMISTRY TRENDS IN INDUSTRY

The first chemical industry to get on board implementing green chemistry principles was the pharmaceutical industry since it was fingered as the one producing the highest E-factors [92]. Despite the advances made in the last 20 years, the same challenges that confronted green chemistry at its infancy still persist today. Tucker, a former employee of Pfizer and the now defunct Bioverdant, has discussed in two short essays the meaning and purpose of green chemistry as well as the current state of affairs in adopting green chemistry thinking in the whole of the chemical industry [98,99]. He suggests that we have surmounted the first hurdle, which is the general acceptance of the philosophy of green chemistry. The second mountain to climb is sustainability. He observes that the application of the 12 green principles has not been widely adopted in every academic and industrial chemistry endeavor, and so the road to climbing the sustainability mountain will not be as straightforward as previously thought. He has identified three areas where opportunities exist to improve the situation: (a) increasing support and engagement of business and academic leaders, (b) entrenching green chemistry education and technical expertise into every facet of the university chemistry curriculum, and (c) "evolving toward proactive and pragmatic regulatory policies." In other words, the subject has to become more rigorous and focused as any area of applied science research. All chemists need to be convinced that practitioners of green chemistry are the most competent of chemists since they know their subject inside out and they can demonstrate innovation with respect to optimization in all its forms. The greatest convincing will be needed among chemists rather than between chemists and other professionals. Increased investment in basic and inspirational research and increased intellectual value need to be assigned to green chemistry endeavors. From the industry side, the practice of outsourcing erodes the drive for efficiency and true sustainability. From the academic side, those who come second in improving existing synthesis plans and processes are not considered novel or even worthy as those who make the initial discovery announcements in the literature. From a practical perspective it is nearly impossible to come up with an optimized or sustainable plan or process with the first hit. Optimization is necessarily an iterative exercise, and so the eventual success of finding a sustainable solution will be on the back of many failures. Those failures have value, and it is this value that granting agencies and investors need to understand. All of science, not just green chemistry, progresses by fault. The models of first-to-the-post getting all the credit from the academic side and high profit margins in every quarterly report from the industrial business side must be eliminated if any modest progress toward the "summit of sustainability" is to be made.

A recent survey of green chemistry trends over the period 1983–2001 in the U.S. patent literature has been published [100]. Over 3200 patents on some aspect of green chemistry were documented with the following demographics: 91% were assigned to companies, universities, and government agencies; 65% came from the United States; 24% came from Europe; 8% came from Japan; and 2% came from the rest of the world. The leading companies in green chemistry patents in the chemicals sector were in order of rank: Proctor & Gamble Company, Bayer AG, BASF, EON AG, Henkel KGAA, Dow Chemical Company, Imperial Chemical Industries PLC, Dow Corning Company, Montedison SPA, Crompton Corporation, E.I. DuPont De Nemours & Company, Rohm & Haas Company, and International Specialty Products. The leading companies in green chemistry patents in the pharmaceuticals sector in order of rank were: GlaxoSmithKline and Novartis AG. A green chemistry relative activity index (AI) was devised as a measure of how various industry sectors actually utilized green chemistry thinking and innovation in their work. This index is calculated by dividing the fraction of green chemistry patents in a given sector by the fraction of all patents in that sector. For example, between 1992 and 2001, there were 43,975 patents issued in the chemicals sector. This number represents 38.3% of all patents granted over this period. Over the same period, 674 green chemistry-related patents were issued in the chemicals sector. This number represents 40.9% of all green chemistry patents granted. The AI value is therefore 40.9/38.3 = 1.07. Sectors with high AI values indicate that green chemistry practices and innovation rank highly in those sectors. Table 2.3 summarizes the AI values for various sectors in the chemical industry. Interestingly, the forest and paper products sector was found to have the highest AI; however, the authors were unable to offer an explanation for this. It may be an artifact of Boolean searches.

The findings in Table 2.3 and the conclusions of the patent survey bear out most of the arguments and observations made by Tucker. The authors made the following points:

a. The AI value for the pharmaceutical industry is significantly lower than for other chemical sectors even though pharmaceutical companies invest three to four times more in research and development than other chemical industries.
b. The higher AI value for the chemical sector suggests that environmental, health, and safety issues matter more to this sector both from public image and profitability points of view.
c. The overall low values for AI across various sectors show that benign chemistry is not yet a primary focus of the chemical industry that produces new products and is highly regulated.
d. Progress in green chemistry technologies is currently slow and is not moving fast enough "toward greener chemistry-based process and manufacturing strategies."

TABLE 2.3

Summary of Green Chemistry Activity Index (AI) for Various Sectors in the Chemical industry

Chemical Industry Sector	AI
Forest and paper products	2.95
Food, beverages, and tobacco	1.22
Chemicals	1.07
Energy	1.05
Electrical	0.96
Health care	0.78
Pharmaceuticals	0.74
Instrument and optical	0.38

On a positive note several examples of green chemistry applications and innovation in chemical manufacturer were recently highlighted by Manley, Anastas, and Cue [101]. These included

a. Semiconductor devices manufactured with condensed CO_2
b. Boat paint without tin (Sea-Nine® versus bis(tributyltin)oxide)
c. Fire extinguisher foam without freons (use of biodegradable foams)
d. Dry cleaning without perchloroethylene (use of supercritical CO_2)
e. Car coatings without lead (use of epoxy electrocoat—e-coat)
f. Lumber without arsenic (use of bivalent copper complex with quaternary ammonium ions)
g. Metal fluid fluids without depleting resources (use of bio-based oils and surfactants)

The authors highlighted the importance of metrics in the subject of green chemistry by pointing out that (a) a set of metrics should be developed to address all 12 green chemistry principles, (b) such a set of metrics should be easy to apply, (c) the metrics should measure progress as well as point chemists in the "green" direction, (d) the metrics should inform and provide momentum for both chemists and corporate managers, and (e) sustainability metrics are valuable since they focus on life cycle assessment and are thus better able to capture a broad evaluation of sustainability.

Concluding recommendations made by these authors, consistent with those of previously cited authors in this discussion, include the following:

a. Green chemistry thinking should begin in the early stages of product development to guarantee enhanced product value throughout its life cycle.
b. The development, dissemination, and application of metrics to all chemical processes is vital to move toward "greener" chemical processes.
c. The high amounts of data generated by metrics analysis need to be managed efficiently to avoid bottlenecks in their interpretation and use.
d. Information transfer from suppliers to manufacturers to end users needs to be made in easy-to-understand terms so that everyone in the chain is on the same page.
e. Green chemistry needs to be made a priority in research and development and funding agencies to guarantee that the drive to develop novel green alternatives to existing chemical syntheses and processes will be purposeful, successful, and profitable.
f. Green chemistry needs to be incorporated into science education since this is the engine that supplies future generations of talented people who will acquire the necessary skills to meet the goals of innovation and sustainability.

An interesting idea of labeling industrially important compounds with "green chemistry" metrics labels alerting buyers and others that those chemicals were made in an environmentally responsible manner may help to convince people of the merits of this endeavor. Imagine the impact this would have if pharmaceuticals were labeled as such. Of course, an independent international governing body such as the International Union of Pure and Applied Chemists, made up of academic and industry members would have to be set up to oversee that all such claims are verifiable and legitimate. This would necessarily require international standards in deciding what metrics were selected and how they should be measured.

Recent examples of pharmaceutical industry innovations recognized by the United States Environmental Protection Agency (EPA) Presidential Green Chemistry Challenge Awards under the greener synthetic pathways category include sitagliptin (Merck, 2006); aprepitant (Merck, 2005); sertraline (Pfizer, 2002); ganciclovir (Roche Colorado Corporation, 2000); and ibuprofen (Boots-Hoechst-Celanese, 1997).

Aprepitant

Ganciclovir

Ibuprofen

Sertraline

Sitagliptin

In this section, it will be shown that metrics are an indispensable tool in justifying the claims made by companies. These examples showcase the power of using metrics properly and it is hoped that this preview will convince the reader that they do reinforce the chemical advances made in the manufacturing of these and other important compounds listed in the synthesis database examined in this book. If not done already, it is suggested that metrics be used by the EPA, and any other similar governing body for that matter, as a key tool to select future worthy candidates for these kinds of awards. The caveat is the disclosure of complete experimental details so that fair and honest assessments can be made. For each case, the citation award claims are itemized and a summary table of the E-factor breakdowns is given (see Tables 2.4 through 2.8). E-kernel refers to waste from by-products, side products, and unreacted starting materials, E-excess refers to waste from excess reagents, and E-aux refers to waste from solvents and other auxiliary materials. Chapter 5 provides a thorough discussion of E-factor calculations and other metrics. Full chemical schemes, graphics, and references for all plans may be found in the accompanying download.

Case I (sitagliptin, an anti-type II diabetic pharmaceutical)

Award citation claims [102]:

- Asymmetric catalytic hydrogenation of unprotected enamines using rhodium salts of a ferrocenyl-based ligand.
- General synthesis of β-amino acids.
- Merck recovers and recycles over 95% of the valuable rhodium.
- Waste reduction by over 80%.
- Completely eliminated aqueous waste streams.
- Elimination of 330 million pounds or more of total waste.
- Elimination of nearly 110 million pounds of aqueous waste.
- Second-generation synthesis creates 220 lb less waste for each pound of sitagliptin manufactured.

TABLE 2.4
Summary of *E*-Factor Breakdown for Various Industrial Syntheses of Sitagliptin Hydrogen Phosphate

Plan	Year	Type	*E*-Kernel	*E*-Excess	*E*-Aux	*E*-Total	Total Mass of Waste (kg/mol)
Merck G3	2009	Linear	7.15	5.24	161.8	174.16	87.9
Merck G2	2005	Convergent	9.4	8.4	239.1	256.86	129.7
Merck G1[a]	2005	Convergent	35.09	>71.72	>1971	>2077.6	>1087

[a] Sitagliptin fumarate.

A comparison of the Merck G2 and Merck G1 plans in Table 2.4 shows a reduction in the *E*-aux contribution from solvents and auxiliary materials of 88%, a reduction in *E*-kernel of 73%, and a reduction in *E*-total of 88%. The difference in *E*-total factors is over 1800 units of waste per unit of sitagliptin produced. These statistics exceed the performance claims made in the citation.

Case II (aprepitant, an antinausea pharmaceutical for patients undergoing chemotherapy)
Award citation claims [103]:

- Shorter synthesis
- Milder reaction conditions
- Reduced energy requirements
- Elimination of all operational hazards encountered in first-generation route including sodium cyanide, dimethyl titanocene, and gaseous ammonia
- New route employs an 80% reduction in raw materials and water usage
- Elimination of approximately 340,000 L of waste per 1,000 kg of aprepitant produced

A comparison of the Merck G2 and Merck G1 plans in Table 2.5 shows a reduction in the *E*-aux contribution from solvents and auxiliary materials of 98%, a reduction of 90% in the *E*-kernel contribution from raw materials, and a reduction in *E*-total of 98%. The difference in *E*-total factors is over 6970 units of waste per unit of aprepitant produced. These statistics exceed the performance claims made in the citation.

Case III (sertraline, an antidepressant pharmaceutical)
Award citation claims [104]:

- Improved safety and material handling
- Reduced energy
- Reduced water usage
- Doubled overall product yield

TABLE 2.5
Summary of *E*-Factor Breakdown for Various Industrial Syntheses of Aprepitant

Plan	Year	Type	*E*-Kernel	*E*-Excess	*E*-Aux	*E*-Total	Total Mass of Waste (kg/mol)
Merck G2	2003	Convergent	3.49	5.98	>101.49	>110.96	>59.3
Merck G1	1998	Convergent	36.35	153.35	>6894.43	>7084.13	>3782.9

TABLE 2.6
Summary of *E*-Factor Breakdown for Various Industrial Syntheses of Sertraline Hydrochloride

Plan	Year	Type	*E*-Kernel	*E*-Excess	*E*-Aux	*E*-Total	Total Mass of Waste (kg/mol)
Pfizer G3	2004	Linear	7.92	55.21	>226.2	>289.76	>99.2
Gedeon Richter	2002	Linear	9.59	42.07	>339.5	>391.19	>133.9
Pfizer G2	1999	Linear	22.34	163.56	>539.7	>725.56	>248.4
Pfizer G1	1984	Linear	103.95	350.3	>4537	>4990.8	>1709

- Three-step sequence in the original manufacturing process was streamlined to a single step.
- Benign solvent ethanol for the combined process.
- Raw material use was cut by 60%, 45%, and 20% for monomethylamine, tetralone, and mandelic acid, respectively.
- Elimination of approximately 310,000 lb per year of titanium tetrachloride.
- Elimination of 220,000 lb per year of 50% sodium hydroxide.
- Elimination of 330,000 lb per year of 35% hydrochloric acid waste.
- Elimination of 970,000 lb per year of solid titanium dioxide waste.

On comparing Pfizer G3 and Pfizer G2 plans, the overall yield increased 2.8-fold, 64% less tetralone and methylamine are used, but 22.5% more (+)-mandelic acid is used. Table 2.6 shows a reduction in the *E*-aux contribution from solvents and auxiliary materials of 58%, a reduction of 65% in the *E*-kernel contribution from raw materials, and a reduction in *E*-total of 60%. The difference in *E*-total factors is over 436 units of waste per unit of sertraline produced.

Case IV (ganciclovir, antiherpes antibiotic)

Award citation claims [105]:

- Guanine triester (GTE) process reduced the number of chemical reagents and intermediates from 22 to 11.
- Overall yield increase of more than 25%.
- 100% increase in production throughput.
- Elimination of 11 different chemicals from the hazardous liquid waste stream.
- Elimination of 1.12 million kg per year of liquid waste.
- Elimination of 25,300 kg per year of solid waste.
- Recycling and reuse four of the five ingredients not incorporated into the final product.

The Roche GTE process has an overall yield of 63.5%, followed by the Merck plan at 23.5% and the Syntex G2 plan at 11.2%. Hexamethydisilazane, propionic anhydride, hydrochloric acid, and one of three equivalents of sodium propionate are sacrificial reagents in the GTE process. On comparing the Roche and Syntex G2 plans, Table 2.7 shows a reduction in the *E*-aux contribution from solvents and auxiliary materials of 77%, a reduction of 75% in the *E*-kernel contribution from raw materials, and a reduction in *E*-total of 81%. The difference in *E*-total factors is over 374 units of waste per unit of ganciclovir produced.

Case V (ibuprofen, nonsteroidal antiinflammatory pharmaceutical)

Award citation claims [106]:

- 80% atom utilization in new process compared to 40% in old process.
- Three catalytic steps in new process compared to six stoichiometric steps.

TABLE 2.7

Summary of *E*-Factor Breakdown for Industrial Syntheses of Ganciclovir

Plan	Year	Type	*E*-Kernel	*E*-Excess	*E*-Aux	*E*-Total	Total Mass of Waste (kg/mol)
Roche	1995	Linear	5.59	3.37	>80.79	>89.79	>22.9
Syntex G2	1983	Convergent	22.33	>95.76	>346.02	>464.11	>118.3
Merck[a]	1988	Convergent	9.41	51.07	>1635.24	>1695.72	>462.9
Syntex G1	1981	Convergent	31.97	>229.82	>1811.21	>2073.01	>528.6
Syntex G3	1988	Convergent	30.70	112.76	>2316.78	>2460.23	>627.4

[a] Ganciclovir hydrate.

TABLE 2.8

Summary of *E*-Factor Breakdown for Industrial Syntheses of (±)-Ibuprofen

Plan	Year	Type	*E*-Kernel	*E*-Excess	*E*-Aux	*E*-Total	Total Mass of Waste (kg/mol)
Upjohn	1977	Linear	>4.78	1.03	>45.9	>51.7	>10.7
Hoechst-Celanese	1988	Linear	1.72	>1.17	>111.9	>114.79	>23.6
Hoechst-Celanese[a]	1988	Linear	1.72	>1.17	>101.41	>104.30	>21.8
Boots	1968	Linear	8.59	>15.42	>96.52	>120.53	>24.8
DuPont	1985	Linear	4.67	61.83	>150.1	>216.61	>44.6

[a] Recycling HF catalyst in step 2.

- Virtually all starting materials are either converted to product or reclaimed by-product, or are completely recovered and recycled in the process.
- Elimination of large volumes of aqueous wastes.
- Anhydrous hydrogen fluoride catalyst/solvent is recovered and recycled with greater than 99.9% efficiency.

The Hoechst-Celanese process has an atom economy of 77%, whereas the original Boots process has an atom economy of 17%. Starting from isobutylbenzene, the Hoechst-Celanese plan involves three linear steps, whereas the Boots plan has four. It is difficult to rank the plans for this target molecule because none of the plans disclose all material consumption. This example is a good case illustrating the frustration in getting the necessary information to make informed decisions about levels of "greenness" of synthesis plans. Perhaps this may have been because it was the first commercial process to be scrutinized in terms of green chemistry principles and so the companies involved were cautious in how much information they disclosed in the early years of green chemistry. It is interesting to note that the details of the Boots process were disclosed in the single French patent issued, but in none of the several British patent documents. The only certainty is that the Hoechst-Celanese process is the most atom economical. Beyond that, it cannot be said that it definitely out-competes the Upjohn plan. Recycling hydrogen fluoride makes a marginal impact on the total *E*-factor as shown in Table 2.8.

REFERENCES

1. Victor, P.A. *Managing without Growth: Slower by Design, Not Disaster*, Edward Elgar Publishing Inc.: Vermont, New England, U.K., 2008.
2. Sherlock, J.P.; Poliakoff, M.; Howdle, S.; Lathbury, D. *Can. Chem. News* 2009, *61*(3), 16.
3. The Chemical Engineer. http://www.tcetoday.com (accessed on June 2010).
4. Poliakoff, M.; Howdle, S.; Sherlock, J.P. *Chem. World* 2009, *6*(1), 42.
5. Cornforth, J. *Chem. Brit.* 1975, *11*, 432.
6. McMurry, J. *Organic Chemistry*, 3rd edn., Brooks/Cole Publishing Co.: Belmont, CA, 1992.
7. Fessenden, R.J.; Fessenden, J.S. *Organic Chemistry*, 4th edn., Brooks/Cole Publishing Co.: Belmont, CA, 1990.
8. Morrison, R.T.; Boyd, R.N. *Organic Chemistry*, Allyn & Bacon: Boston, MA, 1987.
9. Finar, I.L. *Organic Chemistry*, 6th edn., Longman: London, U.K., 1973.
10. Wade, L.G. Jr. *Organic Chemistry*, 5th edn., Pearson Education: Upper Saddle River, NJ, 2003.
11. Ege, S.N. *Organic Chemistry*, D.C. Heath & Co.: Lexington, MA, 1989.
12. Clayden, J.; Greeves, N.; Warren, S.; Wothers, P. *Organic Chemistry*, Oxford University Press: Oxford, U.K., 2001.
13. Brown, R.F. *Organic Chemistry*, Wadsworth: Belmont, CA, 1975.
14. Solomons, T.W.G. *Organic Chemistry*, 6th edn., Wiley: New York, 1996.
15. Carroll, F.A. *Perspectives on Structure and Mechanism in Organic Chemistry*, Brooks/Cole Publishing Co.: Belmont, CA, 1998.
16. Degering, E.F. *An Outline of Organic Chemistry*, 5th edn., Barnes & Noble, Inc.: New York, 1947.
17. Fieser, L.F.; Fieser, M. *Organic Chemistry*, 3rd edn., Reinhold Publishing Corporation: New York, 1956.
18. Karrer, P. *Organic Chemistry*, Elsevier Publishing Co., Inc.: New York, 1950.
19. Noller, C.R. *Chemistry of Organic Compounds*, W.B. Saunders Co.: Philadelphia, PA, 1951.
20. Streitwieser, A. Jr.; Heathcock, C.H. *Introduction to Organic Chemistry*, 3rd edn., Macmillan Publishing Co.: New York, 1985.
21. Vogel, A.I. *Textbook of Practical Organic Chemistry*, Longman: London, U.K., 1948.
22. Crum Brown, A. *Rept. Brit. Assoc. Adv. Sci.* 1874, 45.
23. Tang, P. *Org. Synth.* 2005, *81*, 262.
24. Fieser, L.F. *Org. Synth. Coll.* 1955, *3*, 6.
25. Miller, A.; Solomon, P.H. *Writing Reaction Mechanisms in Organic Chemistry*, Academic Press: San Diego, CA, 2000.
26. Andraos, J. *Can. J. Chem.* 2005, *83*, 1415.
27. Kermack, W.O.; Robinson, R. *J. Chem. Soc.* 1922, *121*, 427.
28. Lapworth, A. *J. Chem. Soc.* 1922, *121*, 416.
29. Cornforth, J.W. *Austr. J. Chem.* 1993, *46*, 157.
30. Baeyer, A.; Villiger, V. *Chem. Ber.* 1904, *37*, 2848.
31. Graebe, C.; Liebermann, C. *Chem. Ber.* 1869, *2*, 332.
32. Sueki, S.; Okamoto, C.; Shimizu, I.; Seto, K.; Furukawa, Y. *Bull. Chem. Soc. Jpn.* 2010, *83*, 385.
33. Wang, X.H.; Cao, X.D.; Tu, S.J.; Zhang, X.H.; Hao, W.J.; Yan, S.; Wu, S.S.; Han, Z.G.; Shi, F.J. *Heterocyclic Chem.* 2009, *46*, 849.
34. Rong, L.; Han, H.; Wang, H.; Jiang, H.; Tu, S.; Shi, D. *J. Heterocyclic Chem.* 2009, *46*, 149.
35. Zhang, X.Y.; Li, X.Y.; Fan, X.S.; Wang, X.; Qu, G.R.; Wang, J.J. *Heterocycles* 2009, *78*, 923.
36. Prelog, V. *My 132 Semesters of Chemistry Studies*, American Chemical Society: Washington, DC, 1991.
37. Anastas, P.T.; Warner, J.C. *Green Chemistry: Theory and Practice*, Oxford University Press: New York, 1998, p. 11.
38. Linthorst, J.A. *Found. Chem.* 2010, *12*, 55.
39. Amato, I. *Science* 1993, *259*, 1538.
40. Walden, P. *Bull. Acad. Imper. Sci. (St. Petersburg)* 1914, 1800.
41. Braun, V. *Chem. Ber.* 1927, *60*, 2557.
42. Calloway, N.O. *Chem. Rev.* 1935, *17*, 327.
43. Wilkes, J.S. *Green Chem.* 2002, *4*, 73.
44. Earle, M.J.; Seddon, K.R. *Pure Appl. Chem.* 2000, *72*, 1391.
45. Gedye, R.N.; Smith, F.; Westaway, K.; Ali, H.; Baldisera, L.; Laberge, L.; Rousell, J. *Tetrahedron Lett.* 1986, *27*, 279.

46. Gedye, R.N.; Smith, F.E.; Westaway, K.C. *Can. J. Chem.* 1988, *66*, 17.
47. Gedye, R.N.; Smith, F.; Westaway, K.C. *Educ. Chem.* 1988, *25*, 55.
48. Gedye, R.N.; Rank, W.; Westaway, K.C. *Can. J. Chem.* 1991, *69*, 706.
49. Ugi, I.; Dömling, A.; Hörl, W. *Endeavour (New Series)* 1994, *18*(3), 115.
50. Kopp, H. *Ann. Chem.* 1856, *98*, 367.
51. Ador, E.; Meyer, V. *Ann. Chem.* 1871, *159*, 1.
52. Ascher, M.; Meyer, V. *Chem. Ber.* 1871, *4*, 323.
53. Meyer, V.; Michler, W. *Chem. Ber.* 1875, *8*, 672.
54. Jones, J.B.; Beck, J.A. *Technol. Chem. NY* 1976, *10*, 107.
55. Jones, J.B. *Tetrahedron* 1986, *42*, 3351.
56. Jones, J.B. *Pure Appl. Chem.* 1990, *62*, 1445.
57. Jones, J.B. *Can. J. Chem.* 1993, *71*, 1273.
58. Lee, T.; Sakowicz, R.; Martichonok, V.; Hogan, J.K.; Gold, M.; Jones, J.B. *Acta Chem. Scand.* 1996, *50*, 697.
59. DeSantis, G.; Jones, J.B. *Curr. Opinion Biotechnol.* 1999, *10*, 324.
60. Jones, J.B.; DeSantis, G. *Acc. Chem. Res.* 1999, *32*, 99.
61. Jones, J.B. *Ann. NY Acad. Sci.* 1987, *501*, 119.
62. Fischer, E. *Chem. Ber.* 1890, *23*, 389.
63. Warburg, O. *Z. Physiol. Chem.* 1906, *48*, 205.
64. Pottevin, H. *Ann. Inst. Pasteur* 1906, *20*, 901.
65. Rosenthaler, L. *Biochem. Z.* 1908, *14*, 238.
66. Prelog, V. *Ciba Foundation Study Group* 1959, *2*, 79.
67. Prelog, V. *Ind. Chim. Belges* 1962, *27*, 1309 (*Chem. Abs.* 59: 10248).
68. Prelog, V. *Colloquium der Gesellschaft fuer Physiologische Chemie* 1964, *14*, 288 (*Chem. Abs.* 62: 30915).
69. Dutler, H.; Coon, M.J.; Kull, A.; Vogel, H.; Waldvogel, G.; Prelog, V. *Eur. J. Biochem.* 1971, *22*, 203.
70. Prelog, V. US 2833694 (Ciba Pharmaceutical Products, Inc., 1958).
71. Ritchie, P.D. *Asymmetric Synthesis and Asymmetric Induction*, Oxford University Press: Oxford, U.K., 1933.
72. Wong, C.H.; Whitesides, G.M. *Enzymes in Synthetic Organic Chemistry*, Pergamon: New York, 1994.
73. Faber, K. *Biotransformations in Preparative Organic Chemistry*, 2nd edn., Springer-Verlag: Heidelberg, Germany, 1995.
74. Manske, R.H.F. US 2471575 (US Rubber Co., 1949).
75. Lea, M.C. *Am. J. Sci.*, 3rd Series, 1892, *43*, 527.
76. Lea, M.C. *Am. J. Sci.*, 3rd Series, 1893, *46*, 413.
77. Lea, M.C. *Phil. Mag. Series 5* 1894, *37*, 470.
78. Takacs, L. *J. Materials Sci.* 2004, *39*, 4987.
79. Takacs, L. *Bull. Hist. Sci.* 2003, *28*, 26.
80. Fernández-Bertran, J.F. *Pure Appl. Chem.* 1999, *71*, 581.
81. Bayer, M.K.; Clausen-Schaumann, H. *Chem. Rev.* 2005, *105*, 2921.
82. Dushkin, A.V. *Chem. Sustainable Dev.* 2004, *12*, 251.
83. Boldyrev, V.V.; Tkacova, K. *J. Mater. Synth. Process.* 2000, *8*, 121.
84. Takacs, L. *J. Miner. Met. Mater. Soc.* 2000, *52*, 12.
85. Wood, R.W.; Loomis, A.L. *Phil. Mag. Series 7* 1927, *4*, 417.
86. Richards, W.T.; Loomis, A.L. *J. Am. Chem. Soc.* 1927, *49*, 3086.
87. Richards, W.T.; Loomis, A.L. *J. Am. Chem. Soc.* 1927, *49*, 1497.
88. Lash, T.D.; Berry, D. *J. Chem. Educ.* 1985, *62*, 85.
89. Clough, S.; Goldman, E.; Williams, S.; George, B. *J. Chem. Educ.* 1986, *63*, 176.
90. Luche, J.L. *Synthetic Organic Sonochemistry*, Plenum Press: New York, 1998.
91. Henglein, A. *Ultrasonics* 1987, *25*, 6.
92. Sheldon, R.A. *Chem Tech.* 1994, *24*(3), 38.
93. Lavoisier, A. Mémoire sur la nature du principe qui se combine avec les métaux pendant leur calcination, et qui en augmente le poids. *Mèmoires de l'Académie Royale des Sciences*, 1778, 1775, 520.
94. Andraos, J. *Org. Process Res. Dev.* 2005, *9*, 149.
95. Trost, B.M. *Science* 1991, *254*, 1471.
96. *The Presidential Green Chemistry Challenge Awards Program Summary of 1998 Award Entries and Recipients*, United States Environmental Protection Agency, EPA744-R-98-001, November 1998, p. 2. http://www.epa.gov/greenchemistry (accessed on June 2010).

97. National Geographic Special Issue: *Water—Our Thirsty World*, 2010, *217*(4), 114.
98. Tucker, J.L. *Org. Process Res. Dev*. 2010, *14*, 328.
99. Tucker, J.L. *Org. Process Res. Dev*. 2006, *10*, 315.
100. Nameroff, T.J.; Garant, R.J.; Albert, M.B. *Research Policy* 2004, *33*, 959.
101. Manley, J.B.; Anastas, P.T.; Cue, B.W. Jr. *J. Cleaner Production* 2008, *16*, 743.
102. *The Presidential Green Chemistry Challenge Award Recipients 1996–2009*, June 2009, 744K09002, United States Environmental Protection Agency, p. 36. http://www.epa.gov (accessed on June 2010).
103. *The Presidential Green Chemistry Challenge Awards Program Summary of 2005 Award Entries and Recipients*, June 2005, EPA744-R-05-002, United States Environmental Protection Agency, p. 6. http://www.epa.gov/greenchemistry (accessed on June 2010).
104. *The Presidential Green Chemistry Challenge Award Recipients 1996–2009*, June 2009, 744K09002, United States Environmental Protection Agency, p. 78. http://www.epa.gov (accessed on June 2010).
105. *The Presidential Green Chemistry Challenge Entries and Recipients 2000*, August 2001, EPA744-R-00-001, United States Environmental Protection Agency, p. 5. http://www.epa.gov/greenchemistry (accessed on June 2010).
106. *The Presidential Green Chemistry Challenge Awards Program Summary of 1997 Award Entries and Recipients*, EPA744-S-97-001, April 1998, United States Environmental Protection Agency, p. 2. http://www.epa.gov/docs/gcc (accessed on June 2010).

3 Problems with Literature Reporting of Synthesis Plans

3.1 INTRODUCTION

The chemistry literature is vast and well organized, but unfortunately, in the world of organic synthesis, is fraught with inconsistencies and errors in the reporting of experimental procedures. How can a chemist trust what is documented? If reproducibility of experimental results is challenged, how will this reflect on the whole of chemistry, and to the wider scientific community? This chapter is dedicated to highlighting some of the most nagging problems. Their impact on the development of green chemistry is also discussed.

There are traditionally four kinds of papers that can be published in the academic literature: communications on original research, full papers on original research, review articles, and essay or opinion pieces. In industry, the only publishable document is the patent. The most notorious of these documents with respect to introduced errors, deliberate withholding of important experimental data, or general sloppy reporting are communications and patents. The purpose of communications is getting results out quickly ahead of competitors. It is all about staking claims and marking territory. Effectively, every communication should begin with the words "keep out" or "hands off." In the world of science what counts is being first—one does not have to be thorough, careful, or even correct in presenting all of the facts. Are reviewers going to check an author's experimental results by trying to duplicate them as part of the standard review process? It would be impractical because of the high volume of submissions and increased review time that it would entail, especially nowadays, with the tsunami of papers published annually online in countless journals. So chemists must rely on trusting their colleagues at their word. Over time any glitches will eventually get picked up and sorted out, but not without paying a price, particularly to the overall reputation of the field. The comments of Nobel Laureate John Cornforth [1] on communications are particularly pertinent: "It has become customary to report a synthesis in a preliminary note or in a sequence of notes, and to defer full publication with proper description of intermediates and methods for years or for ever… It is a poor tribute to chemical synthesis to say that it has been pouring a large volume of unpurified sewage into the chemical literature, but that is too near the truth for comfort. I do not know what can be done about this, though I would support a conspiracy of editors to refuse to publish a preliminary note if the full publication from a previous note was still outstanding after an agreed interval." The situation is much worse for patents since in this situation there is monetary revenue at stake, as opposed to gaining personal reputation and pride as in the world of academia. Patents are written by attorneys whose schooling in chemistry and science is not at the same level as chemists themselves. Patents, too, make claims, mark territory, and protect intellectual property. It is easier to understand and accept lack of details, or even complete falsehoods, in patent documents as routine, rather than in journal articles.

The following ubiquitous practices are particularly problematic:

a. Nondisclosure of reaction yields for all reaction steps, especially in communications.
b. Reporting of yields after a stretch of two or more reactions (even up to five steps!—see Myles [2] plan for (−)-discodermolide) even though all intermediates are actually isolated. This makes for severe assumptions in individual reaction yield performance when determining reaction mass efficiencies (RMEs) and E-factors for a plan. Individual reaction

performance within that sequence of steps is taken as the geometric mean. Estimates of lower limits on *E*-factor (or upper limits on RME) are the only reasonable results that can be obtained for such plans.

c. Secondary references to papers on the syntheses of starting materials are not always the right ones. Those papers may have stated that they were used but the actual syntheses of the materials are not given. Often a separate literature search is necessary to track down the true original citations.

d. Authors tend to reference other extraneous work from their own laboratories with the intent to make the reader more aware of their full body of published work rather than be specific on the relevant reference pertaining to how a particular starting material was made. It also is a clever tactic to artificially increase the citation count for the author.

e. Synthetic chemists excessively use acronyms for reagents without bothering to give their full definitions anywhere in their papers, particularly when they are first mentioned. Authors who do this are not interested in educating novices, such as students. Students should interpret this as arrogance. Prof. K.C. Nicolaou should be commended as having the best laid out plans appearing in the chemistry literature from a powerhouse synthetic organic chemistry research group since all reaction yields and all acronyms for reagents are properly defined.

f. Unnecessarily changing the aspect of chemical structures and thus making visualization and tracking of what bonds are made and broken in reactions very tedious and tiresome. These changes of aspect are imposed on authors by editors and typesetters of journal manuscripts who are more interested in saving space on a page, rather than making things clear for an interested reader to follow. Maintaining the structural aspect is key, as will be demonstrated later in this book, on drawing target bond maps, comparing several target bond maps corresponding to various synthesis plans for the same target molecule, tracking oxidation number changes in bond forming reactions, and in determining the molecular weight fraction of sacrificial reagents used in a synthesis plan.

The following pairwise examples of graphical depictions of chemical reactions illustrate item (f). The reader will find that the second entry in each example is far clearer than the first.

Example 1 [3]

versus

Example 2 [3]

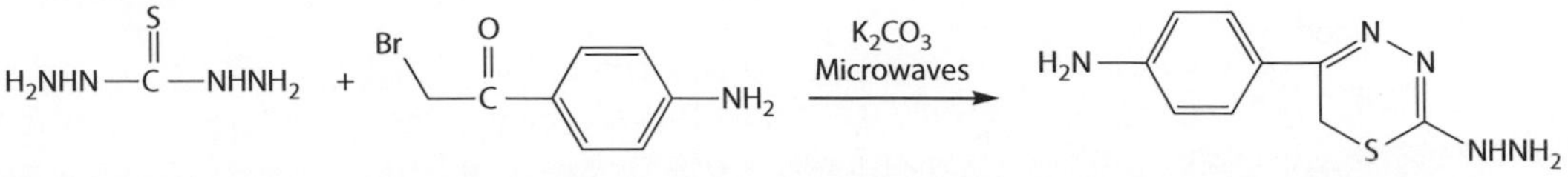

versus

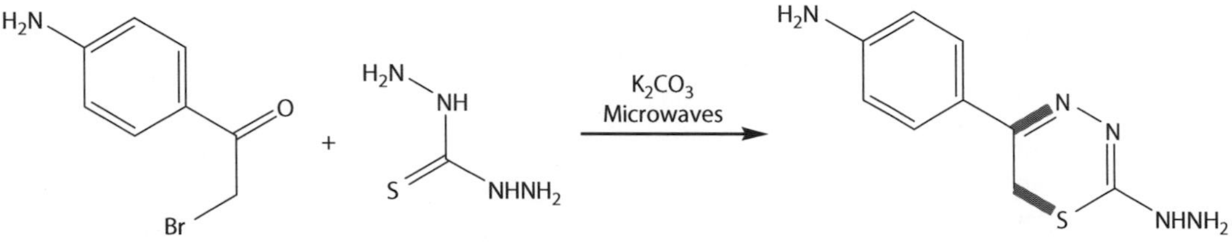

Example 3 [4]

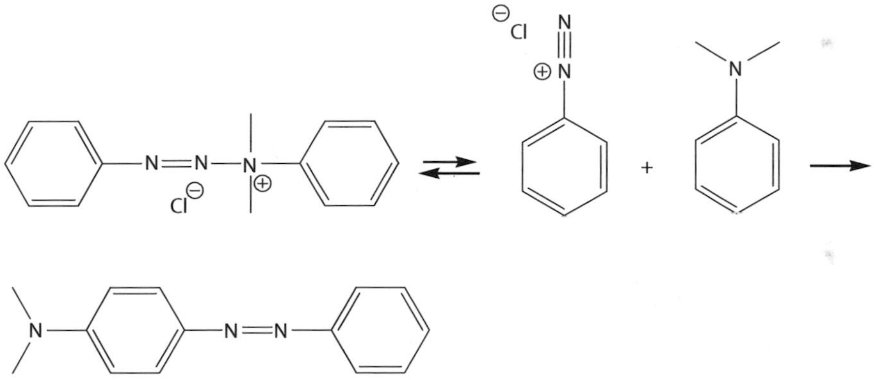

versus

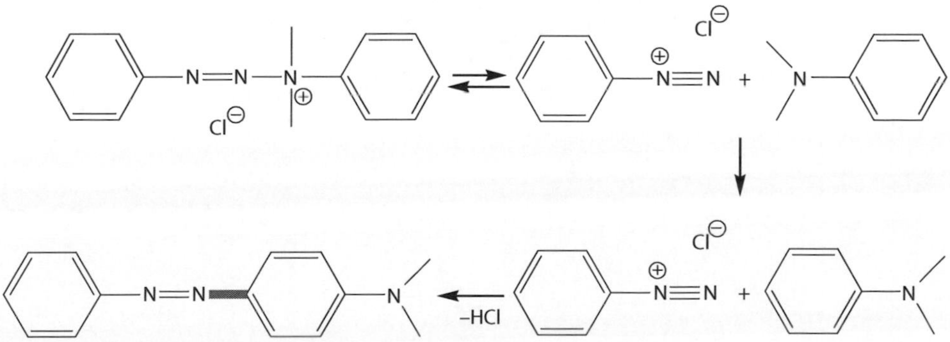

Example 4 [4]

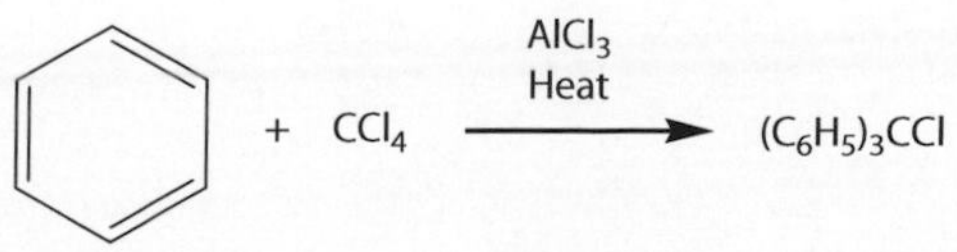

versus

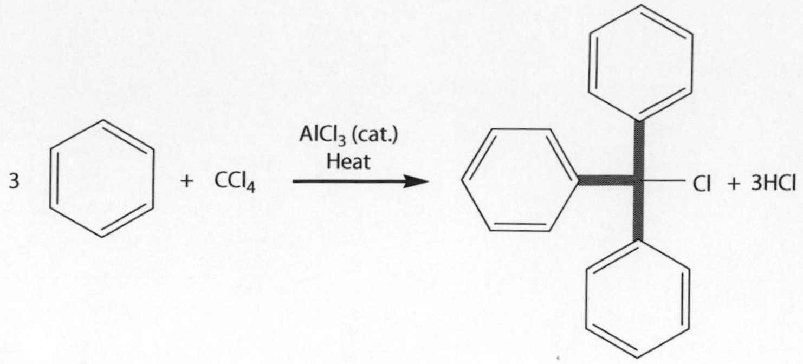

Example 5 [4]

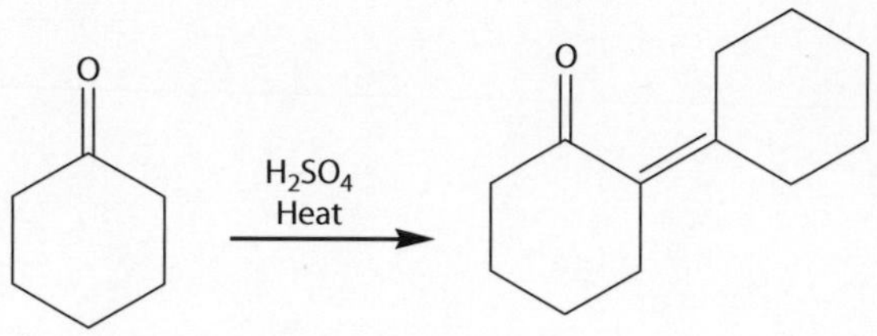

versus

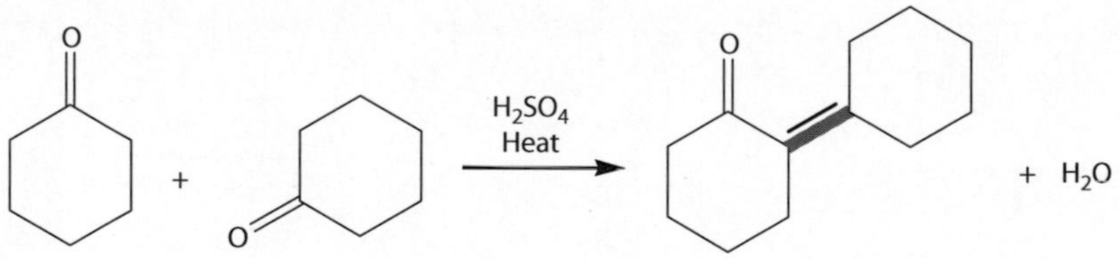

Example 6 [4]

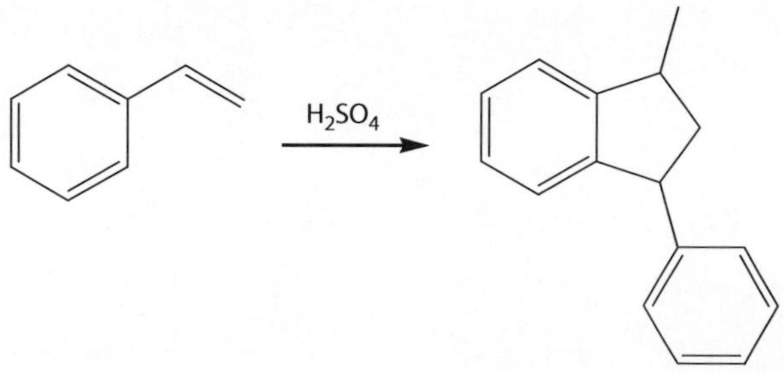

versus

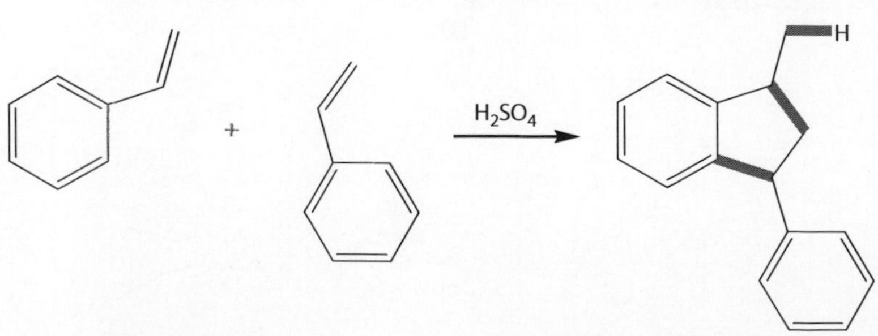

Example 7 [5]

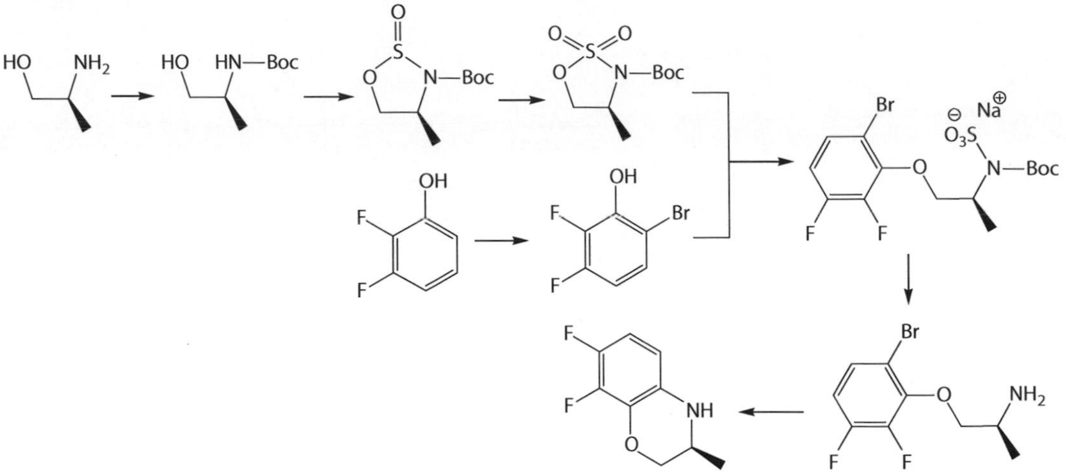

versus

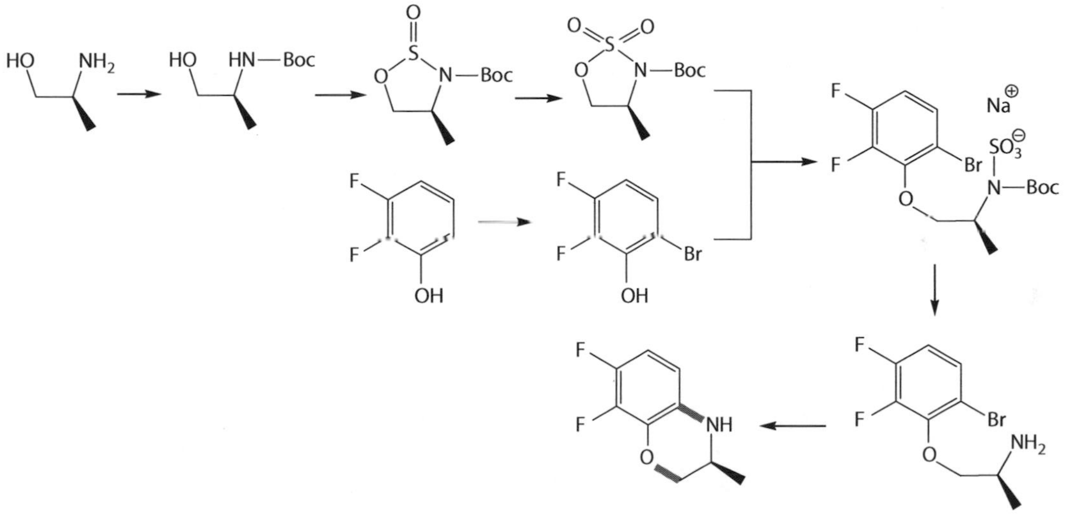

Example 8 [6]

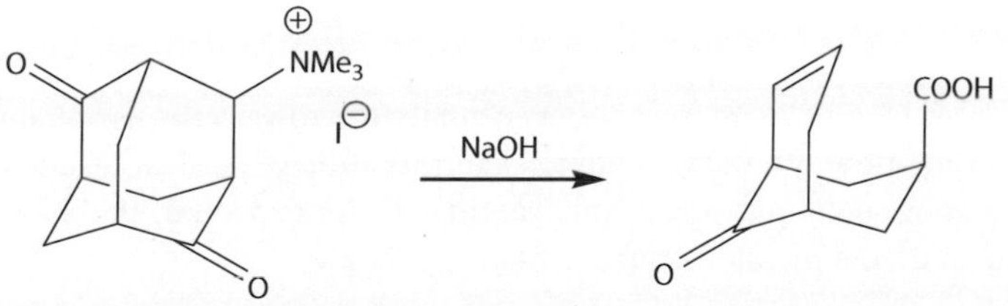

versus

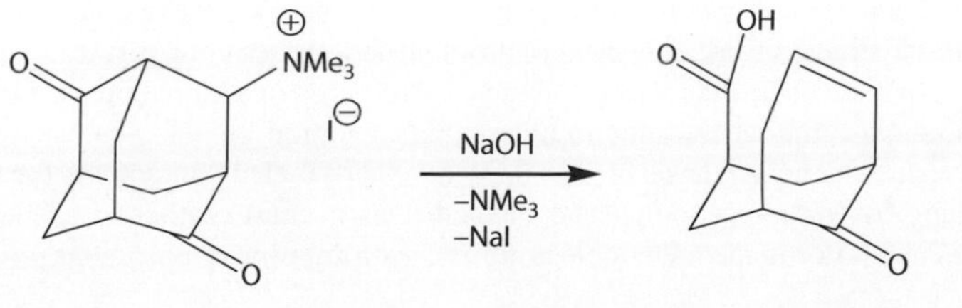

or

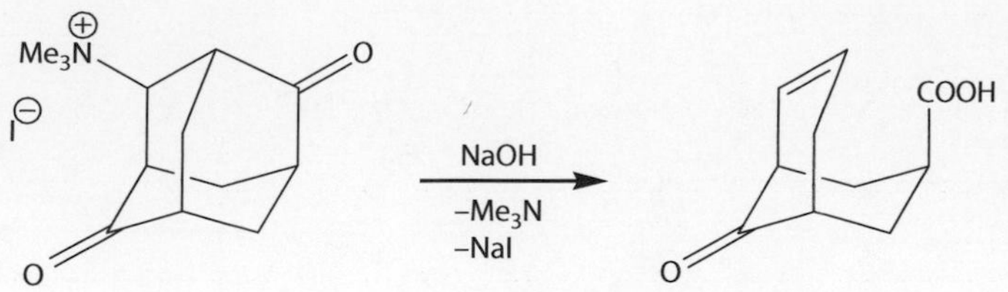

Example 9 [7]

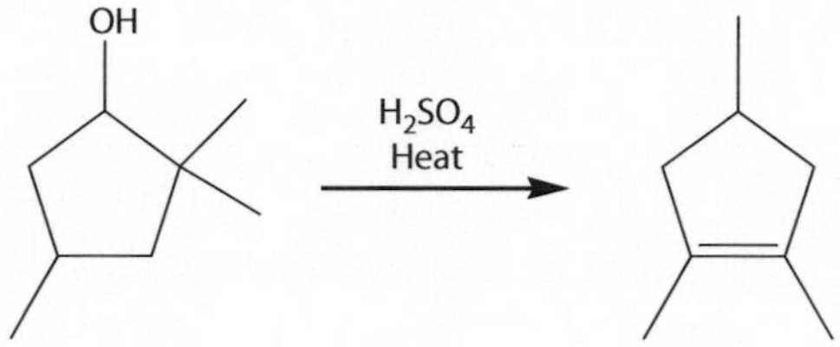

versus

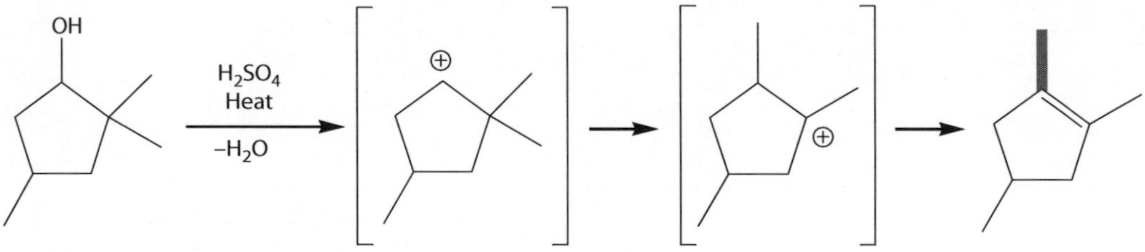

or

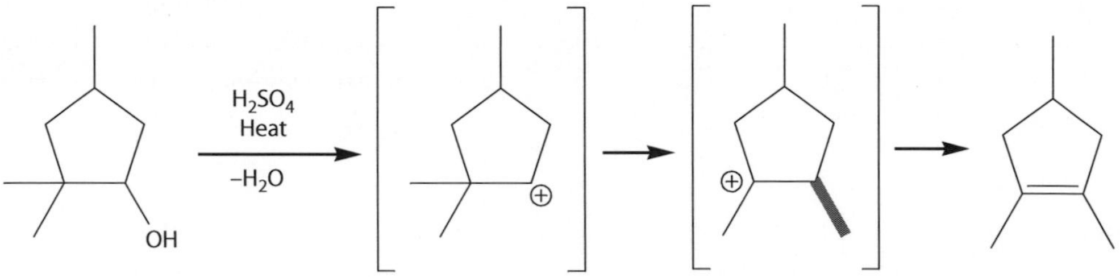

There are still other problems. There is a bias, even contempt, toward the use of any numerical analysis in the world of organic synthesis beyond the determination of percent reaction yield. Following are some reviewer comments that reflect this sentiment. As expected, the anonymous academic referee is more vitriolic than the industrial chemist.

> "It just does not make any sense to apply the so-called "Green Chemistry Metrics" to the academic total synthesis of complex natural products. These syntheses are artworks, major contributions to culture. In Woodward's times syntheses were only used to proof structural assignments. Rapid progress in methodology has now enabled chemists to access complex structures in a more concise fashion, but still not with any ambition to use the schemes for an industrial production. In many cases, synthesis are elaborated only to prepare for the synthesis of new derivatives or labeled compounds for further biological studies. The major driving forces to undertake expeditions in total synthesis include the development of new strategies and tailor-made methodology for the solution of very particular (usually unexpected)

problems, besides the preparation of specific samples for subsequent research. Thus, all attempts to analyze and compare academic total syntheses must be made with great care. The formal application of pseudo-quantitative descriptors not only violates the context and the motivation which have lead to the elaboration of the different syntheses, it also leads to a confusing and needless picture. The paper under consideration has a negligible scientific impact. This paper will only confuse and frustrate those chemists who have a real interest in the development of organic synthesis and in alkaloid chemistry. Actually, the author comes out as someone who has obviously never participated in a serious enterprise in complex total synthesis. He obviously is not sufficiently qualified to recognize the scope and limitation of his metrics approach. As a self-appointed prophet of green chemistry he would be well advised to focus on such processes, which have at least a minor significance with respect to industrial production or other environmental issues.

I strongly suggest the editors of *Molecules* not to accept this manuscript as the journal would endanger its seriousness within the community of chemists interested in total synthesis on a higher academic level."

- An anonymous academic chemist

"The paper contains a very comprehensive set of formulae for calculating green chemistry parameters. They offer a benefit to Green Chemistry experts (rather than your average bench process chemist). But I very much doubt if your average chemist would have the time or perhaps the desire to go near this because it just plain looks far too complex and it actually is, even with the spreadsheets. If progress in Green Chemistry metrics and analysis is to be made at a grass routes level, it needs to be driven down to the scientists and engineers. The spreadsheets are helpful, but in general, still not where we need to be."

- An anonymous industrial chemist

It is amusing to see that chemists sometimes cannot even determine the step count of their own plans correctly! The propensity for making errors increases the more they contract the reporting of reaction yields as mentioned earlier in item (b). There is also a bias toward using computer-assisted synthesis programs—as if they will degrade the task of organic synthesis from being a creative art form to some totally predicable and routine number crunching exercise. Synthetic chemists plan their syntheses according to (a) novelty—showcasing the widespread utility of a key reaction they invented, developing the scope of that reaction and then selecting key natural product target structures where it may be used; (b) the desire to be first to synthesize a target structure (for proof of structure)—prowess and competition are the drivers; (c) the desire to develop a synthesis that is different from prior published plans to the same target molecule; and (d) the desire to develop a plan with the least number of steps and having reactions in which the number of target bond forming events is maximized. There is a bias against ranking plans by any criteria. Here is what a senior professor at the University of Toronto had to say about ranking: "Syntheses should be judged like a judging score in a skating competition. The math is beyond the synthetic community. After all, synthesis is a demonstration of prowess, not a practical exercise."

Obviously, no one wants to be fingered as coming up with the worst plan for a target molecule. Ranking is favored by those coming at the top of the list! This is somewhat like the reaction university presidents have on annual national university rankings by popular news magazines. The most critical are the ones from institutions consistently coming at the bottom of such surveys. One must keep in mind that poor performing plans are needed to invent better ones. The good ones stand on the shoulders of the less efficient ones. No good plan can arise without having made some mistakes in the past. Optimization is an iterative exercise based on comparative analysis. Therefore, every synthesis plan has value. Moreover, the job is never complete as there is always more room for improvement no matter how great the progress. Once all plans are analyzed and ranked for a given target molecule according to some set of criteria, what one would like to see is that the next new plan demonstrates clear improvements over the prior published ones otherwise there is no merit.

In this regard there is the humility factor associated with ranking. Many claims of "conciseness," "efficiency," or "greenness" are made on the basis of one or two criteria like overall yield, or step count, or one of the 12 green principles. Thorough metrics analyses often diminish claims made by authors that are based on a narrow view.

3.2 MISSING INFORMATION IN PLANS

Experimental procedures are often lacking any one or more of the following items:

- Masses of drying agents used (magnesium or sodium sulfate)
- Masses of silica gel or alumina used in flash chromatography
- Masses of chromatographic solvents used as eluents
- Masses of catalysts used
- Masses of reaction solvents
- Masses of wash solvents
- Masses of recrystallization solvents
- Masses of extraction solvents
- Masses of decolorizing charcoal
- Masses of celite
- Mass of dihydrogen used in hydrogenation reactions
- Mass of dioxygen used in oxidation reactions
- Mass of any gaseous reagent

Assumptions used by the Environmental Assessment Tool for Organic Synthesis (EATOS) [8] program when masses of auxiliary materials are not reported in experimental procedures are as follows:

1. Extraction from aqueous media using an organic solvent: 300 mL solvent per L aqueous medium (multiplying factor = 0.3)
2. Aqueous washes of organic solvent: 300 mL water per L organic solvent (multiplying factor = 0.3)
3. Brine: 100 mL brine per L organic solvent (multiplying factor = 0.1)
4. Drying agents: 20 g per L organic solvent (multiplying factor = 0.02)

Unfortunately, these too are limiting as they do not cover all of the items listed earlier.

3.3 MISTAKES IN REPORTED YIELDS AND AMOUNTS OF MATERIALS USED

Faith's book [9] contains many mistakes in amounts of materials. Unfortunately, these have been replicated in the later metric edition [10]. However, these books are the best available references for disclosing details of industrial procedures for making first- and second-generation feedstocks. In compiling the present synthesis database, the syntheses of the following industrial commodity chemicals were assessed using the radial pentagon spreadsheet algorithm (PENTAGON): acetic anhydride, adipic acid, anthraquinone, benzoic acid, bisphenol A, ε-caprolactam, diethyl phthalate, epichlorohydrin, ethylene oxide, glycerol, hexamethylenetetramine, maleic anhydride, melamine, methyl methacrylate, pentaerythritol, phenol, phthalic anhydride, resorcinol, urea, vinyl acetate, and vinyl chloride. All errors were trapped and corrected. This spreadsheet is discussed in detail in Sections 5.2 and 6.2. Reaction yields are supposed to be calculated with respect to the limiting reagent used; however, some authors calculate it with respect to the substrate of interest regardless of whether it is a limiting reagent or not. It is easy to see that such incorrect calculations can result in over- or underestimation of true yield performance. The reader is directed to two recent commentaries on the truthfulness and accuracy of reaction yields and other performance variables reported in the chemistry literature [11,12].

3.4   MISMATCHES IN REPORTED REACTION YIELDS

This case is illustrated by the Jung [13] plan for (–)-kainic acid. The reaction yields reported in the experimental section versus those appearing in the schemes differ by ±5%. Sometimes the yields were higher in the schemes, other times not. In analyzing this plan's material efficiency, the details given in the experimental section were used.

3.5   TACTICS TO ARTIFICIALLY AMPLIFY REACTION PERFORMANCE

When reaction performance is low, that is, less than 50% yield, chemists are reluctant to disclose this directly. Two tactics are used to camouflage poor reaction performance. The first is to report percent conversion and percent selectivity instead of percent yield. The multiplicative product of these fractions is equal to the percent yield (see Section 5.6, for a proof). For example, a reaction in which 71% of the substrate is converted to products and results in a selectivity of 71% to the desired product implies a 50% yield performance toward the desired product. A second tactic is to report the yield based on recovered starting material, often abbreviated as borsm, but defined in a very few papers. In this case, the "yield" includes the desired target product plus unreacted starting material. If we begin with x moles of starting material and after a time t we produce p_1 moles of product P_1 and have remaining r moles of starting material, then the percent yield of product based on recovered starting material is given by

$$\%\text{yield}_{(\text{borsm})} = \left(\frac{p_1 \text{MW}_{P_1} + r\text{MW}_{\text{SM}}}{x\text{MW}_{P_1}} \right) \times 100 \tag{3.1}$$

where it is assumed that the starting material (SM) is the limiting reagent and the variables designated as MW refer to molecular weights of the respective chemical species. Clearly, this numerical value is larger than that for the proper percent yield given by

$$\%\text{yield} = \left(\frac{p_1 \text{MW}_{P_1}}{x\text{MW}_{P_1}} \right) \times 100 \tag{3.2}$$

3.6   REPORTING OF CLASSICAL RESOLUTIONS

Synthesis plans containing a classical resolution step have a sequence of steps as shown in the following.

In step X, a racemic product in the ratio 1:1 is isolated. In step X + 1, a chiral auxiliary is added to produce a diastereomeric salt that preferentially crystallizes out of solution and is isolated. In step X + 2, this salt is exposed to acidic or basic aqueous solution to liberate the free optically pure product that is finally collected. In this kind of resolution, half the mass of racemic product collected in step X is destined for waste. Therefore, the percent reaction yield for step X + 1 should be less than or equal to 50%. The reaction yields for steps X and X + 2 can be as high as 100%. The problem is that papers will report the percent yield for step 2 with a fraction that exceeds 50%. How can this be? What authors have done is the following sleight of hand. The percent yield they report is not with respect to the starting mass of racemic mixture, but with respect to half its mass, that is, the half that will eventually be collected in step X + 2, assuming no losses along the way. Again, chemists have an aversion to reporting any reaction performance having a numerical value below 50%. For example, the Szantay [14] plan for (–)-paroxetine hydrochloride dihydrate involves two classical resolutions. The authors report a yield of 86% for the first resolution instead of the actual value of 43%. A 1:1 mole ratio of racemic amine alcohol to dibenzoyltartrate is used in that step. For the second resolution the authors report a reaction yield of 79%, where the mole ratio of substrate to dibenzoyltartrate used is 1:0.5, not 1:1. The 79% yield is calculated with respect to dibenzoyltartrate as limiting reagent. A value of 35% yield is calculated with respect to the substrate. This artificial doubling of the actual yield performance seems to be commonplace. The following synthesis plans taken from *Organic Syntheses* also apply this practice: (+)-2-octanol [15], (+)-α-phenylethylamine [16], (*S*)(–)-1-phenylethylamine [17], (+)-α-(isopropylideneaminooxy)propionic acid [18], (*S*)-(–)-α-(1-naphthyl)ethylamine [19], (*S*) (+)-binaphthylphosphoric acid [20], (1S,2S)-(–)-1,2-diphenyl-1,2-ethylenediamine [21], (*R*)(+)- BINAP [22], and (R,R)(–)-*N*,*N'*-dimethyl-1,2-diphenylethane-1,2-diamine [23]. The radial pentagons for these cases are shown in Section 6.2. In the present compilation of synthesis plans, whenever classical resolutions were involved the true yield performance was used so that correct computations of overall yield, overall reaction mass efficiency, and overall *E*-factor could be made.

3.7   EXAGGERATED CLAIMS OF EFFICIENCY

A notorious example of mismatches between amounts of materials used and their corresponding mole amounts was found for the Reddy [24] plan for rimonabant. In this paper, the reported number of moles of *every* material used was wrong. Instead of "1 mol of substance X" what the authors actually meant was "1 equivalent of substance X." The giveaway was that the result of dividing the reported mass by the stated number of moles yielded a number that clearly did not correspond to the molecular weight of the substance. The radial pentagon analysis spreadsheet picked up this mistake at once. Imagine if a chemist were to follow these procedures without first checking the correspondence between masses and moles? Having found several discrepancies, which set of figures does one use? Where were the reviewers of *Organic Process Research and Development*? They obviously did not bother to read the experimental section!! Moreover, the paper makes a claim that their new plan is more material and cost efficient than prior published plans. A thorough metrics investigation shows that this claim is not so. In fact, the total *E*-factor for their plan is 438.8, whereas that for the older Sanofi G1 [25] and Sanofi G2 [23] plans were 79.0 and 169.6, respectively.

3.8   GOOD "GREENING" PROGRESSION BUT LITTLE DETAIL
TO SUBSTANTIATE CLAIM

Three plans for isotretinoin show a nice "greening" progression with respect to the kind of chemistry strategy employed. The Roche [26] plan involves a Wittig coupling followed by *cis-trans* isomerization with $Pd(NO_3)_2$ ($E > 239$). The BASF [27] plan involves a Wittig coupling followed by *cis-trans* photoisomerization ($E > 349$). The Abbott [28] plan involves a stereoselective Wittig coupling using $MgCl_2$ catalyst ($E > 253$) with no need for subsequent isomerization. Unfortunately, only lower limit estimates for the total *E*-factor could be made because not all experimental details were given with respect to auxiliary material usage, so it is impossible to verify if the gain made in avoiding the final isomerization step actually translated into an overall more material efficient strategy.

3.9   "READILY AVAILABLE STARTING MATERIALS"

This phrase is used ubiquitously in the organic synthesis literature. Chemists use the phrase to mean commercially available. In fact, commercially available does not mean "readily available" in the sense that it is easy or trivial to make. The most tedious and time-consuming task encountered in compiling the synthesis database for this book was the chasing down in the literature of how such starting materials were actually made. This entailed original literature searching through *Chemical Abstracts* and getting over the hurdles of false leads provided by authors' secondary references. The task was necessary to complete the synthesis analyses properly so that a true picture of waste production could be determined and also to make the plan comparisons and ranking as fair as possible. When one analyzes a plan from a so-called "readily available starting material" it is usually a compound with a fairly advanced structure. If one begins the analysis with this then it artificially reduces the step count and ultimately gives an unfair advantage over other plans. The analogy in amateur or professional sport would be declaring to the world that you won a 100 m sprint race shattering past records, yet you began the race 10 m ahead of the official starting line. This kind of cheating is widespread when making claims of "conciseness" or "efficiency" in publications. Examples of using this tactic found in the present synthesis database are the Trost [29] plan for oseltamivir and the Pfizer G1 [30] plan for sildenafil, where 6-oxa-bicyclo[3.2.1]oct-3-en-7-one and 2-methyl-5-propyl-2*H*-pyrazole-3-carboxylic acid were the advanced starting materials, respectively. The set of starting materials used has another implication. There is the problem of outsourcing portions of synthesis plans to other parties, particularly the early reaction steps of a plan, in order to reduce costs. When this happens there is no control of how the other party makes the required intermediate, particularly the "greenness" of their procedures, unless both parties are in some kind of agreement where both decide an overall green strategy. When a claim is made that an environmentally friendly synthesis was developed for a given target, one should investigate if it pertains to the tail end of the synthesis plan or to the entire range of reaction steps. Often, metrics results reveal disappointments. The fairness of comparisons by metrics analysis is discussed in Sections 4.7, 5.1, and 11.16.

3.10 BIOTRANSFORMATIONS NOT REVEALING REACTION YIELDS IN USUAL WAY

Biotransformations are reactions involving fermentation or chemoenzymatic steps. If a microbe is used it is one enzyme that the microbe produces, which is the active catalyst. Reactions typically take place in aqueous media that contain nutrient broths to keep the microbes alive. If a chemoenzymatic reaction is employed then a particular enzyme extract is used directly without having to deal with a nutrient microbial broth. Instead, the aqueous solvent is carefully controlled by adjusting its pH with appropriate buffers in order to maintain enzyme activity. Upon scale-up, a significant amount of water is used in fermentation reactors, which contributes to the *E*-aux portion of the total *E*-factor. The waste solvent is treated by autoclaving before disposal to destroy any remaining microbes once their activity is spent. There are two main problems in analyzing such reactions: (1) a balanced chemical equation may not be known, and (2) reaction yields are not reported using the usual definition that a chemist would use. As a consequence of not knowing what the balanced chemical equation is with appropriate stoichiometric coefficients, the concept of a limiting reagent is inapplicable. So what is reported, if authors bother do so, is the molar product yield that is obtained by dividing the number of moles of product collected by the number of moles of substrate of interest. Consequently, there are few opportunities to showcase the performances of such plans against conventional chemical methods in head-to-head comparisons. A summary of synthesis plans utilizing biotransformation techniques is discussed in Section 9.6. The reader is directed to the accompanying download for full plan details. These plans were successfully analyzed.

3.11 PAPERS DESCRIBING REACTIONS USING MICROWAVE IRRADIATION

The use of microwave irradiation in organic synthesis has been a great boon for the field of green chemistry. Several books have now been published in this area [31–38]. From a green chemistry perspective several "green" principles are satisfied if microwave irradiation is used. Advantages that have been advertised for this technique are a tremendous savings in reaction time and energy consumption. However, a recent study of efficiency of energy consumption for microwave-mediated reactions compared to conventional heating methods has challenged the energy savings point [39]. A commercially available domestic electricity meter ("Watt-hour meter") was used to measure the consumed energy in kWh and kWh/mol, respectively. The concluding remarks of this paper were as follows: "On the basis of the data presented herein, it is quite clear, however, that it should not be assumed that microwave-assisted synthesis will be more energy-efficient than conventional processing. The widespread general opinion on the relative "greenness" of microwave heating in chemical processing—at least in terms of energy efficiency—therefore needs to be critically questioned." Unfortunately, the experimental section of this paper only gave details on the microwave methods, not the conventional heating ones. Moreover, whatever details were disclosed were skimpy on solvent usage, which makes proper analysis impossible.

From a chemistry perspective a key advantage is that certain transformations occur well under microwave irradiation conditions, whereas no such products are obtained using conventional heating techniques. Sometimes unexpected transformations occur under microwave conditions. One drawback is that very little mechanistic work has been done to explain what is going on in such reactions. The number of synthesis examples dwarfs any publications dealing with reaction mechanisms. Consequently, synthetic organic chemists dominate the field where most publications are one-off synthetic curiosities. Powerful time-resolved kinetic and spectroscopic techniques have not yet been developed to a wide degree for the probing of reaction intermediates in such reactions as is done in physical organic chemistry where such tools as flash photolysis or stopped flow are routinely used, for example. A time-resolved "microwavolysis" apparatus has not yet been invented. However, Leadbeater and coworkers have begun to probe reactions using in situ Raman spectroscopy [40–44] and two reports deal with kinetic studies [45,46].

The first head-to-head comparison of reaction performance based on both material and energy consumption for microwave-mediated reactions was done by Clark [47]. A household electricity meter was also used to measure energy consumption in kWh. Unfortunately, that paper was a communication with sketchy experimental details and no follow-up paper has been published since. A ubiquitous problem in the reporting of such reactions is that both power consumption and reaction time are very often not stated. Often only one of these is stated in experimental sections. The units of power are energy per unit time, such as joules per sec, which is the definition of a watt. Sometimes the power is stated as a dial number on a domestic microwave oven, which is meaningless since every model is unique and no two devices of the same make behave identically. Therefore, the actual energy demand cannot be precisely determined from such reports. Another problem is that authors forget that the measured energy consumption needs to be multiplied by an efficiency factor since microwave ovens typically operate with efficiencies between 50% and 65% [48,49]. This is the energy efficiency of the magnetron in converting electrical energy into microwave irradiation. Andraos has discussed a simple quantitative method of determining energy consumption metrics for individual reactions and synthesis plans [50]. Each reaction in a plan has its own energy demand. The total energy demand for an entire synthesis plan is the sum of the energy demands of all reactions. The underlying assumption is that all reactions are appropriately scaled according to the plan's synthesis tree diagram. For each reaction, the energy that is expended is directed to producing both the intended target product and all other unwanted waste material. What is important is the fraction of the energy expended that is directed to producing the target product for an individual reaction. Clearly, that fraction will be the material efficiency, which is the RME. If a reaction has a high RME because it has a high yield, high atom economy, and does not require much auxiliary material, then most of the energy consumed in producing the target product is channeled in that direction. The overall energy expended to produce the final product in a synthesis plan is given by

$$\Psi_{\text{product}} = \sum_{j} \Psi_j (\text{RME})_j, \tag{3.3}$$

and that expended to produce waste is given by

$$\Psi_{\text{waste}} = \sum_{j} \left[1 - (\text{RME})_j \right] \Psi_j, \tag{3.4}$$

where Ψ_j is the energy expended for reaction j. The total energy input for the entire plan is given by

$$\Psi_{total} = \Psi_{product} + \Psi_{waste}$$

$$= \sum_j \Psi_j (RME)_j + \Psi_j \left[1 - (RME)_j \right]$$

$$= \sum_j \Psi_j \tag{3.5}$$

The respective product forming and waste forming input energy fractions are given by

$$\Phi_{product} = \frac{\sum_j \Psi_j (RME)_j}{\sum_j \Psi_j} \tag{3.6}$$

and

$$\Phi_{waste} = \frac{\sum_j \Psi_j \left[1 - (RME)_j \right]}{\sum_j \Psi_j} \tag{3.7}$$

Note that these fractions are weighted averages with respect to the individual reaction energy inputs. From a green perspective, the idea is to minimize the global energy input (Equation 3.5) and to maximize the fraction of energy input directed toward producing the target product (Equation 3.6). Since that publication, there still are no experimental data available in the literature where these relations can be implemented. The primary obstacle is that no one has yet published a plan showing experimentally determined energy inputs for all reactions in a total synthesis of a target molecule.

3.12 REVIEWING OF SCIENTIFIC PAPERS

Reviewers of scientific papers dealing with organic synthesis typically focus on the abstract, figures, tables, and reaction schemes before passing judgment on the degree of novelty, relevance, importance, and quality of the works. Experimental sections and supplementary material are not as carefully reviewed as one would like or expect. Editors and reviewers are rather casual in their attitude to reviewing these critical parts of a scientific paper. Supplementary material is almost always neglected in the review process. Nowadays, there is even a trend to bury experimental sections into the supplementary material, particularly for communications, which have tight two-page limits. Persons charged with reproducing experimental results for their own research purposes, namely, students and postdoctoral fellows, focus on these neglected sections to carry out their work. They are the first ones who will end up discovering unfortunate errors in publications. As reaction metrics begin to take root as a routine tool for practicing green chemists, they too will discover these errors in experimental sections. Green chemistry demands the disclosure of all experimental details both truthfully and correctly. The spreadsheet algorithms described in this book are excellent proofreading devices that can trap any conceivable error in publications. The radial pentagon spreadsheet in particular should be part of the standard tool package provided by journals for reviewers of manuscripts in synthetic organic chemistry.

As mentioned in the introduction to this chapter, editors and typesetters are too focused on space saving in displaying synthesis plan schemes. This causes authors to contract their schemes thereby increasing the frequency of errors such as missing steps and incorrect yield assignments. Changes in the aspect of structures causes unnecessary slowing down in following the logic of the synthesis progression as too many mental visual flips are needed. This makes the task of reading papers exhausting and can be a real turn-off, especially since the annual output of literature on organic synthesis is huge. Authors have to ask themselves whether or not they want their papers to be read and understood quickly in a sea of thousands published annually in a given journal.

3.13   PATENTS

Common types of patent obfuscations include the following:

a. Omission of catalysts, reaction temperature, and reaction times.
b. Mass-mole mismatches for ingredients.
c. No reaction yields reported.
d. Incorrectly calculated reactions yields reported.
e. Nitrate versus nitrite misspelling.
f. Workups described as done in "usual way" without further elaboration.
g. Classical resolutions exceeding maximum 50% reaction yield are reported—what is actually done is to report the true yield (e.g., 40%) and dividing it by 50%, hence 80% would be the reported result (see Section 3.6).

Any mistake found in a patent should be considered deliberate, not accidental as an innocent typographical error. Something like a change from "nitrate" to "nitrite," for example, should not be taken lightly. The implication of mixing up these kinds of reagents in a reaction could lead to serious accidents. A key disclaimer given in US 5837870 column 17 (Pharmacia and Upjohn plan for (S)-linezolid) [51] about one skilled in the art nicely describes the shift of responsibility onto the reader of the patent not the authors.

"Without further elaboration, it is believed that one skilled in the art, using the preceding description, practice the present invention to its fullest extent. The following detailed examples describe how to prepare the various compounds and/or perform the various processes of the invention and are to be construed as merely illustrative, and not limitations of the preceding disclosure in any way whatsoever. Those skilled in the art will promptly recognize appropriate variations from the procedures both as to reactants and as to reaction conditions and techniques."

A similar one in slightly different language is found in EP 1386607 page 12 (Chemagis plan for donepezil) [52].

"It will be evident to those skilled in the art that the invention is not limited to the details of the foregoing illustrative examples and that the present invention may be embodied in other specific forms without departing from the essential attributes thereof, and it is therefore desired that the present embodiment and examples be considered in all respects as illustrative and not restrictive, reference being made to the appended claims, rather than to the foregoing description, and all changes which come within the meaning and range of equivalency of the claims are therefore intended to be embraced therein."

In compiling the extensive synthesis database presented in this book the following examples illustrate some of the aforementioned obfuscations.

Example 1
The Roche (1953) [53] plan for flunitrazepam: US 3116203 (page 8 line 75 calls for potassium nitrite for carrying out a nitration reaction when the real reagent is potassium nitrate). This was subsequently corrected [54] in Sternbach, L.H.; Fryer, R.I.; Keller, O.; Metlesics, W.; Sach, G.; Steiger, N. *J. Med. Chem.* 1963, *6*, 261. Moreover, the reaction yield was not disclosed.

Example 2
Parke-Davis (1946) [55] plan for phenytoin: all values for the moles of reagents cited in US 2409754 are incorrect.

Example 3
Pfizer-Reddy (2002) [56] plan for sildenafil: hydrazine dihydrochloride salt is used in step 2; US 6444816 gives a mass that corresponds to the monochloride salt, not the dihydrochloride salt.

Example 4
Schwarz Pharma (2005) [57] plan for fesoterodine: US 6713464 patent gives a yield of 75% for a classical resolution step using (S,R)-ephedrine hemihydrate as a chiral salt auxiliary—the correct yield is 46%.

REFERENCES

1. Cornforth, J.W. *Austr. J. Chem.* 1993, *46*, 157.
2. Harried, S.S.; Yang, G.; Strawn, M.A.; Myles, D.C. *J. Org. Chem.* 1997, *62*, 6098.
3. Kidwai, M.; Mohan, R. *Found. Chem.* 2005, *7*, 269.
4. Streitweiser, A. Jr.; Heathcock, C.H. *Introductory Organic Chemistry*, 3rd edn., Macmillan Publishing Company: New York, 1985.
5. Bower, J.F.; Szeto, P.; Gallagher, T. *Org. Lett.* 2007, *9*, 3283.
6. Stetter, H.; Dorsch, U.P. *Liebigs Ann. Chem.* 1976, 1406.
7. Solomons, T.W.G. *Organic Chemistry*, 6th edn., Wiley: New York, 1996.
8. Metzger, J.O.; Eissen, M. (Universität Oldenburg, Germany). http://www.chemie.uni-oldenburg.de/oc/metzger/eatos/
9. Faith, W.L.; Keyes, D.B.; Clark, R.L. *Industrial Chemicals*, 3rd edn., Wiley: New York, 1966.
10. Lowenheim, F.A.; Moran, M.K. *Faith, Keyes, and Clark's Industrial Chemicals*, 4th edn., Wiley: New York, 1975.
11. Wernerova, M.; Hudlicky, T. Synlett, 2010, 2701.
12. Laird, T. *Org. Proc. Res. Dev.*, 2011, *15*, 305.
13. Jung, Y.C.; Yoon, C.H.; Turos, E.; Yoo, K.S.; Jung, K.W. *J. Org. Chem.* 2007, *72*, 10114.
14. Czibula, L.; Nemes, A.; Sebök, F.; Szantay, C. Jr.; Mak, M. *Eur. J. Org. Chem.* 2004, 3336.
15. Kenyon, J. *Org. Synth. Coll.* 1941, *1*, 418.
16. Ingersoll, A.W. *Org. Synth. Coll.* 1943, *2*, 506.
17. Ault, A. *Org. Synth. Coll.* 1973, *5*, 932.
18. Block, P. Jr.; Newman, M.S. *Org. Synth. Coll.* 1973, *5*, 1031.
19. Mohacsi, E.; Leimgruber, W. *Org. Synth. Coll.* 1988, *6*, 826.
20. Jacques, J.; Fouquey, C. *Org. Synth. Coll.* 1993, *8*, 50.
21. Pikul, S.; Corey, E.J. *Org. Synth. Coll.* 1998, *9*, 387.
22. Cai, D.; Hughes, D.L.; Verhoeven, T.R.; Reider, P.J. *Org. Synth. Coll.* 2004, *10*, 93.
23. Alexakis, A.; Aujard, I.; Kanger, T.; Mangensey, P. *Org. Synth. Coll.* 2004, *10*, 312.
24. Kotagiri, V.K.; Suthrapu, S.; Reddy, J.M.; Rao, C.P.; Bollugoddhu, V.; Bhattacharya, A.; Bandichhor, R. *Org. Process Res. Dev.* 2007, *11*, 910.
25. Barth, F.; Casellas, P.; Cong, C.; Martinez, S.; Renaldi, M.; Anne-Archard, G. US 5624941 (Sanofi, 1997).
26. Lucci, R. US 4556518 (Hoffmann La Roche, 1985).
27. John, M.; Paust, J. US 5424465 (BASF, 1995).
28. Wang, X.; Bhatia, A.V.; Hossain, A.; Towne, T.B. WO 9948866 (Abbott Laboratories, 1999).
29. Trost, B.M.; Zhang, T. *Angew. Chem. Int. Ed.* 2008, *47*, 3759.
30. Dale, D.J.; Dunn, P.J.; Golightly, C.; Hughes, M.L.; Levett, P.C.; Pearce, A.K.; Searle, P.M.; Ward, G.; Wood, A.S. *Org. Process Res. Develop.* 2000, *4*, 17.
31. Kingston, H.M.; Haswell, S.J. (eds.) *Microwave-Enhanced Chemistry. Fundamentals, Sample Preparation and Applications*, American Chemical Society: Washington, DC, 1997.
32. Loupy, A. (ed.) *Microwaves in Organic Synthesis*, Wiley-VCH: Weinheim, Germany, 2002.
33. Hayes, B.L. *Microwave Synthesis: Chemistry at the Speed of Light*, CEM Publishing: Matthews, NC, 2002.
34. Lidström, P.; Tierney, J.P. (eds.) *Microwave-Assisted Organic Synthesis*, Blackwell Publishing: Oxford, U.K., 2005.
35. Kappe, C.O.; Stadler, A. *Microwaves in Organic and Medicinal Chemistry*, Wiley-VCH: Weinheim, Germany, 2005.
36. Loupy, A. *Microwaves in Organic Synthesis*, 2nd edn., Wiley-VCH: Weinheim, Germany, 2006.
37. Larhed, M.; Olofsson, K. (eds.) *Microwave Methods in Organic Synthesis*, Springer: Berlin, Germany, 2006.
38. Kappe, C.O.; Dallinger, D.; Murphee, S. *Practical Microwave Synthesis for Organic Synthesis*, Wiley-VCH: Weinheim, Germany, 2009.
39. Razzaq, T.; Kappe, C.O. *Chem. Sus. Chem.* 2008, *1*, 123.
40. Leadbeater, N.E.; Smith, R.J. *Org. Lett.* 2006, *8*, 4589.
41. Leadbeater, N.E.; Smith, R.J.; Barnard, T.M. *Org. Biomol. Chem.* 2007, *5*, 822.
42. Leadbeater, N.E.; Smith, R.J. *Org. Biomol. Chem.* 2007, *5*, 2770.
43. Leadbeater, N.E.; Schmink, J.R. *Nature Protocols* 2008, *3*, 1.

44. Schmink, J.R.; Leadbeater, N.E. *Org. Biomol. Chem.* 2009, *7*, 3842.
45. Schmink, J.R.; Holcomb, J.L.; Leadbeater, N.E. *Chem. Eur. J.* 2008, *14*, 9943.
46. Schmink, J.R.; Holcomb, J.L.; Leadbeater, N.E. *Org. Lett.* 2009, *11*, 365.
47. Gronnow, M.J.; White, R.J.; Clark, J.H.; Macquarrie, D.J. *Org. Process Res. Dev.* 2005, *9*, 516 [corrigendum: Gronnow, M.J.; White, R.J.; Clark, J.H.; Macquarrie, D.J. *Org. Process Res. Dev.* 2007, *11*, 293].
48. Nüchter, M.; Ondruschka, B.; Bonrath, W.; Gum, A. *Green Chem.* 2004, *6*, 128.
49. Nüchter, M.; Müller, U.; Ondruschka, B.; Tied, A.; Lautenschläger, W. *Chem. Eng. Technol.* 2003, *26*, 1207.
50. Andraos, J. In *Green Chemistry Metrics: Measuring and Monitoring Sustainable Processes* (A. Lapkin, D.J.C. Constable, eds.), Wiley-Blackwell: Oxford, U.K., 2008, Chapter 4, Section 4.3.4.
51. Pearlman, B.A.; Perrault, W.R.; Barbachyn, M.R.; Manninen, P.R.; Toops, D.S.; Houser, D.J.; Fleck, T.J. US 5837870 (Pharmacia & Upjohn, 1998).
52. Kaspi, J.; Lerman, O.; Arad, O.; Alnabari, M.; Sery, Y. EP 1386607 (Chemagis Ltd., 2004).
53. Kariss, J.; Newmark, H.L. US 3116203 (Hoffmann-LaRoche, 1963).
54. Sternbach, L.H.; Fryer, R.I.; Keller, O.; Metlesics, W.; Sach, G.; Steiger, N. *J. Med. Chem.* 1963, *6*, 261.
55. Henze, H.R. US 2409754 (Parke-Davis, 1946).
56. Das, S.K.; Purma, P.R.; Akella, V.; Ramanujan, R.; Chakrabarti, R.; Lohray, V.B.; Lohray, B.B.; Paraselli, R.B. US 6444816 (2002-09-03) Reddy's Research Foundation, India.
57. Meese, C.; Sparf, B. US 6713464 (Schwarz Pharma AG, 2004).

4 Problems and Challenges in Synthesis and Green Chemistry

The present chapter outlines some of the ongoing challenges that practicing chemists face in their day-to-day work in synthesizing known and new molecules of academic or commercial interest. Each problem is outlined in brief and where appropriate examples are given.

4.1 PHILOSOPHY AND PRACTICE OF CHEMISTRY

Chemistry is first and foremost an experimental science relying more on empirical work carried out by trial and error than on theory where careful observation of both successes and failures in the laboratory guides the practitioners' thinking on their current research problem. The intervention of theoretical and computational tools has been a recent addition in the evolution of the subject. Other disciplines such as physics, mathematics, computer science, and engineering have played huge roles in developing theories in chemistry and increasing the power of prediction in the subject. This has propelled it to have a firm grounding and it has thus evolved to be a central science that connects with all other scientific fields. One of the great triumphs was quantum mechanics whose practical offspring, spectroscopy, played an immeasurable role in speeding up the process of structure elucidation compared to traditional methods of degradation and analyzing the fragments. Another triumph is the great suite of analytical tools now available to detect molecules and to purify them. Still, despite all these advances, chemists' ability to predict chemical properties of new molecules relies heavily on comparative analysis using the vast database of known compounds. Of course, predictions improve when there is a large enough database available. In the early days when only a few compounds were known, predictive powers were limited. Modern endeavors relying on the existence of large databases, database mining, and pattern recognition are the search for lead pharmaceutical compounds by combinatorial methods and structure determination of large biomolecules such as proteins.

Whatever research problem is investigated, the primary work of a chemist involves synthesizing molecules. Nothing can be done without having a desired compound in hand. If a chemist does not have the training to make it, then he or she will solicit the favors of one who can. This is usually done by purchasing the compound from some chemical company if it is commercially available or by collaborating with an expert colleague if the target compound is unique. Synthesis is a technical term that describes the art of making compounds. More colloquially it is nothing more than cooking. The reader will appreciate that good cooking is an art and so there is no offense in stating that synthesis and cooking are one and the same. For the practicing chemist this entails following known recipes, inventing new ones, and having hands-on dexterity in the use of various laboratory equipment and gadgets to carry out the chemical reactions. Every chemist knows that anyone can assemble glassware and cook up some ingredients in a reaction vessel. However, the real chemist is the one who can fish out the product from that vessel and get it in high yield and in high purity. Few are able to master this skill and one must keep at it in order not to lose the "feel" of it. In doing so, there exists a kind of guild mentality among chemists. In this context, one often hears the phrases "so and so trained with so and so," or "in our hands we were able to...." Apprenticeship is obviously a key attribute of training in chemistry, especially when it comes to synthesis. Associated with this is a "secret" in-house knowledge among schools or laboratories. Modern synthetic organic chemists continue to practice their craft with this traditional mentality. A recent celebration of the 400th anniversary of the earliest professorship at Phillips Universität in Marburg, Germany [1] cited the following house rules:

1. When students of the world's first chemistry professor, Johannes Hartmann, entered the laboratory they were required to leave their swords behind.
2. Students were not allowed to spit on the floor.
3. Students were required to wear linen aprons.
4. Students were obligated to keep whatever experiments they were doing a secret.
5. Students were allowed to ask questions only "with modesty and without bothering the teacher."

Though modern students may find reading these regulations both archaic and amusing, they will find that times have hardly changed. As far as we know, weapons are not permitted in laboratories. Unhygienic practices pose significant health risks when done in a potentially toxic environment filled with laboratory chemicals. Lab coats are still a requirement, though graduate students are well known to slack off on this. Students publicly announce what they are doing in their supervisor's laboratory only after they seek his or her permission. This usually takes the form of conference talks or publications that advisors then write and take credit for the work. Moreover, students are customarily not permitted to pursue their own research questions independently from their advisors. The very few advisors who let their students come up with and pursue their own ideas to completion are special people who do not have insecurities about their own powers and are not afraid or intimidated that that their bright students may surpass them in intellectual ability. The question is whether such an advisor-student interaction is viewed as a battle of egos or an opportunity to make serious discoveries. Students ask questions of their advisors only when they encounter difficulties and hopefully the good ones among them do so after they have done thorough literature searches of their problem beforehand. The more famous their advisor is the more timid the student will be in challenging him or her when a problem is encountered with the research project. In short, the laboratory environment and working relationships of people in it are highly regulated and controlled.

Green chemistry is by its very nature an interdisciplinary subject. It is not a single field of chemistry unto itself. Rather it permeates all fields of chemistry. At its heart is reaction optimization in every sense: material, energy, cost, safety, and life cycle. In effect it is the *cordon bleu* of synthetic chemistry, the core discipline of chemistry. In order to be successful at it, a chemist will require skill development in many fields both from scientific and human resource management points of view. For a start, a green chemist has to know their chemistry inside out. This means synthesis, mechanism, thermodynamics, kinetics, scale-up, engineering, and toxicology to name a few. Since the subject is interdisciplinary this means a green chemist will need to interact with other kinds of scientists. Exposing one's ignorance is part of the territory. So is sharing knowledge. The traditional mentalities discussed earlier challenge these kinds of developments from taking place. Chemical companies are now discussing opportunities to create and share databases of knowledge for their own work [2,3]. This is particularly relevant for the pharmaceutical industry, for example. These are uncharted waters for organizations well known to keep secrets from each other and the general public. It will take a very long time before academic and industrial chemists will put solving problems as the top priority ahead of advancing their own careers and pockets! Perhaps the modern forms of wireless and electronic communication and archival systems will force the issue in the direction of more openness. One immediate hurdle that needs addressing is to cease divorcing the aims of green chemistry from all endeavors in synthesis and process chemistry. This is a completely false notion that has the tendency to segregate these communities into separate silos with little cross talk. It artificially slows down progress in green chemistry and builds up unjustified prejudices among professionals in these fields.

4.2 EDUCATION

The following challenges exist in the education of students in synthetic organic chemistry. These have a negative impact on accepting and adopting green chemistry practices:

1. Balancing of organic chemical equations is not taught in introductory or advanced organic chemistry courses.
2. Synthetic chemists are taught to focus only on the product of interest and its structure while ignoring everything else.
3. A reaction mechanism is only examined when there is a problem in not getting the product of interest the first time around.
4. Synthetic chemists have a phobia against the use of quantitative reasoning in synthesis.
5. Basic skills in arithmetic and mathematics are poor to nonexistent among synthetic chemists.

Minimization of waste is an optimization problem. One cannot succeed in reducing waste for a given reaction if it is not mass balanced and if all products are not identified. Obviously, knowledge of the reaction mechanism will go a long way in getting there. And, yes, computations will be required. One needs to be reminded of the law of conservation of mass discovered by Antoine Lavoisier in 1775 [4]. His mind was wired for balancing things because he was a tax collector—chemistry was just a fun hobby. Tax collectors balance financial books—the credits must match the debits, income, and revenue must at least match expenses to be in the black. So why not apply this thinking to a chemical reaction! Interestingly, the discovery of this law was done on the most difficult kind of reaction possible since one of the reagents and one of the products are both gaseous and at that time were "invisible" to experimental techniques available. The gaseous reagent was of course oxygen (phlogiston) for which he was credited with its discovery along with Joseph Priestley. The gaseous product was carbon dioxide (fixed air). The success of his discovery depended on a new kind of apparatus and setup that allowed the trapping of the respective gases. Further evidence of quantitative reasoning in organic synthesis comes from reading the biographies of the leaders in the field. Robert B. Woodward was proficient in mathematics as an undergraduate [5–7]. Elias J. Corey studied mechanical engineering [8]. K.C. Nicolaou's favorite subject in school was mathematics [9]. August Kekulé was an architect before entering the field of chemistry [10–15]. Would the reader be surprised if Kekulé had the imagination to figure out the structure of benzene having had this background? There is something to be said about those who have skills in quantitative reasoning. The salient skill is the development of logical thinking. Would the reader be surprised to know that the synthesis puzzle problems found in standard organic textbooks involving empirical formulas, filling in the blanks above reaction arrows, and spectroscopic hints about structures of intermediates were pioneered by Woodward? Would Corey's idea of retrosynthesis have come to him without logical thinking learned through mechanical engineering? Jean-Marie Lehn is well known for pioneering the synthesis of supramolecular structures of defined shape [16,17]. Would it be surprising to know that his spouse is a discrete geometer and just perhaps he may get some inspiration from her work?

4.3 CHEMISTRY LITERATURE

With the current exponential explosion of chemistry literature it is imperative to find ways to keep up in an intelligent way. The creation of useful and intelligent database beyond the chronological cataloging of publications as they appear in the literature is desperately needed. This paradigm of chronological cataloging has been going on since 1907 with *Chemical Abstracts* and even earlier in the early nineteenth century with *Poggendorff's* compilation and the *Catalogue of Research Papers published by the Royal Society of Chemistry* (1800–1900). What is needed is a suite of smart databases for reaction performance and synthesis strategy: green syntheses, reaction intermediates, kinetics, and ring construction strategies. The creation of such databases requires significant work beyond trapping keywords in journal article titles and abstracts. A case is made for a ring construction strategy database in Section 10.9. No database exists presently that satisfies this requirement. The onus should be on authors to submit their work in a format suitable to the setup of such databases since such a task would be monumental for any one organization to launch given the current explosion of literature published every year in print and online. The creation of a *Green Organic Syntheses* series in parallel to the venerable *Organic Syntheses* and *Inorganic Syntheses* series would be a significant step in the right direction. Only plans that have been scaled up, checked by independent workers, and proven to be green by some kind of metrics analysis would be included in this elite compilation. The longer the decision to create such databases is delayed, the bigger the task will be. Editors of journals should initiate proof of metrics analysis as a mandatory requirement before authors can profess that their newly announced plans are indeed "green."

4.4   REACTION PERFORMANCE IMPROVEMENTS

In 2005, the American Chemical Society and Green Chemistry Institute in Washington, DC initiated a Pharmaceutical Roundtable comprising process chemists from various companies to discuss their wish lists of green reactions and processes. These were published in the following 2 years [18–20]. The following reactions and solvent methods were targeted as ones for "greening" up:

1. Synthesis of amides
2. Alcohol activation for subsequent nucleophilic substitution reactions
3. Reduction of amides without using hydride reagents
4. Oxidations and epoxidations without use of chlorinated solvents
5. Mitsunobu reaction
6. C–H activation of aromatics for cross-coupling reactions without preparing haloaromatics
7. C–H activation of alkyl groups
8. Asymmetric synthesis of amines from prochiral ketones
9. Asymmetric hydrogenation of unfunctionalized olefins, enamines, and imines
10. Asymmetric hydrolysis of nitriles
11. Asymmetric hydrocyanation
12. Asymmetric hydroformylations
13. Asymmetric hydroaminations of olefins
14. Synthesis of chiral amines
15. Fluorinations
16. Nitrogen-centered chemistry without using azides and hydrazines
17. Bromination
18. Sulfonation
19. Nitration
20. Demethylation
21. Friedel-Crafts reactions on unactivated substrates
22. Ester hydrolysis
23. Wittig chemistry without triphenylphosphine oxide
24. Radial chemistry without tri-*n*-butyltin hydride
25. Use of organocatalysts in catalytic reactions
26. Aromatic etherification reactions
27. Green sources of electrophilic nitrogen
28. Targeted substitution of aromatic rings

29. Replacement of dipolar aprotic solvents such as *N*-methylpyrrolidin-2-one (NMP), *N,N*-dimethylformamide (DMF), and *N,N*-dimethylacetamide (DMAc)
30. Solvent-less reactor cleaning
31. Alternatives to chlorinated solvents

4.5 SYNTHESIS PERFORMANCE

With respect to synthesis performance there is the problem of generalization of a plan that is flexible enough to lead to different targets of related structures versus coming up with optimized plans that are specific to each target molecule. This is important for creating libraries of compounds with a common core structure but having different functional groups. The core structure is called a scaffold and is analogous to a Markush-type structure. Related to this is the problem of reconciling the objectives of the medicinal chemistry route (speed, library construction to identify key pharmacophoric functional groups) with those of the process chemistry route (cost minimization, commercial viability). A fundamental problem of comparing plans to the same target structure is that each plan begins from a different set of starting materials. The question that arises is whether the comparison is fair. Achieving a consensus on a set of readily available materials (first and second generation feedstocks) that all synthesis plans can be traced to will go a long way in eliminating this problem.

4.6 METRICS

The evolution of metrics has paralleled the growth of green chemistry literature over the last two decades. The main problem is that certain metrics are given different names for the same thing thus creating unnecessary confusion. Examples are listed in the following.

Experimental atom economy = kernel reaction mass efficiency (Cann, [21,22])
Experimental atom economy = AE × (1/SF) [23,24]
Actual atom economy = AE × reaction yield [23,24]
Balance yield (bilanzausbeute) = overall RME (Steinbach and Winkenbach) [25]
Mass index = reciprocal of overall RME (Eissen and Metzger) [26]
Mass intensity = 1 + *E*-factor (Constable) [27–29]
Effluent load factor (ELF) = *E*-factor (chemical engineering term) [30]

The green metrics field is plagued by coining names often by repeating those names in articles, book chapters, and conference proceedings. Authors are more interested in wanting to carve out a personal niche with little follow-up in linking ideas into one coherent "big picture." Their motivation is to advance their personal careers rather than to solve pressing scientific problems thoroughly and thoughtfully. Other current problems include:

a. There are too many false claims of "greenness" based on one or two metrics, or one or two of the Anastas-Warner 12 green principles.
b. All relevant metrics parameters need to work together in the same direction in order to make a valid claim of "greenness."
c. Metrics are meaningful in a comparative sense rather than in an absolute sense.
d. Redox economy cannot be parameterized as a fraction between 0 and 1 as all other material efficiency metrics.
e. There is no reporting of energy consumption as part of the standard protocol for writing up experimental procedures for chemical reaction.
f. Reaction yield depends on reaction scale, there is no way to predict it—it can only be found out by experimentation, reaction yields can increase or decrease with increasing scale.
g. Incorporation of toxicity and hazards metrics that are fractions between 0 and 1 for consistency with the majority of other metrics for material and energy efficiency.

4.7 REACTION MECHANISM

When a reaction gives unexpected products from those intended, or when a reaction produces the intended product along with a competing side reaction (usually considered a nuisance), or when a new reaction is discovered, then the consideration of finding out what reaction mechanism is able to explain rationally the product outcome becomes a worthy line of investigation. Often the most useful and attractive types of paired reactions are ones that begin from the same starting structure but under two different reaction conditions result in two different, but useful and interesting product outcomes. Often there is a great lag between the discovery of a reaction and the accompanying experimental evidence that supports its conjectured mechanism. The fraction of named organic reactions that have experimental evidence for their reaction mechanisms versus those that are conjectured is low. Nowadays, the routine is to discover a reaction, conjecture a mechanism with associated intermediates, and then do a computational study to verify the conjectured mechanism. Using computational methods to circumvent experimental verification is risky. The actual experimental evidence is often tedious, takes hard work and care to do, and may require the use of specialized techniques, which the average synthetic organic chemist has no access to or has little expertise in. This is where collaborations with experts who complement the deficiencies of the experimental chemist become highly useful. Publications that marry experiment and theory are particularly attractive and satisfying to both authors and readers.

REFERENCES

1. Everts, S. *Chem. Eng. News* 2009, *87*(44), 32.
2. Wolf, L.K. *Chem. Eng. News* 2009, *87*(37), 36.
3. Jarvis, L. *Chem. Eng. News* 2010, *88*(25), 6.
4. Lavoisier, A. (1778) Mémoire sur la nature du principe qui se combine avec les métaux pendant leur calcination, et qui en augmente le poids. *Mèmoires de l'Académie Royale des Sciences* 1775, 520.
5. Todd, L.; Cornforth, Sir J. *Biog. Memoirs Fellows Roy. Soc.* 1981, *27*, 629.
6. Blout, E. *Biog. Mem. Natl. Acad. Sci.* 2001, *80*, 367.
7. Landry, D.W.; Robert Woodward (1917–1979). In *Nobel Laureates in Chemistry 1901–1992* (L.K. James ed.), American Chemical Society: Washington, DC, 1993, p. 462.
8. Allerson, C.R.; Elias J. Corey (1928–). In *Nobel Laureates in Chemistry 1901–1992* (L.K. James ed.), American Chemical Society: Washington, DC, 1993, p. 750.
9. Bernthsen, A. *Angew. Chem.* 1929, *42*, 891.
10. Nicolaou, K.C. *Angew. Chem. Int. Ed.* 2009, *48*, 5576.
11. Wieland, H. *Angew. Chem.* 1929, *42*, 901.
12. von Weinberg, A. *Angew. Chem.* 1930, *43*, 167.
13. Bernthsen, A. *Angew. Chem.* 1930, *43*, 719.
14. Hafner, K. *Angew. Chem.* 1979, *91*, 685.
15. Gillis, J. *Dictionary of Scientific Biography*, Charles Scribner & Sons: New York, 1981, Vol. 7, p. 279.
16. Dietrich, B. In *Nobel Laureates in Chemistry 1901–1992* (L.K. James ed.), American Chemical Society: Washington, DC, 1993, p. 735.
17. Hargittai, I. *Chem. Intelligencer* 1996, January, 6.
18. Carey, J.S.; Laffan, D.; Thomson, C.; Williams, M.T. *Org. Biomol. Chem.* 2006, *4*, 2337.
19. Tucker, J.L. *Org. Process Res. Dev.* 2006, *10*, 315.
20. Constable, D.J.C.; Dunn, P.J.; Hayler, J.D.; Humphrey, G.R.; Leazer, J.L.; Linderman, R.J.; Lorenz, K.; Manley, J.; Pearlman, B.A.; Wells, A.; Zaks, A.; Zhang, T.Y. *Green Chem.* 2007, *9*, 411.
21. Cann, M.C.; Connelly, M. *Real World Cases in Green Chemistry*, American Chemical Society: Washington, DC, 1998.
22. Cann, M.C. The University of Scranton. http://academic.scranton. edu/faculty/CANNM1/organicmodule. html (accessed March 2007).
23. Cheung, L.L.W.; Styler, S.A.; Dicks, A.P. *J. Chem. Educ.* 2010, *87*, 628.
24. Aktoudianakis, E.; Chan, E.; Edward, A.R.; Jarosz, I.; Lee, V.; Mui, L.; Thatipamala, S.S.; Dicks, A.P. *J. Chem. Educ.* 2009, *86*, 730.
25. Steinbach, A.; Winkenbach, R. *Chemical Engineering*, April 2000, pp. 94–96, 98, 100, 102, 104.

26. Eissen, M.; Metzger, J.O. *Chem. Eur. J.* 2002, *8*, 3580.
27. Constable, D.J.C.; Curzons, A.D.; Freitas dos Santos, L.M.; Geen, G.R.; Hannah, R.E.; Hayler, J.D.; Kitteringham, J.; McGuire, M.A.; Richardson, J.E.; Smith, P.; Webb, R.L.; Yu, M. *Green Chem.* 2001, *3*, 7.
28. Constable, D.J.C.; Curzons, A.D.; Cunningham, V.L. *Green Chem.* 2002, *4*, 521.
29. Curzons, A.D.; Constable, D.J.C.; Mortimer, D.N.; Cunningham, V.L. *Green Chem.* 2001, *3*, 1.
30. Lee, S.; Robinson, G. *Process Development: Fine Chemicals from Grams to Kilograms*, Oxford University Press: Oxford, U.K., 1995, p. 13.

5 Overview of Green Metrics

5.1 ADVANTAGES OF USING GREEN METRICS

A metric is a standard measurement of some parameter or property. The acid dissociation constant is a metric of the propensity of a chemical substance to produce protons in aqueous solution. The bond dissociation energy is a metric of bond strength. The Celsius scale is a metric for temperature. Whenever a metric exists there is always a standardized scale associated with it that has defined units. It is important to know what the reference state is for that standardized scale. For example, the hydrogen electrode is the reference electrode against which the pH scale and oxidation potential scales are defined. A direct consequence of making up scales for things is ranking. In chemistry, it is more meaningful to talk about relative acidity than absolute acidity, for example. Hence, chemists ask the question, is substance A more or less acidic than substance B if the molecular structure of substance A has a particular kind of functional group in it?

Why use metrics at all in organic synthesis? Why use metrics in green chemistry? A number of real-world motivations are listed in the following.

a. You have designed a new synthesis for a known compound and wish to know how it compares with other published plans with respect to material efficiency and cost. Is there a significant improvement? In what areas specifically have improvements been made? Does your new novel strategy pay off in terms of material efficiency?

b. You are acting as a consultant that requires you to evaluate the synthesis performances of past syntheses of an important target molecule your client has assigned you to investigate by mining the available literature.

c. You are a process chemist involved in scaling up a series of experimental procedures for reaction steps to a complex target molecule and wish to identify the bottlenecks in the synthesis plan.

d. You are presenting a course on synthesis for students and wish to highlight, compare, and contrast strategy and material efficiencies of various approaches.

e. You are evaluating various methods of carrying out a transformation and wish to know the best route according to material efficiency, cost, and safety.

f. You are a lecturer and wish to demonstrate to students the utility of quantitative reasoning in a synthesis course.

g. You are an industrial chemist and wish to optimize a procedure or an entire synthesis plan and need a way of tracking progress in a manner that is convincing to chemists and nonchemists.

The idea of incorporating metrics analysis in discussing merits of syntheses to a given target molecule in a comparative sense provides an intelligent in-depth critique of plans beyond the customary recitation in words of what is already evident by looking at a scheme of structures. This is especially beneficial when writing review articles on synthesis methodology or the synthesis of a particular complex natural product. Much deeper insights are gleaned about synthesis strategy if metrics are used. Other benefits include providing proof of novelty in improvements to process patents that are difficult to secure in the first place, and providing concrete proof that a newly reported plan is both different and better by material consumption and strategic design than prior reported plans. The use of metrics is the only meaningful way to convince the chemistry community at large of the "greenness" of any chemical process. Furthermore, it can only be done comparatively. If one process or synthesis exists for some target molecule, then nothing can be said about its "greenness." Hopefully, through a series of well-orchestrated optimizations carried out over time, the so-called "best plan" will surface where all its metrics parameters are optimized in the same direction. Patent attorneys writing patent applications and professors seeking grant funding for their research agenda will find these arguments especially relevant. Standardization of metrics is required so that comparisons between plans can be made in an unbiased way. This needs to be incorporated in patent law regulations and needs to be accepted by all patent claimants. In addition, editors of journals should also incorporate metrics tools as part of their standard protocols in submissions of original research articles. The International Union of Pure and Applied Chemists (IUPAC) should strike a working committee made up of leaders in academia and industry to sort out the details of standardization so that all chemists can benefit. The sooner this is done the quicker green chemistry will be taken seriously by both camps in tracking and advertising substantive progress toward sustainable chemical processes that are backed up by rigorous quantitative data. Finally, and probably most importantly, metrics may be conveniently used as a powerful proofreading tool for patent examiners of patent applications and reviewers of manuscripts to screen for accidental and deliberate errors in experimental procedures as discussed in Chapter 3. The more these errors are allowed to proceed to publication the more chemistry's reputation will be eroded among chemists themselves and other nonchemistry professionals such as regulators, business people, investors, economists, lawyers, engineers, health professionals, and scientists from other disciplines. Irreproducible results in published synthesis plans are a huge turn-off for students who may wish to become prospective chemists.

5.2   MATERIAL EFFICIENCY METRICS AND VISUALS

The literature on green metrics is well documented and is now a mature field. The reader is directed to the review by Calvo-Flores [1] for an introduction to the subject and a more comprehensive monograph reference edited by Lapkin and Constable [2]. A complete listing of publications [3–27] on this subject is given in Section 5.8. To date, only four publications from other groups have utilized green metrics to analyze the performances of new reactions or synthesis plans [28–31]. Clearly, this shows that there is a long way to go before they become accepted as standard practice. In the present work, the following essential metrics were chosen as the best ones to describe succinctly individual reaction and synthesis plan material efficiency performance. All plans in the synthesis database comprising 1050 + plans were evaluated using these parameters. They are listed along with their names, symbols, and definitions as follows.

Reaction yield, ε
Ratio of moles of target product collected to moles of target product expected with respect to the limiting reagent as per the stoichiometry of the given balanced chemical equation.

Atom economy, AE
Ratio of molecular weight of target molecule to sum of molecular weights of reactants assuming a balanced chemical equation.

Kernel reaction mass efficiency, RME_{kernel}
The ratio of the mass of target product collected to the sum of the stoichiometric masses of all reagents used in a chemical reaction or synthesis plan.

Kernel mass of waste per mole of target of molecule
Mass of all stoichiometric reagents used in a synthesis reaction or plan minus the mass of target product collected divided by the mass of 1 mole of the target product.

Kernel E-factor, E-kernel
Contribution to the total or overall E-factor with respect to mass from reaction by-products, reaction side products, and unreacted starting materials (excludes excess reagents).

Excess *E*-factor, *E*-excess

Contribution to the total or overall *E*-factor with respect to mass from excess reagents.

Auxiliary *E*-factor, *E*-aux

Contribution to the total or overall *E*-factor with respect to mass from auxiliary materials such as reaction solvents, work-up, and purification materials.

Total *E*-factor, *E*-total

The ratio of mass of total waste from all sources to mass of target product collected in a given reaction or synthesis plan.

If *E*-factor units are expressed as mass of waste per mole of target product instead of mass of waste per unit mass of target product then one can compare *E*-factors universally for any target compound originating from any synthesis plan.

The following graphics serve as a visual display of the results of the analysis:

a. Percentage kernel waste distribution profile: bar graph showing % kernel waste versus reaction stage
b. Percentage total waste distribution profile: bar graph showing % total waste versus reaction stage
c. Atom economy (AE) profile: bar graph showing AE versus reaction stage
d. Reaction yield profile: bar graph showing reaction yield versus reaction stage
e. Reaction mass efficiency profile: bar graph showing reaction mass efficiency (RME) versus reaction stage
f. Synthesis tree diagram: a diagram showing the following parameters of a synthesis plan: number of branches, number of reaction stages and steps, number of input materials, number of reaction intermediates, reaction yields for each step, the connections between nodes for input materials and the final target product node
g. Radial pentagons for each step: a radial diagram showing reaction yield, AE, excess reagent consumption (1/SF—inverse of stoichiometric factor), auxiliary material consumption (MRP—material recovery parameter), and overall RME for a single reaction step

In writing out the chemical schemes the following convention was used:

1. If a reaction yield was not disclosed by authors then a reasonable value was estimated from secondary literature reports of similar or analogous reaction performance and was designated with an asterisk symbol to reflect this.
2. Parentheses around a chemical structure indicate that the chemical species was not isolated.
3. Curly brackets around a group of chemical structures indicate that those products were isolated as a mixture and not separated—they were used as an entire substrate in the following step.
4. Reagents were written above the reaction arrow, species acting as catalysts and ligands are so indicated.
5. Each reaction is fully balanced and by-products are written below each reaction arrow preceded by minus signs.

The following distinction is made between by-products and side products of a reaction:

By-product of a reaction—a product formed in a reaction between reagents as a direct mechanistic consequence of producing the target product assuming a balanced chemical equation that accounts for the production of that target product.

Side product of a reaction—a product formed in a reaction between reagents, usually undesired, that arises from a competing reaction pathway other than the one that produces the intended target product and its associated byproducts.

The chemical examples taken from Streitwieser and Heathcock's well-known introductory organic chemistry text [32] and shown in the following illustrate the distinction clearly.

Example 1 (p. 755)

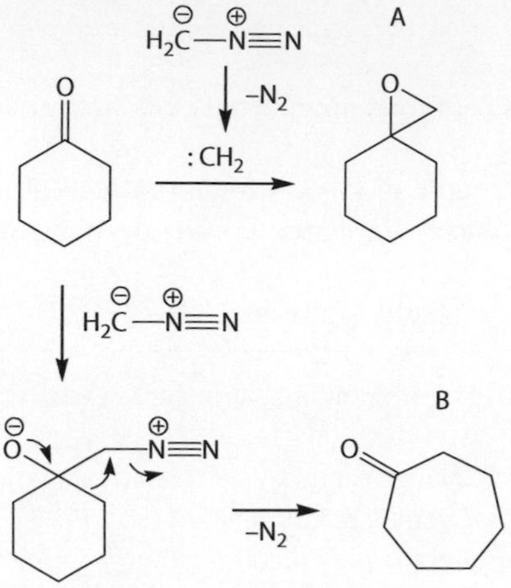

If A is the target product then B is the side product.
If B is the target product then A is the side product.

Nitrogen is a by-product in both cases.

Example 2 (p. 769)

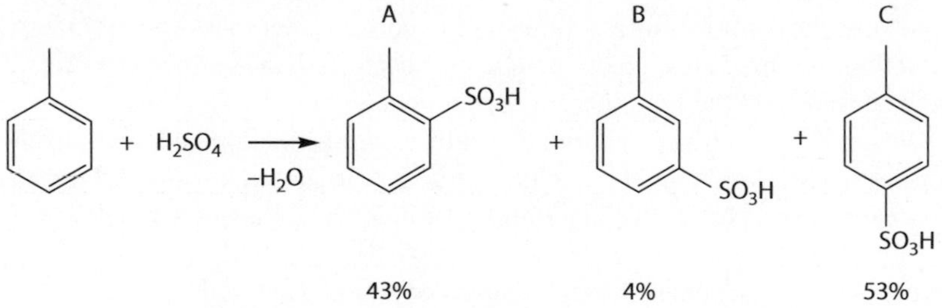

43%   4%   53%

If A is the target product, the B and C are side products.
If B is the target product, then A and C are side products.
If C is the target product, then A and B are side products.

Water is a by-product in all cases.

Example 3 (p. 824)

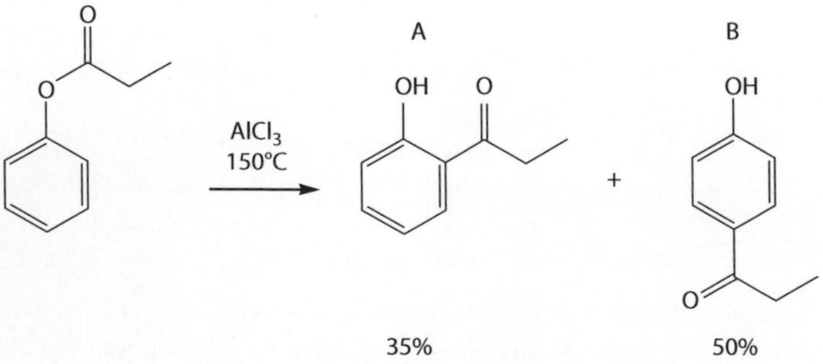

35%   50%

If A is the target product, then B is the side product.
If B is the target product, then A is the side product.
Since the reaction is a rearrangement of the substrate
there are no reaction by-products.

Example 4 (p. 1046)

A   B

CH_3CH_2—C(=O)—CH=CHCl →(NH$_2$OH, EtOH, heat) [isoxazole A] 60% + [isoxazole B] 40%

If [Cl—CH=CH—C(=O)—CH$_2$CH$_3$] + HO—NH$_2$ → [isoxazole] + H_2O + HCl is target reaction

Then (i) A is target product and B is a side product
(ii) H_2O and HCl are by-products

If [Cl—CH=CH—C(=O)—CH$_2$CH$_3$] + H_2N—OH → [isoxazole] + H_2O + HCl is target reaction

Then (i) B is target product and A is a side product
(ii) H_2O and HCl are by-products

The following notations were used for constructing synthesis trees:

1. An asterisk indicates that a reaction yield value is assumed.
2. An open dot indicates an intermediate node.
3. A filled dot indicates a reagent (input material) node.
4. A shaded dot indicates the final product node.
5. [] around a series of reaction yield values indicates that these are calculated as a geometric mean over those steps.
6. The step count for intermediates in branch 1 (main branch) is designated as 1, 2, 3, etc.
7. The step count for intermediates in branch 2 is designated as 1*, 2*, 3*, etc.
8. The step count for intermediates in branch 3 is designated as 1**, 2**, 3**, etc.
9. The step count for intermediates in branch 4 is designated as 1***, 2***, 3***, etc.

The synthesis tree diagrams were highly useful for following material throughput for convergent plans. These diagrams made it easy to spot points of convergence. If two consecutive intermediates of equal molecular weight are connected by a horizontal line in the synthesis tree diagram, then that transformation is a rearrangement. If the molecular weight of the latter intermediate is less than the former then it may be inferred that the transformation is a fragmentation or elimination reaction.

Using Figure 5.1 as an illustrative synthesis tree diagram, we can immediately infer the following about the synthesis tree plan without having seen the formal synthesis scheme for β-ionone:

1. The plan is convergent consisting of three branches, eight reaction stages, and twelve reaction steps.
2. Sixteen input materials are required.

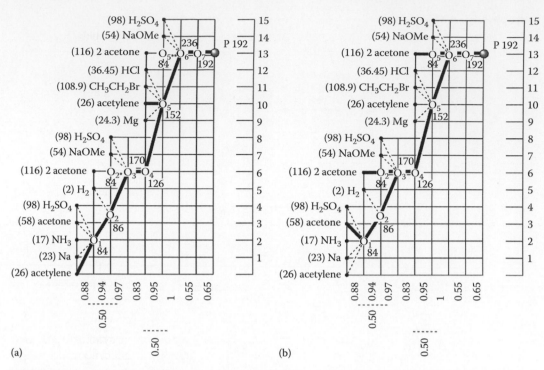

FIGURE 5.1 Synthesis tree diagram for the Roche (1958) synthesis of β-ionone [33–35]. (a) Tracing amount of acetylene used. (b) Tracing amount of acetone used.

3. The eighth step is a rearrangement.
4. Steps 2* and 5** involve the same fragmentation reaction.
5. Steps 4 and 7 involve fragmentation reactions.
6. From Figure 5.1a the stoichiometric mass of acetylene required to make 1 mol of β-ionone is

$$26 \times \left[\frac{1}{0.88 \times 0.94 \times 0.97 \times 0.83 \times 0.95 \times 1 \times 0.55 \times 0.65} + \frac{1}{0.95 \times 1 \times 0.55 \times 0.65} \right].$$

7. From Figure 5.1b the stoichiometric mass of acetone required to make 1 mol of β-ionone is

$$58 \times \left[\frac{1}{0.88 \times 0.94 \times 0.97 \times 0.83 \times 0.95 \times 1 \times 0.55 \times 0.65} \right.$$

$$\left. + \frac{1}{0.50 \times 0.97 \times 0.83 \times 0.95 \times 1 \times 0.55 \times 0.65} + \frac{1}{0.50 \times 1 \times 0.55 \times 0.65} \right]$$

where the first, second, and third terms correspond to the contributions from branches 1, 2, and 3, respectively.

5.3 STRATEGY EFFICIENCY METRICS AND VISUALS

The names, symbols, and definitions of strategy efficiency metrics used are listed in the following.

Type of plan
Linear or convergent.

Number of reaction steps, M
Number of reaction steps in a synthesis plan; a reaction step refers to the interval between a given isolated intermediate and the next consecutive isolated intermediate in a synthesis plan.

Number of reaction stages, N
Number of reaction stages in a synthesis plan; a convergent plan always has $M > N$ and a linear plan always has $M = N$.

Number of input materials, I
Number of input materials in a synthesis plan.

Molecular weight (MW) first moment (building up parameter), μ_1
Parameter that describes the degree of building up going on from the reagent molecules toward the target molecule over the course of a synthesis plan intermediate by intermediate.

Degree of convergence, delta
A parameter determined from the shape of a synthesis tree diagram for a synthesis plan that describes the ratio of the angle subtended at the actual product node vertex to that at a product node vertex corresponding to the hypothetical case of all reaction substrates in a plan reacting in a single step.

Degree of asymmetry, beta
A parameter determined from the shape of a synthesis tree diagram for a synthesis plan that describes the degree of skewness of a triangle whose vertices are the target product node, the origin, and the node for the last reagent projected onto the ordinate axis.

Molecular weight fraction of sacrificial reagents, f(sac)
MW fraction of reagents that absolutely do not end up in the final target molecule structure.

Hypsicity index, HI
A parameter that tracks the oxidation numbers of atoms in target bond forming reactions over the course of a synthesis plan relative to their values in the target molecule.

Number of bonds made per step, B/M
Ratio of sum of number of target bonds made in a synthesis plan to total number of reaction steps.

Number of nonbonding steps per reaction stage, f(nb)
Fraction of number of reaction stages that do not form target bonds.

Oxidation length, |UD|
A parameter that tracks the "oxidation length" traversed over the course of a synthesis plan equal to the absolute value of the sum of all the "ups" and "downs" in a hypsicity profile or bar graph beginning with the zeroth reaction stage.

The following graphics serve as a visual display of the results of the analysis:

a. Target bond maps: a chemical structure drawing of the target product of a synthesis plan showing the target bonds made, the step numbers of each target bond, and the set of atoms that correlate directly with the corresponding reagents that ended up in the target molecule.

b. MW first moment profile: plot of difference between MW of intermediate Pj in step j and MW of final product, Pn, where n is the final step.

c. Target bond forming reaction profile: bar graph of number of target bonds made versus reaction stage.

    d. Hypsicity profile: bar graph of sum of oxidation number changes versus reaction stage.

    e. Radial hexagon: a diagram that depicts the overall atom economy, overall reaction yield along the longest branch, overall kernel RME, MW fraction of reagents in whole or in part that end up in the target molecule, fraction of the kernel waste contribution from target bond forming reactions, and the degree of convergence for a given synthesis plan as a hexagon for easy visualization of the synthesis performance.

5.4 HYPSICITY (OXIDATION LEVEL) ANALYSIS

5.4.1 INTRODUCTION

The concept of oxidation numbers is one that has not received the attention that it should in the education of organic chemists. It has even fallen out of favor in the *Journal of Chemical Education* that has put a moratorium on all submissions on this topic related to balancing equations since 1997 [36]. Though Otis Coe Johnson first introduced it in 1880 as a means of balancing reactions involving inorganic cations and anions [37,38], it is generally confined to high school and university level general chemistry courses. Curiously, the last sentence of Johnson's paper promised to demonstrate the method for organic reactions, but no follow-up paper was ever published. There exists a myth that oxidation numbers are not a useful concept in organic chemistry though clearly every organic textbook has chapters entitled "oxidations" and "reductions." One point that needs to be made is that these titles are misleading because an oxidation cannot occur with out its reduction partner, and vice versa. The two go hand in hand. More correct chapter titles would be "oxidations with respect to the substrate of interest," similarly for reductions. Moreover, none of these chapters balance chemical equations, which makes it difficult to track what is going on in the transformations from a mechanistic point of view. Only a handful of publications discuss balancing organic redox reactions [39–44]. The observation that electron-donating groups (EDG) have atoms of lower oxidation state and electron-withdrawing groups (EWG) have atoms of higher oxidation state is not mentioned in standard texts. For example, in nitrobenzene, the nitrogen atom in the nitro group has an oxidation number of +3, whereas, in aniline, the nitrogen atom in the amino group has an oxidation number of −3. One can classify Hammett-type EDGs and EWGs by the oxidation states of the relevant atoms as shown in Table 5.1.

TABLE 5.1

Classification of Juxtaposed Hammett EDGs and EWGs by Oxidation State[a]

EDG	Oxidation State	EWG	Oxidation State	
$\underline{C}H_3$	−3	$\underline{C}OOH$	+3	
		$\underline{C}OCH_3$	+2	
		$\underline{C}F_3$	+3	
$\underline{C}H_2NH_2$	−1	$\underline{C}N$	+3	
		$\underline{C}ONH_2$	+3	
$\underline{N}H_2$	−3	$\underline{N}O_2$		+3
		$\underline{N}(CH_3)_{3+}$		−3
$\underline{S}H$	−2	$\underline{S}O_3H$		+4
		$\underline{S}O_2CH_3$		+2

[a] Assuming groups are bonded to a phenyl ring.

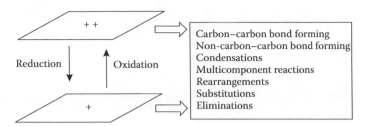

FIGURE 5.2   Redox reactions as central link between various classes of organic reactions.

Another related observation not mentioned is that nucleophilic atoms have lower oxidation states and electrophilic atoms have higher oxidation states. A famous example of a powerful carbon–carbon bond forming reaction is one that can link two fragments having electrophilic carbon centers by means of metallation so that one of the fragments is turned into a nucleophilic center. This change of electron density is of course manifested as a negative change in oxidation state at that carbon atom.

As mentioned earlier, green chemistry demands that the fate of every atom in every reagent used is tracked to its final outcome. It would be impossible to optimize reactions by reducing waste if we do not balance chemical equations and account for the fate of all chemical species in a chemical reaction by some kind of rational mechanism. Redox reactions are a powerful reaction class because they drastically change the electron distributions of atoms. Nature utilizes redox reactions extensively in various biochemical pathways such as glycolysis, citric acid cycle, and photosynthesis that are mediated by electron transfer agents such as nicotinamide adenine dinucleotide (NAD), nicotinamide adenine dinucleotide phosphate (NADP), and metal ion cofactors in enzymes. They may be viewed as a kind of "tunneling" reaction between various classes as suggested by the diagram in Figure 5.2. The Achilles heel is that redox reactions are the most wasteful as a reaction classification since they almost always produce by-products and thus generally have the worst atom economies. Catalytic hydrogenation and oxidation using oxygen gas or hydrogen peroxide are the only notable exceptions. It is also known that balancing redox reactions is the most challenging among all classes of organic reactions. We make the case that tracking the oxidation number changes in atoms forming target bonds is a powerful technique in facilitating balancing chemical equations and in coming up with reaction mechanisms. If these three ideas are tied together coherently, then one can achieve a thorough and deeper understanding of what is going in a given chemical transformation.

This chapter begins by convincing the reader of the following benefits of the concept of oxidation number in organic chemistry through illustrative examples.

1. Tracking oxidation numbers in a reaction mechanism intermediate by intermediate strengthens the link between reaction mechanism and balanced chemical equations that account for all products in a chemical transformation
2. Spotting which atoms are oxidized or reduced in a transformation, especially deep-seated redox transformations
3. Helping to conjecture by-products of a reaction when they are not known

For item (1), the Clemmensen reduction, reduction of nitro aromatics, and the Wolff–Kishner reaction are shown. For each case, a mechanism is given showing the oxidation numbers of key atoms, a redox couple, and an overall balanced chemical equation.

Clemmensen reduction

Reaction:

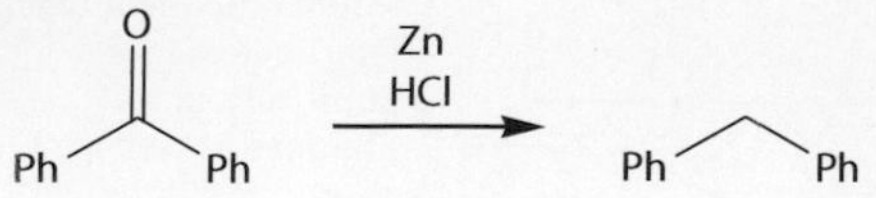

Ph—C(=O)—Ph → (Zn, HCl) → Ph—CH₂—Ph

Possible Mechanism:

$$+2 \quad\quad +2 \xrightarrow{ET} +1 \xrightarrow{ET} 0$$

$$-1 \xleftarrow{ET} 0 \quad\quad 0 \quad\quad 0$$

$$-2 \quad\quad -2$$

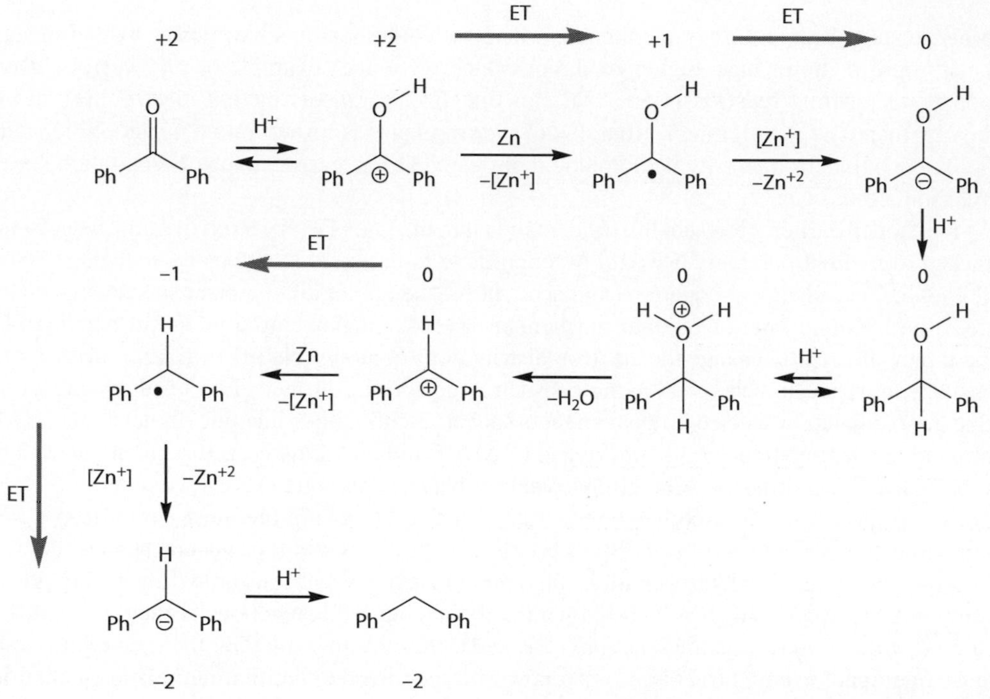

Note that there are four electron transfer steps consistent with the fact that the transformation is a four-electron redox reaction as shown by the redox couple and the difference in oxidation state of the carbon atom from +2 to −2.

Sum of elementary steps:

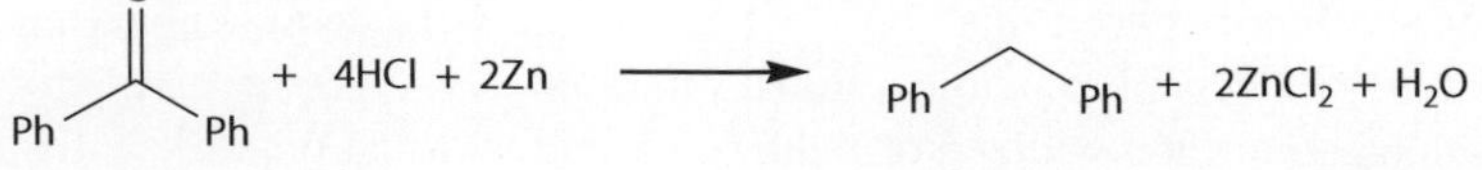

$$Ph\text{-}C(=O)\text{-}Ph + 4HCl + 2Zn \longrightarrow Ph\text{-}CH_2\text{-}Ph + 2ZnCl_2 + H_2O$$

Redox couple:

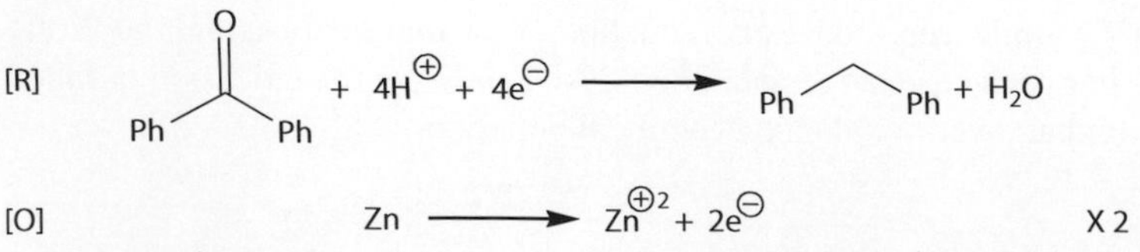

[R] $\quad Ph\text{-}C(=O)\text{-}Ph + 4H^{\oplus} + 4e^{\ominus} \longrightarrow Ph\text{-}CH_2\text{-}Ph + H_2O$

[O] $\quad Zn \longrightarrow Zn^{\oplus 2} + 2e^{\ominus} \quad\quad\quad X\,2$

Reduction of nitro aromatics

Reaction:

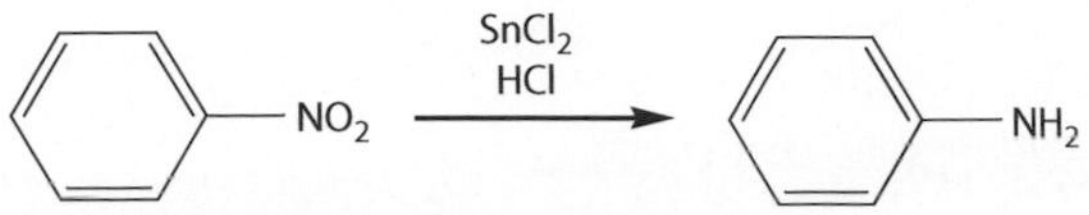

Possible Mechanism:

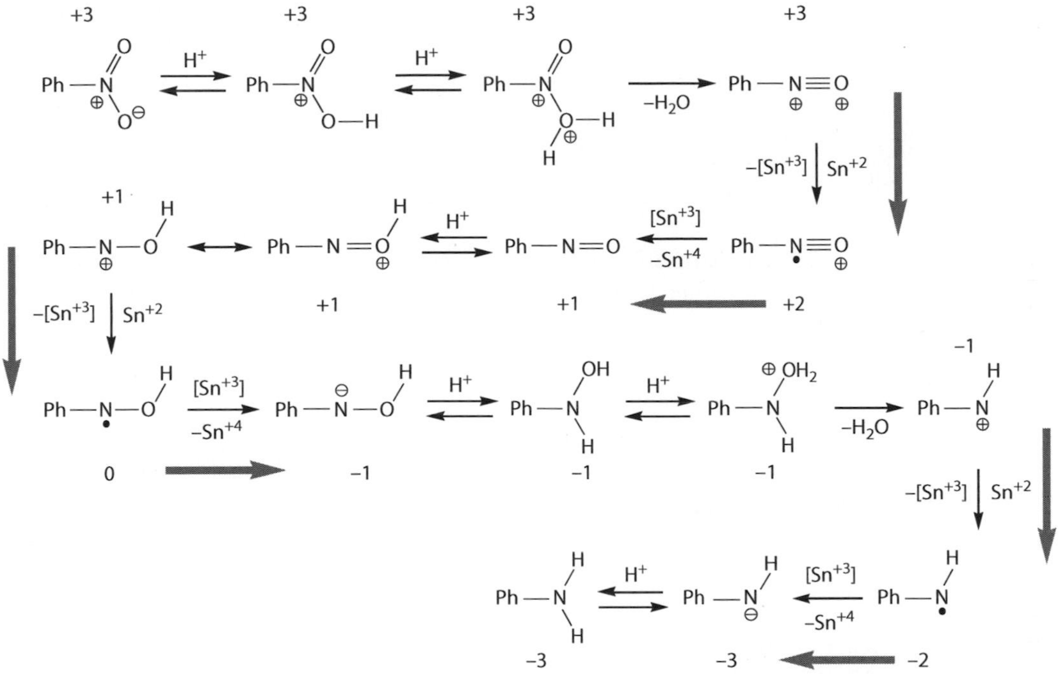

Note that there are six electron transfer steps consistent with the fact that the transformation is a six-electron redox reaction as shown by the redox couple and the difference in oxidation state of the nitrogen atom from +3 to −3.

Summary of elementary steps:

$$PhNO_2 + 6HCl + 3SnCl_2 \rightarrow PhNH_2 + 3SnCl_4 + 2H_2O$$

Redox couple:

[R] $6H^{\oplus} + 6e^{\ominus} + PhNO_2 \longrightarrow PhNH_2 + 2H_2O$

[O] $Sn^{\oplus 2} \longrightarrow Sn^{\oplus 4} + 2e^{\ominus}$ X 3

Wolff–Kishner reduction

Reaction:

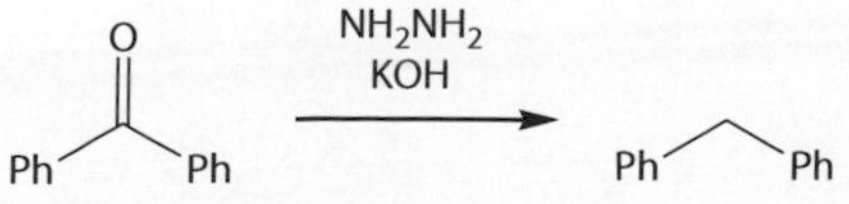

Possible Mechanism:

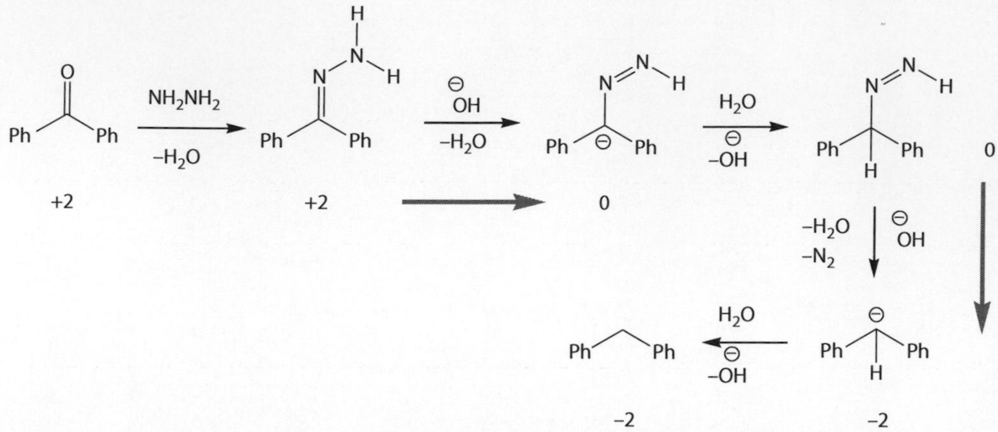

Note that there are two elementary steps each involving a gain of two electrons on the central carbon atom consistent with the fact that the transformation is a four-electron redox reaction as shown by the redox couple and the difference in oxidation state of the carbon atom from +2 to −2.

Summary of elementary steps:

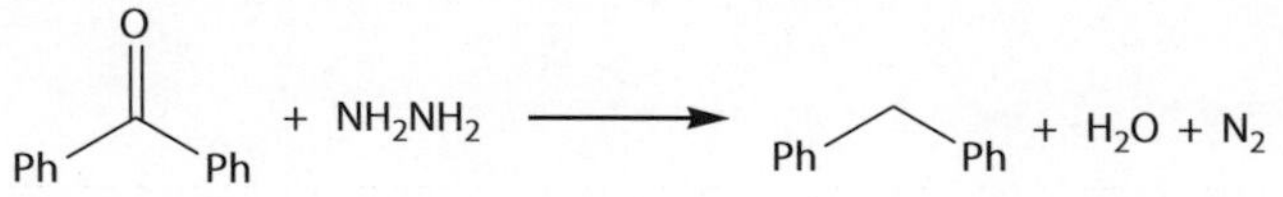

Redox couple:

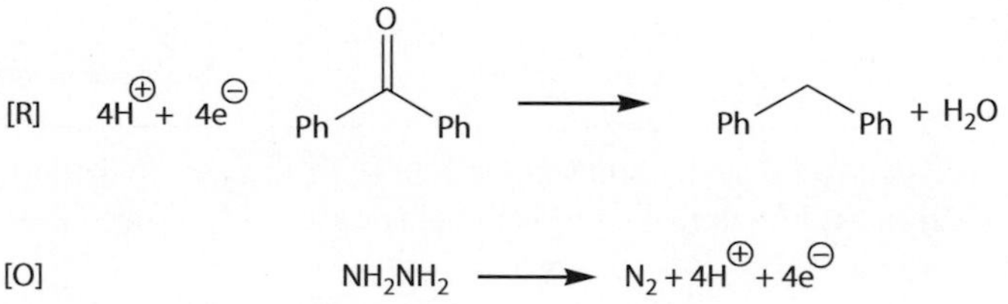

For item (2), the Vogel (1948) [45] and Fierz-David (1949) [46] plans for monastral fast blue B is an example of a deep-seated redox transformation. Two cuprous ions change their oxidation state from +1 to +2 while two nitrile carbons change their oxidation states from +3 to +2.

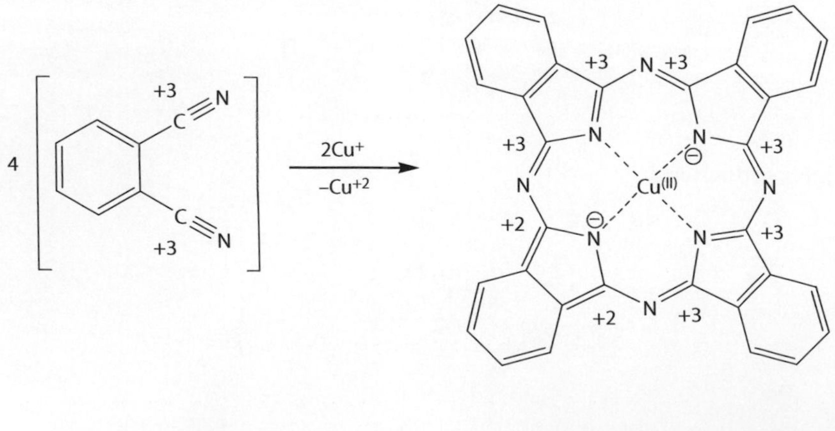

For item (3), the transformation of an oxime to a ketone via exposure to ammonium peroxysulfate begs the question of what is the fate of nitrogen atom. The authors did not bother to question this in their paper [47] as they did not offer a mechanistic explanation.

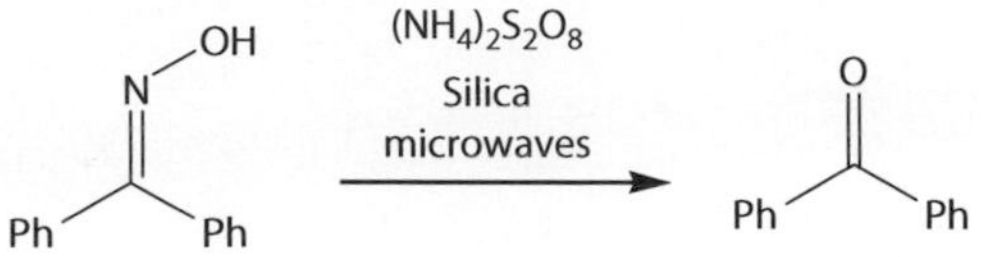

Following is a proposed mechanism showing oxidation numbers of key atoms.

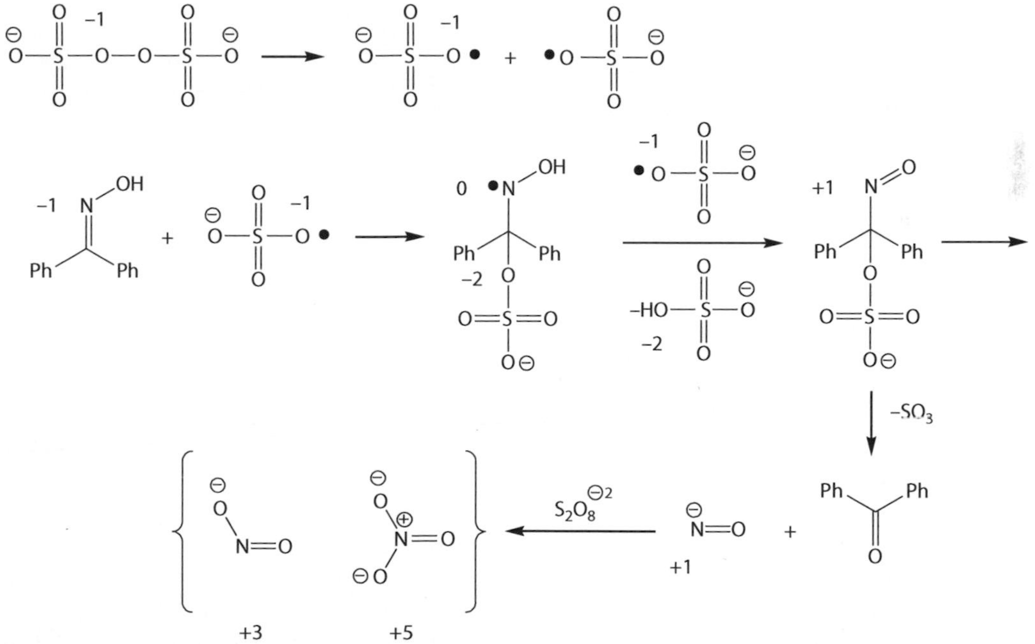

The primary overall transformation is as follows:

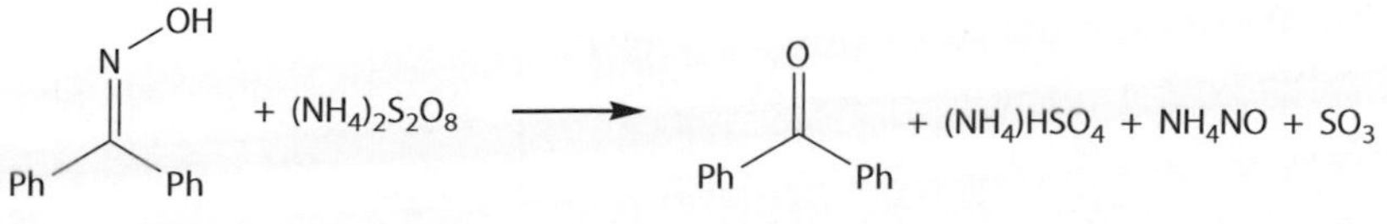

The initially formed NO^- species may be further oxidized to NO_2^- and NO_3^- via secondary reactions with peroxysulfate.

Possibilities for the redox couple include:

$$[R] \quad 2e^{\ominus} + S_2O_8^{\ominus 2} \longrightarrow 2SO_4^{\ominus 2}$$

$$[R] \quad 2e^{\ominus} + 2H^{\oplus} + S_2O_8^{\ominus 2} \longrightarrow SO_3 + SO_4^{\ominus 2} + H_2O$$

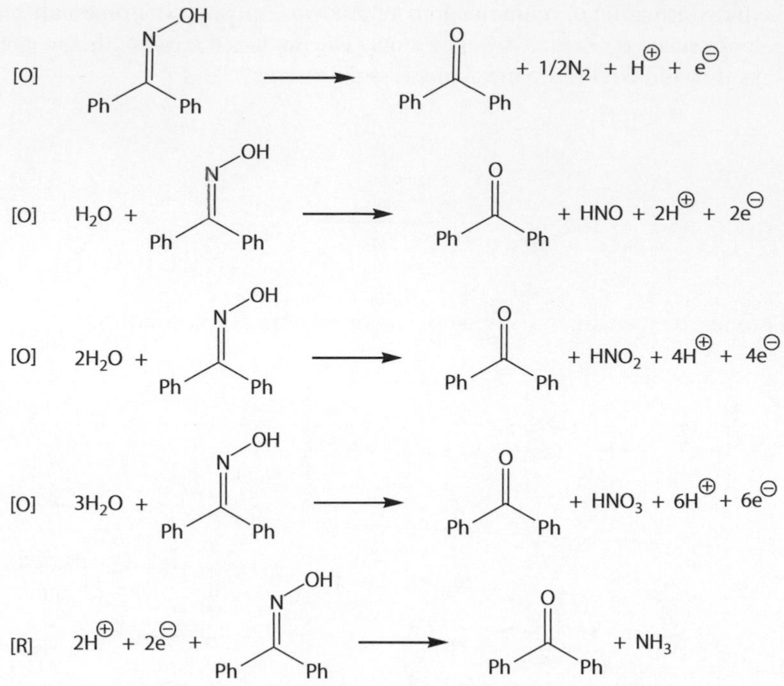

$$[O] \quad \text{(benzophenone oxime)} \longrightarrow \text{(benzophenone)} + 1/2N_2 + H^{\oplus} + e^{\ominus}$$

$$[O] \quad H_2O + \text{(benzophenone oxime)} \longrightarrow \text{(benzophenone)} + HNO + 2H^{\oplus} + 2e^{\ominus}$$

$$[O] \quad 2H_2O + \text{(benzophenone oxime)} \longrightarrow \text{(benzophenone)} + HNO_2 + 4H^{\oplus} + 4e^{\ominus}$$

$$[O] \quad 3H_2O + \text{(benzophenone oxime)} \longrightarrow \text{(benzophenone)} + HNO_3 + 6H^{\oplus} + 6e^{\ominus}$$

$$[R] \quad 2H^{\oplus} + 2e^{\ominus} + \text{(benzophenone oxime)} \longrightarrow \text{(benzophenone)} + NH_3$$

The oxidation couple to nitrogen gas is difficult to rationalize mechanistically, whereas the reduction couple to ammonia can be ruled out since we know that peroxysulfate will be reduced.

At this point the following caveat about using algebraic methods to balance chemical equations is made. It is trivial to work out stoichiometric coefficients using matrix methods especially nowadays since there are many software packages that are available to do this. However, whatever coefficients are obtained must be rationalized with a mechanism. Playing around with integers alone is a dangerous game if it is not couched in the actual chemistry of what is going in a given reaction. Examples in the literature exist of redox reactions that have the curious property of more than one set of stoichiometric coefficients that can satisfy the balancing requirement [43–51]. From a green chemistry point of view, this poses the problem of how to calculate the AE for such reactions. If one uses algebraic or matrix methods to determine coefficients for such reactions, one finds that there exist multiple integer solutions to the simultaneous linear equations. These are so-called diophantine equations that are known since ancient times and make for interesting puzzle problems [52]. Example chemical reactions of this type fall into two groups as follows.

Type 1 (a reactant or product is a diatomic consisting of the same element)

$$\begin{bmatrix} 2 \\ 11 \\ 8 \end{bmatrix} KClO_3 + \begin{bmatrix} 4 \\ 18 \\ 24 \end{bmatrix} HCl \longrightarrow \begin{bmatrix} 2 \\ 11 \\ 8 \end{bmatrix} KCl + \begin{bmatrix} 2 \\ 9 \\ 12 \end{bmatrix} H_2O + \begin{bmatrix} 1 \\ 3 \\ 9 \end{bmatrix} Cl_2 + \begin{bmatrix} 2 \\ 12 \\ 6 \end{bmatrix} ClO_2$$

In this case, chlorine gas is a product.

The associated redox couples for this reaction are given below.

$$[R] \quad 10e^{\ominus} + 12H^{\oplus} + 2KClO_3 \rightarrow Cl_2 + 2K^{\oplus} + 6H_2O$$

$$[R] \quad e^{\ominus} + 2H^{\oplus} + KClO_3 \rightarrow ClO_2 + K^{\oplus} + H_2O$$

$$[R] \quad 6e^{\ominus} + 6H^{\oplus} + KClO_3 \rightarrow KCl + 3H_2O$$

$$[O] \quad Cl^{\ominus} + 2H_2O \rightarrow ClO_2 + 4H^{\oplus} + 5e^{\ominus}$$

$$[O] \quad 2Cl^{\ominus} \rightarrow Cl_2 + 2e^{\ominus}$$

One can verify that various linear combinations of appropriate pairs of these half reactions will lead to the various sets of stoichiometric coefficients given above. For example, the sum of the first reduction half reaction and twice the first oxidation half reaction leads to

$$4H^{\oplus} + 2KClO_3 + 2Cl^{\ominus} \rightarrow Cl_2 + 2K^{\oplus} + 2ClO_2 + 2H_2O$$

Adding two chloride ions to each side gives finally the following balanced equation corresponding to the first set of stoichiometric coefficients.

$$2KClO_3 + 4HCl \rightarrow 2KCl + 2H_2O + Cl_2 + 2ClO_2$$

The same equation is obtained when twice the second reduction half reaction is added to the second oxidation half reaction. If we add all three reduction half reactions and multiply the result by 7, then add both oxidation half reactions and multiply that result by 17, and finally take the overall sum, we obtain the following equation.

$$72H^{\oplus} + 28KClO_3 + 51Cl^{\ominus} \rightarrow 24Cl_2 + 21K^{\oplus} + 36H_2O + 7KCl + 24ClO_2$$

Again, after adding 21 chloride ions to each side to balance the charges and then dividing all coefficients by the common factor 4, we obtain finally an overall equation with yet another set of stoichiometric coefficients.

$$7KClO_3 + 18HCl \rightarrow 7KCl + 9H_2O + 6Cl_2 + 6ClO_2$$

Type 2 (dual ambiguity—reaction is actually a sum of two redox reactions)

$$\begin{bmatrix} 1 \\ 2 \\ 4 \end{bmatrix} NH_4^+ + \begin{bmatrix} 3 \\ 8 \\ 10 \end{bmatrix} NO_2^- + \begin{bmatrix} 2 \\ 6 \\ 6 \end{bmatrix} H^+ \longrightarrow \begin{bmatrix} 1 \\ 1 \\ 5 \end{bmatrix} N_2O + \begin{bmatrix} 2 \\ 8 \\ 4 \end{bmatrix} NO + \begin{bmatrix} 3 \\ 7 \\ 11 \end{bmatrix} H_2O$$

The associated redox couples are given below.

[R] $4e^{\ominus} + 6H^{\oplus} + 2NO_2^{\ominus} \rightarrow N_2O + 3H_2O$

[R] $e^{\ominus} + 2H^{\oplus} + NO_2^{\ominus} \rightarrow NO + H_2O$

[O] $2NH_4^{\oplus} + H_2O \rightarrow N_2O + 10H^{\oplus} + 8e^{\ominus}$

[O] $NH_4^{\oplus} + H_2O \rightarrow NO + 6H^{\oplus} + 5e^{\ominus}$

In this case a redox couple can be written for ammonium ion and nitrite ion going to N_2O, and another can be written for the same pair of reactants going to NO. If we take twice the first reduction half reaction and add it to the first oxidation half reaction we get the first redox couple.

$$2NH_4^{\oplus} + 4NO_2^{\ominus} + 2H^{\oplus} \rightarrow 3N_2O + 5H_2O$$

On the other hand, if we take five times the second reduction half reaction and add it to the second oxidation half reaction we get the second redox couple.

$$NH_4^{\oplus} + 5NO_2^{\ominus} + 4H^{\oplus} \rightarrow 6NO + 4H_2O$$

For these anomalous types of redox reactions, the obvious question is which set of coefficients is the "right" one. The answer is that it is the one that satisfies a rational mechanism made up of the minimum number of elementary reactions whose sum is the overall balanced chemical equation. For the cases cited in the literature, none has had a mechanistic rationale associated with them.

We may conjecture that the combination of stoichiometric coefficients having the least sum is likely the one that can satisfy a reasonable mechanistic explanation. The resultant equation would yield the highest value of the AE with respect to the product of interest for that combination of reagents. What is challenging is to come up with a rational mechanism whose sum of elementary steps is in fact the overall balanced equation that satisfies the above constraint. The following example of a reduction of a carboxylic acid by lithium aluminum hydride is useful in illustrating these points.

$$\begin{bmatrix} 1 \\ 1 \\ 3 \end{bmatrix} RCOOH + \begin{bmatrix} 1 \\ 2 \\ 3 \end{bmatrix} LiAlH_4 + \begin{bmatrix} 3 \\ 7 \\ 10 \end{bmatrix} H_2O \longrightarrow \begin{bmatrix} 1 \\ 1 \\ 2 \end{bmatrix} RCH_2OH + \begin{bmatrix} 1 \\ 2 \\ 3 \end{bmatrix} Al(OH)_3 + \begin{bmatrix} 1 \\ 2 \\ 3 \end{bmatrix} LiOH + \begin{bmatrix} 2 \\ 6 \\ 8 \end{bmatrix} H_2$$

A mechanism that satisfies the first set of coefficients (1, 1, 3, 1, 1, 1, 2) having the least sum of 10 is shown in the following.

The sum of these elementary steps is

$$RCOOH + LiAlH_4 + 3H_2O \rightarrow RCH_2OH + Al(OH)_3 + LiOH + 2H_2.$$

The minimum AE for this reaction when R = H is 0.232 or 23.2%.

Having demonstrated the utility and power of applying oxidation numbers to the balancing of chemical equations in this section, the reader will realize that the current blanket ban in the chemical education literature is unjustified. This state of affairs is an affront to the practice of green chemistry since balancing chemical equations is at the very heart of the subject. All other topics in the field depend on this central point. It is likely that authors and instructors were thrown into a tailspin by the curious reactions mentioned earlier that lead to diophantine equations. Moreover, few of the articles that were published in this area over the years made the link between balancing chemical equations and reaction mechanism. Without this important link there can be no complete and true understanding of any chemical reaction, let alone a redox reaction. Unfortunately, introductory organic chemistry textbooks fail miserably in accomplishing this. In the next section, these concepts are extended to synthesis plan analysis where even more powerful insights are gained.

5.4.2 SYNTHESIS PLANS

By analogy with the tracking of MWs of starting materials and intermediates leading to the final target product in a synthesis plan to gauge how efficiently a target structure is synthesized, it is possible to track the changes in oxidation state, or hypsicity, (Gk: *hypsos*, meaning level or height) of key atoms involved in bond-making and bond-breaking steps. This follows Hendrickson's proposal to aim for *isohypsic* synthesis plans that are characterized by a zero net change in oxidation state of all atoms of starting materials and intermediates involved until the target product is reached [53]. Mathieu [54,55] also wrote about the importance of oxidation numbers in organic synthesis. More recently, Baran has coined the name "redox economy" [56,57] by analogy with the terms "step economy" (coined by Paul Wender) and "atom economy" (coined by Barry Trost) to describe efficient transformations that do not involve net increases or decreases in oxidation number for atoms involved in forming target bonds. All of these treatments dealt with qualitative reasoning; however, preceding Baran's discussions, Andraos in Chapter 4 of reference [2] provided the first formal quantitative analysis of hypsicity in the context of synthesis plan efficiency. Formally, we may define a hypsicity index, HI, as given in Equation 5.1.

$$
\begin{aligned}
\mathrm{HI} &= \frac{\sum_{\text{stages},j}\left[\sum_{\text{atoms},i}\left[(Ox)_{\text{stage},j}^{\text{atom},i} - (Ox)_{\text{stage},N}^{\text{atom},i}\right]\right]}{N+1} \\
&= \frac{\sum_{\text{stages},j}\Delta_j}{N+1} \\
&= \frac{\sum_j \Delta_{\text{row},j}}{N+1}
\end{aligned}
\tag{5.1}
$$

where (Ox) represents the relevant oxidation number of an atom. If HI is zero, then the synthesis is *isohypsic*. If HI is positive valued, then to get to the target molecule, a net reduction is required over the course of the synthesis since an accumulated gain in oxidation level has resulted. Such a condition is termed *hyperhypsic*, by analogy with the term *hyperchromic,* which describes increases in intensities of absorption bands in spectroscopy. Conversely, if HI is negative valued, then to get to the target molecule, a net oxidation is required over the course of the synthesis since an accumulated loss in oxidation level has resulted. Such a condition is termed *hypohypsic*, again by analogy with the term *hypochromic.* The following sequence of steps may be followed to determine HI for a synthesis plan:

1. Enumerate atoms in the target structure that are only involved in the building up process from corresponding starting materials. This set of atoms defines those that are involved in bonding changes occurring in the relevant reaction steps.
2. Work backwards intermediate by intermediate to trace the oxidation numbers of the aforementioned set of atoms back to original starting materials as appropriate following the reaction stages back to the zeroth stage.
3. For each key atom, i, in each reaction stage, j, determine the difference in oxidation number of that atom with respect to what it is in the final target structure. Hence,

$$(Ox)_{\text{stage},j}^{\text{atom},i} - (Ox)_{\text{stage},N}^{\text{atom},i}$$

4. Sum the differences determined in step 3 over all key atoms in stage j. This yields the term

$$\sum_{\text{atoms},i}\left[(Ox)_{\text{stage},j}^{\text{atom},i} - (Ox)_{\text{stage},N}^{\text{atom},i}\right] = \Delta_j$$

5. Finally, take the sum $\sum_{\text{stages},j}\Delta_j$ over the number of stages and divide by $N+1$ accounting for the extra zeroth reaction stage.

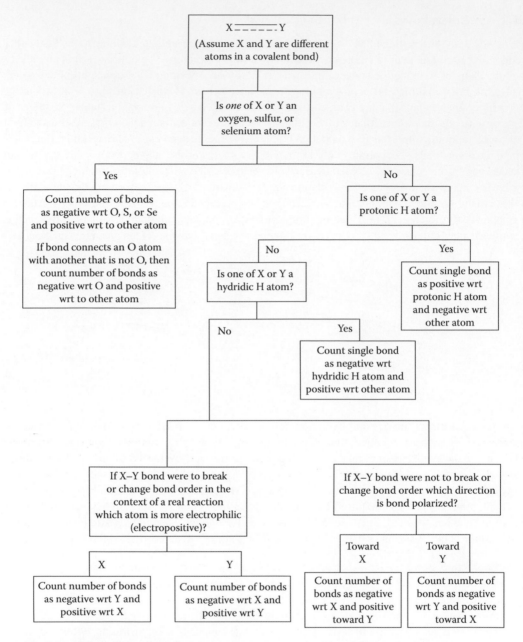

FIGURE 5.3 Flowchart for determining oxidation numbers of atoms in chemical structures.

In order to begin applying these steps, one needs to know how to determine the oxidation numbers of atoms in chemical structures in a systematic way. Figure 5.3 shows a flowchart that can be used as a guide. Equation 5.2 can be used to numerically evaluate the oxidation number of a given atom:

$$Ox_{number} = q - n + p \qquad (5.2)$$

where
    q is the charge on the atom
    n is the number of negative bond contributions
    p is the number of positive bond contributions

Equation 5.3 provides a convenient check calculation.

$$\sum_{j,\text{atoms}} (Ox_{\text{number}})_j = \text{overall charge of molecule} \tag{5.3}$$

Following is a list of useful tips with illustrated examples.

1. If X and Y are identical atoms, then the bond between them does not contribute either way. The oxidation number for each atom will only depend on its formal charge.

   Examples:

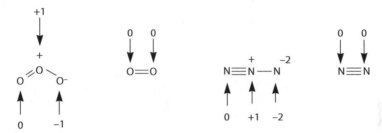

2. Atoms with high oxidation numbers are generally more electrophilic. Atoms with low oxidation numbers are generally more nucleophilic.
3. The more oxygen atoms are bonded to a given atom that is not oxygen, the higher the oxidation number of that atom. The more protonic hydrogen atoms are bonded to a given atom that is not hydrogen, the lower the oxidation number of that atom.

   Examples:

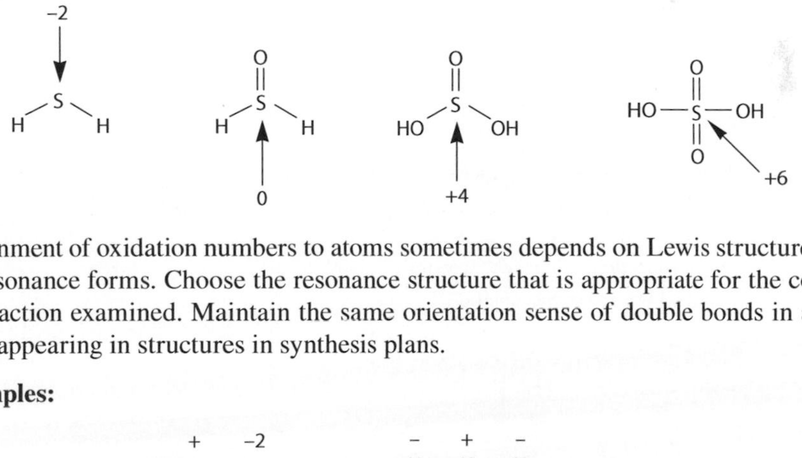

4. Assignment of oxidation numbers to atoms sometimes depends on Lewis structures drawn for resonance forms. Choose the resonance structure that is appropriate for the context of the reaction examined. Maintain the same orientation sense of double bonds in aromatic rings appearing in structures in synthesis plans.

   Examples:

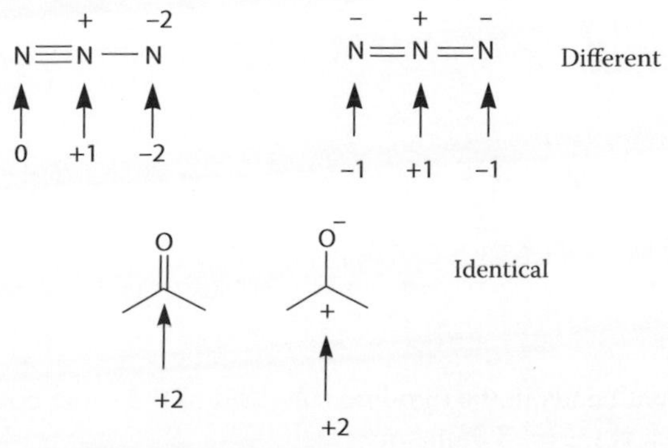

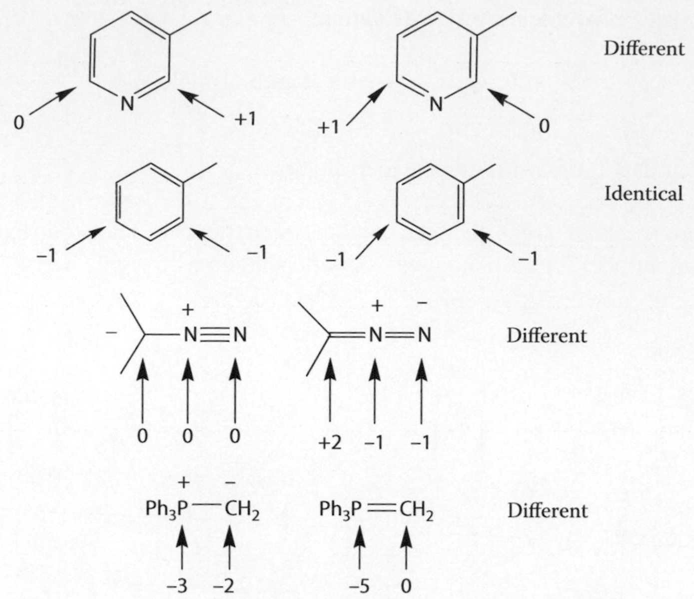

Different

Identical

Different

Different

5. In the context of tracking the oxidation numbers of a given atom from intermediate structure to intermediate structure in a synthesis plan, what is important is the progression of the *change* in oxidation number rather than the absolute values of the oxidation numbers.

Example:

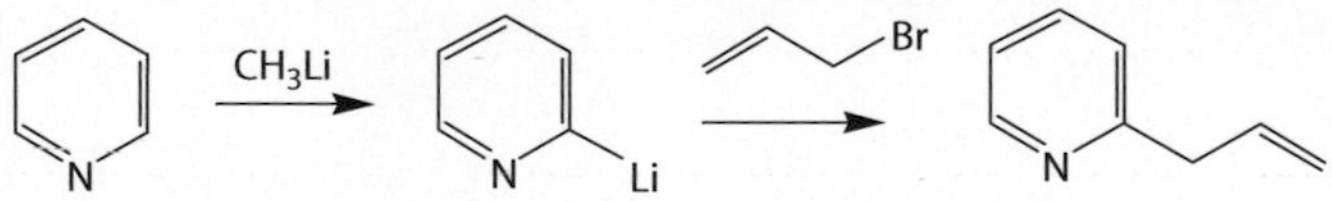

Oxidation state at C2:

    0   0   +1

Change in oxidation state with respect to C2 in target product, Δ_{C2}:

    −1   −1   0

Sum of oxidation number changes at C2:

    $(-1) + (-1) = -2$

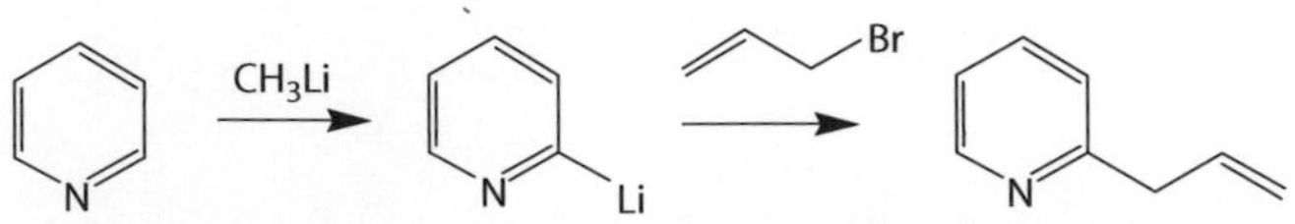

Oxidation state at C2:

    +1   +1   +2

Change in oxidation state with respect to C2 in target product, Δ_{C2}:

    −1   −1   0

Sum of oxidation number changes at C2:

    $(-1) + (-1) = -2$

Both orientations of double bonds in the pyridine ring lead to the same conclusion that the sum of oxidation number changes at C2 is −2 units.

Exercises in assigning oxidation numbers to particular atoms:

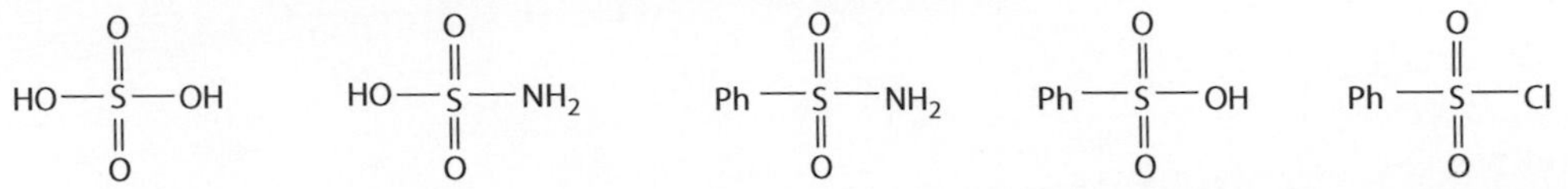

S: +6   +4   +2   +4   +2
N: −1   −1

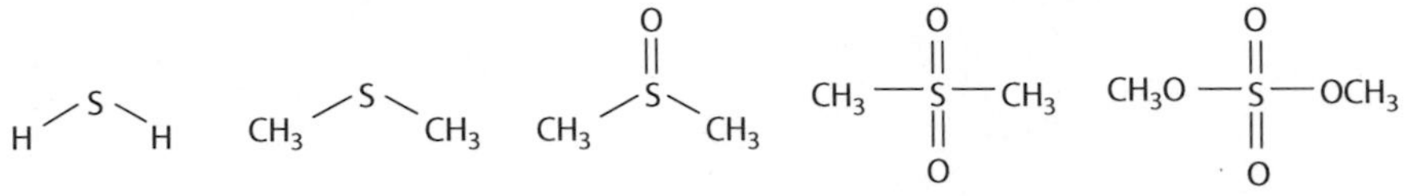

S: −2   −2   0   +2   +6

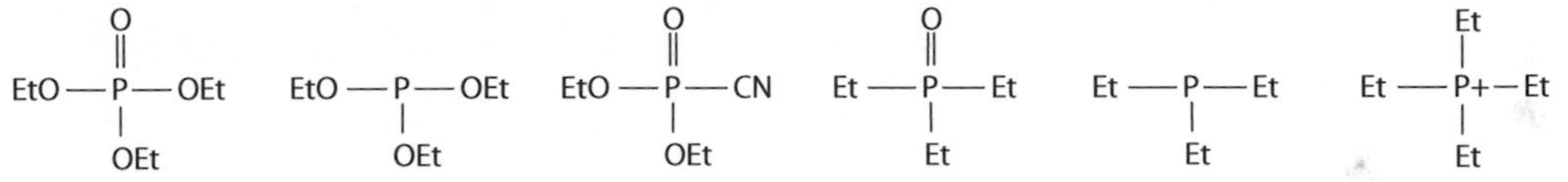

P: +5   +3   +3   −1   −3   −3

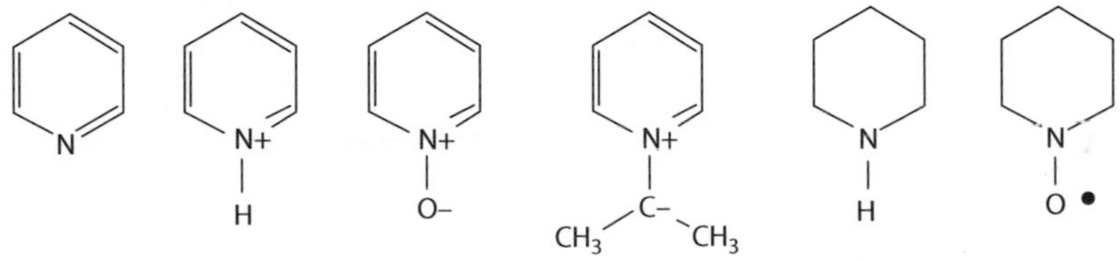

N: −3   −3   −1   −3   −3   −1

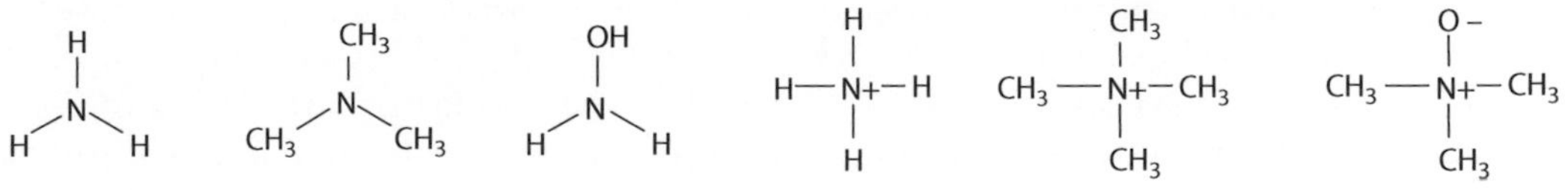

N: −3   −3   −1   −3   −3   −1

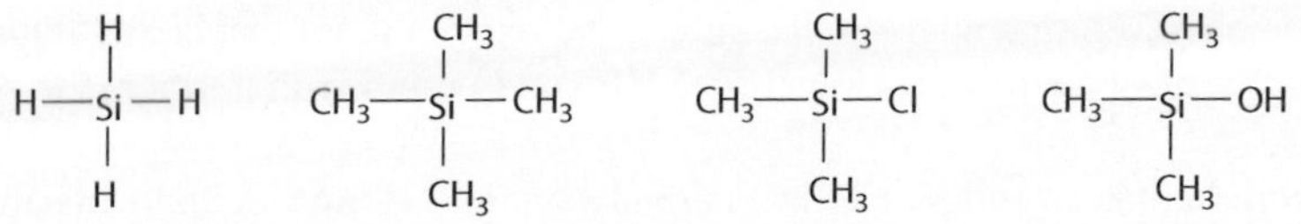

Si: +4   +4   +4   +4

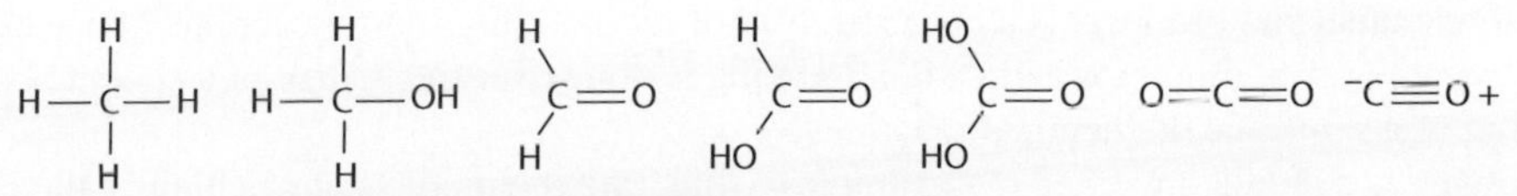

C: −4   −2   0   +2   +4   +4   +2

C: −4   −2   0   +2   +4   +2

I: 0   +1   +7   +5   +3   −1   +3

Cl: −1   0   +1   +3   +5   +7

One can see from this discussion that the concept of assigning oxidation numbers to atoms in a chemical structure is hardly an arbitrary exercise. In fact the following "rules" appearing in high school chemistry books on oxidation numbers that students were forced to memorize are completely rationalized.

1. For any free element the oxidation number is zero. (This is because no electrons are added or subtracted from that atom.)
2. For any monoatomic ion, the oxidation number is equal to the ionic charge. (This follows directly from (1).)
3. For any polyatomic ion, the sum of the oxidation numbers is equal to the ionic charge. (This follows from (1) and (2).)
4. For any neutral compound, the sum of the oxidation numbers is zero. (This is a consequence of algebraic addition.)
5. For hydrogen, the oxidation number in compounds is +1 except in metal hydrides in which it is −1. (The case of +1 corresponds to protic hydrogen and the case of −1 corresponds to hydridic hydrogen.)
6. For oxygen, the oxidation number in compounds is −2 except in peroxides in which it is −1. In the odd example OF_2 it is +2 since fluorine is more electronegative than oxygen and hence, it is observed that nucleophiles attack the oxygen atom displacing fluoride ion. For example, OF_2 reacts with water to produce HF and O_2, and with ammonia to produce N_2O and NH_4F.
7. In their compounds Group IA elements are always +1, Group IIA elements are always +2, and Group IIIA elements are mostly always +3.

By now, the reader will agree that the concept of tracking changes in oxidation numbers for all atoms in reactants and products is a powerful way of recognizing redox reactions in organic chemistry. It becomes even more powerful when tracking is done over the course of a chemical synthesis consisting of a sequence of chemical reactions.

Following is a detailed worked out example to illustrate the method of calculating the HI for the Perkin [58,59] plan for (±)-alpha-terpineol that will convince the reader of the totality and utility of these ideas.

Target bond map

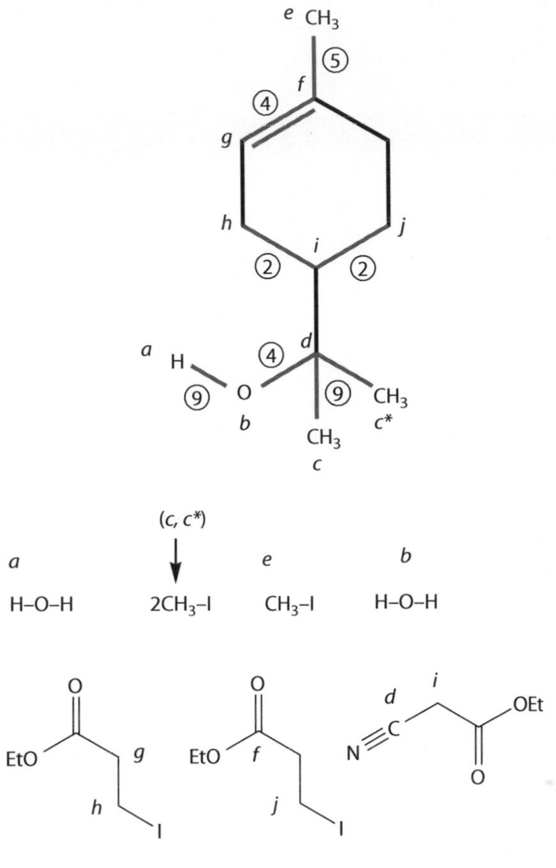

Scheme

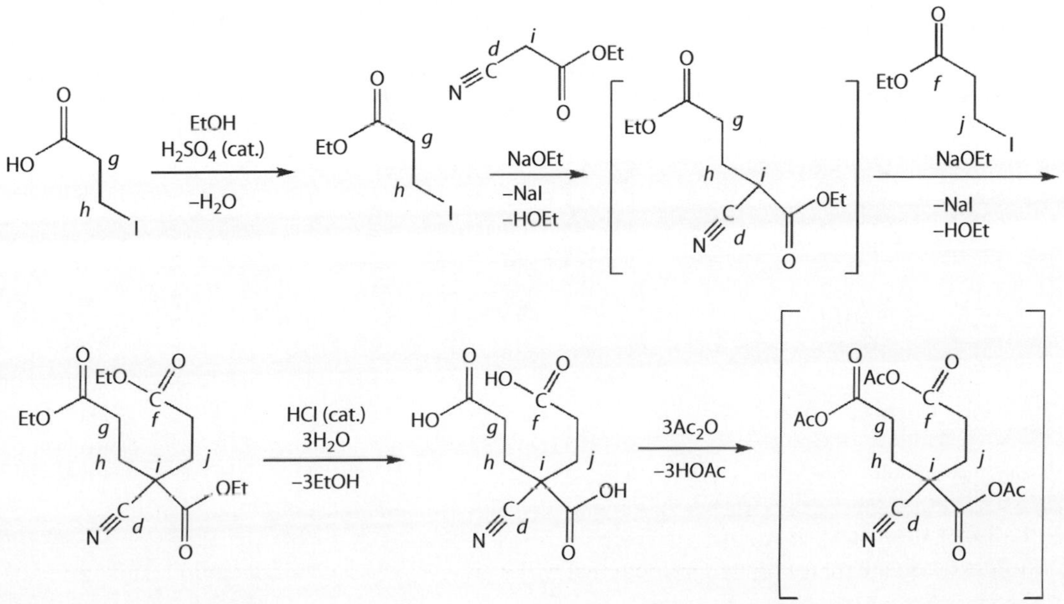

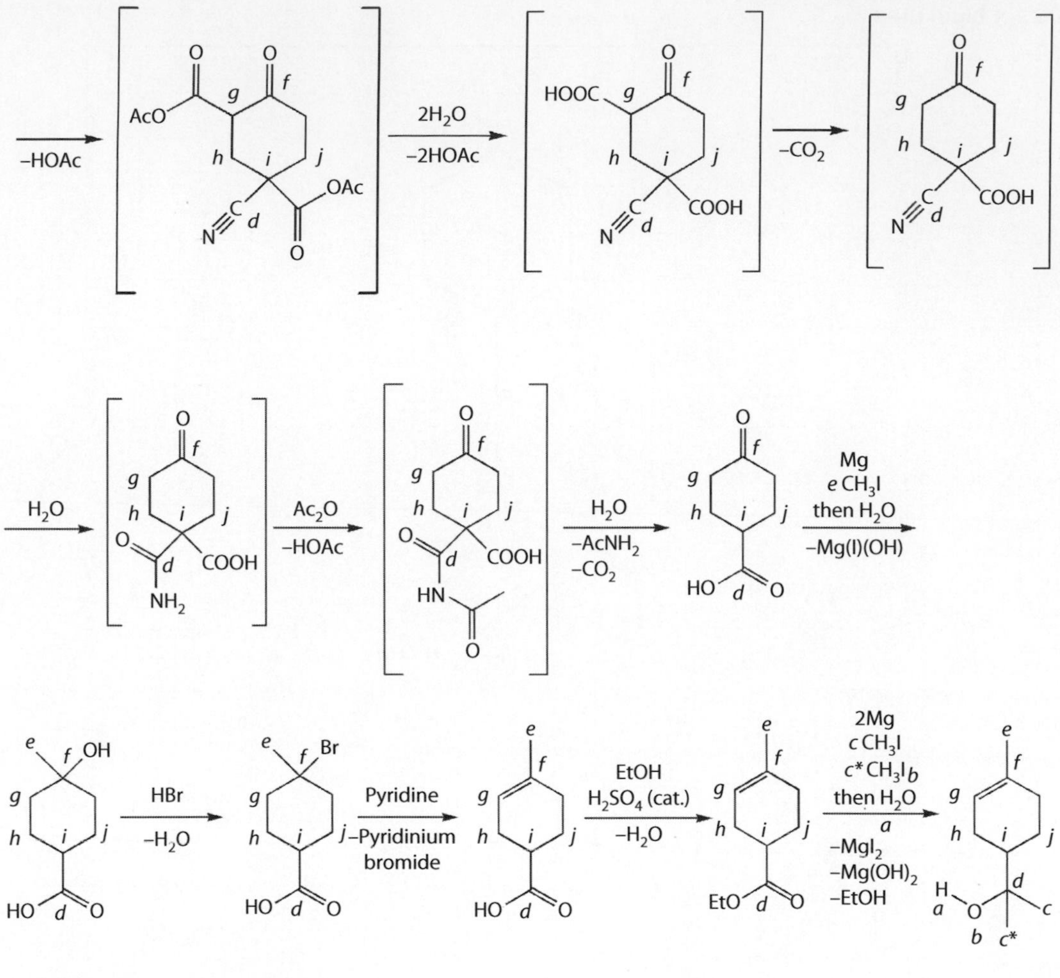

Calculation of hypsicity index

Rxn Stage	a	b	c	c*	d	e	f	g	h	i	j	Delta Row
9	1	−2	−3	−3	1	−3	0	−1	−2	−1	−2	0
8	1	−2	−2	−2	3	−3	0	−1	−2	−1	−2	4
7		−2			3	−3	0	−1	−2	−1	−2	2
6		−2			3	−3	1	−2	−2	−1	−2	2
5		−2			3	−3	1	−2	−2	−1	−2	2
4		−2			3	−2	2	−2	−2	−1	−2	4
3		−2			3		3	−2	−2	0	−2	5
2					3		3	−2	−2	0	−2	5
1					3		3	−2	−1	−2	−1	5
0								−2	−1			0
												29 Sum column
Delta column	0	0	1	1	16	1	13	−7	2	1	1	29 Sum row

$N = 9$

$HI = 29/(9 + 1) = +2.9$

Hypsicity profile

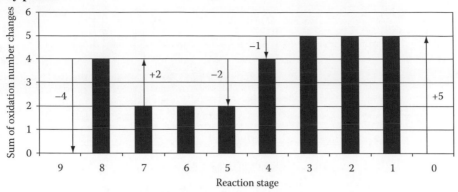

Oxidation length, |UD| (counts the "ups" and "downs" in hypsicity profile)

$$UD = 5 - 1 - 2 + 2 - 4 = 0$$

$$|UD| = 5 + 1 + 2 + 2 + 4 = 14$$

Section 12.7 has a detailed discussion of patterns found in the synthesis database regarding hypsicity trends. "Green" plans with respect to this parameter are highlighted. Appendix A.6 gives a summary of hypsicity trends for reactions contained in the named organic reaction database (see Chapter 7). Examples of each of the possible types of hypsicity profile, taken from the synthesis database, are given in Figures 5.4 and 5.5. The reader is directed to the accompanying download for plan schemes and references.

Some trends with respect to the |UD| parameter are as follows:

1. UD, the algebraic sum of "ups" and "downs" in any hypsicity profile, is always zero.
2. |UD| is always an even number.
3. For monotonic increases (hypohypsic profile) or monotonic decreases (hyperhypsic profile).

|UD| is twice the height of the first bar in the hypsicity profile (see Figure 5.5 for verification).

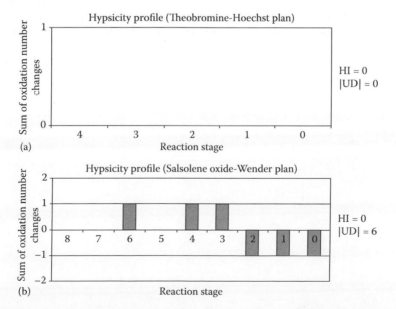

FIGURE 5.4 Isohypsic hypsicity profiles: (a) truly isohypsic and (b) by fortuitous algebraic cancellations of ups and downs.

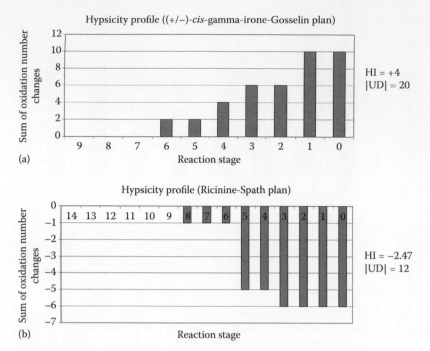

Hypsicity profile ((+/−)-*cis*-gamma-irone-Gosselin plan)

HI = +4
|UD| = 20

Hypsicity profile (Ricinine-Spath plan)

HI = −2.47
|UD| = 12

FIGURE 5.5 (a) Hyperphysic (net reduction) and (b) hypohypsic (net oxidation) hypsicity profiles.

Some interesting curiosities follow with respect to tracking the oxidation numbers of atoms in target bond forming steps in certain reactions.

5.4.2.1 Ozonolysis

Ozone is generated in situ from oxygen gas via an electrical discharge. Ozone then adds to the olefin in a [3 + 2] cycloaddition to form a 1,2,3-trioxane intermediate. This then fragments and rearranges to a 1,2,4-trioxane. The 1,2,4-trioxane can either be worked up in an oxidative work-up with hydrogen peroxide, for example, or in a reductive work-up with dimethylsulfide. The following scheme shows the two possibilities where all atoms are tracked. The problem of assigning oxidation numbers to ozone is bypassed because it originates from oxygen gas whose atoms have an oxidation number of 0. When we calculate the HI for these reactions, we compare the oxidation numbers of the relevant atoms in the product and in the reactants. Formally, ozone is not a reactant since it is made in situ. If we were to use ozone as the reactant, then we would have an ambiguity in the tracking of oxygen atoms because the two resonance forms of ozone would result in two possible sets of labels depending on the orientation of [3 + 2] addition to the olefin.

Oxidative work-up:

$$\Delta = \Delta(a) + \Delta(b) + \Delta(c) + \Delta(d) + \Delta(e) + \Delta(f) + \Delta(g)$$

$$= (1-1) + (-2-0) + (3-(-1)) + (-2-0) + (-2-(-1)) + (3-(-1)) + (-2-0) = +1;$$

$$HI = -\Delta/(N+1) = -\tfrac{1}{2}$$

Reductive work-up:

$$\Delta = \Delta(d) + \Delta(c) + \Delta(f) + \Delta(g) = (-2-0) + (1-(-1)) + (1-(-1)) + (-2-0) = 0; \quad HI = 0$$

Note that in the oxidative work-up case, we obtain a negative value for HI as expected for an oxidation taking place at the olefinic carbon atoms. However, the reductive work-up case yields an apparent isohypsic situation because the newly formed sulfur–oxygen bond in dimethyl sulfoxide is not counted in the HI calculation, as it is not one of the target bonds found in the intended target products.

5.4.2.2 Dihydroxylation

There are two possible ways that osmium tetraoxide can be used to dihydroxylate olefins. One reaction uses it in stoichiometric amounts and the other uses it in catalytic amounts in conjunction with another oxidant that recycles it in situ. In the former case, it is easy to track the origin of the oxygen atoms in the dihydroxy product as shown in the following.

$$\Delta = \Delta(a) + \Delta(b) + \Delta(c) + \Delta(d) + \Delta(e) + \Delta(f) + \Delta(g) + \Delta(h) + \Delta(i) + \Delta(j) + \Delta(k) + \Delta(l)$$

$$= 4 \times (-2-(-2)) + 4 \times (1-1) + 4 \times (0-(-1)) = 4; \quad HI = -\Delta/(N+1) = -2$$

In the latter case, after one cycle of producing dihydroxylated product we would get the same results as the previous one, that is, the oxygen atoms in the diols originate from OsO_4. However, in the second round, we recycle OsO_2 back to OsO_4 via N-methylmorpholine oxide, which contributes two oxygen atoms.

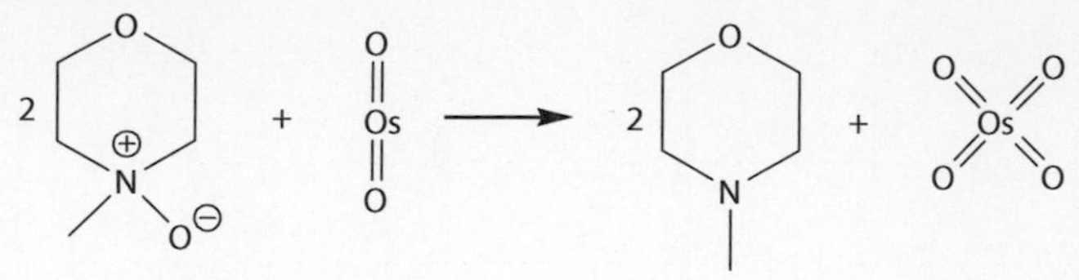

It is easy to see that by the fourth cycle all of the oxygen atoms in OsO_4 will have originated from the morpholine N-oxide, and these in turn will end up in the diol products. So, for this case, as far as tracing the oxygen atoms is concerned for the purpose of calculating the HI, it is as if N-methylmorpholine oxide is directly oxidizing the olefin as shown by the following scheme. Note that one oxygen atom originates from the N-oxide and the other oxygen atom comes from water. The implication is that the olefin initially forms an epoxide that is then ring-opened by water attack. Of course, this is inconsistent with the stereochemistry of the *cis*-addition, which makes sense if OsO_4 were the oxygen transfer agent.

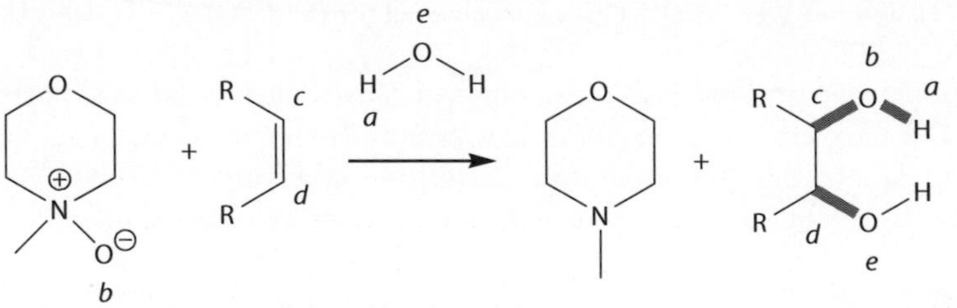

$$\Delta = \Delta(a) + \Delta(b) + \Delta(c) + \Delta(d) + \Delta(e)$$

$$= (1-1) + (-2 - (-2)) + (0 - (-1)) + (0 - (-1)) + (-2 - (-2)) = +2; \quad HI = -\Delta/(N+1) = -1$$

Note that the HI value in this case is -1. The value for the former case is twice this value because two diol products are formed, so really the two results are equivalent. Since OsO_4 is used as a catalyst, it is recovered and it is the N-methylmorpholine oxide that is consumed. This paradoxical situation of using a hypothetical redox reaction in place of the real mechanistically correct one is the only way to satisfy both the tracing of oxygen atoms and the observed net consumption of N-oxide in the overall balanced chemical equation. This will also impact on the determination of f(sac) since the N-oxide will be counted as contributing to the target bond structure map and not osmium tetraoxide.

5.4.2.3   Generation of Diazomethane

When making methyl esters from carboxylic acids, diazomethane is generated first from diazald. Just as in the case of ozone in the ozonolysis reaction, the problem of determining oxidation states for atoms in different resonance structures of diazomethane is avoided since it is generated in situ.

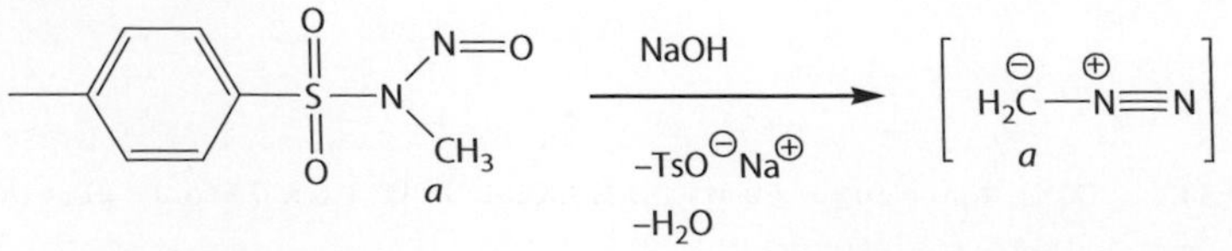

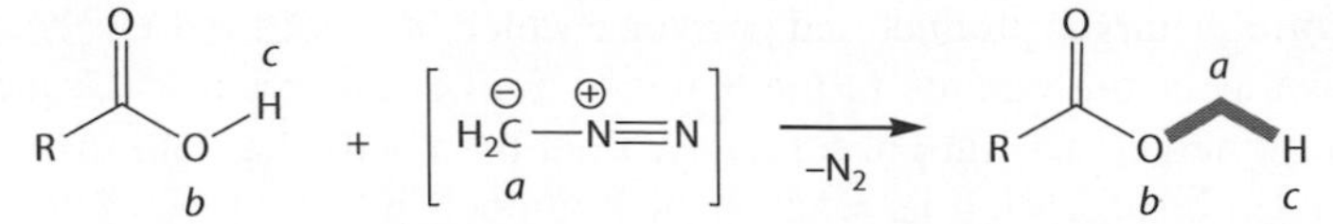

$$\Delta = \Delta(a) + \Delta(b) + \Delta(c) = (-2 - (-2)) + (-2 - (-2)) + (1 - 1) = 0; \quad HI = 0$$

5.4.2.4 Tracking Origin of Sulfur Atom in Saccharin Synthesis

In the Maumee Chemical [60] plan for saccharin, there is a curiosity with respect to the oxidation number of the sulfur atom in the source material as shown in step 1 of the following scheme. It could be either sodium sulfide (oxidation number = −2) or elemental sulfur (oxidation number = 0). In this case, the arithmetic mean of these values was used as the oxidation number of the sulfur atom in the starting material (oxidation number = −1) which, by the way, corresponds to the oxidation number of sulfur in disodium disulfide that is generated in situ from sodium sulfide and elemental sulfur.

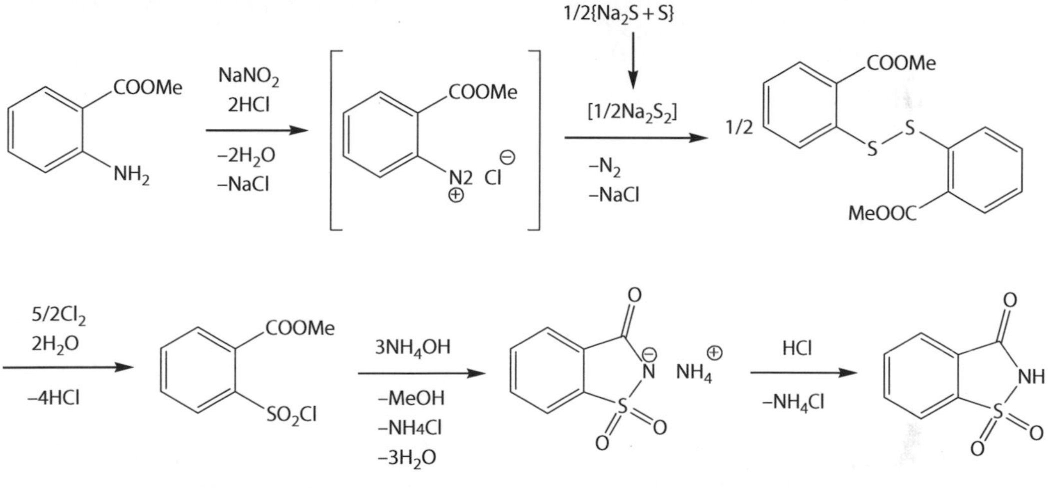

Appendix B gives a listing of challenging redox reactions found in the extensive synthesis database. The reader may use this compilation as a training exercise in assigning oxidation numbers to atoms to identify which are being oxidized and which are being reduced, in writing out appropriate redox half reactions, in writing out rational mechanisms that are consistent with oxidation number changes, half reactions, and stoichiometric balancing, and in determining associated hypsicity indices.

5.5 INSTRUCTIONS ON USING PENTAGON, LINEAR, AND CONVERGENT SPREADSHEETS

Sequence of steps to determine global material efficiency green metrics and synthesis efficiency parameters for any synthesis plan:

1. Obtain detailed experimental procedures for all reactions in the synthesis plan from literature references.
2. Write out set of complete balanced chemical equations for each reaction in the plan showing structures, stoichiometric coefficients, and MWs of all reagents, target product, and byproducts. Note which intermediates are isolated and assign individual reaction yield performances accordingly. Note reaction steps that are telescoped, that is, those that involve sequential chemical transformations without isolation of intermediate products. Note the number of branches involved in the synthesis plan.

3. Draw structure of target product and correlate which atoms in the target structure correspond to which source atoms in the reagents used to construct it. Group the subsets of atoms in the target structure accordingly. Bold the bonds that join the subset groups. For each bolded bond, assign its reaction step number indicating at which reaction step that target bond was made. Beginning with the highest reaction step number and working backward, label the atoms constituting the bond connectivities using letter labels.

4. Construct a synthesis tree for the plan showing structures and MWs for all input materials, and MWs for all intermediates. Input individual reaction yield performances for all steps in the plan. Note the total number of reaction stages, reaction steps, number of branches, and total number of input materials.

INPUTS:

    MW of each reagent
    MW of all intermediates in all branches
    MW of final target product
    Chemical structure of each reagent
    Stoichiometric coefficient of each reagent

OUTPUTS:

    Number of reaction stages
    Number of reaction steps
    Number of input materials
    Number of branches

5. For each balanced chemical equation in the plan, use the RADIAL PENTAGON spreadsheet to obtain its global RME performance as function of AE, reaction yield, excess reagent consumption (inverse of stoichiometric factor (SF)), and auxiliary material consumption (reaction solvents, work-up materials, and purification materials).

    Enter "1" in stoichiometric coefficient (SC) column for each blank reagent entry to preserve line numbers and to ensure that check calculation for SF works properly. Also, delete formula entries under "moles of reagent" (column F) for each blank reagent entry.

    Note the values of reaction yield, reaction scale, SF value, mass of excess reagents, mass of reaction solvents, mass of catalysts, ligands, and impurities, mass of work-up materials, and mass of purification materials.

INPUTS for PENTAGON spreadsheet:

    Reaction literature reference
    Step number
    Reaction classification
    Chemical drawing of complete balanced reaction showing structures
    MW of each reagent
    Stoichiometric coefficient of each reagent
    MW of target product
    Masses, volumes, and densities of all materials used
    Mass of target product collected

OUTPUTS from PENTAGON spreadsheet:

    Number of moles of all reagents, catalysts, and ligands used
    Reaction yield with respect to limiting reagent
    Reaction scale with respect to limiting reagent
    Stoichiometric factor

Mass of excess reagents
Mass of reaction solvents
Mass of work-up materials
Mass of purification materials
Atom economy (AE)
Minimum reaction mass efficiency (RMEmin) when everything but the target product
is destined for waste
Maximum reaction mass efficiency (RMEmax or kernel RME) when only the reaction
byproducts are destined for waste
E-kernel, E-excess, E-aux, and E-total for case where only the target product is the only
material collected and everything else is destined for waste

6. Use LINEAR or CONVERGENT spreadsheet to calculate overall material efficiency performance for plan. Enter data beginning from the last step and working backward toward
the starting materials of each branch following the connectivity of the synthesis tree).

Entering and deleting rows in spreadsheet:

LINEAR spreadsheet is defaulted to a 10-step plan.
CONVERGENT spreadsheet is defaulted to a plan having two 10-step branches.

Insert or delete rows as appropriate anywhere except at the top or bottom rows of each block to avoid
cancellation of embedded cell formulas.

For the following columns, the fidelity of cell formulas is maintained upon inserting or deleting rows:

E(mw) for reaction (column E)
AE for reaction (column F)
Reaction scale in moles defined by synthesis tree (column K)
Kernel RME (column L)
E-kernel (column M)
Kernel mass of reagents (g) (column N)
Kernel mass of waste (g) (column O)
% Kernel waste (column P)
Excess reagents contribution to mass of waste (g) (column R)
Contribution to E-excess (column S)
Total mass of auxiliaries (g) (column X)
Scaling correction factor (column Z)
Corrected auxiliary contribution to mass of waste (g) (column AA)
Contribution to E-auxiliaries (column AB)
Total mass of waste (g) (column AC)
Contribution to E-total (column AD)
% True waste (column AE)
Energy input directed to product based on kernel reaction RME (column AL)

For the following columns the cell formulas are locked with respect to the mole scale of the target
product (cell C5):

Kernel mass of reagents (g) (column N)
Kernel mass of waste (g) (column O)
Excess reagents contribution to mass of waste (g) (column R)
Contribution to E-excess (column S)

Scale correction factor (column Z)
Contribution to E-auxiliaries (column AB)
Contribution to E-total (column AD)

For the following columns it is necessary to copy and paste the *first* cell in the column that contains an embedded formula onto the remaining cells in the column:

Cumulative yield (column J)
Cumulative kernel mass of waste (g) (column Q)
Cumulative true mass of waste (g) (column AF)
Columns to check balancing of chemical equations: BV, BW, and BX

To insert extra branches beyond the two default branches in CONVERGENT spreadsheet, copy entire block of main branch from column headings to subtotals line.

For branches 2 and beyond, two adjustments are required. The first is to adjust the entries in the cumulative yield boxes appearing at the top of each subbranch block. The cell formula will involve multiplying the reaction yield corresponding to the terminal step in the subbranch by the appropriate mole scale in column J that links that subbranch to the main branch. For example, if from the synthesis tree diagram, the subbranch ends at stage X and links with the main branch in a convergent step at stage $X + 1$, then the mole scale in column J that should be used is the one that corresponds to stage $X + 1$ in the main branch. The second change is to adjust cell formulas for overall E-kernel performance for each subbranch in the subtotals line by dividing the mole scale of the final target product (C5) by the mole scale value found in column J that links that branch to the main branch. This J value, which was determined from the first adjustment, corresponds identically to the one used in the cumulative yield box at the top of the block for that branch and represents the mole scale of the terminal node for the given subbranch. Both of these adjustments depend directly on the branch connectivities in the synthesis tree diagram and therefore cannot be preset ahead of time.

Calculation of overall material efficiency performance

INPUTS for LINEAR-COMPLETE or CONVERGENT-COMPLETE spreadsheets:

Target product name
Plan literature reference
Plan type (LINEAR or CONVERGENT).
Number of branches (1 for LINEAR, >1 for CONVERGENT)
Target product scale (set to 1 mol by default)
Number of reaction stages
Number of reaction steps
Number of input materials

For each branch section:

MW of reaction intermediates (column C)
MW of target product (column C)
Sum of MWs of reagents for each step (column B)
Sum of MWs of by-products for each step (column D)

Use check calculation to check correct balancing of chemical equation in each step. Copy and paste columns B, C, and D into columns BS, BT, and BU. Check to see that entries in column BX are all zero for correct balancing of each chemical equation.

From radial pentagon analysis for each step:

Mass of excess reagents (column G)
SF (column H)
Reaction yield (column I)
Mass of reaction solvents (g) (column T)
Mass of work-up materials (g) (column U)
Mass of purification materials (g) (column V)
Mass of catalysts and ligands (g) (column W)
Mole scale of limiting reagent (column Y)

For convergent plans, make sure that the values of the cumulative yield entries for secondary and other branches correctly link with the appropriate reaction steps of the main branch according to the connectivity of the synthesis tree.

OUTPUTS from LINEAR-COMPLETE or CONVERGENT-COMPLETE spreadsheets:

Overall AE
Overall yield along longest branch
Overall kernel RME
E-kernel
E-excess
E-auxiliaries
E-total
Overall true RME
Mass of waste due to by-products and side products (kernel) (g)
Mass of waste due to excess reagents (g)
Mass of waste due to auxiliary materials (g)
Total mass of waste (g)

Calculation of MW first moment parameter

INPUTS:

MW of target product and all reaction intermediate in all branches
MW of starting materials at the beginning of each branch whose structures in whole or in part
    end up in the subsequent intermediate product

OUTPUT:

MW first moment parameter

Copy and paste column C entries in branch blocks into column C block in MW First Moment Calculation section. Also include MW of reagents at beginning of each branch that end up in whole or in part in the next intermediate product.

Check calculation:

INPUTS:

For each branch:

Sum of MW of intermediate products
Number of product nodes
Sum of MW of reagents at start of branch whose structures in whole or in part end up in the
    subsequent intermediate product
Number of starting material nodes

OUTPUT:

MW first moment parameter

Calculation of fraction of sacrificial reagents by MW

INPUTS:

Itemization of reagents ending up in target product according to grouping analysis in Part 3
MW of these reagents

OUTPUT:

f(sac) by MW.

Calculation of degree of asymmetry and degree of convergence
Follow connectivities as per synthesis tree.

INPUTS:

Number of input material dots for each reaction step beginning with the first step in the longest branch of plan. For stage 2 and beyond, the input materials for a given stage are the dark dots for reagents and the open dot for the intermediate product that was obtained in the preceding step. Copy and paste embedded cells in column C as appropriate. Adjust formulas in those cells for each stage according to ordinate values in synthesis tree for immediately preceding intermediate products and input reagent materials.

OUTPUTS:

Centroids of input material dots = ordinates of subsequent intermediate products
Ordinate of dot corresponding to final target product
Degree of convergence
Degree of asymmetry

Ensure that final cell in column C in this section is linked to product ordinate cell in column A under heading title.

Calculation of kernel raw materials cost
For each reaction step enter list of reagents used, the cost per gram of each reagent, and the MW of each reagent. Check that entries in column E for the mole scale of each reagent corresponds to the mole scales given in column K.

Repeat the process for each branch block.

INPUTS:

Reagent name
Cost per gram
MW of reagent

OUTPUTS:

Total kernel raw material cost to produce 1 mol of target product

Calculation of hypsicity (oxidation level) index
From atoms identified in Part 3 involved in the target bond forming reactions, label them with the same letter symbols in each intermediate structure beginning with the final target product and working backward all the way to the respective original input reagents.

INPUTS:

Oxidation states for each atom in the preceding set traced stage by stage from target product to corresponding input reagent.

OUTPUT:

HI

Add extra columns as required for atoms tracked by oxidation state. Copy and paste cells in "delta column" and "delta row" as appropriate. All cell formulas are copied with complete fidelity. Check that numerator in cell formula for HI refers to sum of "delta row" column.

For each column, trace the oxidation state of that atom back to the structure of the source reagent. If the source reagent happens to be introduced at a reaction stage later than the zeroth stage, copy and paste the value of the oxidation state for that atom in the final structure into all the remaining empty cells in that column back to the zeroth stage. Blacken these cells to make it easy to recognize at what reaction stage that atom is first introduced in the synthesis.

Determination of target bond forming reaction profile
Obtained directly from set of bolded bonds in final target product structure found in Part 3.

5.6 MATHEMATICAL DERIVATIONS AND ANALYSIS

All derivations are based on a balanced chemical equation of the form

$$aA + bB \rightarrow pP + qQ$$

where a, b, p, and q are the stoichiometric coefficients of reagents A and B, and target product P and byproduct Q. The MWs of these chemical species are denoted as MW_A, MW_B, MW_P, and MW_Q.

5.6.1 DERIVATION OF EXPRESSION FOR OVERALL E-FACTOR FOR ANY CHEMICAL REACTION

Suppose in the balanced chemical equation that reagent A is the limiting reagent where x moles of A are used, and that an excess of reagent B is used (z moles), where $z > x$. Let the masses of reaction solvent, catalyst, and all auxiliary materials in the work-up and purification stages be s, c, and ω, respectively. Assume that these are not recovered and are destined for waste. Let the moles and mass of target product P collected be y and yMW_P, respectively. Rewrite the balanced chemical equation as

$$A + (b/a)B \rightarrow (p/a)P + (q/a)Q.$$

Beginning with the definition of overall E-factor as the mass ratio of all waste material to target product collected, we have

$$E = \frac{x\mathrm{MW}_A + z\mathrm{MW}_B + c + s + \omega - y\mathrm{MW}_P}{y\mathrm{MW}_P} \tag{5.4}$$

Define the excess moles of B used as $\phi = z - x(b/a)$.
Define the AE as

$$\mathrm{AE} = \frac{(p/a)\mathrm{MW}_P}{\mathrm{MW}_A + (b/a)\mathrm{MW}_B}. \tag{5.5}$$

Since the chemical equation is balanced, we have

$$AE = \frac{(p/a)MW_P}{(p/a)MW_P + (q/a)MW_Q}.$$ (5.6)

We may rewrite Equation 5.4 as

$$E = \frac{xMW_A + (\varphi + x(b/a))MW_B + c + s + \omega}{yMW_P} - 1$$

$$= \frac{x(MW_A + (b/a)MW_B) + \varphi MW_B + c + s + \omega}{yMW_P} - 1$$

$$= \frac{MW_A + (b/a)MW_B + \dfrac{\varphi MW_B}{x} + \dfrac{c + s + \omega}{x}}{\dfrac{yMW_P}{x}} - 1$$ (5.7)

Define reaction yield with respect to limiting reagent A as

$$\varepsilon = \frac{y}{x(p/a)}.$$ (5.8)

Then, E becomes

$$E = \frac{MW_A + (b/a)MW_B + \dfrac{\varphi MW_B}{x} + \dfrac{c + s + \omega}{x}}{(\varepsilon)(p/a)MW_P} - 1.$$ (5.9)

Dividing the numerator and denominator by $MW_A + (b/a)MW_B$, we have

$$E = \frac{1 + \dfrac{\varphi MW_B}{x[MW_A + (b/a)MW_B]} + \dfrac{c + s + \omega}{x[MW_A + (b/a)MW_B]}}{(\varepsilon)(AE)} - 1.$$ (5.10)

Define the stoichiometric factor as

$$SF = 1 + \frac{\varphi MW_B}{x[MW_A + (b/a)MW_B]},$$ (5.11)

where the numerator is the excess mass of reagent B and the denominator is the sum of the stoichiometric masses of reagents A and B.

Then, E becomes

$$E = \frac{SF + \dfrac{c+s+\omega}{x[MW_A + (b/a)MW_B]} - 1}{(\varepsilon)(AE)}$$

$$= \frac{(SF)\left[1 + \dfrac{c+s+\omega}{x[MW_A + (b/a)MW_B](SF)}\right] - 1}{(\varepsilon)(AE)}$$

$$= \frac{(SF)\left[1 + \dfrac{(c+s+\omega)(AE)}{x(p/a)MW_P(SF)}\right] - 1}{(\varepsilon)(AE)}$$

$$= \frac{(SF)\left[1 + \dfrac{(c+s+\omega)(AE)(\varepsilon)}{yMW_P(SF)}\right] - 1}{(\varepsilon)(AE)} \tag{5.12}$$

Define the material recovery parameter (MRP) as

$$MRP = \frac{1}{1 + \dfrac{(c+s+\omega)(AE)(\varepsilon)}{yMW_P(SF)}}. \tag{5.13}$$

Finally, we have

$$E = \frac{(SF)}{(\varepsilon)(AE)(MRP)} - 1 \tag{5.14}$$

5.6.2 CONNECTION BETWEEN STOICHIOMETRIC FACTOR AND MOLE PERCENT EXCESS REAGENT

For the balanced equation given earlier, the mole percent excess reagent used is

$$\%\text{excess} = \frac{\varphi}{x(b/a)} \times 100 = \left(\frac{z - x(b/a)}{x(b/a)}\right) \times 100. \tag{5.15}$$

From Equation 5.11 for the stoichiometric coefficient, we may relate it to the mole percent excess reagent used according to

$$SF = 1 + \frac{\varphi MW_B}{x[MW_A + (b/a)MW_B]}$$

$$= 1 + \frac{(\%\text{excess})x(b/a)MW_B}{x[MW_A + (b/a)MW_B]}$$

$$= 1 + \frac{(\%\text{excess})(b)MW_B}{[MW_A + (b)MW_B]}$$

$$= 1 + \frac{(AE)(\%\text{excess})(b)MW_B}{(p)MW_P} \tag{5.16}$$

We may generalize this result for a balanced chemical equation with more than one reagent used in excess.

$$SF = 1 + \frac{(AE)}{pMW_P} \sum_{j=1} (\%excess)_{B_j} (b_j) MW_{B_j} \tag{5.17}$$

where

b_j is the stoichiometric coefficient of reagent B_j

$(\%excess)_{B_j}$ is the mole percent excess reagent B_j used

5.6.3 Percent Conversion and Percent Selectivity

The percent conversion refers to the fraction of starting material that gets converted to products after a certain reaction time. If we begin with x moles of starting material and after a time t we produce products P1, P2, etc., and have remaining r moles of starting material, then the percent conversion is given by

$$\%conversion = \left(\frac{x-r}{x} \right) \times 100. \tag{5.18}$$

Supposed that p_j moles of product Pj are formed. The percent selectivity for product Pj is the fraction of the amount of starting material that reacted that ends up as product Pj and is given by

$$\%selectivity = \left(\frac{p_j}{x-r} \right) \times 100 \tag{5.19}$$

The percent yield of product Pj formed with respect to the starting material, assuming it is the limiting reagent, is given by

$$\%yield = \left(\frac{p_j}{x} \right) \times 100 \tag{5.20}$$

Therefore, from these relations we can conclude that

$$\%yield = (\%conversion) \times (\%selectivity) \tag{5.21}$$

5.6.4 Derivation of Expressions Relating AE and E(mw), and Overall RME and E-Factor

For a balanced chemical equation, the E-factor based on MWs of chemical species is given by

$$E(mw) = \frac{MW_{byproducts}}{MW_{target\ product}} = \frac{qMW_Q}{pMW_P}. \tag{5.22}$$

From Equation 5.6 for the AE we have

$$AE = \frac{(p/a)MW_P}{(p/a)MW_P + (q/a)MW_Q}$$

$$= \frac{pMW_P}{pMW_P + qMW_Q}$$

$$= \frac{1}{1 + \dfrac{qMW_Q}{pMW_P}}$$

$$= \frac{1}{1 + E(\text{mw})} \tag{5.23}$$

An expression for the overall RME for a chemical reaction can be obtained starting from

$$RME = \frac{yMW_P}{xMW_A + zMW_B + c + s + \omega} \tag{5.24}$$

and following through the same line of sequential steps as before for the derivation of Equation 5.14 for the overall E-factor. The result is

$$RME = (\varepsilon)(AE)\left(\frac{1}{SF}\right)(MRP) \tag{5.25}$$

which is the basis of the radial pentagon analysis. The connecting relationship is then

$$RME = \frac{1}{1 + E} \tag{5.26}$$

which mirrors Equation 5.23.

5.6.5 DERIVATION OF GOLDEN RATIO THRESHOLD FOR "GREEN" REACTIONS

If we set AE equal to $E(\text{mw})$ in Equation 5.23 or RME equal to E in Equation 5.26, we get the expression

$$X = \frac{1}{1 + X} \tag{5.27}$$

which can be solved to give the positive root $\phi = \dfrac{\sqrt{5} - 1}{2} = 0.618...$ identified as the golden ratio. This result can be used to define a threshold boundary above which a chemical reaction could be called "green" as shown by the plot in Figure 5.6. Using this criterion, about 55% of all reactions in the named organic database (see Section 7.11) satisfy this cutoff.

5.6.6 EXPRESSION FOR E-TOTAL FOR ANY CHEMICAL REACTION AS A SUM OF CONTRIBUTORS

For any chemical reaction, E-total is given by

$$E_{\text{total}} = E_{\text{kernel}} + E_{\text{excess}} + E_{\text{aux}} \tag{5.28}$$

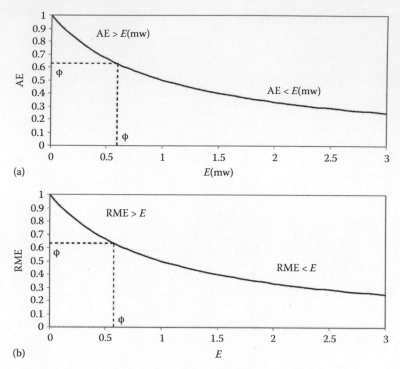

FIGURE 5.6 (a) Plot of AE versus E(mw) and (b) plot of RME versus E showing regions of "greenness."

where

$$E_{\text{kernel}} = \frac{\text{mass of all stoichiometic reagents} - \text{mass of target product}}{\text{mass of target product}}$$

$$E_{\text{excess}} = \frac{\text{mass of excess reagents}}{\text{mass of target product}}$$

$$E_{\text{aux}} = \frac{\text{mass of reaction solvent} + \text{mass of catalysts} + \text{mass of work-up and purification auxiliaries}}{\text{mass of target product}}$$

The kernel waste comprises by-products, side products, and unreacted starting materials (excluding excess reagents). Using the preceding definitions for these terms, we have the following universal relations for any chemical reaction:

$$E_{\text{kernel}} = \frac{1}{(\varepsilon)(\text{AE})} - 1$$

$$E_{\text{excess}} = \frac{(\text{SF}) - 1}{(\varepsilon)(\text{AE})} \tag{5.29}$$

$$E_{\text{aux}} = \frac{(\text{SF})}{(\varepsilon)(\text{AE})} \left[\frac{1}{\text{MRP}} - 1 \right]$$

The reader may verify that summing these three equations leads directly to equation $(1/\text{RME}) - 1$, which is consistent with Equation 5.26.

5.6.7 Expression for *E*-Total for a Synthesis Plan as a Sum of Contributors

The expression for E-total for any synthesis plan is given by

$$E_{\text{total}} = E_{\text{byproducts and unreacted reagents}} + E_{\text{excess reagents}} + E_{\text{auxiliaries}} \tag{5.30}$$

where

$$E_{\text{byproducts and unreacted reagents}} = E_{\text{kernel}} = \frac{1}{p_n} \sum_j \left(\frac{1}{\prod_k^{n \to j} \varepsilon_k} \right) \left(\frac{p_j}{(\text{AE})_j} \right) \left[1 - \varepsilon_j (\text{AE})_j \right] \tag{5.31}$$

$$E_{\text{excess reagents}} = \frac{1}{p_n} \sum_j \left(\frac{1}{\prod_k^{n \to j} \varepsilon_k} \right) \left(\frac{p_j}{(\text{AE})_j} \right) \left[(\text{SF})_j - 1 \right] \tag{5.32}$$

$$E_{\text{auxiliaries}} = \frac{1}{p_n} \sum_j \left(\frac{1}{\prod_k^{n \to j} \varepsilon_k} \right) \left(\frac{c_j + s_j + \omega_j}{x_j^*} \right) \tag{5.33}$$

Symbol definitions:

x is moles of target product in synthesis plan.

$\prod_k^{n \to j} \varepsilon_k$ is the multiplicative chain of reaction yields connecting the target product node to the reactant nodes for step j as per synthesis tree diagram read from right to left (i.e., in the direction $n \to j$).

p_j is the MW of product of step j.

ε_j is the reaction yield with respect to limiting reagent for step j.

$(\text{AE})_j$ is the AE for step j.

$(\text{SF})_j$ is the stoichiometric factor for step j that accounts for excess reagents used in that step.

$c_j + s_j + \omega_j$ is the sum of masses of auxiliary materials used in step j, namely the mass of catalyst, mass of reaction solvent, and mass of all other postreaction materials used in the work-up and purification phases.

x_j^* is the experimental mole scale of limiting reagent in step j as reported in an experimental procedure.

5.6.8 Expression for Raw Material Cost of Target Product for Any Chemical Reaction

$$\text{Cost} = \frac{\sum \$(\text{reagents}) + \$(\text{solvents}) + \$(\text{work-up materials}) + \$(\text{auxiliary materials})}{\text{mass of target product}}$$

This calculation is built in to the radial pentagon analysis.

5.6.9   Expression for Raw Material Cost of Target Product for Any Synthesis Plan

Reagents contribution:

$$\$(\text{reagents}) = \sum_j (\$ \text{ per gram})(\text{MW}_{\text{reagent}})(\text{mole scale of reagent})$$

$$= \sum_j (\$ \text{ per gram})(\text{MW}_{\text{reagent}})\left(\frac{x}{\prod_k^{n \to j} \varepsilon_k}\right)$$

where x is the mole scale of the target product.

Reaction solvents contribution:

$$\$(\text{solvents}) = \sum_j (\$ \text{ per gram})(\text{mass of solvent used in radial pentagon})(\text{scaling correction factor})$$

$$= \sum_j (\$ \text{ per gram})(\text{mass of solvent used in radial pentagon})\left(\frac{x}{x^* \prod_k^{n \to j} \varepsilon_k}\right)$$

Work-up materials contribution:

$$\$(\text{work-up materials}) = \sum_j (\$ \text{ per gram})(\text{mass of work-up material used in radial pentagon})$$
$$\times (\text{scaling correction factor})$$

$$= \sum_j (\$ \text{ per gram})(\text{mass of work-up material used in radial pentagon})$$
$$\times \left(\frac{x}{x^* \prod_k^{n \to j} \varepsilon_k}\right)$$

Auxiliary material contribution:

$$\$(\text{auxiliary materials}) = \sum_j (\$ \text{ per gram})(\text{mass of auxiliary material used in radial pentagon})$$
$$\times (\text{scaling correction factor})$$

$$= \sum_j (\$ \text{ per gram})(\text{mass of auxiliary material used in radial pentagon})$$
$$\times \left(\frac{x}{x^* \prod_k^{n \to j} \varepsilon_k}\right)$$

where x^* is the mole scale of the limiting reagent used in the jth step as per its radial pentagon analysis.

Overall raw material cost is given by

$$\text{Cost} = \frac{\$(\text{reagents}) + \$(\text{solvents}) + \$(\text{work-up materials}) + \$(\text{auxiliary materials})}{\text{mass of target product}}. \quad (5.34)$$

This calculation is built into the LINEAR and CONVERGENT spreadsheets for synthesis plan analysis.

5.6.10 Expression for Finding Centroid of Final Product Node in Synthesis Tree

In a synthesis tree diagram, the coordinates of intermediates are determined as the centroids of the dots corresponding to their preceding reactant input structures according to Equation 5.35 until the final product is reached.

$$\text{centroid} = \frac{1}{2^{n-1}} \left[\sum_{j=0}^{n-1} a_{j+1} \binom{n-1}{j} \right]$$

$$= \frac{1}{2^{n-1}} \left[\sum_{j=0}^{n-1} a_{j+1} \left(\frac{(n-1)!}{(j!)(n-1-j)!} \right) \right] \quad (5.35)$$

where $n \geq 2$, n is the number of points corresponding to the number of reactant input structures, a_{j+1} is the ordinate of the $(j+1)$th input, and $0! = 1$ by definition. Equation 5.35 is well known in classical mechanics where it is used to calculate the center of mass coordinates of equal point masses along a straight line.

5.6.11 Expression for MW First Moment

The MW first moment tracks the degree of building up occurring from starting materials to the final target product going through all intermediates over all branches in the synthesis plan. The general expression is

$$\mu_1 = \frac{1}{N+1} \left[\left(\sum_j (\text{MW}_{Pj} - \text{MW}_{Pn}) + (\text{MW}_{SM(j)} - \text{MW}_{Pn}) \right)_{\text{branch}\#1} \right.$$
$$+ \left(\sum_j (\text{MW}_{Pj} - \text{MW}_{Pn}) + (\text{MW}_{SM(j)} - \text{MW}_{Pn}) \right)_{\text{branch}\#2} + \cdots$$
$$\left. + \left(\sum_j (\text{MW}_{Pj} - \text{MW}_{Pn}) + (\text{MW}_{SM(j)} - \text{MW}_{Pn}) \right)_{\text{branch}\#k} \right] \quad (5.36)$$

where

MW$_{Pj}$, MW$_{Pn}$, and MW$_{SM(j)}$ are the MWs of intermediates, target product, and starting materials that end up in whole or in part in the structure of the first intermediate in each branch

k is the number of branches

N is the number of reaction stages

Another expression for the same thing that gives a quicker computation is given by

$$
\mu_1 = \frac{1}{N+1}\left[\begin{array}{l}\left(\displaystyle\sum_j MW_{Pj}\right)_{\text{branch}\,\#1} + \left(\displaystyle\sum_j MW_{Pj}\right)_{\text{branch}\,\#2} + \cdots + \left(\displaystyle\sum_j MW_{Pj}\right)_{\text{branch}\,\#k} \\[2em] + \left(\displaystyle\sum_j MW_{SM(j)}\right)_{\text{branch}\,\#1} + \left(\displaystyle\sum_j MW_{SM(j)}\right)_{\text{branch}\,\#2} + \cdots + \left(\displaystyle\sum_j MW_{SM(j)}\right)_{\text{branch}\,\#k} \\[2em] - MW_{Pn}\left(\displaystyle\sum_{\text{branches}} \text{intermediate nodes} + \displaystyle\sum_{\text{branches}} \text{starting material nodes}\right)\end{array}\right] \tag{5.37}
$$

5.6.12 Expression for $f(\text{sac})$

The MW fraction of reagents used that do not end up in whole or in part in the final target product structure is given by

$$
f(\text{sac}) = 1 - \frac{\displaystyle\sum_j MW_{\text{reagent},\,j,\,\text{ending up in target product}}}{\displaystyle\sum_j MW_{\text{reagent},\,j}}. \tag{5.38}
$$

5.6.13 Algorithm for LINEAR-COMPLETE and CONVERGENT-COMPLETE Spreadsheets

The algorithm for the computation of overall or global AE, RME, and E-factor for any synthesis plan of any degree of complexity is given by the following sequence of computations beginning with the last step ($j = N$) and working toward the first step ($j = 1$) using the synthesis tree connectivities as a guide:

1. Set the basis mole scale of the target product, x, to be 1 mol.
2. Enter number of branches (b), reaction stages (N), number of reaction steps (M), number of input materials (I), and MW of final target molecule (p_n).
3. For each branch in the plan, enter the sum of the MWs for reagents and by-products in each step, and the sequence of MWs of all intermediate products.
4. For each step in each branch, enter the following parameters from the radial pentagon analysis: mass of excess reagents, SF, reaction yield, mass of reaction solvent, total mass of work-up materials, total mass of purification materials, mass of catalyst and ligand, and mole scale of limiting reagent as reported in the literature procedure.
5. Determine E-factor based on MWs, $(E_{\text{mw}})_j$, using $(E_{\text{mw}})_j = \dfrac{\left(\sum MW_{\text{byproducts}}\right)}{p_j}$.
6. Determine $(AE)_j$ using $(AE)_j = \dfrac{1}{1+(E_{\text{mw}})_j}$.
7. Determine the reaction scales for each reagent as prescribed by synthesis tree using $\dfrac{x}{\prod_k^{n \to j} \varepsilon_k}$.

8. Determine kernel RME and E-kernel (by-products and unreacted starting materials contribution) for any given step using $(\text{RME})_{\text{kernel},j} = (\text{AE})_j \, \varepsilon_j$ and $(E_{\text{kernel}})_j = \dfrac{1}{(\text{RME})_{\text{kernel},j}} - 1$.

9. Determine the kernel mass of reagents used in any given step using

$$\left(\dfrac{x}{\prod\limits_{k}^{n \to j} \varepsilon_k} \right) \left(\sum \text{MW}_{\text{reagents}} \right).$$

10. Determine kernel mass of waste in any given step due to by-products and unreacted starting materials contribution using $\bar{w}_{\text{kernel},j} = \dfrac{p_j}{(\text{AE})_j} \left(\dfrac{x}{\prod\limits_{k}^{n \to j} \varepsilon_k} \right) \left[1 - \varepsilon_j (\text{AE})_j \right].$

11. Determine mass of waste due to excess reagents in any given step using

$$\bar{w}_{\text{excess},j} = \dfrac{p_j}{(\text{AE})_j} \left(\dfrac{x}{\prod\limits_{k}^{n \to j} \varepsilon_k} \right) \left[(\text{SF})_j - 1 \right].$$

12. Determine mass of waste due to auxiliary materials with appropriate scale correction factor for any given step using $\bar{w}_{\text{auxiliary},j} = \left(\dfrac{x}{\prod\limits_{k}^{n \to j} \varepsilon_k} \right) \left(\dfrac{c_j + s_j + \omega_j}{x_j^*} \right).$

13. Determine the total mass of waste in any given step using

$$\bar{w}_{\text{total},j} = \bar{w}_{\text{kernel},j} + \bar{w}_{\text{excess},j} + \bar{w}_{\text{auxiliary},j}$$

$$= \left(\dfrac{p_j}{(\text{AE})_j} \right) \left(\dfrac{x}{\prod\limits_{k}^{n \to j} \varepsilon_k} \right) \left[(\text{SF})_j - \varepsilon_j (\text{AE})_j \right] + \left(\dfrac{x}{\prod\limits_{k}^{n \to j} \varepsilon_k} \right) \left(\dfrac{c_j + s_j + \omega_j}{x_j^*} \right).$$

For the overall performance of a synthesis plan:

14. Determine $(E_{\text{mw}})_{\text{overall}}$ by summing MWs of all by-products in the plan and dividing by the MW of the final target product.

15. Determine $(\text{AE})_{\text{overall}}$ using $(AE)_{\text{overall}} = \dfrac{1}{1 + (E_{\text{mw}})_{\text{overall}}}.$

16. Determine $\bar{w}_{\text{total}} = \sum\limits_{j}^{n} \bar{w}_{\text{total},j}$ by summing all terms found in step (13).

17. Determine overall E-factor using

$$E_{\text{total}} = \dfrac{1}{p_n} \sum\limits_{j} \left(\dfrac{1}{\prod\limits_{k}^{n \to j} \varepsilon_k} \right) \left[\left(\dfrac{p_j}{(\text{AE})_j} \right) \left[(\text{SF})_j - \varepsilon_j (\text{AE})_j \right] + \dfrac{c_j + s_j + \omega_j}{x_j^*} \right].$$

18. Determine overall RME using $(\text{RME})_{\text{total}} = \dfrac{1}{1 + E_{\text{total}}}.$

19. Determine overall yield by multiplying reaction yields along longest branch of synthesis tree.

5.6.14   Note on Augé Method [61]

The author's analysis of individual reactions in the "single reaction" section is essentially a rework-ing of prior work and yields expressions practically identical to what were published previously [3]; so there are no new insights to be gained. However, a common error surfaces in the subsequent sec-tions "linear sequences," "convergent reactions," and "mixed linear and convergent sequences" with respect to the variable definition for the contribution of excess reagents used, namely terms labeled as b. Let us focus on the linear sequence example involving only bimolecular steps (see Scheme 2 in paper) since this is the easiest to analyze. Equation (12) for the mass intensity (MI) of linear sequences given in the paper (see Equation 5.39) is written correctly; however, the accompanying descriptive definition of b_i is incorrect.

$$\mathrm{MI} = \frac{1 + \dfrac{1}{\sum M} \sum_{i=1}^{n} v_{bi} M_{Bi} \left[\phi_i (\varepsilon_1 \varepsilon_2 \varepsilon_3 \ldots \varepsilon_{i-1}) - 1 \right] + s}{(\mathrm{AE}) \left(\prod_{i=1}^{n} \varepsilon_i \right)} = \frac{1 + \sum_{i=1}^{n} b_i + s}{(\mathrm{AE}) \left(\prod_{i=1}^{n} \varepsilon_i \right)} \qquad (5.39)$$

where

$$\varepsilon_0 = 1, \quad \sum M = v_a M_A + \sum_{i=1}^{n} v_{bi} M_{Bi}$$

$$b_i = \frac{v_{bi} M_{Bi} \left[\phi_i (\varepsilon_1 \varepsilon_2 \varepsilon_3 \ldots \varepsilon_{i-1}) - 1 \right]}{\sum M}$$

From Equation 5.39 note that for a single reaction $v_a A + v_{b1} B_1 \to products$ where $n = 1$ with only one excess reagent, $b_1 = \dfrac{(\phi_1 - 1) v_{b1} M_{B1}}{v_a M_A + v_{b1} M_{B1}}$, which corresponds to the ratio of excess mass of B_1 to the stoichiometric mass of A and B_1. When no excess B_1 is used $\phi_1 = 1$ and $b_1 = 0$. However, unlike the case of a single reaction (see equation (3) in paper), the definition of b_i is no longer the ratio of the "mass of excess reactants to the mass of the reactants in stoichiometric amounts" when applied to the analysis of a sequence of steps as stated in the paper. If this were true then under the simplifying conditions of recovery of auxiliary materials and no excess reagents used, $s = 0$ and application of the preceding word definition means that all the b_i terms in Equation 5.39 would be zero, respec-tively. This would lead to the erroneous result that $\mathrm{MI} = \dfrac{1}{(\mathrm{AE}) \left(\prod_{i=1}^{n} \varepsilon_i \right)}$, which would then imply

that $\mathrm{RME} = (\mathrm{AE}) \left(\prod_{i=1}^{n} \varepsilon_i \right)$, a factorable product of overall AE and overall yield. This erroneous conclusion is further discussed in Section 12.3. In fact, the true limiting expression for MI under these conditions is given by

$$\mathrm{MI} = \frac{1 + \dfrac{1}{\sum M} \sum_{i=1}^{n} v_{bi} M_{Bi} \left[(\varepsilon_1 \varepsilon_2 \varepsilon_3 \ldots \varepsilon_{i-1}) - 1 \right]}{(\mathrm{AE}) \left(\prod_{i=1}^{n} \varepsilon_i \right)} \qquad (5.40)$$

where $s = 0$ and all the ϕ_i terms are set to one so that $b_i = \dfrac{v_{bi} M_{Bi} \left[(\varepsilon_1 \varepsilon_2 \varepsilon_3 \ldots \varepsilon_{i-1}) - 1 \right]}{\sum M}$.

For linear sequences of the type shown in Scheme 2 of the paper, the correct expression corresponding to the ratio of the mass of excess reactants to the mass of the reactants in stoichiometric amounts is given by

$$
\frac{\sum_{i=1}^{n} v_{bi} M_{Bi} \left(\varepsilon_1 \varepsilon_2 \varepsilon_3 \ldots \varepsilon_{i-1} \right)\left(\phi_i - 1 \right)}{v_a M_A + \sum_{i=1}^{n} v_{bi} M_{Bi} \left(\varepsilon_1 \varepsilon_2 \varepsilon_3 \ldots \varepsilon_{i-1} \right)} = \frac{\sum_{i=1}^{n} v_{bi} M_{Bi} \left(\varepsilon_1 \varepsilon_2 \varepsilon_3 \ldots \varepsilon_{i-1} \right)\left(\phi_i - 1 \right)}{\sum M - \sum_{i=1}^{n} v_{bi} M_{Bi} + \sum_{i=1}^{n} v_{bi} M_{Bi} \left(\varepsilon_1 \varepsilon_2 \varepsilon_3 \ldots \varepsilon_{i-1} \right)}
$$

$$(5.41)$$

using the author's variable definitions. Now one can see that when all the ϕ_i terms are set to one, the numerator vanishes as it should. This confusion between the descriptive variable definitions and the algebraic expressions applies to all the b terms appearing in equations (18), (20), and (24) in the paper and may inadvertently lead readers to conclude that the author's methodology is incorrect. In fact these parameters are *related* to masses of excess reagents, and can also be negative valued, which is not intuitively translated to a physically meaningful quantity. Unfortunately, it is difficult to come up with a proper word definition for this b parameter. The origin of this odd behavior comes from defining excess reagent consumption as a mass ratio of reagents used to stoichiometric reagents rather than as a mass difference, which would be intuitively obvious.

With respect to the application of the author's algorithm on the synthesis plan for sildenafil, there are the following discrepancies:

1. The masses of aqueous washes in all steps were not accounted for as part of the auxiliary materials used.
2. The ethyl acetate solvent contribution due to the solution of compound P3 (see Scheme 5 in paper) was not taken into account.
3. The excess mass of compound P2′ (see Scheme 5 in paper) was not taken into account in the convergent step.
4. Various scaling errors were made in the number of moles of reagents used in the shorter convergent sequence.

Table 5.2 shows a summary of the results obtained from application of the algorithm given in Section 5.6.13 to this plan. Table 5.3 shows a summary of a check calculation according to a global input analysis where only the sum of all reagent masses are summed using a scale of 9.454 mol

TABLE 5.2

Summary of Green Metrics for Material Efficiency of Sildenafil Plan

Parameter	Value
Overall AE	41.1%
Overall yield (longest branch)	71.9%
E (kernel)	2.49
E (excess reagents)	1.87
E (auxiliaries)	64.75
E (total)	69.11
Overall RME	1.4%
Overall MI	70.11

TABLE 5.3

Comparison of Masses of Reagents Calculated Using Algorithm and Global Input Analysis for a Starting Scale of 9.454 mol for the First Reaction

Parameter	Global Input Analysis	Algorithm
Overall stoichiometric mass of reagents used (g)	11253.57	11248.73
Overall mass of excess reagents used (g)	6045.09	6026.38
Total mass of reagents used (g)	17298.66	17275.11
Mass of target product (g)	3227.36	3227.36
Overall mass of waste (g)	14071.30	14047.75
E-factor	4.36	4.36 (sum of 2.49 for E (kernel) plus 1.87 for E (excess reagents))
MI	5.36	5.36

as the limiting scale of reagents in the first step, as used by the author. It is clear that both methods yield consistent results. The author's result of $E = 4.62$ due to by-products, unreacted starting materials, and excess reagents is numerically close to the actual value of $E = 4.36$. The discrepancy is due to points 3 and 4. The main reasons for this apparent forgiveness in error are that the plan was short and more importantly the reaction yields were very high for each step. The author's result of $E = 26.14$ accounting for all materials used is an underestimation when compared with the true result of $E = 69.11$ because of the neglect of key auxiliary materials as noted in points 1 and 2. When all the aqueous washes are excluded, the present algorithm yields a correct value of $E = 29.92$, still about 4 units above the author's reported value under these conditions. In any event the waste contribution due to auxiliary materials on the overall E-factor is always the greatest. One already knows this even before doing any green metrics calculations. Though a general expression for E-factor as given in Section 5.6.7 is useful to work out overall synthesis performance, it does not give chemists a visual display of performance and hence any real direction to carry out further optimization on a given plan because it does not specifically identify the material bottlenecks. This is where the value of the radial pentagon analysis and preceding algorithm is demonstrated. The radial pentagon analysis identifies, by inspection, potential bottlenecks in the AE, reaction yield, excess reagent consumption, and auxiliary material usage for each reaction in a synthesis plan. The accompanying download contains such an analysis for the sildenafil plan discussed. The spread-sheet algorithm, in addition to giving overall material efficiency performance parameters, also pro-vides graphical profiles by reaction stage of percent waste distribution, AE, reaction yield, kernel RME, and cumulative mass of waste. Any bottlenecks identified by the radial pentagon analysis appear again in such profiles because they have an additive and cumulative effect on waste produc-tion. The accompanying download also contains the results of a spreadsheet analysis with accom-panying histograms of the simple algorithm described here for the sildenafil plan. These should be considered the definitive and correct analyses on this compound based on the best available data revealed in the publications.

Apart from the faults in the Augé paper with respect to an incorrect word definition of the parameter b related to excess reagent consumption and a botched calculation of E-factors for the sildenafil synthesis, there is another more fundamental problem. The treatment of synthesis plans in the forward direction using the mole scale of the limiting reagent appearing in the first step of the longest branch rather than in the reverse direction using the mole scale of the target product as a basis for the entire synthesis plan leads to unwieldy formulas that have no hope of being encoded in any kind of programming language. This is because every time a convergent step is encountered one of the branches needs to be readjusted in scale. This means that the mole scale of every input material pertaining to that branch must be changed accordingly by some factor. If a synthesis plan has N points of convergence originating from $N + 1$ branches, then there will need to be N such adjustments of scale. This leads to unnecessary and tedious calculations that are completely avoided by working backward from the target product node as the basis scale for the entire plan, as no scaling adjustments are required. In summary, the reader will be well advised to use the algorithm described in this chapter, which has been encoded in a spreadsheet format using Microsoft Excel for easy and ready use. The following section illustrates some worked out examples using these spreadsheets. The accompanying download contains the Excel files for these examples. The reader is invited to verify the results for themselves.

5.7   COMPLETE WORKED-OUT EXAMPLES

5.7.1   GUAIFENESIN [62,63]

This example showcases a single-step synthesis. The radial pentagon shows that it has several green attributes including high yield, 100% AE, and no sacrificial reagents used. The transformation is also isohypsic.

Synthesis scheme:

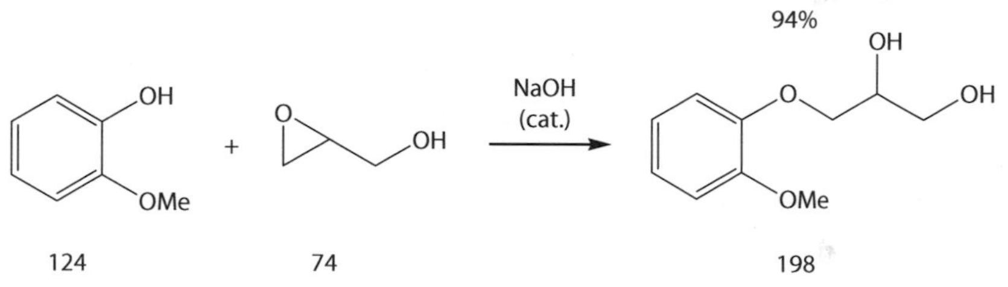

Synthesis tree:

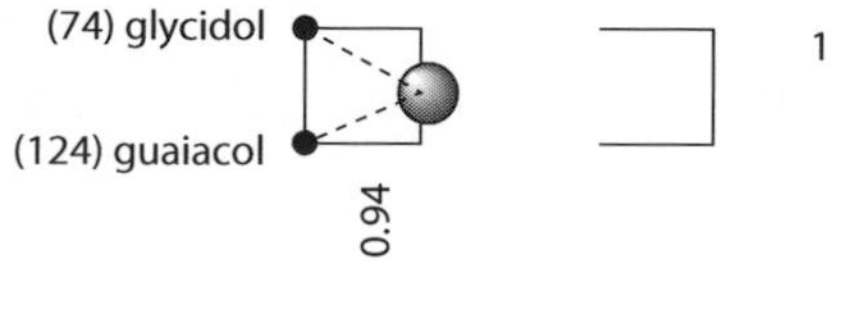

Synthesis metrics:

Plan type = linear
Number of branches = 1
$N = 1$ reaction stage
$M = 1$ reaction
$I = 2$ input materials
Overall AE = 100%
Overall kernel RME = 93.1%
Overall yield = 94%
Kernel mass of waste to produce 1 mol guaifenesin = 12.64 kg
Total mass of waste to produce 1 mol guaifenesin = 15.89 kg
E-kernel = 0.064
E-excess = 0.0096
E-auxiliaries = 0.0053
E-total = 0.080

MW first moment building up parameter $= -99$ g/mol/stage
$f(\text{sac}) = 0$
HI $= 0$
Degree of convergence $= 1$
Degree of asymmetry $= 0$

Skeletal building framework:

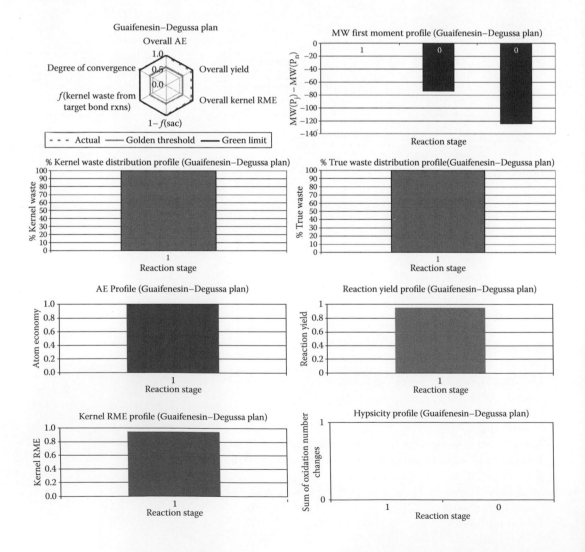

Graphics:

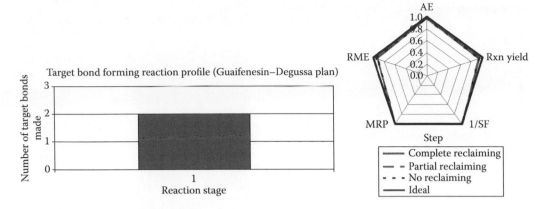

5.7.2 PROBUCANOL [64–68]

This is an example of a three-step linear sequence producing a highly symmetric molecule. The synthesis plan is overall hypohypsic.

Synthesis scheme:

For steps 1 and 2, see: Mueller, E.; Stegmann, H.B.; Scheffler, K. *Ann. Chem.* 1961, *645*, 79.

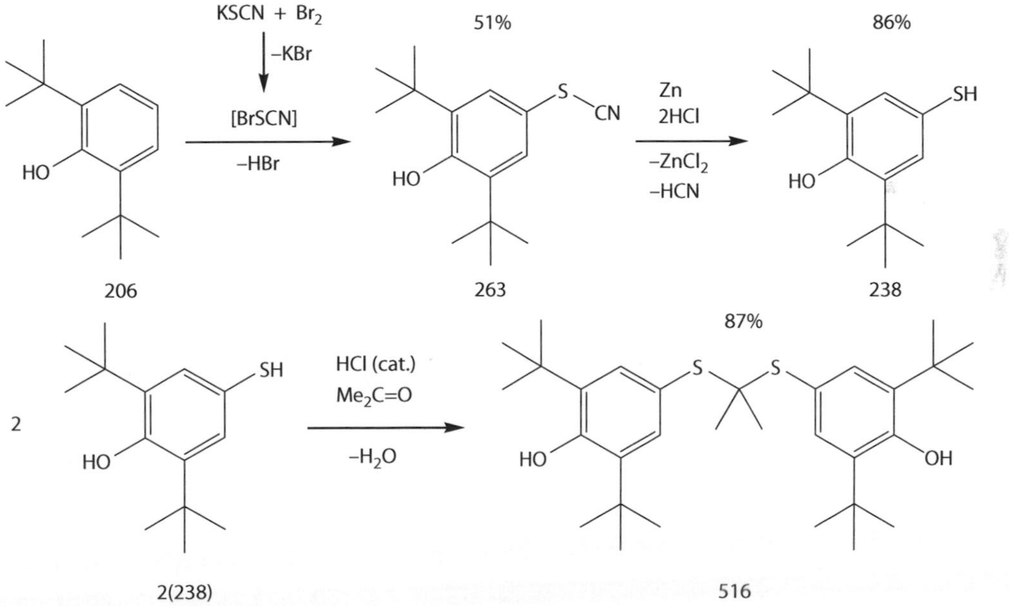

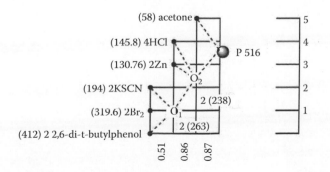

Synthesis tree:

Synthesis metrics:

Plan type = linear
Number of branches = 1
N = 3 reaction stages
M = 3 reactions
I = 6 input materials
Overall AE = 40.9%
Overall kernel RME = 18.0%
Overall yield = 38.2%
Kernel mass of waste to produce 1 mol probucanol = 2.3 kg
Total mass of waste to produce 1 mol probucanol > 32.7 kg
E-kernel = 4.55
E-excess = 7.05
E-auxiliaries > 51.72
E-total > 63.31
MW first moment building up parameter = −114 g/mole/stage
f(sac) = 0.473
HI = −1
Degree of convergence = 0.534
Degree of asymmetry = 0.550

Skeletal building framework:

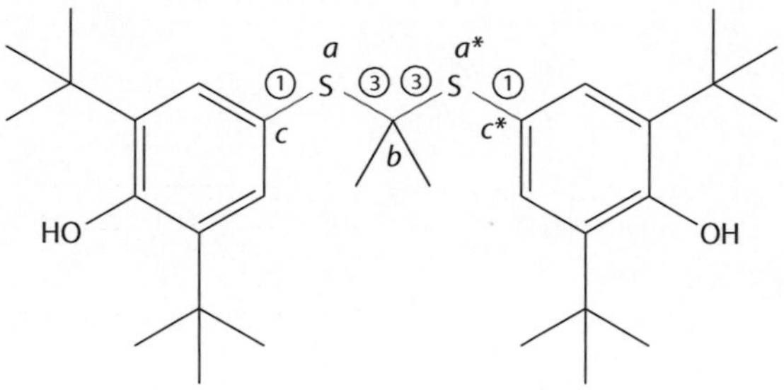

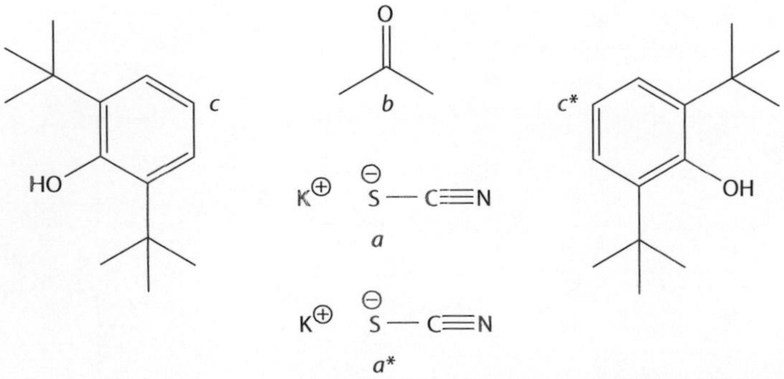

Graphics:

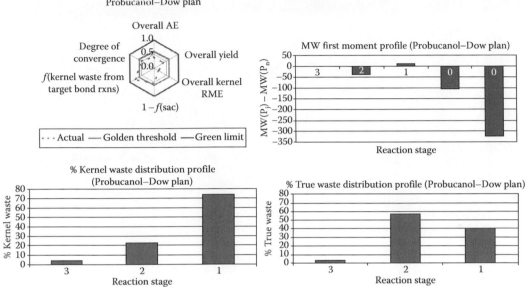

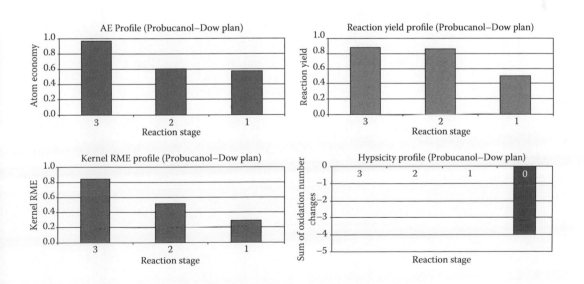

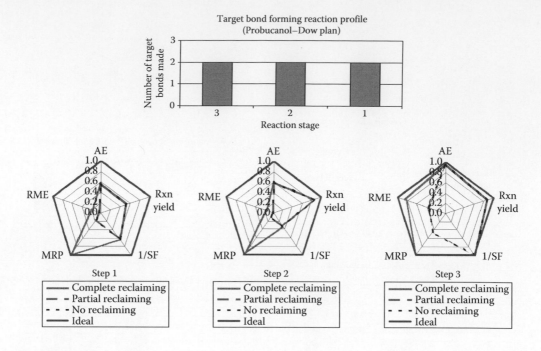

5.7.3 Leonurine [69]

This is an example of a convergent plan with three branches.

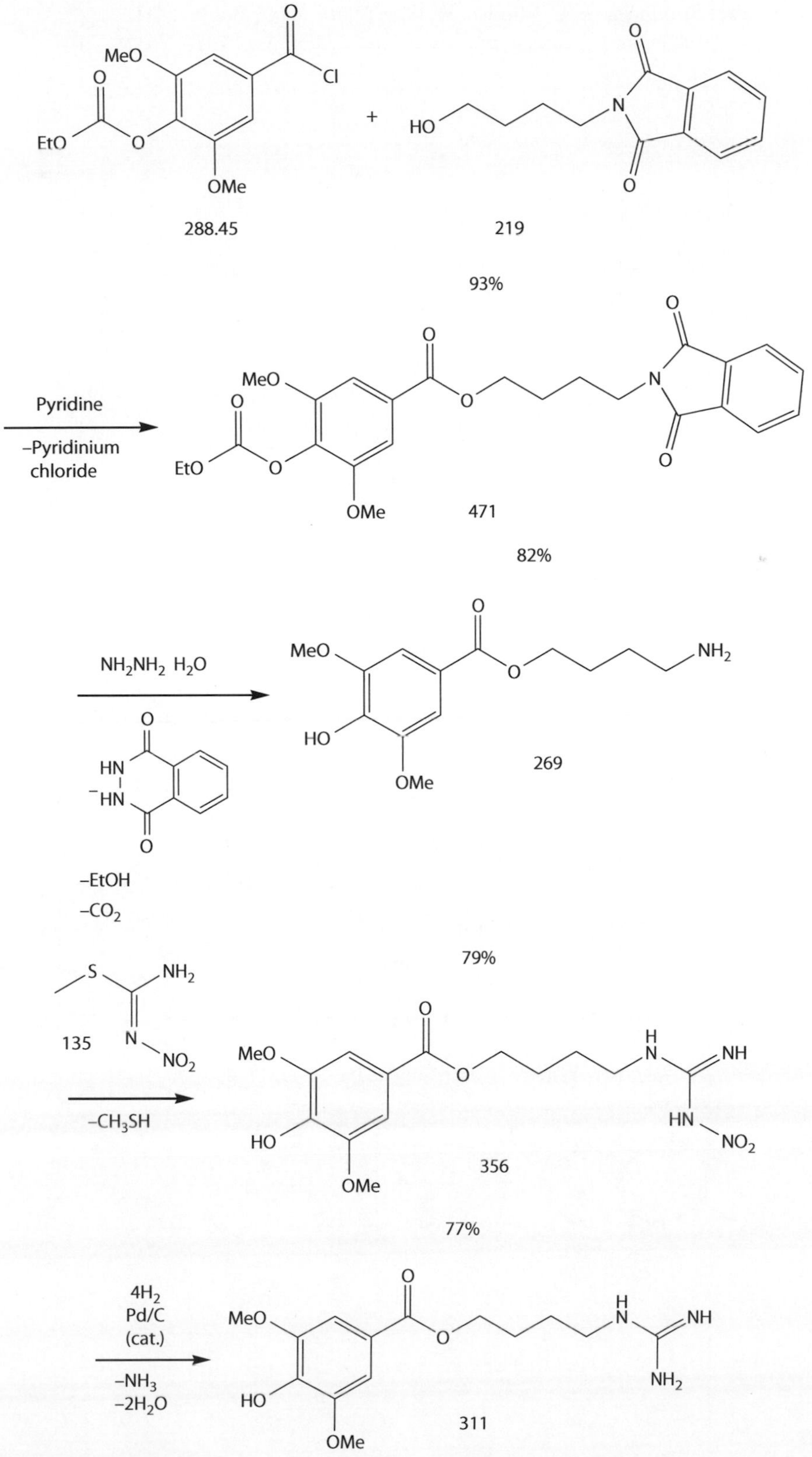

MeO

O

EtO

OMe

288.45

+

HO

O

N

O

219

93%

Pyridine

−Pyridinium chloride

MeO

O

EtO

O

OMe

N

O

O

471

82%

NH$_2$NH$_2$ H$_2$O

HN

N

O

O

−EtOH

−CO$_2$

MeO

HO

OMe

O

O

NH$_2$

269

79%

S

NH$_2$

N

NO$_2$

135

−CH$_3$SH

MeO

HO

OMe

O

O

N
H

NH

HN

NO$_2$

356

77%

4H$_2$

Pd/C

(cat.)

−NH$_3$

−2H$_2$O

MeO

HO

OMe

O

O

N
H

NH

NH$_2$

311

(For step 2**, see: Shildneck, P.R.; Windus, W. *Org. Synth. Coll.* 1943, *2*, 411.)
(For step 3**, see: Fishbein, L.; Gallaghan, J.A. *J. Am. Chem. Soc.* 1954, *76*, 1877.)

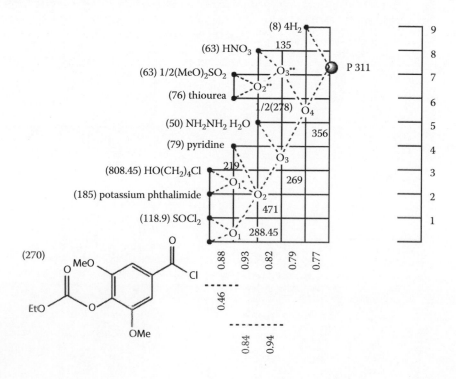

Synthesis tree:

Synthesis metrics:

Plan type = convergent
Number of branches = 3
$N = 5$ reaction stages
$M = 8$ reactions
$I = 10$ input materials
Overall AE = 30.4%
Overall kernel RME = 10.3%
Overall yield = 40.8%
Kernel mass of waste to produce 1 mol leonurine = 2.7 kg
MW first moment building up parameter = −224.03 g/mol/stage
$f(sac) = 0.305$
HI = +0.17
Degree of convergence = 0.483
Degree of asymmetry = 0.608

Skeletal building framework:

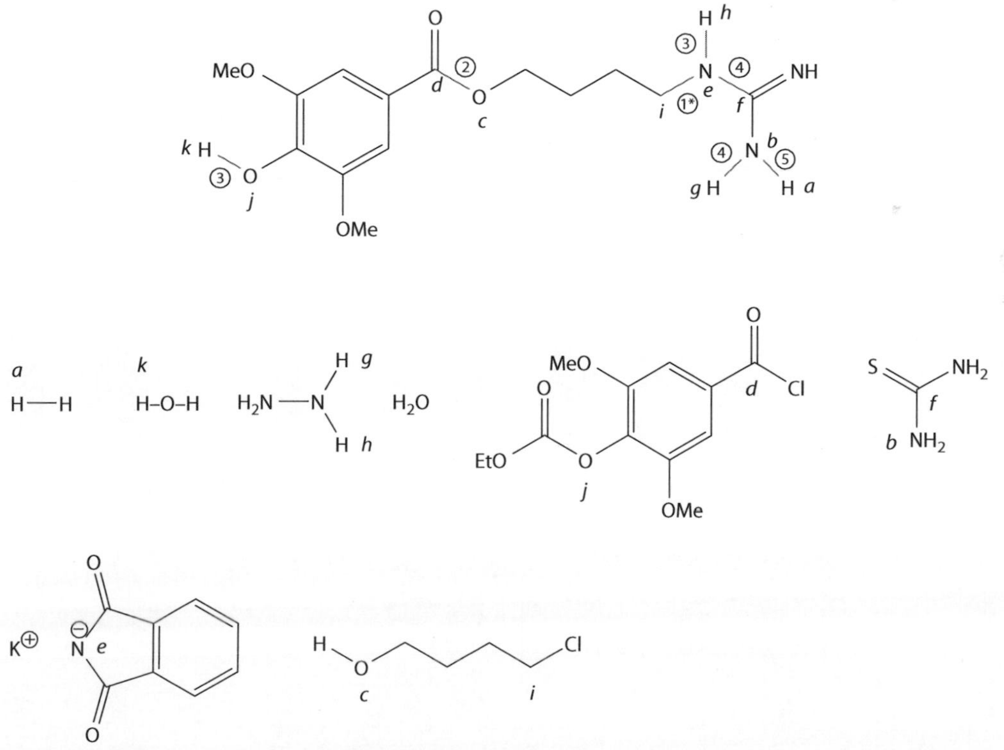

Graphics:

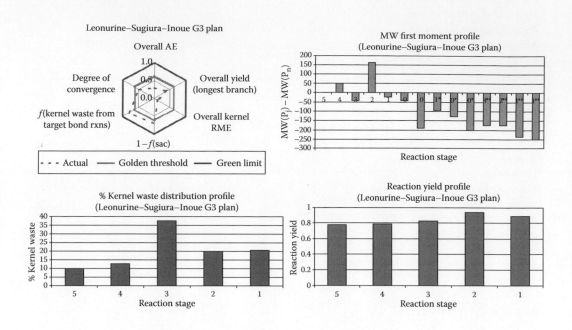

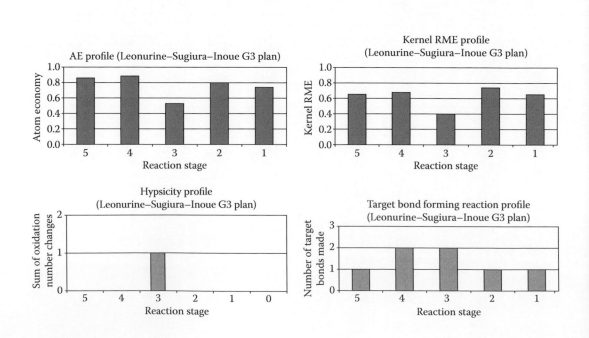

REFERENCES

1. Calvo-Flores, F.G. *ChemSusChem*. 2009, *2*, 905.
2. Lapkin, A.; Constable, D.J.C. *Green Chemistry Metrics: Measuring and Monitoring Sustainable Processes*. Wiley: Chichester, U.K., 2008.
3. Andraos, J. *Org. Process Res. Dev*. 2005, *9*, 149.
4. Andraos, J. *Org. Process Res. Dev*. 2005, *9*, 404.
5. Andraos, J. *Org. Process Res. Dev*. 2006, *10*, 212.
6. Andraos, J.; Izhakova, J. *Chimica Oggi Chem. Today* 2006, *24*(6, Suppl.), 31.
7. Andraos, J.; Sayed, M. *J. Chem. Educ*. 2007, *84*, 1004.
8. Andraos, J. *Can. Chem. News* 2007, *59*(4), 14.
9. Andraos, J. *Org. Process Res. Dev*. 2009, *13*, 161.
10. Trost, B.M. *Science* 1991, *254*, 1471.
11. Trost, B.M. *Acc. Chem. Res*. 2002, *35*, 695.
12. Trost, B.M. *Angew. Chem. Int. Ed*. 1995, *34*, 259.
13. Sheldon, R.A. *Chem. Tech*. 1994, *24*(3), 38.
14. Sheldon, R.A. *Comptes Rendus de l'Académie des Sciences Paris, Série IIc, Chimie* 2000, *3*, 541.
15. Sheldon, R.A. *Pure Appl. Chem*. 2001, *72*, 1233.
16. Sheldon, R.A. *Green Chem*. 2007, *9*, 1273.
17. Sheldon, R.A. *Chem. Commun*. 2008, 3352.
18. Curzons, A.D.; Constable, D.J.C.; Mortimer, D.N.; Cunningham, V.L. *Green Chem*. 2001, *3*, 1.
19. Constable, D.J.C.; Curzons, A.D.; Freitas dos Santos, L.M.; Geen, G.R.; Hannah, R.E.; Hayler, J.D.; Kitteringham, J.; McGuire, M.A.; Richardson, J.E.; Smith, P.; Webb, R.L.; Yu, M. *Green Chem*. 2001, *3*, 7.
20. Steinbach, A.; Winkenbach, R. *Chem. Eng*. April 2000, 94.
21. Constable, D.J.C.; Curzons, A.D.; Cunningham, V.L. *Green Chem*. 2002, *4*, 521.
22. Eissen, M.; Metzger, J.O. *Chem. Eur. J*. 2002, *8*, 3580.
23. Eissen, M.; Hungerbühler, K.; Dirks, S.; Metzger, J. *Green Chem*. 2003, *5*, G25.
24. Metzger, J.O.; Eissen, M. *Comptes Rendus de l'Académie des Sciences Paris, Série IIc, Chimie* 2004, *7*, 569.
25. Eissen, M.; Mazur, R.; Quebbemann, H.G.; Pennemann, K.H. *Helv. Chim. Acta* 2004, *87*, 524.
26. Eissen, M. *Bewertung der Umweltverträglichkeit organisch-chemischer Synthesen*. PhD thesis, 2001, Universität Oldenburg, Oldenburg, Germany.
27. van Aken K.; Strekowski, L.; Patiny, L. *Beilstein J. Org. Chem*. 2006, 2. doi:10.1186/1860-5397-2-3.
28. Kinen, C.O.; Rossi, L.I.; de Rossi, R.H. *Green Chem*. 2009, *11*, 223.
29. Rosini, G.; Borzatta, V.; Paulocci, C.; Righi, P. *Green Chem*. 2008, *10*, 1146.
30. Kuzemko, M.A.; van Arnum, S.D.; Niemczyk, H.J. *Org. Process Res. Dev*. 2007, *11*, 470.
31. Niemczyk, H.J.; van Arnum, S.D. *Green Chem. Lett Rev*. 2008, *1*, 165.
32. Streitweiser, A. Jr.; Heathcock, C.H. *Introductory Organic Chemistry*, 3rd edn., Macmillan Publishing Company: New York, 1985.
33. Froning, J.F.; Hennion, G.F. *J. Am. Chem. Soc*. 1940, *62*, 653.
34. Royals, E.E. *Ind. Eng. Chem*. 1946, *38*, 546.
35. Kimel, W.; Sax, N.W.; Kaiser, S.; Eichmann, G.G.; Chase, G.O.; Ofner, A. *J. Org. Chem*. 1958, *23*, 153.
36. Moore, J.W. *J. Chem. Educ*. 1997, *74*, 1253.
37. Johnson, O.C. *Chem. News* 1880, *42*, 51.
38. Bennett, G.W. *J. Chem. Educ*. 1954, *31*, 157.
39. Burrell, H.P.C. *J. Chem. Educ*. 1959, *36*, 77.
40. Lockwood, K.L. *J. Chem. Educ*. 1968, *45*, 731.
41. Hudlicky, M. *J. Chem. Educ*. 1977, *54*, 100.
42. Cook, D.; Morrison, R.J. *Educ. Chem*. September 1990, 141.
43. Klemm, L.H. *J. Heterocyclic Chem*. 1996, *33*, 569.
44. Anselme, J.P. *J. Chem. Educ*. 1997, *74*, 69.
45. Vogel, A. *Textbook of Practical Organic Chemistry*, Longman: London, U.K., 1948, p. 852.
46. Fierz-David, H.E.; Blangey, L. *The Fundamental Processes of Dye Chemistry*, Interscience Publishing: New York, 1949, p. 338.
47. Chen, F.; Liu, A.; Yan, Q.; Liu, M.; Zhang, D.; Shao, L. *Synth. Commun*. 1999, *29*, 1049.
48. Steinbach, O.F. *J. Chem. Educ*. 1944, *21*, 66.
49. Kolb, D. *J. Chem, Educ*. 1979, *56*, 181.

50. Balasubramanian, K. *J. Math. Chem.* 2001, *30*, 219.
51. Papp, D.; Vizvari, B. *J. Math. Chem.* 2006, *39*, 15.
52. Bashmakova, I.G. *Diophantus and Diophantine Equations*, Mathematical Association of America: Washington, DC, 1997.
53. Hendrickson, J.B. *J. Am. Chem. Soc.* 1971, *93*, 6847.
54. Mathieu, J.; Panico, R.; Weill-Raynal, J. *L'Aménagement Fonctionnel en Synthèse Organique*, Hermann: Paris, France, 1977, Chapter 1.
55. Mathieu, J. *Actualite Chimique* 1975, 20 (*Chem. Abstr.* 88: 36671).
56. Burns, N.Z.; Baran, P.S.; Hoffmann, R.W. *Angew Chem. Int. Ed.* 2009, *48*, 2854.
57. Newhouse, T.; Baran, P.S.; Hoffmann, R.W. *Chem Soc Rev.* 2009, *38*, 3010.
58. Perkin, W.H. Jr. *J. Chem. Soc.* 1904, *85*, 654.
59. Perkin, W.H. Jr. *J. Chem. Soc.* 1904, *85*, 417.
60. Senn, O.F. US 2667503 (Maumee Development Co., 1954).
61. Augé, J. *Green Chem.* 2008, *10*, 225.
62. Merk, W.; Wagner, R.; Werle, P.; Nygren, R.S. DE 3106995 (Degussa, 1982).
63. Merk, W.; Wagner, R.; Werle, P.; Nygren, R.S. US 4390732 (Degussa, 1983).
64. Neuworth, M.B.; Laufer, R.J.; Barnhart, J.W.; Sefranka, J.A.; McIntosh, D.D. *J. Med. Chem.* 1970, *13*, 722.
65. Neuworth, M.B. FR 1561853 (Dow, 1969).
66. Neuworth, M.B. US 3576883 (Dow, 1971).
67. Barnhart, J.W.; Shea, P.J. US 3862332 (Dow, 1975).
68. Laufer, R.J. US 3129262 (Dow, 1964).
69. Sugiura, S.; Inoue, S.; Hayashi, Y.; Kishi, Y.; Goto, T. *Tetrahedron* 1969, *25*, 5155.

6 Optimization

6.1 INTRODUCTION

The concept of optimization in organic synthesis is a multifaceted problem as it is in other disciplines. The common theme is that certain variables need to be first identified as relevant to the problem, and then they are maximized or minimized as appropriate subject to constraints until a threshold criterion is reached. In organic synthesis, the overall goal is to reduce both material waste production and to increase energy efficiency. This has always been the endeavor of process chemists in industry and so the ideas of green chemistry pronounced in the 12 principles are not novel. The new idea behind the implementation of green thinking is to practice this philosophy right at the very beginning of the design stage for the development of a new molecule or process. Traditionally, these were done at the later stages of development after the proof of concept or efficacy was achieved. Barriers or inertias to "green thinking" include the following:

a. The cost of energy and raw materials is kept deliberately low, so the incentive to innovate is weak.
b. Slack or inconsistent enforcement of regulations nationally and internationally with respect to policies on the handling and treatment of waste materials.
c. Lack of sophisticated accounting practices focused on individual processes that includes proper material and energy consumption, life-cycle analysis, and environmental and human resource impact.
d. Difficulty in allocating research and development (R&D) funding for green chemistry in industry.
e. Academics are traditionally taught to ignore optimization altogether as it is thought to be the privy of industry and large-scale operations.
f. Lack of or spotty clean sustainable chemistry examples and topics taught in schools and universities where green chemistry continues to be viewed as a "soft" science.
g. Lack of communication and understanding between chemists and chemical engineers.
h. Lack of rigorous definitions of concepts such as sustainability that are couched in the language of mathematics.
i. First-to-the-post mentality of both business and academia to get to the finish line as quickly as possible before competitors do.
j. Short-term view by industry and investors to show profits in every quarterly report.
k. Optimization by green chemistry methods takes time.
l. Few technically acceptable green substitute products and processes that are success stories.
m. Significant "green washing" and exaggerated claims made in the literature by both academics and industry on the subject of green chemistry making it hard to sift through the chaff to find true examples of successes that can be substantiated by rigorous metrics.
n. Lack of consensus on what metrics to use and how to rank syntheses and processes.

Other points to mention are that there has not been a real demonstration of recycling solvents for second time around usage, even as part of an undergraduate experiment, which has been documented in detail in the literature. Syntheses of specialized catalysts are not mentioned or included in the overall analysis of material efficiency of plans. There is a need to move away from expensive and often toxic organometallic catalysts containing metals such as Pd, Ru, Rh, Os, and Sn for the synthesis of pharmaceutical targets. There is no real demonstration of reusability of organometallic catalysts, even as part of an undergraduate experiment, that is, determining how many times the same catalyst can be used over and over again before its efficacy drops significantly.

Before attempting to minimize waste, one needs to answer the question: what constitutes waste? In order to do this, one needs to know the full chemistry of the process including balancing chemical equations, identifying all products (target, by-products, and side products), probing reaction mechanism, and determining fundamental thermodynamic and kinetic parameters. The old way of determining global waste was to take the mass difference between all input materials and the mass of target product collected. Another way to look at this is that the mass of target product is just the mass difference of goods delivered by delivery vans entering a plant and waste material leaving the plant via waste disposal trucks. This simple arithmetic exercise tells nothing about what are the constituents of the waste and so attempts to minimize it are destined to fail. Consequently, inaccurate and false estimates of waste production will result. For any given chemical reaction, waste is composed of unreacted starting materials, excess reagents, by-products produced as a mechanistic consequence of producing the intended target product, side products produced due to competing side reactions other than the intended target reaction, reaction solvents, auxiliary work-up materials, and auxiliary purification materials. There are generally three levels of waste reduction, which are described as follows:

a. First generation waste reduction: Reducing reaction solvent consumption; reducing solvent consumption in work-up and purification steps; replacing reaction solvents with benign "green" solvents; recycling solvents
b. Second generation waste reduction: Optimizing reaction conditions to increase reaction yields by manipulating reaction parameters such as reaction time, temperature, pressure, choice of solvent, and use of catalysts and additives
c. Third generation waste reduction: Reducing by-products by redesigning reactions; reducing overall waste production by restrategizing a synthesis plan to increase number of target bond forming reactions over number of sacrificial nontarget bond forming reactions

One can see that these generations of waste reduction get progressively harder to do and necessarily take time to achieve. It is not surprising that much effort has been expended in reducing solvent demand since it is well documented that E-aux is by far the largest contribution to E-total. Success depends both on the ingenuity of well-trained professional chemists who know their chemistry and are up-to-date on the latest innovations, and on the patience and vision of business partners and investors. An important realization is that the total number of steps, M, in a synthesis plan comprises two kinds of reaction types: number of target bond forming reactions and number of sacrificial reactions that do not form target bonds. Amazingly, this simple statement appears nowhere in any organic chemistry textbook or monograph on organic synthesis. If the waste contribution from sacrificial reactions exceeds that for target bond forming reactions then the plan is not a synthesis—it is likely a degradation such as the Willstätter [1] plan for cyclooctatetraene from pseudopelletierine, or a poorly designed plan such as the Hamon-Young [2] plan for (±)-twistane. There are fundamentally two kinds of waste: "good" waste that originates from target bond forming reactions and "bad" waste that originates from sacrificial reactions. The objective of material efficiency optimization is then to simultaneously minimize global waste (both "good" and "bad" contributors) and to minimize the waste contribution from sacrificial reactions over target bond forming reactions. Plans may be compared and ranked by examining their overall kernel reaction mass efficiency (RME) or E-kernel values against the vector magnitude ratio (VMR) found from the radial hexagon analysis (see Section 6.3). If the ranking order is preserved by both methods, then "best" plans are those that have minimized overall waste magnitude as well as maximized the "good" waste component of overall waste. Conversely, "poor" performing plans are those that have maximized the overall waste magnitude and minimized the "good" waste contribution. On the other hand, if the ranking order changes by both methods, then this arises from opposing situations such as (a) a plan that minimizes global waste and minimizes "good" waste contribution; or (b) a plan that maximizes global waste and maximizes the "good" waste contribution. Both types of scenarios are examples of mismatches between strategy and material efficiencies, yet these are the very plans that have the greatest chance of being optimized in the right direction where there is a synergy between strategy and material efficiency. The following waste optimization scenarios and options arise when comparing various plan performances to the same target molecule:

Case I: Global waste is high and waste from sacrificial reactions exceeds that from target bond forming reactions ⇒ change design strategy and increase material efficiency for each step.

Case II: Global waste is high and waste from target bond forming reactions exceeds that from sacrificial reactions ⇒ maintain the same strategy and increase material efficiency of each reaction by manipulating reaction parameters such as reaction time, solvent, catalysts, and additives.

Case III: Global waste is low and waste from sacrificial reactions exceeds that from target bond forming reactions ⇒ change design strategy without compromising previous material efficiency gains.

Case IV: Global waste is low and waste from target bond forming reactions exceeds that from sacrificial reactions ⇒ optimum plan is likely reached; maybe some fine tuning opportunities exist.

One can see that the performance order of these cases is I < II < III < IV. An optimized "green" plan should have following metric attributes:

- $f(nb) = 0$, all reactions produce target bonds in final product structure
- $f(sac) = 0$, all reagents in whole or in part end up in final product structure
- The least number of reaction stages (M) and steps (N)
- Reaction yields as high as possible for each step
- Atom economies as high as possible for each step
- E-excess and E-aux as low as possible
- Types of reactions used: *Aufbau*-type reactions and additive redox-type reactions

With respect to strategy and design, the goal is to minimize the overall number of steps to reach the target molecule, to minimize the fraction of reaction steps that produce no target bonds, to minimize overall waste from all reactions, and to minimize the waste contribution from sacrificial reactions. The desired strategy progression over time is that the nth generation plan for a given target reported in year X should aim to have a better set of metric results than the $(n - 1)$th generation plan in year $X - 1$. The desire is to come up with both a different plan demonstrating novelty as well as showing a substantial reduction in overall E-factor. The key take home message is that optimization is an iterative exercise that compares the performances of a set of plans to a common target according to some criteria and that true optimization is achieved when the best possible values of material efficiency and synthetic elegance metrics coincide in the same plan. Often, being different trumps efficiency when disclosing plans in the literature. The words "novel," "concise," and "efficient" are loosely used in titles of journal articles without substantiation by quantitative assessments. It is deflating to find out that claims of efficiency are found to be unsubstantiated when metrics analyses are applied to reported synthesis plans especially when a nifty reaction or sequence of reactions has been implemented. One of the goals of this book is to change the mind-set of chemists and reverse these trends and outcomes; otherwise the quest to legitimize green chemistry as a rigorous branch of chemical science will be significantly hampered.

6.2 WORKED-OUT RADIAL PENTAGON EXAMPLES

In working out the material efficiency of a single chemical reaction, the radial pentagon spreadsheet algorithm was invented to facilitate tedious computations from data written up in experimental procedures. An Excel template spreadsheet of the algorithm called PENTAGON is provided in the accompanying download. Instructions on its use are given in Section 5.4. The spreadsheet lays out experimental procedures in a consistent and easy to read format. The logic of the computations is illustrated in the flowchart shown in Figure 6.1. It turns out that this device is an excellent proofreading tool in identifying and correcting mistakes found in published procedures including incorrect correspondences between quoted masses and moles for reagents, incorrectly reported reaction yields, and incorrect amounts of required ingredients based on the expectations of the balanced chemical equation. As discussed in detail in Chapter 3, the chemistry literature is rife with problem experimental procedures quoting incorrect amounts of materials that jeopardize reproducibility of reaction performance in the laboratory. Even if the reader is not convinced of the merits of green chemistry thinking he or she will find the radial pentagon algorithm to be an indispensable tool in day-to-day laboratory practice. In fact, any deliberate errors the user wishes to introduce in the amounts of reagents used will be trapped because there are built-in check calculations for the reaction yield and excess reagent consumption (stoichiometric factor).

The key points are determining the identity of the limiting reagent and the number of moles used that will define the scale of the given reaction, and setting the stoichiometric coefficient of limiting reagent to unity and adjusting the stoichiometric coefficients of the other reagents accordingly. This is a particularly useful computational trick for handling balanced chemical equations having reagent stoichiometric coefficients that are not equal to unity. For example, if a general balanced chemical equation is of the form

$$aA + bB + cC \rightarrow pP + qQ$$

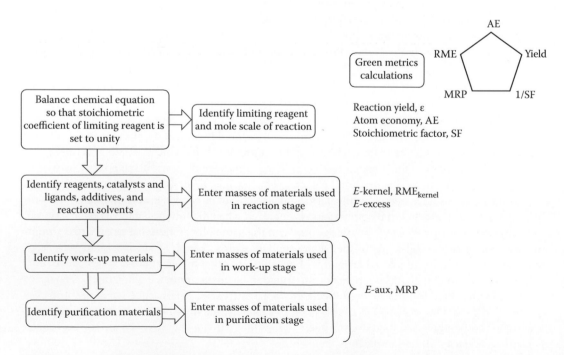

FIGURE 6.1 Flowchart of radial pentagon algorithm and computations.

where $a, b, c, p,$ and q are the stoichiometric coefficients for reagents $A, B,$ and C, target product P, and by-product Q, and if A is the limiting reagent, then rewrite the balanced equation as

$$A + (b/a)B + (c/a)C \rightarrow (p/a)P + (q/a)Q.$$

This becomes the pivotal balanced equation on which all of the key green metrics will be determined. The reaction yield and stoichiometric factor will be pegged against the mole scale of reagent A, x. Thus,

$$\varepsilon = \frac{y}{x(p/a)} \tag{6.1}$$

and

$$SF = 1 + \frac{[z_B - x(b/a)]MW_B + [z_C - x(c/a)]MW_C}{x[MW_A + (b/a)MW_B + (c/a)MW_C]} \tag{6.2}$$

where $y, z_B,$ and z_C are the actual number of moles of P collected, and reagents B and C used, and variables labeled as MW refer to the molecular weights of the respective chemical species. If B were the limiting reagent with mole scale x, the pivot balanced chemical equation is written in the form

$$(a/b)A + B + (c/b)C \rightarrow (p/b)P + (q/b)Q.$$

Then, in this case the reaction yield and stoichiometric factor are

$$\varepsilon = \frac{y}{x(p/b)} \tag{6.3}$$

and

$$SF = 1 + \frac{[z_A - x(a/b)]MW_A + [z_C - x(c/b)]MW_C}{x[(a/b)MW_A + MW_B + (c/b)MW_C]} \tag{6.4}$$

where $y, z_A,$ and z_C are the actual number of moles of P collected, and reagents A and C used. As an exercise, the reader may work out expressions for the reaction yield and stoichiometric factor when reagent C is the limiting reagent.

Assumptions used in the radial pentagon algorithm include the following:

a. For reagents expressed as weight percents; for example, if 37 wt.% HCl were required as a reagent, then the mass used is calculated by multiplying the wt.% by the total mass used—this is entered as reagent amount; the remaining 63 wt.% is counted as water solvent.

b. The densities of mixed solvents were calculated as a linear combination of densities of pure solvents; for example, for x mL of a mixture of solvents A and B in the volume ratio $V_A : V_B$ the overall density is calculated as

$$d = \left[\frac{V_A}{V_A + V_B}\right]d_A + \left[\frac{V_B}{V_A + V_B}\right]d_B$$

where d_A and d_B are the densities of solvents A and B separately.

c. When masses of auxiliary materials are not reported in experimental procedures, then the following guidelines put forward in environmental assessment tool for organic synthesis (EATOS) [3] were followed:
   i. Extraction from aqueous media using an organic solvent: 300 mL solvent per L aqueous medium (multiplying factor = 0.3)
   ii. Aqueous washes of organic solvent: 300 mL water per L organic solvent (multiplying factor = 0.3)
   iii. Brine: 100 mL brine per L organic solvent (multiplying factor = 0.1)
   iv. Drying agents: 20 g per L organic solvent (multiplying factor = 0.02).

The following chemical transformations, collected from the compiled synthesis database, have the curious property of having more than way to write out a balanced chemical equation. These are listed in the following including determinations of E_m and atom economy (AE) for each case. Whichever balanced chemical equation is chosen will affect the calculation of kernel metrics parameters. Basically, the kernel RME will be partitioned differently with respect to reaction yield and AE, and the E-kernel and E-excess contributions will also be affected. When evaluating synthesis trees for kernel waste production where any of these types of transformations were encountered, the ones yielding the lowest AE values were selected so that worst-case scenarios could be determined for the plans. However, global metrics calculated using the radial pentagon analysis will be the same regardless of which chemical equation is chosen. Hence, calculated values for E-total and overall RME will remain unaffected. What matters is that consistency be kept so that meaningful ranking of plans can be made.

1. Sodium sulfate or sodium bisulfate as a by-product in substitution reactions

2. Methylation with dimethyl sulfate

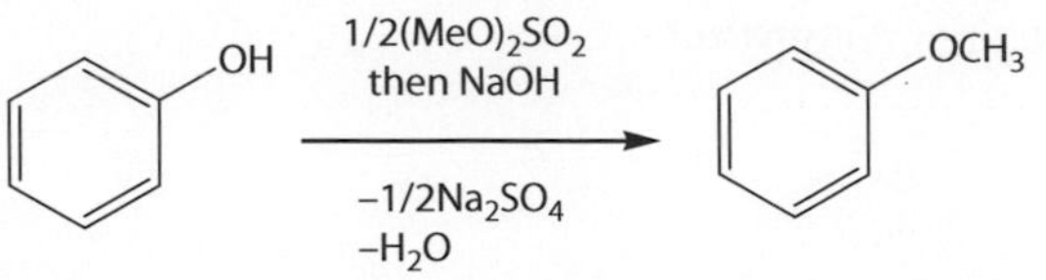

$E_m = 0.824$
$AE = 0.548$

3. Hydrolysis of nitriles

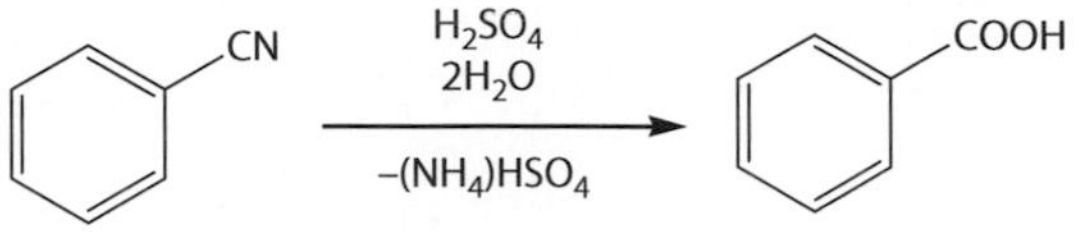

$E_m = 0.943$
$AE = 0.515$

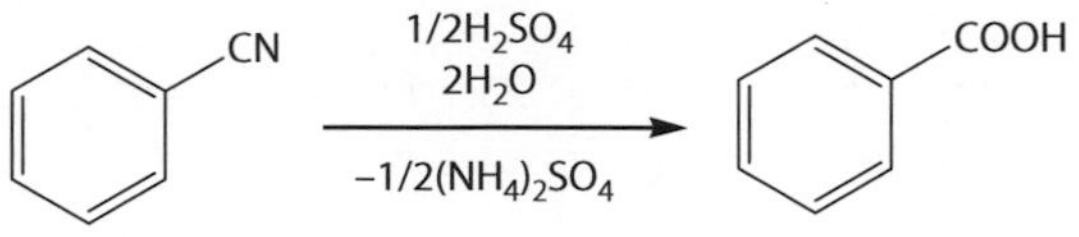

$E_m = 0.541$
$AE = 0.649$

4. Etherification

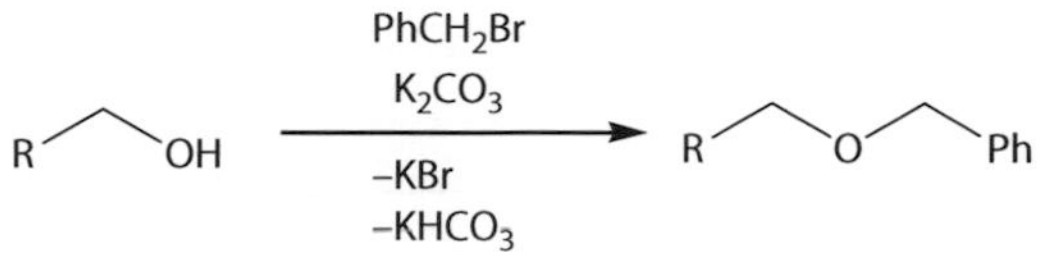

$E_m(max) = 1.794$
$AE(min) = 0.358$

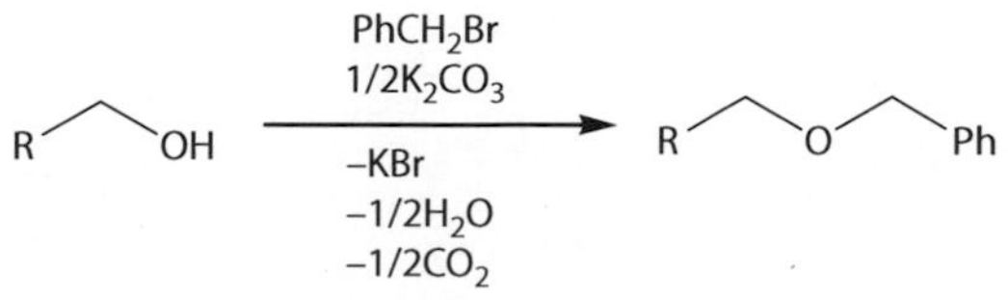

$E_m(max) = 1.229$
$AE(min) = 0.449$

when R = H
5. Friedel–Crafts acylation

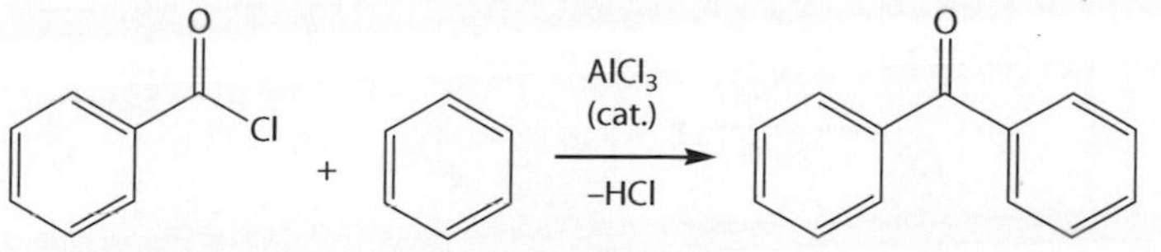

$E_m = 0.200$
$AE = 0.833$

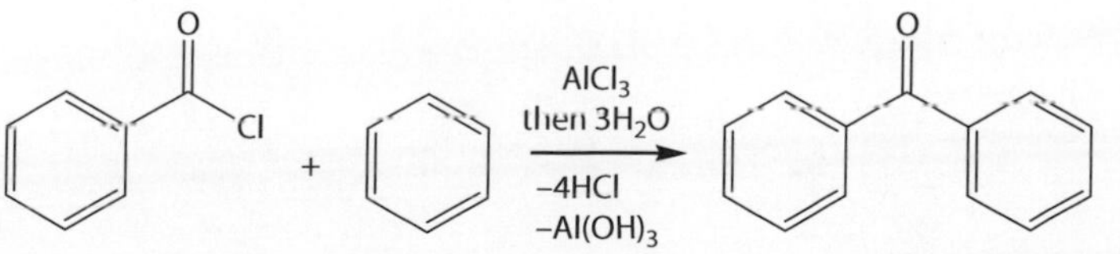

$E_m = 1.230$
$AE = 0.448$

6. Reduction of ketones by sodium borohydride

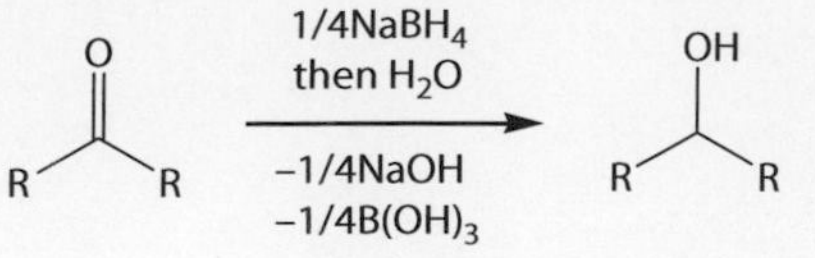

$E_m(max) = 0.797$
$AE(min) = 0.557$

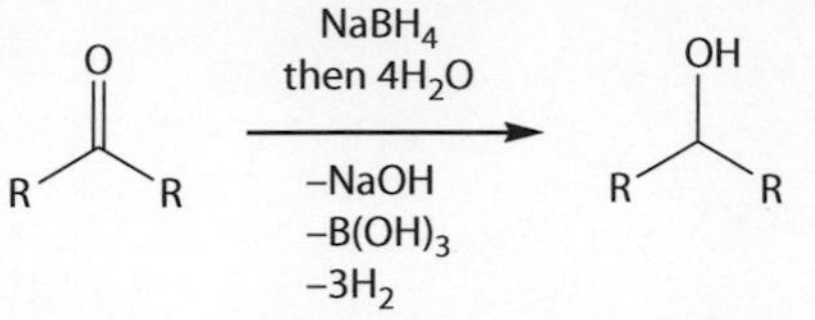

$E_m(max) = 3.375$
$AE(min) = 0.229$

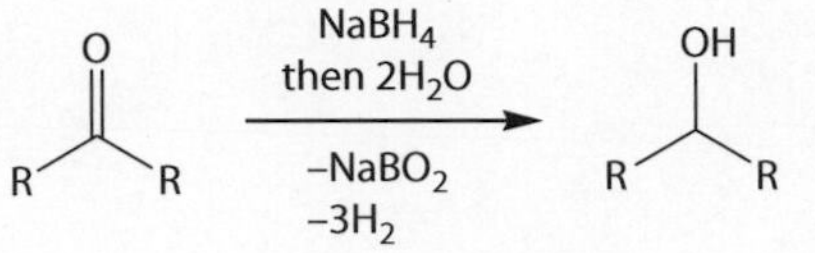

$E_m(max) = 2.250$
$AE(min) = 0.308$

when R = H

7. Reduction of ketones by lithium aluminum hydride

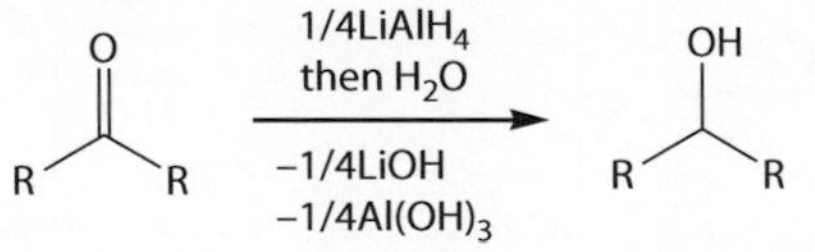

$E_m(max) = 0.797$
$AE(min) = 0.557$

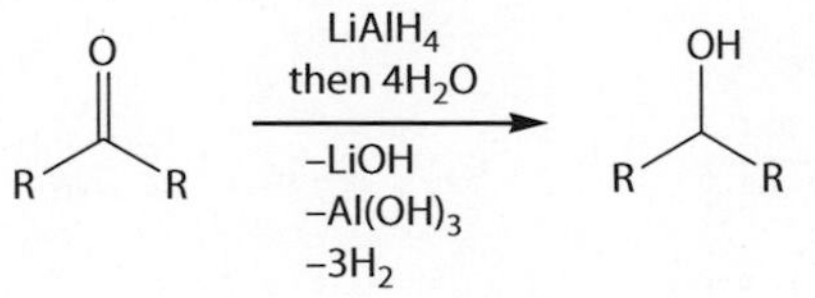

$E_m(max) = 3.375$
$AE(min) = 0.229$

when R = H

8. Elimination

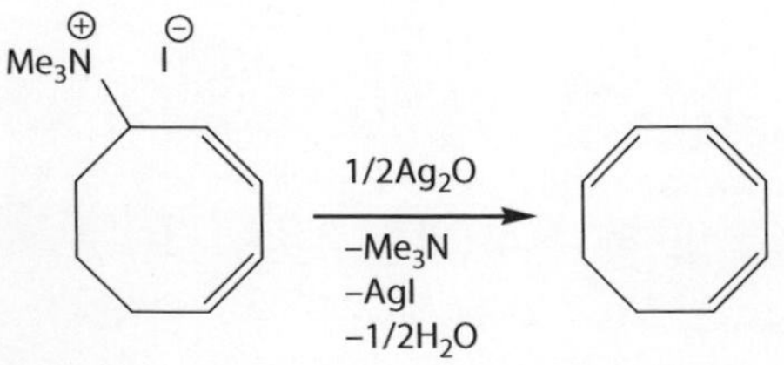

$E_m = 2.858$
$AE = 0.259$

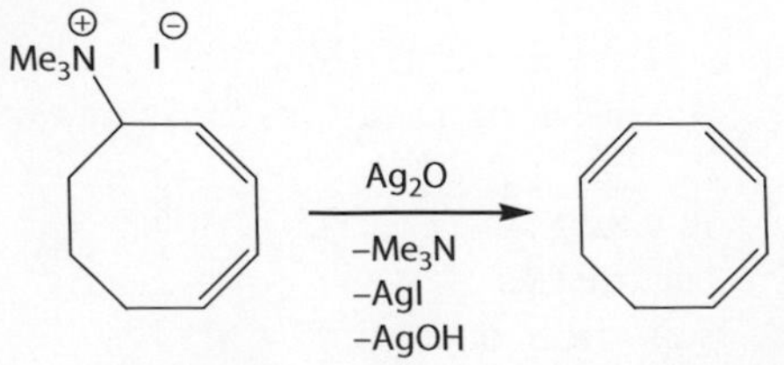

$E_m = 3.951$
$AE = 0.202$

9. Kraus oxidation

$E_m = 0.610$
$AE = 0.621$

$E_m = 0.240$
$AE = 0.807$

$E_m = 1.184$
$AE = 0.456$

10. Dihydroxylation of olefins

$E_m(max) = 1.792$
$AE(min) = 0.358$

$E_m(max) = 1.629$
$AE(min) = 0.380$

when R = H

Examples

Best radial pentagon performances from the synthesis database are shown in Figures 6.2 through 6.4. Note that the experimental radial pentagon is practically at the green limit of all five parameters.

Figures 6.5 and 6.6 show examples of balanced chemical equations with reagent stoichiometric coefficients not equal to one.

The following series of radial pentagons shown in Figures 6.7 through 6.21 are for experimental procedures found in *Organic Syntheses*. The reader should be aware that this is the only resource in the entire chemistry literature that has experimental procedures that have been checked by independent workers and scaled-up. There is no doubt about the reproducibility and veracity of the claims made by the authors of these experiments. The reader is invited to look up the cited references and practice using the PENTAGON spreadsheet algorithm.

1. Two ways of balancing a chemical equation: Friedel–Crafts acylation example using aluminum trichloride as part of balanced chemical equation or as a catalyst. Note that *E*-total has the same value in both cases.
2. Reaction solvent is also a reactant (e.g., thionyl chloride).
3. Classical resolutions using chiral auxiliaries.
4. Reactions advertised as examples of "green chemistry".

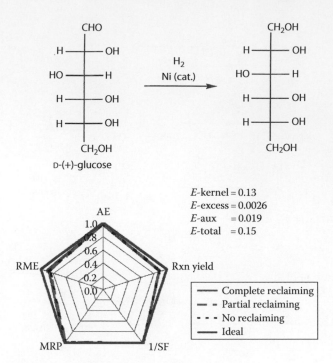

E-kernel = 0.13
E-excess = 0.0026
E-aux   = 0.019
E-total = 0.15

FIGURE 6.2   Radial pentagon for Faith [4] plan for sorbitol.

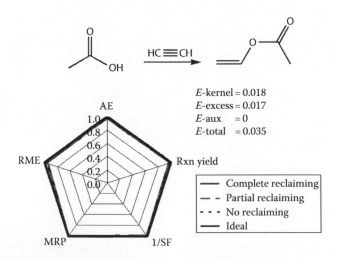

E-kernel = 0.018
E-excess = 0.017
E-aux   = 0
E-total = 0.035

FIGURE 6.3   Radial pentagon for Faith [4] plan for vinyl acetate.

Figure 6.22a and b show radial pentagons for an example reaction that can be analyzed in two ways depending on what is used as the oxidant in the balanced chemical equation. The reaction involves transforming D-fructose to 2,5-diformylfuran [20] via a tandem dehydration-redox sequence. Note that E-total is the same for both analyses. When dioxygen is the oxidant, dimethyl sulfoxide (DMSO) is counted as a solvent; hence, the E-aux contribution is larger. When DMSO is the oxidant, it is counted as a reactant not a solvent. This has the effect of lowering E-aux but increasing E-excess since the number of moles of DMSO is 11 times that of the limiting reagent, D-fructose.

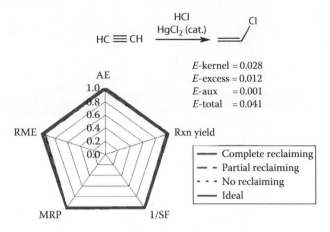

E-kernel = 0.028
E-excess = 0.012
E-aux   = 0.001
E-total = 0.041

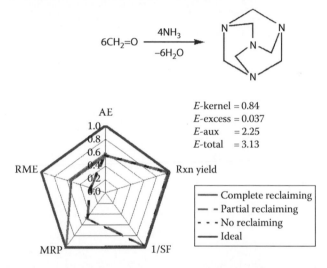

FIGURE 6.4 Radial pentagon for Faith [4] plan for vinyl chloride.

$6CH_2=O \xrightarrow[-6H_2O]{4NH_3}$

E-kernel = 0.84
E-excess = 0.037
E-aux   = 2.25
E-total = 3.13

FIGURE 6.5 Radial pentagon for Faith [4] plan for hexamethylenetetramine (formaldehyde is limiting reagent).

$2NH_3 + CO_2 \longrightarrow$

E-kernel = 0.60
E-excess = 1.30
E-aux   = 0
E-total = 1.9

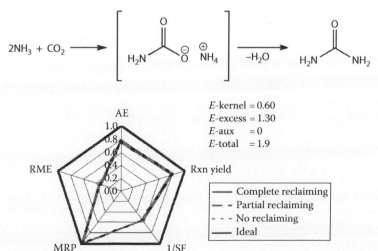

FIGURE 6.6 Radial pentagon for Faith [4] plan for urea (carbon dioxide is limiting reagent).

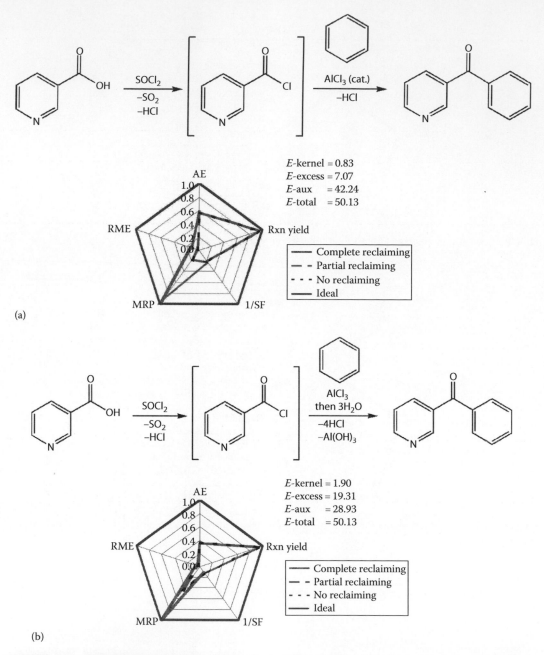

E-kernel = 0.83
E-excess = 7.07
E-aux   = 42.24
E-total = 50.13

E-kernel = 1.90
E-excess = 19.31
E-aux   = 28.93
E-total = 50.13

FIGURE 6.7 Radial pentagon for Friedel–Crafts acylation reaction [5] where aluminum trichloride is treated as (a) a catalyst and (b) a reagent.

6.3 RADIAL HEXAGON ANALYSIS

The creation of the radial hexagon as a visual aid for the easy identification of synthesis plans with "green" attributes follows the same theme as that for the radial pentagon for individual chemical reactions. The six parameters chosen may be categorized into two groups, three under material efficiency (overall yield along the longest branch, overall AE, and overall kernel RME) and the remainder under strategy efficiency (molecular weight fraction of non-sacrificial reagents, $1 - f(\text{sac})$, fraction of the kernel waste coming from target bond forming reactions, fw(tbr), and degree of

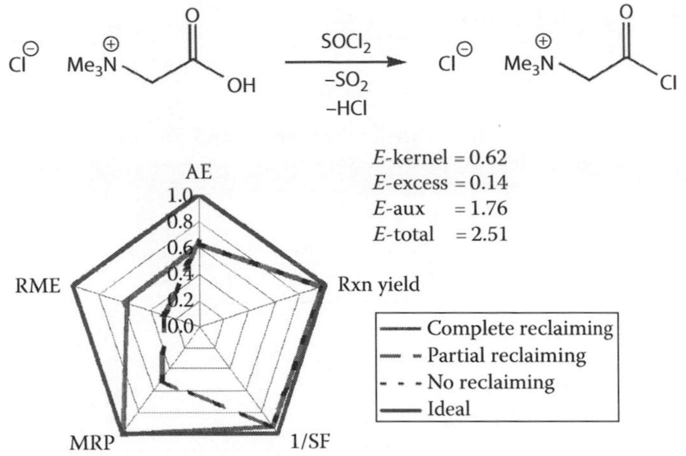

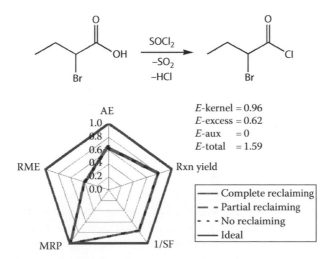

FIGURE 6.8 Radial pentagon for acid chloride synthesis [6]. Thionyl chloride is counted as a reagent not a reaction solvent; however, E-aux is not zero since benzene, toluene, and dichloromethane were used as solvents in the work-up procedure.

FIGURE 6.9 Radial pentagon for acid chloride synthesis [7]. Note that E-aux = 0 since thionyl chloride is counted as a reagent not a reaction solvent and no other work-up or purification materials were used.

convergence, delta). The regular hexagon designated as the "green limit" corresponds to the case when all six parameters have a value of unity, whereas, the one designated as the "golden limit" corresponds to the case when all six parameters have a value of 0.618 ($\varphi = (\sqrt{5} - 1)/2$). Figures 6.23 and 6.24 show radial hexagons for the Givaudan–Roure [21,22] plan for vanillin and Willstätter [23–25] plan for tropinone, respectively. These show the extremes of radial hexagons found in the present synthesis database of 1050+ plans.

A parameter that can describe the overall degree of "greenness," or relative "greenness," of a synthesis plan, with respect to the ideal green synthesis, may be defined as the vector magnitude ratio (VMR) given by Equation 6.5.

$$\text{VMR} = \frac{1}{\sqrt{6}}\left[\varepsilon^2 + (\text{AE})^2 + (\text{RME})^2 + \left(1 - f(\text{sac})\right)^2 + \left(\text{fw(tbr)}\right)^2 + (\text{delta})^2\right]^{1/2}. \tag{6.5}$$

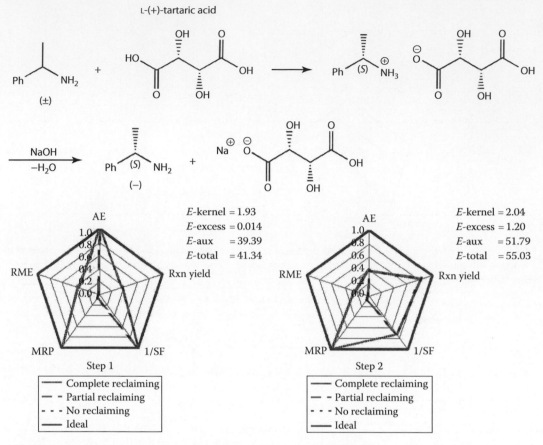

FIGURE 6.10 Radial pentagon for classical resolution of 1-phenylethylamine [8]. Note that the yield of the diastereomeric salt in the first step is less than 50%.

The underlying assumption is that all six parameters are mutually independent. The root 6 factor applies to the magnitude of the vector when all six parameters have a value of 1. Another possible parameter that could have been chosen is the relative area of the actual and green limit hexagons. However, this is problematic because the area of the actual hexagon will depend on the order of the parameters chosen. The vector magnitude does not depend on the order of the coordinates. In the preceding two examples, the VMR values are 0.830 and 0.139, respectively.

6.4 CONNECTIVITY ANALYSIS

Connectivity analysis is rooted in the field of graph theory [26–35] and was adapted by Wender and coworkers for the construction of complex ring containing compounds such as hirsutene [36]. Since his only publication on the subject, there has been no follow-up by that group or others in the field of organic synthesis. The idea that a chemical structure is a graph with atoms represented as vertices and chemical bonds as lines goes back to the work of Arthur Cayley on the problem of enumeration of isomers of hydrocarbons [37–48]. Related to this is the relationship, discovered by Oliver J. Lodge, for determining the number of rings and/or unsaturations for a given molecular formula, which is based on the concept of valence [49–50].

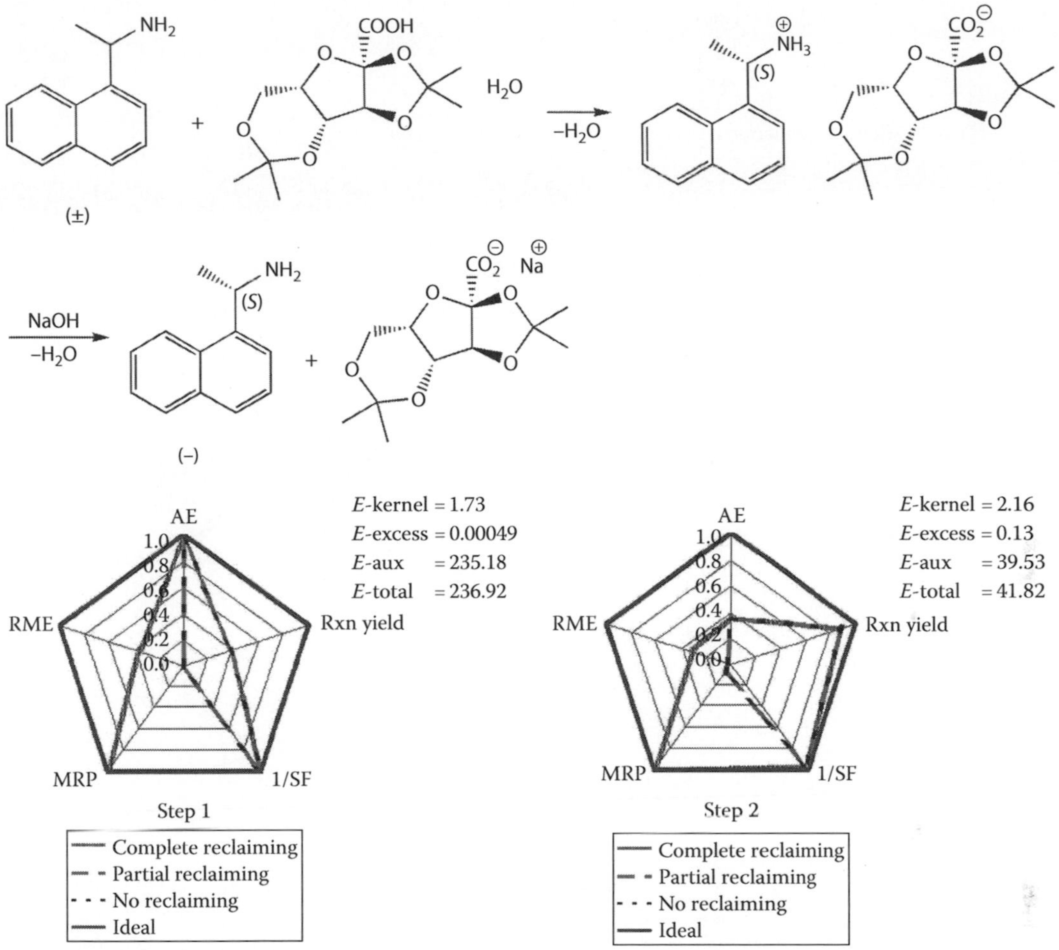

FIGURE 6.11 Radial pentagon for classical resolution of 1-naphthen-1-ylethylamine [9]. Note that the yield of the diastereomeric salt in the first step is less than 50%.

Here we introduce a structure parsing algorithm (SPA) for graphical factorization according to the following steps:

1. Draw the chemical structure as a graph representation.
2. Use heteroatoms as minus sign pivots (e.g., nitrogen or oxygen atoms), and electrophilic atoms as plus sign pivots (e.g., carbonyl groups) for the zeroth generation, G0, graph.
3. For the first generation graph, G1, assign next nearest neighbor vertices (atoms) as plus or minus as appropriate from G0.
4. Continue along chains and rings in this way to come up with G2, G3, etc., graphs.
5. Look for stretches or loops with continuously alternating plus and minus signs. The more plus–minus alternations there are, the easier it is to synthesize the molecule.
6. Identify bond clashes as minus–minus or plus–plus connections between adjacent atoms and atom clashes as dual minus and plus designations for the same atom. The more clashes there are, the more difficult it is to synthesize the molecule. Color code these clashes with gray bonds for easy identification. Examples of bond and atom clashes are given in the following.

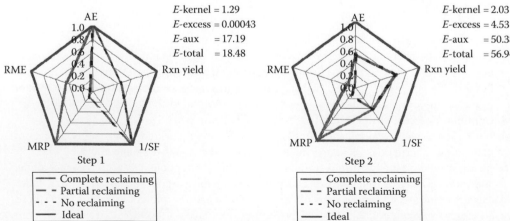

FIGURE 6.12 Radial pentagon for classical resolution of 1,1′-binaphthyl-2,2′-diyl hydrogen phosphate [10]. Note that the yield of the diastereomeric salt in the first step is less than 50%.

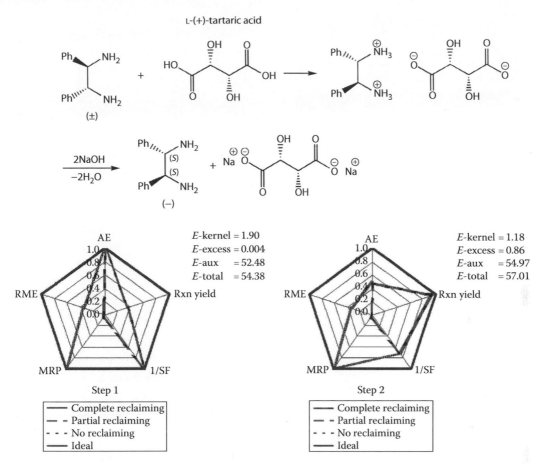

FIGURE 6.13 Radial pentagon for classical resolution of 1,2-diphenylethane-1,2-diamine [11]. Note that the yield of the diastereomeric salt in the first step is less than 50%.

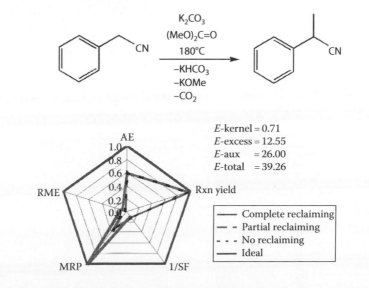

FIGURE 6.14 Radial pentagon for synthesis of 2-phenylpropionitrile [12]. The "green" aspect of this reaction is that dimethylcarbonate was used as a benign methylating agent.

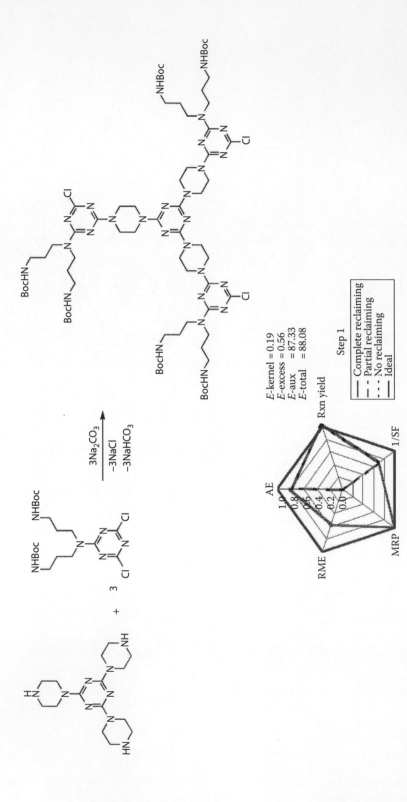

FIGURE 6.15 Radial pentagon for synthesis of a G1 melamine dendrimer [13]. The "green" aspect of this two-step reaction sequence is that it is composed of multi-component reactions.

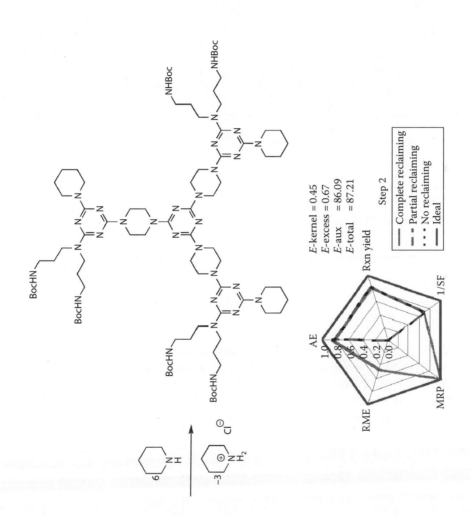

FIGURE 6.15 (continued)

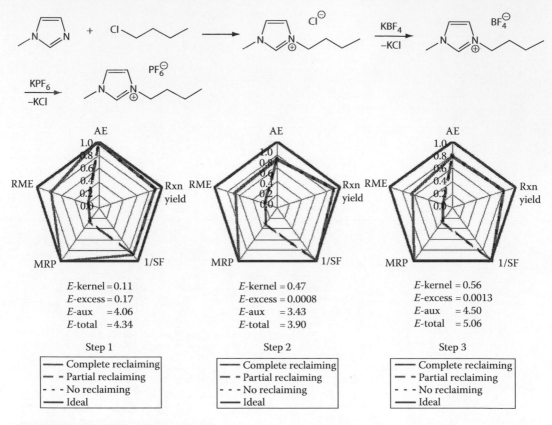

FIGURE 6.16 Radial pentagon for synthesis of an ionic liquid [14]. The "green" aspect of this reaction is that a nonvolatile and potentially recyclable ionic liquid solvent is synthesized for application in other reactions that traditionally use volatile and more toxic solvents.

7. Look for "natural" connections based on functional group transformation reactions or ring construction motifs found in the database of named organic reactions.

8. Look for symmetry elements in the final plus–minus arrangement that may be taken advantage of when proposing a synthesis plan for the target molecule.

9. Any one of the plus–minus connectivities that are identified is a potential target bond in the synthesis plan.

10. Connectivities that are identified as clashes suggest either that those frameworks correspond to starting material fragments, or that those bond connections may be possible through redox reactions where an electrophilic center is converted to a nucleophilic one, or vice versa, before bonding can take place in the usual way between a plus and a minus center. For example,

If there are no nucleophilic or electrophilic pivot atoms in the target structure, such as in a hydrocarbon molecule like hirsutene, then the problem is open-ended with many possible plus–minus connectivities. The presence of these pivots cuts down on the possible ways the skeleton of the target

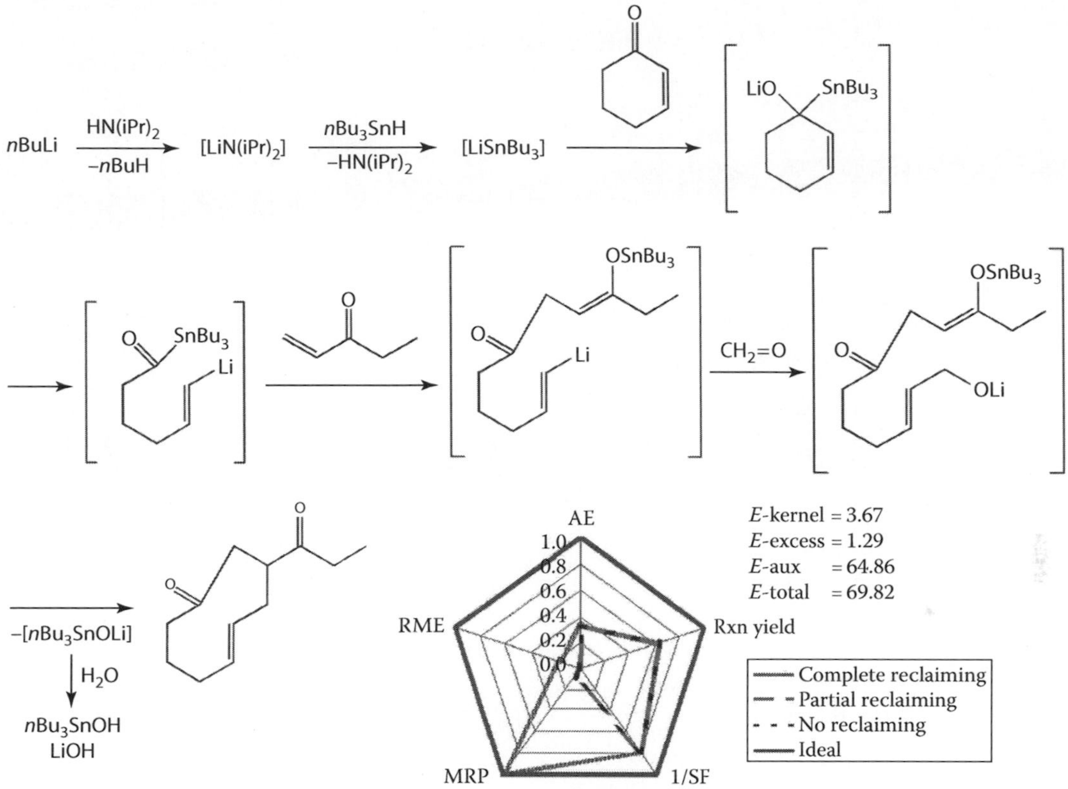

FIGURE 6.17 Radial pentagon for a 3-MCR [15]. The "green" aspect of this reaction is that it is a multi-component reaction.

structure can be put together. The graphical algorithm may be used as a tool in retrosynthetic analysis to posit synthesis plans with the fewest number of steps, hence leading to optimum strategies.

The following example structures help to illustrate the idea of the SPA. Figure 6.25 shows the analysis for crystal violet. The result is a continuous loop of alternating plus–minus connections that is consistent with the synthesis plan for its construction as shown in Figure 6.26.

Figure 6.27 shows the analysis for a heterocyclic compound highlighted in a paper on electronic reaction databases and synthesis network searches [52]. The proposed synthesis plan is shown in Figure 6.28 that features two branches that converge in a three-component coupling reaction.

Figure 6.29 shows the analysis for oseltamivir that results in a number of sign clashes and hence makes it a challenging molecule to synthesize. The four possible orientations corresponding to the G3 graph are shown with gray colored bonds.

Figure 6.30 shows the application of the algorithm to the structure of strychnine. This leads to an interesting proposed ring construction strategy as shown in Figure 6.31.

6.5 PROBABILITY ANALYSIS

At first sight, the idea of introducing probability in the field of organic synthesis may be far fetched. However, when one considers that the oldest measure of reaction performance, reaction yield, is a fraction between 0 and 1, the notion begins to make sense. The reader may recall that all probability values are expressed as fractions between 0 and 1 where values close to zero pertain to low likelihood events and values close to 1 pertain to events that are very likely to occur. If multiple events are mutually

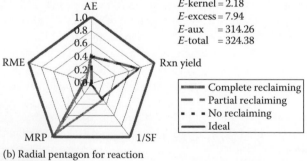

(a) Reaction scheme

(b) Radial pentagon for reaction

E-kernel = 2.18
E-excess = 7.94
E-aux = 314.26
E-total = 324.38

FIGURE 6.18 Radial pentagon for a pseudo 3-MCR [16]. The "green" aspect of this reaction is that it is a multicomponent reaction. It is designated with the prefix pseudo because two of the three reagents are identical.

exclusive, that is, independent from one another, then the probability that they may occur simultaneously is determined as the multiplicative product of the probabilities of each event occurring. Therefore, we may interpret the reaction yield for a given chemical transformation as a probability of reaction occurrence based on experimental factors such as thermodynamic and kinetic considerations, temperature, pressure, choice of solvent, and the presence or absence of catalysts and additives. This probability can only be determined experimentally usually by trial and error. If a reactant is thermodynamically

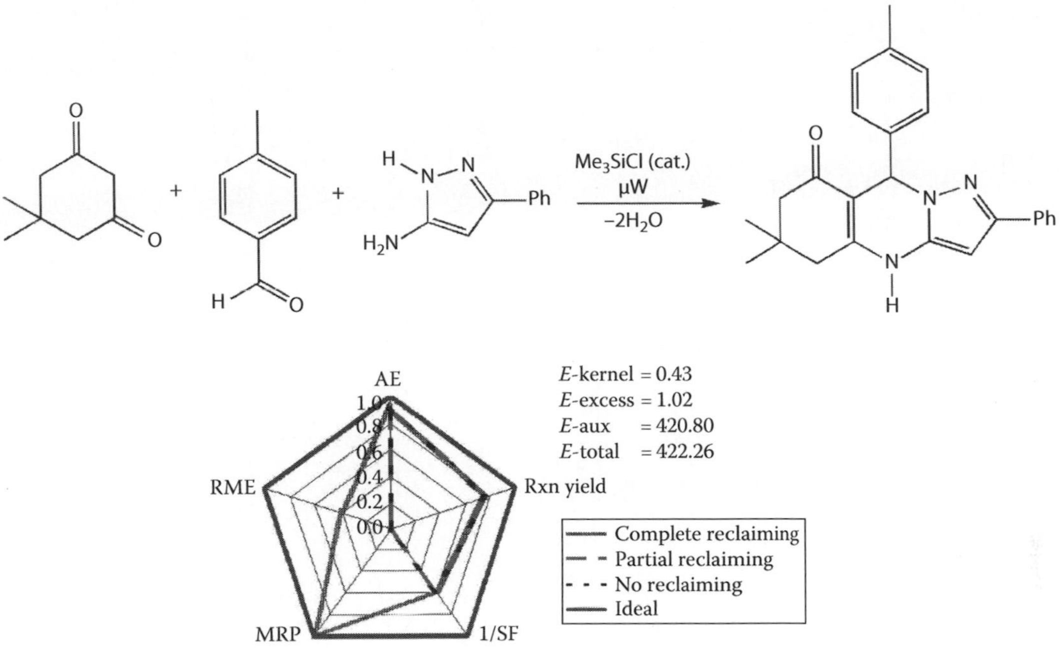

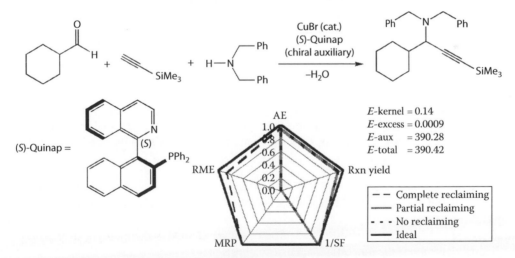

FIGURE 6.19 Radial pentagon for a 3-MCR using microwave reaction [17]. The "green" aspect of this reaction is that it is a multicomponent reaction that requires less energy input than conventional heating methods.

FIGURE 6.20 Radial pentagon for a 3-MCR [18]. The "green" aspect of this reaction is that it is a multi-component reaction.

and kinetically unstable then it will surely react in some way, usually with considerable violence. This dual instability is linked to some structural feature in the molecule's architecture such as a strained ring, or bonds with low bond dissociation energies. On the other hand, if a reactant is thermodynamically and kinetically stable, then it is very unlikely to change its molecular structure under standard temperature and pressure conditions and hence a considerable amount of energy input is required for it to react. Most of the time reactants fall in between these extremes. Chemistry, therefore, is the science of finding the appropriate reaction conditions that control the reactivities of substances so that intended target products may be obtained in reliable and predictable ways. Knowledge of the molecular structures of reactants and products is obviously essential for this endeavor to be successful. This search for

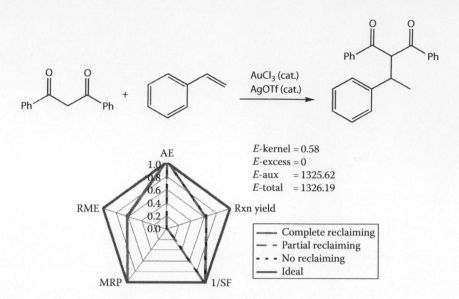

FIGURE 6.21 Radial pentagon for a 100% atom economical reaction using no excess reagents yet having a significant auxiliary material consumption particularly chromatographic solvents [19]. This is a good example of a mismatched scenario with respect to material efficiency.

appropriate reaction conditions is called reaction optimization. So, for any chemical reaction, we may describe the following scenarios where reaction yield, ε, is replaced by, p, for probability:

1. If $p = 0$, this means no reaction occurs.
2. If $p = 1$, this means that the reaction leads to the desired product plus consequential by-products as a result of the reaction mechanism; no starting material is left and no competing side reactions occur.
3. If $0 < p < 1$, this means that the reaction leads to the desired product plus consequential by-products as a result of the reaction mechanism, unreacted starting materials, and any side products if there are other possible reaction pathways other than the one leading to the intended target product.

We may apply these ideas further to a synthesis plan, which is a sequential series of reaction steps leading to a target molecule. If each reaction's outcome is independent from its predecessors, then the overall yield may be interpreted as the overall probability that the sequence of reactions will ultimately lead to the desired product. Recall that the overall yield is the multiplicative product of all reaction yields along the longest branch in a synthesis plan.

The reader will realize by now that all metrics discussed in this book are fractions between 0 and 1. Again, we could interpret all of them as probabilities. We note that overall RME for a given reaction is the multiplicative product of reaction yield, AE, excess reagent consumption (1/SF), and auxiliary material consumption (MRP). Therefore, RME for an individual reaction may be restated as the probability of getting the target product with the least waste produced. Similarly, AE may be restated as the probability of producing only the target structure with no by-products, 1/SF is the probability that the waste produced originates from excess reagents, and MRP is the probability that the waste produced originates from auxiliary materials used in work-up and purification.

Having accepted the interpretation that all metrics with values ranging between 0 and 1 are probabilities, the next idea to introduce is that of thresholds. What we are asking is the following

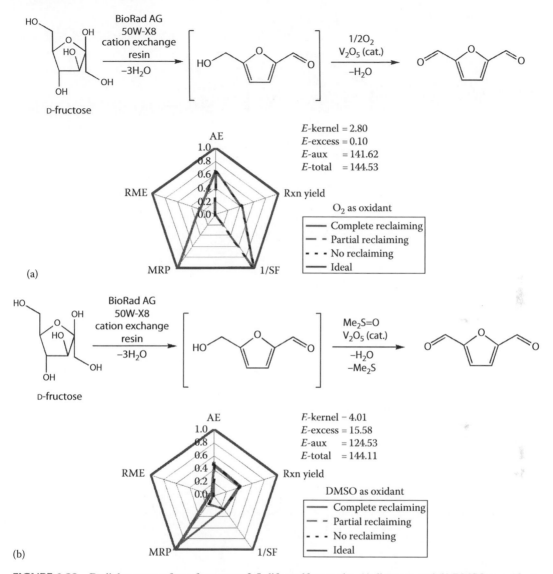

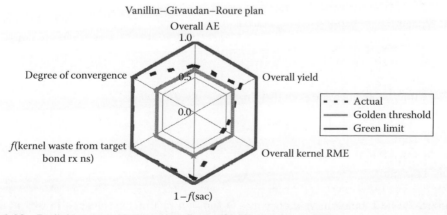

FIGURE 6.22 Radial pentagon for D-fructose to 2,5-diformylfuran using (a) dioxygen and (b) DMSO as oxidant.

FIGURE 6.23 Radial hexagon for Givaudan–Roure [21,22] plan for vanillin.

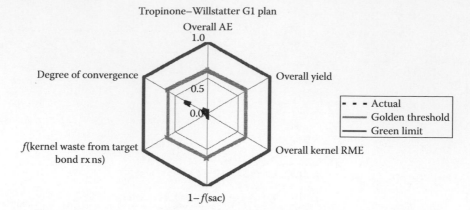

FIGURE 6.24 Radial hexagon for Willstätter [23–25] plan for tropinone.

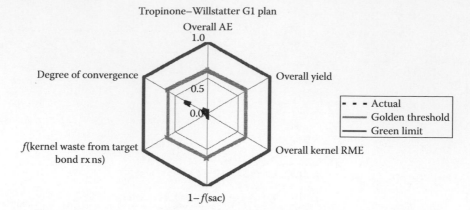

FIGURE 6.25 Structure parsing algorithm applied to crystal violet.

question. What is the probability that a given chemical reaction will have a kernel RME greater than say 50% or 80%? Recalling the expression for kernel RME as

$$RME_{kernel} = (\varepsilon)(AE) \tag{6.6}$$

we impose a threshold value of α such that the inequality (6.7) holds

$$RME_{kernel} = (\varepsilon)(AE) \geq \alpha. \tag{6.7}$$

This may be rewritten as

$$AE \geq \alpha/\varepsilon. \tag{6.8}$$

Since both AE and ε range in value between 0 and 1, a plot of AE versus ε results in a square domain. Figure 6.32 shows such a plot for various threshold values.

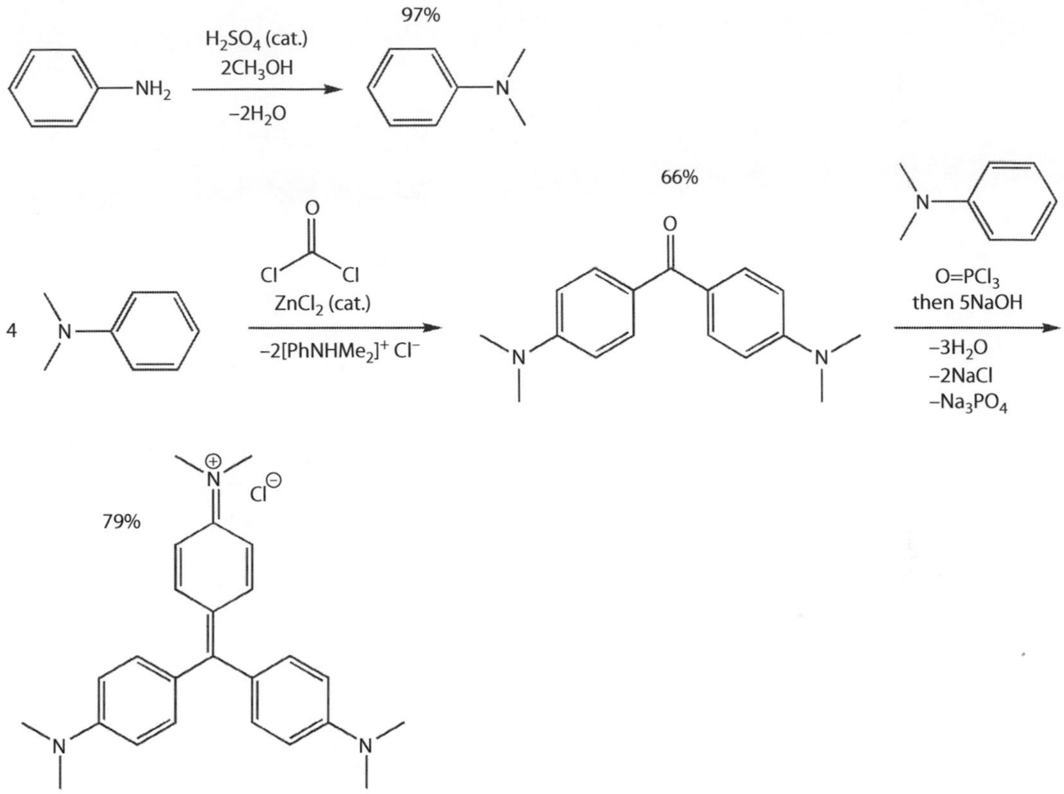

FIGURE 6.26 Synthesis plan for crystal violet [51].

The answer to the question posed earlier may be interpreted graphically as the area of the region above the curve defined by Equation 6.8 relative to the area of the square, which is 1. Hence, we have

$$\text{probability} = \text{region A}$$

$$= 1 - \text{region B} - \text{region C}$$

$$= 1 - \alpha - \alpha \int_{\alpha}^{1} \frac{d\varepsilon}{\varepsilon}$$

$$= 1 - \alpha + \alpha \ln \alpha \qquad (6.9)$$

Figure 6.33 shows a representative plot illustrating graphically the meaning of Equation 6.9 and Figure 6.34 shows a plot of the probability function given by Equation 6.9. We note that as we set the threshold low we increase the probability of meeting that threshold; if we aim for a high threshold then the probability of meeting it diminishes. Hence, we can conclude that achieving a "green" reaction where the kernel RME, based only on reaction yield and AE, is high, appears to be a hard task indeed. If we set α to be equal to the golden ratio 0.618, then $p = 0.0845$ or about 8.5%. Things are much worse if we were to include excess reagent and auxiliary material contributions. The probability would be vanishingly small. This reinforces the notion that if you want to have a decent chance of achieving greenness you have to aim for high atom economical reactions that also have high reaction yields, otherwise there is no hope of meeting any threshold, even of modest value.

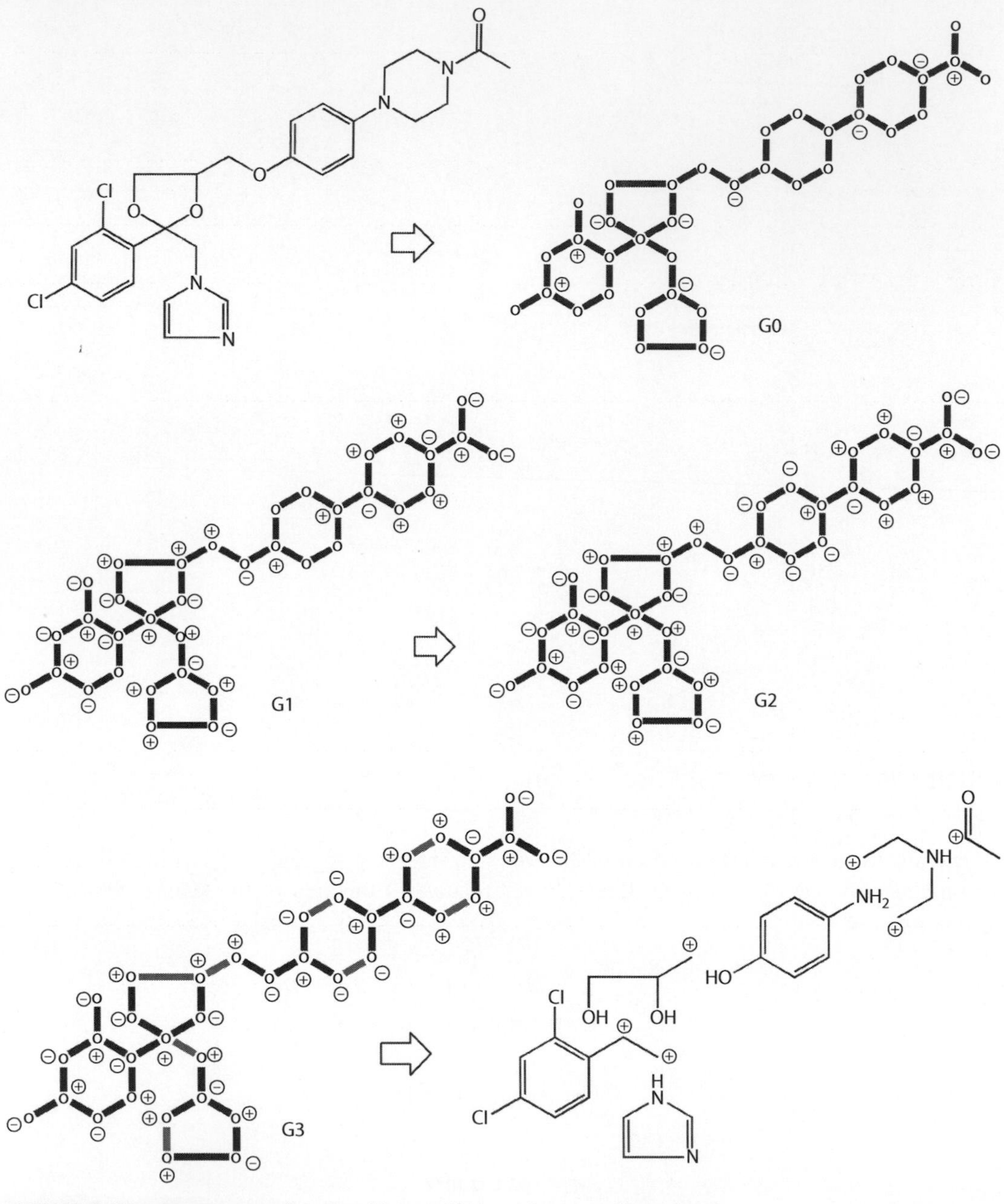

FIGURE 6.27   Structure parsing algorithm applied to a heterocyclic molecule.

We may view the square shown in Figure 6.32 as a kind of dartboard where the goal is to aim for the top right-hand corner if you want a green reaction. At first glance, this might seem like a losing proposition and thus give ammunition to detractors and skeptics of green chemistry thinking. However, things change dramatically in the positive direction when we impose constraints on AE and reaction yield values for a given reaction. This comes as a consequence of writing chemical transformations as general equations using Markush structures. The application of probability analysis on such equations is a very powerful idea that has far-reaching consequences for synthesis planning, especially for template plans that lead to multiple structurally related products as would be the case for building compound libraries based on a core scaffold structure decorated with

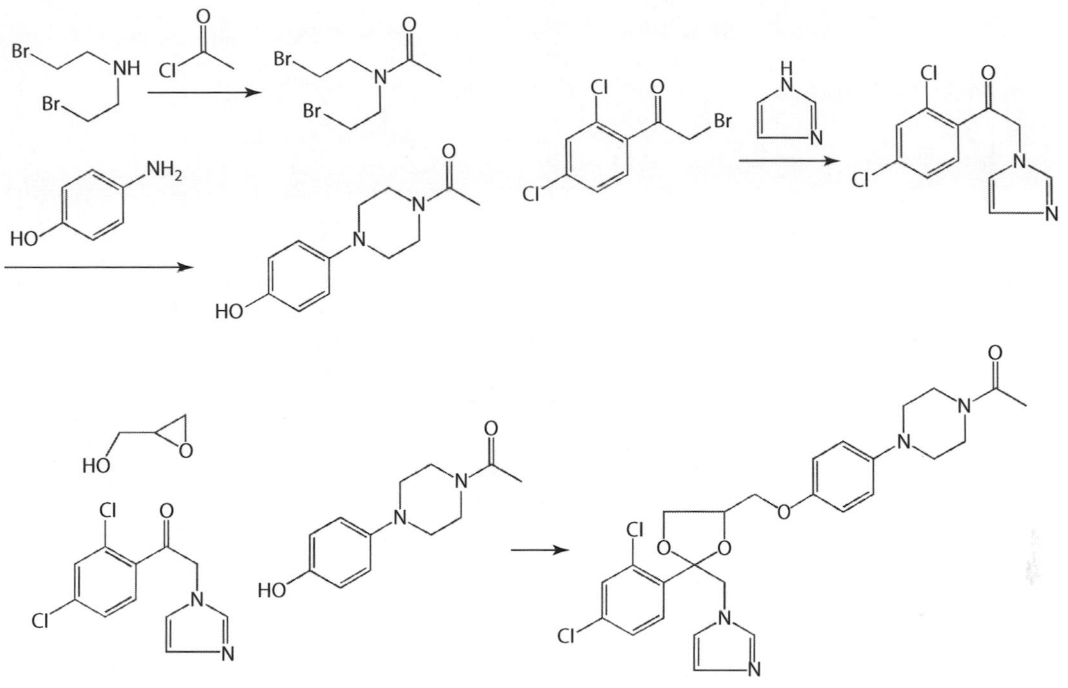

FIGURE 6.28 Proposed convergent synthesis plan.

various R groups. The Biginelli three-component coupling [53–55], which has been rediscovered and studied extensively in the last decade, serves as a good model to illustrate ideas presented here. This reaction has been used to synthesize substituted 3,4-dihydro-1*H*-pyrimidin-2-ones and is shown in the following in general form.

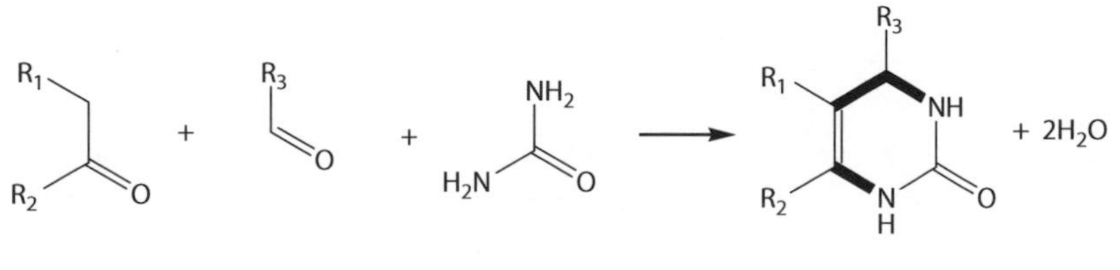

$R_1 + R_2 + 42$ $R_3 + 29$ 60 $R_1 + R_2 + R_3 + 95$ 36

We may pose the following problem:

a. Determine the minimum AE and maximum *E*-factor based on molecular weight (E(mw)) for this reaction.
b. Suppose a minimum target kernel RME of 60% is imposed on this reaction. What is the probability that this target will be met given the minimum AE found in part (a)?
c. Suppose we add another constraint that the reaction yield should have a minimum value of 80%. What is the probability that the kernel RME will be at least 60% given both constraints of a minimum AE found in part (a) and a minimum reaction yield of 80%?

The expression for E(mw) is given by

$$E(\text{mw}) = \frac{36}{R_1 + R_2 + R_3 + 95}. \tag{6.10}$$

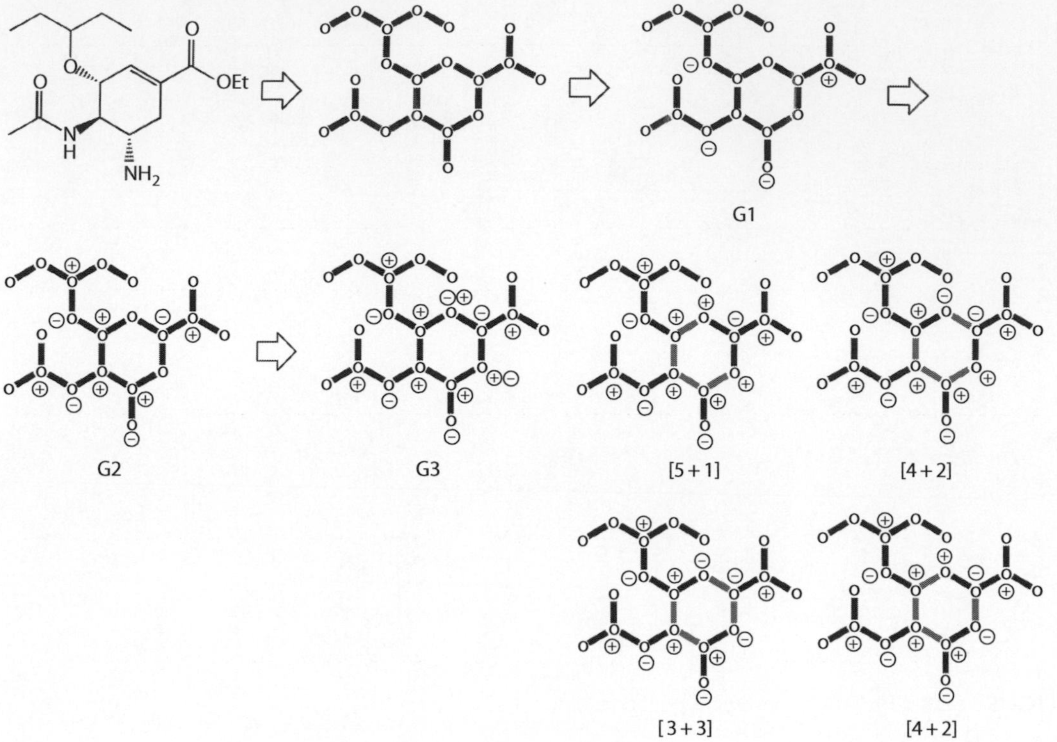

FIGURE 6.29 Structure parsing algorithm applied to oseltamivir.

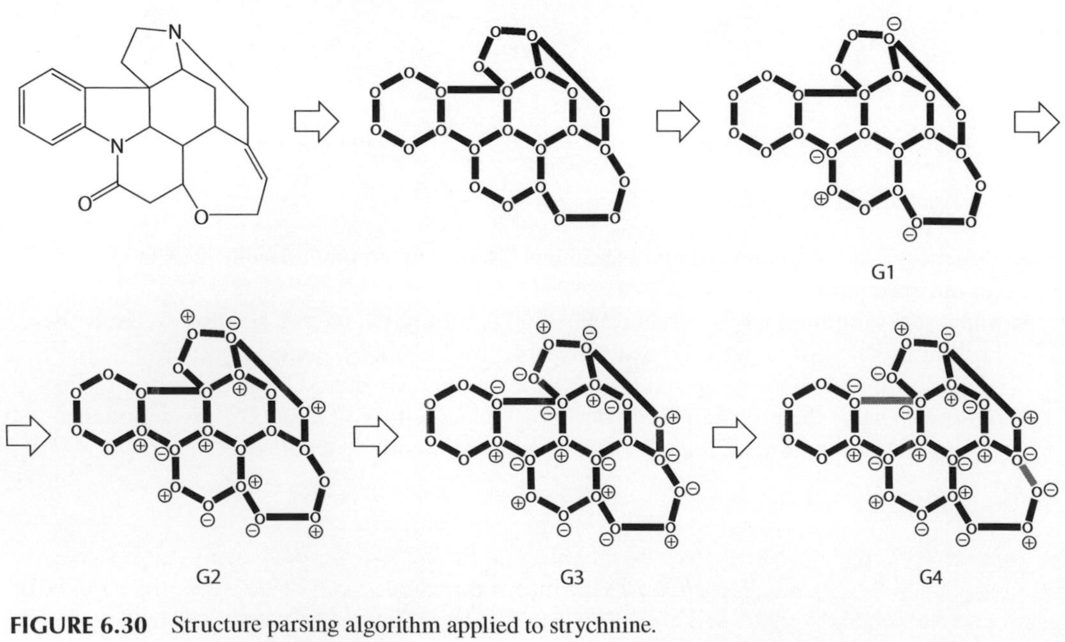

FIGURE 6.30 Structure parsing algorithm applied to strychnine.

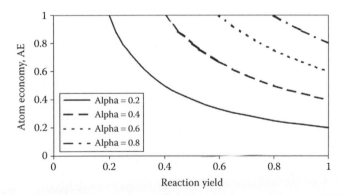

FIGURE 6.31 Proposed plan for constructing strychnine.

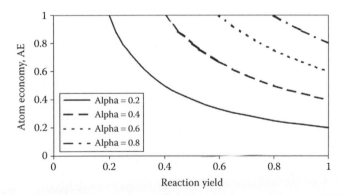

FIGURE 6.32 Plot of AE versus reaction yield according to Equation 6.8 for various values of the threshold, α.

If the R groups are as small as possible, then $E(mw)$ will approach a maximum value. Setting each R group to be a hydrogen atom, we have $E(mw) = 36/98 = 0.367$. Since $E(mw)$ and AE are inversely related by the expression

$$AE = \frac{1}{1 + E(mw)} \tag{6.11}$$

then maximization of $E(mw)$ automatically implies minimization of AE. Hence, AE(min) for this reaction is $1/(1 + 0.367) = 0.731$ or 73.1%. This means that the AE can be no worse than 73.1%. If we impose a threshold kernel RME of 60% ($\alpha = 0.6$), then the dartboard scenario is depicted as in

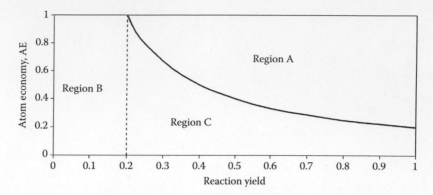

FIGURE 6.33   Relative areas interpreted as a probability according to Equation 6.9 for $\alpha = 0.2$.

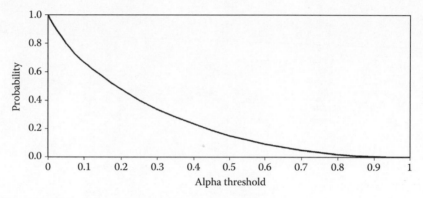

FIGURE 6.34   Plot of the probability function given by Equation 6.9.

Figure 6.35. The probability that the target kernel RME will be met given the constraint AE(min) = 0.731 is given by

$$\text{probability} = \frac{\text{region A}}{\text{region A} + \text{region B}}$$

$$= \frac{\text{area(DIJE)}}{\text{area(DIJE)} + \text{area(ABCD)} + \text{area(DCE)}}$$

$$= \frac{\text{area(DIKH)} - \text{area(DEGH)} - \text{area(EJKG)}}{\text{area(DIKH)} - \text{area(DEGH)} - \text{area(EJKG)} + \text{area(ABCD)} + \text{area(DCE)}} \qquad (6.12)$$

The coordinates of the point E are (0.6/0.731, 0.731) or (0.821, 0.731). Hence, the relevant areas are given by Equation 6.13a through e).

$$\text{area(ABCD)} = (0.6)(1 - 0.731) = 0.161 \qquad (6.13a)$$

$$\text{area(DIKH)} = (1)(1 - 0.6) = 0.4 \qquad (6.13b)$$

$$\text{area(DEGH)} = \alpha \int_{0.6}^{0.821} \frac{d\varepsilon}{\varepsilon} = 0.6\big[\ln(0.821) - \ln(0.6)\big] = 0.188 \qquad (6.13c)$$

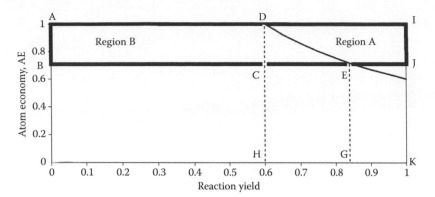

FIGURE 6.35 Dartboard for Biginelli reaction with imposed AE(min) value of 0.731.

$$area(EJKG) = (1 - 0.821)(0.731) = 0.131 \qquad (6.13d)$$

$$area(DCE) = area(DEGH) - area(CEGH)$$

$$= 0.188 - (0.821 - 0.6)(0.731)$$

$$= 0.026 \qquad (6.13e)$$

The required probability is then

$$probability = \frac{0.4 - 0.188 - 0.131}{0.4 - 0.188 - 0.131 + 0.161 + 0.026} = \frac{0.081}{0.268} = 0.302, \qquad (6.14)$$

or 30.2%. This may be compared with the value 9.4% obtained from Equation 6.9 if we had not imposed the minimum AE constraint. When we impose the additional constraint that the reaction yield should not be below 80%, then the dartboard changes to that shown in Figure 6.36. Now, the probability that the target kernel RME will be met given both constraints is given by

$$probability = \frac{region\ A}{region\ A + region\ B}$$

$$= \frac{area(ABCDH)}{area(ABCDH) + area(HDG)}$$

$$= \frac{area(ABCG) - area(HDG)}{area(ABCG) - area(HDG) + area(HDG)}$$

$$= \frac{area(ABCG) - area(HDG)}{area(ABCG)} \qquad (6.15)$$

The coordinates of the point D are (0.6/0.731, 0.731) or (0.821, 0.731). Hence, the relevant areas are given by Equation 6.16a and b.

$$area(ABCG) = (1 - 0.821)(1 - 0.731) = 0.0482 \qquad (6.16a)$$

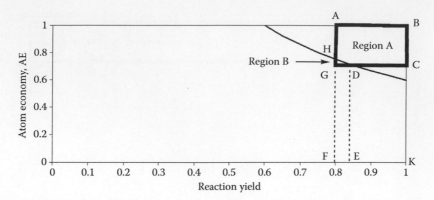

FIGURE 6.36 Dartboard for Biginelli reaction with imposed AE(min) value of 0.731 and ε(min) = 0.80.

$$\text{area(HDG)} = \text{area(HDEF)} - \text{area(GDEF)}$$

$$= \alpha \int_{0.8}^{0.821} \frac{d\varepsilon}{\varepsilon} - (0.821 - 0.8)(0.731)$$

$$= 0.6\big[\ln(0.821) - \ln(0.8)\big] - 0.0154$$

$$= 0.000196 \tag{6.16b}$$

The required probability is then

$$\text{probability} = \frac{0.0482 - 0.000196}{0.0482} = 0.996, \tag{6.17}$$

or 99.6%. We see that our chances have dramatically increased 10-fold from a base probability of 9.4%.

General probability expressions pertaining to various scenarios of minimum atom economies and minimum reaction yields with respect to a threshold kernel RME value of α are summarized in Equations 6.18 through 6.24.

Case I ((AE)min $\geq \alpha$ and $0 \leq \varepsilon$(min) $\leq \alpha$)

$$p = \frac{1 - \text{AE(min)} + \alpha \ln(\text{AE)min}}{\big[1 - \text{AE(min)}\big]\big[1 - \varepsilon(\text{min})\big]} \tag{6.18}$$

Case II ((AE)min $\geq \alpha$ and $\alpha \leq \varepsilon$(min) $\leq \alpha/\text{AE(min)}$)

$$p = \frac{1 - \text{AE(min)} + \alpha - \varepsilon(\text{min}) - \alpha \ln\left[\dfrac{\alpha}{(\text{AE})\min\varepsilon(\text{min})}\right]}{\big[1 - \text{AE(min)}\big]\big[1 - \varepsilon(\text{min})\big]} \tag{6.19}$$

Case III ((AE)min $\geq \alpha$ and $\alpha/\text{AE(min)} \leq \varepsilon$(min) ≤ 1)

$$p = 1 \tag{6.20}$$

Case IV ((AE)min $= \alpha$ and $0 \leq \varepsilon$(min) $\leq \alpha$)

$$p = \frac{1 - \alpha + \alpha \ln \alpha}{\big[1 - \alpha\big]\big[1 - \varepsilon(\text{min})\big]} \tag{6.21}$$

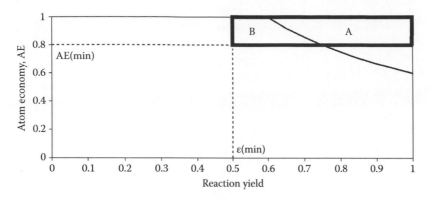

FIGURE 6.37 Dartboard for a reaction with constraints AE(min) = 0.80 and ε(min) = 0.50. (α = 0.60).

Case V ((AE)min = α and α ≤ ε(min) ≤ 1)

$$p = \frac{1 - \varepsilon(\min) + \alpha \ln \varepsilon(\min)}{[1 - \alpha][1 - \varepsilon(\min)]} \tag{6.22}$$

Case VI ((AE)min ≤ α and 0 ≤ ε(min) ≤ α)

$$p = \frac{1 - \alpha + \alpha \ln \alpha}{[1 - AE(\min)][1 - \varepsilon(\min)]} \tag{6.23}$$

Case VII ((AE)min ≤ α and α ≤ ε(min) ≤ 1)

$$p = \frac{1 - \varepsilon(\min) + \alpha \ln \varepsilon(\min)}{[1 - AE(\min)][1 - \varepsilon(\min)]} \tag{6.24}$$

Graphical representations of each of these scenarios are shown in Figures 6.37 through 6.43. These results are arguably the most powerful relationships derived so far in the field of green metrics since they scope out the "green" possibilities for any generalized chemical reaction. The reader may verify the preceding numerical results for the Biginelli example using formulas in Equations 6.18 and 6.19. Note that case III predicts a scenario with 100% certainty. It can be

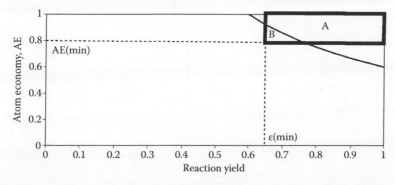

FIGURE 6.38 Dartboard for a reaction with constraints AE(min) = 0.80 and ε(min) = 0.65. (α = 0.60).

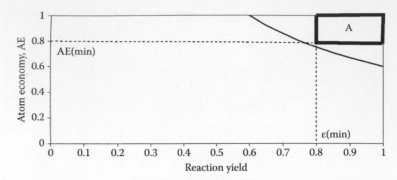

FIGURE 6.39   Dartboard for a reaction with constraints AE(min) = 0.80 and ε(min) = 0.80. (α = 0.60).

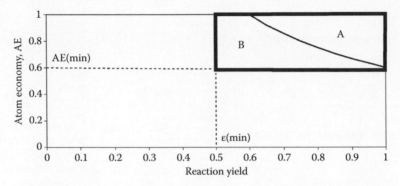

FIGURE 6.40   Dartboard for a reaction with constraints AE(min) = 0.60 and ε(min) = 0.50. (α = 0.60).

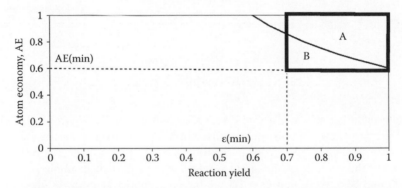

FIGURE 6.41   Dartboard for a reaction with constraints AE(min) = 0.60 and ε(min) = 0.70. (α = 0.60).

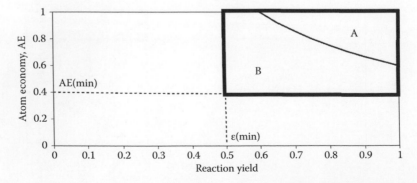

FIGURE 6.42   Dartboard for a reaction with constraints AE(min) = 0.40 and ε(min) = 0.50. (α = 0.60).

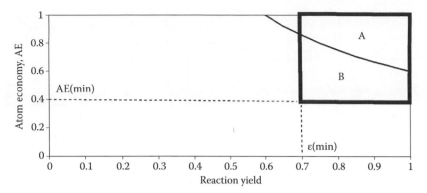

FIGURE 6.43 Dartboard for a reaction with constraints AE(min) = 0.40 and ε(min) = 0.70. (α = 0.60).

reasonably stated that achieving greenness is easily within reach if at the outset reactions have high atom economies and high reaction yields.

6.6 AN EIGHT-COMPONENT COUPLING

A spectacular example of optimization that ties together all of the ideas in this chapter has recently been achieved [56]. In a single one-pot reaction using methanol as the solvent, a tandem [3-MCR – 3-MCR] – Ugi-4-MCR convergent sequence was performed as shown in the generalized scheme (see Figure 6.44). Oxidation states of atoms involved in target forming bonds are indicated as well as a target bond map for the final product. Overall, the reaction is an 8-component coupling that produces 12 target bonds and 10 points of diversity. The minimum AE for this reaction is 0.816 or 81.6% when all the R groups are hydrogen atoms. This highly atom economical performance

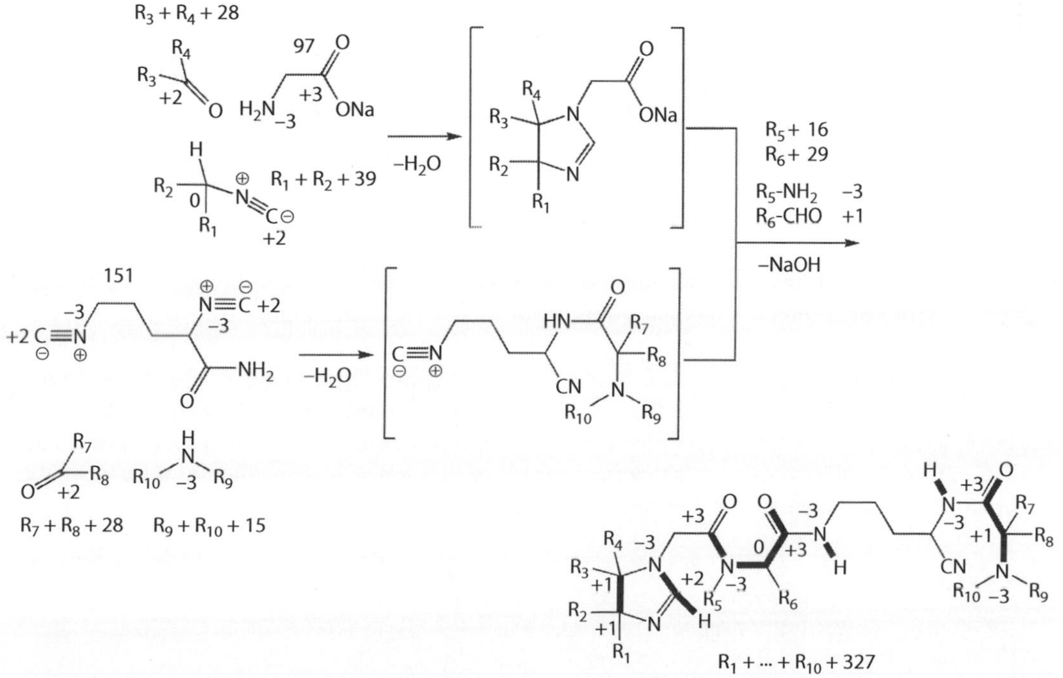

FIGURE 6.44 Generalized scheme for 8-MCR.

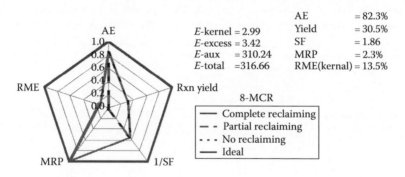

FIGURE 6.45 Scheme for specific 8-MCR reaction.

		AE	= 82.3%
E-kernel = 2.99	Yield	= 30.5%
E-excess = 3.42	SF	= 1.86
E-aux = 310.24	MRP	= 2.3%
E-total = 316.66	RME(kernal) = 13.5%

AE
1.0
0.8
0.6
0.4
0.2
0.0

RME   Rxn yield

8-MCR

——— Complete reclaiming
– – Partial reclaiming
· · · No reclaiming
——— Ideal

MRP   1/SF

FIGURE 6.46 Radial pentagon for specific 8-MCR.

is offset by a reported low yield of only 30.5%. Therefore, the minimum kernel RME is 0.249 or 24.9%. Figure 6.45 shows a specific example for this 8-MCR and Figure 6.46 is the accompanying radial pentagon for its experimental procedure. In terms of synthesis strategy parameters, B/M = 12, *f*(nb) = 0, and *f*(sac) = 0. In terms of changes in oxidation states of atoms involved in forming the target bonds, HI = 0, so the transformation is isohypsic. A synthesis tree would show that the reaction is 100% convergent (delta = 1) and 0% asymmetric (beta = 0). The only drawback as far as optimization is concerned is that the reaction yield is low, which again reinforces the reality that chemistry is a science of compromise and tradeoffs.

REFERENCES

1. Willstätter, R.; Waser, E. *Chem. Ber.* 1911, *44*, 3423.
2. Hamon, D.P.G.; Young, R.N. *Aust. J. Chem.* 1976, *29*, 145.
3. Metzger, J.O.; Eissen, M. (Universität Oldenburg, Germany). http://www.chemie.uni-oldenburg.de/oc/metzger/eatos/

4. Faith, W.L.; Keyes, D.B.; Clark, R.L. *Industrial Chemicals*, 3rd edn., John Wiley & Sons, Inc.: New York, 1966.
5. Villani, F.J.; King, M.S. *Org. Synth. Coll.* 1963, *4*, 88.
6. Vassel, B.; Skelly, W.G. *Org. Synth. Coll.* 1963, *4*, 154.
7. Rutenberg, M.W.; Horning, E.C. *Org. Synth. Coll.* 1963, *4*, 620.
8. Ault, A. *Org. Synth. Coll.* 1973, *5*, 932.
9. Mohacsi, E.; Leimgruber, W. *Org. Synth. Coll.* 1988, *6*, 826.
10. Jacques, J.; Fouquey, C. *Org. Synth. Coll.* 1993, *8*, 50.
11. Pikul, S.; Corey, E.J. *Org. Synth. Coll.* 1998, *9*, 387.
12. Tundo, P.; Selva, M.; Bomben, A. *Org. Synth.* 1999, *76*, 169.
13. Chouai, A.; Venditto, V.J.; Simanek, E.E. *Org. Synth.* 2009, *86*, 151.
14. Dupont, J.; Consorti, C.S.; Suarez, P.A.Z.; de Souza, R.F. *Org. Synth.* 2002, *79*, 236.
15. Webb, K.S.; Asirvatham, E.; Posner, G.H. *Org. Synth. Coll.* 1993, *8*, 562.
16. Morwick, T.; Paquette, L.A. *Org. Synth. Coll.* 1998, *9*, 670.
17. Glasnov, T.N.; Kappe, C.O. *Org. Synth.* 2009, *86*, 252.
18. Gommermann, N.; Knochel, P. *Org. Synth.* 2007, *84*, 1.
19. Li, C.J.; Yao, X. *Org. Synth.* 2007, *84*, 222.
20. Halliday, G.A.; Young, R.J. Jr.; Grushin, V.V. *Org. Lett.* 2003, *5*, 2003.
21. Muheim, A.; Müller, B.; Münch, T.; Wetli, M. EP885968 (Givaudan-Roure, 1998)
22. Muheim, A.; Lerch, K. *Appl. Microbiol. Biotechnol.* 1999, *51*, 456.
23. Willstätter, R. *Chem. Ber.* 1901, *34*, 129.
24. Willstätter, R. *Chem. Ber.* 1901, *34*, 3163.
25. Willstätter, R. *Chem. Ber.* 1896, *29*, 936.
26. Bonchev D.; Rouvray, R.H. *Chemical Graph Theory: Introduction and Fundamentals*, Taylor & Francis: London, U.K., 2003.
27. Andrasfai, B. *Introductory Graph Theory*, Adam Hilger: Bristol, U.K., 1977.
28. Trinajstic, N. *Chemical Graph Theory*; CRC Press: Boca Raton, FL, 1992.
29. Wilson, R.J.; Watkins, J.J. *Graphs: An Introductory Approach*, John Wiley & Sons, Inc.: New York, 1990.
30. Rouvray, D.H. *Endeavour* 1975, *34*, 28.
31. Rouvray, D.H. *Am. Scientist* 1973, *61*, 729.
32. Rouvray, D.H. *CHEMTECH* 1973, *6*, 379.
33. Rouvray, D.H. *Chem. Br.* 1977, *13*, 52.
34. Balaban, A.T.; Kennedy, J.W.; Quintas, L.V. *J. Chem. Educ.* 1988, *65*, 304.
35. Bertz, S.H. *Discrete Appl. Math.* 1988, *19*, 65.
36. Wender, P.; Miller, B.L. Toward the ideal synthesis: Connectivity analysis and multibond-forming processes, in *Organic Synthesis: Theory and Applications* (T. Hudlicky, ed.), Vol. 2, JAI Press: Greenwich, CT, 1993, pp. 27–65.
37. Cayley, A. *Philos. Mag.* 1854, *7*(4), 40.
38. Cayley, A. *Philos. Mag.* 1857, *13*(4), 172.
39. Cayley, A. *Philos. Mag.* 1859, *18*(4), 374.
40. Cayley, A. *Philos. Mag.* 1860, *20*(4), 337.
41. Crum Brown, A. *Trans. R. Soc. Edinburgh* 1864, *23*, 707.
42. Cayley, A. *Philos. Mag.* 1874, *47*(4), 444.
43. Cayley, A. *Chem. Ber.* 1875, *8*, 1056.
44. Cayley, A. *Rep. Br. Assoc. Adv. Sci.* 1875, *45*, 257.
45. Cayley, A. *Philos. Mag.* 1877, *3*(5), 34.
46. Cayley, A. *Am. J. Math.* 1881, *4*, 266.
47. Henze, H.R.; Blair, C.M. *J. Am. Chem. Soc.* 1931, *53*, 3077.
48. Polya, G. *Acta Math.* 1936, *68*, 145.
49. Lodge, O.J. *Philos. Mag.* 1875, *50*(4), 367.
50. Rouvray, D.H. *J. Chem. Educ.* 1975, *52*, 768.
51. Vogel, A. *Textbook of Practical Organic Chemistry*, Longman: London, U.K., 1948, p. 851.
52. Gange, D.M.; D'Antuono, J. *Curr. Opinion Drug Disc. Dev.* 2003, *6*, 362.
53. Biginelli, P. *Gazz. Chim. Ital.* 1893, *23*, 360.
54. Biginelli, P. *Chem. Ber.* 1891, *24*, 1317.
55. Biginelli, P. *Chem. Ber.* 1891, *24*, 2962.
56. Elders, N.; van der Born, D.; Hendrickx, L.J.D.; Timmer, B.J.J.; Krause, A.; Janssen, E.; de Kanter, F.J.J.; Ruijter, E.; Orru, R.V.A. *Angew. Chem. Int. Ed.* 2009, *48*, 5856.

7 Named Organic Reaction Database

This chapter compiles data on 560 named organic reactions that form the database of reactions that synthetic chemists draw upon to plan synthesis routes to complex targets. Sections 7.1 through 7.10 cover the following classes of named organic reactions: carbon–carbon bond forming, condensations, eliminations, multicomponent, non-carbon–carbon bond forming, oxidations with respect to substrate of interest, rearrangements, reductions with respect to substrate of interest, sequences, and substitutions, respectively. Each set of reactions is arranged alphabetically for easy reference. Each reaction is balanced showing all by-products and is generalized using Markush type structures showing those atoms that are specified and those that are variable, which are designated as R groups. From these balanced chemical equations expressions for the minimum atom economy, AE(min), and maximum E-factor with respect to molecular weight, E(max), are determined using the basic definitions given in Chapter 4. These parameters define the worst and best case scenarios of performance at the kernel level for each reaction. Since it is computationally simpler to determine the E-factor over the AE, the following protocol was used. The minimum size R-group is designated as methyl unless otherwise specified and is assigned a molecular weight of 15. The numerical values of E(max) are first determined using R = 15, and then AE(min) is calculated directly using Equation 7.1.

$$AE_{min} = \frac{1}{1+E_{max}} \tag{7.1}$$

Using the acetoacetic acid ester synthesis as an example reaction for illustration, we have the following balanced chemical equation showing the molecular weights below each structure.

$$E = \frac{x+y+44+17+1}{z+57} = \frac{x+y+62}{z+57} \tag{7.2}$$

The group definitions are as follows: R_1 = y; R = z; and X = x(halide). Since carbon dioxide, R_1OH, and HX are by-products, the expression for E is given by Equation 7.2.

If R_1 = R = methyl and X = Cl, then E(max) is given by Equation 7.3.

$$E_{max} = \frac{35.45+15+62}{15+57} = \frac{112.45}{72} = 1.562 \tag{7.3}$$

AE(min) in turn is given by Equation 7.4.

$$AE_{min} = \frac{1}{1+1.562} = 0.390 \tag{7.4}$$

The next item is to determine an expression for the molecular weight fraction of sacrificial reagents used, which is the molecular weight ratio of the sum of all sacrificial reagents to the sum of all reagents used in the balanced chemical equation. In this example, water is the sacrificial reagent since none of the atoms in the water molecule end up in the target product. This parameter is given by Equation 7.5.

$$f(\text{sac}) = \frac{18}{x+y+z+101+18} = \frac{18}{x+y+z+119} \tag{7.5}$$

Again, we can determine the maximum value of this parameter using the minimum definitions of the variable groups given earlier. Hence, $f(\text{sac})(\text{max})$ is given by Equation 7.6.

$$f(\text{sac})(\text{max}) = \frac{18}{35.45+15+15+101+18} = \frac{18}{83.45} = 0.216 \tag{7.6}$$

The final item is the determination of the hypsicity index (HI) for the reaction. In addition to providing a balanced chemical equation for each reaction, the target bonds made are delineated in the target product structure. The corresponding atoms in the product are assigned letter descriptors and these are traced to their original location in the reactant side of the chemical equation. In the foregoing example, only one target bond is made in the reaction and the corresponding two carbon atoms are labeled a and b as shown. The oxidation numbers are assigned for these atoms in the target product and in the reactants. The hypsicity index is then determined according to Equation 7.7 (see Chapters 4 and 6) where $N = 1$ since one reaction step is involved.

$$\begin{aligned} HI &= -\frac{\sum_i (Ox(\text{product}) - Ox(\text{reactant}))_{\text{atom } i}}{N+1} \\ &= -\frac{1}{2}\sum_i (Ox(\text{product}) - Ox(\text{reactant}))_{\text{atom } i} \\ &= -\frac{1}{2}\Delta \end{aligned} \tag{7.7}$$

In the present example, we have

$$HI = -\frac{1}{2}[(-3-(-2))_a + (-2-(-2))_b] = -\frac{1}{2}[-1] = +\frac{1}{2}$$

which indicates that the reaction is hyperhypsic (HI > 0) meaning that a net reduction has occurred in creating the target bond. Indeed, atom b does not change its oxidation number but atom a undergoes a decrease in its oxidation number ongoing from the alkyl halide reagent to the alkylated product.

7.1 CATEGORIZATION OF NAMED C–C COUPLING/ADDITION/ CYCLIZATION REACTIONS BY MINIMUM ATOM ECONOMY AND MAXIMUM ENVIRONMENTAL IMPACT FACTOR, MAXIMUM MOLECULAR WEIGHT FRACTION OF SACRIFICIAL REAGENTS, HYPSICITY INDEX

(Reactions are listed alphabetically.)

Acetoacetic ester synthesis
Simonsen, J.L.; Storey, R. *J. Chem. Soc.* 1910, *95*, 2106.
Simonsen, J.L.; Storey, R. *Proc. Chem. Soc.* 1910, *25*, 290.

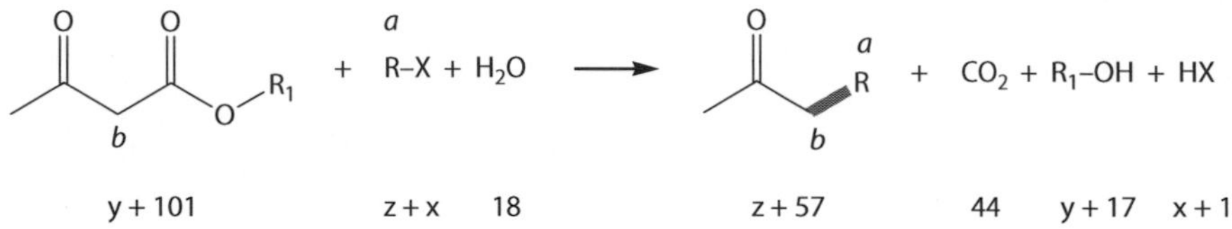

$R_1 = y$; $R = z$; $X = x$(halide)

$$AE = [z+57]/[x+y+z+119]; \quad AE(min) = 58/[x+y+119];$$

$$E = [x+y+62]/[z+57]; \quad E(max) = [x+y+62]/58$$

$$f(sac) = 18/[x+y+z+119]; \quad f(sac)(max) = 18/[x+z+120]$$

X	R_1	E(max)	AE(min)	f(sac)(max)
Cl	Me	1.94	0.34	0.11
Br	Me	2.71	0.27	0.084
I	Me	3.52	0.22	0.069
Cl	Et	2.18	0.31	0.098
Br	Et	2.95	0.25	0.079
I	Et	3.76	0.21	0.065

Set $R = CH_3$

$$\Delta = \Delta(a) + \Delta(b) = (-3 - (-2)) + (-2 - (-2)) = -1; \quad HI = -\Delta/(N+1) = +\frac{1}{2}$$

Alkyne–carbamate–phosphine annulation
Yavari, I.; Bayat, M. *Synth. Commun.* 2002, *32*, 2527.

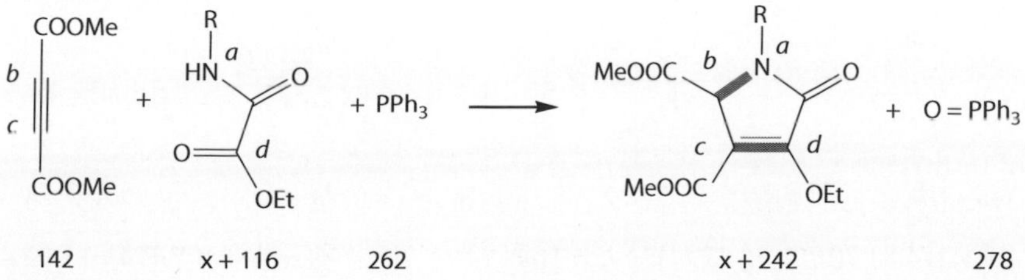

R = x

$$AE = [x + 242]/[x + 520]; \quad \textbf{AE(min)} = \textbf{0.47}; \quad E = 278/[x + 242]; \quad \textbf{E(max)} = \textbf{1.14}$$

$$f(\text{sac}) = 262/[x + 520]; \quad f(\text{sac})(\textbf{max}) = \textbf{0.50}$$

$$\Delta = \Delta(a) + \Delta(b) + \Delta(c) + \Delta(d) = (-3 - (-3)) + (0 - 0) + (0 - 0) + (1 - 3) = -2; \quad \textbf{HI} = -\Delta/(N + 1) = +1$$

Baylis–Hillman reaction
Baylis, A.B.; Hillman, M.E.D., DE 2,155,113 (1972/05/10) Celanese.
Baylis, A.B.; Hillman, M.E.D., US 3,743,669 (1973/07/03) Celanese.

R_1 = EWG (COOR, COR, CN, SO_2R, $CONR_2$)

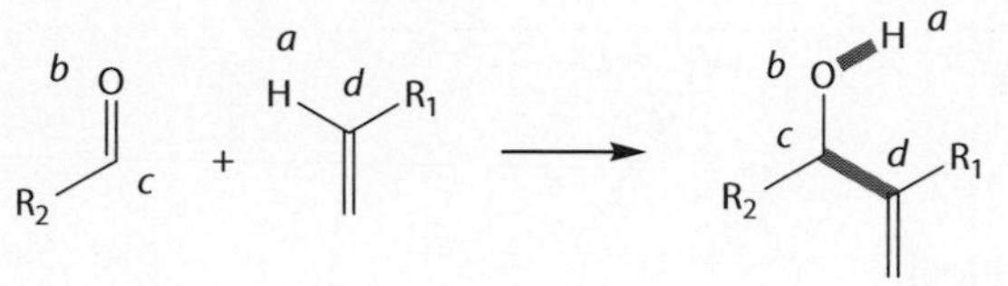

$$AE = 1; E = 0; f(\text{sac}) = 0$$

Set $R_1 = R_2 = CH_3$

$$\Delta = \Delta(a) + \Delta(b) + \Delta(c) + \Delta(d) = (1 - 1) + (-2 - (-2)) + (0 - 1) + (0 - (-1)) = 0; \quad \textbf{HI} = \textbf{0}$$

Bergmann cyclization
Bergman, R.G.; Jones, R.R. *J. Am. Chem. Soc.* 1972, *94*, 660.

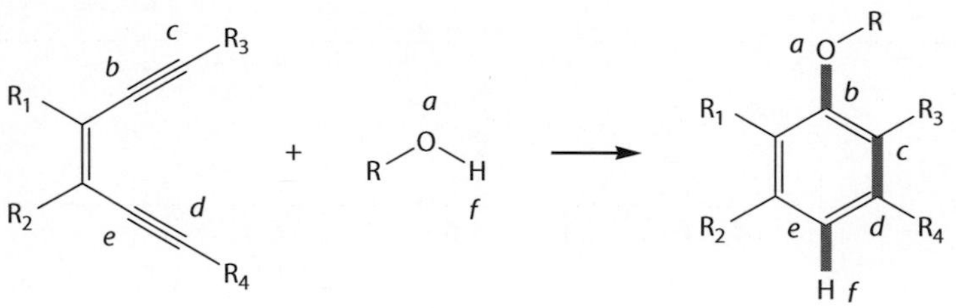

$$AE = 1; E = 0; f(\text{sac}) = 0$$

Set $R_3 = R_4 = CH_3$

$$\Delta = \Delta(a) + \Delta(b) + \Delta(c) + \Delta(d) + \Delta(e) + \Delta(f)$$

$$= (-2 - (-2)) + (1 - 0) + (0 - 0) + (0 - 0) + (-1 - 0) + (1 - 1) = 0; \quad \textbf{HI} = \textbf{0}$$

Blanc reaction
Blanc, H.G. *Compt. Rend.* 1907, *144*, 1356.

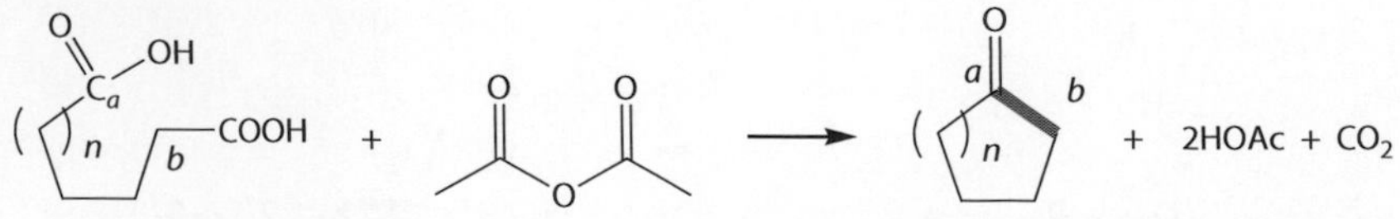

14n + 132   102   14n + 70   120   44

$$AE = [14n + 70]/[14n + 234]; \quad E = 164/[14n + 70]; \quad f(\text{sac}) = 102/[14n + 234]$$

	Ring Size ($n + 1$)		
n	E(max)	AE(min)	f(sac)(max)
1	5	1.95	0.34
2	6	1.67	0.37
3	7	1.46	0.41

$$\Delta = \Delta(a) + \Delta(b) = (2-3) + (-2-(-2)) = -1; \quad \mathbf{HI} = -\Delta/(N+1) = +\frac{1}{2}$$

Cadiot–Chodkiewicz reaction

Chodkiewicz, W. *Ann. Chim. (Paris)* 1957, 2, 819.
Sevin, A.; Chodkiewicz, W.; Cadiot, P. *Bull. Soc. Chim. Fr.* 1974, 913.

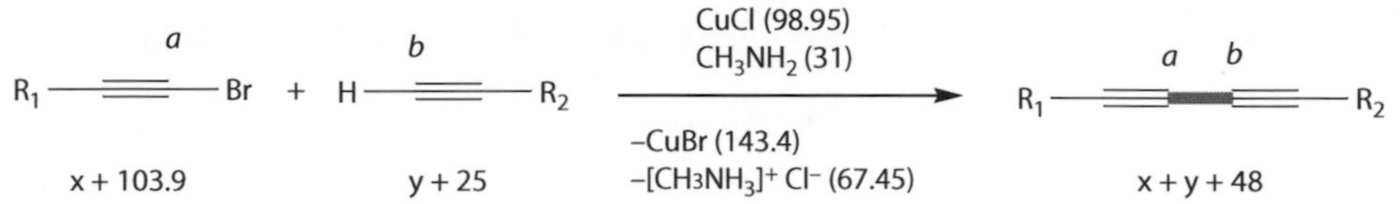

$$R_1 = x; R_2 = y$$

$$AE = [x + y + 48]/[x + y + 258.85]; \quad \mathbf{AE(min) = 0.19}; \quad E = 210.85/[x + y + 48]; \quad \mathbf{E(max) = 4.22}$$

$$f(sac) = 129.95/[x + y + 258.85]; \quad \mathbf{f(sac)(max) = 0.50}$$

$$\Delta = \Delta(a) + \Delta(b) = (0-1) + (0-(-1)) = 0; \quad \mathbf{HI = 0}$$

Ciamician synthesis of pyridines from pyrroles

Ciamician, G.; Dennestedt, M. *Chem. Ber.* 1881, *14*, 1153.

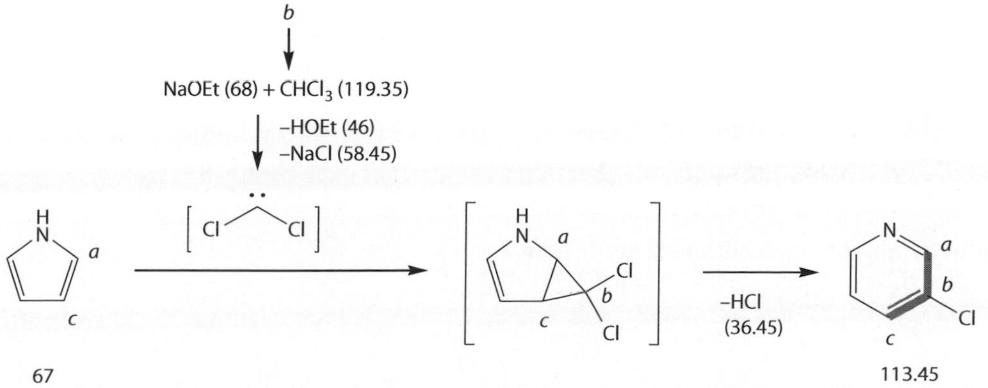

$$\mathbf{AE = 0.45}; \quad \mathbf{E = 140.9/113.45 = 1.24}; \quad \mathbf{f(sac) = 68/254.35 = 0.27}$$

$$\Delta = \Delta(a) + \Delta(b) + \Delta(c) = (1-0) + (1-2) + (-1-(-1)) = 0; \quad \mathbf{HI = 0}$$

Cyclopropanation of olefins with diazomethane

Fischer, H.; Staff, C.E. *Z. Physiol. Chem.* 1935, *234*, 97.
Fischer, H.; Hofmann, H.J. *Z. Physiol. Chem.* 1937, *245*, 139.
Backer, H.J.; Dost, N.; Knotnerus, J. *Rec. Trav. Chim.* 1949, *68*, 237.
De Boer, T.J.; Backer, H.J. *Org. Synth. Coll. Vol. IV* 1963, 250.

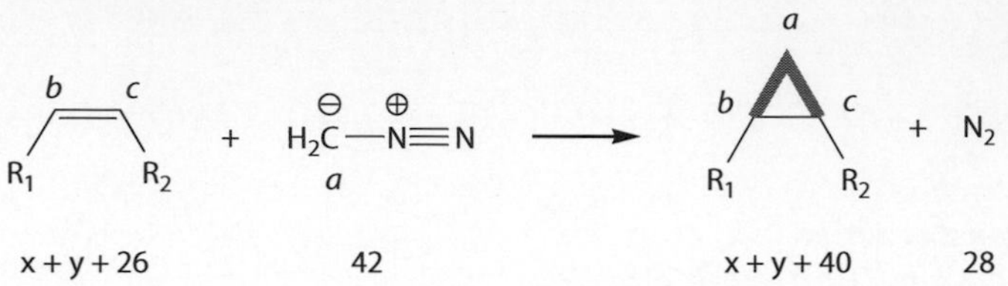

$R_1 = x; R_2 = y$

$$AE = [x+y+40]/[x+y+68]; \quad \textbf{AE(min)} = \textbf{0.60}; \quad E = 28/[x+y+40];$$

$$\textbf{\textit{E}(max)} = \textbf{0.67}; \quad f(\textbf{sac}) = \textbf{0}$$

Set $R_1 = R_2 = CH_3$

$$\Delta = \Delta(a) + \Delta(b) + \Delta(c) = (-2-(-2)) + (-1-(-1)) + (-1-(-1)) = 0; \quad \textbf{HI} = \textbf{0}$$

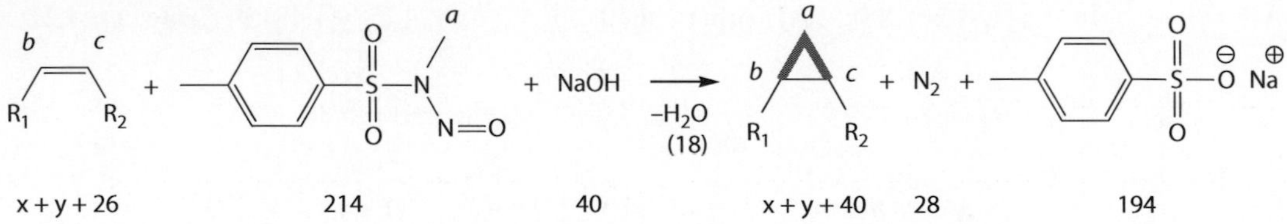

$R = x$

$$AE = [x+y+40]/[x+y+280]; \quad \textbf{AE(min)} = \textbf{0.15}; \quad E = 240/[x+y+40]; \quad \textbf{\textit{E}(max)} = \textbf{5.71}$$

$$f(\textbf{sac}) = 40/[x+y+280]; \quad f(\textbf{sac})(\textbf{max}) = \textbf{0.14}$$

Set $R_1 = R_2 = CH_3$

$$\Delta = \Delta(a) + \Delta(b) + \Delta(c) = (-2-(-2)) + (-1-(-1)) + (-1-(-1)) = 0; \quad \textbf{HI} = \textbf{0}$$

Danheiser alkyne–cyclobutanone cyclization

Danheiser, R.L.; Gee, S.K. *J. Org. Chem.* 1984, *49*, 1672.

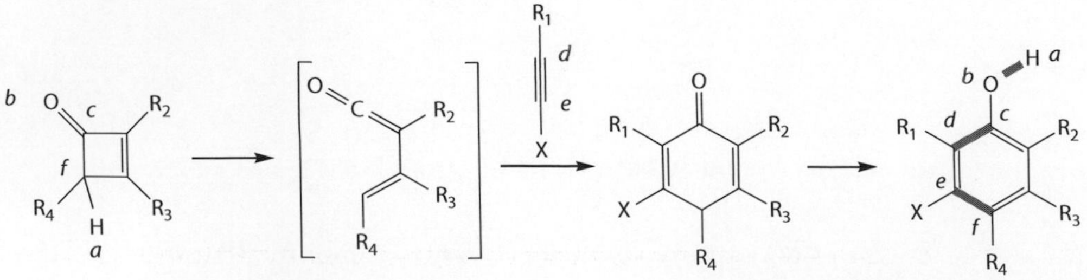

AE = 1; $E = 0$; $f(\text{sac}) = 0$

Set $R_1 = R_4 = X = CH_3$

$$\Delta = \Delta(a) + \Delta(b) + \Delta(c) + \Delta(d) + \Delta(e) + \Delta(f)$$

$$= (1-1) + (-2-(-2)) + (1-2) + (0-0) + (0-0) + (0-(-1)) = 0; \quad \textbf{HI} = \textbf{0}$$

Danheiser [4 + 4] annulation
Danheiser, R.L.; Gee, S.K.; Sard, H. *J. Am. Chem. Soc.* 1982, *104*, 7670.

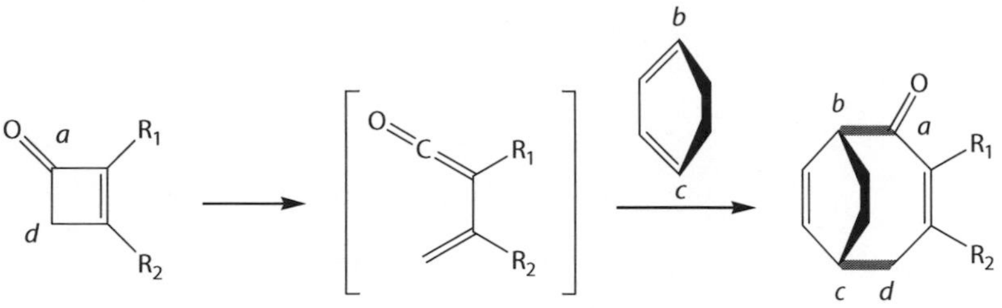

AE = 1; $E = 0$; $f(\text{sac}) = 0$

$$\Delta = \Delta(a) + \Delta(b) + \Delta(c) + \Delta(d) = (2-2) + (-1-(-1)) + (-1-(-1)) + (-2-(-2)) = 0; \quad \textbf{HI} = \textbf{0}$$

Danishefsky reaction
Danishefsky, S.; Kitahura, T. *J. Am. Chem. Soc.* 1974, *96*, 7807.

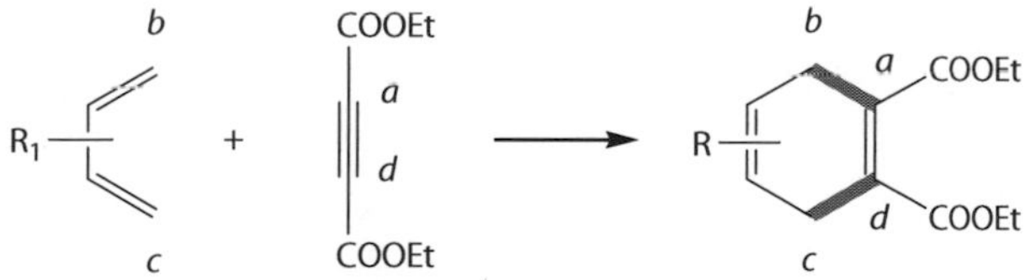

AE = 1; $E = 0$; $f(\text{sac}) = 0$

$$\Delta = \Delta(a) + \Delta(b) + \Delta(c) + \Delta(d) = (0-0) + (-2-(-2)) + (-2-(-2)) + (0-0) = 0; \quad \textbf{HI} = \textbf{0}$$

Diels–Alder reaction
Diels, O.; Alder, K. *Ann. Chem.* 1928, *460*, 98.

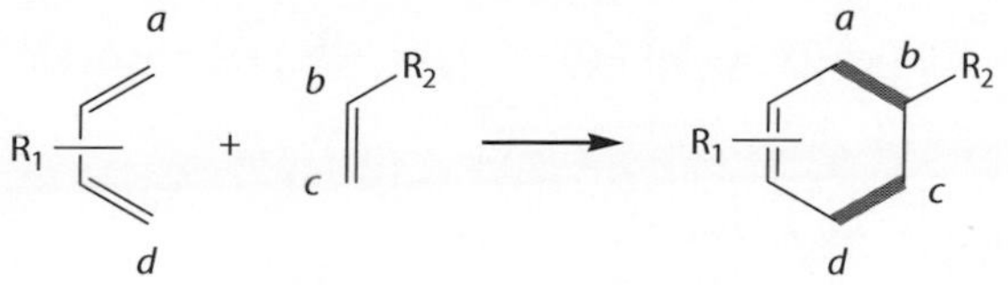

AE = 1; $E = 0$; $f(\text{sac}) = 0$

Set $R_2 = CH_3$

$$\Delta = \Delta(a) + \Delta(b) + \Delta(c) + \Delta(d) = (-2-(-2)) + (-1-(-1)) + (-2-(-2)) + (-2-(-2)) = 0; \quad \textbf{HI} = \textbf{0}$$

Dötz reaction

Dötz, K.H. *Angew. Chem. Int. Engl. Ed.* 1975, *14*, 644.

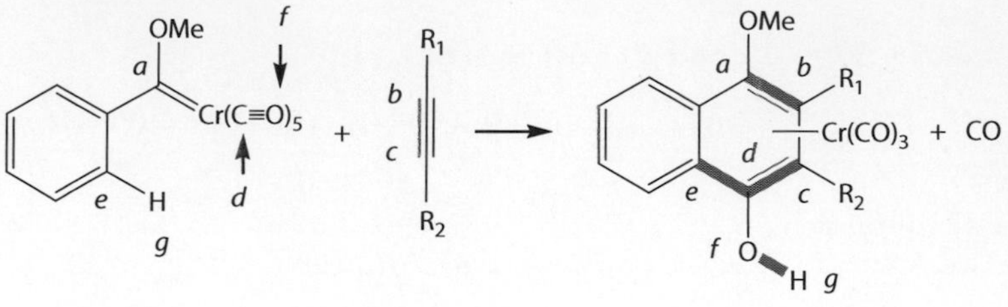

312	x + y + 24	x + y + 308	28

$R_1 = x$; $R_2 = y$

$$AE = [x + y + 308]/[x + y + 336]; \quad \textbf{AE(min)} = \textbf{0.92};$$

$$E = 28/[x + y + 308]; \quad \textbf{E(max)} = \textbf{0.090}; \quad f(\textbf{sac}) = \textbf{0}$$

Set $R_1 = R_2 = CH_3$

$$\Delta = \Delta(a) + \Delta(b) + \Delta(c) + \Delta(d) + \Delta(e) + \Delta(f) + \Delta(g)$$

$$= (1 - (-1)) + (0 - 0) + (0 - 0) + (1 - 2) + (0 - (-1)) + (-2 - (-2)) + (1 - 1) = +2$$

$$\textbf{HI} = -\boldsymbol{\Delta}/(\textbf{N} + \textbf{1}) = -\textbf{1}$$

Eglinton reaction

Eglinton, G.; Jones, E.R.H.; Shaw, B.L.; Whiting, M.C. *J. Chem. Soc.* 1954, 1860.

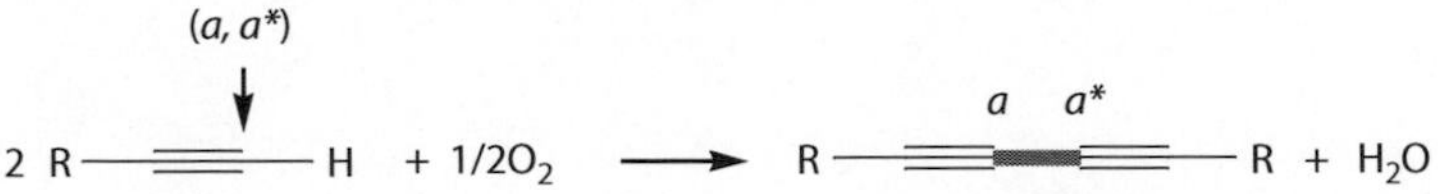

2(x + 25)	16	2x + 48	18

$R = x$

$$AE = [2x + 48]/[2x + 66]; \quad \textbf{AE(min)} = \textbf{0.74}; \quad E = 18/[2x + 48];$$

$$\textbf{E(max)} = \textbf{0.36}; \quad f(\textbf{sac}) = 16/[2x + 66]; \quad f(\textbf{sac})(\textbf{max}) = \textbf{0.24}$$

$$\Delta = \Delta(a) + \Delta(a^*) = (0 - (-1)) + (0 - (-1)) = +2; \quad \textbf{HI} = -\boldsymbol{\Delta}/(\textbf{N}+\textbf{1}) = -\textbf{1}$$

[1,5]-Ene reaction

Alder, K. Pascher, F.; Schmitz, A. *Chem. Ber.* 1943, *76*, 27.

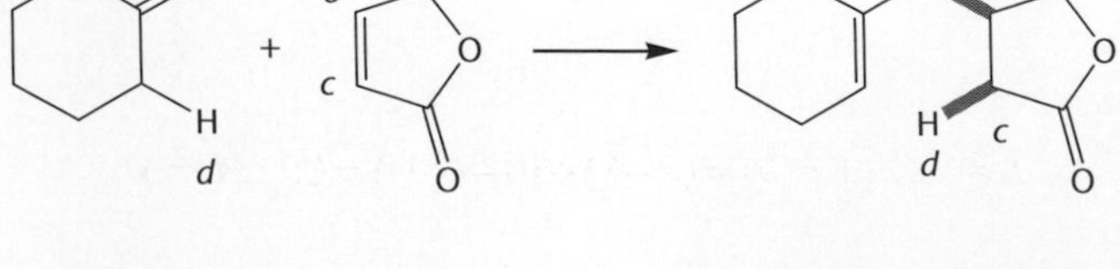

$AE = 1; E = 0; f(sac) = 0$

$$\Delta = \Delta(a) + \Delta(b) + \Delta(c) + \Delta(d) = (-2 - (-2)) + (-1 - (-1)) + (-2 - (-1)) + (1 - 1) = -1;$$

$$HI = -\Delta/(N + 1) = +\tfrac{1}{2}$$

Fischer indole synthesis
Fischer, E.; Jourdan, F. *Chem. Ber.* 1883, *16*, 2241.

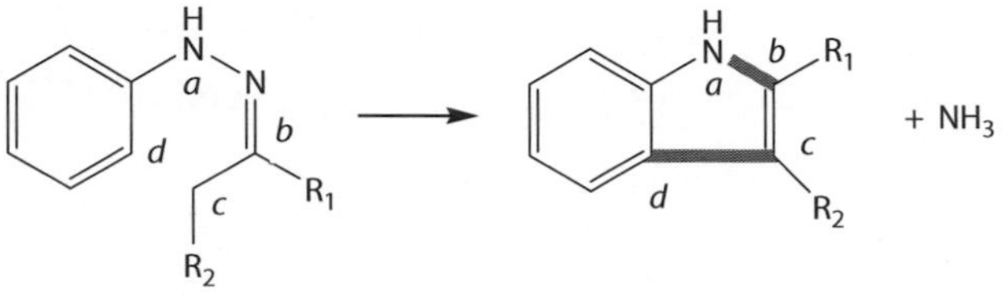

$$x + y + 132 \qquad\qquad x + y + 115 \qquad\qquad 17$$

$R_1 = x; R_2 = y$

$$AE = [x + y + 115]/[x + y + 132]; \quad AE(min) = 0.87;$$

$$E = 17/[x + y + 115]; \quad E(max) = 0.15; \quad f(sac) = 0$$

Set $R_1 = R_2 = CH_3$

$$\Delta = \Delta(a) + \Delta(b) + \Delta(c) + \Delta(d) = (-3 - (-3)) + (1 - 2) + (0 - (-2)) + (0 - (-1)) = +2;$$

$$HI = -\Delta/(N + 1) = -1$$

Friedel–Crafts acylation
Friedel, C.; Crafts, J.M. *Compt. Rend.* 1877, *84*, 1392.
Friedel, C.; Crafts, J.M. *Compt. Rend.* 1877, *84*, 1450.

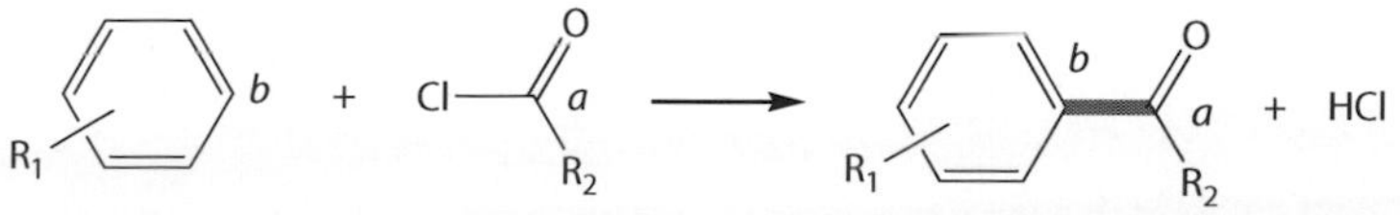

$$x + 73 \qquad\qquad y + 63.45 \qquad\qquad x + y + 100 \qquad\qquad 36.45$$

$R_1 = x$(sum of substituents); $R_2 = y$

$$AE = [x + y + 100]/[x + y + 136.45]; \quad AE(min) = 0.74;$$

$$E = 36.45/[x + y + 100]; \quad E(max) = 0.34; \quad f(sac) = 0$$

Set $R_2 = CH_3$

$$\Delta = \Delta(a) + \Delta(b) = (2 - 3) + (0 - (-1)) = 0; \quad HI = 0$$

Including AlCl₃ catalyst in stoichiometric amounts:

$$x+73 \qquad y+63.45 \quad 133.35 \quad 54 \qquad\qquad x+y+100 \qquad 145.8 \quad 78$$

$$AE = [x+y+100]/[x+y+323.8]; \quad \mathbf{AE(min) = 0.32}; \quad E = 223.8/[x+y+100]; \quad \mathbf{E(max) = 2.11}$$

$$f(\text{sac}) = 187.35/[x+y+323.8]; \quad \mathbf{f(sac)(max) = 0.57}$$

Set $R_2 = CH_3$

$$\Delta = \Delta(a)+\Delta(b) = (2-3)+(0-(-1)) = 0; \quad \mathbf{HI = 0}$$

Friedel–Crafts alkylation
Friedel, C.; Crafts, J.M. *Compt. Rend.* 1877, *84*, 1392.
Friedel, C.; Crafts, J.M. *Compt. Rend.* 1877, *84*, 1450.

$$x+73 \qquad\qquad y+35.45 \qquad\qquad x+y+72 \qquad\qquad 36.45$$

$R_1 = x$(sum of substituents); $R_2 = y$

$$AE = [x+y+72]/[x+y+108.45]; \quad \mathbf{AE(min) = 0.68};$$

$$E = 36.45/[x+y+72]; \quad \mathbf{E(max) = 0.47}; \quad f(\text{sac}) = 0$$

Set $R_2 = CH_3$

$$\Delta = \Delta(a)+\Delta(b) = (-3-(-2))+(0-(-1)) = 0; \quad \mathbf{HI = 0}$$

Including AlCl₃ catalyst in stoichiometric amounts:

$$x+73 \qquad\qquad y+35.45 \quad 133.35 \quad 54 \qquad\qquad x+y+72 \qquad 145.8 \quad 78$$

$$AE = [x+y+72]/[x+y+295.8]; \quad \mathbf{AE(min) = 0.26};$$

$$E = 223.8/[x+y+72]; \quad \mathbf{E(max) = 2.87}$$

$$f(\text{sac}) = 187.35/[x+y+295.8]; \quad \mathbf{f(sac)(max) = 0.62}$$

Set $R_2 = CH_3$

$$\Delta = \Delta(a) + \Delta(b) = (-3 - (-2)) + (0 - (-1)) = 0; \quad \textbf{HI} = \textbf{0}$$

Friedländer synthesis of quinolines
Friedländer, P. *Chem. Ber.* 1882, *15*, 2572.

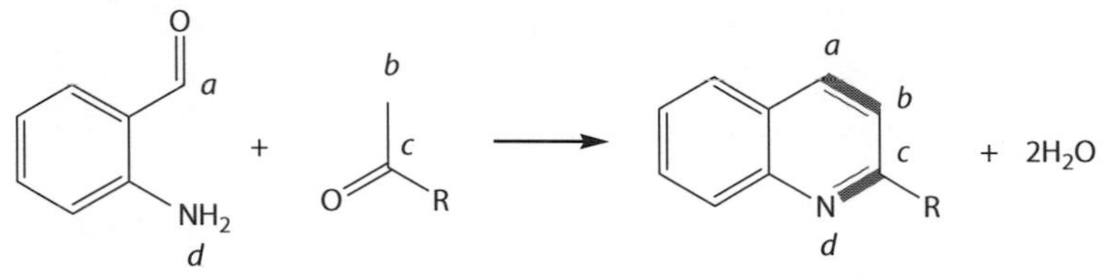

| 121 | x + 43 | x + 128 | 36 |

R = x

$$AE = [x + 128]/[x + 164]; \quad \textbf{AE(min)} = \textbf{0.78}; \quad E = 36/[x + 128]; \quad \textbf{E(max)} = \textbf{0.28}; \quad f(\textbf{sac}) = \textbf{0}$$

Set R = CH_3

$$\Delta = \Delta(a) + \Delta(b) + \Delta(c) + \Delta(d) = (-1 - 1) + (-1 - (-3)) + (2 - 2) + (-3 - (-3)) = 0; \quad \textbf{HI} = \textbf{0}$$

Gattermann–Koch reaction
Gattermann, L.; Koch, J.A. *Chem. Ber.* 1897, *30*, 1622.

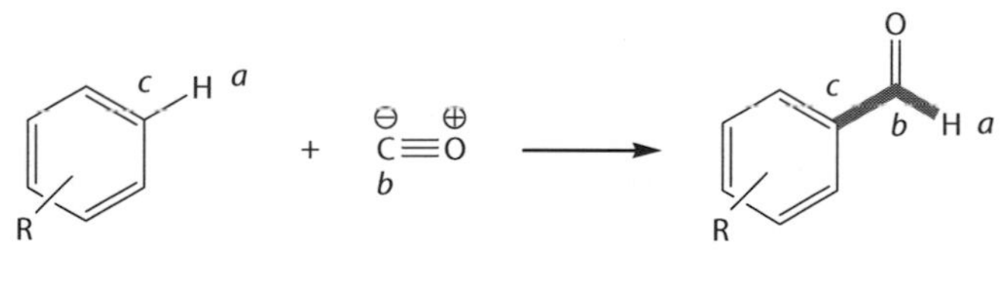

$$\textbf{AE} = \textbf{1}; E = \textbf{0}; f(\textbf{sac}) = \textbf{0}$$

$$\Delta = \Delta(a) + \Delta(b) + \Delta(c) = (1 - 1) + (1 - 2) + (0 - (-1)) = 0; \quad \textbf{HI} = \textbf{0}$$

Glaser coupling
Glaser, C. *Chem. Ber.* 1869, *2*, 422.

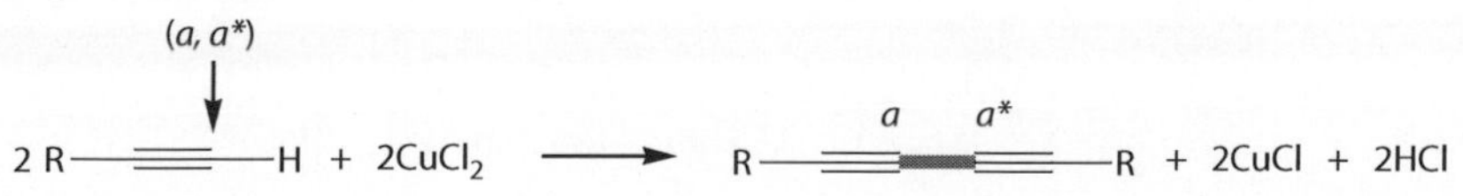

| 2(x + 25) | 268.8 | 2x + 48 | 197.9 | 72.9 |

R = x

$$AE = [2x + 48]/[2x + 318.8]; \quad \textbf{AE(min)} = \textbf{0.16}; \quad E = 270.8/[2x + 48]; \quad \textbf{E(max)} \approx \textbf{5.42}$$

$$f(\text{sac}) = 268.8/[2x + 318.8]; \quad f(\textbf{sac})(\textbf{max}) = \textbf{0.84}$$

$$\Delta = \Delta(a) + \Delta(a^*) = (0 - (-1)) + (0 - (-1)) = +2; \quad \textbf{HI} = -\Delta/(N + 1) = -\textbf{1}$$

Gomberg–Bachmann reaction
Gomberg, M.; Bachmann, W.E. *J. Am. Chem. Soc.* 1924, *46*, 2339.

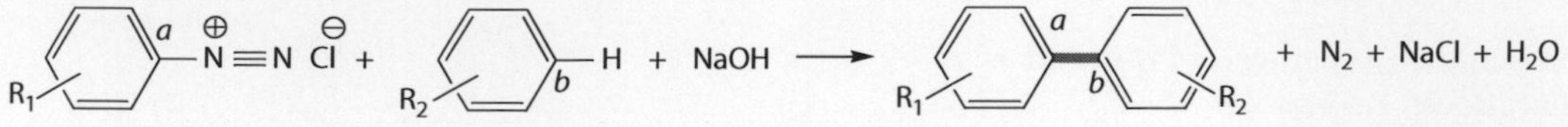

$$x + 135.45 \qquad\qquad y + 73 \qquad\qquad 40 \qquad\qquad x + y + 144 \qquad\qquad 28 \quad 58.45 \quad 18$$

R_1 = x(sum of substituents); R_2 = y(sum of substituents)

$$AE = [x + y + 144]/[x + y + 248.45]; \quad \textbf{AE(min)} = \textbf{0.60};$$

$$E = 104.45/[x + y + 144]; \quad \textbf{E(max)} = \textbf{0.68}$$

$$f(\text{sac}) = 40/[x + y + 248.45]; \quad \textbf{\textit{f}(sac)(max)} = \textbf{0.16}$$

$$\Delta = \Delta(a) + \Delta(b) = (0 - 1) + (0 - (-1)) = 0; \quad \textbf{HI} = \textbf{0}$$

Grignard reaction
Grignard, V. *Compt. Rend.* 1900, *130*, 1322.

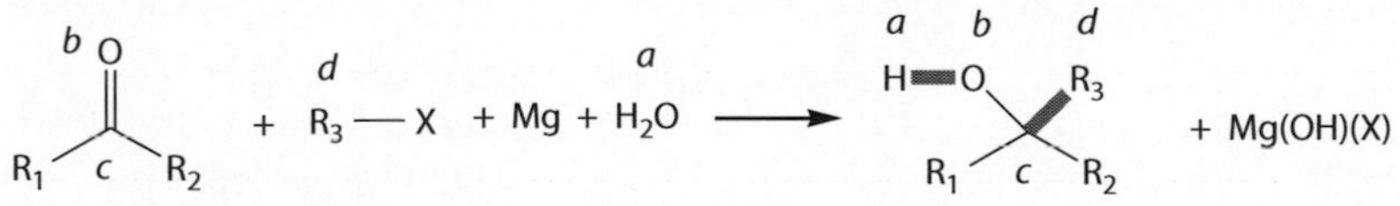

$$y + z + 28 \qquad\qquad w + x \qquad 24.3 \quad 18 \qquad\qquad y + z + w + 29 \qquad x + 41.3$$

R_1 = y; R_2 = z; R_3 = w; X = x(halide)

$$AE = [y + z + w + 29]/[x + y + z + w + 70.3]; \quad E = [x + 41.3]/[y + z + w + 29];$$

$$f(\text{sac}) = 24.3/[x + y + z + w + 70.3]$$

R_3 = **Me, AE(min) = 46/[x + 87.3]; *E*(max) = [x + 41.3]/46; *f*(sac)(max) = 24.3/[x + 87.3]**

X	E(max)	AE(min)	f(sac)(max)
Cl	1.67	0.38	0.20
Br	2.63	0.28	0.15
I	3.66	0.22	0.11

Set $R_1 = R_2 = R_3 = CH_3$

$$\Delta = \Delta(a) + \Delta(b) + \Delta(c) + \Delta(d) = (1 - 1) + (-2 - (-2)) + (1 - 2) + (-3 - (-2)) = -2;$$

$$\textbf{HI} = -\Delta/(N + 1) = +1$$

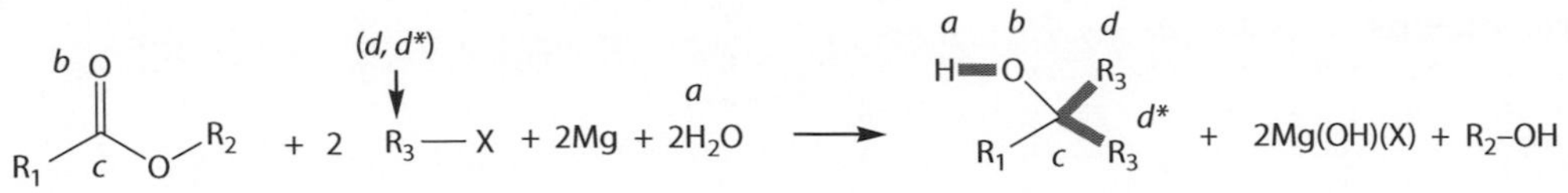

$$y + z + 44 \qquad 2(w + x) \qquad 48.6 \qquad 36 \qquad y + 2w + 29 \qquad 2(x + 41.3) \qquad z + 17$$

$R_1 = y; R_2 = z; R_3 = w; X = x(\text{halide})$

$$AE = [y + 2w + 29]/[2x + y + z + 2w + 128.6];$$

$$E = [2x + z + 99.6]/[y + 2w + 29];$$

$$f(\text{sac}) = 48.6/[2x + y + z + 2w + 128.6]$$

R_2 = Me, R_3 = Me, AE(min) = 60/[2x + 174.6]; E(max) = [2x + 114.6]/60; f(sac)(max) = 48.6/[2x + 174.6]

X	E(max)	AE(min)	f(sac)(max)
Cl	3.09	0.24	0.20
Br	4.57	0.18	0.15
I	6.14	0.14	0.11

Set $R_1 = R_3 = CH_3$

$$\Delta = \Delta(a) + \Delta(b) + \Delta(c) + \Delta(d) + \Delta(d^*) = (1-1) + (-2-(-2)) + (1-3) + 2(-3-(-2)) = -4;$$

$$\mathbf{HI = -\Delta/(N+1) = +2}$$

Hammick reaction
Dyson, P.; Hammick, D.L. *J. Chem. Soc.* 1937, 1724.

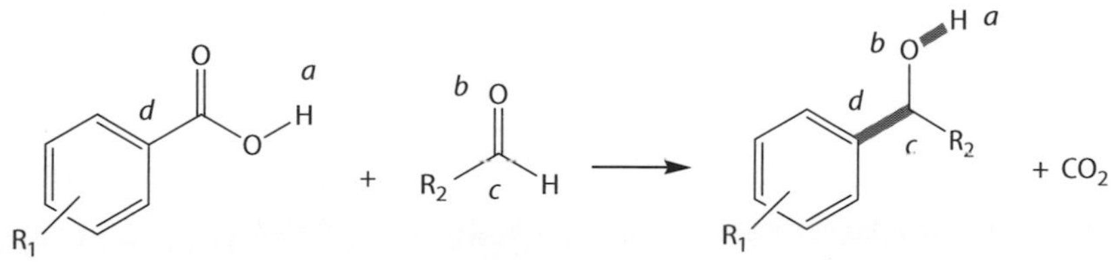

$$x + 117 \qquad\qquad y + 29 \qquad\qquad x + y + 102 \qquad\qquad 44$$

$R_1 = x(\text{sum of substituents}); R_2 = y$

$$AE = [x + y + 102]/[x + y + 146]; \quad \mathbf{AE(min) = 0.71};$$

$$E = 44/[x + y + 102]; \quad \mathbf{E(max) = 0.41; \quad f(sac) = 0}$$

Set $R_2 = CH_3$

$$\Delta = \Delta(a) + \Delta(b) + \Delta(c) + \Delta(d) = (1-1) + (-2-(-2)) + (0-1) + (0-0) = -1;$$

$$\mathbf{HI = -\Delta/(N+1) = +\tfrac{1}{2}}$$

Heck coupling

Heck, R.F.; Nolley, J.P., Jr. *J. Org. Chem.* 1972, *37*, 2320.
Heck, R.F.; Nolley, J.P., Jr. *J. Am. Chem. Soc.* 1968, *90*, 5518.

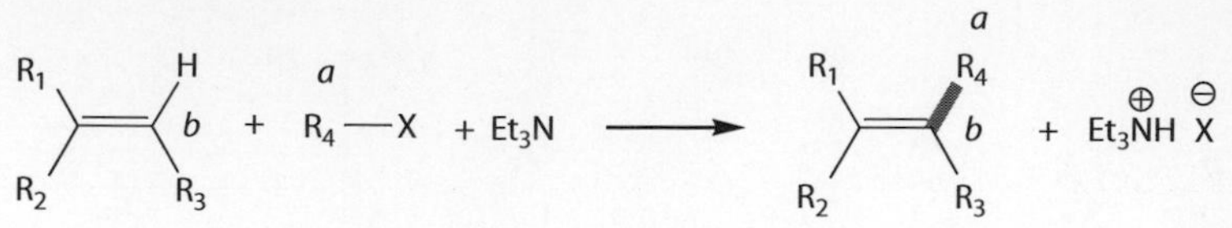

$$R_1 + R_2 + R_3 = y; \quad R_4 = r; \quad X = x(\text{halide})$$

$$AE = [y+r+24]/[y+r+x+126]; \quad \mathbf{AE(min) = 28/[x+130]};$$

$$E = [x+102]/[y+r+24]; \quad \mathbf{E(max) = [x+102]/28}$$

$$\mathbf{\mathit{f}(sac) = 101/[y+r+x+126]; \quad \mathit{f}(sac)(max) = 101/[x+130]}$$

X	E(max)	AE(min)	f(sac)(max)
Cl	4.94	0.17	0.61
Br	6.50	0.13	0.48
I	8.18	0.11	0.39

Set $R_3 = R_4 = CH_3$

$$\Delta = \Delta(a) + \Delta(b) = (-3-(-2)) + (0-(-1)) = 0; \quad \mathbf{HI = 0}$$

Henry reaction (weak acid conditions)

Henry, L. *Compt. Rend.* 1895, *120*, 1265.

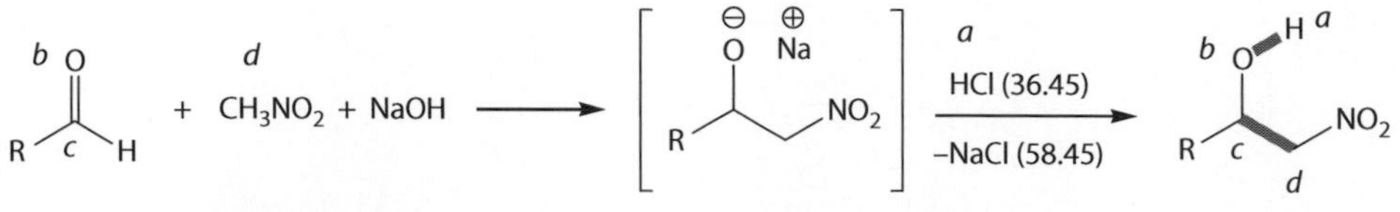

$$AE = [x+90]/[x+166.45]; \quad \mathbf{AE(min) = 0.61};$$

$$E = 58.45/[x+90]; \quad \mathbf{E(max) = 0.64};$$

$$\mathit{f}(sac) = 40/[x+166.45]; \quad \mathbf{\mathit{f}(sac)(max) = 0.24}$$

Set $R = CH_3$

$$\Delta = \Delta(a) + \Delta(b) + \Delta(c) + \Delta(d) = (1-1) + (-2-(-2)) + (0-1) + (-1-(-2)) = 0; \quad \mathbf{HI = 0}$$

Henry reaction (strong acid conditions)

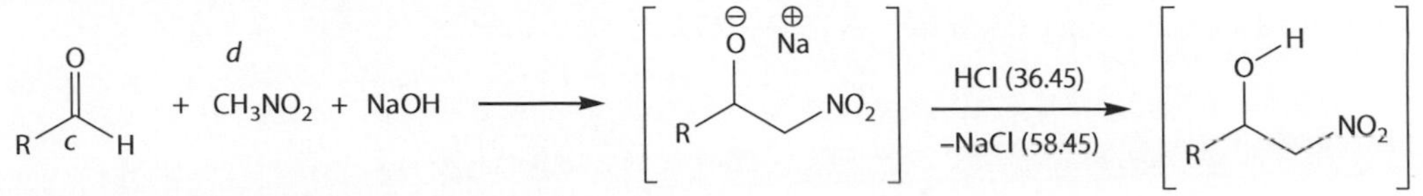

x + 29 61 40

x + 72

R = x

$$AE = [x + 72]/[x + 166.45]; \quad \mathbf{AE(min) = 0.49};$$

$$E = 76.45/[x + 72]; \quad \boldsymbol{E}\mathbf{(max) = 1.05}$$

$$f(sac) = 40/[x + 166.45]; \quad \boldsymbol{f}\mathbf{(sac)(max) = 0.24}$$

Set R = CH_3

$$\Delta = \Delta(c) + \Delta(d) = (-1 - 1) + (0 - (-2)) = 0; \quad \mathbf{HI = 0}$$

Hiyama cross-coupling reaction

Hatanaka, Y.; Fukushima, S.; Hiyama, T. *Heterocycles* 1990, *30*, 303.
Hiyama, T.; Hatanaka, Y. *Pure Appl. Chem.* 1994, *66*, 1471.

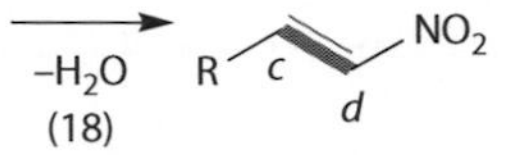

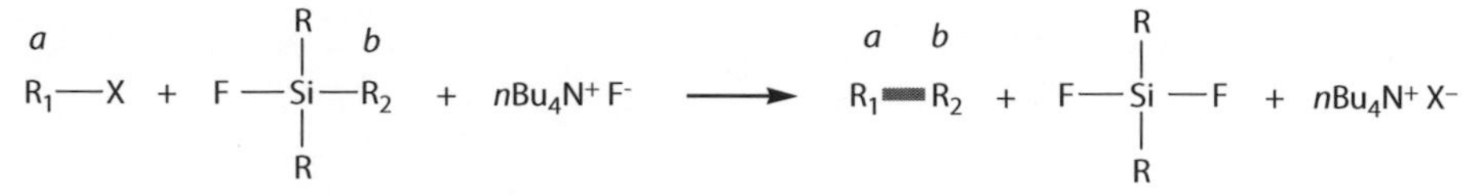

r_1 + x 2r + r_2 + 47 261 r_1 + r_2 2r + 75 x + 242

$R_1 = r_1$, $R_2 = r_2$, R = r, X = x

$$AE = [r_1 + r_2]/[r_1 + r_2 + 2r + x + 308]; \quad E = [2r + x + 317]/[r_1 + r_2];$$

$$f(sac) = 261/[r_1 + r_2 + 2r + x + 308]$$

X	R	*E*(max)	AE(min)
Cl	CH_3	$382.45/[r_1 + r_2]$	$[r_1 + r_2]/[r_1 + r_2 + 373.45]$
Br	CH_3	$426.9/[r_1 + r_2]$	$[r_1 + r_2]/[r_1 + r_2 + 417.9]$
I	CH_3	$474/[r_1 + r_2]$	$[r_1 + r_2]/[r_1 + r_2 + 465]$

Set $R_1 = CH_3$, $R_2 = CH_3$

$$\Delta = \Delta(a) + \Delta(b) = ((-3) - (-2)) + ((-3) - (-4)) = 0; \quad \mathbf{HI = 0}$$

Hosomi–Sakurai reaction

Hosomi, A.; Sakurai, H. *Tetrahedron Lett.* 1976, 1295.

Hosomi, A.; Sakurai. H. *J. Am. Chem. Soc.* 1977, *99*, 1673.

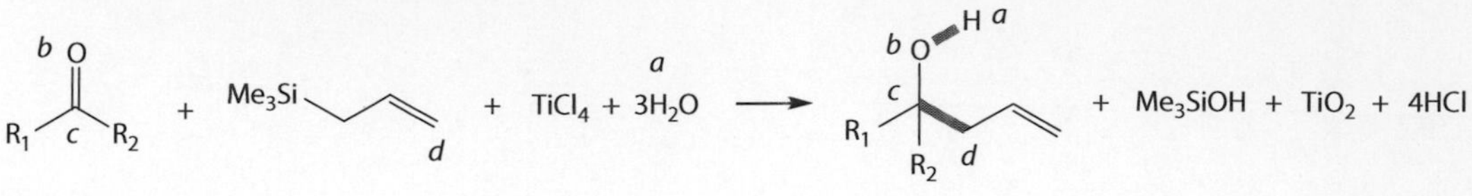

x + y + 28		114		189.7	54	x + y + 70	90	79.9	145.8

$R_1 = x; R_2 = y$

$$AE = [x + y + 70]/[x + y + 385.7]; \quad \textbf{AE(min)} = \textbf{0.19};$$

$$E = 315.7/[x + y + 70]; \quad \textbf{E(max)} = \textbf{4.38}$$

$$f(\text{sac}) = 189.7/[x + y + 385.7]; \quad \textbf{\textit{f}(sac)(max)} = \textbf{0.49}$$

Set $R_1 = R_2 = CH_3$

$$\Delta = \Delta(a) + \Delta(b) + \Delta(c) + \Delta(d) = (1-1) + (-2-(-2)) + (1-2) + (-2-(-2)) = -1;$$

$$\textbf{HI} = -\Delta/(N+1) = +\tfrac{1}{2}$$

Houben–Hoesch reaction

Hoesch, K. *Chem. Ber.* 1915, *48*, 1122.

Houben, J. *Chem. Ber.* 1926, *59*, 2878.

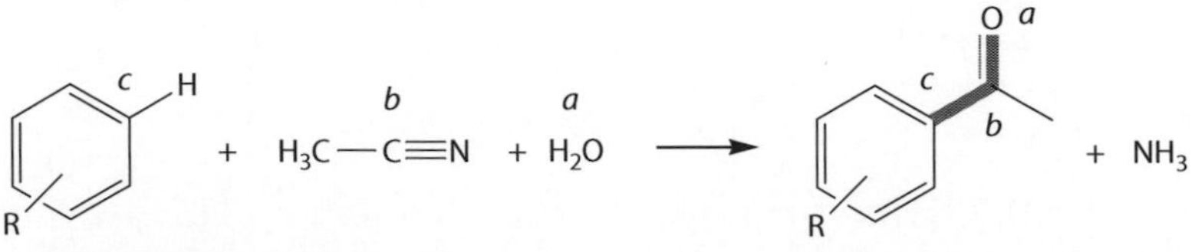

x + 73		41	18	x + 115	17

R = x(sum of 5 substituents)

$$AE = [x + 115]/[x + 132]; \quad \textbf{AE(min)} = \textbf{0.88};$$

$$E = 17/[x + 115]; \quad \textbf{E(max)} = \textbf{0.14}; \quad \textbf{\textit{f}(sac)} = \textbf{0}$$

$$\Delta = \Delta(a) + \Delta(b) + \Delta(c) = (-2-(-2)) + (2-3) + (0-(-1)) = 0; \quad \textbf{HI} = \textbf{0}$$

Jacobs oxidative coupling

Dams, M.; De Vos, D.E.; Celen, S.; Jacobs, P.A. *Angew. Chem. Int. Ed.* 2003, *42*, 3512.

x + 73 | y + z + 54 | 16 | x + y + z + 125 | 18

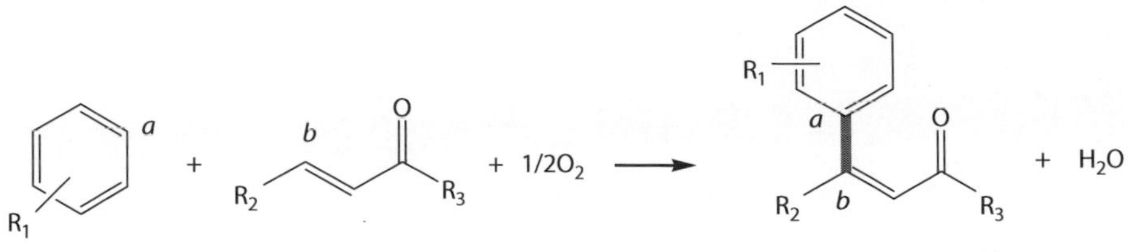

R_1 = x(sum of 5 substituents); R_2 = y; R_3 = z

$$AE = [x + y + z + 125]/[x + y + z + 143];$$

$$E = 18/[x + y + z + 125]; \quad \mathbf{E(max) = 0.14}; \quad \mathbf{AE(min) = 0.88}$$

$$f(sac) = 16/[x + y + z + 143]; \quad \mathbf{f(sac)(max) = 0.11}$$

Set $R_2 = CH_3$

$$\Delta = \Delta(a) + \Delta(b) = (0 - (-1)) + (0 - (-1)) = +2; \quad \mathbf{HI = -\Delta/(N + 1) = -1}$$

Kiliani–Fischer synthesis

Kiliani, H. *Chem. Ber.* 1885, *18*, 3066.
Fischer, E. *Chem. Ber.* 1889, *22*, 2204.

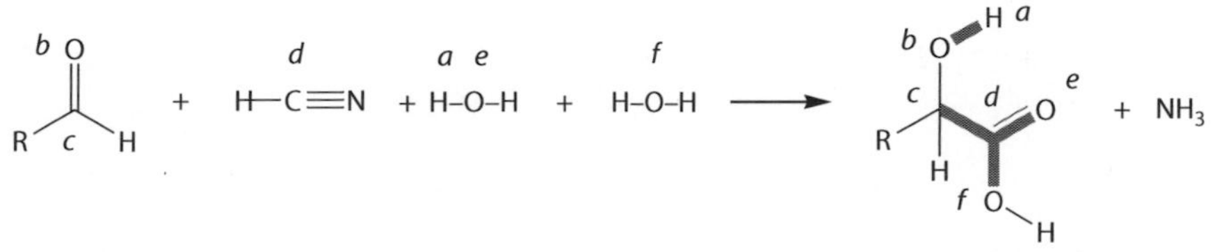

x + 29 | 27 | 18 | 18 | x + 75 | 17

R = x

$$AE = [x + 75]/[x + 92]; \quad \mathbf{AE(min) = 0.82};$$

$$E = 17/[x + 75]; \quad \mathbf{E(max) = 0.22}; \quad \mathbf{f(sac) = 0}$$

Set $R = CH_3$

$$\Delta = \Delta(a) + \Delta(b) + \Delta(c) + \Delta(d) + \Delta(e) + \Delta(f)$$
$$= (1 - 1) + (-2 - (-2)) + (0 - 1) + (3 - 2) + (-2 - (-2)) + (-2 - (-2)) = 0; \quad \mathbf{HI = 0}$$

Koch–Haaf carbonylation reaction
Koch, H. *Brennstoff Chem.* 1955, *36*, 321.
Koch, H.; Haaf, W. *Ann. Chem.* 1958, *618*, 251.

$AE = 1; E = 0; f(sac) = 0$

$$\Delta = \Delta(a) + \Delta(b) + \Delta(c) = (-2-(-1)) + (3-2) + (-2-(-2)) = 0; \quad \textbf{HI} = \textbf{0}$$

Kolbe synthesis
Kolbe, H. *Ann. Chem.* 1860, *113*, 125.
Schmitt, R. *J. Prakt. Chem.* 1885, *31*, 397.

$AE = 1; E = 0; f(sac) = 0$

$$\Delta = \Delta(a) + \Delta(b) + \Delta(c) + \Delta(d) = (1-1) + (-2-(-2)) + (3-4) + (0-(-1)) = 0; \quad \textbf{HI} = \textbf{0}$$

Kulinkovich reaction
Kulinkovich, O.G.; Sviridov, S.V.; Vasilevskii, D.A.; Pritytskaya, T.S. *Zh. Org. Khim.* 1989, *25*, 2244.
Kulinkovich, O.G.; Sviridov, S.V.; Vasilevskii, D.A.; Savchenko, A.I.; Pritytskaya, T.S. *Zh. Org. Khim.* 1991, *27*, 294.
Kulinkovich, O.G.; Vasilevskii, D.A.; Savchenko, A.I.; Sviridov, S.V. *Zh. Org. Khim.* 1991, *27*, 1428.
Kulinkovich, O.G.; Sviridov, S.V.; Vasilevskii, D.A. *Synthesis* 1991, 234.

$x + 2y + 42$ $z + 107.9$ 24.3 $x + 2y + z + 53$ 120.5

$R_1 = x; R_2 = y; R_3 = z$

$$AE = [x + 2y + z + 53]/[x + 2y + z + 173.5]; \quad \textbf{AE(min)} = \textbf{0.33};$$

$$E = 120.5/[x + 2y + z + 53]; \quad \textbf{E(max)} = \textbf{2.04}$$

$$f(sac) = 24.3/[x + 2y + z + 173.5]; \quad \textbf{\textit{f}(sac)(max)} = \textbf{0.14}$$

Set $R_1 = R_3 = CH_3$

$$\Delta = \Delta(a) + \Delta(b) + \Delta(c) = (-2-(-1)) + (1-3) + (-1-(-2)) = -2; \quad \textbf{HI} = -\Delta/(N+1) = \textbf{+1}$$

Kumada cross-coupling

Tamao, K.; Sumitani, K.; Kiso, Y.; Zembayashi, M.; Fujioka, A.; Kodma, S.I.; Nakajima, I.; Minato, A.; Kumada, M. *Bull. Chem. Soc. Jpn.* 1976, *49*, 1958.

$$
\begin{array}{ccc}
a & b & a \quad b \\
R_1\text{---}X \ + \ R_2\text{---}MgX & \longrightarrow & R_1\text{---}R_2 \ + \ MgX_2 \\
r_1 + x \qquad r_2 + x + 24.3 & & r_1 + r_2 \qquad 24.3 + 2x
\end{array}
$$

$R_1 = r, R_2 = r_2, X = x$

$$AE = [r_1 + r_2]/[r_1 + r_2 + 2x + 24.3]; \quad E = [24.3 + 2x]/[r_1 + r_2]; \quad f(\text{sac}) = 0$$

X	E(max)	AE(min)
Cl	$95.2/[r_1 + r_2]$	$[r_1 + r_2]/[r_1 + r_2 + 95.2]$
Br	$184.1/[r_1 + r_2]$	$[r_1 + r_2]/[r_1 + r_2 + 184.1]$
I	$278.3/[r_1 + r_2]$	$[r_1 + r_2]/[r_1 + r_2 + 278.3]$

Set $R_1 = R_2 = CH_3$

$$\Delta = \Delta(a) + \Delta(b) = ((-3) - (-2)) + ((-3) - (-4)) = 0; \quad \textbf{HI} = \textbf{0}$$

Liebeskind–Srogl cross-coupling

Liebeskind, L.S.; Srogl, J. *Org. Lett.* 2002, *4*, 979.

$R_1 = x$(sum of 4 substituents); $R_2 = y$

$$AE = [x + y + 74]/[x + y + 374.55]; \quad \textbf{AE(min)} = \textbf{0.21};$$

$$E = 300.55/[x + y + 74]; \quad \textbf{E(max)} = \textbf{3.80}$$

$$f(\text{sac}) = 208.55/[x + y + 374.55]; \quad \textbf{\textit{f}(sac)(max)} = \textbf{0.55}$$

Set $R_2 = CH_3$

$$\Delta = \Delta(a) + \Delta(b) + \Delta(c) = (-3 - (-4)) + (2 - 3) = 0; \quad \textbf{HI} = \textbf{0}$$

Malonic ester synthesis

Kay, F.W.; Perkin, W.H. Jr. *J. Chem. Soc.* 1907, *89*, 1640.
Avery, S.; Upson, F.W. *J. Am. Chem. Soc.* 1908, *30*, 600.

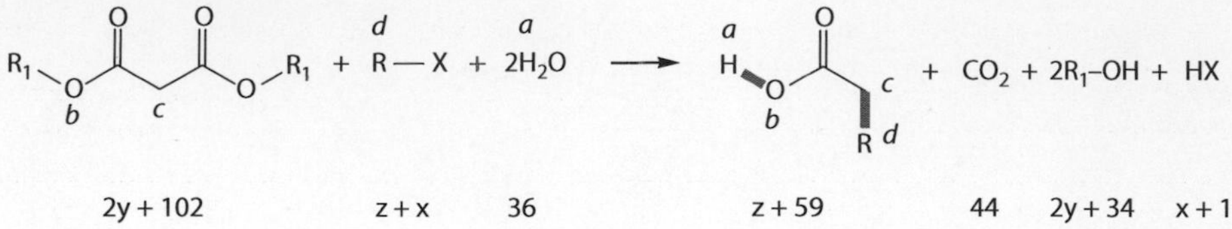

$$2y + 102 \qquad z + x \qquad 36 \qquad\qquad z + 59 \qquad\qquad 44 \quad 2y + 34 \quad x + 1$$

$R_1 = y; \ R = z; \ X = z$

$$AE = [z+59]/[x+2y+z+138]; \quad E = [x+2y+79]/[z+59]$$

$$\mathbf{AE(min) = 60/[x+2y+139]; \quad E(max) = [x+2y+79]/60}$$

$$f(sac) = 18/[x+2y+z+138]; \quad f(sac)(max) = 18/[x+2y+139]$$

X	R_1	E(max)	AE(min)	f(sac)(max)
Cl	Me	2.41	0.29	0.088
Br	Me	3.15	0.24	0.072
I	Me	3.93	0.20	0.061
Cl	Et	2.87	0.26	0.077
Br	Et	3.62	0.22	0.065
I	Et	4.40	0.19	0.056

Set $R = CH_3$

$$\Delta = \Delta(a) + \Delta(b) + \Delta(c) + \Delta(d) = (1-1) + (-2-(-2)) + (-2-(-2)) + (-3-(-2)) = -1;$$

$$\mathbf{HI = -\Delta/(N+1) = +\tfrac{1}{2}}$$

Marschalk reaction

Marschalk, C. *Bull. Soc. Chim. Fr.* 1939, *6*, 655.

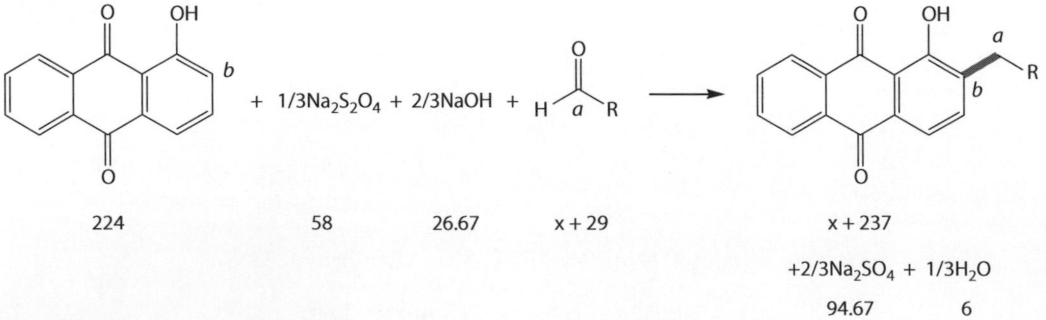

$$224 \qquad\qquad 58 \qquad 26.67 \qquad x+29 \qquad\qquad x+237$$

$$+2/3Na_2SO_4 + 1/3H_2O$$

$$94.67 \qquad\qquad 6$$

$R = x$

$$AE = [x+237]/[x+337.67]; \quad \mathbf{AE(min) = 0.70};$$

$$E = 100.67/[x+237]; \quad \mathbf{E(max) = 0.42}$$

$$f(sac) = 84.67/[x+337.67]; \quad \mathbf{f(sac)(max) = 0.25}$$

Set R = CH_3

$$\Delta = \Delta(a) + \Delta(b) = (-2-1) + (0-(-1)) = 0; \quad \mathbf{HI = 0}$$

McMurry reaction

McMurry, J.E.; Fleming, M.P. *J. Am. Chem. Soc.* 1974, *96*, 4708.

x + y + 28 x + y + 28 2x + 2y + 24

$R_1 = x$; $R_2 = y$

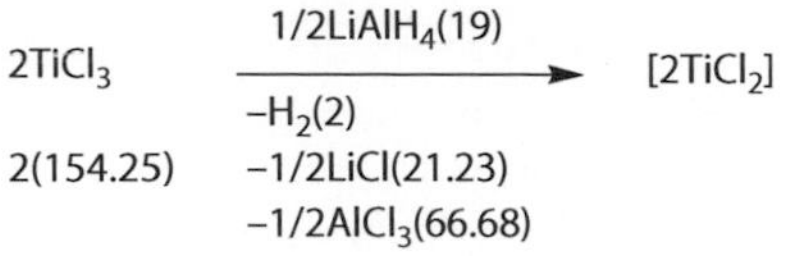

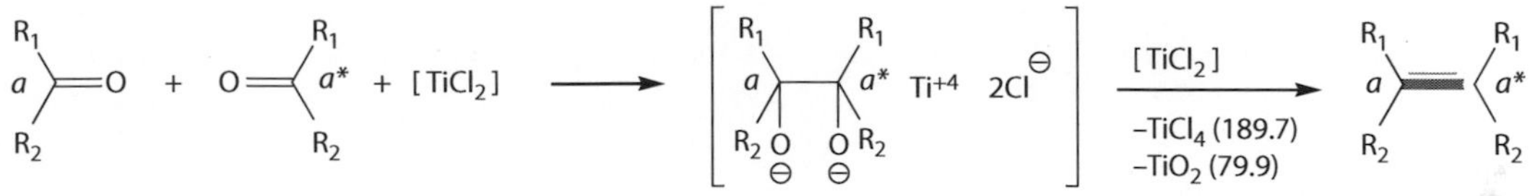

$$AE = [2x + 2y + 24]/[2x + 2y + 383.51]; \quad \mathbf{AE(min) = 0.072};$$

$$E = 359.51/[2x + 2y + 24]; \quad \boldsymbol{E}\mathbf{(max) = 12.84}$$

$$f(\text{sac}) = 327.5/[2x + 2y + 383.51]; \quad \boldsymbol{f}\mathbf{(sac)(max) = 0.85}$$

$$\Delta = \Delta(a) + \Delta(a^*) = (0-2) + (0-2) = -4; \quad \mathbf{HI = -\Delta/(N+1) = +2}$$

Meerwein arylation reaction

Meerwein, H.; Büchner, E.; van Emster, K. *J. Prakt. Chem.* 1939, *152*, 237.

x + 135.45 53 x + 160.45 28

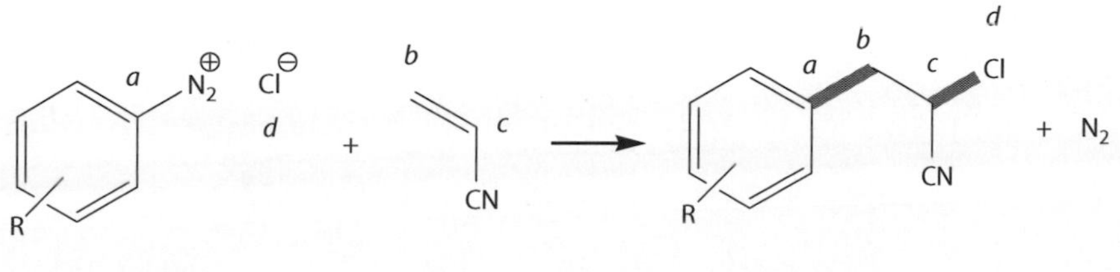

R = x(sum of substituents)

$$AE = [x + 160.45]/[x + 188.45]; \quad \mathbf{AE(min) = 0.86};$$

$$E = 28/[x + 160.45]; \quad \boldsymbol{E}\mathbf{(max) = 0.17}; \quad \boldsymbol{f}\mathbf{(sac) = 0}$$

$$\Delta = \Delta(a) + \Delta(b) + \Delta(c) + \Delta(d) = (0-1) + (-2-(-2)) + (0-(-1)) + (-1-(-1)) = 0; \quad \mathbf{HI = 0}$$

Michael 1,4-conjugate addition
Michael, A. *J. Prakt. Chem.* 1887, *35*, 349.

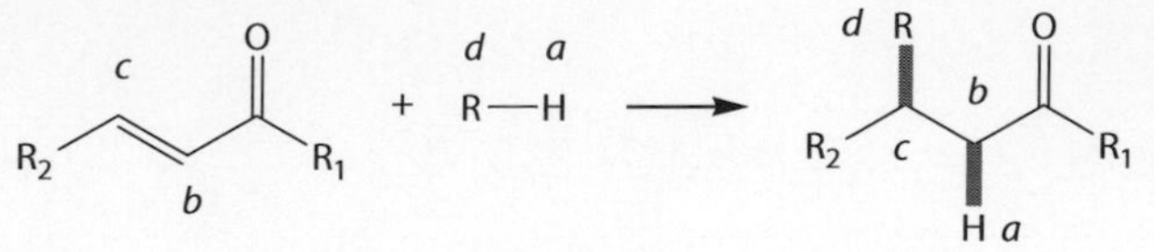

AE = 1; $E = 0$; f(sac) = 0
Set R = R$_2$ = CH$_3$

$$\Delta = \Delta(a) + \Delta(b) + \Delta(c) + \Delta(d) = (1-1) + (-2-(-1)) + (-1-(-1)) + (-3-(-4)) = 0; \quad \mathbf{HI = 0}$$

Mukaiyama aldol reaction
Mukaiyama, T.; Banno, K.; Narasaka, K. *J. Am. Chem. Soc.* 1974, *96*, 7503.
Mukaiyama, T.; Ishihara, I.; Inomata, K. *Chem. Lett.* 1975, 527.
Ishihara, I.; Inomata, K.; Mukaiyama, T. *Chem. Lett.* 1975, 531.
Isawa, T.; Mukaiyama, T. *Chem. Lett.* 1974, 1189.
Narasaka, K.; Soai, K.; Mukaiyama, T. *Chem. Lett.* 1974, 1223.
Mukaiyama, T.; Hayashi, H. *Chem. Lett.* 1974, 15.
Mukaiyama, T.; Ishikawa, H. *Chem. Lett.* 1974, 1077.
Banno, K.; Mukaiyama, T. *Chem. Lett.* 1974, 741.

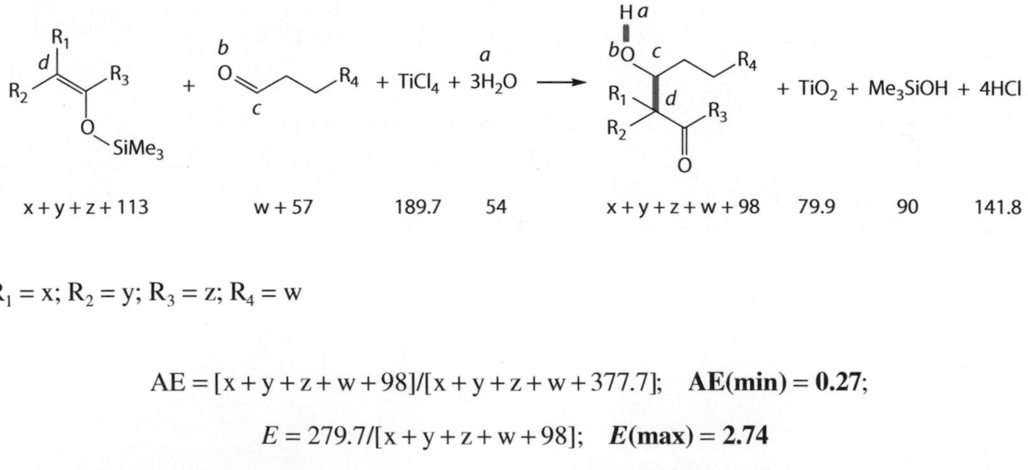

$$\text{R}_1 = \text{x}; \ \text{R}_2 = \text{y}; \ \text{R}_3 = \text{z}; \ \text{R}_4 = \text{w}$$

$$AE = [x + y + z + w + 98]/[x + y + z + w + 377.7]; \quad \mathbf{AE(min) = 0.27};$$

$$E = 279.7/[x + y + z + w + 98]; \quad \mathbf{E(max) = 2.74}$$

$$f(\text{sac}) = 225.7/[x + y + z + w + 377.7]; \quad \mathbf{f(sac)(max) = 0.59}$$

Set R$_1$ = R$_2$ = CH$_3$

$$\Delta = \Delta(a) + \Delta(b) + \Delta(c) + \Delta(d) = (1-1) + (-2-(-2)) + (0-1) + (0-0) = -1;$$

$$\mathbf{HI} = -\Delta/(N+1) = +\tfrac{1}{2}$$

Mukaiyama–Michael reaction

Narasaka, K.; Soai, K.; Aikawa, Y.; Mukaiyama, T. *Bull. Chem. Soc. Jpn.* 1976, *49*, 779.
Mukaiyama, T.; Kobayashi, S. *Heterocycles* 1987, *25*, 245.

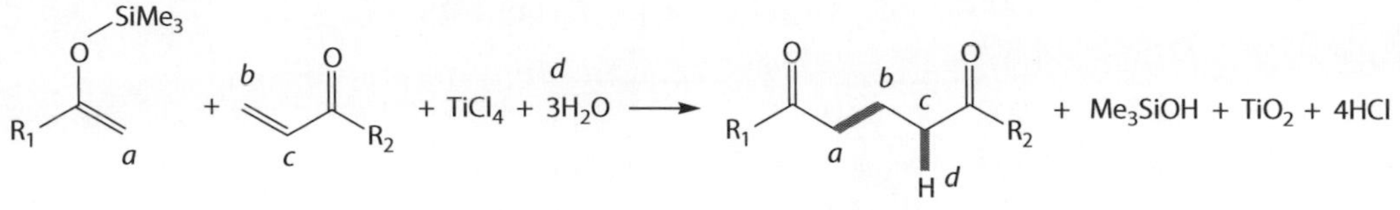

| x + 115 | y + 55 | 189.7 | 54 | x + y + 98 | 90 | 79.9 | 141.8 |

$R_1 = x; R_2 = y$

$$AE = [x + y + 98]/[x + y + 413.7]; \quad \textbf{AE(min)} = \textbf{0.24};$$

$$E = 311.7/[x + y + 98]; \quad \textbf{\textit{E}(max)} = \textbf{3.12}$$

$$f(\text{sac}) = 225.7/[x + y + 413.7]; \quad \textbf{\textit{f}(sac)(max)} = \textbf{0.54}$$

Set $R_2 = CH_3$

$$\Delta = \Delta(a) + \Delta(b) + \Delta(c) + \Delta(d) = (-2 - (-2)) + (-2 - (-2)) + (-2 - (-1)) + (1 - 1) = -1;$$

$$\textbf{HI} = -\Delta/(N+1) = +\tfrac{1}{2}$$

Nazarov cyclization

Nazarov, I.N.; Pinkina, L.N., *Bull. Acad. Sci. USSR, Classe Sci. Chim.* 1946, 633 (see *Chem. Abs.* 1948, 42, 7731i).

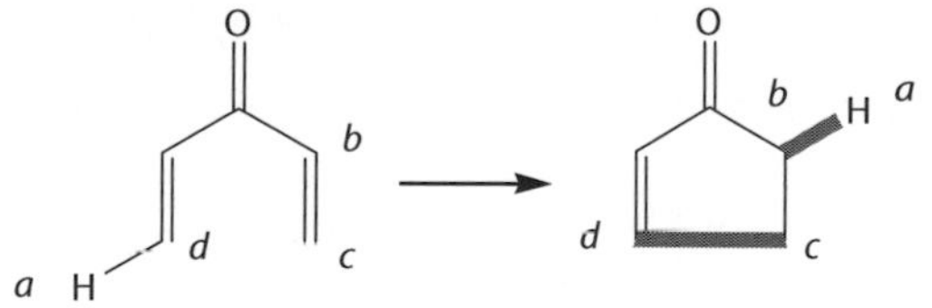

$\textbf{AE} = \textbf{1}; \textit{E} = \textbf{0}; f(\text{sac}) = \textbf{0}$

$$\Delta = \Delta(a) + \Delta(b) + \Delta(c) + \Delta(d) = (1 - 1) + (-2 - (-1)) + (-2 - (-2)) + (-1 - (-2)) = 0; \quad \textbf{HI} = \textbf{0}$$

Negishi cross-coupling reaction

Negishi, E.I.; Baba, S. *Chem. Commun.* 1976, 596.
Negishi, E.I. *Acc. Chem. Res.* 1982, *15*, 340.

| $r_1 + x$ | $r_2 + x + 65.38$ | $r_1 + r_2$ | $65.38 + 2x$ |

$R_1 = r$, $R_2 = r_2$, $X = x$

$$AE = [r_1 + r_2]/[r_1 + r_2 + 2x + 65.38];$$

$$E = [65.38 + 2x]/[r_1 + r_2]; \quad f(sac) = 0$$

X	E(max)	AE(min)
Cl	$136.28/[r_1 + r_2]$	$[r_1 + r_2]/[r_1 + r_2 + 136.28]$
Br	$225.18/[r_1 + r_2]$	$[r_1 + r_2]/[r_1 + r_2 + 225.18]$
I	$319.38/[r_1 + r_2]$	$[r_1 + r_2]/[r_1 + r_2 + 319.38]$

Set $R_1 = R_2 = CH_3$

$$\Delta = \Delta(a) + \Delta(b) = ((-3) - (-2)) + ((-3) - (-4)) = 0; \quad \mathbf{HI = 0}$$

Nieuwland enyne synthesis

Nieuwland, J.A. *J. Am. Chem. Soc.* 1934, *56*, 2120.

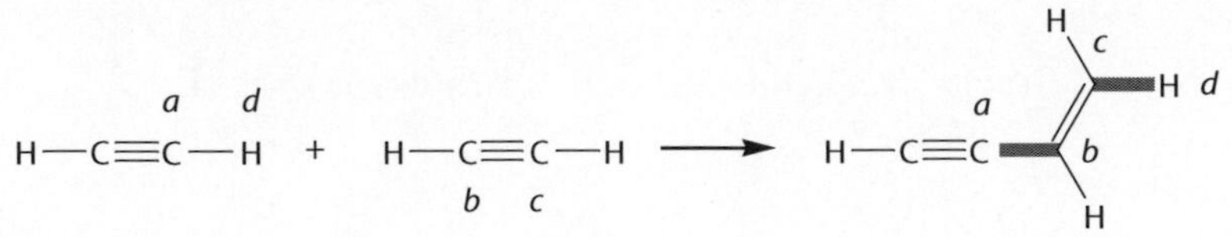

$\mathbf{AE = 1}; \mathbf{\mathit{E} = 0}; \mathbf{\mathit{f}(sac) = 0}$

$$\Delta = \Delta(a) + \Delta(b) + \Delta(c) + \Delta(d) = (0 - (-1)) + (-1 - (-1)) + (-2 - (-1)) + (1 - 1) = 0; \quad \mathbf{HI = 0}$$

Nozaki reaction

Takai, K.; Kimura, K.; Kuroda, T.; Hiyama, T.; Nozaki, H. *Tetrahedron Lett.* 1983, *24*, 5281.

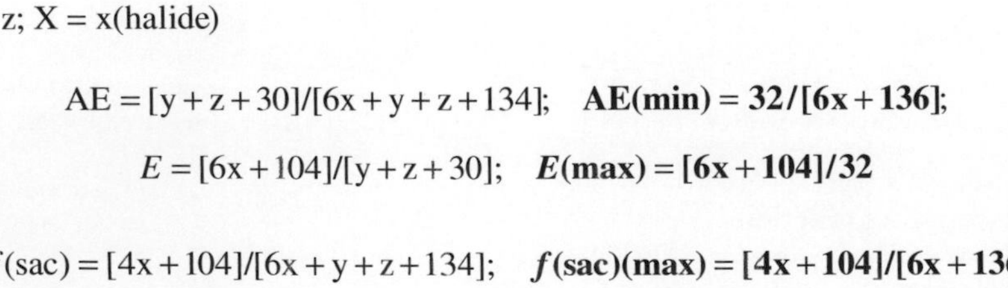

$R_1 = y$; $R_2 = z$; $X = x$(halide)

$$AE = [y + z + 30]/[6x + y + z + 134]; \quad \mathbf{AE(min) = 32/[6x + 136]};$$

$$E = [6x + 104]/[y + z + 30]; \quad \mathbf{E(max) = [6x + 104]/32}$$

$$f(sac) = [4x + 104]/[6x + y + z + 134]; \quad \mathbf{\mathit{f}(sac)(max) = [4x + 104]/[6x + 136]}$$

X	E(max)	AE(min)	f(sac)(max)
Cl	9.90	0.092	0.70
Br	18.23	0.052	0.69
I	27.06	0.036	0.68

Set $R_1 = R_2 = CH_3$

$$\Delta = \Delta(a) + \Delta(b) + \Delta(c) + \Delta(d) = (1-1) + (-2-(-2)) + (0-1) + (-3-(-2)) = -2;$$

$$\mathbf{HI} = -\Delta/(N+1) = +1$$

Organocuprate conjugate addition

Kharasch, M.; Tawney, P.O. *J. Am. Chem. Soc.* 1941, *63*, 2308.
Birch, A.J.; Robinson, R. *J. Chem. Soc.* 1943, 501.
Gilman, H.; Jones, R.G.; Woods, L.A. *J. Am. Chem. Soc.* 1952, *17*, 1630.
Munch-Petersen, J. *J. Org. Chem.* 1957, *22*, 170.
Bjerl-Nielsen, E.; Munch-Petersen, J.; Moller-Jorgensen, P.; Refn, S. *Acta Chem. Scand.* 1959, *13*, 1943.
Bretting, C.; Munch-Petersen, J.; Moller-Jorgensen, P.; Refn, S. *Acta Chem. Scand.* 1960, *14*, 151.
Munch-Petersen, J.; Bretting, C.; Moller-Jorgensen, P.; Refn, S.; Andersen, V.K. *Acta Chem. Scand.* 1961, *15*, 277.
Andersen, V.K.; Munch-Petersen, J. *Acta Chem. Scand.* 1962, *16*, 947.
Kindt-Larsen, T.; Bitsch, V.; Andersen, I.G.K.; Jart, A.; Munch-Petersen, M. *Acta Chem. Scand.* 1963, *17*, 1426.
Jacobsen, S.; Jart, A.; Kindt-Larsen, T.; Andersen, I.G.K.; Munch-Petersen, M. *Acta Chem. Scand.* 1963, *17*, 2423.
House, H.O.; Respess, W.L.; Whitesides, G.M. *J. Org. Chem.* 1966, *31*, 3128.

$R_1 = x$; $R_2 = y$; $R_3 = z$

$$AE = [x + y + z + 55]/[x + y + 2z + 496.55]; \quad E = [z + 441.55]/[x + y + z + 55]$$

$$\mathbf{AE(min) = 58/[2z + 498.55];} \quad \mathbf{E(max) = [z + 441.55]/[z + 57]}$$

$$f(sac) = [z + 423.55]/[2z + 498.55]$$

R_3	E(max)	AE(min)	f(sac)(max)
CH₃ (15)	6.34	0.14	0.83
Ph (77)	3.87	0.21	0.77

Set $R_2 = R_3 = CH_3$

$$\Delta = \Delta(a) + \Delta(b) + \Delta(c) + \Delta(d) = (-3-(-4)) + (-1-(-1)) + (-2-(-1)) + (1-1) = 0; \quad \mathbf{HI = 0}$$

Paterno–Buchi reaction

Paternó, E.; Chietti, G. *Gazz. Chim. Ital*. 1909, *39*, 341.
Büchi, G.; Inman, C.G.; Lipinsky, E.S. *J. Am. Chem. Soc*. 1954, *76*, 4327.

AE = 1; *E* = 0; *f*(sac) = 0

$$\Delta = \Delta(a) + \Delta(b) + \Delta(c) + \Delta(d) = (-2-(-2)) + (0-(-1)) + (-1-(-1)) + (0-1) = 0; \quad \textbf{HI} = \textbf{0}$$

von Pechmann reaction

von Pechmann, H.; Duisberg, C. *Chem. Ber*. 1883, *16*, 2119.

AE = 1; *E* = 0; *f*(sac) = 0

Set $R_1 = R_2 = CH_3$

$$\Delta = \Delta(a) + \Delta(b) + \Delta(c) + \Delta(d) = (1-1) + (-2-0) + (1-0) + (0-0) + (1-(-2)) = 2;$$

$$\textbf{HI} = -\Delta/(N+1) = -1$$

Perkin reaction

Perkin, W.H. *J. Chem. Soc*. 1868, *21*, 53.

x + 101   102   x + 143   60

R = x(sum of substituents)

$$AE = [x+143]/[x+203]; \quad \textbf{AE(min)} = \textbf{0.71}; \quad E = 60/[x+143]; \quad \textbf{E(max)} = \textbf{0.41}; \quad f(sac) = 0$$

$$\Delta = \Delta(a) + \Delta(b) + \Delta(c) + \Delta(d) = (-1-1) + (-1-(-3)) + (-2-(-2)) + (1-1) = 0; \quad \textbf{HI} = \textbf{0}$$

Peterson olefination
Peterson, D.J. *J. Org. Chem.* 1968, *33*, 780.

$$2y + 28 \qquad z + x + 86 \qquad 2y + z + x + 24 \qquad 90$$

$R_1 = y$; $R_2 = z$; $X = x$(halide)

$$AE = [2y + z + x + 24]/[2y + z + x + 114]; \quad E = 90/[2y + z + x + 24]$$

$$\mathbf{AE(min) = [x + 27]/[x + 114]; \quad E(max) = 90/[x + 27]; \quad \mathit{f}(sac) = 0}$$

X	E(max)	AE(min)
Cl	1.44	0.41
Br	0.84	0.54
I	0.58	0.63

Set $R_1 = R_2 = CH_3$

$$\Delta = \Delta(a) + \Delta(b) = (1 - (-1)) + (0 - 2) = 0; \quad \mathbf{HI = 0}$$

Photochemical cyclizations

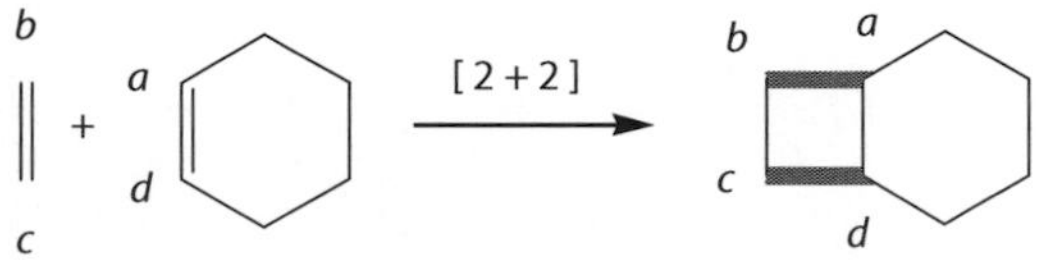

$AE = 1$; $E = 0$; $\mathit{f}(sac) = 0$

$$\Delta = \Delta(a) + \Delta(b) + \Delta(c) + \Delta(d) = (-1 - (-1)) + (-2 - (-2)) + (-2 - (-2)) + (-1 - (-1)) = 0; \quad \mathbf{HI = 0}$$

Pinacol reaction
Fittig, R. *Ann. Chem.* 1860, *114*, 54.

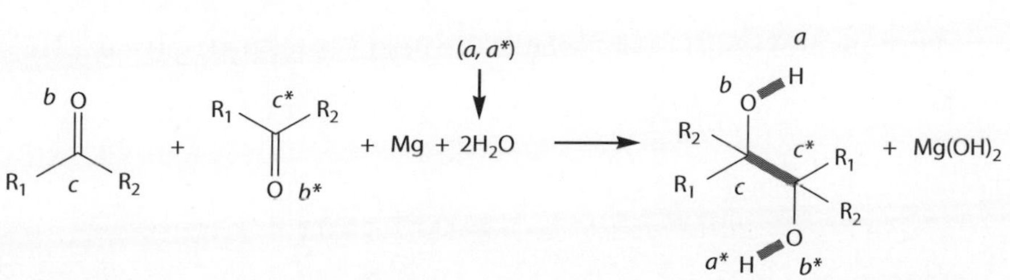

$$x + y + 28 \qquad x + y + 28 \qquad 24.3 \qquad 36 \qquad\qquad 2(x + y + 29) \qquad 58.3$$

$R_1 = x; R_2 = y$

$$AE = [2(x+y+29)][2x+2y+116.3]; \quad \textbf{AE(min)} = \textbf{0.52};$$

$$E = 58.3/[2(x+y+29)]; \quad \textbf{E(max)} = \textbf{0.94}$$

$$f(\text{sac}) = 24.3/[2x+2y+116.3]; \quad \textbf{f(sac)(max)} = \textbf{0.20}$$

Set $R_1 = R_2 = CH_3$

$$\Delta = \Delta(a) + \Delta(b) + \Delta(c) + \Delta(a^*) + \Delta(b^*) + \Delta(c^*) = 2[(1-1)+(-2-(-2))+(1-2)] = -2;$$

$$\textbf{HI} = -\Delta/(N+1) = +\textbf{1}$$

Prins reaction
Prins, H.J., *Chem. Weekblad.* 1917, *14*, 932.
Prins, H.J., *Chem. Weekblad.* 1919, *16*, 1072.
Prins, H.J., *Chem. Weekblad.* 1919, *16*, 1510.
Prins, H.J., *Proc. Acad. Sci. Amsterdam* 1919, *22*, 51.

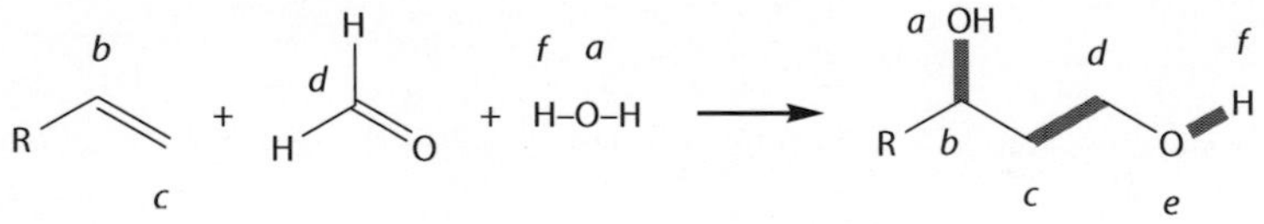

Set $R = CH_3$

$$\Delta = \Delta(a) + \Delta(b) + \Delta(c) + \Delta(d) + \Delta(e) + \Delta(f)$$

$$= (-2-(-2)) + (0-(-1)) + (-2-(-2)) + (-1-0) + (-2-(-2)) + (1-1) = 0; \quad \textbf{HI} = \textbf{0}$$

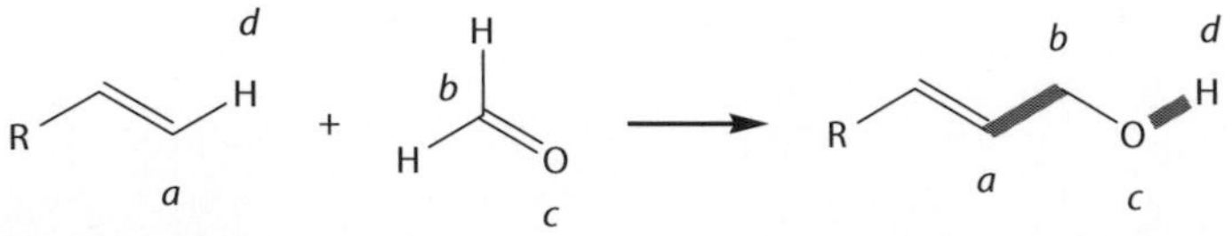

Set $R = CH_3$

$$\Delta = \Delta(a) + \Delta(b) + \Delta(c) + \Delta(d) = (-1-(-2)) + (-1-0) + (-2-(-2)) + (1-1) = 0; \quad \textbf{HI} = \textbf{0}$$

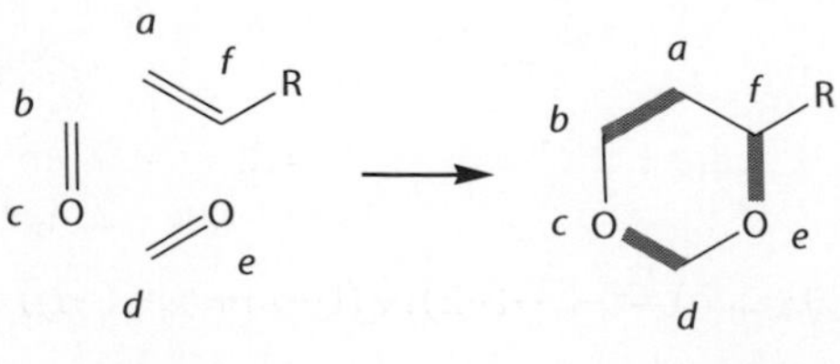

Set R = CH_3

$$\Delta = \Delta(a) + \Delta(b) + \Delta(c) + \Delta(d) + \Delta(e) + (f)$$

$$= (-2 - (-2)) + (-1 - 0) + (-2 - (-2)) + (0 - 0) + (-2 - (-2)) + (0 - (-1)) = 0; \quad \textbf{HI} = \textbf{0}$$

AE = 1; E = 0; f(sac) = 0

Reformatskii reaction

Reformatskii, S. *Chem. Ber.* 1887, *20*, 1210.

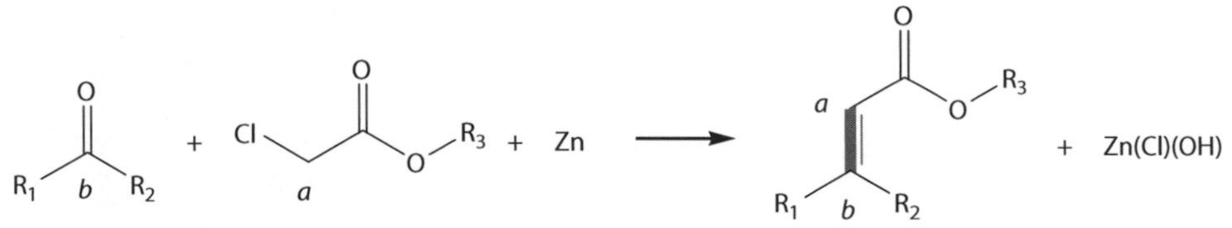

| x + y + 28 | z + 93.45 | 65.38 | x + y + z + 69 | 117.83 |

R_1 = x; R_2 = y; R_3 = z

$$AE = [x + y + z + 69]/[x + y + z + 186.83]; \quad \textbf{AE(min)} = \textbf{0.38};$$

$$E = 117.83/[x + y + z + 69]; \quad \textbf{E(max)} = \textbf{1.64}$$

$$f(\text{sac}) = 65.38/[x + y + z + 186.83]; \quad \textbf{f(sac)(max)} = \textbf{0.34}$$

Set R_1 = R_2 = CH_3

$$\Delta = \Delta(a) + \Delta(b) = (-1 - (-1)) + (0 - 2) = -2; \quad \textbf{HI} = -\Delta/(N + 1) = +1$$

Reimer–Tiemann reaction

Reimer, K.; Tiemann, F. *Chem. Ber.* 1876, *9*, 824.
Reimer, K.; Tiemann, F. *Chem. Ber.* 1876, *9*, 1268.
Reimer, K.; Tiemann, F. *Chem. Ber.* 1876, *9*, 1285.

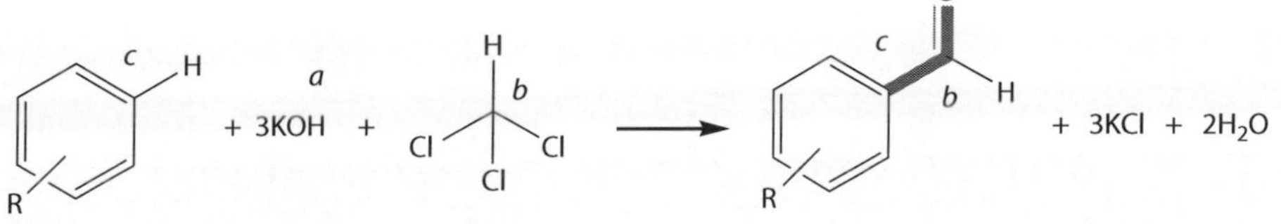

| x + 73 | 168 | 119.35 | x + 101 | 223.35 | 36 |

R = x(sum of 5 substituents)

$$AE = [x + 101]/[x + 360.35]; \quad \textbf{AE(min)} = \textbf{0.29};$$

$$E = 259.35/[x + 101]; \quad \textbf{E(max)} = \textbf{2.45}; \quad \textbf{f(sac)} = \textbf{0}$$

$$\Delta = \Delta(a) + \Delta(b) + \Delta(c) = (-2 - (-2)) + (1 - 2) + (0 - (-1)) = 0; \quad \textbf{HI} = \textbf{0}$$

von Richter reaction

von Richter, V. *Chem. Ber.* 1871, *4*, 21.
von Richter, V. *Chem. Ber.* 1871, *4*, 459.
von Richter, V. *Chem. Ber.* 1871, *4*, 553.

	2KCN	H–O–H	H–O–H		N₂	HCN	2KOH

$$x + 122 \qquad 130 \qquad 18 \qquad 18 \qquad x + 121 \qquad 28 \qquad 27 \qquad 112$$

$R = x$

$$AE = [x + 121]/[x + 288]; \quad \mathbf{AE(min)} = \mathbf{0.42};$$

$$E = 167/[x + 121]; \quad E(\mathbf{max}) = \mathbf{1.37}; \quad f(\mathbf{sac}) = \mathbf{0}$$

$$\Delta = \Delta(a) + \Delta(b) + \Delta(c) + \Delta(d) + \Delta(e) + \Delta(f)$$

$$= (1-1) + (-1-1) + (0-(-1)) + (3-2) + (-2-(-2)) + (-2-(-2)) = 0; \quad \mathbf{HI = 0}$$

Robinson annulation

Rapson, W.S.; Robinson, R. *J. Chem. Soc.* 1935, 1285.
Du Feu, E.C.; McQuillin, F.J., Robinson, R. *J. Chem. Soc.* 1937, 53.

$$x + y + 42 \qquad z + 69 \qquad 68 \quad 36.45 \qquad x + y + z + 93 \qquad 18 \qquad 46 \qquad 58.45$$

$R_1 = x; R_2 = y; R_3 = z$

$$AE = [x + y + z + 93]/[x + y + z + 215.45]; \quad \mathbf{AE(min)} = \mathbf{0.44};$$

$$E = 122.45/[x + y + z + 93]; \quad E(\mathbf{max}) = \mathbf{1.28};$$

$$f(\mathbf{sac}) = 68/[x + y + z + 215.45]; \quad f(\mathbf{sac})(\mathbf{max}) = \mathbf{0.31}$$

Set $R_1 = R_2 = R_3 = CH_3$

$$\Delta = \Delta(a) + \Delta(b) + \Delta(c) + \Delta(d) + \Delta(e) + \Delta(f)$$

$$= (-2-(-2)) + (-1-(-2)) + (0-2) + (0-(-2)) + (-2-(-1)) + (1-1) = 0; \quad \mathbf{HI = 0}$$

Sakurai reaction
Hosomi, A.; Sakurai. H. *J. Am. Chem. Soc.* 1977, *99*, 1673.

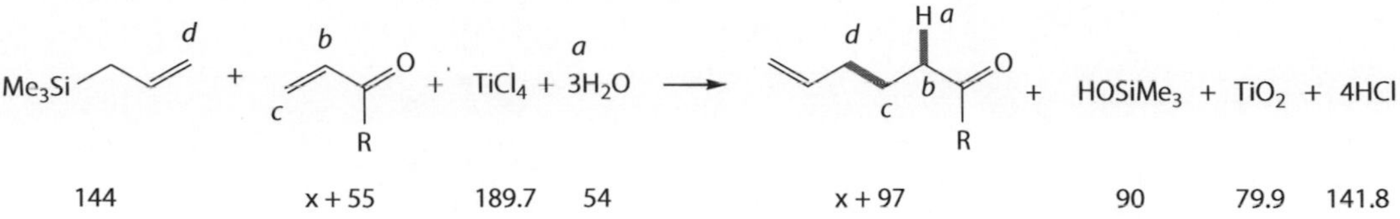

$$R = x$$

$$AE = [x+97]/[x+376.8]; \quad \textbf{AE(min)} = \textbf{0.26}; \quad E = 279.8/[x+97]; \quad \textbf{E(max)} = \textbf{2.86}$$

$$f(\text{sac}) = 189.7/[x+442.7]; \quad \textbf{\textit{f}(sac)(max)} = \textbf{0.43}$$

Set $R = CH_3$

$$\Delta = \Delta(a) + \Delta(b) + \Delta(c) + \Delta(d) = (1-1) + (-2-(-1)) + (-2-(-2)) + (-2-(-2)) = -1;$$

$$\textbf{HI} = -\Delta/(N+1) = +\tfrac{1}{2}$$

Simmons–Smith cyclopropanation reaction
Simmons, H.E.; Smith, R.D. *J. Am. Chem. Soc.* 1958, *80*, 5323.

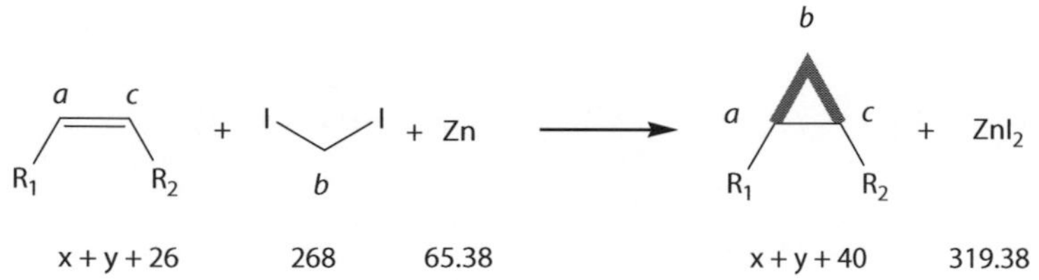

$$R_1 = x; R_2 = y$$

$$AE = [x+y+40]/[x+y+359.38]; \quad \textbf{AE(min)} = \textbf{0.12};$$

$$E = 319.38/[x+y+40]; \quad \textbf{E(max)} = \textbf{7.60}$$

$$f(\text{sac}) = 65.38/[x+y+359.8]; \quad \textbf{\textit{f}(sac)(max)} = \textbf{0.18}$$

Set $R_1 = R_2 = CH_3$

$$\Delta = \Delta(a) + \Delta(b) + \Delta(c) = (-1-(-1)) + (-2-0) + (-1-(-1)) = -2;$$

$$\textbf{HI} = -\Delta/(N+1) = +1$$

Skraup reaction

Skraup, Z.H. *Chem. Ber.* 1880, *13*, 2086.

93 92 324.3 129 253.4 54 72.9

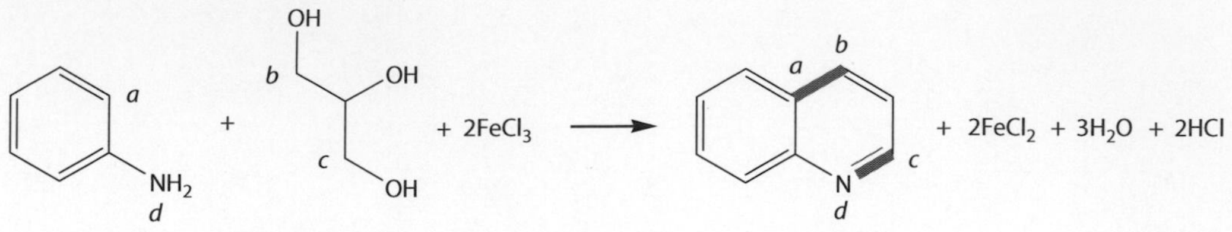

AE = 0.25; *E* = 380.3/129 = 2.95; *f*(sac) = 324.3/509.3 = 0.64

$$\Delta = \Delta(a) + \Delta(b) + \Delta(c) + \Delta(d) = (0-(-1)) + (-1-(-1)) + (1-(-1)) + (-3-(-3)) = +3;$$

$$\mathbf{HI = -\Delta/(N+1) = -3/2}$$

Sonogashira reaction

Sonogashira, K.; Tohda, Y.; Hagihara, N. *Tetrahedron Lett.* 1975, *26*, 4467.

Sonogashira, K.; Yatake, T.; Tohda, Y.; Takahashi, S.; Hagihara, N. *Chem. Commun.* 1977, 291.

r + x + 72 y + 25 r + y + 96 x + 1

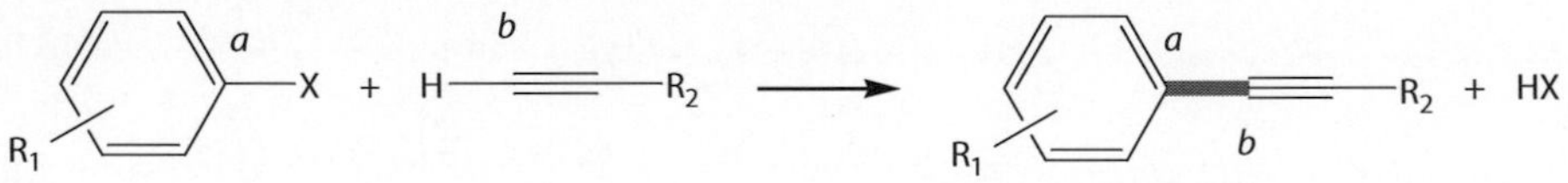

R_1 = r(sum of substituents); X = x(halide); R_2 = y

$$AE = [r+y+96]/[r+y+x+97]; \quad E = [x+1]/[r+y+96]$$

AE(min) = 102/[x + 103]; *E*(max) = [x + 1]/102; *f*(sac) = 0

X	*E*(max)	AE(min)
Cl	0.36	0.74
Br	0.79	0.56
I	1.25	0.44

$\Delta = \Delta(a) + \Delta(b) = (0-1) + (0-(-1)) = 0;$ **HI = 0**

Stetter reaction

Stetter, H. *Angew. Chem. Int. Ed.* 1976, *15*, 639.

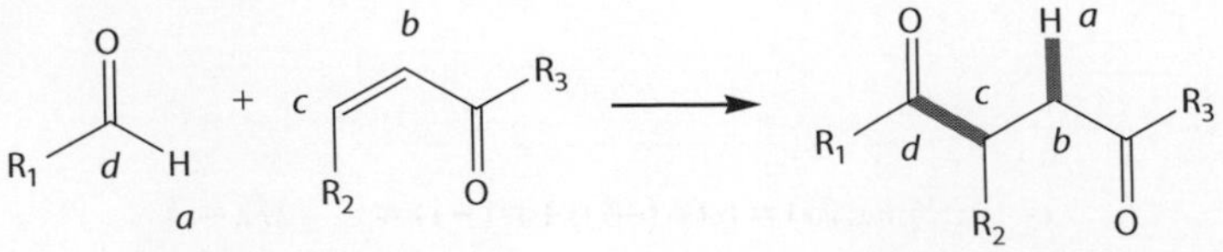

AE = 1; *E* = 0; *f*(sac) = 0

Set $R_1 = R_2 = CH_3$

$$\Delta = \Delta(a) + \Delta(b) + \Delta(c) + \Delta(d) = (1-1) + (-2-(-1)) + (-1-(-1)) + (2-1) = 0; \quad \textbf{HI = 0}$$

Stille coupling
Milstein, D.; Stille, J.K. *J. Am. Chem. Soc.* 1978, *100*, 3636.

r + x	y + 3z + 118.69	r + y	x + 3z + 118.69

$R_1 = r$; $X = x$(halide); $R_2 = y$; $R_3 = z$

$$AE = [r+y]/[r+y+x+3z+119]; \quad E = [x+3z+119]/[r+y]; \quad f(sac) = 0$$

X	R_3	*E*(max)	AE(min)
Cl	*n*Bu	325.14/[r + y]	[r + y]/[r + y + 325.14]
Br	*n*Bu	369.59/[r + y]	[r + y]/[r + y + 369.59]
I	*n*Bu	416.69/[r + y]	[r + y]/[r + y + 416.69]

Set $R_1 = R_2 = CH_3$

$$\Delta = \Delta(a) + \Delta(b) = (-3-(-2)) + (-3-(-4)) = 0; \quad \textbf{HI = 0}$$

Suzuki coupling
Miyaura, N.; Suzuki, A. *Chem. Comm.* 1979, 866.
Miyaura, N.; Yamada, K.; Suzuki, A. *Tetrahedron Lett.* 1979, 3437.

y + 117	z + x + 72	18	y + z + 144	62	x + 1

$R_1 = y$(sum of substituents); $R_2 = z$(sum of substituents); $X = x$(halide)

$$AE = [y+z+144]/[x+y+z+189]; \quad E = [x+45]/[y+z+144]$$

$$f(sac) = 18/[x+y+z+189]; \quad f(sac)(max) = 18/[x+199]$$

$$\textbf{AE(min) = 154/[x+199]}; \quad E(max) = [x+45]/154$$

X	*E*(max)	AE(min)	*f*(sac)(max)
F	0.42	0.71	0.083
Cl	0.52	0.66	0.077
Br	0.81	0.55	0.065
I	1.12	0.47	0.055

$$\Delta = \Delta(a) + \Delta(b) = (0-(-1)) + (0-1) = 0; \quad \textbf{HI = 0}$$

Tebbe olefination

Tebbe, F.N.; Parshell, G.W.; Reddy, G.S. *J. Am. Chem. Soc.* 1978, *100*, 3611.

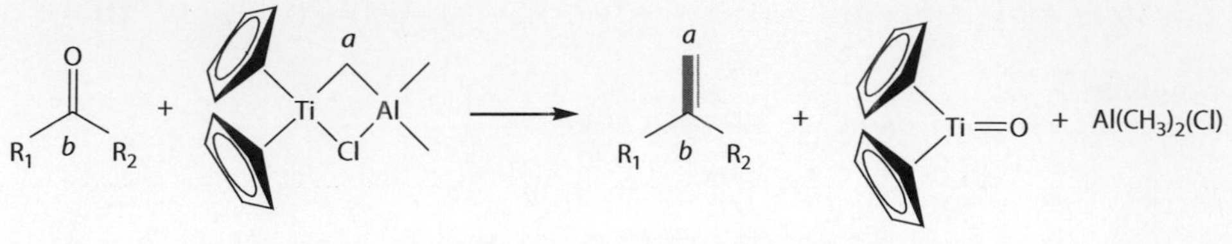

x + y + 28	284.35	x + y + 26	193.9	92.45

$R_1 = x$; $R_2 = y$

$$AE = [x + y + 26]/[x + y + 313.35]; \quad \mathbf{AE(min) = 0.089};$$

$$E = 286.35/[x + y + 26]; \quad \mathbf{E(max) = 10.23}; \quad f(\mathbf{sac}) = \mathbf{0}$$

Set $R_1 = R_2 = CH_3$

$$\Delta = \Delta(a) + \Delta(b) = (-2 - (-4)) + (0 - 2) = 0; \quad \mathbf{HI = 0}$$

Thorpe reaction

Baron, H.; Remfry, F.G.P.; Thorpe, J.F. *J. Chem. Soc.* 1904, *85*, 1726.

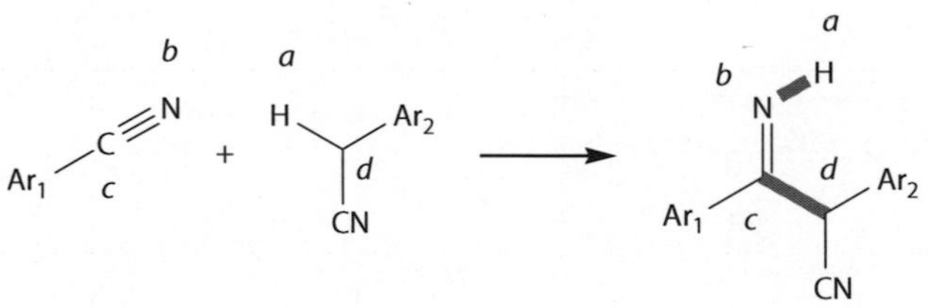

$\mathbf{AE = 1}$; $E = 0$; $f(\mathbf{sac}) = 0$

$$\Delta = \Delta(a) + \Delta(b) + \Delta(c) + \Delta(d) = (1 - 1) + (-3 - (-3)) + (2 - 3) + (-1 - (-2)) = 0; \quad \mathbf{HI = 0}$$

Tishchenko reaction

Tishchenko, V. *J. Russ. Phys. Chem. Soc.* 1906, *38*, 355.
Claisen, L. *Chem. Ber.* 1887, *20*, 646.

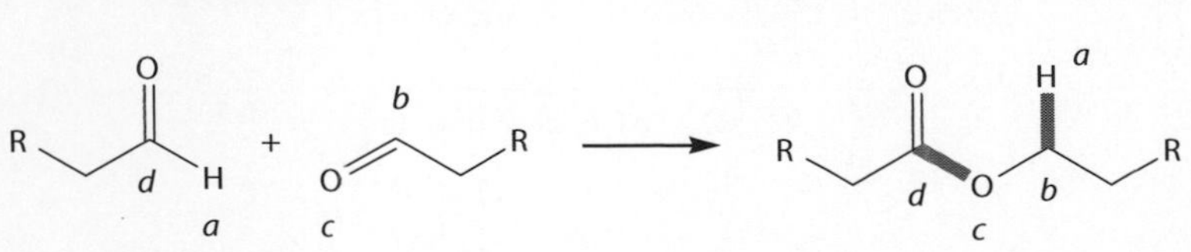

$\mathbf{AE = 1}$; $E = 0$; $f(\mathbf{sac}) = 0$

$$\Delta = \Delta(a) + \Delta(b) + \Delta(c) + \Delta(d) = (1 - 1) + (-1 - 1) + (-2 - (-2)) + (3 - 1) = 0; \quad \mathbf{HI = 0}$$

Trost allene–alkene addition

Trost, B.M.; Pinkerton, A.B. *J. Am. Chem. Soc.* 1999, *121*, 4068.
Trost, B.M.; Pinkerton, A.B.; Seidel, M. *J. Am. Chem. Soc.* 2001, *123*, 12466.

AE = 1; *E* = 0; *f*(sac) = 0

$$\Delta = \Delta(a) + \Delta(b) + \Delta(c) + \Delta(d) = (1-1) + (-2 - (-1)) + (-2 - (-2)) + (0 - 0) = -1;$$

$$\mathbf{HI = -\Delta/(N+1) = +\frac{1}{2}}$$

Ullmann coupling

Ullmann, F. *Ann. Chem.* 1904, *332*, 38.

$$x + y + 72 \qquad x + y + 72 \qquad 63.55 \qquad\qquad 2y + 144 \qquad 2x + 63.55$$

$R_1 = y$(sum of substituents); $X = x$(halide)

$$AE = [2y + 144]/[2x + 2y + 207.55]; \quad E = [2x + 63.55]/[2y + 144]$$

$$\mathbf{AE(min) = 154/[2x + 217.55]; \quad E(max) = [2x + 63.55]/154}$$

$$f(sac) = 63.55/[2x + 2y + 207.55]; \quad f(sac)(max) = \mathbf{63.55/[2x + 217.55]}$$

X	*E*(max)	AE(min)	*f*(sac)(max)
F	0.66	0.60	0.25
Cl	0.87	0.53	0.22
Br	1.45	0.41	0.17
I	2.06	0.33	0.13

$$\Delta = \Delta(a) + \Delta(a^*) = (0-1) + (0-1) = -2; \quad \mathbf{HI = -\Delta/(N+1) = +1}$$

Vilsmeier–Haack–Arnold reaction

Vilsmeier, A.; Haack, A. *Chem. Ber.* 1927, *60*, 119.
Arnold, Z.; Sorm, F. *Collect. Czech. Chem. Commun.* 1958, *23*, 452.
Svoboda, M.; Synácková, M.; Samek, Z.; Fiedler, P.; Arnold, Z. *Collect. Czech. Chem. Commun.* 1978, *43*, 1261.

$$73 \qquad\qquad x + 117 \qquad\quad 153.35 \quad 54 \qquad\quad x + 145 \qquad\qquad 98 \qquad\quad 81.45 \quad 72.9$$

R = x(sum of 4 substituents)

$$AE = [x + 145]/[x + 397.35]; \quad \textbf{AE(min)} = \textbf{0.37}; \quad E = 252.35/[x + 145]; \quad \textbf{\textit{E}(max)} = \textbf{1.69}$$

$$f(\text{sac}) = 207.35/[x + 397.35]; \quad \textbf{\textit{f}(sac)(max)} = \textbf{0.52}$$

$$\Delta = \Delta(a) + \Delta(b) = (0 - (-1)) + (1 - 2) = 0; \quad \textbf{HI} = \textbf{0}$$

Vorbruggen coupling
Niedballa, U.; Vorbrüggen, H. *Angew. Chem. Int. Ed.* 1970, *9*, 461.
Vorbrüggen, H.; Bennua, B. *Tetrahedron Lett.* 1978, 1339.

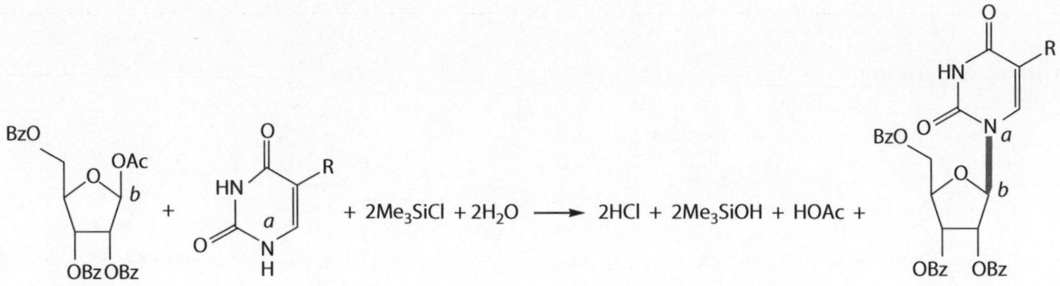

| 504 | x + 111 | 216.9 | 36 | 72.9 | 180 | 60 | x + 555 |

R = x

$$AE = [x + 555]/[x + 867.9]; \quad \textbf{AE(min)} = \textbf{0.64}; \quad E = 312.9/[x + 555]; \quad \textbf{\textit{E}(max)} = \textbf{0.56}$$

$$f(\text{sac}) = 252.9/[x + 867.9]; \quad \textbf{\textit{f}(sac)(max)} = \textbf{0.29}$$

$$\Delta = \Delta(a) + \Delta(b) = (-3 - (-3)) + (1 - 1) = 0; \quad \textbf{HI} = \textbf{0}$$

Wadsworth–Horner–Emmons reaction
Horner, L.; Hoffmann, H.; Wippel, H.G. *Chem. Ber.* 1958, *45*, 61.
Wadsworth, W.S., Jr.; Emmons, W.D. *J. Am. Chem. Soc.* 1961, *83*, 1733.

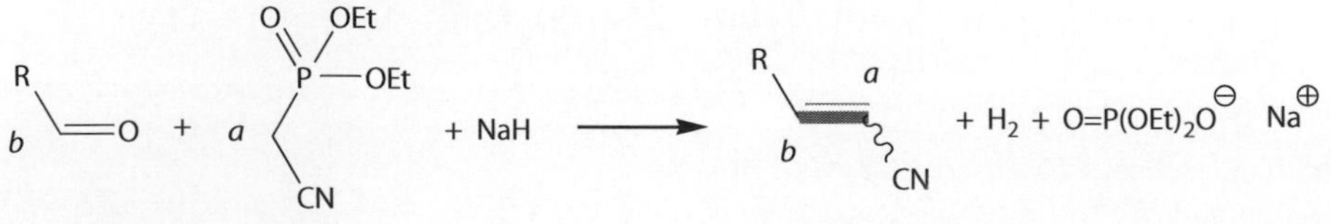

| x + 29 | 177 | 24 | x + 52 | 2 | 176 |

R = x

$$AE = [x + 52]/[x + 230]; \quad \textbf{AE(min) 0.23}; \quad E = 178/[x + 52]; \quad \textbf{\textit{E}(max)} = \textbf{3.36}$$

$$f(\text{sac}) = 24/[x + 230]; \quad \textbf{\textit{f}(sac)(max)} = \textbf{0.10}$$

Set R = CH$_3$

$$\Delta = \Delta(a) + \Delta(b) = (-1 - (-1)) + (-1 - 1) = -2; \quad \textbf{HI} = -\Delta/(N+1) = +1$$

Wender–Trost [5 + 2] cycloaddition

Wender, P.A.; Takahashi, H.; Witulski, B. *J. Am. Chem. Soc.* 1995, *117*, 4720.
Trost, B.M.; Toste, F.D.; Shen, H. *J. Am. Chem. Soc.* 2000, *122*, 2379.

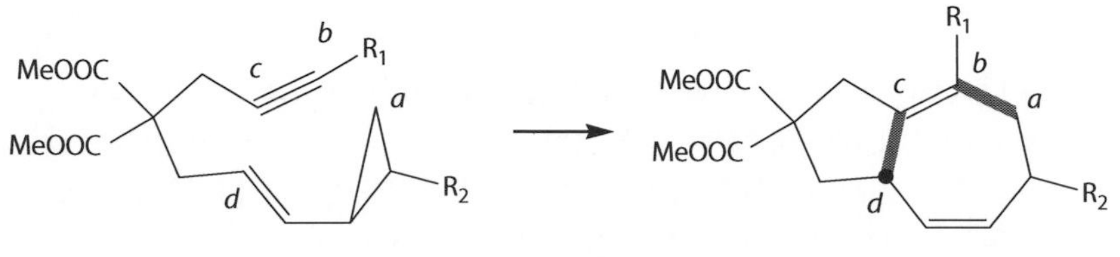

AE = 1; *E* = 0; *f*(sac) = 0
Set R$_1$ = CH$_3$

$$\Delta = \Delta(a) + \Delta(b) + \Delta(c) + \Delta(d) - (-2 - (-2)) + (0 - 0) + (0 - 0) + (-1 - (-1)) = 0; \quad \textbf{HI} = 0$$

Wurtz/Wurtz–Fittig reaction

Wurtz, A. *Ann. Chim. Phys.* 1855, *44*, 275.
Tollens, B.; Fittig, R. *Ann. Chem.* 1864, *131*, 303.

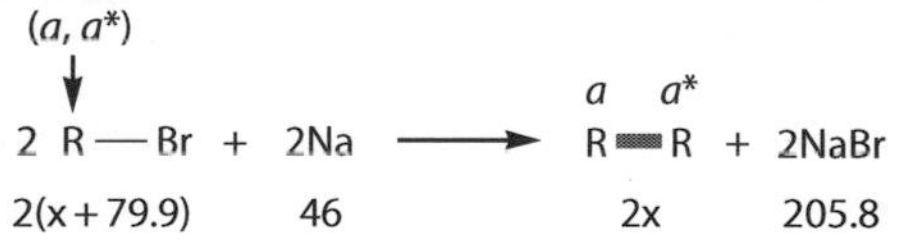

R = x

$$\text{AE} = 2x/[2x + 205.8]; \quad \textbf{AE(min) = 0.010}; \quad E = 205.8/2x; \quad \textbf{\textit{E}(max) = 102.9}$$

$$f(\text{sac}) = 46/[2x + 205.8]; \quad \textbf{\textit{f}(sac)(max) = 0.22}$$

Set R = CH$_3$

$$\Delta = \Delta(a) + \Delta(a^*) = 2(-3 - (-2)) = -2; \quad \textbf{HI} = -\Delta/(N+1) = +1$$

Zincke–Suhl reaction

Zincke, T.; Suhl, R. *Ber.* 1906, *39*, 4148.

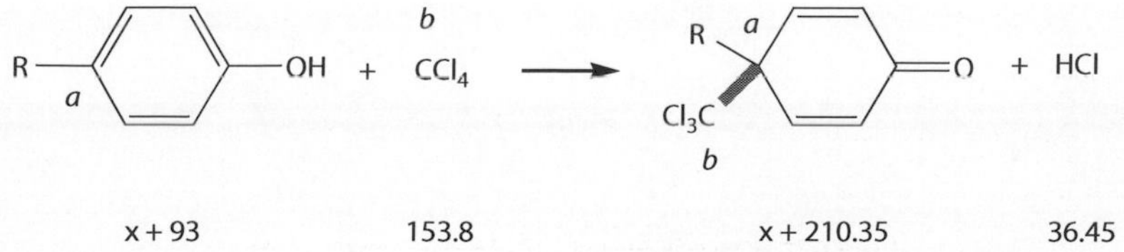

$$AE = [x + 210.35]/[x + 246.8]; \quad \textbf{AE(min)} = \textbf{0.85};$$

$$E = 36.45/[x + 210.35]; \quad \textbf{E(max)} = \textbf{0.17}; \quad f(\textbf{sac}) = \textbf{0}$$

Set R = CH$_3$

$$\Delta = \Delta(a) + \Delta(b) = (0 - 0) + (3 - 4) = -1; \quad \textbf{HI} = -\Delta/(N + 1) = +\tfrac{1}{2}$$

7.2 CATEGORIZATION OF NAMED CONDENSATION REACTIONS BY MINIMUM ATOM ECONOMY AND MAXIMUM ENVIRONMENTAL IMPACT FACTOR, MAXIMUM MOLECULAR WEIGHT FRACTION OF SACRIFICIAL REAGENTS, HYPSICITY INDEX

(Reactions are listed alphabetically.)

Acyloin condensation
Bouveault, L.; Loquin, R. *Compt. Rend.* 1905, *140*, 1593.

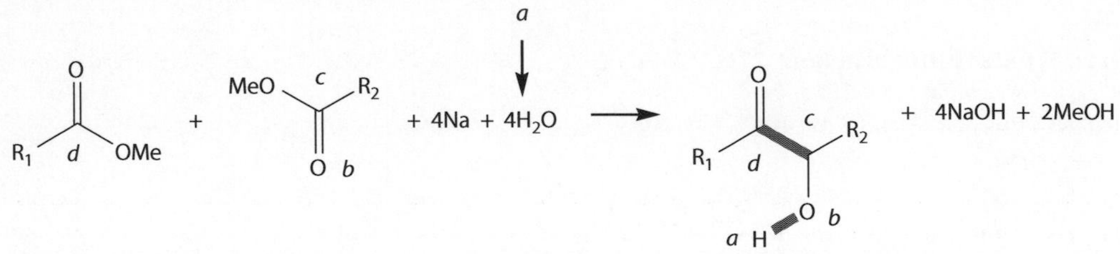

| x + 59 | y + 59 | 92 | 72 | x + y + 58 | 64 | 160 |

R$_1$ = x; R$_2$ = y

$$AE = [x + y + 58]/[x + y + 282]; \quad \textbf{AE(min)} = \textbf{0.21}; \quad E = 224/[x + y + 58]; \quad \textbf{E(max)} = \textbf{3.73}$$

$$f(\textbf{sac}) = 146/[x + y + 282]; \quad f(\textbf{sac})(\textbf{max}) = \textbf{0.51}$$

Set R$_1$ = R$_2$ = CH$_3$

$$\Delta = \Delta(a) + \Delta(b) + \Delta(c) + \Delta(d) = (1 - 1) + (-2 - (-2)) + (0 - 3) + (2 - 3) = -4; \quad \textbf{HI} = -\Delta/(N + 1) = +\textbf{2}$$

Aldol condensation
Kane, R. *Ann. Phys. Chem. Ser. 2* 1838, *44*, 475.
Kane, R. *J. Prakt. Chem.* 1838, *15*, 129.

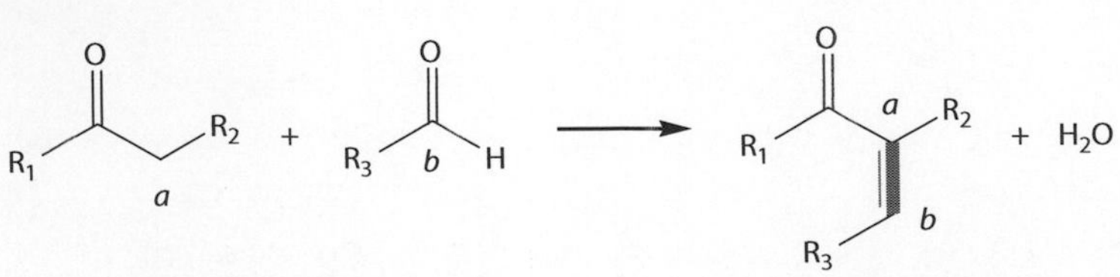

| x + y + 42 | z + 29 | x + y + z + 53 | 18 |

$R_1 = x$; $R_2 = y$; $R_3 = z$

$$AE = [x + y + z + 53]/[x + y + z + 71]; \quad \textbf{AE(min)} = \textbf{0.76};$$

$$E = 18/[x + y + z + 53]; \quad \textbf{E(max)} = \textbf{0.32}; \quad f(\textbf{sac}) = \textbf{0}$$

Set $R_2 = R_3 = CH_3$

$$\Delta = \Delta(a) + \Delta(b) = (0 - (-2)) + (-1 - 1) = 0; \quad \textbf{HI} = \textbf{0}$$

Bamberger–Goldschmidt synthesis of 1,2,4-triazines

Goldschmidt, H.; Rosell, Y. *Chem. Ber.* 1890, *23*, 487.
Goldschmidt, H.; Poltzer, A. *Chem. Ber.* 1891, *24*, 1000.
Bamberger, E.; Dieckmann, W. *Chem. Ber.* 1892, *15*, 534.

| $x + y + 116$ | $z + 29$ | $x + y + z + 127$ | 18 |

$X = x$(sum of substituents); $R_1 = y$; $R_2 = z$

$$AE = [x + y + z + 127]/[x + y + z + 145]; \quad \textbf{AE(min)} = \textbf{0.88};$$

$$E = 18/[x + y + z + 127]; \quad \textbf{E(max)} = \textbf{0.13}; \quad f(\textbf{sac}) = \textbf{0}$$

Set $R_1 = R_2 = CH_3$

$$\Delta = \Delta(a) + \Delta(b) + \Delta(c) = (-3 - (-3)) + (1 - 1) + (-2 - (-1)) = -1; \quad \textbf{HI} = -\Delta/(N+1) = +\tfrac{1}{2}$$

Bamberger–Goldschmidt synthesis of 1,3,5-triazines

Goldschmidt, H.; Rosell, Y. *Chem. Ber.* 1890, *23*, 487.
Goldschmidt, H.; Poltzer, A. *Chem. Ber.* 1891, *24*, 1000.
Bamberger, E.; Dieckmann, W. *Chem. Ber.* 1892, *15*, 534.

| $x + 43$ | $y + 43$ | | $x + y + z + 78$ |

$R_1 = x$; $R_2 = y$; $R_3 = z$

$$AE = [x + y + z + 78]/[x + y + z + 131]; \quad \textbf{AE(min)} = \textbf{0.60};$$

$$E = 53/[x + y + z + 78]; \quad \textbf{E(max)} = \textbf{0.65}; \quad f(\textbf{sac}) = \textbf{0}$$

Set $R_2 = R_3 = CH_3$

$$\Delta = \Delta(a) + \Delta(b) + \Delta(c) + \Delta(d) + \Delta(e)$$

$$= (-3 - (-3)) + (3 - 3) + (-3 - (-3)) + (3 - 3) + (-3 - (-3)) = 0; \quad \mathbf{HI = 0}$$

Benzoin condensation
Lapworth, A.J. *J. Chem. Soc.* 1903, *83*, 995.
Lapworth, A.J. *J. Chem. Soc.* 1904, *85*, 1206.

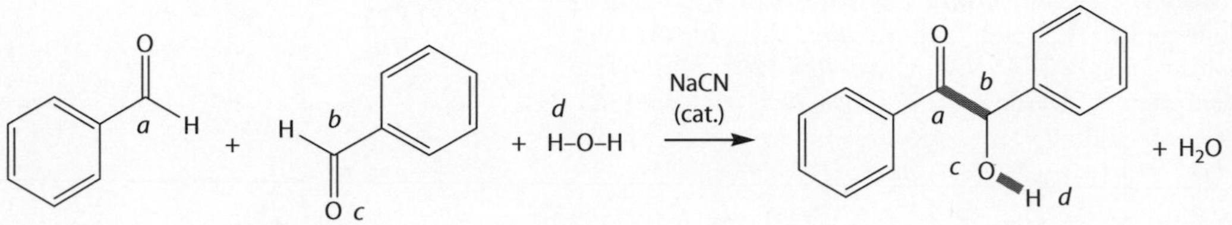

106   106   18   212   18

AE = 1; E = 0; f(sac) = 0

$$\Delta = \Delta(a) + \Delta(b) + \Delta(c) + \Delta(d) = (2-1) + (0-1) + (-2 - (-2)) + (1-1) = 0; \quad \mathbf{HI = 0}$$

Claisen condensation
Claisen, L.; Lowman, O. *Chem. Ber.* 1887, 20, 651.

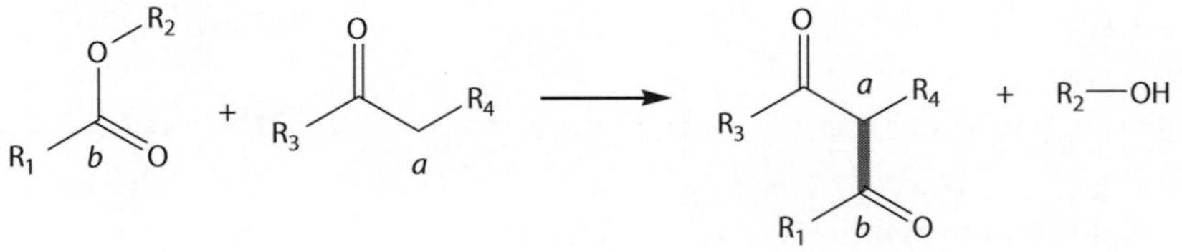

x + y + 44   z + w + 42   x + z + w + 69   y + 17

$R_1 = x;\ R_2 = y;\ R_3 = z;\ R_4 = w$

$$AE = [x + z + w + 69]/[x + y + z + w + 86]; \quad E = [y + 17]/[x + z + w + 69]; \quad f(sac) = 0$$

R_2	E(max)	AE(min)
Me	0.44	0.69
Et	0.64	0.61

Set $R_1 = R_4 = CH_3$

$$\Delta = \Delta(a) + \Delta(b) = (-1 - (-2)) + (2 - 3) = 0; \quad \mathbf{HI = 0}$$

Cyclic ether synthesis

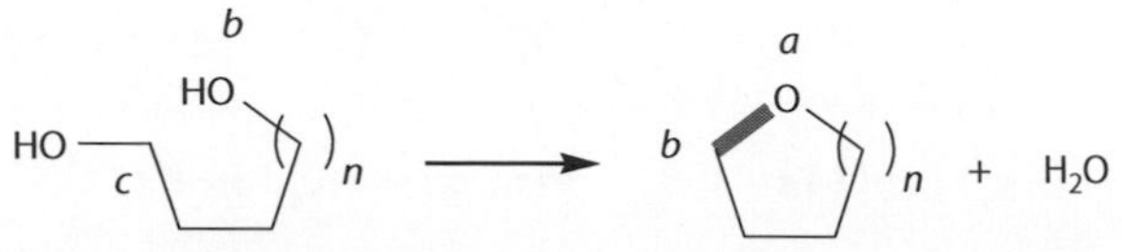

$$14n + 76 \qquad\qquad 14n + 58 \qquad\qquad 18$$

$$AE = [14n + 58]/[14n + 76]; \quad E = 18/[14n + 76]; \quad f(\text{sac}) = 0$$

n	Ring Size ($n + 4$)	E	AE
1	5	0.20	0.83
2	6	0.17	0.85
3	7	0.15	0.87
4	8	0.14	0.88

$$\Delta = \Delta(a) + \Delta(b) = (-2 - (-2)) + (-1 - (-1)) = 0; \quad \mathbf{HI = 0}$$

Darzens condensation
Darzens, G. *Compt. Rend.* 1904, *139*, 1214.

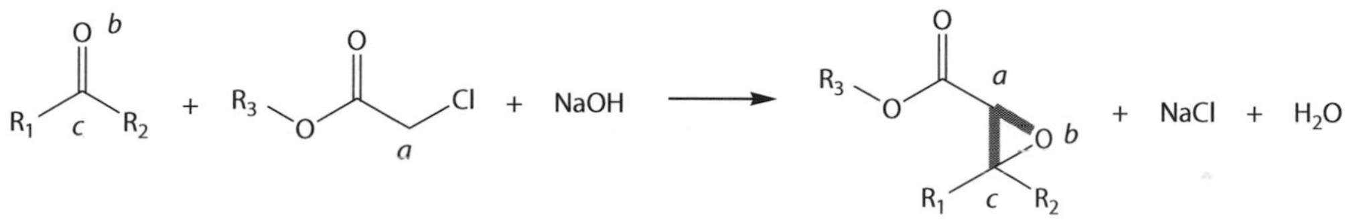

$$x + y + 28 \qquad z + 93.45 \qquad 40 \qquad\qquad x + y + z + 85 \qquad 58.45 \qquad 18$$

$$R_1 = x; R_2 = y; R_3 = z$$

$$AE = [x + y + z + 85]/[x + y + z + 161.45]; \quad \mathbf{AE(min) = 0.54};$$

$$E = 76.45/[x + y + z + 85]; \quad \mathbf{E(max) = 0.87}$$

$$f(\text{sac}) = 40/[x + y + z + 161.45]; \quad \mathbf{f(\text{sac}) = 0.24}$$

Set $R_1 = R_2 = CH_3$

$$\Delta = \Delta(a) + \Delta(b) + \Delta(c) = (0 - (-1)) + (-2 - (-2)) + (1 - 2) = 0; \quad \mathbf{HI = 0}$$

Dieckmann condensation
Dieckmann, W. *Chem. Ber.* 1894, *27*, 102.

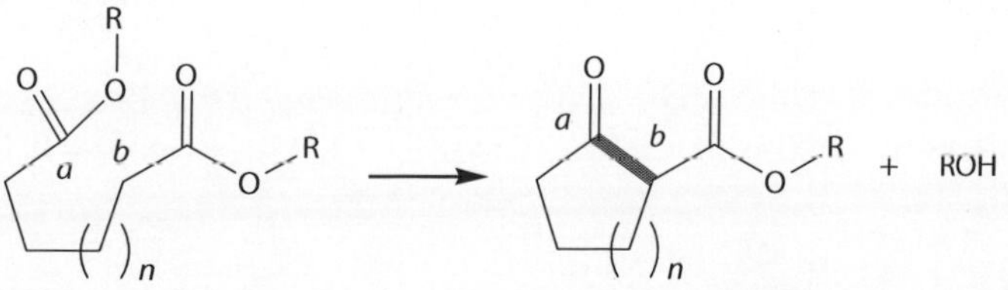

$$14n + x + 130 \qquad\qquad 14n + x + 113 \qquad x + 17$$

R = x

$$AE = [x + 14n + 113]/[x + 14n + 130]; \quad E = [x + 17]/[x + 14n + 113]$$

$$AE(min) = [14n + 114]/[14n + 131]; \quad E(max) = 18/[14n + 114]; \quad f(sac) = 0$$

n	Ring Size (n + 4)	E(max)	AE(min)
1	5	0.14	0.88
2	6	0.13	0.89
3	7	0.12	0.90

$$\Delta = \Delta(a) + \Delta(b) = (2 - 3) + (-1 - (-2)) = 0; \quad \textbf{HI} = \textbf{0}$$

Hinsberg thiophene synthesis
Hinsberg, O. *Chem. Ber.* 1910, *43*, 901.

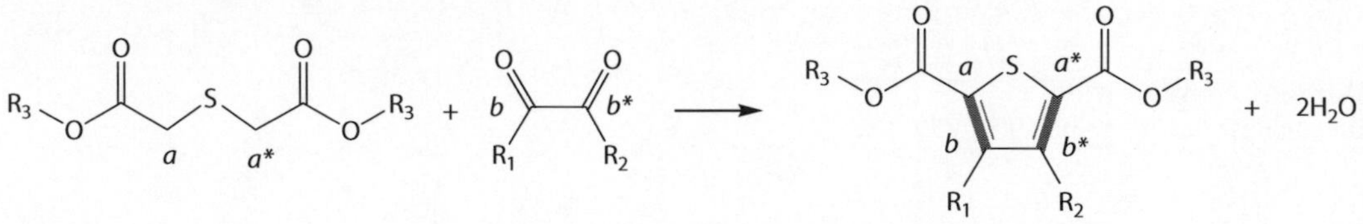

| 2z + 148 | x + y + 56 | x + y + 2z + 168 | 36 |

$R_1 = x; R_2 = y; R_3 = z$

$$AE = [x + y + 2z + 168]/[x + y + 2z + 204]; \quad \textbf{AE} = \textbf{0.82};$$

$$E = 36/[x + y + 2z + 168]; \quad \textbf{E(max)} = \textbf{0.21}; \quad f(sac) = 0$$

Set $R_1 = R_2 = CH_3$

$$\Delta = \Delta(a) + \Delta(a^*) + \Delta(b) + \Delta(b^*) = 2(1 - (-1)) + 2(0 - 2) = 0; \quad \textbf{HI} = \textbf{0}$$

Knoevenagel condensation
Knoevenagel, E. *Chem. Ber.* 1898, *31*, 2596.

| x + 29 | 160 | x + 171 | 18 |

R = x

$$AE = [x + 171]/[x + 189]; \quad \textbf{AE(min)} = \textbf{0.91}; \quad E = 18/[x + 171]; \quad \textbf{E(max)} = \textbf{0.10}; \quad f(sac) = 0$$

Set R = CH_3

$$\Delta = \Delta(a) + \Delta(b) = (0 - (-2)) + (-1 - 1) = 0; \quad \textbf{HI} = \textbf{0}$$

Knorr pyrrole synthesis

Knorr, L. *Ann. Chem.* 1886, *236*, 290.
Knorr, L. *Chem. Ber.* 1884, *17*, 1635.

x + 130   y + 115   x + y + 209   36

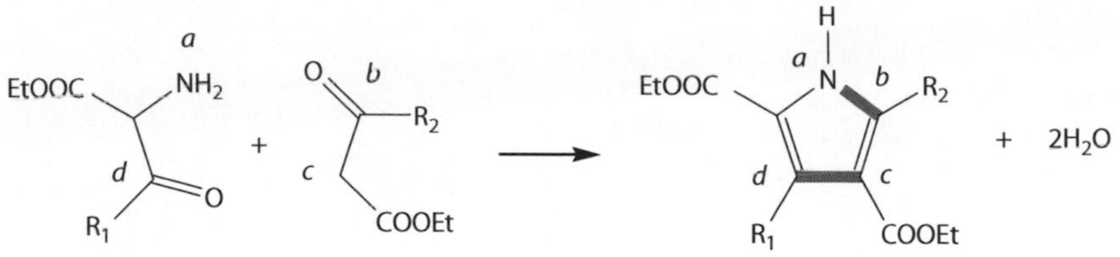

$R_1 = x; R_2 = y$

$$AE = [x + y + 209]/[x + y + 245]; \quad \mathbf{AE(min) = 0.85};$$

$$E = 36/[x + y + 209]; \quad \boldsymbol{E}\mathbf{(max) = 0.17}; \quad \boldsymbol{f}\mathbf{(sac) = 0}$$

Set $R_1 = R_2 = CH_3$

$$\Delta = \Delta(a) + \Delta(b) + \Delta(c) + \Delta(d) = (-3 - (-3)) + (1 - 2) + (0 - (-2)) + (1 - 2) = 0; \quad \mathbf{HI = 0}$$

Mukaiyama aldol reaction

Mukaiyama, T. *Chem. Lett.* 1973, *2*, 1101.
Mukaiyama, T.; Banno, K.; Narasaka, K. *J. Am. Chem. Soc.* 1974, *96*, 7503.

2x + 128   y + 43   189.7   54   2x + y + 99   90   145.8   79.9

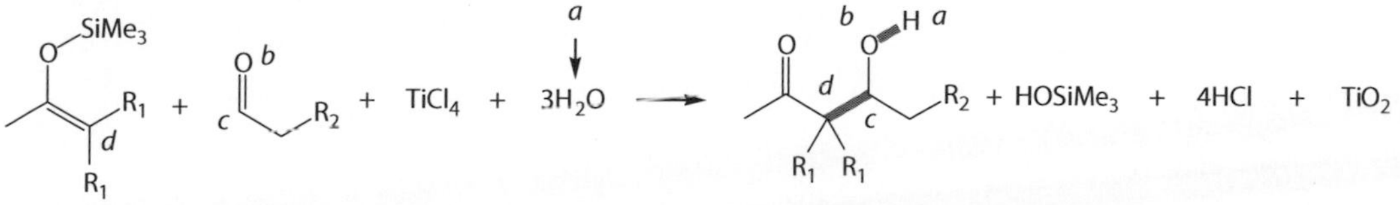

$R_1 = x; R_2 = y$

$$AE = [2x + y + 99]/[2x + y + 414.7]; \quad \mathbf{AE(min) = 0.24};$$

$$E = 315.7/[2x + y + 99]; \quad \boldsymbol{E}\mathbf{(max) = 3.10}$$

$$f(\text{sac}) = 225.7/[2x + y + 414.7]; \quad \boldsymbol{f}\mathbf{(sac)(max) = 0.54}$$

Set $R_1 = CH_3$

$$\Delta = \Delta(a) + \Delta(b) + \Delta(c) + \Delta(d) = (1-1) + (-2-(-2)) + (0-1) + (0-0) = -1;$$

$$\mathbf{HI} = -\Delta/(N+1) = +\tfrac{1}{2}$$

Nenitzescu indole synthesis
Nenitzescu, C.D. *Bull. Soc. Chim. Romania* 1929, *11*, 37.

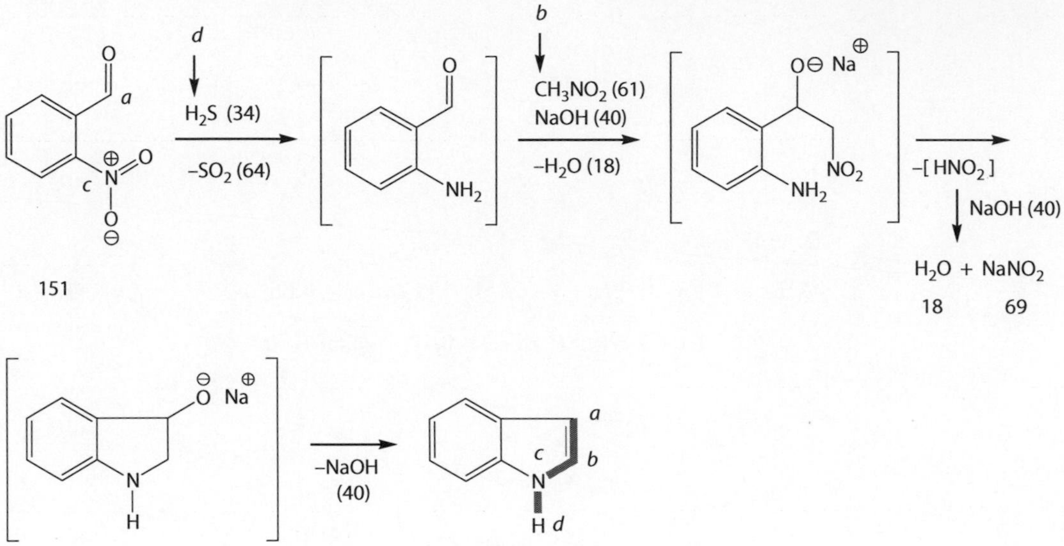

AE = 0.41; *E* = 169/117 = 1.44; *f*(sac) = 40/286 = 0.14

$$\Delta = \Delta(a) + \Delta(b) + \Delta(c) + \Delta(d) = (-1-1) + (0-(-2)) + (-3-3) + (1-1) = -6;$$

$$\mathbf{HI} = -\Delta/(N+1) = +3$$

von Pechmann condensation
von Pechmann, H.; Duisberg, C. *Chem. Ber.* 1883, *16*, 2119.

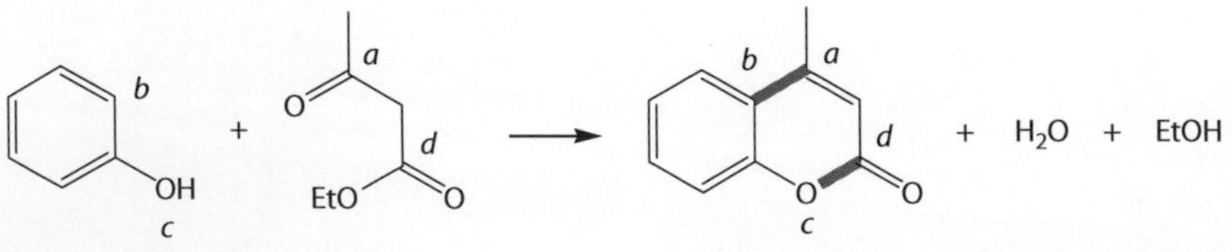

AE = 0.71; *E* = 64/160 = 0.4 ; *f*(sac) = 0

$$\Delta = \Delta(a) + \Delta(b) + \Delta(c) + \Delta(d) = (0-2) + (0-(-1)) + (-2-(-2)) + (3-3) = -1;$$

$$\mathbf{HI} = \Delta/(N+1) = +\tfrac{1}{2}$$

Pellizzari reaction

Pellizzari, G. *Gazz. Chim. Ital*. 1911, *41(II)*, 20.

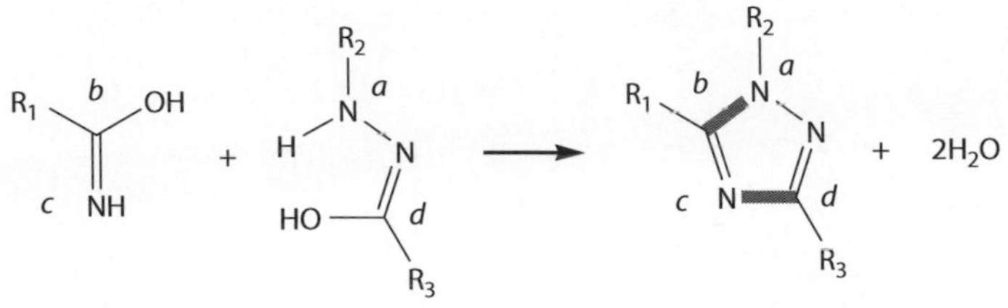

x + 44 x + z + 58 x + y + z + 66 36

$R_1 = x; R_2 = y; R_3 = z$

$$AE = [x+y+z+66]/[x+y+z+102]; \quad \mathbf{AE(min) = 0.66};$$

$$E = 36/[x+y+z+66]; \quad \boldsymbol{E}\mathbf{(max) = 0.52}; \quad \boldsymbol{f}\mathbf{(sac) = 0}$$

Set $R_1 = R_2 = R_3 = CH_3$

$$\Delta = \Delta(a)+\Delta(b)+\Delta(c)+\Delta(d) = (-3-(-3))+(3-3)+(-3-(-3))+(3-3) = 0; \quad \mathbf{HI = 0}$$

Pictet–Spengler isoquinoline synthesis

Pictet, A.; Spengler, T. *Chem. Ber*. 1911, *44*, 2030.

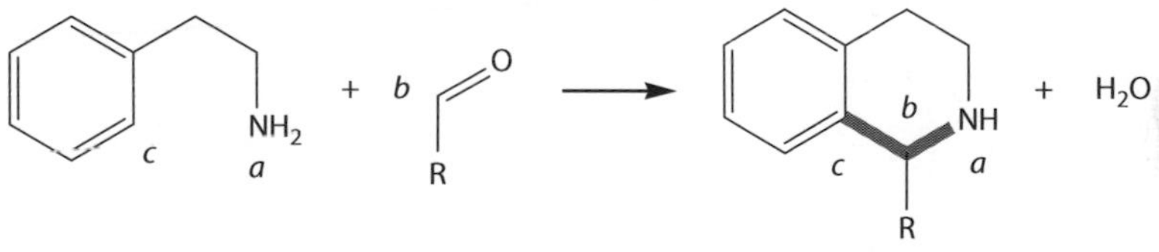

121 x + 29 x + 132 18

R = x

$$AE = [x+132]/[x+150]; \quad \mathbf{AE(min) = 0.88}; \quad E = 18/[x+132]; \quad \boldsymbol{E}\mathbf{(max) = 0.14}; \quad \boldsymbol{f}\mathbf{(sac) = 0}$$

Set $R = CH_3$

$$\Delta = \Delta(a)+\Delta(b)+\Delta(c) = (-3-(-3))+(0-1)+(0-(-1)) = 0; \quad \mathbf{HI = 0}$$

Schiff base imine condensation

Schiff, H. *Ann. Chem*. 1864, *131*, 118.
Schiff, H. *Ann. Chem*. 1866, *140*, 92.

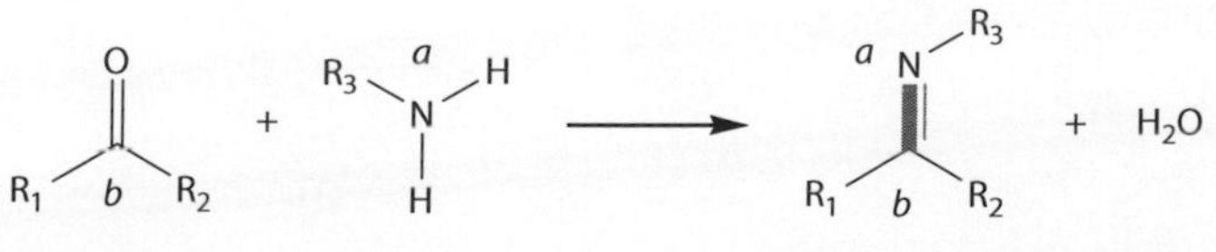

x + y + 28 z + 16 x + y + z + 26 18

$R_1 = x; R_2 = y; R_3 = z$

$$AE = [x + y + z + 26]/[x + y + z + 44]; \quad \textbf{AE(min)} = \textbf{0.62};$$

$$E = 18/[x + y + z + 26]; \quad \textbf{E(max)} = \textbf{0.62}; \quad \textbf{f(sac)} = \textbf{0}$$

Set $R_1 = R_2 = R_3 = CH_3$

$$\Delta = \Delta(a) + \Delta(b) = (-3 - (-3)) + (2 - 2) = 0; \quad \textbf{HI} = \textbf{0}$$

Stobbe condensation
Stobbe, H. *Chem. Ber.* 1893, *26*, 2312.

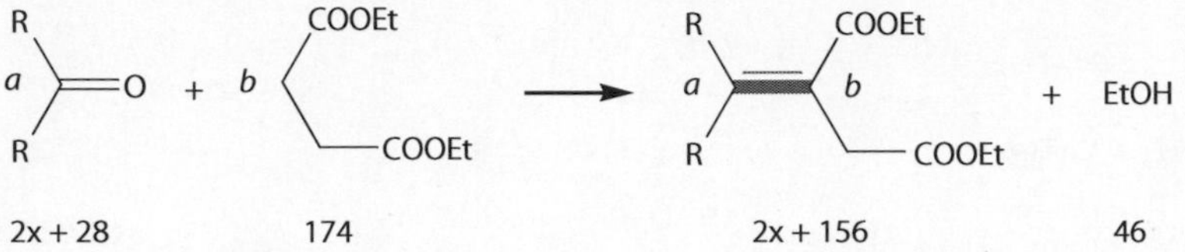

$$AE = [2x + 156]/[2x + 202]; \quad \textbf{AE(min)} = \textbf{0.77};$$

$$E = 46/[2x + 156]; \quad \textbf{E(max)} = \textbf{0.29}; \quad \textbf{f(sac)} = \textbf{0}$$

Set $R = CH_3$

$$\Delta = \Delta(a) + \Delta(b) = (0 - 2) + (0 - (-2)) = 0; \quad \textbf{HI} = \textbf{0}$$

7.3 CATEGORIZATION OF NAMED ELIMINATION/FRAGMENTATION REACTIONS BY MINIMUM ATOM ECONOMY AND MAXIMUM ENVIRONMENTAL IMPACT FACTOR, MAXIMUM MOLECULAR WEIGHT FRACTION OF SACRIFICIAL REAGENTS, HYPSICITY INDEX

(Reactions are listed alphabetically.)

Azetidine synthesis
Howard, C.C.; Marckwald, W. *Chem. Ber.* 1899, *32*, 2031.

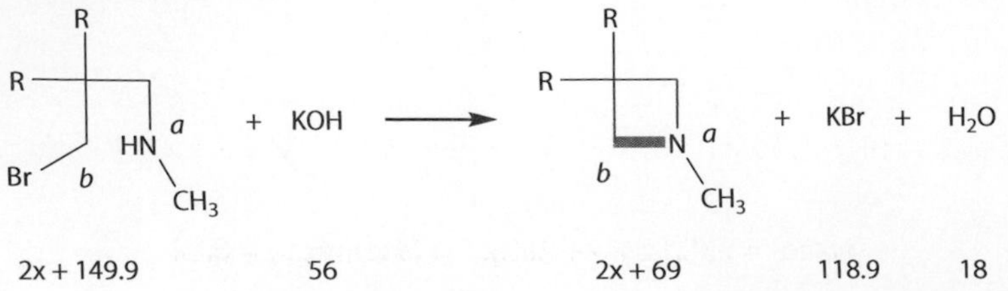

R = x

$$AE = [2x + 69]/[2x + 205.9]; \quad \textbf{AE(min)} = \textbf{0.34}; \quad E = 136.9/[2x + 69]; \quad \textbf{E(max)} = \textbf{1.93}$$

$$f(\text{sac}) = 56/[2x + 205.9]; \quad f(\textbf{sac})(\textbf{max}) = \textbf{0.27}$$

Set R = CH$_3$

$$\Delta = \Delta(a) + \Delta(b) = (-3 - (-3)) + (-1 - (-1)) = 0; \quad \textbf{HI} = \textbf{0}$$

Borodine–Hunsdiecker reaction

Borodine, A. *Ann. Chem.* 1861, *119*, 121.
Hunsdiecker, H.; Hunsdiecker, C. *Chem. Ber.* 1942, *75*, 291.

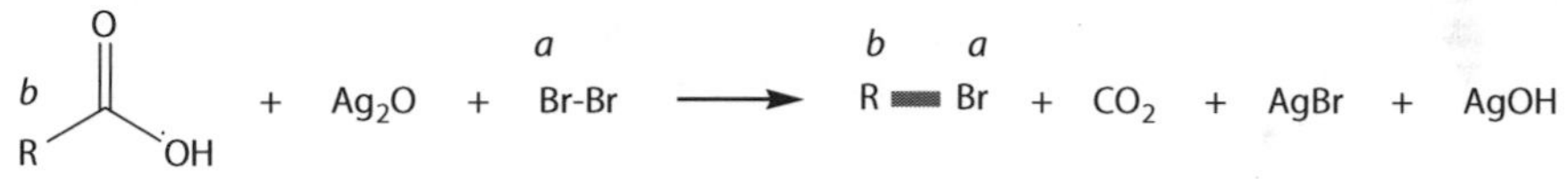

x + 45	232	159.8	x + 79.9	44	187.9	125

R = x

$$AE = [x + 79.9]/[x + 436.8]; \quad \textbf{AE(min)} = \textbf{0.18}; \quad E = 356.9/[x + 79.9]; \quad \textbf{E(max)} = \textbf{4.41}$$

$$f(\text{sac}) = 232/[x + 436.8]; \quad f(\textbf{sac})(\textbf{max}) = \textbf{0.53}$$

Set R = CH$_3$

$$\Delta = \Delta(a) + \Delta(b) = (-1 - 0) + (-2 - (-3)) = 0; \quad \textbf{HI} = \textbf{0}$$

Burgess dehydration

Burgess, E.M. *J. Org. Chem.* 1973, *38*, 26.

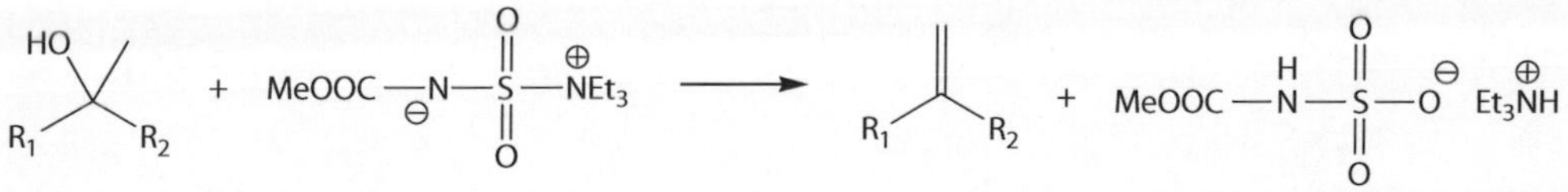

x + y + 44	238	x + y + 26	256

R$_1$ = x; R$_2$ = y

$$AE = [x + y + 26]/[x + y + 282]; \quad \textbf{AE(min)} = \textbf{0.099}; \quad E = 256/[x + y + 26]; \quad \textbf{E(max)} = \textbf{9.14}$$

$$f(\text{sac}) = 238/[x + y + 282]; \quad f(\textbf{sac})(\textbf{max}) = \textbf{0.84}$$

Chugaev reaction

Tschugaeff, L. *Chem. Ber.* 1899, *32*, 3332.

x + y + 134 x + y + 26 60 48

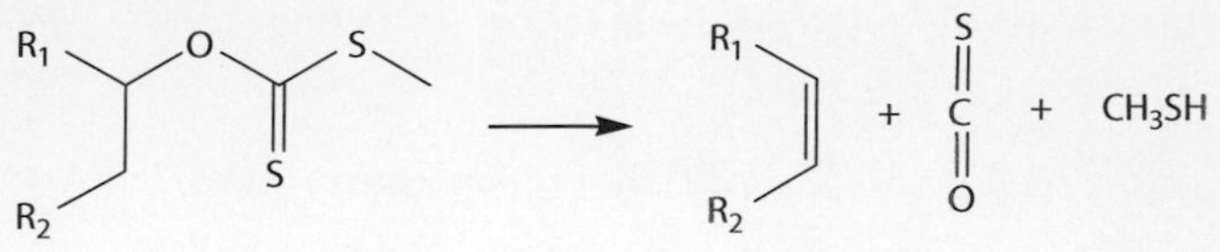

$R_1 = x; R_2 = y$

$$AE = [x + y + 26]/[x + y + 134]; \quad \mathbf{AE(min) = 0.21}; \quad E = 108/[x + y + 26]; \quad \boldsymbol{E}\mathbf{(max) = 3.86}$$

***f*(sac) = 0**

Cope elimination

Cope, A.C.; Foster, T.T.; Toule, P.H. *J. Am. Chem. Soc.* 1949, *71*, 3929.

x + y + z + 86 x + y + z + 25 61

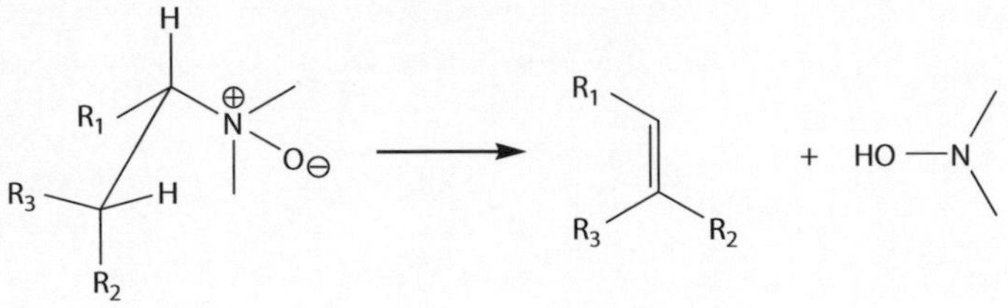

$R_1 = x; R_2 = y; R_3 = z$

$$AE = [x + y + z + 25]/[x + y + z + 86]; \quad \mathbf{AE(min) = 0.31};$$

$$E = 61/[x + y + z + 25]; \quad \boldsymbol{E}\mathbf{(max) = 2.18}$$

***f*(sac) = 0**

Dakin reaction

Dakin, H.D. *Org. Synth. Coll. Vol. I* 1941, *1*, 149.

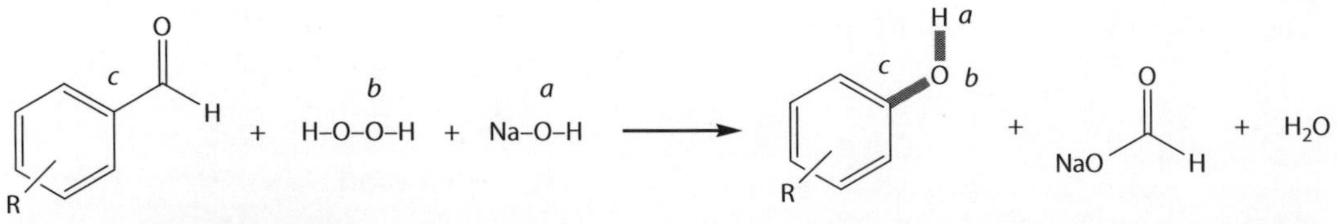

x + 101 34 40 x + 89 68 18

R = x(sum of substituents)

$$AE = [x + 89]/[x + 175]; \quad \mathbf{AE(min) = 0.52}; \quad E = 86/[x + 89]; \quad \boldsymbol{E}\mathbf{(max) = 0.91}$$

***f*(sac) = 0**

$$\Delta = \Delta(a) + \Delta(b) + \Delta(c) = (1 - 1) + (-2 - (-1)) + (1 - 0) = 0; \quad \mathbf{HI = 0}$$

Edman degradation

Edman, P. *Acta Chem. Scand.* 1950, *4*, 283.

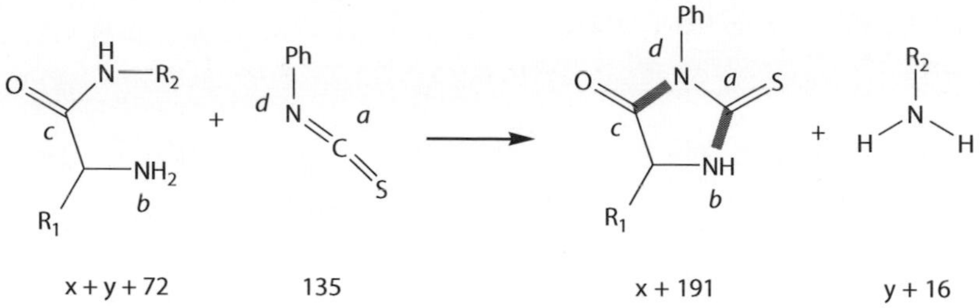

$$\text{AE} = [x + 191]/[x + y + 207]; \quad \textbf{AE(min)} = \textbf{0}; \quad E = [y + 16]/[x + 191]; \quad \textbf{E(max)} = \textbf{infinity}$$

$$f(\text{sac}) = \textbf{0}$$

$$\Delta = \Delta(a) + \Delta(b) + \Delta(c) + \Delta(d) = (4 - 4) + (-3 - (-3)) + (3 - 3) + (-3 - (-3)) = 0; \quad \textbf{HI} = \textbf{0}$$

Grob fragmentation

Grob, C.A., Baumann, W. *Helv. Chim. Acta* 1955, *38*, 594.
Grob, C.A. *Angew. Chem. Int. Engl. Ed.* 1969, *8*, 535.

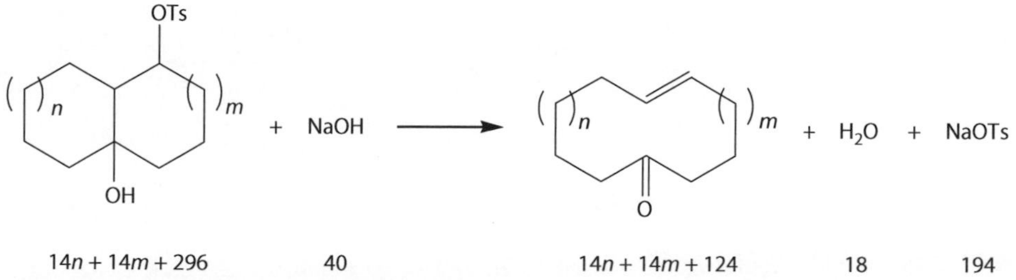

$$\text{AE} = [14n + 14m + 124]/[14n + 14m + 336]; \quad E = 212/[14n + 14m + 124];$$

$$f(\text{sac}) = 40/[14n + 14m + 336]$$

n	m	Ring Size in Product	E	AE	$f(\text{sac})$
1	1	10	1.39	0.42	0.11
1	2	11	1.28	0.44	0.11
1	3	12	1.18	0.46	0.10

Haloform reaction

Gray, A.R.; Walker, J.T.; Fuson, R.C. *J. Am. Chem. Soc.* 1931, *53*, 3494.
Fuson, R.C.; Farlow, M.W.; Stehman, C.J. *J. Am. Chem. Soc.* 1931, *53*, 4097.
Bartlett, P.D. *J. Am. Chem. Soc.* 1934, *56*, 967.
Fuson, R.C.; Bull, B.A. *Chem. Rev.* 1934, *15*, 275.

z + 43   160   6x   z + 67   3(23 + x)   3x + 13   54

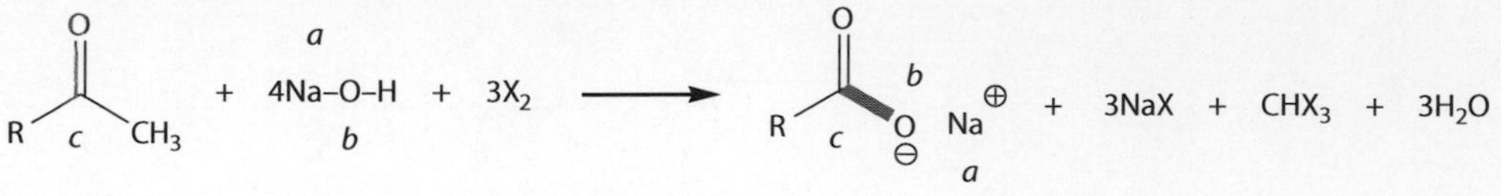

R = z; X = x

$$AE = [z + 67]/[6x + z + 203]; \quad E = [6x + 136]/[z + 67]; \quad f(\text{sac}) = 6x/[6x + z + 203]$$

X	*E*(max)	AE(min)	*f*(sac)(max)
Cl	5.12	0.16	0.51
Br	9.05	0.010	0.70
I	13.21	0.070	0.79

Set R = CH_3

$$\Delta = \Delta(a) + \Delta(b) + \Delta(c) = (1-1) + (-2-(-2)) + (3-2) = +1; \quad \mathbf{HI = -\Delta/(N+1) = -½}$$

Hofmann degradation

Hofmann, A.W. *Chem. Ber.* 1881, *14*, 659.

x + y + 43   426   x + y + 26

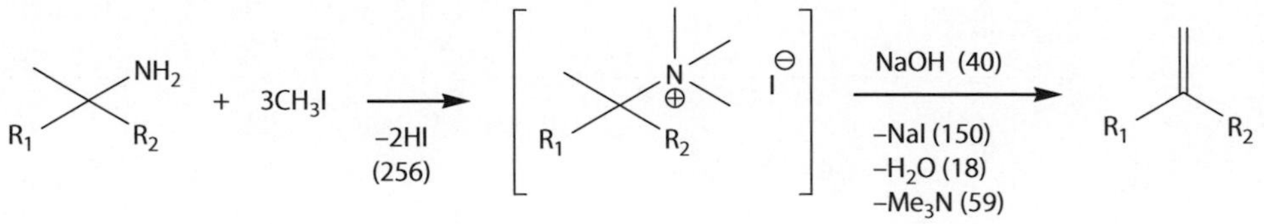

R_1 = x; R_2 = y

$$AE = [x + y + 26]/[x + y + 509]; \quad \mathbf{AE(min) = 0.055}; \quad E = 483/[x + y + 26]; \quad \mathbf{\mathit{E}(max) = 17.25}$$

$$f(\text{sac}) = 466/[x + y + 509]; \quad \mathbf{\mathit{f}(sac)(max) = 0.91}$$

Kochi reaction

Kochi, J.K. *J. Am. Chem. Soc.* 1965, *87*, 2500.

x + 45   443.2   42.45   x + 35.45   325.2   44   60   66

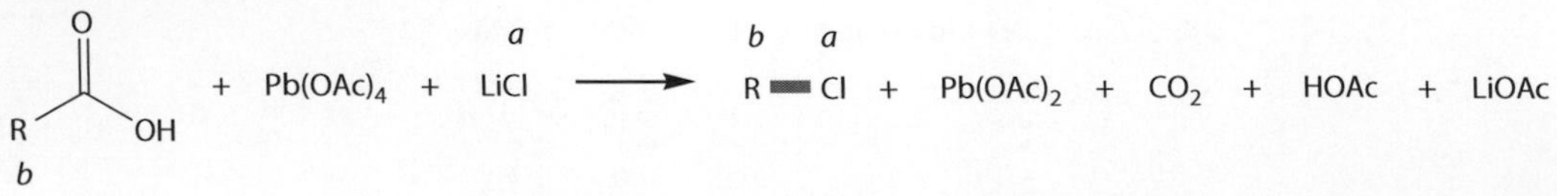

R = x

$$AE = [x + 35.45]/[x + 530.45]; \quad \textbf{AE(min)} = \textbf{0.069}; \quad E = 495/[x + 35.45]; \quad \textbf{E(max)} = \textbf{13.58}$$

$$f(\text{sac}) = 443.2/[x + 530.45]; \quad f(\textbf{sac})(\textbf{max}) = \textbf{0.83}$$

Set R = CH_3

$$\Delta = \Delta(a) + \Delta(b) = (-1 - (-1)) + (-2 - (-3)) = +1; \quad \textbf{HI} = -\Delta/(N+1) = -\tfrac{1}{2}$$

Lossen reaction
Lossen, W. *Ann. Chem.* 1872, *161*, 347.

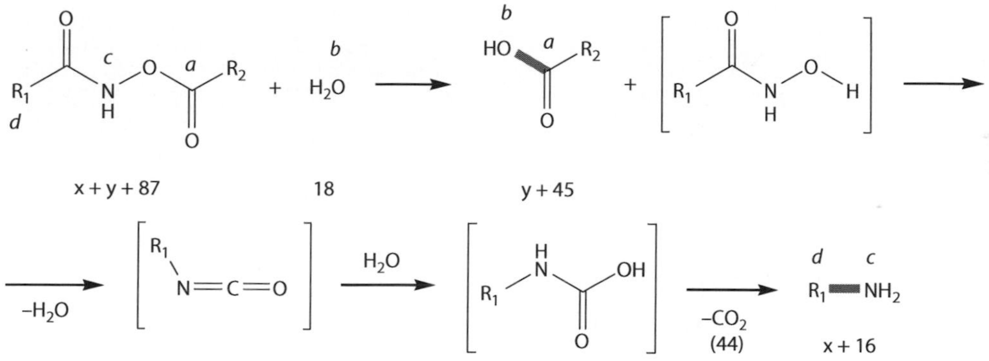

| $x + y + 87$ | | 18 | | $y + 45$ |

$R_1 = x; R_2 = y$

Target product: Carboxylic acid

$$AE = [y + 45]/[x + y + 105]; \quad \textbf{AE(min)} = \textbf{46}/[x + 106];$$

$$E = [x + 60]/[y + 45]; \quad \textbf{E(max)} = [x + 60]/46$$

$f(\textbf{sac}) = \textbf{0}$

R_1	E(max)	AE(min)
H	1.33	0.43
Me	1.63	0.38
Et	1.93	0.34

Set R_2 = CH_3

$$\Delta = \Delta(a) + \Delta(b) = (3 - 3) + (-2 - (-2)) = 0; \quad \textbf{HI} = \textbf{0}$$

Target product: Amine

$$AE = [x + 16]/[x + y + 105]; \quad \textbf{AE(min)} = \textbf{17}/[y + 106];$$

$$E = [y + 89]/[x + 16]; \quad \textbf{E(max)} = [y + 89]/17$$

$$f(\text{sac}) = 18/[x + y + 87]$$

R_2	E(max)	AE(min)	f(sac)(R_1 = Me)
H	5.29	0.16	0.17
Me	6.12	0.14	0.15
Et	6.94	0.13	0.14

Set $R_1 = CH_3$

$$\Delta = \Delta(c) + \Delta(d) = (-3 - (-1)) + (-2 - (-3)) = -1; \quad HI = -\Delta/(N+1) = +\tfrac{1}{2}$$

McFadyen–Stevens reaction
McFadyen, J.S.; Stevens, T.S. *J. Chem. Soc.* 1936, 584.

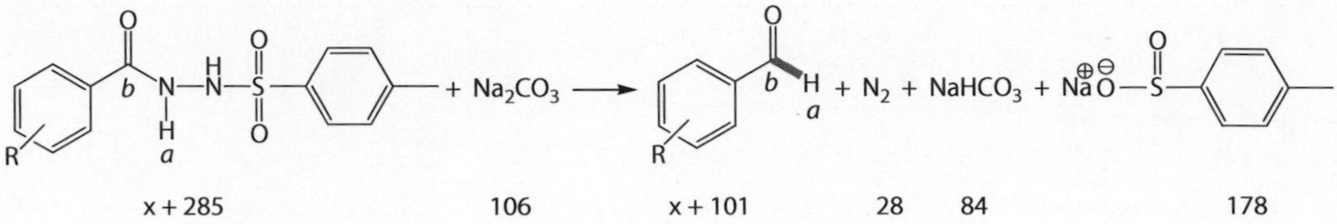

| x + 285 | 106 | x + 101 | 28 | 84 | 178 |

$R = x$(sum of substituents)

$$AE = [x + 101]/[x + 391]; \quad \mathbf{AE(min) = 0.27}; \quad E = 290/[x + 101]; \quad \mathbf{E(max) = 2.74}$$

$$f(sac) = 106/[x + 391]; \quad \mathbf{f(sac)(max) = 0.27}$$

$$\Delta = \Delta(a) + \Delta(b) = (1 - 1) + (1 - 3) = -2; \quad \mathbf{HI = -\Delta/(N+1) = +1}$$

Norrish type II reaction
Bamford, C.H.; Norrish, R.G.W. *J. Chem. Soc.* 1935, 1504.
Norrish, R.G.W.; Bamford, C.H. *Nature* 1936, *138*, 1016.
Norrish, R.G.W.; Bamford, C.H. *Nature* 1937, *140*, 195.
Norrish, R.G.W. *Trans. Faraday Soc.* 1937, *33*, 1521.

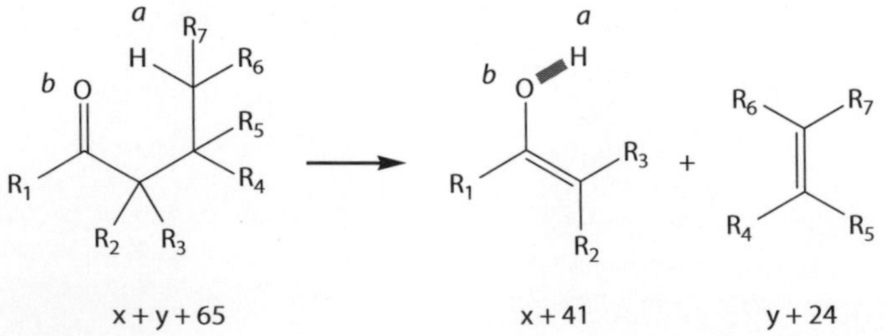

| x + y + 65 | x + 41 | y + 24 |

$R_1 + R_2 + R_3 = x$; $R_4 + R_5 + R_6 + R_7 = y$

f(sac) = 0

Target product: Enol

$$AE = [x + 41]/[x + y + 65]; \quad E = [y + 24]/[x + 41]$$

$$\Delta = \Delta(a) + \Delta(b) = (1 - 1) + (-2 - (-2)) = 0; \quad \mathbf{HI = 0}$$

Target product: Olefin

$$AE = [y + 24]/[x + y + 65]; \quad E = [x + 41]/[y + 24]$$

Pyrolysis of sulfoxides
DePuy, C.H.; King, R.W. *Chem. Rev.* 1960, *60*, 431.
Kingsbury, C.A.; Cram, D.J. *J. Am. Chem. Soc.* 1960, *82*, 1810.

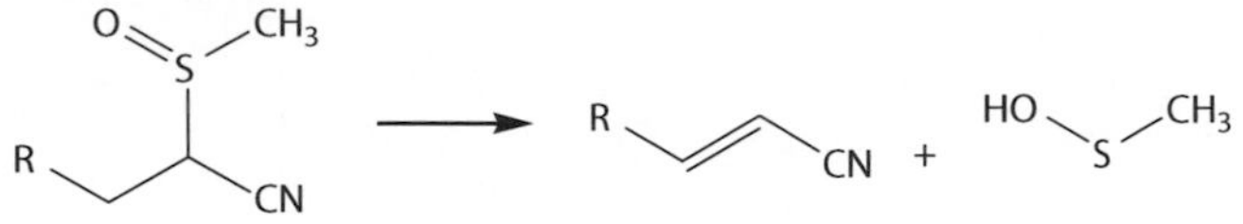

| x + 116 | x + 52 | 64 |

R = x

$$AE = [x + 52]/[x + 116]; \quad \textbf{AE(min)} = \textbf{0.45}; \quad E = 64/[x + 52]; \quad \textbf{E(max)} = \textbf{1.21}$$

f(sac) = 0

Ring closing metathesis reaction
Grubbs, R.H.; Miller, S.J.; Fu, G.C. *Acc. Chem. Res.* 1995, *28*, 446.
Schrock, R.R.; Murdzek, J.S.; Bazan, G.C.; Robbins, J.; DiMare, M.; O'Reagan, M. *J. Am. Chem. Soc.* 1990, *112*, 3875.

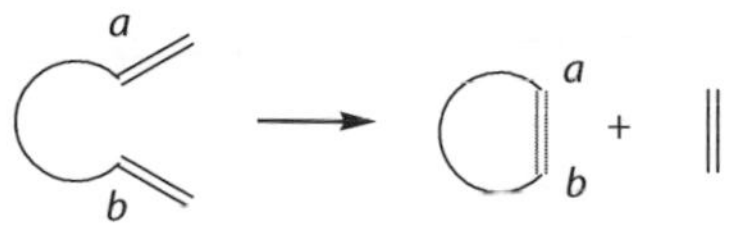

| x + 54 | x + 26 | 28 |

$$AE = [x + 26]/[x + 54]; \quad E = 28/[x + 26]; \quad f(\textbf{sac}) = \textbf{0}$$

X	Ring Size	*E*(max)	AE(min)
CH_2	3	0.70	0.59
$(CH_2)_2$	4	0.52	0.66
$(CH_2)_3$	5	0.41	0.71
$(CH_2)_4$	6	0.34	0.75

$$\Delta = \Delta(a) + \Delta(b) = (-1 - (-1)) + (-1 - (-1)) = 0; \quad \textbf{HI} = \textbf{0}$$

Ruff–Fenton degradation
Ruff, O. *Chem. Ber.* 1898, *31*, 1573.
Fenton, H.J.H. *Proc. Chem. Soc.* 1893, *9*, 113.

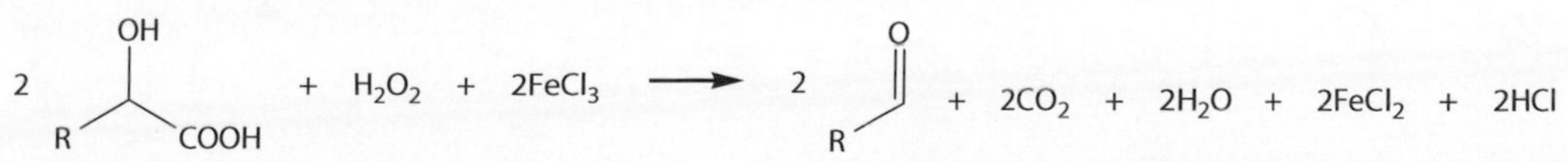

| 2(x + 75) | 34 | 324.7 | 2(x + 29) | 88 | 36 | 253.8 | 72.9 |

R = x

$$AE = [2(x + 29)]/[2(x + 29) + 450.7]; \quad \mathbf{AE(min) = 0.12}; \quad E = 450.7/[2(x + 29)]; \quad \mathbf{E(max) = 7.51}$$

$$f(sac) = 358.7/[2(x + 29) + 450.7]; \quad \mathbf{f(sac)(max) = 0.70}$$

Tiffeneau–Demjanov reaction

Tiffeneau, M.; Weill; Tschoubar, B. *Compt. Rend.* 1937, *205*, 54.
Demjanov, N.J.; Lushnikov, M. *J. Russ. Phys. Chem. Soc.* 1903, *35*, 26.

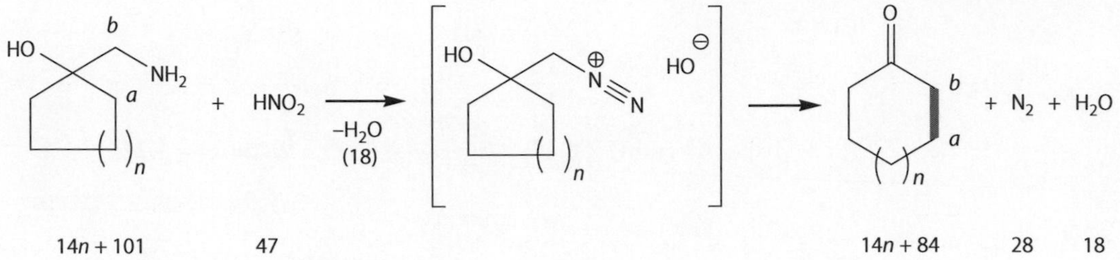

14n + 101   47   14n + 84   28   18

$$AE = [14n + 84]/[14n + 148]; \quad E = 64/[14n + 84]; \quad f(sac) = 47/[14n + 84]$$

n	Ring Size (n + 5)	E(max)	AE(min)	f(sac)
1	6	0.65	0.60	0.48
2	7	0.57	0.64	0.42
3	8	0.51	0.66	0.37

$$\Delta = \Delta(a) + \Delta(b) = (-2 - (-2)) + (-2 - (-1)) = -1; \quad \mathbf{HI} = -\Delta/(N + 1) = +\tfrac{1}{2}$$

7.4 CATEGORIZATION OF MULTICOMPONENT REACTIONS BY MINIMUM ATOM ECONOMY AND MAXIMUM ENVIRONMENTAL IMPACT FACTOR, MAXIMUM MOLECULAR WEIGHT FRACTION OF SACRIFICIAL REAGENTS, HYPSICITY INDEX

(Reactions are listed alphabetically.)

β-Acetoamido carbonyl compound synthesis

Rao, I.N.; Prabhakaran, E.N.; Das, S.K.; Iqbal, J. *J. Org. Chem.* 2003, *68*, 4079.

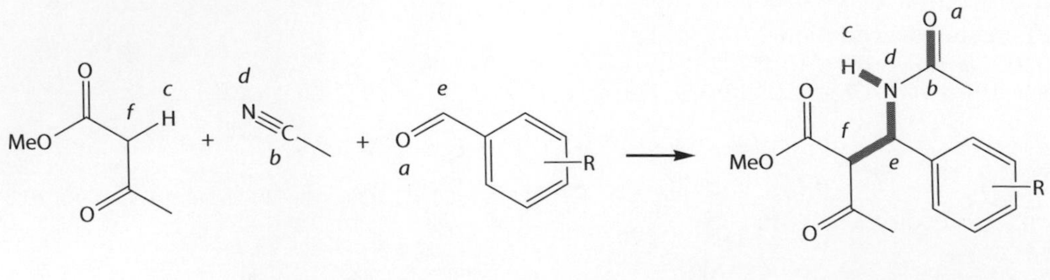

$AE = 1; E = 0; f(sac) = 0$

$$\Delta = \Delta(a) + \Delta(b) + \Delta(c) + \Delta(d) + \Delta(e) + \Delta(f)$$

$$= (-2 - (-2)) + (3 - 3) + (1 - 1) + (-3 - (-3)) + (0 - 1) + (-1 - (-2)) = 0; \quad HI = 0$$

Aldehyde–alkyne–oxadiazoline annulation
Nair, V.; Rajesh, C.; Dhanya, R.; Vinod, A.U. *Tetrahedron Lett.* 2001, *42*, 2043.
Nair, V.; Bindu, S.; Sreekumar, V.; Balagopal, L. *Synthesis* 2003, 1446.

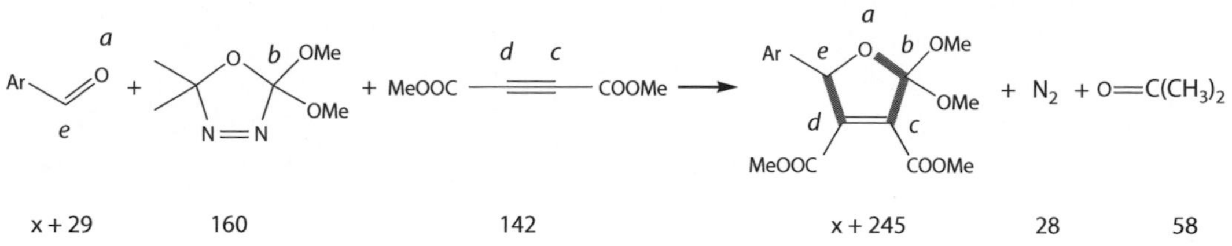

| x + 29 | 160 | 142 | x + 245 | 28 | 58 |

$Ar = x$

$AE = [x + 245]/[x + 331]; \quad AE(min) = 0.79; \quad E = 86/[x + 245]; \quad E(max) = 0.27; \quad f(sac) = 0$

$$\Delta = \Delta(a) + \Delta(b) + \Delta(c) + \Delta(d) + \Delta(e) = (-2 - (-2)) + (3 - 4) + (0 - 0) + (0 - 0) + (0 - 1) = -2;$$

$$HI = -\Delta/(N + 1) = +1$$

Aldehyde–malonylurea–isocyanide condensation
Shaabani, A.; Teimouri, M.B.; Bijanzadeh, H.R. *Tetrahedron Lett.* 2002, *43*, 9097.

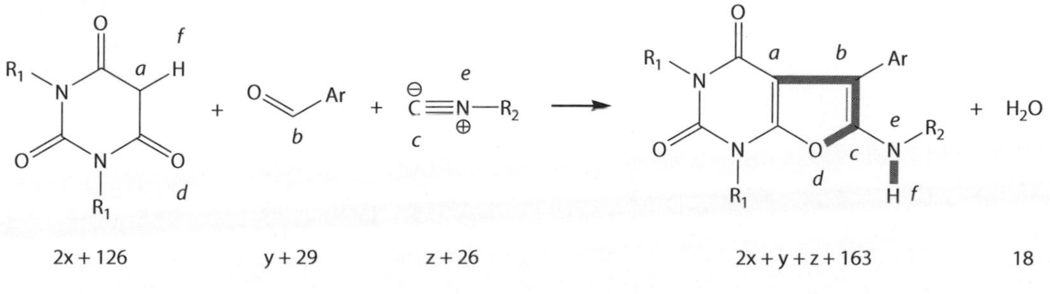

| 2x + 126 | y + 29 | z + 26 | 2x + y + z + 163 | 18 |

$R_1 = x; Ar = y; R_2 = z$

$$AE = [2x + y + z + 163]/[2x + y + z + 181]; \quad AE(min) = 0.93;$$

$$E = 18/[2x + y + z + 163]; \quad E(max) = 0.074$$

$f(sac) = 0$

Set $R_2 = CH_3$

$$\Delta = \Delta(a) + \Delta(b) + \Delta(c) + \Delta(d) + \Delta(e) + \Delta(f)$$

$$= (0 - (-2)) + (0 - 1) + (2 - 2) + (-2 - (-2)) + (-3 - (-3)) + (1 - 1) = +1;$$

$$\mathbf{HI = -\Delta/(N+1) = -1/2}$$

Aldehyde–Meldrum's acid–vinyl ether

Tietze, L.F. *J. Heterocyclic Chem.* 1990, *27*, 47.
Tietze, L.F.; Beifuss, U. *Angew. Chem. Int. Ed.* 1993, *32*, 131.
Tietze, L.F. *Chem. Rev.* 1996, *96*, 115.
Tietze, L.F.; Evers, T.H.; Töpken, E. *Angew. Chem. Int. Ed.* 2001, *40*, 903.

144   x + 29   y + 43   x + y + 198   18

$R_1 = x; R_2 = y$

$$E = 18/[x + y + 198]; \quad E(\text{max}) = 0.09; \quad AE(\text{min}) = 0.92; \quad f(\text{sac}) = 0$$

$$\Delta = \Delta(a) + \Delta(b) + \Delta(c) + \Delta(d) + \Delta(e) = (0 - (-2)) + (-1 - 1) + (-2 - (-2)) + (1 - 0) + (-2 - (-2)) = +1;$$

$$\mathbf{HI = -\Delta/(N+1) = -\frac{1}{2}}$$

Aldehyde–Meldrum's acid–indole

Oikawa, Y.; Hirasawa, H.; Yonemitsu, O. *Tetrahedron Lett.* 1978, 1759.
Oikawa, Y.; Hirasawa, H.; Yonemitsu, O. *Chem. Pharm. Bull.* 1982, *30*, 3092.

144   x + 29   117   x + 272   18

$R_1 = x$

$$E = 18/[x + 272]; \quad E(\text{max}) = 0.066; \quad AE(\text{min}) = 0.94; \quad f(\text{sac}) = 0$$

Set $R_1 = CH_3$

$$\Delta = \Delta(a) + \Delta(b) + \Delta(c) = (-1 - (-2)) + (-1 - 1) + (0 - (-1)) = 0; \quad \mathbf{HI = 0}$$

Aldehyde–Meldrum's acid–ketone
List, B.; Castello, C. *Synlett*. 2001, 1687.

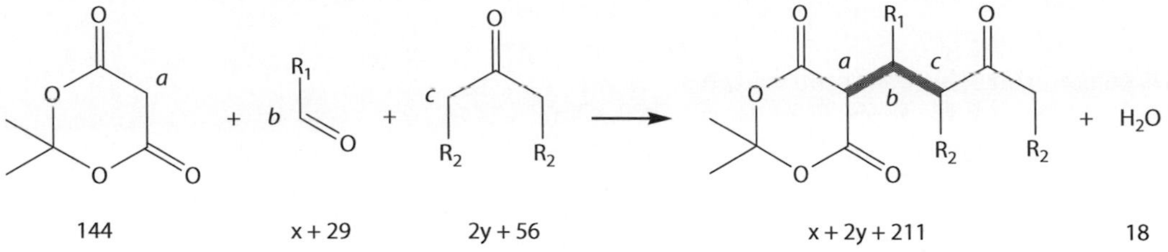

| 144 | x + 29 | 2y + 56 | x + 2y + 211 | 18 |

$R_1 = x; R_2 = y$

$$E = 18/[x + 2y + 211]; \quad \textbf{E(max)} = \textbf{0.084}; \quad \textbf{AE(min)} = \textbf{0.92}; \quad \textbf{\textit{f}(sac)} = \textbf{0}$$

Set $R_1 = R_2 = CH_3$

$$\Delta = \Delta(a) + \Delta(b) + \Delta(c) = (-1 - (-2)) + (-1 - 1) + (-1 - (-2)) = 0; \quad \textbf{HI} = \textbf{0}$$

1,6-Aldol condensation [2 + 2 + 2]
Posner, G.H.; Lu, S.-B.; Arisvatham, E.; Silversmith, E.F.; Shulman, E.M. *J. Am. Chem. Soc.* 1986, *108*, 511.

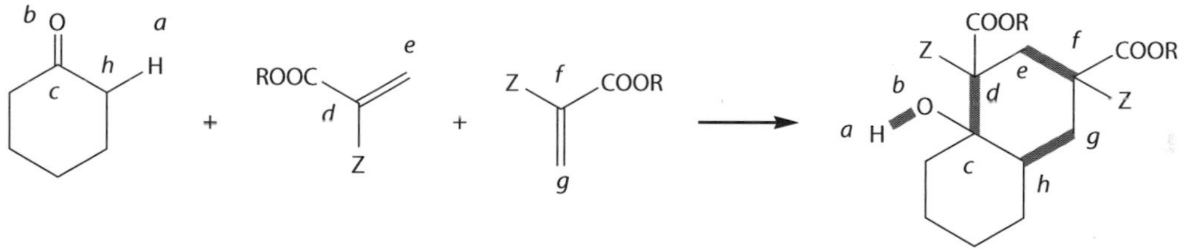

$\textbf{AE} = \textbf{1}; E = \textbf{0}; \textit{f}(\textbf{sac}) = \textbf{0}$

Set $Z = CH_3$

$$\Delta = \Delta(a) + \Delta(b) + \Delta(c) + \Delta(d) + \Delta(e) + \Delta(f) + \Delta(g) + \Delta(h)$$

$$= (1 - 1) + (-2 - (-2)) + (1 - 2) + (0 - 0) + (-2 - (-2)) + (0 - 0) + (-2 - (-2)) + (-1 - (-2)) = 0;$$

$$\textbf{HI} = \textbf{0}$$

Alkene–alkyne annulation
Brown, L.D.; Itoh, K.; Suzuki, H.; Hirai, K.; Ibers, J.A. *J. Am. Chem. Soc.* 1978, *100*, 8232.

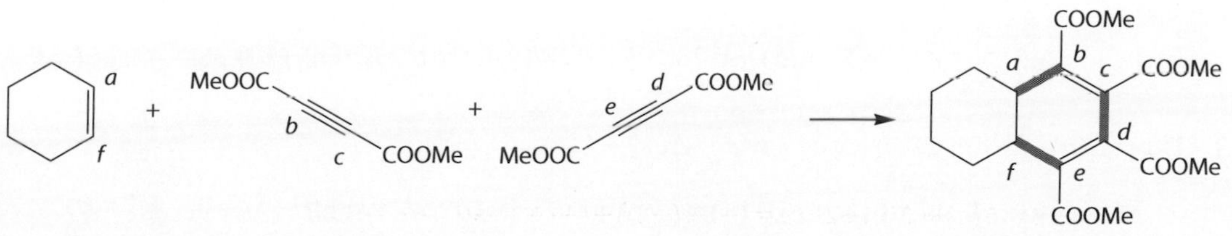

AE = 1; *E* = 0; *f*(sac) = 0

$$\Delta = \Delta(a) + \Delta(b) + \Delta(c) + \Delta(d) + \Delta(e) + \Delta(f) = 2[(-1-(-1)) + (0-0) + (0-0)] = 0; \quad \textbf{HI} = \textbf{0}$$

Alkynes+imines+organoboron reagents

Patel, S.J.; Jamison, J.F. *Angew. Chem. Int. Ed.* 2003, *42*, 1364.

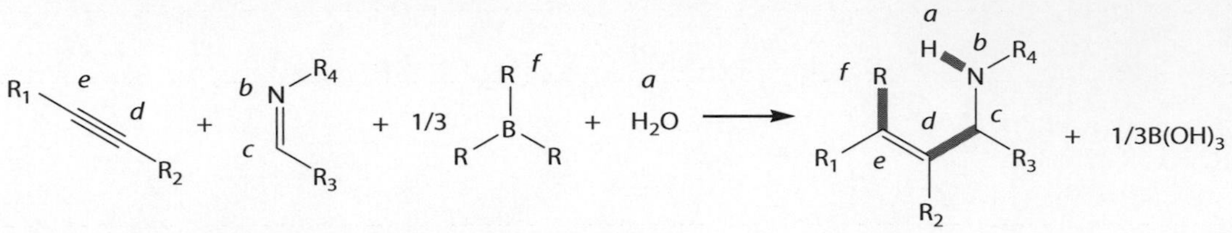

| y + z + 24 | w + v + 27 | 1/3(3x + 11) | 18 | x + y + z + w + v + 52 | 1/3 (62) |

$$R = x; R_1 = y; R_2 = z; R_3 = w; R_4 = v$$

$$AE = [x + y + z + w + v + 52]/[x + y + z + w + v + 72.667]; \quad \textbf{AE(min)} = \textbf{0.73};$$

$$E = (1/3)62/[x + y + z + w + v + 52]; \quad \textbf{E(max)} = \textbf{0.36}$$

***f*(sac) = 0**

Set $R_1 = R_2 = R_3 = R_4 = CH_3$

$$\Delta = \Delta(a) + \Delta(b) + \Delta(c) + \Delta(d) + \Delta(e) + \Delta(f)$$

$$= (1-1) + (-3-(-3)) + (0-1) + (0-0) + (0-0) + (-3-(-4)) = 0; \quad \textbf{HI} = \textbf{0}$$

4-CC allenylation

Grigg, R.; Hodgson, A.; Morris, J.; Sridharan, V. *Tetrahedron Lett.* 2003, *44*, 1023.

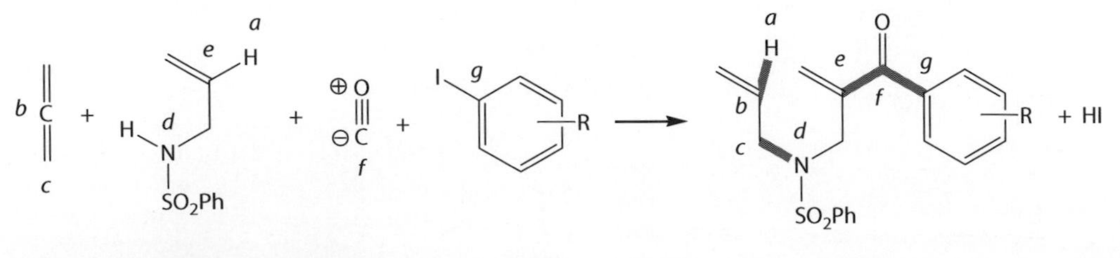

| 40 | 197 | 28 | x + 199 | x + 336 | 128 |

R = x(sum of 5 substituents)

$$AE = [x + 336]/[x + 464]; \quad \textbf{AE(min)} = \textbf{0.73}; \quad E = 128/[x + 336]; \quad \textbf{E(max)} = \textbf{0.38}; \quad f(\text{sac}) = 0$$

$$\Delta = \Delta(a) + \Delta(b) + \Delta(c) + \Delta(d) + \Delta(e) + \Delta(f) + \Delta(g)$$

$$= (1-1) + (-1-0) + (-1-(-2)) + (-1-(-1)) + (0-(-1)) + (2-2) + (0-1) = 0; \quad \textbf{HI} = \textbf{0}$$

Alper carbonylation

Alper, H.; Perera, C.P. *J. Am. Chem. Soc.* 1981, *103*, 1289.
Alper, H.; Mahatantila, C.P. *Organometallics* 1982, *1*, 70.
Alper, H. *Aldrichimica Acta* 1991, *24*, 3.

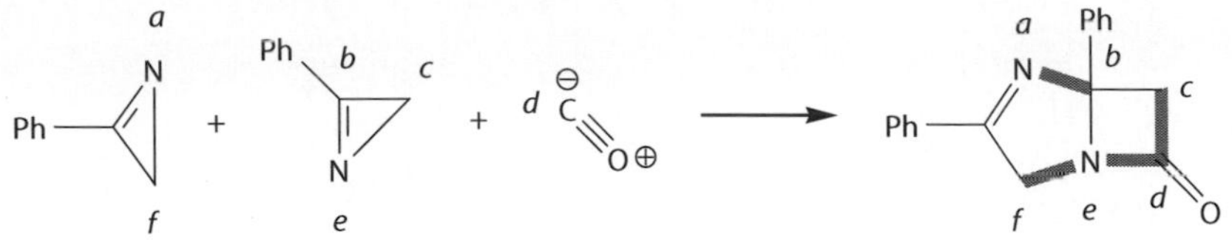

$AE = 1; E = 0; f(sac) = 0$

$$\Delta = \Delta(a) + \Delta(b) + \Delta(c) + \Delta(d) + \Delta(e) + \Delta(f)$$

$$= (-3 - (-3)) + (2 - 2) + (-2 - (-1)) + (3 - 2) + (-3 - (-3)) + (-1 - (-1)) = 0; \quad HI = 0$$

Amide–thioisocyanate–alkylhalide condensation

Graybill, T.L.; Thomas, S.; Wang, M.A. *Tetrahedron Lett.* 2002, *43*, 5305.

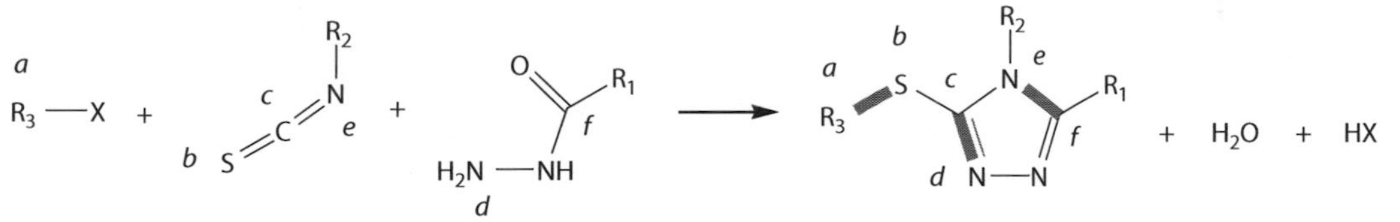

| z + w | y + 58 | x + 59 | x + y + z + 98 | 18 | w + 1 |

$R_1 = x; R_2 = y; R_3 = z; X = w$

$$AE = [x + y + z + 98]/[x + y + z + w + 117]; \quad E = [w + 19]/[x + y + z + 98]; \quad f(sac) = 0$$

X	AE(min)	E(max)
Cl	0.65	0.54
Br	0.51	0.98
I	0.41	1.45

Set $R_1 = R_2 = R_3 = CH_3$

$$\Delta = \Delta(a) + \Delta(b) + \Delta(c) + \Delta(d) + \Delta(e) + \Delta(f)$$

$$= (-2 - (-2)) + (-2 - (-2)) + (4 - 4) + (-2 - (-2)) + (-3 - (-3)) + (3 - 3) = 0; \quad HI = 0$$

α-Aminoacid synthesis via münchnones

Dhawan, R.; Dghaym, R.D.; Arndtsen, B.A. *J. Am. Chem. Soc.* 2003, *125*, 1474.

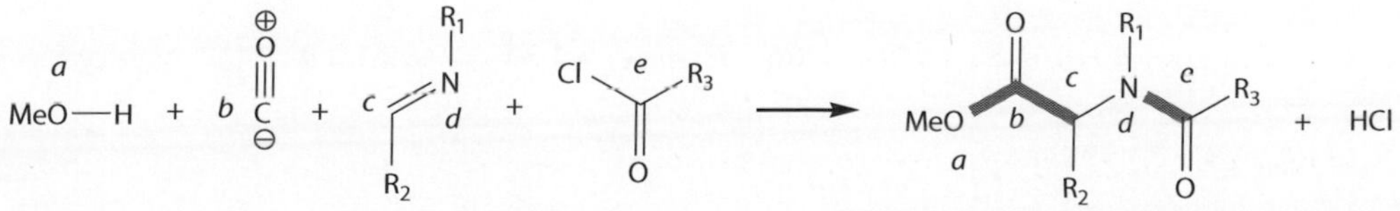

| 32 | 28 | x + y + 27 | z + 63.45 | x + y + z + 114 | 36.45 |

$R_1 = x$; $R_2 = y$; $R_3 = z$

$$AE = [x + y + z + 114]/[x + y + z + 150.45]; \quad \textbf{AE(min)} = \textbf{0.76};$$

$$E = 36.45/[x + y + z + 114]; \quad \textbf{E(max)} = \textbf{0.31}; \quad \textbf{\textit{f}(sac)} = \textbf{0}$$

Set $R_1 = R_2 = R_3 = CH_3$

$$\Delta = \Delta(a) + \Delta(b) + \Delta(c) + \Delta(d) + \Delta(e)$$

$$= (-2 - (-2)) + (3 - 2) + (0 - 1) + (-3 - (-3)) + (3 - 3) = 0; \quad \textbf{HI} = \textbf{0}$$

Aminoalkylation of naphthols with chiral amines
Saidi, M.R.; *Tetrahedron Asymm.* 2003, *14*, 389.

| 121 | x + 29 | 144 | | x + 276 | 18 |

Ar = x

$$AE = [x + 276]/[x + 294]; \quad \textbf{AE(min)} = \textbf{0.95}; \quad E = 18/[x + 276]; \quad \textbf{E(max)} = \textbf{0.051}; \quad \textbf{\textit{f}(sac)} = \textbf{0}$$

$$\Delta = \Delta(a) + \Delta(b) + \Delta(c) = (0 - (-1)) + (0 - 1) + (-3 - (-3)) = 0; \quad \textbf{HI} = \textbf{0}$$

β-Aminocarbonyls synthesis
Adrian, J.C.; Snapper, M.L. *J. Org. Chem.* 2003, *68*, 2143.

| y + 16 | x + 29 | z + 121.9 | x + y + z + 70 | 96.9 |

$R_1 = x$; $R_3 = y$; $R_2 = z$

$$AE = [x + y + z + 70]/[x + y + z + 166.9]; \quad \textbf{AE(min)} = \textbf{0.43};$$

$$E = 96.9/[x + y + z + 70]; \quad \textbf{E(max)} = \textbf{1.33}; \quad \textbf{\textit{f}(sac)} = \textbf{0}$$

Set $R_1 = R_3 = CH_3$

$$\Delta = \Delta(a) + \Delta(b) + \Delta(c) = (-3 - (-3)) + (0 - 1) + (-2 - (-1)) = 0; \quad \textbf{HI} = \textbf{0}$$

Asinger reaction

Asinger, F.; Thiel, M. *Angew. Chem.* 1958, *70*, 667.
Asinger, F.; Offersmanns, H. *Angew. Chem. Int. Ed.* 1967, *6*, 907.

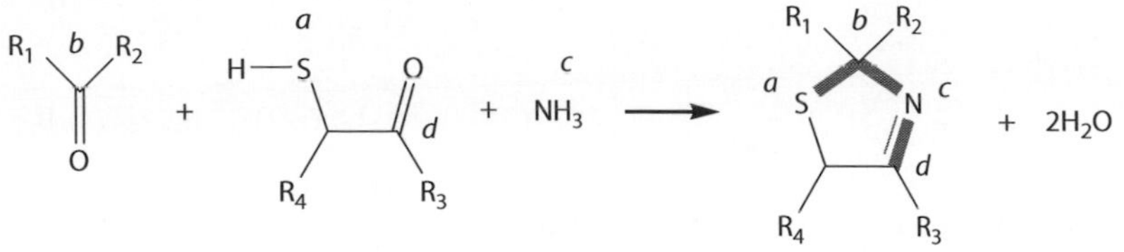

| x + y + 28 | z + w + 74 | 17 | x + y + z + w + 83 | 36 |

$R_1 = x;\ R_2 = y;\ R_3 = w;\ R_4 = z$

$$AE = [x + y + z + w + 83]/[x + y + z + w + 119]; \quad \textbf{AE(min) = 0.71};$$

$$E = 36/[x + y + z + w + 83]; \quad \textbf{E(max) = 0.41}; \quad \textbf{\textit{f}(sac) = 0}$$

Set $R_1 = R_2 = CH_3$

$$\Delta = \Delta(a) + \Delta(b) + \Delta(c) + \Delta(d) = (-2 - (-2)) + (2 - 2) + (-3 - (-3)) + (2 - 2) = 0; \quad \textbf{HI = 0}$$

Asymmetric Mannich-type reaction

Chen, S.L.; Ji, S.J.; Loh, T.P. *Tetrahedron Lett.* 2003, *44*, 2405.

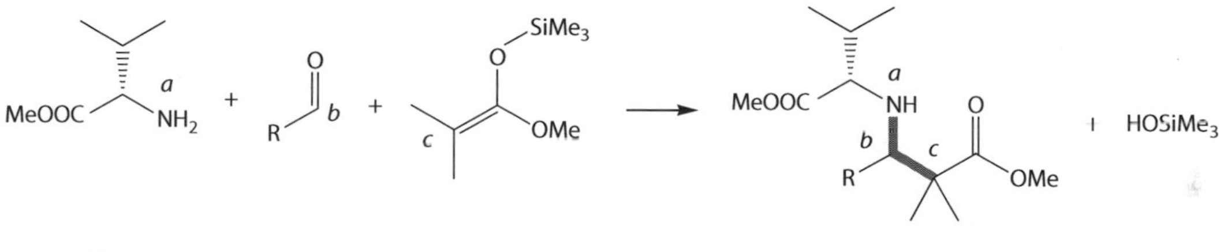

| 131 | x + 29 | 174 | x + 244 | 90 |

$R = x$

$$AE = [x + 244]/[x + 334]; \quad \textbf{AE(min) = 0.73}; \quad E = 90/[x + 244]; \quad \textbf{E(max) = 0.37}; \quad \textbf{\textit{f}(sac) = 0}$$

Set $R = CH_3$

$$\Delta = \Delta(a) + \Delta(b) + \Delta(c) = (-3 - (-3)) + (0 - 1) + (0 - 0) = -1; \quad \textbf{HI} = -\Delta/(N + 1) = +\tfrac{1}{2}$$

Notz, W.; Tanaka, F.; Watanabe, S.; Chowdar, N.S.; Turner, J.M.; Thayumanavan, R.; Barbas, C.F. *J. Org. Chem.* 2003, *68*, 9624.

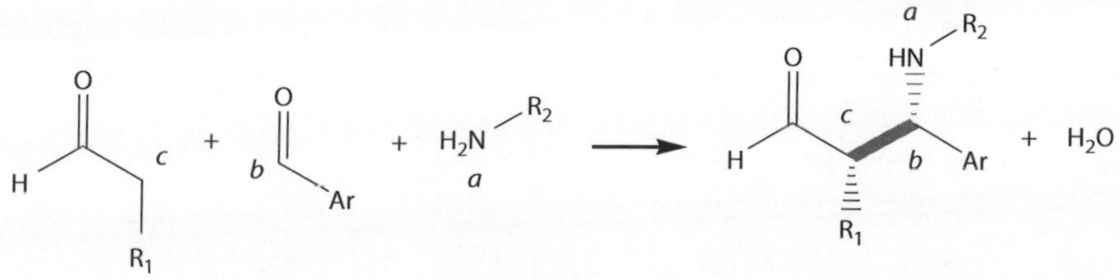

| x + 43 | y + 29 | z + 16 | x + y + z + 142 | 18 |

$R_1 = x$; $R_2 = z$; $y = $ sum of 5 substituents on aromatic ring

$$E = 18/[x + y + z + 142]; \quad E(\text{max}) = 0.12; \quad AE(\text{min}) = 0.89; \quad f(\text{sac}) = 0$$

Set $R_1 = R_2 = CH_3$

$$\Delta = \Delta(a) + \Delta(b) + \Delta(c) = (-3 - (-3)) + (0 - 1) + (-1 - (-2)) = 0; \quad HI = 0$$

Aza-Diels–Alder synthesis of tetrahydroquinolines

Spanedda, M.V.; Hoang, V.D.; Crousse, B.; Bonnet-Delpon, D.; Begue, J.P. *Tetrahedron Lett.* 2003, *44*, 217.

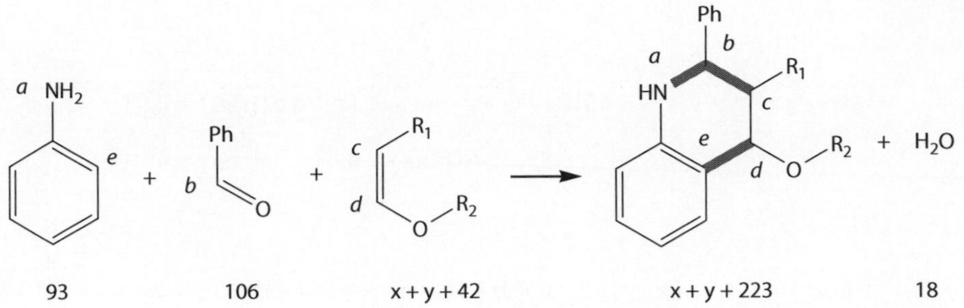

$R_1 = x$; $R_2 = y$

$$AE = [x + y + 223]/[x + y + 241]; \quad AE(\text{min}) = 0.93;$$

$$E = 18/[x + y + 223]; \quad E(\text{max}) = 0.080; \quad f(\text{sac}) = 0$$

Set $R_1 = CH_3$

$$\Delta = \Delta(a) + \Delta(b) + \Delta(c) + \Delta(d) + \Delta(e)$$

$$= (-3 - (-3)) + (0 - 1) + (-1 - (-1)) + (0 - 0) + (0 - (-1)) = 0; \quad HI = 0$$

Annunziata, R.; Cinquini, M.; Cozzi, F.; Molteni, V.; Schupp, O. *Tetrahedron* 1997, *53*, 9715.
Annunziata, R.; Benaglia, M.; Cinquini, M.; Cozzi, F. *Eur. J. Org. Chem.* 2002, 1184.

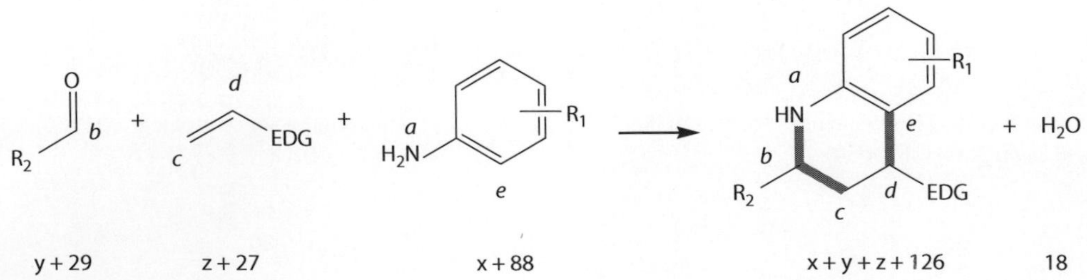

$R_1 = x$(sum of 5 substituents); $R_2 = y$; EDG $= z$

$$AE = [x + y + z + 126]/[x + y + z + 144]; \quad AE(\text{min}) = 0.88;$$

$$E = 18/[x + y + z + 126]; \quad E(\text{max}) = 0.14; \quad f(\text{sac}) = 0$$

Set $R_2 = EDG = CH_3$

$$\Delta = \Delta(a) + \Delta(b) + \Delta(c) + \Delta(d) + \Delta(e)$$

$$= (-3 - (-3)) + (0 - 1) + (-2 - (-2)) + (-1 - (-1)) + (0 - (-1)) = 0; \quad \textbf{HI} = \textbf{0}$$

1,5-Benzodiazepine synthesis

Jarikote, D.V.; Siddiqui, S.A.; Rajagopal, R.; Daniel, T.; Lahoti, R.J.; Srinivasan, K.V. *Tetrahedron Lett.* 2003, *44*, 1835.
Bandgar, B.P.; Bettigeri, S.V.; Joshi, N.S. *Synth. Commun.* 2004, *34*, 1447.
Sivamurugan, V.; Decpa, K.; Palanichamy, M.; Murugesan, V. *Synthesis* 2004, 3833.

| x + 107 | y + 43 | y + 43 | x + 2y + 157 | 36 |

$R_1 = x; R_2 = y$

$$AE = [x + 2y + 157]/[x + 2y + 193]; \quad \textbf{AE(min)} = \textbf{0.82};$$

$$36/[x + 2y + 157]; \quad \textbf{E(max)} = \textbf{0.23}; \quad f(\textbf{sac}) = \textbf{0}$$

Set $R_2 = CH_3$

$$\Delta = \Delta(a) + \Delta(b) + \Delta(c) + \Delta(d) + \Delta(e)$$

$$= (-3 - (-3)) + (2 - 2) + (-2 - (-3)) + (1 - 2) + (-3 - (-3)) = 0; \quad \textbf{HI} = \textbf{0}$$

Betti reaction

Betti, M. *Gazz. Chim. Ital.* 1900, *30(II)*, 301.
Betti, M. *Gazz. Chim. Ital.* 1903, *33(II)*, 2.
Pirrone, F. *Gazz. Chim. Ital.* 1936, *66*, 518.
Pirrone, F. *Gazz. Chim. Ital.* 1937, *67*, 529.

| y + 101 | x + 88 | 154 | x + y + 316 | 18 |

X = x(sum of 5 substituents); Y = y(sum of 5 substituents)

$$AE - [x + y + 316]/[x + y + 334]; \quad \textbf{AE(min)} = \textbf{0.95};$$

$$E = 18/[x + y + 316]; \quad \textbf{E(max)} = \textbf{0.055}; \quad f(\textbf{sac}) = \textbf{0}$$

$$\Delta = \Delta(a) + \Delta(b) + \Delta(c) = (-3 - (-3)) + (0 - 1) + (0 - (-1)) = 0; \quad \textbf{HI} = \textbf{0}$$

Biginelli synthesis

Biginelli, P. *Gazz. Chim. Ital.* 1893, *23*, 360.
Biginelli, P. *Chem. Ber.* 1891, *24*, 1317; 2962.
Sabitha, G.; Reddy, G.S.K.K.; Reddy, K.B.; Yadav, J.S. *Tetrahedron Lett.* 2003, *44*, 6497.
Manjula, A.; Rao, B.V.; Neelakantan, P. *Synth. Commun.* 2004, *34*, 2665.
Srinivas, K.V.N.S.; Das, B. *Synthesis* 2004, *13*, 2091.
Jenner, G. *Tetrahedron Lett.* 2004, *45*, 6195.

$$R_1 = x; R_2 = y; R_3 = z$$

$$AE = [x + y + z + 95]/[x + y + z + 131]; \quad \textbf{AE(min)} = \textbf{0.73};$$

$$E = 36/[x + y + z + 95]; \quad E\textbf{(max)} = \textbf{0.37}; \quad f\textbf{(sac)} = \textbf{0}$$

Set $R_1 = R_2 = R_3 = CH_3$

$$\Delta = \Delta(a) + \Delta(b) + \Delta(c) + \Delta(d) + \Delta(e)$$

$$= (-3 - (-3)) + (0 - 1) + (0 - (-2)) + (1 - 2) + (-3 - (-3)) = 0; \quad \textbf{HI} = \textbf{0}$$

Bis(indolyl)methane synthesis

Bandgar, B.P.; Shaikh, K.A. *Tetrahedron Lett.* 2003, *44*, 1959.
Li, S.J.; Zhou, M.F.; Gu, D.G.; Wang, S.Y.; Loh, T.P. *Synlett* 2003, *13*, 2077.
Sharma, G.V.M.; Reddy, J.J.; Lakshmi, P.S.; Krishna, P.R. *Tetrahedron Lett.* 2004, *45*, 7729.
Xia, M.; Wang, S.; Yuan, W. *Synth. Commun.* 2004, *34*, 3175.

$$R_1 = x; R_2 = y$$

$$AE = [x + y + 244]/[x + y + 262]; \quad \textbf{AE(min)} = \textbf{0.93};$$

$$E = 18/[x + y + 244]; \quad E\textbf{(max)} = \textbf{0.073}; \quad f\textbf{(sac)} = \textbf{0}$$

$$\Delta = \Delta(a) + \Delta(b) + \Delta(c) = (0 - (-1)) + (0 - 2) + (0 - (-1)) = 0; \quad \textbf{HI} = \textbf{0}$$

Bucherer synthesis of hydantoins

Bucherer, H.; Steiner, W. *J. Prakt. Chem.* 1934, *140*, 291.

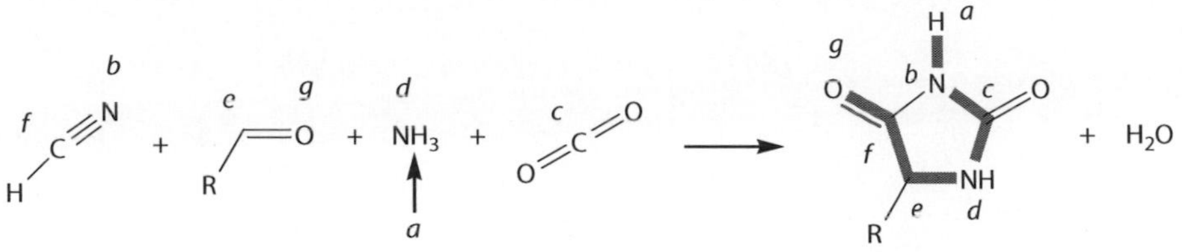

| 27 | x + 29 | 17 | 44 | x + 99 | 18 |

R = x

$$AE = [x+99]/[x+117]; \quad AE(\min) = 0.85;$$

$$E = 18/[x+99]; \quad E(\max) = 0.18; \quad f(\text{sac}) = 0$$

Set R = CH$_3$

$$\Delta = \Delta(a) + \Delta(b) + \Delta(c) + \Delta(d) + \Delta(e) + \Delta(f) + \Delta(g)$$

$$= (1-1) + (-3-(-3)) + (4-4) + (-3-(-3)) + (0-1) + (3-2) + (-2-(-2)) = 0; \quad HI = 0$$

Butadiene trimerization [4 + 4 + 4]

Bogdanovic, B.; Heimbach, P.; Kroner, M.; Wilke, G.; Hoffmann, E.G.; Brandt, J. *Ann. Chem.* 1969, *727*, 143.

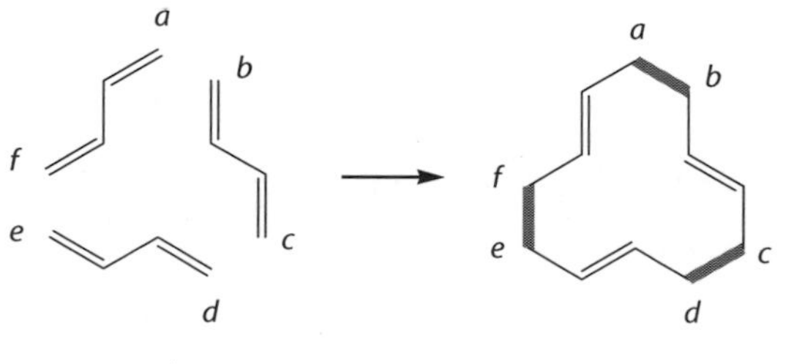

AE = 1; *E* = 0; *f*(sac) = 0

$$\Delta = \Delta(a) + \Delta(b) + \Delta(c) + \Delta(d) + \Delta(e) + \Delta(f) = 6(-2-(-2)) = 0; \quad HI = 0$$

Chichibabin pyridine synthesis

Chichibabin, A.E. *J. Russ. Phys. Chem. Soc.* 1906, *37*, 1229.

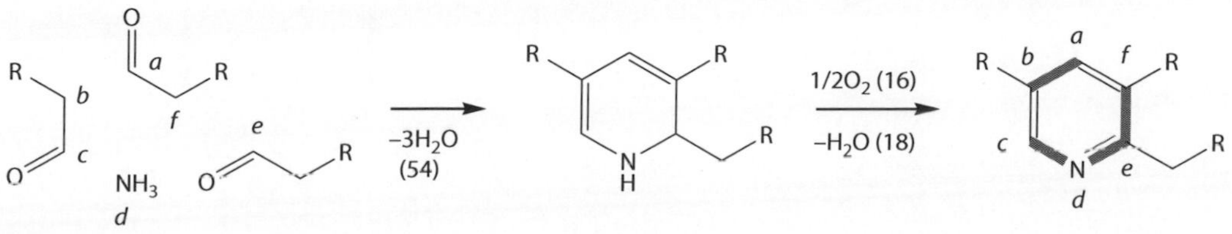

| 17 | 3 (x + 43) | | 3x + 90 |

R = x

$$AE = [3x + 90]/[3x + 162]; \quad \textbf{AE(min)} = \textbf{0.56}; \quad E = 72/[3x + 90]; \quad \textbf{E(max)} = \textbf{0.77}$$

$$f(\text{sac}) = 16/[3x + 162]; \quad \textbf{f(sac)(max)} = \textbf{0.097}$$

Set R = CH_3

$$\Delta = \Delta(a) + \Delta(b) + \Delta(c) + \Delta(d) + \Delta(e) + \Delta(f)$$

$$= (-1 - 1) + (0 - (-2)) + (0 - 1) + (-3 - (-3)) + (2 - 1) + (0 - (-2)) = +2$$

$$\textbf{HI} = -\Delta/(N + 1) = -1$$

Cyclohexanone synthesis [1 + 2 + 3]
Ogura, K.; Yahata, N.; Minoguchi, M.; Ohtsuki, K.; Takahashi, K.; Iida, H. *J. Org. Chem.* 1986, *51*, 508.

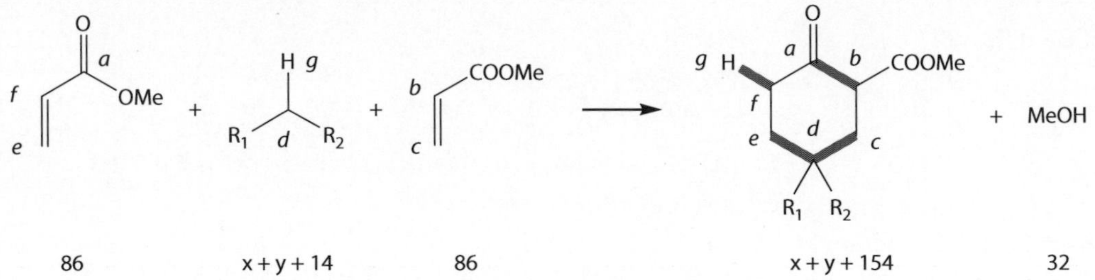

| 86 | x + y + 14 | 86 | x + y + 154 | 32 |

$R_1 = x$; $R_2 = y$

$$AE = [x + y + 154]/[x + y + 171]; \quad \textbf{AE(min)} = \textbf{0.83};$$

$$E = 32/[x + y + 154]; \quad \textbf{E(max)} = \textbf{0.21}; \quad f(\text{sac}) = \textbf{0}$$

Set $R_1 = R_2 = CH_3$

$$\Delta = \Delta(a) + \Delta(b) + \Delta(c) + \Delta(d) + \Delta(e)$$

$$= (2 - 3) + (-1 - (-1)) + (-2 - (-2)) + (0 - (-2)) + (-2 - (-2)) + (-2 - (-1)) + (1 - 1) = 0; \quad \textbf{HI} = \textbf{0}$$

Cyclopropanation [(1 + 2) + (1 + 2)]
Danheiser, R.; Morin, J.M.; Salaski, E.J. *J. Am. Chem. Soc.* 1985, *107*, 8066.

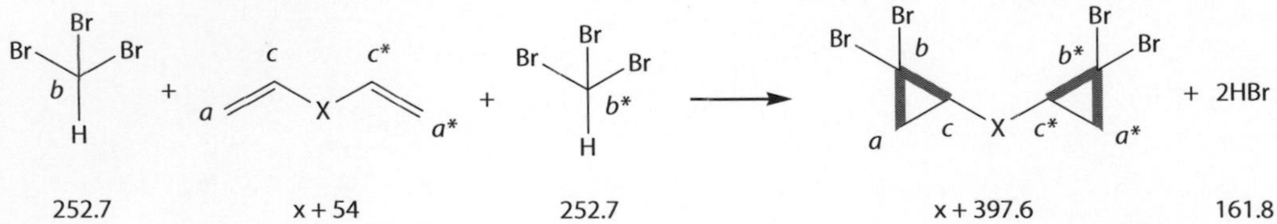

| 252.7 | x + 54 | 252.7 | x + 397.6 | 161.8 |

X = x
AE = [x + 397.6]/[x + 559.4]; E = 161.8/[x + 397.6]

$X = CH_2$ $AE(min) = 0.72$; $E(max) = 0.39$; $f(sac) = 0$

Set $X = CH_2$

$$\Delta = \Delta(a) + \Delta(b) + \Delta(c) + \Delta(a^*) + \Lambda(b^*) + \Delta(c^*)$$

$$= 2[(-2 - (-2)) + (2 - 2) + (-1 - (-1))] = 0; \quad \mathbf{HI = 0}$$

Cyclopropanation [1 + (1 + 2)]

Joucla, M.; Fouchet, B.; LeBrun, J.; Hamelin, J. *Tetrahedron Lett.* 1985, *26*, 1221.

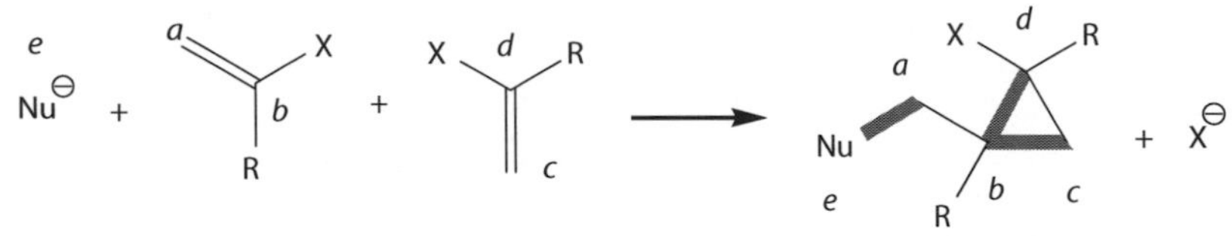

y	r + x + 26	r + x + 26		2r + x + y + 52	x

$R = r$; $X = x$; $Nu = y$

$$AE = [2r + x + y + 52]/[2r + 2x + 52]; \quad E = x/[2r + x + y + 52]; \quad f(sac) = 0$$

Set $R = Nu = CH_3$

X	AE(min)	E(max)
Cl	0.75	0.34
Br	0.65	0.54
I	0.61	0.65

$$\Delta = \Delta(a) + \Delta(b) + \Delta(c) + \Delta(d) + \Delta(e)$$

$$= (-2 - (-2)) + (0 - 1) + (-2 - (-2)) + (1 - 1) + (-3 - (-4)) = 0; \quad \mathbf{HI = 0}$$

1,4-Dihydropyridine synthesis

Yadav, J.S.; Reddy, B.V.S.; Basak, A.K.; Narsaiah, A.V. *Green Chem.* 2003, *5*, 60.

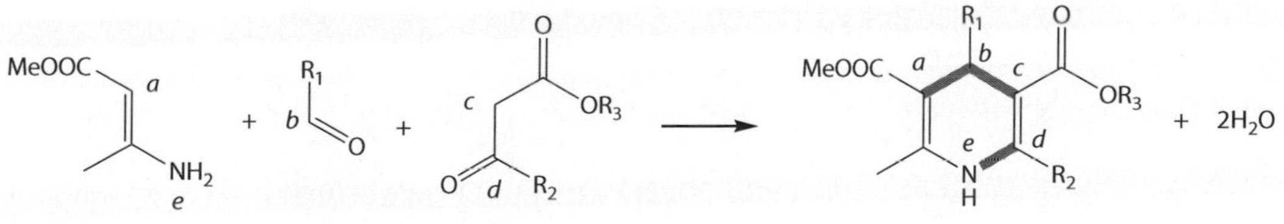

115	x + 29	y + z + 86	x + y + z + 194	36

$R_1 = x$; $R_2 = y$; $R_3 = z$

$$AE = [x + y + z + 194]/[x + y + z + 230]; \quad \mathbf{AE(min) = 0.85};$$

$$E = 36/[x + y + z + 194]; \quad \mathbf{E(max) = 0.18}; \quad f(sac) = 0$$

Set $R_1 = R_2 = CH_3$

$$\Delta = \Delta(a) + \Delta(b) + \Delta(c) + \Delta(d) + \Delta(e)$$

$$= (0 - (-1)) + (-1 - 1) + (0 - (-2)) + (1 - 2) + (-3 - (-3)) = 0; \quad \textbf{HI} = \textbf{0}$$

3,4-Dihydropyrimidin-2-(1H)-one synthesis

Shaabani, A.; Bazgir, A.; Teimouri, F. *Tetrahedron Lett.* 2003, *44*, 857.
Bose, D.L.; Fatima, L.; Merevala, H.B. *J. Org. Chem.* 2003, *68*, 587.
Salehi, P.; Dabir, M.; Zolfigol, M.A.; Fard, M.A.B. *Heterocycles* 2003, *60*, 2435.
Reddy, K.R.; Reddy, C.V.; Mahesh, M.; Raju, P.V.K.; Reddy, V.V.N. *Tetrahedron Lett.* 2003, *44*, 8173.
Gohain, M.; Prajapati, D.; Sandhu, J.S. *Synlett* 2004, 235.
Bose, D.S.; Kumar, R.K.; Fatima, L. *Synlett* 2004, *2*, 279.
Salehi, H.; Guo, Q.X. *Synth. Commun.* 2004, *34*, 171.
Sun, Q.; Wang, Y.Q.; Ge, Z.M.; Cheng, T.M.; Li, R.T. *Synthesis* 2004, *7*, 1047.
Khodaei, M.M.; Khosropour, A.R.; Beygzadeh, M. *Synth. Commun.* 2004, *34*, 1551.
Shailaja, M.; Manjula, A.; Rao, B.V.; Neelakantan, P. *Synth. Commun.* 2004, *34*, 1559.
Yadav, J.S.; Reddy, B.V.S.; Naidu, J.J.; Sadashiv, K. *Chem. Lett.* 2004, *33*, 926.
Narsaiah, A.V.; Basak, A.K.; Nagaiah, K. *Synthesis* 2004, *8*, 1253.
Yadav, J.S.; Kumar, S.P.; Kondaji, G.; Rao, R.S.; Nagaiah, K. *Chem. Lett.* 2004, *33*, 1168.
Mirza-Aghayan, M.; Bolourtchian, M.; Hosseini, M. *Synth. Commun.* 2004, *34*, 3335.
Zhu, Y.; Pan, Y.; Huang, S. *Synth. Commun.* 2004, *34*, 3167.
Reddy, Y.T.; Rajitha, B.; Reddy, P.N.; Kumar, B.S.; Rao, V.P. *Synthesis* 2004, *34*, 3821.
Bhosale, R.S.; Bhosale, S.V.; Bhosale, S.V.; Wang, T.; Zubaidha, P.K. *Tetrahedron Lett.* 2004, *45*, 9111.
Bose, A.K.; Pednekar, S.; Ganguly, S.N.; Chakraborty, G.; Manhas, M.S. *Tetrahedron Lett.* 2004, *45*, 8351.

$$\begin{array}{ccccc}
y + z + 86 & \quad & x + 29 & \quad X + 44 & \qquad X + x + y + z + 123 \qquad 36
\end{array}$$

Use of urea (X = O)

$$AE = [x + y + z + 139]/[x + y + z + 175]; \quad \textbf{AE(min) = 0.80};$$

$$E = 36/[x + y + z + 139]; \quad \textbf{E(max) = 0.25}; \quad f(\textbf{sac}) = \textbf{0}$$

Use of thiourea (X = S):

$$AE = [x + y + z + 155]/[x + y + z + 191]; \quad \textbf{AE(min) = 0.81};$$

$$E = 36/[x + y + z + 155]; \quad \textbf{E(max) = 0.23}; \quad f(\textbf{sac}) = \textbf{0}$$

Set $R_1 = R_2 = CH_3$

$$\Delta = \Delta(a) + \Delta(b) + \Delta(c) + \Delta(d) + \Delta(e)$$

$$= (0 - (-2)) + (0 - 1) + (-3 - (-3)) + (-3 - (-3)) + (1 - 2) = 0; \quad \textbf{HI} = \textbf{0}$$

Doebner reaction
Doebner, O. *Ann. Chem.* 1887, *242*, 256.

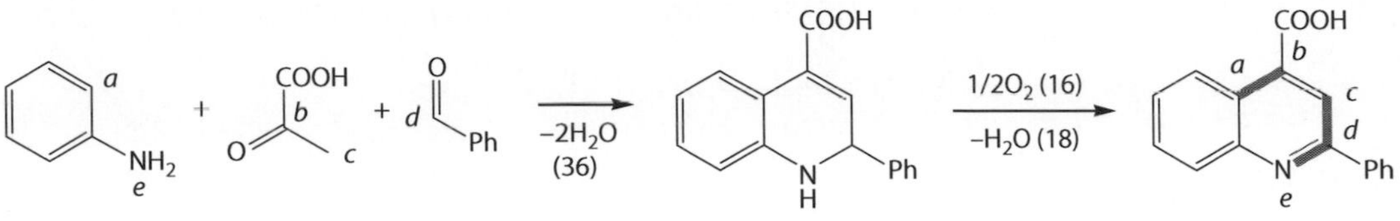

93   88   106   249

AE = 0.82; *E* = 54/249 = 0.22; *f*(sac) = 16/303 = 0.053

$$\Delta = \Delta(a) + \Delta(b) + \Delta(c) + \Delta(d) + \Delta(e)$$

$$= (0-(-1)) + (0-2) + (-1-(-3)) + (2-1) + (-3-(-3)) = 2; \quad \textbf{HI} = \textbf{-1}$$

Dornow–Wiehler isoxazole synthesis
Dornow, A.; Wiehler, G. *Ann. Chem.* 1952, *578*, 113.
Dornow, A.; Wiehler, G. *Ann. Chem.* 1952, *578*, 122.

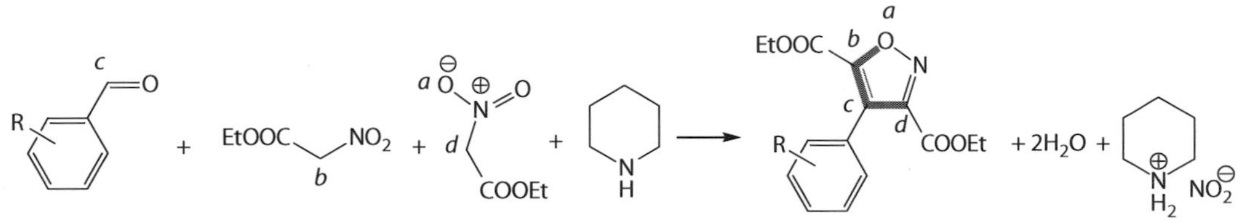

x + 101   133   133   85   x + 284   36   132

R = x(sum of 5 substituents)

$$\text{AE} = [x+284]/[x+452]; \quad \textbf{AE(min)} = \textbf{0.63}; \quad E = 168/[x+284]; \quad \textbf{E(max)} = \textbf{0.58}$$

$$f(\text{sac}) = 85/[x+452]; \quad \textbf{f(sac)(max)} = \textbf{0.19}$$

$$\Delta = \Delta(a) + \Delta(b) + \Delta(c) + \Delta(d)$$

$$= (-2-(-2)) + (1-(-1)) + (0-1) + (2-(-1)) = +4; \quad \textbf{HI} = -\Delta/(N+1) = \textbf{-2}$$

Feist–Benary synthesis of pyrroles
Feist, F. *Chem. Ber.* 1902, *35*, 1537.
Feist, F. *Chem. Ber.* 1902, *35*, 1545.
Benary, F. *Chem. Ber.* 1911, *44*, 489.
Benary, F. *Chem. Ber.* 1911, *44*, 493.

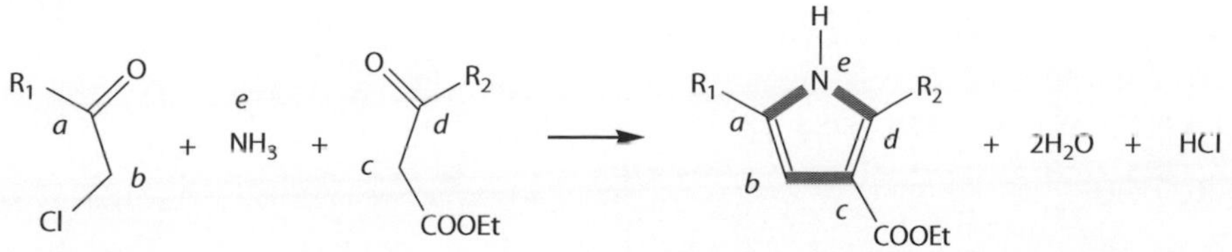

x + 74.45   17   y + 115   x + y + 137   36   36.45

$R_1 = x;\ R_2 = y$

$$AE = [x + y + 137]/[x + y + 209.45]; \quad \textbf{AE(min) = 0.66};$$

$$E = 72.45/[x + y + 137]; \quad \textbf{\textit{E}(max) = 0.52}; \quad \textit{f}\textbf{(sac) = 0}$$

Set $R_1 = R_2 = CH_3$

$$\Delta = \Delta(a) + \Delta(b) + \Delta(c) + \Delta(d) + \Delta(e)$$

$$= (1 - 2) + (-1 - (-1)) + (0 - (-2)) + (1 - 2) + (-3 - (-3)) = 0; \quad \textbf{HI = 0}$$

Formamidine urea synthesis

Ripka, A.S.; Diaz, D.D.; Sharpless, K.B.; Finn, M.G. *Org. Lett.* 2003, 5, 1531.

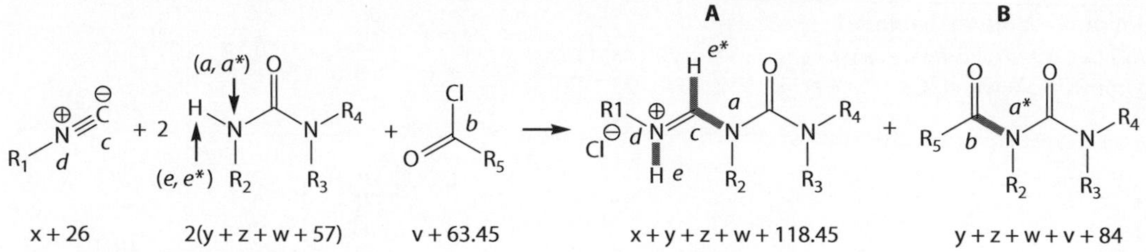

$$\begin{array}{ccccccc}
x + 26 & 2(y + z + w + 57) & v + 63.45 & & x + y + z + w + 118.45 & & y + z + w + v + 84
\end{array}$$

$R_1 = x;\ R_2 = y;\ R_3 = z;\ R_4 = w;\ R_5 = v$

Target product is A:

$$AE = [x + y + z + w + 118.45]/[2(y + z + w) + x + v + 202.45];$$

$$E = [y + z + w + v + 84]/[x + y + z + w + 118.45]$$

AE(min) = 0.58; *E*(max) = 88/122.45 = 0.72; *f*(sac) = 0

Set $R_1 = R_2 = CH_3$

$$\Delta = \Delta(a) + \Delta(c) + \Delta(d) + \Delta(e) + \Delta(e^*)$$

$$= (-3 - (-3)) + (2 - 2) + (-3 - (-3)) + (1 - 1) + (1 - 1) = 0; \quad \textbf{HI = 0}$$

Target product is B:

$$AE = [y + z + w + v + 84]/[2(y + z + w) + x + v + 202.45];$$

$$E = [x + y + z + w + 118.45]/[y + z + w + v + 84]$$

AE(min) = 0.42; *E*(max) = 120.45/88 = 1.37; *f*(sac) = 0

Set $R_2 = R_5 = CH_3$

$$\Delta = \Delta(a^*) + \Delta(b) = (-3 - (-3)) + (3 - 3) = 0; \quad \textbf{HI = 0}$$

Fused 3-aminoimidazoles
Ireland, S.M.; Tye, H.; Whittaker, M. *Tetrahedron Lett.* 2003, *44*, 4369.

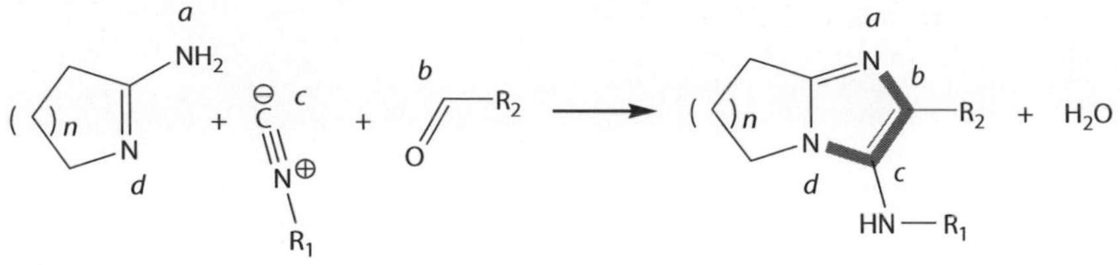

14n + 70   x + 26   y + 29   x + y + 14n + 107   18

$R_1 = x; R_2 = y$

$$AE = [x + y + 14n + 107]/[x + y + 14n + 125]; \quad AE(min) = [14n + 109]/[14n + 127];$$

$$E = 18/[x + y + 14n + 107]; \quad E(max) = 18/[14n + 109]; \quad f(sac) = 0$$

n	Ring size ($n + 4$)	E(max)	AE(min)
1	5	0.15	0.87
2	6	0.13	0.88
3	7	0.12	0.89

Set $R_2 = CH_3$

$$\Delta = \Delta(a) + \Delta(b) + \Delta(c) + \Delta(d) = (-3-(-3)) + (1-1) + (2-2) + (-3-(-3)) = 0; \quad HI = 0$$

Fused benzochromene synthesis
Yavari, I.; Anary-Abbasinejad, M.; Alizadeh, A.; Hossaini, Z. *Tetrahedron* 2003, *59*, 1289.

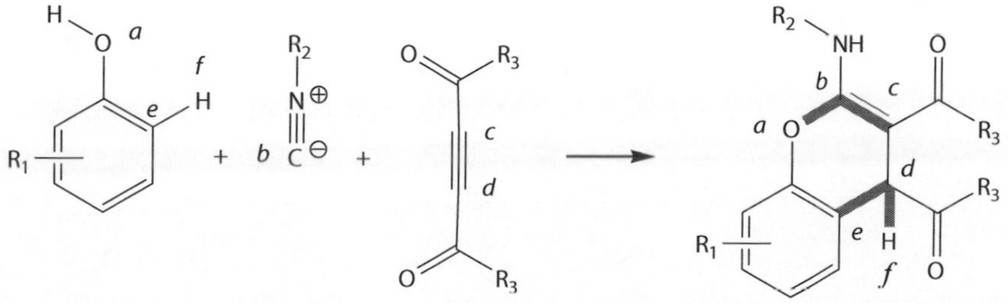

x + 93   y + 26   2z + 80   x + y + z + 199

$AE = 1; E = 0; f(sac) = 0$

$$\Delta = \Delta(a) + \Delta(b) + \Delta(c) + \Delta(d) + \Delta(e)$$

$$= (-2-(-2)) + (2-2) + (0-0) + (-1-0) + (0-(-1)) = 0; \quad HI = 0$$

Gewald aminothiophene synthesis

Gewald, K. *J. Prakt. Chem.* 1966, *31*, 214.
Gewald, K. *J. Prakt. Chem.* 1966, *32*, 26.
Gewald, K.; Schinke, E.; Boettcher, H. *Chem. Ber.* 1966, *99*, 94.
Gewald, K.; Schinke, E. *Chem. Ber.* 1966, *99*, 2712.

$$x + y + 42 \qquad 32 \qquad z + 84 \qquad\qquad x + y + z + 140 \qquad 18$$

$R_1 = x;\ R_2 = y;\ R_3 = z$

$$AE = [x + y + z + 140]/[x + y + z + 158]; \quad \textbf{AE(min)} = \textbf{0.89};$$

$$E = 18/[x + y + z + 140]; \quad \textbf{E(max)} = \textbf{0.13}; \quad \textbf{\textit{f}(sac)} = \textbf{0}$$

Set $R_1 = R_2 = CH_3$

$$\Delta = \Delta(a) + \Delta(b) + \Delta(c) + \Delta(d) + \Delta(e) + \Delta(f) + \Delta(f^*) + \Delta(g)$$

$$= (1 - (-2)) + (-2 - 0) + (2 - 3) + (0 - (-2)) + (0 - 2) + (1 - 1) + (1 - 1) = 0; \quad \textbf{HI} = \textbf{0}$$

Grieco condensation

Grieco, P.A.; Bahsas, A. *Tetrahedron Lett.* 1988, *29*, 5855.

$$x + 136 \qquad\qquad z + w + 26 \qquad y + 29 \qquad\qquad x + y + z + w + 173 \qquad 18$$

$R_1 = x;\ R_2 = y;\ R_3 = z;\ R_4 = w$

$$AE = [x + y + z + w + 173]/[x + y + z + w + 191]; \quad \textbf{AE(min)} = \textbf{0.91};$$

$$E = 18/[x + y + z + w + 173]; \quad \textbf{E(max)} = \textbf{0.10}; \quad \textbf{\textit{f}(sac)} = \textbf{0}$$

Set $R_2 = R_3 = R_4 = CH_3$

$$\Delta = \Delta(a) + \Delta(b) + \Delta(c) + \Delta(d)$$

$$= (0 - (-1)) + (-1 - (-1)) + (-1 - (-1)) + (0 - 1) + (-3 - (-3)) = 0; \quad \textbf{HI} = \textbf{0}$$

Guareschi–Thorpe condensation

Guareschi, I. *Mem. Reale Accad. Sci. Torino II* 1896, *46*, 7; 11; 25.
Baron, H.; Renfry, F.G.P.; Thorpe, J.F. *J. Chem. Soc.* 1904, *85*, 1726.

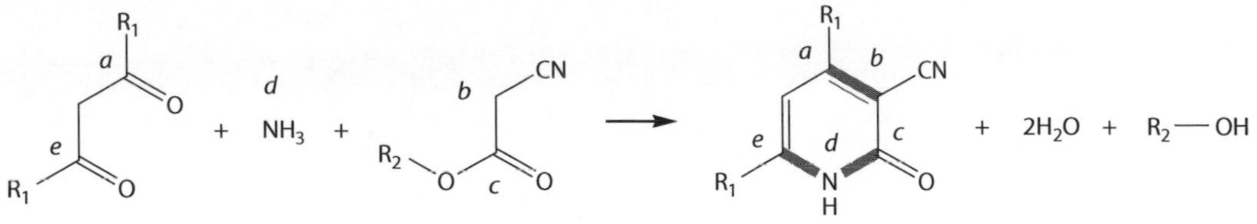

| 2x + 70 | 17 | y + 84 | | 2x + 118 | 36 | y + 17 |

$R_1 = x; R_2 = y$

$$AE = [2x+118]/[2x+y+171]; \quad E = [y+53]/[2x+118]; \quad f(\text{sac}) = 0$$

R_2	$E(\text{max})$	$AE(\text{min})$
Me	0.57	0.64
Et	0.68	0.59

Set $R_1 = CH_3$

$$\Delta = \Delta(a) + \Delta(b) + \Delta(c) + \Delta(d) + \Delta(e)$$

$$= (0-2) + (0-(-2)) + (3-3) + (-3-(-3)) + (1-2) = -1; \quad \mathbf{HI} = -\Delta/(N+1) = +\tfrac{1}{2}$$

Hantzsch synthesis of dihydropyridines

Hantzsch, A. *Ann. Chem.* 1882, *215*, 1.

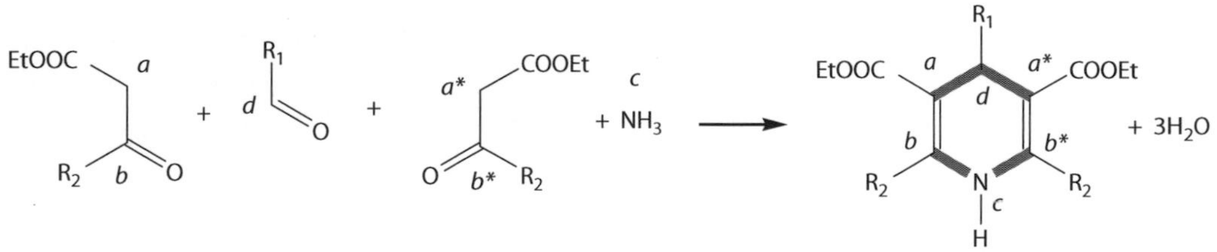

| y + 115 | x + 29 | y + 115 | 17 | 2y + x + 222 | 54 |

$R_1 = x; R_2 = y$

$$AE = [x+2y+222]/[x+2y+276]; \quad \mathbf{AE(min) = 0.81};$$

$$E = 54/[x+2y+222]; \quad \mathbf{E(max) = 0.24}; \quad f(\text{sac}) = 0$$

Set $R_1 = R_2 = CH_3$

$$\Delta = \Delta(a) + \Delta(a^*) + \Delta(b) + \Delta(b^*) + \Delta(c) + \Delta(d)$$

$$= (0-(-2)) + (0-(-2)) + (1-2) + (1-2) + (-3-(-3)) + (-1-1) = 0; \quad \mathbf{HI = 0}$$

Hantzsch synthesis of pyrroles

Hantzsch, A. *Ber.* 1890, *23*, 1474.

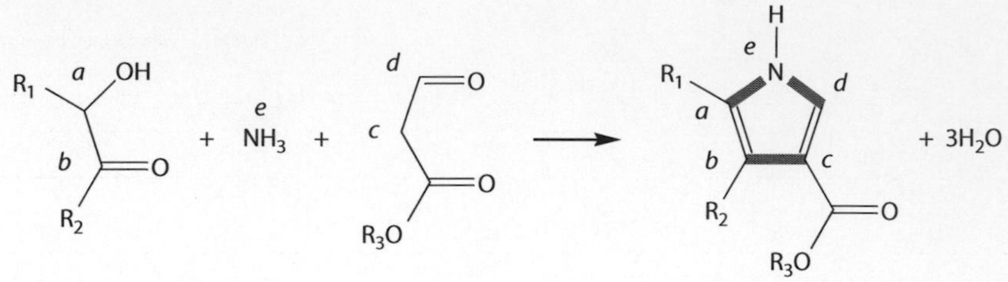

x + y + 58   17   z + 87   x + y + z + 108   54

$R_1 = x; R_2 = y; R_3 = z$

$$AE = [x + y + z + 108]/[x + y + z + 162]; \quad \textbf{AE(min)} = \textbf{0.67};$$

$$E = 54/[x + y + z + 108]; \quad \textit{E}\textbf{(max)} = \textbf{0.49}; \quad \textit{f}\textbf{(sac)} = \textbf{0}$$

Set $R_1 = R_2 = CH_3$

$$\Delta = \Delta(a) + \Delta(b) + \Delta(c) + \Delta(d) + \Delta(e)$$

$$= (1-0) + (0-2) + (0-(-2)) + (0-1) + (-3-(-3)) = 0; \quad \textbf{HI} = \textbf{0}$$

Heck–Diels–Alder reaction

Brown, S.; Grigg, R.; Hinsley, J.; Korn, S.; Sridharan, V.; Uttley, M.D. *Tetrahedron* 2001, *57*, 10347.

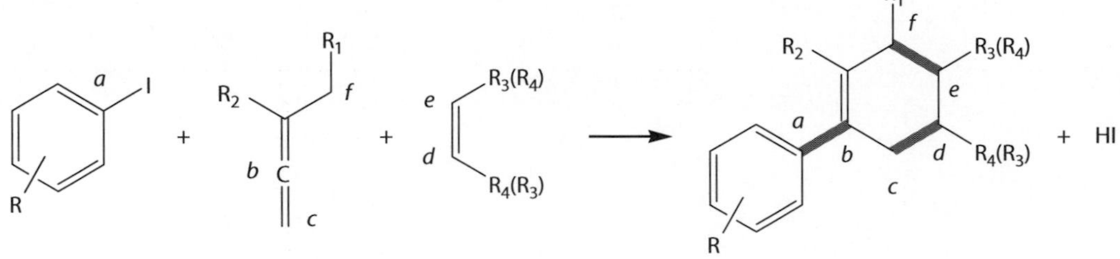

x + 199   y + z + 52   w + v + 26   x + y + z + w + v + 149   128

R = x(sum of 5 substituents); $R_1 = y; R_2 = z; R_3 = w; R_4 = v$

$$AE = [x + y + z + w + v + 149]/[x + y + z + w + v + 277]; \quad \textbf{AE(min)} = \textbf{0.55};$$

$$E = 128/[x + y + z + w + v + 149]; \quad \textit{E}\textbf{(max)} = \textbf{0.81}; \quad \textit{f}\textbf{(sac)} = \textbf{0}$$

Set $R_1 = R_2 = R_3 = R_4 = CH_3$

$$\Delta = \Delta(a) + \Delta(b) + \Delta(c) + \Delta(d) + \Delta(e) + \Delta(f)$$

$$= (0-1) + (0-0) + (-2-(-2)) + (-1-(-1)) + (-1-(-1)) + (-1-(-2)) = 0; \quad \textbf{HI} = \textbf{0}$$

β-Iminoamine synthesis

Cao, C.; Shi, Y.; Odom, A.L. *J. Am. Chem. Soc.* 2003, *125*, 2880.

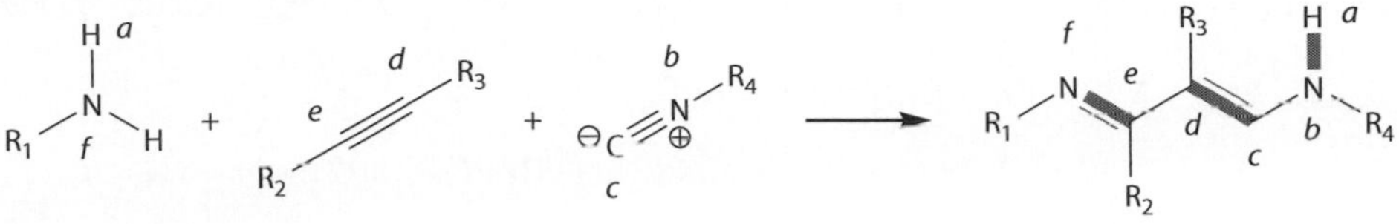

$AE = 1; E = 0; f(sac) = 0$

Set $R_1 = R_2 = R_3 = R_4 = CH_3$

$$\Delta = \Delta(a) + \Delta(b) + \Delta(c) + \Delta(d) + \Delta(e) + \Delta(f)$$

$$= (1-1) + (-3 - (-3)) + (0-2) + (0-0) + (2-0) + (-3 - (-3)) = 0; \quad \mathbf{HI = 0}$$

Indolizine synthesis

Bora, U.; Saikia, A.; Boruah, R.C. *Org. Lett.* 2003, *5*, 435.

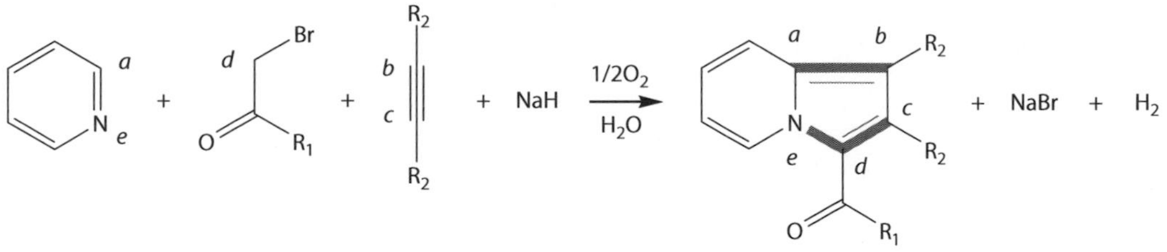

| 79 | x + 121.9 | y + 24 | 24 | x + y + 142 | 102.9 | 2 |

$R_1 = x; R_2 = y$

$$AE = [x+y+142]/[x+y+248.9]; \quad \mathbf{AE(min) = 0.58}; \quad E = 104.9/[x+y+142]; \quad \mathbf{E(max) = 0.73}$$

$$f(sac) = 24/[x+y+248.9]; \quad \mathbf{f(sac)(max) = 0.096}$$

Set $R_2 = CH_3$

$$\Delta = \Delta(a) + \Delta(b) + \Delta(c) + \Delta(d) + \Delta(e)$$

$$= (1-1) + (0-0) + (0-0) + (1-(-1)) + (-3 - (-3)) = +2; \quad \mathbf{HI = -\Delta/(N+1) = -1}$$

Isocyanate–ketone cyclization

Posner, G.H. *Chem. Rev.* 1986, *86*, 831.

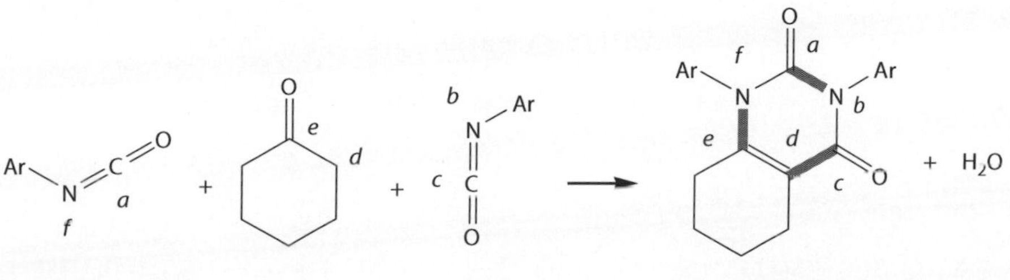

| x + 42 | 98 | x + 42 | 2x + 164 | 18 |

Ar = x

$$AE = [2x + 164]/[2x + 182]; \quad \textbf{AE(min)} = \textbf{0.95}; \quad E = 18/[2x + 164]; \quad \textbf{E(max)} = \textbf{0.057}; \quad f(\textbf{sac}) = \textbf{0}$$

$$\Delta = \Delta(a) + \Delta(b) + \Delta(c) + \Delta(d) + \Delta(e) + \Delta(f)$$

$$= (4 - 4) + (-3 - (-3)) + (3 - 4) + (0 - (-2)) + (1 - 2) + (-3 - (-3)) = 0; \quad \textbf{HI} = \textbf{0}$$

Isoquinoline + DEAD + benzoquinones
Nair, V.; Sreekanth, A.R.; Biju, A.T.; Rath, N.P. *Tetrahedron Lett.* 2003, *44*, 729.

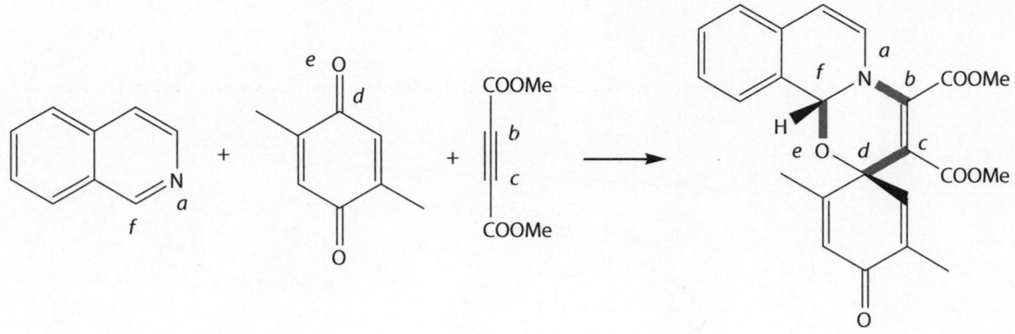

$$\textbf{AE} = \textbf{1}; E = \textbf{0}; f(\textbf{sac}) = \textbf{0}$$

$$\Delta = \Delta(a) + \Delta(b) + \Delta(c) + \Delta(d) + \Delta(e) + \Delta(f)$$

$$= (-3 - (-3)) + (1 - 0) + (0 - 0) + (1 - 2) + (-2 - (-2)) + (1 - 1) = 0; \quad \textbf{HI} = \textbf{0}$$

Isoquinolonic acid synthesis
Yadav, J.S.; Reddy, B.V.S.; Raj, K.S.; Prasad, A.R. *Tetrahedron* 2003, *59*, 1805.

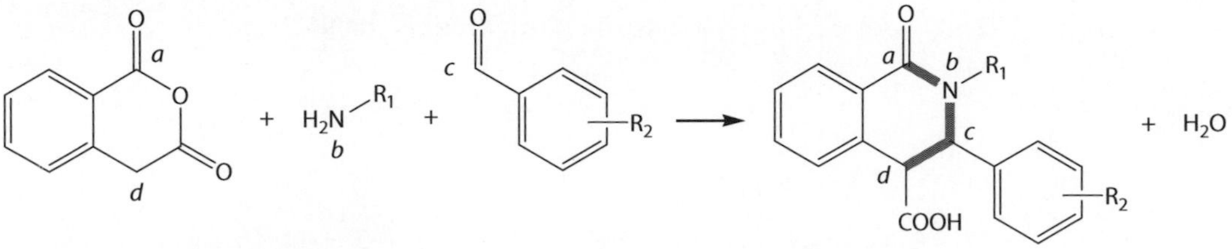

| 162 | y + 16 | x + 101 | x + y + 261 | 18 |

R = x(sum of 5 substituents); R_1 = y

$$AE = [x + y + 261]/[x + y + 279]; \quad \textbf{AE(min)} = \textbf{0.94};$$

$$E = 18/[x + y + 261]; \quad \textbf{E(max)} = \textbf{0.067}; \quad f(\textbf{sac}) = \textbf{0}$$

Set R_1 = CH$_3$

$$\Delta = \Delta(a) + \Delta(b) + \Delta(c) + \Delta(d) = (3 - 3) + (-3 - (-3)) + (0 - 1) + (-1 - (-2)) = 0; \quad \textbf{HI} = \textbf{0}$$

Isoquinolinone synthesis

Grigg, R.; Khamnaen, T.; Rajviroongit, S.; Sridharan, V. *Tetrahedron Lett.* 2002, *43*, 2601.

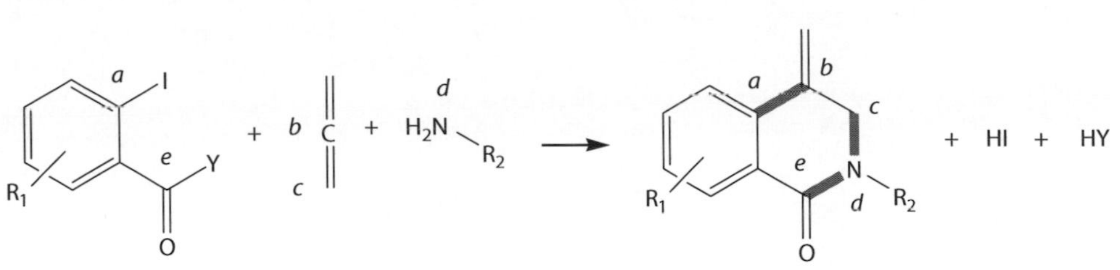

| $x+y+227$ | 40 | $z+16$ | $x+z+152$ | 128 | $y+1$ |

$R = x$(sum of 4 substituents); $Y = y$; $R_2 = z$

$$AE = [x+z+152]/[x+y+z+281]; \quad E = [y+129]/[x+z+152]; \quad f(sac) = 0$$

Y	E(max)	AE(min)
Cl	1.05	0.49
OMe	1.02	0.50

$$\Delta = \Delta(a) + \Delta(b) + \Delta(c) + \Delta(d) + \Delta(e)$$

$$= (0-1) + (0-0) + (-1-(-2)) + (-3-(-3)) + (3-3) = 0; \quad HI = 0$$

Variant of above

Gai, X.; Grigg, R.; Koppen, I.; Marchbank, J.; Sridharan, V. *Tetrahedron Lett.* 2003, *44*, 7445.

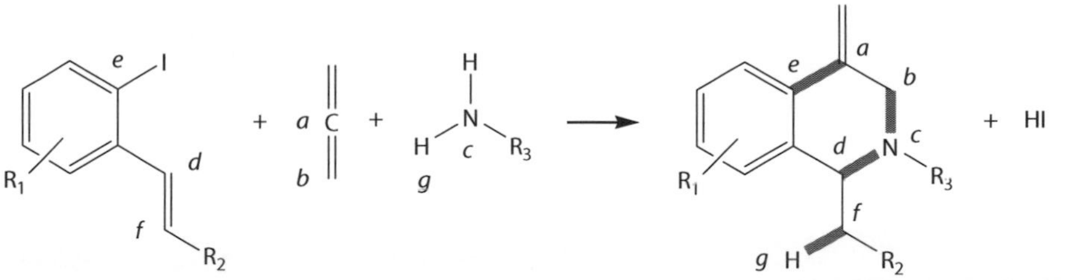

| $x+y+225$ | 40 | $z+16$ | $x+y+z+153$ | 128 |

$R_1 = x$(sum of 4 substituents); $R_2 = y$; $R_3 = z$

$$E = 128/[x+y+z+153]; \quad E(max) = 0.81; \quad AE(min) = 0.55; \quad f(sac) = 0$$

Set $R_2 = R_3 = CH_3$

$$\Delta = \Delta(a) + \Delta(b) + \Delta(c) + \Delta(d) + \Delta(e) + \Delta(f) + \Delta(g)$$

$$= (0-0) + (-1-(-2)) + (-3-(-3)) + (0-(-1)) + (0-1) + (-2-(-1)) + (1-1) = 0; \quad HI = 0$$

Knoevenagel hetero-Diels–Alder reaction
Tietze, L.F.; Evers, H.; Toepken, E. *Helv. Chim. Acta* 2002, *85*, 4200.

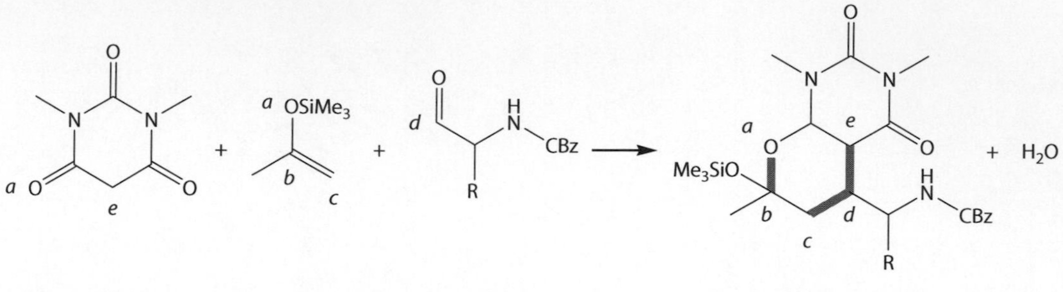

| 156 | 130 | x + 192 | x + 460 | 18 |

R = x

$$AE = [x + 460]/[x + 478]; \quad \mathbf{AE(min) = 0.96}; \quad E = 18/[x + 460]; \quad \mathbf{E(max) = 0.039}; \quad f(\mathbf{sac}) = \mathbf{0}$$

$$\Delta = \Delta(a) + \Delta(b) + \Delta(c) + \Delta(d) + \Delta(e)$$

$$= (-2 - (-2)) + (2 - 1) + (-2 - (-2)) + (-1 - 1) + (-1 - (-2)) = 0; \quad \mathbf{HI = 0}$$

1,6-Mannich condensation
Ziegler, E.; Rüf, W. *Z. Naturforsch. B* 1975, *30B*, 951.

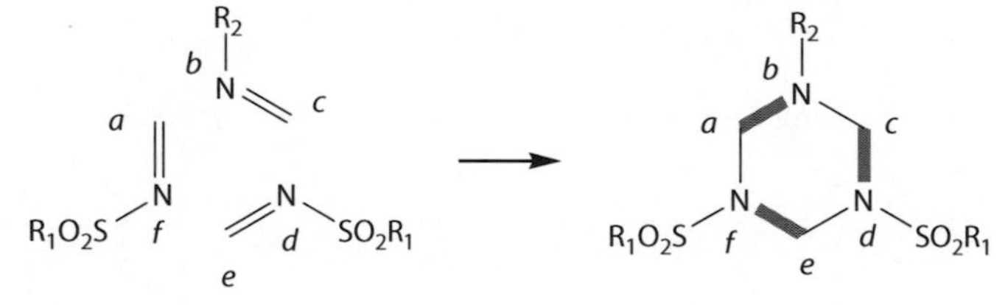

$$\mathbf{AE = 1};\ E = 0;\ f(\mathbf{sac}) = 0$$

Set $R_2 = CH_3$

$$\Delta = \Delta(a) + \Delta(b) + \Delta(c) + \Delta(d) + \Delta(e) + \Delta(f)$$

$$= (0 - 0) + (-3 - (-3)) + (0 - 0) + (-1 - (-1)) + (0 - 0) + (-1 - (-1)) = 0; \quad \mathbf{HI = 0}$$

Mannich condensation
Mannich, C.; Kroesche, I. *Arch. Pharm.* 1912, *250*, 647.

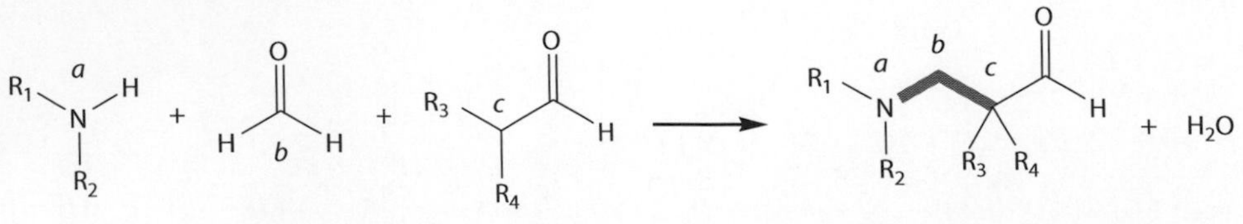

| x + y + 15 | 30 | z + w + 42 | x + y + z + w + 69 | 18 |

$R_1 = x; R_2 = y; R_3 = z; R_4 = w$

$$AE = [x + y + z + w + 69]/[x + y + z + w + 87]; \quad \textbf{AE(min)} = \textbf{0.80};$$

$$E = 18/[x + y + z + w + 69]; \quad \textbf{E(max)} = \textbf{0.25}; \quad f(\text{sac}) = \textbf{0}$$

Set $R_1 = R_2 = R_3 = R_4 = CH_3$

$$\Delta = \Delta(a) + \Delta(b) + \Delta(c) = (-3 - (-3)) + (-1 - 0) + (0 - (-1)) = 0; \quad \textbf{HI} = \textbf{0}$$

Michael–aldol–Horner–Wadsworth–Emmons

Huang, X.; Xiong, Z.C. *Chem. Commun.* 2003, *15*, 1714.

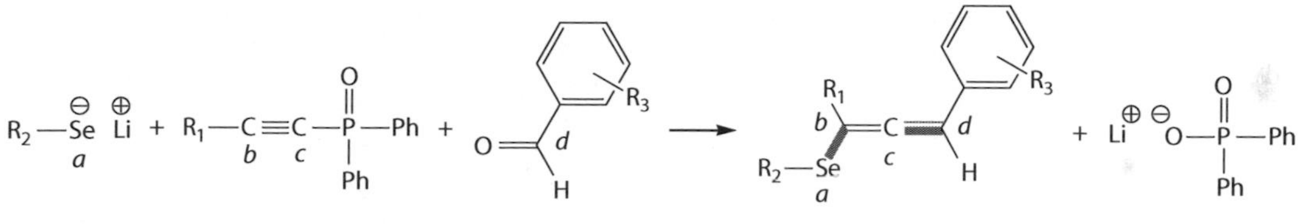

$R_1 = x; R_2 = y; R_3 = z$(sum of 5 substituents)

$$AE = [x + y + z + 187.96]/[x + y + z + 411.96]; \quad \textbf{AE(min)} = \textbf{0.47};$$

$$E = 224/[x + y + z + 187.96]; \quad \textbf{E(max)} = \textbf{1.15}; \quad f(\text{sac}) = \textbf{0}$$

Set $R_1 = R_2 = CH_3$

$$\Delta = \Delta(a) + \Delta(b) + \Delta(c) = (-2 - (-2)) + (1 - 0) + (0 - 1) + (-1 - 1) = -2; \quad \textbf{HI} = -\Delta/(N + 1) = \textbf{+1}$$

Michael–Michael–1,6-Wittig [2 + 2 + 2]

Cory, R.M.; Chan, D.M.T.; Naguib, Y.M.A.; Rastall, M.H.; Renneboog, R.M. *J. Org. Chem.* 1980, *45*, 1852.
Posner, G.H.; Lu, S.-B. *J. Am. Chem. Soc.* 1985, *107*, 1424.

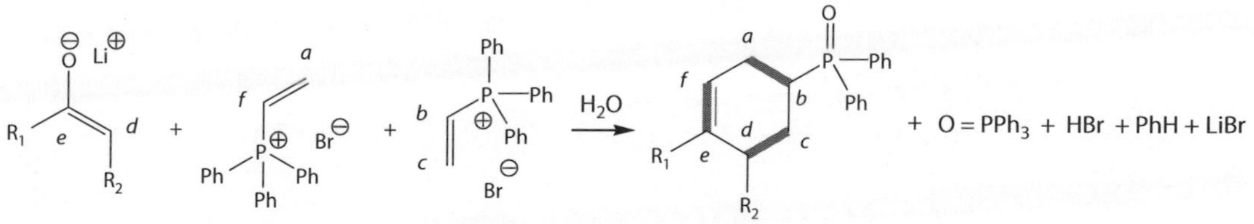

$R_1 = x; R_2 = y$

$$AE = [x + y + 280]/[x + y + 803.8]; \quad \textbf{AE(min)} = \textbf{0.35};$$

$$E = 523.8/[x + y + 280]; \quad \textbf{E(max)} = \textbf{1.86}; \quad f(\text{sac}) = \textbf{0}$$

Set $R_1 = R_2 = CH_3$

$$\Delta = \Delta(a) + \Delta(b) + \Delta(c) + \Delta(d) + \Delta(e) + \Delta(f)$$

$$= (-2 - (-2)) + (0 - 0) + (-2 - (-2)) + (-1 - (-1)) + (0 - 1) + (-1 - 0) = -2$$

$$\mathbf{HI = -\Delta/(N+1) = +1}$$

Nenitzescu–Praill pyrylium salt synthesis

Balaban, A.T.; Nenitzescu, C.D. *Ann. Chem.* 1959, *625*, 74.
Balaban, A.T.; Nenitzescu, C.D. *J. Chem. Soc.* 1961, *4*, 3553.
Balaban, A.T.; Nenitzescu, C.D. *J. Chem. Soc.* 1961, *4*, 3561.
Balaban, A.T.; Nenitzescu, C.D. *J. Chem. Soc.* 1961, 3564.
Praill, P.F.G.; Whitear, A.L. *J. Chem. Soc.* 1961, 3573.
Praill, P.F.G.; Whitear, A.L. *Proc. Chem. Soc.* 1961, 312.

x + y + z + 39   2(2w + 72)   100.45   x + y + z + 2w + 175.45   2w + 90   18

$R_1 = x$; $R_2 = y$; $R_3 = z$; $R_4 = w$

$$AE = [x + y + z + 2w + 175.45]/[x + y + z + 4w + 283.45]; \quad E = [2w + 108]/[x + y + z + 2w + 175.45]$$

$$f(sac) = 100.45/[x + y + z + 4w + 283.45]; \quad f(sac)(max) = \mathbf{100.45/[4w + 286.45]}$$

R_4	E(max)	AE(min)	f(sac)(max)
Me	0.66	0.60	0.29

Set $R_1 = R_3 = R_4 = CH_3$

$$\Delta = \Delta(a) + \Delta(b) + \Delta(c) + \Delta(d) + \Delta(e)$$

$$= (0 - (-2)) + (1 - 3) + (-2 - (-2)) + (2 - 3) + (0 - (-1)) = 0; \quad \mathbf{HI = 0}$$

x + y + z + 39   2w + 126.9   2w + 126.9   100.45   x + y + z + 2w + 175.45   72.9   18

$R_1 = x; R_2 = y; R_3 = z; R_4 = w$

$$AE = [x + y + z + 2w + 175.5]/[x + y + z + 2w + 266.45]; \quad E = 97.8/[x + y + z + 2w + 175.45]$$

$$f(\text{sac}) = 100.45/[x + y + z + 2w + 266.45]; \quad f(\text{sac})(\text{max}) = \mathbf{100.45/[2w + 269.45]}$$

R_4	E(max)	AE(min)	f(sac)(max)
Me	0.47	0.68	0.34

Set $R_1 = R_3 = R_4 = CH_3$

$$\Delta = \Delta(a) + \Delta(b) + \Delta(c) + \Delta(d) + \Delta(e)$$

$$= (0 - (-2)) + (1 - 3) + (-2 - (-2)) + (2 - 3) + (0 - (-1)) = 0; \quad \mathbf{HI = 0}$$

Norbornene synthesis

Bestmann, H.J.; Schobert, R. *Angew. Chem. Int. Ed.* 1985, *24*, 790.

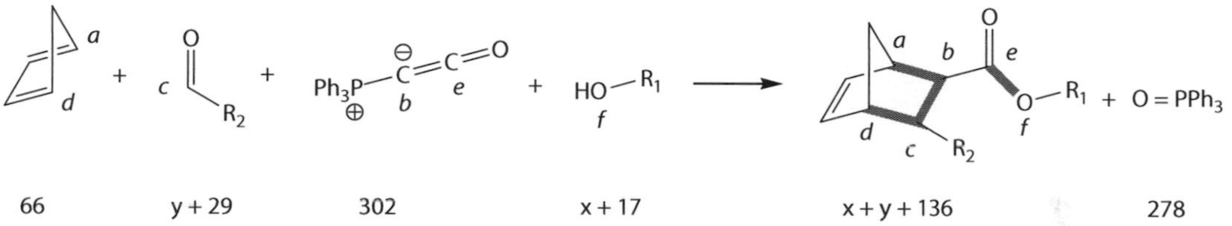

| 66 | $y + 29$ | 302 | $x + 17$ | $x + y + 136$ | 278 |

$$AE = [x + y + 136]/[x + y + 414]; \quad \mathbf{AE(min) = 0.33};$$

$$E = 278/[x + y + 136]; \quad \mathbf{E(max) = 2.01}; \quad f(\text{sac}) = 0$$

Set $R_1 = R_2 = CH_3$

$$\Delta = \Delta(a) + \Delta(b) + \Delta(c) + \Delta(d) + \Delta(e) + \Delta(f)$$

$$= (-1 - (-1)) + (-1 - 0) + (-1 - 1) + (-1 - (-1)) + (3 - 2) + (-2 - (-2)) = -2; \quad \mathbf{HI = +1}$$

Oxazolidin-2-one synthesis

Gabriele, B.; Mancuso, R.; Salerno, G.; Costa, M. *J. Org. Chem.* 2003, *68*, 601.

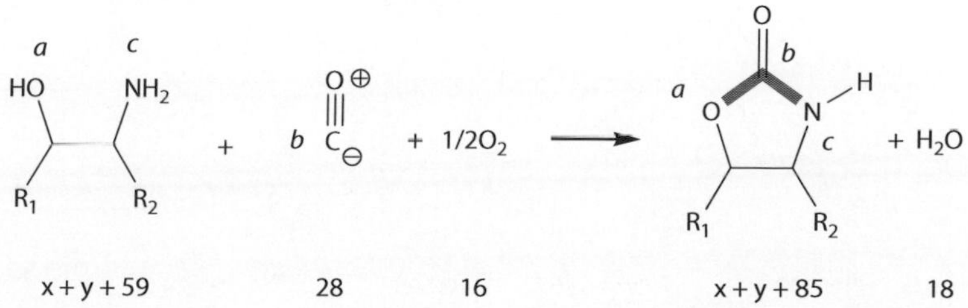

| $x + y + 59$ | 28 | 16 | $x + y + 85$ | 18 |

$R_1 = x; R_2 = y$

$$AE = [x + y + 85]/[x + y + 103]; \quad \textbf{AE(min)} = \textbf{0.83}; \quad E = 18/[x + y + 85]; \quad \textbf{E(max)} = \textbf{0.21}$$

$$f(\text{sac}) = 16/[x + y + 103]; \quad \textbf{\textit{f}(sac)(max)} = \textbf{0.15}$$

$$\Delta = \Delta(a) + \Delta(b) + \Delta(c) = (-2 - (-2)) + (4 - 2) + (-3 - (-3)) = +2; \quad \textbf{HI} = -\Delta/(N+1) = \textbf{-1}$$

Passerini reaction

Passerini, M. *Gazz. Chim. Ital.* 1921, *51(2)*, 126.
Andreana, P.R.; Liu, C.C.; Schreiber, S.L. *Org. Lett.* 2004, *6*, 4231.

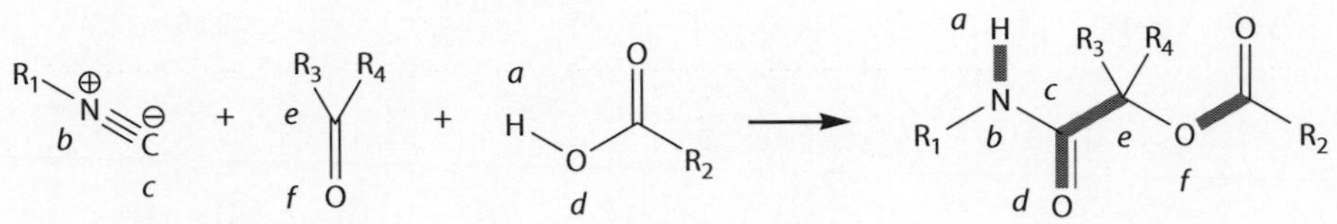

AE = 1; *E* = 0; *f*(sac) = 0

Set $R_1 = R_2 = R_3 = R_4 = CH_3$

$$\Delta = \Delta(a) + \Delta(b) + \Delta(c) + \Delta(d) + \Delta(e) + \Delta(f)$$

$$= (1 - 1) + (-3 - (-3)) + (3 - 2) + (-2 - (-2)) + (1 - 2) + (-2 - (-2)) = 0; \quad \textbf{HI} = \textbf{0}$$

Pauson–Khand reaction

Khand, I.U.; Knox, G.R.; Pauson, P.L. Watts, W.R. *Chem. Commun.* 1971, 36.

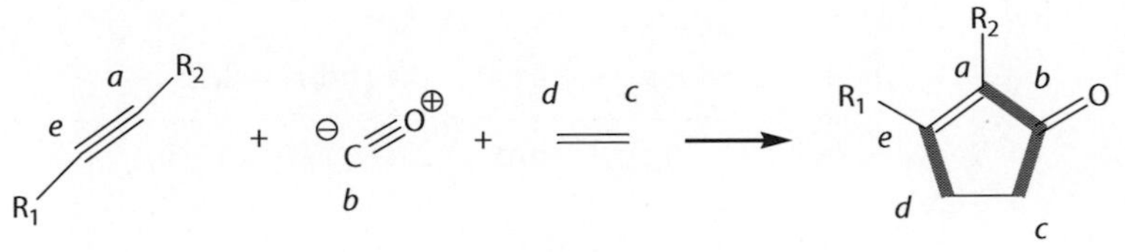

AE = 1; *E* = 0; *f*(sac) = 0

Set $R_1 = R_2 = CH_3$

$$\Delta = \Delta(a) + \Delta(b) + \Delta(c) + \Delta(d) + \Delta(e)$$

$$= (0 - 0) + (2 - 2) + (-2 - (-2)) + (-2 - (-2)) + (0 - 0) = 0; \quad \textbf{HI} = \textbf{0}$$

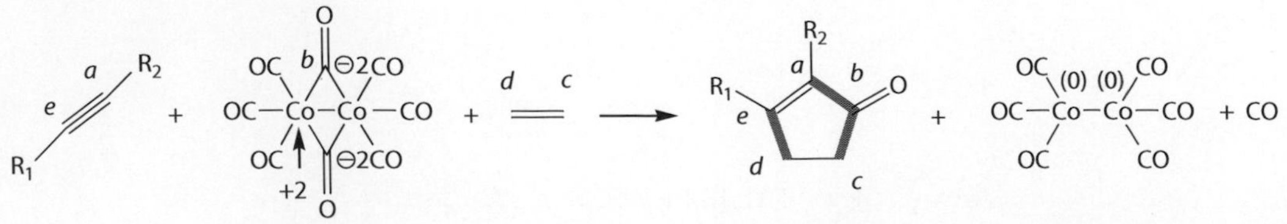

| $x + y + 24$ | 341.86 | 28 | $x + y + 80$ | 285.86 | 28 |

$R_1 = x, R_2 = y$

$$AE = [x + y + 80]/[x + y + 393.86]; \quad \textbf{AE(min)} = \textbf{0.21};$$

$$E = 313.86/[x + y + 80]; \quad \textbf{E(max)} = \textbf{3.83}; \quad f\textbf{(sac)} = \textbf{0}$$

$$\Delta = \Delta(a) + \Delta(b) + \Delta(c) + \Delta(d) + \Delta(e)$$

$$= (0 - 0) + (2 - 0) + (-2 - (-2)) + (-2 - (-2)) + (0 - 0) = 2; \quad \textbf{HI} = -\Delta/(N+1) = -1$$

Petasis condensation
Petasis, N.A.; Goodman, A.; Zavialov, I.A. *Tetrahedron* 1997, *53*, 16463.

| $x + y + 15$ | $z + 45$ | 74 | $x + y + z + 72$ | 62 |

$R_1 = x; R_2 = y; R_3 = z$

$$AE = [x + y + z + 72]/[x + y + z + 134]; \quad \textbf{AE(min)} = \textbf{0.55};$$

$$E = 62/[x + y + z + 72]; \quad \textbf{E(max)} = \textbf{0.83}; \quad f\textbf{(sac)} = \textbf{0}$$

Set $R_1 = R_2 = R_3 = CH_3$

$$\Delta = \Delta(a) + \Delta(b) + \Delta(c) = (-3 - (-3)) + (0 - 1) + (-3 - (-4)) = 0; \quad \textbf{HI} = \textbf{0}$$

Petrenko-Kritschenko reaction
Petrenko-Kritschenko, P.; Zoneff, N. *Chem. Ber.* 1906, *39*, 1358.
Petrenko-Kritschenko, P.; Lewin, N. *Chem. Ber.* 1907, *40*, 2882.
Petrenko-Kritschenko, P.; Petrow, W. *Chem. Ber.* 1908, *41*, 1692.
Petrenko-Kritschenko, P.; Schoettle, S. *Chem. Ber.* 1909, *42*, 2020.

| $y + 16$ | $2(x + 29)$ | 202 | $2x + y + 240$ | 36 |

$R_1 = x; R_2 = y$

$$AE = [2x + y + 240]/[2x + y + 276]; \quad \textbf{AE(min)} = \textbf{0.87};$$

$$E = 36/[2x + y + 240]; \quad \textbf{E(max)} = \textbf{0.15}; \quad f\textbf{(sac)} = \textbf{0}$$

Set $R_1 = R_2 = CH_3$

$$\Delta = \Delta(a) + \Delta(b) + \Delta(c) + \Delta(d) + \Delta(e)$$

$$= (-1-(-2)) + (-1-(-2)) + (0-1) + (-3-(-3)) + (0-1) = 0; \quad \textbf{HI} = \textbf{0}$$

Pinner triazine synthesis
Pinner, A. *Chem. Ber.* 1890, *23*, 2919.

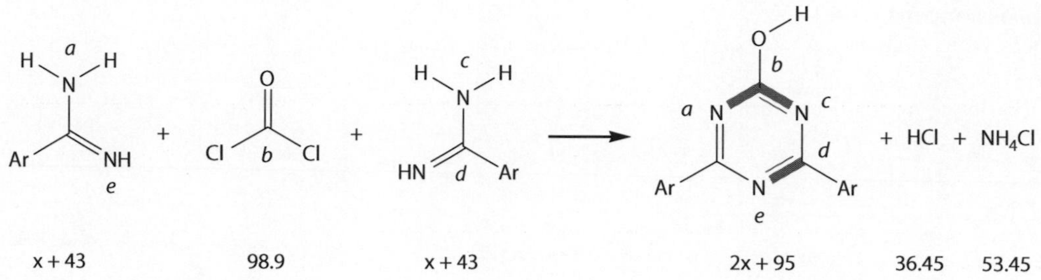

$$\quad x+43 \qquad\qquad 98.9 \qquad\qquad x+43 \qquad\qquad\qquad 2x+95 \qquad\quad 36.45 \quad 53.45$$

Ar = x

$$AE = [2x+95]/[2x+184.9]; \quad \textbf{AE(min)} = \textbf{0.73}; \quad E = 89.9/[2x+95]; \quad \textbf{E(max)} = \textbf{0.36}; \quad f(\text{sac}) = 0$$

$$\Delta = \Delta(a) + \Delta(b) + \Delta(c) + \Delta(d) + \Delta(e)$$

$$= (-3-(-3)) + (4-4) + (-3-(-3)) + (3-3) + (-3-(-3)) = 0; \quad \textbf{HI} = \textbf{0}$$

Pyrano and furanoquinolines
Yadav, J.S.; Reddy, B.V.S.; Reddy, J.S.S.; Rao, R.S. *Tetrahedron* 2003, *59*, 1599.
Ravindranath, N.; Ramesh, C.; Reddy, M.R.; Das, B. *Chem. Lett.* 2003, *32*, 222.

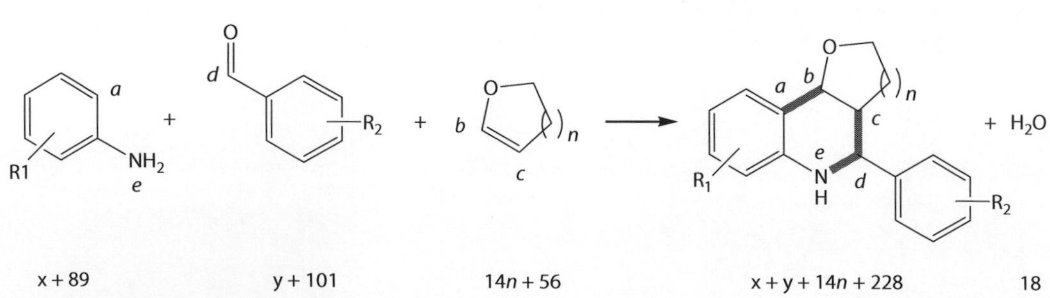

$$\quad x+89 \qquad\qquad y+101 \qquad\qquad 14n+56 \qquad\qquad x+y+14n+228 \qquad\qquad 18$$

R_1 = x(sum of 4 substituents); R_2 = y(sum of 5 substituents)

$$AE = [x+y+14n+228]/[x+y+14n+246]; \quad \textbf{AE(min)} = \textbf{[14n+237]/[14n+255]};$$

$$E = 18/[x+y+14n+228]; \quad \textbf{E(max)} = \textbf{18/[14n+237]}; \quad f(\text{sac}) = 0$$

n	E(max)	AE(min)
1 (furano)	0.072	0.93
2 (pyrano)	0.068	0.94

$$\Delta = \Delta(a) + \Delta(b) + \Delta(c) + \Delta(d) + \Delta(e)$$

$$= (0 - (-1)) + (0 - 0) + (-1 - (-1)) + (0 - 1) + (-3 - (-3)) = 0; \quad \textbf{HI} = \textbf{0}$$

2′,3′-Pyranone(pyrrolidinone)-fused tryptamine synthesis

Dardennes, E.; Kovacs-Kulyassa, A.; Renzetti, A.; Sap, J.; Laronze, J.Y. *Tetrahedron Lett.* 2003, *44*, 221.

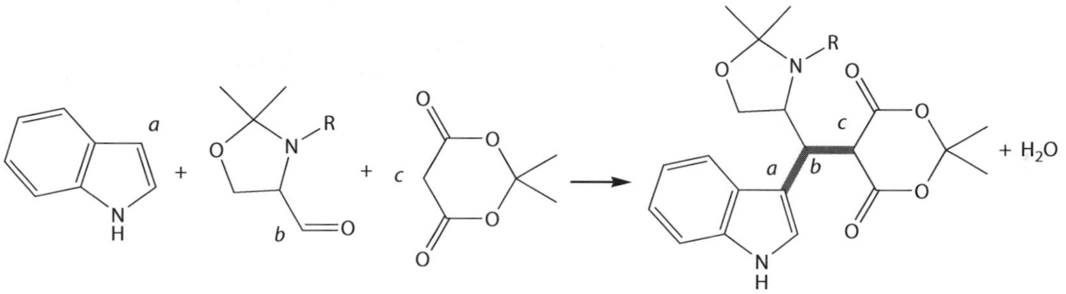

| 117 | x + 128 | 144 | x + 371 | 18 |

R = x

$$AE = [x + 371]/[x + 389]; \quad \textbf{AE(min)} = \textbf{0.95}; \quad E = 18/[x + 371]; \quad \textbf{E(max)} = \textbf{0.048}; \quad f(\text{sac}) = \textbf{0}$$

$$\Delta = \Delta(a) + \Delta(b) + \Delta(c) = (0 - (-1)) + (-1 - 1) + (-1 - (-2)) = 0; \quad \textbf{HI} = \textbf{0}$$

Pyridine synthesis

Kelly, T.R.; Liu, H. *J. Am. Chem. Soc.* 1985, *107*, 4998.

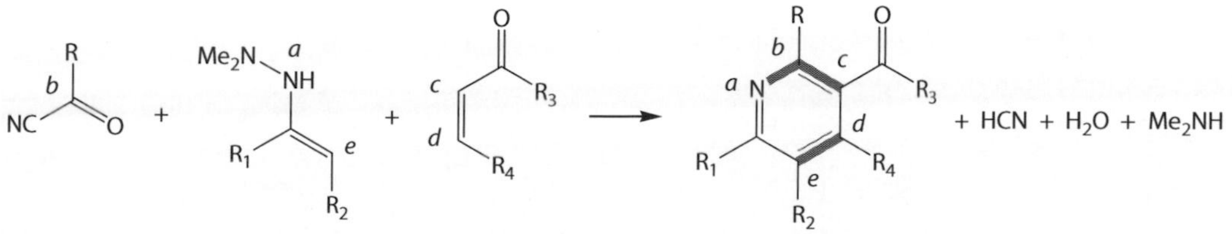

| x + 54 | y + z + 84 | w + r + 54 | x + y + z + w + r + 102 | 27 | 18 | 45 |

R = x; R$_1$ = y; R$_2$ = z; R$_3$ = w; R$_4$ = r

$$AE = [x + y + z + w + r + 102]/[x + y + z + w + r + 192]; \quad \textbf{AE(min)} = \textbf{0.54};$$

$$E = 90/[x + y + z + w + r + 102]; \quad \textbf{E(max)} = \textbf{0.84}; \quad f(\text{sac}) = \textbf{0}$$

Set $R = R_1 = R_2 = R_4 = CH_3$

$$\Delta = \Delta(a) + \Delta(b) + \Delta(c) + \Delta(d) + \Delta(e) = (-3-(-2)) + (1-2) + (0-(-1)) + (0-(-1)) + (0-(-1)) = +1;$$

$$\mathbf{HI = -\Delta/(N+1) = -½}$$

1H-Pyrrolo[3,2e]-1,2,4-triazine synthesis

Charushin, V.N.; Mochulskaya, N.N.; Andreiko, A.A.; Filyakova, V.I.; Kodess, M.I.; Chupakhin, O.N. *Tetrahedron Lett.* 2003, *44*, 2421.

x + 80   102   y + 84   x + y + 206   60

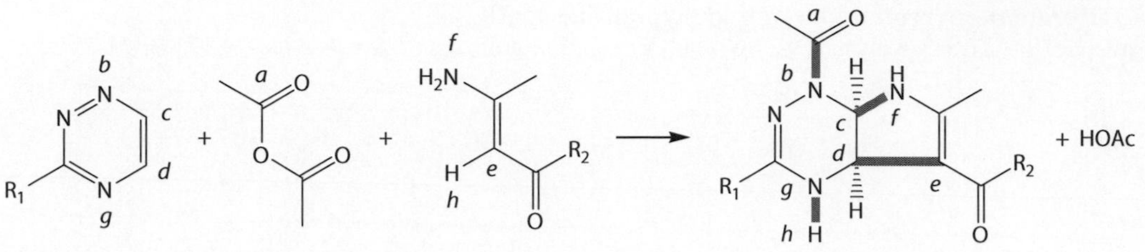

$R_1 = x; R_2 = y$

$$AE = [x + y + 206]/[x + y + 266]; \quad \mathbf{AE(min) = 0.78};$$

$$E = 60/[x + y + 206]; \quad \boldsymbol{E}\mathbf{(max) = 0.29}; \quad \boldsymbol{f}\mathbf{(sac) = 0}$$

$$\Delta = \Delta(a) + \Delta(b) + \Delta(c) + \Delta(d) + \Delta(e) + \Delta(f) + \Delta(g) + \Delta(h)$$

$$= (3-3) + (-2-(-1)) + (1-0) + (0-0) + (0-(-1)) + (-3-(-3)) + (-3-(-3)) + (1-1) = +1;$$

$$\mathbf{HI = -\Delta/(N+1) = -½}$$

Radziszewski reaction

Radziszewski, B. *Chem. Ber.* 1882, *15*, 1493.

z + w + 56   17   y + 29   x + 16   x + y + z + w + 64   54

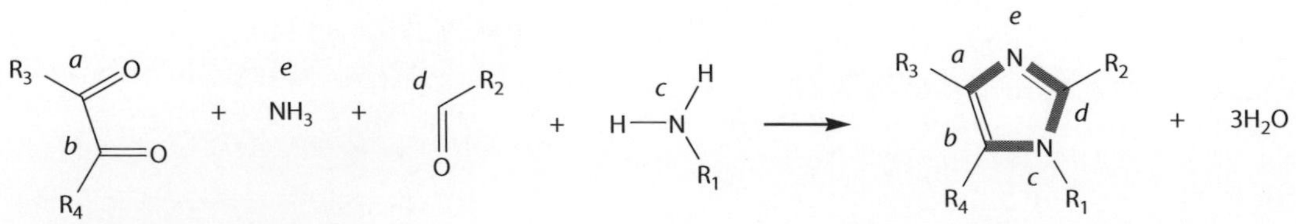

$R_1 = x; R_2 = y; R_3 = z; R_4 = w$

$$AE = [x + y + z + w + 64]/[x + y + z + w + 118]; \quad \mathbf{AE(min) = 0.56};$$

$$E = 54/[x + y + z + w + 64]; \quad \boldsymbol{E}\mathbf{(max) = 0.79}$$

Set $R_1 = R_2 = R_3 = R_4 = CH_3$

$$\Delta = \Delta(a) + \Delta(b) + \Delta(c) + \Delta(d) + \Delta(e)$$

$$= (1-2) + (1-2) + (-3-(-3)) + (3-1) + (-3-(-3)) = 0; \quad \mathbf{HI = 0}$$

Radziszewski-type reaction using microwave irradiation

Balalaie, S.; Hashemi, M.M.; Akhbari, M. *Tetrahedron Lett.* 2003, *44*, 1709.

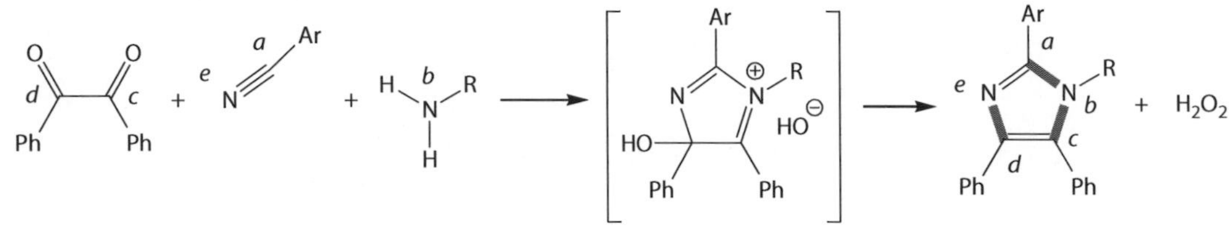

| 210 | x + 98 | y + 16 | | x + y + 290 | 34 |

Ar = x(sum of 5 substituents on phenyl ring); R = y

$$AE = [x + y + 290]/[x + y + 324]; \quad \mathbf{AE(min) = 0.92};$$

$$E = 34/[x + y + 290]; \quad \mathbf{E(max) = 0.092}; \quad f(\mathbf{sac}) = \mathbf{0}$$

Set $R = CH_3$

$$\Delta = \Delta(a) + \Delta(b) + \Delta(c) + \Delta(d) + \Delta(e)$$

$$= (3-3) + (-3-(-3)) + (1-2) + (1-2) + (-3-(-3)) = -2; \quad \mathbf{HI = -\Delta/(N+1) = +1}$$

Robinson–Schoepf synthesis of tropane

Robinson, R. *J. Chem. Soc.* 1917, *111*, 762.
Robinson, R. *J. Chem. Soc.* 1917, *111*, 876.
Schoepf, C. *Angew. Chem.* 1937, *50*, 779.
Schoepf, C. *Angew. Chem.* 1937, *50*, 797.

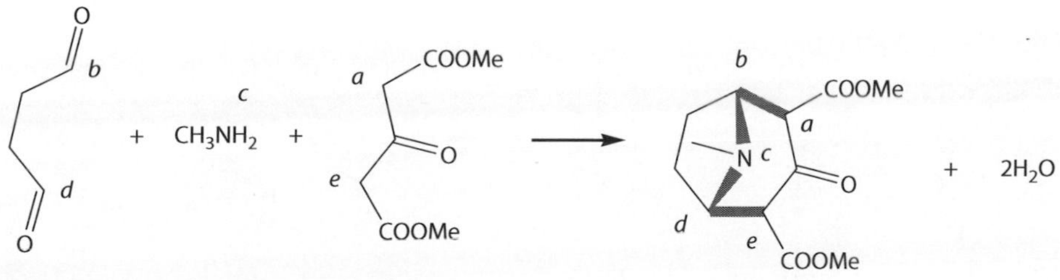

| 86 | 31 | 174 | 255 | 36 |

$AE = 0.88; E = 36/255 = 0.14; f(\mathbf{sac}) = 0$

$$\Delta = \Delta(a) + \Delta(b) + \Delta(c) + \Delta(d) + \Delta(e)$$

$$= (-1-(-2)) + (0-1) + (-3-(-3)) + (0-1) + (-1-(-2)) = 0; \quad \mathbf{HI = 0}$$

Reppe cyclooctatetraene synthesis

Reppe, W.; Schlichting, O.; Klager, K.; Toepel, T. *Ann. Chem.* 1948, *560*, 1.

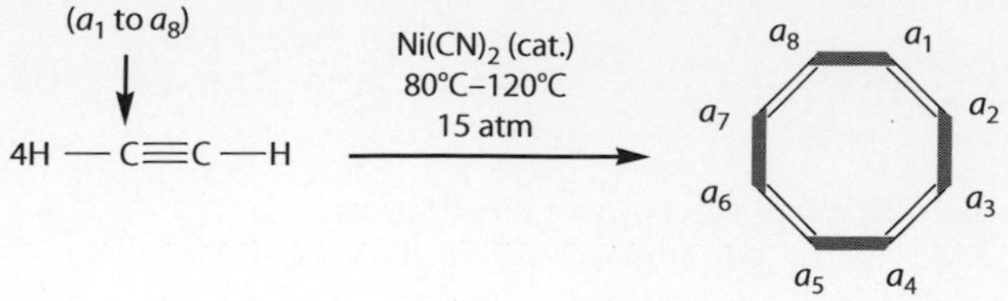

$$AE = 1; \; E = 0; \; f(sac) = 0$$

$$\Delta = \sum_{j=1}^{8} \Delta(a_j) = 8(-1-(-1)) = 0; \quad \mathbf{HI = 0}$$

Reppe synthesis

Reppe, W. *Neue Entwicklungen Aufdem Gebiet der Chemie des Acetylens und Kohlenmonoxyd*, Springer: Berlin, Germany, 1949.

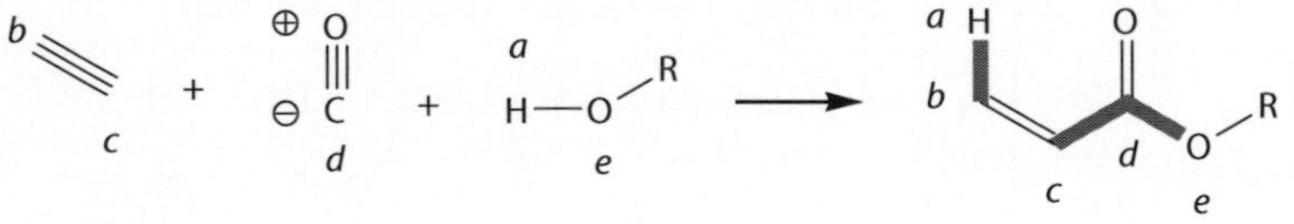

$$AE = 1; \; E = 0; \; f(sac) = 0$$

$$\Delta = \Delta(a) + \Delta(b) + \Delta(c) + \Delta(d) + \Delta(e)$$

$$= (1-1) + (-2-(-1)) + (-1-(-1)) + (3-2) + (-2-(-2)) = 0; \quad \mathbf{HI = 0}$$

Riehm quinoline synthesis

Engler, C.; Riehm, P. *Chem. Ber.* 1885, *18*, 2245.
Levin, J.; Riehm, P. *Chem. Ber.* 1886, *19*, 1394.

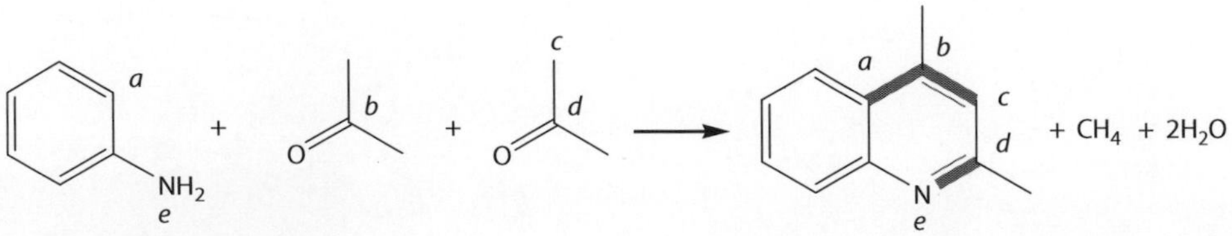

93   58   58   157   16   36

$$AE = 0.75; \; E = 52/157 = 0.33; \; f(sac) = 0$$

$$\Delta = \Delta(a) + \Delta(b) + \Delta(c) + \Delta(d) + \Delta(e)$$

$$= (0-(-1)) + (0-2) + (-1-(-3)) + (2-2) + (-3-(-3)) = +1; \quad \mathbf{HI = -\Delta/(N+1) = -1/2}$$

Roelen synthesis

Roelen, O. *Angew. Chem.* 1948, *60*, 62.
Roelen, O. *Angew. Chem.* 1948, *60*, 213.

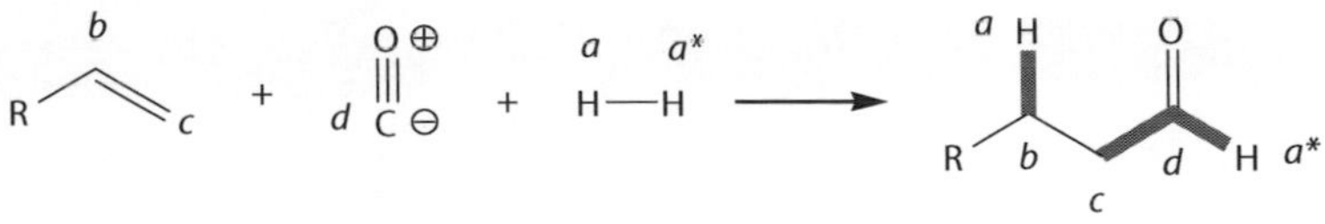

$AE = 1$; $E = 0$; $f(sac) = 0$

Set R = CH$_3$

$$\Delta = \Delta(a) + \Delta(a^*) + \Delta(b) + \Delta(c) + \Delta(d)$$

$$= (1-0) + (1-0) + (-2-(-1)) + (-2-(-2)) + (1-2) = 0; \quad \mathbf{HI = 0}$$

Rothemund reaction

Rothemund, P. *J. Am. Chem. Soc.* 1935, *57*, 2010.
Rothemund, P. *J. Am. Chem. Soc.* 1936, *58*, 625.
Rothemund, P. *J. Am. Chem. Soc.* 1939, *61*, 2912.

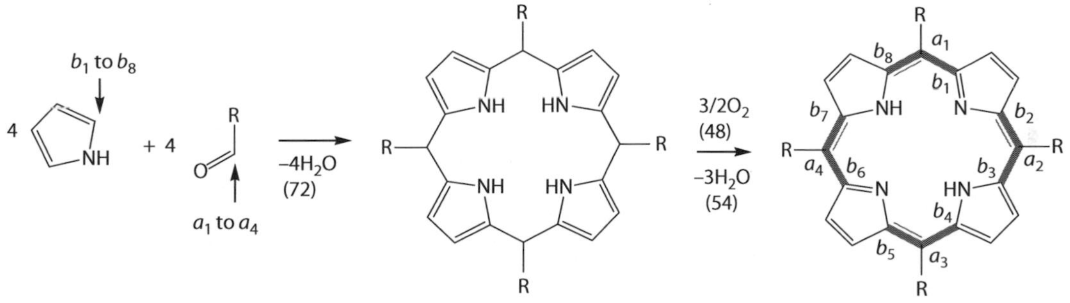

268   4(x + 29)   4x + 306

R = x

$$AE = [4x + 306]/[4x + 432]; \quad \mathbf{AE(min) = 0.71}; \quad E = 126/[4x + 306]; \quad \mathbf{E(max) = 0.41}$$

$$f(sac) = 48/[4x + 432]; \quad \mathbf{f(sac)(max) = 0.11}$$

Set R = CH$_3$

$$\Delta = \sum_{j=1}^{4} a_j + \sum_{j=1}^{8} b_j = 4(0-1) + 2(2-0) + 6(1-0) = +6; \quad \mathbf{HI = -\Delta/(N+1) = -3}$$

Strecker synthesis of α-cyanoamines
Strecker, A. *Ann. Chem.* 1850, *75*, 27.

x + 29	17	27	x + 55	18

R = x

$$AE = [x+55]/[x+73]; \quad \mathbf{AE(min)} = \mathbf{0.76}; \quad E = 18/[x+55]; \quad \mathbf{E(max)} = \mathbf{0.32}; \quad f(sac) = \mathbf{0}$$

Set R = CH$_3$

$$\Delta = \Delta(a) + \Delta(b) + \Delta(c) = (-3 - (-3)) + (0 - 1) + (3 - 2) = 0; \quad \mathbf{HI} = \mathbf{0}$$

Synthesis of 1,7-enynes
Jeganmohan, M.; Shanmugasundaram, M.; Chien-Hong, C. *Org. Lett.* 2003, *5*, 881.

328.69	x + 77	y + w + z + 73.45	x + y + z + w + 154	325.14

R$_1$ = x; R$_2$ = y; R$_3$ = z; R$_4$ = w

$$AE = [x+y+z+w+154]/[x+y+z+w+479.45]; \quad \mathbf{AE(min)} = \mathbf{0.33};$$

$$E = 325.45/[x+y+z+w+154]; \quad \mathbf{E(max)} = \mathbf{2.06}; \quad f(sac) = \mathbf{0}$$

Set R$_1$ = R$_2$ = CH$_3$

$$\Delta = \Delta(a) + \Delta(b) + \Delta(c) + \Delta(d) + \Delta(e)$$

$$= (1 - 1) + (-2 - (-2)) + (-1 - (-1)) + (0 - 0) + (-1 - 0) = -1; \quad \mathbf{HI} = -\Delta/(N+1) = +\tfrac{1}{2}$$

Synthesis of THF derivatives
Liu, G.; Lu, X. *Tetrahedron Lett.* 2003, *44*, 467.

76.45	x + 55	y + w + z + 25	64	x + y + z + w + 120	42.45	58

$R_1 = x$; $R_2 = y$; $E_1 = z$; $E_2 = w$

$$AE = [x + y + z + w + 120]/[x + y + z + w + 220.45]; \quad \textbf{AE(min)} = \textbf{0.55};$$

$$E = 100.45/(x + y + z + w + 120); \quad \textbf{E(max)} = \textbf{0.81}; \quad \textbf{\textit{f}(sac)} = \textbf{0}$$

Set $R_2 = CH_3$, $E_1 = E_2 = CN$

$$\Delta = \Delta(a) + \Delta(b) + \Delta(c) + \Delta(d) + \Delta(e)$$

$$= (-2 - (-2)) + (0 - (-1)) + (0 - 0) + (0 - 0) + (-1 - (-1)) + (-2 - (-2)) = +1;$$

$$\textbf{HI} = -\Delta/(N + 1) = -\frac{1}{2}$$

Tandem Asinger–Ugi condensation (7-CC)

Dömling, A.; Ugi, I. *Angew. Chem. Int. Ed.* 1993, *32*, 563.
Dömling, A.; Herdtweck, E.; Ugi, I. *Acta Chem. Scand.* 1998, *52*, 107.

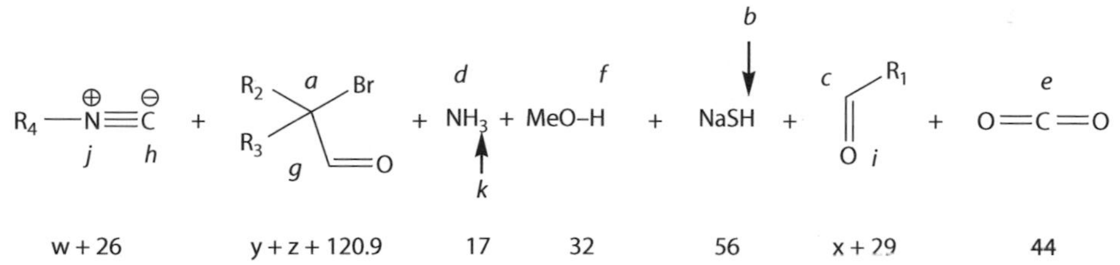

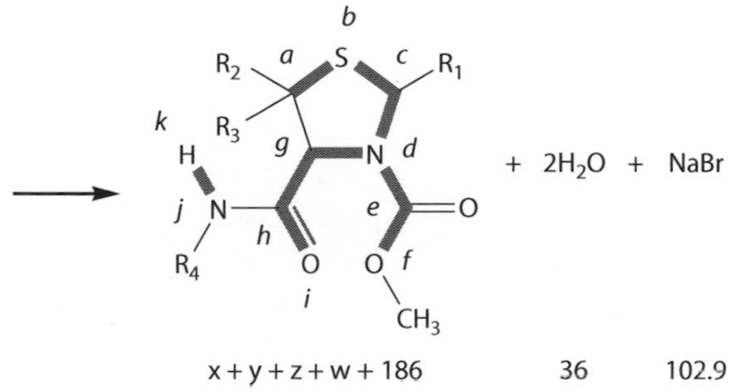

$R_1 = x$; $R_2 = y$; $R_3 = z$; $R_4 = w$

$$E = 138.9/[x + y + z + w + 186]; \quad \textbf{E(max)} = \textbf{0.73};$$

$$AE = [x + y + z + w + 186]/[x + y + z + w + 324.9]; \quad \textbf{AE(min)} = \textbf{0.58}; \quad \textbf{\textit{f}(sac)} = \textbf{0}$$

Set $R_1 = R_2 = R_3 = R_4 = CH_3$

$$\Delta = \Delta(a) + \Delta(b) + \Delta(c) + \Delta(d) + \Delta(e) + \Delta(f) + \Delta(g) + \Delta(h) + \Delta(i) + \Delta(j)$$

$$= (1 - 1) + (-2 - (-2)) + (1 - 1) + (-3 - (-3)) + (4 - 4) + (-2 - (-2)) + (0 - 1)$$

$$+ (3 - 2) + (-2 - (-2)) + (-3 - (-3)) + (1 - 1) = 0; \quad \textbf{HI} = \textbf{0}$$

Tandem Passerini–Wittig reaction

Beck, B.; Magnin-Lachaux, M.; Herdtweck, E.; Dömling, A. *Org. Lett.* 2001, *3*, 2875.

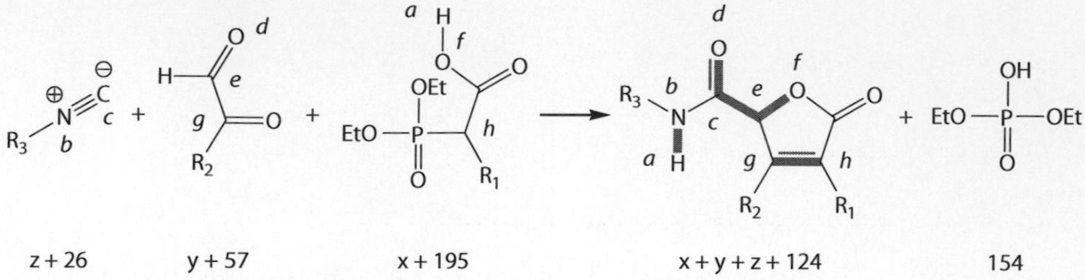

$$z + 26 \qquad y + 57 \qquad x + 195 \qquad\qquad x + y + z + 124 \qquad 154$$

$R_1 = x;\ R_2 = y;\ R_3 = z$

$$E = 154/[x + y + z + 124]; \quad E(\max) = 1.21; \quad AE(\min) = 0.45; \quad f(sac) = 0$$

Set $R_1 = R_2 = R_3 = CH_3$

$$\Delta = \Delta(a) + \Delta(b) + \Delta(c) + \Delta(d) + \Delta(e) + \Delta(f) + \Delta(g) + \Delta(h)$$

$$= (1-1) + (-3-(-3)) + (3-2) + (-2-(-2)) + (0-1) + (-2-(-2)) + (0-2) + (0-0) = -2;$$

$$HI = -\Delta/(N+1) = +1$$

Tandem Petasis–Ugi condensation

Portlock, D.E.; Ostaszewski, R.; Naskar, D.; West, L. *Tetrahedron Lett.* 2003, *44*, 603.

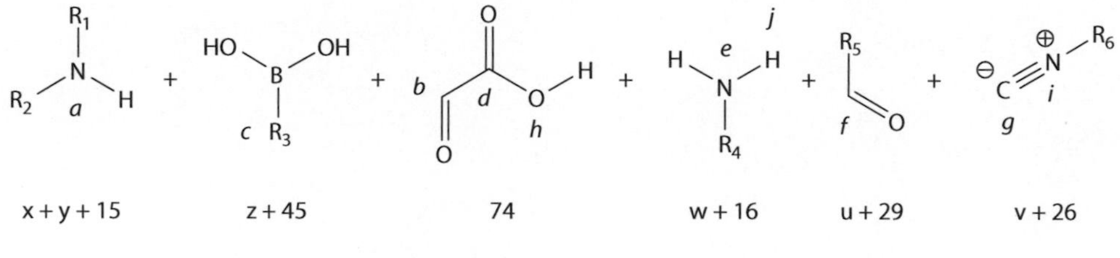

$$x + y + 15 \qquad z + 45 \qquad 74 \qquad w + 16 \qquad u + 29 \qquad v + 26$$

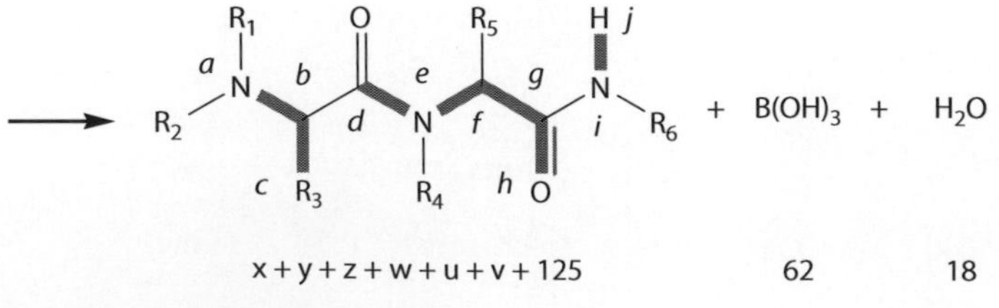

$$x + y + z + w + u + v + 125 \qquad\qquad 62 \qquad\qquad 18$$

$R_1 = x;\ R_2 = y;\ R_3 = z;\ R_4 = w;\ R_5 = u;\ R_6 = v$

$$AE = [x + y + z + w + u + v + 125]/[x + y + z + w + u + v + 205]; \quad AE(\min) = 0.62$$

$$E = 80/[x + y + z + w + u + v + 125]; \quad E(\max) = 0.61; \quad f(sac) = 0$$

Set $R_1 = R_2 = R_3 = R_4 = R_5 = R_6 = CH_3$

$$\Delta = \Delta(a) + \Delta(b) + \Delta(c) + \Delta(d) + \Delta(e) + \Delta(f) + \Delta(g) + \Delta(h) + \Delta(i)$$

$$= (-3 - (-3)) + (0 - 1) + (-3 - (-4)) + (3 - 3) + (-3 - (-3)) + (0 - 1)$$

$$+ (3 - 2) + (-2 - (-2)) + (1 - 1) = 0; \quad \textbf{HI} = \textbf{0}$$

Tetrasubstituted olefins

Zhao, C.; Emrich, D.E.; Larock, R.C. *Org. Lett.* 2003, *5*, 1579.

| x + 199 | y + z + 24 | w + 117 | 18 | x + y + z + w + 168 | 62 | 128 |

$R_1 = $ x(sum of 5 substituents); $R_2 = $ y; $R_3 = $ z; $R_4 = $ w(sum of 5 substituents)

$$AE = [x + y + z + w + 168]/[x + y + z + w + 244]; \quad \textbf{AE(min)} = \textbf{0.51};$$

$$E = 172/[x + y + z + w + 168]; \quad \textbf{E(max)} = \textbf{0.96}$$

$$f(\text{sac}) = 18/[x + y + z + w + 244]; \quad \textbf{\textit{f}(sac)(max)} = \textbf{0.070}$$

$$\Delta = \Delta(a) + \Delta(b) + \Delta(c) + \Delta(d) = (0 - 1) + (0 - 0) + (0 - 0) + (0 - (-1)) = 0; \quad \textbf{HI} = \textbf{0}$$

Thalidomide synthesis

Seijas, J.A.; Vazquez-Tato, M.P.; Gonzalez-Bande, C.; Martinez, M.M.; Beatriz, P.-L. *Synthesis* 2001, 999.

| 147 | 147 | 76 | 258 |

$$\textbf{AE} = \textbf{0.70}; \quad E = 112/258 = \textbf{0.43}; \quad f(\text{sac}) = \textbf{0}$$

$$\Delta = \Delta(a) + \Delta(b) + \Delta(c) + \Delta(d)$$

$$= (-3 - (-3)) + (0 - (-2)) + (3 - 3) + (-3 - (-3)) + (3 - 3) = +2; \quad \textbf{HI} = -\Delta/(N + 1) = \textbf{-1}$$

Thiele reaction
Thiele, J. *Chem. Ber.* 1898, *31*, 1247.

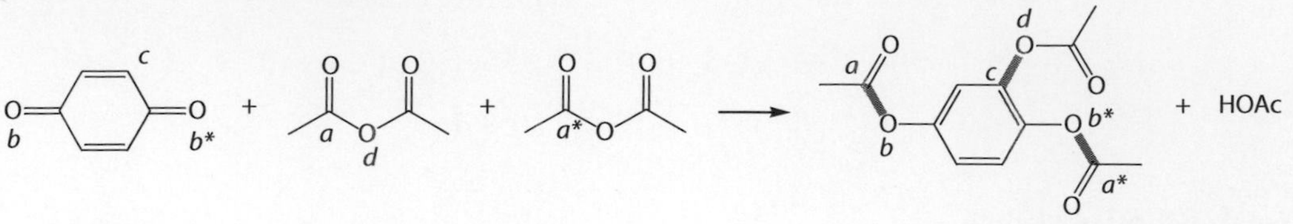

108   102   102   252   60

AE = 0.81; *E* = 60/252 = 0.24; *f*(sac) = 0

$$\Delta = \Delta(a) + \Delta(a^*) + \Delta(b) + \Delta(b^*) + \Delta(c) + \Delta(d)$$

$$= (3-3) + (3-3) + (-2-(-2)) + (-2-(-2)) + (1-(-1)) + (-2-(-2)) = 0; \quad \mathbf{HI = 0}$$

Trimerization of alkynes [2 + 2 + 2]
Kumar, V.G.; Shoba, T.J.; Rao, K.V.C. *Tetrahedron Lett.* 1985, *26*, 6245.

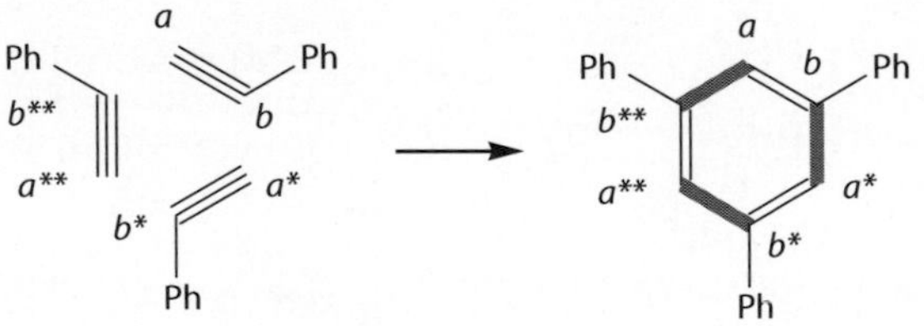

AE = 1; *E* = 0; *f*(sac) = 0

$$\Delta = \Delta(a) + \Delta(a^*) + \Delta(a^{**}) + \Delta(b) + \Delta(b^*) + \Delta(b^{**}) = 3*[(-1-(-1)) + (0-0)] = 0; \quad \mathbf{HI = 0}$$

Tanaka, K.; Shirasaka, K. *Org. Lett.* 2003, *5*, 4697.

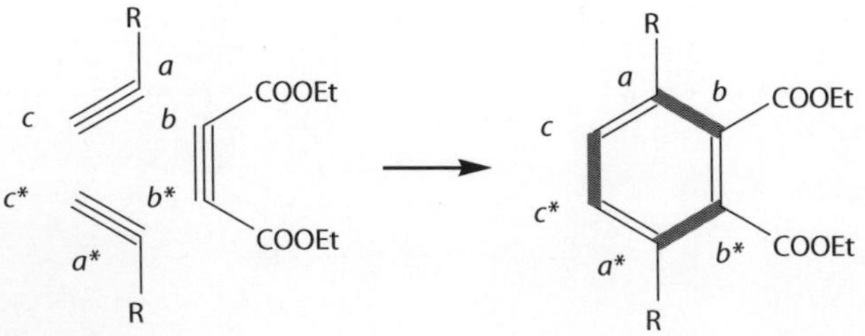

AE = 1; *E* = 0; *f*(sac) = 0

Set R = CH$_3$

$$\Delta = \Delta(a) + \Delta(a^*) + \Delta(b) + \Delta(b^*) + \Delta(c) + \Delta(c^*)$$

$$= (0-0) + (0-0) + (0-0) + (0-0) + (-1-(-1)) + (-1-(-1)) = 0; \quad \mathbf{HI = 0}$$

Trimerization of arylisocyanates [2 + 2 + 2]

Carelli, V.; Liberatore, F.; Moracci, F.M.; Tortorella, S.; Carelli, I.; Inesi, A. *Synth. Commun.* 1985, *15*, 249.
Duong, H.A.; Cross, M.J.; Louie, J. *Org. Lett.* 2004, *6*, 4679.

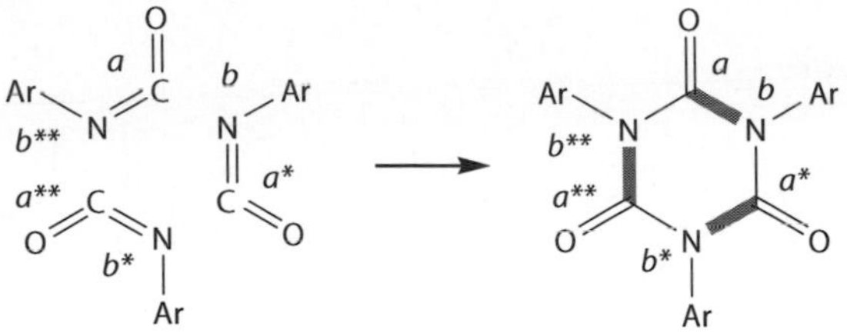

AE = 1; *E* = 0; *f*(sac) = 0

$$\Delta = \Delta(a) + \Delta(a^*) + \Delta(b) + \Delta(b^*) + \Delta(c) + \Delta(c^*) = 3[(4-4) + (-3-(-3))] = 0; \quad \mathbf{HI = 0}$$

Trost's 4-CC synthesis of γ,δ-unsaturated ketones

Trost, B.M.; Pinkerton, A.B. *J. Am. Chem. Soc.* 2000, *122*, 8081.

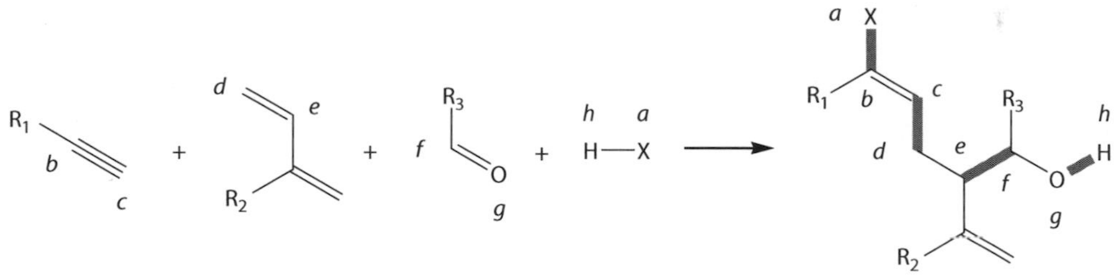

AE = 1; *E* = 0; *f*(sac) = 0

Set $R_1 = R_3 = CH_3$; X = Cl, Br, I

$$\Delta = \Delta(a) + \Delta(b) + \Delta(c) + \Delta(d) + \Delta(e) + \Delta(f) + \Delta(g) + \Delta(h)$$

$$= (-1-(-1)) + (1-0) + (-1-(-1)) + (-2-(-2)) + (-1-(-1)) + (0-1)$$

$$+ (-2-(-2)) + (1-1) = 0; \quad \mathbf{HI = 0}$$

Trost synthesis of 1,5-diketones

Trost, B.M.; Portnoy, M.; Kurihara, H.A. *J. Am. Chem. Soc.* 1997, *119*, 836.

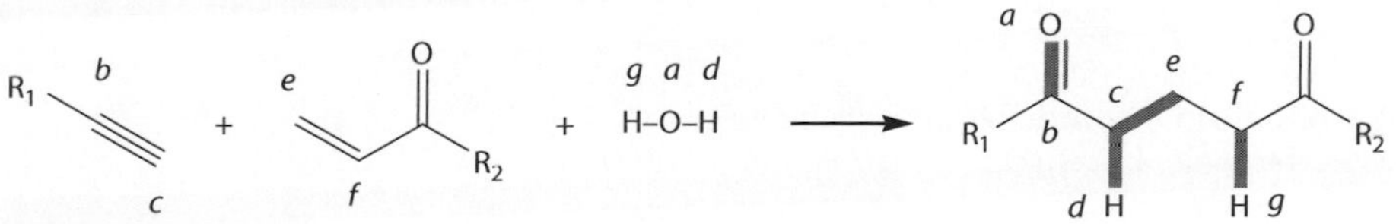

AE = 1; *E* = 0; *f*(sac) = 0

Set $R_1 = CH_3$

$$\Delta = \Delta(a) + \Delta(b) + \Delta(c) + \Delta(d) + \Delta(e) + \Delta(f) + \Delta(g)$$

$$= (-2-(-2)) + (2-0) + (-2-(-1)) + (1-1) + (-2-(-2)) + (-2-(-1)) + (1-1) = 0; \quad \mathbf{HI = 0}$$

Ugi condensation

Ugi, I. *Angew. Chem. Int. Eng. Ed.* 1962, *1*, 8.
Ugi, I.; Steinbrueckner, C. *Chem. Ber.* 1961, *94*, 734; 2802.
Urban, R.; Ugi, I., *Angew. Chem. Int. Engl. Ed.* 1975, *14*, 61.

p + 45   w + 16   y + z + 28   x + 26   x + y + z + w + p + 97   18

$R_1 = x$; $R_2 = y$; $R_3 = z$; $R_4 = w$; $R_5 = p$

$$AE = [x + y + z + w + p + 97]/[x + y + z + w + p + 115]; \quad \textbf{AE(min)} = \textbf{0.85}$$

$$E = 18/[x + y + z + w + p + 97]; \quad \textbf{E(max)} = \textbf{0.18}; \quad f(\textbf{sac}) = \textbf{0}$$

$$\Delta = \Delta(a) + \Delta(b) + \Delta(c) + \Delta(d) + \Delta(e) + \Delta(f) + \Delta(g)$$

$$= (3-3) + (-2-(-2)) + (1-2) + (3-2) + (-3-(-3)) + (-2-(-2)) + (1-1) = 0; \quad \textbf{HI} = \textbf{0}$$

Ugi-3CC MCR

Dömling, A.; Richter, W.; Ugi, I. *Nucleosides & Nucleotides* 1997, *16*, 1753.

AE = 1; *E* = 0; *f*(sac) = 0

Set $R_1 = R_4 = CH_3$

$$\Delta = \Delta(a) + \Delta(b) + \Delta(c) + \Delta(d) + \Delta(e) + \Delta(f)$$

$$= (-3-(-3)) + (0-1) + (3-2) + (-2-(-2)) + (-3-(-3)) + (1-1) = 0; \quad \textbf{HI} = \textbf{0}$$

Vinylphosphonium bromide–ketone cyclization

Posner, G.H.; Mallamo, J.P.; Black, A.Y. *Tetrahedron* 1981, *37*, 3921.

98   160.9   368.9   546.9   80.9

$AE = 0.87; E = 80.9/546.9 = 0.15; f(sac) = 0$

$$\Delta = \Delta(a) + \Delta(b) + \Delta(c) + \Delta(d) + \Delta(e)$$

$$= (-1-(-2)) + (-1-(-1)) + (0-1) + (-2-(-2)) + (0-0) = 0; \quad \mathbf{HI = 0}$$

Weiss reaction
Weiss, U.; Edward, J.M. *Tetrahedron Lett.* 1968, *9*, 4885.

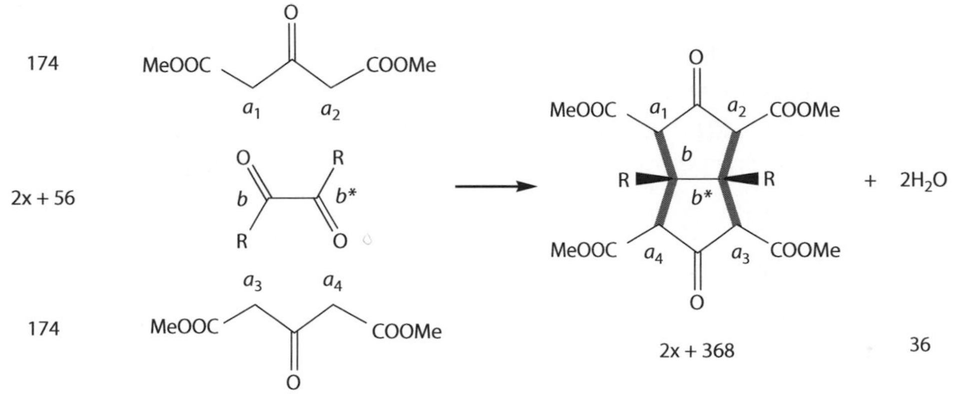

$R = x$

$$AE = [2x + 368]/[2x + 404]; \quad \mathbf{AE(min) = 0.91};$$

$$E = 36/[2x + 368]; \quad \mathbf{E(max) = 0.097}; \quad f(sac) = 0$$

$$\Delta = \Delta(a1) + \Delta(a2) + \Delta(a3) + \Delta(a4) + \Delta(b) + \Delta(b^*) = 4(-1-(-2)) + 2(0-2) = 0; \quad \mathbf{HI = 0}$$

Wulff cyclization
Wulff, W.D.; Yang, D.C. *J. Am. Chem. Soc.* 1984, *106*, 7565.

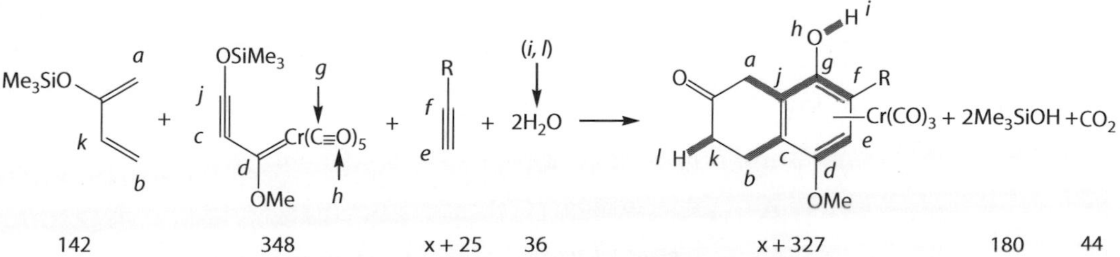

$$AE = [x + 327]/[x + 551]; \quad \mathbf{AE(min) = 0.59};$$

$$E = 208/[x + 327]; \quad \mathbf{E(max) = 0.68}; \quad f(sac) = 0$$

Set $R = CH_3$

$$\Delta = \Delta(a) + \Delta(b) + \Delta(c) + \Delta(d) + \Delta(e) + \Delta(f) + \Delta(g) + \Delta(h) + \Delta(i) + \Delta(j) + \Delta(k) + \Delta(l)$$

$$= (-2-(-2)) + (-2-(-2)) + (0-0) + (1-(-1)) + (-1-(-1)) + (0-0) + (1-2) + (-2-(-2))$$

$$+ (1-1) + (0-1) + (-2-(-1)) + (1-1) = -1; \quad \mathbf{HI = -\Delta/(N+1) = +\tfrac{1}{2}}$$

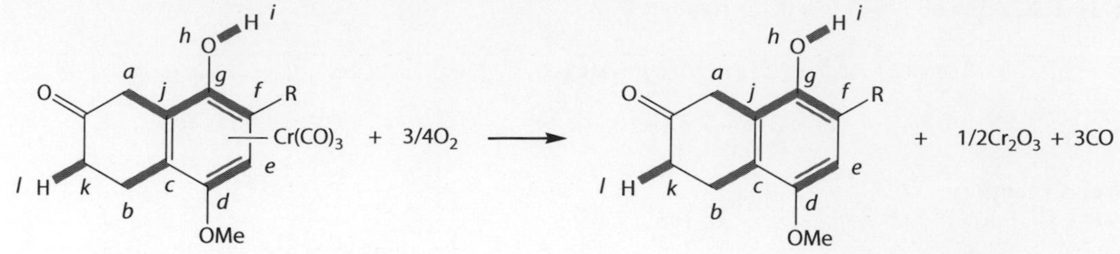

7.5 CATEGORIZATION OF NAMED NON-C–C COUPLING/ADDITION/CYCLIZATION REACTIONS BY MINIMUM ATOM ECONOMY AND MAXIMUM ENVIRONMENTAL IMPACT FACTOR, MAXIMUM MOLECULAR WEIGHT FRACTION OF SACRIFICIAL REAGENTS, HYPSICITY INDEX

(Reactions are listed alphabetically.)

Arbuzov–Michaelis reaction

Michaelis, A.; Kaehne, R. *Chem. Ber.* 1898, *31*, 1048.
Arbuzov, A.E. *J. Russ. Phys. Chem. Soc.* 1906, *38*, 687.

124　　x + y　　y + 109　　x + 15

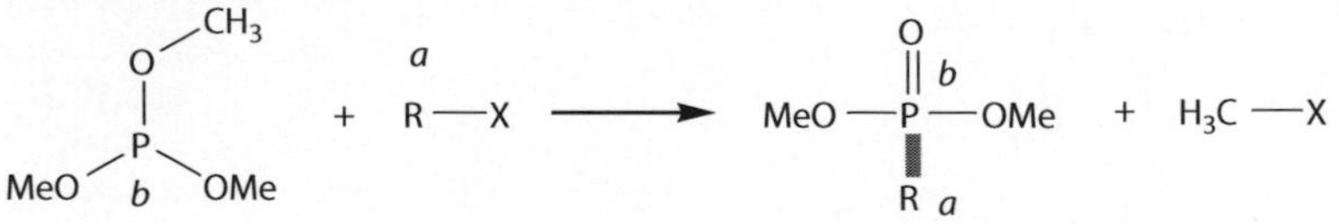

R = y; X = x(halide)

AE = [y + 109]/[x + y + 124]; *E* = [x + 15]/[y + 109]; ***f*(sac) = 0**

X	*E*(max)	AE(min)
Cl	0.47	0.68
Br	0.86	0.54
I	1.29	0.43

Set R = CH_3

$$\Delta = \Delta(a) + \Delta(b) = (-2-(-2)) + (3-3) = 0; \quad \textbf{HI} = \textbf{0}$$

Aziridine synthesis

Naik, K.G.; Triveldi, R.K.; Mehta, S.M. *J. Indian Chem. Soc.* 1943, *20*, 369.
Hassner, A.; Heathcock, C. *Tetrahedron Lett.* 1963, *6*, 393.
Gebelein, C.G.; Swift, G.; Swern, D. *J. Org. Chem.* 1967, *32*, 3314.
Gebelein, C.G.; Swern, D. *J. Org. Chem.* 1968, *33*, 2758.
Gebelein, C.G.; Rosen, S.; Swern, D. *J. Org. Chem.* 1969, *34*, 1677.

x + y + 41　　56　　169　　x + y + 41　　44　　166

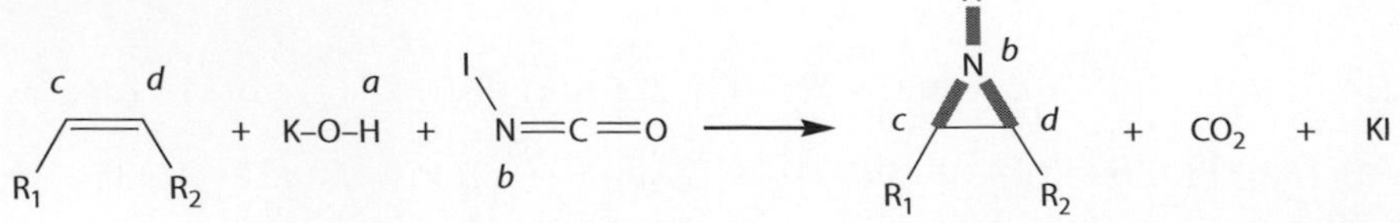

$R_1 = x; R_2 = y$

$$AE = [x + y + 41]/[x + y + 251]; \quad \textbf{AE(min)} = \textbf{0.17}; \quad E = 210/[x + y + 41]; \quad \textbf{E(max)} = \textbf{4.88}$$

$f(\textbf{sac}) = \textbf{0}$

Set $R_1 = R_2 = CH_3$

$$\Delta = \Delta(a) + \Delta(b) + \Delta(c) + \Delta(d) = (1-1) + (-3-(-1)) + (0-(-1)) + (0-(-1)) = 0; \quad \textbf{HI} = \textbf{0}$$

Azo coupling reaction
Griess, P. *Ann. Chem.* 1858, *106*, 123.

x + 135.45   y + 73   x + y + 172   36.45

$$X = x(\text{sum of five substituents}); \quad Y = y(\text{sum offive substituents})$$

$$AE = [x + y + 172]/[x + y + 208.45]; \quad \textbf{AE(min)} = \textbf{0.83};$$

$$E = 36.45/[x + y + 172]; \quad \textbf{E(max)} = \textbf{0.20}$$

$f(\textbf{sac}) = \textbf{0}$

$$\Delta = \Delta(a) + \Delta(b) = (1-(-1)) + (-1-0) = +1; \quad \textbf{HI} = -\Delta/(N+1) = -\tfrac{1}{2}$$

Buchwald–Hartwig cross coupling reaction
Louie, J.; Hartwig, J.F. *Tetrahedron Lett.* 1995, *36*, 3609.
Hartwig, J.F. *Angew. Chem. Int. Ed.* 1998, *37*, 2046.

r + x + 72   y + 16   r + y + 87   x + 1

$$R = r(\text{sum of substituents}); \quad R_1 = y; \quad X = x$$

$$AE = [r + y + 87]/[x + y + r + 88]; \quad E = [x+1]/[[r + y + 87]$$

$$\textbf{AE(min)} = [\textbf{y} + \textbf{92}]/[\textbf{x} + \textbf{y} + \textbf{93}]; \quad \textbf{E(max)} = [\textbf{x} + \textbf{1}]/[\textbf{y} + \textbf{92}]; \quad f(\textbf{sac}) = \textbf{0}$$

X	R_1	E(max)	AE(min)
Br	Me	0.76	0.57
I	Me	1.20	0.46
OSO_2CF_3	Me	1.40	0.42
Br	Ph	0.48	0.68
I	Ph	0.76	0.57
OSO_2CF_3	Ph	0.89	0.53

Set $R_1 = CH_3$

$$\Delta = \Delta(a) + \Delta(b) = (-3-(-3)) + (1-1) = 0; \quad \mathbf{HI = 0}$$

1,3-Dipolar additions

Smith, L.I. *Chem. Rev.* 1938, *23*, 193.
Huisgen, R. *Proc. Chem. Soc.* 1961, 357.
Huisgen, R. *Angew. Chem. Int. Ed.* 1963, *75*, 565.
Huisgen, R. *Angew. Chem. Int. Ed.* 1963, *75*, 633.
Huisgen, R.; Grashey, R.; Steingruber, E. *Tetrahedron Lett.* 1963, *22*, 1441.

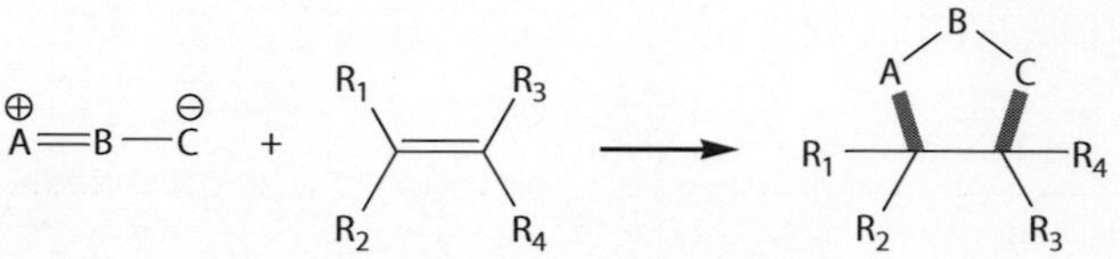

Example:

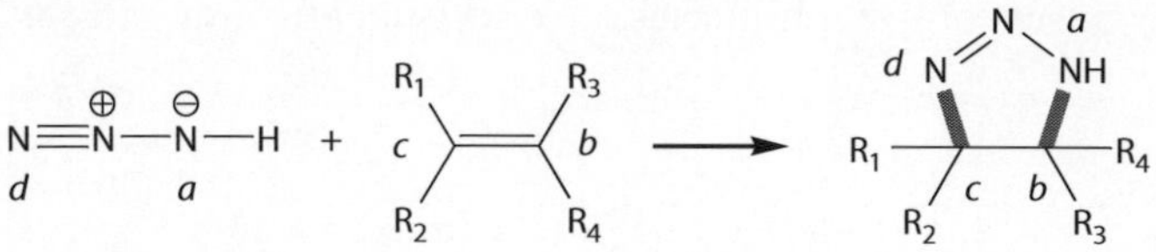

$\mathbf{AE = 1}; \quad \boldsymbol{E} \mathbf{= 0}; \quad \boldsymbol{f}\mathbf{(sac) = 0}$

Set $R_1 = R_2 = R_3 = R_4 = CH_3$

$$\Delta = \Delta(a) + \Delta(b) + \Delta(c) + \Delta(d) = (-2-(-2)) + (1-0) + (1-0) + (-1-0) = +1;$$

$$\mathbf{HI = -\Delta/(\boldsymbol{N}+1) = -½}$$

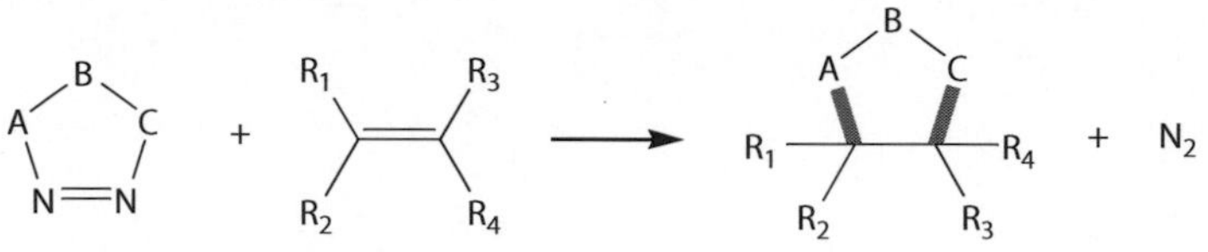

Example:

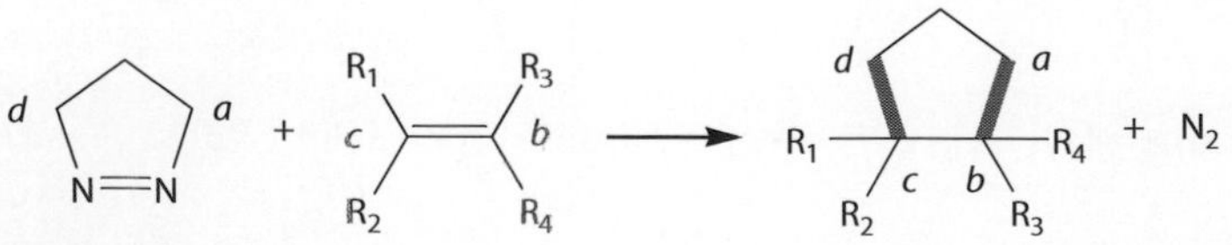

64 y + 24 y + 60 28

$$y = R_1 + R_2 + R_3 + R_4$$

$$AE = [y+60]/[y+88]; \quad \mathbf{AE(min) = 0.70}; \quad E = 28/[y+60]; \quad \boldsymbol{E}\mathbf{(max) = 0.44}$$

$\boldsymbol{f}\mathbf{(sac) = 0}$

Set $R_1 = R_2 = R_3 = R_4 = CH_3$

$$\Delta = \Delta(a) + \Delta(b) + \Delta(c) + \Delta(d) = (-2-(-1)) + (0-0) + (0-0) + (-2-(-1)) = -2;$$

$$\mathbf{HI = -\Delta/(N+1) = +1}$$

Eschweiler–Clarke reaction

Eschweiler, W. *Chem. Ber.* 1905, *38*, 880.
Clarke, H.T.; Gillespie, H.B.; Weisshaus, S.Z. *J. Am. Chem. Soc.* 1933, *55*, 4571.

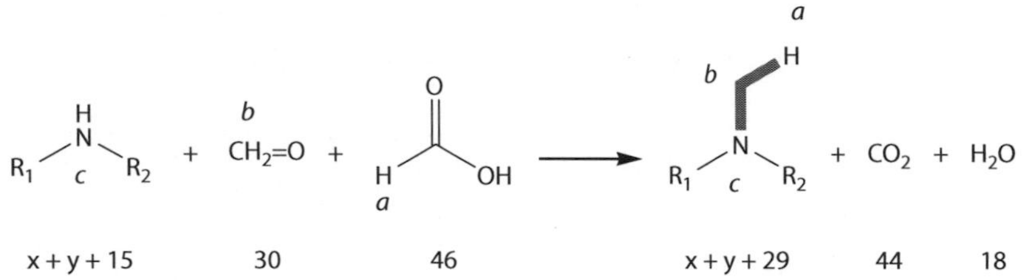

$R_1 = x; \quad R_2 = y$

$$AE = [x+y+29]/[x+y+91]; \quad \mathbf{AE(min) = 0.33}; \quad E = 62/[x+y+29]; \quad \boldsymbol{E}\mathbf{(max) = 2}$$

$\boldsymbol{f}$**(sac) = 0**

Set $R_1 = R_2 = CH_3$

$$\Delta = \Delta(a) + \Delta(b) + \Delta(c) = (1-1) + (-2-0) + (-3-(-3)) = -2; \quad \mathbf{HI = -\Delta/(N+1) = +1}$$

Formation of acetals/ketals

Skrabal, A.; Mirtl, K.H. *Z. Physik. Chem.* 1924, *111*, 98.
Arbusov, A.E. *Z. Physik. Chem.* 1926, *121*, 209.
Broensted, J.N.; Wynne-Jones, W.F.K. *Trans. Faraday Soc.* 1929, *25*, 59.

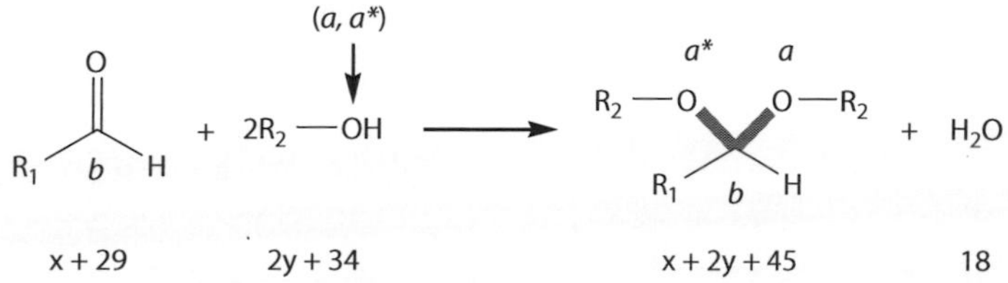

$R_1 = x; \quad R_2 = y$

$$AE = [x+2y+45]/[x+2y+63]; \quad \mathbf{AE(min) = 0.73}; \quad E = 18/[x+2y+45]; \quad \boldsymbol{E}\mathbf{(max) = 0.38}$$

$\boldsymbol{f}$**(sac) = 0**

Set $R_1 = CH_3$

$$\Delta = \Delta(a) + \Delta(a^*) + \Delta(b) = 2(-2-(-2)) + (1-1) = 0; \quad \mathbf{HI = 0}$$

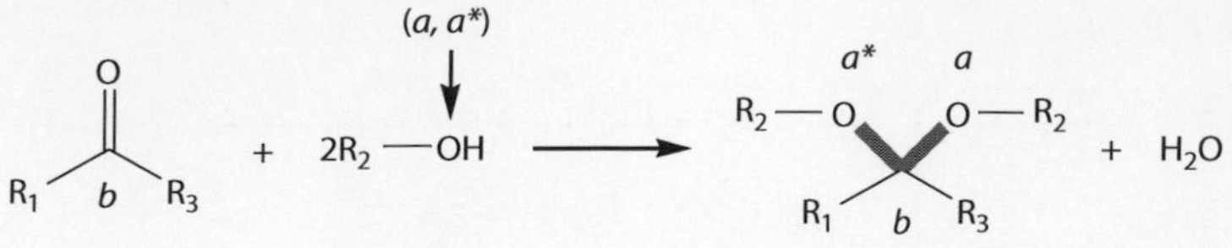

$$x + z + 28 \qquad 2y + 34 \qquad x + z + 2y + 44 \qquad 18$$

$$R_1 = x; \quad R_2 = y; \quad R' = z$$

$$AE = [x + y + 2z + 44]/[x + y + 2z + 62]; \quad \textbf{AE(min)} = \textbf{0.73};$$

$$E = 18/[x + y + 2z + 44]; \quad \textbf{E(max)} = \textbf{0.38}$$

$$f(\textbf{sac}) = \textbf{0}$$

Set $R_1 = R_3 = CH_3$

$$\Delta = \Delta(a) + \Delta(a^*) + \Delta(b) = 2(-2 - (-2)) + (2 - 2) = 0; \quad \textbf{HI} = \textbf{0}$$

Formation of hemiacetals/hemiketals

Jackson, C.L.; MacLaurin, R.D. *Am. Chem. J.* 1907, *38*, 127.
Jackson, C.L.; Flint, H.A. *Am. Chem. J.* 1908, *39*, 80.

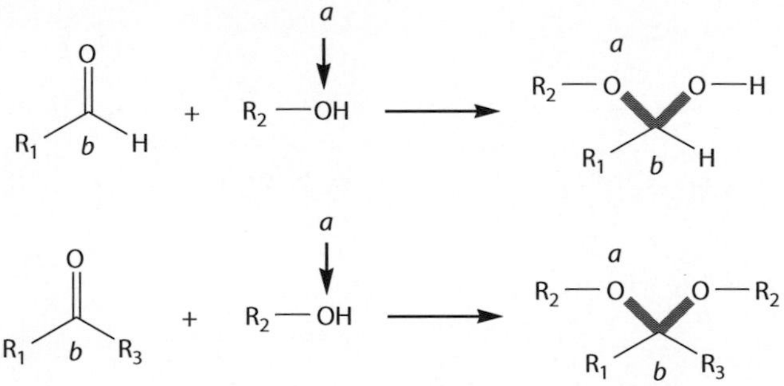

$$\textbf{AE} = \textbf{1}; \quad \textbf{E} = \textbf{0}; \quad f(\textbf{sac}) = \textbf{0}$$

Set $R_1 = R_2 = CH_3$

$$\Delta = \Delta(a) + \Delta(b) = (-2 - (-2)) + (1 - 1) = 0; \quad \textbf{HI} = \textbf{0}$$

$$\Delta = \Delta(a) + \Delta(b) = (-2 - (-2)) + (2 - 2) = 0; \quad \textbf{HI} = \textbf{0}$$

Griess diazotization reaction

Griess, P. *Ann. Chem.* 1858, *106*, 123.

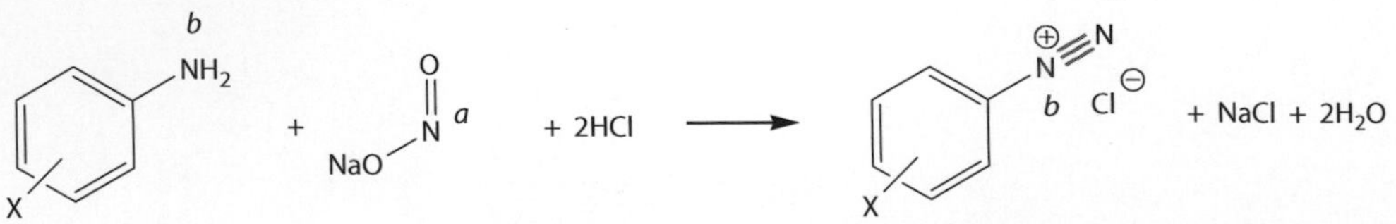

$$x + 88 \qquad\qquad 69 \qquad\qquad 72.9 \qquad\qquad x + 135.45 \qquad\qquad 58.45 \qquad 36$$

X = x(sum of substituents)

$$AE = [x + 135.45]/[x + 229.9]; \quad \textbf{AE(min) = 0.60}; \quad E = 94.45/[x + 135.45]; \quad \textbf{\textit{E}(max) = 0.67}$$

$$f(sac) = 72.9/[x + 229.9]; \quad \textbf{\textit{f}(sac)(max) = 0.32}$$

$$\Delta = \Delta(a) + \Delta(b) = (0 - 3) + (0 - (-3)) = \textbf{0}; \quad \textbf{HI = 0}$$

Hoch–Campbell aziridine synthesis

Hoch, J. *Compt. Rend.* 1934, *198*, 1865.

Campbell, K.N.; McKenna, J.F. *J. Org. Chem.* 1939, *4*, 198.

$$x + y + 57 \qquad\qquad 2(z + 59.75) \qquad 72.9 \qquad\qquad x + y + z + 40 \qquad z + 1 \qquad 190.4 \qquad 18$$

$$R_1 = x; \quad R_2 = y; \quad R_3 = z$$

$$AE = [x + y + z + 40]/[x + y + 2z + 249.4]; \quad \textbf{AE(min) = [z + 42]/[2z + 251.4]}$$

$$E = [z + 209.4]/[x + y + z + 40]; \quad \textbf{\textit{E}(max) = [z + 209.4]/[z + 42]}$$

$$f(sac) = [z + 96.2]/[x + y + 2z + 249.4]; \quad \textbf{\textit{f}(sac)(max) = [z + 96.2]/[2z + 251.4]}$$

Z	E(max)	AE(min)	f(sac)(max)
Me	3.94	0.20	0.40
Ph	2.41	0.29	0.43

Set $R_1 = R_2 = R_3 = CH_3$

$$\Delta = \Delta(a) + \Delta(b) + \Delta(c) + \Delta(d) + \Delta(e)$$

$$= (1 - 1) + (-3 - (-1)) + (0 - (-2)) + (1 - 2) + (-3 - (-4)) = 0; \quad \textbf{HI = 0}$$

Hofmann–Loffler–Freytag reaction

Hofmann, A.W. *Chem. Ber.* 1883, *16*, 558.

Löffler, K.; Freytag, C. *Chem. Ber.* 1909, *42*, 3427.

$$x + y + 14n + 91.45 \qquad\qquad x + y + 14n + 55 \qquad 36.45$$

$$R_1 = x; \quad R_2 = y$$

$$AE = [x + y + 14n + 55]/[x + y + 14n + 91.45]; \quad E = 36.45/[x + y + 14n + 55]; \quad \textbf{\textit{f}(sac) = 0}$$

$$AE(min) = [14n + 57]/[14n + 93.45]$$

$$E(max) = 36.45/[14n + 57]$$

n	Ring Size ($n + 4$)	E(max)	AE(min)
1	5	0.51	0.66
2	6	0.43	0.70
3	7	0.37	0.73

Set $R_1 = R_2 = CH_3$

$$\Delta = \Delta(a) + \Delta(b) = (0 - (-2)) + (-3 - (-1)) = 0; \quad \textbf{HI} = \textbf{0}$$

Leuckart reaction

Leuckart, R. *Chem. Ber.* 1885, *18*, 2341.

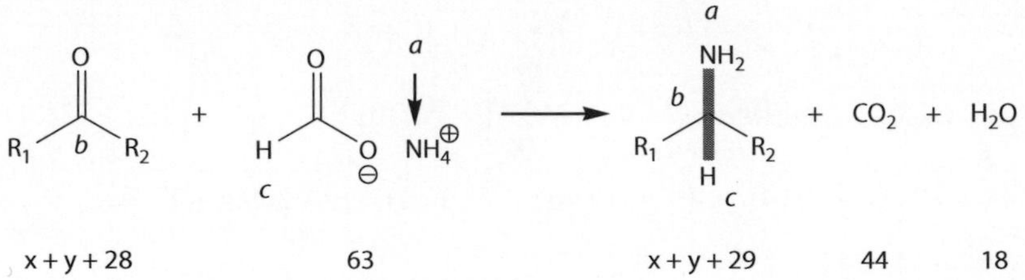

$x + y + 28$ 63 $x + y + 29$ 44   18

$$R_1 = x; \quad R_2 = y$$

$$AE = [x + y + 29]/[x + y + 91]; \quad \textbf{AE(min)} = \textbf{0.33}; \quad E = 62/[x + y + 29]; \quad \textbf{E(max)} = \textbf{2}$$

$$f(sac) = 0$$

Set $R_1 = R_2 = CH_3$

$$\Delta = \Delta(a) + \Delta(b) + \Delta(c) = (-3 - (-3)) + (0 - 2) + (1 - 1) = -2; \quad \textbf{HI} = -\Delta/(N + 1) = +1$$

Menshutkin reaction

Menshutkin, N. *Z. Physik. Chem.* 1890, *5*, 589.

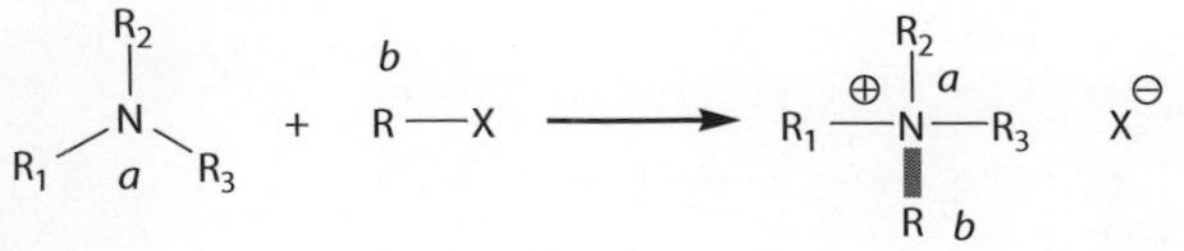

$$\textbf{AE} = \textbf{1}; \quad E = 0; \quad f(sac) = 0$$

Set $R = R_1 = R_2 = R_3 = CH_3$

$$\Delta = \Delta(a) + \Delta(b) = (-3 - (-3)) + (-2 - (-2)) = 0; \quad \textbf{HI} = \textbf{0}$$

Paal–Knorr synthesis of furans, thiophenes, and pyrroles
Paal, C., *Chem. Ber.* 1885, *18*, 367.
Knorr, L. *Chem. Ber.* 1885, *18*, 299.

| 2r + 2x + 52 | 5x + 62 | 2r + x + 50 | 2(3x + 32) |

$$R = r; \quad X = O, S$$

$$AE = [2r + x + 50]/[2r + 7x + 114]; \quad \mathbf{AE(min)} = [\mathbf{x} + \mathbf{52}]/[\mathbf{7x} + \mathbf{116}]$$

$$E = [2(3x + 32)]/[2r + x + 50]; \quad \mathbf{E(max)} = [\mathbf{2(3x} + \mathbf{32)}]/[\mathbf{x} + \mathbf{52}]$$

$$f(sac) = [5x + 62]/[2r + 7x + 114]; \quad \boldsymbol{f}\mathbf{(sac)(max)} = [\mathbf{5x} + \mathbf{62}]/[\mathbf{7x} + \mathbf{116}]$$

X	E(max)	AE(min)	f(sac)(max)
O	2.35	0.30	0.62
S	3.05	0.25	0.65

Set R = CH_3

$$\Delta = \Delta(a) + \Delta(b) = (-2 - (-2)) + (1 - 2) = -1; \quad \mathbf{HI} = -\Delta/(N + 1) = +\tfrac{1}{2}$$

| 2r + 2x + 52 | 96 | 2r + 65 | 79 | 36 |

R = x

$$AE = [2r + 65]/[2r + 180]; \quad \mathbf{AE(min)} = \mathbf{0.37}; \quad E = 115/[2r + 65]; \quad \mathbf{E(max)} = \mathbf{1.72}$$

$\boldsymbol{f}\mathbf{(sac)} = \mathbf{0}$

Set R = CH_3

$$\Delta = \Delta(a) + \Delta(b) + \Delta(c) = (-3 - (-3)) + (1 - 2) + (1 - 2) = -2; \quad \mathbf{HI} = -\Delta/(N + 1) = +\mathbf{1}$$

Polonovski reaction
Polonovski, M.; Polonovski, M. *Bull. Chim. Soc. Fr.* 1926, *39*, 147.
Polonovski, M.; Polonovski, M. *Bull. Chim. Soc. Fr.* 1927, *41*, 1190.

$AE = 0.75$; $E = 0.34$; $f(sac) = 0$

$$\Delta = \Delta(a) + \Delta(b) = (-2 - (-2)) + (1 - (-1)) = +2; \quad \mathbf{HI} = -\Delta/(N+1) = -1$$

Ritter reaction

Ritter, J.J.; Minieri, P.P. *J. Am. Chem. Soc.* 1948, *70*, 4045.

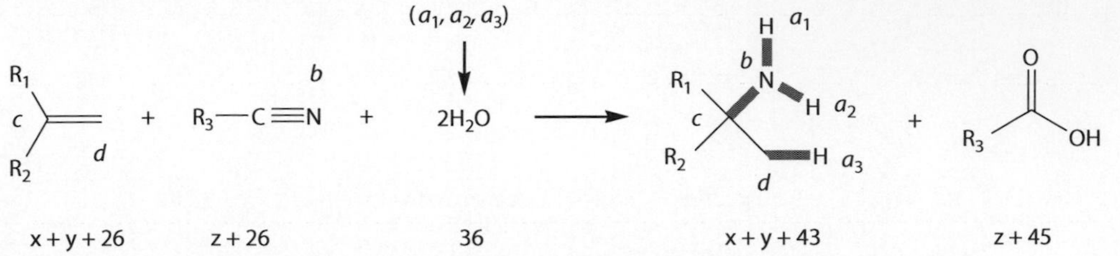

$R_1 = x$; $R_2 = y$; $R_3 = z$

$$AE = [x + y + 43]/[x + y + z + 88]; \quad \mathbf{AE(min)} = 45/[z + 90];$$

$$E = [z + 45]/[x + y + 43]; \quad \mathbf{E(max)} = [z + 45]/45$$

Z	E(max)	AE(min)
H	1.02	0.49
Me	1.33	0.43
Et	1.64	0.38

$f(sac) = 0$

Set $R_1 = R_2 = CH_3$

$$\Delta = \Delta(a_1) + \Delta(a_2) + \Delta(a_3) + \Delta(b) + \Delta(c) + \Delta(d)$$

$$= 3(1-1) + (-3 - (-3)) + (1 - 0) + (-3 - (-2)) = 0; \quad \mathbf{HI = 0}$$

Stahl aerobic oxidative amination

Fix, S.R.; Brice, J.L.; Stahl, S.S. *Angew. Chem. Int. Ed.* 2002, *41*, 164.

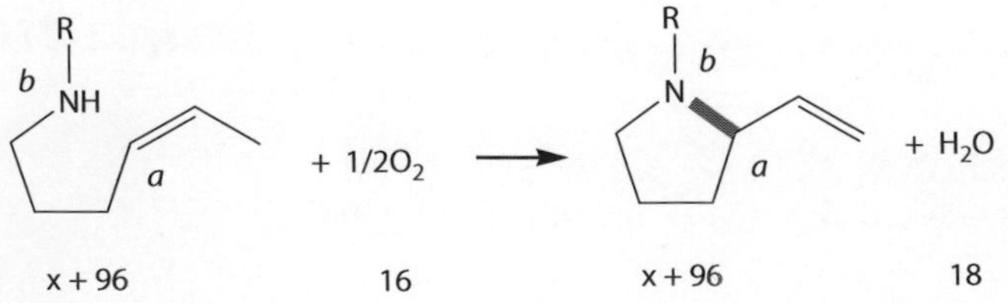

$R = x$

$$AE = [x + 96]/[x + 114]; \quad E = 18/[x + 96]; \quad \mathbf{E(max) = 0.19}; \quad \mathbf{AE(min) = 0.84}$$

$$f(sac) = 16/[x + 114]; \quad \mathbf{f(sac)(max) = 0.14}$$

Set R = CH$_3$

$$\Delta = \Delta(a) + \Delta(b) = (0 - (-1)) + (-3 - (-3)) = +1; \quad \mathbf{HI} = -\Delta/(N+1) = -\tfrac{1}{2}$$

Staudinger reaction

Staudinger, H.; Meyer, J. *Helv. Chim. Acta* 1919, *2*, 635.

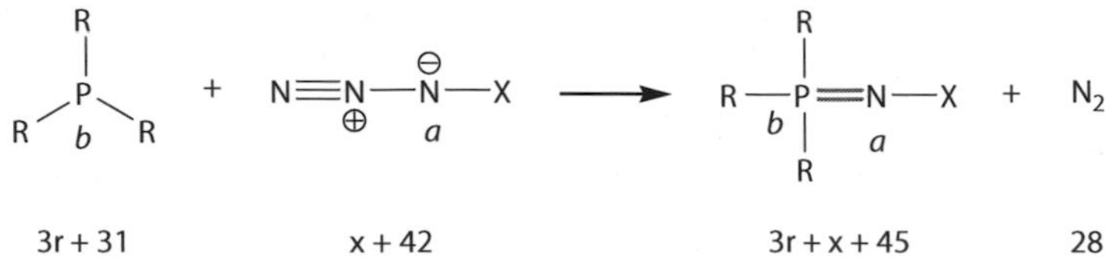

| 3r + 31 | x + 42 | 3r + x + 45 | 28 |

R = r; X = x(halide)

$$AE = [3r + x + 45]/[3r + x + 73]; \quad \mathbf{AE(min)} = [x + 48]/[x + 76];$$

$$E = 28/[3r + x + 45]; \quad \mathbf{E(max)} = 28/[x + 48]; \quad f(\mathbf{sac}) = 0$$

X	E(max)	AE(min)
Cl	0.34	0.75
Br	0.22	0.82
I	0.16	0.86

Set R = CH$_3$

$$\Delta = \Delta(a) + \Delta(b) = (-1 - 0) + (-1 - (-3)) = +1; \quad \mathbf{HI} = -\Delta/(N+1) = -\tfrac{1}{2}$$

Stoltz aerobic oxidative etherification

Trend, R.M.; Ramtohul, E.M.; Ferreira, E.M.; Stoltz, B.M. *Angew. Chem. Int. Ed.* 2003, *42*, 2892.

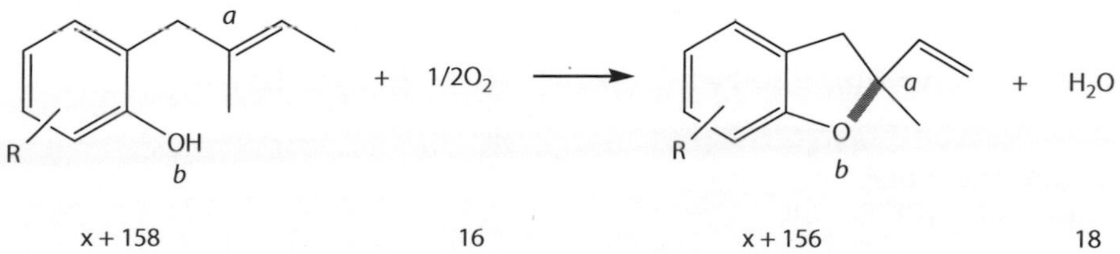

| x + 158 | 16 | x + 156 | 18 |

R = x(sum of 4 substituents)

$$AE = [x + 156]/[x + 174]; \quad E = 18/[x + 156]; \quad \mathbf{E(max)} = \mathbf{0.11}; \quad \mathbf{AE(min)} = \mathbf{0.90}$$

$$f(\mathbf{sac}) = 16/[x + 174]; \quad f(\mathbf{sac})(\mathbf{max}) = \mathbf{0.090}$$

$$\Delta = \Delta(a) + \Delta(b) = (1 - 0) + (-2 - (-2)) = +1; \quad \mathbf{HI} = -\Delta/(N+1) = -\tfrac{1}{2}$$

Wenker synthesis

Wenker, H. *J. Am. Chem. Soc.* 1935, *57*, 2328.

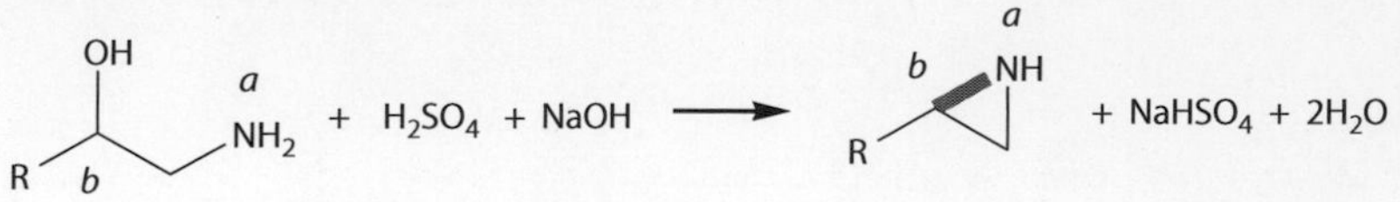

$$x + 60 \qquad\qquad 98 \qquad 40 \qquad\qquad x + 42 \qquad\qquad 120 \qquad 36$$

R = x

$$AE = [x+42]/[x+198]; \quad \textbf{AE(min)} = \textbf{0.22}; \quad E = 156/[x+42]; \quad \textbf{\textit{E}(max)} = \textbf{3.63}$$

$$f(\text{sac}) = 138/[x+198]; \quad f(\text{sac})(\text{max}) = \textbf{0.69}$$

Set R = CH$_3$

$$\Delta = \Delta(a) + \Delta(b) = (-3 - (-3)) + (0 - 0) = 0; \quad \textbf{HI} = \textbf{0}$$

Williamson ether synthesis

Williamson, A.W. *J. Chem. Soc.* 1852, *4*, 229.

$$x + 17 \qquad 40 \qquad y + 79.9 \qquad\qquad x + y + 16 \qquad 118.9 \quad 2$$

R$_1$ = x; R$_2$ = y

$$AE = [x+y+16]/[x+y+136.9]; \quad \textbf{AE(min)} = \textbf{0.13};$$

$$E = 120.9/[x+y+16]; \quad \textbf{\textit{E}(max)} = \textbf{6.72}; \quad f(\text{sac}) = \textbf{0}$$

Set R$_2$ = CH$_3$

$$\Delta = \Delta(a) + \Delta(b) = (-2 - (-2)) + (-2 - (-2)) = 0; \quad \textbf{HI} = \textbf{0}$$

Wöhler synthesis of urea

Wöhler, F. *Ann. Chim.* 1828, *37*, 330.

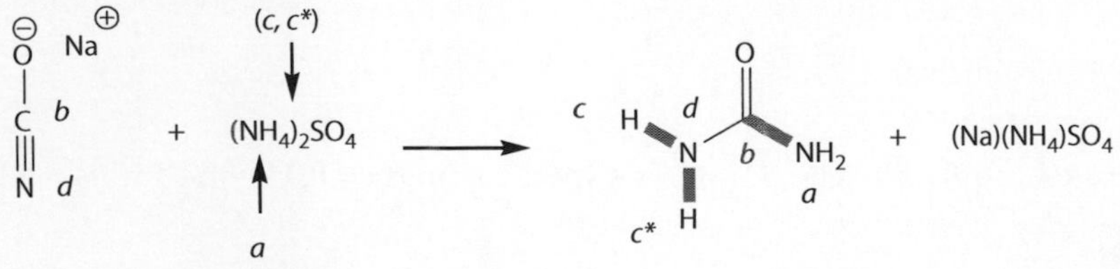

AE = 0.30; *E* = 2.28; *f*(sac) = 0

$$\Delta = \Delta(a) + \Delta(b) + \Delta(c) + \Delta(c^*) + \Delta(d) = (-3 - (-3)) + (4 - 4) + 2(1 - 1) + (-3 - (-3)) = 0; \quad \textbf{HI} = \textbf{0}$$

7.6 CATEGORIZATION OF NAMED OXIDATION REACTIONS (WITH RESPECT TO SUBSTRATE OF INTEREST) BY MINIMUM ATOM ECONOMY, MAXIMUM ENVIRONMENTAL IMPACT FACTOR, MAXIMUM MOLECULAR WEIGHT FRACTION OF SACRIFICIAL REAGENTS, HYPSICITY INDEX

(Reactions are listed alphabetically.)

Baeyer–Villiger oxidation: Additive oxidation
Baeyer, A.; Villiger, V. *Chem. Ber.* 1899, *32*, 3625.
Baeyer, A.; Villiger, V. *Chem. Ber.* 1900, *33*, 858.

$$\begin{array}{cccc} x+y+28 & z+33 & x+y+44 & z+17 \end{array}$$

$R_1 = x; R_2 = y; R' = z$

$$AE = [x+y+44]/[x+y+z+61]; \quad \mathbf{AE(min) = 46/[z+61]};$$

$$E = [z+17]/[x+y+44]; \quad \mathbf{E(max) = [z+17]/46}$$

$$f(\mathbf{sac}) = \mathbf{0}$$

R'	E(max)	AE(min)
H	0.39	0.72
C_7H_4OCl	3.40	0.23

Set $R_1 = R_2 = CH_3$

$$\Delta = \Delta(a) + \Delta(b) + \Delta(c) = (3-2) + (-2-(-1)) + (-2-(-3)) = +1; \quad \mathbf{HI} = -\Delta/(N+1) = -\tfrac{1}{2}$$

Bamford–Stevens oxidation of hydrazones
Bamford, W.R.; Stevens, T.S. *J. Chem. Soc.* 1952, 4735.

Subtractive oxidation:

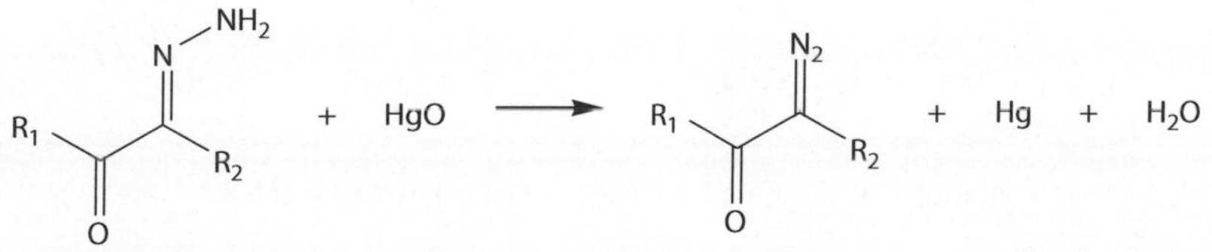

$$\begin{array}{ccccc} x+y+70 & 216.6 & x+y+68 & 200.6 & 18 \end{array}$$

$R_1 = x; R_2 = y$

$$AE = [x + y + 68]/[x + y + 286.6]; \quad \textbf{AE(min)} = \textbf{0.24}; \quad E = 218.6/[x + y + 68]; \quad \textbf{\textit{E}(max)} = \textbf{3.12}$$

$$f(sac) = 216.6/[x + y + 286.6]; \quad \textbf{\textit{f}(sac)(max)} = \textbf{0.75}$$

Additive oxidation with rearrangement:

$$x + y + z + 209 \qquad 54 \qquad\qquad x + y + z + 25 \qquad 28 \quad 32 \qquad\qquad 178$$

$R_1 = x; R_2 = y; R_3 = z$

$$AE = [x + y + z + 25]/[x + y + z + 263]; \quad \textbf{AE(min)} = \textbf{0.11};$$

$$E = 238/[x + y + z + 25]; \quad \textbf{\textit{E}(max)} = \textbf{8.5}$$

$$f(sac) = 54/[x + y + z + 263]; \quad \textbf{\textit{f}(sac)(max)} = \textbf{0.20}$$

Set $R_1 = CH_3$

$$\Delta = \Delta(a) + \Delta(b) = (1 - 1) + (-1 - 2) = -3; \quad \textbf{HI} = -\Delta/(N + 1) = \textbf{+3/2}$$

Boyland–Sims oxidation: Additive oxidation
Boyland, E.; Manson, D.; Sims, P. *J. Chem. Soc.* 1953, 3623.
Boyland, E.; Sims, P. *J. Chem. Soc.* 1954, 980.

$$x + 2r + 87 \qquad\qquad 270 \qquad\qquad 112 \qquad\qquad x + 2r + 103 \qquad 348 \quad 18$$

$X = x$(sum of substituents); $R = r$

$$AE = [x + 2r + 103]/[x + 2r + 469]; \quad \textbf{AE(min)} = \textbf{0.23}; \quad E = 366/[x + 2r + 103]; \quad \textbf{\textit{E}(max)} = \textbf{3.45}$$

$$f(sac) = 112/[x + 2r + 469]; \quad \textbf{\textit{f}(sac)(max)} = \textbf{0.24}$$

$$\Delta = \Delta(a) + \Delta(b) + \Delta(c) = (1 - 1) + (-2 - (-1)) + (1 - (-1)) = +1; \quad \textbf{HI} = -\Delta/(N + 1) = -\textbf{½}$$

Corey γ-lactone synthesis: subtractive oxidation

Corey, E.J.; Kang, M.C. *J. Am. Chem. Soc*. 1984, *106*, 5384.

182 463.8 180 120 345.8

$$\textbf{AE} = \textbf{0.28}; \quad E = 465.8/180 = 2.59; \quad f(\textbf{sac}) = 463.8/645.8 = \textbf{0.72}$$

$$\Delta = \Delta(a) + \Delta(b) + \Delta(c) + \Delta(d) = (-1-(-2)) + (-1-(-1)) + (0-(-1)) + (-2-(-2)) = +2;$$

$$\textbf{HI} = -\Delta/(N+1) = -\textbf{1}$$

Corey–Chaykovsky epoxidation: Additive oxidation

Corey, E.J.; Chaykovsky, M. *J. Am. Chem. Soc*. 1962, *84*, 867.
Corey, E.J.; Chaykovsky, M. *J. Am. Chem. Soc*. 1965, *87*, 1353.

x + y + 28 220 24 x + y + 42 150 78 2

$R_1 = x; R_2 = y$

$$\text{AE} = [x+y+42]/[x+y+272]; \quad \textbf{AE(min)} = \textbf{0.16}; \quad E = 230/[x+y+42]; \quad \textbf{E(max)} = \textbf{5.23}$$

$$f(\text{sac}) = 24/[x+y+272]; \quad f(\textbf{sac})(\textbf{max}) = \textbf{0.088}$$

Set $R_1 = CH_3$

$$\Delta = \Delta(a) + \Delta(b) + \Delta(c) = (-2-(-2)) + (-1-(-2)) + (1-2) = 0; \quad \textbf{HI} = \textbf{0}$$

Corey–Kim oxidation: Subtractive oxidation

Corey, E.J.; Kim, C.U. *J. Am. Chem. Soc*. 1972, *94*, 7586.

x + y + 30 195.45 101 x + y + 28 99 137.45 62

$R_1 = x; R_2 = y$

$$\text{AE} = [x+y+28]/[x+y+326.45]; \quad \textbf{AE(min)} = \textbf{0.091}; \quad E = 298.45/[x+y+28]; \quad \textbf{E(max)} = \textbf{9.95}$$

$$f(\text{sac}) = 296.45/[x+y+326.45]; \quad f(\textbf{sac})(\textbf{max}) = \textbf{0.90}$$

Criegee glycol cleavage reaction: Subtractive oxidation
Criegee, R. *Chem. Ber.* 1931, *64*, 260.

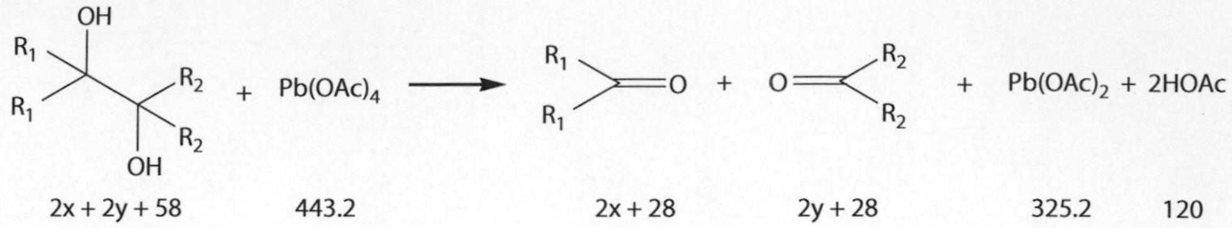

$$R_1 = x; R_2 = y$$

Target product: both ketones

$$AE = [2x + 2y + 56]/[2x + 2y + 501.2]; \quad \mathbf{AE(min) = 0.12};$$

$$E = 445.2/[2x + 2y + 56]; \quad \mathbf{E(max) = 7.42}$$

Target product: $(R_1)_2C{=}O$ ketone

$$AE = [2x + 28]/[2x + 2y + 501.2]; \quad \mathbf{AE(min) = 0.059};$$

$$E = [2y + 473.2]/[2x + 28]; \quad \mathbf{E(max) = 15.83}$$

Target product: $(R_2)_2C{=}O$ ketone

$$AE = [2y + 28]/[2x + 2y + 501.2]; \quad \mathbf{AE(min) = 0.059};$$

$$E = [2x + 473.2]/[2y + 28]; \quad \mathbf{E(max) = 15.83}$$

$$f(\text{sac}) = 443.2/[2x + 2y + 501.2]; \quad \mathbf{f(sac)(max) = 0.88}$$

Dess–Martin oxidation: Subtractive oxidation
Dess, D.B.; Martin, J.C. *J. Org. Chem.* 1983, *48*, 4155.

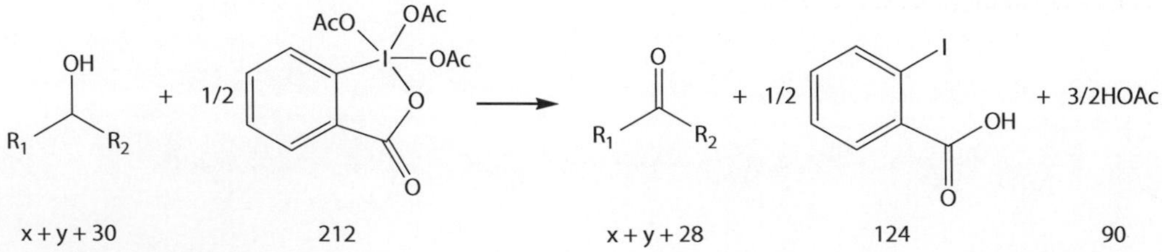

$$R_1 = x; R_2 = y$$

$$AE = [x + y + 28]/[x + y + 242]; \quad \mathbf{AE(min) = 0.12}; \quad E = 214/[x + y + 28]; \quad \mathbf{E(max) = 7.13}$$

$$f(\text{sac}) = 212/[x + y + 242]; \quad \mathbf{f(sac)(max) = 0.87}$$

Elbs reaction: Additive oxidation

Elbs, K. Larsen, E. *Chem. Ber.* 1884, *17*, 2847.
Elbs, K. *J. Prakt. Chem.* 1893, *48*, 179.

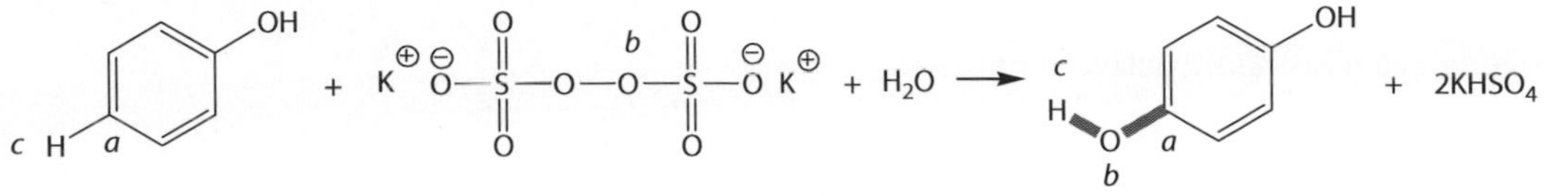

94 270 18 110 272

AE = 0.28; E = 2.47; f(sac) = 0

$$\Delta = \Delta(a) + \Delta(b) + \Delta(c) = (1-(-1)) + (-2-(-1)) + (1-1) = +1; \quad HI = -\Delta/(N+1) = -\tfrac{1}{2}$$

Etard reaction: Additive oxidation

Etard, M.A. *Compt. Rend.* 1880, *90*, 534.
Etard, M.A. *Ann. Chim. Phys.* 1881, *22*, 223.

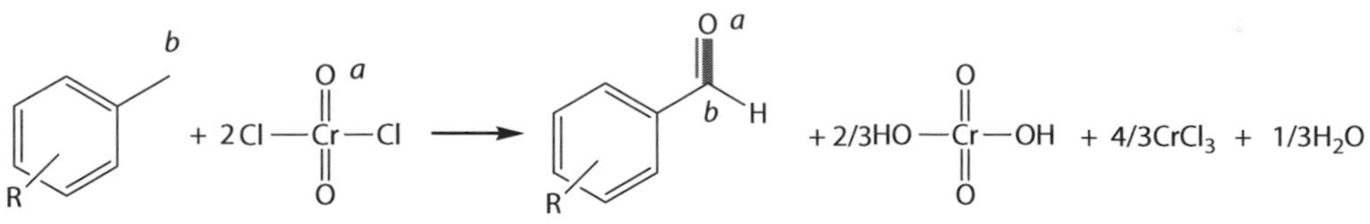

x + 87 317.79 x + 101 78.66 211.13 6

R = x(sum of substituents)

$$AE = [x+101]/[x+404.79]; \quad AE(min) = 0.26; \quad E = 295.79/[x+101]; \quad E(max) = 2.90$$

f(sac) = 0

$$\Delta = \Delta(a) + \Delta(b) = (-2-(-2)) + (1-(-3)) = +4; \quad HI = -\Delta/(N+1) = -2$$

Fehling reaction: Additive oxidation

Fehling, H. *Ann. Chem.* 1849, *72*, 106.

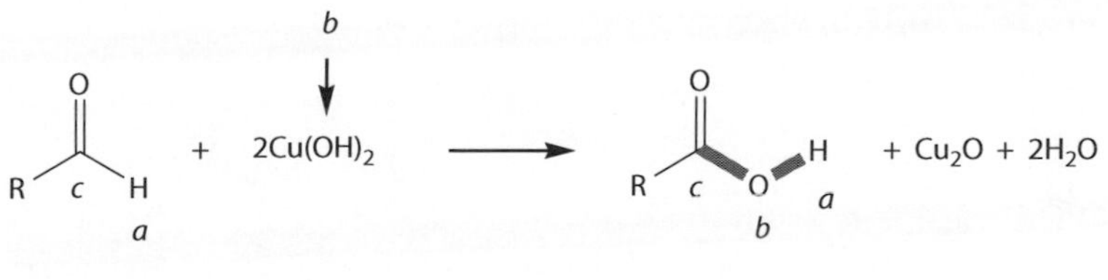

x + 29 195.1 x + 45 143.1 36

R = x

$$AE = [x+45]/[x+224]; \quad AE(min) = 0.20; \quad E = 179/[x+45]; \quad E(max) = 3.89$$

f(sac) = 0

Set R = CH$_3$

$$\Delta = \Delta(a) + \Delta(b) + \Delta(c) = (1-1) + (-2-(-2)) + (3-1) = +2; \quad \mathbf{HI} = -\Delta/(N+1) = \mathbf{-1}$$

Fenton reaction: Subtractive oxidation

Fenton, H.J.H. *Proc. Chem. Soc.* 1893, *9*, 113.

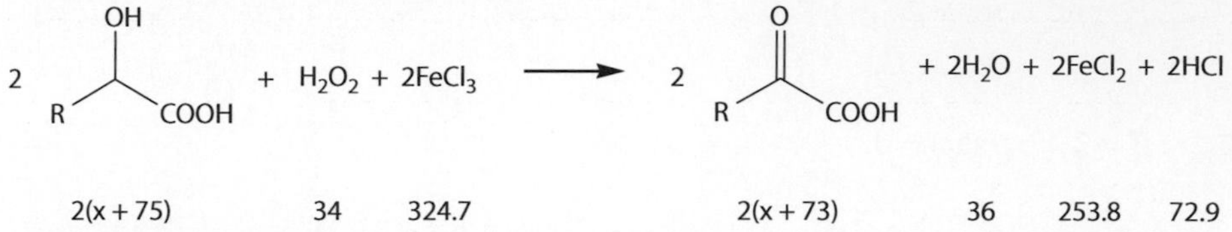

| 2(x + 75) | 34 | 324.7 | 2(x + 73) | 36 | 253.8 | 72.9 |

R = x

$$AE = [2x + 146]/[2x + 508.7]; \quad \mathbf{AE(min) = 0.29}; \quad E = 362.7/[2x + 146]; \quad \mathbf{E(max) = 2.45}$$

$$f(\text{sac}) = 358.7/[2x + 508.7]; \quad \mathbf{f(sac)(max) = 0.70}$$

Fleming oxidation: Additive oxidation

Fleming, I.; Henning, R.; Plaut, H. *Chem. Commun.* 1984, 29.

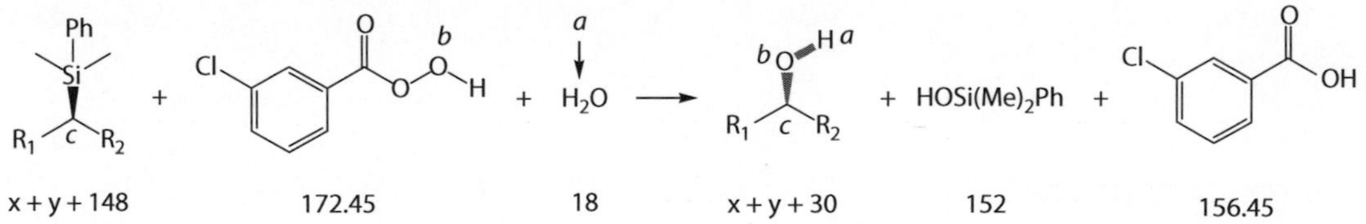

| x + y + 148 | 172.45 | 18 | x + y + 30 | 152 | 156.45 |

$$AE = [x + y + 30]/[x + y + 338.45]; \quad \mathbf{AE(min) = 0.094}; \quad E = 308.45/[x + y + 30]; \quad \mathbf{E(max) = 9.64}$$

$f(\text{sac}) = 0$

Set R$_1$ = R$_2$ = CH$_3$

$$\Delta = \Delta(a) + \Delta(b) + \Delta(c) = (1-1) + (-2-(-1)) + (0-(-2)) = +1; \quad \mathbf{HI} = -\Delta/(N+1) = -\frac{1}{2}$$

Forster reaction: Additive oxidation

Forster, M.O. *J. Chem. Soc.* 1915, *107*, 260.

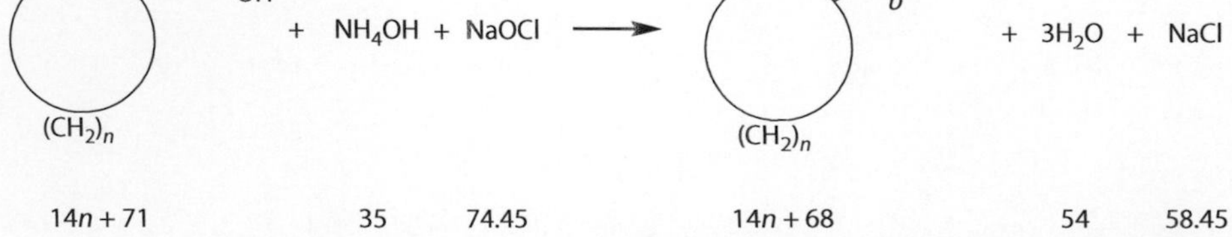

| 14n + 71 | 35 | 74.45 | 14n + 68 | 54 | 58.45 |

$$AE = [14n + 68]/[14n + 180.45]; \quad E = 112.45/[14n + 68]; \quad f(\text{sac}) = 74.45/[14n + 180.45]$$

n	Ring Size	$(n+2)E$	AE	f(sac)
3	5	1.02	0.49	0.33
4	6	0.91	0.52	0.31
5	7	0.81	0.55	0.30
6	8	0.74	0.57	0.28

$$\Delta = \Delta(a) + \Delta(b) = (-1 - (-3)) + (-1 - (-1)) = +2; \quad \mathbf{HI} = -\Delta/(N+1) = -1$$

Graham reaction: Additive oxidation
Graham, W.H. *J. Am. Chem. Soc.* 1965, *87*, 4396.
Graham, W.H. *J. Org. Chem.* 1965, *30*, 2108.

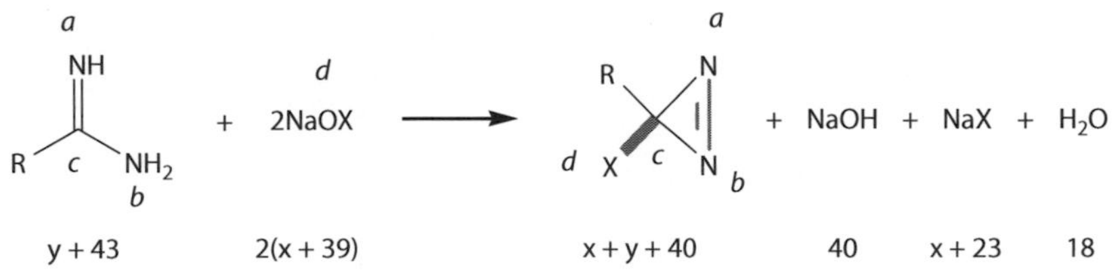

| y + 43 | 2(x + 39) | x + y + 40 | 40 | x + 23 | 18 |

R = y; X = x(halide)

$$AE = [x + y + 40]/[2x + y + 121]; \quad \mathbf{AE(min)} = [x + 41]/[2x + 122]$$

$$E = [x + 81]/[x + y + 40]; \quad \mathbf{E(max)} = [x + 81]/[x + 41]$$

$$f(\text{sac}) = [x + 39]/[2x + y + 121]; \quad \mathbf{f(sac)(max)} = [x + 39]/[2x + 122]$$

X	E(max)	AE(min)	f(sac)(max)
Cl	1.52	0.40	0.39
Br	1.33	0.43	0.42
I	1.23	0.45	0.44

Set R = CH_3

$$\Delta = \Delta(a) + \Delta(b) + \Delta(c) + \Delta(d) = (-1 - (-3)) + (-1 - (-3)) + (3 - 3) + (-1 - 1) = +2;$$

$$\mathbf{HI} = -\Delta/(N+1) = -1$$

Griess diazotization reaction: Additive oxidation
Griess, P. *Ann. Chem.* 1858, *106*, 123.

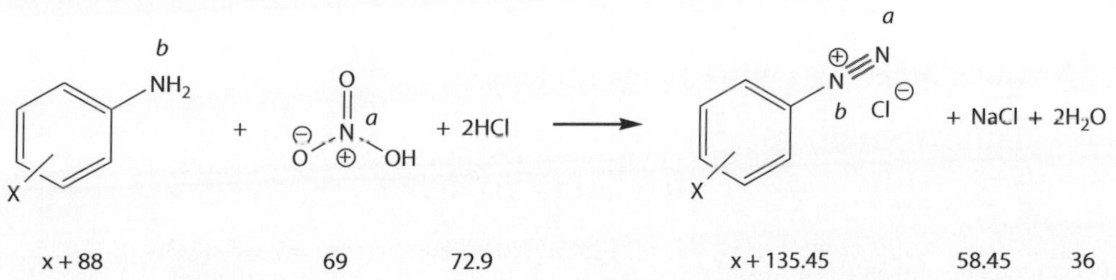

| x + 88 | 69 | 72.9 | x + 135.45 | 58.45 | 36 |

X = x(sum of substituents)

$$AE = [x + 135.45]/[x + 229.9]; \quad \textbf{AE(min) = 0.60}; \quad E = 94.45/[x + 135.45]; \quad \textbf{E(max) = 0.67}$$

$$f(\text{sac}) = 72.9/[x + 229.9]; \quad \textbf{\textit{f}(sac)(max) = 0.32}$$

$$\Delta = \Delta(a) + \Delta(b) = (0 - 5) + (0 - (-3)) = -2; \quad \textbf{HI} = -\Delta/(N + 1) = +1$$

Harries ozonolysis of olefins with oxidative workup: Additive oxidation
Harries, C. *Ann. Chem.* 1905, *343*, 311.

x + y + 26   48

34   x + 45   y + 45   18

$R_1 = x; R_2 = y$

$$AE = [x + y + 90]/[x + y + 108]; \quad \textbf{AE(min) = 0.84}; \quad E = 18/[x + y + 90]; \quad \textbf{E(max) = 0.20}$$

$f(\text{sac}) = 0$

Set $R_1 = R_2 = CH_3$

Using top resonance structure for ozone:

$$\Delta = \Delta(a) + \Delta(b) + \Delta(c) + \Delta(d) + \Delta(e) + \Delta(f) + \Delta(g)$$

$$= (1 - 1) + (-2 - 0) + (3 - (-1)) + (-2 - 1) + (-2 - (-1)) + (3 - (-1)) + (-2 - (-1)) = +1;$$

$$\textbf{HI} = -\Delta/(N + 1) = -\tfrac{1}{2}$$

Using bottom resonance structure for ozone:

$$\Delta = \Delta(a) + \Delta(b) + \Delta(c) + \Delta(d) + \Delta(e) + \Delta(f) + \Delta(g)$$

$$= (1 - 1) + (-2 - (-1)) + (3 - (-1)) + (-2 - 1) + (-2 - (-1)) + (3 - (-1)) + (-2 - 0) = +1;$$

$$\textbf{HI} = -\Delta/(N + 1) = -\tfrac{1}{2}$$

Note: ozone is produced in situ from dioxygen according to overall transformation $3/2 O_2 \rightarrow O_3$.

This implies that oxidation numbers for labeled oxygen atoms b, d, and g are all equal to 0 since they are derived from O_2.

$$\Delta = \Delta(a) + \Delta(b) + \Delta(c) + \Delta(d) + \Delta(e) + \Delta(f) + \Delta(g)$$

$$= (1-1) + (-2-0) + (3-(-1)) + (-2-0) + (-2-(-1)) + (3-(-1)) + (-2-0) = +1;$$

$$\mathbf{HI} = -\Delta/(N+1) = -\tfrac{1}{2} \text{ (as before)}.$$

Harries ozonolysis of olefins with reductive workup: Additive oxidation
Harries, C. *Ann. Chem.* 1905, *343*, 311.

$$R_1 = x; R_2 = y$$

$$AE = [x+y+58]/[x+y+136]; \quad \mathbf{AE(min)} = \mathbf{0.43}; \quad E = 78/[x+y+58]; \quad \mathbf{E(max)} = \mathbf{1.30}$$

$$f(\text{sac}) = 62/[x+y+136]; \quad f(\mathbf{sac})(\mathbf{max}) = \mathbf{0.45}$$

Set $R_1 = R_2 = CH_3$

Using top resonance structure for ozone:

$$\Delta = \Delta(a) + \Delta(b) + \Delta(c) + \Delta(d) = (-2-1) + (1-(-1)) + (1-(-1)) + (-2-(-1)) = 0; \quad \mathbf{HI} = \mathbf{0}$$

Using bottom resonance structure for ozone:

$$\Delta = \Delta(a) + \Delta(b) + \Delta(c) + \Delta(d) = (-2-1) + (1-(-1)) + (1-(-1)) + (-2-0) = -1;$$

$$\mathbf{HI} = \Delta/(N+1) = \tfrac{1}{2}$$

Note: ozone is produced in situ from dioxygen according to overall transformation $3/2O_2 \rightarrow O_3$.

This implies that oxidation numbers for labeled oxygen atoms a and d are all equal to 0 since they are derived from O_2.

$$\Delta = \Delta(a) + \Delta(b) + \Delta(c) + \Delta(d) = (-2-0) + (1-(-1)) + (1-(-1)) + (-2-0) = 0; \quad \mathbf{HI} = \mathbf{0} \text{ (as before)}.$$

Hooker oxidation: Subtractive oxidation with rearrangement
Hooker, S.C. *J. Am. Chem. Soc.* 1936, *58*, 1174.

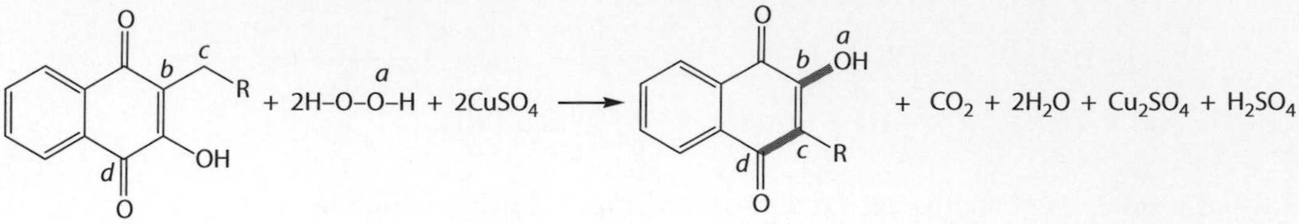

$$x + 187 \qquad\qquad 34 \qquad 319 \qquad\qquad x + 173 \qquad\qquad 44 \qquad 36 \qquad 223 \qquad 98$$

R = x

$$AE = [x+173]/[x+574]; \quad \textbf{AE(min)} = \textbf{0.30}; \quad E = 401/[x+173]; \quad \textbf{E(max)} = \textbf{2.30}$$

$$f(\text{sac}) = 319/[x+574]; \quad \textbf{f(sac)(max)} = \textbf{0.55}$$

Set R = CH_3

$$\Delta = \Delta(a) + \Delta(b) + \Delta(c) + \Delta(d) = (-2 - (-1)) + (1 - 0) + (0 - (-2)) + (2 - 2) = +2;$$

$$\textbf{HI} = -\Delta/(N+1) = \textbf{-1}$$

Jacobsen epoxidation: Additive oxidation
Zhang, W.; Loebach, J.L.; Wilson, S.R.; Jacobsen, E.N. *J. Am. Chem. Soc.* 1990, *112*, 2801.
Jacobsen, E.N.; Zhang, W.; Muci, A.R.; Ecker, J.R.; Deng, L. *J. Am. Chem. Soc.* 1991, *113*, 7063.

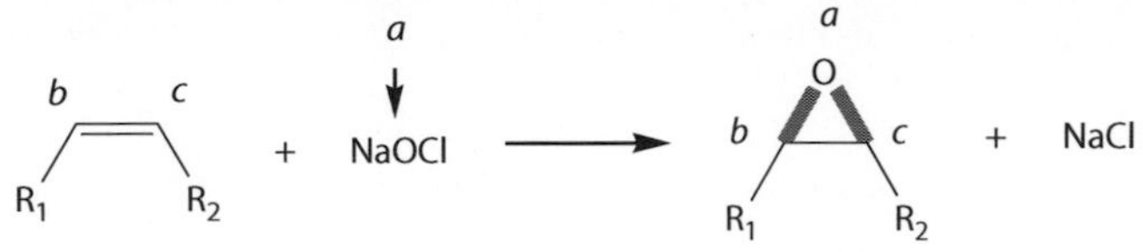

$$x + y + 26 \qquad\qquad 74.45 \qquad\qquad x + y + 42 \qquad\qquad 58.45$$

$$R_1 = x; \quad R_2 = y$$

$$AE = [x+y+42]/[x+y+100.45]; \quad \textbf{AE(min)} = \textbf{0.43}; \quad E = 58.45/[x+y+42]; \quad \textbf{E(max)} = \textbf{1.33}$$

$$\textbf{f(sac)} = \textbf{0}$$

Set $R_1 = R_2 = CH_3$

$$\Delta = \Delta(a) + \Delta(b) + \Delta(c) = (-2 - (-2)) + (0 - (-1)) + (0 - (-1)) = +2; \quad \textbf{HI} = -\Delta/(N+1) = \textbf{-1}$$

Jones oxidation: Subtractive oxidation

Bowden, K.; Heilbron, I.M.; Jones, E.R.H.; Weedon, B.C.L. *J. Chem. Soc.* 1946, 39.
Curtis, R.G.; Heilbron, I.M.; Jones, E.R.H.; Woods, G.F. *J. Chem. Soc.* 1953, 457.
Halsall, T.G.; Hodges, R.; Jones, E.R.H. *J. Chem. Soc.* 1953, 3019.
Bowers, A.; Halsall, T.G.; Jones, E.R.H.; Lemin, A.J. *J. Chem. Soc.* 1953, 2548.

x + y + 30   66.67   98   x + y + 28   130.67   36

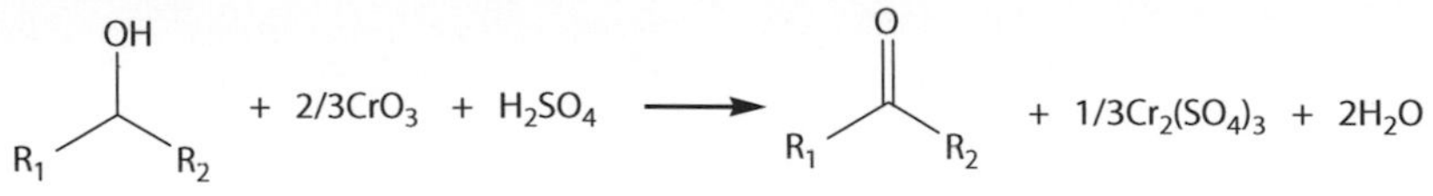

$$R_1 = x; \quad R_2 = y$$

$$AE = [x + y + 28]/[x + y + 194.67]; \quad \mathbf{AE(min) = 0.15}; \quad E = 166.67/[x + y + 28]; \quad \boldsymbol{E}\mathbf{(max) = 5.56}$$

$$f(sac) = 164.67/[x + y + 194.67]; \quad \boldsymbol{f}\mathbf{(sac)(max) = 0.84}$$

Lemieux–Johnson oxidation of olefins to 1,2-diols: Additive oxidation

Pappo, R.; Allen, D.S., Jr.; Lemieux, R.U.; Johnson, W.S. *J. Org. Chem.* 1956, *21*, 478.

x + y + 26   127   18   x + y + 60   111

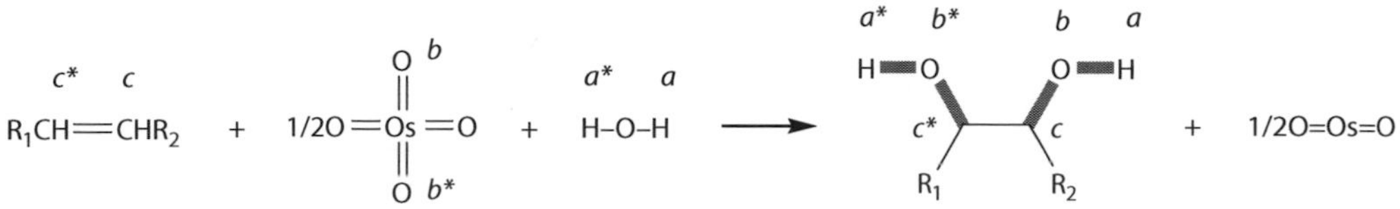

$$R_1 = x; \quad R_2 = y$$

$$AE = [x + y + 60]/[x + y + 171]; \quad \mathbf{AE(min) = 0.37}; \quad E = 111/[x + y + 60]; \quad \boldsymbol{E}\mathbf{(max) = 1.73}$$

$\boldsymbol{f}\mathbf{(sac) = 0}$

Set $R_1 = R_2 = CH_3$

$$\Delta = \Delta(a) + \Delta(b) + \Delta(c) + \Delta(a^*) + \Delta(b^*) + \Delta(c^*) = 2[(1-1) + (-2-(-2)) + (0-(-1))] = +2;$$

$$\mathbf{HI = -\Delta/(\mathit{N} + 1) = -1}$$

Lemieux–Johnson oxidative cleavage of olefins to aldehydes: Additive oxidation

Pappo, R.; Allen, D.S., Jr.; Lemieux, R.U.; Johnson, W.S. *J. Org. Chem.* 1956, *21*, 478.

x + y + 26   127   18   x + y + 60   111

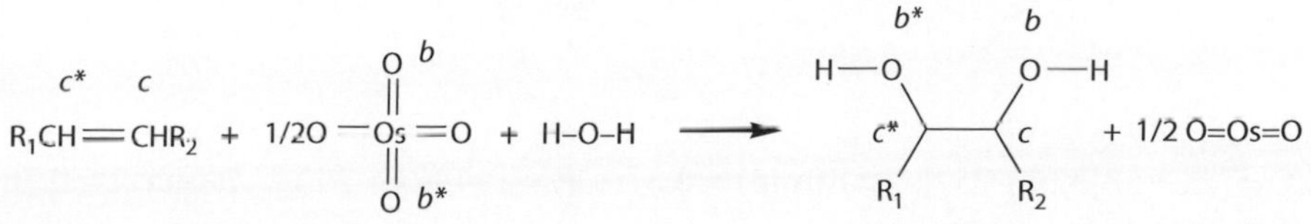

$$R_1 = x; \quad R_2 = y$$

Target product: both aldehydes

$$AE = [x + y + 58]/[x + y + 359]; \quad \mathbf{AE(min) = 0.16}; \quad E = 327/[x + y + 58]; \quad \mathbf{E(max) = 5.45}$$

$$f(sac) = 232/[x + y + 359]; \quad \mathbf{f(sac)(max) = 0.64}$$

$$\Delta = \Delta(b) + \Delta(c) + \Delta(b^*) + \Delta(c^*) = 2[(-2 - (-2)) + (1 - (-1))] = +4; \quad \mathbf{HI} = -\Delta/(N+1) = \mathbf{-2}$$

Target product: $\mathbf{R_1CHO}$

$$AE = [x + 29]/[x + y + 359]; \quad \mathbf{AE(min) = 0.075}; \quad E = [y + 356]/[x + 29]; \quad \mathbf{E(max) = 12.28}$$

$$f(sac) = 232/[x + y + 359]; \quad \mathbf{f(sac)(max) = 0.64}$$

$$\Delta = \Delta(b^*) + \Delta(c^*) = (-2 - (-2)) + (1 - (-1)) = +2; \quad \mathbf{HI} = -\Delta/(N+1) = \mathbf{-1}$$

Target product: $\mathbf{R_2CHO}$

$$AE = [y + 29]/[x + y + 359]; \quad \mathbf{AE(min) = 0.075}; \quad E = [x + 356]/[y + 29]; \quad \mathbf{E(max) = 12.28}$$

$$f(sac) = 232/[x + y + 359]; \quad \mathbf{f(sac)(max) = 0.64}$$

$$\Delta = \Delta(b) + \Delta(c) = (-2 - (-2)) + (1 - (-1)) = +2; \quad \mathbf{HI} = -\Delta/(N+1) = \mathbf{-1}$$

Malaprade oxidation: Subtractive oxidation

Malaprade, L. *Bull. Soc. Chim. Fr.* 1928, *43(4)*, 43.
Malaprade, L. *Bull. Soc. Chim. Fr.* 1928, *43(4)*, 683.
Malaprade, L. *Compt. Rend.* 1928, *186*, 382.

$$R = x$$

$$AE = [2x + 58]/[2x + 252]; \quad \mathbf{AE(min) = 0.57}; \quad E = 194/[2x + 252]; \quad \mathbf{E(max) = 0.76}$$

$$f(sac) = 192/[2x + 252]; \quad \mathbf{f(sac)(max) = 0.76}$$

Oppenauer oxidation: Subtractive oxidation

Oppenauer, R.V. *Rec. Trav. Chim.* 1937, *56*, 137.

| x + y + 30 | 108 | x + y + 28 | 110 |

$$R_1 = x; \quad R_2 = y$$

$$AE = [x + y + 28]/[x + y + 138]; \quad \textbf{AE(min)} = \textbf{0.21}; \quad E = 110/[x + y + 28]; \quad \textbf{E(max)} = \textbf{3.67}$$

$$f(\text{sac}) = 108/[x + y + 138]; \quad f(\textbf{sac})(\textbf{max}) = \textbf{0.77}$$

Oxymercuration of olefins: Additive oxidation

Brook, A.G.; Wright, G.F. *Can. J. Res.* 1950, *28B*, 623.
Wright, G.F. *Chemistry in Canada* 1950, *2(9)*, 29.
Wright, G.F. *Ann. N.Y. Acad. Sci.* 1957, *65*, 436.
Abercrombie, M.J.; Rodgman, A.; Bharucha, K.R.; Wright, G.F. *Can. J. Chem.* 1959, *37*, 1328.

| x + 27 | 319 | 9.5 | 36 | x + 45 | 60 | 15.5 | 201 | 10 |

$$R = x$$

$$AE = [x + 45]/[x + 364.5]; \quad \textbf{AE(min)} = \textbf{0.14}; \quad E = 286.5/[x + 45]; \quad \textbf{E(max)} = \textbf{6.23}$$

$$f(\text{sac}) = 319/[x + 364.5]; \quad f(\textbf{sac})(\textbf{max}) = \textbf{0.87}$$

Set R = CH$_3$

$$\Delta = \Delta(a) + \Delta(b) + \Delta(c) + \Delta(d) = (1 - (-1)) + (-3 - (-2)) + (0 - (-1)) + (-2 - (-2)) = 0; \quad \textbf{HI} = \textbf{0}$$

Permanganate oxidation of olefins to 1,2-diols: Additive oxidation

Wagner, G. *J. Russ. Phys. Chem.* 1895, *27*, 219.
Boeseken, J. *Rec. Trav. Chim.* 1921, *40*, 553.
Boeseken, J.; de Graaff, M.C. *Rec. Trav. Chim.* 1922, *41*, 199.
Boeseken, J. *Rec. Trav. Chim.* 1928, *47*, 683.
Criegee, R. *Ann. Chem.* 1936, *522*, 75.
Criegee, R. *Angew. Chem.* 1938, *51*, 519.
Criegee, R.; Marchand, B.; Wannowius, H. *Ann. Chem.* 1938, *550*, 99.

$$R_1CH = CHR_2 + 2/3KMnO_4 + 4/3H_2O \longrightarrow$$

(b, b*) over the $2/3KMnO_4$, (a, a*) over the $4/3H_2O$, c* c over R_1CH

Product: with labels a*, b*, b, a over $H-O$ ... $O-H$, c* c, R_1, R_2; $+ 2/3MnO_2 + 2/3KOH$

| x + y + 26 | 105.33 | 24 | x + y + 60 | 58 | 37.33 |

$$R_1 = x; \quad R_2 = y$$

$$AE = [x + y + 60]/[x + y + 129.33]; \quad \mathbf{AE(min) = 0.40}; \quad E = 95.33/[x + y + 60]; \quad \mathbf{E(max) = 1.49}$$

$$f(\text{sac}) = 0$$

Set $R_1 = R_2 = CH_3$

$$\Delta = \Delta(a) + \Delta(b) + \Delta(c) + \Delta(a^*) + \Delta(b^*) + \Delta(c^*) = 2[(1-1) + (-2-(-2)) + (0-(-1))] = +2;$$

$$\mathbf{HI = -\Delta/(N+1) = -1}$$

Pfitzner–Moffatt oxidation: Subtractive oxidation

Pfitzner, K.E.; Moffatt, J.G. *J. Am. Chem. Soc.* 1963, *85*, 3027.

$$R \diagup OH + \text{(DMSO)} + Cy-N=C=N-Cy \longrightarrow R \diagup O + \diagup S \diagdown + Cy-NH(C=O)NH-Cy$$

| x + 31 | 78 | 204 | x + 29 | 62 | 222 |

$$R = x$$

$$AE = [x + 29]/[x + 313]; \quad \mathbf{AE(min) = 0.096}; \quad E = 284/[x + 29]; \quad \mathbf{E(max) = 9.47}$$

$$f(\text{sac}) = 282/[x + 313]; \quad \mathbf{f(\text{sac})(max) = 0.90}$$

Prevost reaction: Additive oxidation

Prévost, C. *Compt. Rend.* 1933, *196*, 1129.

$$R_1CH = CHR_2 + I_2 + 2PhCOOAg + 2H_2O \longrightarrow$$

with labels b*, b over $R_1CH=CHR_2$; (a, a*) over; product HO, OH with a*, a and b*, b, R_1, R_2; $+ 2AgI + 2PhCOOH$

| x + y + 26 | 254 | 458 | 36 | x + y + 60 | 470 | 244 |

$$R_1 = x; \quad R_2 = y$$

$$AE = [x + y + 60]/[x + y + 774]; \quad \mathbf{AE(min) = 0.080}; \quad E = 714/[x + y + 60]; \quad \mathbf{E(max) = 11.52}$$

$$f(\text{sac}) = 712/[x + y + 774]; \quad \mathbf{f(\text{sac})(max) = 0.92}$$

Set $R_1 = R_2 = CH_3$

$$\Delta = \Delta(a) + \Delta(a^*) + \Delta(b) + \Delta(b^*) = 2[(-2-(-2)) + (0-(-1))] = +2; \quad \mathbf{HI = -\Delta/(N+1) = -1}$$

Prilezhaev reaction (epoxidation of olefins): Additive oxidation
Prilezhaev, N. *Chem. Ber.* 1909, *42*, 4811.

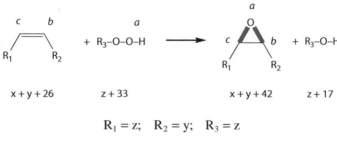

<table>
<tr><td>x + y + 26</td><td>z + 33</td><td>x + y + 42</td><td>z + 17</td></tr>
</table>

$$R_1 = z; \quad R_2 = y; \quad R_3 = z$$

$$AE = [x+y+42]/[x+y+z+59]; \quad \mathbf{AE(min) = 44/[z+59]};$$

$$E = [z+17]/[x+y+42]; \quad \mathbf{E(max) = [z+17]/44}$$

$f(\mathbf{sac}) = \mathbf{0}$

R_3	E(max)	AE(min)
H	0.41	0.71
C_7H_4OCl	3.56	0.22

Set $R_1 = R_2 = CH_3$

$$\Delta = \Delta(a) + \Delta(b) + \Delta(c) = (-2-(-1)) + (0-(-1)) + (0-(-1)) = +1; \quad \mathbf{HI = -\Delta/(N+1) = -\tfrac{1}{2}}$$

Riley oxidation: Additive oxidation
Riley, H.L.; Morley, J.F.; Friend, N.A.C. *J. Chem. Soc.* 1932, 1875.

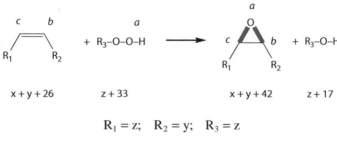

<table>
<tr><td>x + y + 42</td><td>111</td><td>x + y + 56</td><td>79</td><td>18</td></tr>
</table>

$$R_1 = x; \quad R_2 = y$$

$$AE = [x+y+56]/[x+y+153]; \quad \mathbf{AE(min) = 0.37}; \quad E = 97/[x+y+56]; \quad \mathbf{E(max) = 1.67}$$

$f(\mathbf{sac}) = \mathbf{0}$

Set $R_1 = CH_3$

$$\Delta = \Delta(a) + \Delta(b) = (-2-(-2)) + (2-(-2)) = +4; \quad \mathbf{HI = -\Delta/(N+1) = -2}$$

Rubottom reaction: Additive oxidation

Rubottom, G.M.; Vazquez, M.A.; Pelegrina, D.R. *Tetrahedron Lett.* 1974, *15*, 4319.

x + y + 114	172.45	18	x + y + 58	156.45	90

$$R_1 = x; \quad R_2 = y$$

$$AE = [x + y + 58]/[x + y + 304.45]; \quad \mathbf{AE(min) = 0.20}; \quad E = 246.45/[x + y + 58]; \quad \mathbf{\textit{E}(max) = 4.11}$$

$$\mathbf{\textit{f}(sac) = 0}$$

Set $R_2 = CH_3$

$$\Delta = \Delta(a) + \Delta(b) + \Delta(c) = (-2 - (-1)) + (0 - (-1)) + (1 - 1) = 0; \quad \mathbf{HI = 0}$$

Sarett procedure: Subtractive oxidation

Poos, G.I.; Arth, G.E.; Beyler, R.E.; Sarett, L.H. *J. Am. Chem. Soc.* 1953, *75*, 422.

x + 31	258	x + 29	102	158

$$R = x$$

$$AE = [x + 29]/[x + 289]; \quad \mathbf{AE(min) = 0.10}; \quad E = 260/[x + 29]; \quad \mathbf{\textit{E}(max) = 8.67}$$

$$\textit{f}(sac) = 258/[x + 289]; \quad \mathbf{\textit{f}(sac)(max) = 0.89}$$

Sharpless–Jacobsen dihydroxylation: Additive oxidation

Jacobsen, E.N.; Markó, I.; Mungall, W.S.; Schröder, G.; Sharpless, K.B. *J. Am. Chem. Soc.* 1988, *110*, 1968.

2 (x + y + z + 25)	657.7	368.2	2(x + y + z + 27)	222.2	735.7

$$R_1 = x; \quad R_2 = y; \quad R_3 = z$$

$$AE = [2(x + y + z) + 54]/[2(x + y + z) + 1075.9]; \quad \mathbf{AE(min) = 0.059};$$

$$E = 957.9/[2(x + y + z) + 54]; \quad \mathbf{\textit{E}(max) = 15.965}$$

$$f(\text{sac}) = 657.7/[2(x+y+z)+1075.9]; \quad f(\text{sac})(\text{max}) = \mathbf{0.608}$$

$$\Delta = \Delta(a) + \Delta(b) + \Delta(c) + \Delta(d) + \Delta(e) + \Delta(f)$$

$$= (-2-(-2)) + (0-(-1)) - (1-0) + (-2-(-2)) + (1-1) + (1-1) = +2;$$

$$\mathbf{HI} = -\Delta/(N+1) = \mathbf{-1}$$

Including recycling of $K_2[OsO_2(OH)_4]$ by $K_3[Fe(CN)_6]$:

| 2(x+y+z+25) | 1315.4 | 224 | 2(x+y+z+27) | 1471.4 |

$$R_1 = x; \quad R_2 = y; \quad R_3 = z$$

$$AE = [2(x+y+z)+54]/[2(x+y+z)+1539.4]; \quad \mathbf{AE(min)} = \mathbf{0.0392};$$

$$E = 1471.4/[2(x+y+z)+54]; \quad \mathbf{E(max)} = \mathbf{24.523}$$

$$f(\text{sac}) = 1315.4/[2(x+y+z)+1539.4]; \quad \mathbf{f(\text{sac})(\text{max})} = \mathbf{0.851}$$

Set $R_1 = R_2 = R_3 = CH_3$

$$\Delta = \Delta(a) + \Delta(a^*) + \Delta(b) + \Delta(c) = (-2-(-2)) + (-2-(-2)) + (0-(-1)) + (1-0) = +2;$$

$$\mathbf{HI} = -\Delta/(N+1) = \mathbf{-1}$$

Sharpless epoxidation: Additive oxidation

Katsuki, T.; Sharpless, K.B., *J. Am. Chem. Soc.* 1980, *102*, 5974.

| x+y+26 | 90 | x+y+42 | 74 |

$$R_1 = x; \quad R_2 = y$$

$$AE = [x+y+42]/[x+y+116]; \quad \mathbf{AE(min)} = \mathbf{0.37}; \quad E = 74/[x+y+42]; \quad \mathbf{E(max)} = \mathbf{1.68}$$

$$f(\text{sac}) = \mathbf{0}$$

Set $R_1 = R_2 = CH_3$

$$\Delta = \Delta(a) + \Delta(b) + \Delta(c) = (-2-(-1)) + (0-(-1)) + (0-(-1)) = +1; \quad \mathbf{HI} = -\Delta/(N+1) = \mathbf{-\tfrac{1}{2}}$$

Sharpless oxyamination: Additive oxidation

Sharpless, K.B.; Patrick, D.W.; Lonesdale, L.K.; Biller, S.A. *J. Am. Chem. Soc.* 1975, *97*, 2305.

x + y + 26   z + 252   36   x + y + z + 58   256

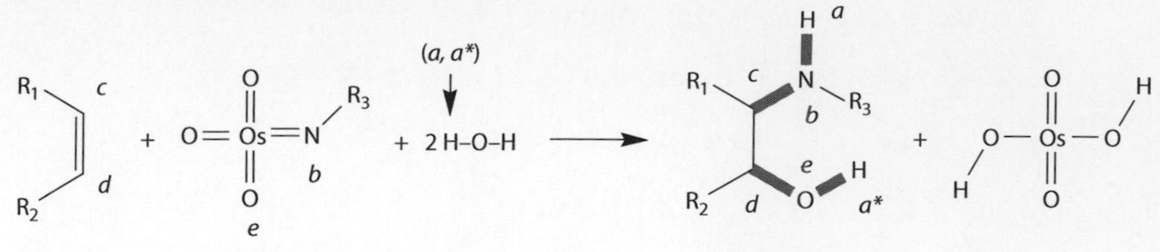

$$R_1 = x; \quad R_2 = y; \quad R_3 = z$$

$$AE = [x + y + z + 58]/[x + y + z + 314]; \quad \mathbf{AE(min) = 0.19};$$

$$E = 256/[x + y + z + 58]; \quad \boldsymbol{E}\mathbf{(max) = 4.20}$$

$$\boldsymbol{f}\mathbf{(sac) = 0}$$

$$R_1 = R_2 = R_3 = CH_3$$

$$\Delta = \Delta(a) + \Delta(b) + \Delta(c) + \Delta(d) + \Delta(e)$$

$$= (1 - 1) + (-3 - (-3)) + (0 - (-1)) + (0 - (-1)) + (-2 - (-2)) = +2;$$

$$\mathbf{HI = -\Delta/(\boldsymbol{N} + 1) = -1}$$

Shi asymmetric epoxidation: Additive oxidation

Wang, Z.X.; Tu, Y.; Frohn, M.; Zhang, J.R.; Shi, Y. *J. Am. Chem. Soc.* 1997, *119*, 11224.
Wang, Z.X.; Shi, Y. *J. Org. Chem.* 1997, *62*, 8622.

x + y + 26   152   56   x + y + 42   174   18

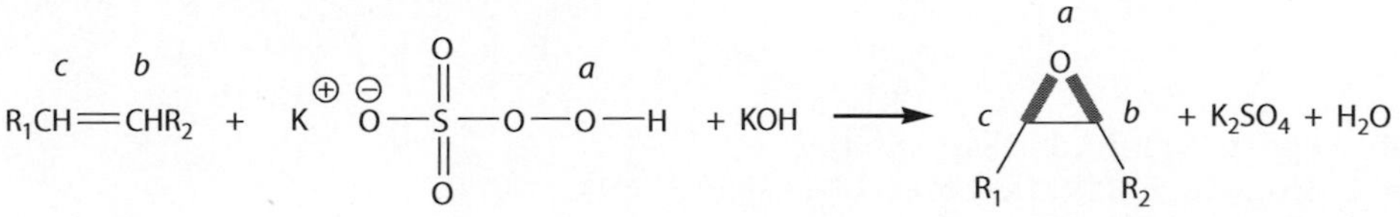

$$R_1 = x; \quad R_2 = y$$

$$AE = [x + y + 42]/[x + y + 234]; \quad \mathbf{AE(min) = 0.19}; \quad E = 192/[x + y + 42]; \quad \boldsymbol{E}\mathbf{(max) = 4.36}$$

$$f(\text{sac}) = 56/[x + y + 234]; \quad \boldsymbol{f}\mathbf{(sac)(max) = 0.24}$$

Set $R_1 = R_2 = CH_3$

$$\Delta = \Delta(a) + \Delta(b) + \Delta(c) = (-2 - (-1)) + (0 - (-1)) + (0 - (-1)) = +1; \quad \mathbf{HI = -\Delta/(\boldsymbol{N} + 1) = -½}$$

Swern oxidation: Subtractive oxidation

Omura, K.; Swern, D. *Tetrahedron* 1978, *34*, 1651.

x + y + 30   78   126.9   202   x + y + 28   62   28   44   274.9

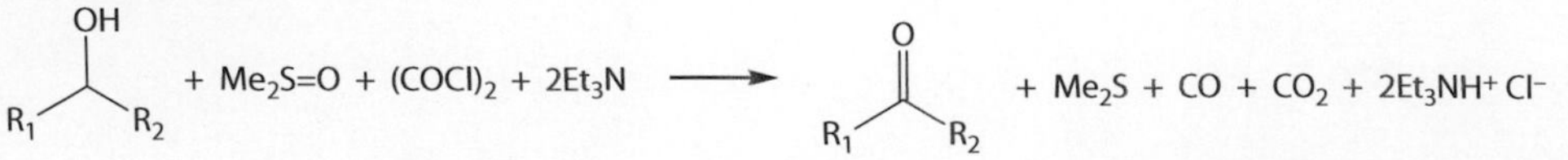

$$R_1 = x; \quad R_2 = y$$

$$AE = [x + y + 28]/[x + y + 436.9]; \quad \textbf{AE(min)} = \textbf{0.068}; \quad E = 408.9/[x + y + 28]; \quad \textbf{E(max)} = \textbf{13.63}$$

$$f(\text{sac}) = 406.9/[x + y + 436.9]; \quad \textbf{f(sac)(max)} = \textbf{0.93}$$

Tollens test: Additive oxidation
Tollens, B. *Chem. Ber.* 1882, *15*, 1635.

| | x + 29 | 375.8 | 18 | x + 45 | 216 | 161.8 |

$$R = x$$

$$AE = [x + 45]/[x + 422.8]; \quad \textbf{AE(min)} = \textbf{0.11}; \quad E = 377.8/[x + 45]; \quad \textbf{E(max)} = \textbf{8.21}$$

$$f(\text{sac}) = 375.8/[x + 422.8]; \quad \textbf{f(sac)(max)} = \textbf{0.89}$$

Set R = CH$_3$

$$\Delta = \Delta(a) + \Delta(b) = (-2 - (-2)) + (3 - 1) = +2; \quad \textbf{HI} = -\Delta/(N + 1) = \textbf{-1}$$

Uemura oxidation: Subtractive oxidation
Nishimura, T.; Onoue, T.; Ohe, K.; Uemura, S. *Tetrahedron Lett.* 1998, *39*, 6011.
Nishimura, T.; Onoue, T.; Ohe, K.; Uemura, S. *J. Org. Chem.* 1999, *64*, 6750.

| x + y + 30 | 16 | x + y + 28 | 18 |

$$R_1 = x; \quad R_2 = y$$

$$AE = [x + y + 28]/[x + y + 46]; \quad E = 18/[x + y + 28]; \quad \textbf{E(max)} = \textbf{0.60}; \quad \textbf{AE(min)} = \textbf{0.63}$$

$$f(\text{sac}) = 16/[x + y + 46]; \quad \textbf{f(sac)(max)} = \textbf{0.33}$$

Wacker–Tsuji oxidation: Additive oxidation
Smidt, J.; Hafner, W.; Jira, R.; Sedlmeier, J.; Sieber, R.; Ruttinger, R.; Kojer, H. *Angew. Chem.* 1959, *71*, 176.
Smidt, J.; Hafner, W.; Jira, R.; Sieber, R.; Sedlmeier, J.; Sabel, J. *Angew. Chem. Int. Ed.* 1962, *1*, 80.
Smidt, J. *Chem. Ind.* 1962, 54.
Tsuji, J. *Pure Appl. Chem.* 1999, *71*, 1539.
Tsuji, J. *New J. Chem.* 2000, *24*, 127.

| 28 | 177.3 | 18 | 44 | 106.4 | 72.9 |

$$\mathbf{AE = 0.20;\quad \mathit{E} = 4.08}$$

$$\mathbf{\mathit{f}(sac) = 177.3/223.3 = 0.79}$$

$$\Delta = \Delta(a)+\Delta(b)+\Delta(c) = (-2-(-2))+(1-(-2))+(-3-(-1)) = +1;\quad \mathbf{HI = -\Delta/(\mathit{N}+1) = -1/2}$$

Wessely oxidation: Additive oxidation

Wessely, F.; Lauterbach-Kiel, G.; Sinwel, F. *Monatsh. Chem.* 1950, *81*, 811.
Wessely, F.; Sinwel, F. *Monatsh. Chem.* 1950, *81*, 1055.

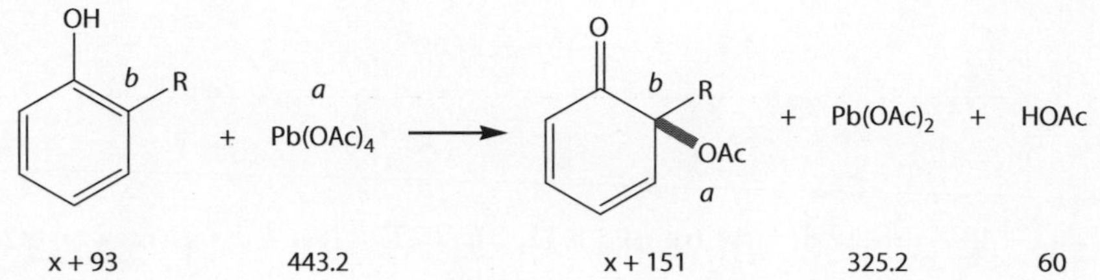

R = x

$$\text{AE} = [\text{x}+151]/[\text{x}+536];\quad \mathbf{AE(min) = 0.28};\quad E = 385/[\text{x}+151];\quad \mathbf{\mathit{E}(max) = 2.53}$$

$$\mathbf{\mathit{f}(sac) = 0}$$

Set R = CH_3

$$\Delta = \Delta(a)+\Delta(b) = (-2-(-2))+(1-0) = +1;\quad \mathbf{HI = -\Delta/(\mathit{N}+1) = -1/2}$$

Willgerodt reaction: Additive oxidation with rearrangement

Willgerodt, C. *Chem. Ber.* 1888, *21*, 534.

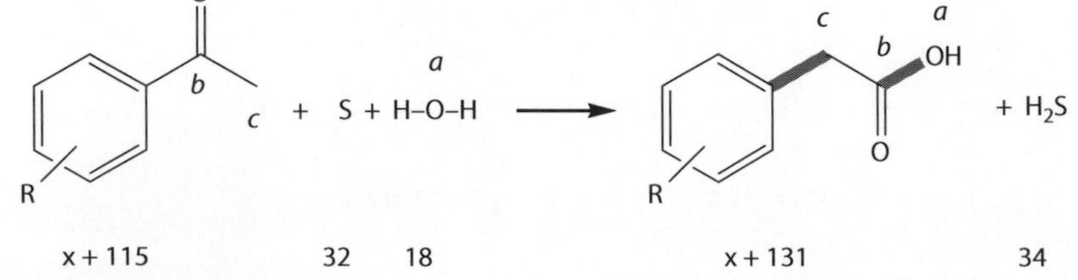

R = x(sum of substituents)

$$\text{AE} = [\text{x}+131]/[\text{x}+165];\quad \mathbf{AE(min) = 0.80};\quad E = 34/[\text{x}+131];\quad \mathbf{\mathit{E}(max) = 0.25}$$

$$f(\text{sac}) = 32/[\text{x}+165];\quad \mathbf{\mathit{f}(sac)(max) = 0.19}$$

$$\Delta = \Delta(a)+\Delta(b)+\Delta(c) = (-2-(-2))+(3-2)+(-2-(-3)) = +2;\quad \mathbf{HI = -\Delta/(\mathit{N}+1) = -1}$$

Woodward *cis*-dihydroxylation: Additive oxidation
Woodward, R.B. *J. Am. Chem. Soc.* 1958, *80*, 209.

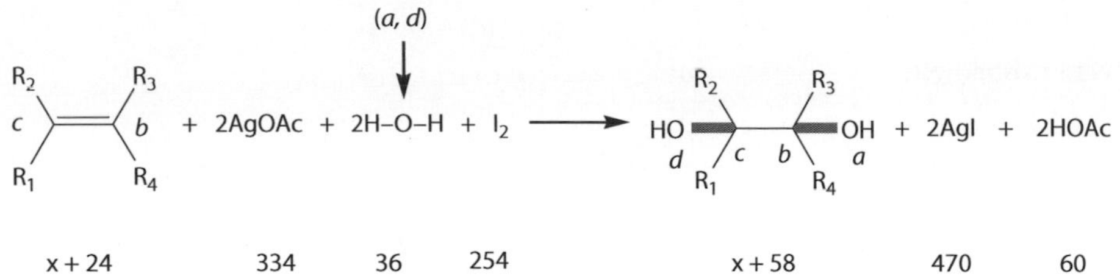

$$R_1 + R_2 + R_3 + R_4 = x$$

$$AE = [x + 58]/[x + 648]; \quad \textbf{AE(min)} = \textbf{0.095}; \quad E = 590/[x + 58]; \quad \textbf{E(max)} = \textbf{9.52}$$

$$f(sac) = 588/[x + 648]; \quad \textbf{f(sac)(max)} = \textbf{0.90}$$

Set $R_1 = R_2 = R_3 = R_4 = CH_3$

$$\Delta = \Delta(a) + \Delta(b) + \Delta(c) + \Delta(d) = (-2 - (-2)) + (1 - 0) + (1 - 0) + (-2 - (-2)) = +2;$$

$$\textbf{HI} = -\Delta/(N + 1) = -\textbf{1}$$

7.7 CATEGORIZATION OF NAMED REARRANGEMENT REACTIONS BY MINIMUM ATOM ECONOMY AND MAXIMUM ENVIRONMENTAL IMPACT FACTOR, MAXIMUM MOLECULAR WEIGHT FRACTION OF SACRIFICIAL REAGENTS, HYPSICITY INDEX

(Reactions are listed alphabetically.)

Acyl rearrangement
Fischer, E. *Chem. Ber.* 1920, *53*, 1621.

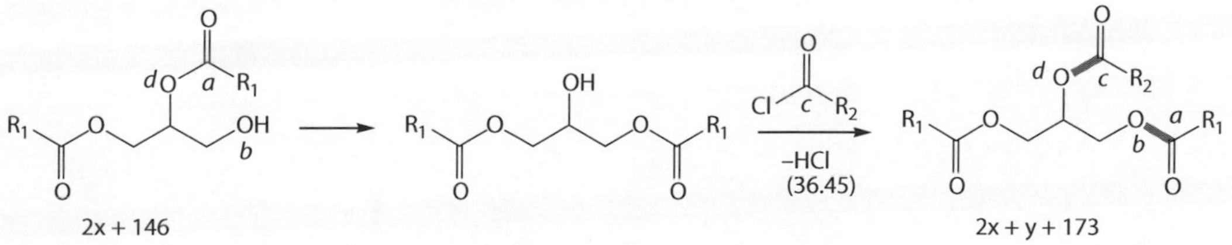

$$R_1 = x; \quad R_2 = y$$

$$AE = [2x + y + 173]/[2x + y + 209.45]; \quad \textbf{AE(min)} = \textbf{0.83};$$

$$E = 36.45/[2x + y + 173]; \quad \textbf{E(max)} = \textbf{0.21}; \quad f(sac) = 0$$

Set $R_1 = R_2 = CH_3$

$$\Delta = \Delta(a) + \Delta(b) + \Delta(c) + \Delta(d) = (3-3) + (-2-(-2)) + (3-3) + (-2-(-2)) = 0; \quad \mathbf{HI = 0}$$

Allylic rearrangement

Claisen, L. *Chem. Ber.* 1912, *45*, 3157.

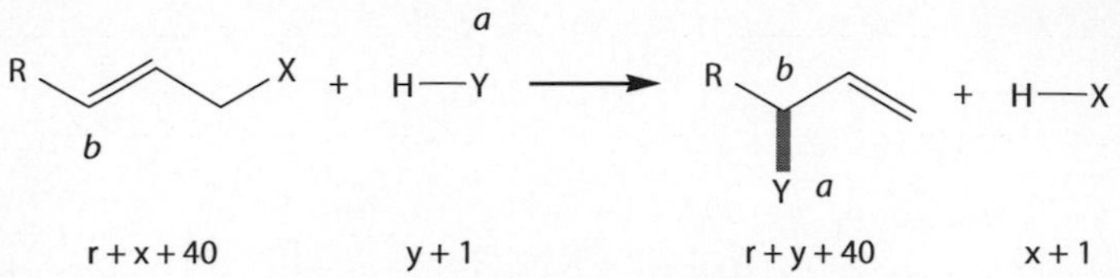

$$r + x + 40 \qquad\qquad y + 1 \qquad\qquad r + y + 40 \qquad\qquad x + 1$$

$$R = r; \quad X = x(\text{halide}); \quad Y = y(\text{halide})$$

$$AE = [r+y+40]/[x+y+r+41]; \quad \mathbf{AE(min)} = [y+41]/[x+y+42];$$

$$E = [x+1]/[r+y+40]; \quad E(\mathbf{max}) = [x+1]/[y+41]; \quad f(\mathbf{sac}) = 0$$

X	Y	E(max)	AE(min)
Cl	Cl	0.48	0.68
Cl	Br	0.30	0.77
Cl	I	0.22	0.82
Br	Cl	1.05	0.49
Br	Br	0.67	0.60
Br	I	0.48	0.68
I	Cl	1.67	0.37
I	Br	1.06	0.49
I	I	0.76	0.57

Set $R = CH_3$

$$\Delta = \Delta(a) + \Delta(b) = (-1-(-1)) + (0-(-1)) = +1; \quad \mathbf{HI} = -\Delta/(N+1) = -\tfrac{1}{2}$$

Aza–Cope [3,3] rearrangement

Klose, H.; Guenther, H. *Chem. Ber.* 1969, *102*, 2230.
Marshall, J.A.; Babler, J.H. *J. Org. Chem.* 1969, *34*, 4186.

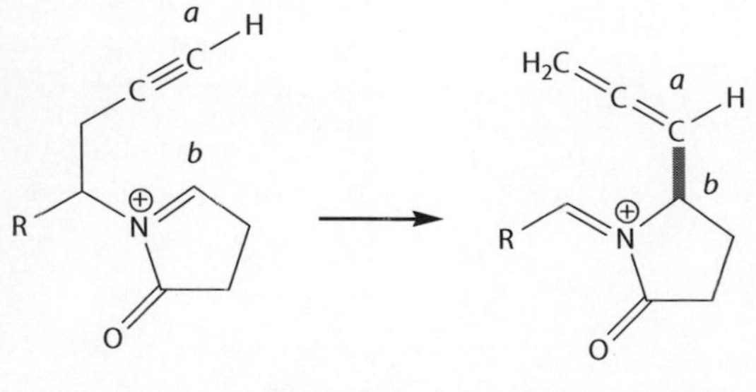

$$\mathbf{AE = 1}; \quad E = 0; \quad f(\mathbf{sac}) = 0$$

$$\Delta = \Delta(a) + \Delta(b) = (-1-(-1)) + (0-1) = -1; \quad \mathbf{HI} = -\Delta/(N+1) = +\tfrac{1}{2}$$

Bamberger rearrangement

Bamberger, E. *Chem. Ber.* 1894, *27*, 1347.
Bamberger, E. *Chem. Ber.* 1894, *27*, 1548.

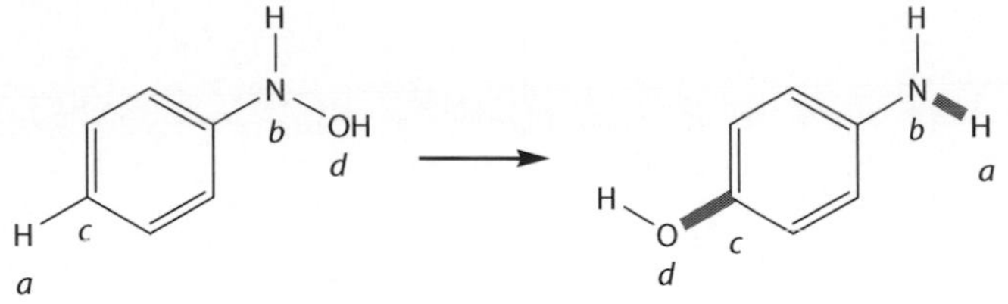

$$AE = 1; \quad E = 0; \quad f(\text{sac}) = 0$$

$$\Delta = \Delta(a) + \Delta(b) + \Delta(c) + \Delta(d) = (1-1) + (-3-(-1)) + (1-(-1)) + (-2-(-2)) = 0; \quad HI = 0$$

Bamberger–Goldschmidt synthesis of isoquinoline

Bamberger, E.; Goldschmidt, C. *Chem. Ber.* 1894, *27*, 1954.

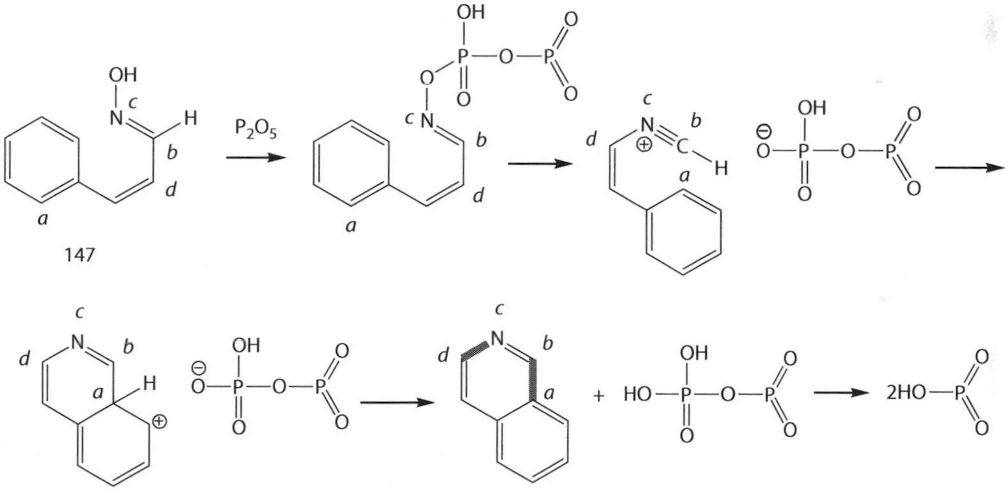

147

129

160

$$AE = 0.45; \quad E = 160/129 = 1.24; \quad f(\text{sac}) = 142/289 = 0.49$$

$$\Delta = \Delta(a) + \Delta(b) + \Delta(c) + \Delta(d) = (0-(-1)) + (1-1) + (-3-(-1)) + (0-(-1)) = 0; \quad HI = 0$$

Barton reaction

Barton, D.H.R.; Beaton, J.M.; Geller, L.E.; Pechet, M.M. *J. Am. Chem. Soc.* 1960, *82*, 2640.

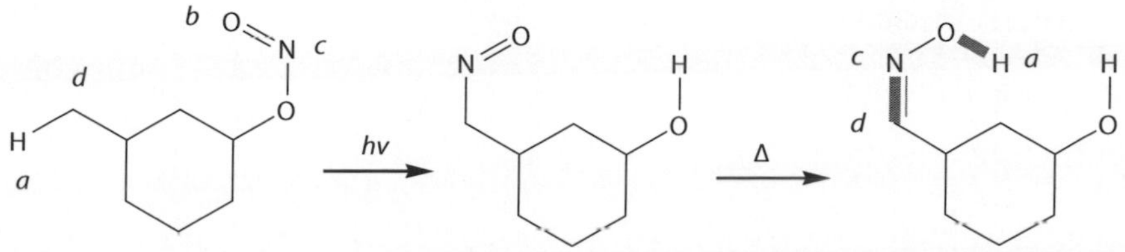

$$AE = 1; \quad E = 0; \quad f(\text{sac}) = 0$$

$$\Delta = \Delta(a) + \Delta(b) + \Delta(c) + \Delta(d) = (1-1) + (-2-(-2)) + (-1-3) + (1-(-3)) = 0; \quad HI = 0$$

Beckmann rearrangement

Beckmann, E. *Chem. Ber.* 1886, *19*, 988.

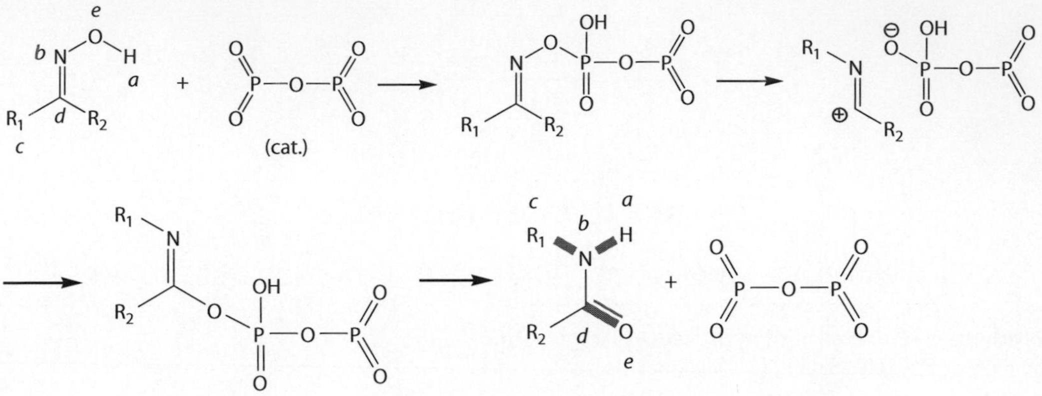

$$AE = 1; \quad E = 0; \quad f(\text{sac}) = 0$$

Set $R_1 = R_2 = CH_3$

$$\Delta = \Delta(a) + \Delta(b) + \Delta(c) + \Delta(d) = (1-1) + (-3-(-1)) + (-2-(-3)) + (3-2) + (-2-(-2)) = 0; \quad \mathbf{HI = 0}$$

Benzidine rearrangement

Hofmann, A.W. *Proc. Roy. Soc. London* 1863, *12*, 576.
Jacobson, P.; Henrich, F.; Klein, J. *Chem. Ber.* 1893, *26*, 688.

$$AE = 1; \quad E = 0; \quad f(\text{sac}) = 0$$

$$\Delta = \Delta(a) + \Delta(b) + \Delta(c) + \Delta(a^*) + \Delta(b^*) + \Delta(c^*) = 2((1-1) + (-3-(-2)) + (0-(-1))) = 0; \quad \mathbf{HI = 0}$$

Benzylic acid rearrangement

Liebig, J. *Ann. Chem.* 1838, *25*, 27.
Zinin, N. *Ann. Chem.* 1839, *31*, 329.

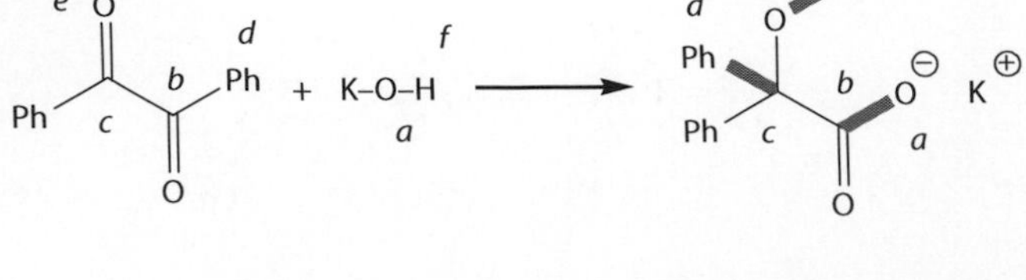

$$AE = 1; \quad E = 0; \quad f(\text{sac}) = 0$$

$$\Delta = \Delta(a) + \Delta(b) + \Delta(c) + \Delta(d) + \Delta(e) + \Delta(f)$$

$$= (-2 - (-2)) + (3 - 2) + (1 - 2) + (0 - 0) + (-2 - (-2)) + (1 - 1) = 0; \quad HI = 0$$

Brook rearrangement

Brook, A.G. *J. Am. Chem. Soc.* 1958, *80*, 1886.

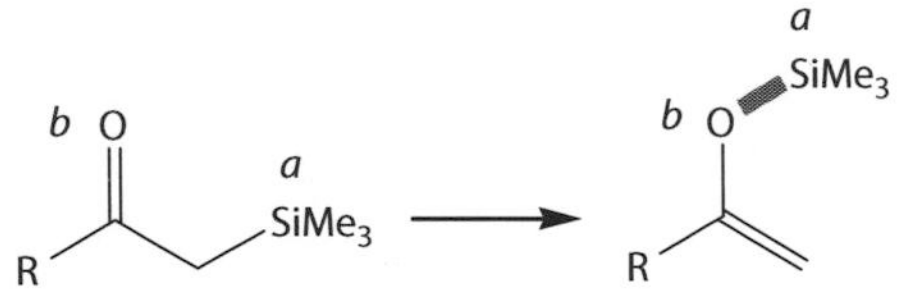

$$AE = 1; \quad E = 0; \quad f(\text{sac}) = 0$$

$$\Delta = \Delta(a) + \Delta(b) = (-2 - (-4)) + (-2 - (-2)) = +2; \quad HI = -\Delta/(N+1) = -1$$

Camphene rearrangement

Wagner, G. *J. Russ. Phys. Chem.* 1899, *31*, 690.
Meerwein, H. *Ann. Chem.* 1914, *405*, 129.

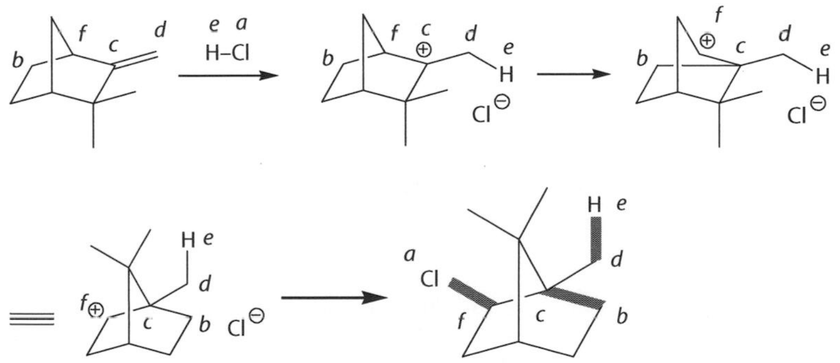

$$AE = 1; \quad E = 0; \quad f(\text{sac}) = 0$$

$$\Delta = \Delta(a) + \Delta(b) + \Delta(c) + \Delta(d) + \Delta(e) + \Delta(f)$$

$$= (-1 - (-1)) + (-2 - (-2)) + (0 - 0) + (-3 - (-2)) + (1 - 1) + (0 - (-1)) = 0; \quad HI = 0$$

Chapman rearrangement

Chapman, A.W. *J. Chem. Soc.* 1925, *127*, 1992.

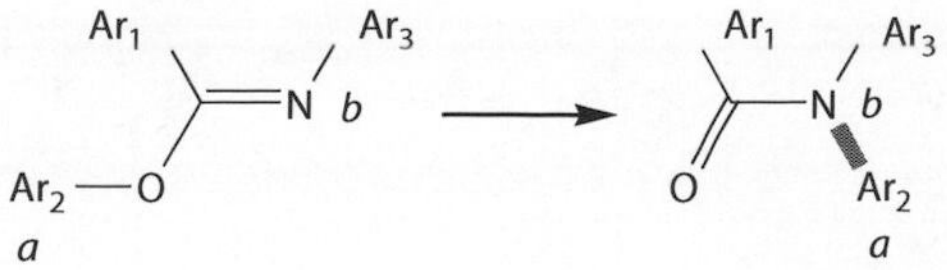

$$AE = 1; \quad E = 0; \quad f(\text{sac}) = 0$$

$$\Delta = \Delta(a) + \Delta(b) = (1-1) + (-3-(-3)) = 0; \quad HI = 0$$

Ciamician photodisproportionation
Ciamician, G.; Silber, P. *Chem. Ber.* 1901, *34*, 2040.

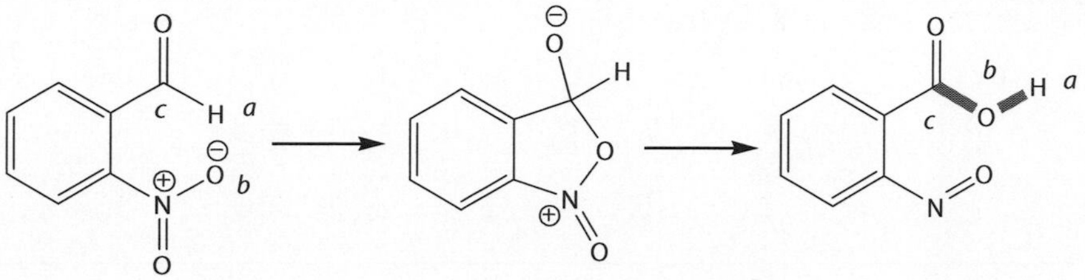

$$AE = 1; \quad E = 0; \quad f(\text{sac}) = 0$$

$$\Delta = \Delta(a) + \Delta(b) + \Delta(c) = (1-1) + (-2-(-2)) + (3-1) = +2; \quad HI = -\Delta/(N+1) = -1$$

Claisen rearrangement
Claisen, L. *Chem. Ber.* 1912, *45*, 3157.

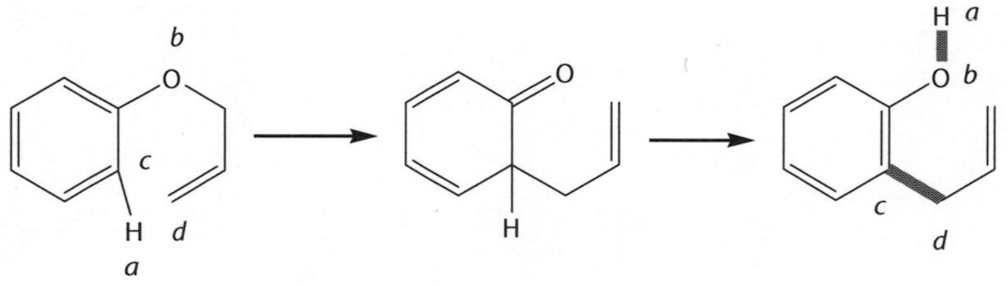

$$AE = 1; \quad E = 0; \quad f(\text{sac}) = 0$$

$$\Delta = \Delta(a) + \Delta(b) + \Delta(c) + \Delta(d)$$

$$= (1-1) + (-2-(-2)) + (0-(-1)) + (-2-(-2)) = +1; \quad HI = -\Delta/(N+1) = -\tfrac{1}{2}$$

Claisen–Ireland rearrangement
Ireland, R.E.; Mueller, R.H. *J. Am. Chem. Soc.* 1972, *94*, 5897.
Ireland, R.E.; Willard, A.K. *Tetrahedron Lett.* 1975, *46*, 3975.

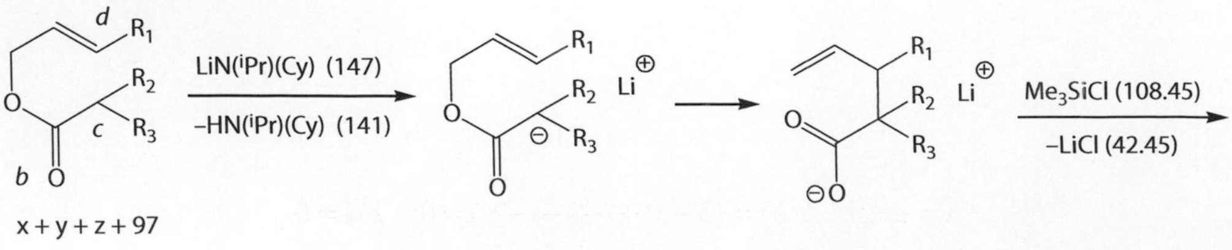

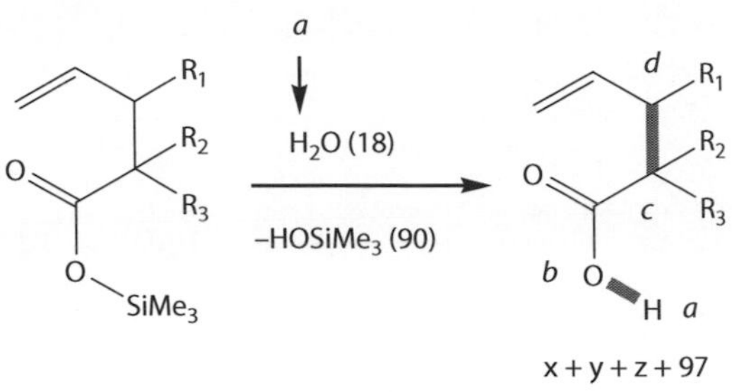

$$R_1 = x; \quad R_2 = y; \quad R_3 = z$$

$$AE = [x+y+z+97]/[x+y+z+370.45]; \quad \mathbf{AE(min) = 0.27};$$

$$E = 273.45/[x+y+z+97]; \quad \mathbf{E(max) = 2.73}$$

$$f(sac) = 255.45/[x+y+z+370.45]; \quad \mathbf{f(sac)(max) = 0.68}$$

Set $R_1 = R_2 = R_3 = CH_3$

$$\Delta = \Delta(a) + \Delta(b) + \Delta(c) + \Delta(d)$$

$$= (1-1) + (-2-(-2)) + (0-(-1)) + (-1-(-1)) = +1; \quad \mathbf{HI} = -\Delta/(N+1) = -\tfrac{1}{2}$$

Cope rearrangement

Cope, A.C.; Hardy, E.M. *J. Am. Chem. Soc.* 1940, *62*, 441.

$$\mathbf{AE = 1}; \quad \mathbf{E = 0}; \quad \mathbf{f(sac) = 0}$$

$$\Delta = \Delta(a) + \Delta(b) = (-2-(-2)) + (-2-(-2)) = 0; \quad \mathbf{HI = 0}$$

Cornforth rearrangement

Cornforth, J.W., *The Chemistry of Penicillin*, Princeton University Press: Princeton, NJ, 1949.

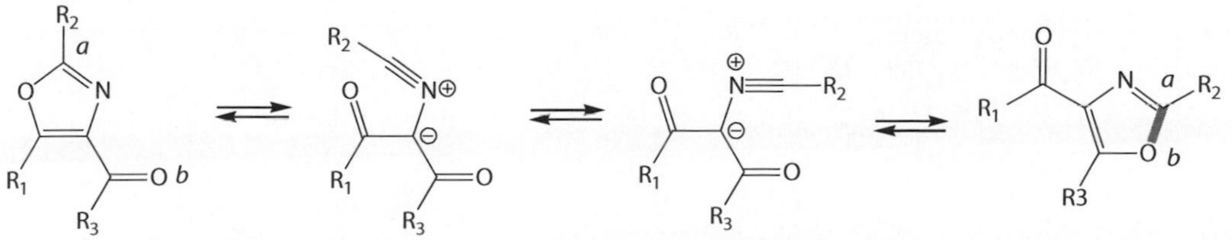

$$\mathbf{AE = 1}; \quad \mathbf{E = 0}; \quad \mathbf{f(sac) = 0}$$

Set $R_2 = CH_3$

$$\Delta = \Delta(a) + \Delta(b) = (3-3) + (-2-(-2)) = 0; \quad \mathbf{HI = 0}$$

Criegee rearrangement
Criegee, R. *Chem. Ber.* 1944, *77*, 722.
Criegee, R.; Kaspar, R. *Ann. Chem.* 1948, *560*, 127.

$$\text{AE} = 1; \quad E = 0; \quad f(\text{sac}) = 0$$

Set $R_1 = R_2 = CH_3$

$$\Delta = \Delta(a) + \Delta(b) + \Delta(c) + \Delta(d) = (0 - (-1)) + (-2 - (-1)) + (2 - 1) + (-2 - (-1)) = 0; \quad \textbf{HI} = \textbf{0}$$

Curtius rearrangement
Curtius, T. *Chem. Ber.* 1890, *23*, 3023.

$$R = x$$

$$\text{AE} = [x + 30]/[x + 160.45]; \quad \textbf{AE(min)} = \textbf{0.19}; \quad E = 130.45/[x + 30]; \quad \textbf{E(max)} = \textbf{4.21}; \quad f(\text{sac}) = 0$$

Set $R = CH_3$

$$\Delta = \Delta(a) + \Delta(b) + \Delta(c) + \Delta(d) = (-1 - (-2)) + (-3 - (-2)) + (1 - 1) + (1 - 1) = 0; \quad \textbf{HI} = \textbf{0}$$

Dieckmann–Thorpe reaction
Dieckmann, W. *Chem. Ber.* 1894, *27*, 102.

$$\text{AE} = 1; \quad E = 0; \quad f(\text{sac}) = 0$$

$$\Delta = \Delta(a) + \Delta(b) + \Delta(c) + \Delta(d) = (1 - 1) + (-3 - (-3)) + (2 - 3) + (-2 - (-3)) = 0; \quad \textbf{HI} = \textbf{0}$$

Dienone-phenol rearrangement

von Auwers, K.; Ziegler, K. *Ann. Chem.* 1921, *425*, 217.

$$AE = 1; \quad E = 0; \quad f(\text{sac}) = 0$$

Set R = CH_3

$$\Delta = \Delta(a) + \Delta(b) + \Delta(c) + \Delta(d) = (1-1) + (-2-(-2)) + (-3-(-3)) - (0-0) = 0; \quad \textbf{HI} = \textbf{0}$$

Di-pi-methane rearrangement

Zimmerman, H.E.; Mariano, P.S. *J. Am. Chem. Soc.* 1969, *91*, 1718.
Hahn, R.C.; Rothman, L.J. *J. Am. Chem. Soc.* 1969, *91*, 2409.
Zimmerman, H.E.; Samuelson, G.E. *J. Am. Chem. Soc.* 1969, *91*, 5307.
Zimmerman, H.E.; Bender, C.O. *J. Am. Chem. Soc.* 1969, *91*, 7516.
Zimmerman, H.E.; Pratt, A.C. *J. Am. Chem. Soc.* 1970, *92*, 1407.
Zimmerman, H.E.; Pratt, A.C. *J. Am. Chem. Soc.* 1970, *92*, 1409.
Zimmerman, H.E.; Epling, G.A. *J. Am. Chem. Soc.* 1970, *92*, 1411.
Zimmerman, H.E.; Bender, C.O. *J. Am. Chem. Soc.* 1970, *92*, 4366.
Zimmerman, H.E.; Pratt, A.C. *J. Am. Chem. Soc.* 1970, *92*, 6259.
Zimmerman, H.E.; Pratt, A.C. *J. Am. Chem. Soc.* 1970, *92*, 6267.
Zimmerman, H.E.; Chen, W. *Chem. Rev.* 1996, *96*, 3065.

$$AE = 1; \quad E = 0; \quad f(\text{sac}) = 0$$

$$\Delta = \Delta(a) + \Delta(b) + \Delta(c) + \Delta(d) = (-1-(-1)) + (0-0) + (-1-(-1)) + (-1-(-1)) = 0; \quad \textbf{HI} = \textbf{0}$$

Epoxide rearrangement

Paul, R. *Compt. Rend.* 1933, *196*, 1409.
Tiffeneau, M.; Ditz, E.; Tchoubar, B. *Compt. Rend.* 1934, *198*, 1039.
Godchot, M.; Mousseron, M. *Compt. Rend.* 1934, *198*, 2003.
Paul, R. *Bull. Soc. Chim. Fr.* 1935, *2(5)*, 745.
Norton, F.H.; Hass, H.B. *J. Am. Chem. Soc.* 1936, *58*, 2147.
Tiffeneau, M.; Gutman, J. *Compt. Rend.* 1936, *203*, 797.

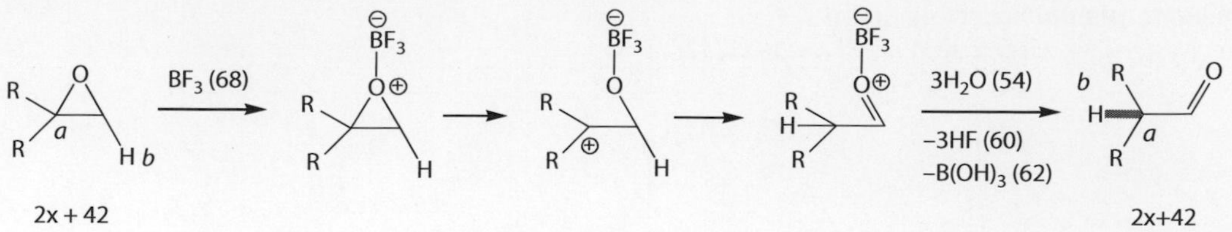

$$2x + 42 \qquad\qquad\qquad\qquad\qquad\qquad\qquad\qquad 2x+42$$

$$R = x$$

$$AE = [2x + 42]/[2x + 164]; \quad \textbf{AE(min) = 0.27}; \quad E = 122/[2x + 42]; \quad \textbf{E(max) = 2.77}$$

$$f(\text{sac}) = 122/[2x + 164]; \quad \textbf{\textit{f}(sac)(max) = 0.73}$$

Set R = CH$_3$

$$\Delta = \Delta(a) + \Delta(b) = (-1-1) + (1-1) = -2; \quad \textbf{HI} = -\Delta/(N+1) = \textbf{+1}$$

Favorskii rearrangement
Favorskii, A.E. *J. Prakt. Chem.* 1913, *88*, 658.

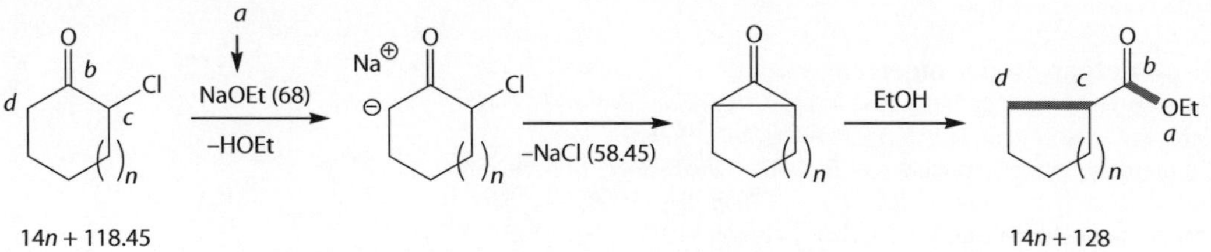

$$14n + 118.45 \qquad\qquad\qquad\qquad\qquad\qquad\qquad\qquad 14n + 128$$

$$AE = [14n + 128]/[14n + 186.45]; \quad E = 58.45/[14n + 128]; \quad f(\text{sac}) = \textbf{0}$$

n	Ring Size ($n + 4$)	E(max)	AE(min)
1	5	0.41	0.71
2	6	0.37	0.73
3	7	0.34	0.74

$$\Delta = \Delta(a) + \Delta(b) + \Delta(c) + \Delta(d) = (-2 - (-2)) + (3 - 2) + (-1 - 0) + (-2 - (-2)) = 0; \quad \textbf{HI = 0}$$

Fischer–Hepp rearrangement
Fischer, O.; Hepp, E. *Ber.* 1886, *19*, 2291.
Fischer, O.; Hepp, E. *Ber.* 1887, *20*, 1247.
Fischer, O.; Hepp, E. *Ber.* 1887, *20*, 2471.
Fischer, O. *Ber.* 1912, *45*, 1098.

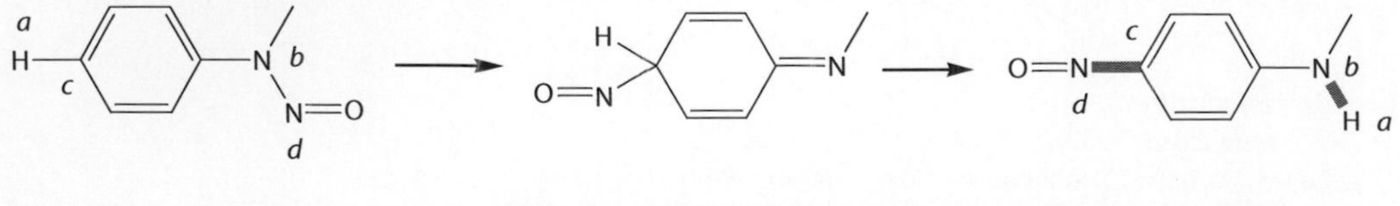

$$\textbf{AE = 1}; \quad \textbf{E = 0}; \quad \textbf{\textit{f}(sac) = 0}$$

$$\Delta = \Delta(a) + \Delta(b) + \Delta(c) + \Delta(d) = (1-1) + (-3-(-2)) + (1-(-1)) + (1-2) = 0; \quad \textbf{HI = 0}$$

Fries rearrangement

Fries, K.; Finck, G. *Chem. Ber.* 1908, *41*, 4271.

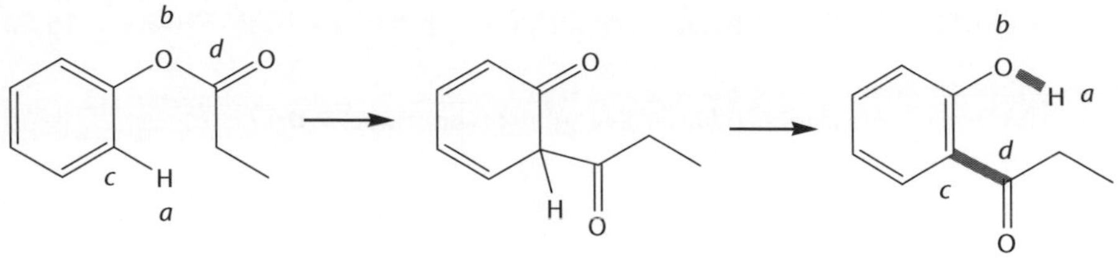

b
O
d
O
c
H
a
O
H
O
b
O
H a
d
c
O

$$\mathbf{AE = 1}; \quad \boldsymbol{E = 0}; \quad \boldsymbol{f}\mathbf{(sac) = 0}$$

$$\Delta = \Delta(a) + \Delta(b) + \Delta(c) + \Delta(d) = (1-1) + (-2-(-2)) + (0-(-1)) + (2-3) = 0; \quad \mathbf{HI = 0}$$

Fritsch–Buttenberg–Wiechell rearrangement

Fritsch, P. *Ann. Chem.* 1894, *279*, 319.
Buttenberg, W.P. *Ann. Chem.* 1894, *279*, 324.
Wiechell, H. *Ann. Chem.* 1894, *279*, 337.

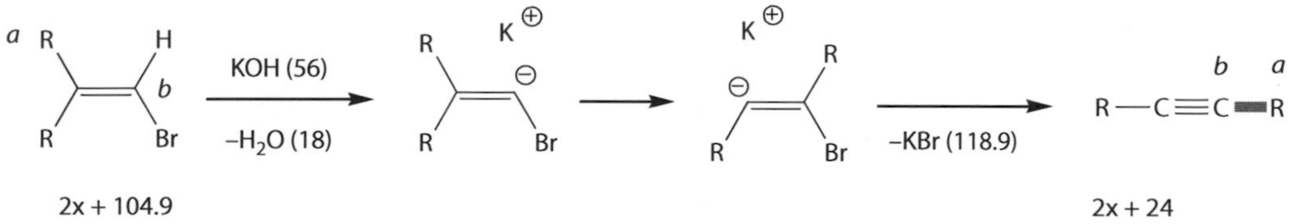

a R
H
b
R
Br
KOH (56)
–H2O (18)
K
R
R
Br
K
R
R
Br
–KBr (118.9)
b a
R—C≡C—R
2x + 104.9
2x + 24

$$\mathrm{R = x}$$

$$\mathrm{AE = [2x + 24]/[2x + 160.9]}; \quad \mathbf{AE(min) = 0.16}; \quad E = \mathrm{136.9/[2x + 24]}; \quad \boldsymbol{E}\mathbf{(max) = 5.27}$$

$$f\mathrm{(sac) = 56/[2x + 160.9]}; \quad \boldsymbol{f}\mathbf{(sac)(max) = 0.34}$$

Set R = CH_3

$$\Delta = \Delta(a) + \Delta(b) = (-3-(-3)) + (0-0) = 0; \quad \mathbf{HI = 0}$$

Hofmann–Martius rearrangement

Hofmann, A.W.; Martius, C.A. *Chem. Ber.* 1871, *4*, 742.
Hofmann, A.W. *Chem. Ber.* 1872, *5*, 704.
Hofmann, A.W. *Chem. Ber.* 1872, *5*, 720.
Hofmann, A.W. *Chem. Ber.* 1874, *7*, 526.

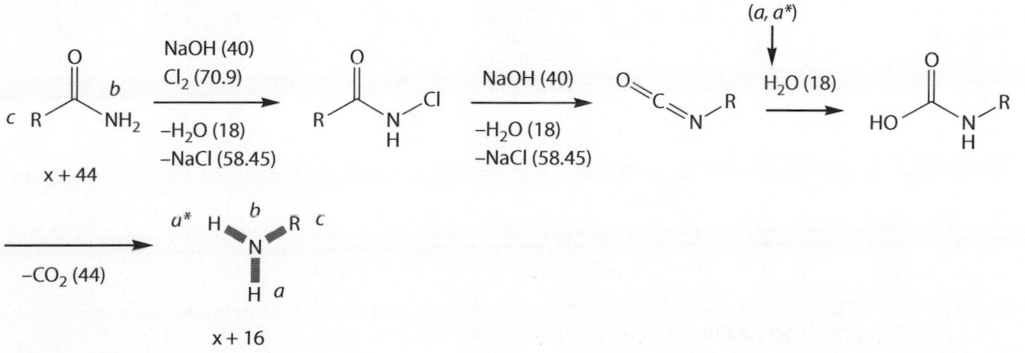

O
b
c R
NH2
x + 44
NaOH (40)
Cl2 (70.9)
–H2O (18)
–NaCl (58.45)
O
R
N
H
Cl
NaOH (40)
–H2O (18)
–NaCl (58.45)
O
C
N
R
(a, a*)
H2O (18)
O
HO
N
H
R
–CO2 (44)
a* H
b
R c
N
H a
x + 16

$$R = x$$

$$AE = [x+16]/[x+194.9]; \quad \textbf{AE(min)} = \textbf{0.087}; \quad E = 178.9/[x+16]; \quad \textbf{E(max)} = \textbf{10.52}$$

$$f(\text{sac}) = 150.9/[x+194.9]; \quad \textbf{\textit{f}(sac)(max)} = \textbf{0.77}$$

Set R = CH$_3$

$$\Delta = \Delta(a) + \Delta(a^*) + \Delta(b) + \Delta(c)$$

$$= (1-1) + (1-1) + (-3-(-3)) + (-2-(-3)) = +1; \quad \textbf{HI} = -\Delta/(N+1) = -\tfrac{1}{2}$$

Hydroboration-borane rearrangement

Brown, H.C.; Subba Rao, B.C. *J. Am. Chem. Soc.* 1956, *78*, 5694.
Brown, H.C.; Subba Rao, B.C. *J. Org. Chem.* 1957, *22*, 1135.
Brown, H.C.; Subba Rao, B.C. *J. Org. Chem.* 1957, *22*, 1136.

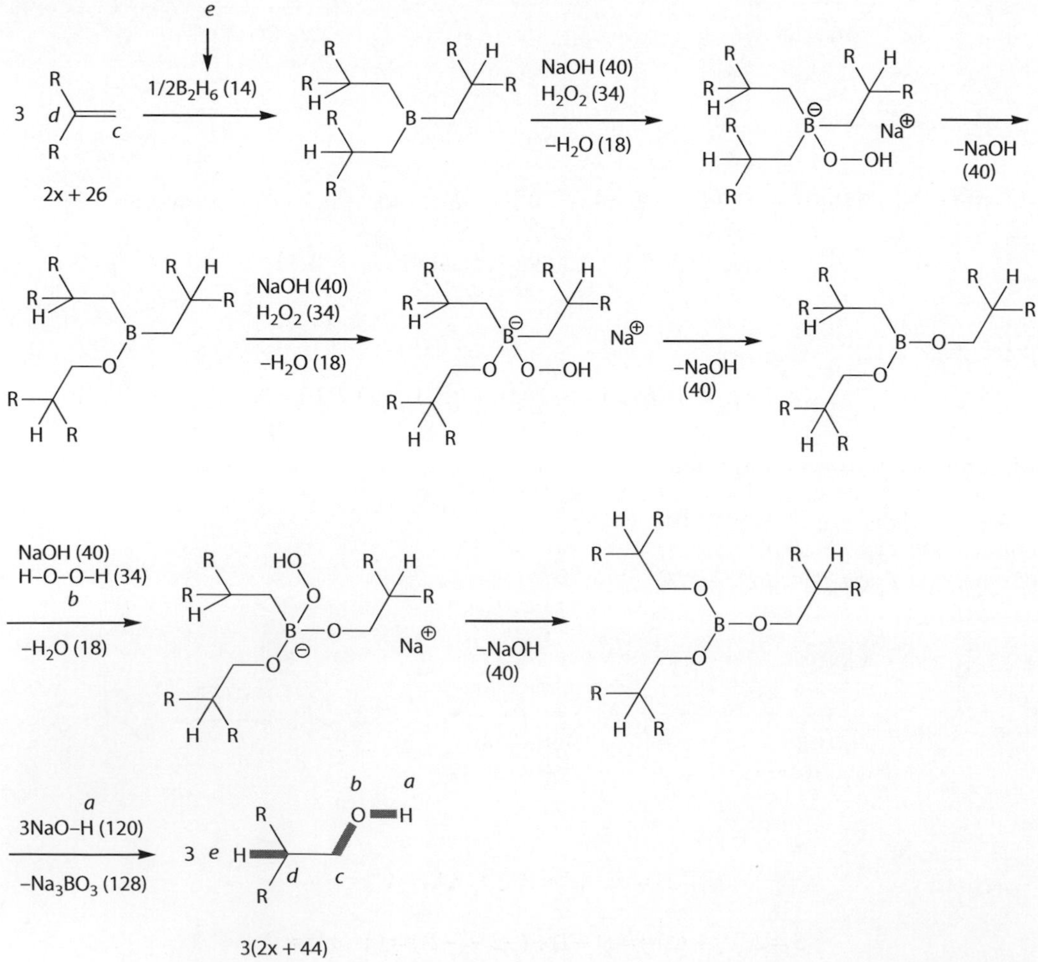

$$R = x$$

$$AE = [3(2x + 44)]/[3(2x + 26) + 236]; \quad \textbf{AE(min)} = \textbf{0.43};$$

$$E = 182/[3(2x + 44)]; \quad \textbf{E(max)} = \textbf{1.32}; \quad f(\textbf{sac}) = \textbf{0}$$

Set R = CH$_3$

$$\Delta = \Delta(a) + \Delta(b) + \Delta(c) + \Delta(d) + \Delta(d) = (1-1) + (-2-(-1)) + (-1-(-2)) + (-1-0) + (1-(-1)) = +1;$$

$$\textbf{HI} = \Delta/(N+1) = -\textbf{½}$$

Hydroperoxide rearrangement

Hock, H.; Lang, S. *Chem. Ber.* 1944, *77B*, 257.
Bartlett, P.D.; Cotman, J.D. Jr. *J. Am. Chem. Soc.* 1950, *72*, 3095.
Hulse, G.E.; Vandenberg, E.J. (Hercules Powder Co.) US Pat. 2,557,968 (1951/06/26).
Kharasch, M.S.; Burt, J.G. *J. Org. Chem.* 1951, *16*, 150.
Vandenberg, E.J., (Distillers Co. Ltd.) GB 676,771 (1952/08/06).
Seubold, F.H. Jr.; Vaughan, W.E. *J. Am. Chem. Soc.* 1953, *75*, 3790.
Bissing, D.E.; Matuszak, C.A.; McEwen, W.E. *Tetrahedron Lett.* 1962, 763.
Bissing, D.E.; Matuszak, C.A.; McEwen, W.E. *J. Am. Chem. Soc.* 1964, *86*, 3824.
Anderson, G.H.; Smith, J.G. *Can. J. Chem.* 1968, *46*, 1553.

3x + 45   2x + 28   x + 17

$$R = x$$

Target product: ketone

$$AE = [2x + 28]/[3x + 45]; \quad \textbf{AE(min)} = \textbf{0.63}; \quad E = [x + 17]/[2x + 28]; \quad \textbf{E(max)} = \textbf{0.60}; \quad f(\textbf{sac}) = \textbf{0}$$

Target product: alcohol

$$AE = [x + 17]/[3x + 45]; \quad \textbf{AE(min)} = \textbf{0.38}; \quad E = [2x + 28]/[x + 17]; \quad \textbf{E(max)} = \textbf{1.67}; \quad f(\textbf{sac}) = \textbf{0}$$

Set R = CH$_3$

$$\Delta = \Delta(a) + \Delta(b) = (-2-(-1)) + (-2-(-3)) = 0; \quad \textbf{HI} = \textbf{0}$$

Kemp elimination

Kemp, D.S. *Tetrahedron* 1967, *23*, 2001.

$$\textbf{AE} = \textbf{1}; \quad E = \textbf{0}; \quad f(\textbf{sac}) = \textbf{0}$$

$$\Delta = \Delta(a) + \Delta(b) = (1-1) + (-2-(-2)) = 0; \quad \textbf{HI} = \textbf{0}$$

Lossen rearrangement

Lossen, W. *Ann. Chem.* 1872, *161*, 347.

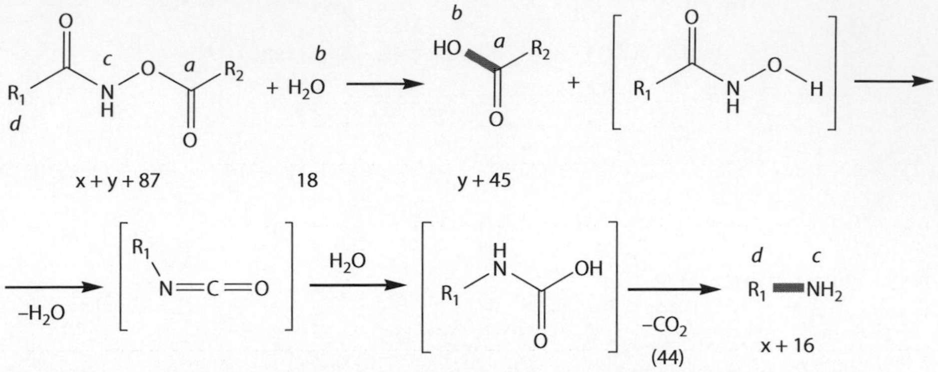

$$x + y + 87 \qquad\qquad 18 \qquad\qquad y + 45$$

$$R_1 = x; \quad R_2 = y$$

Target product: carboxylic acid

$$AE = [y + 45]/[x + y + 105]; \quad AE(\text{min}) = 46/[x + 106];$$

$$E = [x + 60]/[y + 45]; \quad E(\text{max}) = [x + 60]/46$$

$f(\text{sac}) = 0$

R_1	E(max)	AE(min)
H	1.35	0.43
Me	1.65	0.38
Et	1.96	0.34

Set $R_2 = CH_3$

$$\Delta = \Delta(a) + \Delta(b) = (3 - 3) + (-2 - (-2)) = 0; \quad \textbf{HI} = \textbf{0}$$

Target product: Amine

$$AE = [x + 16]/[x + y + 105]; \quad AE(\text{min}) = 17/[y + 106];$$

$$E = [y + 89]/[x + 16]; \quad E(\text{max}) = [y + 89]/17$$

$$f(\text{sac}) = 18/[x + y + 87]$$

R_2	E(max)	AE(min)	f(sac) (R_1 = Me)
H	5.29	0.16	0.17
Me	6.12	0.14	0.15
Et	6.94	0.13	0.14

Set $R_1 = CH_3$

$$\Delta = \Delta(c) + \Delta(d) = (-3 - (-1)) + (-2 - (-3)) = -1; \quad \textbf{HI} = -\Delta/(N + 1) = +\tfrac{1}{2}$$

Martynoff rearrangement

Martynoff, V.F.; Titov, M.I. *J. Gen. Chem. USSR* 1960, *30*, 4072.

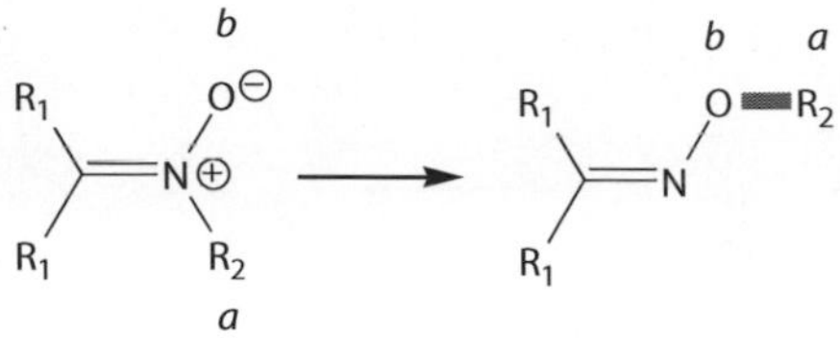

AE =1; $E = 0$; f(sac) = 0

Set $R_2 = CH_3$

$$\Delta = \Delta(a) + \Delta(b) = (-2 - (-2)) + (-2 - (-2)) = 0; \quad \mathbf{HI = 0}$$

McLafferty rearrangement

McLafferty, F.W. *Anal. Chem.* 1959, *31*, 82.

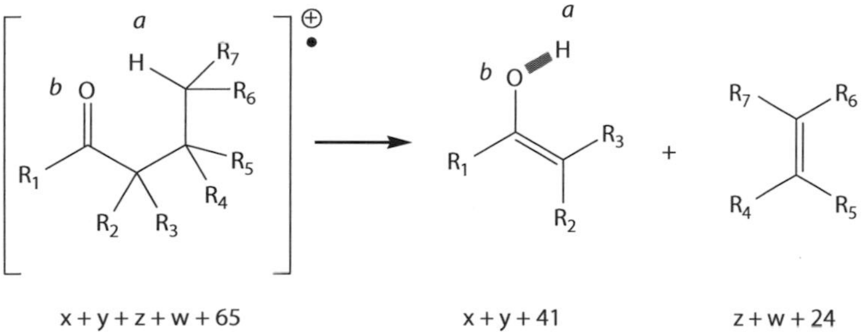

$$x + y + z + w + 65 \qquad x + y + 41 \qquad z + w + 24$$

$R_1 = x; R_2 + R_3 = y; R_4 + R_5 = z; R_6 + R_7 = w$

Target product: Enol

$$AE = [x + y + 41]/[x + y + z + w + 65]; \quad E = [z + w + 24]/[x + y + 41]; \quad f(\text{sac}) = 0$$

$$\Delta = \Delta(a) + \Delta(b) = (1 - 1) + (-2 - (-2)) = 0; \quad \mathbf{HI = 0}$$

Target product: Olefin

$$AE = [z + w + 24]/[x + y + z + w + 65]; \quad E = [x + y + 41]/[z + w + 24]; \quad f(\text{sac}) = 0$$

Meisenheimer rearrangement

Meisenheimer, J. *Chem. Ber.* 1919, *52*, 1667.

AE = 1; $E = 0$; f(sac) = 0

$$\Delta = \Delta(a) + \Delta(b) = (-2 - (-2)) + (-1 - (-1)) = 0; \quad \mathbf{HI = 0}$$

Meyer–Schuster rearrangement

Meyer, K.H.; Schuster, K. *Chem. Ber.* 1922, *55*, 819.

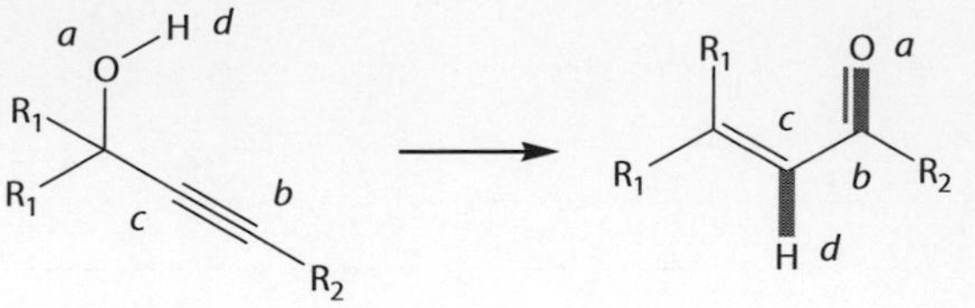

AE = 1; *E* = 0; *f*(sac) = 0

Set $R_2 = CH_3$

$$\Delta = \Delta(a) + \Delta(b) + \Delta(c) + \Delta(d) = (-2 - (-2)) + (2 - 0) + (-1 - 0) + (1 - 1) = +1;$$

$$\mathbf{HI} = -\Delta/(N+1) = -\tfrac{1}{2}$$

Nametkin rearrangement

Nametkin, S.S. *Ann. Chem.* 1923, *432*, 207.

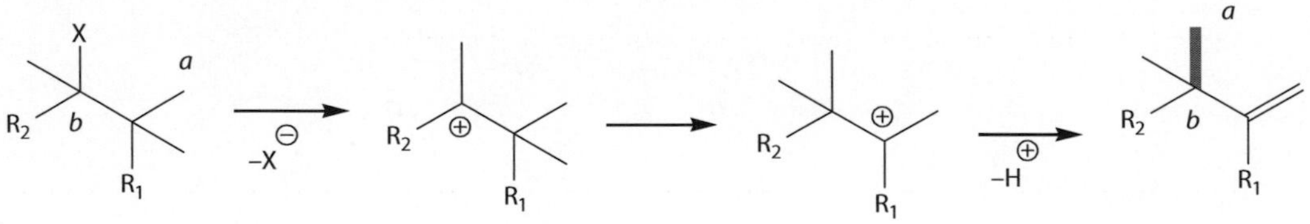

$x + y + z + 69$ $y + z + 68$

$$AE = [y + z + 68]/[x + y + z + 69]; \quad E = [x + 1]/[y + z + 68]; \quad f(\text{sac}) = 0$$

X	*E*(max)	AE(min)
Cl	0.52	0.66
Br	1.16	0.46
I	1.83	0.35

Set $R_2 = CH_3$

$$\Delta = \Delta(a) + \Delta(b) = (-3 - (-3)) + (0 - 1) = -1; \quad \mathbf{HI} = \Delta/(N+1) = +\tfrac{1}{2}$$

Neber rearrangement

Neber, P.W.; Friedolsheim, A. von, *Ann. Chem.* 1926, *449*, 109.

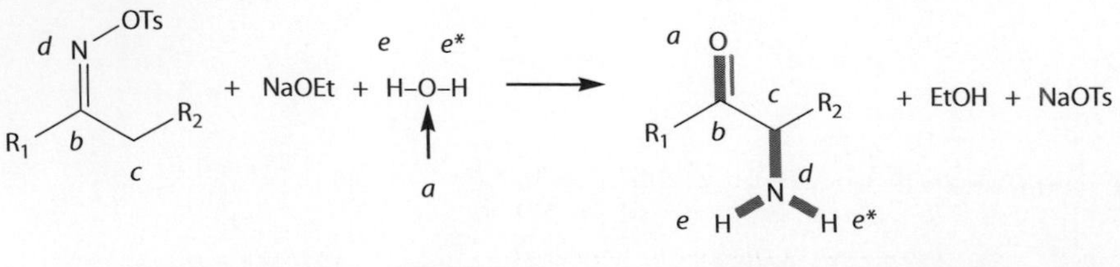

$x + y + 211$ 68   18 $x + y + 57$ 46   194

$R_1 = x; R_2 = y$

$$AE = [x+y+57]/[x+y+297]; \quad \textbf{AE(min)} = \textbf{0.20}; \quad E = 240/[x+y+57]; \quad \textbf{\textit{E}(max)} = \textbf{4.07}$$

$$f(sac) = 68/[x+y+297]; \quad \textbf{\textit{f}(sac)(max)} = \textbf{0.23}$$

Set $R_1 = R_2 = CH_3$

$$\Delta = \Delta(a) + \Delta(b) + \Delta(c) + \Delta(d) + \Delta(e) + \Delta(e^*)$$

$$= (-2-(-2)) + (2-2) + (0-(-2)) + (-3-(-1)) + (1-1) + (1-1) = 0; \quad \textbf{HI} = \textbf{0}$$

Norrish type I reaction

Bamford, C.H.; Norrish, R.G.W. *J. Chem. Soc.* 1935, 1504.
Norrish, R.G.W.; Bamford, C.H. *Nature* 1936, *138*, 1016.
Norrish, R.G.W.; Bamford, C.H. *Nature* 1937, *140*, 195.
Norrish, R.G.W. *Trans. Faraday Soc.* 1937, *33*, 1521.

$AE = 1; E = 0; f(sac) = 0$

$$\Delta = \Delta(a) + \Delta(b) = (1-1) + (-3-(-2)) = -1; \quad \textbf{HI} = -\Delta/(N+1) = +\tfrac{1}{2}$$

Orton rearrangement

Orton, K.J.P. *Proc. Roy. Soc. London* 1903, *71*, 153.
Orton, K.J.P.; Jones, W.J. *Trans. Chem. Soc.* 1909, *95*, 1456.

$AE = 1; E = 0; f(sac) = 0$

$$\Delta = \Delta(a) + \Delta(b) + \Delta(c) + \Delta(d) = (1-1) + (-3-(-1)) + (1-(-1)) + (-1-(-1)) = 0; \quad \textbf{HI} = \textbf{0}$$

Oxy-Cope rearrangement

Berson, J.A.; Jones, M. Jr. *J. Am. Chem. Soc.* 1964, *86*, 5017.
Berson, J.A.; Jones, M. Jr. *J. Am. Chem. Soc.* 1964, *86*, 5019.
Berson, J.A.; Walsh, E.J. Jr. *J. Am. Chem. Soc.* 1968, *90*, 4729.
Berson, J.A.; Walsh, E.J. Jr. *J. Am. Chem. Soc.* 1968, *90*, 4732.

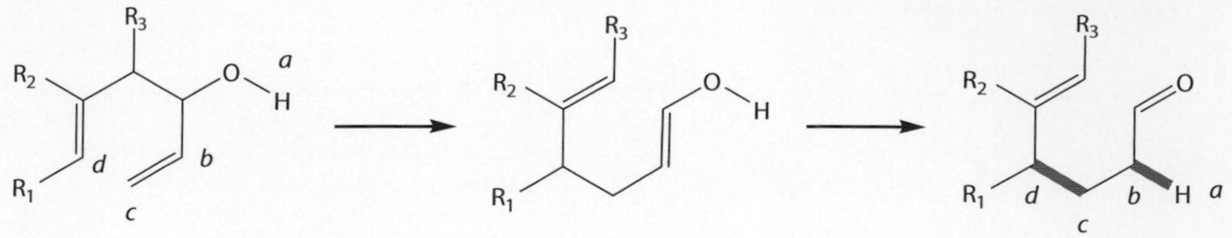

AE = 1; $E = 0$; $f(\text{sac}) = 0$

Set $R_1 = CH_3$

$$\Delta = \Delta(a) + \Delta(b) + \Delta(c) + \Delta(d) = (1-1) + (-2 - (-1)) + (-2 - (-2)) + (-1 - (-1)) = -1;$$

$$\mathbf{HI} = -\Delta/(N+1) = +\tfrac{1}{2}$$

Payne rearrangement
Payne, G.B. *J. Org. Chem.* 1962, *27*, 3819.

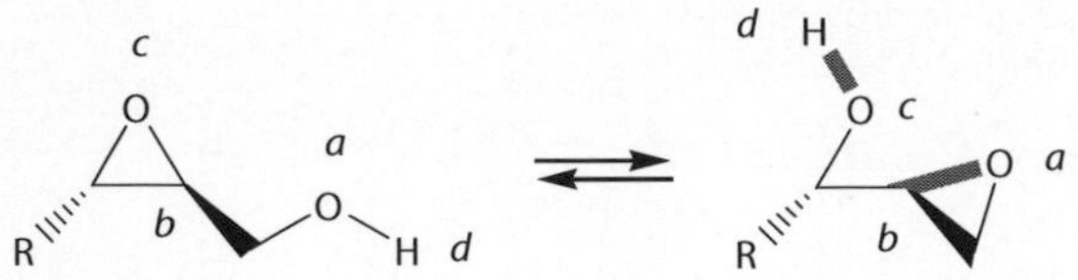

AE = 1; $E = 0$; $f(\text{sac}) = 0$

$$\Delta = \Delta(a) + \Delta(b) + \Delta(c) + \Delta(d) = (-2 - (-2)) + (0 - 0) + (-2 - (-2)) + (1 - 1) = 0; \quad \mathbf{HI} = 0$$

Pentazadiene [1,3] rearrangement
Baines, K.M.; Rourke, T.W.; Vaughan, K.; Hooper, D.L. *J. Org. Chem.* 1981, *46*, 856.

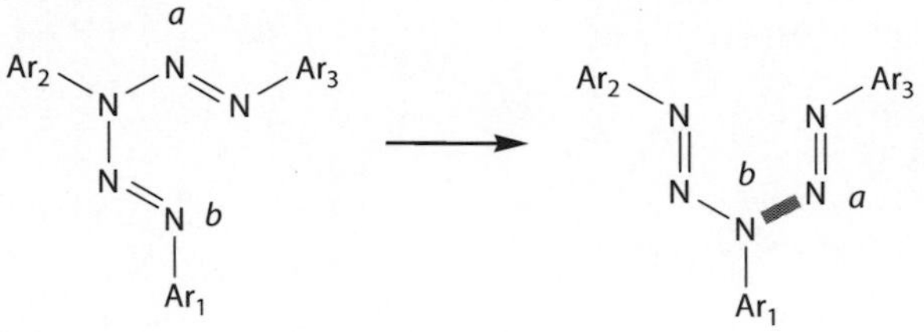

AE = 1; $E = 0$; $f(\text{sac}) = 0$

$$\Delta = \Delta(a) + \Delta(b) = (0 - 0) + (-1 - (-1)) = 0; \quad \mathbf{HI} = 0$$

Perkin rearrangement
Perkin, W.H., *J. Chem. Soc.* 1870, *23*, 368.

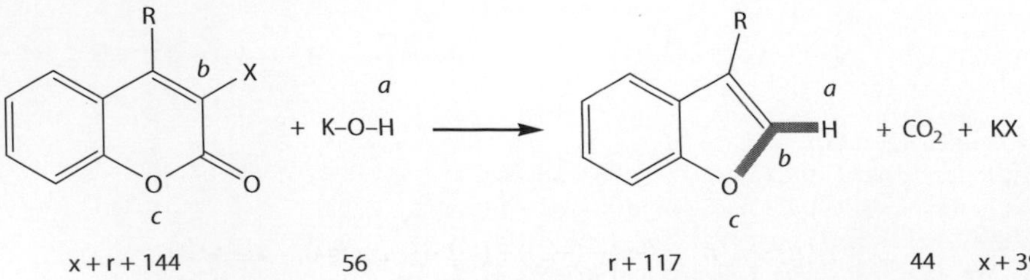

| x + r + 144 | 56 | r + 117 | 44 | x + 39 |

R = r; X = X(halide)

$$AE = [r+117]/[x+r+200]; \quad \mathbf{AE(min) = 118/[x+201]};$$

$$E = [x+83]/[r+117]; \quad \mathbf{E(max) = [x+83]/118}; \quad f(\mathbf{sac}) = \mathbf{0}$$

X	E(max)	AE(min)
Cl	1.00	0.50
Br	1.38	0.42
I	1.78	0.36

$$\Delta = \Delta(a) + \Delta(b) + \Delta(c) = (1-1)+(0-1)+(-2-(-2)) = +1; \quad \mathbf{HI} = -\Delta/(N+1) = -\tfrac{1}{2}$$

Pinacol rearrangement
Fittig, R. *Ann. Chem.* 1860, *114*, 54.

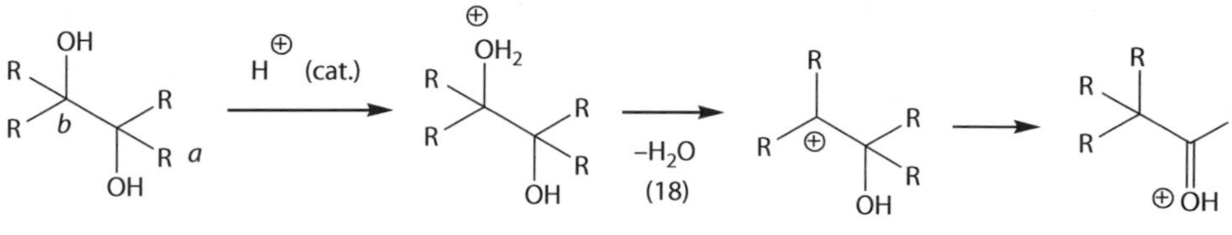

4x + 58

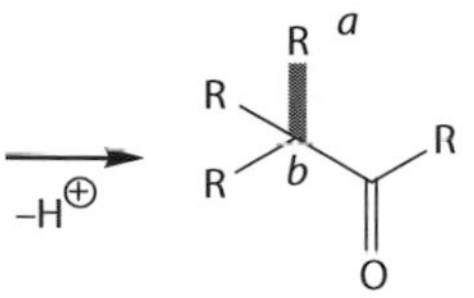

4x + 40

R = x

$$AE = [4x+40]/[4x+58]; \quad \mathbf{AE(min) = 0.71}; \quad E = 18/[4x+40]; \quad \mathbf{E(max) = 0.41}; \quad f(\mathbf{sac}) = \mathbf{0}$$

Set R = CH_3

$$\Delta = \Delta(a) + \Delta(b) = (-3-(-3))+(0-1) = -1; \quad \mathbf{HI} = -\Delta/(N+1) = +\tfrac{1}{2}$$

Polonovski reaction
Polonovski, M.; Polonovski, M. *Bull. Chim. Soc. Fr.* 1926, *39*, 147.
Polonovski, M.; Polonovski, M. *Bull. Chim. Soc. Fr.* 1927, *41*, 1190.

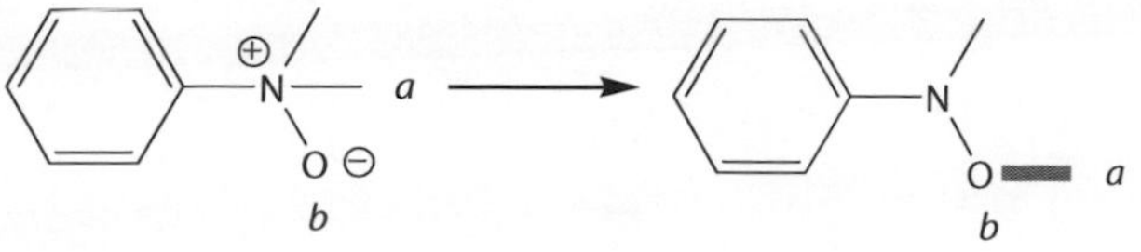

$\mathbf{AE = 1}; E = \mathbf{0}; f(\mathbf{sac}) = \mathbf{0}$

$$\Delta = \Delta(a) + \Delta(b) = (-2-(-2))+(-2-(-2)) = 0; \quad \mathbf{HI} = \mathbf{0}$$

Pummerer rearrangement

Pummerer, R. *Chem. Ber.* 1910, *43*, 1401.

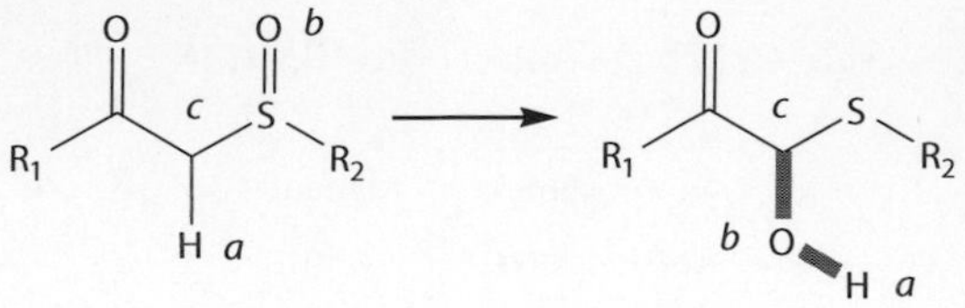

$AE = 1; E = 0; f(sac) = 0$

$$\Delta = \Delta(a) + \Delta(b) + \Delta(c) = (1-1) + (-2-(-2)) + (1-(-1)) = +2;$$

$$\mathbf{HI} = -\Delta/(N+1) = -1$$

Pyridine N-oxide-pyridone [1,4] rearrangement

Boekelheide, V.; Linn, W.J. *J. Am. Chem. Soc.* 1954, *76*, 1286.
Bain, B.M.; Saxton, J.E. *Chem. Ind.* 1960, 402.
Boekelheide, V.; Lehn, W.L. *J. Org. Chem.* 1961, *26*, 428.

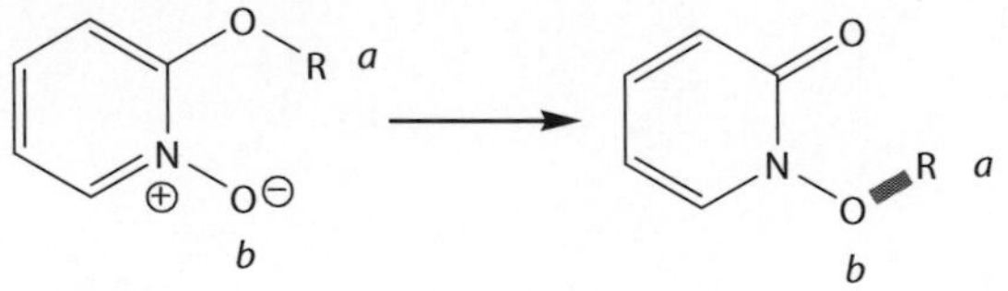

$AE = 1; E = 0; f(sac) = 0$

Set R = CH_3

$$\Delta = \Delta(a) + \Delta(b) = (-2-(-2)) + (-2-(-2)) = 0; \quad \mathbf{HI} = \mathbf{0}$$

Ramberg–Bäckland rearrangement

Ramberg, L.; Bäckland, B., *Arkiv. Kemi Mineral Geol.* 1940, *13A*, 50.

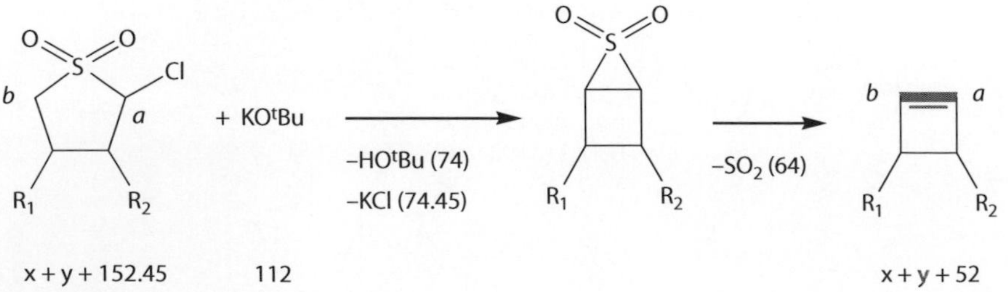

$R_1 = x; R_2 = y$

$$AE = [x + y + 52]/[x + y + 264.45]; \quad \mathbf{AE(min) = 0.20};$$

$$E = 212.45/[x + y + 52]; \quad E(\mathbf{max}) = \mathbf{3.93}$$

$$f(\text{sac}) = 112/[x + y + 264.45]; \quad f(\text{sac})(\text{max}) = 0.42$$

$$\Delta = \Delta(a) + \Delta(b) = (-1 - 1) + (-1 - (-1)) = -2; \quad \text{HI} = -\Delta/(N + 1) = +1$$

Retro Sommelet–Hauser [1,3] rearrangement
Sommelet, M., *Compt. Rend.* 1937, *205*, 56.
Hauser, C.R.; Kantor, S.W., *J. Am. Chem. Soc.* 1951, *73*, 1437.

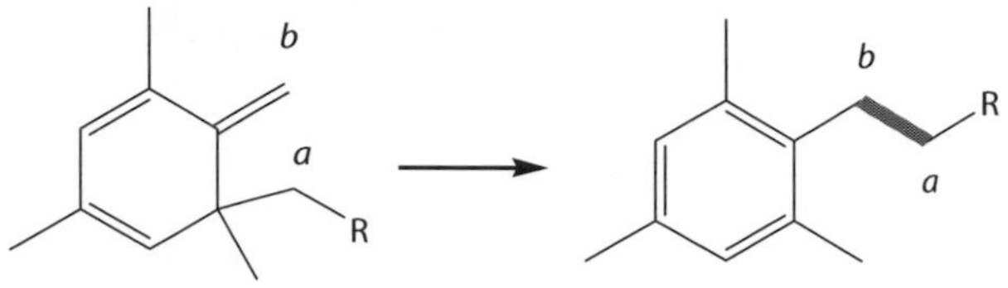

AE = 1; *E* = 0; *f*(sac) = 0

Set R = CH$_3$

$$\Delta = \Delta(a) + \Delta(b) = (-2 - (-2)) + (-2 - (-2)) = 0; \quad \text{HI} = 0$$

Rupe rearrangement
Rupe, H.; Kambli, E., *Helv. Chim. Acta* 1926, *9*, 672.

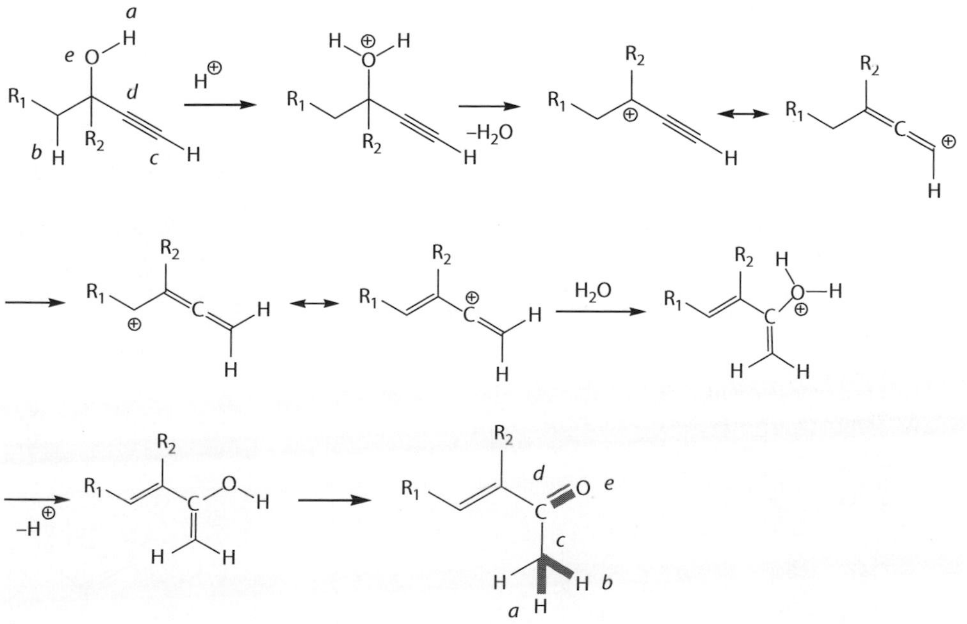

AE = 1; *E* = 0; *f*(sac) = 0

$$\Delta = \Delta(a) + \Delta(b) + \Delta(c) + \Delta(d) + \Delta(e)$$

$$= (1 - 1) + (1 - 1) + (-3 - (-1)) + (2 - 0) + (-2 - (-2)) = 0; \quad \text{HI} = 0$$

Schmidt rearrangement

Claisen, L.; Claparède, A. *Chem. Ber.* 1881, *14*, 2460.
Schmidt, J.G. *Chem. Ber.* 1881, *14*, 1459.

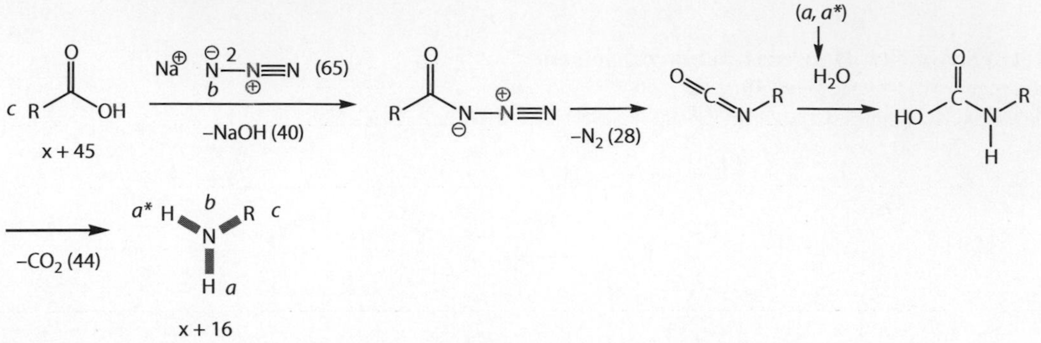

R = x

$$AE = [x+16]/[x+128]; \quad \textbf{AE(min)} = \textbf{0.13}; \quad E = 112/[x+16]; \quad \textbf{E(max)} = \textbf{6.59}; \quad f(\textbf{sac}) = \textbf{0}$$

Set R = CH$_3$

$$\Delta = \Delta(a) + \Delta(a^*) + \Delta(b) + \Delta(c) = (1-1)+(1-1)+(-3-(-2))+(-2-(-3)) = 0; \quad \textbf{HI} = \textbf{0}$$

Sigmatropic rearrangements

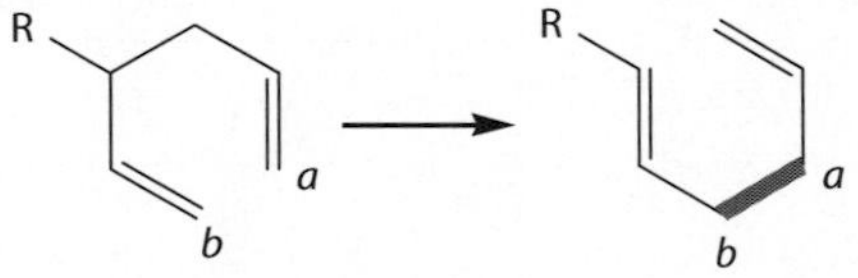

$\textbf{AE} = \textbf{1}; E = \textbf{0}; f(\textbf{sac}) = \textbf{0}$

$$\Delta = \Delta(a) + \Delta(b) = (-2-(-2))+(-2-(-2)) = 0; \quad \textbf{HI} = \textbf{0}$$

Smiles [1,4] rearrangement

Evans, W.J.; Smiles, S. *J. Chem. Soc.* 1935, 181.
Levy, A.A.; Rains, H.C., Smiles, S. *J. Chem. Soc.* 1931, 3264.

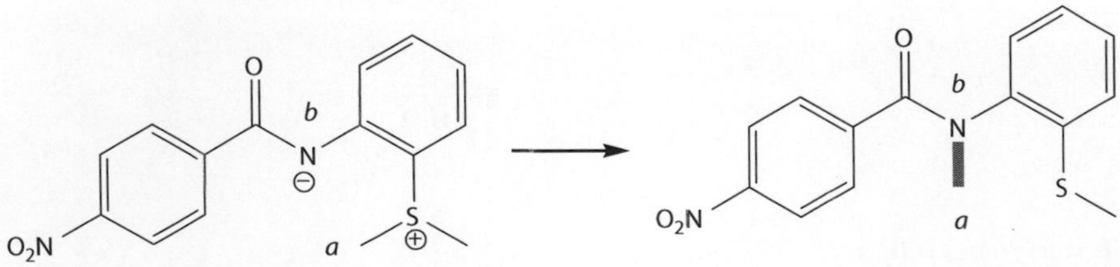

$\textbf{AE} = \textbf{1}; E = \textbf{0}; f(\textbf{sac}) = \textbf{0}$

$$\Delta = \Delta(a) + \Delta(b) = (-2-(-2))+(-3-(-3)) = 0; \quad \textbf{HI} = \textbf{0}$$

Stevens [1,2] rearrangement

Stevens, T.S.; Creighton, E.M.; Gordon, A.B.; MacNicol, M. *J. Chem. Soc.* 1928, 3193.

$$y + 3z + x + 42 \qquad\qquad y + 3z + 41$$

$R_1 = y; R = z; X = x(halide)$

$$AE = [y + 3z + 41]/[x + y + 3z + 81]; \quad \mathbf{AE(min)} = \mathbf{45/[x + 85]};$$

$$E = [x + 40]/[y + 3z + 41]; \quad \mathbf{E(max)} = \mathbf{[x + 40]/45}$$

$$f(sac) = 39/[x + y + 3z + 81]; \quad \mathbf{f(sac)(max)} = \mathbf{39/[x + 85]}$$

X	E(max)	AE(min)	f(sac)
Cl	1.68	0.37	0.32
Br	2.66	0.27	0.24
I	3.71	0.21	0.18

Set $R_1 = R = CH_3$

$$\Delta = \Delta(a) + \Delta(b) = (-3 - (-2)) + (-1 - (-2)) = 0; \quad \mathbf{HI = 0}$$

Sulfenate-sulfoxide [2,3] rearrangement

Carruthers, W.; Entwistle, I.D.; Johnstone, R.A.W.; Millard, B.J. *Chem. Ind.* 1966, 342.
Miller, E.G.; Rayner, D.R.; Mislow, K. *J. Am. Chem. Soc.* 1966, *88*, 3139.
Braverman, S.; Stabinsky, Y. *Chem. Commun.* 1967, 270.
Abbott, D.J.; Stirling, C.J.M. *Chem. Commun.* 1968, 165.
Herriott, A.W.; Mislow, K. *Tetrahedron Lett.* 1968, *25*, 3013.
Bickart, P.; Carson, F.W.; Jacobus, J.; Miller, E.G.; Mislow, K. *J. Am. Chem. Soc.* 1968, *90*, 4869.

$\mathbf{AE = 1; E = 0; f(sac) = 0}$

$$\Delta = \Delta(a) + \Delta(b) = (-2 - (-2)) + (1 - 0) = +1; \quad \mathbf{HI = -\Delta/(N+1) = -\frac{1}{2}}$$

Tiemann rearrangement

Tiemann, F. *Chem. Ber.* 1891, *24*, 4162.

$$x + 59 \qquad\qquad\qquad\qquad\qquad\qquad\qquad\qquad x + 59$$

R = x

$$AE = [x + 59]/[x + 253.45]; \quad \textbf{AE(min) = 0.24}; \quad E = 194.45/[x + 59]; \quad \textbf{E(max) = 3.24}$$

$$f(\text{sac}) = 176.45/[x + 253.45]; \quad f(\text{sac})(\text{max}) = \textbf{0.69}$$

Set R = CH_3

$$\Delta = \Delta(a) + \Delta(b) + \Delta(c) + \Delta(d) + \Delta(e)$$

$$= (-2 - (-2)) + (4 - 3) + (-2 - (-3)) + (-3 - (-1)) + (1 - 1) = 0; \quad \textbf{HI = 0}$$

Vinyl ether rearrangement

Powell, S.G.; Adams, R. *J. Am. Chem. Soc.* 1920, *42*, 646.
Hurd, C.D.; Pollack, M.A. *J. Org. Chem.* 1939, *3*, 550.
Hill, H.S.; Pidgeon, L.M. *J. Am. Chem. Soc.* 1928, *50*, 2718.

$$\textbf{AE = 1; } E \textbf{ = 0; } f\textbf{(sac) = 0}$$

Set R_2 = CH_3

$$\Delta = \Delta(a) + \Delta(b) = (-2 - (-2)) + (-3 - (-2)) = -1; \quad \textbf{HI} = -\Delta/(N+1) = +\tfrac{1}{2}$$

Vinylogous Wolff rearrangement

Smith, A.B. III. *Chem. Commun.* 1974, 695.
Smith, A.B. III; Toder, B.H.; Branca, S.J. *J. Am. Chem. Soc.* 1976, *98*, 7456.
Branca, S.J.; Lock, R.L.; Smith, A.B. III *J. Org. Chem.* 1977, *42*, 3165.

$$R_1 + R_2 + R_3 + R_4 + R_5 = x$$

$$AE = [x + 109]/[x + 137]; \quad \textbf{AE(min) = 0.80}; \quad E = 28/[x + 109]; \quad \textbf{E(max) = 0.25}; \quad f(\text{sac}) = \textbf{0}$$

Set R_2 = R_3 = CH_3

$$\Delta = \Delta(a) + \Delta(b) + \Delta(c) + \Delta(d) + \Delta(e)$$

$$= (-2 - (-2)) + (3 - 2) + (1 - 1) + (-2 - (-1)) + (0 - 0) = 0; \quad \textbf{HI = 0}$$

Wagner–Meerwein rearrangement

Wagner, G. *J. Russ. Phys. Chem.* 1899, *31*, 690.
Meerwein, H. *Ann. Chem.* 1914, *405*, 129.

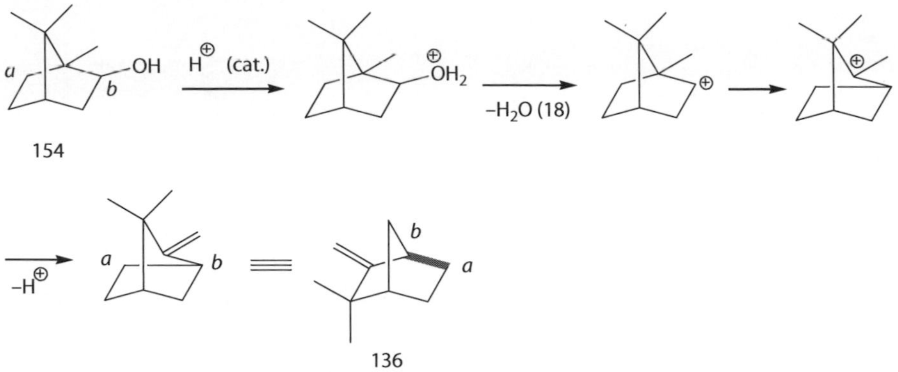

AE = 0.88; *E* **= 0.13;** *f*(sac) **= 0**

$$\Delta = \Delta(a) + \Delta(b) = (-2 - (-2)) + (-1 - 0) = -1; \quad \mathbf{HI} = -\Delta/(N+1) = +\tfrac{1}{2}$$

Wallach rearrangement

Wallach, O.; Belli, L. *Chem. Ber.* 1880, *13*, 525.

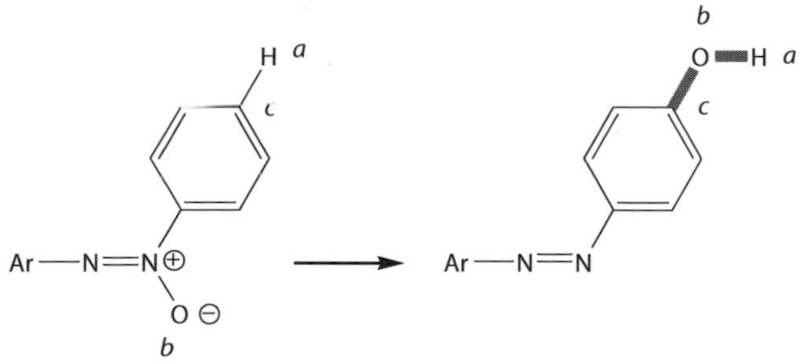

AE = 1; *E* **= 0;** *f*(sac) **= 0**

$$\Delta = \Delta(a) + \Delta(b) + \Delta(c) = (1-1) + (-2-(-2)) + (1-(-1)) = +2; \quad \mathbf{HI} = -\Delta/(N+1) = -1$$

Wawzonek–Yeakey [1,2] rearrangement

Wawzonek, S.; Yeakey, E. *J. Am. Chem. Soc.* 1960, *82*, 5718.

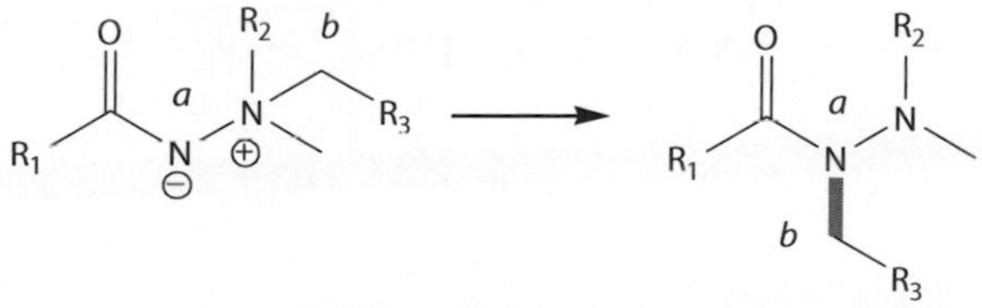

AE = 1; *E* **= 0;** *f*(sac) **= 0**

Set $R_3 = CH_3$

$$\Delta = \Delta(a) + \Delta(b) = (-2 - (-2)) + (-1 - (-1)) = 0; \quad \mathbf{HI} = \mathbf{0}$$

Wittig [1,2] rearrangement

Wittig, G.; Löhmann, L. *Ann. Chem.* 1942, *550*, 260.

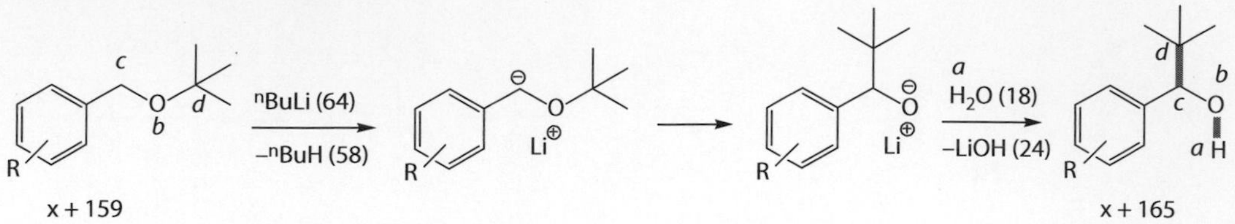

x + 159   x + 165

R = x(sum of substituents)

$$AE = [x + 165]/[x + 223]; \quad \textbf{AE(min)} = \textbf{0.75}; \quad E = 58/[x + 165]; \quad \textbf{E(max)} = \textbf{0.34}$$

$$f(\text{sac}) = 64/[x + 223]; \quad \textbf{f(sac)(max)} = \textbf{0.29}$$

$$\Delta = \Delta(a) + \Delta(b) + \Delta(c) + \Delta(d) = (1-1) + (-2 - (-2)) + (0 - (-1)) + (0 - 1) = 0; \quad \textbf{HI} = \textbf{0}$$

Wolff rearrangement

Wolff, L. *Ann. Chem.* 1912, *394*, 25.

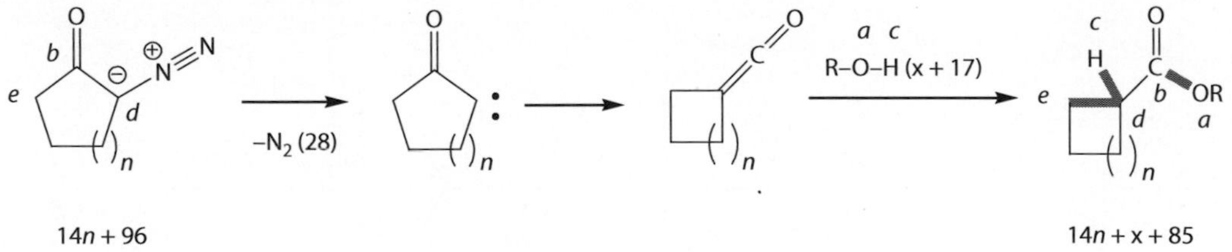

14n + 96   14n + x + 85

R = x

$$AE = [14n + x + 57]/[14n + x + 85]; \quad E = 28/[14n + x + 57]$$

$$\textbf{AE(min)} = [14n + 58]/[14n + 86]; \quad \textbf{E(max)} = 28/[14n + x + 85]; \quad \textbf{f(sac)} = \textbf{0}$$

Set R = CH$_3$

n	Ring Size ($n + 3$)	E(max)	AE(min)
1	4	0.25	0.80
2	5	0.22	0.82
3	6	0.20	0.84
4	7	0.18	0.85
5	8	0.16	0.86

$$\Delta = \Delta(a) + \Delta(b) + \Delta(c) + \Delta(d) + \Delta(e)$$

$$= (-2 - (-2)) + (3 - 2) + (1 - 1) + (-1 - 0) + (-2 - (-2)) = 0; \quad \textbf{HI} = \textbf{0}$$

7.8 CATEGORIZATION OF NAMED REDUCTION REACTIONS (WITH RESPECT TO SUBSTRATE OF INTEREST) BY MINIMUM ATOM ECONOMY, MAXIMUM ENVIRONMENTAL IMPACT FACTOR, MAXIMUM MOLECULAR WEIGHT FRACTION OF SACRIFICIAL REAGENTS, HYPSICITY INDEX

(Reactions are listed alphabetically.)

Benkeser reduction: Additive reduction
Benkeser, R.A.; Robinson, R.E.; Landesman, H. *J. Am. Chem. Soc.* 1952, *74*, 5699.
Benkeser, R.A.; Robinson, R.E.; Sauve, D.M.; Thomas, O.H. *J. Am. Chem. Soc.* 1955, *77*, 3230.

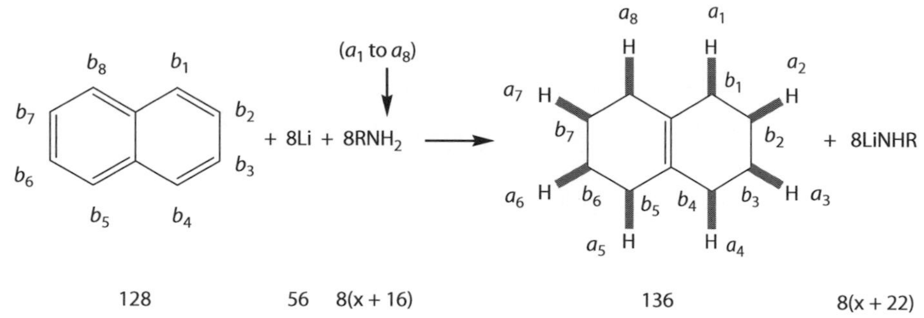

$$R = x$$

$$AE = 136/[8x + 312]; \quad AE(max) = 0.43; \quad E = [8(x+22)]/136; \quad E(min) = 1.35$$

$$f(sac) = 56/[8x+312]; \quad f(sac)(max) = 0.18$$

$$\Delta = \sum_{j=1}^{8}\Delta(a_j) + \sum_{j=1}^{8}\Delta(b_j) = 8(1-1) + 8(-2-(-1)) = -8; \quad HI = -\Delta/(N+1) = +4$$

Birch reduction: Additive reduction
Birch, A.J. *J. Chem. Soc.* 1944, 430.

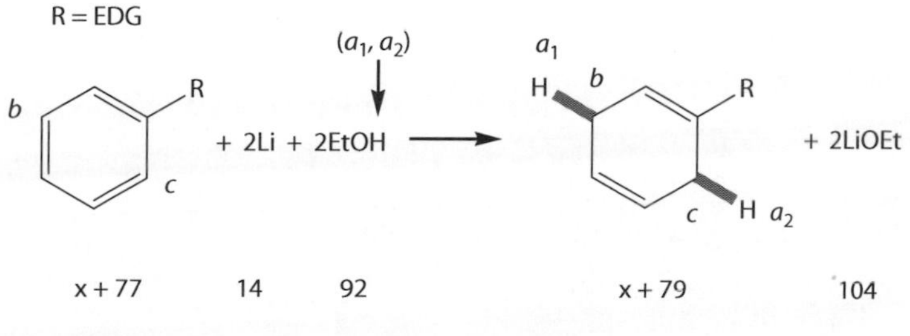

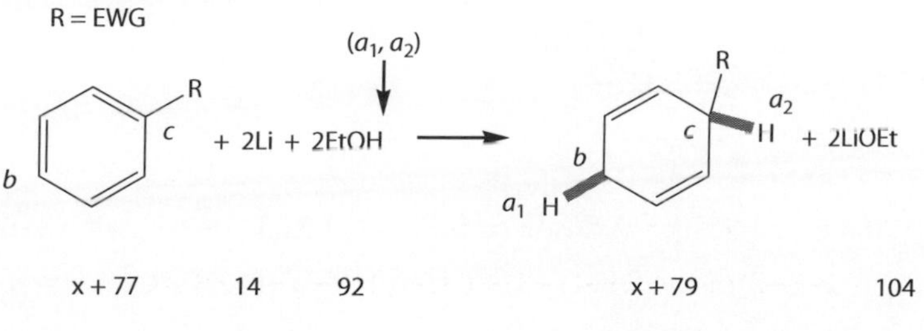

R = x

$$AE = [x + 79]/[x + 183]; \quad \textbf{AE(min)} = \textbf{0.43}; \quad E = 104/[x + 79]; \quad \textbf{\textit{E}(max)} = \textbf{1.30}$$

$$f(\text{sac}) = 14/[x + 183]; \quad f(\textbf{sac})(\textbf{max}) = \textbf{0.076}$$

Set R = CH_3

$$\Delta = \Delta(a_1) + \Delta(a_2) + \Delta(b) + \Delta(c) = (1-1) + (1-1) + (-2 - (-1)) + (-2 - (-1)) = -2;$$

$$\textbf{HI} = -\Delta/(N+1) = \textbf{+1}$$

Borch reduction: Additive reduction

Borch, R.F. *Org. Synth.* 1972, *52*, 124.
Borch, R.F.; Bernstein, M.D.; Durst, H.D. *J. Am. Chem. Soc.* 1971, *93*, 2897.
Borch, R.F.; Hassid, A.I. *J. Org. Chem.* 1972, *37*, 1673.

x + y + 28	21	18	x + y + 30	20.67	16.33

$R_1 = x; R_2 = y$

$$AE = [x + y + 30]/[x + y + 67]; \quad \textbf{AE(min)} = \textbf{0.32}; \quad E = 67/[x + y + 30]; \quad \textbf{\textit{E}(max)} = \textbf{2.09}$$

$f(\textbf{sac}) = \textbf{0}$

Set $R_1 = R_2 = CH_3$

$$\Delta = \Delta(a) + \Delta(b) + \Delta(c) + \Delta(d) = (1-1) + (-2 - (-2)) + (0-2) + (1-(-1)) = 0; \quad \textbf{HI} = \textbf{0}$$

Borohydride reduction: Additive reduction

Wolfrom, M.L.; Wood, H.B. *J. Am. Chem. Soc.* 1951, *73*, 2938.
Woodward, R.B.; Sondheimer, F.; Taub, D. *J. Am. Chem. Soc.* 1951, *73*, 3547.
Woodward, R.B.; Sondheimer, F.; Taub, D. *J. Am. Chem. Soc.* 1951, *73*, 4057.

x + y + 28	9.5	18	x + y + 30	10	15.5

$R_1 = x; R_2 = y$

$$AE = [x + y + 30]/[x + y + 55.5]; \quad \textbf{AE(min)} = \textbf{0.37}; \quad E = 55.5/[x + y + 30]; \quad \textbf{\textit{E}(max)} = \textbf{1.73}$$

$f(\textbf{sac}) = \textbf{0}$

Set $R_1 = R_2 = CH_3$

$$\Delta = \Delta(a) + \Delta(b) + \Delta(c) + \Delta(d) = (1-1) + (-2-(-2)) + (0-2) + (1-(-1)) = 0; \quad \textbf{HI} = \textbf{0}$$

Bouveault–Blanc reaction: Additive reduction
Bouveault, L.; Blanc, G. *Compt. Rend.* 1903, *136*, 1676.

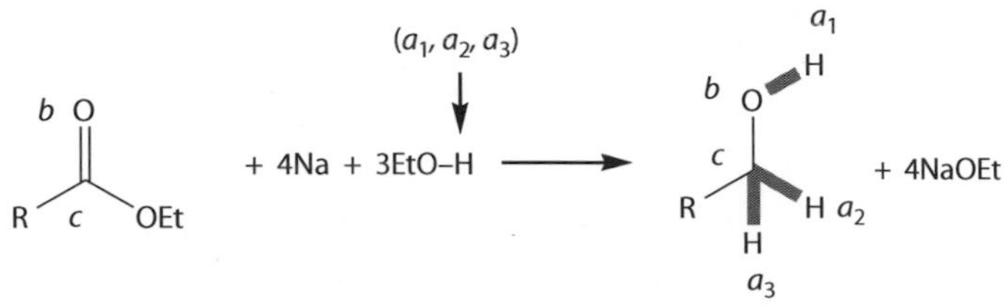

| $x + 73$ | 92 | 138 | $x + 31$ | 272 |

$R = x$

$$AE = [x+31]/[x+303]; \quad \textbf{AE(min)} = \textbf{0.11}; \quad E = 272/[x+31]; \quad \textbf{E(max)} = \textbf{8.50}$$

$$f(\text{sac}) = 92/[x+303]; \quad \textbf{f(sac)(max)} = \textbf{0.30}$$

Set $R = CH_3$

$$\Delta = \sum_{j=1}^{3} \Delta(a_j) + \Delta(b) + \Delta(c) = 3(1-1) + (-2-(-2)) + (-1-3) = -4; \quad \textbf{III} = -\Delta/(N+1) = \textbf{+2}$$

Cannizzaro reaction: Additive reduction
Cannizzaro, S. *Ann. Chem.* 1853, *88*, 129.

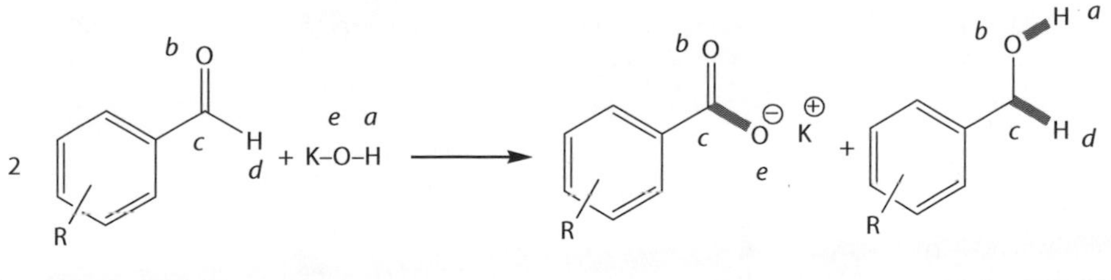

| $2(x + 101)$ | 56 | $x + 155$ | $x + 103$ |

$R = x$(sum of substituents)

Target product: carboxylate salt

$$AE = [x+155]/[x+258]; \quad \textbf{AE(min)} = \textbf{0.60}; \quad E = [x+103]/[x+155]; \quad \textbf{E(max)} = \textbf{0.68}$$

$f(\text{sac}) = \textbf{0}$

$$\Delta = \Delta(b) + \Delta(c) + \Delta(e) = (-2-(-2)) + (3-1) + (-2-(-2)) = +2; \quad \textbf{HI} = -\Delta/(N+1) = \textbf{-1}$$

Target product: alcohol

$$AE = [x+103]/[x+258]; \quad \textbf{AE(min)} = \textbf{0.40}; \quad E = [x+155]/[x+103]; \quad \textbf{E(max)} = \textbf{1.5}$$

$f(\text{sac}) = 0$

$$\Delta = \Delta(a) + \Delta(b) + \Delta(c) + \Delta(d) = (1-1) + (-2-(-2)) + (-1-1) + (1-1) = -2; \quad \mathbf{HI} = -\Delta/(N+1) = +1$$

Clemmensen reduction: Additive reduction with elimination

Clemmensen, E. *Chem. Ber.* 1913, *46*, 1837.
Clemmensen, E. *Chem. Ber.* 1914, *47*, 51.
Clemmensen, E. *Chem. Ber.* 1914, *47*, 681.

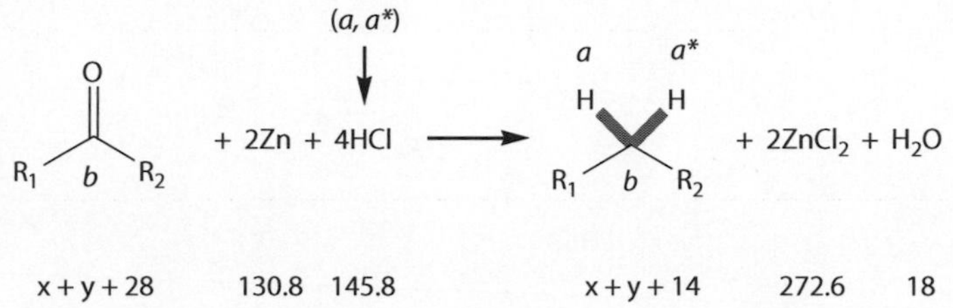

$R_1 = x; R_2 = y$

$$AE = [x+y+14]/[x+y+304.6]; \quad \mathbf{AE(min)} = \mathbf{0.052}; \quad E = 290.6/[x+y+14]; \quad \mathbf{E(max)} = \mathbf{18.16}$$

$$f(\text{sac}) = 130.8/[x+y+304.6]; \quad \mathbf{f(sac)(max)} = \mathbf{0.43}$$

Set $R_1 = R_2 = CH_3$

$$\Delta = \Delta(a) + \Delta(a^*) + \Delta(b) = (1-1) + (1-1) + (-2-2) = -4; \quad \mathbf{HI} = -\Delta/(N+1) = +2$$

Corey–Bakshi–Shibata (CBS) reduction: Additive reduction

Corey, E.J.; Bakshi, R.K.; Shibata, S. *J. Am. Chem. Soc.* 1987, *109*, 5551.
Corey, E.J.; Bakshi, R.K.; Shibata, S.; Chen, C.P.; Singh, V.K. *J. Am. Chem. Soc.* 1987, *109*, 7925.
Corey, E.J.; Bakshi, R.K.; Shibata, S. *J. Org. Chem.* 1988, *53*, 2861.

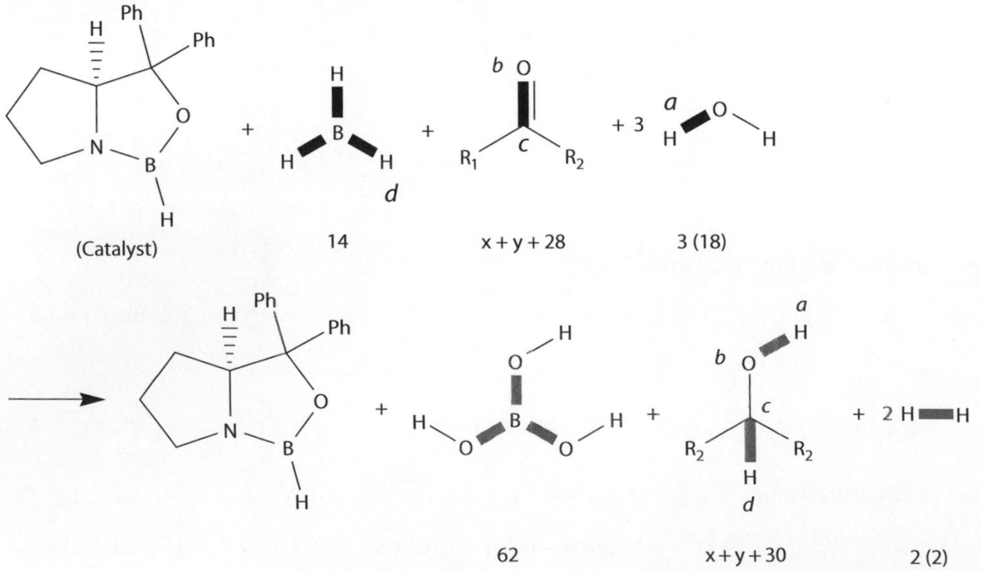

$R_1 = x$; $R_2 = y$

$$AE = [x + y + 30]/[x + y + 96]; \quad \mathbf{AE(min) = 0.33}; \quad E = 64/[x + y + 30]; \quad \mathbf{E(max) = 2}$$

$$f(sac) = 0; \quad \mathbf{f(sac)(max) = 0}$$

Set $R_1 = R_2 = CH_3$

$$\Delta = \Delta(a) + \Delta(b) + \Delta(c) + \Delta(d) = (1-1) + (-2-(-2)) + (0-2) + (1-(-1)) = 0; \quad \mathbf{HI = 0}$$

Corey–Winter reaction: Subtractive reduction
Corey, E.J.; Winter, R.A.E. *J. Am. Chem. Soc.* 1963, *85*, 2677.

| 2x + 60 | 178 | 124 | 2x + 26 | 44 | 156 | 136 |

R = x

$$AE = [2x + 26]/[2x + 362]; \quad \mathbf{AE(min) = 0.077}; \quad E = 336/[2x + 26]; \quad \mathbf{E(max) = 12}$$

$$f(sac) = 302/[2x + 362]; \quad \mathbf{f(sac)(max) = 0.83}$$

Diimide reduction: Additive reduction
Corey, E.J.; Mock, W.L.; Pasto, D.J. *Tetrahedron Lett.* 1961, 347.
Huenig, S.; Mueller, H.R.; Thier, W. *Tetrahedron Lett.* 1961, 353.
Dewey, R.S.; van Tamelen, E.E. *J. Am. Chem. Soc.* 1961, *83*, 3729.
Van Tamelen, E.E.; Dewey, R.S.; Timmons, R.J. *J. Am. Chem. Soc.* 1961, *83*, 3725.
Van Tamelen, E.E.; Timmons, R.J. *J. Am. Chem. Soc.* 1962, *84*, 1067.

| x + y + 26 | 162 | 36 | x + y + 28 | 88 | 80 | 28 |

$R_1 = x$; $R_2 = y$

$$AE = [x + y + 28]/[x + y + 224]; \quad \mathbf{AE(min) = 0.13}; \quad E = 196/[x + y + 28]; \quad \mathbf{E(max) = 6.53}$$

$$f(sac) = 162/[x + y + 224]; \quad \mathbf{f(sac)(max) = 0.72}$$

Set $R_1 = R_2 = CH_3$

$$\Delta = \Delta(a) + \Delta(a^*) + \Delta(b) + \Delta(c) = (1-1) + (1-1) + (-2-(-1)) + (-2-(-1)) = -2;$$

$$\mathbf{HI = -\Delta/(N+1) = +1}$$

$$\underset{x+y+26}{\overset{b \quad\quad c}{R_1CH = CHR_2}} + \underset{32}{\overset{a}{H_2N-NH_2}} + \underset{34}{\overset{a^*}{H_2O_2}} \longrightarrow \underset{x+y+28}{\overset{R_1 \quad\quad R_2}{\underset{a \ H \quad\quad H \ a^*}{\overset{b}{|} \quad\quad \overset{c}{|}}}} + \underset{36}{2H_2O} + \underset{28}{N_2}$$

$R_1 = x; \ R_2 = y$

$$AE = [x + y + 28]/[x + y + 92]; \quad \textbf{AE(min) = 0.32}; \quad E = 64/[x + y + 28]; \quad \textbf{E(max) = 2.13}$$

$$f(\text{sac}) = 34/[x + y + 92]; \quad \textbf{\textit{f}(sac)(max) = 0.36}$$

Set $R_1 = R_2 = CH_3$

$$\Delta = \Delta(a) + \Delta(a^*) + \Delta(b) + \Delta(c) = (1-1) + (1-1) + (-2-(-1)) + (-2-(-1)) = -2;$$

$$\textbf{HI} = -\Delta/(N+1) = \textbf{+1}$$

Gribble reduction of diaryl ketones: Additive reduction
Gribble, G.W.; Leese, R.M.; Evans, B.E. *Synthesis* 1977, 172.

$R_1 = x(\text{sum of substituents}); \ R_2 = y(\text{sum of substituents})$

$$AE = [x + y + 158]/[x + y + 675.5]; \quad \textbf{AE(min) = 0.24}; \quad E = 517.5/[x + y + 158]; \quad \textbf{E(max) = 3.23}$$

$$f(\text{sac}) = 456/[x + y + 675.5]; \quad \textbf{\textit{f}(sac)(max) = 0.67}$$

$$\Delta = \Delta(a) + \Delta(a^*) + \Delta(b) = (1-(-1) + (1-(-1)) + (-2-2) = 0; \quad \textbf{HI = 0}$$

Hydrogenation of olefins: Additive reduction
Lindlar, H. *Helv. Chim. Acta* 1952, *35*, 446.

$$\underset{R_1CH = CHR_2}{\overset{c \quad\quad b \quad\quad a \quad a^*}{}} + \overset{}{H-H} \longrightarrow \underset{a \quad\quad\quad a^*}{\overset{R_1 \quad\quad R_2}{\underset{H \quad\quad H}{\overset{c}{|} \quad\quad \overset{b}{|}}}}$$

$\textbf{AE = 1}; \ \textit{E} = \textbf{0}; \ f(\text{sac}) = \textbf{0}$

Set $R_1 = R_2 = CH_3$

$$\Delta = \Delta(a) + \Delta(a^*) + \Delta(b) + \Delta(c) = (1-0) + (1-0) + (-2-(-1)) + (-2-(-1)) = 0; \quad \mathbf{HI = 0}$$

Hydrogenolysis of benzyl ethers: Additive reduction with elimination

Van Duzee, E.M.; Adkins, H. *J. Am. Chem. Soc.* 1935, *57*, 147.

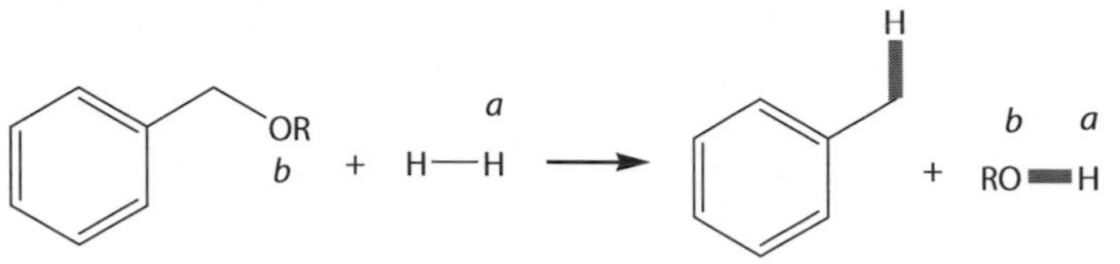

| x + 107 | 2 | 92 | x + 17 |

$R = x$

$$AE = [x + 17]/[x + 109]; \quad \mathbf{AE(min) = 0.16}; \quad E = 92/[x + 17]; \quad \mathbf{E(max) = 5.11}$$

$f(\text{sac}) = 0$

$$\Delta = \Delta(a) + \Delta(b) = (1-0) + (-2-(-2)) = +1; \quad \mathbf{HI} = -\Delta/(N+1) = -\tfrac{1}{2}$$

Lithium aluminum hydride reduction of ketones: Additive reduction

Trevoy, L.W.; Brown, H.G. *J. Am. Chem. Soc.* 1949, *71*, 1675.
Papineau-Couture, G.; Richardson, E.M.; Grant, G.A. *Can. J. Res.* 1949, *27B*, 902.
Lutz, R.E.; Hinkley, D.F. *J. Am. Chem. Soc.* 1950, *72*, 4091.

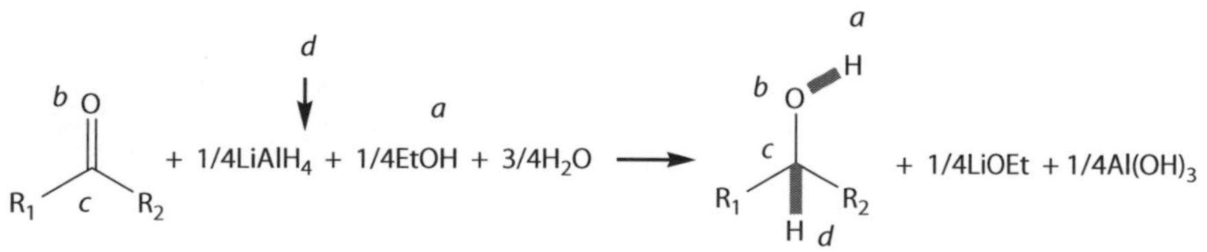

| x + y + 28 | 9.5 | 11.5 | 13.5 | x + y + 30 | 13 | 19.5 |

$R_1 = x; R_2 = y$

$$AE = [x + y + 30]/[x + y + 62.5]; \quad \mathbf{AE(min) = 0.50}; \quad E = 32.5/[x + y + 30]; \quad \mathbf{E(max) = 1.02}$$

$$f(\text{sac}) = 13.5/[x + y + 62.5]; \quad f(\text{sac})(\text{max}) = \mathbf{0.21}$$

Set $R_1 = R_2 = CH_3$

$$\Delta = \Delta(a) + \Delta(b) + \Delta(c) + \Delta(d) = (1-1) + (-2-(-2)) + (0-2) + (1-(-1)) = 0; \quad \mathbf{HI = 0}$$

Lithium aluminum hydride reduction of methyl esters: Additive reduction

Nystrom, R.F.; Weldon, W.G. *J. Am. Chem. Soc.* 1947, *69*, 1197.
Karrer, P.; Portmann, P.; Suter, M. *Helv. Chim. Acta* 1948, *31*, 1617.
Karrer, P.; Banerjea, P. *Helv. Chim. Acta* 1949, *32*, 1692.
Bachmann, W.E.; Dreiding, A.S. *J. Am. Chem. Soc.* 1949, *71*, 3222.
Abdel-Akher, M.A.; Smith, F. *Nature* 1950, *166*, 1037.

$$x + 59 \qquad 19 \qquad 27 \qquad\qquad x + 31 \qquad 16 \qquad 19 \qquad 39$$

$$R_1 = x$$

$$AE = [x + 31]/[x + 105]; \quad \textbf{AE(min)} = \textbf{0.30}; \quad E = 74/[x + 31]; \quad \textbf{E(max)} = \textbf{2.31}$$

$$f(\text{sac}) = 0$$

Set R = CH$_3$

$$\Delta = \Delta(a) + \Delta(b) + \Delta(c) + \Delta(d) + \Delta(d^*) = (1 - 1) + (-2 - (-2)) + (-1 - 3) + 2(1 - (-1)) = 0; \quad \textbf{HI} = \textbf{0}$$

Meerwein–Pondorff–Verley reduction: Additive reduction

Meerwein, H.; Schmidt, R. *Ann. Chem.* 1924, *444*, 221.
Ponndorf, W. *Angew. Chem.* 1926, *39*, 138.
Verley, A. *Bull. Soc. Chim. Fr.* 1925, *37*, 537.

$$x + y + 28 \qquad\qquad 1/3\,(204) \qquad\qquad 18 \qquad\qquad x + y + 30 \qquad 26 \qquad 58$$

$$R_1 = x; R_2 = y$$

$$AE = [x + y + 30]/[x + y + 114]; \quad \textbf{AE(min)} = \textbf{0.28}; \quad E = 84/[x + y + 30]; \quad \textbf{E(max)} = \textbf{2.63}$$

$$f(\text{sac}) = 0$$

Set R$_1$ = R$_2$ = CH$_3$

$$\Delta = \Delta(a) + \Delta(b) + \Delta(c) + \Delta(d) = (1 - 1) + (-2 - (-2)) + (0 - 2) + (1 - 1) = -2; \quad \textbf{HI} = -\Delta/(N + 1) = \textbf{+1}$$

Midland reduction: Additive reduction

Midland, M.M.; Tramontano, A.; Zederic, S.A. *J. Am. Chem. Soc.* 1979, *101*, 2352.

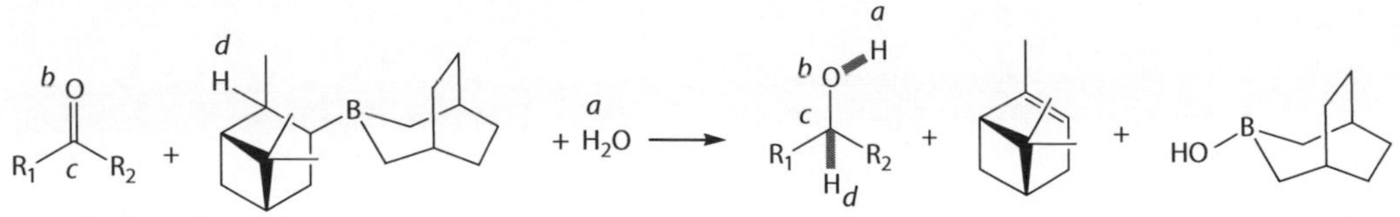

x + y + 28	258	18	x + y + 30	136	138

$R_1 = x; R_2 = y$

$$AE = [x + y + 30]/[x + y + 304]; \quad \mathbf{AE(min) = 0.10}; \quad E = 274/[x + y + 30]; \quad \mathbf{E(max) = 8.56}$$

$f(sac) = 0$

$R_1 = R_2 = CH_3$

$$\Delta = \Delta(a) + \Delta(b) + \Delta(c) + \Delta(d) = (1-1) + (-2-(-2)) + (0-2) + (1-1) = -2; \quad \mathbf{HI} = -\Delta/(N+1) = +1$$

Noyori hydrogenation: Additive reduction

Ikariya, T.; Ishii, Y.; Kawano, H.; Arai, T.; Saburi, M.; Yoshikawa, S.; Akutagawa, S. *Chem. Commun.* 1985, 922.

Noyori, R.; Ohta, M.; Hsiao, Y.; Kitamura, M.; Ohta, T.; Takaya, H. *J. Am. Chem. Soc.* 1986, *108*, 7117.

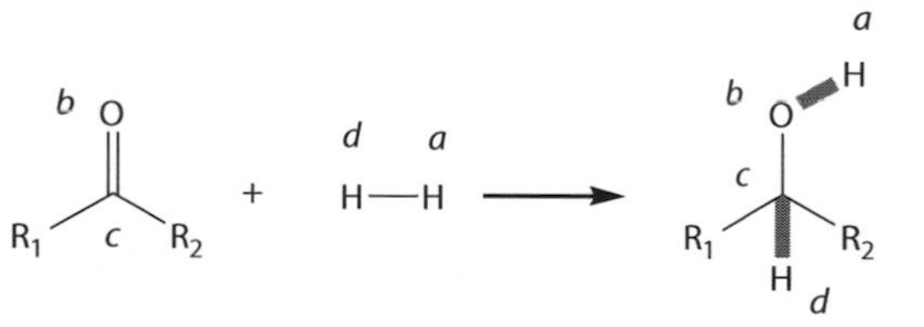

$\mathbf{AE = 1}; E = 0; f(sac) = 0$

Set $R_1 = R_2 = CH_3$

$$\Delta = \Delta(a) + \Delta(b) + \Delta(c) + \Delta(d) = (1-0) + (-2-(-2)) + (0-2) + (1-0) = 0; \quad \mathbf{HI = 0}$$

Radical dehalogenation: Additive reduction with elimination

Strunk, R.J.; DiGiacomo, P.M.; Aso, K.; Kuivila, H.G. *J. Am. Chem. Soc.* 1970, *92*, 2849.

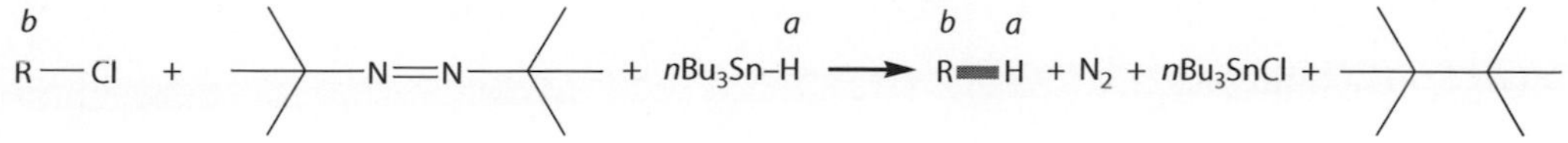

x + 35.45	142	177	x + 1	28	211.45	114

R = x

$$AE = [x + 1]/[x + 354.45]; \quad \mathbf{AE(min) = 0.0056}; \quad E = 353.45/[x + 1]; \quad \mathbf{E(max) = 176.73}$$

$$f(\text{sac}) = 142/[x + 354.45]; \quad f(\text{sac})(\text{max}) = \mathbf{0.40}$$

Set $R = CH_3$

$$\Delta = \Delta(a) + \Delta(b) = (1 - (-1)) + (-4 - (-3)) = +1; \quad \mathbf{HI} = -\Delta/(N + 1) = -\tfrac{1}{2}$$

Reduction of nitroaromatics with hydrogen sulfide: Additive reduction with elimination

Goldschmidt, H.; Larsen, H. *Z. Physik. Chem.* 1910, *71*, 437.

| x + 118 | 34 | x + 88 | 64 |

R = x(sum of substituents)

$$\mathbf{AE} = [x + 88]/[x + 152]; \quad \mathbf{AE(min)} = \mathbf{0.58}; \quad E = 64/[x + 88]; \quad E(\mathbf{max}) = \mathbf{0.72}$$

$$f(\text{sac}) = \mathbf{0}$$

$$\Delta = \Delta(a) + \Delta(b) + \Delta(c) = (1 - 1) + (-3 - 3) + (1 - 1) = -6; \quad \mathbf{HI} = -\Delta/(N + 1) = +\mathbf{3}$$

Reduction of nitroaromatics with iron (Bechamp reduction): Additive reduction with elimination

Bechamp, A.J. *Ann. Chim. Phys.* 1854, *42(3)*, 186.
Neelmeier, W.; Lamberg, W. (General Aniline Works), U.S. Pat. 1,874,581 (1932/08/30).

| x + 118 | 912 | 294 | x + 88 | 1200 | 36 |

R = x(sum of substituents)

$$\mathbf{AE} = [x + 88]/[x + 1324]; \quad \mathbf{AE(min)} = \mathbf{0.067}; \quad E = 1236/[x + 88]; \quad E(\mathbf{max}) = \mathbf{13.89}$$

$$f(\text{sac}) = 912/[x + 1324]; \quad f(\text{sac})(\text{max}) = \mathbf{0.69}$$

$$\Delta = \Delta(a) + \Delta(b) + \Delta(c) = (1 - 1) + (-3 - 3) + (1 - 1) = -6; \quad \mathbf{HI} = -\Delta/(N + 1) = +\mathbf{3}$$

Rosenmund reduction: Additive reduction with elimination

Rosenmund, K.W. *Chem. Ber.* 1918, *51*, 585.

$$\text{x} + 63.45 \qquad 2 \qquad\qquad\qquad \text{x} + 29 \qquad 36.45$$

R = x

$$AE = [x+29]/[x+65.45]; \quad \textbf{AE(min)} = \textbf{0.45}; \quad E = 36.45/[x+29]; \quad \textbf{E(max)} = \textbf{1.22}$$

$f(\textbf{sac}) = \textbf{0}$

Set $R_1 = CH_3$

$$\Delta = \Delta(a) + \Delta(b) = (1-0) + (1-3) = -1; \quad \textbf{HI} = -\Delta/(N+1) = +\tfrac{1}{2}$$

Shapiro reaction: Additive reduction with elimination

Shapiro, R.H.; Heath, M.J. *J. Am. Chem. Soc.* 1967, *89*, 5734.

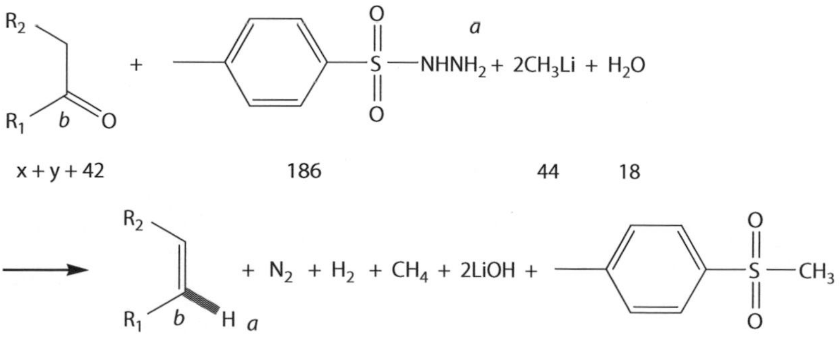

$$\text{x} + \text{y} + 42 \qquad\qquad 186 \qquad\qquad\qquad 44 \qquad 18$$

$$\text{x} + \text{y} + 26 \quad 28 \quad 2 \quad 16 \quad 48 \qquad\qquad 170$$

$R_1 = x; R_2 = y$

$$AE = [x+y+26]/[x+y+290]; \quad \textbf{AE(min)} = \textbf{0.096}; \quad E = 264/[x+y+26]; \quad \textbf{E(max)} = \textbf{9.43}$$

$$f(\text{sac}) = 62/[x+y+290]; \quad f(\textbf{sac})(\textbf{max}) = \textbf{0.21}$$

Set $R_1 = CH_3$

$$\Delta = \Delta(a) + \Delta(b) = (1-1) + (-1-2) = -3; \quad \textbf{HI} = -\Delta/(N+1) = +\tfrac{3}{2}$$

Stephen reduction: Additive reduction with elimination
Stephen, H. *J. Chem. Soc.* 1925, *127*, 1874.

| x + 98 | 72.9 | 189.9 | 18 | x + 101 | 17 | 260.8 |

R = x(sum of substituents)

$$AE = [x+101]/[x+378.8]; \quad \textbf{AE(min)} = \textbf{0.27}; \quad E = 277.8/[x+101]; \quad \textbf{\textit{E}(max)} = \textbf{2.72}$$

$$f(\text{sac}) = 189.9/[x+378.8]; \quad \textbf{\textit{f}(sac)(max)} = \textbf{0.50}$$

$$\Delta = \Delta(a) + \Delta(b) + \Delta(c) = (-2-(-2)) + (1-2) + (1-1) = -1; \quad \textbf{HI} = -\Delta/(N+1) = +\tfrac{1}{2}$$

Thioketal desulfurization: Additive reduction with elimination
Raney, M. US Patent 1,628,190 issued May 10, 1927.
Raney, M. *Ind. Eng. Chem.* 1940, *32*, 1199.

| x + y + 104 | 118 | 145.8 | x + y + 14 | 94 | 259.8 |

$R_1 = x; R_2 = y$

$$AE = [x+y+14]/[x+y+367.8]; \quad \textbf{AE(min)} = \textbf{0.043}; \quad E = 353.8/[x+y+14]; \quad \textbf{\textit{E}(max)} = \textbf{22.11}$$

$$f(\text{sac}) = 118/[x+y+367.8]; \quad \textbf{\textit{f}(sac)(max)} = \textbf{0.32}$$

Set $R_1 = R_2 = CH_3$

$$\Delta = \Delta(a) + \Delta(a^*) + \Delta(b) = (1-1) + (1-1) + (-2-0) = -2; \quad \textbf{HI} = -\Delta/(N+1) = +1$$

Tosylhydrazone reduction: Additive reduction with elimination
Caglioti, L.; Grasselli, P. *Chem. Ind.* 1964, 153.
Caglioti, L.; Grasselli, P. *Politechnico (Milan)* 1964, *46*, 799.
Caglioti, L.; Grasselli, P. *Chim. Ind. (Milan)* 1964, *46*, 1492.
Fischer, M.; Pelah, Z.; Williams, D.H.; Djerassi, C. *Chem. Ber.* 1965, *98*, 3236.
Caglioti, L. *Tetrahedron* 1966, *22*, 487.

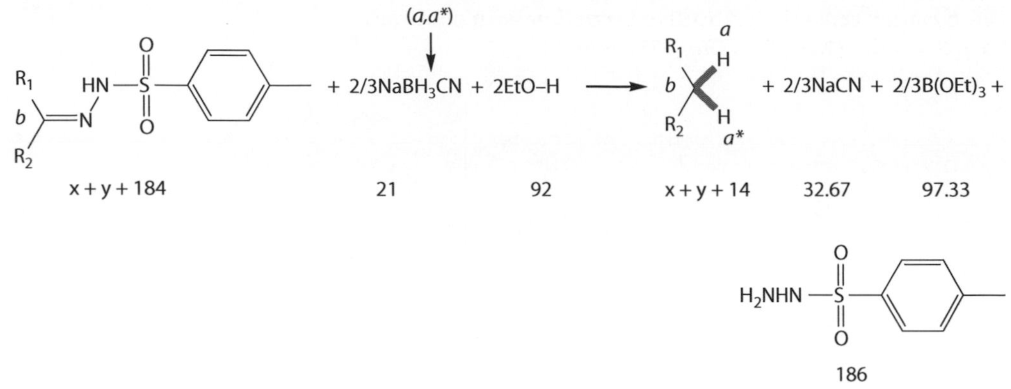

(a,a*)
+ 2/3NaBH3CN + 2EtO-H
+ 2/3NaCN + 2/3B(OEt)3 +
x + y + 184
21
92
x + y + 14
32.67
97.33
H2NHN
186

$R_1 = x$; $R_2 = y$

$$AE = [x + y + 14]/[x + y + 402];\quad \mathbf{AE(min) = 0.038};\quad E = 400/[x + y + 14];\quad \boldsymbol{E}\mathbf{(max) = 25}$$

$$f(\text{sac}) = 92/[x + y + 297];\quad \boldsymbol{f}\mathbf{(sac)(max) = 0.31}$$

Set $R_1 = R_2 = CH_3$

$$\Delta = \Delta(a) + \Delta(a^*) + \Delta(b) = (1 - (-1)) + (1 - (-1)) + (-2 - 2) = 0;\quad \mathbf{HI = 0}$$

Wharton reaction: Additive reduction with elimination

Wharton, P.S.; Bohlen, D.H., *J. Org. Chem.* 1961, *26*, 3615.

Wharton, P.S.; Bohlen, D.H., *J. Org. Chem.* 1961, *26*, 4781.

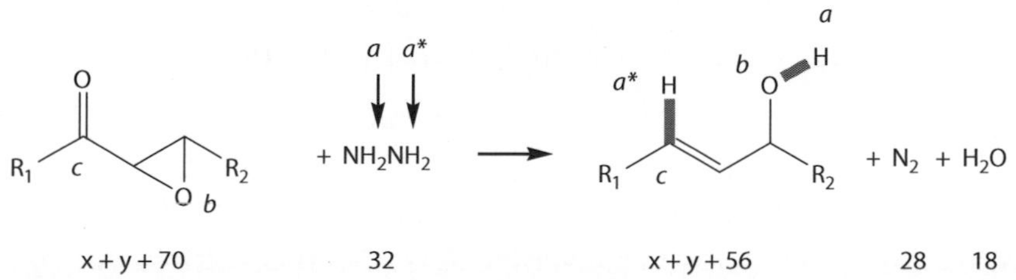

a a*
+ NH2NH2
+ N2 + H2O
x + y + 70
32
x + y + 56
28 18

$R_1 = x$; $R_2 = y$

$$AE = [x + y + 56]/[x + y + 102];\quad \mathbf{AE(min) = 0.56};$$

$$E = 46/[x + y + 56];\quad \boldsymbol{E}\mathbf{(max) = 0.79}$$

***f*(sac) = 0**

Set $R_1 = CH_3$

$$\Delta = \Delta(a) + \Delta(a^*) + \Delta(b) + \Delta(c) = (1 - 1) + (1 - 1) + (-2 - (-2)) + (-1 - 2) = -3;$$

$$\mathbf{HI = -\Delta/(\boldsymbol{N} + 1) = +3/2}$$

Wolff–Kishner reduction: Additive reduction with elimination
Kishner, N. *J. Russ. Phys. Chem. Soc.* 1911, *43*, 582.
Wolff, L. *Ann. Chem.* 1912, *394*, 86.

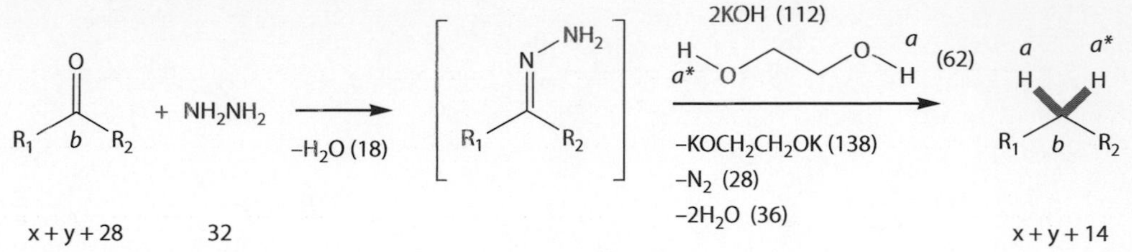

$$R_1 = x; R_2 = y$$

$$AE = [x+y+14]/[x+y+234]; \quad \textbf{AE(min)} = \textbf{0.068}; \quad E = 220/[x+y+14]; \quad \textbf{E(max)} = \textbf{13.75}$$

$$f(\text{sac}) = 144/[x+y+234]; \quad f(\text{sac})(\text{max}) = \textbf{0.61}$$

$$\Delta = \Delta(a) + \Delta(a^*) + \Delta(b) = (1-1) + (1-1) + (-2-2) = -4; \quad \textbf{HI} = -\Delta/(N+1) = \textbf{+2}$$

Zincke disulfide cleavage reaction: Additive reduction
Zincke, T. *Chem. Ber.* 1911, *44*, 769.

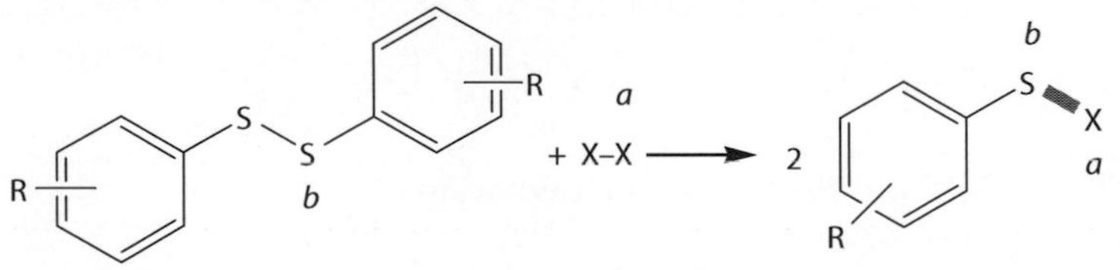

$$\textbf{AE} = \textbf{1};\ E = 0;\ f(\text{sac}) = 0$$

$$\Delta = \Delta(a) + \Delta(b) = (1-0) + (-2-(-1)) = 0; \quad \textbf{HI} = \textbf{0}$$

7.9 CATEGORIZATION OF NAMED SEQUENCES BY MINIMUM ATOM ECONOMY AND MAXIMUM ENVIRONMENTAL IMPACT FACTOR, MAXIMUM MOLECULAR WEIGHT FRACTION OF SACRIFICIAL REAGENTS, HYPSICITY INDEX

(Reactions are listed alphabetically.)

Arndt–Eistert reaction
Arndt, F.; Eistert, B. *Chem. Ber.* 1935, *68*, 200.

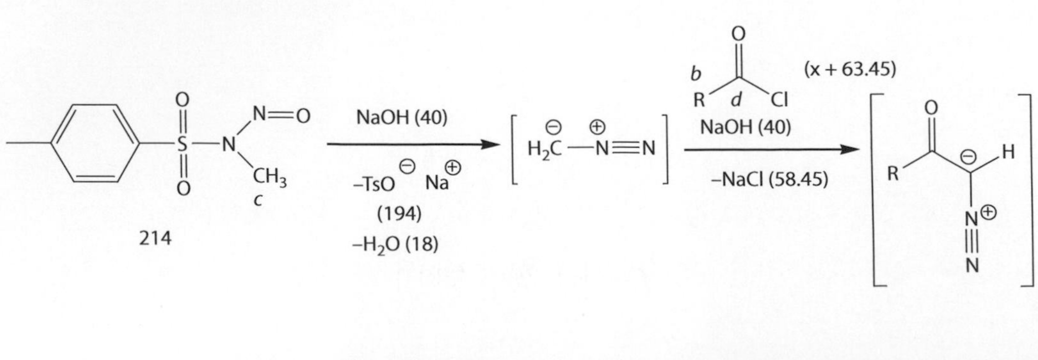

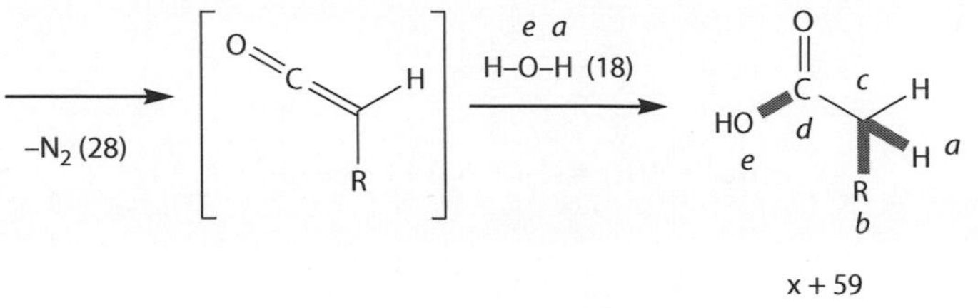

$$R = x$$

$$E = 298.45/[x + 59]$$

E(max) = 4.97

$$AE = [x + 59]/[x + 357.45]$$

AE(min) = 0.17

$$f(sac) = 80/[x + 357.45]$$

f(sac)(max) = 0.22

Set R = CH$_3$

$$\Delta = \Delta(a) + \Delta(b) + \Delta(c) + \Delta(c) + \Delta(d) + \Delta(e)$$

$$= (1-1) + (-3 - (-3)) + (-2 - (-2)) + (3 - 3) + (-2 - (-2)) = 0$$

HI = 0

Barton decarboxylation
Barton, D.H.R.; Crich, D.; Motherwell, W.B. *Chem. Commun.* 1983, 939.

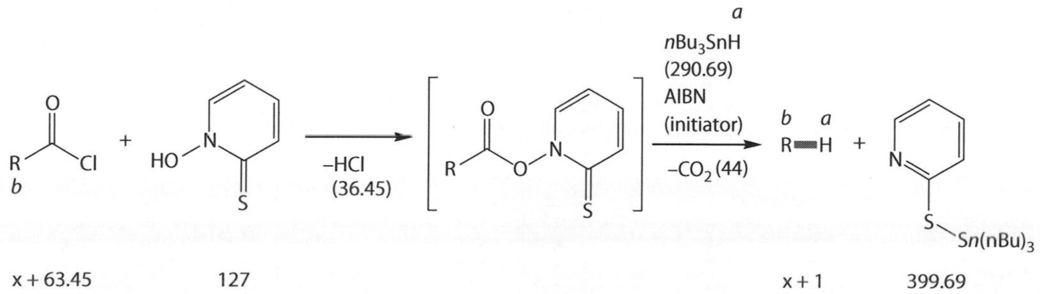

$$R = x$$

$$E = 480.14/[x + 1]$$

E(max) = 30.01

$$AE = [x + 1]/[x + 481.14]$$

AE(min) = 0.032

$$f(sac) = 127/[x + 481.14]$$

$f(\text{sac})(\text{max}) = 0.26$

Set $R = CH_3$

$$\Delta = \Delta(a) + \Delta(b) = (1 - (-1)) + (-4 - (-3)) = +1$$

HI = −1/2

Bischler–Napieralski synthesis

Bischler, A.; Napieralski, B. *Chem. Ber.* 1893, *26*, 1903.

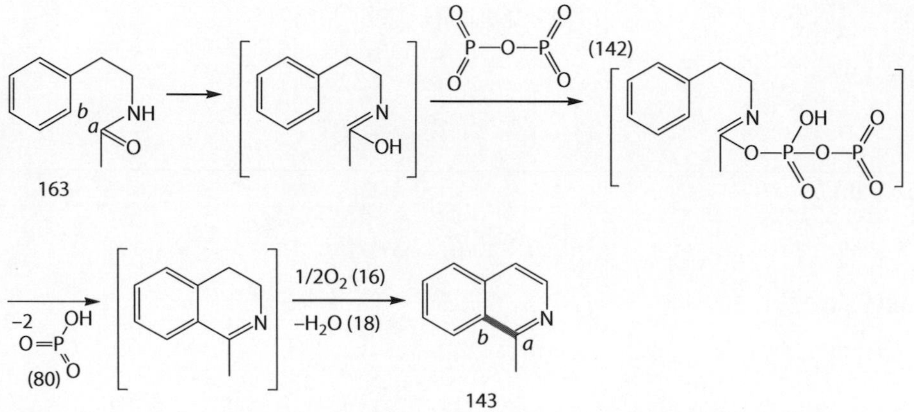

E = 98/143 = 0.69

AE = 0.59

$f(\text{sac}) = 158/321 = 0.49$

$$\Delta = \Delta(a) + \Delta(b) = (2 - 3) + (0 - (-1)) = 0$$

HI = 0

Claisen–Ireland rearrangement

Ireland, R.E.; Mueller, R.H. *J. Am. Chem. Soc.* 1972, *94*, 5897.
Ireland, R.E.; Willard, A.K. *Tetrahedron Lett.* 1975, *46*, 3975.

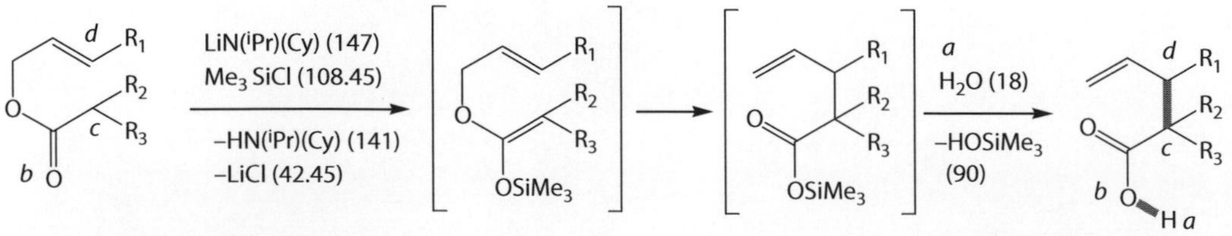

x + y + z + 97   x + y + z + 97

$R_1 = x; R_2 = y; R_3 = z$

$$E = 273.45/[x + y + z + 97]$$

E(max) = 2.73

$$AE = [x + y + z + 97]/[x + y + z + 370.45]$$

AE(min) = 0.27

$$f(\text{sac}) = 255.45/[x + y + z + 370.45]$$

f(sac)(max) = 0.68

Set $R_1 = R_2 = R_3 = CH_3$

HI = −1/2

Doering–LaFlamme allene synthesis

Doering, W. v. E.; LaFlamme, P.M. *Tetrahedron* 1958, *2*, 75.

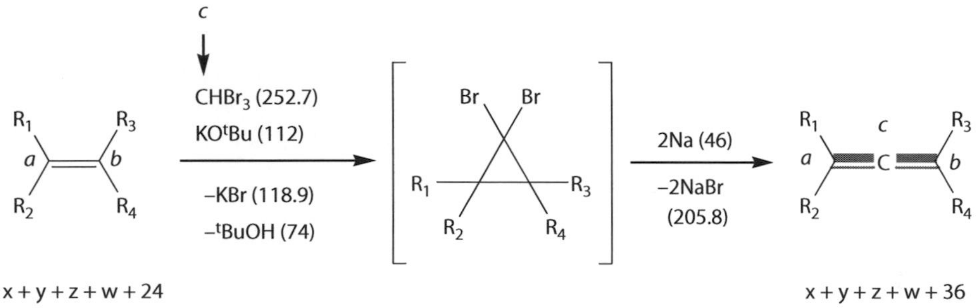

$x + y + z + w + 24$

$x + y + z + w + 36$

$R_1 = x; R_2 = y; R_3 = z; R_4 = w$

$$E = 398.7/[x + y + z + w + 36]$$

E(max) = 9.97

$$AE = [x + y + z + w + 36]/[x + y + z + w + 434.7]$$

AE(min) = 0.091

$$f(\text{sac}) = 158/[x + y + z + w + 434.7]$$

f(sac)(max) = 0.36

Set $R_1 = R_2 = R_3 = R_4 = CH_3$

$$\Delta = \Delta(a) + \Delta(b) + \Delta(c) = (0 - 0) + (0 - 0) + (0 - 2) = -2$$

HI = +1

Japp–Klingemann reaction

Japp, F.R.; Klingemann, F. *Ann. Chem.* 1888, *247*, 190.

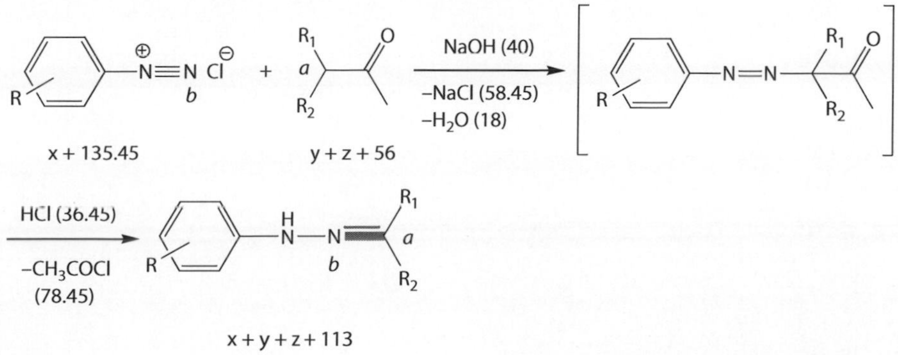

$x + 135.45$

$y + z + 56$

$x + y + z + 113$

R = x(sum of 5 substituents); R_1 = y; R_2 = z

$$E = 154.9/[x + y + z + 113]$$

E(max) = 1.29

$$AE = [x + y + z + 113]/[x + y + z + 267.9]$$

AE(min) = 0.44

$$f(sac) = 76.45/[\ x + y + z + 267.9]$$

f(sac)(max) = 0.28

Set $R_1 = R_2 = CH_3$

$$\Delta = \Delta(a) + \Delta(b) = (2 - (-1)) + (-2 - 0) = +1$$

HI = −1/2

Julia olefination
Julia, M.; Paris, J.M. *Tetrahedron Lett.* 1973, 493.

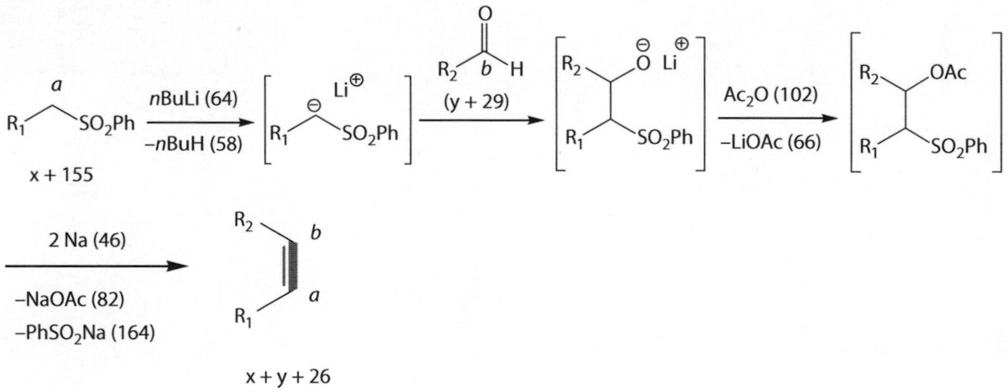

R_1 = x; R_2 = y

$$E = 370/[x + y + 26]$$

E(max) = 13.21

$$AE = [x + y + 26]/[x + y + 396]$$

AE(min) = 0.070

$$f(sac) = 212/[x + y + 396]$$

f(sac)(max) = 0.53

Set $R_1 = R_2 = CH_3$

$$\Delta = \Delta(a) + \Delta(b) = (-1 - (-1)) + (-1 - 1) = -2$$

HI = +1

Martinet dioxindole synthesis

Guyot, A.; Martinet, J. *Compt. Rend.* 1913, *156*, 1625.

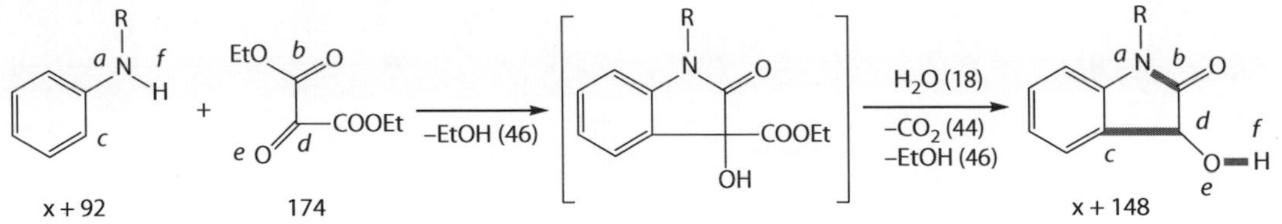

$$E = 136/[x + 148]$$

E(max) = 0.91

$$AE = [x + 148]/[x + 284]$$

AE(min) = 0.52

$$f(sac) = 18/[x + 284]$$

f(sac)(max) = 0.063

Set R = CH$_3$

$$\Delta = \Delta(a) + \Delta(b) + \Delta(c) + \Delta(d) + \Delta(e) + \Delta(f)$$

$$= (-3 - (-3)) + (3 - 3) + (0 - (-1)) + (0 - 2) + (-2 - (-2)) + (1 - 1) = -1$$

HI = +½

Mitsunobu reaction

Mitsunobu, O. *Bull. Chem. Soc. Jpn.* 1967, *40*, 4235.
Mitsunobu, O. *Tetrahedron* 1970, *26*, 5731.

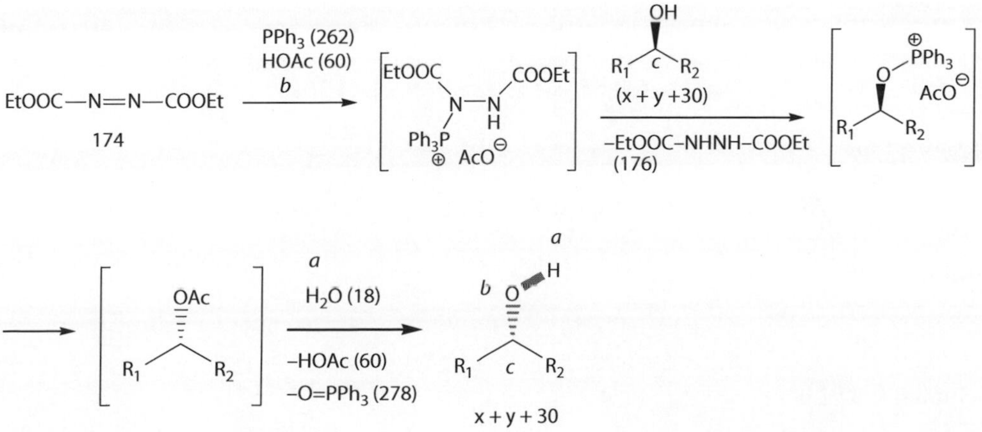

$R_1 = x; R_2 = y$

$$E = 514/[x + y + 30]$$

E(max) = 16.06

$$AE = [x + y + 30]/[x + y + 544]$$

AE(min) = 0.059

$$f(\text{sac}) = 262/[x + y + 544]$$

f(sac)(max) = 0.48

Set $R_1 = R_2 = CH_3$

$$\Delta = \Delta(a) + \Delta(b) + \Delta(c) = (1 - 1) + (-2 - (-2)) + (0 - 0) = 0$$

HI = 0

Murahashi reaction

Tanigawa, Y.; Ohta, H.; Sonoda, A.; Murahashi, S.I. *J. Am. Chem. Soc.* 1978, *100*, 4610.
Tanigawa, Y.; Kanamaru, H.; Sonoda, A.; Murahashi, S.I. *J. Am. Chem. Soc.* 1977, *99*, 2361.

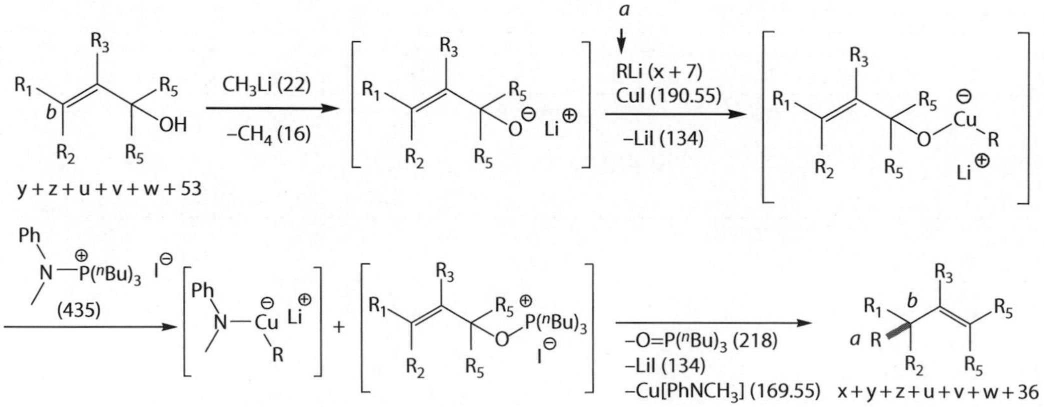

$R = x; R_1 = y; R_2 = z; R_3 = u; R_4 = v; R_5 = w$

$$E = 671.55/[x + y + z + u + v + w + 36]$$

E(max) = 15.99

$$AE = [x + y + z + u + v + w + 36]/[x + y + z + u + v + w + 707.55]$$

AE(min) = 0.059

$$f(\text{sac}) = 647.55/[x + y + z + u + v + w + 707.55]$$

f(sac)(max) = 0.91

Set R = R_1 = R_2 = CH_3

$$\Delta = \Delta(a) + \Delta(b) = (-3 - (-4)) + (0 - 0) = +1$$

HI = −1/2

Nef reaction
Nef, J.U. *Ann. Chem.* 1894, *280*, 263.

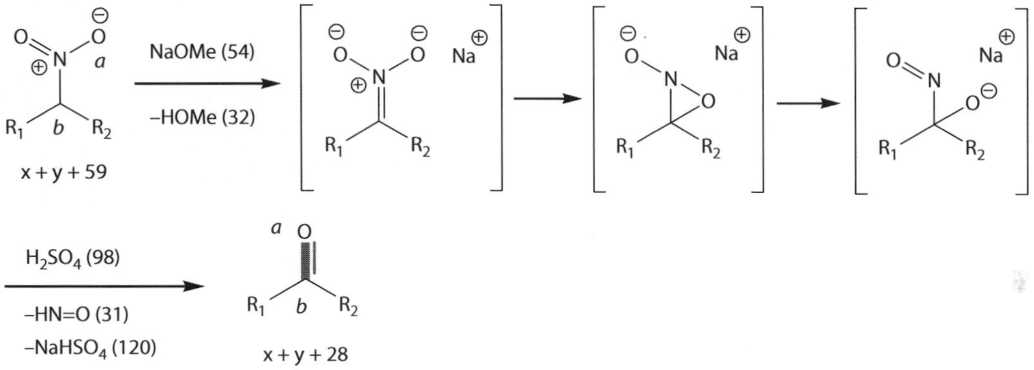

R_1 = x; R_2 = y

$$E = 183/[x + y + 28]$$

***E*(max) = 6.10**

$$AE = [x + y + 28]/[x + y + 211]$$

AE(min) = 0.14

$$f(\text{sac}) = 152/[x + y + 211]$$

***f*(sac)(max) = 0.71**

Set R_1 = R_2 = CH_3

$$\Delta = \Delta(a) + \Delta(b) = (-2 - (-2)) + (2 - 0) = +2$$

HI = −1/2

Nenitzescu–Praill synthesis of substituted pyridines via pyrylium salts
Balaban, A.T.; Nenitzescu, C.D. *Ann. Chem.* 1959, *625*, 74.
Balaban, A.T.; Nenitzescu, C.D. *J. Chem. Soc.* 1961, *4*, 3553.
Balaban, A.T.; Nenitzescu, C.D. *J. Chem. Soc.* 1961, *4*, 3561.
Balaban, A.T.; Nenitzescu, C.D. *J. Chem. Soc.* 1961, *4*, 3564.
Praill, P.F.G.; Whitear, A.L. *J. Chem. Soc.* 1961, 3573.
Praill, P.F.G.; Whitear, A.L. *Proc. Chem. Soc.* 1961, 312.

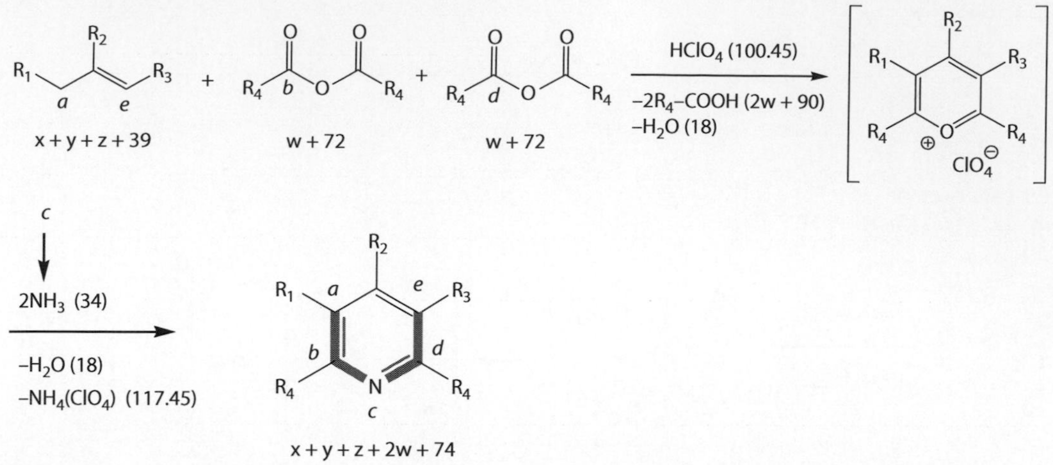

$R_1 = x$; $R_2 = y$; $R_3 = z$; $R_4 = w$

$$E = [2w + 243.45]/[x + y + z + 2w + 74]$$

$$E(max) = [2w + 243.45]/[2w + 77]$$

$$AE = [x + y + z + 2w + 74]/[x + y + z + 2w + 317.45]$$

$$AE(min) = [2w + 74]/[2w + 320.45]$$

$$f(sac) = 117.45/[x + y + z + 2w + 317.45]$$

$$f(sac)(max) = 117.45/[2w + 320.45]$$

R_4	E(max)	AE(min)	f(sac)(max)
CH_3	2.56	0.28	0.34

Set $R_1 = R_2 = CH_3$

$$\Delta = \Delta(a) + \Delta(b) + \Delta(c) + \Delta(d) + \Delta(e) = (0 - (-2)) + (1 - 3) + (-3 - (-3)) + (2 - 3) + (0 - (-1)) = 0$$

HI = 0

Pfitzinger reaction

Pfitzinger, W. *J. Prakt. Chem.* 1886, *33(2)*, 100.
Pfitzinger, W. *J. Prakt. Chem.* 1888, *38*, 582.

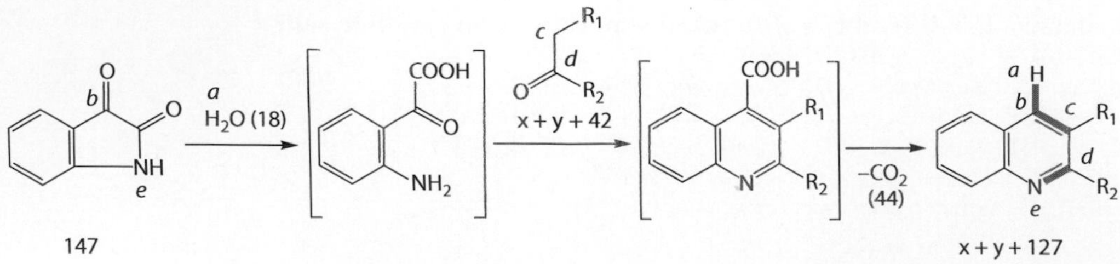

147   x + y + 127

$R_1 = x; R_2 = y$

$$E = 44/[x + y + 127]$$

$E(max) = 0.34$

$$AE = [x + y + 127]/[x + y + 207]$$

$AE(min) = 0.75$

$f(sac) = 0$

Set $R_1 = R_2 = CH_3$

$$\Delta = \Delta(a) + \Delta(b) + \Delta(c) + \Delta(d) + \Delta(e) = (1-1) + (-1-2) + (0-(-2)) + (2-2) + (-3-(-3)) = -1$$

HI = +1/2

Pictet–Spengler isoquinoline synthesis
Pictet, A.; Spengler, T. *Chem. Ber.* 1911, *44*, 2030.

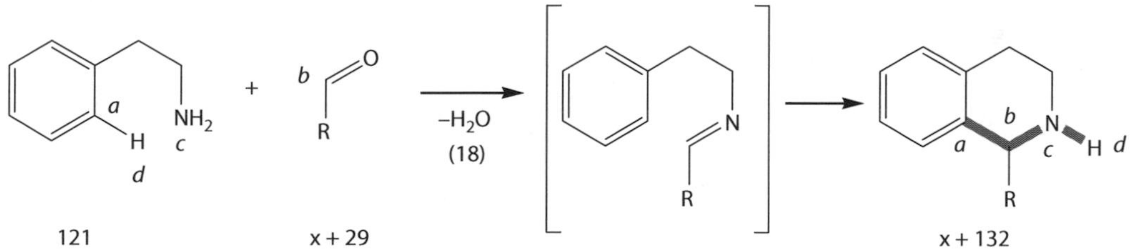

121   x + 29   x + 132

R = x

$$E = 18/[x + 132]$$

$E(max) = 0.14$

$$AE = [x + 132]/[x + 150]$$

$AE(min) = 0.88$

$f(sac) = 0$

Set $R_1 = R_2 = CH_3$

$$\Delta = \Delta(a) + \Delta(b) + \Delta(c) + \Delta(d) = (0-(-1)) + (0-1) + (-3-(-3)) + (1-1) = 0$$

HI = 0

Pinner synthesis
Pinner, A.; Klein, F. *Chem. Ber.* 1877, *10*, 1889.

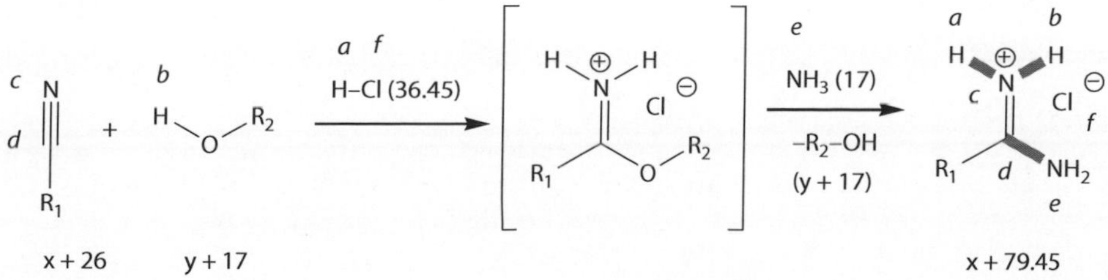

x + 26   y + 17   x + 79.45

$R_1 = x$

$$E = [y + 17]/[x + 79.45]$$

$$\mathbf{\textit{E}(max) = [y + 17]/80.45}$$

$$AE = [x + 79.45]/[x + y + 96.45]$$

AE(min) = 80.45/[y + 97.45]

$f(\text{sac}) = 0$

R_2	E(max)	AE(min)
Me	0.40	0.72
Et	0.57	0.64

Set $R_1 = R_2 = CH_3$

$$\Delta = \Delta(a) + \Delta(b) + \Delta(c) + \Delta(d) + \Delta(e) + \Delta(f)$$

$$= (1 - 1) + (1 - 1) + (-3 - (-3)) + (3 - 3) + (-3 - (-3)) + (-1 - (-1)) = 0$$

HI = 0

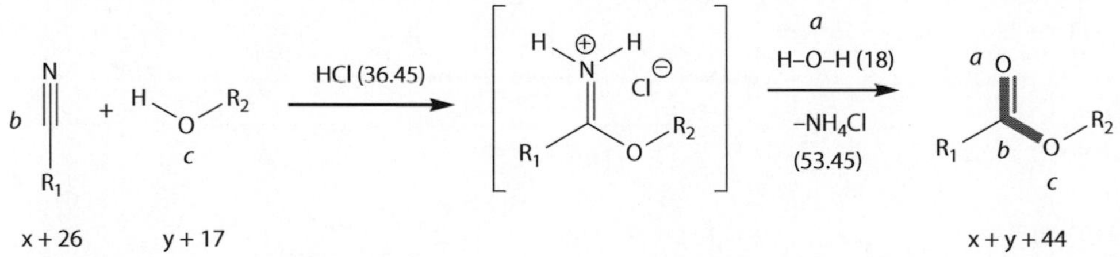

$$x + 26 \qquad y + 17 \qquad\qquad\qquad\qquad\qquad\qquad\qquad\qquad x + y + 44$$

$R_1 = x$; $R_2 = y$

$$E = 53.45/[x + y + 44]$$

E(max) = 1.16

$$AE = [x + y + 44]/[x + y + 97.45]$$

AE(min) = 0.46

$$f(\text{sac}) = 36.45/[x + y + 97.45]$$

f(sac)(max) = 0.37

Set $R_1 = R_2 = CH_3$

$$\Delta = \Delta(a) + \Delta(b) + \Delta(c) = (-2 - (-2)) + (3 - 3) + (-2 - (-2)) = 0$$

HI = 0

Pomeranz–Fritsch reaction
Pomeranz, C. *Monatsch. Chem.* 1893, *14*, 116.
Fritsch, P. *Chem. Ber.* 1893, *26*, 419.

106 133 129

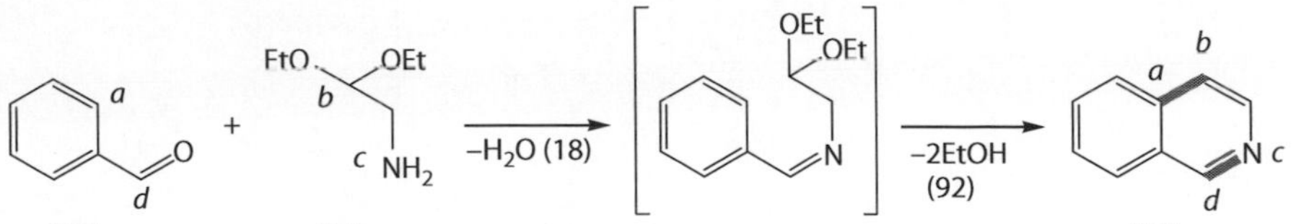

E = 110/129 = 0.85

AE = 0.54

***f*(sac) = 0**

Set $R_1 = R_2 = CH_3$

$$\Delta = \Delta(a) + \Delta(b) + \Delta(c) + \Delta(d) = (0-(-1)) + (-1-1) + (-3-(-3)) + (1-1) = -1$$

HI = +1/2

Schlittler–Mueller modification of Pomeranz–Fritsch reaction
Schlittler, E.; Mueller, J. *Helv. Chim. Acta* 1948, *31*, 914.
Schlittler, E.; Mueller, J. *Helv. Chim. Acta* 1948, *31*, 1119.

x + 106 132 x+128

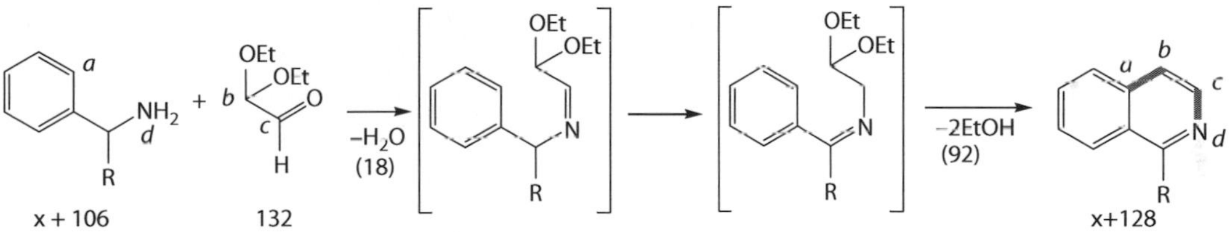

R = x

$$E = 110/[x + 128]$$

***E*(max) = 0.54**

$$AE = [x + 128]/[x + 238]$$

AE(min) = 0.54

***f*(sac) = 0**

Set $R_1 = R_2 = CH_3$

$$\Delta = \Delta(a) + \Delta(b) + \Delta(c) + \Delta(d) = (0-(-1)) + (-1-1) + (0-1) + (-3-(-3)) = -2$$

HI = +1

Reissert reaction
Reissert, A. *Chem. Ber.* 1905, *38*, 1603.
Reissert, A. *Chem. Ber.* 1905, *38*, 3415.

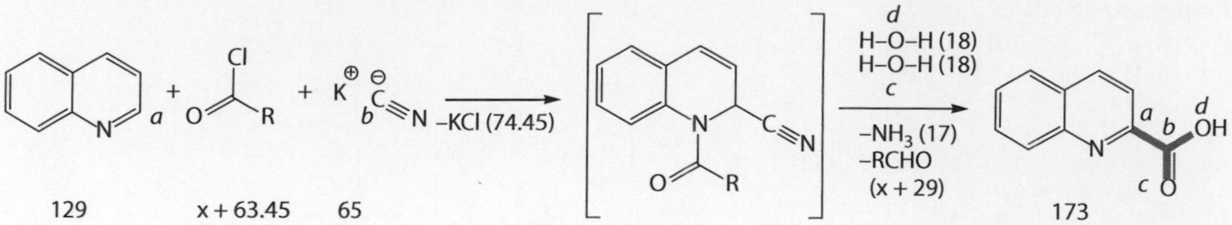

129 x + 63.45 65 173

$$E = [x + 120.45]/173$$

E(max) = 0.78 (R = CH_3)

$$AE = 173/[x + 293.45]$$

AE(min) = 0.56 (R = CH_3)

f(sac) = 0

Set $R_1 = R_2 = CH_3$

$$\Delta = \Delta(a) + \Delta(b) + \Delta(c) + \Delta(d) = (2-1) + (3-2) + (-2-(-2)) + (-2-(-2)) = +2$$

HI = –1

Schlosser modification of Wittig reaction

Schlosser, M.; Christmann, F.K.; Piskala, A.; Coffinet, D. *Synthesis* 1971, 29.
Schlosser, M.; Coffinet, D. *Synthesis* 1971, 380.
Schlosser, M.; Coffinet, D. *Synthesis* 1972, 575.
Schlosser, M.; Christmann, F.K. *Synthesis* 1969, 38.

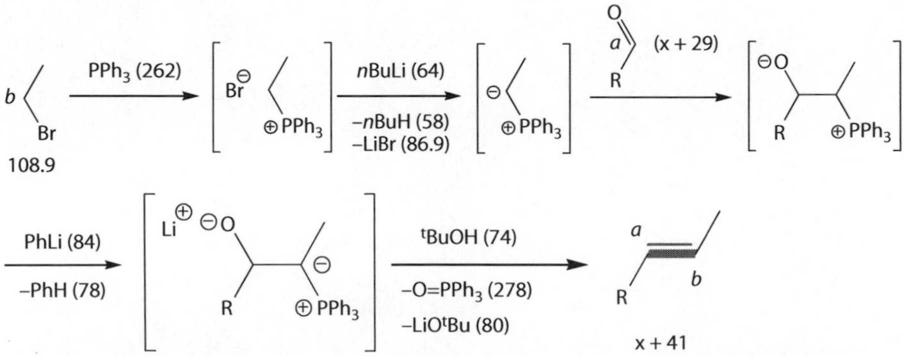

108.9 x + 41

R = x

$$E = 580.9/[x + 41]$$

E(max) = 13.83

$$AE = [x + 41][x + 621.9]$$

AE(min) = 0.067

$$f(\text{sac}) = 484/[x + 621.9]$$

$f(\text{sac})(\text{max}) = 0.78$

Set $R = CH_3$

$$\Delta = \Delta(a) + \Delta(b) = (-1-1) + (-1-(-1)) = -2$$

HI = +1

Stolle synthesis of indoles
Stolle, R. *Chem. Ber.* 1913, *46*, 3915.

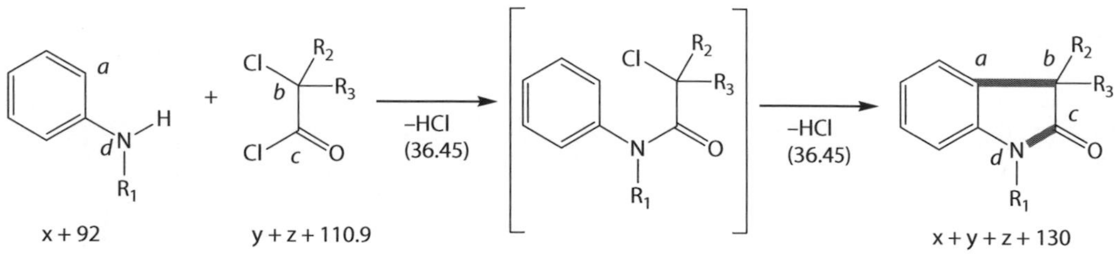

$R_1 = x$; $R_2 = y$; $R_3 = z$

$$E = 72.9/[x + y + z + 130]$$

E(max) = 0.55

$$AE = [x + y + z + 130]/[x + y + z + 202.9]$$

AE(min) = 0.65

f(sac) = 0

Set $R_2 = CH_3$

$$\Delta = \Delta(a) + \Delta(b) + \Delta(c) + \Delta(d) = (0-(-1)) + (0-1) + (3-3) + (-3-(-3)) = 0$$

HI = 0

Stork enamine synthesis
Stork, G.; Terrell, R.; Szmuszkovicz, J. *J. Am. Chem. Soc.* 1954, *76*, 2029.

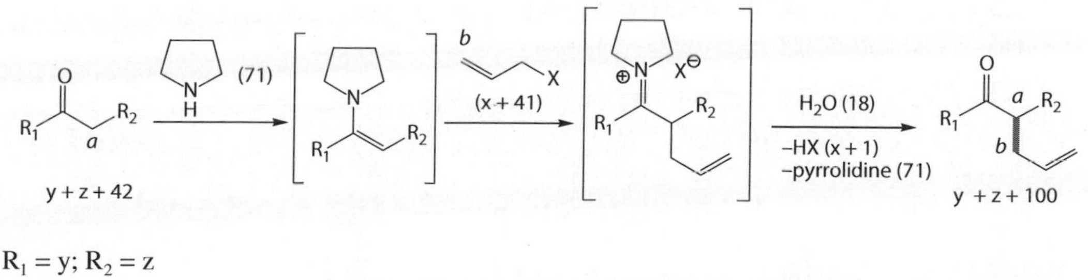

$R_1 = y$; $R_2 = z$

$$E = [x + 72]/[y + z + 100]; \quad E(\text{max}) = [x + 72]/102$$

$$AE = [y + z + 100]/[x + y + z + 172]; \quad AE(\text{min}) = 102/[x + 174]$$

$$f(\text{sac}) = 89/[x + y + z + 172]$$

$f(\text{sac})(\text{max}) = 89/[x + 174]$

X	$E(\text{max})$	$AE(\text{min})$	$f(\text{sac})(\text{max})$
Cl	1.05	0.49	0.42
Br	1.49	0.40	0.35
I	1.95	0.34	0.30

Set $R_2 = CH_3$

$$\Delta = \Delta(a) + \Delta(b) = (-1 - (-2)) + (-2 - (-2)) = +1$$

HI = −1/2

Urech synthesis of hydantoins
Urech, F. *Ann. Chem.* 1873, *165*, 99.

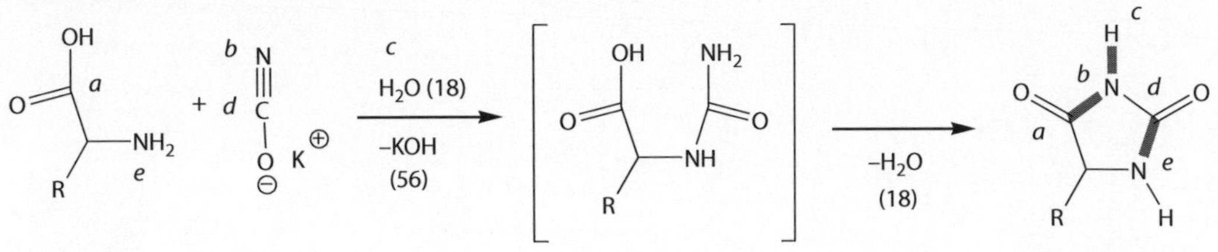

$$x + 74 \qquad\qquad 81 \qquad\qquad\qquad\qquad\qquad\qquad\qquad\qquad\qquad x + 99$$

$R = x$

$E = 74/[x + 99]$

$E(\text{max}) = 0.74$

$$AE = [x + 99]/[x + 173]$$

$AE(\text{min}) = 0.58$

$f(\text{sac}) = 0$

$$\Delta = \Delta(a) + \Delta(b) + \Delta(c) + \Delta(d) + \Delta(e) = (3-3) + (-3-(-3)) + (1-1) + (4-4) + (-3-(-3)) = 0$$

HI = 0

von Richter cinnoline synthesis
von Richter, V. *Chem. Ber.* 1883, *16*, 677.

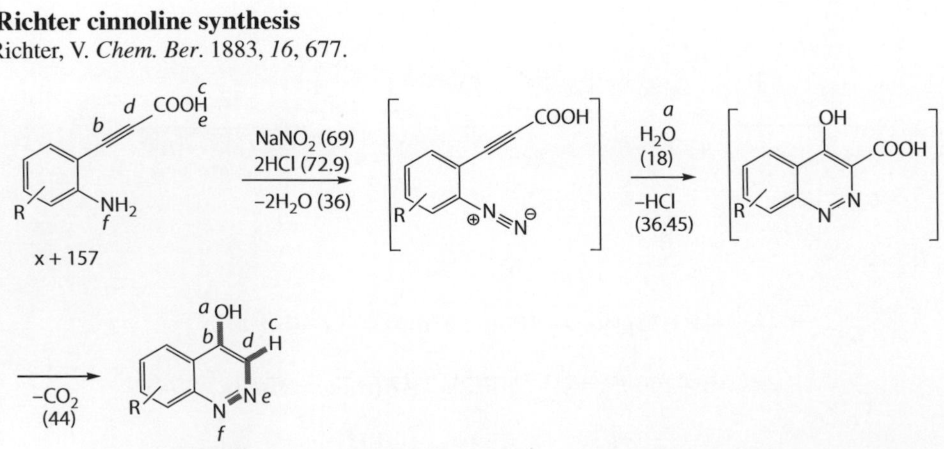

$$x + 157$$

$$x + 142$$

R = x(sum of 4 substituents)

$$E = 116.45/[x + 142]$$

E(max) = 0.80

$$AE = [x + 142]/[x + 316.9]$$

AE(min) = 0.56

$$f(sac) = 72.9/[x + 316.9]$$

f(sac)(max) = 0.23

$$\Delta = \Delta(a) + \Delta(b) + \Delta(c) + \Delta(d) + \Delta(e) + \Delta(f)$$

$$= (-2 - (-2)) + (1 - 0) + (1 - 1) + (0 - 0) + (-1 - 3) + (-1 - (-3)) = -1$$

HI = +1/2

Wittig reaction
Wittig, G.; Schöllkopf, U. *Chem. Ber*. 1954, *87*, 1318.

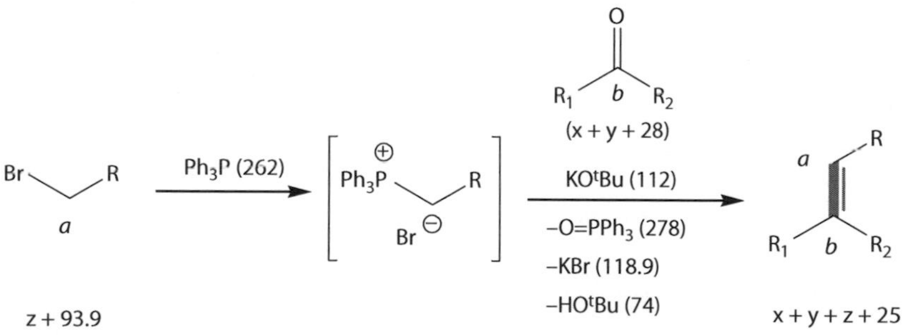

R = z; R_1 = x; R_2 = y

$E = 470.9/[x + y + z + 25]$

E(max) = 16.82

$$AE = [x + y + z + 25]/[x + y + z + 495.9]$$

AE(min) = 0.056

$$f(sac) = 374/[x + y + z + 495.9]$$

f(sac)(max) = 0.75

Set R = R_1 = R_2 = CH_3

$$\Delta = \Delta(a) + \Delta(b) = (-1 - (-1)) + (0 - 2) = -2$$

HI = +1

7.10   CATEGORIZATION OF NAMED SUBSTITUTION REACTIONS BY MINIMUM ATOM ECONOMY AND MAXIMUM ENVIRONMENTAL IMPACT FACTOR, MAXIMUM MOLECULAR WEIGHT FRACTION OF SACRIFICIAL REAGENTS, HYPSICITY INDEX

(Reactions are listed alphabetically.)

Aniline synthesis
Hrutfiord, B.F.; Bunnett, J.F. *J. Am. Chem. Soc.* 1958, *80*, 2021.
Panar, M.; Roberts, J.D. *J. Am. Chem. Soc.* 1960, *82*, 3629.
Bunnett, J.F.; Hrutfriord, B.F. *J. Org. Chem.* 1962, *27*, 4152.
Bunnett, J.F.; Kim, J.K. *J. Am. Chem. Soc.* 1970, *92*, 7464.

$$\text{x} + 107.45 \qquad\qquad 55 \qquad\qquad \text{x} + 88 \qquad\qquad 74.45$$

R = x(sum of substituents)

$$AE = [x+88]/[x+162.45]; \quad \mathbf{AE(min) = 0.56}; \quad E = 74.45/[x+88]; \quad \mathbf{E(max) = 0.80}; \quad \boldsymbol{f}(\mathbf{sac}) = \mathbf{0}$$

$$\Delta = \Delta(a) + \Delta(b) = (-3-(-3)) + (1-1) = 0; \quad \mathbf{HI = 0}$$

Aromatic halogenation
Scheufelen, A. *Ann. Chem.* 1885, *231*, 152.
Schramm, J. *Chem. Ber.* 1885, *18*, 607.
Seelig, E. *Ann. Chem.* 1887, *237*, 129.
Rilliet, A.; Ador, E. *Chem. Ber.* 1875, *8*, 1286.
Mueller, H. *J. Chem. Soc.* 1862, *15*, 41.

$$\text{y} + 73 \qquad\qquad 2\text{x} \qquad\qquad \text{y} + \text{x} + 72 \qquad\qquad \text{x} + 1$$

$$R = y(\text{sum of substituents}); \quad X = x(\text{halogen})$$

$$AE = [x+y+72]/[2x+y+73]; \quad E = [x+1]/[x+y+72]; \quad \boldsymbol{f}(\mathbf{sac}) = \mathbf{0}$$

X	E(max)	AE(min)
Cl	0.32	0.76
Br	0.52	0.66
I	0.63	0.61

$$\Delta = \Delta(a) + \Delta(b) = (-1-0) + (1-(-1)) = +1; \quad \mathbf{HI} = -\Delta/(N+1) = -\tfrac{1}{2}$$

Aromatic nitration
Kopp, H. *Ann. Chem.* 1856, *98*, 367.

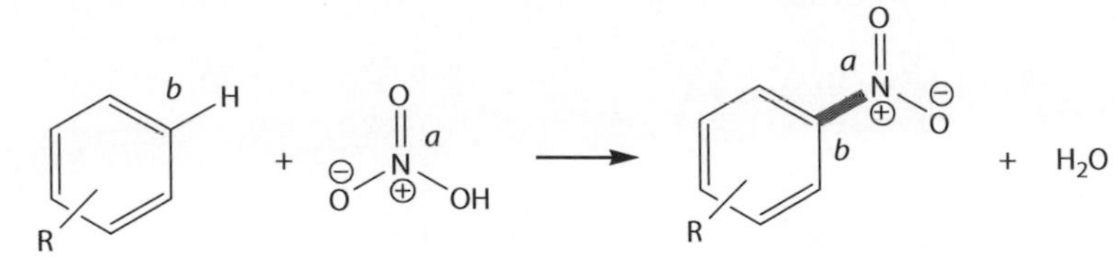

x + 73	63	x + 118	18

R = x(sum of substituents)

$$AE = [x+118]/[x+136]; \quad \mathbf{AE(min) = 0.87}; \quad E = 18/[x+118]; \quad \mathbf{E(max) = 0.15}; \quad \mathbf{\mathit{f}(sac) = 0}$$

$$\Delta = \Delta(a) + \Delta(b) = (3-5) + (1-(-1)) = 0; \quad \mathbf{HI = 0}$$

Aromatic sulphonation
Ador, E.; Meyer, V. *Ann. Chem.* 1871, *159*, 1.
Meyer, V.; Michler, W. *Chem. Ber.* 1875, *8*, 672.
Barth, L.; Senhofer, C. *Chem. Ber.* 1875, *8*, 754.
Ascher, M.; Meyer, V. *Chem. Ber.* 1871, *4*, 323.
Noelting, E. *Chem. Ber.* 1875, *8*, 1091.

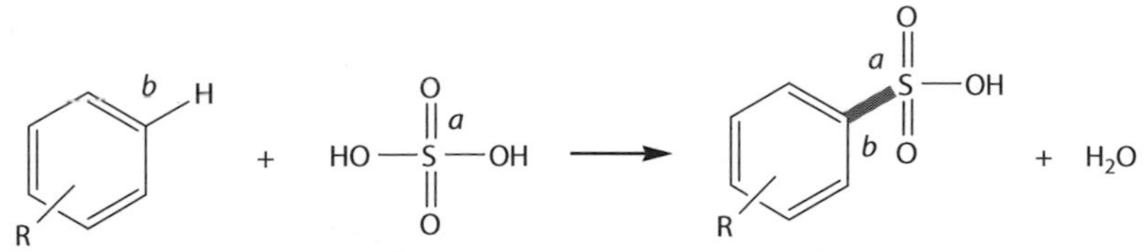

x + 73	98	x + 153	18

X = x(sum of substituents)

$$AE = [x+153]/[x+171]; \quad \mathbf{AE(min) = 0.90}; \quad E = 18/[x+153]; \quad \mathbf{E(max) = 0.11}; \quad \mathbf{\mathit{f}(sac) = 0}$$

$$\Delta = \Delta(a) + \Delta(b) = (4-6) + (1-(-1)) = 0; \quad \mathbf{HI = 0}$$

Bucherer reaction
Bucherer, T. *J. Prakt. Chem.* 1904, *69*, 49.

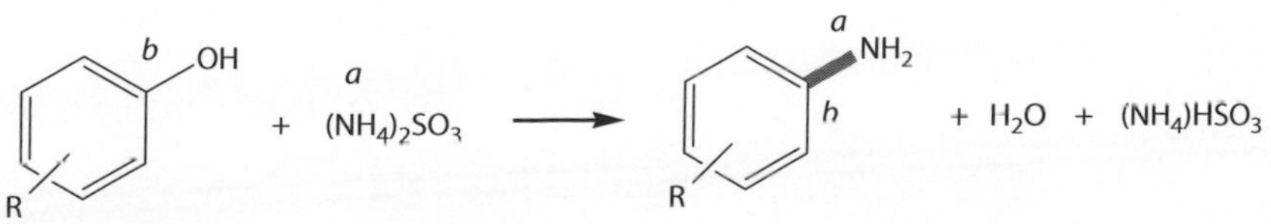

x + 89	116	x + 88	18	99

X = x(sum of substituents)

$$AE = [x+88]/[x+205]; \quad \mathbf{AE(min) = 0.44}; \quad E = 117/[x+88]; \quad \mathbf{E(max) = 1.26}; \quad f(sac) = 0$$

$$\Delta = \Delta(a) + \Delta(b) = (-3-(-3)) + (1-1) = 0; \quad \mathbf{HI = 0}$$

Chichibabin reaction
Chichibabin, A.E.; Zeide, O.A. *J. Russ. Phys. Chem. Soc.* 1914, *46*, 1216.

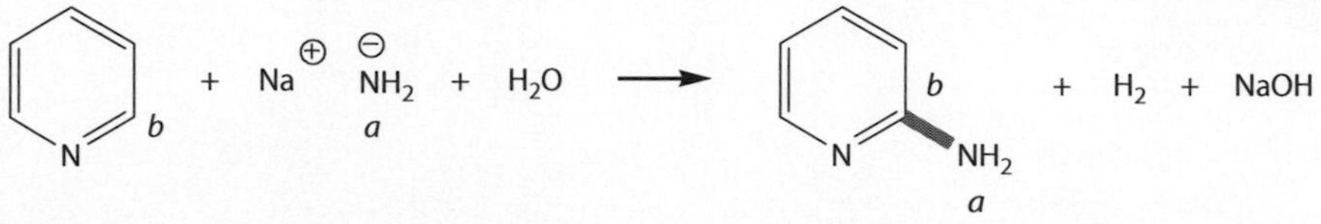

| 79 | 39 | 18 | 94 | 2 | 40 |

AE(min) = 0.69; *E*(max) = 0.45; *f*(sac) = 0.13

$$\Delta = \Delta(a) + \Delta(b) = (-3-(-3)) + (3-1) = +2; \quad \mathbf{HI} = -\Delta/(N+1) = -1$$

Dakin–West reaction
Dakin, H.D.; West, R. *J. Biol. Chem.* 1928, *78*, 91.
Dakin, H.D.; West, R. *J. Biol. Chem.* 1928, *78*, 745.
Dakin, H.D.; West, R. *J. Biol. Chem.* 1928, *78*, 757.

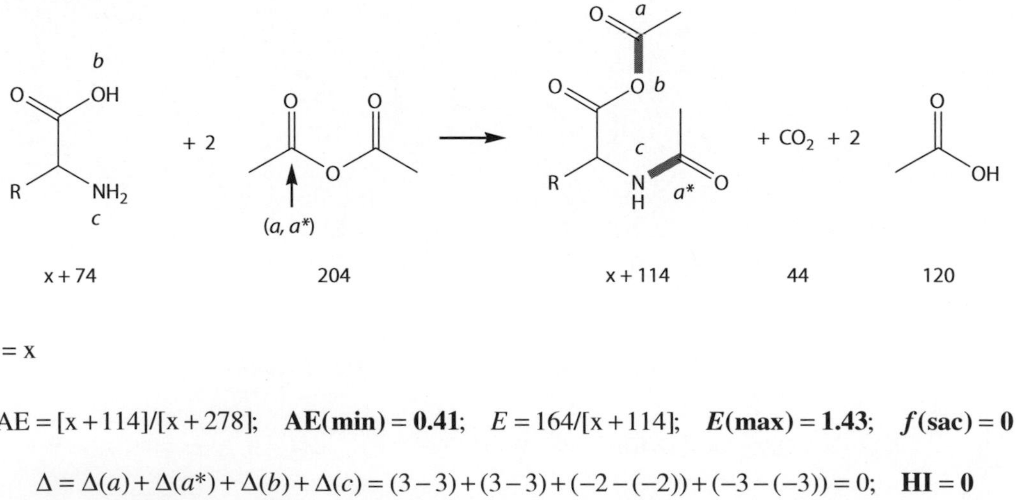

| x + 74 | 204 | x + 114 | 44 | 120 |

R = x

$$AE = [x+114]/[x+278]; \quad \mathbf{AE(min) = 0.41}; \quad E = 164/[x+114]; \quad \mathbf{E(max) = 1.43}; \quad f(sac) = 0$$

$$\Delta = \Delta(a) + \Delta(a^*) + \Delta(b) + \Delta(c) = (3-3) + (3-3) + (-2-(-2)) + (-3-(-3)) = 0; \quad \mathbf{HI = 0}$$

Delepine reaction
Délepine, M. *Bull. Soc. Chim. Fr.* 1895, *13*, 352.

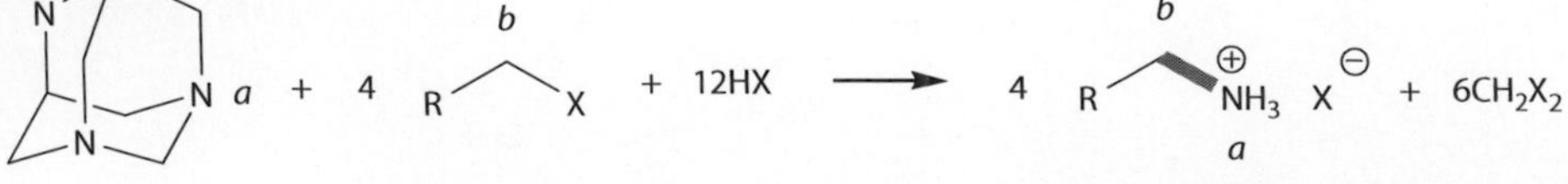

| 140 | 4(x + y + 14) | 12(x + 1) | 4(x + y + 31) | 6(2x + 14) |

X = x(halide); R = y

$$AE = [4(x+y+31)]/[16x+4y+208]; \quad E = [6(2x+14)]/[4(x+y+31)]; \quad f(\text{sac}) = 0$$

X	E(max)	AE(min)
F	1.53	0.40
Cl	1.89	0.35
Br	2.33	0.30
I	2.53	0.28

Set R = CH$_3$

$$\Delta = \Delta(a) + \Delta(b) = (-3-(-3)) + (-2-(-2)) = 0; \quad \textbf{HI} = 0$$

Esterification with diazomethane
Herzig, J.; Pollak, J. *Monatsch. Chem.* 1908, *29*, 263.
Wegschneidler, R.; Gehringer, H. *Monatsch. Chem.* 1909, *29*, 529.
Bouveault, L.; Locquin, R. *Bull. Soc. Chim. Fr.* 1910, *5(4)*, 1136.
De Boer, T.J.; Backer, H.J. *Org. Synth. Coll. Vol. IV* 1963, 250.

R = x

$$AE = [x+59]/[x+87]; \quad \textbf{AE(min)} = \textbf{0.68}; \quad E = 28/[x+59]; \quad \textbf{E(max)} = \textbf{0.47}; \quad f(\text{sac}) = 0$$

$$\Delta = \Delta(a) + \Delta(b) + \Delta(c) = (1-1) + (-2-(-2)) + (-2-(-2)) = 0; \quad \textbf{HI} = 0$$

R = x

$$AE = [x+59]/[x+299]; \quad \textbf{AE(min)} = \textbf{0.20}; \quad E = 240/[x+59]; \quad \textbf{E(max)} = \textbf{4}; \quad f(\text{sac}) = 0$$

$$\Delta = \Delta(a) + \Delta(b) + \Delta(c) = (1-1) + (-2-(-2)) + (-2-(-2)) = 0; \quad \textbf{HI} = 0$$

Finkelstein reaction
Perkin, W.H.; Duppa, B.F. *Ann. Chem.* 1859, *112*, 125.
Finkelstein, H. *Chem. Ber.* 1910, *43*, 1528.

$$AE = [y+r]/[x+y+r+23]; \quad E = [x+23]/[y+r]; \quad f(\text{sac}) = 0$$

X	Y	E(max)	AE(min)
Cl	Cl	1.60	0.38 (degenerate case)
Cl	Br	0.72	0.58
Cl	I	0.46	0.69
Br	Cl	2.82	0.26
Br	Br	1.27	0.44 (degenerate case)
Br	I	0.80	0.55
I	Cl	4.12	0.20
I	Br	1.85	0.35
I	I	1.17	0.46 (degenerate case)

Set $R = CH_3$

$$\Delta = \Delta(a) + \Delta(b) = (-1 - (-1)) + (-2 - (-2)) = 0; \quad \textbf{HI} = \textbf{0}$$

Fischer esterification
Fischer, E.; Speier, A. *Chem. Ber.* 1895, *28*, 3252.

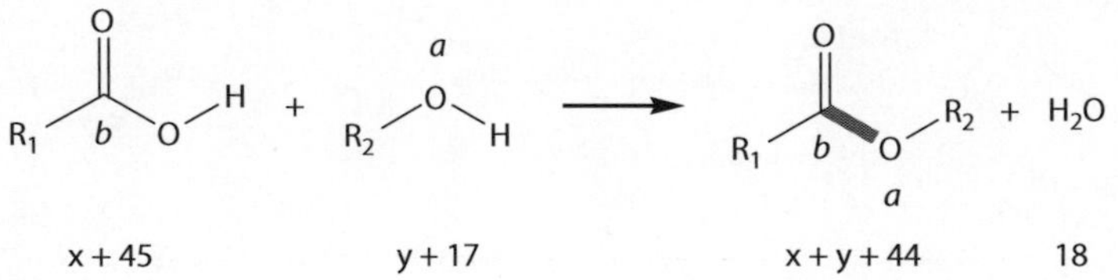

$$x + 45 \qquad\qquad y + 17 \qquad\qquad x + y + 44 \qquad 18$$

$R_1 = x; R_2 = y$

$$AE = [x + y + 44]/[x + y + 62]; \quad \textbf{AE(min)} = \textbf{0.72};$$

$$E = 18/[x + y + 44]; \quad \textbf{E(max)} = \textbf{0.39}; \quad f(\text{sac}) = 0$$

$$\Delta = \Delta(a) + \Delta(b) = (-2 - (-2)) + (3 - 3) = 0; \quad \textbf{HI} = \textbf{0}$$

Gabriel synthesis
Gabriel, S. *Chem. Ber.* 1887, *20*, 2224.

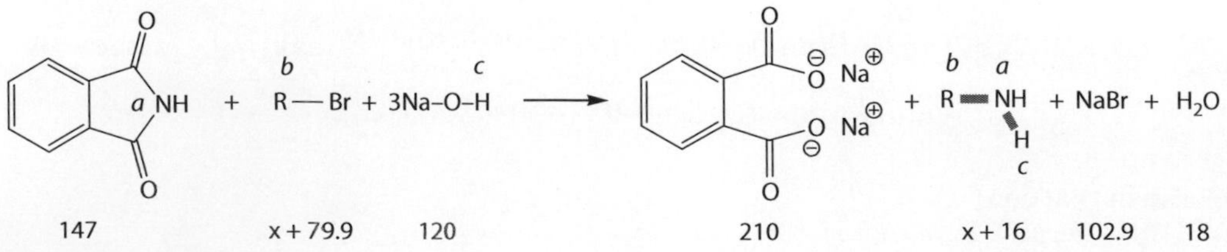

$$147 \qquad\qquad x + 79.9 \quad 120 \qquad\qquad\qquad 210 \qquad\qquad x + 16 \quad 102.9 \quad 18$$

$R = x$

$$AE = [x + 16]/[x + 346.9]; \quad \textbf{AE(min)} = \textbf{0.049}; \quad E = 330.9/[x + 16]; \quad \textbf{E(max)} = \textbf{19.46};$$

$$f(\text{sac}) = 80/[x + 346.9]; \quad f(\textbf{sac})(\textbf{max}) = \textbf{0.23}$$

Set R = CH$_3$

$$\Delta = \Delta(a) + \Delta(b) + \Delta(c) = (-3-(-3)) + (-2-(-2)) + (1-1) = 0; \quad \textbf{HI} = \textbf{0}$$

Gattermann reaction

Gattermann, L. *Chem. Ber.* 1898, *31*, 1149.

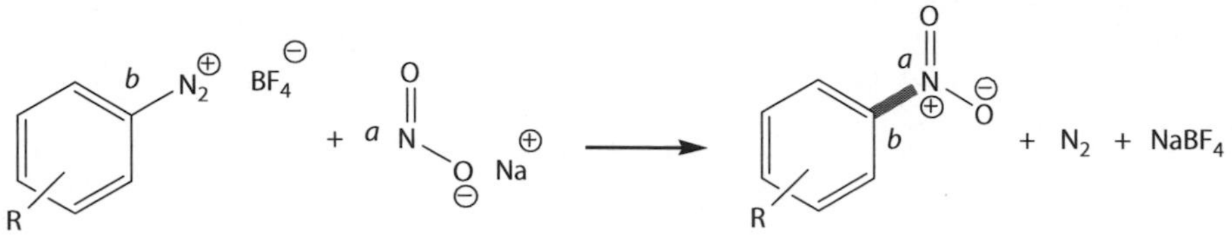

x + 187   69   x + 118   28   110

R = x(sum of substituents)

$AE = [x+118]/[x+256]; \quad \textbf{AE(min)} = \textbf{0.47}; \quad E = 138/[x+118]; \quad \textbf{E(max)} = \textbf{1.12}; \quad f(\textbf{sac}) = \textbf{0}$

$$\Delta = \Delta(a) + \Delta(b) = (3-3) + (1-1) = 0; \quad \textbf{HI} = \textbf{0}$$

Helferich method

Helferich, B.; Schmitz-Hillebrecht, E. *Chem. Ber.* 1933, *66*, 378.

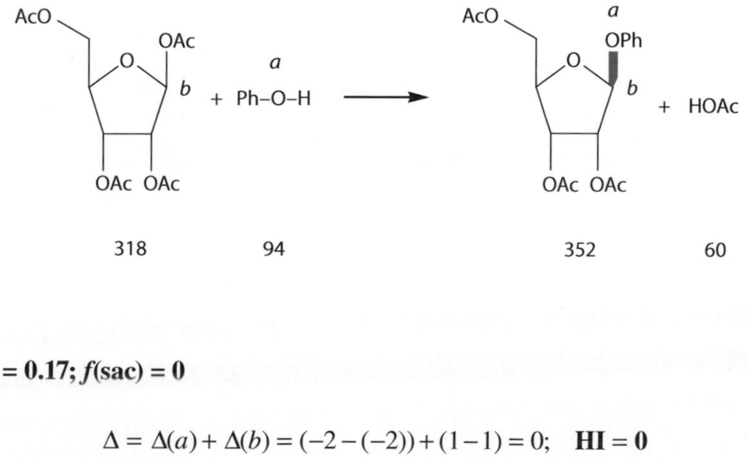

318   94   352   60

5C sugars

$\textbf{AE} = \textbf{0.85}; E = \textbf{0.17}; f(\textbf{sac}) = \textbf{0}$

$$\Delta = \Delta(a) + \Delta(b) = (-2-(-2)) + (1-1) = 0; \quad \textbf{HI} = \textbf{0}$$

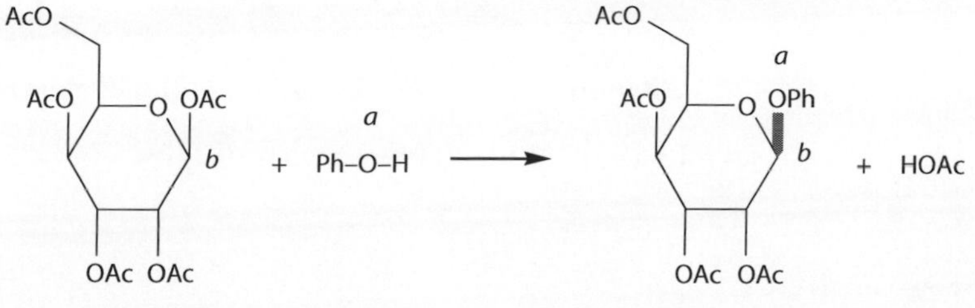

390   94   424   60

6C sugars

$AE = 0.90$; $E = 0.11$; $f(sac) = 0$

$$\Delta = \Delta(a) + \Delta(b) = (-2 - (-2)) + (1 - 1) = 0; \quad HI = 0$$

Hell–Volhard–Zelinsky reaction

Hell, C. *Chem. Be*r. 1881, *14*, 891.
Volhard, J. *Ann. Chem.* 1887, *242*, 141.
Zelinsky, N. *Chem. Ber.* 1887, *20*, 2026.

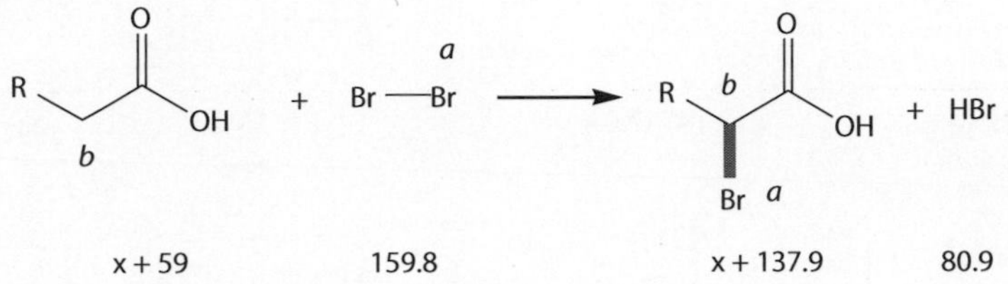

$$\begin{array}{cccc} x + 59 & 159.8 & x + 137.9 & 80.9 \end{array}$$

$R = x$

$$AE = [x + 137.9]/[x + 218.8]; \quad AE(min) = 0.63;$$

$$E = 80.9/[x + 137.9]; \quad E(max) = 0.58; \quad f(sac) = 0$$

Set $R = CH_3$

$$\Delta = \Delta(a) + \Delta(b) = (-1 - 0) + (0 - (-2)) = +1; \quad HI = -\Delta/(N + 1) = -\tfrac{1}{2}$$

Koenigs–Knorr synthesis

Koenigs, W.; Knorr, E. *Chem. Ber.* 1901, *34*, 957.

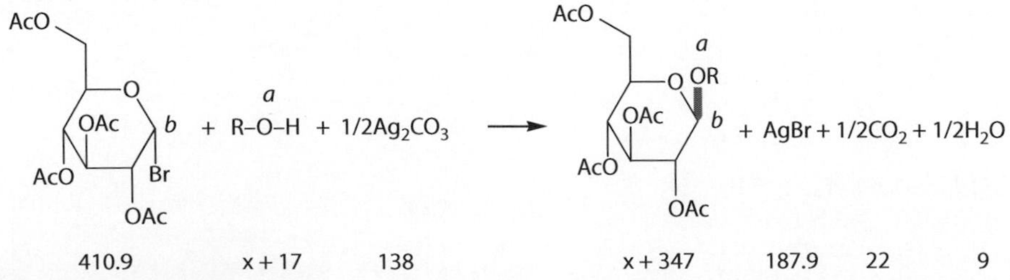

$$\begin{array}{cccccc} 410.9 & x + 17 & 138 & x + 347 & 187.9 & 22 & 9 \end{array}$$

$R = x$

$$AE = [x + 347]/[x + 565.9]; \quad AE(min) = 0.61; \quad E = 218.9/[x + 347]; \quad E(max) = 0.62;$$

$$f(sac) = 138/[x + 565.9]; \quad f(sac)(max) = 0.24$$

Set $R = CH_3$

$$\Delta = \Delta(a) + \Delta(b) = (-2 - (-2)) + (1 - 1) = 0; \quad HI = 0$$

Lapworth reaction

Lapworth, A. *J. Chem. Soc.* 1903, *83*, 995.
Lapworth, A. *J. Chem. Soc.* 1904, *85*, 1206.

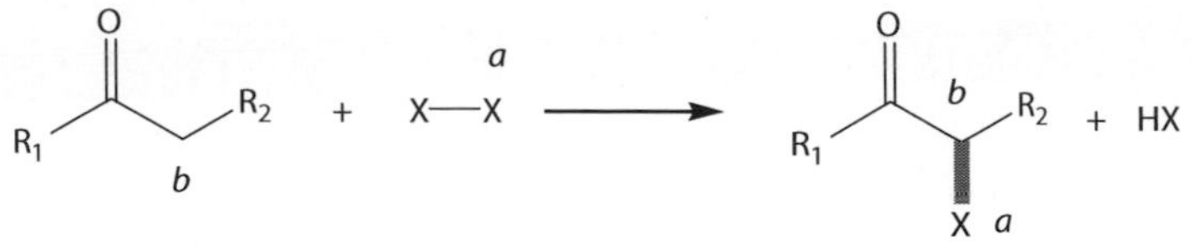

y + z + 42   2x   x + y + z + 41   x + 1

R_1 = y; R_2 = z; X = x(halide)

$$AE = [x+y+z+41]/[2x+y+z+42]; \quad E = [x+1]/[x+y+z+41]; \quad \boldsymbol{f(\text{sac}) = 0}$$

$$\mathbf{AE(min) = [x+43]/[2x+42]; \quad \mathit{E}(max) = [x+1]/[x+43]}$$

X	*E*(max)	AE(min)
Cl	0.46	0.68
Br	0.66	0.60
I	0.75	0.57

Set $R_2 = CH_3$

$$\Delta = \Delta(a) + \Delta(b) = (-1-0) + (0-(-2)) = +1; \quad \mathbf{HI = -\Delta/(\mathit{N}+1) = -1/2}$$

Sandmeyer reaction

Sandmeyer, T. *Chem. Ber.* 1884, *17*, 1633.
Sandmeyer, T. *Chem. Ber.* 1884, *17*, 2650.

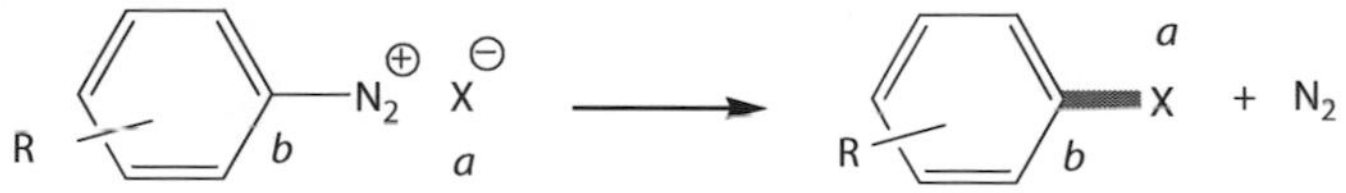

x + y + 100   x + y + 72   28

R = y(sum of substituents); X = x(halide)

$$AE = [x+y+72]/[x+y+100]; \quad E = 28/[x+y+72]; \quad \boldsymbol{f(\text{sac}) = 0}$$

X	*E*(max)	AE(min)
Cl	0.25	0.80
Br	0.18	0.85
I	0.14	0.88

$$\Delta = \Delta(a) + \Delta(b) = (-1-(-1)) + (1-1) = 0; \quad \mathbf{HI = 0}$$

Sanger reaction

Sanger, F. *Biochem. J.* 1945, *39*, 507.

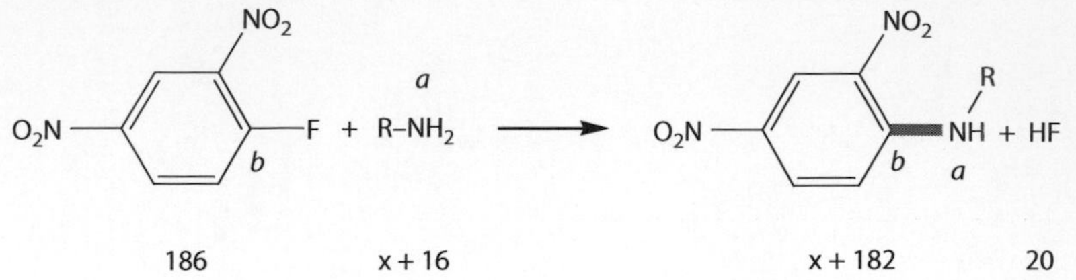

$$R = x$$

$$AE = [x + 182]/[x + 202]; \quad \textbf{AE(min)} = \textbf{0.90}; \quad E = 20/[x + 182]; \quad \textbf{E(max)} = \textbf{0.11}; \quad \textbf{\textit{f}(sac)} = \textbf{0}$$

Set R = CH$_3$

$$\Delta = \Delta(a) + \Delta(b) = (-3 - (-3)) + (1 - 1) = 0; \quad \textbf{HI} = \textbf{0}$$

Schiemann reaction

Balz, G.; Schiemann, G. *Chem. Ber.* 1927, *60*, 1186.

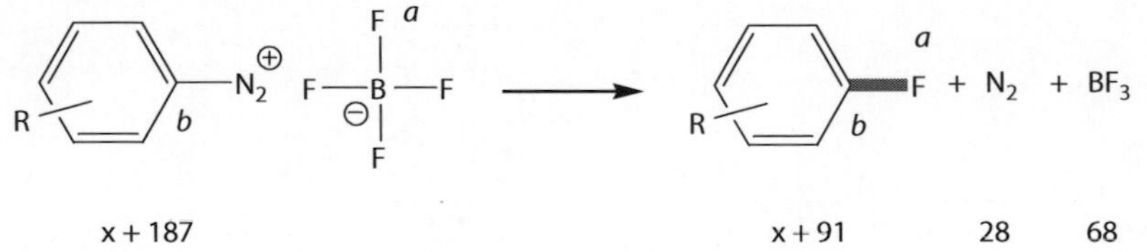

$$R = x(\text{sum of substituents})$$

$$AE = [x + 91]/[x + 187]; \quad \textbf{AE(min)} = \textbf{0.50}; \quad E = 96/[x + 91]; \quad \textbf{E(max)} = \textbf{1}; \quad \textbf{\textit{f}(sac)} = \textbf{0}$$

$$\Delta = \Delta(a) + \Delta(b) = (-1 - (-1)) + (1 - 1) = 0; \quad \textbf{HI} = \textbf{0}$$

Schotten–Baumann reaction

Schotten, C. *Chem. Ber.* 1884, *17*, 2544.
Baumann, E. *Chem. Ber.* 1886, *19*, 3218.

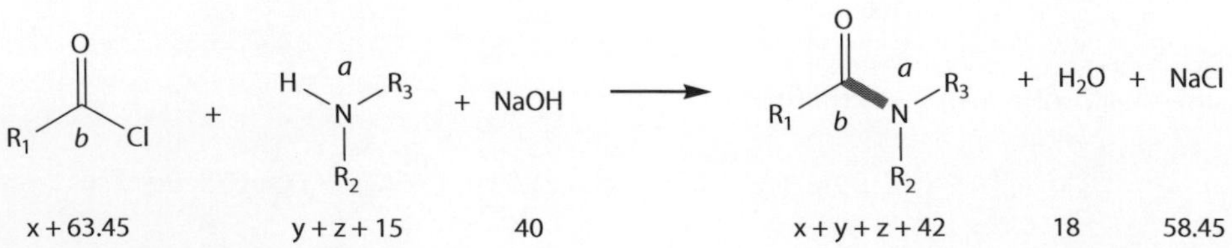

$$R_1 = x; R_2 = y; R_3 = z$$

$$AE = [x + y + z + 42]/[x + y + z + 118.45]; \quad \textbf{AE(min)} = \textbf{0.37};$$

$$E = 76.45/[x + y + z + 42]; \quad \textbf{E(max)} = \textbf{1.70};$$

$$f(\text{sac}) = 40/[x + y + z + 118.45]; \quad \textbf{\textit{f}(sac)(max)} = \textbf{0.33}$$

Set $R_2 = R_3 = CH_3$

$$\Delta = \Delta(a) + \Delta(b) = (-3 - (-3)) + (3 - 3) = 0; \quad \mathbf{HI = 0}$$

Synthesis of acetates via oxymercuration

Brook, A.G.; Wright, G.F. *Can. J. Res.* 1950, *28B*, 623.
Wright, G.F. *Chemistry in Canada* 1950, *2(9)*, 29.
Wright, G.F. *Ann. N.Y. Acad. Sci.* 1957, *65*, 436.
Abercrombie, M.J.; Rodgman, A.; Bharucha, K.R.; Wright, G.F. *Can. J. Chem.* 1959, *37*, 1328.

$R_1 = x; R_2 = y$

$$AE = [x + y + 72]/[x + y + 333]; \quad \mathbf{AE(min) = 0.22};$$

$$E = 261/[x + y + 72]; \quad \mathbf{E(max) = 3.53}; \quad f(\mathbf{sac}) = 0$$

Set $R_1 = R_2 = CH_3$

$$\Delta = \Delta(a) + \Delta(b) = (-2 - (-2)) + (0 - (-2)) = +2; \quad \mathbf{HI} = -\Delta/(N+1) = -\tfrac{1}{2}$$

Synthesis of carbonate diesters, carbamates (urethanes), and ureas

Dumas, J.; Peligot, E. *Ann. Chem.* 1835, *15*, 1.
Röse, B. *Ann. Chem.* 1880, *205*, 227.
Hofmann, A.W. *Ann. Chem.* 1849, *70*, 127.
Vorlaender, D. *Ann. Chem.* 1894, *280*, 167.
Kling, A.; Schumtz, R. *Compt. Rend.* 1919, *168*, 773.
Fosse, R.; DeGraeve, P.; Thomas, P.E. *Compt. Rend.* 1936, *202*, 1544.
Tryon, S.; Benedict, W.S. US 2362865 (General Chemical Co., 1944).
Gehauf, B.; Faber, E.M. US 2806062 (1957).
Lee, J.M. US 2837555 (Dow Chemical, 1958).
Altner, W.; Meisert, E.; Rockstroh, G. DE 1117598 (Bayer, 1961).
Bottenbruch, L.; Schnell, H. DE 1101386 (Bayer, 1961).
Rozsa, L.; Meszaros, L.; Mogyorodi, F. GB 1379977 (1975).
Every, R.L. US 3937728 (Continental Oil Co., 1976).
Schulte-Huermann, W.; Schellmann, E. DE 2847484 (BASF, 1980).

$R = x$

$$AE = [2x + 60]/[2x + 132.9]; \quad \mathbf{AE(min) = 0.46}; \quad E = 72.9/[2x + 60]; \quad \mathbf{E(max) = 1.18}; \quad f(\mathbf{sac}) = 0$$

$$\Delta = \Delta(a) + \Delta(a^*) + \Delta(b) = 2(-2 - (-2)) + (4 - 4) = 0; \quad \mathbf{HI = 0}$$

$$\text{[reaction scheme: acid chloride (98.9) + } R_1\text{-O-H } (x+17) \text{ + amine } (2y+15) \longrightarrow \text{ product } (x+2y+58) + 2HCl \ (72.9)]$$

$R_1 = x;\ R_2 = y$

$$AE = [x + 2y + 58]/[2x + 130.9]; \quad \textbf{AE(min)} = \textbf{0.46};$$

$$E = 72.9/[x + 2y + 58]; \quad \textbf{E(max)} = \textbf{1.20}; \quad \textbf{\textit{f}(sac)} = \textbf{0}$$

Set $R_2 = CH_3$

$$\Delta = \Delta(a) + \Delta(b) + \Delta(c) = (-2 - (-2)) + (4 - 4) + (-3 - (-3)) = 0; \quad \textbf{HI} = \textbf{0}$$

$$\text{[reaction scheme: acid chloride (98.9) + 2 amine } (2(x+15)) \longrightarrow \text{ product } (4x+56) + 2HCl \ (72.9)]$$

$R = x$

$$AE = [4x + 56]/[4x + 128.9]; \quad \textbf{AE(min)} = \textbf{0.45}; \quad E = 72.9/[4x + 56]; \quad \textbf{E(max)} = \textbf{1.22}; \quad \textbf{\textit{f}(sac)} = \textbf{0}$$

Set $R = CH_3$

$$\Delta = \Delta(a) + \Delta(a^*) + \Delta(b) = 2(-3 - (-3)) + (4 - 4) = 0; \quad \textbf{HI} = \textbf{0}$$

Wohl–Ziegler bromination
Wohl, A. *Chem. Ber.* 1919, *52*, 51.
Ziegler, K.; Spath, E.; Schaaf, E.; Schumann, W.; Winkelmann, E. *Ann. Chem.* 1942, *551*, 80.

$$\text{[reaction scheme: N-bromosuccinimide (177.9) + R—H } (x+1) \longrightarrow \text{ succinimide } (99) + R\text{—Br } (x+79.9)]$$

$R = x$

$$AE = [x + 79.9]/[x + 178.9]; \quad \textbf{AE(min)} = \textbf{0.45}; \quad E = 99/[x + 79.9]; \quad \textbf{E(max)} = \textbf{1.22}; \quad \textbf{\textit{f}(sac)} = \textbf{0}$$

Set $R = CH_3$

$$\Delta = \Delta(a) + \Delta(b) = (-1 - (-1)) + (-2 - (-4)) = +2; \quad \mathbf{HI} = -\Delta/(N+1) = -1$$

Zeisel determination
Zeisel, S. *Monatsch. Chem.* 1885, *6*, 989.
Zeisel, S. *Monatsch. Chem.* 1886, *7*, 406.
Zeisel, S.; Fanto, R. *Z. Anal. Chem.* 1903, *42*, 549.

$$R = x(\text{sum of substituents})$$

$$AE = [x + 89]/[x + 231]; \quad \mathbf{AE(min)} = \mathbf{0.40}; \quad E = 142/[x + 89]; \quad \mathbf{E(max)} = \mathbf{1.51}; \quad f(\mathbf{sac}) = \mathbf{0}$$

$$\Delta = \Delta(a) + \Delta(b) = (1 - 1) + (-2 - (-2)) = 0; \quad \mathbf{HI} = \mathbf{0}$$

7.11 SUMMARY OF TRENDS IN NAMED ORGANIC REACTION DATABASE

In this section, Tables 7.1 through 7.10 summarize the AE(min), E(max), and f(sac)(max) parameters calculated for each generalized reaction listed in Sections 7.1 through 7.10. In addition, the difference in oxidation numbers between atoms involved in forming target bonds in the product structure (Δ), the HI, and the hypsicity class are given for each reaction. Since each transformation is a single reaction, $N = 1$ and HI $= -\Delta/2$. Reactions that are isohypsic (HI = 0) do not involve net changes in oxidation numbers of the atoms involved in forming target bonds. Those that are hyperhypsic (HI > 0) imply that the transformations involve net reductions and those that are hypohypsic (HI < 0) imply that the transformations involve net oxidations. What is interesting is that this analysis shows that hyperhypsic and hypohypsic transformations are not restricted to reactions formally classified as oxidations and reductions. The Grignard reaction is hyperhypsic, for example, on account that formation of a Grignard reagent involves formally reducing an electrophilic carbon atom bonded to a halide group by metallation with magnesium, thereby changing it to a nucleophilic center. Formal oxidation and reduction reactions with respect to the substrate of interest are subdivided into additive and subtractive classes. Additive redox reactions are those that involve the addition of atoms from the oxidant or reductant to the substrate. These are good from an atom economical point of view and because the oxidants or reductants employed are not used as sacrificial reagents. An example of an additive reduction is a catalytic hydrogenation reaction of a C=C bond and of an additive oxidation is a dihydroxylation reaction of a C=C bond. Subtractive redox reactions involve the removal of atoms from the substrate. In this case, the oxidants or reductants act as sacrificial reagents and so this class of redox reactions are characterized by low atom economies and high f(sac)(max) values. An example of a subtractive reduction is the Corey–Winter reaction and of a subtractive oxidation is the Swern oxidation. Each table is accompanied by a plot of f(sac)(max) versus AE(min) and one of E(max) versus AE(min). These are given in Figures 7.1 through 7.10.

TABLE 7.1

Summary of Parameters for Named Carbon–Carbon Bond Forming Reactions

Name	AE(min)	E(max)	f(sac)(max)	Δ	HI $= -\Delta/(n + 1)$	Hypsicity Class
Acetoacetic ester synthesis ($X = Cl$, $R_1 = CH_3$)	0.340	1.939	0.106	−1	0.5	Hyperhypsic
Acetoacetic ester synthesis ($X = Br$, $R_1 = CH_3$)	0.270	2.705	0.084	−1	0.5	Hyperhypsic
Acetoacetic ester synthesis ($X = I$, $R_1 = CH_3$)	0.221	3.517	0.069	−1	0.5	Hyperhypsic
Acetoacetic ester synthesis ($X = Cl$, $R_1 = CH_2CH_3$)	0.314	2.180	0.098	−1	0.5	Hyperhypsic
Acetoacetic ester synthesis ($X = Br$, $R_1 = CH_2CH_3$)	0.253	2.947	0.079	−1	0.5	Hyperhypsic
Acetoacetic ester synthesis ($X = I$, $R_1 = CH_2CH_3$)	0.210	3.759	0.065	−1	0.5	Hyperhypsic
Alkyne-carbamate-phosphine annulation	0.466	1.144	0.503	−2	1	Hyperhypsic
Baylis–Hillman	1.000	0	0	0	0	Isohypsic
Bergmann cyclization	1.000	0	0	0	0	Isohypsic
Blanc (product ring size = 5)	0.339	1.952	0.411	−1	0.5	Hyperhypsic
Blanc (product ring size = 6)	0.374	1.673	0.389	−1	0.5	Hyperhypsic
Blanc (product ring size = 7)	0.406	1.464	0.370	−1	0.5	Hyperhypsic
Cadiot–Chodkiewitz	0.192	4.217	0.498	0	0	Isohypsic
Ciamician synthesis of pyridines from pyrroles	0.446	1.242	0.267	0	0	Isohypsic
Cyclopropanation of olefins with diazomethane	0.600	0.667	0	0	0	Isohypsic
Cyclopropanation of olefins with diazald	0.149	5.714	0.142	0	0	Isohypsic
Danheiser alkyne-cyclobutanone cyclization	1.000	0	0	0	0	Isohypsic
Danheiser [4 + 4] annulation	1.000	0	0	0	0	Isohypsic
Danishefsky	1.000	0	0	0	0	Isohypsic
Diels–Alder	1.000	0	0	0	0	Isohypsic
Doetz	0.917	0.090	0	2	−1	Hypohypsic
Eglinton	0.735	0.360	0.235	2	−1	Hypohypsic
[1,5]-ene	1.000	0	0	−1	0.5	Hyperhypsic
Fischer indole synthesis	0.873	0.145	0	2	−1	Hypohypsic
Friedel–Crafts acylation (not including $AlCl_3$)	0.744	0.344	0	0	0	Isohypsic
Friedel–Crafts acylation (including $AlCl_3$)	0.321	2.111	0.568	0	0	Isohypsic
Friedel–Crafts alkylation (not including $AlCl_3$)	0.682	0.467	0	0	0	Isohypsic
Friedel–Crafts alkylation (including $AlCl_3$)	0.258	2.869	0.621	0	0	Isohypsic
Friedlander synthesis	0.782	0.279	0	0	0	Isohypsic
Gattermann–Koch	1.000	0	0	0	0	Isohypsic
Glaser coupling	0.156	5.416	0.838	2	−1	Hypohypsic
Gomberg–Bachmann	0.596	0.678	0.155	0	0	Isohypsic

TABLE 7.1 (continued)
Summary of Parameters for Named Carbon–Carbon Bond Forming Reactions

Name	AE(min)	E(max)	f(sac)(max)	Δ	HI = −Δ/(n + 1)	Hypsicity Class
Grignard (X = Cl, ketone substrate)	0.375	1.668	0.198	−2	1	Hyperhypsic
Grignard (X = Br, ketone substrate)	0.275	2.635	0.145	−2	1	Hyperhypsic
Grignard (X = I, ketone substrate)	0.215	3.659	0.113	−2	1	Hyperhypsic
Grignard (X = Cl, ester substrate)	0.244	3.092	0.198	−4	2	Hyperhypsic
Grignard (X = Br, ester substrate)	0.179	4.573	0.145	−4	2	Hyperhypsic
Grignard (X = I, ester substrate)	0.140	6.143	0.113	−4	2	Hyperhypsic
Hammick	0.711	0.407	0	−1	0.5	Hyperhypsic
Heck coupling (X = Cl)	0.168	4.945	0.607	0	0	Isohypsic
Heck coupling (X = Br)	0.133	6.496	0.481	0	0	Isohypsic
Heck coupling (X = I)	0.109	8.179	0.393	0	0	Isohypsic
Henry (weak acidic conditions)	0.609	0.642	0.239	0	0	Isohypsic
Henry (strong acidic conditions)	0.488	1.047	0.239	0	0	Isohypsic
Hiyama cross coupling (R = CH₃, X = Cl)	0.005	191.225	0.695	0	0	Isohypsic
Hiyama cross coupling (R = CH₃, X = Br)	0.005	213.450	0.622	0	0	Isohypsic
Hiyama cross coupling (R = CH₃, X = I)	0.004	237.000	0.559	0	0	Isohypsic
Hosomi–Sakurai	0.186	4.385	0.489	−1	0.5	Hyperhypsic
Houben–Hoesch	0.874	0.144	0	0	0	Isohypsic
Jacobs oxidative coupling	0.880	0.136	0.107	2	−1	Hypohypsic
Kiliani–Fischer	0.817	0.224	0	0	0	Isohypsic
Koch–Haaf carbonylation	1.000	0	0	0	0	Isohypsic
Kolbe synthesis	1.000	0	0	0	0	Isohypsic
Kulinkovich	0.329	2.042	0.137	−2	1	Hyperhypsic
Kumada cross coupling (X = Cl)	0.021	47.600	0	0	0	Isohypsic
Kumada cross coupling (X = Br)	0.011	92.050	0	0	0	Isohypsic
Kumada cross coupling (X = I)	0.007	139.150	0	0	0	Isohypsic
Liebeskind–Srogl	0.208	3.804	0.549	0	0	Isohypsic
Malonic ester synthesis (X = Cl, R₁ = CH₃)	0.293	2.408	0.088	−1	0.5	Hyperhypsic
Malonic ester synthesis (X = Br, R₁ = CH₃)	0.241	3.148	0.072	−1	0.5	Hyperhypsic
Malonic ester synthesis (X = I, R₁ = CH₃)	0.203	3.933	0.061	−1	0.5	Hyperhypsic
Malonic ester synthesis (X = Cl, R₁ = CH₂CH₃)	0.258	2.874	0.077	−1	0.5	Hyperhypsic
Malonic ester synthesis (X = Br, R₁ = CH₂CH₃)	0.217	3.615	0.065	−1	0.5	Hyperhypsic
Malonic ester synthesis (X = I, R₁ = CH₂CH₃)	0.185	4.400	0.056	−1	0.5	Hyperhypsic
Marshalk	0.703	0.423	0.250	0	0	Isohypsic
McMurry	0.072	12.840	0.845	−4	2	Hyperhypsic
Meerwein arylation	0.855	0.169	0	0	0	Isohypsic
Michael 1,4-conjugate addition	1.000	0	0	0	0	Isohypsic
Mukaiyama aldol	0.267	2.742	0.591	−1	0.5	Hyperhypsic

(continued)

TABLE 7.1 (continued)
Summary of Parameters for Named Carbon–Carbon Bond Forming Reactions

Name	AE(min)	E(max)	f(sac)(max)	Δ	HI = $-\Delta/(n+1)$	Hypsicity Class
Mukaiyama–Michael	0.243	3.117	0.543	−1	0.5	Hyperhypsic
Nazarov Cyclization	1.000	0	0	0	0	Isohypsic
Negishi cross coupling (X = Cl)	0.014	68.14	0	0	0	Isohypsic
Negishi cross coupling (X = Br)	0.009	112.59	0	0	0	Isohypsic
Negishi cross coupling (X = I)	0.006	159.69	0	0	0	Isohypsic
Nieuwland enyne synthesis	1.000	0	0	0	0	Isohypsic
Nozaki (X = Cl)	0.092	9.897	0.705	−2	1	Hyperhypsic
Nozaki (X = Br)	0.052	18.231	0.688	−2	1	Hyperhypsic
Nozaki (X = I)	0.036	27.063	0.682	−2	1	Hyperhypsic
Organocuprate conjugate addition ($R_3 = CH_3$)	0.136	6.341	0.830	0	0	Isohypsic
Organocuprate conjugate addition ($R_3 = Ph$)	0.205	3.870	0.767	0	0	Isohypsic
Paterno–Buchi	1.000	0	0	0	0	Isohypsic
von Pechmann	1.000	0	0	2	−1	Hypohypsic
Perkin	0.712	0.405	0	0	0	Isohypsic
Peterson olefination (X = Cl)	0.410	1.441	0	0	0	Isohypsic
Peterson olefination (X = Br)	0.543	0.842	0	0	0	Isohypsic
Peterson olefination (X = I)	0.631	0.584	0	0	0	Isohypsic
[2 + 2] Photochemical cyclization	1.000	0	0	0	0	Isohypsic
Pinacol	0.515	0.940	0.202	−2	1	Hyperhypsic
Prins	1.000	0	0	0	0	Isohypsic
Reformatskii	0.379	1.637	0.344	−2	1	Hyperhypsic
Reimer–Tiemann	0.290	2.447	0	0	0	Isohypsic
von Richter	0.422	1.369	0	0	0	Isohypsic
Robinson annulation	0.439	1.276	0.311	0	0	Isohypsic
Sakurai	0.259	2.855	0.428	−1	0.5	Hyperhypsic
Simmons–Smith cyclopropanation	0.116	7.604	0.181	−2	1	Hyperhypsic
Skraup	0.253	2.948	0.637	3	−1.5	Hypohypsic
Sonogashira (X = Cl)	0.737	0.357	0	0	0	Isohypsic
Sonogashira (X = Br)	0.558	0.793	0	0	0	Isohypsic
Sonogashira (X = I)	0.443	1.255	0	0	0	Isohypsic
Stetter	1.000	0	0	0	0	Isohypsic
Stille coupling (X = Cl, $R_3 = n$Bu)	0.006	162.570	0.00611	0	0	Isohypsic
Stille coupling (X = Br, $R_3 = n$Bu)	0.005	184.795	0.00538	0	0	Isohypsic
Stille coupling (X = I, $R_3 = n$Bu)	0.005	208.345	0.00478	0	0	Isohypsic
Suzuki coupling (X = F)	0.706	0.416	0.08257	0	0	Isohypsic
Suzuki coupling (X = Cl)	0.657	0.522	0.07678	0	0	Isohypsic
Suzuki coupling (X = Br)	0.552	0.811	0.06454	0	0	Isohypsic
Suzuki coupling (X = I)	0.472	1.117	0.05521	0	0	Isohypsic
Tebbe olefination	0.089	10.227	0	0	0	Isohypsic
Thorpe	1.000	0	0	0	0	Isohypsic
Tishchenko	1.000	0	0	0	0	Isohypsic
Trost allene–alkene addition	1.000	0	0	−1	0.5	Hyperhypsic
Ullmann coupling (X = F)	0.603	0.659	0.249	−2	1	Hyperhypsic
Ullmann coupling (X = Cl)	0.534	0.873	0.220	−2	1	Hyperhypsic
Ullmann coupling (X = Br)	0.408	1.450	0.168	−2	1	Hyperhypsic
Ullmann coupling (X = I)	0.327	2.062	0.135	−2	1	Hyperhypsic

TABLE 7.1 (continued)
Summary of Parameters for Named Carbon–Carbon Bond Forming Reactions

Name	AE(min)	E(max)	f(sac)(max)	Δ	HI = −Δ/(n + 1)	Hypsicity Class
Vilsmeier–Haack–Arnold	0.371	1.694	0.517	0	0	Isohypsic
Vorbruggen coupling	0.640	0.563	0.291	0	0	Isohypsic
Wadsworth–Horner–Emmons	0.229	3.358	0.104	−2	1	Hyperhypsic
Wender–Trost [5 + 2] cycloaddition	1.000	0	0	0	0	Isohypsic
Wurtz–Fittig	0.010	102.900	0.221	−2	1	Hyperhypsic
Zincke–Suhl	0.853	0.172	0	−1	0.5	Hyperhypsic

TABLE 7.2
Summary of Parameters for Named Condensation Reactions

Name	AE(min)	E(max)	f(sac)(max)	Δ	HI = −Δ/(n + 1)	Hypsicity Class
Acyloin	0.211	3.733	0.514	−4	2	Hyperhypsic
Aldol	0.757	0.321	0	0	0	Isohypsic
Bamberger–Goldschmidt synthesis of 1,2,4-triazines	0.882	0.134	0	−1	0.5	Hyperhypsic
Bamberger–Goldschmidt synthesis of 1,3,5-triazines	0.604	0.654	0	0	0	Isohypsic
Benzoin	1.000	0	0	0	0	Isohypsic
Claisen (R_2 = Me)	0.692	0.444	0	0	0	Isohypsic
Claisen (R_2 = Et)	0.610	0.639	0	0	0	Isohypsic
Cyclic ether synthesis (product ring size = 5)	0.833	0.200	0	0	0	Isohypsic
Cyclic ether synthesis (product ring size = 6)	0.852	0.173	0	0	0	Isohypsic
Cyclic ether synthesis (product ring size = 7)	0.868	0.153	0	0	0	Isohypsic
Cyclic ether synthesis (product ring size = 8)	0.880	0.136	0	0	0	Isohypsic
Darzens	0.535	0.869	0.243	0	0	Isohypsic
Dieckmann (product ring size = 5)	0.877	0.141	0	0	0	Isohypsic
Dieckmann (product ring size = 6)	0.888	0.127	0	0	0	Isohypsic
Dieckmann (product ring size = 7)	0.897	0.115	0	0	0	Isohypsic
Hinsberg thiophene synthesis	0.827	0.209	0	0	0	Isohypsic
Knoevenagel	0.905	0.105	0	0	0	Isohypsic
Knorr pyrrole synthesis	0.854	0.171	0	0	0	Isohypsic
Mukaiyama aldol	0.244	3.095	0.540	−1	0.5	Hyperhypsic
Nenitzescu indole synthesis	0.409	1.444	0.140	−6	3	Hyperhypsic
von Pechmann	0.714	0.400	0	−1	0.5	Hyperhypsic
Pellizzari	0.657	0.522	0	0	0	Isohypsic
Pictet–Spengler isoquinoline synthesis	0.881	0.135	0	0	0	Isohypsic
Schiff base imine	0.617	0.621	0	0	0	Isohypsic
Stobbe	0.775	0.291	0	0	0	Isohypsic

TABLE 7.3

Summary of Parameters for Named Elimination Reactions

Name	AE(min)	E(max)	f(sac)(max)	Δ	$HI = -\Delta/(n + 1)$	Hypsicity Class
Azetidine synthesis	0.342	1.928	0.269	0	0	Isohypsic
Borodin–Hundsdiecker	0.185	4.412	0.530	0	0	Isohypsic
Burgess	0.099	9.143	0.838	n/a	n/a	n/a
Chugaev	0.206	3.857	0	n/a	n/a	n/a
Cope elimination	0.315	2.179	0	n/a	n/a	n/a
Dakin	0.522	0.915	0	0	0	Isohypsic
Edman	0.000	infinity	0	0	0	Isohypsic
Grob fragmentation (product ring size = 10)	0.418	1.395	0.110	n/a	n/a	n/a
Grob fragmentation (product ring size = 11)	0.439	1.277	0.106	n/a	n/a	n/a
Grob fragmentation (product ring size = 12)	0.459	1.178	0.102	n/a	n/a	n/a
Haloform reaction (X = Cl)	0.163	5.128	0.510	1	−0.5	Hypohypsic
Haloform reaction (X = Br)	0.100	9.050	0.701	1	−0.5	Hypohypsic
Haloform reaction (X = I)	0.070	13.206	0.789	1	−0.5	Hypohypsic
Hofmann degradation	0.055	17.250	0.912	n/a	n/a	n/a
Kochi	0.069	13.580	0.834	1	−0.5	Hypohypsic
Lossen (product is carboxylic acid, X = Me)	0.380	1.630	0	0	0	Isohypsic
Lossen (product is carboxylic acid, X = H)	0.430	1.326	0	0	0	Isohypsic
Lossen (product is carboxylic acid, X = Et)	0.341	1.935	0	0	0	Isohypsic
Lossen (product is amine, R_1 = Me, R_2 = Me)	0.140	6.118	0.154	−1	0.5	Hyperhypsic
Lossen (product is amine, R_1 = Me, R_2 = H)	0.159	5.294	0.175	−1	0.5	Hyperhypsic
Lossen (product is amine, R_1 = Me, R_2 = Et)	0.126	6.941	0.137	−1	0.5	Hyperhypsic
Norrish type II (product = enol, R's = H)	0.627	0.595	0	0	0	Isohypsic
Norrish type II (product = olefin, R's = H)	0.373	1.680	0	n/a	n/a	n/a
Pyrolysis of sulfoxides	0.453	1.208	0	n/a	n/a	n/a
Ring closing metathesis (X = CH_2)	0.588	0.700	0	0	0	Isohypsic
Ring closing metathesis (X = CH_2CH_2)	0.659	0.519	0	0	0	Isohypsic
Ring closing metathesis (X = $CH_2CH_2CH_2$)	0.708	0.412	0	0	0	Isohypsic
Ring closing metathesis (X = $CH_2CH_2CH_2CH_2$)	0.745	0.341	0	0	0	Isohypsic
Ruff–Fenton	0.117	7.512	0.702	n/a	n/a	n/a
Tiffeneau–Demjanov (product ring size = 6)	0.605	0.653	0.480	−1	0.5	Hyperhypsic
Tiffeneau–Demjanov (product ring size = 7)	0.636	0.571	0.420	−1	0.5	Hyperhypsic
Tiffeneau–Demjanov (product ring size = 8)	0.663	0.508	0.373	−1	0.5	Hyperhypsic

n/a = not applicable.

TABLE 7.4
Summary of Parameters for Named Multicomponent Reactions

Name	AE(min)	E(max)	f(sac)(max)	Δ	HI $= -\Delta/(n + 1)$	Hypsicity Class
Beta-acetoamido carbonyl compound synthesis	1.000	0	0	0	0	Isohypsic
Aldehyde-alkyne-oxadiazoline annulation	0.789	0.267	0	−2	1	Hyperhypsic
Aldehyde-malonylurea-isocyanide condensation	0.931	0.074	0	1	−0.5	Hypohypsic
Aldehyde-Meldrum's acid-vinyl ether	0.917	0.090	0	1	−0.5	Hypohypsic
Aldehyde-Meldrum's acid-indole	0.938	0.066	0	0	0	Isohypsic
Aldehyde-Meldrum's acid-ketone	0.922	0.084	0	0	0	Isohypsic
1,6-Aldol condensation [2 + 2 + 2]	1.000	0	0	0	0	Isohypsic
Alkene–alkyne annulation	1.000	0	0	0	0	Isohypsic
Alkynes-imines-organoboron reagents	0.734	0.363	0	0	0	Isohypsic
4-CC allenylation	0.727	0.375	0	0	0	Isohypsic
Alper carbonylation	1.000	0	0	0	0	Isohypsic
Amide-thioisocyanate-alkylhalide condensation (X = Cl)	0.650	0.539	0	0	0	Isohypsic
Amide-thioisocyanate-alkylhalide condensation (X = Br)	0.505	0.979	0	0	0	Isohypsic
Amide-thioisocyanate-alkylhalide condensation (X = I)	0.409	1.446	0	0	0	Isohypsic
Alpha-aminoacid synthesis via munchnones	0.762	0.312	0	0	0	Isohypsic
Aminoalkylation of naphthols with chiral amines	0.951	0.051	0	0	0	Isohypsic
Beta-aminocarbonyls synthesis	0.430	1.327	0	0	0	Isohypsic
Asinger	0.707	0.414	0	0	0	Isohypsic
Asymmetric Mannich-type	0.731	0.367	0	−1	0.5	Hyperhypsic
Aza-Diels–Alder synthesis of tetrahydroquinolines	0.926	0.080	0	0	0	Isohypsic
1,5-Benzodiazepine synthesis	0.816	0.225	0	0	0	Isohypsic
Betti	0.948	0.055	0	0	0	Isohypsic
Biginelli	0.731	0.367	0	0	0	Isohypsic
bis(indolyl)methane synthesis	0.932	0.073	0	0	0	Isohypsic
Bucherer synthesis of hydantoins	0.847	0.18	0	0	0	Isohypsic
Butadiene trimerization [2 + 2 + 2]	1.000	0	0	0	0	Isohypsic
Chichibabin pyridine synthesis	0.564	0.774	0.0970	2	−1	Hypohypsic
Cyclohexanone synthesis [1 + 2 + 3]	0.830	0.205	0	0	0	Isohypsic
Cyclopropanation [(1 + 2) + (1 + 2)]	0.718	0.393	0	0	0	Isohypsic
Cyclopropanation [1 + (1 + 2)] (X = Cl)	0.747	0.339	0	0	0	Isohypsic
Cyclopropanation [1 + (1 + 2)] (X = Br)	0.651	0.537	0	0	0	Isohypsic
Cyclopropanation [1 + (1 + 2)] (X = I)	0.607	0.648	0	0	0	Isohypsic
1,4-Dihydropyridine synthesis	0.845	0.183	0	0	0	Isohypsic
3,4-Dihydropyrimidin-2-(1H)-one synthesis (X = O)	0.798	0.254	0	0	0	Isohypsic
3,4-Dihydropyrimidin-2-(1H)-one synthesis (X = S)	0.814	0.228	0	0	0	Isohypsic
Doebner	0.822	0.217	0.053	2	−1	Hypohypsic

(continued)

TABLE 7.4 (continued)
Summary of Parameters for Named Multicomponent Reactions

Name	AE(min)	E(max)	f(sac)(max)	Δ	HI $= -\Delta/(n+1)$	Hypsicity Class
Dornow–Wiehler	0.632	0.581	0.186	4	−2	Hypohypsic
Feist–Benary	0.657	0.521	0	0	0	Isohypsic
Formamidine urea synthesis (target product = A)	0.582	0.719	0	0	0	Isohypsic
Formamidine urea synthesis (target product = B)	0.418	1.391	0	0	0	Isohypsic
Fused 3-aminoimidazoles (product ring size = 5)	0.872	0.146	0	0	0	Isohypsic
Fused 3-aminoimidazoles (product ring size = 6)	0.884	0.131	0	0	0	Isohypsic
Fused 3-aminoimidazoles (product ring size = 7)	0.893	0.119	0	0	0	Isohypsic
Fused benzochromene synthesis	1.000	0	0	0	0	Isohypsic
Gewald aminothiophene synthesis	0.888	0.126	0	0	0	Isohypsic
Grieco condensation	0.908	0.102	0	0	0	Isohypsic
Guareschi–Thorpe condensation ($R_2 = CH_3$)	0.638	0.567	0	−1	0.5	Hyperhypsic
Guareschi–Thorpe condensation ($R_2 = CH_2CH_3$)	0.594	0.683	0	−1	0.5	Hyperhypsic
Hantzsch synthesis of dihydropyridines	0.806	0.240	0	0	0	Isohypsic
Hantzsch synthesis of pyrroles	0.673	0.486	0	0	0	Isohypsic
Heck–Diels–Alder	0.552	0.810	0	0	0	Isohypsic
Beta-iminoamine synthesis	1.000	0	0	0	0	Isohypsic
Indolizine synthesis	0.579	0.728	0.0957	2	−1	Hyperhypsic
Isocyanate-ketone cyclization	0.946	0.057	0	0	0	Isohypsic
Isoquinoline-DEAD-benzoquinones	1.000	0	0	0	0	Isohypsic
Isoquinolonic acid synthesis	0.937	0.067	0	0	0	Isohypsic
Isoquinolinone synthesis (Y = Cl)	0.488	1.047	0	0	0	Isohypsic
Isoquinolinone synthesis (Y = OMe)	0.495	1.019	0	0	0	Isohypsic
Knoevenagel hetero-Diels–Alder	0.962	0.039	0	0	0	Isohypsic
1,6-Mannich condensation	1.000	0	0	0	0	Isohypsic
Mannich condensation	0.802	0.247	0	0	0	Isohypsic
Michael–aldol–Horner–Wadsworth–Emmons	0.465	1.149	0	−2	1	Hyperhypsic
Michael–Michael–1,6-Wittig [2 + 2 + 2]	0.350	1.857	0	−2	1	Hyperhypsic
Nenitzescu–Praill pyrylium salt synthesis (with Ac$_2$O)	0.602	0.662	0.290	0	0	Isohypsic
Nenitzescu–Praill pyrylium salt synthesis (with AcCl)	0.681	0.469	0.335	0	0	Isohypsic
Norbornene synthesis	0.332	2.014	0	−2	1	Hyperhypsic
Oxazolidin-2-one synthesis	0.829	0.207	0.152	2	−1	Hypohypsic
Passerini	1.000	0	0	0	0	Isohypsic

TABLE 7.4 (continued)
Summary of Parameters for Named Multicomponent Reactions

Name	AE(min)	E(max)	f(sac)(max)	Δ	HI = −Δ/(n + 1)	Hypsicity Class
Pauson–Khand	1.000	0	0	0	0	Isohypsic
Petasis condensation	0.547	0.827	0	0	0	Isohypsic
Petrenko–Kritschenko	0.871	0.148	0	0	0	Isohypsic
Pinner triazine	0.735	0.361	0	0	0	Isohypsic
Furanoquinoline synthesis	0.933	0.072	0	0	0	Isohypsic
Pyranoquinoline synthesis	0.936	0.068	0	0	0	Isohypsic
2′,3′-pyranone(pyrrolidinone)-fused tryptamine synthesis	0.954	0.048	0	0	0	Isohypsic
Pyridine synthesis	0.543	0.841	0	1	−0.5	Hypohypsic
1H-pyrrolo[3,2e]-1,2,4-triazine synthesis	0.776	0.288	0	1	−0.5	Hypohypsic
Radziszewski	0.557	0.794	0	0	0	Isohypsic
Radziszewski-type reaction using microwaves	0.915	0.092	0	−2	1	Hyperhypsic
Robinson–Schoepf synthesis of tropane	0.876	0.141	0	0	0	Isohypsic
Reppe	1.000	0	0	0	0	Isohypsic
Riehm quinoline synthesis	0.751	0.331	0	1	−0.5	Hypohypsic
Roelen synthesis	1.000	0	0	0	0	Isohypsic
Rothemund	0.711	0.406	0.110	6	−3	Hypohypsic
Strecker synthesis of alpha-cyanoamines	0.757	0.321	0	0	0	Isohypsic
Synthesis of 1,7-enynes	0.327	2.058	0	−1	0.5	Hyperhypsic
Synthesis of THF derivatives	0.552	0.810	0	1	−0.5	Hypohypsic
Tandem Asinger–Ugi condensation	0.578	0.731	0	0	0	Isohypsic
Tandem Passerini–Wittig	0.452	1.213	0	−2	1	Hyperhypsic
Tandem Petasis–Ugi condensation	0.621	0.611	0	0	0	Isohypsic
Tetrasubstituted olefins	0.511	0.956	0.0703	0	0	Isohypsic
Thalidomide synthesis	0.697	0.434	0	2	−1	Hypohypsic
Thiele	0.808	0.238	0	0	0	Isohypsic
Trimerization of alkynes [2 + 2 + 2]	1.000	0	0	0	0	Isohypsic
Trimerization of arylisocyanates [2 + 2 + 2]	1.000	0	0	0	0	Isohypsic
Trost 4-CC synthesis of gamma, delta-unsaturated ketones	1.000	0	0	0	0	Isohypsic
Trost synthesis of 1,5-diketones	1.000	0	0	0	0	Isohypsic
Ugi condensation	0.850	0.176	0	0	0	Isohypsic
Ugi 3-CC condensation	1.000	0	0	0	0	Isohypsic
Vinylphosphonium bromide–ketone cyclization	0.871	0.148	0	0	0	Isohypsic
Weiss	0.911	0.097	0	0	0	Isohypsic
Wulff cyclization	0.594	0.683	0	−1	0.5	Hyperhypsic

TABLE 7.5

Summary of Parameters for Named Non-Carbon-Carbon Bond Forming Reactions

Name	AE(min)	E(max)	f(sac)(max)	Δ	HI $= -\Delta/(n+1)$	Hypsicity Class
Arbuzov–Michaelis (X = Cl)	0.681	0.468	0	0	0	Isohypsic
Arbuzov–Michaelis (X = Br)	0.537	0.863	0	0	0	Isohypsic
Arbuzov–Michaelis (X = I)	0.437	1.291	0	0	0	Isohypsic
Aziridine synthesis (with I-N=C=O)	0.170	4.884	0	0	0	Isohypsic
Azo coupling reaction	0.833	0.200	0	1	−0.5	Hypohypsic
Buchwald–Hartwig cross coupling (X = Br, R_1 = CH_3)	0.569	0.756	0	0	0	Isohypsic
1,3-Dipolar addition (with HN_3)	1.000	0	0	1	−0.5	Hypohypsic
Eschweiler–Clarke	0.333	2	0	−2	1	Hyperhypsic
Acetal synthesis	0.727	0.375	0	0	0	Isohypsic
Ketal synthesis	0.727	0.375	0	0	0	Isohypsic
Hemiacetal synthesis	1.000	0	0	0	0	Isohypsic
Hemiketal synthesis	1.000	0	0	0	0	Isohypsic
Griess diazotization	0.591	0.692	0.316	0	0	Isohypsic
Hoch–Campbell aziridine synthesis (Z = CH_3)	0.203	3.937	0.395	0	0	Isohypsic
Hofmann–Loffler–Freytag (ring size = 5)	0.661	0.513	0	0	0	Isohypsic
Hofmann–Loffler–Freytag (ring size = 6)	0.700	0.429	0	0	0	Isohypsic
Hofmann–Loffler–Freytag (ring size = 7)	0.731	0.368	0	0	0	Isohypsic
Leuckart	0.333	2	0	−2	1	Hyperhypsic
Menshutkin	1.000	0	0	0	0	Isohypsic
Paal–Knorr (X = O) with P_2O_5	0.298	2.353	0.623	−1	0.5	Hyperhypsic
Paal–Knorr (X = S) with P_2S_5	0.247	3.048	0.653	−1	0.5	Hyperhypsic
Paal–Knorr (product = pyrrole, with $(NH_4)_2CO_3$)	0.368	1.716	0	−2	1	Hyperhypsic
Polonovski	0.749	0.335	0	2	−1	Hypohypsic
Ritter (Z = H)	0.495	1.022	0	0	0	Isohypsic
Ritter (Z = Me)	0.429	1.333	0	0	0	Isohypsic
Ritter (Z = Et)	0.378	1.644	0	0	0	isohypsic
Stahl aerobic oxidation amination	0.843	0.186	0.139	1	−0.5	Hypohypsic
Staudinger (X = Cl)	0.749	0.336	0	1	−0.5	Hypohypsic
Staudinger (X = Br)	0.820	0.219	0	1	−0.5	Hypohypsic
Staudinger (X = I)	0.862	0.160	0	1	−0.5	Hypohypsic
Stoltz aerobic oxidative etherification	0.899	0.113	0.090	1	−0.5	Hypohypsic
Wenker	0.216	3.628	0.693	0	0	Isohypsic
Williamson ether synthesis	0.130	6.717	0	0	0	Isohypsic
Wohler urea synthesis	0.305	2.283	0	0	0	Isohypsic

TABLE 7.6
Summary of Parameters for Named Oxidation Reactions
(with respect to Substrate of Interest)

Name	Type	AE(min)	E(max)	f(sac) (max)	Δ	HI = −Δ/ (n + 1)	Hypsicity Class
Baeyer–Villiger (H_2O_2)	Additive	0.719	0.391	0	1	−0.5	Hypohypsic
Baeyer–Villiger (mCPBA)	Additive	0.227	3.401	0	1	−0.5	Hypohypsic
Bamford–Stevens	Subtractive	0.243	3.123	0.751	n/a	n/a	n/a
Bamford–Stevens	Additive with rearrangement	0.105	8.500	0.203	−3	1.5	Hyperhypsic
Boyland–Sims	Additive	0.225	3.453	0.237	1	−0.5	Hypohypsic
Corey gamma-lactone synthesis	Subtractive	0.279	2.588	0.718	2	−1	Hypohypsic
Corey–Chaykovsky	Additive	0.161	5.227	0.088	0	0	Isohypsic
Corey–Kim	Subtractive	0.091	9.948	0.903	n/a	n/a	n/a
Criegee glycol cleavage (product = both ketones)	Subtractive	0.119	7.420	0.877	n/a	n/a	n/a
Dess–Martin	Subtractive	0.130	6.688	0.869	n/a	n/a	n/a
Elbs	Additive	0.288	2.473	0	1	−0.5	Hypohypsic
Etard	Additive	0.256	2.900	0	4	−2	Hypohypsic
Fehling	Additive	0.204	3.891	0	2	−1	Hypohypsic
Fenton	Subtractive	0.290	2.451	0.702	n/a	n/a	n/a
Fleming	Additive	0.094	9.639	0	1	−0.5	Hypohypsic
Forster (ring size = 5)	Additive	0.494	1.022	0.335	2	−1	Hypohypsic
Forster (ring size = 6)	Additive	0.524	0.907	0.315	2	−1	Hypohypsic
Forster (ring size = 7)	Additive	0.551	0.815	0.297	2	−1	Hypohypsic
Forster (ring size = 8)	Additive	0.575	0.740	0.282	2	−1	Hypohypsic
Graham (X = Cl)	Additive	0.396	1.523	0.386	2	−1	Hypohypsic
Graham (X = Br)	Additive	0.429	1.331	0.422	2	−1	Hypohypsic
Graham (X = I)	Additive	0.447	1.238	0.441	2	−1	Hypohypsic
Harries ozonolysis with oxidative workup	Additive	0.836	0.196	0	1	−0.5	Hypohypsic
Harries ozonolysis with reductive workup	Additive	0.435	1.3	0.449	0	0	Isohypsic
Hooker	Additive	0.303	2.305	0.555	2	−1	Hypohypsic
Jacobsen epoxidation	Additive	0.429	1.328	0	2	−1	Hypohypsic
Jones oxidation	Subtractive	0.153	5.556	0.837	n/a	n/a	n/a
Lemieux–Johnson (product = 1,2-diol)	Additive	0.366	1.734	0	2	−1	Hypohypsic
Lemieux–Johnson (product = both aldehydes)	Additive	0.155	5.45	0.643	4	−2	Hypohypsic
Malaprade	Subtractive	0.567	0.764	0.756	n/a	n/a	n/a
Oppenauer	Subtractive	0.214	3.667	0.771	n/a	n/a	n/a
Oxymercuration of olefins	Additive	0.138	6.228	0.873	0	0	Isohypsic
Permanganate oxidation of olefins to 1,2-diols	Additive	0.402	1.490	0	2	−1	Hypohypsic

(continued)

TABLE 7.6 (continued)

Summary of Parameters for Named Oxidation Reactions
(with respect to Substrate of Interest)

Name	Type	AE(min)	E(max)	f(sac) (max)	Δ	HI $= -\Delta/$ $(n + 1)$	Hypsicity Class
Pfitzner–Moffatt	Subtractive	0.096	9.467	0.898	n/a	n/a	n/a
Prevost	Additive	0.080	11.516	0.918	2	−1	Hypohypsic
Prilezhaev (epoxidation of olefins with H_2O_2)	Additive	0.710	0.409	0	1	−0.5	Hypohypsic
Prilezhaev (epoxidation of olefins with mCPBA)	Additive	0.220	3.556	0	1	−0.5	Hypohypsic
Riley	Additive	0.374	1.672	0	4	−2	Hypohypsic
Rubottom	Additive	0.196	4.108	0	0	0	Isohypsic
Sarett procedure	Subtractive	0.103	8.667	0.890	n/a	n/a	n/a
Sharpless–Jacobsen dihydroxylation	Additive	0.059	15.965	0.608	2	−1	Hypohypsic
Sharpless epoxidation	Additive	0.373	1.682	0	1	−0.5	Hypohypsic
Sharpless oxyamination	Additive	0.192	4.197	0	2	−1	Hypohypsic
Shi asymmetric epoxidation	Additive	0.186	4.364	0.237	1	−0.5	Hypohypsic
Swern	Subtractive	0.068	13.63	0.927	n/a	n/a	n/a
Tollens test	Additive	0.109	8.213	0.887	2	−1	Hypohypsic
Uemura	Subtractive	0.625	0.600	0.333	n/a	n/a	n/a
Wacker–Tsuji	Additive	0.197	4.075	0.794	1	−0.5	Hypohypsic
Wessely	Additive	0.283	2.533	0	1	−0.5	Hypohypsic
Willgerodt	Additive with rearrangement	0.795	0.258	0.193	2	−1	Hypohypsic
Woodward *cis*-dihydroxylation	Additive	0.095	9.516	0.902	2	−1	Hypohypsic

n/a = not applicable; HI index does not apply to subtractive oxidation reactions that do not produce target bonds in product.

TABLE 7.7
Summary of Parameters for Named Rearrangement Reactions

Name	AE(min)	E(max)	f(sac) (max)	Δ	HI $= -\Delta/(n + 1)$	Hypsicity Class
Acyl	0.828	0.207	0	0	0	Isohypsic
Allylic (X = Cl, Y = Cl)	0.677	0.477	0	1	−0.5	Hypohypsic
Allylic (X = Cl, Y = Br)	0.768	0.301	0	1	−0.5	Hypohypsic
Allylic (X = Cl, Y = I)	0.822	0.217	0	1	−0.5	Hypohypsic
Allylic (X = Br, Y = Cl)	0.486	1.058	0	1	−0.5	Hypohypsic
Allylic (X = Br, Y = Br)	0.599	0.669	0	1	−0.5	Hypohypsic
Allylic (X = Br, Y = I)	0.675	0.482	0	1	−0.5	Hypohypsic
Allylic (X = I, Y = Cl)	0.374	1.674	0	1	−0.5	Hypohypsic
Allylic (X = I, Y = Br)	0.486	1.059	0	1	−0.5	Hypohypsic
Allylic (X = I, Y = I)	0.568	0.762	0	1	−0.5	Hypohypsic
Aza-Cope	1.000	0	0	−1	0.5	Hyperhypsic
Bamberger	1.000	0	0	0	0	Isohypsic
Bamberger–Goldschmidt synthesis of isoquinoline	0.446	1.240	0.491	0	0	Isohypsic
Barton	1.000	0	0	0	0	Isohypsic
Beckmann	1.000	0	0	0	0	Isohypsic
Benzidine	1.000	0	0	0	0	Isohypsic
Benzylic acid	1.000	0	0	0	0	Isohypsic
Brook	1.000	0	0	2	−1	Hypohypsic
Camphene	1.000	0	0	0	0	Isohypsic
Chapman	1.000	0	0	0	0	Isohypsic
Ciamician photodisproportionation	1.000	0	0	2	−1	Hypohypsic
Claisen	1.000	0	0	1	−0.5	Hypohypsic
Claisen–Ireland	0.268	2.735	0.684	1	−0.5	Hypohypsic
Cope	1.000	0	0	0	0	Isohypsic
Cornforth	1.000	0	0	0	0	Isohypsic
Criegee	1.000	0	0	0	0	Isohypsic
Curtius	0.192	4.208	0	0	0	Isohypsic
Dieckmann–Thorpe	1.000	0	0	0	0	Isohypsic
Dienone-phenol	1.000	0	0	0	0	Isohypsic
Di-pi methane	1.000	0	0	0	0	Isohypsic
Epoxide	0.265	2.773	0.735	−2	1	Hyperhypsic
Favorskii (product ring size = 5)	0.708	0.412	0	0	0	Isohypsic
Favorskii (product ring size = 6)	0.727	0.375	0	0	0	Isohypsic
Favorskii (product ring size = 7)	0.744	0.344	0	0	0	Isohypsic
Fischer–Hepp	1.000	0	0	0	0	Isohypsic
Fries	1.000	0	0	0	0	Isohypsic
Fritsch–Buttenberg–Wiechell	0.160	5.265	0.344	0	0	Isohypsic
Hofmann–Martius	0.087	10.524	0.774	−1	0.5	Hyperhypsic
Hydroboration-borane	0.431	1.319	0	1	−0.5	Hypohypsic
Hydroperoxide (product = ketone)	0.625	0.6	0	n/a	n/a	n/a
Hydroperoxide (product = alcohol)	0.375	1.667	0	0	0	Isohypsic
Kemp	1.000	0	0	0	0	Isohypsic
Lossen (product = carboxylic acid, $R_1 = CH_3$)	0.377	1.652	0	0	0	Isohypsic
Lossen (product = carboxylic acid, $R_1 = H$)	0.426	1.348	0	0	0	Isohypsic

(continued)

TABLE 7.7 (continued)
Summary of Parameters for Named Rearrangement Reactions

Name	AE(min)	E(max)	f(sac) (max)	Δ	HI = $-\Delta/(n + 1)$	Hypsicity Class
Lossen (product = carboxylic acid, $R_1 = CH_2CH_3$)	0.338	1.957	0	0	0	Isohypsic
Lossen (product = amine, $R_2 = H$)	0.159	5.294	0.175	−1	0.5	Hyperhypsic
Lossen (product = amine, $R_2 = CH_3$)	0.140	6.118	0.154	−1	0.5	Hyperhypsic
Lossen (product = amine, $R_2 = CH_2CH_3$)	0.126	6.941	0.137	−1	0.5	Hyperhypsic
Martynoff	1.000	0	0	0	0	Isohypsic
McLafferty (product = enol, R's = H)	0.623	0.605	0	0	0	Isohypsic
McLafferty (product = olefin, R's = H)	0.377	1.654	0	n/a	n/a	n/a
Meisenheimer	1.000	0	0	0	0	Isohypsic
Meyer–Schuster	1.000	0	0	1	−0.5	Hypohypsic
Nametkin (X = Cl)	0.658	0.521	0	−1	0.5	Hyperhypsic
Nametkin (X = Br)	0.464	1.156	0	−1	0.5	Hyperhypsic
Nametkin (X = I)	0.354	1.829	0	−1	0.5	Hyperhypsic
Neber	0.197	4.068	0.227	0	0	Isohypsic
Norrish type I	1.000	0	0	−1	0.5	Hyperhypsic
Orton	1.000	0	0	0	0	Isohypsic
Oxy-Cope	1.000	0	0	−1	0.5	Hyperhypsic
Payne	1.000	0	0	0	0	Isohypsic
Pentazadiene [1,3]	1.000	0	0	0	0	Isohypsic
Perkin (X = Cl)	0.499	1.004	0	1	−0.5	Hypohypsic
Perkin (X = Br)	0.420	1.381	0	1	−0.5	Hypohypsic
Perkin (X = I)	0.360	1.780	0	1	−0.5	Hypohypsic
Pinacol	0.710	0.409	0	−1	0.5	Hyperhypsic
Polonovski	1.000	0	0	0	0	Isohypsic
Pummerer	1.000	0	0	2	−1	Hypohypsic
Pyridine-N-oxide-pyridone [1,4]	1.000	0	0	0	0	Isohypsic
Ramberg–Backlund	0.203	3.934	0.420	−2	1	Hyperhypsic
Retro-Sommelet–Hauser [1,3]	1.000	0	0	0	0	Isohypsic
Rupe	1.000	0	0	0	0	Isohypsic
Schmidt	0.132	6.588	0	0	0	Isohypsic
Sigmatropic [3,3]	1.000	0	0	0	0	Isohypsic
Smiles [1,4]	1.000	0	0	0	0	Isohypsic
Stevens [1,2] (X = Cl)	0.374	1.677	0.324	0	0	Isohypsic
Stevens [1,2] (X = Br)	0.273	2.664	0.237	0	0	Isohypsic
Stevens [1,2] (X = I)	0.212	3.711	0.184	0	0	Isohypsic
Sulfenate-sulfoxide [2,3]	1.000	0	0	1	−0.5	Hypohypsic
Tiemann	0.236	3.241	0.693	0	0	Isohypsic
Vinyl ether	1.000	0	0	−1	0.5	Hyperhypsic
Vinylogous Wolff	0.797	0.255	0	0	0	Isohypsic
Wagner–Meerwein	0.883	0.132	0	−1	0.5	Hyperhypsic
Wallach	1.000	0	0	2	−1	Hypohypsic
Wawzonek–Yeakey [1,2]	1.000	0	0	0	0	Isohypsic
Wittig [1,2]	0.741	0.349	0.286	0	0	Isohypsic
Wolff (product ring size = 4, R = CH_3)	0.803	0.246	0	0	0	Isohypsic
Wolff (product ring size = 5, R = CH_3)	0.821	0.219	0	0	0	Isohypsic
Wolff (product ring size = 6, R = CH_3)	0.835	0.197	0	0	0	Isohypsic
Wolff (product ring size = 7, R = CH_3)	0.848	0.179	0	0	0	Isohypsic
Wolff (product ring size = 8, R = CH_3)	0.859	0.165	0	0	0	Isohypsic

TABLE 7.8
Summary of Parameters for Named Reactions (with respect to Substrate of Interest)

Name	Type	AE(min)	E(max)	f(sac) (max)	Δ	HI = −Δ/ (n + 1)	Hypsicity Class
Benkeser	Additive	0.425	1.353	0.175	−8	4	Hyperhypsic
Birch	Additive	0.435	1.300	0.076	−2	1	Hyperhypsic
Borch	Additive	0.323	2.094	0	0	0	Isohypsic
Borohydride reduction	Additive	0.366	1.734	0	0	0	Isohypsic
Bouveault–Blanc	Additive	0.105	8.500	0.303	−4	2	Hyperhypsic
Cannizzaro reaction (product = alcohol)	Additive	0.400	1.500	0	−2	1	Hyperhypsic
Corey–Bakshi–Shibata	Additive	0.33	2.00	0	0	0	Isohypsic
Corey–Winter	Subtractive	0.077	12.000	0.830	n/a	n/a	n/a
Clemmensen	Additive reduction with elimination	0.052	18.163	0.427	−4	2	Hyperhypsic
Diimide reduction (with NaOOC–N=N–COONa)	Additive	0.133	6.533	0.717	−2	1	Hyperhypsic
Diimide reduction (with H_2O_2 and NH_2NH_2)	Additive	0.319	2.133	0.362	−2	1	Hyperhypsic
Gribble	Additive	0.236	3.234	0.673	0	0	Isohypsic
Hydrogenation of olefins	Additive	1.000	0.000	0	0	0	Isohypsic
Hydrogenolysis of benzyl ethers	Additive reduction with elimination	0.164	5.111	0	1	−0.5	Hypohypsic
$LiAlH_4$ reduction of ketones	Additive	0.496	1.016	0.209	0	0	Isohypsic
$LiAlH_4$ reduction of methyl esters	Additive	0.302	2.313	0	0	0	Isohypsic
Meerwein–Pondorff–Verley	Additive	0.276	2.625	0	−2	1	Hyperhypsic
Midland	Additive	0.105	8.563	0	−2	1	Hyperhypsic
Noyori hydrogenation	Additive	1.000	0.000	0	0	0	Isohypsic
Radical dehalogenation	Additive reduction with elimination	0.006	176.725	0.399	1	−0.5	Hypohypsic
Reduction of nitroaromatics (with H_2S)	Additive reduction with elimination	0.582	0.719	0	−6	3	Hyperhypsic
Reduction of nitroaromatics (with Fe(II) salts)	Additive reduction with elimination	0.067	13.888	0.688	−6	3	Hyperhypsic
Rosenmund	Additive reduction with elimination	0.451	1.215	0	−1	0.5	Hyperhypsic
Shapiro	Additive reduction with elimination	0.096	9.429	0.212	−3	1.5	Hyperhypsic
Stephen	Additive reduction with elimination	0.269	2.724	0.500	−1	0.5	Hyperhypsic
Thioketal desulfurization	Additive reduction with elimination	0.043	22.113	0.319	−2	1	Hyperhypsic
Tosylhydrazone reduction	Additive reduction with elimination	0.038	25.000	0.308	0	0	Isohypsic
Wharton	Additive reduction with elimination	0.558	0.793	0	−3	1.5	Hyperhypsic
Wolff–Kishner	Additive reduction with elimination	0.068	13.750	0.610	−4	2	Hyperhypsic
Zincke disulfide cleavage	Additive	1.000	0.000	0	0	0	Isohypsic

n/a = not applicable; HI index does not apply to subtractive reduction reactions that do not produce target bonds in product.

TABLE 7.9
Summary of Parameters for Named Sequence Reactions

Name	AE(min)	E(max)	f(sac) (max)	Δ	HI = −Δ/ (n + 1)	Hypsicity Class
Arndt–Eistert	0.167	4.974	0.223	0	0	Isohypsic
Barton decarboxylation	0.032	30.009	0.256	1	−0.5	Hypohypsic
Bischler–Napieralski synthesis	0.593	0.685	0.492	0	0	Isohypsic
Claisen–Ireland rearrangement	0.268	2.735	0.684	1	−0.5	Hypohypsic
Doering–LaFlamme allene synthesis	0.091	9.968	0.360	−2	1	Hyperhypsic
Japp–Klingemann	0.437	1.291	0.278	1	−0.5	Hypohypsic
Julia olefination	0.070	13.214	0.533	−2	1	Hyperhypsic
Martinet dioxindole synthesis	0.523	0.913	0.063	−1	0.5	Hyperhypsic
Mitsunobu	0.059	16.063	0.480	0	0	Isohypsic
Murahashi	0.059	15.989	0.908	1	−0.5	Hypohypsic
Nef	0.141	6.100	0.714	2	−1	Hypohypsic
Nenitzescu–Praill synthesis of substituted pyridines	0.281	2.556	0.335	0	0	Isohypsic
Pfitzinger	0.746	0.341	0	−1	0.5	Hyperhypsic
Pictet–Spengler isoquinoline synthesis	0.881	0.135	0	0	0	Isohypsic
Pinner (product is guanidinium chloride salt, $R_2 = CH_3$)	0.715	0.398	0	0	0	Isohypsic
Pinner (product is guanidinium chloride salt, $R_2 = CH_2CH_3$)	0.636	0.572	0	0	0	Isohypsic
Pinner (product is ester)	0.463	1.162	0.367	0	0	Isohypsic
Pomeranz–Fritsch	0.540	0.853	0	−1	0.5	Hyperhypsic
Schlitter–Mueller modification of Pomeranz–Fritsch reaction	0.540	0.853	0	−2	1	Hyperhypsic
Reissert (R = CH_3)	0.561	0.783	0	2	−1	Hypohypsic
Schlosser modification of Wittig reaction	0.067	13.831	0.777	−2	1	Hyperhypsic
Stolle synthesis of indoles	0.646	0.548	0	0	0	Isohypsic
Stork enamine synthesis (X = Cl)	0.487	1.053	0.425	1	−0.5	Hypohypsic
Stork enamine synthesis (X = Br)	0.402	1.489	0.351	1	−0.5	Hypohypsic
Stork enamine synthesis (X = I)	0.339	1.951	0.296	1	−0.5	Hypohypsic
Urech synthesis of hydantoins	0.575	0.740	0	0	0	Isohypsic
von Richter cinnoline synthesis	0.556	0.798	0.227	−1	0.5	Hyperhypsic
Wittig	0.056	16.818	0.750	1	−0.5	Hypohypsic

TABLE 7.10
Summary of Parameters for Named Substitution Reactions

Name	AE(min)	E(max)	f(sac) (max)	Δ	HI = −Δ/(n + 1)	Hypsicity Class
Aniline synthesis	0.555	0.801	0	0	0	Isohypsic
Aromatic halogenation (X = Cl)	0.755	0.324	0	1	−0.5	Hypohypsic
Aromatic halogenation (X = Br)	0.660	0.516	0	1	−0.5	Hypohypsic
Aromatic halogenation (X = I)	0.614	0.627	0	1	−0.5	Hypohypsic
Aromatic nitration	0.872	0.146	0	0	0	Isohypsic
Aromatic sulfonation	0.898	0.114	0	0	0	Isohypsic
Bucherer	0.443	1.258	0	0	0	Isohypsic
Chichibabin	0.691	0.447	0.132	2	−1	Hypohypsic
Dakin–West	0.412	1.426	0	0	0	Isohypsic
Delepine (X = F)	0.395	1.529	0	0	0	Isohypsic
Delepine (X = Cl)	0.346	1.888	0	0	0	Isohypsic
Delepine (X = Br)	0.300	2.330	0	0	0	Isohypsic
Delepine (X = 127)	0.283	2.528	0	0	0	Isohypsic
Esterification with diazomethane	0.200	4.000	0	0	0	Isohypsic
Finkelstein (X = Cl, Y = Cl)	0.384	1.604	0	0	0	Isohypsic
Finkelstein (X = Cl, Y = Br)	0.581	0.722	0	0	0	Isohypsic
Finkelstein (X = Cl, Y = I)	0.687	0.457	0	0	0	Isohypsic
Finkelstein (X = Br, Y = Cl)	0.262	2.823	0	0	0	Isohypsic
Finkelstein (X = Br, Y = Br)	0.440	1.272	0	0	0	Isohypsic
Finkelstein (X = Br, Y = I)	0.554	0.804	0	0	0	Isohypsic
Finkelstein (X = I, Y = Cl)	0.195	4.115	0	0	0	Isohypsic
Finkelstein (X = I, Y = Br)	0.350	1.854	0	0	0	Isohypsic
Finkelstein (X = I, Y = I)	0.460	1.172	0	0	0	Isohypsic
Fischer esterification	0.719	0.391	0	0	0	Isohypsic
Gabriel synthesis	0.049	19.465	0.230	0	0	isohypsic
Gattermann	0.471	1.122	0	0	0	Isohypsic
Helferich (5 carbon sugars)	0.854	0.170	0	0	0	Isohypsic
Helferich (6 carbon sugars)	0.876	0.142	0	0	0	Isohypsic
Hell–Volhard–Zelinsky	0.632	0.582	0	1	−0.5	Hypohypsic
Koenigs–Knorr	0.614	0.629	0.243	0	0	Isohypsic
Lapworth (X = Cl)	0.683	0.465	0	0	0	Isohypsic
Lapworth (X = Br)	0.603	0.658	0	0	0	Isohypsic
Lapworth (X = I)	0.570	0.753	0	0	0	Isohypsic
Sandmeyer (X = Cl)	0.801	0.249	0	0	0	Isohypsic
Sandmeyer (X = Br)	0.849	0.178	0	0	0	Isohypsic
Sandmeyer (X = I)	0.879	0.137	0	0	0	Isohypsic
Sanger	0.901	0.109	0	0	0	Isohypsic
Schiemann	0.500	1.000	0	0	0	Isohypsic
Schotten–Baumann	0.371	1.699	0.329	0	0	Isohypsic
Synthesis of acetates via oxymercuration	0.219	3.575	0	2	−1	Hypohypsic
Synthesis of carbonate diesters	0.460	1.176	0	0	0	Isohypsic
Synthesis of carbamates (urethanes)	0.456	1.195	0	0	0	Isohypsic
Synthesis of ureas	0.451	1.215	0	0	0	Isohypsic
Wohl–Ziegler	0.450	1.224	0	2	−1	Hypohypsic
Zeisel determination	0.398	1.511	0	0	0	Isohypsic

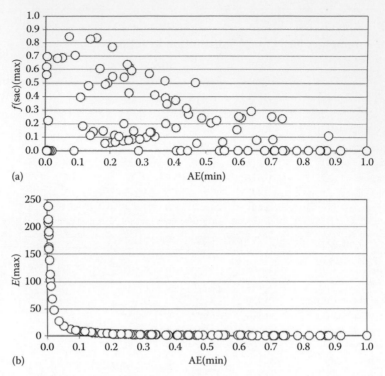

FIGURE 7.1 (a) Correlation between f(sac)(max) with AE(min); and (b) correlation between E(max) and AE(min) for named carbon-carbon bond forming reactions.

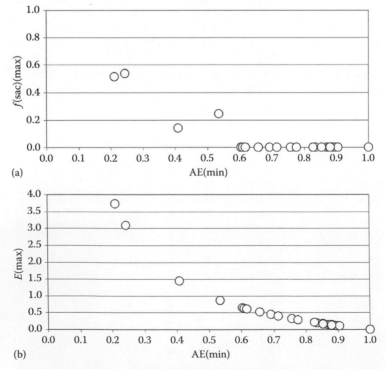

FIGURE 7.2 (a) Correlation between f(sac)(max) with AE(min); and (b) correlation between E(max) and AE(min) for named condensation reactions.

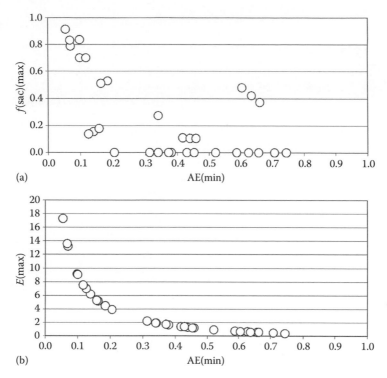

FIGURE 7.3 (a) Correlation between f(sac)(max) with AE(min); and (b) correlation between E(max) and AE(min) for named elimination reactions.

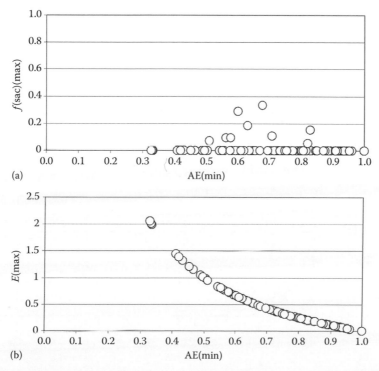

FIGURE 7.4 (a) Correlation between f(sac)(max) with AE(min); and (b) correlation between E(max) and AE(min) for named multicomponent reactions.

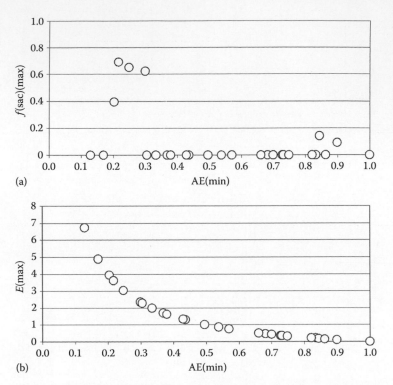

FIGURE 7.5 (a) Correlation between $f(sac)(max)$ with AE(min); and (b) correlation between $E(max)$ and AE(min) for named non-carbon–carbon bond forming reactions.

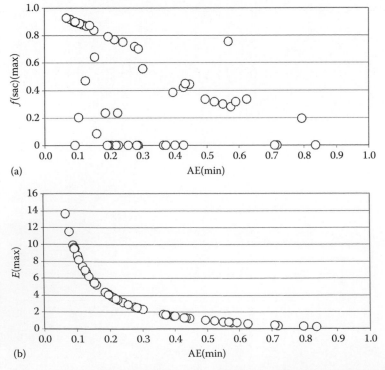

FIGURE 7.6 (a) Correlation between $f(sac)(max)$ with AE(min); and (b) correlation between $E(max)$ and AE(min) for named oxidation reactions (with respect to substrate of interest).

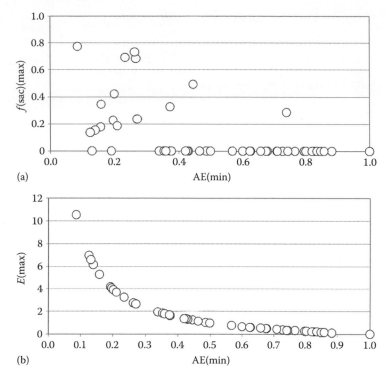

FIGURE 7.7   (a) Correlation between f(sac)(max) with AE(min); and (b) correlation between E(max) and AE(min) for named rearrangement reactions.

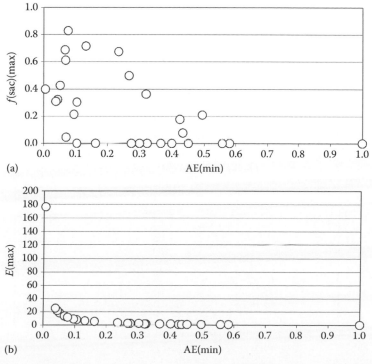

FIGURE 7.8   (a) Correlation between f(sac)(max) with AE(min); and (b) correlation between E(max) and AE(min) for named reduction reactions (with respect to substrate of interest).

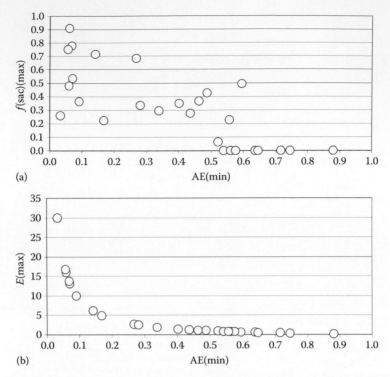

FIGURE 7.9 (a) Correlation between f(sac)(max) with AE(min); and (b) correlation between E(max) and AE(min) for named sequence reactions.

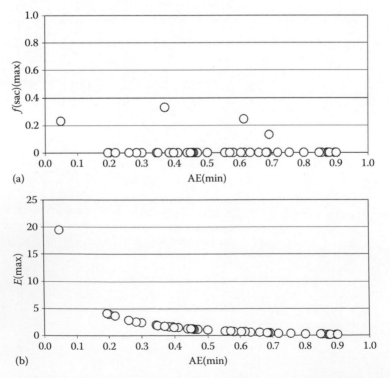

FIGURE 7.10 (a) Correlation between f(sac)(max) with AE(min); and (b) correlation between E(max) and AE(min) for named substitution reactions.

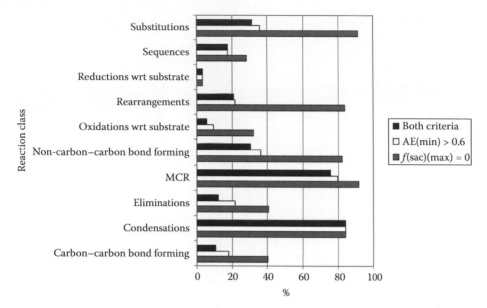

FIGURE 7.11 Distribution of reactions in each classification that satisfy the criteria AE(min) > 0.6 (minimum atom economies above 60%), f(sac)(max) = 0 (no sacrificial reagents used), and both.

Good performing reactions have high values for AE(min), low values for f(sac)(max), and low values for E(max), whereas poor performing reactions have exactly the opposite trend. When examining these figures, the good reactions are the ones depicted by data points appearing in the lower right hand corner and the poor reactions are depicted by data points appearing in the top left hand corner. E(max) and AE(min) are related by the simple expression given in Equation 7.8a and b

$$AE_{min} = \frac{1}{1 + E_{max}} \tag{7.8a}$$

$$E_{max} = \frac{1}{AE_{min}} - 1 \tag{7.8b}$$

Figure 7.11 shows the distribution of reactions in each classification that satisfy the criteria AE(min) > 0.6 (minimum atom economy above 60%), f(sac)(max) = 0 (no sacrificial reagents used), and both constraints. Not surprisingly, the best performing reaction classifications are multicomponent reactions and condensations. The worst performers are the redox reactions and eliminations. Most rearrangements, non-carbon–carbon bond forming reactions, and substitutions satisfy the condition f(sac)(max) = 0.

8 Example Transformations Illustrating Sacrificial Reagents

Examples are given of single- and two-step reactions involving sacrificial reagents whose atoms in whole or in part do not become incorporated into the target product structure. Each reaction is generalized as far as possible using Markush structures and balanced to show all by-products. Then, general expressions for and estimates of minimum atom economy (AE(min)) are determined according to Equations 8.1 and 8.2 where the minimum size R group is taken to be a methyl group.

$$E_{max} = \frac{\sum MW_{byproducts}}{MW_{target\ product}} \qquad (8.1)$$

$$AE_{min} = \frac{1}{1 + E_{max}} \qquad (8.2)$$

These kinds of reactions necessarily have low minimum atom economies particularly if the sacrificial reagents are of high molecular weight compared to the substrate of interest. A brief description of the role of the sacrificial reagents is also given.

8.1 SINGLE-STEP REACTIONS

Reaction 1

$$E_{max} = \frac{64 + 2(36.45)}{R + 59} \rightarrow \frac{136.9}{74} = 1.850; \quad AE_{min} = 0.351$$

This reaction is formally a substitution reaction of a hydroxyl group for a methoxy group. The thionyl chloride is used as a sacrificial reagent to convert the acid into an acid chloride, which has a more electrophilic carbonyl group, thereby facilitating the substitution. Hence, this example illustrates that the purpose of using a sacrificial reagent is to improve the electronics of the nucleophile and electrophile interactions.

Reaction 2

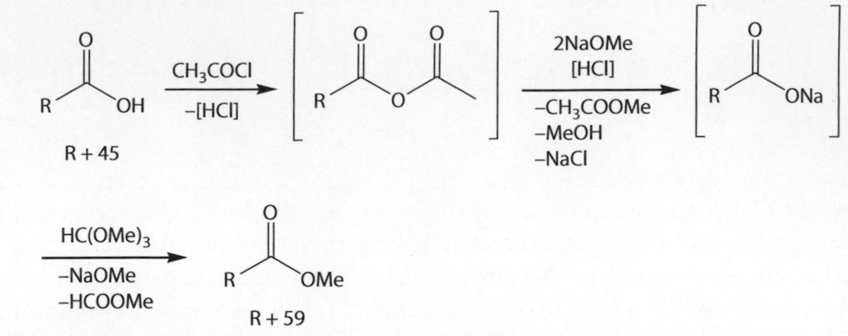

$$E_{max} = \frac{74 + 32 + 58.45 + 60}{R + 59} \rightarrow \frac{224.45}{74} = 3.033; \quad AE_{min} = 0.248$$

In this second example again, the purpose of using sacrificial reagents is to alter the electronics of the reaction; however, the roles of reacting partners are reversed. The methyl group is introduced in the form of an electrophilic reagent, trimethoxyorthoformate.

Reaction 3

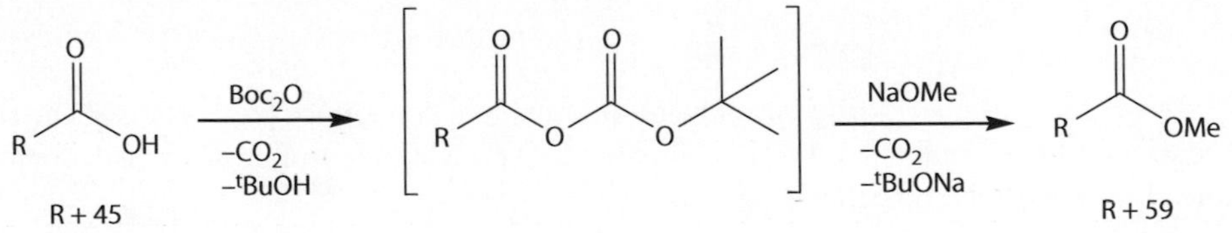

$$E_{max} = \frac{2(44) + 74 + 96}{R + 59} \rightarrow \frac{258}{74} = 3.486; \quad AE_{min} = 0.223$$

This reaction is electronically analogous to the first reaction and is facilitated by the production of carbon dioxide and *tert*-butanol as by-products.

Reaction 4

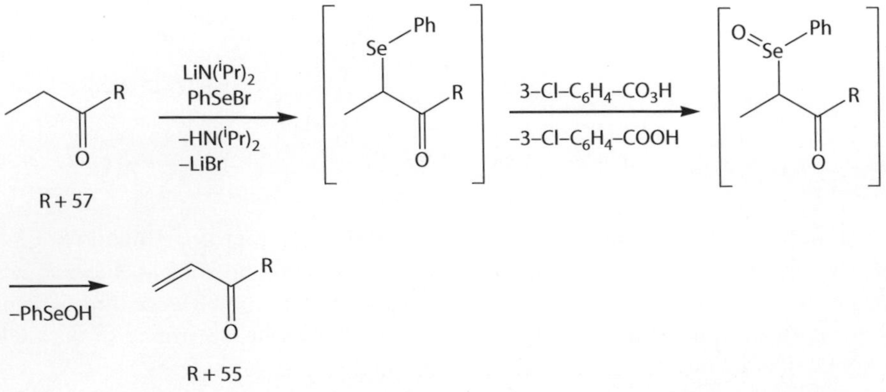

$$E_{max} = \frac{101 + 86.9 + 156.45 + 172.96}{R + 57} \rightarrow \frac{517.31}{72} = 7.185; \quad AE_{min} = 0.122$$

This reaction formally introduces a carbon–carbon double bond next to a carbonyl group. Again, the sacrificial reagents facilitate the electronic requirements of the reaction.

Reaction 5

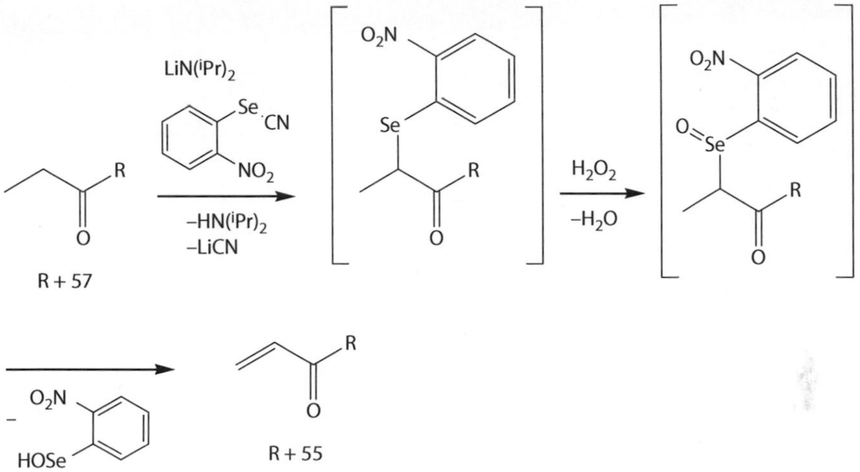

$$E_{max} = \frac{101 + 33 + 18 + 217.96}{R + 57} \rightarrow \frac{369.96}{72} = 5.138; \quad AE_{min} = 0.163$$

This example is analogous to reaction 4.

Reaction 6

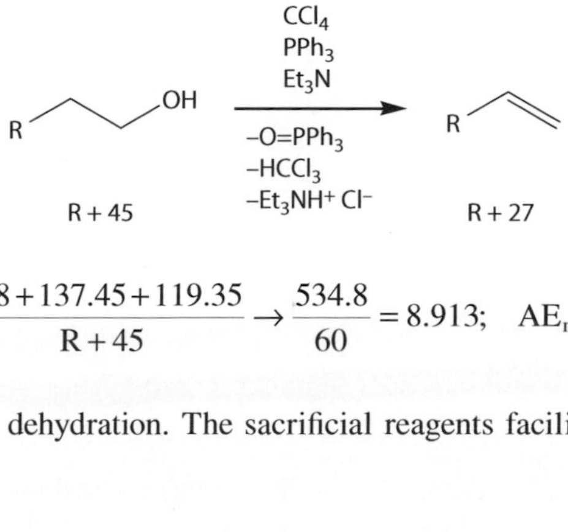

$$E_{max} = \frac{278 + 137.45 + 119.35}{R + 45} \rightarrow \frac{534.8}{60} = 8.913; \quad AE_{min} = 0.101$$

This reaction is formally a dehydration. The sacrificial reagents facilitate the electronic requirements of the reaction.

Reaction 7

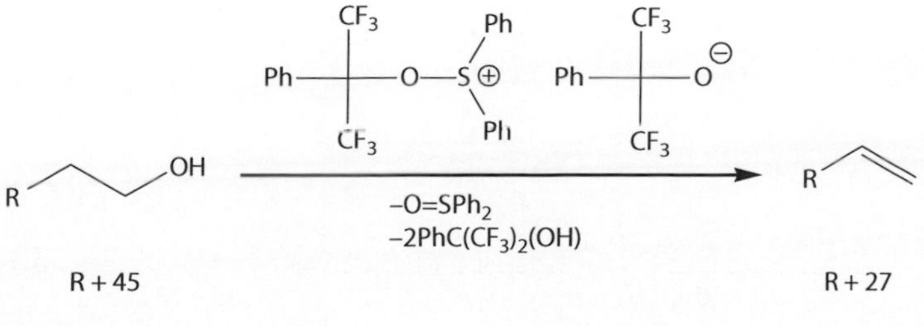

$$E_{max} = \frac{2(244) + 202}{R + 45} \rightarrow \frac{690}{60} = 11.500; \quad AE_{min} = 0.080$$

This example is analogous to reaction 6, except that a very bulky sacrificial reagent is used.

Reaction 8

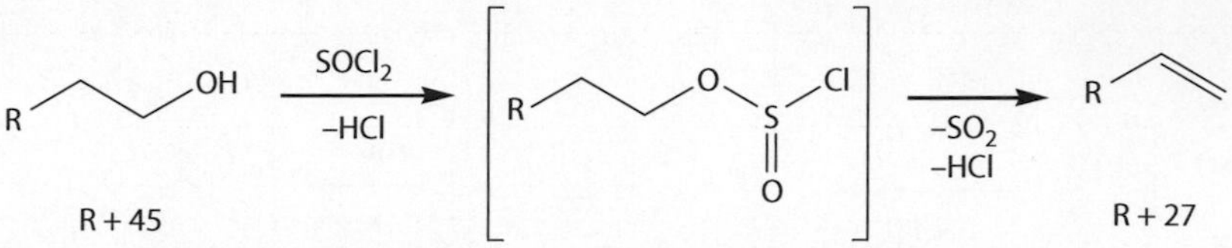

$$E_{max} = \frac{2(36.45) + 64}{R + 45} \rightarrow \frac{136.9}{60} = 2.282; \quad AE_{min} = 0.305$$

This example is analogous to reactions 6 and 7 except that a lower molecular weight sacrificial reagent is used and hence this leads to the highest minimum atom economy for this transformation.

Reaction 9

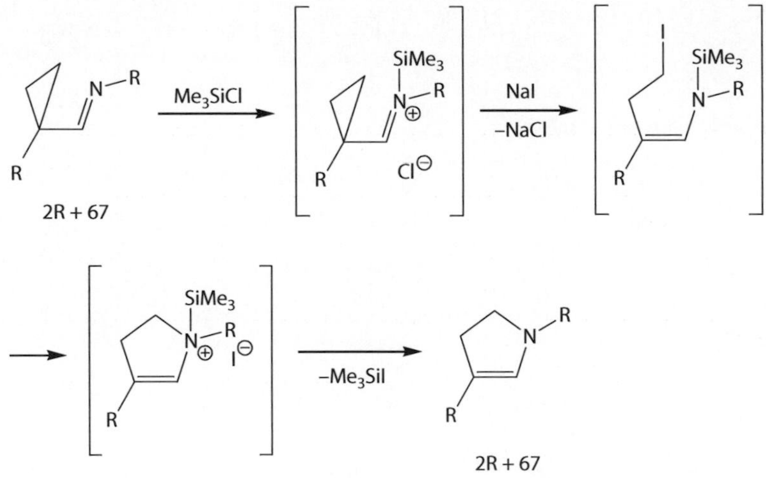

$$E_{max} = \frac{58.45 + 200}{2R + 67} \rightarrow \frac{258.45}{97} = 2.664; \quad AE_{min} = 0.273$$

This reaction is formally a rearrangement. The sacrificial reagents facilitate the electronic requirements.

Reaction 10

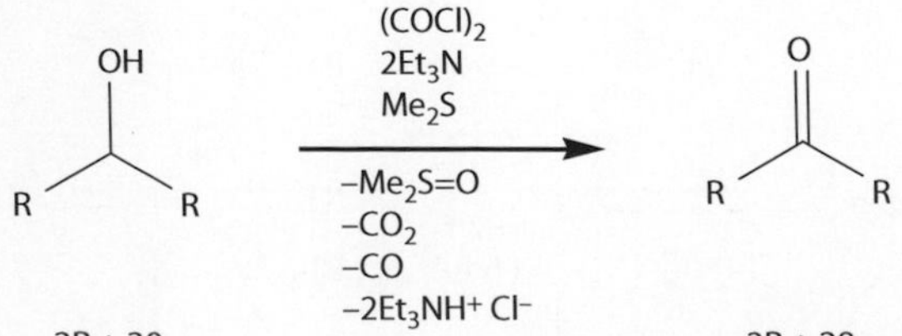

$$E_{max} = \frac{78 + 44 + 28 + 2(137.45)}{2R + 28} \rightarrow \frac{424.9}{58} = 7.326; \quad AE_{min} = 0.120$$

The Swern oxidation involves three sacrificial reagents to formally remove two hydrogen atoms from a secondary alcohol. The sacrificial reagents facilitate the electronic requirements.

Reaction 11

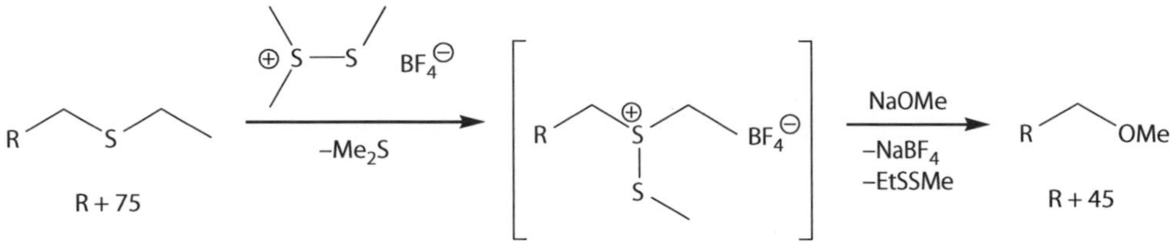

$$E_{max} = \frac{62 + 110 + 108}{R + 45} \rightarrow \frac{280}{60} = 4.667; \quad AE_{min} = 0.176$$

This reaction is formally a substitution. The sacrificial reagent is used to improve the leaving group ability of the thioethyl group.

Reaction 12

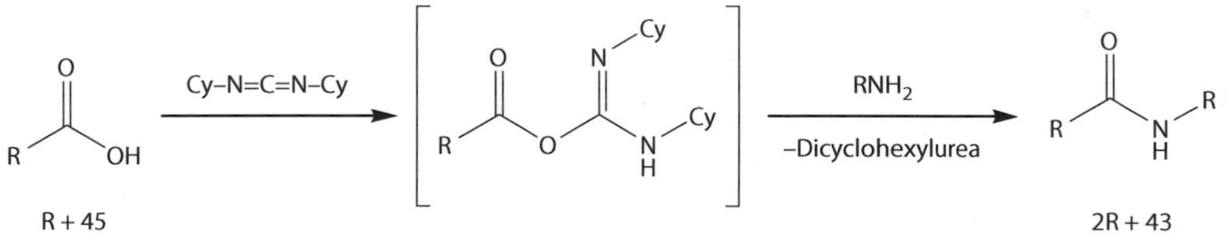

$$E_{max} = \frac{224}{2R + 43} \rightarrow \frac{224}{73} = 3.068; \quad AE_{min} = 0.246$$

This reaction is formally a condensation in which dicyclohexylcarbodiimide (DCC) acts as a sacrificial reagent. The amidation proceeds under energetically favorable conditions driven by the production of a thermodynamically stable by-product, dicyclohexylurea, that precipitates out of solution.

Reaction 13

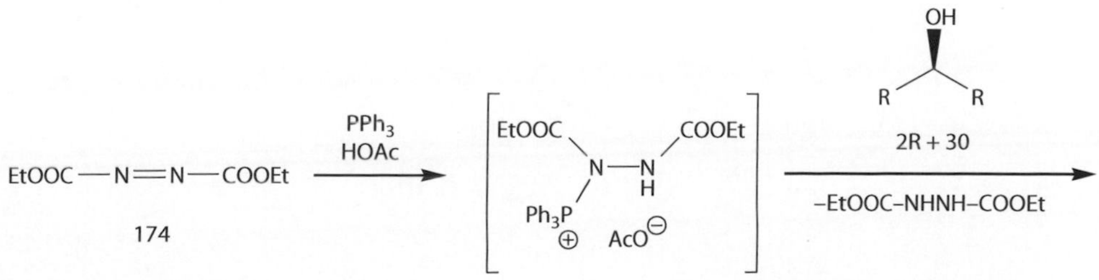

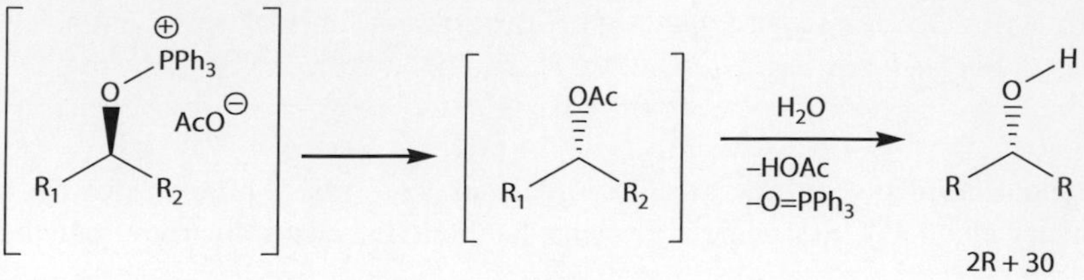

$$2R + 30$$

$$E_{max} = \frac{176 + 60 + 278}{2R + 30} \to \frac{514}{60} = 8.567; \quad AE_{min} = 0.105$$

In the Mitsunobu reaction, diethylazodicarboxylate (DEAD) and triphenylphosphine are sacrificial reagents.

8.2   TWO-STEP REACTIONS

Reaction 1

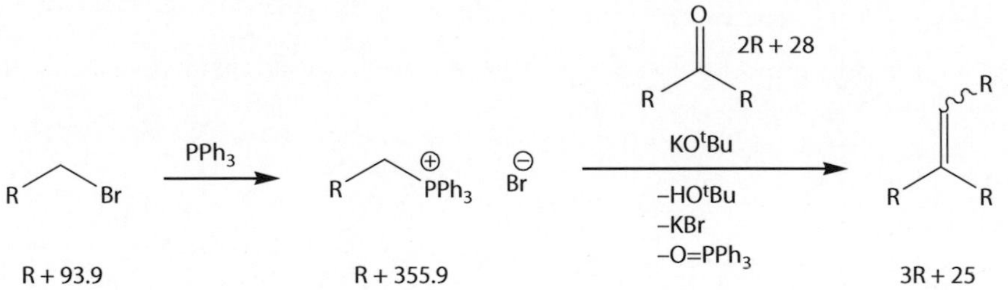

$$R + 93.9 \qquad\qquad R + 355.9 \qquad\qquad 3R + 25$$

Step 1: $E_{max} = 0; \quad AE_{min} = 1$

Step 2: $E_{max} = \dfrac{278 + 118.9 + 74}{3R + 25} \to \dfrac{470.9}{70} = 6.727; \quad AE_{min} = 0.129$

In the Wittig reaction, the sacrificial reagents are triphenylphosphine and potassium t-butoxide. The formation of the triphenylphosphonium bromide is 100% atom economical.

Reaction 2

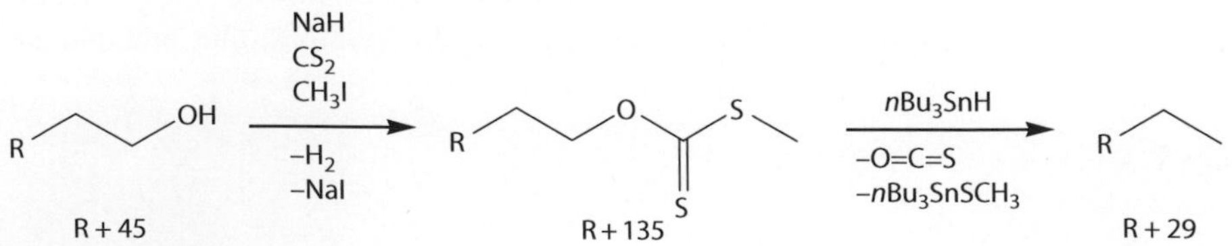

$$R + 45 \qquad\qquad R + 135 \qquad\qquad R + 29$$

Step 1: $E_{max} = \dfrac{2 + 150}{R + 135} \to \dfrac{152}{150} = 1.013; \quad AE_{min} = 0.497$

Step 2: $E_{max} = \dfrac{60 + 336.69}{R + 29} \to \dfrac{396.69}{44} = 9.016; \quad AE_{min} = 0.010$

The overall transformation is substitution of a hydroxyl group for a hydrogen atom. The sacrificial reagents are sodium hydride, carbon disulfide, and iodomethane in the first step to prepare the xanthate ester. The source of the hydrogen atom in the product is tri-*n*-butyltin hydride.

Reaction 3

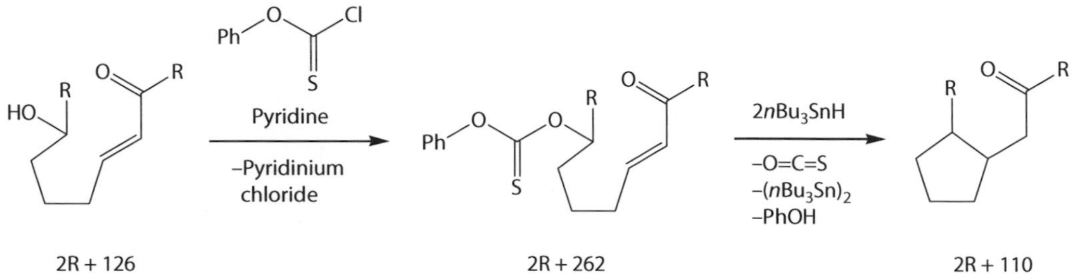

$$2R + 126 \qquad\qquad\qquad 2R + 262 \qquad\qquad\qquad 2R + 110$$

Step 1: $E_{max} = \dfrac{115.45}{2R+262} \rightarrow \dfrac{115.45}{292} = 0.395; \quad AE_{min} = 0.717$

Step 2: $E_{max} = \dfrac{94+60+579.38}{2R+110} \rightarrow \dfrac{733.38}{140} = 5.238; \quad AE_{min} = 0.160$

In this transformation, one of the tri-*n*-butyltin hydride molecules is acting as a sacrificial reagent in donating a hydrogen atom to the departing phenoxyl radical.

Reaction 4

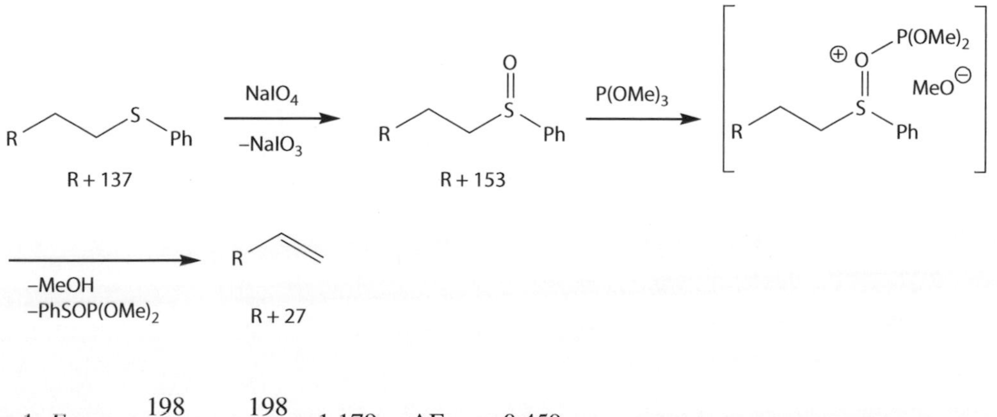

$$R + 137 \qquad\qquad\qquad R + 153$$

$$R + 27$$

Step 1: $E_{max} = \dfrac{198}{R+153} \rightarrow \dfrac{198}{168} = 1.179; \quad AE_{min} = 0.459$

Step 2: $E_{max} = \dfrac{32+218}{R+27} \rightarrow \dfrac{250}{42} = 5.952; \quad AE_{min} = 0.144$

Sodium periodate acts as a sacrificial oxidant in the first step and in the second step trimethylphosphite electronically improves the leaving group ability of the departing group.

Reaction 5

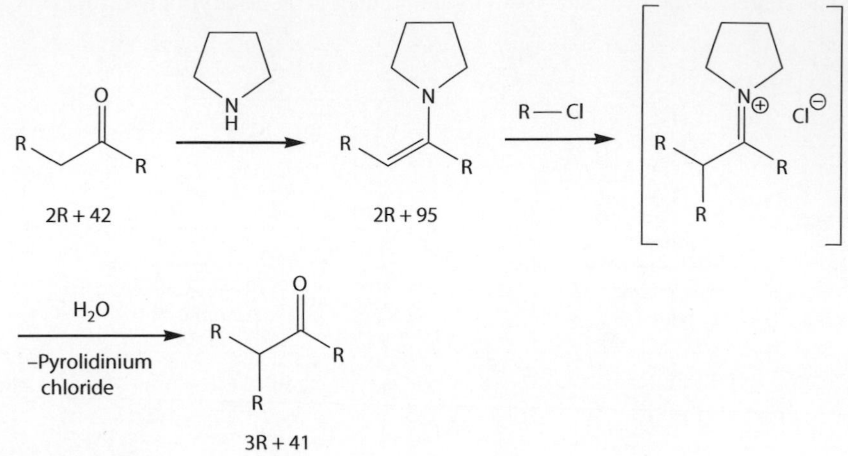

$$2R + 42$$

$$2R + 95$$

$$3R + 41$$

Step 1: $E_{max} = 0$; $AE_{min} = 1$

Step 2: $E_{max} = \dfrac{107.45}{3R + 41} \rightarrow \dfrac{107.45}{86} = 1.249$; $AE_{min} = 0.445$

In the Stork enamine synthesis, the pyrrolidine acts as a sacrificial reagent to electronically facilitate the alpha-carbon alkylation. The formation of the enamine is 100% atom economical.

Reaction 6

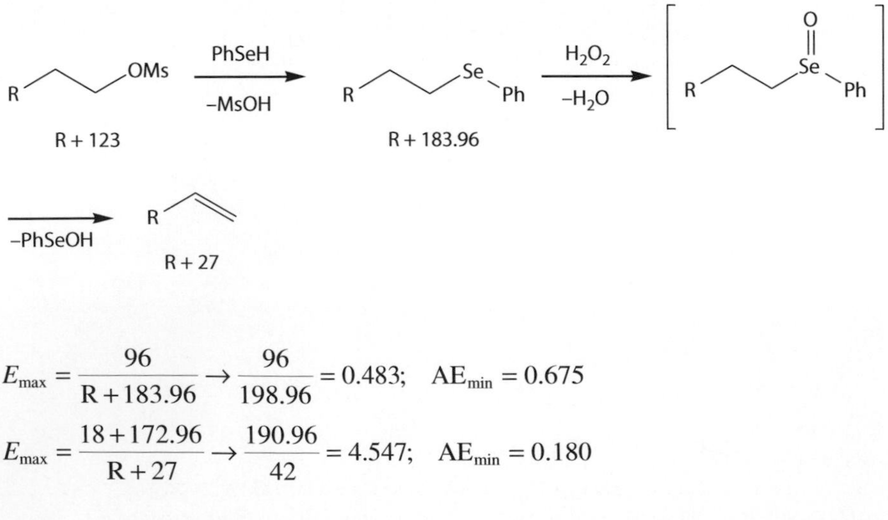

$$R + 123$$

$$R + 183.96$$

$$R + 27$$

Step 1: $E_{max} = \dfrac{96}{R + 183.96} \rightarrow \dfrac{96}{198.96} = 0.483$; $AE_{min} = 0.675$

Step 2: $E_{max} = \dfrac{18 + 172.96}{R + 27} \rightarrow \dfrac{190.96}{42} = 4.547$; $AE_{min} = 0.180$

The sequence is a formal elimination of methanesulfonic acid. The swapping of the methanesulfonyl group for a benzeneselenyl group followed by oxidation with hydrogen peroxide as sacrificial oxidant facilitates the electronics for elimination to occur.

Reaction 7

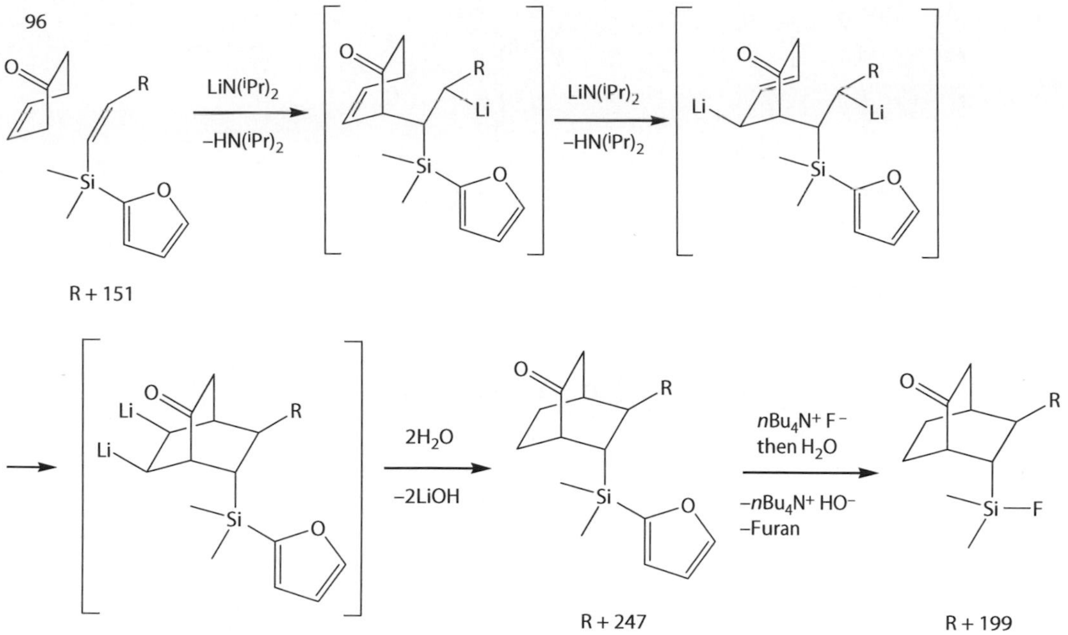

96

R + 151

R + 247

R + 199

Step 1: $E_{max} = \dfrac{2(101) + 2(24)}{R + 247} \rightarrow \dfrac{250}{262} = 0.954; \quad AE_{min} = 0.512$

Step 2: $E_{max} = \dfrac{68 + 259}{R + 199} \rightarrow \dfrac{327}{214} = 1.528; \quad AE_{min} = 0.396$

In this sequence, the furan ring acts sacrificially as a chiral directing group.

Reaction 8

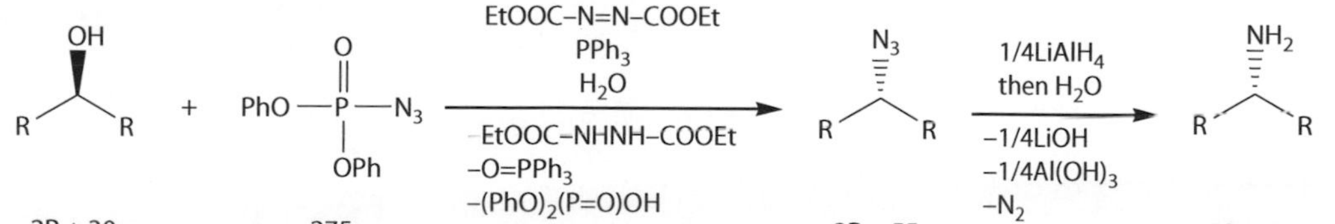

2R + 30   275   2R + 55   2R + 29

Step 1: $E_{max} = \dfrac{176 + 278 + 250}{2R + 55} \rightarrow \dfrac{704}{85} = 8.282; \quad AE_{min} = 0.108$

Step 2: $E_{max} = \dfrac{0.25(24 + 78) + 28}{2R + 29} \rightarrow \dfrac{53.5}{59} = 0.907; \quad AE_{min} = 0.524$

In this two-step Mitsunobu azidation-hydrogenation sequence, DEAD and triphenylphosphine are sacrificial reagents as in reaction 13 in the previous section. This may be compared with the two-step sequence shown below for the same transformation involving substitution of a hydroxyl group for an amino group with inversion of configuration. In this second method, the amino nitrogen atom originates from phthalimide and the hydrogen atoms originate from hydrazine, so both reagents are not sacrificial in this case.

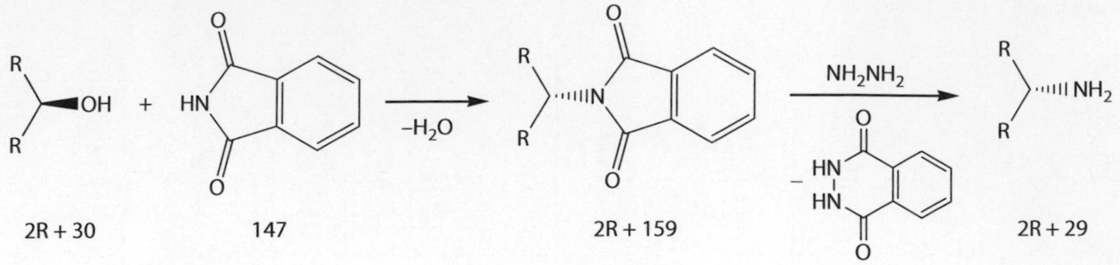

$$\text{Step 1: } E_{max} = \frac{18}{R+159} \rightarrow \frac{18}{174} = 0.103; \quad AE_{min} = 0.906$$

$$\text{Step 2: } E_{max} = \frac{162}{2R+29} \rightarrow \frac{162}{59} = 2.746; \quad AE_{min} = 0.267$$

Reaction 9 [1,2]

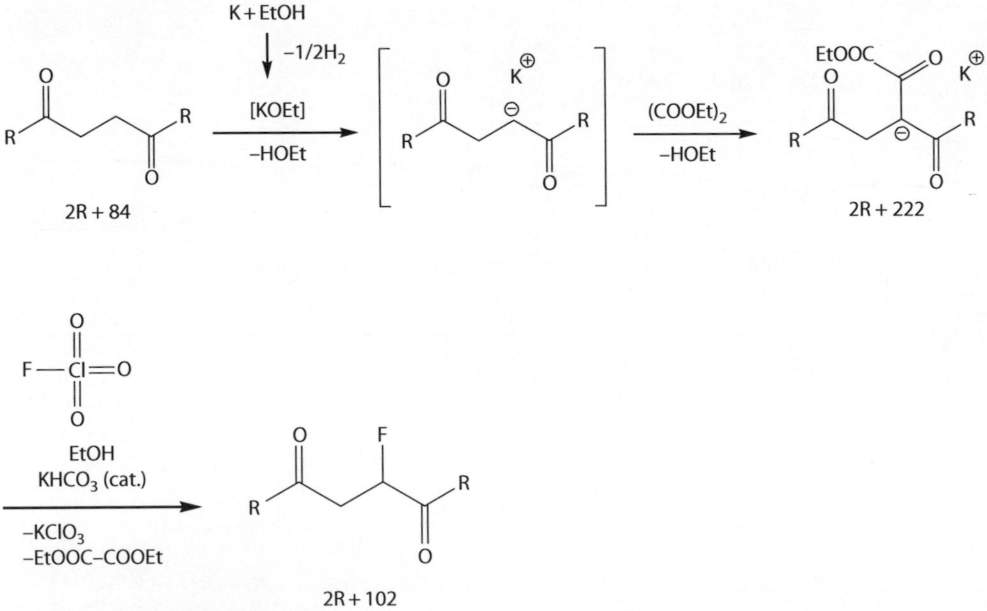

$$\text{Step 1: } E_{max} = \frac{0.5(2)+2(46)}{2R+222} \rightarrow \frac{93}{252} = 0.369; \quad AE_{min} = 0.730$$

$$\text{Step 2: } E_{max} = \frac{146+122.45}{2R+102} \rightarrow \frac{268.45}{132} = 2.034; \quad AE_{min} = 0.330$$

In this alpha-fluorination reaction sequence, diethyl oxalate is a sacrificial reagent that facilitates the electrophilic transfer of a fluorine atom from perchlorylfluoride.

Reaction 10

$$\text{Step 1: } E_{max} = \frac{139.65}{R+310} \rightarrow \frac{139.65}{325} = 0.430; \quad AE_{min} = 0.699$$

$$\text{Step 2: } E_{max} = \frac{316}{2R+44} \rightarrow \frac{316}{74} = 4.270; \quad AE_{min} = 0.190$$

TABLE 8.1
Summary of Sacrificial Reactions Appearing in the Compiled Synthesis Database

Target Compound	Plan	Sacrificial Reaction
Alizarin	Graebe-Liebermann (1869)	Salt switching step to aid in crystallization and purification (sodium to calcium) (step 5)
Astaxanthin	Roche G2 (1981)	Sacrificial chiral directing groups in step 3 (methyl) and step 8 (dibromination)
(±)-cocaine	Tufariello (1979)	Methyl acrylate in step 3, then removed in step 6
(−)-codeine	Overman (1993)	5-aminodibenzosuberane ring system used to deliver N-atom in step 6*; ring removed by hydrogenation in step 12
(S)-coniine	Gramain (2000)	(R)-1-phenylethylamine chiral directing group introduced in step 3 to effect a stereo-selective reduction of an iminium ion
(S)-coniine	Hurvois (2006)	(S)-1-phenylethylamine chiral directing group introduced in step 4 to effect a stereo-selective reductive elimination of a nitrile moiety
(S)-coniine	Shipman (1996)	S-1-phenylethylamine chiral directing group introduced in step 3 to effect a stereo-selective reduction of an iminium ion
(−)-kainic acid	Anderson (2005)	(−)-8-phenylmenthol used as chiral directing group in step 8
(−)-kainic acid	Lautens (2005)	(−)-menthol chiral directing group in step 2*
(−)-kainic acid	Ogasawara G1 (1997)	Cyclopentadiene used in step 4, then removed in step 12
(−)-kainic acid	Poisson G1 (2005)	9-phenyl-9-bromofluorene introduced in step 14*, then removed as 9-phenylfluorene in step 21
(−)-paroxetine	Ferrosan (1975)	(−)-menthol used as chiral resolving group in step 4
Platensimycin	Nicolaou G4 (2007)	(S,S)-pseudoephedrine used as chiral auxiliary in step 1 to effect stereoselective C-alkylation in step 2
Quinine	Aggarwal (2010)	(−)-isothiocineole used as a chiral directing group in step 14
(−)-swainsonine	Trost (1995)	Anthracene used in step 1, then removed in step 8
Vitamin E	B. Breit (2007)	2-diphenylphosphanyl benzoic acid chiral auxiliary introduced in step 10** used to direct stereoselectivity in steps 11** and 16

* refers to second branch in plan.
** refers to third branch in plan.

In this example, the sacrificial (+)-B-chlorodiisopinocampheylborane acts as a chiral directing group to effect a stereoselective Grignard reaction.

8.3   SACRIFICIAL REACTIONS FROM SYNTHESIS DATABASE

In this section, various sacrificial reactions found in the compiled synthesis database are summarized in Table 8.1. The actual reactions may be found by looking up the plans in the accompanying download.

REFERENCES

1. Friedman, L.; Kosower, E. *Org. Synth. Coll.* 1955, *3*, 510.
2. Dean, F.H.; Pattison, F.L.M. *Can. J. Chem.* 1963, *41*, 1833.

9 Synthesis Strategies

9.1 BRAINSTORMING EXERCISES

The concept of synthesis strategy is the hallmark of organic chemistry. After one has learned about and amassed a library of organic reaction types, one uses this database to develop and invent synthesis plans composed of various combinations of these reactions in some kind of organized sequence to a complex target structure. Though it is an idea that is difficult to define and parameterize, it is often best illustrated by examples. This has been the tradition of any type of compilation on organic chemistry. But what are the mechanics of how to come up with a synthesis strategy to a given target molecule? Retrosynthetic analysis provides a path for achieving this. When faced with a difficult problem, an effective method of solution is to think and work backward in the reverse sense. This idea, well known in mathematical science as inverse problem solving, provides elegant and beautiful solutions, especially when complementary algebraic and geometric solutions to a problem are found. A good mathematical analogy is factorization of polynomials or factorization of natural numbers into their constituent prime numbers. Retrosynthesis may be thought of in the same way. In effect, retrosynthetic analysis is the solution to the inverse problem of coming up with a route or plan to a target molecule. We may state the problem as follows: find a set of building block starting materials that can assemble to form the complex target molecule. When we add green chemistry thinking to this problem, the problem may be posed as follows: find an optimal set of building block starting materials that can assemble to form the complex target molecule, where optimal refers to building blocks that are used in their entirety and that are benign.

The exercise of analyzing any published synthesis to come up with a target bond map of the target structure as shown in Chapter 5 and Section 6.3 is in fact the solution to the structural "factorization" problem. The map tells us which bonds were made over the course of the synthesis, when those bonds were made via the step count number, and which parts of the starting materials ended up in the final target structure. At the heart of parsing a given chemical structure by connectivity analysis are the following points:

1. Bonds may be thought of as the algebraic sum of electrophilic (pluses) and nucleophilic (minuses) centers, or the sum of homophilic centers (radicals).
2. The problem is fundamentally a combinatorial one with many possible ways or solutions of assembly.
3. Any appropriate combination of pluses and minuses can theoretically work, that is to say, any kind of chemical reaction with appropriate combinations of reacting centers is possible; the key of course is to find the right reaction conditions for this to happen; in a practical sense this usually means finding the magic catalyst that will do the trick, that is overcoming the thermodynamic constraints of enthalpy and entropy that constitute what chemists call the energy barrier for a reaction.
4. Open-minded brainstorming without being biased by any constraints leads to paths and solutions that are unexpected and therefore result in new discoveries.

The modern literature of organic synthesis methodology shows quite clearly that chemists are moving in the direction of finding all possible ways to make any given type of functional group transformation. This is the bedrock of new reaction discovery. Often, in the process, plus–plus or minus–minus algebraic bond addition becomes possible using appropriate metal-mediated catalysts or reagents. This leads to bond making combinations that were once thought to be inconceivable and impossible. A classic example of this is the Grignard reaction that involves the coupling of two electrophilic (plus) centers via elemental magnesium: a carbonyl center and an alkyl or aryl halide center. As a consequence of this same sign addition there must be an in situ change in oxidation number in one of the coupling centers for this to happen. In the Grignard reaction, the alkyl or aryl halide carbon center decreases its oxidation number in order for it to be a nucleophilic center so that it can couple with the carbonyl carbon partner of higher oxidation state. Hence, the original plus–plus combination is changed to a minus–plus combination so that a chemical bond is made. The same arguments can be made about any modern cross-coupling reaction mediated by metal, especially palladium, catalysts that are so ubiquitous today. In effect, modern day organic chemists are increasing the magnitude of the set of possible reactions, that is, increasing the number of fundamental reactions forming the database upon which synthetic chemists can draw upon to use in their quest to come up with viable synthetic routes to useful target molecules. Green chemists try to turn these into practical, safe, and optimal reactions.

Here we present four simple brainstorming exercises that illustrate the ideas discussed earlier. These are especially useful for students to expand their minds in thinking about ways how structures could be built by as many different strategies as possible.

9.1.1 NITRILE TO CARBOXYLIC ACID TRANSFORMATION

Acid catalyzed route:

$$R-C\equiv N \ + \ 2H_2O + HCl \ \longrightarrow \ R\overset{O}{\underset{}{\diagup}}OH \ + \ NH_4Cl$$

$$R+26 \qquad 2(18) \quad 36.45 \qquad\qquad R+45 \qquad\qquad 53.45$$

$$E = \frac{53.45}{R+45}; \quad E_{max} = \frac{53.45}{15+45} = 0.891; \quad AE_{min} = 0.529$$

$$f(sac) = \frac{36.45}{R+26+2(18)+36.45}; \quad f(sac)(max) = \frac{36.45}{15+98.45} = 0.321$$

	a	b	c	delta row	
1	3	−2	−2	0	
0	3	−2	−2	0	
				0	sum row
delta column	0	0	0	0	sum column

$$HI = \frac{\sum_j \Delta_j}{N+1} = \frac{1}{2}(0) = 0 \text{ (isohypsic)}$$

Note that though this reaction is acid catalyzed where the concentration of acid exceeds that of the nitrile, the acid will react with the liberated ammonia by-product in a secondary acid-base neutralization reaction. Hence, the acid appears as a reactant in the overall balanced chemical equation. In choosing this route, the R group would need to be resistant to acid. The transformation does not involve any changes in oxidation state of atoms forming the target bonds.

Oxidation—diazotization—substitution sequence:

$$E = \frac{120}{R+44}; \quad E_{max} = \frac{120}{15+44} = 2.034; \quad AE_{min} = 0.330$$

$$E = \frac{53.45 + 2(18)}{R+45}; \quad E_{max} = \frac{89.45}{15+45} = 1.203; \quad AE_{min} = 0.454$$

$$f(sac) = \frac{104 + 69 + 36.45}{R + 26 + 34 + 104 + 69 + 36.45 + 18} = \frac{209.45}{R + 287.45}; \quad f(sac)(max) = \frac{209.45}{15 + 287.45} = 0.693$$

	a	b	c	delta row	
2	3	-2	-2	0	
1	3	-2	-2	0	
0	3		-1	1	
				1	sum row
delta column	0	0	1	1	sum column

$$HI = \frac{\sum_j \Delta_j}{N+1} = \frac{1}{3}(1) = 0.333 \text{ (hyperhypsic)}$$

This is a two-step sequence than can be used when R is an acid-sensitive group. The price of carrying out this methodology is the overall low minimum atom economy (AE) compared with the acid catalyzed option, which is a more direct synthesis. Note that in the second step, the water molecule added after the formation of the diazonium salt is not cancelled with the two water molecules formed prior since part of this water molecule ends up in the final product structure. In the first step, the water and sodium hydroxide molecules cancel each other out in the overall balanced chemical equation.

Since the entire transformation is hyperhypsic, a net reduction has occurred. This corresponds to one of the oxygen atoms in hydrogen peroxide that is being reduced. The sulfur atom in sodium bisulfite is oxidized to sodium bisulfate, but since the sodium bisulfite reagent is a sacrificial one the oxidation number change at sulfur does not count in the calculation of the hypsicity index.

9.1.2 MAKING A METHYL ESTER

Acid catalyzed route (Fischer esterification):

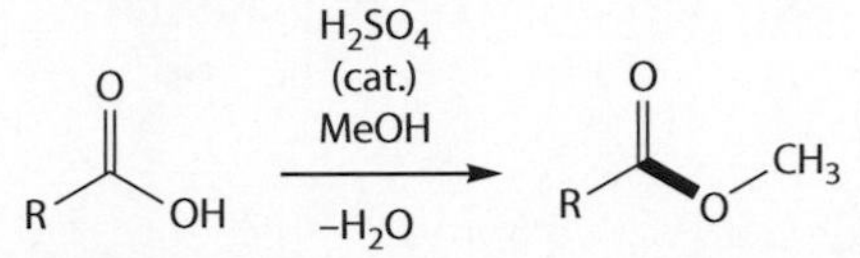

$$E = \frac{18}{R+59}; \quad E_{max} = \frac{18}{15+59} = 0.243; \quad AE_{min} = 0.804$$

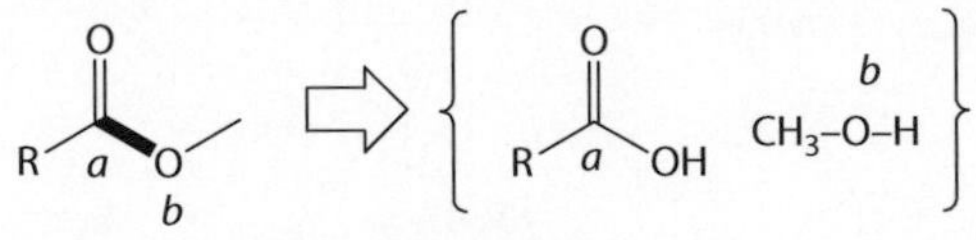

$$f(sac) = 0$$

	a	b	delta row	
1	3	−2	0	
0	3	−2	0	
			0	sum row
delta column	0	0	0	sum column

$$HI = \frac{\sum_j \Delta_j}{N+1} = \frac{1}{2}(0) = 0 \text{ (isohypsic)}$$

Saponification—Substitution sequence:

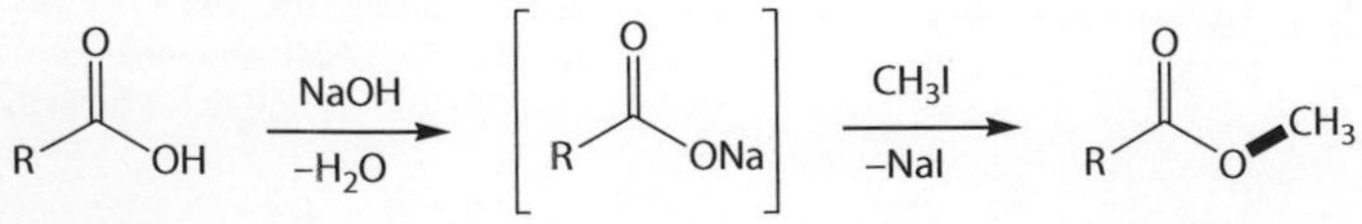

$$E = \frac{18+150}{R+59}; \quad E_{max} = \frac{168}{15+59} = 2.270; \quad AE_{min} = 0.306$$

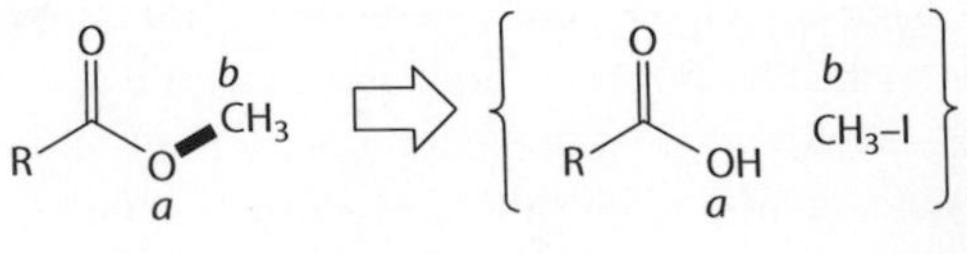

$$f(\text{sac}) = \frac{40}{R+45+40+142}; \quad f(\text{sac})(\max x) = \frac{40}{15+227} = 0.165$$

	a	b	delta row	
1	–2	–2	0	
0	–2	–2	0	
			0	sum row
delta column	0	0	0	sum column

$$HI = \frac{\sum_j \Delta_j}{N+1} = \frac{1}{2}(0) = 0 \text{ (isohypsic)}$$

Diazomethane method:

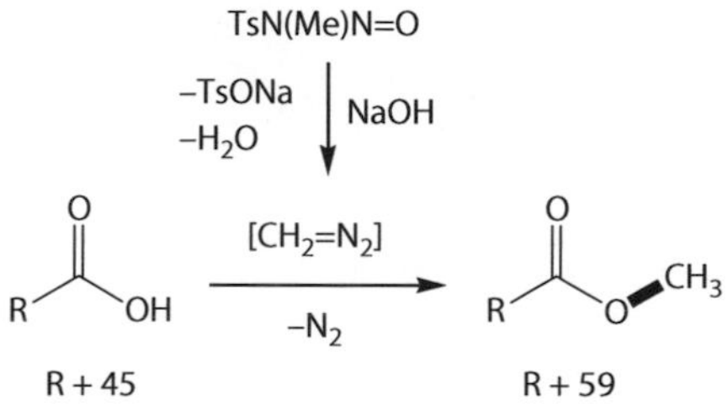

TsN(Me)N=O

–TsONa | NaOH
–H₂O ↓

R + 45   [CH₂=N₂]   –N₂   R + 59

$$E = \frac{194+18+28}{R+59}; \quad E_{max} = \frac{240}{15+59} = 3.243; \quad AE_{min} = 0.236$$

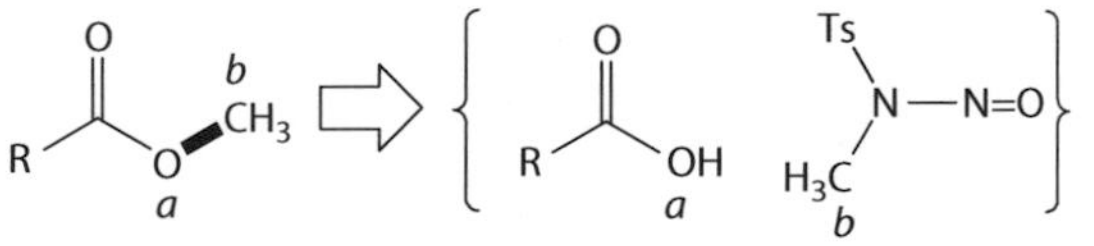

$$f(\text{sac}) = \frac{40}{R+45+214+40}; \quad f(\text{sac})(\max) = \frac{40}{15+299} = 0.127$$

	a	b	delta row	
1	–2	–2	0	
0	–2	–2	0	
			0	sum row
delta column	0	0	0	sum column

$$HI = \frac{\sum_j \Delta_j}{N+1} = \frac{1}{2}(0) = 0 \text{ (isohypsic)}$$

Carbonylation:

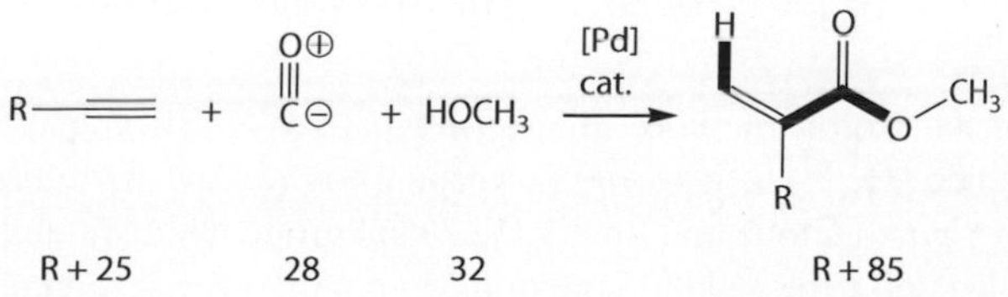

R + 25   28   32   R + 85

$$E = 0; \quad AE = 1$$

$$f(\text{sac}) = 0$$

	a	b	c	d	e	delta row	
1	1	−2	0	3	−2	0	
0	1	−1	0	2	−2	0	
						0	sum row
delta column	0	1	0	−1	0	0	sum column

$$\text{HI} = \frac{\sum_j \Delta_j}{N+1} = \frac{1}{2}(0) = 0 \text{ (isohypsic)}$$

Cyanohydrin–oxidation–substitution sequence (Corey method) [1]:

$$E = \frac{86.94 + 27}{R + 59}; \quad E_{\max} = \frac{113.94}{15 + 59} = 1.540; \quad AE_{\min} = 0.394$$

$$f(\text{sac}) = \frac{27}{R + 29 + 27 + 86.94 + 32}; \quad f(\text{sac})(\max x) = \frac{27}{15 + 174.94} = 0.142$$

	a	b	delta row	
1	−2	3	0	
0	−2	1	−2	
			−2	sum row
delta column	0	−2	−2	sum column

$$\text{HI} = \frac{\sum_j \Delta_j}{N+1} = \frac{1}{2}(-2) = -1 \text{ (hypohypsic)}$$

With the exception of the Corey method, all transformations are isohypsic. The Corey method involves hydrogen cyanide (HCN) as a sacrificial reagent to produce the intermediate cyanohydrin that is then oxidized to a cyanoketone and finally the cyano group is substituted by a methoxy group. The Fischer esterification and carbonylation methods do not involve sacrificial reagents. The former has an AE(min) of 80% and the latter is 100% atom economical. However, the carbonylation method is specific to making esters with olefinic R groups that originate from acetylenic precursors. The diazomethane method is the least atom economical and involves the use of a hazardous reagent.

9.1.3 Making an Amide

Acid chloride—Schotten–Baumann sequence:

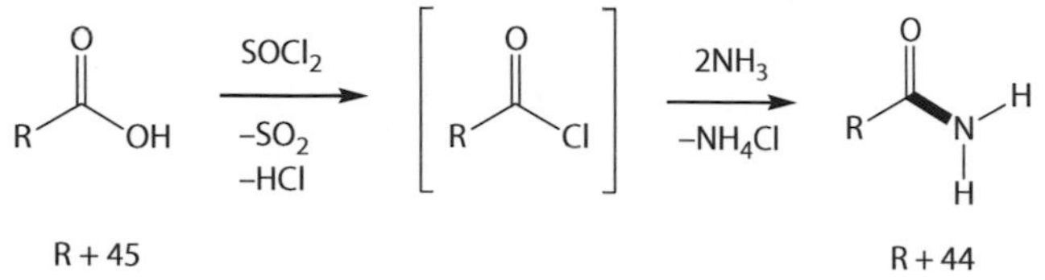

$$E = \frac{64 + 36.45 + 53.45}{R + 44}; \quad E(\max) = \frac{153.9}{15 + 44} = 2.608; \quad AE(\min) = 0.277$$

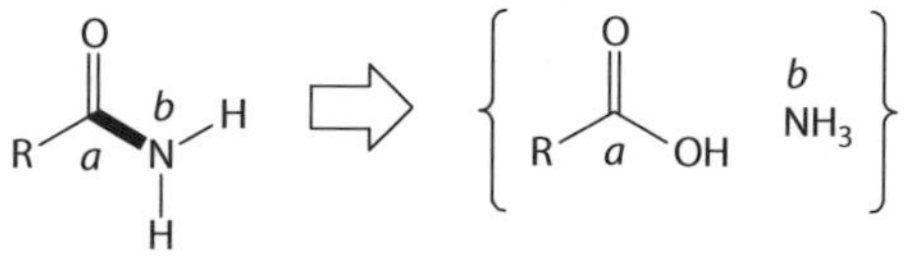

$$f(\mathrm{sac}) = \frac{118.9 + 17}{R + 45 + 118.9 + 2(17)}; \quad f(\mathrm{sac})(\max) = \frac{135.9}{15 + 197.9} = 0.638$$

	a	b	delta row	
1	3	−3	0	
0	3	−3	0	
			0	sum row
delta column	0	0	0	sum column

$$\mathrm{HI} = \frac{\sum_j \Delta_j}{N + 1} = \frac{1}{2}(0) = 0 \text{ (isohypsic)}$$

Carbodiimide method:

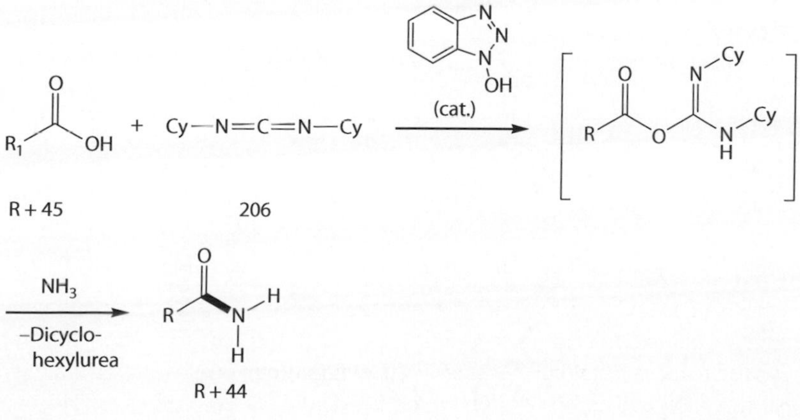

$$E = \frac{224}{R+44}; \quad E_{max} = \frac{224}{15+44} = 3.797; \quad AE_{min} = 0.208$$

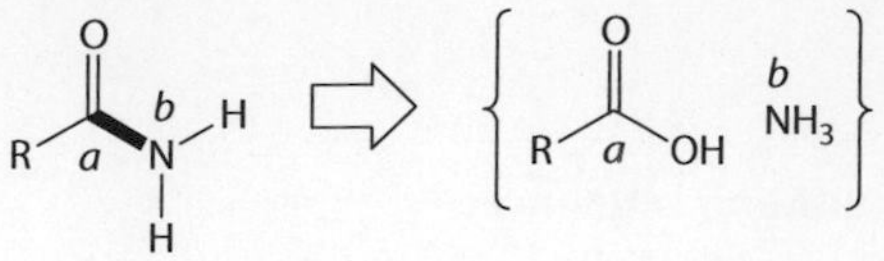

$$f(sac) = \frac{206}{R+45+206+17}; \quad f(sac)(max) = \frac{206}{15+268} = 0.728$$

	a	b	delta row	
1	3	-3	0	
0	3	-3	0	
			0	sum row
delta column	0	0	0	sum column

$$HI = \frac{\sum_j \Delta_j}{N+1} = \frac{1}{2}(0) = 0 \text{ (isohypsic)}$$

Carbonyldiimidazole method:

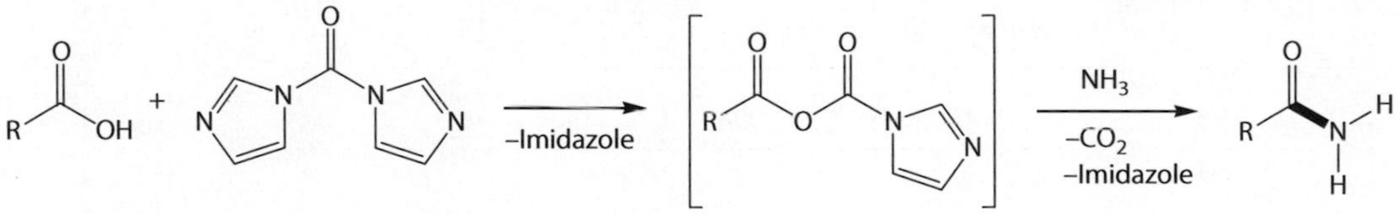

R + 45   162   R + 44

$$E = \frac{2(68)+44}{R+44}; \quad E_{max} = \frac{180}{15+44} = 3.051; \quad AE_{min} = 0.247$$

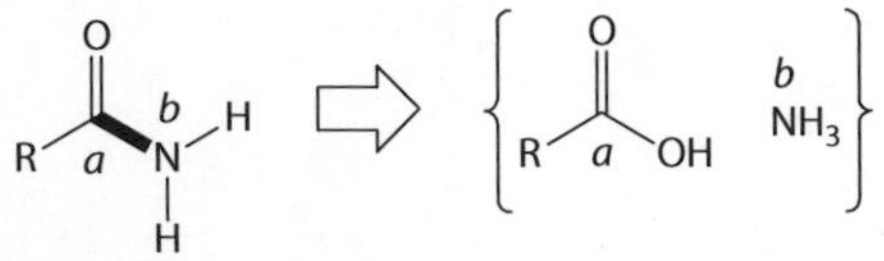

$$f(sac) = \frac{162}{R+45+162+17}; \quad f(sac)(max) = \frac{162}{15+224} = 0.678$$

	a	b	delta row	
1	3	-3	0	
0	3	-3	0	
			0	sum row
delta column	0	0	0	sum column

$$HI = \frac{\sum_j \Delta_j}{N+1} = \frac{1}{2}(0) = 0 \text{ (isohypsic)}$$

Nitrile hydrolysis:

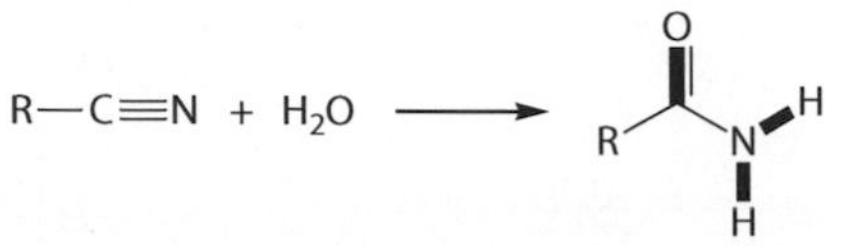

$$R-C\equiv N \; + \; H_2O \longrightarrow \underset{R}{\overset{O}{\big|\big|}}{\diagdown}N{\diagup}^{H}_{H}$$

$$E = 0; \quad AE = 1$$

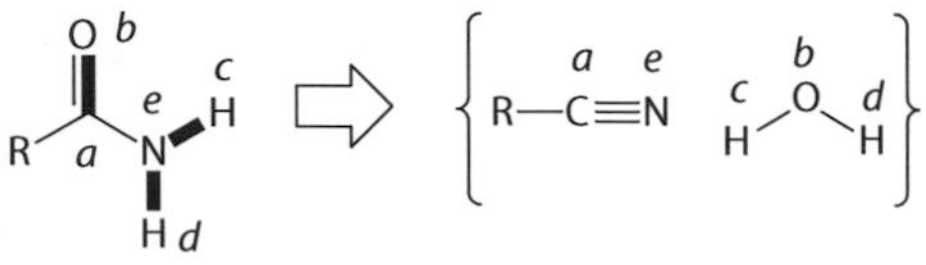

$$f(\text{sac}) = 0$$

	a	b	c	d	e	delta row	
1	3	−2	1	1	−3	0	
0	3	−2	1	1	−3	0	
						0	sum row
delta column	0	0	0	0	0	0	sum column

$$HI = \frac{\sum_j \Delta_j}{N+1} = \frac{1}{2}(0) = 0 \text{ (isohypsic)}$$

Oxidation of nitriles:

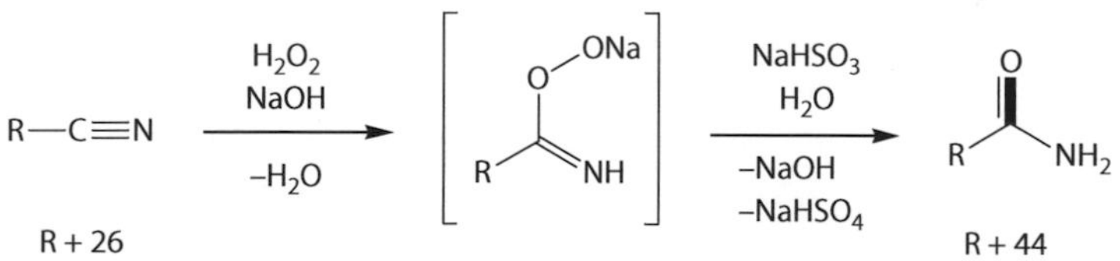

$$R + 26 \qquad\qquad\qquad R + 44$$

$$E = \frac{120}{R+44}; \quad E_{max} = \frac{120}{15+44} = 2.034; \quad AE_{min} = 0.330$$

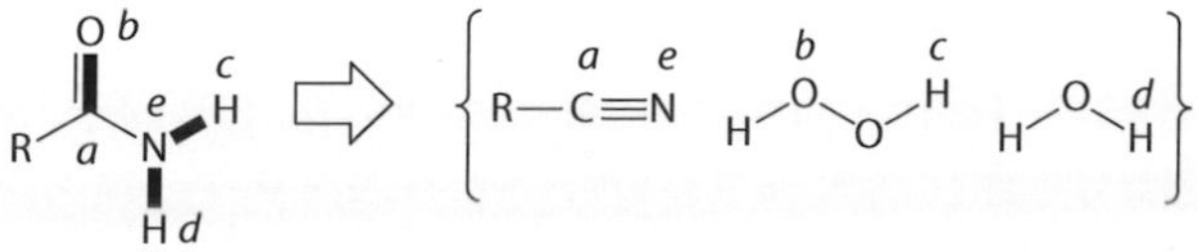

$$f(\text{sac}) = \frac{104}{R+26+34+104}; \quad f(\text{sac})(\text{max}) = \frac{104}{15+164} = 0.581$$

	a	b	c	d	e	delta row	
1	3	−2	1	1	−3	0	
0	3	−1	1	1	−3	1	
						1	sum row
delta column	0	1	0	0	0	1	sum column

$$HI = \frac{\sum_j \Delta_j}{N+1} = \frac{1}{2}(1) = \frac{1}{2} \text{ (hyperhypsic)}$$

Umpolung method [2]:

(1) $\quad R_1 \diagdown NO_2 \xrightarrow[\substack{-H_2O \\ -KBr}]{\substack{KOH \\ Br_2}} R_1 \diagdown NO_2$ (Br)

$R + 60 \qquad\qquad\qquad R + 138.9$

$$E = \frac{18 + 118.9}{R + 60}; \quad E_{max} = \frac{136.9}{15 + 60} = 1.825; \quad AE_{min} = 0.354$$

–Succinimide $\Big|$ NH_3

(2) $\quad R_1 \diagdown NO_2$ (Br) $\xrightarrow[\substack{-KHCO_3 \\ -KI}]{\substack{[NH_2I] \\ K_2CO_3}} \left[R_1 \diagdown NO_2 \text{ (Br, NH}_2) \right] \xrightarrow[\substack{-KBr \\ -KHCO_3}]{\substack{H_2O \\ K_2CO_3}} \left[R_1 \diagdown NO_2 \text{ (OH, NH}_2) \right] \xrightarrow[-HNO_2]{} R \diagdown N{-}H \text{ (O, H)}$

$R + 138.9 \qquad\qquad\qquad\qquad\qquad\qquad\qquad\qquad\qquad\qquad\qquad\qquad R + 44$

$$E = \frac{99 + 100 + 166 + 118.9 + 100 + 47}{R + 44}; \quad E_{max} = \frac{630.9}{15 + 44} = 10.693; \quad AE_{min} = 0.0855$$

$$f(sac) = \frac{56 + 159.8 + 225 + 2(138)}{R + 60 + 56 + 159.8 + 225 + 17 + 2(138) + 18}; \quad f(sac)(max) = \frac{716.8}{15 + 811.8} = 0.867$$

	a	b	c	delta row	
2	3	–3	–2	0	
1	1	–3	–2	–2	
0	–1			–4	
				–6	sum row
delta column	–6	0	0	–6	sum column

$$HI = \frac{\sum_j \Delta_j}{N + 1} = \frac{1}{3}(-6) = -2 \text{ (hypohypsic)}$$

Not surprisingly, the hydrolysis of nitriles is the most atom economical reaction. The isohypsic transformations are the acid chloride–Schotten–Baumann sequence, carbodiimide and carbonyl-diimidazole methods, and the hydrolysis of nitriles. The oxidation of nitriles with hydrogen peroxide is hyperhypsic for the reasons mentioned in Section 9.1.1 for the nitrile to carboxylic acid transformation. The umpolung method, which electronically reverses the roles of the reacting partners, is hypohypsic since a net oxidation has occurred at the reactive carbon atom. This is a novel transformation and is useful for substituted amines having chiral substituents since no epimerization occurs under the reaction conditions, unlike the other methods. However, this novelty comes at a high price. The transformation has the lowest minimum AE and uses the highest fraction of sacrificial reagents. This example clearly illustrates the kind of trade-offs chemists face in manipulating the electronics of functional groups in chemical transformations and in developing efficient reactions. Usually, gains made in one direction are offset by the other.

9.1.4   CYCLOHEXANONE RING CONSTRUCTION

Ring construction is most challenging and is of high interest to synthetic organic chemists. The main hurdles are designing building blocks with the correct positive and negative ends and overcoming thermodynamic barriers, namely ring strain for small rings and entropy for large rings. Chapter 10 is dedicated to this important endeavor. Here we present a simple exercise in thinking about novel and alternative ways of constructing a six-membered ring compound. In exploring options for the synthesis of cyclohexanone, it is useful to first scope out all possible one-bond and two-bond disconnections. The target bonds are bolded for clarity. Noting the plane of symmetry in the molecule to eliminate redundancies the possibilities are outlined in the following.

One-bond disconnection combinations:

$$[6+0] \qquad [6+0] \qquad [6+0]$$

Two-bond disconnection combinations:

$$[5+1]$$

$$[4+2]$$

$$[3+3]$$

Plus-minus options for one-bond disconnections

A   B   C

Possible schemes that would satisfy the plus–minus combinations are given in the following.

Option A

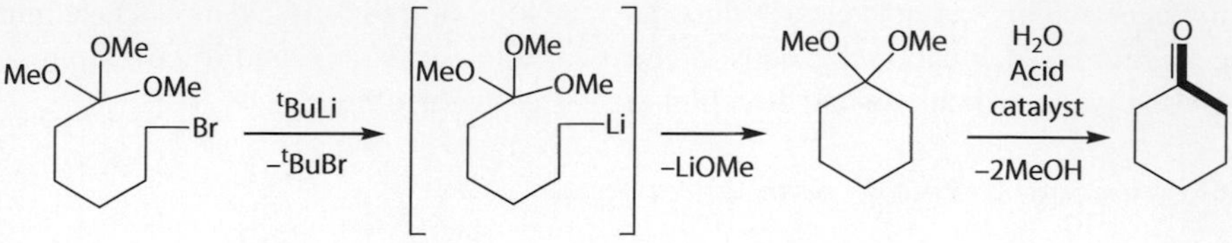

Option B

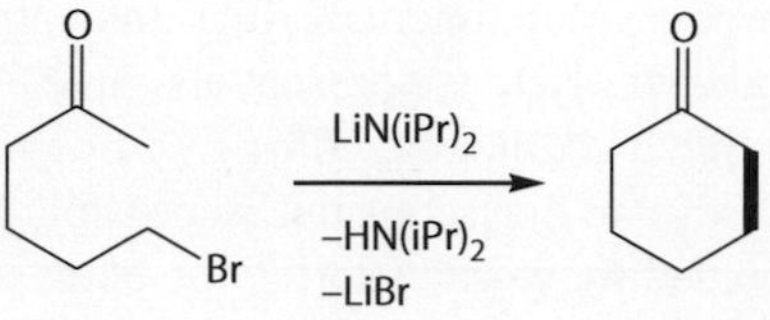

Option C

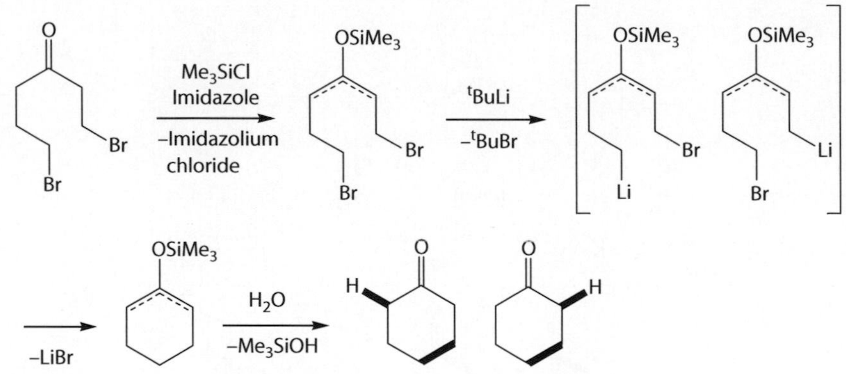

Plus-minus options for two-bond disconnections

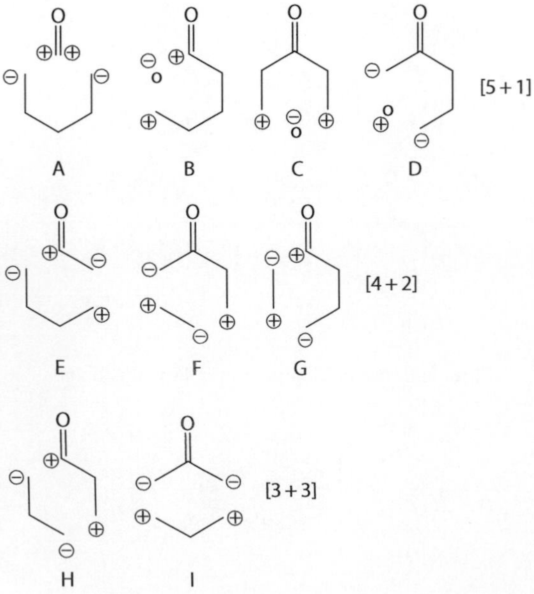

A   B   C   D   [5 + 1]

E   F   G   [4 + 2]

H   I   [3 + 3]

Possible schemes that would satisfy the plus–minus combinations are given in the following.

Option A

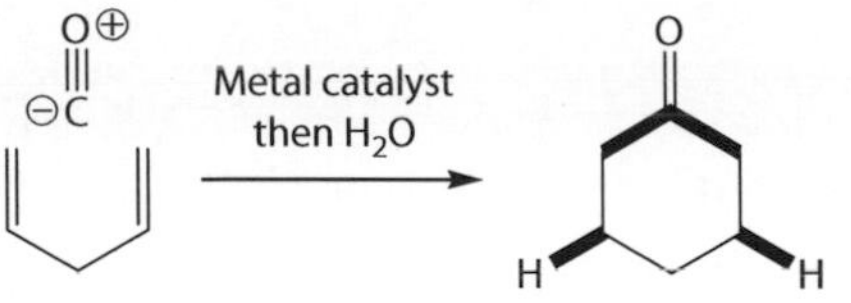

Option B

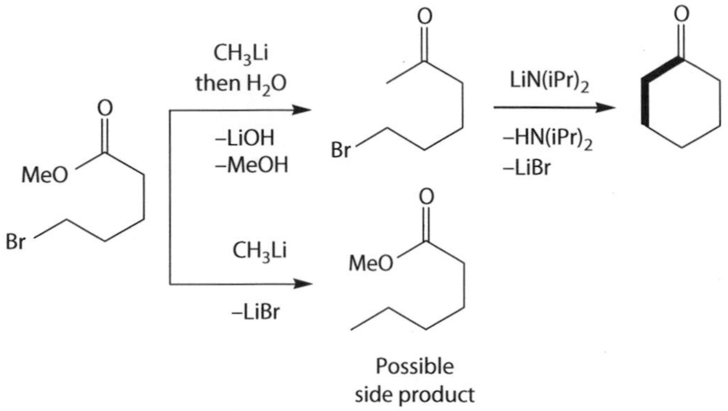

Possible
side product

Option C

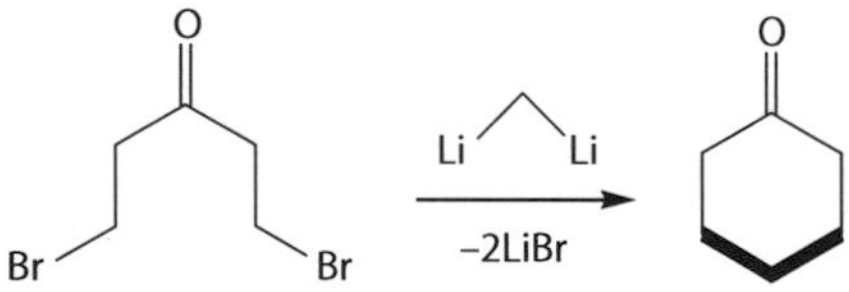

Option D

Possible side
product

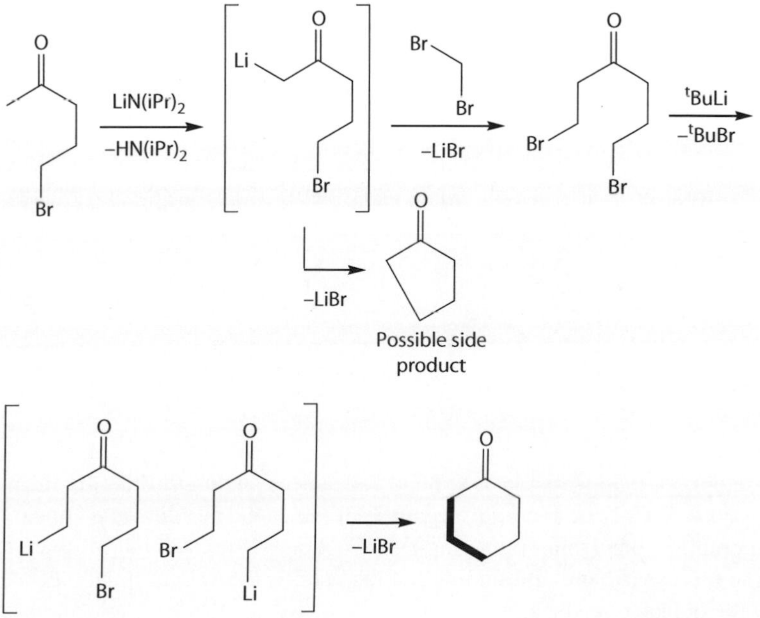

Option E

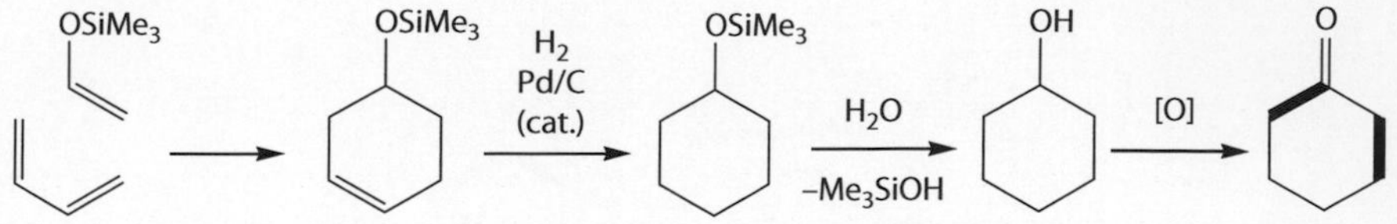

Option F

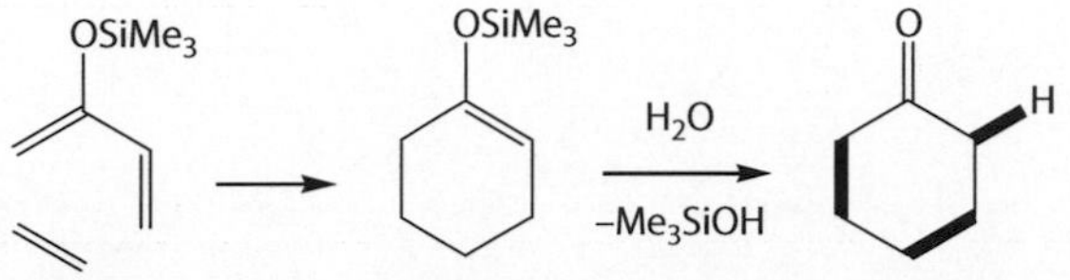

Option G

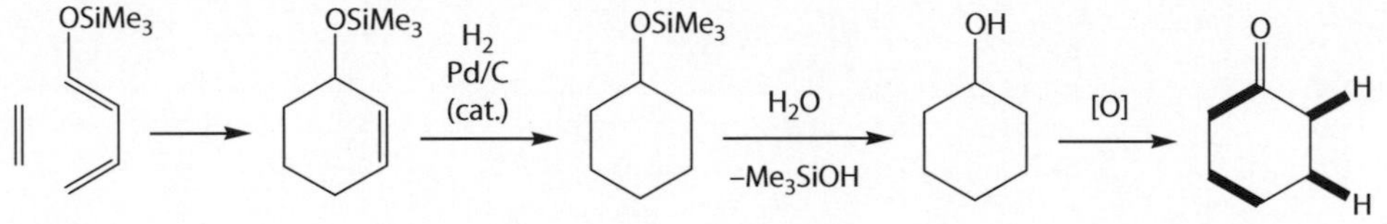

Option H

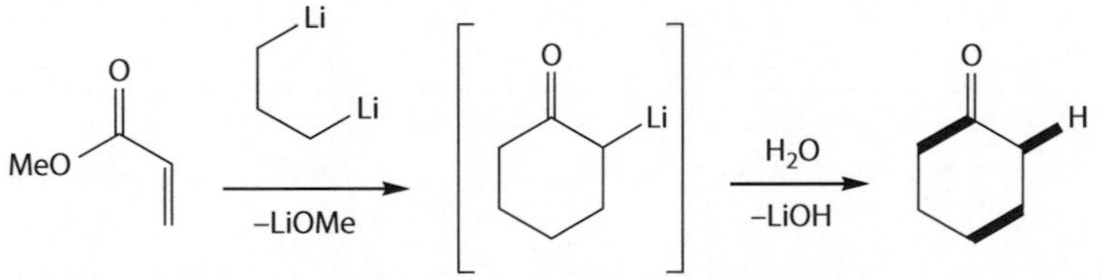

Option I

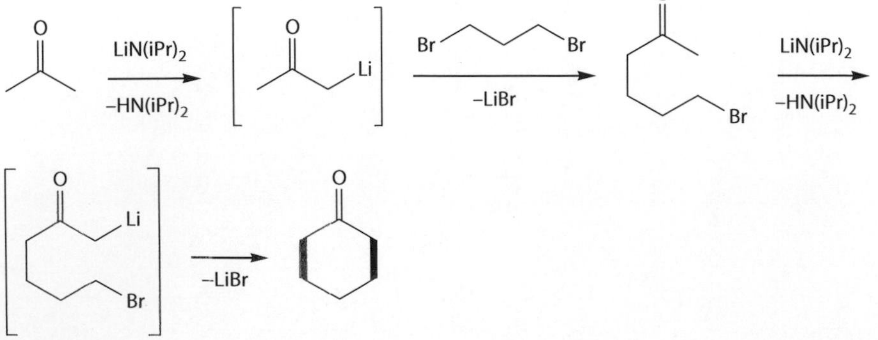

Looking at these options it appears that two-bond disconnection possibilities are the most attractive. In particular, options A, C, F, H, and I are elegant solutions to the problem and appear to be practical with the least complications from competing side reactions. However, the ultimate test of validity is actually doing the experiment and finding out the results. Many a chemist has been humbled or exhilarated by the outcome as the case may be.

In summary, the following steps help to streamline retrosynthetic analysis toward direct and efficient synthetic routes, especially for ring construction:

1. Identify symmetry elements in the final target structure.
2. Identify electrophilic and nucleophilic centers.
3. Identify heteroatoms and assign these as nucleophilic pivots.
4. Parse the target structure into all possible one-bond disconnections.
5. From each of the one-bond disconnection options deduce a second bond disconnection to come up with an overall range of two-bond disconnection possibilities.
6. From each of the two-bond disconnection options deduce a third bond disconnection to come up with an overall range of three-bond disconnection possibilities; these will result in possible three-component coupling reactions.
7. For each of the disconnected structure diagrams assign pluses and minuses as appropriate to come up with the combinations of pieces that will come together in an additive sense to build up the target ring structure.
8. Assign appropriate functional groups to these pieces identified in the one-bond, two-bond, and three-bond disconnection diagrams that will serve as electron deficient or electron rich groups consistent with the plus and minus assignments.
9. Write out the synthesis plans in the forward sense for each case found in (8).
10. Rank these plans according to step length, AE, and possibilities of complications such as side reactions, etc. to identify best candidate routes.

Steps (4) to (6) were applied to various monocyclic and bicyclic structures in compiling the possible disconnection graphs listed in Appendix A7.

9.2   DIRECT SYNTHESES

Direct syntheses are characterized as being composed mainly of reactions that are additive in the target bond forming sense. They incorporate appropriate positive (electrophilic) and negative (nucleophilic) units to build up the target structures without the need for adjusting redox states of atoms involved in forming new bonds. They are thus associated with aufbau-type or building up type reactions such as carbon-carbon-additions, non-carbon-carbon additions, additive redox, substitutions, and multicomponent reactions. As a consequence of this repertoire of reaction types, direct synthesis plans usually result in the shortest number of steps. This is why these types of reactions are the most powerful building block reactions in the database of named organic reactions. Direct syntheses use appropriate starting materials whose skeletons remain intact as they are elaborated toward the more complex target molecule; for example, starting materials with required stereocenters selected from the chiral pool such as menthol, carvone, amino acids, and carbohydrates. The required skeleton in the starting material usually has a privileged ring system. There is no need to change bond types between atoms that are already connected; that is, a starting C–C linkage remains a C–C linkage, etc. Direct syntheses require no need for protecting groups and no need for oxidation level changes for atoms involved in target bond forming steps. Examples of reactions appearing in direct syntheses are cross-coupling reactions to make substituted olefins with the required orientation of groups about the C=C bond (see, e.g., Section 7.1). An excellent example of a direct synthesis is the Campos [3] synthesis of (R)-nicotine that involves a Negishi coupling of a pyrrolidine and pyridine ring precursors. Not surprisingly this is the number one ranking plan to make this target compound albeit it is the nonnatural stereoisomer that is formed. Presumably, if (+)-sparteine were used as the chiral ligand, then the natural (S)-nicotine could be formed. However, another wrinkle in this strategy is that (+)-sparteine is the unnatural enantiomer of the natural (−)-sparteine. Pursuing this idea means that a total synthesis of (+)-sparteine is required, which is even more complex than the target (*S*)-nicotine. This example shows how insidious Nature is in thwarting the best laid plans of chemists!

Appropriate application of connectivity analysis especially for ring constructions can go a long way in coming up with direct synthetic routes to such targets as was done in the cyclohexanone example given in Section 9.1.3.

9.3   INDIRECT SYNTHESES

Indirect syntheses are the antithesis of direct syntheses. They are associated with sacrificial-type reactions such as those itemized in Chapter 8. Reaction types involved are eliminations or fragmentations, subtractive redox reactions, and to some extent rearrangements.

Plans that are stacked with these kinds of reactions suggest that "wrong" starting materials were probably used. As a result indirect syntheses necessarily have an increased number of steps to the final target structure. The following "corrections" are usually found in indirect syntheses:

1. Elimination reactions required to change bond type between already connected atoms, for example, C–C bond changes to a C=C bond.
2. Epimerization reactions required to change the configuration of a stereogenic center.
3. *Cis-trans* isomerization reactions are required—for example, Inhoffen [4] and Roche G1 [5] plans for beta-carotene require heating in the last steps, respectively; Al-Hassan [6] plan for tamoxifen requires photoisomerization in the final step.
4. Some kind of resolution required to obtain correct stereoisomer of compound before proceeding to the next step.
5. Use of protecting groups to momentarily cap a specific functional group thereby preventing unwanted potential side reactions.
6. Required change of ring size via rearrangement; rearrangements may be viewed as a kind of "tunneling" between structure types and are often discovered accidentally.
7. Employing a strategy that begins with an aromatic ring starting material in order to make a nonaromatic ring or vice versa; these require redox changes; for example, Tanaka-Bersohn [7] plan for atiprimod that employs a Birch reduction to dearomatize an aromatic ring in order to eventually make a substituted cyclohexane ring (this example will be revisited in Section 9.9).
8. Degradation reactions that pear down or prune unwanted elements of a starting material via oxidative degradation or eliminations before proceeding with the real productive synthesis steps, for example, Hamon-Young [8] plan for (±)-twistane; Firmenich [9] plan for (−)-patchouli alcohol where only a portion of the alpha-pinene starting material is retained in the final product structure; Merck [10], Gates [11], and Martin [12] plans for quinine begin with 3-ethyl-4-methylpyridine that needs to be dearomatized via hydrogenation, classically resolved to get the correct stereochemistry, and the 3-ethyl group needs to be changed to a vinyl group in four steps via photochlorination of the terminal carbon atom and elimination to create the C=C double bond, which conforms to the quinuclidine moiety found in the final structure.

In the last example, a more direct approach to make the vinyl group may be via a Wittig strategy from a nicotinic aldehyde or via a direct aromatic substitution strategy. It should be noted that the Wittig and Heck reactions were known at the time the Merck, Gates, and Martin plans of quinine were disclosed.

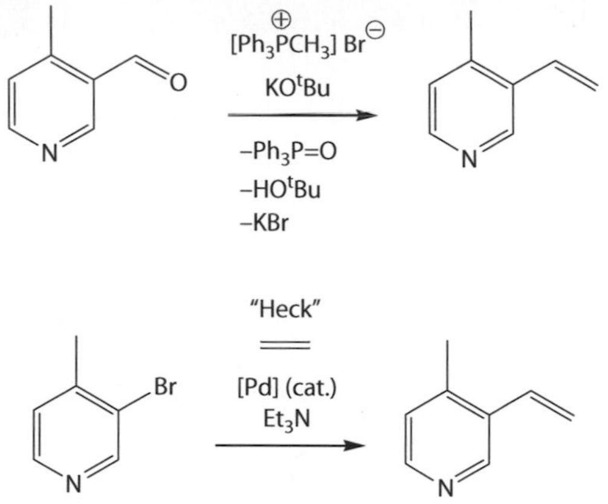

We have seen that target bond maps readily illustrate which reagents end up in the final product structure and thus are skeletally preserved. The greater the preservation of skeletal structure of the starting material in the final target, the more direct the synthesis strategy is; the less the preservation is the more indirect is the strategy. This is also reinforced by the target bond making profile—more gaps appearing in such a profile mean indirect strategies were followed using sacrificial type reactions. This is further linked with the hypsicity profile that typically shows significant changes in oxidation changes over the course of the synthesis and thus results in a high oxidation length, |UD| characterized by multiple "ups" and "downs."

Examples of indirect syntheses taken from the synthesis database are the following.

The Cocker [13,14] plan for (+)-fenchone shows redundancy in bond making and bond breaking between C1 and C2 as indicated in the following scheme.

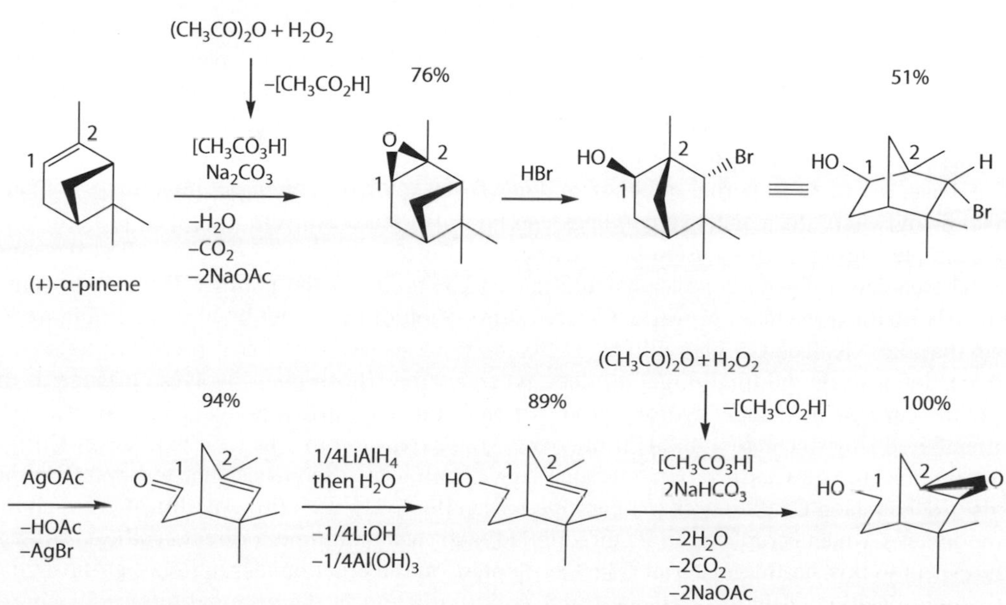

The Carretero [15] plan for (–)-swainsonine involves tandem kinetic resolutions in steps 5 and 6.

The question arises if it is possible to bypass the central intermediate by acylating the unwanted stereoisomer of the starting alcohol with another kind of lipase according to the following.

A nice comparison of carvone usage as a starting material to make (–)-platensimycin is illustrated by the Nicolaou G5 [16] plan beginning with (R)-carvone and the Ghosh [17] plan beginning with (S)-carvone. The former plan involves 20 reaction stages, produces 203 kg waste per mole (E-kernel = 460.3), and has f(sac) = 0.754. The latter plan involves 31 reaction stages, produces 345 kg waste per mole (E-kernel = 786.4), and has f(sac) = 0.732. The target bond maps for each plan showing where the respective carvones end up in (–)-platensimycin are given in Figure 9.1. This example illustrates the tension between being different, novel, and efficient all at the same time. The choice of (S)-carvone as starting material lengthened the plan by 11 reaction stages and nearly doubled the mass of waste. Closer examination of the target bond maps in Figure 9.1 shows that the Nicolaou G5 plan utilizes (R)-carvone in such a way that it preserves its stereo-chemical integrity in the final target product; whereas, the Ghosh plan involves cleavage of the ring in (S)-carvone by tandem hydroxy bromination of the isopropenyl moiety, ring closure to the six-membered ring to create a [3.2.1] framework, ring expansion to a lactone via Baeyer-Villiger oxidation, and in situ lactonization—lactone ring opening (see sequence that follows). In other words, the Nicolaou G5 plan uses (R)-carvone more effectively with little modification as shown by the intact six-membered ring in its target bond map thus making it a direct synthesis strategy with respect to this starting material. The Ghosh plan, on the other hand, requires significant cor-rective modification of the (S)-carvone framework to make it fit the eventual tetracyclic core of (–)-platensimycin,

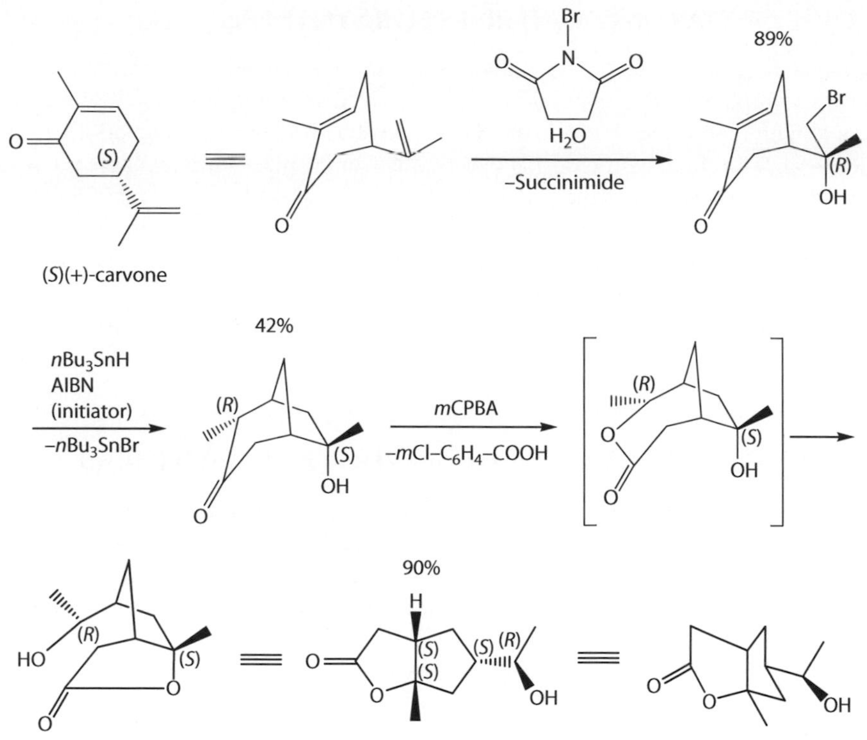

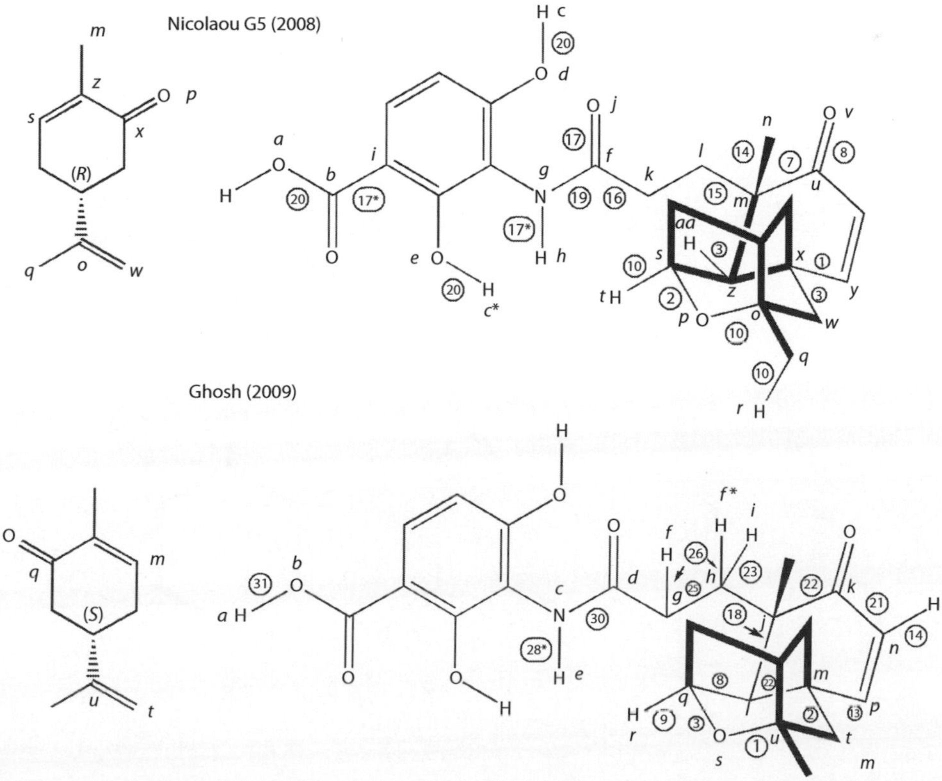

FIGURE 9.1 Target bond maps for Nicolaou G5 and Ghosh plans for (−)-platensimycin showing superposition of (R)-carvone and (S)-carvone structures denoted by bolded bonds.

9.4 CHOICE OF STARTING MATERIALS FOR SYNTHESIS

The more remote in structure the starting materials are to the intended target product, the more indirect is the synthesis plan. Such a plan will be characterized as having a sparse looking target bond making profile with high values for f(nb) and low values for B/M. An example of this is the Tatsuta [18] plan for (−)-tetracycline hydrochloride that begins from D-glucosamine. Table 9.1 summarizes key metrics for this plan as well as the plan's ranking among others for this target. Figure 9.2 shows the target bond map where the structure of D-glucosamine is superimposed on the (−)-tetracycline structural framework. The only stereocenter that is preserved is that of the amino group. Four out of five stereocenters in D-glucosamine are not utilized.

Versatile starting materials for syntheses found in the present synthesis database are summarized in Figure 9.3.

9.5 COMPARISON OF TOTAL SYNTHESIS VERSUS SEMISYNTHESIS

An excellent example of the great advantage of employing a semisynthetic strategy that begins from a natural product whose structure resembles the final target product is that of using 10-deacetylbaccatin III as a starting material to synthesize (−)-paclitaxel. This material can be obtained from the dried needles of the *Taxus baccata* yew tree (1000 g dried needles yield 1 g

TABLE 9.1

Summary of Plan Performances for (−)-Tetracycline Hydrochloride

Plan	Year	N	M	I	% Yield	% AE	% Kernel RME	Kernel Mass of Waste (kg/mol)	B/M	f(nb)
Myers G2	2007	17	24	51	2.8	8.3	0.4	131.9	1	0.35
Myers G1	2005	18	27	46	1.3	10.1	0.3	175.9	0.93	0.39
Tatsuta	2000	36	40	73	0.06	5.6	0.02	2041.5	0.6	0.67

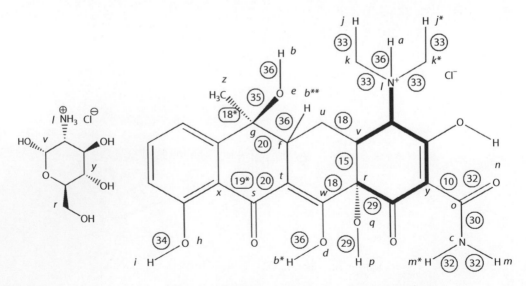

FIGURE 9.2   Target bond map for Tatsuta (2000) plan for (−)-tetracycline hydrochloride.

10-deacetylbaccatin III). Such a strategy effectively gives a tremendous head start over a true total synthesis and thus is the most effective way of cutting down the step count and reducing overall waste production. These conclusions are confirmed by the data in Table 9.2 where the highest and lowest ranking plans for the total and semisynthesis of (−)-paclitaxel are summarized. The radial hexagons shown in Figure 9.4 give a visual representation of the respective plan performances.

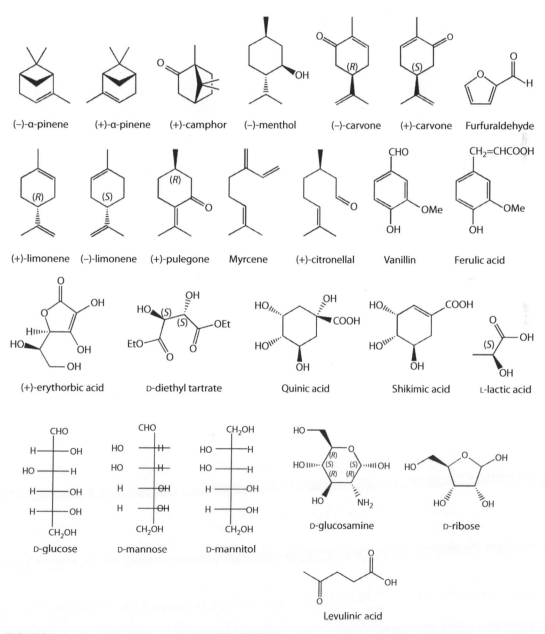

FIGURE 9.3 Versatile starting materials for synthesis.

(continued)

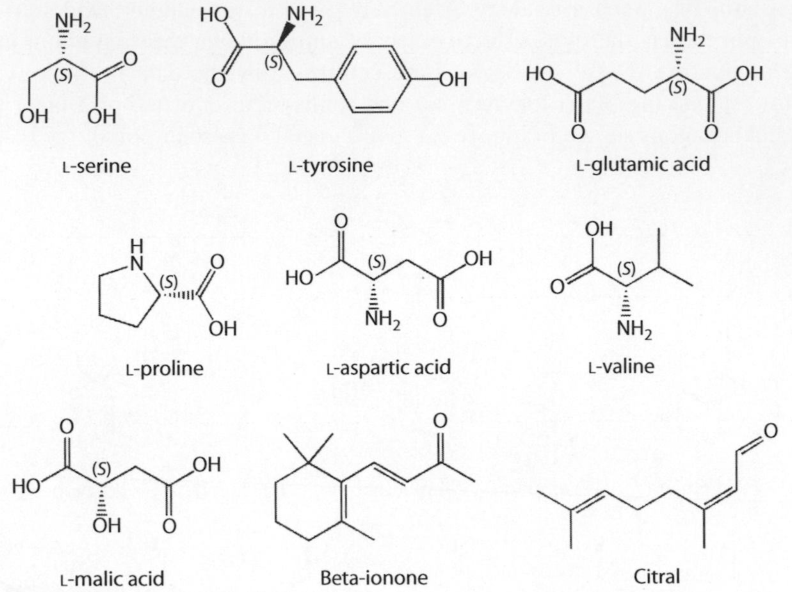

L-serine

L-tyrosine

L-glutamic acid

L-proline

L-aspartic acid

L-valine

L-malic acid

Beta-ionone

Citral

FIGURE 9.3 (continued)

TABLE 9.2
Summary of Plan Best and Worst Performances for (−)-Paclitaxel
Total and Semisyntheses

Plan	Year	N	M	I	% Overall Yield	% AE	Kernel Mass of Waste (kg/mol)
Total syntheses							
Wender [19,20]	1997	38	40	87	0.03	8.9	3428.2
Takahashi [21]	2006	48	62	109	0.002	6.9	90441.4
Semisyntheses							
Greene G2 [22]	1994	6	8	16	45.2	35.2	3.1
Commerçon [23]	1992	11	15	27	18.5	25.2	10.6

9.6 BIO-INSPIRED STRATEGIES

Synthesis plans falling in this category are those that involve chemoenzymatic steps, microbial fermentation steps, use biomass as a starting material, or begin from renewable biofeedstocks that are obtained from natural sources by some method of extraction. Excellent reviews of the subject have appeared in the literature with respect to transformation of biomass and agricultural waste into commodity chemicals [24], biocatalysis applications for the synthesis of pharmaceuticals [25–27], transformation of carbohydrate feedstocks to O- and N-heterocyclic compounds [28], and biocatalytic processes for the synthesis of aromatic commodity chemicals [29]. Many of these versatile biomass-derived starting materials appear in Figure 9.3. The use of biofeedstocks is not a new idea. In the older literature, it is known under the name *chemurgy*, a word coined by William Jay Hale in 1934 [30,31] that was formed by joining the word *chemistry* with the Greek word for work, *ergon*. One of the industrial leaders who advocated using agricultural materials, particularly soybeans, to make commodity chemicals, biofuels,

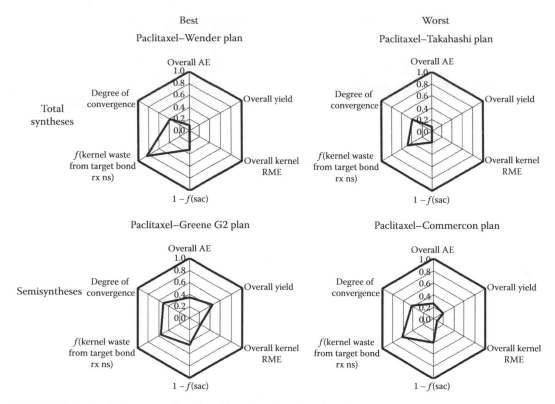

FIGURE 9.4   Radial hexagons for (–)-paclitaxel total and semisyntheses given in Table 9.2.

and automotive parts was Henry Ford way back in the 1930s. Other supporters of using renewable feedstocks from agricultural waste for the chemical industry were George W. Carver, Irenee DuPont, Thomas Edison, and Charles Kettering [32–37]. Those who know the history of biofuels will know that the first biofuels were peanut and vegetable oils used by Otto Diesel to power his diesel engine horseless carriage vehicle. *Organic Syntheses* has a procedure to make furfuraldehyde from corn cobs [38] (E-total = 43.2) and another to make levulinic acid from cane sugar [39] (E-total = 31.2). Figures 9.5 through 9.7 show example bio-inspired syntheses of a variety of valuable target molecules.

Despite the rosy outlook of these examples, there are serious problems that need addressing from a green chemistry point of view. These are itemized as follows:

a. It is disappointing to not be able to verify claims because there is a lack of hard data in experimental procedures to do an assessment of material efficiency; for example, proper yields and quantities of materials used are not stated explicitly.

b. Literature papers dealing with bioengineered microbes are heavily weighted toward the microbiology and genetics side of things with very little relevant information that a chemist could sink his or her teeth into.

c. One needs to consider extraction processes from natural sources in the overall assessment; this means there will be a significant solvent demand—often this is completely overlooked.

d. There is a serious problem with balancing chemical equations mainly because the actual understanding of the mechanisms of the biotransformations is poor or nonexistent.

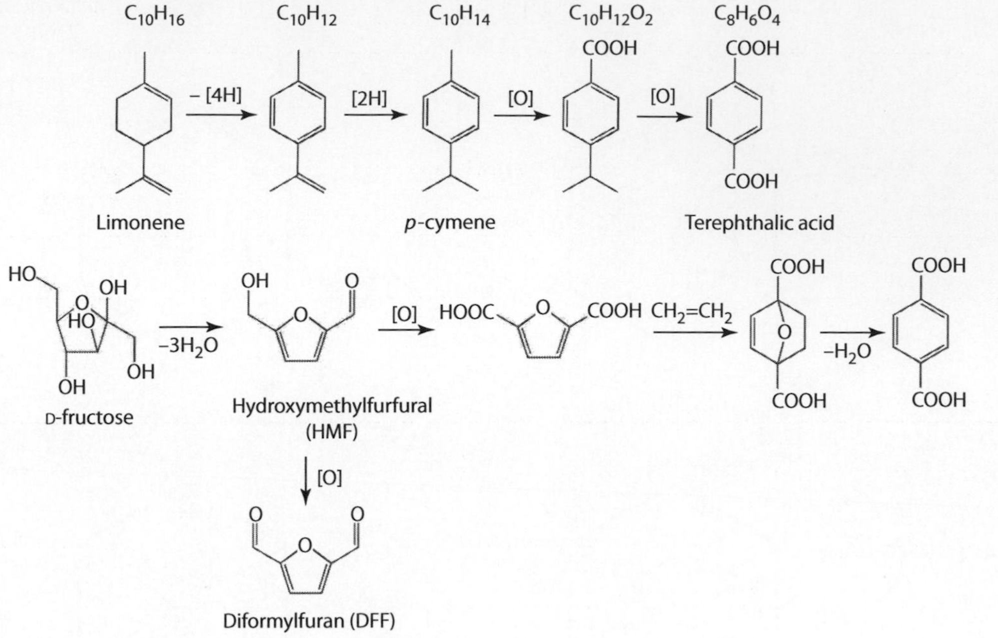

FIGURE 9.5 Example syntheses using biofeedstock materials.

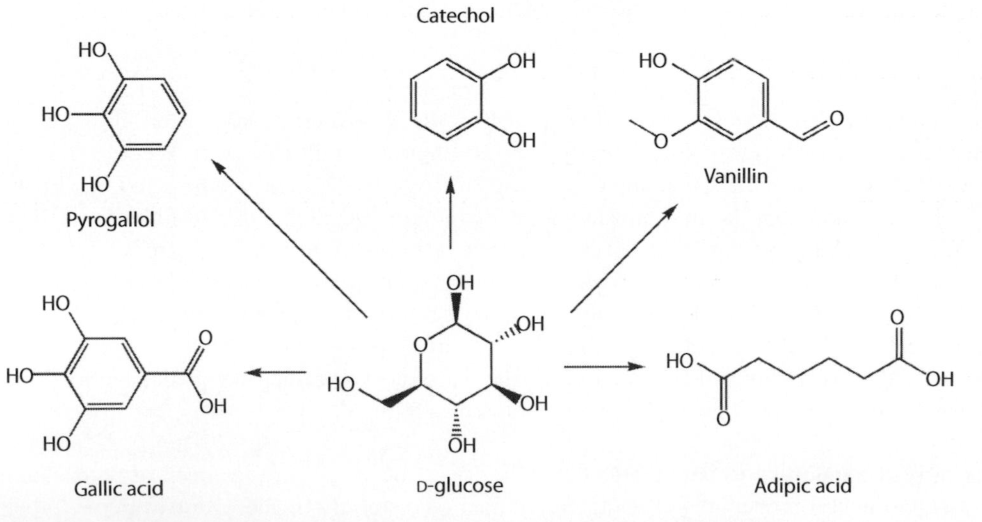

FIGURE 9.6 D-glucose as a biofeedstock for a variety of target molecules [40–44].

e. No head-to-head comparisons have been made between conventional and biomass-derived syntheses with respect to rigorous metrics analysis.

f. On the face of it some claims appear to be sensational.

The reader can verify some of these deficiencies in the schemes given in Figures 9.5 through 9.7 particularly the synthesis of indigo from D-glucose. One reaction encountered in the present

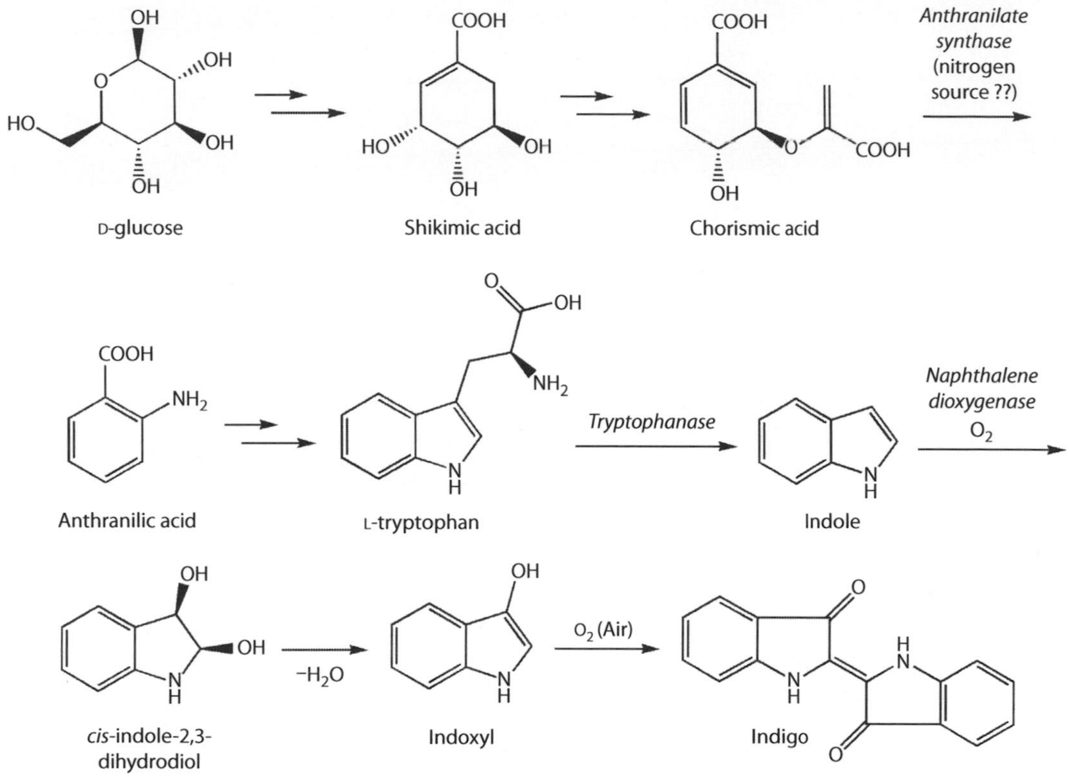

FIGURE 9.7 Bio-inspired synthesis of indigo [45,46].

synthesis database compilation, which required an assumption to be made with respect to balancing the chemical equation, is the *cis*-dihydroxylation of aromatic rings by dioxygen in the presence of bioengineered *Pseudomonas putida*. This reaction was used in the synthesis of vitamin C, (−)-tetracycline, and oseltamivir phosphate (see Table 9.3). The following simplified mechanism was invoked, which results in the production of hydrogen peroxide as a by-product.

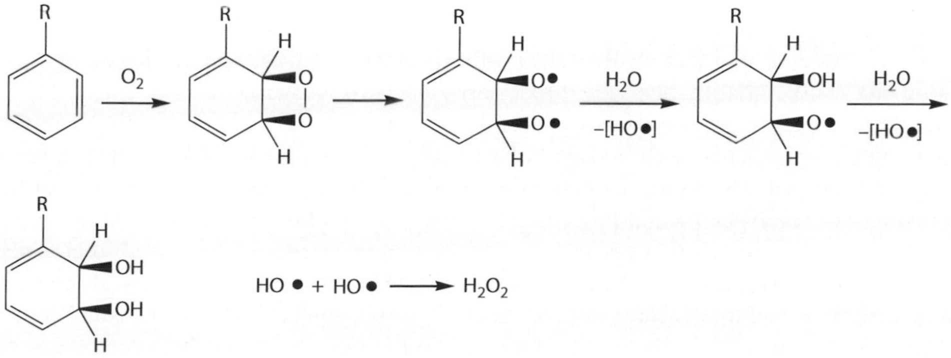

The real mechanism may involve metal cofactors in a dioxygenase enzyme according to the following scheme using iron as an arbitrarily chosen metal ion.

At this time, since the mechanism is unknown the simpler mechanism was used to calculate metrics. What can be said with certainty is that until all of the problems itemized earlier are resolved by serious fundamental research there can be no credible way to prove the "greenness" of these kinds of transformations from either a material or energy consumption point of view.

Table 9.3 lists example bio-inspired plans found in the present synthesis database compilation along with their ranking with respect to kernel waste production among other traditional chemosynthetic plans. Though these types of plans fulfill one of the 12 green principles of green chemistry, it does not automatically follow that they are indeed "greener" overall in plan performance over total synthesis plans. Many claims appearing in the literature are exaggerated and not backed up by any metrics analysis. What would be more convincing is if all steps in a plan were of this type rather than one or two key reactions. In fact, the Givaudan–Roure [47,48] plan for vanillin is a good example of a single step plan that fulfills this criterion. The examples listed in Table 9.3 are the first in the literature that have been backed up by a rigorous metrics analysis. The reader is referred to the accompanying download for full plan details and references.

9.7 CHEMOSELECTIVITY VERSUS CLASSICAL RESOLUTION COMPARISON

An interesting performance comparison between plans to stereoisomeric target molecules is a comparison between a plan that involves the synthesis of a racemic mixture followed by classical resolution and another that involves the direct stereoselective synthesis of the required stereoisomeric target molecule. The resolution of a racemic mixture is usually done by adding a chiral discriminating molecule such as a chiral acid or base that produces two diastereomeric salt products one of which preferentially crystallizes out of solution and the other remains in the mother liquor. The remainder of the experimental procedure involves the isolation of whichever diastereomeric salt is wanted. If it is the one that crystallizes out, then the procedure is easy—all that is needed to do is to collect the product by filtration, then liberate the free stereoisomer by adding acid or base as appropriate, and finally recrystallize the crude stereoisomer. If the desired diastereomeric salt is the one that remains in the mother liquor, then the unwanted salt that precipitates out of solution is first filtered off as before and the solvent is removed from the remaining mother liquor by evaporation. The crude diastereomeric salt is then dissolved in acidic or basic solution to liberate the free stereoisomer of opposite orientation and finally it is subjected to purification by recrystallization. Either of these sequences may be repeated if the optical purity of the collected target stereoisomer is not satisfactory. Stereoselective syntheses, in contrast, involve one of the

TABLE 9.3

List of Chemoenzymatic Plans in Synthesis Database

Target Product	Plan	Rank with respect to Kernel Mass of Waste Produced	Key Bio-Inspired Steps
Astaxanthin	Ito G1	4 out of 5	Step 3: Stereoselective hydrogenation with *Saccharomyces cerevisiae* and sucrose
(+)-codeine[a]	Hudlicky	14 out of 15	Step 1: *cis*-dihydroxylation of (2-Bromoethyl) benzene with *E. coli* JM 109 (pDTG601A)
(*S*)-dapoxetine	Gotor	2 out of 4	Step 4: Selective amidation with *Candida antarctica* lipase A
(*S*)-fluoxetine	A. Kumar	11 out of 15	Step 4: Ketone reduction to secondary alcohol with *Saccharomyces cerevisiae* and glucose
Glycerol	Faith G3	3 out of 3	Single step: Starting material is a triglyceride with three stearic acid side chains
(−)-hirsutene[a]	Banwell	26 out of 30	Step 1: *cis*-dihydroxylation of toluene with *Pseudomonas putida*
(*S*)-metolachlor	Zhang	4 out of 5	Step 2: Stereoselective ester hydrolysis with *Candida antarctica* (CAL-B) lipase
Oseltamivir phosphate[a]	Banwell	21 out of 29	Step 1: *cis*-dihydroxylation of bromobenzene with *Pseudomonas putida*
Oseltamivir phosphate[a]	Hudlicky G1	16 out of 29	Step 1: *cis*-dihydroxylation of ethyl benzoate with *E. coli* JM 109 (pDTG601A)
Oseltamivir phosphate[a]	Fang G2	4 out of 29	Step 1: *cis*-dihydroxylation of bromobenzene with *Pseudomonas putida*
Oseltamivir phosphate[a]	Fang G3	5 out of 29	Step 1: *cis*-dihydroxylation of bromobenzene with *Pseudomonas putida*
Oseltamivir phosphate[a]	Hudlicky G2	14 out of 29	Step 1: *cis*-dihydroxylation of ethyl benzoate with *E. coli* JM 109 (pDTG601A)
(−)-tetracycline HCl[a]	Myers G1	2 out of 4	Step 1*: *cis*-dihydroxylation of benzoic acid with *Alcaligenes eutrophus*
Vanillin	Givaudan–Roure	2 out of 14	Step 1: Oxidative cleavage with O_2 by *Streptomyces setonii* ATCC 39116
Vanillin	Frost	14 out of 14	Step 1: D-glucose to vanillic acid via glycolytic, pentose phosphate, and shikimic acid pathways followed by methylation with L-methionine
Vitamin C[a]	Banwell	4 out of 5	Step 1: *cis*-dihydroxylation of chlorobenzene with *Pseudomonas putida*
Vitamin C	Biotechnical resources	2 out of 5	Single step: D-glucose to L-ascorbic acid in the presence of *Chlorella pyrenoidosa* UV101-158
Vitamin C	Reichstein	3 out of 5	Step 2: Hydroxyl group oxidation with dioxygen in the presence of *Acetobacter suboxydans*
Vitamin C	Roche	1 out of 5	Step 1: D-sorbitol to 2-keto-L-gulonic acid in the presence of *Gluconobacter suboxydans IFO 3255* and the strain DSM 4025
(+/−)-welwitindolinone A[a]	Wood	1 out of 2	Step 1: *cis*-dihydroxylation of benzene with *Pseudomonas putida*

[a] Green metrics worked out under the assumption that the byproduct of the *cis*-dihydroxylation reaction of aromatic precursors in the presence of dioxygen and bio-engineered *Pseudomonas putida* is hydrogen peroxide since the mechanism of the reaction is unknown.

reaction steps in the sequence as the key step that utilizes either a chiral auxiliary or catalyst to direct the addition of a specific reagent to one of the chiral faces of the substrate, or uses an enzyme as a chiral discriminator, or uses a natural product starting material from the chiral pool that has the appropriate orientation of groups that map on to those found in the final target product structure, or uses a chiral directing group in a sacrificial sense. Various resolution strategies may be employed including kinetic resolution (one enantiomer is collected, at least half the starting material is destined for waste), parallel kinetic resolution (one enantiomer is collected, the other is not destined for waste but is shunted toward a different reaction path to another useful product), and dynamic kinetic resolution (the unwanted enantiomer is isomerized in situ to the desired one so that in the end potentially all of the starting material is directed to producing a product that originates from the desired enantiomer). The key question that may be asked from

TABLE 9.4
Summary of Metrics for the Synthesis of (+)-Cocaine by Various Strategies

Plan	Year	Type	N	M	I	f(sac)	% Yield	% AE	Kernel Mass of Waste (kg/mol)	Strategy
Cha	2000	Linear	7	7	20	0.716	36.4	12.2	4.4	Stereoselective
Casale	1987	Convergent	8	9	15	0.564	7.7	21.5	10.9	Resolution by differential crystallization of diastereomeric tartrate salts
Pearson	2004	Convergent	15	17	35	0.763	1.7	8.4	84.7	Resolution by differential crystallization of N-boc protected diester product

TABLE 9.5
Summary of Metrics for the Synthesis of (S)-Dapoxetine by Various Strategies

Plan	Year	Type	N	M	I	f(sac)	% Yield	% AE	Kernel Mass of Waste (kg/mol)	Strategy
Lilly	1988	Linear	3	3	6	0.414	35.7	46.1	1.4	Resolution by differential crystallization of diastereomeric tartrate salts
Gotor	2006	Linear	7	7	14	0.579	12.6	21.4	4.7	Stereoselective using *Candida antartica* lipase A (CAL-A)
Srinivasan	2007	Linear	10	10	21	0.648	13.4	12.9	8.9	Stereoselective using chiral auxiliary (DHQ)$_2$(PHAL)
Deshmukh	2009	Linear	14	14	23	0.786	1.6	11.6	49.5	Stereoselective using (R,R)-diethyl tartrate template starting material

a green chemistry point of view is which of these many strategies is more material efficient. The classical resolution method has the drawback of high solvent demand in the isolation and purification of the target stereoisomeric product due to repeated recrystallizations, but has the advantage of usually fewer reaction steps to get to the racemic product. Direct stereoselective routes have the opposite scenario: more reaction steps but may gain by using less solvents in work-up and purification procedures overall.

Here we present the results of head-to-head comparisons where chemoselective routes are found to be better than classical resolution routes and vice versa. This exercise shows that one cannot make statements about plan performance based simply on generalizations. Each plan needs to be examined on its own by a thorough metrics analysis and both its strengths and weaknesses are identified in relation to other competing plans. Tables 9.4 and 9.5 summarize results for the syntheses of (+)-cocaine and (−)-dapoxetine, respectively. Full details of plans are given in the accompanying download.

In the Cha [49] plan, the key step is step 3:

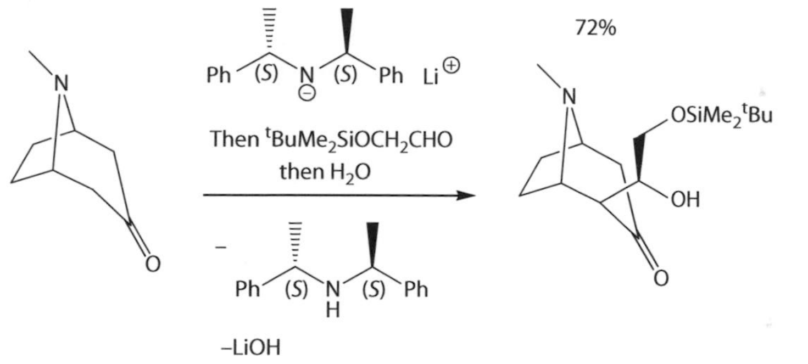

In the Casale [50] plan the key steps are steps 3–5:

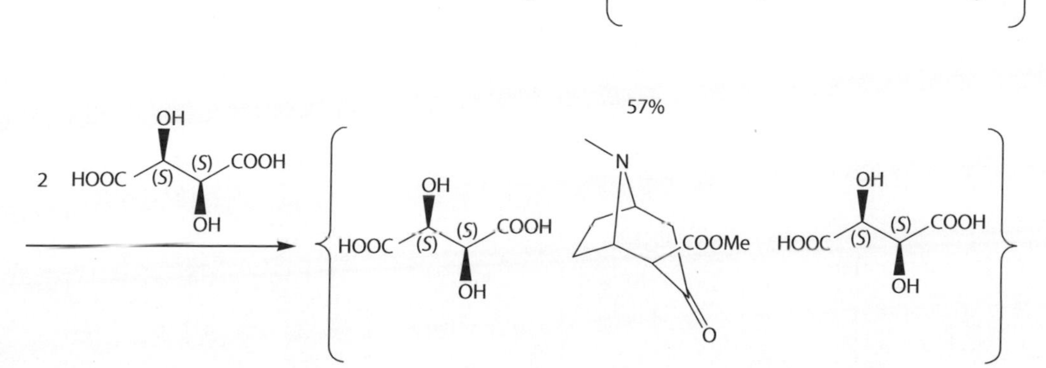

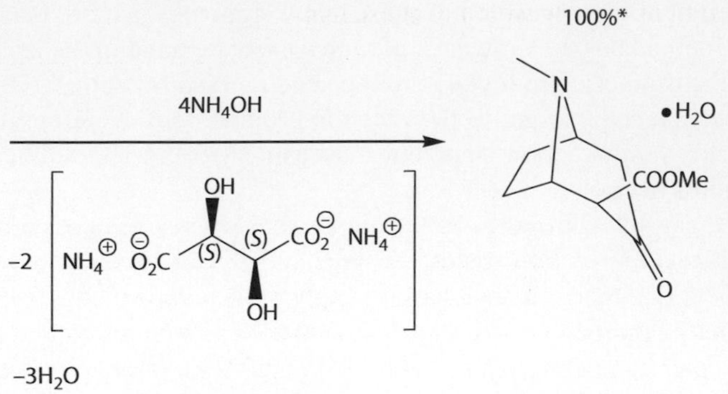

In the Pearson [51] plan the key step is step 15:

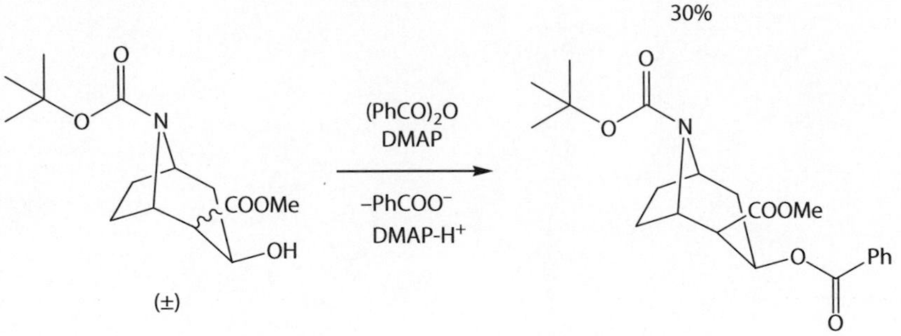

In the Lilly [52] plan, racemic dapoxetine was first prepared, then (+)-dapoxetine tartrate was made by treating (±)-dapoxetine with (+)-tartaric acid. Treatment of (±)-dapoxetine with (−)-tartaric acid yielded (−)-dapoxetine tartrate. No details on reaction performance were given for either of these steps, hence the figures in Table 9.5 for this plan pertain to the synthesis of (±)-dapoxetine.

In the Gotor [53] plan, the key step is step 4:

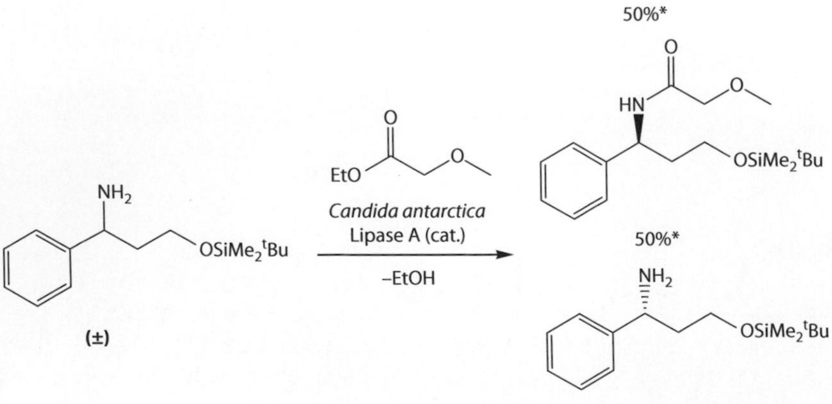

In the Srinivasan [54] plan, the key steps are steps 1–3:

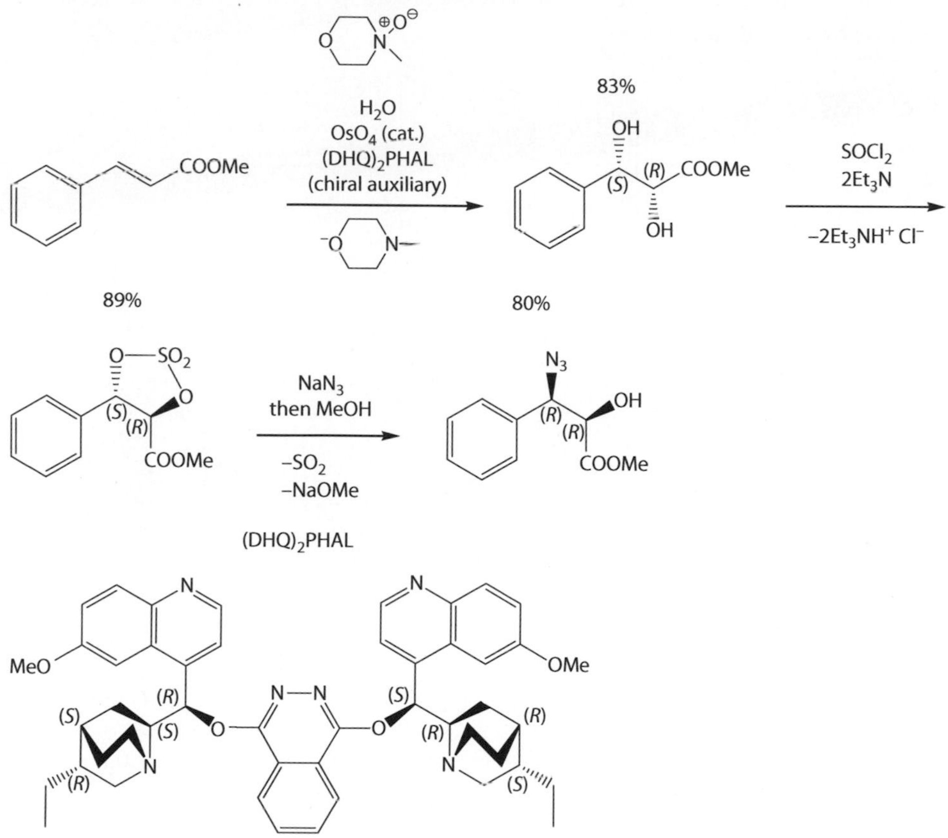

In the Deshmukh [55] plan the key step is step 3:

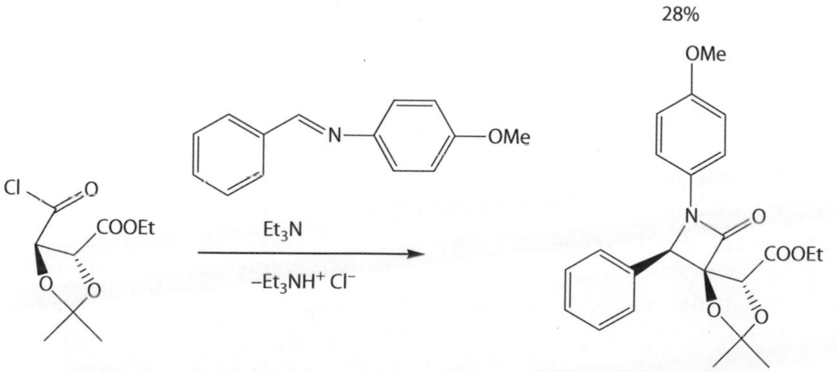

It should be noted that the Deshmukh [59] stereoselective plan for (S)-dapoxetine is far more wasteful than all prior plans for this target. This is a good example of a case where authors came up with a new plan but had not checked to see if substantive improvements were actually achieved in terms of synthesis efficiency and waste reduction. This is where the usefulness of metrics analysis comes in at the design stage. An analysis based on kernel metrics would have easily alerted authors at the outset that their proposed plan could never be competitive with what was already known.

What is more interesting is that no one to date has utilized a stereoselective imine catalytic hydrogenation strategy to make (S)-dapoxetine directly according to a four-step reaction sequence such as the one proposed in the following.

Such a scheme has the added advantage that (R)-dapoxetine could also be prepared by using the opposite stereoisomer of the chiral auxiliary in the catalytic hydrogenation step. Note that steps 1 and 2 are telescoped. In the chemical industry, catalytic asymmetric hydrogenation of imines and ketones is by far the most implemented type of stereoselective reaction to make chiral fine chemicals, pharmaceuticals, and agrichemicals. In terms of popularity among strategies, classical resolution using optically active mandelic acid, tartaric acid, cinchona alkaloids, or 1-phenylethyl-amine is used more often than devising stereoselective routes. The biggest gain is the few number of reaction steps required. The high solvent demand is offset by low cost solvents and their ready availability, and also because the resolution steps are usually done in aqueous or alcoholic media that are simpler to treat in the waste stream. Enzymatic stereoselective routes are usually more often chosen than chemo-stereoselective routes because synthetic chiral auxiliaries and catalysts need to be synthesized themselves, which often is more complex and expensive than the original target molecule. Furthermore, enzymes are more exquisite in their stereoselectivity performance compared to synthetic chiral auxiliaries. Choosing starting materials from the chiral pool such as carbohydrates, amino acids, and simple terpenes is also widely used and more effective than the use of synthetic chiral auxiliaries.

9.8   MISMATCH BETWEEN STRATEGY AND WASTE PRODUCTION

There are two statements that abound in organic textbooks, treatises, reviews, or discussions on organic synthesis. One is that convergent plans are more efficient than linear ones. The other is that plans with fewer steps are also more efficient. Both of these statements are generalizations based on a very simplistic "mathematical" model in which the compared plans have identical reaction yield performances for all steps. This is the only condition that makes them true. Moreover, the statement of synthesis efficiency is solely based on the overall yield criterion without taking into account the other three factors determining reaction material efficiency performance: AE (accounts for by-products), excess reagent consumption, and auxiliary material consumption. By now the reader can clearly see that these assumptions are unrealistic. The simple-minded argument usually goes like this. If we take the reaction yield for all steps as 90%, then a five-step plan would have an overall yield of $(0.90)^5 = 0.59\%$ or 59%. A two-step plan on the other hand would have an overall yield of $(0.90)^2 = 81\%$. So, yes, the shorter plan has the higher overall yield. Now, let us remove the constraint that all reaction yields should have the same value so that our analysis conforms to reality. We could easily come up with a combination of two fractions whose product is less than that of a combination of five fractions. An example would be plan A with reaction yields 50% and 60% for steps 1 and 2 giving an overall yield of 30%. Compare this with plan B with reaction yields 90%, 80%, 85%, 70%, and 93% for steps 1–5 with an overall yield of 40%. Here we have an apparent "counterintuitive" result that the shorter plan is less material efficient than a longer one. When this numbers game is applied to comparing convergent plans with linear ones, the simplistic approach based on equal reaction yields takes on an even more unrealistic turn because multiple branches are constrained to have identical performances. Again, using our 90% value for all reaction yields, a linear plan of six steps would have an overall yield of 53%. Arranging the six reaction steps into two branches is shown as follows

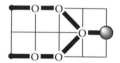

results in an overall yield with respect to either branch of $(0.90)^4 = 0.66$ or 66%. As before, if we lift the constraint of equal reaction yields we can easily come up with a combination of four fractions referring to either branch whose product is less than that of a combination of six fractions. In fact, since the two branches are of equal length we have the added complication that there are two possible values one could calculate for overall yield depending on which branch is chosen. The upshot of all this is that the original two statements are really naïve generalizations based on unrealistic scenarios. In the context of green chemistry principles they are not useful. Zhang [56] has highlighted the inadequacy of using step count and overall yield as determinants of overall synthesis efficiency and also that a global overall yield cannot be applied to convergent plans.

Here we present real examples from the synthesis database that substantiate the conclusions of the present discussion. Table 9.6 summarizes head-to-head pairwise comparisons of plans to various target molecules according to type (linear or convergent), number of steps, overall yield, AE, and E-kernel. The reader is directed to the accompanying download for plan details and references. When metrics analyses reveal apparent conflicting results between plans for the same target mol-ecule such as a shorter plan has a lower overall yield than a longer one, or that a longer plan is more atom economical than a shorter one, or that a more atom economical plan has a lower overall yield than a higher atom economical one, then these are clear signs that plan optimization has not been achieved. True synthesis plan optimization is achieved when it can be proved by metrics that the most material efficient plan to a specific target product among many candidate plans is indeed the one that has the fewest number of steps, the highest AE, the highest overall yield, requires the least input materials, requires the least molecular weight fraction of sacrificial reagents, has the highest degree of building up over the course of the synthesis (most negative molecular weight first moment parameter), has every reaction stage producing at least one target bond, has the highest number of target bonds made per reaction step, and has been made by the most direct strategy using additive

TABLE 9.6
Summary of Head-to-Head Pairwise Comparisons of Plans Highlighting Mismatches in Kernel Metrics Parameters

Target	Plan	Year	Type	N	M	I	% Yield	% AE	Kernel Mass of Waste (kg/mol)	E-Kernel
(±)-hirsutene	Cohen	1993	Linear	7	7	19	37.4	9.5	3.9	19.1
(±)-hirsutene	Wender	1982	Linear	7	7	8	4.2	23.3	11.0	53.9
(−)-kainic acid	Fukuyama G3	2008	Linear	22	22	47	3	3.6	74.4	349.3
(−)-kainic acid	Cossy	1999	Convergent	22	24	44	0.06	3.6	2402.7	11280.3
(S)-naproxen	Noyori	1987	Linear	6	6	10	76.1	27.5	0.8	3.5
(S)-naproxen	Syntex G3	1986	Convergent	6	8	15	15.7	16.6	5.1	22.2
Oseltamivir phosphate	Roche G3	1999	Linear	13	13	19	39	21	3.1	7.6
Oseltamivir phosphate	Roche G4	2000	Linear	9	9	17	1.1	23.9	27.4	66.8
Papaverine	Redel–Bouteville	1949	Linear	8	8	12	7.6	30.7	8.1	23.9
Papaverine	Pictet–Gams	1909	Convergent	8	11	22	5	11.9	43.4	128.0
(−)-physostigmine	Nakada	2008	Linear	21	21	38	5.9	8.3	28.4	103.3
(−)-physostigmine	Takano G1	1982	Linear	12	12	25	0.4	13.9	237.3	862.9
(−)-salsolene oxide	Paquette	1997	Linear	15	15	37	6.8	5.4	18.2	82.7
(±)-salsolene oxide	Wender	2006	Linear	8	8	13	1.7	18.0	21.8	99.1
(−)-swainsonine	Blechert	2002	Linear	14	14	30	22.4	4.2	7.3	42.2
(−)-swainsonine	Kibayashi	1994	Convergent	20	21	36	1.9	4.3	70.9	409.8
Tamoxifen	Larock	2005	Linear	3	3	11	33.5	32.4	2.3	6.2
Tamoxifen	Armstrong	1997	Convergent	3	6	22	16.4	14	13.8	37.2
Vanillin	Mottern	1934	Linear	4	4	3	40.2	43.6	0.6	3.9
Vanillin	Geigy	1899	Linear	1	1	5	30	22.3	2.1	13.8
(±)-welwitindolinone A	Wood	2004	Linear	25	25	46	2.3	6.5	84.0	238.3
(+)-welwitindolinone A	Baran	2005	Linear	11	11	21	0.3	12.1	287.2	814.9

type reactions (CC bond forming, non–CC–bond forming, additive redox, multicomponent, tandem, cascade, domino). One of the great disappointments to avoid is finding a plan that satisfies all of the aforementioned criteria only to lose its prestigious rank because it has a high auxiliary material consumption.

9.9 ROUTE SELECTION AND REACTION NETWORKS

A reaction network is a concise representation of several synthesis plans in a single diagram. It is constructed from a retrosynthetic map beginning from one or more target structures and going backward to find the progenitor precursor structures until readily available feedstocks or natural product starting materials are reached. In graph theory, a network is called a directed graph or digraph. We first denote features of digraph representations of reaction networks and outline a convention for drawing them. Then we demonstrate how one can use these digraphs to enumerate all possible routes to a given target molecule. From these enumerations of routes, one constructs synthesis tree diagrams for each route as was introduced in Chapter 5. In turn, for each of these synthesis tree diagrams, one then evaluates the kernel and global metrics to evaluate its material and strategy performance using the radial pentagon and linear or convergent spreadsheet algorithms introduced in Chapters 5 and 6. Using all of these data, the most optimal route is selected. This represents a simple and systematic method of route selection that is rapid, robust, and effective in identifying potential "green" routes to a specified target product.

To assist the reader in following the thread of the ideas, we introduce the route selection process for aniline, which is an important industrial commodity chemical. An example retrosynthetic map for aniline is shown in Figure 9.8. A corresponding synthetic map for aniline is shown in Figure 9.9.

A digraph representation of aforementioned synthetic plan is shown in Figure 9.10.

Key features of digraphs are nodes and arrows. The three kinds of nodes are source nodes (S_i), target product nodes (P_i), and intermediate nodes (I_i). Arrows connect the nodes to each other. They correspond to conventional reaction arrows in the forward sense from structure to structure in a sequence. The following conventions are followed in drawing digraphs:

1. Source nodes are represented as filled dots with (S_i) labels.
2. Target product nodes are represented as shaded dots with (P_i) labels.

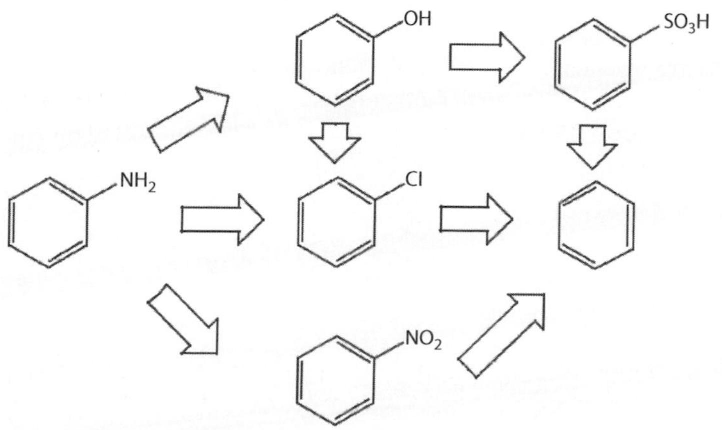

FIGURE 9.8   Retrosynthetic map for aniline.

SO3H
NaOH
3008C
OH
H2SO4
NaOH
NH3/catalyst
Cl
Cl2
NH3
NH2
FeCl3
Catalyst
3H2
HNO3
NO2
Pd C

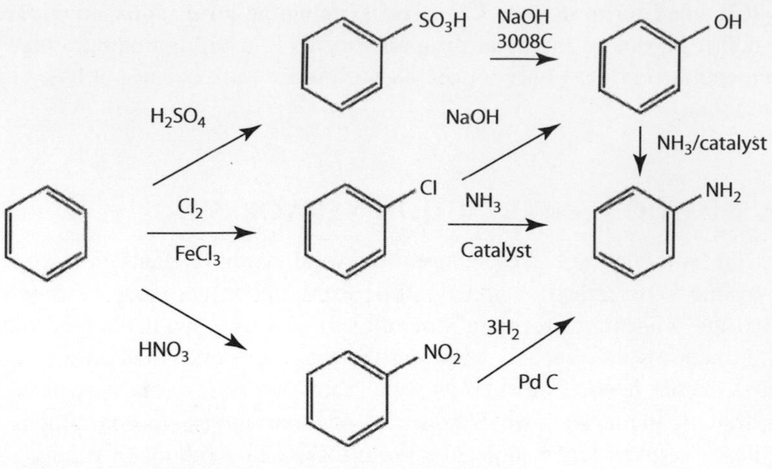

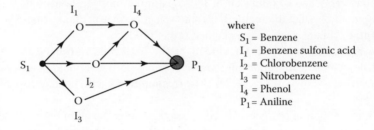

FIGURE 9.9 Synthetic map for aniline.

I1
I4
S1
P1
I2
I3
where
S1 = Benzene
I1 = Benzene sulfonic acid
I2 = Chlorobenzene
I3 = Nitrobenzene
I4 = Phenol
P1 = Aniline

FIGURE 9.10 Digraph for aniline reaction network.

3. Intermediate nodes are represented as open dots with (I_i) labels.
4. Reaction paths.

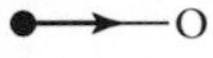

S1
I1

corresponds to a reaction $S_1 \rightarrow I_1$ where S_1 is the substrate of interest and I_1 is the immediate product of the reaction. Note that other reagent structures, solvents, catalysts, and by-products do not appear. If there are multiple reactions connecting S_1 to I_1 then the networks are represented as shown.

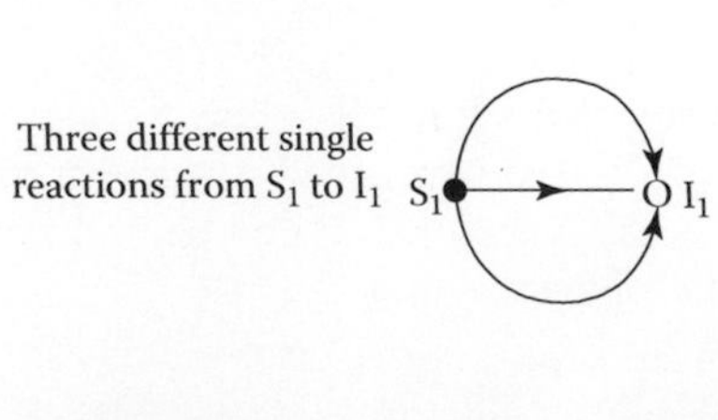

Two different single
reactions from S1 to I1
S1
I1
Three different single
reactions from S1 to I1
S1
I1

Each arrow exiting a node is counted as an outdegree with a + 1 designation with respect to that node. If no arrows exit a node then the outdegree at that node is 0. Examples of outdegree designations for various network nodes are shown as follows.

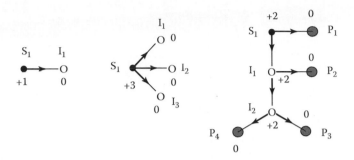

Each arrow entering a node is counted as an indegree with a − 1 designation with respect to that node. If no arrows enter a node then the indegree at that node is 0. Examples of indegree designations for various network nodes are shown as follows.

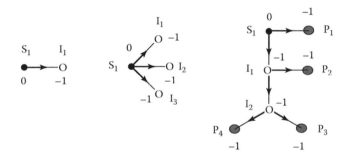

Rules:

1. At any given node the degree at that node is the sum of the indegrees and the outdegrees.

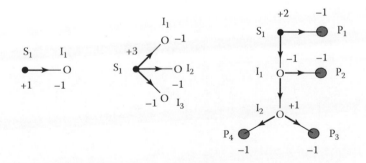

2. The sum of all outdegrees equals the sum of all indegrees for any network.
3. The sum of all degrees of all nodes in any reaction network is zero. The sum of all positive degrees equals the sum of all negative degrees in any network. This is known as the handshaking di-lemma.

In earlier examples, we have

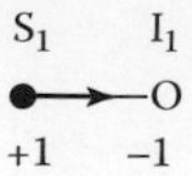

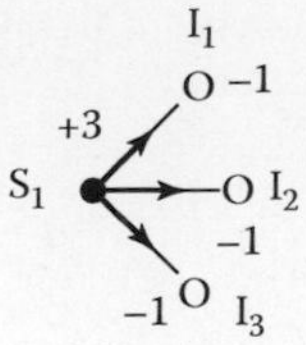

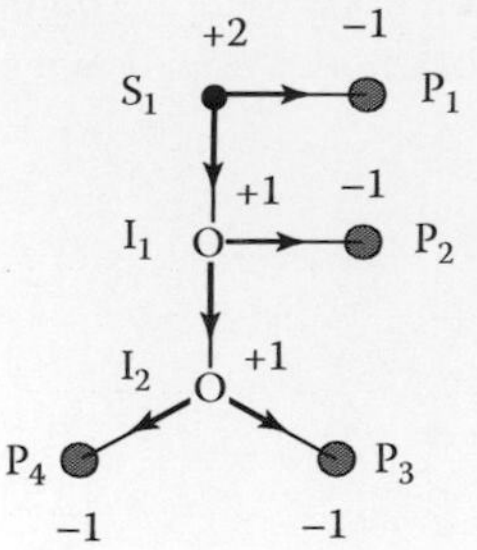

$$+1-1=0 \qquad\qquad +3-1-1-1=0 \qquad\qquad +2-1+1-1+1-1-1=0$$

Types of networks:

Type I

Single source to single target product via multiple routes or paths; this is usually used to represent many ways to make a given target product from the same starting material via different routes.

Example: benzene to aniline synthesis plan shown in Figure 9.10.

Type II

Multiple sources to single target product (usually used to represent many ways to make a given target product from different starting materials via different routes).

Example:

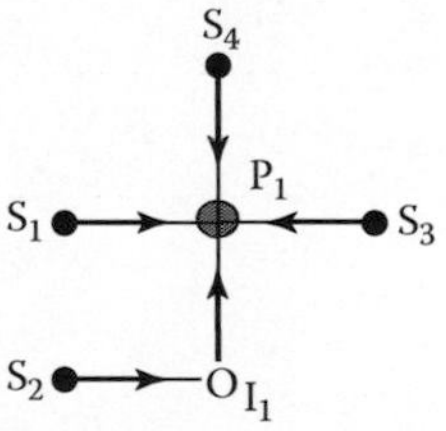

This digraph could correspond to the synthesis plan shown in Figure 9.11.

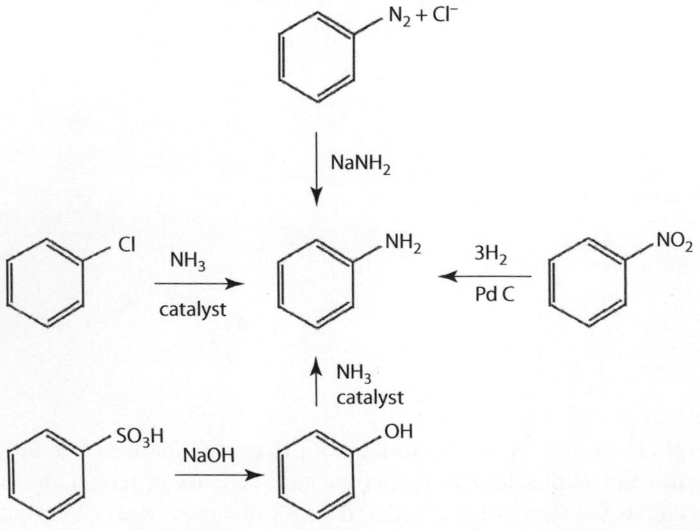

FIGURE 9.11 Type II digraph for aniline.

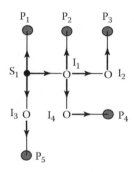

FIGURE 9.12 Type III digraph for ethylene oxide.

Type III
Single source to multiple target products; usually used to represent versatility of a particular starting material in preparing other useful products.

Example:

This digraph could correspond to the synthesis plan shown in Figure 9.12.

Counting pathways:

a. Begin with reaction network diagram.
b. Construct its corresponding reduced network diagram by eliminating intermediates with a single inpath and a single outpath.
c. Determine type of network, that is, single source to single target (Type I), multiple sources to single target (Type II), or single source to multiple targets (Type III).
d. For Type II networks, draw reduced networks of Types I or III for each source starting material. For Type I and Type III networks go to (e).
e. From (d), unpack reduced network to find and enumerate total number of paths by drawing a divergent tree. Begin with source node and draw outdegree arrows from it to all intermediates in the first tier. For each intermediate node reached in the first tier, draw corresponding outdegree arrows to all intermediates in the second tier. Keep repeating this process until all target product nodes are reached.

From the divergent tree, the following relation holds.

Total number of paths = sum of positive degrees = sum of negative degrees.

Once paths are enumerated, synthesis trees may be constructed and the material, energy, and cost efficiencies may be readily determined.

Example 1 (from digraph in Figure 9.9)

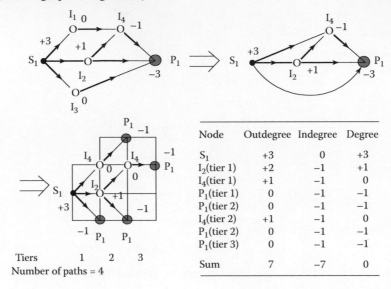

Node	Outdegree	Indegree	Degree
S_1	+3	0	+3
I_2(tier 1)	+2	−1	+1
I_4(tier 1)	+1	−1	0
P_1(tier 1)	0	−1	−1
P_1(tier 2)	0	−1	−1
I_4(tier 2)	+1	−1	0
P_1(tier 2)	0	−1	−1
P_1(tier 3)	0	−1	−1
Sum	7	−7	0

Tiers   1   2   3

Number of paths = 4

Example 2 (from Figure 9.10)

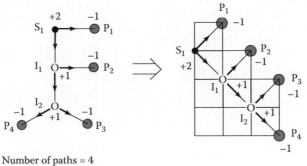

Number of paths = 4

Example 3 (from Figure 9.11)

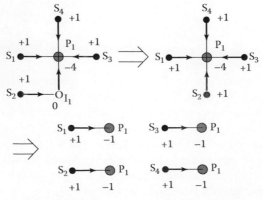

Number of paths = 4

Example 4 (from Figure 9.12)

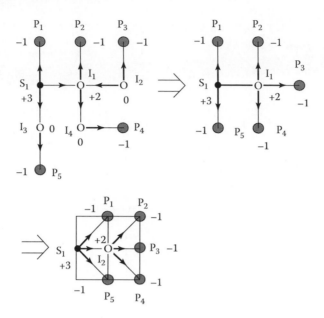

Number of paths = 5

We now apply the whole process of route selection to the reaction network for 4-aminophenol shown in Figure 9.13. Reaction yields were taken from Faith's and Shreve's books on industrial chemical synthesis [57,58].

FIGURE 9.13 Reaction network for 4-aminophenol.

The reduced digraph, divergent tree, and enumeration of routes are as follows.

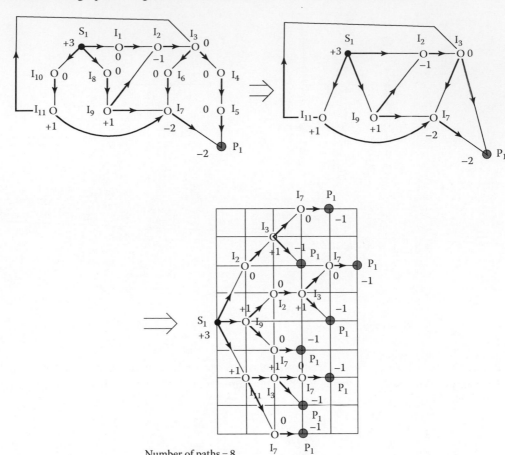

Number of paths = 8

The eight synthesis routes along with their synthesis trees are shown in Figures 9.14 through 9.21.

Table 9.7 summarizes the kernel metrics for the eight plans to 4-aminophenol. From these data the meritorious plans are the following:

1. Route 1 has the highest overall yield.
2. Route 2 has the highest overall yield and involves the greatest degree of building up.
3. Route 4 has the least oxidation changes and is hypohypsic.
4. Route 5 has the least number of steps, the highest number of bonds made per step, and the least number of sacrificial steps, and produces the lowest kernel mass of waste.
5. Route 8 has the least number of steps, the least molecular weight fraction of sacrificial reagents, the highest atom economy, the highest number of bonds made per step, and the least number of sacrificial steps, and produces the lowest kernel mass of waste.

It appears then that the "winning" plan is route 8 closely followed by route 5. These would be the ones worth pursuing. Inclusion of auxiliary material demand is likely to separate them if a global metrics analysis is done. Note the mismatches between plans with the highest overall yield (routes 1 and 2) and the highest atom economy (route 8). The shorter plans are not the ones with the highest overall yield though route 5 and routes 1 and 2 are separated by a 2% margin with respect to this parameter. These observations indicate that further optimization is still possible. Since all plans have a common starting material and end product their comparison is quite fair.

(i) S1 ---> I2 ---> I3 ---> I7 ---> P1

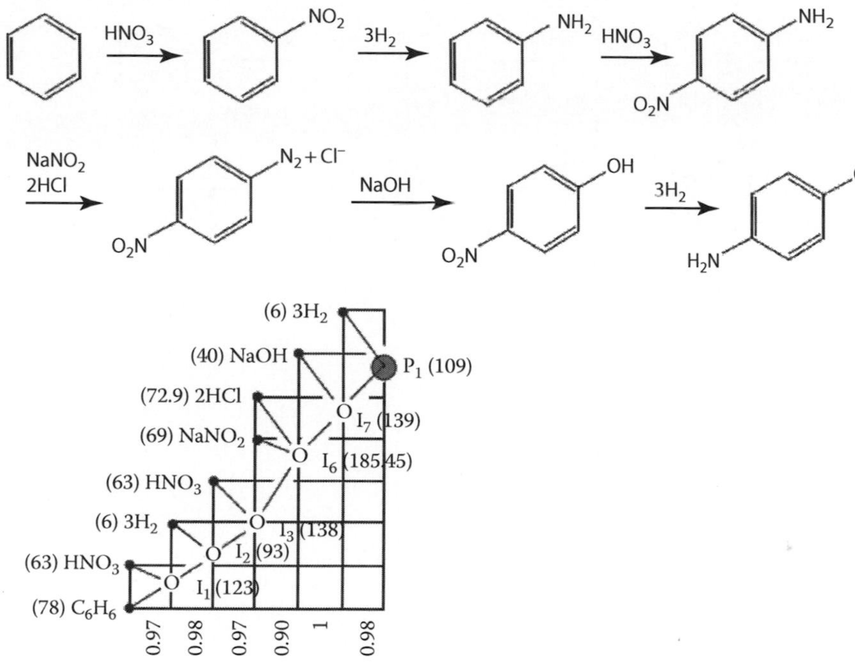

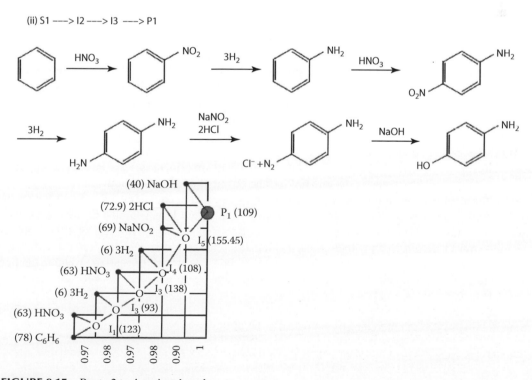

FIGURE 9.14 Route 1 to 4-aminophenol.

(ii) S1 ----> I2 ----> I3 ----> P1

FIGURE 9.15 Route 2 to 4-aminophenol.

(iii) S1 ----> I9 ----> I2 ----> I3 ----> I7 ----> P1

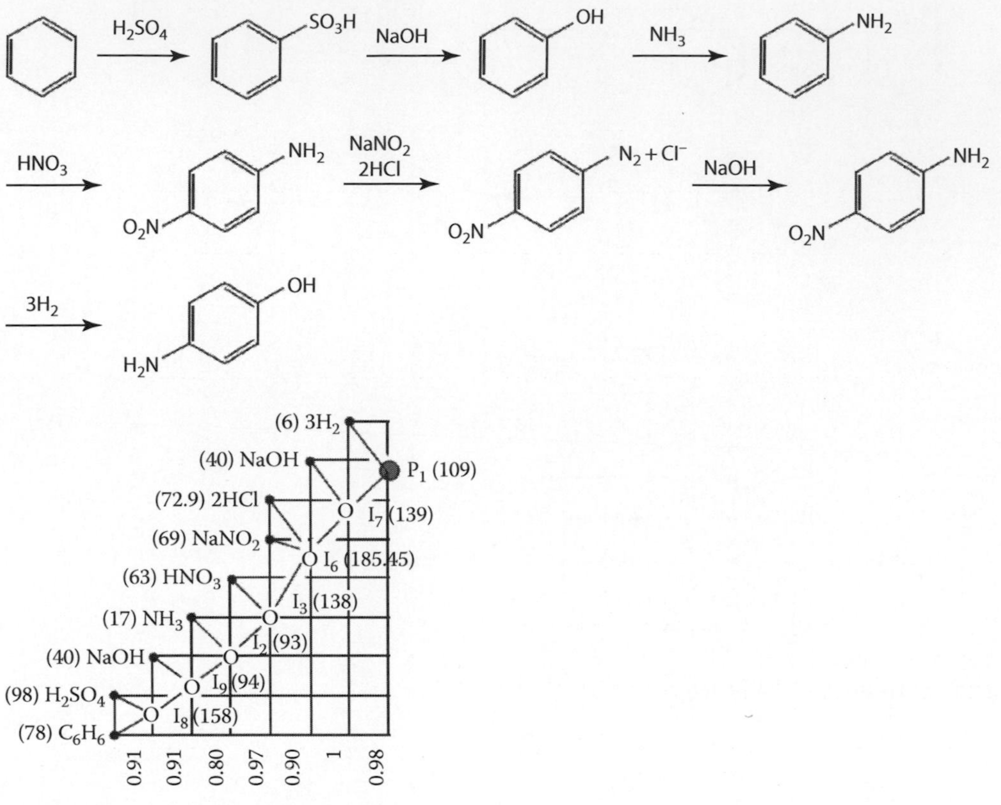

FIGURE 9.16   Route 3 to 4-aminophenol.

The next example taken from the synthesis database is a reaction network for various routes to aspirin [57,58]. Figures 9.22 through 9.24 show reaction networks for phenol, acetic anhydride, and aspirin, respectively. The digraph in Figure 9.25 shows the combination of two subnetworks for phenol and acetic anhydride that converge to produce the final target product.

By inspection it can be easily seen that there are five routes to salicylic acid and two routes to acetic anhydride. Therefore, the combined number of routes to aspirin is $5 \times 2 = 10$. The digraph shows that the convergent step is the one linking the terminal node of one network (I_2—salicylic acid) to the terminal node of the other (I_3—acetic anhydride). The point of convergence is depicted by the coming together of the two arrows emanating from I_2 and I_3 as a double arrow toward the final product node P. Hence, the indegree at P is -2 and this makes the overall sum of all indegrees and outdegrees zero consistent with the handshaking di-lemma condition. Figures 9.26 through 9.35 show the synthesis trees for each route. The first five routes begin with the carbide route to acetic anhydride starting from calcium carbide and the latter five begin with the Wacker route starting from ethylene.

Tables 9.8 and 9.9 summarize the kernel and global metrics for the possible routes, respectively. Route numbers correspond to the chronological order of the synthesis tree diagrams.

(iv) S1 ---> I9 ---> I2 ---> I3 ---> P1

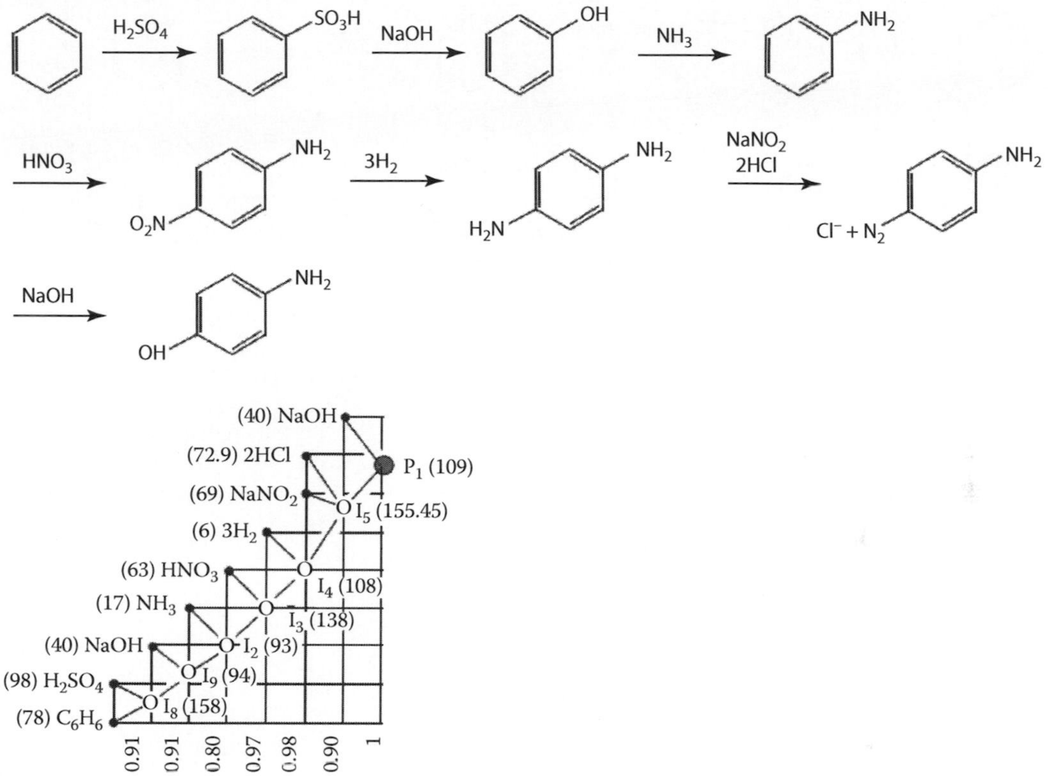

FIGURE 9.17   Route 4 to 4-aminophenol.

(v) S1 ---> I9 ---> I7 ---> P1

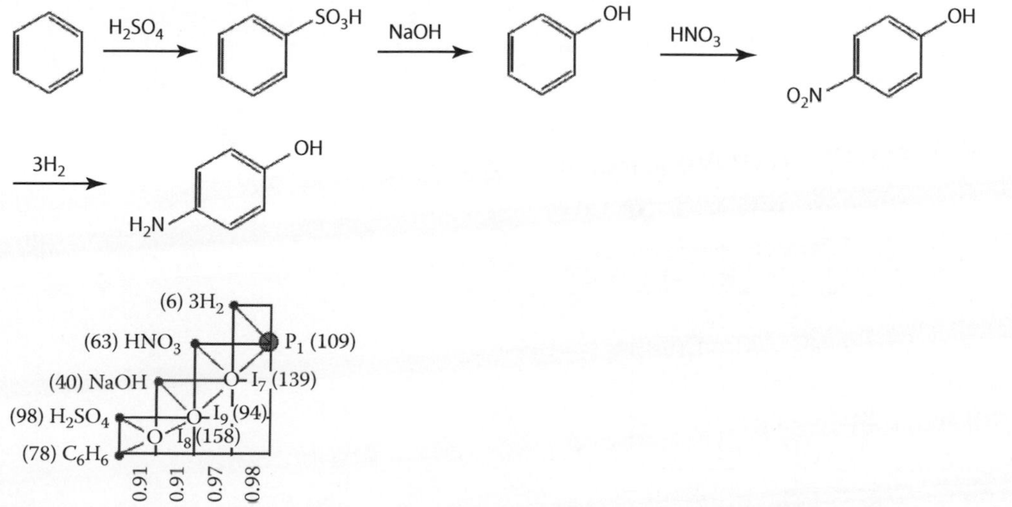

FIGURE 9.18   Route 5 to 4-aminophenol.

(vi) S1 ----> I11 ----> I3 ----> I7 ----> P1

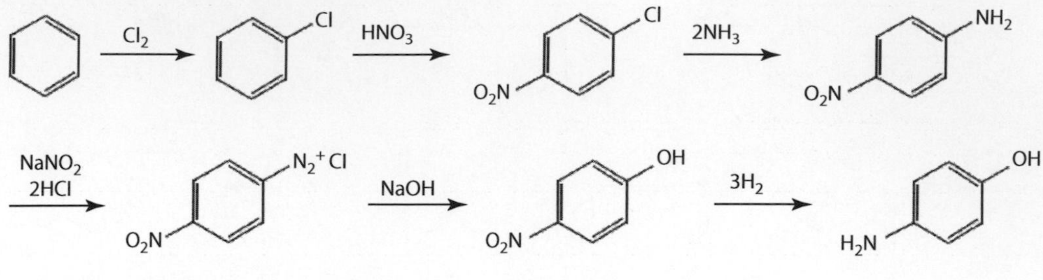

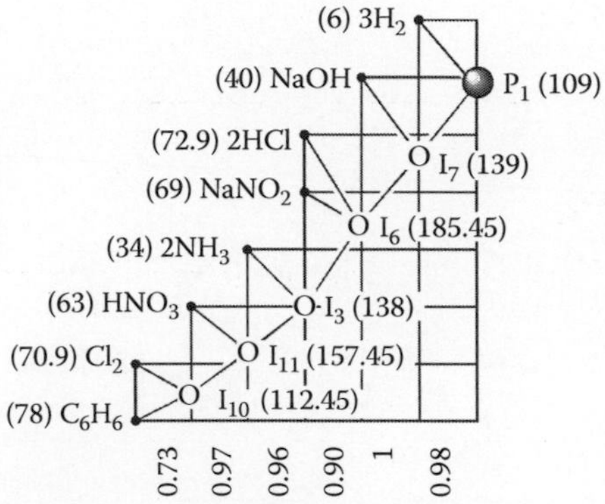

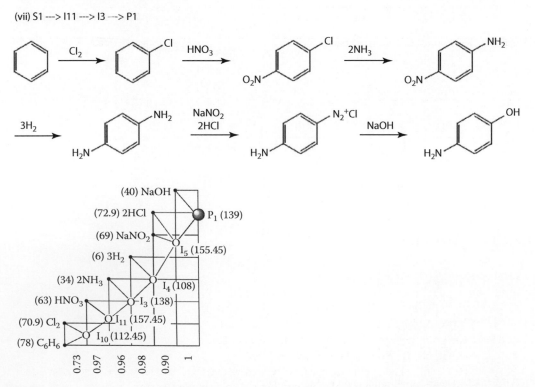

FIGURE 9.19   Route 6 to 4-aminophenol.

(vii) S1 ---> I11 ---> I3 ---> P1

FIGURE 9.20   Route 7 to 4-aminophenol.

(vii) S1 ---> I11 ---> I7 ---> P1

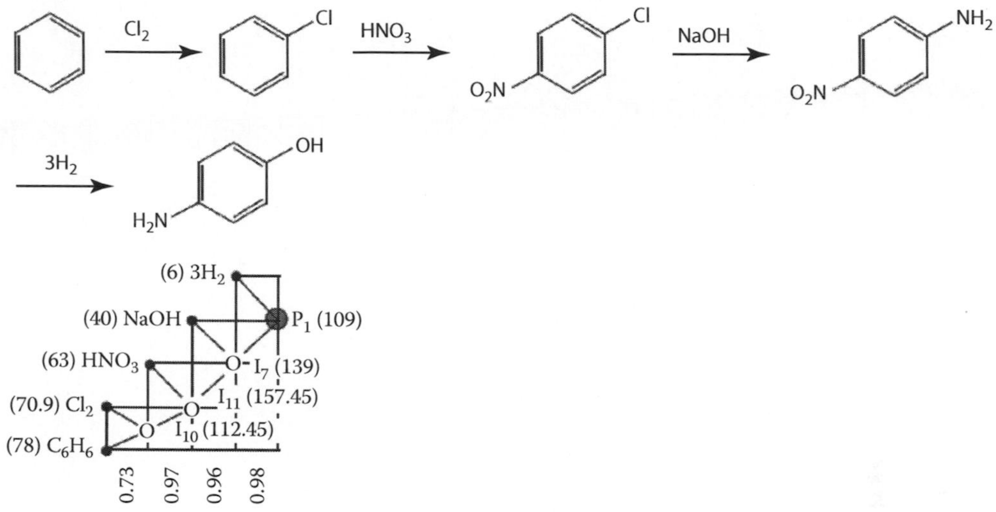

FIGURE 9.21 Route 8 to 4-aminophenol.

TABLE 9.7

Summary of Kernel Metrics for the Eight Linear Plans to 4-Aminophenol

Plan	N	M	I	μ1	f(sac)	% Yield	% AE	Kernel Mass of Waste (kg/mol)	B/M	f(nb)	|UD|
Route 1	6	6	8	8.06	0.535	81.3	27.4	0.35	0.67	0.50	20
Route 2	6	6	8	−0.65	0.535	81.3	27.4	0.35	0.67	0.50	12
Route 3	7	7	9	13.93	0.618	56.7	22.5	0.57	0.57	0.57	20
Route 4	7	7	9	6.31	0.721	56.7	22.5	0.57	0.29	0.71	8
Route 5	4	4	5	4.4	0.351	78.7	38.2	0.23	1.00	0.25	20
Route 6	6	6	8	16.89	0.574	60	25.1	0.46	0.67	0.50	20
Route 7	6	6	7	8.18	0.574	60	25.1	0.46	0.67	0.50	20
Route 8	4	4	5	2.56	0.283	66.6	42.3	0.23	1.00	0.25	20

The meritorious routes are as follows at the kernel level:

1. Routes 6–10 have the least number of reaction stages.
2. Routes 6 and 7 have the least number of reaction steps.
3. Routes 8 and 9 have the least number of input materials.
4. Route 7 has the highest degree of building up toward the target product.
5. Route 8 has the least molecular weight fraction of sacrificial reagents.
6. Routes 6–10 have the highest overall yield.
7. Routes 6, 8, and 9 have the highest atom economies and the lowest kernel masses of waste.
8. Routes 1, 2, 6, and 7 have the highest number of bonds made per step.

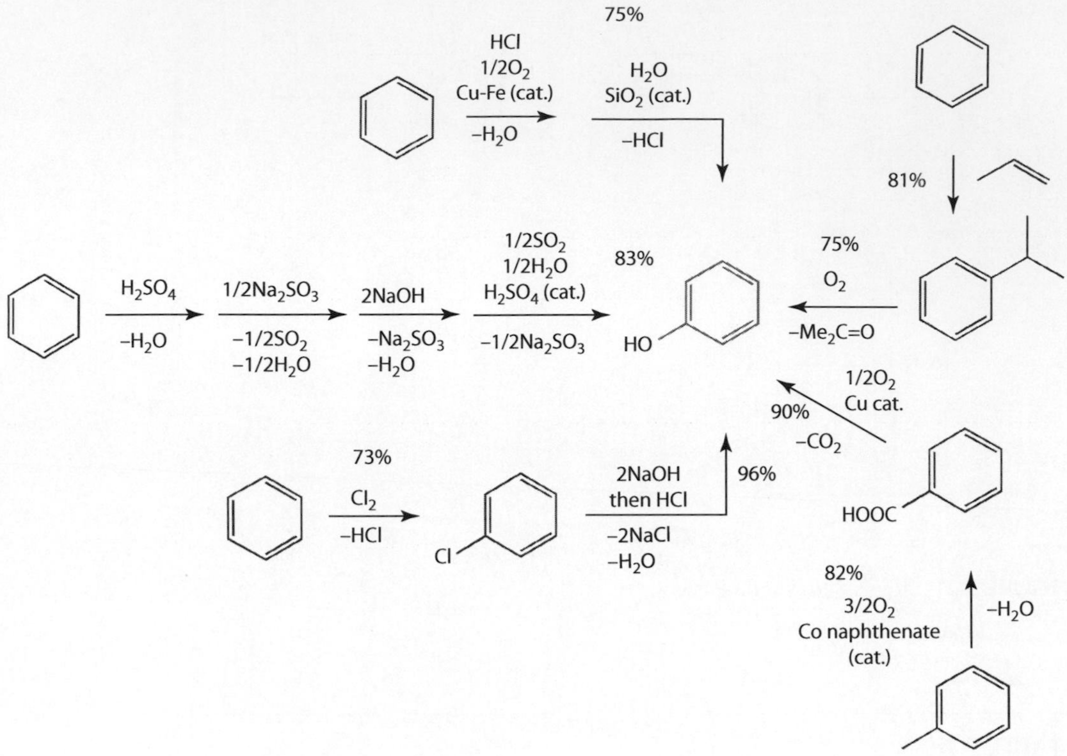

FIGURE 9.22   Reaction network for phenol synthesis.

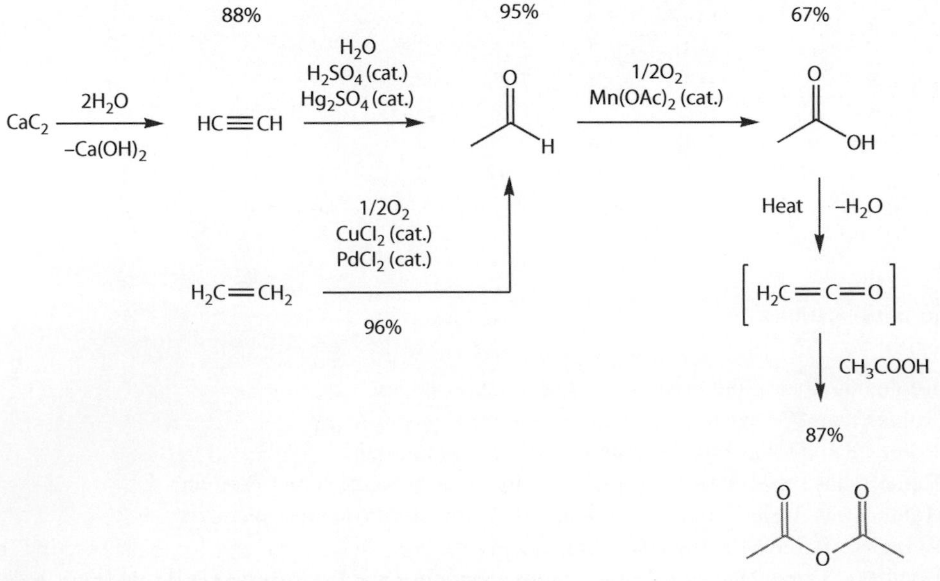

FIGURE 9.23   Reaction network for acetic anhydride synthesis.

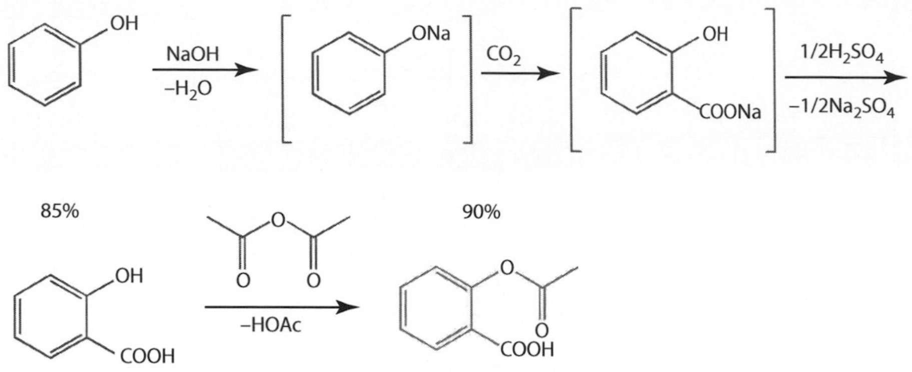

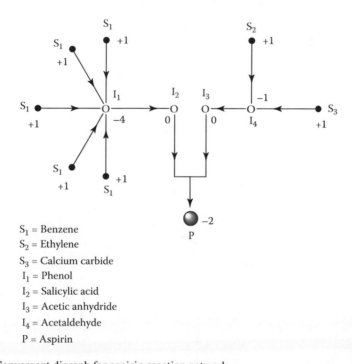

FIGURE 9.24 Reaction network for aspirin synthesis.

FIGURE 9.25 Convergent digraph for aspirin reaction network.

S_1 = Benzene
S_2 = Ethylene
S_3 = Calcium carbide
I_1 = Phenol
I_2 = Salicylic acid
I_3 = Acetic anhydride
I_4 = Acetaldehyde
P = Aspirin

All routes have every reaction stage producing one target bond. At the global metrics level, routes 6 and 9 have the least E-kernel, routes 8 and 9 have the least E-excess, and routes 6–10 have the least E-auxiliaries. Overall, routes 8 and 9 have the least E-total. On balance, routes 6 (Wacker–Raschig process) and 9 (Wacker-toluene oxidation process) appear to have the most positive counts based on all metrics parameters.

This section closes with two reaction network exercises for glycerol and trimex shown in Figures 9.36 and 9.37 that the reader can practice drawing digraphs, enumerate synthesis trees, and rank plans according to kernel metrics.

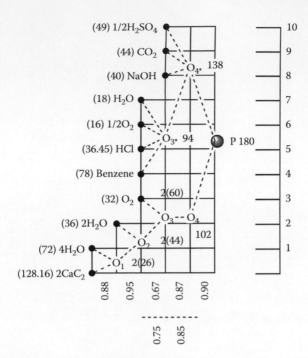

FIGURE 9.26 Synthesis tree for carbide-Raschig process.

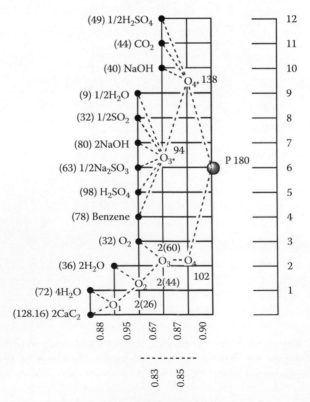

FIGURE 9.27 Synthesis tree for carbide-benzenesulfonate process.

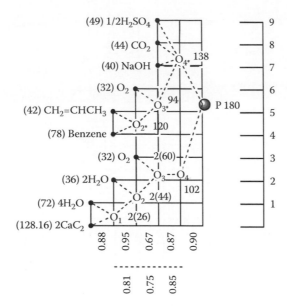

FIGURE 9.28 Synthesis tree for carbide-cumene peroxidation process.

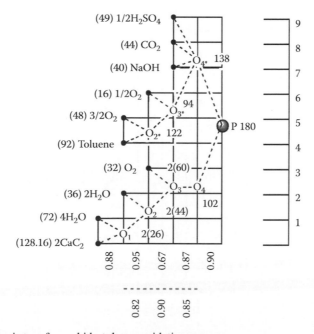

FIGURE 9.29 Synthesis tree for carbide-toluene oxidation process.

9.10 SCHEDULING

The utility of synthesis tree diagrams has been well demonstrated for keeping track of branches, input starting materials, intermediates, and reaction yield chains, especially when dealing with multiconvergent synthesis plans. It can also be transformed into a Gantt diagram to aid in scheduling batch processes. Gantt invented a number of charts to streamline the scheduling of work and hence this resulted in increasing the efficiency of production management [59,60]. The concept of Gantt diagrams has been applied to the chemical industry [61,62]. In this section, a simple example illustrates the idea. Figure 9.38 shows a reduced form of the synthesis tree diagram for a convergent plan

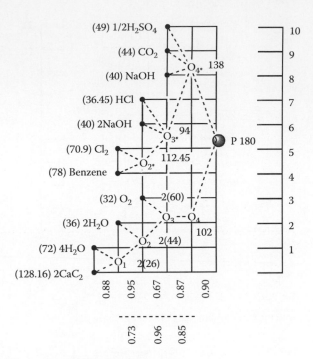

FIGURE 9.30 Synthesis tree for carbide-caustic process.

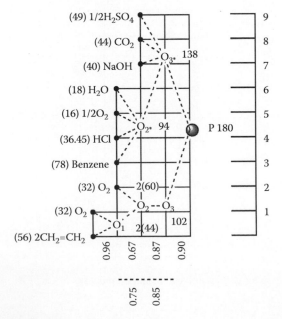

FIGURE 9.31 Synthesis tree for Wacker–Raschig process.

consisting of four branches, three points of convergence, 10 reaction stages, and 17 reaction steps. The diagram shows only the nodes for intermediates, the final product, and the starting materials at the beginning of each branch. For clarity, all other starting materials are not shown. Instead of having reaction yields across the bottom of the diagram, the reduced diagram shows the cycle time for each reaction step in hours. The cycle time for a given reaction step is the length of time required for all unit operations relevant to carrying out that chemical transformation. This includes reaction

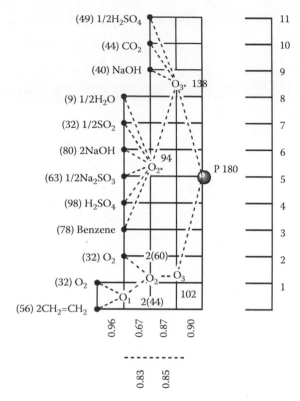

FIGURE 9.32   Synthesis tree for Wacker-benzenesulfonate process.

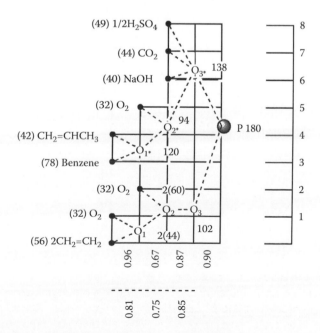

FIGURE 9.33   Synthesis tree for Wacker-cumene peroxidation process.

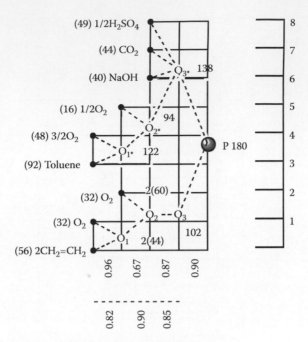

FIGURE 9.34   Synthesis tree for Wacker-toluene oxidation process.

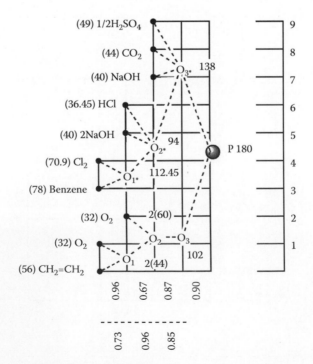

FIGURE 9.35   Synthesis tree for Wacker-caustic process.

TABLE 9.8

Summary of Kernel Metrics for the 10 Plans to Aspirin

Plan	N	M	I	$\mu 1$	f(sac)	% Yield	% AE	Kernel Mass of Waste (kg/mol)	B/M	f(nb)	|UD|
Route 1	5	7	11	−151.64	0.474	43.9	32.8	0.84	1	0	10
Route 2	5	7	13	−176.47	0.591	43.9	23.6	1.15	1	0	10
Route 3	5	8	10	−157.64	0.452	43.9	32.5	0.9	0.875	0	6
Route 4	5	8	10	−156.64	0.485	43.9	32.3	0.85	0.875	0	6
Route 5	5	8	11	−154.08	0.533	43.9	27	1.02	0.875	0	10
Route 6	4	6	10	−178.8	0.42	50.4	44.8	0.49	1	0	10
Route 7	4	6	13	−208.6	0.584	50.4	29.4	0.8	1	0	10
Route 8	4	7	9	−186	0.39	50.4	44.4	0.54	0.857	0	6
Route 9	4	7	9	−187.6	0.435	50.4	44	0.5	0.857	0	6
Route 10	4	7	10	−181.73	0.508	50.4	34.7	0.67	0.857	0	10

TABLE 9.9

Summary of Global Metrics for the 10 Plans to Aspirin

Plan	E-Kernel	E-Excess	E-Aux	E-Total
Route 1	4.68	22.66	2.25	29.59
Route 2	6.39	22.71	2.25	31.35
Route 3	4.99	20.62	2.25	27.85
Route 4	4.75	20.65	2.25	27.65
Route 5	5.67	21.72	2.25	29.64
Route 6	2.72	4.7	4.32	11.74
Route 7	4.43	4.75	4.32	13.5
Route 8	3.02	2.66	4.32	10
Route 9	2.78	2.69	4.32	9.79
Route 10	3.7	3.76	4.32	11.78

time, work-up time, and purification time. From this diagram, we can deduce that 36 h are needed to complete the entire synthesis assuming no stoppages along the way. This is found by taking the sum of the cycle times for branch 1 (B1). The corresponding Gantt diagram is shown in Figure 9.39 where the four branches are displayed as separate time blocks. Each unit square corresponds to a 1 h interval. The black arrows indicate the times when production of intermediates is complete. The light gray arrows indicate the start times of each branch. The start times for the four branches are as follows: T for B1, T + 2 for B2, T + 9 for B3, and T + 18 for B4. The dark gray arrow shows the time when the final product is complete. From the Gantt diagram, it is possible to write out the complete schedule of events for carrying out the entire synthesis plan in chronological order as shown in Table 9.10. The diagram tells us when to start certain operations for each branch so that several operations may be run in parallel, thus saving time. A parameter often used by process chemists to measure synthesis production efficiency is space-time-yield, STY, given by Equation 9.1.

$$STY = \frac{\text{mass-of-product}}{(\text{total-process-time})(\text{total-volume-of-input-materials})} \tag{9.1}$$

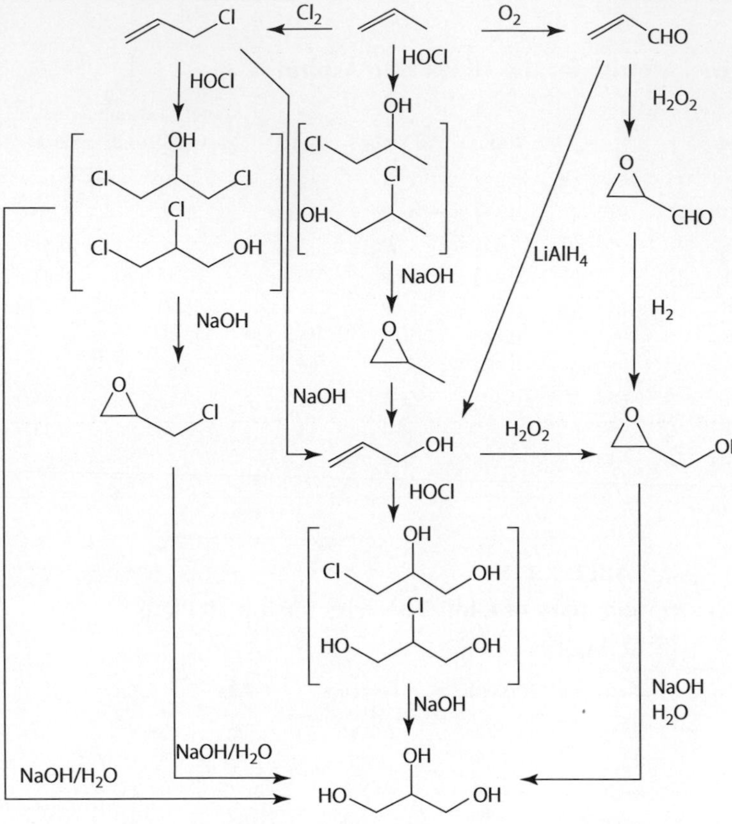

FIGURE 9.36   Reaction network for glycerol.

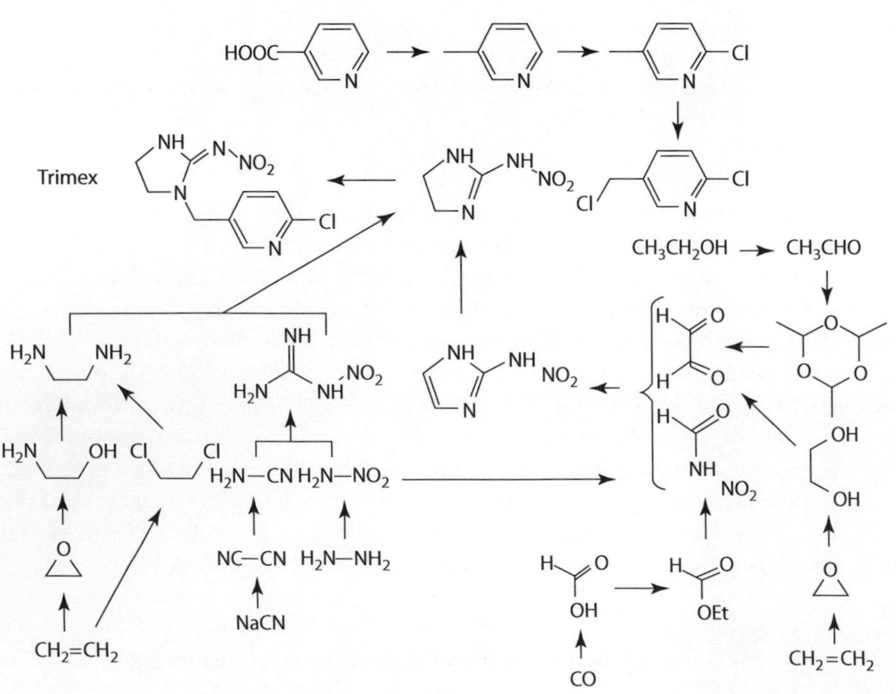

FIGURE 9.37   Reaction network for trimex.

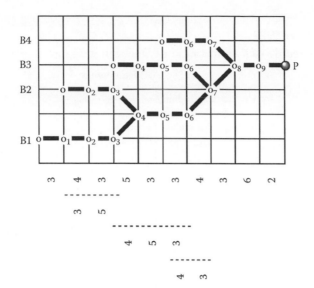

FIGURE 9.38 Reduced synthesis tree diagram for a convergent plan showing branches, intermediates, and cycle times in hours for each reaction step.

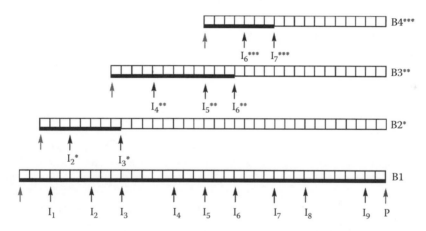

FIGURE 9.39 Gantt diagram for reduced synthesis tree diagram shown in Figure 9.38.

The total process time is the sum of the cycle times for each reaction step along the longest branch of the synthesis plan. In carrying out a synthesis, the sum of all input materials will be transformed to producing both the target product and waste products. The overall reaction mass efficiency accounts for the mass of target product produced per total mass of input materials. The STY for a synthesis plan may be written more uniformly in mass units per unit time as shown in Equation 9.2.

$$STY = \frac{RME_{overall}}{total\text{-}process\text{-}time} \tag{9.2}$$

So, for this example, if 36 h were needed to produce 1000 kg of target product P, which required 10,000 kg of total input materials (reagents, solvents, and other auxiliaries), then overall RME = 1,000/10,000 = 0.1 and STY = 0.1/36 = 0.00278 kg product/kg inputs/h.

TABLE 9.10

Schedule of Events for Gantt Diagram in Figure 9.39

Time	Description
T	Start branch 1
T + 2	Start branch 2
T + 3	Intermediate 1 complete
T + 5	Intermediate 2* complete
T + 7	Intermediate 2 complete
T + 9	Start branch 3
T + 10	Intermediates 3 and 3* complete
T + 13	Intermediate 4** complete
T + 15	Intermediate 4 complete
T + 18	Intermediates 5 and 5** complete, start branch 4
T + 21	Intermediates 6 and 6** complete
T + 22	Intermediate 6*** complete
T + 25	Intermediates 7 and 7*** complete
T + 28	Intermediate 8 complete
T + 34	Intermediate 9 complete
T + 36	Product complete

9.11   WHAT CAN GO WRONG WITH COMPUTER-ASSISTED SYNTHESIS SOFTWARE

In this section, we demonstrate that attempts to come up with novel plans using computer-assisted software do not always coincide with those plans being material efficient. This frustratingly mismatched situation is illustrated by the example anti-arthritic compound atiprimod developed by Smith Kline & French in 1988 [63,64]. The details of the synthesis plans are given in the accompanying download along with references. A computer-assisted program called SYNSUP by Bersohn and Tanaka was used to come up with 10 routes to this compound [12,65–67]. These are summa-rized in Figures 9.40 through 9.46. The plan selected for experimentation by the authors followed a

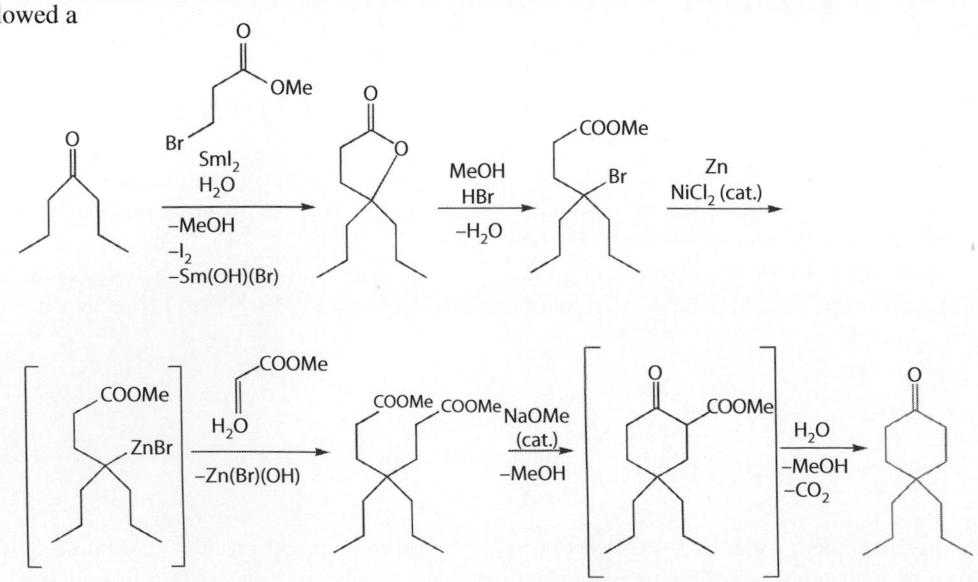

FIGURE 9.40   SYNSUP route to 4,4'-di-n-propylcyclohexanone (Route A).

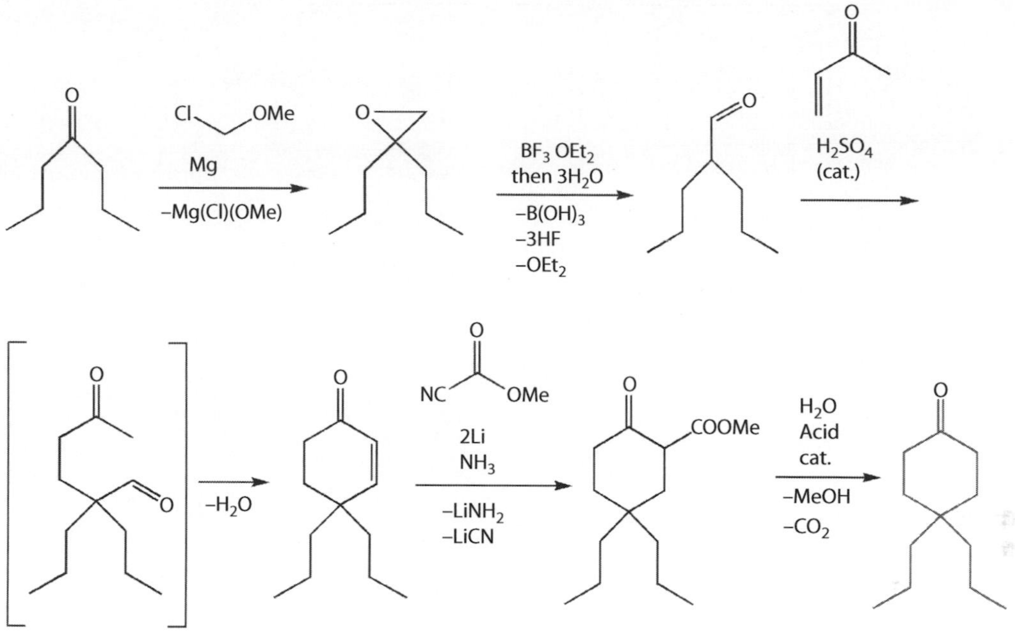

FIGURE 9.41 SYNSUP route to 4,4′-di-*n*-propylcyclohexanone (Route B).

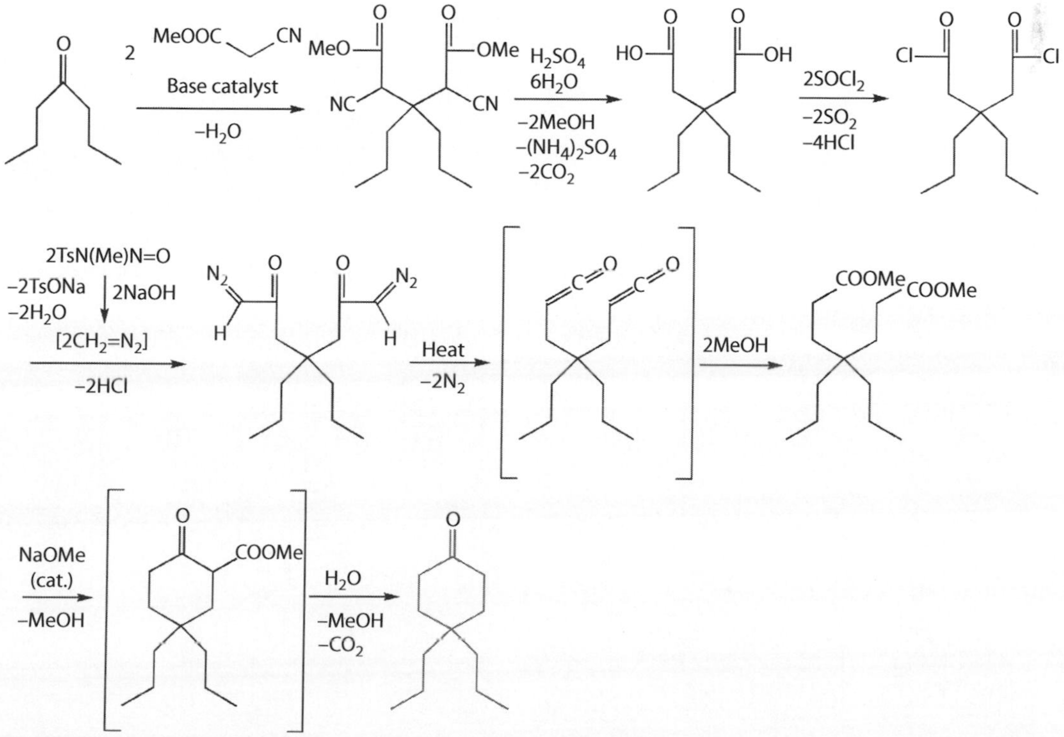

FIGURE 9.42 SYNSUP route to 4,4′-di-*n*-propylcyclohexanone (Route C).

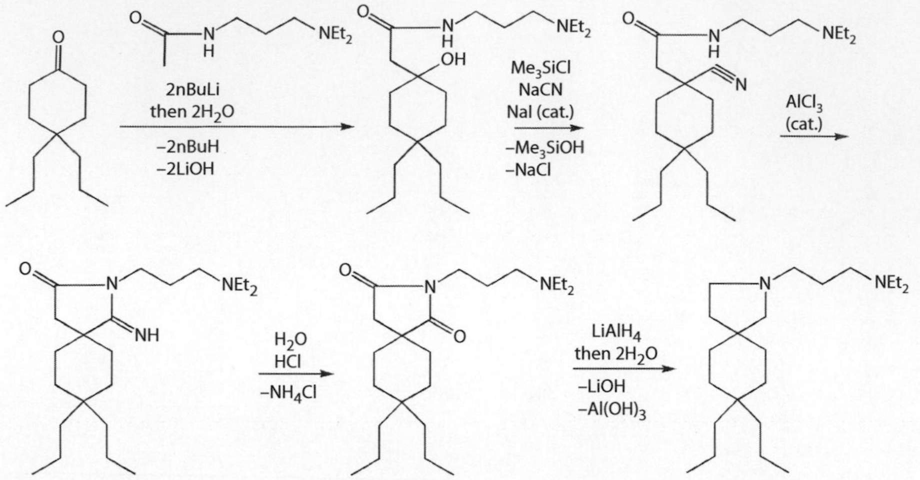

FIGURE 9.43 SYNSUP route to atiprimod (Route A*).

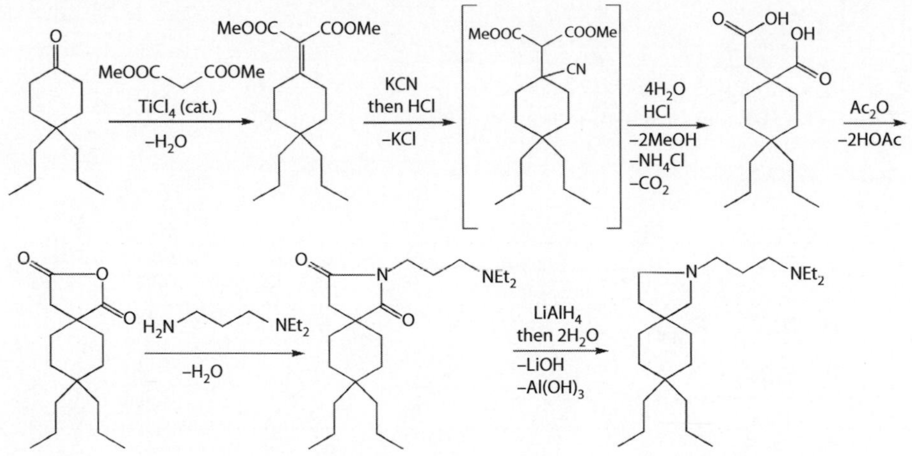

FIGURE 9.44 SYNSUP route to atiprimod (Route B*).

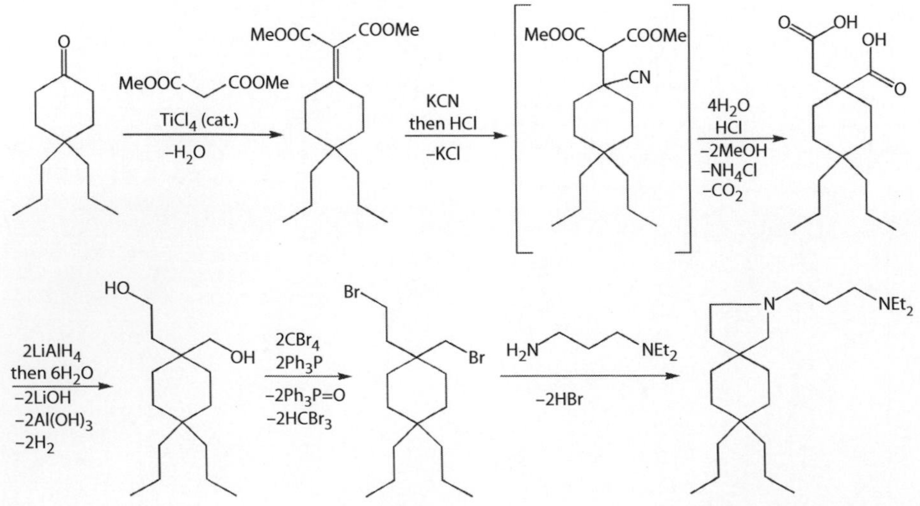

FIGURE 9.45 SYNSUP route to atiprimod (Route C*).

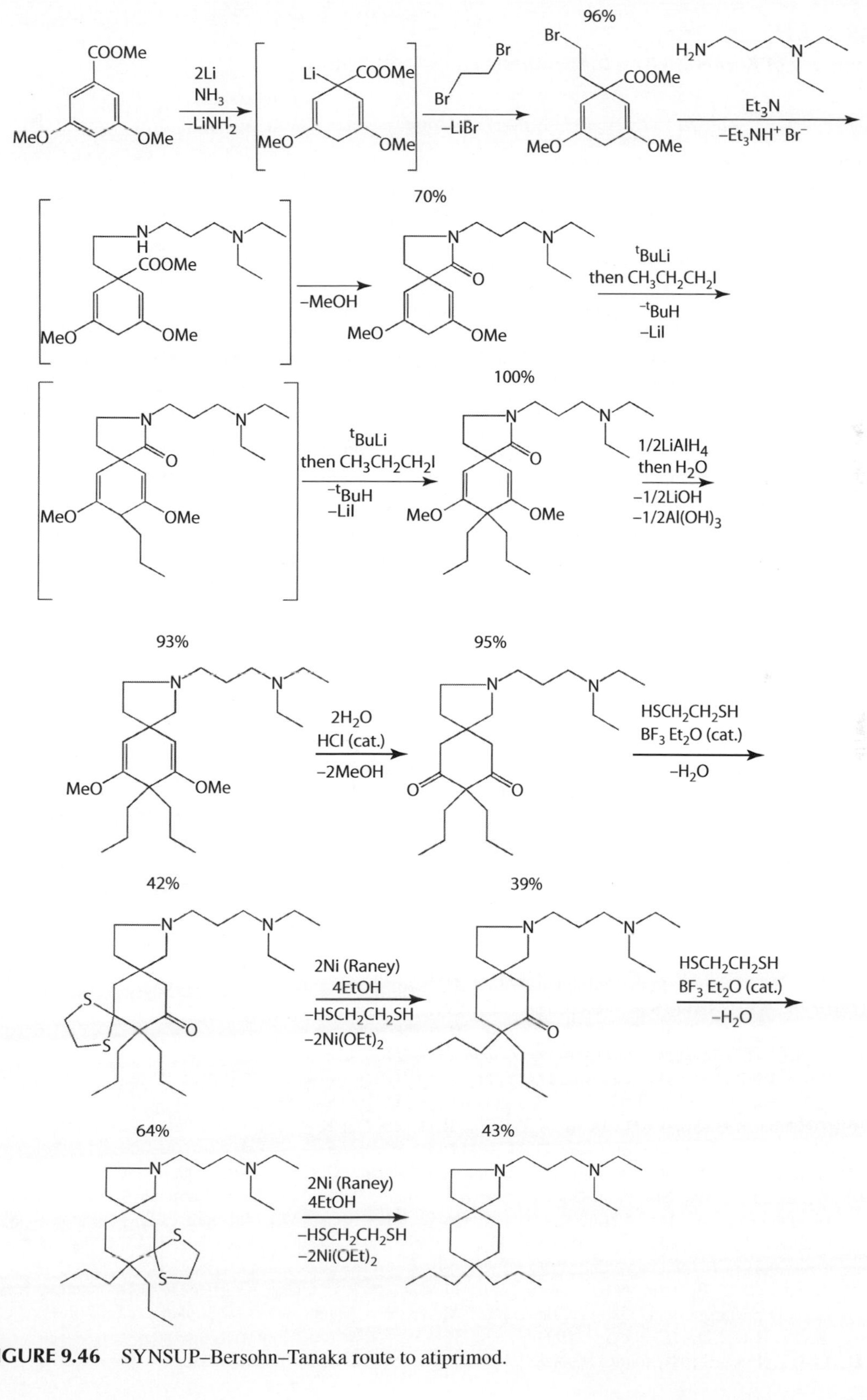

FIGURE 9.46 SYNSUP–Bersohn–Tanaka route to atiprimod.

TABLE 9.11
Summary of Kernel Metrics for Synthesis Plans to Atiprimod

Plan	N	M	I	μ1	f(sac)	% Yield	% AE	Kernel Mass of Waste (kg/mol)	B/M	f(nb)	|UD|
SmithKline Beecham	8	9	19	−158.78	0.695	47.8	16.5	2.6	1.56	0.13	26
SmithKline and French	9	9	15	−124.5	0.384	26.9	25.7	2.7	1.78	0.22	22
Rice	14	14	23	−141.5	0.622	1.4	14.8	40.3	1.43	0.36	30
Bersohn–Tanaka	9	9	17	−29.03	0.447	0.03	17	42.9	1.44	0.22	28
SYNSUP[a]	10	11	20	−163.95	0.525	N/A	25.8	>0.96	1.27	0.20	22

[a] Best overall atom economical route: route B from 4-heptanone to 4,4′-di-n-propylcyclohexanone and route A* from 4,4′-di-n-propylcyclohexanone to atiprimod.

Birch reduction strategy that converts a starting aromatic material to a substituted cyclohexane ring that links to a five-membered ring in a spiro fashion (see Figure 9.42). When compared with prior routes to this target compound, the Birch reduction strategy was indeed novel as the other plans involved the same strategy of cyclizing acyclic fragments to make the six-membered cyclohexane ring. The unfortunate outcome of this novel strategy is that it ranks dead last in terms of kernel material efficiency as shown by the data in Table 9.11. Another unfortunate situation is that the experimental procedure documented by Bersohn and Tanaka had not enough detail to carry out a global metrics analysis.

REFERENCES

1. Corey, E.J.; Gilman, N.W.; Ganem, B.E. *J. Am. Chem. Soc.* 1968, *90*, 5616.
2. Chen, B.; Makley, D.M.; Johnston, J.N. *Nature* 2010, *465*, 1027.
3. Campos, K.R.; Klapers, A.; Waldman, J.H.; Dormer, P.G.; Chen, C.Y. *J. Am. Chem. Soc.* 2006, *128*, 3538.
4. Inhoffen, H.H.; Bohlmann, F.; Bartram, K.; Rummert, G.; Pommer, H. *Ann. Chem.* 1950, *570*, 54.
5. Surmatis, J.D.; Ofner, A. *J. Org. Chem.* 1961, *26*, 1171.
6. Al-Hassan, M.I. *Synthesis* 1987, 816.
7. Tanaka, A.; Kawai, T.; Takabatake, T.; Oka, N.; Okamoto, H.; Bersohn, M. *Tetrahedron Lett.* 2006, *47*, 6733.
8. Näf, F.; Decorzant, R.; Giersch, W.; Ohloff, G. *Helv. Chim. Acta.* 1981, *64*, 1387.
9. Gutzwiller, J.; Uskokovic, M. *J. Am. Chem. Soc.* 1970, *92*, 204.
10. Gates, M.; Sugavanam, R.; Schreiber, W.L. *J. Am. Chem. Soc.* 1970, *92*, 205.
11. Taylor, E.C.; Martin, S.F. *J. Am. Chem. Soc.* 1972, *94*, 6218.
12. Hamon, D.P.G.; Young, R.N. *Austr. J. Chem.* 1976, *29*, 145.
13. Boyle, P.H.; Cocker, W.; Grayson, D.H.; Shannon, P.V.R. *Chem. Commun.* 1971, 395.
14. Boyle, P.H.; Cocker, W.; Grayson, D.H.; Shannon, P.V.R. *J. Chem. Soc. (C)* 1971, 2136.
15. de Vicente, J.; Arrayas, R.G.; Canada, J.; Carretero, J.C. *Synlett.* 2000, 53.
16. Nicolaou, K.C.; Pappo, D.; Tsang, K.Y.; Gibe, R.; Chen, D.Y.K. *Angew. Chem. Int. Ed.* 2008, *47*, 944.
17. Ghosh, A.K.; Xi, K. *J. Org. Chem.* 2009, *74*, 1163.
18. Tatsuta, K.; Yoshimoto, T.; Gunji, H.; Okado, Y.; Takahashi, M. *Chem. Lett.* 2000, *6*, 646.
19. Wender, P.A.; Badham, N.F.; Conway, S.P.; Floreancig, P.E.; Glass, T.E.; Gränicher, C.; Houze, J.B.; Jänichen, J.; Lee, D.; Marquess, D.G.; McCrane, P.L.; Meng, W.; Mucciaro, T.P.; Mühlebach, M.; Natchus, M.G.; Paulsen, H.; Rawlins, D.B.; Satkofsky, J.; Shuker, A.J.; Sutton, J.C.; Taylor, R.E.; Tomooka, K. *J. Am. Chem. Soc.* 1997, *119*, 2755.
20. Wender, P.A.; Badham, N.F.; Conway, S.P.; Floreancig, P.E.; Glass, T.E.; Houze, J.B.; Krauss, N.E.; Lee, D.; Marquess, D.G.; McCrane, P.L.; Meng, W.; Natchus, M.G.; Shuker, A.J.; Sutton, J.C.; Taylor, R.E. *J. Am. Chem. Soc.* 1997, *119*, 2757.
21. Doi, T.; Fuse, S.; Miyamoto, S.; Nakai, K.; Sasuga, D.; Takahashi, T. *Chem. Asian J.* 2006, *1*, 370.

22. Kanazawa, A.M.; Denis, J.N.; Greene, A.E. *Chem. Commun.* 1994, 2591.
23. Commerçon, A.; Bezard, D.; Bernard, F.; Bourzat, J.D. *Tetrahedron Lett.* 1992, *33*, 5185.
24. Corma, A.; Iborra, S.; Velty, A. *Chem. Rev.* 2007, *107*, 2411.
25. Tao, J.; Zhao, L.; Ran, N. *Org. Process Res. Dev.* 2007, *11*, 259.
26. Ran, N.; Zhao, L.; Chen, Z.; Tao, J. *Green Chem.* 2008, *10*, 361.
27. Sheldon, R.A.; van Rantwijk, F. *Aust. J. Chem.* 2004, *57*, 281.
28. Lichtenthaler, F.W. *Acc. Chem. Res.* 2002, *35*, 728.
29. Frost, J.W.; Lievense, J. *New J. Chem.* 1994, *18*, 341.
30. Hale, W.J. *The Farm Chemurgic: Farmward the Star of Destiny Lights Our Way*, Stratford Company: Boston, MA, 1934, p. 141.
31. Hale, W.J. *Ind. Eng. Chem.* 1930, *22*, 1311.
32. McMillen, W. *Ind. Eng. Chem.* 1939, *31*, 540.
33. Kovarik, B. *Autom. Hist. Rev.*1998, *32*, 7.
34. Kovarik, B. http://www.radford.edu/~wkovarik/papers/fuel.html (accessed July 12, 2010).
35. Beeman, R. *Agri. Hist.* 1994, *68*, 23.
36. Bennett, S. *Chemistry World* 2007, October, *10*, 66.
37. Finlay, M.R. *J. Ind. Ecol.* 2004, *7*, 33.
38. Adams, R.; Voorhees, V. *Org. Synth. Coll.* 1943, *1*, 280.
39. McKenzie, B.F. *Org. Synth. Coll.* 1941, *1*, 335.
40. Draths, K.M.; Frost, J.W. *J. Am. Chem. Soc.* 1994, *116*, 399.
41. Niu, W.; Draths, K.M.; Frost, J.W. *Biotechnol. Prog.* 2002, *18*, 201.
42. Kambourakis, S.; Draths, K.M.; Frost, J.W. *J. Am. Chem. Soc.* 2000, *122*, 9042.
43. Li, K.; Frost, J.W. *J. Am. Chem. Soc.* 1998, *120*, 10545.
44. Draths, K.M.; Frost, J.W. *J. Am. Chem. Soc.* 1995, *117*, 2395.
45. Ensley, B.D.; Ratzkin, B.J.; Osslund, T.D.; Simon, M.J.; Wackett, L.P.; Gibson, D.T. *Science* 1983, *222*, 167.
46. Berry, A.; Dodge, T.C.; Pepsin, M.; Weyler, W. *J. Industrial Microbiol. Biotechnol.* 2002, *28*, 127.
47. Muheim, A.; Müller, B.; Münch, T.; Wetli, M. EP885968 (Givaudan-Roure, 1998).
48. Muheim, A.; Lerch, K. *Appl. Microbiol. Biotechnol.* 1999, *51*, 456.
49. Lee, J.C.; Lee, K.I.; Cha, J.K. *J. Org. Chem.* 2000, *65*, 4773.
50. Casale, J.F. *Forensic Sci. Int.* 1987, *33*, 275.
51. Mans, D.M.; Pearson, W.H. *Org. Lett.* 2004, *6*, 3305.
52. Robertson, D.W.; Thompson, D.C.; Wong, D.T. EP 288188 (October 26, 1988) Eli Lilly, Inc.
53. Torre, O.; Gotor-Fernandez, V.; Gotor, V. *Tetrahedron Asymm.* 2006, *17*, 860.
54. Siddiqui, S.A.; Srinivasan, K.V. *Tetrahedron Asymm.* 2007, *18*, 2099.
55. Chincholkar, P.M.; Kale, A.S.; Gumaste, V.K.; Deshmukh, A.R.A.S. *Tetrahedron* 2009, *65*, 2605.
56. Zhang, T.Y. *Chem. Rev.* 2006, *106*, 2583.
57. Faith, W.L.; Keyes, D.B.; Clark, R.L. *Industrial Chemicals*, 3rd edn., Wiley: New York, 1966.
58. Shreve, R.N. *Chemical Process Industries*, 3rd edn., McGraw-Hill Book Co: New York, 1967.
59. Gantt, H.L. *Trans. Am. Soc. Mechanical Eng.* 1903, *24*, 1322.
60. Gantt, H.L. *Organizing for Work,* Harcourt, Brace, and Howe: New York, 1919.
61. Ku, H.; Rajagopalan, D.; Karimi, I. *Chem. Eng. Progress* 1987, August, *83*, 35.
62. Musier, R.F.H.; Evans, L.B. *Chem. Eng. Progress* 1990, June, *86*, 66.
63. Badger, A.M.; DiMartino, M.J.; Mirabelli, C.K.; Cheeseman, E.N.; Dorman, J.W.; Picker, D.H.; Schwartz, D.A. EP 310321 (SmithKline & French, 1988).
64. Badger, A.M.; Schwartz, D.A.; Picker, D.H.; Dorman, J.W.; Bradley, F.C.; Cheeseman, E.N.; DiMartino, M.J.; Hanna, N.; Mirabelli, C.K. *J. Med. Chem.* 1990, *33*, 2963.
65. Tanaka, A.; Kawai, T.; Takabatake, T.; Oka, N.; Okamoto, H.; Bersohn, M. *Tetrahedron* 2007, *63*, 10226.
66. Dogane, I.; Takabatake, T.; Bersohn, M. *Recl. Trav. Chim. Pays-Bas* 1992, *111*, 291.
67. Bersohn, M. *Bull. Chem. Soc. Jpn.* 1972, *45*, 1897.

10 Ring Construction Strategies

The feature of rings in organic molecules is a top draw to make a molecule a worthy target of synthesis for synthetic chemists. It is here where a chemist can display his or her greatest ingenuity and prowess. It is also arguably the single most important structural motif that has driven the development of synthetic methodology in organic chemistry. There is a direct link between overall molecular complexity and ring system complexity. The more exotic the ring system is the more synthetic chemists are attracted to take up the challenge of coming up with a viable synthesis plan for that target structure. Of course, the greatest architect of such ring systems is Nature. Rote combinatorial chemical approaches have no hope of matching the exquisite ring systems found in Nature. This is why natural products chemistry always inspires chemists, drives the discovery of novel chemistry development, and has had the most success in delivering lead compounds that become future medicines.

As the synthetic organic chemistry literature grows exponentially and haphazardly in every direction, it is imperative that chemists create intelligent databases and search tools to mine them for ring construction strategies. This chapter along with Appendix A.7 is the first attempt to focus attention on this task. We begin by introducing a simple universal code or graphical algebra for enumerating the construction of monocyclic and bicyclic rings. This facilitates the codification of the nebulous concept of strategy as well as the sorting of strategies according to patterns. The following sections summarize the results for the synthesis database described in this book. Ultimately, such information may be used as a powerful tool in synthesis optimization so that the "greenest" plan to any given target may be found.

We illustrate the basic ideas using the six-membered monocycle and the [3.3.0] and [2.2.1] bicyclic ring systems as examples. The rings are displayed as graphs with lines and nodes to show the framework of connectivity. New bonds made in the transformation are shown in gray scale. Next, we explore all possible ways such frameworks can be put together using one-bond, two-bond, etc., disconnections. For example, for the six-membered ring, we have the following graphical equation for a one-bond disconnection:

which can be abbreviated as $\xrightarrow{[6+0]} A$. If this transformation were to occur in the nth step in a synthesis plan then the code would be $\xrightarrow{[6+0]} A_n$. Similarly, for a [3 + 3] transformation involving two-bond disconnections occurring in step 11, for example, we have $\xrightarrow{[3+3]} A_{11}$.

Note that the number of components reacting to form the ring is equal to the number of elements appearing in brackets. Hence, for a [3 + 3] cycloaddition there are two elements, so it is a two-component coupling reaction. The number of bond disconnections equals the number of components reacting. Note also that the sum of the elements is 6, which corresponds to the monocycle ring size.

For the [6 + 0] transformation, since one of the elements is zero it does not count as a component and so this transformation is a unimolecular reaction. A [2 + 2 + 2] transformation to form a six-membered ring involves three-bond disconnections and hence is a three-component coupling reaction.

Conversely, if we have a graph that looks like

then it means that the transformation is [3 + 2 + 1] three-component coupling to form ring C.

For a [3.3.0] framework, we may have the following transformation occurring in step 8, for example.

The abbreviated code is $B_0 \xrightarrow{[3+2]} [3.3.0] \, A_8$, which means that the plan began with a five-membered ring B, and in the 8th step, a [3 + 2] two-component coupling occurred to form the other five-membered ring A. For the case,

abbreviated as $\xrightarrow{[(5+0)_A + (5+0)_B]} [3.3.0]_{10}$, we have a transformation occurring in step 10 that makes both five-membered rings A and B via [5 + 0] ring closures. For a [2.2.1] framework, we may have the following transformation occurring in step 5, for example.

The abbreviated code is $\xrightarrow{[(6+0)+(5+0)]} [2.2.1]_5$. Note that the new bond made can be parsed in two ways depending on the point of view of which ring is closed. Another possibility may be

which is abbreviated as $\xrightarrow{[(5+0)+(5+0)]} [2.2.1]_5$. The following would be a two-component coupling reaction occurring in step 5.

This is abbreviated as $\xrightarrow{[4+2]}$ [2.2.1]$_5$.

Appendix A.7 has a complete listing of all possible combinations for constructing 3–10-membered monocycles, [n.m.0] fused bicyclic ring systems, and the following common bicyclics: [1.1.1], [2.1.1], [2.2.1], [2.2.2], and [3.2.2]. Ring construction strategies found in the synthesis database are summarized in Section 10.1 using the aforementioned graphical algebra notation. The compilation is given according to the target compounds listed in alphabetical order. References to the plans are given at the beginning of each subsection. A summary of the ring construction strategies arranged according to patterns follows each section pertaining to the target compound. Ring expansions and contractions and bicyclic ring formation found in the named organic reaction database are summarized in Section 10.2 through 10.4. The use of sacrificial rings is covered in Section 10.5. The use of the versatile furan ring is covered in Section 10.6. Various novel [x + y + z] strategies mimicking the Pauson-Khand reaction that were pioneered by Wender are given in Section 10.7. Finally, the chapter closes with Section 10.8 on spectacular ring forming reactions found in the synthesis database that showcase numerous rings being made in a single reaction step.

10.1 RING CONSTRUCTION STRATEGIES IN SYNTHESIS DATABASE

Absinthin ([4.4.0] to [5.3.0] rearrangement; six-membered ring via [4 + 2] cycloaddition)
Zhang, W.; Luo, S.; Fang, F.; Chen, Q.; Hu, H.; Jia, X.; Zhai, H. *J. Am. Chem. Soc.* 2005, *127*, 18.

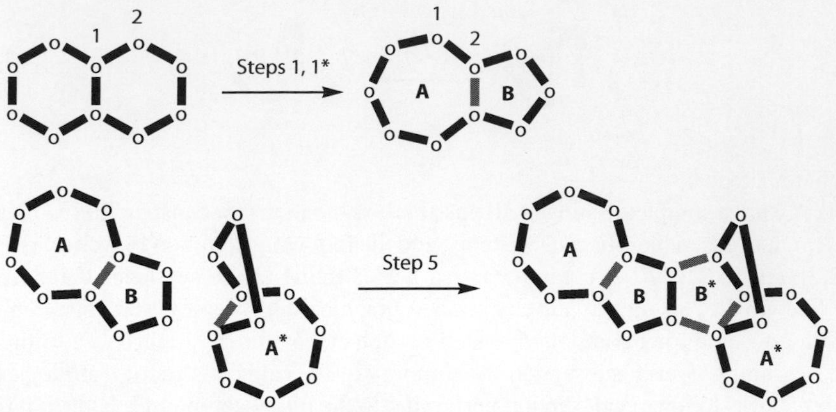

Sequence:
$$[4.4.0]_0 \rightarrow (A+B)_1 [5.3.0]$$
$$[4.4.0]_0 \rightarrow (A^* + B^*)_{1*} [5.3.0]$$
$$\left.\right\} \xrightarrow{[4+2]} B_5^*$$

Alizarin (six-membered ring via [6 + 0] cycloaddition)

Fierz-David, H.E.; Blangey, L. *The Fundamental Processes of Dye Chemistry*, Interscience Publishing: New York, 1949, pp. 224, 225, 228, 314.

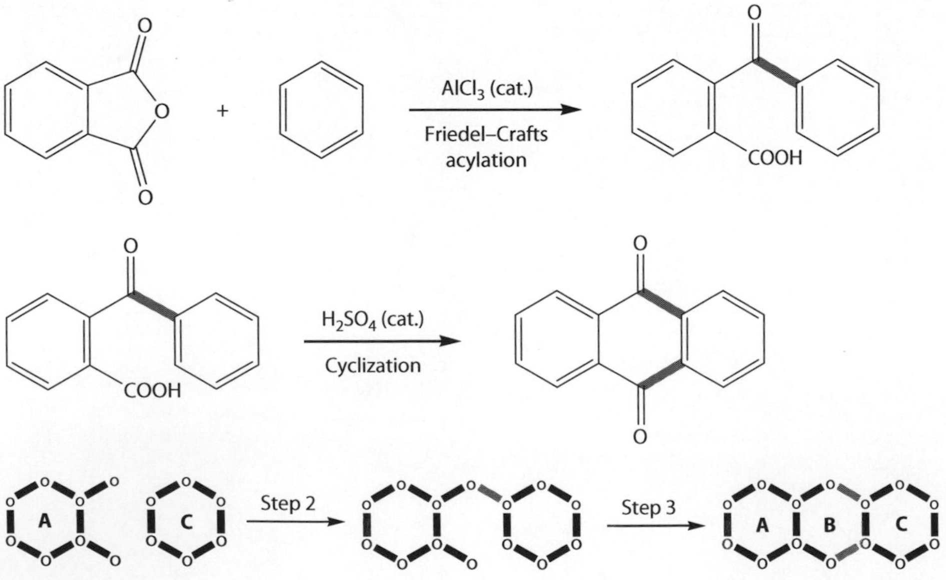

Sequence: $(A + C)_0 \xrightarrow{[6+0]} B_3$

Alizarin (six-membered ring via [4 ± 2] cycloaddition)

Graebe, C.; Liebermann, C. *Chem. Ber.* 1869, *2*, 332.

Fierz-David, H.E. *The Fundamental Processes of Dye Chemistry*, J. & A. Churchill: London, 1921, p. 167.

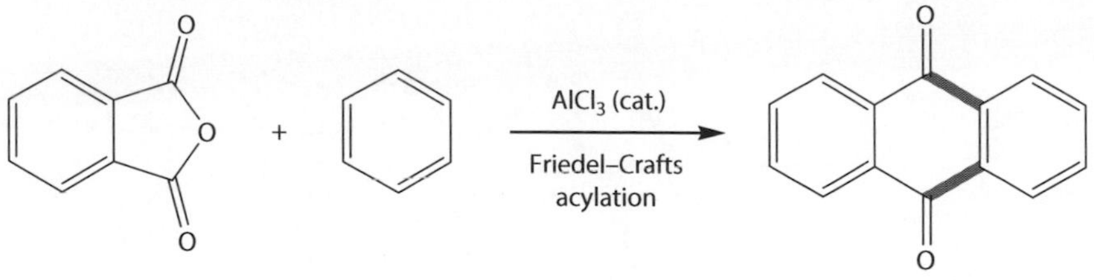

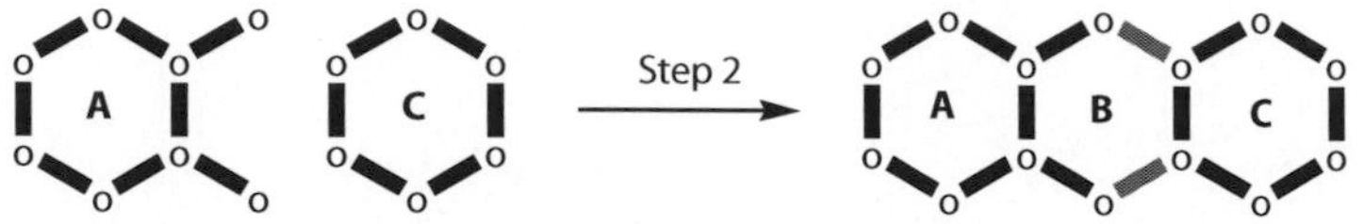

Sequence: $(A + C)_0 \xrightarrow{[4+2]} B_2$

(±)-α-Santonin (six-membered ring via [6 ± 0] cycloaddition; five-membered ring via [5 ± 0] cycloaddition)

Marshall, J.A.; Wuts, P.G.M. *J. Org. Chem.* 1978, *43*, 1086.

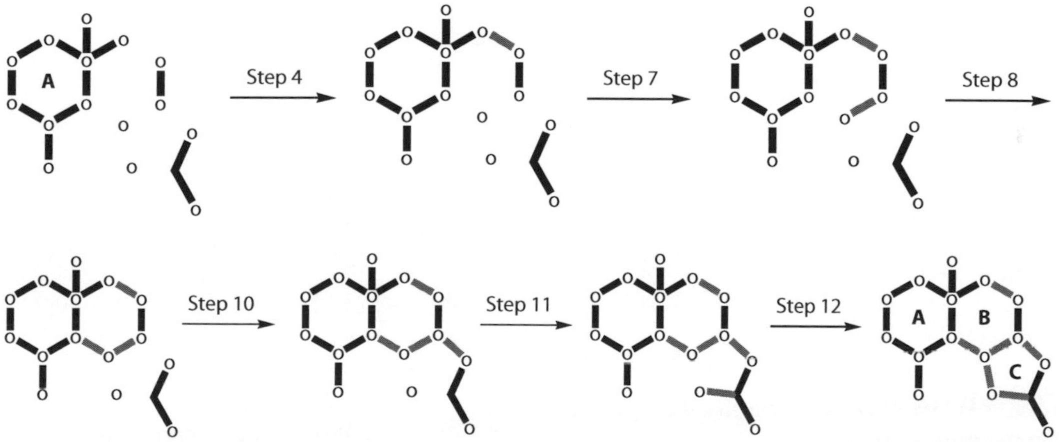

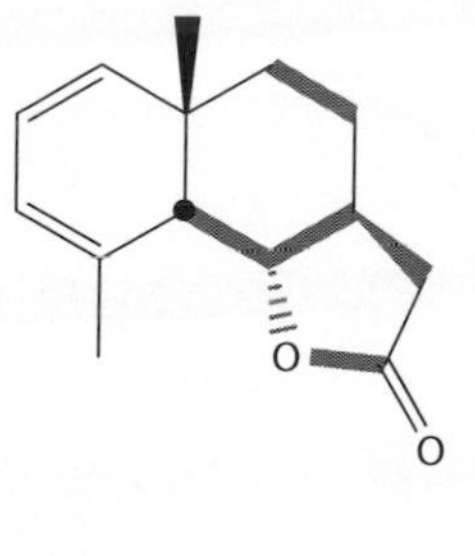

Sequence: $A_0 \xrightarrow{[6+0]} B_8 \xrightarrow{[5+0]} C_{12}$

(±)-α-Santonin (six-membered ring via [4 ± 2] cycloaddition; five-membered ring via [5 ± 0] cycloaddition)

Abe, Y.; Harukawa, T.; Ishikawa, H.; Miki, T.; Sumi, M.; Toga, T. *J. Am. Chem. Soc.* 1956, *78*, 1416.
Abe, Y.; Harukawa, T.; Ishikawa, H.; Miki, T.; Sumi, M.; Toga, T. *J. Am. Chem. Soc.* 1956, *78*, 1422.

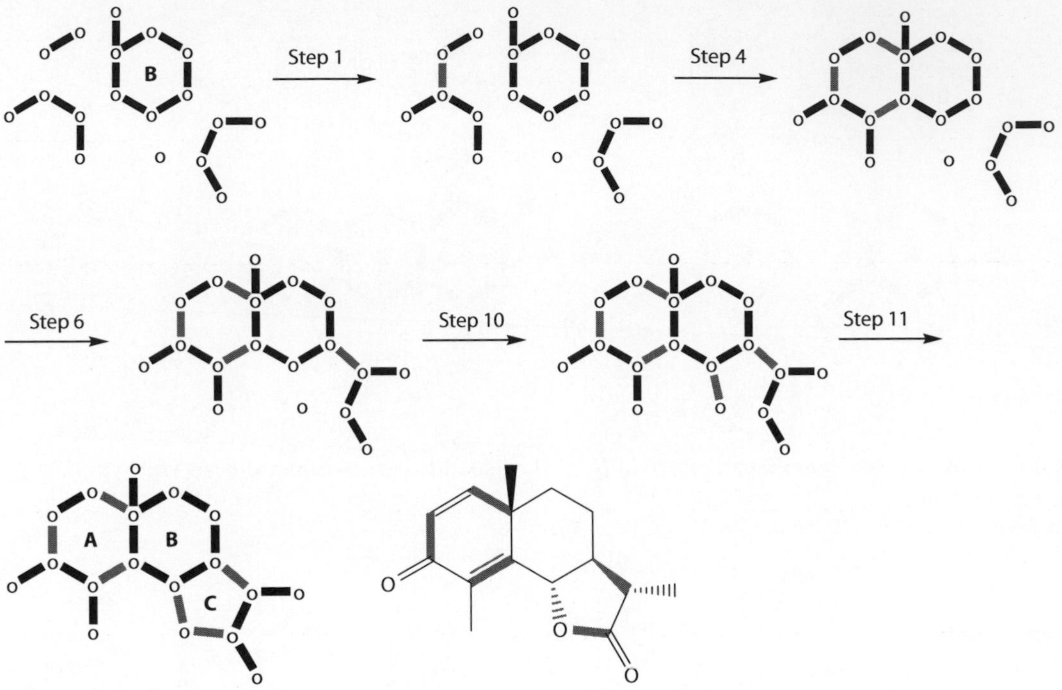

Sequence: $B_0 \xrightarrow{\ [4+2]\ } A_4 \xrightarrow{\ [5+0]\ } C_{11}$

Amphidinolide P

(3-membered ring via [2 + 1] cycloaddition; 6-membered ring via [6 + 0] cycloaddition; 13-membered ring via [10 + 3] cycloaddition)

Trost, B.M.; Papillon, J.P.N. *J. Am. Chem. Soc.* 2004, *126*, 13618.
Trost, B.M.; Papillon, J.P.N.; Nussbaumer, T. *J. Am. Chem. Soc.* 2005, *127*, 17921.

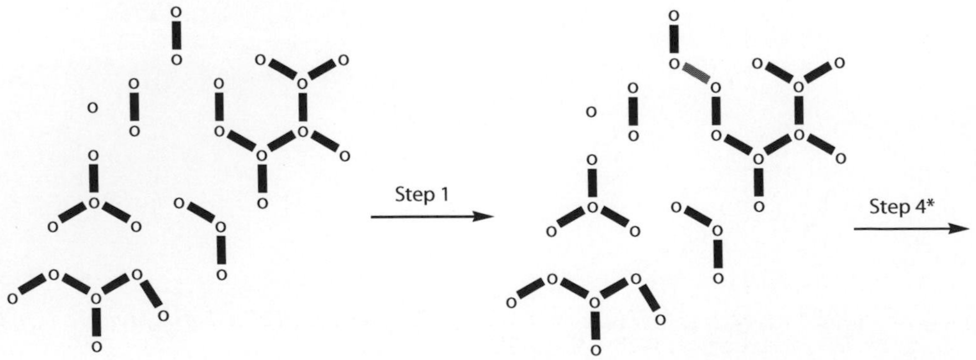

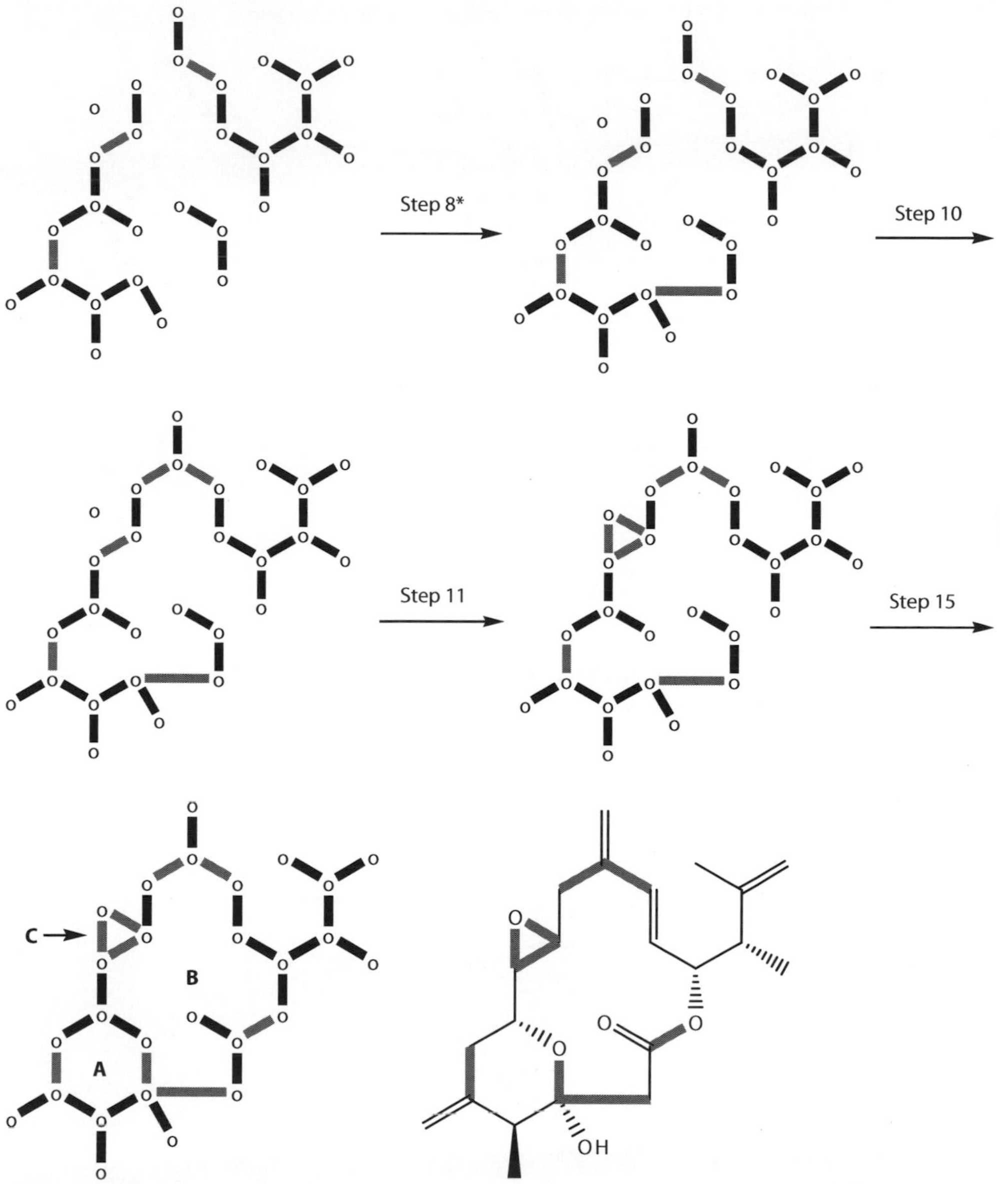

Sequence: $\xrightarrow{[2+1]} C_{11} \xrightarrow{[(6+0)_A+(10+3)_B]} (A+B)_{15}$

Antipyrine (five-membered via [3 + 2] cycloaddition)
Furniss, B.S.; Hannaford, A.J.; Rogers, V.; Smith, P.W.G.; Tatchell, A.R. *Vogel's Textbook of Practical Organic Chemistry*, Longman: London, 1978, pp. 540, 727, 882.

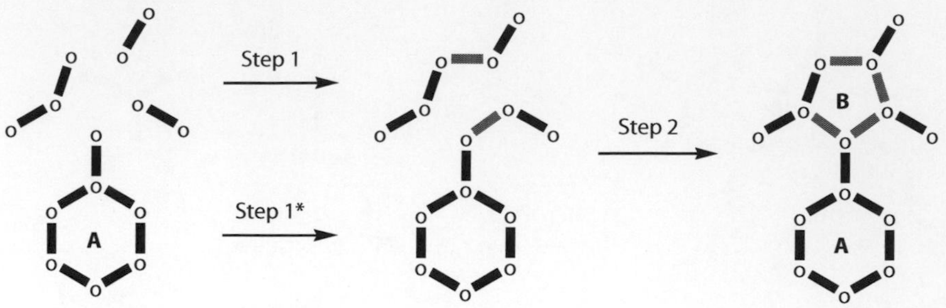

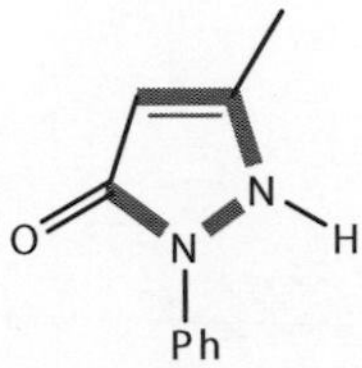

Sequence: $A_0 \xrightarrow{[3+2]} B_2$

Aspidophytine
(five-membered ring via [5 + 0] cycloaddition; tricyclic ring cascade via [(4 + 2) + (4 + 1) + (5 + 1)] cycloaddition; five-membered ring via [4 + 1] cycloaddition)

He, F.; Bo, Y.; Altom, J.D.; Corey, E.J. *J. Am. Chem. Soc.* 1999, *121*, 6771.

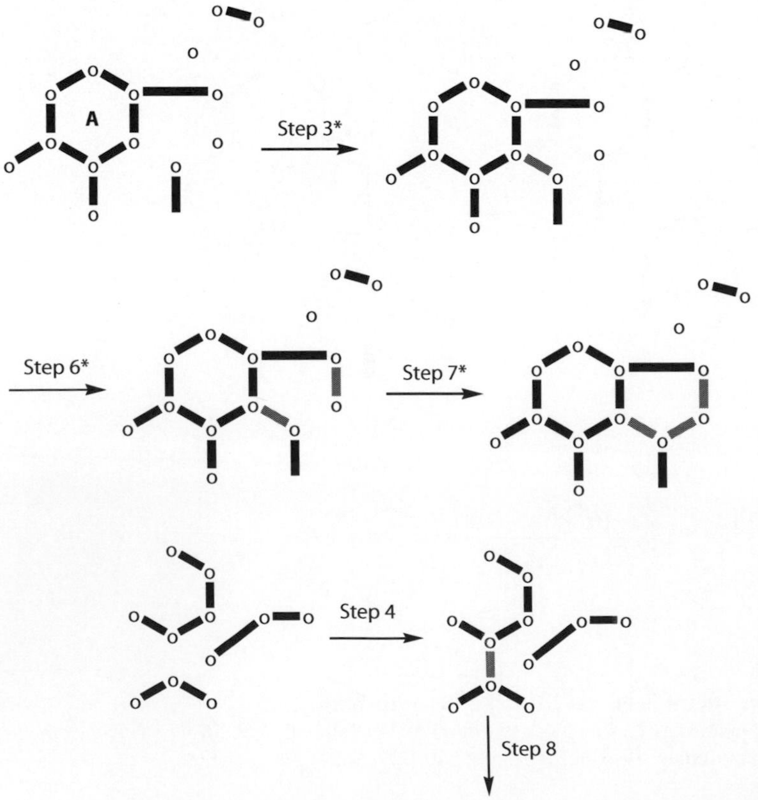

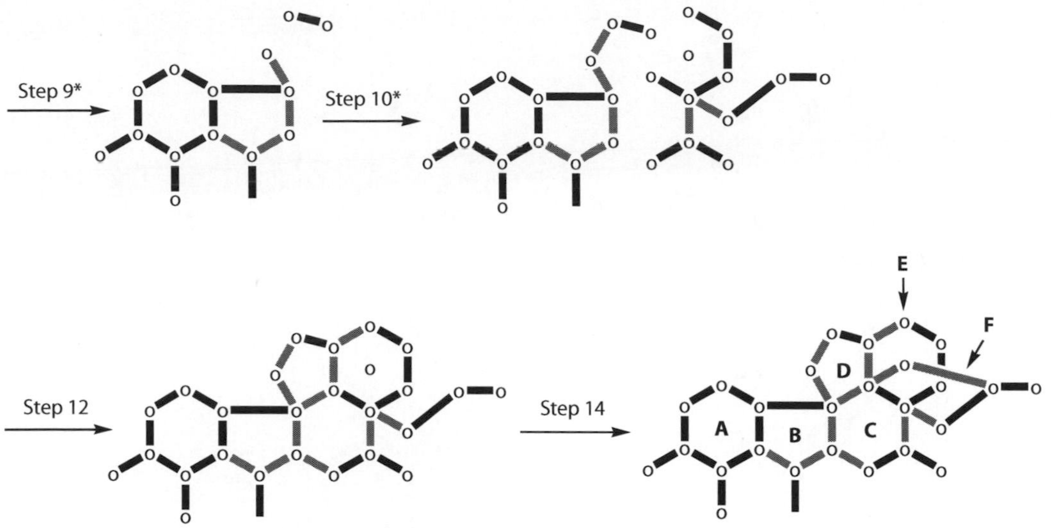

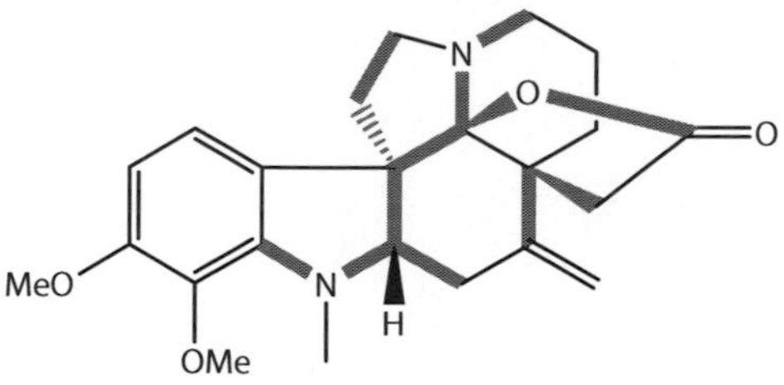

Sequence: $A_0 \xrightarrow{[5+0]} B_{7*} \xrightarrow{[(4+2)_C + (4+1)_D + (5+1)_g]} (C+D+E)_{12} \xrightarrow{[4+1]} F_{14}$

Azadirachtin

Ley, S.V.; Abad-Somovilla, A.; Anderson, J.C.; Ayats, C.; Bänteli, R.; Beckmann, E.; Boyer, A.; Brasca, M.G.; Brice, A.; Broughton, H.B.; Burke, B.J.; Cleator, E.; Craig, D.; Denholm, A.A.; Denton, R.M.; Durand-Reville, T.; Gobbi, L.B.; Göbel, M.; Gray, B.L.; Grossmann, R.B.; Gutteridge, C.E.; Hahn, N.; Harding, S.L.; Jennens, D.C.; Jennens, L.; Lovell, P.J.; Lovell, H.J.; de la Puente, M.L.; Kolb, H.C.; Koot, W.J.; Maslen, S.L.; McCusker, C.F.; Mattes, A.; Pape, A.R.; Pinto, A.; Santafianos, D.; Scott, J.S.; Smith, S.C.; Somers, A.Q.; Spilling, C.D.; Stelzer, F.; Toogood, P.L.; Turner, R.M.; Veitch, G.E.; Wood, A.; Zumbrunn, C. *Chem. Eur. J.* 2008, *14*, 10683.

Veitch, G.E.; Beckmann, E.; Burke, B.J.; Boyer, A.; Ayats, C.; Ley, S.V. *Angew. Chem. Int. Ed.* 2007, *46*, 7633.

Veitch, G.E.; Beckmann, E.; Burke, B.J.; Boyer, A.; Maslen, S.L.; Ley, S.V. *Angew. Chem. Int. Ed.* 2007, *46*, 7629.

Veitch, G.E.; Boyer, A.; Ley, S.V. *Angew. Chem. Int. Ed.* 2008, *47*, 9402.

Left-hand side of molecule:

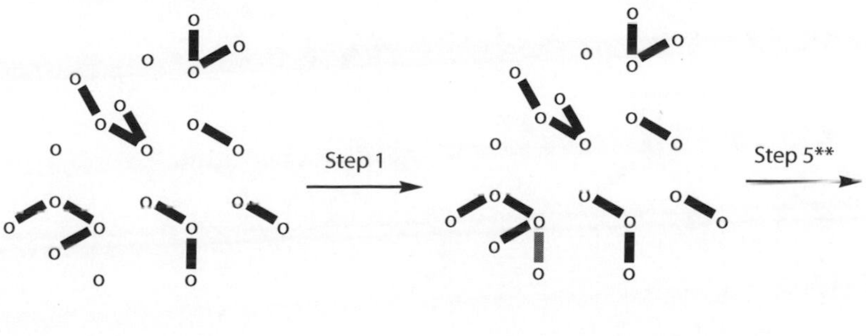

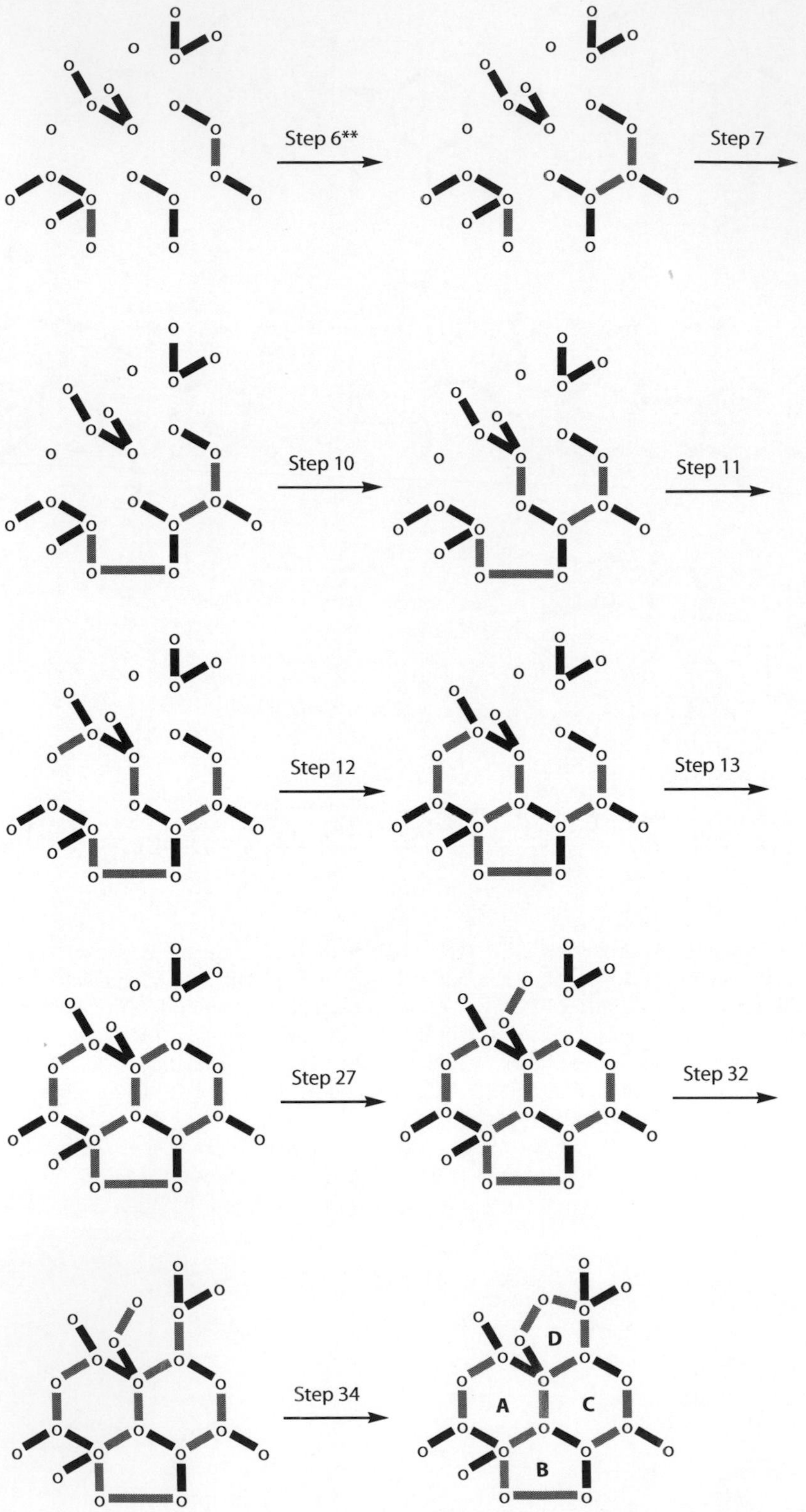
Step 6**
Step 7
Step 10
Step 11
Step 12
Step 13
Step 27
Step 32
Step 34
A
B
C
D

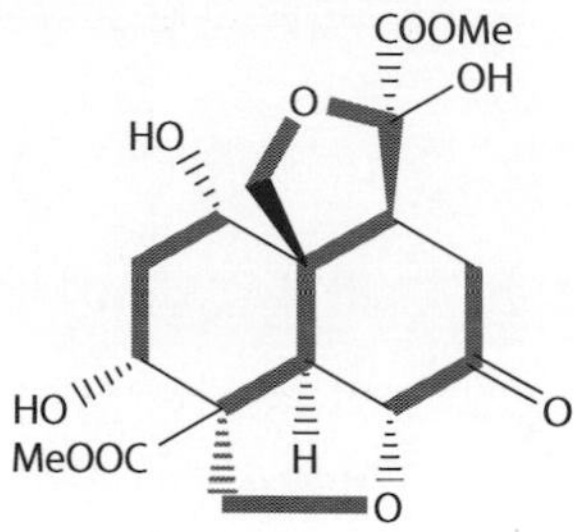

Sequence: $\xrightarrow{[4+2]_A +[5+0]_B} [4.3.0](A+B)_{12} \xrightarrow{[6+0]} C_{13} \xrightarrow{[5+0]} D_{34}$

Right-hand side of molecule:

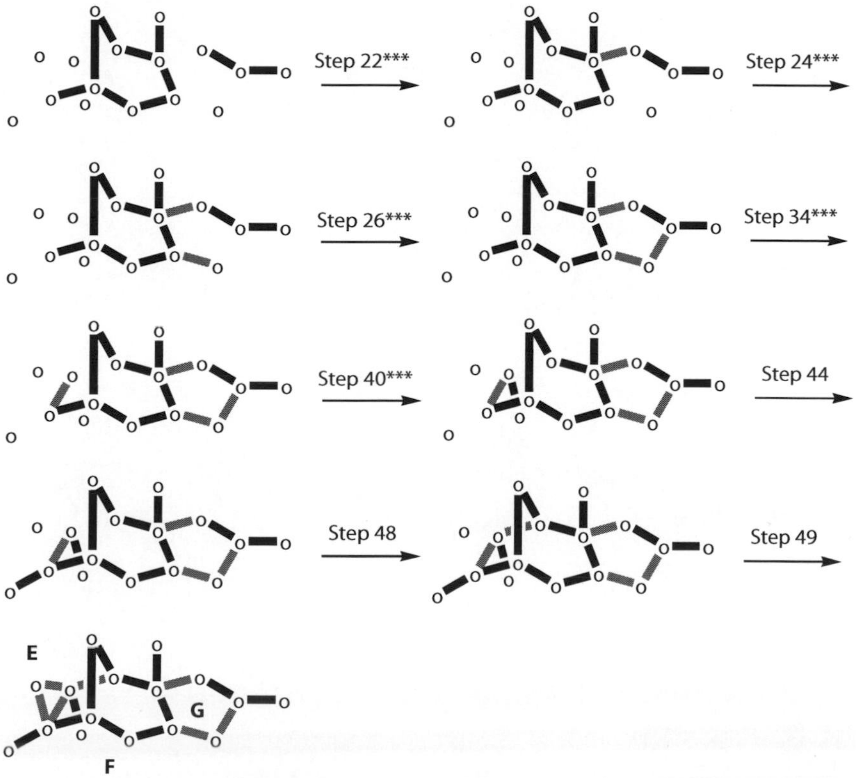

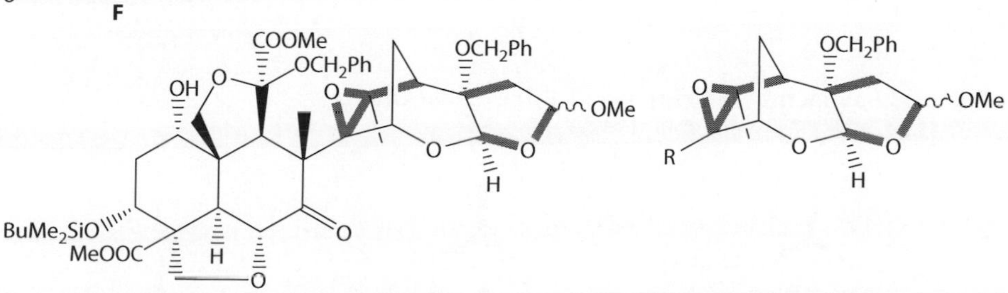

Sequence: $\xrightarrow{[5+0]} G_{26***} \xrightarrow{[(7+0)+(5+0)]} [3.2.1] \, F_{48} \xrightarrow{[2+1]} E_{49}$

Boyer, A.; Veitch, G.E.; Beckmann, E.; Ley, S.V. *Angew. Chem. Int. Ed.* 2009, *48*, 1317.

Left-hand side of molecule: as mentioned earlier

Right-hand side of molecule:

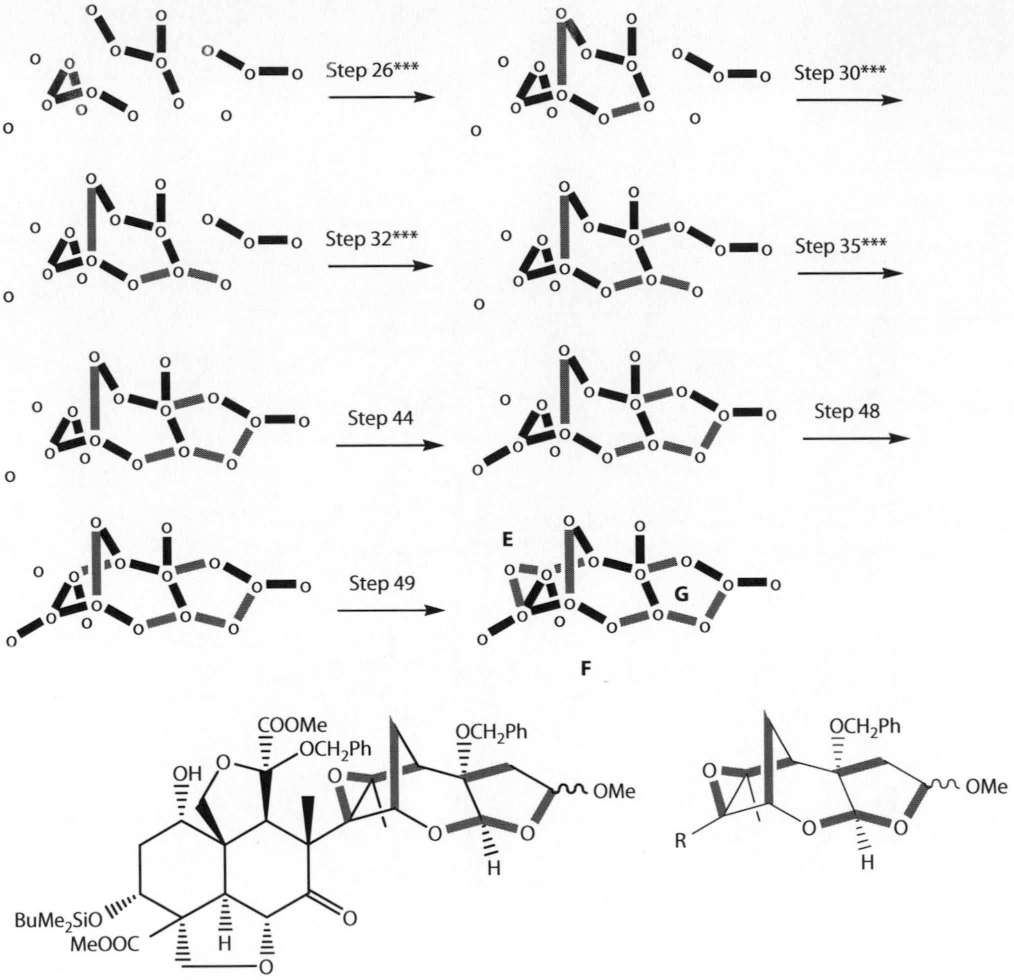

Sequence: $\xrightarrow{[4+2]} [6]_{26^{***}} \xrightarrow{[5+0]} G_{35^{***}} \xrightarrow{[(7+0)+(5+0)]} [3.2.1]F_{48} \xrightarrow{[2+1]} E_{49}$

Barbituric acid (six-membered ring via [3 + 3] cycloaddition)
Dickey, J.B.; Gray, A.R. *Org. Synth. Coll.* 1943, *2*, 60.

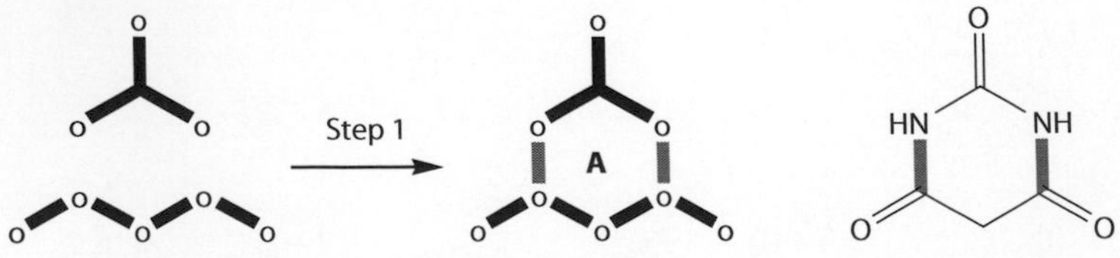

Sequence: $\xrightarrow{[3+3]} A_1$

Barrelene ([2.2.2] bicyclic ring via [4 + 2 + 2] cycloaddition)

Zimmerman, H.E.; Grunewald, G.L.; Paufler, R.M.; Sherwin, M.A. *J. Am. Chem. Soc.* 1969, *91*, 2330.

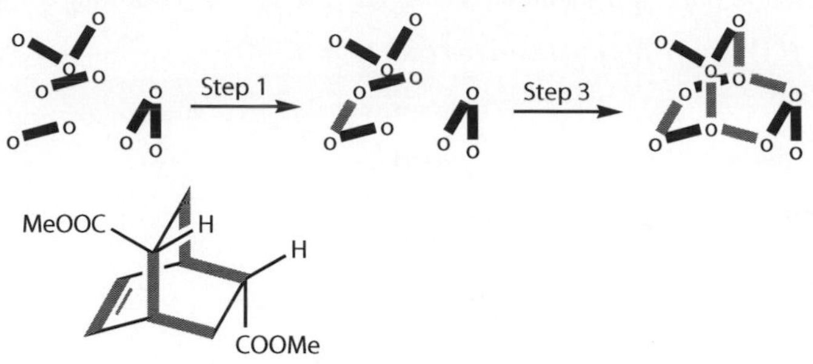

Sequence: $\xrightarrow{[(4+2) + (4+2)]}$ $[2.2.2]_3$

Basketene

(Eight-membered ring via [2 + 2 + 2 + 2] cycloaddition; [8] to [4.2.0] rearrangement; six-membered ring via [4 + 2] cycloaddition; four-membered ring via [2 + 2] cycloaddition)

Dauben, W.G.; Whalen, D.L. *Tetrahedron Lett.* 1966, *7*, 3743.

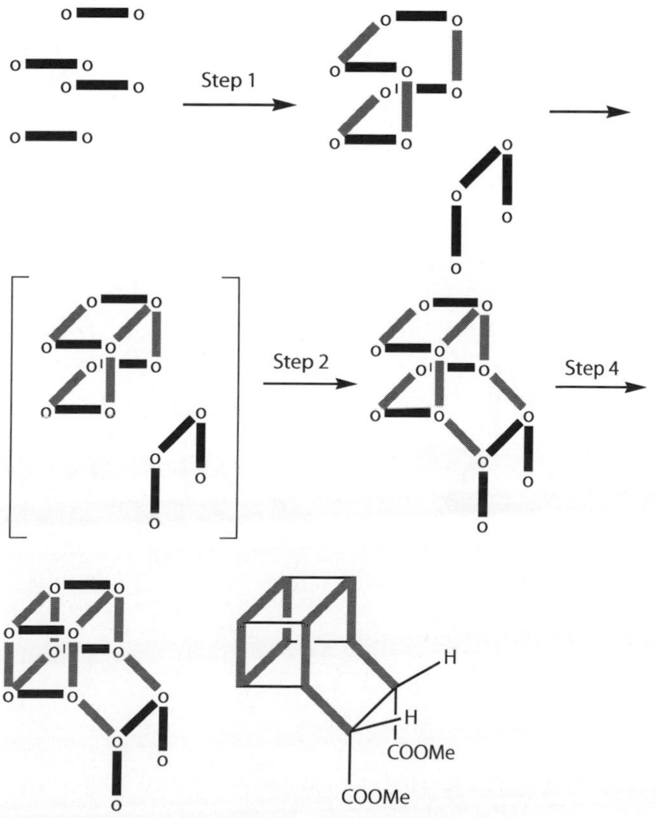

Sequence: $\xrightarrow{[2+2+2+2]_1}$ $[8] \rightarrow [4.2.0]$ $\xrightarrow{[4+2]_2}$ $\xrightarrow{[2+2]_4}$

Basketene

(Eight-membered ring via [2 + 2 + 2 + 2] cycloaddition; [8] to [4.2.0] rearrangement; six-membered ring via [4 + 2] cycloaddition; four-membered ring via [2 + 2] cycloaddition)

Masamune, S.; Cuts, H.; Hogben, M.G. *Tetrahedron Lett.* 1966, *7*, 1017.

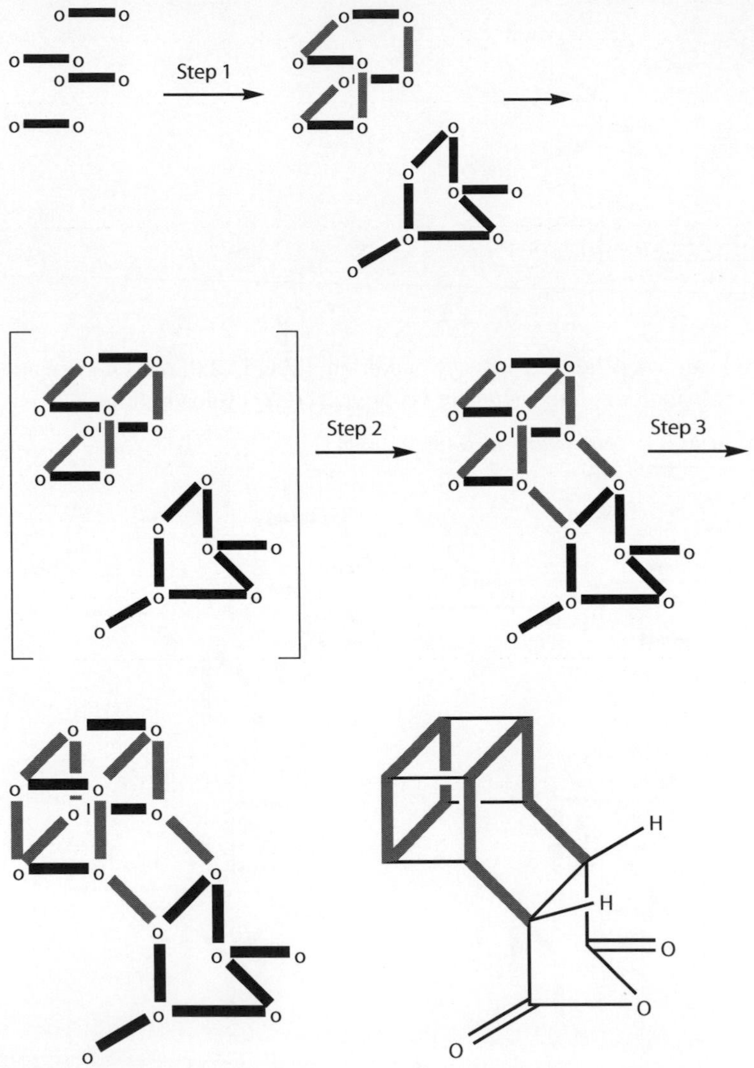

Sequence: $\xrightarrow{[2+2+2+2]}$ [8] → [4.2.0] $\xrightarrow{[4+2]_2}$ $\xrightarrow{[2+2]_3}$

Bullvalene

(Seven-membered ring via [6 + 1] cycloaddition; three-membered ring via [2 + 1] cycloaddition; seven-membered ring via [6 + 1] cycloaddition)

von Doering, W.; Ferrier, B.M.; Fossel, E.T.; Harnenstein, J.H.; Jones, M. Jr.; Klumpp, G.; Rubin, R.M.; Saunders, M. *Tetrahedron* 1967, *23*, 3943.

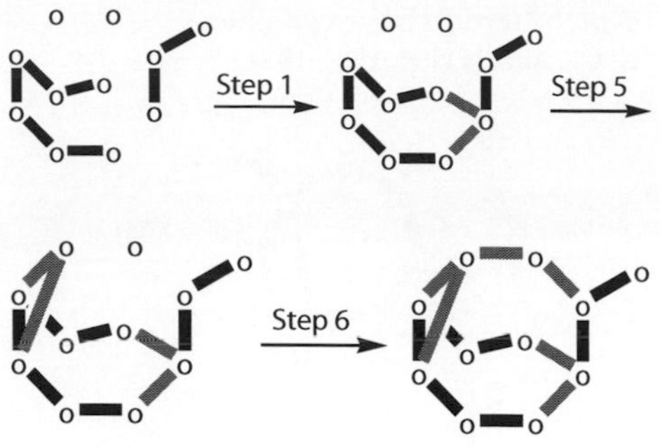

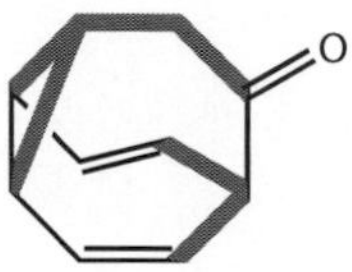

Sequence: $\xrightarrow{[6+1]_1}$ $\xrightarrow{[2+1]_5}$ $\xrightarrow{[6+1]_6}$

Bullvalene

(Eight-membered ring via [2 + 2 + 2 + 2] cycloaddition; seven-membered ring via [5 + 2] cycloaddition)

Schröder, G. *Chem. Ber.* 1964, *97*, 3131.
Schröder, G. *Chem. Ber.* 1964, *97*, 3140.

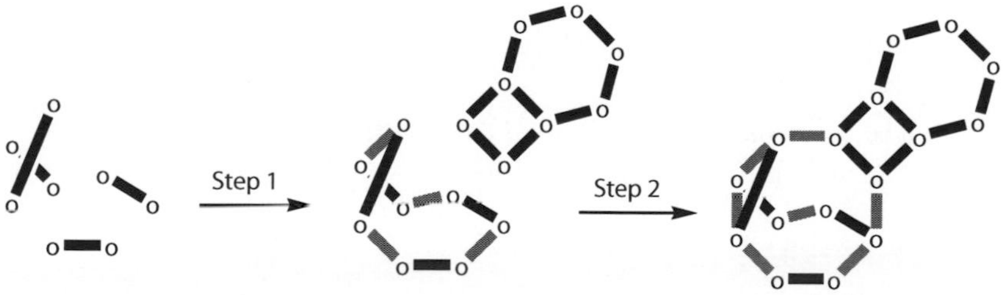

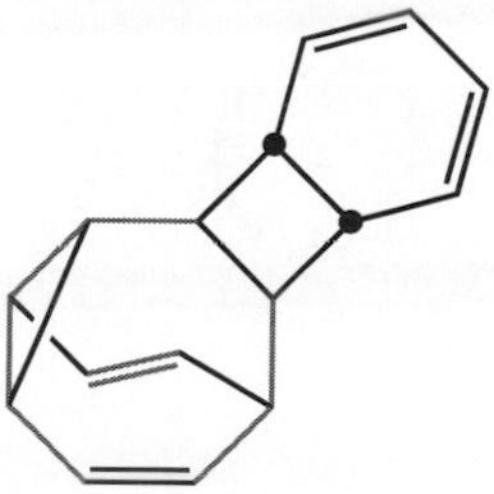

Sequence: $\xrightarrow{[2+2+2+2]_1}$ $\xrightarrow{[5+2]_2}$

ε-Caprolactam (six- to seven-membered ring expansion)

Hawkins, E.G.E. US 3947406 (BP Chemicals U.K. Ltd., 1976).

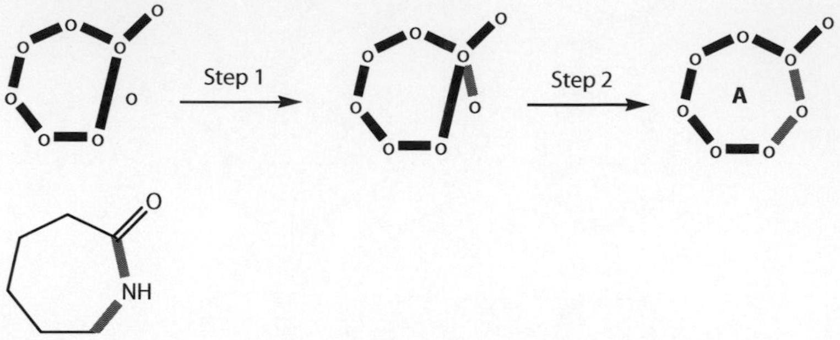

Sequence: $[6] \rightarrow A_2[7]$

ε-Caprolactam (six- to seven-membered ring expansion)

Nelson, J.D.; Modi, D.P.; Evans, P.A. *Org. Synth.* 2002, *79*, 165.

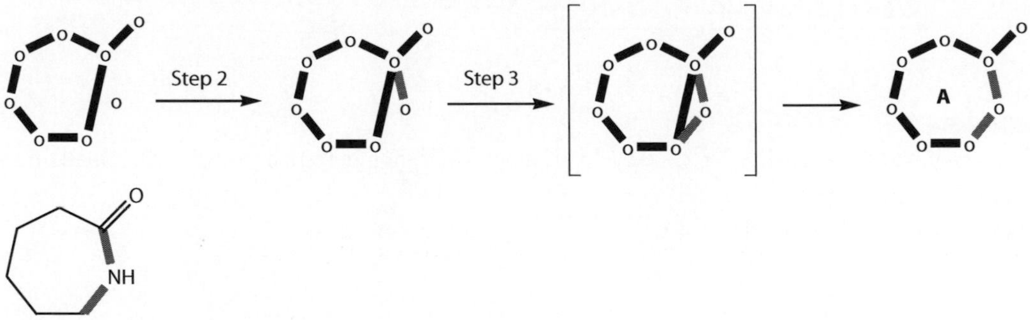

Sequence: $[6] \rightarrow [4.1.0] \rightarrow A_3[7]$

ε-Caprolactam (six- to seven-membered ring expansion)

Lowenheim, F.A.; Moran, M.K. *Faith, Keyes, and Clark's Industrial Chemicals*, 4th edn., Wiley: New York, 1975, p. 201.

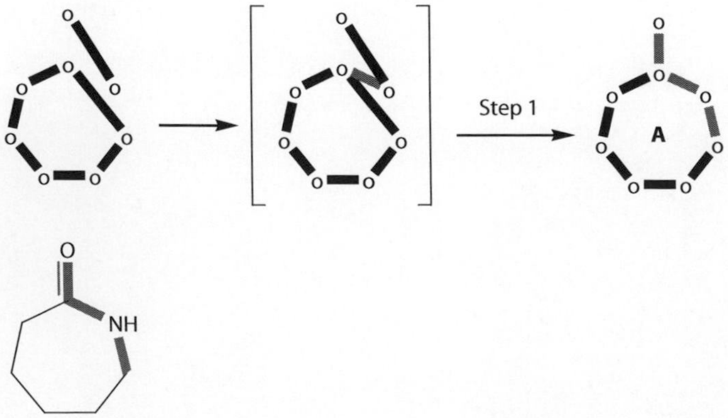

Sequence: $[6] \rightarrow A_1[7]$

ε-Caprolactam (six- to seven-membered ring expansion)
Yamamoto, S.; Sakaguchi, S.; Ishii, Y. *Green Chem.* 2003, *5*, 300.

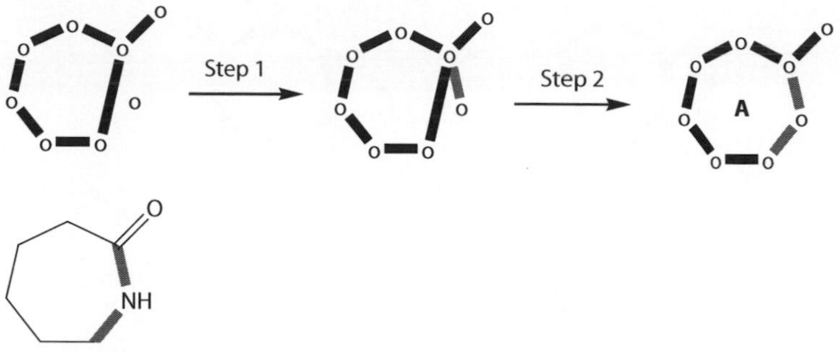

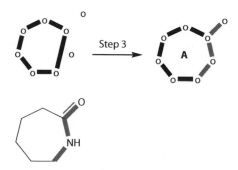

Sequence: [6] → A_2[7]

ε-Caprolactam (six- to seven-membered ring expansion)
Ishii, Y.; Sakaguchi, S.; Iwahama, T. *Adv. Synth. Catal.* 2001, *343*, 393.
Sakaguchi, S.; Nishiwaki, Y.; Kitamura, T.; Ishii, Y. *Angew. Chem. Int. Ed.* 2001, *40*, 222.

Sequence: [6] → A_3[7]

ε-Caprolactam (six- to seven-membered ring expansion)
Marvel, C.S.; Eck, J.C. *Org. Synth. Coll.* 1943, *2*, 371.
Eck, J.C.; Marvel, C.S. *Org. Synth. Coll.* 1943, *2*, 76.

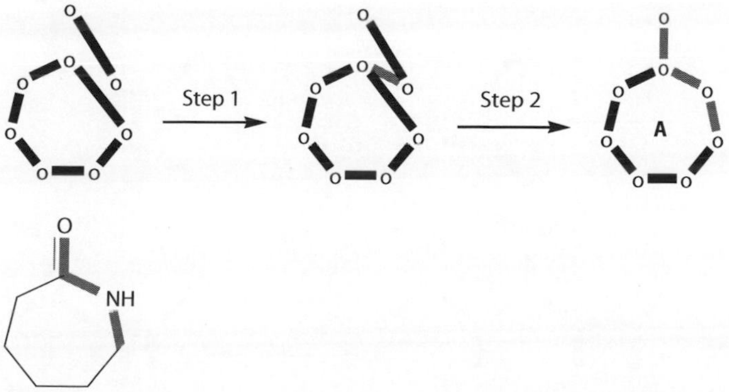

Sequence: [6] → A_2[7]

ε-Caprolactam (six- to seven-membered ring expansion)

Thomas, J.M.; Raja, R. *Proc. Nat. Acad. Sci.* 2005, *102*, 13732.

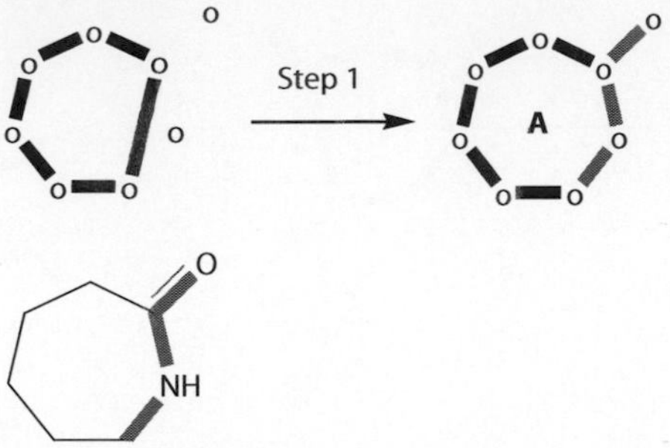

Sequence: [6] → A_1[7]

Summary:
Strategy #1: BP, Ishii G1
Strategy #2: Evans
Strategy #3: Ishii G2, Thomas
Strategy #4: Faith, Marvel

Cephalosporin C

(Four-membered ring via [4 + 0] cycloaddition; six-membered ring via [6 + 0] cycloaddition)

Woodward, R.B.; Heusler, K.; Gosteli, J.; Naegeli, P.; Oppolzer, W.; Ramage, R.; Ranganathan, S.; Vorbrüggen, H.
        J. Am. Chem. Soc. 1966, *88*, 852.
Woodward, R.B. *Science* 1966, *153*, 487.

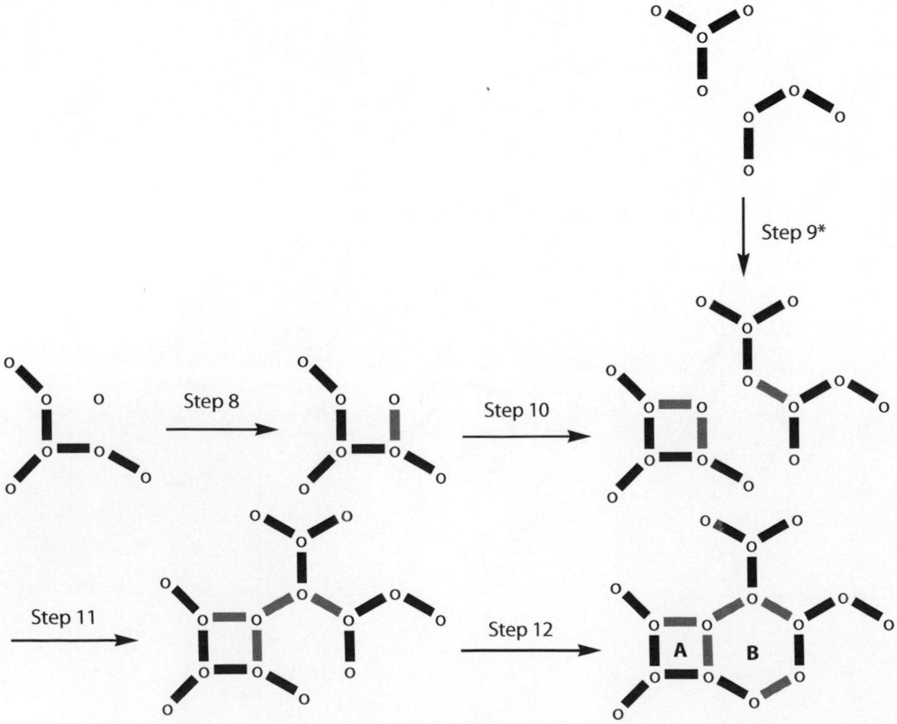

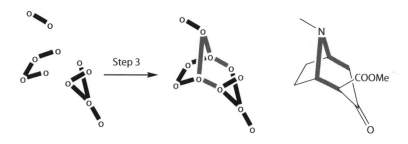

Sequence: $\xrightarrow{[4+0]} A_{10} \xrightarrow{[6+0]} B_{12}$

(±)-Cocaine

Casale, J.F. *Forensic Sci. Int.* 1987, *33*, 275.

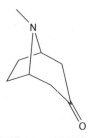

Sequence: $\xrightarrow{[(4+1)+(4+3)]} [3.2.1]_3$

(±)-Cocaine

Lee, J.C.; Lee, K.L.; Cha, J.K. *J. Org. Chem.* 2000, *65*, 4773.

(±)-Cocaine

Mans, D.M.; Pearson, W.H. *Org. Lett.* 2004, *6*, 3305.

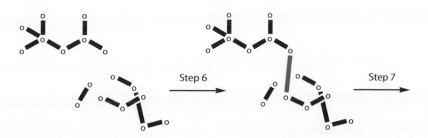

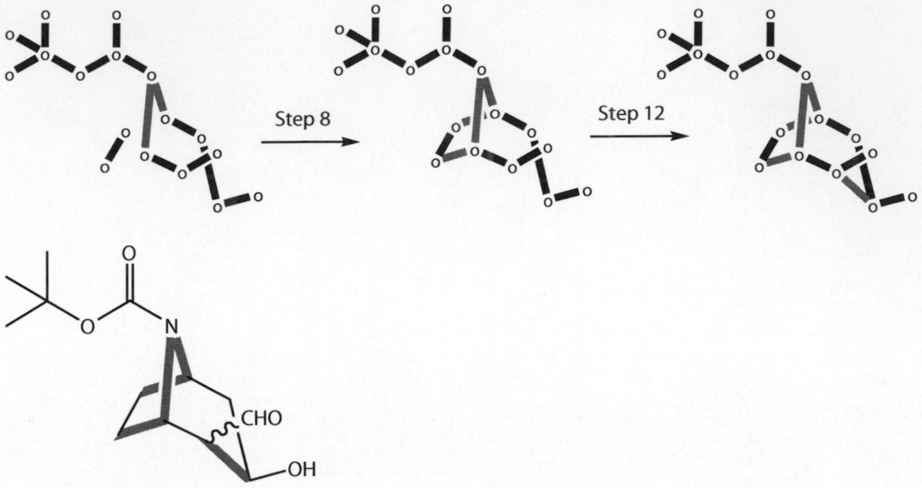

Sequence: $\xrightarrow{[3+2]}$ $[5]_8$ $\xrightarrow{[(7+0) + (6+0)]}$ $[3.2.1]_{12}$

(–)-Cocaine

Lin, R.; Castells, J.; Rapoport, H. *J. Org. Chem.* 1998, *63*, 4069.

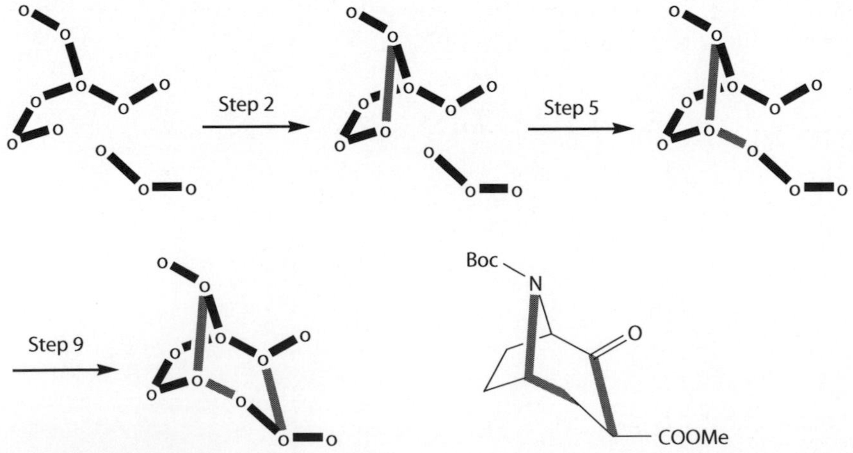

Sequence: $\xrightarrow{[5+0]}$ $[5]_2$ $\xrightarrow{[(7+0)+(6+0)]}$ $[3.2.1]_9$

(±)-Cocaine

Tufariello, J.J.; Mullen, G.B.; Tegeler, J.J.; Trybulski, E.J.; Wong, S.C.; Ali, S.A. *J. Am. Chem. Soc.* 1979, *101*, 2435.

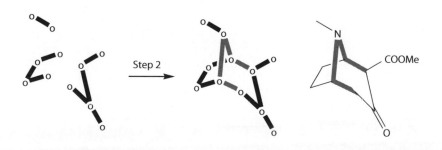

Sequence: $\xrightarrow{[(7+0) + (6+0)]}$ $[3.2.1]_6$

(±)-Cocaine

Willstätter, R.; Wolfes, O.; Mäder, H. *Ann. Chem.* 1923, *434*, 111.

Sequence: $\xrightarrow{[(4+1) +(4+3)]}$ $[3.2.1]_2$

Summary:

Strategy #1:
[(4 + 1) + (4 + 3)]: Casale, Willstätter

Strategy #2:
[(7 + 0) + (6 + 0)]: Pearson, Rapoport, Tufariello

Strategy #3:
[3.2.1] bicyclic ring precursor: Cha

(−)-Codeine
Tanimoto, H.; Saito, R.; Chida, N. *Tetrahedron Lett.* 2008, *49*, 358.

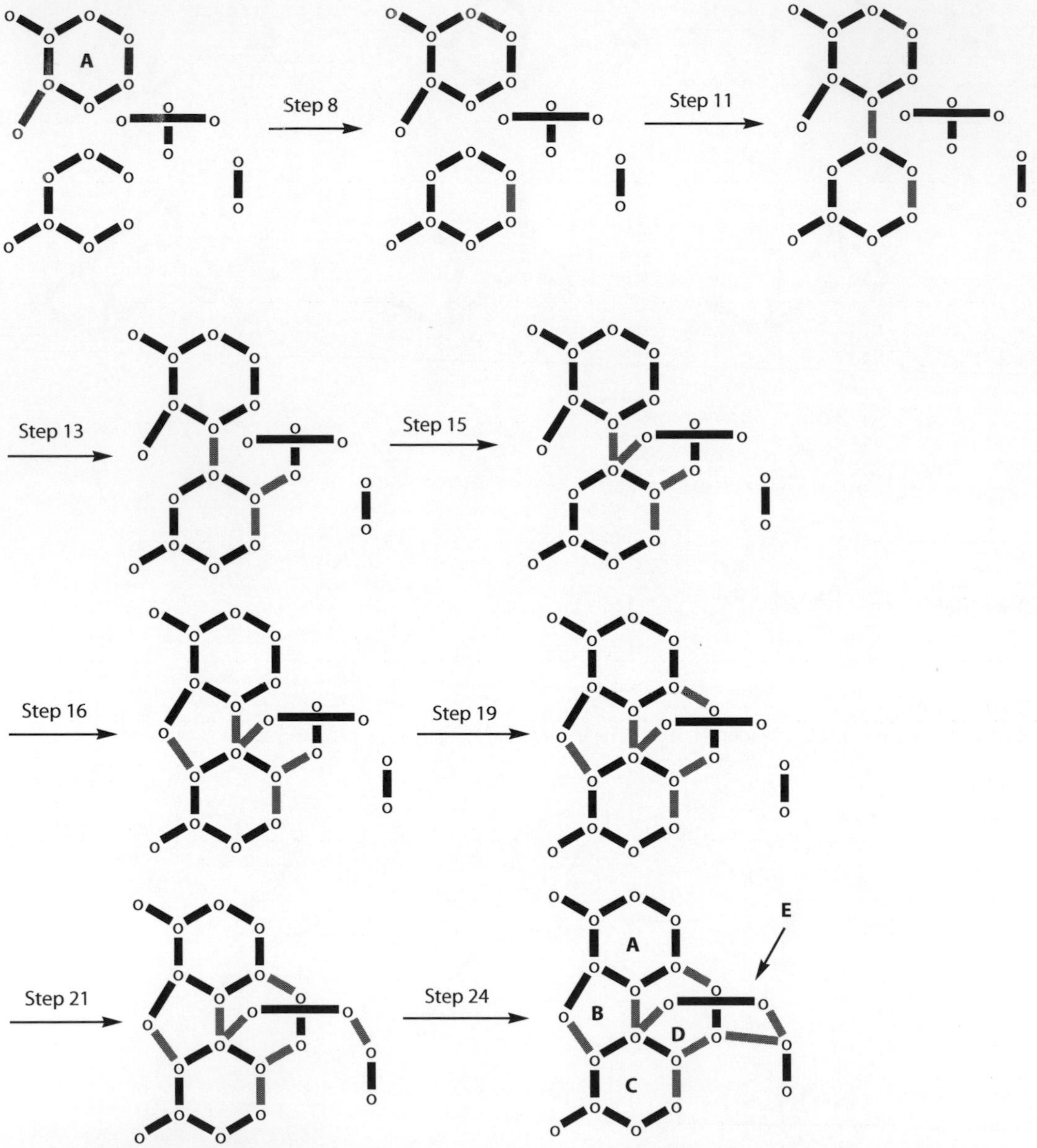

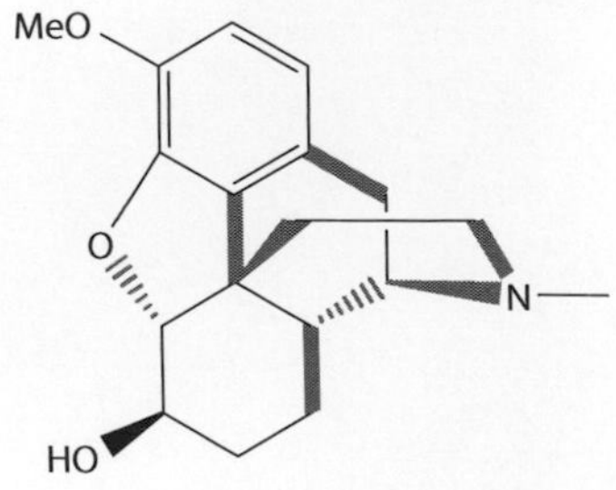

Sequence: $A_0 \xrightarrow{[6+0]} C_8 \xrightarrow{[5+0]} B_{16} \xrightarrow{[6+0]} D_{19} \xrightarrow{[6+0]} E_{24}$

(±)-Codeine

Uchida, K.; Yokoshima, S.; Kan, T.; Fukuyama, T. *Org. Lett.* 2006, *8*, 5311.

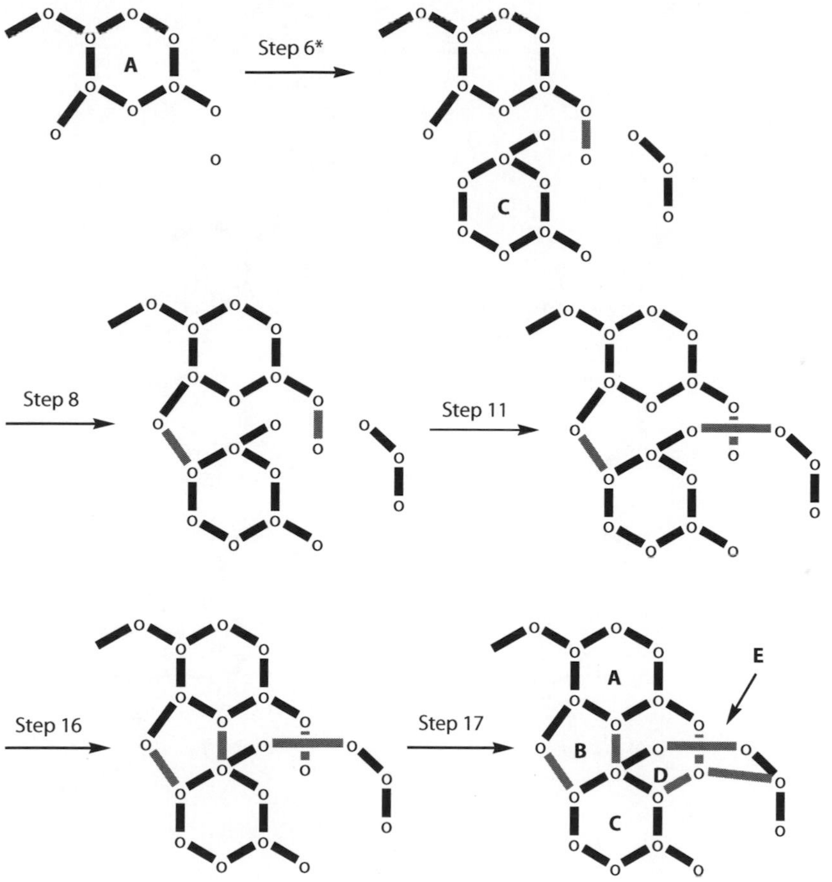

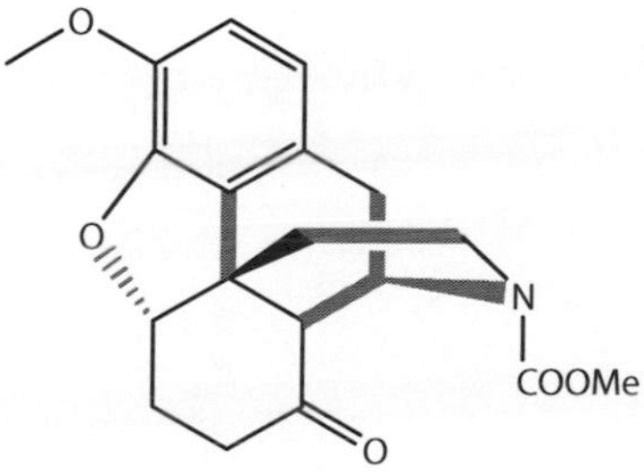

Sequence: $A_{1*} + C_0 \xrightarrow{[5+0]} B_{16} \xrightarrow{[6+0]_D + [5+1]_E} (D+E)_{17}$

(−)-Codeine

Gates, M. *J. Am. Chem. Soc.* 1950, *72*, 228.
Gates, M.; Tschudi, G. *J. Am. Chem. Soc.* 1956, *78*, 1380.

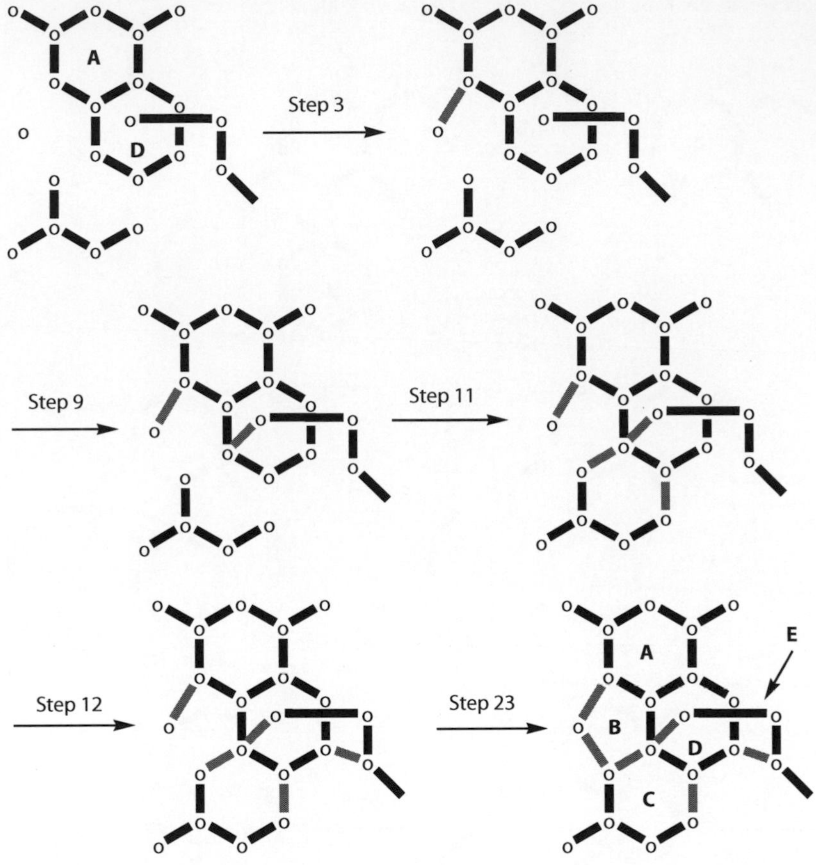

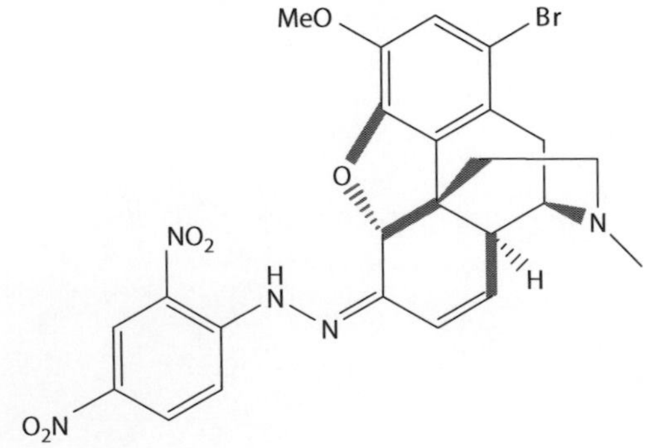

Sequence: $(A + D)_0 \xrightarrow{[4+2]} C_{11} \xrightarrow{[6+0]} E_{12} \xrightarrow{[5+0]} B_{23}$

(±)-Codeine

Omori, A.T.; Finn, K.J.; Leisch, H.; Carroll, R.J.; Hudlicky, T. *Synlett* 2007, 2859.

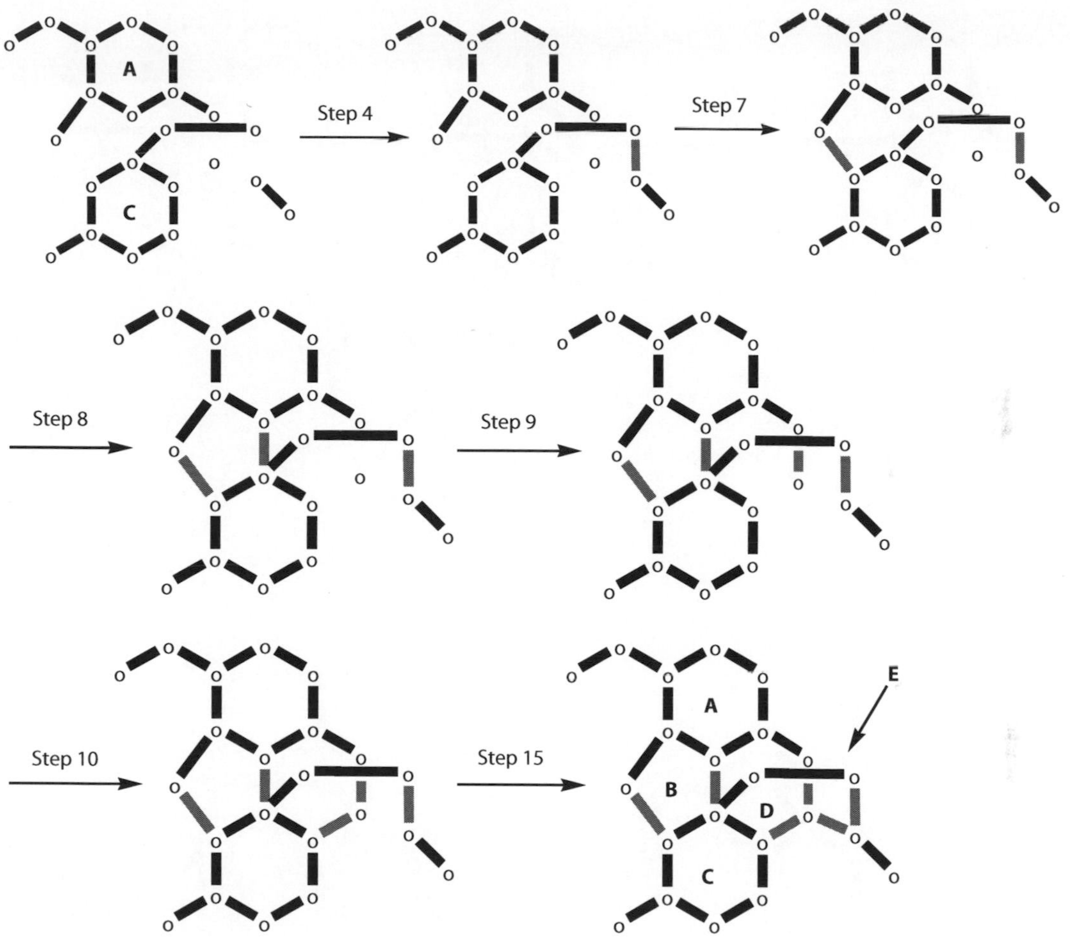

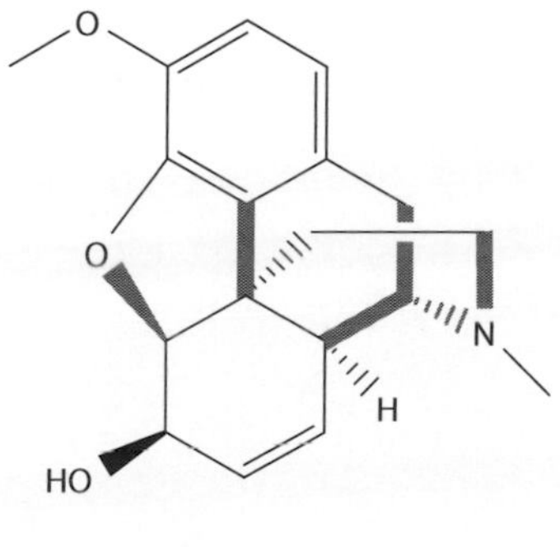

Sequence: $A_6 + C_0 \xrightarrow{[5+0]} B_8 \xrightarrow{[6+0]} D_{10} \xrightarrow{[6+0]} E_{15}$

(±)-Codeine

Varin, M.; Barre, E.; Iorga, B.; Guillou, C. *Chem. Eur. J.* 2008, *14*, 6606.

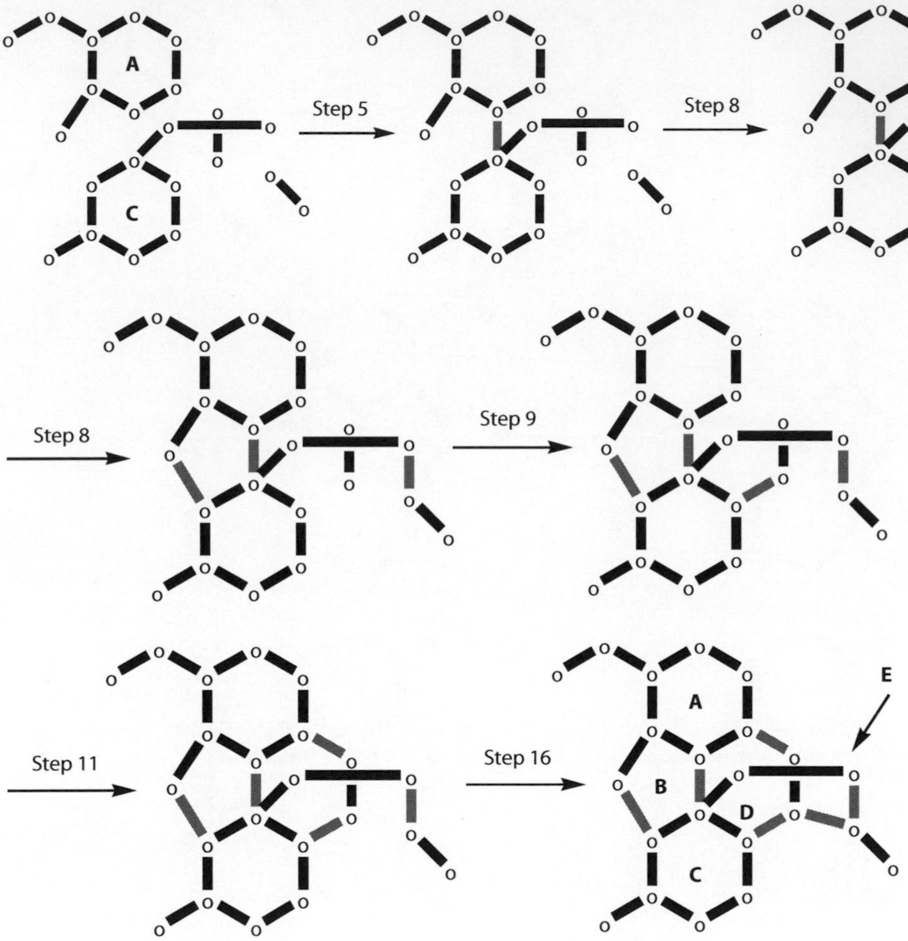

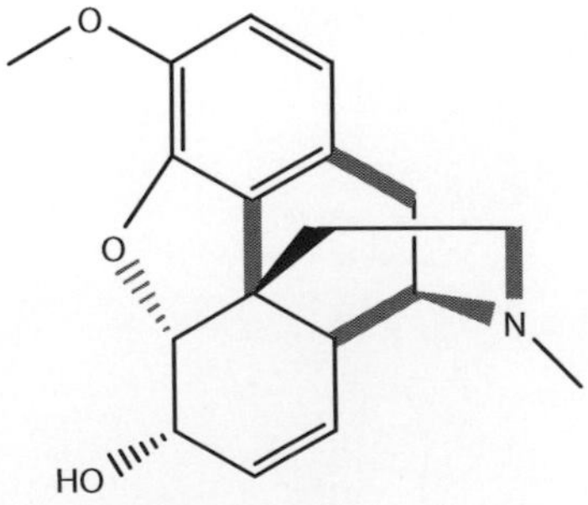

Sequence: $A_0 + C_{1*} \xrightarrow{[5+0]} B_8 \xrightarrow{[6+0]} D_{11} \xrightarrow{[6+0]} E_{16}$

(–)-Codeine

Trauner, D.; Bats, J.W.; Werner, A.; Mulzer, J. *J. Org. Chem.* 1998, *63*, 5908.

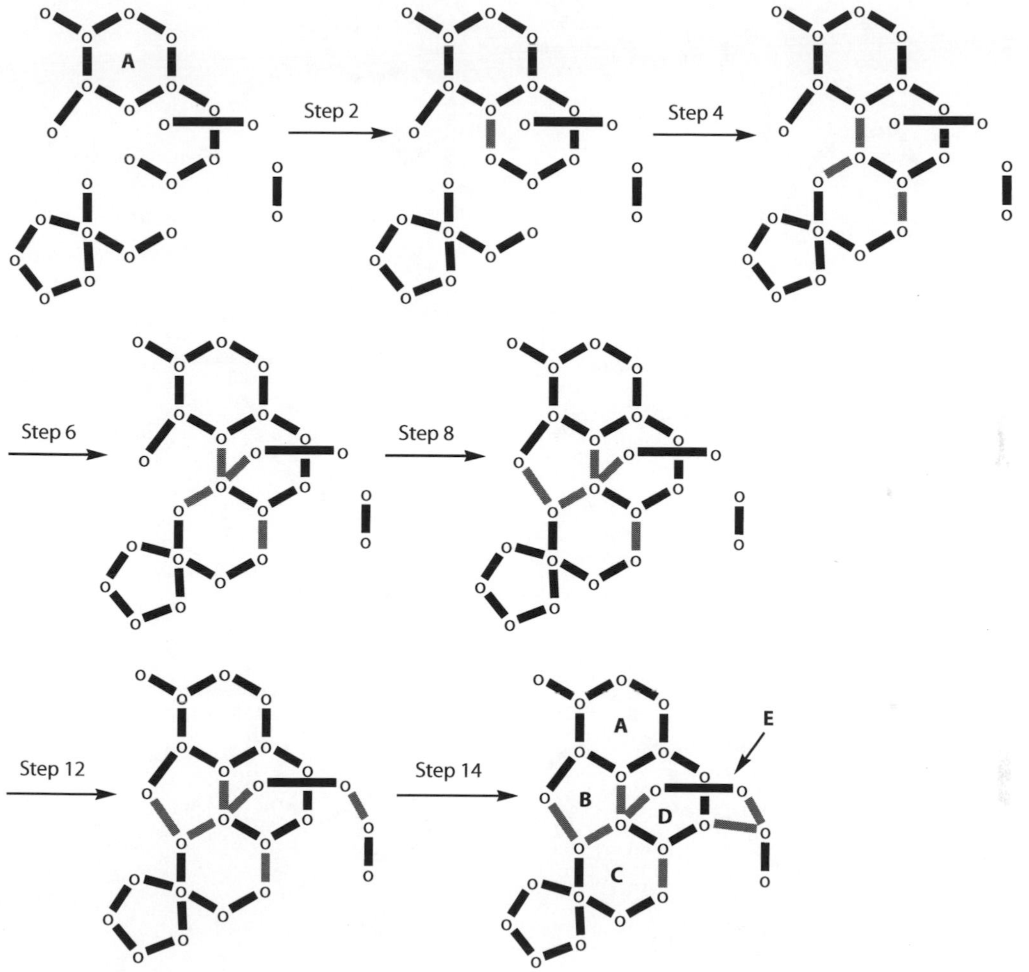

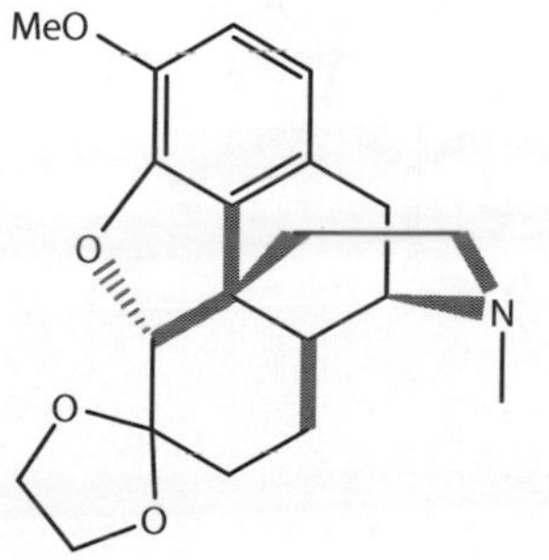

Sequence: $A_0 \xrightarrow{[6+0]} D_2 \xrightarrow{[4+2]} C_4 \xrightarrow{[5+0]} B_8 \xrightarrow{[6+0]} E_{14}$

(–)-Codeine

Nagata, H.; Miyazawa, N.; Ogasawara, K. *Chem. Commun.* 2001, 1094.

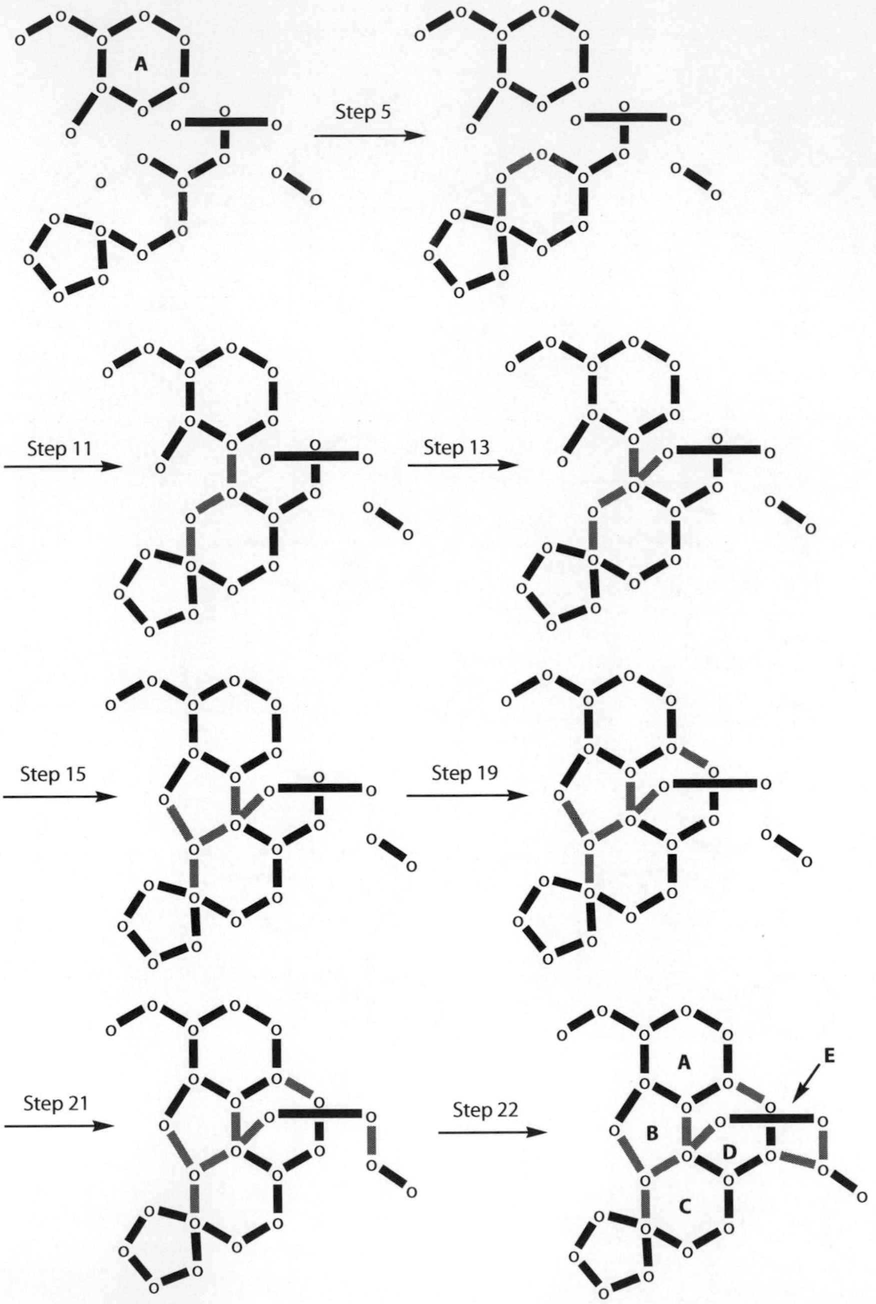
A
Step 5
Step 11
Step 13
Step 15
Step 19
Step 21
Step 22
A
B
C
D
E

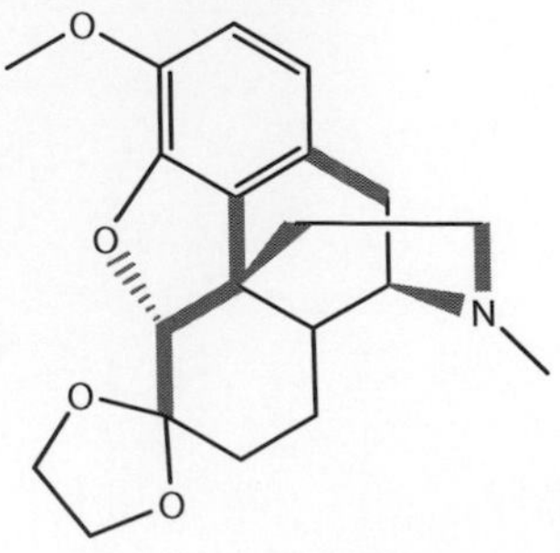
O
O
N
O
O

Sequence: $A_{10} \xrightarrow{[5+1]} C_5 \xrightarrow{[5+0]} B_{15} \xrightarrow{[6+0]} D_{19} \xrightarrow{[6+0]} E_{22}$

(−)-Codeine

Hong, C.Y.; Kado, N.; Overman, L.E. *J. Am. Chem. Soc.* 1993, *115*, 11028.

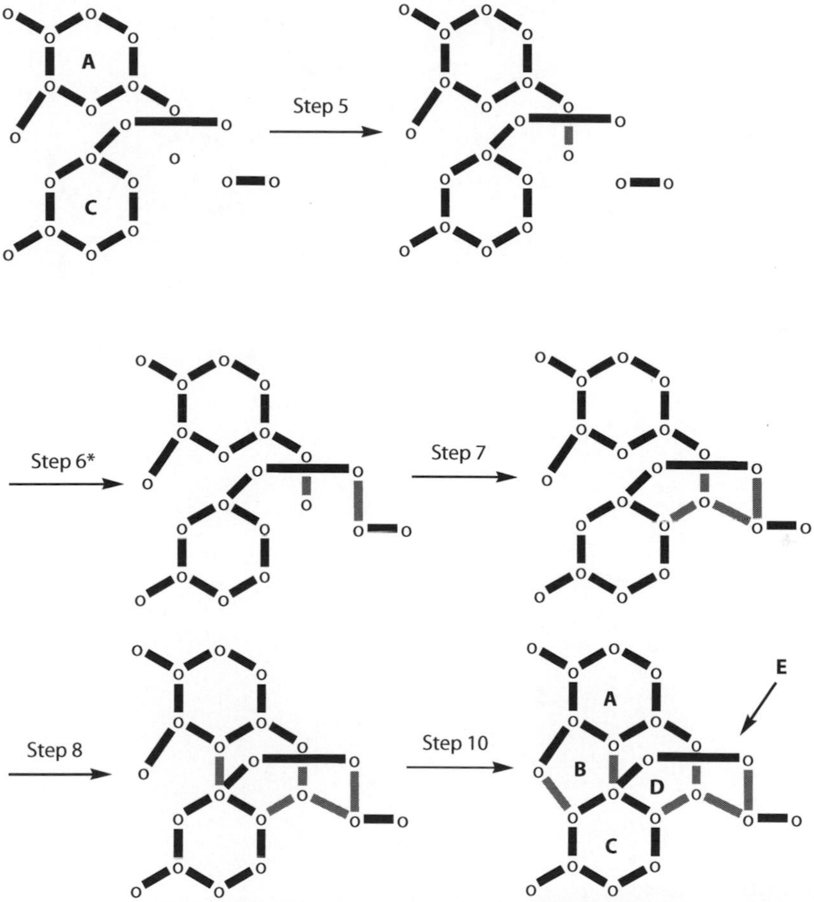

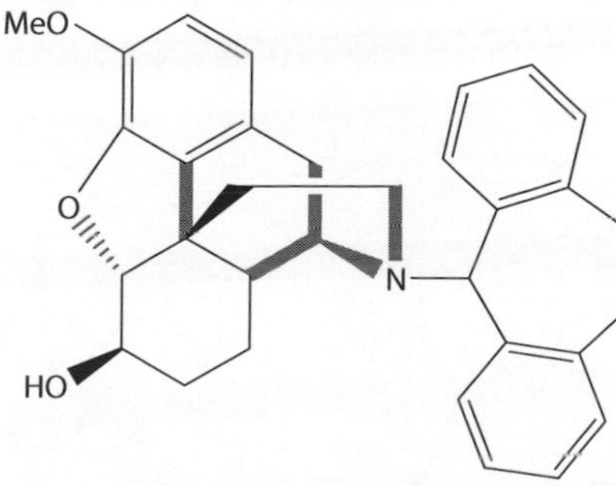

Sequence: $A_0 + C_{1*} \xrightarrow{[5+1]} E_7 \xrightarrow{[6+0]} D_8 \xrightarrow{[5+0]} B_{10}$

(–)-Codeine

Parker, K.A.; Fokas, D. *J. Org. Chem.* 2006, *71*, 449.

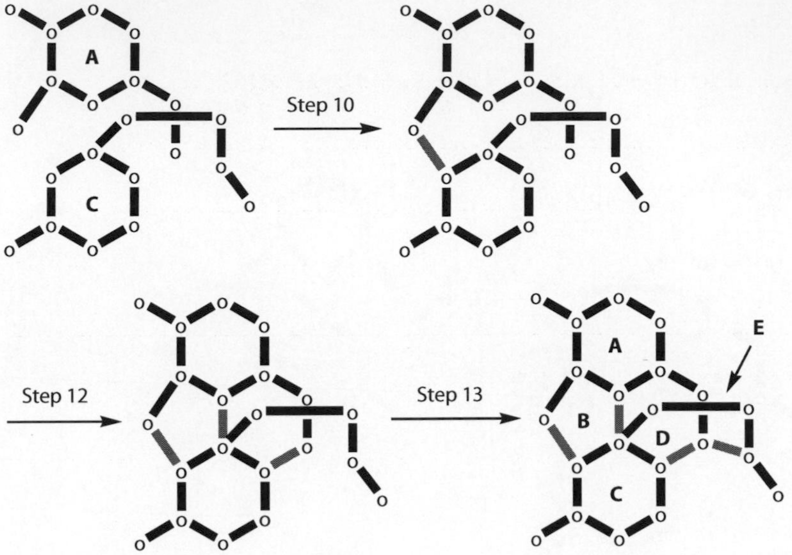
A
C
Step 10
Step 12
Step 13
A
B
C
D
E

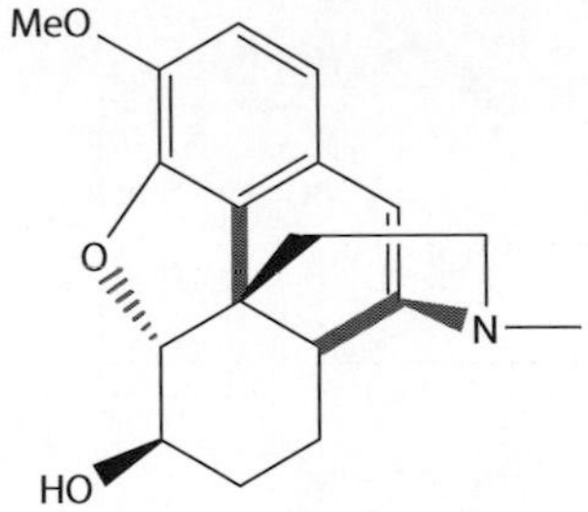
MeO
O
N
HO

Sequence: $A_9 + C_0 \xrightarrow{[5+0]_B+[4+2]_D} (B+D)_{12} \xrightarrow{[6+0]} E_{13}$

(±)-Codeine

Iijima, I.; Minamikawa, J.; Jacobson, A.E.; Brossi, A.; Rice, K.C. *J. Org. Chem.* 1978, *43*, 1462.

Rice, K.C. *J. Org. Chem.* 1980, *45*, 3137.

Rice, K.C. *J. Med. Chem.* 1977, *20*, 164.

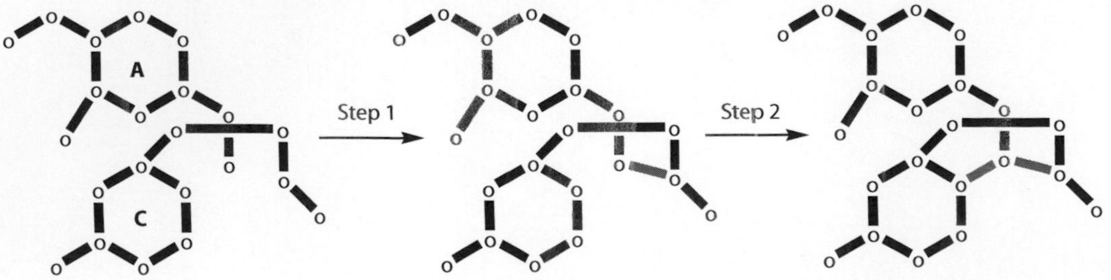
A
C
Step 1
Step 2

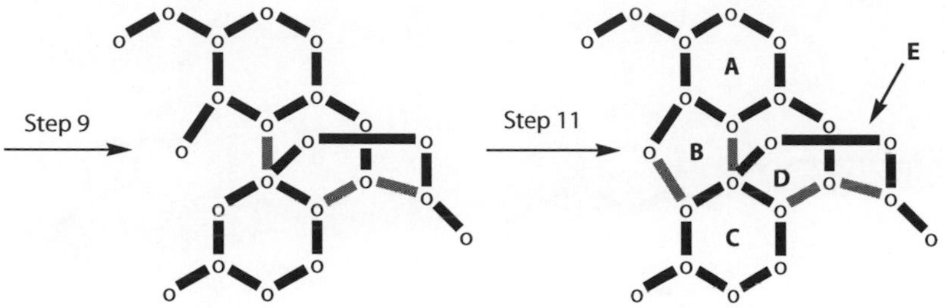

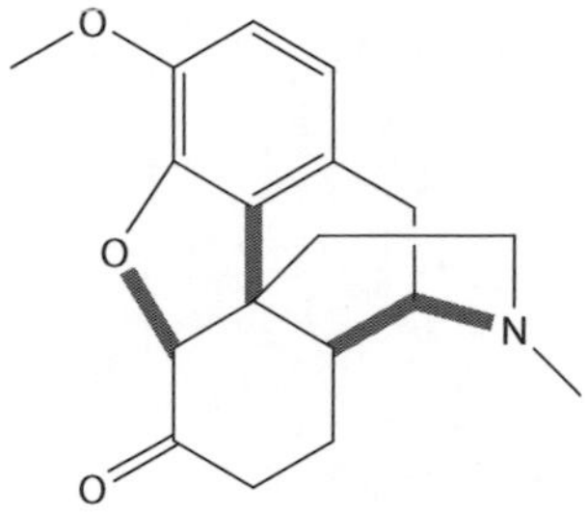

Sequence: $A_0 + C_0 \xrightarrow{[6+0]} E_2 \xrightarrow{[6+0]} D_9 \xrightarrow{[5+0]} B_{11}$

(−)-Codeine

Taber, D.F.; Neubert, T.D.; Rheingold, A.L. *J. Am. Chem. Soc.* 2002, *124*, 12416.

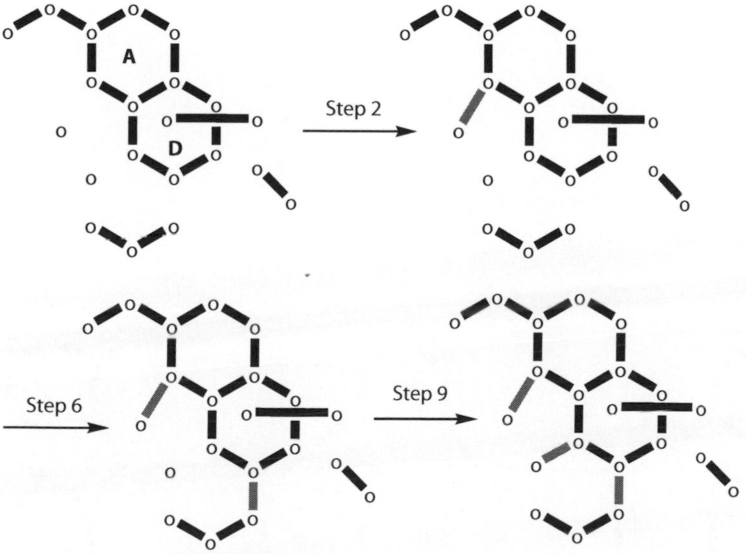

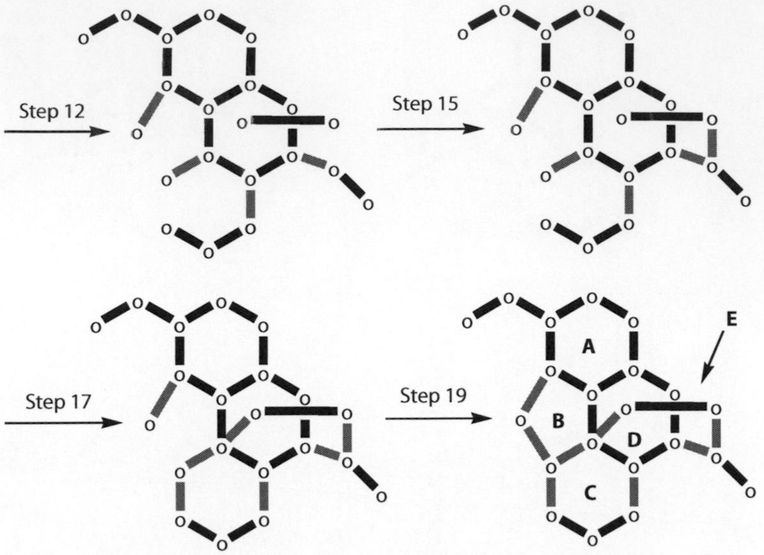

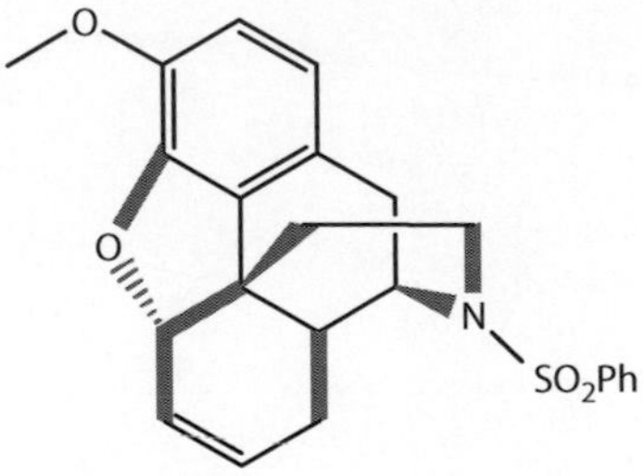

Sequence: $(A+D)_0 \xrightarrow{[6+0]_C + [6+0]_E} (C+E)_{17} \xrightarrow{[5+0]} B_{19}$

(−)-Codeine

Trost, B.M.; Tang, W. *J. Am. Chem. Soc.* 2002, *124*, 14542.
Trost, B.M.; Tang, W.; Toste, F.D. *J. Am. Chem. Soc.* 2005, *127*, 14785.

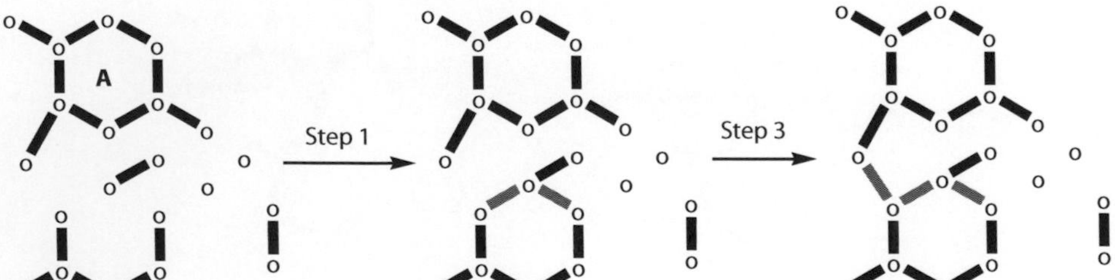

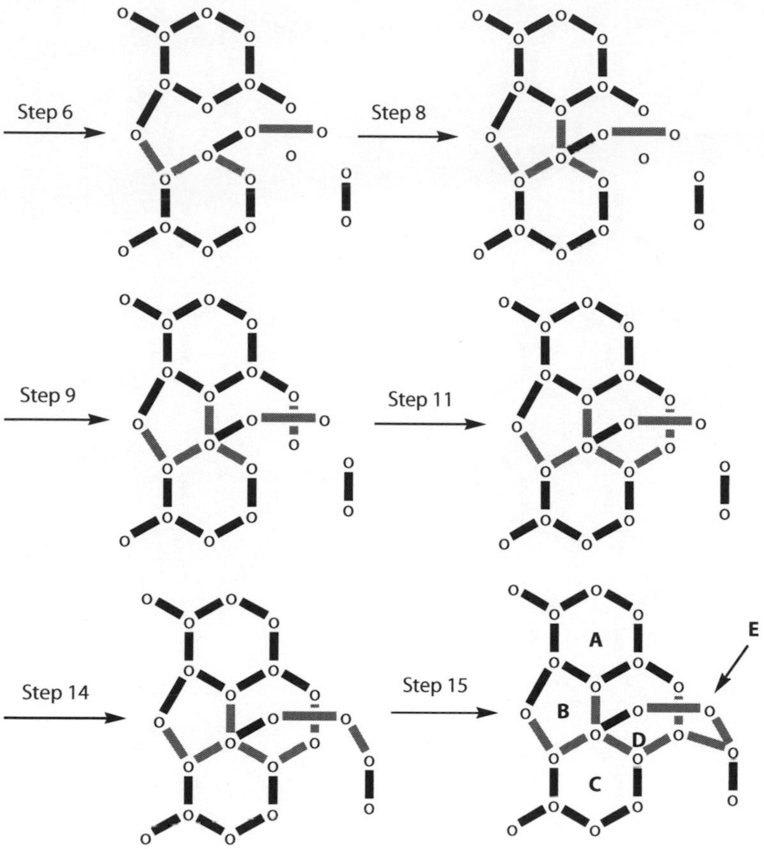

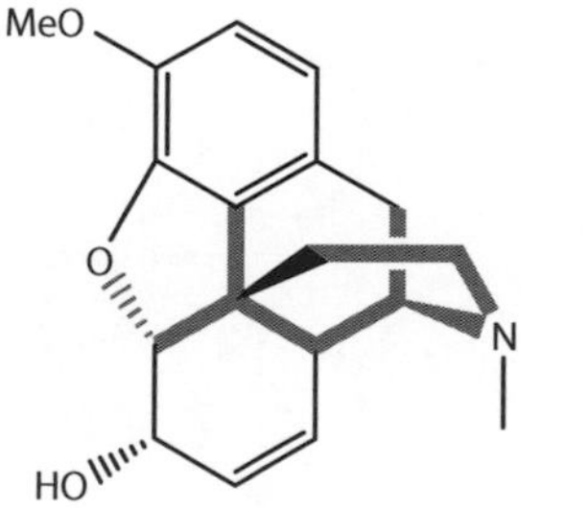

Sequence: $A_{1*} \xrightarrow{[5+1]} C_1 \xrightarrow{[5+0]} B_8 \xrightarrow{[6+0]} D_{11} \xrightarrow{[6+0]} E_{15}$

(±)-Codeine

White, J.D.; Hrnciar, P.; Stappenbeck, F. *J. Org. Chem.* 1997, *62*, 5250.

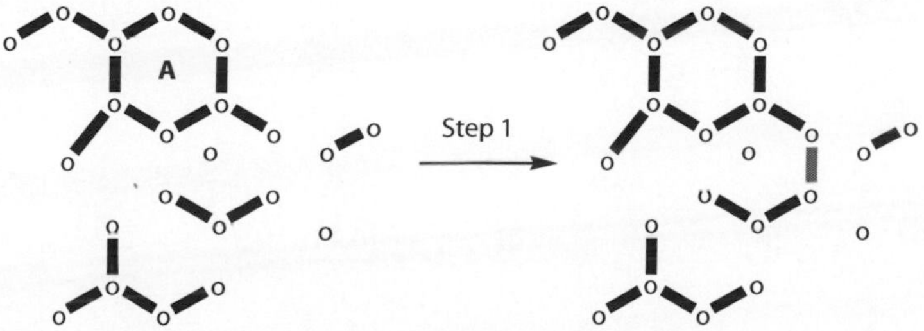

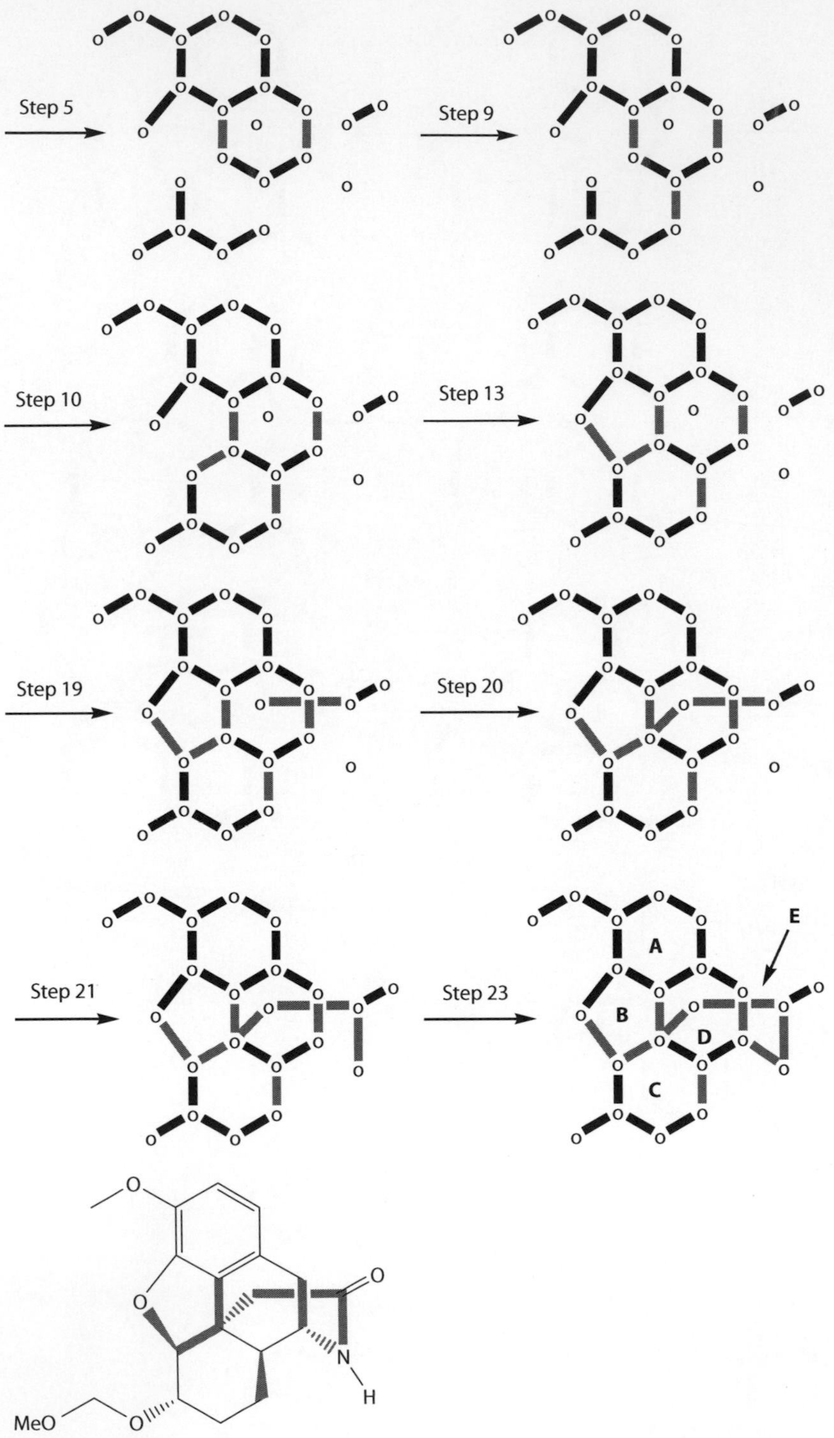

Sequence: $A_0 \xrightarrow{[6+0]} D_5 \xrightarrow{[6+0]} C_{10} \xrightarrow{[5+0]} B_{13} \xrightarrow{[6+0]} E_{23}$

(±)-Codeine

Stork, G.; Yamashita, A.; Adams, J.; Schulte, G.R.; Chesworth, R.; Miyazaki, Y.; Farmer, J.J. *J. Am. Chem. Soc.* 2009, *131*, 11402.

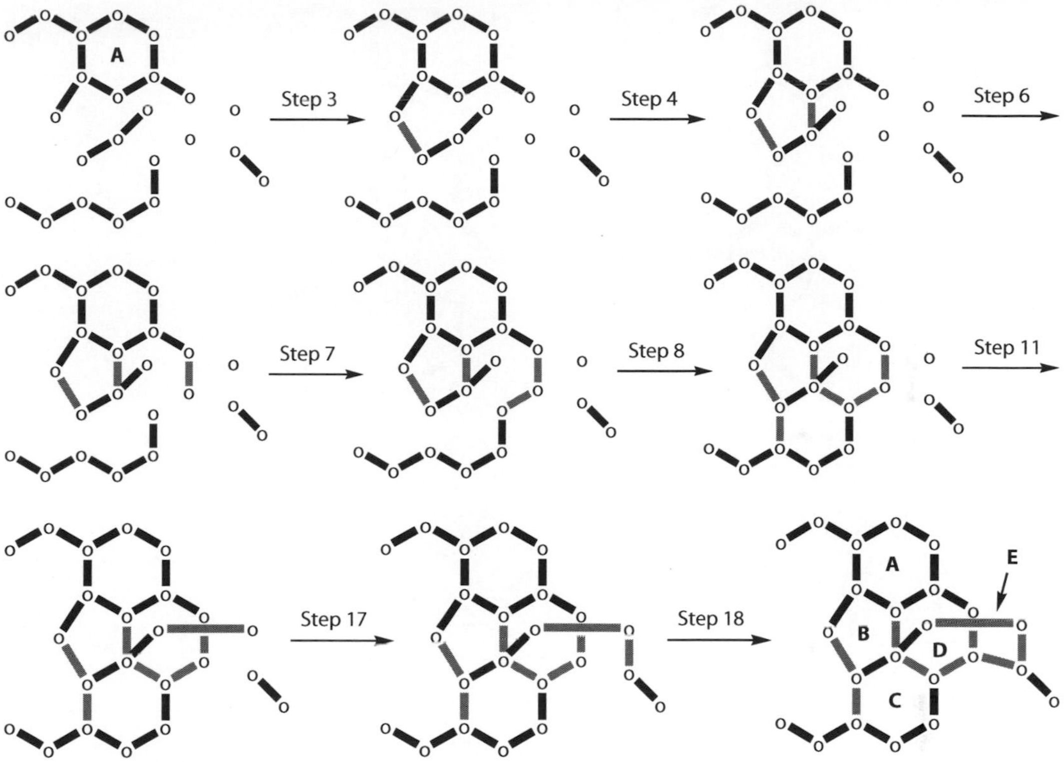

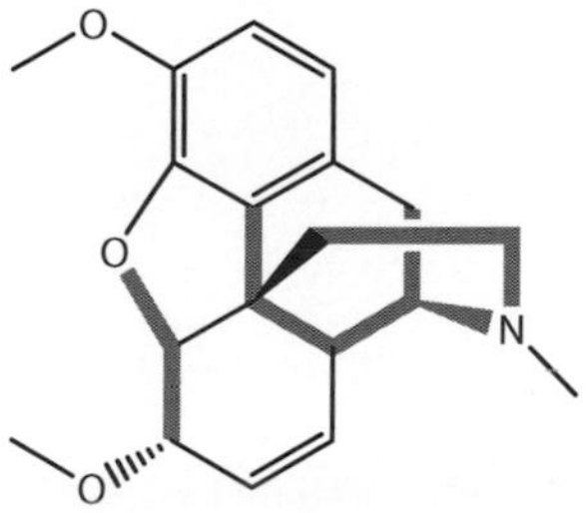

Sequence: $A_0 \xrightarrow{[5+0]} B_4 \xrightarrow{[(4+2)_C + (6+0)_D]} (C+D)_8 \xrightarrow{[6+0]} E_{18}$

(±)-Codeine

Magnus, P.; Sane, N.; Fauber, B.P.; Lynch, V. *J. Am. Chem. Soc.* 2009, *131*, 16045.

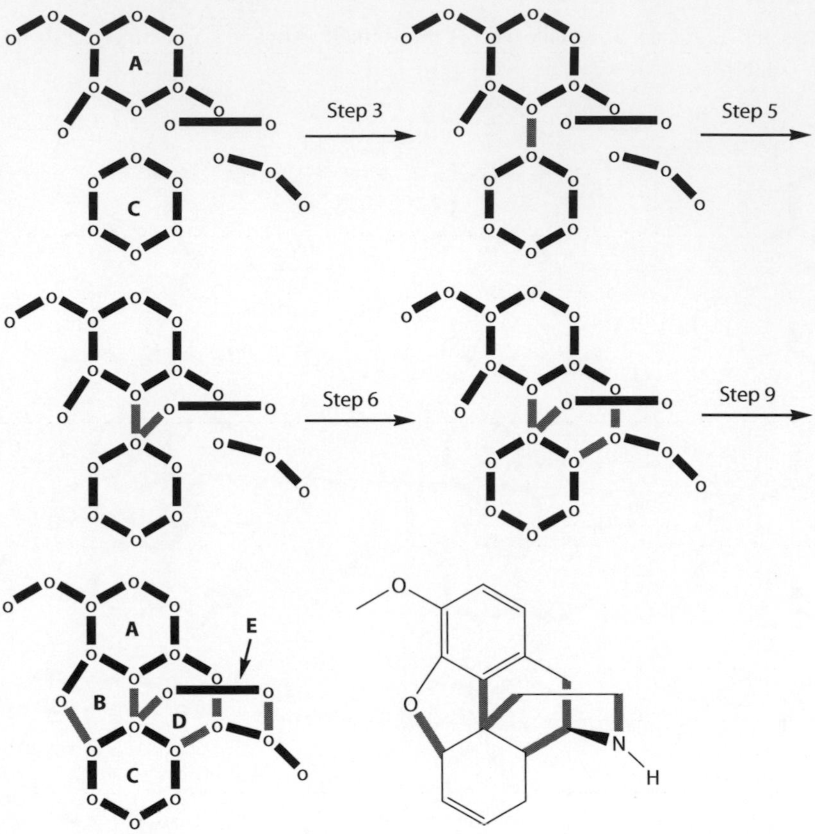

Sequence: $(A+C)_0 \xrightarrow{[5+1]} D_6 \xrightarrow{[(5+0)_B + (6+0)_E]} (B+E)_9$

Summary:

Strategy A-C-B-D-E: Chida, Fukuyama, Hudlicky, Iorga, Ogasawara, Parker, Trost
Strategy (A + D)-C-E-B: Gates, Taber
Strategy A-D-C-B-E: Mulzer, White
Strategy (A + C)-E-D-B: Overman, Rice
Strategy A-B-(C + D)-E: Stork
Strategy (A + C)-D-(B + E): Magnus

Colchicine
Banwell, M.G. *Pure Appl. Chem.* 1996, *68*, 539.

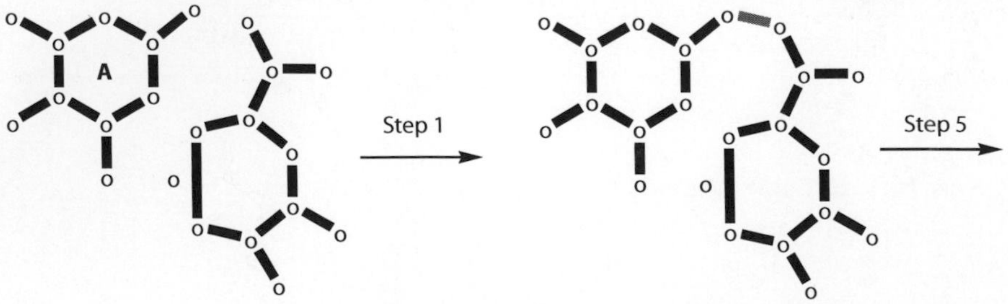

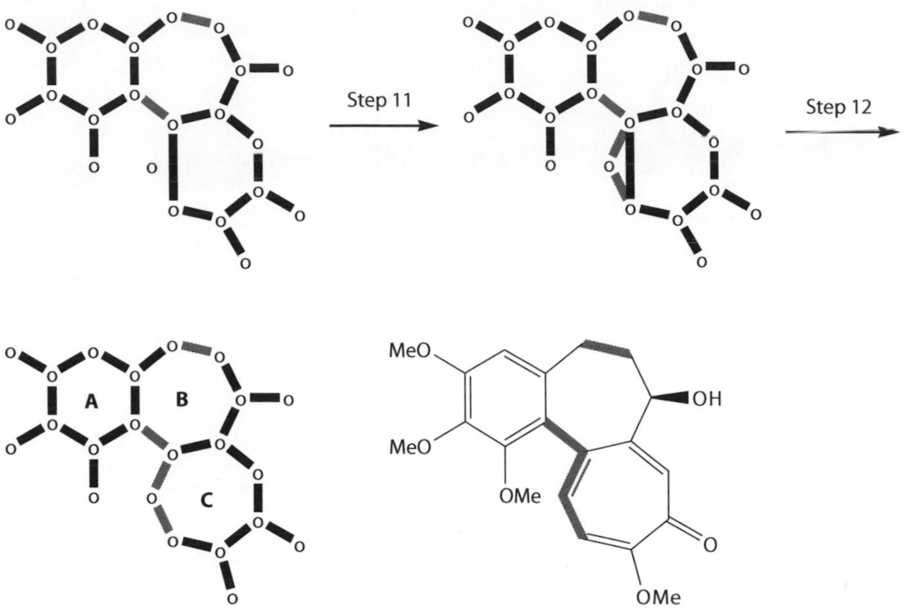

Sequence: $A_0 \xrightarrow{[7+0]} B_5 \xrightarrow{[2+1]} [4.1.0]_{11} \rightarrow [7]C_{12}$

Colchicine

Lee, J.C.; Jin, S.; Cha, J.K. *J. Org. Chem.* 1998, *63*, 2804.
Lee, J.C.; Cha, J.K. *Tetrahedron* 2000, *56*, 10175.

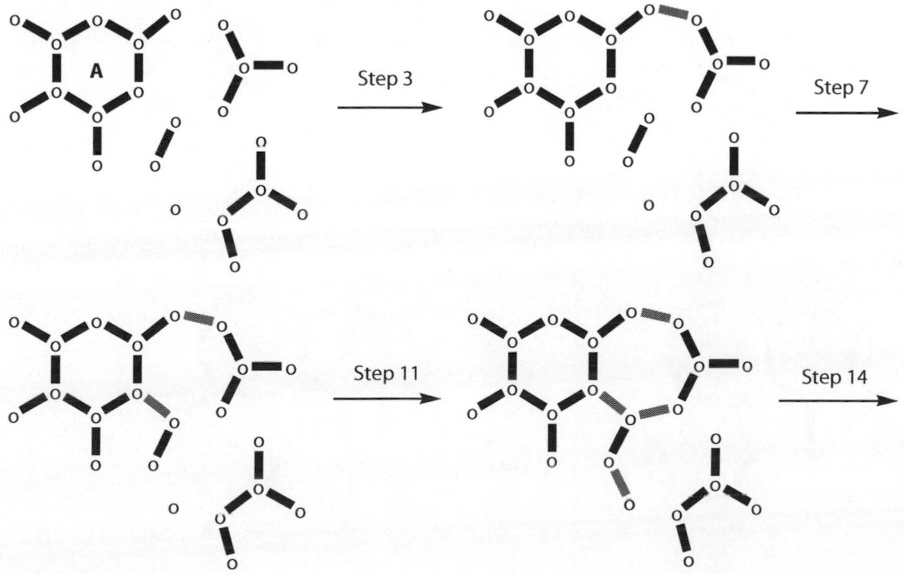

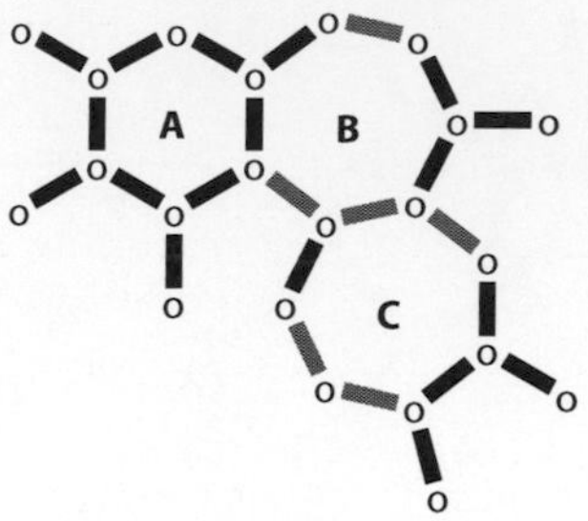

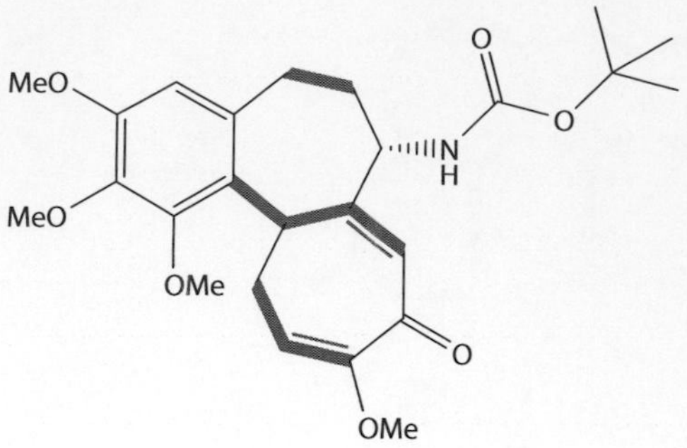

Sequence: $A_0 \xrightarrow{[7+0]} B_{11} \xrightarrow{[4+3]} C_{14}$

Colchicine

Schreiber, J.; Leimgruber, W.; Pesaro, M.; Schudel, P.; Eschenmoser, A. *Angew. Chem.* 1959, *71*, 637.
Schreiber, J.; Leimgruber, W.; Pesaro, M.; Schudel, P.; Eschenmoser, A. *Helv. Chim. Acta.* 1961, *44*, 540.

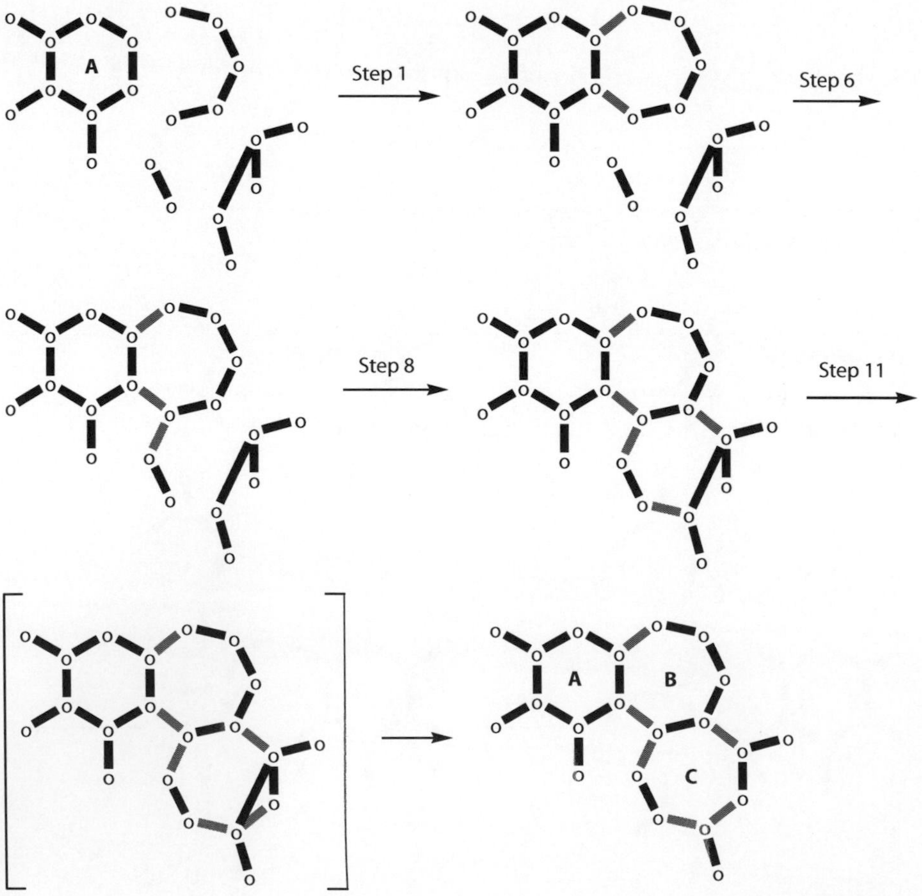

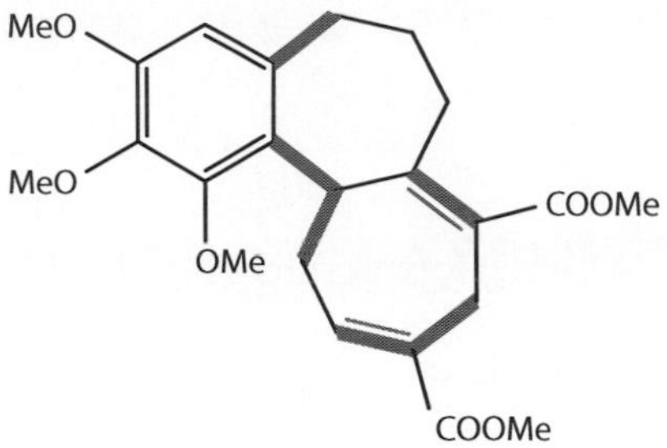

Sequence: $A_0 \xrightarrow{[5+2]} B_1 \xrightarrow{[4.1.0] \rightarrow [7]} C_{11}$

Colchicine

Evans, D.A.; Tanis, S.P.; Hart, D.J. *J. Am. Chem. Soc.* 1981, *103*, 5813.

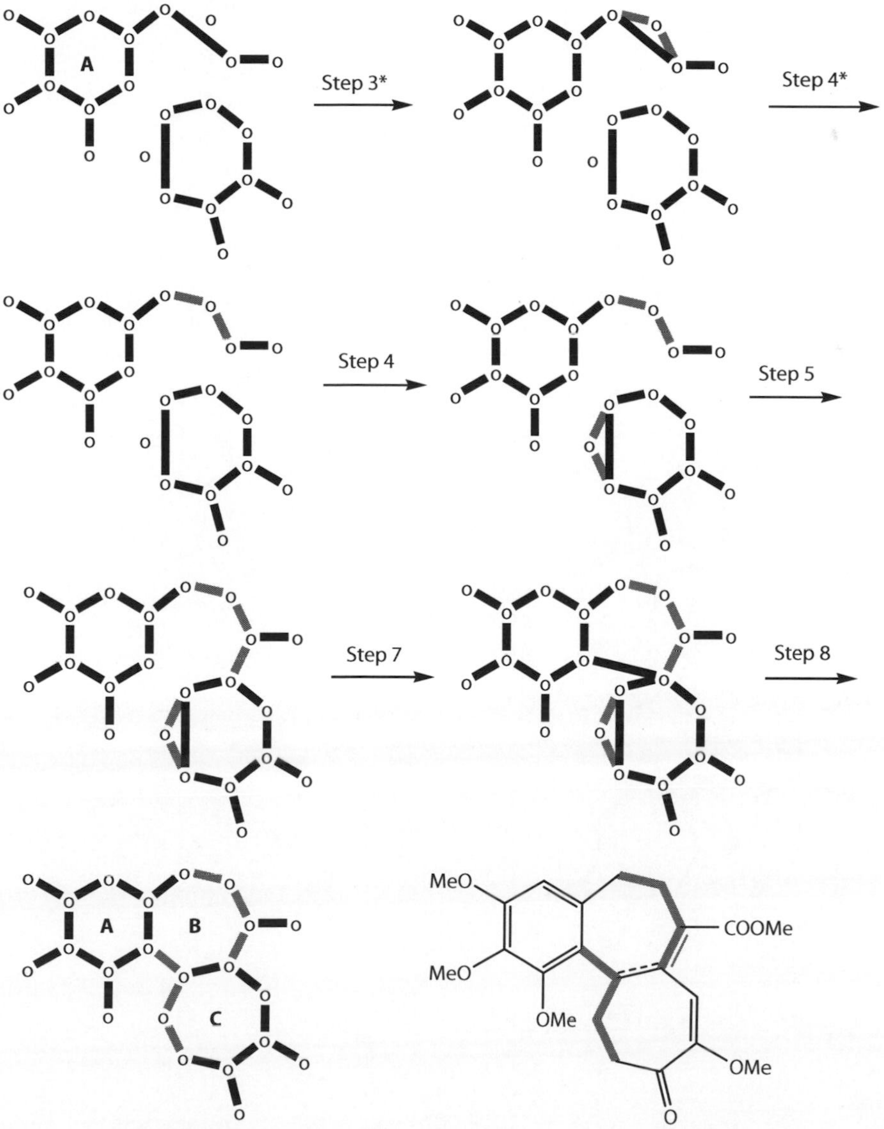

Sequence: $A_0 \xrightarrow{\ spiro[6,[4.2.0]]\ \rightarrow[5.5.0]\ } (B+C)_8$

Colchicine

Martel, J.; Toromanoff, E.; Huynh, C. *J. Org. Chem.* 1965, *30*, 1752.

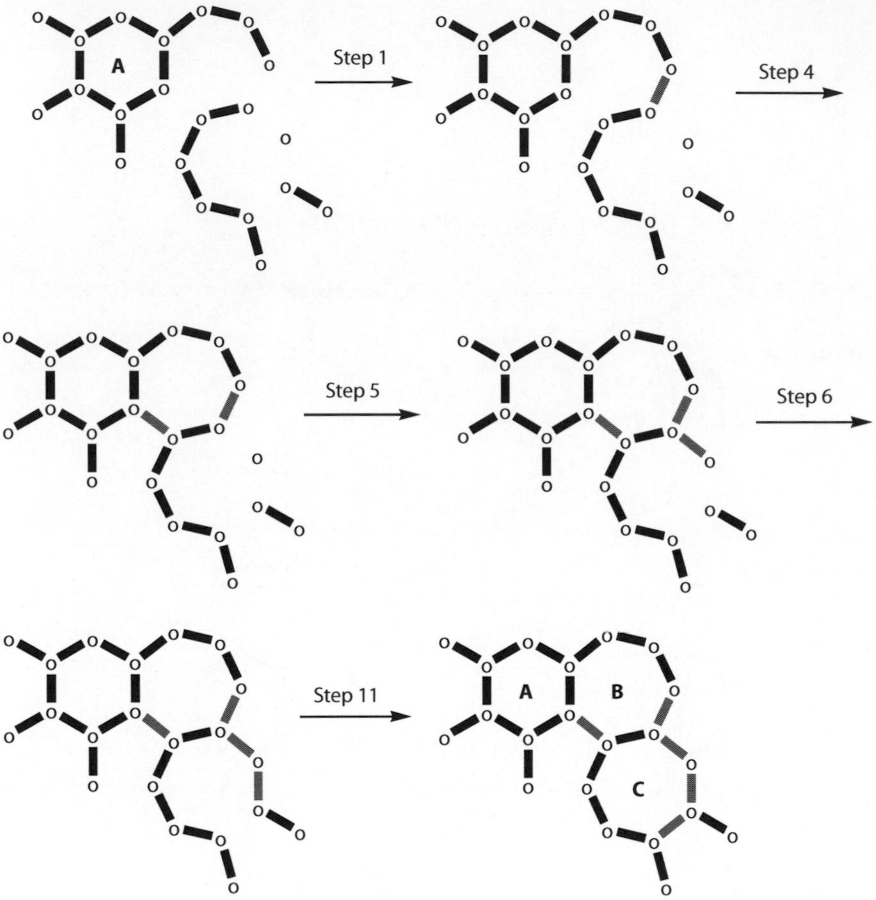

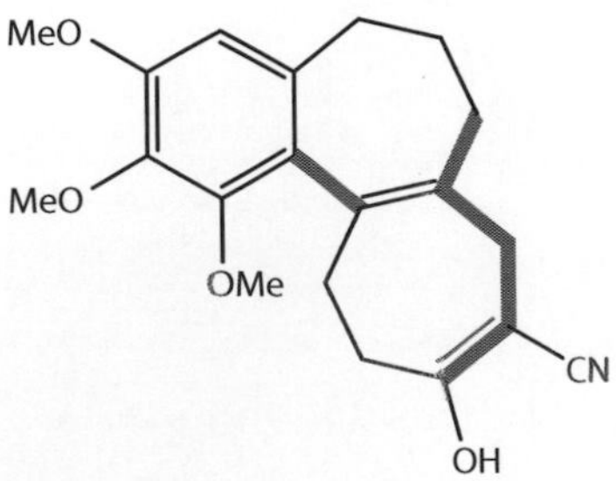

Sequence: $A_0 \xrightarrow{\ [7+0]\ } B_4 \xrightarrow{\ [7+0]\ } C_{11}$

Colchicine

Graening, T.; Bette, V.; Neudörfl, J.; Lex, J.; Schmalz, H.G. *Org. Lett.* 2005, *7*, 4317.

Graening, T.; Friedrichsen, W.; Lex, J.; Schmalz, H.G. *Angew. Chem. Int. Ed.* 2002, *41*, 1524.

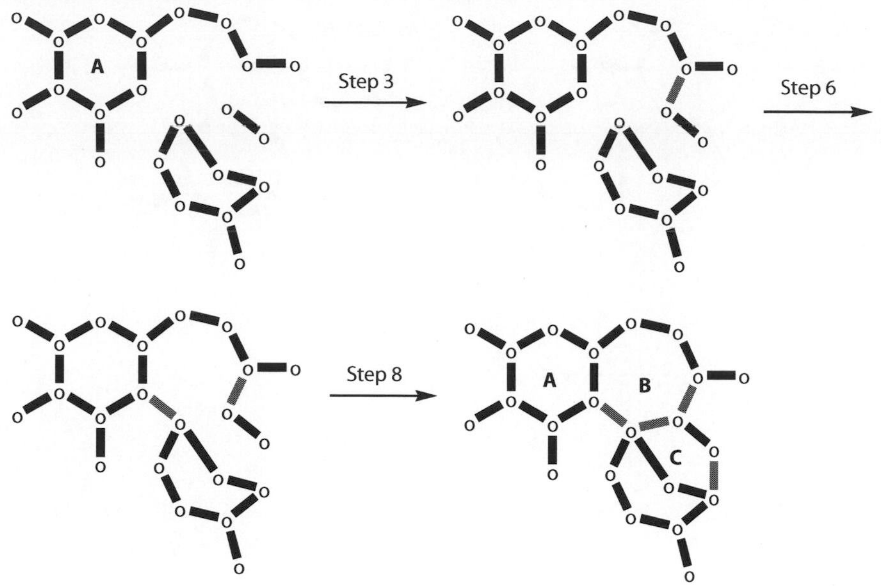

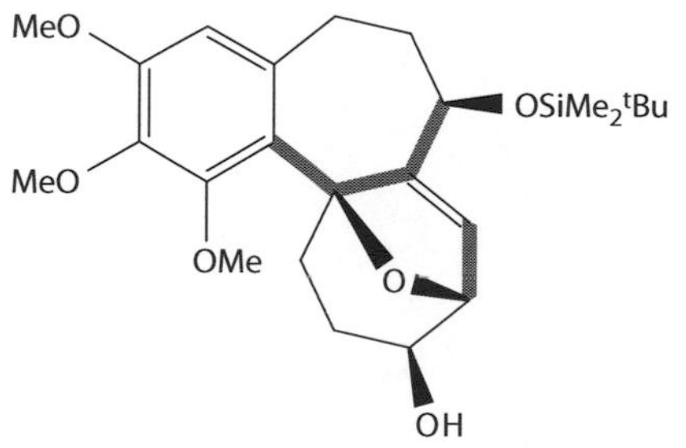

Sequence: $A_0 \xrightarrow{[(7+0)_B + (5+2)_C]} [5.5.0]\,(B+C)_8$

Colchicine

Scott, A.I.; McCapra, F.; Buchanan, R.L.; Day, A.C.; Young, D.W. *Tetrahedron* 1965, *21*, 3605.

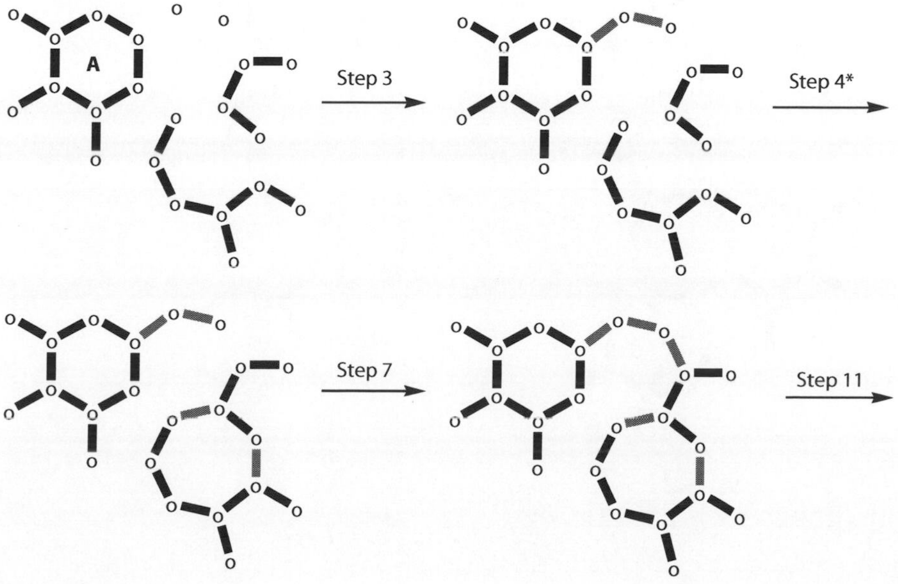

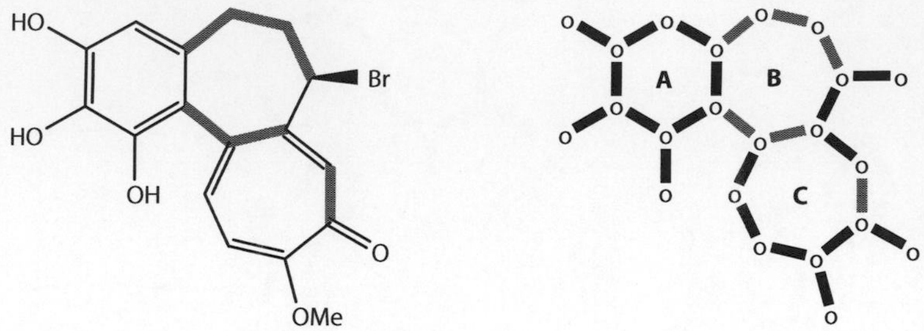

Sequence: $A_0 \xrightarrow{[5+2]} B_{4*} \xrightarrow{[7+0]} C_{11}$

Colchicine

van Tamelen, E.E.; Spencer, T.A. Jr.; Allen, D.S. Jr.; Ovis, R.L. *J. Am. Chem. Soc.* 1959, *81*, 6341.
van Tamelen, E.E.; Spencer, T.A. Jr.; Allen, D.S. Jr.; Ovis, R.L. *Tetrahedron* 1961, *14*, 8.

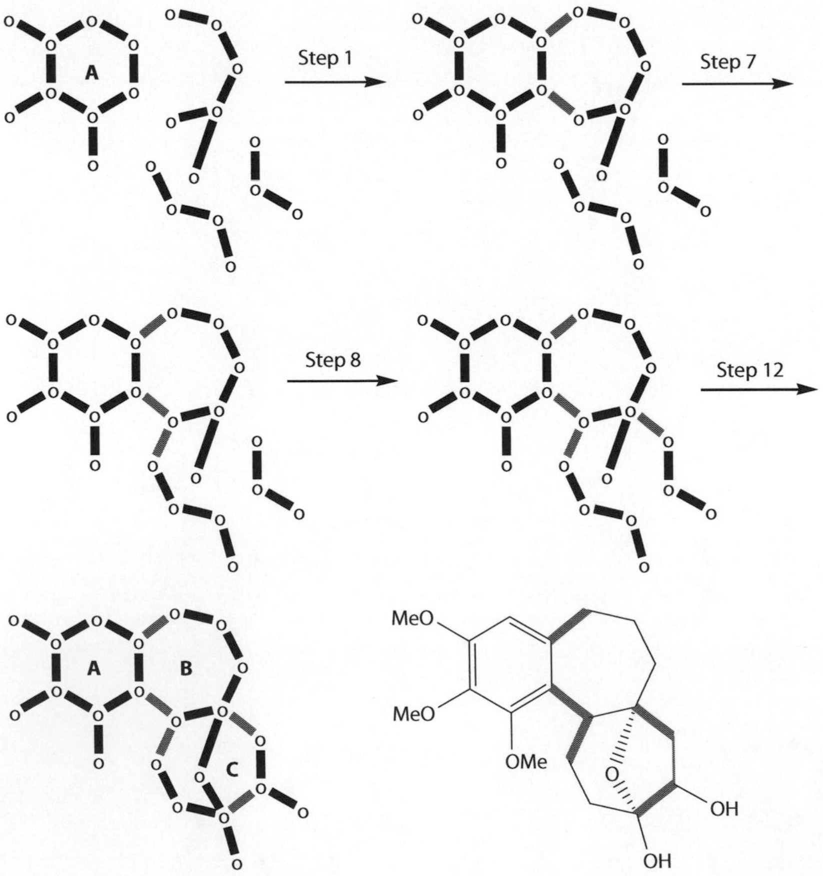

Sequence: $A_0 \xrightarrow{[5+2]} B_1 \xrightarrow{[7+0]} C_{12}$

Colchicine
Woodward, R.B. *Harvey Lectures* 1963, *59*, 31.

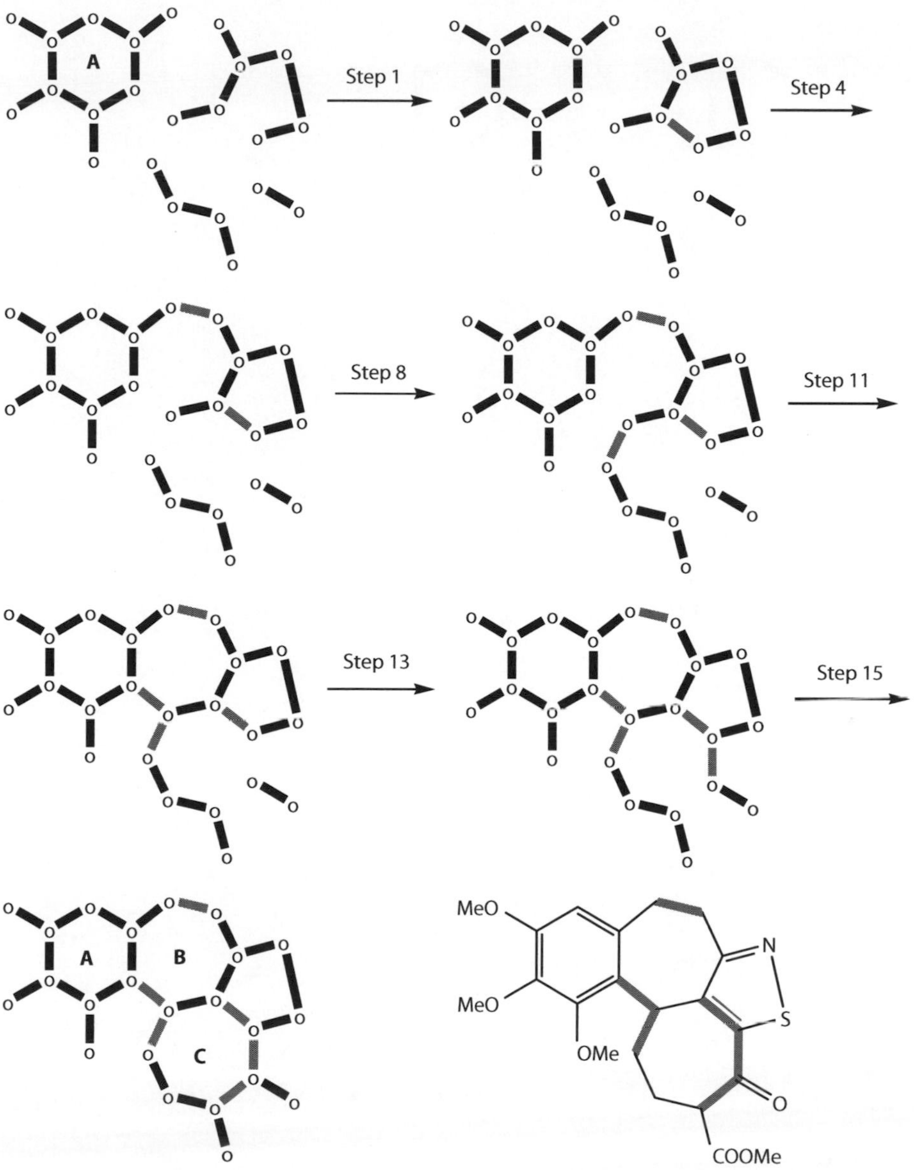

Sequence: $A_0 \xrightarrow{[7+0]} B_{11} \xrightarrow{[7+0]} C_{15}$

Summary:
Strategy A-B-C #1: Banwell
Strategy A-B-C #2: Roussel, Woodward
Strategy A-B-C #3: Scott, van Tamelen
Strategy A-B-C #4: Eschenmoser
Strategy A-B-C #5: Cha
Strategy A-(B + C) #1: Schmalz
Strategy A-(B + C) #2: Evans

(*S*)-Coniine
Aketa, K.I.; Terashima, S.; Yamada, S.I. *Chem. Pharm. Bull.* 1976, *24*, 621.

Sequence: $\xrightarrow{[6+0]} A_1$

(*S*)-Coniine
Reding, M.T.; Buchwald, S.L. *J. Org. Chem.* 1998, *63*, 6344.

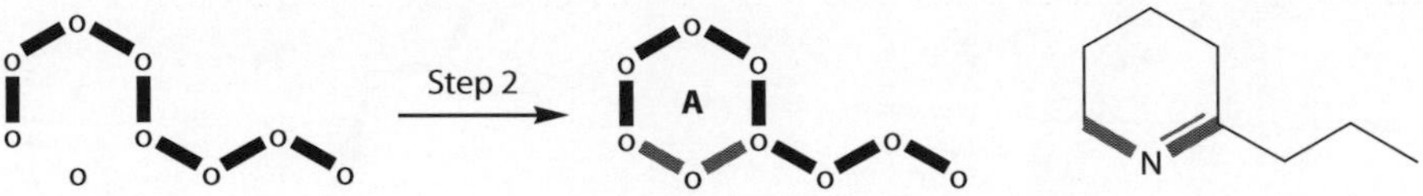

Sequence: $\xrightarrow{[5+1]} A_2$

(*S*)-Coniine
Lebrun, S.; Couture, A.; Deniau, E.; Grandclaudon, P. *Org. Lett.* 2007, *9*, 2473.

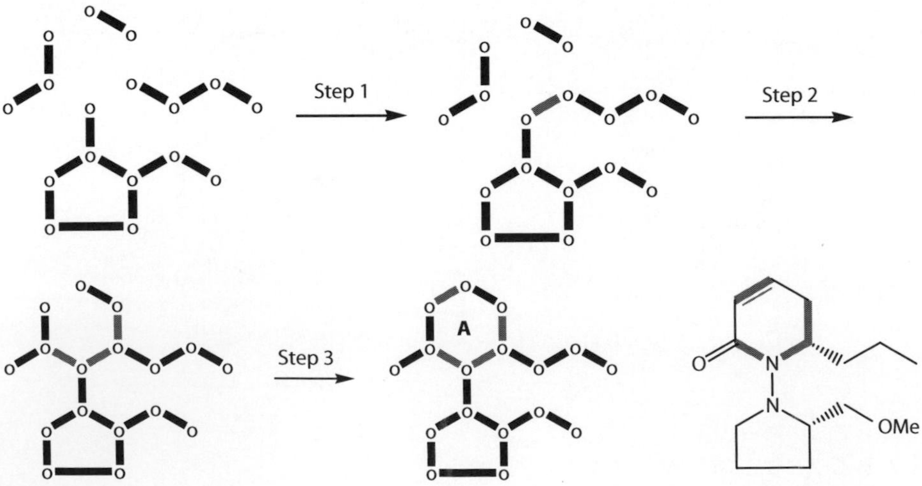

Sequence: $\xrightarrow{[6+0]} A_3$

(R)-Coniine
Lathbury, D.; Gallagher, T. *Chem. Commun.* 1986, 114.

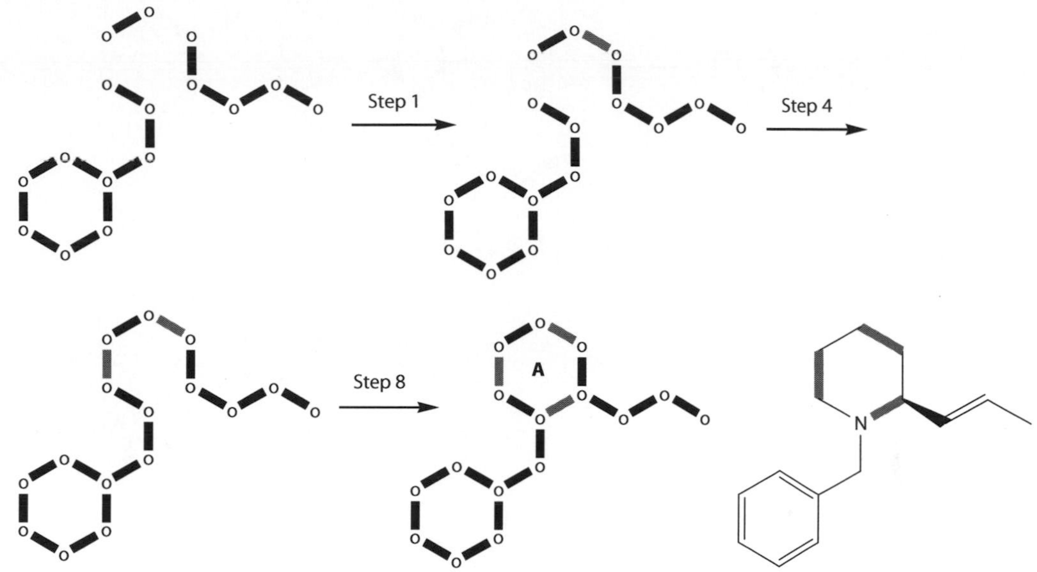

Sequence: $\xrightarrow{[6+0]} A_8$

(S)-Coniine
Bois, F.; Gardette, D.; Gramain, J.C. *Tetrahedron Lett.* 2000, *41*, 8769.

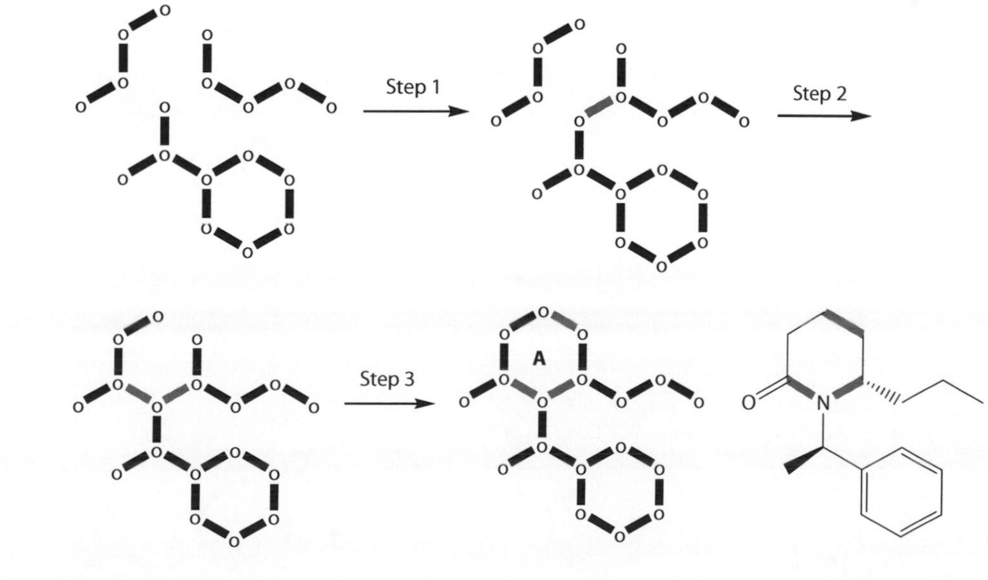

Sequence: $\xrightarrow{[6+0]} A_3$

(S)-Coniine
Girard, N.; Pouchain, L.; Hurvois, J.P.; Moinet, C. *Synlett* 2006, 1679.

Sequence: $\xrightarrow{[5+1]} A_1$

(S)-Coniine

Guerrier, L.; Royer, J.; Grierson, D.S.; Husson, H.P. *J. Am. Chem. Soc.* 1983, *105*, 7754.

Sequence: $\xrightarrow{[5+1]} A_1$

(R)-Coniine

Ito, M.; Maeda, M.; Kibayashi, C. *Tetrahedron Lett.* 1992, *33*, 3765.

(S)-Coniine

Gommermann, N.; Knochel, P. *Chem. Commun.* 2004, 2324.
Gommermann, N.; Knochel, P. *Chem. Eur. J.* 2006, *12*, 4380.

Sequence: $\xrightarrow{[6+0]} A_6$

(±)-Coniine
Ladenburg, A. *Chem. Ber.* 1886, *19*, 439.

(*S*)-Coniine
Munchhof, M.J.; Meyers, A.I. *J. Org. Chem.* 1995, *60*, 7084.

Sequence: $\xrightarrow{[5+1]} A_3$

(±)-Coniine
Nagasaka, T.; Hayashi, H.; Hamaguchi, F. *Heterocycles* 1988, *27*, 1685.
Nagasaka, T.; Tamano, H.; Hamaguchi, F. *Heterocycles* 1986, *24*, 1231.

(*R*)-Coniine
Pandey, G.; Das, P. *Tetrahedron Lett.* 1997, *38*, 9073.

(S)-Coniine

Hayes, J.F.; Shipman, M.; Twin, H. *Chem. Commun.* 2001, 1784.
Ince, J.; Ross, T.M.; Shipman, M.; Slawin, A.M.Z. *Tetrahedron* 1996, *52*, 7037.

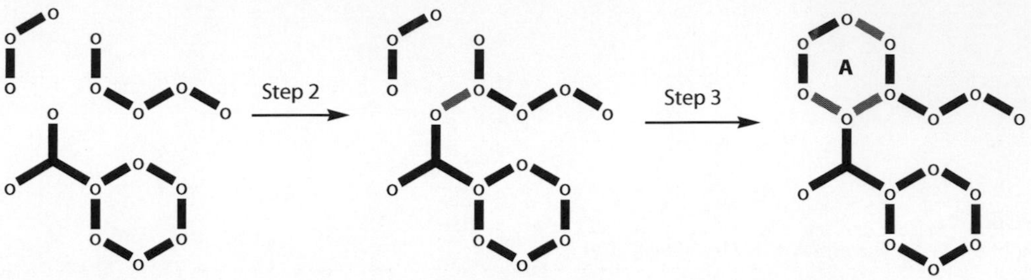

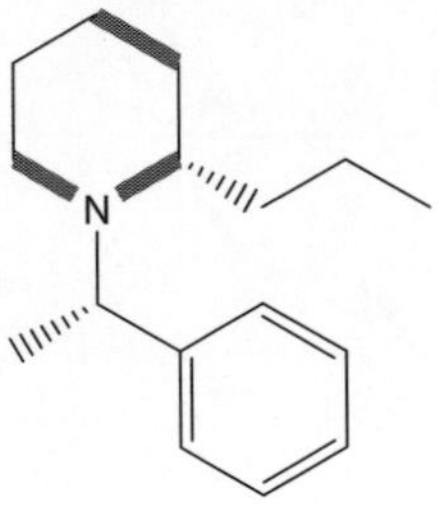

Sequence: $\xrightarrow{[3+3]} A_3$

(S)-Coniine

Takahata, H.; Kubota, M.; Takahashi, S.; Momose, T. *Tetrahedron Asymm.* 1996, *7*, 3047.

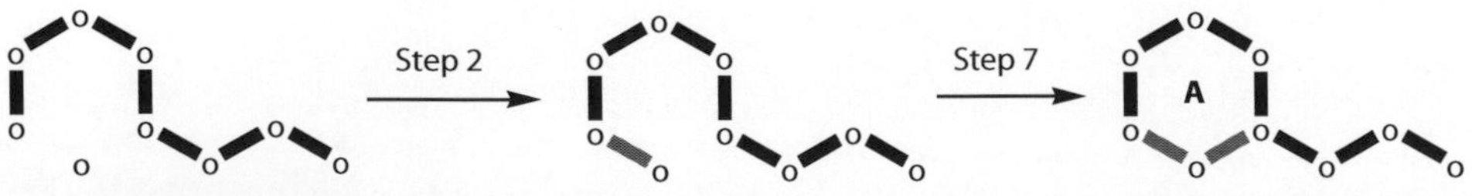

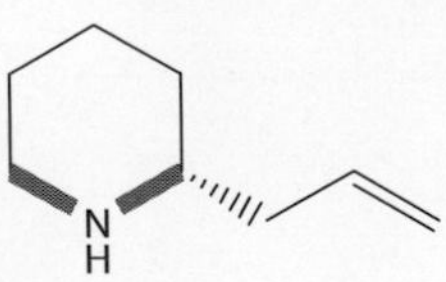

Sequence: $\xrightarrow{[6+0]} A_7$

(±)-Coniine

Bourgeois, J.; Dion, I.; Cebrowski, P.H.; Loiseau, F.; Bédard, A.C.; Beauchemin, A.M. *J. Am. Chem. Soc.* 2009, *131*, 874.

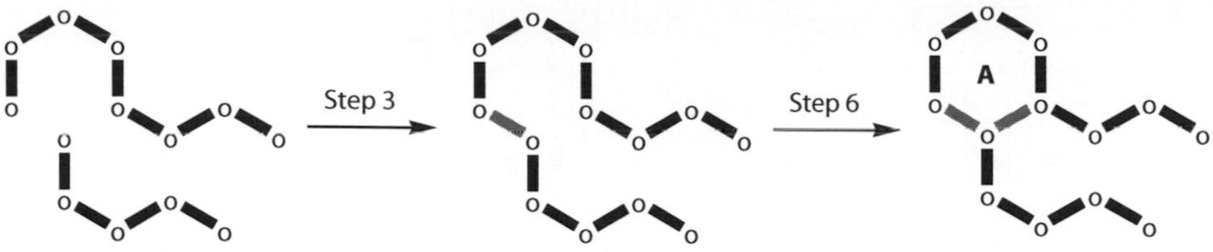

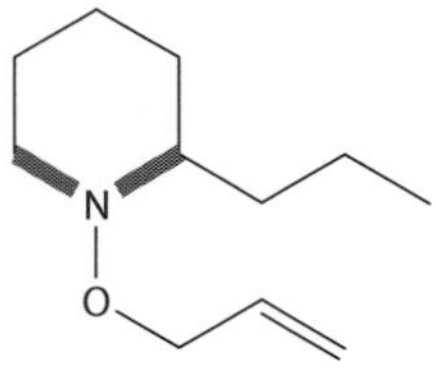

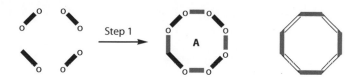

Sequence: $\xrightarrow{[6+0]} A_6$

Summary:

Strategy [6 + 0]: Aketa, Beauchemin, Couture, Gallagher, Gramain, Knochel, Takahata
Strategy [5 + 1]: Buchwald, Hurvois, Husson, Meyers
Strategy [3 + 3]: Shipman
Strategy with precursors containing six-membered ring:
Kibayashi, Ladenburg, Nagasaka, Pandey

Cyclooctatetraene

Reppe, W.; Schlichting, O.; Klager, K.; Toepel, T. *Ann. Chem.* 1948, *560*, 1.

Sequence: $\xrightarrow{[2+2+2+2]} A_1$

Cyclooctatetraene

Willstatter, R.; Waser, E. *Chem. Ber.* 1911, *44*, 3423.
Willstatter, R.; Heidelberger, M. *Chem. Ber.* 1913, *46*, 517.
Cope, A.C.; Overberger, C.G. *J. Am. Chem. Soc.* 1948, *70*, 1433.

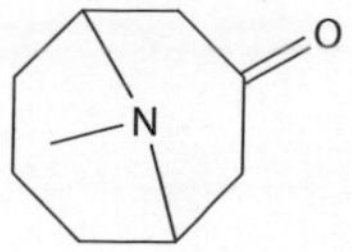

(−)-Cylindricine C
Molander, G.A.; Rönn, M. *J. Org. Chem.* 1999, *64*, 5183.

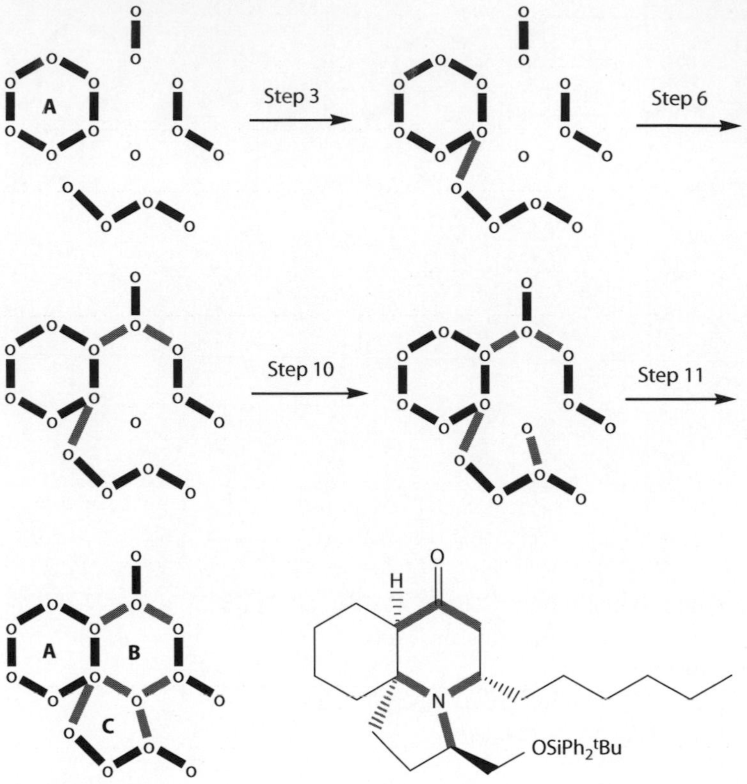

Sequence: $A_0 \xrightarrow{[(5+1)_B + (5+0)_C]} (B+C)_{11}$

(±)-Cyclindricine C
Trost, B.M.; Rudd, M.T. *Org. Lett.* 2003, *5*, 4599.

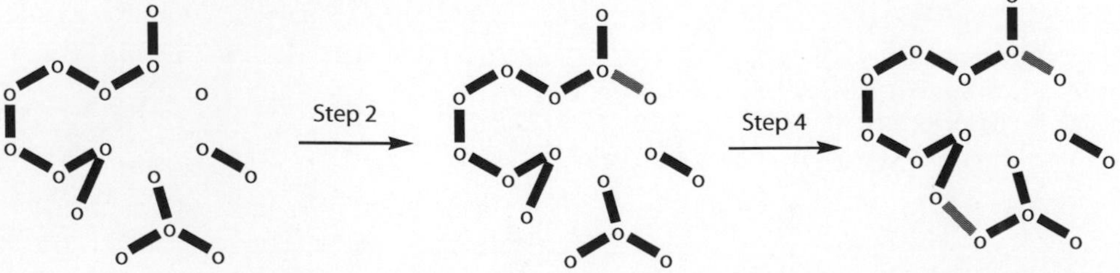

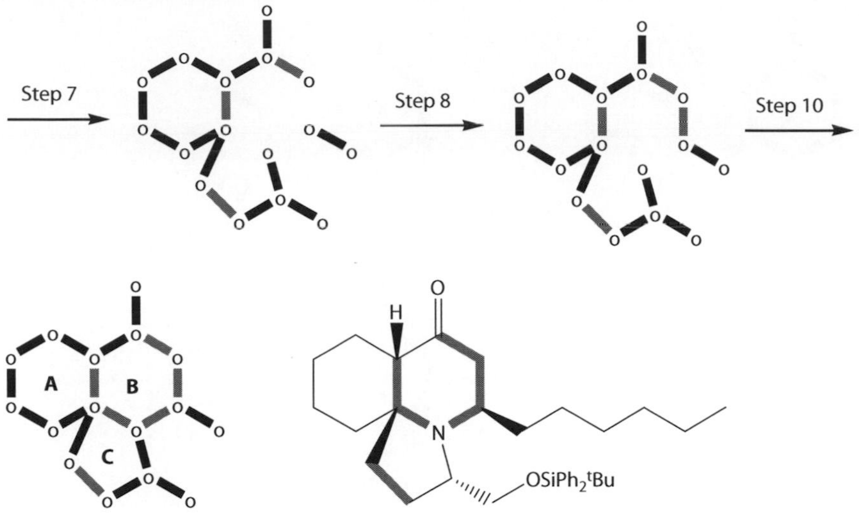

Sequence: $\xrightarrow{[6+0]} A_7 \xrightarrow{[(5+1)_B + (5+0)_C]} (B+C)_{10}$

Dewar benzene

van Tamelen, E.E.; Pappas, S.P. *J. Am. Chem. Soc.* 1963, *85*, 3297.
van Tamelen, E.E.; Pappas, S.P.; Kirk, K.L. *J. Am. Chem. Soc.* 1971, *93*, 6092.

Sequence: $\xrightarrow{[6] \rightarrow [2.2.0]} (A+B)_2$

Eleutherobin

Castoldi, D.; Caggiano, L.; Panigada, L.; Sharon, O.; Costa, A.M.; Gennari, C. *Chem. Eur. J.* 2006, *12*, 51.
Ceccarelli, S.; Piarulli, U.; Gennari, C. *Tetrahedron Lett.* 1999, *40*, 153.

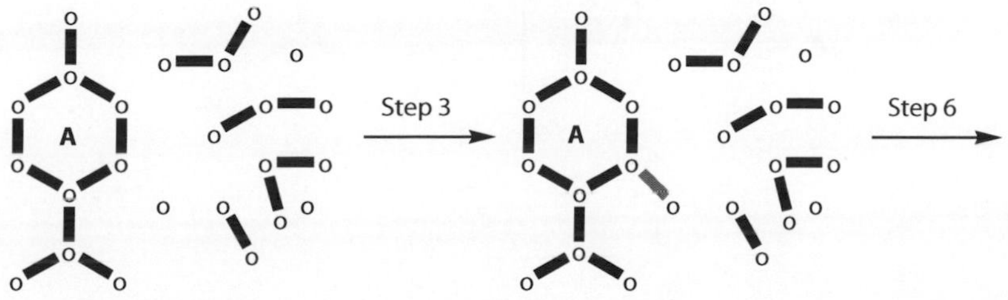

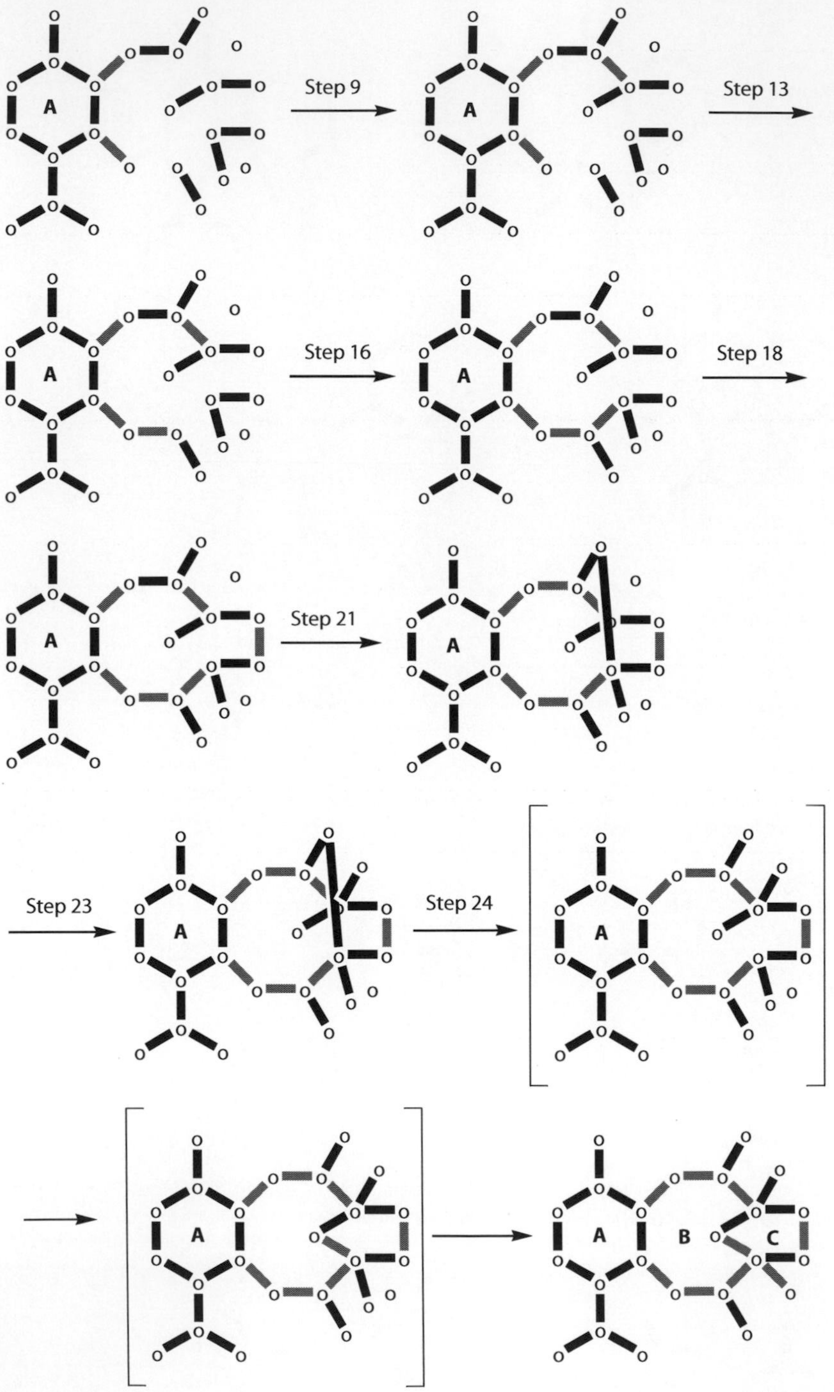
Step 9
Step 13
Step 16
Step 18
Step 21
Step 23
Step 24
A
B
C

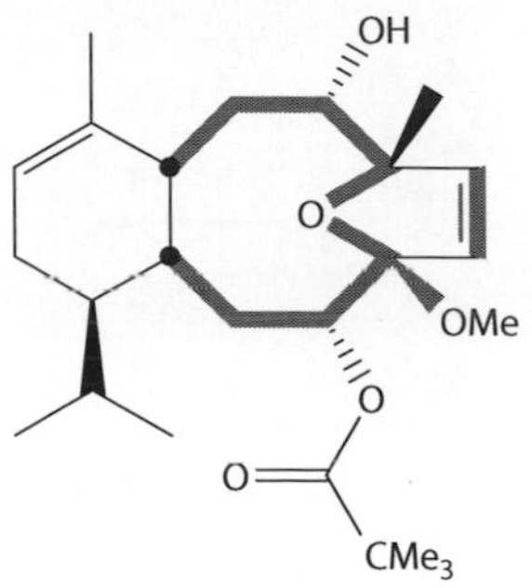

Sequence: $A_0 \xrightarrow{[9+0]_B + [5+0]_C} [6.2.1](B+C)_{24}$

Chen, X.T.; Bhattacharya, S.K.; Zhou, B.; Gutteridge, C.E.; Pettus, T.R.R.; Danishefsky, S.J. *J. Am. Chem. Soc.* 1999, *121*, 6563.

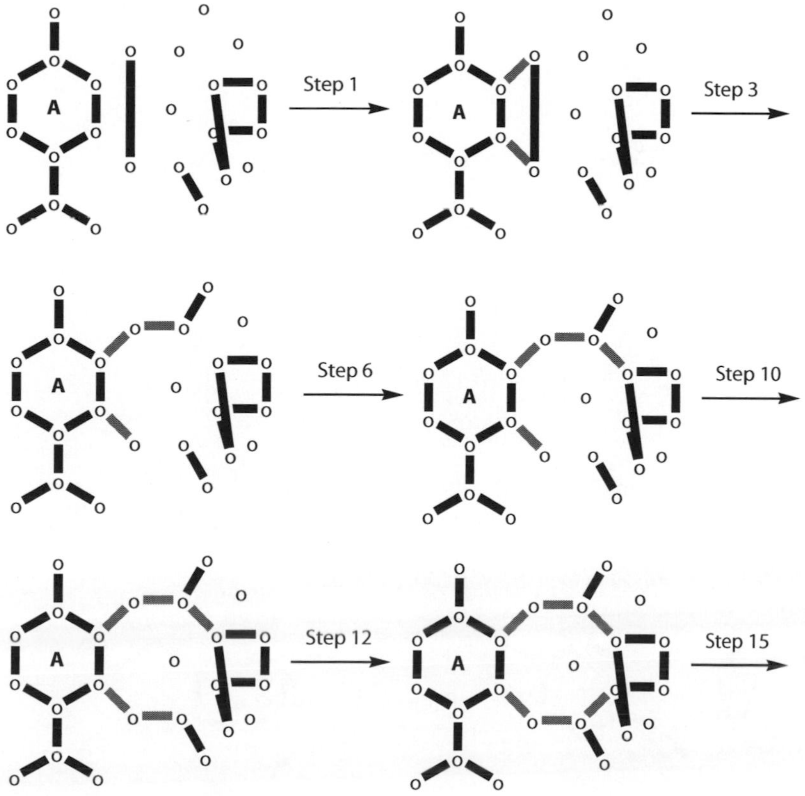

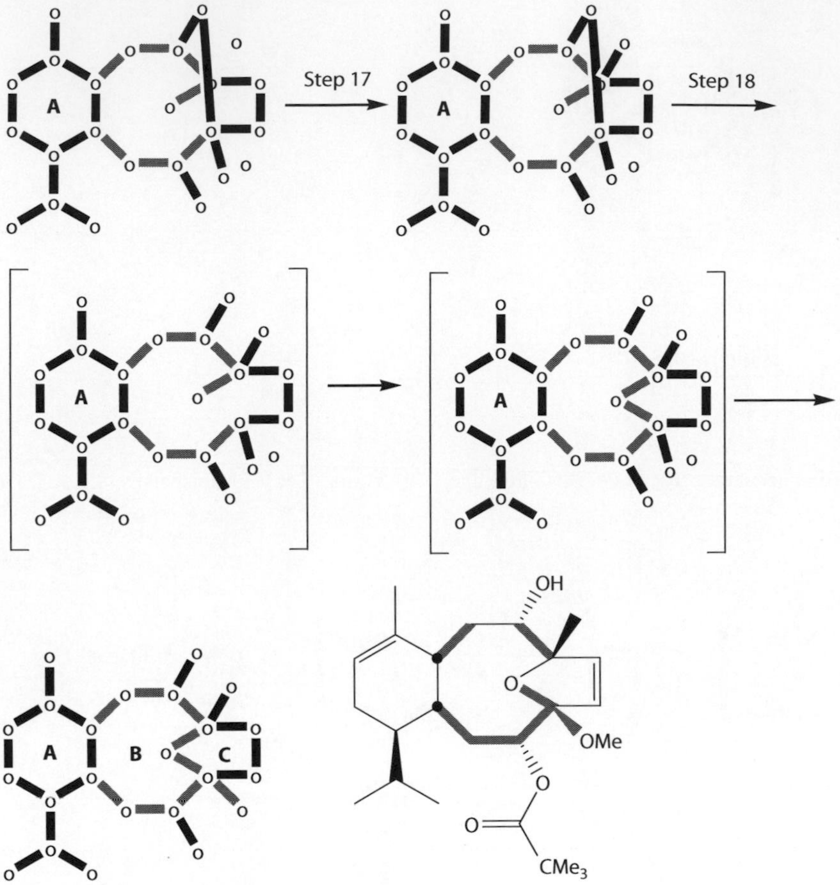

Sequence: $A_0 \xrightarrow{[9+0]_B + [5+0]_C} [6.2.1](B+C)_{18}$

Summary:
Strategy: $A \rightarrow [6.2.1](B+C)$ via $[9+0] + [5+0]$

(−)-Epibatidine
Aggarwal, V.K.; Olofsson, B. *Angew. Chem. Int. Ed.* 2005, *44*, 5516.

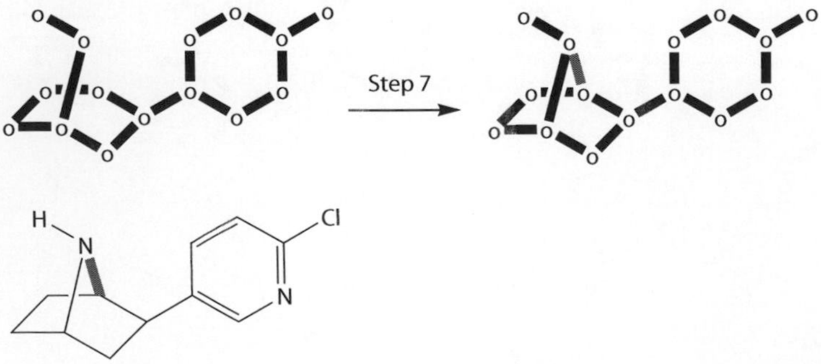

Sequence: $[6] \xrightarrow{[(5+0)+(5+0)]} [2.2.1]_7$

(±)-Epibatidine
Xu, R.; Chu, G.; Bai, D. *Tetrahedron Lett.* 1996, *37*, 1463.

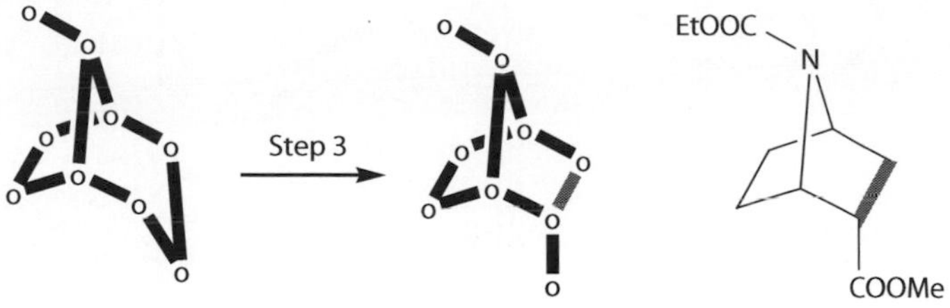

Sequence: $[3.2.1] \rightarrow [2.2.1]_3$

(−)-Epibatidine
Broka, C. *Tetrahedron Lett.* 1993, *34*, 3251.

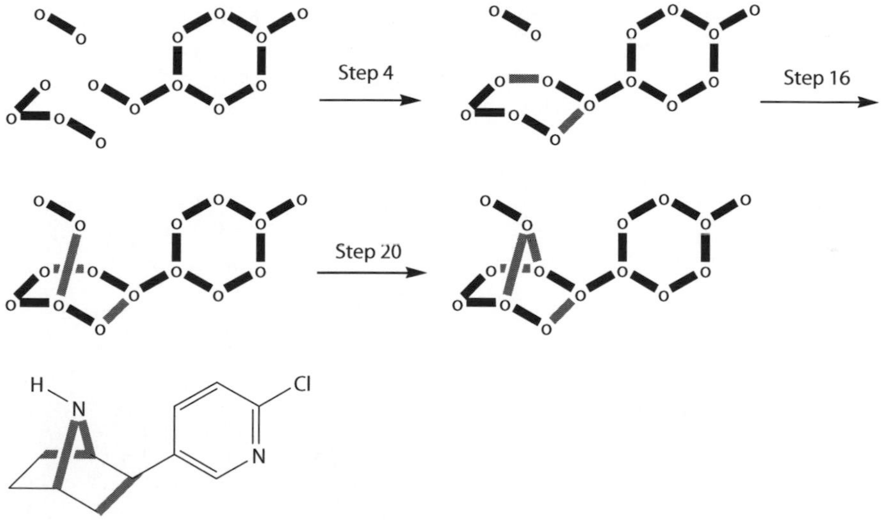

Sequence: $\xrightarrow{[4+2]}[6]_4 \xrightarrow{[(5+0)+(5+0)]}[2.2.1]_{20}$

(−)-Epibatidine
Kotian, P.L.; Carroll, F.I. *Synth. Commun.* 1995, *25*, 63.

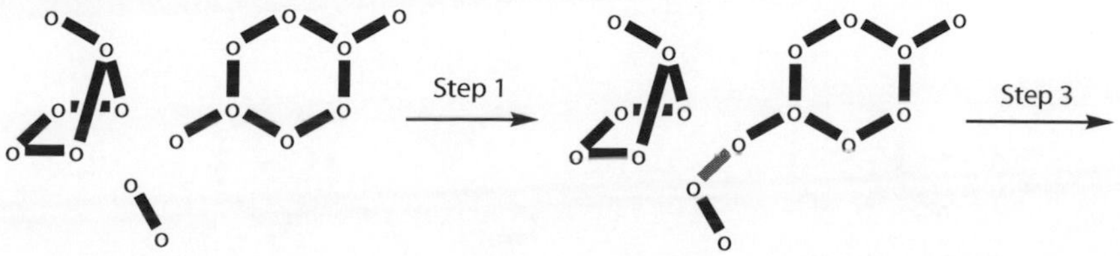

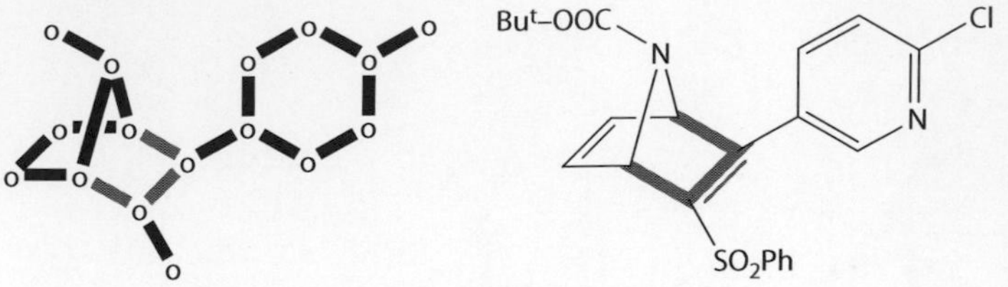

Sequence: $\xrightarrow{[(4+2)+(3+2)]}$ [2.2.1]₃

(–)-Epibatidine
Corey, E.J.; Loh, T.P.; AchyuthaRao, S.; Daley, D.C.; Sarshar, S. *J. Org. Chem.* 1993, *58*, 5600.

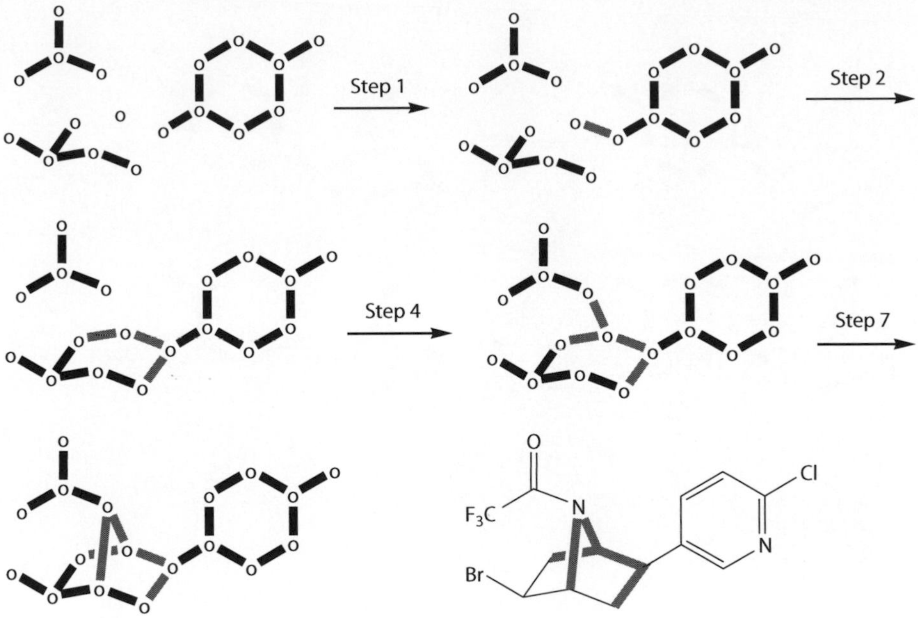

Sequence: $\xrightarrow{[4+2]}$ [6]₂ $\xrightarrow{[(5+0)+(5+0)]}$ [2.2.1]₇

(–)-Epibatidine
Evans, D.A.; Scheidt, K.A.; Downey, C.W. *Org. Lett.* 2001, *3*, 3009.

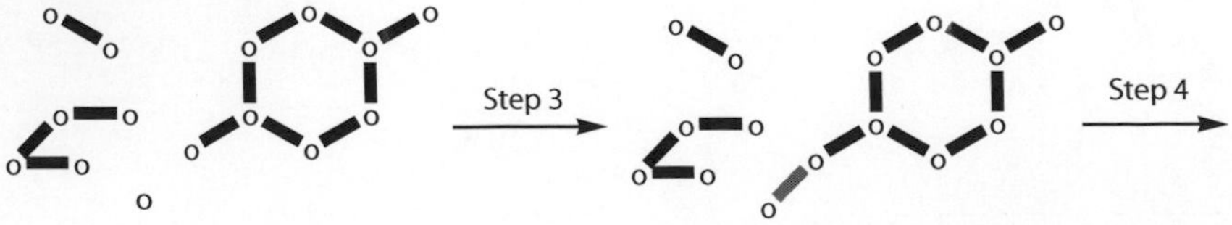

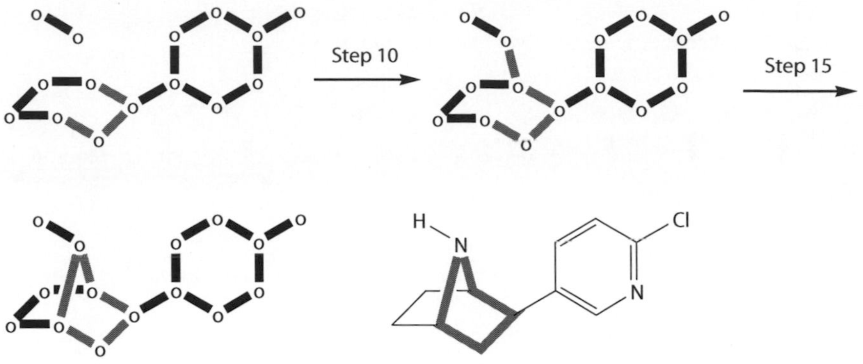

Sequence: $\xrightarrow{[4+2]}$ $[6]_4$ $\xrightarrow{[(5+0)+(5+0)]}$ $[2.2.1]_{15}$

(−)-Epibatidine

Fletcher, S.R.; Baker, R.; Chambers, M.S.; Hobbs, S.C.; Mitchell, P.J. *Chem. Commun.* 1993, 1216.

Fletcher, S.R.; Baker, R.; Chambers, M.S.; Herbert, R.H.; Hobbs, S.C.; Thomas, S.R.; Verrier, H.M.; Watt, A.P.; Ball, R.G. *J. Org. Chem.* 1994, *59*, 1771.

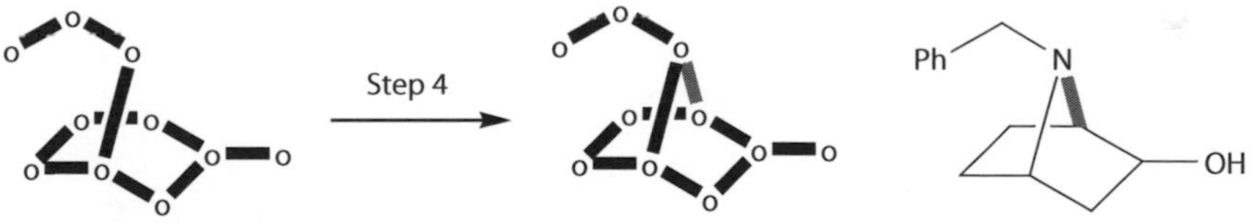

Sequence: $\xrightarrow{[(5+0)+(5+0)]}$ $[2.2.1]_4$

(±)-Epibatidine

Giblin, G.M.P.; Jones, C.D.; Simpkins, N.S. *Synlett* 1997, 589.

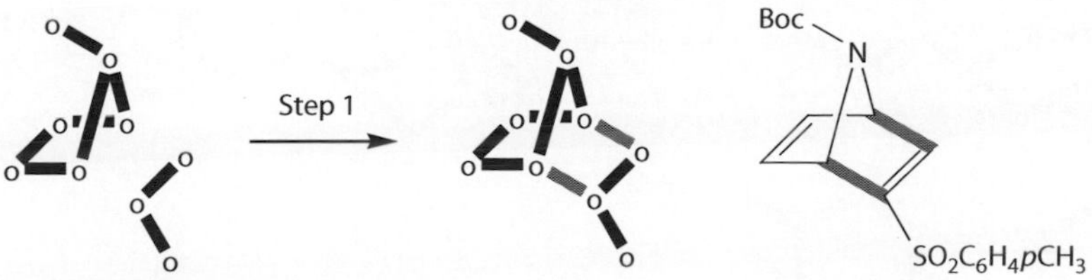

Sequence: $\xrightarrow{[(4+2)+(3+2)]}$ $[2.2.1]_1$

(−)-Epibatidine
Hoashi, Y.; Yabuta, T.; Takemoto, Y. *Tetrahedron Lett.* 2004, *45*, 9185.

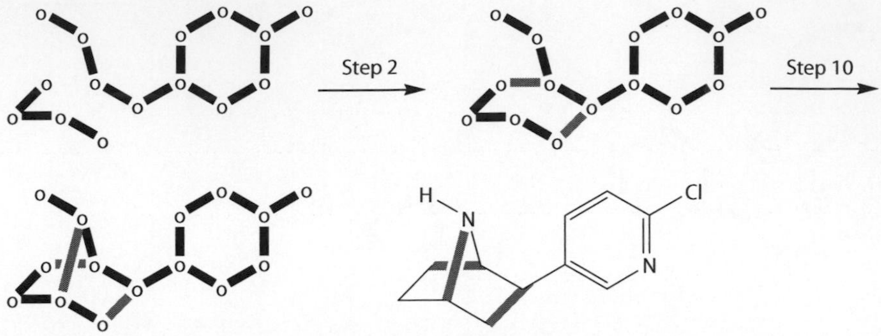

Sequence: $\xrightarrow{[4+2]} [6]_2 \xrightarrow{[(5+0)+(5+0)]} [2.2.1]_{10}$

(−)-Epibatidine
Aoyagi, S.; Tanaka, R.; Naruse, M.; Kibayashi, C. *J. Org. Chem.* 1998, *63*, 8397.

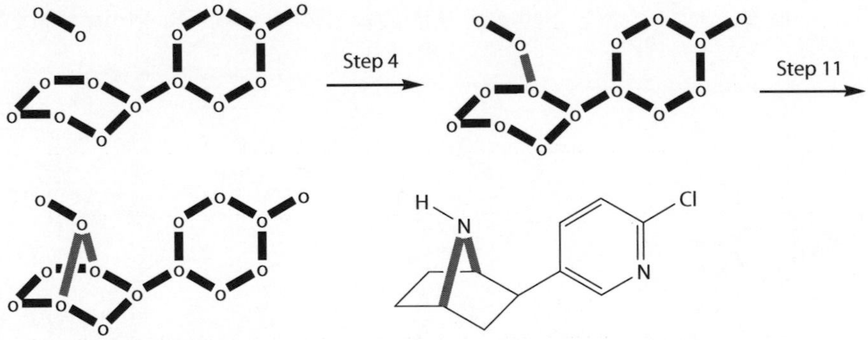

Sequence: $\xrightarrow{[(5+0)+(5+0)]} [2.2.1]_{11}$

(−)-Epibatidine
Kosugi, H.; Abe, M.; Hatsuda, R.; Uda, H.; Kato, M. *Chem. Commun.* 1997, 1857.

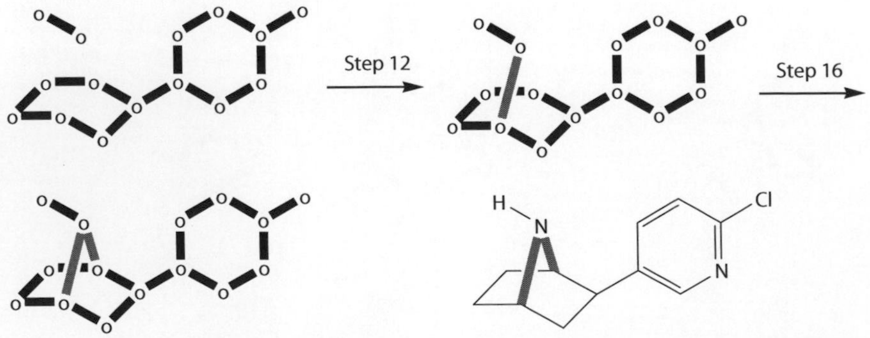

Sequence: $\xrightarrow{[(5+0)+(5+0)]} [2.2.1]_{16}$

(–)-Epibatidine
Lee, C.L.; Loh, T.P. *Org. Lett.* 2005, *7*, 2965.

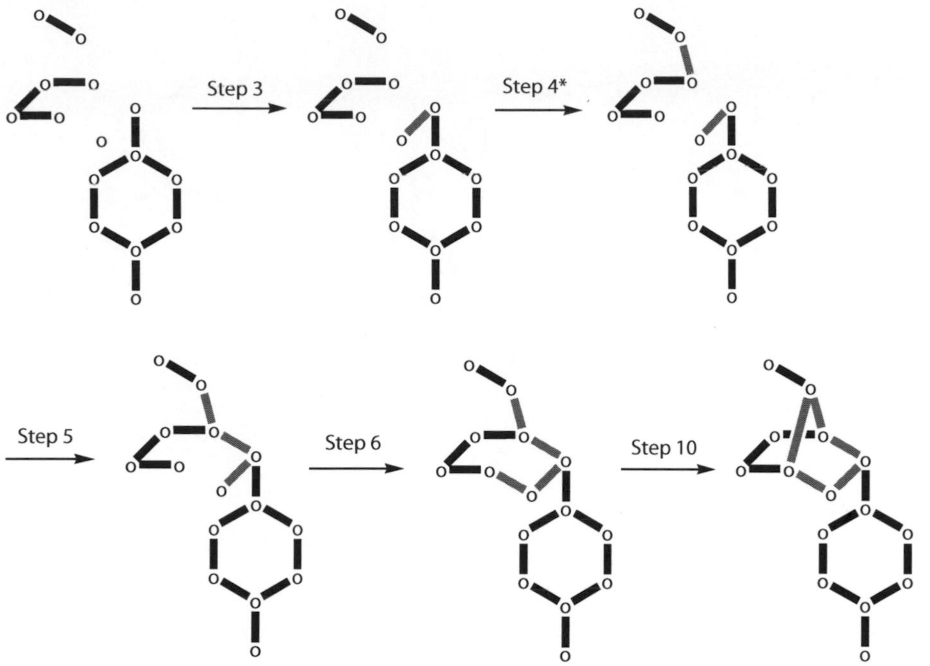

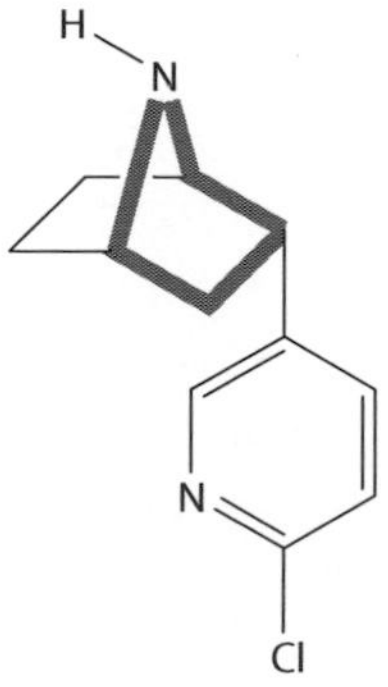

Sequence: $\xrightarrow{[6+0]}$ [6]$_6$ $\xrightarrow{[(5+0)+(5+0)]}$ [2.2.1]$_{10}$

(±)-Epibatidine
Olivo, H.F.; Hemenway, M.S. *J. Org. Chem.* 1999, *64*, 8968.

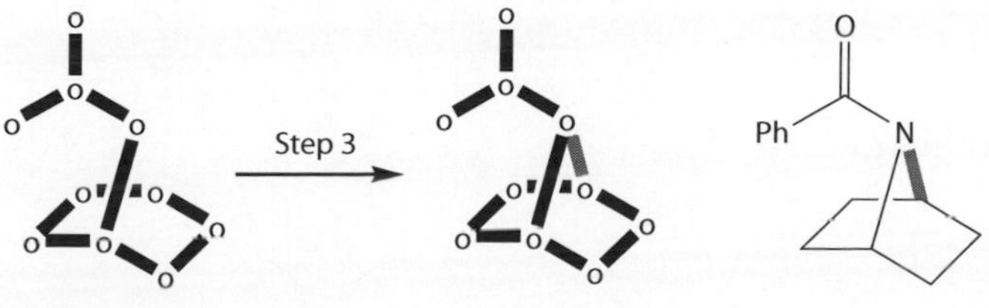

Sequence: $\xrightarrow{[(5+0)+(5+0)]}$ [2.2.1]$_3$

(±)-Epibatidine

Pandey, G.; Bagul, T.D.; Sahoo, A.K. *J. Org. Chem.* 1998, *63*, 760.

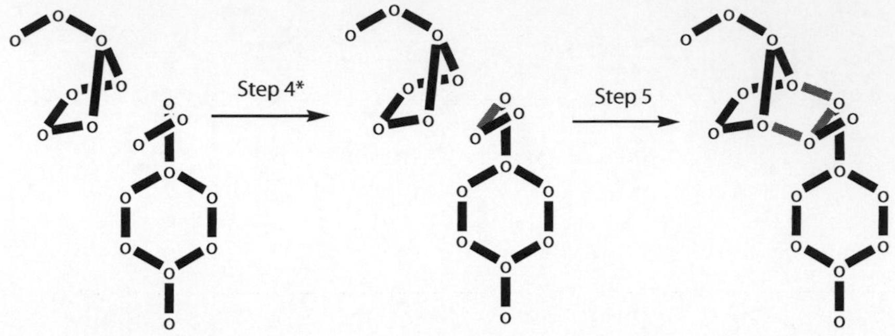

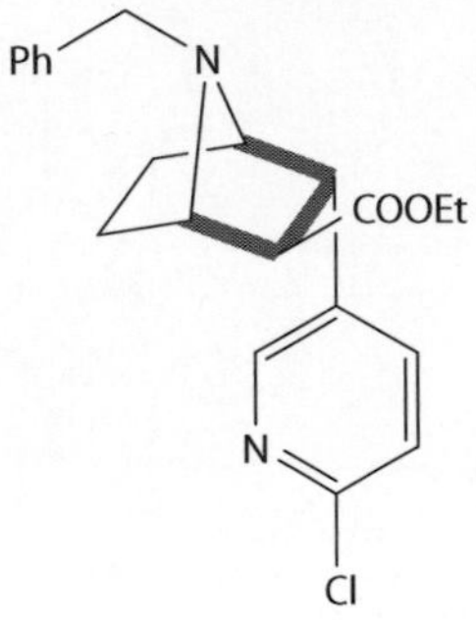

Sequence: $\xrightarrow{[(4+2)+(3+2)]}$ $[2.2.1]_5$

(±)-Epibatidine

Clayton, S.; Regan, A.C. *Tetrahedron Lett.* 1993, *34*, 7493.

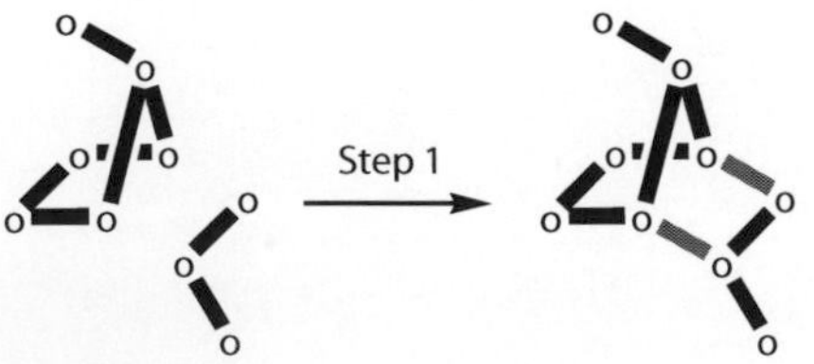

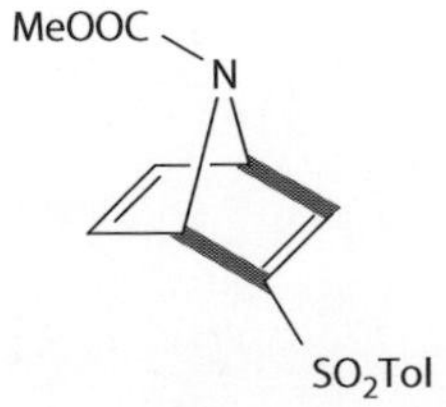

Sequence: $\xrightarrow{[(4+2)+(3+2)]}$ $[2.2.1]_1$

(±)-Epibatidine

Huang, D.F.; Shen, T.Y. *Tetrahedron Lett.* 1993, *34*, 4477.

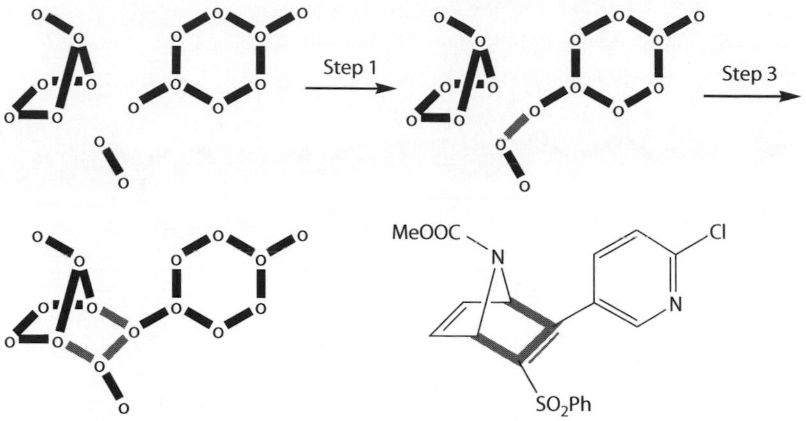

Sequence: $\xrightarrow{[(4+2)+(3+2)]}$ $[2.2.1]_3$

(–)-Epibatidine

Szantay, C.; Kardos-Balogh, Z.; Moldvai, I.; Szantay, C.; Temesvari-Major, E.; Blasko, G. *Tetrahedron* 1996, *52*, 11053.

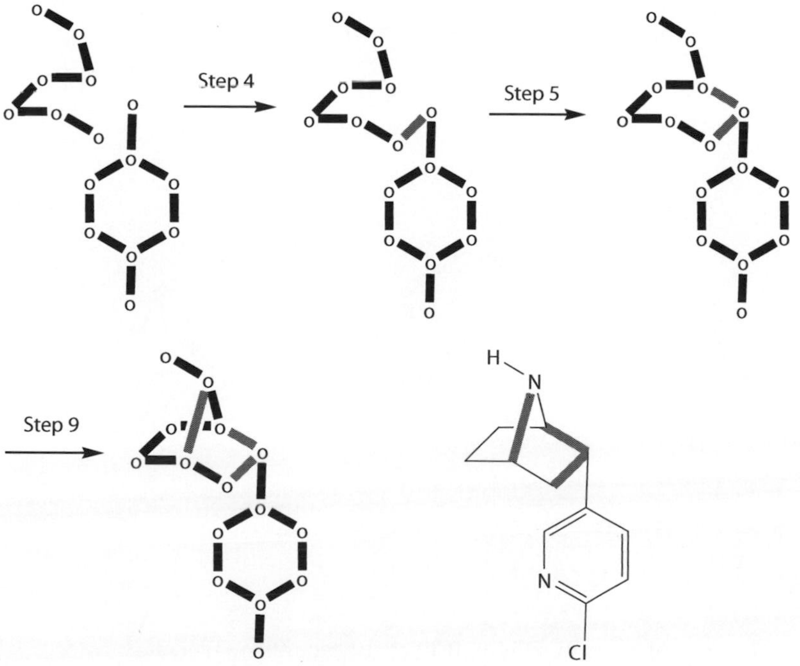

Sequence: $\xrightarrow{[6+0]}$ $[6]_5$ $\xrightarrow{[(5+0)+(5+0)]}$ $[2.2.1]_9$

(−)-Epibatidine

Szantay, C.; Kardos-Balogh, Z.; Moldvai, I.; Szantay, C.; Temesvari-Major, E.; Blasko, G. *Tetrahedron* 1996, *52*, 11053.

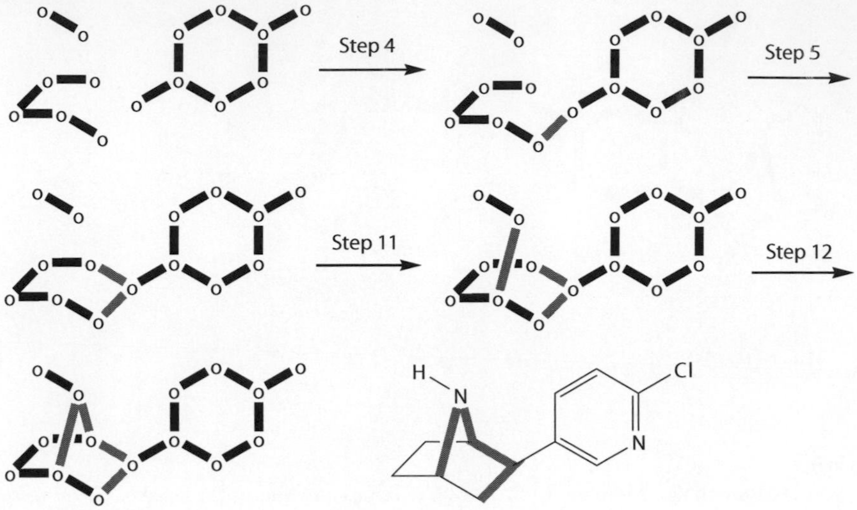

Sequence: $\xrightarrow{[6+0]} [6]_4 \xrightarrow{[(5+0)+(5+0)]} [2.2.1]_{12}$

(±)-Epibatidine

Zhang, C.; Trudell, M.L. *J. Org. Chem.* 1996, *61*, 7189.

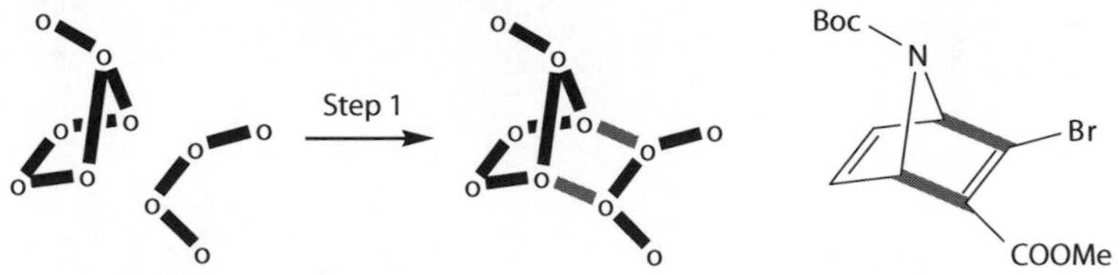

Sequence: $\xrightarrow{[(4+2)+(3+2)]} [2.2.1]_1$

Summary:

Strategy [(4 + 2) + (3 + 2)]: Carroll, Giblin, Pandey, Regan, Shen, Trudell
Strategy [(5 + 0) + (5 + 0)]: Aggarwal, Fletcher, Kibayashi, Kosugi, Olivo
Strategy [6 + 0] + [(5 + 0) + (5 + 0)]: Loh, Szantay G1, Szantay G2
Strategy [4 + 2] + [(5 + 0) + (5 + 0)]: Broka, Corey, Evans, Hoashi
Strategy [3.2.1] to [2.2.1] rearrangement: Bai

Estrone

Bachmann, W.E.; Kushner, S.; Stevenson, A.C. *J. Am. Chem. Soc.* 1942, *64*, 974.

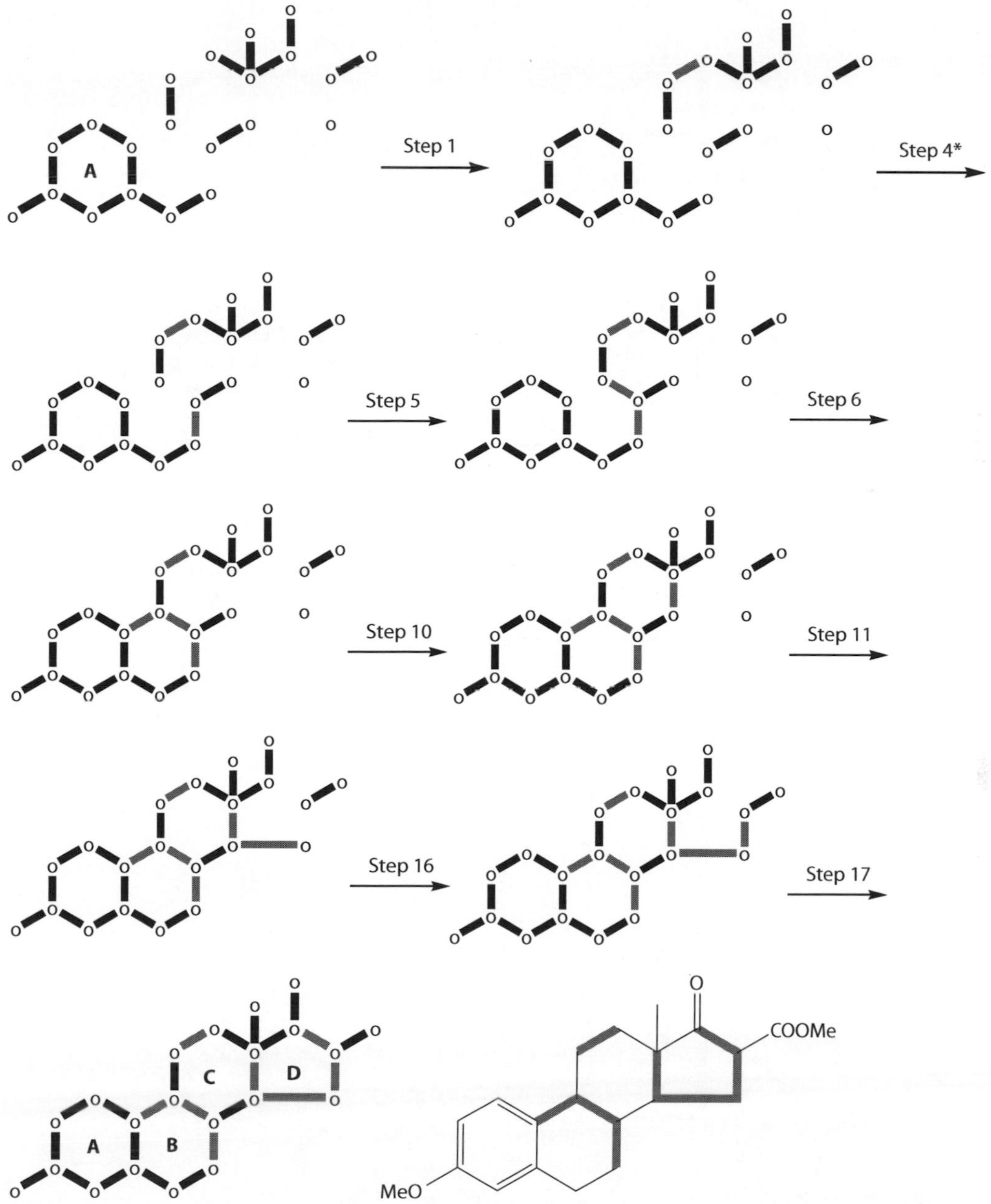

Sequence: $A_0 \xrightarrow{[6+0]} B_6 \xrightarrow{[6+0]} C_{10} \xrightarrow{[5+0]} D_{17}$

(±)-Estrone
Yeung, Y.Y.; Chein, R.J.; Corey, E.J. *J. Am. Chem. Soc.* 2007, *129*, 10346.

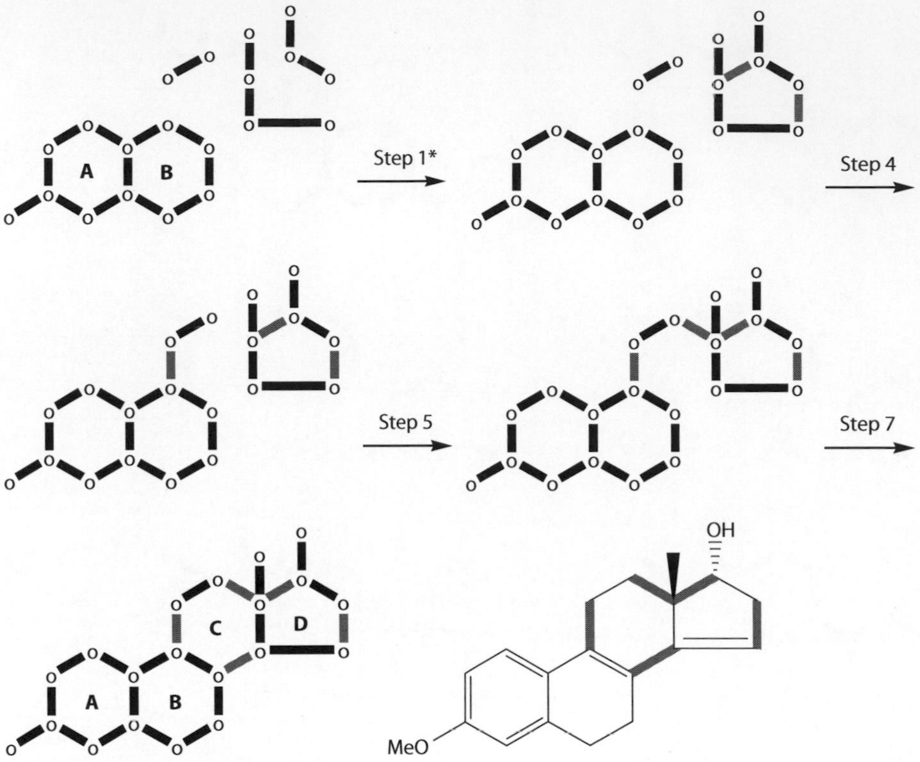

Sequence: $(A+B)_0 \xrightarrow{\quad [3+2] \quad} D_{1*} \xrightarrow{\quad [6+0] \quad} C_7$

(±)-Estrone
Danishefksy, S.; Cain, P. *J. Am. Chem. Soc.* 1976, *98*, 4975.

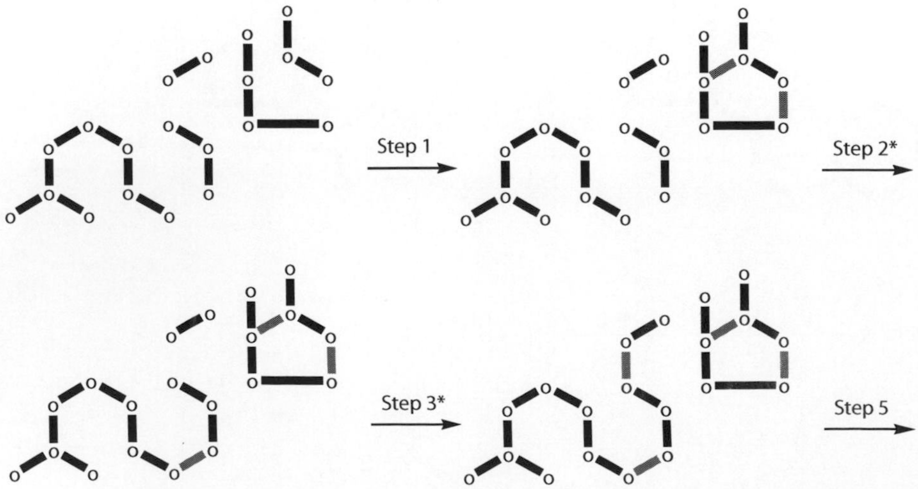

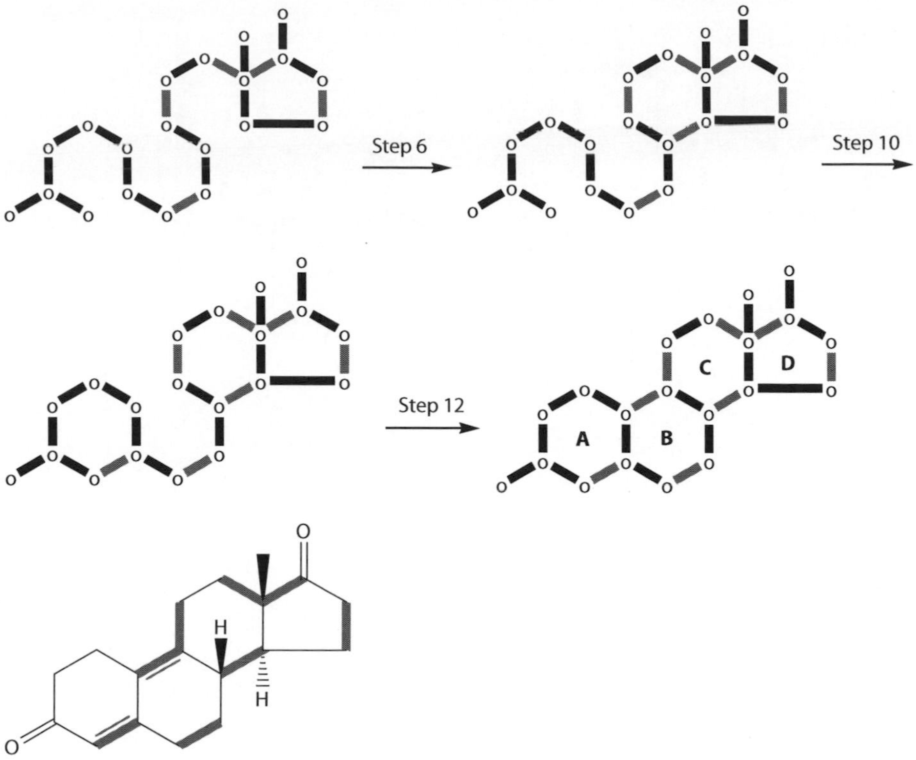

Sequence: $\xrightarrow{[3+2]} D_1 \xrightarrow{[6+0]} C_6 \xrightarrow{[6+0]} A_{10} \xrightarrow{[6+0]} B_{12}$

(±)-Estrone

Grieco, P.A.; Takigawa, T.; Schillinger, W.J. *J. Org. Chem.* 1980, *45*, 2257.

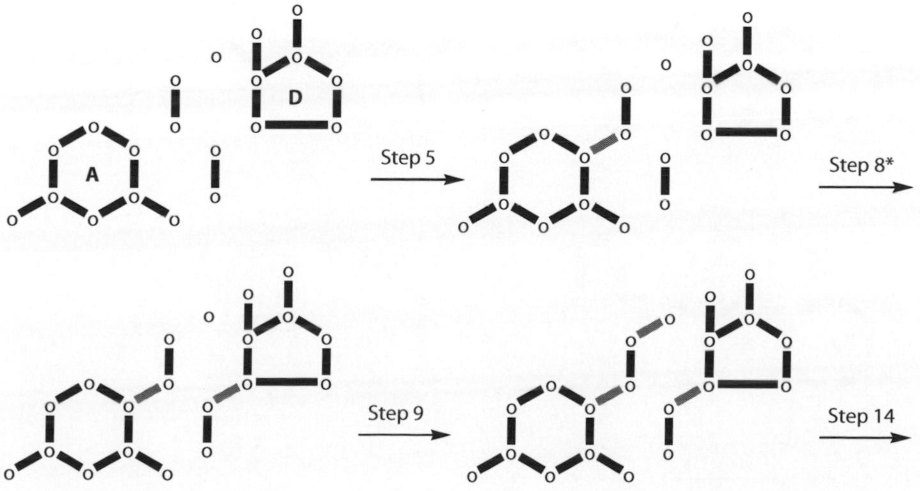

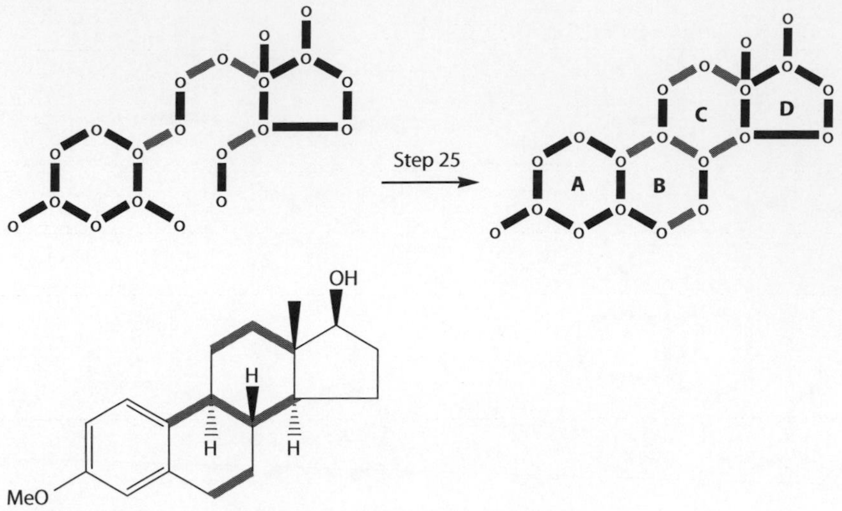

Sequence: $(A + D)_0 \xrightarrow{[(4+2)_B + (6+0)_C]} [4.4.0] (B + C)_{25}$

(±)-Estrone

Johnson, W.S.; Banerjee, D.K.; Schneider, W.P.; Gutsche, C.D.; Shelberg, W.E.; Chinn, L.J. *J. Am. Chem. Soc.* 1952, *74*, 2832.

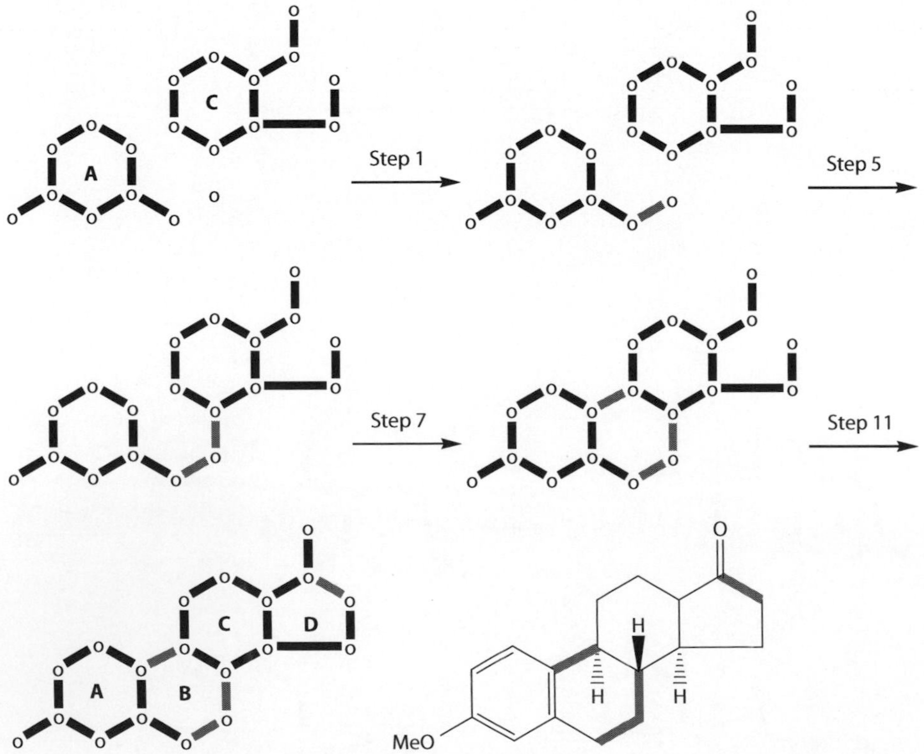

Sequence: $(A + C)_0 \xrightarrow{[6+0]} B_7 \xrightarrow{[5+0]} D_{11}$

(±)-Estrone

Bartlett, P.A.; Johnson, W.S. *J. Am. Chem. Soc.* 1973, *95*, 7501.

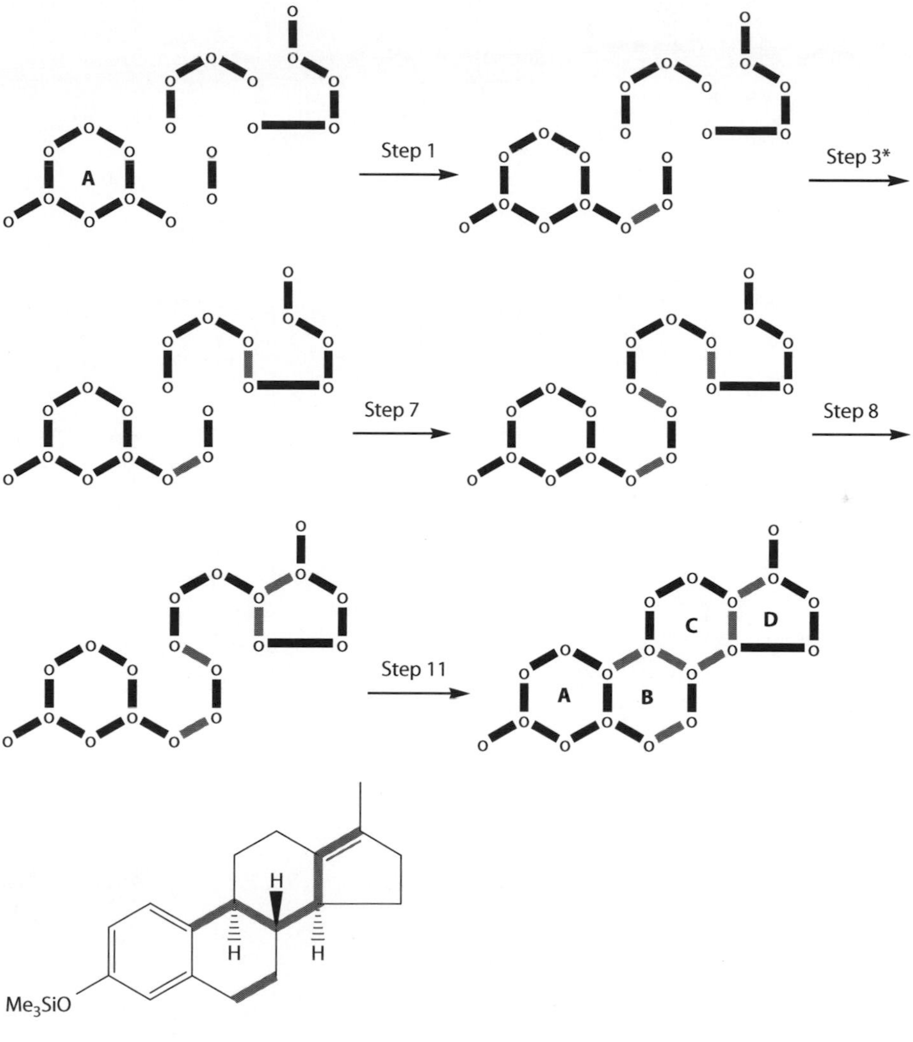

Sequence: $A_0 \xrightarrow{[5+0]} D_8 \xrightarrow{[(6+0)_B + (6+0)_C]} [4.4.0] (B+C)_{11}$

(±)-Estrone

Kametani, T.; Nemoto, H.; Ishikawa, H.; Shiroyama, K.; Hiroo, M.; Fukumoto, K. *J. Am. Chem. Soc.* 1977, *99*, 3461.

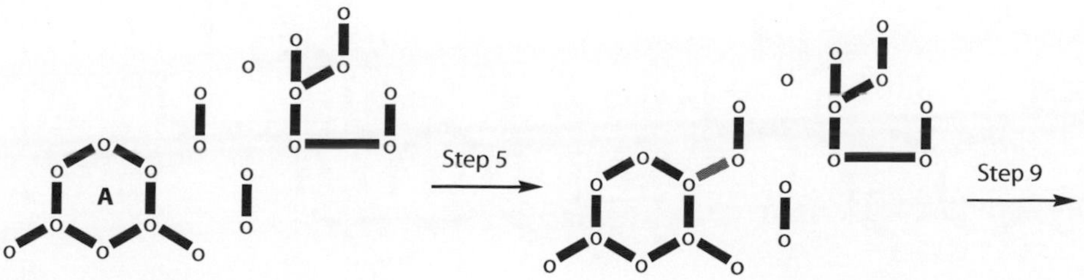

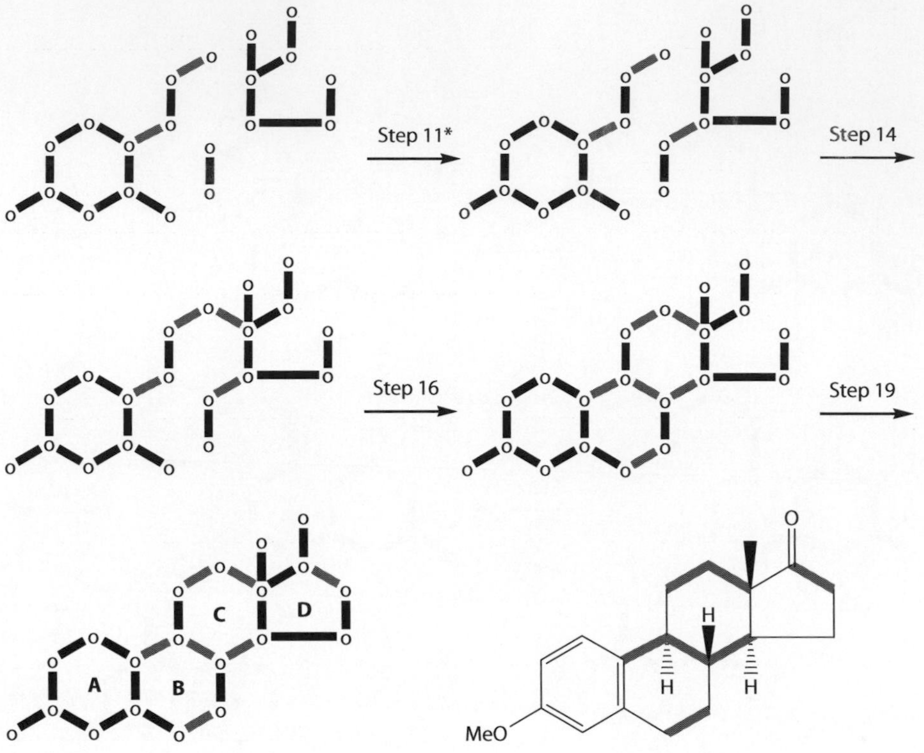

Sequence: $A_0 \xrightarrow{[(4+2)_B+(6+0)_C]} [4.4.0]\,(B+C)_{16} \xrightarrow{[5+0]} D_{19}$

(±)-Estrone
Pattenden, G.; Gonzalez, M.A.; McCulloch, S.; Walter, A.; Woodhead, S.J. *Proc. Natl. Acad. Sci. USA* 2004, *101*, 12024.

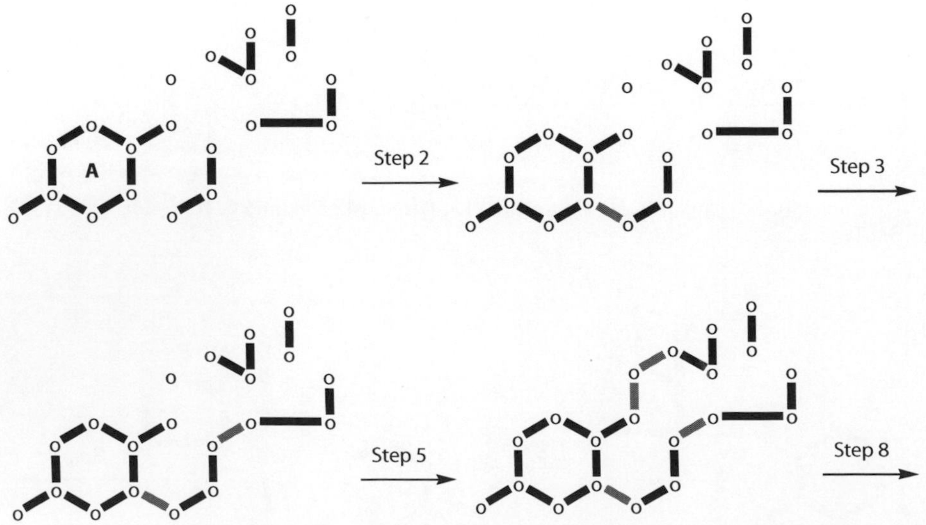

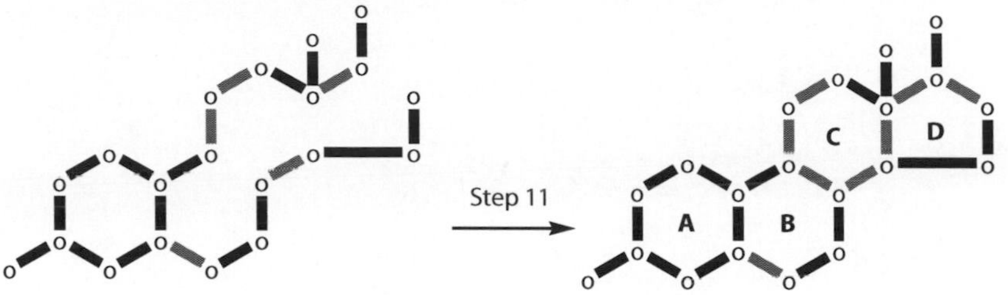

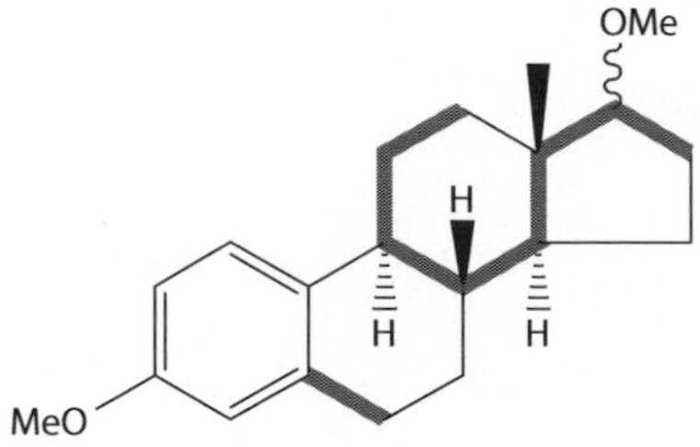

Sequence: $A_0 \xrightarrow{\ [(6+0)_B + (4+2)_C + (3+2)_D]\ } ([4.4.0]_{BC} + [4.3.0]_{CD}) \, (B+C+D)_1$

(±)-Estrone

Douglas, G.H.; Graves, J.M.H.; Hartley, D.; Hughes, G.A.; McLoughlin, B.J.; Siddall, J.; Smith, H. *J. Chem. Soc.* 1963, 5072.

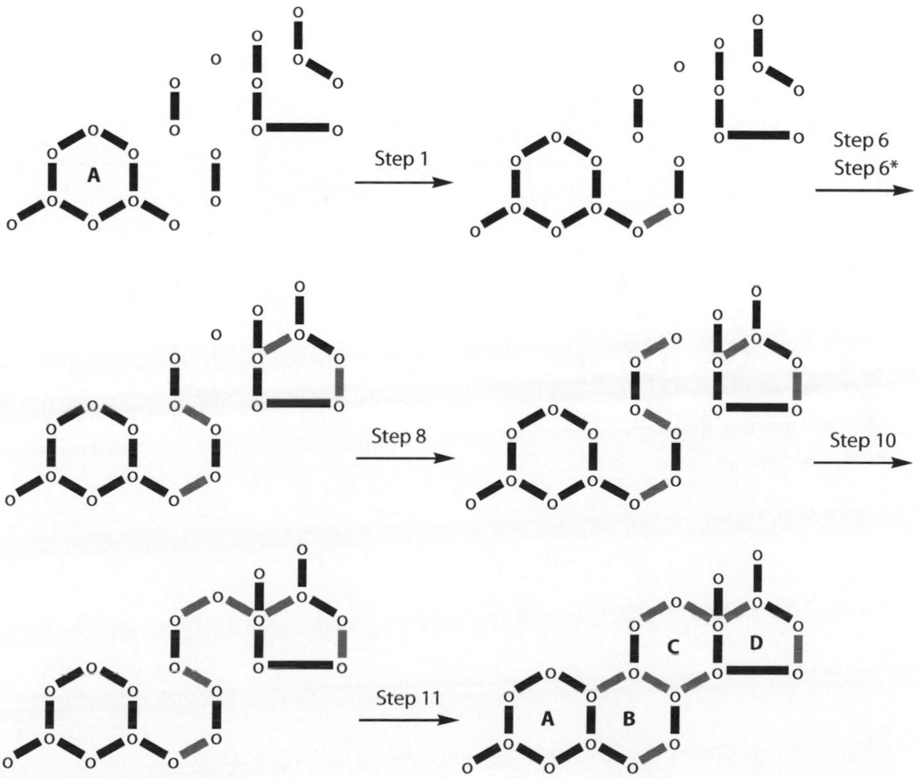

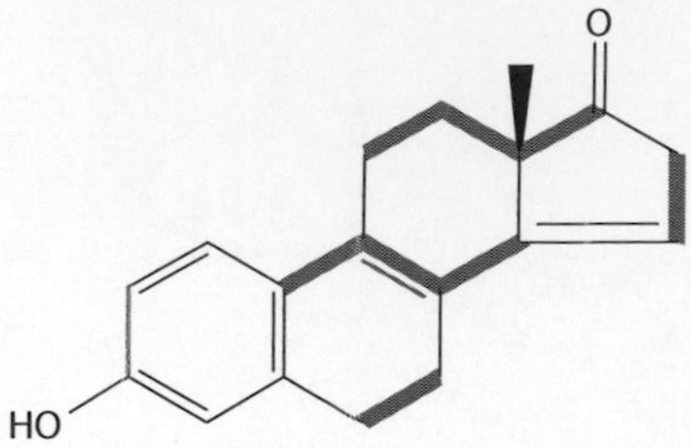

Sequence: $A_0 \xrightarrow{[3+2]} D_{6*} \xrightarrow{[(6+0)_B + (6+0)_C]} [4.4.0]\,(B+C)_{11}$

(±)-Estrone

Ananchenko, S.N.; Torgov, I.V. *Tetrahedron Lett*. 1963, 1553.
Ananchenko, S.N.; Limanov, V.Y.; Leonov, V.N.; Rzheznikov, V.N.; Torgov, I.V. *Tetrahedron* 1962, *18*, 1355.

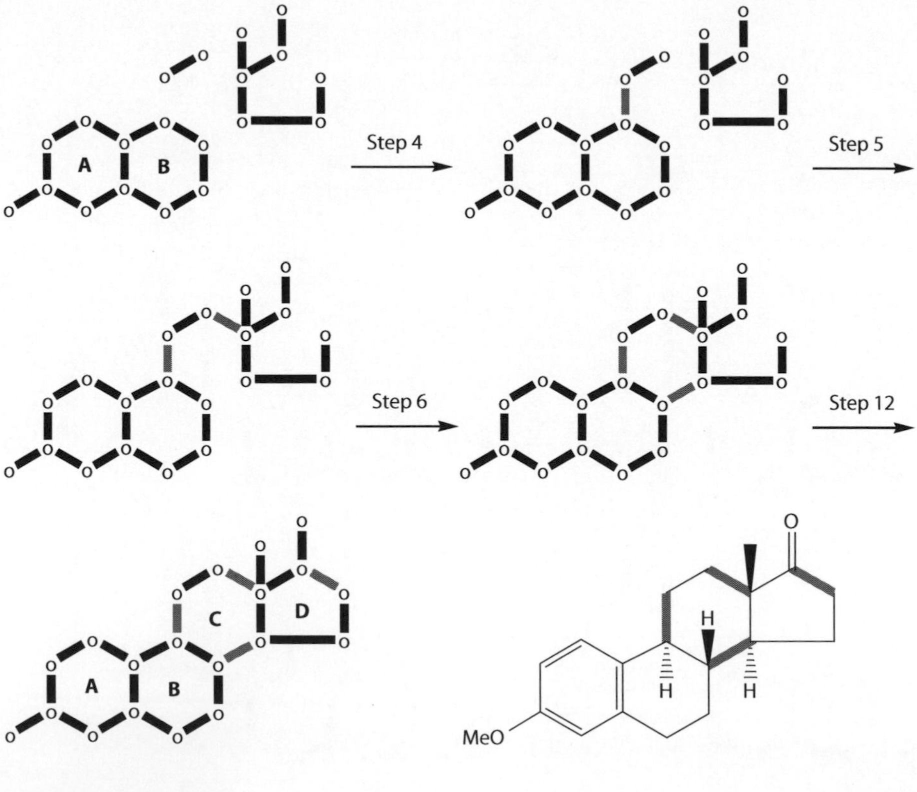

Sequence: $(A+B)_0 \xrightarrow{[6+0]} C_6 \xrightarrow{[5+0]} D_{12}$

Summary:

Strategy #1:

$$A_0 \xrightarrow{[6+0]} B_6 \xrightarrow{[6+0]} C_{10} \xrightarrow{[5+0]} D_{17}: \text{Bachmann}$$

Strategy #2:

$$(A+B)_0 \xrightarrow{[3+2]} D_{1*} \xrightarrow{[6+0]} C_7: \text{Corey}$$

Strategy #3:

$$\xrightarrow{[3+2]} D_1 \xrightarrow{[6+0]} C_6 \xrightarrow{[6+0]} A_{10} \xrightarrow{[6+0]} B_{12}: \text{Danishefsky}$$

Strategy #4:

$$(A+D)_0 \xrightarrow{[(4+2)_B+(6+0)_C]} [4.4.0]\,(B+C)_{25}: \text{Grieco}$$

Strategy #5:

$$(A+C)_0 \xrightarrow{[6+0]} B_7 \xrightarrow{[5+0]} D_{11}: \text{Johnson G1}$$

Strategy #6:

$$A_0 \xrightarrow{[5+0]} D_g \xrightarrow{[(6+0)_B+(6+0)_C]} [4.4.0]\,(B+C)_{11}: \text{Johnson G2}$$

$$A_0 \xrightarrow{[3+2]} D_{6*} \xrightarrow{[(6+0)_B+(6+0)_C]} [4.4.0]\,(B+C)_{11}: \text{Smith}$$

Strategy #7:

$$A_0 \xrightarrow{[(4+2)_B+(6+0)_C]} [4.4.0]\,(B+C)_{16} \xrightarrow{[5+0]} D_{19}: \text{Kametani}$$

Strategy #8:

$$A_0 \xrightarrow{[(6+0)_B+(4+2)_C+(3+2)_D]} ([4.4.0]_{BC}+[4.3.0]_{CD})\,(B+C+D)_{11}: \text{Pattenden}$$

Strategy #9:

$$(A+B)_0 \xrightarrow{[6+0]} C_6 \xrightarrow{[5+0]} D_{12}: \text{Torgov}$$

(±)-Fenchone

Buchbauer, G.; Rohner, H.C. *Ann. Chem.* 1981, 2093.

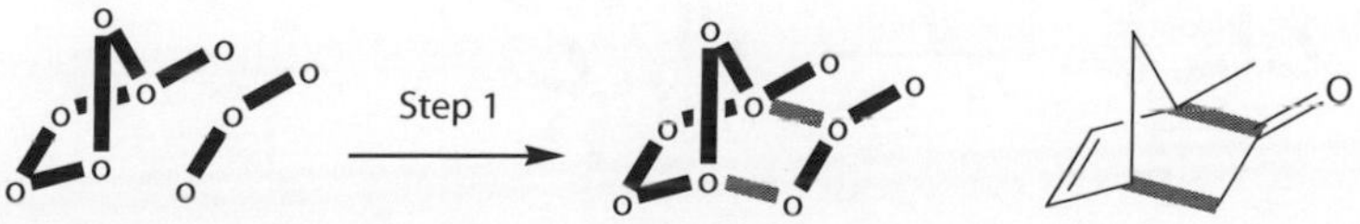

Sequence: $\xrightarrow{[(4+2)+(3+2)]} [2.2.1]_1$

Komppa, G.; Klami, A. *Chem. Ber.* 1935, *68*, 2001.

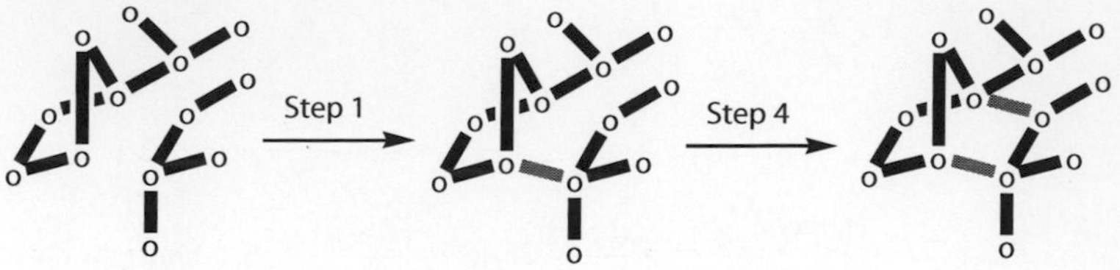

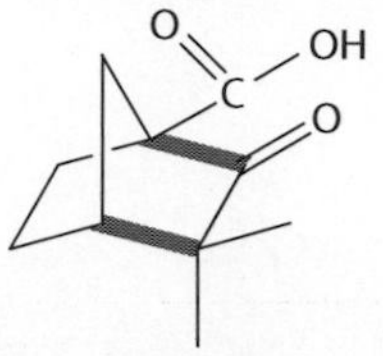

Sequence: $\xrightarrow{[(6+0)+(5+0)]} [2.2.1]_4$

Ruzicka, L. *Chem. Ber.* 1917, *50*, 1362.

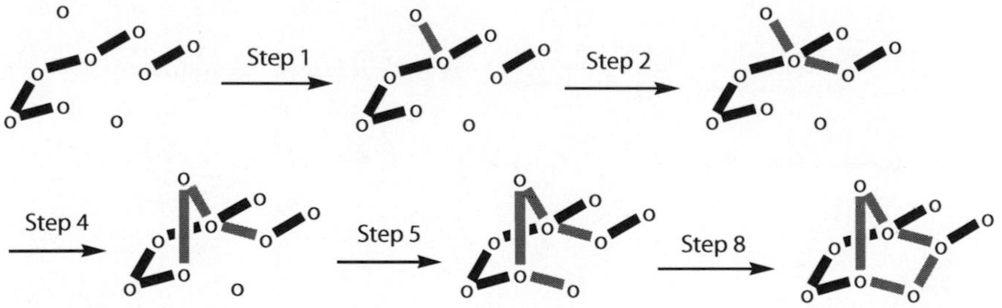

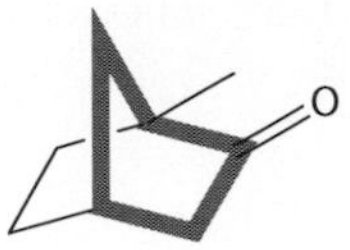

Sequence: $\xrightarrow{[5+0]} [5]_4 \xrightarrow{[(6+0)+(5+0)]} [2.2.1]_8$

(±)-Fenchone
Boyle, P.H.; Cocker, W.; Grayson, D.H.; Shannon, P.V.R. *Chem. Commun.* 1971, 395.
Boyle, P.H.; Cocker, W.; Grayson, D.H.; Shannon, P.V.R. *J. Chem. Soc. (C)* 1971, 2136.

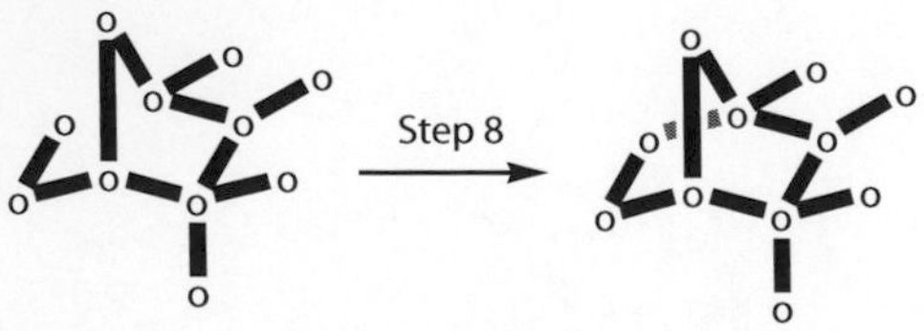

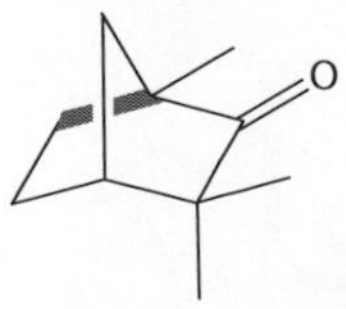

Sequence: $\xrightarrow{[(6+0)+(5+0)]} [2.2.1]_8$

Summary:

Strategy $\xrightarrow{[(4+2)+(3+2)]}$ $[2.2.1]_1$: Buchbauer

Strategy $\xrightarrow{[(6+0)+(5+0)]}$ $[2.2.1]_8$: Cocker

Strategy $\xrightarrow{[(6+0)+(5+0)]}$ $[2.2.1]_4$: Komppa

Strategy $\xrightarrow{[5+0]}$ $[5]_4$ $\xrightarrow{[(6+0)+(5+0)]}$ $[2.2.1]_8$: Ruzicka

α-Himachalene

Wenkert, E.; Naemura, K. *Synth. Commun.* 1973, *3*, 45.

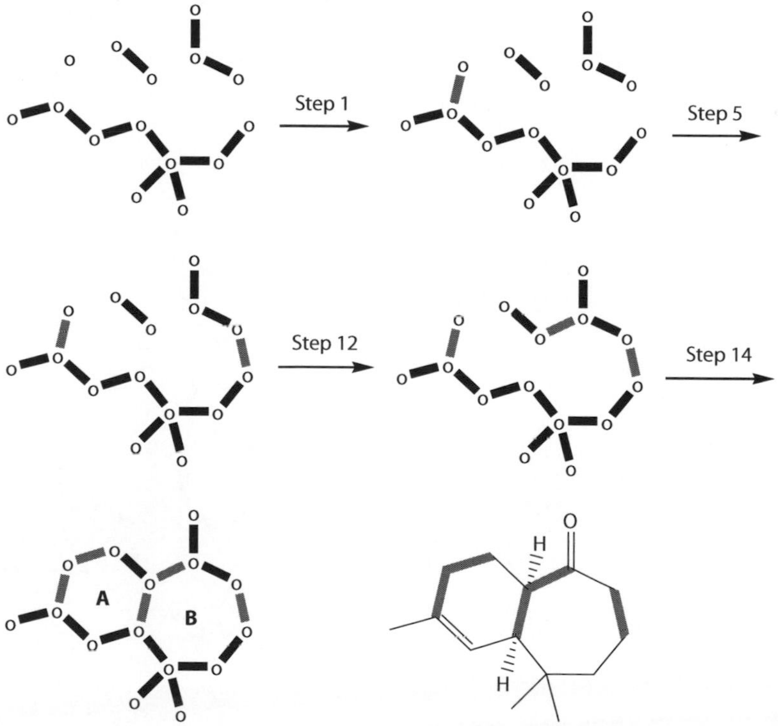

Sequence: $\xrightarrow{[(4+2)_A+(7+0)_B]}$ $[5.4.0]_{14}$

Oppolzer, W.; Snowden, R.L. *Helv. Chim. Acta* 1981, *64*, 2592.

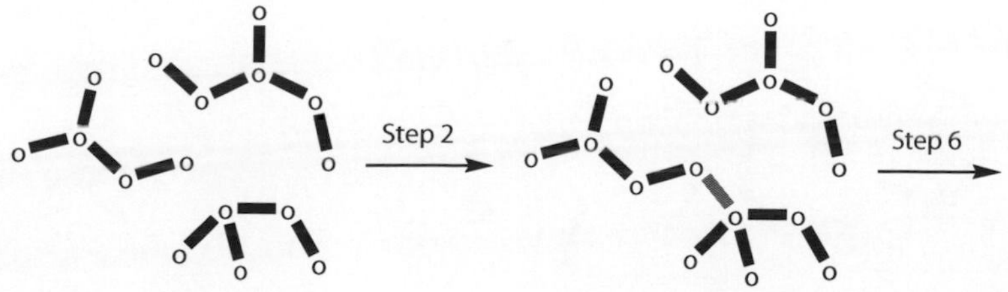

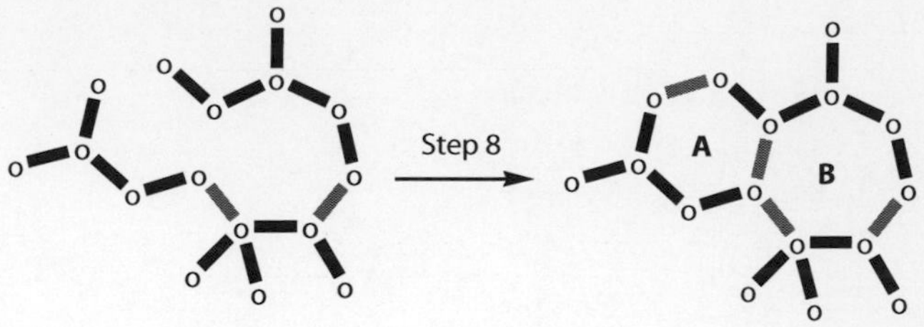

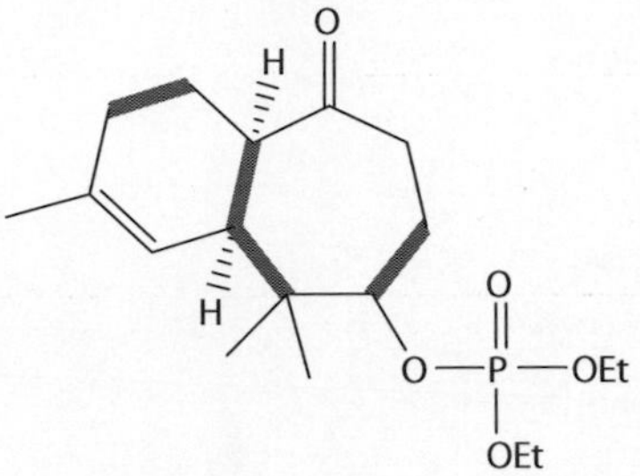

Sequence: $\xrightarrow{\;[(4+2)_A + (7+0)_B]\;}$ $[5.4.0]_8$

Evans, D.A.; Ripin, D.H.B.; Johnson, J.S.; Shaughnessy, E.A. *Angew. Chem. Int. Ed.* 1997, *36*, 2117.

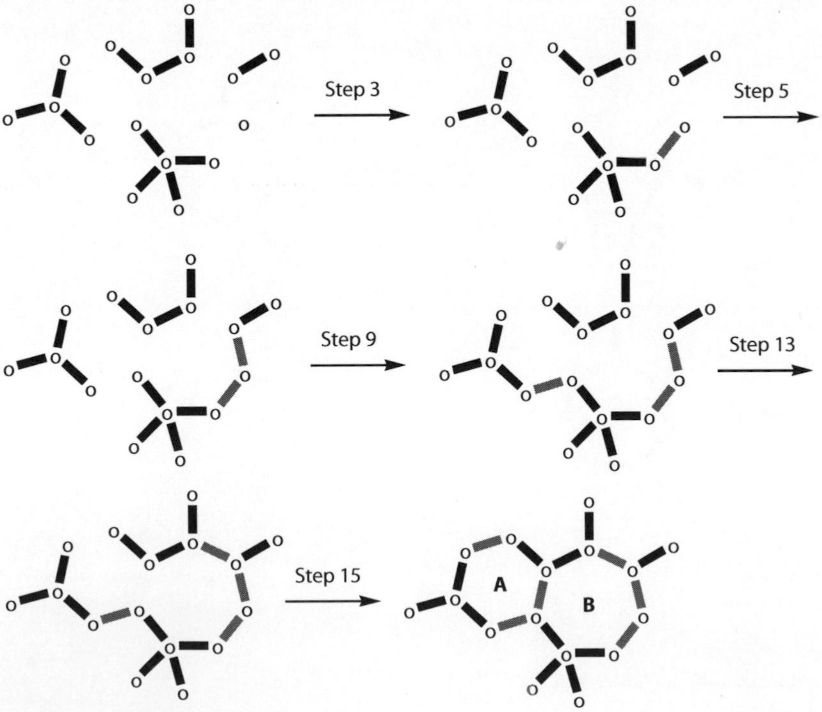

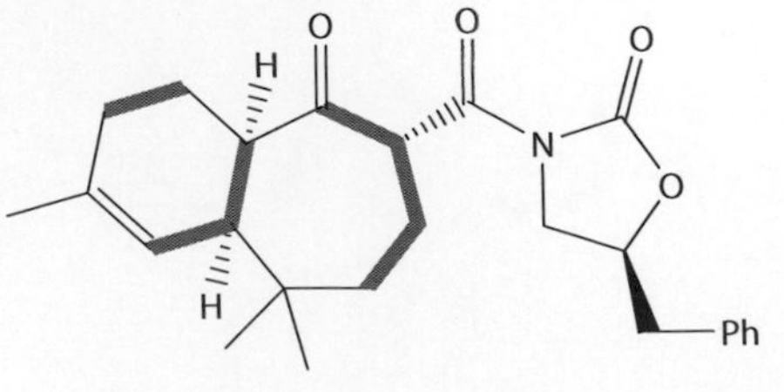

Sequence: $\xrightarrow{[(4+2)_A+(7+0)_B]}$ $[5.4.0]_{15}$

Summary:
Strategy [4 + 2] + [7 + 0] → [5.4.0]: Evans, Oppolzer, Wenkert

(−)-Hirsutene

Banwell, M.G.; Edwards, A.J.; Harfoot, G.J.; Joliffe, K.A. *J. Chem. Soc. Perkin Trans. I* 2002, 2439.
Banwell, M.G.; Edwards, A.J.; Harfoot, G.J.; Joliffe, K.A. *Tetrahedron* 2004, *60*, 535.

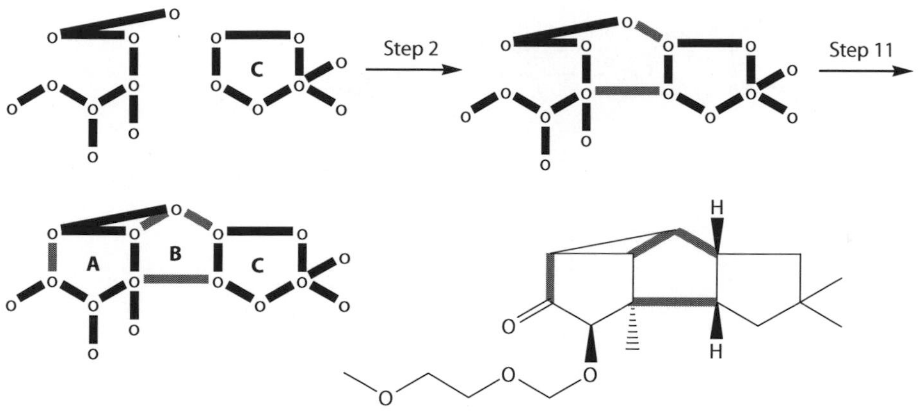

Sequence: $C_2 \xrightarrow{[4+2]} [6]_2 \xrightarrow{[(5+0)_A+(5+0)_B]} [3.3.0](A+B)_{11}$

(±)-Hirsutene

Cohen, T.; McNamara, K.; Kuzemko, M.A.; Ramig, K.; Landi, J.J. Jr.; Dong, Y. *Tetrahedron* 1993, *49*, 7931.
Ramig, K.; Kuzemko, M.A.; McNamara, K.; Cohen, T. *J. Org. Chem.* 1992, *57*, 1968.

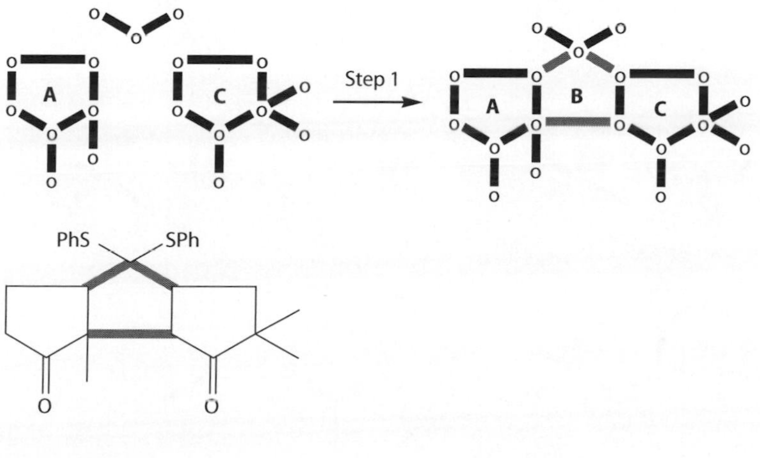

Sequence: $(A+C)_0 \xrightarrow{[2+2+1]} B_1$

(±)-Hirsutene

Cossy, J.; Belotti, D.; Pete, J.P. *Tetrahedron* 1990, *46*, 1859.
Cossy, J.; Belotti, D.; Pete, J.P. *Tetrahedron Lett*. 1987, *28*, 4547.

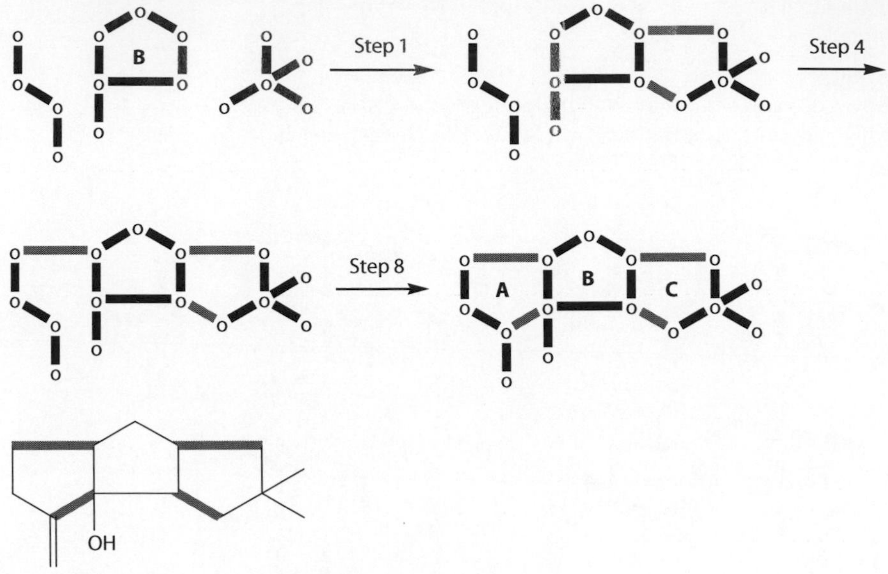

Sequence: $B_0 \xrightarrow{[3+2]} C_1 \xrightarrow{[5+0]} A_8$

(±)-Hirsutene

Curran, D.P.; Rakiewicz, D.M. *Tetrahedron* 1985, *41*, 3943.
Curran, D.P.; Rakiewicz, D.M. *J. Am. Chem. Soc*. 1985, *107*, 1448.

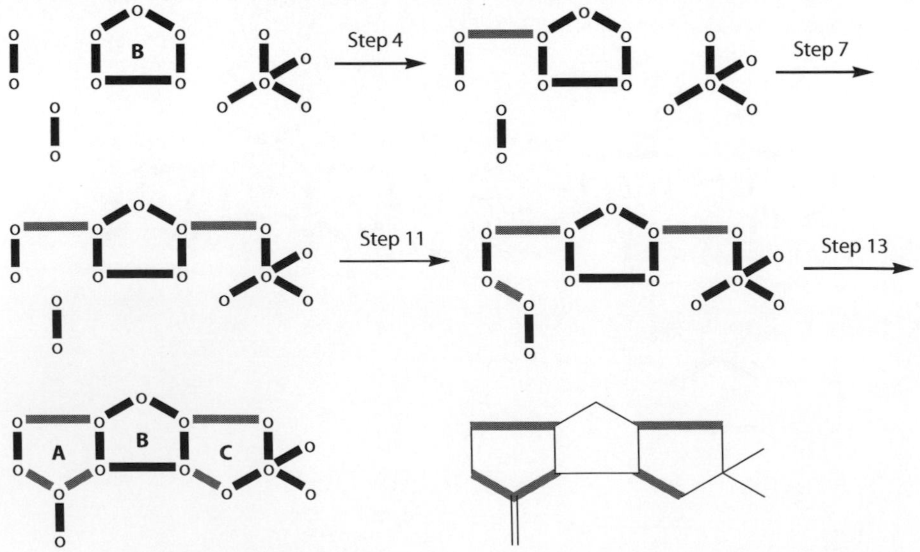

Sequence: $B_0 \xrightarrow{[(5+0)_A + (5+0)_C]} (A+C)_{13}$

(±)-Hirsutene

Anger, T.; Graalmann, O.; Schroeder, H.; Gerke, R.; Kaiser, U.; Fitjer, L.; Noltemeyer, M. *Tetrahedron* 1998, *54*, 10713.

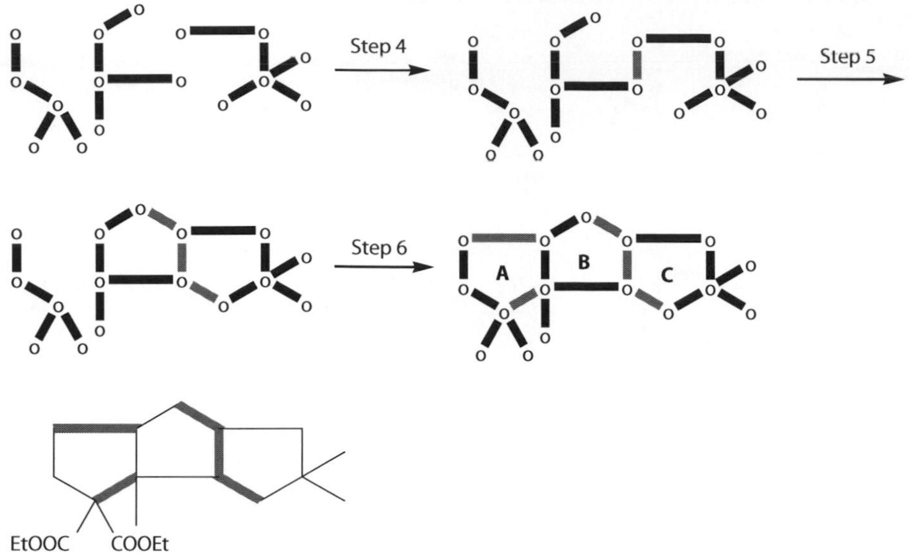

Sequence: $\xrightarrow{[(5+0)+(5+0)]}$ $[3.3.0]\,(B+C)_5$ $\xrightarrow{[3+2]}$ A_6

(±)-Hirsutene

Franck-Neumann, M.; Miesch, M.; Lacroix, E.; Metz, B.; Kern, J.M. *Tetrahedron* 1992, *48*, 1911.
Franck-Neumann, M.; Miesch, M.; Lacroix, E. *Tetrahedron Lett.* 1989, *30*, 3529.

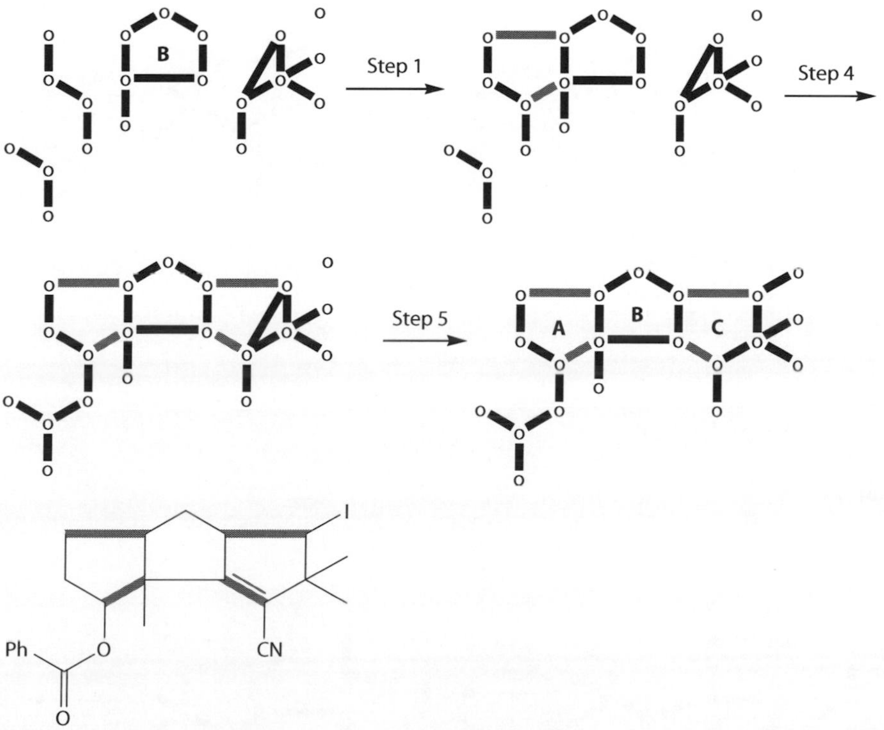

Sequence: B_0 $\xrightarrow{[3+2]}$ A_1 $\xrightarrow{[2+2]}$ $[2.1.0]_4$ $\xrightarrow{[2.1.0]\,\rightarrow[5]}$ C_5

(±)-Hirsutene

Toyota, M.; Nishikawa, Y.; Motoki, K.; Yoshida, N.; Fukumoto, K. *Tetrahedron* 1993, *49*, 11189.
Toyota, M.; Nishikawa, Y.; Motoki, K.; Yoshida, N.; Fukumoto, K. *Tetrahedron Lett.* 1993, *34*, 6099.

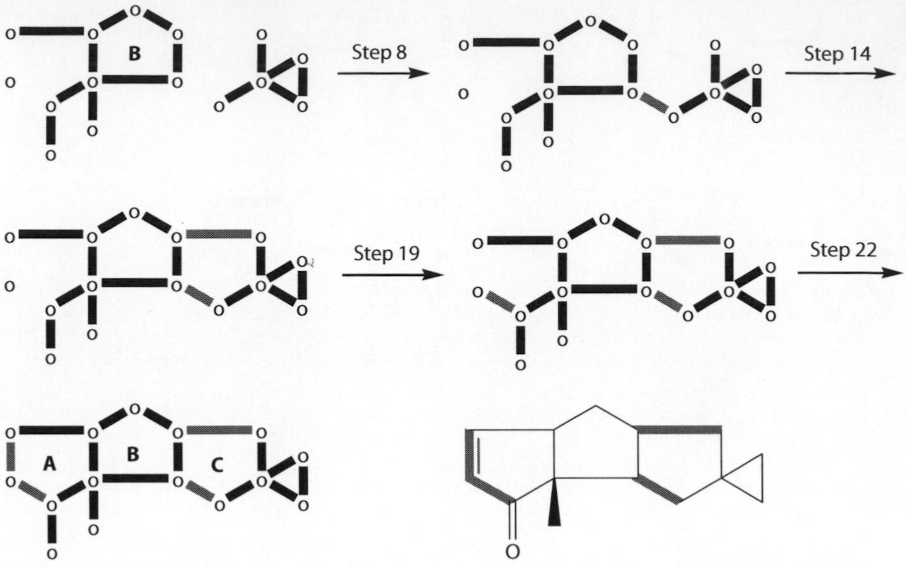

Sequence: $B_0 \xrightarrow{[5+0]} C_{14} \xrightarrow{[5+0]} A_{22}$

(±)-Hirsutene

Funk, R.L.; Bolton, G.L. *J. Org. Chem.* 1984, *49*, 5021.

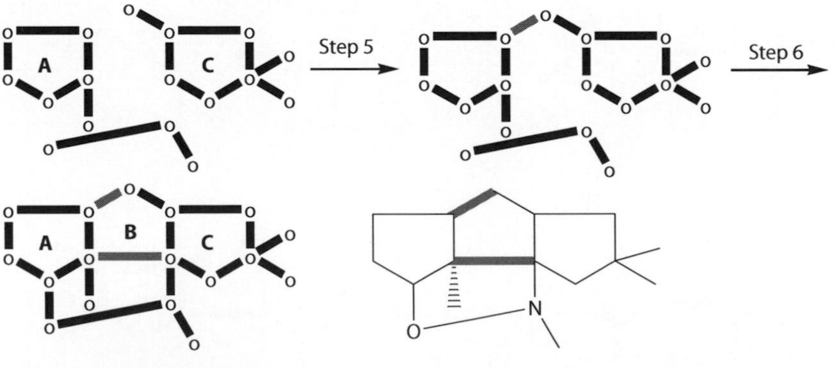

Sequence: $A_5 + C_0 \xrightarrow{[5+0]} B_6$

(±)-Hirsutene

Castro, J.; Sorensen, H.; Riera, A.; Morin, C.; Moyano, A.; Pericas, M.A.; Greene, A.E. *J. Am. Chem. Soc.* 1990, *112*, 9388.

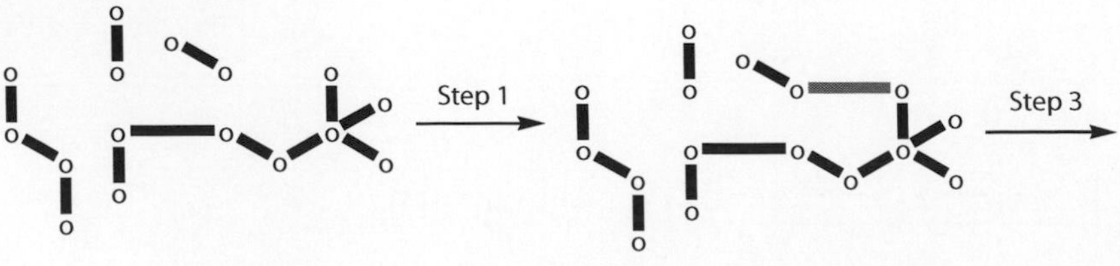

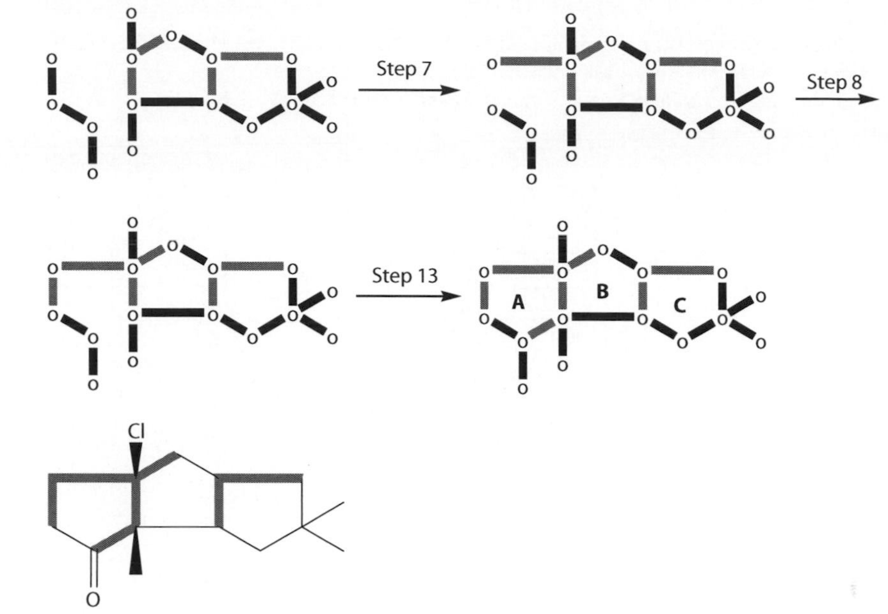

Sequence: $\xrightarrow{[2+2+1]_B + [5+0]_C} [3.3.0]\,(B+C)_3 \xrightarrow{[5+0]} A_{13}$

(±)-Hirsutene

Hewson, A.T.; MacPherson, D.T. *J. Chem. Soc. Perkin Trans. I* 1985, 2625.

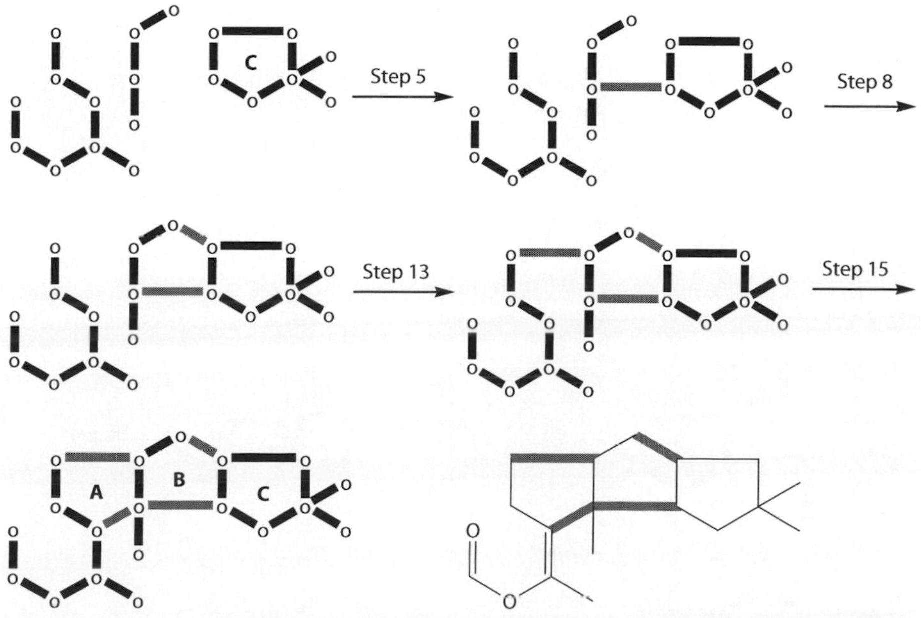

Sequence: $C_0 \xrightarrow{[5+0]} B_8 \xrightarrow{[5+0]} A_{15}$

(±)-Hirsutene

Hua, D.H.; Venkataraman, S.; Ostrander, R.A.; Sinai, G.Z.; McCann, P.J.; Coulter, M.J.; Xu, M.R. *J. Org. Chem.* 1988, *53*, 507.
Hua, D.H.; Sinai-Zingde, G.; Venkataraman, S. *J. Am. Chem. Soc.* 1985, *107*, 4088.

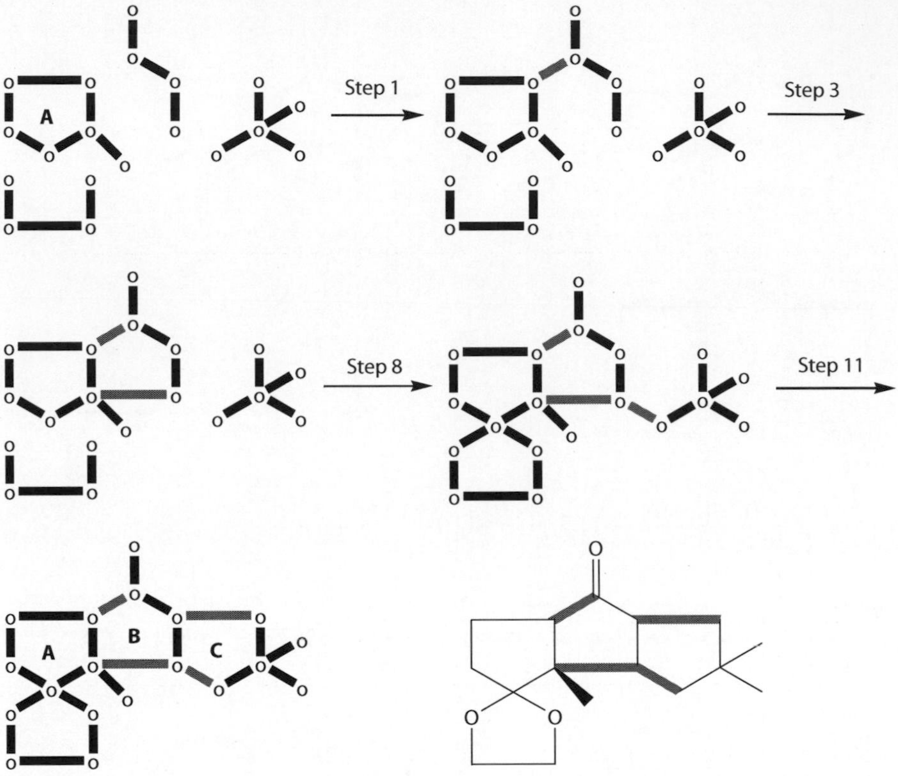

Sequence: $A_0 \xrightarrow{[5+0]} B_3 \xrightarrow{[5+0]} C_{11}$

(±)-Hirsutene

Hudlicky, T.; Kutchan, T.M. *J. Am. Chem. Soc.* 1980, *102*, 6353.

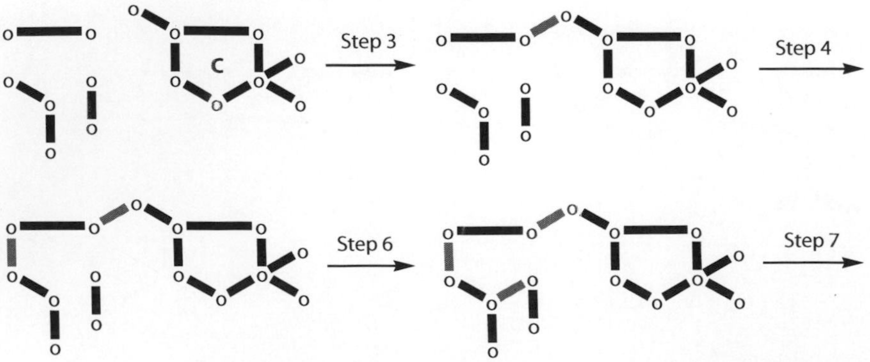

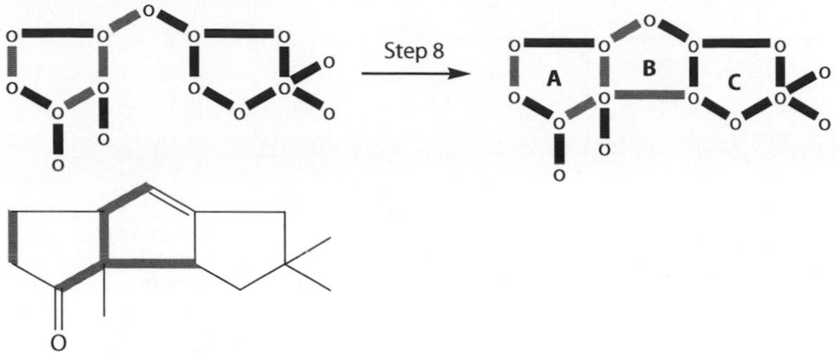

Sequence: $C_0 \xrightarrow{[5+0]} A_7 \xrightarrow{[5+0]} B_8$

(±)-Hirsutene

Iyoda, M.; Kushida, T.; Kitami, S.; Oda, M. *Chem. Commun.* 1986, 1049.

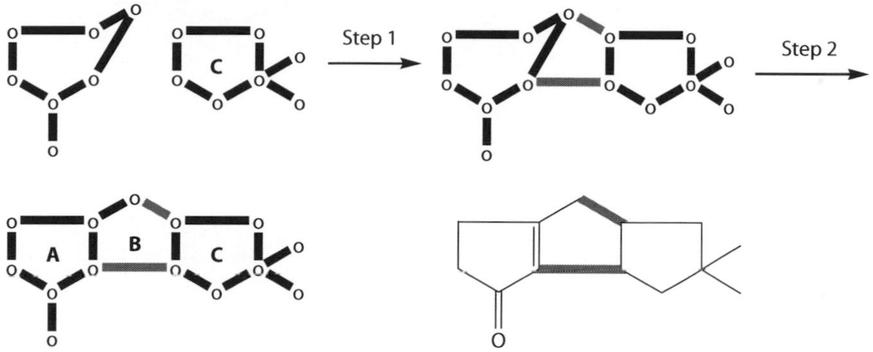

Sequence: $C_0 \xrightarrow{[2+2]} [4]_1 \xrightarrow{[4.2.0] \rightarrow [3.3.0]} (A+B)$

(±)-Hirsutene

Wang, J.C.; Krische, M.J. *Angew. Chem. Int. Ed.* 2003, *42*, 5855.

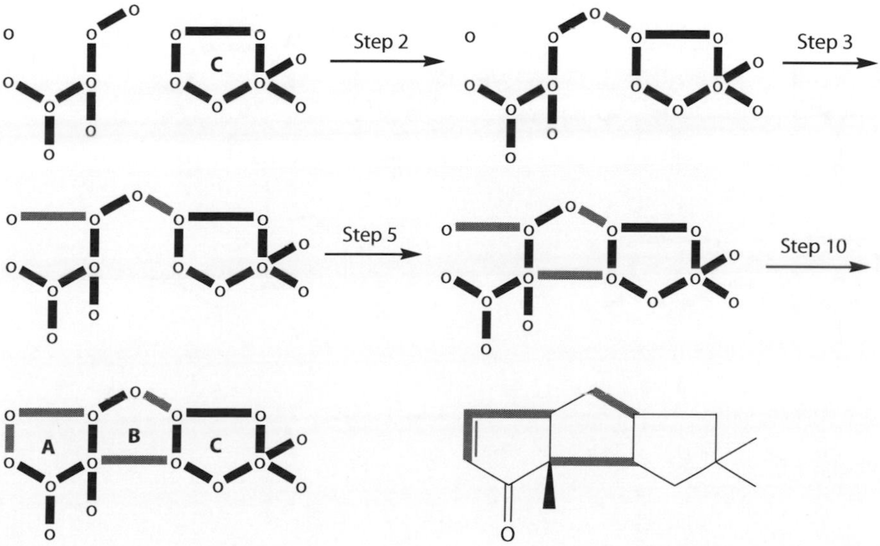

Sequence: $C_0 \xrightarrow{[5+0]} B_5 \xrightarrow{[5+0]} A_{10}$

(±)-Hirsutene
Leonard, J.; Bennett, L.; Mahmood, A. *Tetrahedron Lett.* 1999, *40*, 3965.

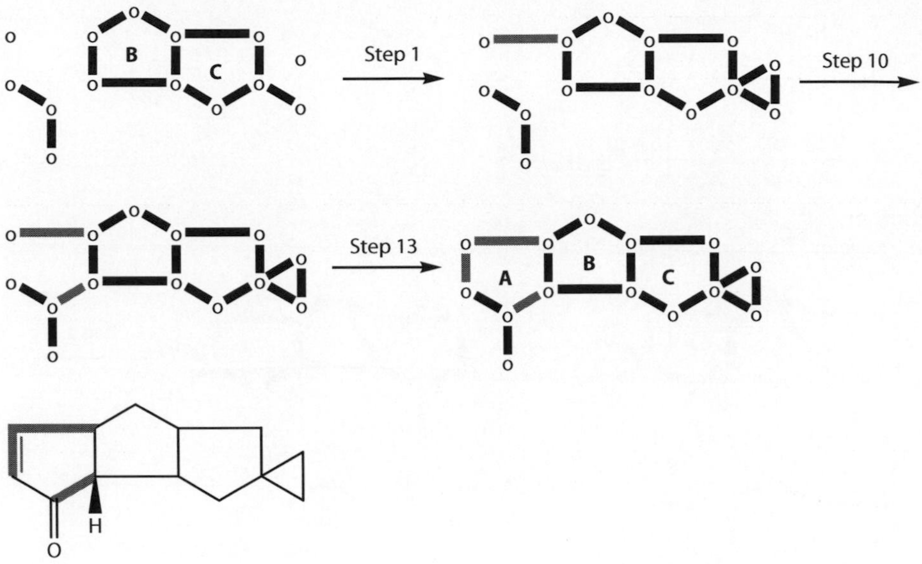

Sequence: $(B+C)_0 \xrightarrow{[5+0]} A_{13}$

(±)-Hirsutene
Ley, S.V.; Murray, J.; Palmer, B.D. *Tetrahedron* 1985, *41*, 4765.

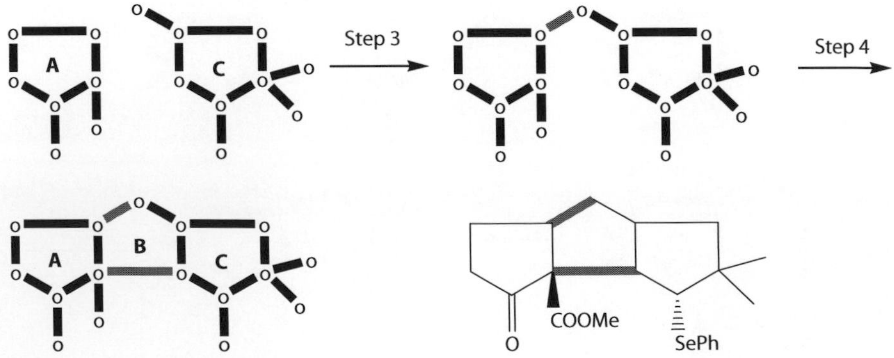

Sequence: $A_{0*} + C_0 \xrightarrow{[5+0]} B_4$

(±)-Hirsutene

Chandler, C.L.; List, B. *J. Am. Chem. Soc.* 2008, *130*, 6737.

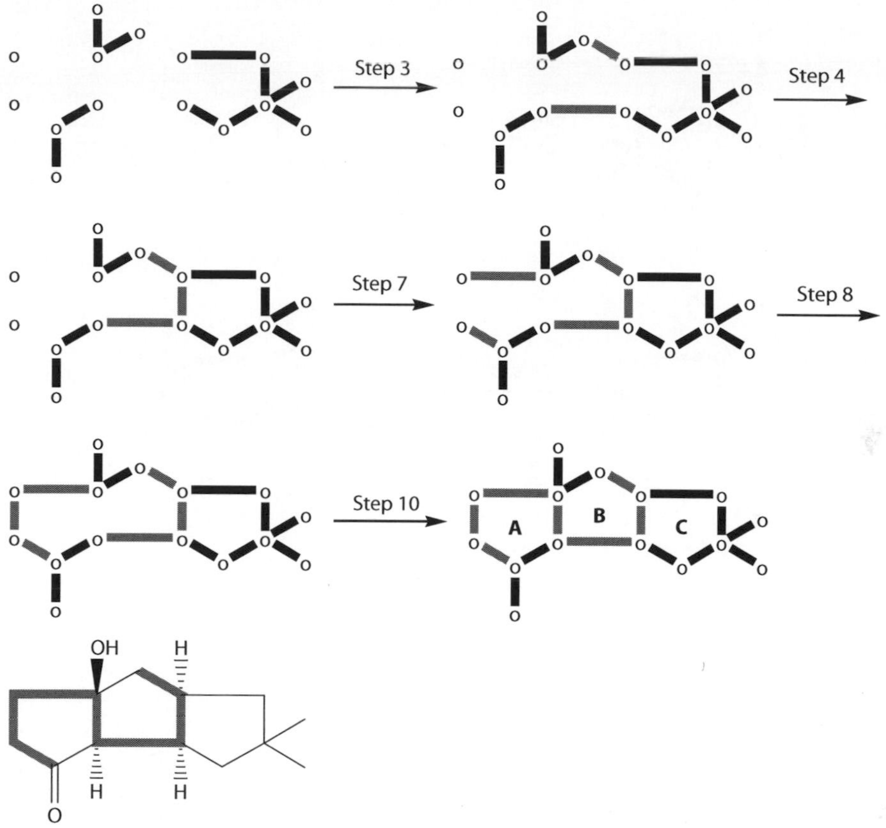

Sequence: $\xrightarrow{[5+0]} C_4 \xrightarrow{[8]\rightarrow[3.3.0]} (A+B)_{10}$

(±)-Hirsutene

Little, R.D.; Higby, R.G.; Moeller, K.D. *J. Org. Chem.* 1983, *48*, 3139.

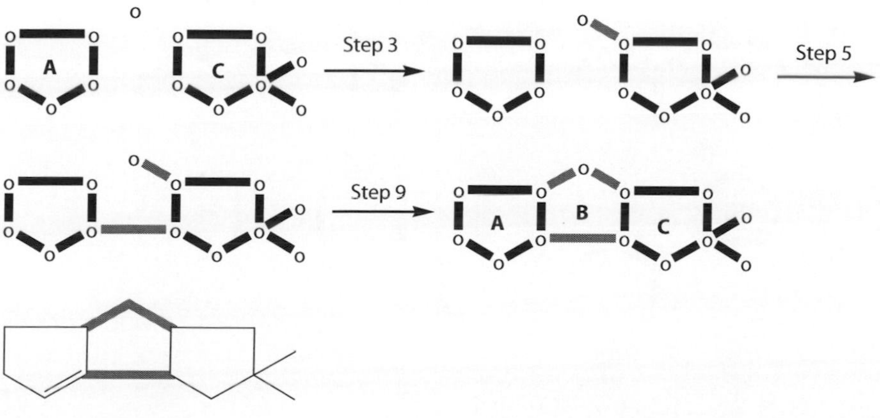

Sequence: $A_5 + C_0 \xrightarrow{[5+0]} B_9$

558

(±)-Hirsutene

Magnus, P.; Quagliato, D. *J. Org. Chem.* 1985, *50*, 1621.

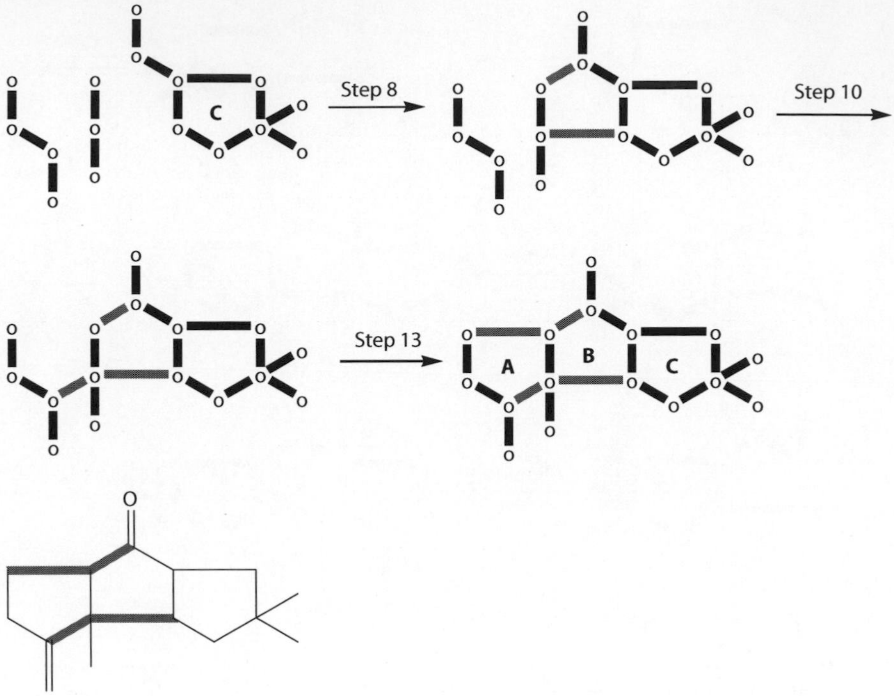

Sequence: $C_0 \xrightarrow{[3+2]} B_8 \xrightarrow{[5+0]} A_{13}$

(±)-Hirsutene

Mehta, G.; Murthy, A.N.; Reddy, D.S.; Reddy, A.V. *J. Am. Chem. Soc.* 1986, *108*, 3443.
Mehta, G.; Reddy, A.V. *Chem. Commun.* 1981, 757.

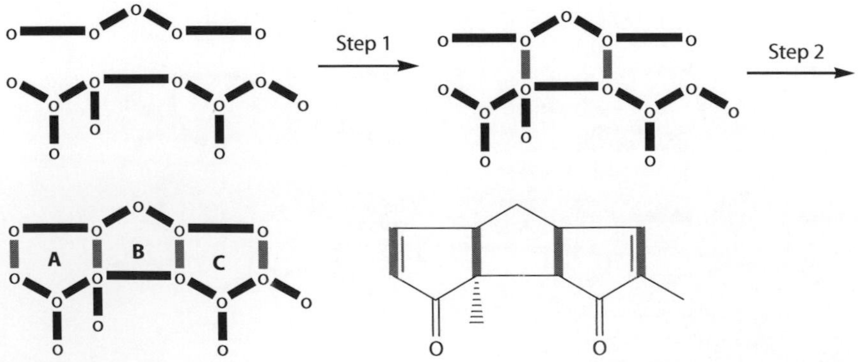

Sequence: $\xrightarrow{[3+2]} B_1 \xrightarrow{[(5+0)_A + (5+0)_C]} (A+C)_2$

(±)-Hirsutene

Nozoe, S.; Furukawa, J.; Sankawa, U.; Shibata, S. *Tetrahedron Lett*. 1976, 195.

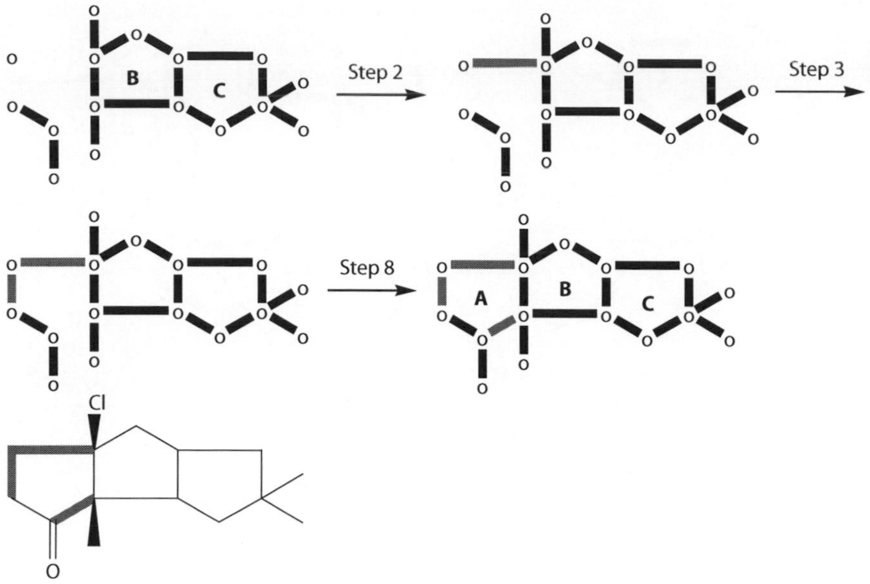

Sequence: $(B+C)_0 \xrightarrow{[5+0]} A_8$

(±)-Hirsutene

Oppolzer, W.; Robyr, C. *Tetrahedron* 1994, *50*, 415.

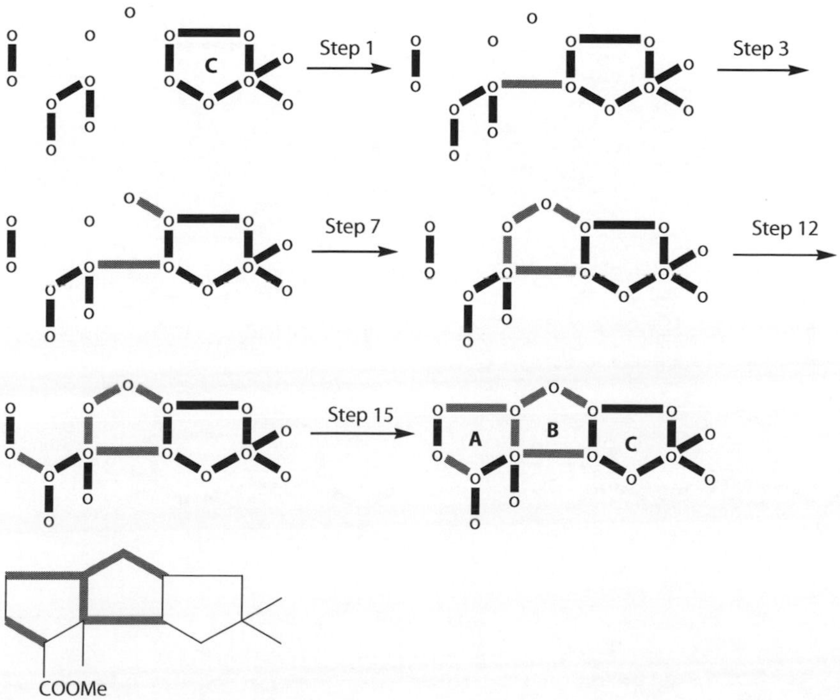

Sequence: $C_0 \xrightarrow{[4+1]} B_7 \xrightarrow{[5+0]} A_{15}$

(±)-Hirsutene

Paquette, L.A.; Moriarty, K.J.; Shen, C.C. *Isr. J. Chem.* 1991, *31*, 195.

Sequence: $C_0 \xrightarrow{[(5+0)_A + (5+0)_B]} [3.3.0] (A+B)_8$

(±)-Hirsutene

Sarkar, T.K.; Ghosh, S.K.; Rao, P.S.V.S.; Satapathi, T.K.; Mamdapur, V.R. *Tetrahedron* 1992, *48*, 6897.
Sarkar, T.K.; Ghosh, S.K.; Rao, P.S.V.S.; Mamdapur, V.R. *Tetrahedron Lett.* 1990, *31*, 3465.

Sequence: $\xrightarrow{[5+0]} C_7 \xrightarrow{[5+0]} B_{11} \xrightarrow{[5+0]} A_{14}$

(±)-Hirsutene

Singh, V.; Vedantham, P.; Sahu, P.K. *Tetrahedron Lett.* 2002, *43*, 519.

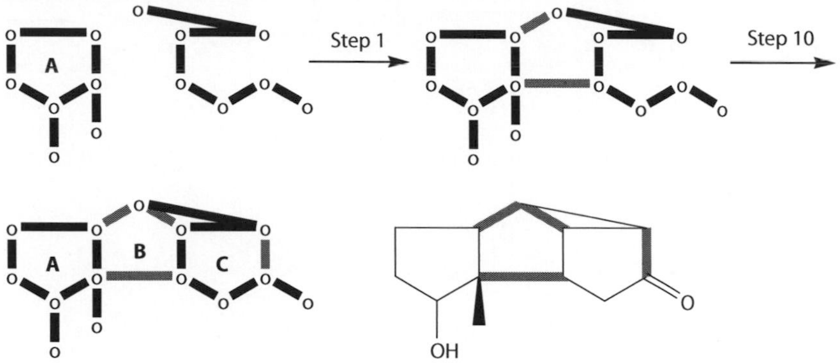

Sequence: $A_0 \xrightarrow{[(5+0)_B + (5+0)_C]} [3.3.0] (B+C)_{10}$

(±)-Hirsutene

Sternbach, D.D.; Ensinger, C.L. *J. Org. Chem.* 1990, *55*, 2725.

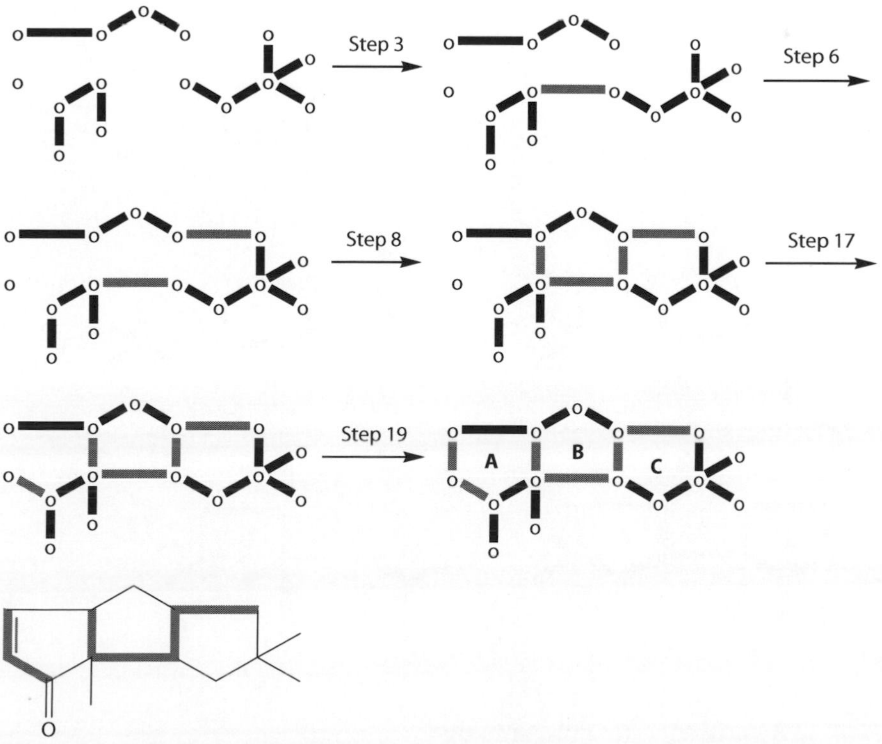

Sequence: $\xrightarrow{[(3+2)_B + (5+0)_C]} [3.3.0] (B+C)_8 \xrightarrow{[5+0]} A_{19}$

(±)-Hirsutene
Disanayaka, B.W.; Weedon, A.C. *Chem. Commun.* 1985, 1282.

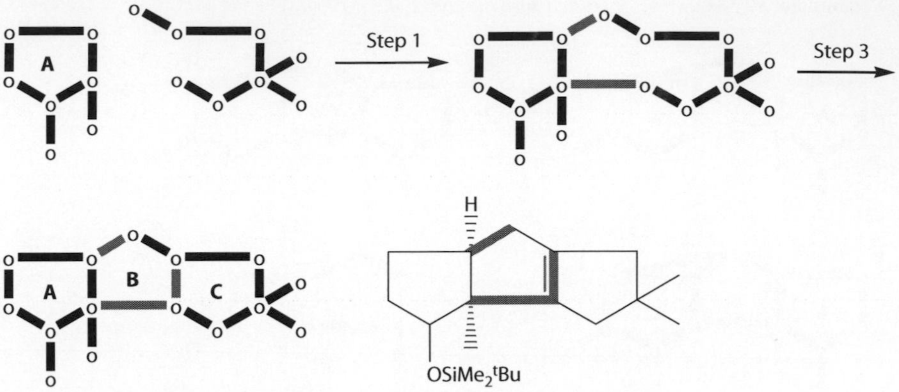

Sequence: $A_0 \xrightarrow{[8] \to [3.3.0]} (B+C)_3$

(−)-Hirsutene
Weinges, K.; Reichert, H.; Huber-Patz, U.; Irngartinger, H. *Liebigs Ann. Chem.* 1993, 403.

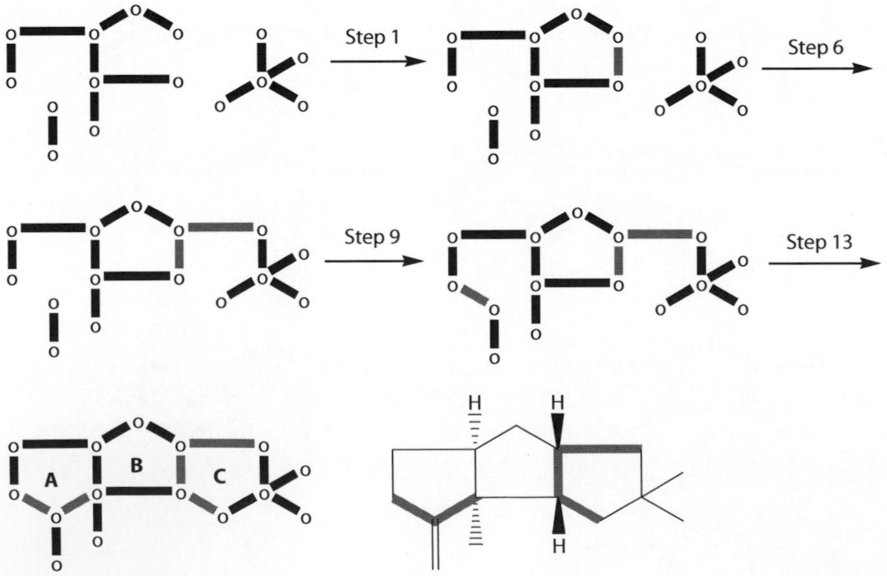

Sequence: $\xrightarrow{[5+0]} B_1 \xrightarrow{[(5+0)_A + (5+0)_C]} (A+C)_{13}$

(±)-Hirsutene

Wender, P.A.; Howbert, J.J. *Tetrahedron Lett.* 1982, *23*, 3983.

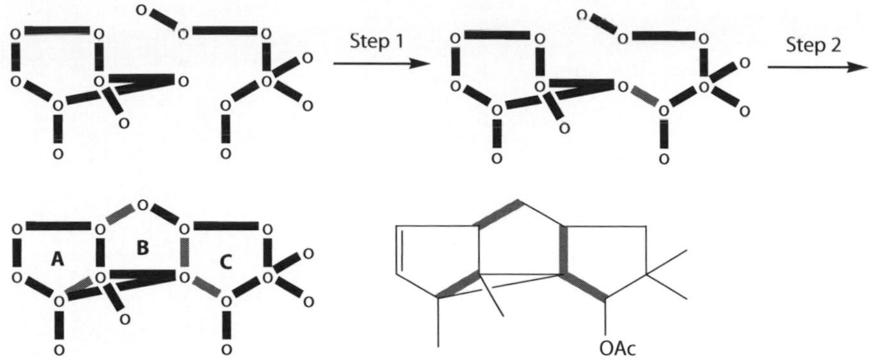

Sequence: $\xrightarrow{[(5+0)_A + (3+2)_B + (5+0)_C]} (A + B + C)_2$

(±)-Hirsutene

Jiao, L.; Yuan, C.; Yu, Z.X. *J. Am. Chem. Soc.* 2008, *130*, 4421.

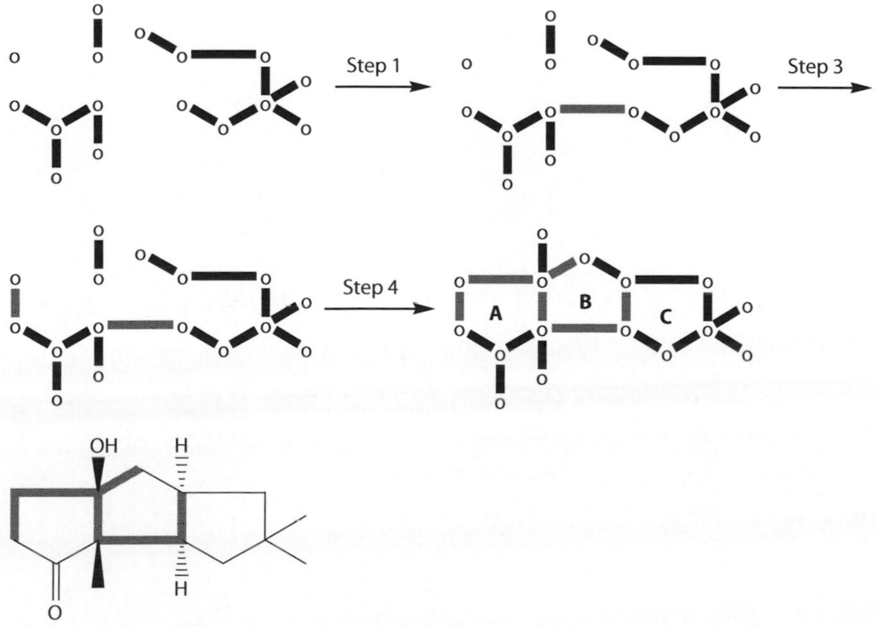

Sequence: $\xrightarrow{[(4+1)_A + (2+2+1)_B + (5+0)_C]} (A + B + C)_4$

Plan	Strategy
Banwell	$C_2 \xrightarrow{[4+2]} [6]_2 \xrightarrow{[(5+0)_A + (5+0)_B]} [3.3.0]\,(A+B)_{11}$
Cohen	$(A+C)_0 \xrightarrow{[2+2+1]} B_1$
Cossy	$B_0 \xrightarrow{[3+2]} C_1 \xrightarrow{[5+0]} A_8$
Curran	$B_0 \xrightarrow{[(5+0)_A + (5+0)_C]} (A+C)_{13}$
Fitjer	$\xrightarrow{[(5+0)+(5+0)]} [3.3.0]\,(B+C)_5 \xrightarrow{[3+2]} A_6$
Franck–Neumann	$B_0 \xrightarrow{[3+2]} A_1 \xrightarrow{[2+2]} [2.1.0]_4 \xrightarrow{[2.1.0] \rightarrow [5]} C_5$
Fukumoto	$B_0 \xrightarrow{[5+0]} C_{14} \xrightarrow{[5+0]} A_{22}$
Funk	$A_5 + C_0 \xrightarrow{[5+0]} B_6$
Greene–Nozoe	$\xrightarrow{[2+2+1]_B + [5+0]_C} [3.3.0]\,(B+C)_3 \xrightarrow{[5+0]} A_{13}$
Hewson	$C_0 \xrightarrow{[5+0]} B_8 \xrightarrow{[5+0]} A_{15}$
Hua	$A_0 \xrightarrow{[5+0]} B_3 \xrightarrow{[5+0]} C_{11}$
Hudlicky	$C_0 \xrightarrow{[5+0]} A_7 \xrightarrow{[5+0]} B_8$
Iyoda	$C_0 \xrightarrow{[2+2]} [4]_1 \xrightarrow{[4.2.0] \rightarrow [3.3.0]} (A+B)_2$
Krische	$C_0 \xrightarrow{[5+0]} B_5 \xrightarrow{[5+0]} A_{10}$
Leonard	$(B+C)_0 \xrightarrow{[5+0]} A_{13}$
Ley	$A_{0^*} + C_0 \xrightarrow{[5+0]} B_4$
List	$\xrightarrow{[5+0]} C_4 \xrightarrow{[8] \rightarrow [3.3.0]} (A+B)_{10}$
Little	$A_5 + C_0 \xrightarrow{[5+0]} B_9$
Magnus	$C_0 \xrightarrow{[3+2]} B_8 \xrightarrow{[5+0]} A_{13}$
Mehta	$\xrightarrow{[3+2]} B_1 \xrightarrow{[(5+0)_A + (5+0)_C]} (A+C)_2$
Nozoe	$(B+C)_0 \xrightarrow{[5+0]} A_8$
Oppolzer	$C_0 \xrightarrow{[4+1]} B_7 \xrightarrow{[5+0]} A_{15}$
Paquette	$C_0 \xrightarrow{[(5+0)_A + (5+0)_B]} [3.3.0]\,(A+B)_8$
Sarkar	$\xrightarrow{[5+0]} C_7 \xrightarrow{[5+0]} B_{11} \xrightarrow{[5+0]} A_{14}$
Singh	$A_0 \xrightarrow{[(5+0)_B + (5+0)_C]} [3.3.0]\,(B+C)_{10}$

(continued)

Plan	Strategy
Sternbach	$\xrightarrow{[(3+2)_B+(5+0)_C]} [3.3.0] (B+C)_8 \xrightarrow{[5+0]} A_{19}$
Weedon	$A_0 \xrightarrow{[8] \to [3.3.0]} (B+C)_3$
Weinges	$\xrightarrow{[5+0]} B_1 \xrightarrow{[(5+0)_A+(5+0)_C]} (A+C)_{13}$
Wender	$\xrightarrow{[(5+0)_A+(3+2)_B+(5+0)_C]} (A+B+C)_2$
Yu	$\xrightarrow{[(4+1)_A+(2+2+1)_B+(5+0)_C]} (A+B+C)_4$

$A \to B \to C$: Hua

$A \to (B+C)$: Singh, Weedon

$(A+C) \to B$: Cohen, Funk, Ley, Little

$B \to C \to A$: Cossy, Fukumoto

$B \to A \to C$: Franck-Neumann

$B \to (A+C)$: Curran

$(B+C) \to A$: Leonard, Nozoe

$C \to B \to A$: Hewson, Krische, Magnus, Oppolzer

$C \to A \to B$: Hudlicky

$C \to (A+B)$: Banwell, Iyoda, Paquette

$\to C \to B \to A$: Sarkar

$\to C \to (A+B)$: List

$\to B \to (A+C)$: Mehta, Weinges

$\to (B+C) \to A$: Fitjer, Greene–Nozoe, Sternbach

$\to (A+B+C)$: Wender, Yu

(±)-Hypoglycin A
Carbon, J.A.; Martin, W.B.; Swett, L.R. *J. Am. Chem. Soc.* 1958, *80*, 1002.

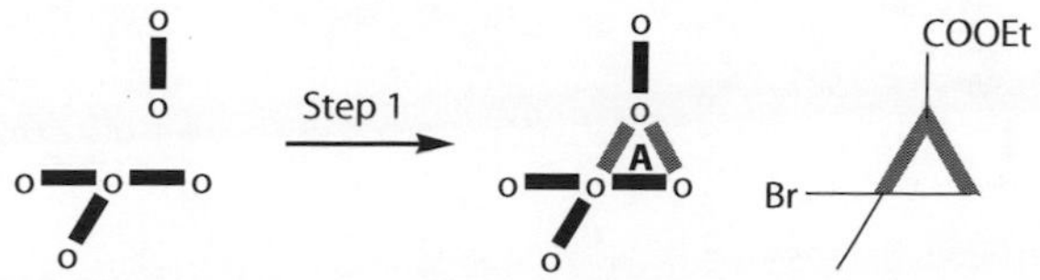

Sequence: $\xrightarrow{[2+1]} A_1$

(2S,3R)-Hypoglycin A

Baldwin, J.E.; Adlington, R.M.; Bebbington, D.; Russell, A.T. *Chem. Commun.* 1992, 1249.
Baldwin, J.E.; Adlington, R.M.; Bebbington, D.; Russell, A.T. *Tetrahedron* 1994, *50*, 12015.

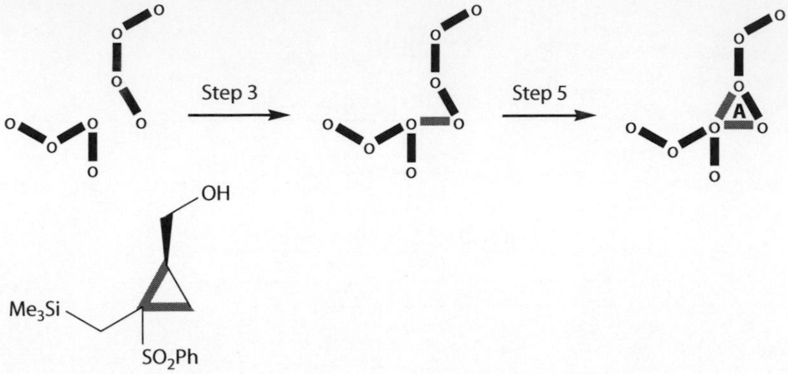

Sequence: $\xrightarrow{[3+0]} A_5$

(±)-Hypoglycin A

Black, D.K.; Landor, S.R. *Tetrahedron Lett.* 1963, 1065.

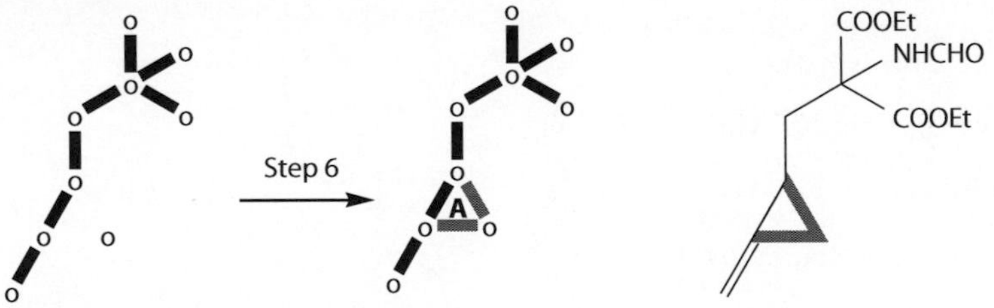

Sequence: $\xrightarrow{[2+1]} A_6$

(±)-Ibogamine

Imanishi, T.; Yagi, N.; Shin, H.; Hanaoka, M. *Tetrahedron Lett.* 1981, *22*, 4001.
Imanishi, T.; Shin, H.; Yagi, N.; Hanaoka, M. *Tetrahedron Lett.* 1980, *21*, 3285.

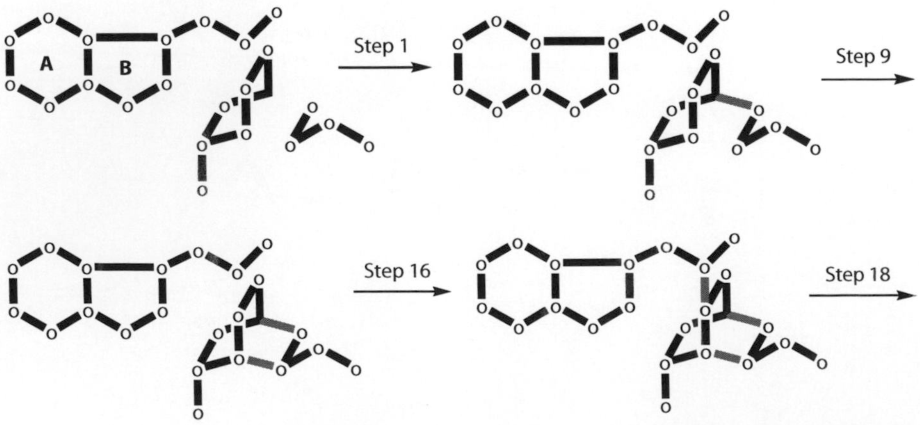

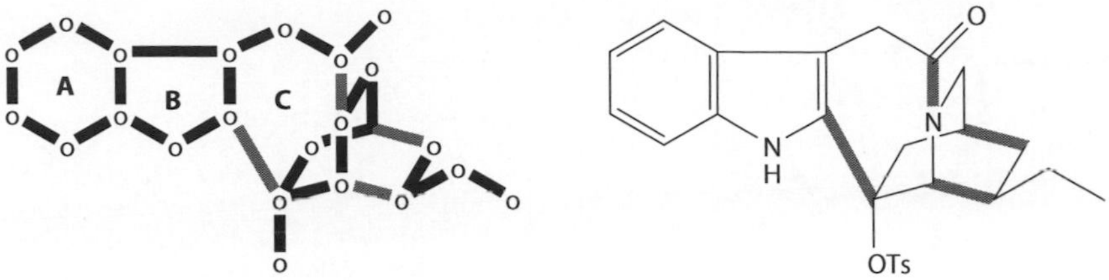

Sequence: $(A + B)_0 \xrightarrow{[(6+0) + (6+0)]} [2.2.2]_9 \xrightarrow{[7+0]} C_{18}$

(±)-Ibogamine

Hodgson, D.M.; Galano, J.M. *Org. Lett.* 2005, 7, 2221.

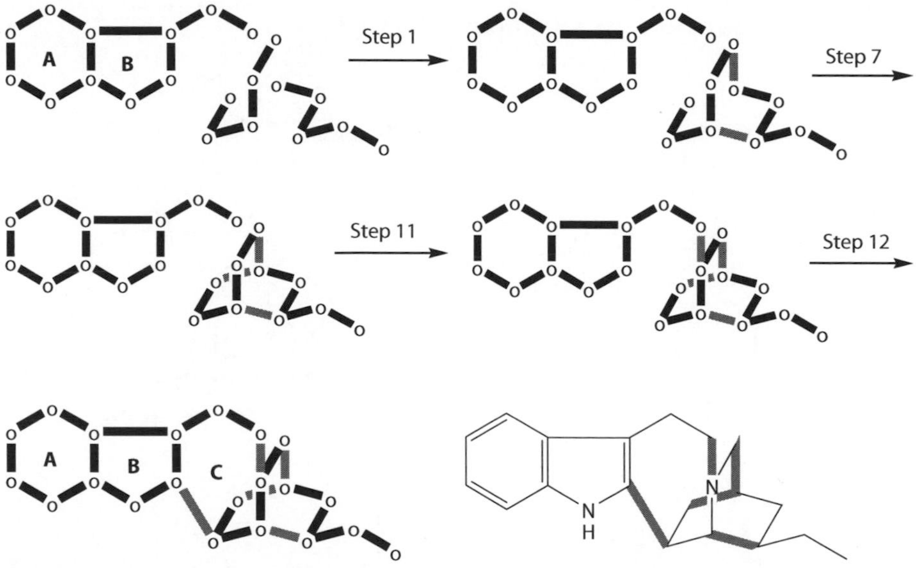

Sequence: $(A + B)_0 \xrightarrow{[(6+0) + (6+0)]} [2.2.2]_7 \xrightarrow{[7+0]} C_{12}$

(±)-Ibogamine

Nagata, W.; Hirai, S.; Okumura, T.; Kawata, K. *J. Am. Chem. Soc.* 1968, *90*, 1650.

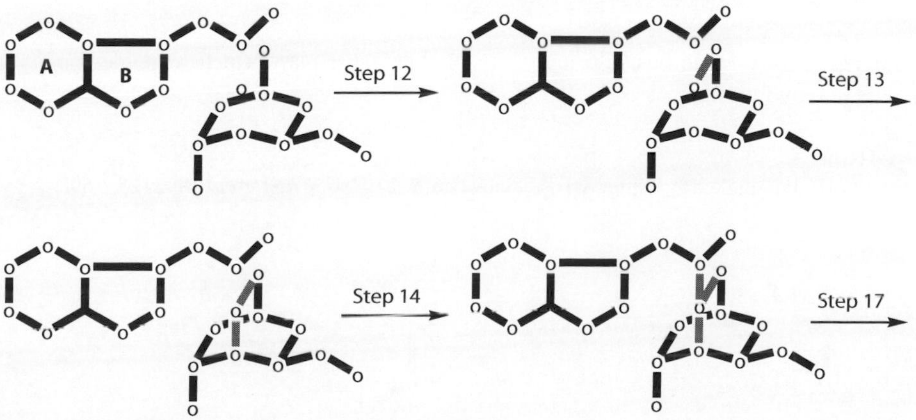

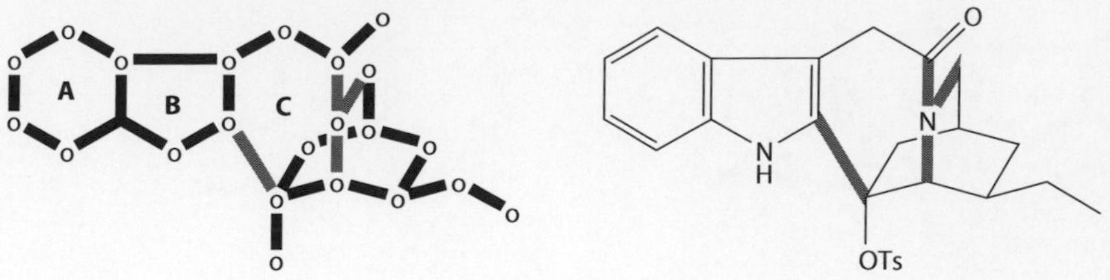

Sequence: $(A+B)_0 \xrightarrow{[(6+0)+(6+0)]} [2.2.2]_{13} \xrightarrow{[7+0]} C_{17}$

(±)-Ibogamine

Trost, B.M.; Godleski, S.A.; Genet, J.P. *J. Am. Chem. Soc.* 1978, *100*, 3930.

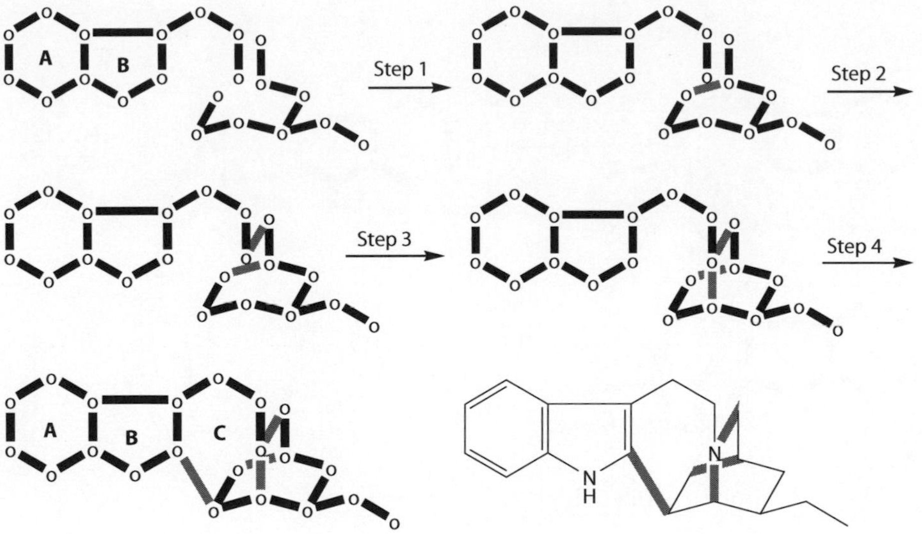

Sequence: $(A+B)_0 \xrightarrow{[(6+0)+(6+0)]} [2.2.2]_3 \xrightarrow{[7+0]} C_4$

Note: Four plans have same basic strategy except that Trost plan attaches indole moiety before making [2.2.2] bicyclic ring.

(±)–*cis*-α-Irone

Schulte-Elte, K.H.; Pamingle, H.; Uijttewaal, A.; Snowden, R.L. *Helv. Chim. Acta* 1992, *75*, 759.

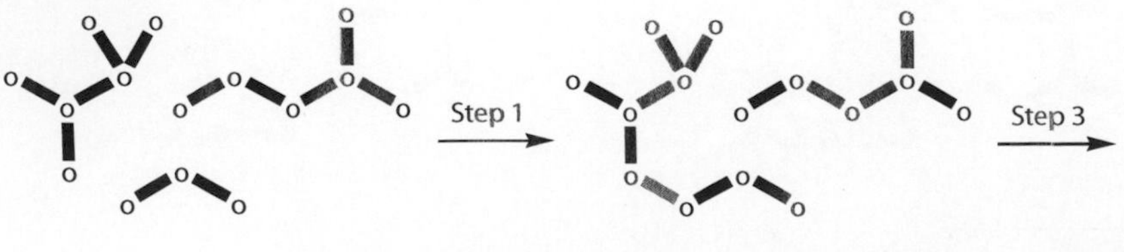

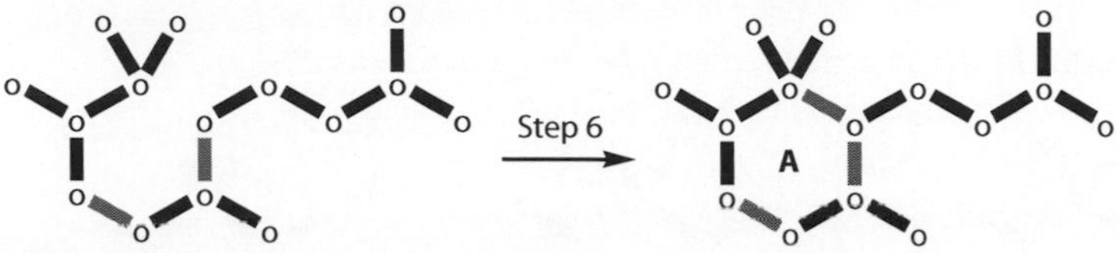

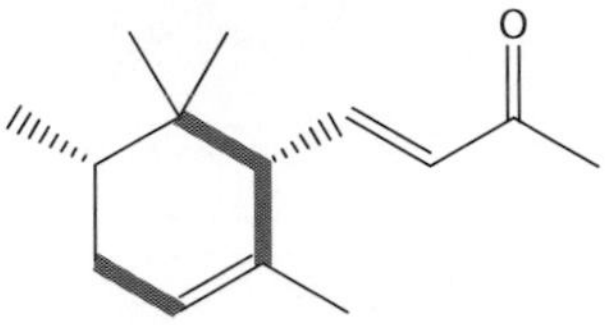

Sequence: $\xrightarrow{[6+0]}$ A_6

Nussbaumer, C.; Fráter, G. *J. Org. Chem.* 1987, *52*, 2096.

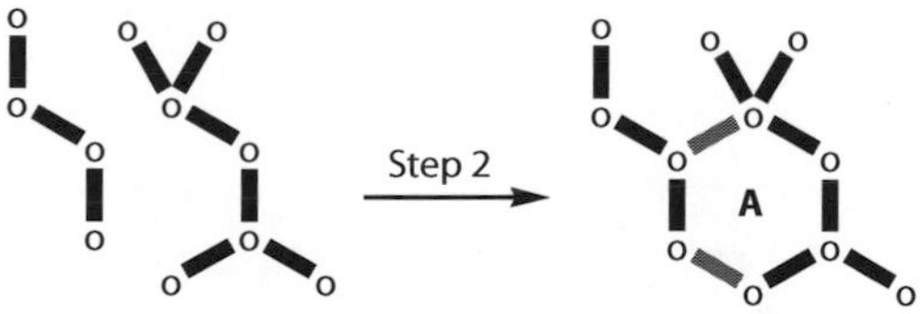

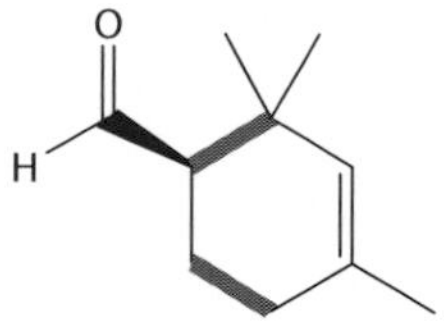

Sequence: $\xrightarrow{[4+2]}$ A_2

(−)-(2S,6R)-*cis*-α-Irone
Inoue, T.; Kiyota, H.; Oritani, T. *Tetrahedron Asymm.* 2000, *11*, 3807.

As mentioned earlier.

(±)-(2R,6S)-*cis*-α-Irone
Ohtsuka, Y.; Itoh, F.; Oishi, T. *Chem. Pharm. Bull.* 1991, *39*, 2540.

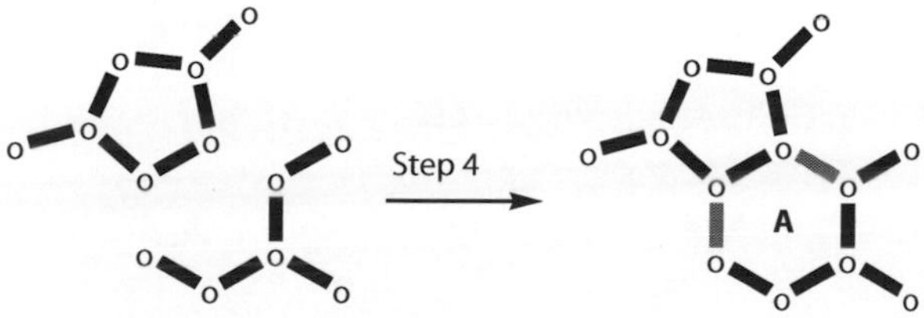

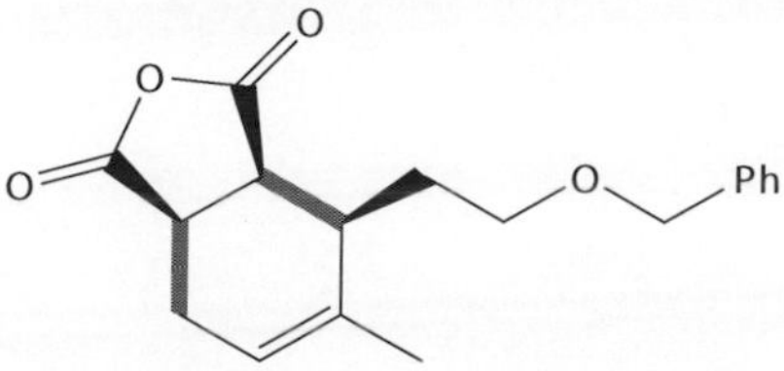

Sequence: $\xrightarrow{[4+2]}$ A_4

(±)-(2S,6S)-*trans*-α-Irone
Rautenstrauch, V.; Willhalm, B.; Thommen, W.; Ohloff, G. *Helv. Chim. Acta* 1984, *67*, 325.

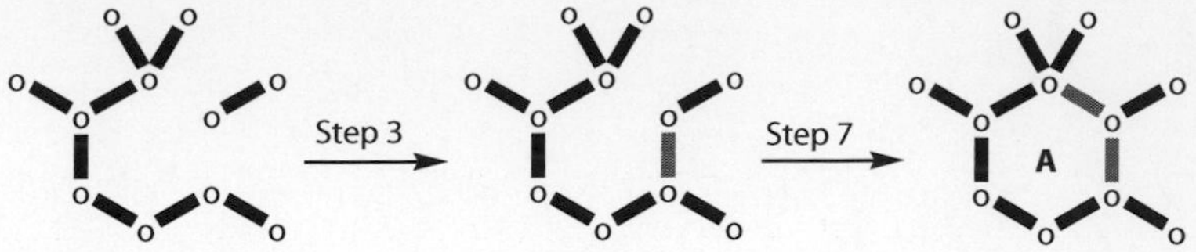

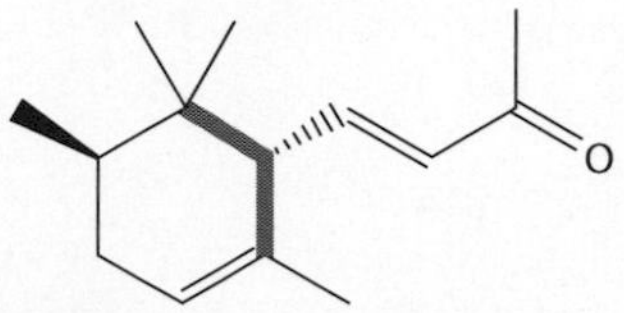

Sequence: $\xrightarrow{[6+0]} A_7$

(±)-*trans*-α-Irone
Eschinazi, H.E. *J. Am. Chem. Soc.* 1959, *81*, 2905.

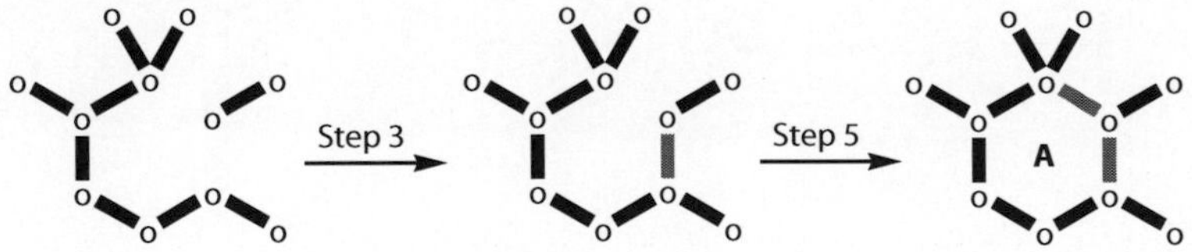

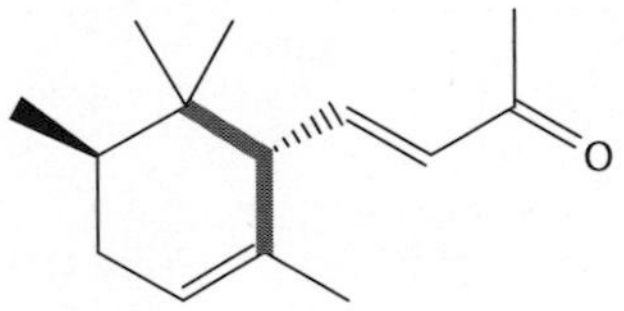

Sequence: $\xrightarrow{[6+0]} A_5$

(±)-(2S,6S)-*trans*-α-Irone
Helmlinger, D.; Fráter, G. *Helv. Chim. Acta* 1989, *72*, 1515.

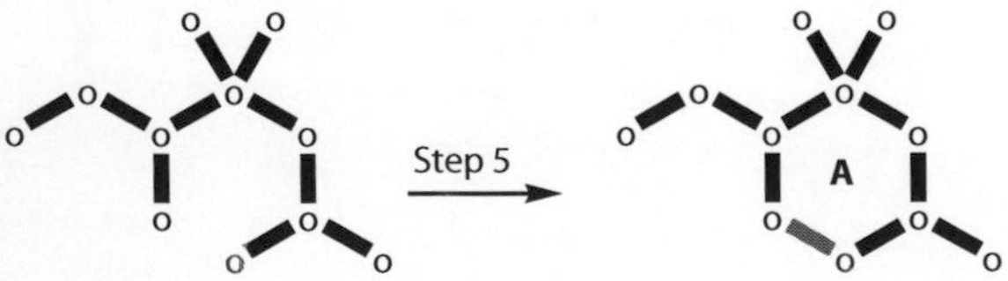

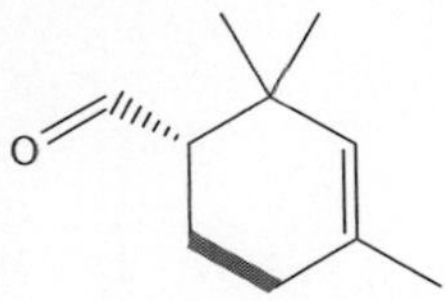

Sequence: $\xrightarrow{[6+0]} A_5$

Summary:
Strategy [4 + 2]: Givaudan G2, Kiyota, Ohtsuka
Strategy [6 + 0]: Givaudan G1, Givaudan G3, Firmenich G1, Firmenich G2

(±)-*trans*-γ-Irone
Brenna, E.; Fuganti, C.; Ronzani, S.; Serra, S. *Helv. Chim. Acta* 2001, *84*, 3650.

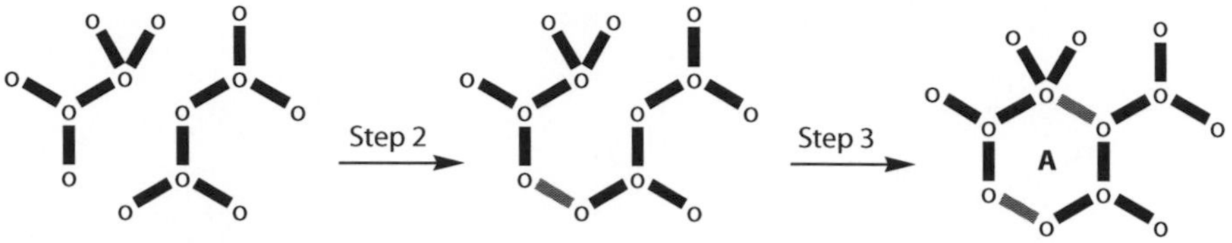

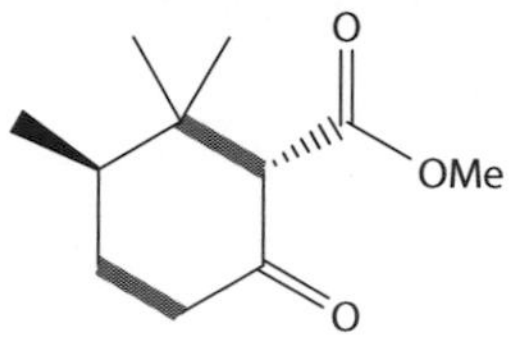

Sequence: $\xrightarrow{[6+0]} A_3$

Takazawa, O.; Kogami, K.; Hayashi, K. *Bull. Chem. Soc. Jpn.* 1985, *58*, 389.

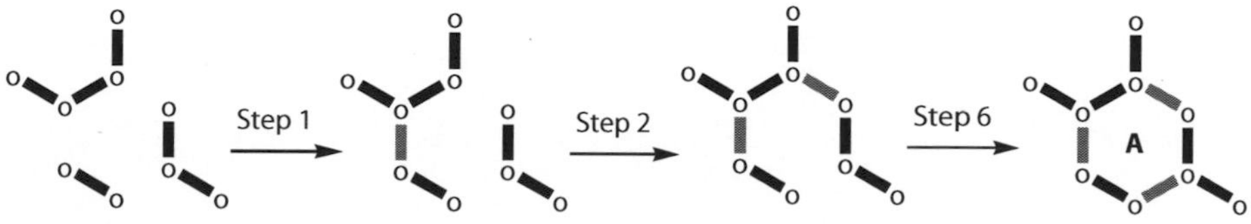

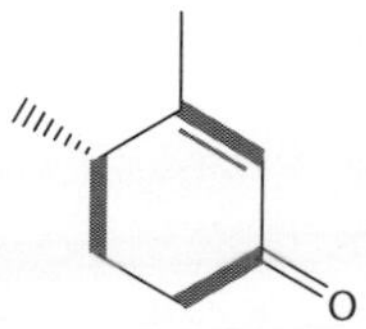

Sequence: $\xrightarrow{[6+0]} A_6$

(±)-(2R,6R)-*trans*-γ-Irone
Brenna, E.; Fuganti, C.; Ronzani, S.; Serra, S. *Helv. Chim. Acta* 2001, *84*, 3650.

Preformed ring precursor: (+)-*trans*-α-irone

Monti, H.; Andran, G.; Monti, J.P.; Leandri, G. *J. Org. Chem.* 1996, *61*, 6021.

Preformed ring precursor: (−)-carvone

(−)-(2S,6S)-*trans*-γ-Irone

Bezant, S.; Giannini, E.; Zanoni, G.; Vidari, G. *Eur. J. Org. Chem.* 2003, 3958.

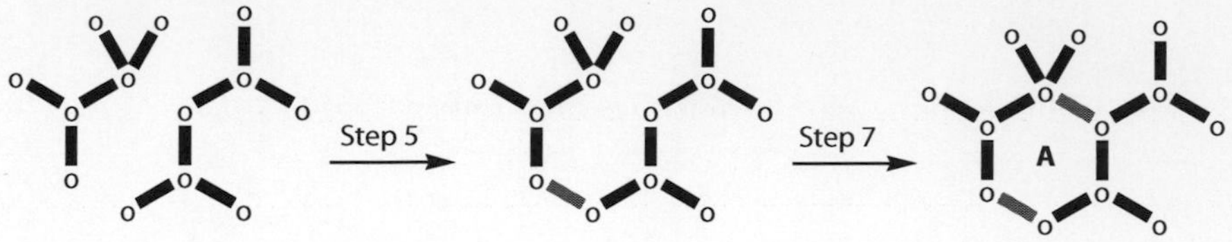

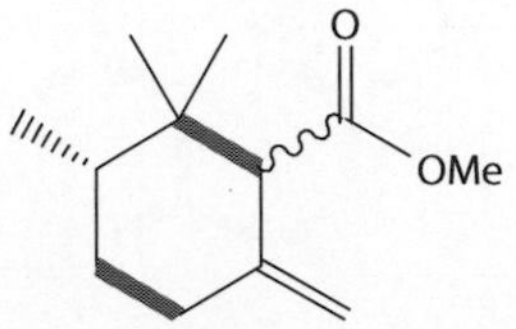

Sequence: $\xrightarrow{[6+0]} A_7$

(±)-*cis*-γ-Irone

Garnero, J.; Joulain, D. *Bull. Soc. Chim. Fr.* 1979, 15.

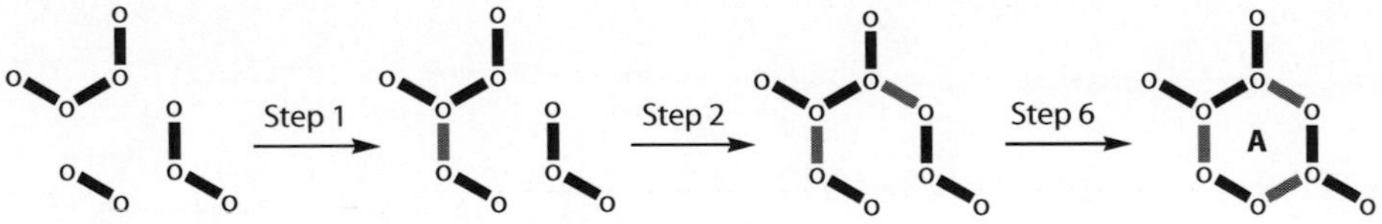

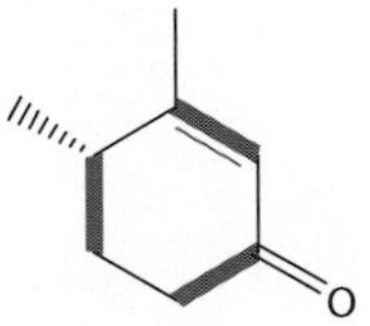

Sequence: $\xrightarrow{[6+0]} A_6$

Nussbaumer, C.; Fráter, G. *Helv. Chim. Acta* 1988, *71*, 619.

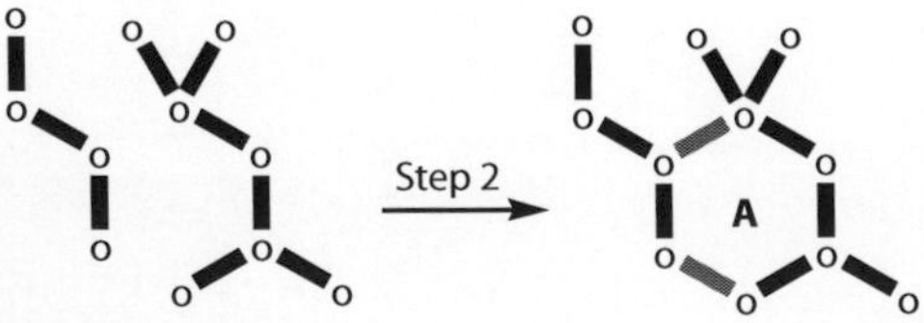

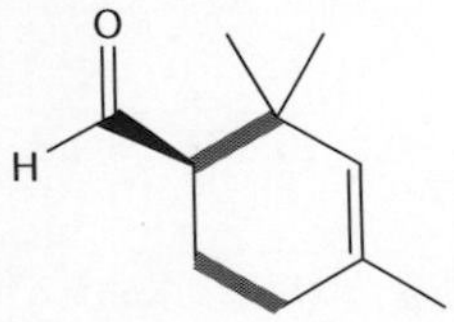

Sequence: $\xrightarrow{[4+2]} A_2$

Gosselin, P.; Perrotin, A.; Mille, S. *Tetrahedron* 2001, *57*, 733.

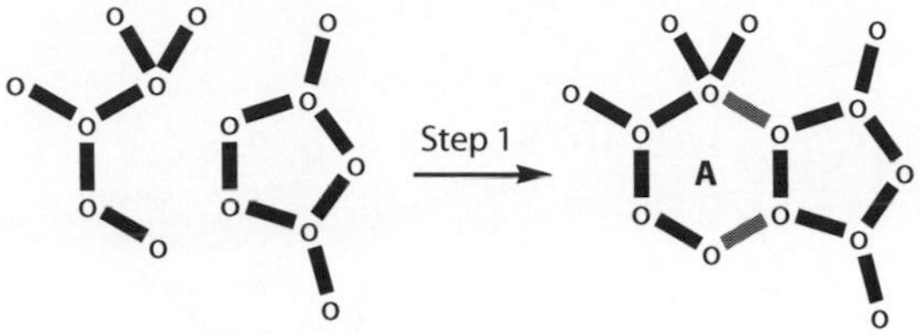

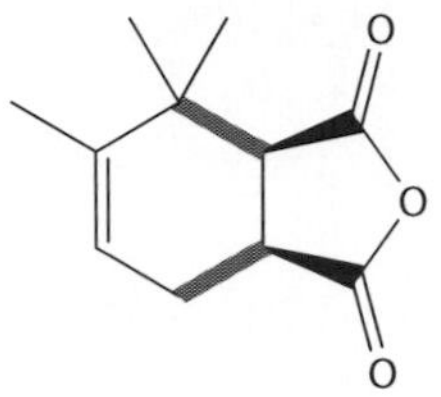

Sequence: $\xrightarrow{[4+2]} A_1$

Leyendecker, F.; Comte, M.T. *Tetrahedron* 1987, *43*, 85.

Preformed ring precursor: (±)-carvone

Monti, H.; Laval, G.; Feraud, M. *Eur. J. Org. Chem.* 1999, 1825.

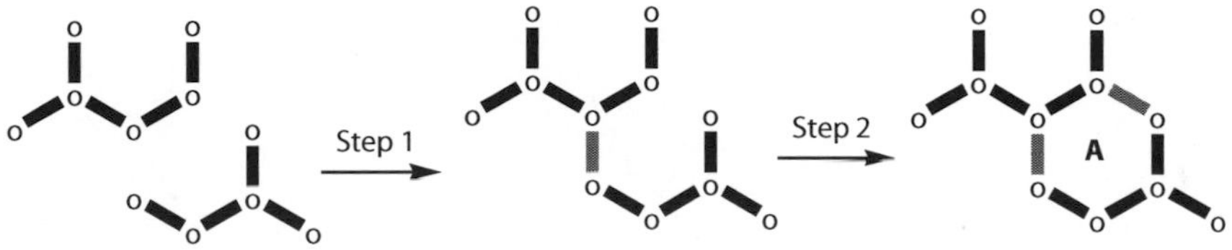

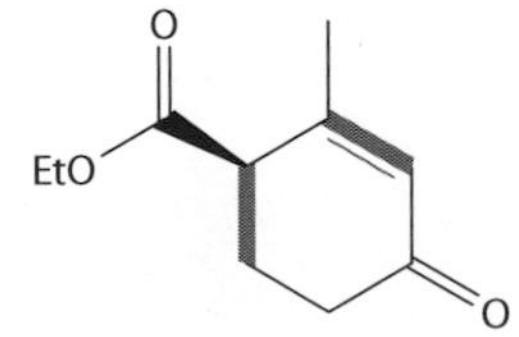

Sequence: $\xrightarrow{[6+0]} A_2$

Kitahara, T.; Tanida, K.; Mori, K. *Agric. Biol. Chem.* 1983, *47*, 581.

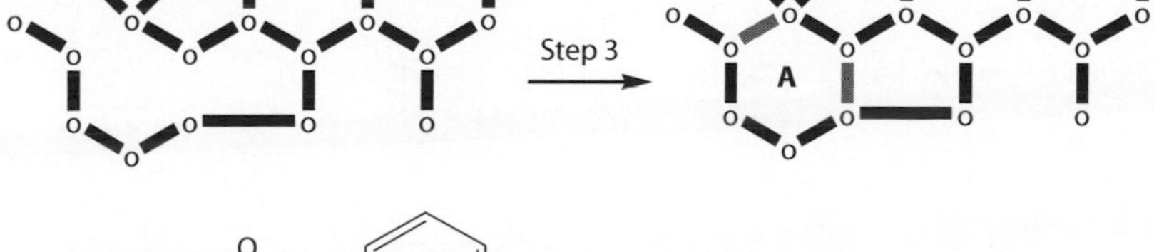

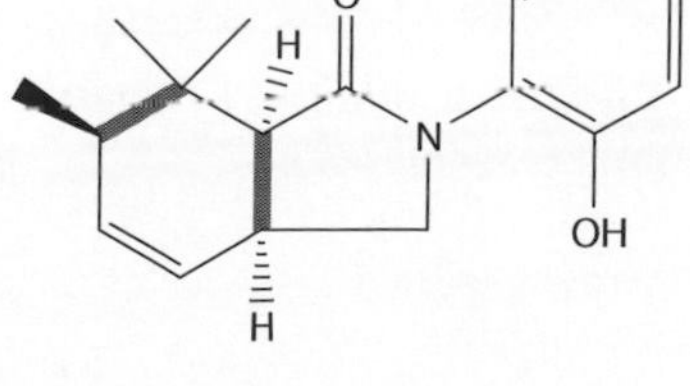

Sequence: $\xrightarrow{[4+2]} A_3$

(–)-(2S,6R)-*cis*-γ-Irone
Inoue, T.; Kiyota, H.; Oritani, T. *Tetrahedron Asymm.* 2000, *11*, 3807.

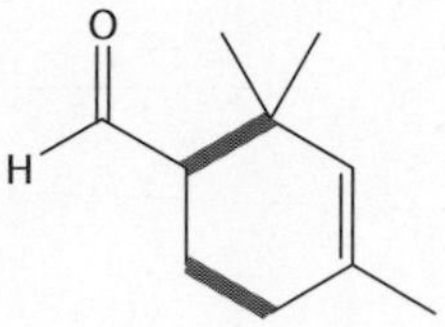

Sequence: $\xrightarrow{[4+2]} A_2$

Bezant, S.; Giannini, E.; Zanoni, G.; Vidari, G. *Eur. J. Org. Chem.* 2003, 3958.

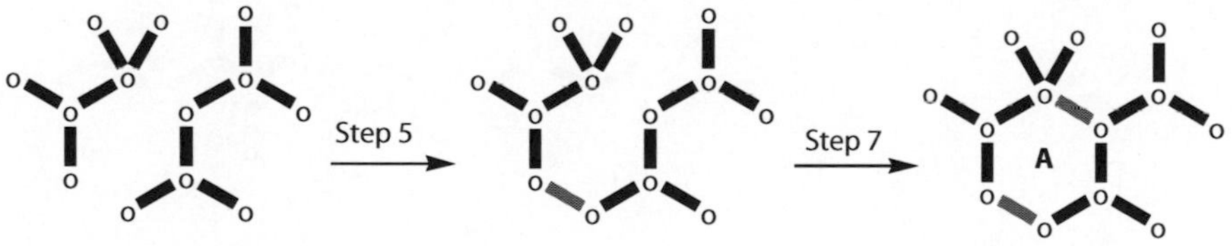

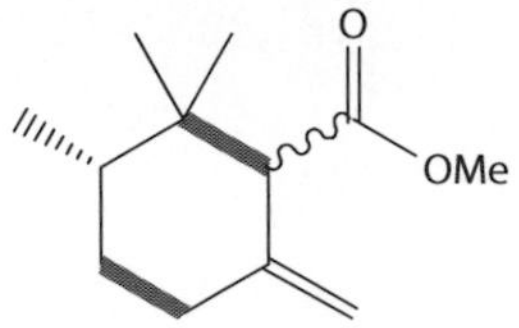

Sequence: $\xrightarrow{[6+0]} A_7$

(±)-(2R,6S)-*cis*-γ-Irone
Laval, G.; Andran, G.; Galano, J.M.; Monti, H. *J. Org. Chem.* 2000, *65*, 3551.

Preformed ring precursor: (–)-carvone

Summary:
Strategy [4+2]: Givaudan, Gosselin, Kiyota, Mori
Strategy [6+0]: Brenna G1, Garnero-Joulain, Monti G1, Takazawa, Vidari G1, Vidari G2
Preformed ring precursors: Brenna G2, Leyendecker, Monti G2, Monti G3

(–)-Kainic acid
Anderson, J.C.; O'Loughlin, J.M.A.; Tornos, J.A. *Org. Biomol. Chem.* 2005, *3*, 2741.

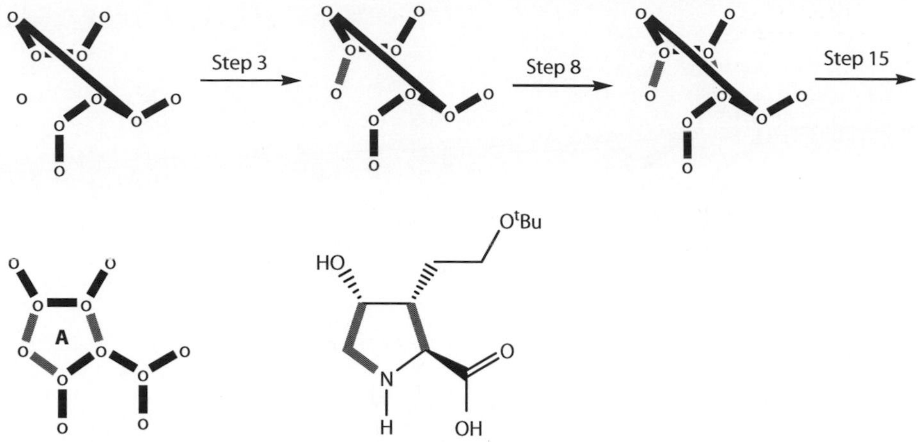

Sequence: $\xrightarrow{[5]\rightarrow[5+0]} A_{15}$

Bachi, M.D.; Melman, A. *J. Org. Chem.* 1997, *62*, 1896.
Bachi, M.D.; Melman, A. *Pure Appl. Chem.* 1998, *70*, 259.

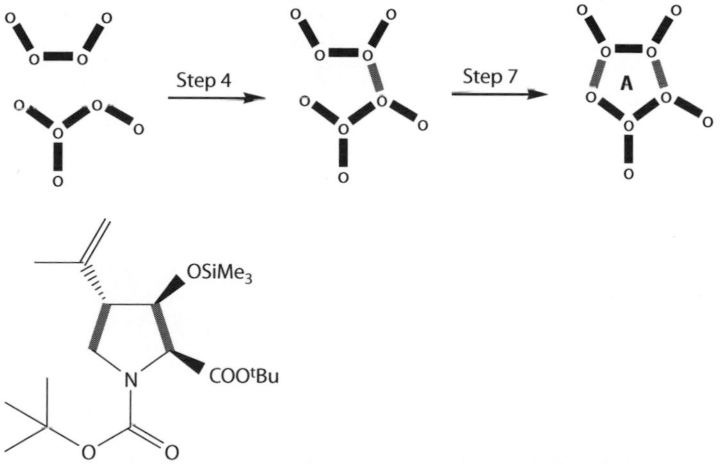

Sequence: $\xrightarrow{[5+0]} A_7$

Baldwin, J.E.; Moloney, M.G.; Parsons, A.F. *Tetrahedron* 1990, *46*, 7263.

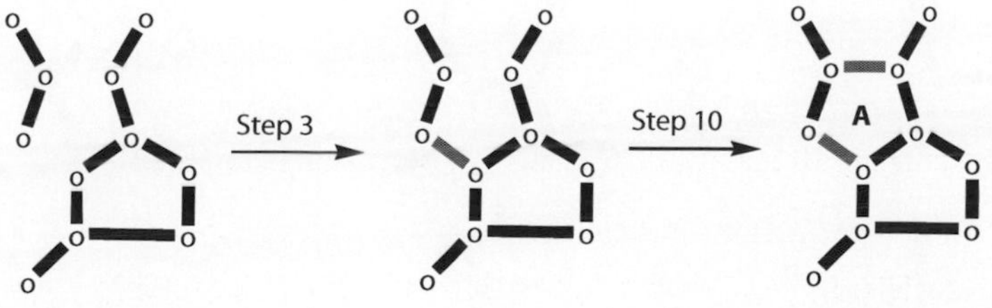

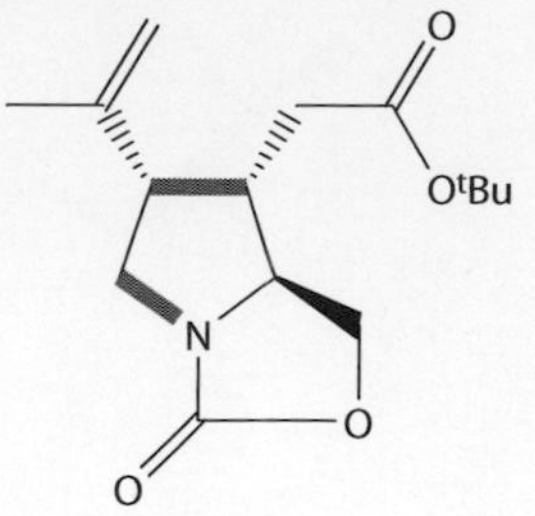

Sequence: $\xrightarrow{[5+0]} A_{10}$

Barco, A.; Benetti, S.; Pollini, G.P.; Spalluto, G.; Zanirato, V. *Chem. Commun.* 1991, 390.
Barco, A.; Benetti, S.; Spalluto, G.; Casolari, A.; Pollini, G.P.; Zanirato, V. *J. Org. Chem.* 1992, *57*, 6279.

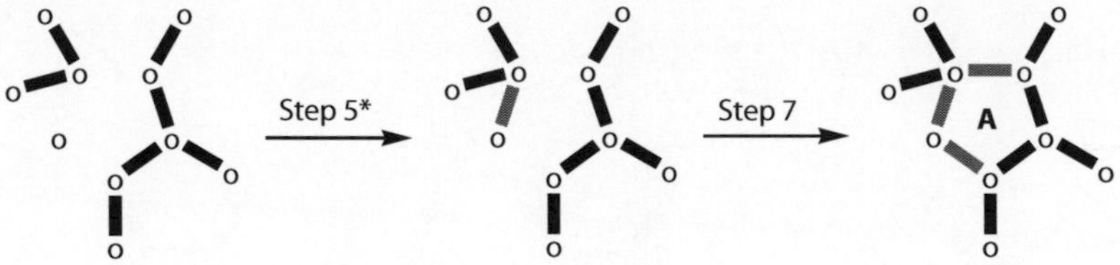

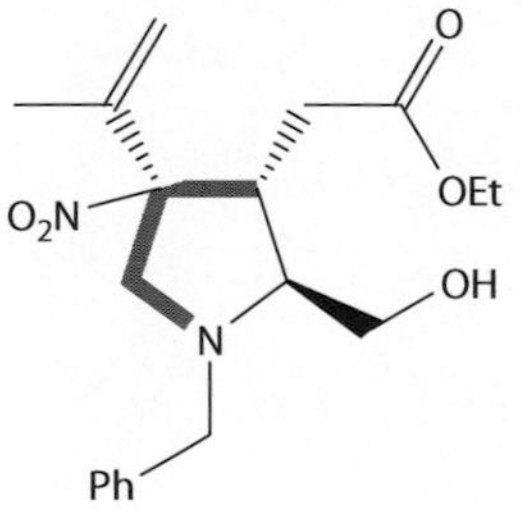

Sequence: $\xrightarrow{[3+2]} A_7$

Clayden, J.; Menet, C.J.; Tchabanenko, K. *Tetrahedron* 2002, *58*, 4727.
Clayden, J.; Menet, C.J.; Mansfield, D.J. *Chem. Commun.* 2002, 38.

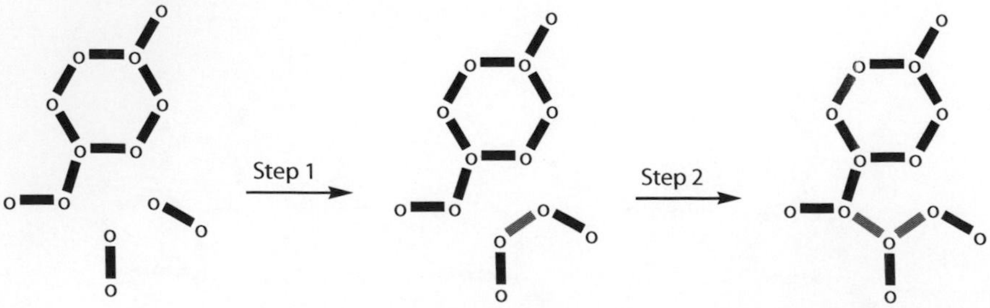

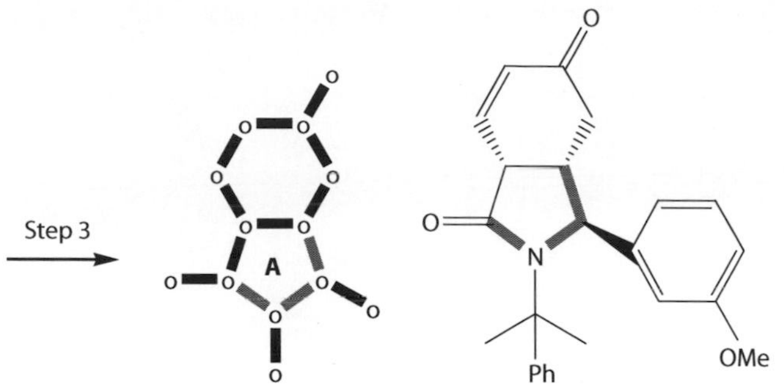

Sequence: $\xrightarrow{[5+0]} A_3$

Cossy, J.; Cases, M.; Pardo, D.G. *Tetrahedron* 1999, *55*, 6153.
Cossy, J.; Cases, M.; Pardo, D.G. *Synlett* 1998, 507.

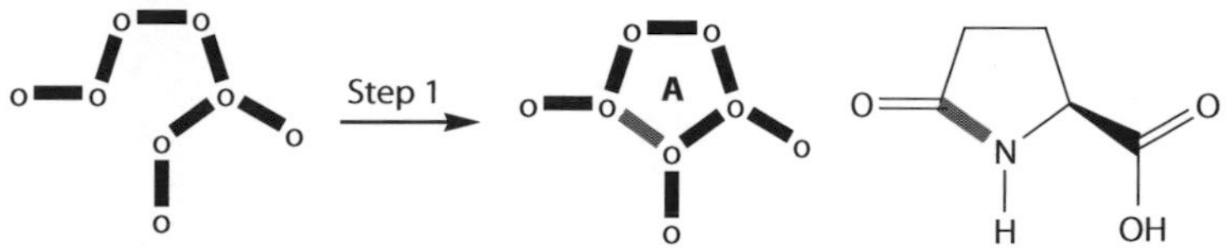

Sequence: $\xrightarrow{[5+0]} A_1$

Morita, Y.; Tokuyama, H.; Fukuyama, T. *Org. Lett.* 2005, *7*, 4337.

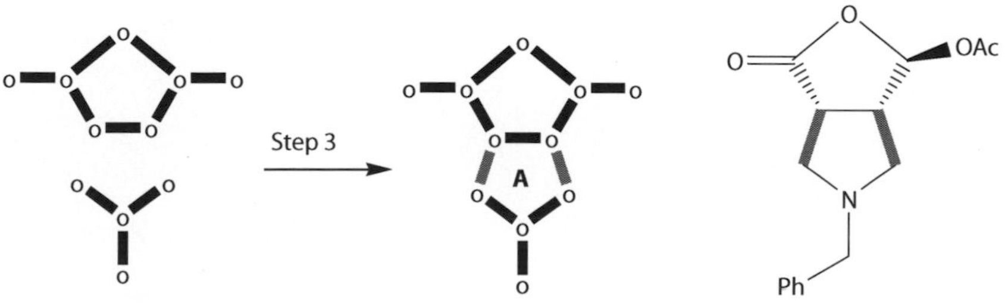

Sequence: $\xrightarrow{[3+2]} A_3$

Sakaguchi, H.; Tokuyama, H.; Fukuyama, T. *Org. Lett.* 2007, *9*, 1635.

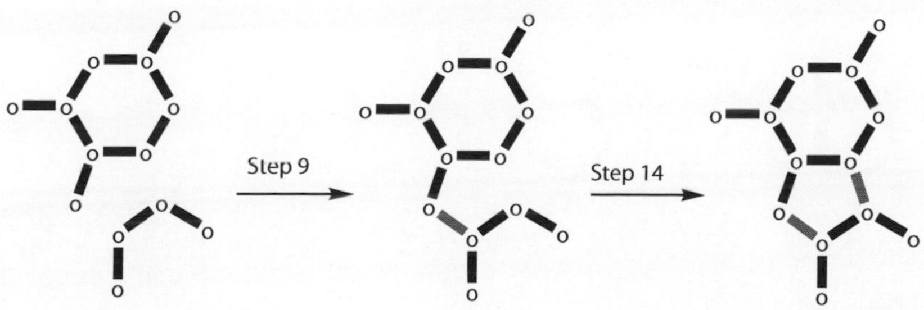

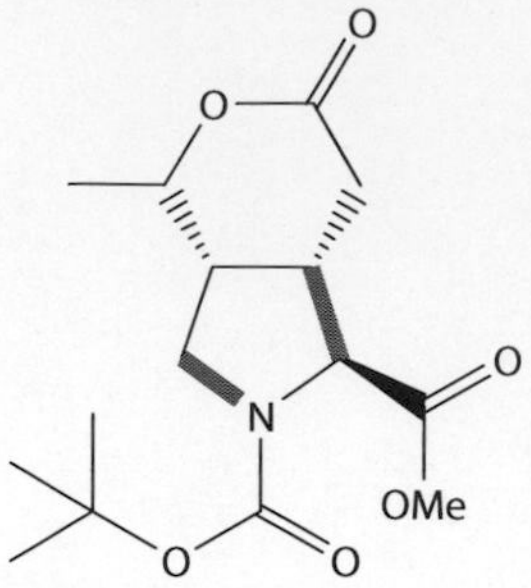

Sequence: $\xrightarrow{[5+0]} A_{14}$

Sakaguchi, H.; Tokuyama, H.; Fukuyama, T. *Org. Lett.* 2008, *10*, 1711.

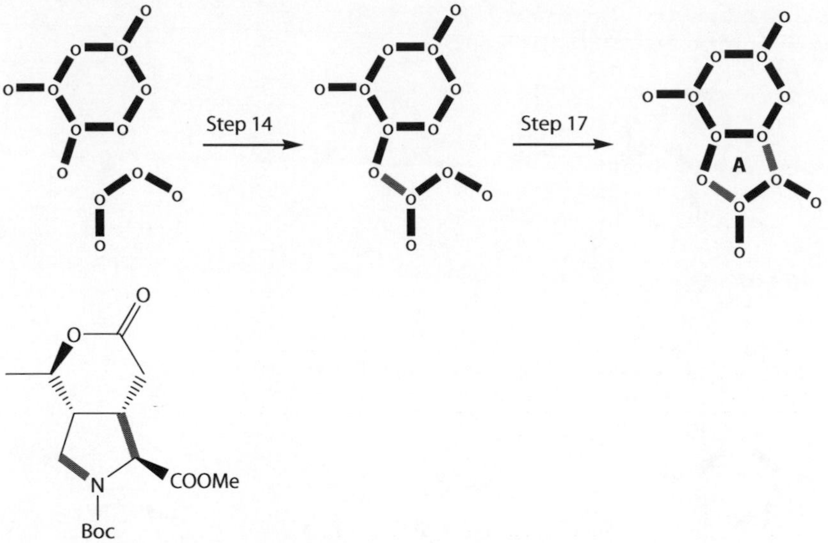

Sequence: $\xrightarrow{[5+0]} A_{17}$

Xia, Q.; Ganem, B. *Org. Lett.* 2001, *3*, 485.

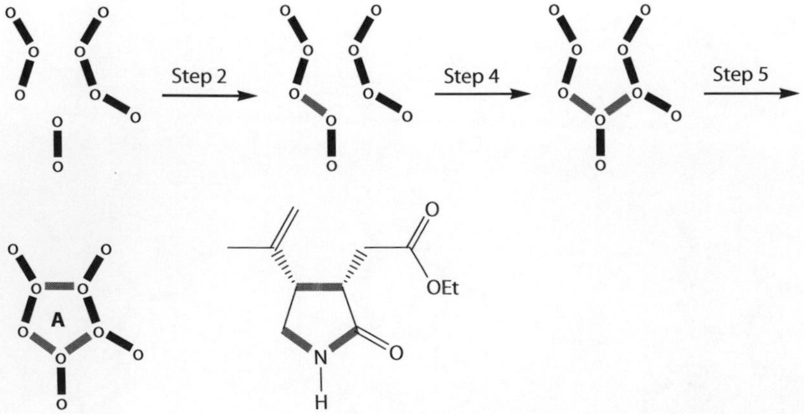

Sequence: $\xrightarrow{[5+0]} A_5$

Hanessian, S.; Ninkovic, S. *J. Org. Chem.* 1996, *61*, 5418.

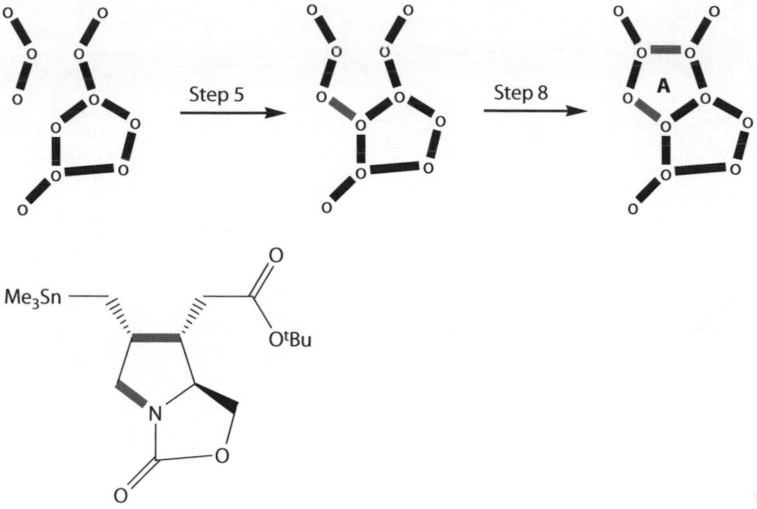

Sequence: $\xrightarrow{[5+0]} A_8$

Martinez, M.M.; Hoppe, D. *Eur. J. Org. Chem.* 2005, 1427.

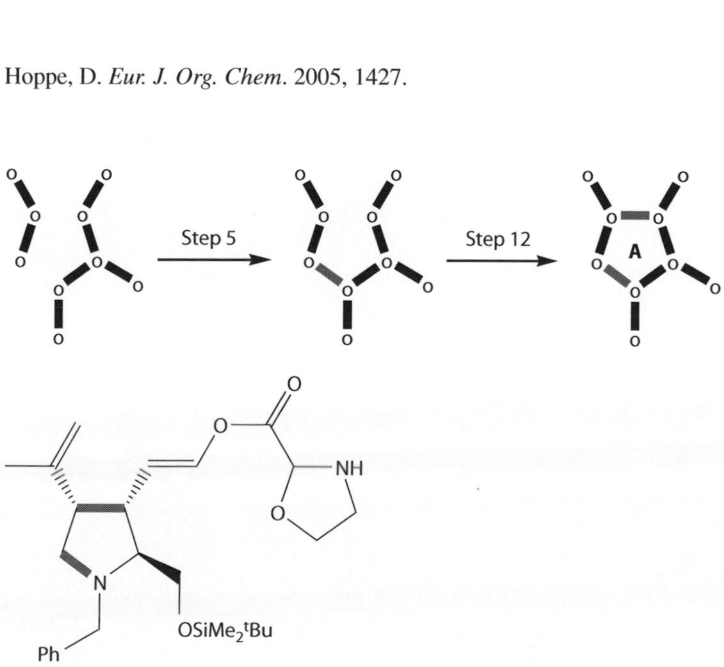

Sequence: $\xrightarrow{[5+0]} A_{12}$

Jung, Y.C.; Yoon, C.H.; Turos, E.; Yoo, K.S.; Jung, K.W. *J. Org. Chem.* 2007, *72*, 10114.

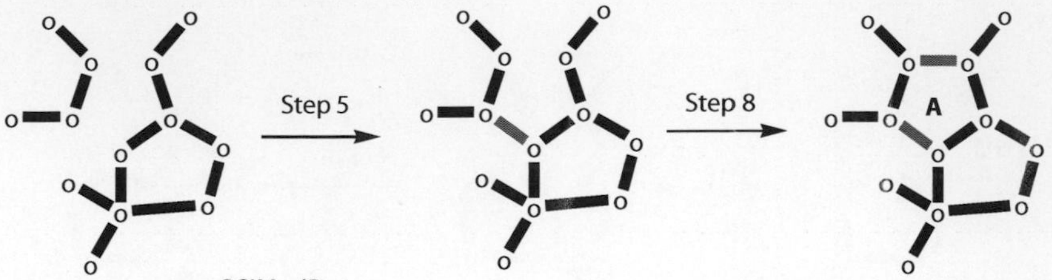

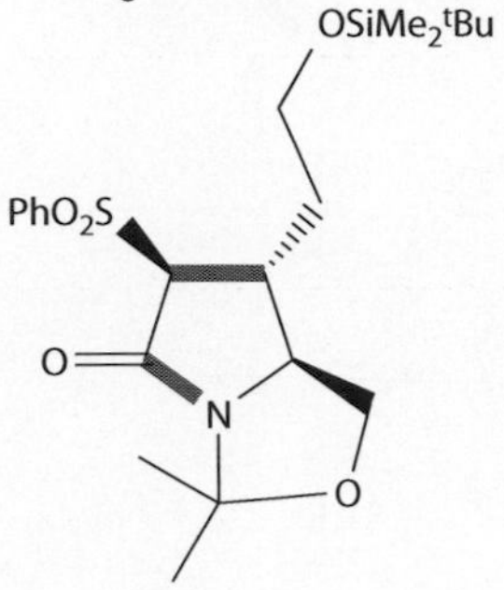

Sequence: $\xrightarrow{[5+0]} A_8$

Cooper, J.; Knight, D.W.; Gallagher, P.T. *Chem. Commun.* 1987, 1220.
Cooper, J.; Knight, D.W.; Gallagher, P.T. *J. Chem. Soc. Perkin Trans. I* 1992, 553.

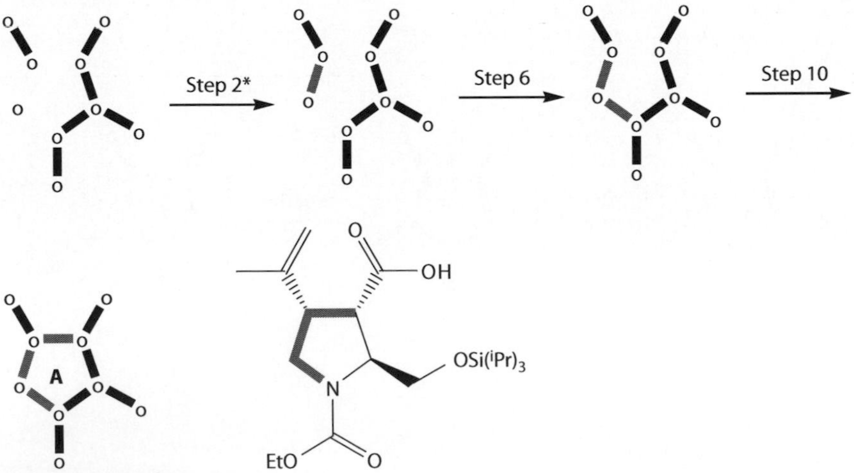

Sequence: $\xrightarrow{[5+0]} A_{10}$

Scott, M.E.; Lautens, M. *Org. Lett.* 2005, *7*, 3045.

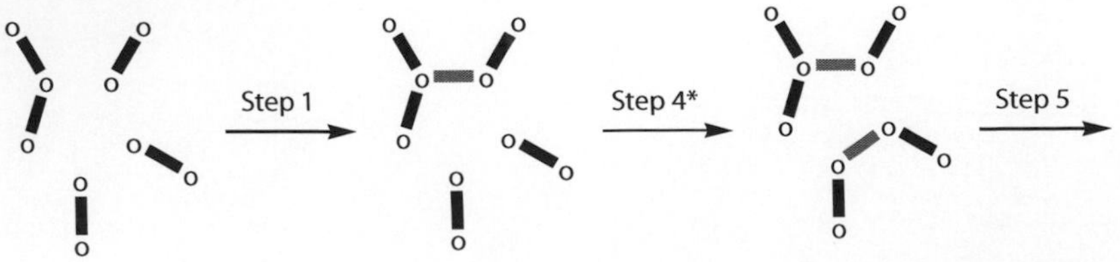

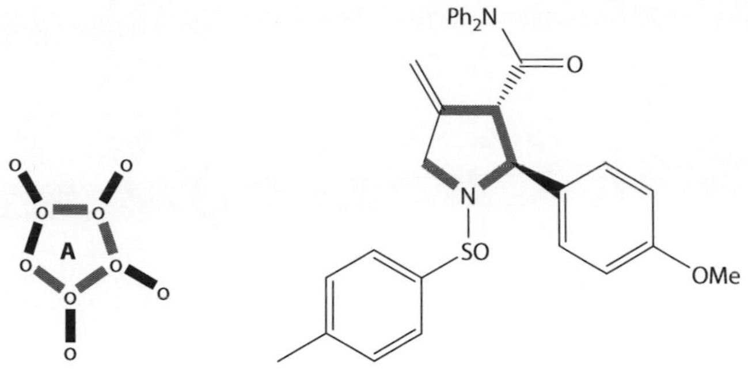
A
Ph2N
O
N
SO
OMe

Sequence: $\xrightarrow{[3+2]} A_5$

Chevliakov, M.V.; Montgomery, J. *J. Am. Chem. Soc.* 1999, *121*, 11139.

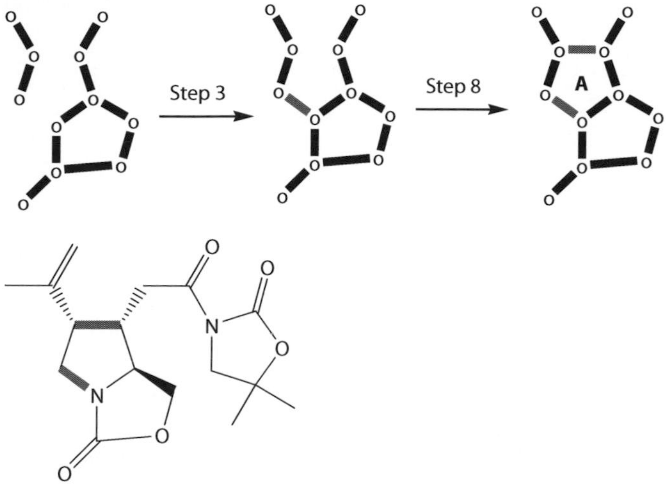
Step 3
Step 8
A

Sequence: $\xrightarrow{[5+0]} A_8$

Miyato, O.; Ozawa, Y.; Ninomiya, I.; Naito, T. *Synlett* 1997, 275.
Miyato, O.; Ozawa, Y.; Ninomiya, I.; Naito, T. *Tetrahedron* 2000, *56*, 6199.

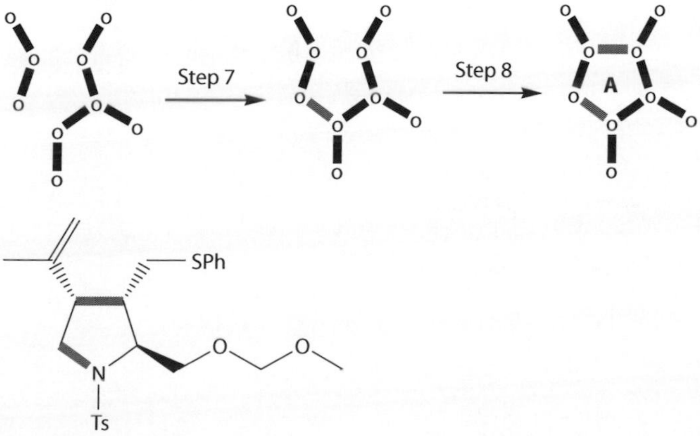
Step 7
Step 8
A
SPh
N
Ts

Sequence: $\xrightarrow{[5+0]} A_8$

Nakada, Y.; Sugahara, T.; Ogasawara, K. *Tetrahedron Lett.* 1997, *38*, 857.

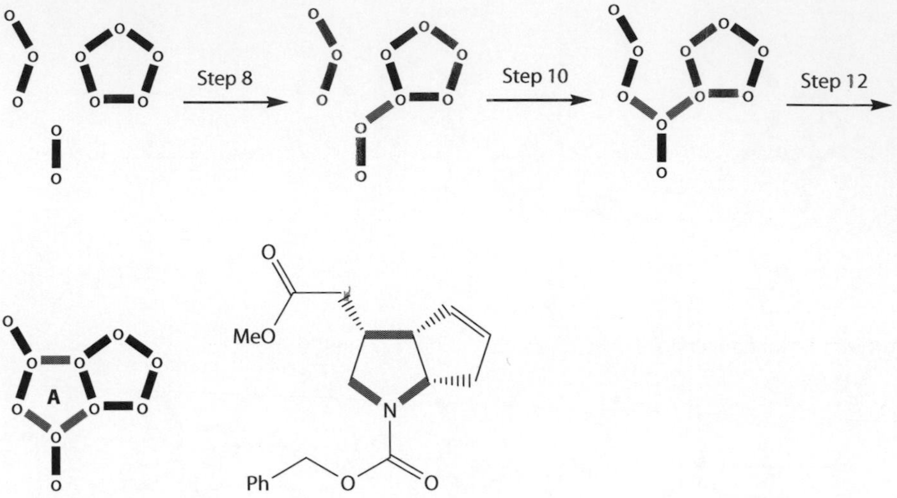

Sequence: $\xrightarrow{[5+0]}$ A_{12}

Nakagawa, H.; Sugahara, T.; Ogasawara, K. *Org. Lett.* 2000, *2*, 3181.

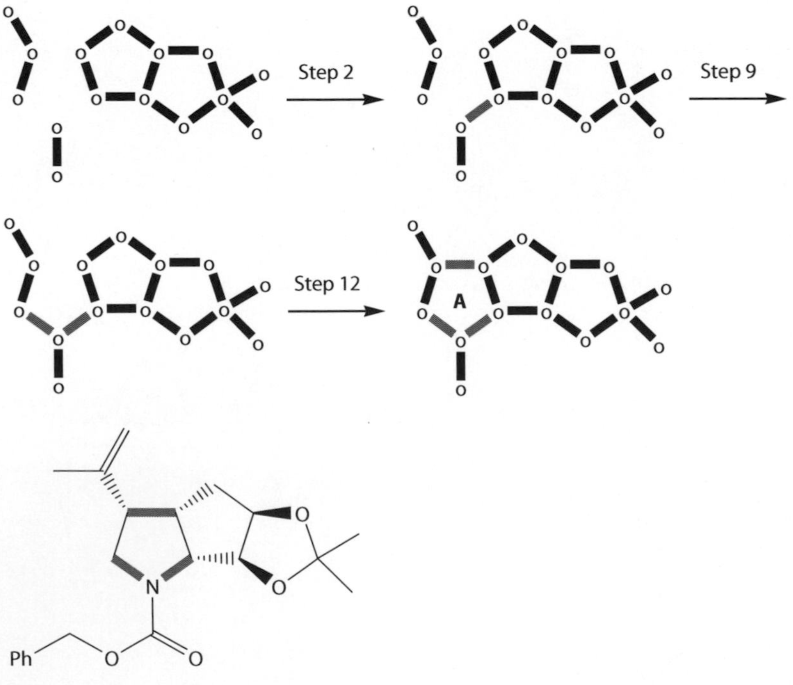

Sequence: $\xrightarrow{[5+0]}$ A_{12}

Hirasawa, H.; Taniguchi, T.; Ogasawara, K. *Tetrahedron Lett.* 2001, *42*, 7587.

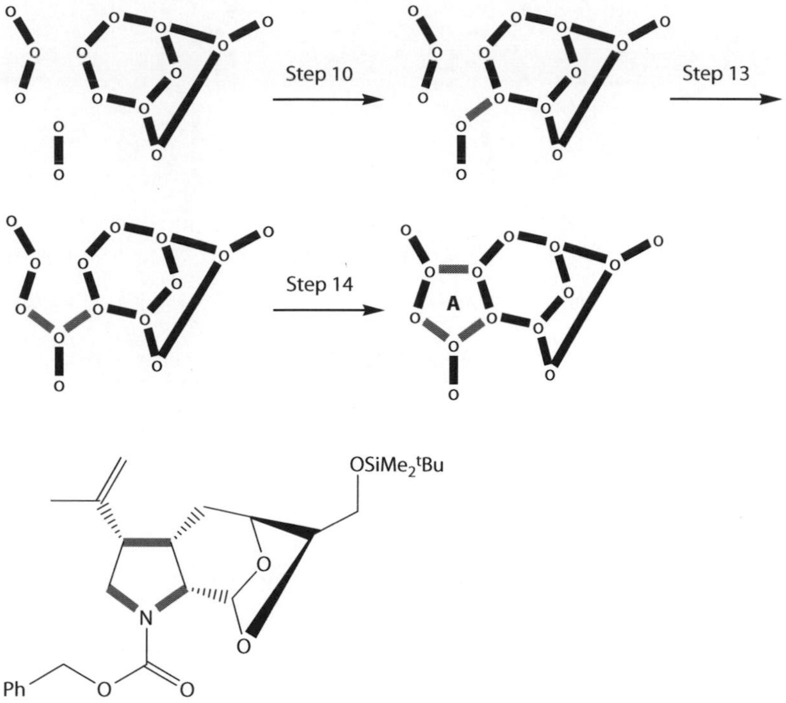

Sequence: $\xrightarrow{[5+0]} A_{14}$

Oppolzer, W.; Thirring, K. *J. Am. Chem. Soc.* 1982, *104*, 4978.

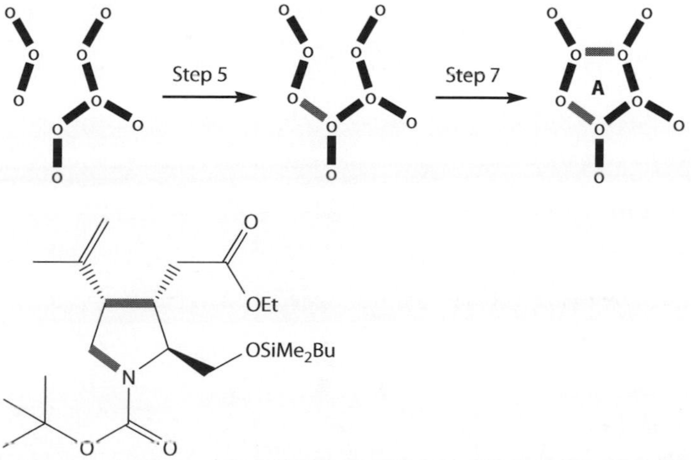

Sequence: $\xrightarrow{[5+0]} A_7$

Poisson, J.F.; Orellana, A.; Greene, A.E. *J. Org. Chem.* 2005, *70*, 10860.

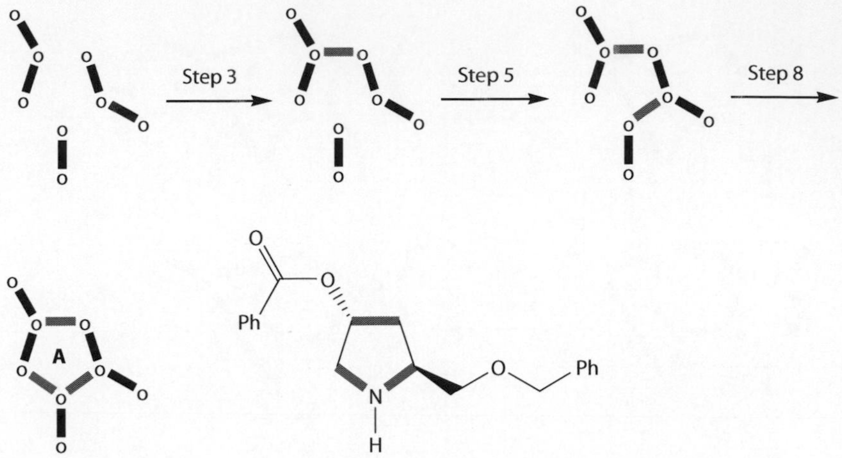

Sequence: $\xrightarrow{[5+0]} A_8$

Pandey, S.K.; Orellana, A.; Greene, A.E.; Poisson, J.F. *Org. Lett.* 2006, *8*, 5665.

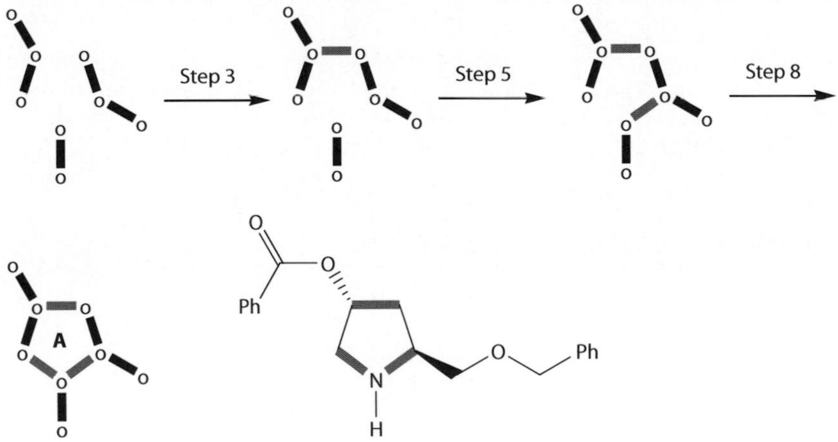

Sequence: $\xrightarrow{[5+0]} A_8$

Rubio, A.; Ezquerra, J.; Escribano, A.; Remuifion, M.J.; Vaquero, J.J. *Tetrahedron Lett.* 1998, *39*, 2171.

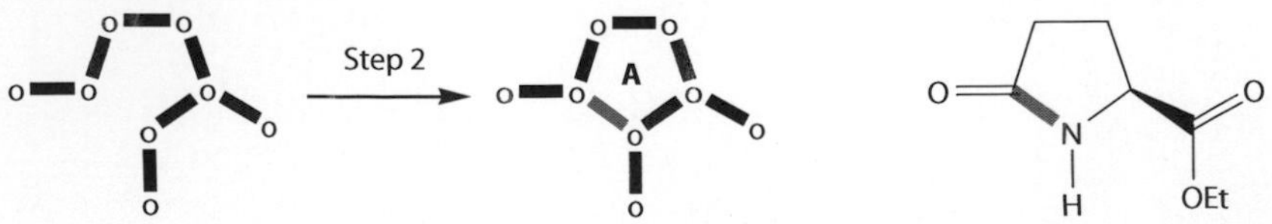

Sequence: $\xrightarrow{[5+0]} A_2$

Takano, S.; Iwabuchi, Y.; Ogasawara, K. *Chem. Commun.* 1988, 1204.

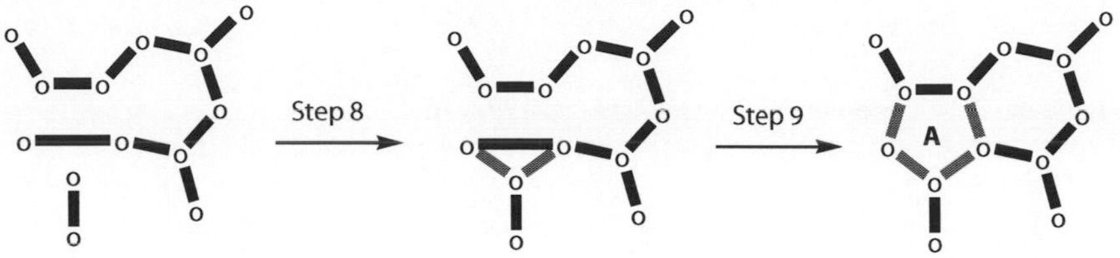

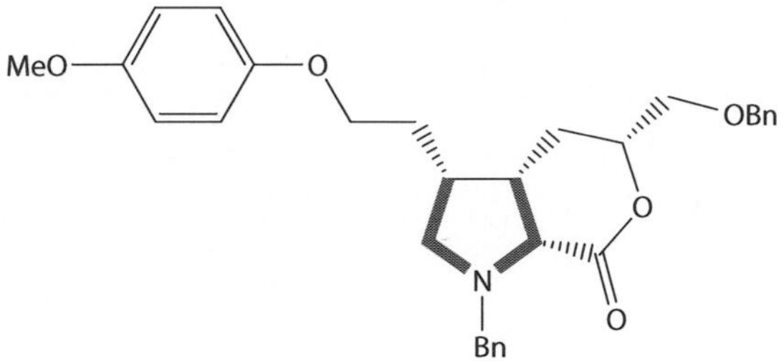

Sequence: $\xrightarrow{[3+2]} A_9$

Takano, S.; Sugihara, T.; Satoh, S.; Ogasawara, K. *J. Am. Chem. Soc.* 1988, *110*, 6467.

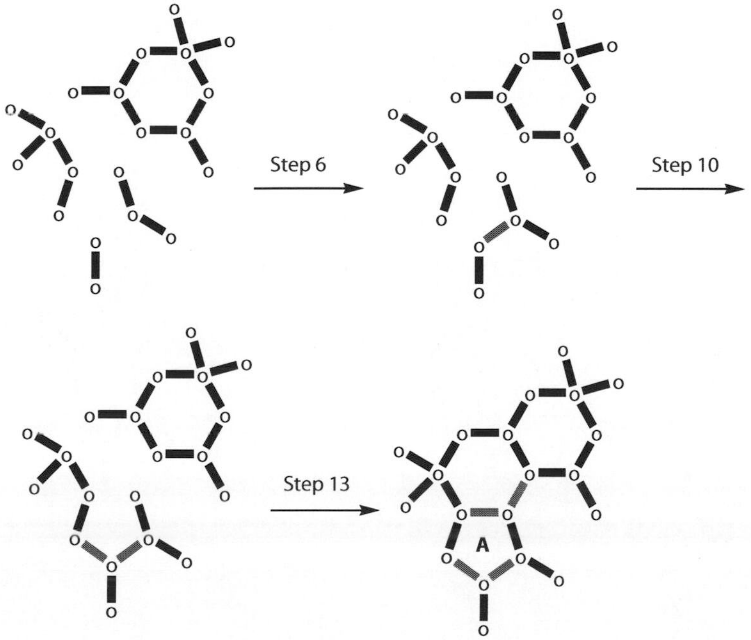

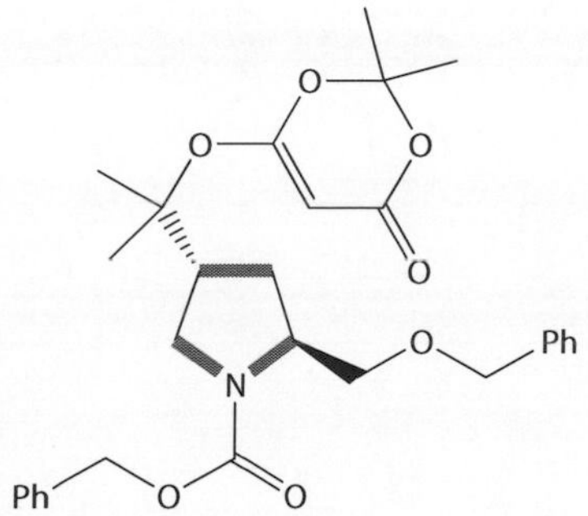

Sequence: $\xrightarrow{[5+0]_A + [3+2+1]}$ $[4.3.0]A_{13}$

Takano, S.; Inomata, K.; Ogasawara, K. *Chem. Commun.* 1992, 169.

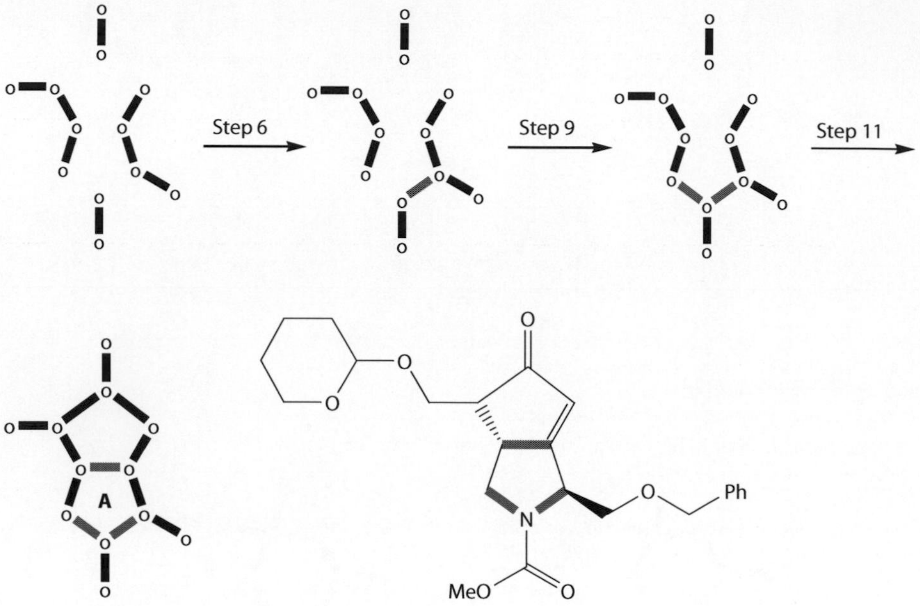

Sequence: $\xrightarrow{[5+0]_A + [2+2+1]}$ $[3.3.0]A_{11}$

Hatakeyama, S.; Sugawara, K.; Takano, S. *Chem. Commun.* 1993, 125.

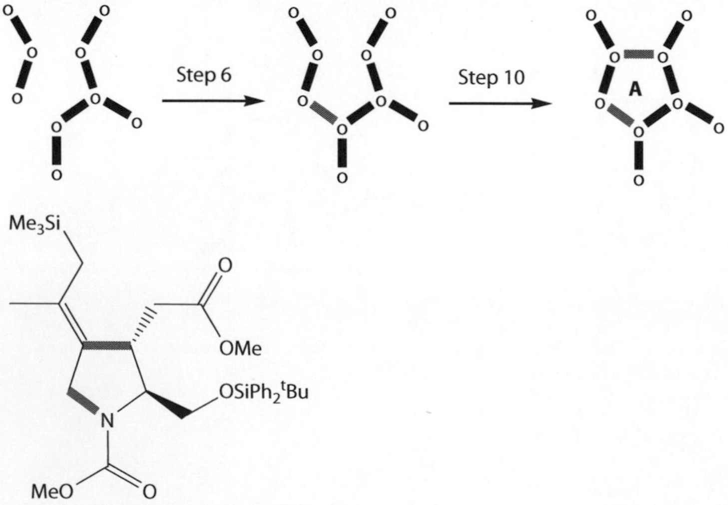

Sequence: $\xrightarrow{[5+0]}$ A_{10}

Campbell, A.D.; Raynham, T.M.; Taylor, R.J.K. *Chem. Commun.* 1999, 245.
Campbell, A.D.; Raynham, T.M.; Taylor, R.J.K. *J. Chem. Soc. Perkin Trans. 1*, 2000, 3194.

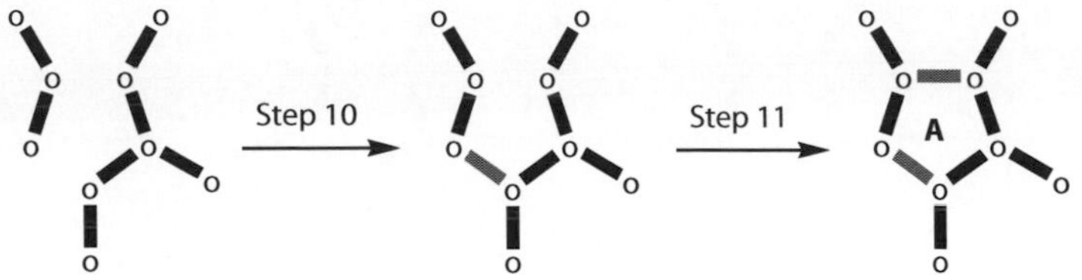

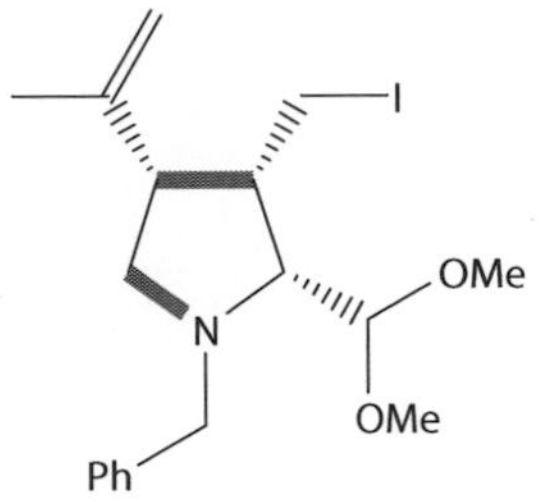

Sequence: $\xrightarrow{[5+0]} A_{11}$

Yoo, S.; Lee, S.H. *J. Org. Chem.* 1994, *59*, 6968.

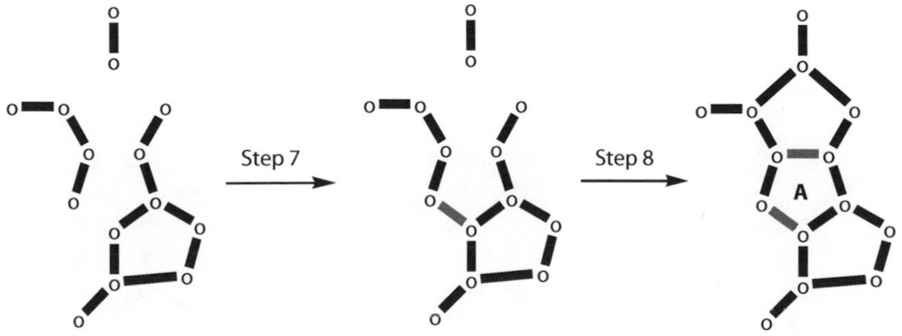

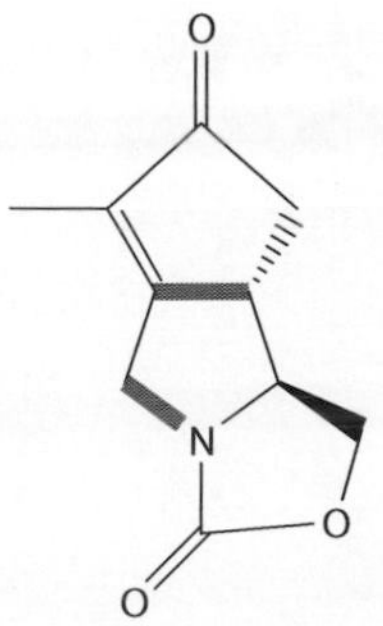

Sequence: $\xrightarrow{[5+0]_A + [2+2+1]} [3.3.0]A_8$

(±)-Kainic acid
Trost, B.M.; Rudd, M.T. *Org. Lett.* 2003, *5*, 1467.

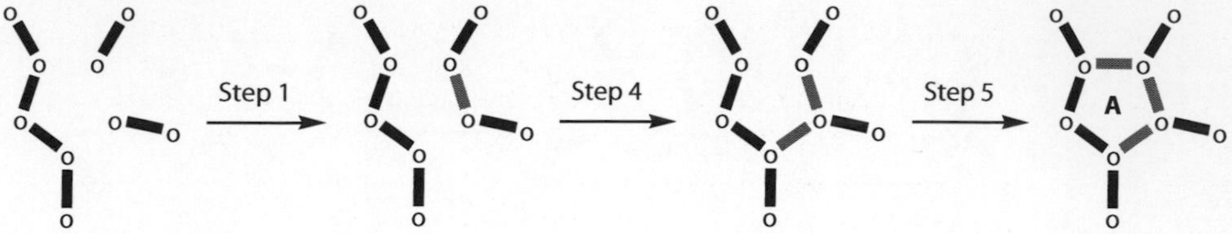

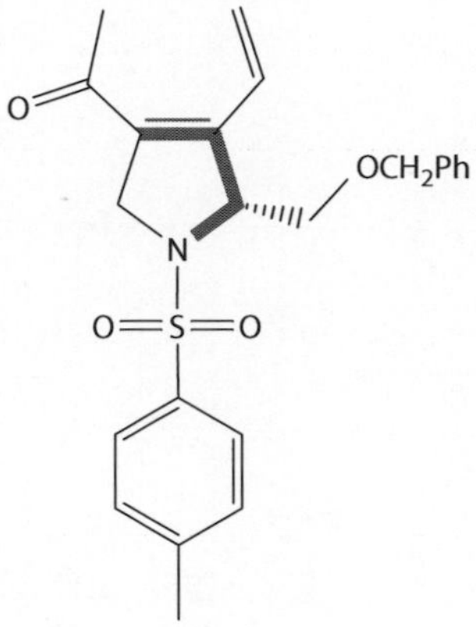

Sequence: $\xrightarrow{[5+0]} A_5$

(±)-Kainic acid
Monn, J.A.; Valli, M.J. *J. Org. Chem.* 1994, *59*, 2773.

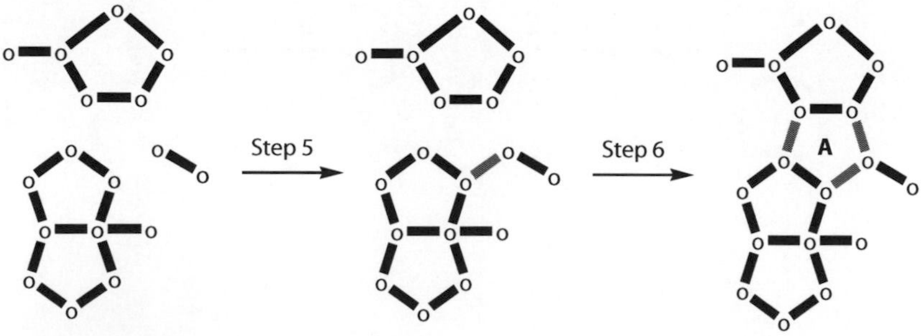

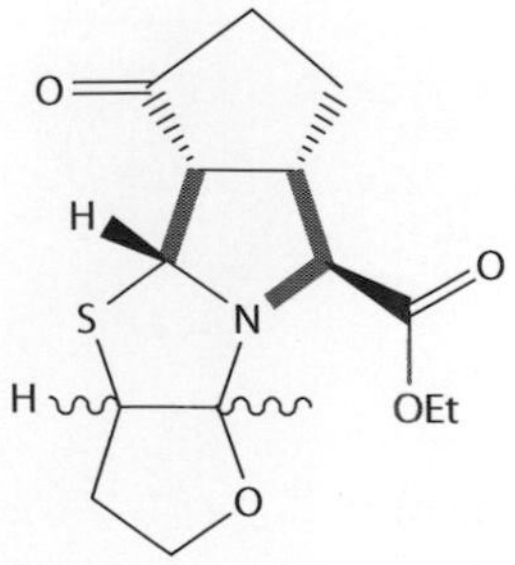

Sequence: $\xrightarrow{[3+2]} A_6$

Majik, M.S.; Parameswaran, P.S.; Tilve, S.G. *J. Org. Chem.* 2009, *74*, 3591.

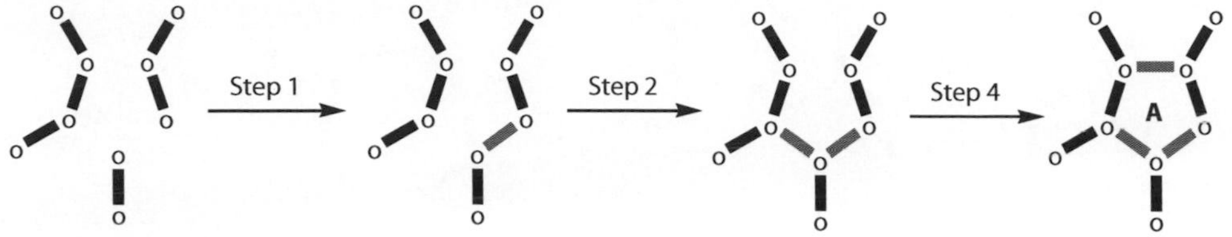

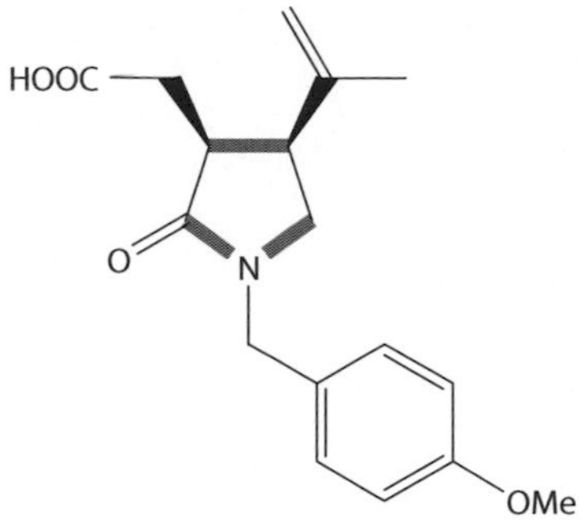

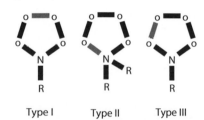

Sequence: $\xrightarrow{[5+0]} A_4$

Summary:

Strategy #1: $\xrightarrow{[5] \rightarrow [5+0]} A_{15}$ Anderson

Type I   Type II   Type III

Strategy #2a: [5 + 0] Type I: Baldwin, Ganem, Hanessian, Hoppe, Jung, Knight, Montgomery, Naito, Ogasawara G1, Ogasawara G2, Ogasawara G3, Oppolzer, Takano G4, Taylor, Tilve, Trost

Strategy #2b: [5 + 0] Type II: Cossy, Poisson G1, Poisson G2, Rubio

Strategy #2c: [5 + 0] Type III: Bachi, Clayden, Fukuyama G2, Fukuyama G3

Strategy #3: [3 + 2] Benetti, Fukuyama G1, Lautens, Monn, Takano G1

Strategy #4: $\xrightarrow{[5+0]_A + [3+2+1]} [4.3.0]A_{13}$ Takano G2

Strategy #5: $\xrightarrow{[5+0]_A + [2+2+1]} [3.3.0]A_{11}$ Takano G3

$\xrightarrow{[5+0]_A + [2+2+1]} [3.3.0]A_8$ Yoo

(±)-Laurallene

Crimmins, M.T.; Tabet, E.A. *J. Am. Chem. Soc.* 2000, *122*, 5473.

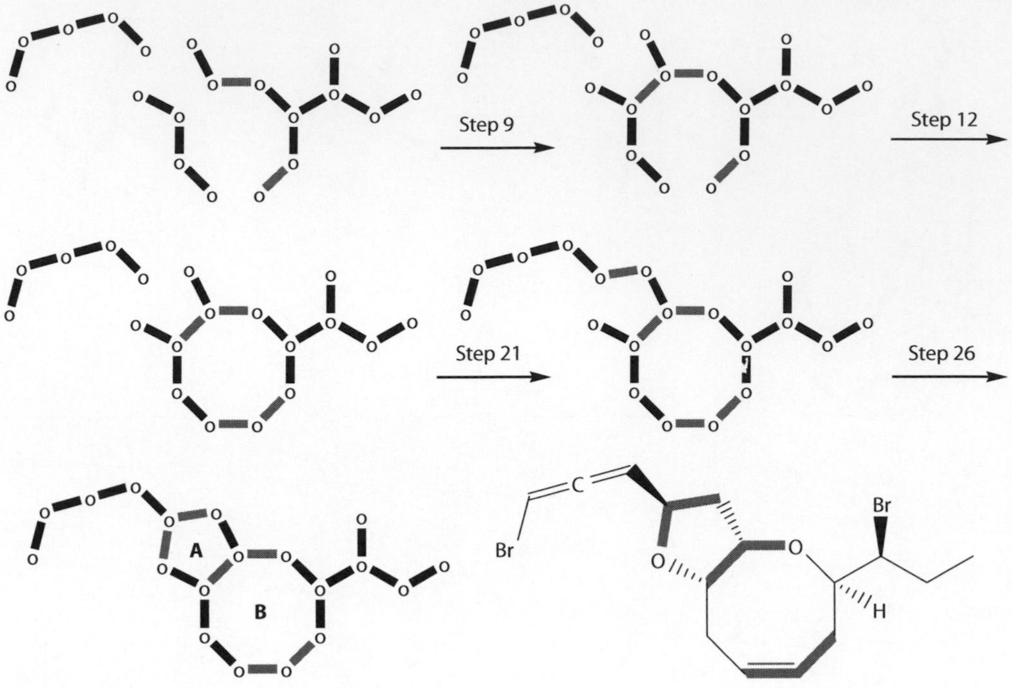

Sequence: $\xrightarrow{[8+0]} B_{12} \xrightarrow{[5+0]} A_{26}$

Sasaki, M.; Hashimoto, A.; Tanaka, K.; Kawahata, M.; Yamaguchi, K.; Takeda, K. *Org. Lett.* 2008, *10*, 1803.

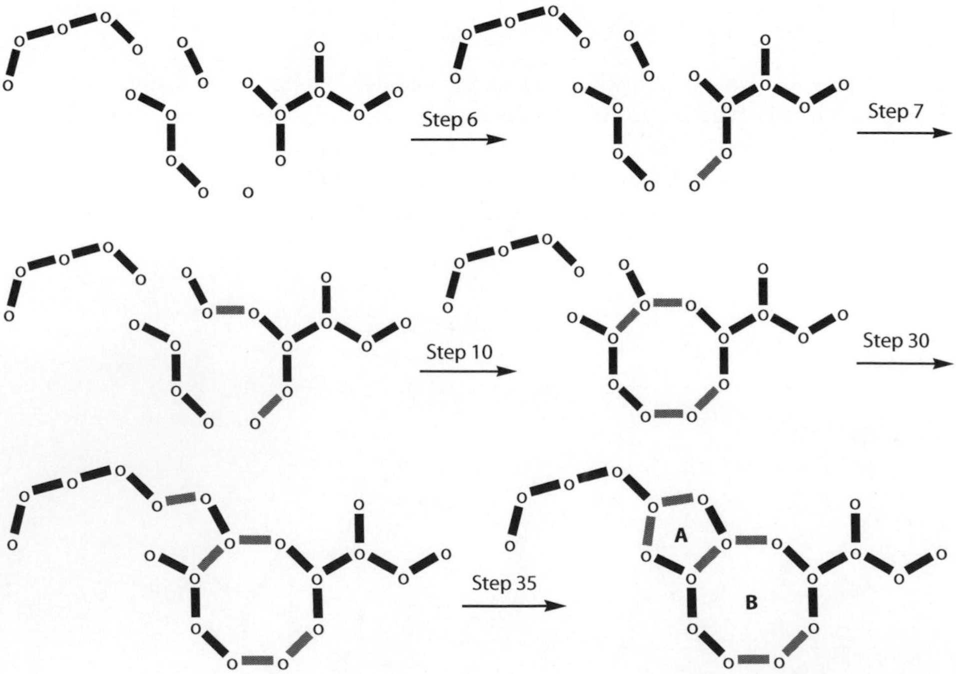

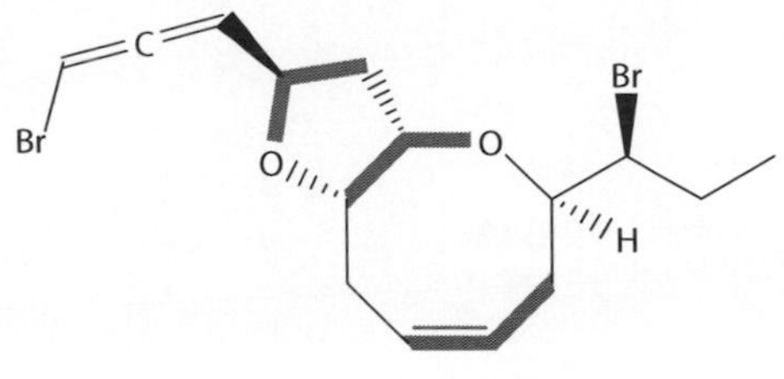

Sequence: $\xrightarrow{[5+3]} B_{10} \xrightarrow{[5+0]} A_{35}$

Saitoh, T.; Suzuki, T.; Sugimoto, M.; Hagiwara, H.; Hoshi, T. *Tetrahedron Lett.* 2003, *44*, 3175.

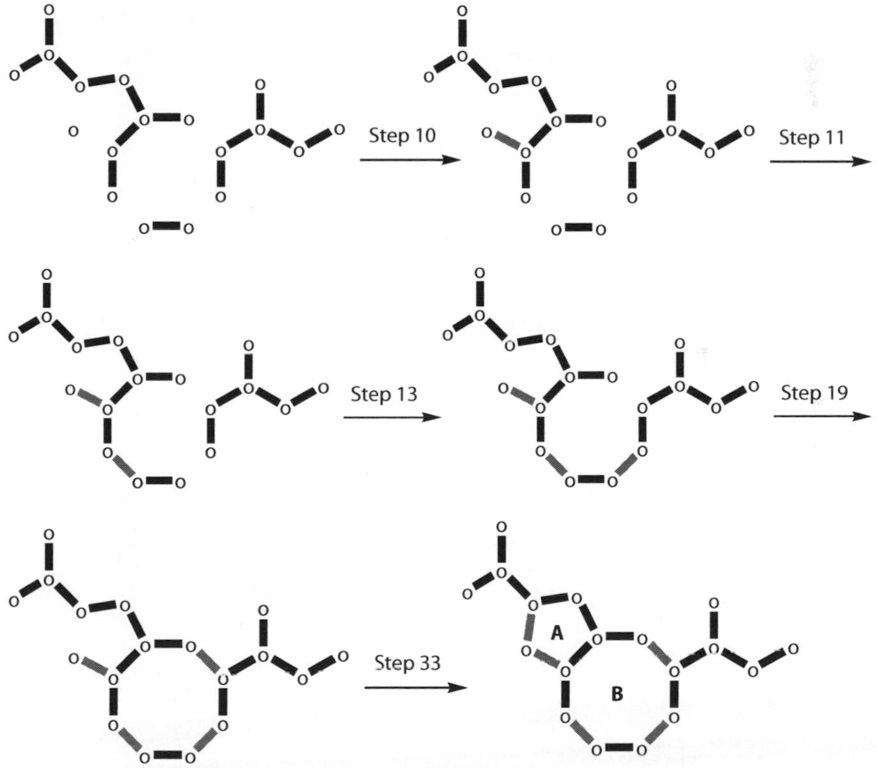

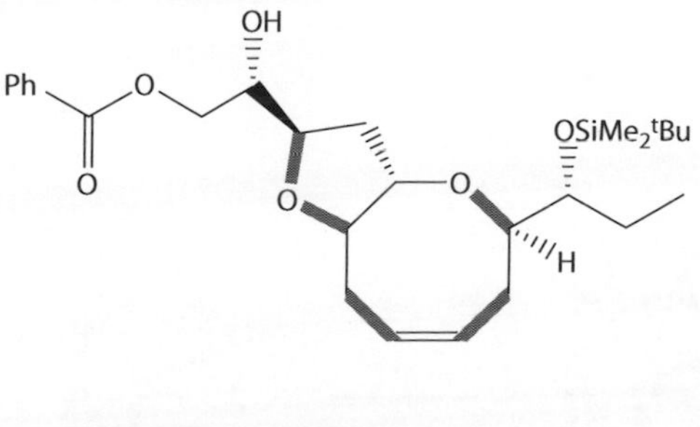

Sequence: $\xrightarrow{[8+0]} B_{19} \xrightarrow{[5+0]} A_{33}$

Summary:
Strategy #1

$$\xrightarrow{[8+0]} B_{12} \xrightarrow{[5+0]} A_{26}: \text{Crimmins}$$

$$\xrightarrow{[8+0]} B_{19} \xrightarrow{[5+0]} A_{33}: \text{Suzuki}$$

Strategy #2

$$\xrightarrow{[5+3]} B_{10} \xrightarrow{[5+0]} A_{35}: \text{Takeda}$$

(±)-Longifolene

Corey, E.J.; Ohno, M.; Vatakencherry, P.A.; Mitra, R.B. *J. Am. Chem. Soc.* 1961, *83*, 1251.
Corey, E.J.; Ohno, M.; Mitra, R.B.; Vatakencherry, P.A. *J. Am. Chem. Soc.* 1964, *86*, 478.

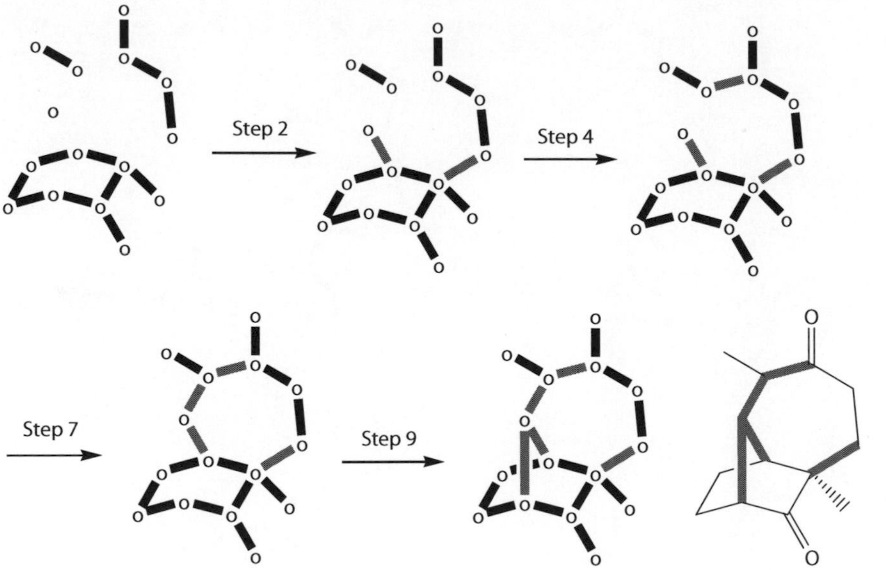

Sequence: $[6]_0 \xrightarrow{[7+0]} [7]_7 \xrightarrow{[(5+0)+(5+0)]} [2.2.1]_9$

(±)-Longifolene

Lei, B.; Fallis, A.G. *J. Org. Chem.* 1993, *58*, 2186.

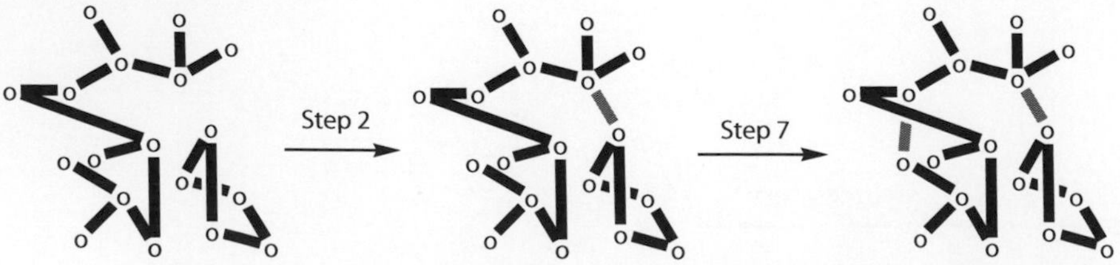

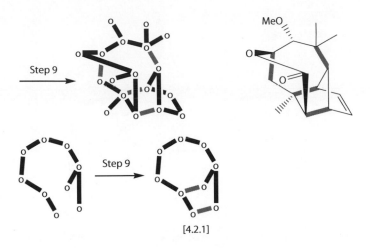

Sequence: $[5]_0 \xrightarrow{[6+0]} [6]_7 \xrightarrow{[(7+0)_{[7]}+[(4+2)+(3+2)]_{[2.2.1]}]} [4.2.1]([7]+[2.2.1])_9$

(±)-Longifolene

Volkmann, R.A.; Andrews, G.C.; Johnson, W.S. *J. Am. Chem. Soc.* 1975, *97*, 4777.

Sequence: $[5]_0 \xrightarrow{[(7+0)_{[7]}+[(4+2)+(3+2)_{[2.2.1]}]} [4.2.1]([7]+[2.2.1])_5$

(±)-Longifolene

Karimi, S.; Tavares, P. *J. Nat. Prod.* 2003, *66*, 520.
Karimi, S. *J. Nat. Prod.* 2001, *64*, 406.

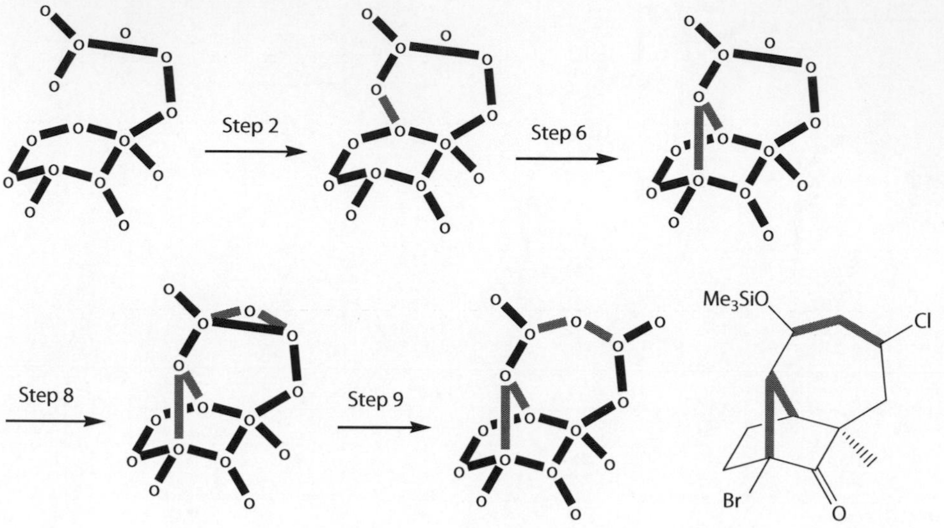

Sequence:

$$[6]_0 \xrightarrow{[6+0]} [6]_2 \xrightarrow{[(5+0)+(5+0)]} [2.2.1]_6 \xrightarrow{[2+1]} [4.1.0]_8 \xrightarrow{[4.1.0] \rightarrow [7]} [7]_9$$

(±)-Longifolene

Ho, T.L.; Yeh, W.L.; Yule, J.; Liu, H.J. *Can. J. Chem.* 1992, *70*, 1375.

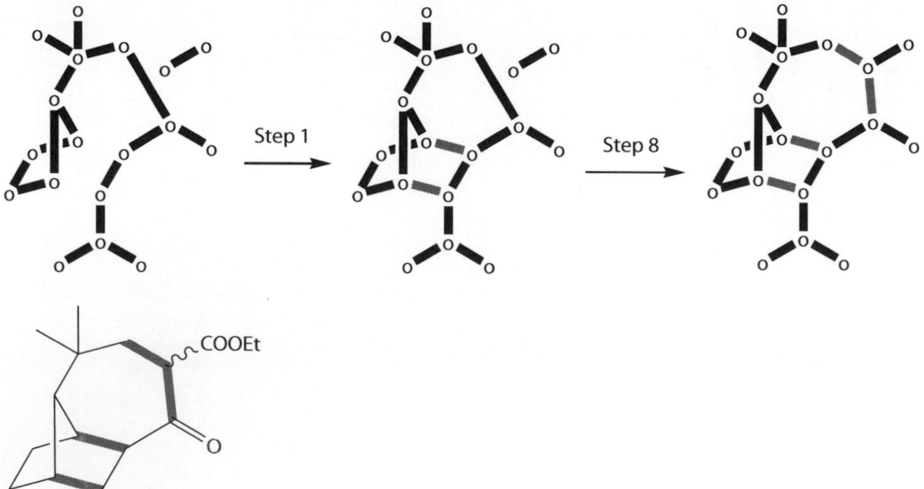

Sequence: $[5]_0 \xrightarrow{[(4+2)+(3+2)]} [2.2.1]_1 \xrightarrow{[3.2.1]+[1] \rightarrow [4.2.1]} [4.2.1]_8$

(±)-Longifolene
McMurry, J.E.; Isser, S.J. *J. Am. Chem. Soc.* 1972, *94*, 7132.

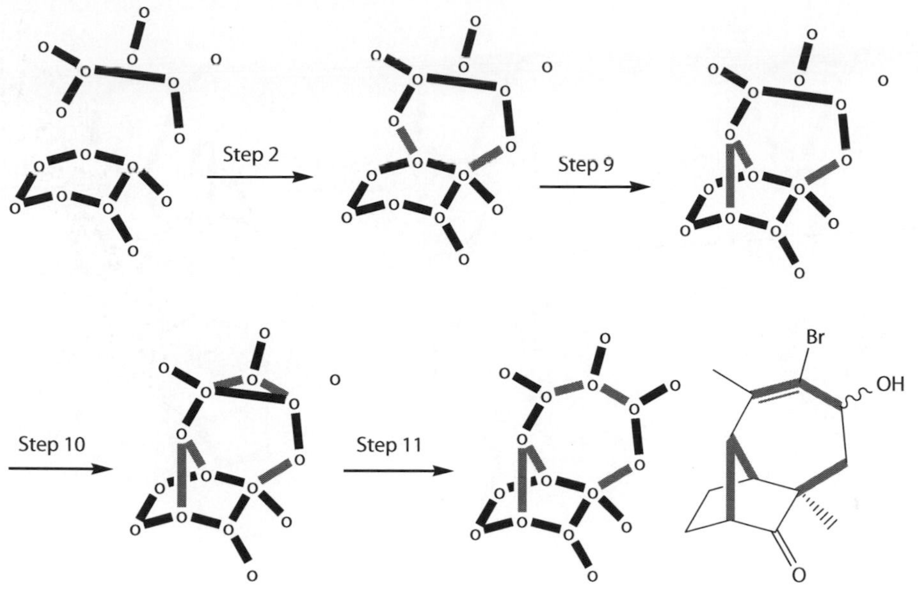

Sequence:

$$[6]_0 \xrightarrow{[4+2]} [6]_2 \xrightarrow{[(5+0)+(5+0)]} [2.2.1]_9 \xrightarrow{[2+1]} [4.1.0]_{10} \xrightarrow{[4.1.0] \rightarrow [7]} [7]_{11}$$

(±)-Longifolene
Kuo, D.L.; Money, T. *Can. J. Chem.* 1988, *66*, 1794.

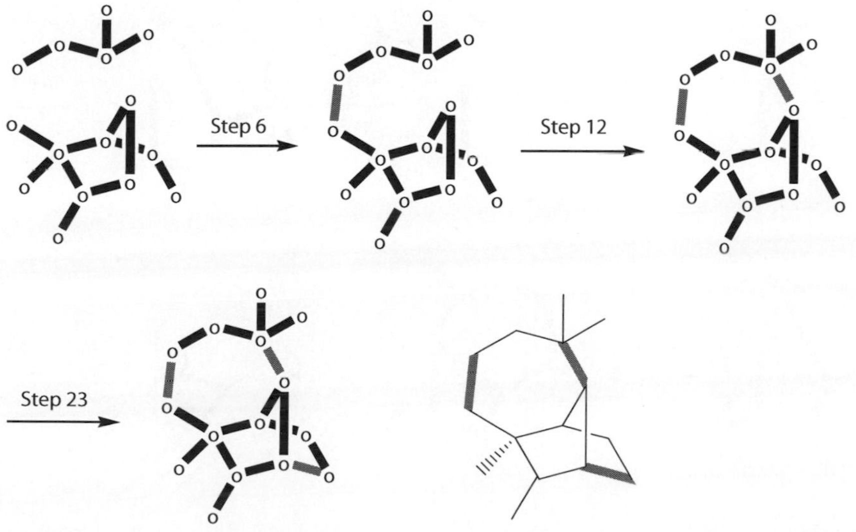

Sequence: $\xrightarrow{[7+0]} [7]_{12} \xrightarrow{[(6+0)+(5+0)]} [2.2.1]_{23}$

(±)-Longifolene

Oppolzer, W.; Godel, T. *J. Am. Chem. Soc.* 1978, *100*, 2583.

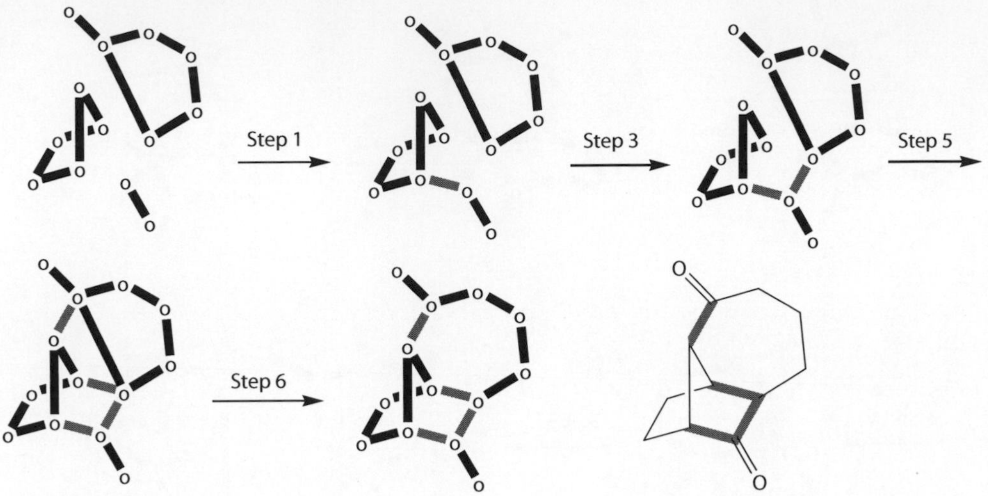

Sequence: $[5]_0 \xrightarrow{[(6+0)+(5+0)]_{[2.2.1]} + [2+2]_{[3.2.0]}} ([2.2.1] + [3.2.0])_5 \xrightarrow{[3.2.0] \rightarrow [7]} [7]_6$

(±)-Longifolene

Oppolzer, W.; Godel, T. *Helv. Chim. Acta* 1984, *67*, 1184.

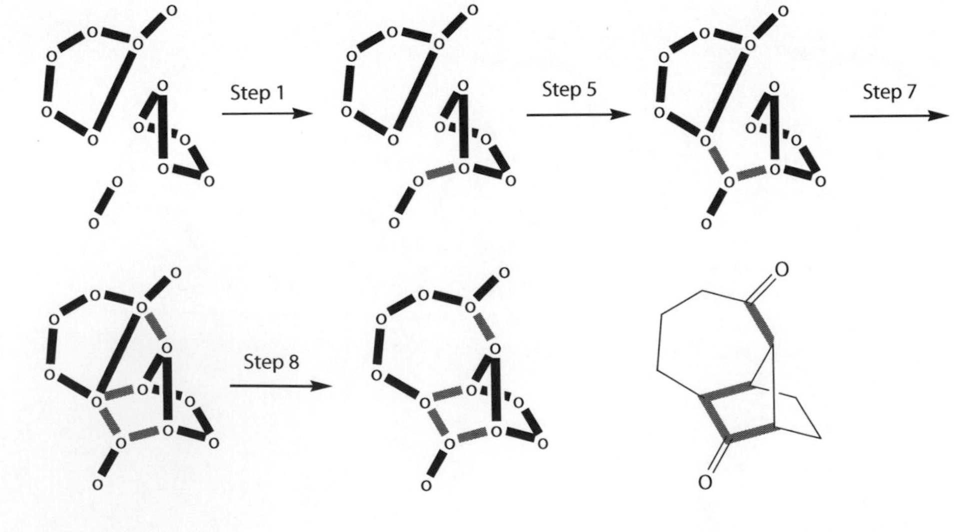

Sequence: $[5]_0 \xrightarrow{[(6+0)+(5+0)]_{[2.2.1]} + [2+2]_{[3.2.0]}} ([2.2.1]+[3.2.0])_7 \xrightarrow{[3.2.0] \rightarrow [7]} [7]_8$

(±)-Longifolene

Schultz, A.G.; Puig, S. *J. Org. Chem.* 1985, *50*, 915.

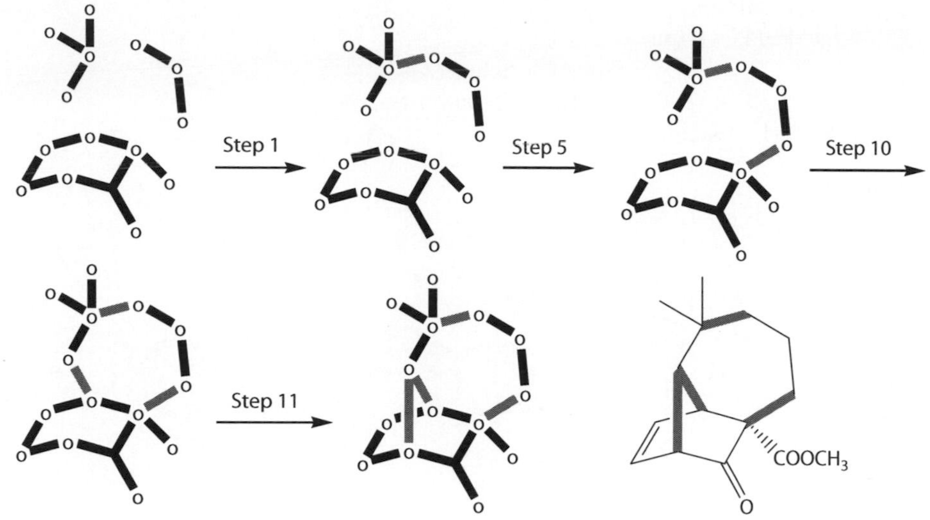

Sequence: $[6]_0 \xrightarrow{[7+0]} [7]_{10} \xrightarrow{[(5+0)+(5+0)]} [2.2.1]_{11}$

(−)-Longifolene

Schultz, A.G.; Puig, S. *J. Org. Chem.* 1985, *50*, 915.

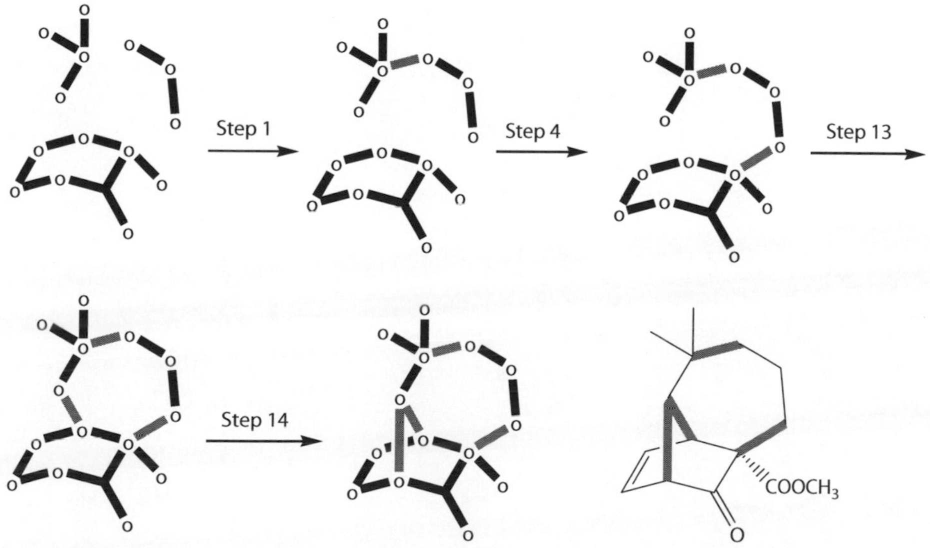

Sequence: $[6]_0 \xrightarrow{[7+0]} [7]_{13} \xrightarrow{[(5+0)+(5+0)]} [2.2.1]_{14}$

Summary:
Strategy [5]→([2.2.1] + [3.2.0])→[7]: Oppolzer G1, Oppolzer G2
Strategy [5]→[6]→[4.2.1]: Fallis
Strategy [5]→[4.2.1]: Johnson
Strategy [5]→[2.2.1]→[4.2.1]: Liu
Strategy [6]→[7]→[2.2.1]: Corey, Schultz G1, Schultz G2
Strategy [6]→[6]→[2.2.1]→[4.1.0]→[7]: Karimi, McMurry
Strategy →[7]→[2.2.1]: Money

(±)-Lycopodine
Ayer, W.A.; Bowman, W.R.; Joseph, T.C.; Smith, P. *J. Am. Chem. Soc.* 1968, *90*, 1648.
Ayer, W.A.; Bowman, W.R.; Cooke, G.A.; Soper, A.C. *Tetrahedron Lett.* 1966, 2021.

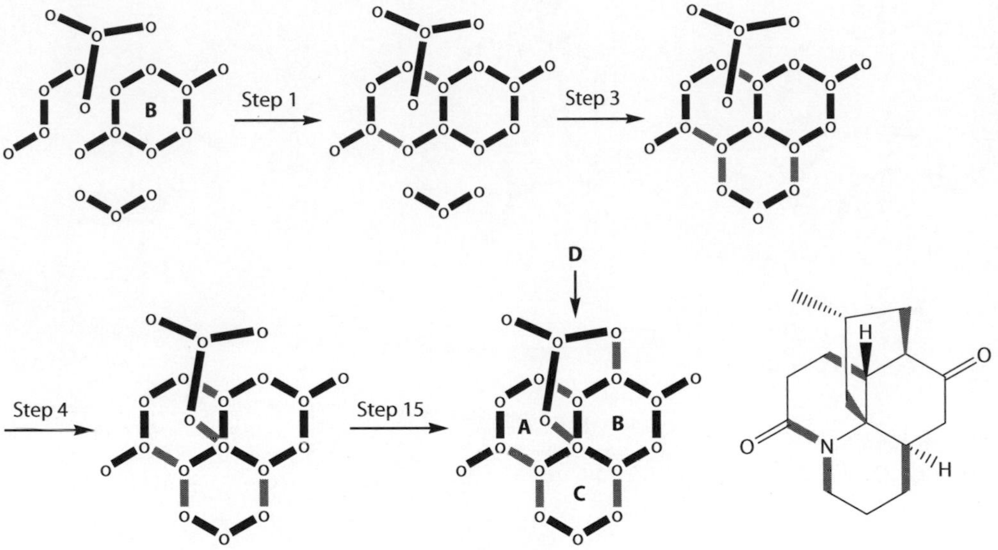

Sequence: $B_0 \xrightarrow{[3+3]} A_1 \xrightarrow{[3+3]} C_3 \xrightarrow{[5+0]} D_{15}$

(±)-Lycopodine
Stork, G.; Kretchmer, R.A.; Schlessinger, R.A. *J. Am. Chem. Soc.* 1968, *90*, 1647.

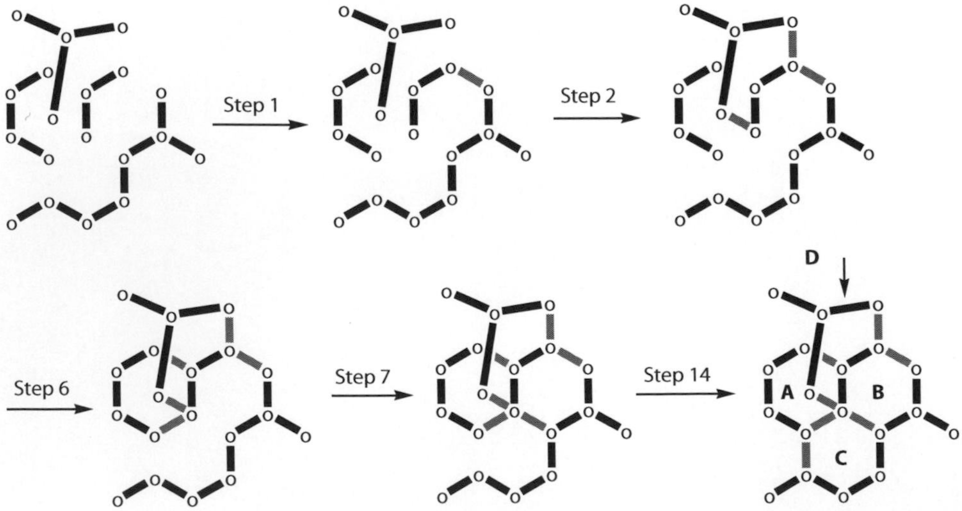

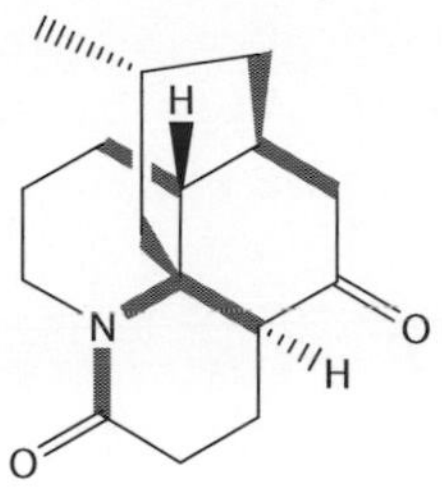

Sequence: $\xrightarrow{[3+3]} D_2 \xrightarrow{[4+2]} A_6 \xrightarrow{[6+0]} B_7 \xrightarrow{[6+0]} C_{14}$

Lysergic acid

Inoue, T.; Yokoshima, S.; Fukuyama, T. *Heterocycles* 2009, *79*, 373.

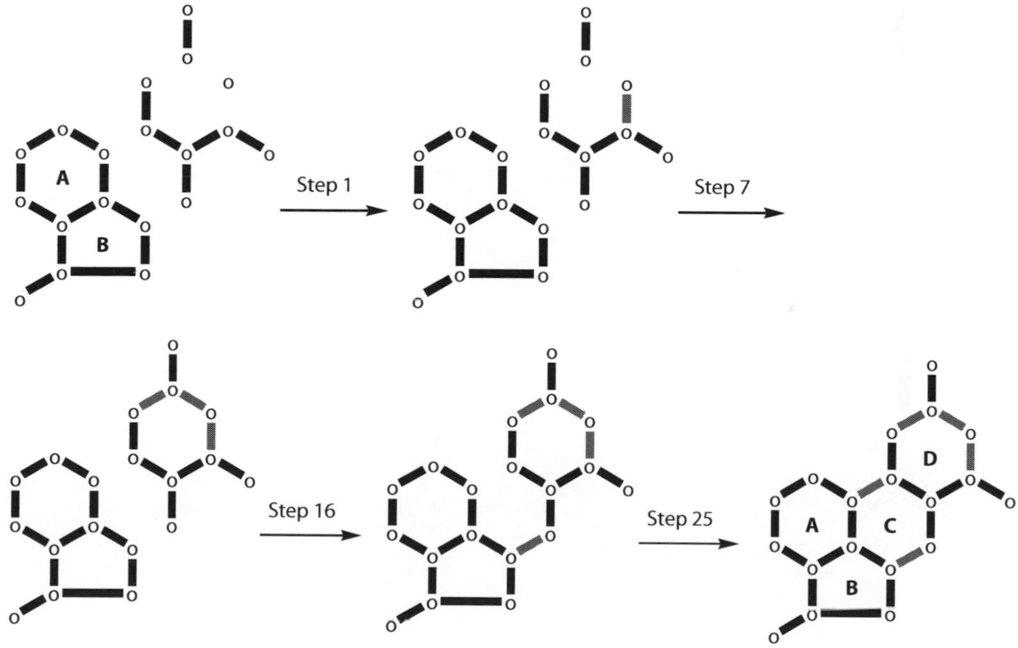

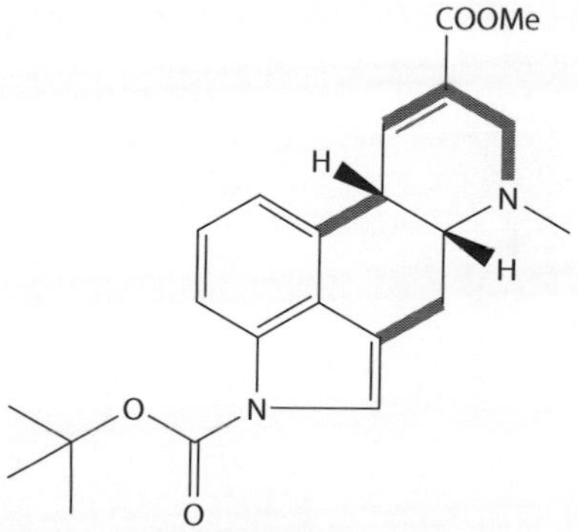

Sequence: $(A + B)_0 \xrightarrow{[5+1]} D_7 \xrightarrow{[6+0]} C_{25}$

Lysergic acid

Kurokawa, T.; Isomura, M.; Tokuyama, H.; Fukuyama, T. *Synlett* 2009, 775.

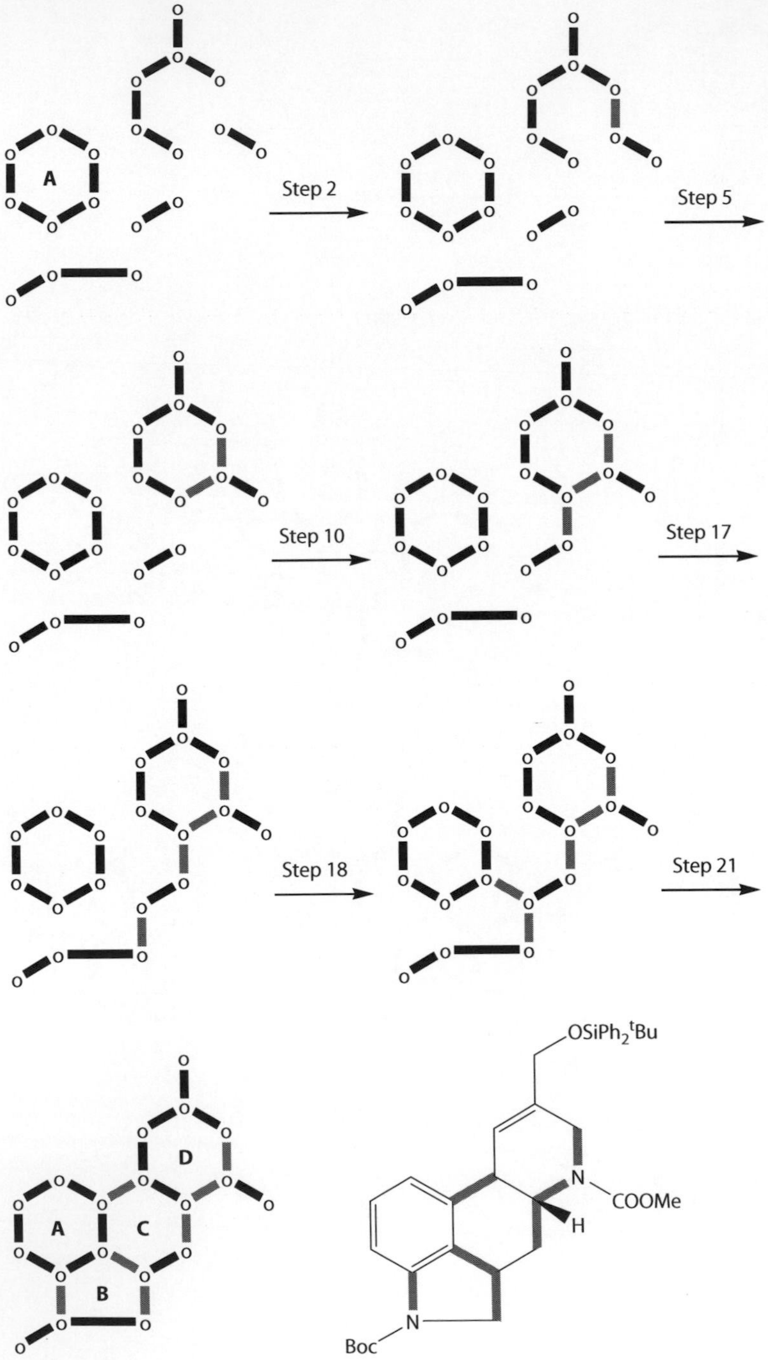

Sequence: $A_0 \xrightarrow{[6+0]} D_5 \xrightarrow{[(5+0)_B + (6+0)_C]} (B+C)_{21}$

Lysergic acid

Hendrickson, J.B.; Wang, J. *Org. Lett.* 2004, *6*, 3.

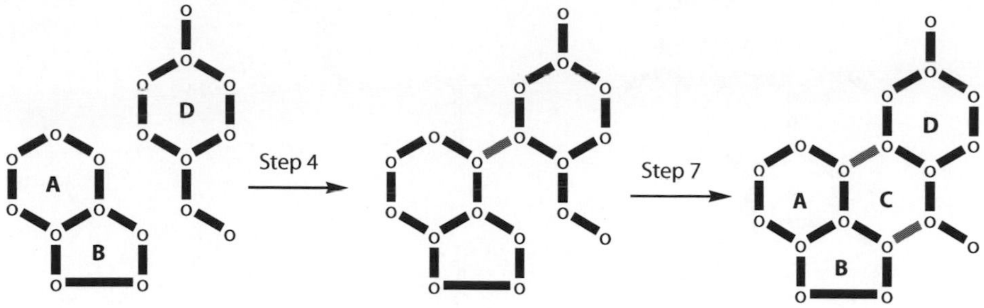

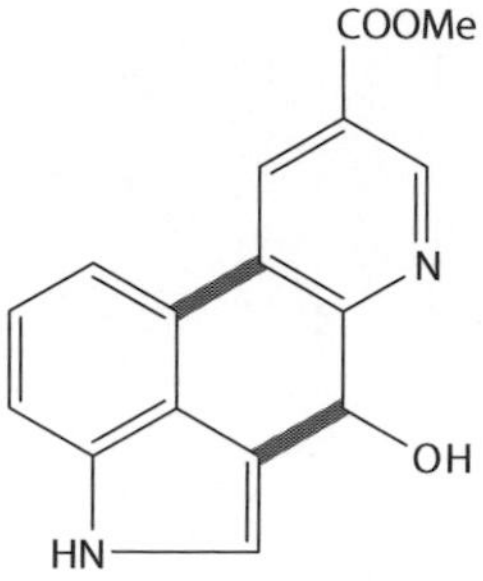

Sequence: $(A + B + D)_0 \xrightarrow{[6+0]} C_7$

Lysergic acid

Kurihara, T.; Terada, T.; Harusawa, S.; Yoneda, R. *Chem. Pharm. Bull.* 1987, *35*, 4793.
Kurihara, T.; Terada, T.; Yoneda, R. *Chem. Pharm. Bull.* 1986, *34*, 442.

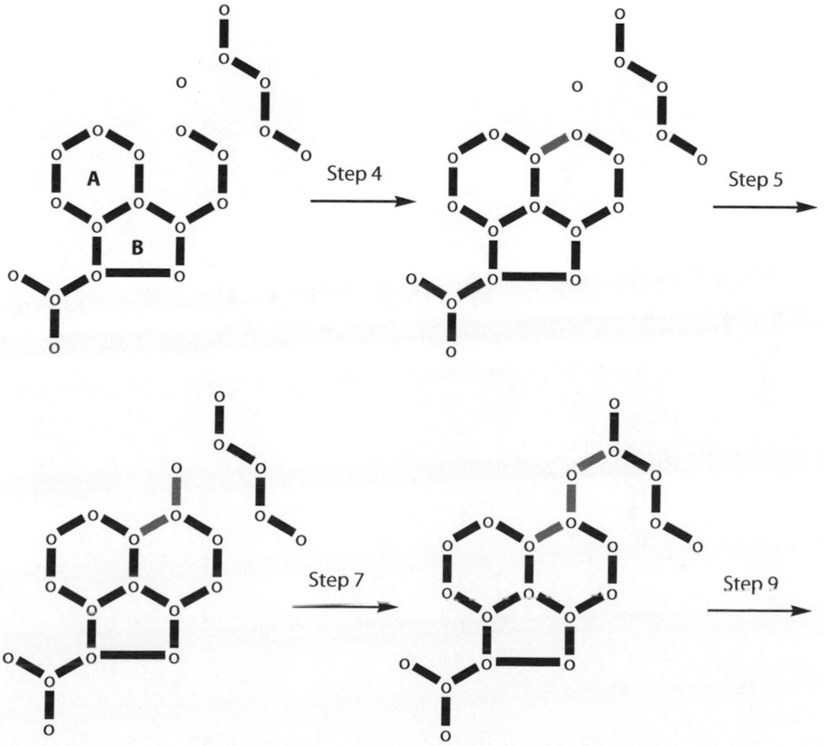

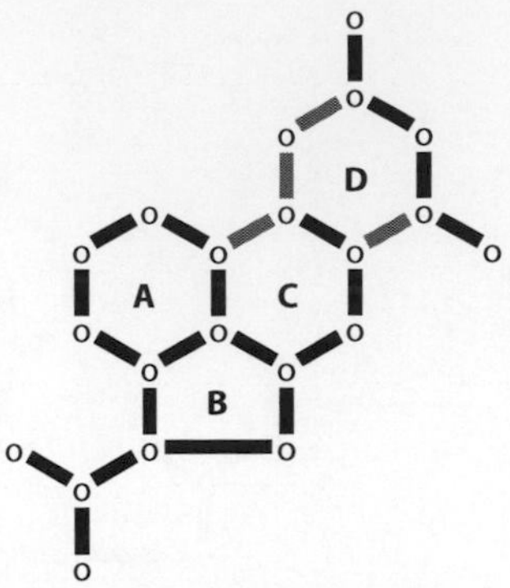

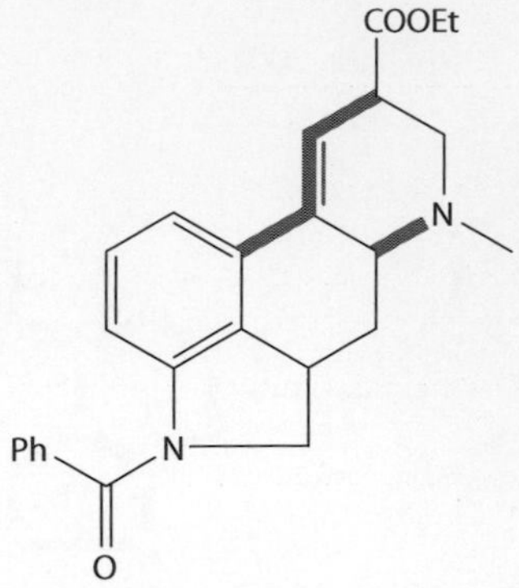

Sequence: $(A+B)_0 \xrightarrow{[6+0]} C_4 \xrightarrow{[6+0]} D_9$

Lysergic acid

Ninomiya, I.; Hashimoto, C.; Kiguchi, T.; Naito, T. *J. Chem. Soc. Perkin Trans. I* 1985, 941.
Kiguchi, T.; Hashimoto, C.; Naito, T.; Ninomiya, I. *Heterocycles* 1982, *19*, 2279.

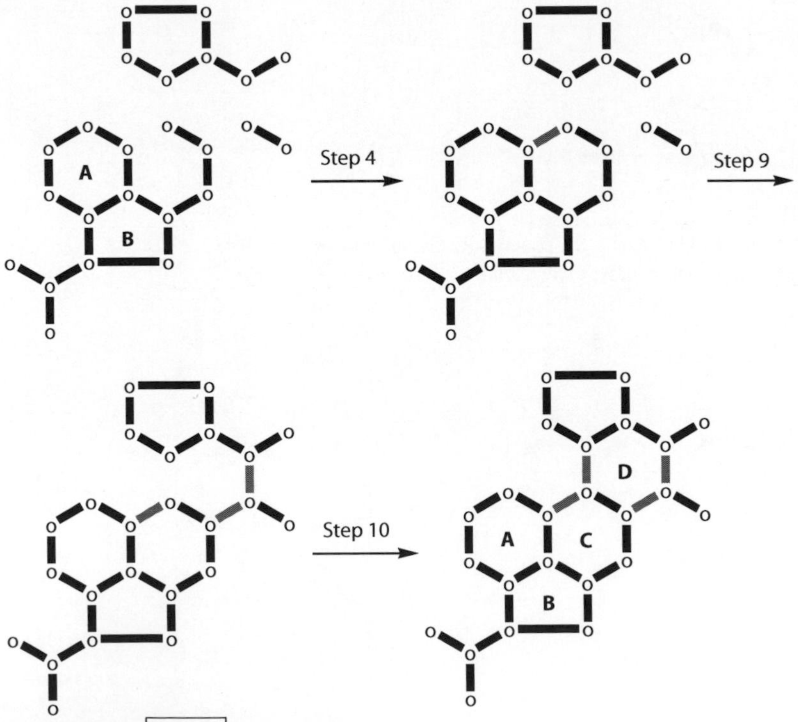

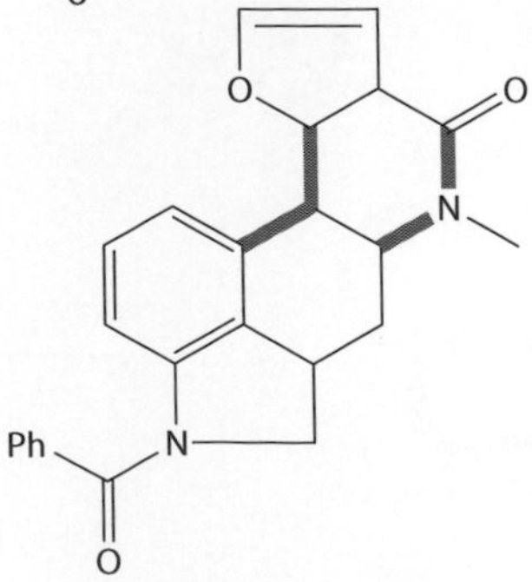

Step 4

Step 9

Step 10

Sequence: $(A + B)_0 \xrightarrow{[6+0]} C_4 \xrightarrow{[6+0]} D_{10}$

Lysergic acid
Oppolzer, W.; Francotte, E.; Battig, K. *Helv. Chim. Acta* 1981, *64*, 478.

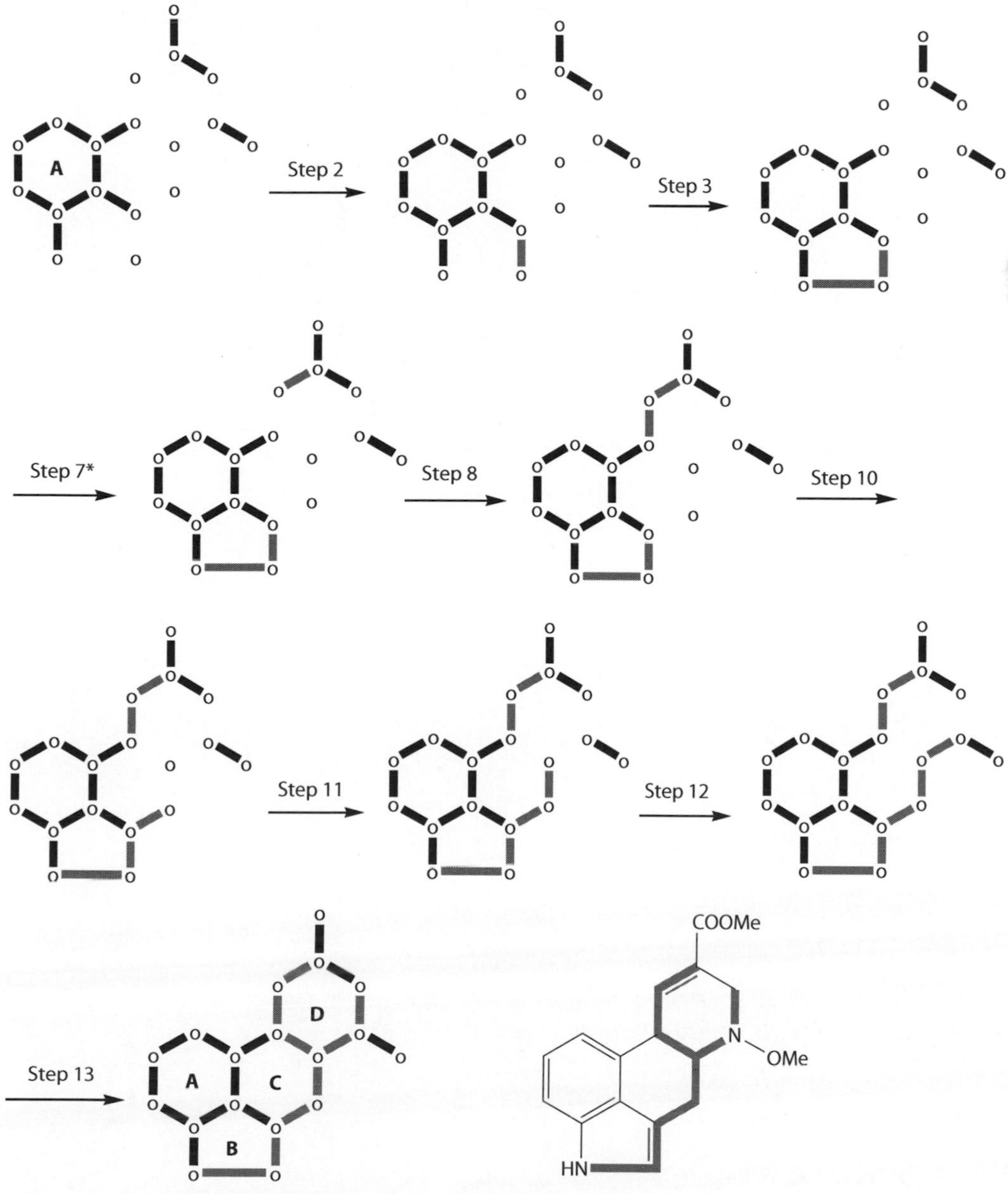

Sequence: $A_0 \xrightarrow{[5+0]} B_3 \xrightarrow{[(6+0)_C + (4+2)_D]} [4.4.0](C+D)_{13}$

Lysergic acid

Cacchi, S.; Ciattini, P.G.; Morera, E.; Ortar, G. *Tetrahedron Lett.* 1988, *29*, 3117.

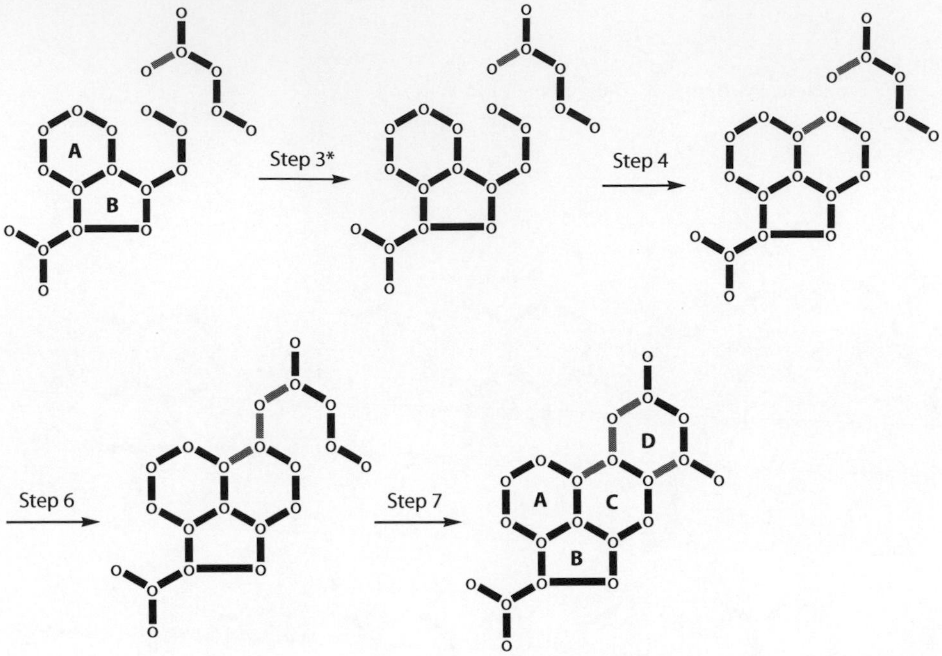

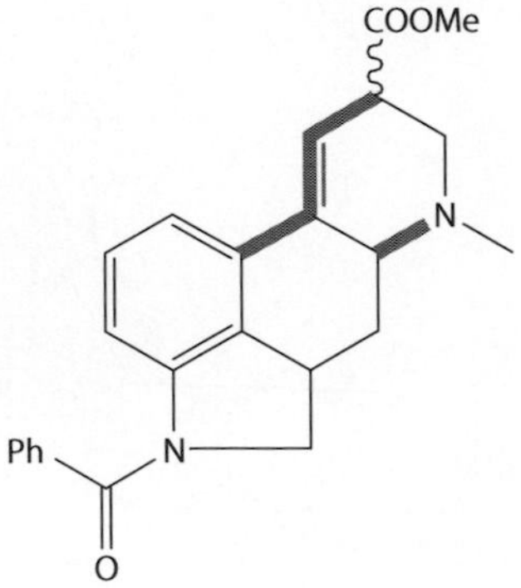

Sequence: $(A + B)_0 \xrightarrow{[6+0]} C_4 \xrightarrow{[6+0]} D_7$

Lysergic acid

Ramage, R.; Armstrong, V.W.; Coulton, S. *Tetrahedron* 1981, *37*, Suppl. 1, 157.
Armstrong, V.W.; Coulton, S.; Ramage, R. *Tetrahedron Lett.* 1976, 4311.

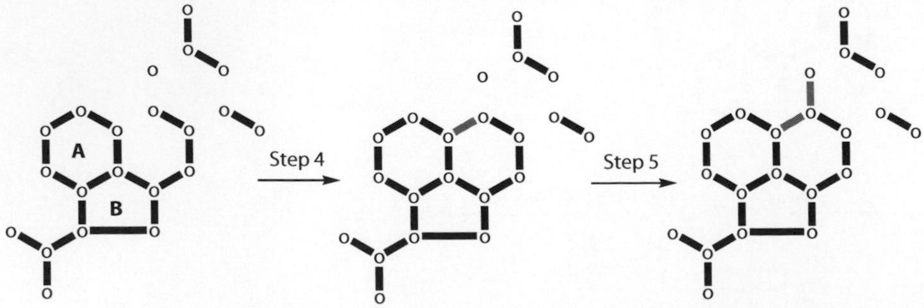

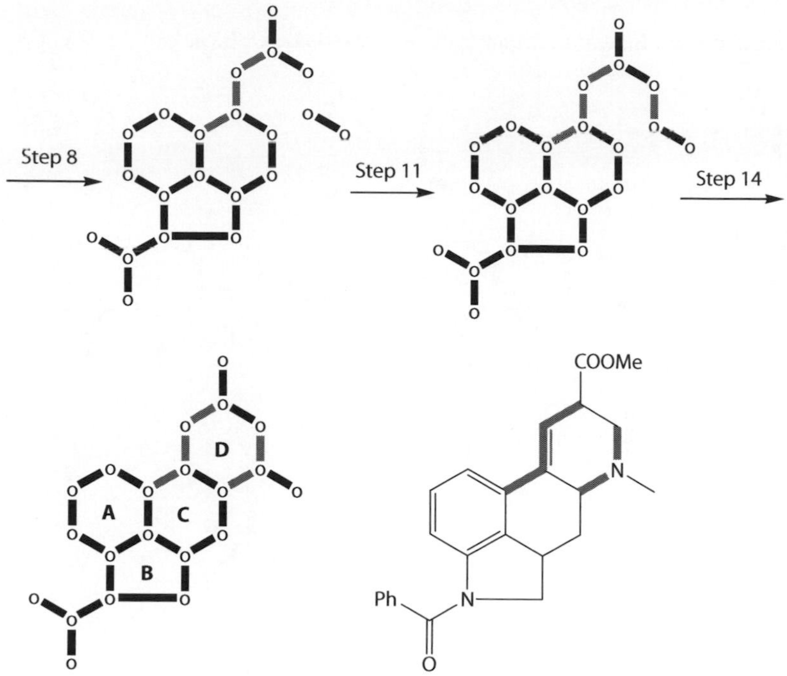

Sequence: $(A+B)_0 \xrightarrow{[6+0]} C_4 \xrightarrow{[6+0]} D_{14}$

Lysergic acid

Rebek Jr., J; Tai, D.F. *Tetrahedron Lett.* 1983, *24*, 859.

Rebek Jr., J.; Tai, D.F.; Shue, Y.K. *J. Am. Chem. Soc.* 1984, *106*, 1813.

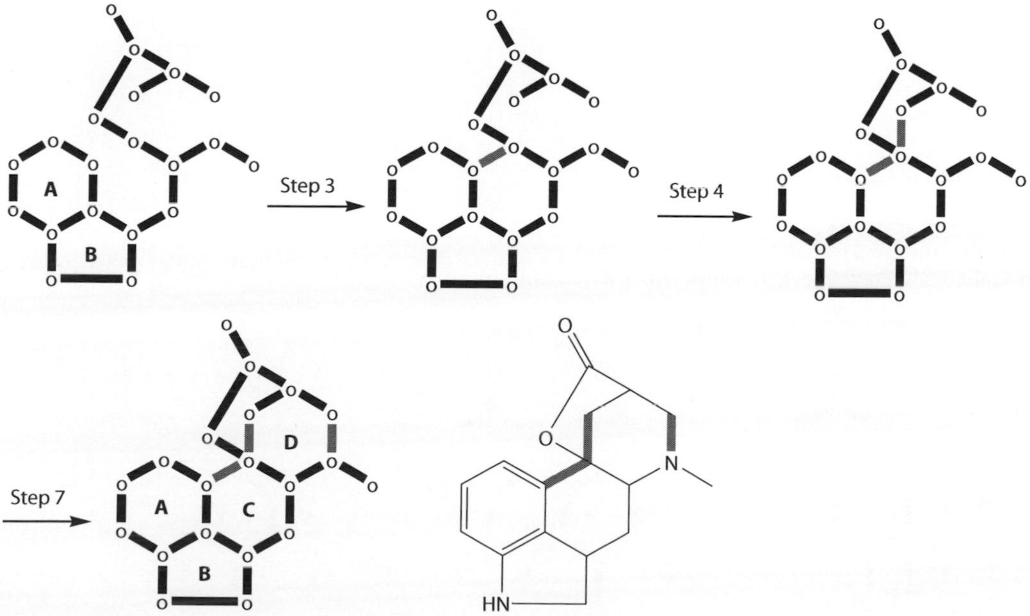

Sequence: $(A+B)_0 \xrightarrow{[6+0]} C_3 \xrightarrow{[6+0]} D_7$

Lysergic acid

Moldvai, I.; Temesvari-Major, E.; Incze, M.; Szentirmay, E.; Gacs-Baitz, E.; Szantay, C. *J. Org. Chem.* 2004, *69*, 5993.

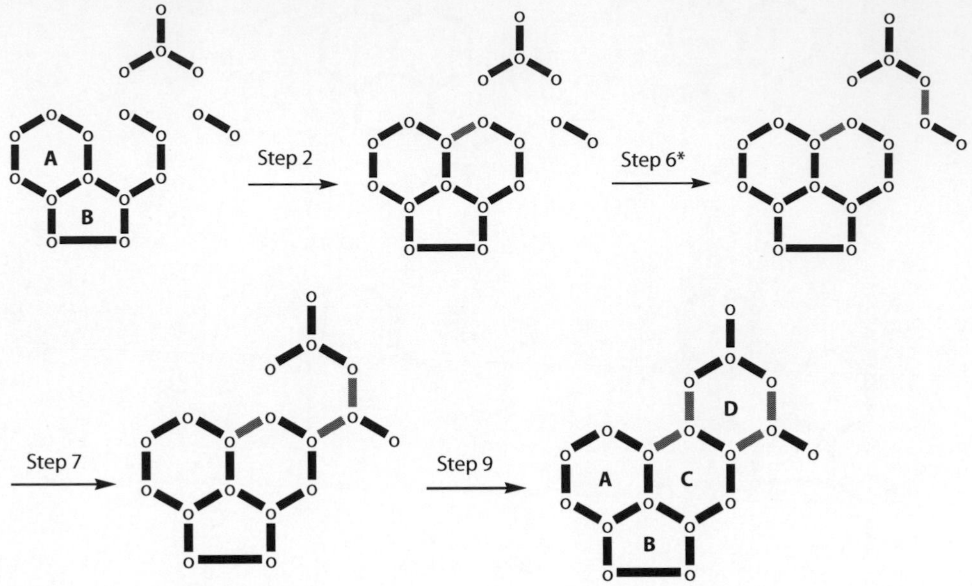

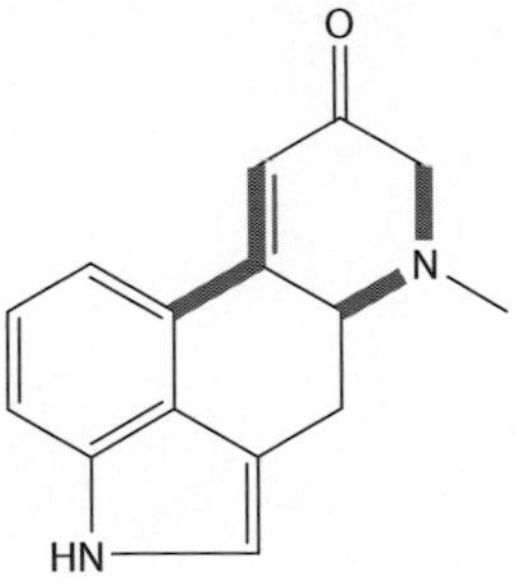

Sequence: $(A+B)_0 \xrightarrow{[6+0]} C_2 \xrightarrow{[6+0]} D_9$

Lysergic acid

Kornfeld, E.C.; Fornefeld, E.J.; Kline, G.B.; Mann, M.J.; Morrison, D.E.; Jones, R.G.; Woodward, R.B. *J. Am. Chem. Soc.* 1956, *78*, 3087.

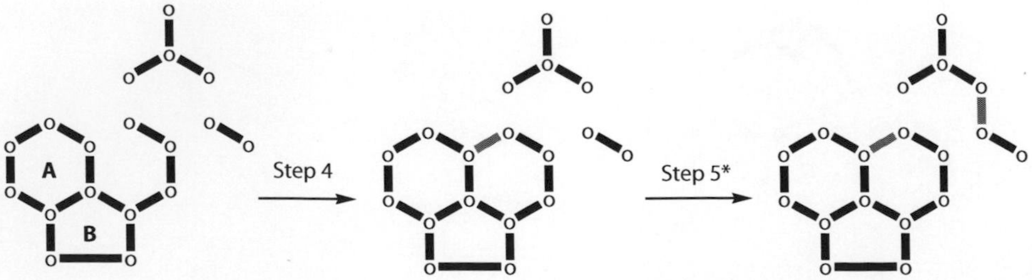

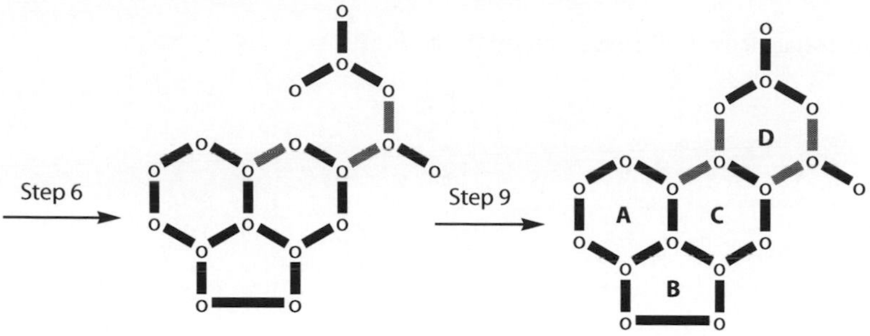

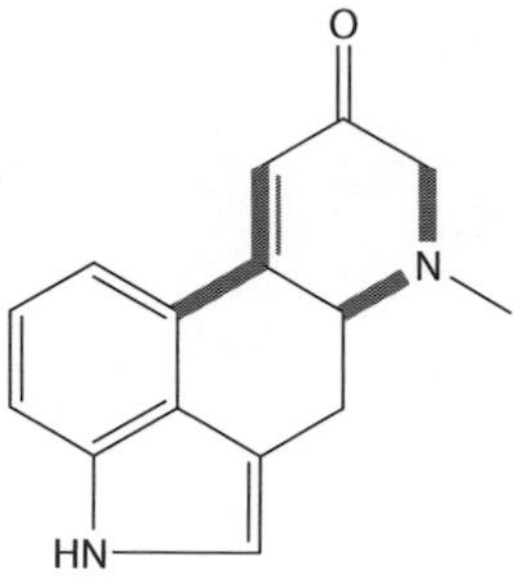

Sequence: $(A+B)_0 \xrightarrow{[6+0]} C_4 \xrightarrow{[6+0]} D_9$

Summary:

Strategy: $(A + B) \rightarrow C \rightarrow D$:
Kurihara, Ninomiya, Ortar, Ramage, Rebek, Szantay, Woodward
Strategy: $A \rightarrow B \rightarrow [4.4.0](C + D)$: Oppolzer
Strategy: $(A + B + D) \rightarrow C$: Hendrickson
Strategy: $(A + B) \rightarrow D \rightarrow C$: Fukuyama G1
Strategy: $A \rightarrow D \rightarrow (B + C)$: Fukuyama G2

(±)-Menthol

Friederang, A.; Pasedach, H. DE 2203807 (BASF, 1973).

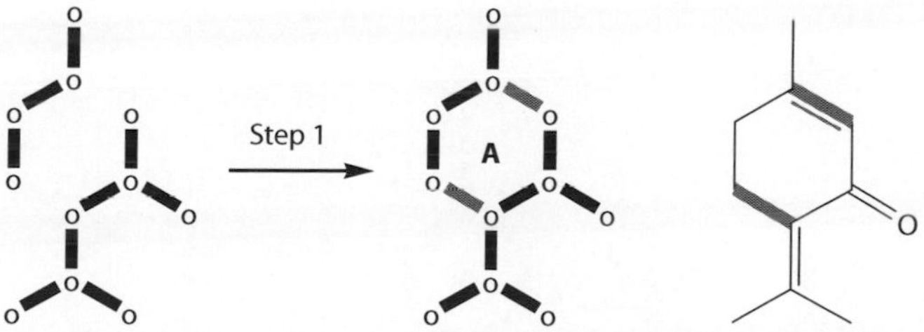

Sequence: $\xrightarrow{[3+3]} A_1$

(±)-Menthol

Brode, W.R.; van Dolah, R.W. *Ind. Eng. Chem.* 1947, *39*, 1157.

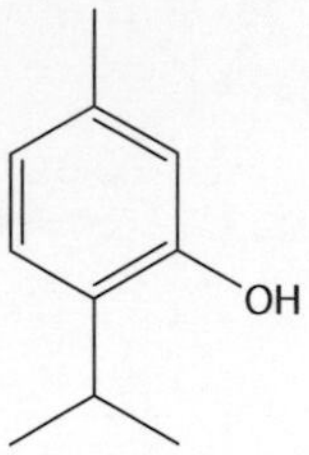

(–)-Menthol

Ohshima, T.; Tadaoka, H.; Hori, K.; Sayo, N.; Mashima, K. *Chem. Eur. J.* 2008, *14*, 2060.

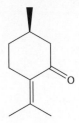

(–)-Menthol

GB 1209396 (RJ Reynolds Tobacco, 1970).

Leffingwell, J.C.; Shackelford, R.E. (RJ Reynolds Tobacco, TCRC Presentation Oct. 1973), personal
    communication

(–)-Menthol

Sayo, N.; Matsumoto, T. US 2002007094 (2002).

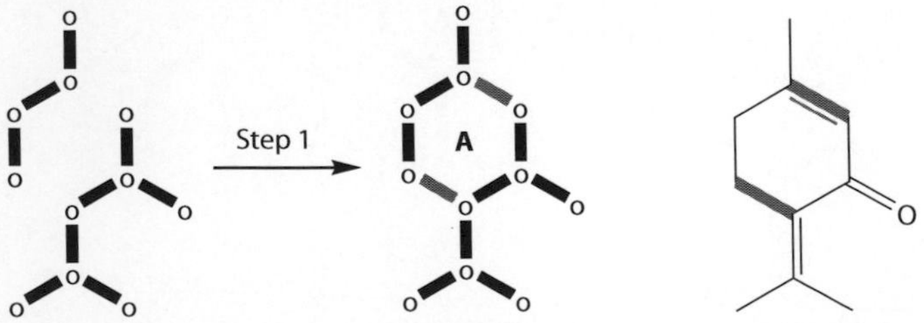

Sequence: $\xrightarrow{[3+3]} A_1$

(–)-Menthol

Takaya, H.; Akutagawa, S.; Noyori, R. *Org. Synth.* 1989, *67*, 20.
Tani, K.; Yamagata, T.; Otsuka, S.; Kumobayashi, H.; Akutagawa, S. *Org. Synth.* 1989, *67*, 33.
Akutagawa, S. *Topics in Catalysis* 1997, *4*, 271.

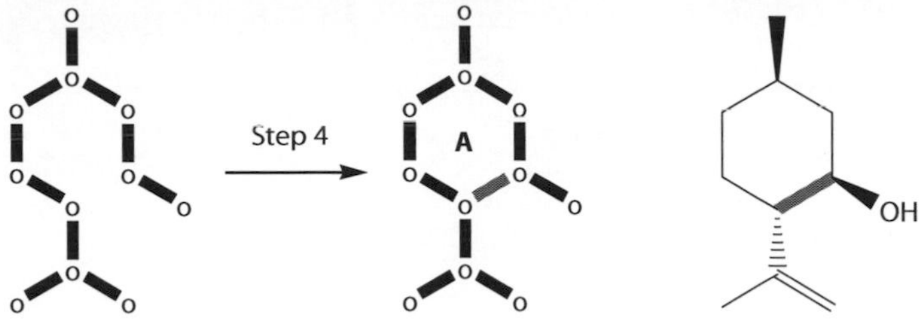

Sequence: $\xrightarrow{[6+0]} A_4$

(–)-Menthol

Iwata, T.; Okeda, Y.; Hori, Y. EP 1225163 (Takasago International Corp., 2002).

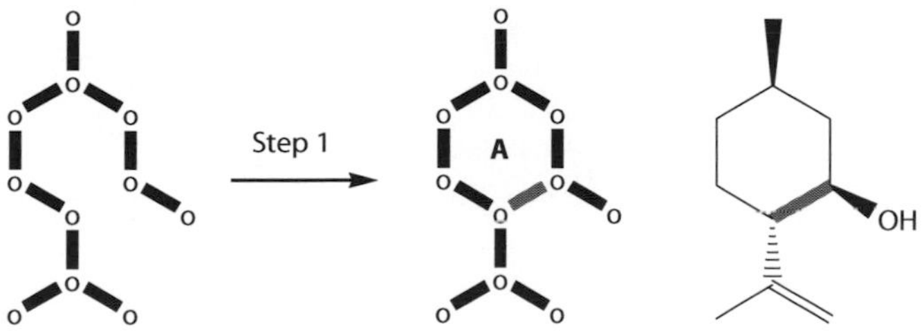

Sequence: $\xrightarrow{[6+0]} A_1$

(±)-Menthol

Ohshiro, S.; Doi, K. *Yakugaku Zasshi* 1968, *88*, 417 (CA 70: 4289).

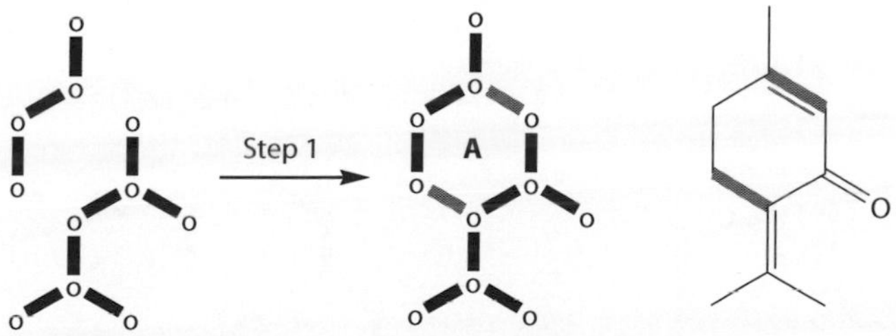

Sequence: $\xrightarrow{[3+3]} A_1$

Summary:

Strategy [3 + 3]: BASF, Sayo-Matsumoto, Tanabe-Seiyaku
Strategy [6 + 0]: Takasago G1, Takasago G2
Precursors containing ring: Brode, Mashima, RJ Reynolds Tobacco

(±)-Nakadomarin A

Young, I.S.; Kerr, M.A. *J. Am. Chem. Soc.* 2007, *129*, 1465.

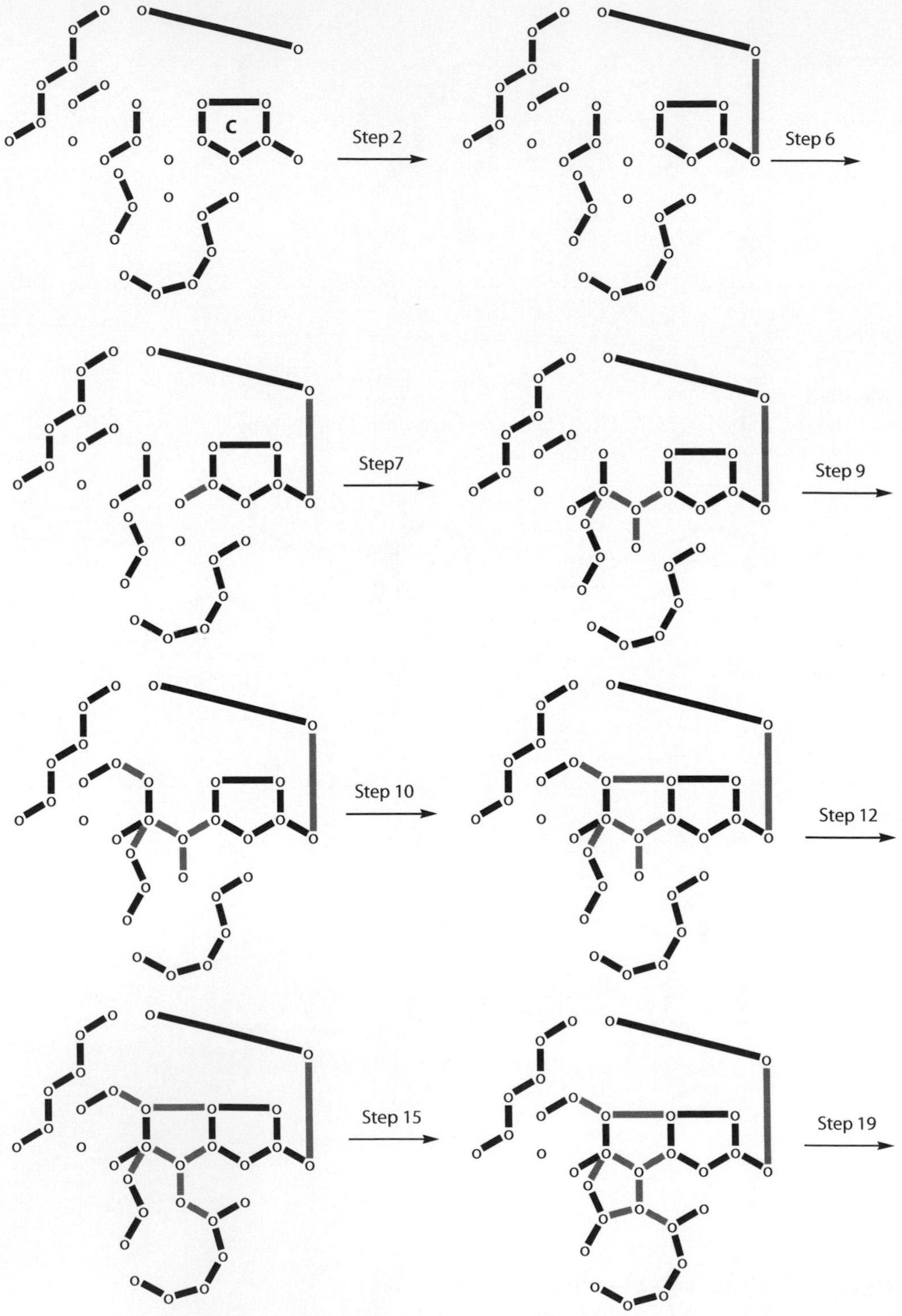

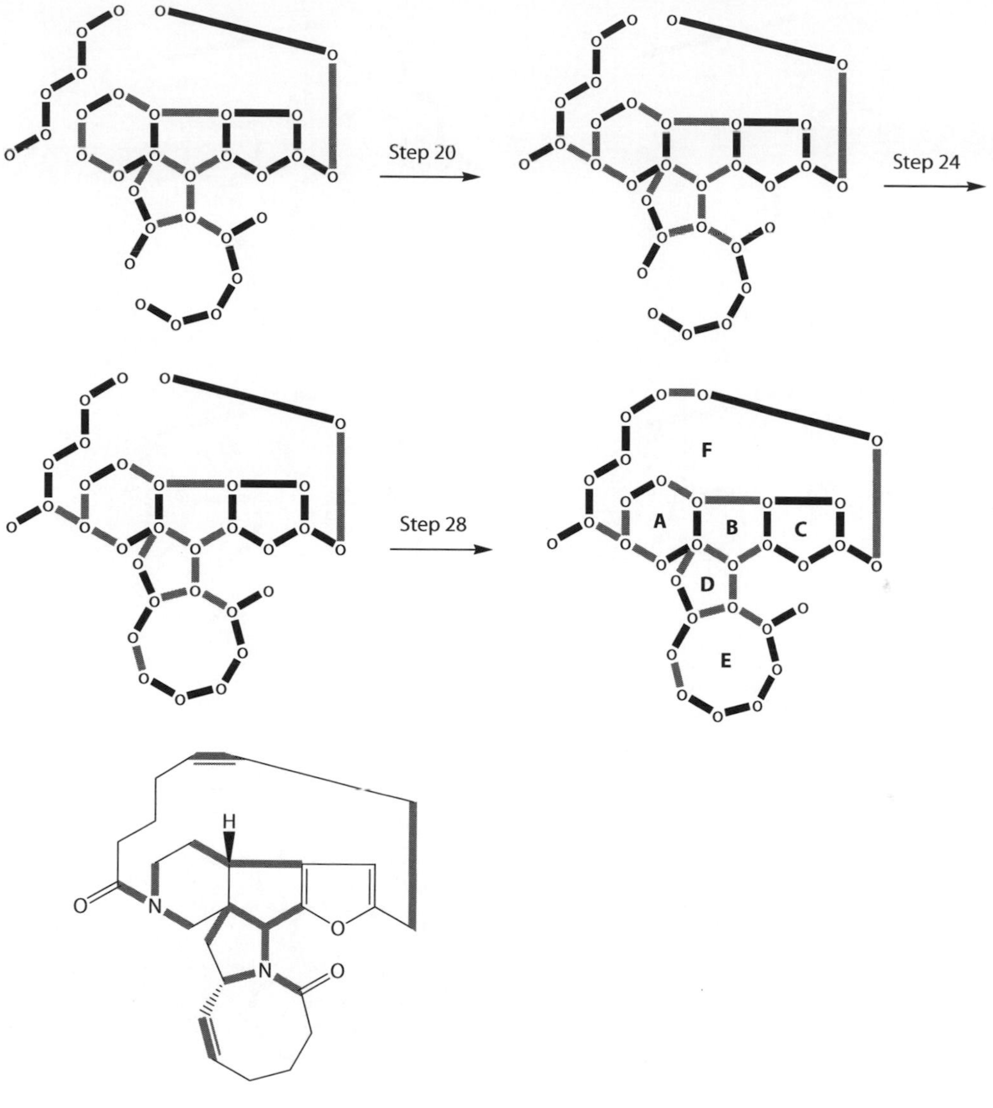

Sequence: $C_0 \xrightarrow{[5+0]} B_{10} \xrightarrow{[5+0]} D_{15} \xrightarrow{[5+1]} A_{19} \xrightarrow{[8+0]} E_{24} \xrightarrow{[15+0]} F_{28}$

Nagata, T.; Nakagawa, M.; Nishida, A. *J. Am. Chem. Soc.* 2003, *125*, 7484.
Nagata, T.; Nakagawa, M.; Nishida, A. *J. Am. Chem. Soc.* 2003, *125*, 13618.

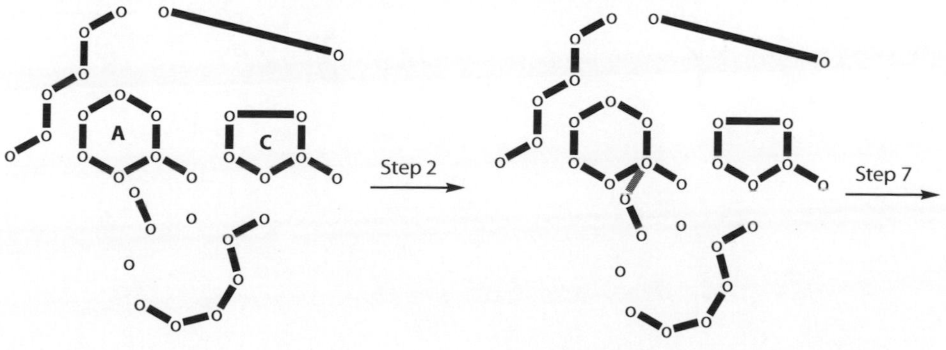

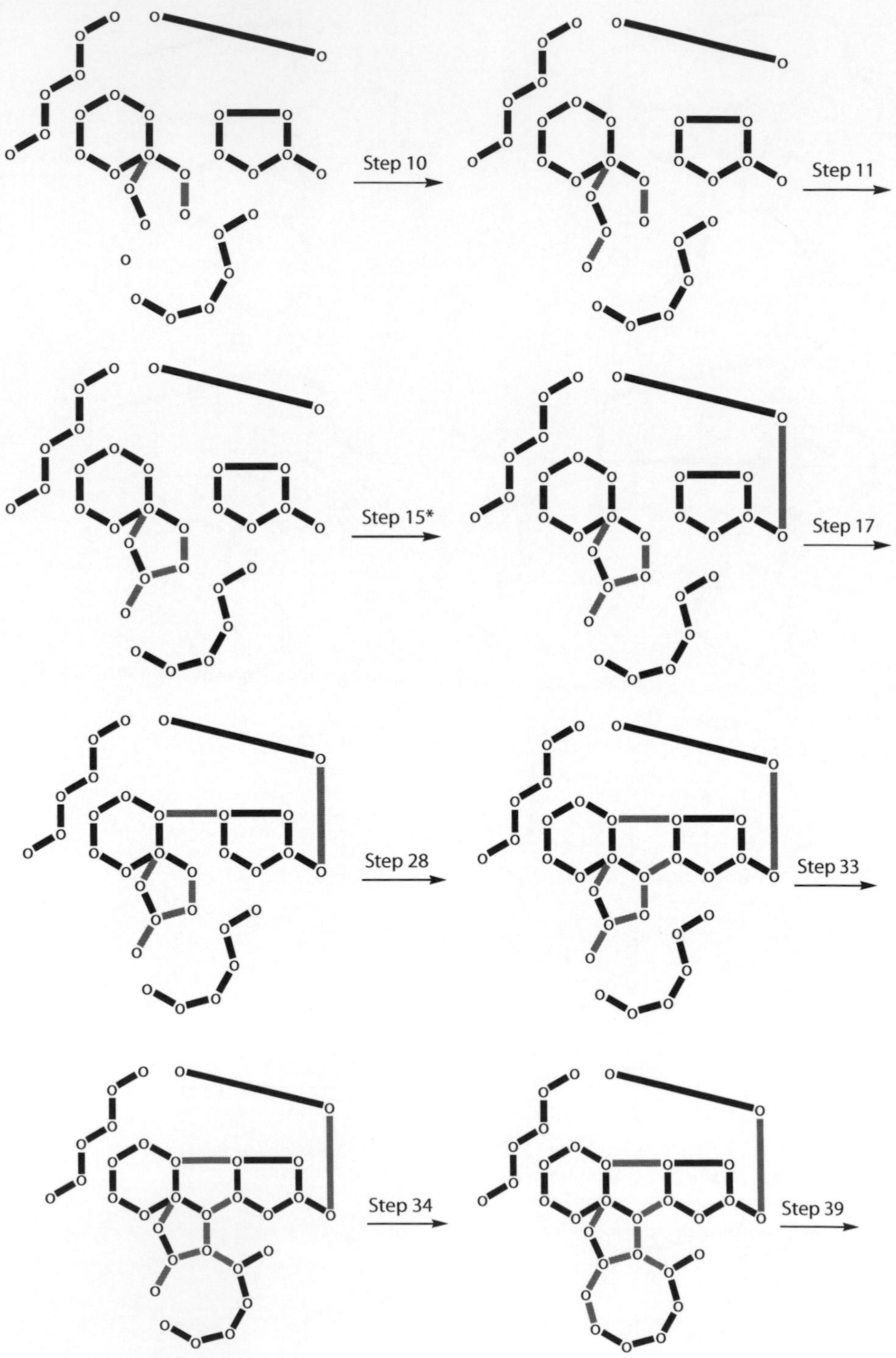
Step 10
Step 11
Step 15*
Step 17
Step 28
Step 33
Step 34
Step 39

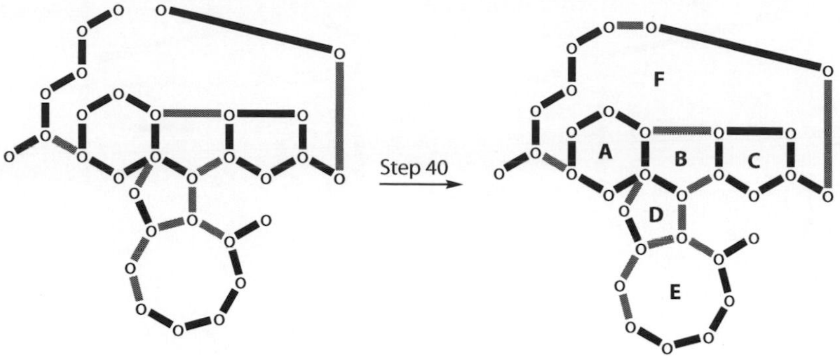

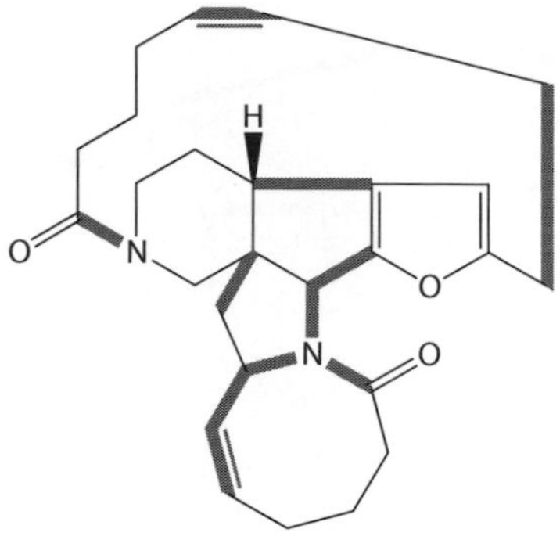

Sequence: $A_0 + C_{14*} \xrightarrow{[5+0]} D_{11} \xrightarrow{[5+0]} B_{28} \xrightarrow{[8+0]} E_{34} \xrightarrow{[15+0]} F'_{40}$

(−)-Nakadomarin A

Ono, K.; Nakagawa, M.; Nishida, A. *Angew. Chem. Int. Ed.* 2004, *43*, 2020.

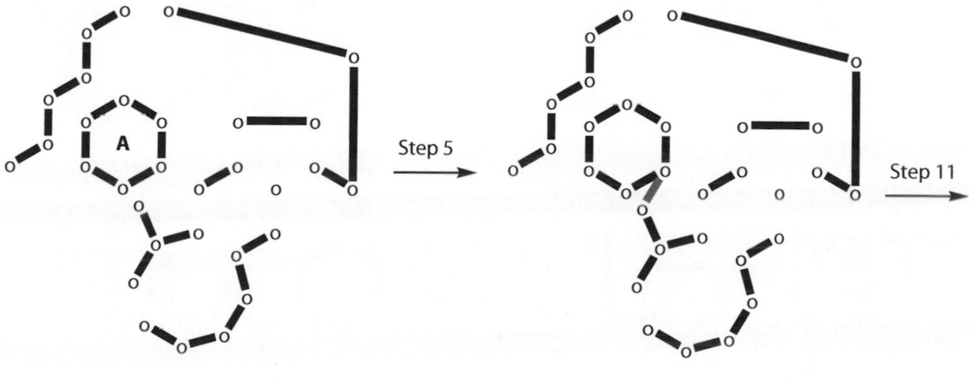

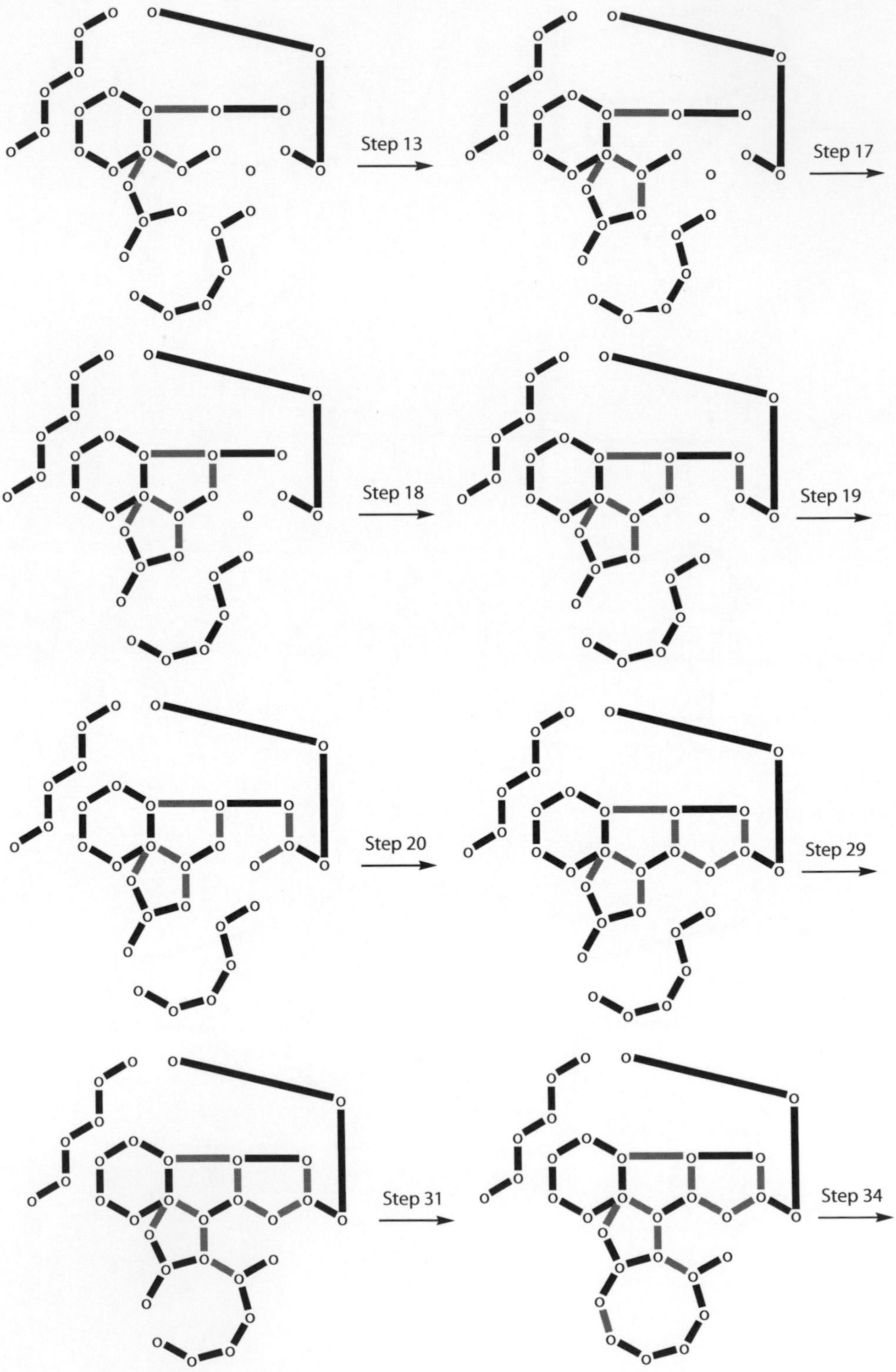
Step 13
Step 17
Step 18
Step 19
Step 20
Step 29
Step 31
Step 34

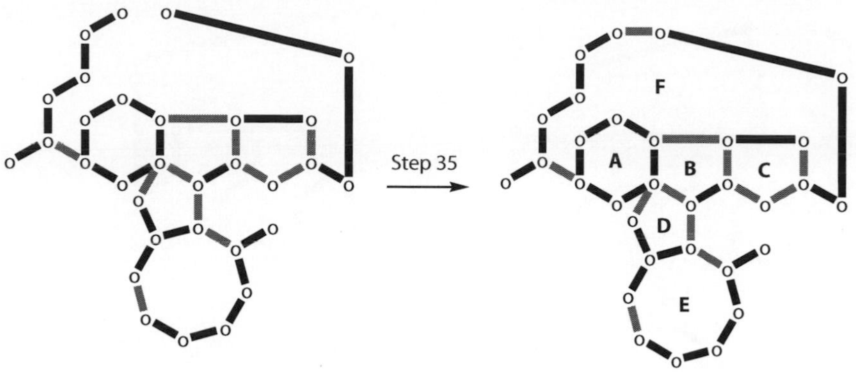

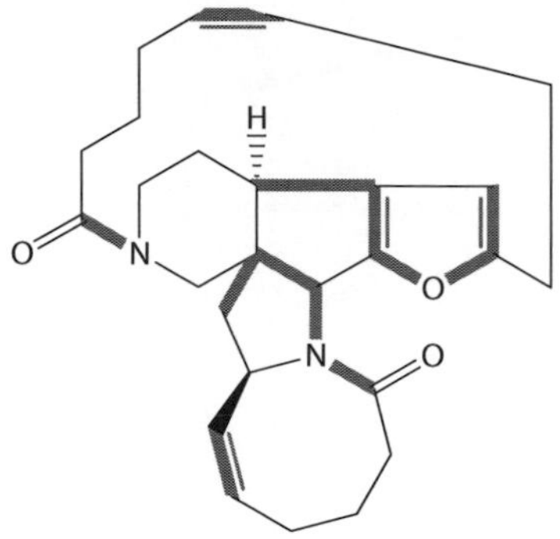

Sequence: $A_0 \xrightarrow{[5+0]} D_{13} \xrightarrow{[5+0]} B_{17} \xrightarrow{[5+0]} C_{20} \xrightarrow{[8+0]} E_{31} \xrightarrow{[15+0]} F_{35}$

(–)-Nakadomarin A

Jakubec, P.; Cockfield, D.M.; Dixon, D.J. *J. Am. Chem. Soc.* 2009, *131*, 16632.

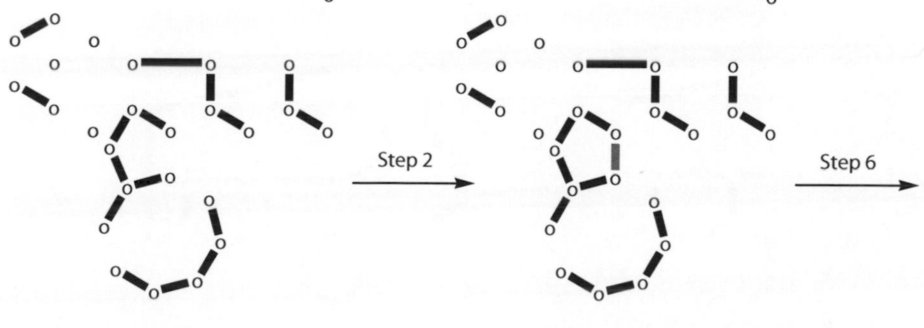

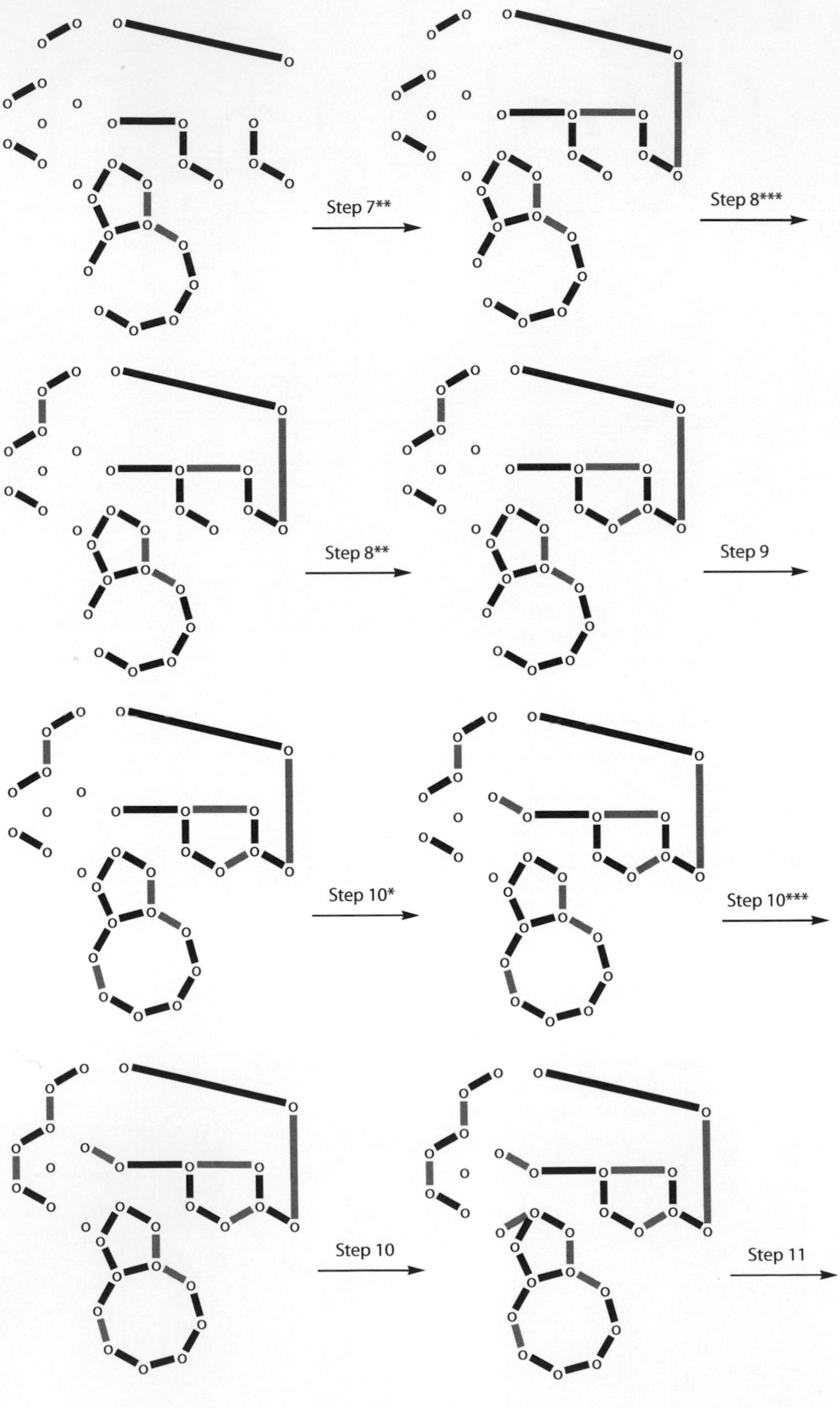
Step 7**
Step 8***
Step 8**
Step 9
Step 10*
Step 10***
Step 10
Step 11

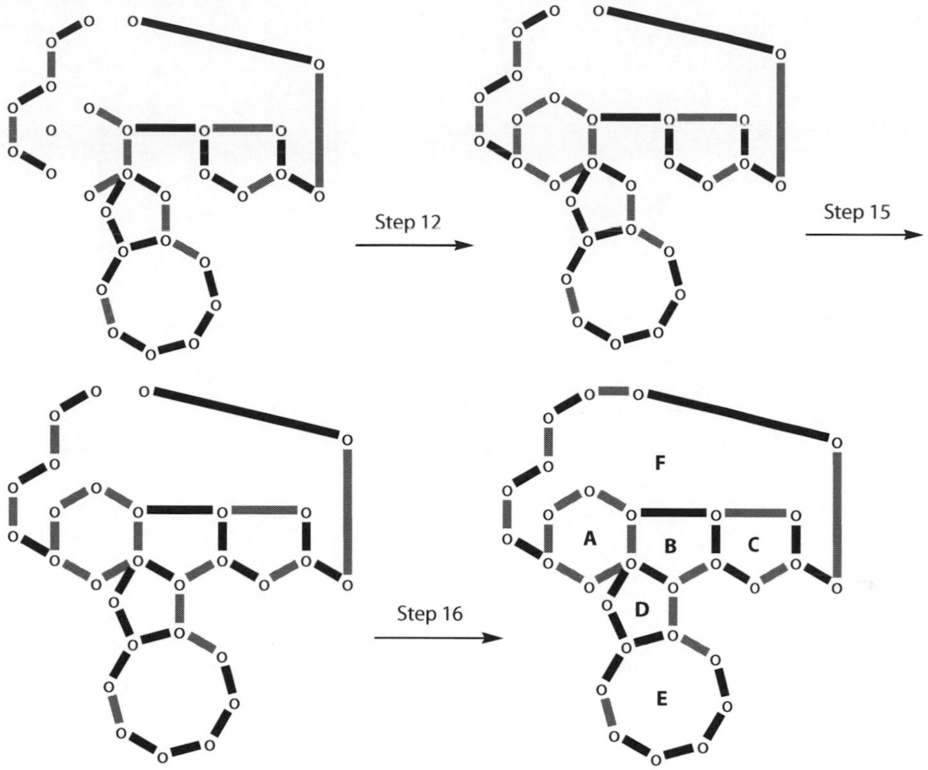

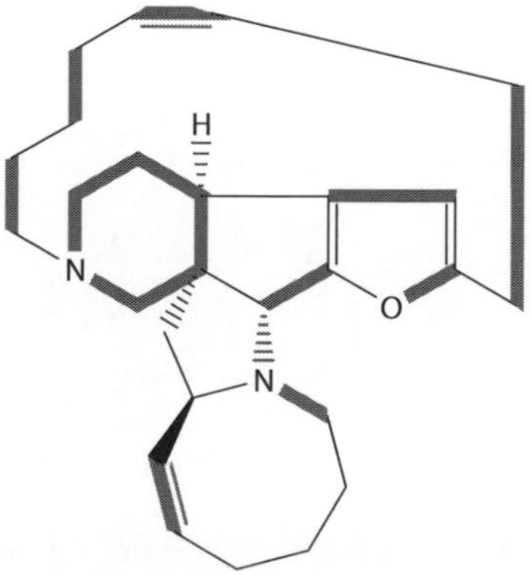

Sequence:

$$\xrightarrow{[5+0]} D_2 \xrightarrow{[5+0]} C_{8**} \xrightarrow{[8+0]} E_9 \xrightarrow{[4+1+1]} A_{12} \xrightarrow{[5+0]} B_{15} \xrightarrow{[15+0]} F_{16}$$

Summary:

A + B → D → B → E → F: Nishida G1

A → D → B → C → E → F: Nishida G2

C → B → D → A → E → F: Kerr

→ D → C → E → A → B → F: Dixon

(−)-Nepetalactone

Achmad, S.A.; Cavill, G.W.K. *Proc. Chem. Soc.* 1968, 166.

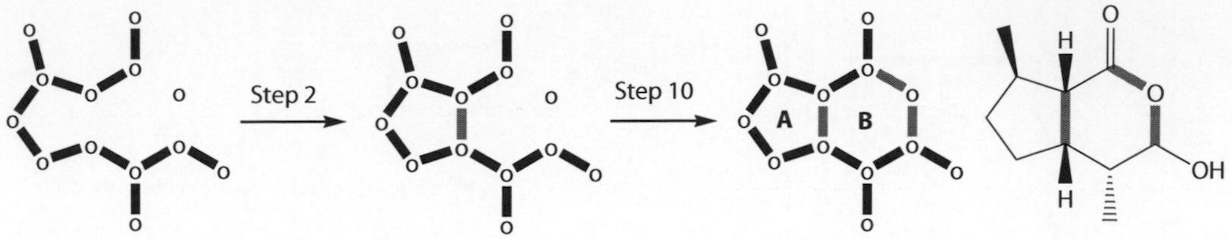

Sequence: $\xrightarrow{[5+0]} A_2 \xrightarrow{[5+1]} B_{10}$

(±)-Nepetalactone

Sakurai, K.; Ikeda, K.; Mori, K. *Agric. Biol. Chem.* 1988, *52*, 2369.

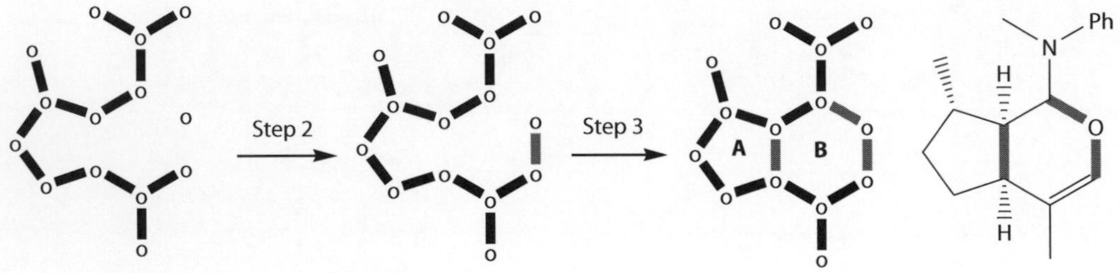

Sequence: $\xrightarrow{[(5+0)_A +(4+2)_B]} [4.3.0]_3$

(±)-Nepetalactone

Dawson, G.W.; Pickett, J.A.; Smiley, D.W.M. *Bioorg. Med. Chem.* 1996, *4*, 351.

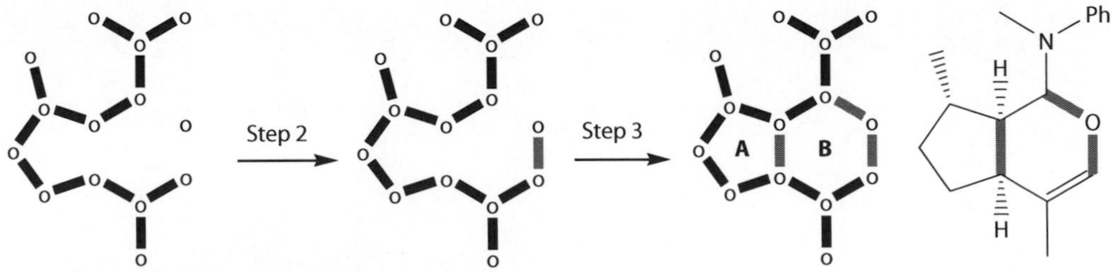

Sequence: $\xrightarrow{[(5+0)_A +(4+2)_B]} [4.3.0]_3$

(−)-Nepetalactone

Dawson, G.W.; Pickett, J.A.; Smiley, D.W.M. *Bioorg. Med. Chem.* 1996, *4*, 351.

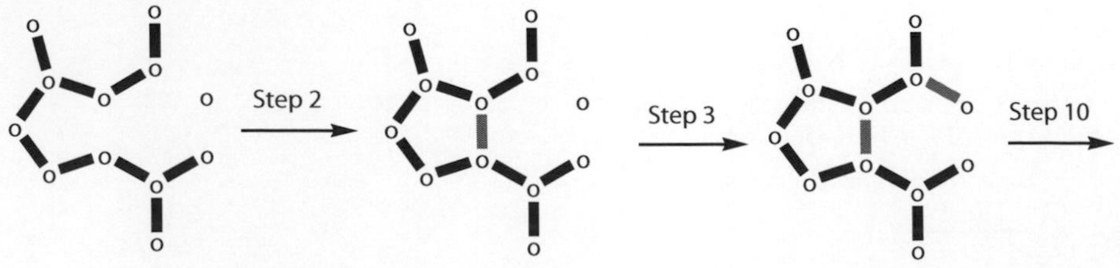

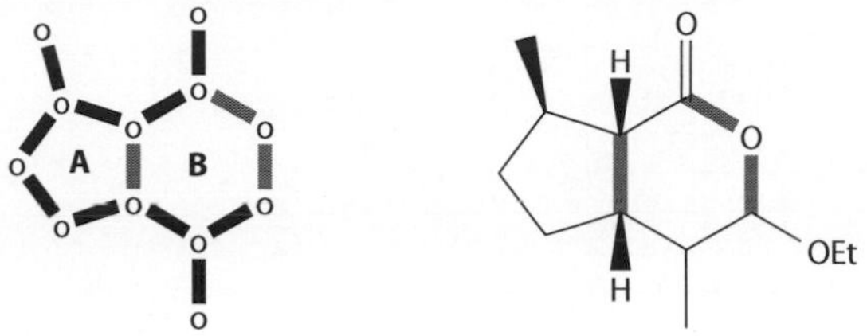

Sequence: $\xrightarrow{[5+0]} A_2 \xrightarrow{[6+0]} B_{10}$

(±)-Nepetalactone

Suemune, H.; Oda, K.; Saeki, S.; Sakai, K. *Chem. Pharm. Bull.* 1988, *36*, 172.

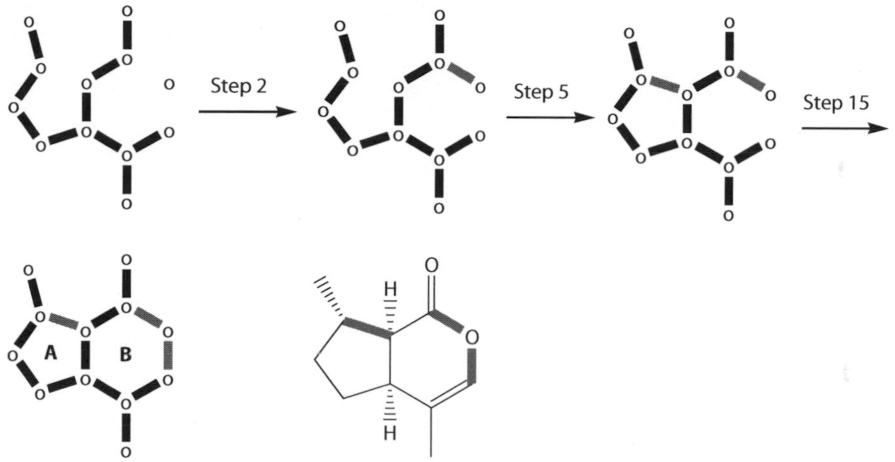

Sequence: $\xrightarrow{[5+0]} A_5 \xrightarrow{[6+0]} B_{15}$

(±)-Nepetalactone

Sakan, T.; Fujino, A.; Murai, F.; Suzui, A.; Butsugan, Y. *Bull. Chem. Soc. Jpn.* 1960, *33*, 1737.

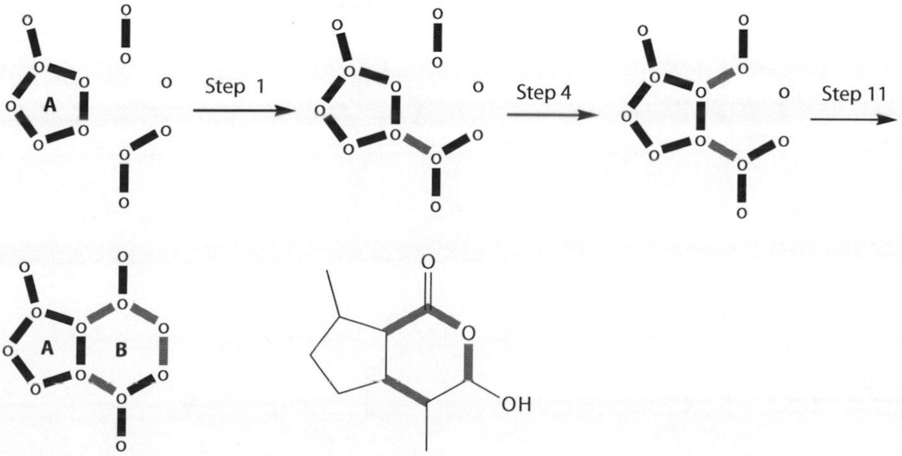

Sequence: $A_0 \xrightarrow{[5+1]} B_{11}$

Summary:
Strategy #1

$$\xrightarrow{[5+0]} A_2 \xrightarrow{[5+1]} B_{10}: \text{Cavill}$$

$$\xrightarrow{[5+0]} A_5 \xrightarrow{[6+0]} B_{15}: \text{Sakai}$$

$$\xrightarrow{[5+0]} A_2 \xrightarrow{[6+0]} B_{10}: \text{Pickett G2}$$

Strategy #2

$$\xrightarrow{[(5+0)_A + (4+2)_B]} [4.3.0]_3: \text{Pickett G1}$$

$$\xrightarrow{[(5+0)_A + (4+2)_B]} [4.3.0]_3: \text{Mori}$$

Strategy #3

$$A_0 \xrightarrow{[5+1]} B_{11}: \text{Sakan}$$

(R)-Nicotine
Campos, K.R.; Klapers, Waldman, J.H.; Dormer, P.G.; Chen, C.Y. *J. Am. Chem. Soc.* 2006, *128*, 3538.

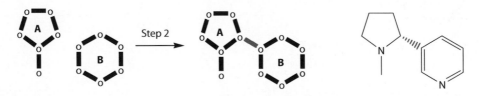

Sequence: $A_0 + B_0 \rightarrow (A + B)_2$

(S)-Nicotine
Chavdarian, C.G.; Sanders, E.B.; Bassfield, R.L. *J. Org. Chem.* 1982, *47*, 1069.

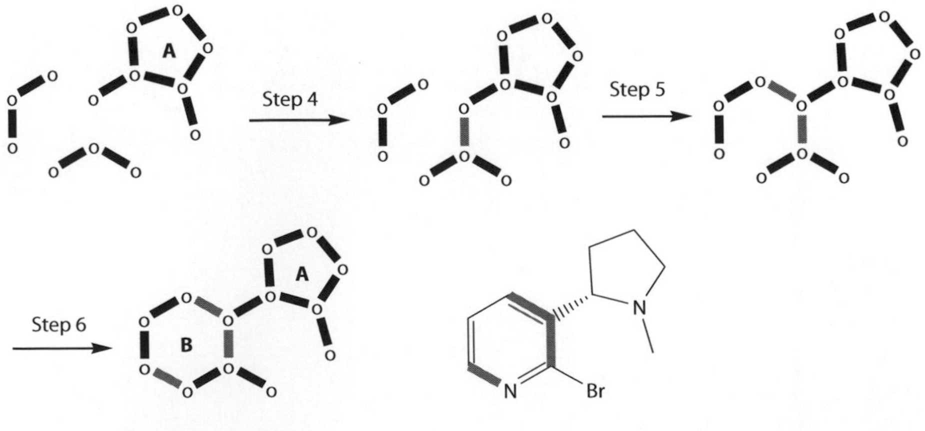

Sequence: $A_0 \xrightarrow{[6+0]} B_6$

(±)-Nicotine

Craig, L.C. *J. Am. Chem. Soc.* 1933, *55*, 2854.

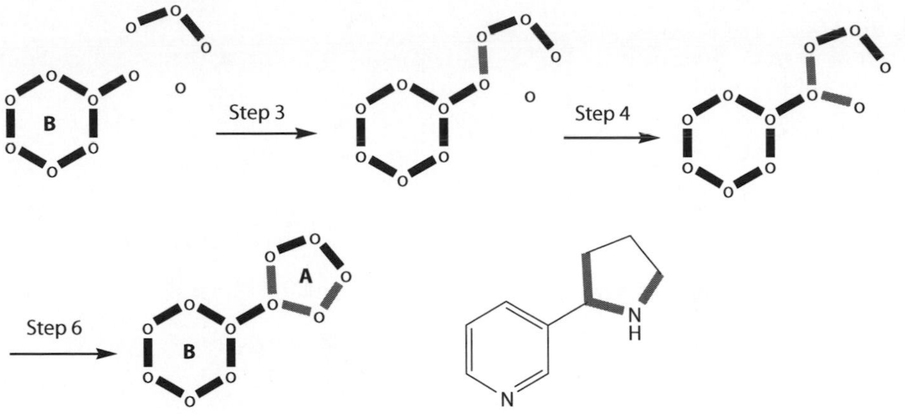

Sequence: $B_0 \xrightarrow{[5+0]} A_6$

(*S*)-Nicotine

Marquez, F.; Llebaria, A.; Delgado, A. *Tetrahedron Asymmetry* 2001, *12*, 1625.

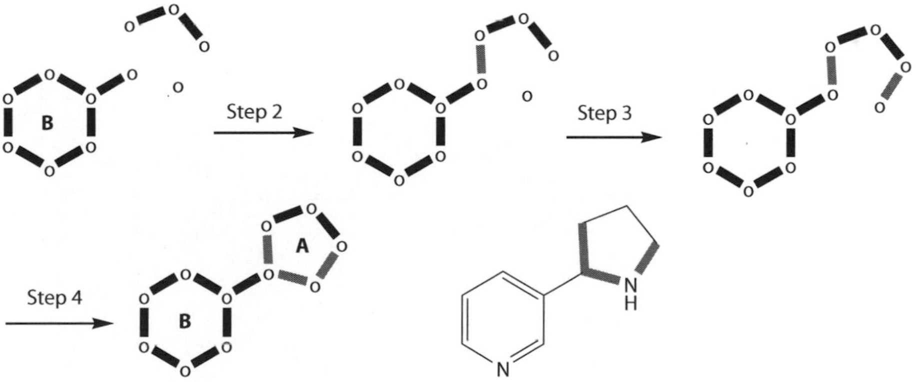

Sequence: $B_0 \xrightarrow{[5+0]} A_4$

(*S*)-Nicotine

Welter, C.; Moreno, R.M.; Streiff, S.; Helmchen, G. *Org. Biomol. Chem.* 2005, *3*, 3266.

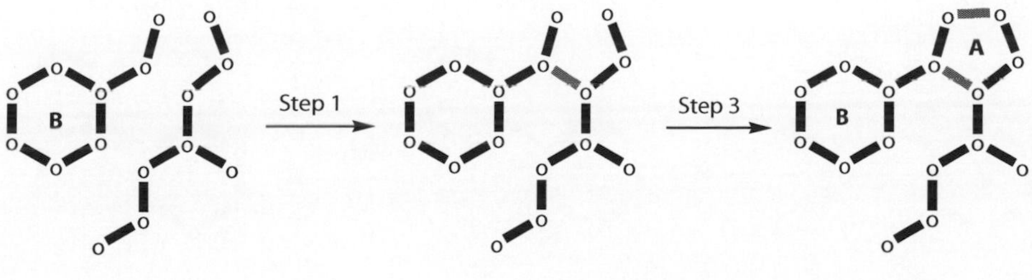

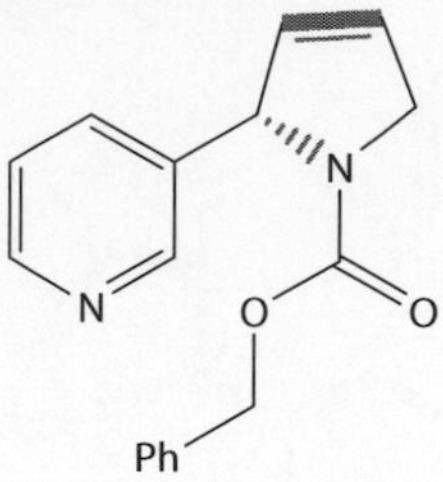

Sequence: $B_0 \xrightarrow{\text{[5+0]}} A_3$

(S)-Nicotine
Jacob, P. III *J. Org. Chem.* 1982, *47*, 4167.

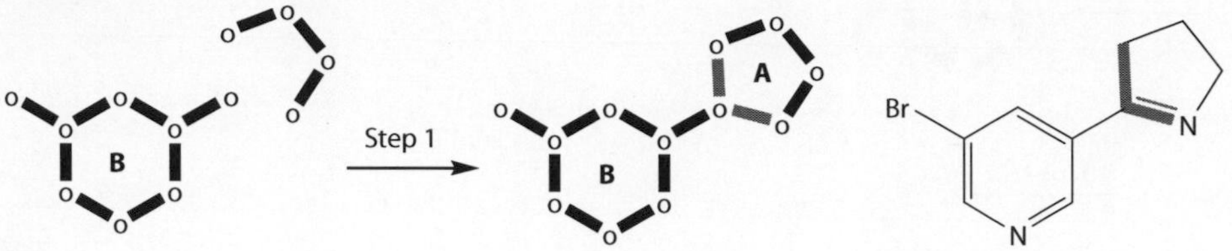

Sequence: $B_0 \xrightarrow{\text{[4+1]}} A_1$

(S)-Nicotine
Felpin, F.X.; Girard, S.; Vo-Thanh, G.; Robins, R.J.; Villieras, J.; Lebreton, J. *J. Org. Chem.* 2001, *66*, 6305.

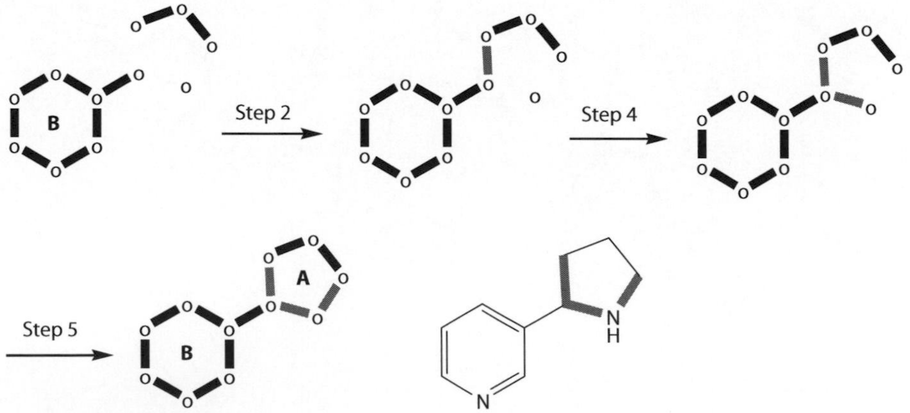

Sequence: $B_0 \xrightarrow{\text{[5+0]}} A_5$

(±)-Nicotine
Baxendale, I.R.; Bruscotti, G.; Matsuoka, M.; Ley, S.V. *J. Chem. Soc. Perkin Trans. I* 2002, 143.

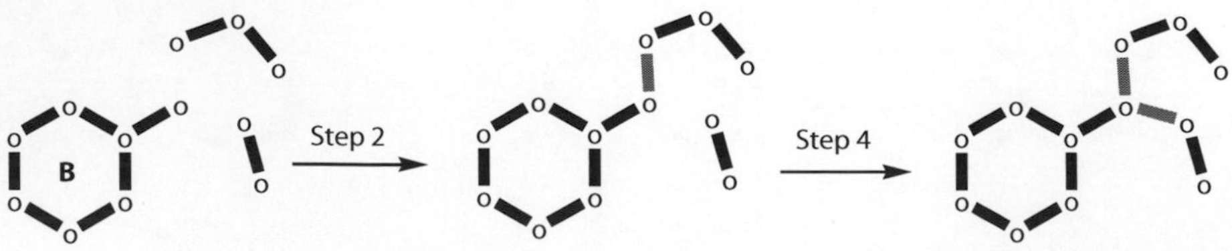

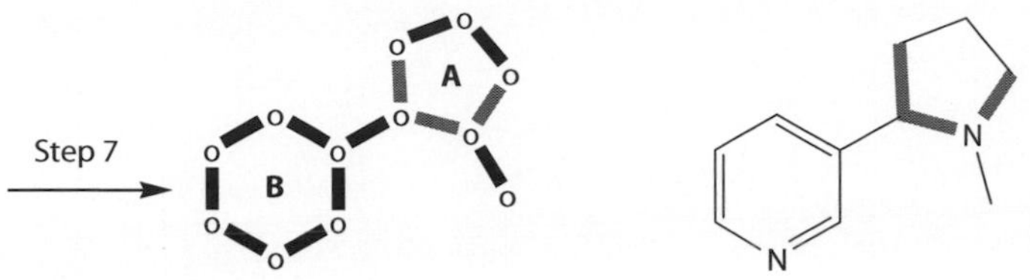
Step 7
A
B
N
N

Sequence: $B_0 \xrightarrow{[5+0]} A_7$

(*S*)-Nicotine

Loh, T.P.; Zhou, J.R.; Li, X.R.; Sim, K.Y. *Tetrahedron Lett.* 1999, *40*, 7847.

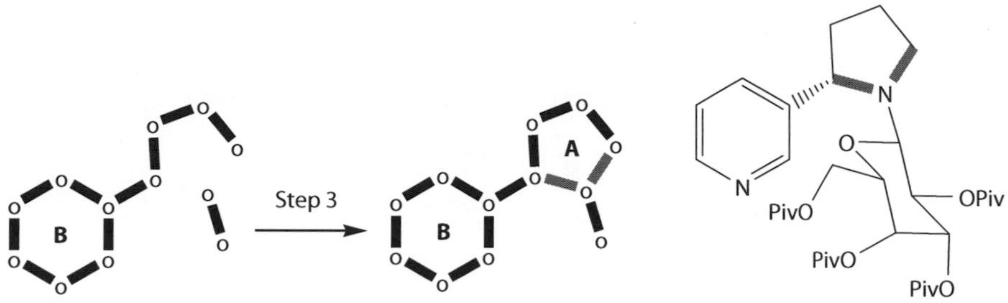
B
Step 3
A
B
N
N
O
OPiv
PivO
PivO
PivO

Sequence: $B_0 \xrightarrow{[4+1]} A_3$

(±)-Nicotine

Pictet, A.; Rotschy, A. *Chem. Ber.* 1904, *37*, 1225.
Pictet, A. *Chem. Ber.* 1900, *33*, 2355.
Pictet, A.; Crépieux, P. *Chem. Ber.* 1895, *28*, 1904.

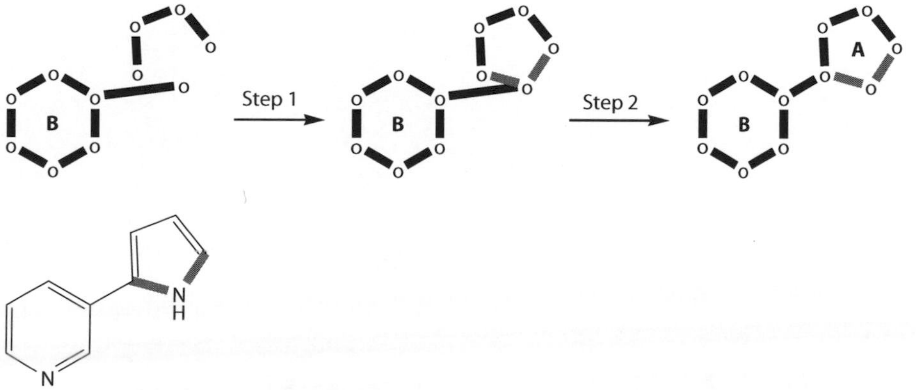
B
Step 1
B
Step 2
A
B
N
H
N

Sequence: $B_0 \xrightarrow{[4+1]} A_1 \xrightarrow{1,2-shift} A_2$

(±)-Nicotine

Späth, E.; Bretschneider, H. *Chem. Ber.* 1928, *61*, 327.

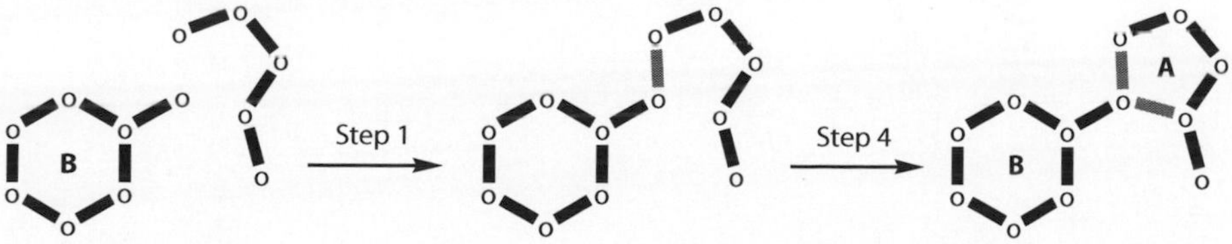
B
Step 1
Step 4
A
B

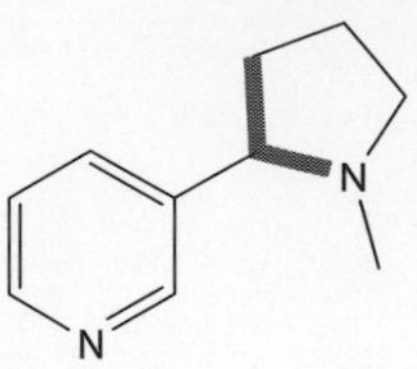

Sequence: $B_0 \xrightarrow{\;[5+0]\;} A_4$

Summary:

Strategy B → A via [4 + 1]: Jacob, Loh

Strategy B → A via [4 + 1] and 1,2-rearrangement: Pictet

Strategy B → A via [5 + 0]: Craig, Delgado, Helmchen, Lebreton, Ley, Spath

Strategy A → B: Chavdarian

Strategy A + B → (A + B): Campos

Oseltamivir phosphate

Matveenko, M.; Willis, A.C.; Banwell, M.G. *Tetrahedron Lett*. 2008, *49*, 7018.

Shie, J.J.; Fang, J.M.; Wong, C.H. *Angew. Chem. Int. Ed*. 2008, *47*, 5788.

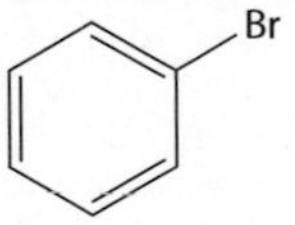

Sullivan, B.; Carrera, I.; Drouin, M.; Hudlicky, T. *Angew. Chem. Int. Ed*. 2009, *48*, 4229.

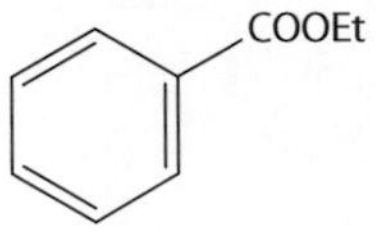

Yeung, Y.Y.; Hong, S.; Corey, E.J. *J. Am. Chem. Soc*. 2006, *128*, 6310.

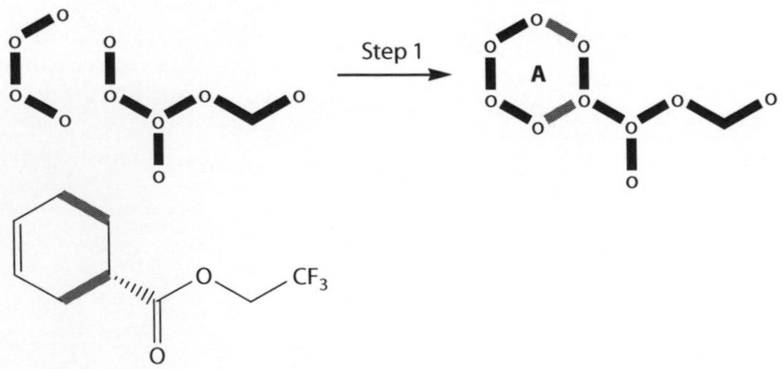

Sequence: $\xrightarrow{\;[4+2]\;} A_1$

Shie, J.J.; Fang, J.M.; Wang, S.Y.; Tsai, K.C.; Cheng, Y.S.E.; Yang, A.S.; Hsaio, S.C.; Su, C.Y.; Wong, C.H. *J. Am. Chem. Soc.* 2007, *129*, 11892.

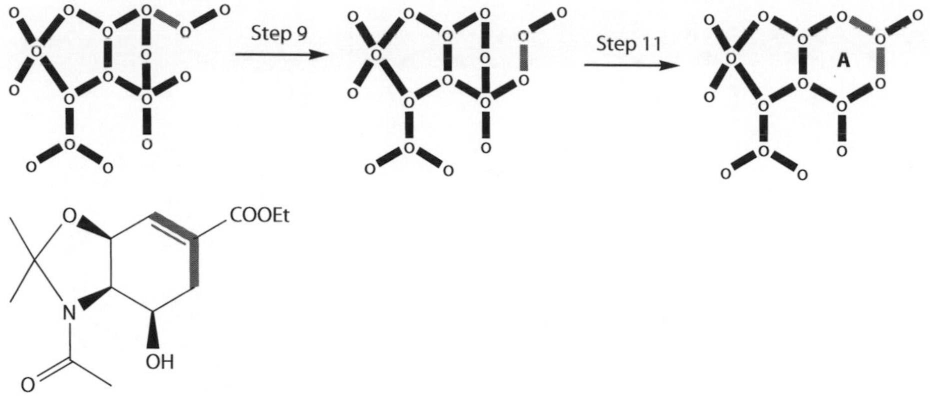

Sequence: $\xrightarrow{[6+0]} A_{11}$

Satoh, N.; Akiba, T.; Yokoshima, S.; Fukuyama, T. *Angew. Chem. Int. Ed.* 2007, *46*, 5734.

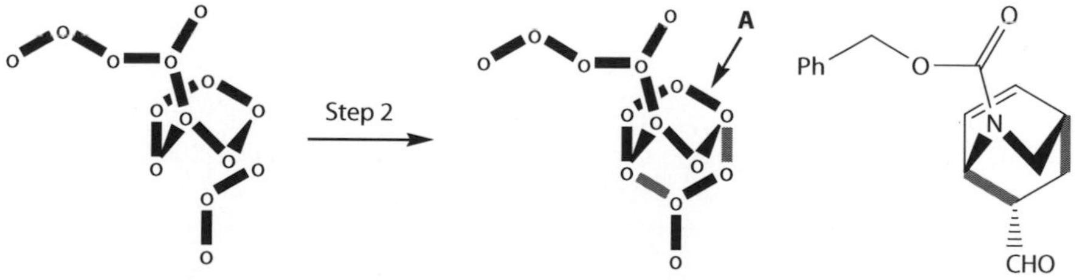

Sequence: $\xrightarrow{[4+2]} A_2$

Rohloff, J.C.; Kent, K.M.; Postich, M.J.; Becker, M.W.; Chapman, H.H.; Kelly, D.E.; Lew, W.; Louie, M.S.; McGee, L.R.; Prisbe, E.J.; Schultze, L.M.; Yu, R.H.; Zhang, L. *J. Org. Chem.* 1998, *63*, 4545.
Federspiel, M.; Fischer, R.; Hennig, M.; Mair, H.J.; Oberhauser, T.; Rimmler, G.; Albiez, T.; Bruhin, J.; Estermann, H.; Gandert, C.; Göckel, V.; Götzö, S.; Hoffmann, U.; Huber, G.; Janatsch, G.; Lauper, S.; Röckel-Stäbler, O.; Trussardi, R.; Zwahlen, A.G. *Org. Process Res. Develop.* 1999, *3*, 266.
Harrington, P.J.; Brown, J.D.; Foderaro, T.; Hughes, R.C. *Org. Process Res. Develop.* 2004, *8*, 86.
Karpf, M.; Trussardi, R. *J. Org. Chem.* 2001, *66*, 2044.

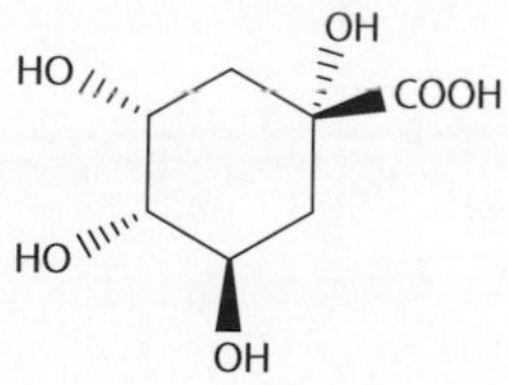

Ishikawa, H.; Suzuki, T.; Hayashi, Y. *Angew. Chem. Int. Eng. Ed.* 2009, *48*, 1304.

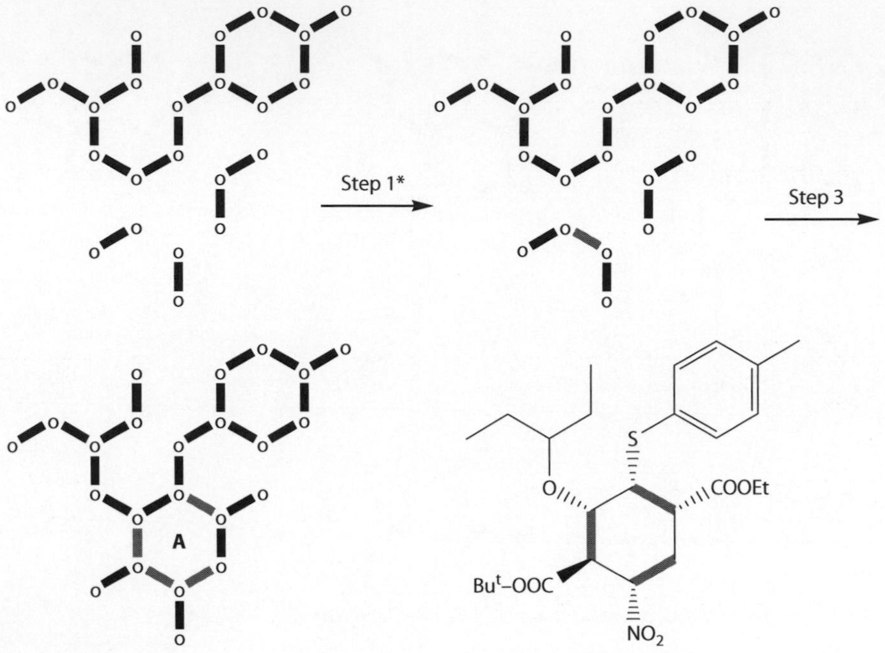

Sequence: $\xrightarrow{[2+2+2]} A_3$

Bromfield, K.M.; Gradén, H.; Hagberg, D.P.; Olsson, T.; Kann, N. *Chem. Commun.* 2007, 3183.

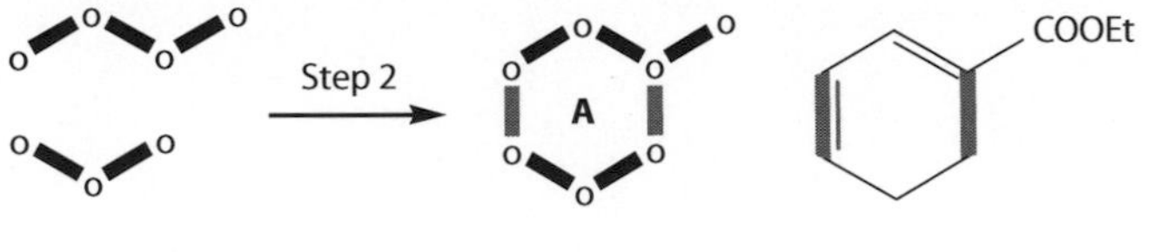

Sequence: $\xrightarrow{[3+3]} A_2$

Mandai, T.; Oshitari, T. *Synlett* 2009, 783.

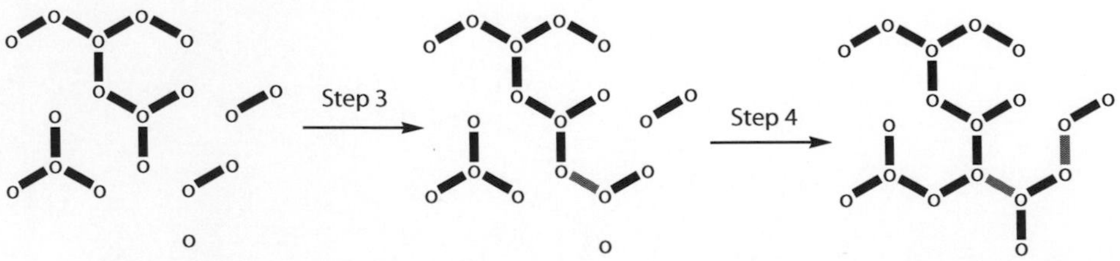

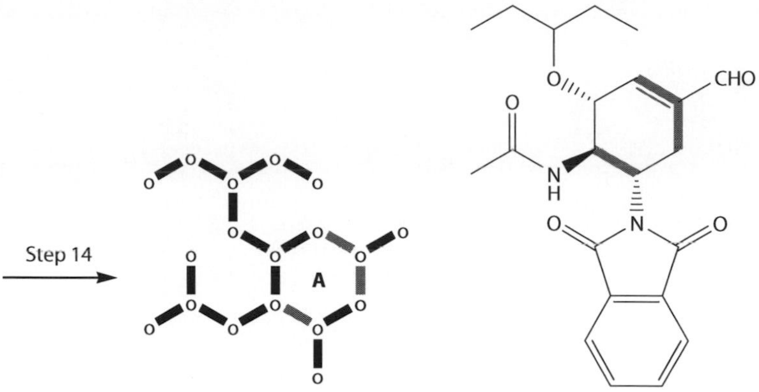

Step 14

Sequence: $\xrightarrow{[6+0]} A_{14}$

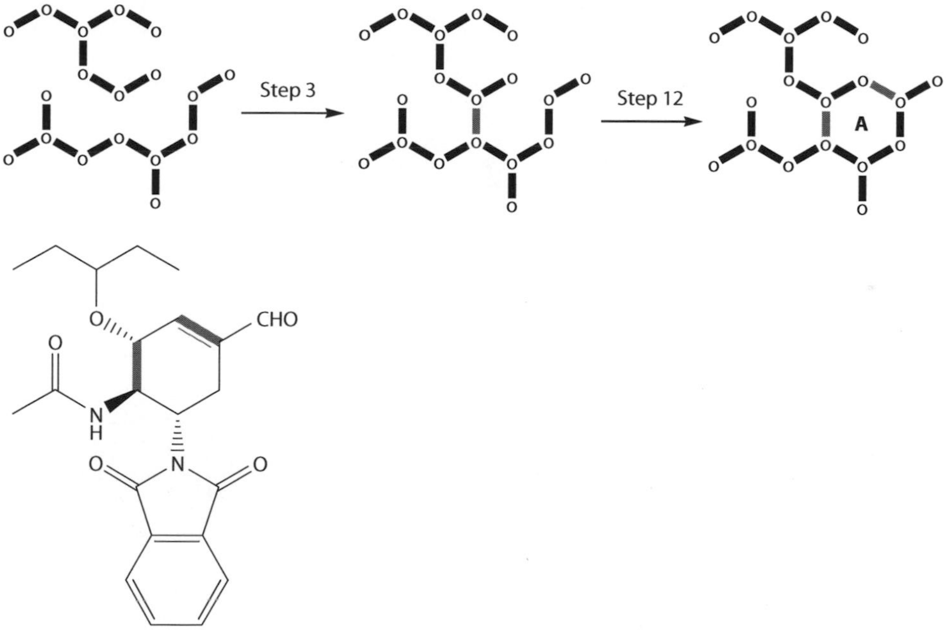

Step 3

Step 12

Sequence: $\xrightarrow{[6+0]} A_{12}$

Kipassa, N.T.; Okamura, H.; Kina, K.; Hamada, T.; Iwagawa, T. *Org. Lett.* 2008, *10*, 815.

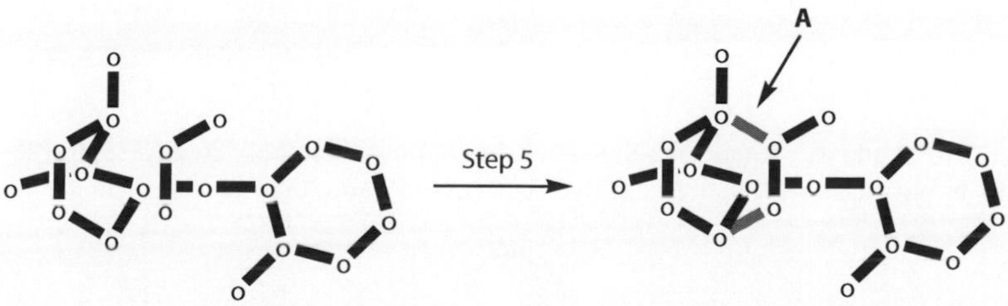

Step 5

Sequence: $\xrightarrow{\ [4+2]\ }$ A_5

Federspiel, M.; Fischer, R.; Hennig, M.; Mair, H.J.; Oberhauser, T.; Rimmler, G.; Albiez, T.; Bruhin, J.; Estermann, H.; Gandert, C.; Göckel, V.; Götzö, S.; Hoffmann, U.; Huber, G.; Janatsch, G.; Lauper, S.; Röckel-Stäbler, O.; Trussardi, R.; Zwahlen, A.G. *Org. Process Res. Develop.* 1999, *3*, 266.
Harrington, P.J.; Brown, J.D.; Foderaro, T.; Hughes, R.C. *Org. Process Res. Develop.* 2004, *8*, 86.
Nie, L.D.; Shi, X.X.; Ko, K.H.; Lu, W.D. *J. Org. Chem.* 2009, *74*, 3970.
Karpf, M.; Trussardi, R. *Angew. Chem. Int. Ed.* 2009, *48*, 5760.

Abrecht, S.; Karpf, M.; Trussardi, R.; Wirz, B. EP 1127872 Tamiflu via Diels-Alder (F. Hoffmann-La Roche AG, 2000).

Sequence: $\xrightarrow{\ [4+2]\ }$ A_1

Zutter, U.; Iding, H.; Wirz, EP 1146036 (F. Hoffmann-La Roche AG, 2000).

Fukuta, Y.; Mita, T.; Fukuda, N.; Kanai, M.; Shibasaki, M. *J. Am. Chem. Soc.* 2006, *128*, 6312.
Mita, T.; Fukuda, N.; Roca, F.X.; Kanai, M.; Shibasaki, M. *Org. Lett.* 2007, *9*, 259.

Yamatsugu, K.; Kamijo, S.; Suto, Y.; Kanai, M.; Shibasaki, M. *Tetrahedron Lett*. 2007, *48*, 1403.

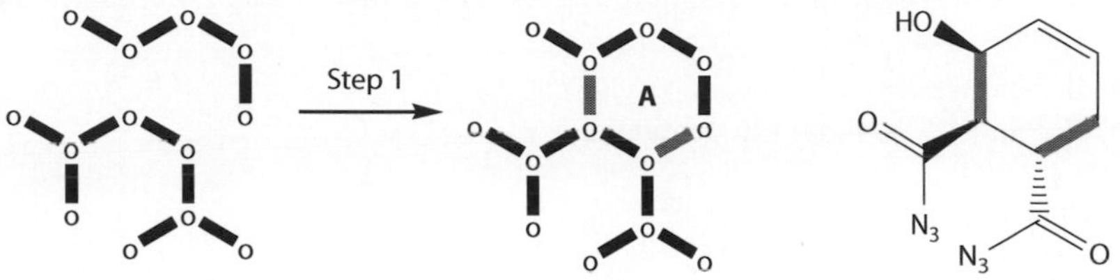

Sequence: $\xrightarrow{[4+2]}$ A_1

Yamatsugu, K.; Kanai, M.; Shibasaki, M. *Tetrahedron* 2009, *65*, 6017.
Yamatsugu, K.; Yin, L.; Kamijo, S.; Kimura, Y.; Kanai, M.; Shibasaki, M. *Angew. Chem. Int. Ed*. 2009, *48*, 1070.

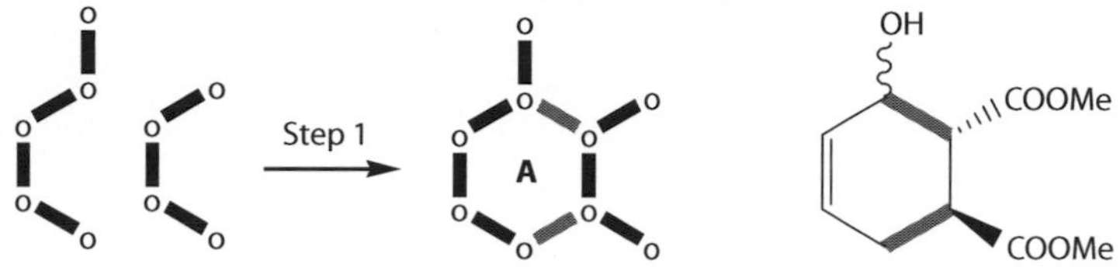

Sequence: $\xrightarrow{[4+2]}$ A_1

Trost, B.M.; Zhang, T. *Angew. Chem. Int. Ed*. 2008, *47*, 3759.

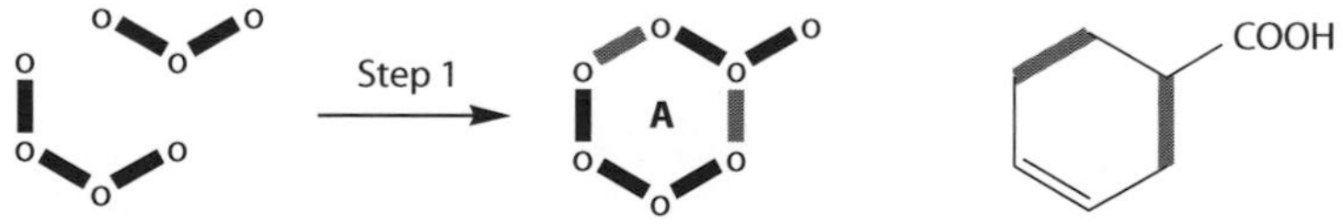

Sequence: $\xrightarrow{[4+2]}$ A_1

Ma, J.; Zhao, Y.; Ng, S.; Zhang, J.; Zeng, J.; Than, A.; Chen, P.; Liu, X.W. *Chem. Eur. J*. 2010, *16*, 4533.

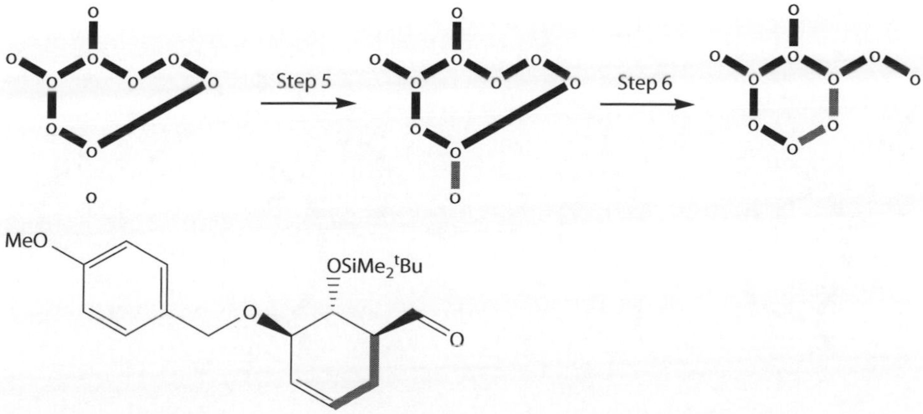

Sequence: [6] $\xrightarrow{[6+0]}$ [6]

Summary:
Strategy [4 + 2]: Corey, Fukuyama, Okamura, Roche G4, Shibasaki G3, Shibasaki G4, Shibasaki
     G5, Trost
Strategy [3 + 3]: Kann
Strategy [6 + 0]: Fang G1, Mandai G1, Mandai G2
Strategy [2 + 2 + 2]: Hayashi
Strategy [6] → [6]: Chen-Liu

Precursors containing ring:

Banwell, Fang G2, Fang G3, Gilead, Hudlicky G1, Roche G1, Roche G2, Roche G3, Roche G5, Roche G6,
     Shi, Shibasaki G1, Shibasaki G2

(–)-Paclitaxel
Commercon, A.; Bezard, D.; Bernard, F.; Bourzat, J.D. *Tetrahedron Lett.* 1992, *33*, 5185.
Gennari, C.; Carcano, M.; Donghi, M.; Mongelli, N.; Vanotti, E.; Vulpetti, A. *J. Org. Chem.* 1997, *62*, 4746.
Denis, J.N.; Corres, A.; Greene, A.E. *J. Org. Chem.* 1991, *56*, 6939.
Kanazawa, A.M.; Denis, J.N.; Greene, A.E. *Chem. Commun.* 1994, 2591.
Kingston, D.G.I.; Chaudhury, A.G.; Gunatilaka, A.A.L.; Middleton, M.L. *Tetrahedron Lett.* 1994, *35*, 4483.

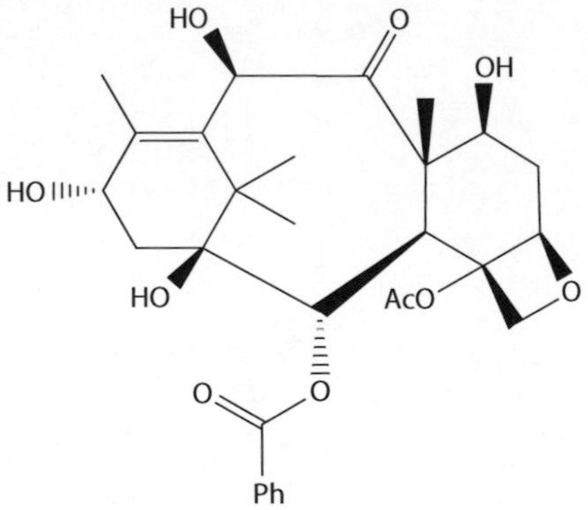

(–)-Paclitaxel
Masters, J.J.; Link, J.T.; Snyder, L.B.; Young, W.B.; Danishefsky, S.J. *Angew. Chem. Int. Ed.* 1995, *34*, 1723.
Danishefsky, S.J.; Masters, J.J.; Young, W.B.; Link, J.T.; Snyder, L.B.; Magee, T.V.; Jung, D.K.;
     Isaacs, R.C.A.; Bornmann, W.G.; Alaimo, C.A.; Coburn, C.A.; Di Grandi, M.J. *J. Am. Chem. Soc.*
     1996, *118*, 2843.

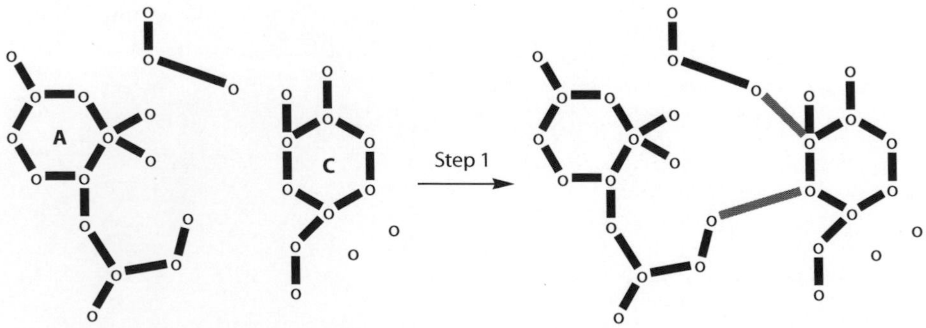

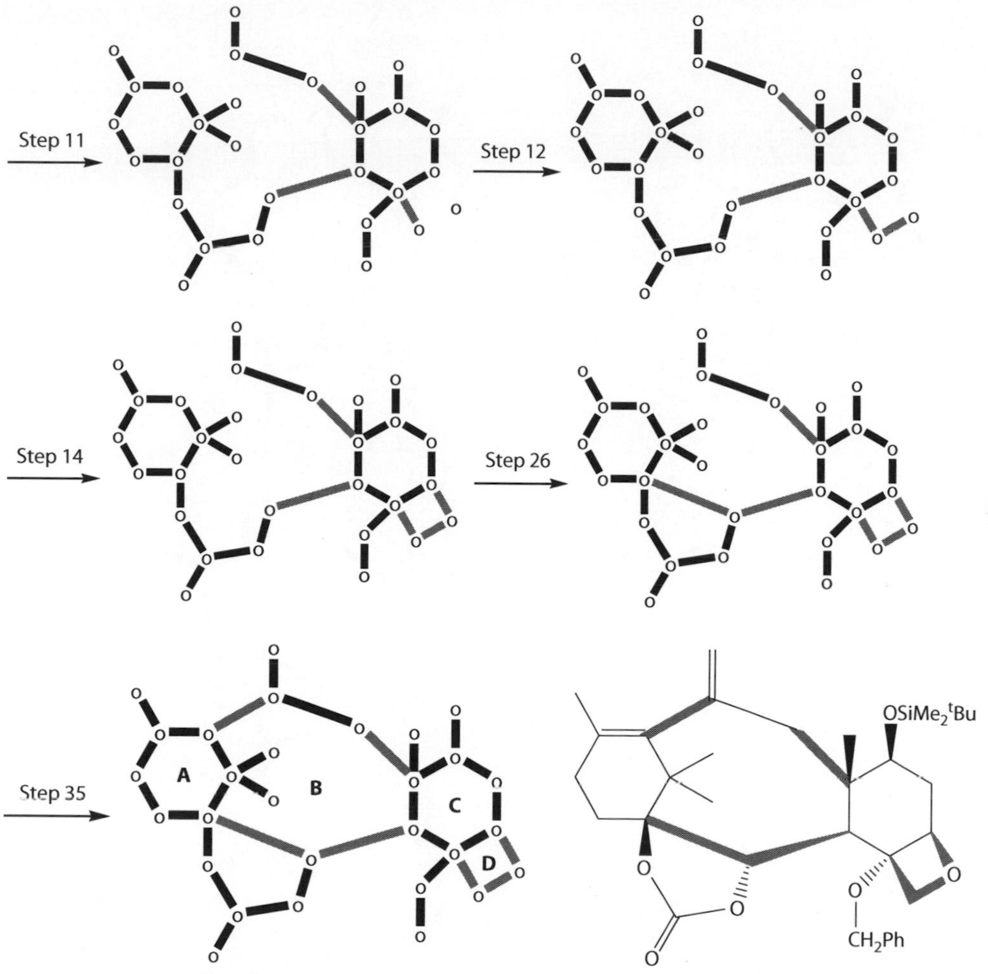

Sequence: $A_{22*} + C_0 \xrightarrow{[4+0]} D_{14} \xrightarrow{[8+0]} B_{35}$

(±)-Paclitaxel

Holton, R.A.; Juo, R.R.; Kim, H.B.; Williams, A.D.; Harusawa, S.; Lowenthal, R.E.; Yogai, S. *J. Am. Chem. Soc.* 1988, *110*, 6558.

Holton, R.A.; Somoza, C.; Kim, H.B.; Liang, F.; Biedliger, R.J.; Boatman, P.D.; Shindo, M.; Smith, C.C.; Kim, S.; Nadizadeh, H.; Suzuki, Y.; Tao, C.; Vu, P.; Tang, S.; Zhang, P.; Murthi, K.K.; Gentile, L.N.; Liu, J.H. *J. Am. Chem. Soc.* 1994, *116*, 1597.

Holton, R.A.; Somoza, C.; Kim, H.B.; Liang, F.; Biedliger, R.J.; Boatman, P.D.; Shindo, M.; Smith, C.C.; Kim, S.; Nadizadeh, H.; Suzuki, Y.; Tao, C.; Vu, P.; Tang, S.; Zhang, P.; Murthi, K.K.; Gentile, L.N.; Liu, J.H. *J. Am. Chem. Soc.* 1994, *116*, 1599.

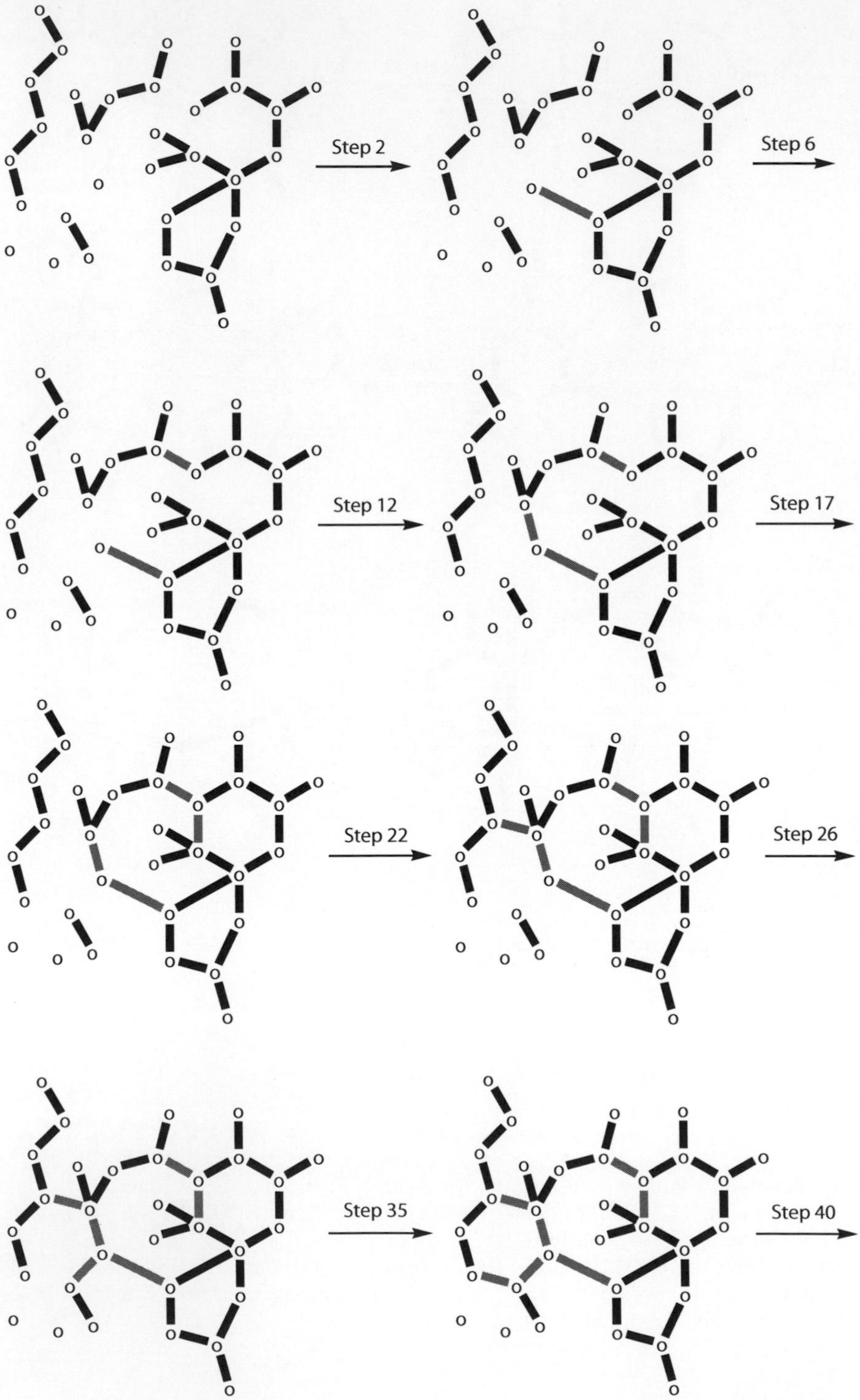
Step 2
Step 6
Step 12
Step 17
Step 22
Step 26
Step 35
Step 40

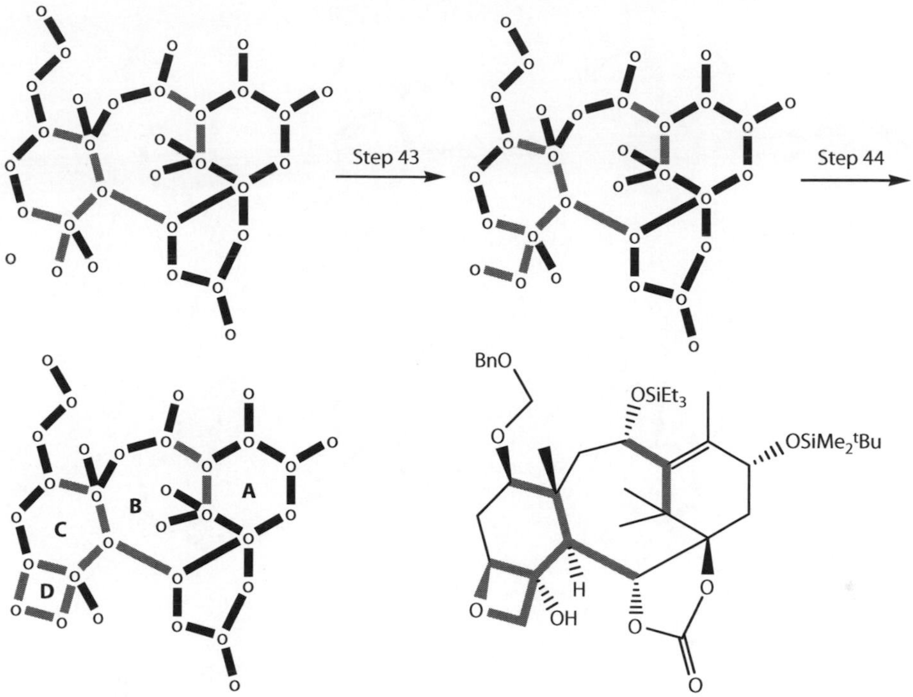

Sequence: $\xrightarrow{[(8+0)+(6+0)]}$ $[5.3.1](A+B)_{17}$ $\xrightarrow{[6+0]}$ C_{35} $\xrightarrow{[4+0]}$ D_{44}

(−)-Paclitaxel

Kusama, H.; Hara, R.; Kawahara, S.; Nishimori, T.; Kashima, H.; Nakamura, N.; Morihara, K.; Kuwajima, I. *J. Am. Chem. Soc.* 2000, *122*, 3811.

Morihara, K.; Hara, R.; Kawahara, S.; Nishimori, T.; Nakamura, N.; Kusama, H.; Kuwajima, I. *J. Am. Chem. Soc.* 1998, *120*, 12980.

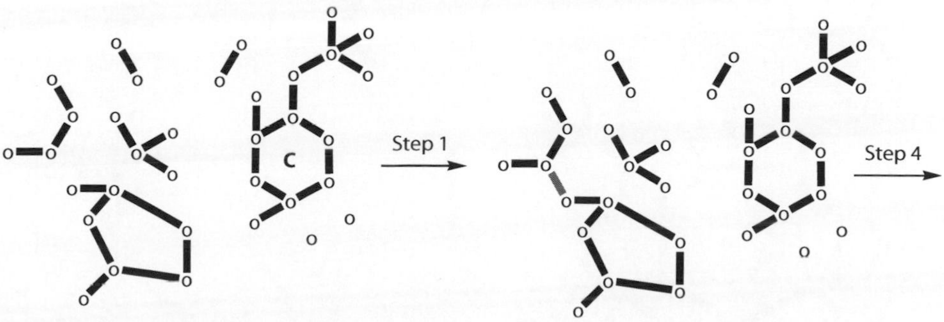

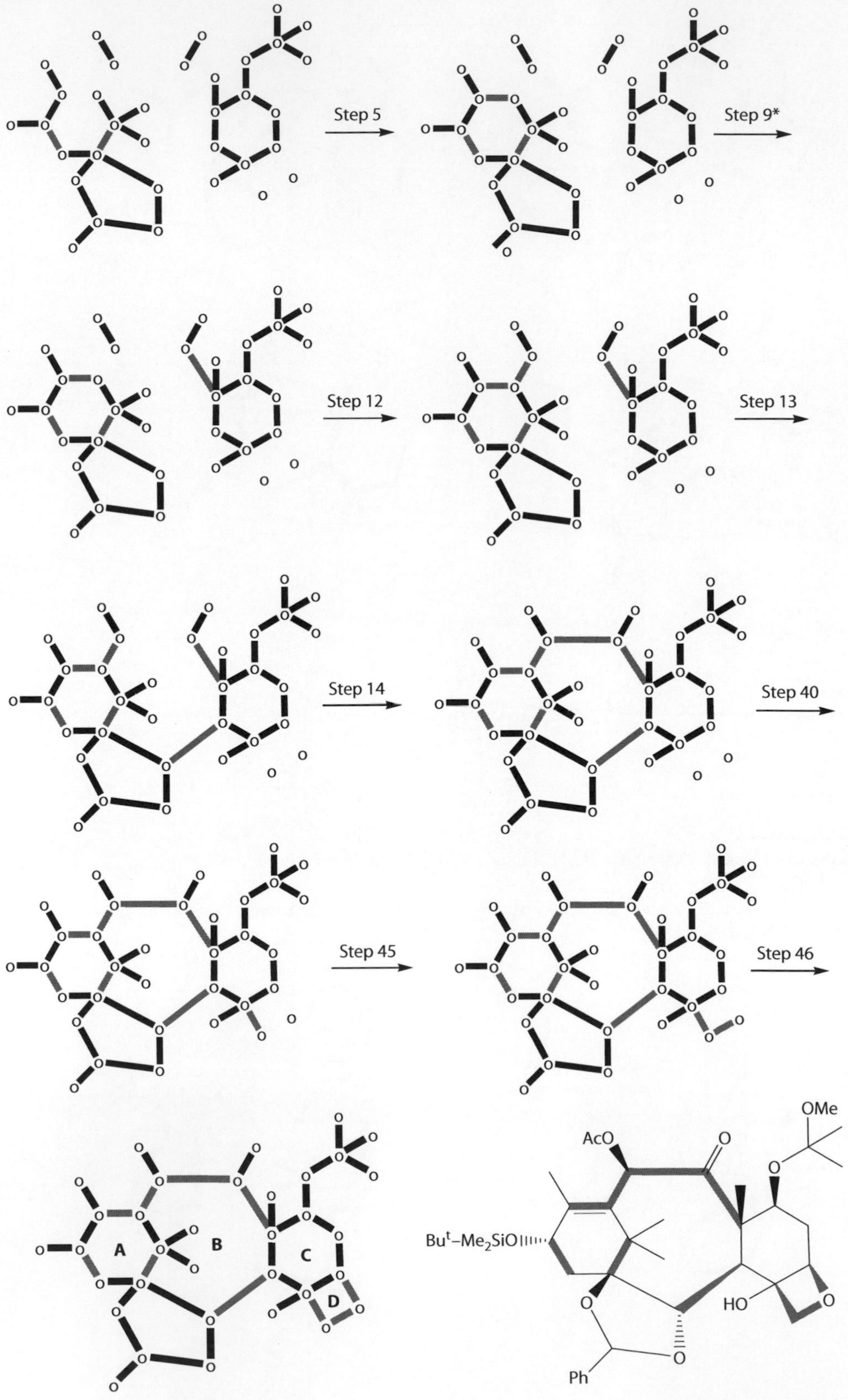

Step 5
Step 9*
Step 12
Step 13
Step 14
Step 40
Step 45
Step 46
A
B
C
D
AcO
O
OMe
Bu^t–Me$_2$SiO
HO
O
O
O
Ph

Sequence: $C_{6*} \xrightarrow{[6+0]} A_5 \xrightarrow{[8+0]} B_{13} \xrightarrow{[4+0]} D_{46}$

(–)-Paclitaxel

Mukaiyama, T.; Shiina, I.; Iwadare, H.; Saitoh, M.; Nishimura, T.; Ohkawa, N.; Sakoh, H.; Nishimura, K.; Tani, Y.; Hasegawa, M.; Yamada, K.; Saitoh, K. *Chem. Eur. J.* 1999, *5*, 121.

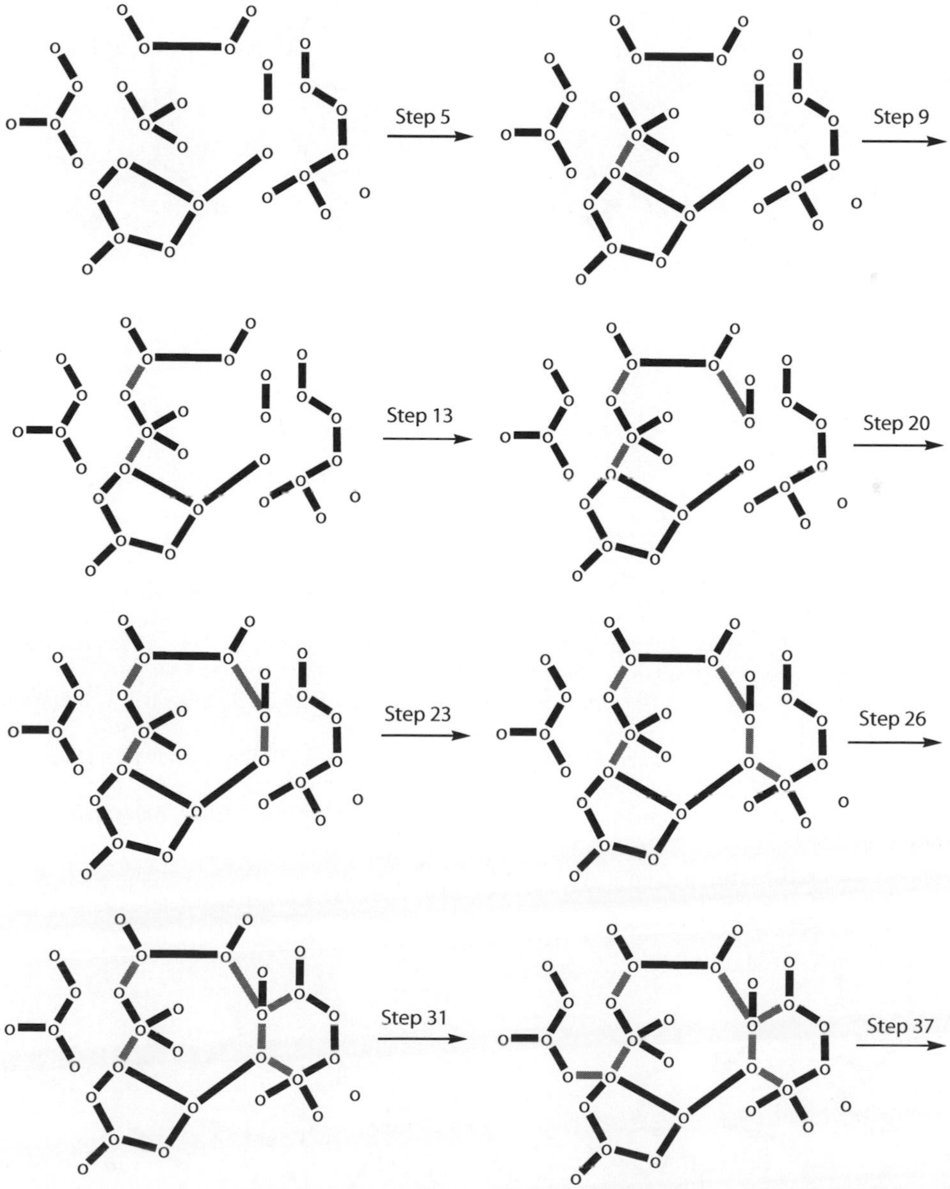

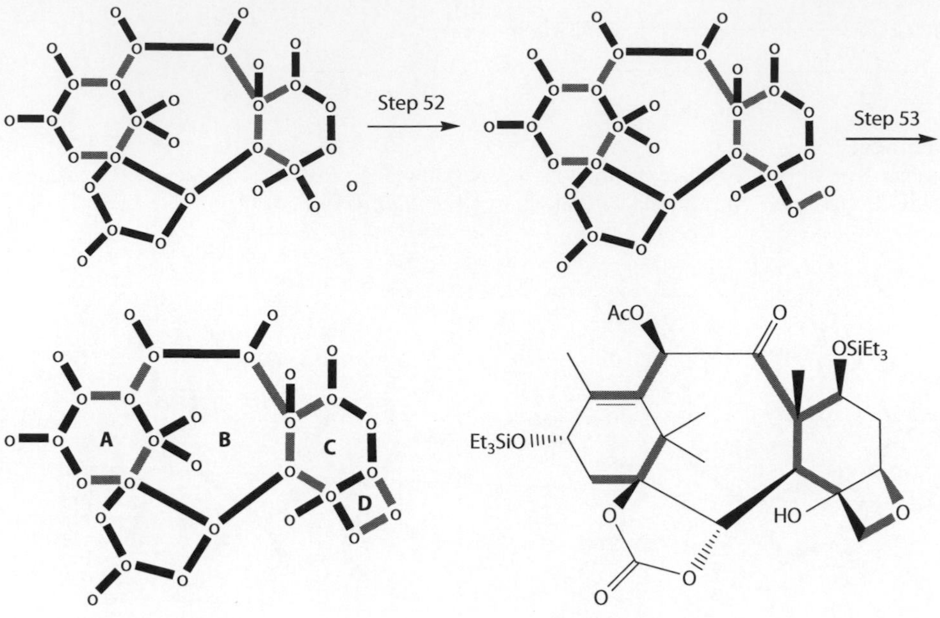

Sequence: $\xrightarrow{[8+0]} B_{20} \xrightarrow{[6+0]} C_{26} \xrightarrow{[6+0]} A_{37} \xrightarrow{[4+0]} D_{53}$

(−)-Paclitaxel

Nicolaou, K.C.; Nantermet, P.G.; Ueno, H.; Guy, R.K.; Couladouros, E.A.; Sorensen, E.J. *J. Am. Chem. Soc.* 1995, *117*, 624.

Nicolaou, K.C.; Liu, J.J.; Yang, Z.; Ueno, H.; Sorensen, E.J.; Claiborne, C.F.; Guy, R.K.; Hwang, C.K.; Nakada, M.; Nantermet, P.G. *J. Am. Chem. Soc.* 1995, *117*, 634.

Nicolaou, K.C.; Yang, Z.; Liu, J.J.; Nantermet, P.G.; Claiborne, C.F.; Renaud, J.; Guy, R.K.; Shibayama, K. *J. Am. Chem. Soc.* 1995, *117*, 645.

Nicolaou, K.C.; Ueno, H.; Liu, J.J.; Nantermet, P.G.; Yang, Z.; Renaud, J.; Paulvannan, K.; Chadha, R. *J. Am. Chem. Soc.* 1995, *117*, 653.

Nicolaou, K.C.; Claiborne, C.F.; Nantermet, P.G.; Couladouros, E.A.; Sorensen, E.J. *J. Am. Chem. Soc.* 1994, *116*, 1591.

Nicolaou, K.C.; Guy, R.K. *Angew. Chem. Int. Ed.* 1995, *34*, 2079.

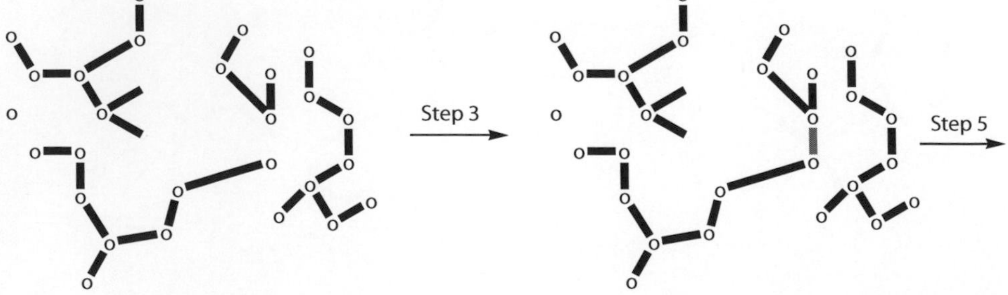

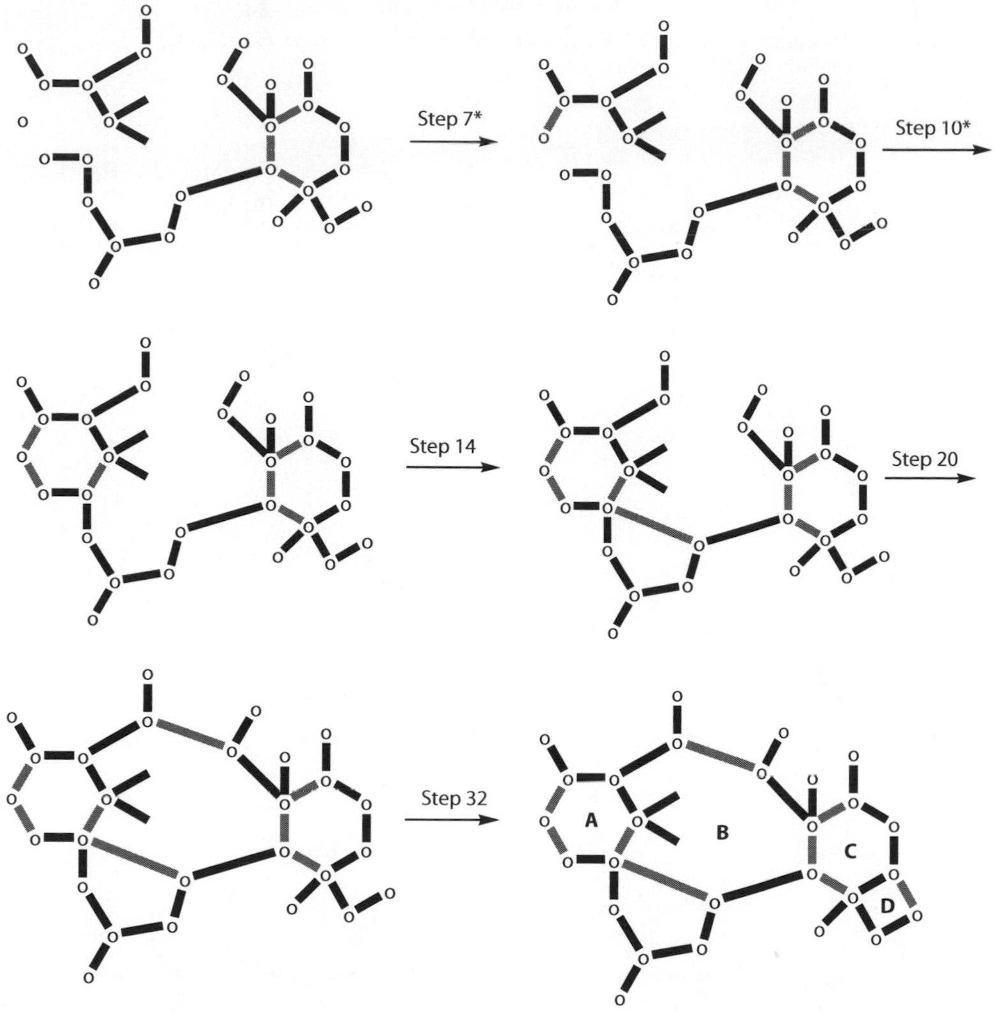

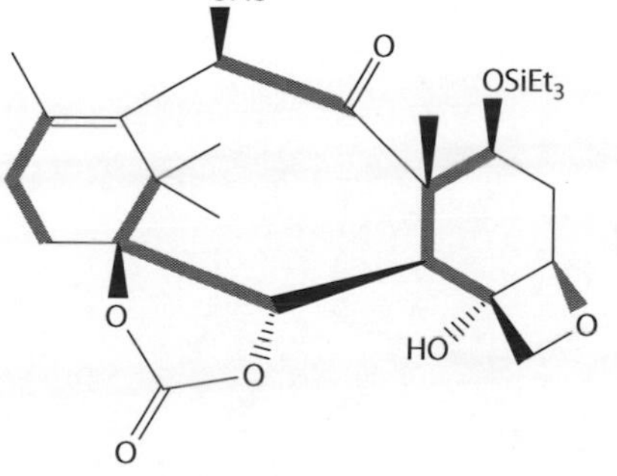

Sequence: $\xrightarrow{[4+2]} C_5 \xrightarrow{[4+2]} A_{10*} \xrightarrow{[8+0]} B_{20} \xrightarrow{[4+0]} D_{32}$

(−)-Paclitaxel
Doi, T.; Fuse, S.; Miyamoto, S.; Nakai, K.; Sasuga, D.; Takahashi, T. *Chem. Asian J.* 2006, *1*, 370.

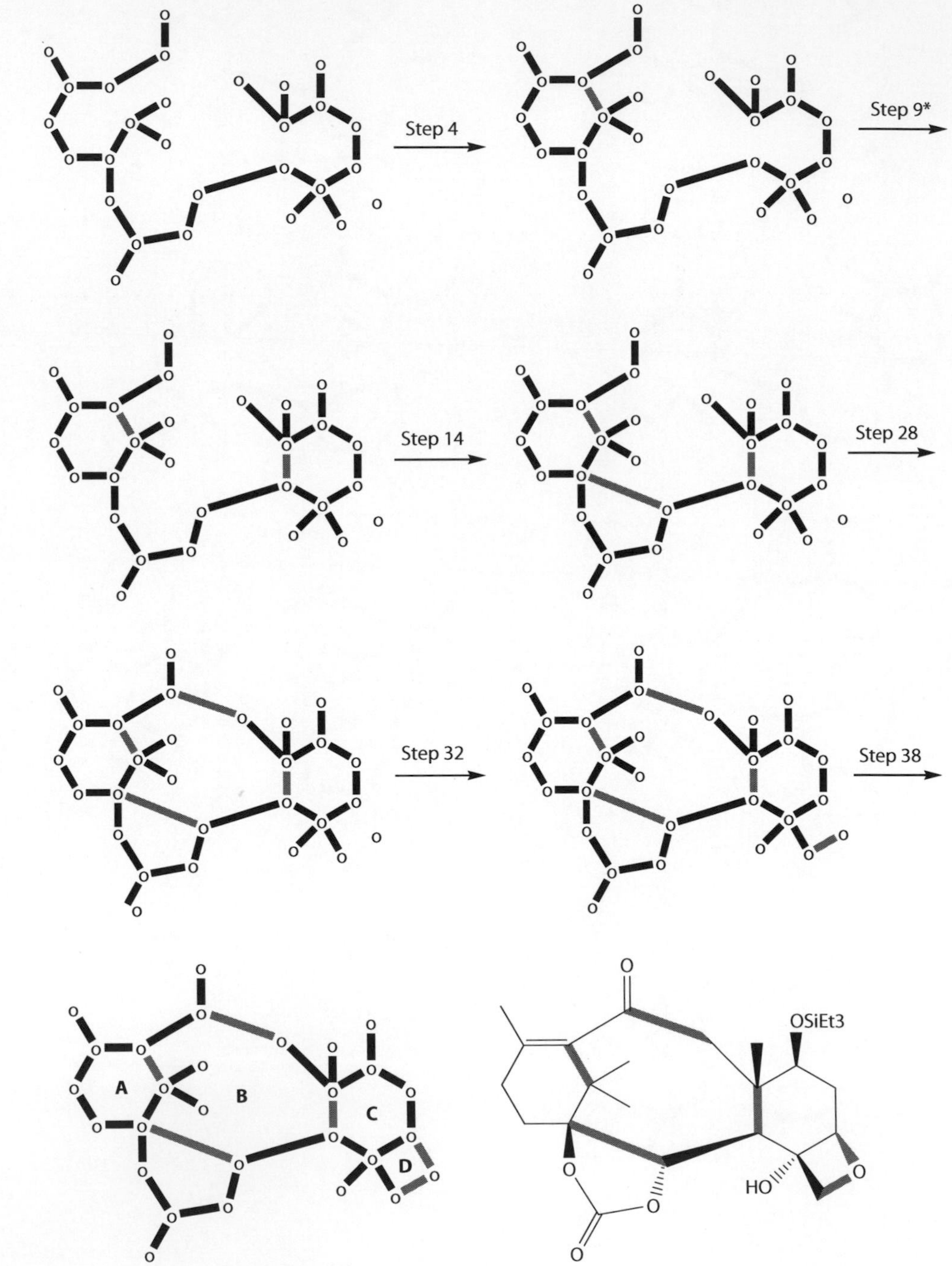

Sequence: $\xrightarrow{[6+0]} A_4 \xrightarrow{[6+0]} C_{9*} \xrightarrow{[8+0]} B_{28} \xrightarrow{[4+0]} D_{38}$

(–)-Paclitaxel

Wender, P.A.; Badham, N.F.; Conway, S.P.; Floreancig, P.E.; Glass, T.E.; Gränicher, C.; Houze, J.B.; Jänichen, J.; Lee, D.; Marquess, D.G.; McCrane, P.L.; Meng, W.; Mucciaro, T.P.; Mühlebach, M.; Natchus, M.G.; Paulsen, H.; Rawlins, D.B.; Satkofsky, J.; Shuker, A.J.; Sutton, J.C.; Taylor, R.E.; Tomooka, K. *J. Am. Chem. Soc.* 1997, *119*, 2755.

Wender, P.A.; Badham, N.F.; Conway, S.P.; Floreancig, P.E.; Glass, T.E.; Houze, J.B.; Krauss, N.E.; Lee, D.; Marquess, D.G.; McCrane, P.L.; Meng, W.; Natchus, M.G.; Shuker, A.J.; Sutton, J.C.; Taylor, R.E. *J. Am. Chem. Soc.* 1997, *119*, 2757.

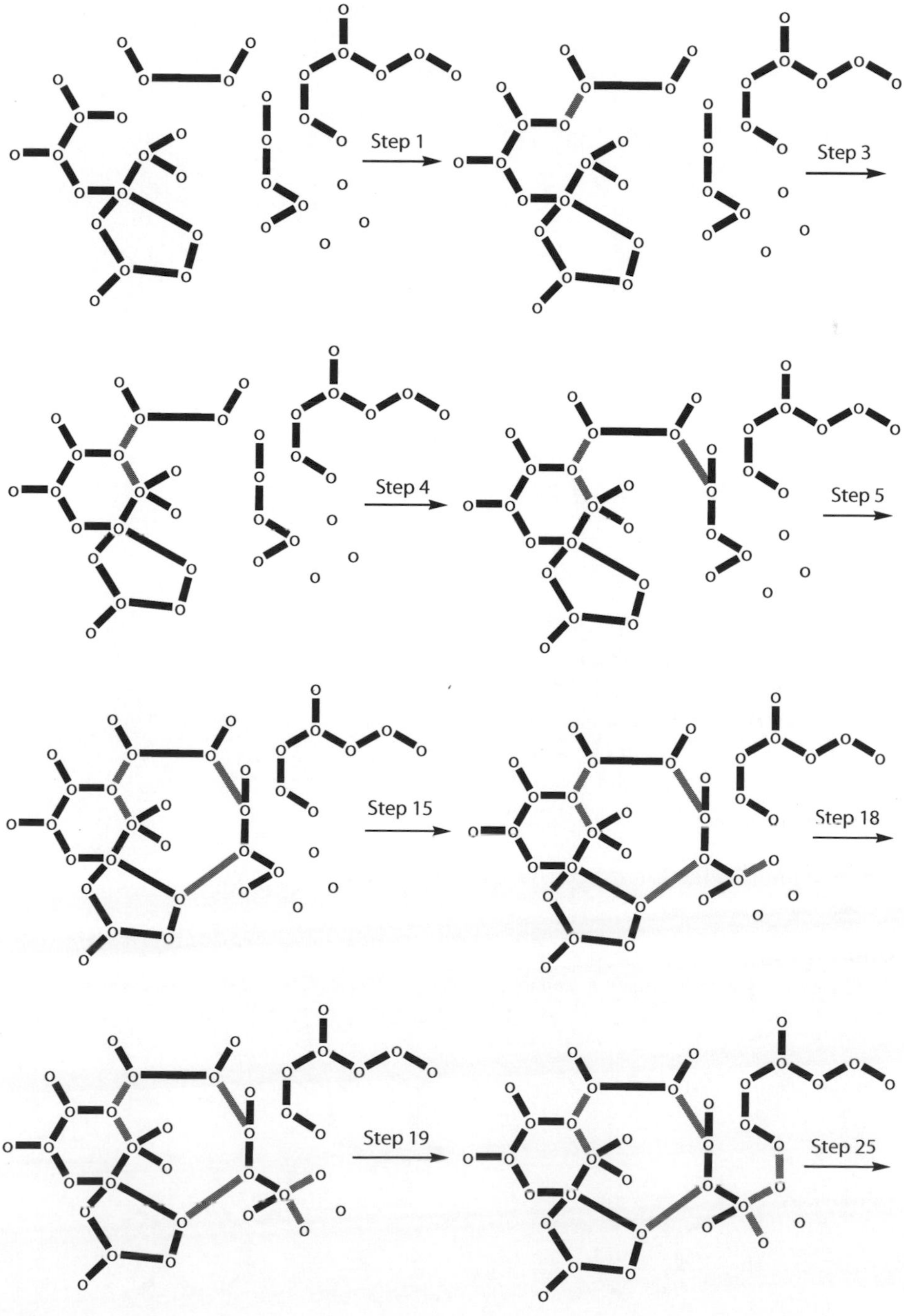

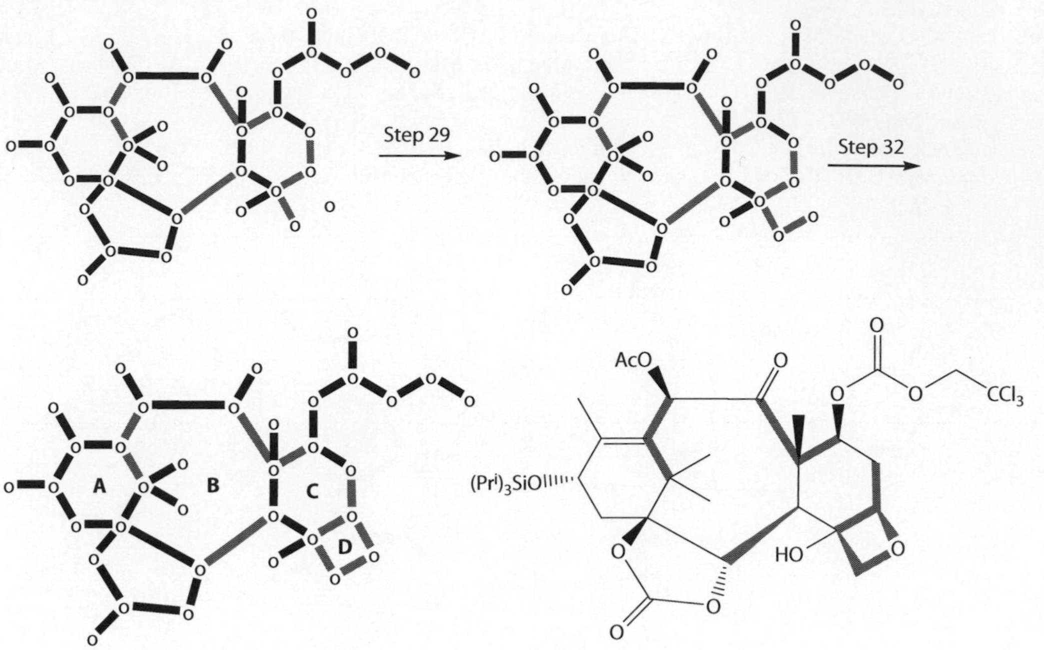

Sequence: $\xrightarrow{[6+0]} A_3 \xrightarrow{[8+0]} B_5 \xrightarrow{[6+0]} C_{25} \xrightarrow{[4+0]} D_{32}$

Summary:
Strategy → A → B → C → D: Wender
Strategy → A → C → B → D: Takahashi
Strategy → C → A → B → D: Nicolaou
Strategy C → A → B → D: Kuwajima
Strategy → B → C → A → D: Mukaiyama
Strategy A + C → D → B: Danishefsky
Strategy → [5.3.1] (A + B) → C → D: Holton

Precursors containing ring: Commercon, Gennari, Greene G1, Greene G2, Kingston

Papaverine
Dean, F.H. Ontario Research Foundation, unpublished results, 1974.

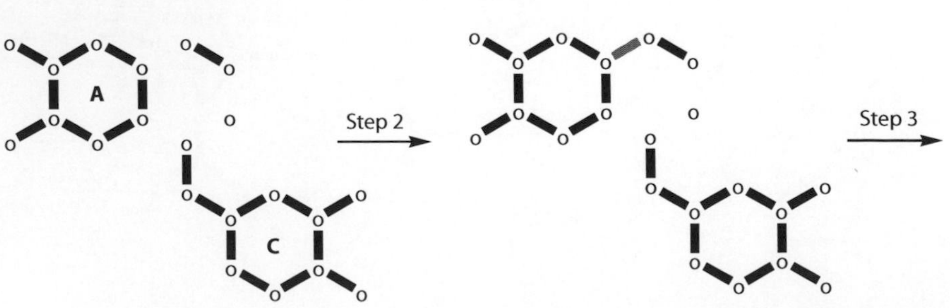

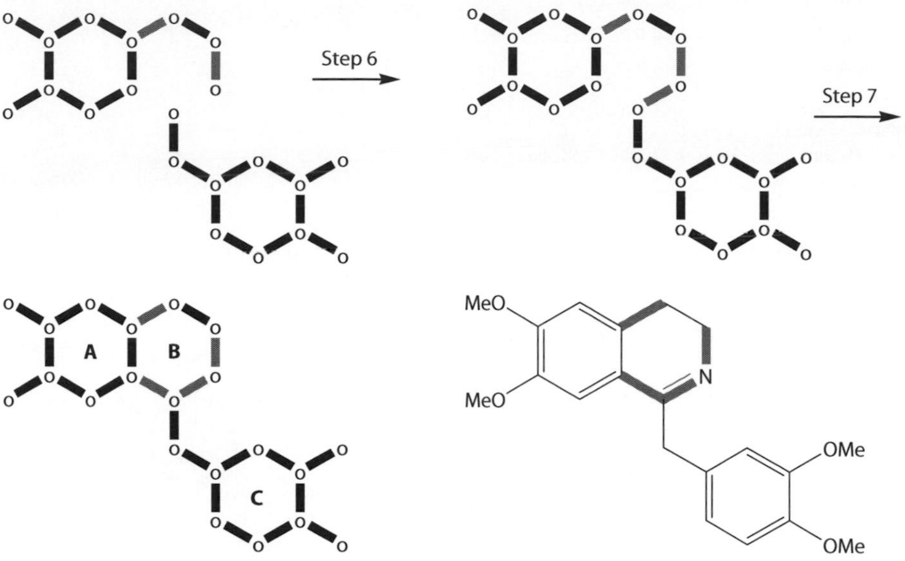

Sequence: $A_0 + C_{3*} \xrightarrow{\;[6+0]\;} B_7$

Papaverine

Kropp, W.; Decker, H. *Chem. Ber.* 1909, *42*, 1184.
Wahl, H. *Bull. Chim. Soc. Fr.* 1950, 680.

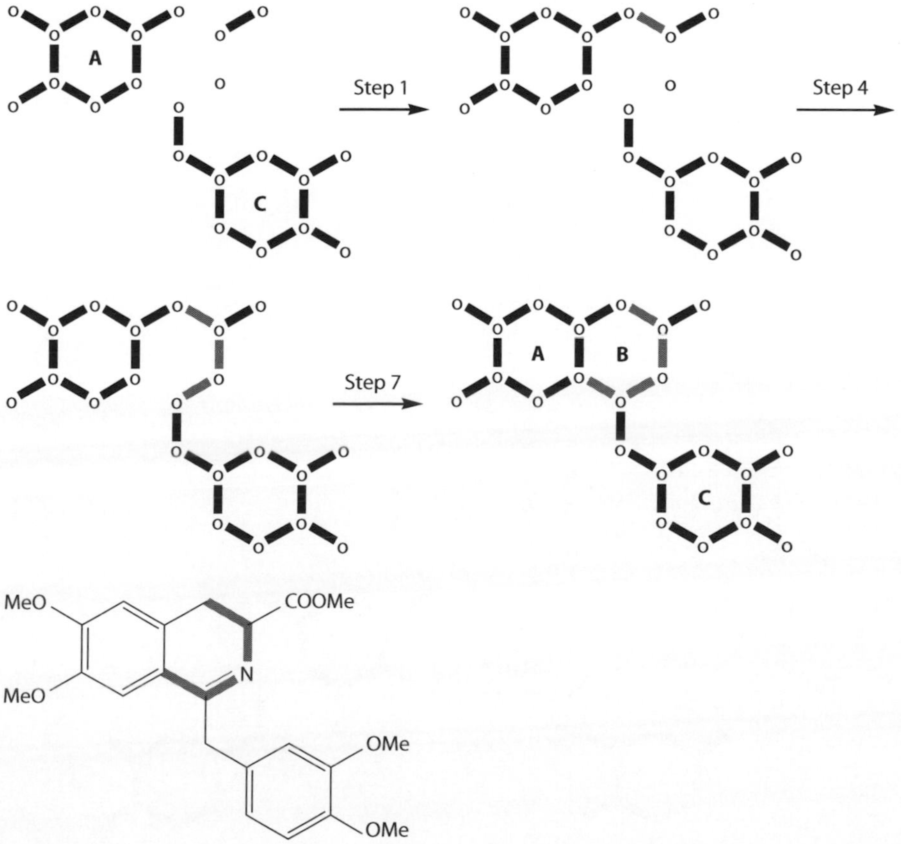

Sequence: $A_0 + C_{0*} \xrightarrow{[6+0]} B_7$

Papaverine
Kindler, K.; Peschke, K. *Arch. Pharm.* 1934, *272*, 236.
Pal, B.C. *J. Sci. Ind. Res.* 1958, *17A*, 270.

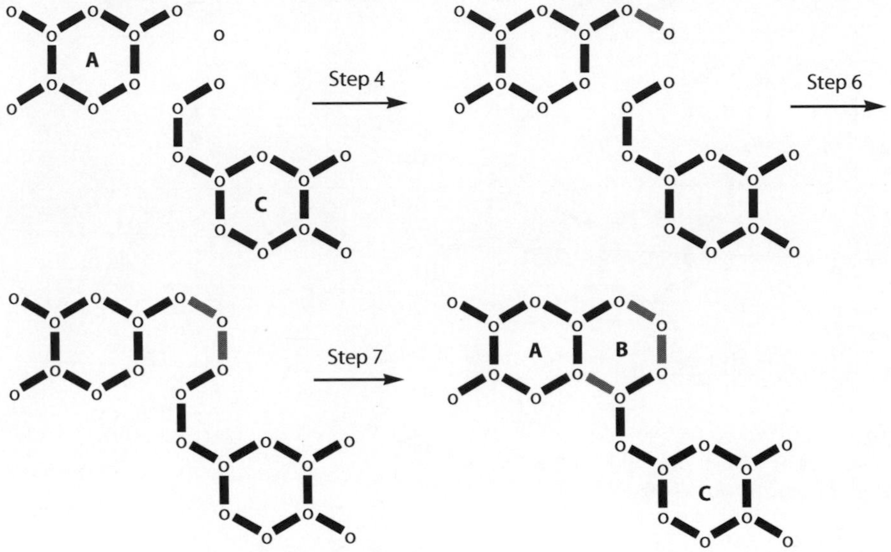

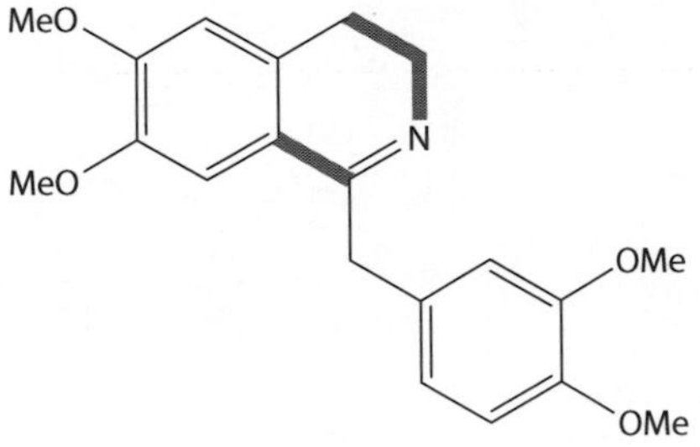

Sequence: $A_0 + C_{0*} \xrightarrow{[6+0]} B_7$

Papaverine
Pictet, A.; Gams, A. *Compt. Rend.* 1909, *149*, 210.
Pictet, A.; Gams, A. *Chem. Ber.* 1909, *42*, 2943.

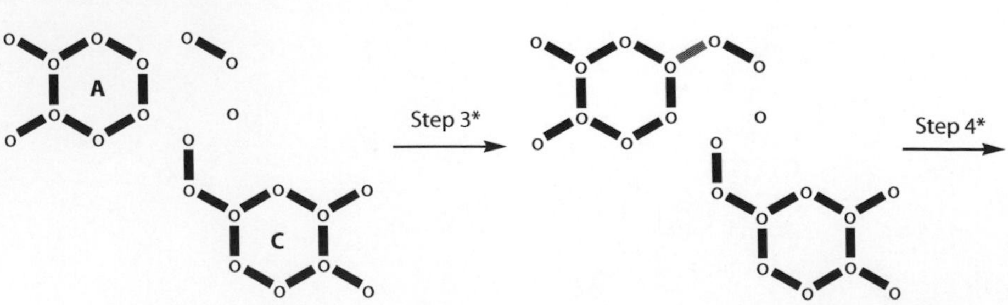

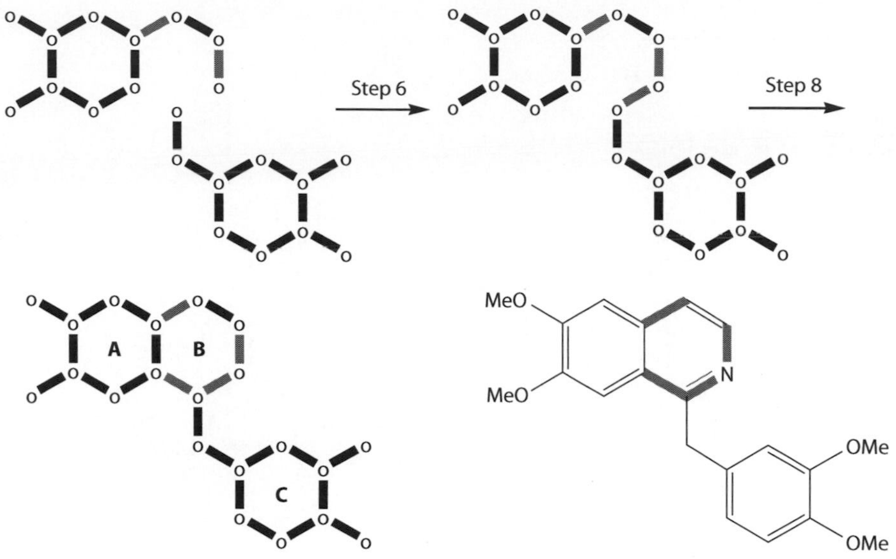

Sequence: $A_{2*} + C_0 \xrightarrow{[6+0]} B_8$

Papaverine

Redel, J.; Bouteville, A. *Bull. Soc. Chim. Fr.* 1949, 443.

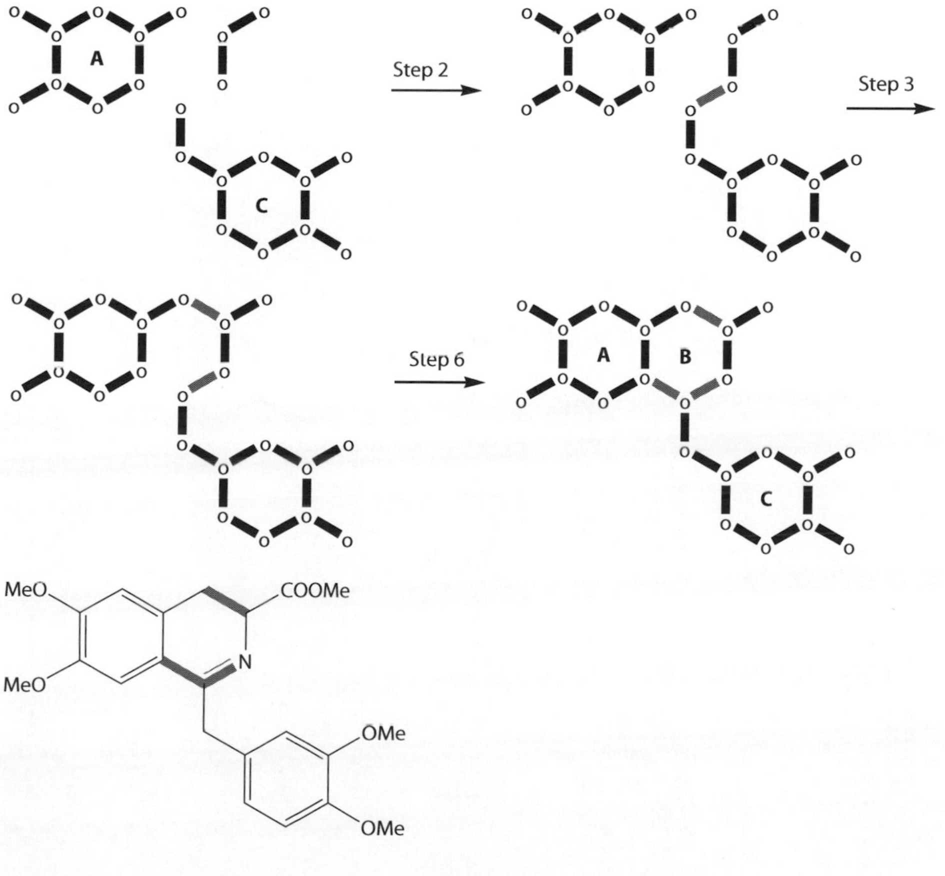

Sequence: $A_2 + C_0 \xrightarrow{[6+0]} B_6$

Summary:
Strategy $(A + C) \rightarrow B$: all plans

(−)-Paroxetine
Amat, M.; Bosch, J.; Hidalgo, J.; Canto, M.; Perez, M.; Llor, N.; Molins, E.; Miravitlles, Orozco, M.; Luque, J. *J. Org. Chem.* 2000, *65*, 3074.

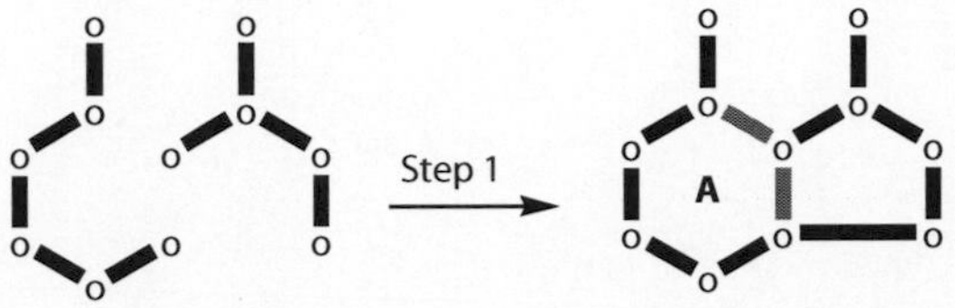

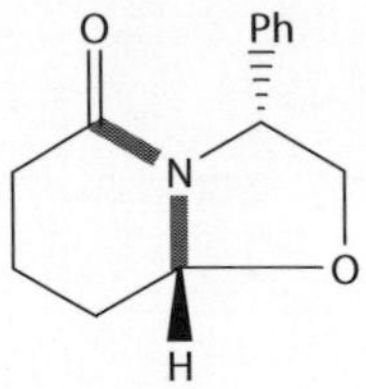

Sequence: $\xrightarrow{[5+1]} A_1$

Johnson, T.A.; Curtis, M.D.; Beak, P. *J. Am. Chem. Soc.* 2001, *123*, 1004.

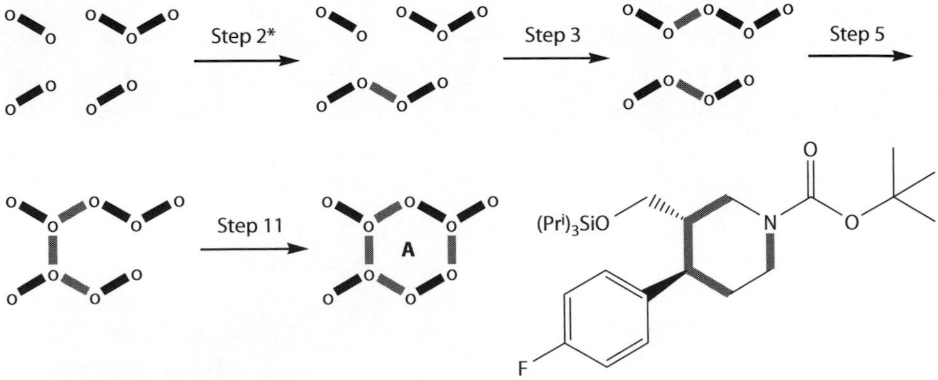

Sequence: $\xrightarrow{[6+0]} A_{11}$

Hughes, G.; Kimura, M.; Buchwald, S.L. *J. Am. Chem. Soc.* 2003, *125*, 11253.

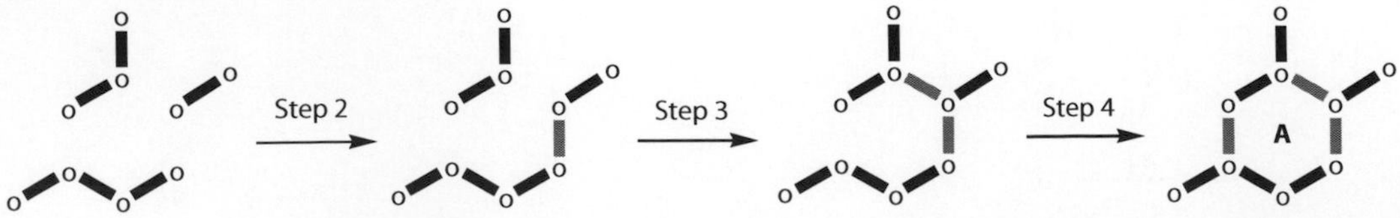

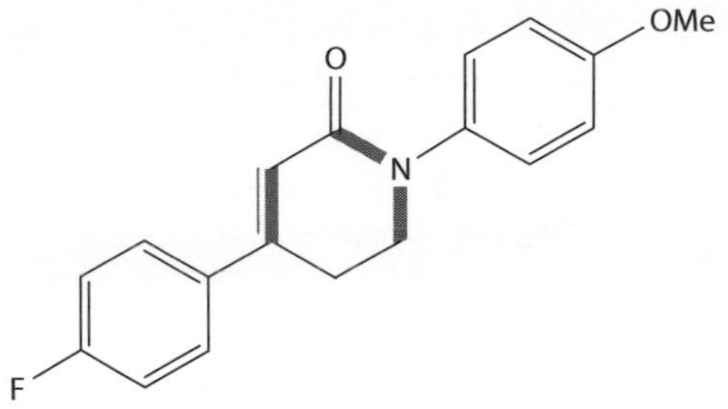

Sequence: $\xrightarrow{[6+0]} A_4$

Chen, C.Y.; Chang, B.R.; Tsai, M.R.; Chang, M.Y.; Chang, N.C. *Tetrahedron* 2003, *59*, 9383.

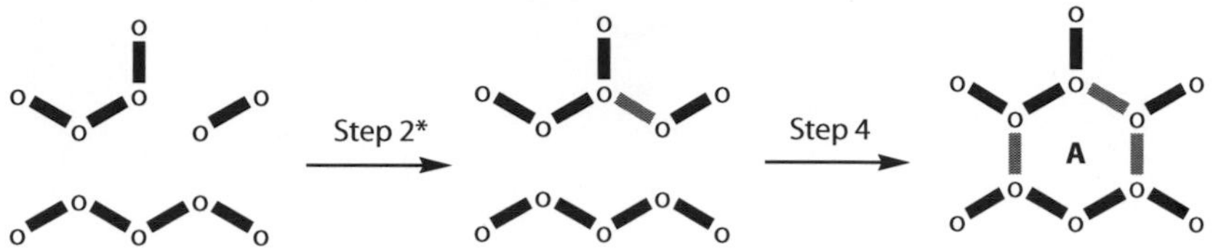

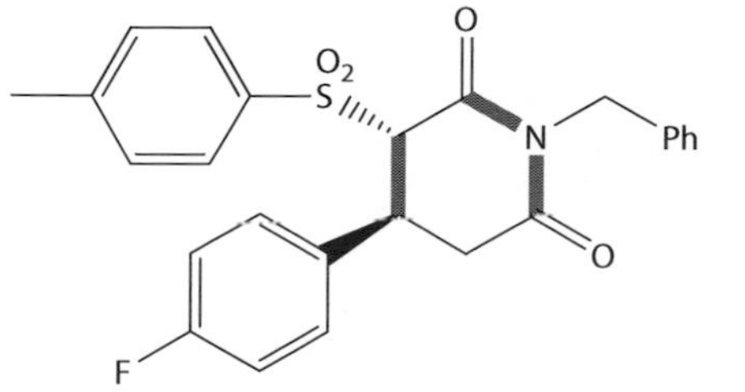

Sequence: $\xrightarrow{[3+3]} A_4$

Cossy, J.; Mirguet, O.; Pardo, D.G.; Desmurs, J.R. *Eur. J. Org. Chem.* 2002, 3543.

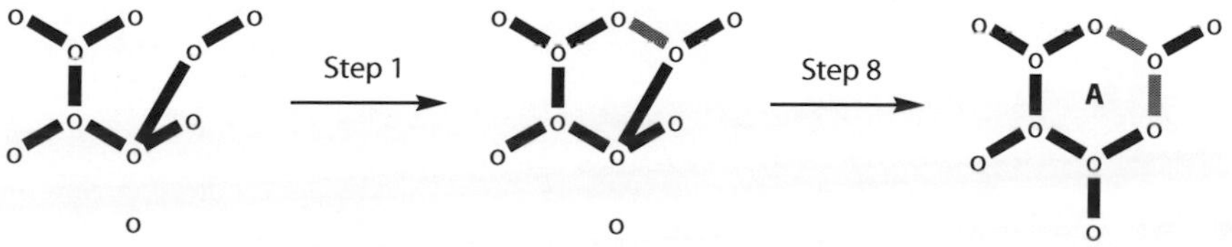

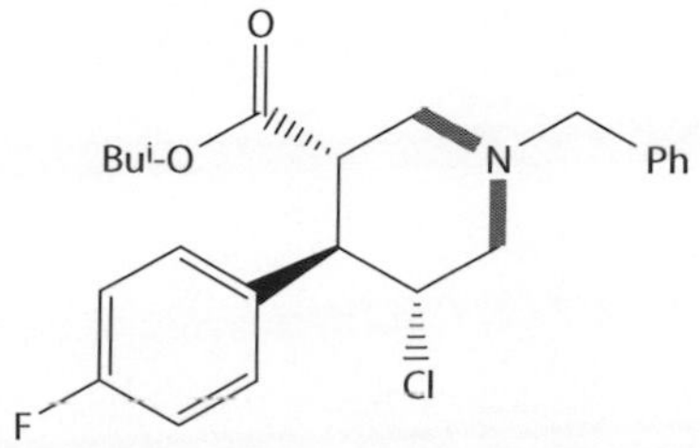

Sequence: $\xrightarrow{[5]\rightarrow[6]} A_8$

Hynes, P.S.; Stupple, P.A.; Dixon, D.J. *Org. Lett.* 2008, *10*, 1389.

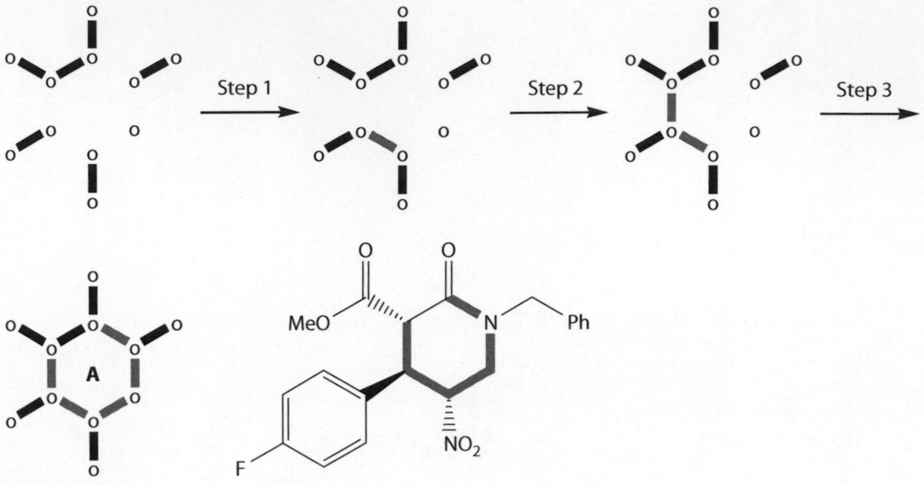

Sequence: $\xrightarrow{[4+1+1]} A_3$

Christensen, J.A.; Squires, R.F. US 4007196 (A/S Ferrosan, 1977).
Christensen, J.A.; Squires, R.F. US 3912743 (A/S Ferrosan, 1975).

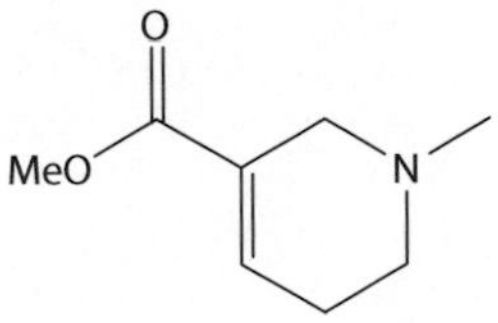

Bower, J.F.; Riis-Johannessen, T.; Szeto, P.; Whitehead, A.J.; Gallagher, T. *Chem. Commun.* 2007, 728.

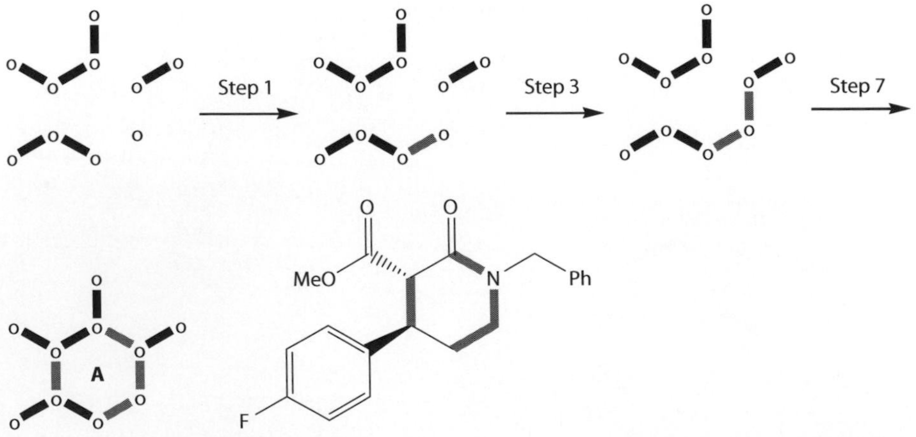

Sequence: $\xrightarrow{[4+2]} A_7$

De Gonzalo, G.; Brieva, R.; Sanchez, V.M.; Bayod, M.; Gotor, V. *J. Org. Chem.* 2001, *66*, 8947.

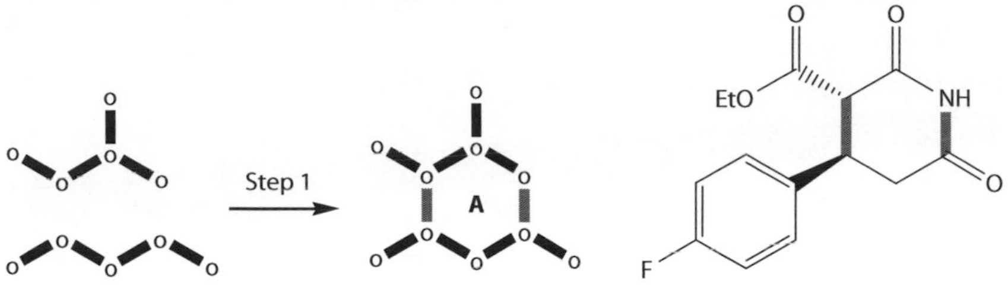

Sequence: $\xrightarrow{[3+3]} A_1$

Senda, T.; Ogasawara, M.; Hayashi, T. *J. Org. Chem.* 2001, *66*, 6852.

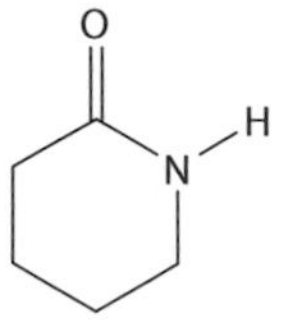

Brandau, S.; Landa, A.; Franzén, J.; Marigo, M.; Jørgensen, K.A. *Angew. Chem. Int. Ed.* 2006, *45*, 4305.

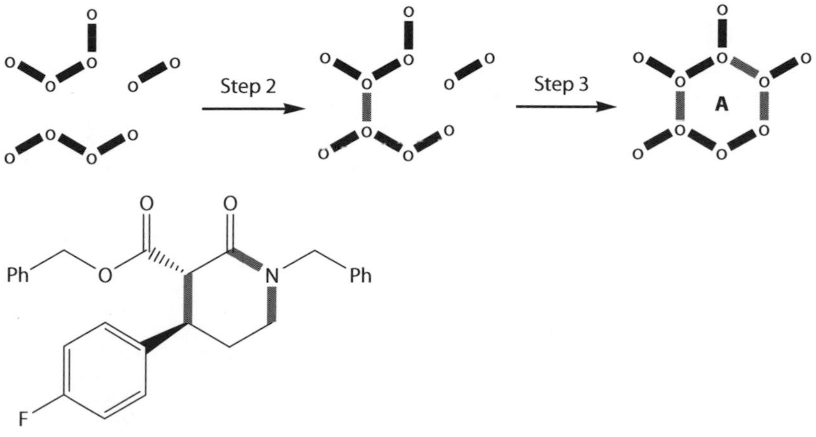

Sequence: $\xrightarrow{[5+1]} A_3$

Koech, P.K.; Krische, M.J. *Tetrahedron* 2006, *62*, 10594.

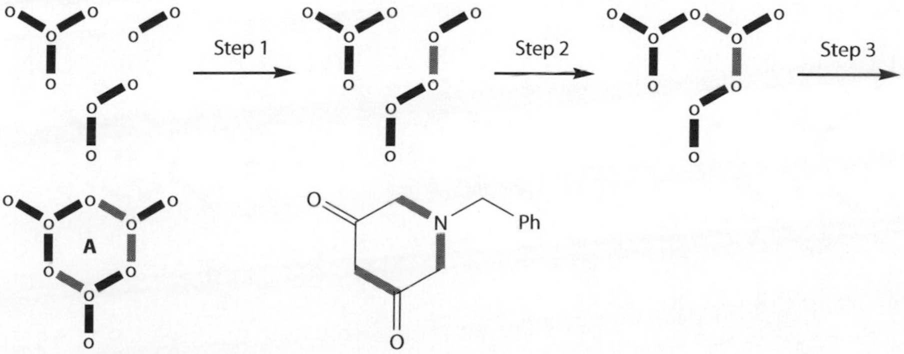

Sequence: $\xrightarrow{[6+0]} A_3$

Liu, L.T.; Hong, P.C.; Huang, H.L.; Chen, S.F.; Wang, C.L.J.; Wen, Y.S. *Tetrahedron Asymm.* 2001, *12*, 419.

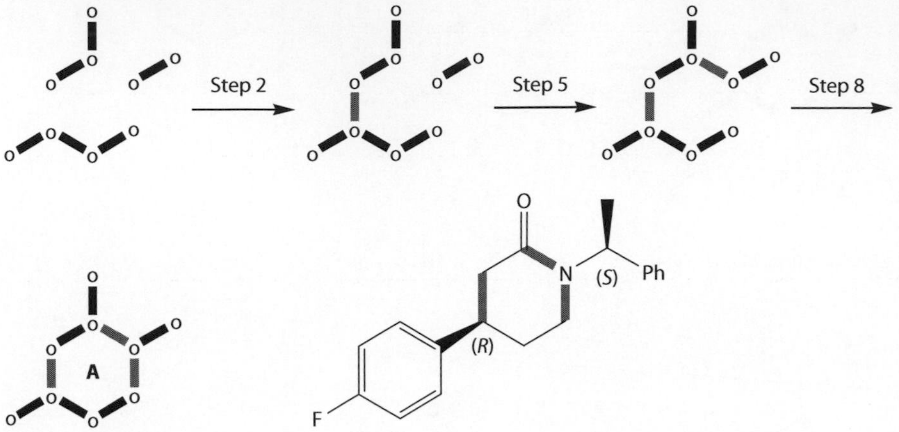

Sequence: $\xrightarrow{[6+0]} A_8$

Greenhalgh, D.A.; Simpkins, N.S. *Synlett* 2002, 2074.
Gill, C.D.; Greenhalgh, D.A.; Simpkins, N.S. *Tetrahedron* 2003, *59*, 9213.

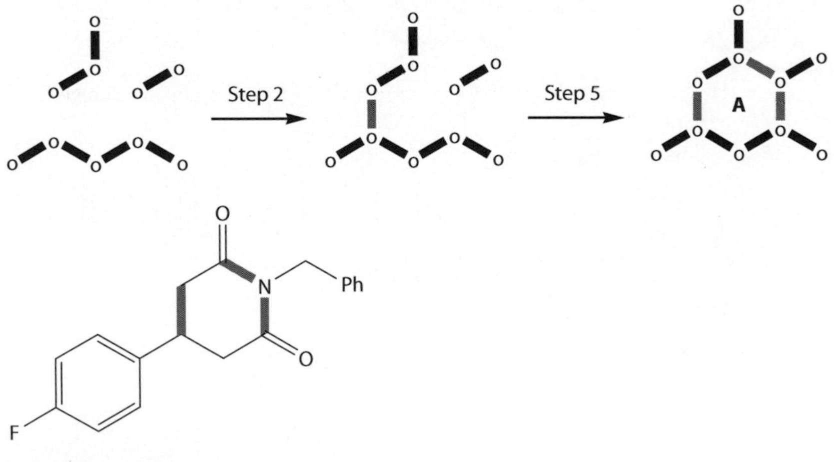

Sequence: $\xrightarrow{[5+1]} A_5$

Czibula, L.; Nemes, A.; Sebök, F.; Szantay, C. Jr.; Mak, M. *Eur. J. Org. Chem.* 2004, 3336.

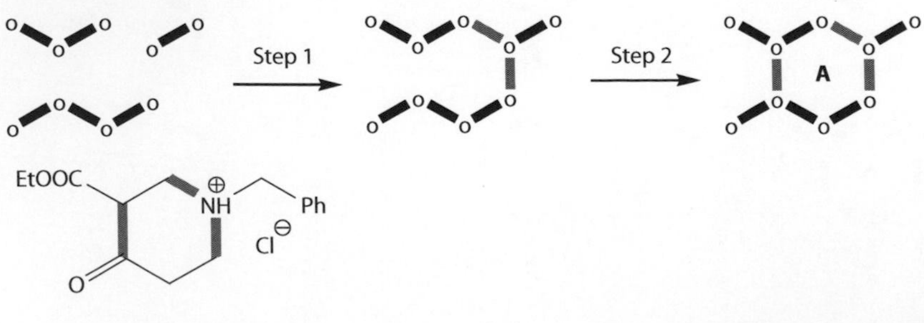

Sequence: $\xrightarrow{[6+0]} A_2$

Valero, G.; Schimer, J.; Cisarova, I.; Vesely, J.; Moyano, A.; Rios, R. *Tetrahedron Lett.* 2009, *50*, 1943.

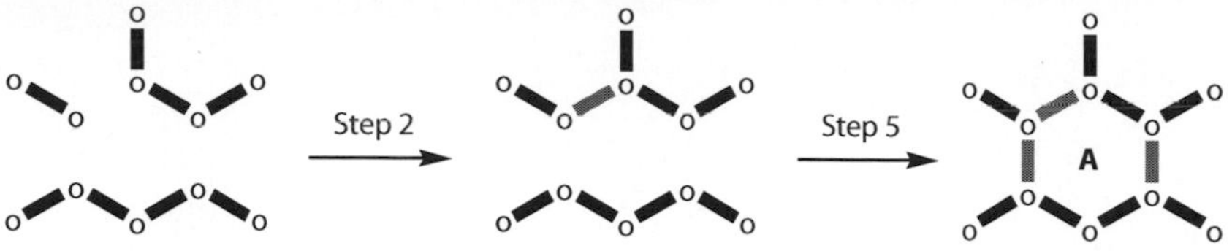

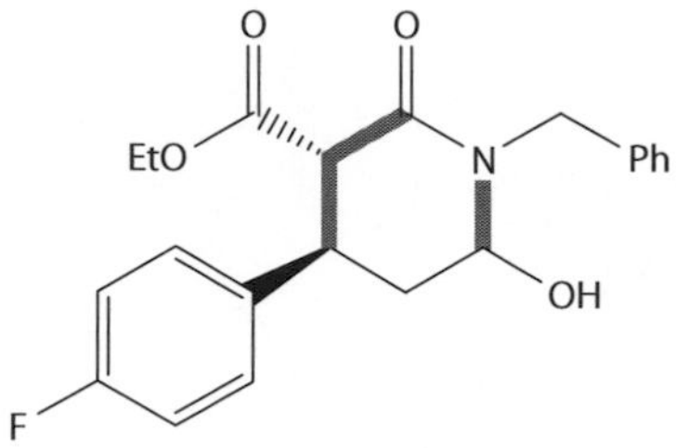

Sequence: $\xrightarrow{[3+3]} A_5$

(+)-Paroxetine

Takasu, K.; Nishida, N.; Tomimura, A.; Ihara, M. *J. Org. Chem.* 2005, *70*, 3957.

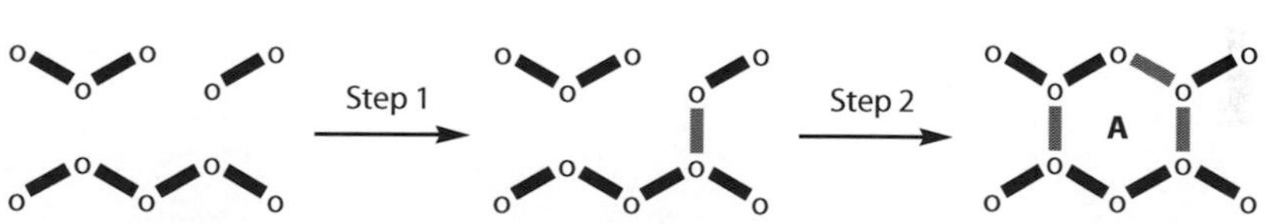

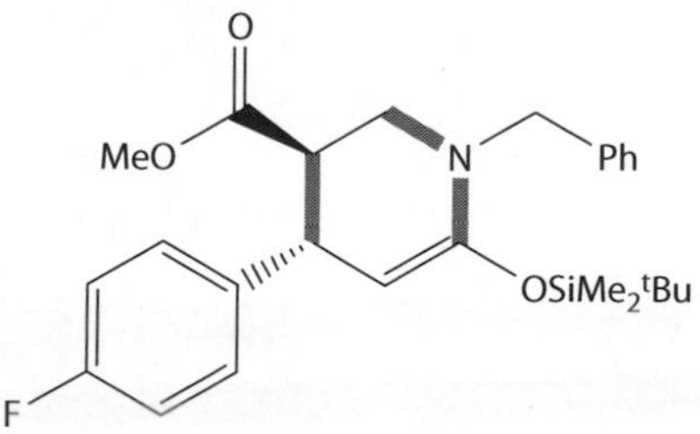

Sequence: $\xrightarrow{[4+2]} A_2$

Yu, M.S.; Lantos, I.; Peng, Z.Q.; Yu, J.; Cacchio, T. *Tetrahedron Lett.* 2000, *41*, 5647.

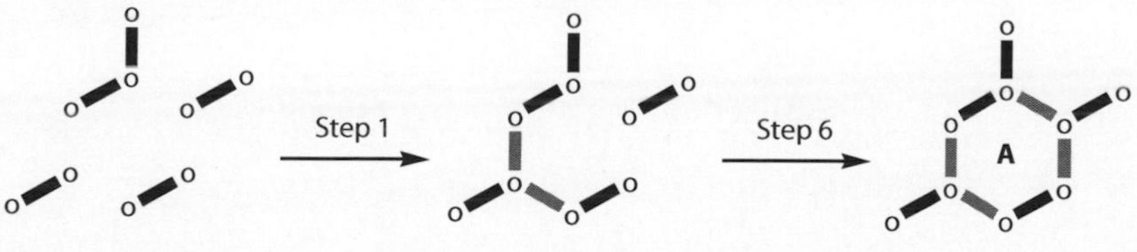

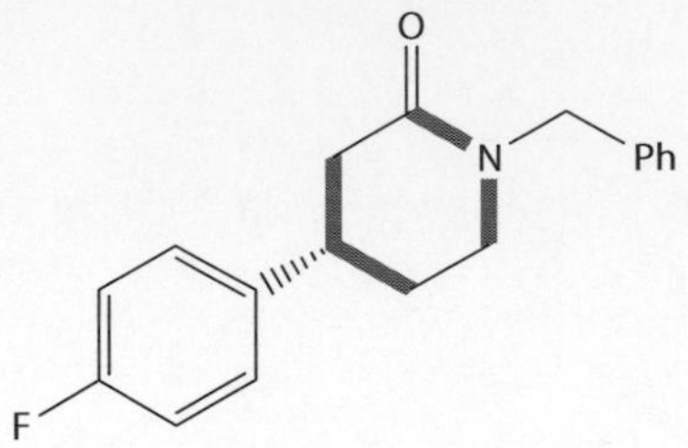

Sequence: $\xrightarrow{[5+1]} A_6$

Summary:

Strategy #1: [5 + 1] Amat G1, Amat G2, Jorgensen, Simpkins, Yu

Strategy #2: [6 + 0] Beak, Buchwald, Krische, Liu, Szantay

Strategy #3: [3 + 3] Chang, Gotor, Vesely-Moyano-Rios

Strategy #4: [4 + 2] Gallagher, Takasu-Ihara

Strategy #5: [5]→[6] Cossy

Strategy #6: [4 + 1 + 1] Dixon

Strategy #7: beginning with six-membered ring starting material (arecoline—Ferrosan; δ-lactam—Hayashi)

(±)-Patchouli alcohol

Bertrand, M.; Teisseire, P.; Pelerin, G. *Tetrahedron Lett.* 1980, *21*, 2055.

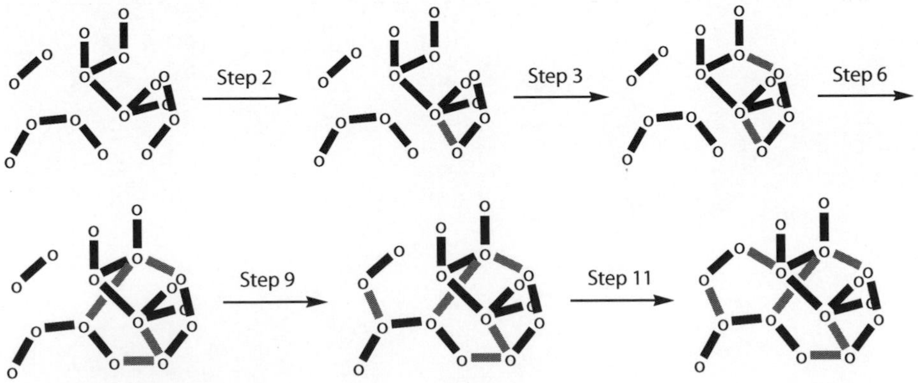

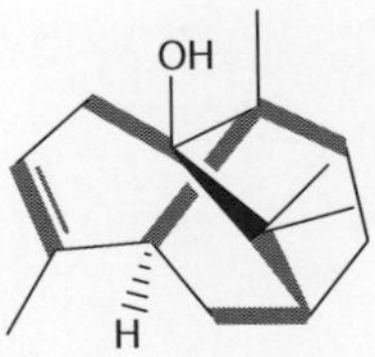

Sequence: $\xrightarrow{[6+0]} [6]_3 \xrightarrow{[(4+2)+(4+2)]} [2.2.2]_6 \xrightarrow{[(6+0)+(8+0)]} [4.4.0]_{11}$

Danishefsky, S.; Dumas, D. *Chem. Commun.* 1968, 1287.

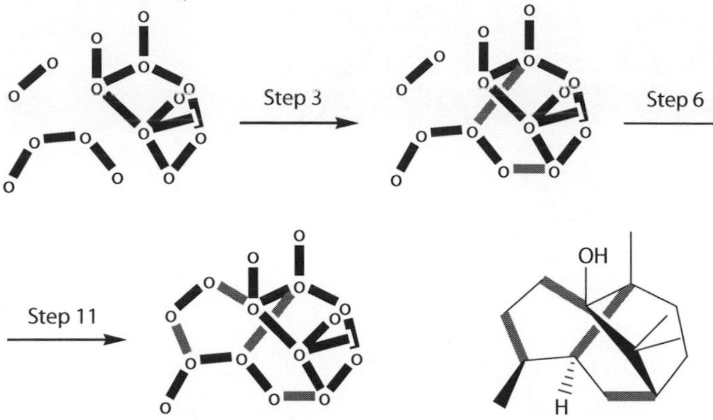

Sequence: $\xrightarrow{[(4+2)+(4+2)]} [2.2.2]_3 \xrightarrow{[6+0]} [4.4.0]_{11}$

Magee, T.V.; Stork, G.; Fludzinski, P. *Tetrahedron Lett.* 1995, *36*, 7607.

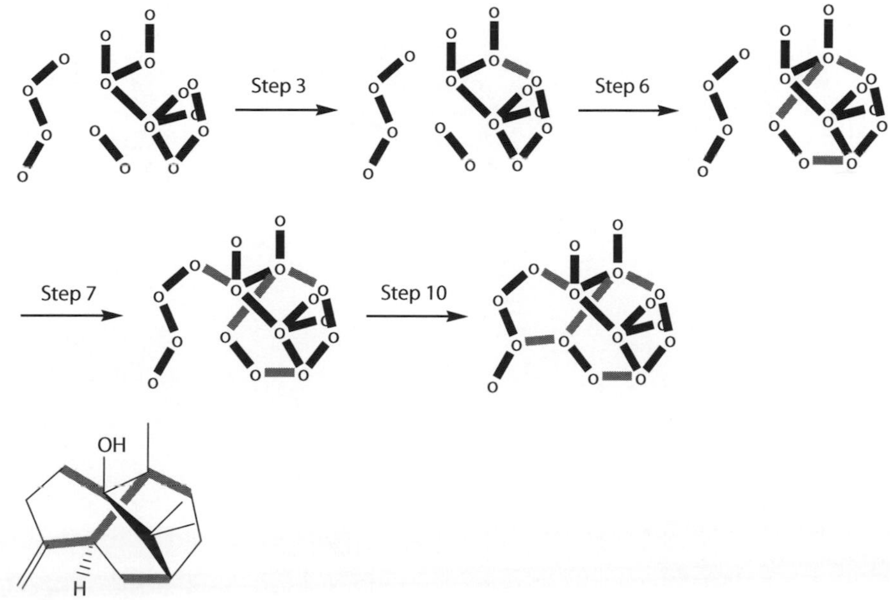

Sequence: $\xrightarrow{[(4+2)+(4+2)]} [2.2.2]_6 \xrightarrow{[6+0]} [4.4.0]_{10}$

Mirrington, R.N.; Schmalzl, K.J. *J. Org. Chem.* 1972, *37*, 2871.

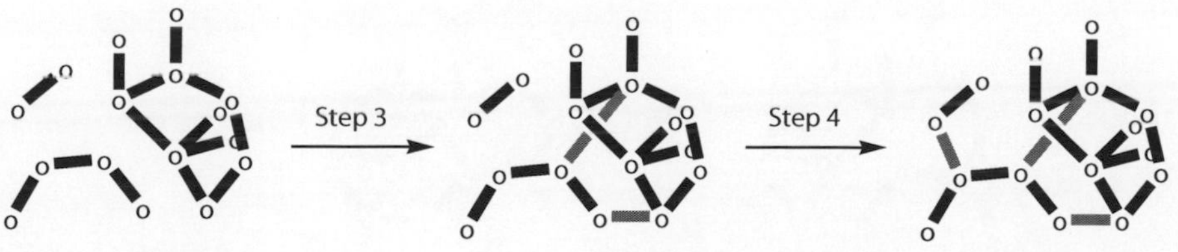

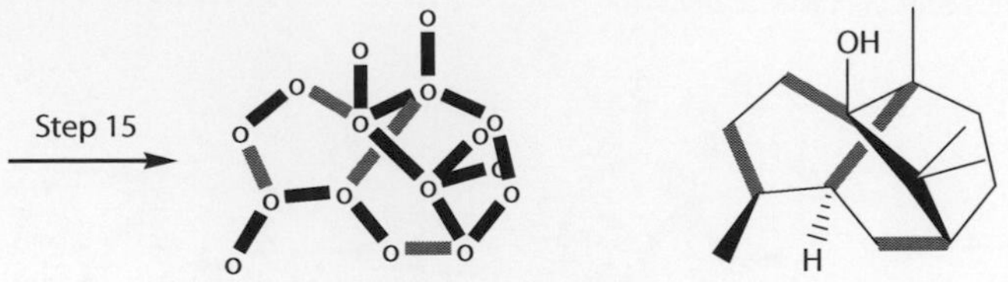

Sequence: $\xrightarrow{[(4+2)+(4+2)]} [2.2.2]_3 \xrightarrow{[6+0]} [4.4.0]_{15}$

(−)-Patchouli alcohol

Büchi, G.; MacLeod, W.D. Jr.; Padilla, J. *J. Am. Chem. Soc.* 1964, *86*, 4438.
Büchi, G.; Erickson, R.E.; Wakabayashi, N. *J. Am. Chem. Soc.* 1961, *83*, 927.

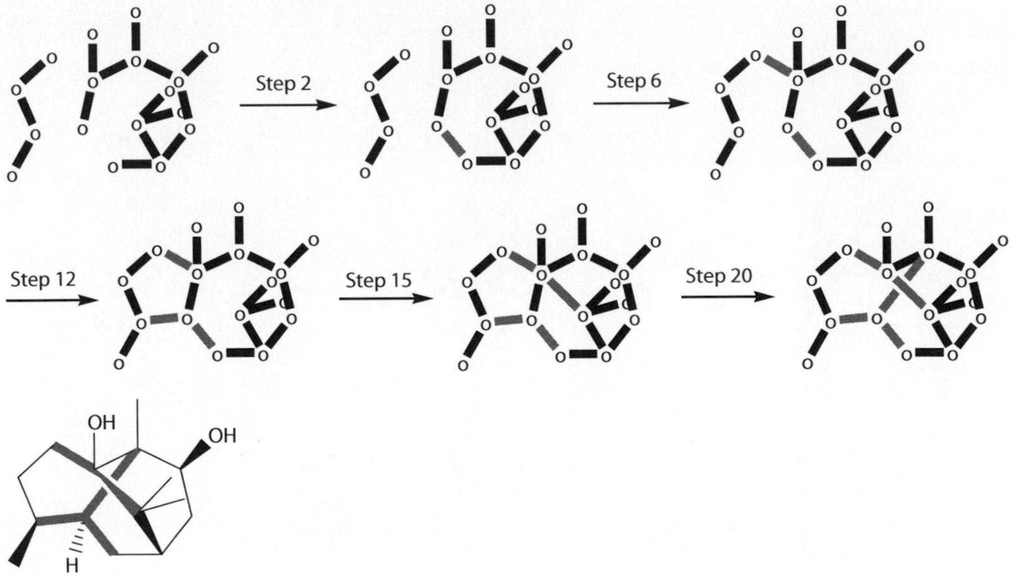

Sequence: $\xrightarrow{[8+0]} [8]_2 \xrightarrow{[5+0]} [5.3.0]_{12} \xrightarrow{[5+0]} [3.2.1]_{15} \xrightarrow{[5] \to [6]} [4.4.0]_{20}$

Näf, F.; Decorzant, R.; Giersch, W.; Ohloff, G. *Helv. Chim. Acta* 1981, *64*, 1387.

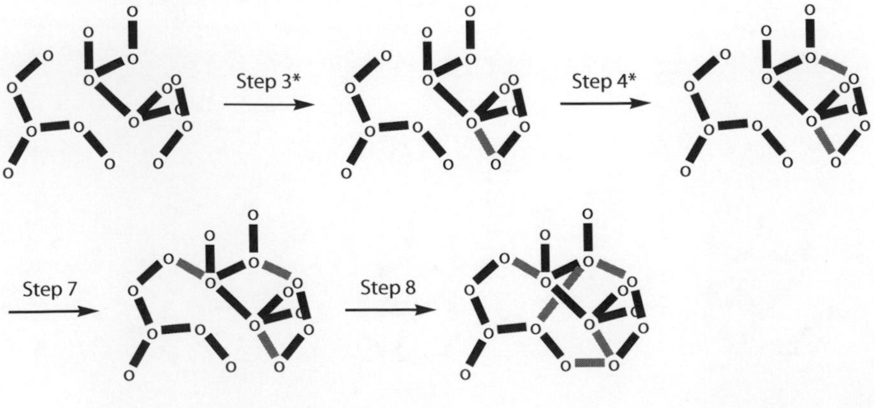

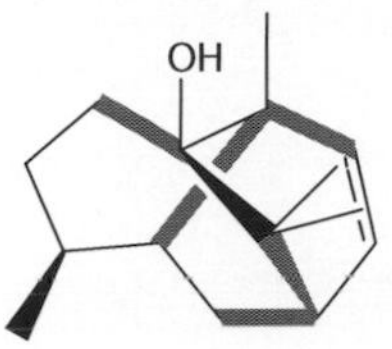

Sequence: $\xrightarrow{[6+0]} [6]_{4*} \xrightarrow{[(6+0)+(4+2)]} [4.4.0]_8$

Kaliappan, K.P.; Rao, G.S.R.S. *J. Chem. Soc. Perkin Trans. I* 1997, 1385.

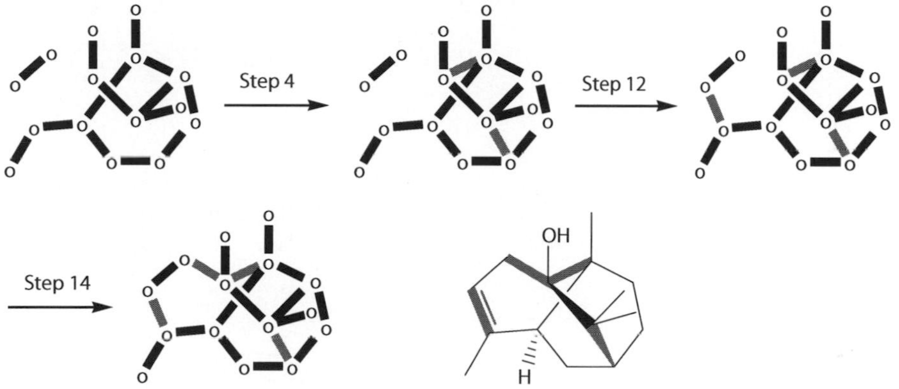

Sequence: $\xrightarrow{[(4+2)+(4+2)]} [2.2.2]_4 \xrightarrow{[6+0]} [4.4.0]_{14}$

Srikrishna, A.; Satyanarayana, G. *Tetrahedron Asymm.* 2005, *16*, 3992.

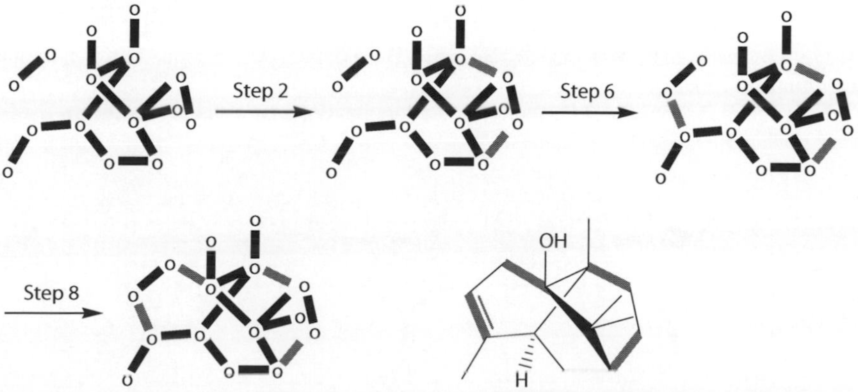

Sequence: $\xrightarrow{[(4+2)+(4+2)]} [2.2.2]_2 \xrightarrow{[6+0]} [4.4.0]_8$

Summary:
Strategy #1

$$\xrightarrow{[(4+2)+(4+2)]} [2.2.2]_2 \xrightarrow{[6+0]} [4.4.0]_8: \text{Srikrishna}$$

$$\xrightarrow{[(4+2)+(4+2)]} [2.2.2]_3 \xrightarrow{[6+0]} [4.4.0]_{11}: \text{Danishefsky}$$

$$\xrightarrow{[(4+2)+(4+2)]} [2.2.2]_3 \xrightarrow{[6+0]} [4.4.0]_{15}: \text{Mirrington}$$

$$\xrightarrow{[(4+2)+(4+2)]} [2.2.2]_4 \xrightarrow{[6+0]} [4.4.0]_{14}: \text{Rao}$$

$$\xrightarrow{[(4+2)+(4+2)]} [2.2.2]_6 \xrightarrow{[6+0]} [4.4.0]_{10}: \text{Magee}$$

Strategy #2

$$\xrightarrow{[6+0]} [6]_{4*} \xrightarrow{[(6+0)+(4+2)]} [4.4.0]_8: \text{Firmenich}$$

Strategy #3

$$\xrightarrow{[8+0]} [8]_2 \xrightarrow{[5+0]} [5.3.0]_{12} \xrightarrow{[5+0]} [3.2.1]_{15} \xrightarrow{[5]\to[6]} [4.4.0]_{20}: \text{Buchi}$$

Strategy #4

$$\xrightarrow{[6+0]} [6]_3 \xrightarrow{[(4+2)+(4+2)]} [2.2.2]_6 \xrightarrow{[(6+0)+(8+0)]} [4.4.0]_{11}: \text{Bertrand}$$

Phenytoin

Biltz, H. *Chem. Ber.* 1908, *41*, 1379.

Biltz, H.; Seydel, K. *Chem. Ber.* 1911, *44*, 411.

Muccioli, G.G.; Poupaert, J.H.; Wouters, J.; Norberg, B.; Poppitz, W.; Scriba, G.K.E.; Lambert, D.M. *Tetrahedron* 2003, *59*, 1301.

Muccioli, G.G.; Wouters, J.; Poupaert, J.H.; Norberg, B.; Poppitz, W.; Scriba, G.K.E.; Lambert, D.M. *Org. Lett.* 2003, *5*, 3599.

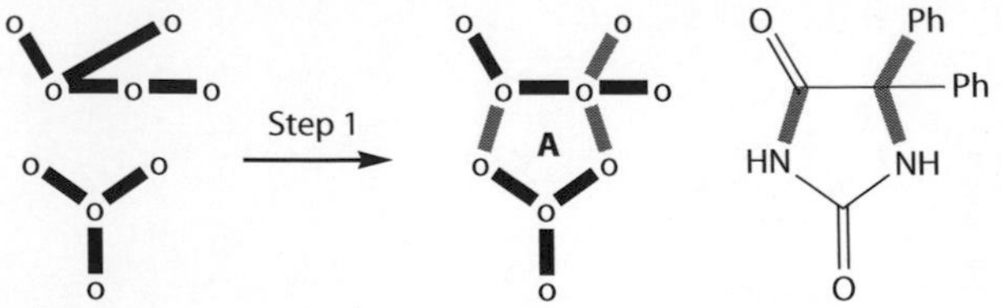

Sequence: $\xrightarrow{[3+2]} A_1$

Klumpp, D.A.; Yeung, K.Y.; Prakash, G.K.S.; Olah, G.A. *Synlett* 1998, 918.

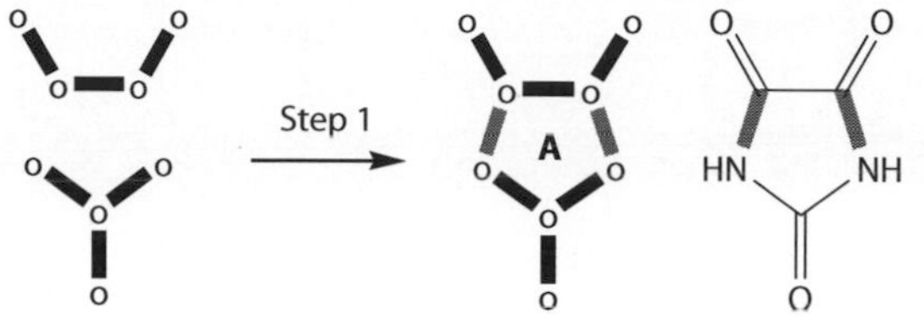

Sequence: $\xrightarrow{[3+2]} A_1$

Henze, H.R. US 2409754 (Parke-Davis, 1946).

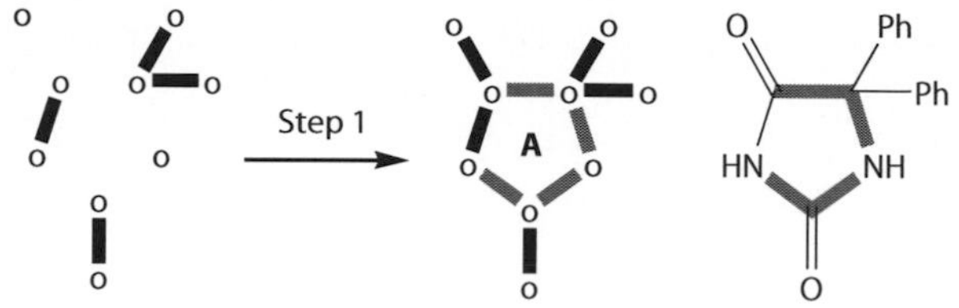

Sequence: $\xrightarrow{[2+1+1+1]} A_1$

Summary:
Strategy [3 + 2]: Biltz, Lambert G1, Lambert G2, Olah
Strategy [2 + 1 + 1 + 1]: Parke-Davis

(−)-Physostigmine
Node, M.; Hao, X.; Fuji, K. *Chem. Lett.* 1991, *20*, 57.
Fuji, K.; Node, M.; Nagasawa, H.; Naniwa, Y.; Taga, T.; Machida, K.; Snatzke, G. *J. Am. Chem. Soc.* 1989, *111*, 7921.

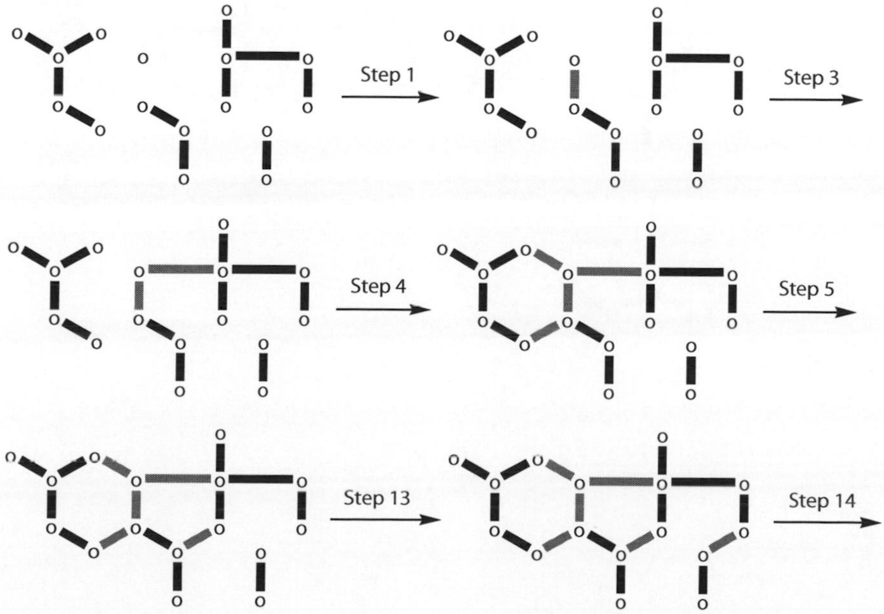

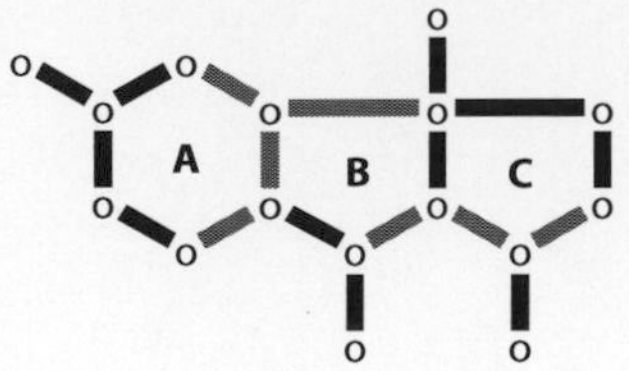

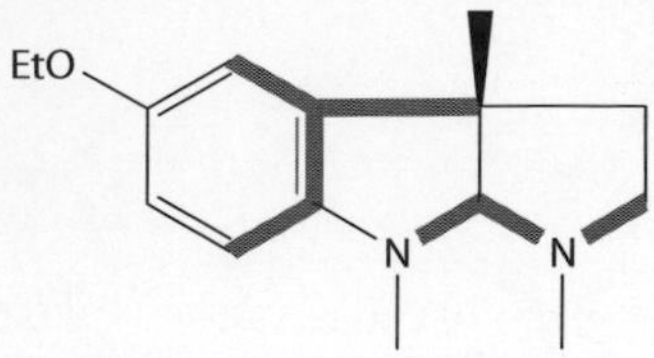

Sequence: $\xrightarrow{[4+2]} A_4 \xrightarrow{[5+0]} B_5 \xrightarrow{[5+0]} C_{14}$

(±)-Physostigmine

Shishido, K.; Shitara, E.; Komatsu, H.; Hiroya, K.; Fukumoto, K.; Kametani, T. *J. Org. Chem.* 1986, *51*, 3007.
Kametani, T.; Kajiwara, M.; Fukumoto, K. *Tetrahedron* 1974, *30*, 1053.

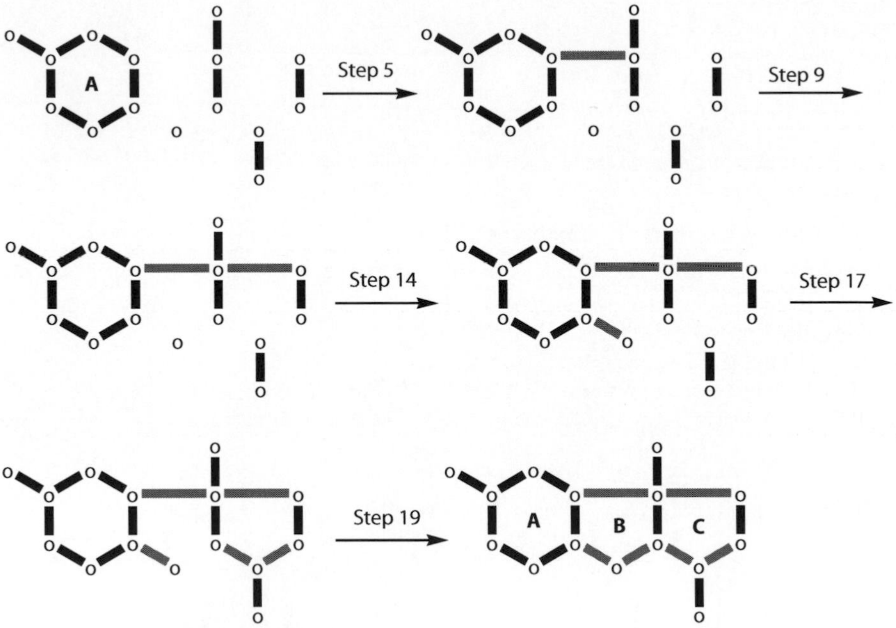

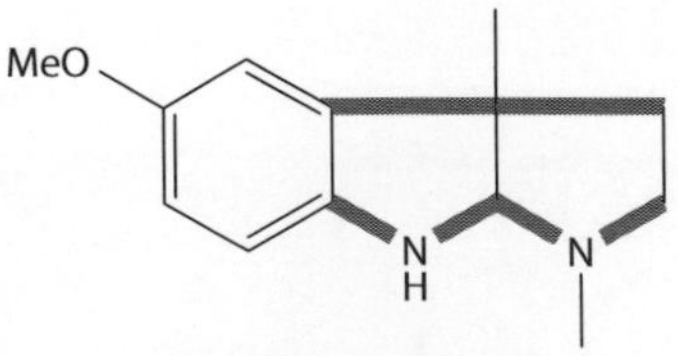

Sequence: $A_0 \xrightarrow{[4+1]} C_{17} \xrightarrow{[5+0]} B_{19}$

(±)-Physostigmine

Rege, P.D.; Johnson, F. *J. Org. Chem.* 2003, *68*, 6133.

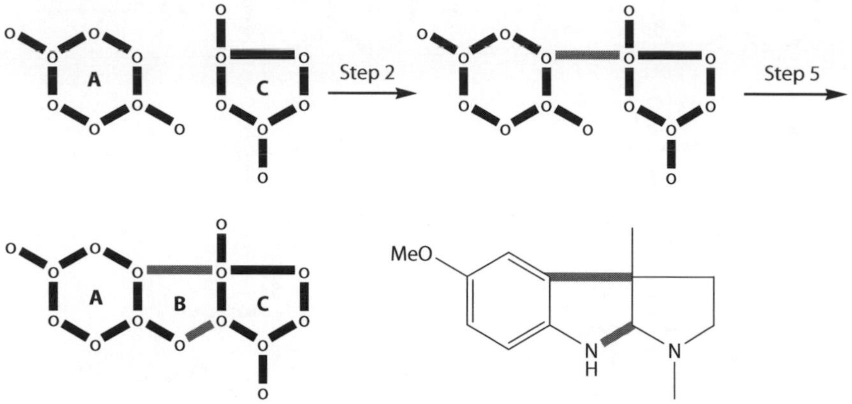

Sequence: $A_1 + C_0 \xrightarrow{\;[5+0]\;} B_5$

(±)-Physostigmine

Julian, P.L.; Pikl, J. *J. Am. Chem. Soc.* 1935, *57*, 563.
Julian, P.L.; Pikl, J. *J. Am. Chem. Soc.* 1935, *57*, 755.
Julian, P.L.; Pikl, J. *J. Am. Chem. Soc.* 1935, *57*, 539.
Julian, P.L.; Pikl, J.; Boggess, D. *J. Am. Chem. Soc.* 1934, *56*, 1797.

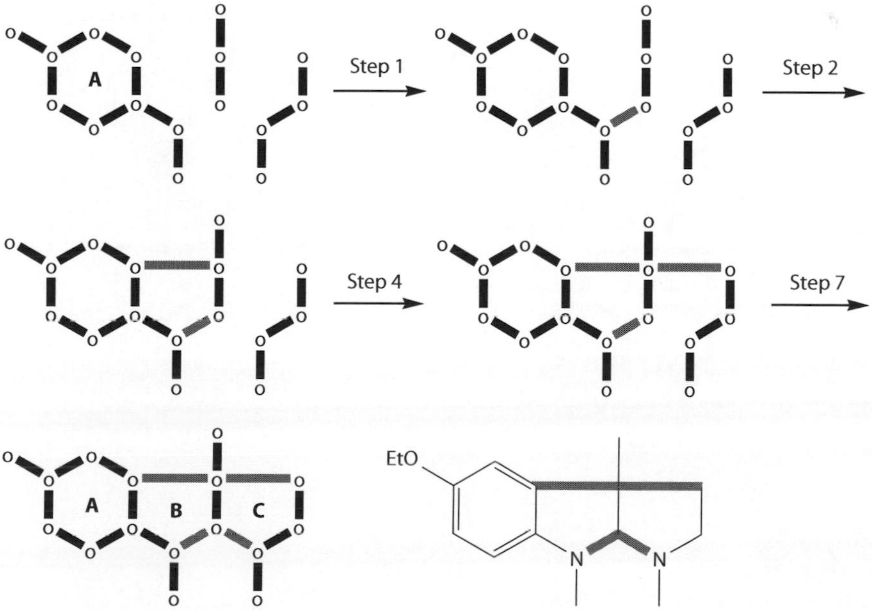

Sequence: $A_0 \xrightarrow{\;[5+0]\;} B_2 \xrightarrow{\;[5+0]\;} C_7$

(±)-Physostigmine

Kulkarni, M.G.; Dhondge, A.P.; Borhade, A.S.; Gaikwad, D.D.; Chavhan, S.W.; Shaikh, Y.B.; Ningdale, V.B.; Desai, M.P.; Birhade, D.R.; Shinde, M.P. *Tetrahedron Lett.* 2009, *50*, 2411.

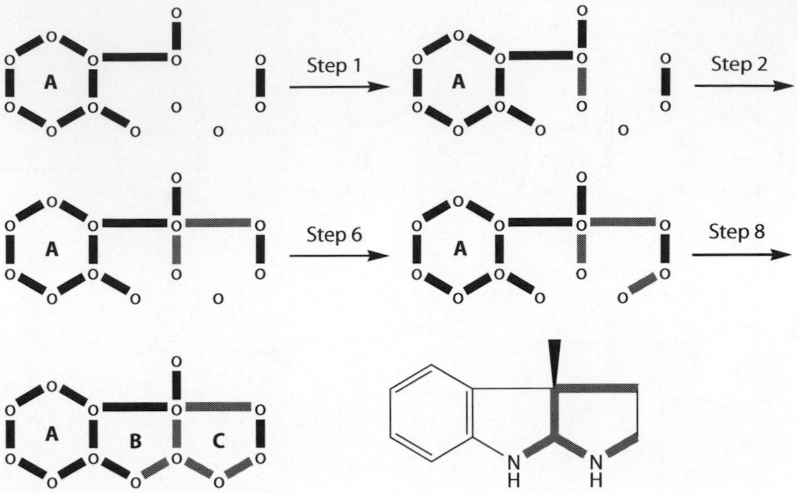

Sequence: $A_0 \xrightarrow{[(5+0)_B + (5+0)_C]} [3.3.0](B+C)_8$

(−)-Physostigmine
Marino, J.P.; Bogdan, S.; Kimura, K. *J. Am. Chem. Soc.* 1992, *114*, 5566.

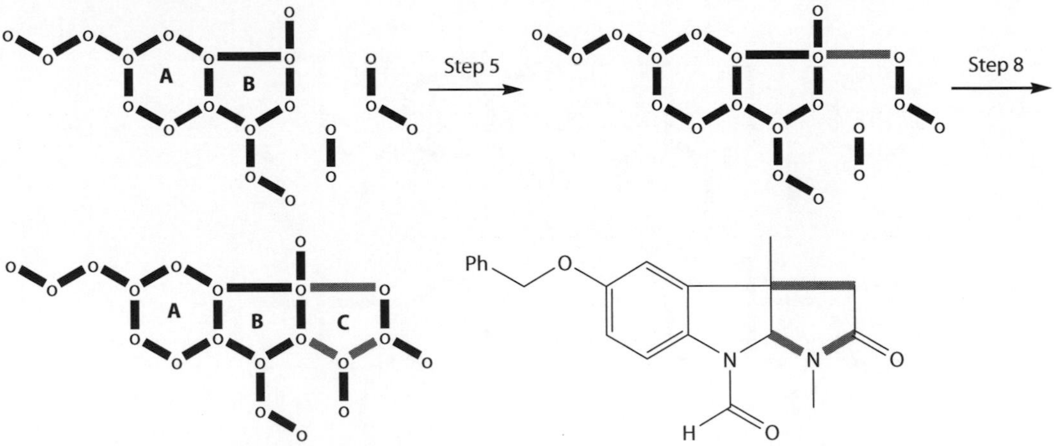

Sequence: $(A+B)_0 \xrightarrow{[4+1]} C_8$

(±)-Physostigmine
Mukai, C.; Yoshida, T.; Sorimachi, M.; Odani, A. *Org. Lett.* 2006, *8*, 83.

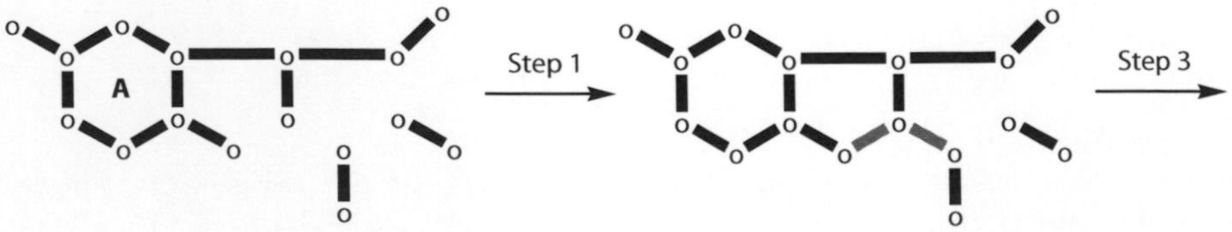

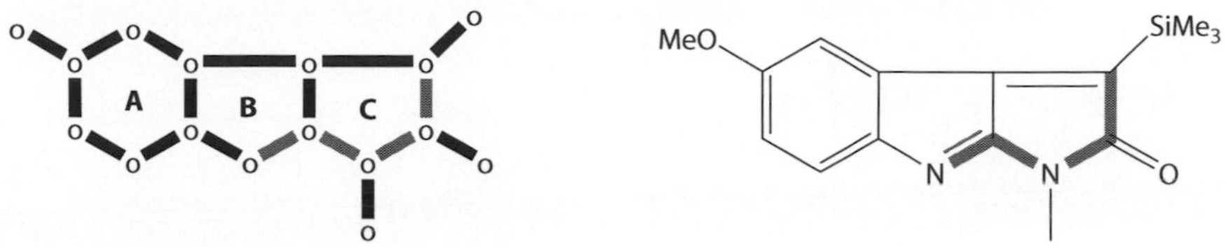

Sequence: $A_0 \xrightarrow{[5+0]} B_1 \xrightarrow{[4+1]} C_3$

(−)-Physostigmine
Asakawa, K.; Noguchi, N.; Takashima, S.; Nakada, M. *Tetrahedron Asymmetry* 2008, *19*, 2304.

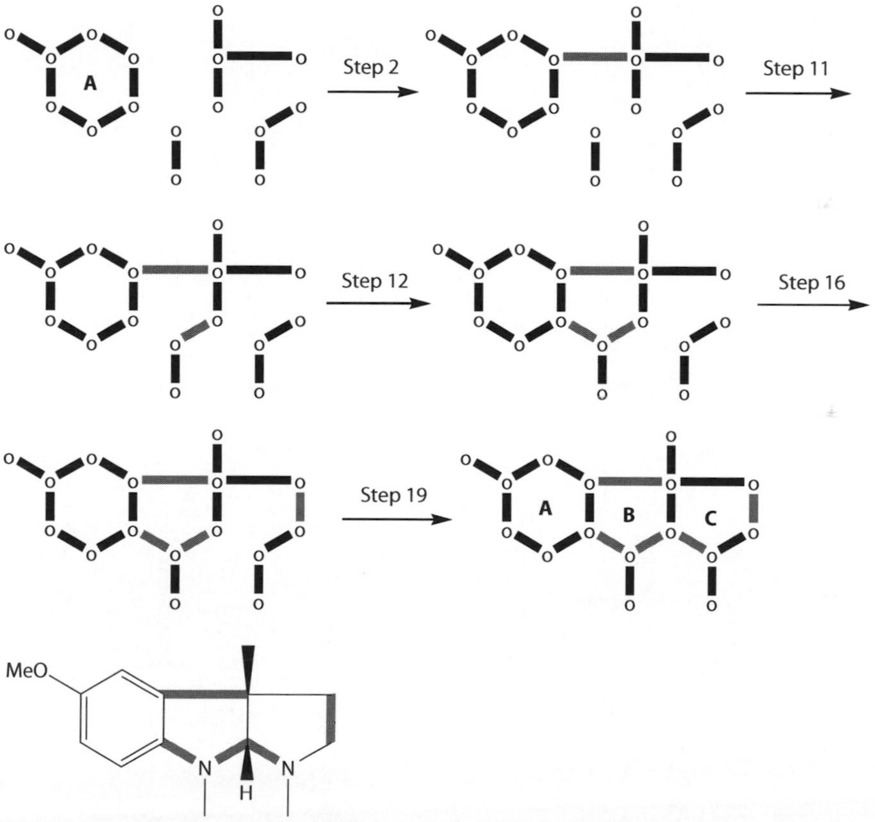

Sequence: $A_0 \xrightarrow{[5+0]} B_{12} \xrightarrow{[5+0]} C_{19}$

(−)-Physostigmine
Kawahara, M.; Nishida, A.; Nakagawa, M. *Org. Lett.* 2000, *2*, 675.

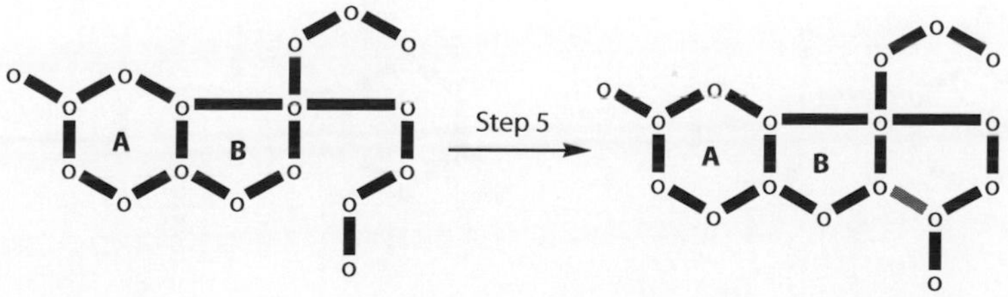

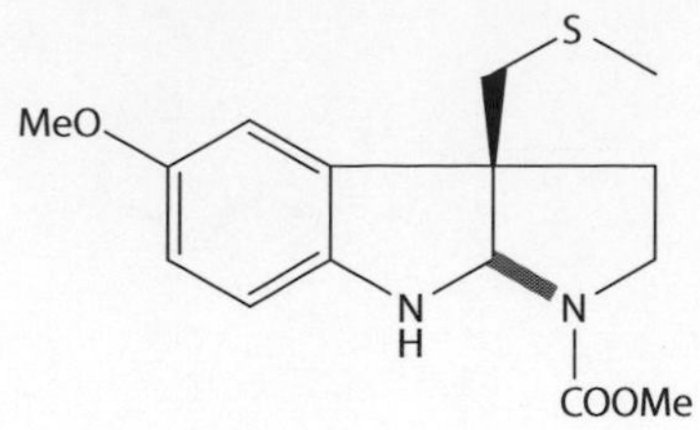

Sequence: $(A + B)_0 \xrightarrow{[5+0]} C_5$

(−)-Physostigmine

Matsuura, T.; Overman, L.E.; Poon, D.J. *J. Am. Chem. Soc.* 1998, *120*, 6500.

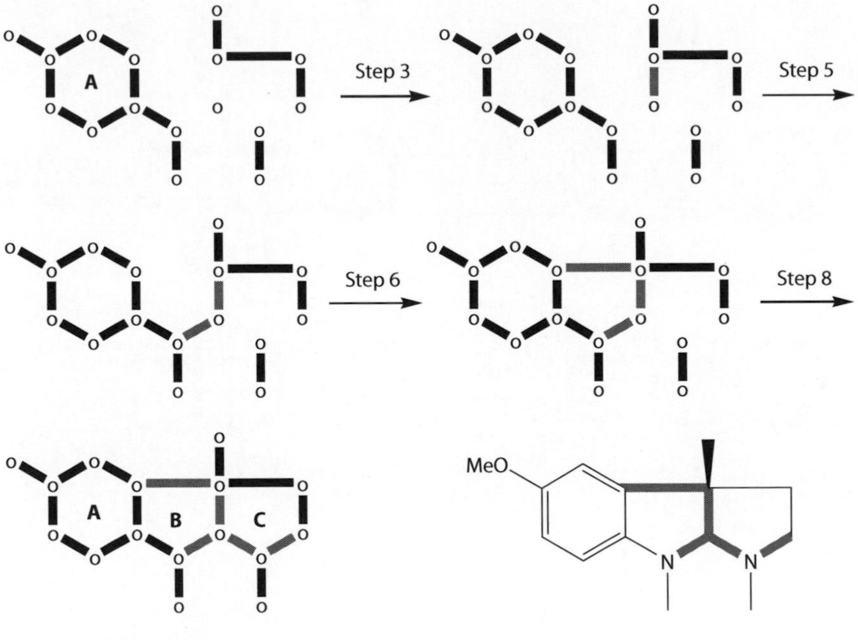

Sequence: $A_4 \xrightarrow{[5+0]} B_6 \xrightarrow{[4+1]} C_8$

(±)-Physostigmine

Wunberg, J.B.P.A.; Speckamp, W.N. *Tetrahedron* 1978, *34*, 2399.

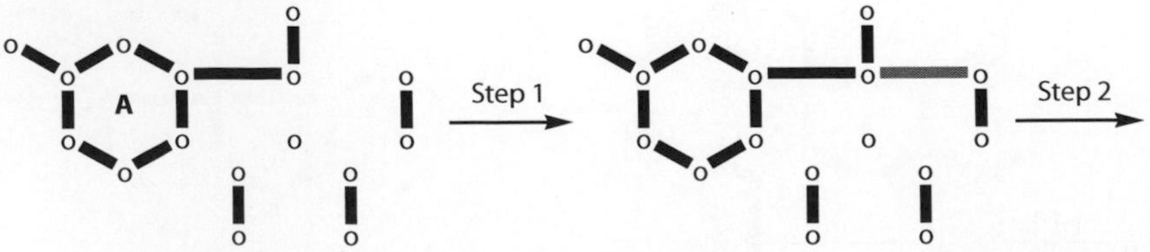

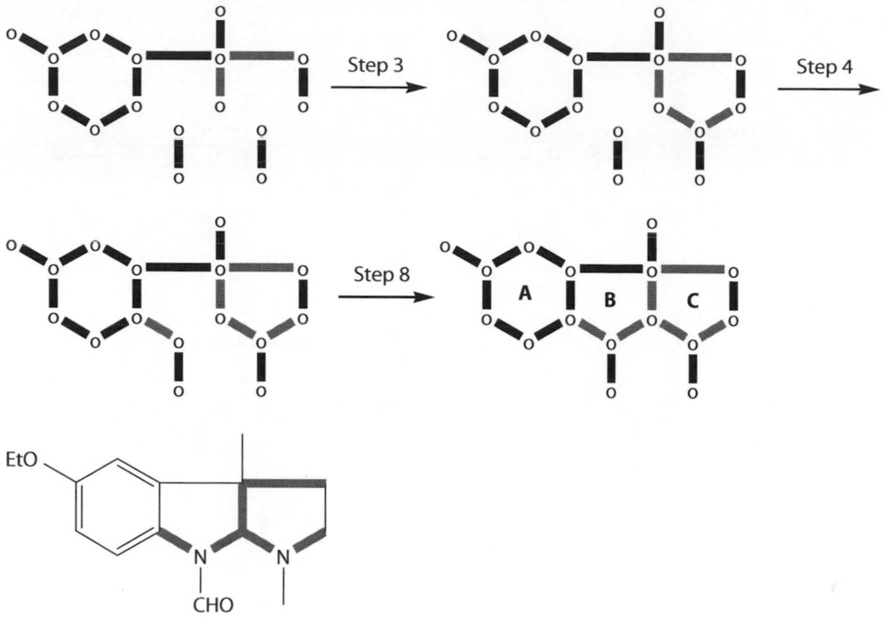

Sequence: $A_0 \xrightarrow{[4+1]} C_3 \xrightarrow{[5+0]} B_8$

(–)-Physostigmine

Takano, S.; Moriya, M.; Ogasawara, K. *J. Org. Chem.* 1991, *56*, 5982.

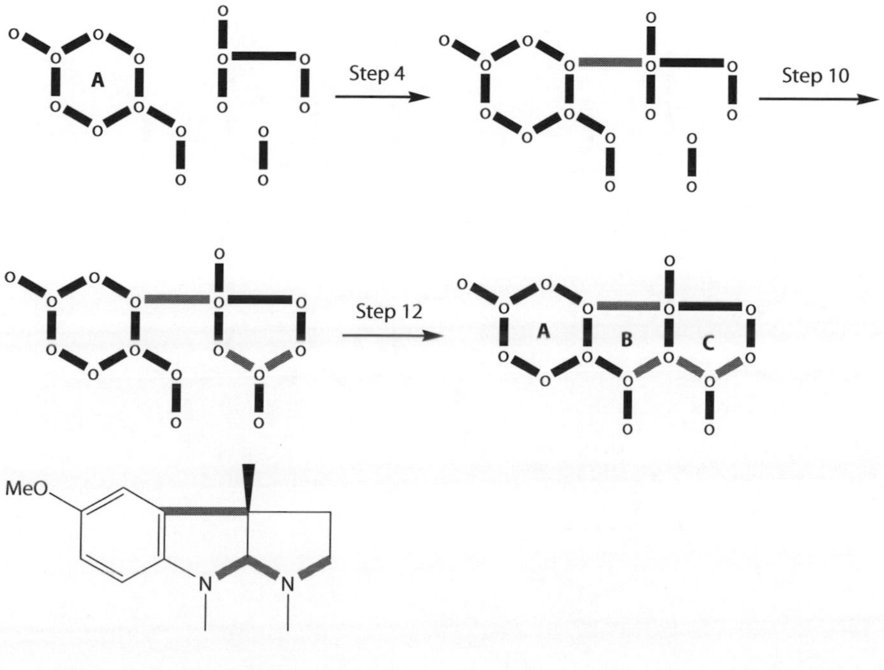

Sequence: $A_0 \xrightarrow{[4+1]} C_{10} \xrightarrow{[5+0]} B_{12}$

(–)-Physostigmine

Takano, S.; Goto, E.; Hirama, M.; Ogasawara, K. *Chem. Pharm. Bull.* 1982, *30*, 2641.

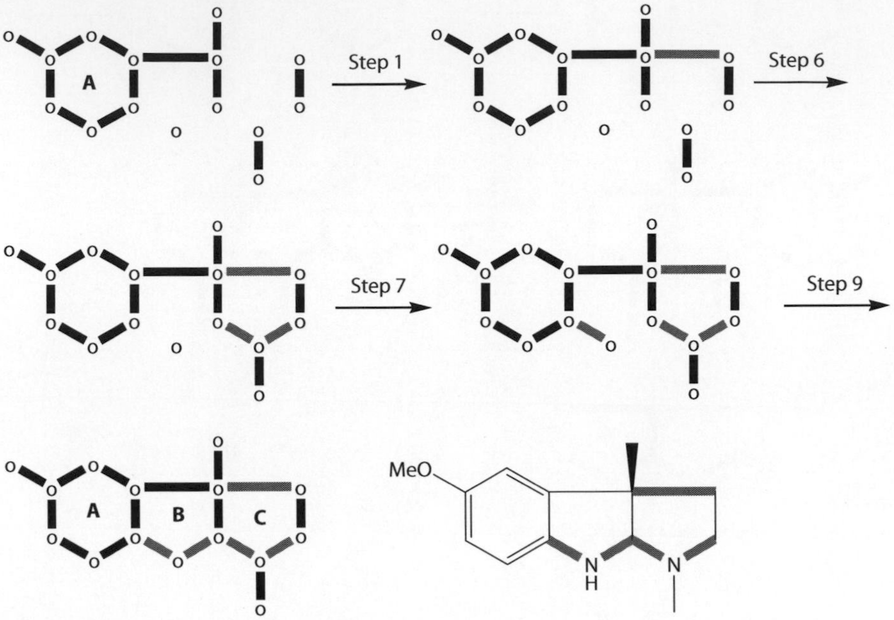

Sequence: $A_0 \xrightarrow{[4+1]} C_6 \xrightarrow{[5+0]} B_9$

(–)-Physostigmine

Trost, B.M.; Zhang, Y. *J. Am. Chem. Soc.* 2006, *128*, 4590.

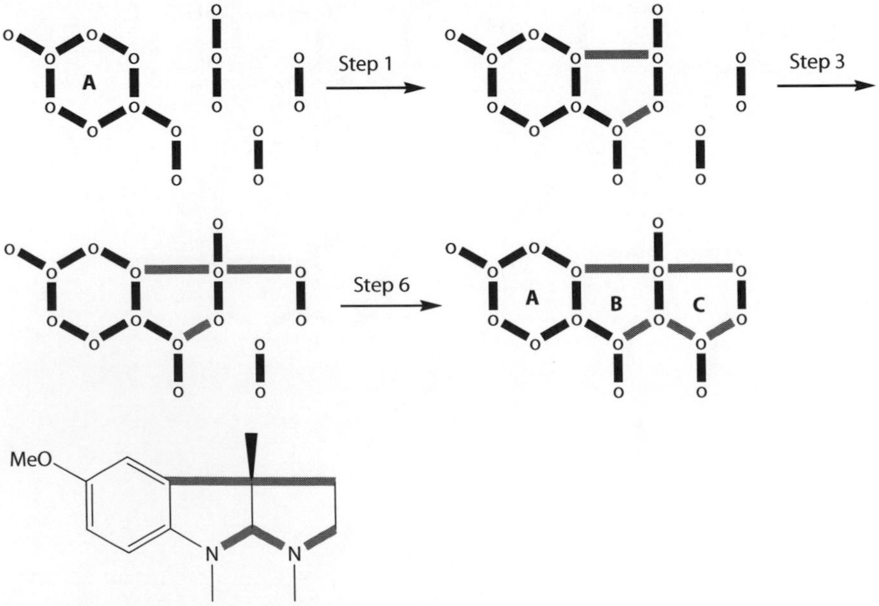

Sequence: $A_0 \xrightarrow{[3+2]} B_1 \xrightarrow{[4+1]} C_6$

(–)-Physostigmine

Yu, Q.S.; Lu, B.Y. *Heterocycles* 1994, *39*, 519.

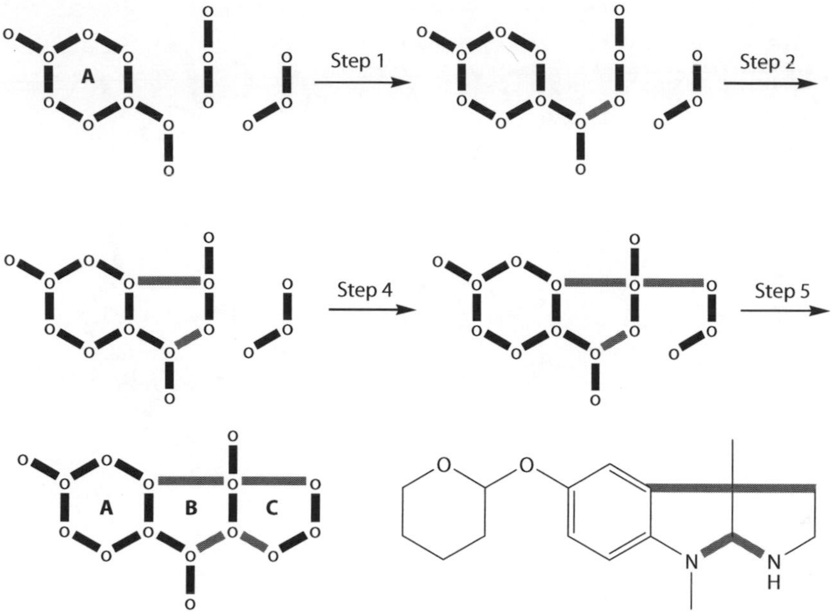

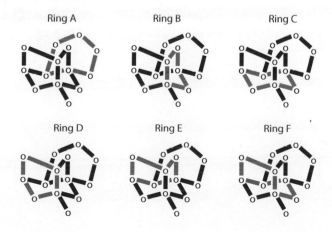

Sequence: $A_0 \xrightarrow{\text{[5+0]}} B_1 \xrightarrow{\text{[5+0]}} C_5$

Summary:

Strategy A → B → C [5 + 0]-[5 + 0] sequence: Julian, Nakada, Yu
Strategy A → B → C [5 + 0]-[4 + 1] sequence: Mukai, Overman
Strategy A → B → C [3 + 2]-[4 + 1] sequence: Trost
Strategy A → C → B: Fukumoto, Speckamp, Takano G1, Takano G2
Strategy → A → B → C: Fuji
Strategy (A + C) → B: Johnson
Strategy (A + B) → C: Marino, Nakagawa
Strategy A → (B + C): Kulkarni

(±)-Platensimycin

Ring nomenclature:

Ring A Ring B Ring C

Ring D Ring E Ring F

Nicolaou, K.C.; Li, A.; Edmonds, D.J. *Angew. Chem. Int. Ed.* 2006, *45*, 7086.

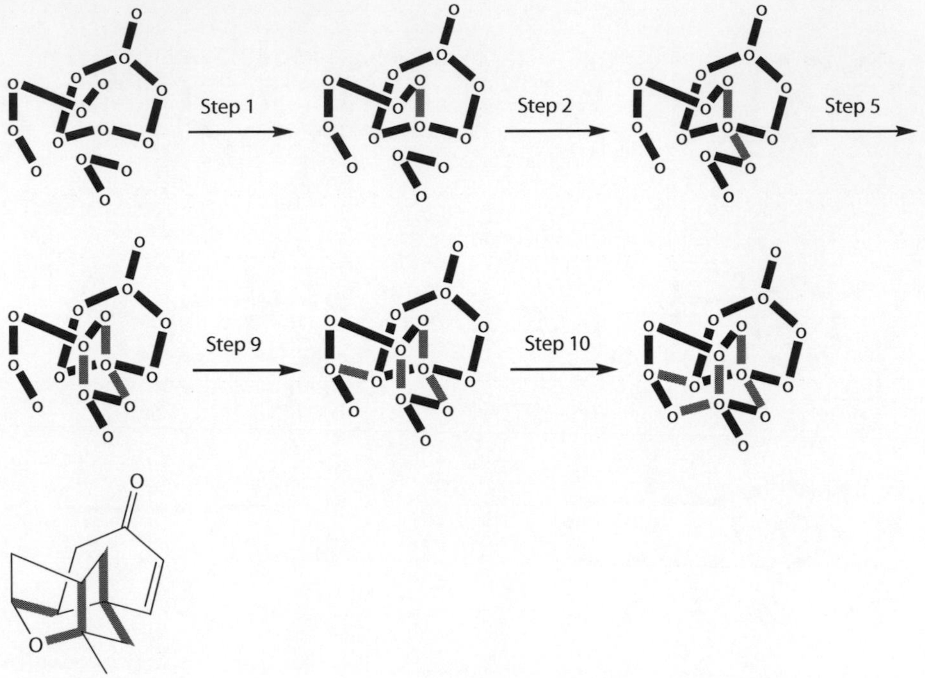

Sequence: $A_0 \xrightarrow{[5+0]} B_5 \xrightarrow{[6+0]} E_9 \xrightarrow{[(6+0)_C + (5+0)_D]} (C + D)_{10}$

Nicolaou, K.C.; Tang, Y.; Wang, J. *Chem. Commun.* 2007, 1922.

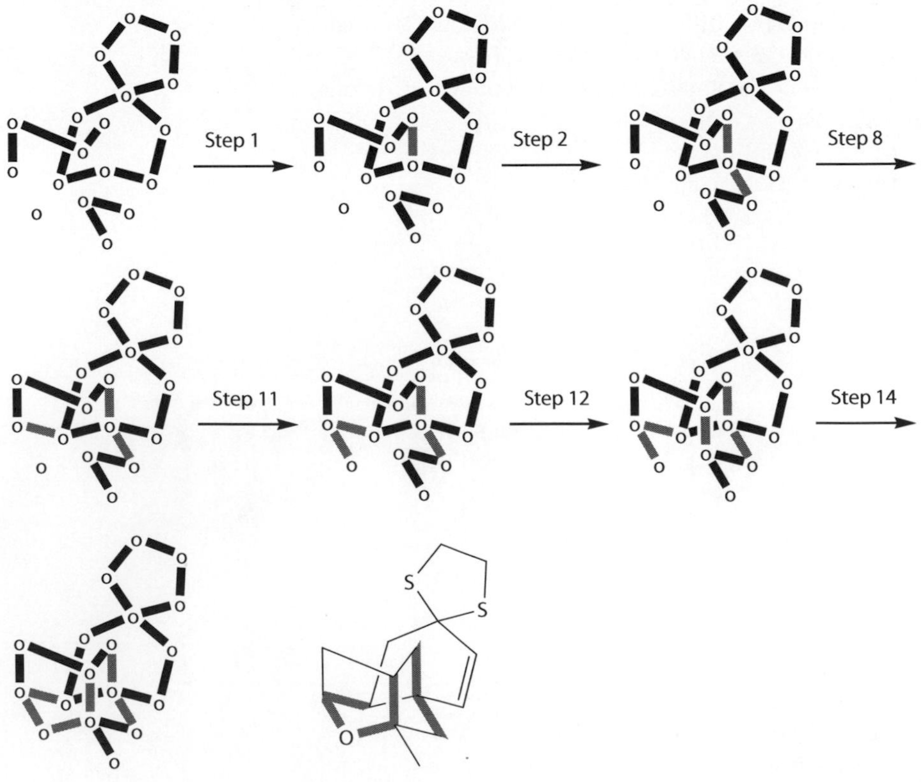

Sequence: $A_0 \xrightarrow{[6+0]} E_8 \xrightarrow{[5+0]} B_{12} \xrightarrow{[(6+0)_C + (5+0)_D]} (C+D)_{14}$

Zou, Y.; Chen, C.H.; Taylor, C.D.; Foxman, B.M.; Snider, B.B. *Org. Lett.* 2007, *9*, 1825.

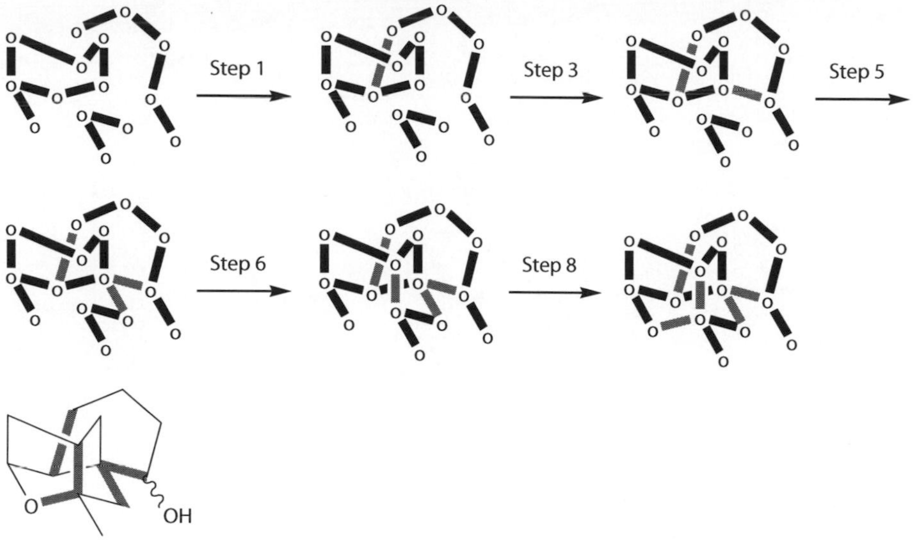

Sequence: $E_0 \xrightarrow{[6+0]} A_3 \xrightarrow{[5+0]} B_6 \xrightarrow{[(6+0)_C + (5+0)_D]} (C+D)_8$

McGrath, N.A.; Bartlett, E.S.; Sittihan, S.; Njardarson, J.T. *Angew. Chem. Int. Ed.* 2009, *48*, 8543.

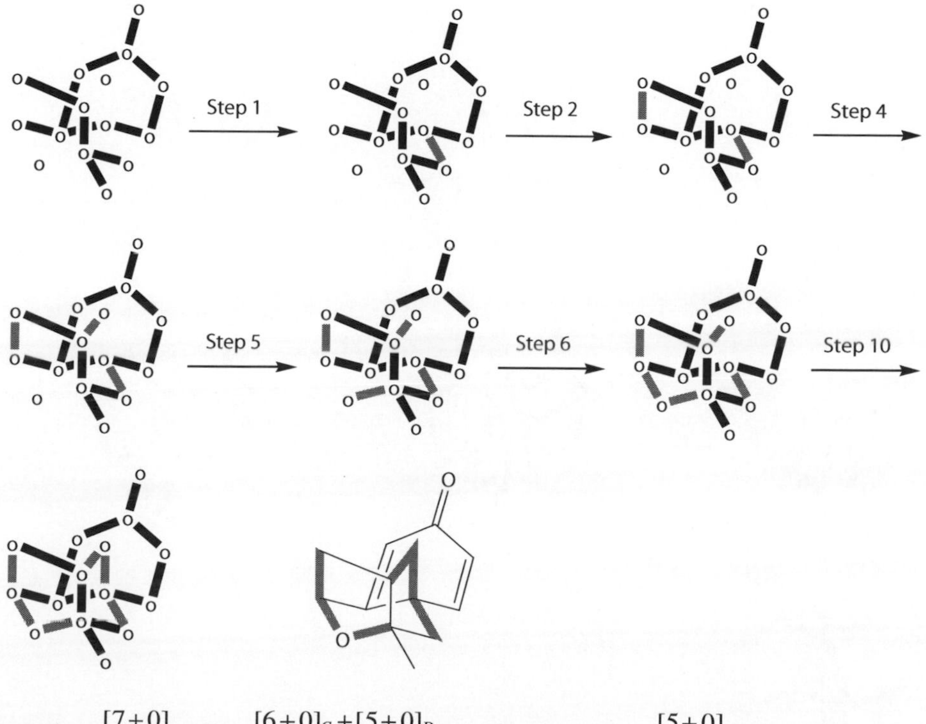

Sequence: $A_0 \xrightarrow{[7+0]} F_2 \xrightarrow{[6+0]_C + [5+0]_D} [3.2.1](C+D)_6 \xrightarrow{[5+0]} B_{10}$

(–)-Platensimycin
Lalic, G.; Corey, E.J. *Org. Lett.* 2007, *9*, 4921.

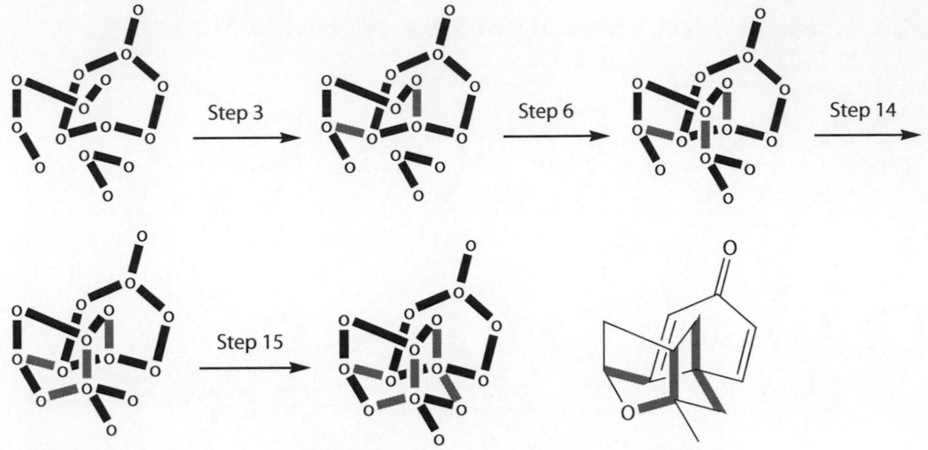

Sequence: $A_0 \xrightarrow{[4+2]} E_3 \xrightarrow{[5+0]} D_{14} \xrightarrow{[(6+0)_C +(5+0)_D]} (C+D)_{15}$

Ghosh, A.K.; Xi, K. *J. Org. Chem.* 2009, *74*, 1163.

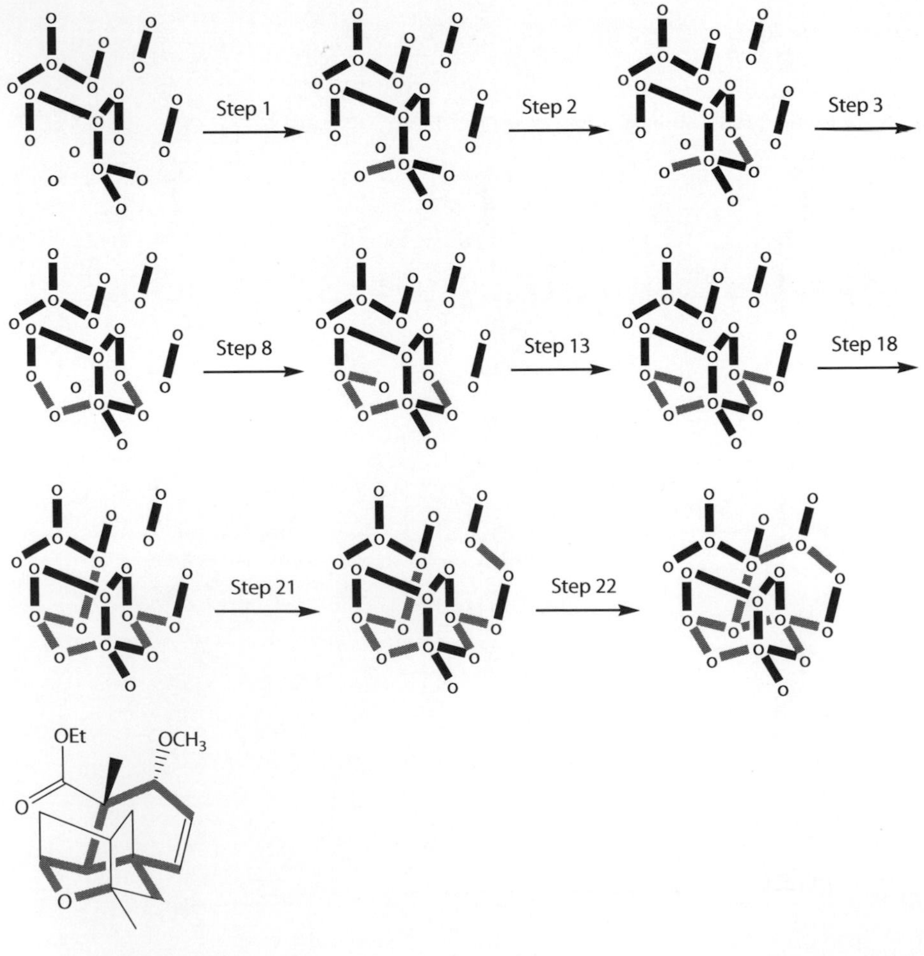

Sequence: $\xrightarrow{[5+0]} B_2 \xrightarrow{[5+0]} D_3 \xrightarrow{[(4+2)_A + (6+0)_C + (6+0)_E]} (A + C + E)_{22}$

Kim, C.H.; Jang, K.P.; Choi, S.Y.; Chung, Y.K.; Lee, E. *Angew. Chem. Int. Ed.* 2008, *47*, 4009.

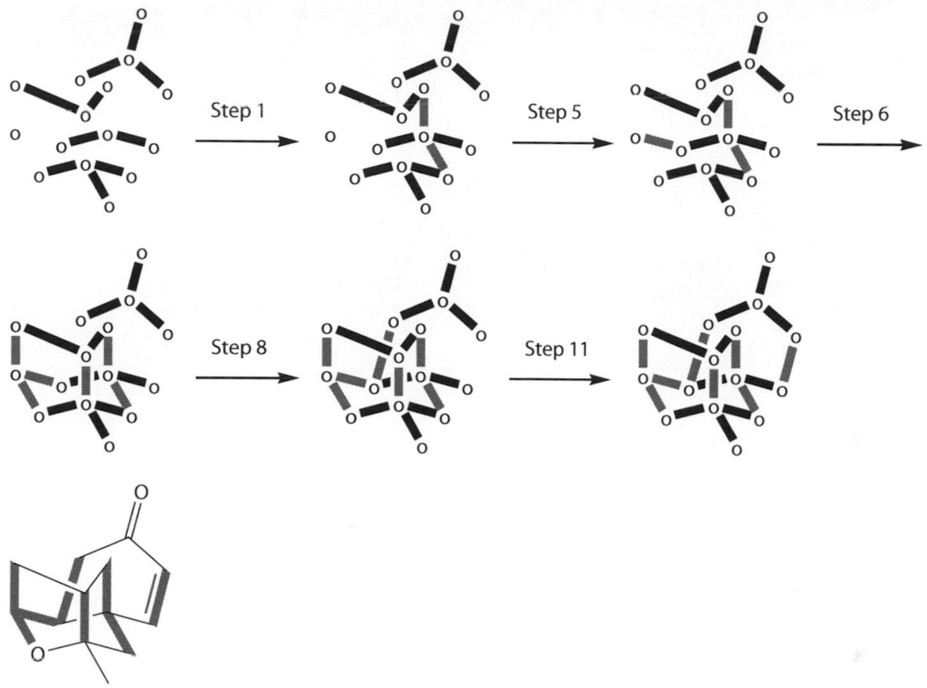

Sequence: $\xrightarrow{[(5+0)_B + (6+0)_C + (2+2+1)_D + (6+0)_E]} (B + C + D + E)_6 \xrightarrow{[6+0]} A_{11}$

Tiefenbacher, K.; Mulzer, J. *Angew. Chem. Int. Ed.* 2007, *46*, 8074.

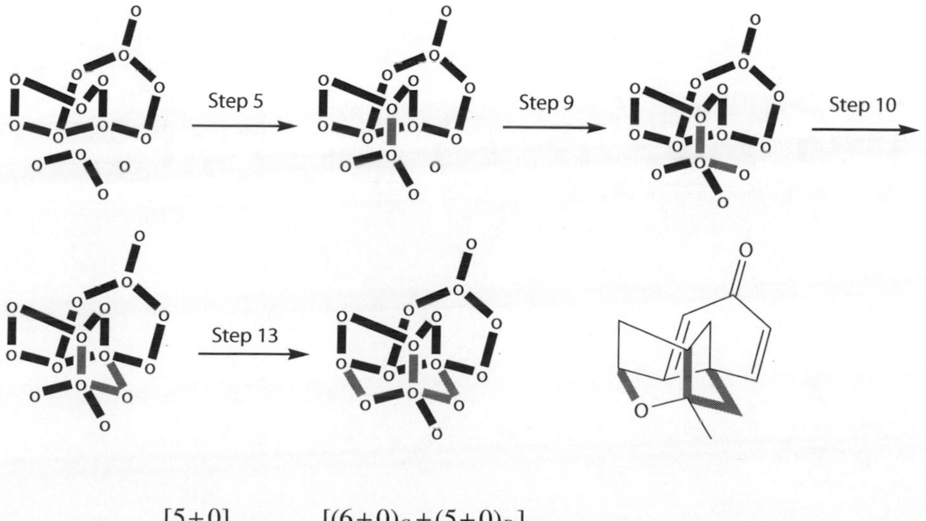

Sequence: $(A + E)_0 \xrightarrow{[5+0]} B_{10} \xrightarrow{[(6+0)_C + (5+0)_D]} (C + D)_{13}$

Nicolaou, K.C.; Edmonds, D.J.; Li, A.; Tria, G.S. *Angew. Chem. Int. Ed.* 2007, *46*, 3942.

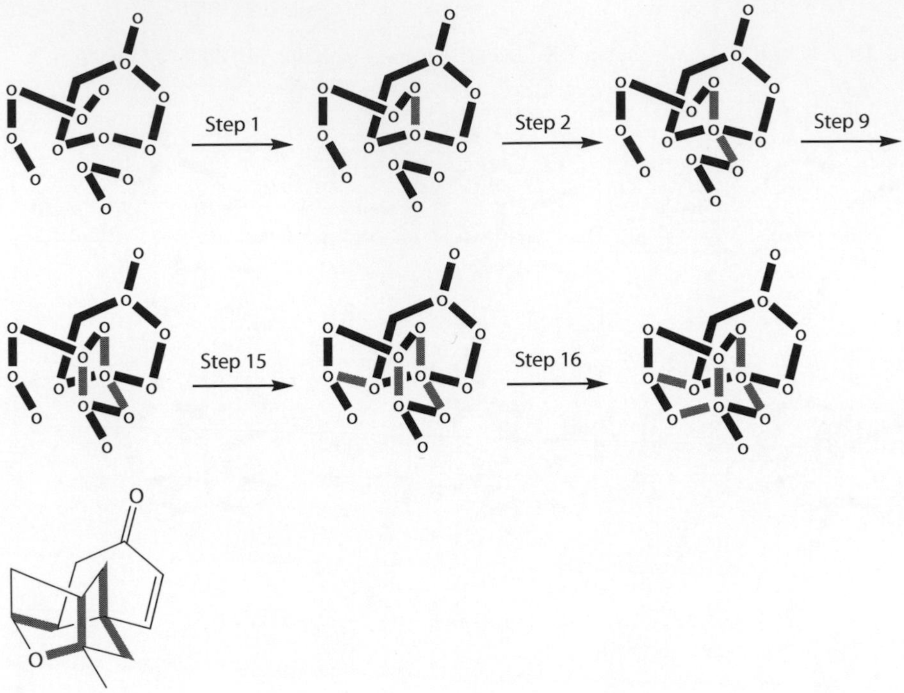

Sequence: $A_0 \xrightarrow{[5+0]} B_9 \xrightarrow{[6+0]} E_{15} \xrightarrow{[(6+0)_C + (5+0)_D]} (C+D)_{16}$

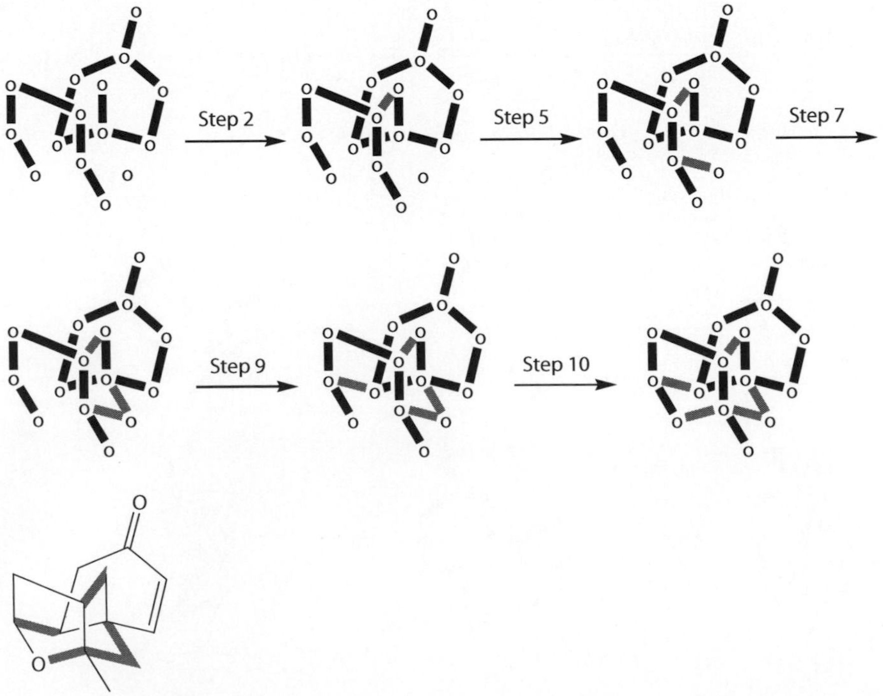

Sequence: $A_0 \xrightarrow{[5+0]} B_7 \xrightarrow{[6+0]} E_9 \xrightarrow{[(6+0)_C + (5+0)_D]} (C+D)_{10}$

Nicolaou, K.C.; Pappo, D.; Tsang, K.Y.; Gibe, R.; Chen, D.Y.K. *Angew. Chem. Int. Ed.* 2008, *47*, 944.

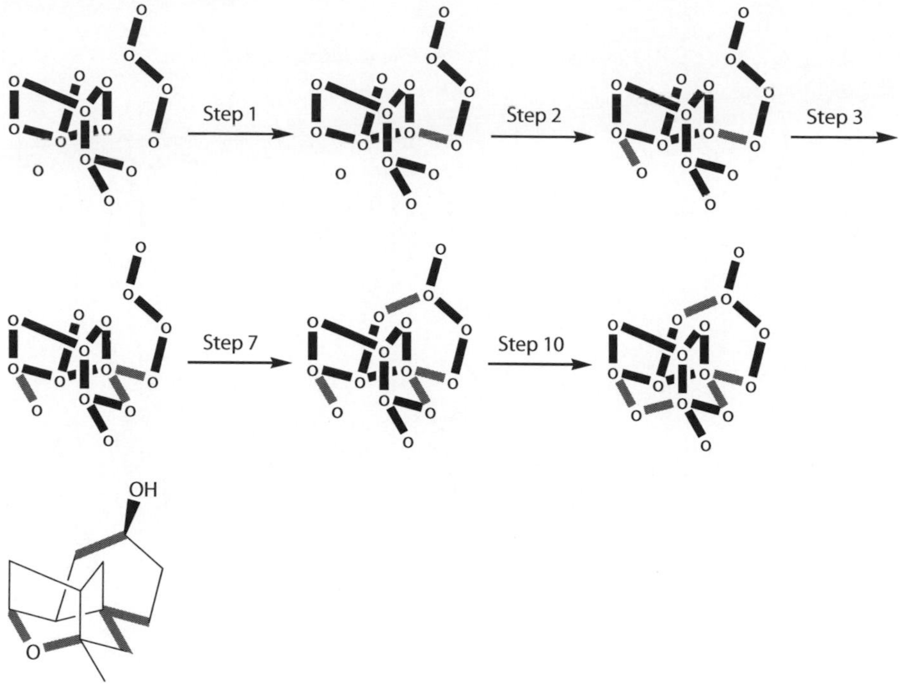

Sequence: $E_0 \xrightarrow{[5+0]} B_3 \xrightarrow{[6+0]} A_7 \xrightarrow{[(6+0)_C + (5+0)_D]} (C+D)_{10}$

Li, P.; Pavette, J.N.; Yamamoto, H. *J. Am. Chem. Soc.* 2007, *129*, 9534.

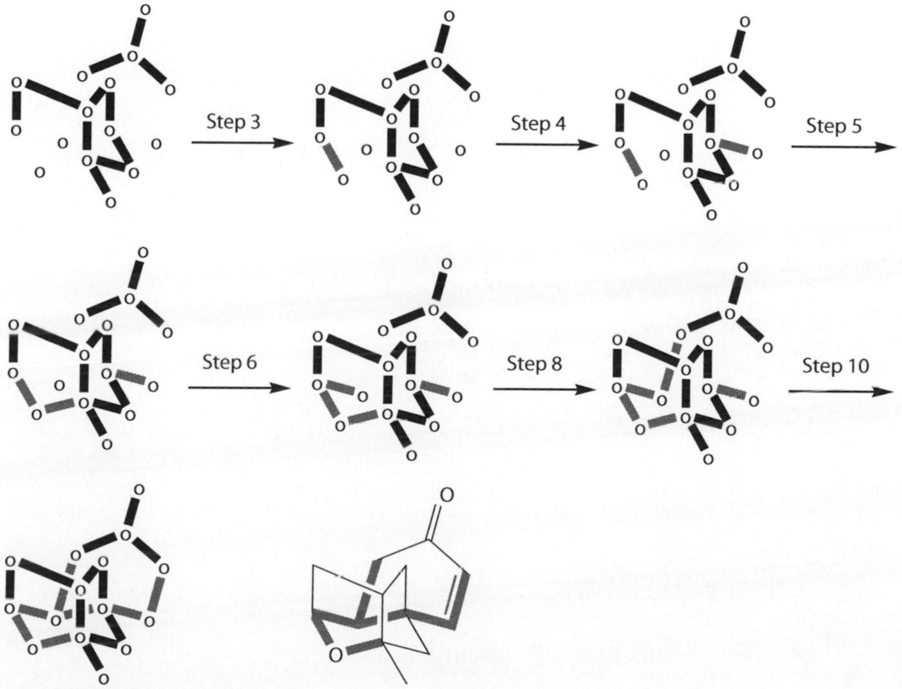

Sequence: $B_0 \xrightarrow{[5+0]} D_5 \xrightarrow{[(6+0)_A + (6+0)_C + (6+0)_E]} (A + C + E)_{10}$

Nicolaou, K.C.; Li, A.; Ellery, S.P.; Edmonds, D.J. *Angew. Chem. Int. Ed.* 2009, *48*, 6293.

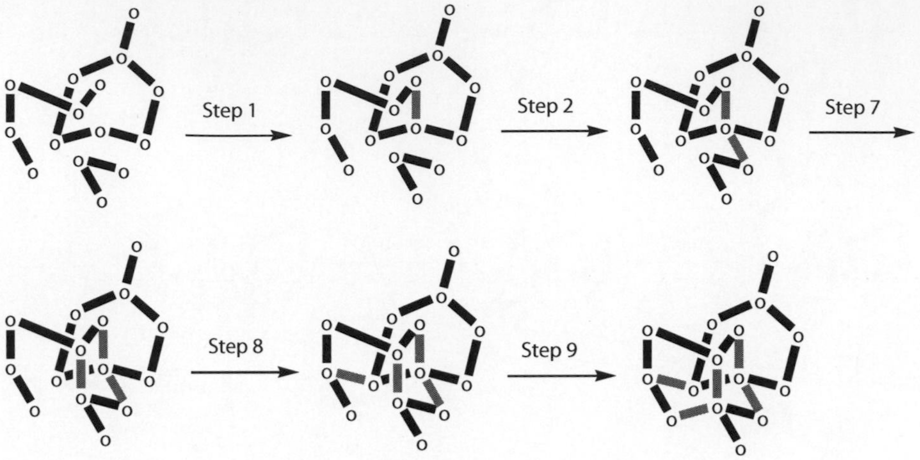

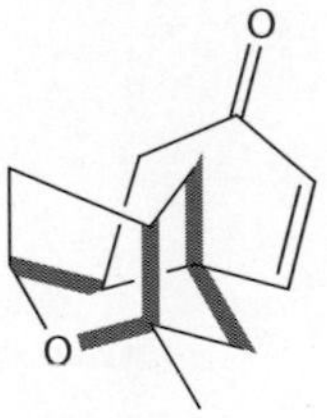

Sequence: $A_0 \xrightarrow{[5+0]} B_7 \xrightarrow{[6+0]} E_8 \xrightarrow{[(6+0)_C + (5+0)_D]} (C + D)_9$

Summary:

Strategy #1

$B_0 \xrightarrow{[5+0]} D_5 \xrightarrow{[(6+0)_A + (6+0)_C + (6+0)_E]} (A + C + E)_{10}$: Yamamoto

Strategy #2

$E_0 \xrightarrow{[5+0]} B_3 \xrightarrow{[6+0]} A_7 \xrightarrow{[(6+0)_C + (5+0)_D]} (C + D)_{10}$: Nicolaou G5

Strategy #3

$A_0 \xrightarrow{[5+0]} B_7 \xrightarrow{[6+0]} E_8 \xrightarrow{[(6+0)_C + (5+0)_D]} (C + D)_9$: Nicolaou G6

$A_0 \xrightarrow{[5+0]} B_7 \xrightarrow{[6+0]} E_9 \xrightarrow{[(6+0)_C + (5+0)_D]} (C + D)_{10}$: Nicolaou G4

$A_0 \xrightarrow{[5+0]} B_9 \xrightarrow{[6+0]} E_{15} \xrightarrow{[(6+0)_C + (5+0)_D]} (C + D)_{16}$: Nicolaou G3

$A_0 \xrightarrow{[5+0]} B_5 \xrightarrow{[6+0]} E_9 \xrightarrow{[(6+0)_C + (5+0)_D]} (C + D)_{10}$: Nicolaou G1

Strategy #4

$$A_0 \xrightarrow{[4+2]} E_3 \xrightarrow{[5+0]} D_{14} \xrightarrow{[(6+0)_C + (5+0)_D]} (C+D)_{15}: \text{Corey}$$

Strategy #5

$$A_0 \xrightarrow{[6+0]} E_8 \xrightarrow{[5+0]} B_{12} \xrightarrow{[(6+0)_C + (5+0)_D]} (C+D)_{14}: \text{Nicolaou G2}$$

Strategy #6

$$E_0 \xrightarrow{[6+0]} A_3 \xrightarrow{[5+0]} B_6 \xrightarrow{[(6+0)_C + (5+0)_D]} (C+D)_8: \text{Snider}$$

Strategy #7

$$(A+E)_0 \xrightarrow{[5+0]} B_{10} \xrightarrow{[(6+0)_C + (5+0)_D]} (C+D)_{13}: \text{Mulzer}$$

Strategy #8

$$\xrightarrow{[(5+0)_B + (6+0)_C + (3+2)_D + (6+0)_E]} (B+C+D+E)_6 \xrightarrow{[6+0]} A_{11}: \text{Lee}$$

Strategy #9

$$\xrightarrow{[5+0]} B_2 \xrightarrow{[5+0]} D_3 \xrightarrow{[(4+2)_A + (6+0)_C + (6+0)_E]} (A+C+E)_{22}: \text{Ghosh}$$

Strategy #10

$$A_0 \xrightarrow{[7+0]} F_2 \xrightarrow{[6+0]_C + [5+0]_D} [3.2.1](C+D)_6 \xrightarrow{[5+0]} B_{10}: \text{Njardarson}$$

[1.1.1]-Propellane
Wiberg, K.B. *Acc. Chem. Res.* 1984, *17*, 379.
Wiberg, K.K.; Walker, F.H. *J. Am. Chem. Soc.* 1982, *104*, 5239.

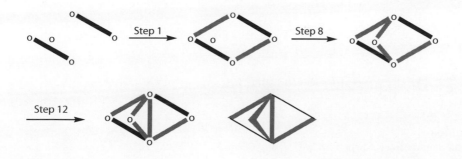

Sequence:

$$\xrightarrow{[2+2]} [4]_1 \xrightarrow{[3+1]+[3+1]} [1.1.1]_8 \xrightarrow{[3+0]+[3+0]+[3+0]} [1.1.1.0]_{12}$$

(±)-Quinine
Igarashi, J.; Katsukawa, M.; Wang, Y.G.; Acharya, H.P.; Kobayashi, Y. *Tetrahedron Lett.* 2004, *45*, 3783.

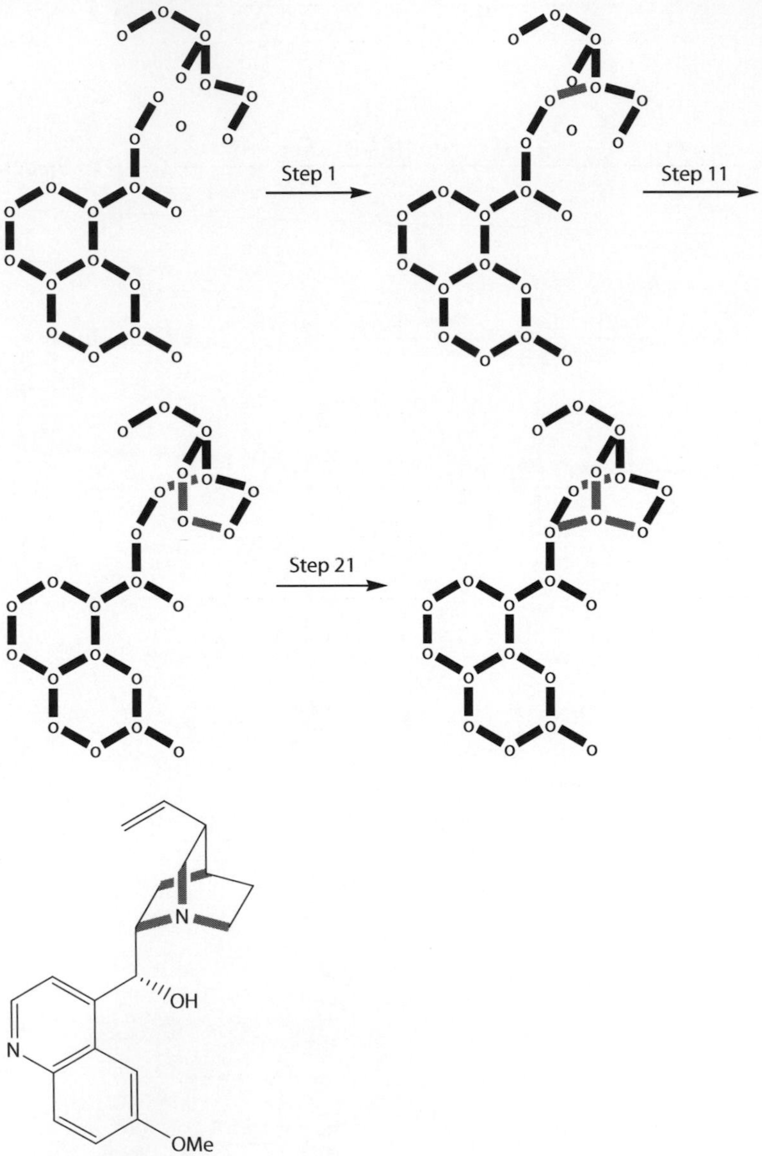

Sequence: $\xrightarrow{[5+1]} [6]_{11} \xrightarrow{[(6+0)+(6+0)]} [2.2.2]_{21}$

(−)-Quinine
Gates, M.; Sugavanam, R.; Schreiber, W.L. *J. Am. Chem. Soc.* 1970, *92*, 205.

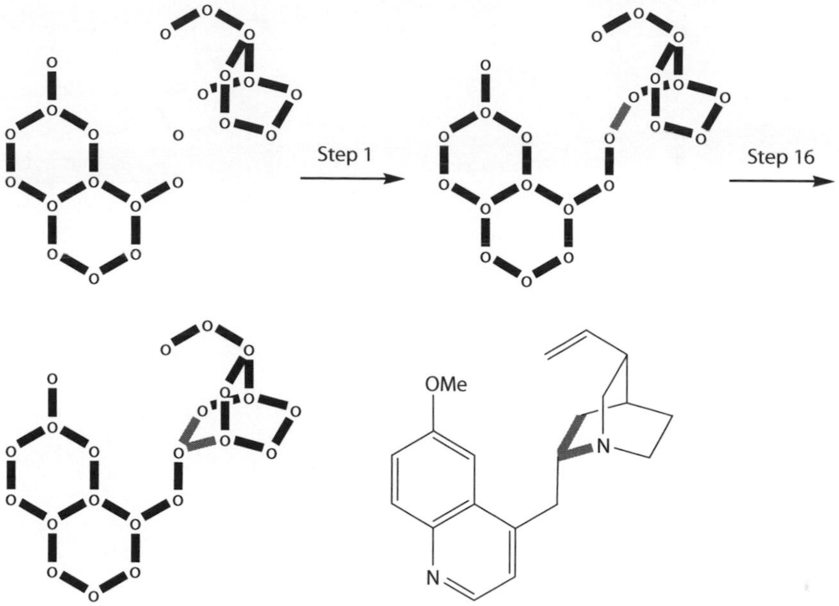

Step 1

Step 16

Sequence: $[6]_0 \xrightarrow{[(6+0)+(6+0)]} [2.2.2]_{16}$

Gutzwiller, J.; Uskokovic, M. *J. Am. Chem. Soc.* 1970, *92*, 204.

As mentioned earlier.

Sequence: $[6_0] \xrightarrow{[(6+0)+(6+0)]} [2.2.2]_{15}$

Taylor, E.C.; Martin, S.F. *J. Am. Chem. Soc.* 1972, *94*, 6218.

As mentioned earlier.

Sequence: $[6]_0 \xrightarrow{[(6+0)+(6+0)]} [2.2.2]_{14}$

(±)-Quinine

Raheem, I.T.; Goodman, S.N.; Jacobsen, E.N. *J. Am. Chem. Soc.* 2004, *126*, 706.

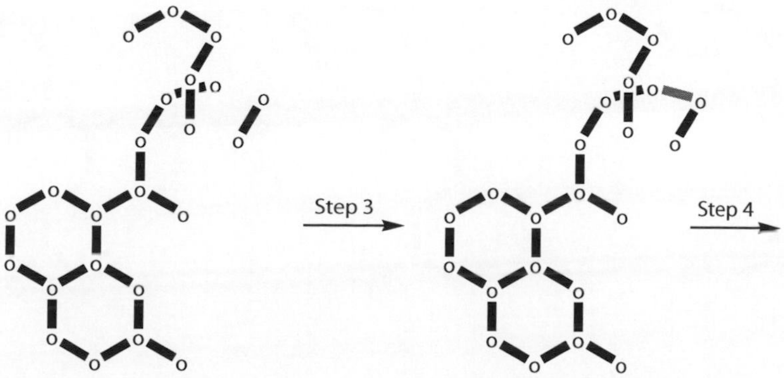

Step 3

Step 4

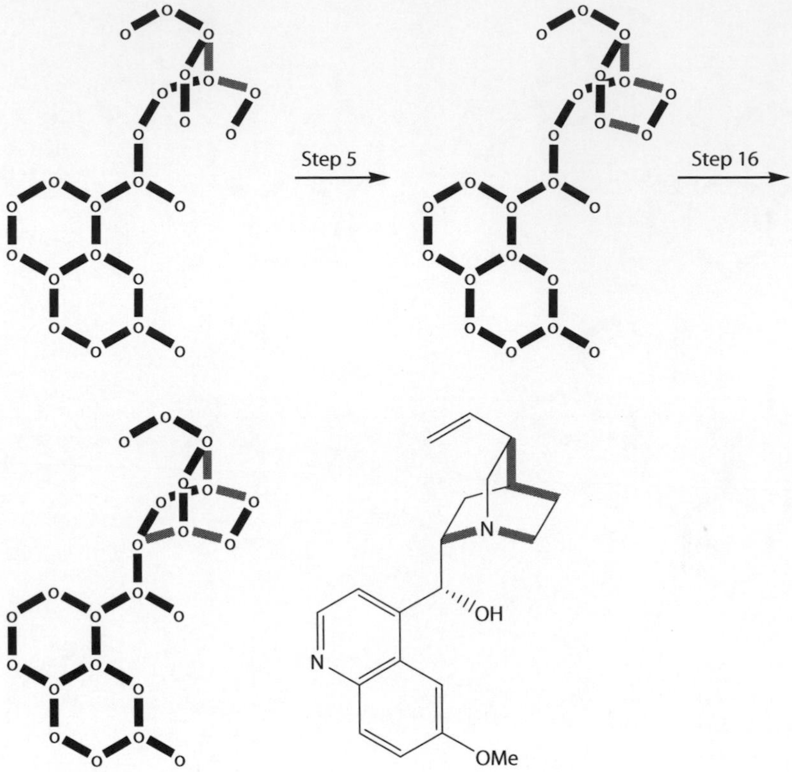

Sequence: $\xrightarrow{[6+0]} [6]_5 \xrightarrow{[(6+0)+(6+0)]} [2.2.2]_{16}$

(±)-Quinine

Webber, P.; Krische, M.J. *J. Org. Chem.* 2008, *73*, 9379.

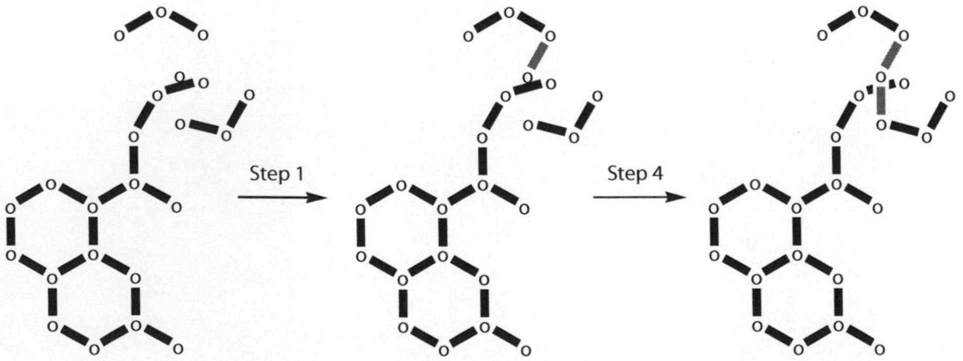

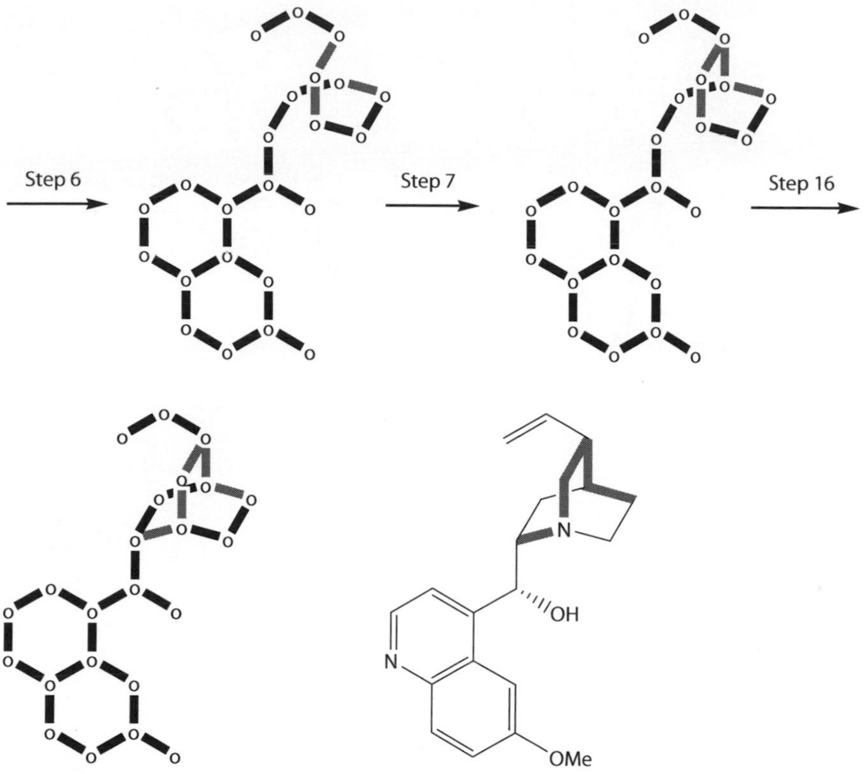

Sequence: $\xrightarrow{[6+0]}$ $[6]_7$ $\xrightarrow{[(6+0)+(6+0)]}$ $[2.2.2]_{16}$

(−)-Quinine

Stork, G.; Niu, D.; Fujimoto, A.; Koft, E.R.; Balkovec, J.M.; Tata, J.R.; Dake, G.R. *J. Am. Chem. Soc.* 2001, *123,* 3239.

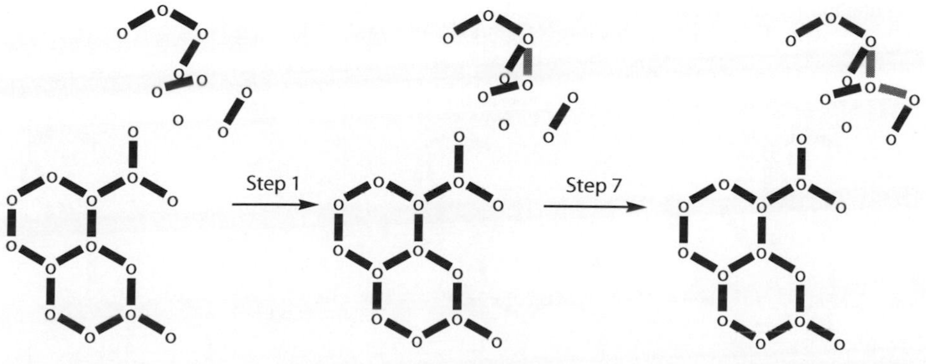

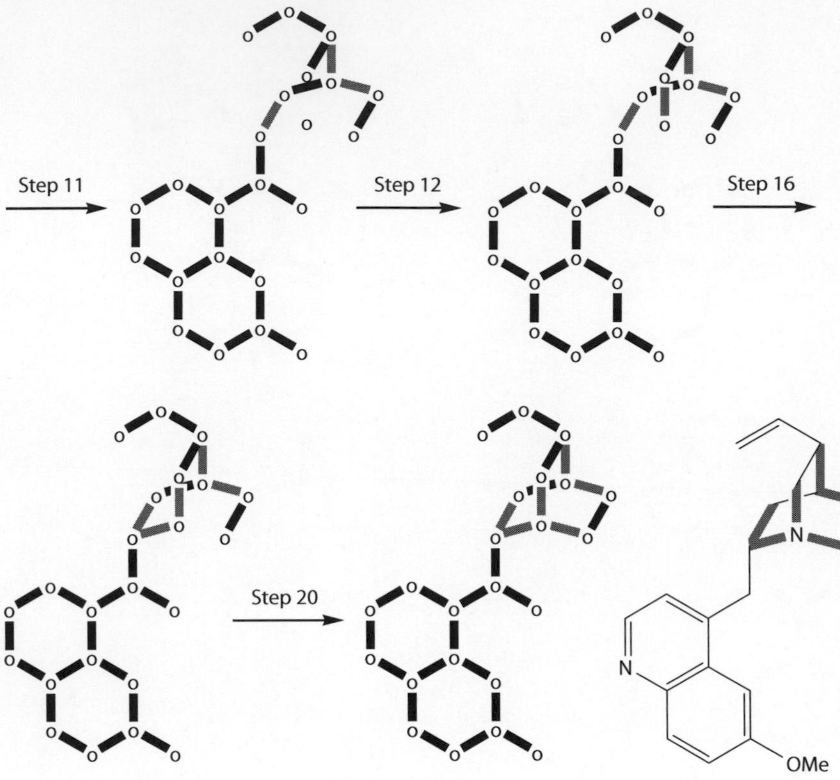

Sequence: $\xrightarrow{\;[6+0]\;}$ $[6]_{16}$ $\xrightarrow{\;[(6+0)+(6+0)]\;}$ $[2.2.2]_{20}$

(±)-Quinine

Woodward, R.B.; Doering, W.E. *J. Am. Chem. Soc.* 1945, *67*, 860.
Rabe, P. *Chem. Ber.* 1911, *44*, 2088.
Rabe, P.; Kindler, K. *Chem. Ber.* 1918, *51*, 466.
Rabe, P.; Huntenburg, W.; Schultze, A.; Volger, G. *Chem. Ber.* 1931, *64*, 2487.
Smith, A.C.; Williams, R.M. *Angew. Chem. Int. Ed.* 2008, *47*, 1736.

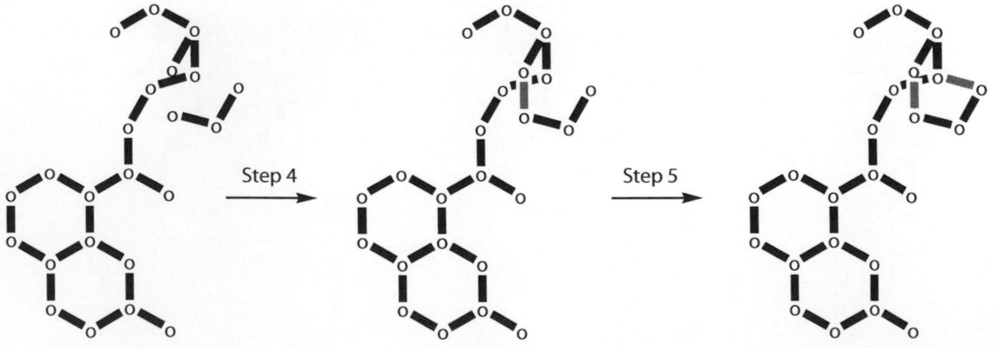

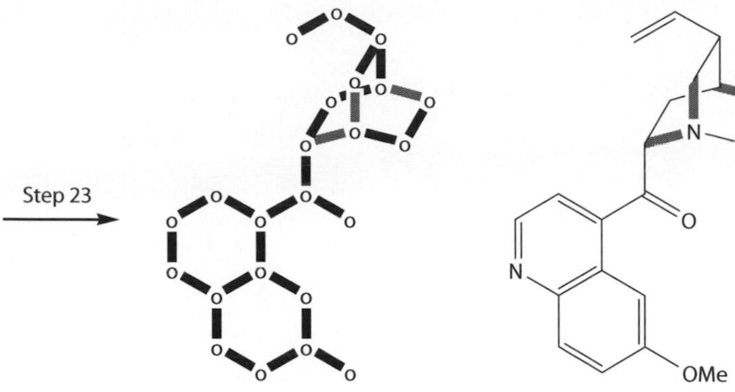

Sequence: $\xrightarrow{[6+0]}$ $[6]_5$ $\xrightarrow{[(6+0)+(6+0)]}$ $[2.2.2]_{23}$

(−)-Quinine
Illa, O.; Arshad, M.; Ros, A.; McGarrigle, E.M.; Aggarwal, V.K. *J. Am. Chem. Soc.* 2010, *132*, 1828.

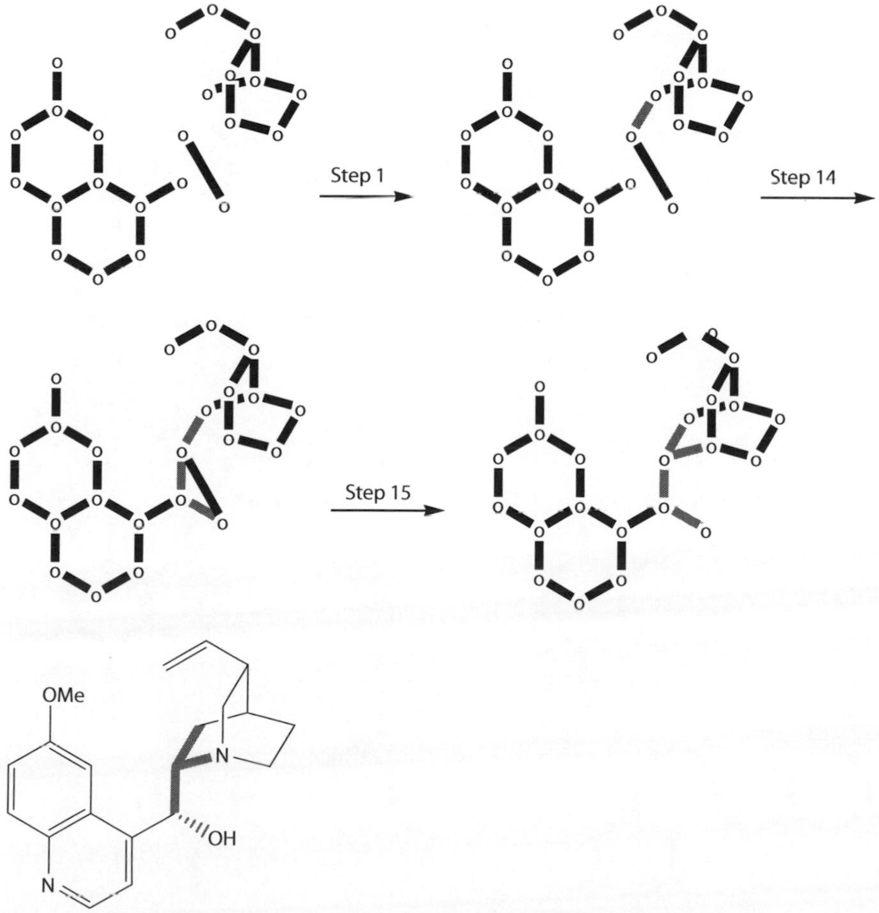

Sequence: $[6]_0$ $\xrightarrow{[(6+0)+(6+0)]}$ $[2.2.2]_{15}$: Aggawal

Summary:

Strategy $\xrightarrow{[5+1]}$ $[6]_{11}$ $\xrightarrow{[(6+0)+(6+0)]}$ $[2.2.2]_{21}$: Kobayashi

Strategy $\xrightarrow{[6+0]}$ $[6]_5$ $\xrightarrow{[(6+0)+(6+0)]}$ $[2.2.2]_{16}$: Jacobsen

Strategy $\xrightarrow{[6+0]}$ $[6]_7$ $\xrightarrow{[(6+0)+(6+0)]}$ $[2.2.2]_{16}$: Krische

Strategy $\xrightarrow{[6+0]}$ $[6]_{16}$ $\xrightarrow{[(6+0)+(6+0)]}$ $[2.2.2]_{20}$: Stork

Strategy $\xrightarrow{[6+0]}$ $[6]_5$ $\xrightarrow{[(6+0)+(6+0)]}$ $[2.2.2]_{23}$: Woodward

Strategy $[6]_0$ $\xrightarrow{[(6+0)+(6+0)]}$ $[2.2.2]_{16}$: Gates

Strategy $[6]_0$ $\xrightarrow{[(6+0)+(6+0)]}$ $[2.2.2]_{15}$: Merck

Strategy $[6]_0$ $\xrightarrow{[(6+0)+(6+0)]}$ $[2.2.2]_{14}$: Martin

Strategy $[6]_0$ $\xrightarrow{[(6+0)+(6+0)]}$ $[2.2.2]_{15}$: Aggarwal

Reserpine

Gomez, A.M.; Lopez, J.C.; Fraser-Reid, B. *J. Org. Chem.* 1994, *59*, 4048.
Gomez, A.M.; Lopez, J.C.; Fraser-Reid, B. *J. Org. Chem.* 1995, *60*, 3859.

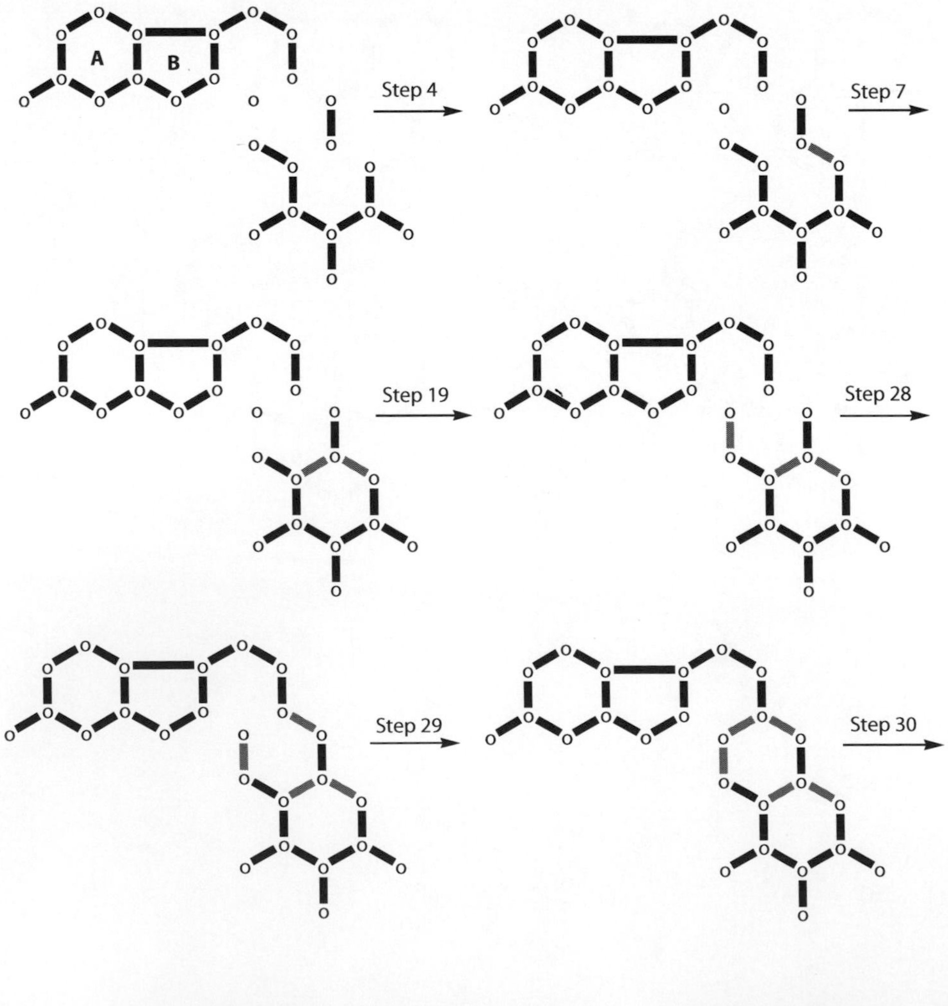

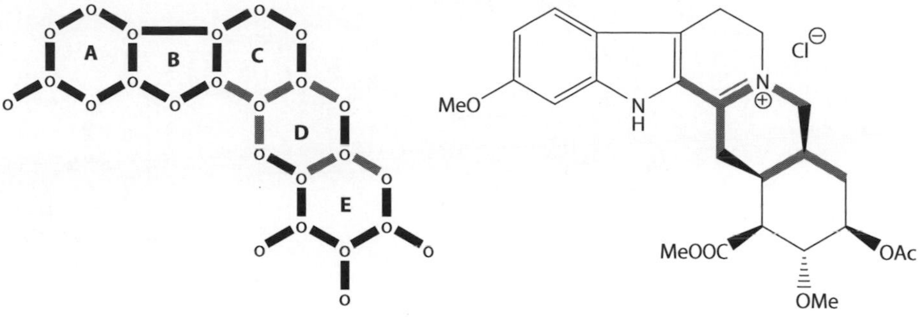

Sequence: $(A + B)_{27} \xrightarrow{[6+0]} E_7 \xrightarrow{[6+0]} D_{29} \xrightarrow{[6+0]} C_{30}$

Reserpine

Hanessian, S.; Pan, J.; Carnell, A.; Bouchard, H.; Lesage, L. *J. Org. Chem.* 1997, *62*, 465.

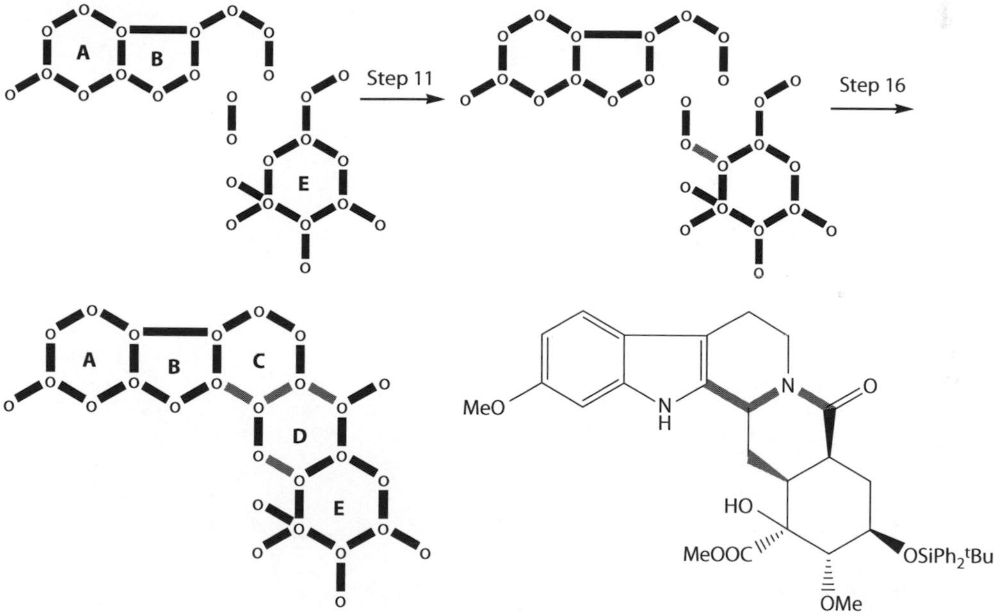

Sequence: $(A + B)_{15} + E_0 \xrightarrow{[(5+1)_C + (5+1)_D]} [4.4.0] (C + D)_{16}$

Reserpine

Chu, C.S.; Liao, C.C.; Rao, P.D. *Chem. Commun.* 1996, 1537.

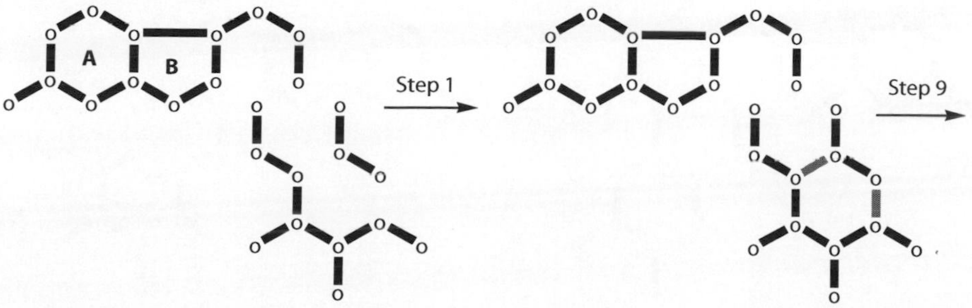

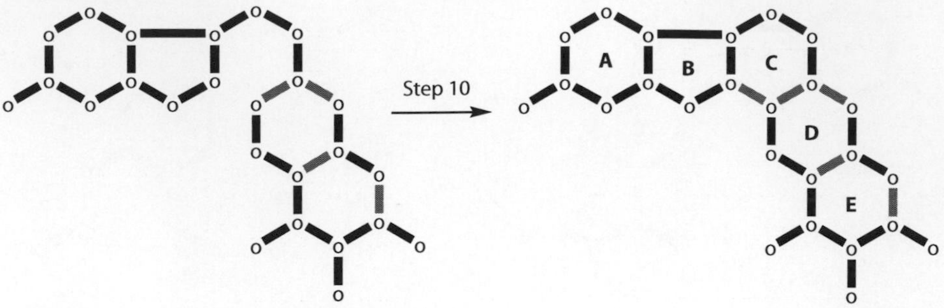

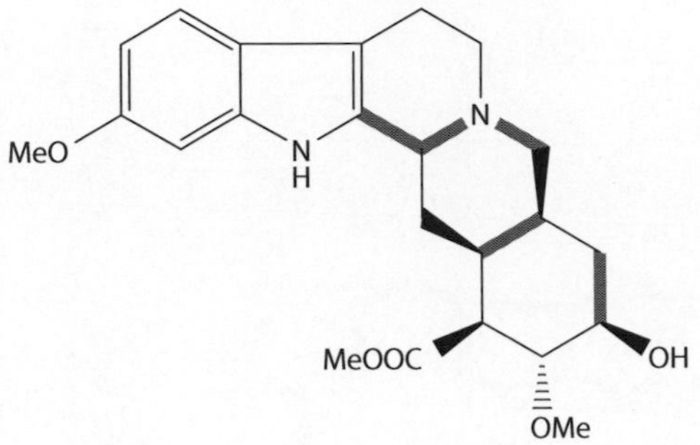

Sequence: $(A+B)_8 \xrightarrow{[4+2]} E_1 \xrightarrow{[5+1]} D_9 \xrightarrow{[6+0]} C_{10}$

Reserpine

Martin, S.F.; Grzejszczak, S.; Rueger, H.; Williamson, S.A. *J. Am. Chem. Soc.* 1985, *107*, 4072.
Martin, S.F.; Rueger, H.; Williamson, S.A.; Grzejszczak, S. *J. Am. Chem. Soc.* 1987, *109*, 6124.
Martin, S.F.; Williamson, S.A.; Gist, R.P.; Smith, K.M. *J. Org. Chem.* 1983, *48*, 5170.

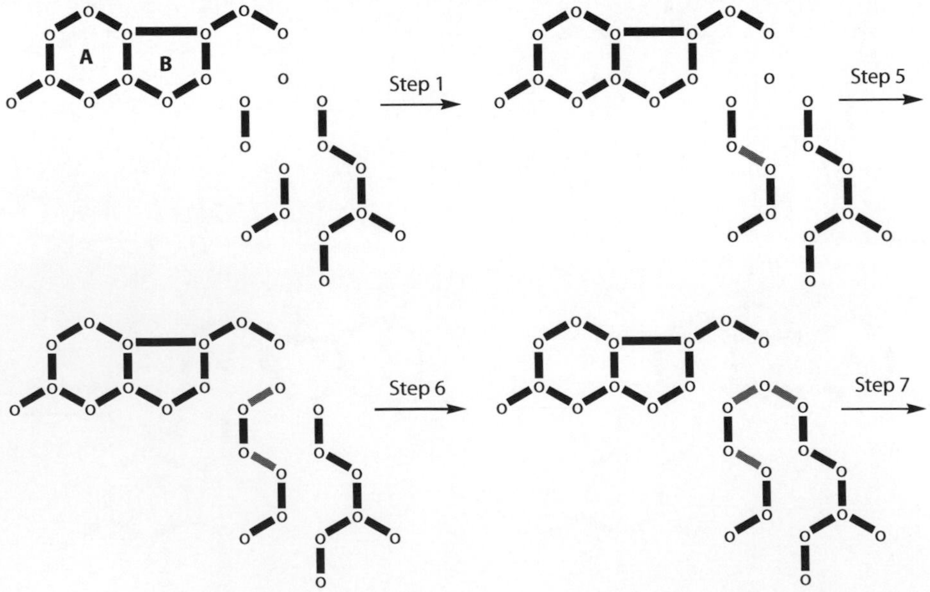

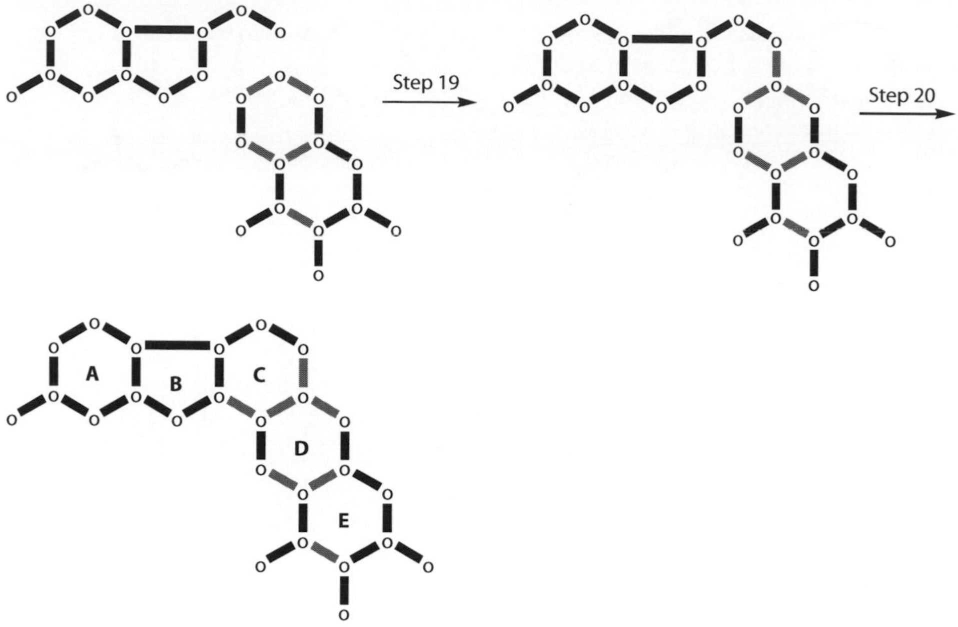
Step 19
Step 20
A
B
C
D
E

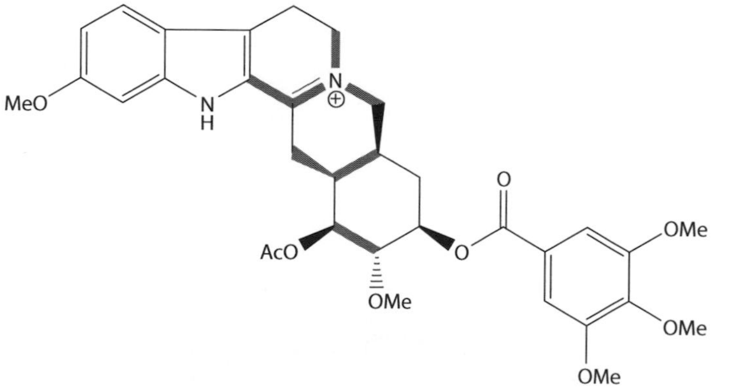
MeO
N
H
N
AcO
OMe
O
O
OMe
OMe
OMe

Sequence: $(A+B)_{18} \xrightarrow{[(6+0)_D+(4+2)_E]} [4.4.0]\,(D+E)_7 \xrightarrow{[6+0]} C_{20}$

Reserpine

Mehta, G.; Reddy, D.S. *J. Chem. Soc. Perkin Trans. I* 1998, 2125.
Mehta, G.; Reddy, D.S. *J. Chem. Soc. Perkin Trans. I* 2000, 1399.

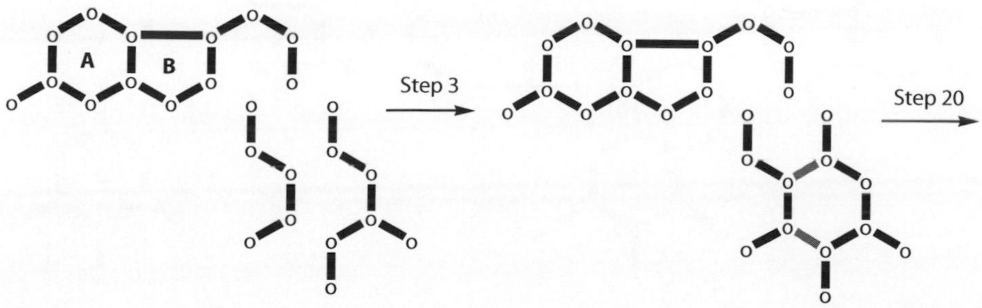
A
B
Step 3
Step 20

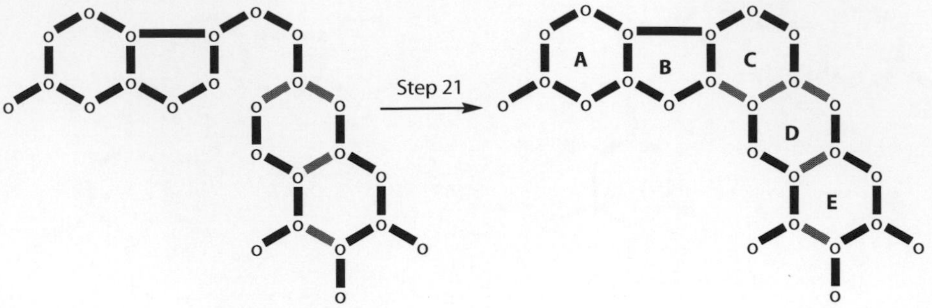

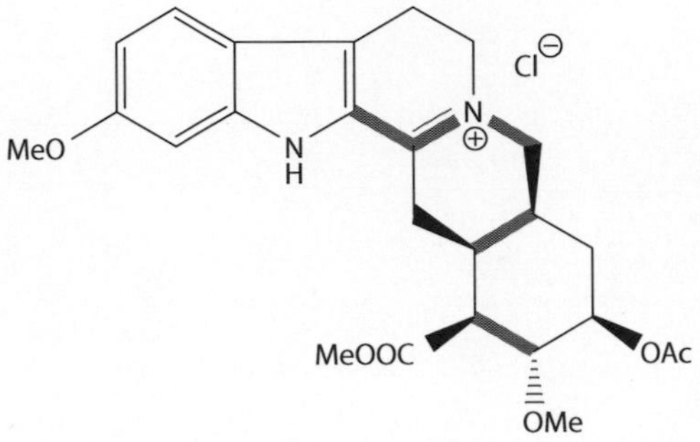

Sequence: $(A + B)_{19} \xrightarrow{[4+2]} E_3 \xrightarrow{[5+1]} D_{20} \xrightarrow{[6+0]} C_{21}$

Reserpine

Pearlman, B. *J. Am. Chem. Soc.* 1979, *101*, 6404.

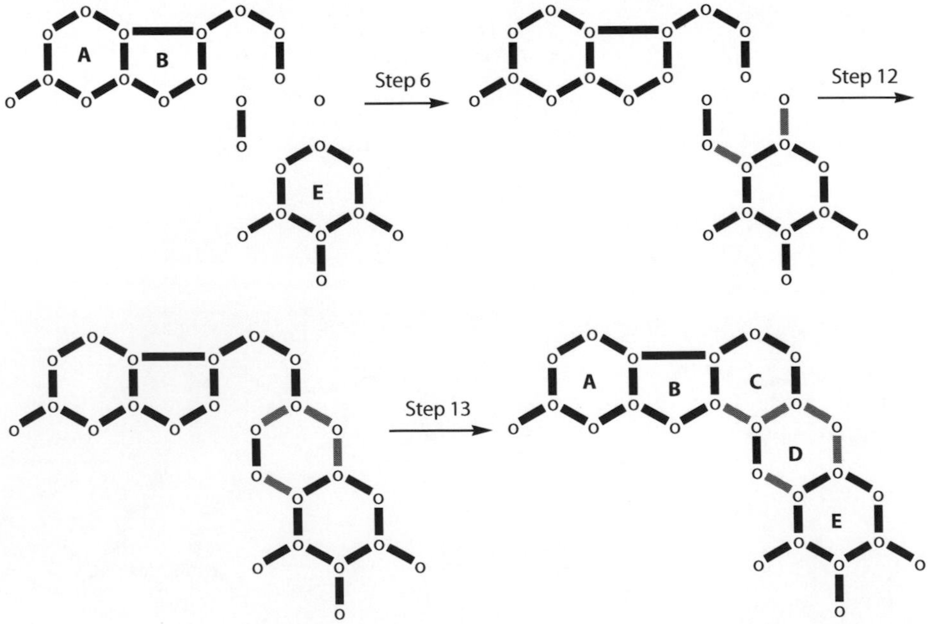

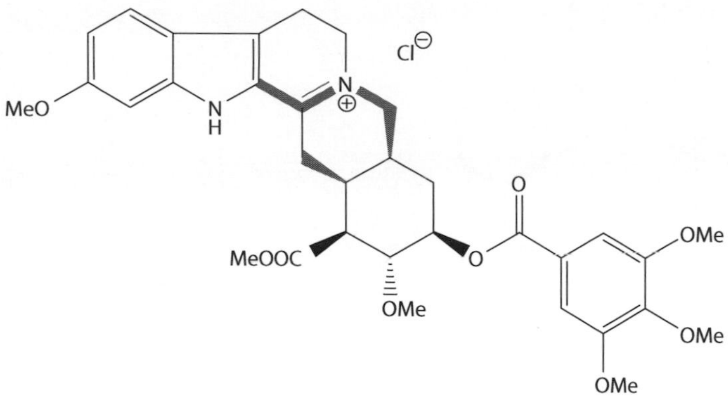

Sequence: $(A + B)_{11} + E_0 \xrightarrow{[5+1]} D_{12} \xrightarrow{[6+0]} C_{13}$

Reserpine

Sparks, S.M.; Gutierrez, A.J.; Shea, K.J. *J. Org. Chem.* 2003, *68*, 5274.

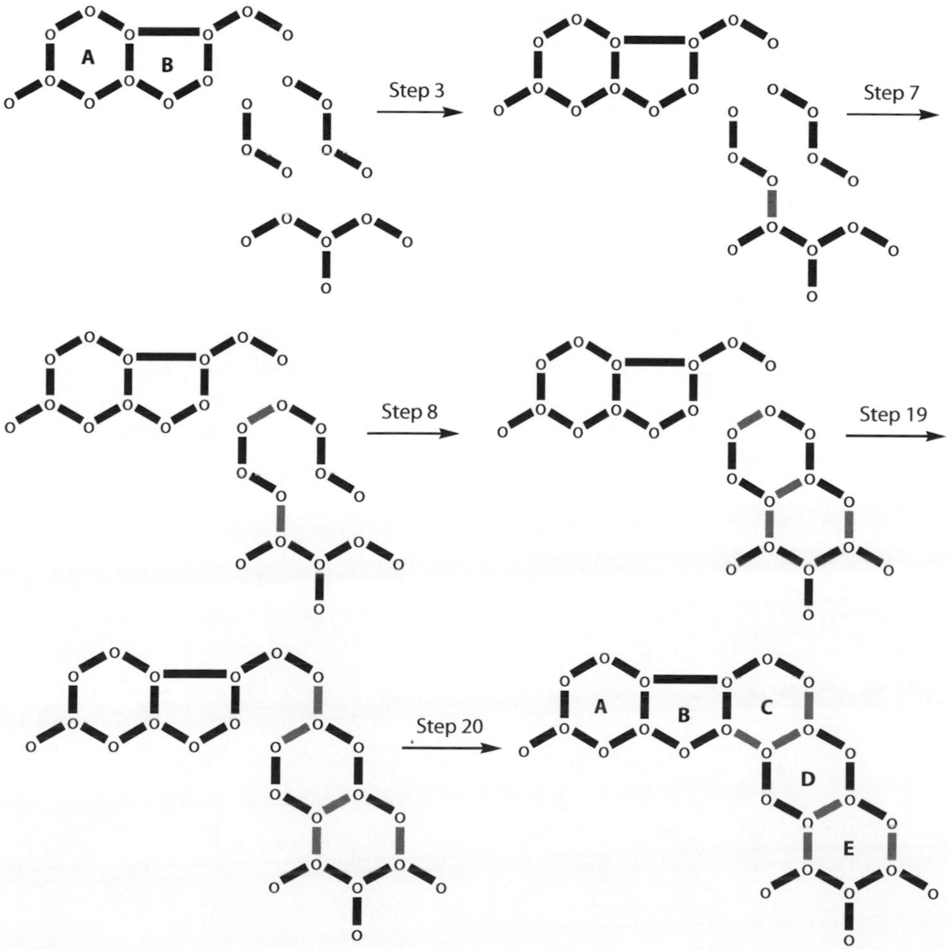

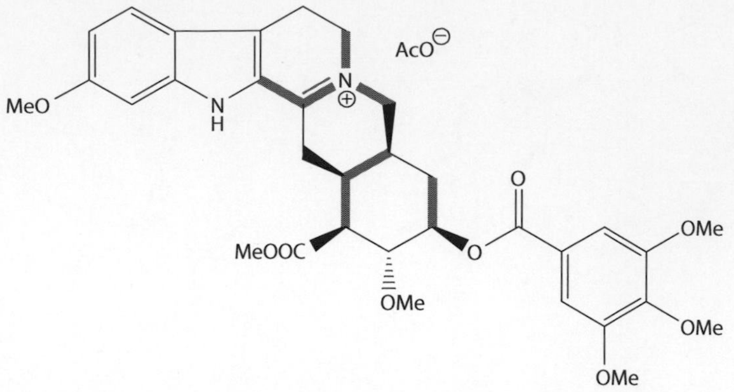

Sequence: $(A+B)_{18} \xrightarrow{[(6+0)_D + (4+2)_E]} [4.4.0] (D+E)_8 \xrightarrow{[6+0]} C_{20}$

Reserpine
Stork, G.; Tang, P.C.; Casey, M.; Goodman, B.; Toyota, M. *J. Am. Chem. Soc.* 2005, *127*, 16255.

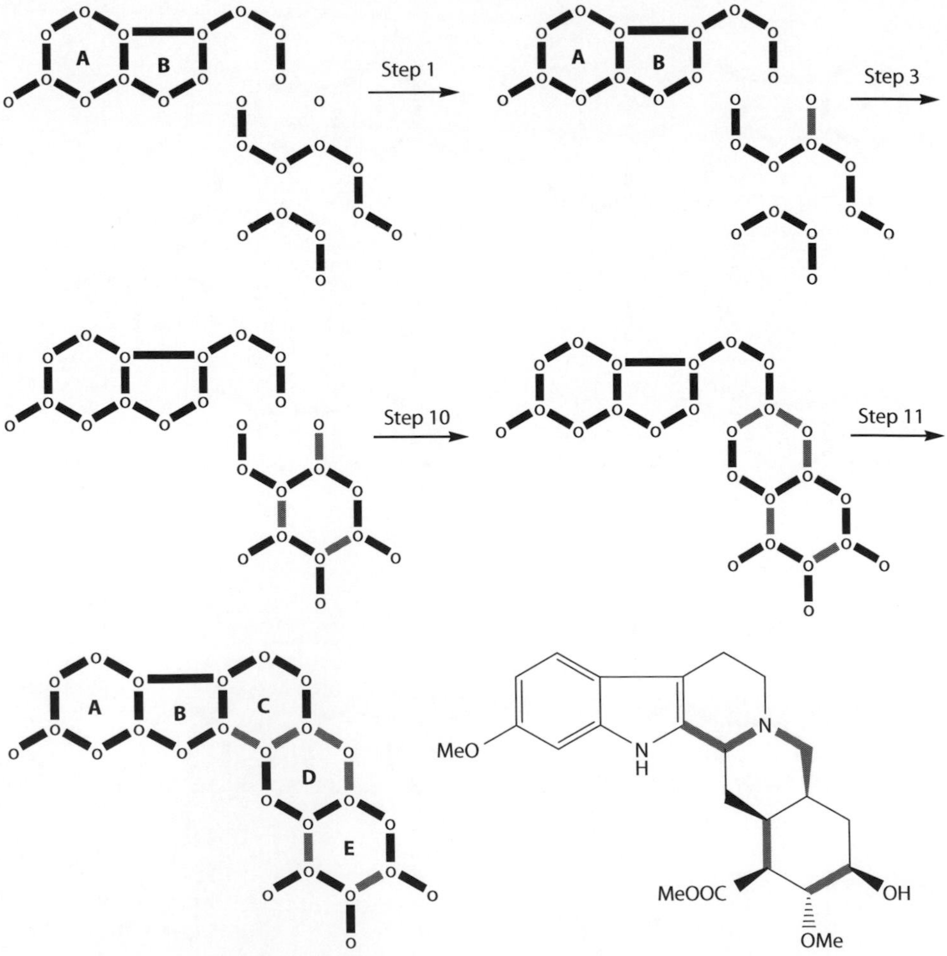

Sequence: $(A+B)_9 \xrightarrow{[4+2]} E_3 \xrightarrow{[5+1]} D_{10} \xrightarrow{[6+0]} C_{11}$

Reserpine

Wender, P.A.; Schaus, J.M.; White, A.M. *J. Am. Chem. Soc.* 1980, *102*, 6159.
Wender, P.A.; Schaus, J.M.; White, A.M. *Heterocycles* 1987, *25*, 263.

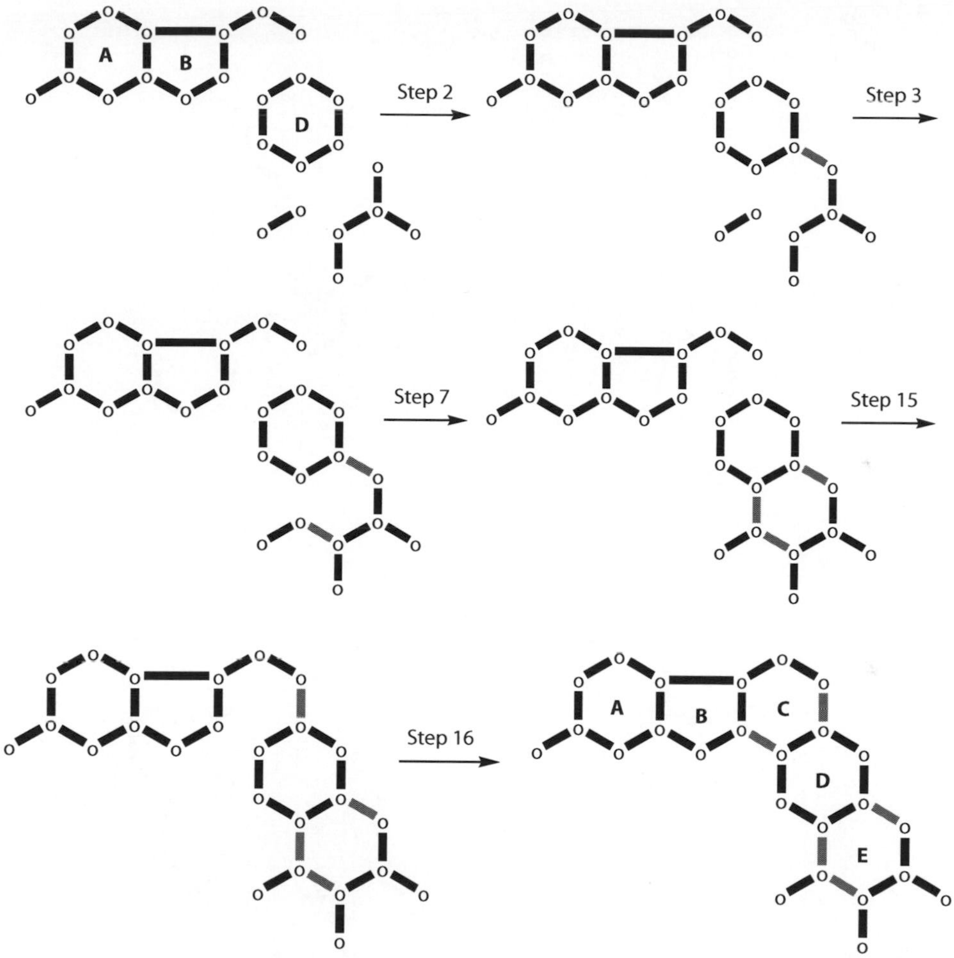

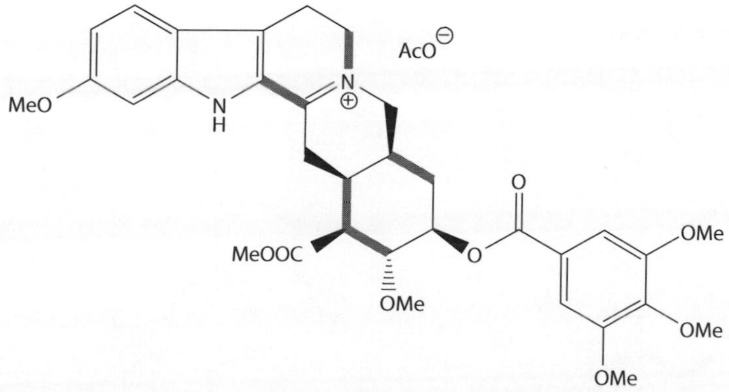

Sequence: $(A+B)_{14} + D_0 \xrightarrow{\;[6+0]\;} E_7 \xrightarrow{\;[6+0]\;} C_{16}$

Reserpine
Woodward, R.B.; Bader, F.E.; Bickel, H.; Frey, A.J.; Kierstead, R.W. *Tetrahedron* 1958, *2*, 1.

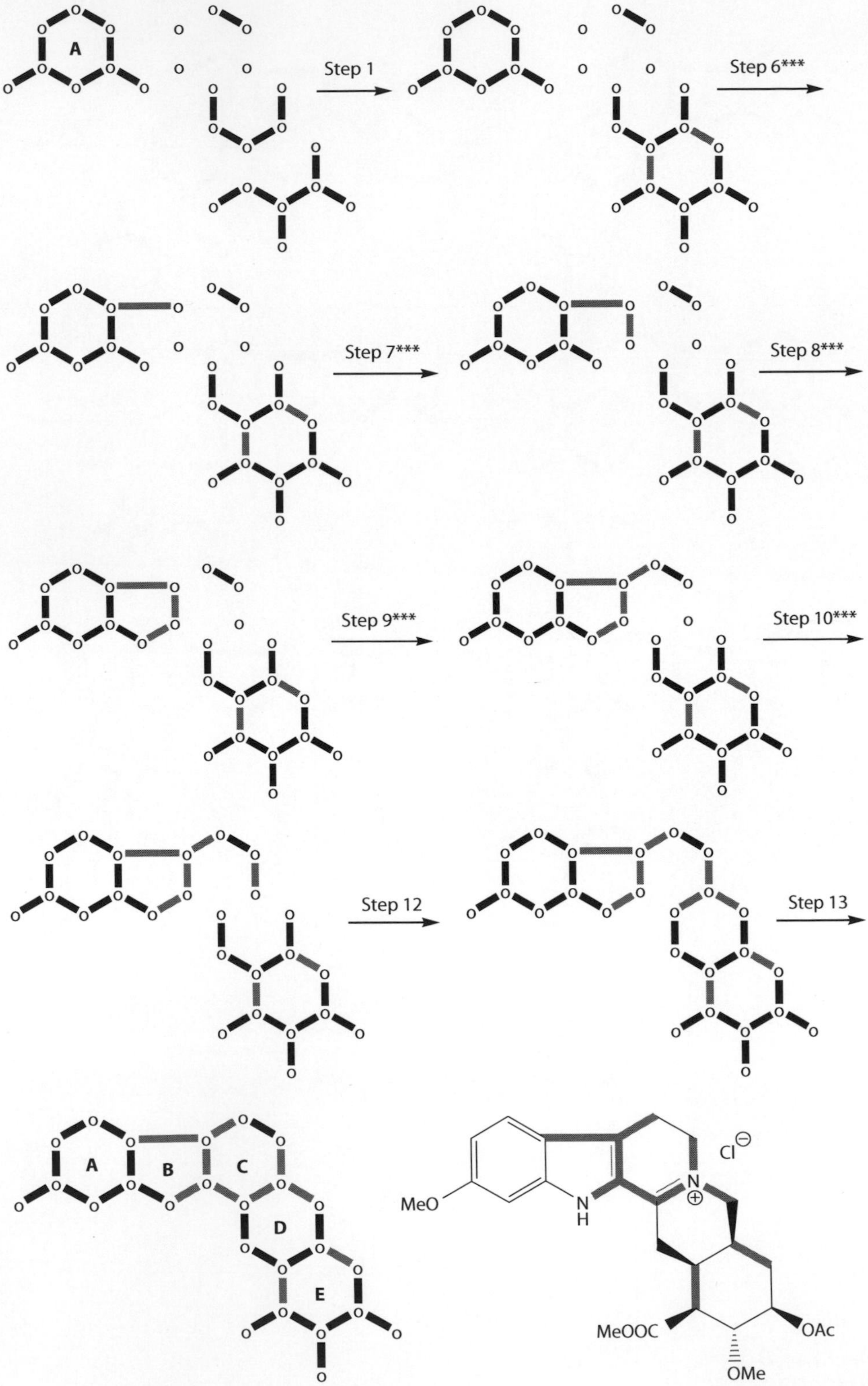

Sequence: $A_{5***} \xrightarrow{[4+2]} E_1 \xrightarrow{[5+0]} B_{8***} \xrightarrow{[5+1]} D_{12} \xrightarrow{[6+0]} C_{13}$

Summary:

$(A+B)_{15} + E_0 \xrightarrow{[(5+1)_C + (5+1)_D]} [4.4.0](C+D)_{16}$: Hanessian

$(A+B)_{11} + E_0 \xrightarrow{[5+1]} D_{12} \xrightarrow{[6+0]} C_{13}$: Pearlman

$(A+B)_{14} + D_0 \xrightarrow{[6+0]} E_7 \xrightarrow{[6+0]} C_{16}$: Wender

$(A+B)_{27} \xrightarrow{[6+0]} E_7 \xrightarrow{[6+0]} D_{29} \xrightarrow{[6+0]} C_{30}$: Fraser-Reid

$(A+B)_8 \xrightarrow{[4+2]} E_1 \xrightarrow{[5+1]} D_9 \xrightarrow{[6+0]} C_{10}$: Liao

$(A+B)_9 \xrightarrow{[4+2]} E_3 \xrightarrow{[5+1]} D_{10} \xrightarrow{[6+10]} C_{11}$: Stork

$(A+B)_{19} \xrightarrow{[4+2]} E_3 \xrightarrow{[5+1]} D_{20} \xrightarrow{[6+0]} C_{21}$: Mehta

$(A+B)_{18} \xrightarrow{[(6+0)_D + (4+2)_E]} [4.4.0](D+E)_7 \xrightarrow{[6+0]} C_{20}$: Martin

$(A+B)_{18} \xrightarrow{[(6+0)_D + (4+2)_E]} [4.4.0](D+E)_8 \xrightarrow{[6+0]} C_{20}$: Shea

$A_{5***} \xrightarrow{[4+2]} E_1 \xrightarrow{[5+0]} B_{8***} \xrightarrow{[5+1]} D_{12} \xrightarrow{[6+0]} C_{13}$: Woodward

Resveratrol

Moreno-Manas, M.; Pleixats, R. *An. Quim. Ser. C* 1985, *81*, 157.

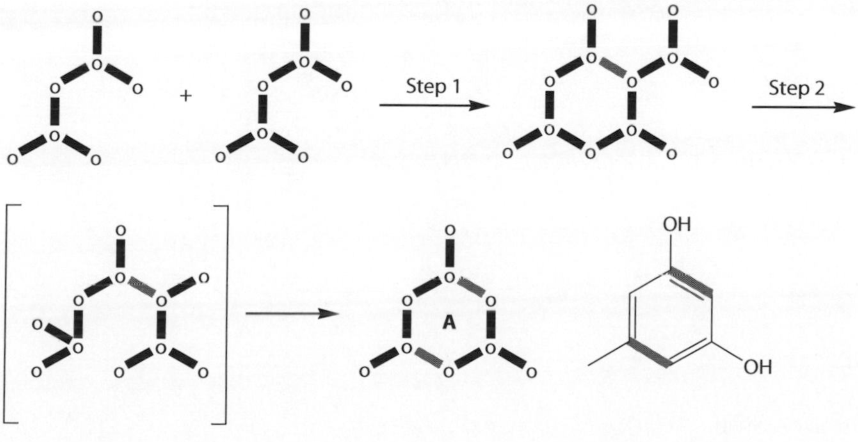

Sequence: $\xrightarrow{[6+0]} A_2$

Ricinine
Schroeter, G.; Seidler, C.; Sulzbacher, M.; Kanitz, R. *Chem. Ber.* 1932, *65*, 432.

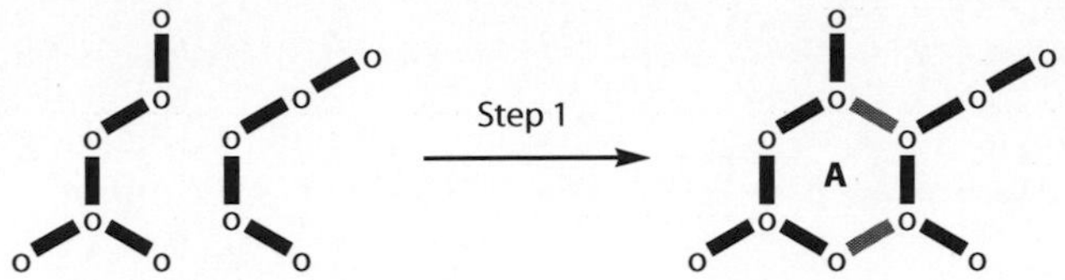

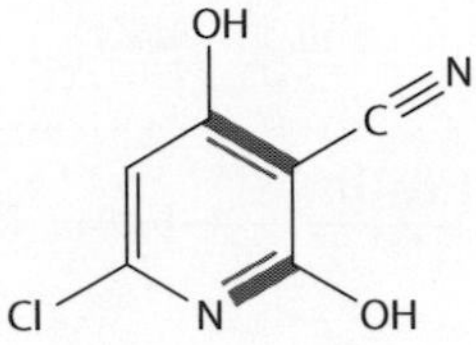

Sequence: $\xrightarrow{[4+2]} A_1$

Späth, E.; Koller, G. *Chem. Ber.* 1923, *56*, 2454.

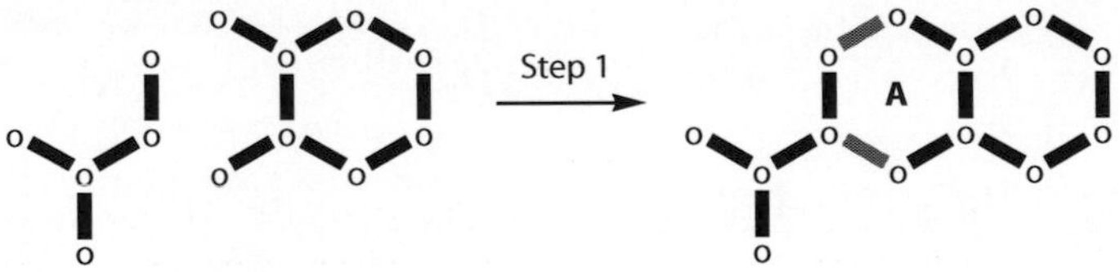

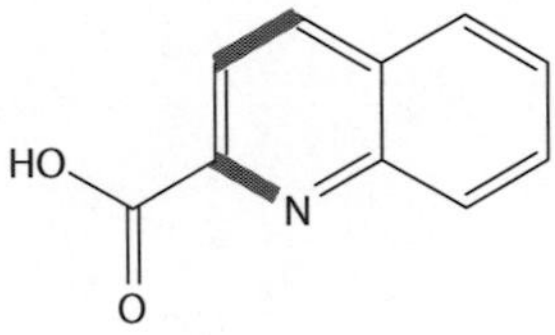

Sequence: $\xrightarrow{[4+2]} A_1$

Sugasawa, T.; Sasakura, K.; Toyoda, T. *Chem. Pharm. Bull.* 1974, *22*, 763.

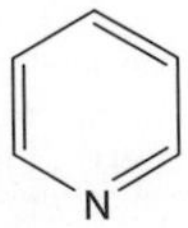

Taylor, E.C. Jr.; Crovetti, A.J. *J. Am. Chem. Soc.* 1956, *78*, 214.

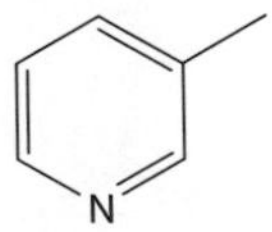

Summary:
Strategy #1:
[4 + 2]: Schroeter, Spath

Strategy #2:
Preformed ring precursors: Sugasawa, Taylor

Rosefuran

Barco, A.; Benetti, S.; De Risi, C.; Pollini, G.P.; Zanirato, V. *Tetrahedron* 1995, *51*, 7721.

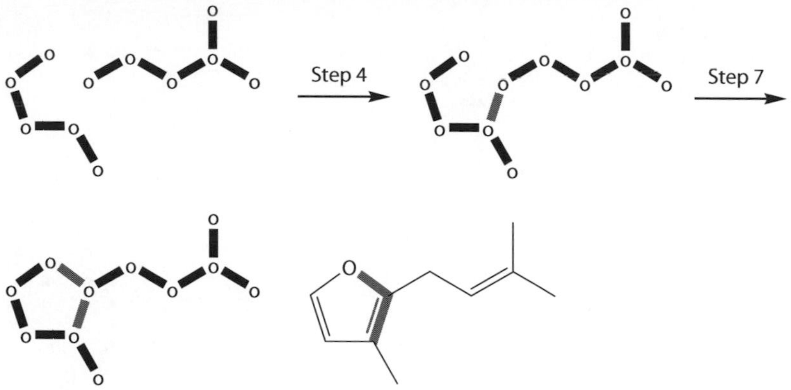

Sequence: $\xrightarrow{[5+0]}$ $[5]_7$

Birch, A.J.; Slobbe, J. *Tetrahedron Lett.* 1976, 2079.
Büchi, G.; Kovats, E.; Enggist, P.; Uhde, G. *J. Org. Chem.* 1968, *33*, 1227.

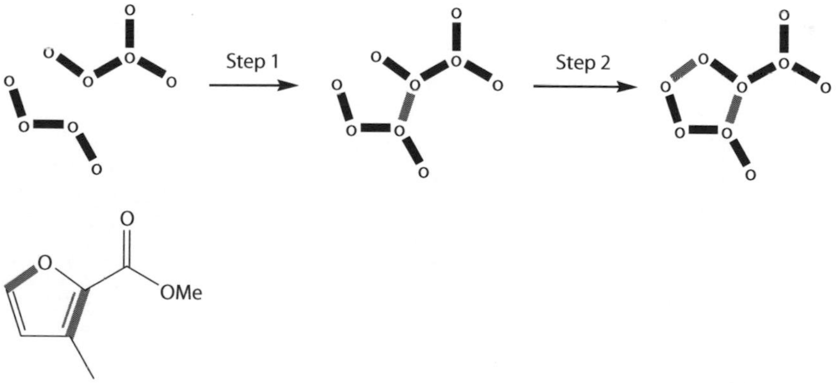

Sequence: $\xrightarrow{[5+0]}$ $[5]_2$

Araki, S.; Jin, S.J.; Butsugan, Y. *J. Chem. Soc. Perkin Trans. I* 1995, 549.

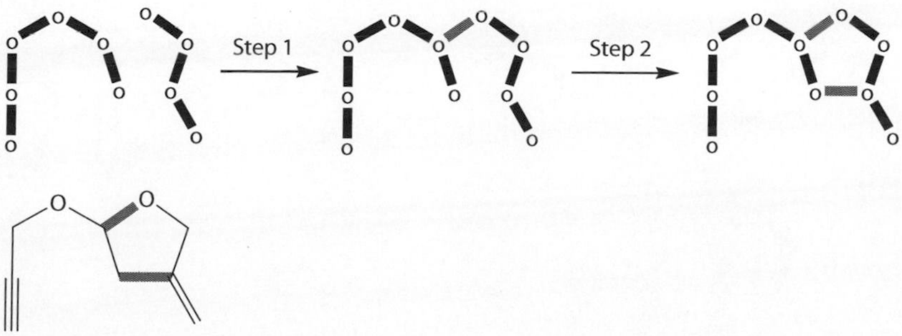

Sequence: $\xrightarrow{[5+0]}$ $[5]_2$

Iriye, R.; Uno, T.; Ohwa, I.; Konishi, A. *Agric. Biol. Chem.* 1990, *54*, 1841.
Iriye, R.; Yorifuji, T.; Takeda, N.; Tatematsu, A. *Agric. Biol. Chem.* 1984, *48*, 2923.

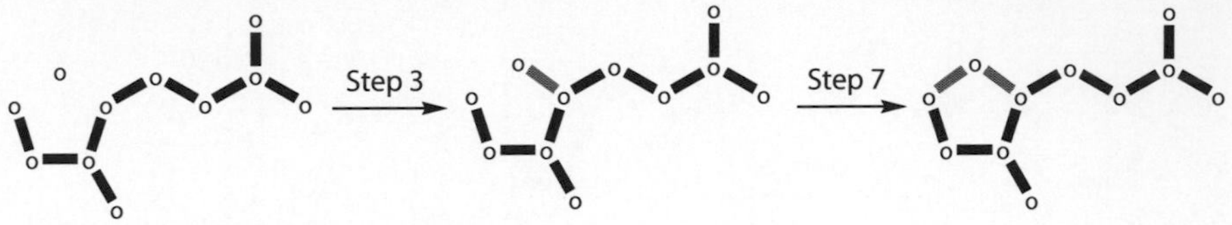

or

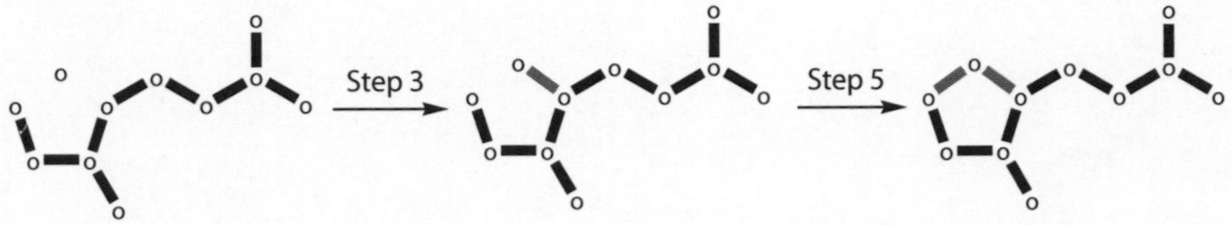

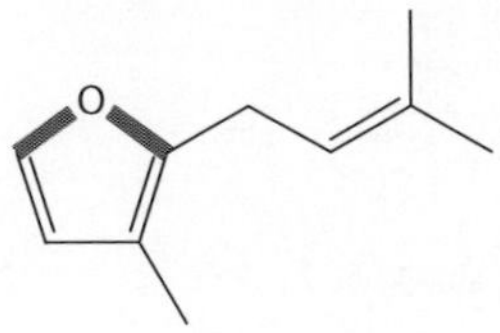

Sequence: $\xrightarrow{[5+0]}$ $[5]_7$ or $\xrightarrow{[5+0]}$ $[5]_5$

Marshall, J.A.; DuBay, W.J. *J. Org. Chem.* 1993, *58*, 3602.

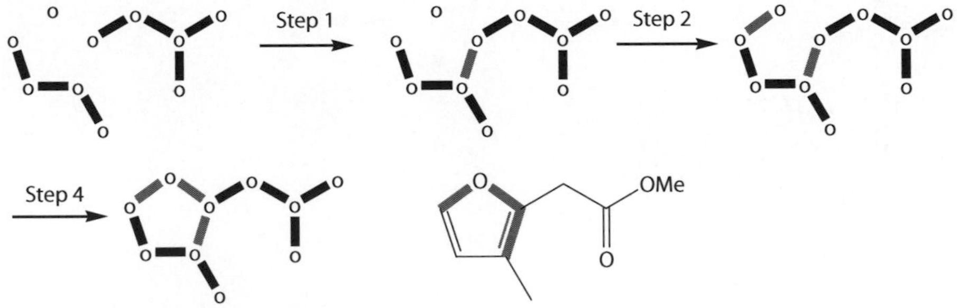

Sequence: $\xrightarrow{[5+0]}$ $[5]_4$

Okazaki, R.; Negishi, Y.; Inamoto, N. *J. Org. Chem.* 1984, *49*, 3819.

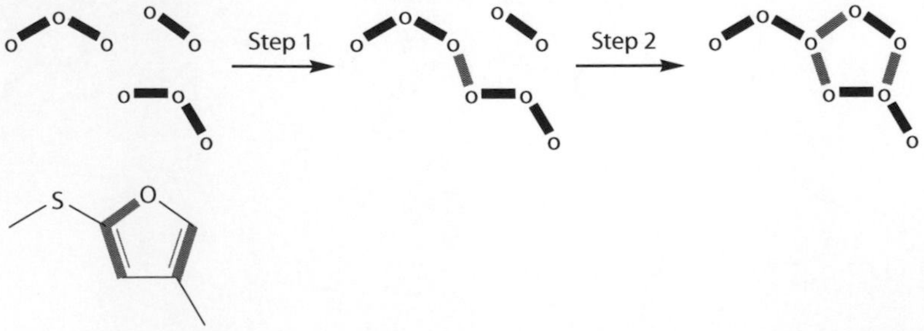

Sequence: $\xrightarrow{[3+2]}$ $[5]_2$

Gedge, D.R.; Pattenden, G. *Tetrahedron Lett.* 1977, 4443.

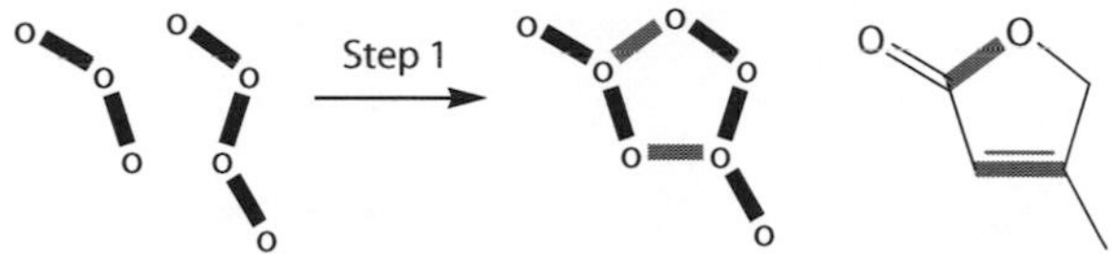

Step 1

Sequence: $\xrightarrow{[3+2]}$ $[5]_1$

Gabriele, B.; Salerno, G. *Chem. Commun.* 1997, 1083.

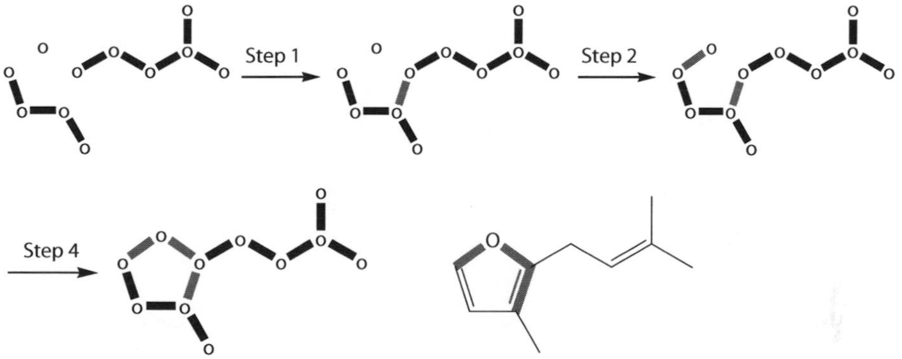

Step 1

Step 2

Step 4

Sequence: $\xrightarrow{[5+0]}$ $[5]_4$

Meier, L.; Scharf, H.D. *Liebigs Ann. Chem.* 1986, 731.

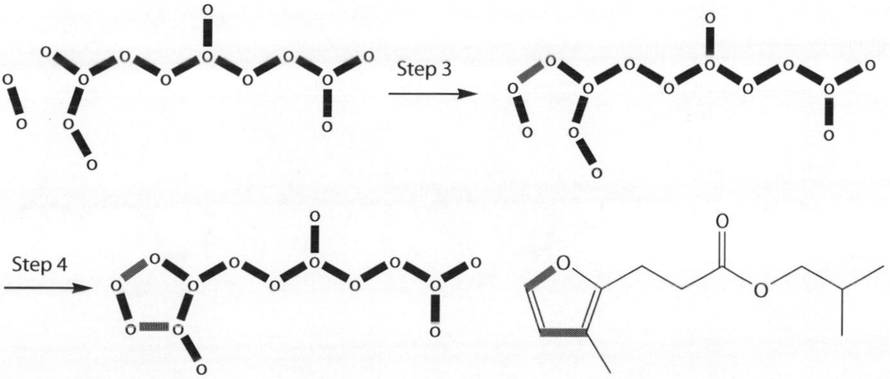

Step 3

Step 4

Sequence: $\xrightarrow{[5+0]}$ $[5]_4$

Takano, S.; Morimoto, M.; Satoh, S.; Ogasawara, K. *Chem. Lett.* 1984, 1261.

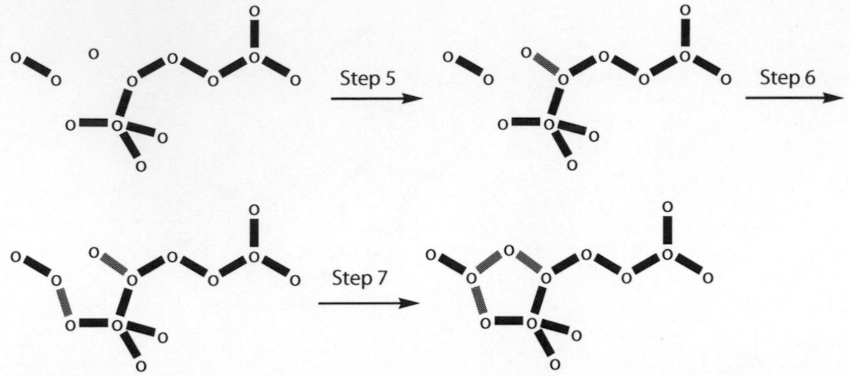

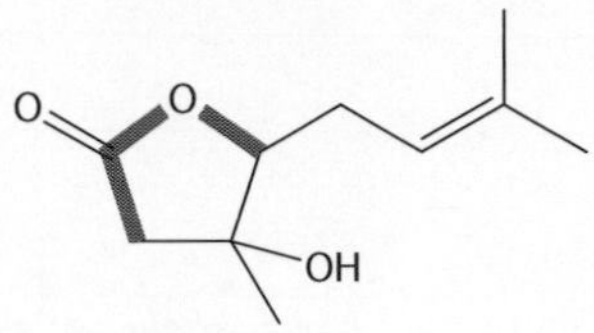

Sequence: $\xrightarrow{[5+0]}$ $[5]_7$

Takeda, A.; Shinhama, K.; Tsuboi, S. *Bull. Chem. Soc. Jpn.* 1977, *50*, 1903.

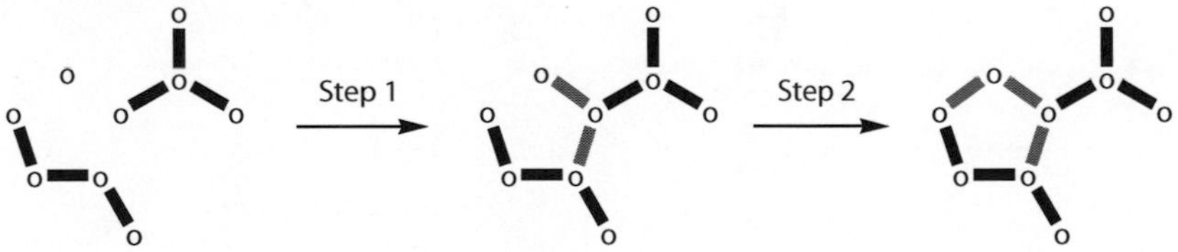

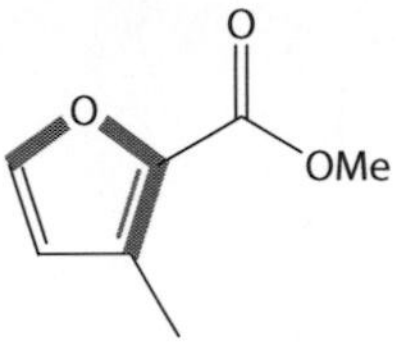

Sequence: $\xrightarrow{[5+0]}$ $[5]_2$

Trost, B.M.; Flygare, J.A. *J. Org. Chem.* 1994, *59*, 1078.

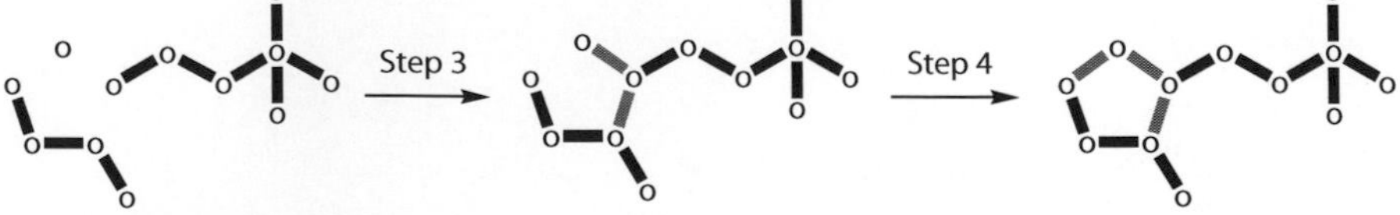

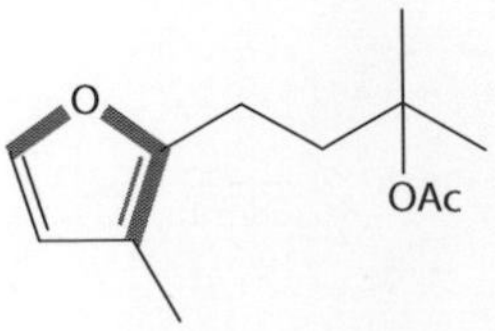

Sequence: $\xrightarrow{[5+0]}$ $[5]_4$

Tsukasa, H. *Agric. Biol. Chem.* 1989, *53*, 3091.

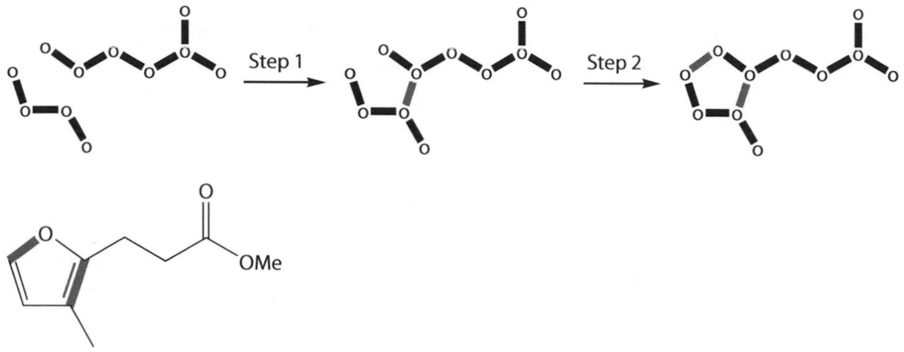

Sequence: $\xrightarrow{[5+0]}$ $[5]_2$

Vig, O.P.; Vig, A.K.; Handa, V.K.; Sharma, S.D. *Ind. J. Chem.* 1974, *51*, 900.

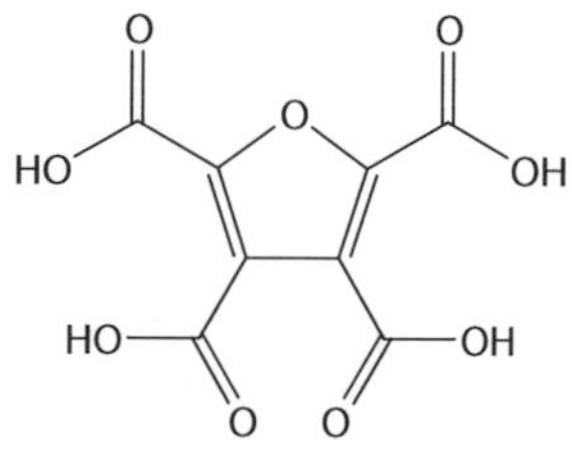

Wenkert, E.; Marsaioli, A.J.; Moeller, P.D.R. *J. Chromatography* 1988, *440*, 449.

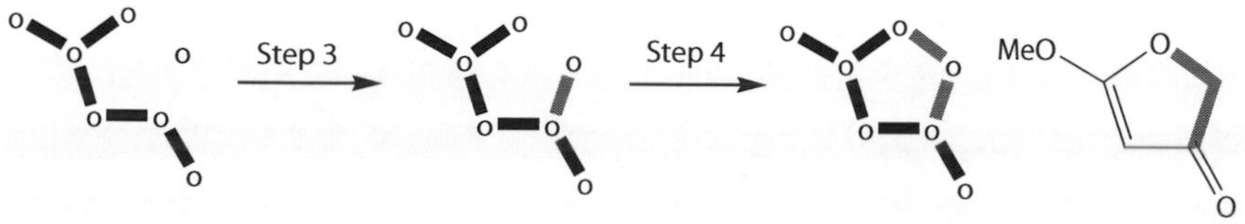

Sequence: $\xrightarrow{[5+0]}$ $[5]_4$

Wong, M.K.; Leung, C.Y.; Wong, H.N.C. *Tetrahedron* 1997, *53*, 3497.

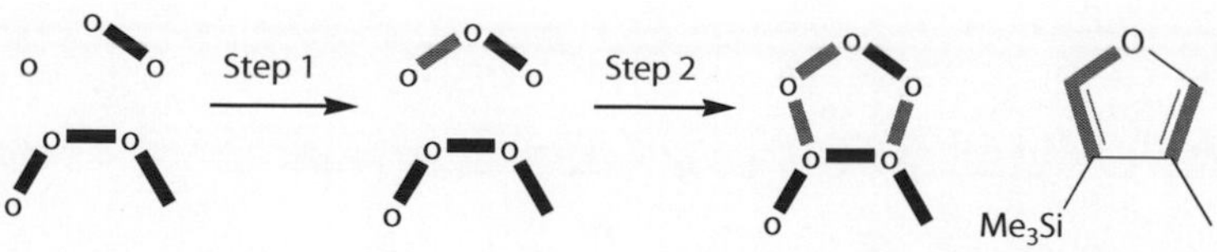

Sequence: $\xrightarrow{[3+2]}$ $[5]_2$

Summary:

I II III IV V

Strategy [5 + 0] type I: Barco, Wenkert
Strategy [5 + 0] type II: Birch, Tsukasa
Strategy [5 + 0] type III: Botsugan, Scharf
Strategy [5 + 0] type IV: Iriye G1, Iriye G2
Strategy [5 + 0] type V: Marshall, Salerno, Takano, Takeda, Trost
Strategy [3 + 2]: Okazaki, Pattenden, Wong

Precursor containing ring: Vig

(±)-Salsolene oxide
Wender, P.A.; Croatt, M.P.; Witulski, B. *Tetrahedron* 2006, *62*, 7505.

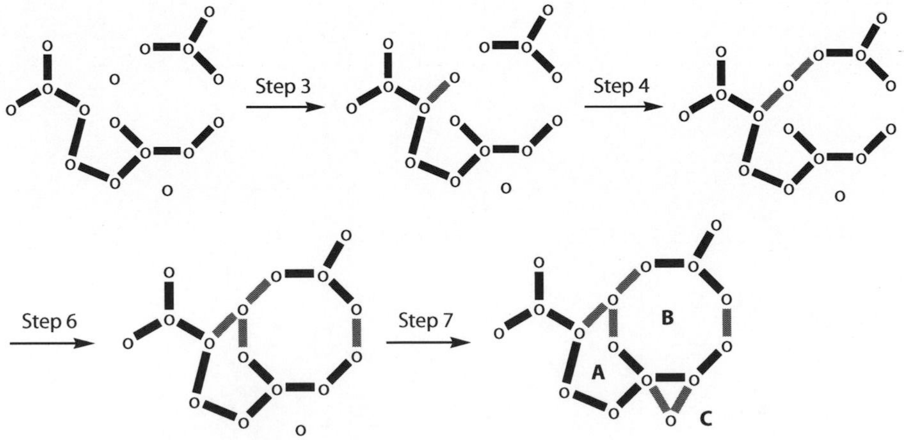

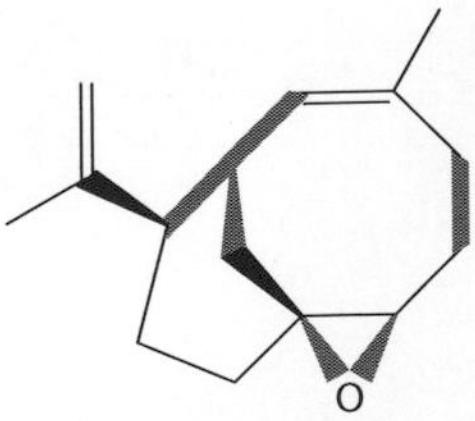

Sequence: $\xrightarrow{[(6+0) + (4+4)]}$ $[5.3.1](A+B)_6$ $\xrightarrow{[2+1]}$ $[6.1.0]C_7$

(−)-Salsolene oxide

Paquette, L.A.; Sun, L.Q.; Watson, T.J.N.; Friedrich, D.; Freeman, B.T. *J. Am. Chem. Soc.* 1997, *119*, 2767.

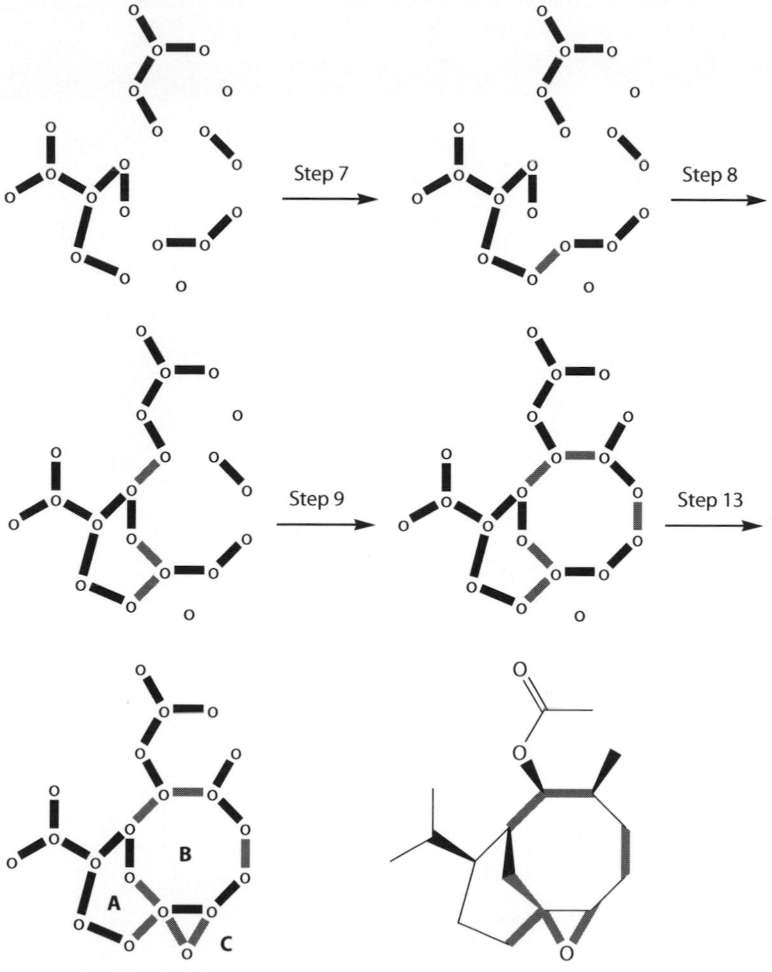

Sequence: $\xrightarrow{[6+0]} A_8 \xrightarrow{[6+2]} [5.3.1]B_9 \xrightarrow{[2+1]} [6.1.0]C_{13}$

Serotonin

Harley-Mason, J.; Jackson, A.H. *J. Chem. Soc.* 1954, 1165.

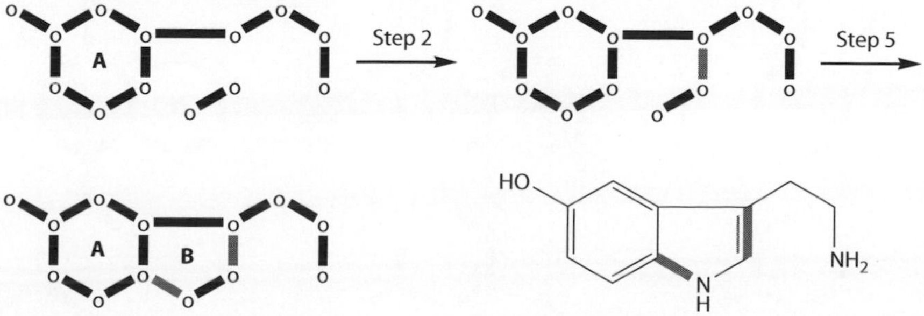

Sequence: $\xrightarrow{[5+0]} B_5$

(±)-Sertraline

Yun, J.; Buchwald, S.L. *J. Org. Chem.* 2000, *65*, 767.
Quallich, G.J.; Williams, M.T.; Friedmann, R.C. *J. Org. Chem.* 1999, *55*, 4971.
Taber, G.P.; Pfisterer, D.M.; Colberg, J.C. *Org. Process Res. Develop.* 2004, *8*, 385.

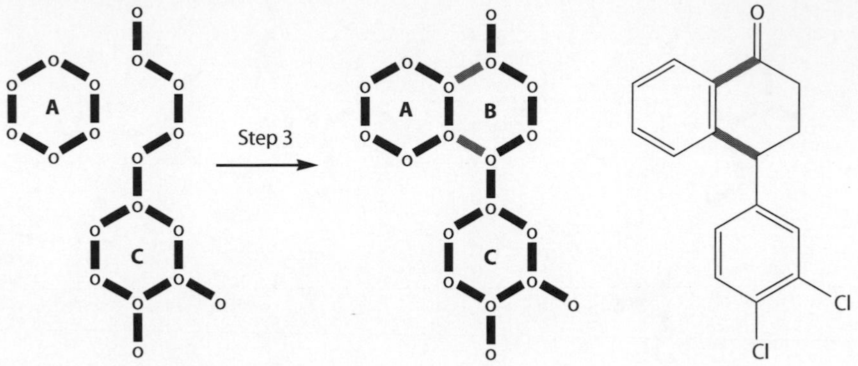

Sequence: $A_0 + C_0 \xrightarrow{[4+2]} B_3$

Vukics, K.; Fodor, T.; Fischer, J.; Fellegvári, I.; Lévai, S. *Org. Process Res. Develop.* 2002, *6*, 82.
Wang, G.; Zheng, C.; Zhao, G. *Tetrahedron Asymm.* 2006, *17*, 2074.

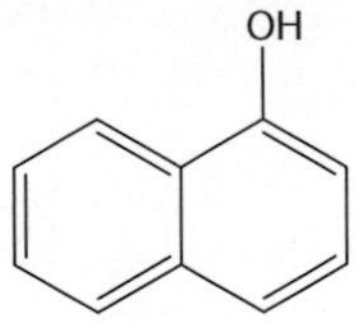

Lautens, M.; Rovis, T. *Tetrahedron* 1999, *55*, 8967.

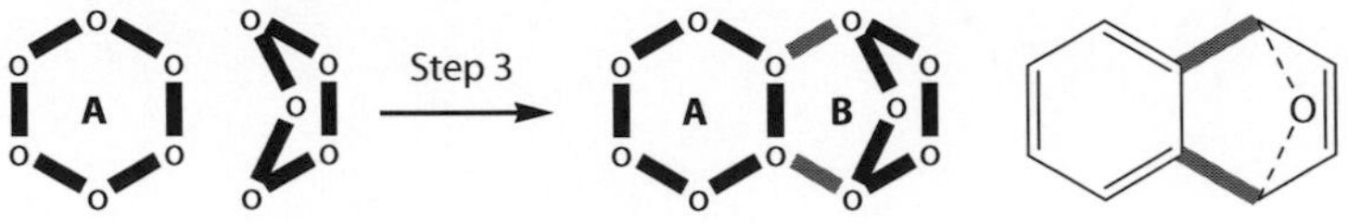

Sequence: $A_0 \xrightarrow{[4+2]} B_3$

Welch, W.M.; Kraska, A.R.; Sarges, R.; Koe, B.K. *J. Med. Chem.* 1984, *27*, 1508.

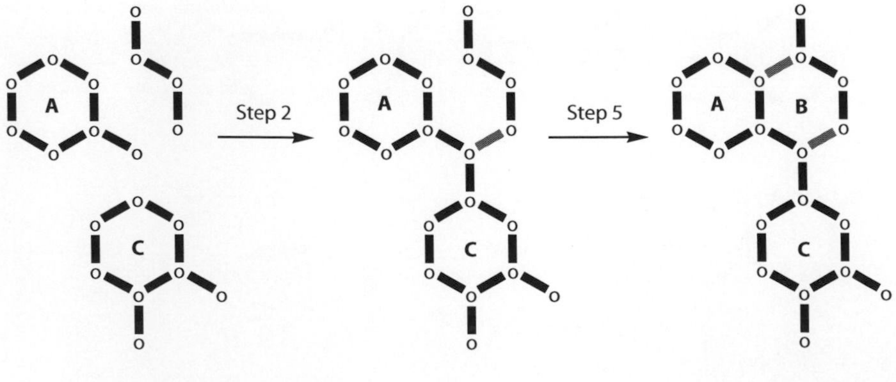

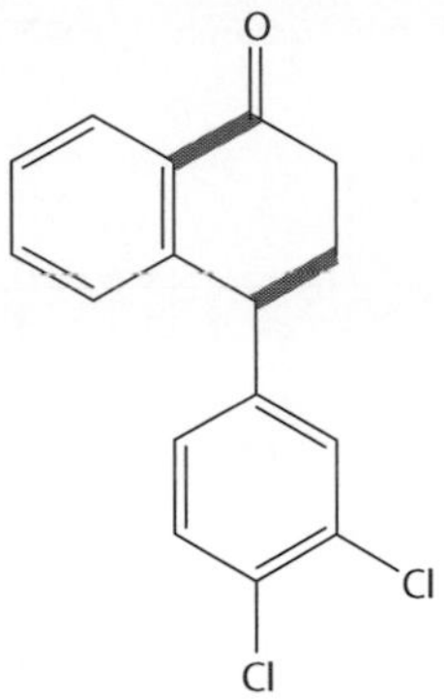

Sequence: $A_0 + C_0 \xrightarrow{[6+0]} B_5$

Summary:
Strategy #1

$$A_0 + C_0 \xrightarrow{[4+2]} B_3: \text{ Buchwald, Pfizer G2, Pfizer G3}$$

Strategy #2

$$A_0 \xrightarrow{[4+2]} B_3: \text{ Lautens}$$

Strategy #3

$$A_0 + C_0 \xrightarrow{[6+0]} B_5: \text{ Pfizer G1}$$

Strategy #4
Precursor containing ring: Gedeon-Richter, Zhao

Silphinene
Crimmins, M.T.; Mascarella, S.W. *J. Am. Chem. Soc.* 1986, *108*, 3435.

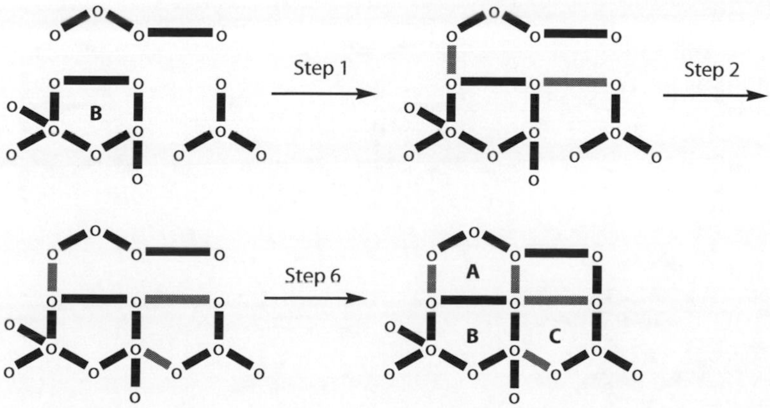

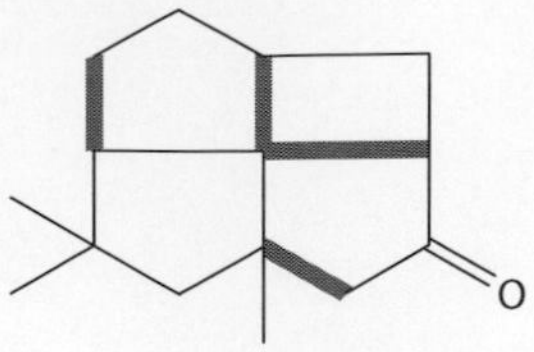

Sequence: $B_0 \xrightarrow{[5+0]} C_2 \xrightarrow{[5+0]} A_6$

Franck-Neumann, M.; Miesch, M.; Gross, L. *Tetrahedron Lett.* 1991, *32*, 2135.
Miesch, M.; Miesch, L.; Franck-Neumann, M. *Tetrahedron* 1997, *53*, 2103.

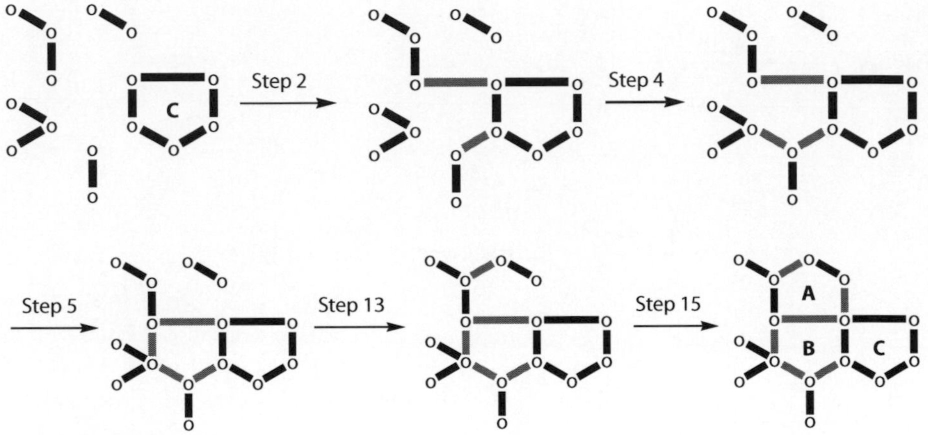

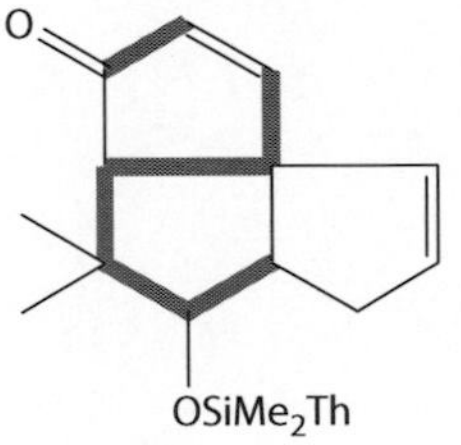

Sequence: $C_0 \xrightarrow{[5+0]} B_5 \xrightarrow{[5+0]} A_{15}$

Tsunoda, T.; Komdama, M.; Ito, S. *Tetrahedron Lett.* 1983, *24*, 83.

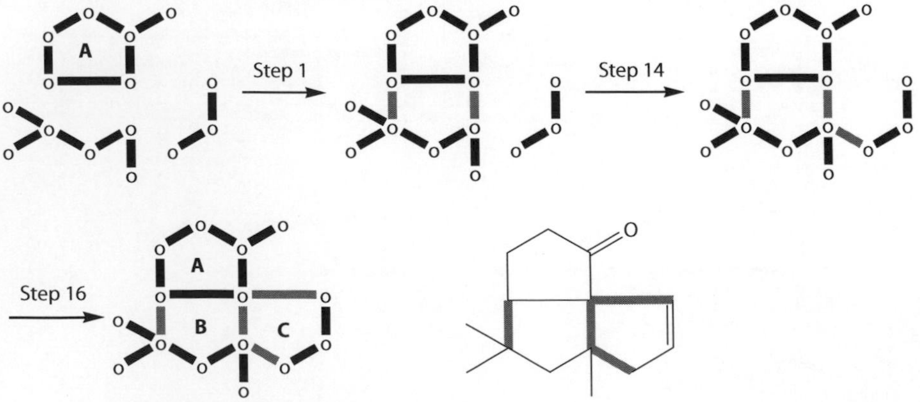

Sequence: $A_0 \xrightarrow{[3+2]} B_1 \xrightarrow{[5+0]} C_{16}$

Koteswar, R.Y.; Nagarajan, M. *Tetrahedron Lett*. 1988, *29*, 107.

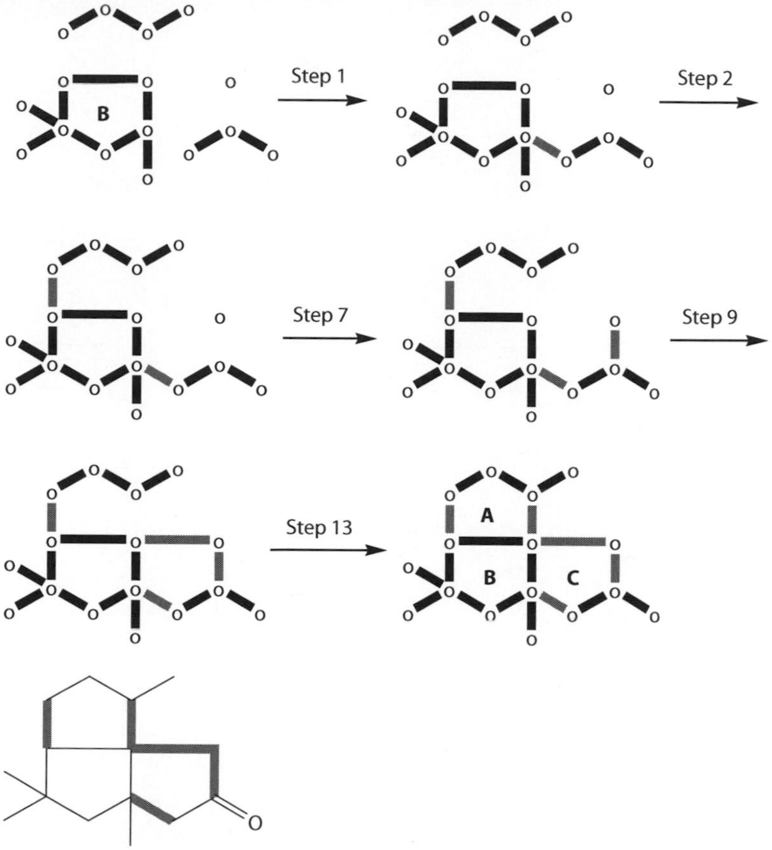

Sequence: $B_0 \xrightarrow{[5+0]} C_9 \xrightarrow{[5+0]} A_{13}$

Paquette, L.A.; Leone-Bay, A. *J. Am. Chem. Soc*. 1983, *105*, 7352.

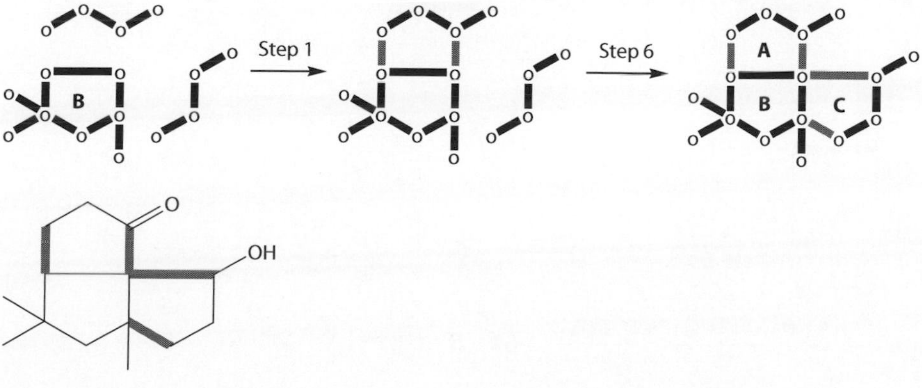

Sequence: $B_0 \xrightarrow{[3+2]} A_1 \xrightarrow{[3+2]} C_6$

Sternbach, D.D.; Hughes, J.W.; Burdi, D.F.; Banks, B.A. *J. Am. Chem. Soc.* 1985, *107*, 2149.

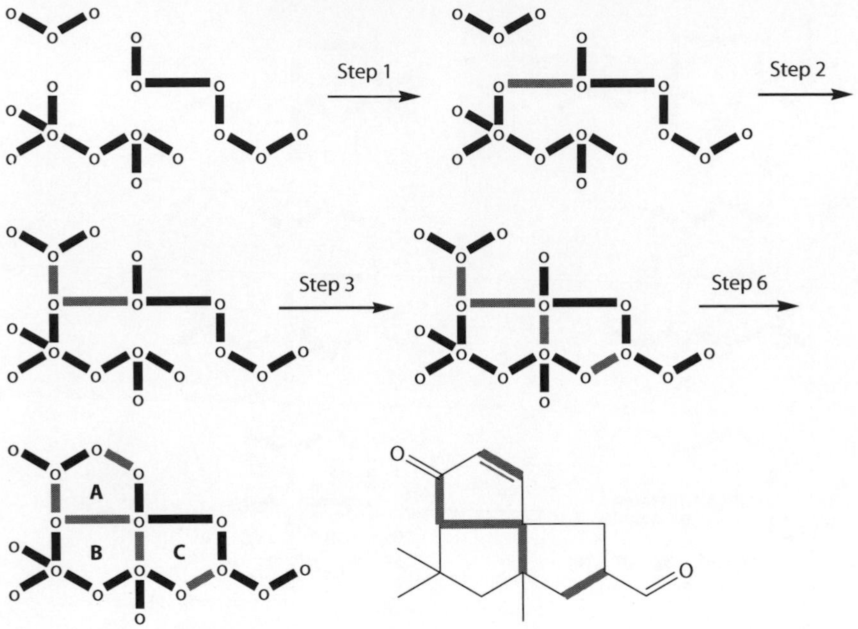

Sequence: $\xrightarrow{[(5+0)_B + (3+2)_C]} [3.3.0] (B+C)_3 \xrightarrow{[5+0]} A_6$

Wender, P.A.; Ternansky, R.J. *Tetrahedron Lett.* 1985, *26*, 2625.

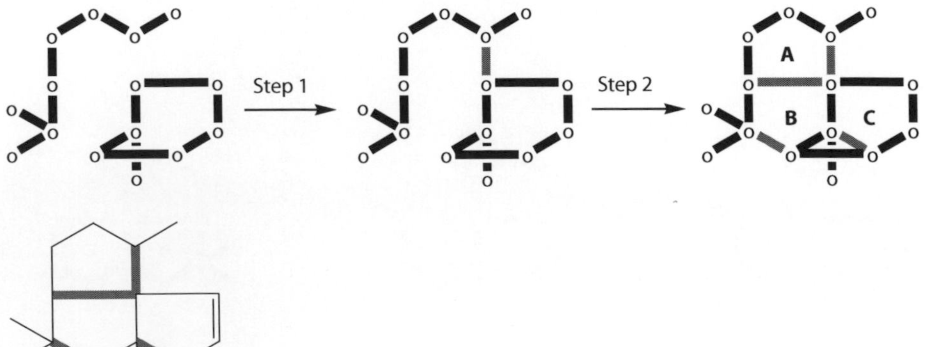

Sequence: $\xrightarrow{[(5+0)_A + (3+2)_B + (5+0)_C]} (A+B+C)_2$

Yamamura, S.; Ohkubo, M.; Shizuri, Y. *Tetrahedron Lett.* 1989, *30*, 3798.

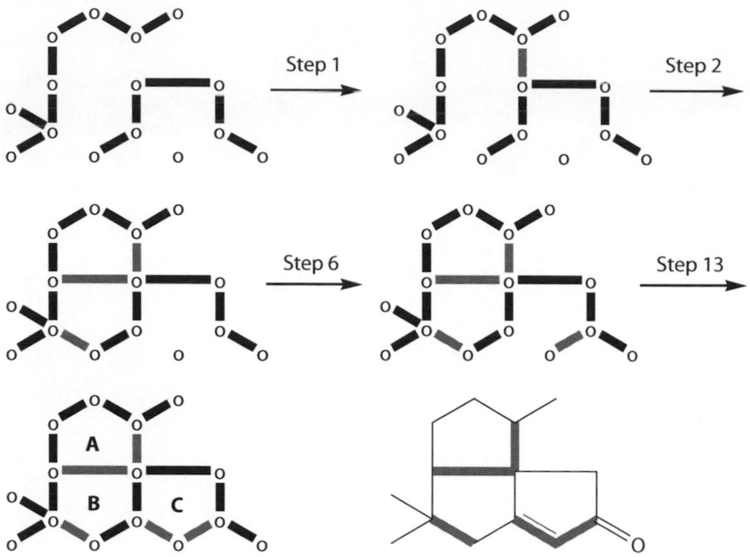

Sequence: $\xrightarrow{[(5+0)_A+(3+2)_B]}$ $[3.3.0]\,(A+B)_2$ $\xrightarrow{[5+0]}$ C_{13}

Summary:

$\xrightarrow{[(5+0)_A+(3+2)_B]}$ $[3.3.0]\,(A+B)_2$ $\xrightarrow{[5+0]}$ C_{13}: Yamamura

$\xrightarrow{[(5+0)_B+(3+2)_C]}$ $[3.3.0]\,(B+C)_3$ $\xrightarrow{[5+0]}$ A_6: Sternbach

$\xrightarrow{[(5+0)_A+(3+2)_B+(5+0)_C]}$ $(A+B+C)_2$: Wender

B_0 $\xrightarrow{[5+0]}$ C_9 $\xrightarrow{[5+0]}$ A_{13}: Nagarajan

B_0 $\xrightarrow{[5+0]}$ C_2 $\xrightarrow{[5+0]}$ A_6: Crimmins

C_0 $\xrightarrow{[5+0]}$ B_5 $\xrightarrow{[5+0]}$ A_{15}: Franck-Neumann

A_0 $\xrightarrow{[3+2]}$ B_1 $\xrightarrow{[5+0]}$ C_{16}: Ito

B_0 $\xrightarrow{[3+2]}$ A_1 $\xrightarrow{[3+2]}$ C_6: Paquette

(±)-Sparteine
Anet, E.F.L.J.; Hughes, G.K.; Ritchie, E. *Austr. J. Chem.* 1950, *3*, 635.

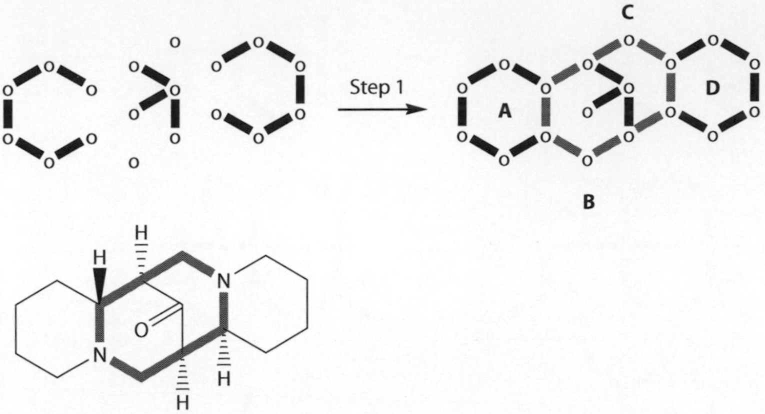

Sequence: $\dfrac{[(6+0)_A + (3+2+1)_B + (3+2+1)_C + (6+0)_D]}{} \longrightarrow (A + B + C + D)_1$

(±)-Sparteine
Smith, B.T.; Wendt, J.A.; Aubé, J. *Org. Lett.* 2002, *4*, 2577.

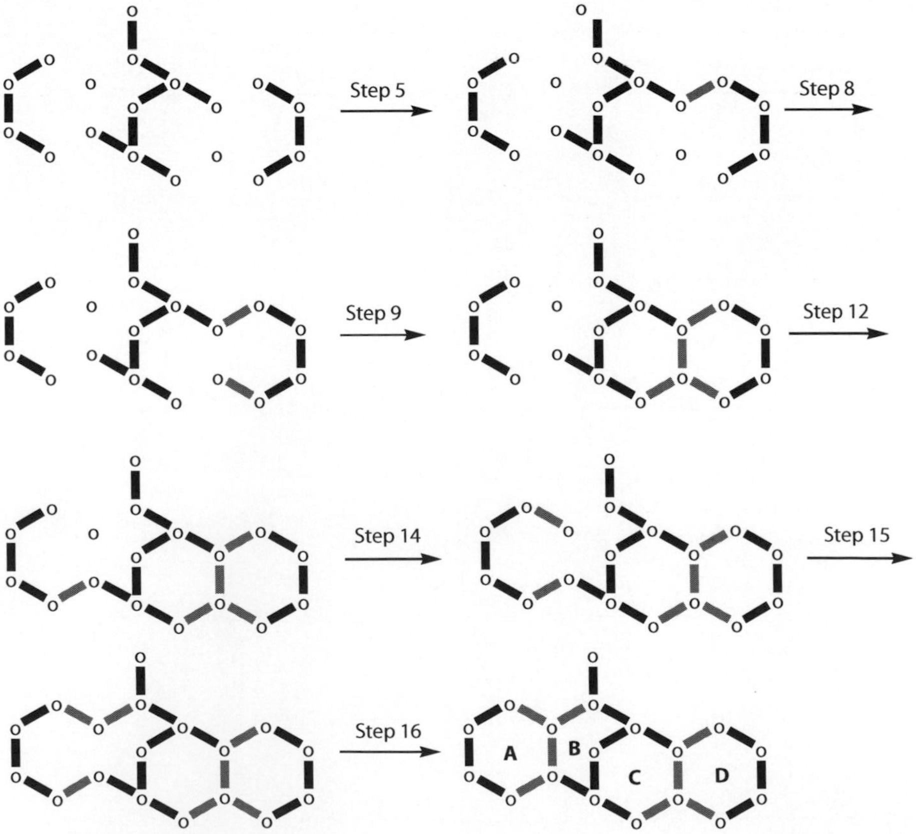

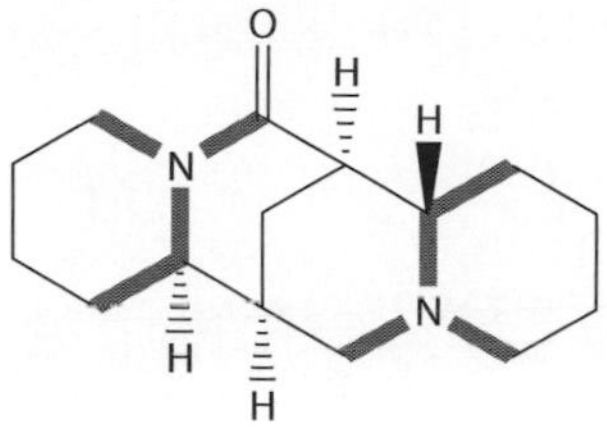

Sequence: $\xrightarrow{[(5+1)_C+(6+0)_D]}$ $[4.4.0]\,(C+D)_9$ $\xrightarrow{[(6+0)_A+(6+0)_B]}$ $[4.4.0]\,(A+B)_{16}$

(±)-Sparteine

Norcross, N.R.; Melbardis, J.P.; Solera, M.F.; Sephton, M.A.; Kilner, C.; Zakharov, L.N.; Astles, P.C.; Warriner, S.L.; Blakemore, P.R. *J. Org. Chem.* 2008, *73*, 7939.

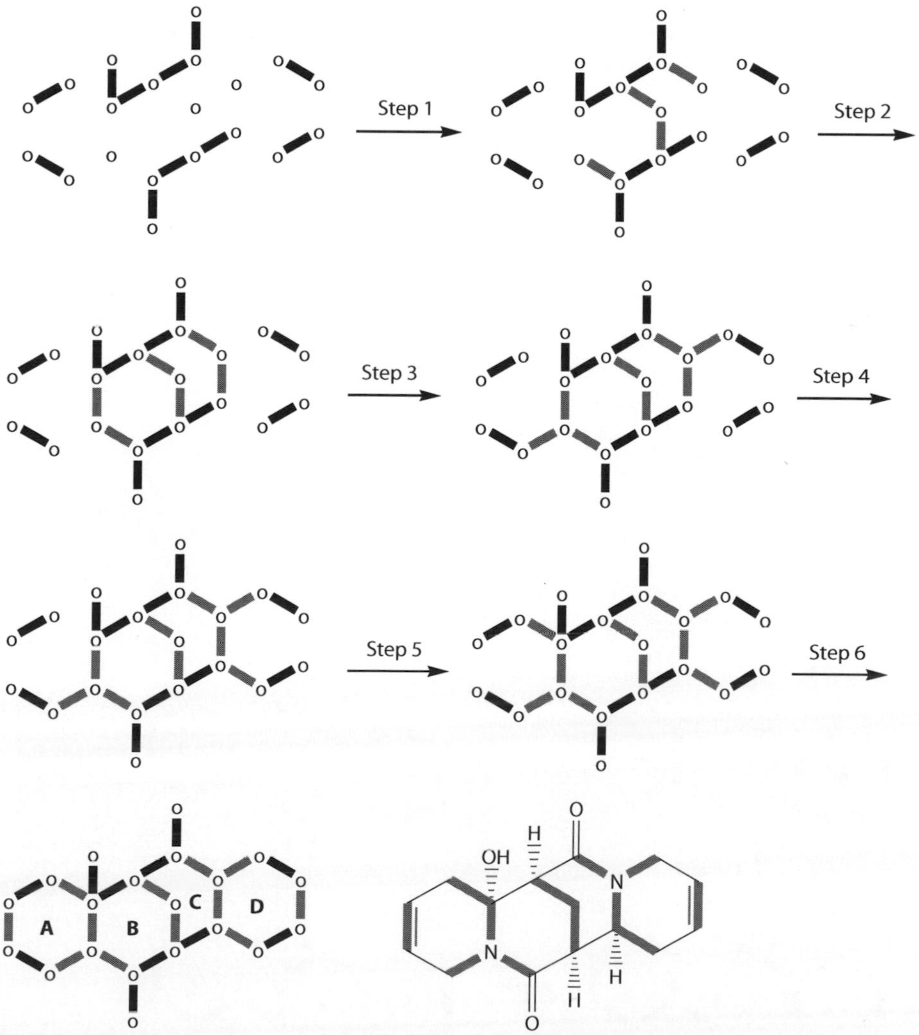

Sequence: $\xrightarrow{[(6+0)_B+(6+0)_C]}$ $[3.3.1]\,(B+C)_2$ $\xrightarrow{[(6+0)_A+(6+0)_D]}$ $(A+D)_6$

(±)-Sparteine

Bohlmann, F.; Müller, H.J.; Schumann, D. *Chem. Ber.* 1973, *106*, 3026.
Gray, D.; Gallagher, T. *Angew. Chem. Int. Ed.* 2006, *45*, 2419.

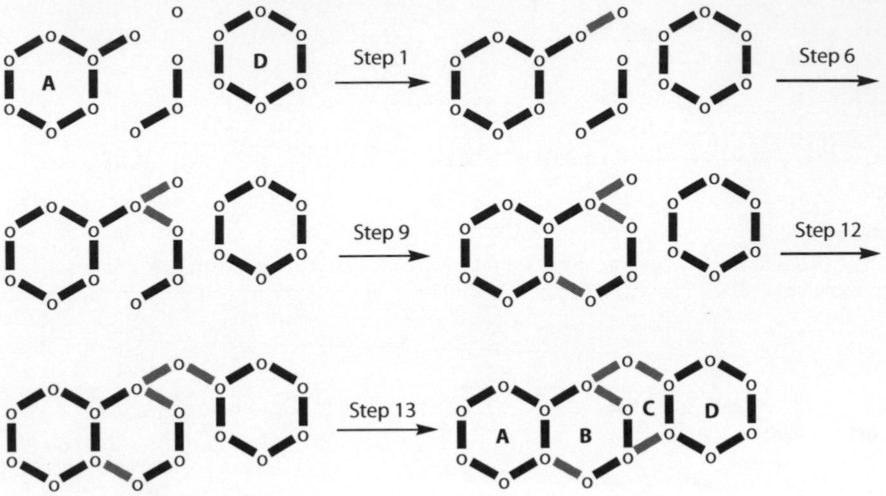

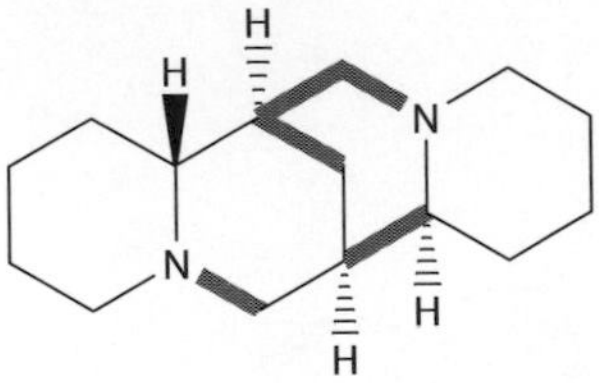

Sequence: $A_0 + D_{11} \xrightarrow{\ [6+0]\ } B_9 \xrightarrow{\ [6+0]\ } C_{13}$

(±)-Sparteine

Clemo, G.R.; Raper, R.; Short, W.S. *J. Chem. Soc.* 1949, 663.
Clemo, G.R.; Morgan, W.M.; Raper, R. *J. Chem. Soc.* 1936, 1025.

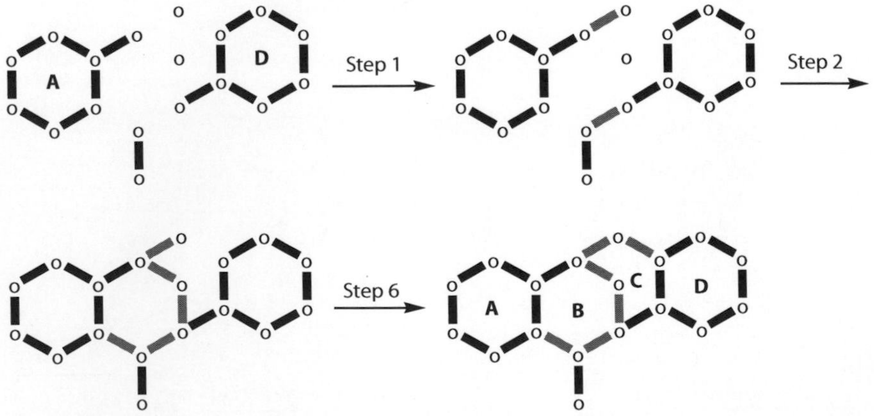

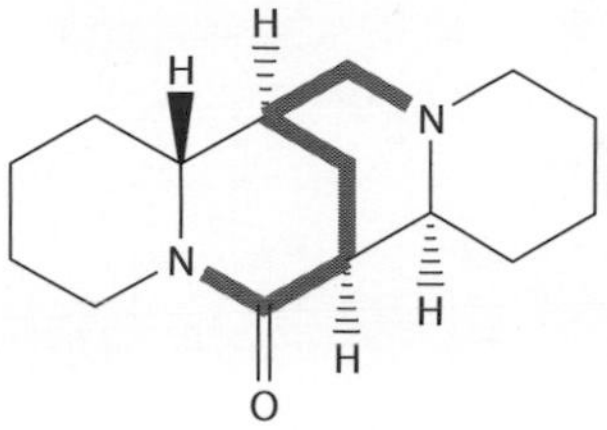

Sequence: $(A+D)_0 \xrightarrow{[3+2+1]} B_2 \xrightarrow{[6+0]} C_6$

(±)-Sparteine
Buttler, T.; Fleming, I. *Chem. Commun.* 2004, 2404.

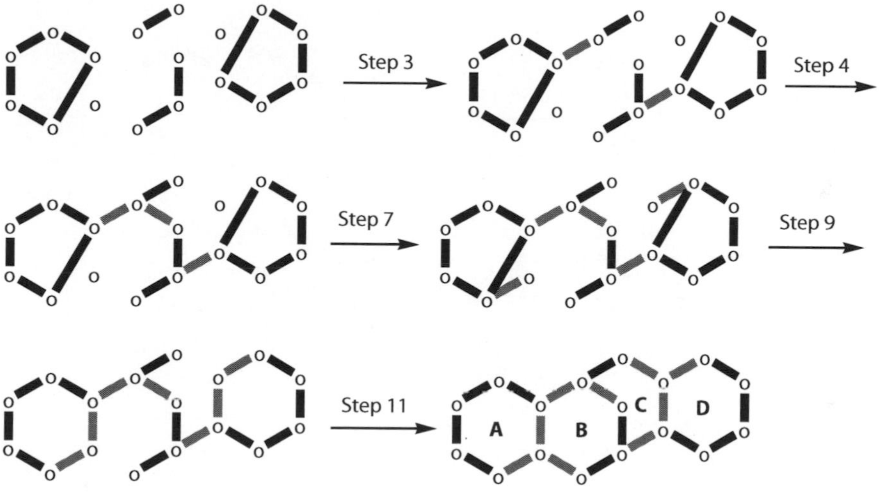

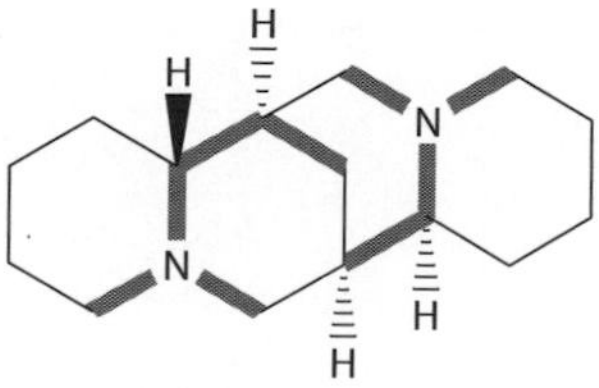

Sequence: $\xrightarrow{[([5]\rightarrow[6])_A + ([5]\rightarrow[6])_D]} A_9 + D_9 \xrightarrow{[(6+0)_B + (6+0)_C]} [3.3.1] (B+C)_{11}$

(±)-Sparteine
Wanner, M.J.; Koomen, G.J. *J. Org. Chem.* 1996, *61*, 5581.

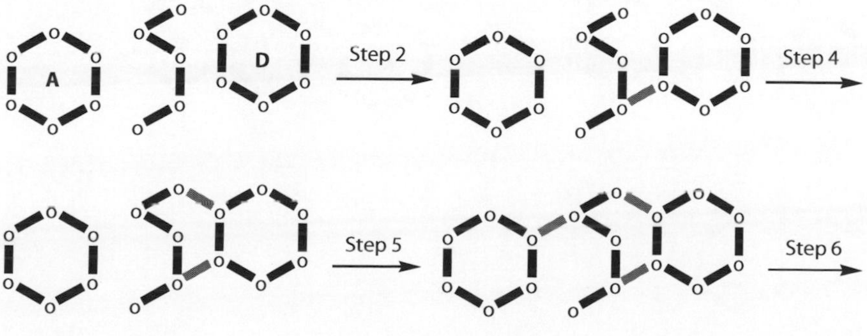

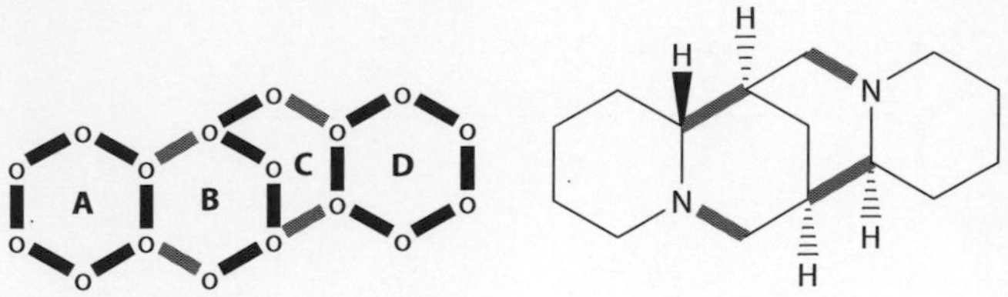

Sequence: $(A+D)_0 \xrightarrow{[6+0]} C_4 \xrightarrow{[6+0]} B_6$

(±)-Sparteine

Leonard, N.J.; Beyler, R.E. *J. Am. Chem. Soc.* 1948, *70*, 2298.
Leonard, N.J.; Beyler, R.E. *J. Am. Chem. Soc.* 1950, *72*, 1316.

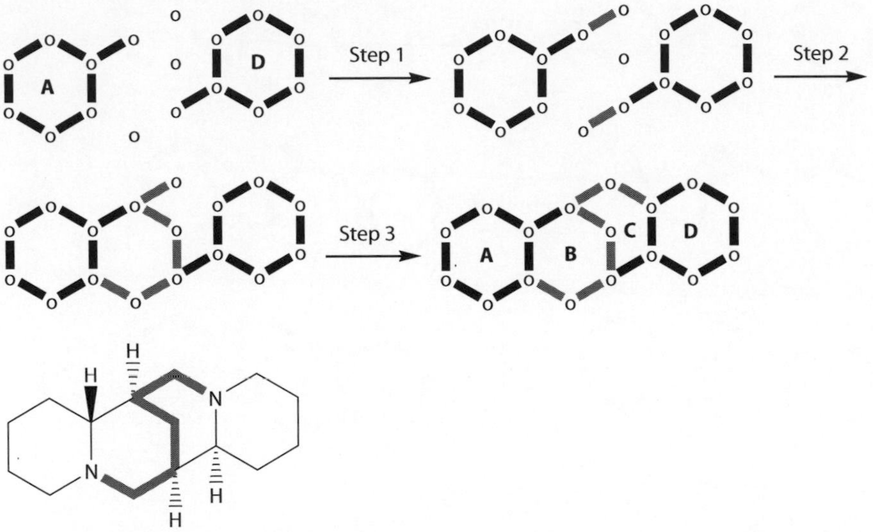

Sequence: $(A+D)_0 \xrightarrow{[3+2+1]} B_2 \xrightarrow{[6+0]} C_3$

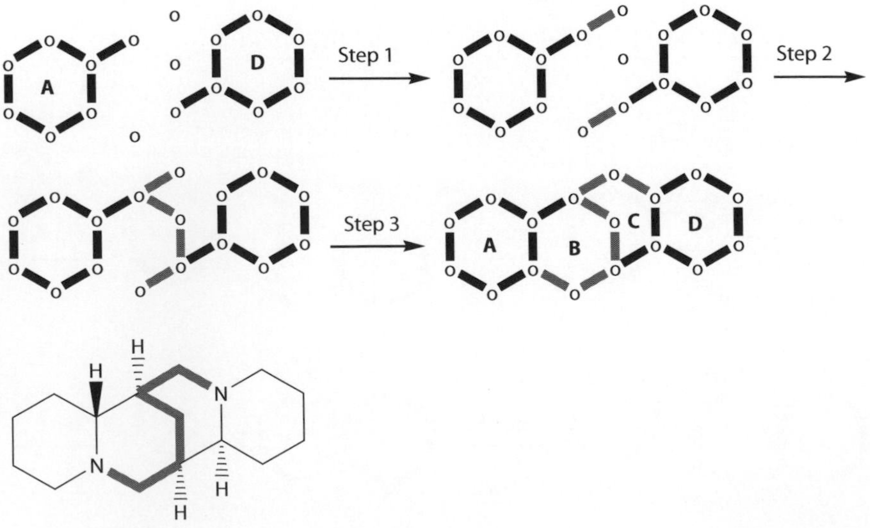

Sequence: $(A + D)_0 \xrightarrow{[(6+0)_B + (6+0)_C]} [3.3.1] (B + C)_3$

(±)-Sparteine

Takatsu, N.; Ohmiya, S.; Otomasu, H. *Chem. Pharm. Bull.* 1987, *35*, 891.
Takatsu, N.; Noguchi, M.; Ohmiya, S.; Otomasu, H. *Chem. Pharm. Bull.* 1987, *35*, 4990.

Sequence: $A_0 + D_{4*} \xrightarrow{[6+0]} B_3 \xrightarrow{[5+1]} C_6$

(±)-Sparteine

Van Tamelen, E.E.; Foltz, R.L. *J. Am. Chem. Soc.* 1960, *82*, 2400.
Van Tamelen, E.E.; Foltz, R.L. *J. Am. Chem. Soc.* 1969, *91*, 7372.

Sequence: $(A + D)_0 \xrightarrow{[(6+0)_B + (6+0)_C]} [3.3.1] (B + C)_2$

Summary:

$$(A + D)_0 \xrightarrow{[(6+0)_B + (6+0)_C]} [3.3.1] (B + C)_2: \text{van Tamelen}$$

$$(A + D)_0 \xrightarrow{[(6+0)_B + (6+0)_C]} [3.3.1] (B + C)_3: \text{Leonard G2}$$

$$A_0 + D_{11} \xrightarrow{[6+0]} B_9 \xrightarrow{[6+0]} C_{13}: \text{Bohlmann-Gallagher}$$

$$A_0 + D_{4*} \xrightarrow{[6+0]} B_3 \xrightarrow{[5+1]} C_6: \text{Takatsu}$$

$$(A+D)_0 \xrightarrow{[3+2+1]} B_2 \xrightarrow{[6+0]} C_6: \text{Clemo-Raper}$$

$$(A+D)_0 \xrightarrow{[3+2+1]} B_2 \xrightarrow{[6+0]} C_3: \text{Leonard G1}$$

$$(A+D)_0 \xrightarrow{[6+0]} C_4 \xrightarrow{[6+0]} B_6: \text{Koomen}$$

$$\xrightarrow{[(6+0)_B + (6+0)_C]} [3.3.1](B+C)_2 \xrightarrow{[(6+0)_A + (6+0)_D]} (A+D)_6: \text{Blakemore}$$

$$\xrightarrow{[(5+1)_C + (6+0)_D]} [4.4.0](C+D)_9 \xrightarrow{[(6+0)_A (6+0)_B]} [4.4.0](A+B)_{16}: \text{Aube}$$

$$\xrightarrow{[([5]\to[6])_A + ([5]\to[6])_D]} A_9 + D_9 \xrightarrow{[(6+0)_B + (6+0)_C]} [3.3.1](B+C)_{11}: \text{Fleming}$$

$$\xrightarrow{[(6+0)_A + (3+2+1)_B + (3+2+1)_C + (6+0)_D]} (A+B+C+D)_1: \text{Anet-Hughes-Ritchie}$$

(±)-Stenine
Zeng, Y.; Aube, J. *J. Am. Chem. Soc.* 2005, *127*, 15712.

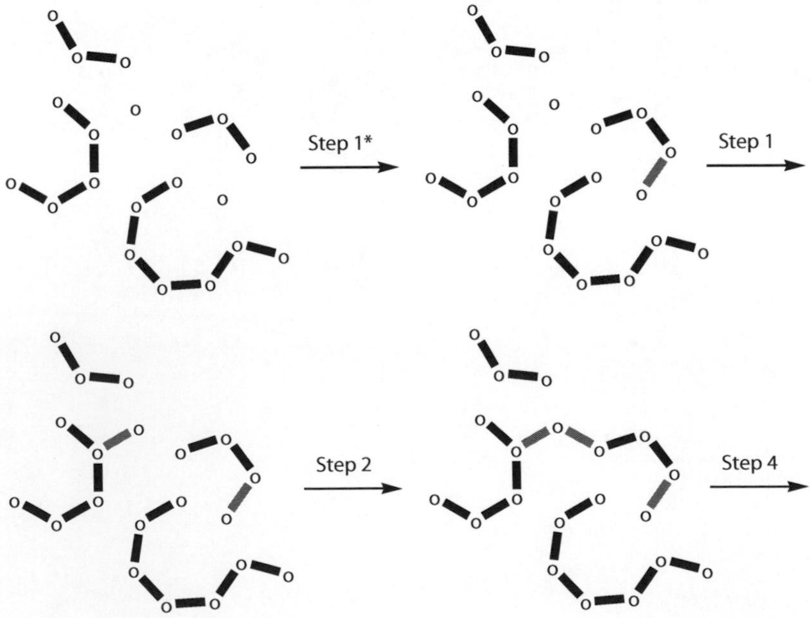

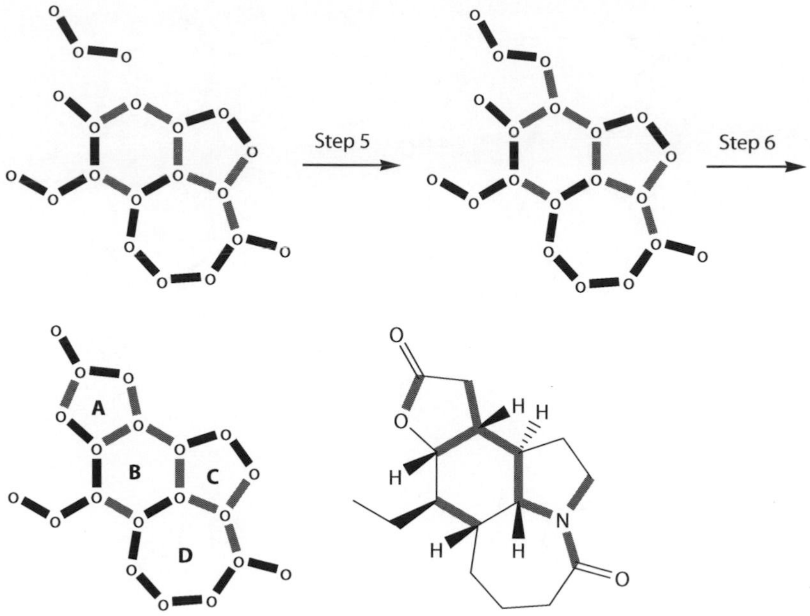

Sequence: $\xrightarrow{([4+2]_B + [4+1]_C + [6+1]_D)} (B+C+D)_4 \xrightarrow{[5+0]} A_6$

(−)-Stenine

Wipf, P.; Kim, Y.; Goldstein, D.M. *J. Am. Chem. Soc.* 1995, *117*, 11106.

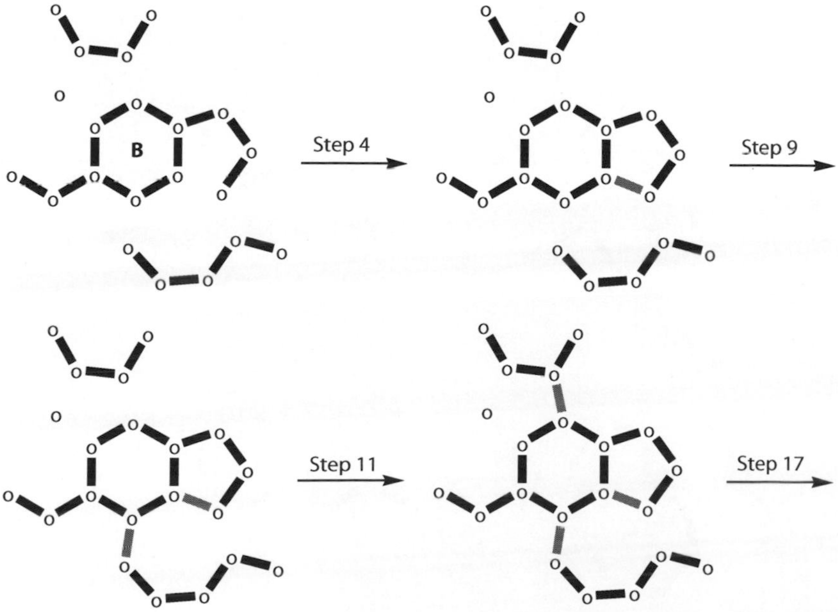

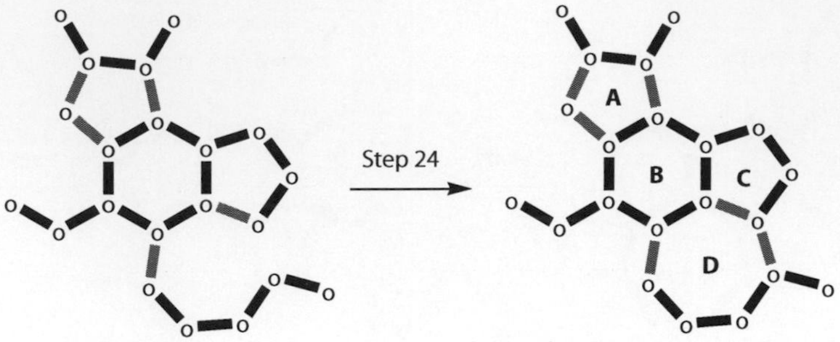

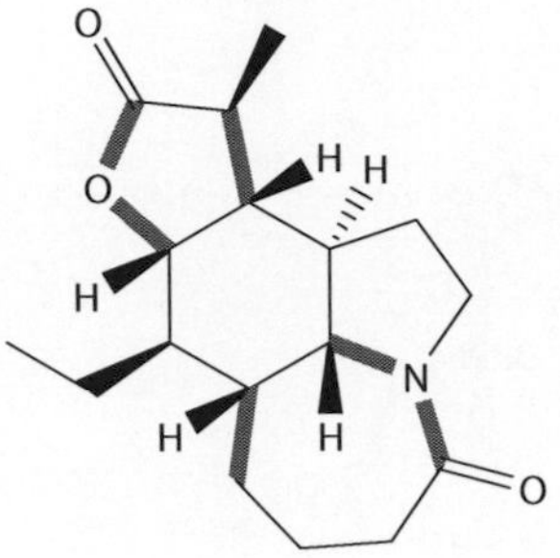

Sequence: $B_0 \xrightarrow{[5+0]} C_4 \xrightarrow{[4+1]} A_{17} \xrightarrow{[7+0]} D_{24}$

(±)-Stenine

Chen, C.; Hart, D.J. *J. Org. Chem.* 1990, *55*, 6236.
Chen, C.; Hart, D.J. *J. Org. Chem.* 1993, *58*, 3840.

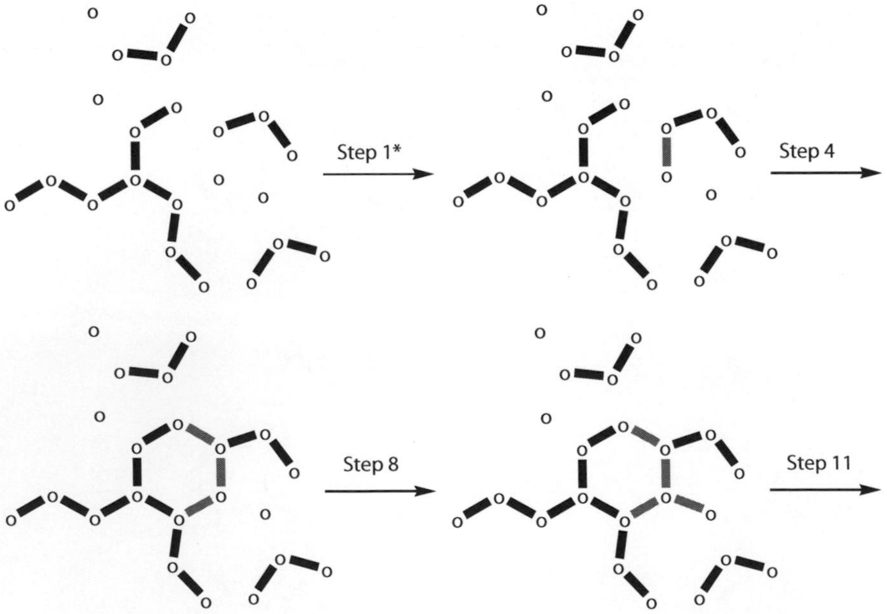

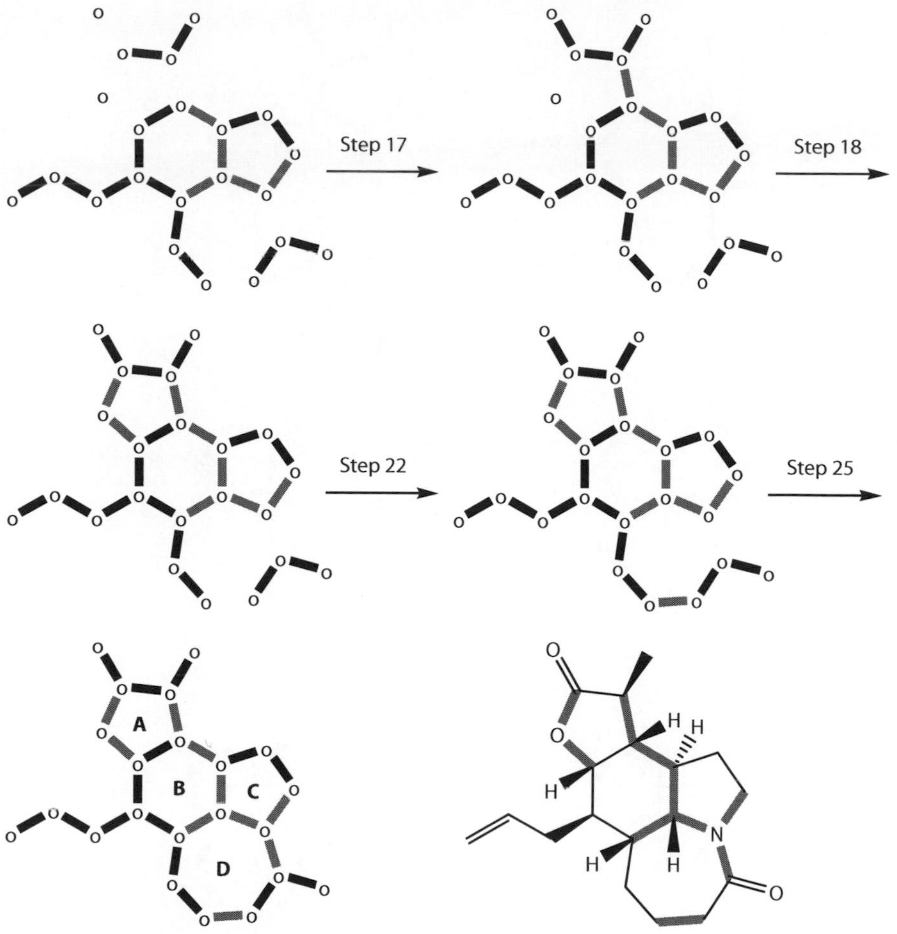

Sequence: $\xrightarrow{[4+2]} B_4 \xrightarrow{[5+0]} C_{11} \xrightarrow{[4+1]} A_{18} \xrightarrow{[7+0]} D_{25}$

(–)-Stenine

Morimoto, Y.; Iwahashi, M.; Nishida, K.; Hayashi, Y.; Shirahama, H. *Angew. Chem. Int. Ed.* 1996, *35*, 904.
Morimoto, Y.; Iwahashi, M.; Kinoshita, T.; Nishida, K. *Chem. Eur. J.* 2001, *7*, 4107.

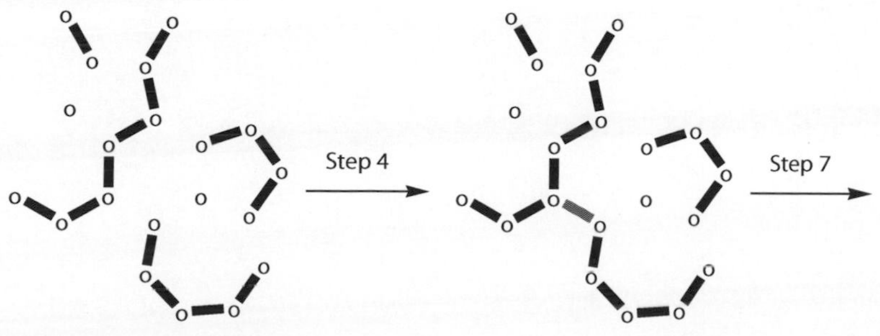

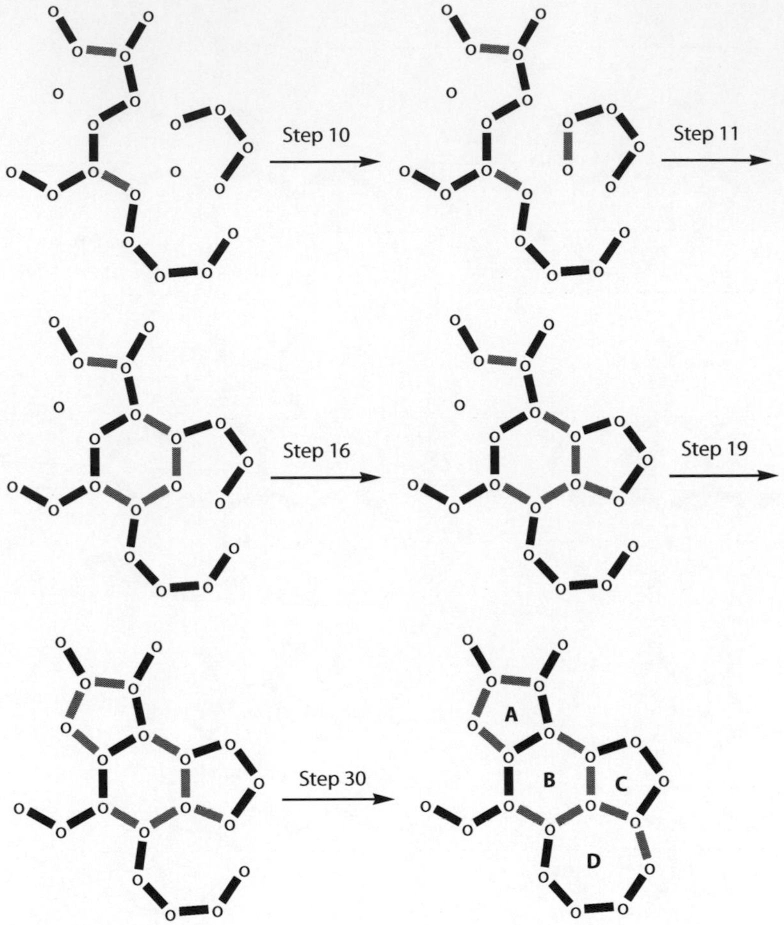

Step 10

Step 11

Step 16

Step 19

Step 30

A
B   C
D

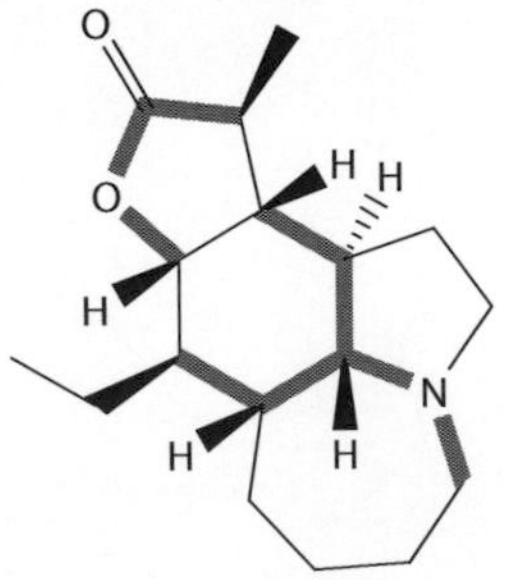

Sequence: $\xrightarrow{[4+2]} B_{11} \xrightarrow{[5+0]} C_{16} \xrightarrow{[4+1]} A_{19} \xrightarrow{[7+0]} D_{30}$

(±)-Stenine

Padwa, A.; Ginn, J.D. *J. Org. Chem.* 2005, *70*, 5197.
Ginn, J.D.; Padwa, A. *Org. Lett.* 2002, *4*, 1515.

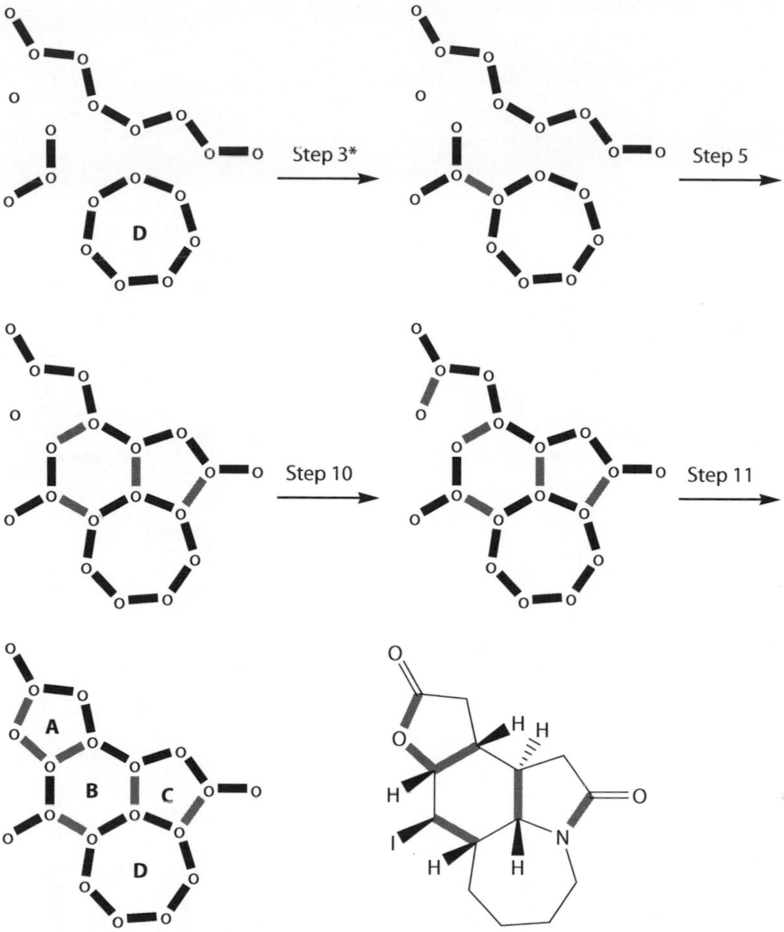

Sequence: $D_0 \xrightarrow{([4+2]_B + [3+2]_C)} (B+C)_5 \xrightarrow{[5+0]} A_{11}$

Summary:

Strategy #1:

$$\xrightarrow{([4+2]_B + [4+1]_C + [6+1]_D)} (B+C+D)_4 \xrightarrow{[5+0]} A_6 : \text{Aube}$$

Strategy #2:

$$B_0 \xrightarrow{[5+0]} C_4 \xrightarrow{[4+1]} A_{17} \xrightarrow{[7+0]} D_{24} : \text{Wipf}$$

Strategy #3:

$$\xrightarrow{[4+2]} B_4 \xrightarrow{[5+0]} C_{11} \xrightarrow{[4+1]} A_{18} \xrightarrow{[7+0]} D_{25} : \text{Hart}$$

$$\xrightarrow{[4+2]} B_{11} \xrightarrow{[5+0]} C_{16} \xrightarrow{[4+1]} A_{19} \xrightarrow{[7+0]} D_{30} : \text{Morimoto}$$

Strategy #4:

$$D_0 \xrightarrow{([4+2]_B + [3+2]_C)} [4.3.0]\,(B+C)_5 \xrightarrow{[5+0]} A_{11}: \text{Padwa}$$

Strychnine
Bodwell, G.J.; Li, *J. Angew. Chem. Int. Ed.* 2002, *41*, 3261.

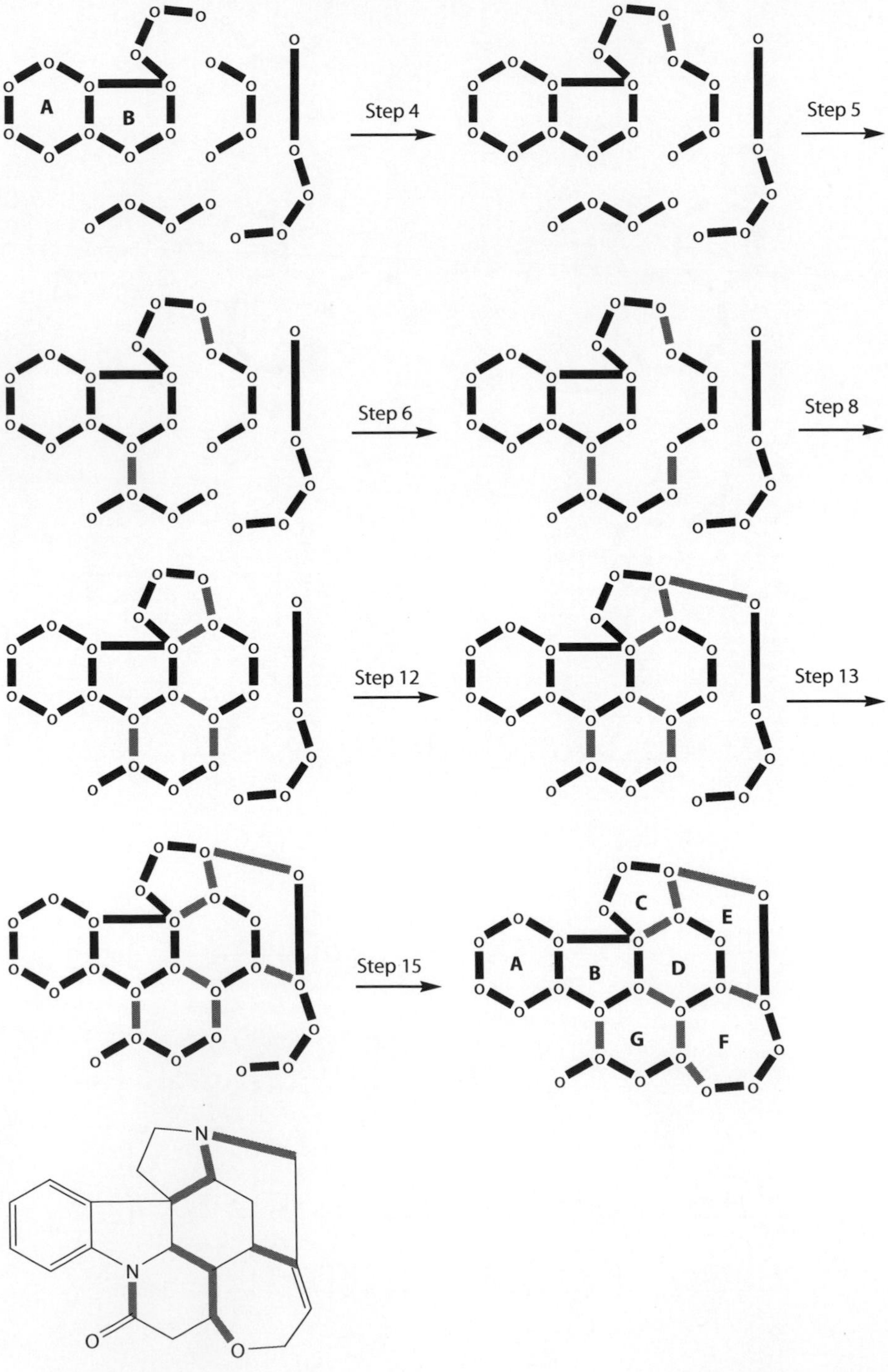

Sequence: $(A+B)_3 \xrightarrow{[(5+0)_C + (4+2)_D + (6+0)_G]} (C+D+G)_8 \xrightarrow{[6+0]} E_{13} \xrightarrow{[7+0]} F_{15}$

Sole, D.; Bonjoch, J.; Garcia-Rubio, S.; Peidro, E.; Bosch, J. *Chem. Eur. J.* 2000, *6*, 655.
Sole, D.; Bonjoch, J.; Garcia-Rubio, S.; Peidro, E.; Bosch, J. *Angew. Chem. Int. Ed.* 1999, *38*, 395.

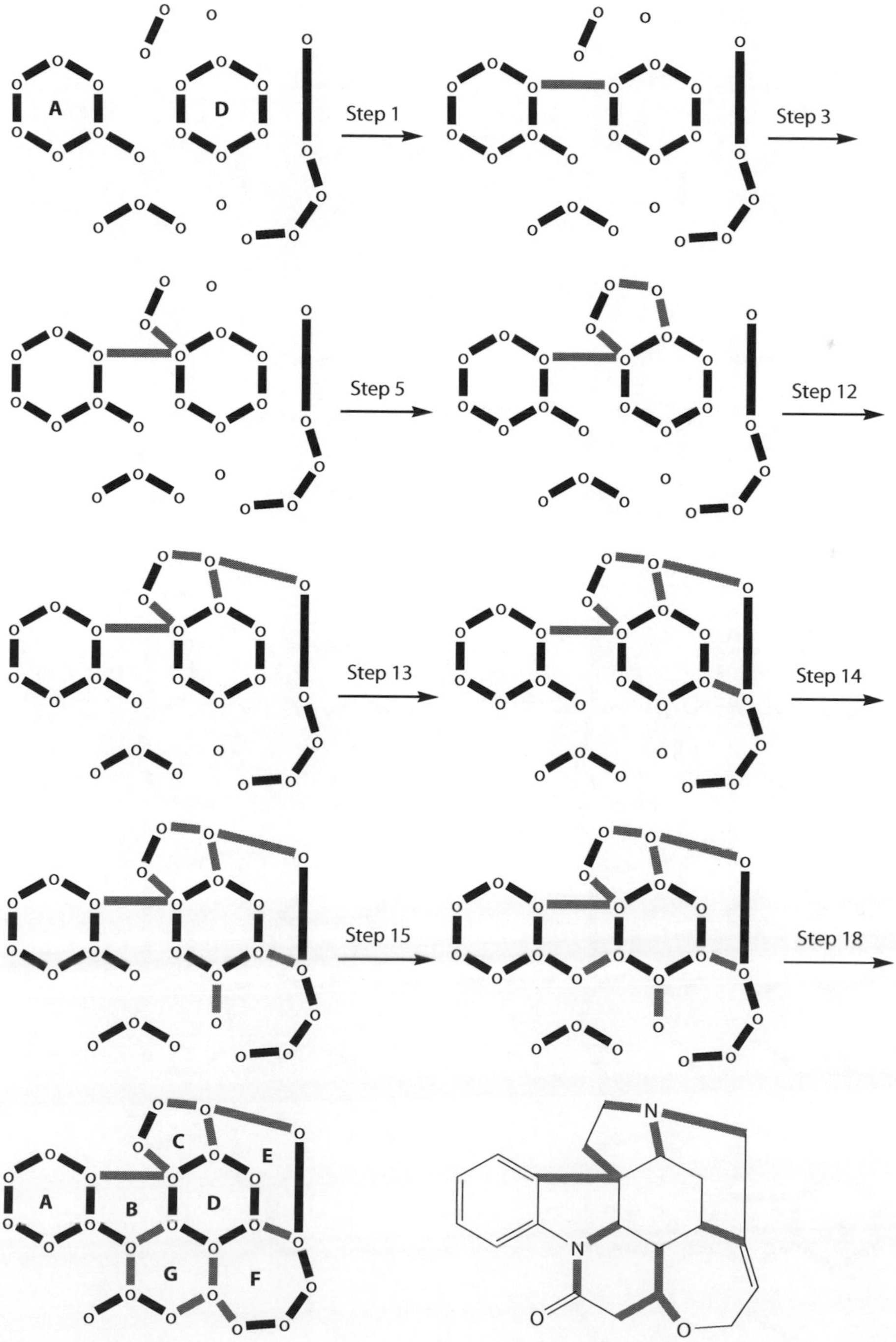

Sequence:

$$A_0 + D_0 \xrightarrow{[4+1]} C_5 \xrightarrow{[6+0]} E_{13} \xrightarrow{[5+0]} B_{15} \xrightarrow{[(7+0)_F + (4+2)_G]} [5.4.0]\,(F+G)_{18}$$

Kaburagi, Y.; Tokuyama, H.; Fukuyama, T. *J. Am. Chem. Soc.* 2004, *126*, 10246.

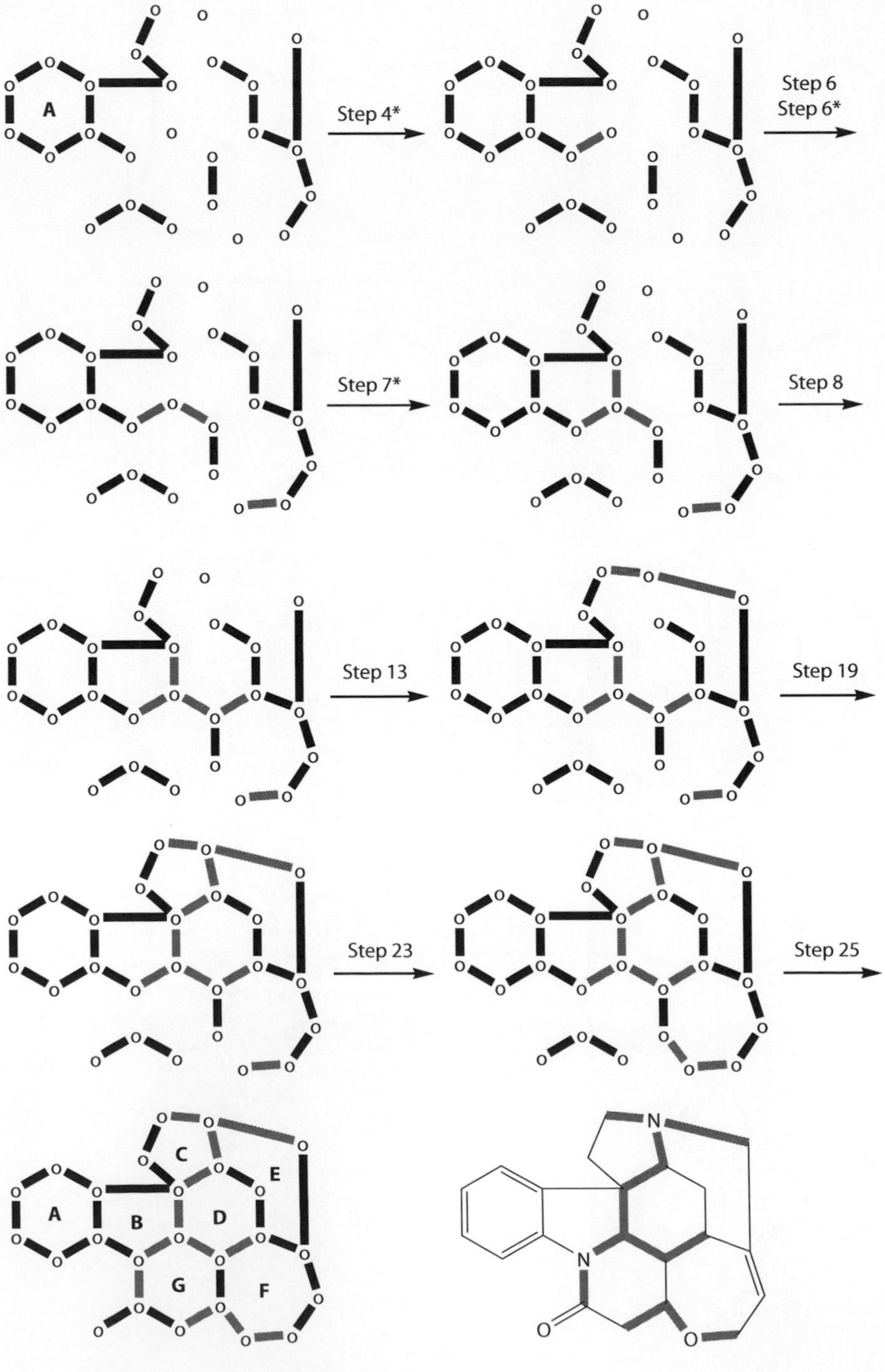

Sequence:

$$A_{3*} \xrightarrow{[5+0]} B_{7*} \xrightarrow{[(4+1)_C + (6+0)_D + (6+0)_E]} (C+D+E)_{19} \xrightarrow{[7+0]} F_{23} \xrightarrow{[4+2]} G_{25}$$

Kuehne, M.E.; Xu, F. *J. Org. Chem.* 1993, *58*, 7490.
Parsons, R.L.; Berk, J.D.; Kuehne, M.E. *J. Org. Chem.* 1993, *58*, 7482.

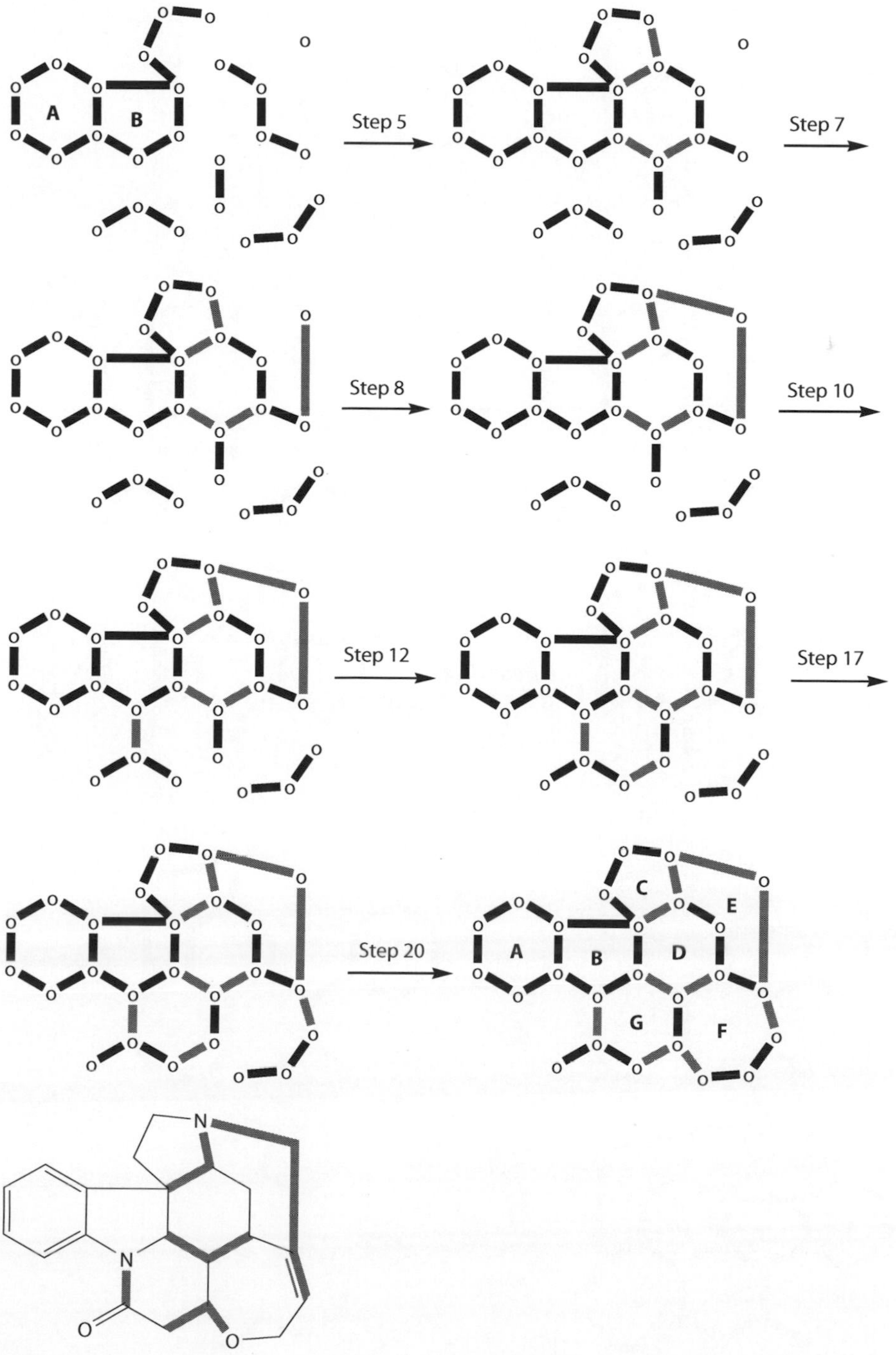

Sequence:

$$(A+B)_0 \xrightarrow{[(4+1)_C + (3+2+1)_D]} [4.3.0](C+D)_5 \xrightarrow{[6+0]} E_8 \xrightarrow{[6+0]} G_{12} \xrightarrow{[7+0]} F_{20}$$

Magnus, P.; Giles, M.; Bonnert, R.; Kim, C.S.; McQuire, L.; Merritt, A.; Vicker, N. *J. Am. Chem. Soc.* 1992, *114*, 4403.
Magnus, P.; Giles, M.; Bonnert, R.; Johnson, G.; McQuire, L.; Deluca, M.; Merritt, A.; Kim, C.S.; Vicker, N.
    J. Am. Chem. Soc. 1993, *115*, 8116.
Magnus, P.; Ladlow, M.; Kim, C.S.; Boniface, P. *Heterocycles* 1989, *28*, 951.

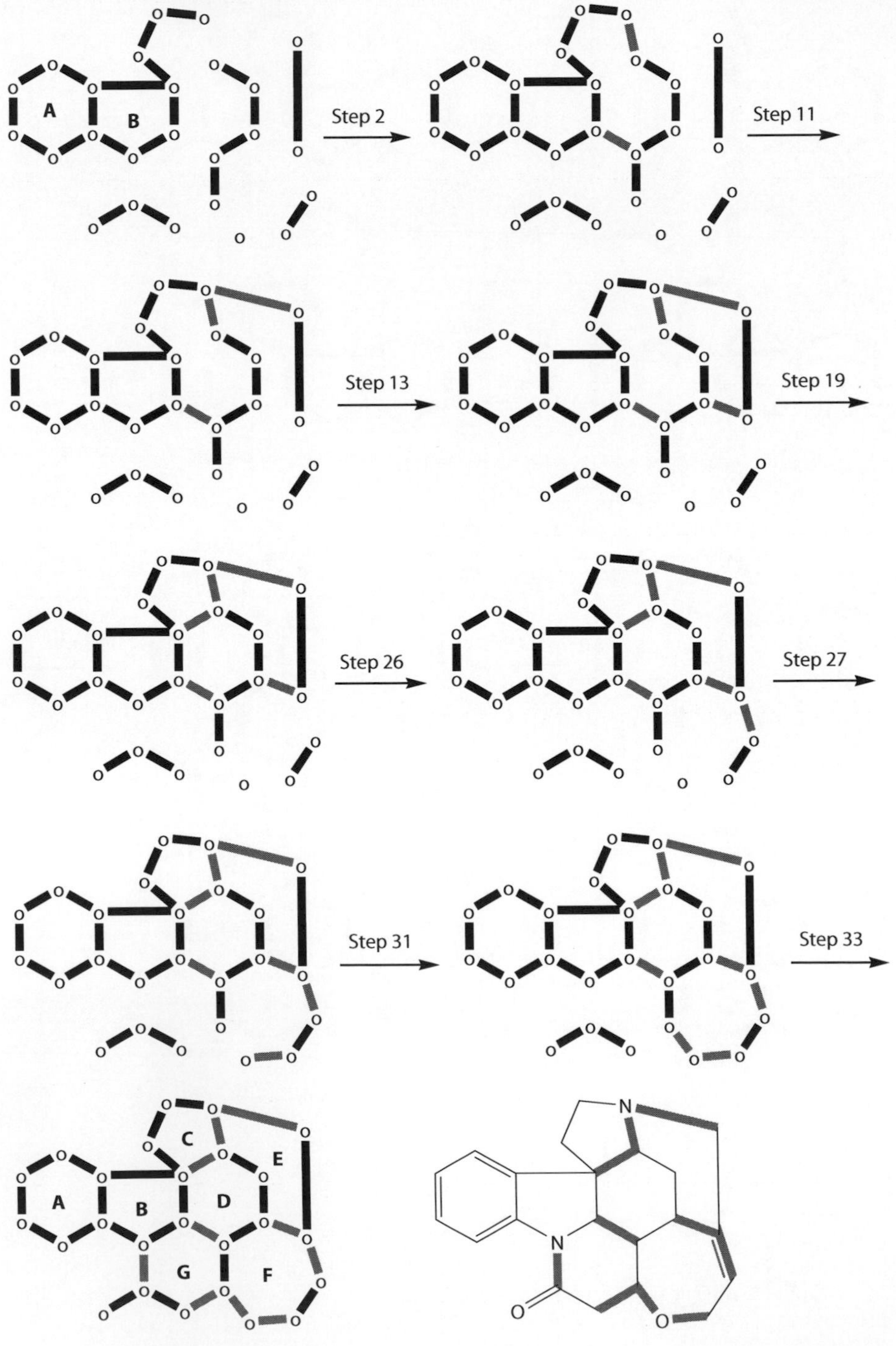

Sequence:

$$(A+B)_0 \xrightarrow{[6+0]} E_{13} \xrightarrow{[(5+0)_C+(6+0)_D]} [4.3.0]\,(C+D)_{19} \xrightarrow{[7+0]} F_{31} \xrightarrow{[4+2]} G_{33}$$

Ito, M.; Clark, C.W.; Mortimore, M.; Goh, J.B.; Martin, S.F. *J. Am. Chem. Soc.* 2001, *123*, 8003.

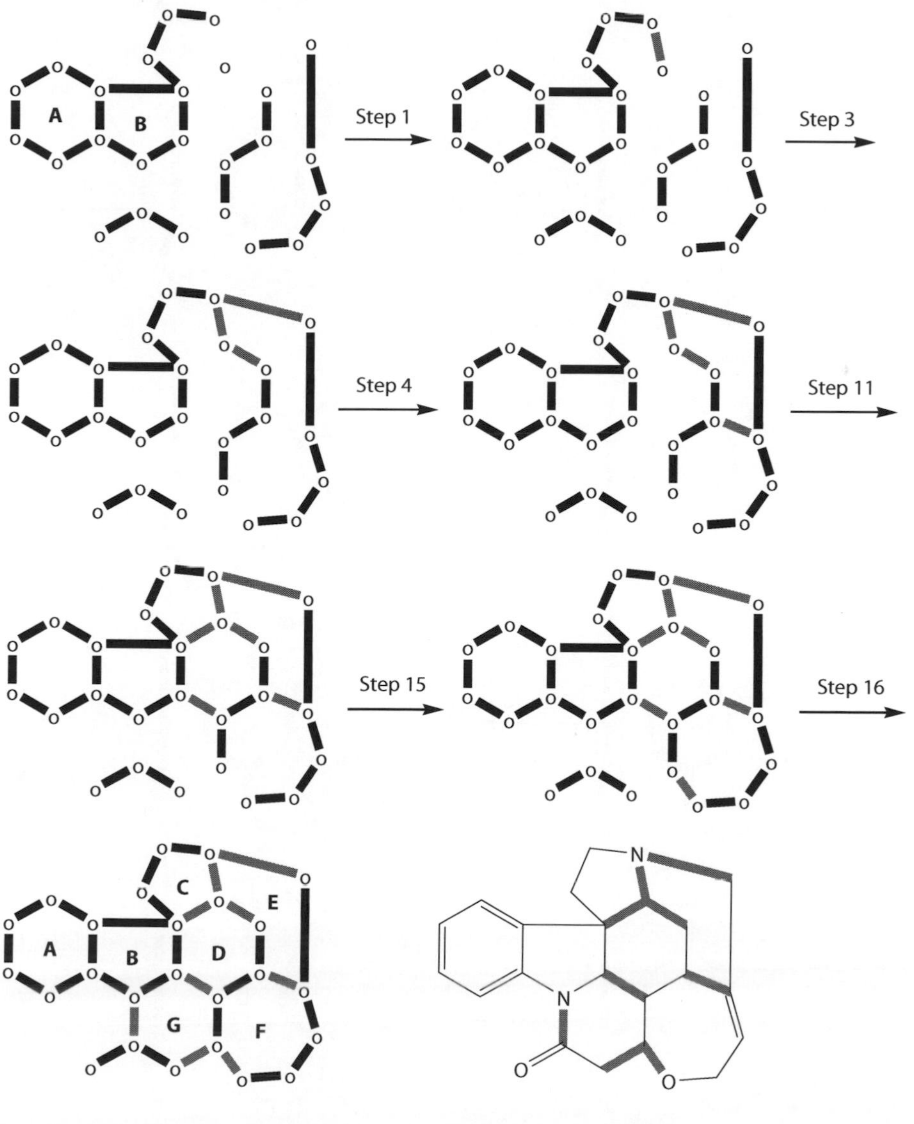

Sequence:

$$(A+B)_0 \xrightarrow{[6+0]} E_4 \xrightarrow{[(5+0)_C+(4+2)_D]} [4.3.0]\,(C+D)_{11} \xrightarrow{[7+0]} F_{15} \xrightarrow{[4+2]} G_{16}$$

Mori, M.; Nakanishi, M.; Kajishima, D.; Sato, Y. *Org. Lett.* 2001, *3*, 1913.
Nakanishi, M.; Mori, M. *Angew. Chem. Int. Ed.* 2002, *41*, 1934.
Mori, M.; Nakanishi, M.; Kajishima, D.; Sato, Y. *J. Am. Chem. Soc.* 2003, *125*, 9801.

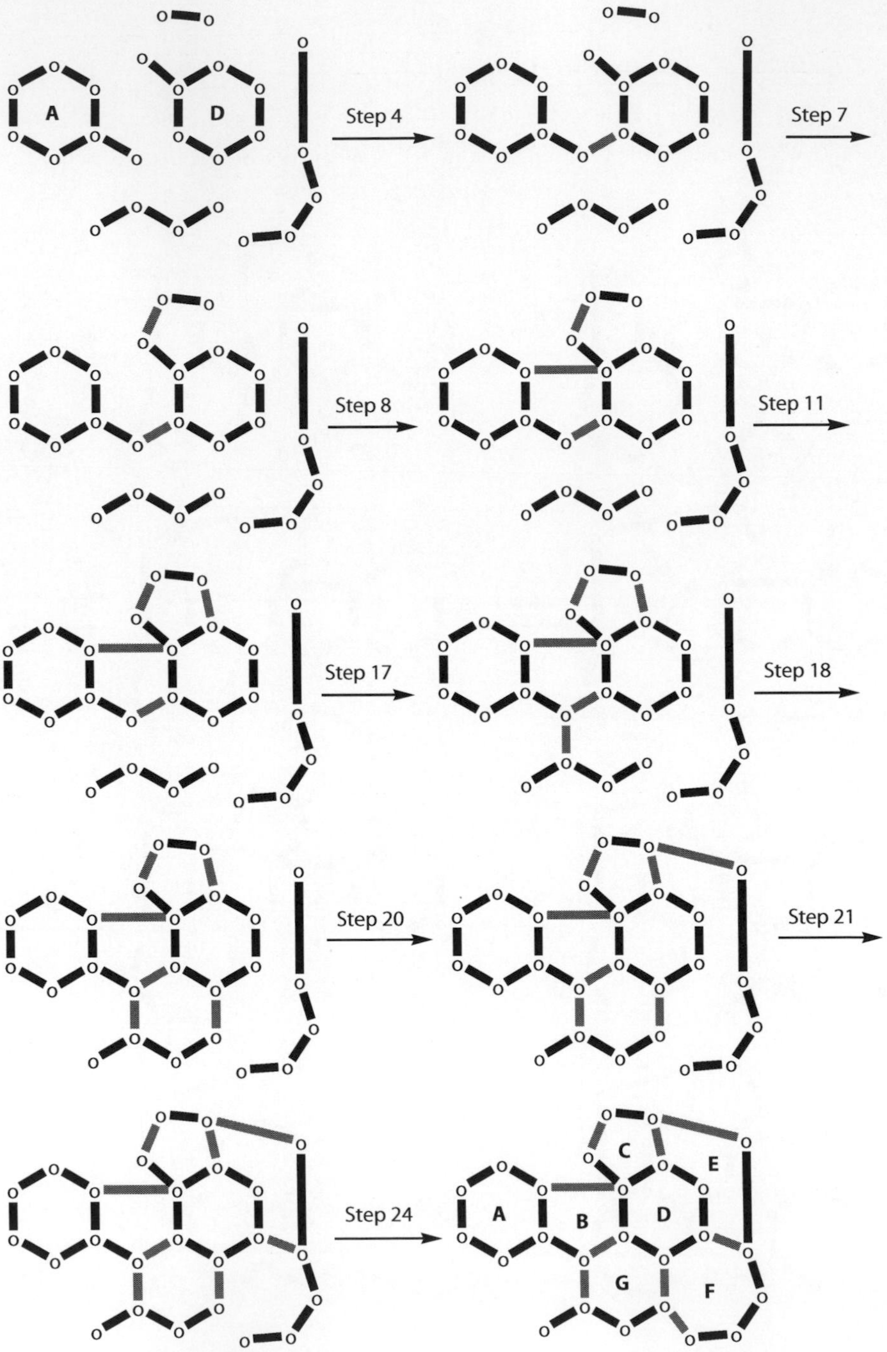
A
D
Step 4
Step 7
Step 8
Step 11
Step 17
Step 18
Step 20
Step 21
Step 24
A
B
C
D
E
F
G
N
N
O
O

Sequence: $A_3 + D_0 \xrightarrow{[5+0]} B_8 \xrightarrow{[5+0]} C_{11} \xrightarrow{[6+0]} G_{18} \xrightarrow{[6+0]} E_{21} \xrightarrow{[7+0]} F_{24}$

Knight, S.D.; Overman, L.E.; Pairaudeau, G. *J. Am. Chem. Soc.* 1993, *115*, 9293.
Knight, S.D.; Overman, L.E.; Pairaudeau, G. *J. Am. Chem. Soc.* 1995, *117*, 5776.

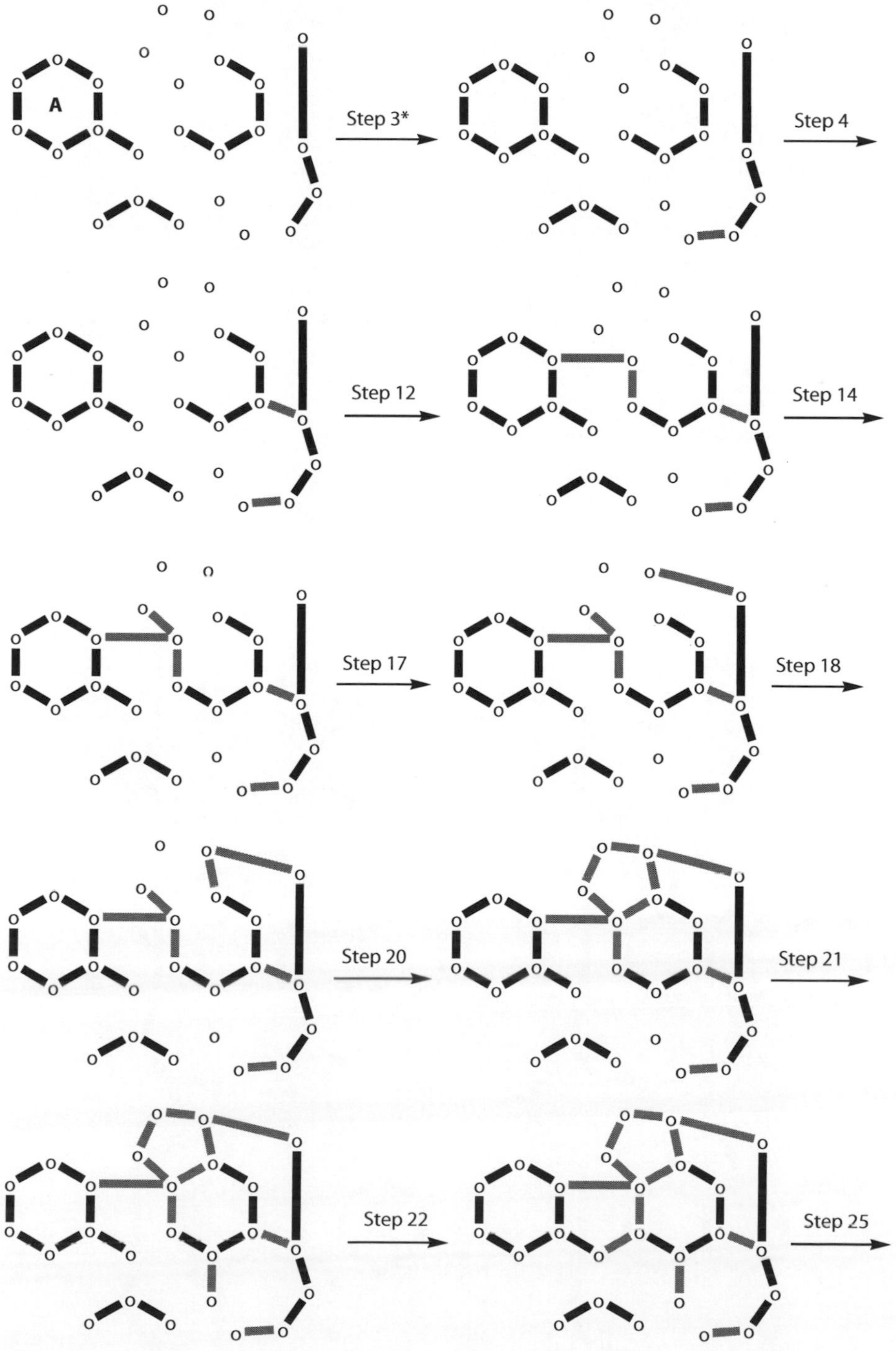

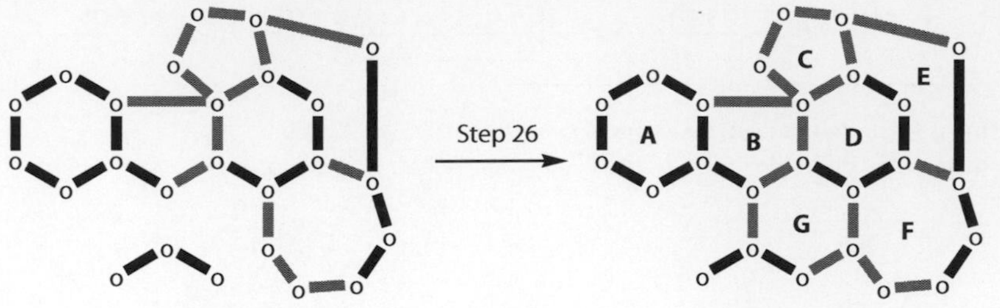

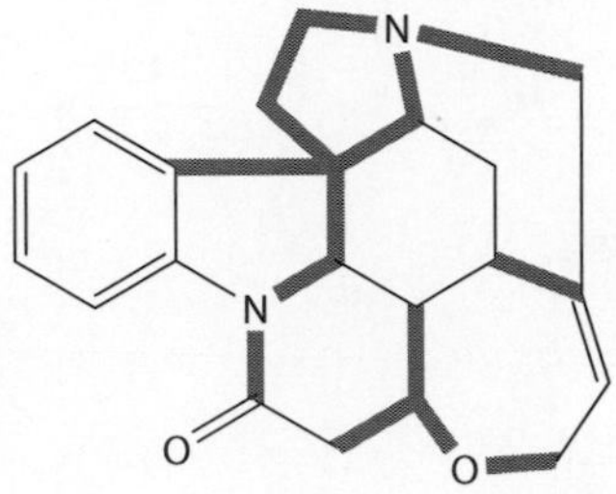

Sequence:

$$A_{10*} \xrightarrow{[6+0]} E_{18} \xrightarrow{[(2+2+1)_C + (6+0)_D]} [4.3.0] (C+D)_{20} \xrightarrow{[5+0]} B_{22}$$

$$\xrightarrow{[7+0]} F_{25} \xrightarrow{[4+2]} G_{26}$$

Zhang, H.; Boonsombat, J.; Padwa, A. *Org. Lett.* 2007, *9*, 279.

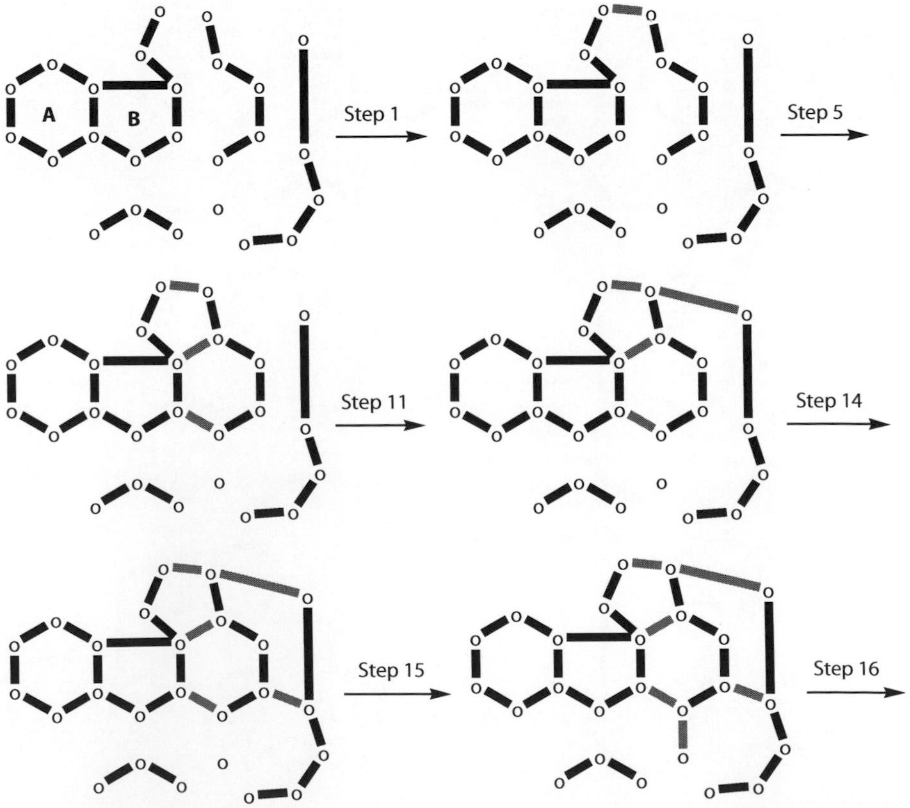

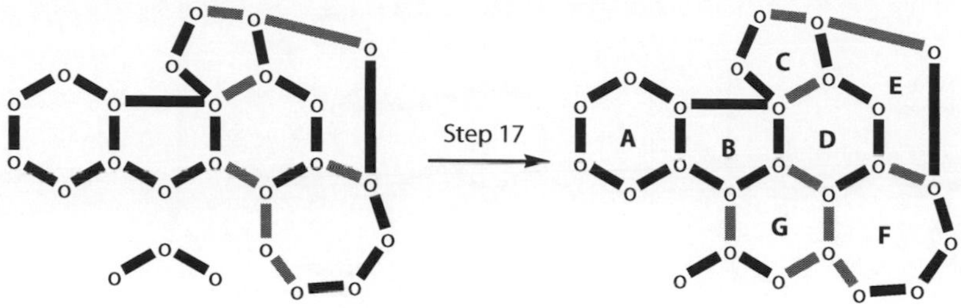

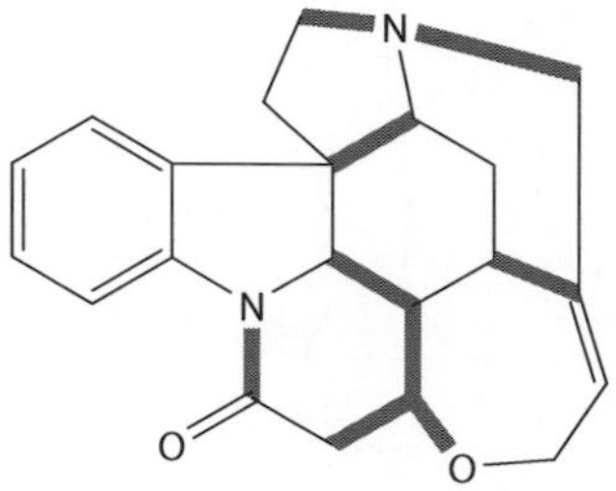

Sequence:

$$(A + B)_0 \xrightarrow{[(5+0)_C + (4+2)_D]} [4.3.0](C+D)_5 \xrightarrow{[6+0]} E_{14} \xrightarrow{[7+0]} F_{16} \xrightarrow{[4+2]} G_{17}$$

Rawal, V.H.; Iwasa, S. *J. Org. Chem.* 1994, *59*, 2685.
Sole, D.; Bonjoch, J.; Garcia-Rubio, S.; Peidro, E.; Bosch, J. *Chem. Eur. J.* 2000, *6*, 655.

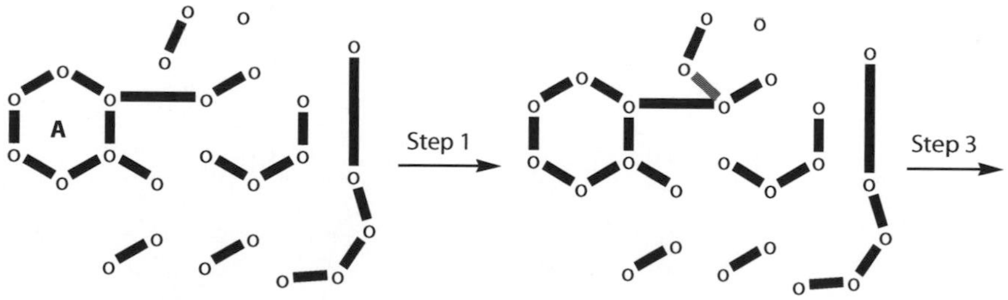

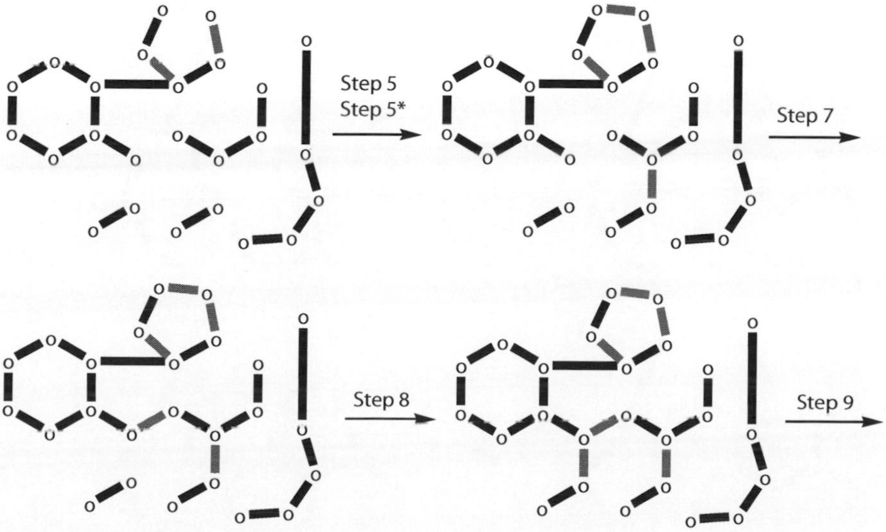

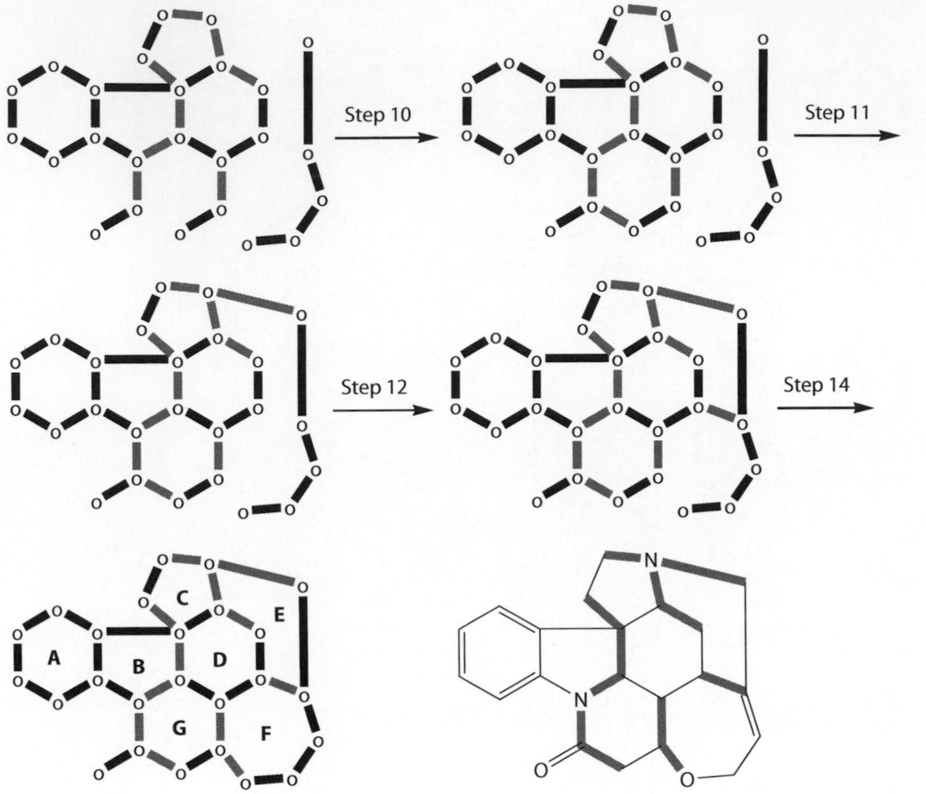

Sequence:

$$A_0 \xrightarrow{[5+0]} C_3 \xrightarrow{[(5+0)_B + (4+2)_D]} [4.3.0](B+D)_9 \xrightarrow{[6+0]} G_{10} \xrightarrow{[6+0]} E_{12}$$

$$\xrightarrow{[7+0]} F_{14}$$

Beemalmanns, C.; Reissig, H.U. *Angew. Chem. Int. Ed.* 2010, doi: 10.1002/anie.201003320.

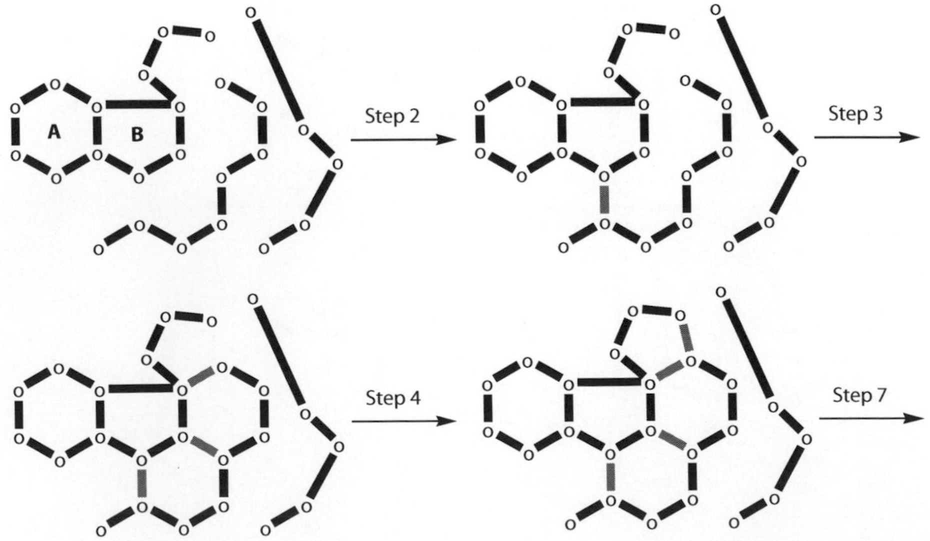

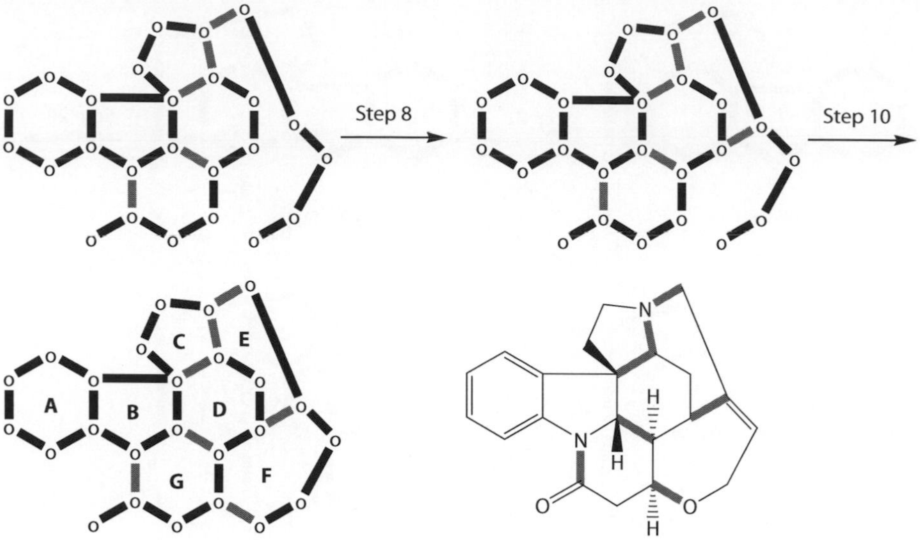

Sequence:

$$(A+B)_0 \xrightarrow{[(4+2)_D + (6+0)_G]} [4.4.0]\,(D+G)_3 \xrightarrow{[5+0]} C_4 \xrightarrow{[6+0]} E_8 \xrightarrow{[7+0]} F_{10}$$

Ohshima, T.; Xu, Y.; Takita, R.; Shibasaki, M. *Tetrahedron* 2004, *60*, 9569.
Ohshima, T.; Xu, Y.; Takita, R.; Shimizu, S.; Zhong, D.; Shibasaki, M. *J. Am. Chem. Soc.* 2002, *124*, 14546.

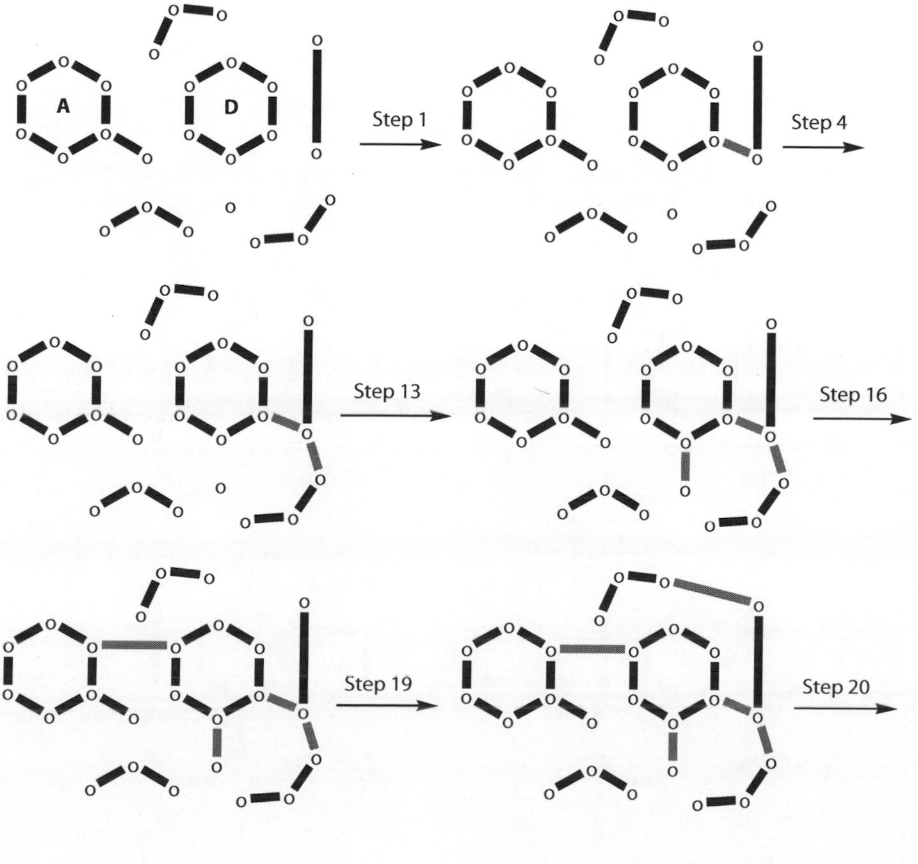

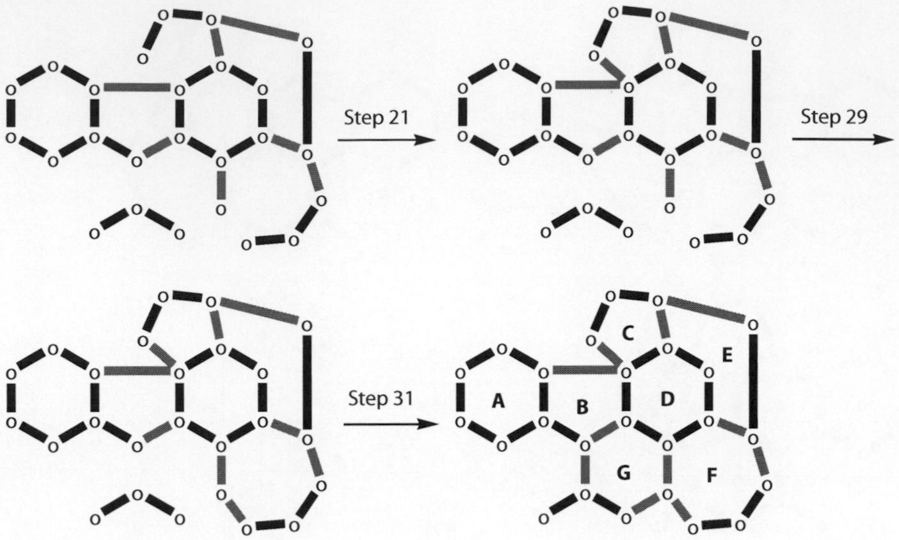

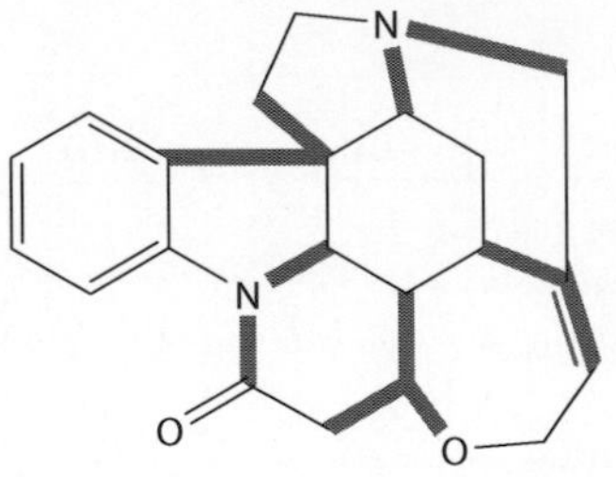

Sequence: $A_{15} + D_0 \xrightarrow{[5+0]_B + [6+0]_E} B_{20} + E_{20} \xrightarrow{[5+0]} C_{21} \xrightarrow{[7+0]} F_{29} \xrightarrow{[4+2]} G_{31}$

Stork, G. Ischia Advanced School of Organic Chemistry, Ischia Porto, Italy, September 21, 1992.
Bonjoch, J.; Sole, J. *Chem. Rev.* 2000, *100*, 3455.

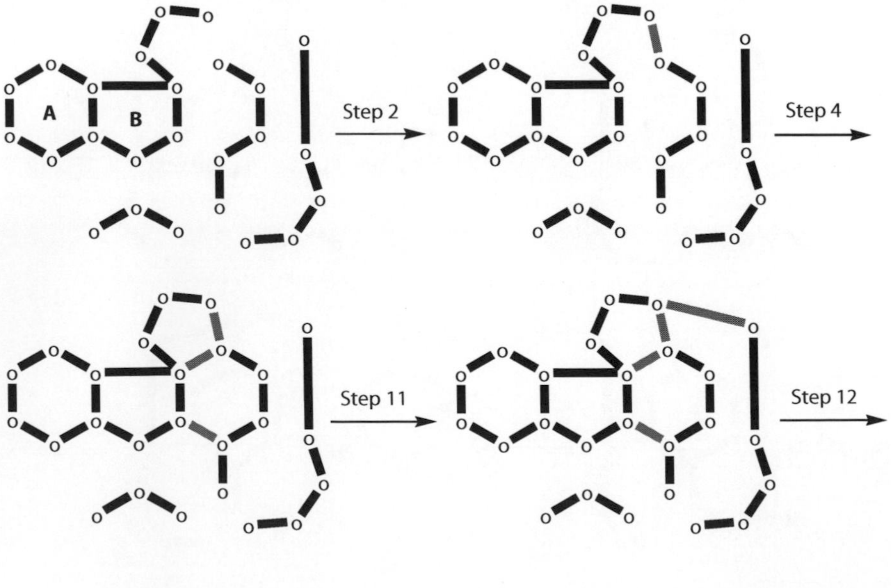

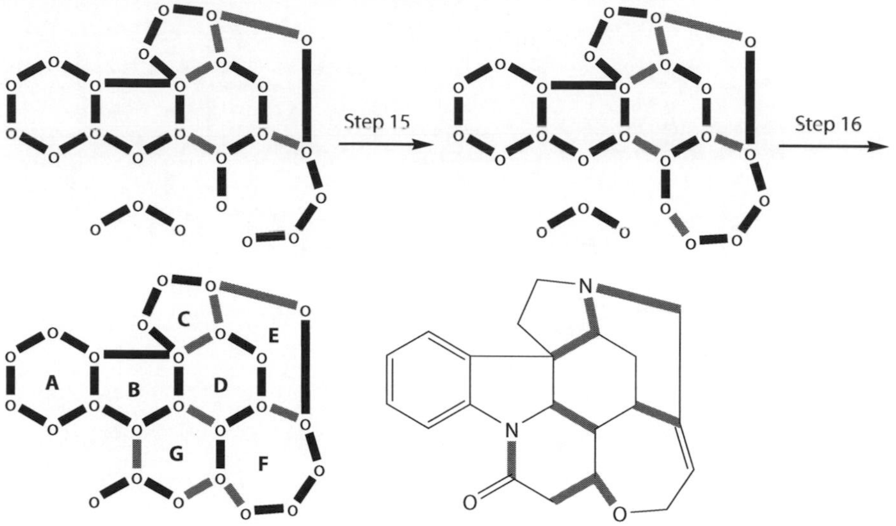

Sequence:

$$(A+B)_0 \xrightarrow{[(5+0)_C+(4+2)_D]} [4.3.0] (C+D)_4 \xrightarrow{[6+0]} E_{12} \xrightarrow{[7+0]} F_{15} \xrightarrow{[4+2]} G_{16}$$

Eichberg, M.J.; Dorta, R.L.; Lamottke, K.; Vollhardt, K.P.C. *Org. Lett.* 2000, *2*, 2479.
Eichberg, M.J.; Dorta, R.L.; Grotjahn, D.B.; Lamottke, K.; Schmidt, M.; Vollhardt, K.P.C. *J. Am. Chem. Soc.* 2001, *123*, 9324.

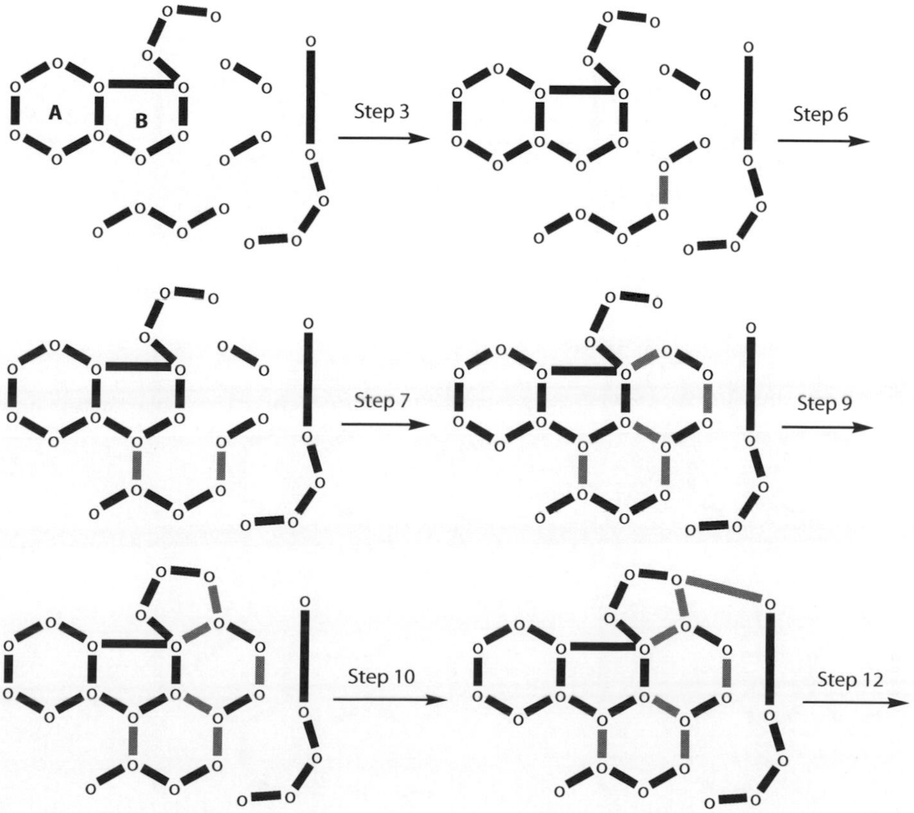

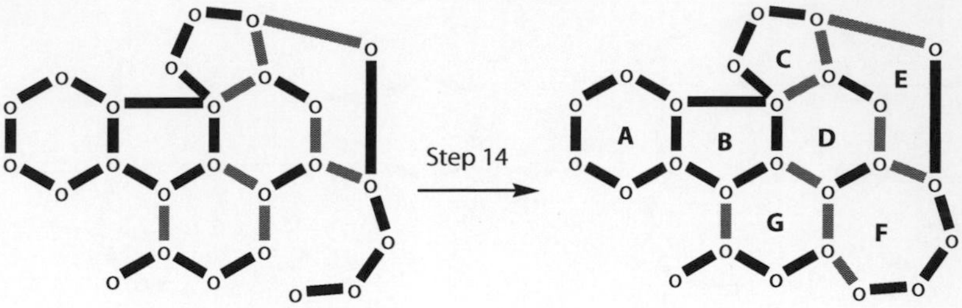

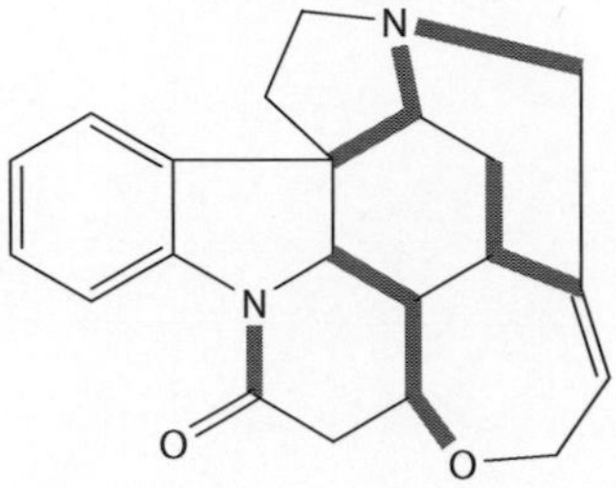

Sequence:

$$(A+B)_5 \xrightarrow{[(2+2+2)_D+(6+0)_G]} [4.4.0] (D+G)_7 \xrightarrow{[5+0]} C_9 \xrightarrow{[6+0]} E_{12} \xrightarrow{[7+0]} F_{14}$$

Woodward, R.B.; Cava, M.P.; Ollis, W.D.; Hunger, A.; Daeniker, H.U.; Schenker, K. *J. Am. Chem. Soc.* 1954, *76*, 4749.
Woodward, R.B.; Cava, M.P.; Ollis, W.D.; Hunger, A.; Daeniker, H.U.; Schenker, K. *Tetrahedron* 1963, *19*, 247.

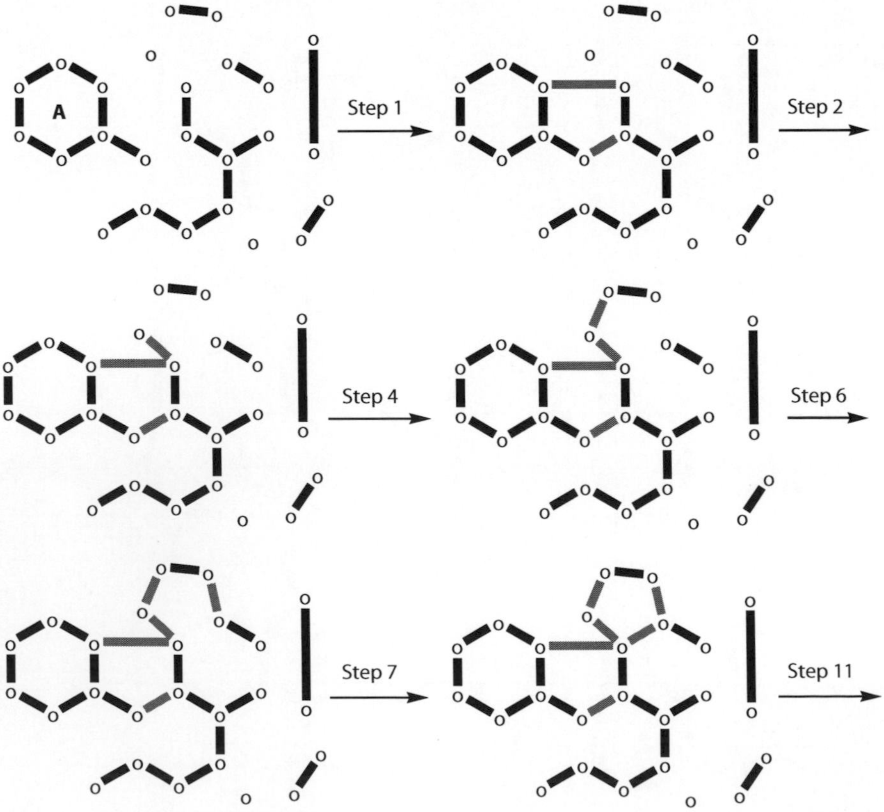

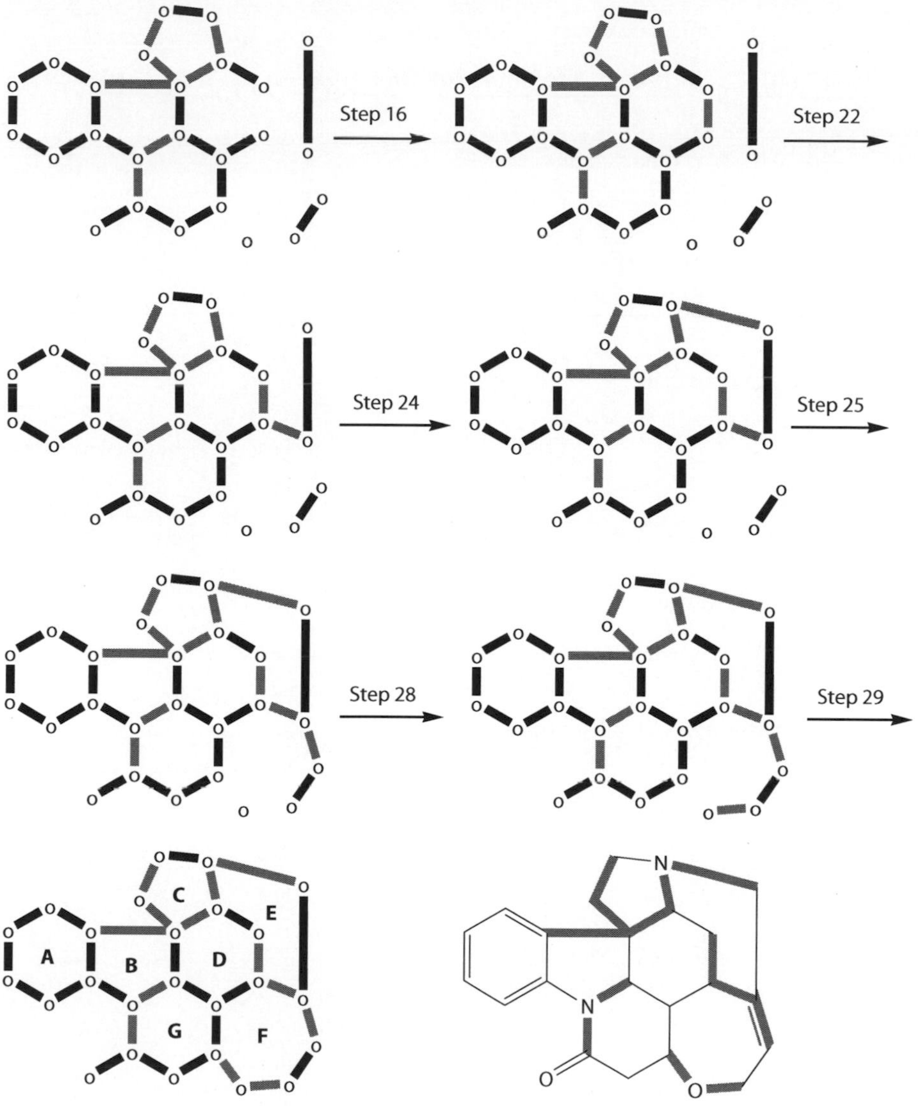

Sequence:

$$A_0 \xrightarrow{[3+2]} B_1 \xrightarrow{[5+0]} C_7 \xrightarrow{[6+0]} G_{11} \xrightarrow{[6+0]} D_{16} \xrightarrow{[6+0]} E_{24} \xrightarrow{[7+0]} F_{29}$$

Summary:
Strategy #1

$$A_0 \xrightarrow{[3+2]} B_1 \xrightarrow{[5+0]} C_7 \xrightarrow{[6+0]} G_{11} \xrightarrow{[6+0]} D_{16} \xrightarrow{[6+0]} E_{24}$$

$$\xrightarrow{[7+0]} F_{29} : \text{Woodward}$$

Strategy #2

$$A_{3*} \xrightarrow{[5+0]} B_{7*} \xrightarrow{[(4+1)_C + (6+0)_D + (6+0)_E]} (C+D+E)_{19} \xrightarrow{[7+0]} F_{23}$$

$$\xrightarrow{[4+2]} G_{25}: \text{Fukuyama}$$

Strategy #3

$$A_{10*} \xrightarrow{[6+0]} E_{18} \xrightarrow{[(2+2+1)_C + (6+0)_D]} [4.3.0](C+D)_{20} \xrightarrow{[5+0]} B_{22} \xrightarrow{[7+0]} F_{25}$$

$$\xrightarrow{[4+2]} G_{26}: \text{Overman}$$

Strategy #4

$$A_0 \xrightarrow{[5+0]} C_3 \xrightarrow{[(5+0)_B + (4+2)_D]} [4.3.0](B+D)_9 \xrightarrow{[6+0]} G_{10} \xrightarrow{[6+0]} E_{12}$$

$$\xrightarrow{[7+0]} F_{14}: \text{Rawal}$$

Strategy #5

$$(A+B)_0 \xrightarrow{[6+0]} E_{13} \xrightarrow{[(5+0)_C + (6+0)_D]} [4.3.0](C+D)_{19} \xrightarrow{[7+0]} F_{31}$$

$$\xrightarrow{[4+2]} G_{33}: \text{Magnus}$$

$$(A+B)_0 \xrightarrow{[6+0]} E_4 \xrightarrow{[(5+0)_C + (4+2)_D]} [4.3.0](C+D)_{11} \xrightarrow{[7+0]} F_{15}$$

$$\xrightarrow{[4+2]} G_{16}: \text{Martin}$$

Strategy #6

$$(A+B)_0 \xrightarrow{[(5+0)_C + (4+2)_D]} [4.3.0](C+D)_5 \xrightarrow{[6+0]} E_{14} \xrightarrow{[7+0]} F_{16} \xrightarrow{[4+2]} G_{17}: \text{Padwa}$$

$$(A+B)_0 \xrightarrow{[(5+0)_C + (4+2)_D]} [4.3.0](C+D)_4 \xrightarrow{[6+0]} E_{12} \xrightarrow{[7+0]} F_{15} \xrightarrow{[4+2]} G_{16}: \text{Stork}$$

Strategy #7

$$(A+B)_0 \xrightarrow{[(4+1)_C + (3+2+1)_D]} [4.3.0](C+D)_5 \xrightarrow{[6+0]} E_8 \xrightarrow{[6+0]} G_{12}$$

$$\xrightarrow{[7+0]} F_{20}: \text{Kuehne}$$

Strategy #8

$$(A+B)_5 \xrightarrow{[(2+2+2)_D + (6+0)_G]} [4.4.0](D+G)_7 \xrightarrow{[5+0]} C_9 \xrightarrow{[6+0]} E_{12}$$

$$\xrightarrow{[7+0]} F_{14}: \text{Vollhardt}$$

Strategy #9

$$(A + B)_3 \xrightarrow{[(5+0)_C + (4+2)_D + (6+0)_G]} (C + D + G)_8 \xrightarrow{[6+0]} E_{13} \xrightarrow{[7+0]} F_{15} : \text{Bodwell}$$

Strategy #10

$$A_{15} + D_0 \xrightarrow{[5+0]_B + [6+0]_E} B_{20} + E_{20} \xrightarrow{[5+0]} C_{21} \xrightarrow{[7+0]} F_{29} \xrightarrow{[4+2]} G_{31} : \text{Shibasaki}$$

Strategy #11

$$A_3 + D_0 \xrightarrow{[5+0]} B_8 \xrightarrow{[5+0]} C_{11} \xrightarrow{[6+0]} G_{18} \xrightarrow{[6+0]} E_{21} \xrightarrow{[7+0]} F_{24} : \text{Mori}$$

Strategy #12

$$A_0 + D_0 \xrightarrow{[4+1]} C_5 \xrightarrow{[6+0]} E_{13} \xrightarrow{[5+0]} B_{15} \xrightarrow{[(7+0)_F + (4+2)_G]} [5.4.0] (F + G)_{18} : \text{Bonjoch}$$

Strategy #13

$$(A + B)_0 \xrightarrow{[(4+2)_D + (6+0)_G]} [4.4.0] (D + G)_3 \xrightarrow{[5+0]} C_4 \xrightarrow{[6+0]} E_8 \xrightarrow{[7+0]} F_{10} : \text{Reissig}$$

(±)-Swainsonine
Martin-Lopez, M.J.; Rodriguez, R.; Bermejo, F. *Tetrahedron* 1998, *54*, 11623.

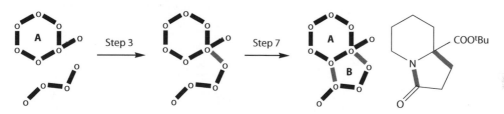

Sequence: $A_0 \xrightarrow{[5+0]} B_7$

(−)-Swainsonine
Buschmann, N.; Ruckert, A.; Blechert, S. *J. Org. Chem.* 2002, *67*, 4325.

Sequence: $\xrightarrow{[5+0]} B_6 \xrightarrow{[6+0]} A_{10}$

(−)-Swainsonine
de Vicente, J.; Arrayas, R.G.; Canada, J.; Carretero, J.C. *Synlett* 2000, 53.

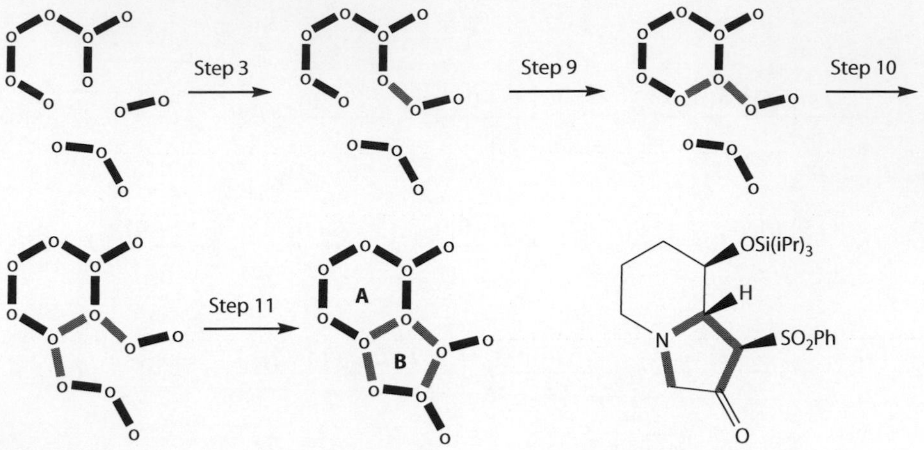

Sequence: $\xrightarrow{[6+0]} A_9 \xrightarrow{[5+0]} B_{11}$

(−)-Swainsonine
Bennett, R.B. III; Choi, J.R.; Montgomery, W.D.; Cha, J.K. *J. Am. Chem. Soc.* 1989, *111*, 2580.

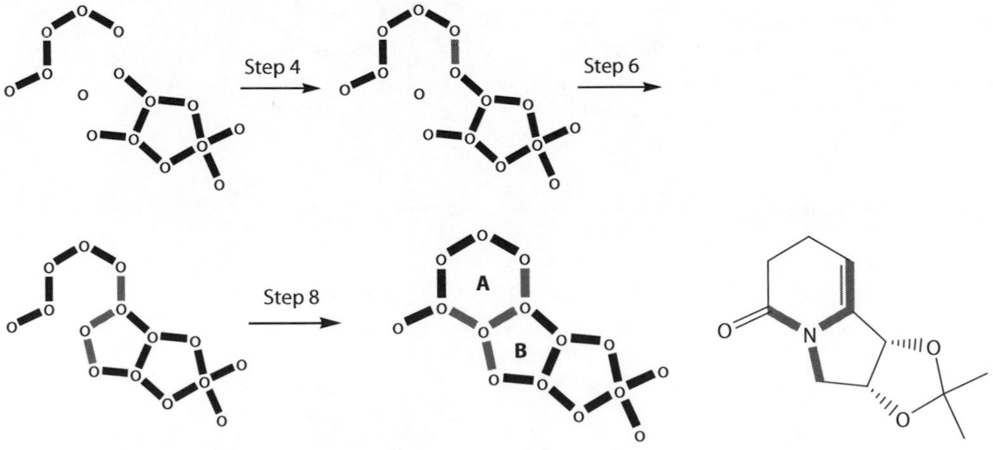

Sequence: $\xrightarrow{[4+1]} B_6 \xrightarrow{[6+0]} A_8$

(−)-Swainsonine
Dechamps, I.; Pardo, D.G.; Cossy, J. *Tetrahedron* 2007, *63*, 9082.

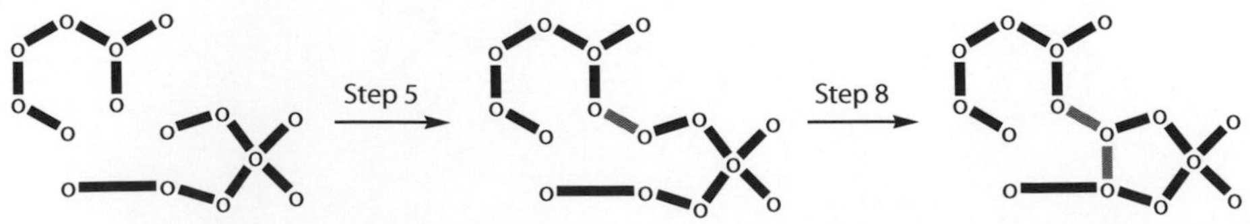

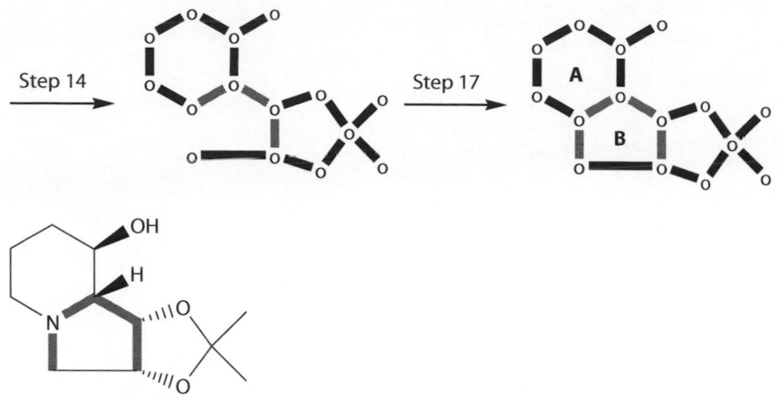

Step 14
Step 17
A
B
OH
H
N
O
O

Sequence: $\xrightarrow{[6+0]} A_{14} \xrightarrow{[5+0]} B_{17}$

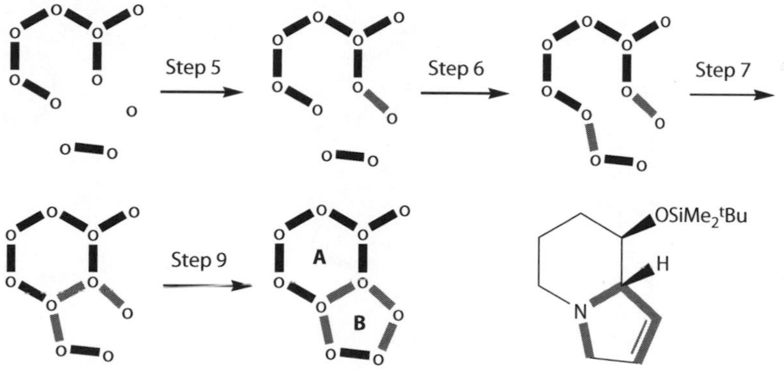

Step 5
Step 6
Step 7
Step 9
A
B
OSiMe2tBu
H
N

Sequence: $\xrightarrow{[6+0]} A_{7} \xrightarrow{[5+0]} B_{9}$

(–)-Swainsonine

Ferreira, F.; Greck, C.; Genet, J.P. *Bull. Soc. Chim. Fr.* 1997, *134*, 615.

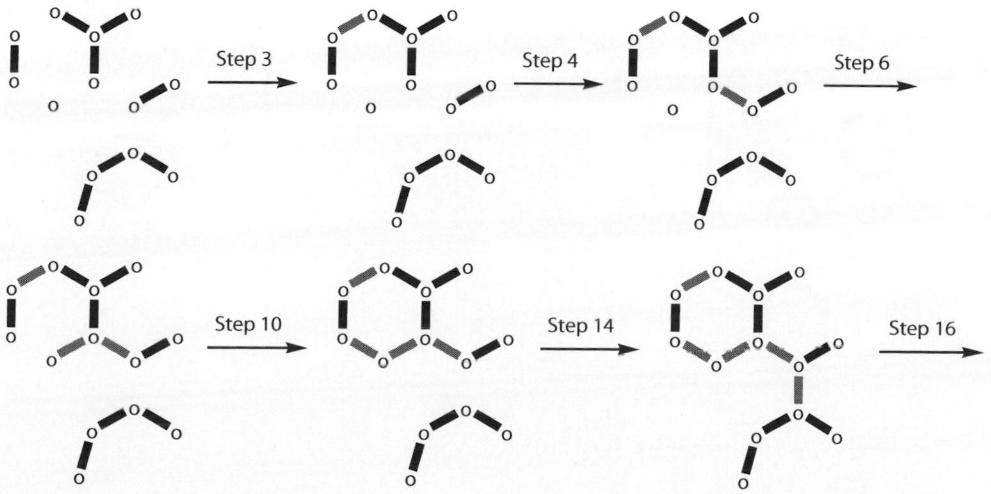

Step 3
Step 4
Step 6
Step 10
Step 14
Step 16

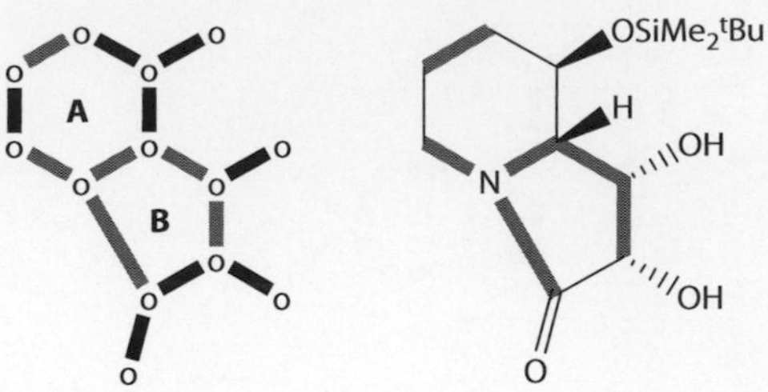

Sequence: $\xrightarrow{[6+0]} A_{10} \xrightarrow{[5+0]} B_{16}$

(–)-Swainsonine
Fleet, G.W.J.; Gough, M.J.; Smith, P.W. *Tetrahedron Lett.* 1984, *25*, 1853.

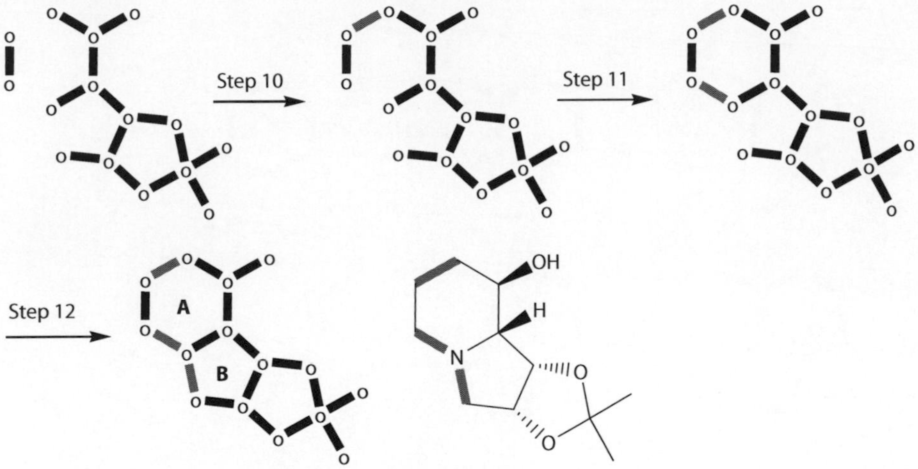

Sequence: $\xrightarrow{[6+0]} A_{11} \xrightarrow{[5+0]} B_{12}$

(–)-Swainsonine hydrochloride
Sharma, P.K.; Shah, R.N.; Carver, J.P. *Org. Process. Res. Develop.* 2008, *12*, 831.

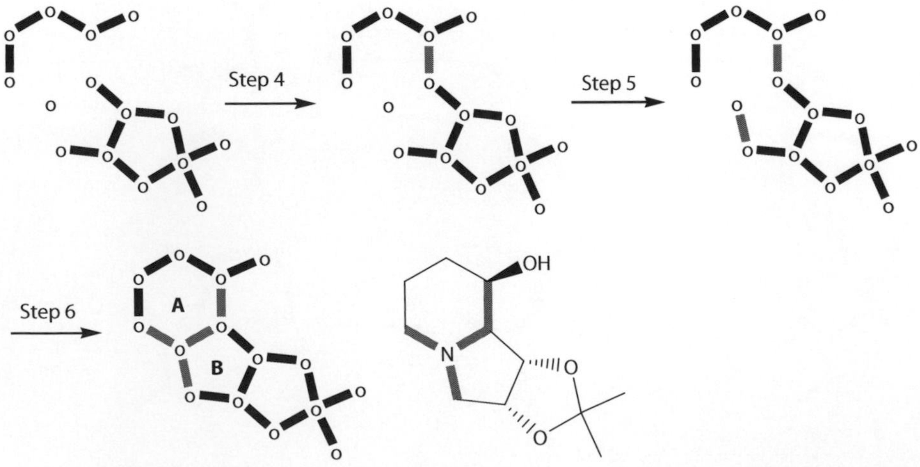

Sequence: $\xrightarrow{[(5+1)_A + (5+0)_B]} [4.3.0] (A + B)_6$

(−)-Swainsonine

Tian, Y.S.; Joo, J.E.; Kong, B.S.; Pham, V.T.; Lee, K.Y.; Ham, W.H. *J. Org. Chem.* 2009, *74*, 3962.

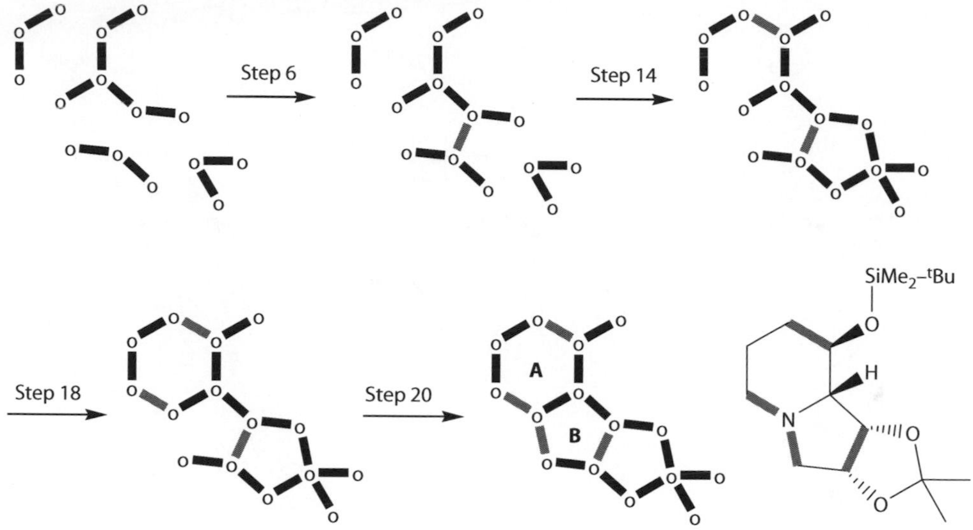

Sequence: $\xrightarrow{[6+0]} A_{14} \xrightarrow{[5+0]} B_{20}$

(−)-Swainsonine

Setoi, H.; Takeno, H.; Hashimoto, M. *J. Org. Chem.* 1985, *50*, 3950.

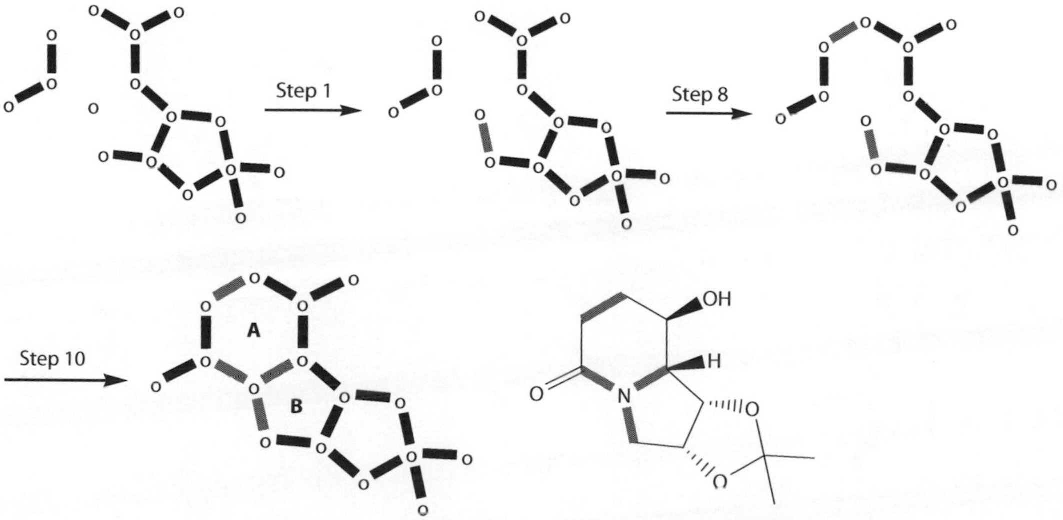

Sequence: $\xrightarrow{[(5+1)_A + (5+0)_B]} [4.3.0] (A + B)_{10}$

(±)-Swainsonine
Oishi, T.; Iwakuma, T.; Hirama, M.; Ito, S. *Synlett* 1995, 404.

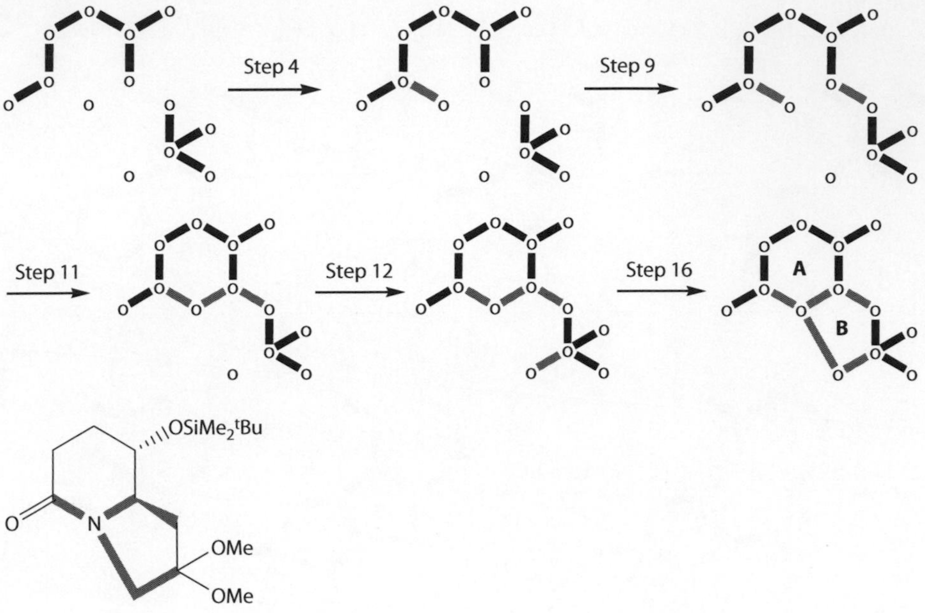

Sequence: $\xrightarrow{[6+0]} A_{11} \xrightarrow{[5+0]} B_{16}$

(−)-Swainsonine
Ikota, N.; Hanaki, A. *Chem. Pharm. Bull.* 1990, *38*, 2712.

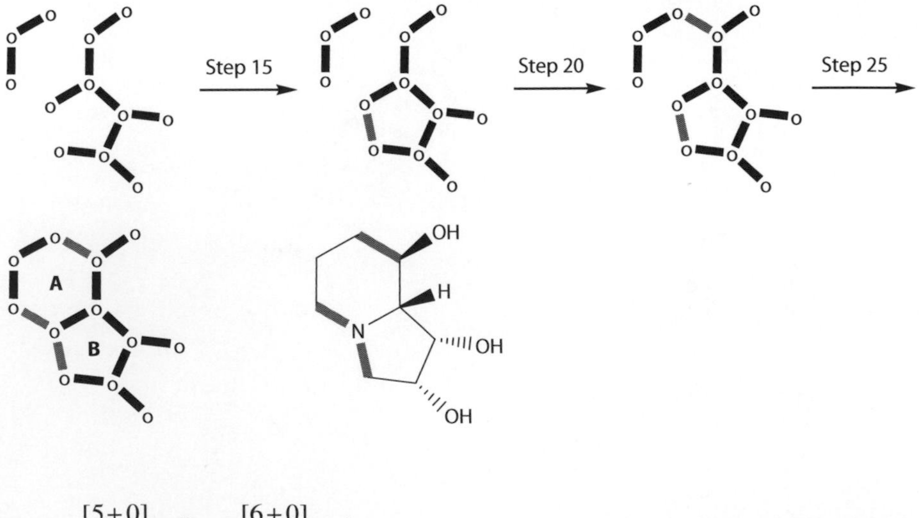

Sequence: $\xrightarrow{[5+0]} B_{15} \xrightarrow{[6+0]} A_{25}$

(−)-Swainsonine
Kang, S.H.; Kim, G.T. *Tetrahedron Lett.* 1995, *36*, 5049.

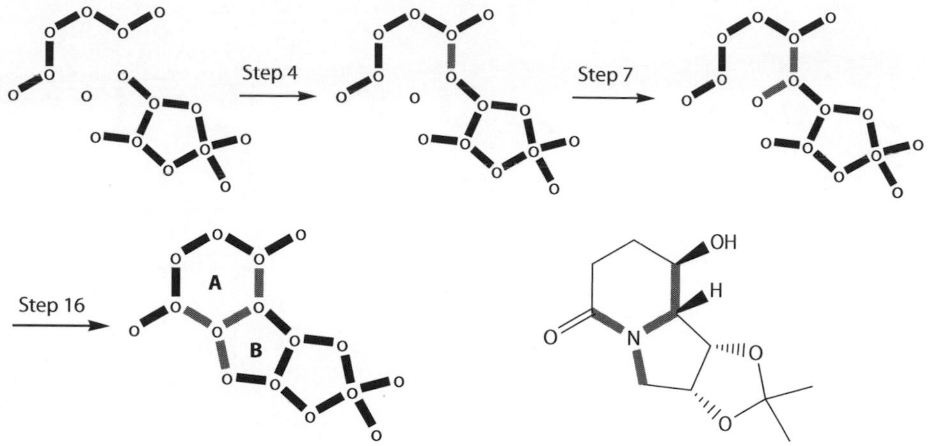

Sequence: $\xrightarrow{[(6+0)_A + (5+0)_B]} [4.3.0]\,(A+B)_{16}$

(−)-Swainsonine
Naruse, M.; Aoyagi, S.; Kibayashi, C. *J. Org. Chem.* 1994, *59*, 1358.

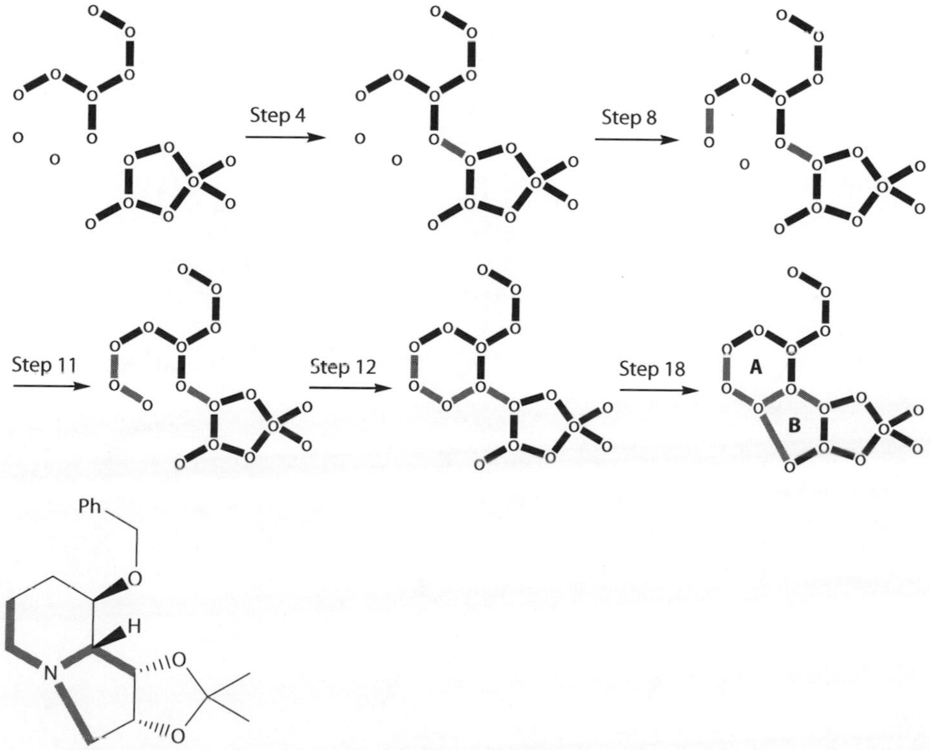

Sequence: $\xrightarrow{[6+0]} A_{12} \xrightarrow{[5+0]} B_{18}$

(−)-Swainsonine
Zhao, H.; Hans, S.; Cheng, X.; Mootoo, D.R. *J. Org. Chem.* 2001, *66*, 1761.

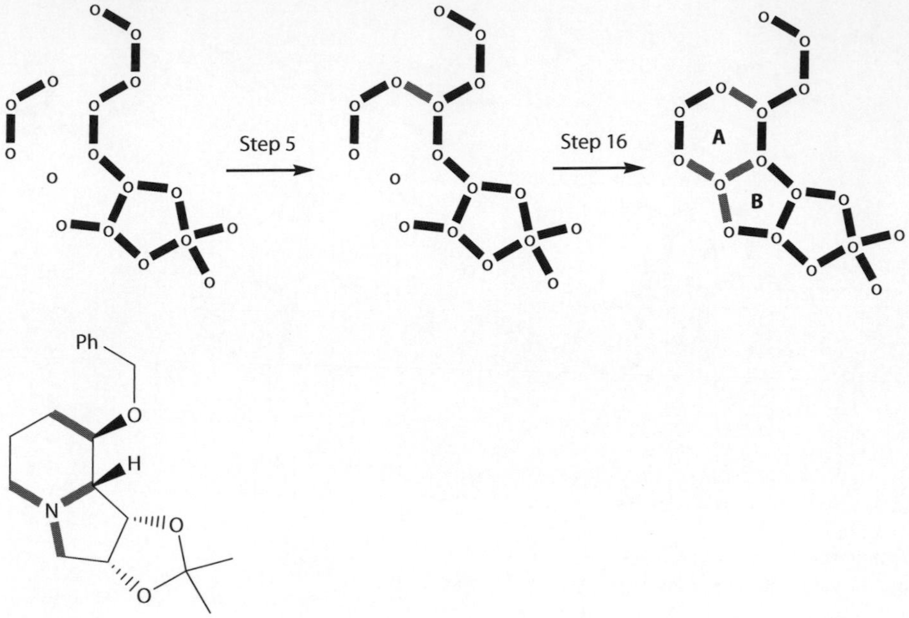

Sequence: $\xrightarrow{[(5+1)_A+(4+1)_B]}$ [4.3.0] $(A+B)_{16}$

(±)-Swainsonine
Guo, H.; O'Doherty, G.A. *Org. Lett.* 2006, *8*, 1609.

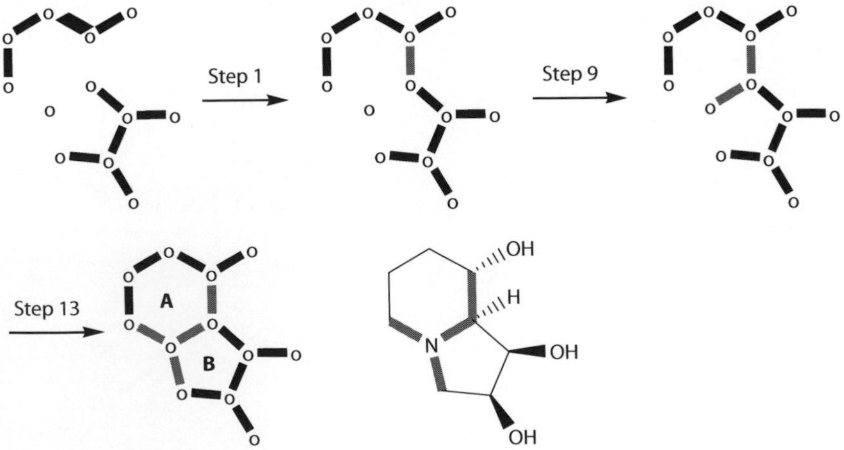

Sequence: $\xrightarrow{[(6+0)_A+(5+0)_B]}$ [4.3.0] $(A+B)_{13}$

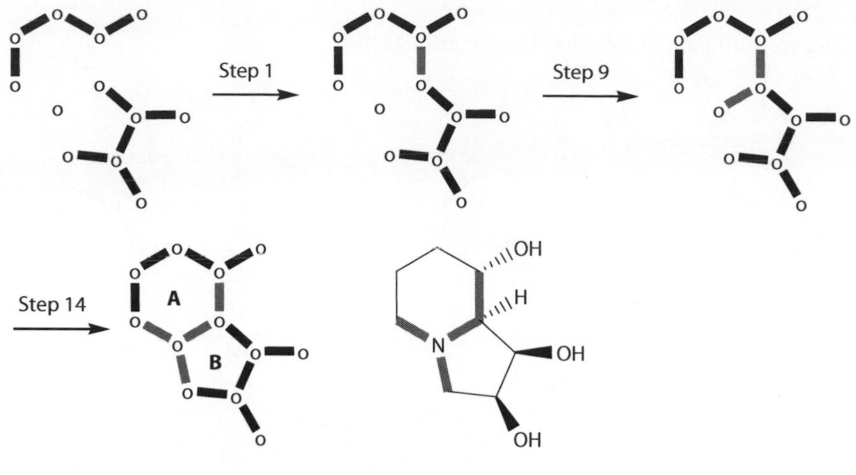

Sequence: $\xrightarrow{[(6+0)_A + (5+0)_B]} [4.3.0]\,(A+B)_{14}$

(–)-Swainsonine

Pearson, W.H.; Hembre, E.J. *J. Org. Chem.* 1996, *61*, 7217.

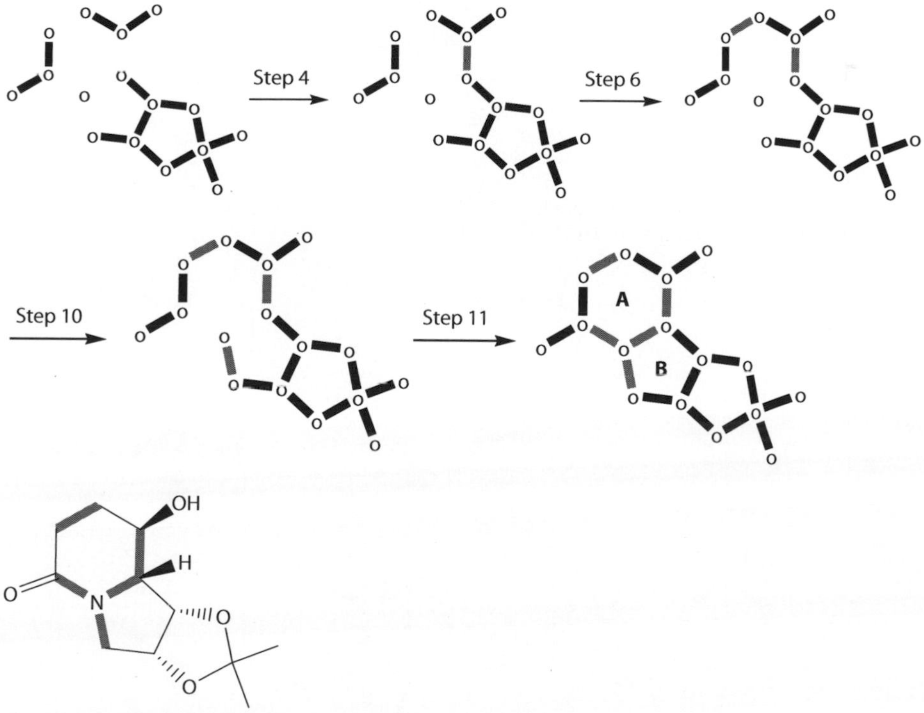

Sequence: $\xrightarrow{[(5+1)_A + (5+0)_B]} [4.3.0]\,(A+B)_{11}$

(–)-Swainsonine
Ceccon, J.; Greene, A.E.; Poisson, J.F. *Org. Lett.* 2006, *8*, 4739.

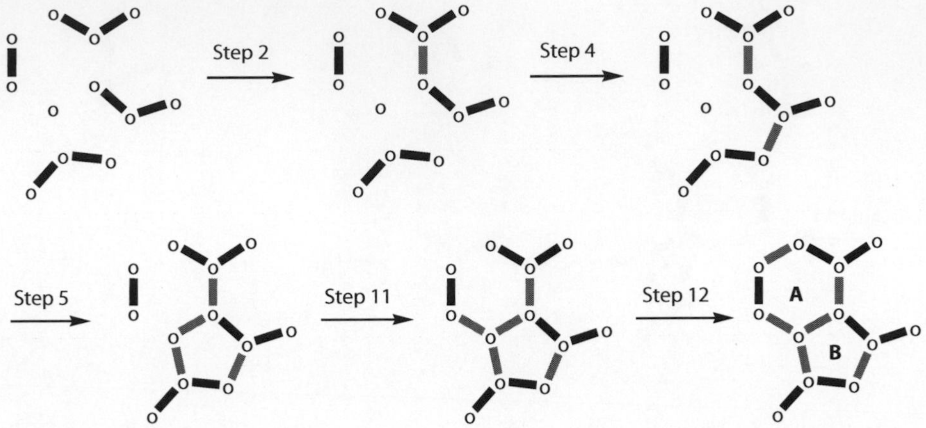

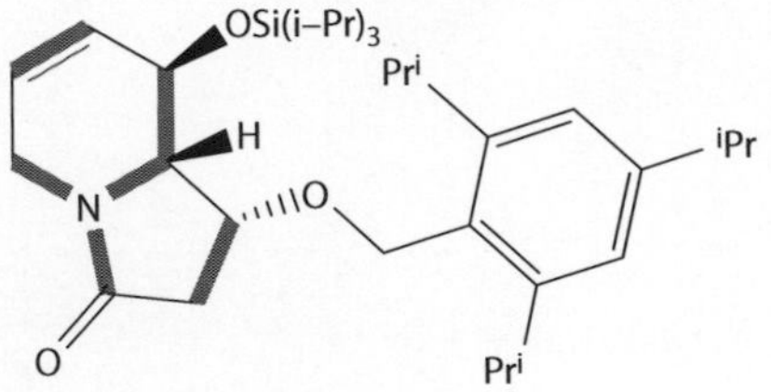

Sequence: $\xrightarrow{[4+1]} B_5 \xrightarrow{[6+0]} A_{12}$

(–)-Swainsonine
Lindsay, K.B.; Pyne, S.G. *Austr. J. Chem.* 2004, *57*, 669.
Lindsay, K.B.; Pyne, S.G. *J. Org. Chem.* 2002, *67*, 7774.

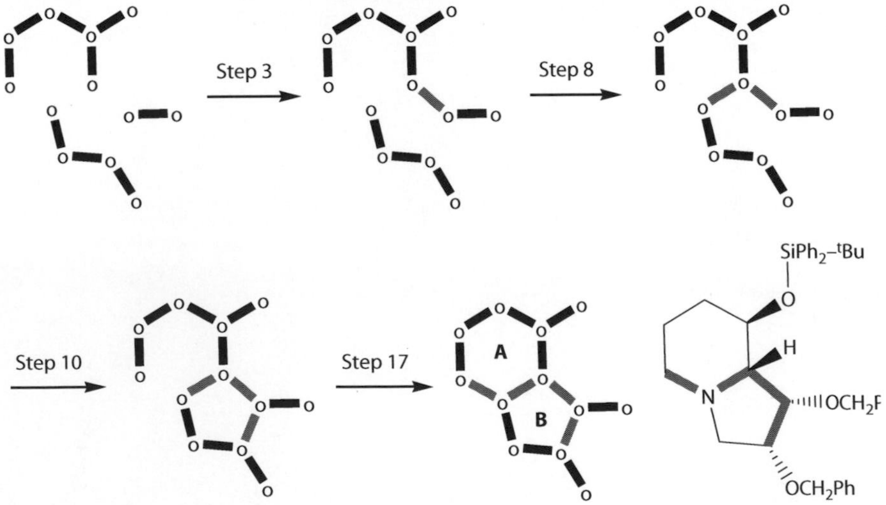

Sequence: $\xrightarrow{[5+0]} B_{10} \xrightarrow{[6+0]} A_{17}$

Au, C.W.G.; Pyne, S.G. *J. Org. Chem.* 2006, *71*, 7097.

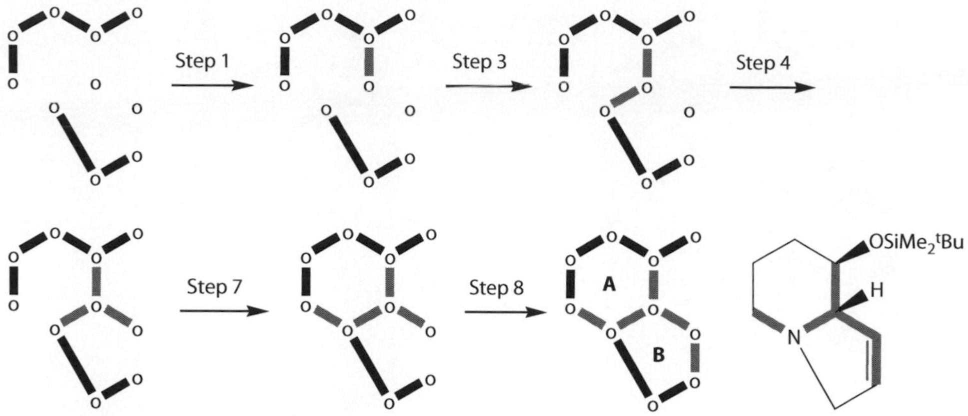

Sequence: $\xrightarrow{[6+0]} A_7 \xrightarrow{[5+0]} B_8$

(−)-Swainsonine

Heimgartner, G.; Raatz, D.; Reiser, O. *Tetrahedron* 2005, *61*, 643.

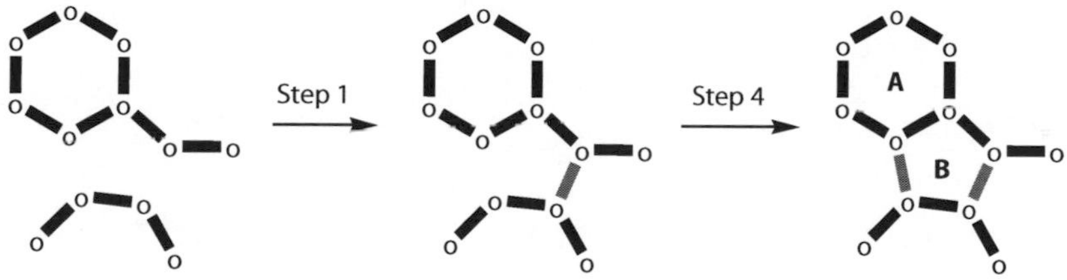

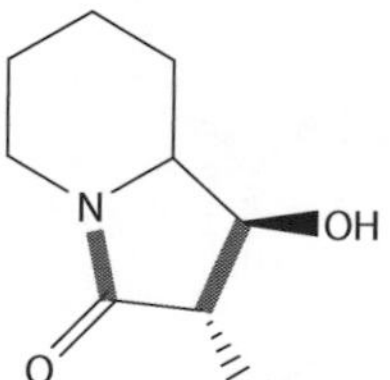

Sequence: $A_0 \xrightarrow{[5+0]} B_4$

(−)-Swainsonine

Ali, M.H.; Hough, L.; Richardson, A.C. *Carbohydrate Res.* 1985, *136*, 225.
Ali, M.H.; Hough, L.; Richardson, A.C. *Chem. Commun.* 1984, 447.

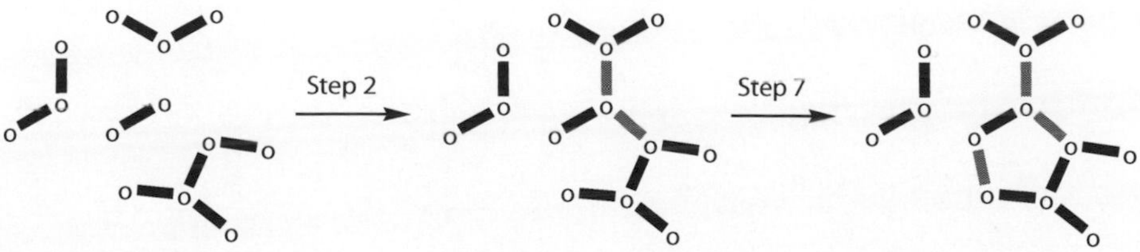

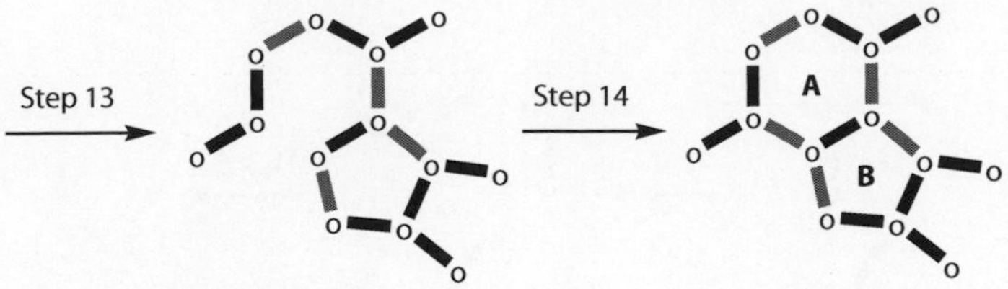

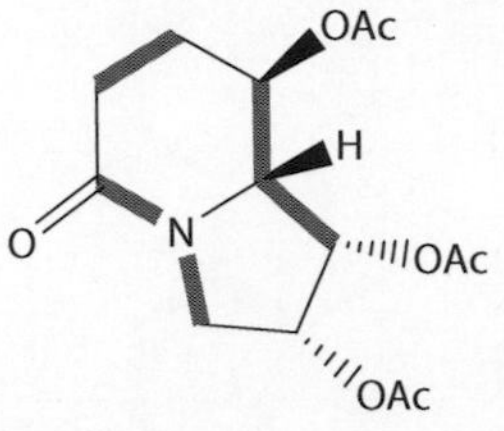

Step 13
Step 14
A
B

OAc
H
O
N
OAc
OAc

Sequence: $\xrightarrow{[5+0]} B_7 \xrightarrow{[6+0]} A_{14}$

(–)-Swainsonine

Martin, R.; Murruzzu, C.; Pericas, M.A.; Riera, A. *J. Org. Chem.* 2005, *70*, 2325.

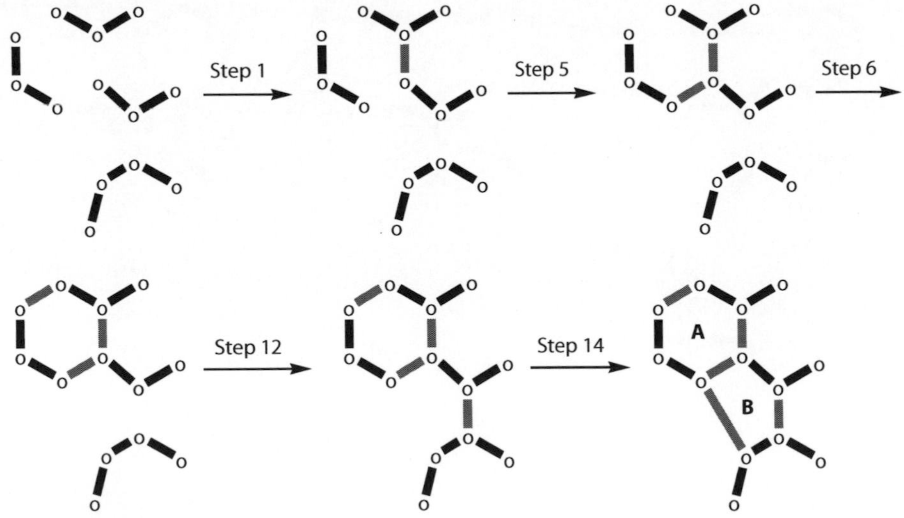

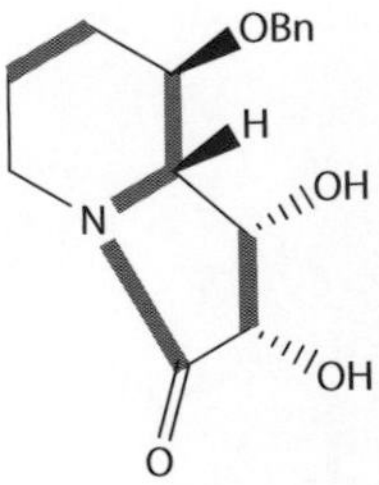

Step 1
Step 5
Step 6
Step 12
Step 14
A
B

OBn
H
N
OH
OH
O

Sequence: $\xrightarrow{[6+0]} A_6 \xrightarrow{[5+0]} B_{14}$

(−)-Swainsonine

Hunt, J.A.; Roush, W.R. *Tetrahedron Lett*. 1995, *36*, 501.

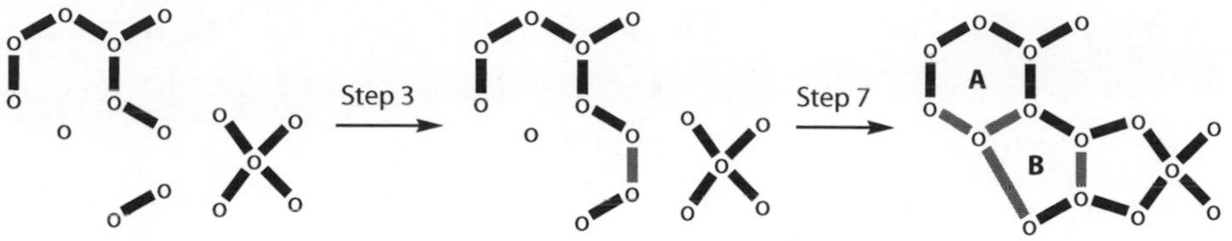

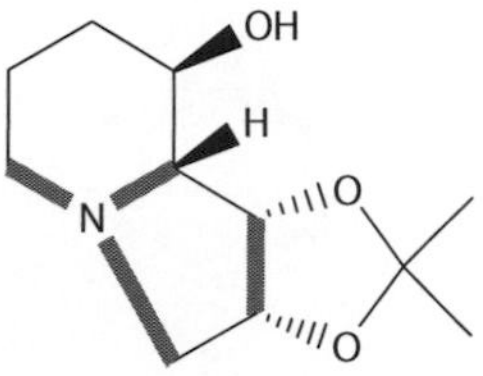

Sequence: $\xrightarrow{[(5+1)_A + (4+1)_B]}$ $[4.3.0] (A+B)_7$

(−)-Swainsonine

Adams, C.E.; Walker, F.J.; Sharpless, K.B. *J. Org. Chem*. 1985, *50*, 420.

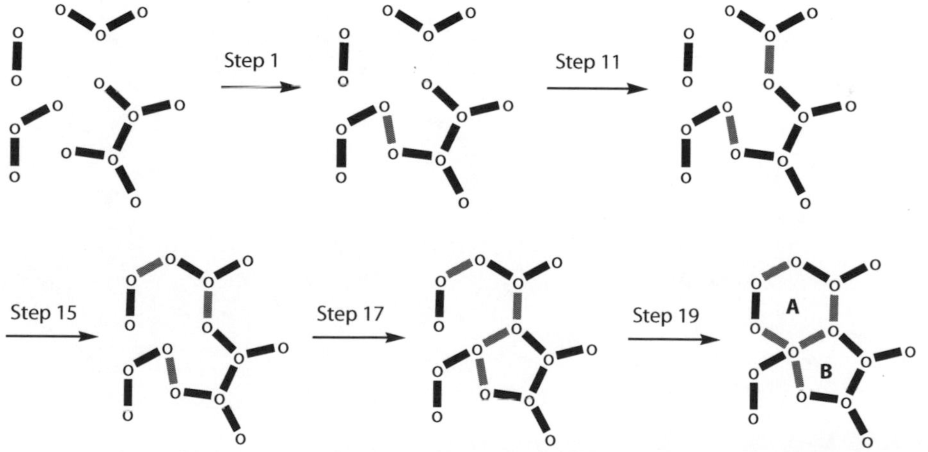

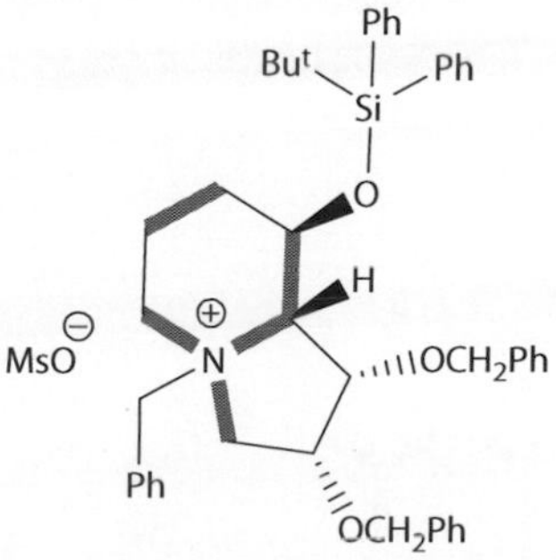

Sequence: $\xrightarrow{[5+0]} B_{17} \xrightarrow{[6+0]} A_{19}$

(–)-Swainsonine
Suami, T.; Tadano, K.; Iimura, Y. *Chem. Lett*. 1984, 513.

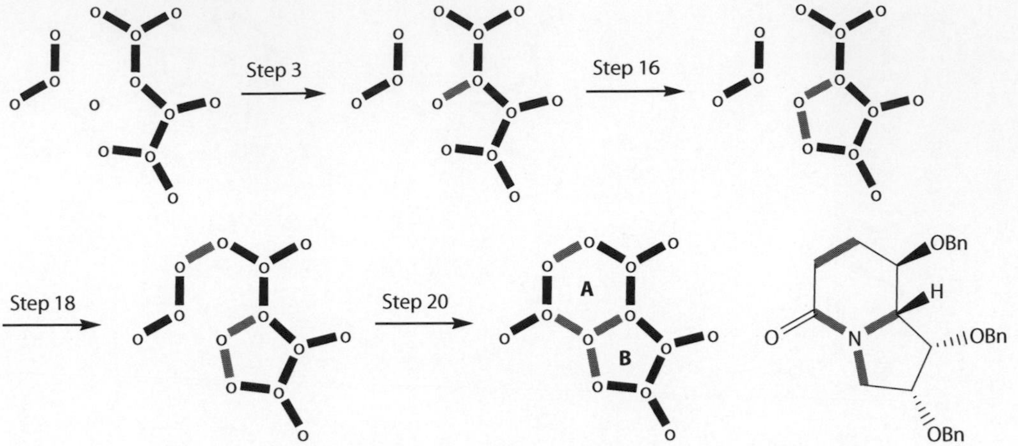

Sequence: $\xrightarrow{[5+0]} B_{16} \xrightarrow{[6+0]} A_{20}$

Suami, T.; Tadano, K.; Iimura, Y. *Carbohydrate Res*. 1985, *136*, 67.

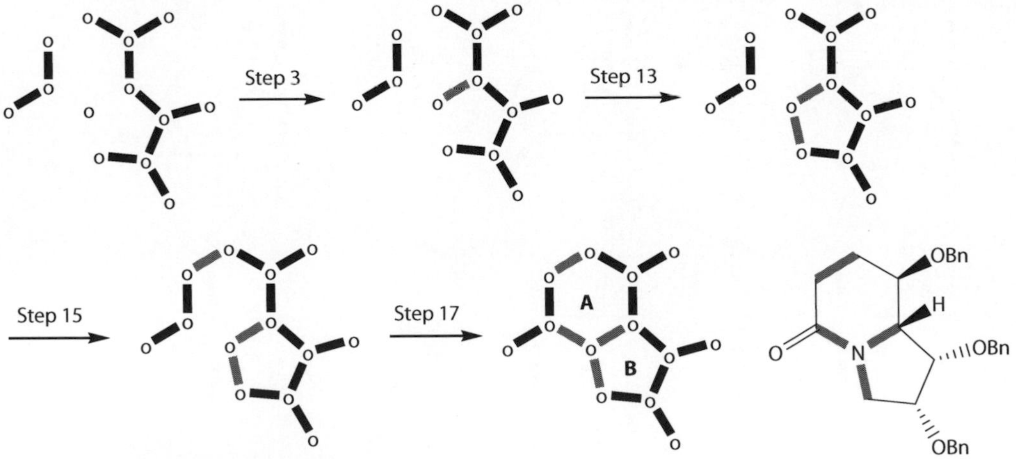

Sequence: $\xrightarrow{[5+0]} B_{13} \xrightarrow{[6+0]} A_{17}$

(–)-Swainsonine
Yasuda, N.; Tsutsumi, H.; Takaya, T. *Chem. Lett*. 1984, 1201.

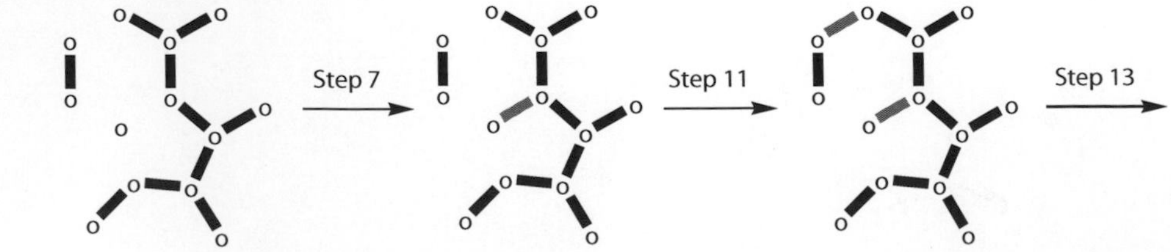

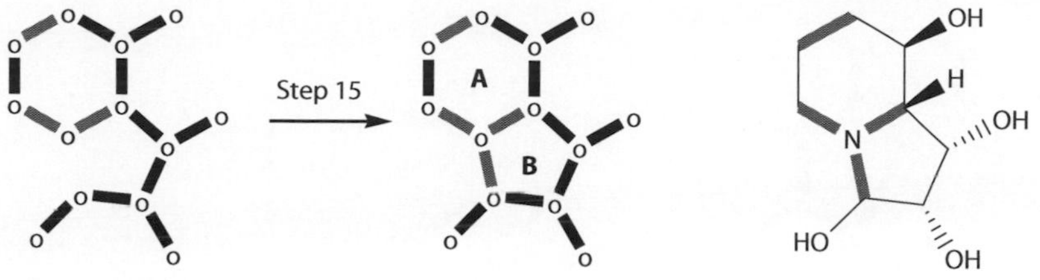

Sequence: $\xrightarrow{[6+0]} A_{13} \xrightarrow{[5+0]} B_{15}$

(−)-Swainsonine
Trost, B.M.; Patterson, D.E. *Chem. Eur. J.* 1995, 5, 3279.

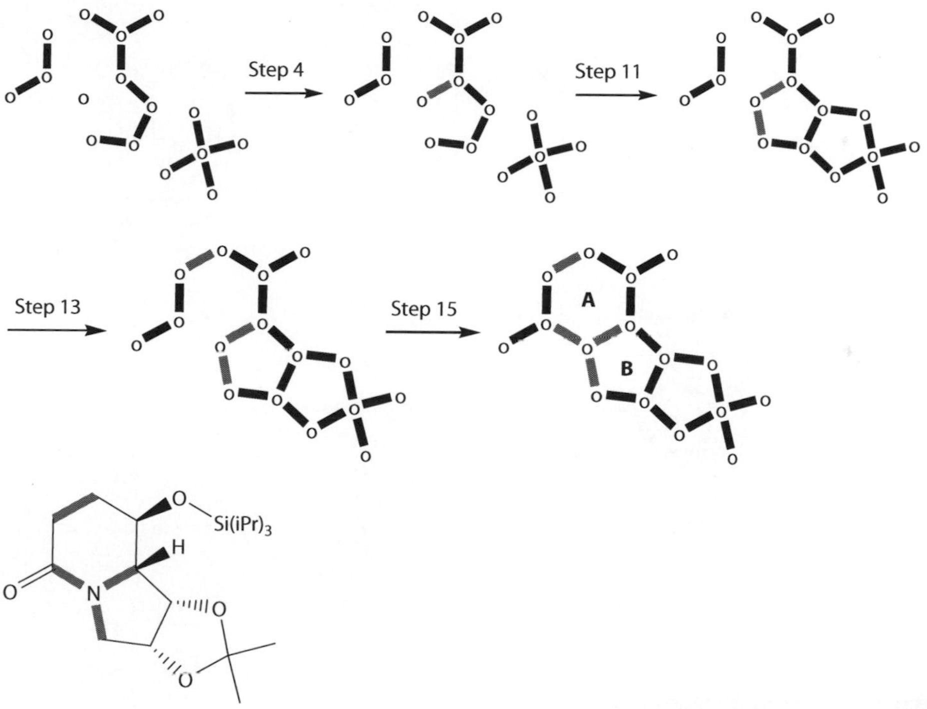

Sequence: $\xrightarrow{[5+0]} B_{11} \xrightarrow{[6+0]} A_{15}$

(±)-Swainsonine
Alam, M.A.; Kumar, A.; Vankar, Y.D. *Eur. J. Org. Chem.* 2009, 4972.

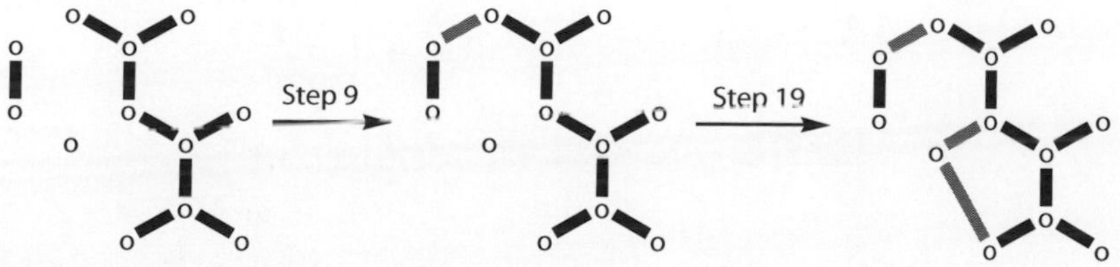

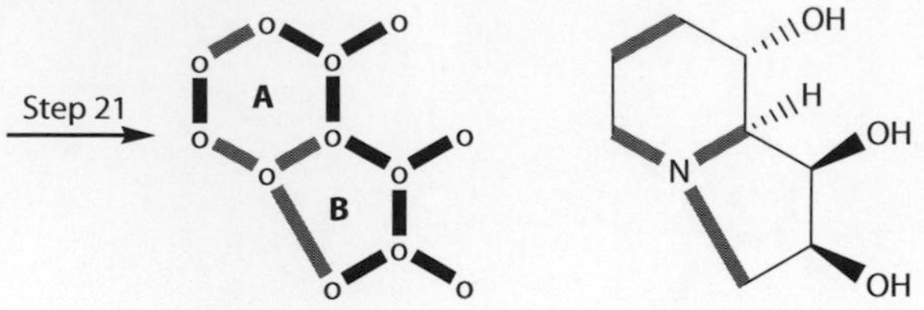

Sequence: $\xrightarrow{[4+1]} B_{19} \xrightarrow{[6+0]} A_{21}$

(−)-Swainsonine

Zhou, W.S.; Xie, W.G.; Lu, Z.H.; Pan, X.F. *Tetrahedron Lett.* 1995, *36*, 1291.
Zhou, W.S.; Xie, W.G.; Lu, Z.H.; Pan, X.F. *J. Chem. Soc. Perkin Trans. I* 1995, 2599.

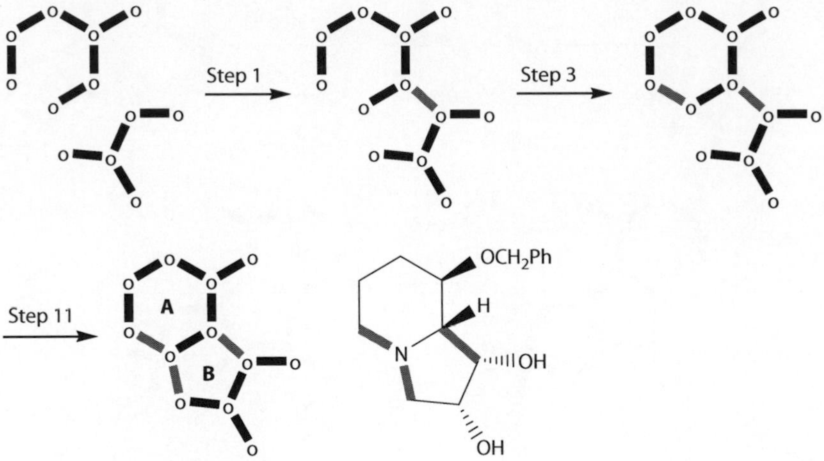

Sequence: $\xrightarrow{[6+0]} A_3 \xrightarrow{[5+0]} B_{11}$

Summary:
Strategy #1

$\xrightarrow{[6+0]} A_9 \xrightarrow{[5+0]} B_{11}$: Carretero (19.6)

$\xrightarrow{[6+0]} A_{14} \xrightarrow{[5+0]} B_{17}$: Cossy G1 (635.4)

$\xrightarrow{[6+0]} A_7 \xrightarrow{[5+0]} B_9$: Cossy G2 (14.7)

$\xrightarrow{[6+0]} A_{10} \xrightarrow{[5+0]} B_{16}$: Ferreira (21.1)

$$\xrightarrow{[6+0]} A_{11} \xrightarrow{[5+0]} B_{12}: \textbf{Fleet (8.3)}$$

$$\xrightarrow{[6+0]} A_{11} \xrightarrow{[5+0]} B_{16}: \text{Hirama (151.1)}$$

$$\xrightarrow{[6+0]} A_{12} \xrightarrow{[5+0]} B_{18}: \text{Kibayashi (70.9)}$$

$$\xrightarrow{[6+0]} A_{7} \xrightarrow{[5+0]} B_{8}: \text{Pyne G2 (45.8)}$$

$$\xrightarrow{[6+0]} A_{6} \xrightarrow{[5+0]} B_{14}: \text{Riera (28.7)}$$

$$\xrightarrow{[6+0]} A_{13} \xrightarrow{[5+0]} B_{15}: \text{Takaya (1248.5)}$$

$$\xrightarrow{[6+0]} A_{3} \xrightarrow{[5+0]} B_{11}: \text{Zhou (38.9)}$$

$$\xrightarrow{[6+0]} A_{14} \xrightarrow{[5+0]} B_{20}: \text{Ham (21.0)}$$

Strategy #2a

$$\xrightarrow{[5+0]} B_{6} \xrightarrow{[6+0]} A_{10}: \textbf{Blechert (7.3)}$$

$$\xrightarrow{[5+0]} B_{15} \xrightarrow{[6+0]} A_{25}: \text{Ikota (495.4)}$$

$$\xrightarrow{[5+0]} B_{10} \xrightarrow{[6+0]} A_{17}: \text{Pyne G1 (23.0)}$$

$$\xrightarrow{[5+0]} B_{7} \xrightarrow{[6+0]} A_{14}: \text{Richardson (181.9)}$$

$$\xrightarrow{[5+0]} B_{17} \xrightarrow{[6+0]} A_{19}: \text{Sharpless (29.4)}$$

$$\xrightarrow{[5+0]} B_{16} \xrightarrow{[6+0]} A_{20}: \text{Suami G1 (296.0)}$$

$$\xrightarrow{[5+0]} B_{13} \xrightarrow{[6+0]} A_{17}: \text{Suami G2 (375.3)}$$

$$\xrightarrow{[5+0]} B_{11} \xrightarrow{[6+0]} A_{15}: \text{Trost (10.7)}$$

Strategy #2b

$$\xrightarrow{[4+1]} B_6 \xrightarrow{[6+0]} A_g: \text{Cha (10.4)}$$

$$\xrightarrow{[4+1]} B_5 \xrightarrow{[6+0]} A_{12}: \text{Poisson (53.0)}$$

$$\xrightarrow{[4+1]} B_{19} \xrightarrow{[6+0]} A_{21}: \textbf{Vankar (9.7)}$$

Strategy #3a

$$\xrightarrow{[(5+1)_A + (5+0)_B]} [4.3.0] (A+B)_6: \textbf{GlycoDesign (6.8)}$$

$$\xrightarrow{[(5+1)_A + (5+0)_B]} [4.3.0] (A+B)_{10}: \text{Hashimoto (47.6)}$$

$$\xrightarrow{[(5+1)_A + (5+0)_B]} [4.3.0] (A+B)_{11}: \text{Pearson (18.4)}$$

Strategy #3b

$$\xrightarrow{[(5+1)_A + (4+1)_B]} [4.3.0] (A+B)_{16}: \text{Mootoo (17.4)}$$

$$\xrightarrow{[(5+1)_A + (4+1)_B]} [4.3.0] (A+B)_7: \textbf{Roush (7.1)}$$

Strategy #3c

$$\xrightarrow{[(6+0)_A + (5+0)_B]} [4.3.0] (A+B)_{16}: \text{Kang (15.7)}$$

$$\xrightarrow{[(6+0)_A + (5+0)_B]} [4.3.0] (A+B)_{13}: \textbf{O'Doherty G1 (16.0)}$$

$$\xrightarrow{[(6+0)_A + (5+0)_B]} [4.3.0] (A+B)_{14}: \text{O'Doherty G2 (7.1)}$$

Strategy #4

$$A_0 \xrightarrow{[5+0]} B_7: \textbf{Bermejo (23.6)}$$

$$A_0 \xrightarrow{[5+0]} B_4: \text{Reiser (35.6)}$$

kernel waste in parentheses (kg per mole target product)

(−)-Tetracycline hydrochloride

Charest, M.G.; Siegel, D.R.; Myers, A.G. *J. Am. Chem. Soc.* 2005, *127*, 8292.
Charest, M.G.; Lerner, C.D.; Brubaker, J.D.; Siegel, D.R.; Myers, A.G. *Science* 2005, *308*, 395.

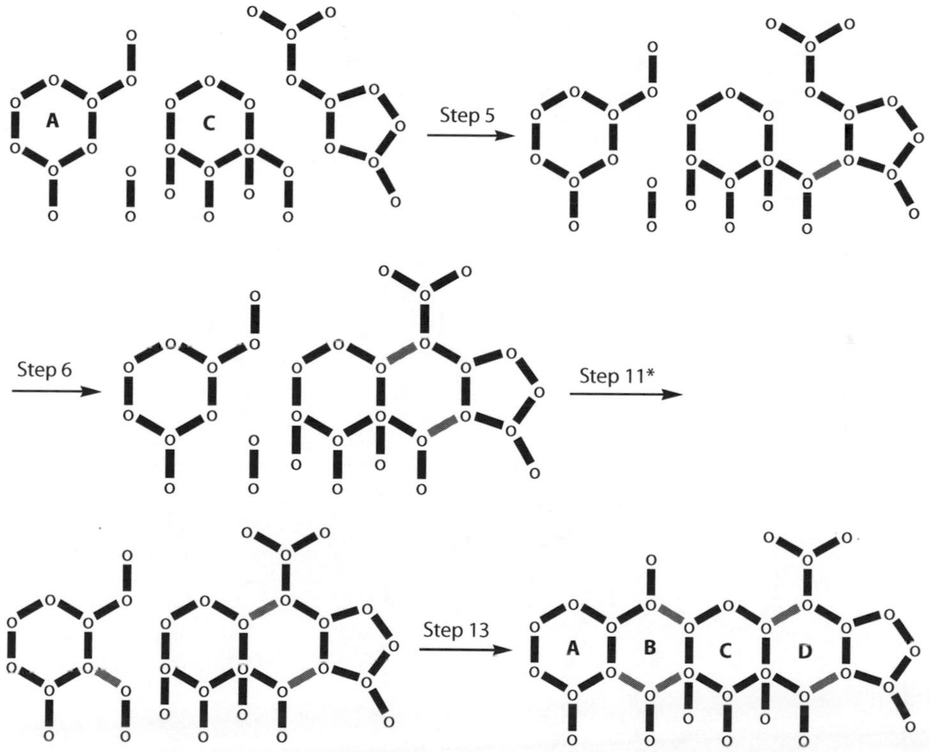

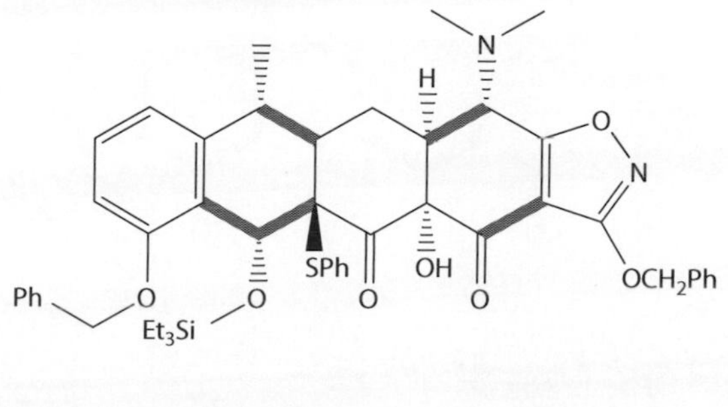

Sequence: $A_{7*} + C_{0*} \xrightarrow{[6+0]} D_6 \xrightarrow{[4+2]} B_{13}$

(−)-Tetracycline hydrochloride

Brubaker, J.D.; Myers, A.G. *Org. Lett.* 2007, *9*, 3523.
Charest, M.G.; Siegel, D.R.; Myers, A.G. *J. Am. Chem. Soc.* 2005, *127*, 8292.
Charest, M.G.; Lerner, C.D.; Brubaker, J.D.; Siegel, D.R.; Myers, A.G. *Science* 2005, *308*, 395.

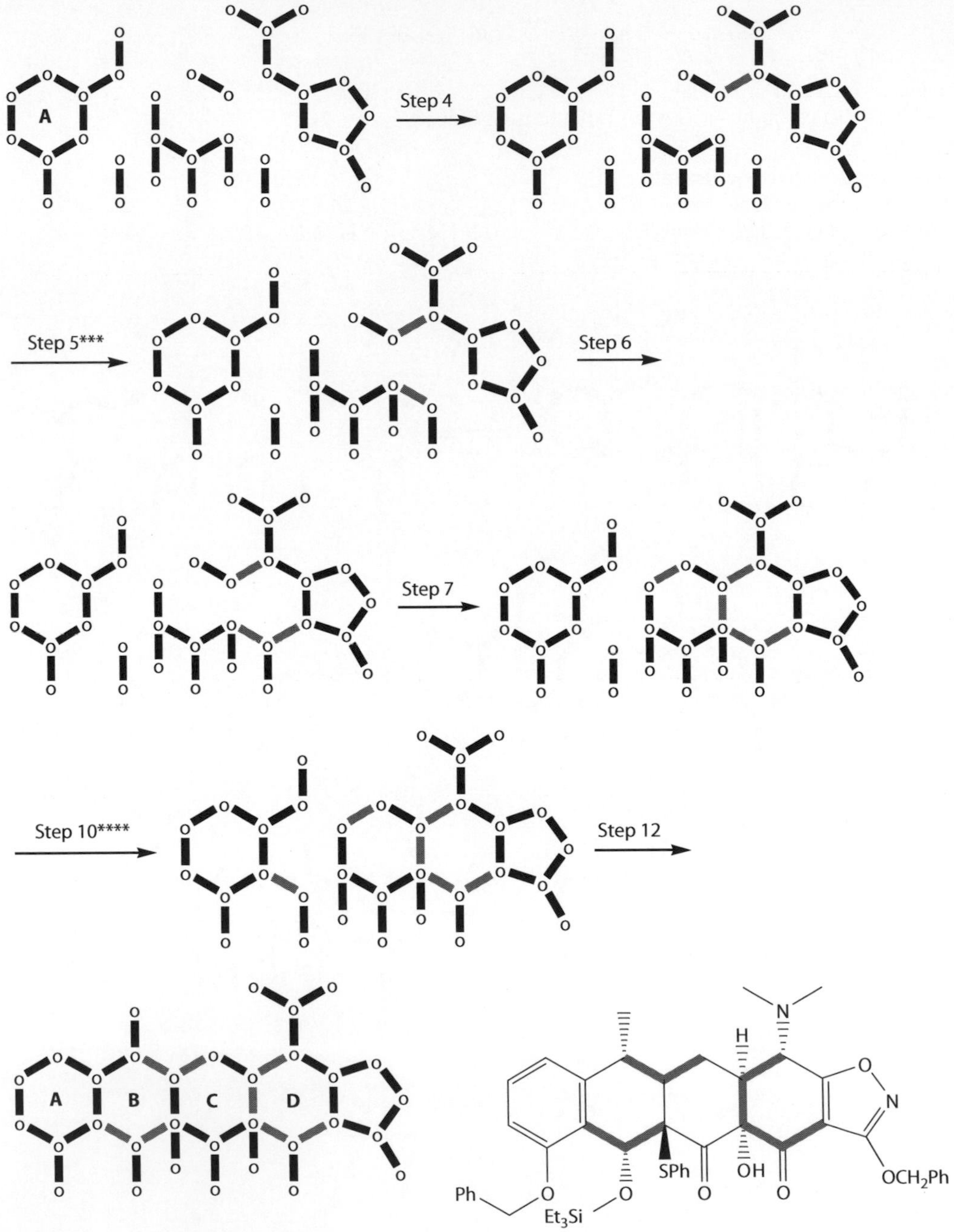

Sequence: $A_{6****} \xrightarrow{[(4+2)_C + (6+0)_D]} [4.4.0](C+D)_7 \xrightarrow{[4+2]} B_{12}$

(−)-Tetracycline hydrochloride

Kolosov, M.N.; Popravko, S.A.; Shemyakin, M.M. *Ann. Chem.* 1963, *668*, 86.

Gurevich, A.I.; Karapetyan, M.G.; Kolosov, M.N.; Korobko, V.G.; Onoprienko, V.V.; Popravko, S.A.; Shemyakin, M.M. *Tetrahedron Lett*. 1967, 131.
Kolosov, M.N.; Vasina, I.V.; Shemyakin, M.M. *Zh. Obsh. Khim*. 1964, *34*, 2534 (CA 61:84024).

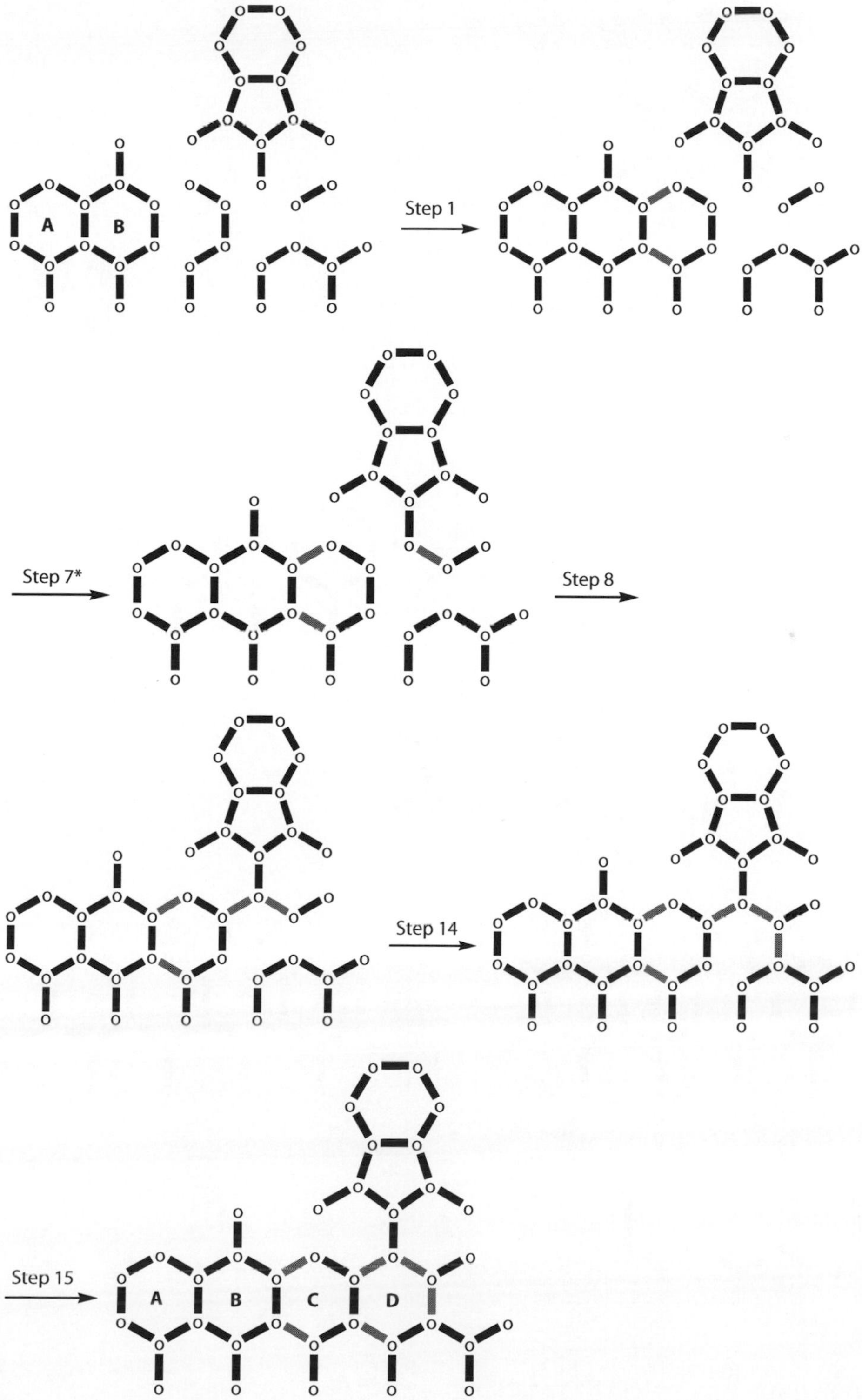

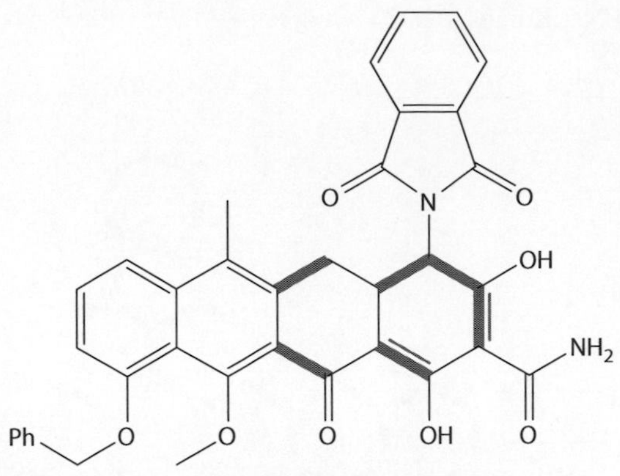

Sequence: $(A+B)_0 \xrightarrow{[4+2]} C_1 \xrightarrow{[6+0]} D_{15}$

(−)-Tetracycline hydrochloride

Tatsuta, K.; Yoshimoto, T.; Gunji, H.; Okado, Y.; Takahashi, M. *Chem. Lett.* 2000, 646.

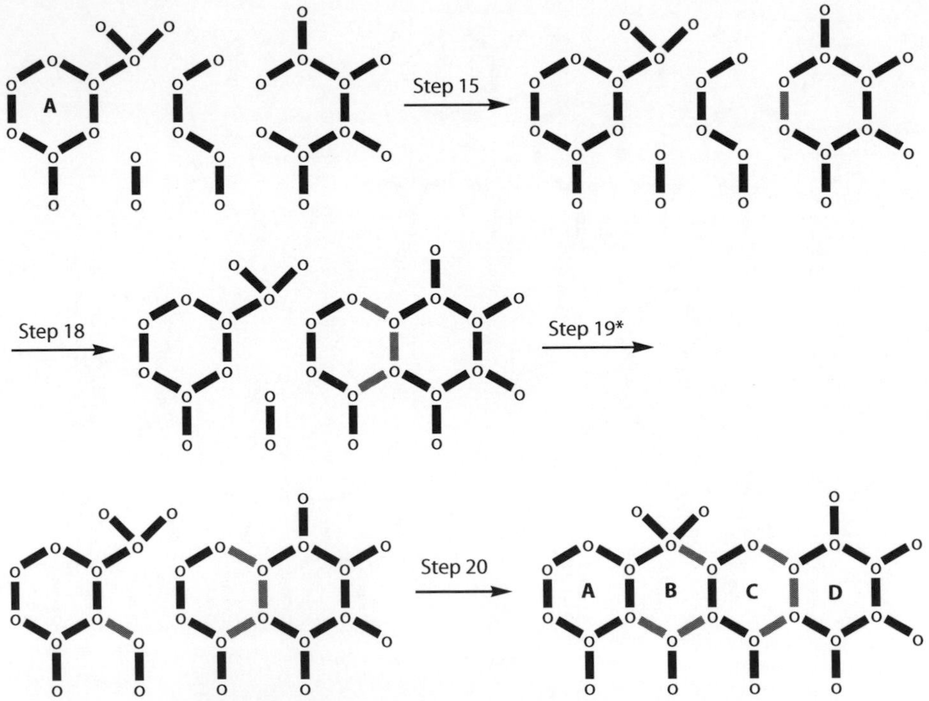

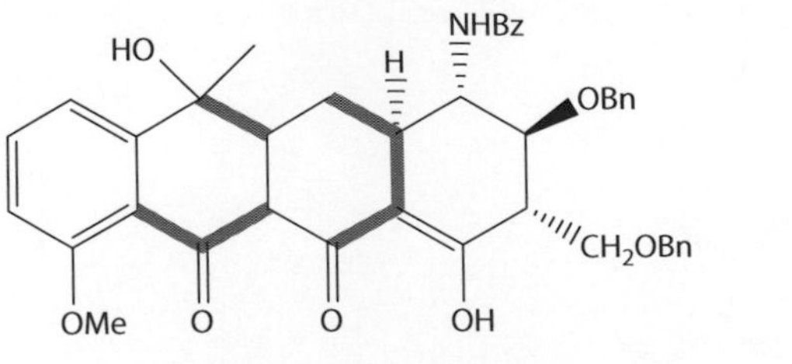

Sequence: $A_{16*} \xrightarrow{[6+0]} D_{15} \xrightarrow{[4+2]} C_{18} \xrightarrow{[4+2]} B_{20}$

Summary:

$$A_{7*} + C_{0*} \xrightarrow{[6+0]} D_6 \xrightarrow{[4+2]} B_{13} : \text{Myers G1}$$

$$A_{6****} \xrightarrow{[(4+2)_C + (6+0)_D]} [4.4.0] (C+D)_7 \xrightarrow{[4+2]} B_{12} : \text{Myers G2}$$

$$(A+B)_0 \xrightarrow{[4+2]} C_1 \xrightarrow{[6+0]} D_{15} : \text{Shemyakin}$$

$$A_{16*} \xrightarrow{[6+0]} D_{15} \xrightarrow{[4+2]} C_{18} \xrightarrow{[4+2]} B_{20} : \text{Tatsuta}$$

(–)-α-Thujone

Oppolzer, W.; Pimm, A.; Stammen, B.; Hume, W.E. *Helv. Chim. Acta* 1997, *80*, 623.

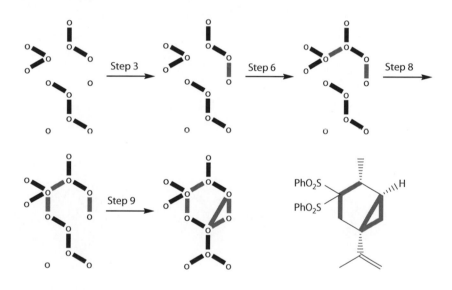

Sequence: $\xrightarrow{[(5+0)+(2+1)]} [3.1.0]_9$

Tropinone

Bazilevskaya, G.I.; Gura, D.V.; Bainova, M.S.; Dyumaev, K.M.; Sarycheva, I.K.; Preobrazhenskii, N.A. *Zh. Obsh. Khim.* 1958, *28*, 1097.

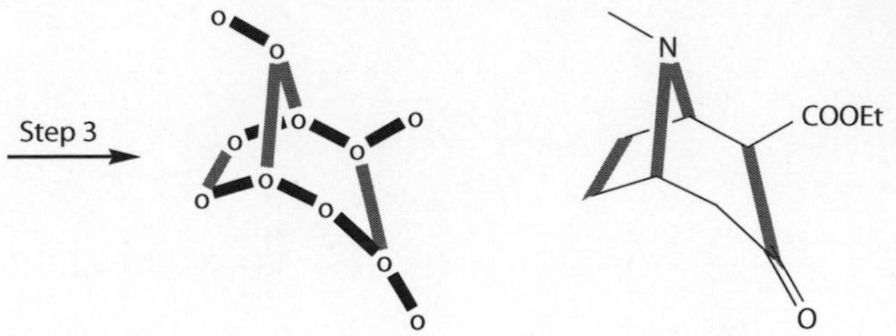

Sequence: $\xrightarrow{[4+1]} [5]_2 \xrightarrow{[(7+0)+(6+0)]} [3.2.1]_3$

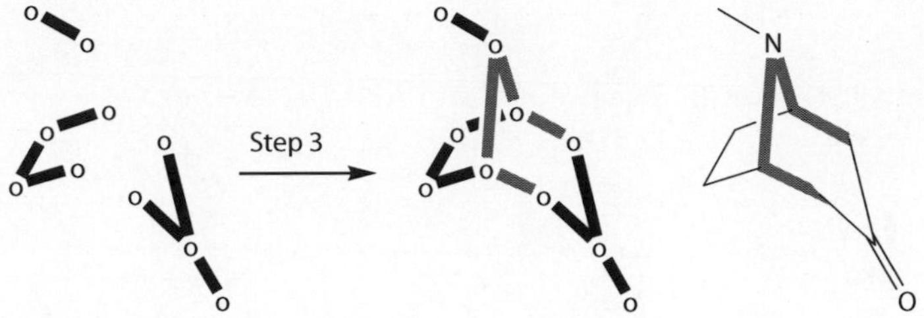

Sequence: $\xrightarrow{[(4+1)+(4+3)]} [3.2.1]_3$

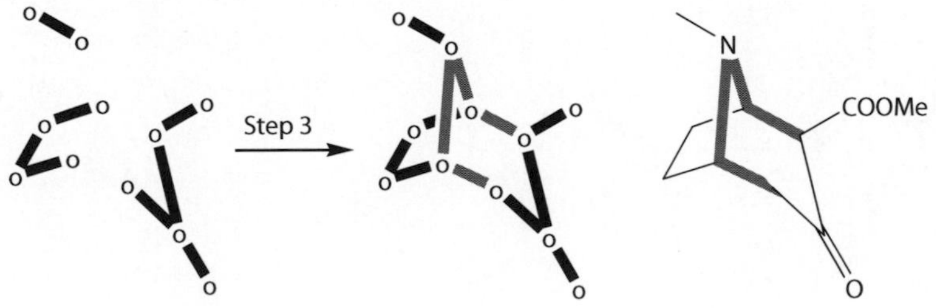

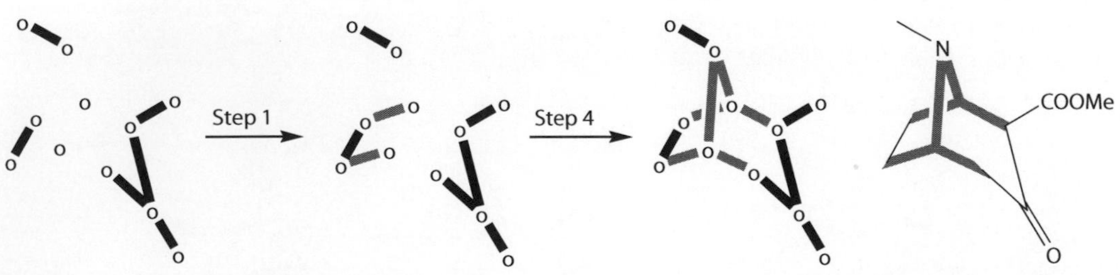

Sequence: $\xrightarrow{[(4+1)+(4+3)]} [3.2.1]_3$

Sequence: $\xrightarrow{[(4+1)+(4+3)]} [3.2.1]_4$

Karrer, P.; Alagil, H. *Helv. Chim. Acta* 1947, *30*, 1776.

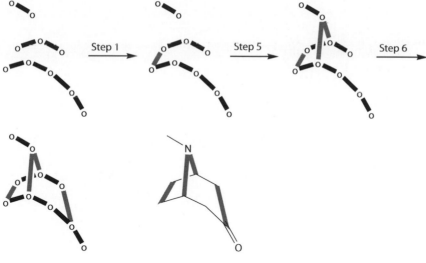

Sequence: $\xrightarrow{[4+1]}$ $[5]_5$ $\xrightarrow{[(7+0)+(6+0)]}$ $[3.2.1]_6$

Parker, W.; Raphael, R.A.; Wilkinson, D.I. *J. Chem. Soc.* 1959, 2433.

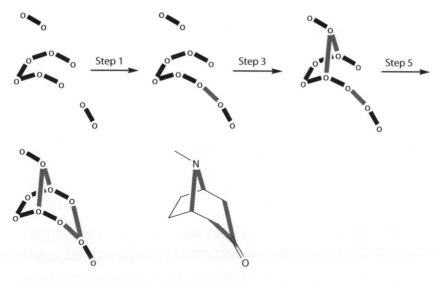

Sequence: $\xrightarrow{[4+1]}$ $[5]_3$ $\xrightarrow{[(7+0)+(6+0)]}$ $[3.2.1]_5$

Robinson, R. *J. Chem. Soc.* 1917, 762.

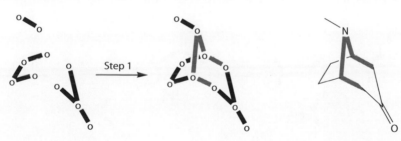

Sequence: $\xrightarrow{[(4+1)+(4+3)]}$ $[3.2.1]_1$

Willstätter, R. *Chem. Ber.* 1901, *34*, 129.
Willstätter, R. *Chem. Ber.* 1901, *34*, 3163.
Willstäatter, R. *Chem. Ber.* 1896, *29*, 936.

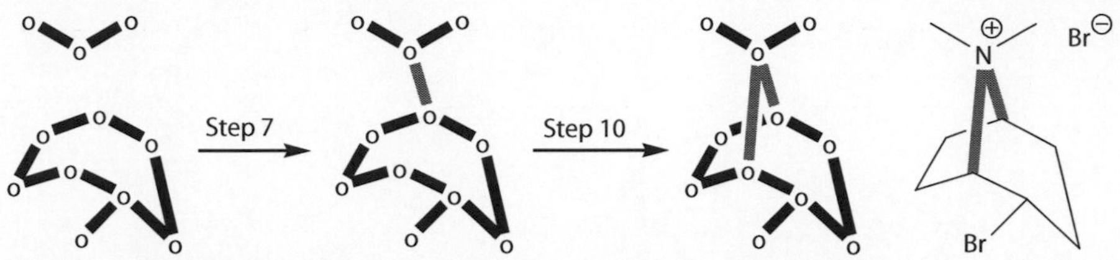

Sequence: $[7_0]$ $\xrightarrow{[(6+0)+(5+0)]}$ $[3.2.1]_{10}$

Willstätter, R.; Racke, F. *Ann. Chem.* 1921, *425*, 1.
Willstätter, R.; Bommer, M. *Ann. Chem.* 1921, *422*, 15.

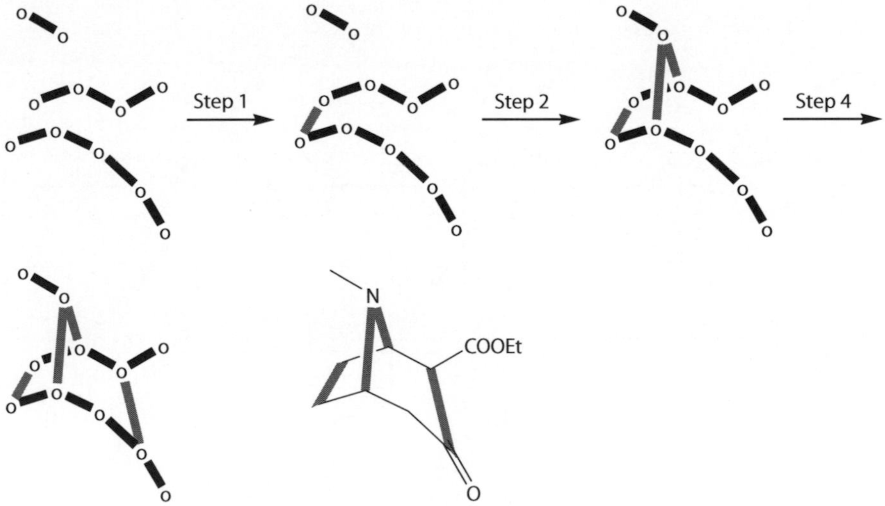

Sequence: $\xrightarrow{[4+1]}$ $[5]_2$ $\xrightarrow{[(7+0)+(6+0)]}$ $[3.2.1]_4$

(±)-Thienamycin

Hanessian, S.; Desilets, D.; Bennani, Y.L. *J. Org. Chem.* 1990, *55*, 3098.
Hanessian, S.; Bedeschi, A.; Battistini, C.; Mongelli, N. *J. Am. Chem. Soc.* 1985, *107*, 1438.

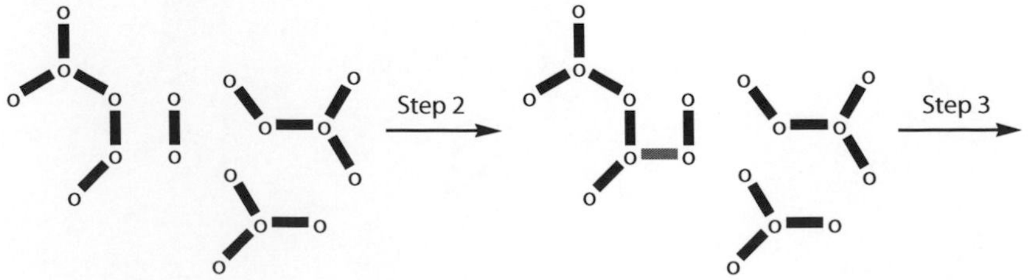

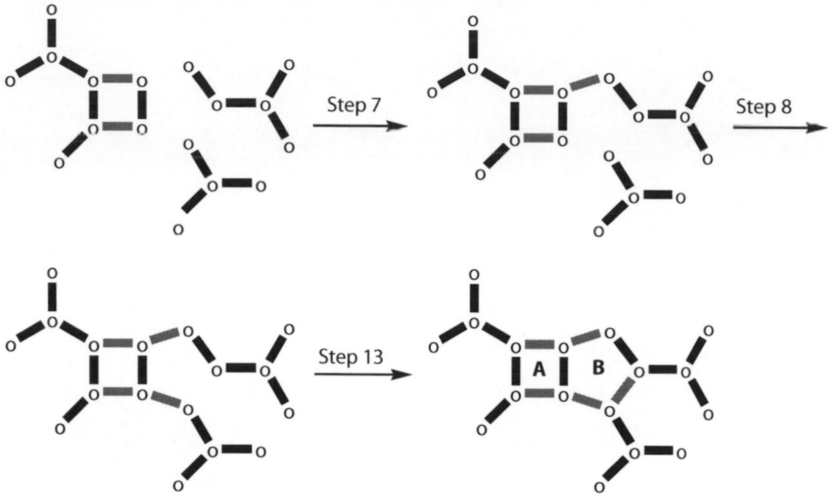

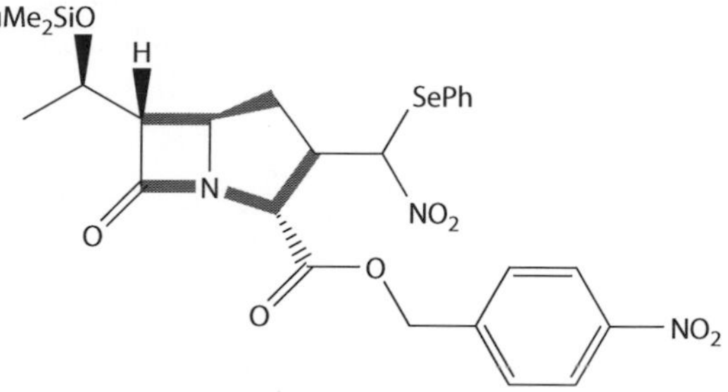

Sequence: $\xrightarrow{[4+0]} A_3 \xrightarrow{[5+0]} B_{13}$

(±)-Thienamycin

Melillo, D.G.; Liu, T.; Ryan, K.; Sletzinger, M.; Shinkai, T. *Tetrahedron Lett.* 1981, *22*, 913.
Melillo, D.G.; Shinkai, I.; Liu, T.; Ryan, K.; Sletzinger, M. *Tetrahedron Lett.* 1980, *21*, 2783.

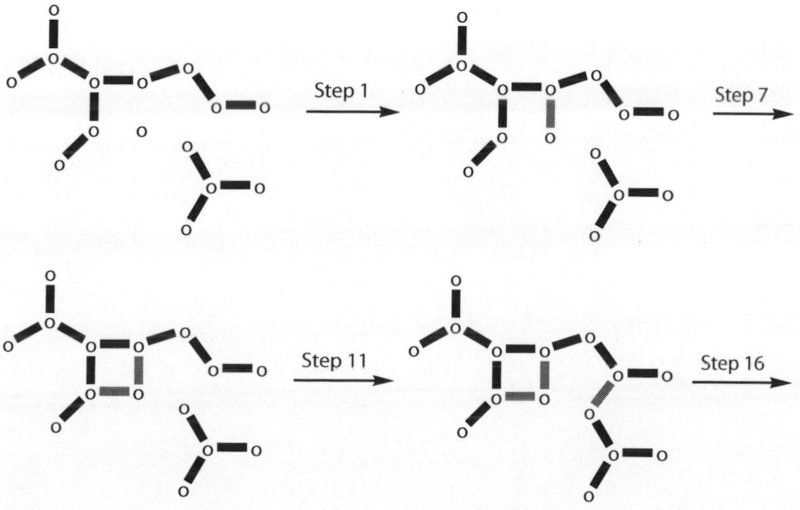

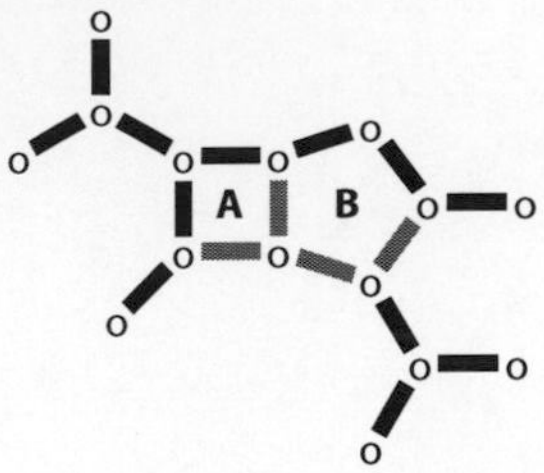

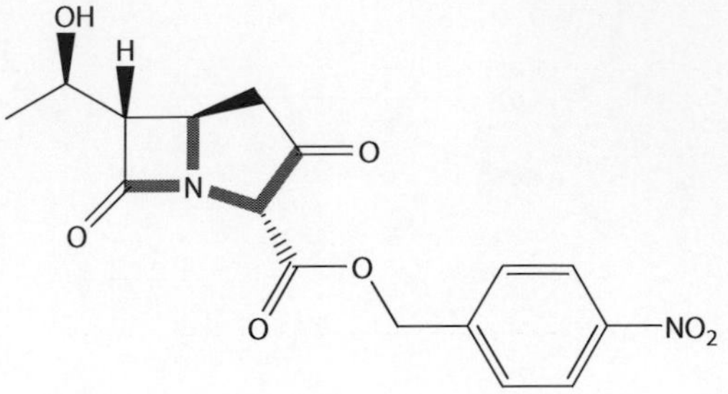

Sequence: $\xrightarrow{[4+0]} A_7 \xrightarrow{[5+0]} B_{16}$

(±)-Thienamycin

Reider, P.J.; Grabowski, E.J.J. *Tetrahedron Lett.* 1982, *23*, 2293.

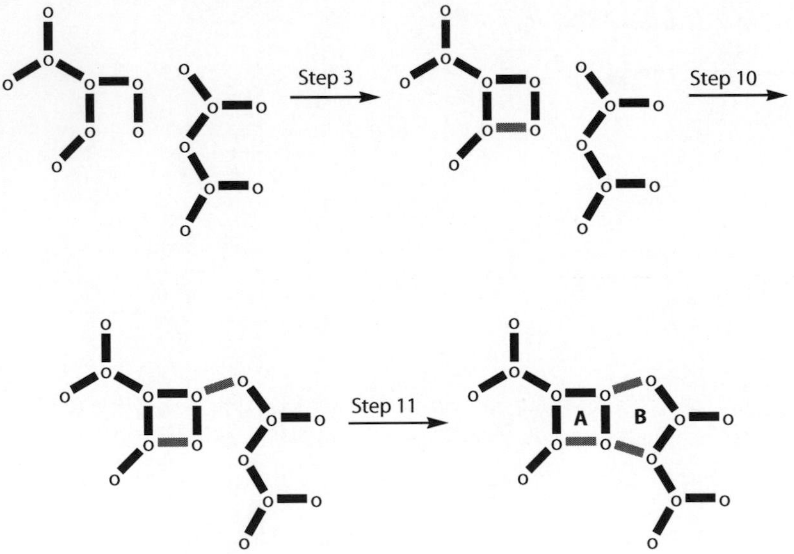

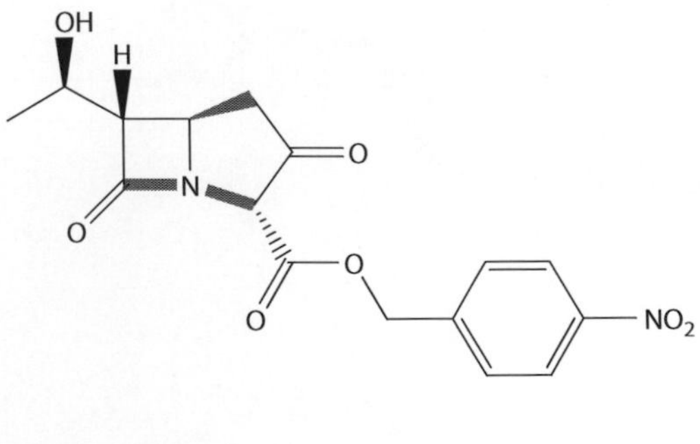

Sequence: $\xrightarrow{[4+0]} A_3 \xrightarrow{[5+0]} B_{11}$

(±)-Thienamycin

Salzmann, T.N.; Ratcliffe, R.W.; Christensen, B.G.; Bouffard, F.A. *J. Am. Chem. Soc.* 1980, *102*, 6161.

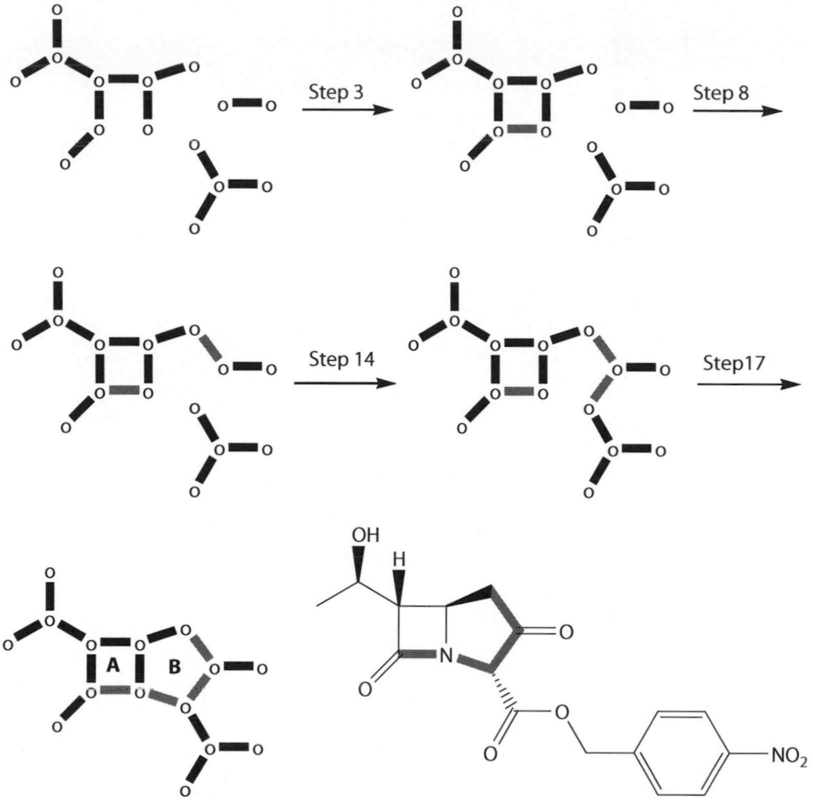

Sequence: $\xrightarrow{[4+0]} A_3 \xrightarrow{[5+0]} B_{17}$

(±)-Thienamycin

Johnston, D.B.R.; Schmitt, S.M.; Bouffard, F.A.; Christensen, B.G. *J. Am. Chem. Soc.* 1978, *100*, 313.
Bouffard, F.A.; Johnson, D.B.R.; Christensen, B.G. *J. Org. Chem.* 1980, *45*, 1130.
Schmitt, S.M.; Johnston, D.B.R.; Christensen, B.G. *J. Org. Chem.* 1980, *45*, 1142.

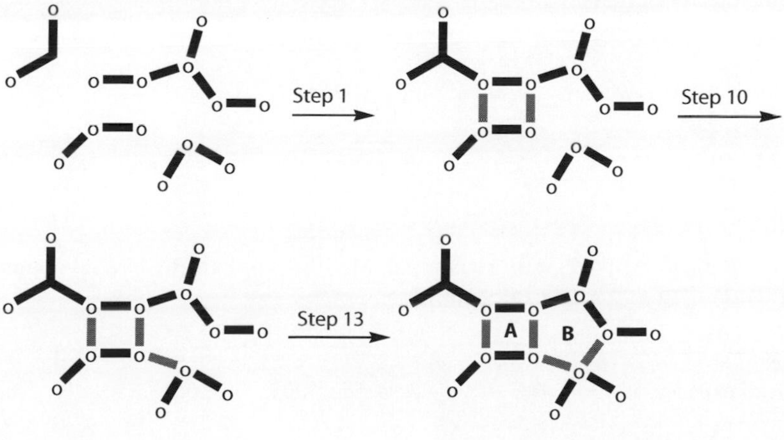

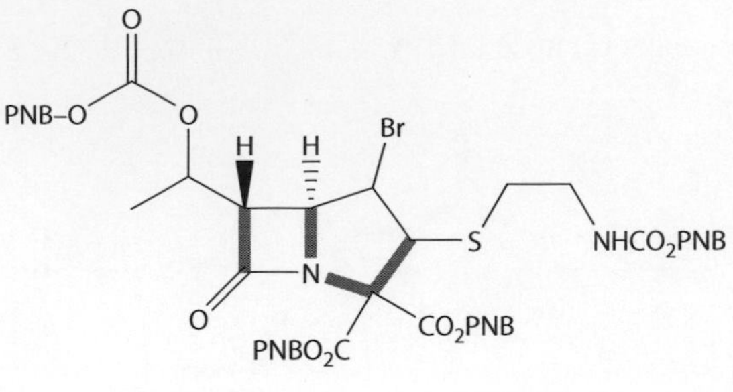

Sequence: $\xrightarrow{[2+2]} A_1 \xrightarrow{[5+0]} B_{13}$

(±)-Thienamycin

Tatsuta, K. *Adv. Synth. Catal.* 2001, *343*, 143.

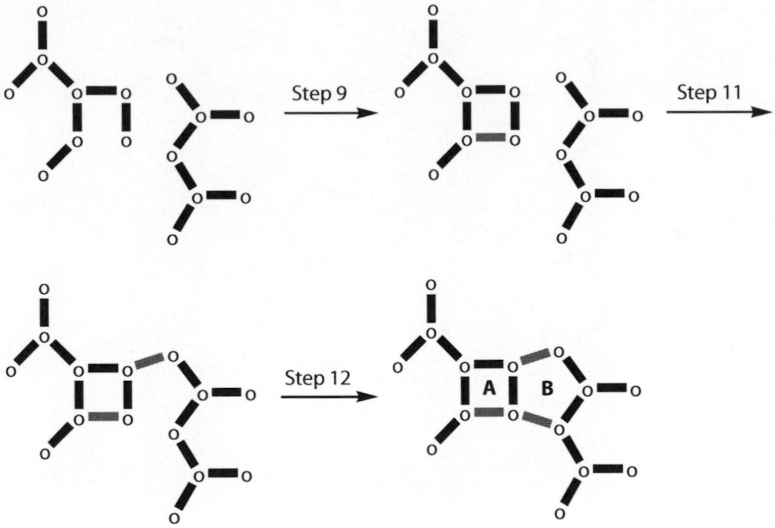

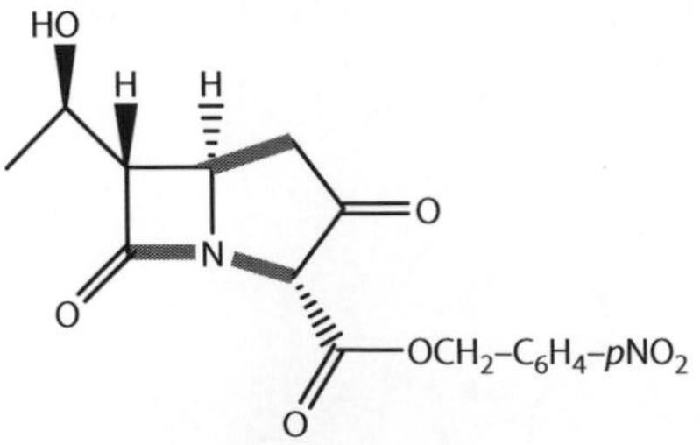

Sequence: $\xrightarrow{[4+0]} A_9 \xrightarrow{[5+0]} B_{12}$

Summary:

Strategy → A → B via [4 + 0] + [5 + 0]: Hanessian, Melillo-Shinkai, Reider, Salzmann Tatsuta
Strategy → A → B via [2 + 2] + [5 + 0]: Schmitt

Triquinacene

Woodward, R.B.; Fukunaga, T.; Kelly, R.C. *J. Am. Chem. Soc.* 1964, *86*, 3162.

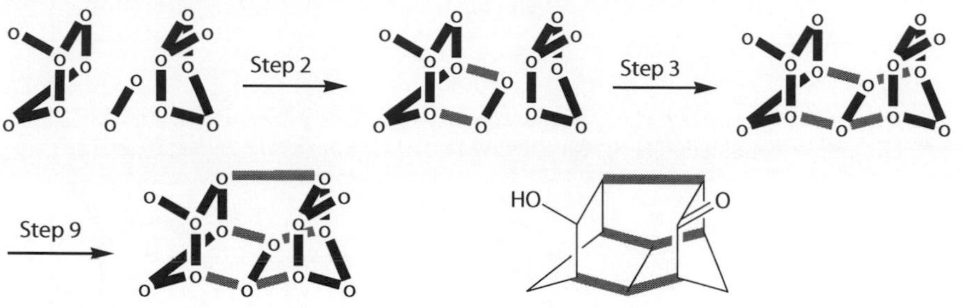

Sequence: $\xrightarrow{[3+2]} [5]_2 \xrightarrow{[(3+2)+(4+2)]} [3.3.0]_3 \xrightarrow{[5+0]} [5]_9$

Tropolone

Chapman, O.L.; Fitton, P. *J. Am. Chem. Soc.* 1963, *85*, 41.

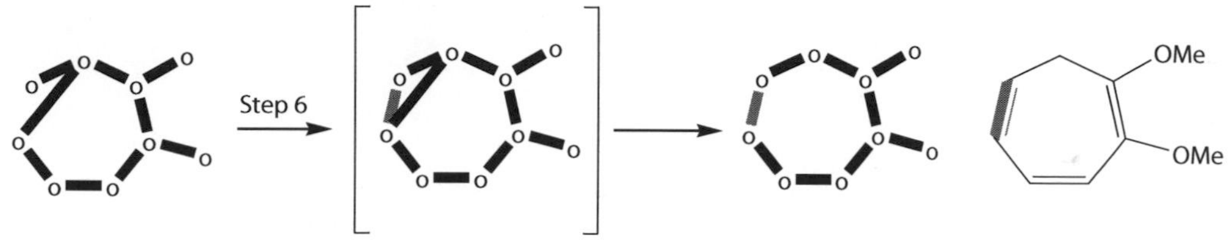

Sequence: $[6] \xrightarrow{[6] \rightarrow [7]} [7]_6$

Cook, J.W.; Gibb, A.R.; Raphael, R.A.; Somerville, A.R. *J. Chem. Soc.* 1951, 503.

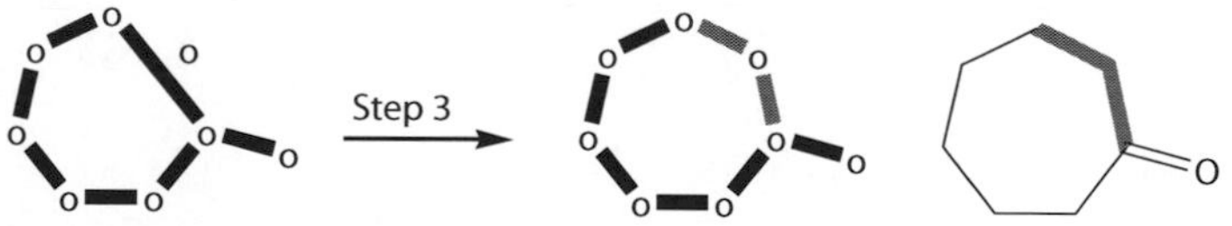

Sequence: $[6] \xrightarrow{[6]+1} [7]_3$

Knight, J.D.; Cram, D.J. *J. Am. Chem. Soc.* 1951, *73*, 4136.

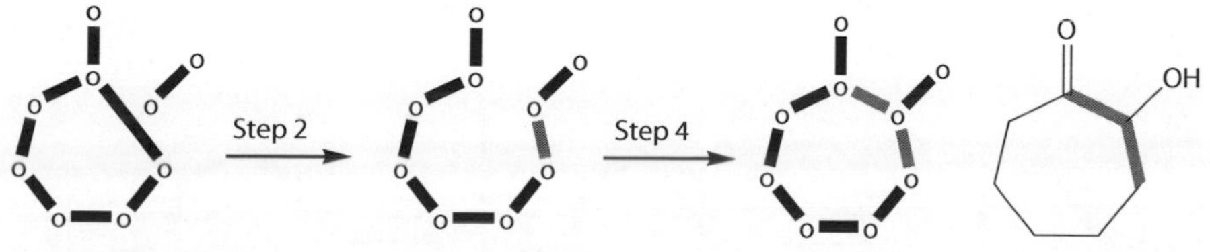

Sequence: $[6] \xrightarrow{[7+0]} [7]_4$

von Doering, W.E.; Knox, L.H. *J. Am. Chem. Soc.* 1951, *73*, 828.
von Doering, W.E.; Knox, L.H. *J. Am. Chem. Soc.* 1950, *72*, 2305.

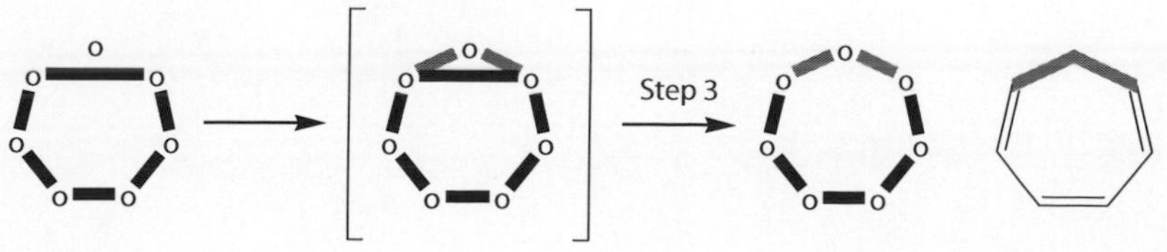

Sequence: $[6] \xrightarrow{[6]+1} [7]_3$

Drysdale, J.J.; Gilbert, W.W.; Sinclair, H.K.; Sharkey, W.H. *J. Am. Chem. Soc.* 1958, *80*, 245.
Drysdale, J.J.; Gilbert, W.W.; Sinclair, H.K.; Sharkey, W.H. *J. Am. Chem. Soc.* 1958, *80*, 3672.

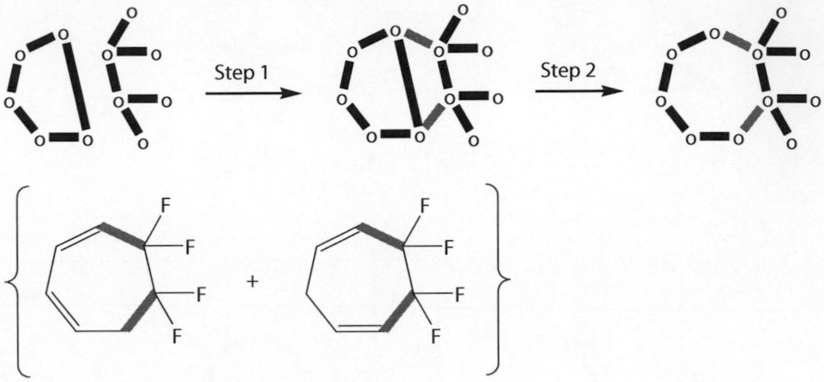

Sequence: $\xrightarrow{[2+2]} [3.2.0]_1 \xrightarrow{[3.2.0] \rightarrow [7]} [7]_2$

Minns, R.A. *Org. Synth. Coll.* 1988, *6*, 1037.

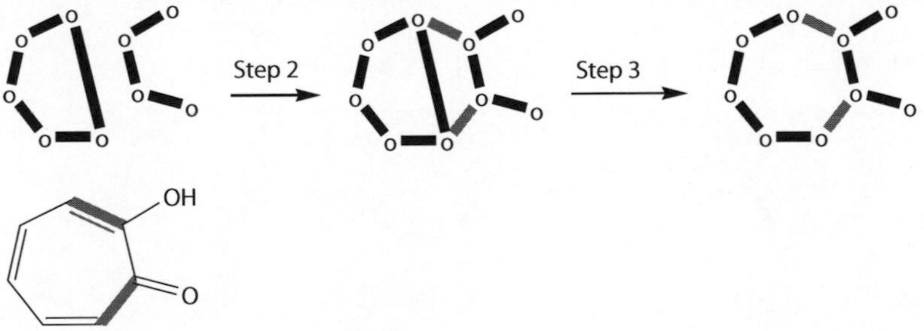

Sequence: $\xrightarrow{[2+2]} [3.2.0]_2 \xrightarrow{[3.2.0] \rightarrow [7]} [7]_3$

Oda, M.; Kitahara, Y. *Tetrahedron Lett.* 1969, 3295.

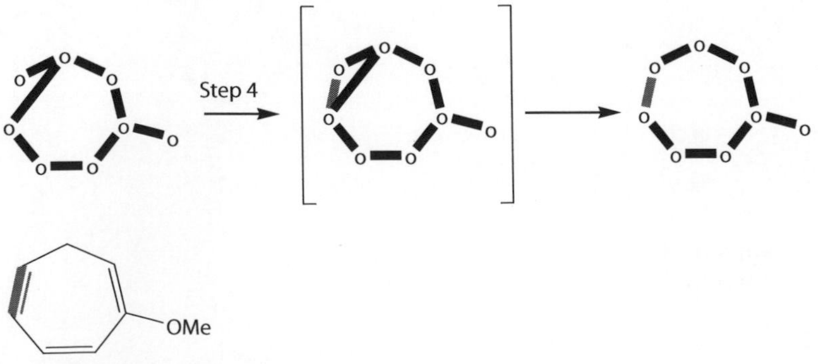

Sequence: $[6] \xrightarrow{[6] \rightarrow [7]} [7]_4$

Stevens, H.C.; Reich, D.A.; Brandt, D.R.; Fountain, K.R.; Gaughan, E.J. *J. Am. Chem. Soc.* 1965, *87*, 5257.

Sequence: $\xrightarrow{[2+2]} [3.2.0]_2 \xrightarrow{[3.2.0] \rightarrow [7]} [7]_3$

van Tamelen, E.E.; Hildahl, G.T. *J. Am. Chem. Soc.* 1956, *78*, 4405.

Sequence: $[6] \xrightarrow{[6]+1} [7]_5$

ter Borg, A.P.; van Helden, R.; Bickel, A.F. *Rec. Trav. Chim. Pays Bas* 1962, *81*, 177.
ter Borg, A.P.; van Helden, R.; Bickel, A.F. *Rec. Trav. Chim. Pays Bas* 1962, *81*, 164.

Sequence: $[6] \xrightarrow{[6]+1} [7]_3$

Summary:
Strategy #1

$$[6] \xrightarrow{[6]+1} [7]_5 \text{: van Tamelen}$$

Strategy #2

$$[6] \xrightarrow{[6]+1} [7]_3 \text{: ter Borg}$$

$$[6] \xrightarrow{[6]+1} [7]_3 \text{: von Doering}$$

Strategy #3

$$\xrightarrow{[2+2]} [3.2.0]_2 \xrightarrow{[3.2.0]\to[7]} [7]_3 : \text{Stevens}$$

$$\xrightarrow{[2+2]} [3.2.0]_2 \xrightarrow{[3.2.0]\to[7]} [7]_3 : \text{Minns}$$

$$\xrightarrow{[2+2]} [3.2.0]_1 \xrightarrow{[3.2.0]\to[7]} [7]_2 : \text{DuPont}$$

Strategy #4

$$[6] \xrightarrow{[6]\to[7]} [7]_4 : \text{Oda}$$

$$[6] \xrightarrow{[6]\to[7]} [7]_6 : \text{Chapman}$$

Strategy #5

$$[6] \xrightarrow{[7+0]} [7]_4 : \text{Cram}$$

Strategy #6

$$[6] \xrightarrow{[6]+1} [7]_3 : \text{Cook}$$

(±)-Twistane

Gauthier, J.; Deslongchamps, P. *Can. J. Chem.* 1967, *45*, 297.

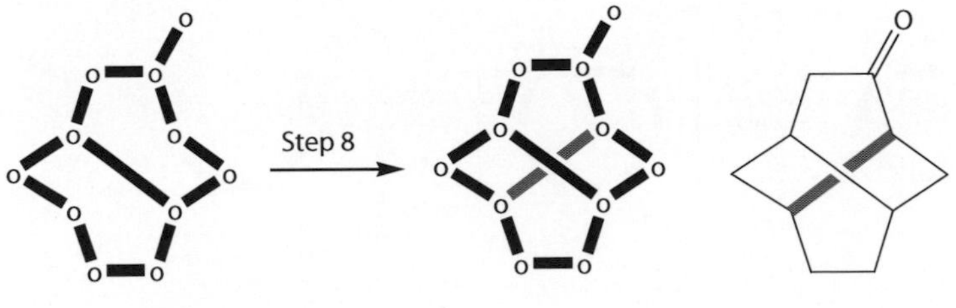

Sequence: $[4.4.0] \xrightarrow{[(6+0)+(6+0)]} [4.4.0]_8$

Belanger, A.; Lambert, Y.; Deslongchamps, P. *Can. J. Chem.* 1969, *47*, 795.
Belanger, A.; Poupart, J.; Deslongchamps, P. *Tetrahedron Lett.* 1968, 2127.

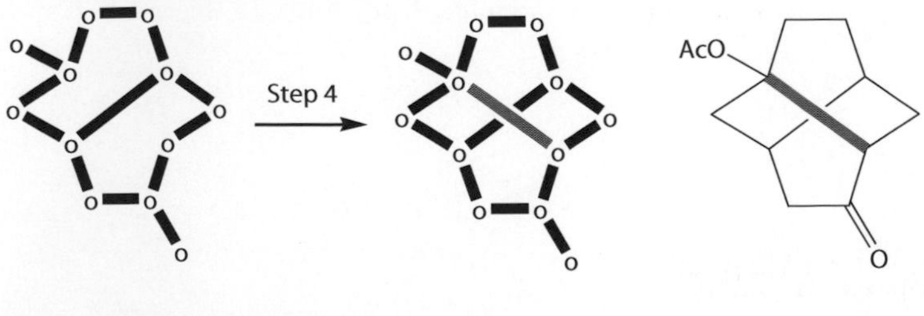

Sequence: $[4.4.0] \xrightarrow{[(6+0)+(6+0)]} [4.4.0]_4$

Hamon, D.P.G.; Young, R.N. *Austr. J. Chem.* 1976, *29*, 145.

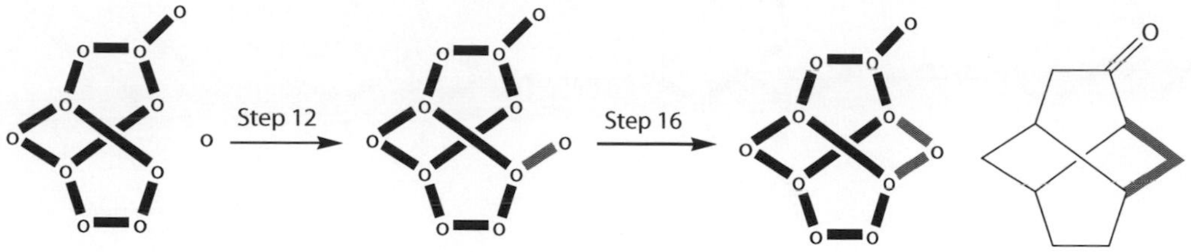

Sequence: $\xrightarrow{[(6+0)+(6+0)]} P_{16}$

Whitlock, H.W. Jr.; Siefken, N.W. *J. Am. Chem. Soc.* 1968, *90*, 4929.
Whitlock, H.W. Jr. *J. Am. Chem. Soc.* 1962, *84*, 314.

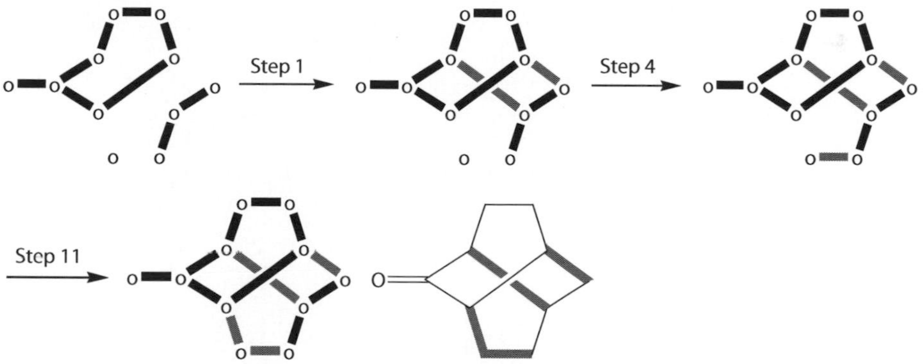

Sequence: $[6] \xrightarrow{[(6+0)+(6+0)]} [2.2.2]_1 \xrightarrow{[(6+0)+(6+0)]} P_{11}$

(±)-Twistane
Dodds, D.R.; Jones, J.B. *Chem. Commun.* 1982, 1080.
Dodds, D.R.; Jones, J.B. *J. Am. Chem. Soc.* 1988, *110*, 577.

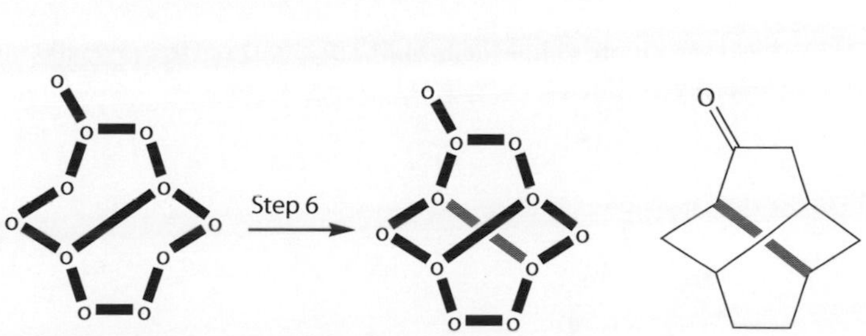

Sequence: $[4.4.0] \xrightarrow{[(6+0)+(6+0)]} [4.4.0]_6$

Tichy, M.; Sicher, J. *Tetrahedron Lett.* 1969, 4609.

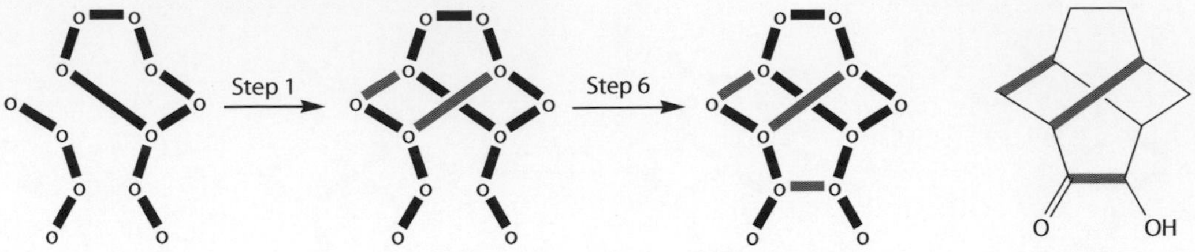

Sequence: $[6] \xrightarrow{[(6+0)+(6+0)]} [2.2.2]_1 \xrightarrow{[(6+0)+(6+0)]} P_6$

Summary:
Strategy #1

$[4.4.0] \xrightarrow{[(6+0)+(6+0)]} [4.4.0]_8$: Deslongschamps G1

$[4.4.0] \xrightarrow{[(6+0)+(6+0)]} [4.4.0]_4$: Deslongschamps G2

$[4.4.0] \xrightarrow{[(6+0)+(6+0)]} [4.4.0]_6$: Jones

Strategy #2

$[6] \xrightarrow{[(6+0)+(6+0)]} [2.2.2]_1 \xrightarrow{[(6+0)+(6+0)]} P_6$: Tichy

$[6] \xrightarrow{[(6+0)+(6+0)]} [2.2.2]_1 \xrightarrow{[(6+0)+(6+0)]} P_{11}$: Whitlock

Strategy #3

$\xrightarrow{[(6+0)+(6+0)]} P_{16}$: Hamon-Young

(−)-Vindoline
Choi, Y.; Ishikawa, H.; Velcicky, J.; Elliott, G.I.; Miller, M.M.; Boger, D.L. *Org. Lett.* 2005, 7, 4539.
Ishikawa, H.; Elliott, G.I.; Velcicky, J.; Choi, Y.; Boger, D.L. *J. Am. Chem. Soc.* 2006, *128*, 10596.

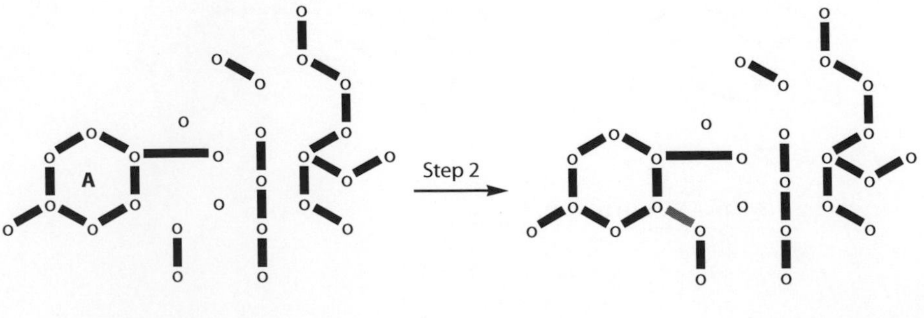

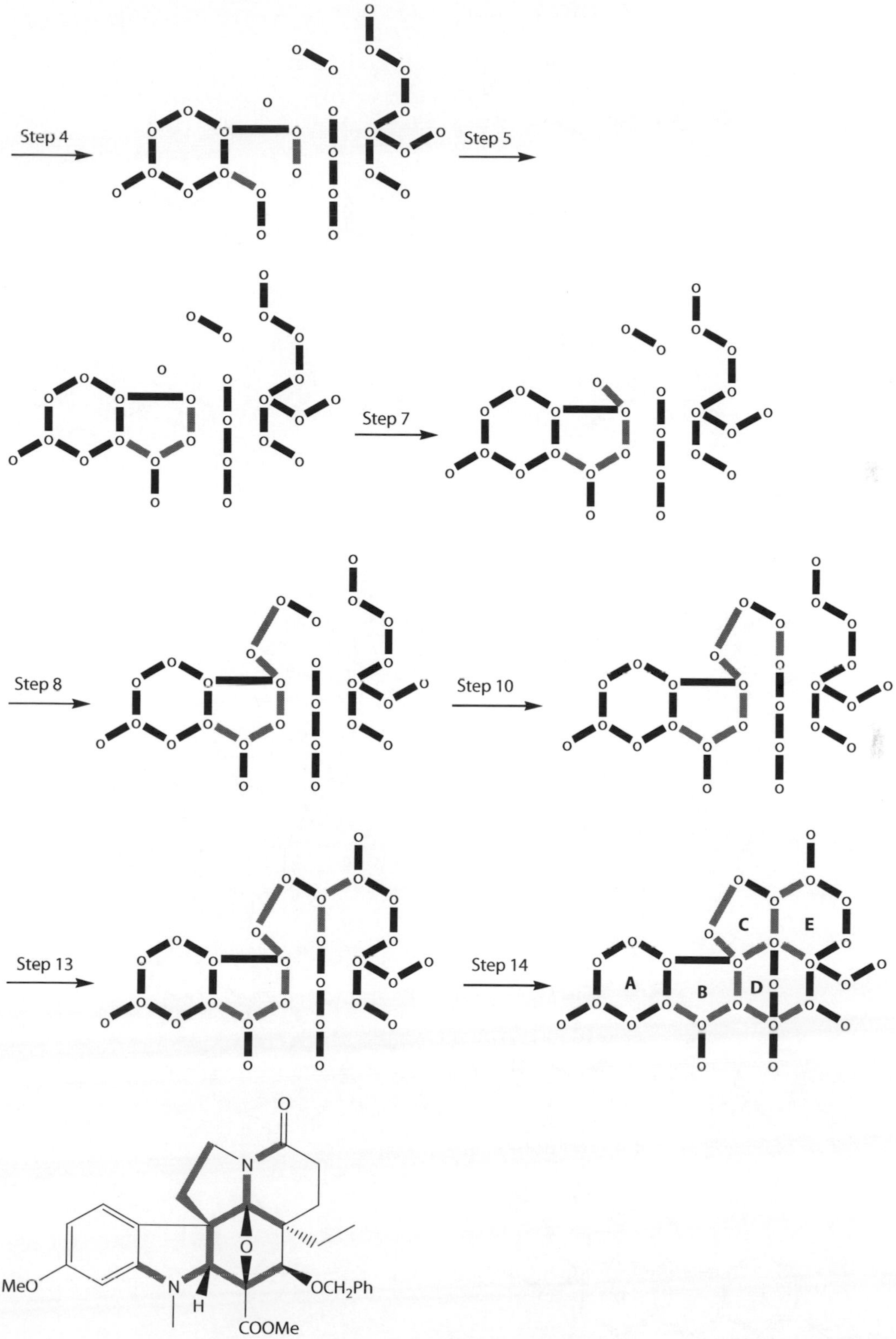
Step 4
Step 5
Step 7
Step 8
Step 10
Step 13
Step 14
A
B
C
D
E
O
N
MeO
N
H
COOMe
OCH2Ph

Sequence: $A_0 \xrightarrow{[5+0]} B_5 \xrightarrow{[(5+0)+(6+0)+(2+2+1+1)]} (C+D+E)_{14}$

Kobayashi, S.; Ueda, T.; Fukuyama, T. *Synlett* 2000, 883.

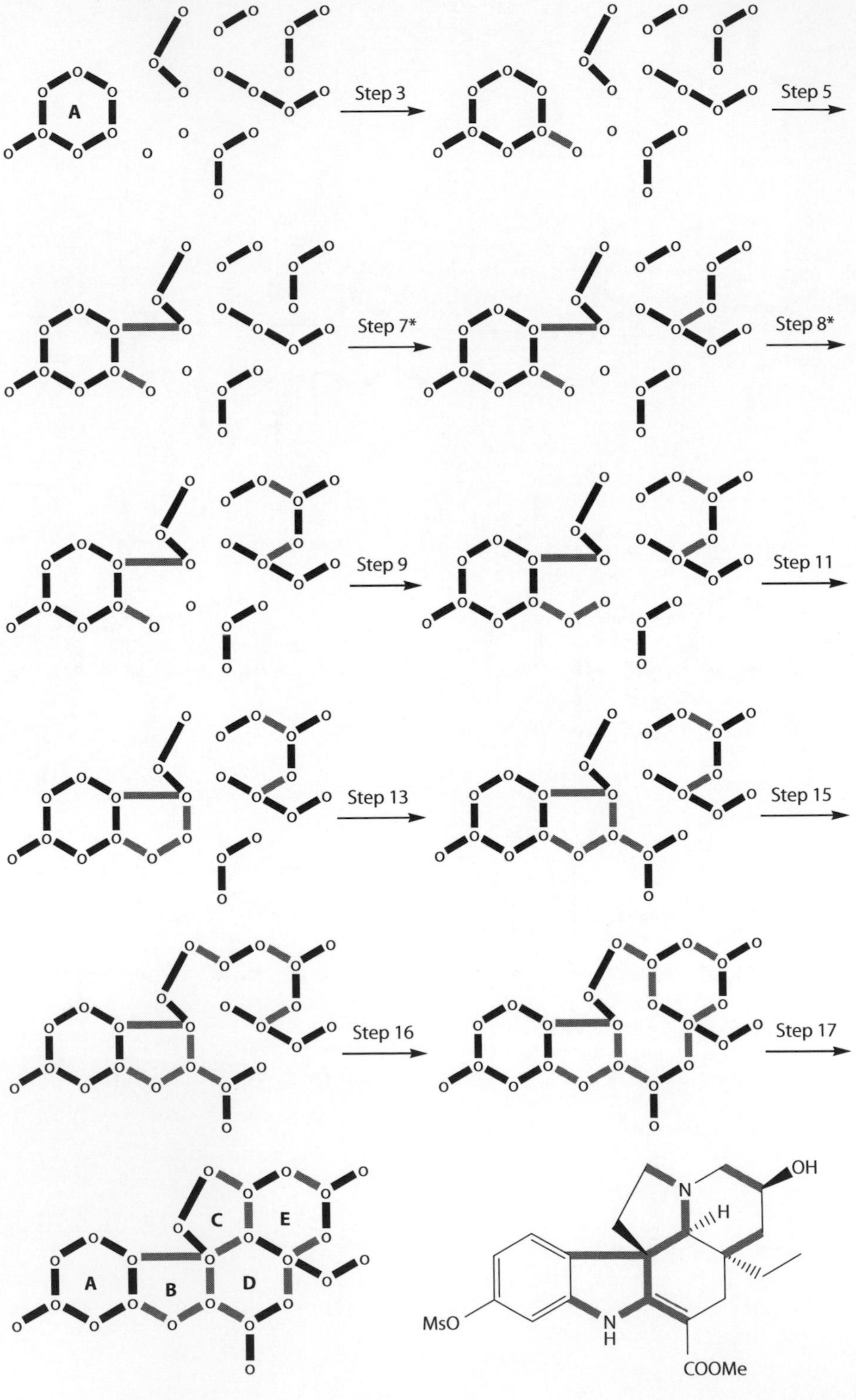

Sequence: $A_0 \xrightarrow{[5+0]} B_{11} \xrightarrow{[6+0]} E_{16} \xrightarrow{[(6+0)+(5+0)]} [4.3.0]\,(C+D)_{17}$

Yokoshima, S.; Ueda, T.; Kobayashi, S.; Sato, A.; Kuboyama, T.; Tokuyama, H.; Fukuyama, T. *J. Am. Chem. Soc.* 2002, *124*, 2137.

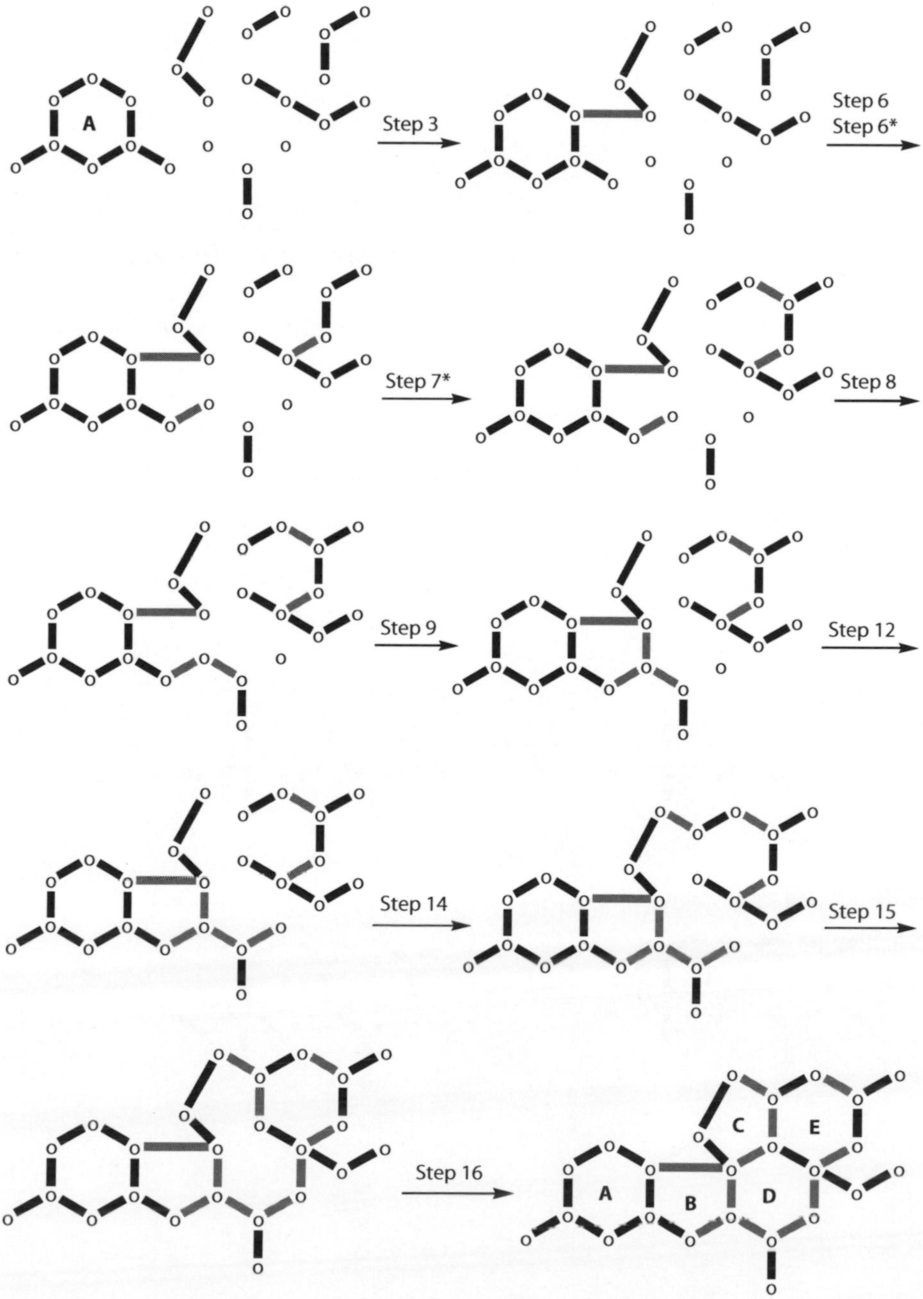

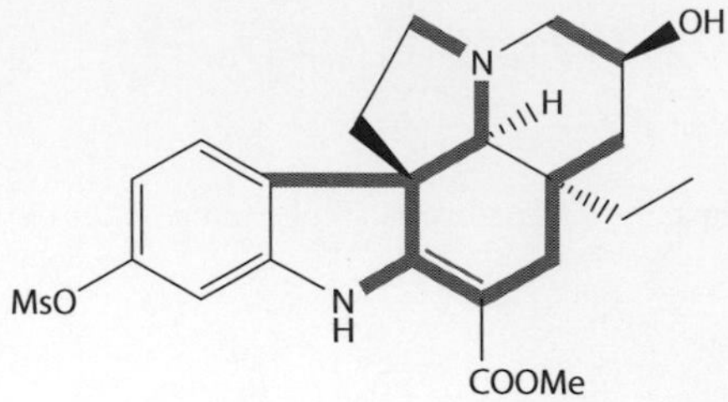

Sequence: $A_0 \xrightarrow{[5+0]} B_9 \xrightarrow{[6+0]} E_{15} \xrightarrow{[(6+0)+(5+0)]} [4.3.0]\,(C+D)_{16}$

Summary:
Strategy #1

$A_0 \xrightarrow{[5+0]} B_{11} \xrightarrow{[6+0]} E_{16} \xrightarrow{[(6+0)+(5+0)]} [4.3.0]\,(C+D)_{17}$: Fukuyama G1

$A_0 \xrightarrow{[5+0]} B_9 \xrightarrow{[6+0]} E_{15} \xrightarrow{[(6+0)+(5+0)]} [4.3.0]\,(C+D)_{16}$: Fukuyama G2

Strategy #2

$A_0 \xrightarrow{[5+0]} B_5 \xrightarrow{[(5+0)+(6+0)+(2+2+1+1)]} (C+D+E)_{14}$: Bodger

Vitamin E
Rein, C.; Demel, P.; Outten, R.A.; Netscher, T.; Breit, B. *Angew. Chem. Int. Ed.* 2007, *46*, 8670.

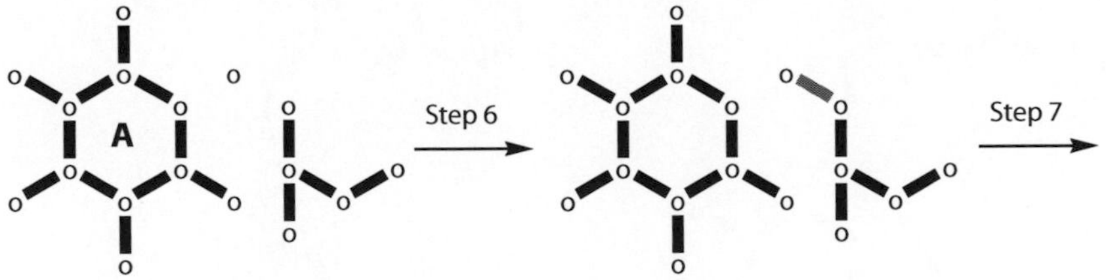

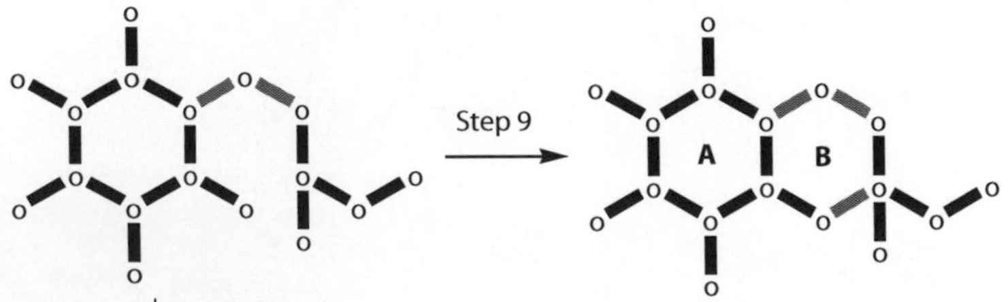

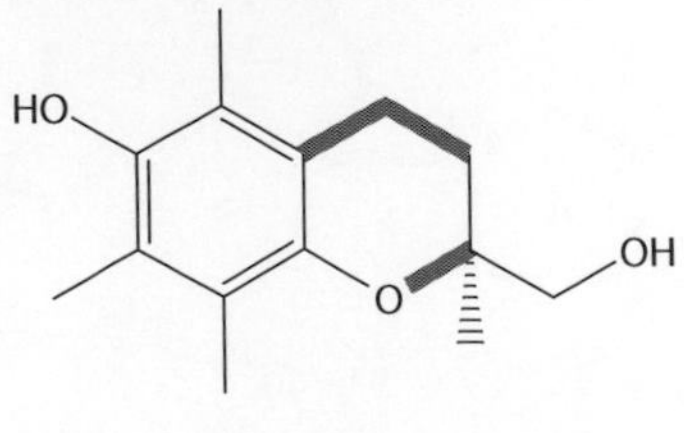

Sequence: $A_{5*} \xrightarrow{[6+0]} B_9$

Karrer, P.; Ringier, B.H. *Helv. Chim. Acta* 1939, *22*, 610.

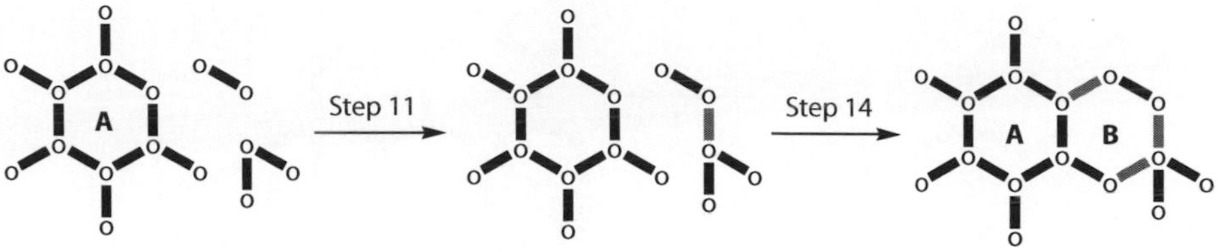

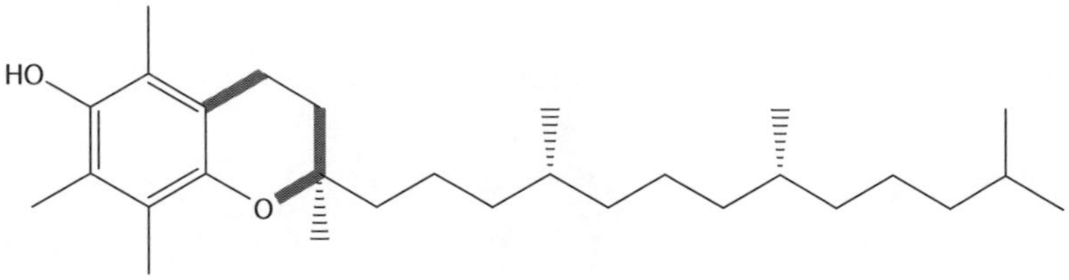

Sequence: $A_{13} \xrightarrow{[3+3]} B_{14}$

Knight, D.W.; Qing, X. *Tetrahedron Lett.* 2009, *50*, 3534.

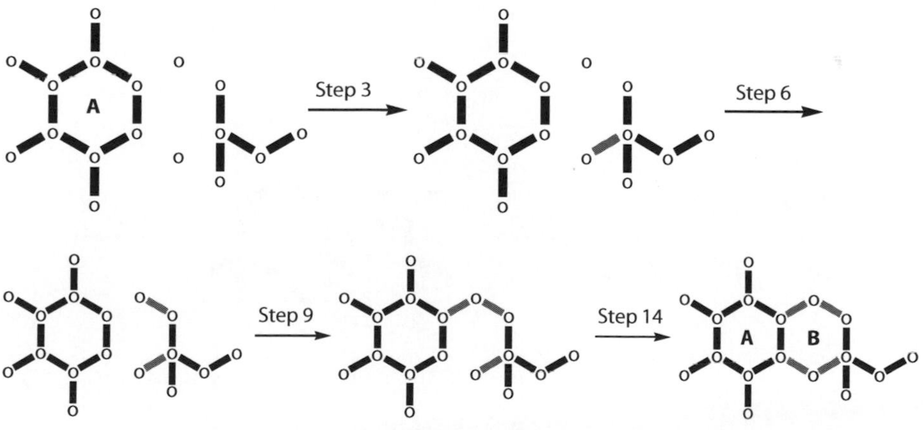

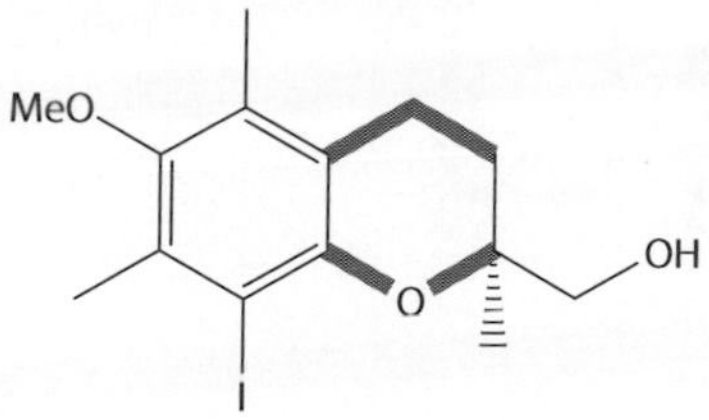

Sequence: $A_{3^*} \xrightarrow{[6+0]} B_{14}$

Mayer, H.; Schudel, P.; Rüegg, R.; Isler, O. *Helv. Chim. Acta* 1963, *67*, 650.

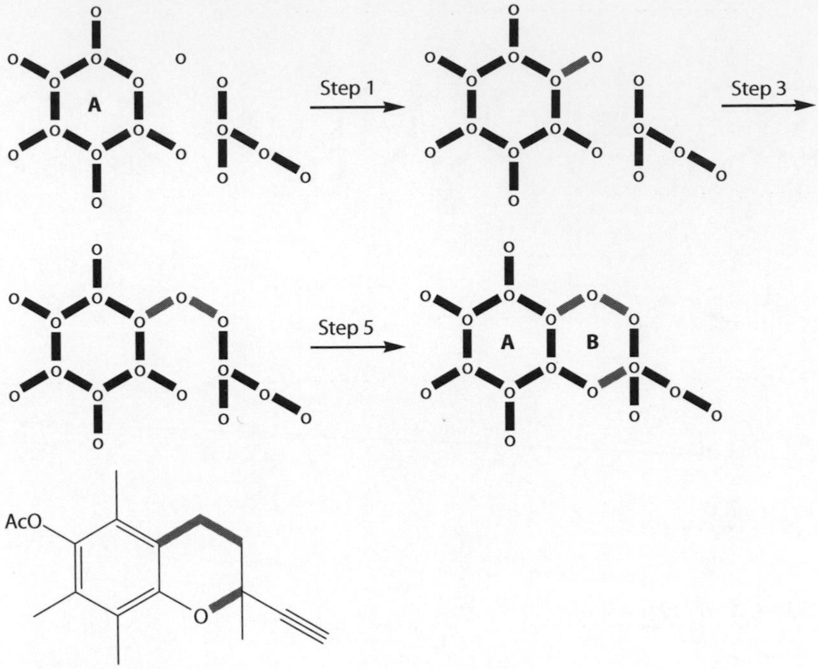

Sequence: $A_0 \xrightarrow{[6+0]} B_5$

Cohen, N.; Lopresti, R.J.; Saucy, G. *J. Am. Chem. Soc.* 1979, *101*, 6710.

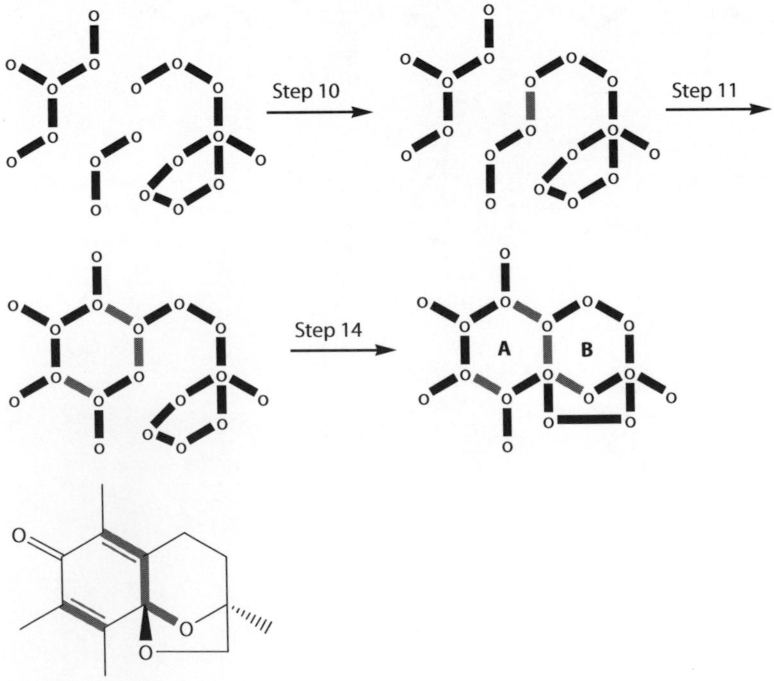

Sequence: $\xrightarrow{[3+3]} A_{11} \xrightarrow{[6+0]} B_{14}$

Tietze, L.F.; Stecker, F.; Zinngrebe, J.; Sommer, K.M. *Chem. Eur. J.* 2006, *12*, 8770.
Tietze, L.F.; Kinzel, T. *Pure Appl. Chem.* 2007, *79*, 629.

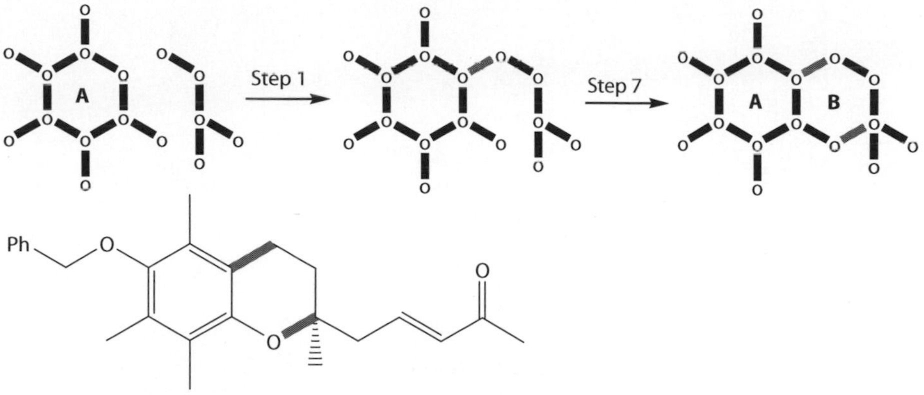

Sequence: $A_0 \xrightarrow{\text{[6+0]}} B_7$

Liu, K.; Chougnet, A.; Woggon, W.D. *Angew. Chem. Int. Ed.* 2008, *47*, 5827.

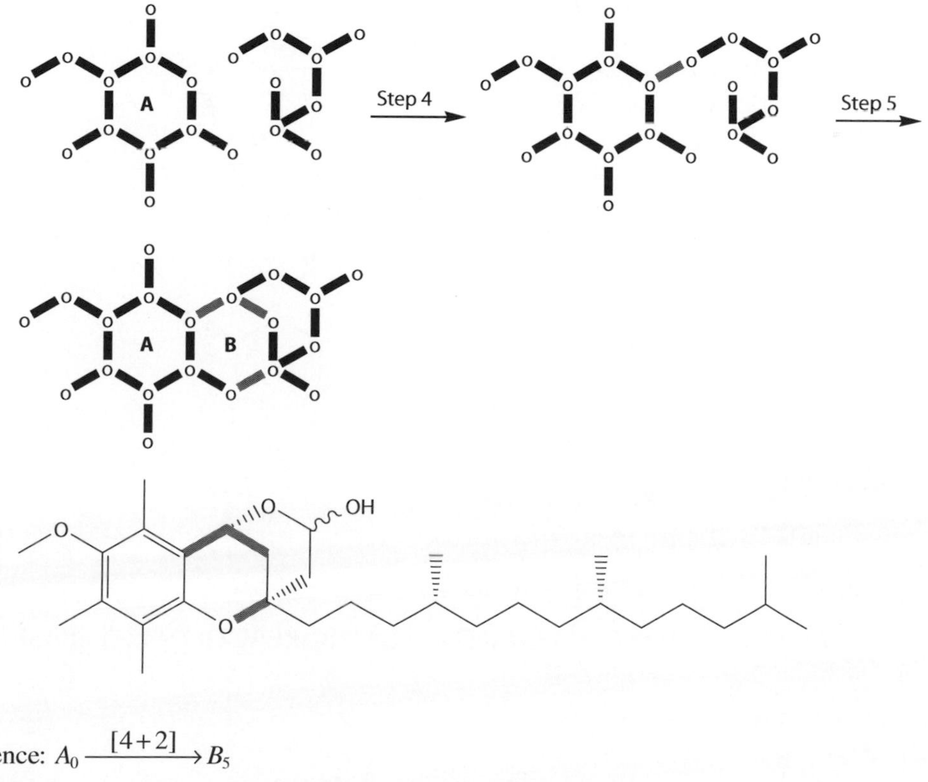

Sequence: $A_0 \xrightarrow{\text{[4+2]}} B_5$

Summary:
Strategy #1

$$A_0 \xrightarrow{\text{[4+2]}} B_5: \text{Woggon}$$

Strategy #2

$$A_0 \xrightarrow{[6+0]} B_7: \text{Tietze}$$

$$A_0 \xrightarrow{[6+0]} B_5: \text{Roche G1}$$

$$A_{3*} \xrightarrow{[6+0]} B_{14}: \text{Knight}$$

$$A_{5*} \xrightarrow{[6+0]} B_9: \text{Breit}$$

Strategy #3

$$A_{13} \xrightarrow{[3+3]} B_{14}: \text{Karrer}$$

Strategy #4

$$\xrightarrow{[3+3]} A_{11} \xrightarrow{[6+0]} B_{14}: \text{Roche G2}$$

Zamifenacin
Cossy, J.; Dumas, C.; Pardo, D.G. *Bioorg. Med. Chem. Lett.* 1997, 7, 1343.

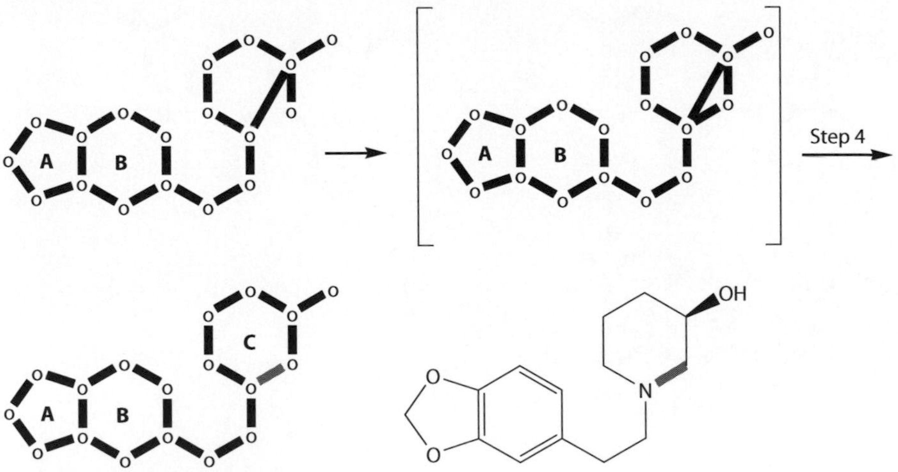

Sequence: $[5] \xrightarrow{[5]+1} [6]_4$

10.2 RING CONTRACTIONS IN NAMED ORGANIC REACTION DATABASE

Favorskii rearrangement
Favorskii, A.E. *J. Prakt. Chem.* 1913, 88, 658.

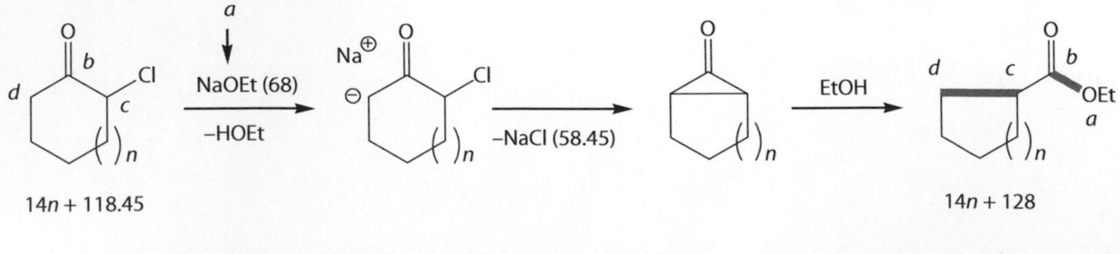

$14n + 118.45$

$14n + 128$

Perkin rearrangement
Perkin, W.H. *J. Chem. Soc.* 1870, *23*, 368.

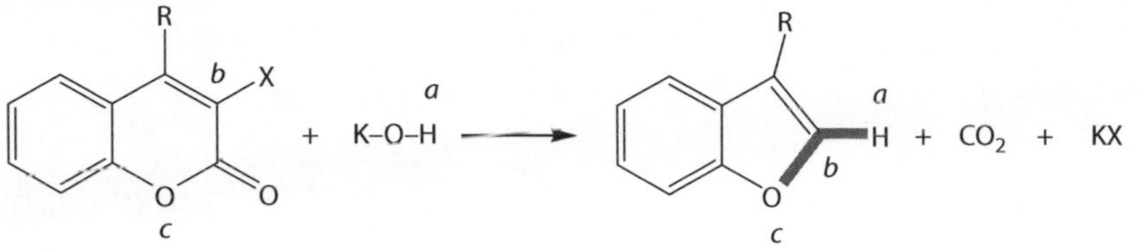

$$x + r + 144 \qquad 56 \qquad\qquad r + 117 \qquad 44 \qquad x + 39$$

Ramberg–Bäckland rearrangement
Ramberg, L.; Bäckland, B. *Arkiv. Kemi Mineral Geol.* 1940, *13A*, 50.

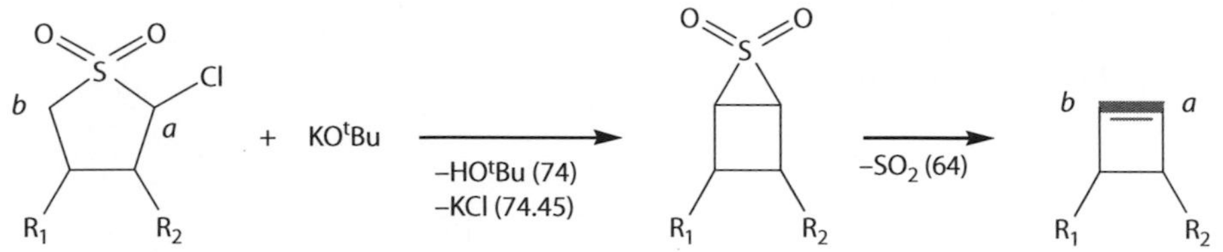

$$x + y + 152.45 \qquad 112 \qquad\qquad\qquad\qquad x + y + 52$$

Wolff rearrangement
Wolff, L. *Ann. Chem.* 1912, *394*, 25.

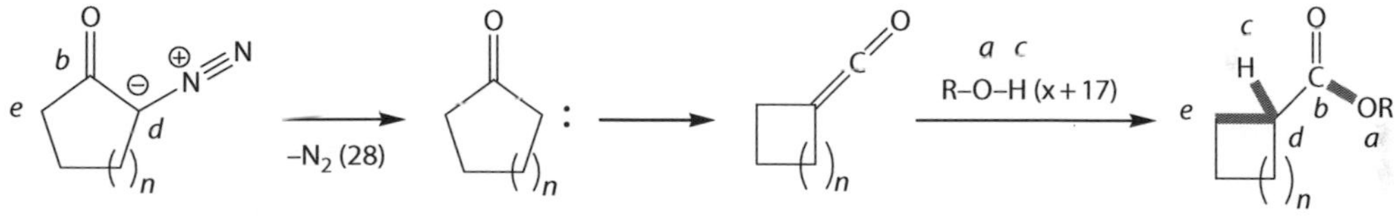

$$14n + 96 \qquad\qquad\qquad\qquad\qquad\qquad 14n + x + 85$$

10.3 RING EXPANSIONS IN NAMED ORGANIC REACTION DATABASE

Ciamician synthesis of pyridines from pyrroles
Ciamician, G.; Dennestedt, M. *Chem. Ber.* 1881, *14*, 1153.

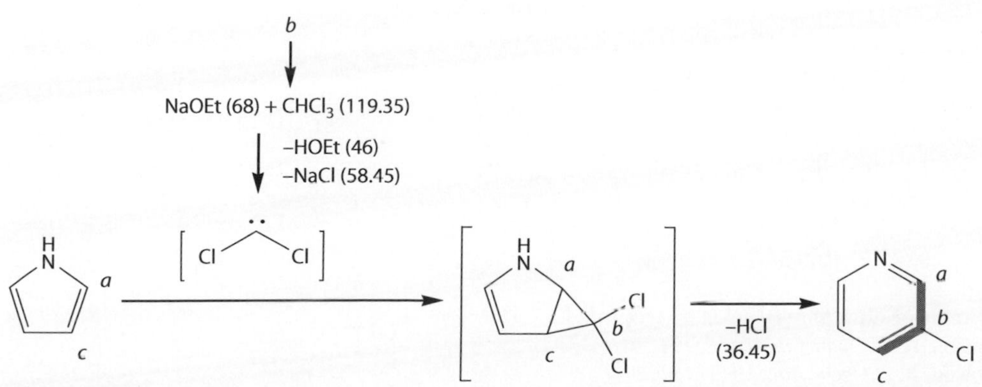

$$\text{NaOEt (68)} + \text{CHCl}_3 \text{ (119.35)}$$

−HOEt (46)
−NaCl (58.45)

−HCl
(36.45)

67

113.45

Grob fragmentation

Grob, C.A.; Baumann, W. *Helv. Chim. Acta* 1955, *38*, 594.
Grob, C.A. *Angew. Chem. Int. Engl. Ed.* 1969, *8*, 535.

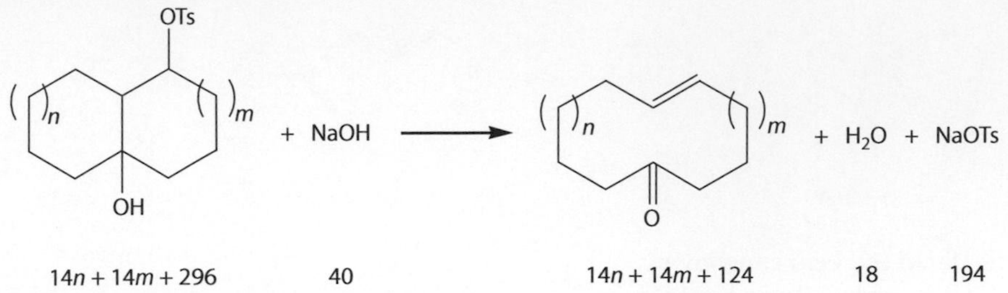

OTs
OH
+ NaOH
O
+ H2O + NaOTs
14n + 14m + 296
40
14n + 14m + 124
18
194

Tiffeneau–Demjanov reaction

Tiffeneau, M.; Weill; Tschoubar, B. *Compt. Rend.* 1937, *205*, 54.
Demjanov, N.J.; Lushnikov, M. *J. Russ. Phys. Chem. Soc.* 1903, *35*, 26.

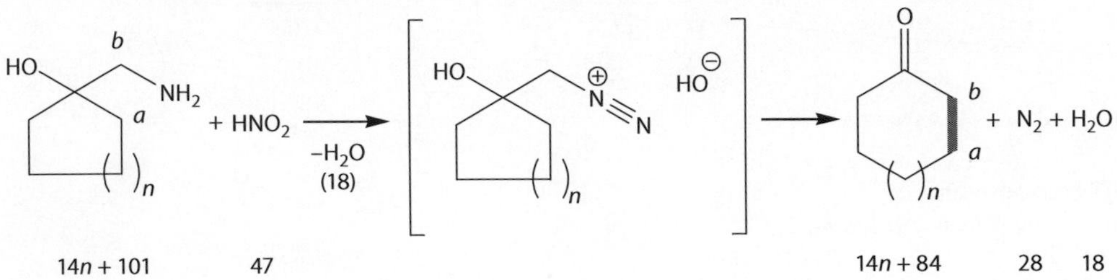

HO
NH2
+ HNO2
–H2O
(18)
HO
+ N2 + H2O
14n + 101
47
14n + 84
28
18

10.4 BICYCLIC FORMATION REACTIONS IN NAMED ORGANIC REACTION DATABASE

Alper carbonylation [3.2.0]

Alper, H.; Perera, C.P. *J. Am. Chem. Soc.* 1981, *103*, 1289.
Alper, H.; Mahatantila, C.P. *Organometallics* 1982, *1*, 70.
Alper, H. *Aldrichimica Acta* 1991, *24*, 3.

Ph
Ph
Ph
Ph
Ph

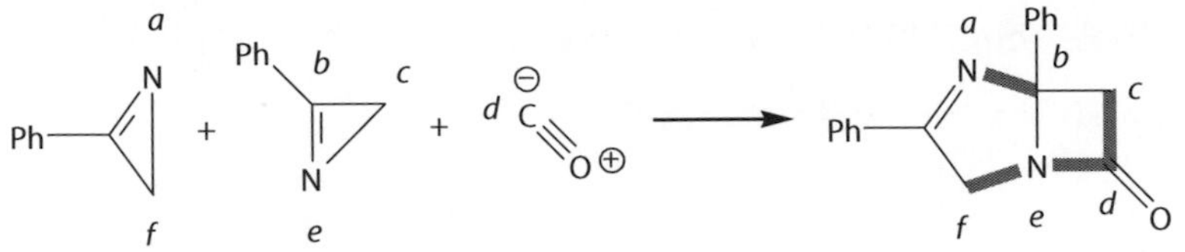

Norbornene synthesis [2.2.1]

Bestmann, H.J.; Schobert, R. *Angew. Chem. Int. Ed.* 1985, *24*, 790.

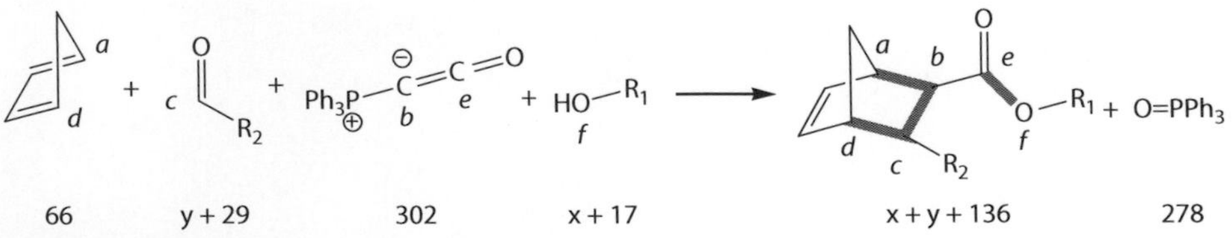

Ph3P
+ HO—R1
+ O=PPh3
66
y + 29
302
x + 17
x + y + 136
278

Robinson–Schoepf synthesis of tropane [3.2.1]

Robinson, R. *J. Chem. Soc.* 1917, *111*, 762.
Robinson, R. *J. Chem. Soc.* 1917, *111*, 876.
Schoepf, C. *Angew. Chem.* 1937, *50*, 779.
Schoepf, C. *Angew. Chem.* 1937, *50*, 797.

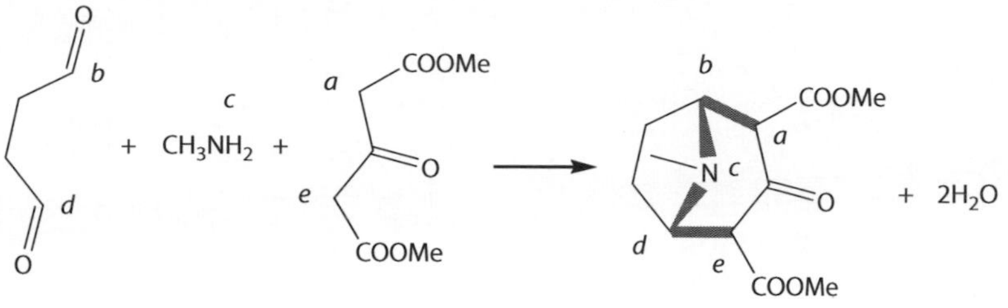

86   31   174   255   36

Weiss reaction [3.3.0]
Weiss, U.; Edward, J.M. *Tetrahedron Lett*. 1968, 4885.

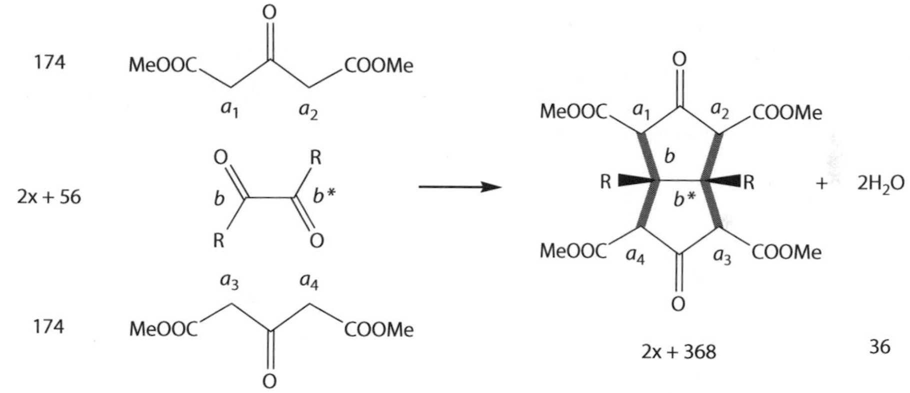

174

$2x + 56$

174

$2x + 368$ 36

Wulff cyclization [4.4.0]
Wulff, W.D.; Yang, D.C. *J. Am. Chem. Soc*. 1984, *106*, 7565.

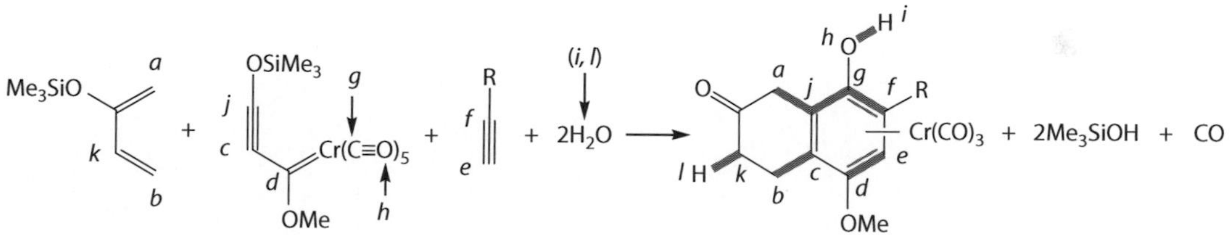

142   348 $x + 25$ 36 $x + 327$ 180   28

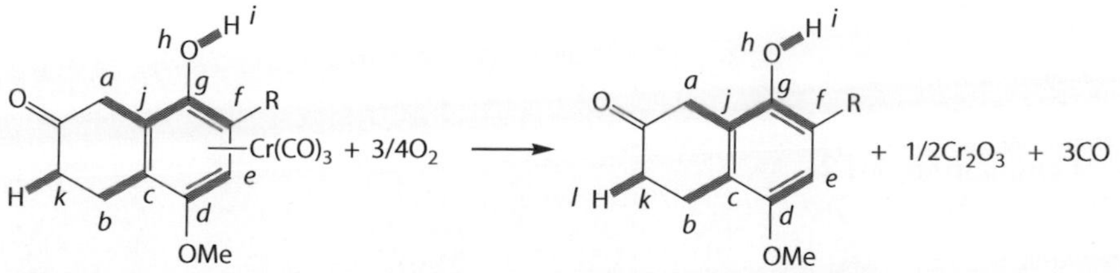

Camphene rearrangement [2.2.1]
Wagner, G. *J. Russ. Phys. Chem*. 1899, *31*, 690.
Meerwein, H. *Ann. Chem*. 1914, *405*, 129.

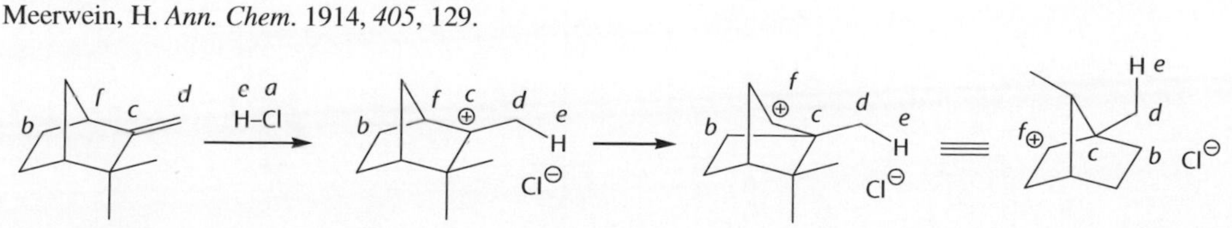

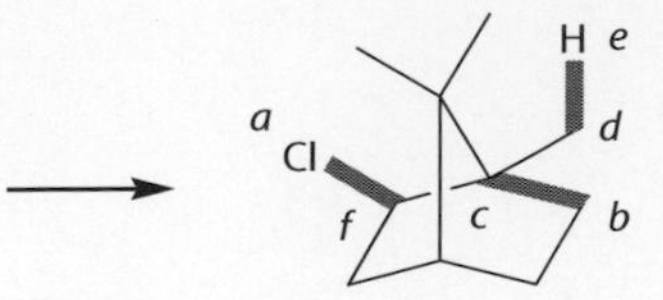

Di-pi-methane rearrangement [3.1.0]

Zimmerman, H.E.; Mariano, P.S. *J. Am. Chem. Soc.* 1969, *91*, 1718.
Hahn, R.C.; Rothman, L.J. *J. Am. Chem. Soc.* 1969, *91*, 2409.
Zimmerman, H.E.; Samuelson, G.E. *J. Am. Chem. Soc.* 1969, *91*, 5307.
Zimmerman, H.E.; Bender, C.O. *J. Am. Chem. Soc.* 1969, *91*, 7516.
Zimmerman, H.E.; Pratt, A.C. *J. Am. Chem. Soc.* 1970, *92*, 1407.
Zimmerman, H.E.; Pratt, A.C. *J. Am. Chem. Soc.* 1970, *92*, 1409.
Zimmerman, H.E.; Epling, G.A. *J. Am. Chem. Soc.* 1970, *92*, 1411.
Zimmerman, H.E.; Bender, C.O. *J. Am. Chem. Soc.* 1970, *92*, 4366.
Zimmerman, H.E.; Pratt, A.C. *J. Am. Chem. Soc.* 1970, *92*, 6259.
Zimmerman, H.E.; Pratt, A.C. *J. Am. Chem. Soc.* 1970, *92*, 6267.
Zimmerman, H.E.; Chen, W. *Chem. Rev.* 1996, *96*, 3065.

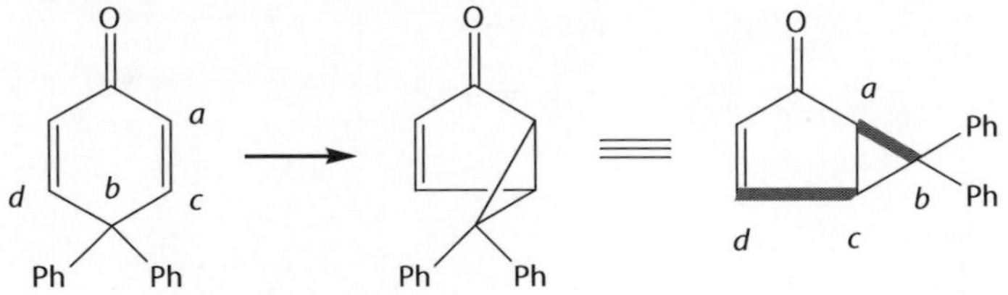

Wagner–Meerwein rearrangement [2.2.1]

Wagner, G. *J. Russ. Phys. Chem.* 1899, *31*, 690.
Meerwein, H. *Ann. Chem.* 1914, *405*, 129.

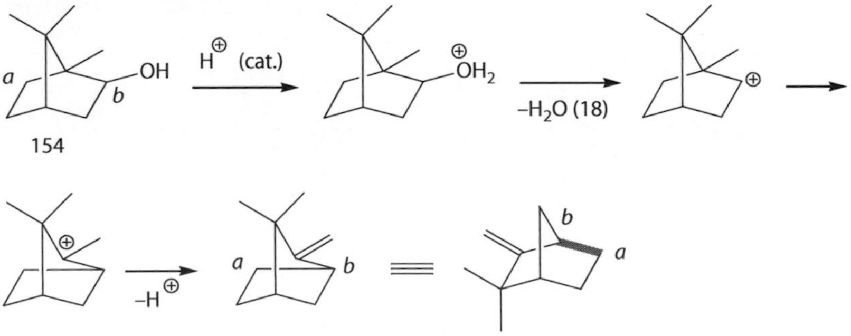

Danheiser [4 + 4] annulation [4.2.2]

Danheiser, R.L.; Gee, S.K.; Sard, H. *J. Am. Chem. Soc.* 1982, *104*, 7670.

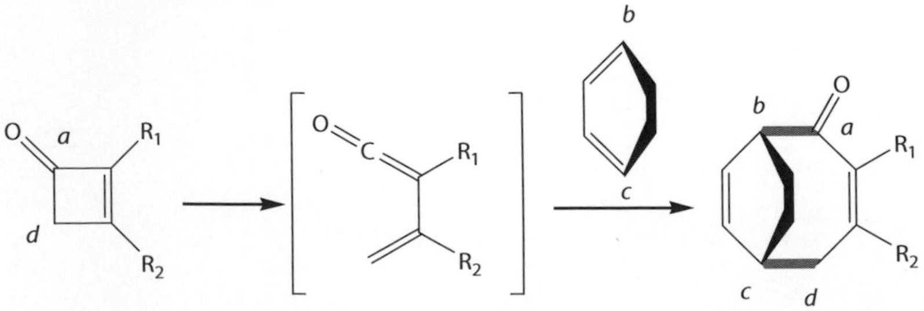

Diels–Alder reaction [2.2.1]
Diels, O.; Alder, K. *Ann. Chem.* 1928, *460*, 98.

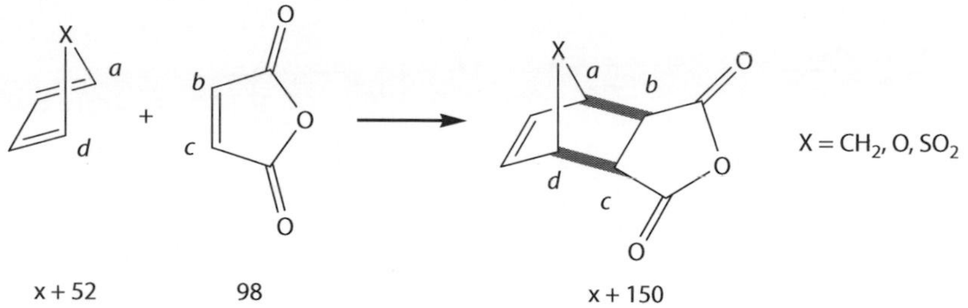

X = CH$_2$, O, SO$_2$

x + 52   98   x + 150

Corey γ-lactone synthesis—Subtractive oxidation [3.3.0]
Corey, E.J.; Kang, M.C. *J. Am. Chem. Soc.* 1984, *106*, 5384.

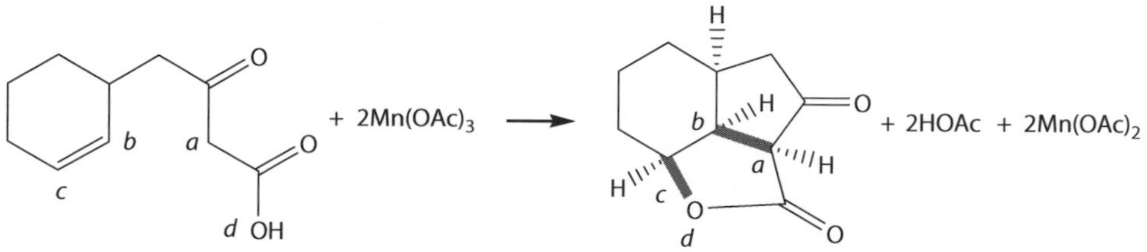

+ 2Mn(OAc)$_3$ + 2HOAc + 2Mn(OAc)$_2$

182   463.8   180   120   345.8

10.5   SACRIFICIAL RINGS

In this section are given examples excerpted from the synthesis database of strategies that involve using sacrificial rings. References to the plans are given in the accompanying download.

G. Schröder (1964) plan for bullvalene: benzocyclobutane ring cleaved as benzene (step 3).

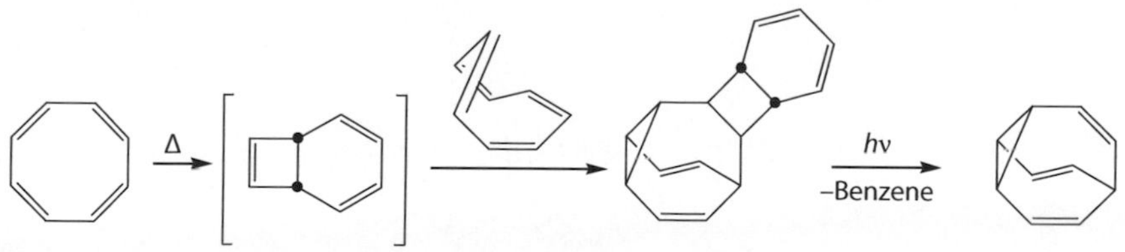

Taber (2002) plan for (−)-codeine: oxidative cleavage of cyclopentene ring by ozonolysis followed by recyclization to form cyclohexone ring (steps 16 and 17).

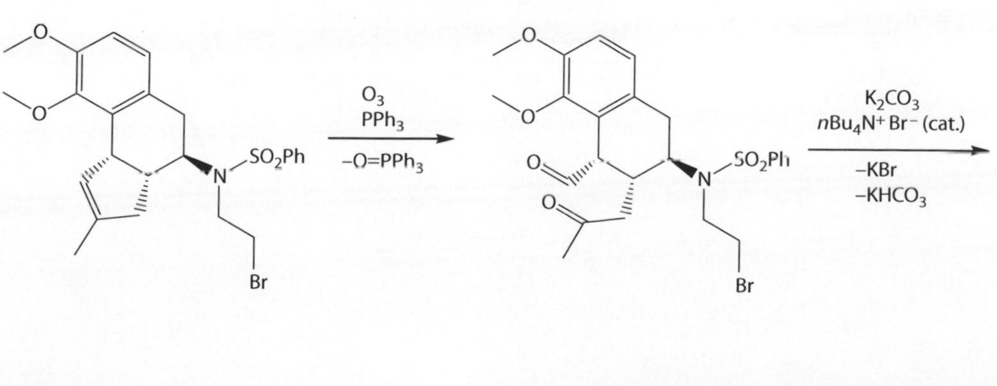

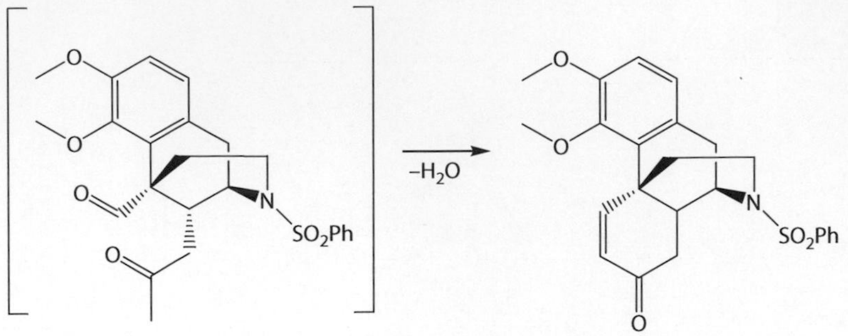

Banwell (1996) plan for colchicine: cyclopropane ring opening that results in a ring expansion of a [4.1.0] bicyclic system to a six-membered ring in step 12.

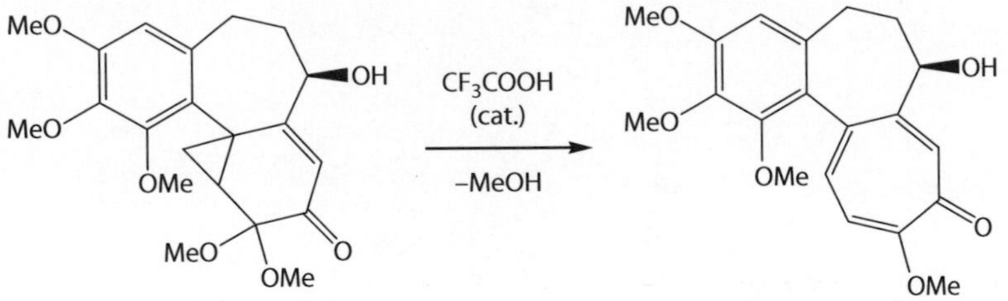

Eschenmoser (1959) plan for colchicine: tropolone ring B is reduced (step 3), tropolone ring C is built up by creating a six-membered ring, which is then ring expanded via a cyclopropane ring (steps 6 to 11).

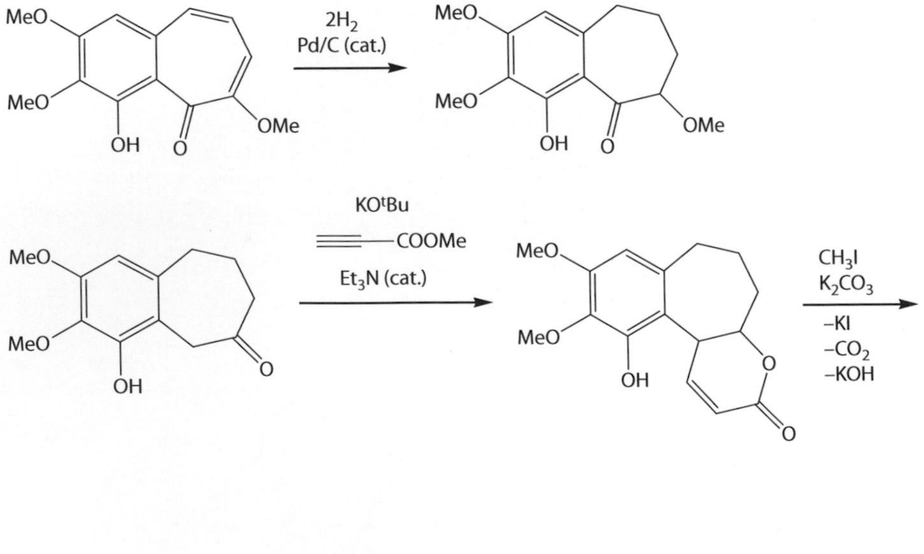

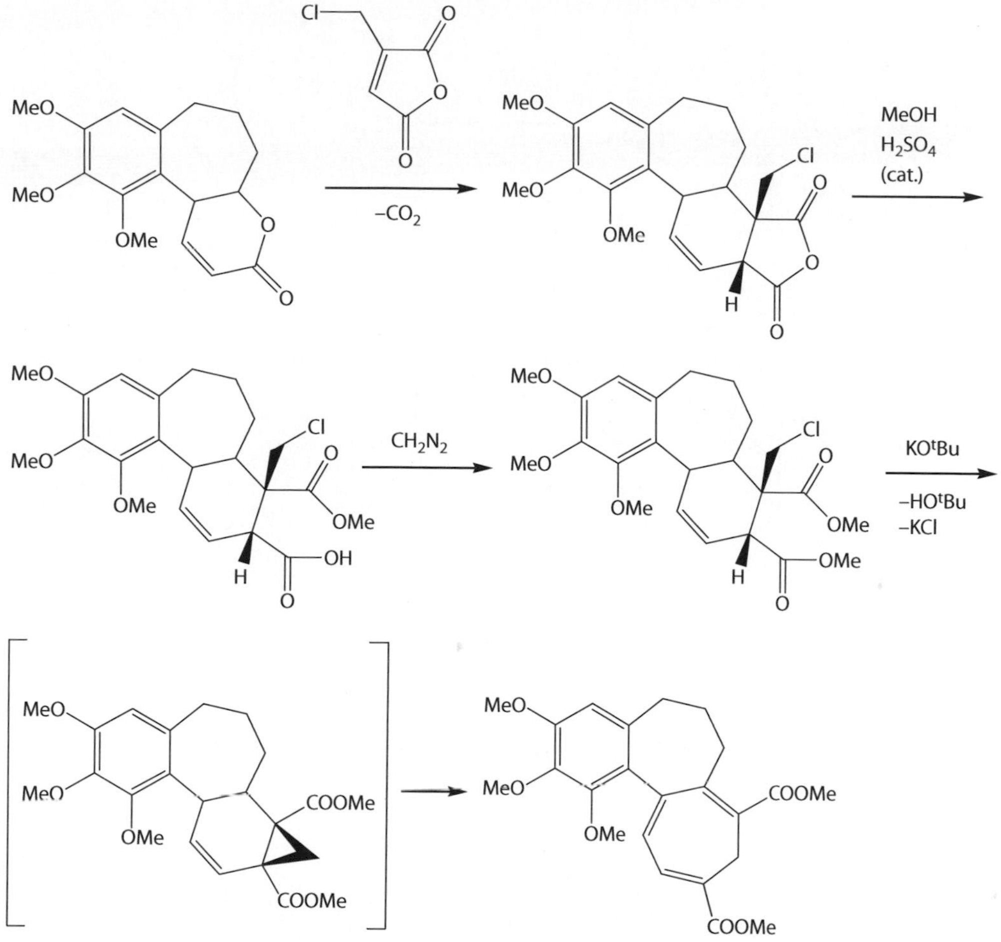

Evans (1981) plan for colchicine: cyclopropane ring opening that results in simultaneous ring expansion of two spiro-six-membered rings in step 8.

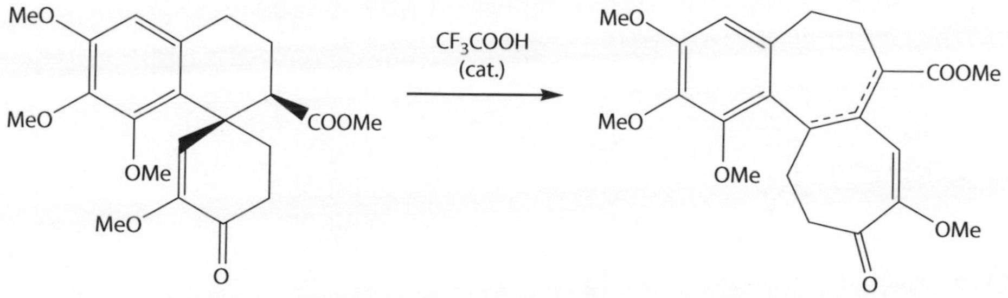

Scott (1965) plan for colchicine: decarboxylation in step 8.

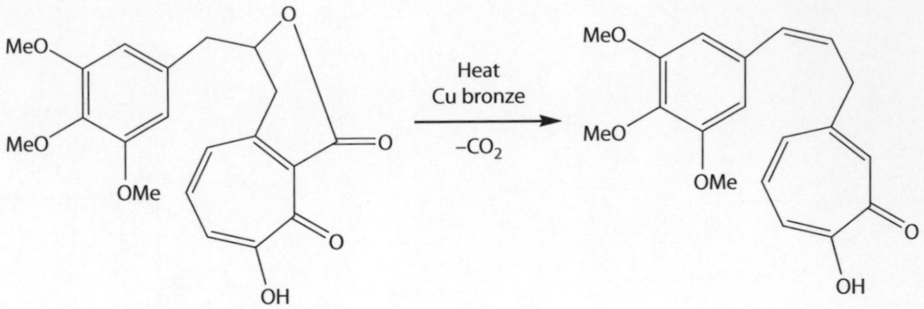

Woodward (1963) plan for colchicine: isothiazole ring opened via reduction with Raney Ni (step 22).

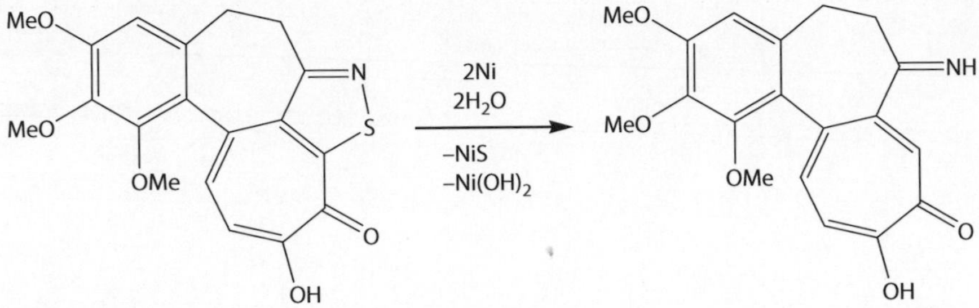

Wender (1982) plan for (±)-hirsutene: cyclopropane ring opening via elimination (step 3).

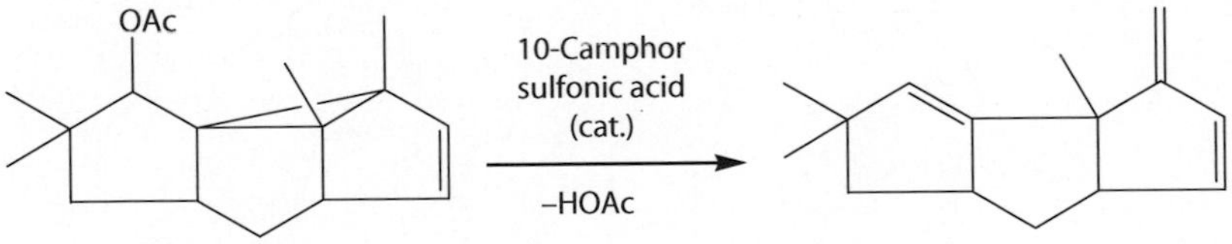

Yu (2004) plan for (±)-hirsutene: cyclopropane ring opening (step 4).

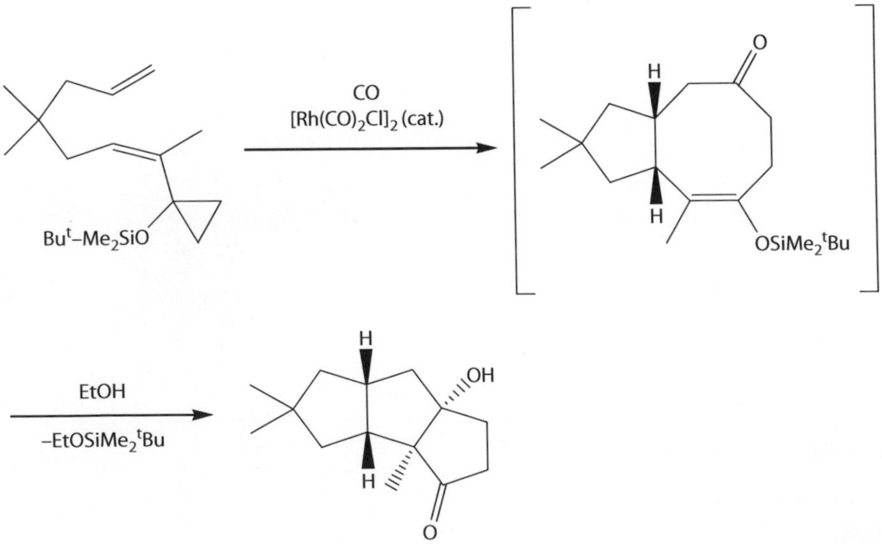

Pattenden (2004) plan for (+)-estrone: cyclopropane radical ring opening and zipper cyclization (step 11).

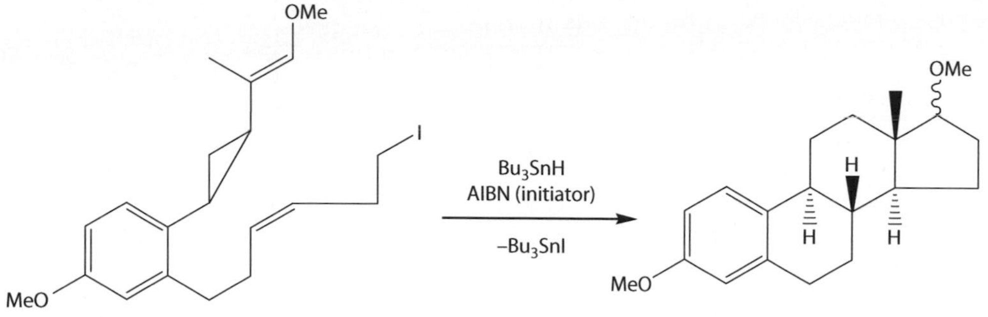

Takano G1 (1988) plan for (−)-kainic acid: aziridine ring opening to affect an intramolecular [3 + 2] cycloaddition in step 9.

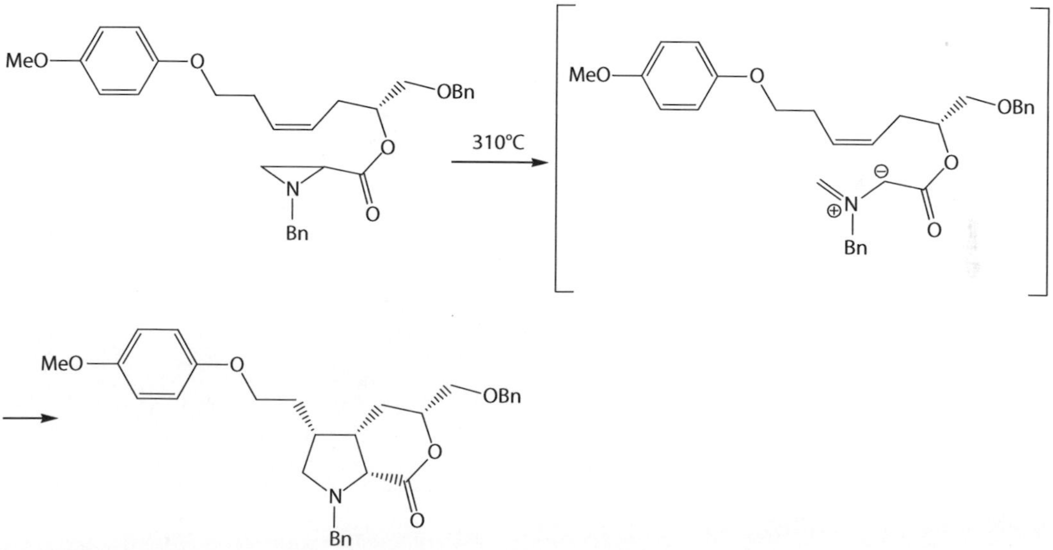

Takano G3 (1992) plan for (−)-kainic acid: Pausan-Khand reaction to make top ring in step 11, which is then ring expanded by Baeyer-Villiger oxidation (step 17) and opened up in steps 23.

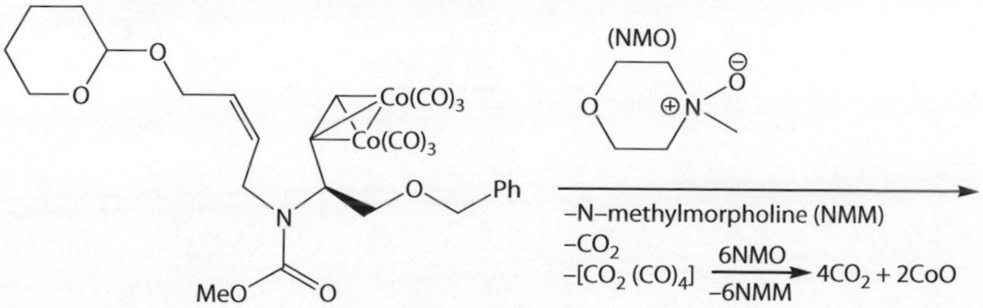

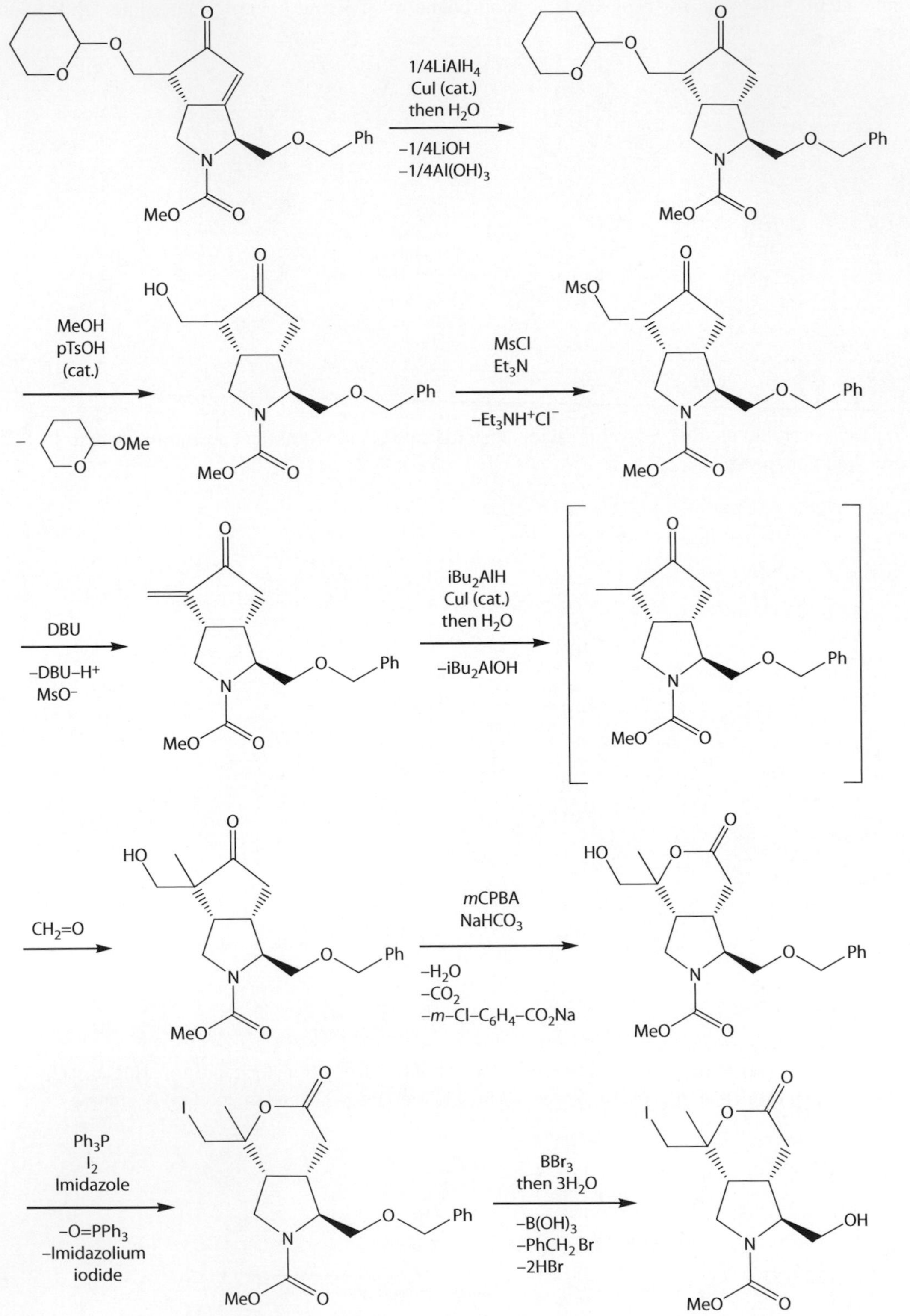
1/4LiAlH4
CuI (cat.)
then H2O
–1/4LiOH
–1/4Al(OH)3
MeOH
pTsOH
(cat.)
OMe
MsCl
Et3N
–Et3NH+Cl–
DBU
–DBU–H+
MsO–
iBu2AlH
CuI (cat.)
then H2O
–iBu2AlOH
CH2=O
mCPBA
NaHCO3
–H2O
–CO2
–m–Cl–C6H4–CO2Na
Ph3P
I2
Imidazole
–O=PPh3
–Imidazolium
iodide
BBr3
then 3H2O
–B(OH)3
–PhCH2 Br
–2HBr
HO
MsO
MeO
Ph
OH
I

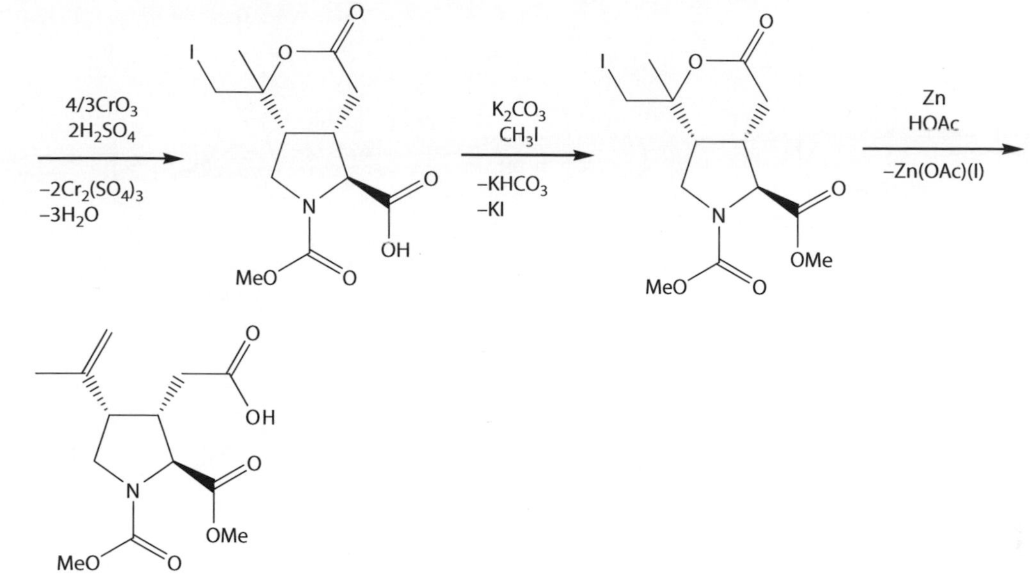

Yoo (1994) plan for (−)-kainic acid: Pausan–Khand reaction to make top ring in step 8, which is then ring opened in step 13 by ozonolysis

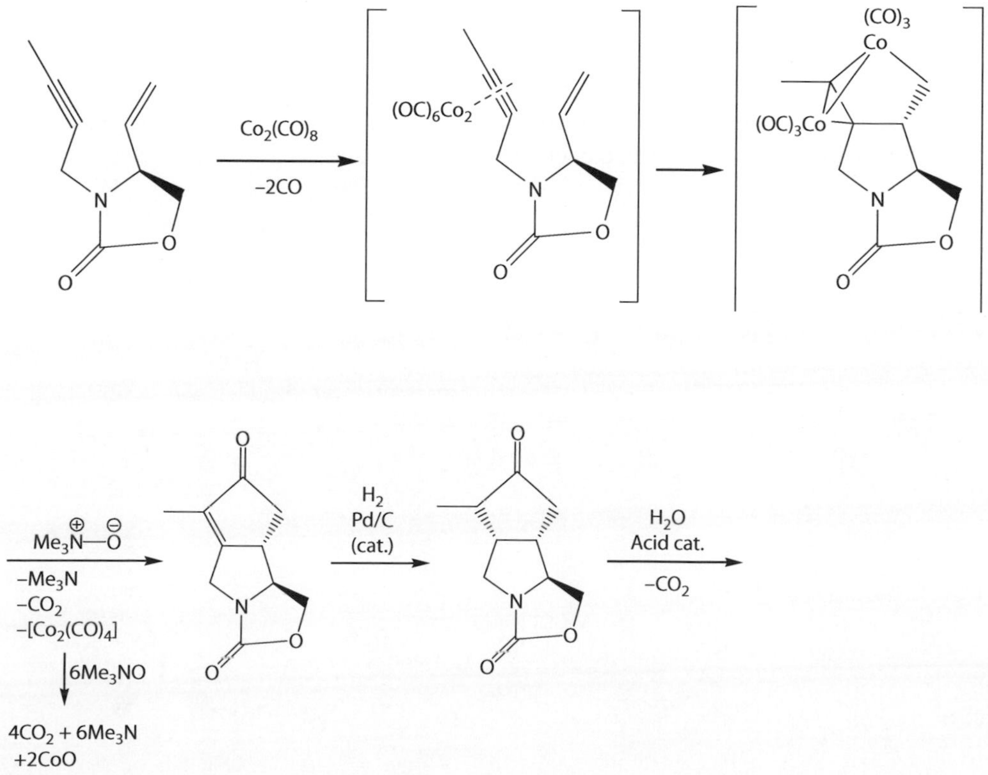

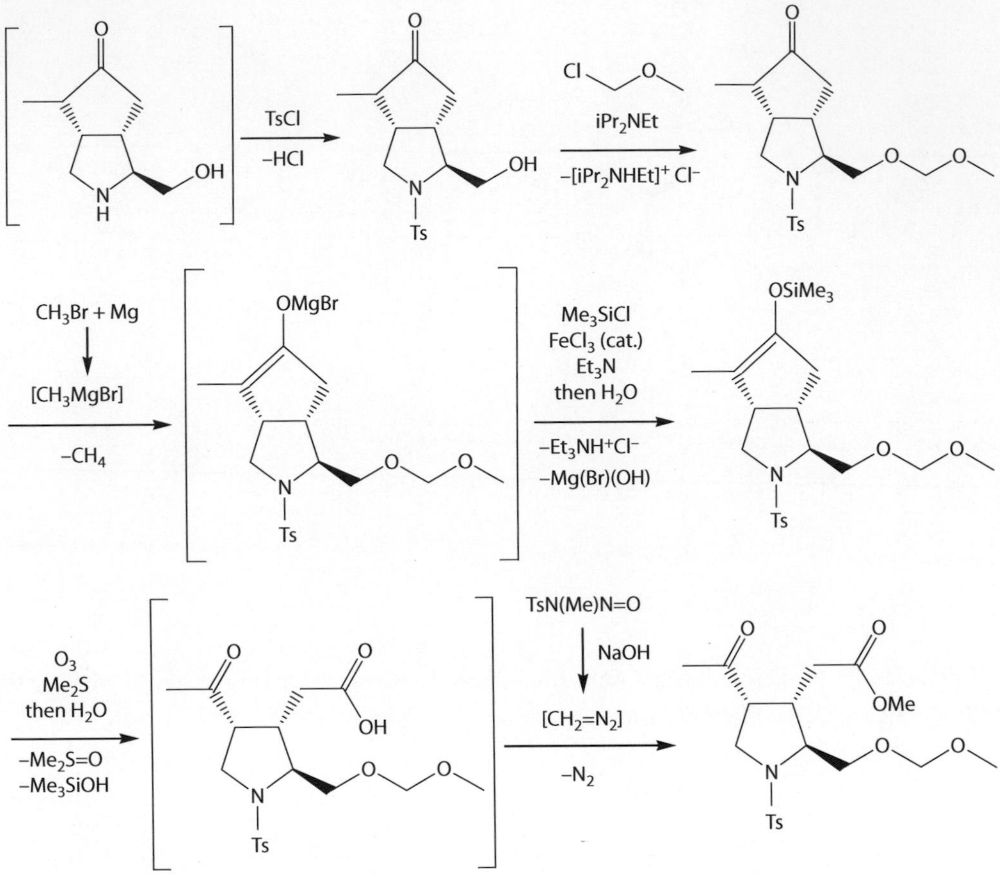

Jung (2007) plan for (−)-kainic acid: cyclohexenone top ring is made in step 13, ring expanded by Baeyer–Villiger oxidation in step 20, and opened up in step 21.

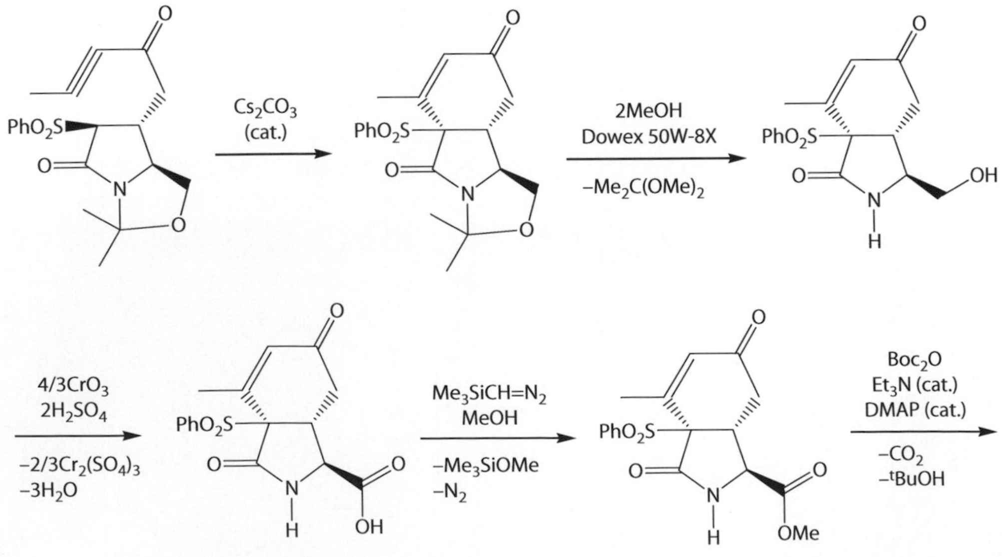

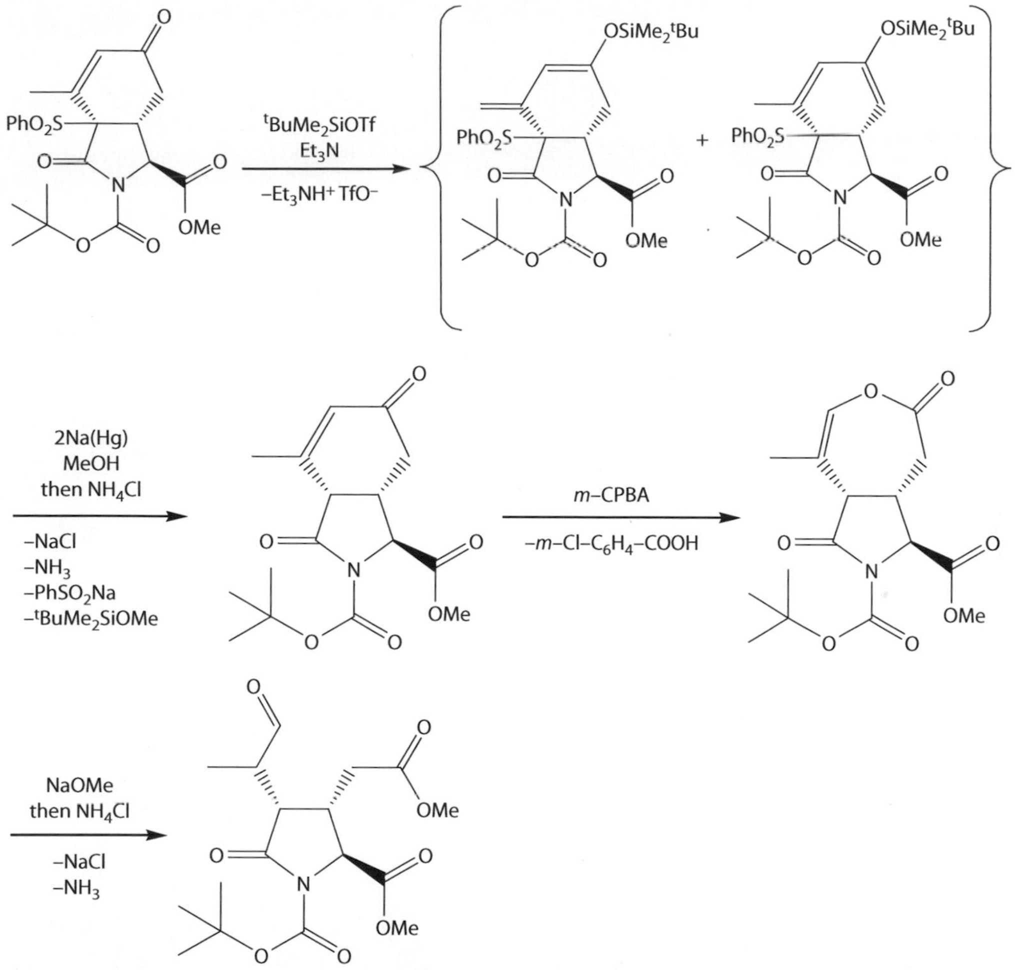

Clayden (2002) plan for (−)-kainic acid: oxidative degradation of aromatic ring leaving one carbon atom as a carboxylic acid (step 7); cyclohexanone top ring is made in step 3, ring expanded by Baeyer–Villiger oxidation (step 9), and opened up in step 10.

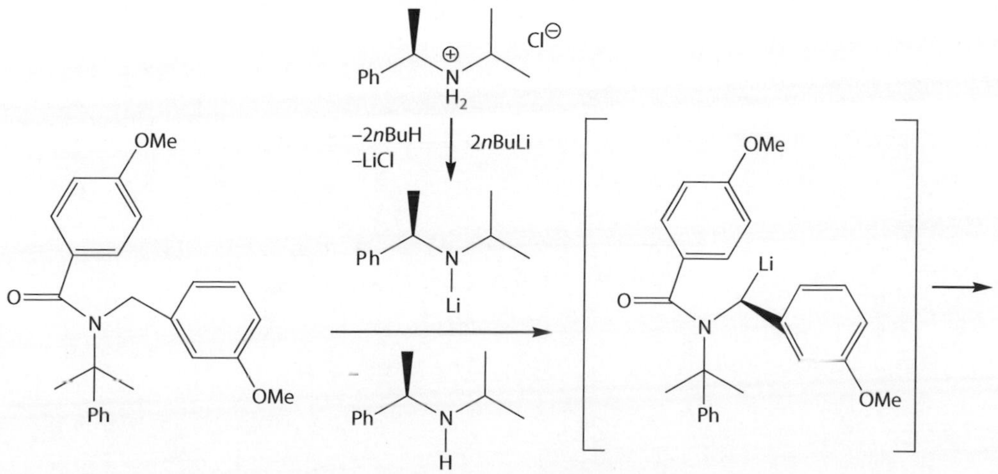

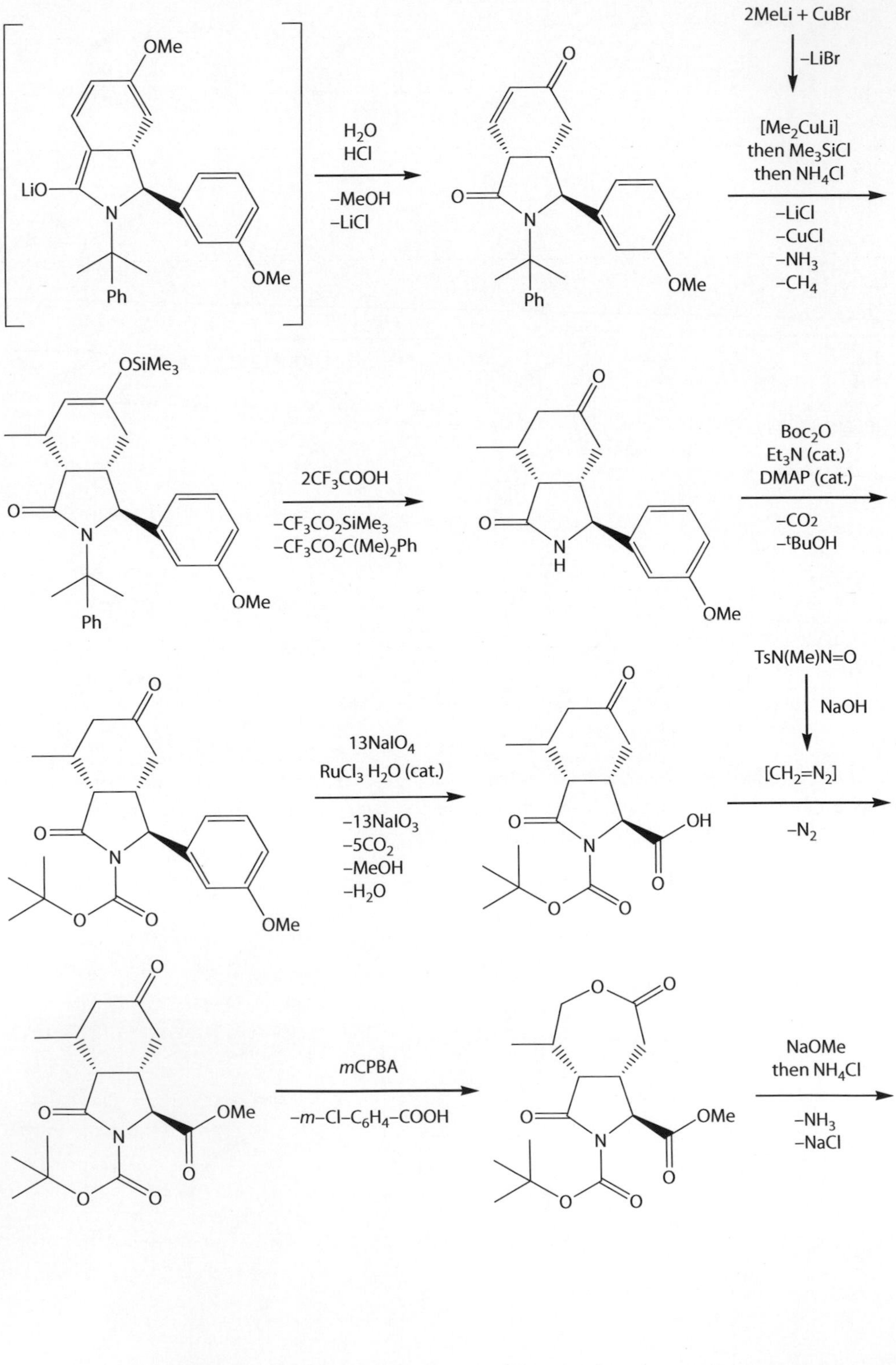
OMe
LiO
N
Ph
OMe
H2O
HCl
–MeOH
–LiCl
O
O
N
Ph
OMe
2MeLi + CuBr
–LiBr
[Me2CuLi]
then Me3SiCl
then NH4Cl
–LiCl
–CuCl
–NH3
–CH4
OSiMe3
O
N
Ph
OMe
2CF3COOH
–CF3CO2SiMe3
–CF3CO2C(Me)2Ph
O
O
N
H
OMe
Boc2O
Et3N (cat.)
DMAP (cat.)
–CO2
–tBuOH
O
O
N
O
O
OMe
13NaIO4
RuCl3 H2O (cat.)
–13NaIO3
–5CO2
–MeOH
–H2O
O
O
N
OH
O
O
O
TsN(Me)N=O
NaOH
[CH2=N2]
–N2
O
O
N
OMe
O
O
O
mCPBA
–m–Cl–C6H4–COOH
O
O
O
N
OMe
O
O
O
NaOMe
then NH4Cl
–NH3
–NaCl

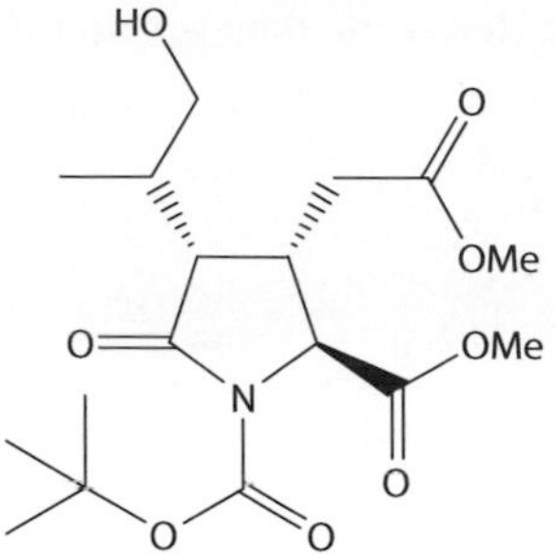

Cossy (1998) plan for (–)-kainic acid: cyclopentene top ring is made in step 14 and oxidatively opened up in step 16.

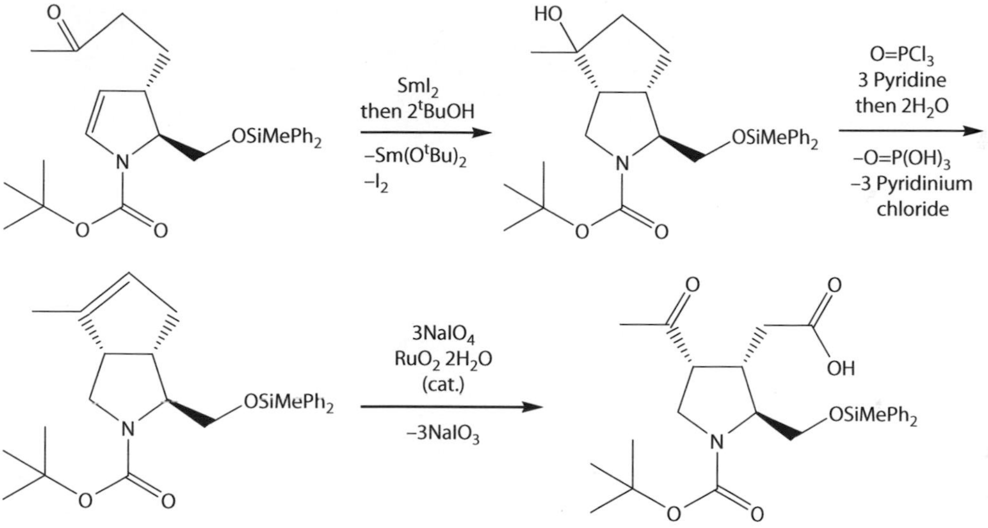

Bachi (1997) plan for (–)-kainic acid: sulfone top ring is made in step 12 and then is reductively opened up in step 13.

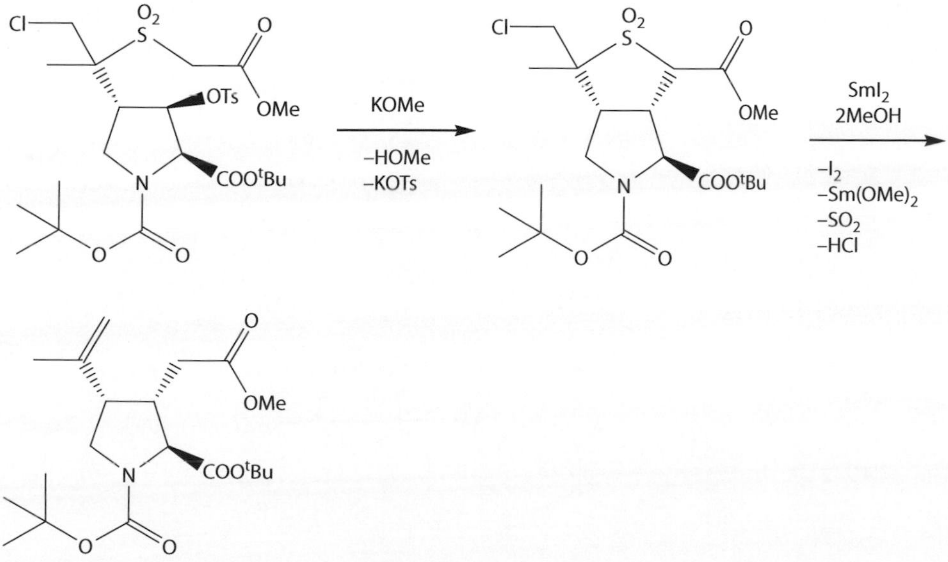

Monn (1994) plan for (±)-kainic acid: cyclopentene top ring is made in step 6 and then is oxidatively opened up in step 10.

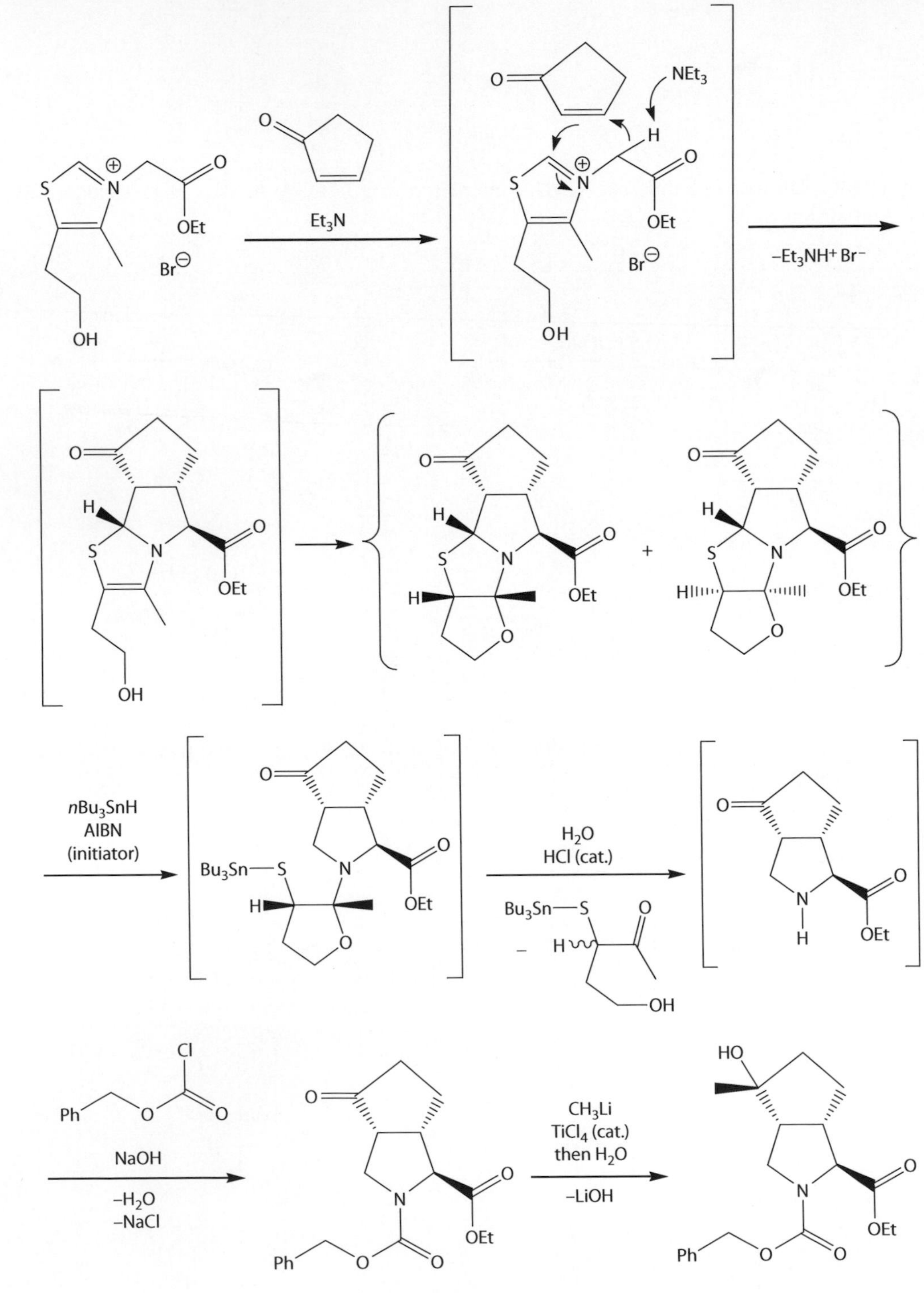

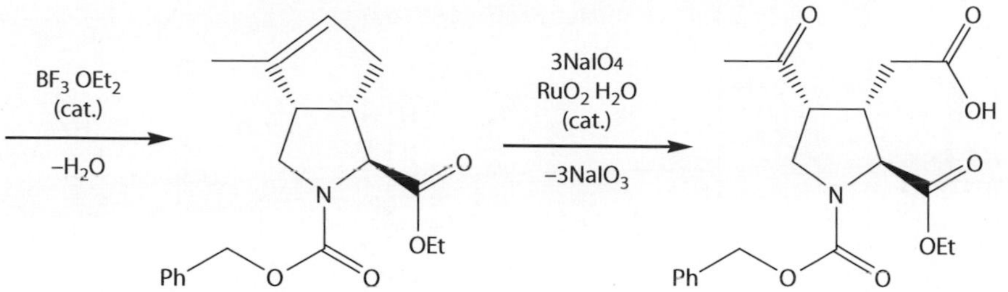

Oppolzer (1981) plan for lysergic acid: cyclopentadiene is eliminated in a tandem retro-Diels–Alder–Diels–Alder reaction (step 13).

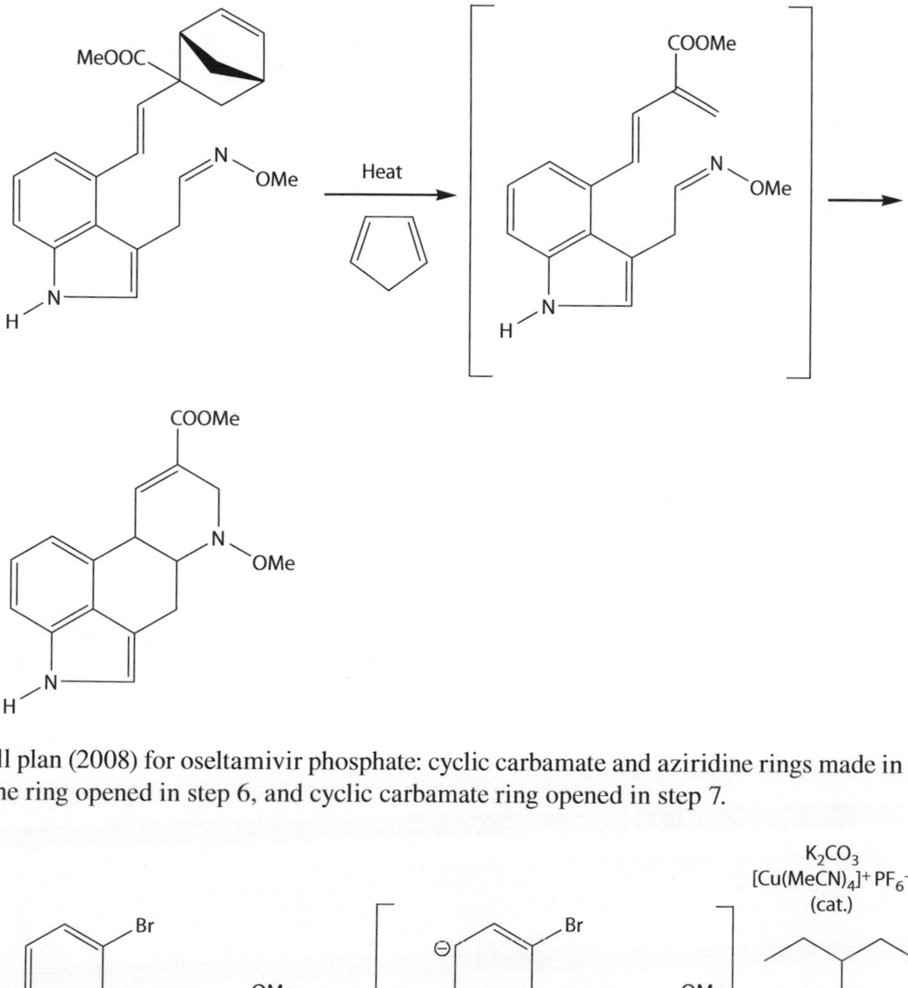

Banwell plan (2008) for oseltamivir phosphate: cyclic carbamate and aziridine rings made in step 5, aziridine ring opened in step 6, and cyclic carbamate ring opened in step 7.

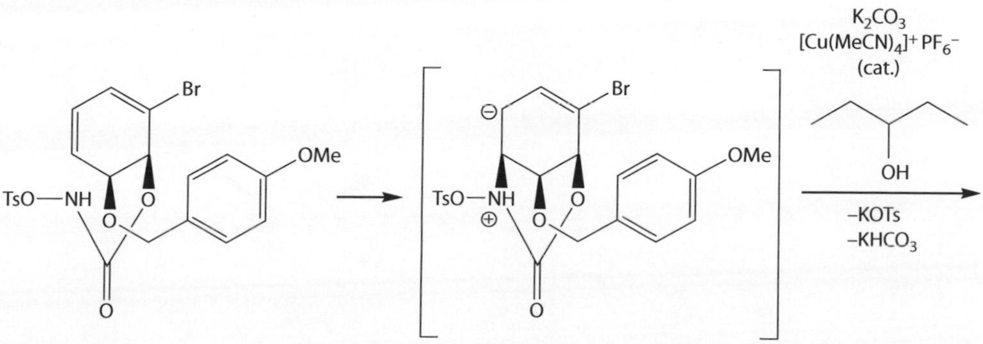

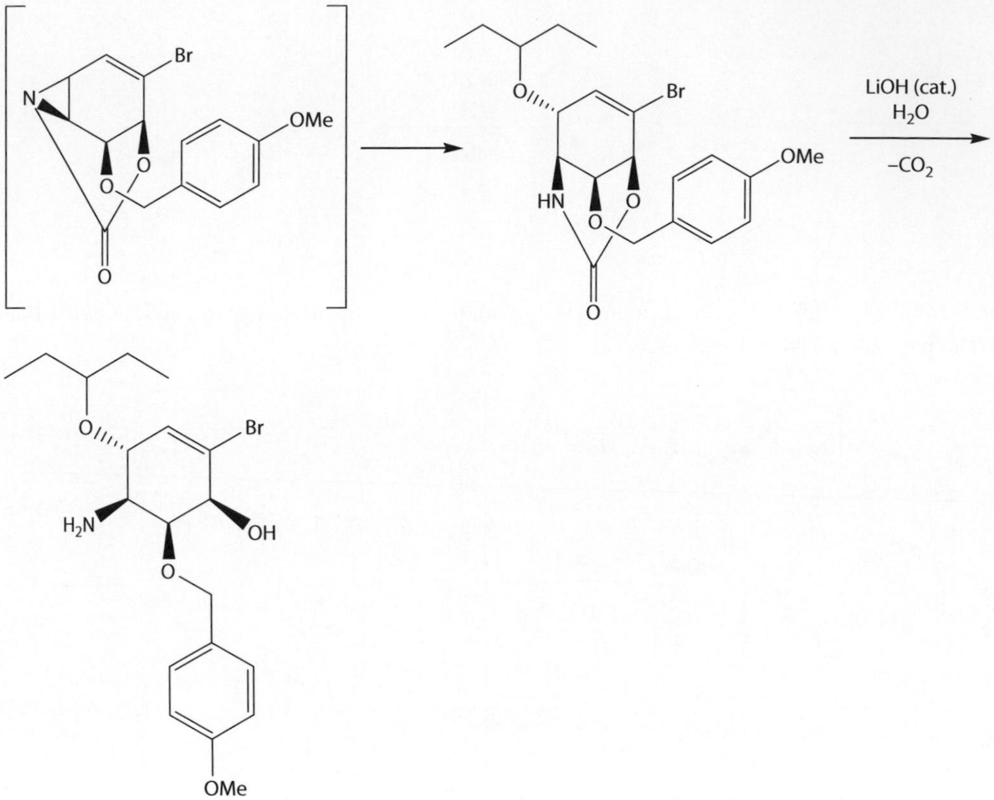

Chen–Liu (2010) plan for oseltamivir phosphate: Claisen rearrangement in step 6; cyclic carbamate made in step 12 and ring opened in step 15.

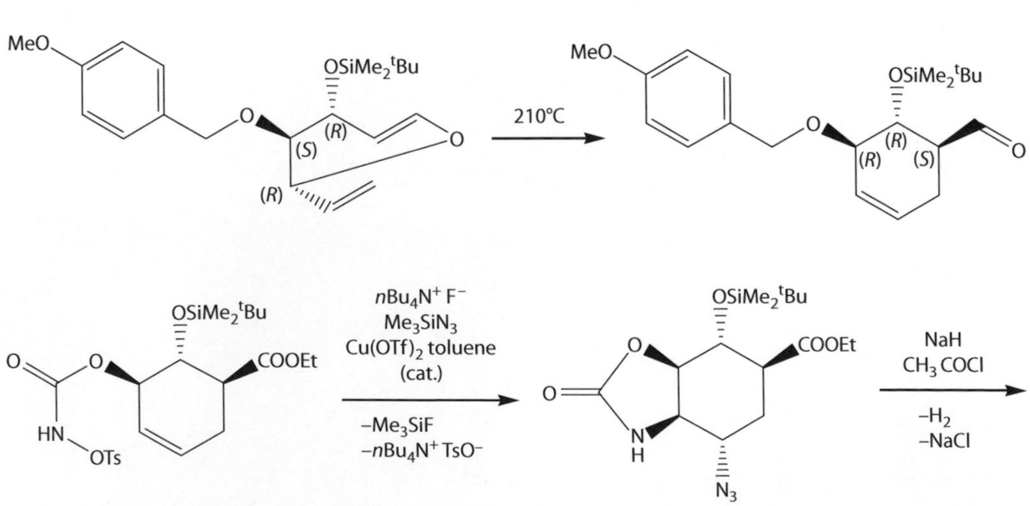

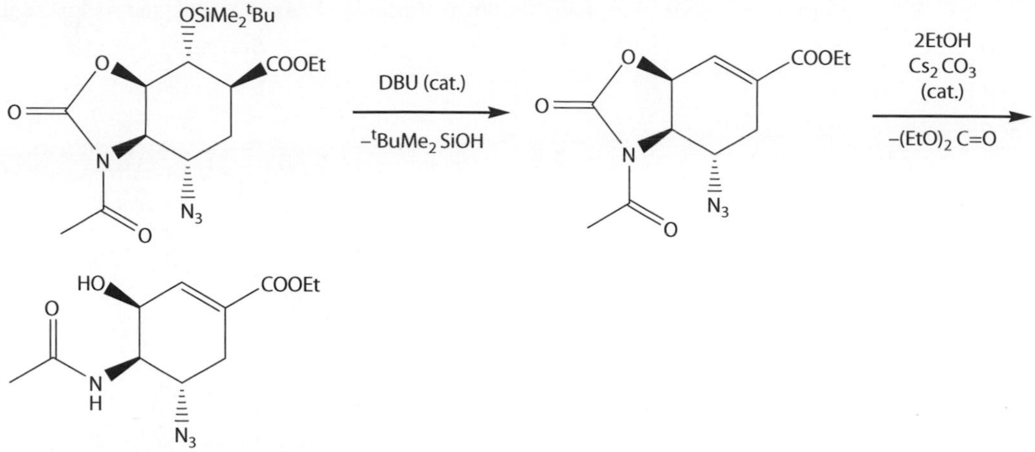

Corey (2006) plan for oseltamivir phoshate: lactam formation in step 3, lactam opened in step 7, aziridine ring made in step 9 then opened in step 10.

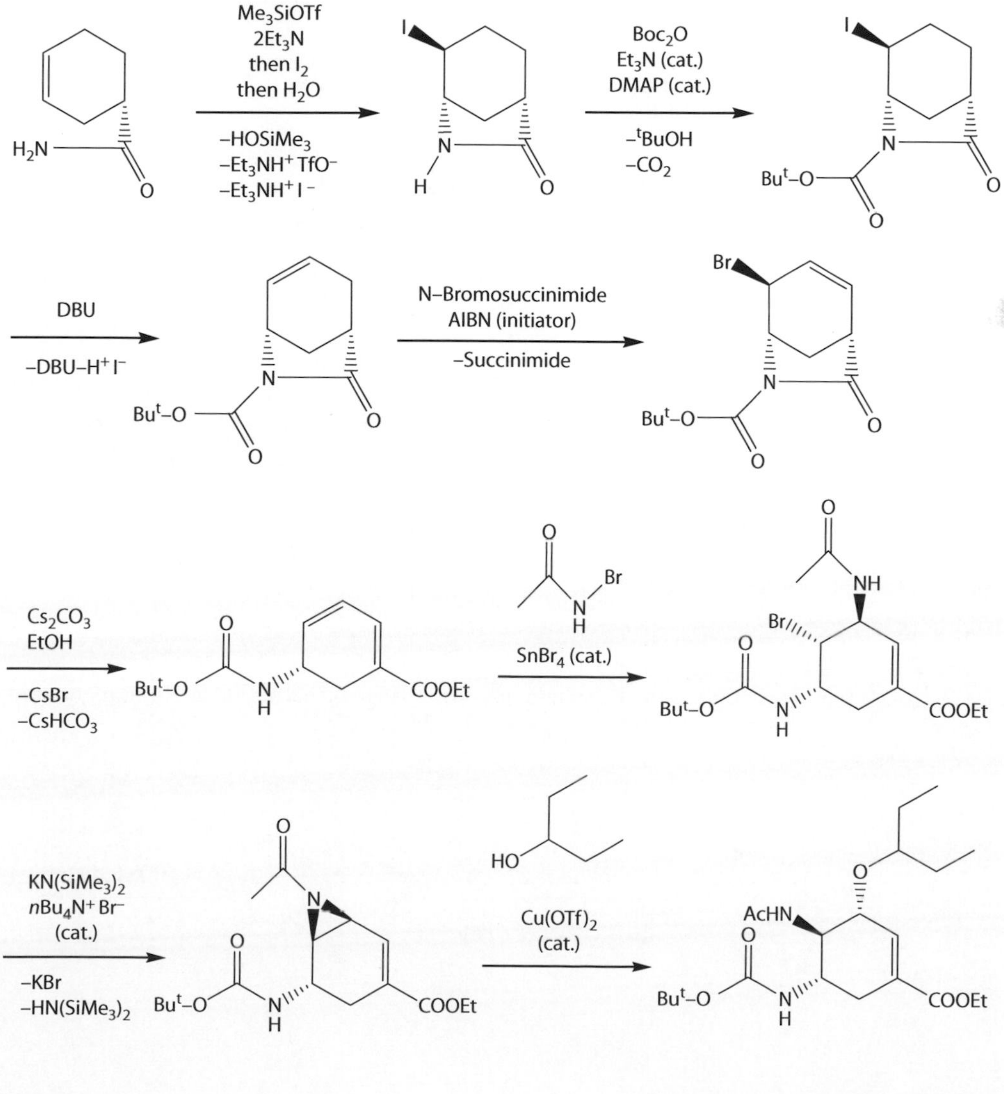

Fang G1 (2007) plan for oseltamivir phosphate: tetrahydrofuran-2-ol ring opening and reclosure in step 11.

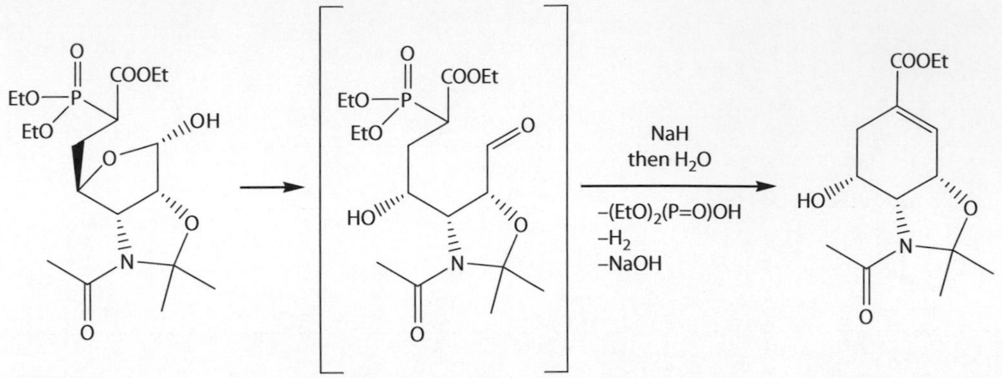

Fang G2 (2008) plan for oseltamivir phosphate: aziridine ring formation in step 4 and opening in step 5.

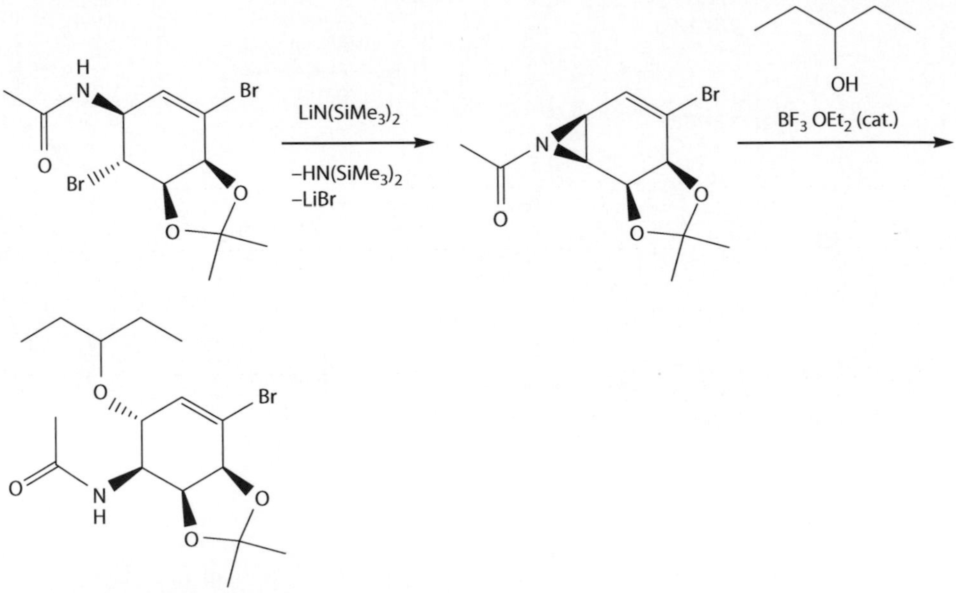

Fang G3 (2008) plan for oseltamivir phosphate: aziriding ring formation in step 4 and opening in step 5.

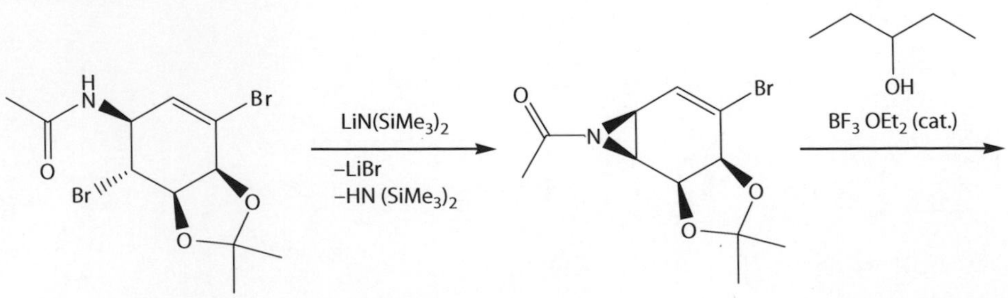

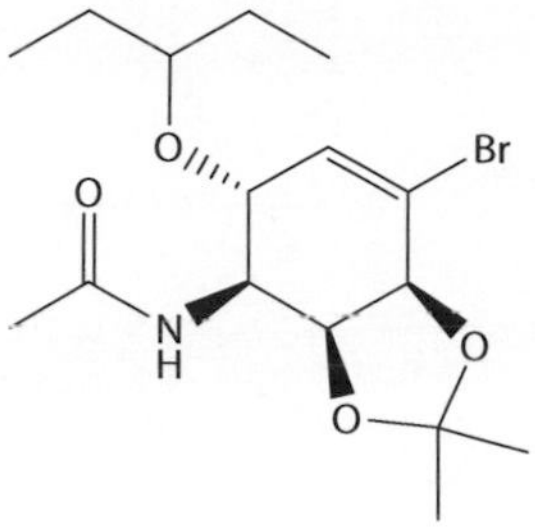

Fukuyama (2007) plan for oseltamivir phosphate: tandem [2.2.1] bicyclic ring opening and aziridine ring formation in step 9 and aziridine ring opening in step 10.

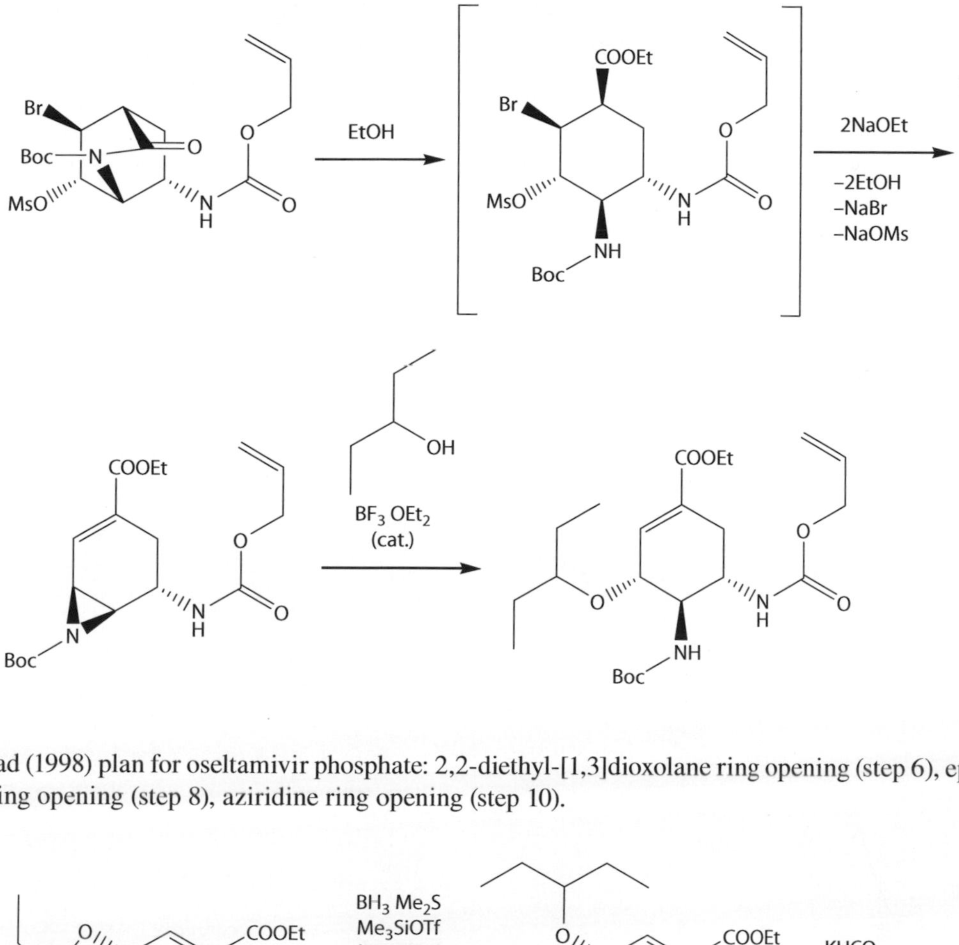

Gilead (1998) plan for oseltamivir phosphate: 2,2-diethyl-[1,3]dioxolane ring opening (step 6), epoxide ring opening (step 8), aziridine ring opening (step 10).

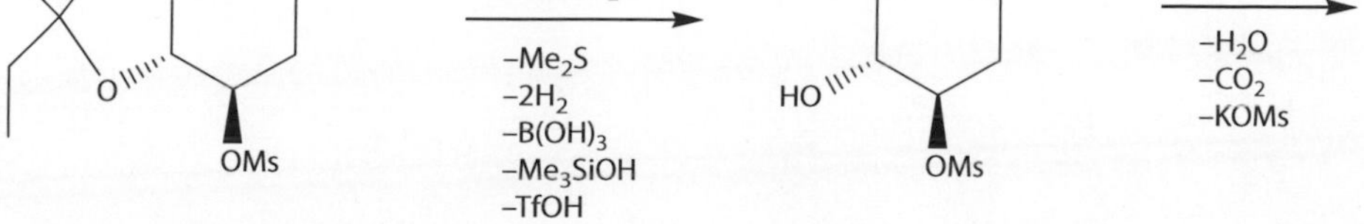

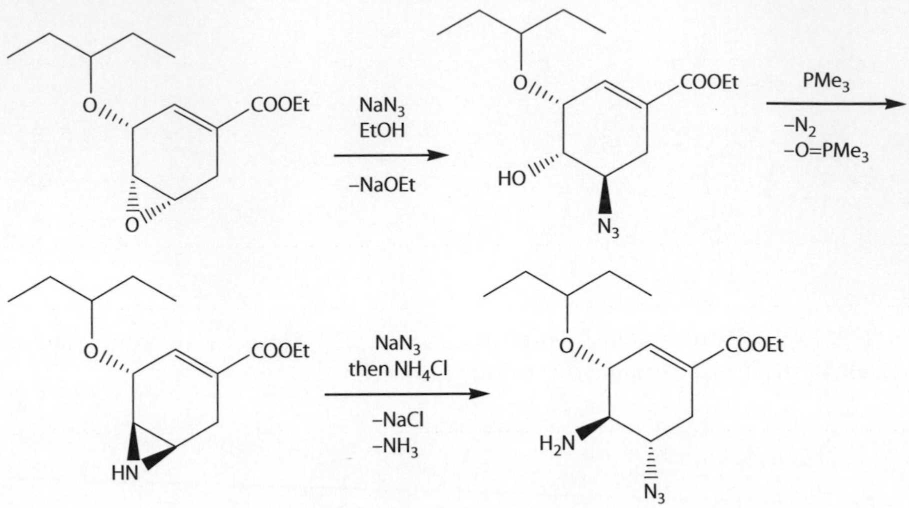

Hudlicky G1 (2009) plan for oseltamivir phosphate: [2.2.2] bicyclic ring made in step 2 and opened in step 3, dihydrooxazole ring formation in step 4 and opened in step 5.

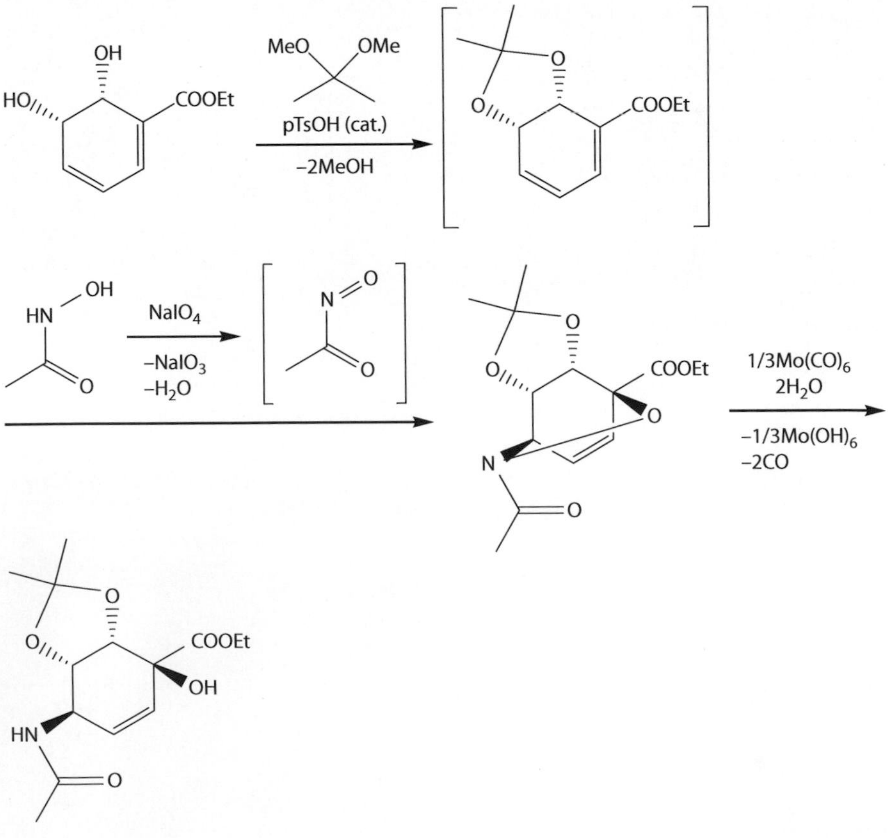

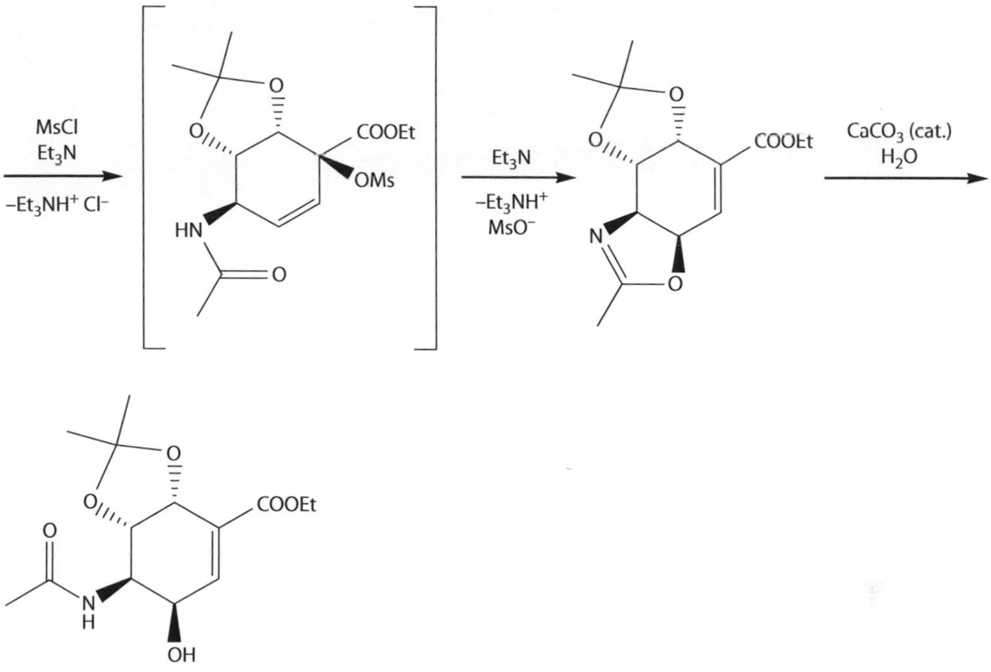

Kann (2007) plan for oseltamivir phosphate: epoxidation in step 10 and opening in step 11, aziridine ring formation in step 13 and opening in step 14.

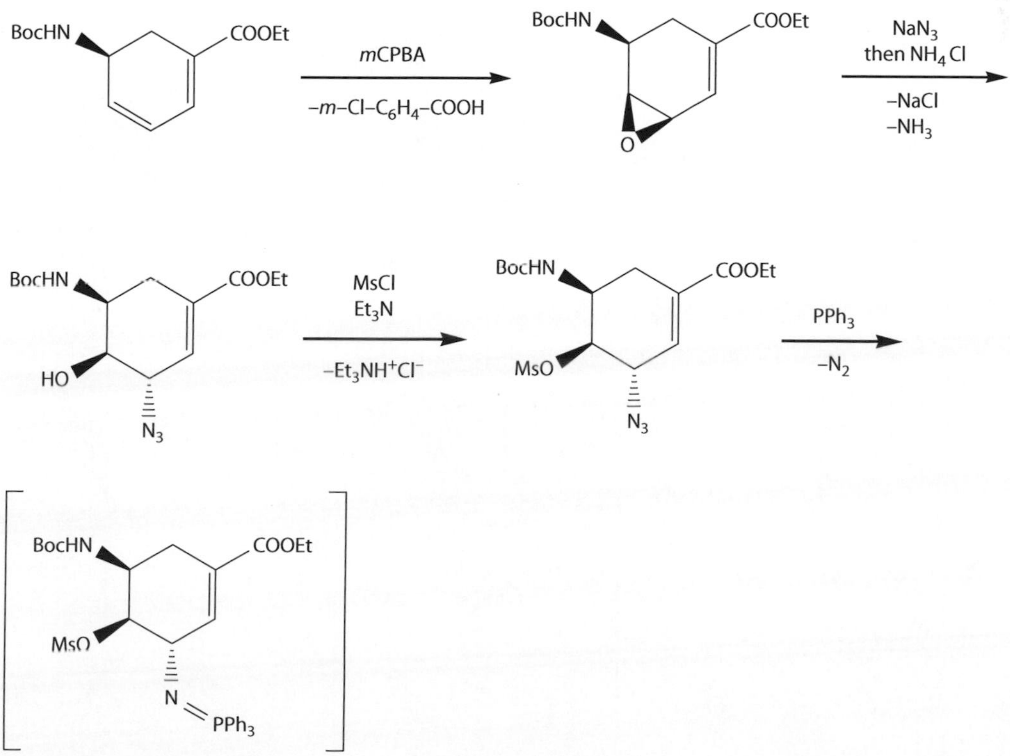

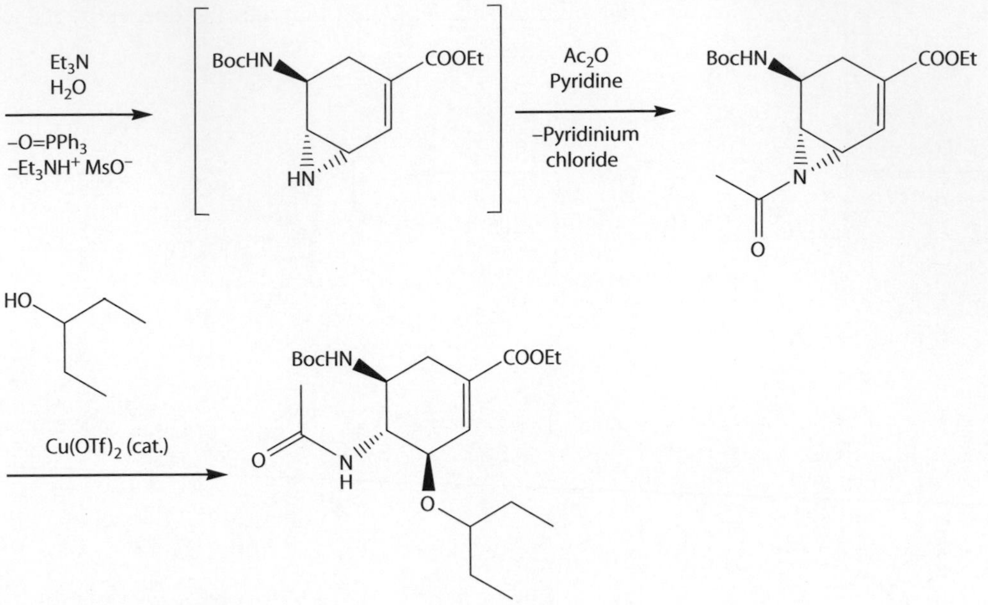

Mandai G2 (2009) plan for oseltamivir phosphate: β-lactam formation in step 3 via [2 + 2] cycloaddition and opened in step 9.

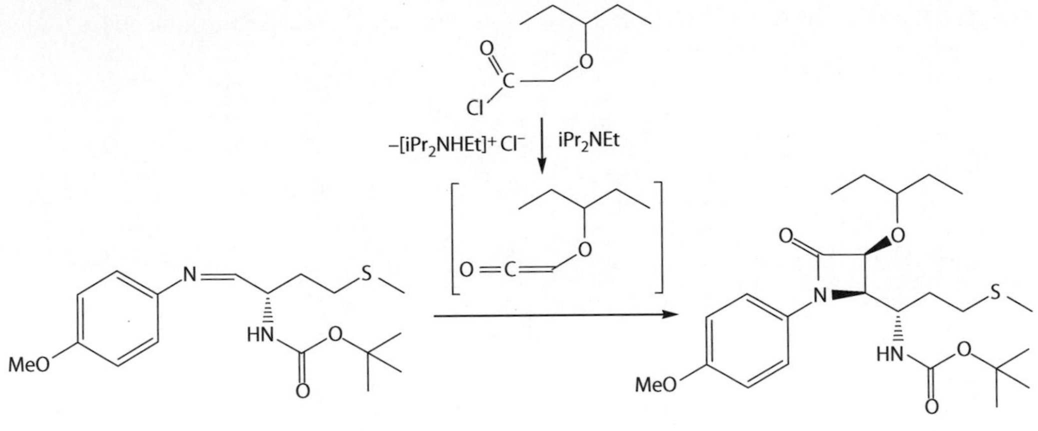

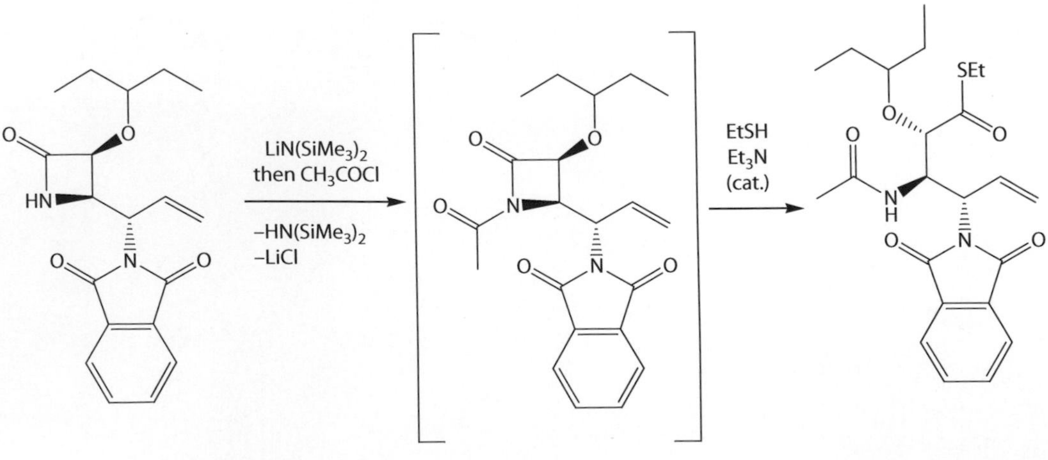

Okamura–Corey (2008) plan for oseltamivir phosphate: [2.2.2] bicyclic ring opened by reduction (step 6); aziridine ring opening in step 12.

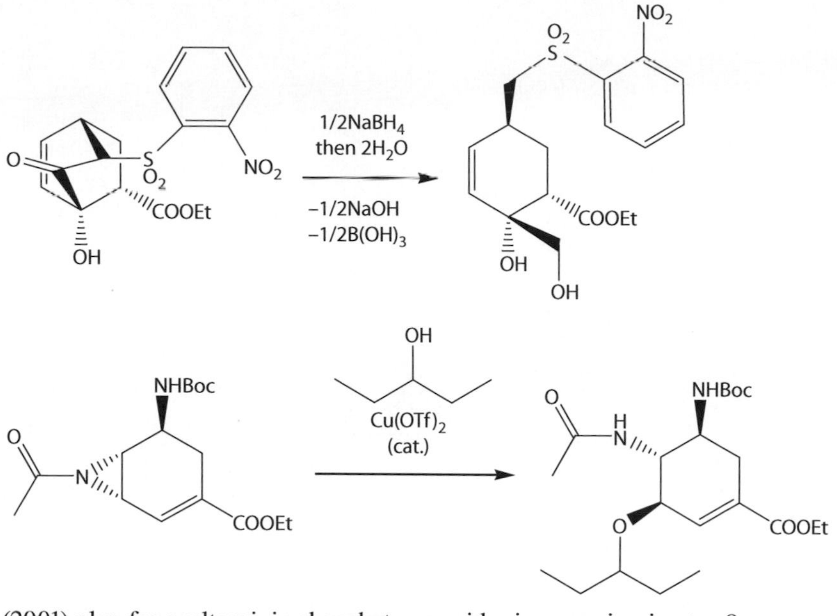

Roche G1 (2001) plan for oseltamivir phosphate: epoxide ring opening in step 8.

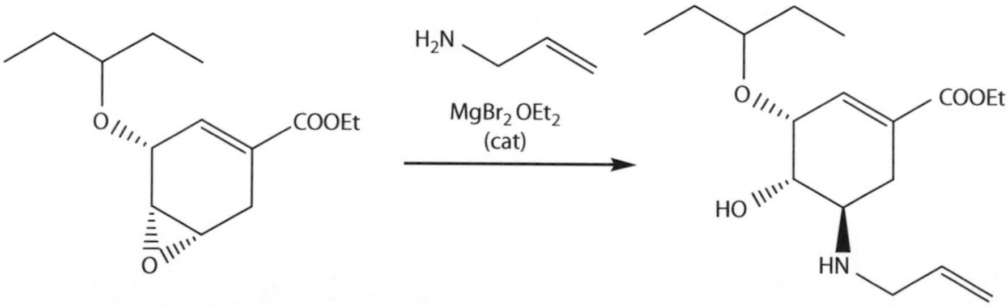

Roche G2 (1999, 2004) plan for oseltamivir phosphate: epoxide ring opening (step 8), aziridine formation (step 9), and aziridine ring opening (step 10).

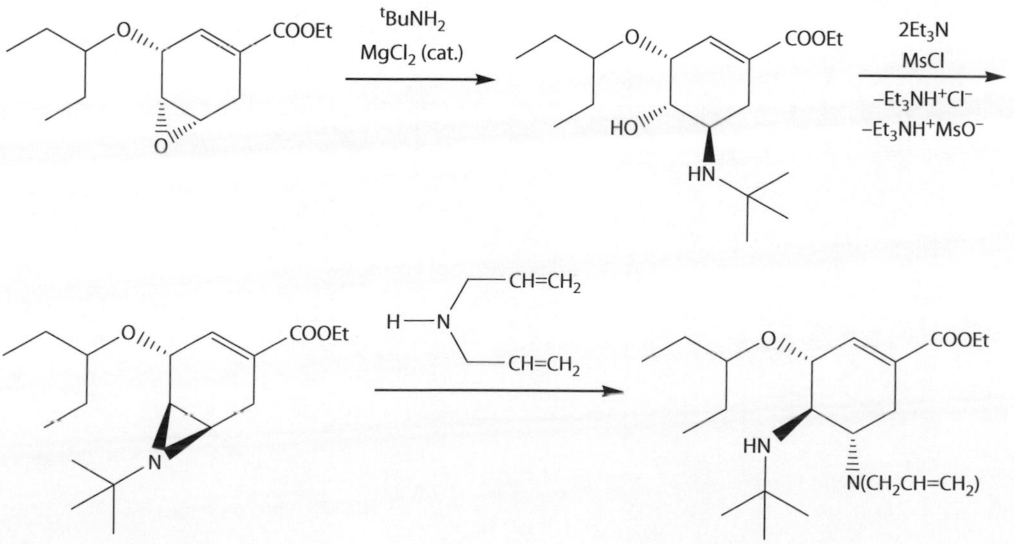

Roche G3 (1999, 2004) plan for oseltamivir phosphate: epoxide ring opening (step 7), aziridine formation (step 8), and aziridine ring opening (step 9).

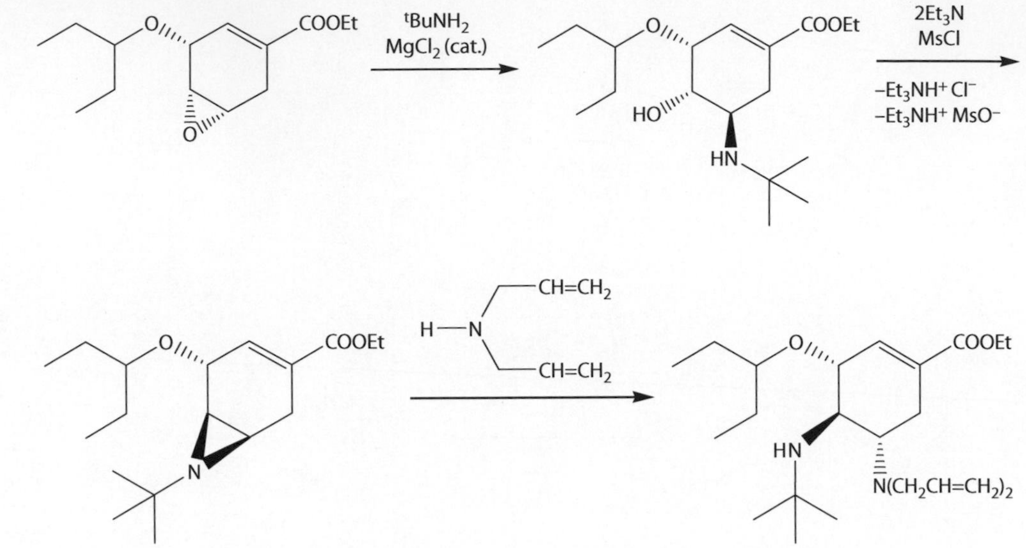

Roche G4 (2000) plan for oseltamivir phosphate: aziridine ring made in step 3 and ring opened in step 5.

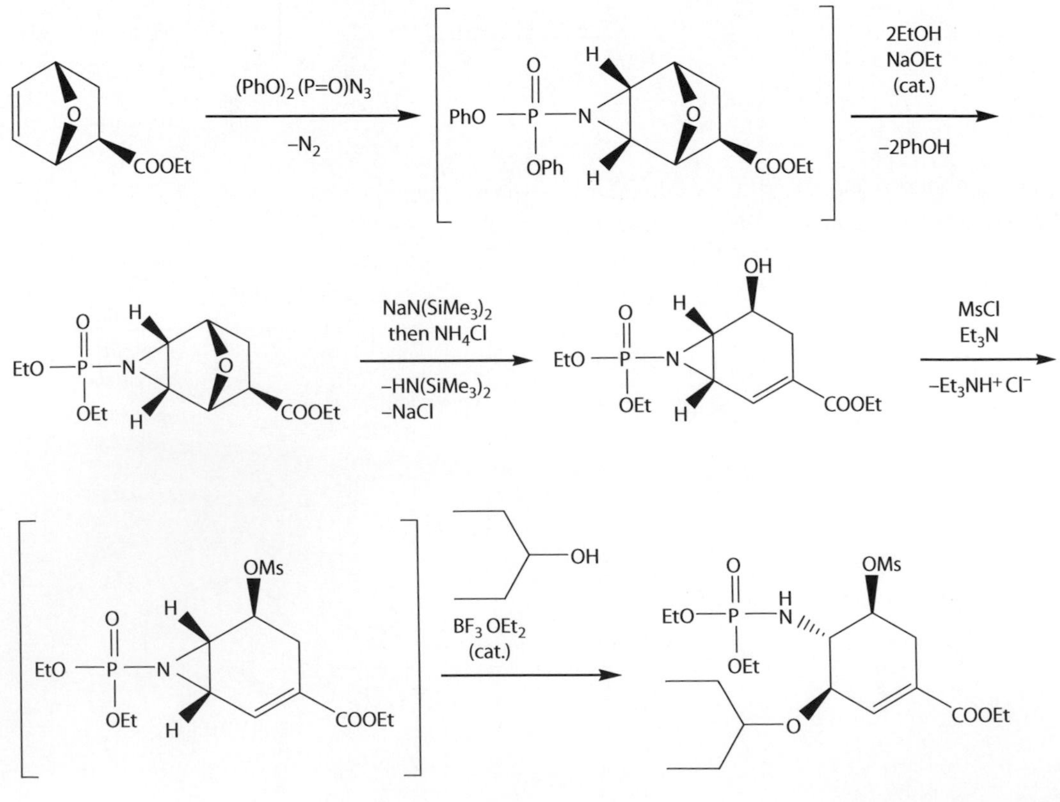

Roche G5 (2008) plan for oseltamivir phosphate: cyclic carbamate ring opening (step 9).

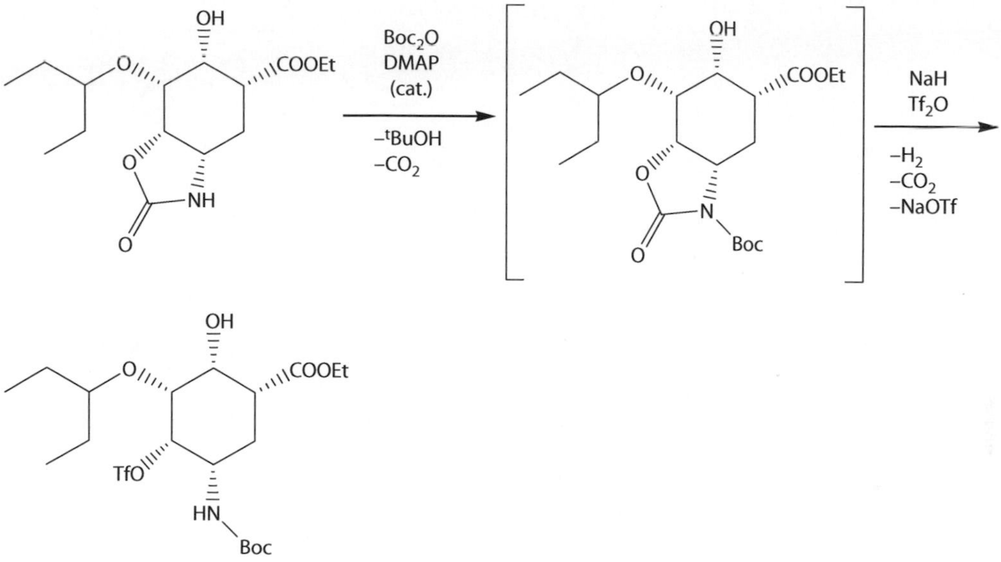

Roche G6 (2009) plan for oseltamivir phosphate: aziridine ring opening (step 5).

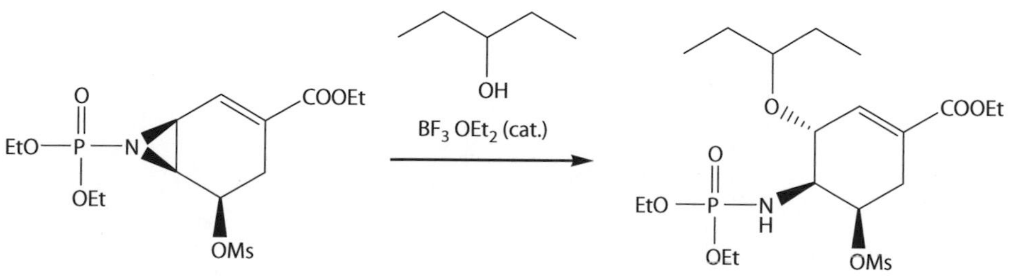

Shi (2009) plan for oseltamivir phosphate: aziridine ring opening (step 6).

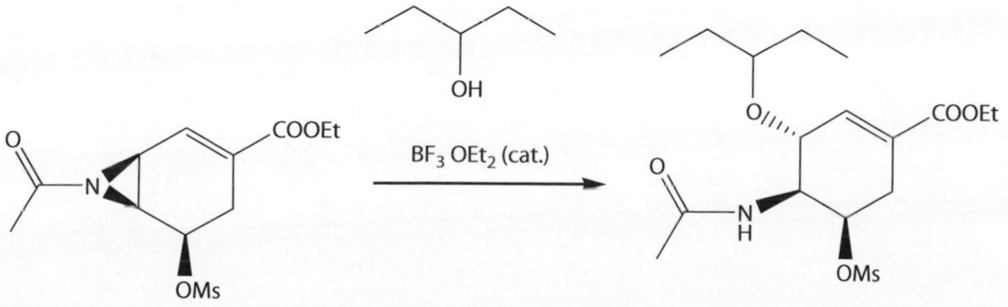

Shibasaki G1 (2006) plan for oseltamivir phosphate: epoxide ring opening in step 2, aziridine ring opening in step 4, second aziridine ring opening in step 11.

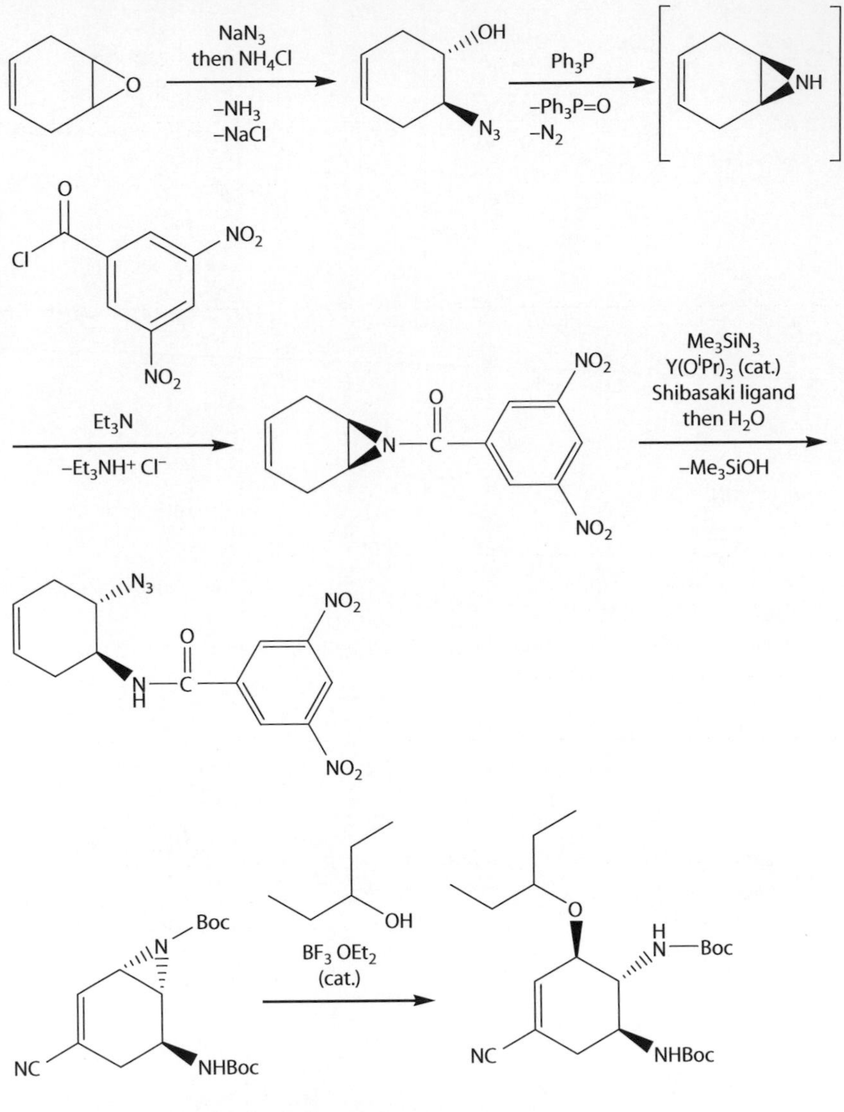

Shibasaki G2 (2007) plan for oseltamivir phosphate: epoxide ring opening in step 2, aziridine ring opening in step 4, cyclic carbamate ring opening in step 9, second aziridine ring opening in step 14.

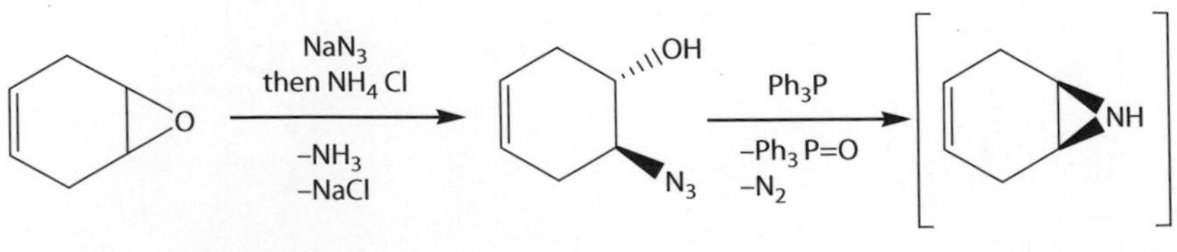

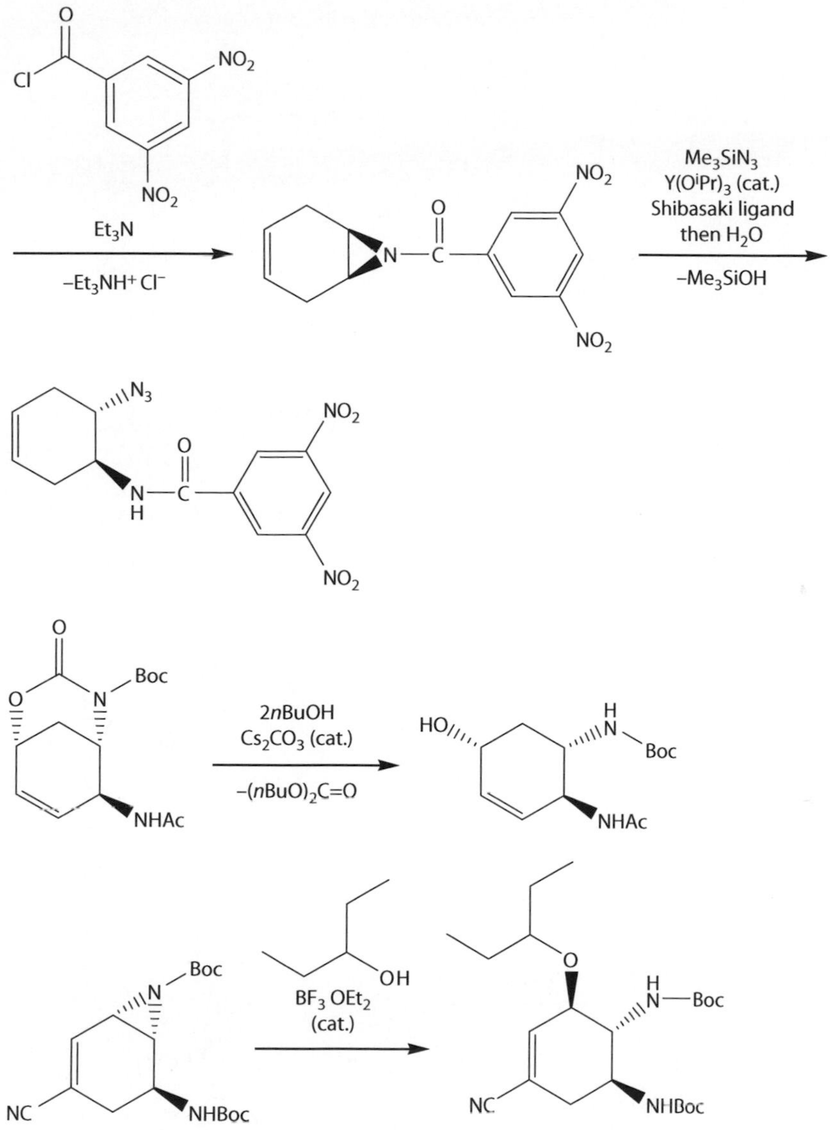

Shibasaki G3 (2007) plan for oseltamivir phosphate: cyclic carbamate ring opening in step 3, aziridine ring opening in step 9, aziridine ring opening in step 10.

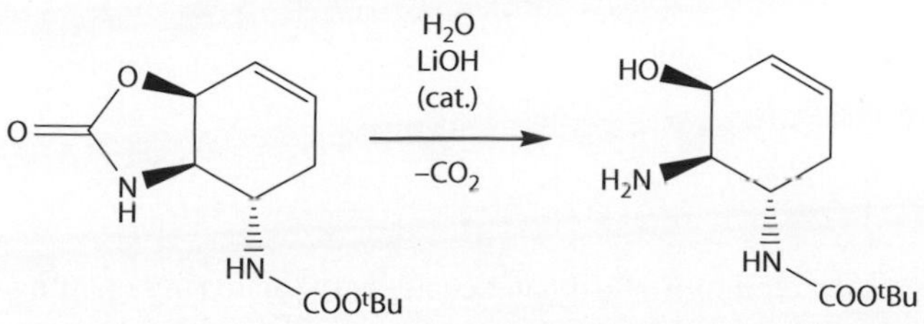

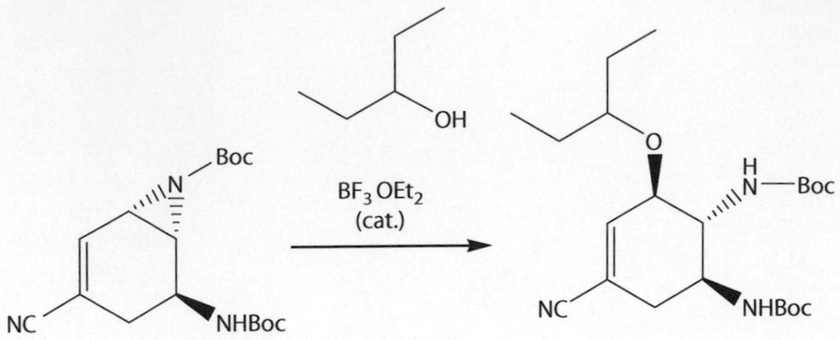

Shibasaki G4 (2009) plan for oseltamivir phosphate: cyclic carbamate ring opening in step 5, epoxide ring opening in step 7,

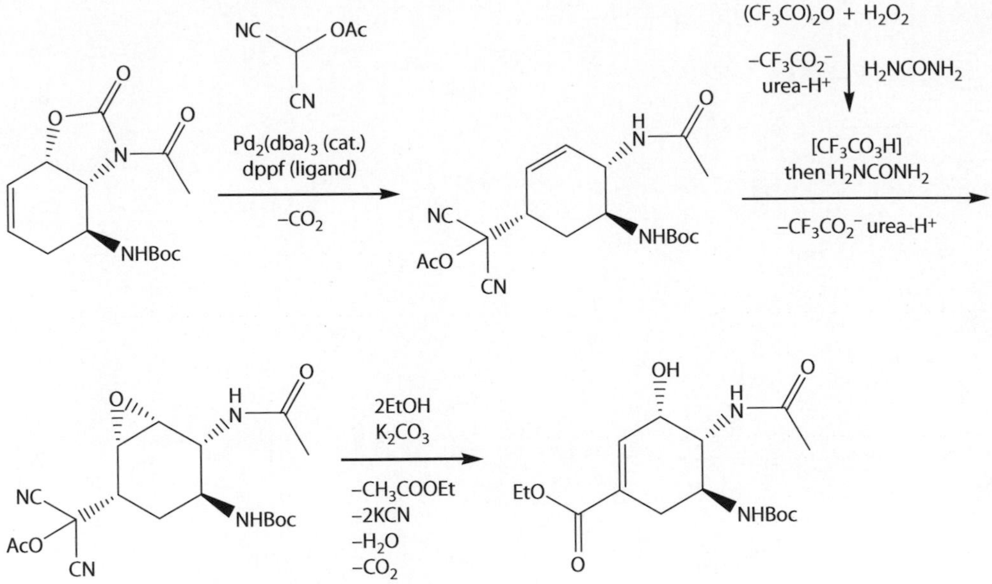

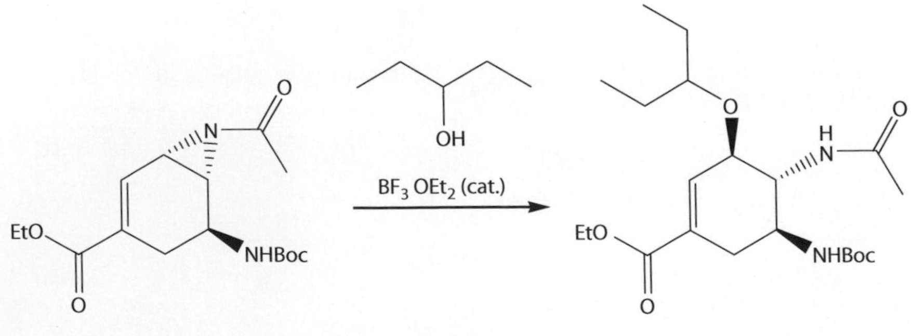

Shibasaki G4 (2009) plan for oseltamivir phosphate: cyclic carbamate ring opening in step 5, aziridine ring opening in step 10.

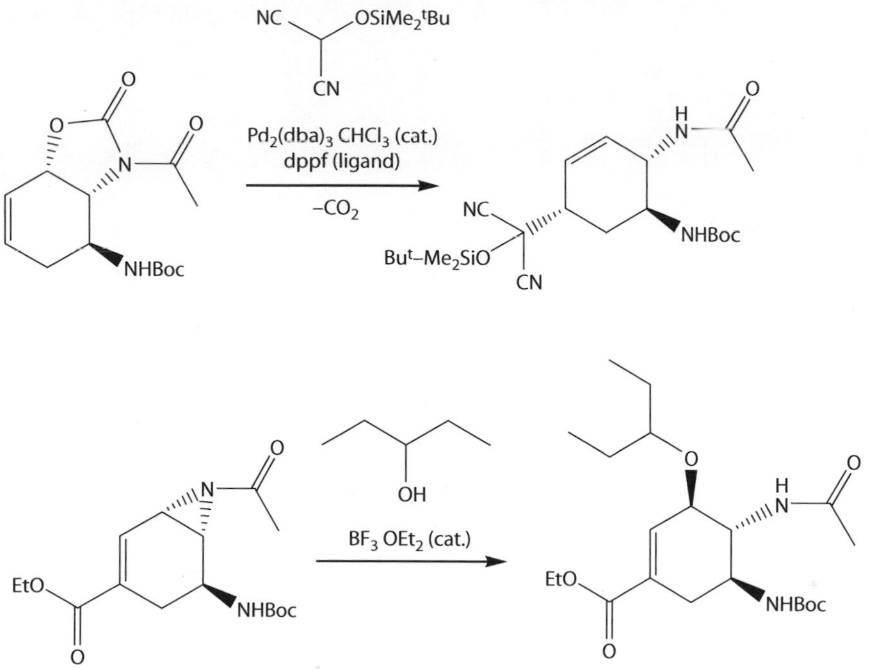

Trost (2008) plan for oseltamivir phosphate: lactone ring opening in step 4, aziridine ring opening in step 8.

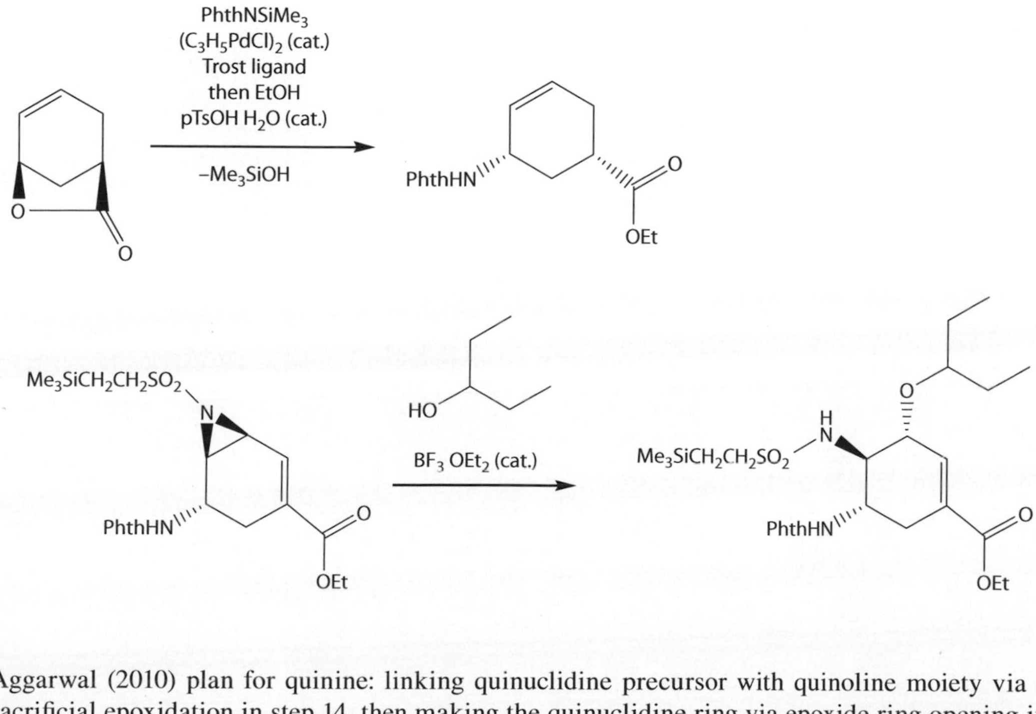

Aggarwal (2010) plan for quinine: linking quinuclidine precursor with quinoline moiety via a sacrificial epoxidation in step 14, then making the quinuclidine ring via epoxide ring opening in step 15.

Späth (1923) plan for ricinine: benzene ring in 4-chloroquinoline is oxidatively cleaved in step 6.

Fleming (2004) plan for (±)-sparteine: cyclopropane ring radical opening with concomitant ring expansion of a fused six-membered ring into a seven-membered ring in step 5.

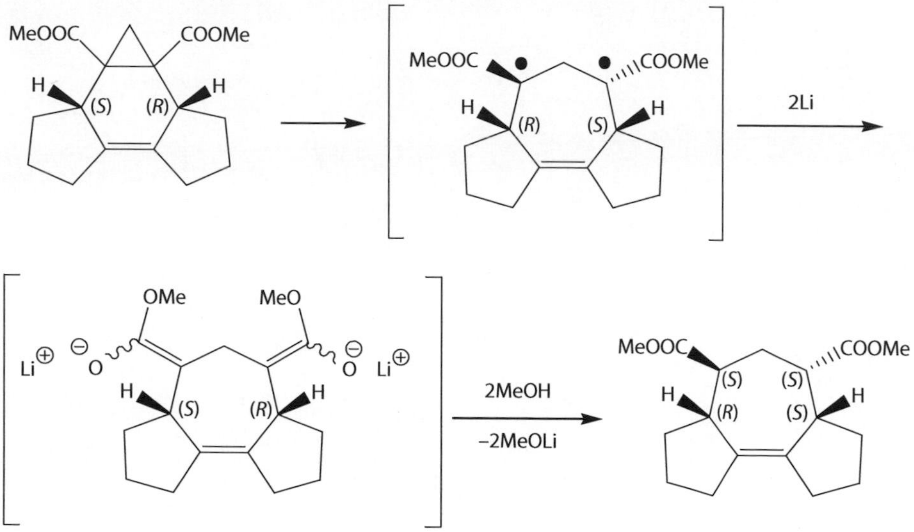

von Doering (1950) plan for tropolone: step 3.

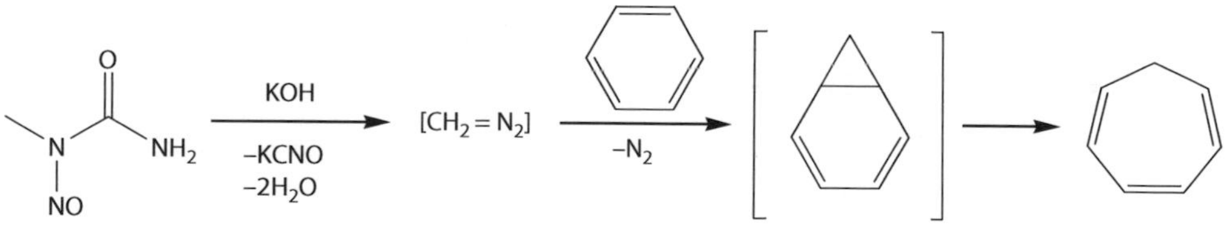

DuPont (1958) plan for tropolone: step 2.

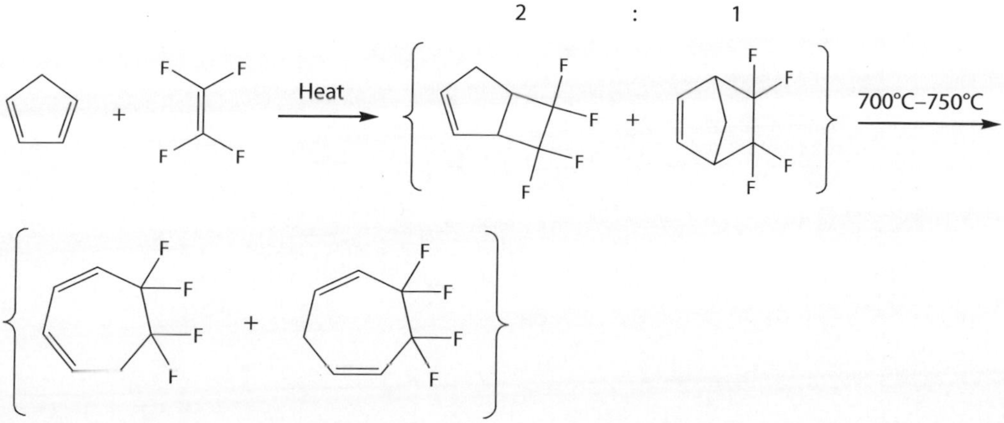

Minns (1988) plan for tropolone: step 3.

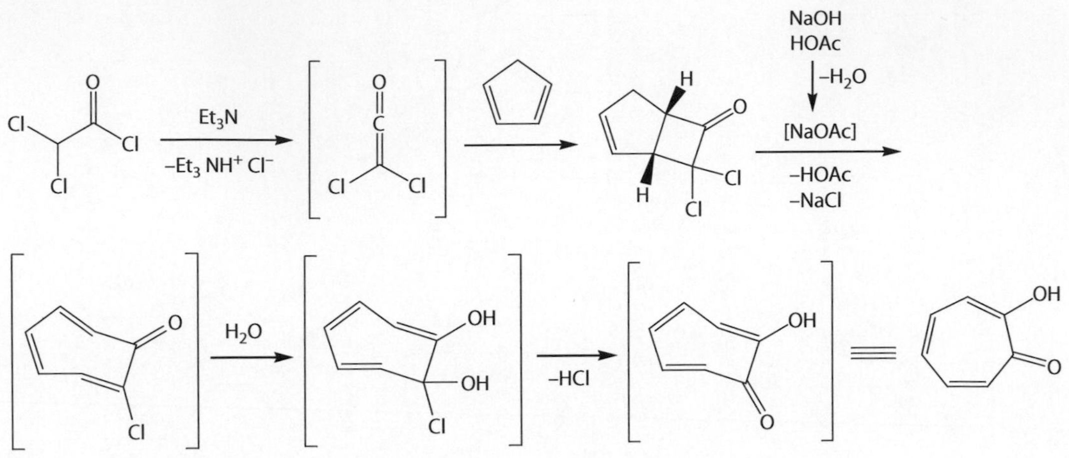

Oda (1969) plan for tropolone: step 4 and step 7.

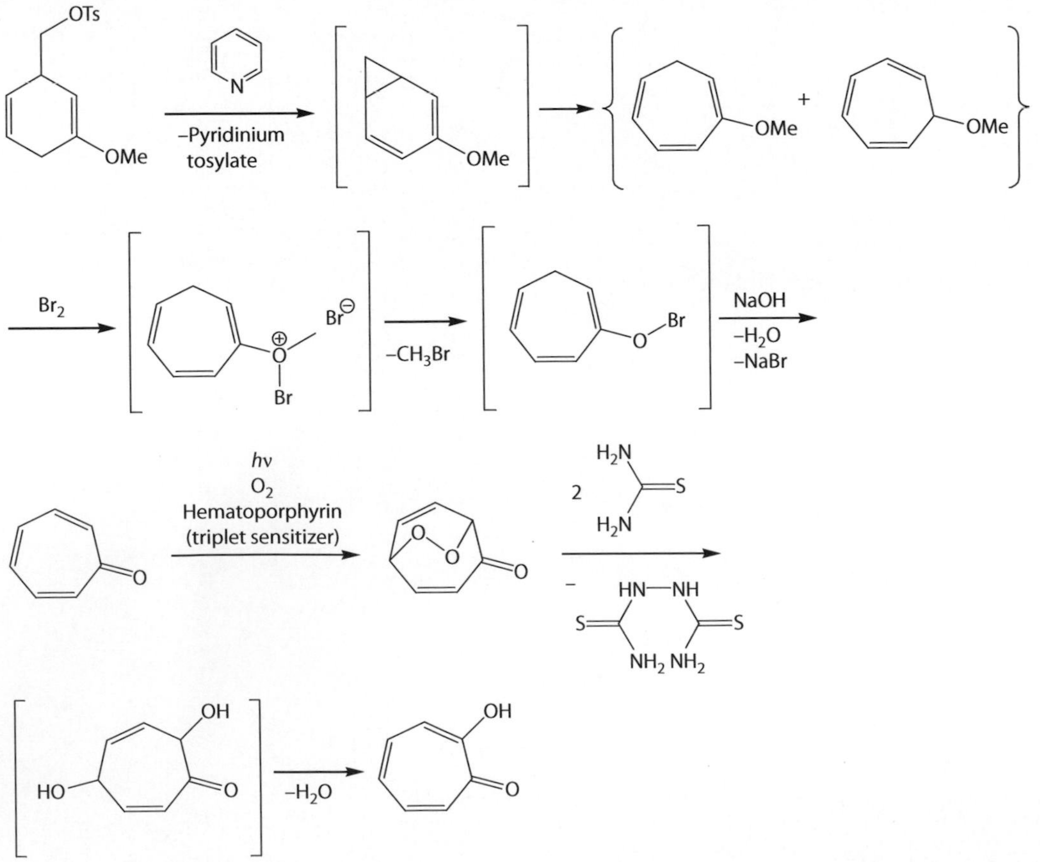

Stevens (1965) plan for tropolone: step 3.

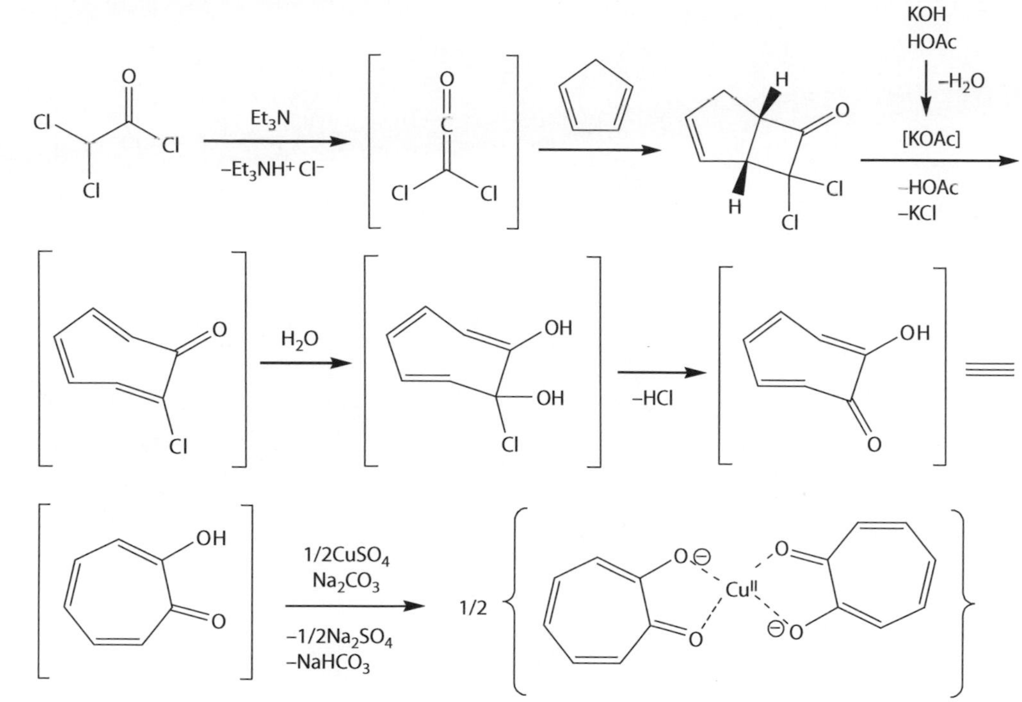

ter Borg (1962) plan for tropolone: step 3.

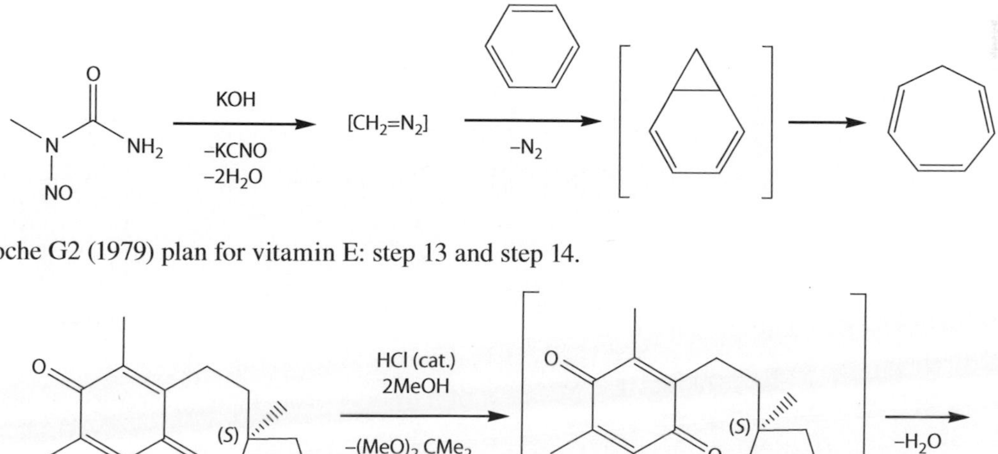

Roche G2 (1979) plan for vitamin E: step 13 and step 14.

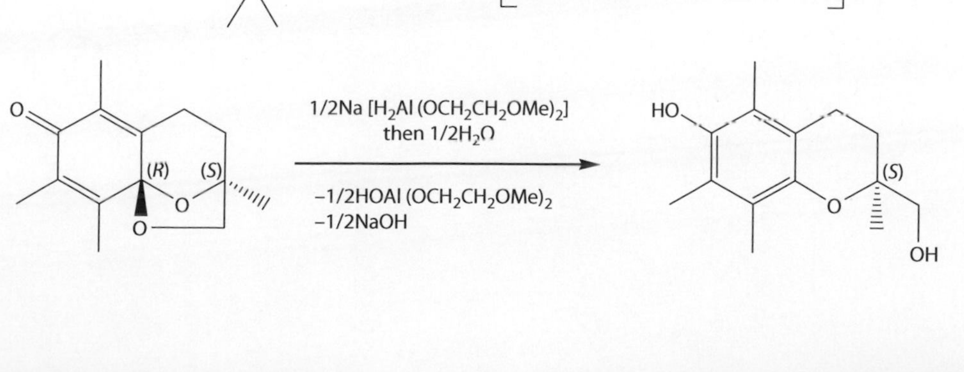

Woggon (2008) plan for vitamin E: key ring forming step in step 5 then hydrogenation to break sacrificial ring in step 7.

O
(R)
(R)
(R)
N
H
OSiEt3
Ar
Ar
(cat.)
−H2O
Ar = 3', 5'-Ditrifluoromethylphenyl
O
OH
OH
H2O
−
Et3SiO
NH

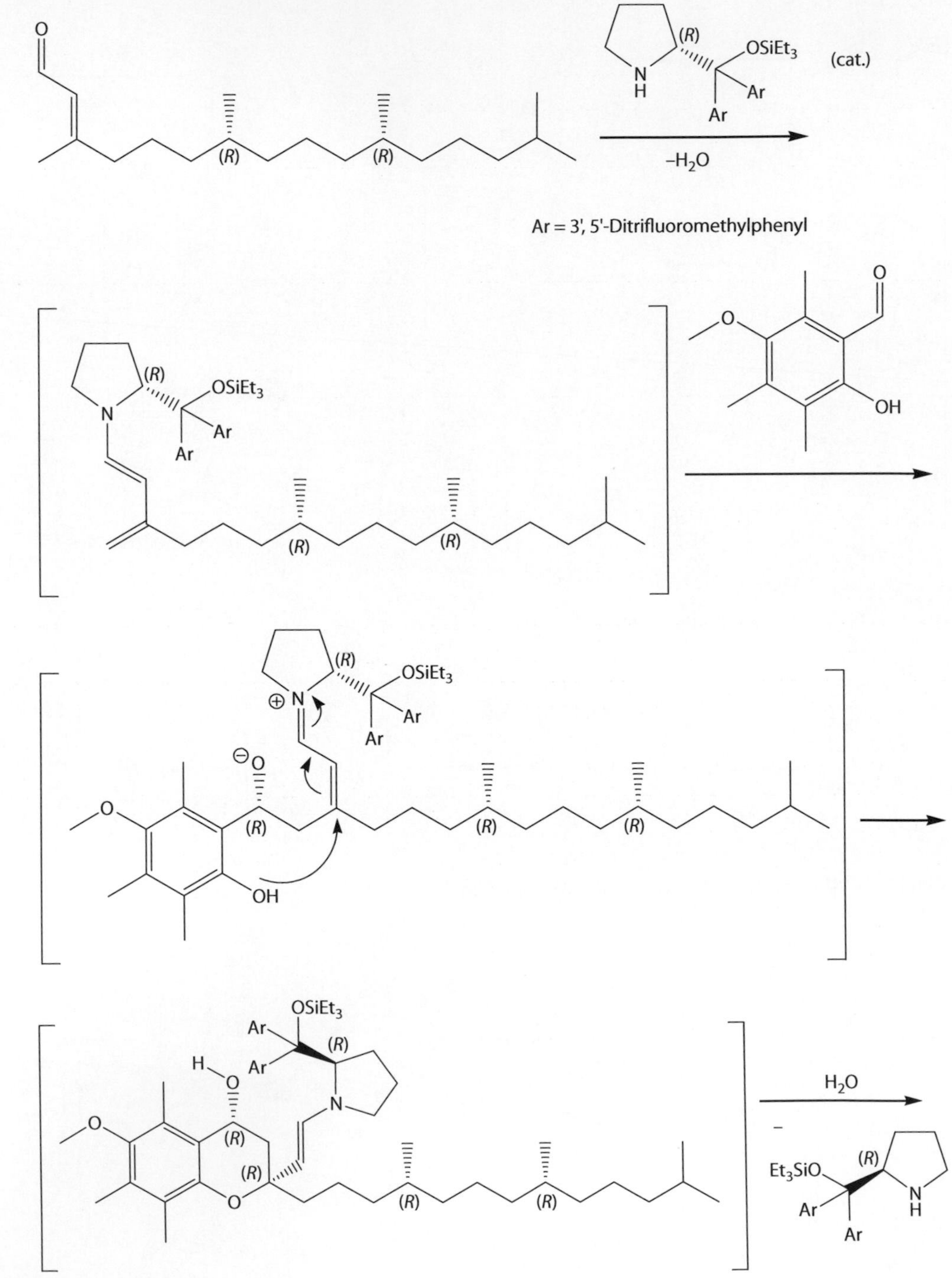

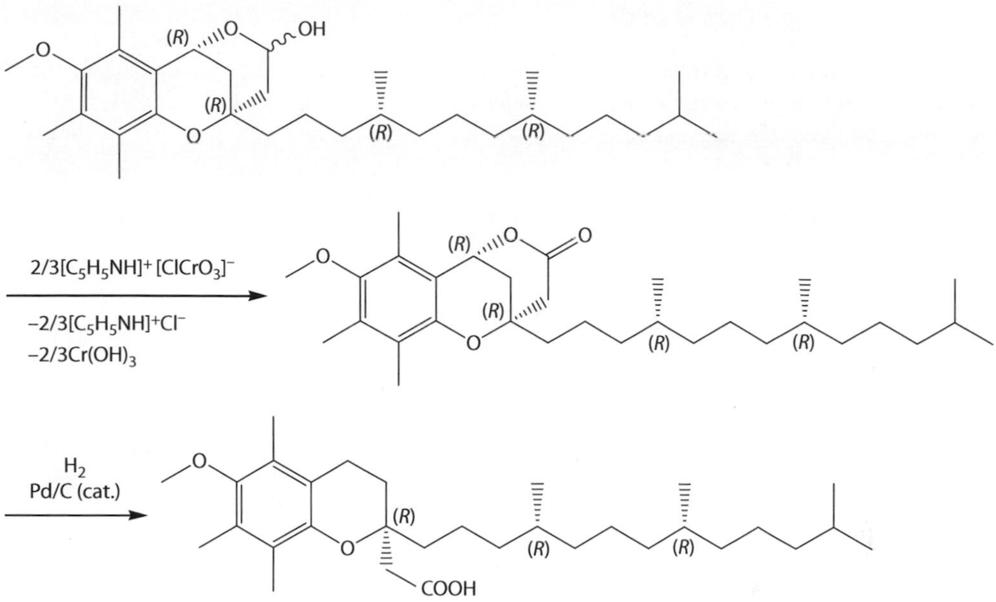

Cossy (1997) plan for zamifenacin: three-membered ring opened in step 4 via a [3.1.0] to [6] rearrangement

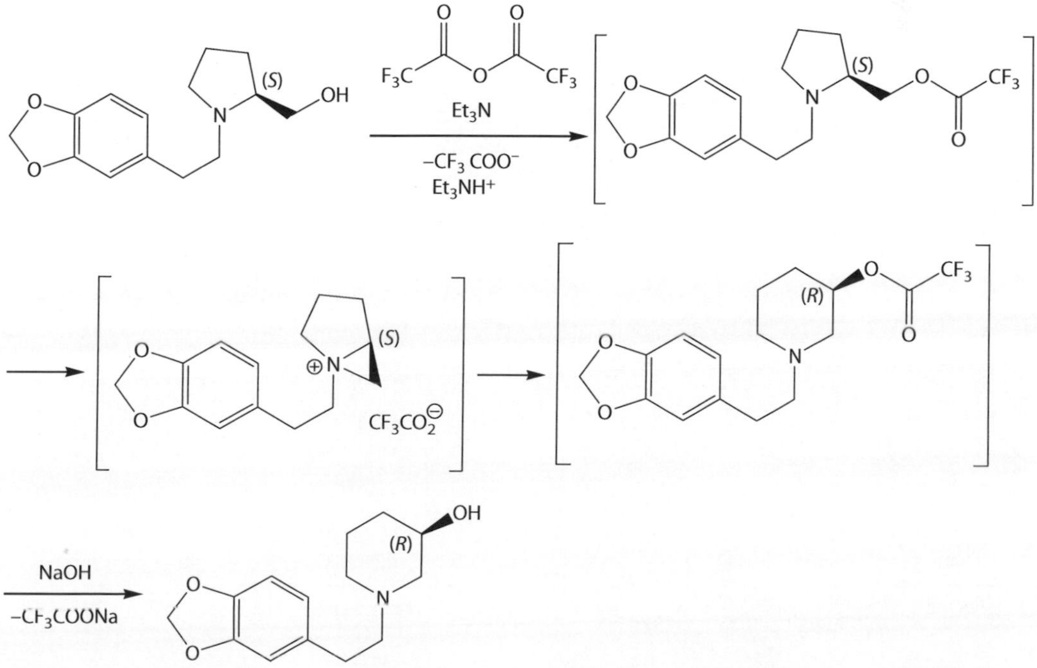

10.6   USE OF THE FURAN RING

The following examples excerpted from the synthesis database highlight various uses of the furan ring as a sacrificial ring: as a diene in Diels–Alder reactions, in ring expansions, and in oxidations of the ring followed by expansion. References to the plans are given in the accompanying download.

Stork (2009) plan for (±)-codeine: benzofuran intermediate in Diels–Alder reaction (step 8).

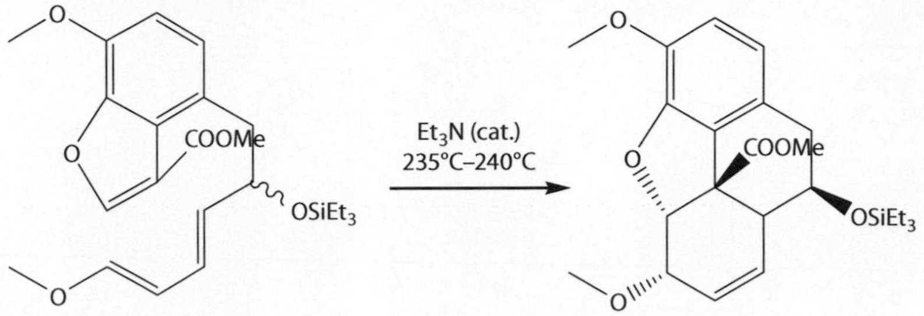

Cha (1998) plan for colchicine: tandem Diels–Alder—HCN elimination (step 11); Diels–Alder reaction (step 14).

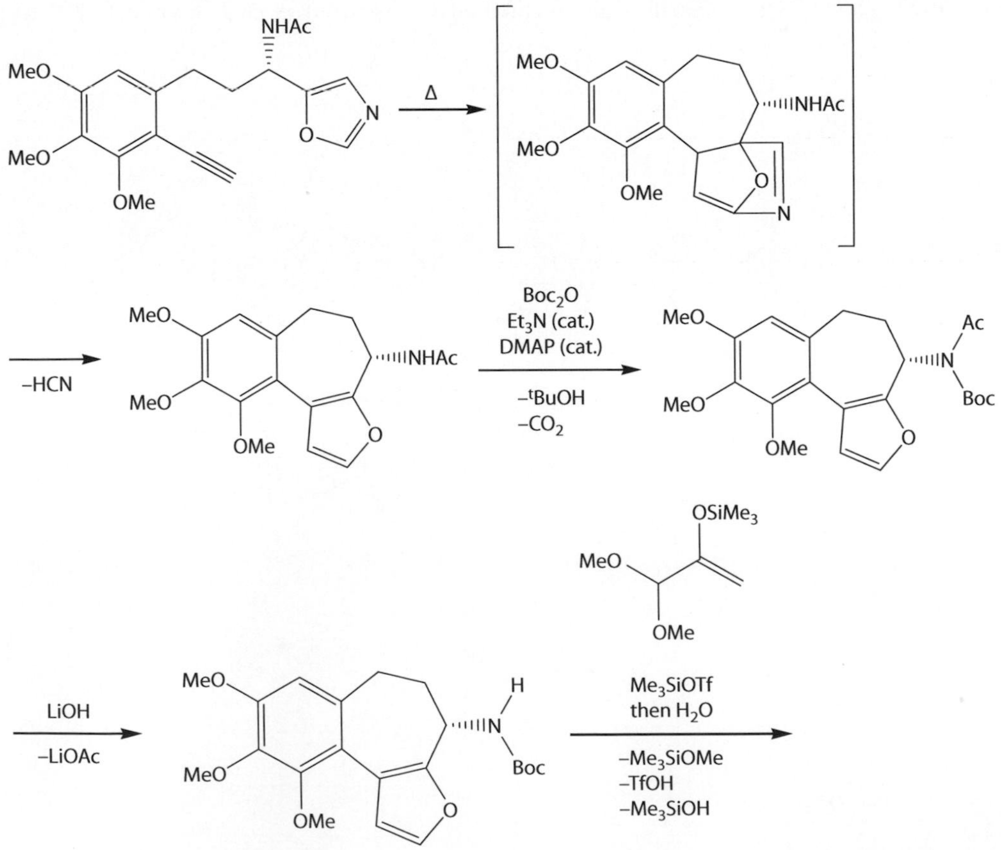

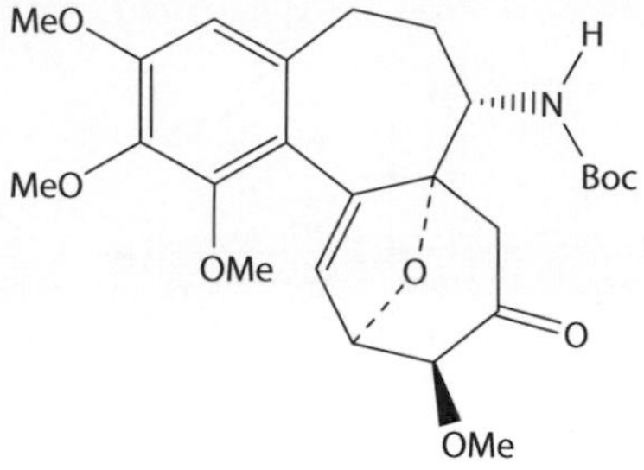

Danishefsky (1999) plan for eleutherobin: transformation in step 15.

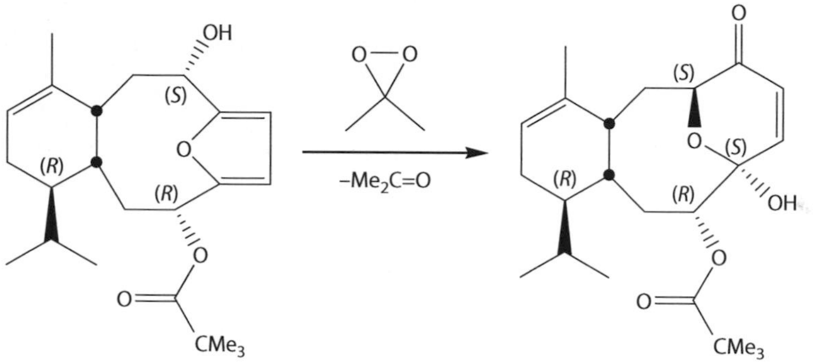

Johnson G2 (1973) plan for (±)-estrone: ring opening in step 2.

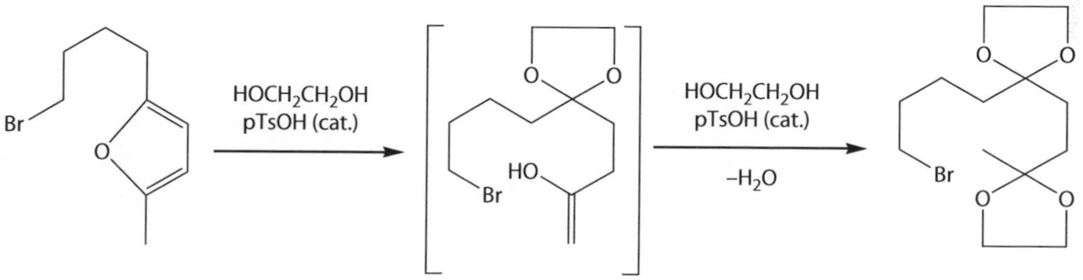

Ogasawara G3 (2001) plan for (−)-kainic acid: oxidation of furfural and ring opening (step 5).

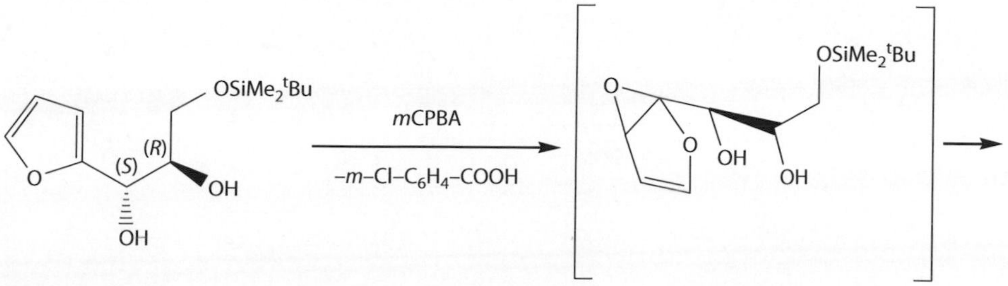

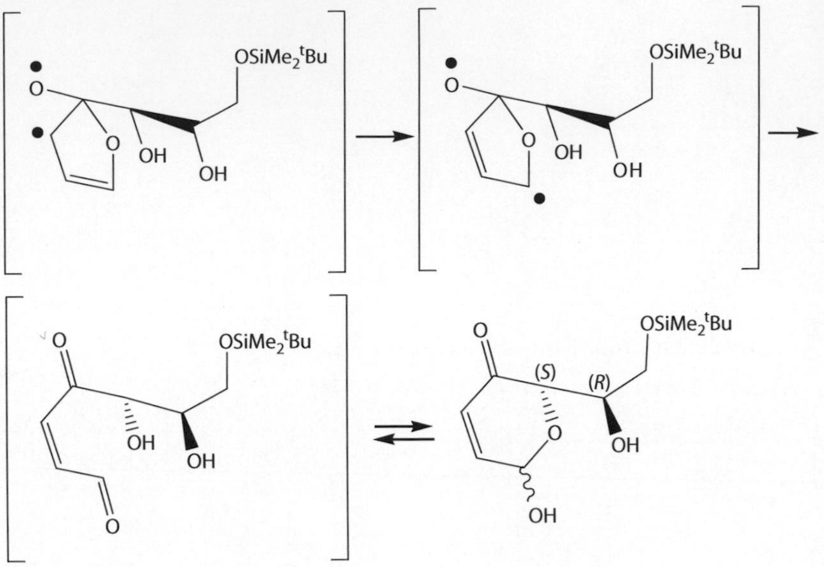

Fukuyama G1 (2005) plan for (–)-kainic acid: oxidation of furfural (step 1) and ring opening (step 6).

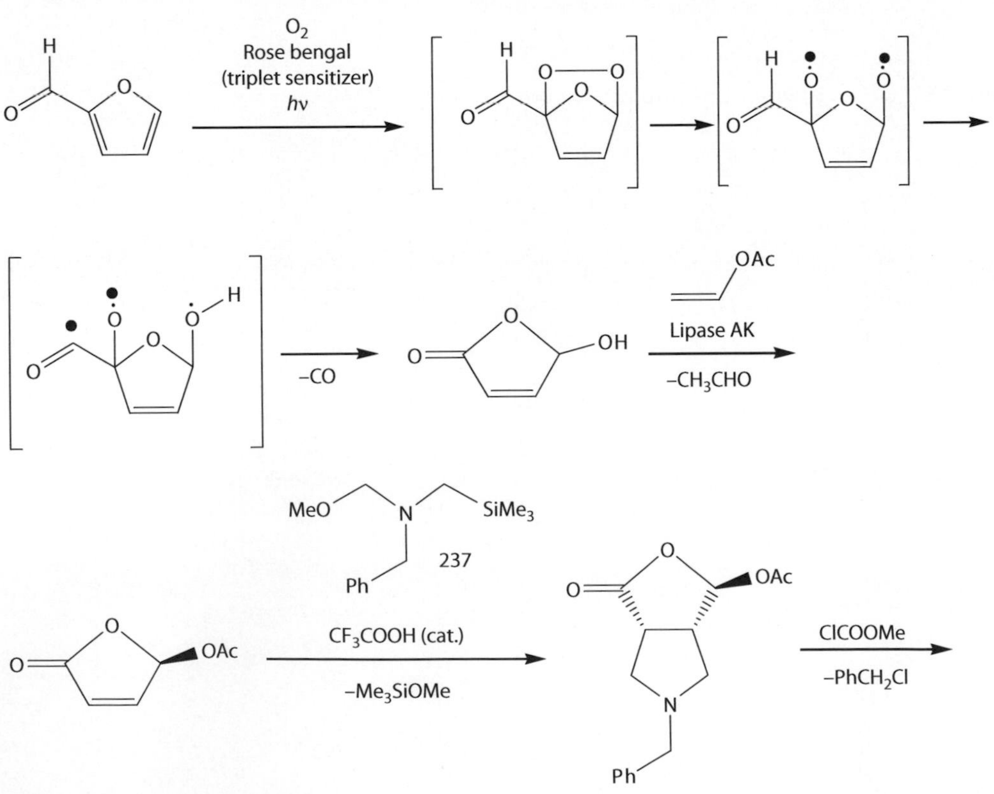

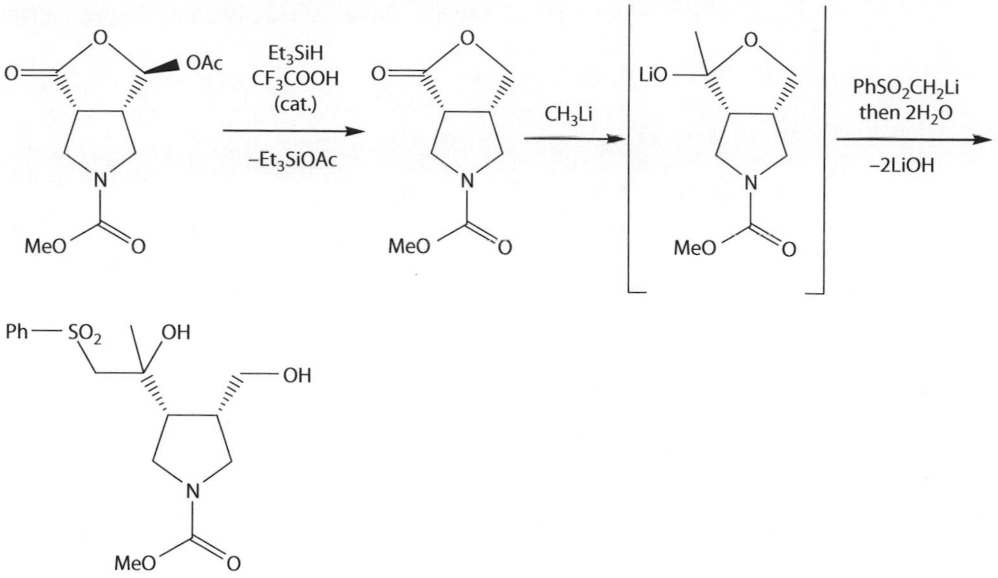

Okamura–Corey (2008) plan for oseltamivir phosphate: ring opening and pyridine formation (step 1).

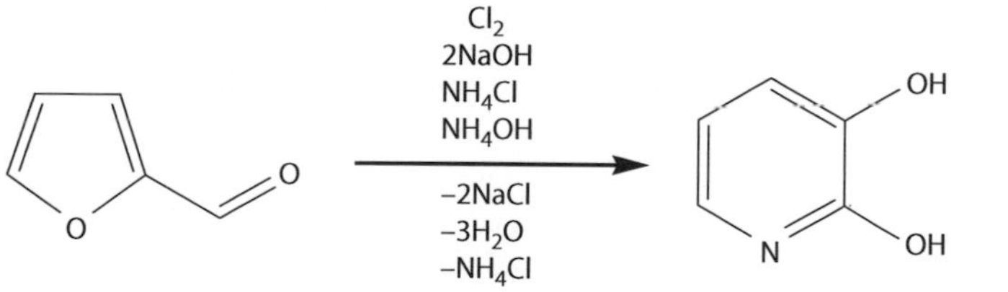

Roche G4 (2000) plan for oseltamivir phosphate: Diels–Alder reaction in step 1 and ring opening in step 4.

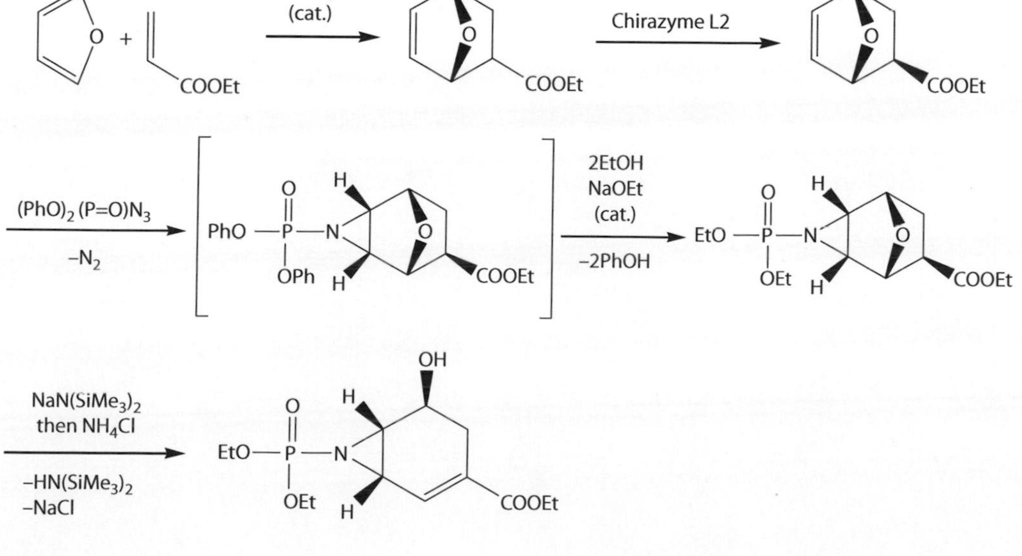

Lautens G1 (1997) plan for (+)-sertraline: Diels–Alder—reductive ring opening sequence (steps 3 and 4).

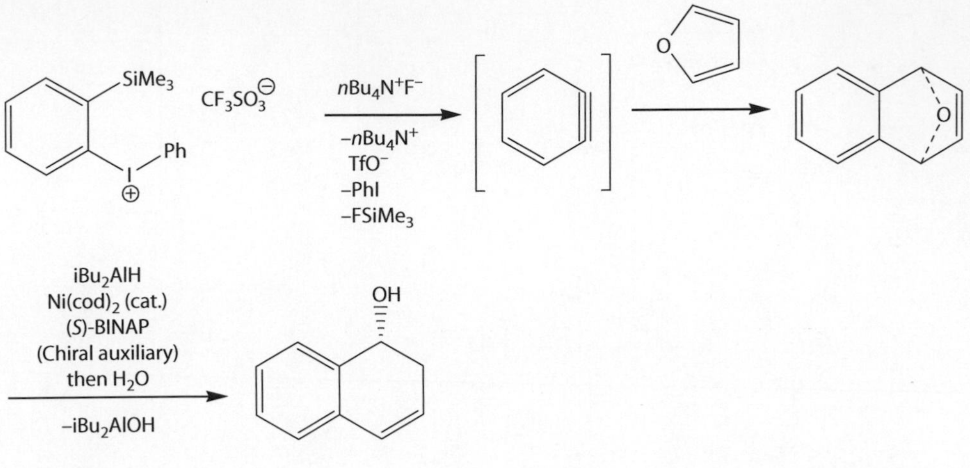

Padwa (2002) plan for (±)-stenine: furan ring intermediate undergoes Diels–Alder reaction then thermal ring opening and recyclization sequence (step 5).

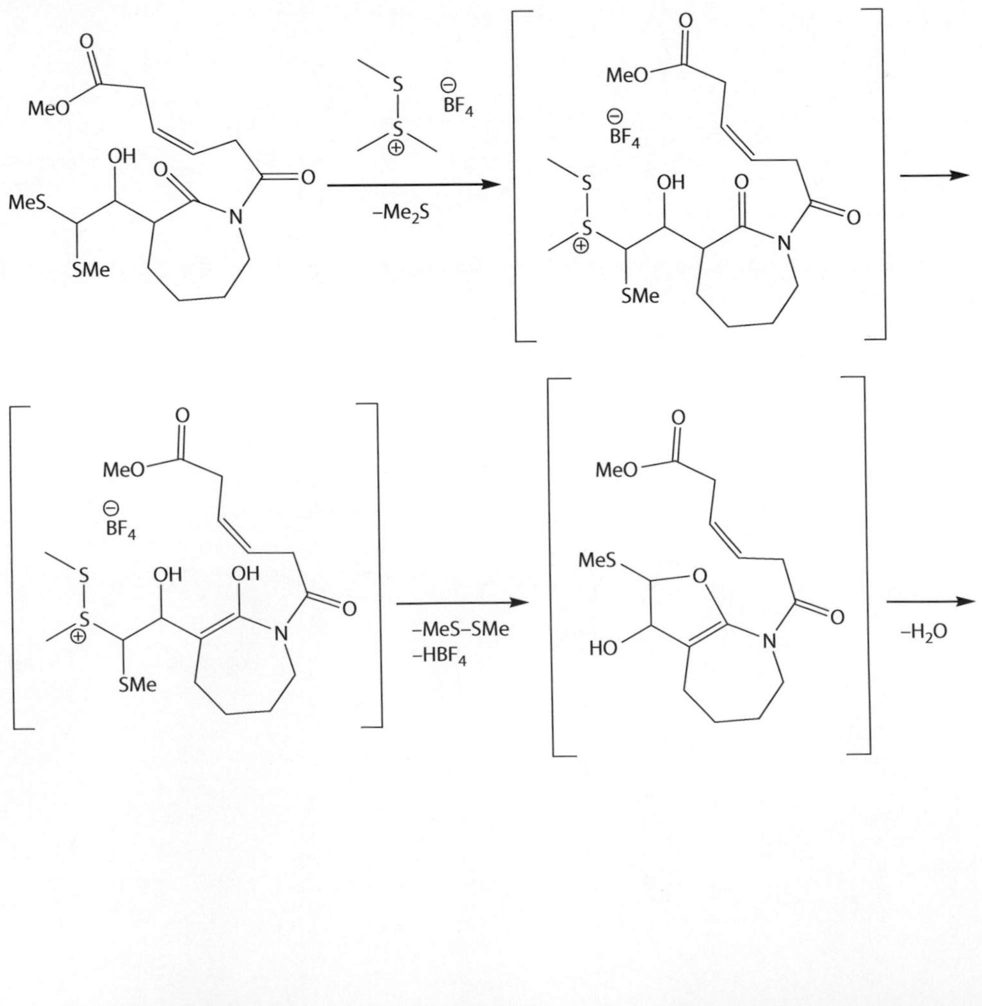

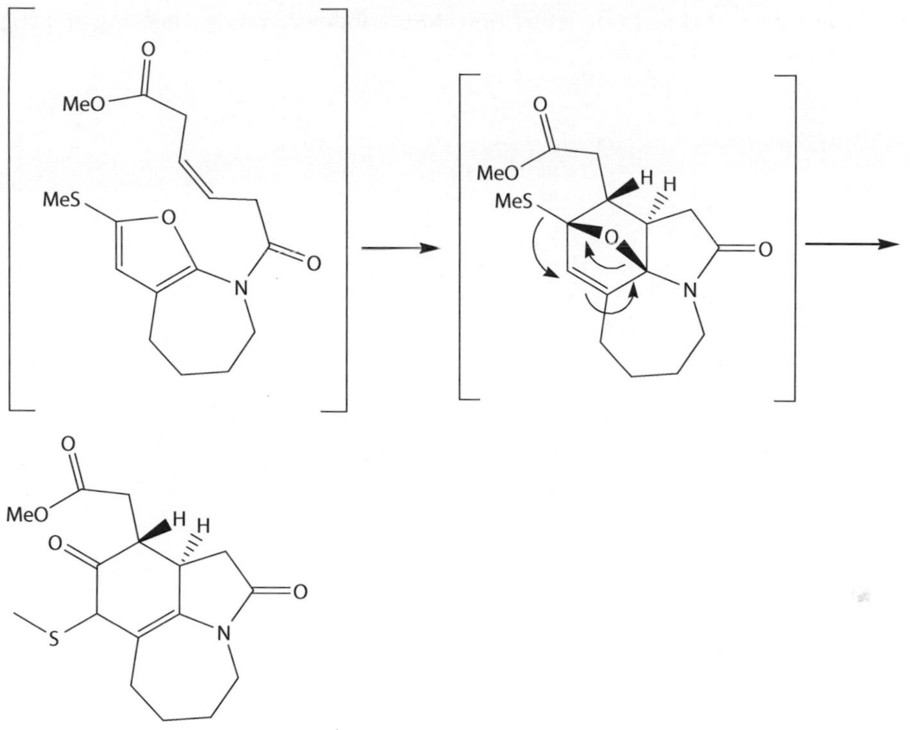

Kuehne (1993) plan for strychnine: furan ring opened to 1,4-diacetal (step 2* and 3*)

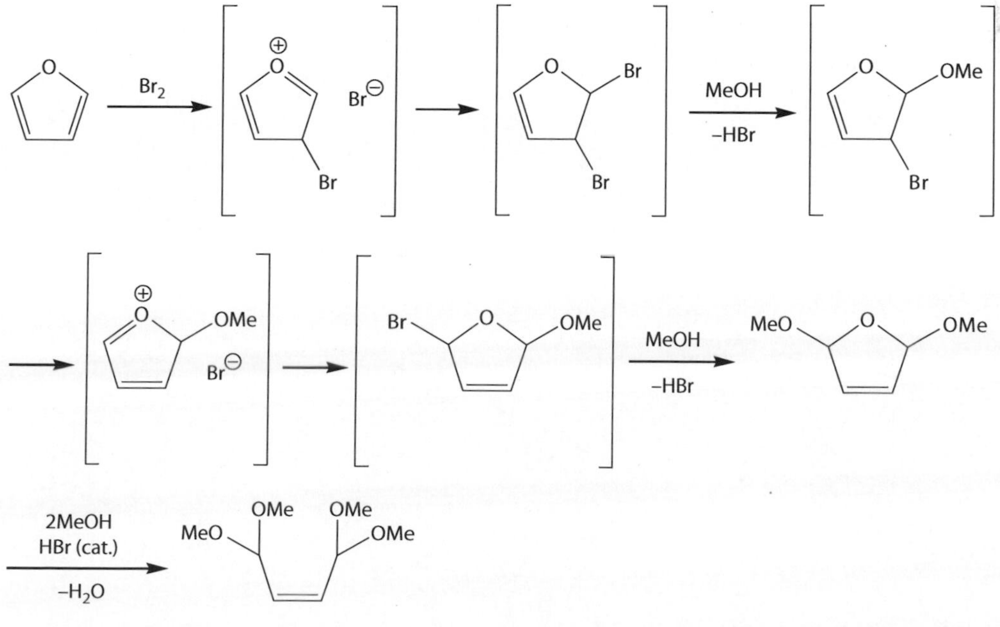

Padwa (2007) plan for strychnine: Diels–Alder ring closure followed by ring opening and [1,2]-H transfer (step 5).

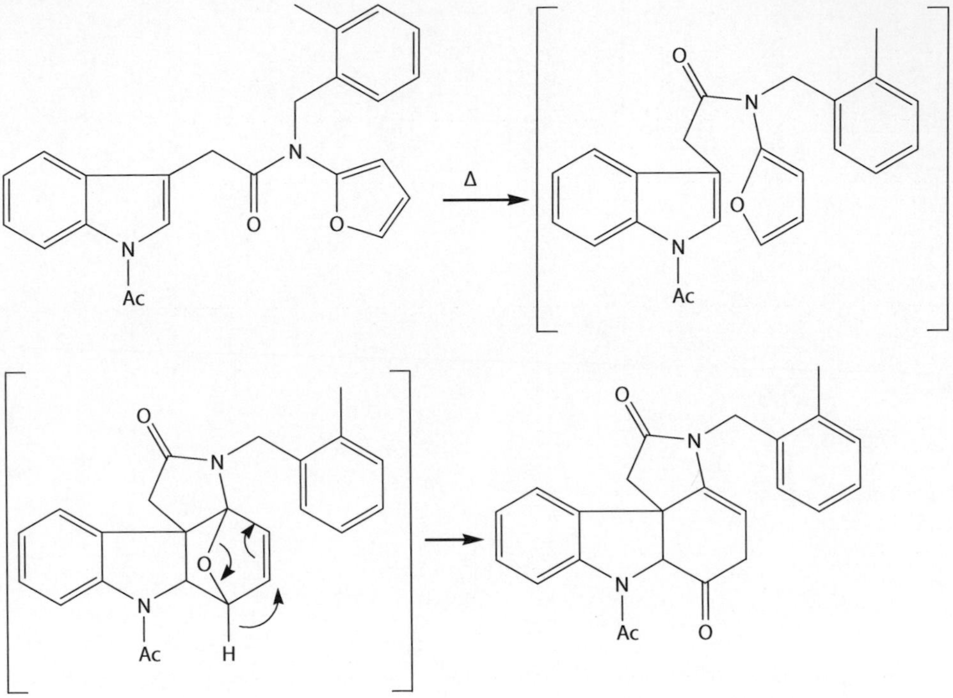

Myers G2 (2007) plan for (–)-tetracycline HCl: Diels–Alder reaction in step 7 and ring opening to a cyclohexone derivative in step 9.

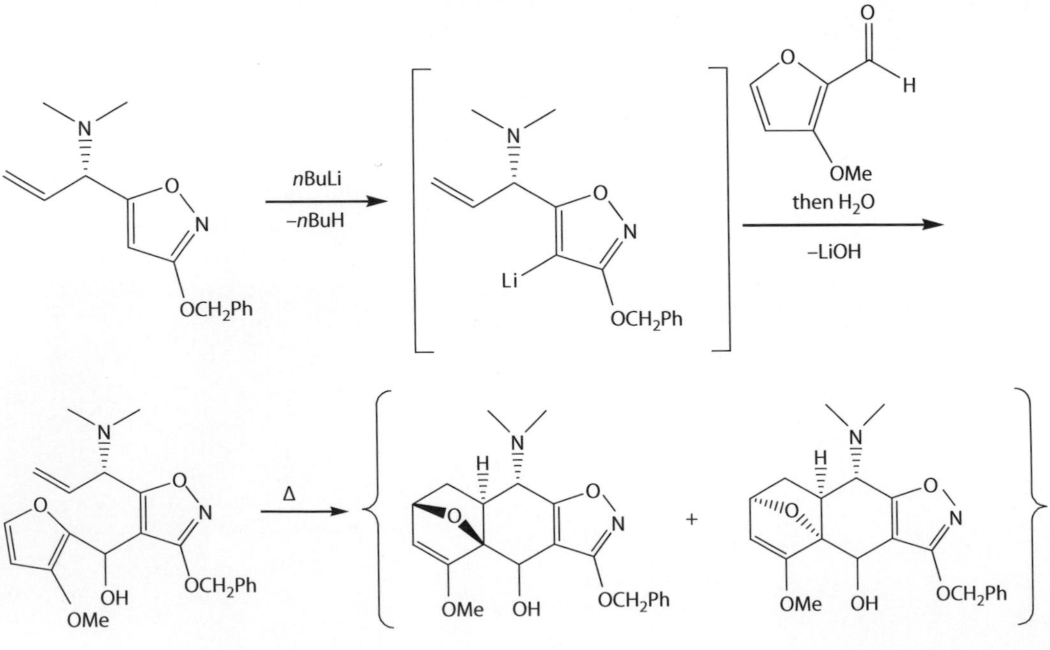

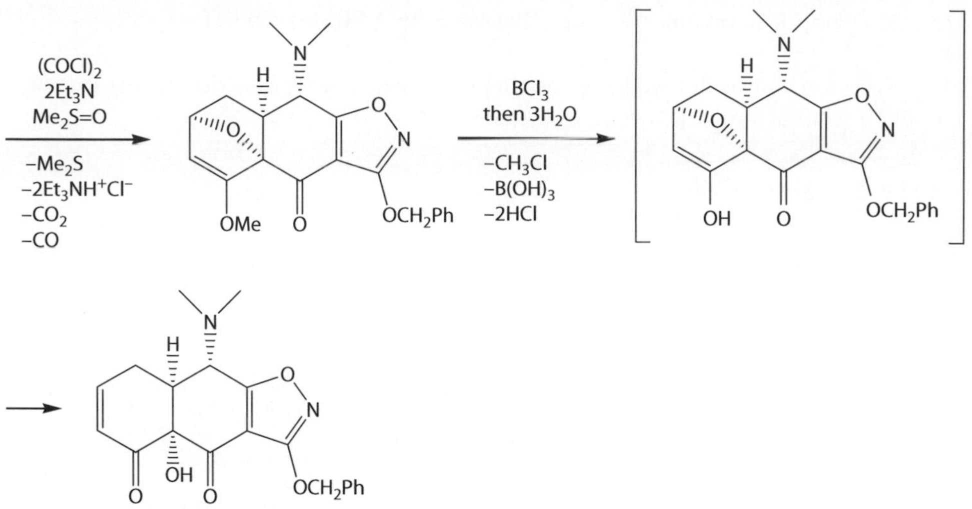

10.7 WENDER [X + Y + Z] RING CONSTRUCTION STRATEGIES

Wender [4 + 2] cycloaddition

Wender, P.A.; Jenkins, T.E. *J. Am. Chem. Soc.* 1989, *111*, 6432.

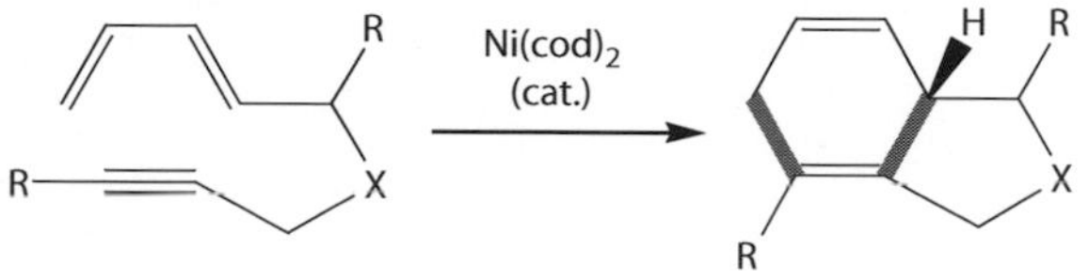

AE = 1; *E* = 0; *f*(sac) = 0; HI = 0

Wender [5 + 2] cycloaddition

Intramolecular:

Wender, P.A.; Takahashi, H.; Witulski, B. *J. Am. Chem. Soc.* 1995, *117*, 4720.

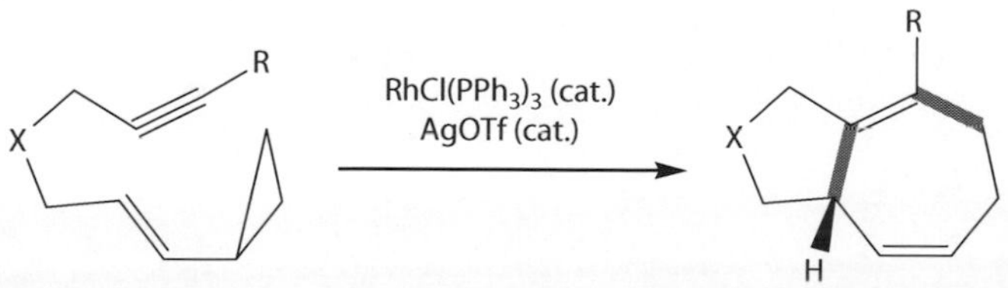

AE = 1; *E* = 0; *f*(sac) = 0; HI = 0

Intermolecular:

Wender, P.A.; Rieck, H.; Fuji, M. *J. Am. Chem. Soc.* 1998, *120*, 10976.

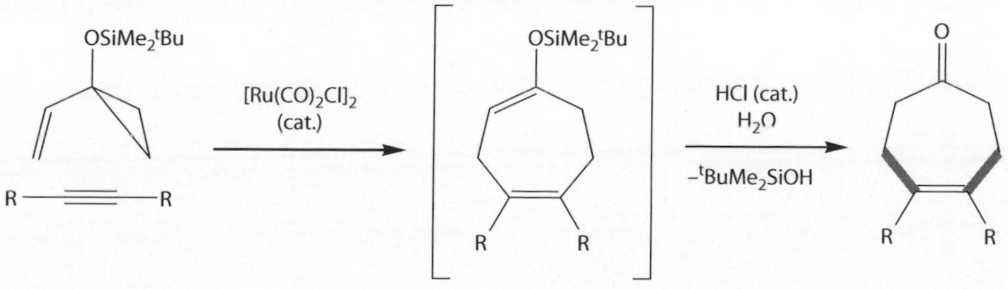

$E = 132/[2R + 108]$; $E(\text{max}) = 0.957$ (when R = methyl); $AE(\text{min}) = 0.511$
$HI = 0$

Note that in these examples the cyclopropane ring in the substrate is a sacrificial ring that acts as a surrogate for a 1,3-dipolarophile.

Wender [2 + 2 + 1] cycloaddition

Two-component coupling:
Wender, P.A.; Croatt, M.P.; Deschamps, N.M. *J. Am. Chem. Soc.* 2004, *126*, 5948.
Wender, P.A.; Deschamps, N.M.; Gamber, G.G. *Angew. Chem. Int. Ed.* 2003, *42*, 1853.

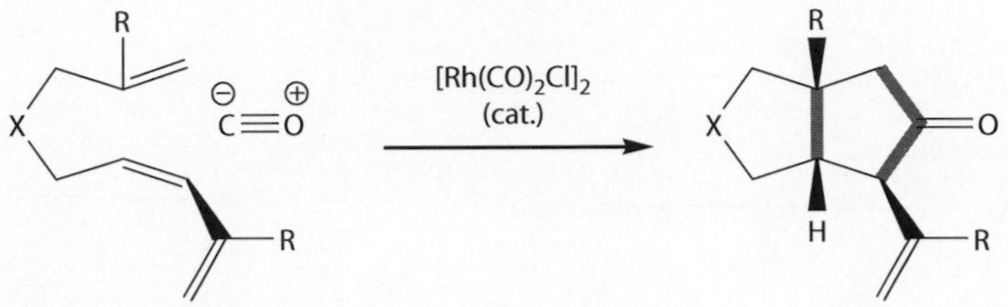

$AE = 1$; $E = 0$; $f(\text{sac}) = 0$; $HI = 0$

Three-component coupling:
Wender, P.A.; Deschamps, N.M.; Williams, T.J. *Angew. Chem. Int. Ed.* 2004, *43*, 3076.

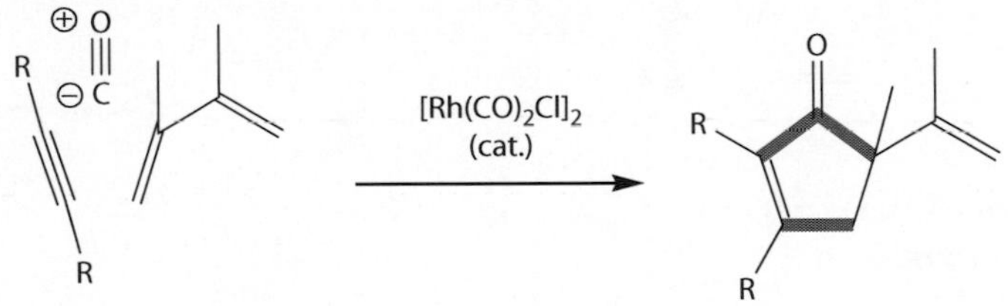

$AE = 1$; $E = 0$; $f(\text{sac}) = 0$; $HI = 0$

Wender [5 + 2 + 1] cycloaddition
Wender, P.A.; Gamber, G.G.; Hubbard, R.D.; Zhang, L. *J. Am. Chem. Soc.* 2002, *124*, 2876.

Three-component coupling:

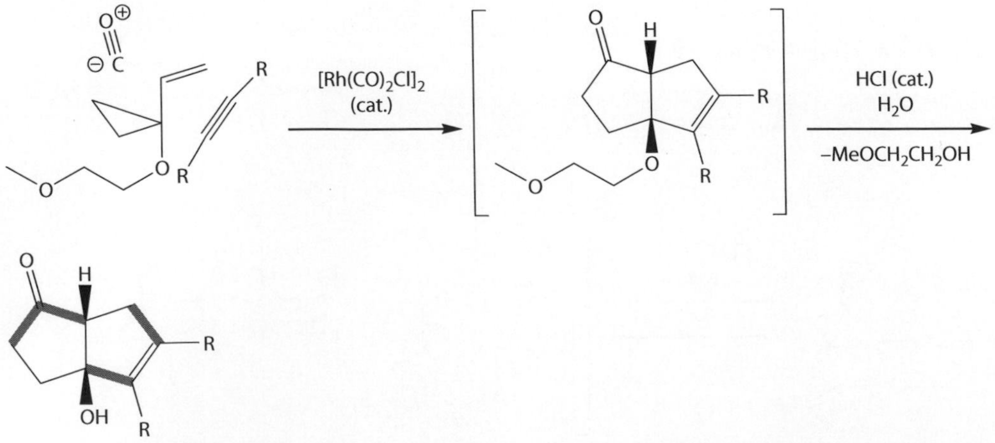

E = 76/[2R + 136]; *E*(max) = 0.458 (when R = methyl); AE(min) = 0.686
HI = 0

Wegner, H.A.; de Meijere, A.; Wender, P.A. *J. Am. Chem. Soc.* 2005, *127*, 6530.

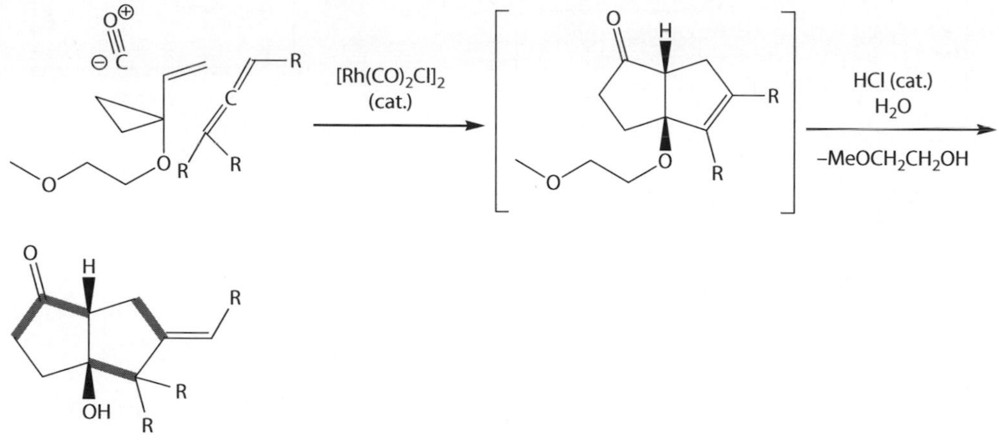

E = 76/[3R + 149]; *E*(max) = 0.392 (when R = methyl); AE(min) = 0.719
HI = 0

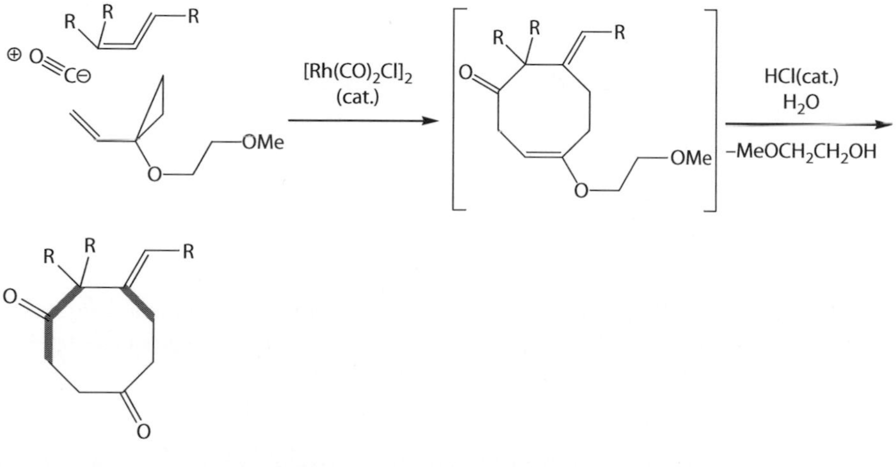

E = 76/[3R + 149]; *E*(max) = 0.392 (when R = methyl); AE(min) = 0.719
HI = 0

Note that in these examples the cyclopropane ring in the substrate is a sacrificial ring that acts as a surrogate for a 1,3-dipolarophile.

Wender [5 + 1 + 2 + 1] cycloaddition

Wender, P.A.; Gamber, G.G.; Hubbard, R.D.; Pham, S.M.; Zhang, L. *J. Am. Chem. Soc.* 2004, *127*, 2836.

Four-component coupling:

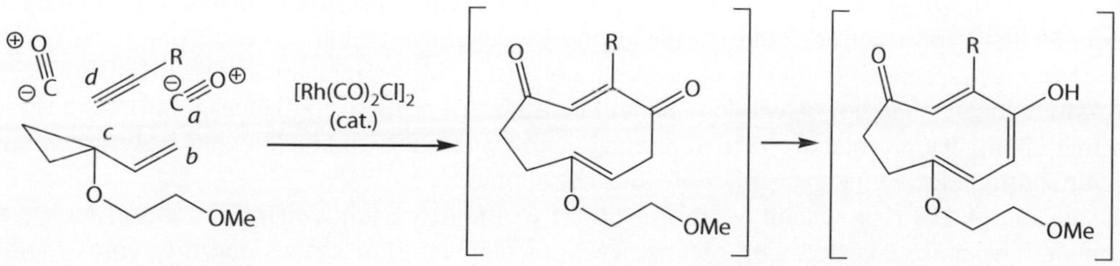

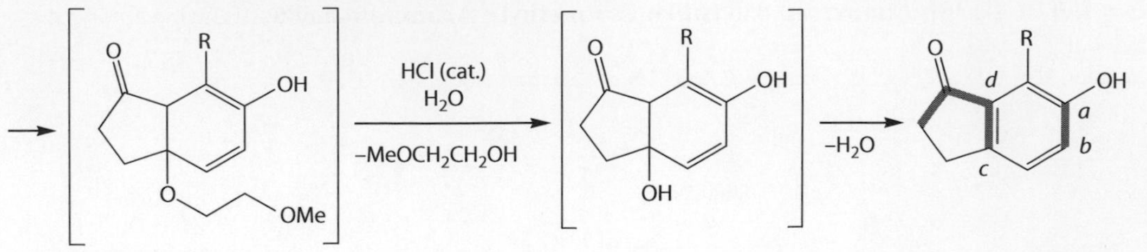

$E = [76 + 18]/[R + 147]$; $E(\text{max}) = 0.580$ (when R = methyl); $AE(\text{min}) = 0.633$
$HI = -(1/2) [(1 - 2)_a + (-1 - (-2))_b + (0 - 1)_c + (0 - (-1))_d] = 0$

Seven-component coupling:

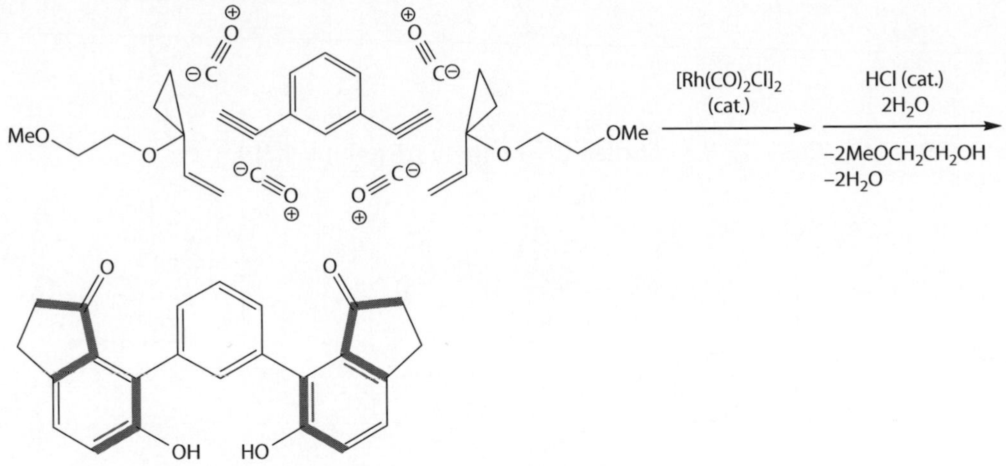

Note that in these examples the cyclopropane ring in the substrate is a sacrificial ring that acts as a surrogate for a 1,3-dipolarophile. The seven-component coupling reaction is an example of a two-directional synthesis due to the high symmetry of the target product.

10.8   SPECTACULAR RING CONSTRUCTION STRATEGIES

The following compilation, taken from the synthesis database, showcases spectacular ring constructions in a single reaction step. These illustrate the feasibility of increasing molecular complexity in reactions that produce multiple target bonds in a single step. Alternative strategies to make the very same complex targets usually require several steps. Despite this success, there are some general problems associated with these reactions: (a) they are usually low yield transformations; (b) there is a trade-off between making many target bonds in a single step to build a complex ring structure and a low conversion of starting materials to products; (c) thermodynamics, namely, entropy and enthalpy, are the biggest hurdles working against this from happening efficiently, especially for intermolecular type reactions; and (d) for an intramolecular reaction to occur, one must operate under low concentration conditions to prevent competing side biomolecular reactions from occurring, which means that the solvent demand will be high. All of these challenges reinforce the assertion that chemistry is a science of compromise, and particularly, that achieving truly green atom economical transformations is neither obvious nor simple.

Using the [4.4.0] ring system as an illustrative example, certain general structural motifs are pertinent. These are codified with nicknames signifying familiar shapes and are represented by the following graphical representations. These may be correlated directly with the combinations of possible monocyclic and bicyclic transformations shown in Appendix A.7.

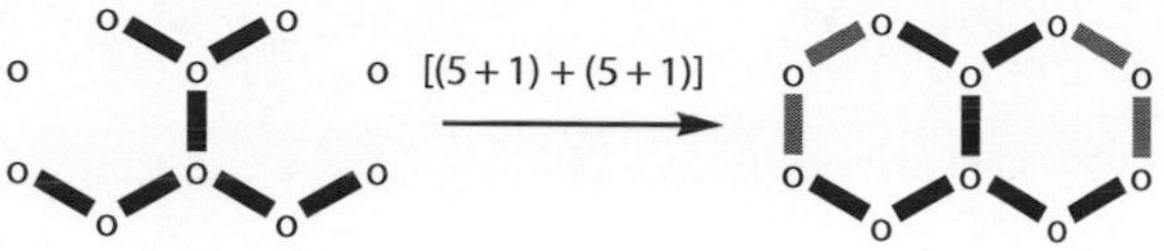

Anchor

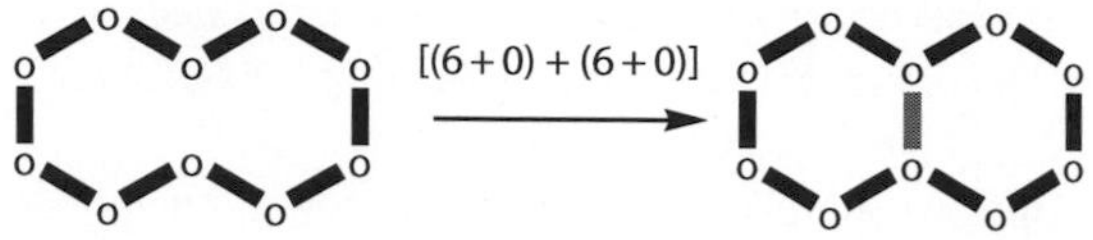

$[(5+1)+(5+1)]$

Bubble

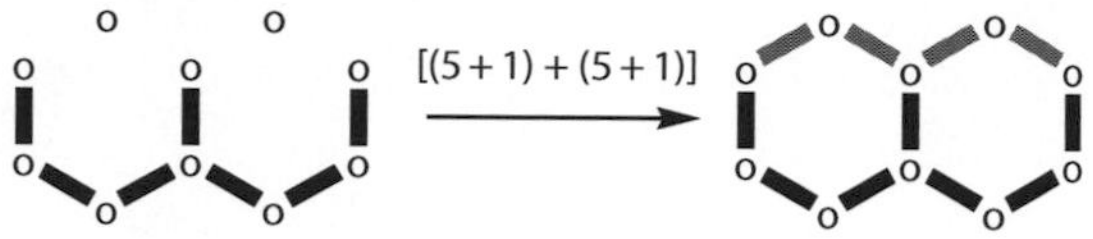

$[(6+0)+(6+0)]$

Claw

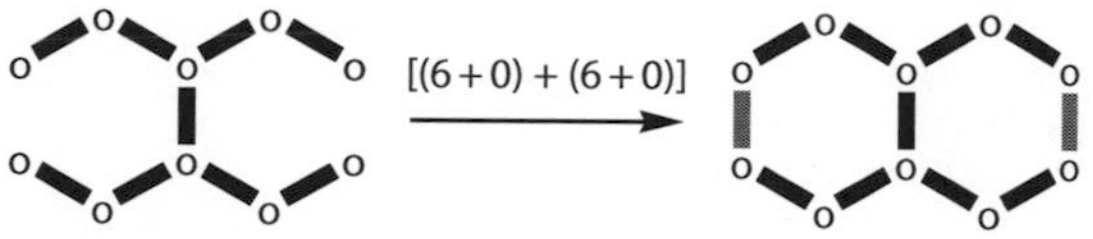

$[(5+1)+(5+1)]$

Pincer

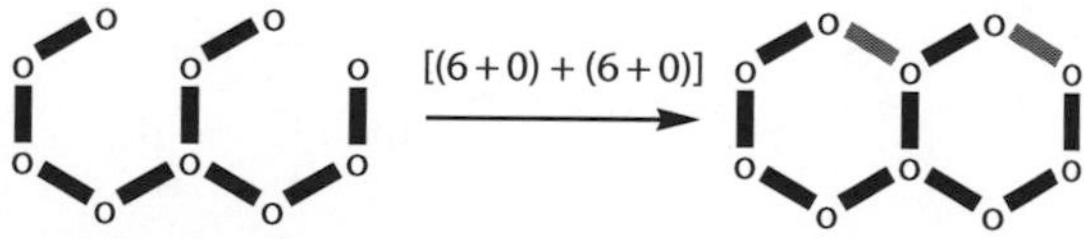

$[(6+0)+(6+0)]$

Swaying hands (right)

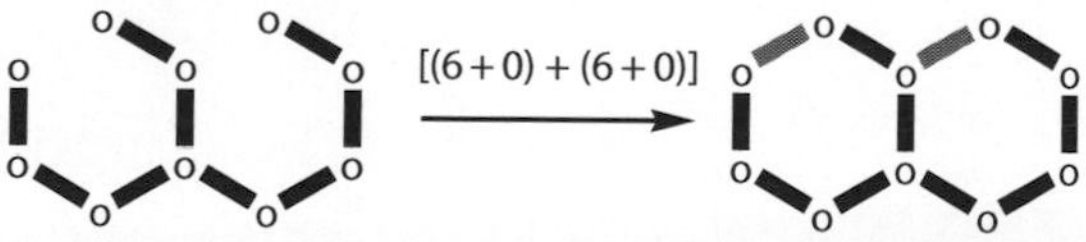

$[(6+0)+(6+0)]$

Swaying hands (left)

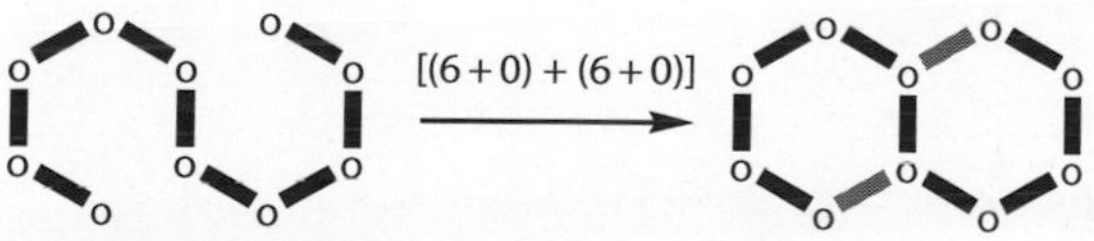

$[(6+0)+(6+0)]$

S-wave

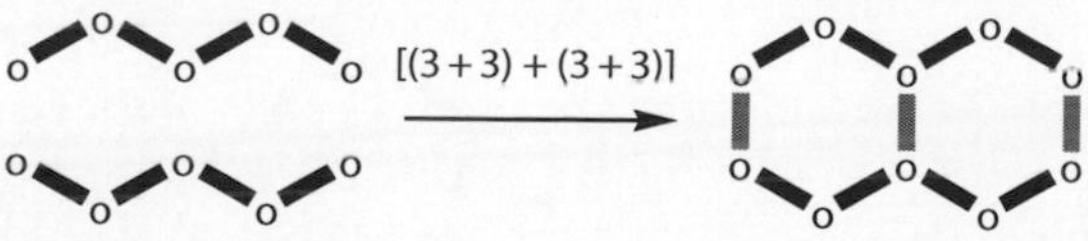

$[(6+0)+(6+0)]$

Zipper (intermolecular)

$[(3+3)+(3+3)]$

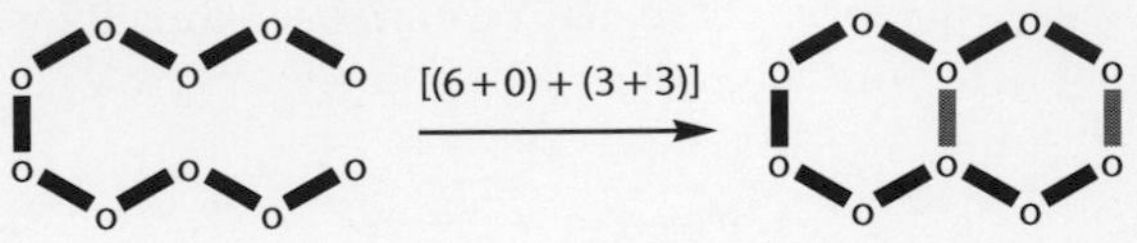

Zipper (intramolecular)

For each case, the chemical transformation is given followed by a simplified graphical equation of the formal transformation to form the relevant ring system using the notation presented at the beginning of this chapter. References for the plans are given in the accompanying download

Zhai (2005) plan for absinthin: photo-rearrangement in step 1; Diels–Alder reaction in step 5.

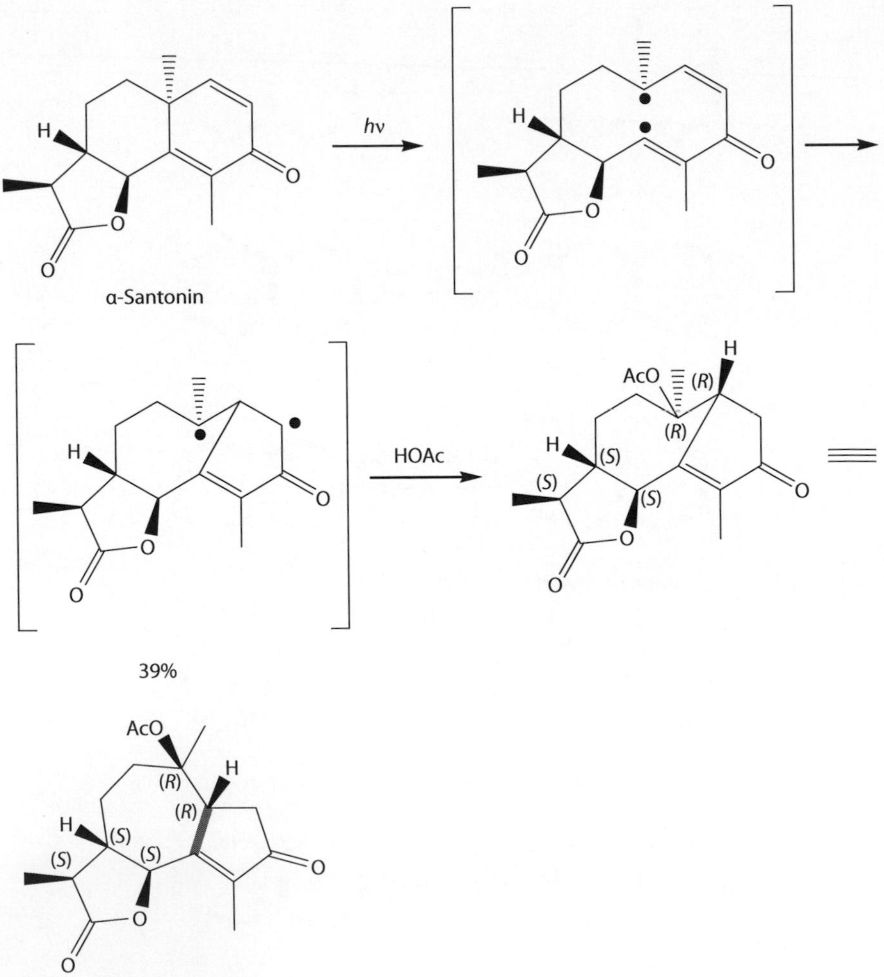

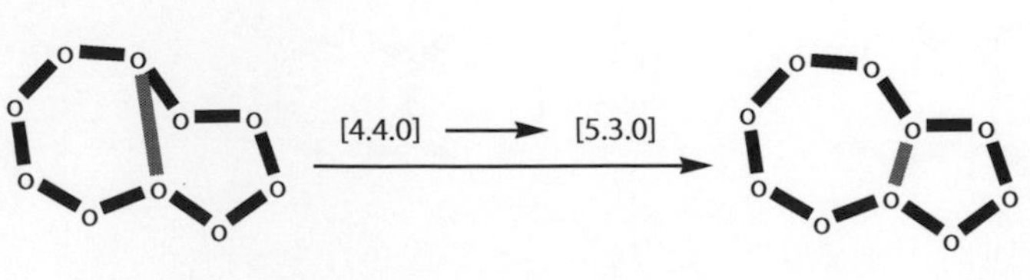

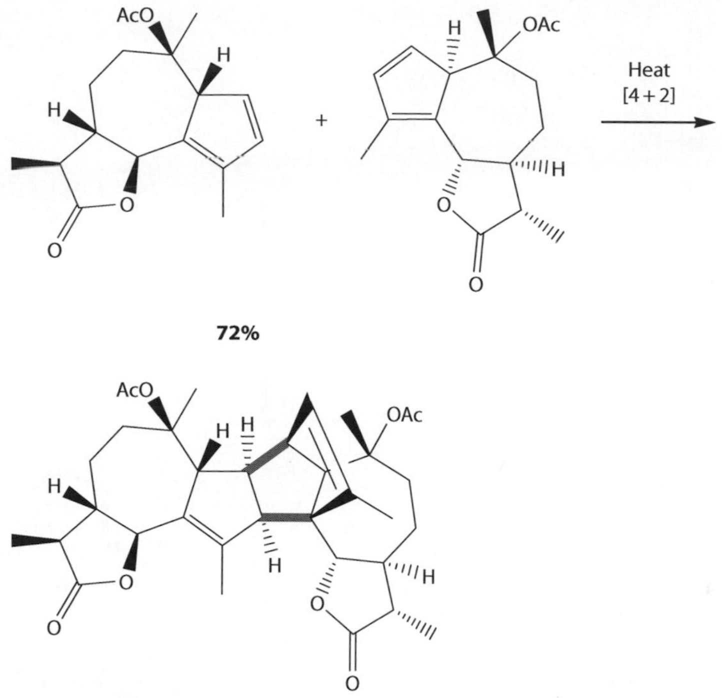

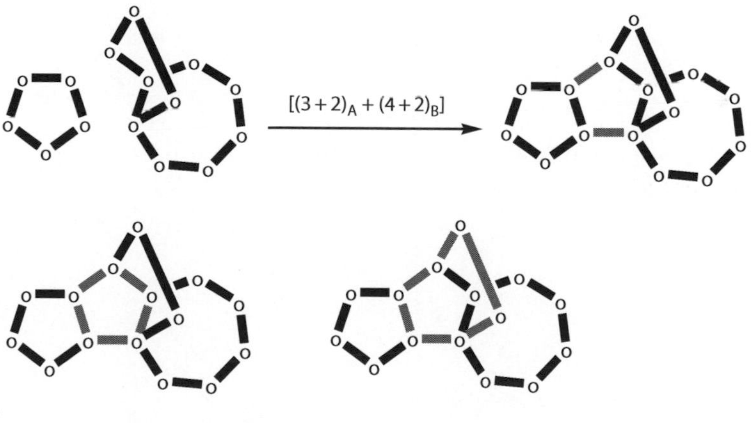

72%

Ring A   Ring B

Trost (2004) plan for amphidinolide P: step 15 (macrocyclization-rearrangement)

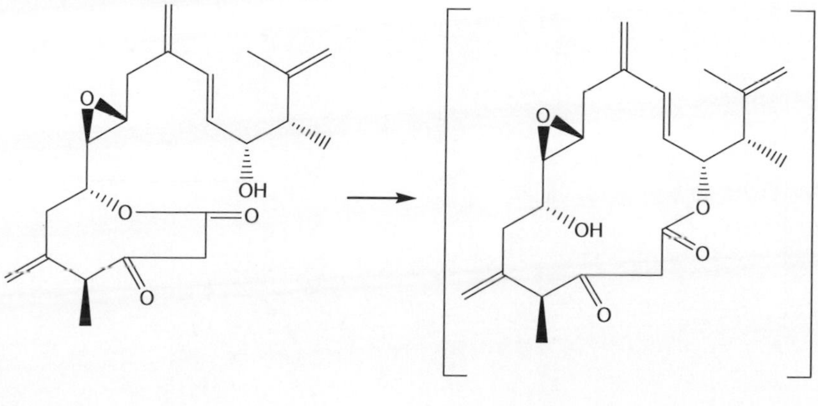

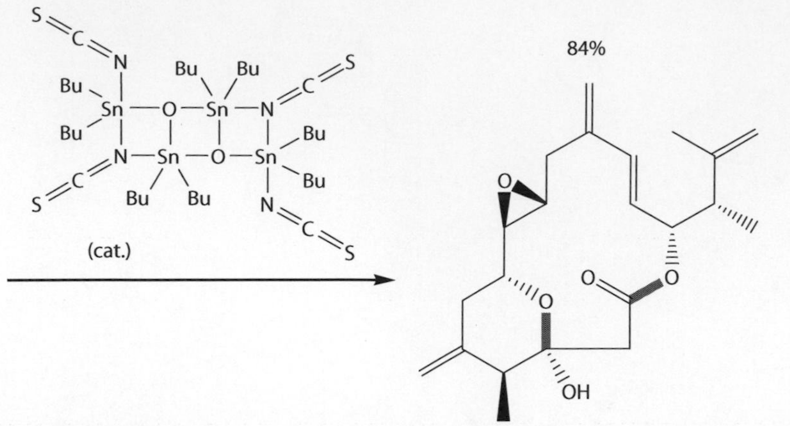

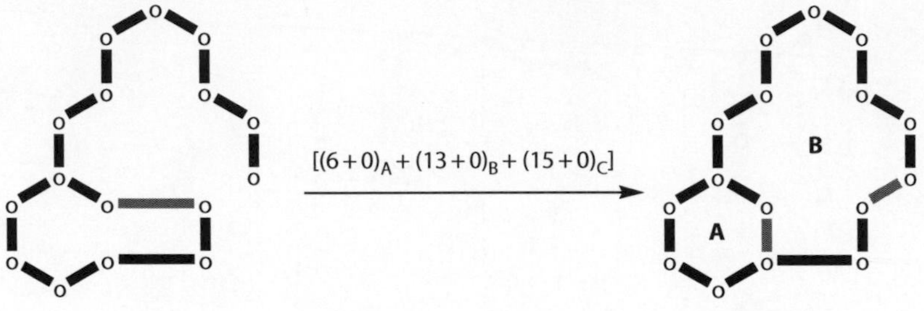

Note that the lactone ring in the substrate is a sacrificial ring in this transformation.
Corey (1999) plan for aspidophytine: step 12.

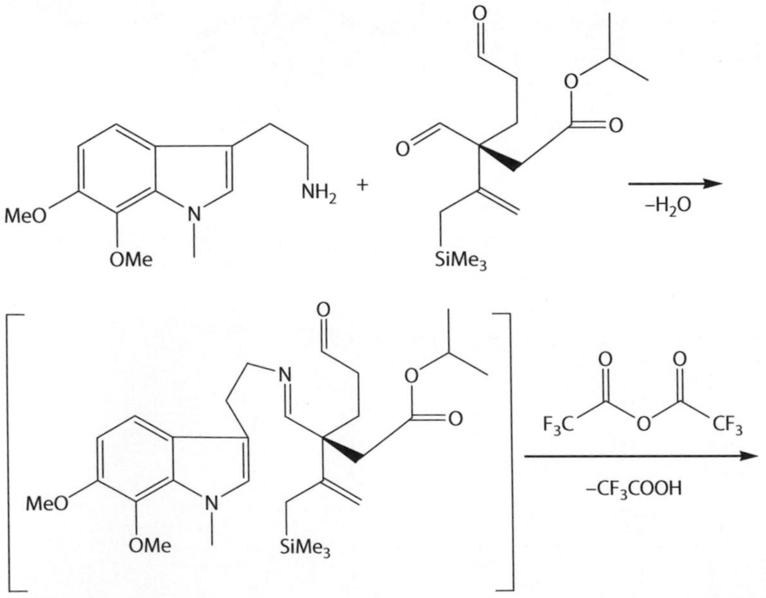

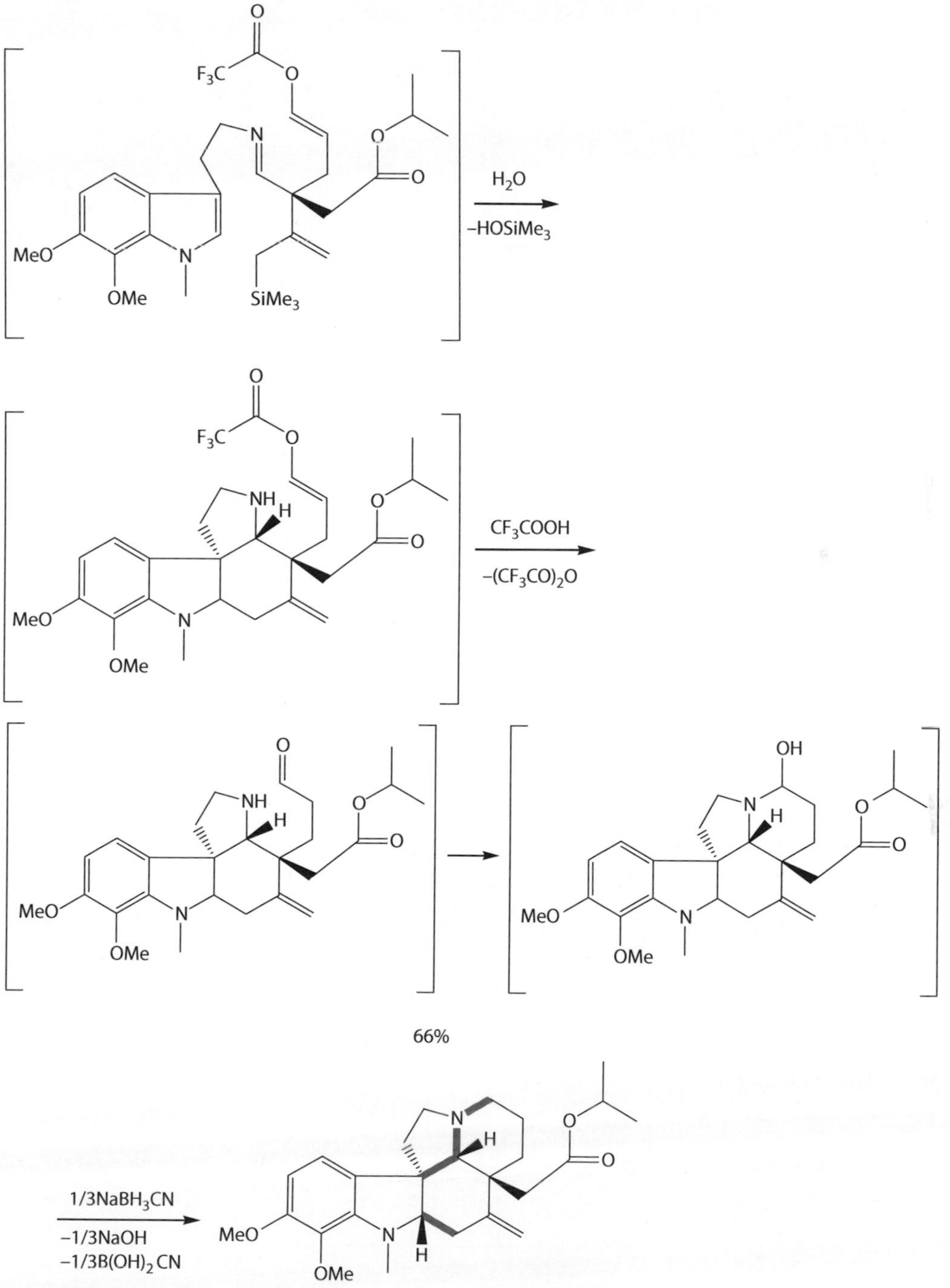
F3C
O
N
MeO
OMe
SiMe3
H2O
–HOSiMe3
NH
H
CF3COOH
–(CF3CO)2O
OH
66%
1/3NaBH3CN
–1/3NaOH
–1/3B(OH)2CN

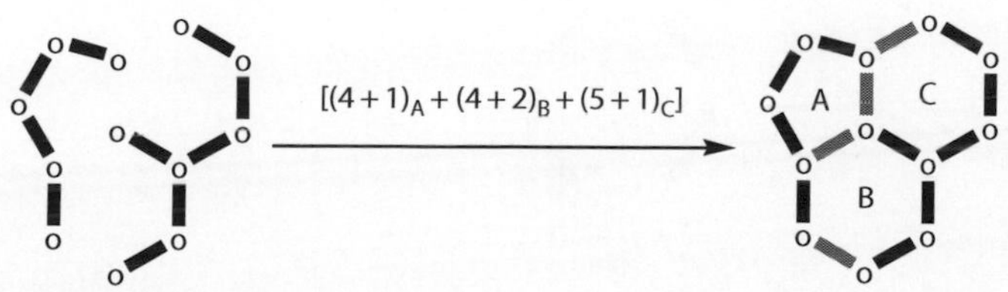
[(4 + 1)A + (4 + 2)B + (5 + 1)C]
A
B
C

Schmalz (2002) plan for colchichine: step 8.

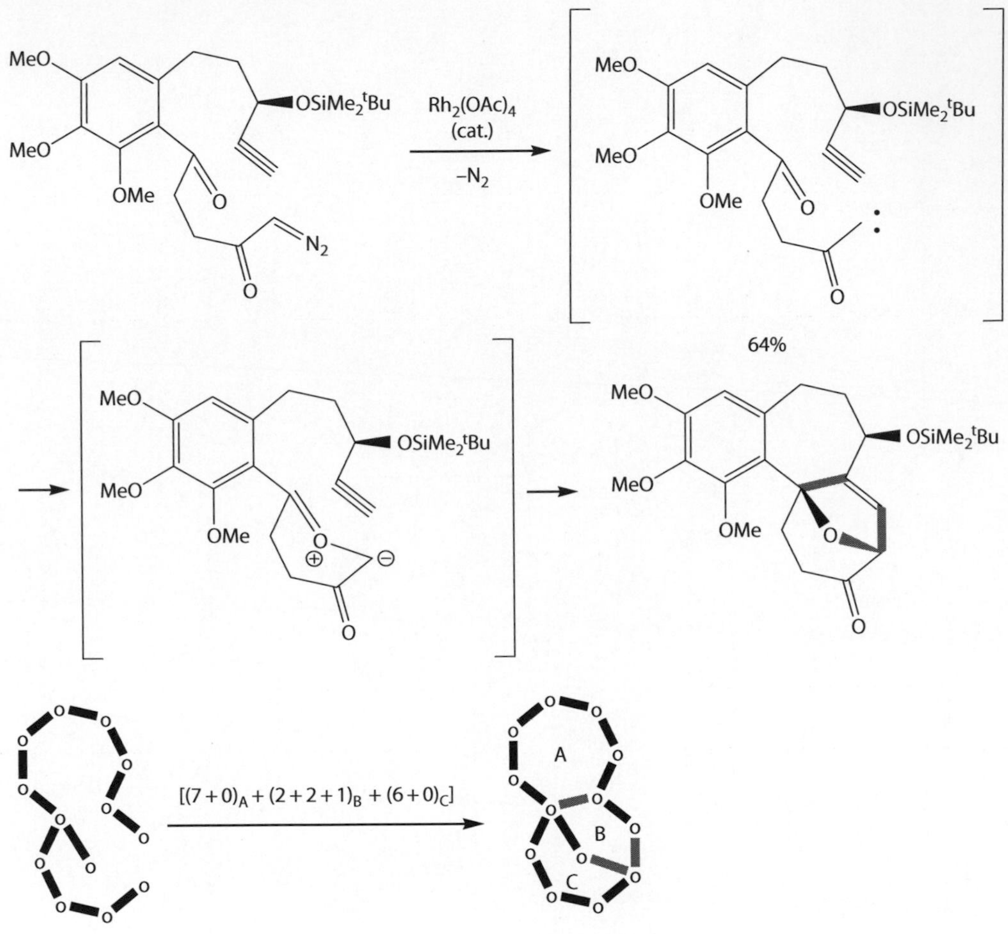

MeO
MeO
OMe
O
O
N2
OSiMe2tBu
Rh2(OAc)4
(cat.)
–N2
64%
A
B
C
[(7+0)A + (2+2+1)B + (6+0)C]

Reppe (1948) plan for cyclooctatetraene: step 1.

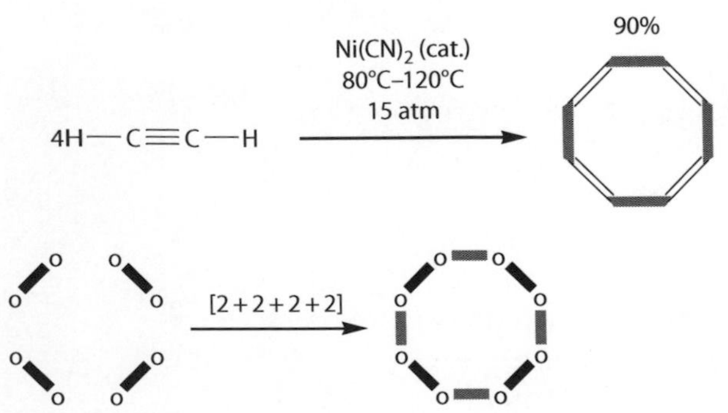

90%
Ni(CN)2 (cat.)
80°C–120°C
15 atm
4H—C≡C—H
[2+2+2+2]

Pattenden (2004) plan for (+)-estrone: step 11 (zipper)

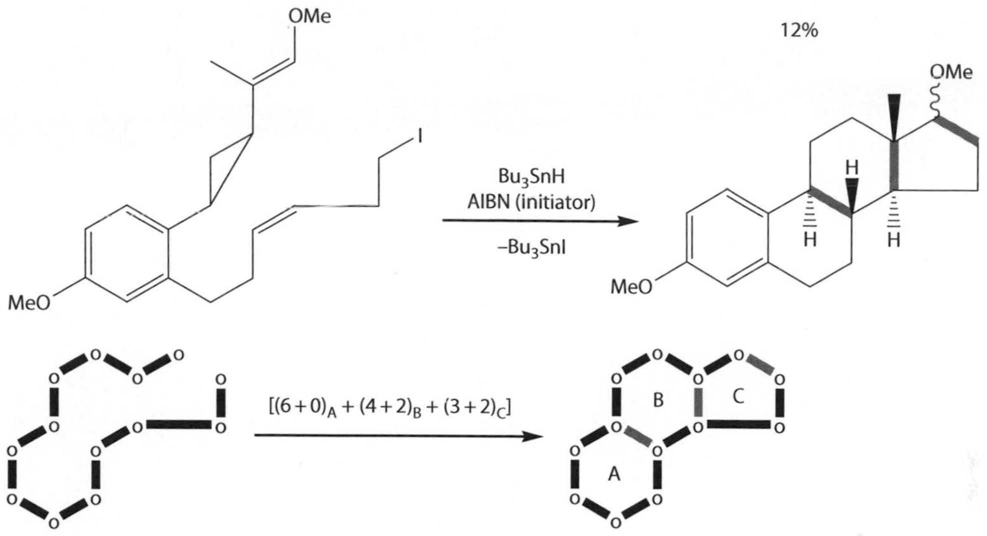

Trost (1978) plan for (+)-ibogamine: step 3.

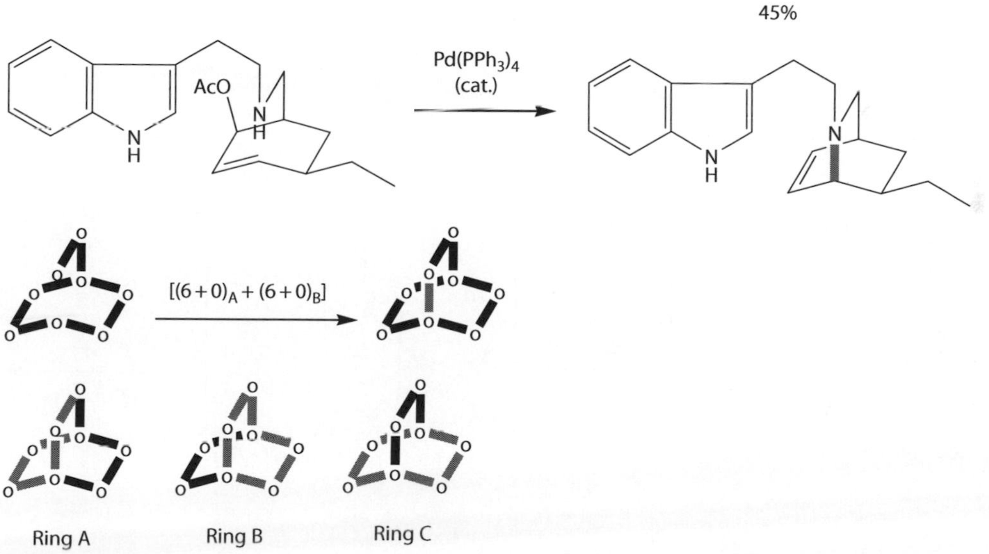

Ring A   Ring B   Ring C

Wenkert (1973) plan for (±)-alpha-himachalene: step 14 (intramolecular zipper)

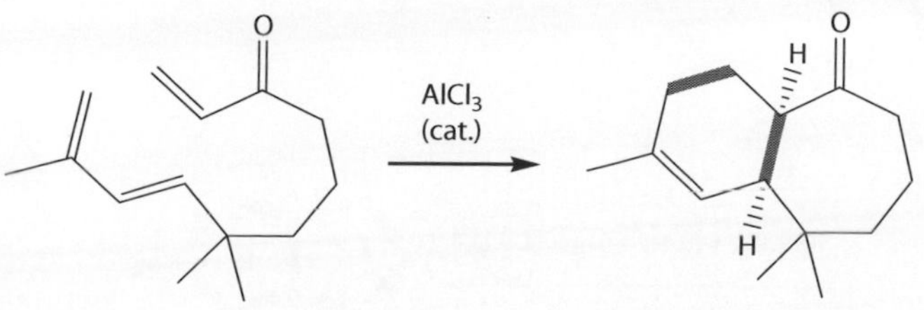

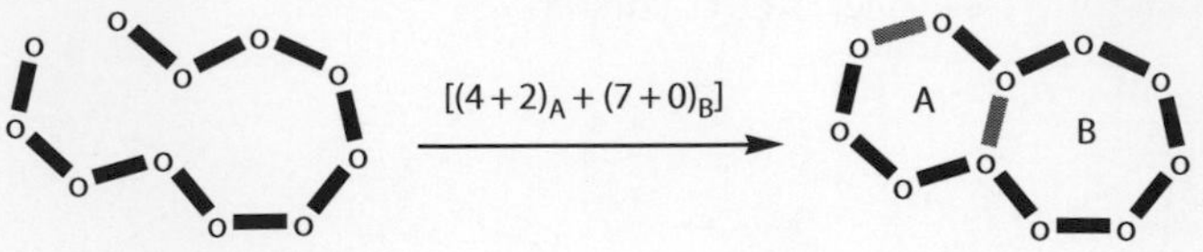
[(4 + 2)A + (7 + 0)B]
A
B

Wender (1982) plan for (±)-hirsutene: step 2.

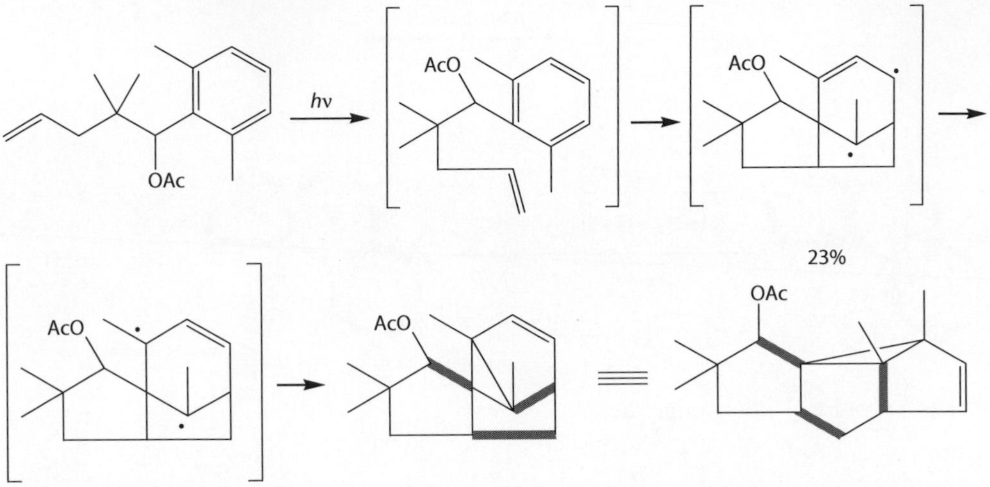
OAc
hν
AcO
AcO
AcO
AcO
23%
OAc

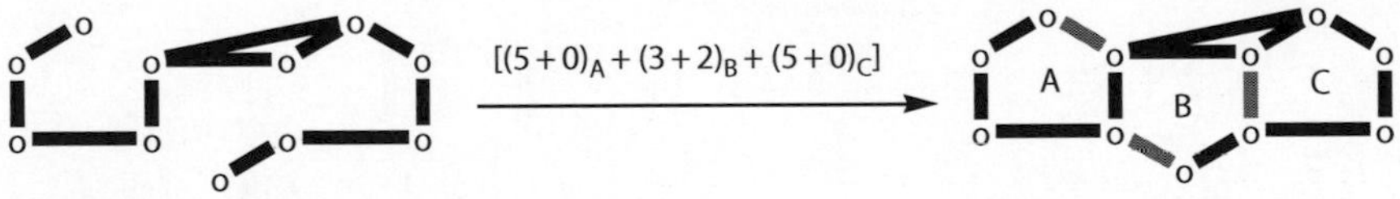
[(5 + 0)A + (3 + 2)B + (5 + 0)C]
A
B
C

Yu (2008) plan for (±)-hirsutene: step 4.

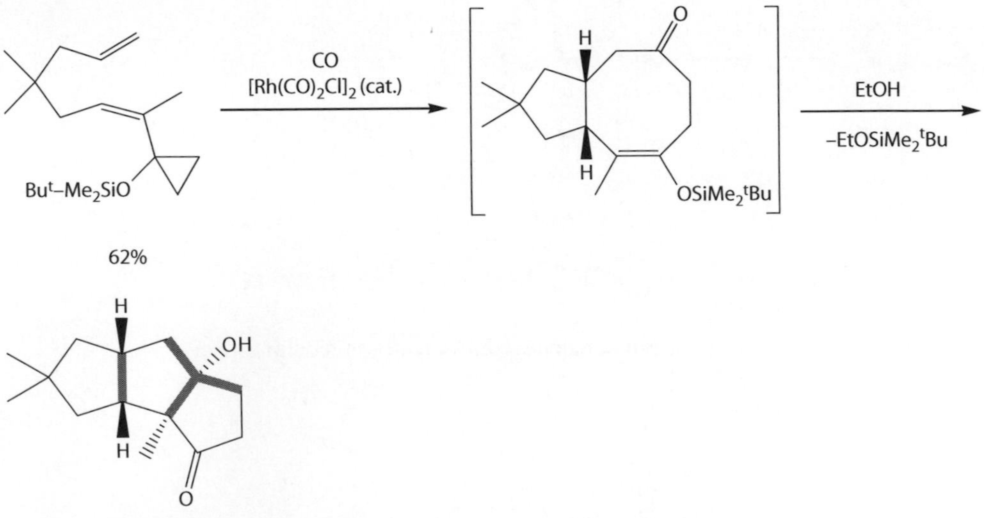
Bu^t–Me2SiO
CO
[Rh(CO)2Cl]2 (cat.)
H
H
O
OSiMe2tBu
EtOH
–EtOSiMe2tBu
62%
H
H
OH
O

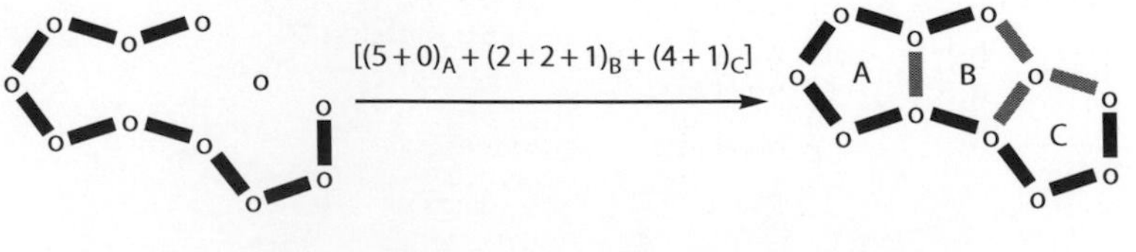
[(5 + 0)A + (2 + 2 + 1)B + (4 + 1)C]
A
B
C

Curran (1985) plan for (±)-hirsutene: step 13.

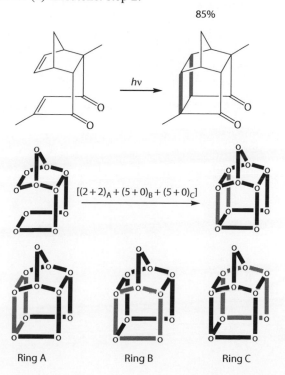

Mehta (1981, 1986) plan for (±)-hirsutene: step 2.

Ring A   Ring B   Ring C

Weinges (1993) plan for (–)-hirsutene: step 13.

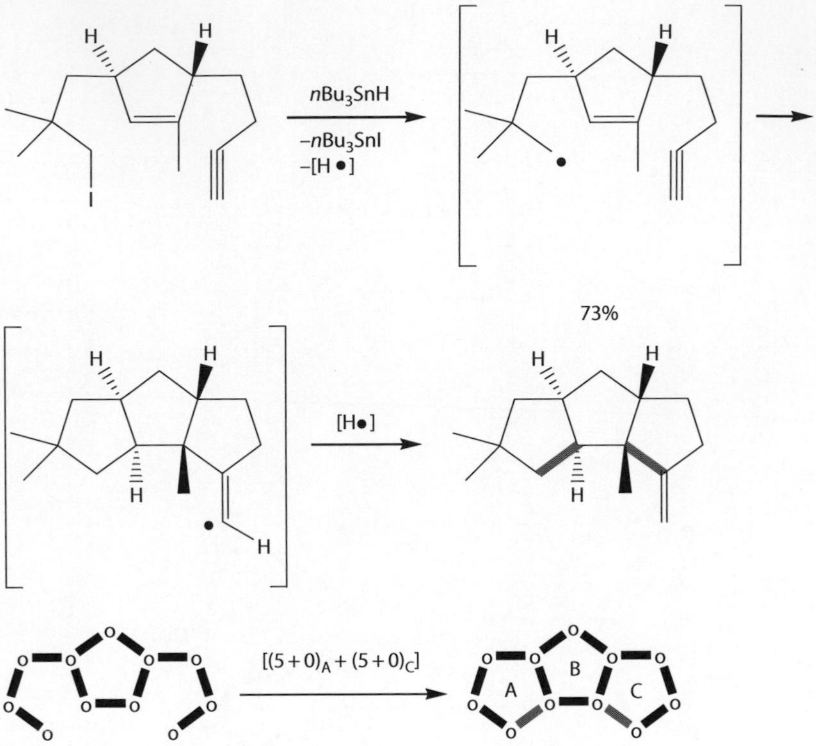

73%

$[(5+0)_A + (5+0)_C]$

Oppolzer (1981) plan for lysergic acid: step 13 (intramolecular zipper)

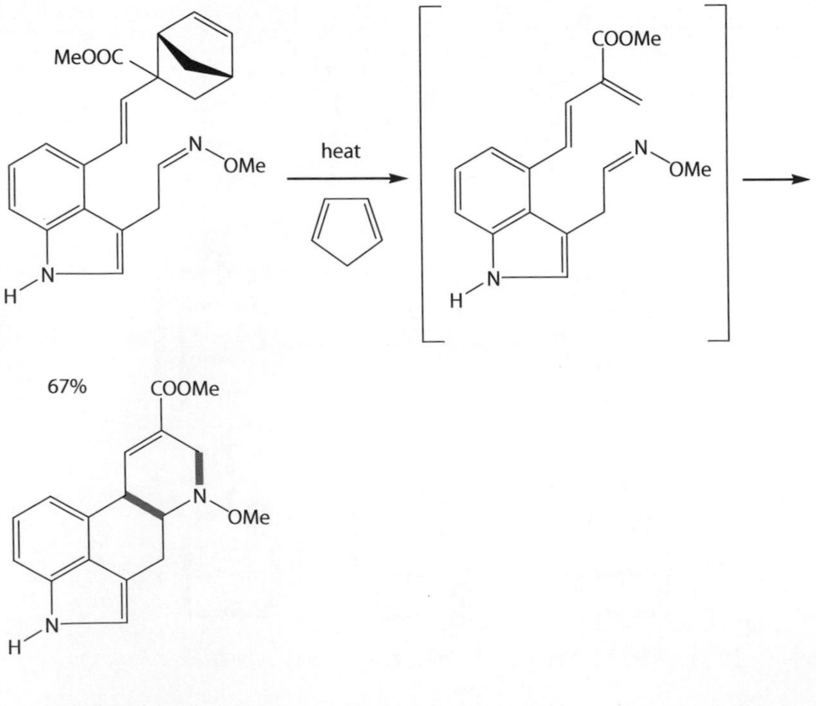

67%

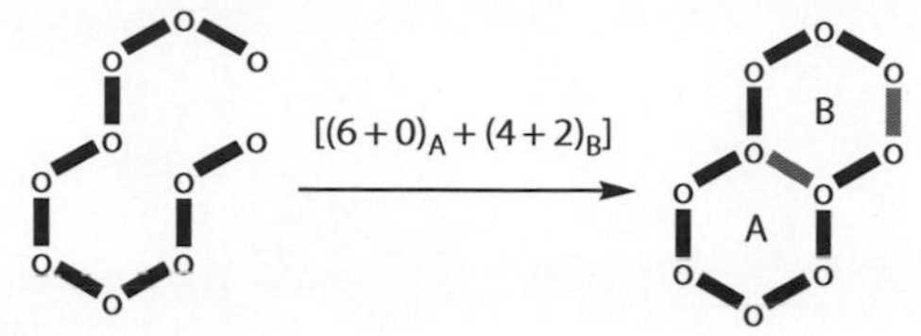

[4.4.0]

Fukuyama G2 (2009) plan for (+)-lysergic acid: double Heck reaction (step 21).

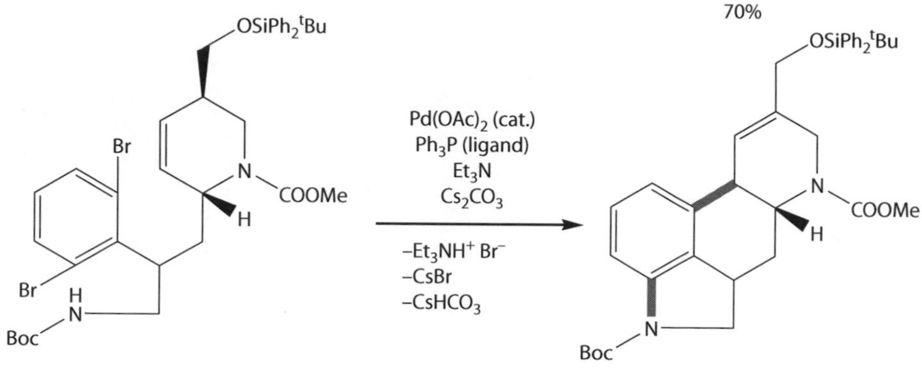

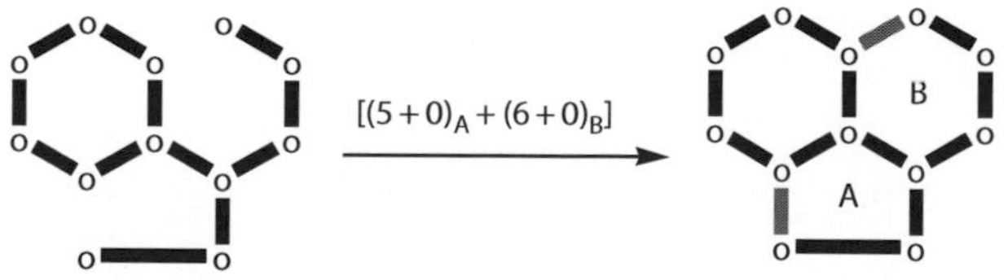

Mori (1988) plan for (+)-nepetalactone: step 3 (intramolecular zipper)

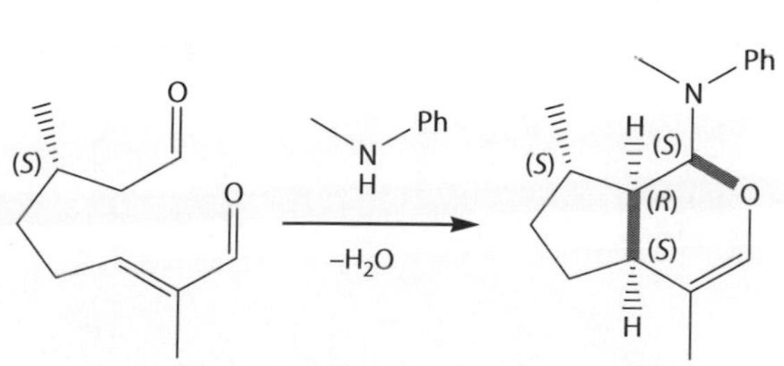

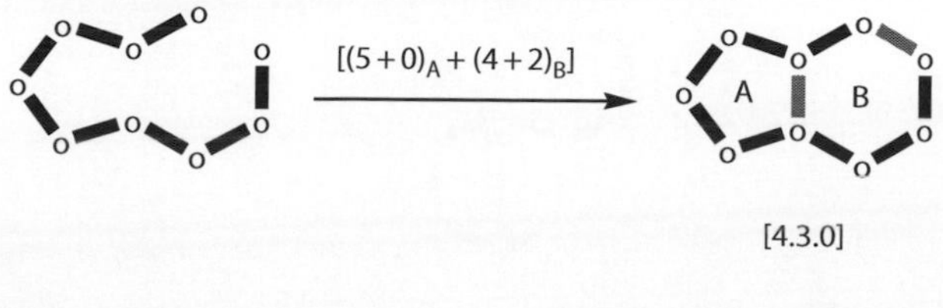

[4.3.0]

Pickett G1 (1996) plan for (+)-nepetalactone: step 3 (intramolecular zipper as mentioned earlier)
Baran (2010) plan for palau'amine: tandem lactamization—transannular reaction in step 21.

17%

[4.3.0]

Wiberg (1982, 1984) plan for [1.1.1]-propellane: step 12.

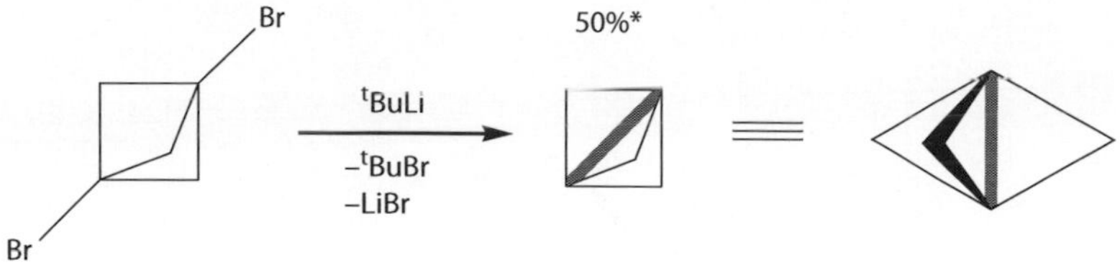
Br
Br
^tBuLi
$-^tBuBr$
–LiBr
50%*

*assumed reaction yield

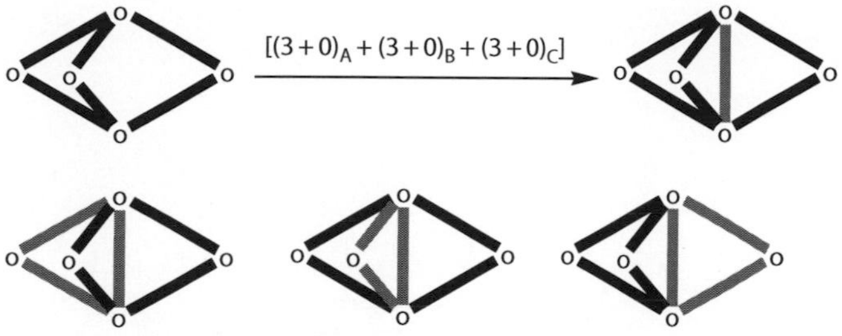
$[(3+0)_A + (3+0)_B + (3+0)_C]$

Ring A Ring B Ring C

Kulkarni (2009) plan for (±)-physostigmine: step 8 (claw—swaying hands)

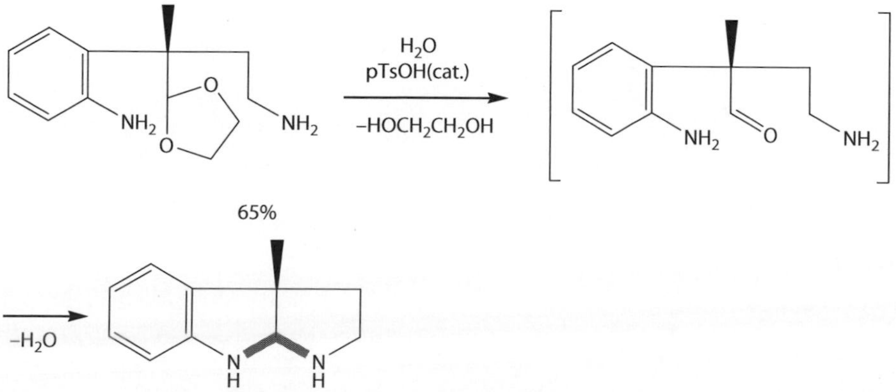
NH_2
NH_2
H_2O
pTsOH(cat.)
$-HOCH_2CH_2OH$
NH_2
NH_2
65%
$-H_2O$
N H
N H

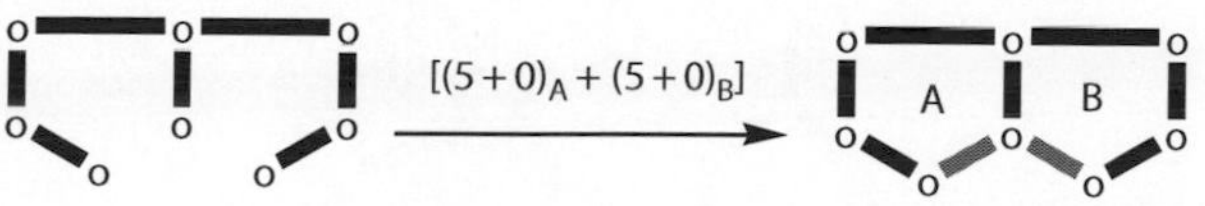
$[(5+0)_A + (5+0)_B]$
A
B

[3.3.0]

Aggarwal (2010) plan for quinine: linking quinuclidine precursor with quinoline moiety via a sacrificial epoxidation in step 14, then making the quinuclidine ring via epoxide ring opening in step 15.

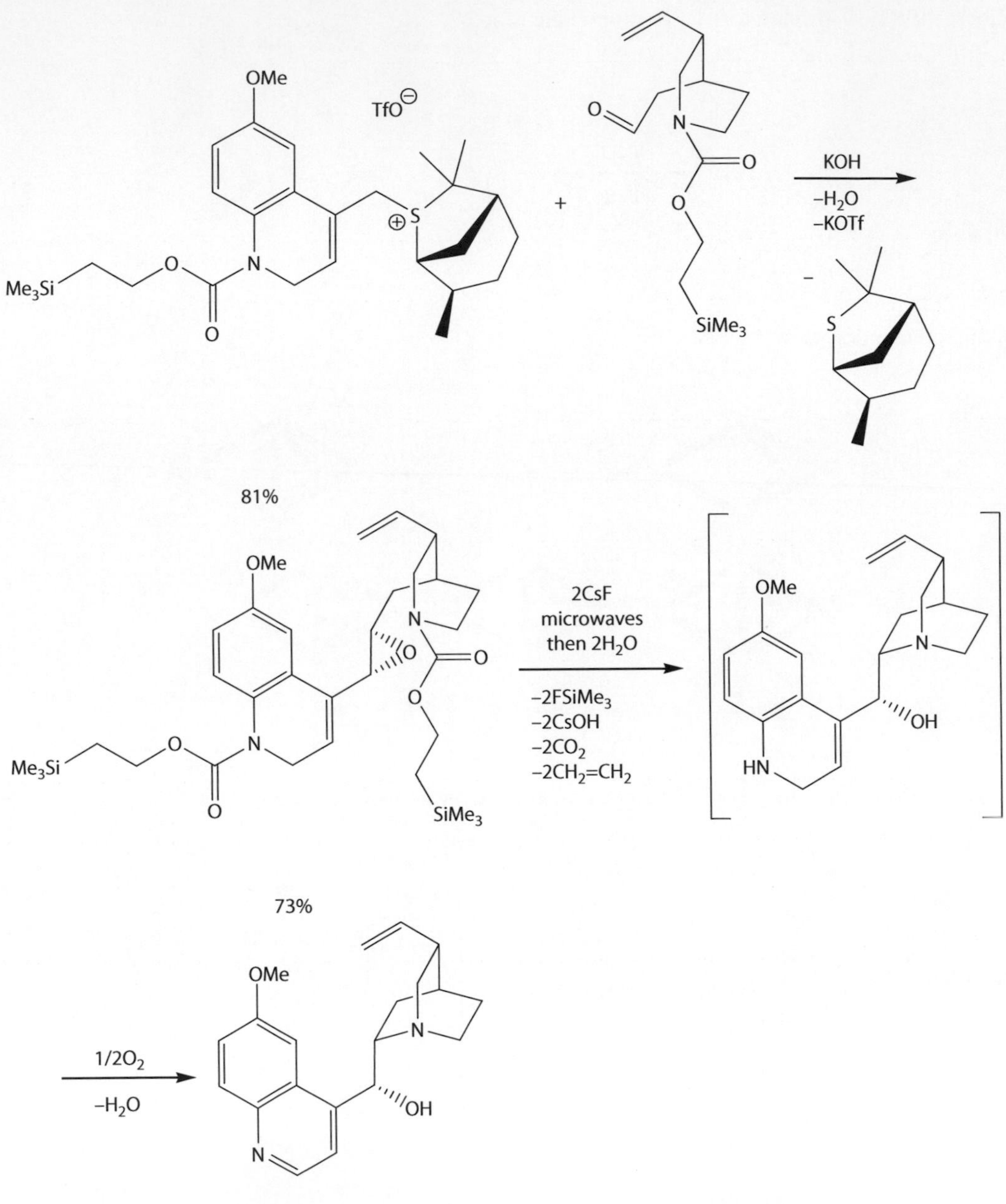
OMe
TfO
S
Me3Si
O
N
+
O
N
SiMe3
KOH
–H2O
–KOTf
S
81%
OMe
N
O
2CsF
microwaves
then 2H2O
–2FSiMe3
–2CsOH
–2CO2
–2CH2=CH2
HN
OH
73%
1/2O2
–H2O
OMe
N
OH

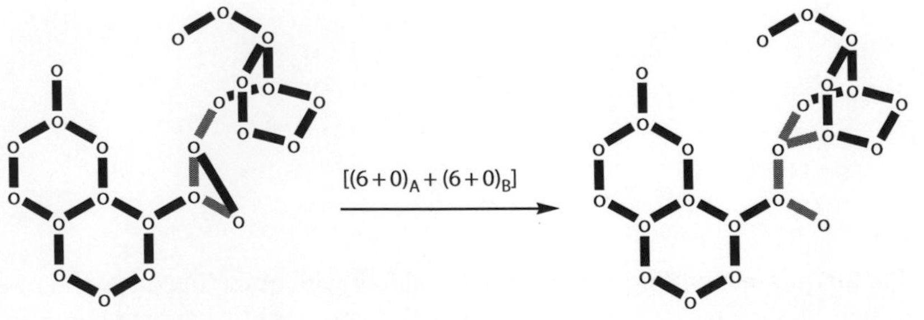
[(6 + 0)A + (6 + 0)B]

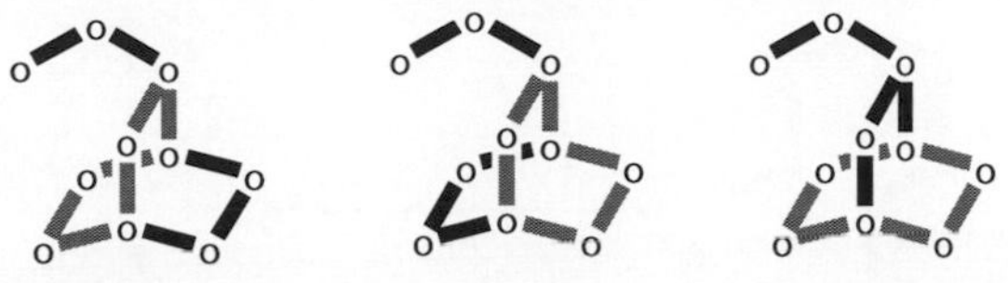

Ring A   Ring B   Ring C

Martin (1983) plan for reserpine: tandem Diels–Alder—retro-Diels–Alder reaction in step 7 (intramolecular zipper)

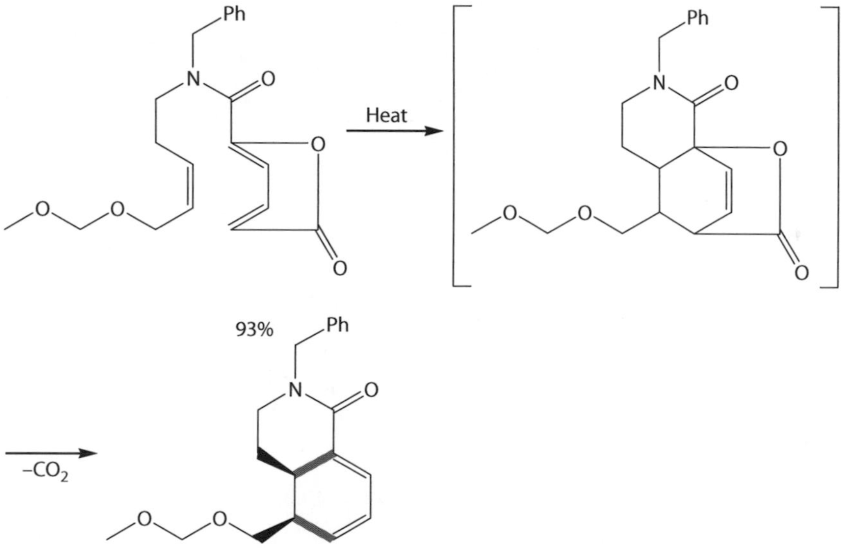

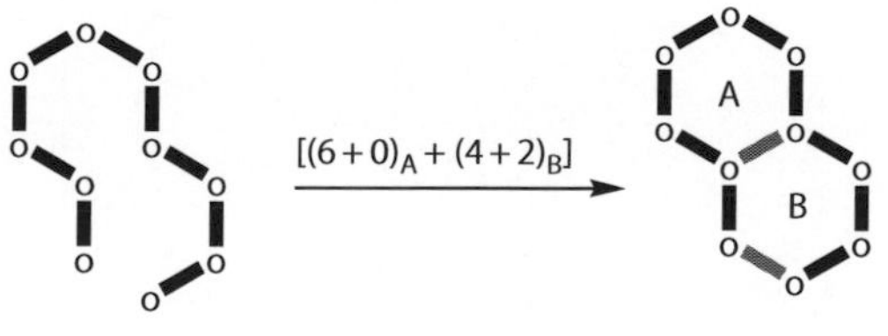

93%

$-CO_2$

$[(6+0)_A + (4+2)_B]$

[4.4.0]

Shea (2003) plan for reserpine: Diels–Alder reaction in step 8.

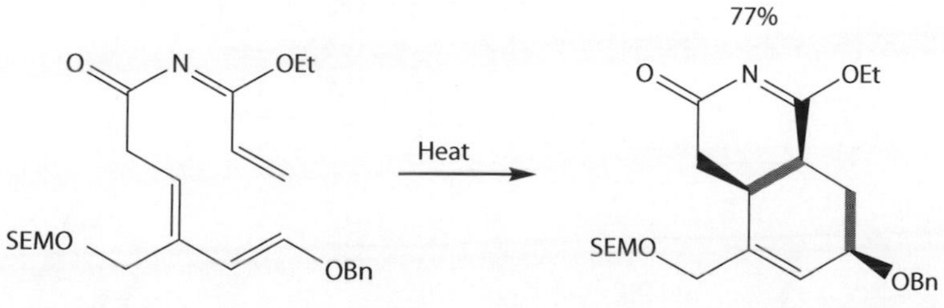

77%

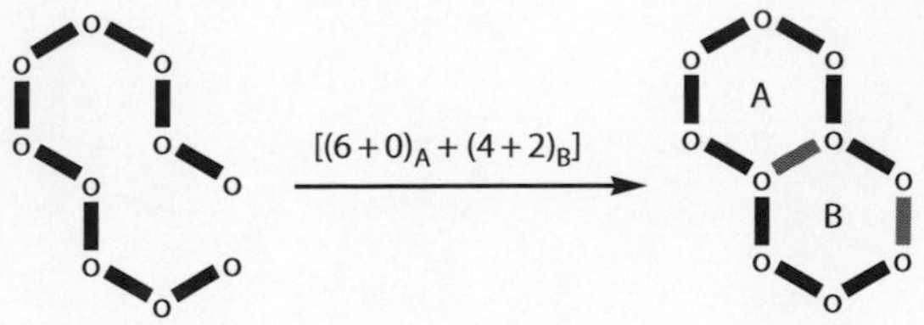
A
B
[(6 + 0)A + (4 + 2)B]

[4.4.0]

Wender (2006) plan for (±)-salsolene oxide: step 6 (intramolecular zipper)

24%

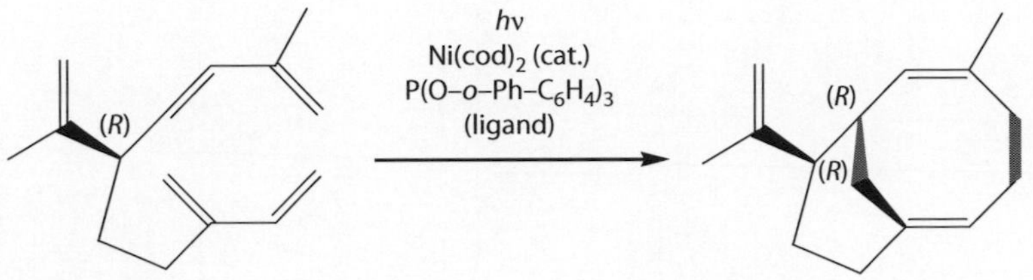
hν
Ni(cod)2 (cat.)
P(O–o–Ph–C6H4)3
(ligand)
(R)
(R)
(R)

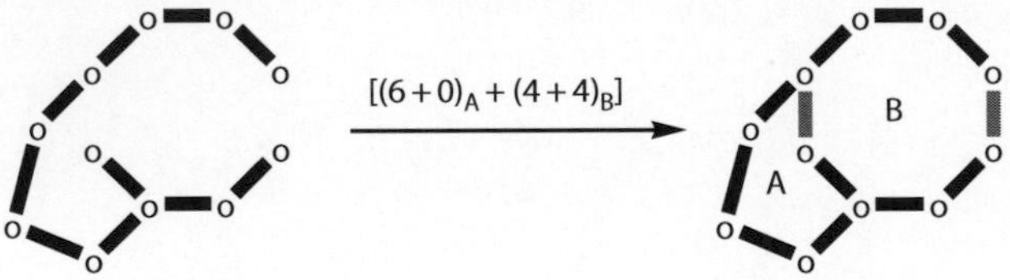
[(6 + 0)A + (4 + 4)B]
A
B

[5.3.1]

Wender (1985) plan for silphinene: step 2 (intramolecular zipper)

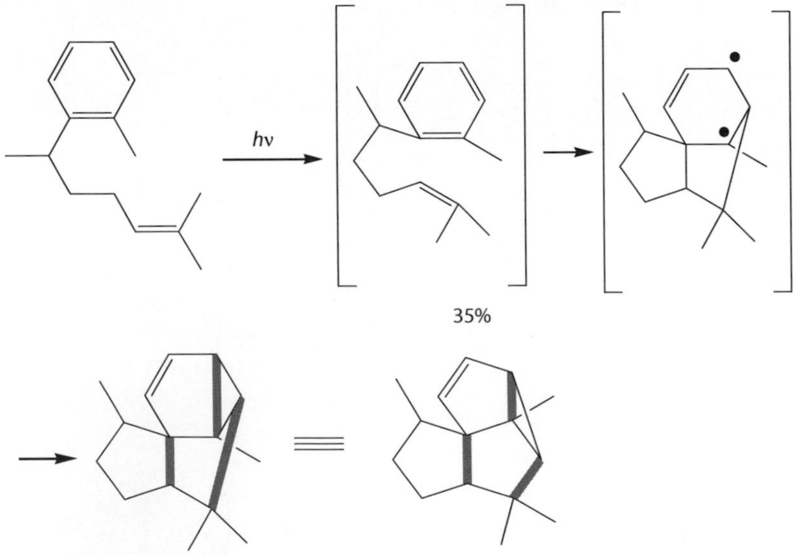
hν
35%

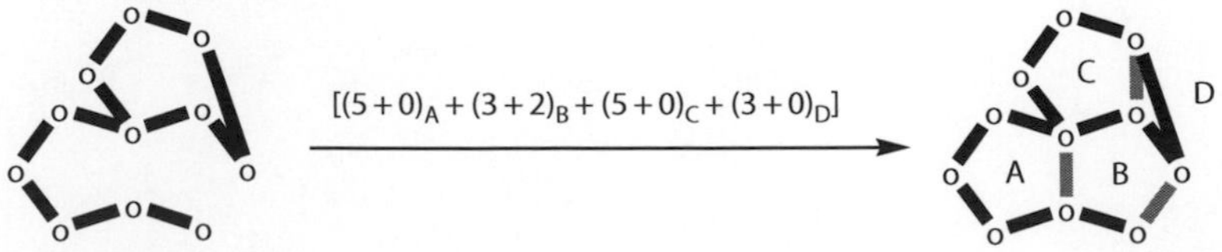
[(5 + 0)A + (3 + 2)B + (5 + 0)C + (3 + 0)D]
A
B
C
D

Anet (1950) plan for (±)-sparteine: step 1 (idea is analogous to strategy employed in Robinson (1917) plan for tropinone)

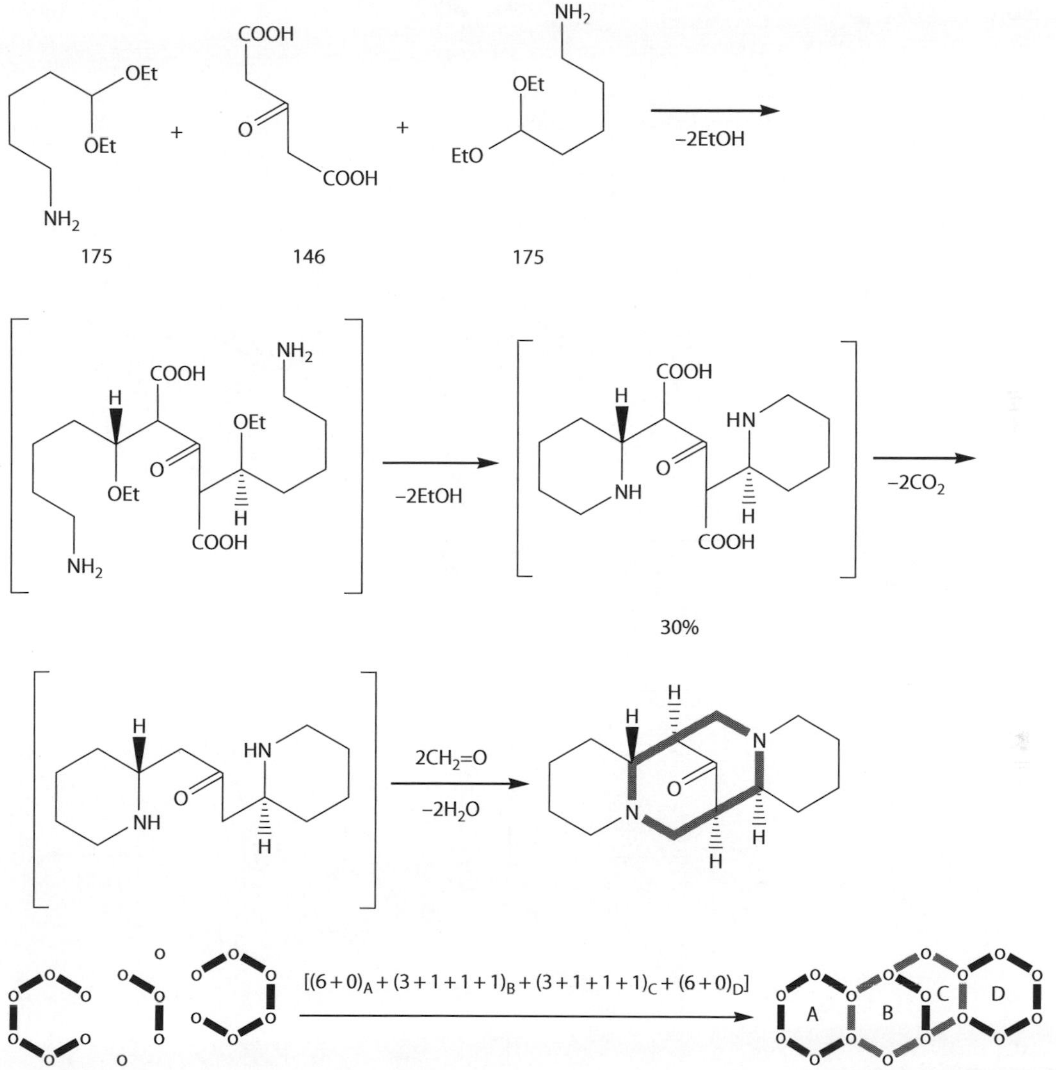
175
146
175
−2EtOH
−2EtOH
−2CO2
30%
2CH2=O
−2H2O
[(6 + 0)A + (3 + 1 + 1 + 1)B + (3 + 1 + 1 + 1)C + (6 + 0)D]
A
B
C
D

Blakemore (2008) plan for (±)-sparteine: step 2 (pincer)

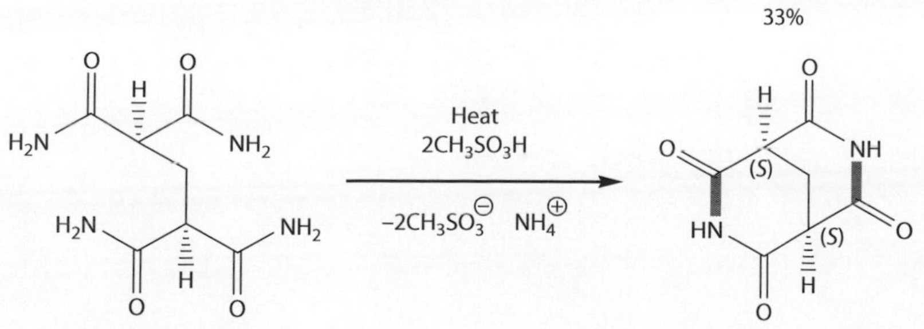
33%
Heat
2CH3SO3H
−2CH3SO3⊖ NH4⊕
(S)
(S)

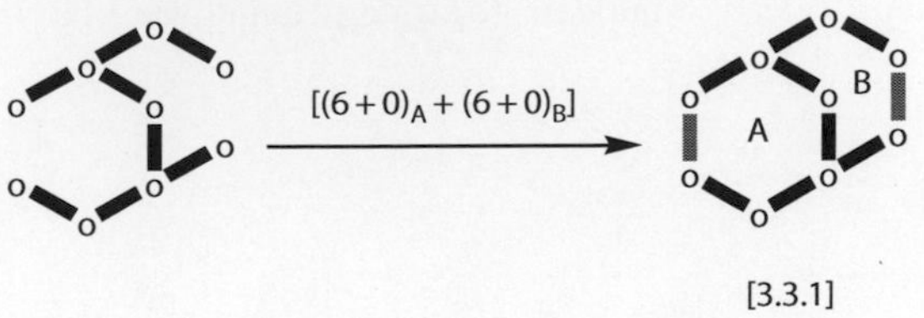

[3.3.1]

Fleming (2004) plan for (±)-sparteine: step 11 (pincer)

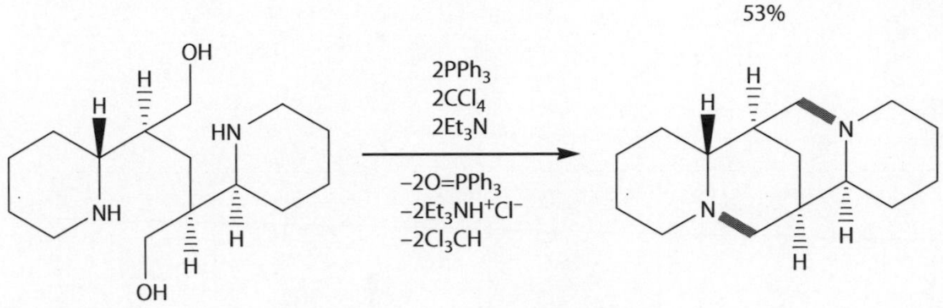

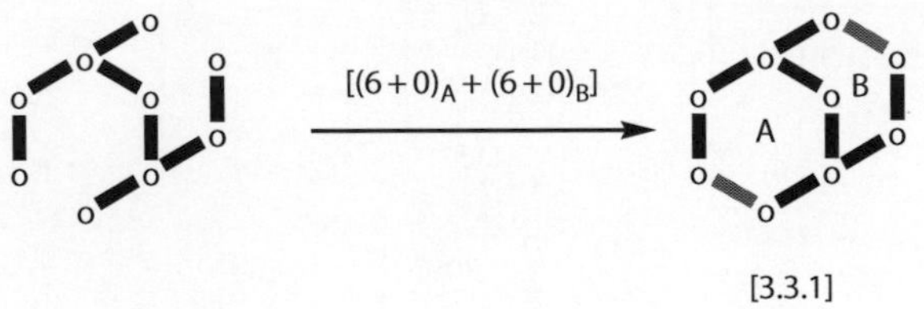

[3.3.1]

Leonard G2 (1950) plan for (±)-sparteine: step 3 (pincer)

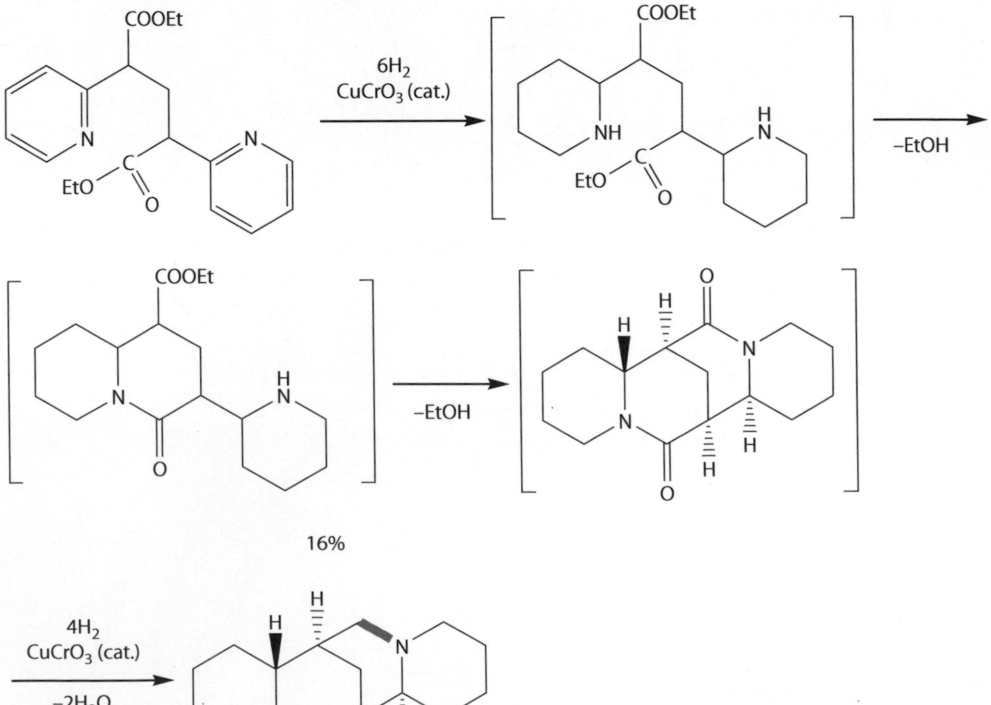

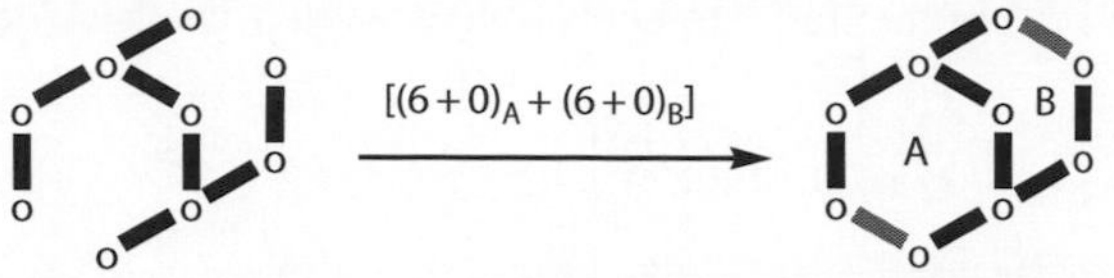

$$[(6+0)_A + (6+0)_B]$$

[3.3.1]

van Tamelen (1960) plan for (±)-sparteine: step 4 (S-wave)

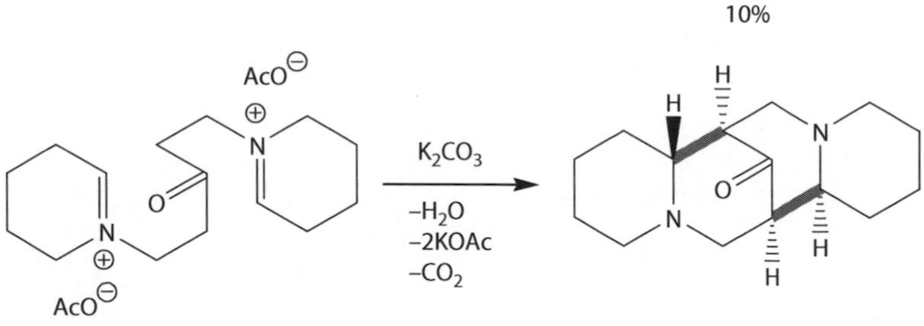

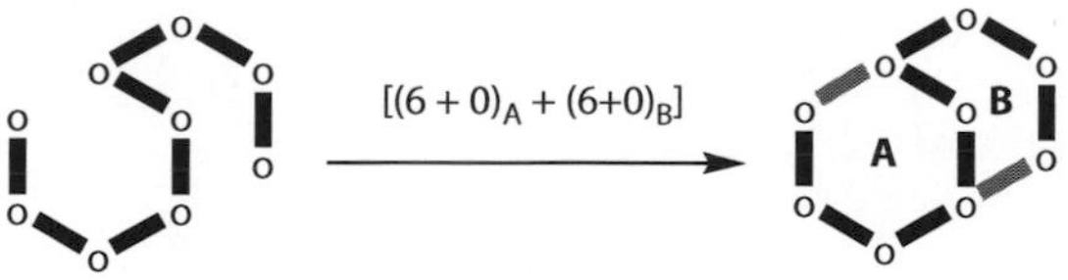

10%

$$[(6+0)_A + (6+0)_B]$$

[3.3.1]

Aube (2005) plan for (±)-stenine: step 4.

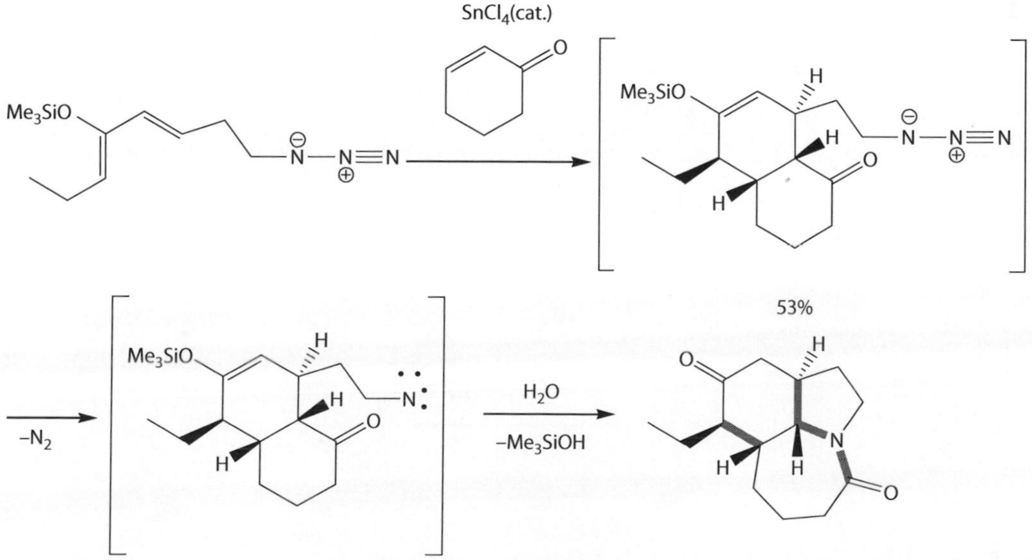

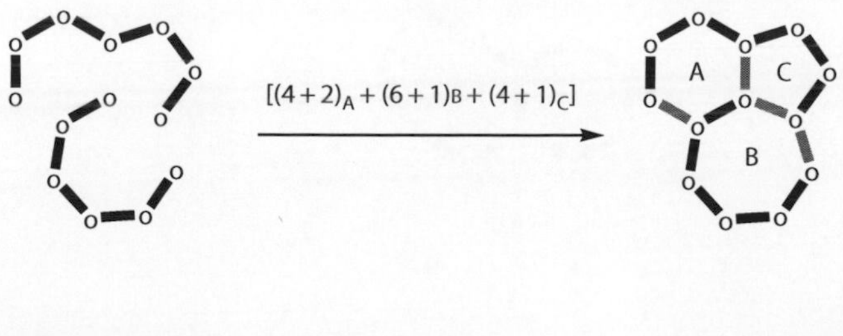

53%

$$[(4+2)_A + (6+1)_B + (4+1)_C]$$

Bodwell (2002) plan for strychnine: tandem Diels–Alder—retro-Diels–Alder reaction in step 8.

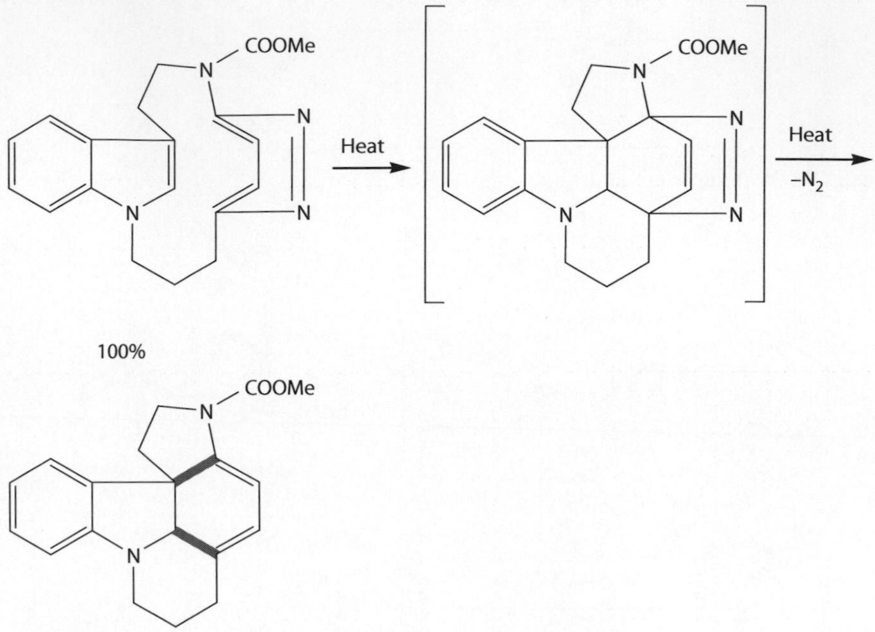

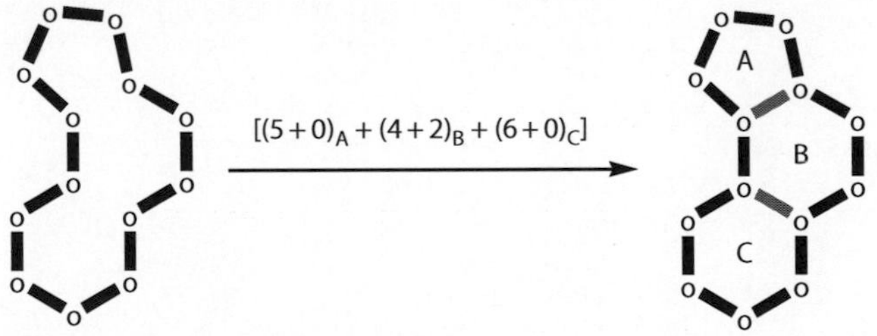

100%

$[(5+0)_A + (4+2)_B + (6+0)_C]$

Note that the pyridazine ring in the substrate is a sacrificial ring in this transformation.

Bonjoch (2000) plan for strychnine: step 18.

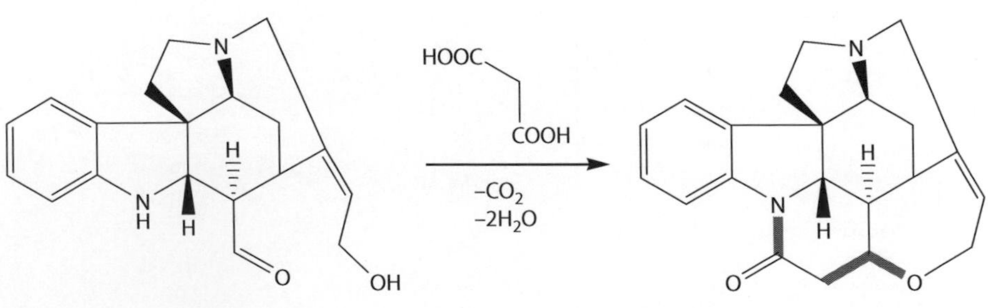

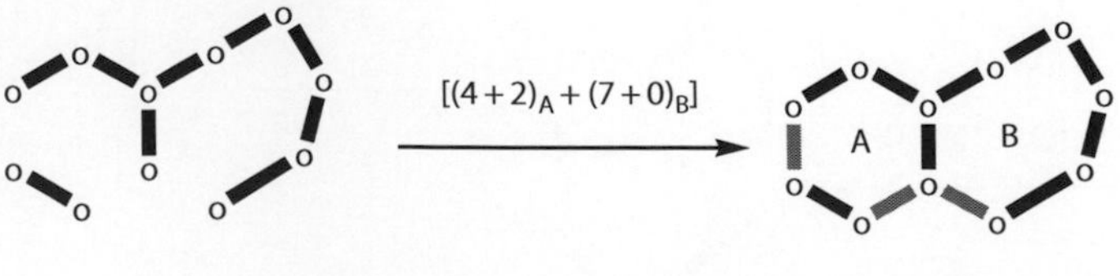

49%

$[(4+2)_A + (7+0)_B]$

[5.4.0]

Fukuyama (2004) plan for strychnine: step 19.

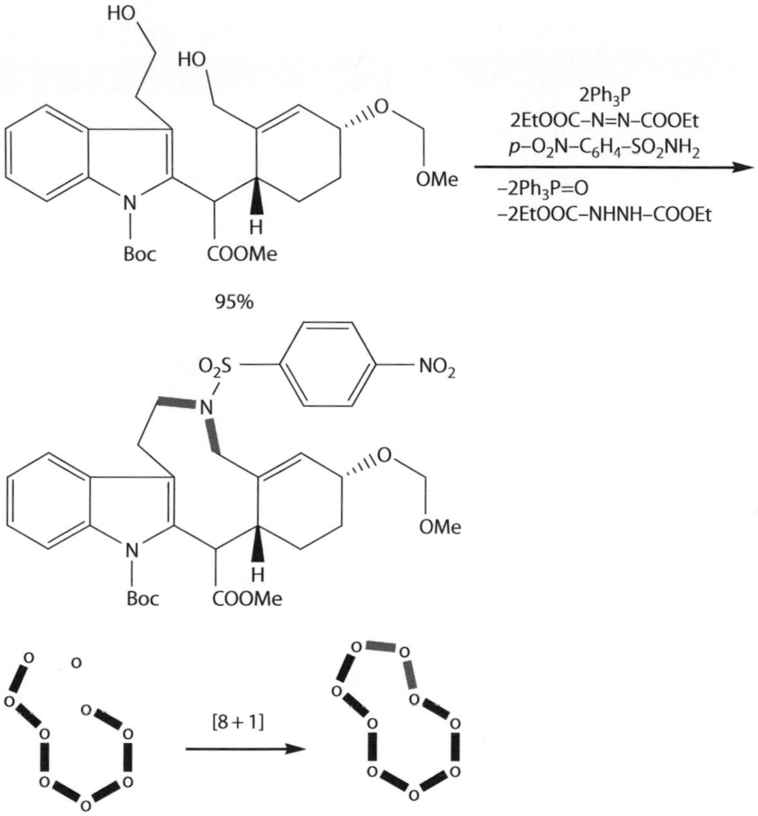

95%

Martin (2001) plan for strychnine: step 11 (intramolecular zipper)

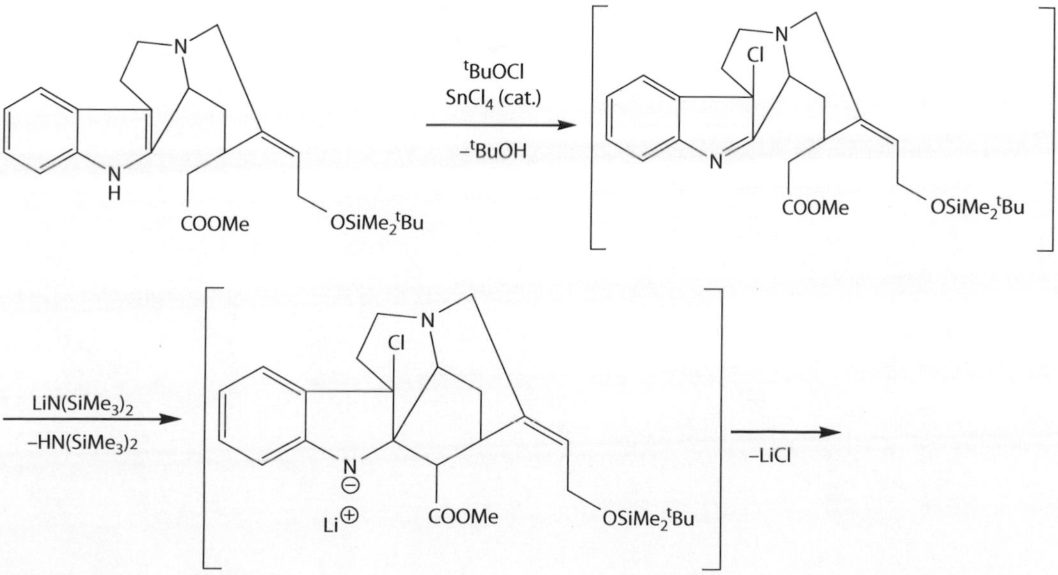

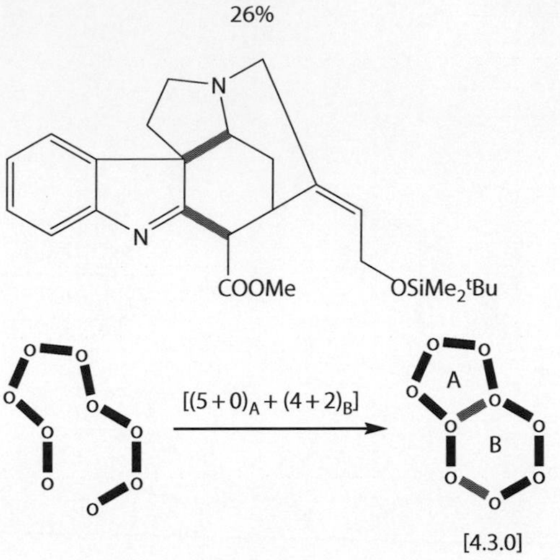
26%
N
N
COOMe
OSiMe2tBu
[(5+0)A+(4+2)B]
A
B
[4.3.0]

Reissig (2010) plan for strychnine: step 3.

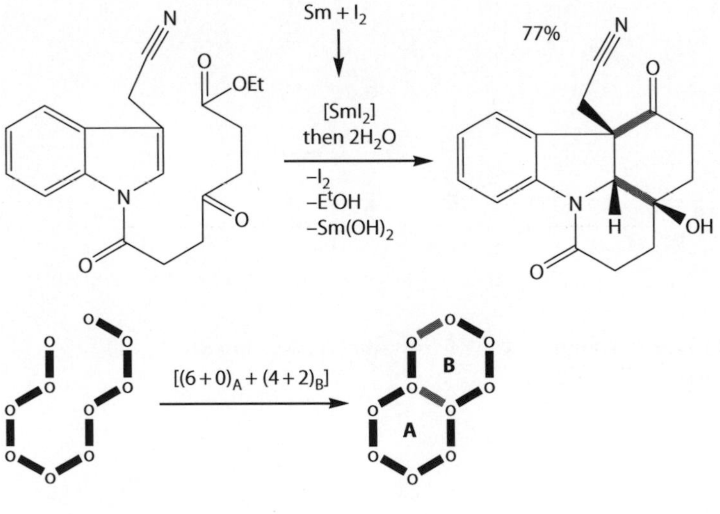
N
O
OEt
N
O
O
Sm + I2
[SmI2]
then 2H2O
–I2
–EtOH
–Sm(OH)2
77%
N
O
N
H
OH
O
[(6+0)A+(4+2)B]
B
A

Vollhardt (2000) plan for strychnine: step 7.

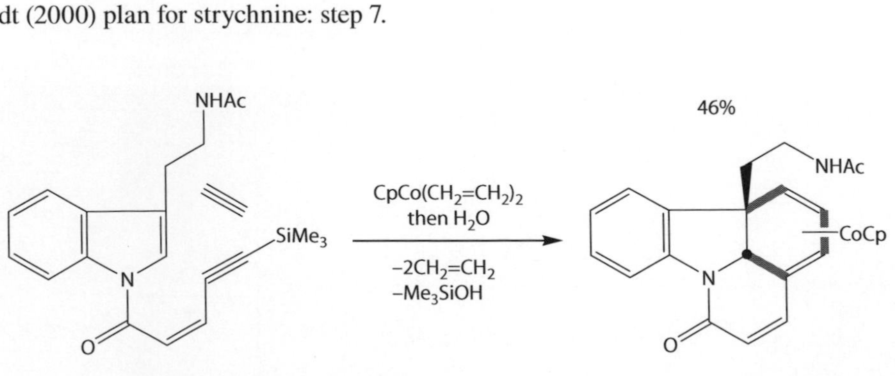
NHAc
SiMe3
N
O
CpCo(CH2=CH2)2
then H2O
–2CH2=CH2
–Me3SiOH
46%
NHAc
CoCp
N
O

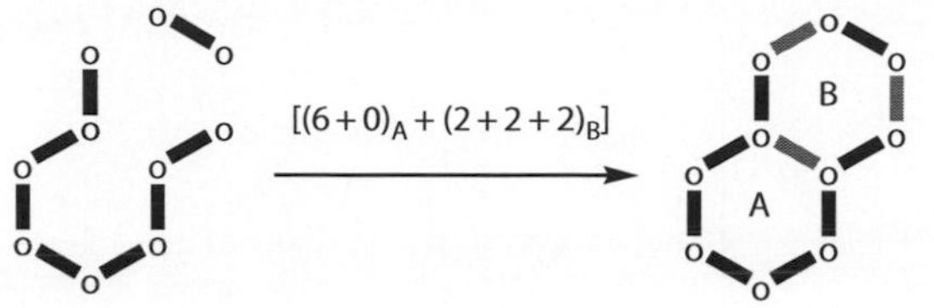

[4.4.0]

Kang (1995) plan for (–)-swainsonine: step 16 (claw)

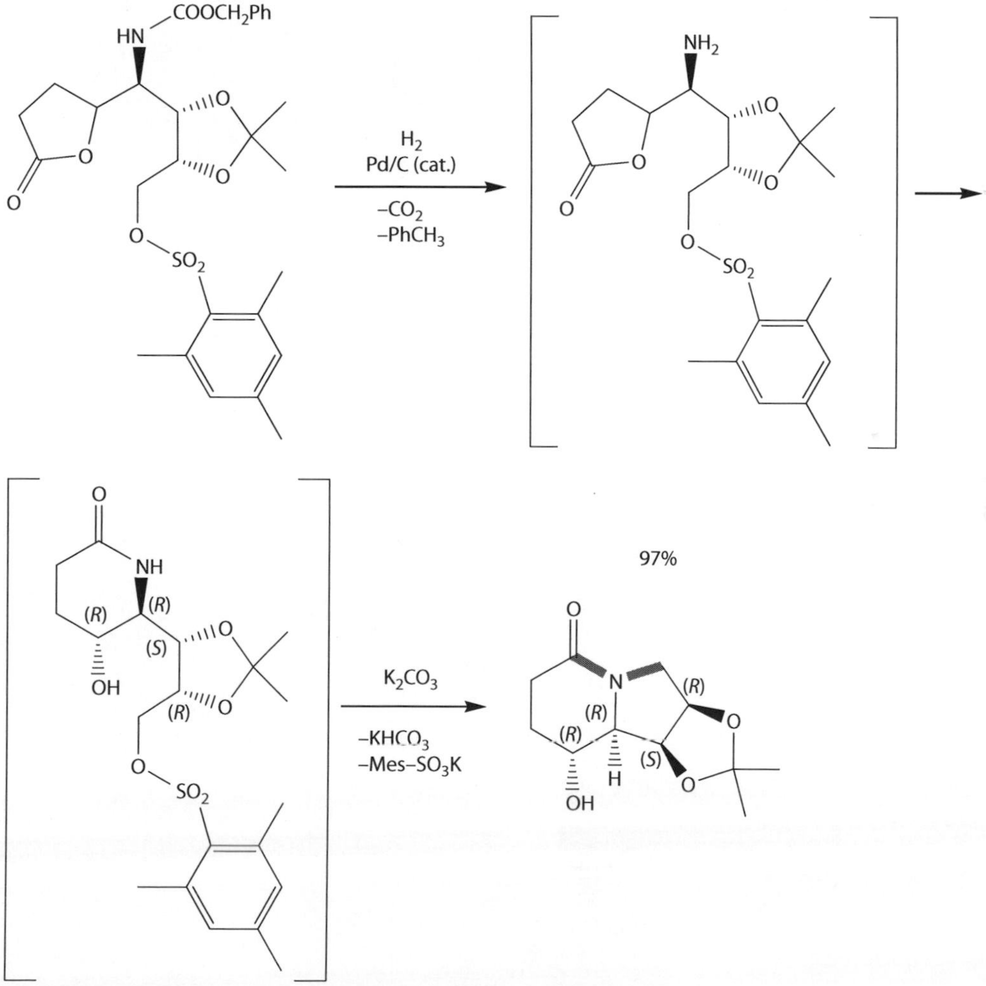

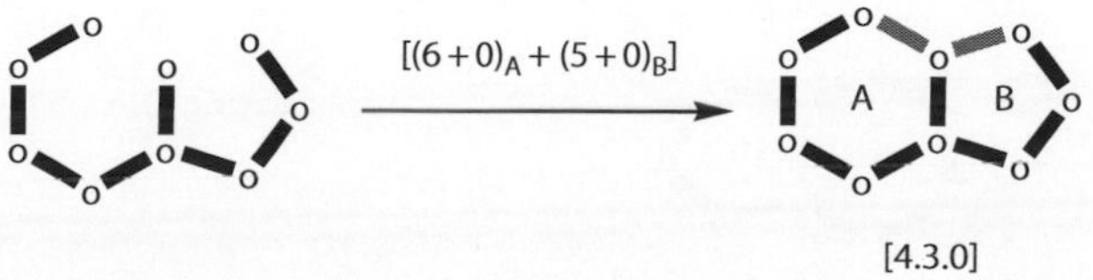

[4.3.0]

Mootoo (2001) plan for (–)-swainsonine: step 12.

69%

[4.3.0]

Note that both rings comprising the spiro ring moiety in the substrate are sacrificial rings in this transformation.

O'Doherty G1 (2006) plan for (+)-swainsonine: step 13 (claw)

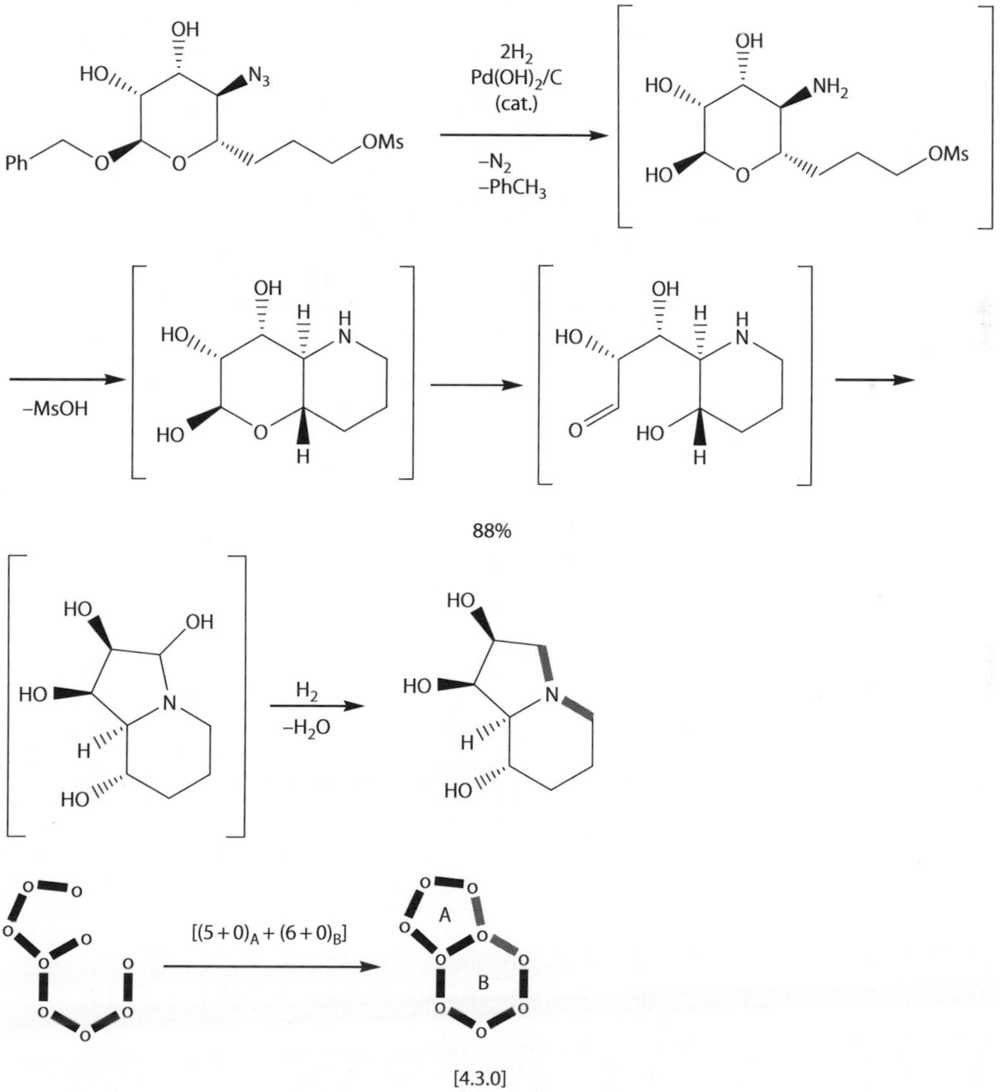

88%

[4.3.0]

Note that the tetrahydropyran ring in the substrate is a sacrificial ring in this transformation.

O'Doherty G2 (2006) plan for (+) swainsonine: step 14 (as mentioned earlier)
Pearson (1996) plan for (−)-swainsonine: step 11.

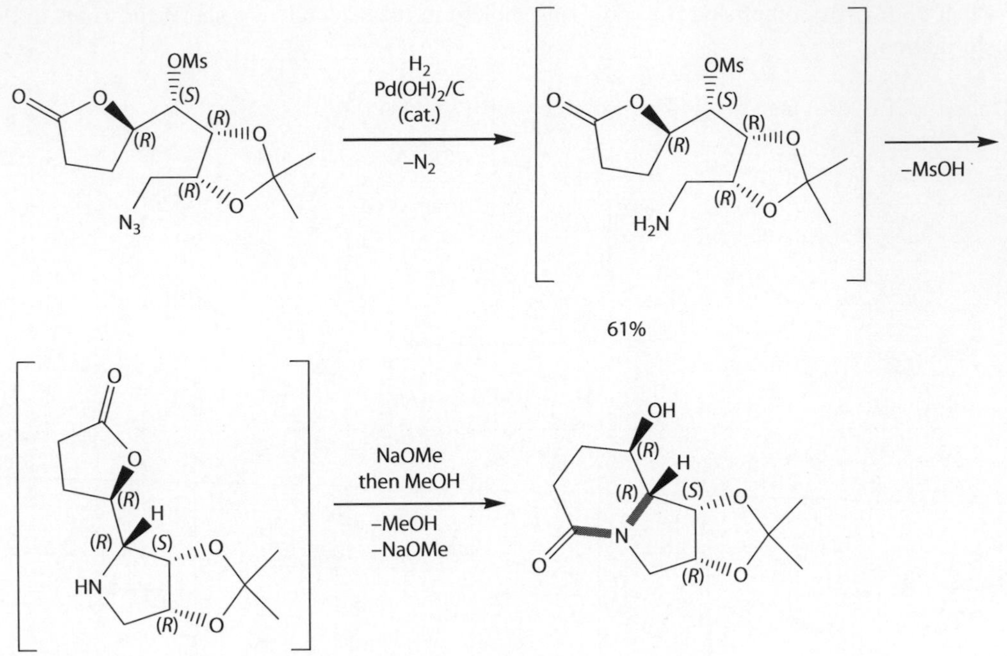

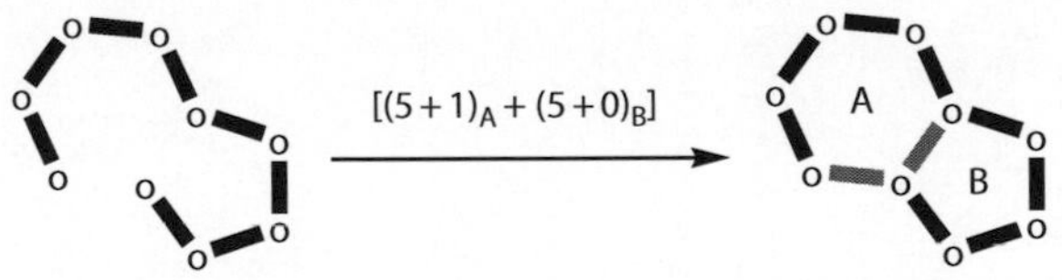

Note that the lactone ring in the substrate is a sacrificial ring in this transformation.

Roush (1995) plan for (−)-swainsonine: step 7.

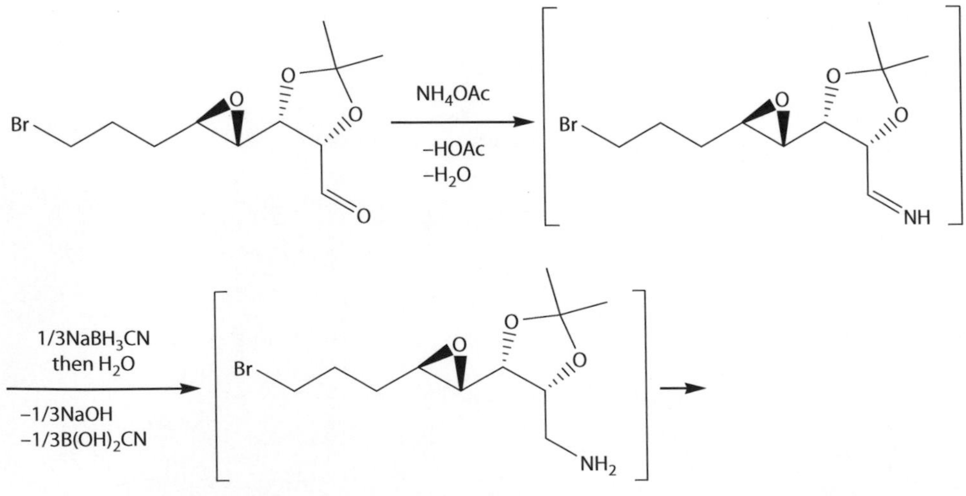

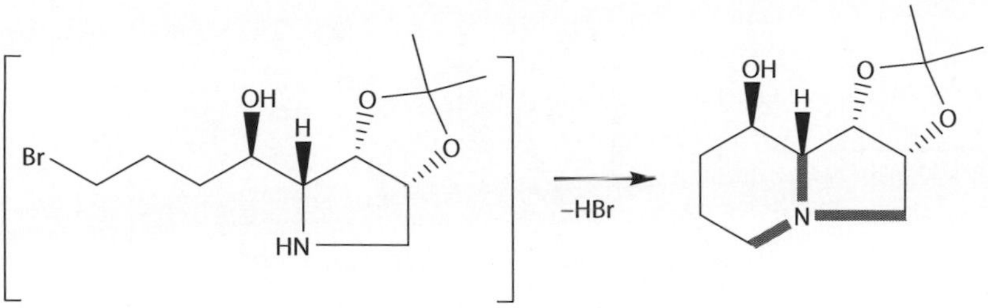

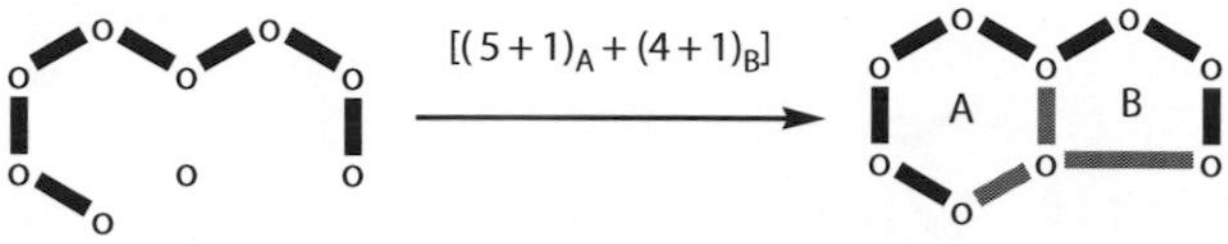

[4.3.0]

Myers G2 (2007) plan for (–)-tetracycline hydrochloride: step 7 (intramolecular zipper)

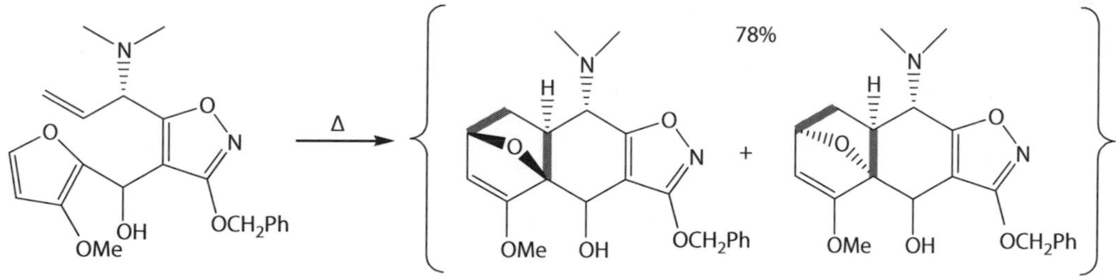

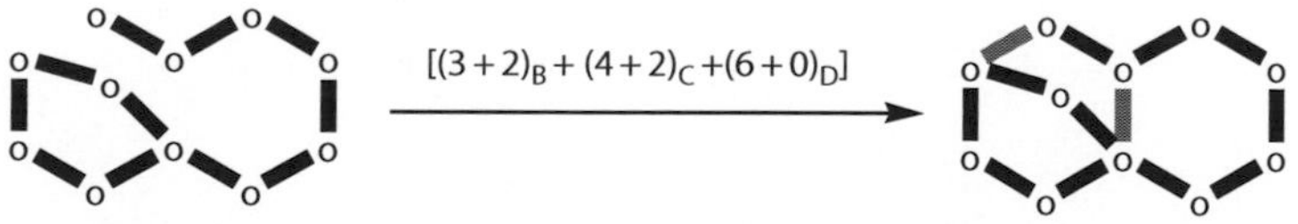

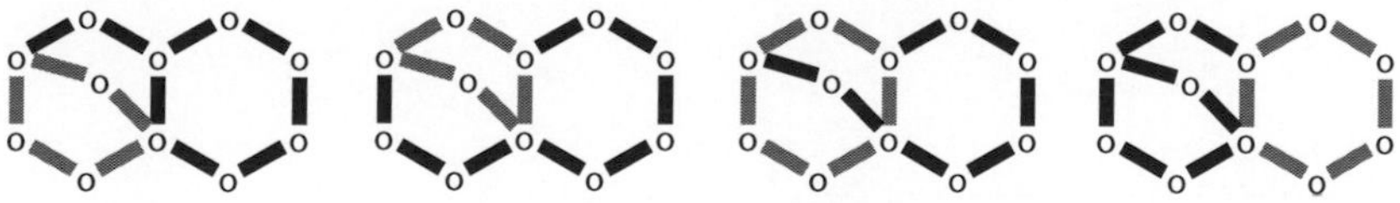

Ring A   Ring B   Ring C   Ring D

Oppolzer (1997) plan for (–)-alpha-thujone: step 9.

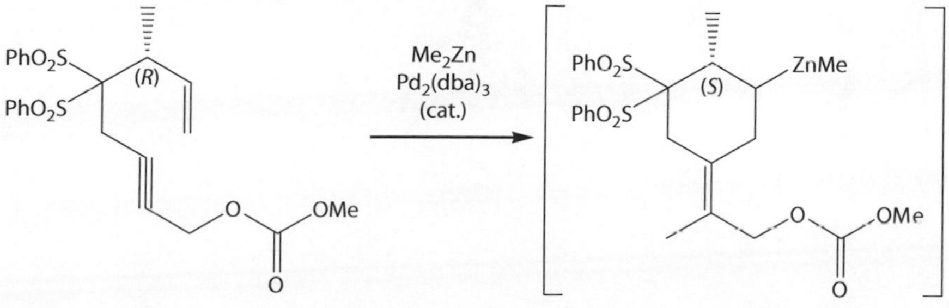

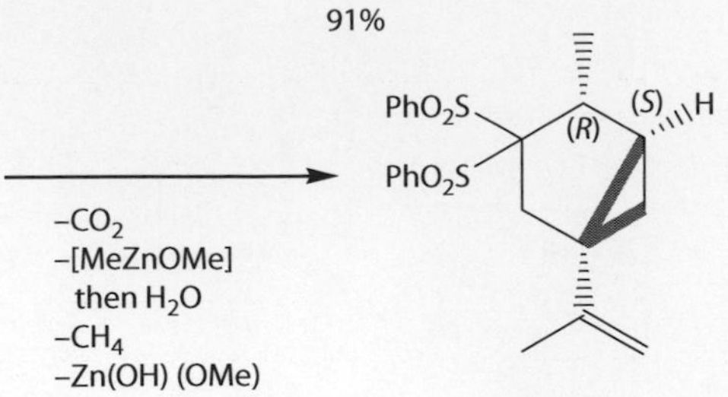
91%
PhO2S
PhO2S
(R)
(S)
H
-CO2
-[MeZnOMe]
then H2O
-CH4
-Zn(OH) (OMe)

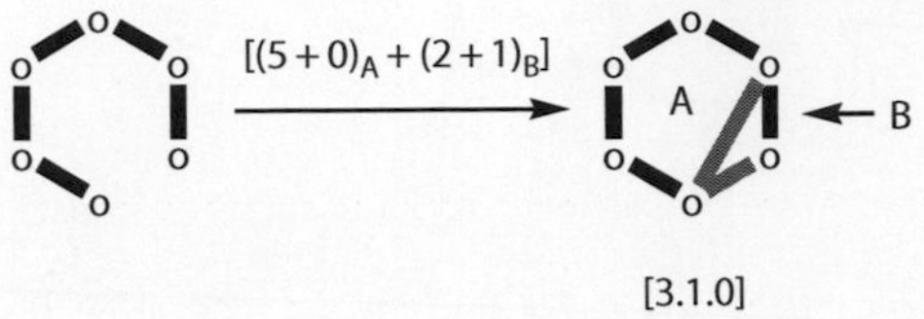
[(5 + 0)A + (2 + 1)B]
A
B
[3.1.0]

Bazilevskaya G4 (1958) plan for tropinone: step 4.

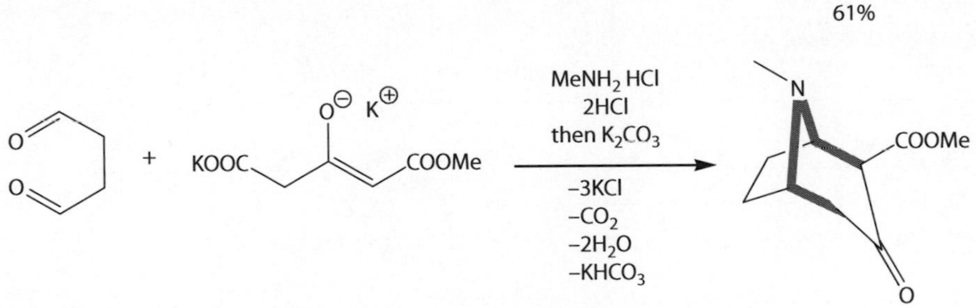
61%
O
O
+
KOOC
O
K
COOMe
MeNH2 HCl
2HCl
then K2CO3
-3KCl
-CO2
-2H2O
-KHCO3
N
COOMe
O

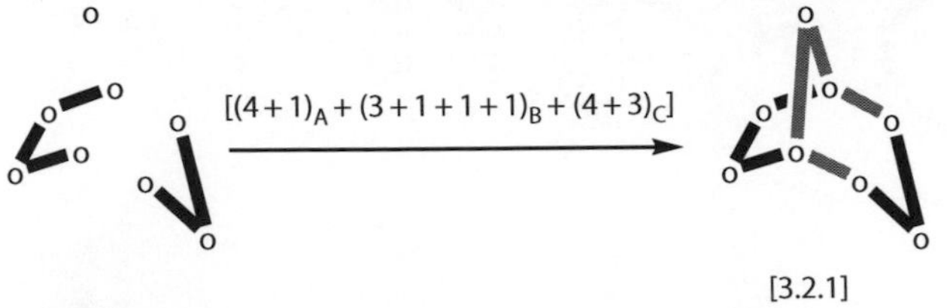
[(4 + 1)A + (3 + 1 + 1 + 1)B + (4 + 3)C]
[3.2.1]

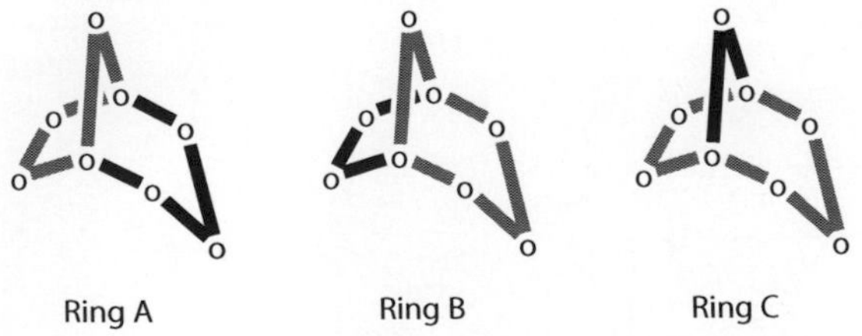
Ring A
Ring B
Ring C

R. Robinson (1917) plan for tropinone: step 1.

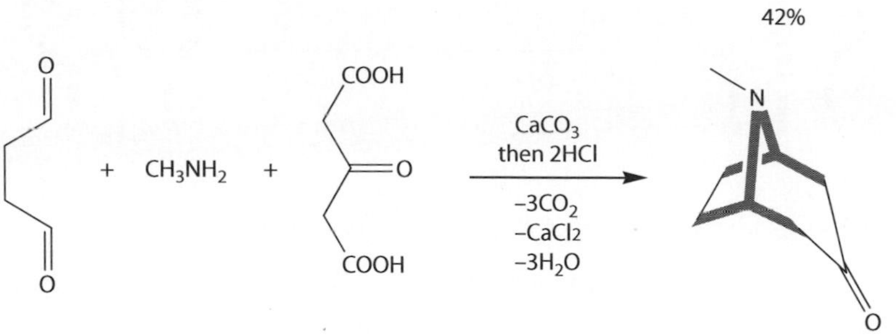

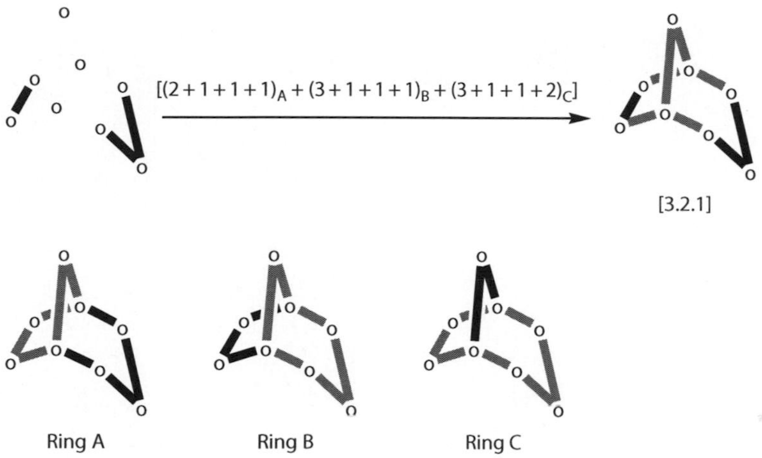

Ring A   Ring B   Ring C

Boger (2005) plan for (−)-vindoline: step 14.

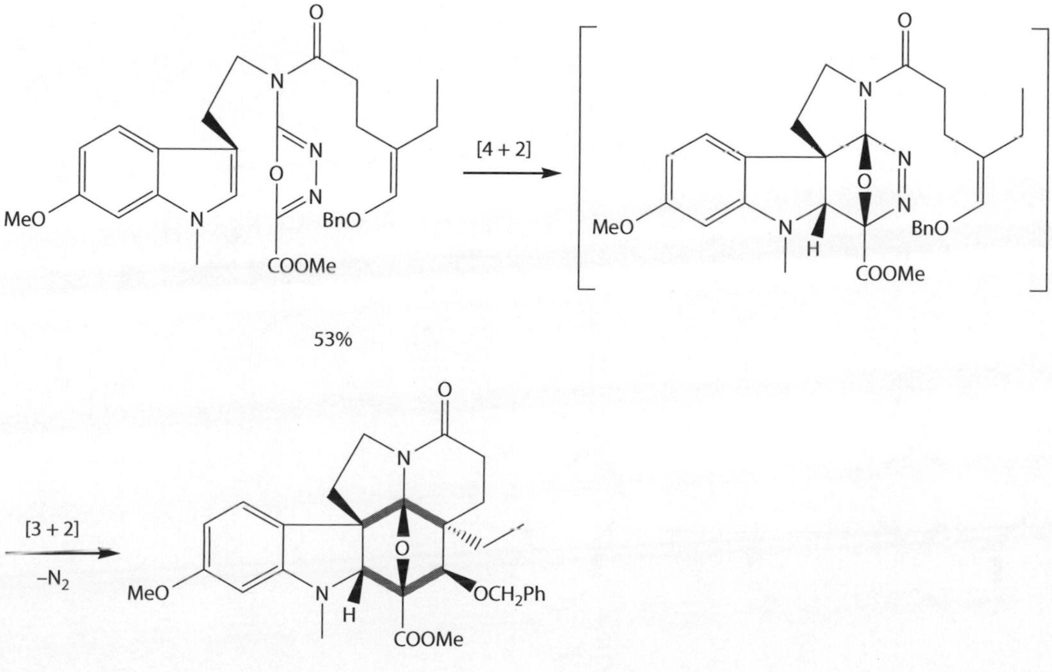

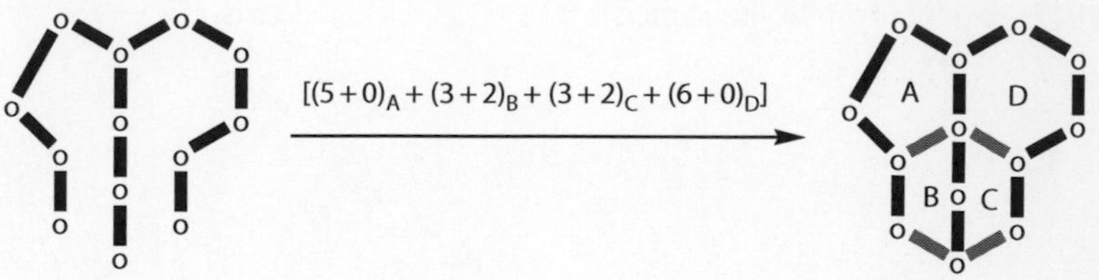

$$[(5+0)_A + (3+2)_B + (3+2)_C + (6+0)_D]$$

Note that the [1,3,4]-oxadiazole ring is sacrificial in this transformation.

Fukuyama G1 (2000) plan for (–)-vindoline: step 16 (six- and nine-membered rings are made simultaneously)

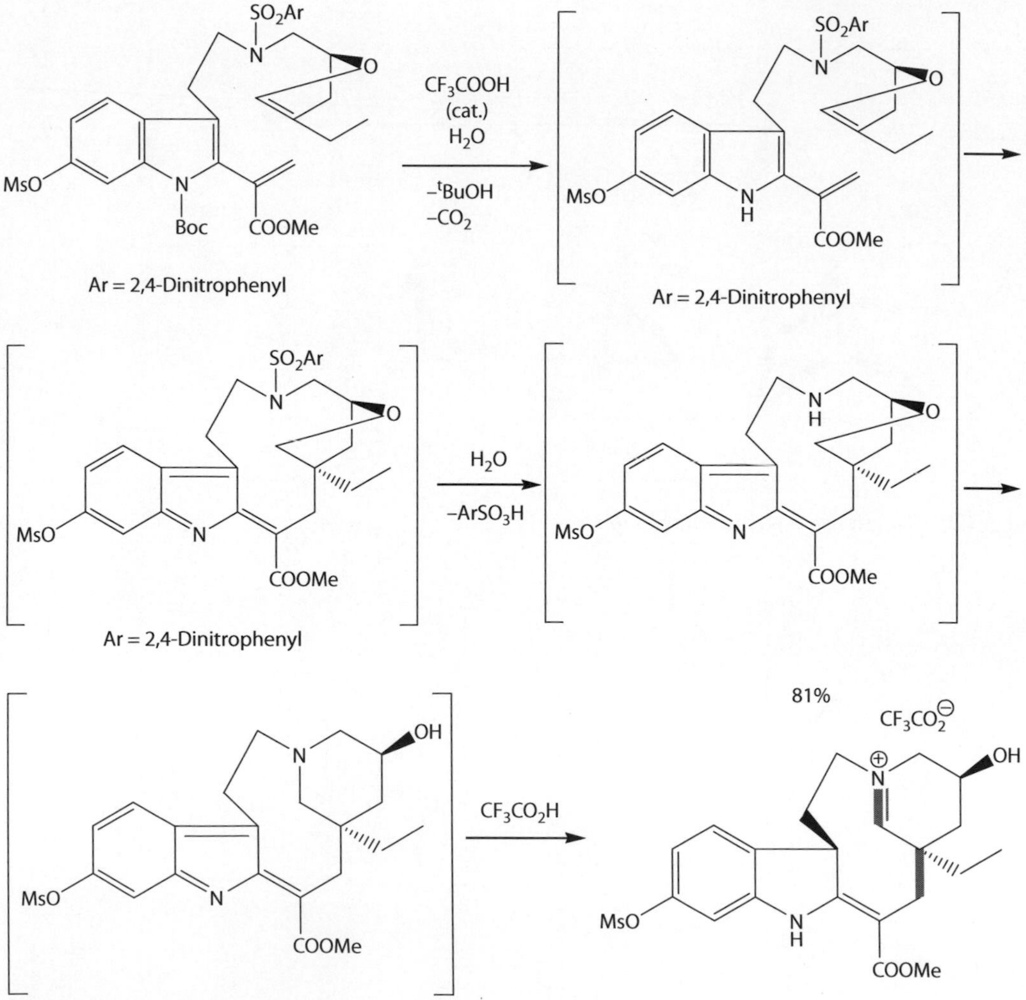

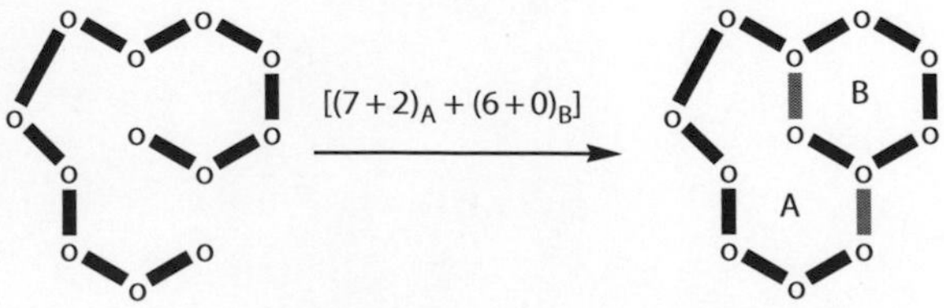

Ar = 2,4-Dinitrophenyl

CF₃COOH (cat.)
H₂O
–ᵗBuOH
–CO₂

Ar = 2,4-Dinitrophenyl

Ar = 2,4-Dinitrophenyl

H₂O
–ArSO₃H

81%

CF₃CO₂H

$$[(7+2)_A + (6+0)_B]$$

[6.3.1]

Fukuyama G2 (2002) plan for (–)-vindoline: step 15 (as mentioned earlier)
Woggon (2008) plan for vitamin E: step 5.

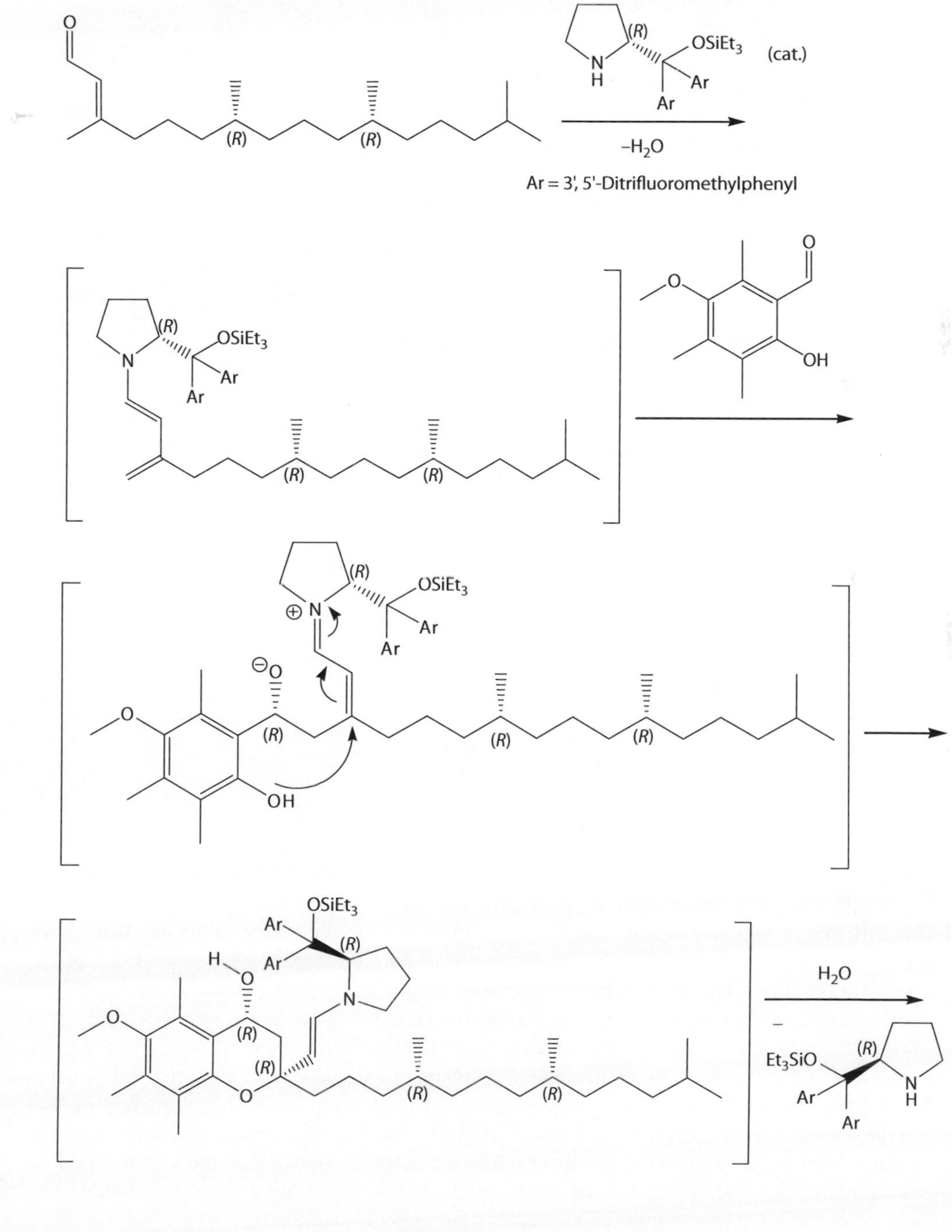

Ar = 3', 5'-Ditrifluoromethylphenyl

58%

$[(4+2)_A + (4+2)_B]$

[3.3.1]

10.9 PROPOSAL FOR A COMPUTER-SEARCHABLE RING CONSTRUCTION DATABASE

Suppose a chemist wishes to propose a new synthesis for quinine that was different strategically from prior published plans. The goal is to come up with a plan that is different from past published plans that demonstrates novelty in building up the ring system in a different way than earlier published plans, and hopefully that this new methodology translates into a more material efficient plan.

In order to accomplish this efficiently and to guarantee the novelty criterion, a searchable ring construction database would be required. For a given published synthesis plan to a particular target molecule this would involve entering its target bond map that highlights the target bonds made in building up the ring system. For the purpose of this discussion these special bonds are highlighted in red. In addition, the aspect of all structure diagrams and target bond maps must be preserved for easy comparisons. So, for the quinine example there are nine published plans for this target structure. Figure 10.1 shows the respective target bond maps as described earlier.

A chemist would carry out a new retrosynthetic analysis of quinine showing which bonds are targeted for disconnection. The goal is obviously to have a different set of bonds than those highlighted in red in the aforementioned target bond maps. A pairwise similarity score may be assigned for the set of red bonds in the proposed structure and each of the aforementioned structures corresponding to literature plans. Strategic novelty is characterized by a low similarity score. Such a similarity score may be described by the Tanimoto index [1], T, given by Equation 10.1.

$$T = \frac{R(AB)}{R(A) + R(B) - R(AB)} \tag{10.1}$$

where

R(AB) is the number of common gray scale bonds found in both structures A and B

R(A) is the number of gray scale bonds in structure A

R(B) is the number of gray scale bonds in structure B

If $T = 1$, then both structures A and B highlight the same set of gray scale bonds. If $T = 0$, then there are no common gray scale bonds found in structures A and B. Table 10.1 summarizes the Tanimoto indices for the 28 pairwise comparisons possible for the eight published quinine plans. From this table we find that the Gates and Martin plans have identical sets of target bonds ($T = 1$) and the Krische—Jacobsen pair are the most dissimilar ($T = 0.240$). A second criterion to distinguish strategies is the target bond profile as a function of reaction stage. This is represented as a bar graph showing the number of target bonds in each of the reaction stages over the course of the synthesis plan. This diagram can distinguish between two plans that have identical Tanimoto similarity indices. Even though the same set of target bonds appears in both structures, the order in which they are made may be different. This difference will be reflected in the bar graph. The target bond profiles for the nine published quinine plans appear in Figure 10.2. These profiles would also accompany the target bond maps shown earlier in the strategy database. Note that profiles for the Gates and Martin plans are slightly different even though both have $T = 1$ values. The Martin plan is two steps shorter and has more target bonds made in the final stages. We may also infer from these profiles that the Martin, Gates, Merck, and Aggarwal plans have a high proportion of sacrificial reactions as shown by the significant gaps where no target bonds are made. In contrast, the Stork, Kobayashi, Jacobsen, and Krische plans have a high proportion of target bond forming reactions.

Thirdly, it is possible to employ the graphical algebraic code and diagrams introduced in Section 10.1 to describe various strategies used in building up the quinuclidine ring system. These are summarized in the following and Figure 10.3 shows the key transformations in graphical form.

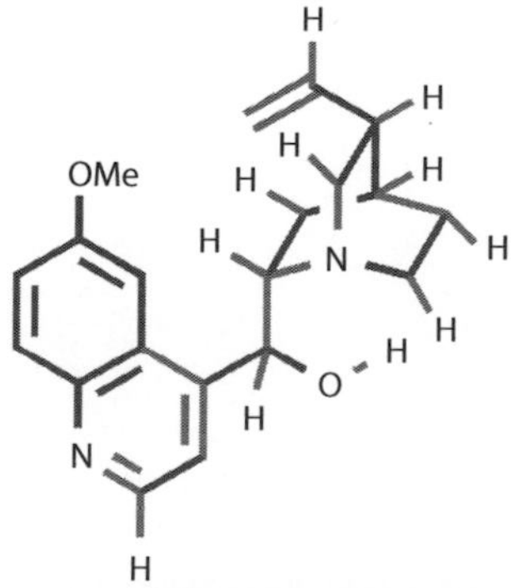

Woodward–Doering–Rabe (1948)

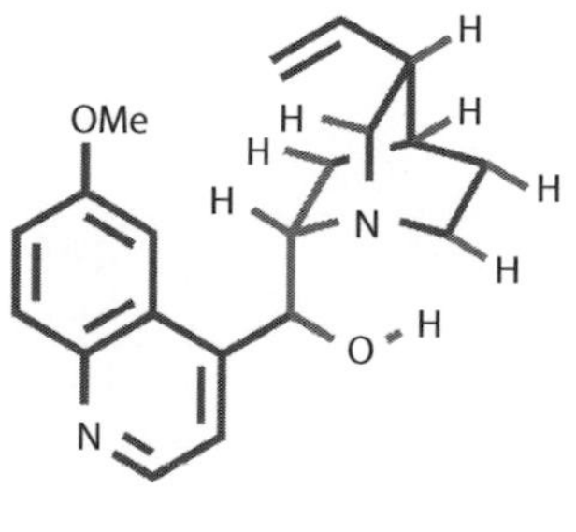

Gates (1970)

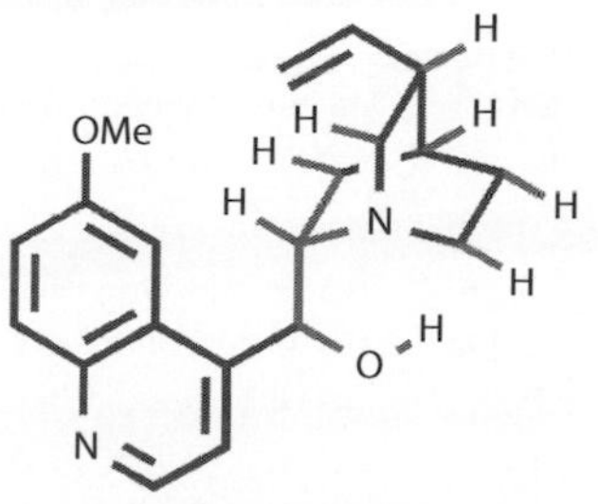

Merck (1970)

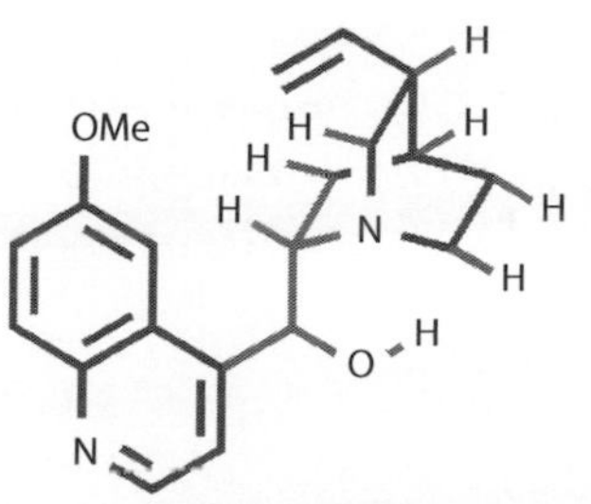

Martin (1972)

FIGURE 10.1 Target bond maps for various synthesis plans for quinine.

(continued)

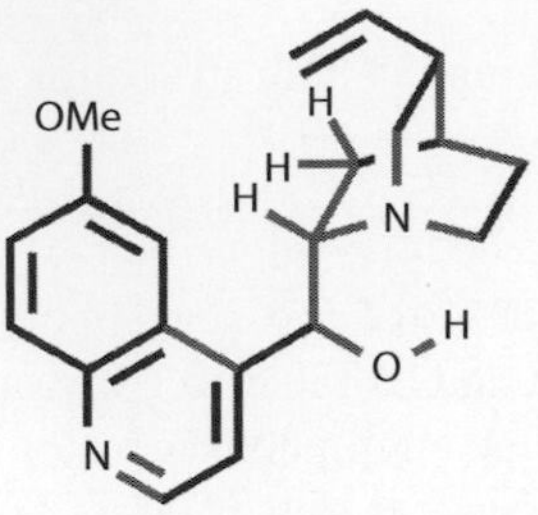

Stork (2001) 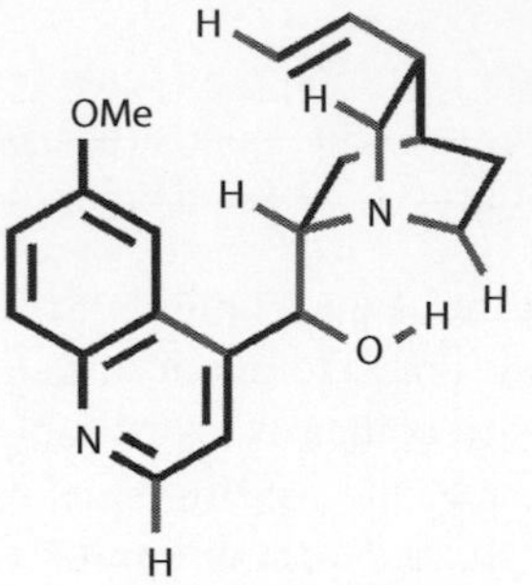   Kobayashi (2004)

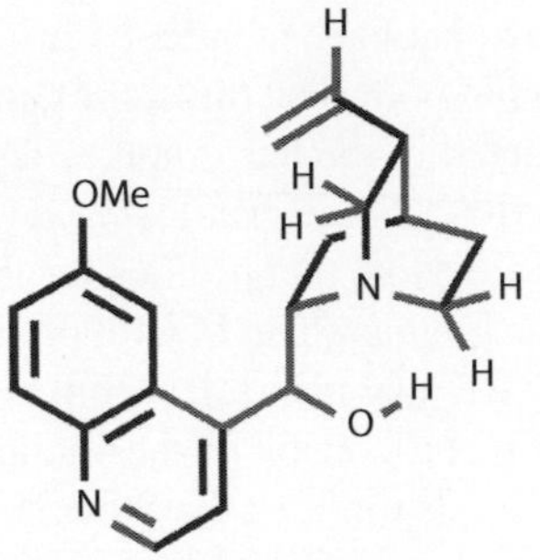

Jacobsen (2004) 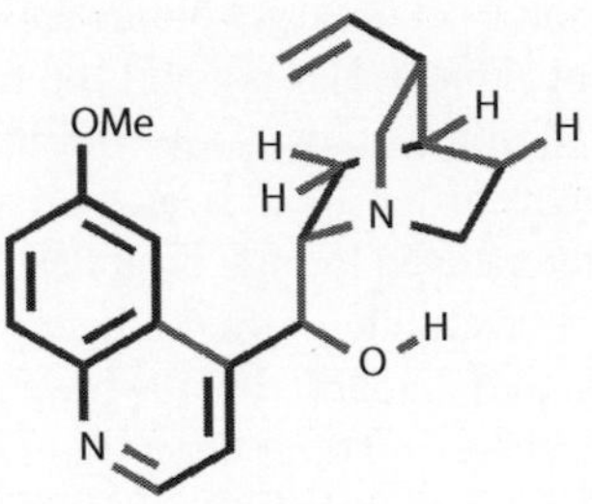   Krische (2008)

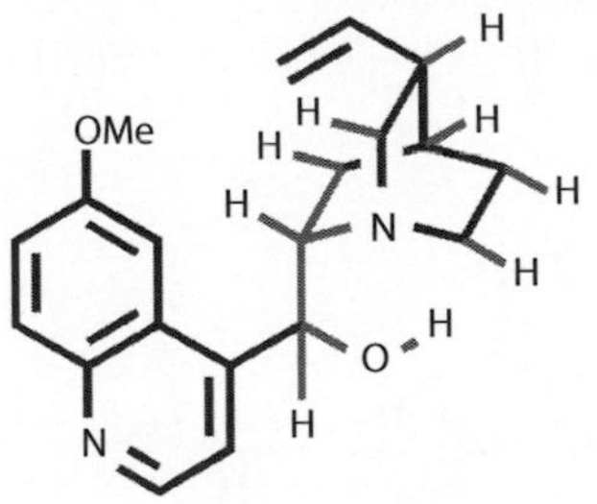

Aggarwal (2010)

FIGURE 10.1 (continued)

TABLE 10.1
Summary of Tanimoto Similarity Indices for Nine Quinine Plans

Plan	Aggarwal	Gates	Jacobsen	Kobayashi	Krische	Martin	Merck	Stork	Woodward–Doering–Rabe
Aggarwal		0.923	0.261	0.350	0.333	0.857	0.923	0.350	0.524
Gates			0.273	0.368	0.350	0.923	1.000	0.368	0.476
Jacobsen				0.364	0.240	0.318	0.273	0.429	0.346
Kobayashi					0.208	0.350	0.368	0.400	0.269
Krische						0.333	0.350	0.450	0.259
Martin							0.923	0.350	0.455
Merck								0.368	0.476
Stork									0.269
Woodward–Doering–Rabe									

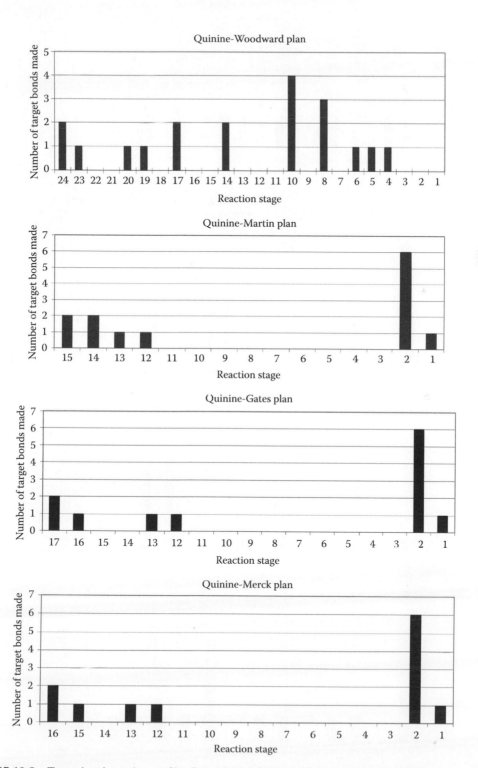

FIGURE 10.2 Target bond reaction profiles for various published quinine total synthesis plans.

(continued)

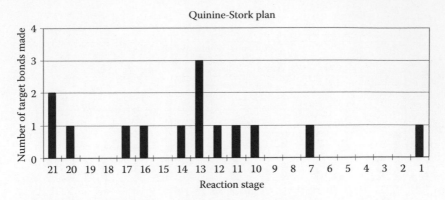

Quinine-Stork plan

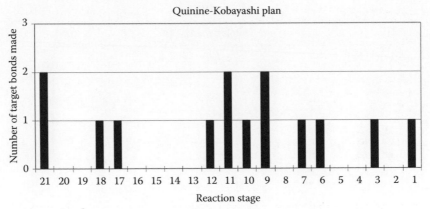

Quinine-Kobayashi plan

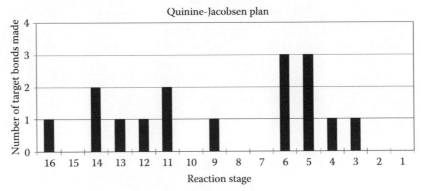

Quinine-Jacobsen plan

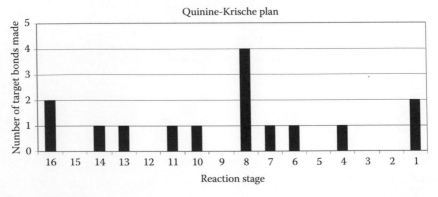

Quinine-Krische plan

FIGURE 10.2 (continued)

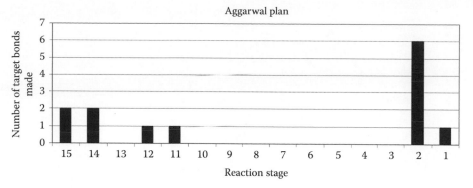

FIGURE 10.2 (continued)

FIGURE 10.3 Graphs showing the ring construction strategies employed in building up the quinuclidine ring moiety in quinine.

(*continued*)

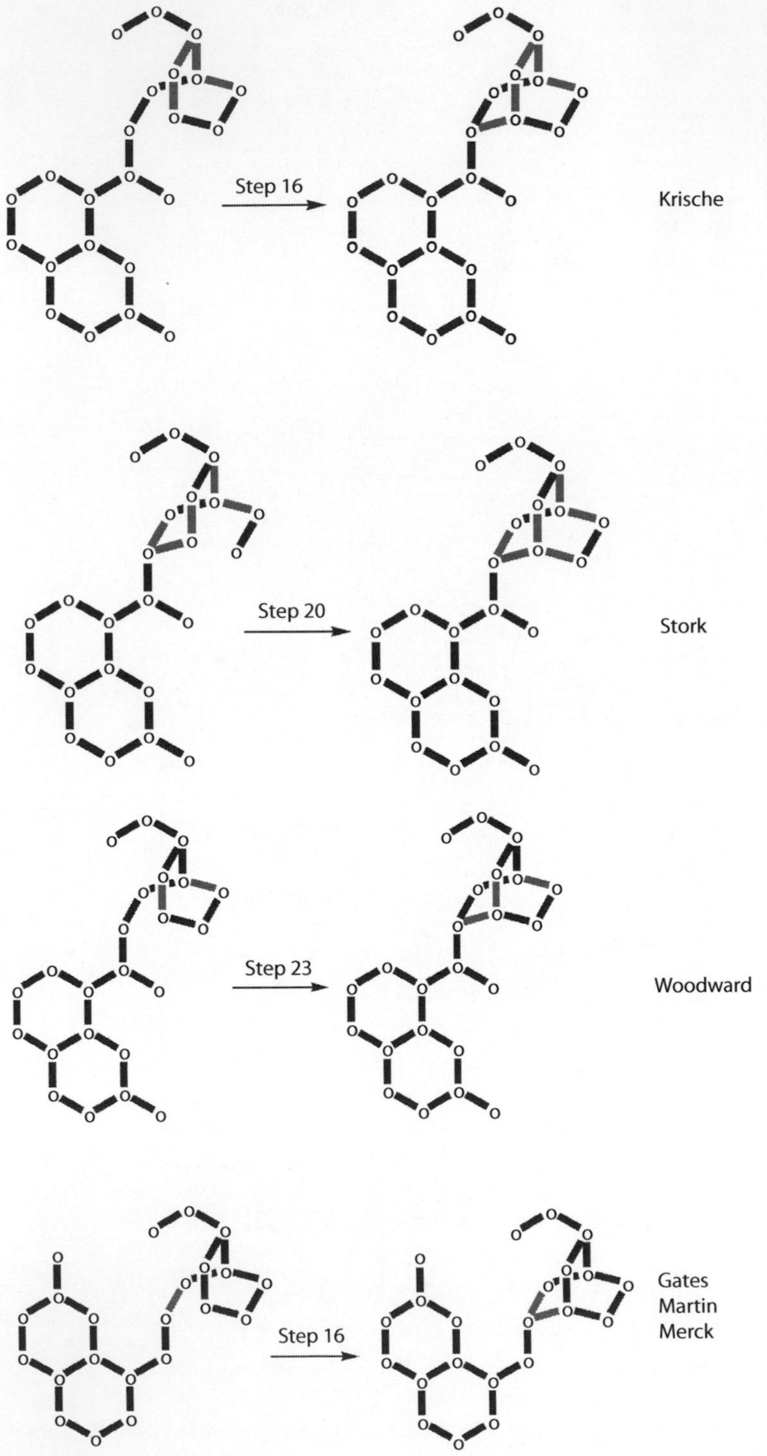
Step 16
Krische
Step 20
Stork
Step 23
Woodward
Step 16
Gates
Martin
Merck

FIGURE 10.3 (continued)

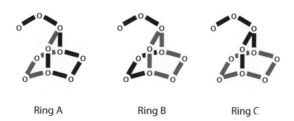

FIGURE 10.3 (continued)

Strategy $\xrightarrow{[5+1]_B}$ $[6]_{11}$ $\xrightarrow{[(6+0)_A+(6+0)_C]}$ $[2.2.2]_{21}$: Kobayashi

Strategy $\xrightarrow{[6+0]_B}$ $[6]_5$ $\xrightarrow{[(6+0)_A+(6+0)_C]}$ $[2.2.2]_{16}$: Jacobsen

Strategy $\xrightarrow{[6+0]_B}$ $[6]_7$ $\xrightarrow{[(6+0)_A+(6+0)_C]}$ $[2.2.2]_{16}$: Krische

Strategy $\xrightarrow{[6+0]_B}$ $[6]_{16}$ $\xrightarrow{[(6+0)_A+(6+0)_C]}$ $[2.2.2]_{20}$: Stork

Strategy $\xrightarrow{[6+0]_B}$ $[6]_5$ $\xrightarrow{[(6+0)_A+(6+0)_C]}$ $[2.2.2]_{23}$: Woodward

Strategy B_0 $\xrightarrow{[(6+0)_A+(6+0)_C]}$ $[2.2.2]_{16}$: Gates

Strategy B_0 $\xrightarrow{[(6+0)_A+(6+0)_C]}$ $[2.2.2]_{15}$: Merck

Strategy B_0 $\xrightarrow{[(6+0)_A+(6+0)_C]}$ $[2.2.2]_{14}$: Martin

Strategy B_0 $\xrightarrow{[(6+0)_A+(6+0)_C]}$ $[2.2.2]_{15}$: Aggarwal

Ring A Ring B Ring C

Rings A and C in all plans are made by a [6 + 0] cycloaddition as the last step in making the [2.2.2] ring system. The Kobayashi plan makes ring B via a [5 + 1] cycloaddition, whereas the Jacobsen, Krische, Stork, and Woodward plans make it via a [6 + 0] cycloaddition. The Gates, Merck, Martin, and Aggarwal plans begin with precursors that have this ring preformed. The Aggarwal plan is unique in linking the quinoline ring moiety to the quinuclidine ring precursor via a sacrificial epoxidation in step 14 and then making the quinuclidine ring and opening up the epoxide ring in step 15.

REFERENCE

Rogers, D.J.; Tanimoto, T.T. *Science* 1960, *132*, 1115.

11 Example Highlights from Database

11.1 WHICH PLAN TO CHOOSE FOR SCALE-UP?

The example chosen for this is the synthesis of the polyketide natural product (+)-discodermolide discovered in the marine sponge *Discodermia dissoluta*. Since the compound showed promise as an anticancer agent, it was required to develop a scaled-up total synthesis to provide enough quantities of this material for clinical trials. The natural source amount is 7 mg from 434 g of frozen sponge or 0.002% w/w [1]. With a molecular weight of 593 g/mol, this works out to an *E*-factor of 62,000. This is a minimum estimate since it does not include the solvent demand required in the extraction process. In 2000, there were two total syntheses reported in the literature, which we refer to in this compilation as the Smith G2 [1,2] and Paterson G1 [3,4] plans. The Tanimoto similarity index for the two plans with respect to the target bonds selected in the plans is 42%. Novartis [5–8] decided to choose the Smith G2 plan for scale-up and not surprisingly both plans had a 93% Tanimoto similarity index. By contrast, the Paterson G1-Novartis pair has a 58% similarity in the target bonds made. The data presented in Table 11.1 show that this was not a wise choice since even at the level of kernel metrics (reaction yield and atom economy) the Paterson G1 plan produced six times less waste than the Smith G2 plan. The decision made by Novartis was done without the benefit of a thorough metrics evaluation of the synthesis performance. Moreover, probably because of scale-up issues, the Novartis plan turned out to produce 1.5 times more waste than the Smith G2 plan. Since that time, Smith produced two more synthesis generations with the fourth announced in 2005 [9] producing 79.1 kg waste per mole, a 4.5-fold improvement over the Smith G2 plan. In 2004, Paterson also improved his strategy to 44.3 kg waste per mole in Paterson G3 [10] and this represents so far the most material efficient plan at the kernel metrics level. This example illustrates to process chemists contemplating a scaled-up plan for a given target from a set of published procedures, that a well thought out and executed metrics analysis can go a long way in making good decisions about which plans to choose. The automatic consequence of selecting and carrying out material efficient and cost effective plans is the achievement of greener syntheses.

11.2 SPARSE TARGET BOND MAKING PROFILES

The target bond making profile can be a very informative visual aid in showing how strategically efficient a synthesis plan is. If every reaction stage produces a target bond, then we may conclude that all of the waste produced arises from target bond reactions. If on the other hand, there are more empty gaps in the profile than bars, then these gaps are directly correlated with sacrificial reactions that do not produce target bonds. A synthesis plan producing the majority of its waste from sacrificial reactions is a poorly strategized plan. In this section, we highlight outstanding examples of such disastrous plans. For each case, the corresponding target bond profile is given as well as the target bond map (Figures 11.1 through 11.5). The reaction schemes are given in Figures 11.1c through 11.5c where the target bond forming steps are color coded in gray scale.

TABLE 11.1

Comparison of Kernel Metrics for Plans to Discodermolide

Plan	Number of Steps	%AE	E-Kernel	Kernel Mass of Waste (kg/mol)
Paterson G1 (2000)	47	4.3	106.9	63.4
Smith G2 (2000)	45	4.4	605.1	358.8
Novartis (2004)	48	4.2	878.4	520.9

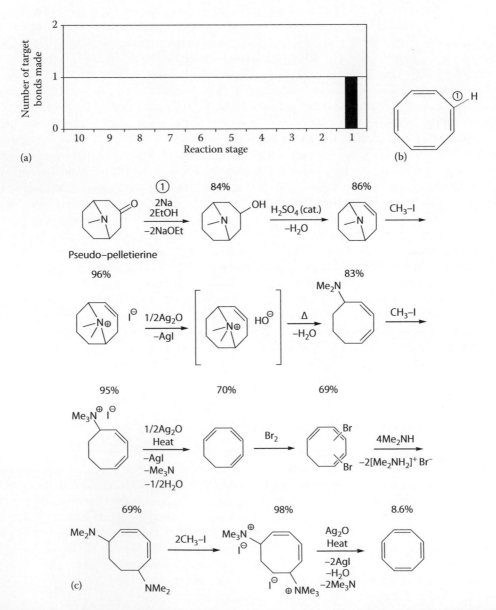

FIGURE 11.1 (a) Target bond profile for Willstätter (1911, 1913) plan for cyclooctatetraene. (b) Target bond map for Willstätter (1911, 1913) plan for cyclooctatetraene. (c) Scheme for Willstätter (1911, 1913) plan for cyclooctatetraene.

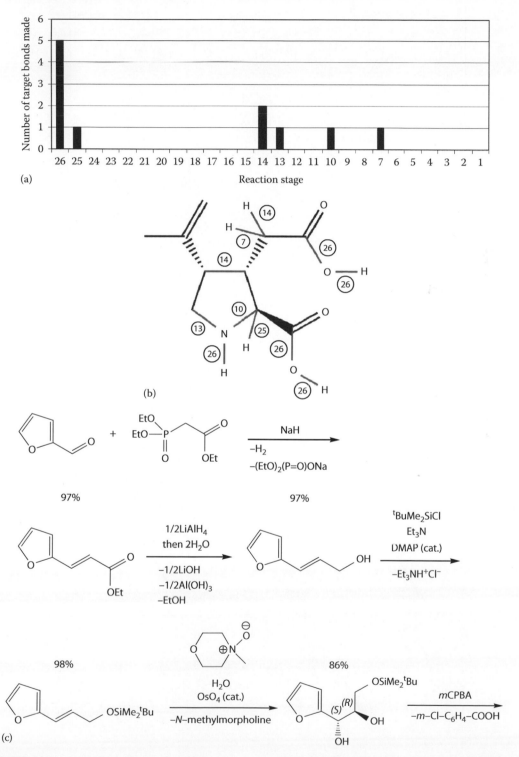

FIGURE 11.2 (a) Target bond profile for Ogasawara G3 (2001) plan for (−)-kainic acid. (b) Target bond map for Ogasawara G3 (2001) plan for (−)-kainic acid. (c) Scheme for Ogasawara G3 (2001) plan for (−)-kainic acid.

(continued)

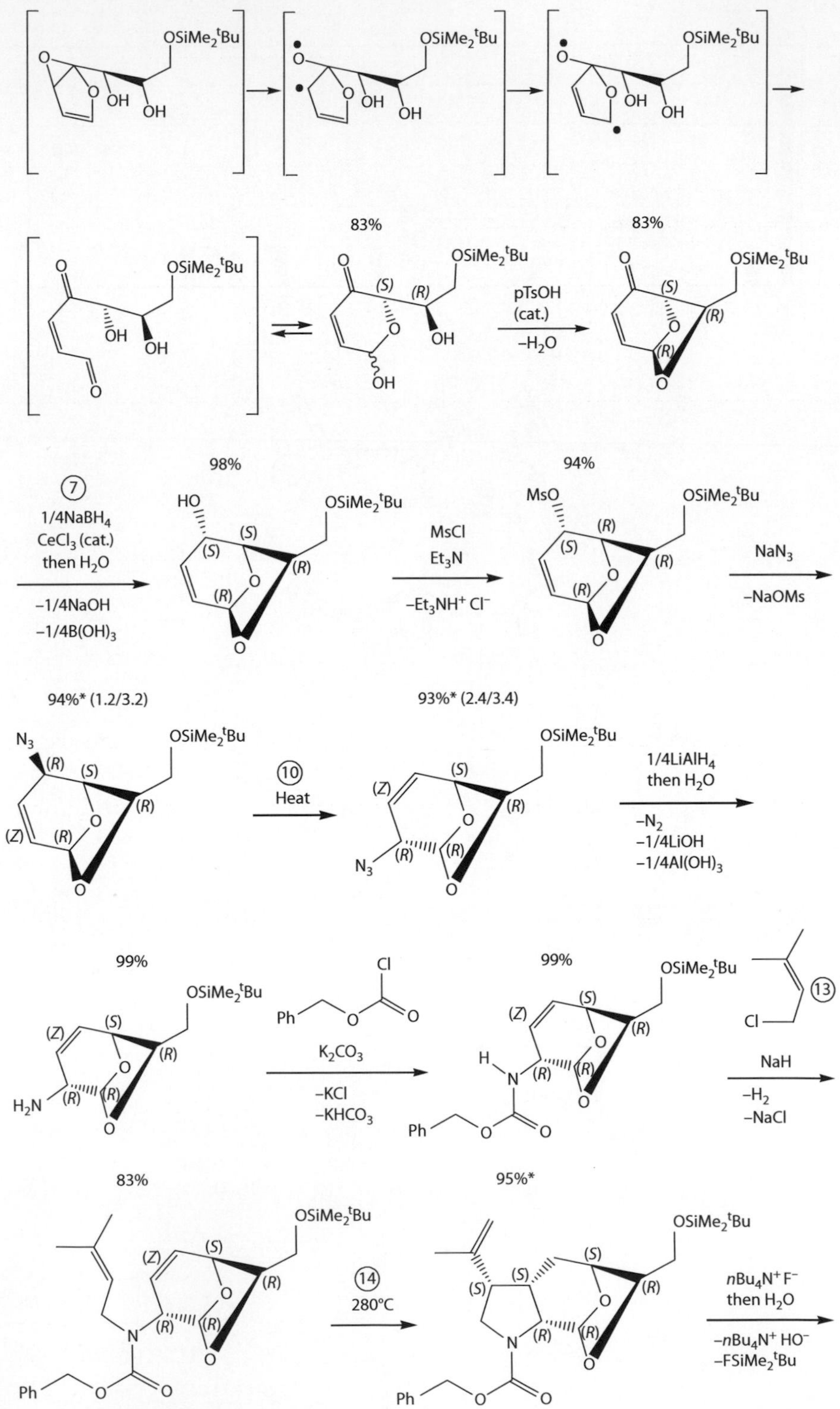

FIGURE 11.2c (continued)

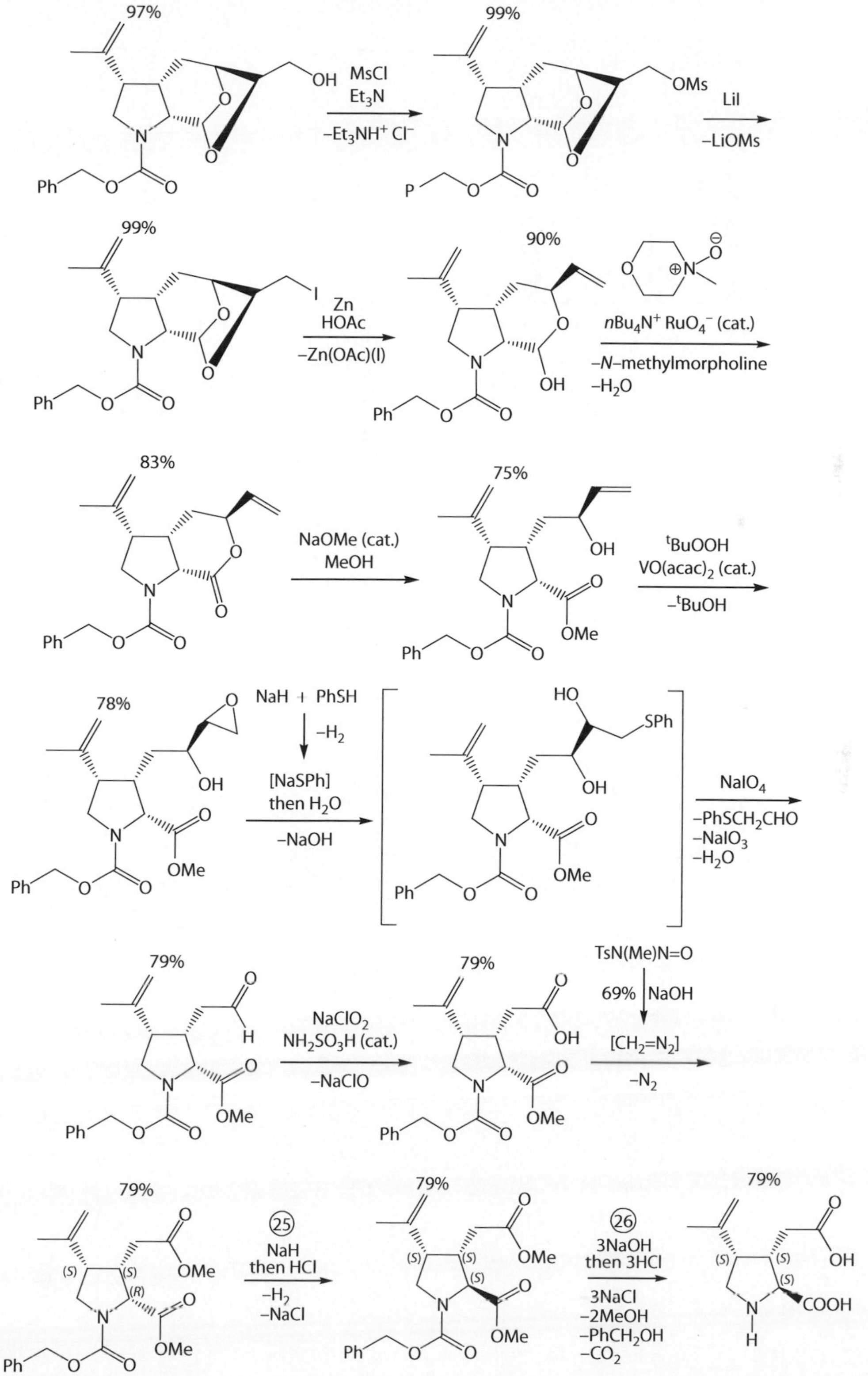

FIGURE 11.2c (continued)

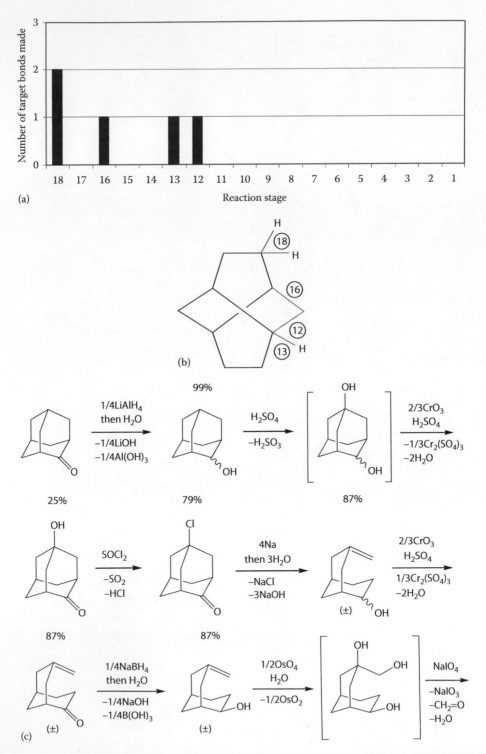

FIGURE 11.3 (a) Target bond profile for Hamon–Young (1976) plan for (±)-twistane. (b) Target bond map for Hamon–Young (1976) plan for (±)-twistane. (c) Scheme for Hamon–Young (1976) plan for (±)-twistane.

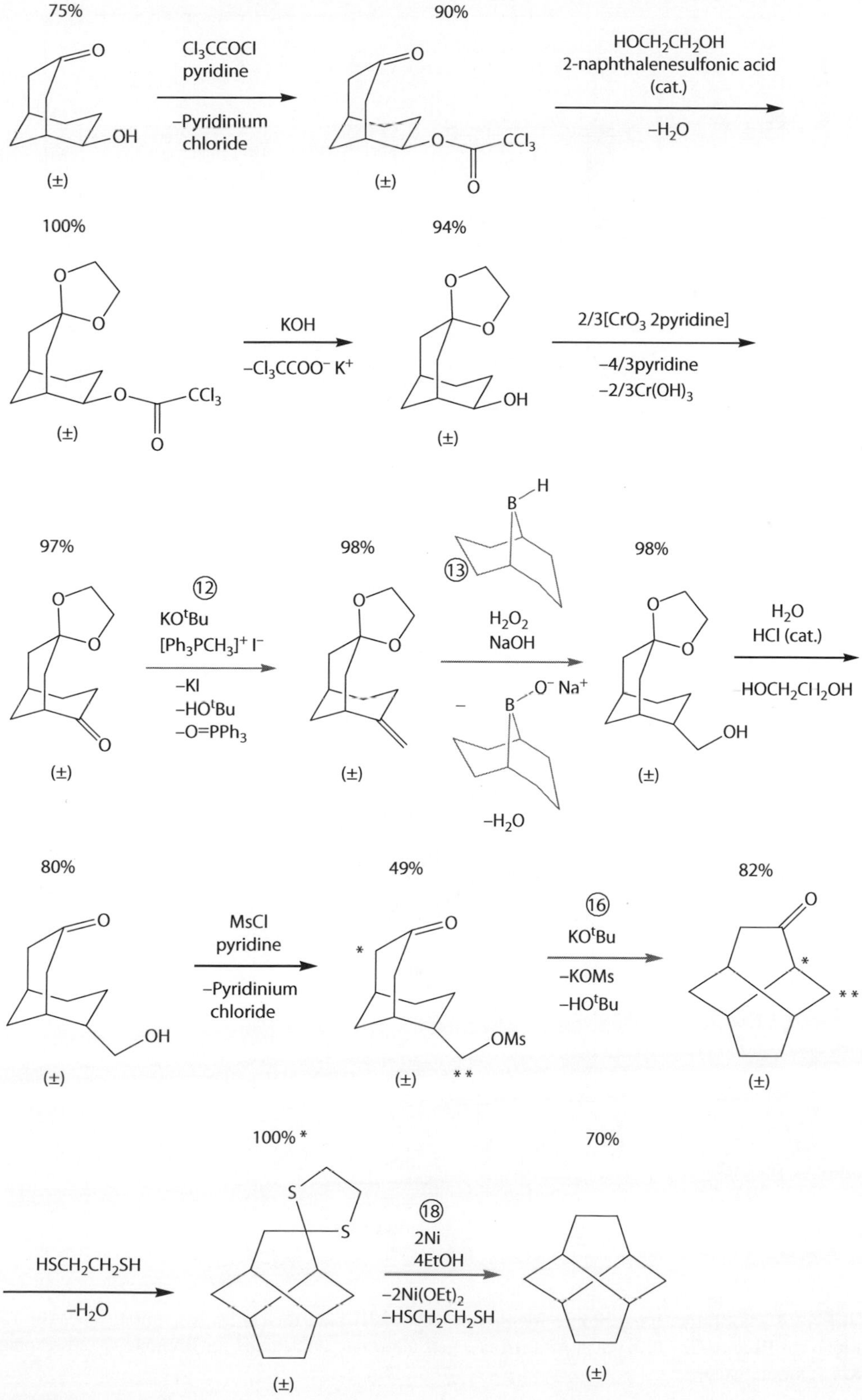

FIGURE 11.3c (continued)

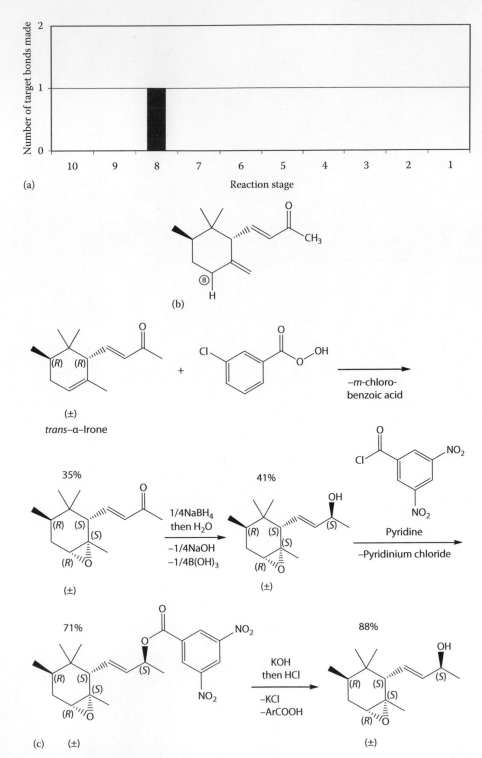

FIGURE 11.4 (a) Target bond profile for Brenna G2 (2001) plan for (+)-*trans*-gamma-irone. (b) Target bond map for Brenna G2 (2001) plan for (+)-*trans*-gamma-irone. (c) Scheme for Brenna G2 (2001) plan for (+)-*trans*-gamma-irone.

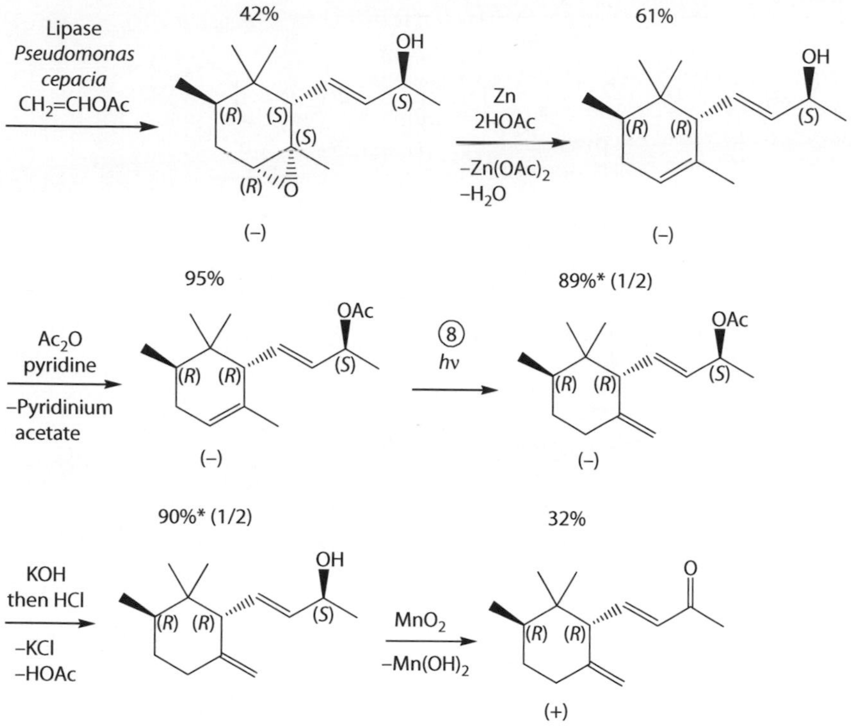

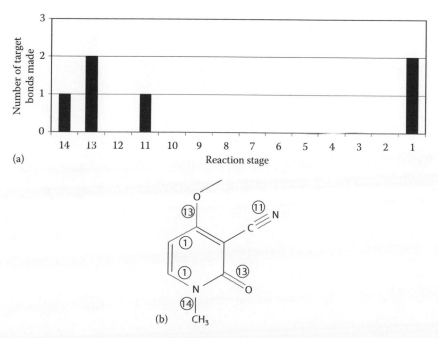

FIGURE 11.4c (continued)

FIGURE 11.5 (a) Target bond profile for Späth (1923) plan for ricinine. (b) Target bond map for Späth (1923) plan for ricinine.

(*continued*)

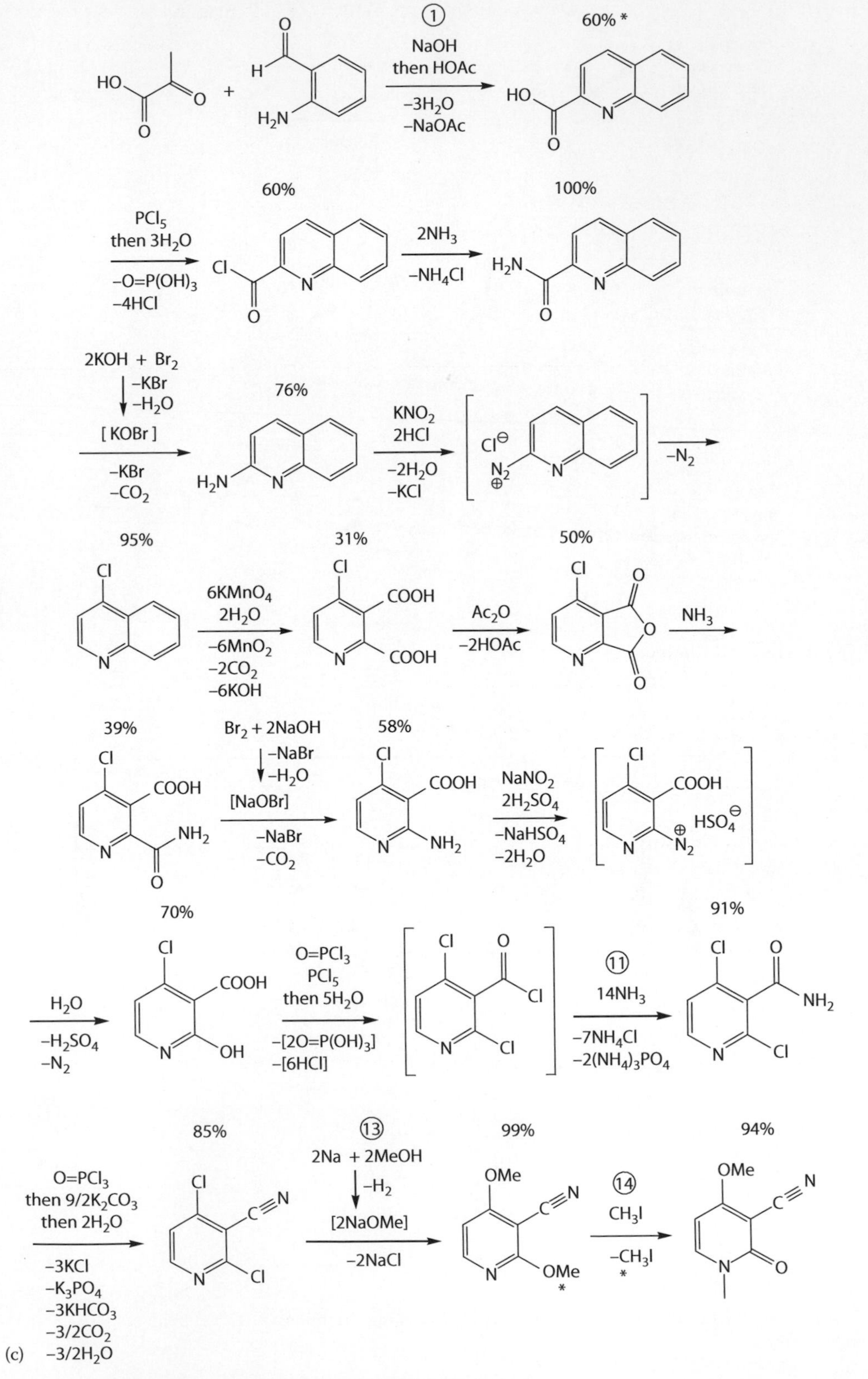

FIGURE 11.5 (continued)   (c) Scheme for Späth (1923) plan for ricinine.

a. Willstätter [11–13] plan for cyclooctatetraene: 9/10 = 90% of reaction stages produce no target bonds; 1/10 = 10% of reactions produce target bonds; 19.6% of the mass of total kernel waste produced arises from target bond forming reactions; 80.4% of the mass of total kernel waste produced arises from sacrificial reactions. This plan is in fact not a synthesis at all–it is a degradation of the natural product pseudo-pelletierine.

b. Ogasawara G3 [14] plan for (–)-kainic acid: 20/26 = 77% of reaction stages produce no target bonds; 11/27 = 41% of reactions produce target bonds; 5% of the mass of total kernel waste produced arises from target bond forming reactions; 95% of the mass of total kernel waste produced arises from sacrificial reactions.

c. Hamon and Young [15] plan for (±)-twistane: 14/18 = 78% of reaction stages produce no target bonds; 5/18 = 28% of reaction produce target bonds; 10% of the mass of total kernel waste produced arises from target bond forming reactions; 90% of the mass of total kernel waste produced arises from sacrificial reactions.

d. Brenna G2 [16] plan for (+)-*trans*-gamma-irone: 9/10 = 90% of reaction stages produce no target bonds; 1/10 = 10% of reaction produce target bonds; 0.7% of the mass of total kernel waste produced arises from target bond forming reactions; 99.3% of the mass of total kernel waste produced arises from sacrificial reactions.

e. Späth [17] plan for ricinine: 10/14 = 67% of reaction stages produce no target bonds; 6/14 = 43% of reaction produce target bonds; 23% of the mass of total kernel waste produced arises from target bond forming reactions; 77% of the mass of total kernel waste produced arises from sacrificial reactions.

11.3 RECYCLING OPTIONS

In this section, we present analyses of the impact of recycling by-products back to reagents in the case of the Wittig reaction [18] and recycling stereoisomeric side products produced in the syntheses of (–)-epibatidine [19] and penicillin V [20–22].

The scheme for the Wittig reaction producing methylenecyclohexane as the target product and triphenylphosphine oxide as the by-product is shown in Figure 11.6.

Using a basis scale of 1 mol of methylenecyclohexane target product produced this two-step sequence would result in the production of 0.096 kg methylenecyclohexane and 0.278 kg triphenylphosphine oxide. The LINEAR-COMPLETE spreadsheet algorithm calculates, using the experimental data provided in the *Organic Syntheses* procedure, that this scheme produces 14.90 kg total waste per mole methylenecyclohexane (includes the 0.278 kg of triphenylphosphine oxide by-product) and has the following E-factor breakdown: E-kernel = 12.61; E-excess = 1.27; E-auxiliaries = 141.37; E-total = 155.24. Figure 11.7 shows the accompanying radial pentagons for the two steps.

If we change our point of view and say that triphenylphosphine oxide is the target product and methylenecyclohexane is a by-product, then the corresponding scheme would be written as shown in Figure 11.8.

Again, using a basis scale of 1 mol of triphenylphosphine oxide target product produced this two-step sequence would result in the production of 0.096 kg methylenecyclohexane and 0.278 kg

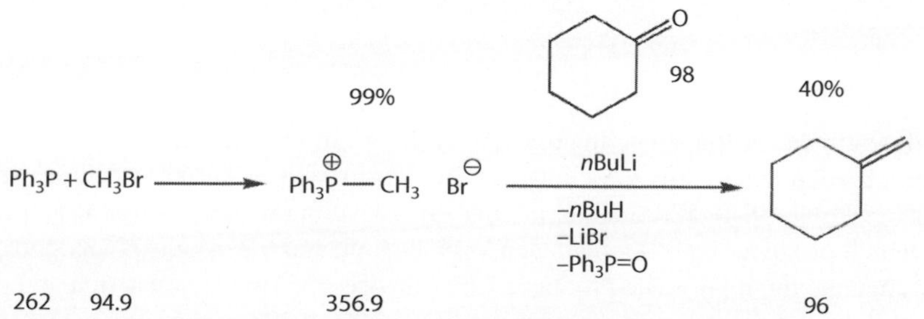

FIGURE 11.6 Wittig reaction scheme with methylenecyclohexane as target product.

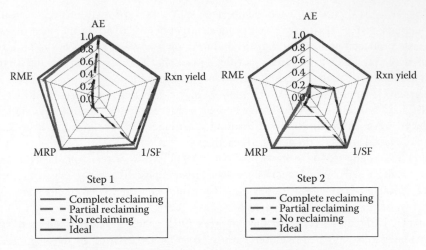

FIGURE 11.7   Radial pentagons for the two-step sequence shown in Figure 11.6.

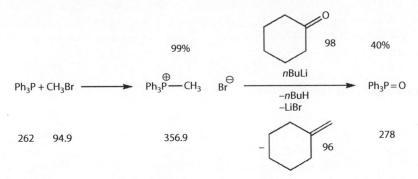

FIGURE 11.8   Wittig reaction scheme with triphenylphosphine oxide as target product.

triphenylphosphine oxide as before. The LINEAR-COMPLETE spreadsheet algorithm calculates, using the experimental data provided in the *Organic Syntheses* procedure, that this scheme produces 14.70 kg total waste per mole triphenylphosphine oxide (includes the 0.096 kg of methylenecyclohexane by-product) and has the following E-factor breakdown: E-kernel = 3.70; E-excess = 0.44; E-auxiliaries = 48.82; E-total = 52.96. Figure 11.9 shows the accompanying radial pentagons for the two steps.

Suppose now that we wish to recycle the triphenylphosphine oxide produced in the two-step sequence back into triphenylphosphine. Clearly, this will involve a redox type reaction to carry out this transformation. Such a reaction has been developed by BASF [23] as shown in the following.

$$
\underset{278}{Ph_3P=O} \quad \xrightarrow[-CO_2]{Cl_2C=O} \quad [Ph_3PCl_2] \quad \xrightarrow[-2/3AlCl_3]{2/3Al} \quad \underset{262}{\overset{96\%}{Ph_3P}}
$$

The radial pentagon for this recycling reaction is shown in Figure 11.10.

The E-factor breakdown for this recycling reaction is as follows: E-kernel = 0.57; E-excess = 0.046; E-auxiliaries = 7.74; and E-total = 8.35. If we link this recycling reaction with the scheme shown in Figure 11.6, which produces 0.096 kg methylenecyclohexane and 0.278 kg triphenylphosphine oxide, then we can calculate the total waste produced from the original two-step reaction and the recycling reaction. From the first two-step sequence, we produce 14.90−0.278 = 14.622 kg net waste. If we take the recovered 0.278 kg of triphenylphosphine oxide and subject it to the recycling reaction we will

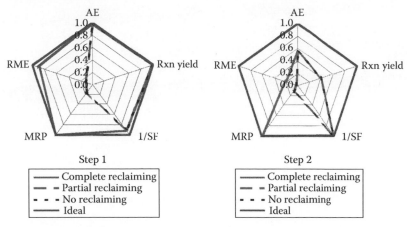

FIGURE 11.9   Radial pentagons for the two-step sequence shown in Figure 11.8.

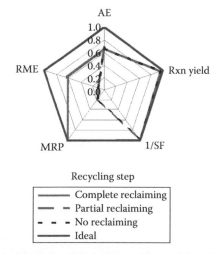

FIGURE 11.10   Radial pentagon for triphenylphosphine oxide to triphenylphosphine recycling reaction.

produce 8.35 kg waste per kg recovered triphenylphosphine. The radial pentagon analysis shows that 23.21 kg triphenylphosphine oxide produce 21 kg of triphenylphosphine and 175.4 kg waste. Note that 175.4/21 = 8.35, which corresponds to E-total for this reaction. So, assuming no change in performance upon change of reaction scale, 0.278 kg triphenylphosphine oxide would produce 0.278 × (21/23.21) = 0.252 kg triphenylphosphine and 0.278 × (175.4/23.21) = 2.10 kg of additional waste. Hence, the option of including a recycling reaction yields 0.096 kg of target methylenecyclohexane and 14.622 + 2.10 = 16.722 kg of total waste, which may be compared with 14.90 kg waste if we chose not to recycle triphenylphosphine oxide and kept only methylenecyclohexane. In order to produce 0.096 kg (1 mol) methylenecyclohexane, the synthesis tree diagram for the two-step sequence indicates that 262/(0.40 × 0.99) = 0.662 kg of triphenylphosphine is required. Applying the recycling reaction means that (0.252/0.662) × 100 = 38% of this reagent is recovered. Most of the triphenylphosphine is lost as unreacted Wittig reagent (methyl triphenylphosphonium bromide) in step 2. If the reaction yield of step 2 had been 90% instead of 40%, then a similar analysis would show that 85% of triphenylphosphine could be recovered by applying the recycling reaction. A 100% yield for the Wittig reaction would translate to a 95% maximum recovery of triphenylphosphine. These reaction yield improvements translate to E-total values of 68.44 and 61.60, respectively (cf. 155.24). Alternatively, the corresponding values for total mass of waste per mol methylenecyclohexane are 6.6 and 5.9 kg, respectively (cf. 14.90 kg).

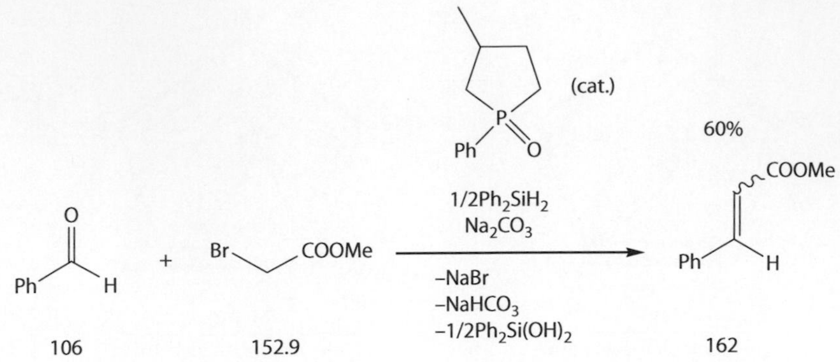

FIGURE 11.11   Catalytic Wittig reaction applied to the synthesis of methyl cinnamate.

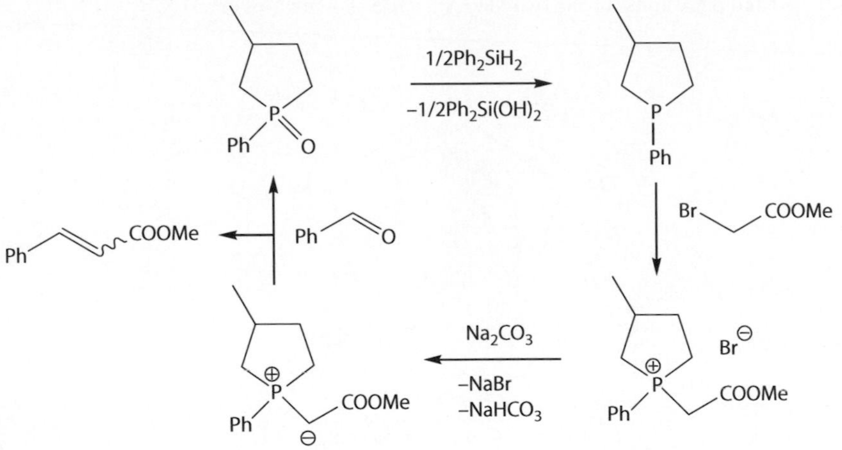

FIGURE 11.12   Mechanism of catalytic Wittig reaction.

A catalytic version of the Wittig reaction was disclosed in 2009 [24] in which 3-methyl-1-phenylphospholane-1-oxide functions as a precatalyst. The active catalyst, 3-methyl-1-phenylphospholane, is produced via reduction with diphenylsilane. Figure 11.11 shows a scheme to produce methyl cinnamate in this way and Figure 11.12 shows the proposed mechanism. This new methodology imparts a number of advantages: (a) the phosphine is not used in stoichiometric amounts; (b) the reaction operates under catalytic conditions and hence no separate recycling reaction is necessary; and (c) the reaction occurs in a single step. The atom economy for this transformation is 35.5%. A radial pentagon analysis reveals the following E-factor breakdown: E-kernel = 3.71; E-excess = 1.84; E-auxiliaries = 3.17; and E-total = 8.73. Under the assumptions that triphenylphosphine is used in the same molar amount in place of diphenylsilane, the same solvent demand is applied, the formation of the Wittig reagent is quantitative, and the yield of methyl cinnamate is unchanged, an analogous calculation for the conventional Wittig protocol reveals that it has an atom economy of 25.8% and an E-factor breakdown of E-kernel = 5.46; E-excess = 0.97; E-auxiliaries = 2.97; and E-total = 9.41. We see that the catalytic method has a 9.7% higher atom economy and 7% lower E-total. When we compare the catalytic method performance against conventional Wittig methods using green technologies such as microwave irradiation [25] and using water as a reaction solvent [26], the catalytic method appears to be quite superior. Figure 11.13 shows the radial pentagons for each of these methods side by side for easy visual comparison along with their E-factor breakdowns.

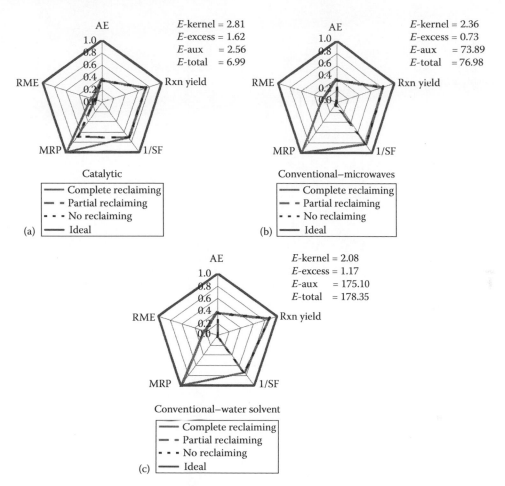

FIGURE 11.13 Radial pentagons for the synthesis of methyl cinnamate under the following conditions: (a) Catalytic Wittig, (b) Conventional Wittig reaction using microwave irradiation, and (c) Conventional Wittig using water as a solvent.

Synthesis tree diagrams facilitate calculations of the impact of recycling unwanted isomers produced in reactions via isomerization operations, for example. The synthesis database reveals two such examples for the synthesis of penicillin V and (−)-epibatidine. Figures 11.14 and 11.15 show abbreviated schemes for each plan indicating the structures of the key isomers produced in each case. The full plans are given in the accompanying download. One mole of (±)-valine produces 0.02492 mol of the alpha-isomer and 0.03115 mol of the gamma-isomer intermediate in the convergent step. From this point 0.02492 mol of the alpha-isomer leads to 0.000678 mol of penicillin V. On the other hand, 0.03115 mol of the gamma-isomer leads to 0.01184 mol of the alpha-isomer after isomerization, which in turn leads to an additional 0.000322 mol of penicillin V. Hence, the total moles of penicillin V obtained is 0.000678 + 0.000322 = 0.001000 mol. Therefore, recovery of the gamma-isomer increases the yield of target product by 47.5%. These calculations confirm that it is advantageous to do this since the gamma-isomer is the major one produced in the key convergent step (gamma to alpha ratio = 1.25:1).

In the case of the Carroll plan to (−)-epibatidine, 1 mol of starting 6-chloronicotinoyl chloride produces 0.1207 mol of stereoisomer A and 0.0618 mol of stereoisomer B. From this point 0.0618 mol of stereoisomer B leads to 0.0168 mol of (−)-epibatidine. On the other hand, 0.1207 mol

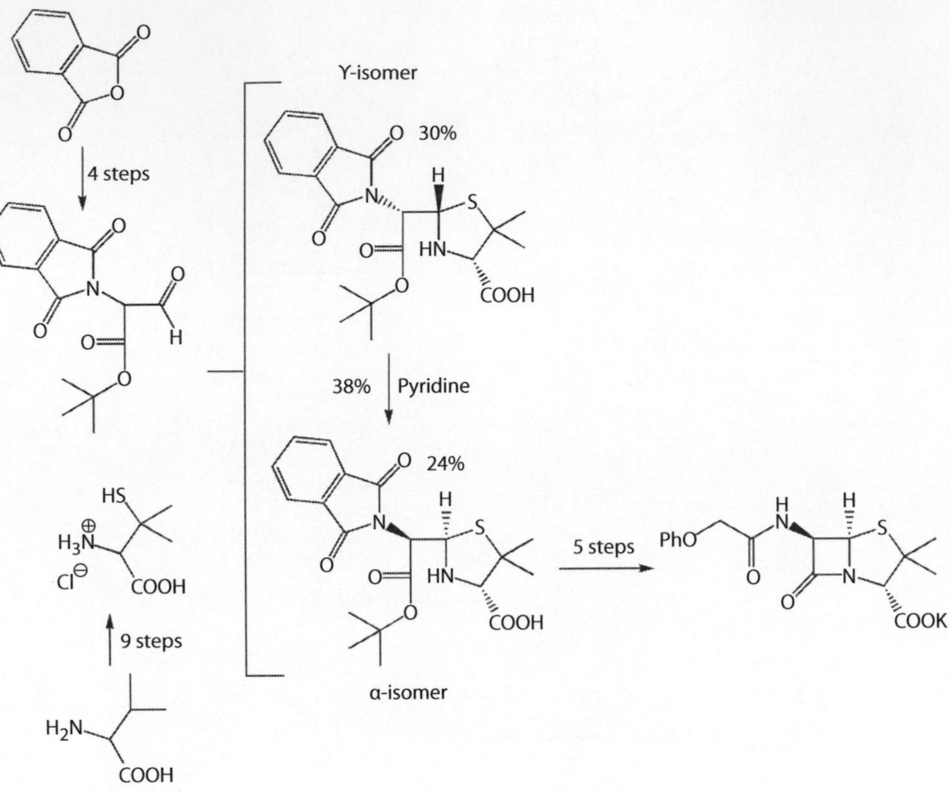

FIGURE 11.14   Abbreviated scheme for Sheehan (1959) synthesis of penicillin V.

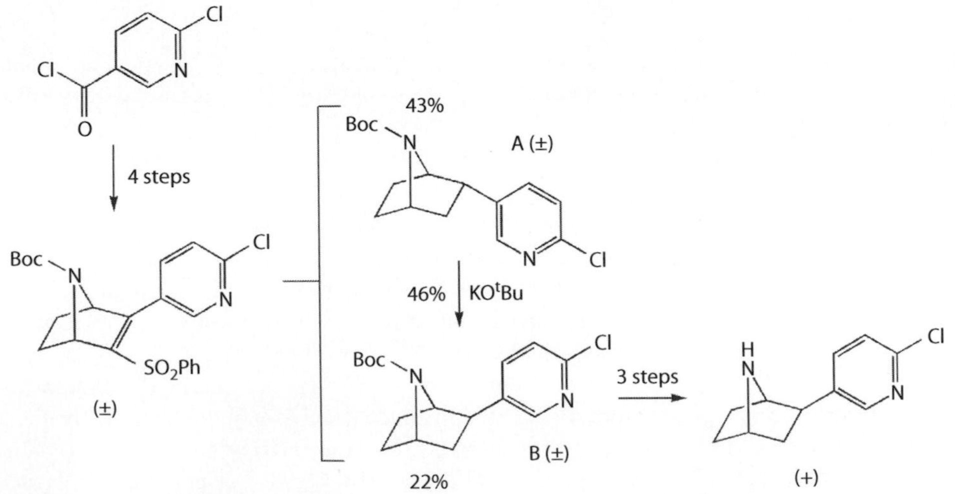

FIGURE 11.15   Abbreviated scheme for Carroll (1995) synthesis of (−)-epibatidine.

of stereoisomer A leads to 0.0556 mol of stereoisomer B after isomerization, which in turn leads to an additional 0.0151 mol of (−)-epibatidine. Hence, the total moles of (−)-epibatidine obtained is 0.0168 + 0.0151 = 0.0319 mol. Therefore, recovery of stereoisomer A increases the yield of target product by 90%. In this case, it is even more advantageous to recycle the unwanted isomer since it is the major one produced in the key convergent step (A to B ratio = 2:1).

11.4 IMPROVEMENT IN WASTE REDUCTION

A nice demonstration of overall kernel waste reduction is the Johnson G1 [27] and Johnson G2 [28] plans for (±)-estrone. The plans produce 260.6 and 22.2 kg of waste per mole (±)-estrone, respectively. Therefore, the G2 plan cuts down the kernel waste by 260.6/22.2 = 11.7 fold! This sort of progression in the green direction with time is rare as most of the time different plans using different strategies do not translate into improved material efficient plans as claimed by authors. In fact, the words "concise" and "efficient" are used loosely in titles of papers without rigorous checking of the metrics. The usual arguments based only on overall yield and number of steps are wholly inadequate in making claims of efficiency. The estrone example is again instructive in this regard as the data in Table 11.2 clearly show.

The more efficient G2 plan has more reaction steps and stages, uses more input materials, and has a lower overall atom economy. It also utilizes a higher molecular weight fraction of sacrificial reagents. These apparent counterintuitive results are rationalized when one examines the high reaction yield efficiencies of each step compared with the G1 plan. The main bottlenecks in the G1 plan are a 20% reaction yield for an intramolecular Friedel–Crafts alkylation in step 7,

a 30% yield for an oxidation of a 1,5-diol to the corresponding 1,5-diketone in step 4*,

and a 50% yield for a coupling reaction in step 5.

TABLE 11.2

Summary of Metrics for the Johnson G1 and G2 Plans for (±)-Estrone

Plan	N	M	I	% AE	% Overall Yield	$f(sac)$
G1	12	14	19	13.3	0.6	0.656
G2	16	20	26	10	5.9	0.710

By contrast the lowest yield reaction in the G2 plan is a Wittig coupling occurring in 65% yield.

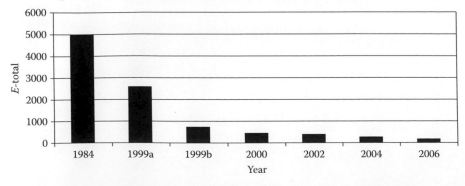

This mismatch in metrics parameters is a clear sign, despite the great improvement in waste reduction, that true optimization has still not yet been achieved since all parameters are not optimally gravitating in the same plan.

The case of various syntheses of (+)-sertraline hydrochloride is a beautiful example where indeed this goal has been achieved over time as shown by the E-factor profile in Figure 11.16. These data correspond to the Pfizer G1 [29], Lautens G1 [30,31], Pfizer G2 [32], Buchwald [33], Gedeon Richter [34], Pfizer G3 [35], and Zhao [36] plans, which can be found in the accompanying download. The expo-nential reduction in waste produced is stunning. Such an observation makes the Presidential Green Chemistry Challenge Award bestowed on Pfizer in 2002 particularly meritorious. However, this trend was spoiled in 2010 by the announcement of the Bäckvall stereoselective plan [37] that produces an E-factor of 2240. Not surprisingly, the killer culprit despite the use of lipases to effect a stereoselective synthesis is the high contribution of solvents and washes that yield E-auxiliaries = 2234 and account for 99.7% of the total waste! One point of caution in these comparisons is that the industrial plans were done on large scales, whereas, the academic ones were done on milligram scales and so the merit of the academic plans on large scale remains untested and unproven. The safest comparison that can be made is between the G1, G2, and G3 Pfizer plans and the Gedeon Richter plan.

Exactly the opposite trend in waste profile is found for synthesis plans for the natural product (+)-laurallene. Figure 11.17 shows that the best plan is the Crimmins plan of 2000 [38], the first

FIGURE 11.16 Time profile of E-total for various syntheses of (+)-sertaline hydrochloride.

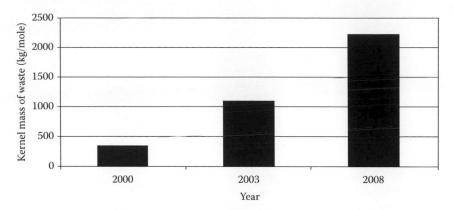

FIGURE 11.17 Time profile of kernel mass of waste produced per mole for various syntheses of (+)-laurallene.

synthesis plan disclosed for this target molecule. The Suzuki [39] and Takeda [40] plans of 2003 and 2008 progressively increase waste production. Despite the novelties of the synthesis strategies employed, they unfortunately did not translate into material efficient plans. The reader is directed to the accompanying download for a full analysis of these plans.

11.5 SPECTACULAR EXAMPLES OF SYNERGY BETWEEN STRATEGY AND MATERIAL EFFICIENCY METRICS

Rice [41–43] plan for (±)-codeine (rank = 1)

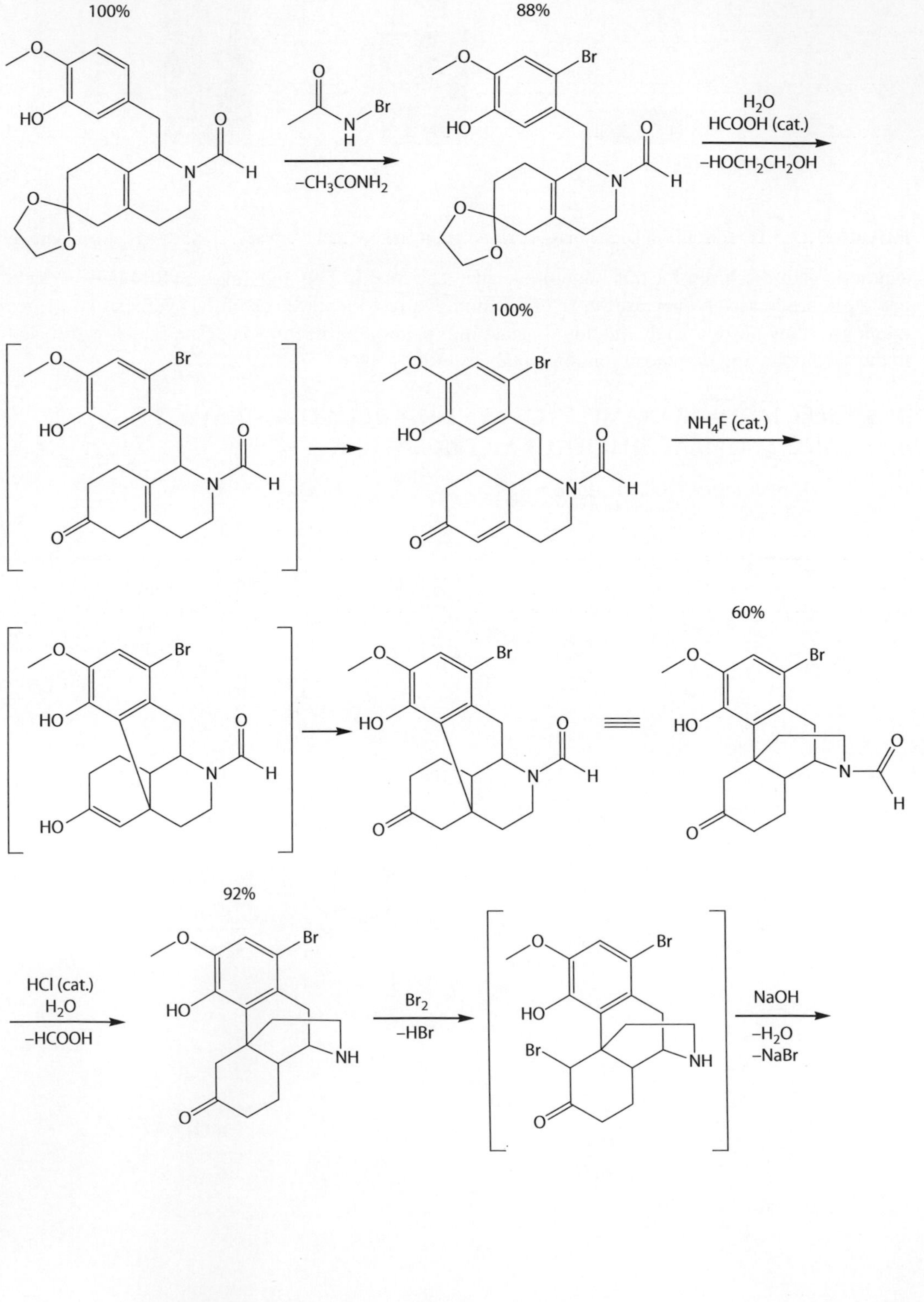
100%
88%
Br
HO
O
N
H
–CH3CONH2
H2O
HCOOH (cat.)
–HOCH2CH2OH
100%
NH4F (cat.)
60%
92%
HCl (cat.)
H2O
–HCOOH
Br2
–HBr
NH
NaOH
–H2O
–NaBr

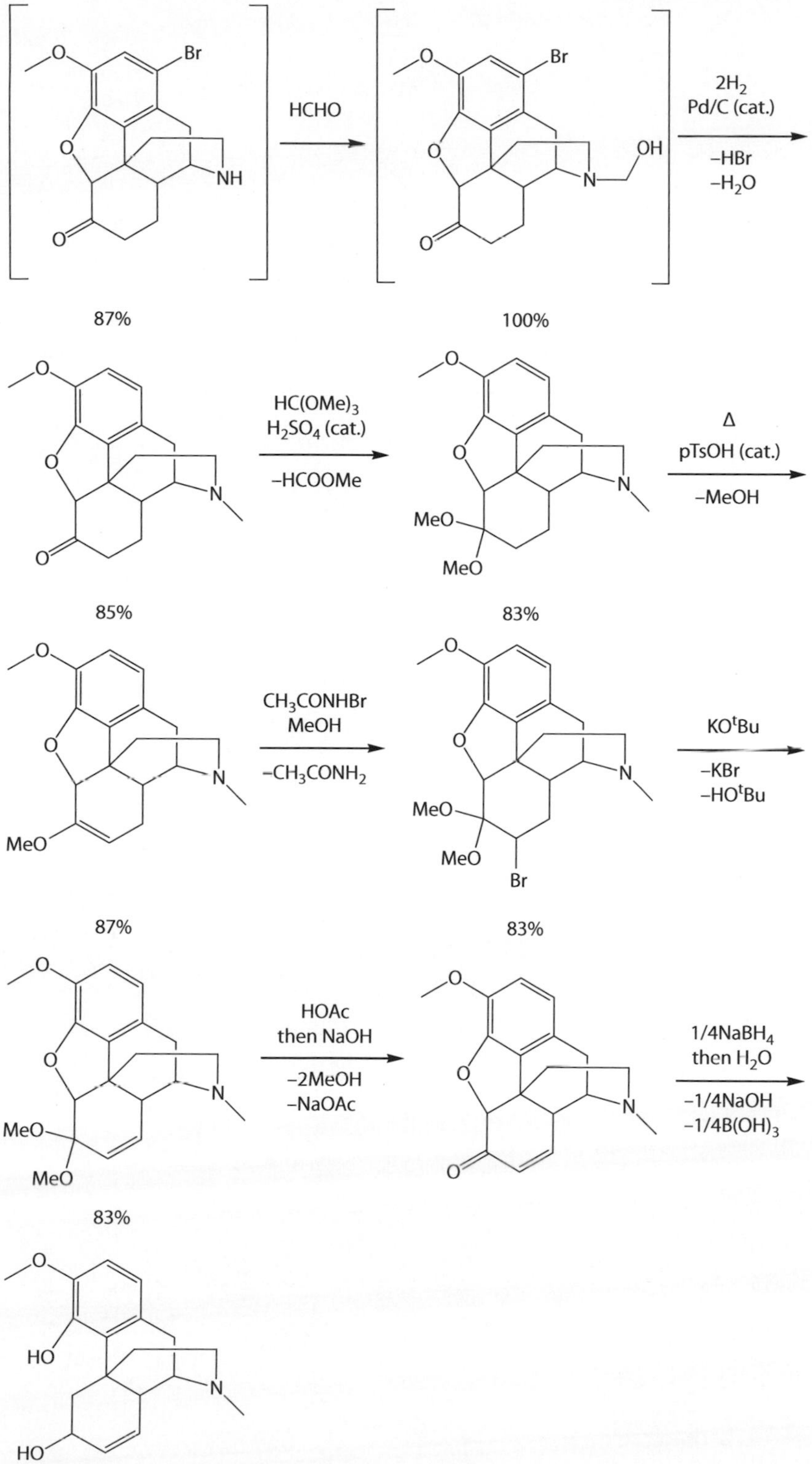
Br
NH
O
HCHO
Br
OH
N
2H$_2$
Pd/C (cat.)
–HBr
–H$_2$O
87%
100%
N
HC(OMe)$_3$
H$_2$SO$_4$ (cat.)
–HCOOMe
MeO
MeO
Δ
pTsOH (cat.)
–MeOH
85%
83%
MeO
CH$_3$CONHBr
MeOH
–CH$_3$CONH$_2$
Br
KO^{t}Bu
–KBr
–HO^{t}Bu
87%
83%
HOAc
then NaOH
–2MeOH
–NaOAc
1/4NaBH$_4$
then H$_2$O
–1/4NaOH
–1/4B(OH)$_3$
83%
HO
HO

Magnus [44] plan for (±)-codeine (rank = 2)

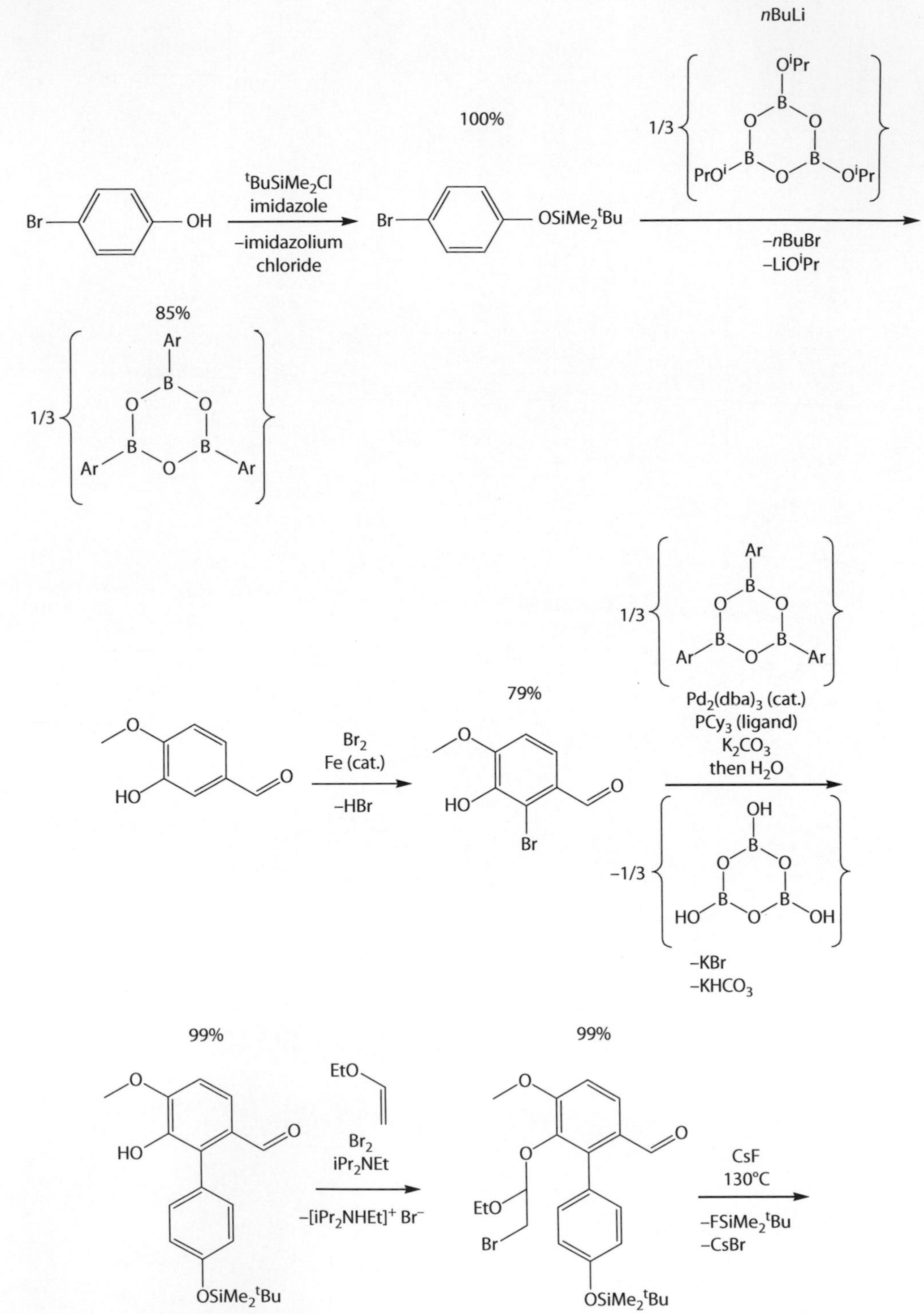

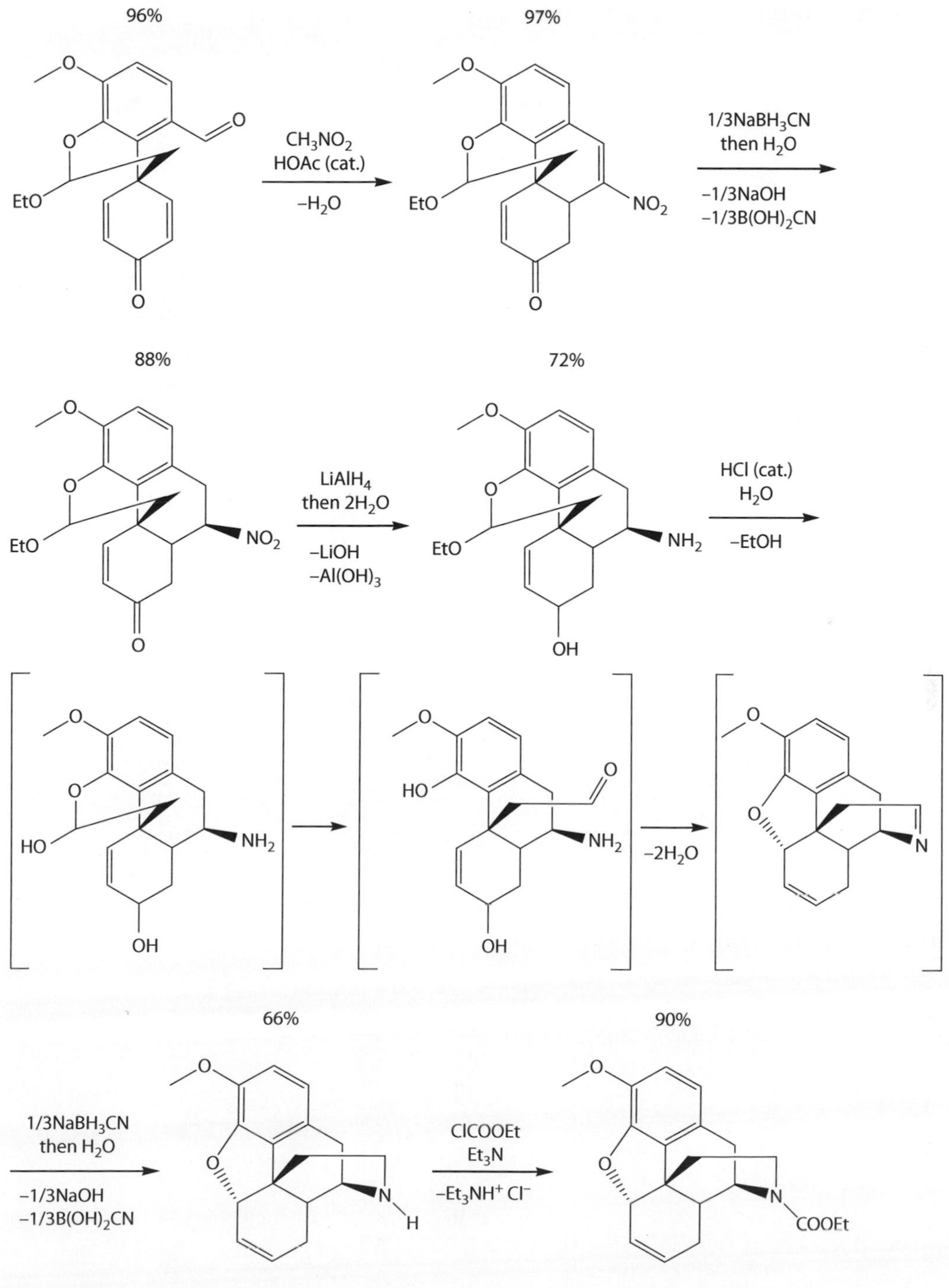

96%
97%
CH3NO2
HOAc (cat.)
–H2O
EtO
NO2
1/3NaBH3CN
then H2O
–1/3NaOH
–1/3B(OH)2CN
88%
72%
LiAlH4
then 2H2O
–LiOH
–Al(OH)3
NH2
OH
HCl (cat.)
H2O
–EtOH
HO
–2H2O
N
66%
90%
ClCOOEt
Et3N
–Et3NH+ Cl–
H
COOEt

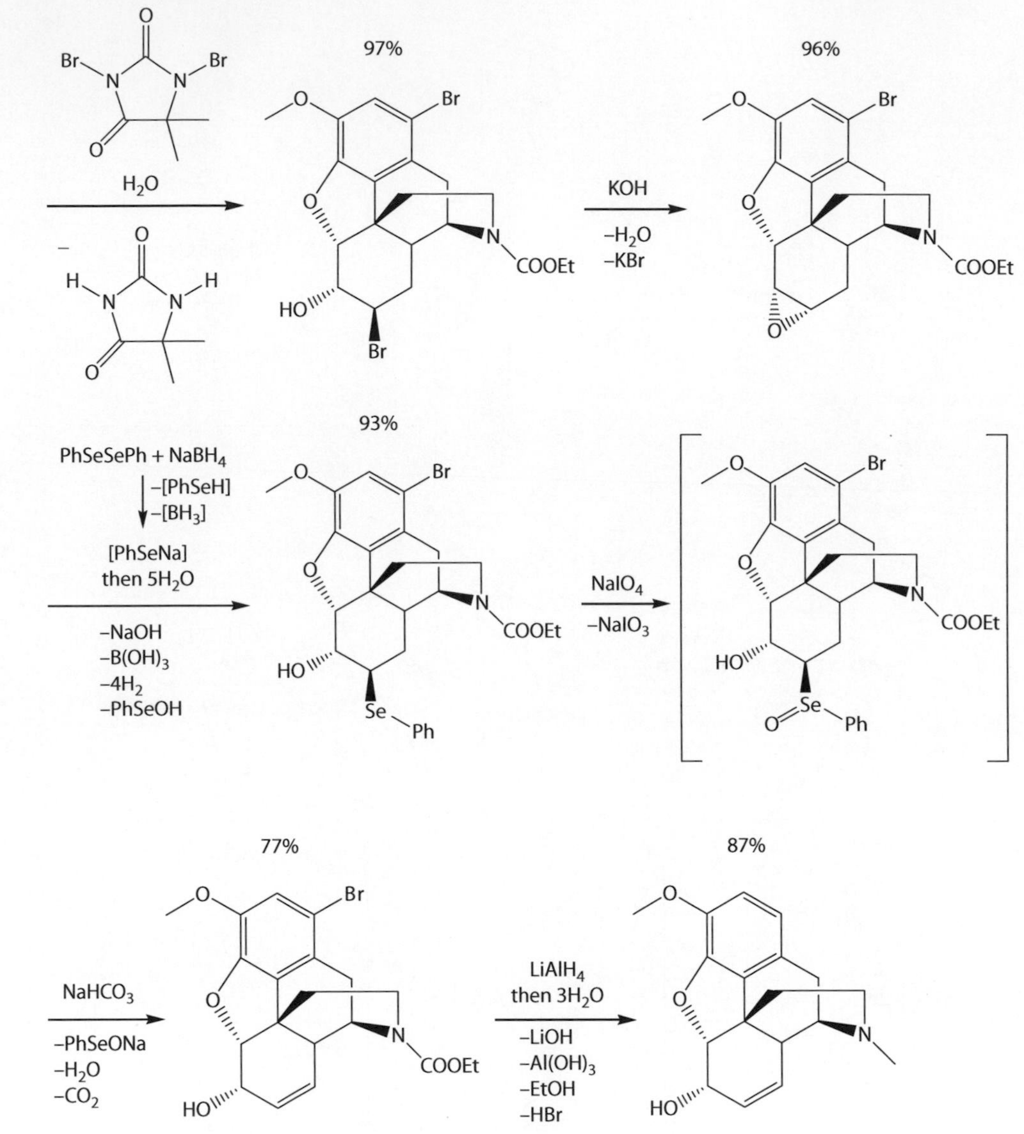

Reppe [45] plan for cyclooctatetraene (rank = 1)

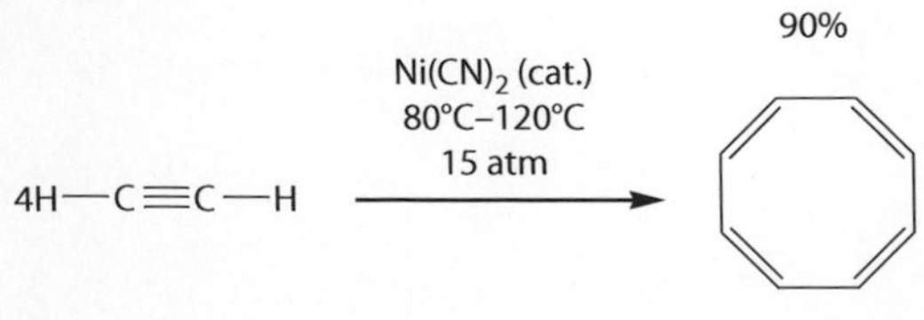

Pfizer [46] plan for fesoterodine fumarate (rank = 1)

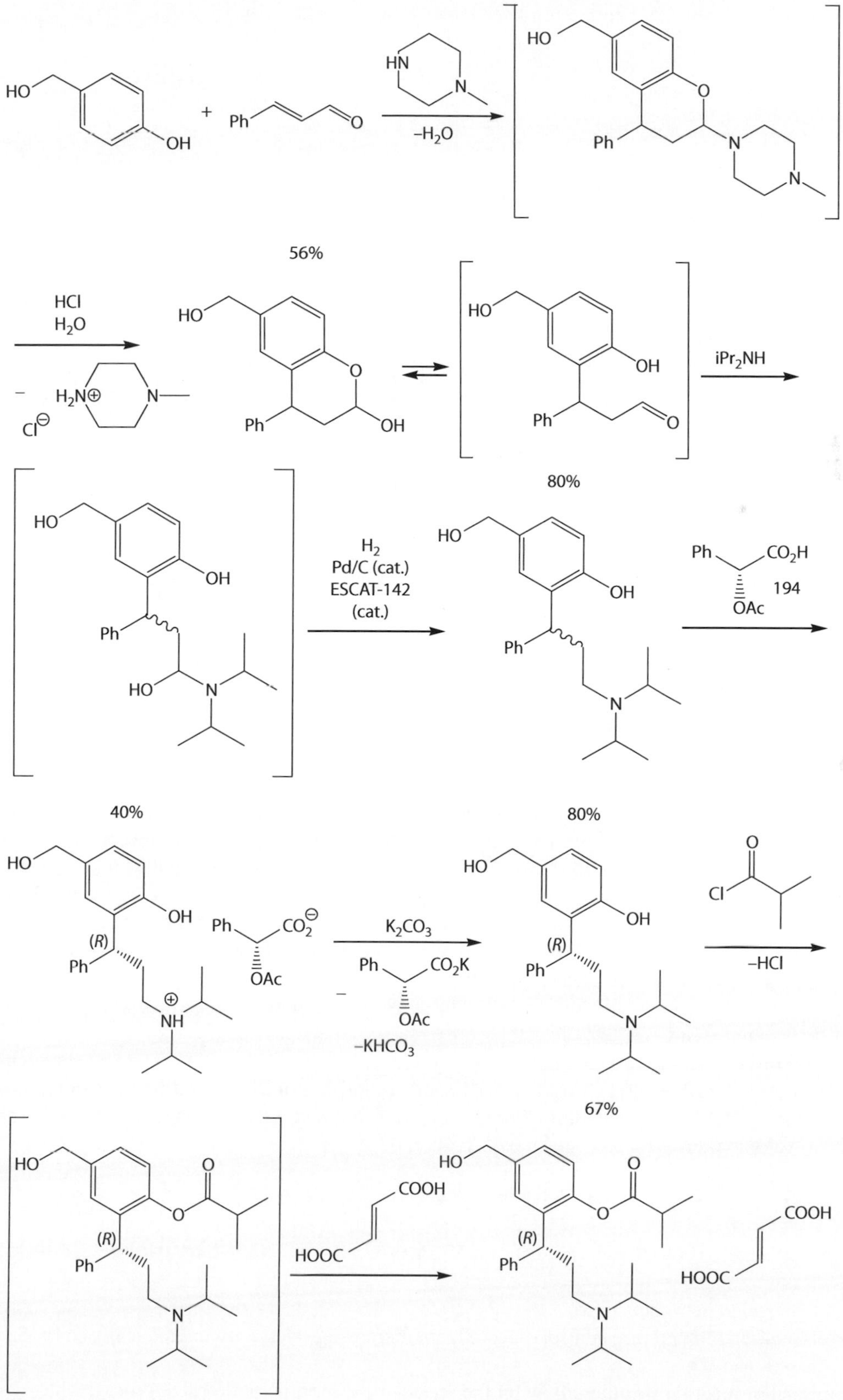
HO
OH
+
Ph
O
HN
N
$-H_2O$
HO
O
Ph
N
N
56%
HCl
H_2O
$H_2N^{\oplus}$
N
$Cl^{\ominus}$
HO
O
Ph
OH
HO
OH
Ph
O
iPr_2NH
80%
HO
OH
Ph
HO
N
H_2
Pd/C (cat.)
ESCAT-142
(cat.)
HO
OH
Ph
N
Ph
CO_2H
OAc
194
40%
80%
HO
OH
(R)
Ph
NH
Ph
$CO_2^{\ominus}$
OAc
K_2CO_3
Ph
CO_2K
OAc
$-KHCO_3$
HO
OH
(R)
Ph
N
O
Cl
$-HCl$
67%
HO
O
O
(R)
Ph
N
COOH
HOOC
HO
O
O
(R)
Ph
N
COOH
HOOC

Campos [47] plan for (*R*)-nicotine (rank = 1)

* assumed yield

Robinson [48] plan for tropinone (rank = 1)

11.6   MICROCHANNEL AND FLOW TECHNIQUE STRATEGIES

A new emerging technology in fine chemical synthesis is microchannel or microreactor flow chemistry [49–71]. Authors have heavily advertised this technique as satisfying green chemistry principles. Rather than carrying out reactions in flasks and batch reactors, reactions are run through capillary tubes in fabricated devices that look like circuit boards. Solutions of each reagent are prepared separately and then flowed at controlled rates by means of pumps reminiscent of high pressure liquid chromatography (HPLC) apparatuses. Reactions take place inside flow tubes filled with solid supported reagents or coated with reagents. Tubes may be heated conventionally or subjected to pyrolysis, photolysis, or microwave irradiation. The entire apparatus is fully automated and computer controlled. This flow methodology has a number of advantages over conventional ways of carrying out reactions. These include the following:

a. Efficient mixing.
b. Improved surface area to volume ratio; $200\,cm^2/mL$ for microreactors. This may be compared with values of $1\,cm^2/mL$ for a $100\,mL$ round bottom flask and $0.06\,cm^2/mL$ for a $1000\,L$ batch reactor. Assuming the shape of a reaction vessel is a sphere, the surface area to volume ratio (SAVR) is given by SAVR $= 3/r$.

$$100\,mL\,flask, \quad r = 2.9\ cm, \quad SAVR = 1.034$$

$$1000\,L\,batch\ reactor, \quad r = 27\ in. = 68.58\ cm, \quad SAVR = 0.044$$

c. Controlled heat transfer.
d. Increased temperature control especially for increasing selectivity in kinetically controlled reactions [72].
e. Smaller reaction volumes allow for the replacement of large scale batch processes.

f. Reduced consumption of solvents and reagents.

g. Utilization of continuous flow can achieve large amounts of product; "scaling out" and "numbering up" paradigms are used instead of "scaling up"; simple and inexpensive replication of reactions with multiple microreactors running in parallel rather than increasing the size of any one reaction unit.

h. Governed by laminar flow (low Reynolds numbers), characterized as smooth constant fluid motion.

i. Time resolution is manifested as space resolution, that is, reaction time is controlled by tube length.

j. High heat exchanging efficiency permits highly exothermic reactions to be performed under isothermal conditions so that no hot spots occur.

k. Prevention of competing side reactions.

l. Allows for safer protocols when using hazardous materials since reactants are compartmentalized until they are brought together to react.

m. Less chance for "runaway" reactions occurring.

n. Faster optimization to "right" reaction conditions is possible.

o. Less energy intensive [73].

p. Precise control of various reaction parameters is possible.

q. Technique is directly applicable to combinatorial and multistep synthesis.

r. Real-time spectroscopic monitoring of reaction progress by IR (infrared) and nuclear magnetic resonance (NMR).

s. Full automation of apparatus.

t. Significant laboratory space savings.

u. Purification protocols can be incorporated into the flow system.

v. Elimination of process optimization requirements.

w. Catalysis operates under favorable kinetic conditions.

x. On-the-fly changes in reaction conditions are possible.

As with any technique in chemistry there are always limitations. Some disadvantages include:

a. Flow cannot be applied to all reactions types though great strides are being made to increase the repertoire of reactions: hydrogenations, ozonolysis, heterogeneous reactions, liquid phase reactions, etc.

b. Microreactors are incompatible with solid reagents.

c. Mainly useful for fast reactions.

d. Technology is still expensive.

e. Clogging of flow lines by reagents or by-products that may not be soluble in the reaction solvent.

f. Solubilities and viscosities of starting materials, target product of reaction, and by-products of reaction need to be carefully evaluated.

g. Dilute conditions are necessary for in-line monitoring by spectroscopic means.

Experimental procedures documented in the literature using flow techniques are even worse than those using conventional methods in terms of properly mentioning reaction conditions and amounts of materials used (see Chapter 2). Of particular note is the inconsistent reporting of both flow rate and flow time to get an estimate of the volume of solvent used. Often procedures cite one of these parameters, not both. This shortcoming makes it impossible to estimate actual solvent usage. In fact, a key caveat with this technique is the rather high solvent demand as will be shown for the following examples cited from the synthesis database presented in this work. Another point that is overlooked is the solvent used to clean the lines, which is reminiscent of the same oversight in cleaning batch reactors and other equipment between operations.

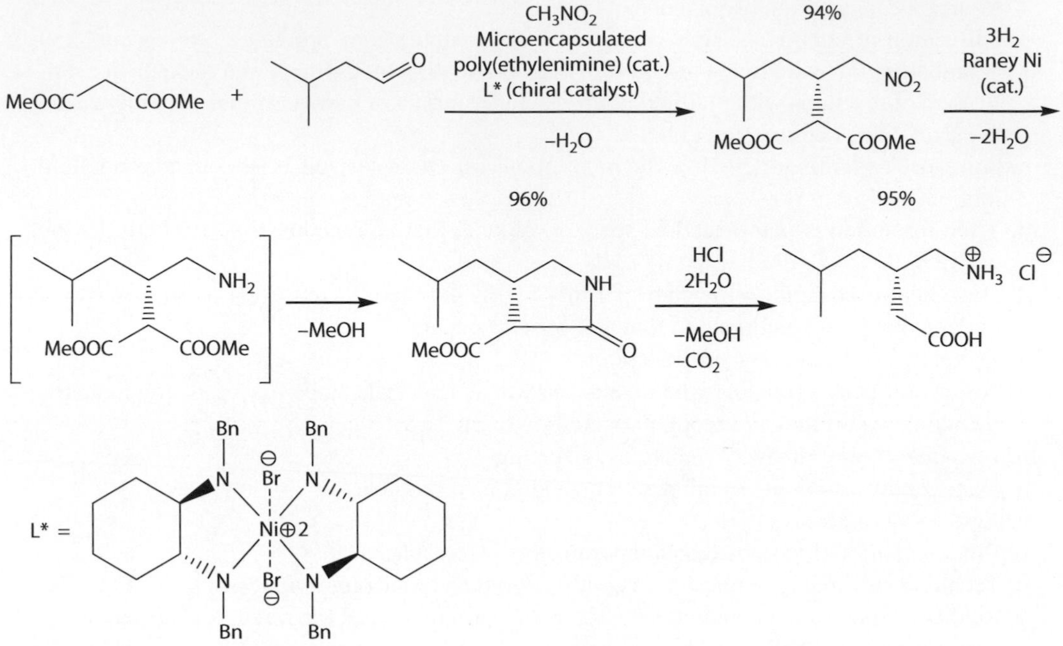

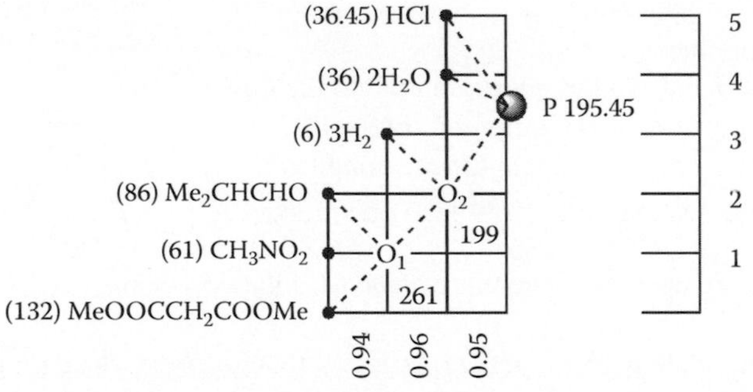

FIGURE 11.18 Synthesis tree for McQuade (2007) plan for (S)-pregabalin hydrochloride.

These problems detract from the claims of greenness especially when it is already well established that the contribution of E-auxiliaries to E-total is often well over 80% for plans run using conventional batch reactors and reaction vessels. Despite the number of examples already appearing in the literature, what has not yet been done is a head-to-head comparison of conventional and flow techniques for a given reaction that proves the green merits of flow by a thorough metrics analysis. The examples highlighted here are the very first steps in this direction though they are hampered by limited disclosed data.

The first example is the McQuade [74] plan for (S)-pregabalin hydrochloride shown in the following and as a synthesis tree in Figure 11.18.

The reaction solvent to limiting reagent mass ratio for each step is as follows:

Step 1 (19.4:1), step 2 (7.5:1), and step 3 (0:1). In the third step, water is a reagent, so it is not formally counted as a solvent. This plan ranks 15th out of 24 in terms of E-total as shown by the data in Table 11.3.

TABLE 11.3
E-Factor Breakdowns for Synthesis Plans for (S)-Pregabalin

Plan	Year	Type	E-Kernel	E-Excess	E-Aux	E-Total	Total Mass of Waste (kg)
Pfizer G1	2008	Linear	4.27	0.99	36.35	41.61	6.6
Pfizer G2	2008	Linear	9.37	4.88	52.9	67.15	10.7
Dowpharma	2003	Linear	7.46	9.57	179.8	196.81	31.3
Warner-Lambert G1	1996	Linear	25.8	24.36	166.5	216.63	34.4
Parke-Davis G4	1997	Linear	23	28.77	176.3	228.02	36.3
Hamersak G2	2007	Linear	13.53	17.3	230.2	260.98	41.5
Silverman G1	1989	Linear	2.4	28.34	236.6	267.3	42.5
Parke-Davis G2	1997	Linear	28.83	22.18	227.6	278.56	44.3
Silverman G2	1990	Linear	29.07	60.85	216.3	306.18	48.7
Warner-Lambert G2	2001	Linear	8.85	7.47	289.7	306.02	48.7
Hayashi	2007	Linear	14.8	16.86	284.7	316.37	50.3
Parke-Davis G1	1997	Convergent	22.54	15.15	290.1	327.77	52.1
Koskinen	2009	Linear	10.53	4.06	338.9	353.5	69.1
Hamersak G1	2007	Linear	20.88	24.92	448.9	494.69	78.7
McQuade	2007	Linear	1.09	46.98	522.6	570.7	111.5
Shibasaki	2005	Linear	4.58	15.03	744.3	763.87	149.3
Felluga	2008	Linear	20.97	67.95	1131	1219.6	238.4
Jacobsen	2003	Linear	6.61	11.56	1202	1220.5	238.5
Lee	2007	Linear	12.7	17.49	2121	2151.5	342.1
Sartillo-Piscil	2007	Linear	30.24	664.4	5586	6280.2	998.5
Armstrong	2006	Linear	7.71	N/A	N/A	N/A	N/A
Hu	2004	Linear	15.82	N/A	N/A	N/A	N/A
Ortuno	2008	Linear	24.06	N/A	N/A	N/A	N/A
Parke-Davis G3	1997	Linear	4.77	N/A	N/A	N/A	N/A

N/A = not available.

The second example is the McQuade [75] plan for (±)-ibuprofen shown in the following and as a synthesis tree in Figure 11.19.

The reaction solvent to limiting reagent mass ratio for each step is as follows:

Step 1 (1.1:1) and step 2 (84.7:1). This plan ranks last out of 10 in terms of E-total as shown by the data in Table 11.4.

The third example is the Ley [76] plan for (±)-oxomaritidine shown in the following and its synthesis tree in Figure 11.20.

This plan had a sketchy documentation of experimental procedures and so unfortunately no full metrics could be evaluated beyond atom economy and reaction yield.

The fourth example is the Ley [77] plan for imatinib shown in the following and as a synthesis tree in Figure 11.21.

This plan had the best-documented experimental procedure using the microreactor technique so far encountered. The reaction solvent to limiting reagent mass ratio for each step is as follows:

Step 1 (560:1), step 2 (428:1), step 3 (375:1), step 1* (2:1), and step 2* (6.8:1). This plan ranks last out of six in terms of E-total as shown by the data in Table 11.5.

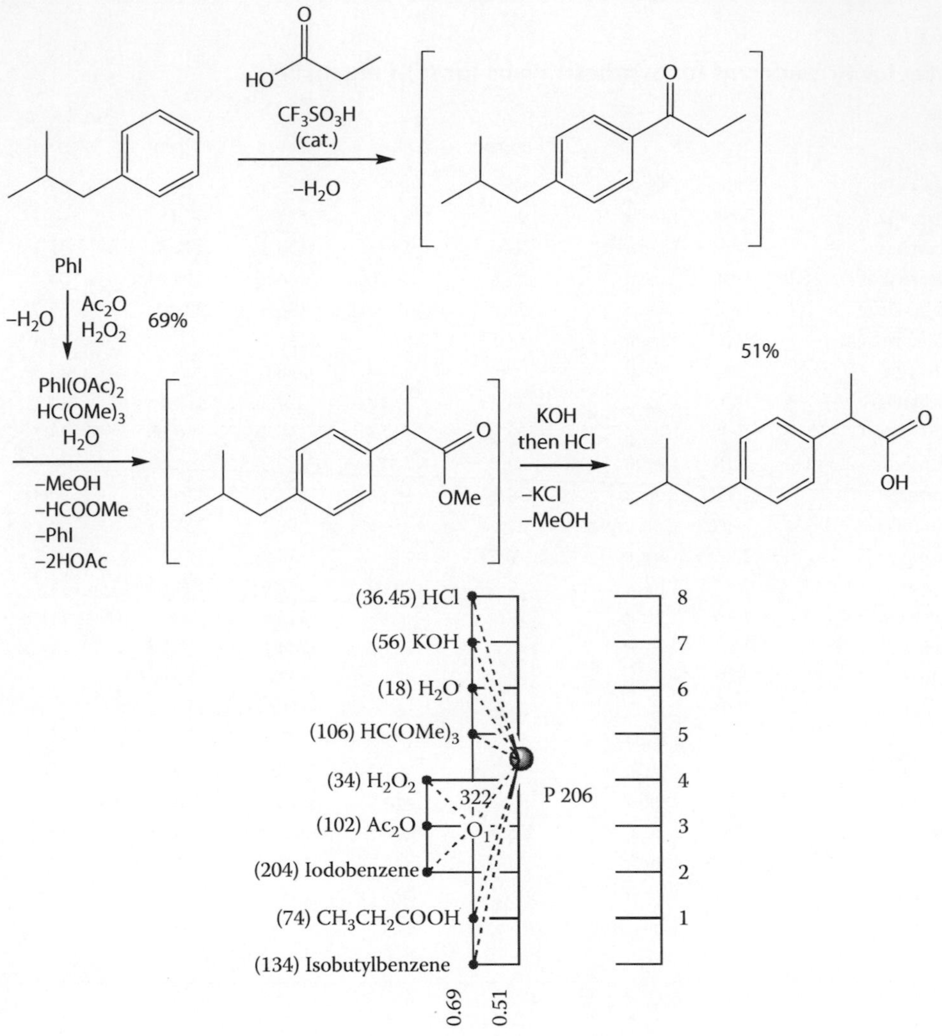

FIGURE 11.19 Synthesis tree for McQuade (2009) plan for (±)-ibuprofen.

TABLE 11.4
***E*-Factor Breakdowns for Synthesis Plans for (±)-Ibuprofen**

Plan	Year	Type	*E*-Kernel	*E*-Excess	*E*-Aux	*E*-Total	Total Mass of Waste (kg)
Upjohn	1977	Linear	4.78	1.03	45.9	51.7	10.7
DD113889	1975	Linear	9.17	15.55	30.47	55.19	11.4
Hoechst-Celanese	1988	Linear	1.72	1.17	111.9	114.79	23.6
Boots	1968	Linear	8.59	15.42	96.52	120.53	24.8
Ruchardt	1991	Linear	12.62	52.06	79.98	144.66	29.8
DuPont	1985	Linear	4.67	61.83	150.1	216.61	44.6
Pinhey	1984	Convergent	7.8	152.2	540.3	700.33	144.3
RajanBabu	2009	Linear	3.91	21.81	1888	1913.8	394.2
Furstoss, (*S*)-isomer	1999	Linear	17.8	12.13	2286	2316.2	477.1
McQuade	2009	Linear	7.73	346.6	41968	42322	8718

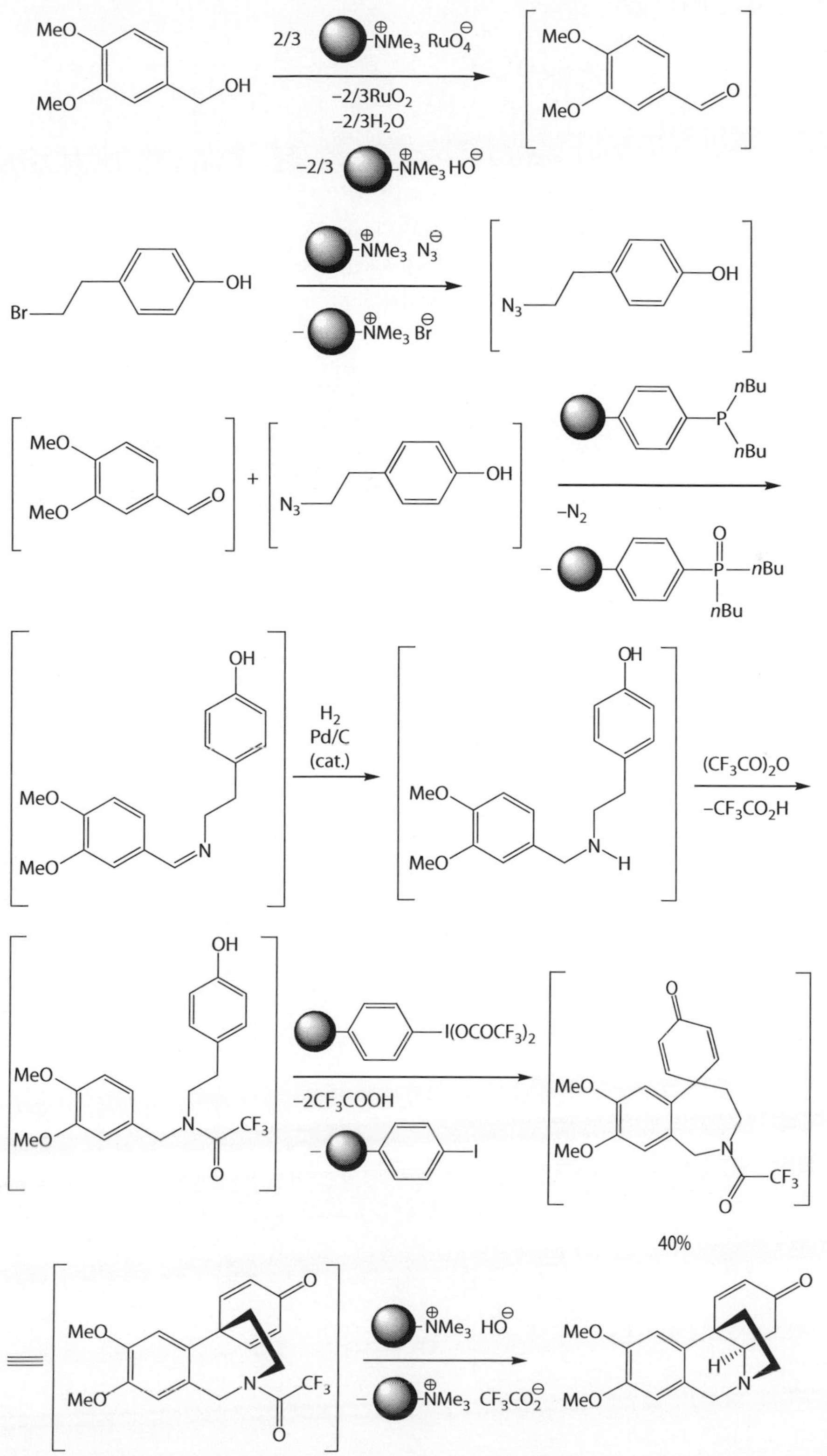

FIGURE 11.20 Synthesis tree for Ley (2006) plan for (±)-oxomaritidine.

(continued)

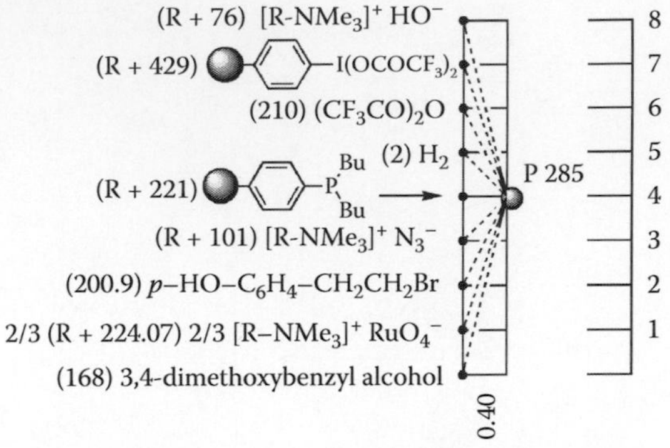

FIGURE 11.20 (continued)

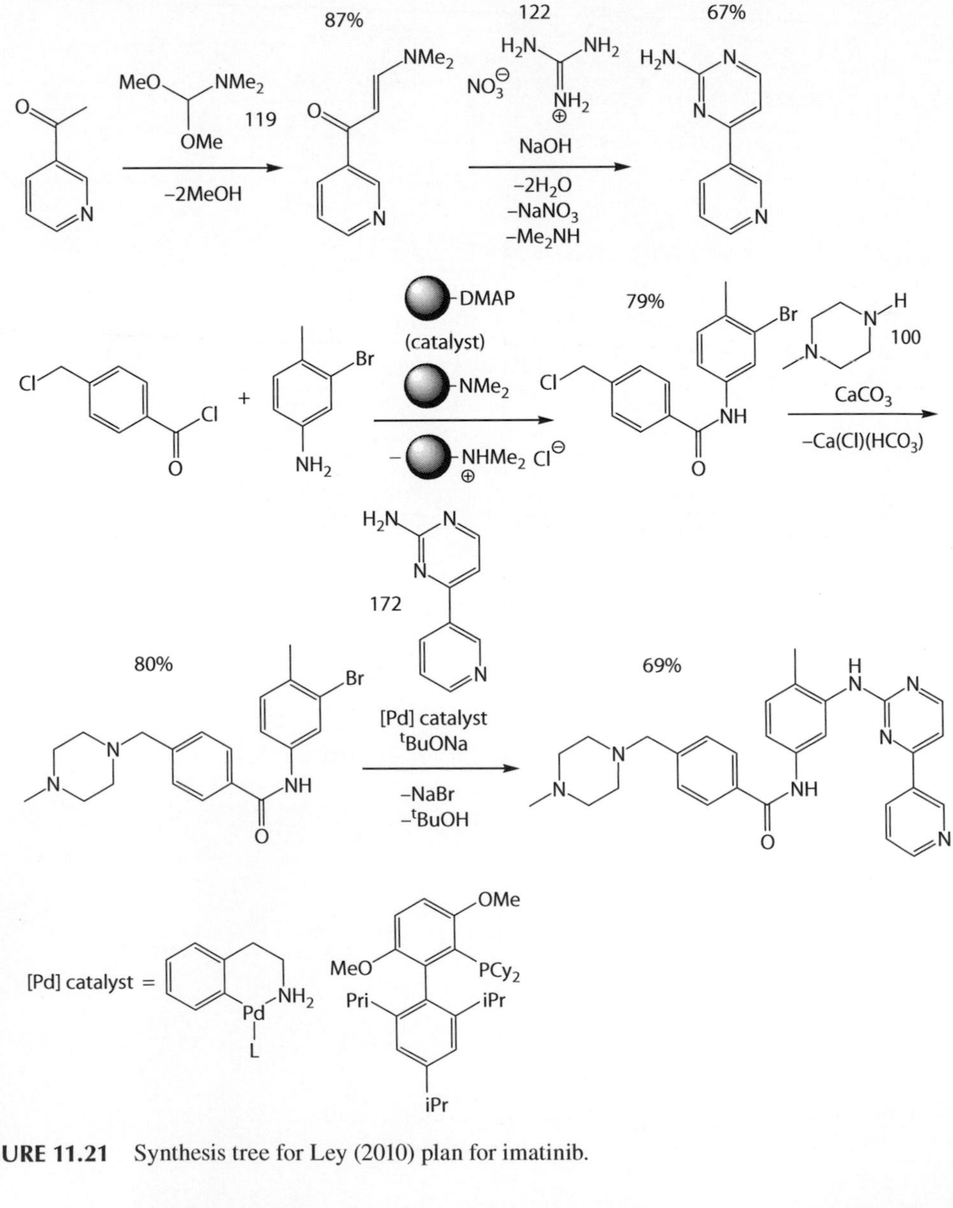

FIGURE 11.21 Synthesis tree for Ley (2010) plan for imatinib.

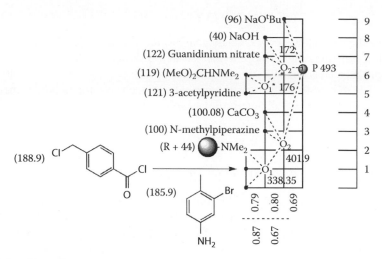

FIGURE 11.21 (continued)

TABLE 11.5

***E*-Factor Breakdowns for Synthesis Plans for Imatinib**

Plan	Year	Type	*E*-Kernel	*E*-Excess	*E*-Aux	*E*-Total	True Mass of Waste (kg)*
Ivanov	2009	Convergent	4.45	5.96	151.2	161.63	79.7
Novartis	2003	Linear	1.2	0.8	330	332	163.7
Noszal	2005	Convergent	13.6	37.3	436.6	487.5	240.3
Natco Pharma	2004	Convergent	17.7	55.3	569.1	642.1	316.5
Wang	2008	Convergent	4.3	30.8	618.3	653.4	322.1
Ley	2010	Convergent	3.89	111.7	2210	2325.6	1147

* Refers to second branch in plan.

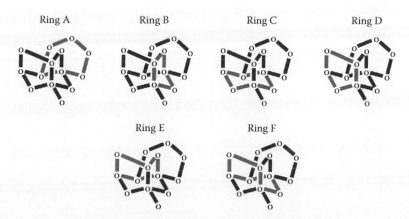

FIGURE 11.22 Enumeration of rings in platensimycin tetracyclic core structure.

11.7   RING CONSTRUCTION NOVELTY

The synthesis of the natural product platensimycin has recently attracted the attention of synthetic chemists because of its unusual tetracyclic core structure shown graphically in Figure 11.22. The Njardarson plan [78] stands out from the other reported syntheses in terms of novelty because it is the only one to date that targeted the construction of the hidden seven-membered ring F. The reader is referred to Section 10.1 for a complete summary of the ring construction strategies employed for this compound.

11.8   WRONG STARTING MATERIALS USED

The following are two examples illustrating the problem when an inappropriately chosen starting material is used in a synthesis plan. The first is the Danishefsky plan [79] for (+)-estrone in which a pyridine containing starting material is used. This plan requires excision of the nitrogen atom from the pyridine ring and then reconnection to form the phenolic ring of estrone. Clearly, the pyridine is a sacrificial ring in this synthesis. The key sequence of steps is shown in the following.

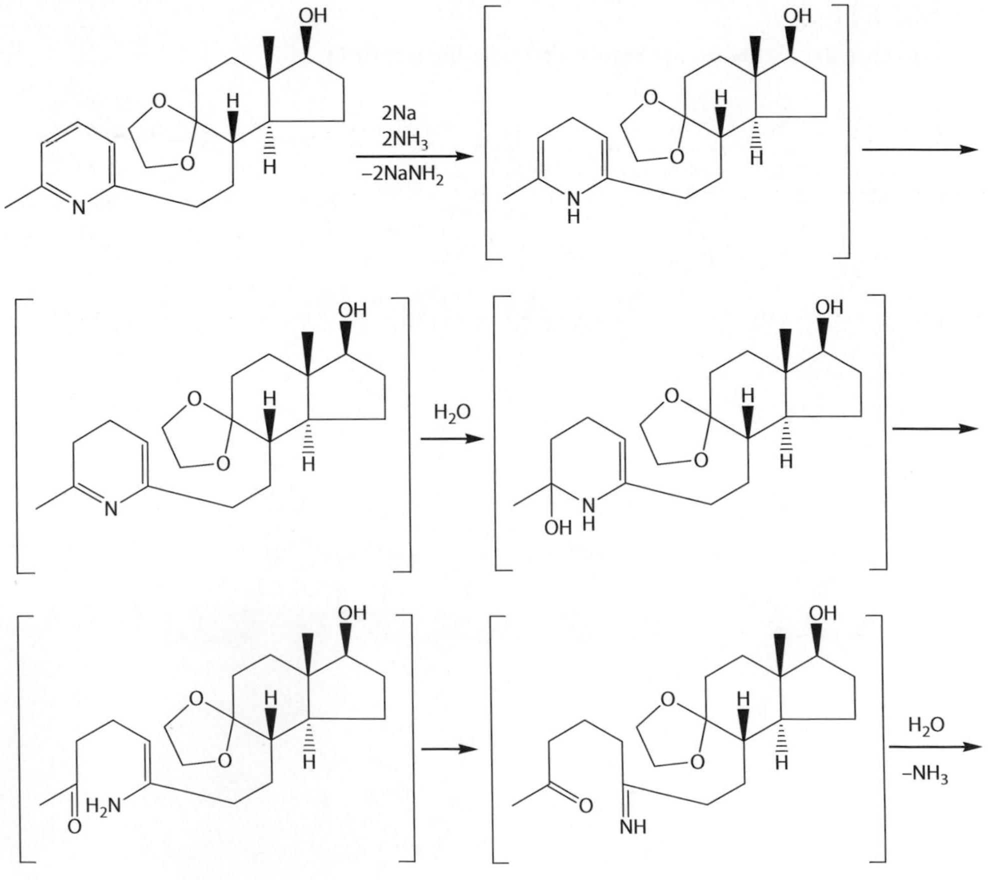

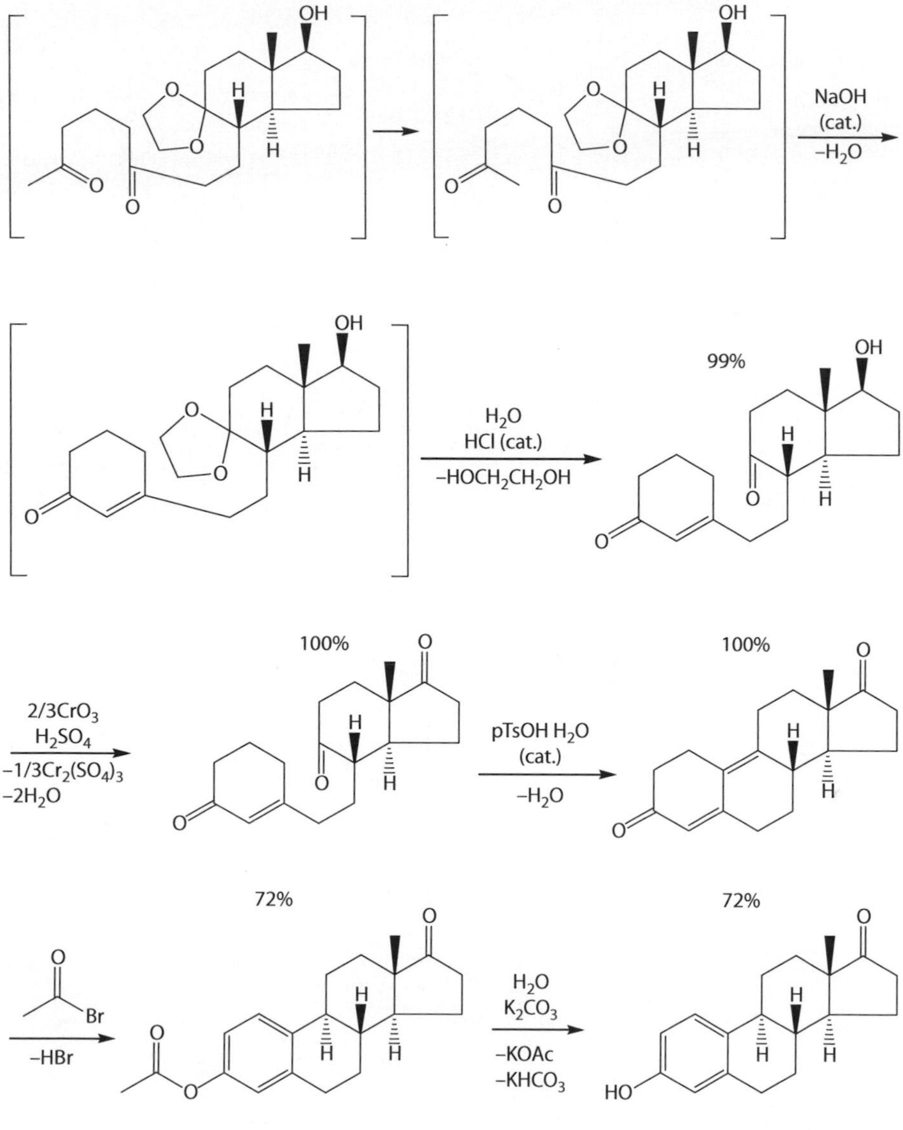

The second example is one already seen before in Figure 11.3a–c, which refers to the Hamon and Young [15] plan for (±)-twistane. The first 11 steps produce no target bonds as indicated by the target bond profile. The plan requires excision of a methylene group from the adamantanone ring system, so the real starting material in this scheme is bicyclo[3.3.1]nonane-2,7-dione. This compound has been made by the following sequence [80–82].

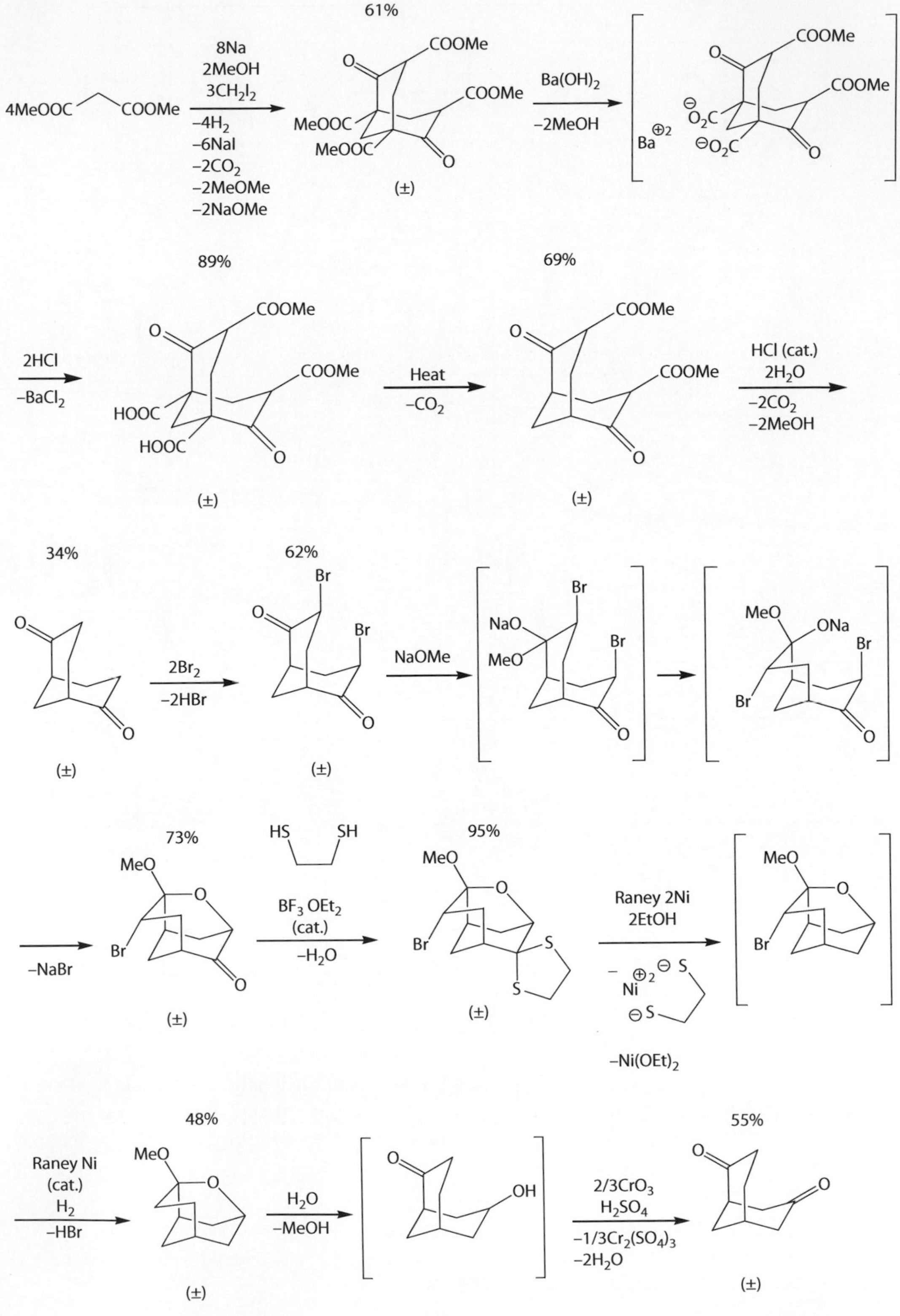

61%

4MeOOC⌒COOMe

$\xrightarrow[\begin{array}{c}-4H_2\\-6NaI\\-2CO_2\\-2MeOMe\\-2NaOMe\end{array}]{\begin{array}{c}8Na\\2MeOH\\3CH_2I_2\end{array}}$

$\xrightarrow[-2MeOH]{Ba(OH)_2}$

89%

$\xrightarrow[-BaCl_2]{2HCl}$

69%

$\xrightarrow[-CO_2]{Heat}$

$\xrightarrow[\begin{array}{c}-2CO_2\\-2MeOH\end{array}]{\begin{array}{c}HCl\ (cat.)\\2H_2O\end{array}}$

34%

(±)

62%

$\xrightarrow[-2HBr]{2Br_2}$

(±)

$\xrightarrow{NaOMe}$

73%

$\xrightarrow{-NaBr}$

(±)

HS⌒SH

$\xrightarrow[\begin{array}{c}(cat.)\\-H_2O\end{array}]{BF_3\ OEt_2}$

95%

(±)

$\xrightarrow[\begin{array}{c}Ni^{\oplus 2}\ ^{\ominus}S\\{}^{\ominus}S\end{array}]{\begin{array}{c}Raney\ 2Ni\\2EtOH\end{array}}$

$-Ni(OEt)_2$

48%

$\xrightarrow[-HBr]{\begin{array}{c}Raney\ Ni\\(cat.)\\H_2\end{array}}$

(±)

$\xrightarrow[-MeOH]{H_2O}$

55%

$\xrightarrow[\begin{array}{c}-1/3Cr_2(SO_4)_3\\-2H_2O\end{array}]{\begin{array}{c}2/3CrO_3\\H_2SO_4\end{array}}$

(±)

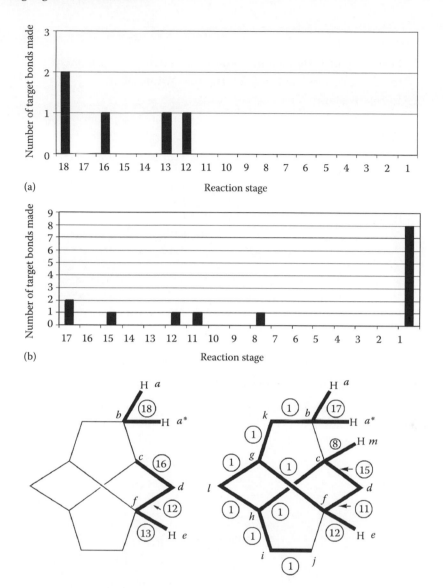

FIGURE 11.23 Comparison of target bond forming profiles and target bond maps for original (a) and modified (b) Hamon-Young plans to (±)-twistane.

When this sequence is linked to the rest of the Hamon–Young plan for (±)-twistane the target bond forming profile for the resultant modified plan shows a significant degree of building up in step 1 where 8 bonds are made in a single step. Figure 11.23 shows a head-to-head comparison of both plans in terms of strategy efficiency. Though there are significant gains made in terms of strategy efficiency, the material efficiency performance is offset compared to the original plan as shown by the data in Table 11.6. The large mass of by-products produced in step 1 (AE = 24%) contributes significantly to the poor material performance in the modified plan. Again, what we have here is a trade-off between material and synthesis efficiency.

TABLE 11.6
Summary of Material and Synthesis Efficiency Performances
for the Original and Modified Hamon–Young Plans
to (±)-Twistane

Parameter	Original Plan	Modified Plan	Trend with respect to Modified Plan
Number of steps	18	17	Decrease
Number of input materials	31	29	Decrease
MW first moment (g/mol)	48.35	147.98	Increase
$f(\text{sac})$	0.761	0.526	Decrease
HI	+3.47	+6.33	
% Yield	1.7	0.3	Decrease
% Atom economy	4.2	3.2	Decrease
% Kernel RME	0.3	0.02	Decrease
Kernel mass of waste (kg/mol)	40.9	648.4	Increase
Number of bonds made per step	0.28	0.82	Increase
$f(\text{nb})$	0.78	0.65	Decrease
|UD|	30	40	Increase

11.9   PLAN WITH NO TARGET BONDS MADE

This example is more of a curiosity than anything else. The Sandborn plan [83] for (–)-menthone involves a subtractive oxidation of (–)-menthol since this one-step transformation requires only the removal of two hydrogen atoms. No hypsicity or target bond making profiles are applicable for this plan.

11.10   OLD REACTIONS USING ONE OF 12 PRINCIPLES OF GREEN CHEMISTRY

Though green chemistry appears to be a new and emerging field, the foundation principles on which it is based are well rooted in the past literature. Any chemist who is engaged in improving reaction performance or redesigning synthesis plans to obtain the intended target product cost effectively is actually doing green chemistry. This activity has been ongoing since the dawn of organic chemistry. Following are some examples taken from the synthesis database presented in this work that justify this point.

Example #1
The Manske plan [84] for 2,4-D or 2,4-diphenoxyacetic acid, a constituent of the herbicide Agent Orange employs a solvent free protocol. The total *E*-factor for this two-step plan is only 2.21 or

0.49 kg waste produced per mole of product. This example is somewhat ironic since a "green" method is used to make a compound with a color-coded nickname for the purpose of getting rid of green foliage! A rather catchy phrase for a title of a paper could be dreamed up if it had been developed in modern times.

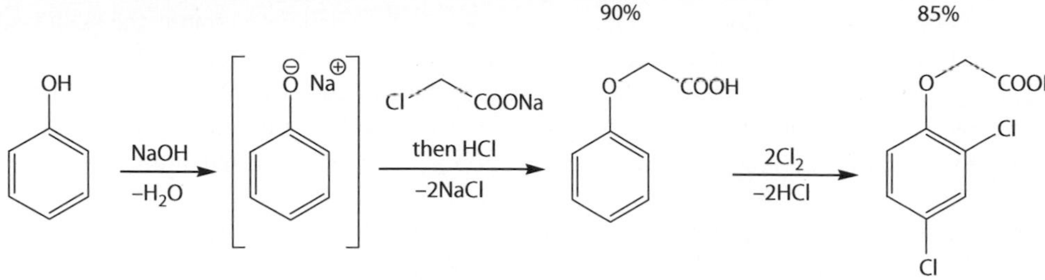

Example #2

The Merck G3 plan [85] for indomethacin involves a 3-component multicomponent reaction (MCR) for a single step synthesis.

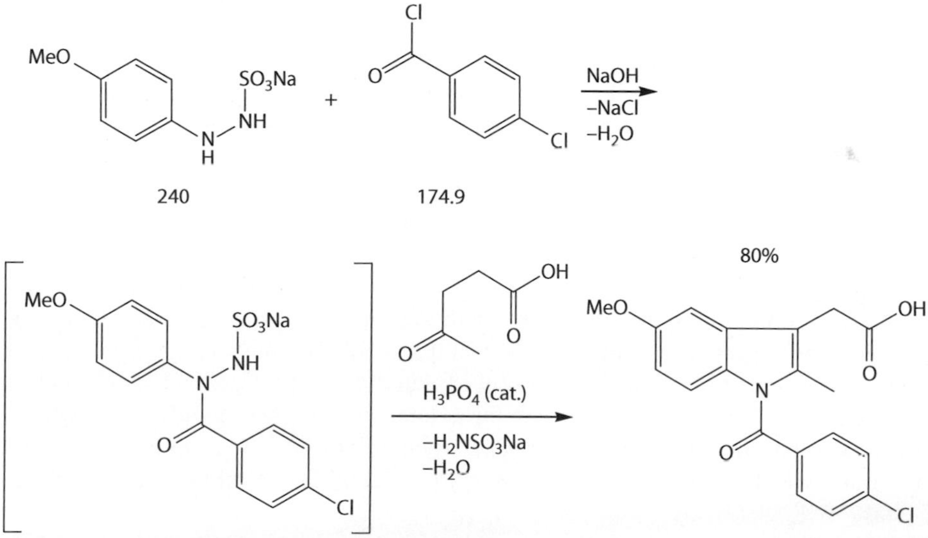

Example #3

The Parke-Davis plan [86] for phenytoin is another example of a 3-component MCR for a single step synthesis.

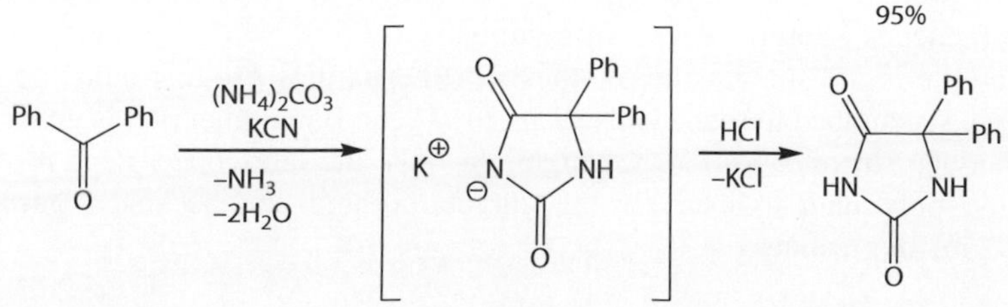

11.11   TELESCOPING STEPS

Telescoping or concatenation of steps is an effective method of cutting down the step count in a synthesis plan. It is a favorite option used by industrial process chemists. The idea involves carrying out the first reaction in a reaction vessel, then once it is complete, the reaction solvent is evaporated and replaced with another for the second reaction to take place in the same vessel. This is called solvent switching. If one is lucky enough that the solvent of the second reaction is the same as the first, then no evaporation is required. In either case, the intermediate product from the first reaction is not formally isolated, so there is a significant savings in the consumption of work-up and purification materials. Such telescoping has implications on the way a synthesis tree diagram is drawn as is shown by the following examples taken from the synthesis database. Intermediates not isolated are shown in square brackets in the schemes.

The first example is the Merck G3 [87] plan for sitagliptin hydrogen phosphate, an antidiabetic pharmaceutical used to treat juvenile type II diabetes. The synthesis scheme is shown in the following and the corresponding synthesis tree diagram is given in Figure 11.24.

Telescoping is achieved in step 3: tandem esterification—aldol condensation—fragmentation to a ketene intermediate—amidation—imination—isomerization. The synthesis tree diagram shows all input materials in order for these steps without showing the intermediates. The net effect is the vertical elongation of the diagram that results in a high value for the degree of convergence ($\delta = 0.749$) and a low value for the degree of asymmetry (beta $= 0.505$). With respect to waste production, the Merck G3 plan outperforms the Merck G1 plan that does not use telescoping. Accordingly, the kernel mass waste per mole of sitagliptin hydrogen phosphate is reduced 5-fold and the E-total is reduced 12-fold.

The second example is the Hayashi [88] plan for oseltamivir phosphate, a neuraminidase inhibitor used to treat influenza. The synthesis scheme is shown in the following and the corresponding synthesis tree diagram is given in Figure 11.25.

Telescoping takes place in step 3 (tandem organocatalyzed Michael addition—[4 + 2] cyclo-addition—Michael addition (thiolation)), step 4 (tandem ester hydrolysis—azidation), and step 5 (tandem Curtius rearrangement—acylation—nitro group reduction—elimination). This plan ranks third in a list of 28 plans for this target product. However, this high ranking needs to be qualified by two important points. One is that plan feasibility has been demonstrated on a millimole scale and has not yet been scaled up to a level that would be convincing for process chemists to employ. The second point is that authors of this plan did not disclose the masses of chromatographic solvents used in purification steps, so the E-total value of 597 should be considered a very conservative lower limit. There is no doubt that when all solvent consumption is accounted for this plan's rank will drop significantly in relation to the number one ranking Roche G3 plan [89] with E-total $= 231$.

The next two examples come from academic groups on syntheses of natural products.

The Organ G2 plan [90] for bupleurynol is shown in the following and its synthesis tree is given in Figure 11.26.

The entire plan is essentially an 8-component coupling, though from a practical point of view, the first four input materials are assembled in one flask, the next two components are assembled in a second flask, and then the contents of these two flasks are mixed along with the addition of the last two reagents. From the shape of the synthesis tree diagram, one can infer immediately that the degree of convergence is 1 and the degree of asymmetry is 0.

The last example is the spectacular synthesis of the natural product palau'amine that after a 20 year odyssey succumbed to total synthesis in 2010. The Baran plan [91] is given in its entirety in the accompanying download. Here we highlight only the telescoped steps in the following. Since this is the only plan to date for this target, nothing can be said regarding synthesis efficiency or degree of greenness.

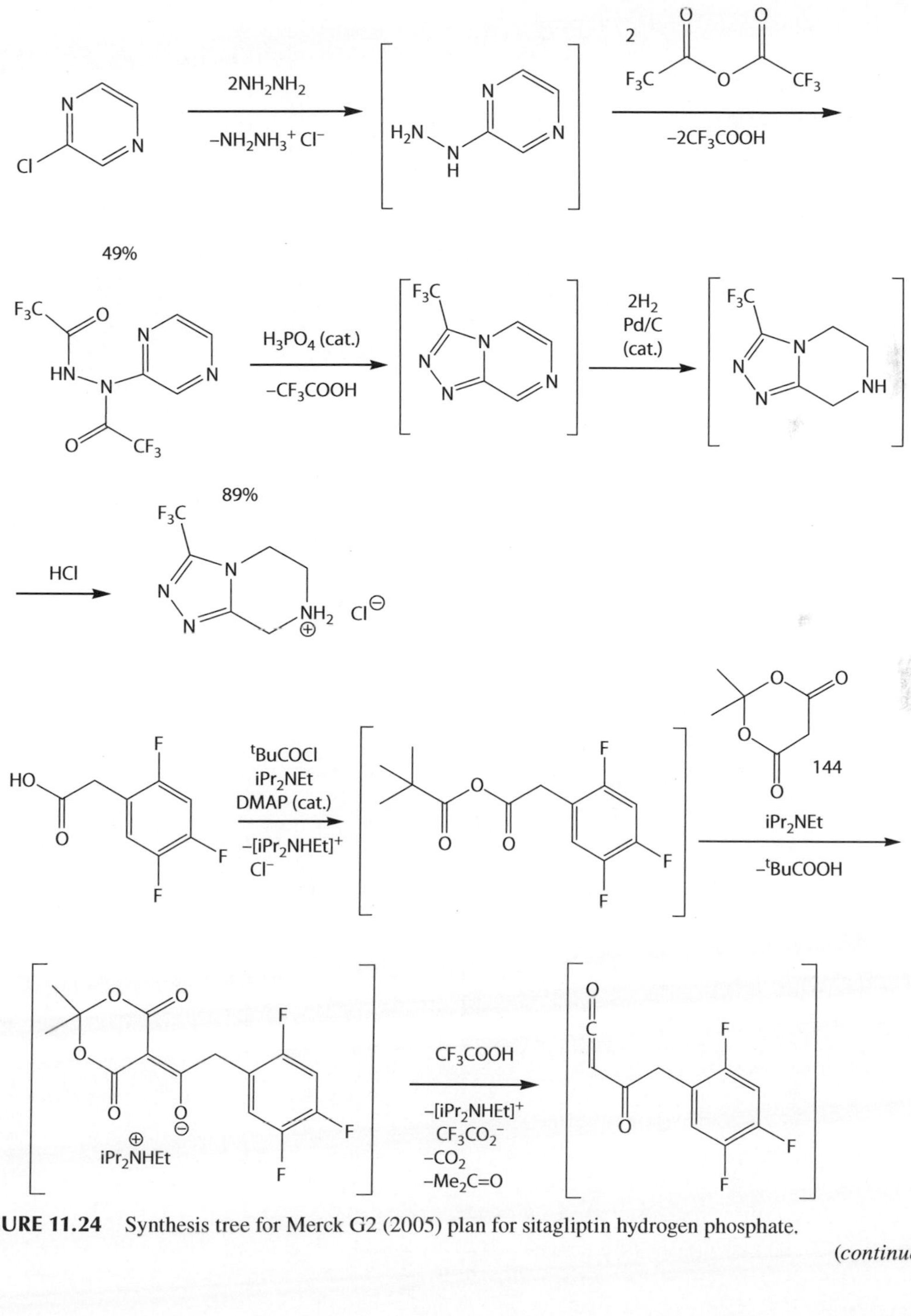

FIGURE 11.24 Synthesis tree for Merck G2 (2005) plan for sitagliptin hydrogen phosphate.

(*continued*)

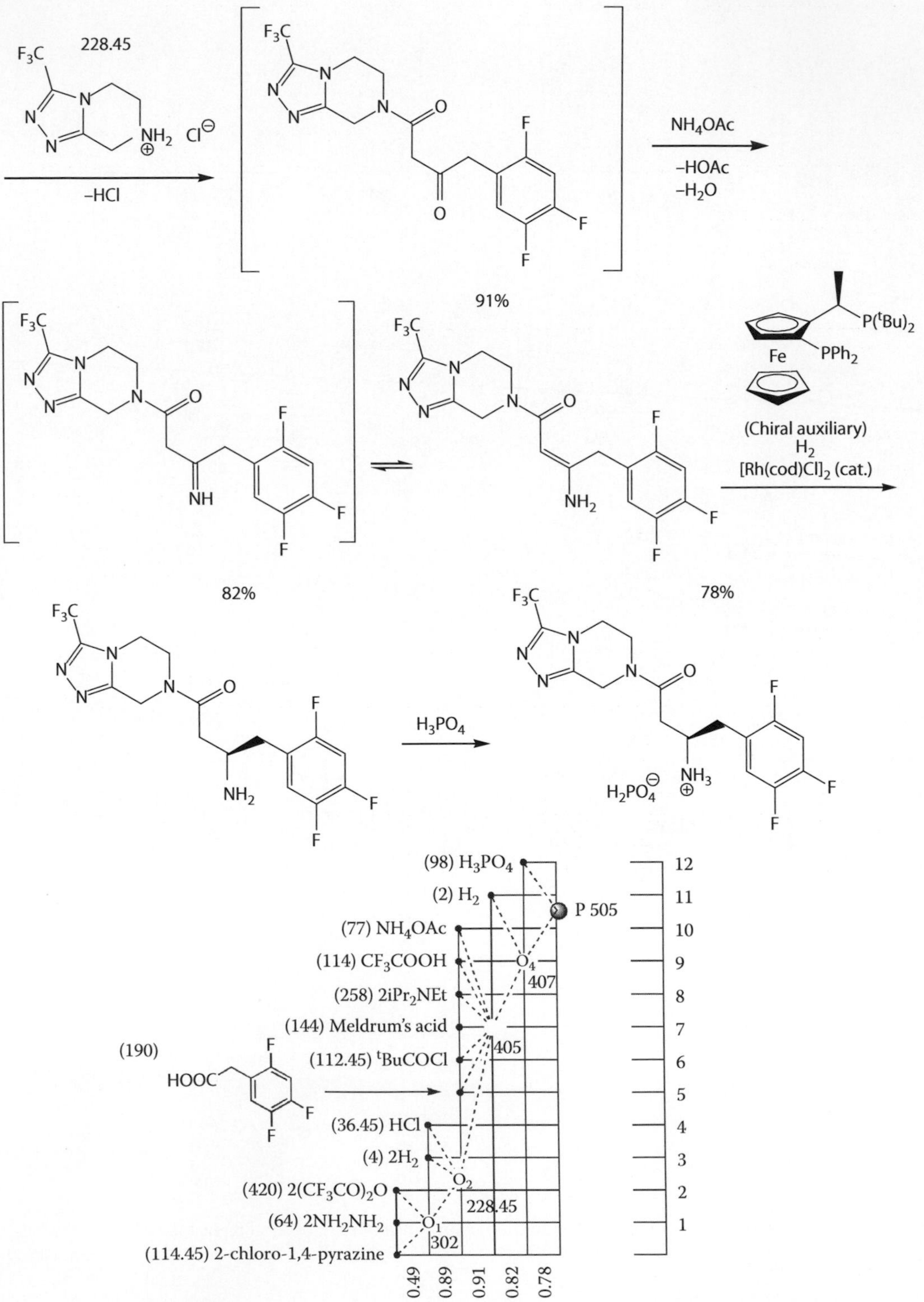

FIGURE 11.24 (continued)

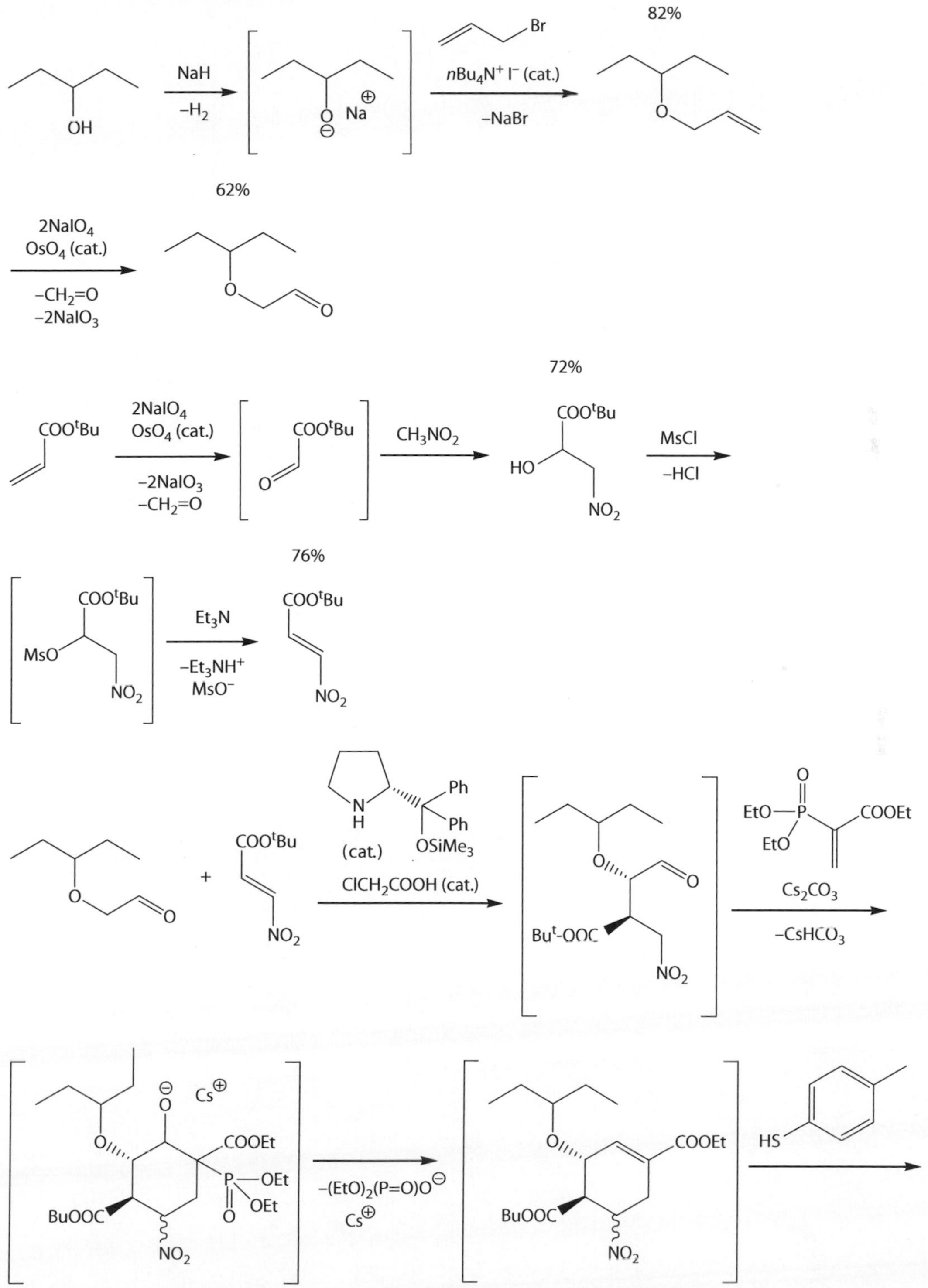

FIGURE 11.25 Synthesis tree for Hayashi (2009) plan for oseltamivir phosphate.

(*continued*)

FIGURE 11.25 (continued)

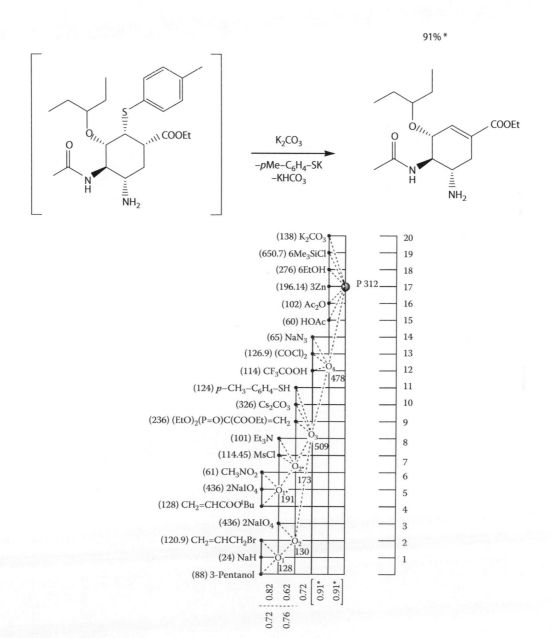

FIGURE 11.25 (continued)

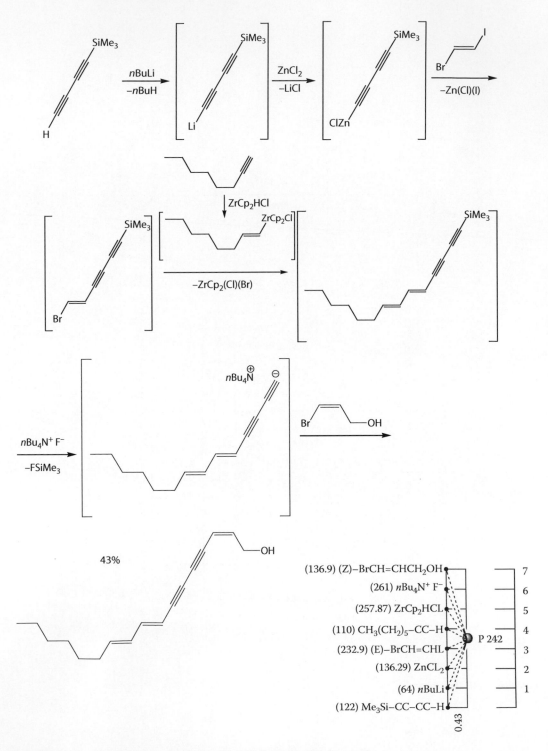

FIGURE 11.26 Synthesis tree for Organ G2 (2004) plan for bupleurynol.

Step 17

64%

Step 21

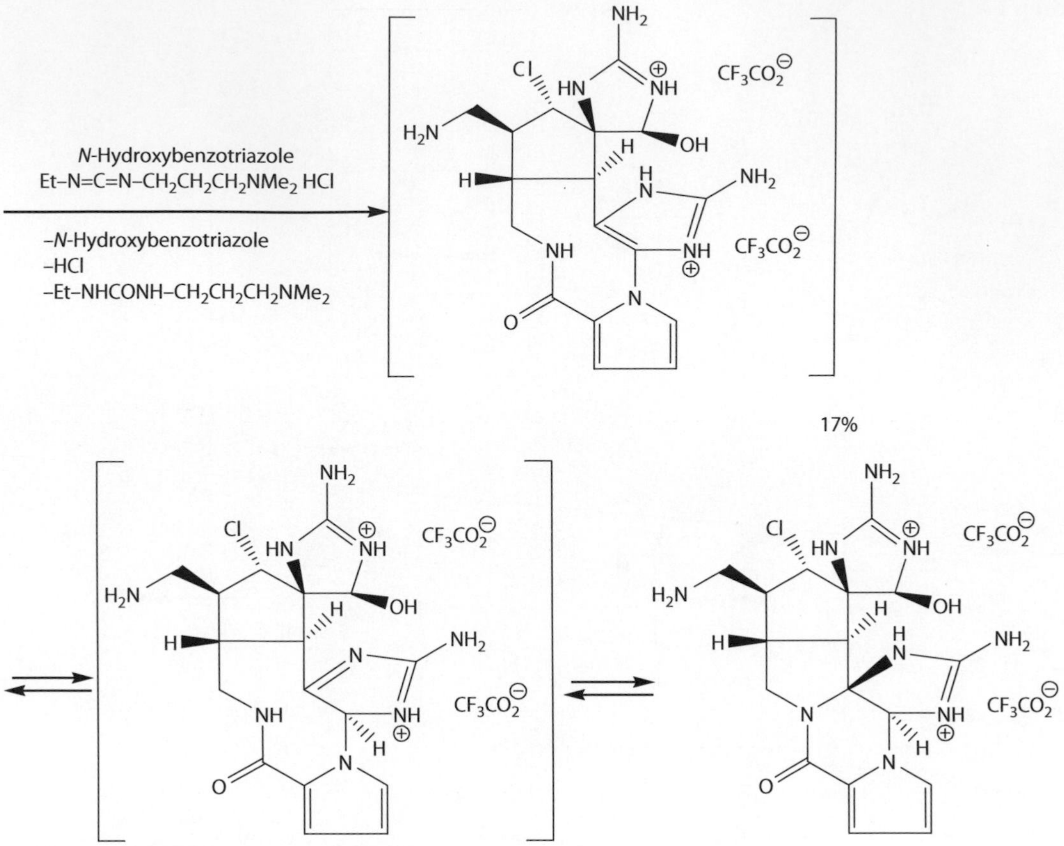

N-Hydroxybenzotriazole
Et–N=C=N–CH₂CH₂CH₂NMe₂ HCl

–N-Hydroxybenzotriazole
–HCl
–Et–NHCONH–CH₂CH₂CH₂NMe₂

17%

11.12 EXTREME CONVERGENCE IN PLANS

The following plans in the synthesis database exemplify extreme degrees of convergence:

Khorana–Todd [92] plan for coenzyme A (six branches), Myles [93] plan for (−)-discodermolide (six branches); Panek [94,95] plan for (+)-discodermolide (7 branches), and Paterson G1 plan [3,4] for (+)-discodermolide (5 branches). Figures 11.27 through 11.30 show abbreviated synthesis tree diagrams for these plans along with the reaction yield chains, color-coded branches, and points of convergence. Since these plans have multiple branches, the term overall yield can only be applied with respect to a specified branch. Taking the Khorana–Todd plan as an example, the overall yield values with respect to each of the six branches are as follows:

1. B1: $0.7 \times 0.31 \times 0.84 \times 0.95 \times 1 \times 0.5 \times 0.7 \times 0.66 \times 0.7 \times 0.9 \times 0.65 = 0.0164$ (1.6%)
2. B2*: $0.28 \times 1 \times (0.5 \times 0.7 \times 0.66 \times 0.7 \times 0.9 \times 0.65) = 0.0265$ (2.7%)
3. B3**: $0.8 \times (0.7 \times 0.9 \times 0.65) = 0.3276$ (32.8%)
4. B4***: $0.72 \times 0.76 \times 0.6 \times 0.88 \times 0.88 \times 1 \times (0.65) = 0.1653$ (16.5%)
5. B5****: $0.87 \times (0.6 \times 0.88 \times 0.88 \times 1) \times (0.65) = 0.2628$ (26.3%)
6. B6*****: $0.8 \times (0.88 \times 0.88 \times 1) \times (0.65) = 0.4027$ (40.3%)

The conventional practice is to state the overall yield with respect to the longest branch, which in this case is B1. Note that the chain of reaction yields is easily linked using these synthesis tree diagrams: B2* links to B1, B3** links to B1, B4*** links to B1, B5**** links to B4*** which in turn links to B1, and B6***** links to B4*** which in turn links to B1. A general rule is that the number of branches is equal to the number of points of convergence plus one. The reader can easily verify that the

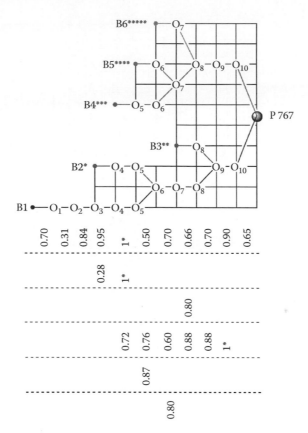

FIGURE 11.27 Abbreviated synthesis tree diagram for the Khorana–Todd (1961) plan for coenzyme A.

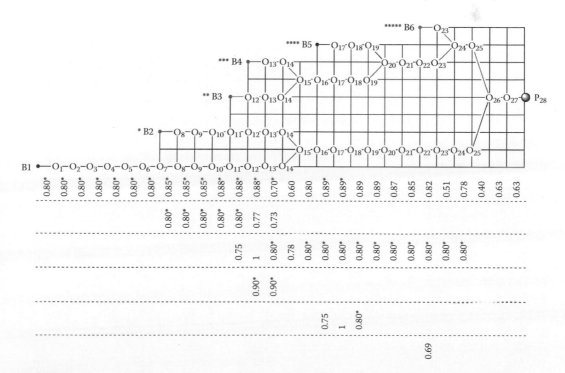

FIGURE 11.28 Abbreviated synthesis tree diagram for the Myles (1997) plan for (–)-discodermolide.

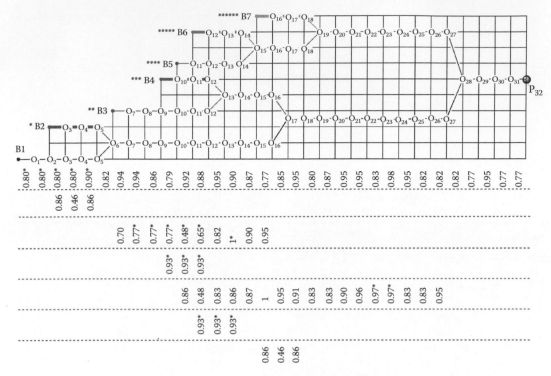

FIGURE 11.29 Abbreviated synthesis tree diagram for the Panek (2005) plan for (+)-discodermolide.

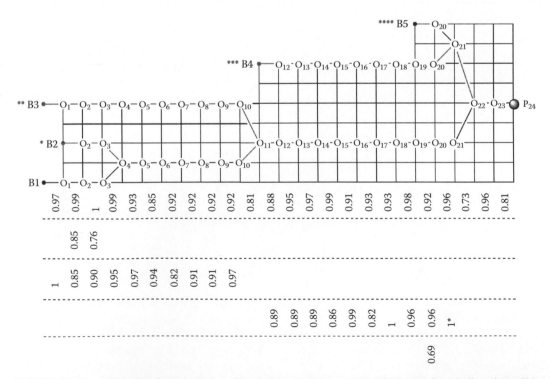

FIGURE 11.30 Abbreviated synthesis tree diagram for the Paterson G1 (2000) plan for (+)-discodermolide.

Khorana–Todd plan has six branches and five points of convergence: B1 and B2* converge, B3** and B1 converge, B4*** and B1 converge, B5**** and B4*** converge, and B6***** and B4*** converge.

11.13 USE OF A HUB COMMON INTERMEDIATE

When a complex target molecule has a common substructure that is repeated, this can be exploited to great advantage in planning a synthesis. To gain overall synthesis efficiency, the linear sequence to that common or hub intermediate must be as efficient as possible because it will appear in multiple branches in the overall synthesis tree diagram and thus it will be counted by that number of times in the determination of the overall kernel reaction mass efficiency. Plans incorporating such common sequences are called reflexive plans. The parsing of a complex structure in this way in addition to looking for symmetry elements are two effective techniques when applying retrosynthetic analysis. Synthesis plans for discodermolide illustrate this strategy well.

a. Smith G4 [9] plan for (+)-discodermolide: 5 step stretch (49% yield) to hub intermediate A is repeated twice

A

b. Smith G3 [96] plan for (+)-discodermolide: 5 step stretch (49% yield) to hub intermediate A is repeated three times
c. Smith G2 [1,2] plan for (+)-discodermolide: 5 step stretch (49% yield) to hub intermediate A is repeated three times
d. Smith G1 [97] plan for (−)-discodermolide: 5 step stretch (58% yield) to hub intermediate A is repeated three times
e. Novartis [5–8] plan for (+)-discodermolide: 7 step stretch (52% yield) to hub intermediate A is repeated three times
f. Paterson G2 [98] plan for discodermolide: 5 step stretch (53% yield) to hub intermediate B is repeated three times

B

11.14 COMPROMISE BETWEEN STRATEGY AND MATERIAL EFFICIENCIES

A good example illustrating this kind of compromise is the Panek [94,95] plan for (+)-discodermolide. Here the gains made in achieving stereoselectivity trump waste production as shown by the schemes in the following. The key reactions involve successive stereoselective additions of C6 fragments that are then pruned by three carbon atoms by ozonolysis in later steps. The net result is the addition of a C3 fragment in a stereoselective fashion by virtue of the chiral dimethylphenylsilyl directing group attached to the original C6 fragment.

Steps 6 to 8:

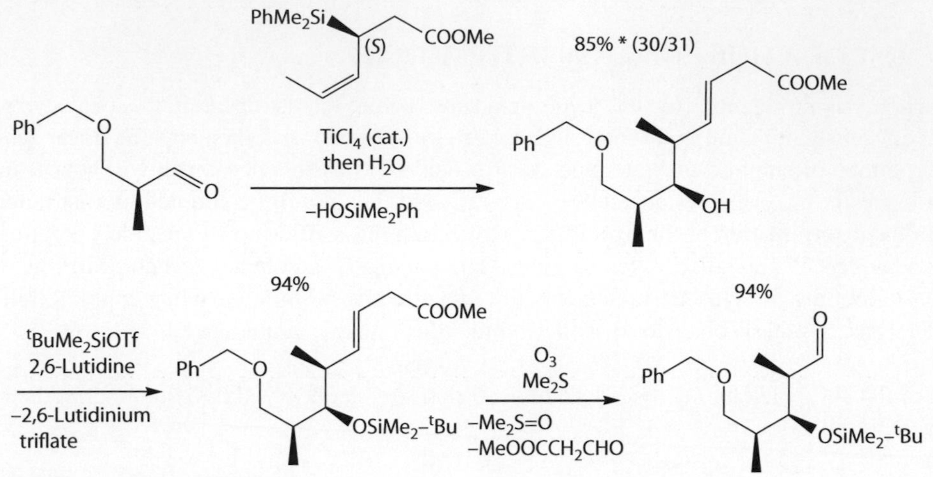

Steps 13 to 16**:**

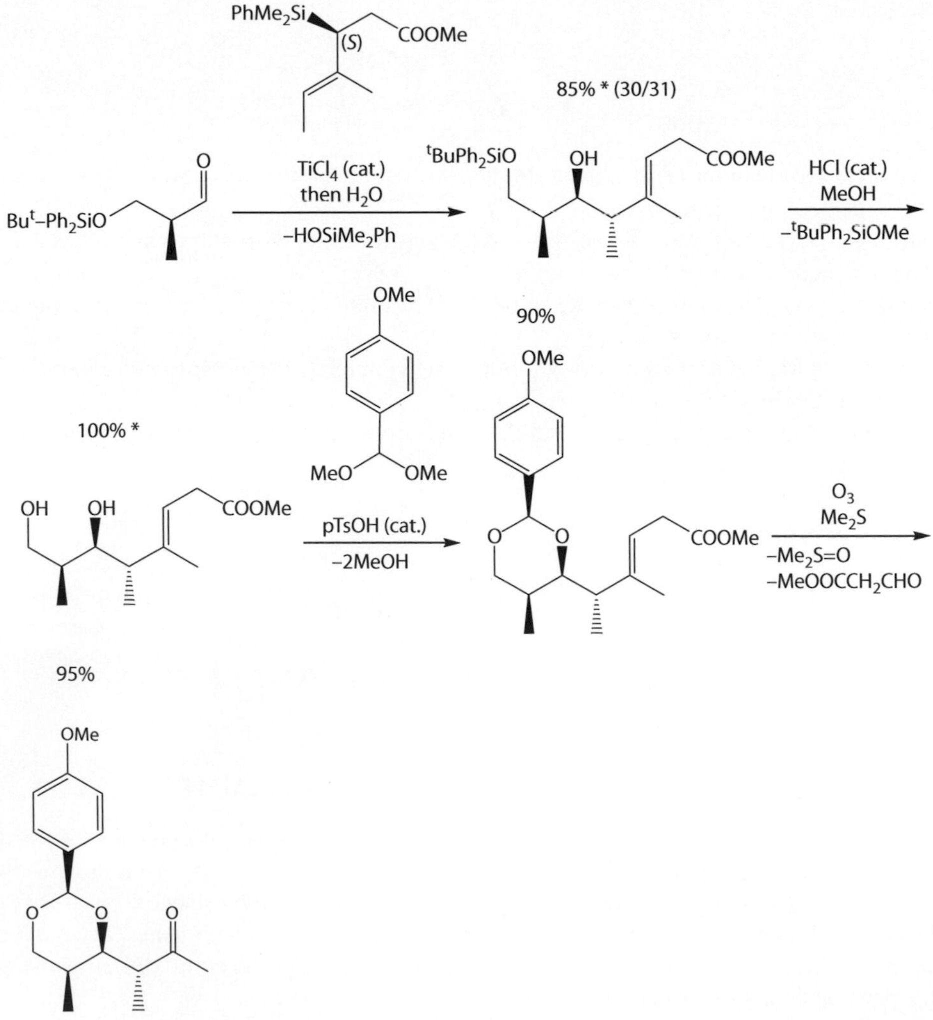

Steps 15**** to 23****:

11.15 INCLUSION OF SYNTHESES OF CATALYSTS AND LIGANDS FOR COMPLETE ANALYSIS

When a reaction step in a synthesis plan to a particular target compound requires an auxiliary material that itself needs to be synthesized, then the metrics analysis for the target compound synthesis plan must also incorporate the synthesis plan for the auxiliary material. Often the structures of these auxiliary materials are more complex than the final target compounds whose synthesis plans require their use. There is no evidence in the literature that these auxiliary materials are actually recovered and reused again once the intended target compound is synthesized. Nevertheless, in order to do a proper metrics analysis, the mole scale of the auxiliary material in the key step needs to be determined in relation to the mole scale of the target compound. The Takasago G1 [99–102] plan for (–)-menthol is worked out here in detail for illustrative purposes. The synthesis plan for (–)-menthol is given in the following.

The corresponding synthesis tree diagram for this plan is shown in Figure 11.31.

The basis mole scale of (–)-menthol is taken to be 1 mol. The second step involves a chiral 2,2′-bis(diphenylphosphino)-1,1′-binaphthyl (BINAP) catalyst to carry out the isomerization. From Figure 11.31, the scale of the second step is $1/(0.68 \times 0.7 \times 0.91 \times 0.94) = 2.456$ mol. The experimental

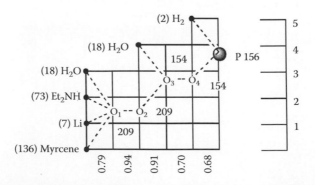

FIGURE 11.31 Synthesis tree diagram for Takasago G1 (1989) plan for (–)-menthol.

procedure for this step indicates that the catalyst loading is 0.1 mol% with respect to the limiting reagent geranylamine. Therefore, the mole scale of chiral BINAP catalyst required to aim for a 1 mol scale of (−)-menthol is 2.456 × (0.1/100) = 0.002456 mol. The E-factor breakdown for the synthesis of (−)-menthol as calculated using the LINEAR COMPLETE spreadsheet algorithm is: E-kernel = 3.95, E-excess = 16.20, E-auxiliaries = 290.40, and E-total = 310.55 or 48.4 kg waste per mole (−)-menthol. The E-total value includes the 0.002456 mol of BINAP catalyst as part of the total accumulated waste material. The BINAP catalyst is synthesized according to the scheme [100,103,104] shown in the following.

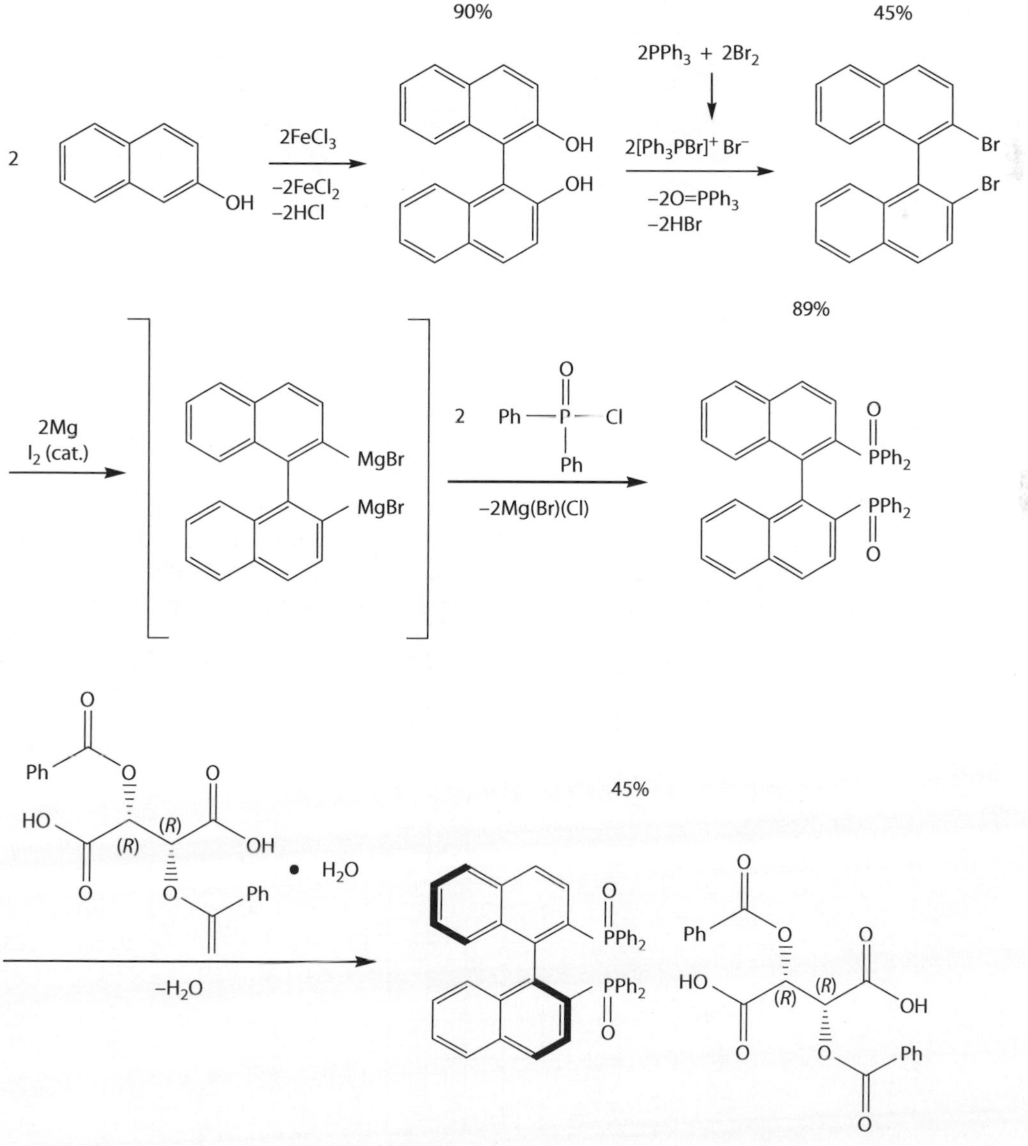

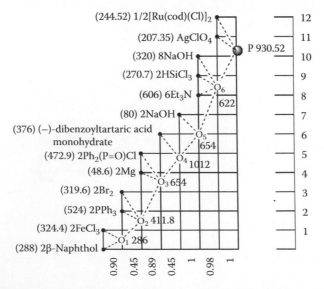

The corresponding synthesis tree diagram is shown in Figure 11.32.

The E-factor breakdown as calculated using the LINEAR COMPLETE spreadsheet algorithm is as follows:

E-kernel = 12.50, E-excess = 6.16, E-auxiliaries = 753.31, and E-total = 753.31 or 718.3 kg/mole chiral BINAP catalyst. Since 0.002456 mol of BINAP catalyst are required, then 718.3 × 0.002456 = 1.76 kg of additional waste will be produced. In total, then, the synthesis of 1 mol of (−)-menthol by the Takasago G1 plan will produce 48.4 + 1.76 = 50.16 kg of waste material.

FIGURE 11.32 Synthesis tree diagram for synthesis plan chiral BINAP catalyst.

From its molecular weight of 156 g/mol, the overall E-total will be 50.16/0.156 = 321.57, which may be compared with 310.55 calculated earlier without including the synthesis of the chiral BINAP catalyst.

In a similar fashion the following plans are analyzed:

a. Trost [105] plan for oseltamivir phosphate including synthesis of Trost ligand and DuBois catalyst
b. Corey [106] plan for oseltamivir phosphate including synthesis of Corey-Shibata-Lee catalyst
c. Shibasaki G1 [107] plan for oseltamivir phosphate including synthesis of Shibasaki ligand
d. Shibasaki G2 [108] plan for oseltamivir phosphate including synthesis of Shibasaki ligand
e. Shibasaki G4 [109] plan for oseltamivir phosphate including synthesis of Shibasaki ligand
f. Shibasaki G5 [110,111] plan for oseltamivir phosphate including synthesis of Shibasaki ligand
g. Fukuyama [112] plan for oseltamivir phosphate including synthesis of Macmillan catalyst
h. Ciba-Syngenta [113] plan for metolachlor including synthesis of chiral ferrocenyl xylophos auxiliary.

The structures of these auxiliary materials are shown in Figure 11.33.

Table 11.7 summarizes the results of these calculations. The adjusted E-total values increased in magnitude between 1.2% and 9.8% when the syntheses of auxiliary materials were taken into account.

11.16 SYNTHESIS PLANS TRACING BACK TO COMMON STARTING MATERIALS FOR FAIREST COMPARISONS

A key problem in comparing several synthesis plans to a common target molecule is that the comparison is one-sided since different sets of starting materials lead to the same product node. It can therefore be argued that the whole exercise of comparing plans inherently can never be a fair exercise. This is where a recommendation to get the chemistry community to come to a consensus on itemizing a set of starting first and second generation feedstocks from which all synthesis plans originate becomes important. If such a set of starting materials could be agreed upon, then the one-sided problem would vanish. However, at the present moment we can only make do with what authors of plans say they used as starting materials–usually they are commercially available chemicals. In several instances while developing the present database, these materials were found to be advanced structures. For these cases, literature searches were done to trace how these molecules were made and then these reactions were linked with the remainder of the plans so that fair comparisons could be made as best as possible. Of the 200+ target molecules examined only three had plans that could be traced to exactly the same set of starting materials. These were ibuprofen, warfarin, and (+)-paroxetine. Figure 11.34 shows the structures of these pharmaceuticals and their progenitor starting materials. The statistics on these plans may be found in Chapter 13.

11.17 MOST CHALLENGING MOLECULE TO SYNTHESIZE: COLCHICINE

Of the 1050+ plans examined in this database, the poorest performing ones refer to the synthesis of colchicine. Seven out of nine published plans to date for this target compound fall into the same trap of the isocolchicine—colchicines equilibrium in the late stages of the synthesis plans as shown in Figure 11.35. The Cha and the Banwell plans are the exceptions. The final structure

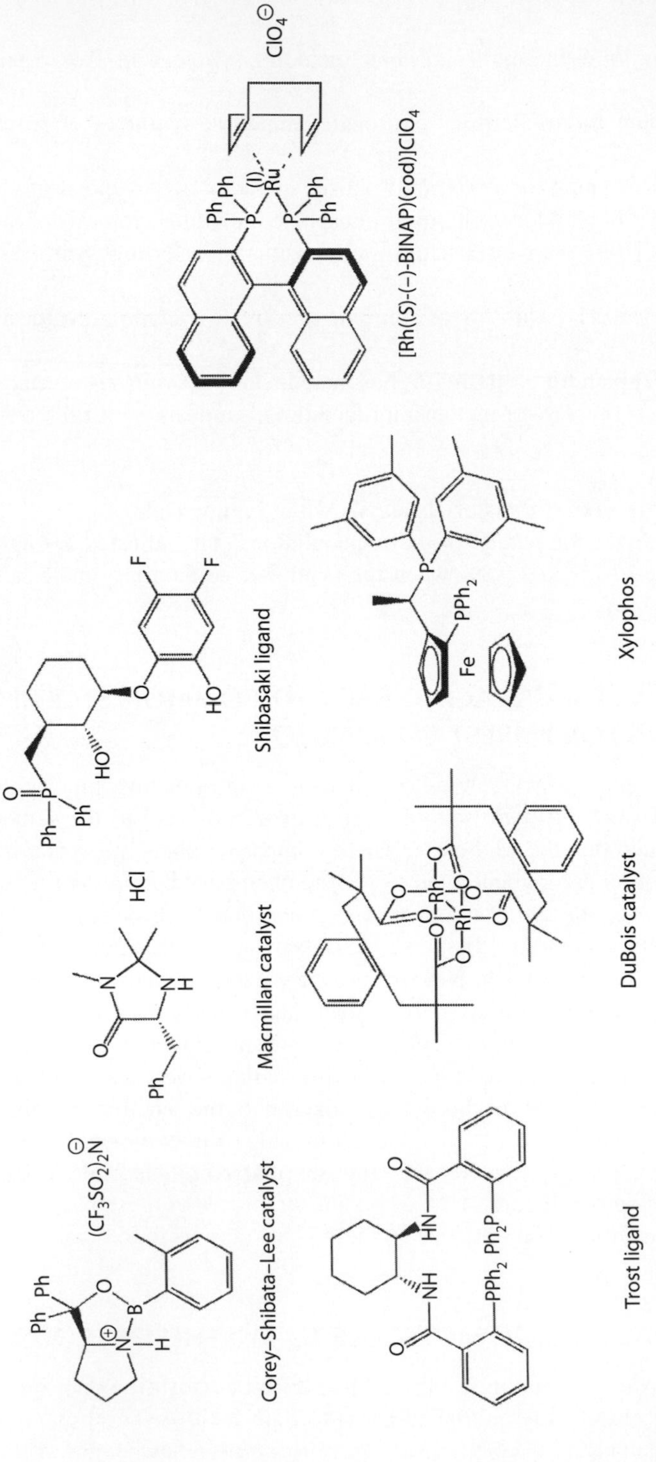

FIGURE 11.33   Structures of auxiliary materials.

TABLE 11.7
Summary of Calculations for Plans Involving Auxiliary Materials That Need to Be Synthesized in Addition to Target Compound

Plan	Step Number in Target Compound Synthesis Tree	Mole Scale of Step	Mol% of Auxiliary Required in Step	Adjusted Mole Scale of Auxiliary	Raw Mass of Waste (kg) to Produce 1 mol of Target Compound	Raw E-Total for Target Compound	Mass of Waste (kg) to Produce 1 mol Auxiliary Material	Additional Mass of Waste (kg) Produced in Making Auxiliary	Adjusted E-Total for Target Compound
Takasago G1 (1989)	2	2.456	0.1	0.002456	48.4	310.55	718.3	1.76	321.57
Trost (2008)	4[a]	3.34[a]	7.5[a]	0.25[a]	1138.9	2778	101.6[a]	25.5[a]	2859
	7[b]	2.24[b]	2.0[b]	0.045[b]			170.3[b]	7.6[b]	
Corey (2006)	1	4.45	10.0	0.45	1295.4	3159	113.5	50.6	3283
Shibasaki G1 (2006)	4	45.3	4.0	1.81	6047.3	14749	326.9	592.4	16194
Shibasaki G2 (2007)	4	14.4	4.0	0.57	8100.2	19756	327.0	187.8	20214
Shibasaki G4 (2009)	1	10.2	2.5	0.26	1947.7	4751	321.9	83.7	4955
Shibasaki G5 (2009)	1	15.0	2.5	0.37	2151.2	5247	327.0	122.4	5545
Fukuyama (2007)	2	17.4	10.2	1.77	1638.8	3997	11.4	20.3	4046
Ciba-Syngenta (1999)	1	1.111	0.01	1.111e-4	1.05	3.72	280.1	0.031	3.83

[a] Trost ligand.
[b] DuBois catalyst.

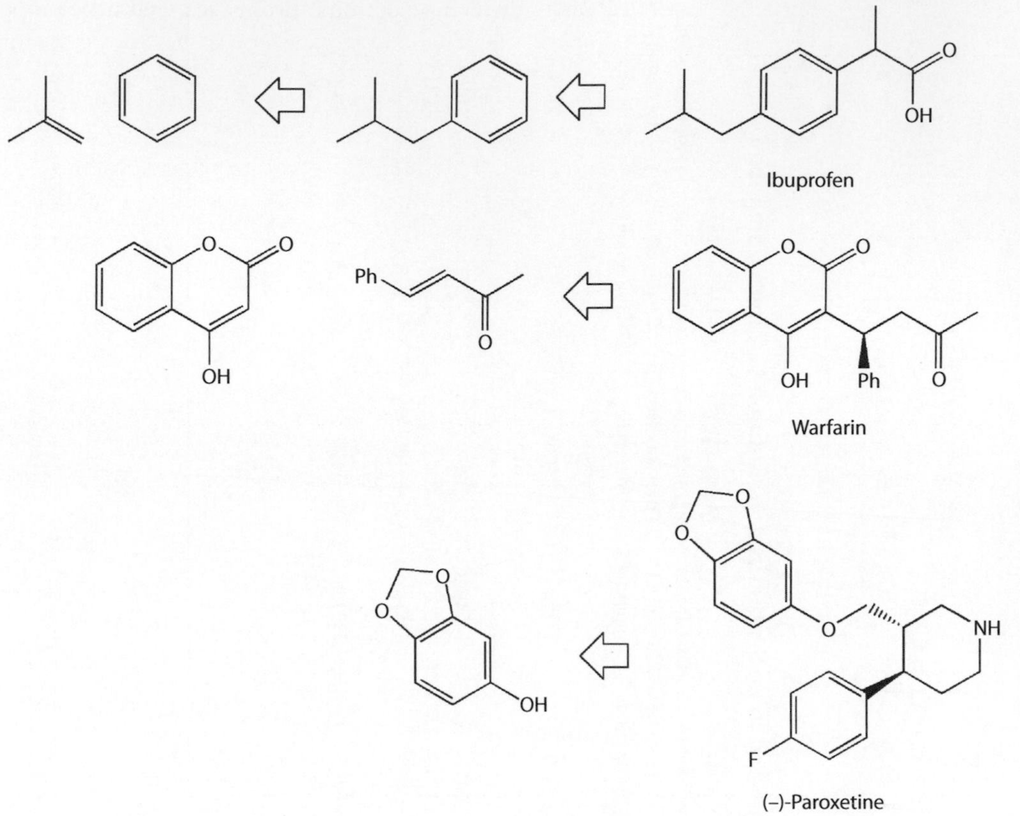

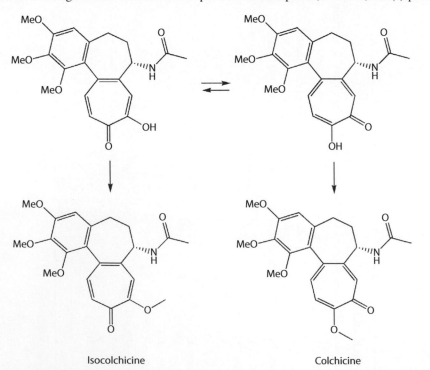

FIGURE 11.34 Starting materials used in various plans to make ibuprofen, warfarin, and (+)-paroxetine.

FIGURE 11.35 Isocolchicine—colchicines equilibrium system.

is made of three fused rings that have vastly different electronic properties, and hence this is the source of the challenging task in constructing such a system. The best performing plan to date is the one by Cha [114,115] with 1.6% overall yield over 16 steps and producing 79.1 kg kernel waste per mole colchicine. The worst performing plan is that of van Tamelen [116,117] with 0.00001% overall yield over 22 steps and producing an astonishing 12,307,680 kg of kernel waste per mole colchicine. Details of the plans are given in the accompanying download and a summary of the plan statistics is given in Chapter 13. Ring construction strategies are given in Section 10.1.

11.18 TARGET COMPOUND WHERE FUNDAMENTAL NAMED ORGANIC REACTIONS WERE DISCOVERED

β-carotene is an exceptional target molecule because three named organic reactions were discovered or developed in conjunction with its total synthesis. These are the McMurry reaction [118], the Wittig reaction as exemplified by the BASF plan [119], and the hydrogenation of acetylenes to cis-olefins using Lindlar's catalyst as exemplified by the Roche G1 [120,121] plan.

McMurry coupling:

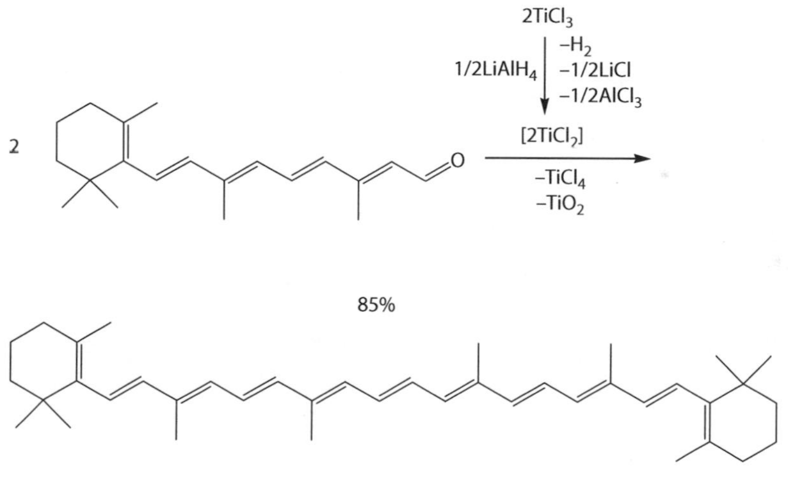

Wittig reaction:

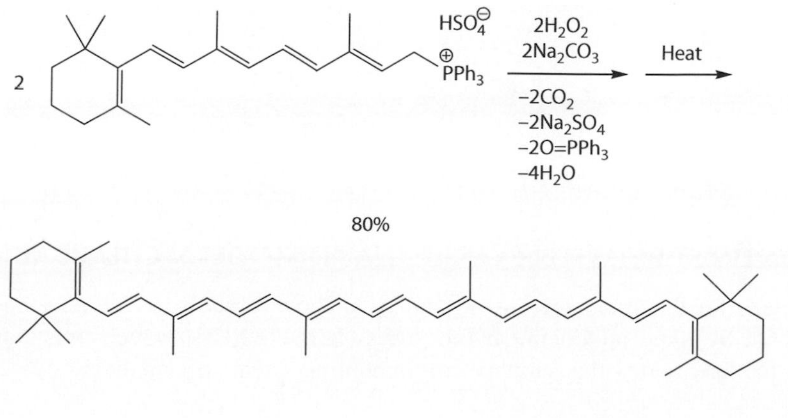

Lindlar catalyst for hydrogenation of acetylenes to *cis*-olefins:

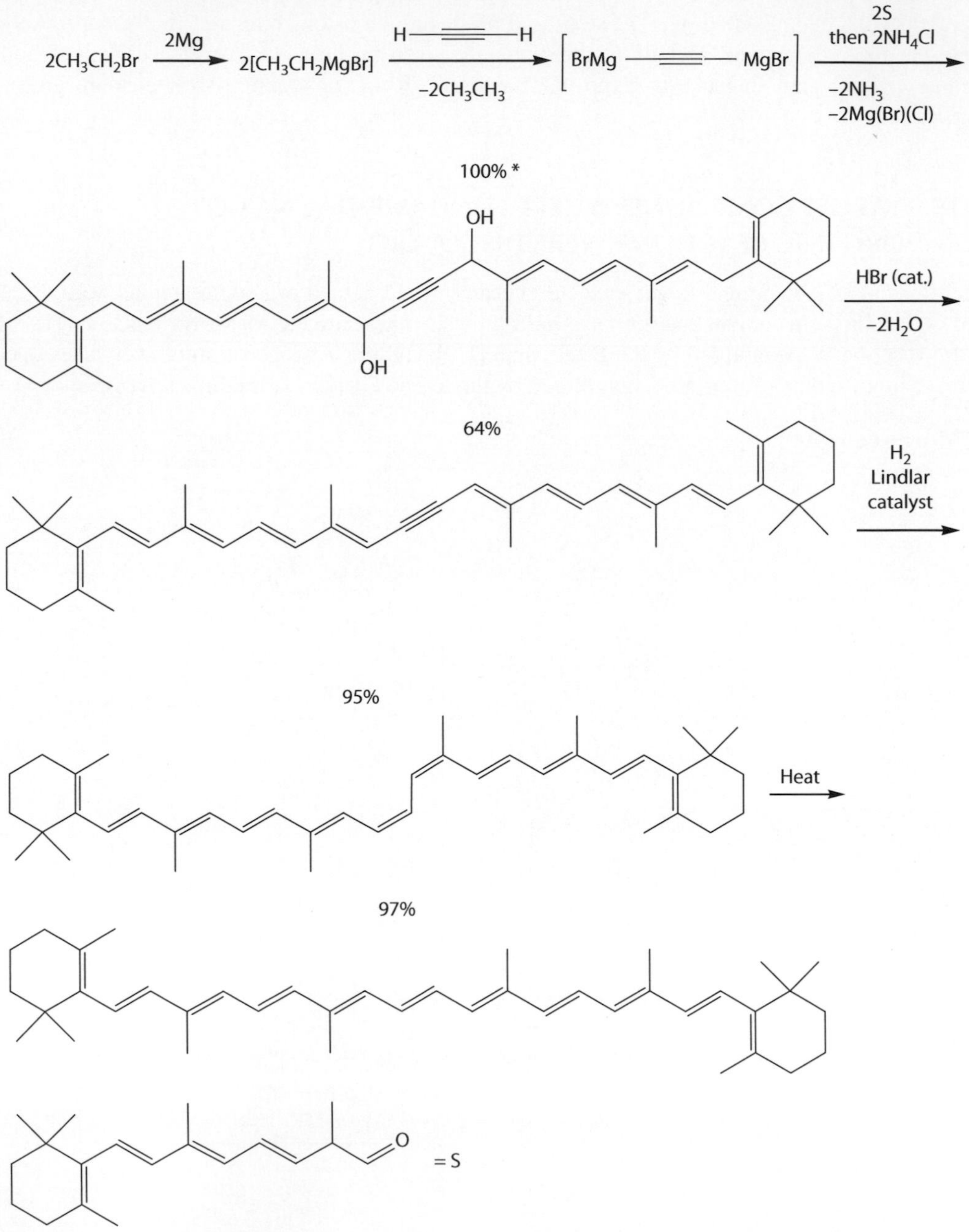

$$2CH_3CH_2Br \xrightarrow{2Mg} 2[CH_3CH_2MgBr] \xrightarrow[-2CH_3CH_3]{H\!\!=\!\!=\!\!=\!\!H} [BrMg\!\!=\!\!=\!\!=\!\!MgBr] \xrightarrow[\substack{-2NH_3\\-2Mg(Br)(Cl)}]{\substack{2S\\ \text{then } 2NH_4Cl}}$$

100% *

HBr (cat.)

−2H₂O

64%

H₂
Lindlar
catalyst

95%

Heat

97%

= S

11.19   CONTRAST BETWEEN CLASSICAL AND MODERN CHEMICAL ROUTES

The set of plans for the anticancer agent tamoxifen show a nice contrast between the original discovery route by Imperial Chemical Industries (ICI) [122,123], which uses classical chemistry versus modern routes that employ cross coupling chemistry mediated by palladium salt catalysts. The classical route is shown in the following. The key carbon–carbon bond forming reactions are a benzoin condensation (step 1), alpha-carbon alkylation (step 3), and a Grignard reaction (step 4).

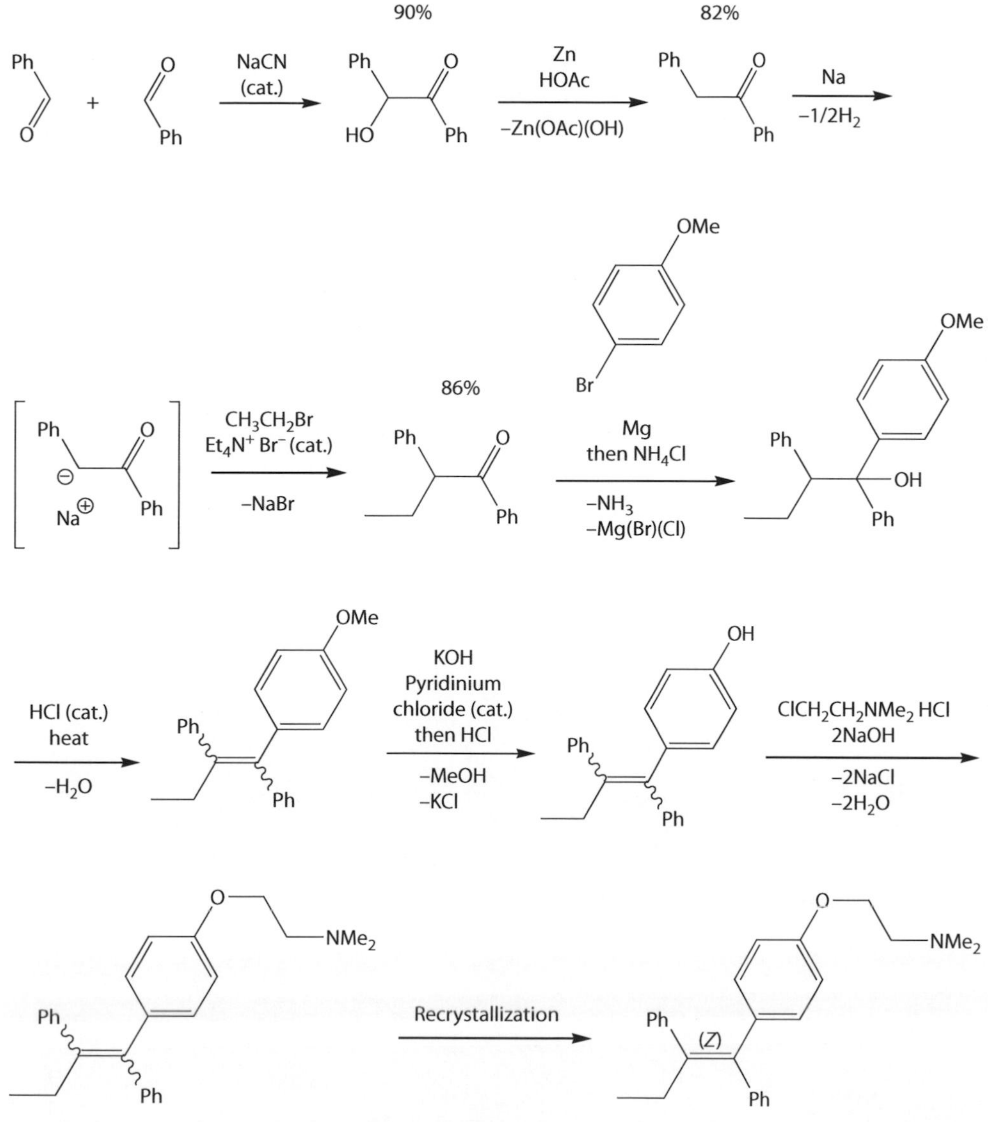

90%   82%

NaCN (cat.)

Zn
HOAc
−Zn(OAc)(OH)

Na
−1/2H$_2$

CH$_3$CH$_2$Br
Et$_4$N$^+$ Br$^-$ (cat.)
−NaBr

86%

Mg
then NH$_4$Cl
−NH$_3$
−Mg(Br)(Cl)

HCl (cat.)
heat
−H$_2$O

KOH
Pyridinium chloride (cat.)
then HCl
−MeOH
−KCl

ClCH$_2$CH$_2$NMe$_2$ HCl
2NaOH
−2NaCl
−2H$_2$O

Recrystallization

(Z)

The reader is directed to the accompanying download to view the remaining 19 plans and to Chapter 13 for a summary of the plan statistics. Unfortunately, because the ICI plan did not disclose reaction yields for all steps including experimental procedures, little can be done to compare its performance against the modern cross-coupling strategies. The National Research Development Corporation (NRDC) plan [124,125], which uses McMurry coupling chemistry, is so far the best performer with the least E-total of 179. The worst performer is the Katzenellenbogen [126] plan with an E-total of 27,138.

11.20   RESOLUTION WITH LIPASE

In this section, two synthetic plans using lipases as chiral discriminators are presented where it is used in two different roles. The first case is the Bracher plan [127] for (R)-fluoxetine shown in the plan as follows. In step 2, the acetylated product is unwanted and the nonacetylated chiral alcohol is the desired compound, which is taken on in the next step.

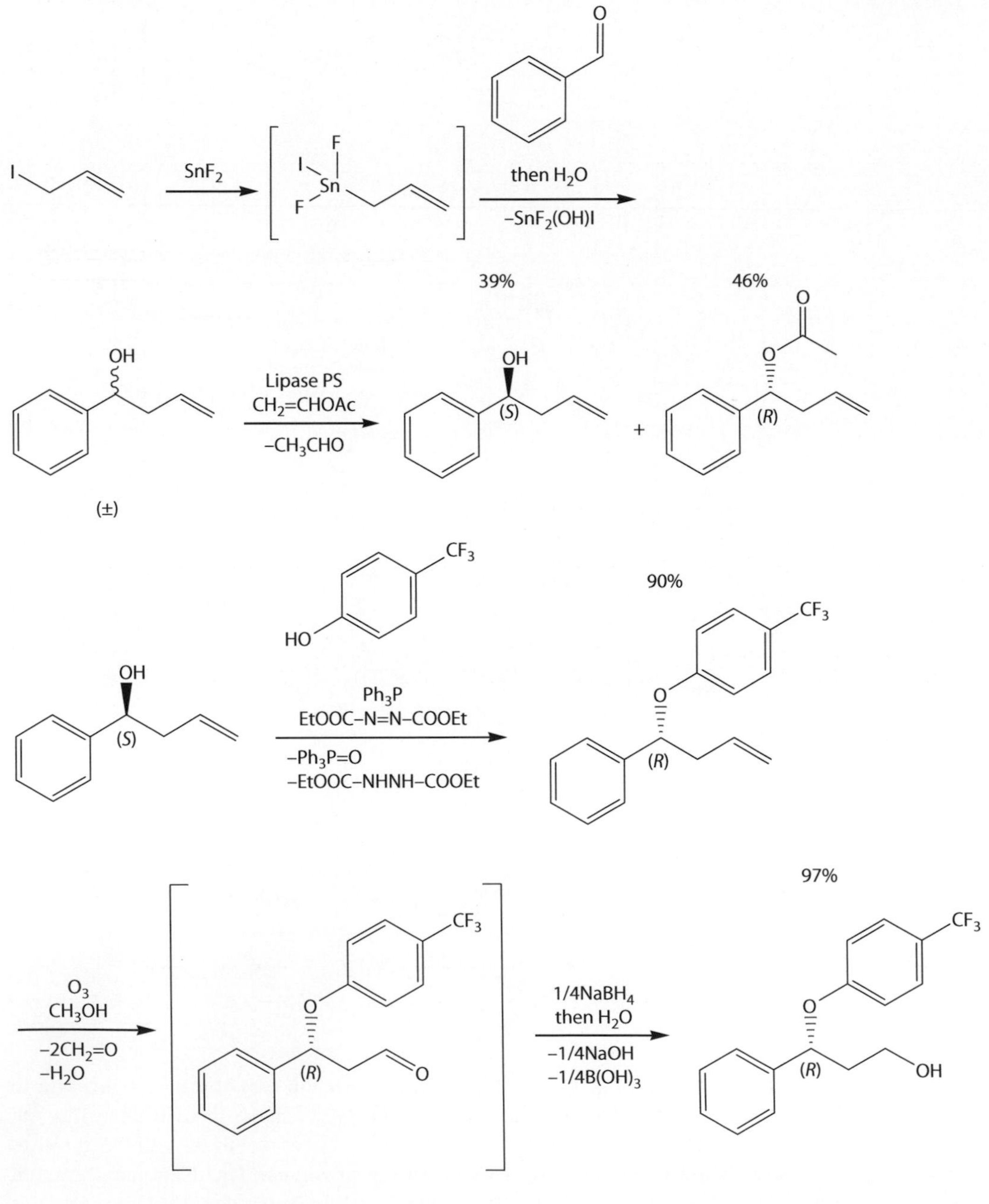

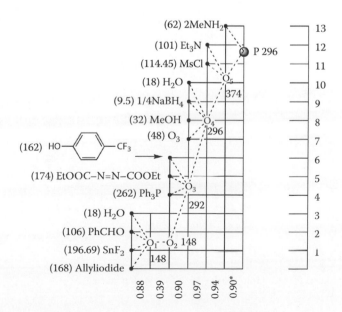

Figure 11.36 shows the synthesis tree for the Bracher plan. Note that the vinyl acetate reagent is not included as a reagent in the diagram because it operates on the unwanted stereoisomer and therefore does not enter into the balanced chemical equation for that step. However, in the calculation of the E-factor for that step in the radial pentagon analysis and in E-total for the entire plan it is of course counted as an input material.

The second case is the Gotor [128] plan for (S)-dapoxetine whose synthesis scheme is shown in the following. In this case, step 4 shows that the chiral amide product is the desired one and the free amine product is unwanted. Figure 11.37 shows the synthesis tree for this plan.

Note that in step 4 ethyl methoxyacetate is included as a reagent in the tree diagram because it is part of the balanced chemical equation leading to the desired amide product.

FIGURE 11.36 Synthesis tree for Bracher (1996) plan for (R)-fluoxetine.

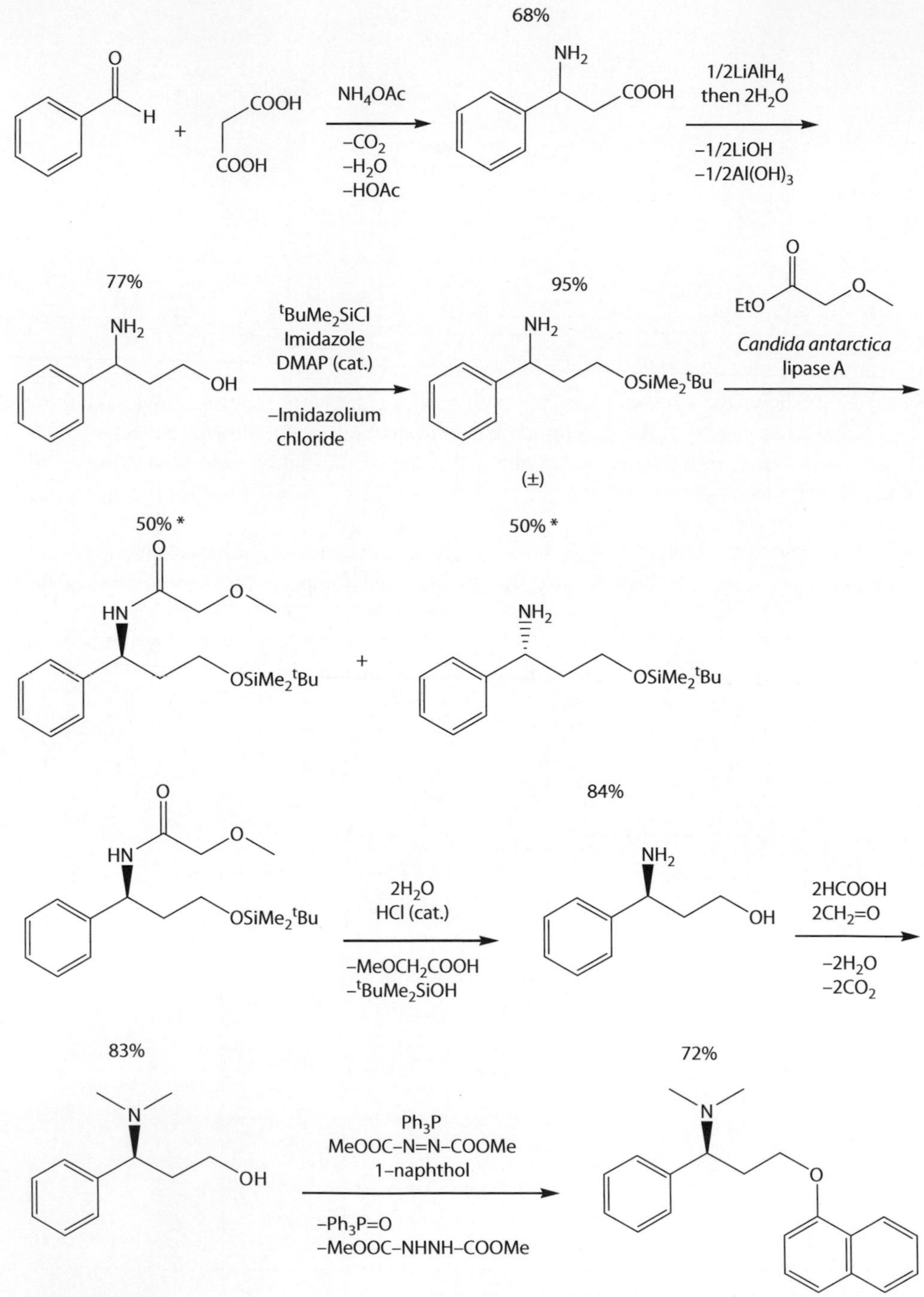

FIGURE 11.37 Synthesis tree for Gotor (2006) plan for (*S*)-dapoxetine.

*Assumed yield

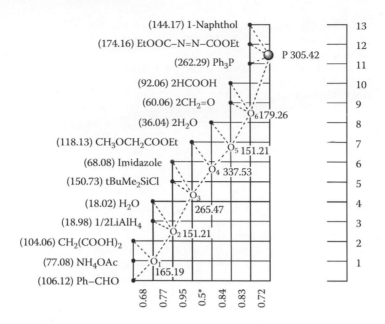

FIGURE 11.37 (continued)

REFERENCES

1. Smith, A.B. III; Kaufman, M.D.; Beauchamp, T.J.; LaMarche, M.J.; Arimoto, H. *Org. Lett.* 1999, *1*, 1823.
2. Smith, A.B. III; Beauchamp, T.J.; LaMarche, M.J.; Kaufman, M.D.; Qiu, Y.; Arimoto, H.; Jones, D.R.; Kobayashi, K. *J. Am. Chem. Soc.* 2000, *122*, 8654.
3. Paterson, I.; Florence, G.J.; Gerlach, K.; Scott, J.P. *Angew. Chem. Ind. Ed.* 2000, *39*, 377.
4. Paterson, I.; Florence, G.J.; Gerlach, K.; Scott, J.P.; Sereinig, N. *J. Am. Chem. Soc.* 2001, *123*, 9535.
5. Mickel, S.J.; et al. *Org. Process Res. Develop.* 2004, *8*, 92.
6. Mickel, S.J.; et al. *Org. Process Res. Develop.* 2004, *8*, 101.
7. Mickel, S.J.; et al. *Org. Process Res. Develop.* 2004, *8*, 107.
8. Mickel, S.J.; et al. *Org. Process Res. Develop.* 2004, *8*, 113.
9. Smith, A.B. III; Freeze, B.S.; Xian, M.; Hirose, T. *Org. Lett.* 2005, *7*, 1825.
10. Paterson, I.; Lyothier, I. *Org. Lett.* 2004, *6*, 4933.
11. Willstätter, R.; Waser, E. *Chem. Ber.* 1911, *44*, 3423.
12. Willstätter, R.; Heidelberger, M. *Chem. Ber.* 1913, *46*, 517.
13. Cope, A.C.; Overberger, C.G. *J. Am. Chem. Soc.* 1948, *70*, 1433.
14. Hirasawa, H.; Taniguchi, T.; Ogasawara, K. *Tetrahedron Lett.* 2001, *42*, 7587.
15. Hamon, D.P.G.; Young, R.N. *Austr. J. Chem.* 1976, *29*, 145.
16. Brenna, E.; Fuganti, C.; Ronzani, S.; Serra, S. *Helv. Chim. Acta* 2001, *84*, 3650.
17. Späth, E.; Koller, G. *Chem. Ber.* 1923, *56*, 2454.
18. Wittig, G.; Schoellkopf, U. *Org. Synth. Coll.* 1973, *5*, 751.
19. Kotian, P.L.; Carroll, F.I. *Synth. Commun.* 1995, *25*, 63.
20. Sheehan, J.C.; Henery-Logan, K.R. *J. Am. Chem. Soc.* 1957, *79*, 1262.
21. Sheehan, J.C.; Henery-Logan, K.R. *J. Am. Chem. Soc.* 1959, *81*, 3089.
22. Sheehan, J.C.; Henery-Logan, K.R. *J. Am. Chem. Soc.* 1962, *84*, 2983.
23. Hermeling, D.; Bassler, P.; Hammes, P.; Hugo, R.; Lechtken, P.; Siegel, H. US 5527966 (BASF, 1996).
24. O'Brien, C.J.; Tellez, J.L.; Nixon, Z.S.; Kang, L.J.; Carter, A.L.; Kunkel, S.R.; Przeworski, K.C.; Chass, G.A. *Angew. Chem. Int. Ed.* 2009, *48*, 6836.
25. Mahajana, R.P.; Patila, S.L.; Malia, R.S. *Org. Prep. Proced. Int.* 2005, *37*, 286.
26. El-Batta, A.; Jiang, C.; Zhao, W.; Anness, R.; Cooksy, A.L.; Bergdahl, M. *J. Org. Chem.* 2007, *72*, 5244.
27. Johnson, W.S.; Banerjee, D.K.; Schneider, W.P.; Gutsche, C.D.; Shelberg, W.E.; Chinn, L.J. *J. Am. Chem. Soc.* 1952, *74*, 2832.
28. Bartlett, P.A.; Johnson, W.S. *J. Am. Chem. Soc.* 1973, *95*, 7501.

29. Welch, W.M.; Kraska, A.R.; Sarges, R.; Koe, B.K. *J. Med. Chem.* 1984, *27*, 1508.
30. Lautens, M.; Rovis, T. *Tetrahedron* 1999, *55*, 8967.
31. Lautens, M.; Rovis, T. *J. Org. Chem.* 1997, *62*, 5246.
32. Quallich, G.J.; Williams, M.T.; Friedmann, R.C. *J. Org. Chem.* 1999, *55*, 4971.
33. Yun, J.; Buchwald, S.L. *J. Org. Chem.* 2000, *65*, 767.
34. Vukics, K.; Fodor, T.; Fischer, J.; Fellegvári, I.; Lévai, S. *Org. Process Res. Develop.* 2002, *6*, 82.
35. Taber, G.P.; Pfisterer, D.M.; Colberg, J.C. *Org. Process Res. Develop.* 2004, *8*, 385.
36. Wang, G.; Zheng, C.; Zhao, G. *Tetrahedron Asymm.* 2006, *17*, 2074.
37. Krumlinde, P.; Bogar, K.; Bäckvall, J.E. *Chem. Eur. J.* 2010, *16*, 4031.
38. Crimmins, M.T.; Tabet, E.A. *J. Am. Chem. Soc.* 2000, *122*, 5473.
39. Saitoh, T.; Suzuki, T.; Sugimoto, M.; Hagiwara, H.; Hoshi, T. *Tetrahedron Lett.* 2003, *44*, 3175.
40. Sasaki, M.; Hashimoto, A.; Tanaka, K.; Kawahata, M.; Yamaguchi, K.; Takeda, K. *Org. Lett.* 2008, *10*, 1803.
41. Iijima, I.; Minamikawa, J.; Jacobson, A.E.; Brossi, A.; Rice, K.C. *J. Org. Chem.* 1978, *43*, 1462.
42. Rice, K.C. *J. Org. Chem.* 1980, *45*, 3137.
43. Rice, K.C. *J. Med. Chem.* 1977, *20*, 164.
44. Magnus, P.; Sane, N.; Fauber, B.P.; Lynch, V. *J. Am. Chem. Soc.* 2009, *131*, 16045.
45. Reppe, W.; Schlichting, O.; Klager, K.; Toepel, T. *Ann. Chem.* 1948, *560*, 1.
46. Dirat, O. *Organic Process Research & Development Conference*, Lisbon, Portugal, September 2009.
47. Campos, K.R.; Klapers, W.J.H.; Dormer, P.G.; Chen, C.Y. *J. Am. Chem. Soc.* 2006, *128*, 3538.
48. Robinson, R. *J. Chem. Soc.* 1917, 762.
49. Baxendale, I.R.; Hayward, J.J.; Lanners, S.; Ley, S.V.; Smith, C.D. Organic chemistry in microreactors: Heterogeneous reactions, in *Microreactors in Organic Synthesis and Catalysis* (T. Wirth, ed.), Wiley-VCH: Weinheim, Germany, 2008, Chapter 4.2, pp. 84–122.
50. Baxendale, I.R.; Ley, S.V. Solid supported reagents in multi-step flow synthesis, in *New Avenues to Efficient Chemical Synthesis Emerging Technologies* (P.H. Seeberger, T. Blume, eds.), Springer: Berlin, Germany, 2007, pp. 151–185.
51. Ehrfeld, W.; Hessel, V.; Löwe, H. *Microreactors—New Technology for Modern Chemistry*, Wiley-VCH: Weinheim, Germany, 2000.
52. Geyer, K.; Codée, J.D.C.; Seeberger. P.H. *Chem. Eur. J.* 2006, *12*, 8434.
53. Ahmed-Omer, B.; Brandt, J.C.; Wirth, T. *Org. Biomol. Chem.* 2007, *5*, 733.
54. Pennemann, H.; Watts, P.; Haswell, S.J.; Hessel, V.; Löwe, H. *Org. Process Res. Dev.* 2004, *8*, 422.
55. Comer, E.; Organ, M.G. *Chem. Eur. J.* 2005, *11*, 7223.
56. Comer, E.; Organ, M.G. *J. Am. Chem. Soc.* 2005, *127*, 8160.
57. Hornung, C.H.; Mackley, M.R.; Baxendale, I.R.; Ley. S.V. *Org. Process Res. Dev.* 2007, *11*, 399 (micro-capillary flow disk microreactor).
58. Watts, P.; Haswell, S.J. *Drug Discovery Today* 2003, *8*, 586.
59. Watts, P. *Chem. Ingen. Technik* 2004, *76*, 555.
60. Fletcher, P.D.I.; Haswell, S.J.; Pombo-Villar, E.; Warrington, B.H.; Watts, P.; Wong, S.Y.F.; Zhang, X. *Tetrahedron* 2002, *58*, 4735.
61. Haswell, S.J.; Middleton, R.J.; O'Sullivan, B.; Skelton, V.; Watts, P.; Styring, P. *Chem. Commun.* 2001, 391.
62. Hessel, V.; Löwe, H. *Chem. Eng. Technol.* 2005, *28*, 267.
63. Watts, P.; Haswell, S.J. *Chem. Eng. Technol.* 2005, *28*, 290.
64. Markowz, G.; Schirrmeister, S.; Albrecht, J.; Becker, F.; Schütte, R.; Caspary, K.J.; Klemm, E. *Chem. Eng. Technol.* 2005, *28*, 459.
65. Roberge, D.M.; Ducry, L.; Bieler, N.; Cretton, P.; Zimmermann, B. *Chem. Eng. Technol.* 2005, *28*, 318.
66. Haswell, S.J.; Watts, P. *Green Chem.* 2003, *5*, 240.
67. Schwalbe, T.; Kursawe, A.; Sommer, J. *Chem. Eng. Technol.* 2005, *28*, 408.
68. Zhang, X.; Stefanick, S.; Villani, F.J. *Org. Process Res. Dev.* 2004, *8*, 455.
69. Doku, G.N.; Verboom, W.; Reinhoudt, D.N.; van den Berg, A. *Tetrahedron* 2005, *61*, 2733.
70. Watts, P.; Wiles, C. *Org. Biomol. Chem.* 2007, *5*, 727.
71. Wheeler, R.C.; Benali, O.; Deal, M.; Farrant, E.; MacDonald, S.J.F.; Warrington, B.H. *Org. Process Res. Dev.* 2007, *11*, 704.
72. Schwalbe, T.; Autze, V.; Hohmann, M.; Stirner, W. *Org. Process Res. Dev.* 2004, *8*, 440.
73. Kralisch, D.; Kreisel, G. *Chem. Eng. Sci.* 2007, *62*, 1094.
74. Poe, S.L.; Kobaslija, M.; McQuade, D.T. *J. Am. Chem. Soc.* 2007, *129*, 9216.
75. Bogdan, A.R.; Poe, S.L.; Kubis, D.C.; Broadwater, S.J.; McQuade, D.T. *Angew. Chem. Int. Ed.* 2009, *48*, 8547.

76. Baxendale, I.R.; Deeley, J.; Griffiths-Jones, C.M.; Ley, S.V.; Saaby, S.; Tranmer, G.K. *Chem. Commun.* 2006, 2566.
77. Hopkin, M.D.; Baxendale, I.R.; Ley, S.V. *Chem. Commun.* 2010, 2450.
78. McGrath, N.A.; Bartlett, E.S.; Sittihan, S.; Njardarson, J.T. *Angew. Chem. Int. Ed.* 2009, *48*, 8543.
79. Danishefksy, S.; Cain, P. *J. Am. Chem. Soc.* 1976, *98*, 4975.
80. Meerwein, H.; Schürmann, W. *Ann. Chem.* 1913, *398*, 196.
81. Butkus, E.; Stoncius, S.; Zilinskas, A. *Chirality* 2001, *13*, 694.
82. Prelog, V.; Seiwerth, R. *Chem. Ber.* 1941, *74*, 1644.
83. Sandborn, L.T. *Org. Synth. Coll.* 1941, *1*, 340.
84. Manske, R.H.F. US 2471575 (US Rubber Co., 1949).
85. Sletzinger, M.; Chemerda, J.M. DE 1643463 (Merck & Co., 1967).
86. Henze, H.R. US 2409754 (Parke-Davis, 1946).
87. Hansen, K.B.; Hsiao, Y.; Xu, F.; Rivera, N.; Clausen, A.; Kubryk, M.; Krska, S.; Rosner, T.; Simmons, B.; Balsells, J.; Ikemoto, N.; Sun, Y.; Spindler, F.; Malan, C.; Grabowskii, E.J.J.; Armstrong, J.D. III, *J. Am. Chem. Soc.* 2009, *131*, 8798.
88. Ishikawa, H.; Suzuki, T.; Hayashi, Y. *Angew. Chem. Int. Eng. Ed.* 2009, *48*, 1304.
89. Federspiel, M.; Fischer, R.; Hennig, M.; Mair, H.J.; Oberhauser, T.; Rimmler, G.; Albiez, T.; Bruhin, J.; Estermann, H.; Gandert, C.; Göckel, V.; Götzö, S.; Hoffmann, U.; Huber, G.; Janatsch, G.; Lauper, S.; Röckel-Stäbler, O.; Trussardi, R.; Zwahlen, A.G. *Org. Process Res. Dev.* 1999, *3*, 266.
90. Ghasemi, H.; Antunes, L.M.; Organ, M.G. *Org. Lett.* 2004, *6*, 2913.
91. Seiple, I.B.; Su, S.; Young, I.S.; Lewis, C.A.; Yamaguchi, J.; Baran, P.S. *Angew. Chem. Int. Ed.* 2010, *49*, 1095.
92. Moffatt, J.G.; Khorana, H.G. *J. Am. Chem. Soc.* 1961, *83*, 663.
93. Harried, S.S.; Yang, G.; Strawn, M.A.; Myles, D.C. *J. Org. Chem.* 1997, *62*, 6098.
94. Arefolov, A.; Panek, J.S. *J. Am. Chem. Soc.* 2005, *127*, 5596.
95. Arefolov, A.; Panek, J.S. *Org. Lett.* 2002, *4*, 2397.
96. Smith, A.B., III; Freeze, B.S.; Brouard, I.; Hirose, T. *Org. Lett.* 2003, *5*, 4405.
97. Smith, A.B. III; Qiu, Y.; Jones, D.R.; Kobayashi, K. *J. Am. Chem. Soc.* 1995, *117*, 12011.
98. Paterson, I.; Delgado, O.; Florence, G.J.; Lyothier, I.; Scott, J.P.; Sereinig, N. *Org. Lett.* 2003, *5*, 35.
99. Takabe, K.; Katagiri, T.; Tanaka, J.; Fujita, T.; Watanabe, S.; Suga, K. *Org. Synth.* 1989, *67*, 44.
100. Tani, K.; Yamagata, T.; Otsuka, S.; Kumobayashi, H.; Akutagawa, S. *Org. Synth.* 1989, *67*, 33.
101. Nakatani, Y.; Kawashima, K. *Synlett* 1978, 147.
102. Ravasio, N.; Poli, N.; Psaro, R.; Saba, M.; Zaccheria, F. *Top. Catal.* 2000, *13*, 195.
103. Pummerer, R.; Prell, E.; Rieche, A. *Chem. Ber.* 1926, *59*, 2159.
104. Takaya, H.; Akutagawa, S.; Noyori, R. *Org. Synth.* 1989, *67*, 20.
105. Trost, B.M.; Zhang, T. *Angew. Chem. Int. Ed.* 2008, *47*, 3759.
106. Yeung, Y.Y.; Hong, S.; Corey, E.J. *J. Am. Chem. Soc.* 2006, *128*, 6310.
107. Fukuta, Y.; Mita, T.; Fukuda, N.; Kanai, M.; Shibasaki, M. *J. Am. Chem. Soc.* 2006, *128*, 6312.
108. Mita, T.; Fukuda, N.; Roca, F.X.; Kanai, M.; Shibasaki, M. *Org. Lett.* 2007, *9*, 259.
109. Yamatsugu, K.; Yin, L.; Kamijo, S.; Kimura, Y.; Kanai, M.; Shibasaki, M. *Angew. Chem. Int. Ed.* 2009, *48*, 1070.
110. Yamatsugu, K.; Kanai, M.; Shibasaki, M. *Tetrahedron* 2009, *65*, 6017.
111. Satoh, N.; Akiba, T.; Yokoshima, S.; Fukuyama, T. *Angew. Chem. Int. Ed.* 2007, *46*, 5734.
112. Dorta, R.; Broggini, D.; Stoop, R.; Rüegger, H.; Spindler, F.; Togni, A. *Chem. Eur. J.* 2004, *10*, 267.
113. Blaser, H.U.; Buser, H.P.; Jalett, H.P.; Pugin, B.; Spindler, F. *Synlett* 1999, 867.
114. Lee, J.C.; Jin, S.; Cha, J.K. *J. Org. Chem.* 1998, *63*, 2804.
115. Lee, J.C.; Cha, J.K. *Tetrahedron* 2000, *56*, 10175.
116. van Tamelen, E.E.; Spencer, T.A. Jr.; Allen, D.S. Jr.; Ovis, R.L. *J. Am. Chem. Soc.* 1959, *81*, 6341.
117. van Tamelen, E.E.; Spencer, T.A. Jr.; Allen, D.S. Jr.; Ovis, R.L. *Tetrahedron* 1961, *14*, 8.
118. McMurry, J.E.; Fleming, M.P. *J. Am. Chem. Soc.* 1974, *96*, 4708.
119. Nürrenbach, A.; Paust, J.; Pommer, H.; Schneider, J.; Schulz, B. *Ann. Chem.* 1977, 1146.
120. Isler, O.; Huber, W.; Ronco, A.; Kofler, M. *Helv. Chim. Acta* 1947, *30*, 1911.
121. Isler, O.; Lindlar, H.; Montavon, M.; Ruegg, R.; Zeller, P. *Helv. Chim. Acta* 1956, *39*, 249.
122. GB 1013907 (Imperial Chemical Industries, Ltd., 1965).
123. GB 1064629 (Imperial Chemical Industries, Ltd., 1966).
124. Coe, P.L.; Scriven, C.E.; Percy, R.K. EP 168175 (National Research Development Corp., 1987).
125. Coe, P.L.; Scriven, C.E. *J. Chem. Soc. Perkin Trans. I* 1986, 475.
126. Robertson, D.W.; Katzenellenbogen, J.A. *J. Org. Chem.* 1985, *47*, 2387.
127. Bracher, F.; Litz, T. *Bioorg. Med. Chem.* 1996, *4*, 877.
128. Torre, O.; Gotor-Fernandez, V.; Gotor, V. *Tetrahedron Asymm.* 2006, *17*, 860.

12 Summary of Overall Trends in Synthesis Database

This chapter presents several plots of correlations between key metrics parameters for the 1050 plans documented in the extensive synthesis database. Each trend is discussed in the context of achieving optimum and green syntheses. Examples of synthesis plans that achieve some measure of greenness are highlighted. The reader is referred to Chapters 5 and 6 regarding the definitions of metrics parameters.

12.1 OVERALL ATOM ECONOMY TRENDS

Figure 12.1 shows the trend between overall atom economy (AE) and molecular weight fraction of sacrificial reagents, f(sac). The "greener" plans are the ones with low values for f(sac) and high values for % AE. One can see that the density of plans is rather sparse in the left-hand side of the plot where the "greener" plans are located. Another observation is that there are a set of plans with f(sac) = 0 and yet have AEs less than or equal to the golden limit cutoff of 61.8%. Even though all reagents end up in whole or in part in the target product structures, they do so at the price of producing significant by-products along the way. These plans are as follows where the % AEs are given in parentheses: (a) Faith G3 [1] plan for glycerol (9.7); (b) Stolz G1 [2–6] plan for (–)-ephedrine hydrochloride (12.1); (c) Faith [7] plan for vinyl acetate (36.7); (d) Faith G2 [8,9] plan for glycerol (40.7); (e) Vogel [10] plan for hexamethylenetetramine (43.8); (f) Faith [11] plan for vinyl chloride (45.7); (g) DuPont [12] plan for ferrocene (56.7); (h) Dickey–Gray [13] plan for barbituric acid (58.2); (i) Vogel [14] plan for monastral fast blue B (61.5); and (j) Faith G4 [15,16] plan for phenol (61.8).

Figure 12.2 shows a plot of the % overall AE versus vector magnitude ratio (VMR) obtained from the radial hexagon analysis based on 587 plans for compounds with multiple plans as noted in Chapter 13. "Greener" plans are those with high AEs and high values of VMR. Again, as in Figure 12.1, the greatest density of points is in the opposite direction, that is, in the lower left-hand corner of the plot. Plans exceeding the golden limit for AE are as follows with the % AE values in parentheses: (a) Givaudan–Roure [17,18] plan for vanillin (67.3); (b) Späth [19] plan for resveratrol (73.1); (c) Roesler [20] plan for vanillin (74.0); (d) Hoechst-Celanese [21,22] plan for ibuprofen (77.4); (e) Dominion Rubber [22–24] plan for DEET (84.1); and (f) Pfizer [25,26] plan for DEET (84.1).

12.2 OVERALL YIELD TRENDS

Figure 12.3 shows the trend between % overall yield and VMR obtained from the radial hexagon analysis based on 587 plans for compounds with multiple plans as noted in Chapter 13. Most plans gravitate to the lower left-hand corner of the plot. The four plans in the top right corner are as follows with the % overall yields given in parentheses: (a) Roesler plan [20] for vanillin (70.0); (b) Givaudan–Roure [17,18] plan for vanillin (75.0); (c) McQuade [27] plan for (S)-(+)-pregabalin hydrochloride (85.7); and (d) Mayer [28] plan for vanillin (92.0).

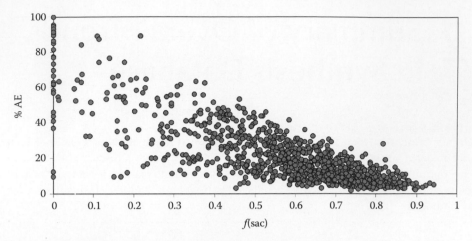

FIGURE 12.1   Correlation between % overall AE and f(sac).

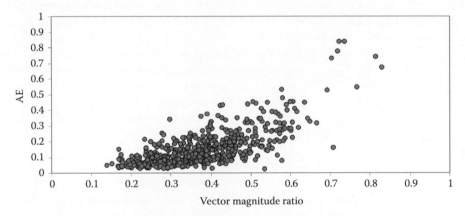

FIGURE 12.2   Correlation between AE and VMR from radial hexagon analysis.

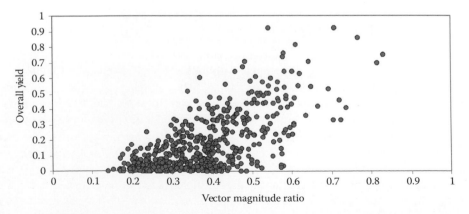

FIGURE 12.3   Correlation between % overall yield and VMR from radial hexagon analysis.

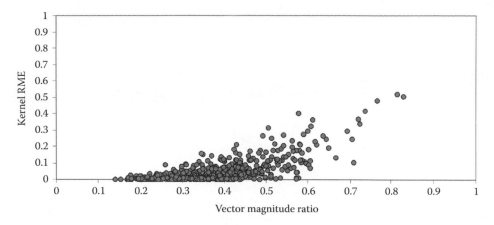

FIGURE 12.4   Correlation between overall RME and VMR from radial hexagon analysis.

12.3   KERNEL OVERALL REACTION MASS EFFICIENCY TRENDS

Figure 12.4 shows a plot of overall reaction mass efficiency (RME) versus and VMR obtained from the radial hexagon analysis based on 587 plans for compounds with multiple plans as noted in Chapter 13. No plans had overall RMEs exceeding the golden threshold limit of 61.8%. When comparing this plot with Figure 12.2 for the AE trend with VMR, one notes that the RME plot shows a significant compression in the y-axis. Again, the majority of plans gravitate in the left-hand corner of the plot. The top five plans in order of RME values are as follows: (a) Roesler plan [20] for vanillin (51.8); (b) Givaudan–Roure [17,18] plan for vanillin (50.4); (c) McQuade [27] plan for (S)-(+)-pregabalin hydrochloride (47.9); (d) Dominion Rubber [22,23] plan for DEET (41.4); and (e) Volynkin [29] plan for vanillin (40.2).

Proof that % overall kernel RME for a plan is not equal to the product of the overall AE and overall yield (Y) is given by the plots shown in Figure 12.5a through c. The first plot shows the full range of values where a straight line is fit to the data giving an apparent high correlation coefficient of 97%. However, this is deceptive as the second plot shows an expansion of the lower left-hand corner of the former. It is obvious that the greatest deviation between the two parameters is in the low range of both parameters as expected. This observation is confirmed by the deviation plot shown in Figure 12.5c. As both AE and overall yield increase in value, their multiplicative product numerically gets closer to the true value of RME. Of course, when $AE = 1$ and $Y = 1$, then indeed RME = 1. The six points in Figure 12.5a that are outliers refer to the following plans: (a) Fierz-David [30] plan for picric acid; (b) Fierz-David [31] plan for congo red; (c) Faith [32] plan for maleic anhydride; (d) Faith G3 [33] plan for phenol; (e) Thomas [34] plan for ε-caprolactam; and (f) Fierz-David [35] plan for Michler's hydrol. The first five of these plans involve a single-step sequence and the last is a three-step sequence.

12.4   DEGREES OF CONVERGENCE AND ASYMMETRY TRENDS

The two parameters that define the shapes of synthesis tree diagrams are degree of convergence (delta) and degree of asymmetry (beta). Figures 12.6 and 12.7 show the trends of these parameters with the number of reaction steps in a plan for all plans in the database. Thirty-five percent of the plans in the database are convergent plans and the remaining 65% are linear. The delta versus N plot shows that as the number of steps increases the degree of convergence increases, whereas the beta versus N plot shows that as the number of steps increases the degree of asymmetry increases. These correlations may be predicted directly from Equations 12.1 and 12.2 that relate these parameters to N for linear sequences made up of successive bimolecular reactions.

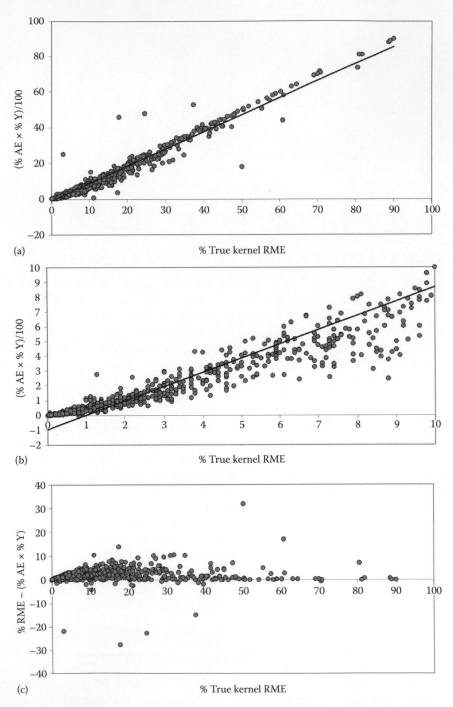

FIGURE 12.5 (continued)   (a) Correlation between (% AE × % Y)/100 and % true kernel RME: Slope = 0.96; intercept = 0.96; r^2 = 0.97. (b) Correlation between (% AE × % Y)/100 and % true kernel RME: slope = 0.96; intercept = 0.96; r^2 = 0.97 (same as [a] but with a smaller scale for the x-axis). (c) Deviation plot showing difference between (% AE × % Y)/100 and % true kernel RME.

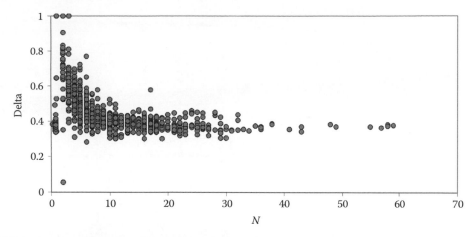

FIGURE 12.6 Correlation between degree of convergence and number of steps.

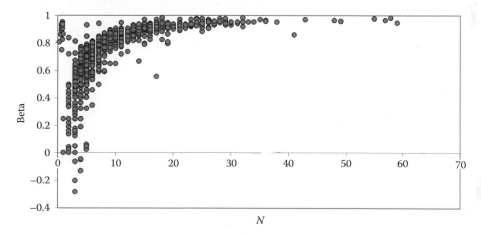

FIGURE 12.7 Correlation between degree of symmetry and number of steps.

$$\text{delta} = \frac{\arctan\left(\dfrac{N-1+2^{-N}}{N}\right) + \arctan\left(\dfrac{1-2^{-N}}{N}\right)}{2\arctan\left(\dfrac{N}{2}\right)} \qquad (12.1)$$

$$\text{beta} = \frac{2(N-1)+2^{1-N}}{N} - 1 \qquad (12.2)$$

Figure 12.8 shows graphically what these functions look like. Superposition of these functions on the observed data points in Figures 12.6 and 12.7 shows good agreement with respect to shape. For purely linear plans with large values of N, Figure 12.8 predicts that delta approaches 0.25 and beta approaches 1 (completely asymmetric). The limiting value of delta in Figure 12.6 is about 0.4 instead of 0.25 since this plot covers both linear and convergent plans of various combinations of reaction types involving any number of input materials. Figure 12.9 shows the correlation between delta and VMR obtained from the radial hexagon analysis based on 587 plans for compounds with multiple plans as noted in Chapter 13. Table 12.1 summarizes the plans with 100% degree of convergence. Note that all but two plans are single step reactions. The reader is directed to the accompanying download for plan schemes and references.

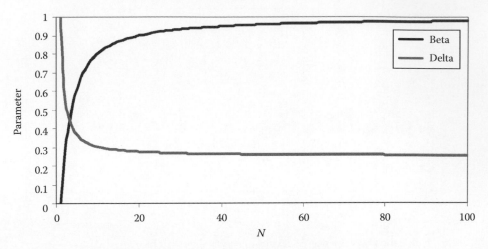

FIGURE 12.8   Plots of delta and beta versus N according to Equations 12.1 and 12.2.

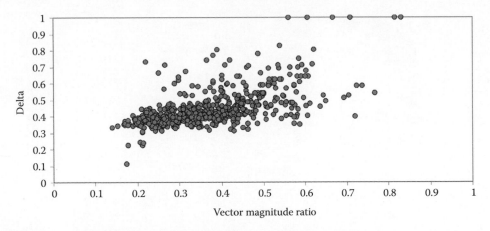

FIGURE 12.9   Correlation between delta and VMR from radial hexagon analysis.

Another observation with respect to Figure 12.7 is that nine data points take on negative values of beta. Indeed, Equation (12.2) for linear plans predicts beta < 0 when N < 1. Though the fact that N < 1 has no physical sense, the real reason why nine plans have apparent negative values of beta is because of the unusual shapes of their respective synthesis tree diagrams. Figures 12.10 and 12.11 show the synthesis tree diagrams for the bisphenol A and aspirin plans, respectively. All plans are convergent and exhibit unusual branching shapes. Negative beta values arise because the ordinate value of the final target node is lower than the value of the centroid of the input nodes arranged in a vertical line. The normal situation is for the ordinate of the target node to be equal to (beta = 0) or greater than this centroid value (beta > 0).

12.5   TARGET BOND FORMING PROFILE PARAMETERS

The two key parameters obtained from the target bond profiles are the number of target bonds made per step, B/M, where M is the total number of reaction steps, and the fraction of reaction stages producing no target bonds, f(nb). Figure 12.12 shows the correlation between B/M and f(sac). The "greener" plans are those with small f(sac) values and produce a high number of target bonds per step. The Vogel [10] plan to hexamethylenetetramine produces 12 target bonds in a single step where both formaldehyde and ammonium hydroxide reagents end up in the final target structure. Figure 12.13 shows the correlation between f(nb) and f(sac). The density of points is highest toward the right-hand

TABLE 12.1

Summary of Plans in Database with 100% Degree of Convergence (Delta = 1)

N	Target	Plan	Year
1	Acetaminophen	Warner-Lambert	1961
1	Adipic acid	Vogel	1978
1	Allantoin	Grimaux	1877
1	Allantoin	Hartman	1943
1	Barbituric acid	Dickey-Gray	1943
1	Benzoic acid	Faith G4	1966
3	Bullvalene	Schroeder	1964
1	Bupleurynol	Organ G2	2004
1	Cyclooctatetraene	Reppe	1948
1	DDT	Vogel	1978
1	ε-Caprolactam	Thomas	2005
1	ε-Caprolactam	Faith	1975
2	ε-Caprolactam	BP	1976
1	Epichlorohydrin	Faith	1966
1	Ethylene oxide	Faith G1	1966
1	Ethylene oxide	Faith G2	1966
1	Ferrocene	BASF G1	1963
1	Ferrocene	Wilkinson G1	1963
1	Ferrocene	BASF G2	1963
1	Ferrocene	Wilkinson G2	1963
1	Ferrocene	DuPont	1957
1	Ferrocene	Pauson	1951
1	Fluorescein	Vogel	1948
1	Glycerol	Faith G3	1966
1	Hexamethylenetetramine	Vogel	1948
1	Indomethacin	Merck G3	1967
1	Maleic anhydride	Faith	1966
1	Menthone	Sandborn	1941
1	Methamphetamine	Ogata	1919
1	Oxomaritidine	Ley G2	2006
1	Phenol	Faith G1	1966
1	Phenol	Faith G3	1966
1	Phenolphthalein	Vogel	1948
1	Phenytoin	Biltz	1908
1	Phenytoin	Lambert G1	2003
1	Phenytoin	Parke-Davis	1946
1	Phthalic anhydride	Faith G1	1966
1	Phthalic anhydride	Faith G2	1966
1	Picric acid	Fierz-David	1949
1	Picric acid	Vogel	1948
1	Saccharin	BASF	1984
1	Safranine	Fierz-David	1921
1	Sorbitol	Faith	1966
1	Tropinone	Robinson	1917
1	Urea	Faith	1966
1	Vanillin	Mayer	1949
1	Vanillin	Roesler	1907
1	Vanillin	Geigy	1899
1	Vanillin	Givaudan–Roure	1998
1	Vitamin C	Biotechnical Resources	1991
1	Warfarin	Jorgensen	2003

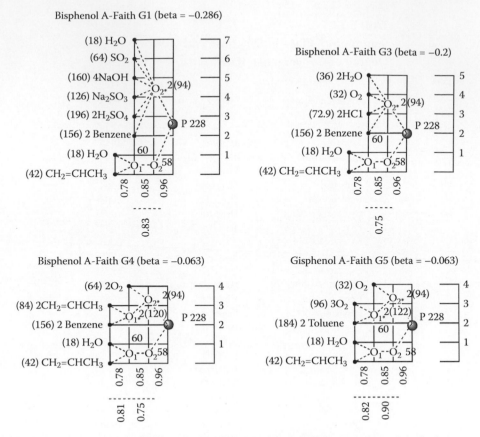

FIGURE 12.10 Synthesis tree diagrams for bisphenol A plans exhibiting negative values for beta.

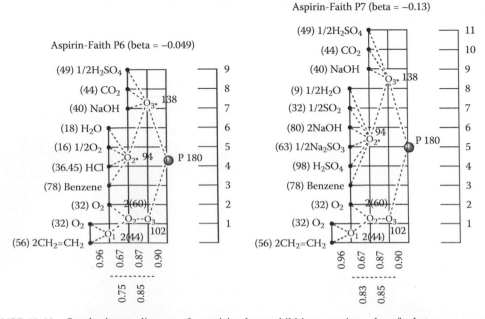

FIGURE 12.11 Synthesis tree diagrams for aspirin plans exhibiting negative values for beta.

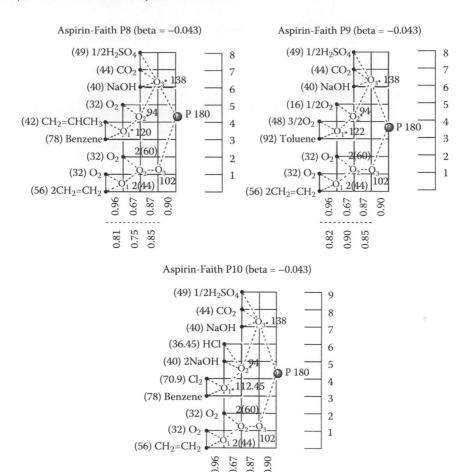

FIGURE 12.11 (continued)

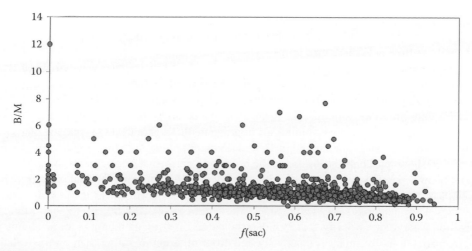

FIGURE 12.12 Correlation between B/M and f(sac).

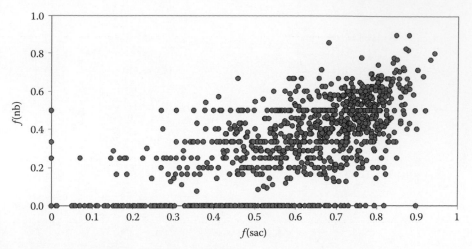

FIGURE 12.13   Correlation between $f(nb)$ and $f(sac)$.

corner where both $f(nb)$ and $f(sac)$ are high. Again, "greener" plans are scarce. An unusual feature of this profile is that there are 197 plans with $f(nb) = 0$ and yet have a wide range of values for $f(sac)$. In these plans, every step produces at least one target bond but at the expense of using sacrificial reagents. Thirty-two plans have both $f(nb) = 0$ and $f(sac) = 0$. These plans are especially "green" because every reaction produces at least one target bond and all reagents in whole or in part end up in the target bond structure. Since these plans have this unusually rare characteristic, they are identified and summarized in Table 12.2. Figures 12.14 and 12.15 show the correlations of B/M and $f(nb)$ with respect to number of reaction stages, N. As expected, as the number of reaction stages increases the lower is the number of bonds made per step. The $f(nb)$ versus N plot is far more scattered. For completeness, Figure 12.16 shows the direct correlation between B/M and $f(nb)$. There are 229 plans with $f(nb) = 0$ and various values of B/M. An adequate parameter that captures and separates the different synthesis strategies is the ratio $f(nb)$ to B/M. Using this discriminating parameter, we can write the following four general strategy outcomes in descending order of performance:

1. "Green" syntheses have $f(nb)/(B/M) = 0$, $f(nb) = 0$, and $f(sac) = 0$.
2. Direct syntheses have $f(nb)/(B/M) = 0$, $f(nb) = 0$, and $0 < f(sac) < 1$.
3. "In between" syntheses have $0 < f(nb)/(B/M) < 1.5$, $0 < f(nb) < 1$, and $0 < f(sac) < 1$.
4. Indirect syntheses have $f(nb)/(B/M) > 1.5$, $f(nb) > 0.6$, and $f(sac) > 0.6$.

Figures 12.17 and 12.18 show the correlation of this ratio parameter with $f(sac)$ and $f(nb)$, respectively. The outlier points in the two plots correspond to the Willstätter [36–38] plan for cyclooctatetraene and the Brenna G2 [39] plan for (+)-(2R,6R)-*trans*-gamma-irone. The reader is directed to Section 11.2 on discussions of plans with sparse target bond forming profiles where these plans were mentioned. The "green" syntheses are exactly those itemized in Table 12.2. Figure 12.19a shows a pie chart highlighting the proportion of each type of plan found in the entire database. Though the "green" plans are in the minority, the majority of plans fall in the "in between" category where optimistically there are great opportunities for improvement. The good news is that the plans falling in the type D category and summarized in Table 12.3 do not outnumber the "green" plans by much. Figure 12.19b shows pie charts according to the preceding performance categories for academic and industrial plans separately. This representation is more revealing in that industrial plans have higher proportions of "green plans" and direct syntheses and lower proportions of "in between" and indirect syntheses as compared to academic plans. These data are consistent with the expectation that optimization is more of a priority goal in industrial synthesis strategy than it is for academic plans which, instead, put more emphasis on novelty and proof of structure rather than on true efficiency. It may be concluded that industrial plans are generally "greener" than academic ones.

TABLE 12.2

Summary of "Green" Plans with Both
f(nb) = 0 and f(sac) = 0

Target	Plan	Year
Acetic anhydride	Faith G2	1966
Allantoin	Grimaux	1877
Anthraquinone	Vogel	1948
Barbituric acid	Dickey–Gray	1943
Bisphenol A	Faith G4	1966
Cyclooctatetraene	Reppe	1948
DDT	Vogel	1948
ε-Caprolactam	Thomas	2005
(−)-ephedrine hcl	Stolz G1	1931
Ethylene oxide	Faith G2	1966
Ferrocene	DuPont	1957
Fluorescein	Vogel	1948
Glycerol	Faith G2	1975
Glycerol	Faith G3	1966
Hexamethylenetetramine	Vogel	1948
Ibuprofen	Hoechst-Celanese	1988
Malathion	American Cyanamid	1951
Menthol	Takasago G2	2002
Menthol	Sayo–Matsumoto	2002
Phenolphthalein	Vogel	1948
Phenytoin	Biltz	1908
Phenytoin	Lambert G1	2003
Phenytoin	Parke-Davis	1946
Phthalic anhydride	Faith G2	1966
Resveratrol	Spath	1941
Sorbitol	Faith	1966
Thalidomide	Seijas	2001
Urea	Faith	1966
Vinyl acetate	Faith	1966
Vinyl chloride	Faith	1966
Warfarin	Jorgensen	2003
Zingerone	Nomura	1917

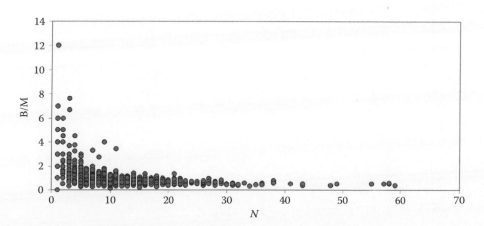

FIGURE 12.14 Correlation between B/M and N.

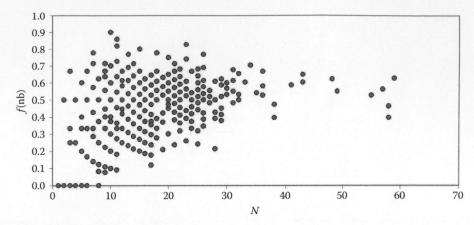

FIGURE 12.15   Correlation between $f(nb)$ and N.

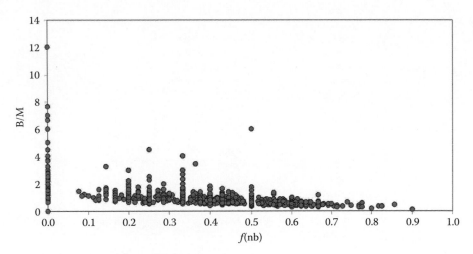

FIGURE 12.16   Correlation between B/M and $f(nb)$.

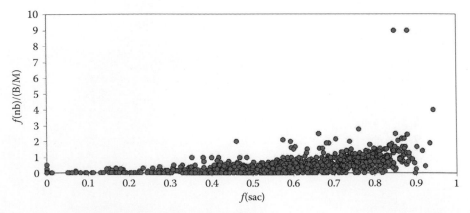

FIGURE 12.17   Correlation between $f(nb)/(B/M)$ and $f(sac)$.

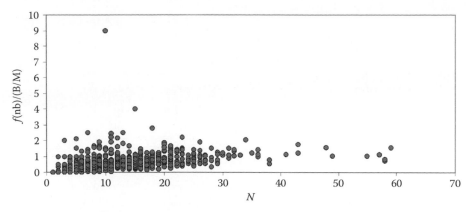

FIGURE 12.18 Correlation between $f(\text{nb})/(\text{B/M})$ and N.

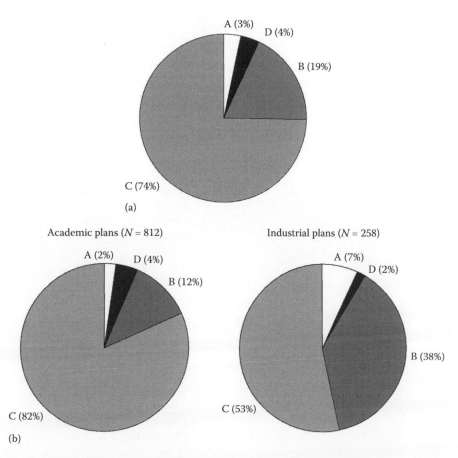

FIGURE 12.19 (a) Pie chart showing the breakdown of general synthesis strategy types according to magnitudes of $f(\text{nb})/(\text{B/M})$ for all plans in database. (b) Pic chart showing the breakdown of general synthesis strategy types according to magnitudes of $f(\text{nb})/(\text{B/M})$ according to academic and industrial plans.

TABLE 12.3

Summary of Plans with $f(nb)/(B/M) > 1.5$, $f(nb) > 0.6$, and $f(sac) > 0.6$

$f(nb)/(B/M)$	Target	Plan	Years
1.500	Epibatidine	Olivo	1999
1.500	Rosefuran	Iriye G2	1990
1.500	Codeine	Chida	2008
1.500	Ricinine	Taylor	1956
1.500	(−)-Kainic acid	Takano G4	1993
1.518	(−)-Kainic acid	Poisson G1	2005
1.526	Colchicine	Roussel–Uclaf	1965
1.550	Paclitaxel	Takahashi	2006
1.556	Paclitaxel	Mukaiyama	1999
1.563	Vitamin A	Roche G6c	1976
1.563	Isotretinoin	Roche	1985
1.569	Vitamin E	Knight	2009
1.600	Epibatidine	Szantay G2	1909, 1996
1.600	Papaverine	Decker–Wahl	1950
1.667	Barrelene	Zimmerman G2	1969
1.667	Fenchone	Cocker	1971
1.667	Morphine	Parker	2006
1.667	Papaverine	Redel–Bouteville	1949
1.667	Ricinine	Spath	1923
1.733	Longifolene	Schultz G2	1985
1.800	Epibatidine	Fletcher	1994
1.875	Codeine	Parker	2006
1.888	(−)-Kainic acid	Ogasawara G3	2001
2.000	Carvone	Royals	1951
2.000	Carvone	A.M. Todd	1958
2.000	Vanillin	Volynkin	1938
2.059	Lysergic acid	Fukuyama G2	2009
2.100	Naproxen	Brunner	2000
2.133	Juvenile hormone	Schulz–Sprung	1969
2.182	Nepetalactone	Pickett G2	1996
2.203	Triquinacene	Woodward	1964
2.455	Coniine	Pandey	1997
2.500	Rosefuran	Iriye G1	1990
2.500	Naproxen	Zambon G2	1989
2.800	Twistane	Hamon–Young	1976
4.000	Tropinone	Willstatter G1	1901
9.000	Cyclooctatetraene	Willstatter–Cope	1913, 1948
9.000	(+)-(2R,6R)-*trans*-gamma-irone	Brenna G2	2001

12.6   SACRIFICIAL REAGENTS AND SACRIFICIAL REACTION TRENDS

The following graphics refer to the set of compounds with multiple plans as noted in Chapter 13 that were analyzed using radial hexagons, which capture both material and strategy efficiency metrics. The key parameter is the VMR, which is correlated with $f(\text{sac})$, ($f(\text{non-sac})$) molecular weight fraction of non-sacrificial reagents, fraction of kernel waste produced from target bond reactions (fw(tbr)), and fraction of kernel waste produced from sacrificial reactions (fw(sr)) as shown in Figures 12.20 through 12.24. Figure 12.20a and b shows that most plans have high $f(\text{sac})$ values. The plans with fw(tbr) = 1 are summarized in Table 12.4.

Figure 12.22a and b shows two views of how the difference fw(tbr) – fw(sr) correlates with VMR. The "greener" plans are the ones with a positive difference. The dots arranged in a vertical column, where fw(tbr) – fw(sr) = 1, refer to the plans listed in Table 12.4 where 100% of the kernel waste is coming from target bond forming reactions. 223 plans have a negative difference (38%) and 364 plans have a positive difference (62%).

Figure 12.23a and b shows two views of how the difference $f(\text{non-sac})$ – $f(\text{sac})$ correlates with VMR. The "greener" plans are the ones with a positive difference. In this case, the "greener" plans are sparse represented by only 18% (105/587) of all plans examined.

From Figure 12.24 the plans represented by dots across the top of the graph when fw(tbr) = 1 are those "greener" plans listed in Table 12.4.

Figure 12.25 shows a summary radial hexagon for all 587 plans where the best performing all round plan, labeled as "maximum" is the Givaudan–Roure [17,18] plan for vanillin, and the worst plan, labeled as "minimum" is the Willstätter G1 [40–42] plan for tropinone. Recall that this latter plan is in fact a degradation, not a synthesis, as mentioned in Section 11.2.

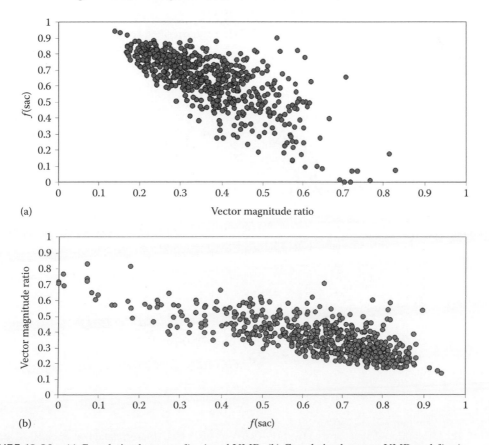

FIGURE 12.20   (a) Correlation between $f(\text{sac})$ and VMR. (b) Correlation between VMR and $f(\text{sac})$.

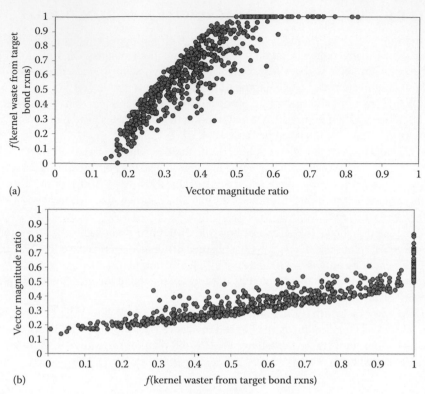

FIGURE 12.21 (a) Correlation between fw(tbr) and VMR. (b) Correlation between VMR and fw(tbr).

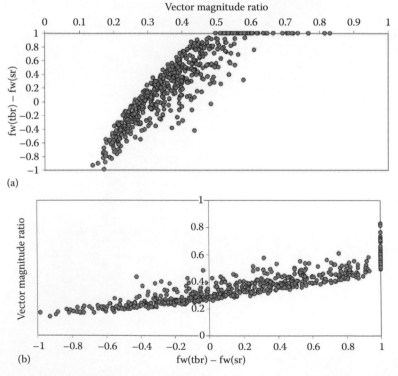

FIGURE 12.22 (a) Correlation between fw(tbr) – fw(sr) versus VMR. (b) Correlation between VMR versus fw(tbr) – fw(sr).

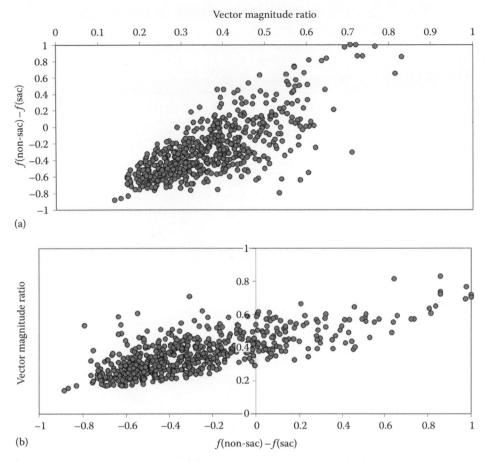

FIGURE 12.23 (a) Correlation between f(non-sac) – f(sac) versus VMR. (b) Correlation between VMR versus f(non-sac) – f(sac).

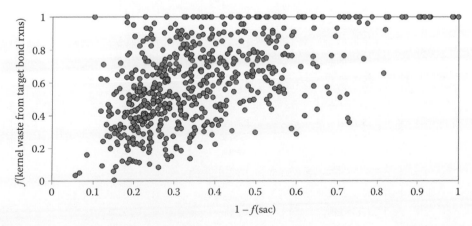

FIGURE 12.24 Correlation between fw(tbr) versus $1 - f$(sac).

TABLE 12.4
Summary of Plans with fw(tbr) = 1

Target	Plan	Year
Aspirin	Vogel	1978
Coniine	Buchwald	1998
Coniine	Couture	2007
Coniine	Husson	1983
DEET	Bhattacharya	2006
DEET	Dominion Rubber	1962
DEET	Khalafi-Nezhad	2005
DEET	LeFevre	1990
DEET	Neeland	1998
DEET	Pfizer	1979
DEET	Showa Denko	1976
DEET	Virginia Chemicals	1971
DEET	Wang	1974
Epibatidine	Giblin	1997
Fluoxetine	Panunzio	2004
Fluoxetine	Shibasaki	2004
Fluoxetine	Wang	2005
Ibuprofen	Boots	1968
Ibuprofen	DuPont	1985
Ibuprofen	Hoechst-Celanese	1988
Ibuprofen	Upjohn	1977
Indigo	Harley-Mason	1950
Naproxen	DuPont	1985
Naproxen	Noyori	1987
Nicotine	Campos	2006
Nicotine	Delgado	2001
Pregabalin	Felluga	2008
Pregabalin	McQuade	2007
Pregabalin	Pfizer G1	2008
Pregabalin	Pfizer G2	2008
Pregabalin	Shibasaki	2005
Pregabalin	Silverman G1	1989
Resveratrol	Spath	1941
Rosefuran	Okazaki	1984
Rosefuran	Pattenden	1977
Rosefuran	Salerno	1997
Rosefuran	Wong	1997
Sertraline	Zhao	2006
Sertraline	Pfizer G3	2004
Silphinene	Wender	1985
Sparteine	Anet	1950
Sparteine	Blakemore	2008
Sparteine	Leonard G1	1950
Sparteine	Leonard G2	1950
Tropinone	Bazilevskaya G1	1958
Tropinone	Robinson	1917
Tropinone	Willstatter G2	1921

TABLE 12.4 (continued)
Summary of Plans with fw(tbr) = 1

Target	Plan	Year
Tyrian purple	Gerlach	1989
Tyrian purple	Rottig	1935
Vanillin	Frost	1998
Vanillin	Geigy	1899
Vanillin	Givaudan–Roure	1998
Vanillin	Lampman G1	1977
Vanillin	Mayer	1949
Vanillin	Roesler	1907
Vanillin	Zasosov	1959

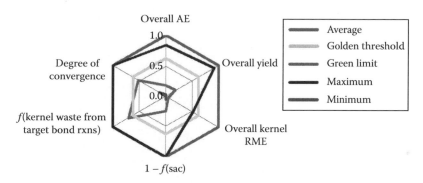

FIGURE 12.25 Overall radial hexagon for compounds with multiple plans as noted in Chapter 13.

12.7 HYPSICITY TRENDS

With respect to tracking oxidation number changes in the making of target bond forming bonds in synthesis plans, several interesting results and patterns emerged. The key parameters are hypsicity index (HI), oxidation length or number of "ups" and "downs" in the hypsicity profiles (|UD|), and delta row. Figure 12.26 shows a scatter plot of HI versus f(sac). There is a mild upward positive correlation between |UD| versus f(sac) as shown in Figure 12.27. One expects a greater spread in delta row for plans with high f(sac) values and this is confirmed in Figure 12.28. When |UD| is plotted against HI, a V-shaped plot results depending on whether HI is positive or negative for a given synthesis plan as shown in Figure 12.29. An analogous plot of N versus HI is less well defined in shape as shown by Figure 12.30. A weak linear correlation is found between delta row and HI (see Figure 12.31). For completeness, Figures 12.32 through 12.35 show the remaining pairwise comparisons between variables. The reader is directed to Appendix A6 for tables of plans that are identified as isohypsic (HI = 0), hyperhypsic (HI > 0), and hypohypsic (HI < 0). Figure 12.36a through c shows modest linear correlations for hyperhypsic plans and similarly Figure 12.37a through c shows those for hypohypsic plans. "Greenness" with respect to hypsicity is exemplified by a plan that has f(sac) = 0 and shows either of the following situations: (a) HI = 0 as consequence of not having any redox type reactions in the plan; (b) HI > 0 where the hypsicity profile shows a steady decrease in oxidation level from the zeroth to the last reaction stage and where additive reduction reactions have been used; or (c) HI < 0 where the hypsicity profile shows a steady increase in oxidation level from the zeroth to the last reaction stage and where additive oxidation reactions have been used. Examples of plans in each category are shown in the following along with the respective hypsicity profiles in Figure 12.38.

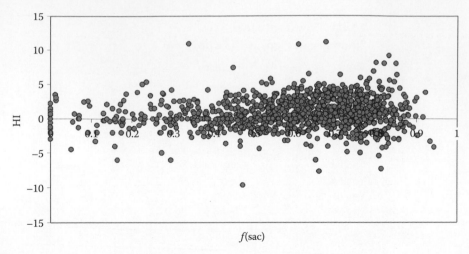

FIGURE 12.26 Correlation between HI and f(sac).

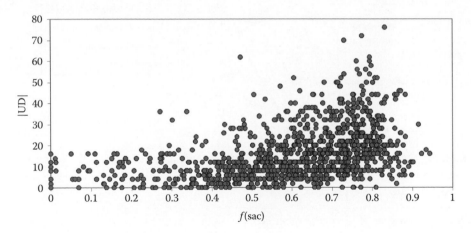

FIGURE 12.27 Correlation between |UD| and f(sac).

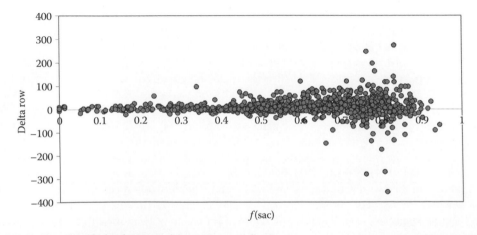

FIGURE 12.28 Correlation between delta row versus f(sac).

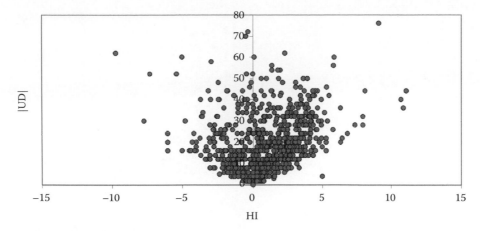

FIGURE 12.29 Correlation between |UD| versus HI.

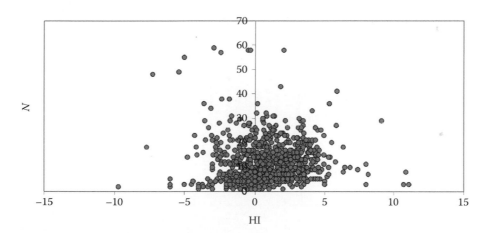

FIGURE 12.30 Correlation between N and HI.

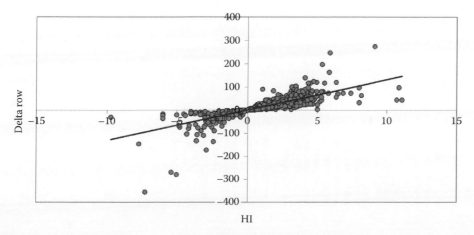

FIGURE 12.31 Correlation between delta row and HI. Linear fit yields slope = 13.10; intercept = 0.24; $r^2 = 0.61$.

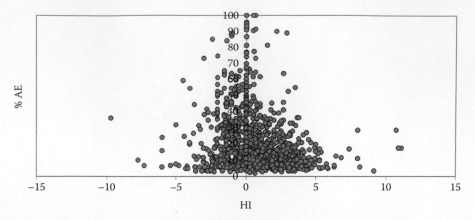

FIGURE 12.32   Correlation between % AE versus HI.

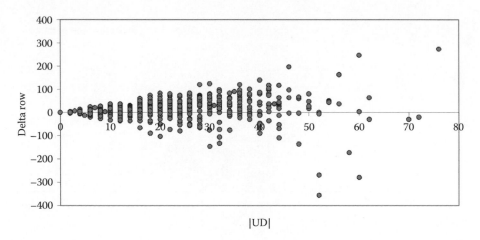

FIGURE 12.33   Correlation between delta row versus |UD|.

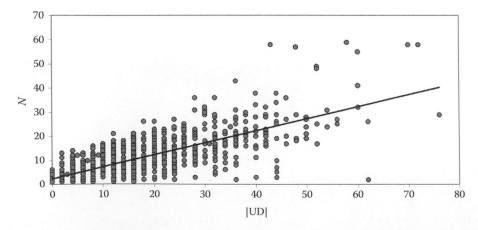

FIGURE 12.34   Correlation between N versus |UD|. Linear fit with slope = 0.50, intercept = 2.08, r^2 = 0.58.

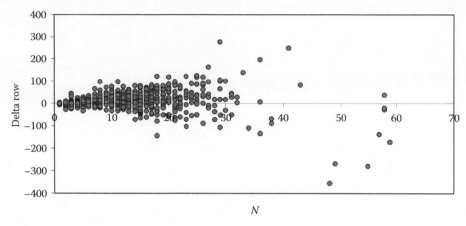

FIGURE 12.35   Correlation between delta row versus N.

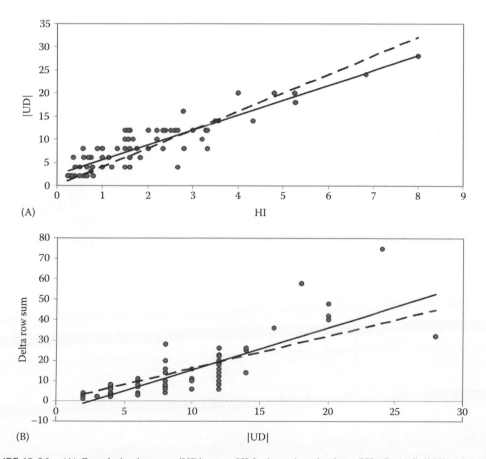

FIGURE 12.36   (A) Correlation between |UD| versus HI for hyperhypsic plans (HI > 0). (a) Solid line: Linear fit with slope = 3.22, intercept = 2.32, r^2 = 0.81; (b) Broken line: Linear fit with slope = 3.93, intercept = 0, r^2 = 0.77. (B) Correlation between delta row sum versus |UD| for hyperhypsic plans (HI > 0). (a) Solid line: Linear fit with slope = 2.08, intercept = −5.41, r^2 = 0.71. (b) Broken line: Linear fit with slope = 1.66, intercept = 0, r^2 = 0.65.

(*continued*)

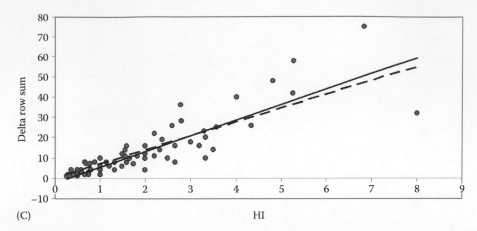

FIGURE 12.36 (continued)   (C) Correlation between delta row sum versus HI for hyperhypsic plans (HI > 0). (a) Solid line: Linear fit with slope = 7.70, intercept = −2.36, r^2 = 0.76; (b) Broken line: Linear fit with slope = 7.00, intercept = 0, r^2 = 0.75.

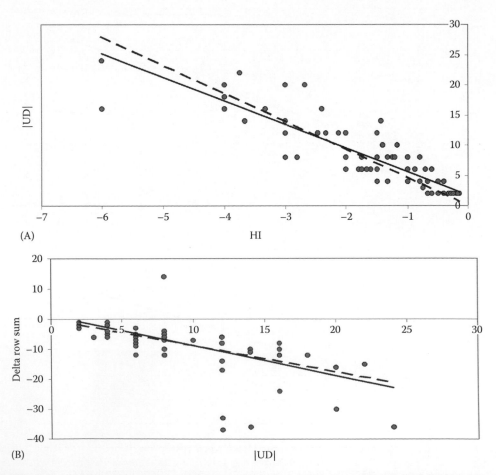

FIGURE 12.37   (A) Correlation between |UD| versus HI for hypohypsic plans (HI < 0). (a) Solid line: Linear fit with slope = −3.90, intercept = 1.77, r^2 = 0.77. (b) Broken line: Linear fit with slope = −4.67, intercept = 0, r^2 = 0.72. (B) Correlation between delta row sum versus |UD| for hypohypsic plans (HI < 0). (a) Solid line: Linear fit with slope = −1.00, intercept = 1.35, r^2 = 0.47. (b) Broken line: Linear fit with slope = −0.90, intercept = 0, r^2 = 0.44.

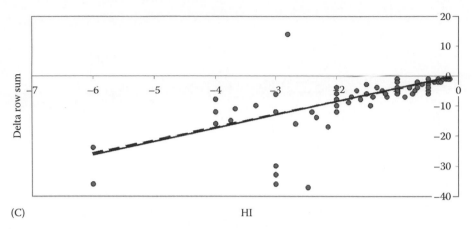

(C)   HI

FIGURE 12.37 (continued) (C) Correlation between delta row sum versus HI for hypohypsic plans (HI < 0). (a) Solid line: Linear fit with slope = 4.44, intercept = 0.35, r^2 = 0.46. (b) Broken line: Linear fit with slope = 4.28, intercept = 0, r^2 = 0.46.

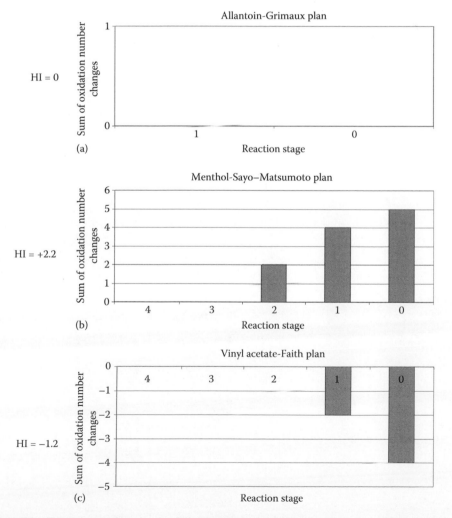

FIGURE 12.38 Hypsicity profiles for the following plans: (a) Allantoin (Grimaux, 1877), (b) (−)-menthol (Sayo–Matsumoto, 2002), (c) Vinyl acetate (Faith, 1966).

Isohypsic: Grimaux [43] plan for allantoin

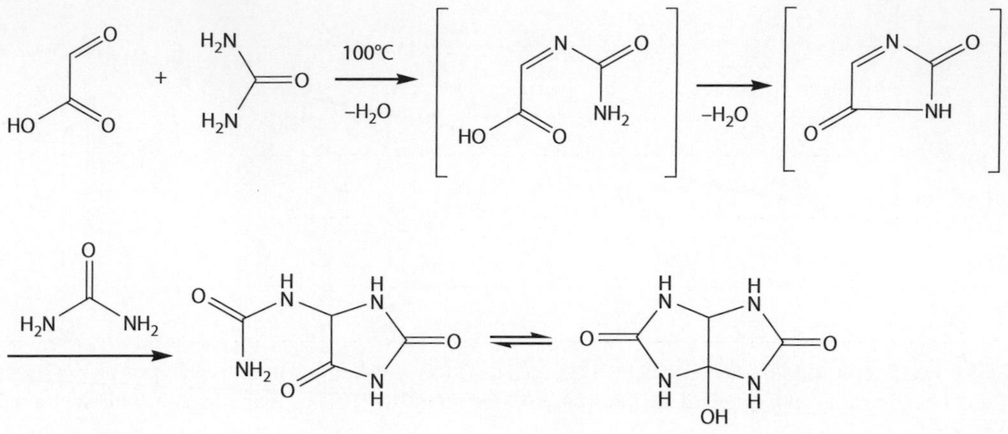

Hyperhypsic: Sayo–Matsumoto [44] plan for (–)-menthol

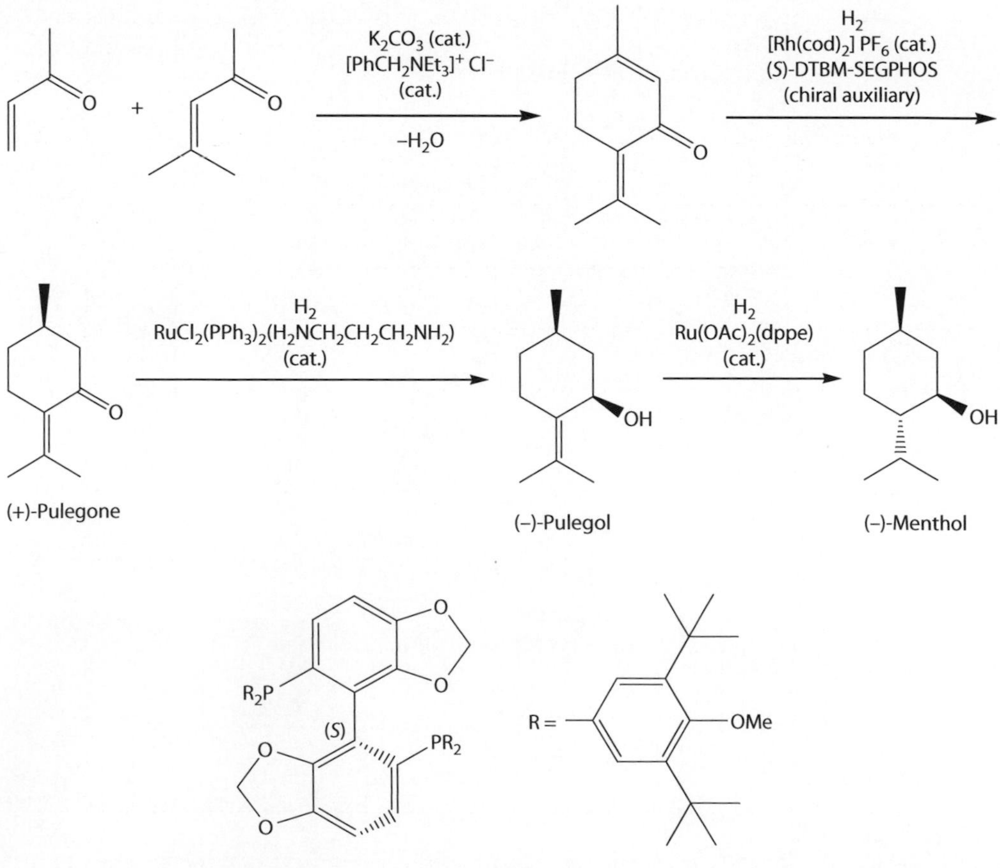

(+)-Pulegone   (–)-Pulegol   (–)-Menthol

(S)-DTBM-SEGPHOS

Hypohypsic: Faith [45] plan for vinyl acetate

REFERENCES

1. Faith, W.L.; Keyes, D.B.; Clark, R.L. *Industrial Chemicals*, 3rd edn., Wiley: New York, 1966, p. 726.
2. Stolz, F.; Flaecher, F.; Krohs, W.; Hallensleben, J. DE 538455, 1931.
3. GB 297385 (I.G. Farben, 1929).
4. GB 318488 (I.G. Farben, 1930).
5. GB 365541 (I.G. Farben, 1932).
6. GB 365535 (I.G. Farben, 1932).
7. Faith, W.L.; Keyes, D.B.; Clark, R.L. *Industrial Chemicals*, 3rd edn., Wiley: New York, 1966, pp. 1, 12, 28, 800.
8. Lowenheim, F.A.; Moran, M.K. *Faith, Keyes, and Clark's Industrial Chemicals*, 4th edn., Wiley: New York, 1975, p. 430.
9. Faith, W.L.; Keyes, D.B.; Clark, R.L. *Industrial Chemicals*, 3rd edn., Wiley: New York, 1966, pp. 458, 472.
10. Vogel, A. *Textbook of Practical Organic Chemistry,* Longman: London, U.K., 1948, p. 324.
11. Faith, W.L.; Keyes, D.B.; Clark, R.L. *Industrial Chemicals*, 3rd edn., Wiley: New York, 1966, pp. 28, 805.
12. Anzilotti, W.F.; Weinmayr, V. US 2791597 (DuPont, 1957).
13. Dickey, J.B.; Gray, A.R. *Org. Synth. Coll.* 1943, *2*, 60.
14. Vogel, A. *Textbook of Practical Organic Chemistry*, Longman: London, U.K., 1948, p. 852.
15. Faith, W.L.; Keyes, D.B.; Clark, R.L. *Industrial Chemicals*, 3rd edn., Wiley: New York, 1966, p. 583.
16. Lowenheim, F.A.; Moran, M.K. *Faith, Keyes, and Clark's Industrial Chemicals*, 4th edn., Wiley: New York, 1975, p. 294.
17. Muheim, A.; Müller, B.; Münch, T.; Wetli, M. EP885968 (Givaudan-Roure, 1998).
18. Muheim, A.; Lerch, K. *Appl. Microbiol. Biotechnol.* 1999, *51*, 456.
19. Späth, E.; Kromp, K. *Chem. Ber.* 1941, *74*, 867.
20. Roeslcr, A. DE 189037, 1907.
21. Elango, V.; Murphy, M.A.; Smith, B.L.; Davenport, K.G.; Mott, G.N.; Moss, G.L. EP 284310 (Hoechst-Celanese Corp., USA, 1988).
22. Elango, V.; Davenport, K.G.; Murphy, M.A.; Mott, G.N.; Zey, E.G.; Smith, B.L.; Moss, G.L. EP 400892 (Hoechst Celanese Corp, USA, 1990).
23. Van Stryk, F.G. CA 716609 (Dominion Rubber, 1962).
24. Van Stryk, F.G. US 3198831 (US Rubber Company, 1965).
25. Hull, E.H. US 4133833 (Pfizer, 1979).
26. Hull, E.H. DE 2900231 (Pfizer, 1979).
27. Poe, S.L.; Kobaslija, M.; McQuade, D.T. *J. Am. Chem. Soc.* 2007, *129*, 9216.
28. Mayer, E. *Oesterreichische Chem. Ztg.* 1949, *50*, 40.
29. Volynkin, N.I. *Zh. Prikladnoi Khimii* 1938, *11*, 423 (*Chem. Abs.* **32**: 41767).
30. Fierz-David, H.E.; Blangey, L. *The Fundamental Processes of Dye Chemistry*, Interscience Publishing: New York, 1949, p. 150.
31. Fierz-David, H.E.; Blangey, L. *The Fundamental Processes of Dye Chemistry*, Interscience Publishing: New York, 1949, pp. 124, 175, 180, 261, 293.
32. Faith, W.L.; Keyes, D.B.; Clark, R.L. *Industrial Chemicals*, 3rd edn., Wiley: New York, 1966, p. 497.

33. Faith, W.L.; Keyes, D.B.; Clark, R.L. *Industrial Chemicals*, 3rd edn., Wiley: New York, 1966, p. 583.
34. Thomas, J.M.; Raja, R. *Proc. Nat. Acad. Sci.* 2005, *102*, 13732.
35. Fierz-David, H.E.; Blangey, L. *The Fundamental Processes of Dye Chemistry*, Interscience Publishing: New York, 1949, pp. 133, 137, 138.
36. Willstätter, R.; Waser, E. *Chem. Ber.* 1911, *44*, 3423.
37. Willstätter, R.; Heidelberger, M. *Chem. Ber.* 1913, *46*, 517.
38. Cope, A.C.; Overberger, C.G. *J. Am. Chem. Soc.* 1948, *70*, 1433.
39. Brenna, E.; Fuganti, C.; Ronzani, S.; Serra, S. *Helv. Chim. Acta* 2001, *84*, 3650.
40. Willstätter, R. *Chem. Ber.* 1901, *34*, 129.
41. Willstätter, R. *Chem. Ber.* 1901, *34*, 3163.
42. Willstätter, R. *Chem. Ber.* 1896, *29*, 936.
43. Grimaux, E. *Ann. Chim. Phys.* 1877, *11*(5), 356.
44. Sayo, N.; Matsumoto, T. US 2002007094, 2002.
45. Faith, W.L.; Keyes, D.B.; Clark, R.L. *Industrial Chemicals*, 3rd edn., Wiley: New York, 1966, pp. 1, 12, 28, 800.

13 Compounds with Multiple Plans

This chapter compiles those compounds in the database having at least six plans each. The list of compounds in alphabetical order followed by the number of plans in parentheses is as follows: aspirin (11), bombykol (14), codeine (15), colchicine (9), coniine (16), *N,N*-diethyl-meta-toluamide (DEET) (11), discodermolide (13), epibatidine (19), estrone (10), fluoxetine (13), gamma-irone (14), hirsutene (30), ibuprofen (10), indigo (12), kainic acid (33), longifolene (11), lysergic acid (11), morphine (12), naproxen (11), nicotine (10), oseltamivir phosphate (29), paclitaxel (7), paroxetine (19), physostigmine (15), platensimycin (13), pregabalin (24), quinine (9), reserpine (10), resveratrol (10), rosefuran (18), sertraline (9), silphinene (8), sparteine (11), strychnine (15), swainsonine (33), tamoxifen (19), tropinone (8), tropolone (10), twistane (6), tyrian purple (10), vanillin (14), and vitamin A (9). The accompanying CD-ROM to this book contains the complete schemes, synthesis trees, target bond maps, green metrics analyses, and full graphical displays. In this chapter, summary tables (Tables 13.1 through 13.50) are given showing the performances of the plans according to the following parameters: number of steps (N), number of stages (M), number of input materials used (I), molecular weight building up parameter (μ_1), degree of asymmetry (beta), degree of convergence (delta), molecular weight fraction of sacrificial reagents used (f(sac)), hypsicity index (HI), percent atom economy (%AE), percent kernel reaction mass efficiency (%RME), and kernel mass of waste produced per mole of target product. For pharmaceutical compounds only, a second table of *E*-factor breakdowns is also given. Each table is accompanied by a figure showing the fractional waste contributions from sacrificial and building-up reactions (Figures 13.1 through 13.42). Plans are ranked according to the kernel mass of waste produced per mole of target product.

Aspirin

TABLE 13.1
Summary of Green Metrics Parameters for Aspirin Syntheses

Plan	Year	Type	N	M	I	μ_1	Beta	Delta	$f(sac)$	HI	% Yield	% AE	% Kernel RME	Kernel Mass of Waste (kg) per Mole
Faith P1[a]	1966	Convergent	5	7	11	−151.64	0.044	0.572	0.474	−1.5	43.9	32.8	17.6	0.84
Faith P2[b]	1966	Convergent	5	7	13	−176.47	0.026	0.623	0.591	−1.5	43.9	23.6	13.5	1.15
Faith P3[c]	1966	Convergent	5	8	10	−157.64	0.059	0.541	0.452	−1.17	43.9	32.5	16.7	0.9
Faith P4[d]	1966	Convergent	5	8	10	−156.64	0.059	0.541	0.485	−1.17	43.9	32.3	17.4	0.85
Faith P5[e]	1966	Convergent	5	8	11	−154.08	0.048	0.571	0.533	−1.5	43.9	27	15	1.02
Faith P6[f]	1966	Convergent	4	6	10	−178.8	−0.05	0.624	0.42	−1.4	50.4	44.8	26.9	0.49
Faith P7[g]	1966	Convergent	4	6	13	−208.6	−0.13	0.695	0.584	−1.4	50.4	29.4	18.4	0.8
Faith P8[h]	1966	Convergent	4	7	9	−186	−0.04	0.592	0.39	−1	50.4	44.4	24.9	0.54
Faith P9[i]	1966	Convergent	4	7	9	−187.6	−0.04	0.592	0.435	−1	50.4	44	26.4	0.5
Faith P10[j]	1966	Convergent	4	7	10	−181.73	−0.04	0.624	0.508	−1.4	50.4	34.7	21.3	0.67
Vogel[k]	1978	Linear	2	2	6	−123.7	0.4	0.712	0.387	0	22.7	39.9	10.8	1.5

[a] Carbide-Raschig.
[b] Carbide-benzenesulfonate.
[c] Carbide-cumene peroxidation.
[d] Carbide-toluene oxidation.
[e] Carbide-caustic.
[f] Wacker–Raschig.
[g] Wacker-benzenesulfonate.
[h] Wacker-cumene peroxidation.
[i] Wacker-toluene oxidation.
[j] Wacker-caustic.
[k] Carboxylation-acetylation process.

TABLE 13.2

Summary of *E*-Factor Breakdown for Aspirin Syntheses

Plan	Year	Type	E-Kernel	E-Excess	E-Aux	E-Total
Faith P1[a]	1966	Convergent	4.68	22.66	2.25	29.59
Faith P2[b]	1966	Convergent	6.39	22.71	2.25	31.35
Faith P3[c]	1966	Convergent	4.99	20.62	2.25	27.85
Faith P4[d]	1966	Convergent	4.75	20.65	2.25	27.65
Faith P5[e]	1966	Convergent	5.67	21.72	2.25	29.64
Faith P6[f]	1966	Convergent	2.72	4.7	4.32	11.74
Faith P7[g]	1966	Convergent	4.43	4.75	4.32	13.5
Faith P8[h]	1966	Convergent	3.02	2.66	4.32	10
Faith P9[i]	1966	Convergent	2.78	2.69	4.32	9.79
Faith P10[j]	1966	Convergent	3.7	3.76	4.32	11.78
Vogel[k]	1978	Linear	8.22	2.48	24.57	35.27

[a] Carbide-Raschig.
[b] Carbide-benzenesulfonate.
[c] Carbide-cumene peroxidation.
[d] Carbide-toluene oxidation.
[e] Carbide-caustic.
[f] Wacker–Raschig.
[g] Wacker-benzenesulfonate.
[h] Wacker-cumene peroxidation.
[i] Wacker-toluene oxidation.
[j] Wacker-caustic.
[k] Carboxylation-acetylation process.

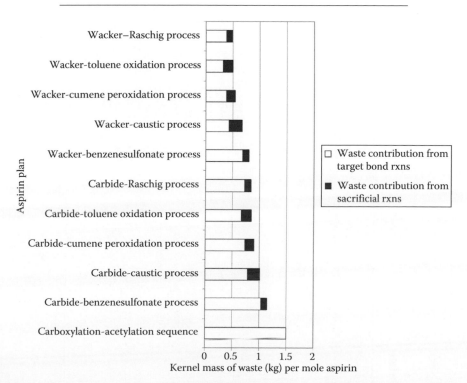

FIGURE 13.1 Bar graph showing sacrificial and building-up reaction proportions for each synthesis plan for aspirin. Plans are ranked according to kernel mass of waste produced per mole aspirin.

Bombykol

TABLE 13.3
Summary of Green Metrics Parameters for Bombykol Syntheses

Plan	Year	Type	N	M	l	μ_1	Beta	Delta	f(sac)	HI	% Yield	% AE	% Kernel RME	Kernel Mass of Waste (kg) per Mole
Alexakis G1	1984	Convergent	7	11	27	−184.58	0.793	0.551	0.723	3.38	32.1	9.5	4.7	4.8
Alexakis G2	1989	Convergent	8	9	23	46.17	0.864	0.467	0.76	0.44	24.5	7.9	3.2	7.2
Bayer G1	1962	Linear	9	9	21	−18.56	0.882	0.426	0.695	2.7	14.6	15.2	3.4	6.8
Bayer G2	1962	Linear	7	7	13	−35.5	0.811	0.412	0.359	2.5	15.2	14	3.7	6.3
Bestmann	1977	Linear	4	4	7	−50.4	0.583	0.47	0.277	3.6	26.5	20.1	7.4	3
Butenandt G1	1961	Linear	6	7	15	−100.47	0.789	0.478	0.657	3.14	13.5	17.2	3.1	7.5
Butenandt G2	1961	Linear	6	6	10	−23.14	0.762	0.407	0.51	0.71	18.2	35.9	7.3	3
Dasaradhi	1991	Linear	5	5	8	64.05	0.627	0.428	0.275	3.67	17.4	19.2	5.4	4.1
Naso	1989	Convergent	6	11	23	−73.44	0.554	0.648	0.75	3.29	33.4	9.6	4.4	5.2
Negishi	1973	Linear	6	6	16	−34.22	0.816	0.48	0.595	0.57	11.2	14	2.9	7.8
Normant	1980	Convergent	6	10	23	−154.09	0.725	0.585	0.519	5.14	27.4	15.5	4	5.7
Suginome	1983	Convergent	6	8	13	−115.47	0.596	0.496	0.585	−0.71	40.3	18.2	9.2	2.4
Trost	1980	Linear	6	6	12	−18.43	0.539	0.77	0.441	−0.14	21.4	27	8	2.7
Uenishi	2000	Linear	11	11	22	44.06	0.874	0.4	0.587	2	11.5	7.6	2.2	10.5

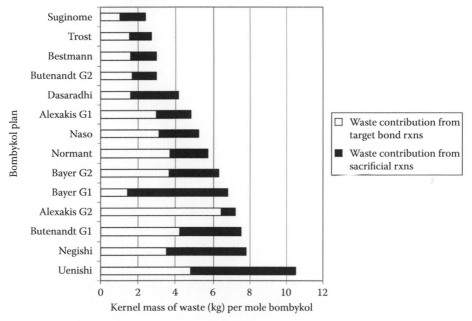

FIGURE 13.2   Bar graph showing sacrificial and building-up reaction proportions for each synthesis plan for bombykol. Plans are ranked according to kernel mass of waste produced per mole bombykol.

Codeine

TABLE 13.4
Summary of Green Metrics Parameters for Codeine Syntheses

Plan	Year	Type	N	M	I	μ_1	Beta	Delta	f(sac)	HI	% Yield	% AE	% Kernel RME	Kernel Mass of Waste (kg) per Mole
Rice	1980	Linear	17	17	25	25.75	0.89	0.338	0.758	1.06	11.1	17.4	3.4	8.5
Magnus	2009	Convergent	15	16	33	32.83	0.915	0.4	0.775	4.94	16.9	9.5	2.6	11.4
Mulzer	1998	Linear	21	21	40	26.83	0.932	0.37	0.806	0.68	2.9	9.2	0.7	41.7
Trost	2005	Convergent	15	16	30	−18.05	0.879	0.394	0.608	4.31	3.1	7	0.7	42.2
White	1997	Convergent	29	31	37	0.057	0.93	0.303	0.765	3.3	2.2	8.3	0.5	58.6
Overman	1993	Convergent	18	23	39	23.57	0.932	0.394	0.763	0.42	4.2	7.7	0.5	65.7
Parker	2006	Linear	20	20	37	59.88	0.927	0.367	0.841	0.67	2.6	7	0.4	69.5
Fukuyama	2006	Convergent	25	31	56	61.54	0.883	0.413	0.808	4.5	1.9	4.3	0.4	70.2
Stork	2009	Linear	21	21	36	56.66	0.927	0.355	0.638	3.59	1.3	6.6	0.3	100.5
Chida	2008	Linear	31	31	54	71.95	0.95	0.35	0.802	0.84	1.3	4.6	0.2	127.2
Iorga	2008	Convergent	16	18	34	−9.1	0.902	0.402	0.762	−0.41	0.5	7.9	0.09	346.7
Ogasawara	2001	Linear	29	29	51	−8.37	0.947	0.352	0.74	0.6	0.6	5.9	0.08	365.5
Taber	2002	Linear	27	27	47	54.89	0.945	0.352	0.74	−0.43	0.3	5.4	0.06	523.8
Hudlicky	2007	Linear	15	15	28	44.4	0.878	0.383	0.774	0.75	0.2	7	0.05	547.5
Gates	1956	Linear	25	25	46	20.65	0.943	0.361	0.882	1.42	0.01	6	0.003	11737

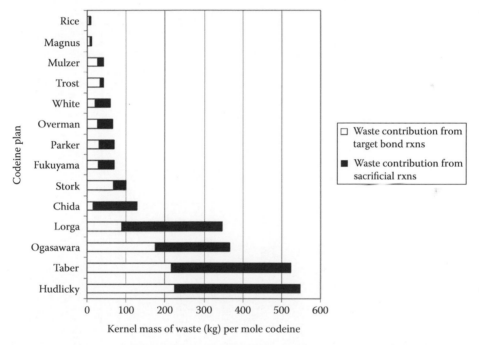

FIGURE 13.3 Bar graph showing sacrificial and building-up reaction proportions for each synthesis plan for codeine. Plans are ranked according to kernel mass of waste produced per mole codeine.

Colchicine

TABLE 13.5
Summary of Green Metrics Parameters for Colchicine Syntheses

Plan	Year	Type	N	M	I	μ_1	Beta	Delta	$f(sac)$	HI	% Yield	% AE	% Kernel RME	Kernel Mass of Waste (kg) per Mole
Cha	1998	Linear	16	16	30	−37.62	0.884	0.381	0.72	2.82	1.6	10.7	0.5	79.1
Evans	1984	Convergent	13	16	25	−180.2	0.877	0.376	0.38	−1.27	0.75	16.2	0.29	136.2
Banwell	1996	Linear	15	15	23	−33.5	0.88	0.356	0.646	−0.13	0.87	11.4	0.24	165.9
Schmalz	2005	Convergent	16	18	39	−63.29	0.933	0.408	0.781	0.53	0.94	8	0.15	258.3
Scott	1965	Convergent	16	20	23	−187.66	0.886	0.335	0.626	0	0.026	10.8	0.0041	97868
Roussel–Uclaf	1965	Convergent	23	27	36	−127.4	0.917	0.341	0.634	−1.79	0.00092	10.9	2.90E-04	138514
Eschenmoser	1959	Convergent	25	28	35	−114.84	0.913	0.322	0.608	−1.92	3.40E-05	11.2	1.90E-05	2146210
Van Tamelen	1959	Convergent	22	26	34	−135.35	0.932	0.337	0.651	−0.22	1.00E-05	9.6	3.20E-06	12307680
Woodward	1963	Convergent	26	28	42	−79.64	0.911	0.354	0.812	3.11	<100	8	<7.4	>5

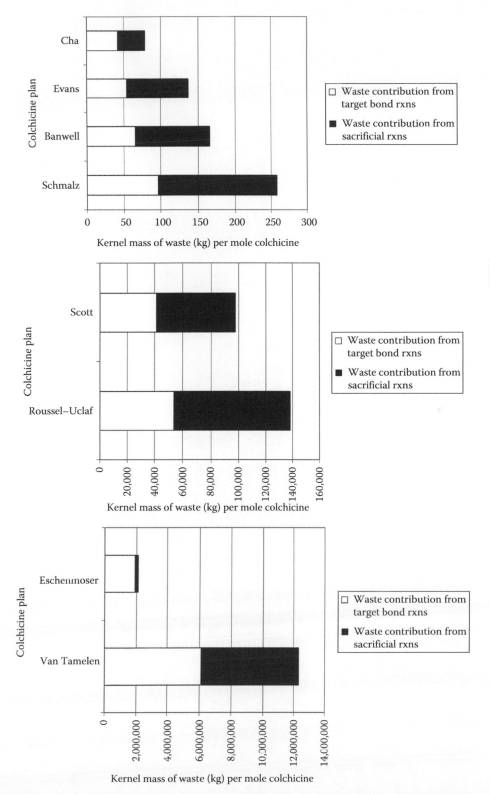

FIGURE 13.4 Bar graphs showing sacrificial and building-up reaction proportions for each synthesis plan for colchicine. Plans are ranked according to kernel mass of waste produced per mole colchicine.

Coniine

TABLE 13.6
Summary of Green Metrics Parameters for Coniine Syntheses

Plan	Year	Type	N	M	I	μ_1	Beta	Delta	f(sac)	HI	% Yield	% AE	% Kernel RME	Kernel Mass of Waste (kg) per Mole
Husson	1983	Linear	4	4	8	68.2	0.67	0.484	0.361	0.6	46.6	21.1	12.9	0.9
Couture	2007	Linear	5	5	7	68.33	0.578	0.403	0.092	2.5	30.9	32.4	11.4	1
Nagasaka	1988	Linear	6	6	11	25.49	0.743	0.429	0.573	1.14	26.3	18.3	7.3	1.6
Ladenburg	1886	Linear	4	4	5	3.8	0.406	0.407	0.225	5	35.7	15.5	6.1	1.9
Buchwald	1998	Linear	3	3	10	−22.3	0.625	0.626	0.5	1.75	27.3	14.7	5.9	2
Gallagher	1986	Linear	10	10	15	48.9	0.834	0.359	0.497	2.36	31.1	11.5	5	2.4
Meyers	1995	Linear	6	6	12	74.76	0.789	0.437	0.351	2.71	35.8	11.3	5	2.4
Hurvois	2006	Linear	5	5	11	79.8	0.765	0.468	0.483	−0.5	34.5	12.3	4.7	2.6
Gramain	2000	Linear	6	6	11	64.71	0.767	0.424	0.459	1.57	14.5	18.1	4.6	2.6
Knochel	2004	Linear	6	6	13	128.57	0.778	0.454	0.75	1.29	40.7	7.6	4.2	2.9
Shipman	2001	Linear	4	4	9	63.14	0.617	0.521	0.184	1	22.6	11.6	3.4	3.6
Takahata	1996	Linear	8	8	16	40.83	0.853	0.412	0.866	1.33	25.5	5.8	2.3	5.4
Pandey	1997	Convergent	11	12	23	64.17	0.878	0.406	0.878	0.58	14	6.3	2.1	6.1
Kibayashi	1992	Convergent	12	13	24	105.62	0.9	0.392	0.685	3.15	18.3	4.6	1.7	7.4
Beauchemin	2009	Linear	7	7	14	86.23	0.794	0.43	0.712	1.88	4.4	6.2	0.4	28.6
Aketa	1976	Convergent	12	13	23	159.96	0.894	0.387	0.796	−1.85	1	6.1	0.1	115.8

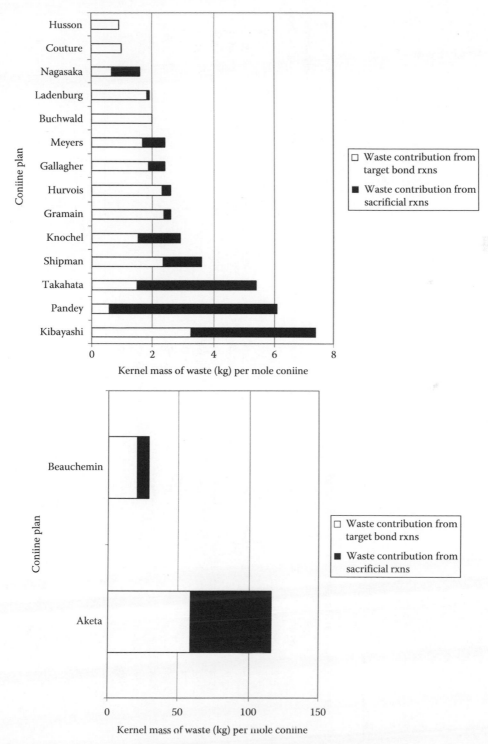

FIGURE 13.5 Bar graphs showing sacrificial and building-up reaction proportions for each synthesis plan for coniine. Plans are ranked according to kernel mass of waste produced per mole coniine.

DEET

TABLE 13.7
Summary of Green Metrics Parameters for DEET Syntheses

Plan	Year	Type	N	M	I	μ_1	Beta	Delta	f(sac)	HI	% Yield	% AE	% Kernel RME	Kernel Mass of Waste (kg) per Mole
Dominion Rubber	1962	Linear	2	2	3	−99.67	0.25	0.584	0.07	−1.33	41.2	84.1	41.4	0.27
Pfizer	1979	Linear	2	2	3	−99.67	0.25	0.584	0.07	−1.33	33.3	84.1	33.5	0.38
Neeland	1998	Linear	2	2	4	−99.67	0.25	0.644	0.496	−1.33	47.5	45.6	32	0.41
U.S. Dept. Agriculture	1953	Linear	3	3	4	−83.89	0.417	0.458	0.496	−1	46.1	45.6	31	0.43
Wang	1974	Linear	2	2	4	−99.67	0.25	0.644	0.496	−1.33	43.6	45.6	29.3	0.46
Virginia Chemicals	1971	Linear	2	2	4	−99.67	0.25	0.644	0.518	−1.33	41.7	43.6	27.2	0.51
Showa Denko	1976	Linear	2	2	7	−100	0.5	0.719	0.494	0	55.3	39.1	22.9	0.64
Bhattacharya	2006	Linear	2	2	10	−100	0.431	0.805	0.624	0	63	29	20.8	0.7
Khalafi-Nezhad	2005	Linear	2	2	5	−99.67	0.219	0.699	0.643	−1.33	37.2	32.3	19.3	0.8
LeFevre	1990	Linear	2	2	4	−99.67	0.25	0.644	0.471	−1.33	25	47.9	17.4	0.91
Snieckus	1989	Linear	5	5	9	13.67	0.791	0.44	0.551	0.33	20.3	24	5.9	3.1

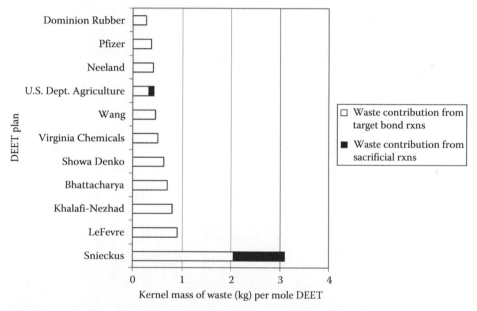

FIGURE 13.6 Bar graph showing sacrificial and building-up reaction proportions for each synthesis plan for DEET. Plans are ranked according to kernel mass of waste produced per mole DEET.

Discodermolide

TABLE 13.8
Summary of Green Metrics Parameters for Discodermolide Syntheses

Plan	Year	Type	N	M	I	μ_1	Beta	Delta	f(sac)	HI	% Yield	% AE	% Kernel RME	Kernel Mass of Waste (kg) per Mole
Paterson G3	2004	Convergent	23	42	93	−290.14	0.97	0.446	0.751	1.67	13	5.5	1.3	44.3
Paterson G1	2000	Convergent	24	47	114	−441.86	0.974	0.457	0.793	2	13.5	4.3	0.9	63.4
Smith G4	2005	Convergent	17	41	90	−429.66	0.971	0.47	0.776	2.61	10.5	4.7	0.7	79.1
Paterson G2	2003	Convergent	25	47	109	−365.08	0.973	0.45	0.764	1.42	5.5	4.3	0.5	129.2
Marshall	1998	Convergent	29	49	98	−227.76	0.976	0.425	0.754	−1.73	1.3	5.2	0.3	223.1
Ardisson	2007	Convergent	21	39	84	−262.97	0.969	0.446	0.752	−0.68	1.6	5.8	0.2	238.2
Smith G2	2000	Convergent	24	45	97	−251.54	0.971	0.445	0.786	1.88	1.6	4.4	0.2	358.8
Panek	2005	Convergent	32	71	125	−584.5	0.98	0.435	0.793	0.12	0.8	4	0.2	384.1
Smith G3	2003	Convergent	24	44	92	−283.2	0.97	0.441	0.768	1.84	1.4	4.7	0.2	379.6
Novartis	2004	Convergent	26	48	108	−409.66	0.972	0.447	0.794	2.33	0.9	4.2	0.1	520.9
Schreiber	1993	Convergent	25	49	113	−352.7	0.979	0.449	0.723	0.038	0.8	4	0.1	590.1
Smith G1	1995	Convergent	28	51	103	−183.38	0.977	0.432	0.801	0.62	1.3	4.1	0.07	832
Myles	1997	Convergent	28	55	103	−440.52	0.95	0.447	0.824	−0.38	0.07	5.1	0.03	1749.2

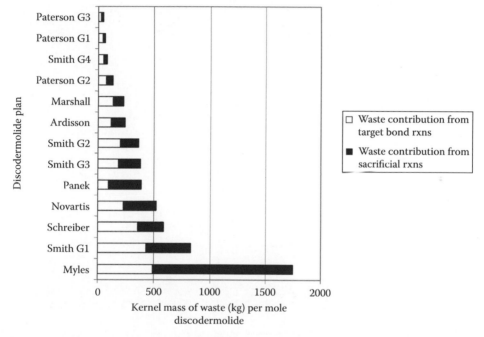

FIGURE 13.7 Bar graph showing sacrificial and building-up reaction proportions for each synthesis plan for discodermolide. Plans are ranked according to kernel mass of waste produced per mole discodermolide.

Epibatidine

TABLE 13.9
Summary of Green Metrics Parameters for Epibatidine Syntheses

Plan	Year	Type	N	M	I	μ_1	Beta	Delta	f(sac)	HI	% Yield	% AE	% Kernel RME	Kernel Mass of Waste (kg) per Mole
Giblin	1997	Linear	6	6	12	83.19	0.757	0.444	0.244	2	24.1	19.9	6.6	3
Hoashi	2004	Linear	10	10	14	50.88	0.794	0.351	0.636	6.46	18.6	12.5	4.4	4.5
Corey	1993	Linear	9	9	15	87.03	0.853	0.377	0.52	0.1	35.5	9	4.2	4.7
Trudell	1996	Linear	9	9	19	67.41	0.872	0.413	0.606	2.6	17.6	10.6	3.7	5.4
Aggarwal	2005	Linear	7	7	12	112.28	0.803	0.399	0.648	0.38	4.3	11.7	1.9	10.6
Bai	1996	Linear	9	9	17	39.92	0.823	0.406	0.739	0.7	10.4	8.8	1.8	11.4
Olivo	1999	Linear	10	10	16	43.57	0.847	0.368	0.742	0.46	7.2	12	1.7	12.3
Regan	1993	Convergent	5	7	12	9.9	0.717	0.498	0.549	1.5	3.7	20.6	1.5	13.4
Kosugi	1997	Linear	16	16	30	87.5	0.908	0.376	0.806	1.71	12	6.1	1.1	18.5
Evans	2001	Convergent	15	16	26	104.16	0.917	0.368	0.799	1.06	7.1	5.7	0.9	22.3
Shen	1993	Linear	6	6	10	55	0.76	0.407	0.471	2.86	2.5	20.6	0.9	24.2
Szantay G1	1996	Linear	10	10	13	62.12	0.778	0.338	0.807	5.27	2.6	10.2	0.8	24.6
Loh	2005	Convergent	12	13	23	33.92	0.882	0.389	0.757	1.31	7	5.6	0.8	27.4
Kibayashi	1998	Convergent	11	12	25	111.83	0.903	0.412	0.8	1	2.2	7	0.4	54.1
Carroll	1995	Linear	8	8	15	115.22	0.848	0.402	0.724	2	1.7	8.3	0.4	55.4
Pandey	1998	Convergent	10	13	26	-0.1	0.907	0.432	0.724	-0.18	3.1	6.4	0.3	59.7
Szantay G2	1996	Linear	12	12	15	132.33	0.816	0.322	0.84	1.62	1	9	0.3	71.6
Fletcher	1994	Linear	13	13	18	49.06	0.872	0.333	0.787	0.93	1.2	9.8	0.2	85
Broka	1993	Linear	20	20	35	113.71	0.922	0.36	0.867	0.33	1.4	4	0.2	86.9

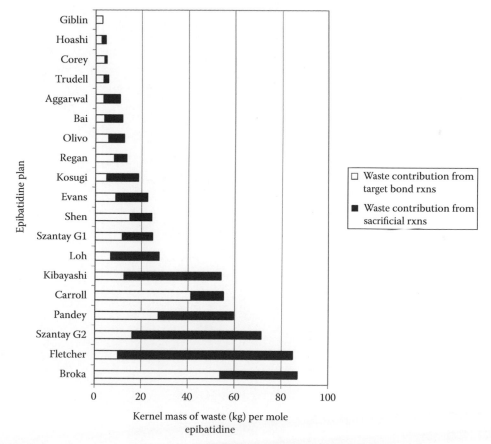

FIGURE 13.8   Bar graph showing sacrificial and building-up reaction proportions for each synthesis plan for epibatidine. Plans are ranked according to kernel mass of waste produced per mole epibatidine.

Estrone

TABLE 13.10
Summary of Green Metrics Parameters for Estrone Syntheses

Plan	Year	Type	N	M	I	μ_1	Beta	Delta	f(sac)	HI	% Yield	% AE	% Kernel RME	Kernel Mass of Waste (kg) per Mole
Corey	2007	Convergent	10	14	24	−112.82	0.876	0.431	0.661	3.64	12.4	11.4	1.8	15.1
Johnson G2	1973	Convergent	16	20	26	−4.47	0.907	0.353	0.71	0.94	5.9	10	1.2	22.2
Danishefsky	1976	Convergent	14	17	30	−60.5	0.909	0.398	0.697	2.87	3.7	11.9	1.1	23.5
Torgov	1963	Convergent	13	14	23	−74.86	0.894	0.372	0.613	1.29	1.9	14.3	0.6	45.9
Grieco	1980	Convergent	27	35	56	−52.21	0.952	0.376	0.867	3.43	1.5	5.6	0.4	61.1
Smith	1963	Convergent	14	18	24	−94.27	0.821	0.38	0.64	3.73	1.8	15.1	0.4	66.3
Pattenden	2004	Linear	13	13	26	73.57	0.893	0.393	0.483	2.14	0.9	5.8	0.2	157.2
Bachmann	1942	Convergent	19	22	31	−24.51	0.927	0.348	0.551	3.6	0.6	10.6	0.1	192.4
Johnson G1	1952	Convergent	12	14	19	−57.55	0.871	0.359	0.656	0.62	0.6	13.3	0.1	260.6
Kametani	1977	Convergent	20	23	39	−55.68	0.938	0.373	0.769	2.24	0.1	8.9	0.04	686.6

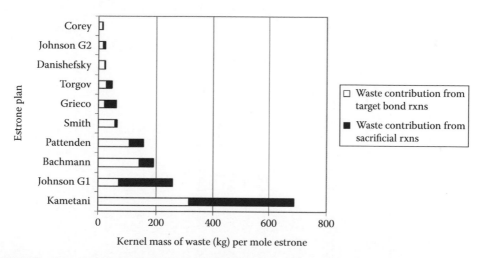

FIGURE 13.9 Bar graph showing sacrificial and building-up reaction proportions for each synthesis plan for estrone. Plans are ranked according to kernel mass of waste produced per mole estrone.

Fluoxetine

TABLE 13.11
Summary of Green Metrics Parameters for Fluoxetine Syntheses

Plan	Year	Type	N	M	I	μ_1	Beta	Delta	f(sac)	HI	% Yield	% AE	% Kernel RME	Kernel Mass of Waste (kg) per Mole
Corey	1989	Linear	5	5	10	−183.65	0.637	0.477	0.469	0.33	81.3	40.9	35.9	0.62
Wang	2005	Linear	5	5	14	−214.62	0.729	0.526	0.272	2.67	70.9	34.4	24.3	0.9
Shibasaki	2004	Linear	6	6	12	−203.74	0.697	0.457	0.097	1.86	36.3	45.1	26.5	0.96
Sharpless	1988	Linear	8	8	12	−187.45	0.792	0.371	0.293	0.78	21.8	35.1	15.9	1.8
Lilly G1	1982	Linear	7	7	15	−101.91	0.765	0.45	0.611	0.5	32.3	28.9	11.8	2.3
Brown	1988	Convergent	5	6	12	−216.06	0.738	0.492	0.654	0.33	40.2	24.3	12.8	2.3
Lilly G2	1994	Linear	6	6	14	−169.67	0.677	0.495	0.462	0.57	18.5	37.8	12.1	2.5
P Kumar	2004	Linear	7	7	16	−161.45	0.794	0.455	0.692	0.5	61	17.9	11.8	2.6
Bracher	1996	Linear	6	6	14	−77.14	0.806	0.46	0.676	0	25.3	20.1	9.3	2.9
Sepracor	2001	Linear	8	8	17	−162.17	0.835	0.427	0.659	3.78	16.1	20.5	7.1	4.1
A Kumar	1991	Linear	9	9	19	−116.51	0.843	0.421	0.785	2.3	18.4	16.9	8.4	4.2
Miles	2001	Linear	6	6	14	−146.45	0.738	0.479	0.810	1.71	35.1	11.9	7.5	4.3
Panunzio	2004	Linear	6	6	14	−49.08	0.802	0.462	0.432	1.57	17.5	20	17.5	5.5

TABLE 13.12

Summary of _E_-Factor Breakdown for Fluoxetine Syntheses

Plan	Year	Type	_E_-Kernel	_E_-Excess	_E_-Aux	_E_-Total	Total Mass of Waste (kg) per Mole
A Kumar	1991	Linear	10.84	5.02	22.59	38.45	13.3
Wang[a]	2005	Linear	3.12	6.56	149.1	158.82	54.9
P Kumar	2004	Linear	7.5	33.73	176.8	218.05	75.3
Brown	1988	Convergent	6.79	8	262.5	277.26	95.8
Bracher	1996	Linear	9.78	105	404.1	518.84	153.6
Panunzio	2004	Linear	17.67	70.13	588.3	676.06	208.9
Sharpless	1988	Linear	5.31	11.9	1451	1467.7	507
Lilly G1	1982	Linear	7.51	181.2	1990	2178.9	673.1
Miles	2001	Linear	12.38	26.78	1928	1967.1	679.5
Shibasaki	2004	Linear	2.71	>6.58	>723.89	>733.18	>253.3
Sepracor	2001	Linear	13.13	>24.77	>59.9	>97.8	>30.2
Lilly G2	1994	Linear	7.28	>0	>0	>7.28	>2.5
Corey	1989	Linear	1.79	>1.79	>0.81	>4.39	>1.52

[a] Includes synthesis of chiral ligand resin.

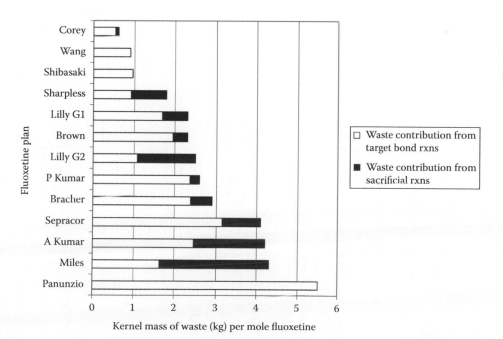

FIGURE 13.10 Bar graph showing sacrificial and building-up reaction proportions for each synthesis plan for fluoxetine. Plans are ranked according to kernel mass of waste produced per mole fluoxetine.

Gamma-Irone

TABLE 13.13

Summary of Green Metrics Parameters for Gamma-Irone Syntheses

Plan	Year	Type	N	M	I	μ_1	Beta	Delta	f(sac)	HI	% Yield	% AE	% Kernel RME	Kernel Mass of Waste (kg) per Mole
Gosselin	2001	Linear	9	9	17	−13	0.74	0.423	0.573	4	14	12.7	3.2	6.3
Monti G1	1999	Linear	13	13	23	24.43	0.878	0.376	0.67	3.5	12.7	9.8	2.8	7.1
Brenna G1	2001	Linear	5	5	13	−41.2	0.68	0.508	0.413	2.17	8.6	13.7	2.6	7.9
Mori	1983	Linear	10	10	20	−27.95	0.781	0.424	0.628	2.73	7.9	11.2	1.9	10.6
Givaudan	1988	Linear	10	10	19	−17.27	0.851	0.398	0.659	2.27	3.8	12.8	1.7	12.2
Monti G2	1996	Linear	11	11	27	−21.67	0.89	0.428	0.794	1.58	10	7.5	1.6	12.7
Leyendecker	1987	Convergent	11	14	31	−46.65	0.908	0.442	0.794	1.25	8.5	6	1.2	16.8
Monti G3	2000	Linear	14	14	35	−24.93	0.916	0.419	0.801	2.27	3.9	4.8	0.6	34.8
Vidari G1	2003	Linear	12	12	27	−27.4	0.891	0.413	0.736	2.62	3.1	6.4	0.5	39.8
Takazawa	1985	Linear	13	13	30	−27.9	0.856	0.426	0.713	2.57	0.6	7.4	0.4	46.8
Vidari G2	2003	Linear	12	12	25	−27.4	0.893	0.401	0.719	2.62	2.4	6.8	0.4	50
Garnero–Joulain	1979	Linear	12	12	24	−34.97	0.888	0.395	0.588	2.31	0.4	7.5	0.3	69.2
Brenna G2	2001	Linear	10	10	15	30.95	0.833	0.359	0.848	1.18	0.1	15.2	0.06	324.7
Kiyota	2000	Convergent	17	18	26	−6.18	0.893	0.344	0.8	3	0.08	8.2	0.03	646

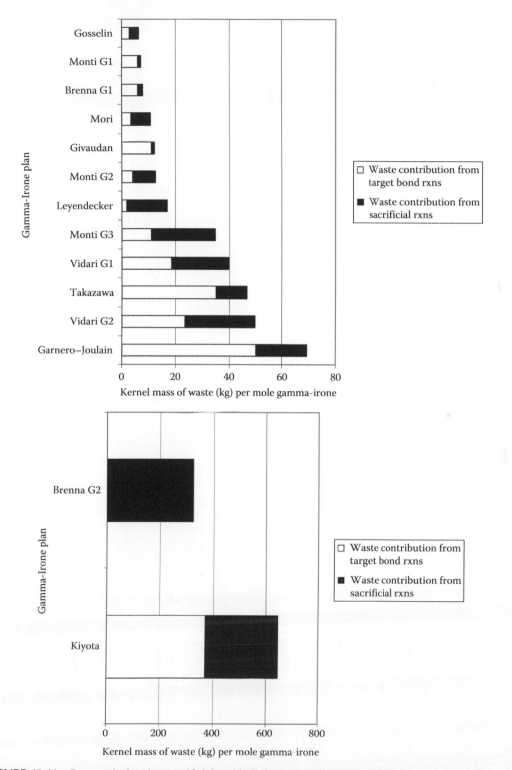

FIGURE 13.11 Bar graph showing sacrificial and building-up reaction proportions for each synthesis plan for gamma-irone. Plans are ranked according to kernel mass of waste produced per mole gamma-irone.

Hirsutene

TABLE 13.14
Summary of Green Metrics Parameters for Hirsutene Syntheses

Plan	Year	Type	N	M	I	μ_1	Beta	Delta	f(sac)	HI	% Yield	% AE	% Kernel RME	Kernel Mass of Waste (kg) per Mole
Iyoda	1986	Linear	4	4	9	−42.4	0.647	0.514	0.43	3.2	34.3	16.1	8.3	2.2
Hudlicky	1980	Linear	10	10	14	−3.23	0.802	0.35	0.514	3.18	25.1	12.5	4.9	4.0
Cohen	1993	Linear	7	7	19	62.75	0.841	0.47	0.544	5.25	37.4	9.5	5.0	3.9
Funk	1984	Linear	12	12	19	7.77	0.861	0.361	0.671	2.78	13.3	9.8	3.0	6.6
Krische	2003	Linear	12	12	21	6.34	0.87	0.377	0.606	4.31	9.7	9.6	2.4	8.2
Yu	2008	Linear	7	7	15	41	0.815	0.437	0.424	2.38	15.5	8.5	2.1	9.6
Paquette	1991	Linear	9	9	19	98.1	0.852	0.419	0.595	2.3	10.5	7.6	2.1	9.8
Franck-Neumann	1992	Linear	10	10	18	58.81	0.849	0.389	0.621	2.73	8.7	9.7	1.9	10.8
Sternbach	1990	Linear	21	21	37	20.88	0.927	0.359	0.701	3.86	19.5	5.2	1.8	10.9
Wender	1982	Linear	7	7	8	29.61	0.705	0.33	0.531	2.38	4.2	23.3	1.8	11.0
Mehta	1986	Linear	10	10	12	−5.45	0.758	0.324	0.235	5.27	3.2	16.9	1.5	13.6
Hua	1988	Convergent	16	18	39	49.95	0.926	0.411	0.811	3.41	16.3	4.5	1.3	15.8
Weedon	1985	Linear	7	7	15	17	0.81	0.439	0.766	2.75	8.2	8.0	1.1	18.6
Cossy	1990	Convergent	9	10	18	−40.11	0.808	0.42	0.544	2	5.2	11.9	1.1	19.0
Curran	1985	Convergent	13	14	26	28.3	0.907	0.389	0.735	3	7.9	5.2	0.8	24.9
Fitjer	1998	Linear	9	9	16	94.49	0.822	0.395	0.781	0.7	5.8	6.0	0.8	25.0
Weinges	1993	Linear	13	13	30	34.92	0.925	0.406	0.83	3.21	4.4	4.0	0.7	30.6
List	2008	Linear	13	13	18	−4.75	0.834	0.338	0.473	4.07	6.5	6.4	0.7	30.7
Ley	1985	Convergent	8	10	32	40.01	0.825	0.528	0.85	6.33	4.4	5.6	0.6	31.6
Nozoe	1976	Linear	11	11	17	−2.55	0.839	0.361	0.723	1.75	2.2	9.1	0.5	41.1
Oppolzer	1994	Convergent	18	20	38	46.71	0.936	0.386	0.723	4.42	5.4	4.3	0.4	46.6
Fukumoto	1993	Linear	25	25	41	21.38	0.937	0.345	0.628	3.31	3.3	4.2	0.4	49.2
Magnus	1985	Convergent	16	17	37	7.07	0.926	0.403	0.779	4.35	2.0	5.4	0.4	53.9
Hewson	1985	Linear	17	17	30	56.17	0.909	0.36	0.586	3.11	2.7	5.0	0.4	54.5
Little	1983	Linear	15	15	25	−4.02	0.889	0.361	0.756	2.19	1.6	6.8	0.3	62.3
Banwell	2002	Linear	18	18	28	41.58	0.906	0.344	0.467	3.47	1.0	5.4	0.2	129.7
Greene–Nozoe	1990	Linear	16	16	27	40.22	0.901	0.36	0.609	2.77	0.4	5.7	0.1	296.5
Singh	2002	Linear	16	16	24	−16.82	0.914	0.338	0.453	3.29	0.3	5.6	0.1	307.4
Sarkar	1992	Convergent	20	21	35	37.8	0.919	0.356	0.685	0.095	0.2	5.5	0.1	391.6
Leonard	1999	Linear	20	20	39	17.95	0.93	0.375	0.617	2.43	0.4	4.5	0.1	420.0

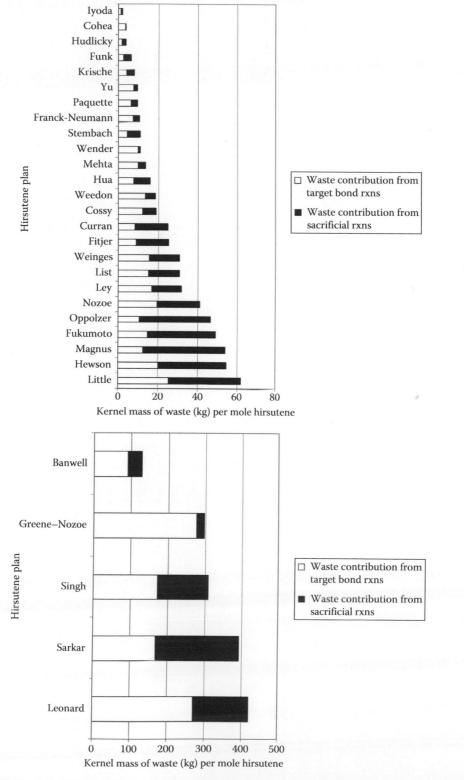

FIGURE 13.12 Bar graphs showing sacrificial and building-up reaction proportions for each synthesis plan for hirsutene. Plans are ranked according to kernel mass of waste produced per mole hirsutene.

Ibuprofen

TABLE 13.15
Summary of Green Metrics Parameters for Ibuprofen Syntheses

Plan	Year	Type	N	M	I	μ_1	Beta	Delta	f(sac)	HI	% Yield	% AE	% Kernel RME	Kernel Mass of Waste (kg) per Mole
Hoechst-Celanese	1988	Linear	4	4	5	−81.6	0.531	0.399	0	0.6	45.7	77.4	36.8	0.4
RajanBabu	2009	Linear	7	7	11	−61	0.77	0.387	0.408	−0.75	55	31.4	20.4	0.8
DuPont	1985	Linear	3	3	8	−50.41	0.458	0.587	0.476	0.75	41.1	31.9	17.6	0.96
Upjohn	1977	Linear	6	6	10	−48.57	0.739	0.41	0.443	−0.14	57.7	27	17.3	1
McQuade	2009	Linear	2	2	9	3.33	0.137	0.831	0.542	−0.67	35.2	26.9	11.5	1.6
Pinhey	1984	Convergent	5	6	8	15.07	0.603	0.431	0.689	−0.5	45.4	23.7	11.4	1.6
Boots	1968	Linear	5	5	13	−57.67	0.718	0.514	0.631	−0.33	48.9	17.1	10.4	1.8
DD113889	1975	Linear	8	8	13	−34.44	0.791	0.386	0.564	−0.56	34.9	22.1	9.8	1.9
Ruchardt	1991	Linear	6	6	11	−59.86	0.717	0.434	0.665	0.57	27.1	18.4	7.3	2.6
Furstoss	1999	Linear	6	6	10	−60.86	0.711	0.415	0.305	−0.86	17.3	22.7	5.3	3.7

TABLE 13.16
Summary of E-Factor Breakdown for Ibuprofen Syntheses

Plan	Year	Type	E-Kernel	E-Excess	E-Aux	E-Total	Total Mass of Waste (kg) per Mole
Upjohn	1977	Linear	4.78	1.03	45.9	51.7	10.7
DD113889	1975	Linear	9.17	15.55	30.47	55.19	11.4
Hoechst-Celanese	1988	Linear	1.72	1.17	111.9	114.79	23.6
Boots	1968	Linear	8.59	15.42	96.52	120.53	24.8
Ruchardt	1991	Linear	12.62	52.06	79.98	144.66	29.8
DuPont	1985	Linear	4.67	61.83	150.1	216.61	44.6
Pinhey	1984	Convergent	7.8	152.2	540.3	700.33	144.3
RajanBabu	2009	Linear	3.91	21.81	1888	1913.8	394.2
Furstoss	1999	Linear	17.8	12.13	2286	2316.2	477.1
McQuade	2009	Linear	7.73	346.6	41968	42322	8718

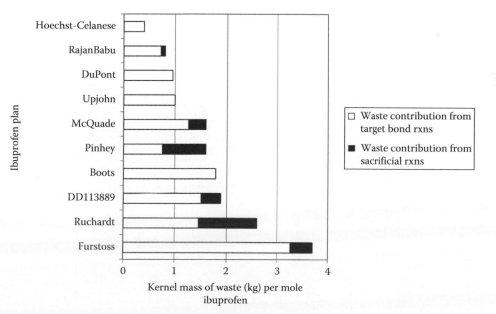

FIGURE 13.13 Bar graph showing sacrificial and building-up reaction proportions for each synthesis plan for ibuprofen. Plans are ranked according to kernel mass of waste produced per mole ibuprofen.

Indigo

TABLE 13.17
Summary of Green Metrics Parameters for Indigo Syntheses

Plan	Years	Type	N	M	I	μ₁	Beta	Delta	f(sac)	HI	% Yield	% AE	% Kernel RME	Kernel Mass of Waste (kg) per Mole
Ziegler G2	1965	Linear	4	4	7	67.18	0.453	0.487	0.559	-3.2	70.8	21.3	16.3	0.7
Ziegler G1	1964	Linear	6	6	6	69.99	0.588	0.317	0.558	-2	60.6	21.4	14	0.8
Gosteli G2	1977	Linear	4	4	6	35.98	0.489	0.445	0.53	-2	40.7	28.6	13.7	0.8
Madelung G1	1914	Linear	4	4	8	4.69	0.688	0.481	0.813	-2.4	92	14	13.6	0.83
Gosteli G3	1977	Linear	4	4	7	29.18	0.62	0.464	0.622	-2	37	22.9	10.8	1.1
Gosteli G1	1977	Linear	3	3	6	35.75	0.5	0.542	0.45	2.5	24.9	26.8	8.5	1.4
Baeyer G1	1879	Linear	4	4	8	3.65	0.634	0.491	0.517	1.6	34.8	17.4	7	1.7
Heumann–Fierz-David	1891, 1949	Linear	5	5	13	6.5	0.649	0.534	0.699	-1.33	32.5	13.6	6.4	1.9
Sandmeyer	1903	Linear	5	5	12	67	0.539	0.542	0.818	3.33	32.8	9.4	3.4	3.7
Harley-Mason	1950	Linear	3	3	9	29.4	0.645	0.598	0.72	2.5	35.1	8.3	3	4.2
Vogel	1978	Convergent	4	5	11	-32.13	0.653	0.557	0.761	-2.4	12.8	12.9	2.3	5.7
Madelung G2	1914	Linear	7	7	12	-0.66	0.788	0.401	0.772	-2.5	11.1	13.3	3.8	6.6

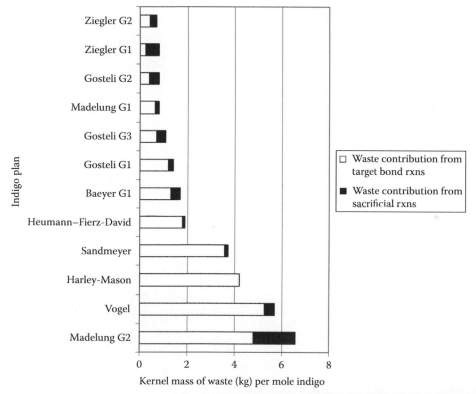

FIGURE 13.14 Bar graph showing sacrificial and building-up reaction proportions for each synthesis plan for indigo. Plans are ranked according to kernel mass of waste produced per mole indigo.

Kainic Acid

TABLE 13.18
Summary of Green Metrics Parameters for Kainic Acid Syntheses

Plan	Year	Type	N	M	l	μ_1	Beta	Delta	f(sac)	HI	% Yield	% AE	% Kernel RME	Kernel Mass of Waste (kg) per Mole
Ganem	2001	Convergent	9	11	15	−94.27	0.81	0.384	0.322	1.7	6.7	17.7	3.7	1.4
Tilve	2009	Linear	7	7	18	44.48	0.752	0.49	0.629	1.88	15.1	7.4	2.3	9.1
Trost	2003	Linear	14	14	29	182.6	0.901	0.359	0.729	−4.8	15.1	6.9	1.9	11
Yoo	1994	Convergent	19	21	39	59.27	0.928	0.383	0.702	−2.6	10.2	4.5	1.2	17.8
Bachi	1997	Linear	14	14	25	134.65	0.89	0.374	0.736	1.8	7.3	6.4	1.3	16.6
Benetti	1992	Convergent	14	18	33	13.26	0.932	0.405	0.758	−1.53	11.6	6	1.1	19.4
Oppolzer	1982	Linear	11	11	18	85.75	0.872	0.366	0.761	−2.42	3.4	10.5	0.9	22.3
Monn	1994	Linear	13	13	23	12.56	0.86	0.379	0.59	−0.29	2.2	7.1	0.8	28
Fukuyama G2	2007	Linear	19	19	34	75.94	0.924	0.363	0.695	0.8	4.2	6	0.7	28.7
Naito	1997	Convergent	16	21	31	70.83	0.909	0.381	0.775	−2.12	3.8	6.8	0.6	35.5
Ogasawara G2	2000	Convergent	18	19	31	104.31	0.909	0.361	0.825	−4.16	2.7	6.3	0.5	38.9
Fukuyama G1	2005	Linear	19	19	30	35.65	0.922	0.344	0.692	1.75	0.7	6.5	0.4	51.1
Clayden	2002	Convergent	14	16	29	86.36	0.907	0.394	0.786	0.4	6.3	3.4	0.4	51.7
Jung	2007	Linear	26	26	42	124.42	0.937	0.342	0.785	0	4.4	4.3	0.4	53.7
Takano G4	1993	Convergent	21	27	47	166.4	0.936	0.393	0.856	−3	4.6	3.7	0.4	55.4
Fukuyama G3	2008	Linear	22	22	47	119.6	0.946	0.386	0.803	3.09	3	3.6	0.3	74.4
Ogasawara G3	2001	Convergent	26	27	42	102.3	0.866	0.343	0.933	−3.3	1.2	5.7	0.2	106.1
Taylor	1999	Convergent	17	21	38	7.03	0.925	0.398	0.692	−0.78	1.5	4.6	0.2	108.1
Montgomery	1999	Convergent	12	18	36	−51.12	0.933	0.438	0.803	−1.85	1.4	5.8	0.2	108.6
Takano G1	1988	Convergent	21	23	43	104.75	0.937	0.38	0.763	−1.14	0.9	4.4	0.2	110.8
Poisson G2	2006	Convergent	31	34	53	43.5	0.825	0.371	0.823	−1.63	0.8	3.4	0.2	112.1
Baldwin	1990	Convergent	15	17	32	48.06	1.023	0.365	0.752	−2.06	0.8	5.7	0.2	130.4
Poisson G1	2005	Convergent	28	30	54	71.75	0.948	0.366	0.799	−2.52	0.7	3.6	0.2	138.2
Knight	1987	Convergent	17	23	38	77.4	0.937	0.394	0.734	−2.72	0.7	5.8	0.1	157.4
Lautens	2005	Convergent	18	22	39	138.2	0.927	0.392	0.791	0.32	0.8	2.8	0.1	188.8
Anderson	2005	Convergent	23	24	44	159.12	0.937	0.369	0.814	−4.29	0.7	4.6	0.1	192.3
Rubio	1998	Linear	21	21	32	172	0.925	0.336	0.816	1.55	0.8	5.1	0.1	198.7
Hanessian	1996	Convergent	20	23	42	21.03	0.939	0.383	0.806	−2.76	1.6	3.6	0.1	155.8
Takano G2	1988	Convergent	23	24	39	131.33	0.922	0.353	0.802	−3.08	0.4	4.2	0.09	246.1
Hoppe	2005	Convergent	17	21	43	142	0.939	0.414	0.824	−2.44	0.6	4.5	0.07	294.9
Ogasawara G1	1997	Convergent	29	30	56	92.47	0.952	0.365	0.777	−3.57	5.5	3.5	0.06	355.8
Takano G3	1992	Linear	24	24	46	108.33	0.94	0.368	0.801	−2.24	0.4	3	0.04	594.6
Cossy	1999	Convergent	22	24	44	155.16	0.935	0.377	0.763	−2.7	0.06	3.6	0.009	2402.7

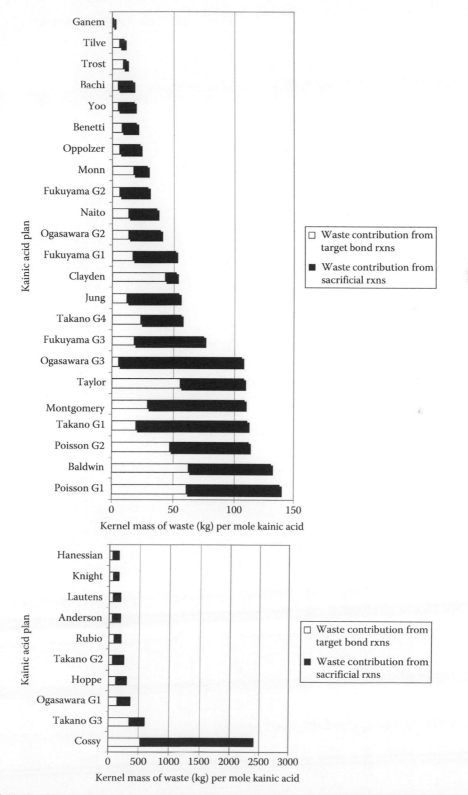

FIGURE 13.15 Bar graphs showing sacrificial and building-up reaction proportions for each synthesis plan for kainic acid. Plans are ranked according to kernel mass of waste produced per mole kainic acid.

Longifolene

TABLE 13.19
Summary of Green Metrics Parameters for Longifolene Syntheses

Plan	Year	Type	N	M	I	μ_1	Beta	Delta	$f(sac)$	HI	% Yield	% AE	% Kernel RME	Kernel Mass of Waste (kg) per Mole
Oppolzer G1	1978	Convergent	12	13	22	−57.9	0.873	0.384	0.522	2.46	9.2	9.6	2.5	8
Schultz G1	1985	Linear	17	19	30	40.66	0.908	0.366	0.778	3.39	14.6	7.3	3.2	8.5
Johnson	1975	Linear	11	11	28	−26.5	0.901	0.43	0.834	1.25	8.9	6.3	1.8	10.9
Fallis	1993	Convergent	15	16	30	−0.39	0.791	0.415	0.673	4.94	0.7	7.4	0.2	11.8
McMurry	1972	Linear	20	20	38	−3.97	0.929	0.371	0.746	3.29	5.6	6.5	1.4	14.2
Schultz G2	1985	Convergent	20	24	38	77.03	0.928	0.371	0.838	4.05	12.5	5.2	1.3	16
Liu	1992	Linear	22	22	32	19.91	0.914	0.33	0.782	1.13	3	8.3	1.1	17.7
Oppolzer G2	1984	Convergent	14	15	25	−54.38	0.889	0.374	0.572	2.6	1.7	8.6	0.9	23.6
Money	1988	Linear	23	23	45	48.16	0.938	0.369	0.779	2.29	4.7	4.2	0.6	35.1
Karimi	2003	Linear	17	17	37	31.24	0.927	0.394	0.64	4.22	2.9	4.8	0.6	36.3
Corey	1964	Linear	16	16	34	8.47	0.846	0.407	0.685	3.77	0.4	7.2	0.1	179.1

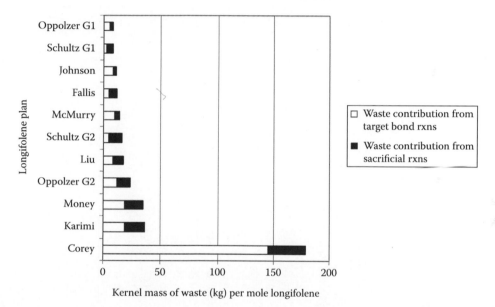

FIGURE 13.16 Bar graph showing sacrificial and building-up reaction proportions for each synthesis plan for longifolene. Plans are ranked according to kernel mass of waste produced per mole longifolene.

Lysergic Acid

TABLE 13.20
Summary of Green Metrics Parameters for Lysergic Acid Syntheses

Plan	Year	Type	N	M	I	μ_1	Beta	Delta	f(sac)	HI	% Yield	% AE	% Kernel RME	Kernel Mass of Waste (kg) per Mole
Hendrickson	2004	Convergent	10	12	26	−68.64	0.887	0.439	0.763	4.18	6.3	10.2	1.1	23.9
Rebek	1984	Linear	11	11	22	59.82	0.896	0.395	0.755	0.67	1.2	10	0.46	58.3
Ramage	1981	Linear	16	16	30	57.97	0.911	0.375	0.682	0.18	1.7	7.9	0.3	88.8
Kurihara	1987	Linear	13	13	35	40.32	0.929	0.425	0.796	0.43	1	7.2	0.18	148.4
Ortar	1988	Convergent	9	13	28	−15.66	0.897	0.463	0.859	−0.1	1.3	9.7	0.17	160.9
Woodward	1956	Convergent	15	17	29	−19.78	0.971	0.367	0.789	1.19	0.69	11.2	0.16	168.6
Oppolzer	1981	Convergent	16	17	32	18.05	0.913	0.385	0.817	1.59	0.96	6.1	0.14	193.3
Szantay	2004	Convergent	15	17	33	−10.26	0.911	0.401	0.764	1.31	0.53	9.1	0.12	228
Ninomiya	1985	Linear	19	19	36	27.67	0.933	0.37	0.835	−0.95	0.45	7.7	0.1	261.6
Fukuyama G2	2009	Convergent	34	35	58	143.35	0.951	0.346	0.856	−3.11	0.45	3.4	0.057	470.6
Fukuyama G1	2009	Convergent	27	29	51	76.87	0.949	0.363	0.751	1.5	0.13	5.1	0.024	1095.6

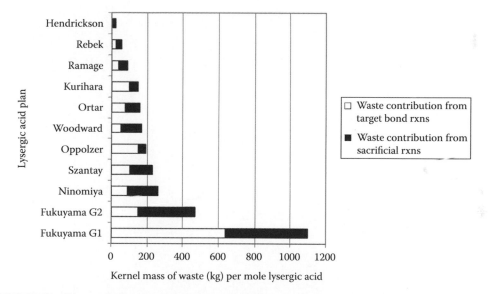

FIGURE 13.17   Bar graph showing sacrificial and building-up reaction proportions for each synthesis plan for lysergic acid. Plans are ranked according to kernel mass of waste produced per mole lysergic acid.

Morphine

TABLE 13.21
Summary of Green Metrics Parameters for Morphine Syntheses

Plan	Year	Type	N	M	I	μ_1	Beta	Delta	f(sac)	HI	% Yield	% AE	% Kernel RME	Kernel Mass of Waste (kg) per Mole
Rice	1980	Linear	18	18	27	38.39	0.898	0.339	0.786	1	10.1	14.1	2.8	9.7
Mulzer	1998	Linear	22	22	42	39.66	0.935	0.37	0.818	0.65	2.7	8	0.6	46.2
Trost	2005	Convergent	16	17	32	−0.51	0.907	0.387	0.631	4.06	2.8	6.2	0.6	46.6
Overman	1993	Convergent	19	24	41	42	0.935	0.392	0.777	0.4	3.9	6.9	0.4	70
Parker	2006	Linear	21	21	39	71.16	0.93	0.367	0.848	0.64	2.4	6.3	0.4	76.7
White	1997	Convergent	30	32	39	15.86	0.931	0.306	0.78	3.19	1.6	7.3	0.4	78.6
Fukuyama	2006	Convergent	26	32	58	77.41	0.953	0.385	0.814	4.33	1.4	3.9	0.3	91.2
Stork	2009	Linear	22	22	38	54.19	0.929	0.355	0.658	3.44	1	6.2	0.2	116.3
Ogasawara	2001	Linear	30	30	53	6.35	0.949	0.353	0.751	0.58	0.5	5.3	0.07	402
Chida	2008	Linear	32	32	56	83.77	0.952	0.35	0.809	0.82	1.2	4.2	0.2	140.1
Taber	2002	Linear	28	28	49	67	0.945	0.352	0.751	−0.41	0.2	4.8	0.05	609.5
Gates	1956	Linear	26	26	47	33.89	0.95	0.357	0.862	1.37	0.004	5.6	0.0008	34521

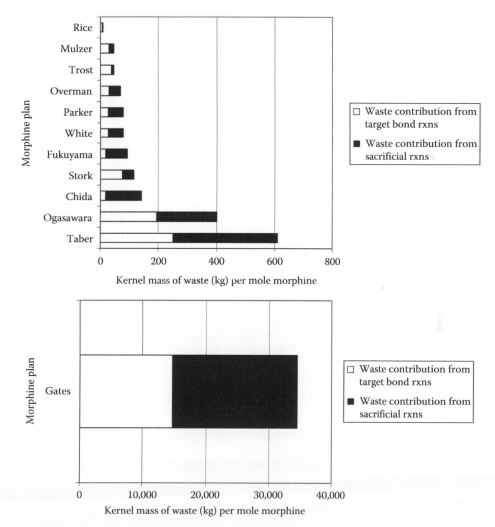

FIGURE 13.18 Bar graphs showing sacrificial and building-up reaction proportions for each synthesis plan for morphine. Plans are ranked according to kernel mass of waste produced per mole morphine.

Naproxen

TABLE 13.22
Summary of Green Metrics Parameters for Naproxen Syntheses

Plan	Year	Type	N	M	I	μ_1	Beta	Delta	f(sac)	HI	% Yield	% AE	% Kernel RME	Kernel Mass of Waste (kg) per Mole
Noyori	1987	Linear	6	6	10	−58.34	0.741	0.41	0.611	1.43	76.1	27.5	21.6	0.8
DuPont	1985	Linear	5	5	12	−72.23	0.623	0.522	0.672	0.5	59.8	20.7	13.9	1.4
Brunner	2000	Linear	5	5	10	−14.57	0.7	0.464	0.573	0.33	44.1	21.9	11.2	1.8
Zambon G1	1987	Linear	6	6	13	76.86	0.785	0.453	0.789	0.29	51.7	15.6	10.6	1.9
Chan	1995	Linear	4	4	8	−47.91	0.616	0.495	0.431	1.8	21.9	41	10.3	2
Effenberger	1997	Linear	6	6	12	−23.65	0.792	0.436	0.627	1.14	19.3	22.7	6.5	3.3
Syntex G3	1986	Convergent	6	8	15	−76.5	0.787	0.478	0.735	0.71	15.7	16.6	4.4	5.1
Syntex G2	1972	Linear	6	6	12	−6.63	0.799	0.434	0.673	0.29	18	16.7	3.7	6
Ruchardt	1991	Linear	9	9	15	−21.56	0.81	0.385	0.716	0.7	7.7	18.8	2.8	7.9
Syntex G1	1970	Linear	7	7	15	5.06	0.69	0.468	0.631	−0.88	6.4	17.9	2.2	10.1
Zambon G2	1989	Linear	13	13	22	142.31	0.89	0.366	0.851	0.5	10.7	11.9	2.2	10.3

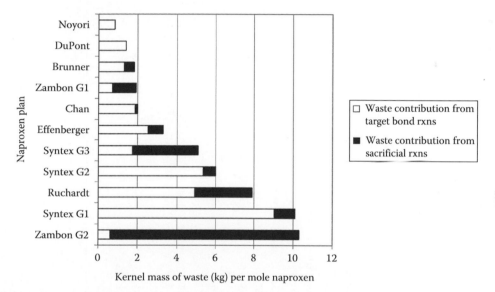

FIGURE 13.19 Bar graph showing sacrificial and building-up reaction proportions for each synthesis plan for naproxen. Plans are ranked according to kernel mass of waste produced per mole naproxen.

Nicotine

TABLE 13.23
Summary of Green Metrics Parameters for Nicotine Syntheses

Plan	Year	Type	N	M	I	μ_1	Beta	Delta	f(sac)	HI	% Yield	% AE	% Kernel RME	Kernel Mass of Waste (kg) per Mole
Campos	2006	Linear	2	2	6	−62.67	0.722	0.35	0.498	−0.33	46.8	21.9	11.5	1.2
Lebreton	2001	Linear	7	7	16	51.71	0.776	0.459	0.694	1	67.1	11.1	8.3	1.8
Helmchen	2005	Linear	5	5	9	33.33	0.671	0.447	0.295	4.17	45.3	11.4	7.5	2
Jacob	1982	Linear	5	5	12	71.6	0.743	0.491	0.507	1.67	9.7	16.1	2.5	6.3
Delgado	2001	Linear	5	5	11	43	0.766	0.468	0.79	2.17	11	10.6	2	7.9
Loh	1999	Linear	5	5	13	65.65	0.778	0.496	0.38	1	11.7	10.6	1.7	9.3
Chavdarian	1982	Linear	7	7	14	−23.4	0.821	0.424	0.599	0.75	3.5	15.9	0.8	20.7
Spath	1928	Linear	4	4	13	0.4	0.762	0.552	0.718	1.6	1	15.4	0.3	61.7
Craig	1933	Linear	7	7	17	−25	0.833	0.454	0.588	1.75	0.4	11.5	0.2	92.8
Ley	2002	Linear	7	7	13	10.06	0.821	0.41	0.633	2.38	24	14.4	<4.4	>3.5
Pictet	1895–1904	Linear	8	8	15	48.3	0.814	0.409	0.728	3.67	<100	6.6	<6.6	>2.3

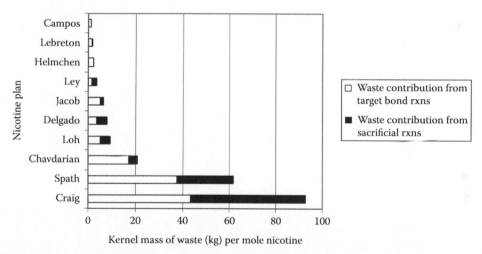

FIGURE 13.20 Bar graph showing sacrificial and building-up reaction proportions for each synthesis plan for nicotine. Plans are ranked according to kernel mass of waste produced per mole nicotine.

Oseltamivir Phosphate

TABLE 13.24
Summary of Green Metrics Parameters for Oseltamivir Phosphate Syntheses

Plan	Years	Type	N	M	l	μ_1	Beta	Delta	$f(sac)$	HI	% Yield	% AE	% Kernel RME	Kernel Mass of Waste (kg) per Mole
Shi	2009	Linear	8	8	14	−128.44	0.818	0.409	0.661	0.22	47.6	21.7	11.9	3
Roche G3 (shikimic acid route)	1999, 2004	Linear	13	13	19	−102.68	0.861	0.345	0.457	0.143	39	21	11.5	3.1
Roche G2 (quinic acid route)	1999, 2004	Linear	14	14	20	−93.57	0.868	0.338	0.467	0.133	21.9	20.3	9	4.2
Fang G2	2008	Linear	12	12	20	−154.09	0.875	0.367	0.452	1.39	26.5	14.5	7.2	5.3
Fang G3	2008	Linear	12	12	21	−130.86	0.864	0.378	0.557	1.39	22.5	17.1	6.8	5.6
Trost-short	2008	Linear	9	9	17	−35.4	0.865	0.397	0.63	−0.2	29.9	16.1	5.6	6.9
Corey	2006	Linear	11	11	17	−153.68	0.843	0.361	0.743	−3	22.4	17.2	5.5	7.2
Roche G5 (desymmetrization)	2000	Linear	11	11	24	−110.48	0.779	0.44	0.677	1	25.6	13.8	5.3	7.3
Hayashi	2009	Convergent	5	7	21	−185.52	0.697	0.622	0.756	4	29.3	8	4.1	7.3
Roche G6	2009	Linear	8	8	15	−26.44	0.762	0.42	0.659	0.67	15.2	21.9	4.6	8.5
Trost-long	2008	Linear	12	12	21	−99.92	0.892	0.372	0.69	−1.54	16.2	13.4	4	9.8
Roche G1 (quinic acid route)	2001	Linear	12	12	21	−131.23	0.862	0.378	0.625	0.154	7.6	18.5	3.2	12.5

Fang G1	2007	Convergent	17	18	35	−167.02	0.950	0.379	0.671	1.444	13.4	12	3.1	12.7
Hudlicky G2	2010	Linear	9	9	21	−175.60	0.871	0.43	0.705	0.3	5.3	19.9	3.1	12.7
Gilead	1998	Linear	12	12	21	−135.85	0.867	0.377	0.607	0.385	6.3	20.5	2.7	15
Hudlicky G1	2009	Convergent	12	13	23	−262.72	0.855	0.396	0.741	0.39	3.5	20.6	2.4	16.5
Fukuyama	2007	Linear	13	13	22	−133.48	0.875	0.369	0.638	−2.64	5.5	15.9	2.4	16.4
Mandai G2	2009	Convergent	16	18	26	26.43	0.907	0.353	0.493	1.29	8.2	12.6	1.8	17.1
Mandai G1	2009	Linear	18	18	30	33.16	0.92	0.353	0.714	−2.79	7	7.8	1.5	20.4
Shibasaki G4	2009	Linear	11	11	25	−124.67	0.89	0.416	0.584	−1.58	9.6	11.5	1.9	21.1
Banwell	2008	Linear	14	14	29	−84.69	0.903	0.395	0.656	1.47	4.6	13.1	1.9	21.3
Roche G4 (Diels–Alder)	2000	Linear	9	9	17	−142.56	0.756	0.432	0.505	−0.5	1.1	23.9	1.5	27.4
Shibasaki G5	2009	Linear	11	11	21	−96.17	0.871	0.393	0.509	−1.75	6.7	13	1.4	28
Okamura–Corey	2008	Linear	13	13	25	−124.76	0.893	0.386	0.754	−0.43	2.6	16.8	1.3	32
Kann	2007	Linear	15	15	25	−61.26	0.88	0.363	0.788	−0.56	3.4	11.8	0.9	47.4
Shibasaki G2	2007	Linear	16	16	32	−163.86	0.914	0.385	0.837	−0.71	4.5	10	0.8	47.9
Shibasaki G3	2007	Linear	11	11	23	−167.26	0.88	0.406	0.481	−0.33	1.4	16.1	0.6	73.6
Chen–Liu	2010	Linear	21	21	33	−73.91	0.928	0.341	0.651	−1.59	1.7	9.4	0.5	86.8
Shibasaki G1	2006	Linear	15	15	34	−132.10	0.919	0.403	0.766	−1.19	1.4	9.6	0.3	150.3

TABLE 13.25
Summary of *E*-Factor Breakdowns for Oseltamivir Phosphate Syntheses

Plan	Years	Type	*E*-Kernel	*E*-Excess	*E*-Aux	*E*-Total	True Mass of Waste (kg) per Mole
Roche G3 (shikimic acid route)	1999, 2004	Linear	7.7	24.6	198.62	230.9	94.7
Roche G2 (quinic acid route)	1999, 2004	Linear	10.1	30	267.74	307.9	126.2
Hayashi	2009	Convergent	23.39	177.79	395.66	596.85	186.2
Shi	2009	Linear	7.37	28.19	429.2	464.76	190.5
Roche G1 (quinic acid route)	2001	Linear	30.6	71.1	755.5	857.2	351.4
Roche G5 (desymmetrization)	2000	Linear	17.8	68.4	847.4	933.6	382.8
Gilead	1998	Linear	36.7	91.5	808.6	936.7	384
Fang G2	2008	Linear	12.89	59.16	1656.7	1728.7	708.7
Roche G6	2009	Linear	20.85	84.9	1784.1	1889.8	774.8
Fang G3	2008	Linear	13.7	104.3	1858.6	1976.6	810.3
Fang G1	2007	Convergent	31	274.8	2275.1	2580.9	1058
Trost-short[a]	2008	Linear	16.8	141.5	2527.1	2685.4	1101
Hudlicky G1[b]	2009	Convergent	40.22	133.41	2593	2766.6	1134
Trost-long[a]	2008	Linear	23.8	144.4	2690.5	2858.7	1172
Corey[c]	2006	Linear	17.5	208.8	3056.5	3282.9	1346
Hudlicky G2[b]	2010	Linear	30.97	220.33	3359.2	3610.5	1480
Fukuyama[d]	2007	Linear	40	163.4	3843	4046.5	1659
Shibasaki G4[e]	2009	Linear	51.39	1162.46	3740.9	4954.7	2031
Roche G4 (Diels–Alder)	2000	Linear	66.8	181.2	4855.6	5103.5	2092
Shibasaki G5[e]	2009	Linear	68.29	1476.91	4000.2	5545.4	2274
Hudlicky G1	2009	Convergent	40.22	6174.08	5792.4	12007	4923
Banwell	2008	Linear	21.3	838.66	11683	12573	5155
Kann	2007	Linear	115.5	285.9	13238	13640	5592
Chen–Liu	2010	Linear	211.64	337.9	14945	15494	6353
Shibasaki G1[a]	2006	Linear	366.6	3772.8	12055	16194	6640
Shibasaki G2[a]	2007	Linear	116.8	1279.9	18818	20215	8288
Okamura–Corey	2008	Linear	78	439.8	21926	22444	9202
Shibasaki G3	2007	Linear	179.5	1554.1	24806	26539	10881

[a] Includes synthesis of Trost ligand and DuBois catalyst.
[b] Excludes excess water and fermentation medium for step 1 in waste calculation.
[c] Includes synthesis of Corey–Shibata–Lee catalyst.
[d] Includes synthesis of MacMillan catalyst.
[e] Includes synthesis of Shibasaki ligand.

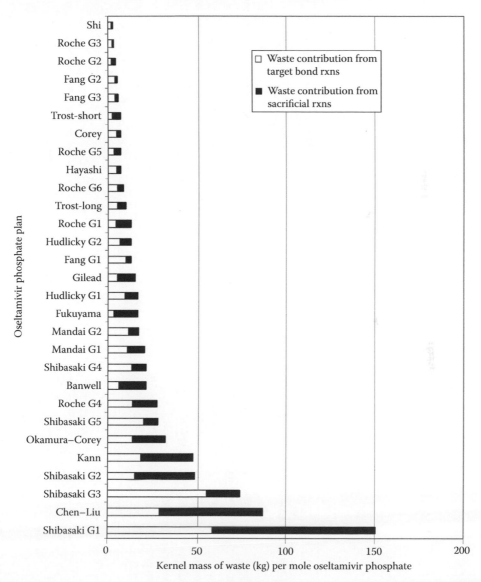

FIGURE 13.21 Bar graph showing sacrificial and building-up reaction proportions for each synthesis plan for oseltamivir phosphate. Plans are ranked according to kernel mass of waste produced per mole oseltamivir phosphate.

Paclitaxel

TABLE 13.26
Summary of Green Metrics Parameters for Paclitaxel Syntheses

Plan	Year	Type	N	M	l	μ_1	Beta	Delta	$f(sac)$	HI	% Yield	% AE	% Kernel RME	Kernel Mass of Waste (kg) per Mole
Holton	1994	Convergent	55	59	117	−420.34	0.975	0.37	0.761	−5	0.1	7.5	0.07	1175.4
Wender	1997	Convergent	38	40	87	−300.49	0.956	0.387	0.678	−2.33	0.03	8.9	0.02	3428.2
Nicolaou	1995	Convergent	38	47	85	−508.14	0.955	0.384	0.751	−1.77	0.08	8.9	0.02	3654.5
Mukaiyama	1999	Convergent	59	67	126	−341.65	0.945	0.379	0.798	−2.9	0.02	5.6	0.01	8254.6
Danishefsky	1995	Convergent	49	55	98	−472.64	0.961	0.366	0.806	−5.38	0.01	6.5	0.008	10435.7
Kuwajima	2000	Convergent	57	65	114	−405.03	0.967	0.364	0.774	−2.38	0.009	5.5	0.003	27205
Takahashi	2006	Convergent	48	62	109	−469.54	0.965	0.382	0.813	−7.27	0.002	6.9	0.0009	90441.4

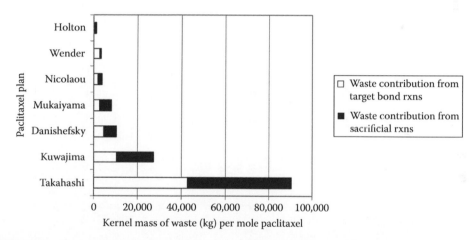

FIGURE 13.22   Bar graph showing sacrificial and building-up reaction proportions for each synthesis plan for paclitaxel. Plans are ranked according to kernel mass of waste produced per mole paclitaxel.

Paroxetine

TABLE 13.27
Summary of Green Metrics Parameters for Paroxetine Syntheses

Plan	Year	Type	N	M	I	μ_1	Beta	Delta	f(sac)	HI	% Yield	% AE	% Kernel RME	Kernel Mass of Waste (kg) per Mole
Amat G1	2000	Linear	9	9	17	−28	0.854	0.399	0.554	3.4	9.9	16.8	3	10.5
Amat G2	2000	Linear	10	10	17	−35.64	0.854	0.378	0.561	3.46	13.4	16.6	4.1	7.7
Beak	2001	Convergent	14	17	35	−53.12	0.931	0.414	0.82	2.93	10.8	9.8	2.9	11.1
Buchwald	2003	Linear	11	11	17	−55.3	0.858	0.358	0.459	2.42	13.1	15.1	4.2	7.5
Chang	2003	Convergent	11	13	24	−79.72	0.899	0.408	0.497	4.08	7.1	16.4	2.9	11.1
Cossy	2002	Linear	13	13	27	−37.83	0.91	0.394	0.686	4.5	11.9	10	2.1	15.2
Dixon	2008	Linear	9	9	15	−53.4	0.833	0.381	0.358	2.4	23.2	19.4	8.5	3.5
Ferrosan	1975	Linear	9	9	16	−87.09	0.823	0.406	0.638	2.2	13.6	20.6	4.6	6.8
Gallagher	2007	Linear	11	11	22	−97.45	0.873	0.4	0.568	2.67	19.5	22.5	6.1	5.6
Gotor	2001	Linear	9	9	14	−45.9	0.827	0.369	0.587	0.4	3	17.2	1.5	21
Hayashi	2001	Linear	9	9	18	−62.56	0.863	0.407	0.607	1.8	27.4	22.1	9.2	3.3
Jorgensen	2006	Linear	7	7	15	−67.76	0.833	0.433	0.301	2.5	28.4	21.2	8.7	3.5
Krische	2006	Convergent	14	16	31	−81.21	0.886	0.41	0.754	2.87	1.9	9.2	0.8	47.5
Liu	2001	Linear	13	13	29	−83.07	0.89	0.411	0.684	3.64	3.6	15.5	1.3	27.9
Simpkins	2003	Linear	10	10	22	−68	0.841	0.426	0.675	4.18	4.4	14.7	1.4	23.9
Szantay	2004	Linear	14	14	26	−46.06	0.887	0.381	0.731	1.33	2.1	14.8	1.1	35.7
Takasu–Ihara	2005	Linear	7	7	13	−26.63	0.806	0.413	0.579	2.38	34.8	24.3	11.4	2.6
Vesely–Moyano–Rios	2009	Linear	9	10	16	−155.68	0.844	0.39	0.431	3.8	14.9	23.2	7.4	4.1
Yu	2000	Linear	11	11	20	−70.33	0.877	0.383	0.482	3.25	30.9	18.4	8.5	3.5

TABLE 13.28

Summary of *E*-Factor Breakdowns for Oseltamivir Phosphate Syntheses

Plan	Year	Type	*E*-Kernel	*E*-Excess	*E*-Aux	*E*-Total	Total Mass of Waste (kg) per Mole
Amat G1	2000	Linear	31.85	125.1	4828	4984.7	1640
Amat G2	2000	Linear	23.34	80.21	3575	3678.2	1210
Beak	2001	Convergent	33.72	388.7	7672	8094.7	2663
Buchwald	2003	Linear	22.83	243	3067	3333.1	1097
Chang[a]	2003	Convergent	33.63	25.91	2135	2194.6	722
Cossy	2002	Convergent	47.21	528.9	6901	7477.6	2460
Dixon[a]	2008	Linear	10.79	118.4	415.4	544.64	179.2
Ferrosan[a]	1975	Linear	20.78	38.37	550.8	609.93	200.7
Gallagher	2007	Linear	15.28	15.02	8546	8576.8	3134
Gotor	2001	Linear	63.69	285.2	9403	9751.9	3208
Hayashi[a]	2001	Linear	9.88	1.47	272.1	293.44	95.5
Jorgensen[a]	2006	Linear	10.5	479.7	14777	15267	5023
Krische[a]	2006	Convergent	130.1	198.4	4851	5187.8	1896
Liu	2001	Linear	76.43	435	3391	3902.6	1426
Simpkins[a]	2003	Linear	72.63	308.9	12959	13340	4389
Szantay[a]	2004	Linear	89.01	133.3	1345	1567.1	629.1
Takasu–Ihara[b]	2005	Linear	7.81	0	0	0	2.6
Vesely–Moyano–Rios[a]	2009	linear	12.55	50.51	966.2	1029.3	338.6
Yu[b]	2000	linear	10.72	0	0	0	3.5

[a] Lower limit.

[b] No experimental details given.

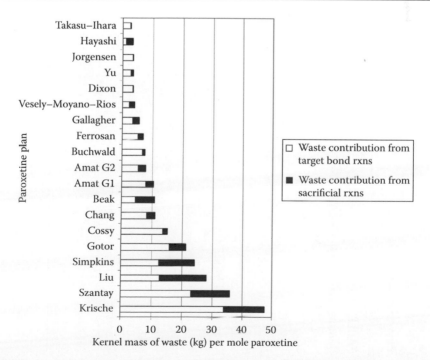

FIGURE 13.23 Bar graph showing sacrificial and building-up reaction proportions for each synthesis plan for paroxetine. Plans are ranked according to kernel mass of waste produced per mole paroxetine.

Physostigmine

TABLE 13.29
Summary of Green Metrics Parameters for Physostigmine Syntheses

Plan	Year	Type	N	M	I	μ_1	Beta	Delta	f(sac)	HI	% Yield	% AE	% Kernel RME	Kernel Mass of Waste (kg) per Mole
Yu	1994	Linear	7	7	14	−44.91	0.737	0.442	0.247	1.5	64.2	30.1	21.1	1
Julian	1935	Linear	9	9	20	−67.1	0.824	0.435	0.565	1.7	27	17.6	7	3.6
Trost	2006	Linear	8	8	17	−81.39	0.79	0.438	0.646	0.67	22.6	18.3	6.7	3.9
Nakagawa	2000	Linear	9	9	19	−54.16	0.815	0.428	0.507	1.9	13.9	15.2	4.6	5.7
Overman	1998	Linear	10	10	22	−6.64	0.84	0.427	0.735	2.09	19.7	11.1	3.2	8.3
Johnson	2003	Linear	8	8	17	−75.62	0.804	0.445	0.669	2.44	10.1	15.6	2.3	11.7
Speckamp	1978	Convergent	11	12	22	−90.96	0.841	0.408	0.531	4.92	4.8	17.7	1.8	15.2
Mukai	2006	Linear	9	9	24	−20.39	0.855	0.456	0.727	3.4	11.7	8.8	1.7	16.4
Kulkarni	2009	Linear	13	13	23	−33.9	0.849	0.381	0.67	1.79	5.6	12.6	1.3	21.2
Marino	1992	Convergent	10	11	22	−1.92	0.801	0.437	0.505	3.73	4.7	9.8	1.1	24.7
Nakada	2008	Linear	21	21	38	−11.42	0.91	0.367	0.697	4.09	5.9	8.3	1	28.4
Takano G2	1991	Linear	14	14	26	−49.4	0.867	0.385	0.688	0.27	3	11.6	0.7	36.4
Fuji	1991	Linear	16	16	34	−53.41	0.899	0.398	0.696	2.71	4.1	7.4	0.7	40.4
Fukumoto	1986	Linear	22	22	41	−43.85	0.915	0.367	0.761	−0.26	1.3	7.3	0.3	107.3
Takano G1	1982	Linear	12	12	25	−37.54	0.861	0.409	0.565	1.54	0.4	13.9	0.1	237.3

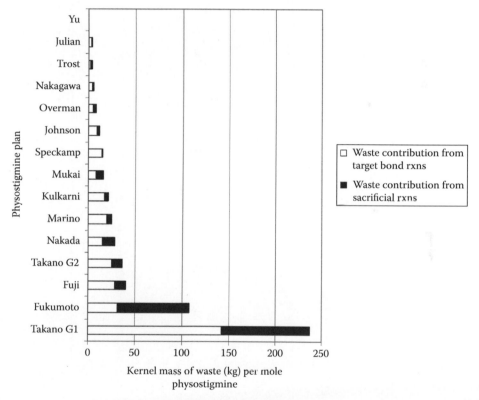

FIGURE 13.24 Bar graph showing sacrificial and building-up reaction proportions for each synthesis plan for physostigmine. Plans are ranked according to kernel mass of waste produced per mole physostigmine.

Platensimycin

TABLE 13.30
Summary of Green Metrics Parameters for Platensimycin Syntheses

Plan	Year	Type	N	M	I	μ₁	Beta	Delta	f(sac)	HI	% Yield	% AE	% Kernel RME	Kernel Mass of Waste (kg) per Mole
Lee	2008	Convergent	18	24	48	−280.37	0.918	0.424	0.572	3.11	7.9	9	1.8	23.8
Nicolaou G6	2009	Convergent	16	21	36	−233.13	0.89	0.409	0.636	−0.76	7.5	10	1.8	24.5
Nicolaou G1	2006	Convergent	17	22	40	−224.57	0.902	0.412	0.67	−1	6.5	9.1	1.4	31.3
Njardarson	2009	Convergent	20	25	43	−225.1	0.905	0.396	0.651	0.71	4.5	7.9	1.4	32
Nicolaou G4	2007	Convergent	17	22	42	−208.01	0.906	0.418	0.715	−0.06	6.8	8.4	1.2	36.3
Corey	2007	Convergent	25	30	55	−211.02	0.929	0.391	0.777	0.62	6	6.5	1.1	39
Nicolaou G2	2007	Convergent	22	27	45	−169.7	0.913	0.385	0.591	−1	4.3	7.7	1.1	40.5
Yamamoto	2007	Convergent	17	22	43	−300.23	0.909	0.421	0.621	−0.22	2	9.4	0.9	49.7
Nicolaou G3	2007	Convergent	23	28	47	−199.3	0.917	0.384	0.659	−0.33	2.4	8	0.5	81.2
Mulzer	2007	Convergent	22	27	47	−275.58	0.917	0.391	0.722	−0.57	1	8.9	0.4	104.9
Nicolaou G5	2008	Convergent	20	25	50	−243.06	0.922	0.413	0.754	−2.38	0.8	6.7	0.2	203
Snider	2007	Convergent	17	22	46	−274.13	0.915	0.428	0.606	−0.44	0.2	8	0.1	315.9
Ghosh	2009	Convergent	31	33	54	−176.07	0.942	0.352	0.732	1.41	0.4	5.5	0.1	346.8

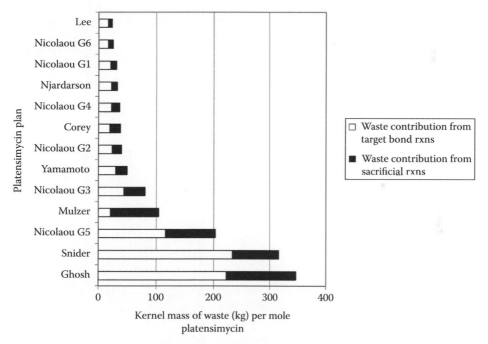

FIGURE 13.25 Bar graph showing sacrificial and building-up reaction proportions for each synthesis plan for platensimycin. Plans are ranked according to kernel mass of waste produced per mole platensimycin.

Pregabalin

TABLE 13.31

Summary of Green Metrics Parameters for Pregabalin Syntheses

Plan	Year	Type	N	M	I	μ_1	Beta	Delta	f(sac)	HI	% Yield	% AE	% Kernel RME	Kernel Mass of Waste (kg) per Mole
McQuade	2007	Linear	3	3	6	−59.56	0.5	0.542	0.011	3.5	85.7	54.7	47.9	0.2
Silverman G1	1989	Linear	3	3	5	−11.25	0.406	0.51	0.013	3	53.4	52.5	29.4	0.4
Pfizer G1	2008	Linear	4	4	8	23.4	0.623	0.494	0.082	2.4	54.4	32.5	19	0.7
Shibasaki	2005	Linear	5	5	10	−13.28	0.717	0.461	0.338	2	64.2	26.8	17.9	0.9
Dowpharma	2003	Linear	5	5	12	−16.67	0.69	0.506	0.657	4.33	38.8	27.1	11.8	1.2
Jacobsen	2003	Linear	6	6	10	0.89	0.714	0.415	0.583	1.86	41.4	22.9	13.1	1.3
Warner-Lambert G2	2001	Linear	5	5	12	−16.67	0.69	0.506	0.585	4.33	31.7	27.1	10.2	1.4
Armstrong	2006	Linear	6	6	9	−36.81	0.685	0.396	0.444	1.43	28.1	32.6	11.5	1.5
Pfizer G2	2008	Linear	4	4	9	23.8	0.647	0.514	0.171	2	28.1	28.3	9.6	1.5
Lee	2007	Linear	7	7	13	19.42	0.825	0.409	0.537	1	31.7	14.4	7.3	2
Koskinen	2009	Linear	5	5	10	−68.19	0.745	0.455	0.598	4.5	32.1	22.6	8.7	2.1

Hamersak G2	2007	Linear	5	5	11	88.17	0.769	0.467	0.412	-0.17	48.7	10.7	6.9	2.2
Hayashi	2007	Linear	5	5	9	-26.33	0.732	0.436	0.271	2	18.1	28.4	6.4	2.3
Hu	2004	Linear	8	8	11	24.86	0.6	0.375	0.62	-0.33	29.8	15.2	5.9	3.1
Parke-Davis G3	1997	Linear	9	9	19	53.17	0.872	0.413	0.655	2.4	33.9	10.6	4.8	3.2
Hamersak G1	2007	Linear	6	6	12	91	0.804	0.433	0.445	-0.29	30.5	11.6	4.6	3.3
Parke-Davis G1	1997	Convergent	9	10	17	65.44	0.87	0.396	0.712	1.7	32.5	10.4	4.2	3.6
Parke-Davis G4	1997	Linear	7	7	14	18.5	0.656	0.442	0.589	-0.13	20.9	14.6	4.2	3.7
Ortuno	2008	Linear	11	11	14	32.25	0.808	0.328	0.501	3.25	22.2	9.6	4	3.8
Felluga	2008	Linear	4	4	7	-21.05	0.583	0.47	0.213	3.8	11.3	36	4.6	4.1
Warner-Lambert G1	1996	Linear	4	4	9	18.6	0.616	0.522	0.285	1.8	14.3	21.8	3.7	4.1
Parke-Davis G2	1997	Linear	4	4	9	18.6	0.616	0.522	0.361	1.6	12.9	21.8	3.4	4.6
Silverman G2	1990	Linear	7	7	13	20.5	0.809	0.413	0.503	0.25	11.8	11.9	3.3	4.6
Sartillo-Piscil	2007	Linear	7	7	13	113.08	0.78	0.419	0.476	0.5	28.1	9.3	3.2	4.8

TABLE 13.32

Summary of *E*-Factor Breakdowns for Pregabalin Syntheses

Plan	Year	Type	*E*-Kernel	*E*-Excess	*E*-Aux	*E*-Total	Total Mass of Waste (kg) per Mole
Pfizer G1	2008	Linear	4.27	0.99	36.35	41.61	6.6
Pfizer G2	2008	Linear	9.37	4.88	52.9	67.15	10.7
Dowpharma	2003	Linear	7.46	9.57	179.8	196.81	31.3
Warner-Lambert G1	1996	Linear	25.8	24.36	166.5	216.63	34.4
Parke-Davis G4	1997	Linear	23	28.77	176.3	228.02	36.3
Hamersak G2	2007	Linear	13.53	17.3	230.2	260.98	41.5
Silverman G1	1989	Linear	2.4	28.34	236.6	267.3	42.5
Parke-Davis G2	1997	Linear	28.83	22.18	227.6	278.56	44.3
Silverman G2	1990	Linear	29.07	60.85	216.3	306.18	48.7
Warner-Lambert G2	2001	Linear	8.85	7.47	289.7	306.02	48.7
Hayashi	2007	Linear	14.8	16.86	284.7	316.37	50.3
Parke-Davis G1	1997	Convergent	22.54	15.15	290.1	327.77	52.1
Koskinen	2009	Linear	10.53	4.06	338.9	353.5	69.1
Hamersak G1	2007	Linear	20.88	24.92	448.9	494.69	78.7
McQuade	2007	Linear	1.09	46.98	522.6	570.7	111.5
Shibasaki	2005	Linear	4.58	15.03	744.3	763.87	149.3
Felluga	2008	Linear	20.97	67.95	1131	1219.6	238.4
Jacobsen	2003	Linear	6.61	11.56	1202	1220.5	238.5
Lee	2007	Linear	12.7	17.49	2121	2151.5	342.1
Sartillo-Piscil	2007	Linear	30.24	664.4	5586	6280.2	998.5
Armstrong	2006	Linear	7.71	N/A	N/A	N/A	N/A
Hu	2004	Linear	15.82	N/A	N/A	N/A	N/A
Ortuno	2008	Linear	24.06	N/A	N/A	N/A	N/A
Parke-Davis G3	1997	Linear	4.77	N/A	N/A	N/A	N/A

N/A = not available since no experimental details given.

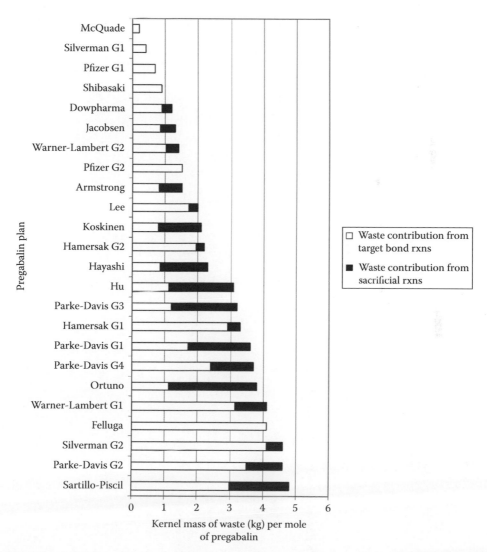

FIGURE 13.26 Bar graph showing sacrificial and building-up reaction proportions for each synthesis plan for pregabalin. Plans are ranked according to kernel mass of waste produced per mole pregabalin.

Quinine

TABLE 13.33
Summary of Green Metrics Parameters for Quinine Syntheses

Plan	Year	Type	N	M	I	μ_1	Beta	Delta	f(sac)	HI	% Yield	% AE	% Kernel RME	Kernel Mass of Waste (kg) per Mole
Aggarwal	2010	Convergent	15	18	44	−177.68	0.897	0.448	0.873	2.5	4.9	6.2	1.1	28
Jacobsen	2004	Convergent	16	21	39	−114.71	0.909	0.416	0.753	2	3.2	5.6	0.79	40.6
Merck	1970	Convergent	16	17	33	−108.7	0.893	0.395	0.797	1.94	1.6	12.2	0.6	57.9
Kobayashi	2004	Convergent	21	28	53	−57.14	0.944	0.407	0.825	−0.41	2.4	4.6	0.47	69.2
Stork	2001	Convergent	21	24	32	−4.03	0.921	0.384	0.735	1.05	1.1	6.4	0.43	75
Martin	1972	Convergent	15	17	32	−147.96	0.893	0.401	0.731	2.44	1.1	9.9	0.4	90
Krische	2008	Convergent	16	20	51	−62.28	0.931	0.444	0.807	1.94	2.1	4.6	0.3	102.8
Gates	1970	Convergent	17	18	34	−119.44	0.874	0.394	0.802	1.72	0.8	11	0.3	116.9
Woodward–Doering–Rabe	1945	Convergent	24	30	44	−95.24	0.937	0.362	0.815	1.24	0.012	7.6	0.0026	12492

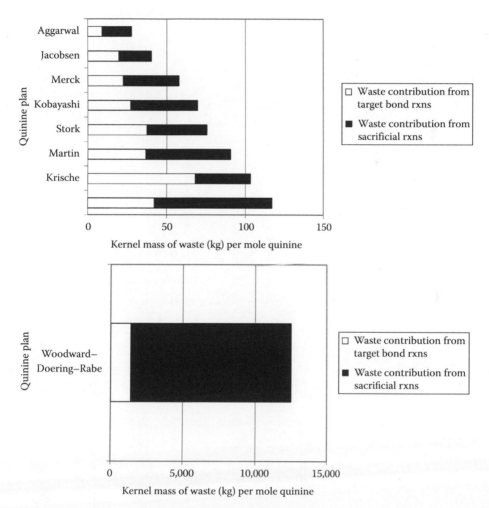

FIGURE 13.27 Bar graphs showing sacrificial and building-up reaction proportions for each synthesis plan for quinine. Plans are ranked according to kernel mass of waste produced per mole quinine.

Reserpine

TABLE 13.34
Summary of Green Metrics Parameters for Reserpine Syntheses

Plan	Year	Type	N	M	I	μ_1	Beta	Delta	$f(sac)$	HI	% Yield	% AE	% Kernel RME	Kernel Mass of Waste (kg) per Mole
Stork	2005	Convergent	12	14	26	-419.39	0.893	0.407	0.526	0.39	15.2	22.4	4.4	13.1
Liao	1996	Linear	11	11	20	-309.75	0.86	0.387	0.661	1.75	3.4	19.8	2.2	27.4
Martin	1987	Convergent	21	22	32	-289.69	0.917	0.337	0.468	-1.18	3.2	17	1.5	39.1
Wender	1987	Convergent	17	19	28	-365.81	0.906	0.354	0.576	0.83	1.9	15.8	1.3	44
Shea	2003	Convergent	21	22	34	-230.09	0.923	0.346	0.538	-1.73	2	13.2	0.8	78.8
Hanessian	1997	Convergent	23	24	35	-162.4	0.926	0.34	0.6	4.96	1.9	10.9	0.6	108.6
Pearlman	1979	Convergent	14	15	24	-399.66	0.887	0.374	0.481	2.53	0.1	27.7	0.3	243.5
Woodward	1958	Convergent	19	25	49	-564.55	0.946	0.412	0.712	2.9	0.09	14.4	0.1	464.4
Fraser-Reid	1995	Linear	36	36	70	-225.63	0.961	0.363	0.731	0.22	0.4	8	0.094	642.6
Mehta	2000	Convergent	27	28	49	-310.87	0.944	0.358	0.712	1	0.003	12.2	0.002	28932

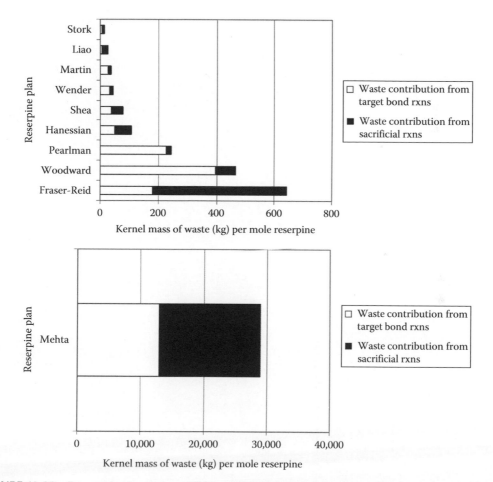

FIGURE 13.28 Bar graphs showing sacrificial and building-up reaction proportions for each synthesis plan for quinine. Plans are ranked according to kernel mass of waste produced per mole quinine.

Resveratrol

TABLE 13.35
Summary of Green Metrics Parameters for Resveratrol Syntheses

Plan	Year	Type	N	M	I	μ_1	Beta	Delta	f(sac)	HI	% Yield	% AE	% Kernel RME	Kernel Mass of Waste (kg) per Mole
Spath	1941	Linear	2	2	2	−33.33	0	0.528	0	0.67	33.1	73.1	24.2	0.7
Marra	2006	Linear	5	5	9	12.98	0.703	0.522	0.475	0	57.9	14.8	9.5	2.2
Yus G3	2009	Linear	4	4	7	77.16	0.477	0.536	0.751	0	51	13.4	8.2	2.6
Yus G2	2009	Linear	4	4	7	47.38	0.477	0.536	0.771	0.4	30.6	13.8	6.4	3.3
Yus G1	1997	Linear	4	4	9	−13.11	0.574	0.531	0.615	0.33	20.2	18.9	6.2	3.4
Guiso	2002	Linear	5	5	12	24.48	0.757	0.487	0.607	1	47.4	10.5	6.2	3.5
Meier	1997	Convergent	5	6	12	−80.02	0.714	0.499	0.715	2.33	42.9	11	5.9	3.7
Shen	2002	Convergent	5	7	13	−40.23	0.727	0.512	0.876	1.14	31.7	7.1	3.5	6.2
Cushman	1993	Linear	6	6	13	128.96	0.723	0.468	0.844	−0.63	29.8	8	3.7	5.9
Moreno-Manas	1985	Linear	7	7	9	54.09	0.594	0.364	0.672		10.1	14.7	3	7.4

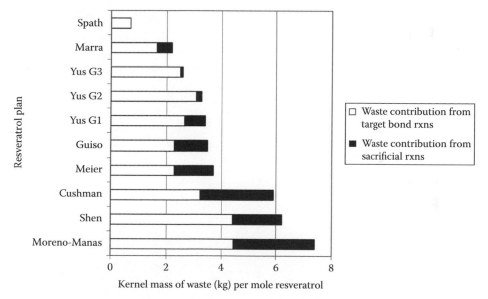

FIGURE 13.29   Bar graph showing sacrificial and building-up reaction proportions for each synthesis plan for resveratrol. Plans are ranked according to kernel mass of waste produced per mole resveratrol.

Rosefuran

TABLE 13.36
Summary of Green Metrics Parameters for Rosefuran Syntheses

Plan	Year	Type	N	M	l	μ_1	Beta	Delta	f(sac)	HI	% Yield	% AE	% Kernel RME	Kernel Mass of Waste (kg) per Mole
Salerno	1997	Linear	4	4	8	−81.71	0.5	0.482	0.284	−0.4	20.1	37.8	11.6	1.1
Wong	1997	Convergent	4	5	10	−44.66	0.677	0.529	0.383	0.6	30.5	17.7	7.6	1.8
Butsugan	1995	Convergent	4	5	13	−44.12	0.621	0.604	0.76	0.4	32	14.5	7.2	1.9
Birch	1976	Convergent	5	6	11	−24.29	0.702	0.484	0.622	0.5	34.2	14.6	6.9	2
Trost	1994	Linear	6	6	10	−25.3	0.747	0.409	0.619	−1.14	22.8	23	6.2	2.3
Wenkert	1988	Convergent	7	9	12	−38.97	0.751	0.408	0.325	2.63	46.6	10.8	6	2.3
Iriye G2	1990	Linear	5	5	7	25.82	0.578	0.403	0.6	−0.67	16.1	25.6	5.7	2.5
Tsukasa	1989	Linear	4	4	6	−8.4	0.188	0.466	0.18	2.6	9.8	26.2	5.3	2.7
Vig	1974	Linear	5	5	9	−8	0.574	0.463	0.316	2.5	23.6	13.4	5.2	2.7
Marshall	1993	Linear	6	6	13	−40.08	0.743	0.463	0.658	1.14	13.4	13.4	4.6	3.1
Pattenden	1977	Convergent	3	4	8	−54.83	0.589	0.589	0.254	1.5	14.4	20	4.2	3.4
Buchi	1968	Convergent	6	7	12	0.94	0.674	0.461	0.576	0.29	19.6	16.3	4.1	3.5
Takeda	1977	Convergent	6	7	10	−35.55	0.717	0.414	0.461	0.57	13	20.7	3.7	3.9
Scharf	1986	Linear	7	7	10	22	0.637	0.385	0.412	1.13	7.8	15.1	1.9	3.9
Iriye G1	1990	Linear	7	7	8	40.61	0.712	0.329	0.663	−0.5	4.6	21.6	1.6	9.4
Takano	1984	Linear	9	9	14	28.2	0.821	0.37	0.719	−0.7	7.2	9.2	1.4	10.7
Okazaki	1984	Convergent	4	5	11	−29.64	0.708	0.54	0.584	1.6	5	12.7	0.9	16.7
Barco	1995	Linear	7	7	12	7.42	0.94	0.373	0.741	−0.5	5.2	8	0.6	24.8

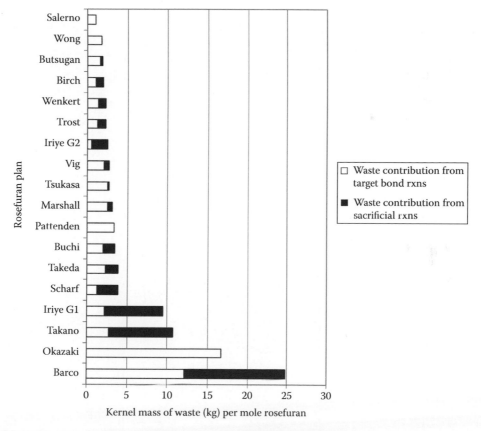

FIGURE 13.30 Bar graph showing sacrificial and building-up reaction proportions for each synthesis plan for rosefuran. Plans are ranked according to kernel mass of waste produced per mole rosefuran.

Sertraline

TABLE 13.37
Summary of Green Metrics Parameters for Sertraline Syntheses

Plan	Year	Type	N	M	I	μ_1	Beta	Delta	f(sac)	HI	% Yield	% AE	% Kernel RME	Kernel Mass of Waste (kg) per Mole
Zhao	2006	Linear	5	5	9	−60.98	0.738	0.435	0.495	1	35	43	18.1	1.4
Pfizer G3	2004	Linear	5	5	14	−96.77	0.781	0.508	0.54	2	25	39	11.2	2.7
Gedeon Richter	2002	Linear	5	5	11	−58.69	0.734	0.476	0.521	1.33	20.5	36.6	9.4	3.3
Lautens G2	2005	Convergent	8	9	16	−55.73	0.84	0.415	0.478	0.89	14.5	16	4.1	7.1
Buchwald	2000	Linear	5	5	12	−85.98	0.76	0.486	0.358	2.33	19.2	38.9	8.6	3.2
Pfizer G2	1999	Linear	8	8	17	−72.83	0.839	0.426	0.562	1.56	8.7	34.9	4.3	7.6
Backvall	2010	Linear	12	12	24	−18.1	0.737	0.43	0.671	1.77	18.8	8.9	3.8	7.8
Lautens G1	1999	Convergent	11	13	28	−15.81	0.776	0.471	0.764	1	14.1	7.1	1.9	15.9
Pfizer G1	1984	Linear	10	10	18	−50.04	0.848	0.389	0.579	1.64	1.8	26.7	1	35.6

TABLE 13.38
Summary of *E*-Factor Breakdowns for Sertraline Syntheses

Plan	Year	Type	*E*-Kernel	*E*-Excess	*E*-Aux	*E*-Total	Total Mass of Waste (kg) per Mole
Zhao	2006	Linear	4.54	89.21	64.64	158.38	48.4
Pfizer G3	2004	Linear	7.92	55.21	226.2	289.76	99.2
Gedeon Richter	2002	Linear	9.59	42.07	339.5	391.19	133.9
Buchwald	2000	Linear	10.56	90.25	352.2	453.05	138.6
Pfizer G2	1999	Linear	22.34	163.56	539.7	725.56	248.4
Backvall	2010	Linear	25.39	180.54	2234	2440.4	746.5
Lautens G1	1999	Convergent	51.99	114.58	2425	2591.6	792.8
Lautens G2	2005	Convergent	23.14	29.72	4117	4169.7	1276
Pfizer G1	1984	Linear	103.95	350.3	4537	4990.8	1709

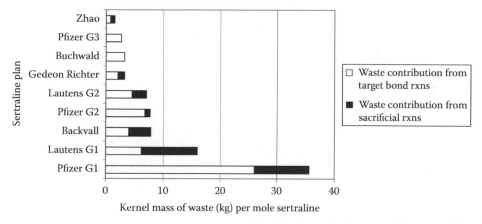

FIGURE 13.31 Bar graph showing sacrificial and building-up reaction proportions for each synthesis plan for sertraline. Plans are ranked according to kernel mass of waste produced per mole sertraline.

Silphinene

TABLE 13.39
Summary of Green Metrics Parameters for Silphinene Syntheses

Plan	Year	Type	N	M	I	μ_1	Beta	Delta	f(sac)	HI	% Yield	% AE	% Kernel RME	Kernel Mass of Waste (kg) per Mole
Wender	1985	Linear	3	3	10	−28.8	0.583	0.64	0.34	2.5	20.1	34.3	7.6	2.5
Sternbach	1985	Linear	10	10	20	−1.64	0.825	0.413	0.74	2.55	18.2	12.6	5.1	3.8
Paquette	1983	Linear	13	13	25	−0.08	0.861	0.394	0.713	2.5	18.9	10.2	3.5	5.6
Crimmins	1986	Linear	10	10	27	37.54	0.651	0.522	0.589	2.91	18.2	8.4	2.9	6.9
Yamamura	1989	Linear	16	16	30	52.99	0.825	0.394	0.778	3	3.6	7.7	0.8	26.1
Nagarajan	1988	Convergent	14	15	29	64.33	0.819	0.416	0.845	5.27	4.4	5.5	0.6	28.8
Ito	1983	Convergent	21	22	35	−14.9	0.902	0.355	0.858	4.36	5	6.8	0.6	33.4
Franck-Neumann	1991	Linear	20	20	34	85.16	0.899	0.36	0.832	5.52	1.9	5.4	0.3	81

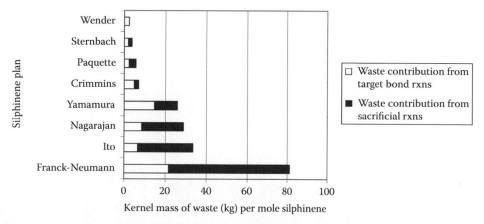

FIGURE 13.32 Bar graph showing sacrificial and building-up reaction proportions for each synthesis plan for silphinene. Plans are ranked according to kernel mass of waste produced per mole silphinene.

Sparteine

TABLE 13.40
Summary of Green Metrics Parameters for Sparteine Syntheses

Plan	Years	Type	N	M	I	μ_1	Beta	Delta	$f(sac)$	HI	% Yield	% AE	% Kernel RME	Kernel Mass of Waste (kg) per Mole
Anet	1950	Linear	2	2	5	−44	0.375	0.678	0.245	3.33	24	28.1	8.8	2.4
Takatsu	1987	Convergent	11	12	17	−84.83	0.839	0.361	0.688	3.33	13.6	18	8.8	5.4
Leonard G1	1950	Linear	3	3	9	−45	0.672	0.589	0.676	11.13	7.3	17.2	1.7	13.6
Leonard G2	1950	Linear	3	3	8	−33	0.607	0.584	0.61	10.75	3.6	28.2	1.1	21.9
Blakemore	2008	Linear	8	8	15	−8.44	0.809	0.41	0.338	10.89	1.9	17.2	0.7	32.3
Aube	2002	Linear	17	17	36	15.88	0.926	0.39	0.818	2.17	7.1	4.5	0.7	32.6
Fleming	2004	Convergent	11	12	25	26.47	0.862	0.425	0.849	8	3.9	5.8	0.6	38.6
Koomen	1996	Convergent	6	7	23	−48.31	0.764	0.567	0.842	0	2.8	11.1	0.6	41.1
Van Tamelen	1969	Linear	5	5	9	−14.03	0.594	0.46	0.78	1.33	0.2	11	0.07	323.3
Bohlmann–Gallagher	1973, 2006	Linear	13	13	24	−23.08	0.819	0.395	0.617	4.21	0.7	8.4	0.06	422
Clemo–Raper	1936, 1949	Linear	7	7	17	4.36	0.837	0.453	0.799	8.13	0.3	8.6	0.05	482

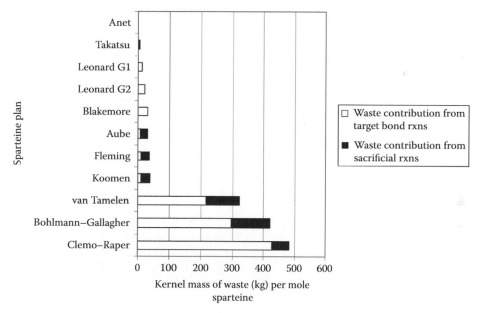

FIGURE 13.33 Bar graph showing sacrificial and building-up reaction proportions for each synthesis plan for sparteine. Plans are ranked according to kernel mass of waste produced per mole sparteine.

Strychnine

TABLE 13.41
Summary of Green Metrics Parameters for Strychnine Syntheses

Plan	Year	Type	N	M	I	μ_1	Beta	Delta	$f(sac)$	HI	% Yield	% AE	% Kernel RME	Kernel Mass of Waste (kg) per Mole
Rawal	1994	Convergent	14	20	34	−126.25	0.893	0.422	0.699	1.67	33.1	7.8	2.2	15.1
Reissig	2010	Convergent	10	14	28	−81.27	0.869	0.457	0.817	1.91	9.2	10.4	1.4	22.8
Bodwell	2002	Convergent	15	19	30	−43.45	0.88	0.394	0.691	0.81	2.9	9.8	1	33.4
Kuehne	1993	Convergent	20	23	38	−34.96	0.927	0.372	0.683	2.53	2.3	7.6	0.4	84.9
Padwa	2007	Convergent	17	21	36	−6.26	0.938	0.386	0.657	3.89	1.6	9	0.37	89.8
Overman	1995	Convergent	26	28	50	−0.5	0.952	0.365	0.653	3.37	1.5	6	0.3	111.6
Fukuyama	2004	Convergent	25	29	50	59.2	0.952	0.371	0.753	4.73	1.1	5.2	0.26	128
Martin	2001	Convergent	16	18	29	−63.55	0.917	0.369	0.719	3.35	0.79	12.5	0.2	166.8
Vollhardt	2001	Convergent	14	19	28	−90.92	0.871	0.397	0.676	0.6	0.56	10.2	0.15	218.1
Shibasaki	2004	Linear	31	31	53	128.53	0.955	0.346	0.776	1.31	0.46	5.2	0.1	333.8
Mori	2003	Convergent	24	28	50	71.21	0.945	0.379	0.77	0.08	0.32	5.4	0.043	785.3
Bonjoch	2000	Convergent	18	22	39	−6.54	0.938	0.389	0.713	6.32	0.14	6.9	0.041	810.3
Magnus	1993	Linear	33	33	60	97.54	0.955	0.355	0.83	4.09	0.0033	4.4	0.00085	39232
Woodward	1963	Convergent	29	30	45	38.18	0.939	0.334	0.768	3.47	4.6E-05	9.8	0.000019	1732663
Stork	1992	Convergent	16	20	33	26.64	0.874	0.4	0.755	2.53	<100	10.4	<6	>5.2

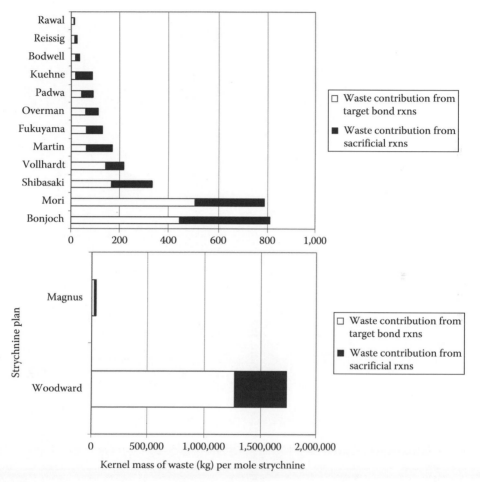

FIGURE 13.34 Bar graphs showing sacrificial and building-up reaction proportions for each synthesis plan for strychnine. Plans are ranked according to kernel mass of waste produced per mole strychnine.

Swainsonine

TABLE 13.42
Summary of Green Metrics Parameters for Swainsonine Syntheses

Plan	Year	Type	N	M	I	μ_1	Beta	Delta	$f(sac)$	HI	% Yield	% AE	% Kernel RME	Kernel Mass of Waste (kg) per Mole
O'Doherty G1	2006	Linear	13	13	25	124.57	0.903	0.384	0.78	1.5	17.3	8.2	2.8	6
GlycoDesign	2008	Linear	7	7	21	−30.2	0.836	0.488	0.642	2.25	25.5	7.9	3	6.8
Roush	1995	Convergent	8	10	20	102.08	0.86	0.447	0.724	−1.33	17.3	8.6	2.4	7.1
O'Doherty G2	2006	Linear	15	15	27	142.13	0.917	0.368	0.784	1.75	14.8	7.7	2.4	7.1
Blechert	2002	Linear	14	14	30	88.4	0.917	0.396	0.786	−0.07	22.4	4.2	2.3	7.3
Fleet	1984	Linear	13	13	20	221.29	0.894	0.349	0.72	3.36	10	8.3	2	8.3
Vankar	2009	Linear	21	21	37	246.73	0.936	0.357	0.876	1.59	17	5.4	1.8	9.7
Cha	1989	Linear	10	10	19	31.91	0.828	0.404	0.608	2.55	14.7	7.6	1.6	10.4
Trost	1999	Linear	17	17	26	264.94	0.904	0.342	0.721	2.5	14.2	6.2	1.6	10.7
Cossy G2	2007	Linear	13	13	32	64.64	0.922	0.415	0.836	−0.29	14.1	3.7	1.2	14.7
Kang	1995	Convergent	18	21	37	236.77	0.926	0.385	0.864	1.47	17.3	3.8	1.1	15.7
Mootoo	2001	Convergent	14	16	28	219.48	0.92	0.385	0.86	1.93	8.9	6.5	1.1	17.4
Pearson	1996	Linear	13	13	27	64.5	0.912	0.394	0.775	2.5	11.8	4.5	0.9	18.4
Carretero	2000	Linear	17	17	26	185.33	0.909	0.342	0.733	−0.72	6.9	5.3	0.9	19.6
Ham	2009	Linear	21	21	38	215.05	0.941	0.36	0.784	−0.64	7.5	4.8	0.8	21

Ferreira	1997	Linear	19	19	37	116.05	0.935	0.374	0.717	2	5.7	5	0.8	21.1
Pyne G1	2004	Linear	19	19	33	170.04	0.925	0.358	0.861	-0.95	3.9	4.6	0.7	23
Bermejo	1998	Convergent	15	16	30	97.74	0.894	0.39	0.81	-0.5	5.6	6.4	0.7	23.6
Riera	2005	Linear	18	18	27	43.95	0.917	0.336	0.622	0.68	2.7	7.5	0.6	28.7
Sharpless	1985	Linear	21	21	41	354.83	0.945	0.371	0.664	0.59	5.2	3.3	0.6	29.4
Reiser	2005	Linear	11	11	24	38.42	0.867	0.416	0.726	3.17	3.9	5.7	0.5	35.6
Zhou	1995	Linear	14	14	27	168.2	0.916	0.38	0.779	0.27	4	5.7	0.4	38.9
Pyne G2	2006	Linear	12	12	26	178	0.904	0.404	0.77	-1.08	5.1	3.1	0.4	45.8
Hashimoto	1985	Linear	12	12	19	132.85	0.873	0.359	0.591	1.85	2.2	10.5	0.4	47.6
Poisson	2006	Linear	20	20	35	194.48	0.929	0.358	0.756	1.19	3.1	2.9	0.3	53
Kibayashi	1994	Convergent	20	21	36	94.49	0.933	0.362	0.783	0.48	1.9	4.3	0.2	70.9
Hirama	1995	Linear	24	24	47	120.2	0.927	0.375	0.793	3.52	0.6	3.5	0.1	151.1
Richardson	1985	Linear	16	16	24	135.67	0.886	0.343	0.76	3.88	0.6	4.9	0.1	181.9
Suami G1	1984	Linear	22	22	36	247.87	0.934	0.346	0.811	3	0.2	4.1	0.06	296
Suami G2	1985	Linear	19	19	38	241.65	0.937	0.377	0.829	3.05	0.2	3.7	0.05	375.3
Ikota	1990	Linear	25	25	51	176.53	0.956	0.373	0.915	0.15	0.3	3.5	0.03	495.4
Cossy G1	2007	Linear	18	18	34	107.55	0.931	0.371	0.778	0.84	0.2	5.4	0.03	635.4
Takaya	1984	Linear	16	16	23	111.06	0.895	0.334	0.607	3.82	0.07	8.4	0.01	1248.5

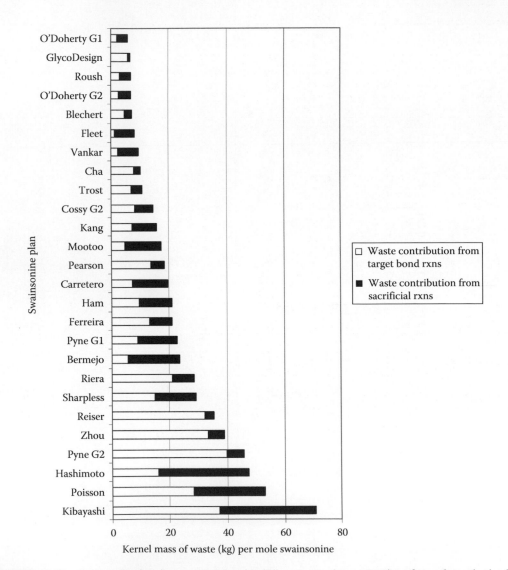

FIGURE 13.35 Bar graphs showing sacrificial and building-up reaction proportions for each synthesis plan for swainsonine. Plans are ranked according to kernel mass of waste produced per mole swainsonine.

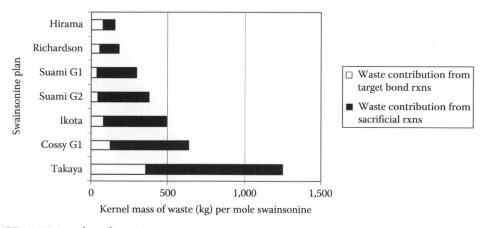

FIGURE 13.35 (continued)

Tamoxifen

TABLE 13.43
Summary of Green Metrics Parameters for Tamoxifen Syntheses

Plan	Year	Type	N	M	I	μ_1	Beta	Delta	$f(sac)$	HI	% Yield	% AE	% Kernel RME	Kernel Mass of Waste (kg) per Mole
Larock	2005	Linear	3	3	11	−178.58	0.574	0.664	0.432	0.5	33.5	32.4	13.9	2.3
Bristol-Myers	1984	Linear	4	4	11	−223.96	0.492	0.599	0.578	2	26.2	34.8	12.1	2.7
O'Shea	2006	Convergent	3	4	13	−465.75	0.559	0.706	0.663	0.25	40	26.8	11.7	2.8
Knochel	1997	Convergent	3	4	13	−384.06	0.587	0.696	0.635	0.75	37.2	23.3	11.7	3.1
NRDC	1987	Linear	5	5	11	−165.18	0.729	0.477	0.537	1.67	23.2	31.9	11.1	3.1
McCague	1990	Convergent	8	9	19	−262.72	0.849	0.442	0.718	1.33	23.6	18.5	8.6	3.9
Taniguchi	2009	Convergent	4	5	16	−195.35	0.751	0.595	0.738	0.6	32.9	19.3	8.2	4.1
Fallis	2003	Linear	7	7	16	−106.25	0.844	0.441	0.571	1.38	27.5	18.9	7.9	4.3
Miller	1985	Linear	8	8	21	−120.37	0.864	0.453	0.697	−0.33	32.8	17.7	7.4	4.6
Dow Chemical	1995	Convergent	3	4	13	−287.48	0.151	0.782	0.701	0.75	26.6	18.9	5.7	6
Mori–Sato	2006	Convergent	9	10	20	−124.4	0.878	0.42	0.518	2	22.8	14.5	5.6	6.3
Nishihara	2007	Convergent	3	5	18	−333	0.815	0.39	0.658	5	25.4	19.6	5.1	6.9
Shiina	2004	Linear	4	4	9	−178.8	0.656	0.512	0.527	0	9.7	34.3	4.9	7.2
Al-Hassan	1987	Linear	4	4	12	−167.86	0.579	0.578	0.554	0.4	11.1	31.3	4.5	7.9
Yus	2003	Convergent	5	7	15	−388.68	0.605	0.58	0.685	1.67	13.1	22	4.3	8.3
Shindo	2005	Linear	7	7	14	−51.21	0.792	0.43	0.463	0.38	14.4	20.1	4.1	8.7
Itami	2003	Convergent	5	7	20	−253.37	0.846	0.538	0.682	0.33	16.8	15.7	3.7	9.6
Armstrong	1997	Convergent	3	6	22	−369.82	0.564	0.807	0.688	1.5	16.4	14	2.6	13.8
Katzenellenbogen	1985	Linear	7	7	17	−70.99	0.809	0.461	0.433	2.38	5.2	26.9	2.2	16.8
ICI	1965	Linear	8	8	12	−89.67	0.754	0.376	0.379	0.78	<63.5	35.4	<29.8	>0.9

TABLE 13.44

Summary of *E*-Factor Breakdowns for Tamoxifen Syntheses

Plan	Year	Type	*E*-Kernel	*E*-Excess	*E*-Aux	*E*-Total	Total Mass of Waste (kg) per Mole
NRDC	1987	Linear	8.27	27.93	143.2	179.35	66.5
Bristol-Myers	1984	Linear	7.29	15.67	423.1	446.07	165.5
Knochel	1997	Convergent	7.57	96.84	631.2	735.63	299.7
Shiina	2004	Linear	19.29	808.3	17.86	845.42	313.7
Nishihara	2007	Convergent	18.5	9.15	903.3	930.9	345.4
O'Shea	2006	Convergent	7.54	123.2	949.6	1080.4	400.8
Taniguchi	2009	Convergent	11.17	63.02	1339	1413.2	524.3
Dow–Chemical	1995	Convergent	16.08	159.2	1545	1720.6	638.3
Yus	2003	Convergent	22.48	67.35	1698	1787.7	663.2
Larock	2005	Linear	6.19	514.9	1567	2087.7	774.5
Fallis	2003	Linear	11.71	30.75	2382	2425	899.7
Mori–Sato	2006	Convergent	16.92	19.85	2418	2454.6	910.6
Miller	1985	Linear	12.47	36.79	2631	2679.9	994.2
Itami	2003	Convergent	25.9	113.9	2868	3007.9	1116
Al-Hassan	1987	Linear	21.39	294.3	2796	3112.2	1155
Shindo	2005	Linear	23.36	224.7	3122	3370.3	1250
McCague	1990	Convergent	10.59	378.9	5850	6239.4	2315
Armstrong	1997	Convergent	37.14	422.8	6953	7413.1	2750
Katzenellenbogen	1985	Linear	45.34	280.9	26812	27138	10068

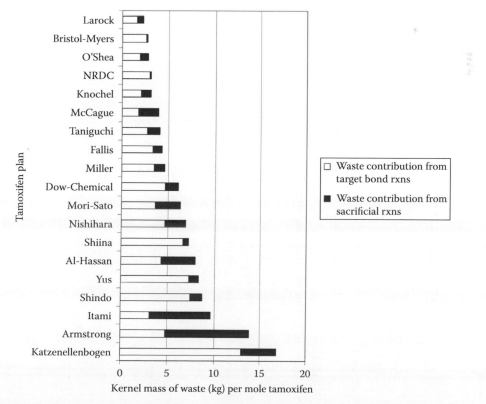

FIGURE 13.36 Bar graph showing sacrificial and building-up reaction proportions for each synthesis plan for tamoxifen. Plans are ranked according to kernel mass of waste produced per mole tamoxifen.

Tropinone

TABLE 13.45
Summary of Green Metrics Parameters for Tropinone Syntheses

Plan	Year	Type	N	M	I	μ_1	Beta	Delta	f(sac)	HI	% Yield	% AE	% Kernel RME	Kernel Mass of Waste (kg) per Mole
Robinson	1917	Linear	1	1	5	−77	0	1	0.397	1	42	31.9	13.4	0.9
Bazilevskaya G3	1958	Convergent	4	5	11	0.2	0.673	0.551	0.638	1.2	40.3	15.3	7.3	1.8
Bazilevskaya G2	1958	Linear	3	3	9	−40.5	0.48	0.644	0.724	1.5	41	13.5	6.9	1.9
Bazilevskaya G1	1958	Linear	4	4	4	104.76	0.417	0.358	0.131	0.8	4	27.8	1.2	11.2
Parker	1959	Linear	5	5	10	38.33	0.715	0.461	0.802	2.67	6.2	15.7	1.2	11.4
Willstatter G2	1921	Linear	5	5	7	118	0.63	0.397	0.139	2	2.4	25.1	0.7	19.2
Bazilevskaya G4	1958	Convergent	5	6	15	36.67	0.765	0.526	0.668	3.33	2.1	7.9	0.3	47.4
Willstatter G1	1901	Linear	15	15	21	8.84	0.899	0.33	0.943	−4.13	0.1	4.9	0.02	566.2
Karrer	1947	Linear	6	6	9	108.94	0.687	0.396	0.498	1.14	<4.9	12.3	<0.8	>16.3

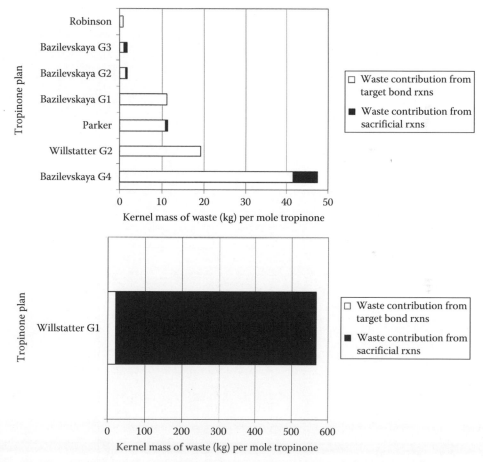

FIGURE 13.37 Bar graphs showing sacrificial and building-up reaction proportions for each synthesis plan for tropinone. Plans are ranked according to kernel mass of waste produced per mole tropinone.

Tropolone

TABLE 13.46
Summary of Green Metrics Parameters for Tropolone Syntheses

Plan	Year	Type	N	M	I	μ_1	Beta	Delta	f(sac)	HI	% Yield	% AE	% Kernel RME	Kernel Mass of Waste (kg) per Mole
Minns	1988	Linear	3	3	7	42.01	0.453	0.587	0.655	0.75	64.3	19.8	14	0.7
Stevens	1965	Linear	4	4	10	39.77	0.607	0.548	0.725	0	33.2	14.6	5.9	1.9
DuPont	1958	Linear	3	3	4	2.5	0.25	0.467	0.36	0.75	8.2	20.5	4.6	2.5
Ter Borg	1962	Linear	10	10	14	50.65	0.749	0.356	0.845	-3	8.2	8.6	1.1	10.9
Van Tamelen	1956	Linear	7	7	11	26.5	0.746	0.391	0.733	-1.5	4.5	10.1	1	11.7
Chapman	1963	Linear	7	7	16	52.13	0.78	0.459	0.881	1.5	5.4	7.4	0.8	14.4
Oda	1969	Linear	7	7	13	9.81	0.821	0.41	0.755	-0.5	2.8	10.4	0.6	21.5
Von Doering	1951	Linear	5	5	10	-44.63	0.703	0.463	0.465	-3.33	3	12	0.5	22.2
Cram	1951	Linear	5	5	9	37.33	0.703	0.442	0.76	0.83	2.5	9.6	0.4	28.4
Cook	1951	Linear	7	7	10	2.77	0.776	0.368	0.716	-1.25	1.7	10.9	0.4	33.4

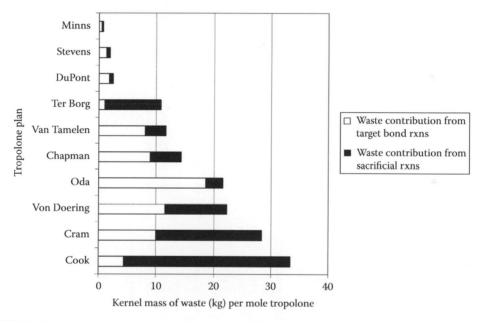

FIGURE 13.38   Bar graphs showing sacrificial and building-up reaction proportions for each synthesis plan for tropolone. Plans are ranked according to kernel mass of waste produced per mole tropolone.

Twistane

TABLE 13.47
Summary of Green Metrics Parameters for Twistane Syntheses

Plan	Year	Type	N	M	I	μ_1	Beta	Delta	f(sac)	HI	% Yield	% AE	% Kernel RME	Kernel Mass of Waste (kg) per Mole
Whitlock	1968	Linear	12	12	23	40.85	0.848	0.397	0.813	4.15	8.8	8.5	1.7	8.1
Tichy	1969	Linear	10	10	13	76.73	0.827	0.332	0.793	6.82	5.5	11.4	1.3	10.2
Deslongchamps G2	1969	Linear	9	9	15	55.65	0.765	0.393	0.707	4.3	4.3	8.9	1	13
Jones	1982	Linear	8	8	12	52.67	0.77	0.374	0.661	4.67	4.2	9.6	1	13.1
Deslongchamps G1	1967	Linear	10	10	17	76	0.839	0.381	0.739	4.55	1.5	7.9	0.4	36.6
Hamon–Young	1976	Linear	18	18	31	48.35	0.982	0.346	0.761	3.47	1.7	4.2	0.3	40.9

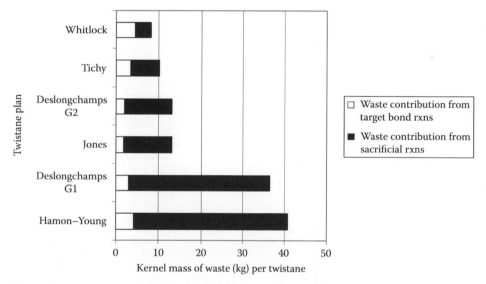

FIGURE 13.39 Bar graphs showing sacrificial and building-up reaction proportions for each synthesis plan for twistane. Plans are ranked according to kernel mass of waste produced per mole twistane.

Tyrian Purple

TABLE 13.48
Summary of Green Metrics Parameters for Tyrian Purple Syntheses

Plan	Year	Type	N	M	I	μ_1	Beta	Delta	f(sac)	HI	% Yield	% AE	% Kernel RME	Kernel Mass of Waste (kg) per Mole
Gerlach	1989	Linear	3	3	8	−7.53	0.516	0.608	0.818	8	48.6	28.4	15.9	1.1
Cooksey G1	1994	Convergent	7	8	15	−39.94	0.699	0.466	0.701	3.5	21.7	17	6.5	3.2
Imming	2001	Linear	5	5	11	−28.93	0.734	0.476	0.682	3	9.9	19.5	3.2	6.3
Rottig	1935	Linear	4	4	11	−56.34	0.665	0.695	0.544	1.6	5.9	25.8	1.9	10.7
Cooksey G2	2005	Linear	5	5	11	−28.93	0.734	0.476	0.71	3.5	6.8	17.8	1.9	10.9
Sachs	1903	Linear	8	8	14	−22.03	0.821	0.395	0.782	4.44	6.6	10.2	1.2	18
Tanoue	2001	Linear	8	8	17	23.98	0.839	0.426	0.635	−1.78	2.4	6.1	0.7	28
Majima	1930	Linear	5	5	9	8.57	0.715	0.439	0.653	−6	<15.6	24.2	<4.8	>4.1
Friedlander	1912	Linear	7	7	13	−21.21	0.795	0.43	0.738	−1.5	<4.3	8.6	<0.7	>27.9
Grandmougin	1914	Linear	9	9	18	−38.7	0.804	0.421	0.842	−0.4	<2.2	6.3	<0.5	>39.5

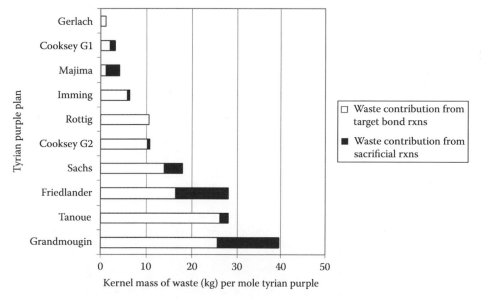

FIGURE 13.40   Bar graphs showing sacrificial and building-up reaction proportions for each synthesis plan for tyrian purple. Plans are ranked according to kernel mass of waste produced per mole tyrian purple.

Vanillin

TABLE 13.49
Summary of Green Metrics Parameters for Vanillin Syntheses

Plan	Year	Type	N	M	I	μ_1	Beta	Delta	f(sac)	HI	% Yield	% AE	% Kernel RME	Kernel Mass of Waste (kg) per Mole
Roesler	1907	Linear	1	1	4	−143.5	0	1	0.177	0	70	74	51.8	0.1
Givaudan–Roure	1998	Linear	1	1	2	−47	0	1	0.071	0	75	67.3	50.4	0.15
Volynkin	1938	Linear	3	3	6	−98.5	0.488	0.544	0.46	−0.25	73.7	53.3	40.2	0.2
Mottern	1934	Linear	4	4	3	−1.2	0.25	0.311	0.352	0.8	40.2	43.6	21.2	0.6
Harries–Haarmann	1915	Linear	2	2	4	8	0	0.655	0.467	0.33	52.5	38.3	23.5	0.49
Givaudan	1921	Linear	3	3	7	19.5	0.161	0.624	0.613	−1.75	57.8	16.5	12	1.1
Guyot	1910	Linear	4	4	6	26.2	0.5	0.444	0.726	−0.4	55.9	11.4	7.5	1.9
Delvaux	1947	Linear	2	2	4	28.9	0.25	0.644	0.4	−0.67	39	33.6	15.2	0.8
Mayer	1949	Linear	1	1	7	−66.25	0	1	0.654	−1.5	92	16.2	10.4	0.9
Zasosov	1959	Linear	2	2	5	28.9	0.219	0.699	0.563	−0.67	54.1	20.7	11.6	1.2
Lampman G1	1977	Linear	2	2	6	8	0	0.753	0.453	−1.67	27	31.7	10.3	1.3
Geigy	1899	Linear	1	1	5	−75	0	1	0.774	−1	30	22.3	6.7	2.1
Riedel	1927	Linear	4	4	11	−9.6	0.492	0.599	0.55	−1.8	16.3	18.4	4.2	3.5
Frost	1998	Linear	2	2	6	33.33	0.4	0.712	0.897	0.33	44.2	2.3	1	14.4

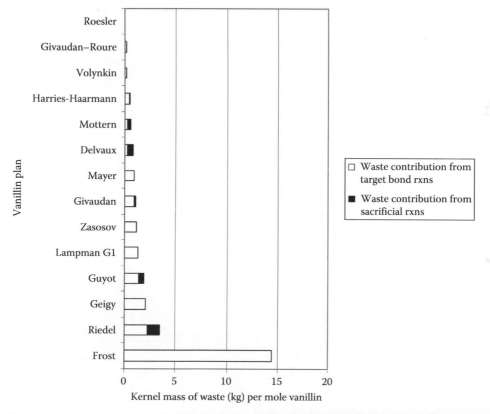

FIGURE 13.41 Bar graphs showing sacrificial and building-up reaction proportions for each synthesis plan for vanillin. Plans are ranked according to kernel mass of waste produced per mole vanillin.

Vitamin A

TABLE 13.50
Summary of Green Metrics Parameters for Vitamin A Syntheses

Plan	Year	Type	N	M	l	μ_1	Beta	Delta	f(sac)	HI	% Yield	% AE	% Kernel RME	Kernel Mass of Waste (kg) per Mole
Roche G4	1976	Convergent	6	7	15	−235.74	0.763	0.486	0.651	0	36.9	26.3	16.9	1.4
Roche G6c	1976	Convergent	8	10	16	−165.46	0.658	0.455	0.673	1.67	40	22.7	15.8	1.5
Roche G6b	1976	Convergent	7	9	14	−175.94	0.604	0.47	0.633	1.5	29.6	25.5	15.2	1.6
Roche G3	1976	Convergent	4	6	12	−251.95	0.712	0.556	0.58	0	39	30	12.3	2
Roche G1	1947	Convergent	7	8	16	−132.19	0.836	0.443	0.609	1.25	14.9	25.1	8.8	2.9
Roche G6a	1976	Convergent	6	8	15	−201.65	0.633	0.522	0.691	1.14	15.2	22.9	7.1	3.7
Roche G2	1949	Convergent	7	8	19	−132.19	0.851	0.467	0.732	1.25	13.9	17.2	6.5	4.1
Roche G7	1978	Convergent	6	8	16	−153.84	0.774	0.494	0.831	−1.29	21.9	12.2	4.6	5.9
Merck	1951	Linear	8	8	15	−32.57	0.807	0.411	0.43	3.11	6.5	25.9	3.1	8.99
Roche G5	1976	Convergent	9	11	22	−202.31	0.872	0.437	0.672	1.8	0.8	18.5	0.5	69.71

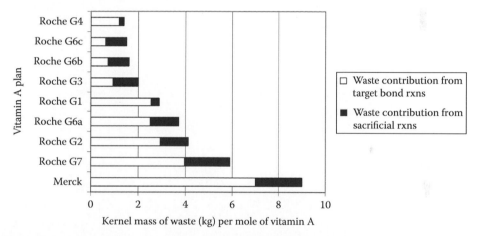

FIGURE 13.42   Bar graphs showing sacrificial and building-up reaction proportions for each synthesis plan for vitamin A. Plans are ranked according to kernel mass of waste produced per mole vitamin A.

Appendix A

A.1 TABLES OF DENSITIES OF SOLVENTS AND WORK-UP SOLUTIONS USED IN RADIAL PENTAGON ANALYSES

Solvent	Density (g/mL)[a]
1,2-Dichloroethane	1.256
1,4-Dioxane	1.034
95% EtOH	0.816
Acetic acid (glacial)	1.0498
Acetic anhydride	1.081
Acetone	0.791
Acetonitrile	0.78
Ammonia (liquid)	0.6818
Benzene	0.879
Carbitol	0.999
CCl_4	1.594
CF_3COOH	1.48
CH_2Cl_2	1.336
Chloroform	1.492
$ClCH_2CH_2Cl$	1.253
Cyclohexane	0.779
Diethyl ether	0.708
DMF	0.944
DMSO	1.101
Ethanol	0.789
Ethylene glycol	1.113
Ethylene glycol dimethyl ether	0.867
EtOAc	0.901
37 wt.% formaldehyde	1.09
HOAc	1.049
i-PrOH	0.785
Isooctane	0.692
Isopropyl acetate	0.874
Isopropyl ether	0.725
Ligroin	0.656
MeOAc	0.932
Methanol	0.792
Methyl ethyl ketone	0.805
MTBE	0.74

(continued)

(continued)

Solvent	Density (g/mL)[a]
n-Butanol	0.811
n-Heptane	0.684
n-Hexane	0.659
Nitrobenzene	1.203
N-Methylpyrrolidone	1.028
n-Pentane	0.626
Petroleum ether	0.64
Pyridine	0.983
Sulfolane	1.261
t-Amyl alcohol	0.805
t-Butanol	0.779
THF	0.888
Thionyl chloride	1.631
Toluene	0.866
Trichloroethylene	1.463
Water	1
Xylene-*m*	0.864
Xylene-*o*	0.88
Xylene-*p*	0.862
Xylenes	0.869

[a] Lange, N.A., *Handbook of Chemistry*, 10th edn., McGraw-Hill Book Co., Inc., New York, 1961, pp. 1118–1171.

Work-Up Solution	Density (g/mL)[a]
0.1 M HCl	1
1 M H_2SO_4	1.06
1 M HCl	1.018
1 M NaOH	1.033
1 wt.% NaOH	1.0095
10 wt.% Citric acid	1.0392
10 wt.% HCl	1.0474
10 wt.% HOAc	1.0125
10 wt.% K_2CO_3	1.0904
10 wt.% Na_2CO_3	1.1029
10 wt.% Na_2SO_3	1.0948
10 wt.% NaCl	1.0707
10 wt.% NaOH	1.1089
10 wt.% NH_4Cl	1.0286
28 wt.% NH_4OH	0.898
30 wt.% H_2O_2	1.1122
37 wt.% HCl	1.1837
48 wt.% HBr	1.49
5 wt.% $NaHCO_3$	1.0354
5 wt.% NaOH	1.0538
50 wt.% H_2SO_4	1.3951
50 wt.% KOH	1.5143
50 wt.% NaOH	1.5253
70 wt.% HNO_3	1.4134
85 wt.% H_3PO_4	1.689
98 wt.% H_2SO_4	1.8361
Glacial HOAc (99.8 wt.%)	1.049
Sat'd K(Na)tartrate (Rochelle salt)	1.4001
Sat'd Na_2CO_3	1.1463
Sat'd $Na_2S_2O_3$	1.3827
Sat'd Na_2SO_3	1.1755
Sat'd NaCl	1.1804
Sat'd $NaHCO_3$	1.0581
Sat'd NaOH	1.5253
Sat'd NH_4Cl	1.0674

[a] Lange, N.A., *Handbook of Chemistry*, 10th ed., McGraw-Hill Book Co., Inc., New York, 1961, pp. 1118–1171.

A.2   LIST OF ABBREVIATIONS FOR REAGENTS

Acronym	Full Name	Structure
ACES	N-(2-Acetamido)-2-aminoethanesulfonic acid	
ACN	Acetonitrile	CH_3CN
ADA	N-(2-Acetamido)-iminodiacetic acid	
Aq	Aqueous	H_2O
AIBN	Azobisisobutyronitrile	
Aldrithiol®	2,2'-Dipyridyldisulphide	
ATP	Adenosinetriphosphate	
9-BBN	9-Borabicyclo[3.3.1]nonane	
BDSC	t-Butyldimethylchlorosilane	

(continued)

Acronym	Full Name	Structure
BES	*N,N*-Bis (2-hydroxyethyl)-2-aminoethanesulfonic acid	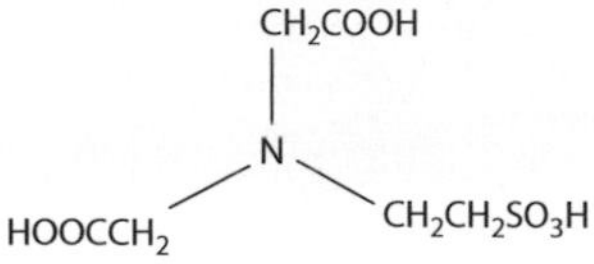
Bicine	*N,N*-Bis(2-hydroxyethyl)glycine	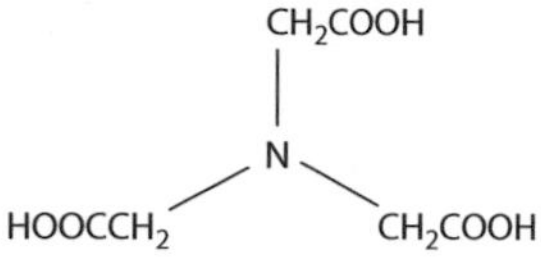
BINAP	2,2′-Bis(diphenylphosphino)-1,1′-Binaphthyl	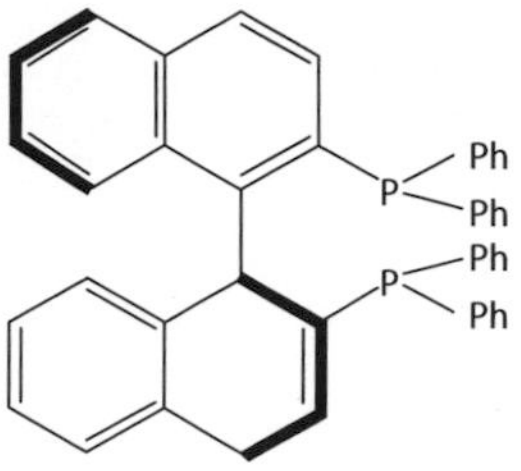 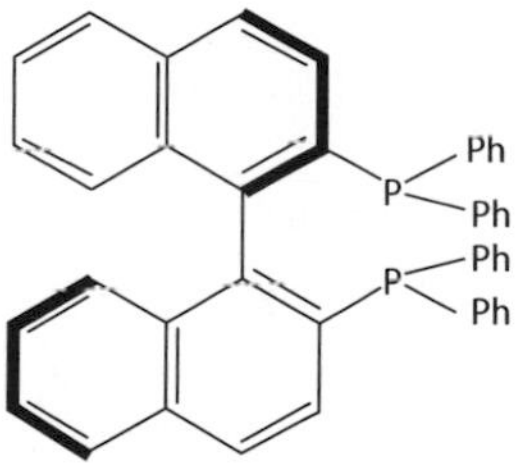
BINAPO	Phospinous acid, diphenyl-[1,1′-binaphthalene]-2,2′-dilyl ester	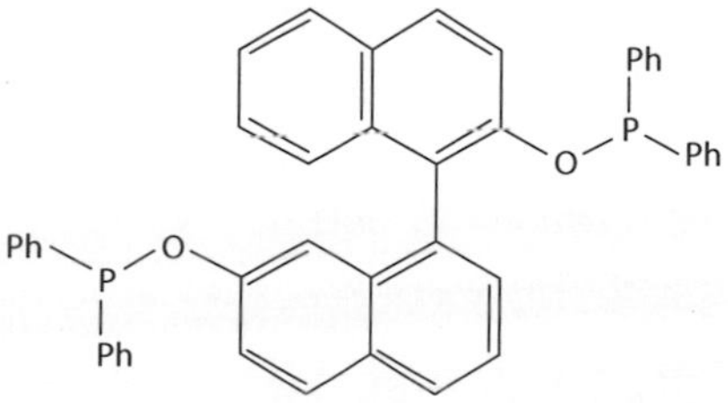

(*continued*)

(continued)

Acronym	Full Name	Structure
BINOL	2,2'-Dihydroxy-1,1'-binaphthyl (binaphthyl alcohol)	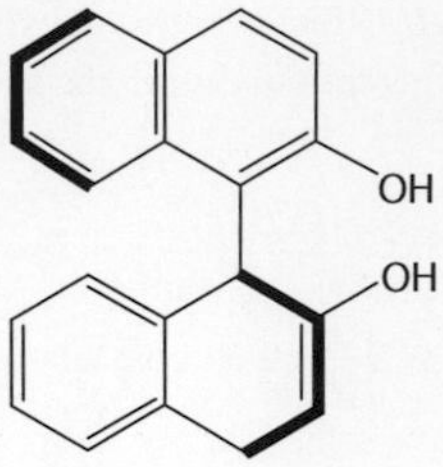 (S)-(−) 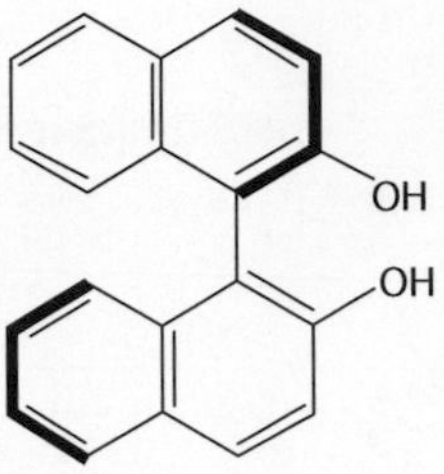 (R)-(+)
BMB	Bisamylborane	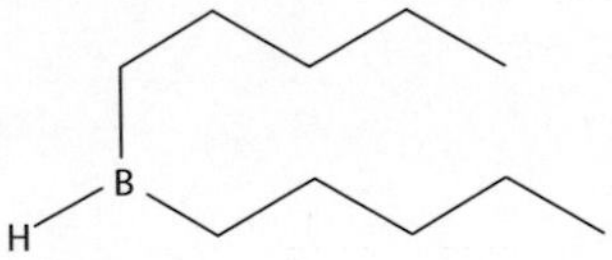
BMS	Borane dimethylsulfide complex	B_2H_6–$(CH_3)_2S$
BOM	Benzyloxymethyl	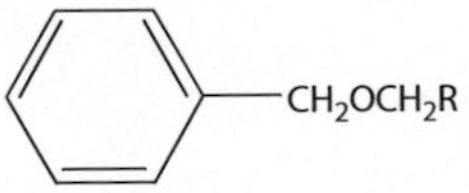 –CH_2OCH_2R
BOOB	Di-t-butylperoxide	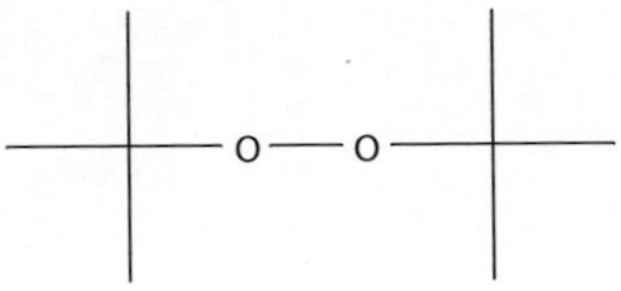
BOP–Cl	Bis-(2-oxo-3-oxazolidinyl)phosphinic acid	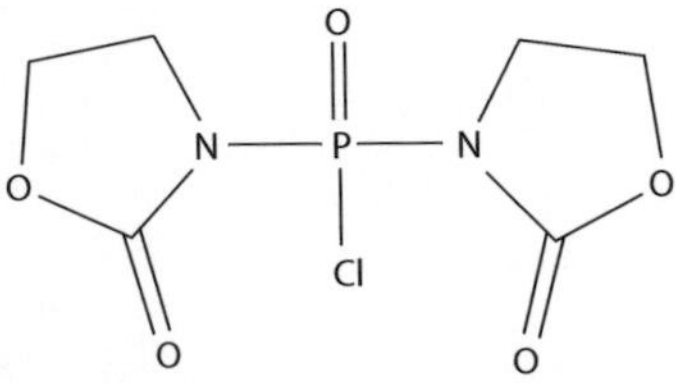
BQ	Benzoquinone	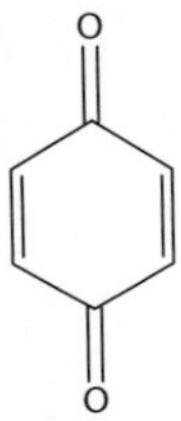 p-Quinone 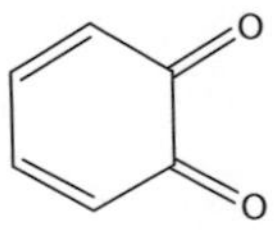 o-Quinone

(continued)

Acronym	Full Name	Structure
BSA	*N*,O-Bis(trimethylsilyl)acetamide benzeneseleninic acid	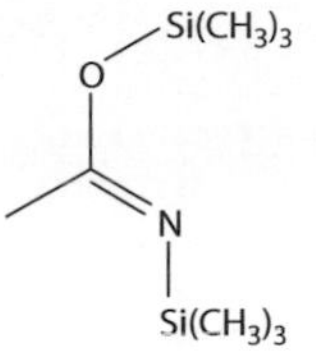 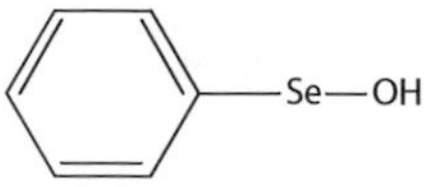
BTAF	Benzyltrimethylammonium fluoride	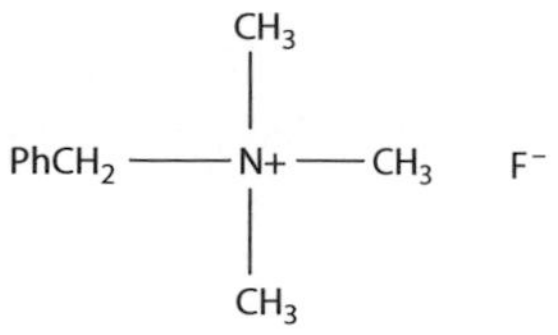
BTMSA	*N*,*N*-Bis(trimethylsilyl)acetamide	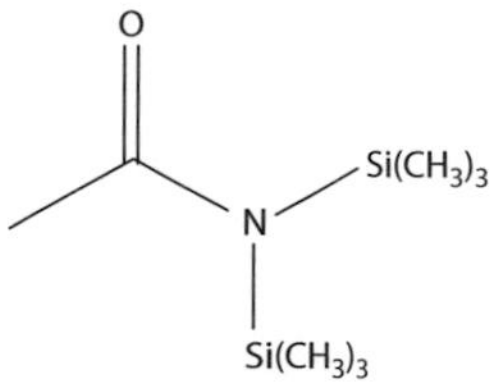
Bz	Benzene	
CAN	Ceric ammonium nitrate	$(NH_4)_2\ Ce(NO_3)_6$
CAPS	3-(Cyclohexylamino)-1- propanesulfonic acid	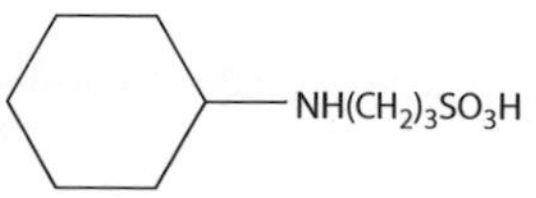 $NH(CH_2)_3SO_3H$
CHES	2-(Cyclohexylamino)-1- ethanesulfonic acid	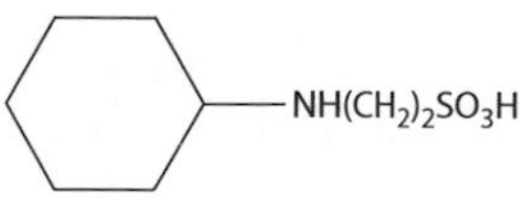 $NH(CH_2)_2SO_3H$
CSA	Camphor-10-sulfonic acid	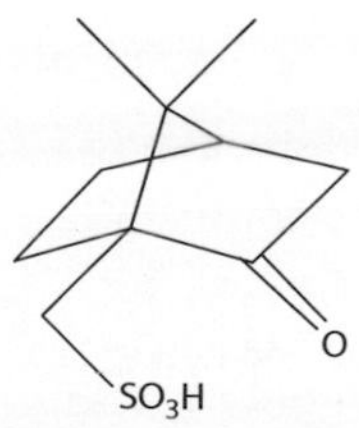 SO_3H
CSI	Chlorosulfonyl isocyanate	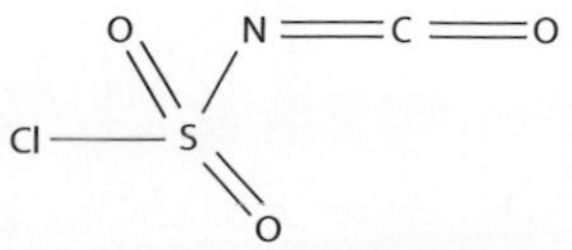

(*continued*)

(continued)

Acronym	Full Name	Structure
DABCO	1,4-Diazabicyclo[2.2.2]octane	
DAST	(Diethylamino)sulfur trifluoride	
DBN	1,5-Diazabicyclo[4.3.0]non-5-ene	
DBQ	2,3-Dichloro-5,6-dicyano-1,4-dibenzoquinone	
DBU	1,8-Diazabicyclo[5.4.0]undec-7-ene	
DCC	1,3-Dicyclohexylcarbodiimide	
DDQ	2,3-Dichloro-5,6-dicyano-1,4-benzoquinone	
DEAD	Diethylazodicarboxylate	
DIBAH (DIBAL)	Diisobutylaluminum hydride	

(continued)

Acronym	Full Name	Structure
DHP	Dihydropyran	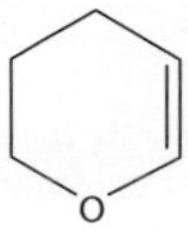
DHQD	Dihydroquinidine	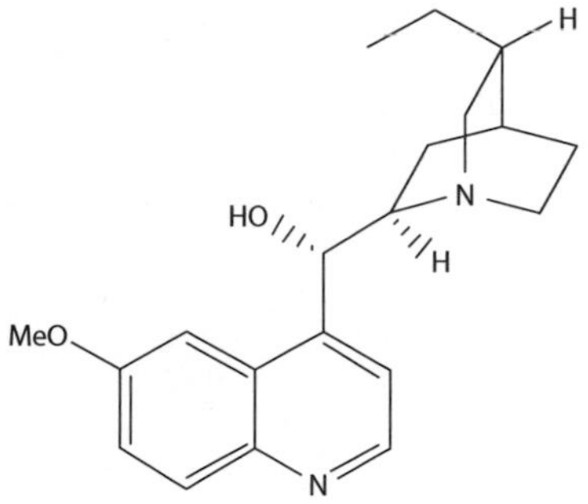
DIC	Diisopropylcarbodiimide	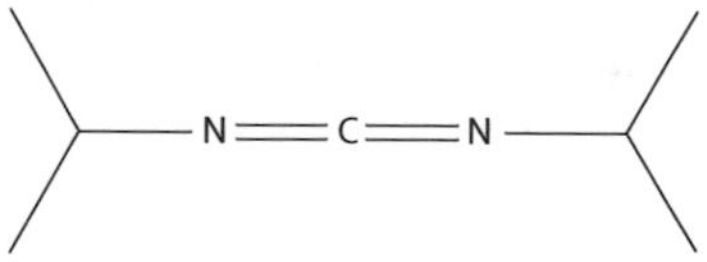
DME	1,2-Dimethoxyethane (glyme)	$CH_3OCH_2CH_2OCH_3$
DMAP	4-Dimethylaminopyridine	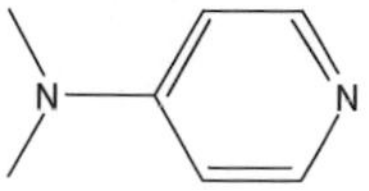
DMF	Dimethylformamide	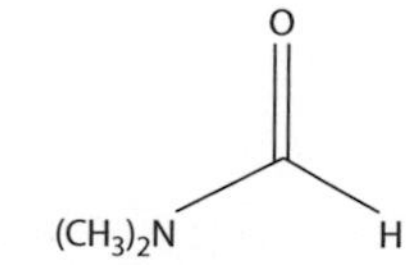
DMP	Dess–Martin periodinane	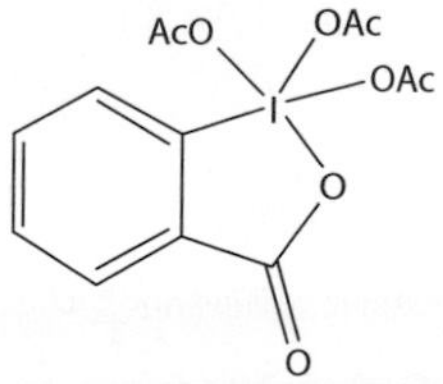
DMS	Dimethylsulfide	Me_2S
DMSO	Dimethylsulfoxide	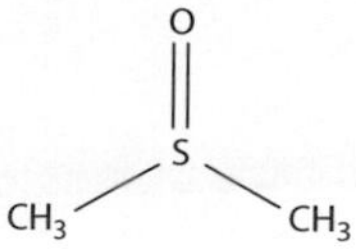

(*continued*)

(continued)

Acronym	Full Name	Structure
DNA	Deoxyribonucleic acid	
DNFB	2,4-Dinitrofluorobenzene	
DNQ	Diazonaphthoquinone	
DPK	Diphenylketone	
DPPA	Diphenylphosphoryl azide	
DPPE	1,2-Bis(diphenylphosphine)ethane	
DPPP	1,3-Bis(diphenylphosphine)propane	

(continued)

Acronym	Full Name	Structure
EDA	Ethylenediamine	$H_2N\text{-}CH_2 \text{---} CH_2\text{-}NH_2$
EDDA	Ethylene diammonium acetate	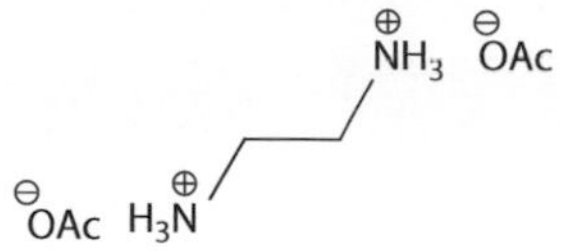
EDTA	Ethylenediaminetetraacetic acid	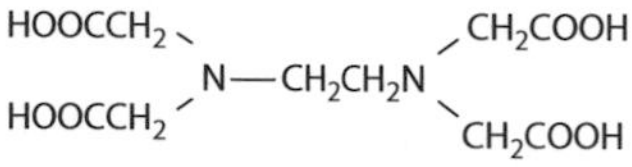
EEA	Ethyl acetoacetate	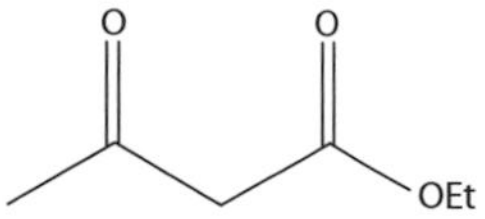
FDPP	Pentafluorophenyl diphenylphosphinate	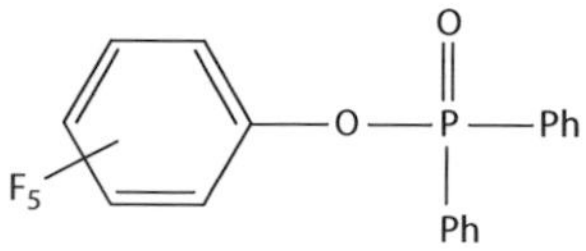
HEPES	4-(2-Hydroxyethyl)-1-piperazineethanesulfonic acid	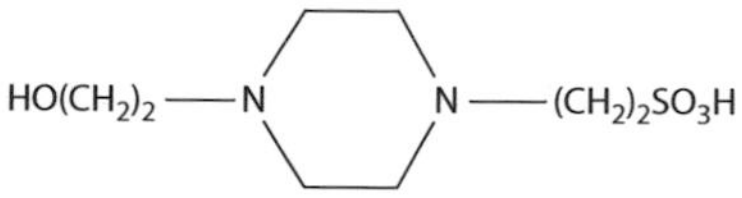
HEPPS	4-(2-Hydroxyethyl)-piperazine-1-propanesulfonic acid	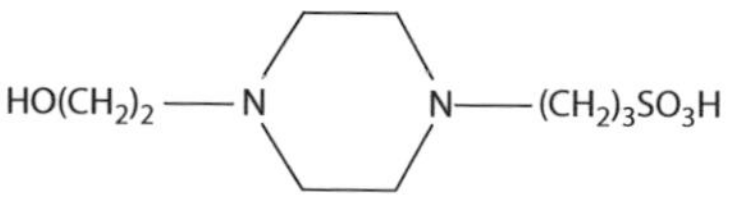
HFIP	Hexafluoroisopropyl alcohol	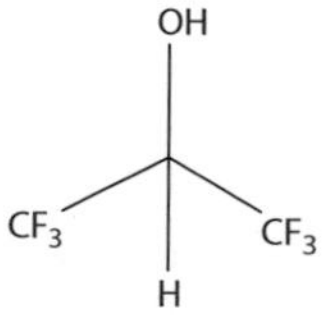
HMDS	Hexamethyldisilazane or bis(trimethylsilyl)amine	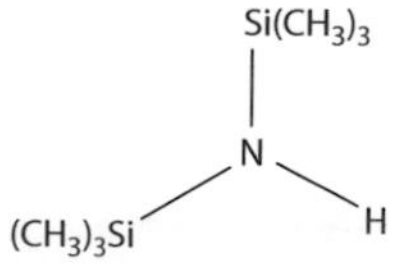
HMPA	Hexamethylphosphoramide	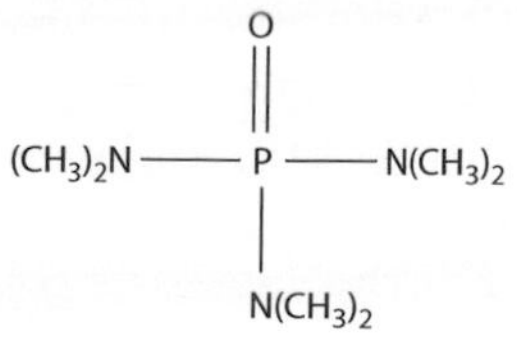
HMPT	Hexamethylphosphoric triamide	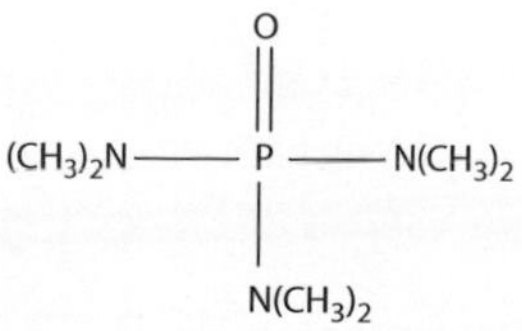

(continued)

(continued)

Acronym	Full Name	Structure
KDA	Potassium diisopropylamide	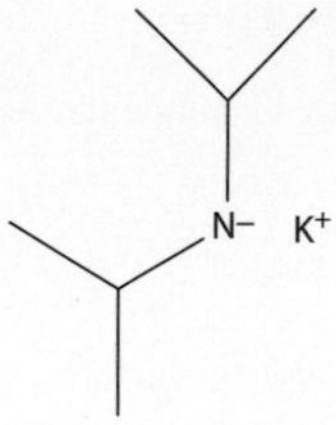
K-Selectride®	Potassium tri-*sec*-butylborohydride	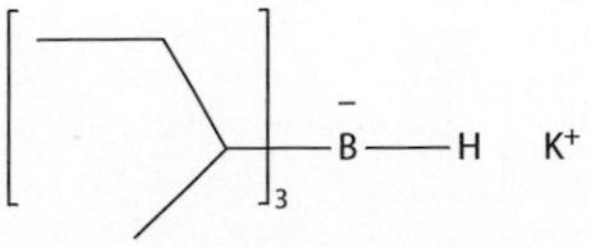
LAH	Lithium aluminum hydride	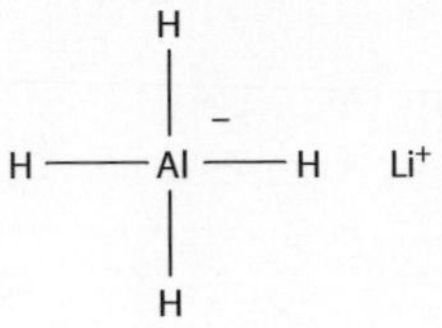
LDA	Lithium diisopropylamide	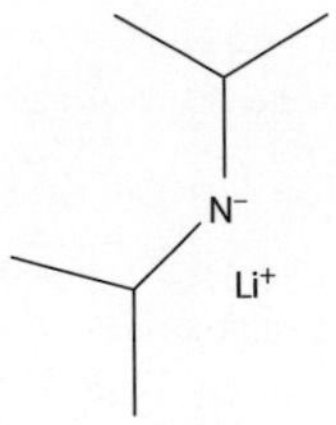
LDEA	Lithium diethylamide	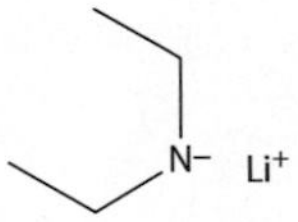
LHMDS	Lithium bis(trimethylsilyl)amide	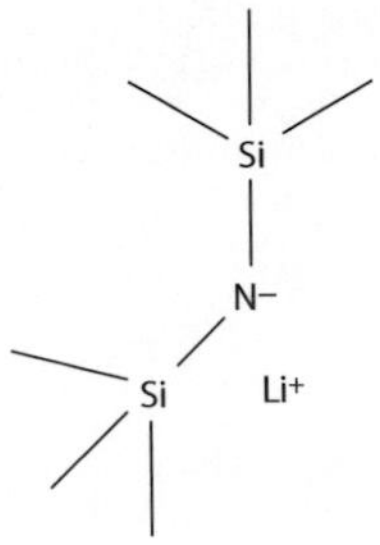
MCPBA	*Meta*-chloroperoxybenzoic acid	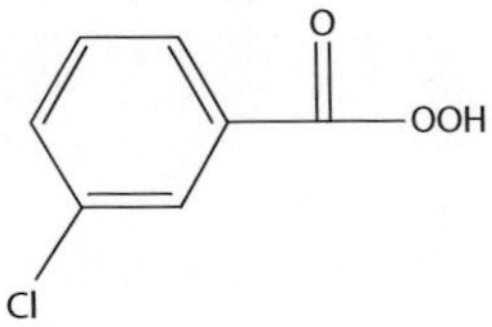
MES	2-Morpholinoethanesulfonic acid	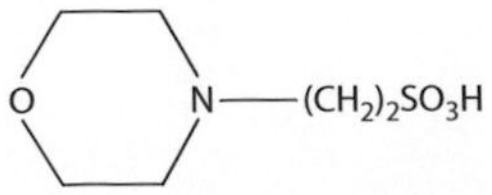

(continued)

Acronym	Full Name	Structure
MET	Methyl ethyl ketone	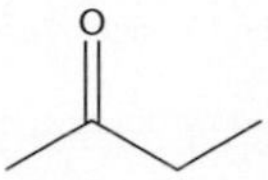
MMTS	Methyl methylthiomethylsulfoxide	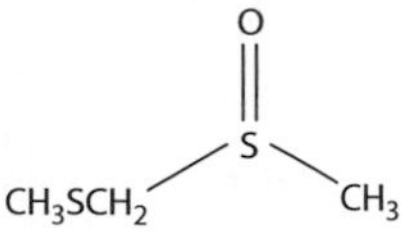 CH_3SCH_2 — S — CH_3
MOPS	3-Morpholinopropanesulfonic acid	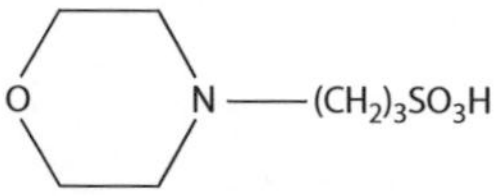 N — $(CH_2)_3SO_3H$
MVK	Methyl vinyl ketone	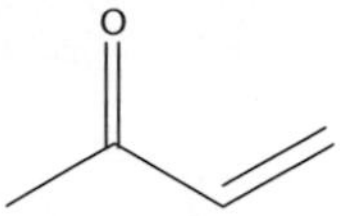
NBS	*N*-Bromosuccinamide	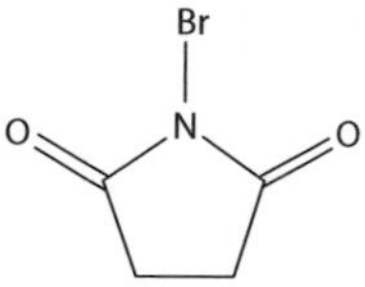
NCS	*N*-Chlorosuccinamide	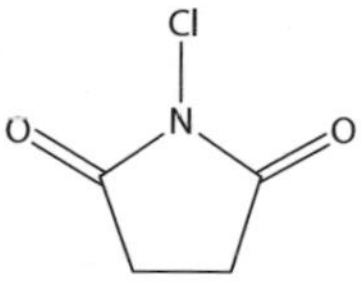
NMM	*N*-Methylmorpholine	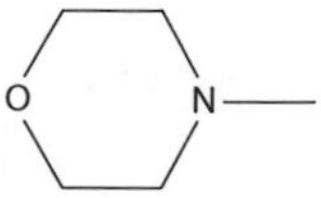
NMO	*N*-Methylmorpholine N-oxide	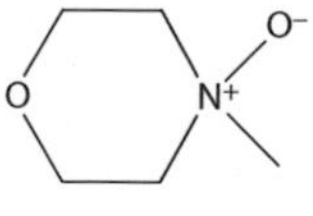
NMP	*N*-Methyl-2-pyrrolidone	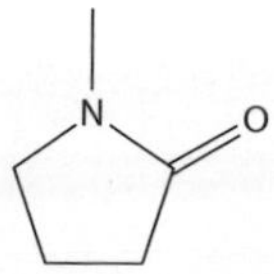
N-PSP	*N*-Phenylselenophthalimide	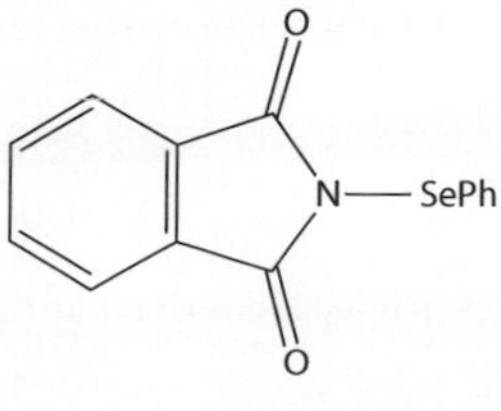 N — SePh

(continued)

(continued)

Acronym	Full Name	Structure
PCC	Pyridinium chlorochromate	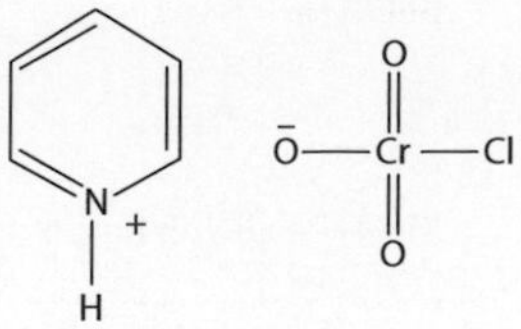
PDC	Pyridinium dichromate	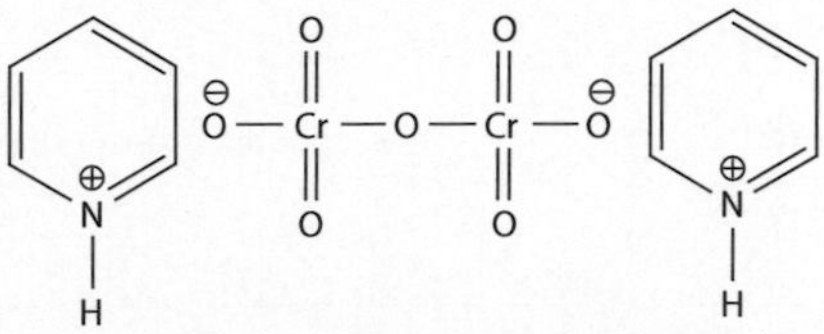
PIPES	Piperazine-1,4-bis(2-ethanesulfonic acid)	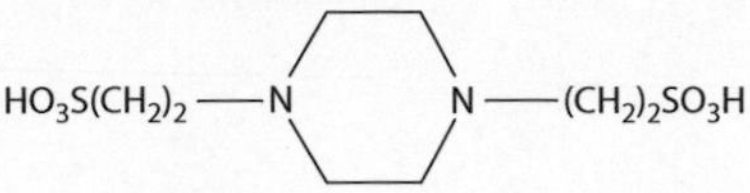
PPA	Polyphosphoric acid	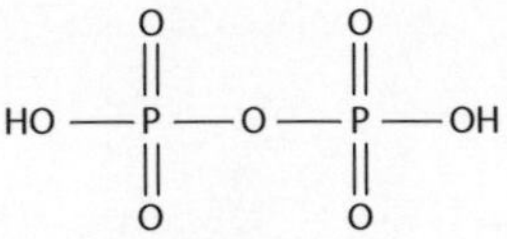 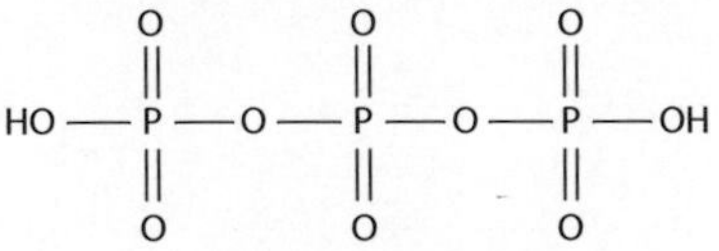
PPE	Polyphosphoric ester	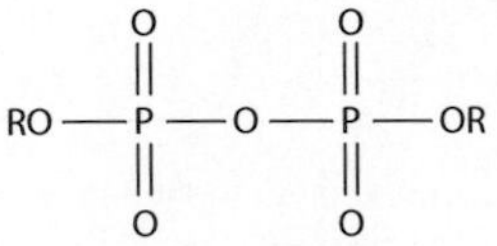 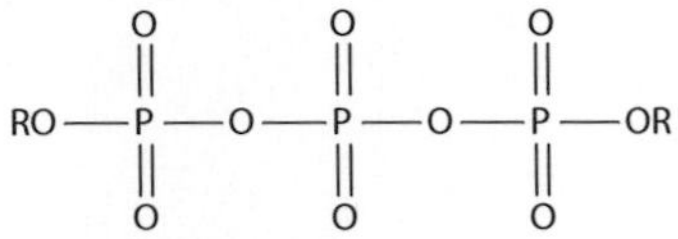
PPTS	Pyridinium p-toluenesulfonate	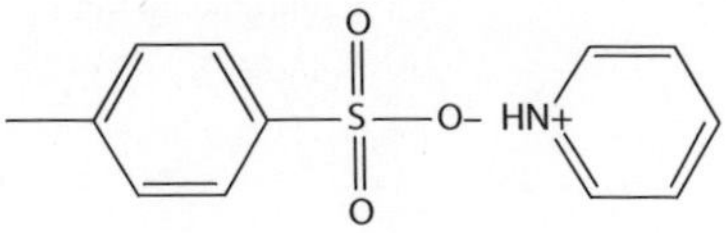
PTSA	p-Toluenesulfonic acid	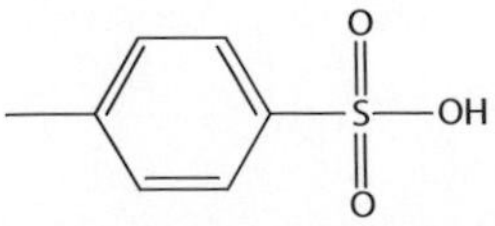

(continued)

Acronym	Full Name	Structure
Pyr or Py	Pyridine	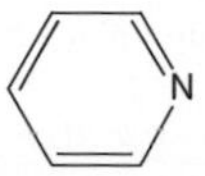
PyBroP	Bromotripyrrolidinophosphonium hexafluorophosphate	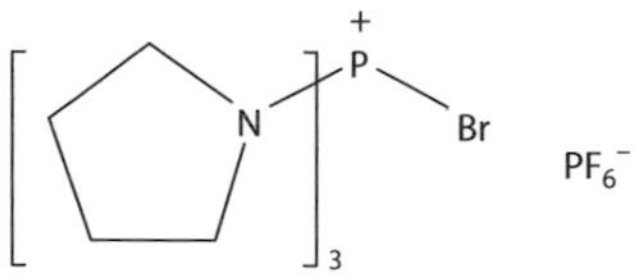
Red-Al®	Sodium bis(2-methoxyethoxy) aluminum hydride	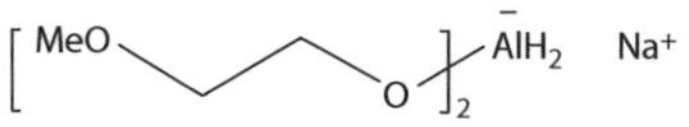
RNA	Ribonucleic acid	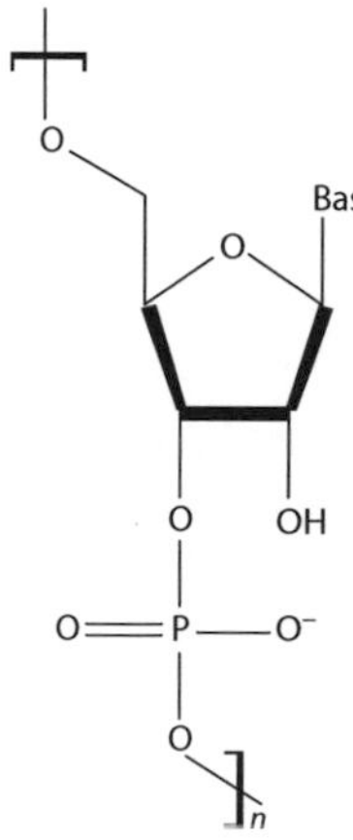
SALEN	N,N'-Disalicylidenethylenediamine, N,N'-bis(salicylidene) ethylenediamine	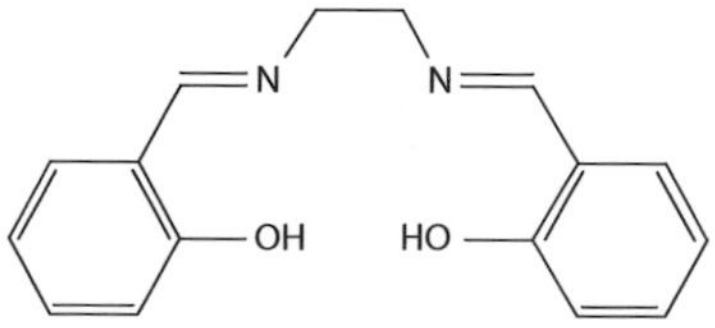
SEMCl	[2-(Trimethylsilyl)ethoxy]methyl chloride	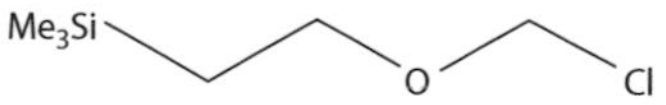
SMEAH	Sodium bis(2-methoxyethoxy) aluminum hydride	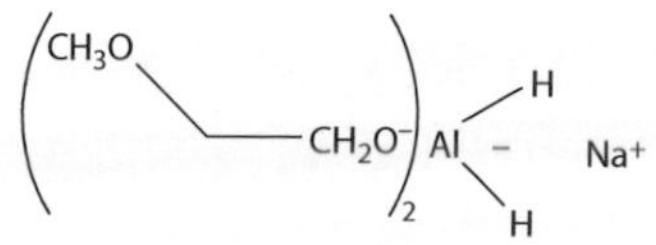
TAPS	N-[Tris(hydroxymethyl)-methyl]-3-aminopropanesulfonic acid	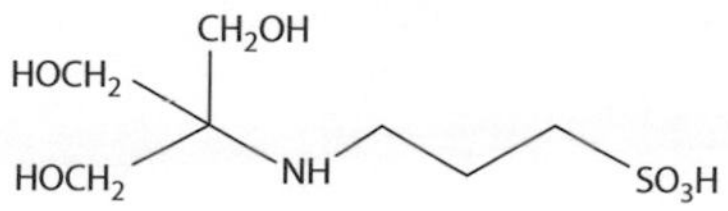
TBAF	Tetrabutylammonium fluoride	$(n\text{-Bu})_4 \overset{+}{N} \ F^-$

(*continued*)

(continued)

Acronym	Full Name	Structure
TCDI	Thiocarbonyldiimidazole	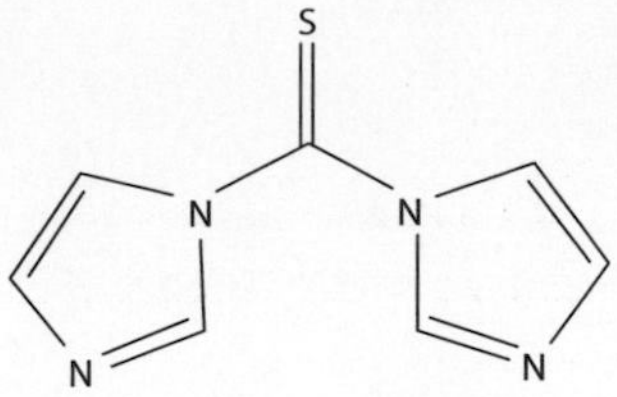
TCNE	Tetracyanoethylene	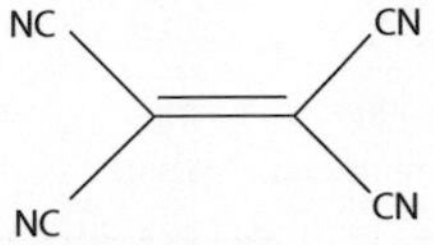
TCNQ	Tetracyanoquinodimethane	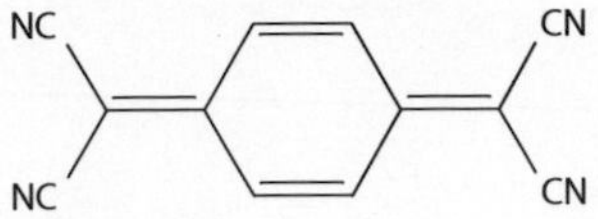
TEA	Triethylamine	Et_3N
TED	Triethylenediamine	$NH_2(CH_2CH_2)_3NH_2$
TEMPO	2,2,6,6-Tetramethylpiperidinooxy, free radical or 2,2,6,6-tetramethylpiperidine N-oxide	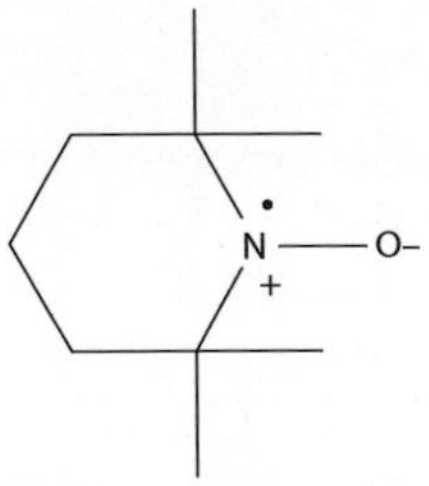
TES	N-[Tris(hydroxymethyl)-methyl]-2-aminoethanesulfonic acid	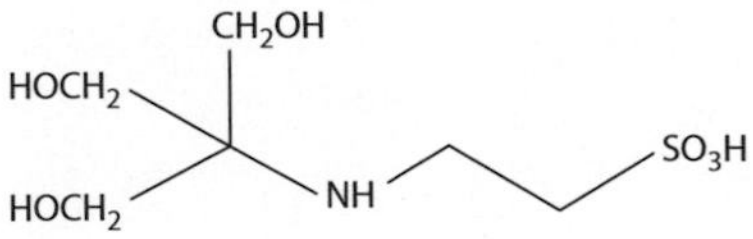
TFA	Trifluoroacetic acid	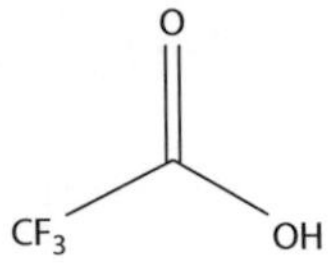
TFAA	Trifluoroacetic anhydride	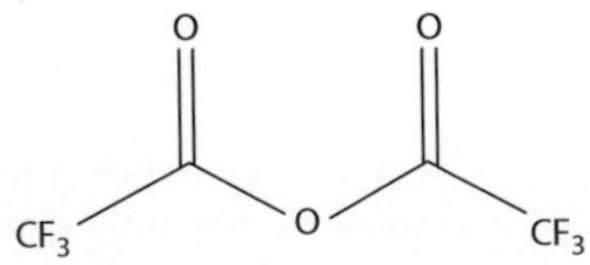
TFE	Trifluoroethanol	CF_3CH_2OH
THF	Tetrahydrofuran	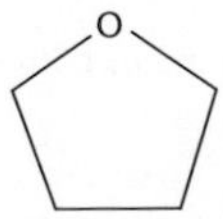

(continued)

Acronym	Full Name	Structure
TMEDA	*N,N,N′,N′-* Tetramethylethylenediamine	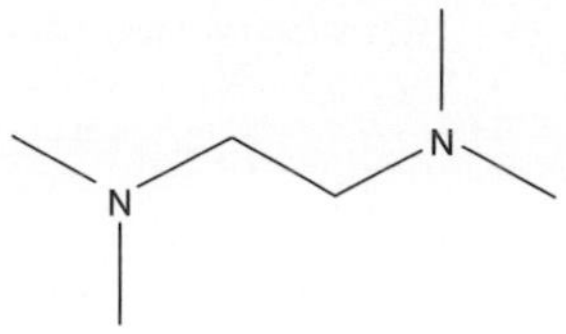
TMCS	Chlorotrimethylsilane	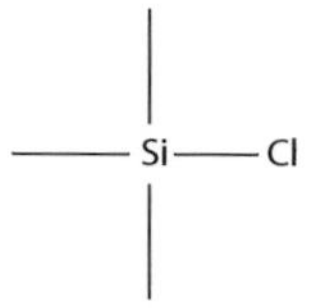
TMG	Tetramethylguanidine	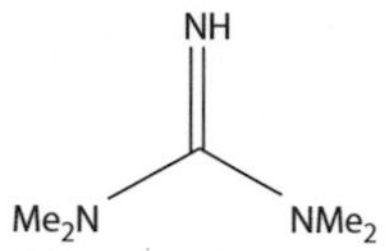
TMNO	Trimethylamine N-oxide monohydrate	$(CH_3)_3 \overset{+}{N} \!-\! \overset{-}{O} \;\bullet\; H_2O$
TMS	Tetramethylsilane	$(CH_3)_4 Si$
TMSCl	Trimethylchlorosilane	$(CH_3)_3 SiCl$
TMSOTf	Trimethylsilyl trifluoromethanesulfonate	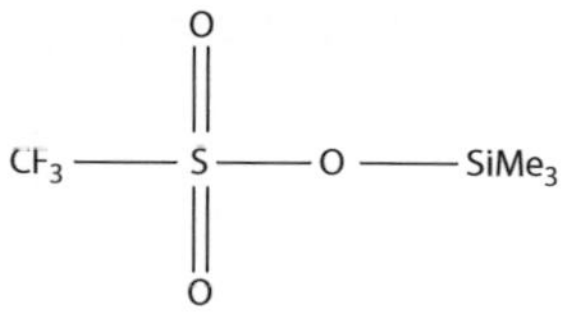
TNT	Trinitrotoluene	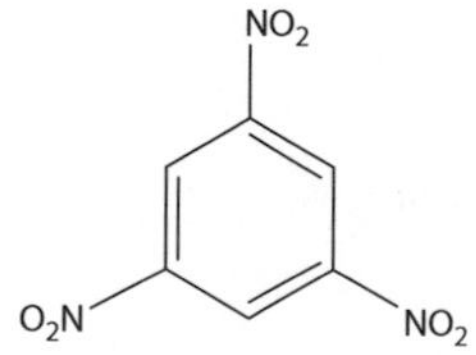
TPAP	Tetra-*n*-propylammonium perruthenate	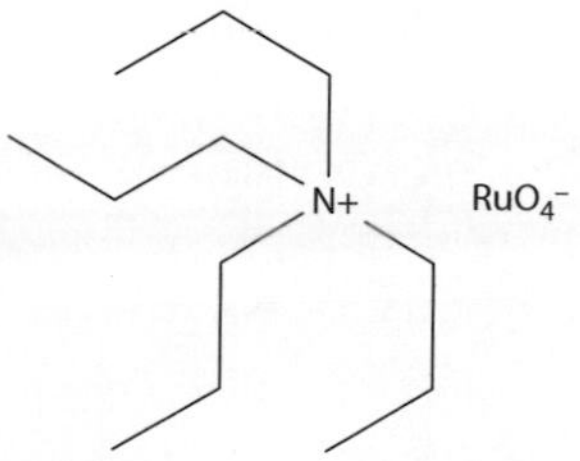
TPP	Triphenylphosphine	$Ph_3 P$
TPP-DEAD	Triphenylphosphine-DEAD	$Ph_3 P - EtOOCN = NCOOEt$
TPPO	Triphenylphosphine oxide	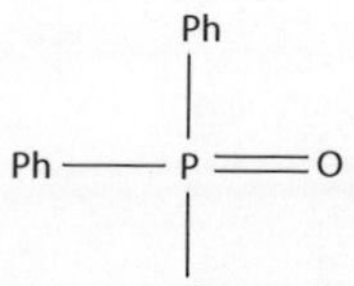

(*continued*)

(continued)

Acronym	Full Name	Structure
Tricine	*N*-[Tris(hydroxymethyl)-methyl]-glycine	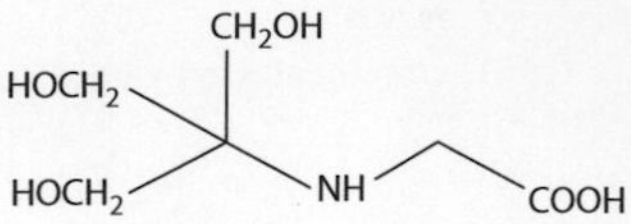 CH₂OH, HOCH₂, HOCH₂, NH, COOH structure
TRIS	Tris(hydroxymethyl)aminomethane	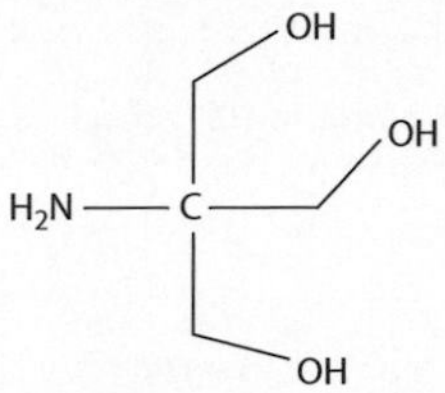 H₂N—C with three CH₂OH / OH groups
Troc-Cl	2,2,2-Trichloroethylchloroformate	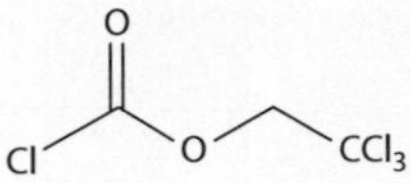 Cl—C(=O)—O—CH₂—CCl₃
Viltride®	Sodium bis(2-methoxyethoxy) aluminum hydride	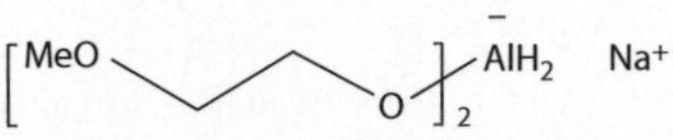 [MeO—CH₂CH₂—O]₂AlH₂⁻ Na⁺

A.3 LIST OF ABBREVIATIONS FOR FUNCTIONAL GROUPS

Acronym	Full Name	Structure
Ac	Acyl, acetyl	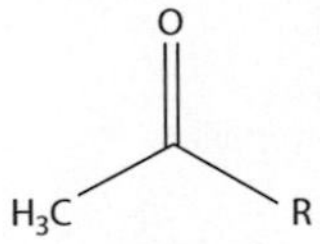 H₃C—C(=O)—R
AcAc	Acetoacetyl	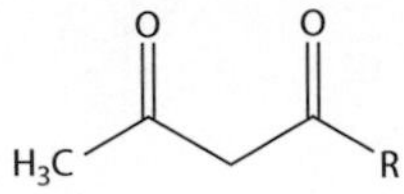 H₃C—C(=O)—CH₂—C(=O)—R
All	Allyl	 CH₂=CH—CH₂—R
Alloc	Allyloxycarbonyl	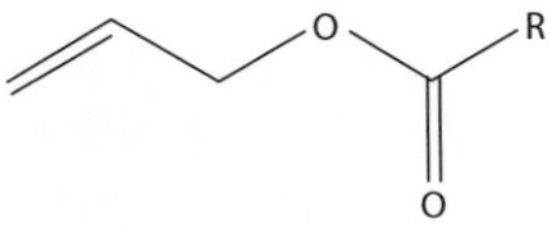 CH₂=CH—CH₂—O—C(=O)—R
Ar	Aryl	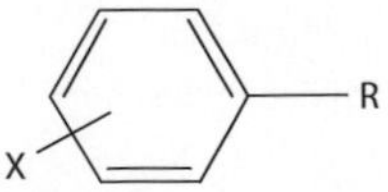 aromatic ring with X and R
Bn	Benzyl	PhCH₂R
Boc	*t*-Butoxycarbonyl	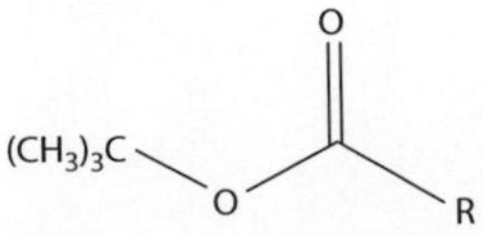 (CH₃)₃C—O—C(=O)—R

(continued)

Acronym	Full Name	Structure
Bs	Brosyl	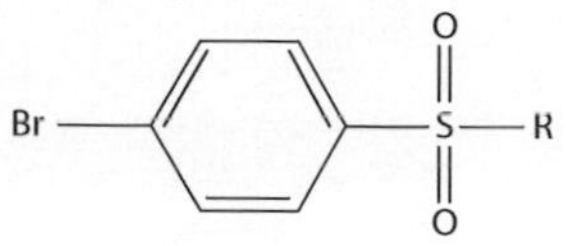
n-Bu	Butyl	$CH_3CH_2CH_2CH_2R$
Bz	Benzoyl	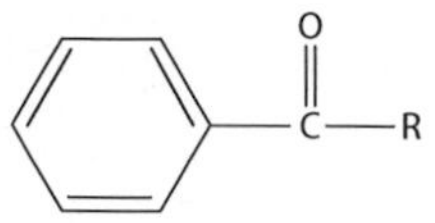
Cbz	Benzyloxycarbonyl	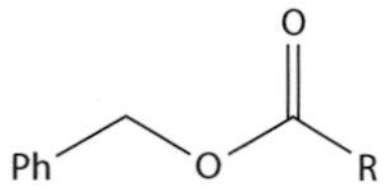
Cp	Cyclopentadienyl	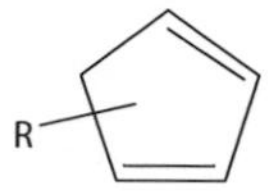
Dan	Dansyl	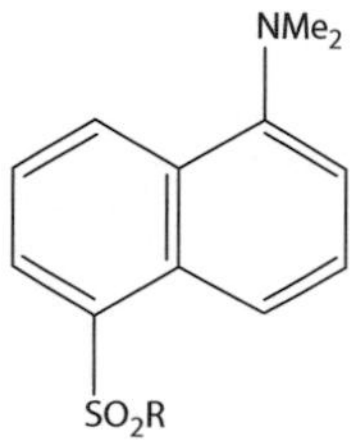
Dip	2,6-Diisopropylphenyl	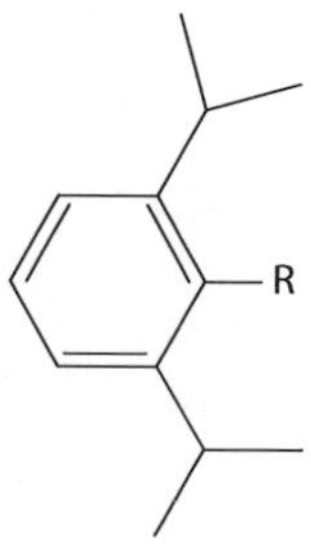
DMPM	3,4-Dimethoxybenzyl	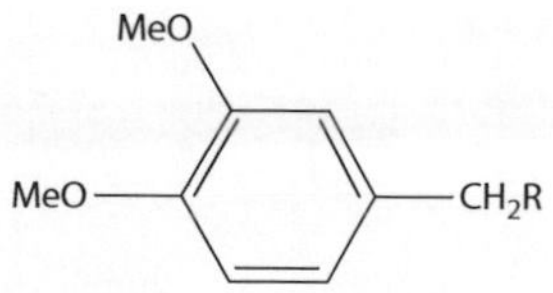
DNP	2,4-Dinitrophenyl	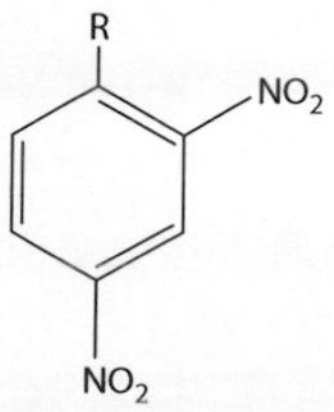

(*continued*)

(continued)

Acronym	Full Name	Structure
EE	1-Ethoxyethyl	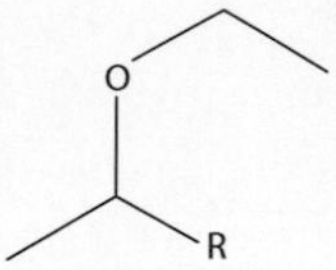
Et	Ethyl	CH_3CH_2R
FMOC	Fluorenylmethyloxycarbonyl	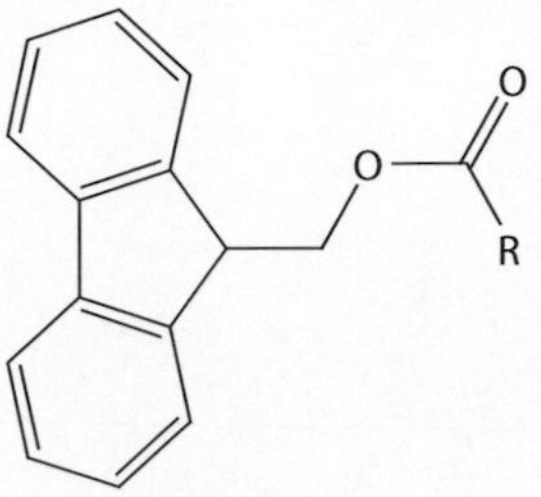
i-	*iso-* or *ipso*	
i-Bu	*iso*-butyl	$(CH_3)_2CH\ CH_2R$
i-Pr	*iso*-propyl	$(CH_3)_2CHR$
m-	*Meta-*	
Me	Methyl	CH_3R
MEM	2-Methoxyethoxymethyl	$CH_3OCH_2CH_2OCH_2R$
Mes	Mesityl	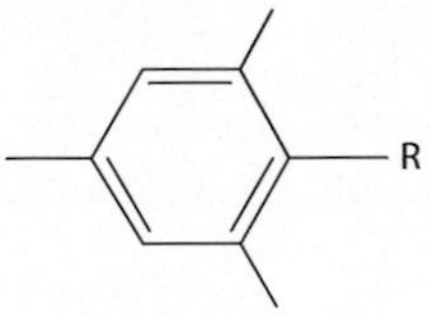
Ms	Mesyl or methanesulfonyl	CH_3SO_2R
MOM	Methoxymethyl	CH_3OCH_2R
MPM	*p*-Methoxybenzyl	
o-	*Ortho-*	
n-	*Normal-*	
p-	*Para-*	
Ns	*p*-Nitrophenylsulfonyl	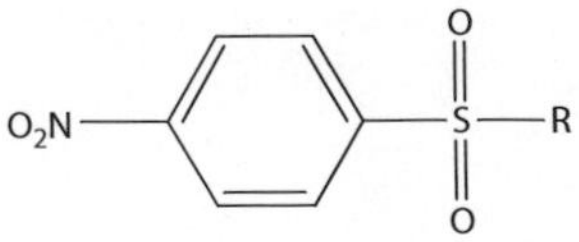

(continued)

Acronym	Full Name	Structure
PE	Polyethylene	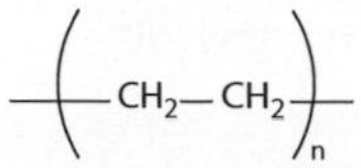 $-(CH_2-CH_2)_n-$
PEG	Polyethyleneglycol	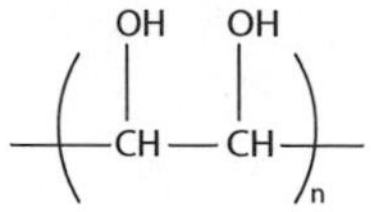 $-(CH(OH)-CH(OH))_n-$
Ph	Phenyl	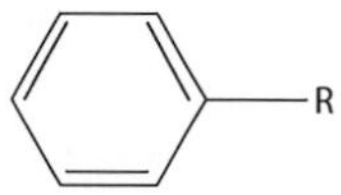 C_6H_5-R
Piv, Pv	Pivaloyl	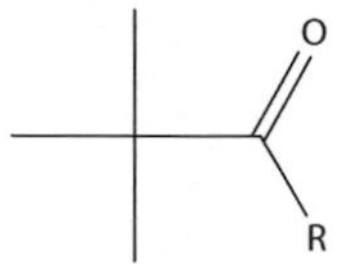 $(CH_3)_3C-C(=O)-R$
PMB	*p*-Methoxybenzyl	 $MeO-C_6H_4-CH_2R$
PMP	*Para*-methoxyphenyl	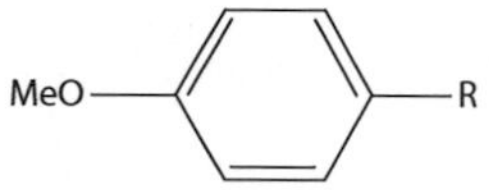 $MeO-C_6H_4-R$
Pr	Propyl	$CH_3CH_2CH_2$
PVA	Polyvinylalcohol	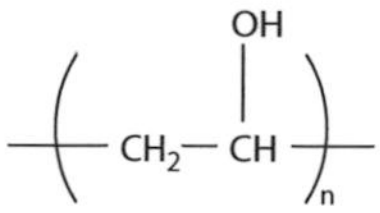 $-(CH_2-CH(OH))_n-$
PVC	Polyvinylchloride	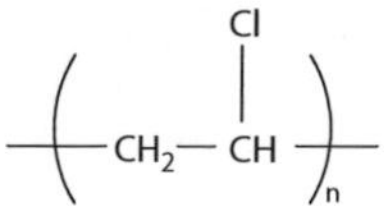 $-(CH_2-CH(Cl))_n-$
R	Any group	
s-Bu	*Sec*-butyl	$CH_3CH_2CH(CH_3)R$
SEM	[2-(Trimethylsilyl) ethyl]	 $Me_3Si-CH_2CH_2-R$
t-	*Tert*-	
t-Boc	*Tert*-butoxycarbonyl	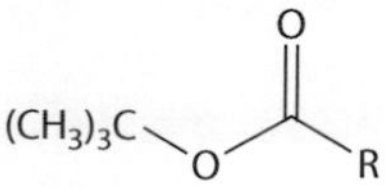 $(CH_3)_3C-O-C(=O)-R$
t-Bu	*Tert*-butyl	$(CH_3)_3CR$

(*continued*)

(continued)

Acronym	Full Name	Structure
TBDPS	*t*-Butyldiphenylsilyl	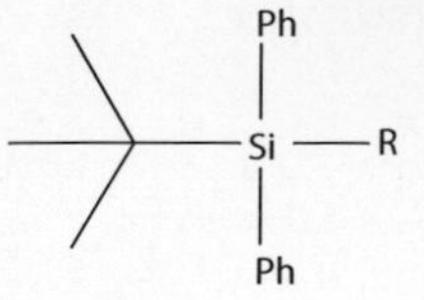
TBS	*t*-Butyldimethylsilyl	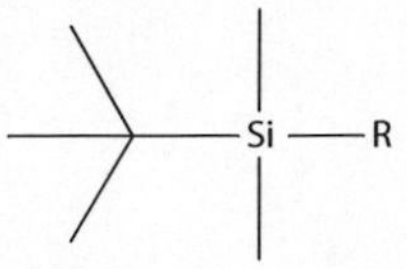
Tbt	2,4,6-Tris[bis(trimethylsilyl)methyl]phenyl	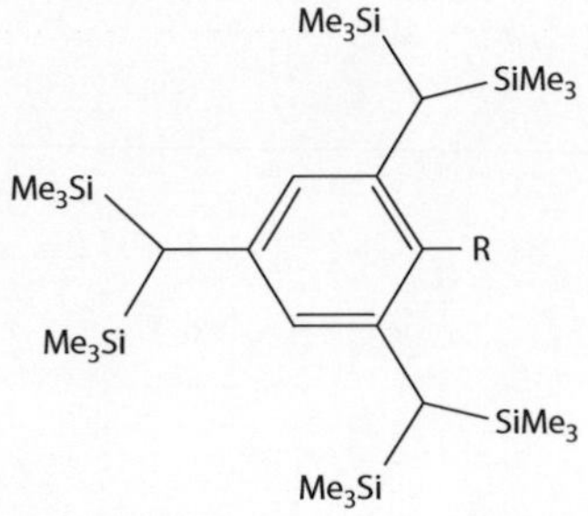
Tf	Triflate, triflyl, trifluoromethylsulfonyl	CF_3SO_2R
Thex	Thexyl	$(CH_3)_2CHC(CH_3)_2R$
THP	Tetrahydropyranyl	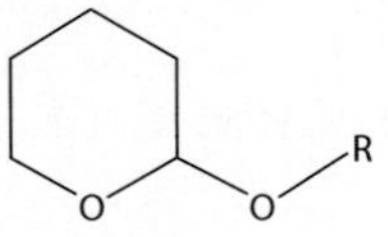
TIPS	Triisopropylsilyl	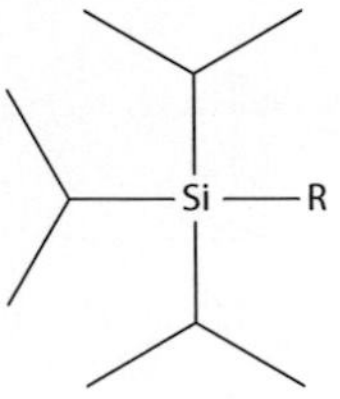
TMS	Trimethylsilyl	$(CH_3)_3 SiR$
Tpt	Tryptyl	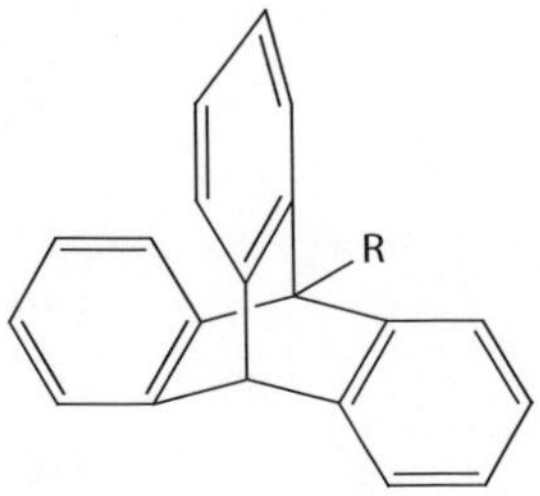
Troc	2,2,2-Trichloroethylformate	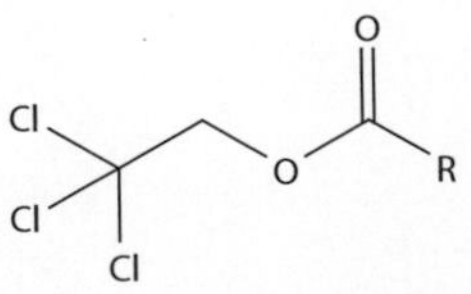

(continued)

Acronym	Full Name	Structure
Try, Tr	Trityl, triphenylmethyl	Ph$_3$CR
Ts	Tosyl	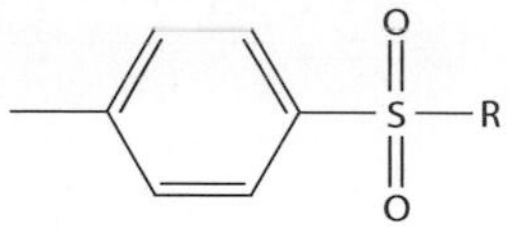
X	Halide group	F, Cl, Br, I
Z	Benzyloxycarbonyl	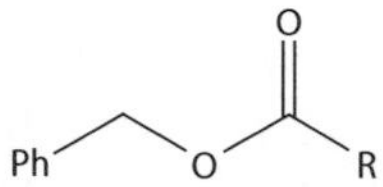

A.4   LIST OF CHIRAL AUXILIARIES USED IN SYNTHESIS PLANS

Target Product	Plan (Step Number)	Structure of Chiral Auxiliary
(−)-Adociacetylene B	Trost (step 4)	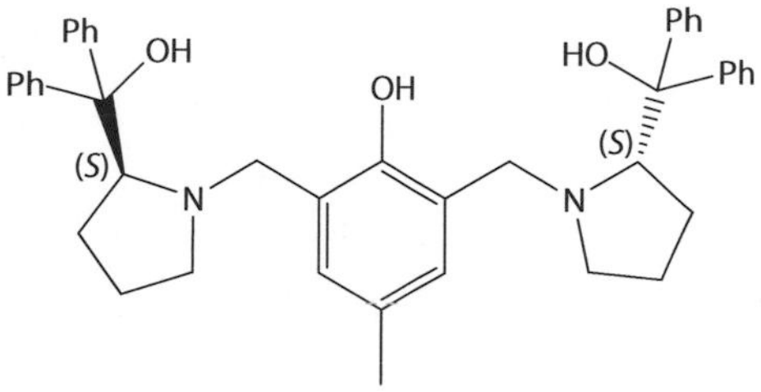
(−)-Codeine, (−)-Morphine	Trost (step 3)	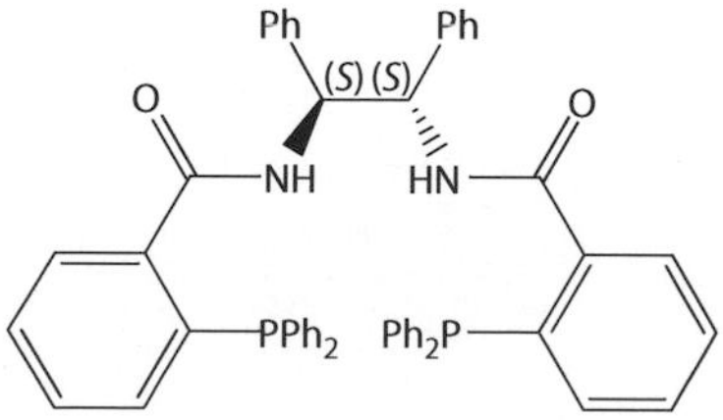
(+)-Codeine	White (step 3)	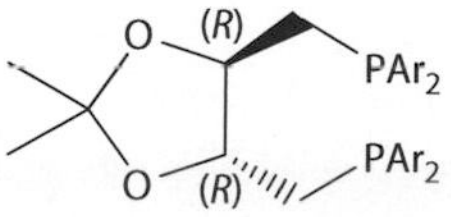 Ar = 3,4,5-Trimethoxyphenyl
(S)-Dapoxetine	Srinivasan (step 1)	(DHQ)$_2$PHAL 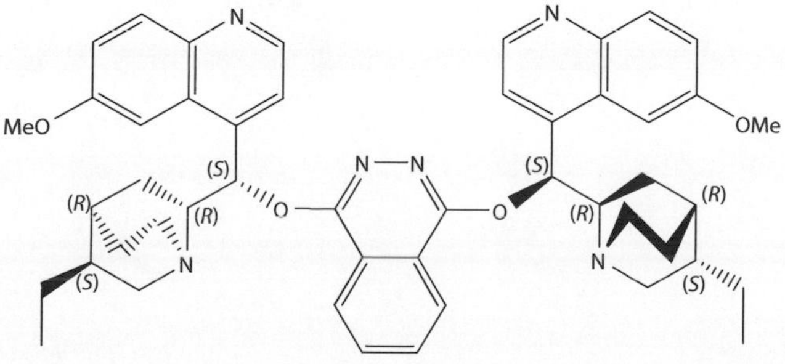

(continued)

(continued)

Target Product	Plan (Step Number)	Structure of Chiral Auxiliary
(−)-Epinephrine	Carreira (step 1)	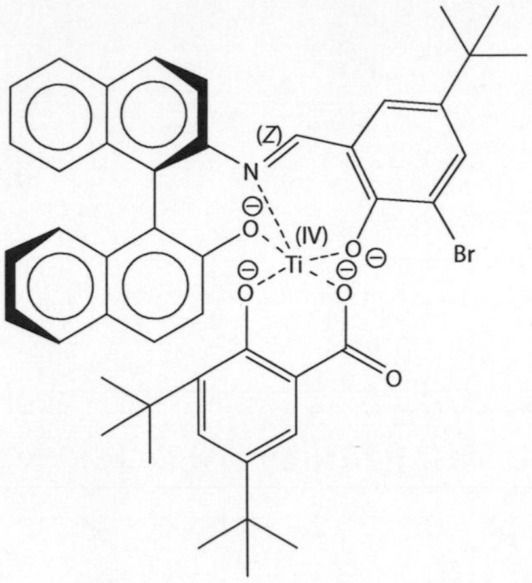
(+)-Estrone	Corey (step 6)	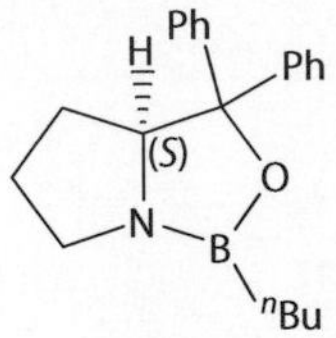
(R)-Fluoxetine hydrochloride	Sharpless (step 3)	(+)-Diisopropyltartrate
(−)-Kainic acid	Bachi (step 4)	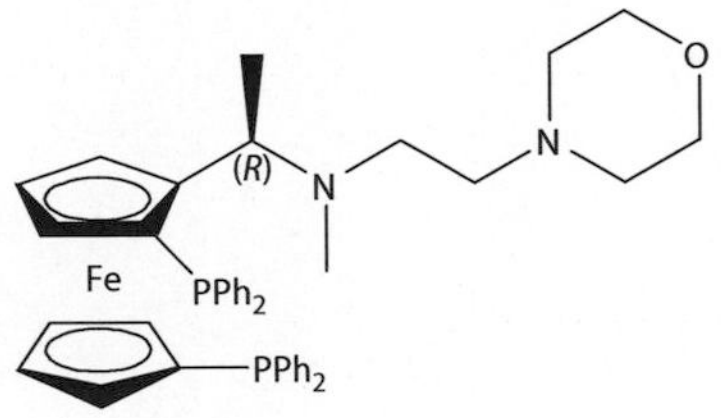
(−)-Kainic acid	Fukuyama G1 (step 11)	(+)-Sparteine
(−)-Kainic acid	Poisson G2 (step 1)	(+)-Diisopropyltartrate
(−)-Kainic acid	Takano G3 (step 3)	(+)-Diisopropyltartrate
(+)-Kainic acid	Trost (step 3)	(R)-BINOL
(−)-Menthol	Takasago G1 (step 2)	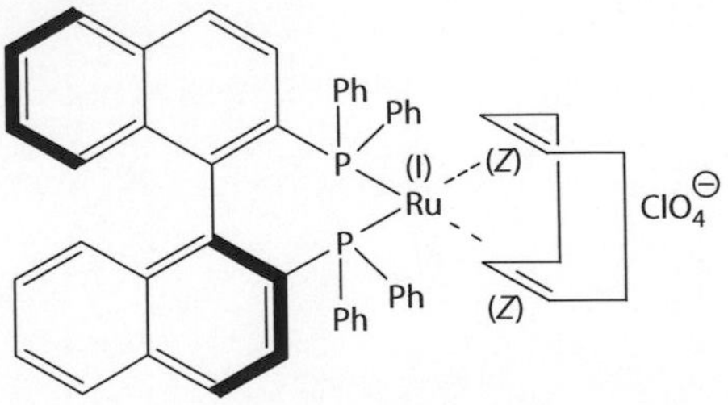

[Rh((S)-(−)-BINAP)(cod)]ClO₄

(continued)

Target Product	Plan (Step Number)	Structure of Chiral Auxiliary
(S)(−)-Metolachlor	Ciba-Syngenta (step 1)	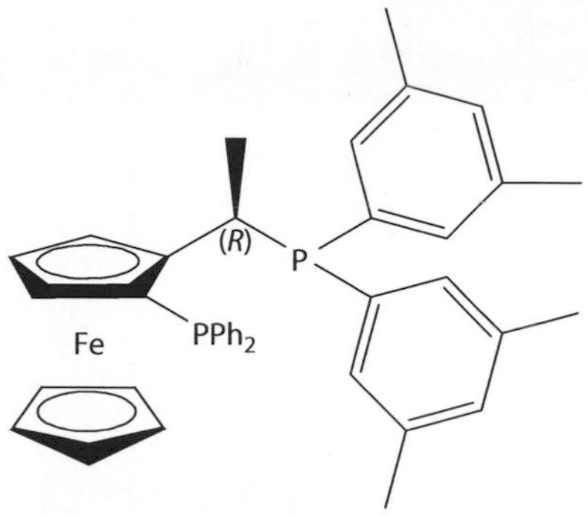

(R)-Nicotine	Campos (step2)	(−)-Sparteine (by conjecture)
Oseltamivir phosphate	Corey (step 1)	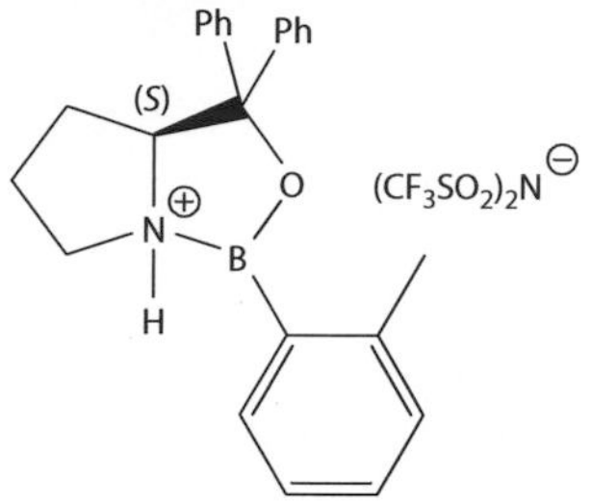

Oseltamivir phosphate	Fukuyama (step 2)	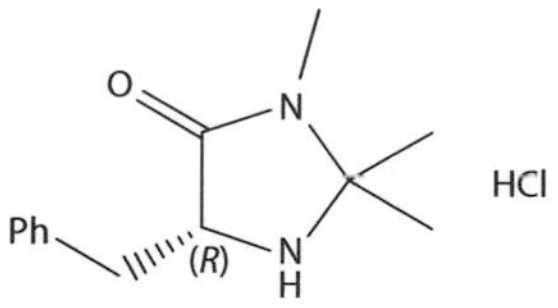

Oseltamivir phosphate	Hayashi (step 3)	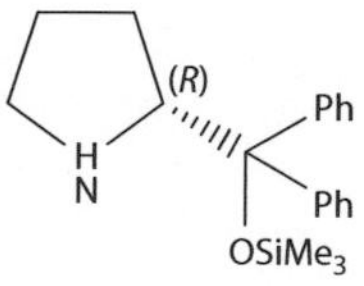

Oseltamivir phosphate	Mandai G2 (step 10)	

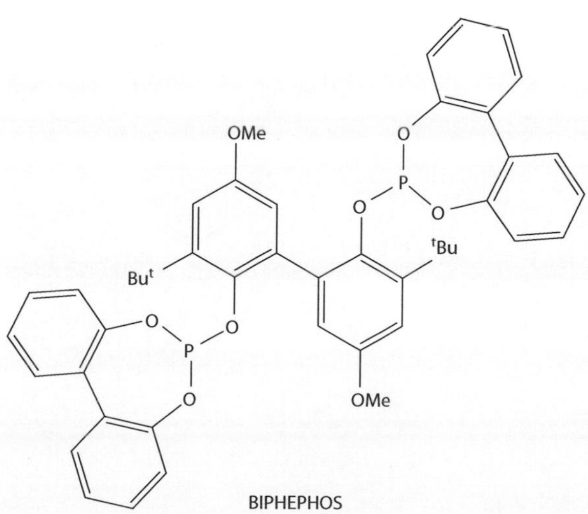

BIPHEPHOS

(continued)

(continued)

Target Product	Plan (Step Number)	Structure of Chiral Auxiliary
Oseltamivir phosphate	Shibasaki G1 (step 4)	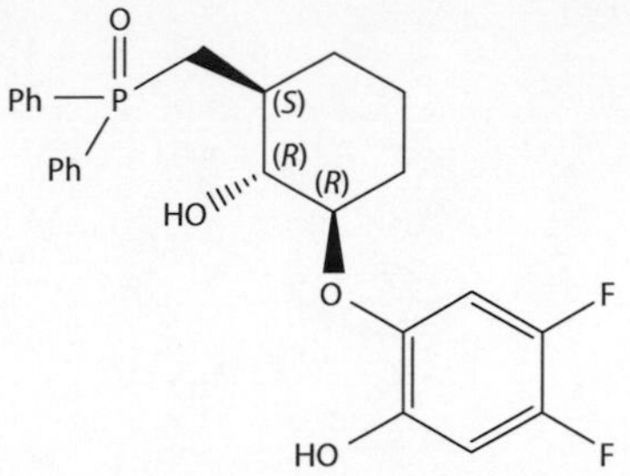
Oseltamivir phosphate	Shibasaki G4 (step 1)	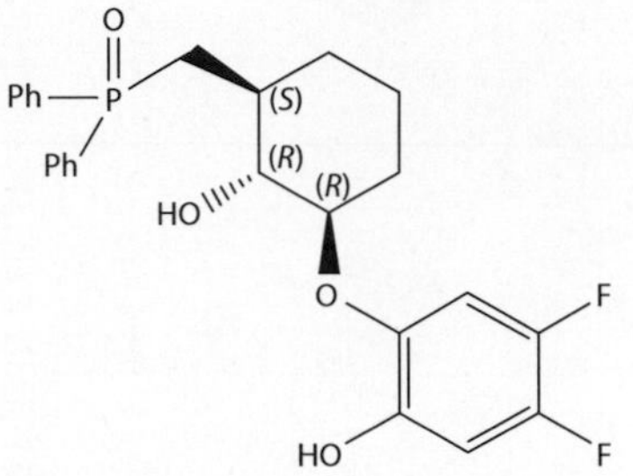
Oseltamivir phosphate	Shibasaki G5 (step 1)	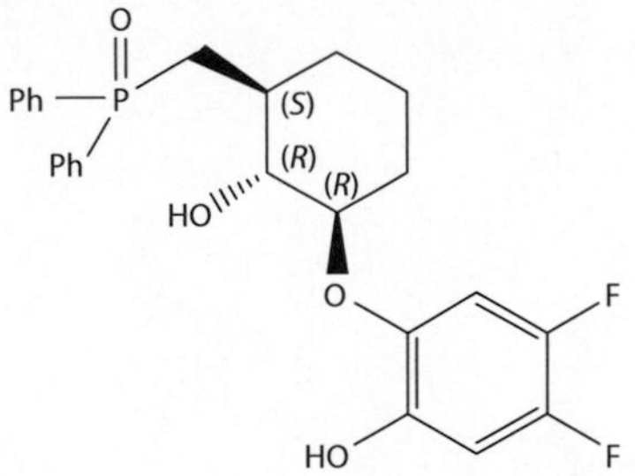
Oseltamivir phosphate	Trost (step 4)	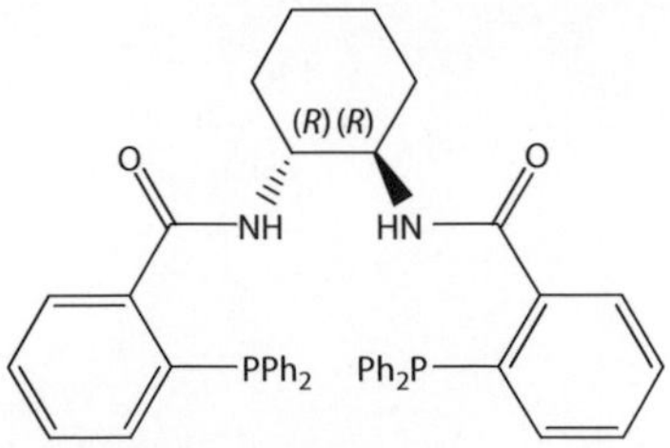
Oseltamivir phosphate	Trost (step 7)	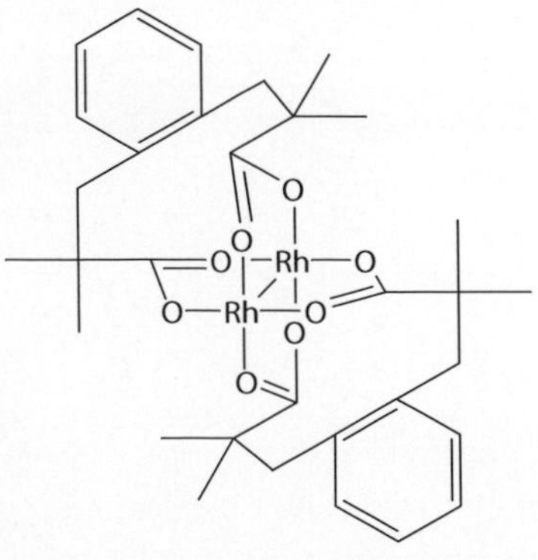

(continued)

Target Product	Plan (Step Number)	Structure of Chiral Auxiliary
(−)-Paroxetine	Beak (step 5)	(−)-Sparteine
(−)-Paroxetine	Buchwald (step 5)	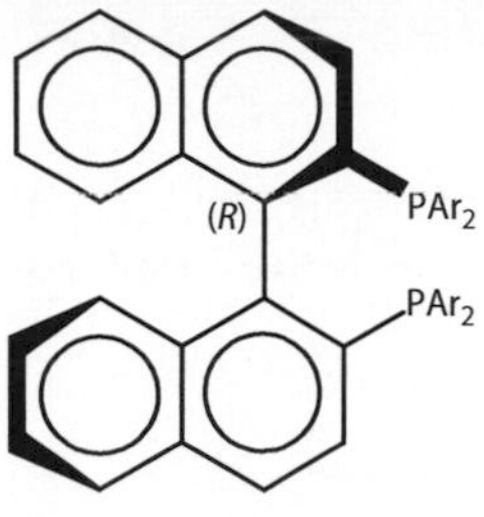

Ar = p–CH₃–C₆H₄–

(−)-Paroxetine	Dixon (step 2)	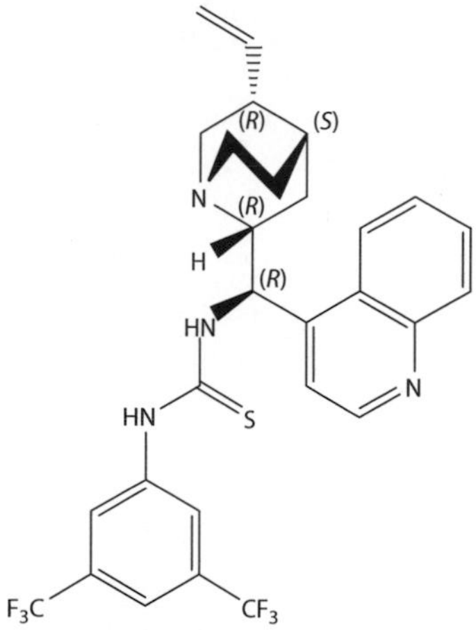

(−)-Paroxetine hydrochloride	Gallagher (step 2)	

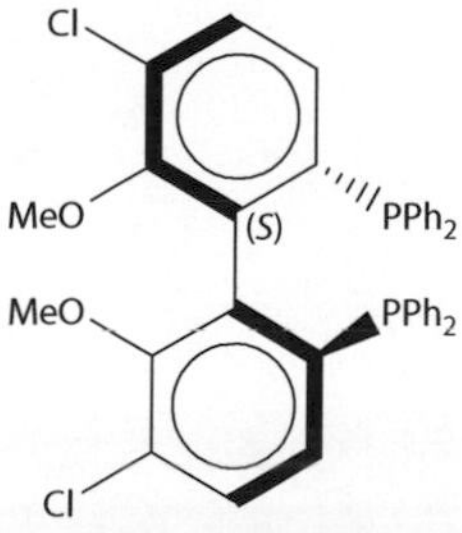

(S)–Cl–MeO–BIPHEP

(−)-Paroxetine	Hayashi (step 4)	

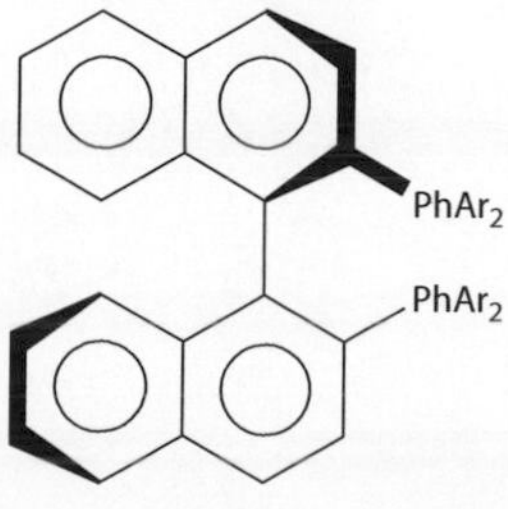

Ar = 3,5-Dimethyl-4-methoxyphenyl

(continued)

(continued)

Target Product	Plan (Step Number)	Structure of Chiral Auxiliary
(−)-Paroxetine	Jorgensen (step 2)	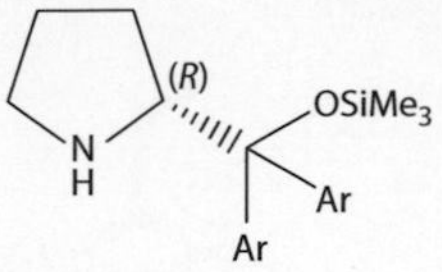 Ar = 3,5-Bis(trifluoromethyl)phenyl
(−)-Paroxetine hydrochloride	Krische (step 8)	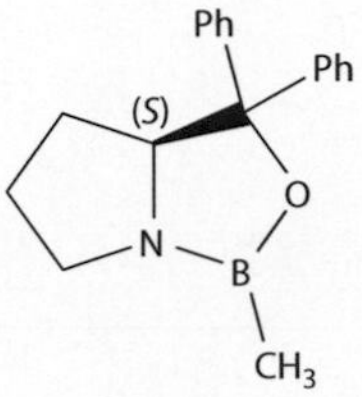
(−)-Paroxetine	Simpkins (step 6)	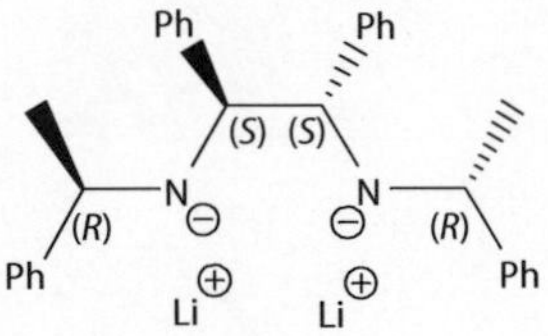
(−)-Paroxetine	Vesely–Moyano–Rios (step 5)	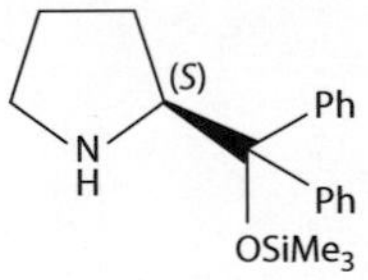
Physostigmine	Trost (step 3)	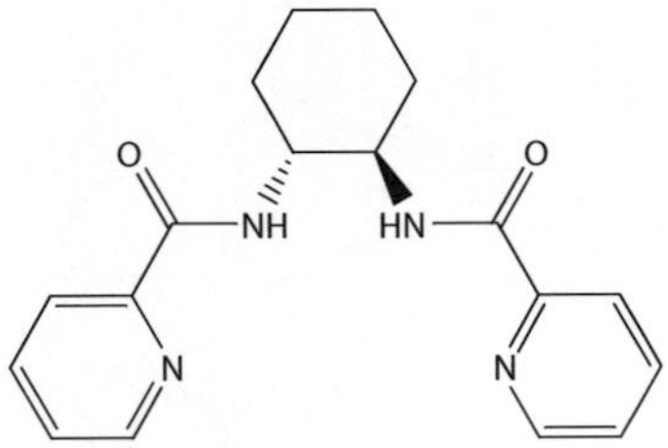
(−)-Platensimycin	Corey (step 16)	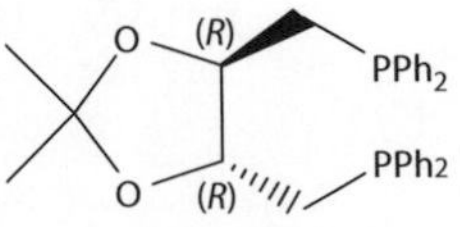 (R,R)–DIOP

(continued)

Target Product	Plan (Step Number)	Structure of Chiral Auxiliary
(−)-Platensimycin	Yamamoto (step 1)	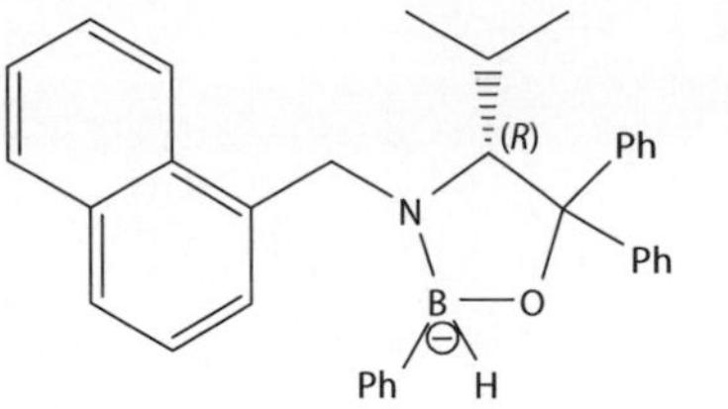
(S)-Pregabalin	DowPharma-Pfizer (step 5)	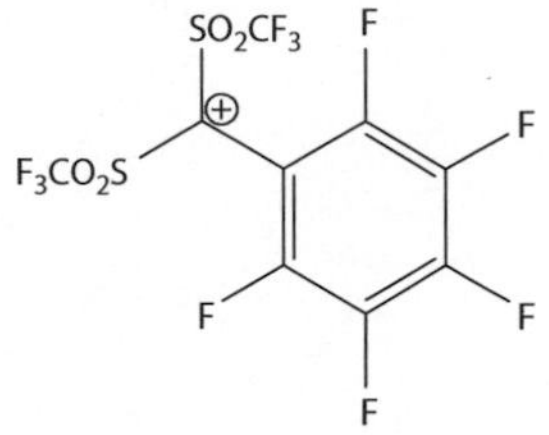 [(R,R)–(Me–DuPHOS)Rh(cod)]⁺ BF₄⁻
(S)-Pregabalin	Jacobsen (step 4)	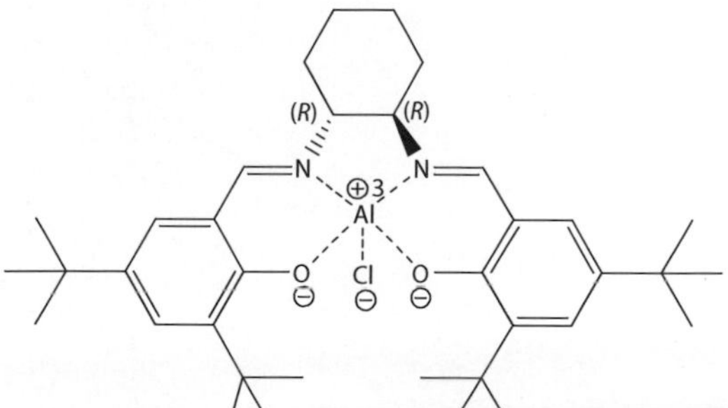

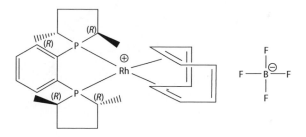

[(R,R)–(Me–DuPHOS)Rh(cod)]⁺ BF₄⁻

(*continued*)

(continued)

Target Product	Plan (Step Number)	Structure of Chiral Auxiliary
(S)-Pregabalin	Koskinen (step 3)	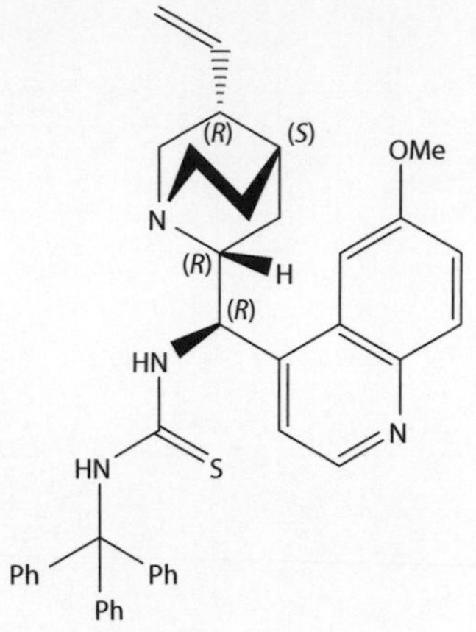
(S)-Pregabalin	Shibasaki (step 3)	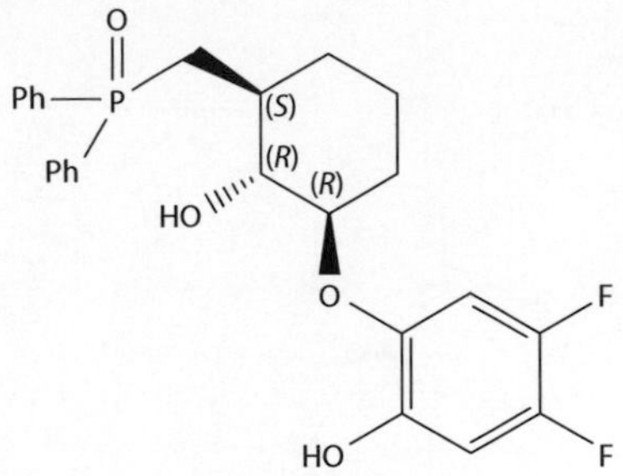
(+)-Sparteine	Aube (step 1)	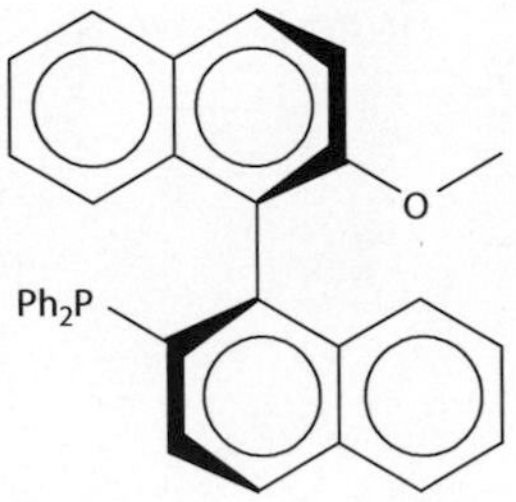 (S)–MeO–MOP
(−)-Swainsonine	Blechert (step 2)	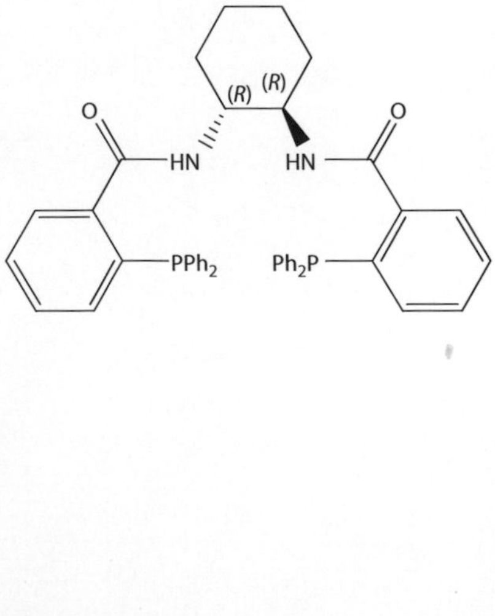

(S)–MeO–MOP

(continued)

Target Product	Plan (Step Number)	Structure of Chiral Auxiliary
(−)-Swainsonine	Trost (step 4)	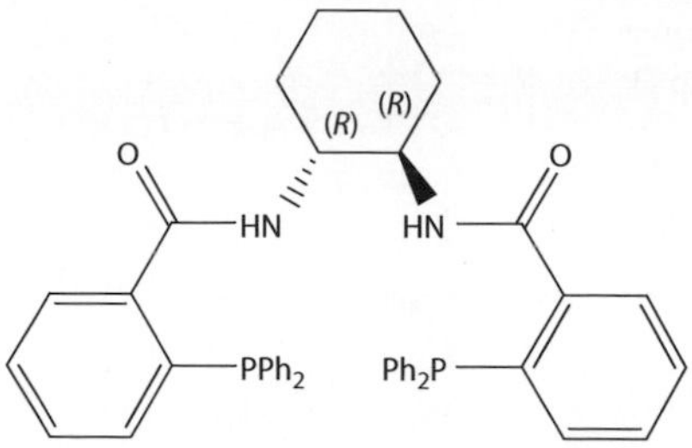

(−)-Swainsonine	Zhou (step 3)	(−)-Diisopropyltartrate
Vitamin E	Tietze (step 7)	

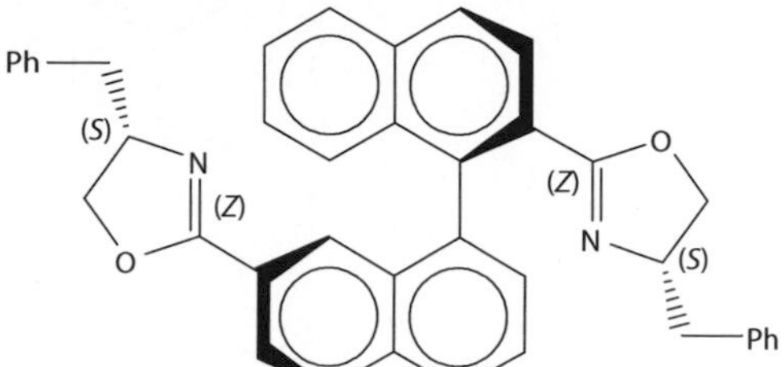

(S,S)–Bn–BOXAX

Vitamin E	Woggon (step 5)	

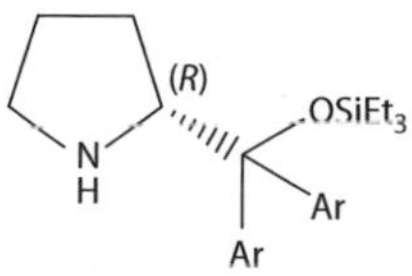

Ar = 3′,5′-Ditrifluoromethylphenyl

(S)-Warfarin	DuPont-Merck G1 (step 4)	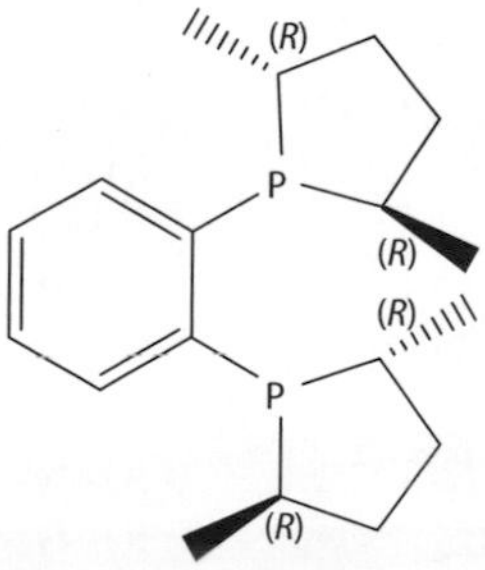

(S)-Warfarin	DuPont-Merck G2 (step 4)	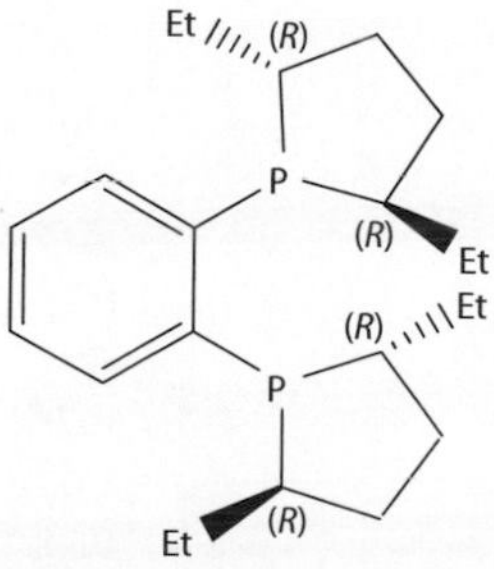

(*continued*)

(continued)

Target Product	Plan (Step Number)	Structure of Chiral Auxiliary
(S)-Warfarin	Jorgensen (step 1)	
(S)-Warfarin	Sodeoka (step 4)	

A.5 LIST OF PLANS WITH STEADY BUILDING UP MW PROFILE

Target	Plan	Year	Type	μ_1 (g/mol/Stage)
(±)-*Trans*-gamma-irone	Brenna G1	2001	Linear	−41.2
(±)-Fenchone	Buchbauer	1981	Linear	−72.64
(R)-Fluoxetine	Sepracor	2001	Linear	−162.17
(R)-Fluoxetine hydrochloride	P Kumar	2004	Linear	−161.45
(R)-Fluoxetine hydrochloride	Brown	1988	Convergent	−216.06
(R)-Fluoxetine hydrochloride	Corey	1989	Linear	−183.65
(R)-Fluoxetine hydrochloride	Lilly G2	1994	Linear	−169.67
(R)-Fluoxetine hydrochloride	Sharpless	1988	Linear	−187.45
(R)-Fluoxetine hydrochloride	Shibasaki	2004	Linear	−203.74
(S)-Fluoxetine hydrochloride	Wang	2005	Linear	−214.62
2,4-D	Dolge	1941	Linear	−88
2,4-D	Manske	1949	Linear	−161.57
2,4-D	Haskelberg G2	1947	Linear	−161.57
2,4-D	ICI	1945	Linear	−88
Acetaminophen	Vogel	1978	Linear	−39.03
Acetaminophen	Warner-Lambert	1961	Linear	−113
Acetaminophen	Pearson	1953	Linear	−41
Adipic acid	Faith G1	1975	Linear	−90
Adipic acid	Faith G2	1975	Linear	−90
Adipic acid	Ellis	1941	Linear	−79
Adipic acid	Vogel	1978	Linear	−33
Allantoin	Grimaux	1877	Linear	−61
Aniline yellow	Vogel	1948	Linear	−46.33
Anthraquinone	Faith G1	1966	Linear	−84
Anthraquinone	Faith G2	1966	Linear	−91.33

(continued)

Target	Plan	Year	Type	μ_1 (g/mol/Stage)
Antipyrine	Vogel	1978	Linear	−111.5
Aspirin	Faith P1	1966	Convergent	−151.64
Aspirin	Faith P2	1966	Convergent	−176.47
Aspirin	Faith P3	1966	Convergent	−157.64
Aspirin	Faith P4	1966	Convergent	−156.64
Aspirin	Faith P5	1966	Convergent	−154.08
Aspirin	Faith P6	1966	Convergent	−178.8
Aspirin	Faith P7	1966	Convergent	−208.6
Aspirin	Faith P8	1966	Convergent	−186
Aspirin	Faith P9	1966	Convergent	−187.6
Aspirin	Faith P10	1966	Convergent	−181.73
Aspirin	Vogel	1978	Linear	−123.7
Atrazine	Geigy	1960	Linear	30.07
Barbituric acid	Dickey-Gray	1943	Linear	−18
Benfluralin	Lilly	1966	Linear	−127.45
Bifenazate	CK Witco	2001	Linear	−116.16
Bifenazate	Felpin	2008	Linear	−175.44
Bifenazate	Sajiki	2007	Linear	−77.98
Bifenazate	Uniroyal G1	1993	Linear	−102.19
Bisphenol A	Faith G1	1966	Convergent	−248.5
Bisphenol A	Faith G2	1966	Convergent	−213.05
Bisphenol A	Faith G3	1966	Convergent	−259.5
Bisphenol A	Faith G4	1966	Convergent	−244.5
Bisphenol A	Faith G5	1966	Convergent	−233.5
Caffeine	Bredereck	1950	Linear	−147.82
Carbofuran	Union Carbide	1984	Linear	−87.89
Carbofuran	FMC G3	1969	Linear	−88.89
Carbofuran	FMC G1	1966	Linear	−63.93
Cis-alpha-irone	Firmenich G2	1992	Linear	−45.43
Cocaine	Willstatter	1923	Linear	−119.4
Coenzyme A	Khorana–Todd	1961	Convergent	−1319.8
Congo red	Fierz-David	1949	Convergent	−642.67
Congo red	Vogel	1948	Convergent	−813.2
DDT	Faith	1966	Convergent	−398.92
DDT	Vogel	1948	Linear	−159.13
DEET	Wang	1974	Linear	−99.67
DEET	Virginia Chemicals	1971	Linear	−99.67
DEET	U.S. Dept. Agriculture	1953	Linear	−83.89
DEET	Showa Denko	1976	Linear	−100
DEET	Pfizer	1979	Linear	−99.67
DEET	Neeland	1998	Linear	−99.67
DEET	LeFevre	1990	Linear	−99.67
DEET	Khalafi-Nezhad	2005	Linear	−99.67
DEET	Bhattacharya	2006	Linear	−100
DEET	Dominion Rubber	1962	Linear	−99.67
Dimethoate	American Cyanamid	1950	Convergent	−251.1
Dimethoate	Montecatini	1958	Convergent	−251.1
e-Caprolactam	Thomas	2005	Linear	−104
e-Caprolactam	Marvel	1943	Linear	−51.33

(continued)

(continued)

Target	Plan	Year	Type	μ_1 (g/mol/Stage)
e-Caprolactam	Faith	1975	Linear	−143.5
Epichlorohydrin	Faith	1966	Linear	−62.23
Eriochrome blue black R	Shreve	1967	Convergent	−545.67
Ethylene oxide	Faith G1	1966	Linear	−21
Ethylene oxide	Faith G2	1966	Linear	−22
Fesoterodine	Pfizer	2009	Linear	−323.67
Fluorescein	Vogel	1948	Linear	−148
Fluridone	Hooker	1980	Linear	−68.69
Glycerol	Faith G1	1975	Linear	−21.55
Glycerol	Faith G2	1975	Convergent	−137.5
Glyphosate	Monsanto	1972	Linear	−52.71
Hirsutene	Iyoda	1986	Linear	−42.4
Ibuprofen	Hoechst-Celanese	1988	Linear	−81.6
Ibuprofen	Furstoss	1999	Linear	−60.86
Ibuprofen	RajanBabu	2009	Linear	−61
Ibuprofen	DuPont	1985	Linear	−50.41
Ibuprofen	Ruchardt	1991	Linear	−59.86
Imatinib	Novartis	2003	Linear	−387.28
Imatinib	Natco Pharma	2004	Convergent	−586.48
Imatinib	Noszal	2005	Convergent	−647.81
Imatinib	Wang	2008	Convergent	−530.88
Imatinib	Ivanov	2009	Convergent	−649.95
Imatinib	Ley	2010	Convergent	−560.24
Indomethacin	Sumitomo	1968	Linear	−165.18
Indomethacin	Merck G3	1967	Linear	−270.73
Indoxacarb	DuPont G1	1999	Convergent	−634.76
Indoxacarb	DuPont G2	1999	Convergent	−717.39
Isotretinoin	Salman	2005	Linear	−76.8
Juvenile hormone	Hoffmann	1969	Convergent	−176.6
Juvenile hormone	Trost	1967	Convergent	−152.15
Linalool	Ruzicka	1919	Linear	−42
Malachite green	Muhlhauser	1887	Convergent	−375.51
Malachite green	Fierz-David	1921	Convergent	−375.51
Malathion	American Cyanamid	1951	Linear	−228.33
Maleic anhydride	Faith	1966	Linear	−11
MDMA	Daeid G2	2008	Linear	−45.44
MDMA	Braun	1980	Linear	−45.44
MDMA	Daeid G1	2008	Linear	−45.44
MDMA	Daeid G3	2008	Linear	−45.44
Menthol	BASF	1973	Linear	−33
Menthol	Takasago G2	2002	Linear	−1.33
Menthol	Mashima	2008	Linear	−40
Menthol	Sayo–Matsumoto	2002	Linear	−31.2
Methamphetamine	Ogata	1919	Linear	−224.18
Methyl methacrylate	Faith	1975	Convergent	−82.4
Methyl red	Vogel	1978	Convergent	−275.75
Methylene blue	Fierz-David	1921	Convergent	−454.57
Methylene blue	Muhlhauser G1	1887	Linear	−134.12
Methylene blue	Caro	1878	Linear	−134.12
Methylene blue	American Cyanamid	1980	Convergent	−384.74

(continued)

Target	Plan	Year	Type	μ_1 (g/mol/Stage)
Methylene blue	Kehrmann	1916	Linear	−224.9
Methylene blue	Muhlhauser G1	1887	Linear	−195.67
Metolachlor	Ciba-Syngenta	1999	Linear	−233.93
Metolachlor	Ciba-Geigy G1	1972	Linear	−118.48
Metolachlor	Cho	1992	Linear	−127.4
Metolachlor	Zhang	2006	Linear	−84.73
Mischler's hydrol	Fierz-David	1949	Linear	−53.5
Monastral fast blue B	Fierz-David	1949	Convergent	−471.1
Naproxen	DuPont	1985	Linear	−72.23
Nelfinavir	Agouron-Lilly	1997	Convergent	−526.72
Nelfinavir	Japan Tobacco-Agouron	1998	Convergent	−523.8
Nicotine	Campos	2006	Linear	−62.67
Orange I	Muhlhauser	1887	Convergent	−474.67
Orange II	Fierz-David	1949	Convergent	−444.39
Orange II	Muhlhauser	1887	Convergent	−444.39
Orange II	Vogel	1978	Convergent	−599.39
Orange III	Vogel	1978	Convergent	−440
Orange IV	Fierz-David	1949	Convergent	−530.18
Oseltamivir	Shibasaki G2	2007	Linear	−163.86
Oseltamivir	Shibasaki G3	2007	Linear	−167.26
Oseltamivir	Okamura–Corey	2008	Linear	−124.76
Oseltamivir	Corey	2006	Linear	−153.68
Oseltamivir	Gilead	1998	Linear	−135.85
Oseltamivir	Roche G1 (quinic acid route)	2001	Linear	−131.23
Oseltamivir	Shibasaki G1	2006	Linear	−132.1
Oseltamivir	Shibasaki G2	2007	Linear	−102.68
Oseltamivir	Fukuyama	2007	Linear	−133.48
Oseltamivir	Roche G2 (quinic acid route)	1999, 2004	Linear	−93.57
Oseltamivir	Roche G4 (Diels–Alder)	2000	Linear	−142.56
Oseltamivir	Roche G5 (desymmetrization)	2000	Linear	−110.48
Oseltamivir	Fang G1	2007	Convergent	−167.02
Oseltamivir	Fang G2	2008	Linear	−154.09
Oseltamivir	Fang G3	2008	Linear	−130.86
Oseltamivir	Shi	2009	Linear	−128.44
Oseltamivir	Hudlicky G1	2009	Convergent	−262.72
Oseltamivir	Shibasaki G4	2009	Linear	−124.67
Papaverine	Kindler–Peschke–Pal	1934, 1958	Convergent	−280.23
Papaverine	Pictet–Gams	1909	Convergent	−223.18
Papaverine	Dean	1974	Convergent	−212.89
Parathion	American Cyanamid	1948	Linear	−144.96
Patchouli alcohol	Magee	1995	Linear	−60.26
Patchouli alcohol	Firmenich	1981	Convergent	−144.22
Pentaerythritol	Faith	1966	Convergent	−99.48
Permethrin	Kuraray	1978	Convergent	−270.79
Phenacetin	Vogel	1978	Linear	−58.33
Phenol	Faith G1	1966	Linear	−77.5
Phenol	Faith G3	1966	Linear	−46
Phenolphthalein	Vogel	1948	Linear	−150
Phenolsulfonphthalein	Vogel	1948	Linear	−188.87

(*continued*)

(continued)

Target	Plan	Year	Type	μ_1 (g/mol/Stage)
Phenytoin	Lambert G1	2003	Linear	−117
Phenytoin	Biltz	2003	Linear	−117
Phenytoin	Parke-Davis	1946	Linear	−314.28
Phenytoin	Olah	1998	Linear	−205.03
Phorate	Bayer	1956	Convergent	−307.03
Phthalic anhydride	Faith G1	1966	Linear	−36
Phthalic anhydride	Faith G2	1966	Linear	−47
Physostigmine	Yu	1994	Linear	−44.91
Physostigmine	Speckamp	1978	Convergent	−90.96
Physostigmine	Trost	2006	Linear	−81.39
Physostigmine	Nakagawa	2000	Linear	−54.16
Physostigmine	Johnson	2003	Linear	−75.62
Picric acid	Fierz-David	1949	Linear	−87.5
Picric acid	Vogel	1948	Linear	−87.5
Pirate	American Cyanamid	1992	Linear	−299.94
Quinapril	Warner-Lambert-Parke-Davis	1986	Convergent	−383.36
Raltegravir	Merck	2006	Convergent	−367.5
Reserpine	Stork	2005	Convergent	−419.39
Reserpine	Pearlman	1979	Convergent	−399.66
Reserpine	Fraser-Reid	1995	Linear	−225.63
Reserpine	Liao	1996	Linear	−309.75
Reserpine	Mehta	2000	Convergent	−310.87
Reserpine	Martin	1987	Convergent	−289.69
Reserpine	Wender	1987	Convergent	−365.81
Reserpine	Shea	2003	Convergent	−230.09
Reserpine	Woodward	1958	Convergent	−564.55
Ricinine	Sugasawa	1974	Linear	−51
Rimonabant	Sanofi G1	1997	Linear	−183.83
Rimonabant	Reddy	2007	Linear	−220.79
Rimonabant	Sanofi G2	1997	Linear	−200.6
Risperidone	Janssen	1989	Convergent	−357.21
Rosefuran	Salerno	1997	Linear	−81.71
Rynaxypyr	DuPont G3	2007	Linear	−225.14
Rynaxypyr	DuPont G2	2007	Linear	−209.39
Rynaxypyr	DuPont G1	2007	Convergent	−429.29
Saccharin	BASF	1984	Linear	−295.68
Safranine	Fierz-David	1921	Linear	−622.45
Sancycline	Woodward	1968	Linear	−149.1
Sertraline	Buchwald	2000	Linear	−85.98
Sertraline	Zhao	2006	Linear	−60.98
Sildenafil	Pfizer-Reddy	2000	Convergent	−380.52
Sildenafil	Pfizer	2000	Convergent	−423.1
Silphinene	Wender	1985	Linear	−28.78
Sitagliptin fumarate	Merck G1	2005	Convergent	−479.8
Sitagliptin hydrogen phosphate	Merck G3	2009	Linear	−306.02
Sitagliptin hydrogen phosphate	Merck G2	2005	Convergent	−558.87
Sorbitol	Faith	1966	Linear	−91

(continued)

Target	Plan	Year	Type	μ_1 (g/mol/Stage)
Tamoxifen	Al-Hassan	1987	Linear	−167.86
Tamoxifen	Bristol-Myers	1984	Linear	−223.96
Tamoxifen	Dow-Chemical	1995	Convergent	−287.48
Tamoxifen	Fallis	2003	Linear	−106.25
Tamoxifen	ICI	1965	Linear	−89.67
Tamoxifen	Knochel	1997	Convergent	−384.06
Tamoxifen	Larock	2005	Linear	−178.58
Tamoxifen	McCague	1990	Convergent	−262.72
Tamoxifen	Nishihara	2007	Convergent	−333
Tamoxifen	NRDC	1987	Linear	−165.18
Tamoxifen	O'Shea	2006	Convergent	−465.75
Tamoxifen	Shiina	2004	Linear	−178.8
Tamoxifen	Yus	2003	Convergent	−388.68
Tebufenozide	Rohm and Haas	1994	Linear	−181.03
Thalidomide	Gruenthal G1	1957	Linear	−64
Thalidomide	Celgene	1999	Linear	−93
Theobromine	Bredereck G2	1950	Linear	−73
Thyroxine	Harington–Barger	1927	Linear	−327.79
Thyroxine	Chalmers	1949	Linear	−421.44
Trans-alpha-irone	Givaudan G1	1959	Linear	−62
Trans-alpha-irone	Firmenich G1	1984	Linear	−46.5
Triclosan	Lourens	1997	Convergent	−171.03
Trifluralin	Lilly	1963	Linear	−150.16
Tropinone[a]	Bazilevskaya G2	1958	Linear	−40.5
Urea	Faith	1966	Linear	−21
Uric acid	Traube	1900	Linear	−18.4
Vanillin	Volynkin	1938	Linear	−98.5
Vanillin	Roesler	1907	Linear	−143.5
Vanillin	Geigy	1899	Linear	−75
Vitamin A	Roche G6a	1976	Convergent	−201.65
Vitamin A	Roche G6b	1976	Convergent	−175.94
Vitamin A	Roche G6c	1976	Convergent	−165.46
Vitamin C	Biotechnical Resources	1991	Linear	−165
Vitamin E	Karrer	1939	Linear	−186.96
Warfarin	Jorgensen	2003	Linear	−154
Zamifenacin	Cossy	1997	Linear	−199.28
Zingerone	Nomura	1917	Linear	−60
Zingerone	Mukaiyama	1976	Linear	−36
Zafirlukast	ICI	1990	Linear	−295.79
Zafirlukast	MSN	2004	Linear	−242.88
Zafirlukast	Reddy	2009	Convergent	−461.01

[a] Approximate value since reducing agent not specified.

A.6   SUMMARY TABLES FOR TRENDS IN REACTION DATABASE

A.6.1   LIST OF PLANS THAT ARE ISOHYPSIC (HI = 0 WITH NO ALGEBRAIC CANCELLATIONS OF "UPS" AND "DOWNS")

Target	Plan	Year	Type	f(sac)
Adamantane	Schleyer	1957	Linear	0
Adamantane	DuPont	1960	Linear	0
Atrazine	Geigy	1960	Linear	0.281
Allantoin	Grimaux	1877	Linear	0
Aniline yellow	Vogel	1948	Linear	0.345
Anthraquinone	Vogel	1948	Linear	0
Aspirin	Vogel	1978	Linear	0.387
Barbituric acid	Dickey–Gray	1943	Linear	0
Barbituric acid	Shimo	1959	Linear	0.347
Basketene	Dauben	1966	Linear	0.808
Basketene	Masamune	1966	Linear	0.754
Bullvalene	Schroeder	1964	Linear	0.375
ε-Caprolactam	Thomas	2005	Linear	0
ε-Caprolactam	Faith	1975	Linear	0
Carbaryl	Union Carbide	1959	Linear	0.206
Cyclooctatetraene	Reppe	1948	Linear	0
DDT	Vogel	1948	Linear	0
DEET	Bhattacharya	2006	Linear	0.624
DEET	Showa Denko	1976	Linear	0.494
Dimethoate	American cyanamid	1950	Convergent	0.221
Dimethoate	Montecatini	1958	Convergent	0.23
(−)-Ephedrine HCl	Manske–Johnson	1929	Linear	0.499
(+)-Ephedrine HCl	Manske–Johnson	1929	Linear	0.293
Ethylene oxide	Faith G2	1966	Linear	0
Ganciclovir	Ogilvie G1	1982	Linear	0.35
Ganciclovir	Ogilvie G2	1982	Convergent	0.341
Ganciclovir	Ogilvie G3	1982	Convergent	0.641
Ganciclovir	Roche	1995	Linear	0.441
Ganciclovir	Syntex G1	1982	Convergent	0.338
Ganciclovir	Syntex G2	1983	Convergent	0.298
Ganciclovir	Merck	1988	Convergent	0.282
Glycerol	Faith G3	1966	Linear	0
Glyphosate	Monsanto	1972	Linear	0.292
Hexamethylenetetramine	Vogel	1948	Linear	0
Imatinib	Ley	2010	Convergent	0.251
Indomethacin	Merck G3	1967	Linear	0.07
Indomethacin	Merck G1	1964	Linear	0.567

(continued)

Target	Plan	Year	Type	f(sac)
Malathion	American Cyanamid	1951	Linear	0
MDMA	Daeid G2	2008	Linear	0.322
MDMA	Braun	1980	Linear	0.335
MDMA	Daeid G1	2008	Linear	0.366
Melamine	Faith	1975	Linear	0.5
Menthol	Brode	1947	Linear	0
Metolachlor	Ciba-Geigy G1	1972	Linear	0.26
Paclitaxel	Gennari G1	1997	Convergent	0.504
Phenacetin	Vogel	1978	Linear	0.158
Phenolphthalein	Vogel	1948	Linear	0
Phenytoin	Parke-Davis	1946	Linear	0
Phenytoin	Biltz	1908	Linear	0
Phenytoin	Lambert G1	2003	Linear	0
Phorate	Bayer	1956	Convergent	0.372
Picric acid	Fierz-David	1949	Linear	0.409
Picric acid	Vogel	1948	Linear	0.409
Resveratrol	Yus G3	2009	Linear	0.751
Resveratrol	Yus G2	2009	Linear	0.771
Rimonabant	Sanofi G1	1997	Linear	0.366
Rimonabant	Reddy	2007	Linear	0.491
Saccharin	BASF	1984	Linear	0.382
Safranine	Fierz-David	1921	Linear	0.561
Sorbitol	Faith	1966	Linear	0
Tamoxifen	Shiina	2004	Linear	0.527
Tebufenozide	Rohm and Haas	1994	Linear	0.444
Thalidomide	Grunenthal G1	1957	Linear	0.423
Thalidomide	Grunenthal G2	1960	Linear	0.537
Thalidomide	Grunenthal G3	1970	Linear	0.433
Thalidomide	Seijas	2001	Linear	0
Thalidomide	Celgene	1999	Linear	0.666
Theobromine	Hoechst	1994	Linear	0.559
Theobromine	Bredereck G2	1950	Linear	0.561
Tropinone	Stevens	1965	Linear	0.725
Urea	Faith	1966	Linear	0
Uric acid	Bills	1962	Linear	0
Vanillin	Roesler	1907	Linear	0.177
Vanillin	Givaudan–Roure	1998	Linear	0.071
Vinyl chloride	Faith	1966	Linear	0
Vitamin A	Roche G3	1976	Convergent	0.581
Vitamin C	Biotechnical Resources	1991	Linear	0.139
Warfarin	Jorgensen	2003	Linear	0

A.6.2 List of Plans That Are Isohypsic (HI = 0 as a Consequence of Algebraic Cancellations of "Ups" and "Downs")

Target	Plan	Year	Type	f(sac)	$|UD|$
Benzoic acid	Faith G1	1966	Linear	0.386	4
Colchicine	Scott	1965	Convergent	0.626	24
Crystal violet	Vogel	1948	Convergent	0.415	6
12a-Deoxytetracycline	Stork	1996	Convergent	0.745	16
R-Fluoxetine	Bracher	1996	Linear	0.676	8
Ganciclovir	Reddy Laboratories	2009	Convergent	0.486	8
Glycerol	Faith	1975	Convergent	0	8
Hipposudoric acid	Hashimotot	2006	Convergent	0.822	14
(−)-Kainic acid	Jung	2007	Linear	0.785	26
Methyl methacrylate	Faith	1975	Convergent	0.59	4
Permethrin	Kuraray	1978	Convergent	0.547	4
Phenol	Faith G5	1966	Linear	0.308	4
Resveratrol	Marra	2006	Linear	0.475	4
Salsolene oxide	Wender	2006	Linear	0.604	6
Sparteine	Koomen	1996	Convergent	0.842	14
Vitamin A	Roche G4	1976	Convergent	0.651	8

A.6.3 List of Plans That Are Hyperhypsic (HI > 0)

Target	Plan	Years	Type	f(sac)	HI	$|UD|$
Adamantane	Prelog G1	1941	Linear	0.342	8	24
Adamantane	Stetter	1956	Linear	0.651	7.11	24
Acetaminophen	Warner-Lambert	1961	Linear	0.637	2	8
Acetaminophen	Vogel	1978	Linear	0.417	1	8
Actarit	Mitsubishi Chemical	1988	Linear	0.185	3.57	14
Aniline	Shreve–Faith G2	1966, 1967	Linear	0.014	2.67	4
Antipyrine	Vogel	1978	Convergent	0.546	1.5	10
Beta-carotene	Roche G1	1947, 1956	Linear	0.544	2.8	8
Bullvalene	von Doering	1964	Convergent	0.733	1.6	8
Bupleurynol	Organ G2	2004	Linear	0.349	1	4
Caffeine	Traube	1900	Linear	0.44	0.57	8
Caffeine	Bredereck	1950	Linear	0.623	1.6	8
ε-Caprolactam	Marvel	1943	Linear	0.683	1.33	8
Congo red	Vogel	1948	Convergent	0.565	2	8
Coniine	Buchwald	1998	Linear	0.5	1.75	8
Cyclooctatetraene	Willstatter–Cope	1913, 1948	Linear	0.88	0.36	6
Dewar benzene	van Tamelen	1963, 1971	Linear	0.728	0.5	4
(−)-Ephedrine HCl	Stolz G1	1931	Linear	0	1.33	8
Epibatidine	Trudell	1996	Linear	0.606	2.6	12
(±)-Epinephrine	Dakin	1905	Linear	0.641	1.5	4
(±)-Epinephrine	Stolz	1904	Linear	0.641	1.5	4
Fenchone	Komppa	1935	Linear	0.717	4.8	20
(±)-Fluoxetine	Lilly G1	1982	Linear	0.611	0.5	4
Fluorescein	Vogel	1948	Linear	0	0.5	2
Fesoterodine	Pfizer	2009	Linear	0.456	0.5	4

(continued)

Target	Plan	Years	Type	f(sac)	HI	|UD|
Hirsutene	Cohen	1993	Linear	0.544	5.25	20
Hirsutene	Iyoda	1986	Linear	0.43	3.2	10
Hirsutene	Funk	1984	Linear	0.671	2.78	16
Hirsutene	Wender	1982	Linear	0.531	2.38	12
Hirsutene	Mehta	1986	Linear	0.235	5.27	18
Hypoglycin A	Abbott	1958	Convergent	0.512	3.29	12
Ibogamine	Trost	1978	Linear	0.223	0.6	6
Ibuprofen	DuPont	1985	Linear	0.476	0.75	6
Imatinib	Noszal	2005	Convergent	0.611	3	12
Imatinib	Novartis	2003	Linear	0.161	0.25	2
Indigo	Baeyer G4	1882	Linear	0.675	1.5	4
Indigo	Sandmeyer	1903	Linear	0.818	3.33	8
Indomethacin	Sumitomo	1968	Linear	0.421	0.6	2
(±)-*Cis*-gamma-gamma-irone	Gosselin	2001	Linear	0.573	4	20
Isodrin	Shell development corp.	1955	Linear	0.071	0.4	4
Isotretinoin	Salman	2005	Linear	0.396	1.6	12
L-DOPA	Roche	1973	Linear	0.513	0.4	4
Linezolid	Upjohn G2	1998	Linear	0.62	1.5	12
Linezolid	Gotor	2008	Convergent	0.481	0.89	8
Lysergic acid	Rebek	1984	Linear	0.755	0.67	4
Menthol	Tanabe–Seiyaku	1968	Linear	0.703	1.57	10
Menthol	BASF	1973	Linear	0	1.5	12
Menthol	Takasago G2	2002	Linear	0	0.67	4
Menthol	Mashima	2008	Linear	0	0.5	4
Menthol	Sayo–Matsumoto	2002	Linear	0	2.2	10
Methamphetamine	Ogata	1919	Linear	0.404	1	4
Metolachlor	Ciba-Syngenta	1999	Linear	0.23	0.33	2
Metolachlor	Cho	1992	Linear	0.266	0.5	4
Metolachlor	Zhang	2006	Linear	0.336	1.33	8
Metolachlor	Ciba-Geigy G2	1982	Linear	0.437	2	8
Naproxen	Effenberger	1997	Linear	0.627	1.14	6
Naproxen	Ruchardt	1991	Linear	0.716	0.7	6
Nelfinavir	Agouron-Lilly	1997	Convergent	0.687	1	6
Nicotine	Spath	1928	Linear	0.718	1.6	6
Nicotine	Jacob	1982	Linear	0.507	1.67	10
Nicotine	Loh	1999	Linear	0.38	1	8
Parathion	American Cyanamid	1948	Linear	0.307	0.25	2
Paroxetine	Ferrosan	1975	Linear	0.615	2.2	12
Phenytoin	Olah	1998	Linear	0.67	0.67	4
Pregabalin	Warner-Lambert G2	2001	Linear	0.585	4.33	14
Pregabalin	Dowpharma	2003	Linear	0.657	4.33	14
Pregabalin	McQuade	2007	Linear	0.011	3.5	14
Resveratrol	Spath	1941	Linear	0	0.67	2
Resveratrol	Shen	2002	Convergent	0.876	2.33	8
(*R*)–fluoxetine HCl	Sharpless	1988	Linear	0.293	0.78	4
Ricinine	Schroeter	1932	Linear	0.394	1.83	6
Rosefuran	Pattenden	1977	Convergent	0.254	1.5	8

(*continued*)

(continued)

Target	Plan	Years	Type	f(sac)	HI	|UD|
Rosefuran	Buchi	1968	Convergent	0.576	0.29	2
Rosefuran	Okazaki	1984	Convergent	0.584	1.6	8
Sertraline	Pfizer G2	1999	Linear	0.562	1.56	12
Sertraline	Pfizer G3	2004	Linear	0.54	2	12
Sertraline	Buchwald	2000	Linear	0.358	2.33	12
(S)–fluoxetine HCl	Wang	2005	Linear	0.272	2.67	12
Sildenafil	Pfizer	2000	Convergent	0.453	2.67	12
Silphinene	Wender	1985	Linear	0.34	2.5	12
Sparteine	Anet	1950	Linear	0.245	3.33	12
Tamoxifen	Al-Hassan	1987	Linear	0.554	0.4	2
Tamoxifen	Armstrong	1997	Convergent	0.688	1.5	8
Tamoxifen	Larock	2005	Linear	0.432	0.5	2
Tamoxifen	O'Shea	2006	Convergent	0.663	0.25	2
Tropinone	Robinson	1917	Linear	0.397	1	4
Tropinone	Bazilevskaya G4	1958	Convergent	0.668	3.33	12
Tropinone	Bazilevskaya G3	1958	Convergent	0.638	1.2	4
Tropinone	Bazilevskaya G2	1958	Linear	0.724	1.5	4
Tropinone	Karrer	1947	Linear	0.498	1.14	6
Tropolone	DuPont	1958	Linear	0.36	0.75	3
Twistane	Tichy	1969	Linear	0.793	6.82	24
Tyrian purple	Gerlach	1989	Linear	0.818	8	28
Tyrian purple	Rottig	1935	Linear	0.544	1.6	4
Vanillin	Mottern	1934	Linear	0.352	0.8	2
Vanillin	Harries–Haarmann	1915	Linear	0.467	0.33	2
Zingerone	Nomura	1917	Linear	0	0.67	4
Zingerone	Mukaiyama	1976	Linear	0.06	0.67	4

A.6.4 LIST OF PLANS THAT ARE HYPOHYPSIC (HI < 0)

Target	Plan	Years	Type	f(sac)	HI	|UD|
2,4-D	Dolge	1941	Linear	0.261	−0.67	4
2,4-D	Haskelberg G1	1947	Linear	0.055	−0.5	4
2,4-D	Manske	1949	Linear	0.093	−2	8
2,4-D	Haskelberg G2	1947	Linear	0.093	−2	8
2,4-D	ICI	1945	Linear	0.261	−0.67	4
Acetaminophen	Pearson	1953	Linear	0.544	−0.5	4
Acetic anhydride	Faith G1	1966	Linear	0.268	−2	8
Acetic anhydride	Faith G2	1966	Linear	0	−1.5	8
Adipic acid	Faith G1	1975	Linear	0.11	−3.33	16
Adipic acid	Ellis	1941	Linear	0.164	−6	16
Adipic acid	Vogel	1978	Linear	0.157	−4	16
Alizarin	Fierz-David G1	1949	Linear	0.466	−4	20
Alizarin	Vogel	1978	Linear	0.345	−2.8	8
Allantoin	Hartman	1943	Linear	0.532	−0.5	2
Aniline	Shreve–Faith G3	1966, 1967	Linear	0.484	−0.67	4
Anthraquinone	Faith G1	1966	Linear	0.229	−0.67	4
Anthraquinone	Faith G2	1966	Linear	0.114	−2	12

(continued)

Target	Plan	Years	Type	f(sac)	HI	|UD|
Benfluralin	Lilly	1966	Linear	0.375	−2.67	20
Benzoic acid	Faith G3	1966	Linear	0.624	−2	12
Benzoic acid	Faith G4	1966	Linear	0.114	−1	4
Bifenazate	Chee	2006	Linear	0.472	−0.6	6
Bifenazate	Uniroyal G3	2004	Linear	0.53	−0.67	4
Bifenazate	Uniroyal G2	2000	Linear	0.491	−0.4	4
Bisphenol A	Faith G2	1966	Convergent	0.375	−2	12
Bisphenol A	Faith G4	1966	Convergent	0	−1.5	12
ε-Caprolactam	Evans	2002	Linear	0.589	−1.5	6
ε-Caprolactam	Ishii G1	2003	Linear	0.526	−2	8
ε-Caprolactam	BP	1976	Linear	0.498	−1.33	4
Carbofuran	FMC G1	1966	Linear	0.412	−0.33	2
Carbofuran	FMC G2	1967	Linear	0.13	−0.25	2
Carbofuran	FMC G3	1969	Linear	0.349	−0.5	2
Carbofuran	Union Carbide	1984	Linear	0.396	−1.75	8
Carvone	Royals	1951	Linear	0.592	−1.25	8
Carvone	A.M.Todd	1958	Linear	0.738	−0.75	3
Clothianidin	Novartis G1	2000	Convergent	0.514	−2.13	12
Dapoxetine	Lilly	1988	Linear	0.414	−1	4
DDT	Faith	1966	Convergent	0.174	−2	12
DEET	Dominion Rubber	1962	Linear	0.07	−1.33	8
DEET	Khalafi-Nezhad	2005	Linear	0.643	−1.33	8
DEET	LeFevre	1990	Linear	0.471	−1.33	8
DEET	Neeland	1998	Linear	0.496	−1.33	8
DEET	Pfizer	1979	Linear	0.07	−1.33	8
DEET	U.S. Dept. Agriculture	1953	Linear	0.496	−1	8
DEET	Virginia Chemicals	1971	Linear	0.518	−1.33	8
DEET	Wang	1974	Linear	0.496	−1.33	8
Epichlorohydrin	Faith	1966	Linear	0.368	−1.5	6
Eriochrome blue black R	Shreve	1967	Convergent	0.583	−2.33	12
Ethylene oxide	Faith G1	1966	Linear	0.683	−1	4
Ferrocene	DuPont	1957	Linear	0	−2	8
Ferrocene	Pauson	1951	Linear	0.475	−0.5	2
Ferrocene	BASF G2	1963	Linear	0.294	−1	4
Ferrocene	BASF G1	1963	Linear	0.294	−2	8
Ferrocene	Wilkinson G2	1963	Linear	0.361	−2	8
Ferrocene	Wilkinson G1	1963	Linear	0.151	−2	8
Fluorescein	Muhlhauser	1877	Convergent	0.63	−3.75	22
Ganciclovir	Syntex G3	1988	Convergent	0.58	−0.33	4
Glycerol	Faith G1	1975	Linear	0.602	−1.5	8
Ibuprofen	Pinhey	1984	Convergent	0.689	−0.5	2
Indigo	Heumann G1	1890	Linear	0.752	−3	8
Indigo	Vogel	1978	Convergent	0.761	−2.4	16
Indoxacarb	DuPont G1	1999	Convergent	0.385	−1.43	14
Malachite green	Muhlhauser	1887	Convergent	0.553	−1.4	10
Malachite green	Fierz-David	1921	Convergent	0.553	−1.4	10
Maleic anhydride	Faith	1966	Linear	0.216	−1	4
Methylene blue	Muhlhauser G1	1887	Linear	0.641	−4	18
Methylene blue	Caro	1878	Linear	0.641	−4	18

(continued)

(continued)

Target	Plan	Years	Type	f(sac)	HI	|UD|
Methylene blue	Kehrmann	1916	Linear	0.429	−3.67	14
Methylene blue	Muhlhauser G1	1887	Linear	0.682	−4	18
Mischler's hydrol	Fierz-David	1949	Linear	0.521	−1.5	4
Mischler's ketone	Fierz-David	1949	Linear	0.432	−0.67	4
Naproxen	Syntex G1	1970	Linear	0.631	−0.88	6
Nelfinavir	Japan Tobacco-Agouron	1998	Convergent	0.681	−0.5	4
Nicotinamide	Vogel	1978	Linear	0.416	−1.5	12
Nicotine	Campos	2006	Linear	0.498	−0.33	2
Orange I	Muhlhauser	1887	Convergent	0.342	−2	12
Orange II	Fierz-David	1949	Convergent	0.408	−2	12
Orange II	Muhlhauser	1887	Convergent	0.465	−2	12
Orange II	Vogel	1978	Convergent	0.427	−2	12
Orange IV	Fierz-David	1949	Convergent	0.279	−1.33	8
Oseltamivir	Roche G4 (Diels–Alder)	2000	Linear	0.505	−0.5	4
Oseltamivir	Corey	2006	Linear	0.743	−3	14
Oxomaritidine	Ley G2	2006	Linear	0.697	−1	4
Oxomaritidine	Holton	1970	Linear	0.494	−2	6
Paclitaxel	Greene G2	1994	Convergent	0.554	−0.14	2
Phenobarbital	Chamberlain	1953	Linear	0.303	−0.2	2
Phenobarbital	Nelson	1928	Linear	0.39	−0.6	2
Phenobarbital	Rising G1	1927	Linear	0.351	−0.29	2
Phenobarbital	Stieglitz	1918	Convergent	0.487	−1	4
Phenol	Faith G1	1966	Linear	0.647	−1	4
Phenol	Faith G2	1966	Linear	0.418	−0.67	4
Phenol	Faith G3	1966	Linear	0.353	−1	4
Phenol	Faith G4	1966	Linear	0	−0.67	4
Phthalic anhydride	Faith G1	1966	Linear	0.176	−1	4
Phthalic anhydride	Faith G2	1966	Linear	0	−3	12
Resorcinol	Muhlhauser	1887	Linear	0.681	−1.33	8
Resorcinol	Fierz-David	1949	Linear	0.726	−1.33	8
Ricinine	Spath	1923	Linear	0.902	−2.47	12
Ricinine	Taylor	1956	Linear	0.828	−3	20
Rimonabant	Sanofi G2	1997	Linear	0.405	−0.17	2
Rosefuran	Salerno	1997	Linear	0.284	−0.4	2
Rosefuran	Iriye G1	1990	Linear	0.663	−0.5	4
Rosefuran	Iriye G2	1990	Linear	0.6	−0.67	4
Rynaxypyr	DuPont G3	2007	Linear	0.466	−0.67	4
Rynaxypyr	DuPont G2	2007	Linear	0.461	−0.71	6
Saccharin	Maumee-Sherwin-Williams	1980s	Linear	0.728	−1.17	10
Saccharin	Maumee	1954	Linear	0.554	−1.6	6
Sulfapyridine	Vogel	1978	Convergent	0.258	−0.8	4
Triclosan	Lourens	1997	Convergent	0.528	−1	8
Triclosan	Geigy	1970	Convergent	0.567	−0.8	8
Triclosan	Ciba G1	1998	Linear	0.307	−1.4	10
Triclosan	Ciba G2	1998	Convergent	0.395	−1	8
Trifluralin	Lilly	1963	Linear	0.206	−1.5	8

(continued)

Target	Plan	Years	Type	f(sac)	HI	|UD|
Tyrian purple	Majima	1930	Linear	0.653	−6	24
Uric acid	Traube	1900	Linear	0.794	−0.67	4
Vanillin	Lampman G1	1977	Linear	0.453	−1.67	6
Vanillin	Mayer	1949	Linear	0.654	−1.5	6
Vanillin	Zasosov	1959	Linear	0.563	−0.67	2
Vanillin	Geigy	1899	Linear	0.774	−1	4
Vanillin	Delvaux	1947	Linear	0.4	−0.67	2
Vanillin	Riedel	1927	Linear	0.55	−1.8	6
Vanillin	Givaudan	1921	Linear	0.613	−1.75	6
Veronal	Vogel	1978	Linear	0.576	−1	6
Vinyl acetate	Faith	1966	Convergent	0	−1.2	8
Vitamin C	Roche	1992	Linear	0.351	−0.67	4
Vitamin C	Banwell	1998	Linear	0.656	−3	12
Warfarin	Cravotto	2001	Linear	0.359	−0.25	2
Warfarin	Link	1961	Linear	0.514	−0.25	2

A.6.5 LIST OF PLANS THAT INVOLVE CHEMO-ENZYMATIC STEPS

Target Product	Plan
Astaxanthin	Ito G1
(+)-Codcine	Hudlicky
(S)-Dapoxetine	Gotor
(S)-Fluoxetine	A.Kumar
(−)-Hirsutene	Banwell
(S)-Metolachlor	Zhang
Oseltamivir	Banwell
Oseltamivir	Hudlicky G1
Oseltamivir	Fang G2
Oseltamivir	Fang G3
Oseltamivir	Hudlicky G2
(−)-Tetracycline	Myers G1
Vanillin	Givaudan–Roure
Vanillin	Frost
Vitamin C	Banwell
Vitamin C	Biotechnical Resources
Vitamin C	Rcichstein
Vitamin C	Roche
(±)-Welwitindolinone A	Wood

A.6.6 LIST OF PLANS BY FINAL PRODUCT THAT TRANSFORM A STARTING NATURAL PRODUCT INTO ANOTHER TARGET NATURAL PRODUCT

Target Natural Product	Starting Natural Product	Plan
(−)-Cocaine	(±)-Malic acid	Rapoport
(−)-Cocaine	(R)-Glutamic acid	Rapoport
(−)-Ephedrine HCl	Amygdalin	Freudenberg
(−)-Hirsutene	(R)-(−)-Carvone	Weinges
(−)-Kainic acid	L-Aspartic acid	Fukuyama G2
(−)-Kainic acid	D-Serine	Baldwin
(−)-Kainic acid	L-Serine	Taylor
(−)-Kainic acid	D-Serine	Benetti
(−)-Kainic acid	L-Glutamic acid	Cossy
(−)-Kainic acid	L-Phenylalanine	Fukuyama G2
(−)-Kainic acid	L-Aspartic acid	Fukuyama G3
(−)-Kainic acid	L-Serine	Hanessian
(−)-Kainic acid	D-Serine	Hoppe
(−)-Kainic acid	L-Glutamic acid	Jung
(−)-Kainic acid	L-Aspartic acid	Knight
(−)-Kainic acid	D-Serine	Montgomery
(−)-Kainic acid	L-Methionine	Naito
(−)-Kainic acid	L-Glutamic acid	Oppolzer
(−)-Kainic acid	L-Glutamic acid	Rubio
(−)-Kainic acid	L-Glutamic acid	Yoo
(−)-Kainic acid	L-Serine	Takano G4
(−)-Kainic acid	Furfuraldehyde	Ogasawara G3
(−)-Kainic acid	Furfuraldehyde	Fukuyama G1
(−)-Menthol	(R)-(+)-Pulegone	Mashima
(−)-Menthol	Myrcene	Takasago G1
(−)-Menthol	(+)-Citronellal	Takasago G2
(−)-Nakadomarin A	L-Serine	Nishida G2
(−)-Nakadomarin A	D-Glutamic acid	Dixon
(−)-Nepetalactone	(R)-(+)-Pulegone	Cavill
(−)-Nepetalactone	(R)-(+)-pulegone	Pickett G1
(−)-Oleocanthal	D-Ribose	Smith G1
(−)-Oleocanthal	D-Ribose	Smith G2
(−)-Paclitaxel	Verbenone	Wender
(−)-Paclitaxel	(+)-Camphor	Holton
(−)-Paclitaxel	L-Serine	Mukaiyama
(−)-Paclitaxel	10-Deacetylbaccatin III	Commercon
(−)-Paclitaxel	10-Deacetylbaccatin III	Gennari
(−)-Paclitaxel	10-Deacetylbaccatin III	Greene G1
(−)-Paclitaxel	10-Deacetylbaccatin III	Greene G2
(−)-Paclitaxel	10-Deacetylbaccatin III	Kingston
(−)-Patchouli alcohol	(+)-Camphor	Buchi
(−)-Patchouli alcohol	(+)-Alpha-pinene	Firmenich
(−)-Patchouli alcohol	(R)-(−)-Carvone	Srikrishna
(−)-Platensimycin	(S)-(+)-Carvone	Ghosh
(−)-Platensimycin	(R)-(−)-Carvone	Nicolaou G5
(−)-Stenine	L-Tyrosine	Wipf
(−)-Swainsonine	D-(+)-Glucose	Vankar

(continued)

Target Natural Product	Starting Natural Product	Plan
(−)-Swainsonine	D-(+)-Glucose	Fleet
(−)-Swainsonine	(+)-Erythorbic acid	GlycoDesign
(−) Swainsonine	(+)-Erythorbic acid	Pearson
(−)-Swainsonine	(+)-Erythorbic acid	Cha
(−)-Swainsonine	(+)-Erythorbic acid	Kang
(−)-Swainsonine	L-(S)-Proline	Cossy G1
(−)-Swainsonine	L-(S)-Proline	Cossy G2
(−)-Swainsonine	L-Serine	Ham
(−)-Swainsonine	D-Mannose	Hashimoto
(−)-Swainsonine	D-Mannose	Mootoo
(−)-Swainsonine	Menthofuran	Roush
(−)-Swainsonine	(R)-Malic acid	Kibayashi
(−)-Tetracycline HCl	Juglone	Shemyakin
(−)-Tetracycline HCl	D-Glucosamine	Tatsuta
(−)-*Trans*-alpha-irone	(+)-Citronellal	Yoshikoshi
(+)-(2R,6R)-*Trans*-gamma-irone	(±)-Alpha-irone	Brenna G2
(+)-(2R,6R)-*Trans*-gamma-irone	(R)-(−)-Carvone	Monti G2
(+)-(2R,6S)-*Cis*-gamma-irone	(R)-(−)-Carvone	Monti G1
(±)-*Cis*-gamma-irone	(±)-Carvone	Leyendecker
(+)-Fenchone	(+)-Alpha-pinene	Cocker
(+)-Longifolene	(+)-Camphor	Money
(+)-Nepetalactone	(S)-Citronellol	Mori
(+)-Nepetalactone	(S)-Citronellol	Pickett G1
(+)-Nepetalactone	(S)-(−)-Limonene	Sakai
(+)-Swainsonine	L-Glutamic acid	Ikota
(+)-Swainsonine	L-Glutamic acid	Hirama
(+)-Thienamycin	L-Threonine	Hanessian
(+)-Thienamycin	L-Aspartic acid	Reider
(+)-Thienamycin	L-Aspartic acid	Salzmann
(+)-*Trans*-alpha-irone	(−)-Beta-pinene	Givaudan G2
(+)-*Trans*-alpha-irone	(−)-Alpha-pinene	Firmenich G1
(+)-Vinblastine	Catharanthine	Boger
(+)-Vinblastine	Catharanthine	Kutney
(±)-Epibatidine	Tropinone	Bai
(±)-Menthol	Thymol	Brode
(±)-Nicotine	Mucic acid	Pictet
(±)-Terramycin	Juglone	Muxfeldt
(±)-*Trans*-alpha-irone	(±)-Alpha-pinene	Givaudan G1
(R)-(−)-Carvone	(R)-(+)-Limonene	Royals
(R)-(−)-Carvone	(R)-(+)-Limonene	A.M. Todd
(S)-Nicotine	L-(S)-Proline	Chavdarian
(S)-Thyroxine	(S)-Tyrosine	Chalmers
12a-Deoxytetracycline	Juglone	Stork
6-Aminopenicillanic acid	(±)-Valine	Sheehan
Absinthin	Alpha-santonin	Zhai
Cephalosporin C	L-(+)-Cysteine	Woodward
Coenzyme A	Uric acid	Khorana–Todd
Curcumin	Ferulic acid	Lampe
Eleutherobin	(R)-(−)-Carvone	Gennari

(continued)

(continued)

Target Natural Product	Starting Natural Product	Plan
Eleutherobin	Beta-phellandrene	Danishefsky
Helminthosporal	(S)-(+)-Carvone	Corey
Laurallene	D-Diethyl tartrate	Suzuki
L-DOPA	Vanillin	Roche
Lysergic acid	L-(S)-Tryptophan	Rebek
Menthone	(−)-Menthol	Sandborn
Oseltamivir	D-Ribose	Fang G1
Oseltamivir	D-Mannitol	Mandai G1
Oseltamivir	Furfuraldehyde	Okamura
Oseltamivir	Quinic acid	Gilead
Oseltamivir	Quinic acid	Roche G1
Oseltamivir	Quinic acid	Roche G2
Oseltamivir	Shikimic acid	Roche G3
Oseltamivir	Shikimic acid	Shi
Oseltamivir	D-Glucal	Chen–Liu
Penicillin V	(±)-Valine	Sheehan
Quisqualic acid	L-Serine	Baldwin
Reserpine	Quinic acid	Hanessian
Rosefuran	Citral	Iriye G1
Rosefuran	Citral	Iriye G2
Rosefuran	Furfuraldehyde	Scharf
Salsolene oxide	Myrcene	Wender
Salsolene oxide	(R)-(−)-Carvone	Paquette
Sorbitol	D-(+)-Glucose	Faith
Theobromine	Guanine	Bredereck G2
Vanillin	Ferulic acid	Givaudan–Roure
Vitamin C	D-(+)-Glucose	Biotechnical Resources
Vitamin C	D-(+)-Galactose	Haworth
Vitamin C	D-(+)-Glucose	Reichstein
Vitamin E	Phytol	Woggon
Vitamin E	Phytol	Knight
Vitamin E	Phytol	Roche G1
Vitamin E	Phytol	Roche G2
Zingerone	Vanillin	Nomura
Zingerone	Vanillin	Mukaiyama
(−)-Alpha-pinene	(+)-*Trans*-alpha-irone	Firmenich G1
(−)-Beta-pinene	(+)-*Trans*-alpha-irone	Givaudan G2
(−)-Menthol	Menthone	Sandborn
(+)-Alpha-pinene	(+)-Fenchone	Cocker
(+)-Alpha-pinene	(−)-Patchouli alcohol	Firmenich
(+)-Camphor	(−)-Paclitaxel	Holton
(+)-Camphor	(−)-Patchouli alcohol	Buchi
(+)-Camphor	(+)-Longifolene	Money
(+)-Citronellal	(−)-Menthol	Takasago G2
(+)-Citronellal	(−)-*Trans*-alpha-irone	Yoshikoshi
(+)-Erythorbic acid	(−)-Swainsonine	GlycoDesign
(+)-Erythorbic acid	(−)-Swainsonine	Pearson
(+)-Erythorbic acid	(−)-Swainsonine	Cha
(+)-Erythorbic acid	(−)-Swainsonine	Kang
(±)-Alpha-irone	(+)-(2R,6R)-*Trans*-gamma-irone	Brenna G2

(continued)

Target Natural Product	Starting Natural Product	Plan
(±)-Alpha-pinene	(±)-*Trans*-alpha-irone	Givaudan G1
(±)-Carvone	(±)-*Cis*-gamma-irone	Leyendecker
(±)-Malic acid	(−)-Cocaine	Rapoport
(±)-Valine	6-Aminopenicillanic acid	Sheehan
(±)-Valine	Penicillin V	Sheehan
(*R*)-(−)-Carvone	(−)-Platensimycin	Nicolaou G5
(*R*)-(−)-Carvone	(−)-Hirsutene	Weinges
(*R*)-(−)-Carvone	(−)-Patchouli alcohol	Srikrishna
(*R*)-(−)-Carvone	Eleutherobin	Gennari
(*R*)-(−)-Carvone	Salsolene oxide	Paquette
(*R*)-(−)-Carvone	(+)-(2R,6R)-*Trans*-gamma-irone	Monti G2
(*R*)-(−)-Carvone	(+)-(2R,6S)-*Cis*-gamma-irone	Monti G1
(*R*)-(+)-Limonene	(*R*)-(−)-Carvone	Royals
(*R*)-(+)-Limonene	(*R*)-(−)-Carvone	A.M. Todd
(*R*)-(+)-Pulegone	(−)-Menthol	Mashima
(*R*)-(+)-Pulegone	(−)-Nepetalactone	Cavill
(*R*)-(+)-Pulegone	(−)-Nepetalactone	Pickett G1
(*R*)-Glutamic acid	(−)-Cocaine	Rapoport
(*R*)-Malic acid	(−)-Swainsonine	Kibayashi
(*S*)-(−)-Limonene	(+)-Nepetalactone	Sakai
(*S*)-(+)-Carvone	(−)-Platensimycin	Ghosh
(*S*)-(+)-Carvone	Helminthosporal	Corey
(*S*)-Citronellol	(+)-Nepetalactone	Mori
(*S*)-Citronellol	(+)-Nepetalactone	Pickett G1
(*S*)-Tyrosine	(*S*)-Thyroxine	Chalmers
10-Deacetylbaccatin III	(−)-Paclitaxel	Commercon
10-Deacetylbaccatin III	(−)-Paclitaxel	Gennari
10-Deacetylbaccatin III	(−)-Paclitaxel	Greene G1
10-Deacetylbaccatin III	(−)-Paclitaxel	Greene G2
10-Deacetylbaccatin III	(−)-Paclitaxel	Kingston
Alpha-santonin	Absinthin	Zhai
Amygdalin	(−)-Ephedrine HCl	Freudenberg
Beta-phellandrene	Eleutherobin	Danishefsky
Catharanthine	(+)-Vinblastine	Boger
Catharanthine	(+)-Vinblastine	Kutney
Citral	Rosefuran	Iriye G1
Citral	Rosefuran	Iriye G2
D-(+)-Galactose	Vitamin C	Haworth
D-(+)-Glucose	(−)-Swainsonine	Vankar
D-(+)-Glucose	(−)-Swainsonine	Fleet
D-(+)-Glucose	Sorbitol	Faith
D-(+)-Glucose	Vitamin C	Biotechnical Resources
D-(+)-Glucose	Vitamin C	Reichstein
D-Diethyl tartrate	Laurallene	Suzuki
D-Glucal	Oseltamivir	Chen–Liu
D-Glucosamine	(−)-Tetracycline HCl	Tatsuta
D-Glutamic acid	(−)-Nakadomarin A	Dixon
D-Mannitol	Oseltamivir	Mandai G1
D-Mannose	(−)-Swainsonine	Hashimoto

(continued)

(continued)

Target Natural Product	Starting Natural Product	Plan
D-Mannose	(−)-Swainsonine	Mootoo
D-Ribose	(−)-Oleocanthal	Smith G1
D-Ribose	(−)-Oleocanthal	Smith G2
D-Ribose	Oseltamivir	Fang G1
D-Serine	(−)-Kainic acid	Baldwin
D-Serine	(−)-Kainic acid	Benetti
D-Serine	(−)-Kainic acid	Hoppe
D-Serine	(−)-Kainic acid	Montgomery
Ferulic acid	Vanillin	Givaudan–Roure
Ferulic acid	Curcumin	Lampe
Furfuraldehyde	Oseltamivir	Okamura
Furfuraldehyde	Rosefuran	Scharf
Furfuraldehyde	(−)-Kainic acid	Ogasawara G3
Furfuraldehyde	(−)-Kainic acid	Fukuyama G1
Guanine	Theobromine	Bredereck G2
Juglone	(−)-Tetracycline HCl	Shemyakin
Juglone	(±)-Terramycin	Muxfeldt
Juglone	12a-Deoxytetracycline	Stork
L-(+)-Cysteine	Cephalosporin C	Woodward
L-(S)-Proline	(−)-Swainsonine	Cossy G1
L-(S)-Proline	(−)-Swainsonine	Cossy G2
L-(S)-Proline	(S)-Nicotine	Chavdarian
L-(S)-Tryptophan	Lysergic acid	Rebek
L-Aspartic acid	(−)-Kainic acid	Fukuyama G2
L-Aspartic acid	(+)-Thienamycin	Reider
L-Aspartic acid	(+)-Thienamycin	Salzmann
L-Aspartic acid	(−)-Kainic acid	Fukuyama G3
L-Aspartic acid	(−)-Kainic acid	Knight
L-Glutamic acid	(+)-Swainsonine	Ikota
L-Glutamic acid	(+)-Swainsonine	Hirama
L-Glutamic acid	(−)-Kainic acid	Cossy
L-Glutamic acid	(−)-Kainic acid	Jung
L-Glutamic acid	(−)-Kainic acid	Oppolzer
L-Glutamic acid	(−)-Kainic acid	Rubio
L-Glutamic acid	(−)-Kainic acid	Yoo
L-Methionine	(−)-Kainic acid	Naito
L-Phenylalanine	(−)-Kainic acid	Fukuyama G2
L-Serine	(−)-Nakadomarin A	Nishida G2
L-Serine	(−)-Paclitaxel	Mukaiyama
L-Serine	(−)-Swainsonine	Ham
L-Serine	Quisqualic acid	Baldwin
L-Serine	(−)-Kainic acid	Taylor
L-Serine	(−)-Kainic acid	Hanessian
L-Serine	(−)-Kainic acid	Takano G4
L-Threonine	(+)-Thienamycin	Hanessian

(continued)

Target Natural Product	Starting Natural Product	Plan
L-Tyrosine	(−)-Stenine	Wipf
Menthofuran	(−)-Swainsonine	Roush
Mucic acid	(±)-Nicotine	Pictet
Myrcene	(−)-Menthol	Takasago G1
Myrcene	Salsolene oxide	Wender
Phytol	Vitamin E	Woggon
Phytol	Vitamin E	Knight
Phytol	Vitamin E	Roche G1
Phytol	Vitamin E	Roche G2
Quinic acid	Oseltamivir	Gilead
Quinic acid	Oseltamivir	Roche G1
Quinic acid	Oseltamivir	Roche G2
Quinic acid	Reserpine	Hanessian
Shikimic acid	Oseltamivir	Roche G3
Shikimic acid	Oseltamivir	Shi
Thymol	(±)-Menthol	Brode
Tropinone	(±)-Epibatidine	Bai
Uric Acid	Coenzyme A	Khorana–Todd
Vanillin	L-DOPA	Roche
Vanillin	Zingerone	Nomura
Vanillin	Zingerone	Mukaiyama
Verbenone	(−)-Paclitaxel	Wender

A.7   MONOCYCLIC AND BICYCLIC RING ENUMERATION DATABASE

A.7.1   LIST OF PLANS INVOLVING CONSTRUCTING THREE-MEMBERED MONOCYCLIC RINGS

Type	Target	Step	Plan
[0 + 3]	(2S,4R)-Hypoglycin A	5	Baldwin
[1 + 2]	Colchicine	11	Banwell
[1 + 2]	Amphidinolide P	11	Trost
[1 + 2]	Bullvalene	5	Von Doering
[1 + 2]	(±)-Hypoglycin A	1	Abbott
[1 + 2]	(±)-Hypoglycin A	6	Black
[1 + 2]	(±)-Longifolene	8	Karimi
[1 + 2]	(±)-Longifolene	10	McMurry
[1 + 2]	Salsolene oxide	7	Wender
[1 + 2]	Salsolene oxide	13	Paquette
[1 + 2]	Azadirachtin	49	Ley G1
[1 + 2]	Azadirachtin	49	Ley G2

A.7.2   List of Plans Involving Constructing Four-Membered Monocyclic Rings

Type	Target	Step	Plan
[0 + 4]	Cephalosporin C	10	Woodward
[0 + 4]	(−)-Paclitaxel	14	Danishefsky
[0 + 4]	(+)-Paclitaxel	44	Holton
[0 + 4]	(−)-Paclitaxel	46	Kuwajima
[0 + 4]	(−)-Paclitaxel	53	Mukaiyama
[0 + 4]	(−)-Paclitaxel	32	Nicolaou
[0 + 4]	(−)-Paclitaxel	38	Takahashi
[0 + 4]	(−)-Paclitaxel	32	Wender
[0 + 4]	(+)-Thienamycin	3	Hanessian
[0 + 4]	(+)-Thienamycin	7	Melillo–Shinkai
[0 + 4]	(+)-Thienamycin	3	Reider
[0 + 4]	(+)-Thienamycin	3	Salzmann
[0 + 4]	(+)-Thienamycin	9	Tatsuta
[1 + 3]	No entries		
[2 + 2]	Basketene	4	Dauben
[2 + 2]	Basketene	3	Masamune
[2 + 2]	(±)-Hirsutene	2	Franck-Neumann
[2 + 2]	(±)-Hirsutene	1	Iyoda
[2 + 2]	(±)-Longifolene	4	Oppolzer G1
[2 + 2]	[1.1.1]-Propellane	1	Wiberg
[2 + 2]	(±)-Thienamycin	1	Schmitt
[2 + 2]	Tropolone	1	DuPont
[2 + 2]	Tropolone	2	Minns

A.7.3   LIST OF PLANS INVOLVING CONSTRUCTING FIVE-MEMBERED MONOCYCLIC RINGS

Type	Target	Step	Plan
[0 + 5]	(−)-Cocaine	2	Rapoport
[0 + 5]	(−)-Codeine	16	Chida
[0 + 5]	(−)-Codeine	23	Gates
[0 + 5]	(−)-Codeine	8	Mulzer
[0 + 5]	(−)-Codeine	15	Ogasawara
[0 + 5]	(−)-Codeine	10	Overman
[0 + 5]	(−)-Codeine	19	Taber
[0 + 5]	(−)-Codeine	8	Trost
[0 + 5]	(−)-Hirsutene	1	Weinges
[0 + 5]	(−)-Kainic acid	15	Anderson
[0 + 5]	(−)-Kainic acid	7	Bachi
[0 + 5]	(−)-Kainic acid	10	Baldwin
[0 + 5]	(−)-Kainic acid	3	Clayden
[0 + 5]	(−)-Kainic acid	1	Cossy
[0 + 5]	(−)-Kainic acid	14	Fukuyama G2
[0 + 5]	(−)-Kainic acid	17	Fukuyama G3
[0 + 5]	(−)-Kainic acid	5	Ganem
[0 + 5]	(−)-Kainic acid	8	Hanessian
[0 + 5]	(−)-Kainic acid	12	Hoppe
[0 + 5]	(−)-Kainic acid	8	Jung
[0 + 5]	(−)-Kainic acid	10	Knight
[0 + 5]	(−)-Kainic acid	8	Montgomery
[0 + 5]	(−)-Kainic acid	8	Naito
[0 + 5]	(−)-Kainic acid	12	Ogasawara G1
[0 + 5]	(−)-Kainic acid	12	Ogasawara G2
[0 + 5]	(−)-Kainic acid	14	Ogasawara G3
[0 + 5]	(−)-Kainic acid	7	Oppolzer
[0 + 5]	(−)-Kainic acid	8	Poisson G1
[0 + 5]	(−)-Kainic acid	8	Poisson G2
[0 + 5]	(−)-Kainic acid	2	Rubio
[0 + 5]	(−)-Kainic acid	10	Takano G4
[0 + 5]	(−)-Kainic acid	11	Taylor
[0 + 5]	(−)-Nakadomarin A	13	Nishida G2
[0 + 5]	(−)-Nakadomarin A	17	Nishida G2
[0 + 5]	(−)-Nakadomarin A	20	Nishida G2
[0 + 5]	(−)-Nakadomarin A	2	Dixon
[0 + 5]	(−)-Nakadomarin A	8**	Dixon
[0 + 5]	(−)-Nepetalactone	2	Cavill
[0 + 5]	(−)-Nepetalactone	2	Pickett G2
[0 + 5]	(−)-Physostigmine	5	Fuji
[0 + 5]	(−)-Physostigmine	14	Fuji
[0 + 5]	(−)-Physostigmine	12	Nakada
[0 + 5]	(−)-Physostigmine	19	Nakada
[0 + 5]	(−)-Physostigmine	5	Nakagawa
[0 + 5]	(−)-Physostigmine	6	Overman
[0 + 5]	(−)-Physostigmine	12	Takano G1
[0 + 5]	(−)-Physostigmine	9	Takano G2
[0 + 5]	(−)-Physostigmine	1	Yu

(continued)

(continued)

Type	Target	Step	Plan
[0 + 5]	(–)-Physostigmine	5	Yu
[0 + 5]	(–)-Platensimycin	7	Nicolaou G4
[0 + 5]	(–)-Platensimycin	9	Nicolaou G3
[0 + 5]	(–)-Platensimycin	14	Corey
[0 + 5]	(–)-Platensimycin	7	Nicolaou G6
[0 + 5]	(–)-Stenine	4	Wipf
[0 + 5]	(–)-Stenine	16	Morimoto
[0 + 5]	(–)-Swainsonine	6	Blechert
[0 + 5]	(–)-Swainsonine	11	Carretero
[0 + 5]	(–)-Swainsonine	17	Cossy G1
[0 + 5]	(–)-Swainsonine	9	Cossy G2
[0 + 5]	(–)-Swainsonine	16	Ferreira
[0 + 5]	(–)-Swainsonine	12	Fleet
[0 + 5]	(–)-Swainsonine	15	Ikota
[0 + 5]	(–)-Swainsonine	18	Kibayashi
[0 + 5]	(–)-Swainsonine	10	Pyne G1
[0 + 5]	(–)-Swainsonine	8	Pyne G2
[0 + 5]	(–)-Swainsonine	4	Reiser
[0 + 5]	(–)-Swainsonine	7	Richardson
[0 + 5]	(–)-Swainsonine	14	Riera
[0 + 5]	(–)-Swainsonine	17	Sharpless
[0 + 5]	(–)-Swainsonine	16	Suami G1
[0 + 5]	(–)-Swainsonine	13	Suami G2
[0 + 5]	(–)-Swainsonine	15	Takaya
[0 + 5]	(–)-Swainsonine	11	Trost
[0 + 5]	(–)-Swainsonine	11	Zhou
[0 + 5]	(–)-Swainsonine	20	Ham
[0 + 5]	(–)-Vindoline	5	Boger
[0 + 5]	(–)-Vindoline	11	Fukuyama G1
[0 + 5]	(–)-Vindoline	9	Fukuyama G2
[0 + 5]	(+)-Codeine	8	Hudlicky
[0 + 5]	(+)-Codeine	13	White
[0 + 5]	(+)-Hirsutene	13	Greene
[0 + 5]	(+)-Hirsutene	4	List
[0 + 5]	(+)-Hirsutene	8	Nozoe
[0 + 5]	(+)-Kainic acid	5	Trost
[0 + 5]	(+)-Laurallene	26	Crimmins
[0 +5]	(+)-Laurallene	35	Takeda
[0 + 5]	(+)-Laurallene	33	Suzuki
[0 + 5]	(+)-Nakadomarin A	10	Kerr
[0 + 5]	(+)-Nakadomarin A	15	Kerr
[0 + 5]	(+)-Nakadomarin A	11	Nishida G1
[0 + 5]	(+)-Nakadomarin A	28	Nishida G1
[0 + 5]	(+)-Nakadomarin A	15	Dixon
[0 + 5]	(+)-Nepetalactone	5	Sakai
[0 + 5]	(+)-Swainsonine	16	Hirama
[0 + 5]	(+)-Thienamycin	13	Hanessian
[0 + 5]	(+)-Thienamycin	16	Melillo–Shinkai
[0 + 5]	(+)-Thienamycin	11	Reider
[0 + 5]	(+)-Thienamycin	17	Salzmann

(continued)

Type	Target	Step	Plan
[0 + 5]	(+)-Thienamycin	12	Tatsuta
[0 + 5]	(±)-Codeine	9	Magnus
[0 + 5]	(±)-Codeine	16	Fukuyama
[0 + 5]	(±)-Codeine	8	Iorga
[0 + 5]	(±)-Codeine	11	Rice
[0 + 5]	(±)-Codeine	4	Stork
[0 + 5]	(±)-Hirsutene	8	Cossy
[0 + 5]	(±)-Hirsutene	14	Fukumoto
[0 + 5]	(±)-Hirsutene	22	Fukumoto
[0 + 5]	(±)-Hirsutene	6	Funk
[0 + 5]	(±)-Hirsutene	8	Hewson
[0 + 5]	(±)-Hirsutene	15	Hewson
[0 + 5]	(±)-Hirsutene	3	Hua
[0 + 5]	(±)-Hirsutene	11	Hua
[0 + 5]	(±)-Hirsutene	7	Hudlicky
[0 + 5]	(±)-Hirsutene	8	Hudlicky
[0 + 5]	(±)-Hirsutene	5	Krische
[0 + 5]	(±)-Hirsutene	10	Krische
[0 + 5]	(±)-Hirsutene	13	Leonard
[0 + 5]	(±)-Hirsutene	4	Ley
[0 + 5]	(±)-Hirsutene	9	Little
[0 + 5]	(±)-Hirsutene	13	Magnus
[0 + 5]	(±)-Hirsutene	15	Oppolzer
[0 + 5]	(±) Hirsutene	7	Sarkar
[0 + 5]	(±)-Hirsutene	11	Sarkar
[0 + 5]	(±)-Hirsutene	14	Sarkar
[0 + 5]	(±)-Hirsutene	19	Sternbach
[0 + 5]	(±)-Kainic acid	4	Tilve
[0 + 5]	(±)-Lycopodine	15	Ayer
[0 + 5]	(±)-Nicotine	6	Craig
[0 + 5]	(±)-Nicotine	7	Ley
[0 + 5]	(±)-Nicotine	4	Spath
[0 + 5]	(±)-Physostigmine	19	Fukumoto
[0 + 5]	(±)-Physostigmine	5	Johnson
[0 + 5]	(±)-Physostigmine	2	Julian
[0 + 5]	(±)-Physostigmine	7	Julian
[0 + 5]	(±)-Physostigmine	1	Mukai
[0 + 5]	(±)-Physostigmine	8	Speckamp
[0 + 5]	(±)-Platensimycin	12	Nicolaou G2
[0 + 5]	(±)-Platensimycin	5	Nicolaou G1
[0 + 5]	(±)-Platensimycin	6	Njardarson
[0 + 5]	(±)-Platensimycin	10	Njardarson
[0 + 5]	(±)-Stenine	6	Aube
[0 + 5]	(±)-Stenine	11	Padwa
[0 + 5]	(±)-Stenine	11	Hart
[0 + 5]	(±)-Swainsonine	7	Bermejo
[0 + 5]	(±)-Thienamycin	13	Schmitt
[0 + 5]	(S)-Nicotine	4	Delgado
[0 + 5]	(S)-Nicotine	3	Helmchen

(*continued*)

(continued)

Type	Target	Step	Plan
[0 + 5]	(S)-Nicotine	5	Lebreton
[0 + 5]	Alpha-santonin	12	Marshall
[0 + 5]	Alpha-santonin	11	Abe
[0 + 5]	Aspidophytine	7*	Corey
[0 + 5]	Azadirachtin	26***	Ley G1
[0 + 5]	Azadirachtin	35***	Ley G2
[0 + 5]	Estrone	17	Bachmann
[0 + 5]	Estrone	11	Johnson G1
[0 + 5]	Estrone	8	Johnson G2
[0 + 5]	Estrone	12	Torgov
[0 + 5]	Lysergic acid	3	Oppolzer
[0 + 5]	Patchouli alcohol	12	Buchi
[0 + 5]	Patchouli alcohol	15	Buchi
[0 + 5]	Reserpine	8***	Woodward
[0 + 5]	Rosefuran	7	Barco
[0 + 5]	Rosefuran	2	Birch
[0 + 5]	Rosefuran	2	Buchi
[0 + 5]	Rosefuran	2	Botsugan
[0 + 5]	Rosefuran	5	Iriye G2
[0 + 5]	Rosefuran	7	Iriye G1
[0 + 5]	Rosefuran	4	Marshall
[0 + 5]	Rosefuran	4	Salerno
[0 + 5]	Rosefuran	4	Scharf
[0 + 5]	Rosefuran	7	Takano
[0 + 5]	Rosefuran	2	Takeda
[0 + 5]	Rosefuran	4	Trost
[0 + 5]	Rosefuran	2	Tsukasa
[0 + 5]	Rosefuran	4	Wenkert
[0 + 5]	Serotonin	5	Harley-Mason
[0 + 5]	Silphinene	13	Yamamura
[0 + 5]	Silphinene	6	Sternbach
[0 + 5]	Silphinene	9	Nagarajan
[0 + 5]	Silphinene	13	Nagarajan
[0 + 5]	Silphinene	2	Crimmins
[0 + 5]	Silphinene	6	Crimmins
[0 + 5]	Silphinene	5	Franck-Neumann
[0 + 5]	Silphinene	15	Franck-Neumann
[0 + 5]	Silphinene	16	Ito
[0 + 5]	Strychnine	15	Bonjoch
[0 + 5]	Strychnine	7*	Fukuyama
[0 + 5]	Strychnine	8	Mori
[0 + 5]	Strychnine	11	Mori
[0 + 5]	Strychnine	22	Overman
[0 + 5]	Strychnine	3	Rawal
[0 + 5]	Strychnine	20	Shibasaki
[0 + 5]	Strychnine	21	Shibasaki
[0 + 5]	Strychnine	9	Vollhardt
[0 + 5]	Strychnine	7	Woodward

(continued)

Type	Target	Step	Plan
[0 + 5]	Triquinacene	9	Woodward
[1 + 4]	(−)-Physostigmine	8	Marino
[1 + 4]	(−)-Physostigmine	8	Overman
[1 + 4]	(−)-Physostigmine	10	Takano G1
[1 + 4]	(−)-Physostigmine	6	Takano G2
[1 + 4]	(−)-Physostigmine	6	Trost
[1 + 4]	(−)-Stenine	17	Wipf
[1 + 4]	(−)-Stenine	19	Morimoto
[1 + 4]	(−)-Swainsonine	6	Cha
[1 + 4]	(−)-Swainsonine	5	Poisson
[1 + 4]	(+)-Swainsonine	19	Vankar
[1 + 4]	(±)-Hirsutene	7	Oppolzer
[1 + 4]	(±)-Physostigmine	17	Fukumoto
[1 + 4]	(±)-Physostigmine	3	Mukai
[1 + 4]	(±)-Physostigmine	3	Speckamp
[1 + 4]	(±)-Stenine	18	Hart
[1 + 4]	(S)-Nicotine	1	Jacob
[1 + 4]	(S)-Nicotine	3	Loh
[1 + 4]	Aspidophytine	14	Corey
[1 + 4]	Strychnine	5	Bonjoch
[1 + 4]	Tropinone	2	Bazilevskaya G1
[1 + 4]	Tropinone	5	Karrer
[1 + 4]	Tropinone	3	Parker
[1 + 4]	Tropinone	2	Willstatter G2
[2 + 3]	Antipyrine	2	Vogel
[2 + 3]	(−)-Kainic acid	7	Benetti
[2 + 3]	(−)-Kainic acid	3	Fukuyama G1
[2 + 3]	(−)-Kainic acid	5	Lautens
[2 + 3]	(−)-Kainic acid	9	Takano G1
[2 + 3]	(−)-Physostigmine	1	Trost
[2 + 3]	(+)-Cocaine	8	Pearson
[2 + 3]	(±)-Hirsutene	1	Cossy
[2 + 3]	(±)-Hirsutene	6	Fitjer
[2 + 3]	(±)-Hirsutene	1	Franck-Neumann
[2 + 3]	(±)-Hirsutene	8	Magnus
[2 + 3]	(±)-Hirsutene	1	Mehta
[2 + 3]	(±)-Kainic acid	6	Monn
[2 + 3]	Estrone	1*	Corey
[2 + 3]	Estrone	1	Danishefsky
[2 + 3]	Estrone	6*	Smith
[2 + 3]	Phenytoin	1	Biltz
[2 + 3]	Phenytoin	1	Lambert G1
[2 + 3]	Phenytoin	1	Lambert G2
[2 + 3]	Phenytoin	1	Olah
[2 + 3]	Rosefuran	2	Okazaki
[2 + 3]	Rosefuran	1	Pattenden
[2 + 3]	Rosefuran	2	Wong
[2 + 3]	Silphinene	1	Ito

(continued)

(continued)

Type	Target	Step	Plan
[2 + 3]	Silphinene	1	Paquette
[2 + 3]	Silphinene	6	Paquette
[2 + 3]	Strychnine	1	Woodward
[2 + 3]	Triquinacene	2	Woodward
[1 + 1 + 3]	No entries		
[1 + 2 + 2]	(±)-Hirsutene	1	Cohen
[1 + 1 + 1 + 2]	Phenytoin	1	Parke-Davis

A.7.4   LIST OF PLANS INVOLVING CONSTRUCTING SIX-MEMBERED MONOCYCLIC RINGS

Type	Target	Step	Plan
[0 + 6]	Alizarin	3	Fierz-David
[0 + 6]	(−)-*Cis*-gamma-irone	7	Vidari G1
[0 + 6]	(−)-Codeine	8	Chida
[0 + 6]	(−)-Codeine	19	Chida
[0 + 6]	(−)-Codeine	24	Chida
[0 + 6]	(−)-Codeine	12	Gates
[0 + 6]	(−)-Codeine	2	Mulzer
[0 + 6]	(−)-Codeine	14	Mulzer
[0 + 6]	(−)-Codeine	19	Ogasawara
[0 + 6]	(−)-Codeine	22	Ogasawara
[0 + 6]	(−)-Codeine	8	Overman
[0 + 6]	(−)-Codeine	13	Parker
[0 + 6]	(−)-Codeine	11	Trost
[0 + 6]	(−)-Codeine	15	Trost
[0 + 6]	(−)-Epibatidine	6	Loh
[0 + 6]	(−)-Epibatidine	5	Szantay G1
[0 + 6]	(−)-Epibatidine	4	Szantay G2
[0 + 6]	(−)-Menthol	4	Takasago G1
[0 + 6]	(−)-Menthol	1	Takasago G2
[0 + 6]	(−)-Nepetalactone	10	Pickett G2
[0 + 6]	(−)-Paclitaxel	5	Kuwajima
[0 + 6]	(−)-Paclitaxel	26	Mukaiyama
[0 + 6]	(−)-Paclitaxel	37	Mukaiyama
[0 + 6]	(−)-Paclitaxel	4	Takahashi
[0 + 6]	(−)-Paclitaxel	9*	Takahashi
[0 + 6]	(−)-Paclitaxel	3	Wender
[0 + 6]	(−)-Paclitaxel	25	Wender
[0 + 6]	(−)-Paroxetine	11	Beak
[0 + 6]	(−)-Paroxetine	4	Buchwald
[0 + 6]	(−)-Paroxetine	3	Krische
[0 + 6]	(−)-Paroxetine	8	Liu
[0 + 6]	(−)-Paroxetine	2	Szantay
[0 + 6]	(−)-Platensimycin	6	Lee
[0 + 6]	(−)-Quinine	16	Stork
[0 + 6]	(−)-Swainsonine	10	Blechert
[0 + 6]	(−)-Swainsonine	9	Carretero
[0 + 6]	(−)-Swainsonine	8	Cha
[0 + 6]	(−)-Swainsonine	14	Cossy G1

(continued)

Type	Target	Step	Plan
[0 + 6]	(−)-Swainsonine	7	Cossy G2
[0 + 6]	(−)-Swainsonine	10	Ferreira
[0 + 6]	(−)-Swainsonine	11	Fleet
[0 + 6]	(−)-Swainsonine	25	Ikota
[0 + 6]	(−)-Swainsonine	12	Kibayashi
[0 + 6]	(−)-Swainsonine	12	Poisson
[0 + 6]	(−)-Swainsonine	17	Pyne G1
[0 + 6]	(−)-Swainsonine	7	Pyne G2
[0 + 6]	(−)-Swainsonine	14	Richardson
[0 + 6]	(−)-Swainsonine	6	Riera
[0 + 6]	(−)-Swainsonine	19	Sharpless
[0 + 6]	(−)-Swainsonine	20	Suami G1
[0 + 6]	(−)-Swainsonine	17	Suami G2
[0 + 6]	(−)-Swainsonine	13	Takaya
[0 + 6]	(−)-Swainsonine	15	Trost
[0 + 6]	(−)-Swainsonine	14	Ham
[0 + 6]	(−)-Swainsonine	3	Zhou
[0 + 6]	(−)-Tetracycline HCl	6	Myers G1
[0 + 6]	(−)-Tetracycline HCl	15	Shemyakin
[0 + 6]	(−)-Tetracycline HCl	15	Tatsuta
[0 + 6]	(−)-Vindoline	16	Fukuyama G1
[0 + 6]	(−)-Vindoline	15	Fukuyama G2
[0 + 6]	(+)-Codeine	10	Hudlicky
[0 + 6]	(+)-Codeine	15	Hudlicky
[0 + 6]	(+)-Codeine	5	White
[0 + 6]	(+)-Codeine	10	White
[0 + 6]	(+)-Codeine	23	White
[0 + 6]	(+)-Cylindricine	7	Trost
[0 + 6]	(+)-Longifolene	7	Fallis
[0 + 6]	(+)-Nepetalactone	15	Sakai
[0 + 6]	(+)-Paclitaxel	35	Holton
[0 + 6]	(+)-Sertraline	5	Pfizer G1
[0 + 6]	(+)-Swainsonine	11	Hirama
[0 + 6]	(+)-Swainsonine	21	Vankar
[0 + 6]	(+)-*Trans*-alpha-irone	7	Firmenich G1
[0 + 6]	(+)-*Trans*-gamma-irone	7	Vidari G2
[0 + 6]	(±)-*Cis*-alpha-irone	6	Firmenich G2
[0 + 6]	(±)-*Cis*-gamma-irone	6	Garnero–Joulain
[0 + 6]	(±)-*Cis*-gamma-irone	2	Monti G1
[0 + 6]	(±)-Codeine	11	Iorga
[0 + 6]	(±)-Codeine	16	Iorga
[0 + 6]	(±)-Codeine	2	Rice
[0 + 6]	(±)-Codeine	9	Rice
[0 + 6]	(±)-Codeine	18	Stork
[0 + 6]	(±)-Codeine	9	Magnus
[0 + 6]	(±)-Coniine	6	Beauchemin
[0 + 6]	(±)-Longifolene	2	Karimi
[0 + 6]	(±)-Lycopodine	7	Stork
[0 + 6]	(±)-Lycopodine	14	Stork

(continued)

(continued)

Type	Target	Step	Plan
[0 + 6]	(±)-Platensimycin	6	Njardarson
[0 + 6]	(±)-Quinine	5	Jacobsen
[0 + 6]	(±)-Quinine	7	Krische
[0 + 6]	(±)-Quinine	5	Woodward
[0 + 6]	(±)-Sparteine	6	Blakemore
[0 + 6]	(±)-Sparteine	9	Bohlmann–Gallagher
[0 + 6]	(±)-Sparteine	13	Bohlmann–Gallagher
[0 + 6]	(±)-Sparteine	4	Koomen
[0 + 6]	(±)-Sparteine	6	Koomen
[0 + 6]	(±)-Sparteine	3	Leonard G1
[0 + 6]	(±)-Sparteine	3	Takatsu
[0 + 6]	(±)-*Trans*-alpha-irone	5	Givaudan G1
[0 + 6]	(±)-*Trans*-alpha-irone	5	Givaudan G3
[0 + 6]	(±)-*Trans*-gamma-irone	3	Brenna G1
[0 + 6]	(±)-*Trans*-gamma-irone	6	Takazawa
[0 + 6]	(*R*)-Coniine	8	Gallagher
[0 + 6]	(*S*)-Coniine	1	Aketa
[0 + 6]	(*S*)-Coniine	3	Couture
[0 + 6]	(*S*)-Coniine	3	Gramain
[0 + 6]	(*S*)-Coniine	6	Knochel
[0 + 6]	(*S*)-Coniine	7	Takahata
[0 + 6]	(*S*)-Nicotine	6	Chavdarian
[0 + 6]	Alpha-santonin	8	Marshall
[0 + 6]	Cephalosporin C	12	Woodward
[0 + 6]	Estrone	6	Bachmann
[0 + 6]	Estrone	10	Bachmann
[0 + 6]	Estrone	7	Corey
[0 + 6]	Estrone	6	Danishefsky
[0 + 6]	Estrone	10	Danishefsky
[0 + 6]	Estrone	12	Danishefsky
[0 + 6]	Estrone	7	Johnson G1
[0 + 6]	Estrone	6	Torgov
[0 + 6]	Lysergic acid	7	Hendrickson
[0 + 6]	Lysergic acid	4	Kurihara
[0 + 6]	Lysergic acid	9	Kurihara
[0 + 6]	Lysergic acid	4	Ninomiya
[0 + 6]	Lysergic acid	10	Ninomiya
[0 + 6]	Lysergic acid	4	Ortar
[0 + 6]	Lysergic acid	7	Ortar
[0 + 6]	Lysergic acid	4	Ramage
[0 + 6]	Lysergic acid	14	Ramage
[0 + 6]	Lysergic acid	3	Rebek
[0 + 6]	Lysergic acid	7	Rebek
[0 + 6]	Lysergic acid	2	Szantay
[0 + 6]	Lysergic acid	9	Szantay
[0 + 6]	Lysergic acid	25	Fukuyama G1
[0 + 6]	Lysergic acid	5	Fukuyama G2
[0 + 6]	Oseltamivir	14	Mandai G1
[0 + 6]	Oseltamivir	12	Mandai G2
[0 + 6]	Oseltamivir phosphate	11	Fang G1

(continued)

Type	Target	Step	Plan
[0 + 6]	Papaverine	7	Dean
[0 + 6]	Papaverine	7	Decker–Wahl
[0 + 6]	Papaverine	7	Kindlcr–Peschke–Pal
[0 + 6]	Papaverine	8	Pictet–Gams
[0 + 6]	Papaverine	6	Redel–Bouteville
[0 + 6]	Patchouli alcohol	4*	Firmenich
[0 + 6]	Patchouli alcohol	3	Bertrand
[0 + 6]	Patchouli alcohol	8	Srikrishna
[0 + 6]	Patchouli alcohol	11	Danishefsky
[0 + 6]	Patchouli alcohol	15	Mirrington
[0 + 6]	Patchouli alcohol	14	Rao
[0 + 6]	Patchouli alcohol	10	Magee
[0 + 6]	Reserpine	7	Fraser-Reid
[0 + 6]	Reserpine	29	Fraser-Reid
[0 + 6]	Reserpine	30	Fraser-Reid
[0 + 6]	Reserpine	10	Liao
[0 + 6]	Reserpine	20	Martin
[0 + 6]	Reserpine	21	Mehta
[0 + 6]	Reserpine	13	Pearlman
[0 + 6]	Reserpine	20	Shea
[0 + 6]	Reserpine	11	Stork
[0 + 6]	Reserpine	7	Wender
[0 + 6]	Reserpine	16	Wender
[0 + 6]	Reserpine	13	Woodward
[0 + 6]	Resveratrol	2	Moreno-Manas
[0 + 6]	Salsolene oxide	8	Paquette
[0 + 6]	Strychnine	13	Bodwell
[0 + 6]	Strychnine	13	Bonjoch
[0 + 6]	Strychnine	8	Kuehne
[0 + 6]	Strychnine	12	Kuehne
[0 + 6]	Strychnine	13	Magnus
[0 + 6]	Strychnine	4	Martin
[0 + 6]	Strychnine	18	Mori
[0 + 6]	Strychnine	21	Mori
[0 + 6]	Strychnine	18	Overman
[0 + 6]	Strychnine	14	Padwa
[0 + 6]	Strychnine	10	Rawal
[0 + 6]	Strychnine	12	Rawal
[0 + 6]	Strychnine	20	Shibasaki
[0 + 6]	Strychnine	12	Stork
[0 + 6]	Strychnine	12	Vollhardt
[0 + 6]	Strychnine	11	Woodward
[0 + 6]	Strychnine	16	Woodward
[0 + 6]	Strychnine	24	Woodward
[0 + 6]	Vitamin E	14	Knight
[0 + 6]	Vitamin E	5	Roche G1
[0 + 6]	Vitamin E	14	Roche G2
[0 + 6]	Vitamin E	7	Tietze
[1 + 5]	(−)-Codeine	5	Ogasawara

(continued)

(continued)

Type	Target	Step	Plan
[1 + 5]	(−)-Codeine	7	Overman
[1 + 5]	(−)-Codeine	1	Trost
[1 + 5]	(±)-Codeine	6	Magnus
[1 + 5]	(S)-Coniine	2	Buchwald
[1 + 5]	(S)-Coniine	1	Hurvois
[1 + 5]	(S)-Coniine	1	Husson
[1 + 5]	(S)-Coniine	3	Meyers
[1 + 5]	Lysergic acid	7	Fukuyama G1
[1 + 5]	(+)-Nakadomarin A	19	Kerr
[1 + 5]	(−)-Nepetalactone	10	Cavill
[1 + 5]	(±)-Nepetalactone	11	Sakan
[1 + 5]	(−)-Paroxetine	1	Amat G1
[1 + 5]	(+)-Paroxetine	1	Amat G2
[1 + 5]	(−)-Paroxetine	3	Jorgensen
[1 + 5]	(−)-Paroxetine	5	Simpkins
[1 + 5]	(+)-Paroxetine	6	Yu
[1 + 5]	(±)-Quinine	11	Acharya
[1 + 5]	Reserpine	9	Liao
[1 + 5]	Reserpine	20	Mehta
[1 + 5]	Reserpine	12	Pearlman
[1 + 5]	Reserpine	10	Stork
[1 + 5]	Reserpine	12	Woodward
[1 + 5]	(±)-Sparteine	6	Takatsu
[2 + 4]	Absinthin	5	Zhai
[2 + 4]	Alizarin	2	Fierz-David
[2 + 4]	Azadirachtin	26***	Ley G2
[2 + 4]	(±)-Cis-alpha-irone	2	Givaudan G2
[2 + 4]	(−)-Cis-alpha-irone	2	Kiyota
[2 + 4]	(+)-Cis-alpha-irone	4	Ohtsuka
[2 + 4]	Alpha-santonin	4	Abe
[2 + 4]	Basketene	2	Dauben
[2 + 4]	Basketene	2	Masamune
[2 + 4]	(−)-Codeine	11	Gates
[2 + 4]	(−)-Codeine	4	Mulzer
[2 + 4]	(−)-Epibatidine	4	Broka
[2 + 4]	(−)-Epibatidine	2	Corey
[2 + 4]	(−)-Epibatidine	4	Evans
[2 + 4]	(−)-Epibatidine	2	Hoashi
[2 + 4]	(−)-Hirsutene	2	Banwell
[2 + 4]	(±)-Cis-gamma-irone	2	Givaudan
[2 + 4]	(±)-Cis-gamma-irone	1	Gosselin
[2 + 4]	(−)-Cis-gamma-irone	2	Kiyota
[2 + 4]	(±)-Cis-gamma-irone	3	Mori
[2 + 4]	(±)-Longifolene	2	McMurry
[2 + 4]	(±)-Lycopodine	6	Stork
[2 + 4]	Oseltamivir phosphate	1	Banwell
[2 + 4]	Oseltamivir phosphate	2	Fukuyama
[2 + 4]	Oseltamivir phosphate	5	Okamura
[2 + 4]	Oseltamivir phosphate	1	Roche G4
[2 + 4]	Oseltamivir phosphate	1	Trost

(continued)

Type	Target	Step	Plan
[2 + 4]	Oseltamivir phosphate	1	Shibasaki G4
[2 + 4]	Oseltamivir phosphate	1	Shibasaki G5
[2 + 4]	(−)-Paclitaxel	5	Nicolaou
[2 + 4]	(−)-Paclitaxel	10*	Nicolaou
[2 + 4]	(−)-Physostigmine	4	Fuji
[2 + 4]	(−)-Paroxetine	7	Gallagher
[2 + 4]	(+)-Paroxetine	2	Takasu–Ihara
[2 + 4]	Reserpine	1	Liao
[2 + 4]	Reserpine	3	Mehta
[2 + 4]	Reserpine	3	Stork
[2 + 4]	Reserpine	1	Woodward
[2 + 4]	(+)-Sertraline	3	Buchwald
[2 + 4]	(+)-Sertraline	3	Pfizer G2
[2 + 4]	(+)-Sertraline	3	Pfizer G3
[2 + 4]	(+)-Sertraline	3	Lautens
[2 + 4]	(±)-Stenine	4	Hart
[2 + 4]	(−)-Stenine	11	Morimoto
[2 + 4]	Strychnine	25	Fukuyama
[2 + 4]	Strychnine	33	Magnus
[2 + 4]	Strychnine	16	Martin
[2 + 4]	Strychnine	26	Overman
[2 + 4]	Strychnine	17	Padwa
[2 + 4]	Strychnine	31	Shibasaki
[2 + 4]	Strychnine	16	Stork
[2 + 4]	(−)-Tetracycline HCl	13	Myers G1
[2 + 4]	(−)-Tetracycline HCl	12	Myers G2
[2 + 4]	(−)-Tetracycline HCl	1	Shemyakin
[2 + 4]	(−)-Tetracycline HCl	18	Tatsuta
[2 + 4]	(−)-Tetracycline HCl	20	Tatsuta
[2 + 4]	Vitamin E	5	Woggon
[2 + 4]	Ricinine	1	Schroeter
[2 + 4]	Ricinine	1	Spath
[3 + 3]	Barbituric acid	1	Dickey–Gray
[3 + 3]	(S)-Coniine	3	Shipman
[3 + 3]	(±)-Lycopodine	1	Ayer
[3 + 3]	(±)-Lycopodine	3	Ayer
[3 + 3]	(±)-Lycopodine	2	Stork
[3 + 3]	(±)-Menthol	1	BASF
[3 + 3]	(−)-Menthol	1	Sayo–Matsumoto
[3 + 3]	(±)-Menthol	1	Tanabe–Seiyaku
[3 + 3]	Oseltamivir phosphate	2	Kann
[3 + 3]	(−)-Paroxetine	4	Chang
[3 + 3]	(−)-Paroxetine	1	Gotor
[3 + 3]	(−)-Paroxetine	5	Vesely–Moyano–Rios
[3 + 3]	Vitamin E	11	Roche G2
[1 + 1 + 4]	(−)-Paroxetine	3	Dixon
[1 + 1 + 4]	(−)-Nakadomarin A	12	Dixon

(continued)

(continued)

Type	Target	Step	Plan
[1 + 2 + 3]	(±)-Sparteine	2	Clemo–Raper
[1 + 2 + 3]	(±)-Sparteine	2	Leonard G1
[2 + 2 + 2]	Oseltamivir	3	Hayashi
[1 + 1 + 1 + 3]	No entries		
[1 + 1 + 2 + 2]	No entries		
[1 + 2 + 1 + 2]	No entries		

A.7.5   LIST OF PLANS INVOLVING CONSTRUCTING SEVEN-MEMBERED MONOCYCLIC RINGS

Type	Target	Step	Plan
[0 + 7]	Colchicine	11	Cha
[0 + 7]	Colchicine	5	Banwell
[0 + 7]	Colchicine	4	Roussel
[0 + 7]	Colchicine	11	Roussel
[0 + 7]	Colchicine	11	Scott
[0 + 7]	Colchicine	12	Van Tamelen
[0 + 7]	Colchicine	11	Woodward
[0 + 7]	Colchicine	15	Woodward
[0 + 7]	(±)-Ibogamine	18	Hanaoka
[0 + 7]	(+)-Ibogamine	12	Hodgson
[0 + 7]	(±)-Ibogamine	17	Nagata
[0 + 7]	(±)-Ibogamine	4	Trost
[0 + 7]	(±)-Longifolene	7	Corey
[0 + 7]	(+)-Longifolene	12	Money
[0 + 7]	(±)-Longifolene	10	Schultz G1
[0 + 7]	(±)-Longifolene	13	Schultz G2
[0 + 7]	(±)-Platensimycin	2	Njardarson
[0 + 7]	(−)-Stenine	24	Wipf
[0 + 7]	(±)-Stenine	25	Hart
[0 + 7]	(−)-Stenine	30	Morimoto
[0 + 7]	Strychnine	15	Bodwell
[0 + 7]	Strychnine	23	Fukuyama
[0 + 7]	Strychnine	20	Kuehne
[0 + 7]	Strychnine	31	Magnus
[0 + 7]	Strychnine	15	Martin
[0 + 7]	Strychnine	24	Mori
[0 + 7]	Strychnine	25	Overman
[0 + 7]	Strychnine	16	Padwa
[0 + 7]	Strychnine	14	Rawal
[0 + 7]	Strychnine	29	Shibasaki
[0 + 7]	Strychnine	15	Stork
[0 + 7]	Strychnine	14	Vollhardt
[0 + 7]	Strychnine	29	Woodward
[0 + 7]	Tropolone	4	Cram
[1 + 6]	Bullvalene	1	Von Doering
[1 + 6]	Bullvalene	6	Von Doering
[2 + 5]	Bullvalene	2	Schroeder
[2 + 5]	Colchicine	1	Eschenmoser

(continued)

Type	Target	Step	Plan
[2 + 5]	Colchicine	4*	Scott
[2 + 5]	Colchicine	1	Van Tamelen
[3 + 4]	Colchicine	14	Cha
[1 + 1 + 5]	No entries		
[1 + 2 + 4]	No entries		
[1 + 3 + 3]	No entries		
[2 + 2 + 3]	No entries		
[1 + 1 + 1 + 4]	No entries		
[1 + 1 + 2 + 3]	No entries		
[1 + 2 + 1 + 3]	No entries		
[1 + 2 + 2 + 2]	No entries		

A.7.6 LIST OF PLANS INVOLVING CONSTRUCTING EIGHT-MEMBERED MONOCYCLIC RINGS

Type	Target	Step	Plan
[0 + 8]	(−)-Paclitaxel	35	Danishefsky
[0 + 8]	(−)-Paclitaxel	13	Kuwajima
[0 + 8]	(−)-Paclitaxel	20	Mukaiyama
[0 + 8]	(−)-Paclitaxel	20	Nicolaou
[0 + 8]	(−)-Paclitaxel	28	Takahashi
[0 + 8]	(−)-Paclitaxel	5	Wender
[0 + 8]	(+)-Nakadomarin A	24	Kerr
[0 + 8]	(+)-Nakadomarin A	34	Nishida G1
[0 + 8]	(−)-Nakadomarin A	31	Nishida G2
[0 + 8]	(−)-Nakadomarin A	9	Dixon
[0 + 8]	Patchouli alcohol	2	Buchi
[0 + 8]	(+)-Laurallene	12	Crimmins
[0 + 8]	(+)-Laurallene	19	Suzuki
[1 + 7]	No entries		
[2 + 6]	Salsolene oxide	9	Paquette
[3 + 5]	(+)-Laurallene	10	Takeda
[4 + 4]	No entries		
[1 + 1 + 6]	No entries		
[1 + 2 + 5]	No entries		
[1 + 3 + 4]	No entries		
[2 + 2 + 4]	No entries		
[2 + 3 + 3]	No entries		
[1 + 1 + 1 + 5]	No entries		
[1 + 1 + 2 + 4]	No entries		
[1 + 1 + 3 + 3]	No entries		
[1 + 2 + 1 + 4]	No entries		
[1 + 2 + 2 + 3]	No entries		
[1 + 2 + 3 + 2]	No entries		
[1 + 3 + 1 + 3]	No entries		
[2 + 2 + 2 + 2]	Basketene	1	Dauben
[2 + 2 + 2 + 2]	Basketene	1	Masamune
[2 + 2 + 2 + 2]	Bullvalene	1	Schroeder
[2 + 2 + 2 + 2]	Cyclooctatetraene	1	Reppe

A.7.7 LIST OF PLANS INVOLVING CONSTRUCTING NINE-MEMBERED MONOCYCLIC RINGS

Type	Target	Step	Plan
[0 + 9]	No entries		
[1 + 8]	No entries		
[2 + 7]	No entries		
[3 + 6]	No entries		
[4 + 5]	No entries		
[1 + 1 + 7]	No entries		
[1 + 2 + 6]	No entries		
[1 + 3 + 5]	No entries		
[1 + 4 + 4]	No entries		
[2 + 2 + 5]	No entries		
[2 + 3 + 4]	No entries		
[3 + 3 + 3]	No entries		
[1 + 1 + 1 + 6]	No entries		
[1 + 1 + 2 + 5]	No entries		
[1 + 1 + 3 + 4]	No entries		
[1 + 2 + 1 + 5]	No entries		
[1 + 2 + 2 + 4]	No entries		
[1 + 2 + 3 + 3]	No entries		
[1 + 2 + 4 + 2]	No entries		
[1 + 3 + 1 + 4]	No entries		
[1 + 3 + 2 + 3]	No entries		
[2 + 2 + 2 + 3]	No entries		

A.7.8 LIST OF PLANS INVOLVING CONSTRUCTING 10-MEMBERED MONOCYCLIC RINGS

Type	Target	Step	Plan
[0 + 10]	No entries		
[1 + 9]	No entries		
[2 + 8]	No entries		
[3 + 7]	No entries		
[4 + 6]	No entries		
[5 + 5]	No entries		
[1 + 1 + 8]	No entries		
[1 + 2 + 7]	No entries		
[1 + 3 + 6]	No entries		
[1 + 4 + 5]	No entries		
[2 + 2 + 6]	No entries		
[2 + 3 + 5]	No entries		
[2 + 4 + 4]	No entries		
[3 + 3 + 4]	No entries		
[1 + 1 + 1 + 7]	No entries		
[1 + 1 + 2 + 6]	No entries		
[1 + 1 + 3 + 5]	No entries		
[1 + 1 + 4 + 4]	No entries		
[1 + 2 + 1 + 6]	No entries		
[1 + 2 + 2 + 5]	No entries		
[1 + 2 + 3 + 4]	No entries		

(continued)

Type	Target	Step	Plan
[1 + 2 + 4 + 3]	No entries		
[1 + 2 + 5 + 2]	No entries		
[1 + 3 + 1 + 5]	No entries		
[1 + 3 + 3 + 3]	No entries		
[1 + 4 + 1 + 4]	No entries		
[1 + 4 + 2 + 3]	No entries		
[2 + 2 + 2 + 4]	No entries		
[2 + 2 + 3 + 3]	No entries		
[2 + 3 + 2 + 3]	No entries		

A.7.9 LIST OF PLANS INVOLVING CONSTRUCTING 15-MEMBERED MONOCYCLIC RINGS

Type	Target	Step	Plan
[0 + 15]	(+)-Nakadomarin A	28	Kerr
[0 + 15]	(+)-Nakadomarin A	40	Nishida G1
[0 + 15]	(−)-Nakadomarin A	35	Nishida G2
[0 + 15]	(−)-Nakadomarin A	16	Dixon

A.7.10 LIST OF PLANS INVOLVING CONSTRUCTING BICYCLIC RINGS

Bicyclic Ring	Type	Target	Step	Plan
[1.1.1]	[(3 + 1) + (3 + 1)]	[1.1.1]-Propellane	8	Wiberg
[10.3.1]	[(6 + 0) + (10 + 3)]	Amphidinolide P	15	Trost
[2.1.0]	[(5 + 0) + (4 + 2)]	(−)-Codeine	12	Parker
[2.2.1]	[(5 + 0) + (5 + 0)]	(−)-Epibatidine	7	Aggarwal
[2.2.1]	[(5 + 0) + (5 + 0)]	(−)-Epibatidine	20	Broka
[2.2.1]	[(4 + 2) + (3 + 2)]	(−)-Epibatidine	3	Carroll
[2.2.1]	[(5 + 0) + (5 + 0)]	(−)-Epibatidine	7	Corey
[2.2.1]	[(5 + 0) + (5 + 0)]	(−)-Epibatidine	15	Evans
[2.2.1]	[(5 + 0) + (5 + 0)]	(−)-Epibatidine	4	Fletcher
[2.2.1]	[(4 + 2) + (3 + 2)]	(±)-Epibatidine	1	Giblin
[2.2.1]	[(5 + 0) + (5 + 0)]	(−)-Epibatidine	10	Hoashi
[2.2.1]	[(5 + 0) + (5 + 0)]	(−)-Epibatidine	11	Kibayashi
[2.2.1]	[(5 + 0) + (5 + 0)]	(−)-Epibatidine	16	Kosugi
[2.2.1]	[(5 + 0) + (5 + 0)]	(−)-Epibatidine	10	Loh
[2.2.1]	[(5 + 0) + (5 + 0)]	(−)-Epibatidine	3	Olivo
[2.2.1]	[(4 + 2) + (3 + 2)]	(±)-Epibatidine	5	Pandey
[2.2.1]	[(4 + 2) + (3 + 2)]	(±)-Epibatidine	1	Regan
[2.2.1]	[(4 + 2) + (3 + 2)]	(±)-Epibatidine	3	Shen
[2.2.1]	[(5 + 0) + (5 + 0)]	(−)-Epibatidine	9	Szantay G1
[2.2.1]	[(5 + 0) + (5 + 0)]	(−)-Epibatidine	12	Szantay G2
[2.2.1]	[(4 + 2) + (3 + 2)]	(±)-Epibatidine	1	Trudell
[2.2.1]	[(5 + 0) + (5 + 0)]	(±)-Longifolene	9	Corey
[2.2.1]	[(5 + 0) + (5 + 0)]	(±)-Longifolene	6	Karimi
[2.2.1]	[(4 + 2) + (3 + 2)]	(±)-Longifolene	1	Liu
[2.2.1]	[(5 + 0) + (5 + 0)]	(±)-Longifolene	9	McMurry
[2.2.1]	[(6 + 0) + (5 + 0)]	(+)-Longifolene	23	Money

(*continued*)

(continued)

Bicyclic Ring	Type	Target	Step	Plan
[2.2.1]	[(6 + 0) + (5 + 0)]	(±)-Longifolene	5	Oppolzer G1
[2.2.1]	[(6 + 0) + (5 + 0)]	(±)-Longifolene	7	Oppolzer G2
[2.2.1]	[(5 + 0) + (5 + 0)]	(±)-Longifolene	11	Schultz G1
[2.2.1]	[(5 + 0) + (5 + 0)]	(±)-Longifolene	14	Schultz G2
[2.2.1]	[(5 + 0) + (6 + 0)]	(±)-Fenchone	8	Ruzicka
[2.2.1]	[(4 + 2) + (3 + 2)]	(±)-Fenchone	1	Buchbauer
[2.2.1]	[(6 + 0) + (5 + 0)]	(±)-Fenchone	4	Komppa
[2.2.1]	[(6 + 0) + (5 + 0)]	(+)-Fenchone	8	Cocker
[2.2.2]	[(4 + 2) + (4 + 2)]	Barrelene	3	Zimmerman
[2.2.2]	[(6 + 0) + (6 + 0)]	(±)-Ibogamine	9	Hanaoka
[2.2.2]	[(6 + 0) + (6 + 0)]	(+)-Ibogamine	7	Hodgson
[2.2.2]	[(6 + 0) + (6 + 0)]	(±)-Ibogamine	13	Nagata
[2.2.2]	[(6 + 0) + (6 + 0)]	(±)-Ibogamine	3	Trost
[2.2.2]	[(6 + 0) + (6 + 0)]	(±)-Quinine	21	Acharya
[2.2.2]	[(6 + 0) + (6 + 0)]	(±)-Quinine	16	Jacobsen
[2.2.2]	[(6 + 0) + (6 + 0)]	(±)-Quinine	16	Krische
[2.2.2]	[(6 + 0) + (6 + 0)]	(−)-Quinine	20	Stork
[2.2.2]	[(6 + 0) + (6 + 0)]	(±)-Quinine	23	Woodward
[2.2.2]	[(6 + 0) + (6 + 0)]	(−)-Quinine	15	Aggawal
[2.2.2]	[(6 + 0) + (6 + 0)]	(+)-Twistane	1	Tichy
[2.2.2]	[(6 + 0) + (6 + 0)]	(±)-Twistane	1	Whitlock
[2.2.2]	[(6 + 0) + (6 + 0)]	(−)-Quinine	16	Gates
[2.2.2]	[(6 + 0) + (6 + 0)]	(−)-Quinine	15	Merck
[2.2.2]	[(6 + 0) + (6 + 0)]	(−)-Quinine	14	Martin
[2.2.2]	[(4 + 2) + (4 + 2)]	Patchouli alcohol	2	Srikrishna
[2.2.2]	[(4 + 2) + (4 + 2)]	Patchouli alcohol	3	Danishefsky
[2.2.2]	[(4 + 2) + (4 + 2)]	Patchouli alcohol	3	Mirrington
[2.2.2]	[(4 + 2) + (4 + 2)]	Patchouli alcohol	4	Rao
[2.2.2]	[(4 + 2) + (4 + 2)]	Patchouli alcohol	6	Magee
[2.2.2]	[(4 + 2) + (4 + 2)]	Patchouli alcohol	6	Bertrand
[3.1.0]	[(5 + 0) + (2 + 1)]	(−)-Alpha-thujone	9	Oppolzer
[3.2.1]	[(7 + 0) + (6 + 0)]	Tropinone	3	Bazilevskaya G1
[3.2.1]	[(4 + 1) + (4 + 3)]	Tropinone	3	Bazilevskaya G2
[3.2.1]	[(4 + 1) + (4 + 3)]	Tropinone	3	Bazilevskaya G3
[3.2.1]	[(4 + 1) + (4 + 3)]	Tropinone	4	Bazilevskaya G4
[3.2.1]	[(7 + 0) + (6 + 0)]	Tropinone	6	Karrer
[3.2.1]	[(7 + 0) + (6 + 0)]	Tropinone	5	Parker
[3.2.1]	[(4 + 1) + (4 + 3)]	Tropinone	1	Robinson
[3.2.1]	[(5 + 0) + (6 + 0)]	Tropinone	10	Willstatter G1
[3.2.1]	[(7 + 0) + (6 + 0)]	Tropinone	4	Willstatter G2
[3.2.1]	[5 + 0]	Patchouli alcohol	15	Buchi
[3.2.1]	[(4 + 1) + (4 + 3)]	(+)-Cocaine	3	Casale
[3.2.1]	[(7 + 0) + (6 + 0)]	(+)-Cocaine	12	Pearson
[3.2.1]	[(7 + 0) + (6 + 0)]	(−)-Cocaine	9	Rapoport
[3.2.1]	[(7 + 0) + (6 + 0)]	(±)-Cocaine	6	Tufariello
[3.2.1]	[(4 + 1) + (4 + 3)]	(±)-Cocaine	2	Willstatter
[3.2.1]	[(6 + 0) + (5 + 0)]	(±)-Platensimycin	6	Njardarson
[3.2.1]	[(7 + 0) + (5 + 0)]	Azadirachtin	48	Ley G1
[3.2.1]	[(7 + 0) + (5 + 0)]	Azadirachtin	48	Ley G2
[3.3.0]	[(5 + 0) + (5 + 0)]	(−)-Hirsutene	11	Banwell

(continued)

Bicyclic Ring	Type	Target	Step	Plan
[3.3.0]	[(5 + 0) + (5 + 0)]	(±)-Hirsutene	5	Fitjer
[3.3.0]	[(2 + 2 + 1) + (5 + 0)]	(+)-Hirsutene	3	Greene
[3.3.0]	[(5 + 0) + (5 + 0)]	(±)-Hirsutene	8	Paquette
[3.3.0]	[(3 + 2) + (5 + 0)]	(±)-Hirsutene	8	Sternbach
[3.3.0]	[(5 + 0) + (3 + 2)]	Silphinene	2	Yamamura
[3.3.0]	[(5 + 0) + (3 + 2)]	Silphinene	3	Sternbach
[3.3.0]	[(3 + 2) + (4 + 2)]	Triquinacene	3	Woodward
[3.3.0]	[5 + 0]	(−)-Platensimycin	5	Yamamoto
[3.3.0]	[(5 + 0) + (2 + 2 + 1)]	(−)-Platensimycin	6	Lee
[3.3.0]	[5 + 0]	(−)-Platensimycin	3	Ghosh
[3.3.0]	[(5 + 0) + (2 + 2 + 1)]	(−)-Kainic acid	11	Takano G3
[3.3.0]	[(5 + 0) + (2 + 2 + 1)]	(−)-Kainic acid	8	Yoo
[3.3.0] + [4.4.0]	[6 + 0]	(−)-Platensimycin	10	Nicolaou G5
[3.3.0] + [4.4.0]	[6 + 0]	(−)-Platensimycin	10	Nicolaou G4
[3.3.0] + [4.4.0]	[6 + 0]	(−)-Platensimycin	16	Nicolaou G3
[3.3.0] + [4.4.0]	[6 + 0]	(−)-Platensimycin	13	Mulzer
[3.3.0] + [4.4.0]	[5 + 0]	(−)-Platensimycin	15	Corey
[3.3.0] + [4.4.0]	[5 + 0]	(±)-Platensimycin	8	Snider
[3.3.0] + [4.4.0]	[5 + 0]	(±)-Platensimycin	14	Nicolaou G2
[3.3.0] + [4.4.0]	[6 + 0]	(±)-Platensimycin	10	Nicolaou G1
[3.3.0] + [4.4.0]	[6 + 0]	(−)-Platensimycin	9	Nicolaou G6
[3.3.1]	[(6 + 0) + (5 + 1)]	(±)-Codeine	17	Fukuyama
[3.3.1]	[(6 + 0) + (6 + 0)]	(±)-Sparteine	2	Blakemore
[3.3.1]	[(6 + 0) + (6 + 0)]	(±)-Sparteine	11	Fleming
[3.3.1]	[(6 + 0) + (6 + 0)]	(±)-Sparteine	3	Leonard G2
[3.3.1]	[(6 + 0) + (6 + 0)]	(±)-Sparteine	2	van Tamelen
[4.2.1]	[(7 + 0) + (3 + 2)]	(+)-Longifolene	9	Fallis
[4.2.1]	[(7 + 0) + (3 + 2)]	(±)-Longifolene	5	Johnson
[4.3.0]	[(5 + 1) + (5 + 0)]	(−)-Cylindricine C	11	Molander
[4.3.0]	[(5 + 1) + (5 + 0)]	(+)-Cylindricine C	10	Trost
[4.3.0]	[(4 + 1) + (3 + 2 + 1)]	Srychnine	5	Kuehne
[4.3.0]	[(5 + 0) + (6 + 0)]	Strychnine	33	Magnus
[4.3.0]	[(5 + 0) + (4 + 2)]	Strychnine	11	Martin
[4.3.0]	[(2 + 2 + 1) + (6 + 0)]	Strychnine	20	Overman
[4.3.0]	[(5 + 0) + (4 + 2)]	Strychnine	5	Padwa
[4.3.0]	[(5 + 0) + (4 + 2)]	Strychnine	9	Rawal
[4.3.0]	[(5 + 0) + (4 + 2)]	Strychnine	4	Stork
[4.3.0]	[(6 + 0) + (5 + 0)]	(−)-Vindoline	17	Fukuyama G1
[4.3.0]	[(6 + 0) + (5 + 0)]	(−)-Vindoline	16	Fukuyama G2
[4.3.0]	[(5 + 1) + (5 + 0)]	(−)-Swainsonine	6	GlycoDesign
[4.3.0]	[(5 + 1) + (5 + 0)]	(−)-Swainsonine	10	Hashimoto
[4.3.0]	[(6 + 0) + (5 + 0)]	(−)-Swainsonine	16	Kang
[4.3.0]	[(5 + 1) + (4 + 1)]	(−)-Swainsonine	16	Mootoo
[4.3.0]	[(6 + 0) + (5 + 0)]	(−)-Swainsonine	13	O'Doherty G1
[4.3.0]	[(6 + 0) + (5 + 0)]	(−)-Swainsonine	14	O'Doherty G2
[4.3.0]	[(5 + 1) +(5 + 0)]	(−)-Swainsonine	11	Pearson
[4.3.0]	[(5 + 1) + (4 + 1)]	(−)-Swainsonine	7	Roush
[4.3.0]	[(5 + 0) + (4 + 2)]	(+)-Nepetalactone	3	Mori
[4.3.0]	[(5 + 0) + (6 + 0)]	Lysergic acid	21	Fukuyama G2

(continued)

(continued)

Bicyclic Ring	Type	Target	Step	Plan
[4.3.0]	[5 + 0]	(−)-Platensimycin	3	Nicolaou G5
[4.3.0]	[5 + 0]	(−)-Platensimycin	10	Mulzer
[4.3.0]	[(4 + 2) + (3 + 2)]	(±)-Stenine	5	Padwa
[4.3.0]	[(5 + 0) + (3 + 2 + 1)]	(−)-Kainic acid	13	Takano G2
[4.3.1]	[(6 + 0) + (6 + 0)]	(−)-Codeine	17	Taber
[4.4.0]	[(6 + 0) + (4 + 2)]	Lysergic acid	13	Oppolzer
[4.4.0]	[(5 + 1) + (5 + 1)]	Reserpine	16	Hanessian
[4.4.0]	[(6 + 0) + (4 + 2)]	Reserpine	7	Martin
[4.4.0]	[(6 + 0) + (4 + 2)]	Reserpine	8	Shea
[4.4.0]	[(5 + 1) + (6 + 0)]	(+)-Sparteine	9	Aube
[4.4.0]	[(6 + 0) + (6 + 0)]	(+)-Sparteine	16	Aube
[4.4.0]	[(2 + 2 + 2) + (6 + 0)]	Strychnine	7	Vollhardt
[4.4.0]	[(4 + 2) + (6 + 0)]	(−)-Tetracycline HCl	7	Myers G2
[4.4.0]	[(6 + 0) + (6 + 0)]	(±)-Twistane	8	Deslongschamps G1
[4.4.0]	[(6 + 0) + (6 + 0)]	(±)-Twistane	4	Deslongschamps G2
[4.4.0]	[(6 + 0) + (6 + 0)]	(+)-Twistane	6	Jones
[4.4.0]	[6 + 0]	Patchouli alcohol	8	Srikrishna
[4.4.0]	[6 + 0]	Patchouli alcohol	11	Danishefsky
[4.4.0]	[6 + 0]	Patchouli alcohol	15	Mirrington
[4.4.0]	[6 + 0]	Patchouli alcohol	14	Rao
[4.4.0]	[6 + 0]	Patchouli alcohol	10	Magee
[4.4.0]	[(6 + 0) + (4 + 2)]	Patchouli alcohol	8	Firmenich
[4.4.0]	[(6 + 0) + (8 + 0)]	Patchouli alcohol	11	Bertrand
[4.4.0]	[(6 + 0) + (4 + 2)]	(−)-Platensimycin	10	Yamamoto
[4.4.0]	[6 + 0]	(−)-Platensimycin	7	Nicolaou G5
[4.4.0]	[6 + 0]	(−)-Platensimycin	11	Lee
[4.4.0]	[(6 + 0) + (4 + 2)]	(−)-Platensimycin	22	Ghosh
[4.4.0]	[4 + 2]	(−)-Platensimycin	3	Corey
[4.4.0]	[6 + 0]	(±)-Platensimycin	3	Snider
[4.4.0]	[6 + 0]	(±)-Platensimycin	8	Nicolaou G2
[4.4.0]	[6 + 0]	(±)-Platensimycin	9	Nicolaou G1
[4.4.0]	[(4 + 2) + (6 + 0)]	(±)-Codeine	8	Stork
[4.4.0]	[6 + 0]	(−)-Platensimycin	8	Nicolaou G6
[4.4.0]	[(4 + 2) + (6 + 0)]	Estrone	25	Grieco
[4.4.0]	[(6 + 0) + (6 + 0)]	Estrone	11	Johnson G2
[4.4.0]	[(6 + 0) + (6 + 0)]	Estrone	11	Smith
[4.4.0]	[(4 + 2) + (6 + 0)]	Estrone	16	Kametani
[5.3.0]	[5 + 0]	Patchouli alcohol	12	Buchi
[5.3.1]	[(8 + 0) + (6 + 0)]	(+)-Paclitaxel	17	Holton
[5.3.1]	[(6 + 0) + (4 + 4)]	Salsolene oxide	6	Wender
[5.3.1]	[6 + 2]	Salsolene oxide	9	Paquette
[5.4.0]	[(7 + 0) + (4 + 2)]	Strychnine	18	Bonjoch
[5.4.0]	[(4 + 2) + (7 + 0)]	(±)-Alpha-himachalene	14	Wenkert
[5.4.0]	[(4 + 2) + (7 + 0)]	(±)-Alpha-himachalene	8	Oppolzer
[5.4.0]	[(4 + 2) + (7 + 0)]	(−)-Alpha-himachalene	15	Evans
[5.5.0]	[(7 + 0) + (5 + 2)]	Colchicine	8	Schmalz
[6.1.0]	[2 + 1]	Salsolene oxide	7	Wender
[6.1.0]	[2 + 1]	Salsolene oxide	13	Paquette
[6.2.1]	[(9 + 0) + (5 + 0)]	Eleutherobin	24	Gennari
[6.2.1]	[(9 + 0) + (5 + 0)]	Eleutherobin	18	Danishefsky

A.7.11 LIST OF PLANS INVOLVING CONSTRUCTING TRICYCLIC RINGS

Type	Target	Step	Plan
[(3 + 0) + (3 + 0) + (3 + 0)]	[1.1.1]-Propellane	12	Wiberg
[(4 + 1) + (2 + 2 + 1) + (5 + 0)]	(±)-Hirsutene	4	Yu
[(4 + 1) + (6 + 0) + (6 + 0)]	Strychnine	19	Fukuyama
[(4 + 2) + (4 + 1) + (5 + 1)]	Aspidophytine	12	Corey
[(5 + 0) + (3 + 2) + (5 + 0)]	(±)-Hirsutene	2	Wender
[(5 + 0) + (3 + 2) + (5 + 0)]	Silphinene	2	Wender
[(5 + 0) + (4 + 2) + (6 + 0)]	Strychnine	8	Bodwell
[(5 + 0) + (6 + 0) + (2 + 2 + 1 + 1)]	(−)-Vindoline	14	Boger
[(5 + 0) + [5] + (5 + 0)]	(±)-Hirsutene	13	Curran
[(5 + 0) + [5] + (5 + 0)]	(±)-Hirsutene	2	Mehta
[(5 + 0) + [5] + (5 + 0)]	(−)-Hirsutene	13	Weinges
[(6 + 0) + (4 + 2) + (3 + 2)]	Estrone	11	Pattenden
[(4 + 2) + (4 + 1) + (6 + 1)]	(±)-Stenine	4	Aube

A.7.12 LIST OF PLANS INVOLVING CONSTRUCTING QUADRICYCLIC RINGS

Type	Target	Step	Plan
[(6 + 0) + (3 + 2 + 1) + (3 + 2 + 1) + (6 + 0)]	(±)-Sparteine	1	Anet–Hughes–Ritchie

A.7.13 LIST OF PLANS INVOLVING REARRANGEMENTS FROM ONE RING TYPE TO ANOTHER

From	To	Target	Step	Plan
[2.1.0]	[5]	(±)-Hirsutene	5	Franck-Neumann
[3.2.0]	[7]	(±)-Longifolene	6	Oppolzer G1
[3.2.0]	[7]	(±)-Longifolene	8	Oppolzer G2
[3.2.0]	[7]	Tropolone	2	DuPont
[3.2.0]	[7]	Tropolone	3	Minns
[3.2.0]	[7]	Tropolone	3	Stevens
[3.2.1]	[2.2.1]	(±)-Epibatidine	3	Bai
[3.2.1] + [1]	[4.2.1]	(±)-Longifolene	8	Liu
[4.1.0]	[7]	Colchicine	12	Banwell
[4.1.0]	[7]	Colchicine	11	Eschenmoser
[4.1.0]	[7]	(±)-Longifolene	9	Karimi
[4.1.0]	[7]	(±)-Longifolene	11	McMurry
[4.2.0]	[3.3.0]	(±)-Hirsutene	2	Iyoda
[4.4.0]	[5.3.0]	Absinthin	1	Zhai
[5]	[6]	(±)-Sparteine	9	Fleming
[5]	[6]	(−)-Paroxetine	8	Cossy
[5] + 1	[6]	Zamifenacin	4	Cossy
[5]→[6]	[4.4.0]	Patchouli alcohol	20	Buchi
[6]	[2.2.0]	Dewar benzene	2	Van Tamelen
[6]	[7]	Tropolone	6	Chapman
[6]	[7]	Tropolone	4	Oda
[6]	[6]	Oseltamivir	6	Chen–Liu
[6] + 1	[7]	Caprolactam	2	BP

(continued)

(continued)

From	To	Target	Step	Plan
[6] + 1	[4.1.0] → [7]	Caprolactam	3	Evans
[6] + 1	[7]	Caprolactam	1	Faith
[6] + 1	[7]	Caprolactam	2	Ishii G1
[6] + 1	[7]	Caprolactam	3	Ishii G2
[6] + 1	[7]	Caprolactam	2	Marvel
[6] + 1	[7]	Caprolactam	1	Thomas
[6] + 1	[7]	Tropolone	3	Cook
[6] + 1	[7]	Tropolone	3	Von Doering
[6] + 1	[7]	Tropolone	5	Van Tamelen
[6] + 1	[7]	Tropolone	3	ter Borg
[8]	[4.2.0]	Basketene	2	Dauben
[8]	[4.2.0]	Basketene	2	Masamune
[8]	[3.3.0]	(+)-Hirsutene	10	List
[8]	[3.3.0]	(±)-Hirsutene	3	Weedon
1,2-shift		(±)-Nicotine	3	Pictet
spiro[6,[4.2.0]]	[5.5.0]	Colchicine	8	Evans

A.7.14   List of Named Organic Reactions Involving Ring Constructions

Ring Size	Fragmentation Type	Reaction
3	[0 + 3]	Graham reaction—additive oxidation
3	[0 + 3]	Hoch–Campbell aziridine synthesis
3	[0 + 3]	Payne rearrangement
3	[0 + 3]	Wenker synthesis
3	[1 + 2]	Aziridine synthesis
3	[1 + 2]	Corey–Chaykovsky epoxidation—additive oxidation
3	[1 + 2]	Cyclopropanation [(1 + 2) + (1 + 2)]
3	[1 + 2]	Cyclopropanation [1 + (1 + 2)]
3	[1 + 2]	Cyclopropanation of olefins with diazomethane
3	[1 + 2]	Darzens condensation
3	[1 + 2]	Jacobsen epoxidation—additive oxidation
3	[1 + 2]	Kulinkovich reaction
3	[1 + 2]	Prilezhaev reaction (epoxidation of olefins)—additive oxidation
3	[1 + 2]	Sharpless epoxidation—additive oxidation
3	[1 + 2]	Shi asymmetric epoxidation—additive oxidation
3	[1 + 2]	Simmons–Smith cyclopropanation reaction
3	[1 + 2]	Vinylphosphonium bromide-ketone cyclization
4	[0 + 4]	None
4	[1 + 3]	Azetidine synthesis
4	[2 + 2]	Paterno-Buchi reaction
4	[2 + 2]	Photochemical cyclizations
5	[0 + 5]	Blanc reaction ($n = 1$)
5	[0 + 5]	Cornforth rearrangement
5	[0 + 5]	Cyclic ether synthesis ($n = 1$)
5	[0 + 5]	Dieckmann condensation ($n = 1$)
5	[0 + 5]	Dieckmann–Thorpe reaction
5	[0 + 5]	Edman degradation
5	[0 + 5]	Hofmann–Loffler–Freytag reaction ($n = 1$)
5	[0 + 5]	Nazarov cyclization

(continued)

Ring Size	Fragmentation Type	Reaction
5	[0 + 5]	Paal–Knorr synthesis of furans, thiophenes, and pyrroles
5	[0 + 5]	Ring closing metathesis reaction
5	[0 + 5]	Stahl aerobic oxidative amination
5	[0 + 5]	Stoltz aerobic oxidative etherification
5	[1 + 4]	Nenitzescu indole synthesis
5	[1 + 4]	Oxazolidin-2-one synthesis
5	[1 + 4]	Paal–Knorr synthesis of furans, thiophenes, and pyrroles
5	[2 + 3]	[2,3]-Radical addition
5	[2 + 3]	1,3-Dipolar additions
5	[2+ 3]	1H-Pyrrolo[3,2e]-1,2,4-triazine synthesis
5	[2 + 3]	Alkyne–carbamate–phosphine annulation
5	[2 + 3]	Amide–thioisocyanate-alkylhalide condensation
5	[2 + 3]	Fischer indole synthesis
5	[2 + 3]	Hinsberg thiophene synthesis
5	[2 + 3]	Knorr pyrrole synthesis
5	[2 + 3]	Martinet dioxindole synthesis
5	[2 + 3]	Pellizzari reaction
5	[2 + 3]	Stolle synthesis of indoles
5	[2 + 3]	Synthesis of THF derivatives
5	[2 + 3]	Tandem Passerini–Wittig reaction
5	[2 + 3]	Urech synthesis of hydantoins
5	[2 + 3]	Von Pechmann reaction
5	[1 + 2 + 2]	Aldehyde–alkyne–oxadiazoline annulation
5	[1 + 2 + 2]	Feist–Benary synthesis of pyrroles
5	[1 + 2 + 2]	Gewald aminothiophene synthesis
5	[1 + 2 + 2]	Hantzsch synthesis of pyrroles
5	[1 + 2 + 2]	Indolizine synthesis
5	[1 + 2 + 2]	Pauson–Khand reaction
5	[1 + 2 + 2]	Radziszewski-type reaction using microwave irradiation
5	[1 + 1 + 3]	Aldehyde–malonylurea-isocyanide condensation
5	[1 + 1 + 3]	Asinger reaction
5	[1 + 1 + 3]	Bucherer synthesis of hydantoins
5	[1 + 1 + 3]	Dornow–Wiehler isoxazole synthesis
5	[1 + 1 + 3]	Fused 3-aminoimidazoles
5	[1 + 1 + 3]	Radziszewski reaction
5	[1 + 1 + 3]	Tandem Asinger–Ugi condensation (7-CC)
5	[1 + 1 + 1 + 2]	None
6	[0 + 6]	Aldol condensation (intramolecular)
6	[0 + 6]	Bergmann cyclization
6	[0 + 6]	Bischler–Napieralski synthesis
6	[0 + 6]	Blanc reaction ($n = 2$)
6	[0 + 6]	Cyclic ether synthesis ($n = 2$)
6	[0 + 6]	Dieckmann condensation ($n = 2$)
6	[0 + 6]	Friedel–Crafts acylation (intramolecular)
6	[0 + 6]	Friedel–Crafts alkylation (intramolecular)
6	[0 + 6]	Hofmann–Loffler–Freytag reaction ($n = 2$)
6	[0 + 6]	Mukaiyama aldol reaction (intramolecular)
6	[0 + 6]	Mukaiyama–Michael reaction (intramolecular)

(*continued*)

(continued)

Ring Size	Fragmentation Type	Reaction
6	[0 + 6]	Ring closing metathesis reaction
6	[1 + 5]	Bamberger–Goldschmidt synthesis of 1,2,4-triazines
6	[1 + 5]	Pictet–Spengler isoquinoline synthesis
6	[1 + 5]	Thalidomide synthesis
6	[1 + 5]	Von Richter cinnoline synthesis
6	[2 + 4]	Bamberger–Goldschmidt synthesis of isoquinoline
6	[2 + 4]	Danheiser alkyne-cyclobutanone cyclization
6	[2 + 4]	Danishefsky reaction
6	[2 + 4]	Diels–Alder reaction
6	[2 + 4]	Friedländer synthesis of quinolines
6	[2 + 4]	Heck–Diels–Alder reaction
6	[2 + 4]	Pfitzinger reaction
6	[2 + 4]	Robinson annulation
6	[2 + 4]	Schlittler–Mueller modification of Pomeranz–Fritsch reaction
6	[3 + 3]	Pomeranz-Fritsch reaction
6	[3 + 3]	Skraup reaction
6	[3 + 3]	Von Pechmann condensation
6	[2 + 2 + 2]	1,6-Aldol condensation [2 + 2 + 2]
6	[2 + 2 + 2]	1,6-Mannich condensation
6	[2 + 2 + 2]	Alkene–alkyne annulation
6	[2 + 2 + 2]	Isocyanate–ketone cyclization
6	[2 + 2 + 2]	Isoquinoline + DEAD + benzoquinones
6	[2 + 2 + 2]	Michael–Michael–1,6-Wittig [2 + 2 + 2]
6	[2 + 2 + 2]	Prins reaction
6	[2 + 2 + 2]	Trimerization of alkynes [2 + 2 + 2]
6	[2 + 2 + 2]	Trimerization of arylisocyanates [2 + 2 + 2]
6	[1 + 2 + 3]	1,4-Dihydropyridine synthesis
6	[1 + 2 + 3]	3,4-Dihydropyrimidin-2-(1H)-one synthesis
6	[1 + 2 + 3]	Aldehyde–Meldrum's acid–vinyl ether
6	[1 + 2 + 3]	Aza-Diels–Alder synthesis of tetrahydroquinolines
6	[1 + 2 + 3]	Bamberger-Goldschmidt synthesis of 1,3,5-triazines
6	[1 + 2 + 3]	Biginelli synthesis
6	[1 + 2 + 3]	Cyclohexanone synthesis [1 + 2 + 3]
6	[1 + 2 + 3]	Doebner reaction
6	[1 + 2 + 3]	Doetz reaction
6	[1 + 2 + 3]	Fused benzochromene synthesis
6	[1 + 2 + 3]	Grieco condensation
6	[1 + 2 + 3]	Guareschi–Thorpe condensation
6	[1 + 2 + 3]	Isoquinolinone synthesis
6	[1 + 2 + 3]	Knoevenagel hetero-Diels–Alder reaction
6	[1 + 2 + 3]	Nenitzescu–Praill pyrylium salt synthesis
6	[1 + 2 + 3]	Pinner triazine synthesis
6	[1 + 2 + 3]	Pyrano and furanoquinolines
6	[1 + 2 + 3]	Pyridine synthesis
6	[1 + 2 + 3]	Riehm quinoline synthesis
6	[1 + 1 + 4]	Isoquinolonic acid synthesis
6	[1 + 2 + 1 + 2]	Hantzsch synthesis of dihydropyridines
6	[1 + 1 + 2 + 2]	Chichibabin pyridine synthesis
6	[1 + 1 + 1 + 2]	Nenitzescu–Praill pyridine synthesis via pyrylium salts
6	[1 + 1 + 1 + 2]	Petrenko–Kritschenko reaction

(continued)

Ring Size	Fragmentation Type	Reaction
7	[0 + 7]	Blanc reaction ($n = 3$)
7	[0 + 7]	Cyclic ether synthesis ($n = 3$)
7	[0 + 7]	Dieckmann condensation ($n = 3$)
7	[0 + 7]	Hofmann–Loffler–Freytag reaction ($n = 3$)
7	[0 + 7]	Ring closing metathesis reaction
7	[1 + 6]	None
7	[2 + 5]	Wender–Trost [5 + 2] cycloaddition
7	[3 + 4]	None
7	[1 + 2 + 4]	1,5-Benzodiazepine synthesis
7	[1 + 1 + 5]	None
7	[1 + 3 + 3]	None
7	[2 + 2 +3]	None
8	[0 + 8]	Cyclic ether synthesis ($n = 4$)
8	[0 + 8]	Ring closing metathesis reaction
8	[1 + 7]	None
8	[2 + 6]	None
8	[3 + 5]	None
8	[4 + 4]	Danheiser [4 + 4] annulation
8	[1 + 1 + 6]	None
8	[1 + 2 + 5]	None
8	[1 + 3 + 4]	None
8	[2 + 2 + 4]	None
8	[2 + 3 + 3]	None
8	[2 + 2 + 2 + 2]	Reppe cyclooctatetraene synthesis
8	[1 + 1 + 1 + 5]	None
8	[1 + 1 + 2 + 4]	None
8	[1 + 1 + 3 + 3]	None
8	[1 + 2 + 1 + 4]	None
8	[1 + 2 + 2 + 3]	None
8	[1 + 2 + 3 + 2]	None
8	[1 + 3 + 1 + 3]	None
12	[4 + 4 + 4]	Butadiene trimerization [4 + 4 + 4]
16	[1 + 3 + 1 + 3 + 1 + 3 + 1 + 3]	Rothemund reaction (porphyrins)

A.7.15   ENUMERATION OF BOND DISCONNECTIONS BY RING SIZE FOR MONOCYCLIC RINGS

	Disconnections			
Ring Size	1-Bond	2-Bond	3-Bond	4-Bond
3	1	1		
4	1	2		
5	1	2	2	1
6	1	3	3	3
7	1	3	4	4
8	1	4	5	8
9	1	4	7	10
10	1	5	8	16

A.7.16 ENUMERATION OF MONOCYCLIC RING CONSTRUCTION TEMPLATES

Ring Type	Fragmentation Type	Number of Bond Disconnections	Graphic
	[0 + 3]	1	
	[1 + 2]	2	
	[0 + 4]	1	
	[1 + 3]	2	
	[2 + 2]	2	
	[1 + 1 + 2]	3	
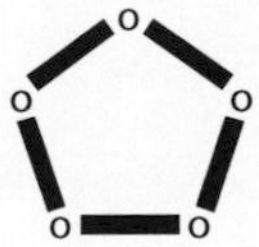	[0 + 5]	1	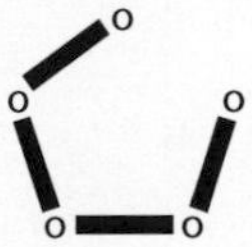
	[1 + 4]	2	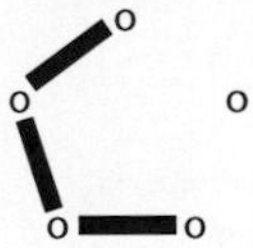
	[2 + 3]	2	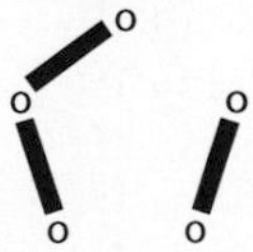
	[1 + 1 + 3]	3	
	[1 + 2 + 2]	3	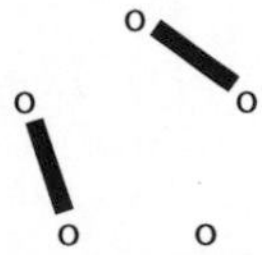
	[1 + 1 + 1 + 2]	4	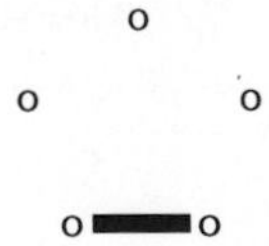

(continued)

Ring Type	Fragmentation Type	Number of Bond Disconnections	Graphic
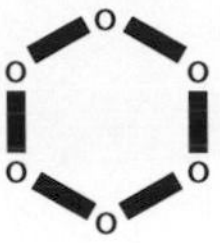	[0 + 6]	1	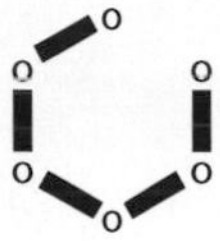
	[1 + 5]	2	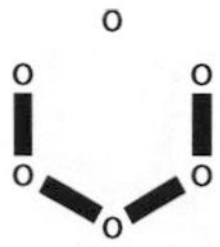
	[2 + 4]	2	
	[3 + 3]	2	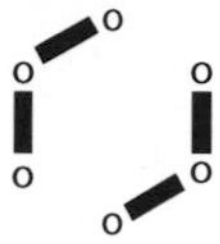
	[1 + 1 + 4]	3	
	[1 + 2 + 3]	3	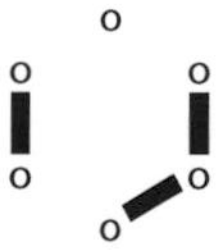
	[2 + 2 + 2]	3	
	[1 + 1 + 1 + 3]	4	
	[1 + 1 + 2 + 2]	4	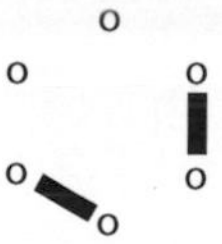

(continued)

(continued)

Ring Type	Fragmentation Type	Number of Bond Disconnections	Graphic
	[1 + 2 + 1 + 2]	4	
	[0 + 7]	1	
	[1 + 6]	2	
	[2 + 5]	2	
	[3 + 4]	2	
	[1 + 1 + 5]	3	
	[1 + 2 + 4]	3	
	[1 + 3 + 3]	3	
	[2 + 2 + 3]	3	

(continued)

Ring Type	Fragmentation Type	Number of Bond Disconnections	Graphic
	[1 + 1 + 1 + 4]	4	
	[1 + 1 + 2 + 3]	4	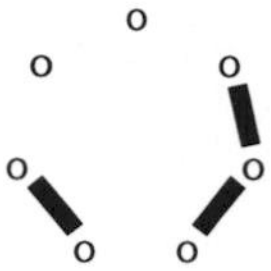
	[1 + 2 + 1 + 3]	4	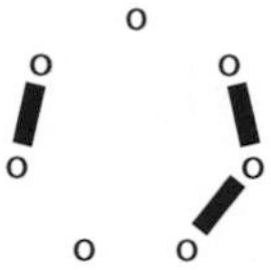
	[1 + 2 + 2 + 2]	4	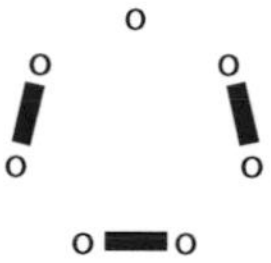
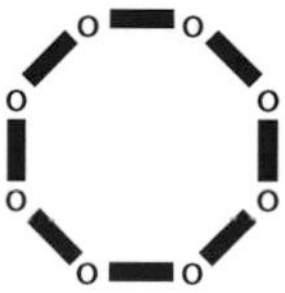	[0 + 8]	1	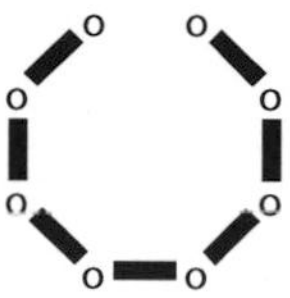
	[1 + 7]	2	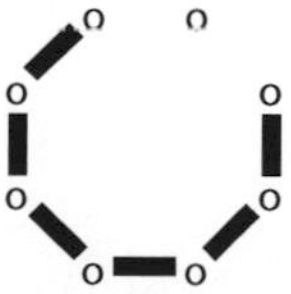
	[2 + 6]	2	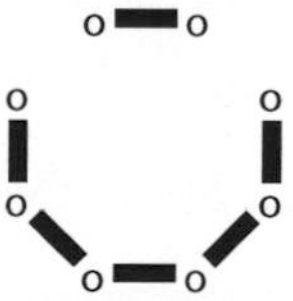
	[3 + 5]	2	
	[4 + 4]	2	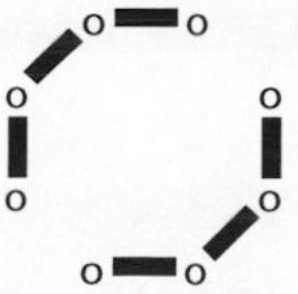

(continued)

(continued)

Ring Type	Fragmentation Type	Number of Bond Disconnections	Graphic
	[1 + 1 + 6]	3	
	[1 + 2 + 5]	3	
	[1 + 3 + 4]	3	
	[2 + 2 + 4]	3	
	[2 + 3 + 3]	3	
	[1 + 1 + 1 + 5]	4	
	[1 + 1 + 2 + 4]	4	
	[1 + 1 + 3 + 3]	4	
	[1 + 2 + 1 + 4]	4	

(continued)

Ring Type	Fragmentation Type	Number of Bond Disconnections	Graphic
	[1 + 2 + 2 + 3]	4	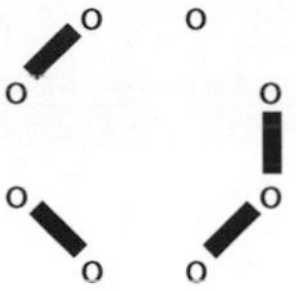
	[1 + 2 + 3 + 2]	4	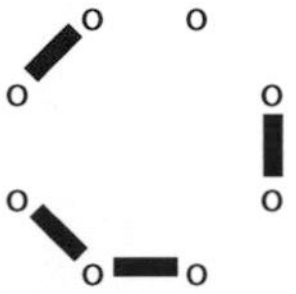
	[1 + 3 + 1 + 3]	4	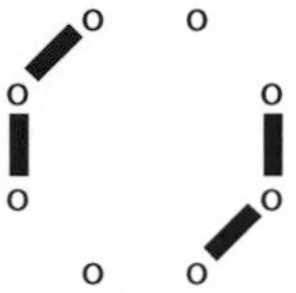
	[2+ 2 + 2 + 2]	4	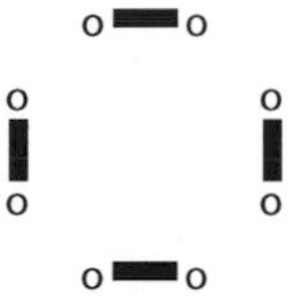
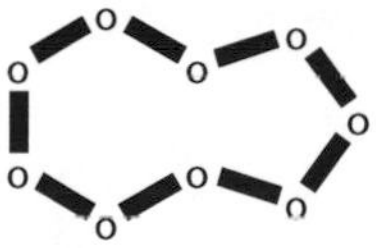	[0 + 9]	1	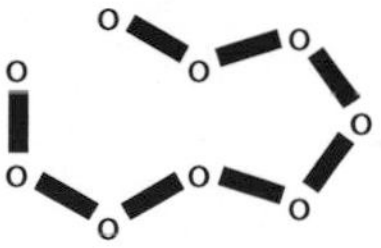
	[1 + 8]	2	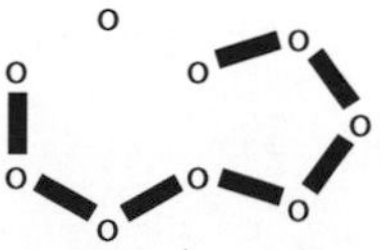
	[2 + 7]	2	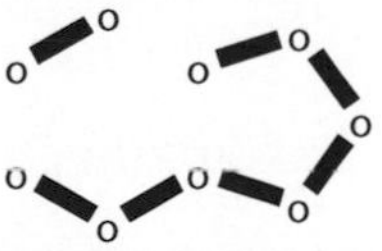
	[3 + 6]	2	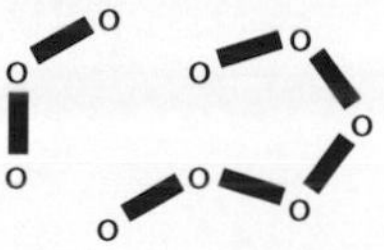
	[4 + 5]	2	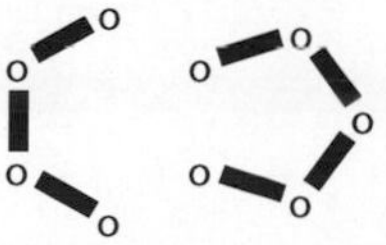

(continued)

(continued)

Ring Type	Fragmentation Type	Number of Bond Disconnections	Graphic
	[1 + 1 + 7]	3	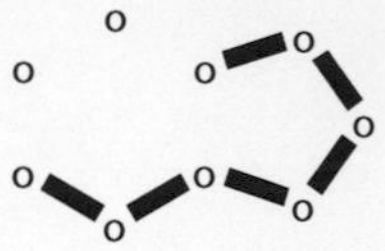
	[1 + 2 + 6]	3	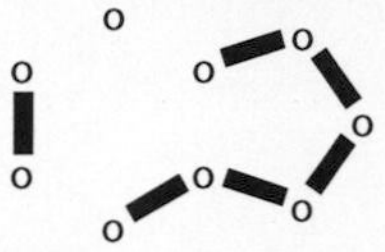
	[1 + 3 + 5]	3	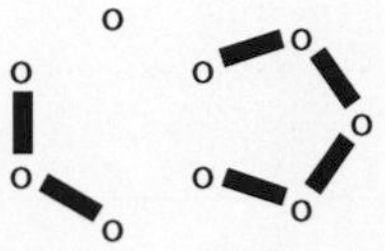
	[1 + 4 + 4]	3	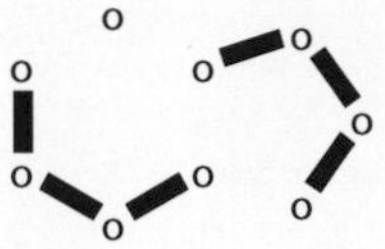
	[2 + 2 + 5]	3	
	[2 + 3 + 4]	3	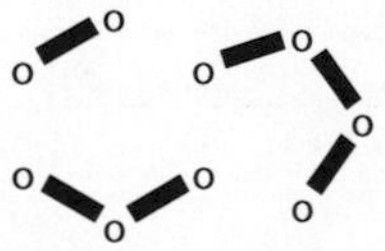
	[3 + 3 + 3]	3	
	[1 + 1 + 1 + 6]	4	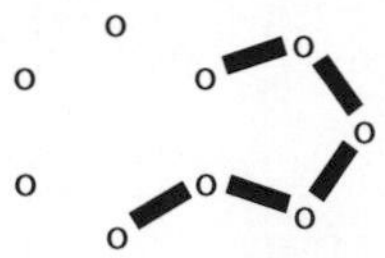
	[1 + 1 + 2 + 5]	4	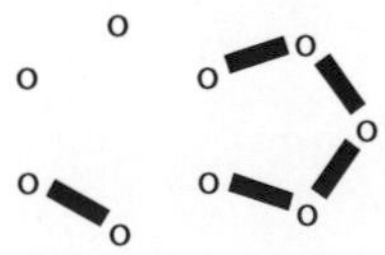
	[1 + 1 + 3 + 4]	4	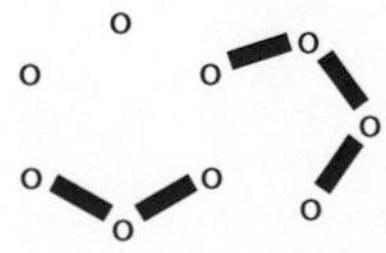

(continued)

Ring Type	Fragmentation Type	Number of Bond Disconnections	Graphic
	[1 + 2 + 1 + 5]	4	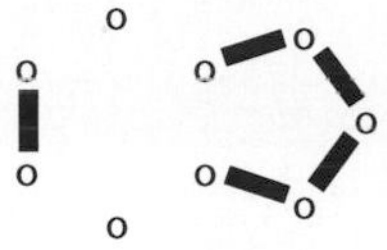
	[1 + 2 + 2 + 4]	4	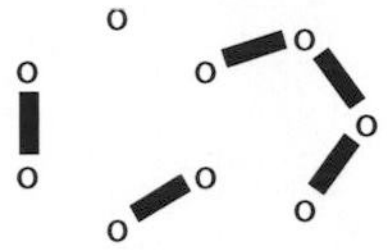
	[1 + 2 + 3 + 3]	4	
	[1 + 2 + 4 + 2]	4	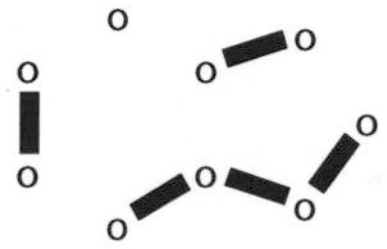
	[1 + 3 + 1 + 4]	4	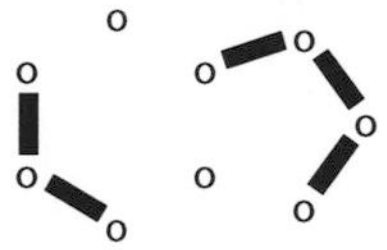
	[1 + 3 + 2 + 3]	4	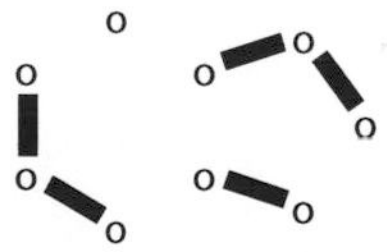
	[2 + 2 + 2 + 3]	4	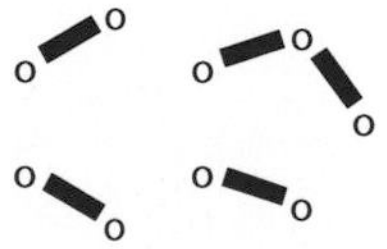
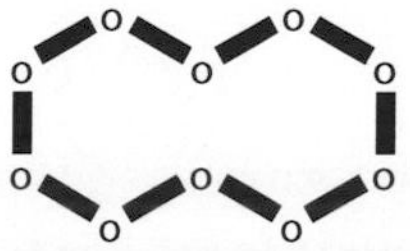	[0 + 10]	1	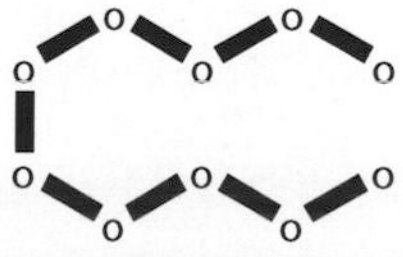
	[1 + 9]	2	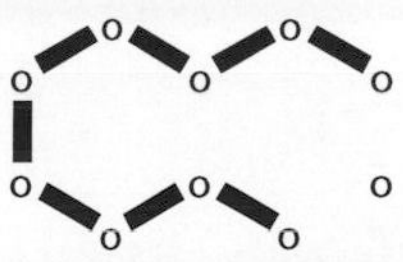
	[2 + 8]	2	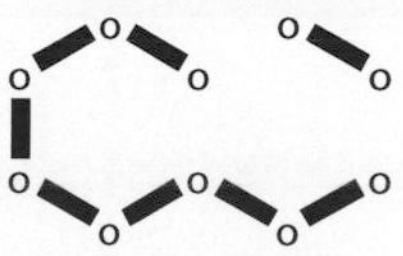

(*continued*)

(continued)

Ring Type	Fragmentation Type	Number of Bond Disconnections	Graphic
	[3 + 7]	2	
	[4 + 6]	2	
	[5 + 5]	2	
	[1 + 1 + 8]	3	
	[1 + 2 + 7]	3	
	[1 + 3 + 6]	3	
	[1 + 4 + 5]	3	
	[2 + 2 + 6]	3	
	[2 + 3 + 5]	3	
	[2 + 4 + 4]	3	

(continued)

Ring Type	Fragmentation Type	Number of Bond Disconnections	Graphic
	[3 + 3 + 4]	3	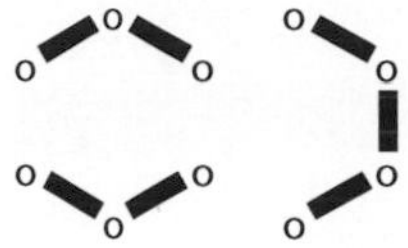
	[1 + 1 + 1 + 7]	4	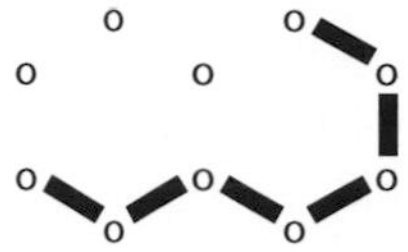
	[1 + 1 + 2 + 6]	4	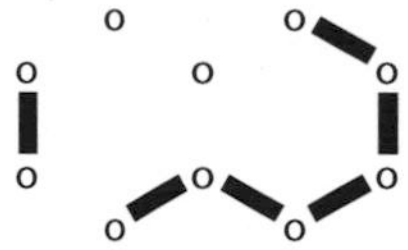
	[1 + 1 + 3 + 5]	4	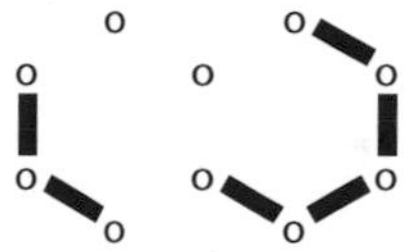
	[1 + 1 + 4 + 4]	4	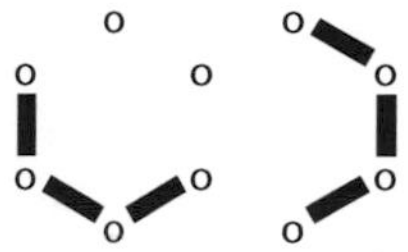
	[1 + 2 + 1 + 6]	4	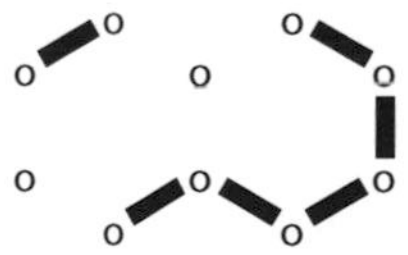
	[1 + 2 + 2 + 5]	4	
	[1 + 2 + 3 + 4]	4	
	[1 + 2 + 4 + 3]	4	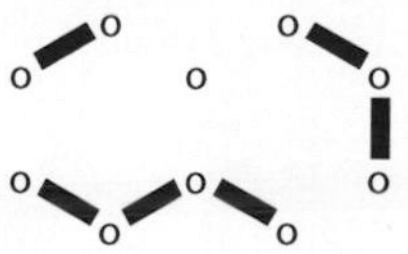
	[1 + 2 + 5 + 2]	4	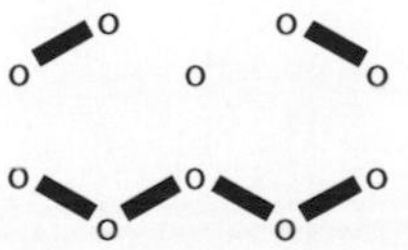

(continued)

(continued)

Ring Type	Fragmentation Type	Number of Bond Disconnections	Graphic
	[1 + 3 + 1 + 5]	4	
	[1 + 4 + 1 + 4]	4	
	[1 + 4 + 2 + 3]	4	
	[2 + 2 + 2 + 4]	4	
	[2 + 2 + 3 + 3]	4	
	[2 + 3 + 2 + 3]	4	
	[1 + 3 + 3 + 3]	4	

A.7.17 ENUMERATION OF BOND DISCONNECTIONS BY RING SIZE FOR FUSED BICYCLIC RINGS

Fused Bicyclic Ring System	Disconnections		2-Bond Disconnections	
	1-Bond	2-Bond	1-Ring Formed	2-Rings Formed
[2.2.0]	3	10	3	7
[3.2.0]	4	13	6	7
[4.2.0]	6	21	8	13
[5.2.0]	6	25	11	14
[3.3.0]	3	12	4	8
[4.3.0]	6	25	10	15
[5.3.0]	6	30	13	17
[4.4.0]	4	20	7	13
[5.4.0]	7	37	15	22
[5.5.0]	4	24	9	15

A.7.18 Enumeration of Fused Bicyclic Ring Construction Templates

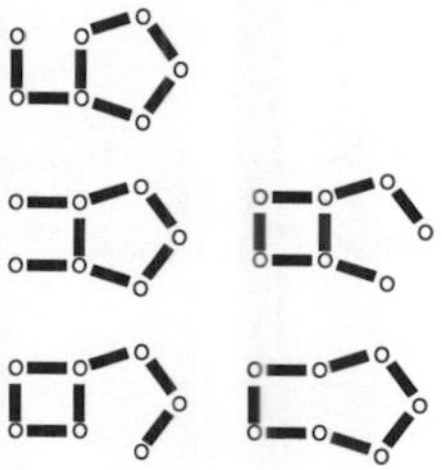

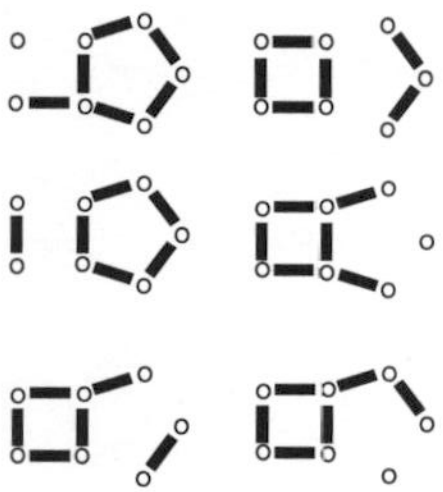

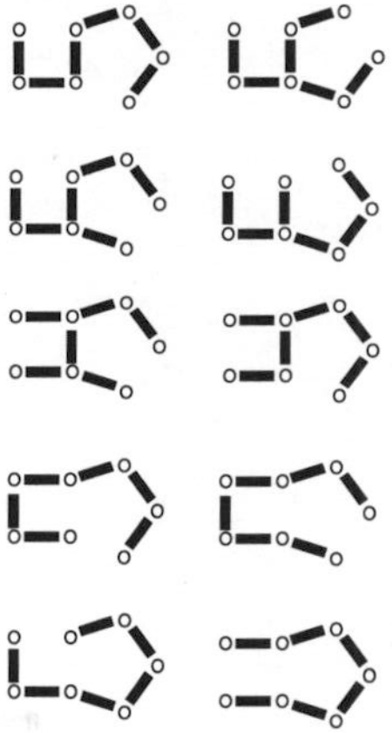

(continued)

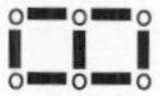

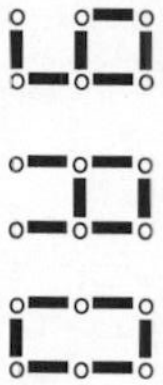

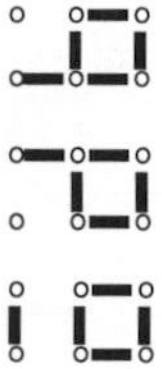

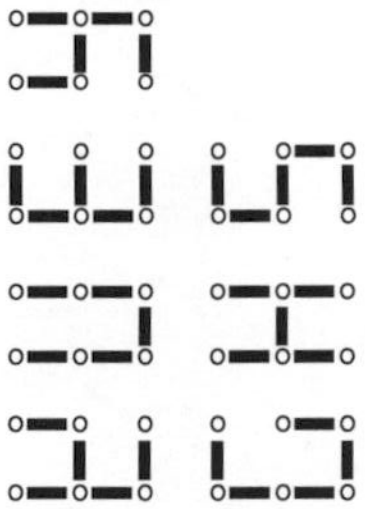

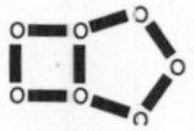

(continued)

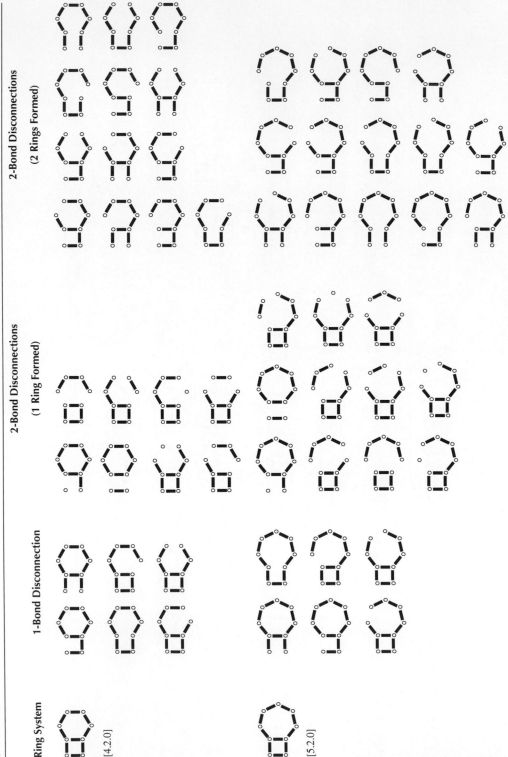

Ring System

[4.2.0]

[5.2.0]

1-Bond Disconnection

2-Bond Disconnections (1 Ring Formed)

2-Bond Disconnections (2 Rings Formed)

(continued)

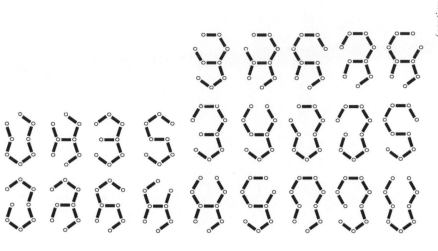

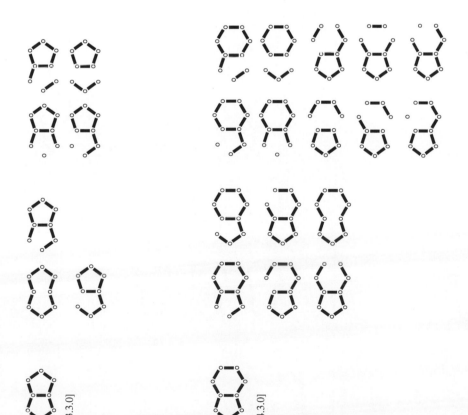

[3.3.0]

[4.3.0]

(continued)

Ring System

1-Bond Disconnection

2-Bond Disconnections
(1 Ring Formed)

2-Bond Disconnections
(2 Rings Formed)

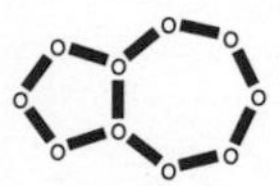

[5.3.0]

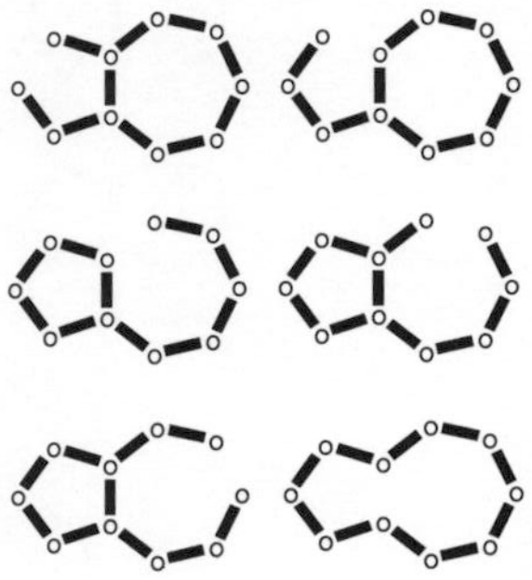

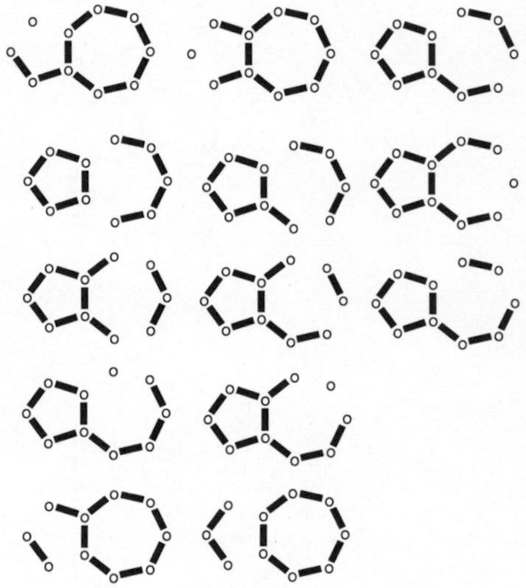

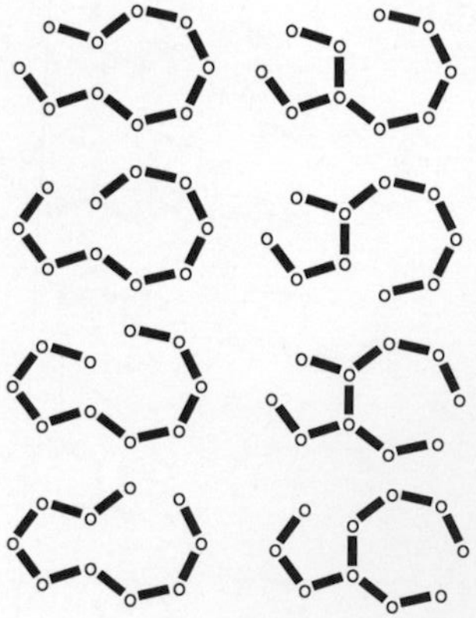

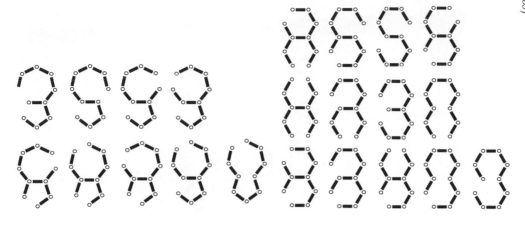

(continued)

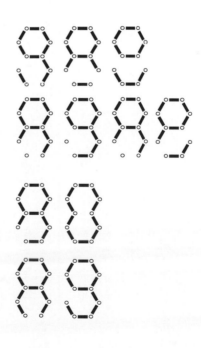

[5.3.0]

[4.4.0]

(continued)

Ring System	1-Bond Disconnection	2-Bond Disconnections (1 Ring Formed)	2-Bond Disconnections (2 Rings Formed)

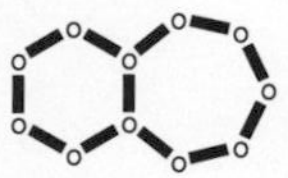

[5.4.0]

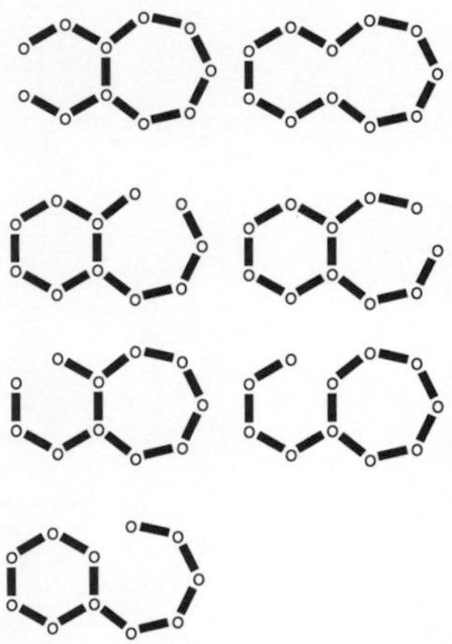

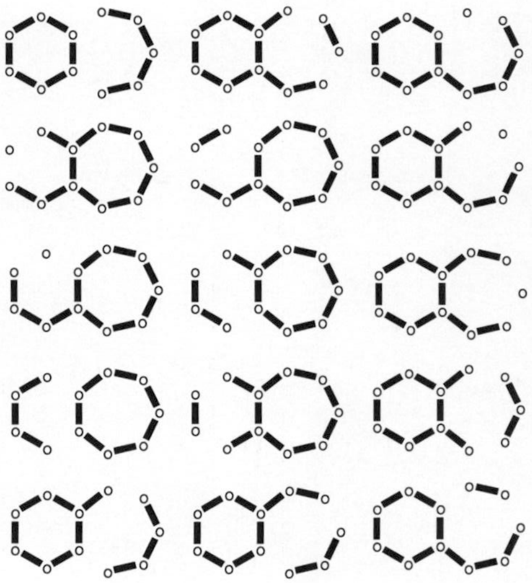

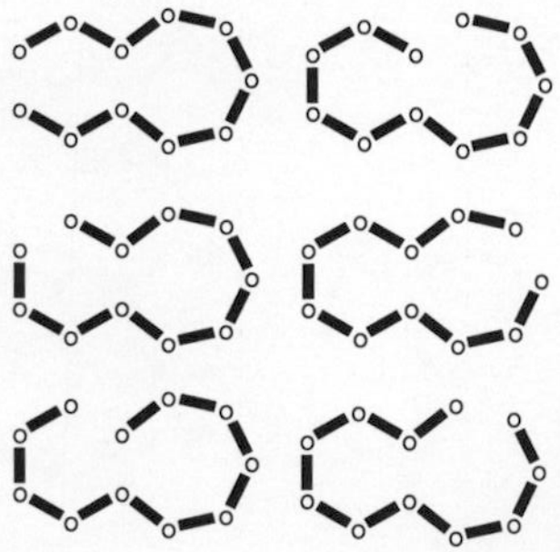

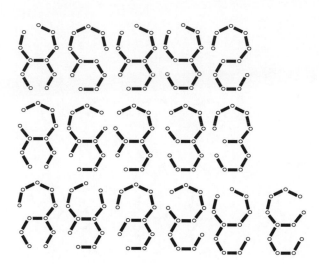

(continued)

[5.4.0]

[5.4.0]

(continued)

Ring System

1-Bond Disconnection

**2-Bond Disconnections
(1 Ring Formed)**

**2-Bond Disconnections
(2 Rings Formed)**

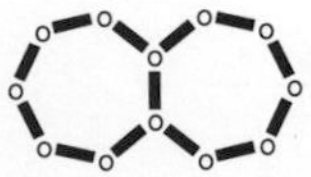

[5.5.0]

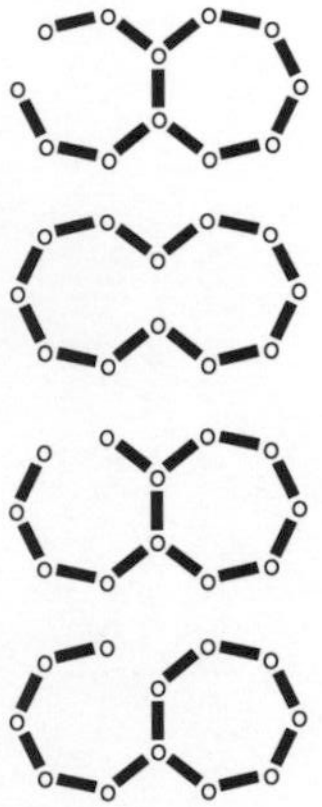

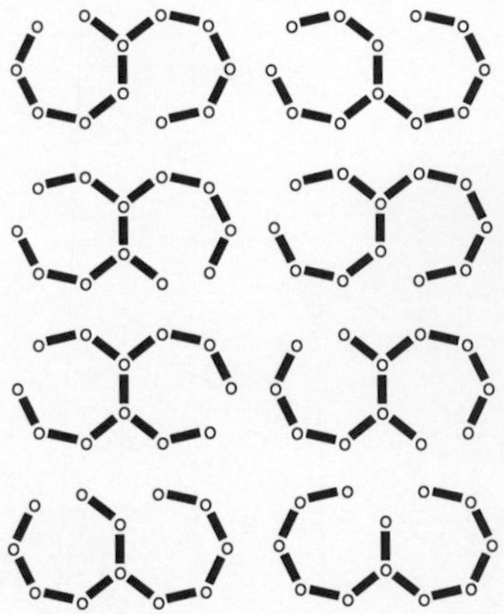

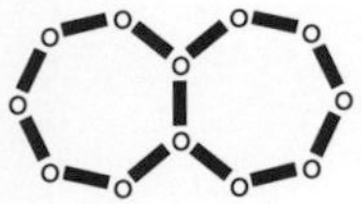

[5.5.0]

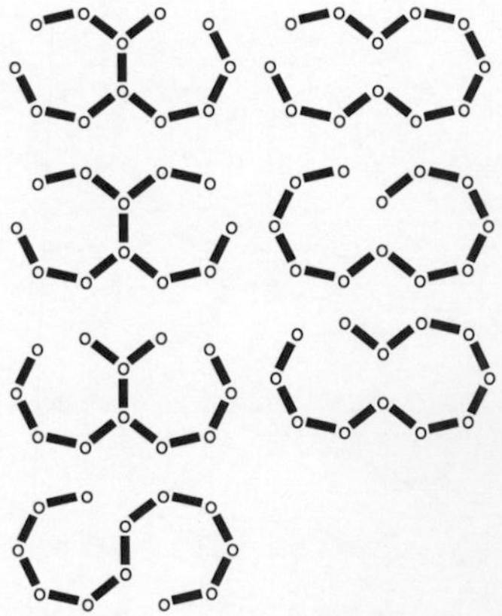

A.7.19   Enumeration of Bond Disconnections by Ring Size for Bicyclic Rings

Bicyclic	Disconnections		
Ring System	1-Bond	2-Bonds	3-Bonds
[1.1.1]	1	3	4
[2.1.1]	3	6	16
[2.2.1]	3	11	18
[2.2.2]	2	6	18
[3.2.2]	5	23	58

[1.1.1]

1-Bond disconnection

2-Bond disconnections

3-Bond disconnections

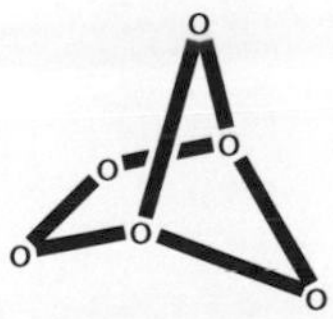

[2.2.1]

1-Bond disconnection

2-Bond disconnections

3-Bond disconnections

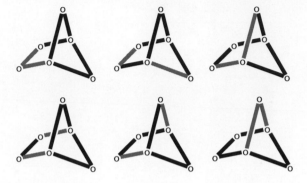

[2.2.1]

1-Bond disconnection

2-Bond disconnections

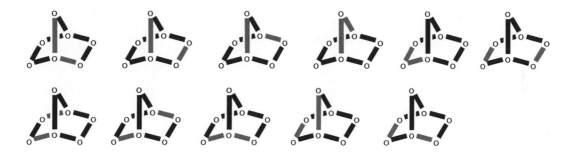

3-Bond disconnections

[2.2.2]

1-Bond disconnection

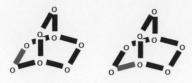

2-Bond disconnections

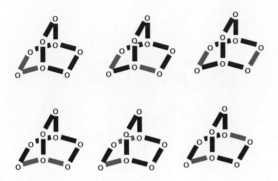

3-Bond disconnections

[3.2.1]

1-Bond disconnection

2-Bond disconnections

3-Bond disconnections

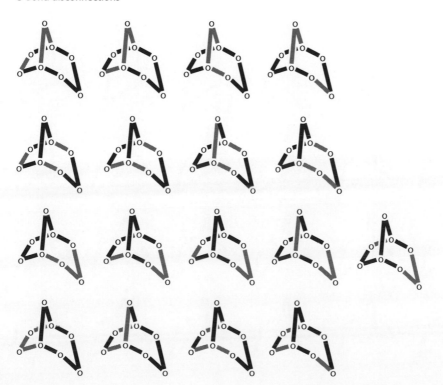

3-Bond disconnections

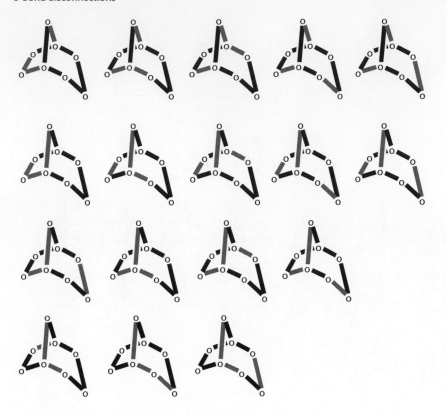

3-Bond disconnections

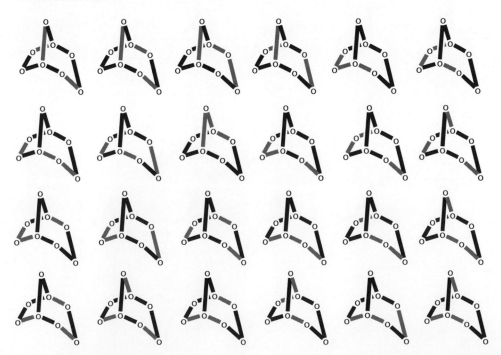

A.8 DI-SUBSTITUTED AROMATIC COMBINATIONS USING DONOR AND ACCEPTOR GROUPS

[D] = donor group (EDG, electron source); [A] = acceptor group (EWG, electron sink)

Constraints:
[D] groups are ortho- and para-directing.
[A] groups are meta-directing.

(I) Para orientation combinations.

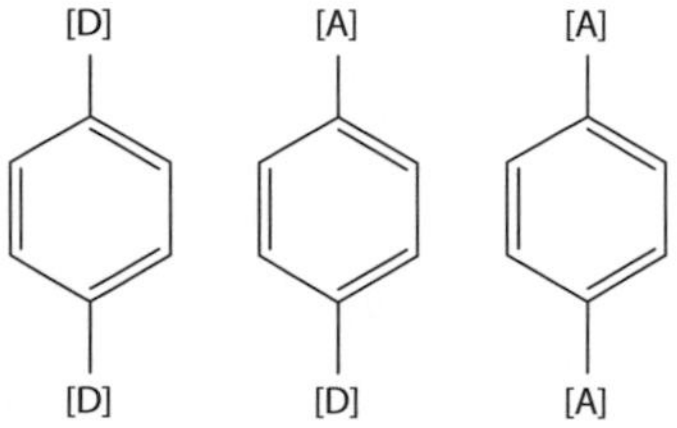

Synthesis strategies with fewest steps:

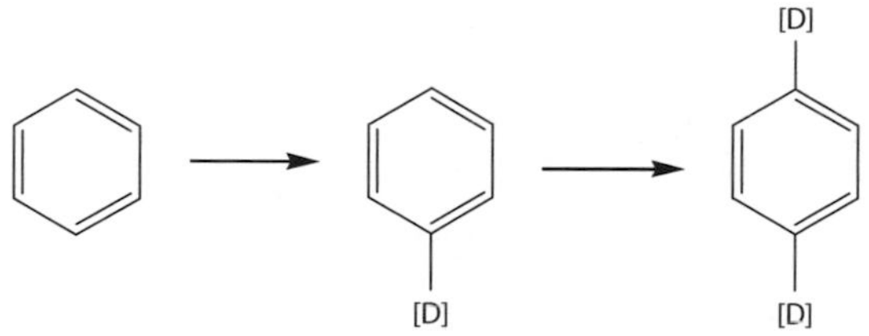

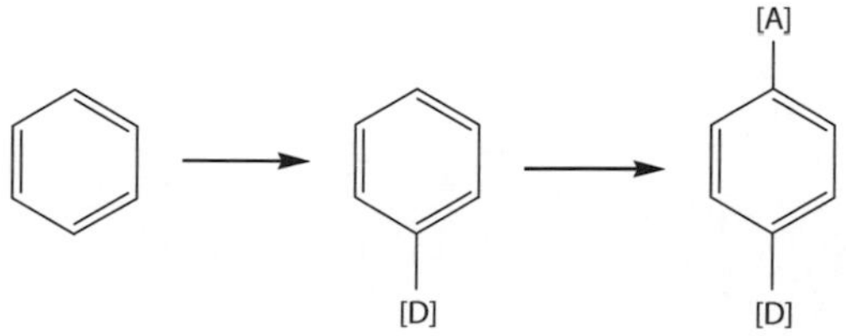

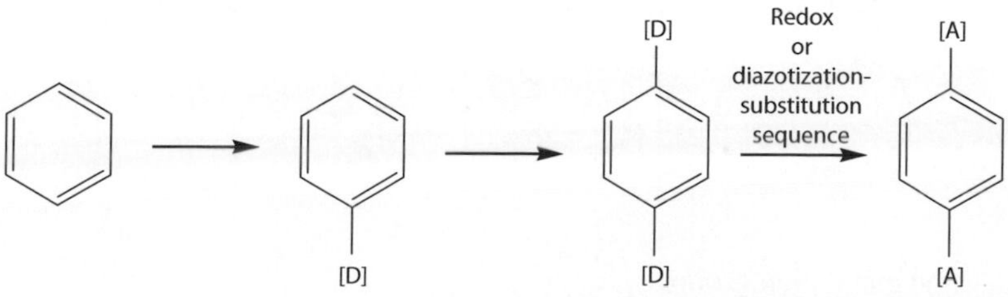

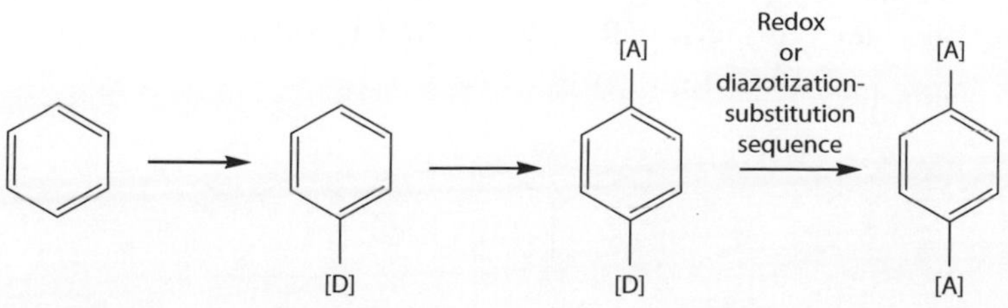

(II) Ortho orientation combinations

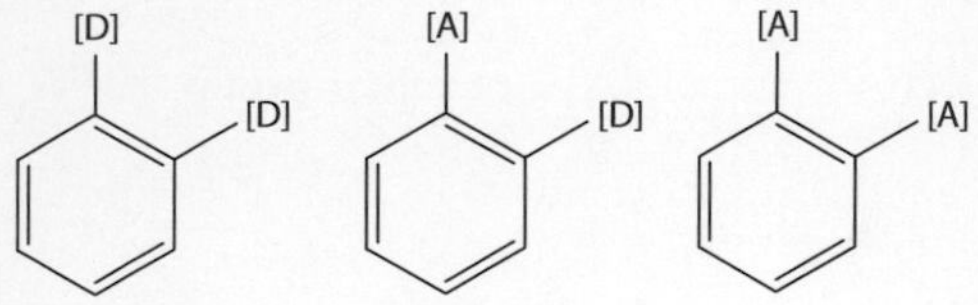

Synthesis strategies with fewest steps:

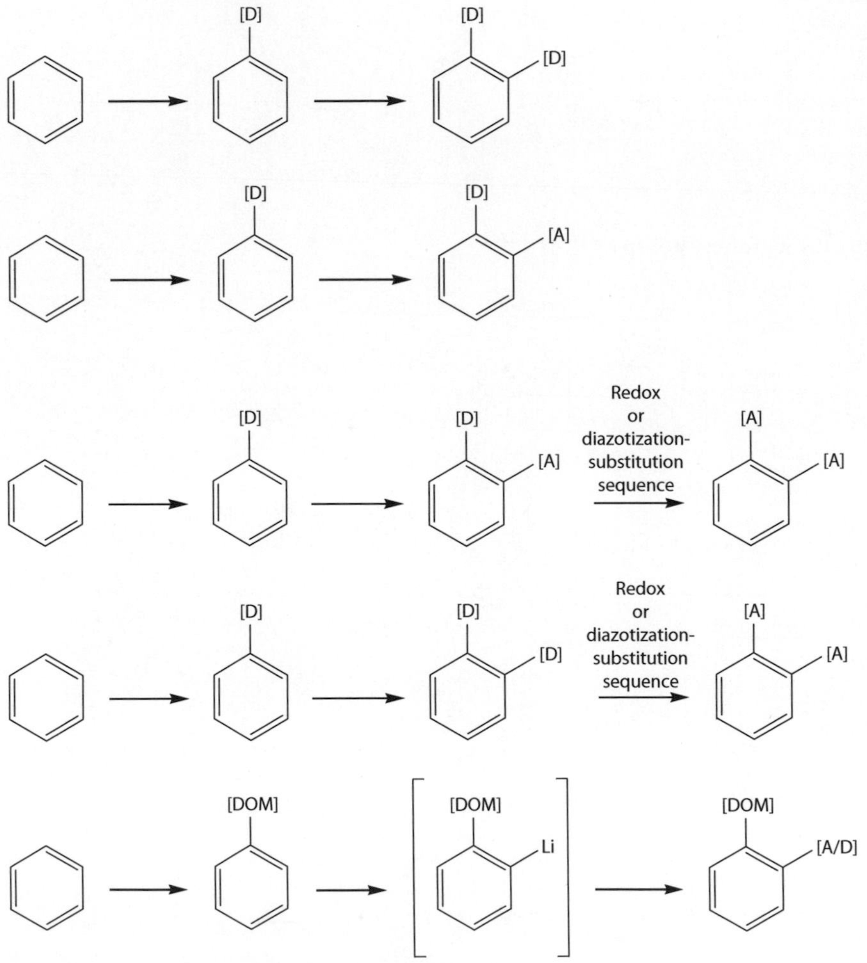

[DOM] = directed ortho metalation group.

Some DOM groups are acceptors (e.g., SO_2NR_2, SO_2R, SO_3R, $CONR_2$).
Some DOM groups are donors (e.g., SR, $NHCO_2R$, OCH_2OCH_3, $OCONR_2$).

(III) Meta orientation combinations

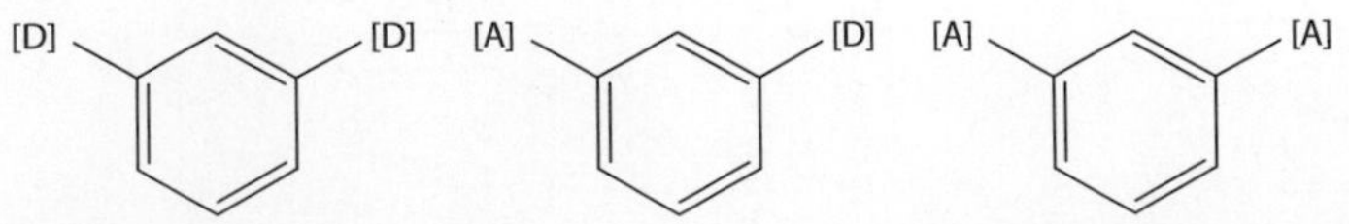

Synthesis strategies with fewest steps:

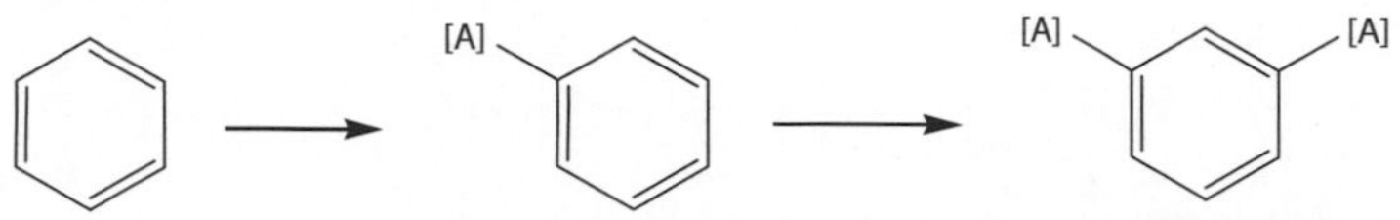

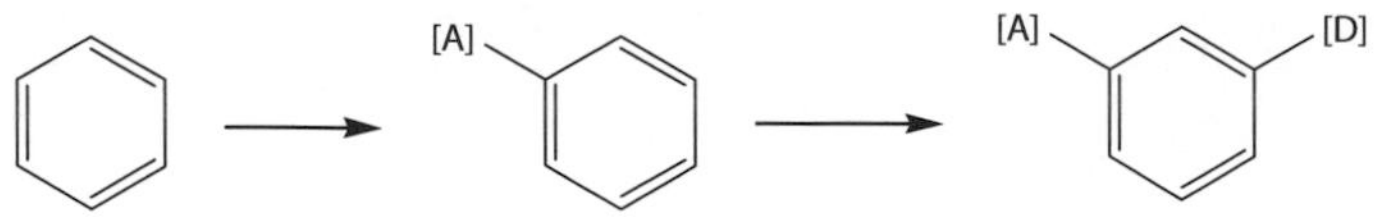

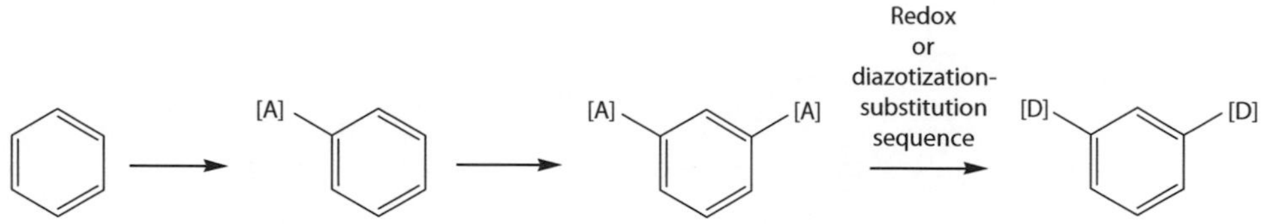

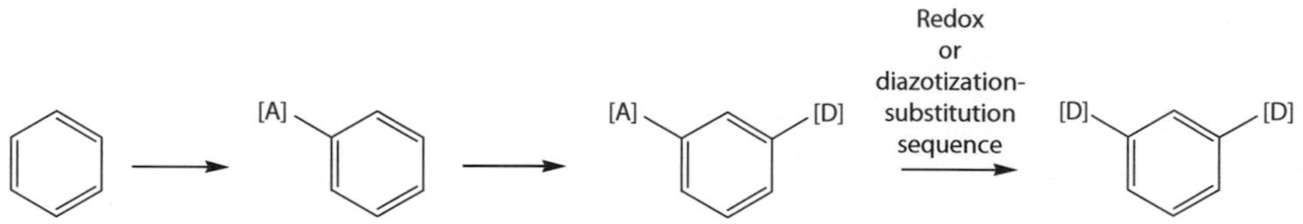

A.9 TRI-SUBSTITUTED AROMATIC COMBINATIONS USING DONOR AND ACCEPTOR GROUPS

[D] = donor group (EDG, electron source); [A] = acceptor group (EWG, electron sink)

Constraints:
[D] groups are ortho- and para-directing.
[A] groups are meta-directing.

(I) 1-2-4 Combinations.

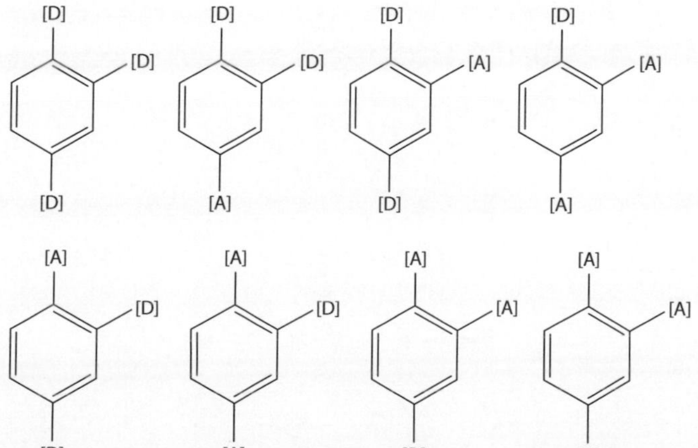

Synthesis strategies with fewest steps:

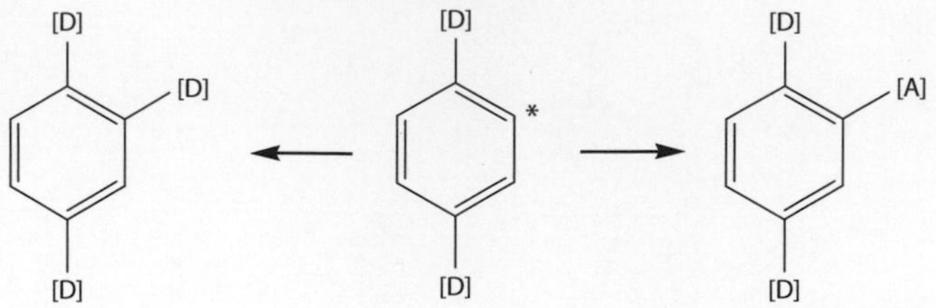

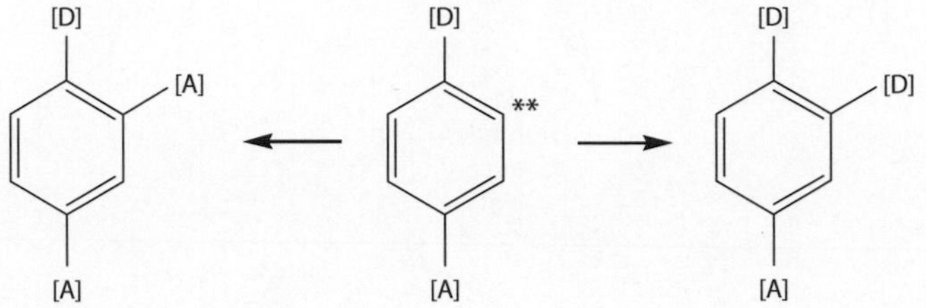

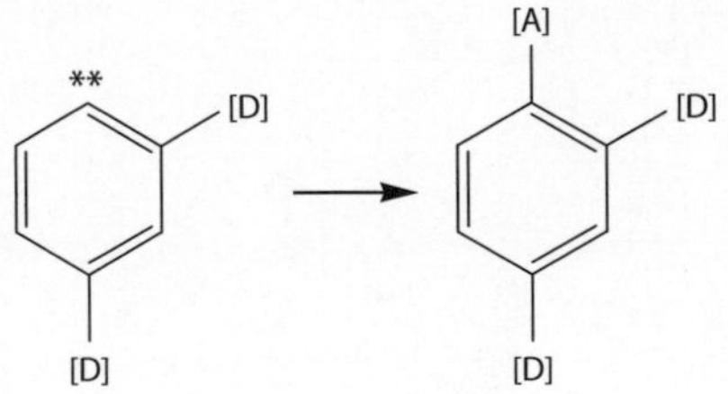

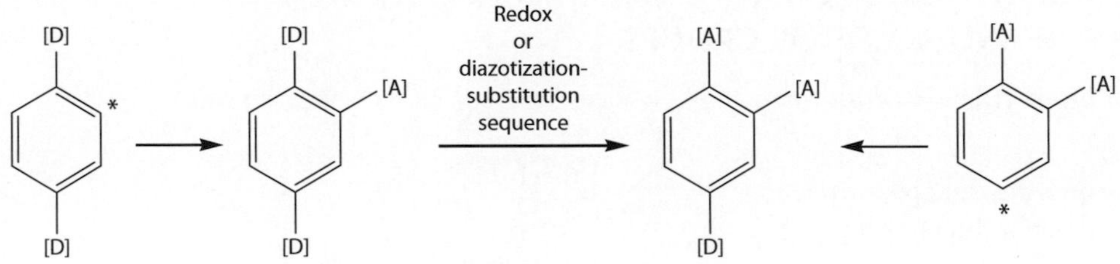

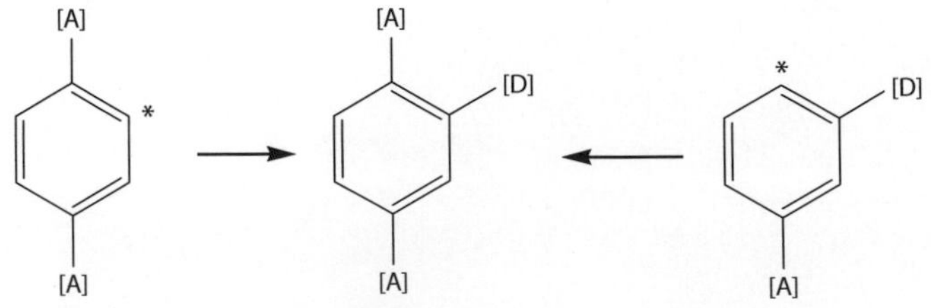

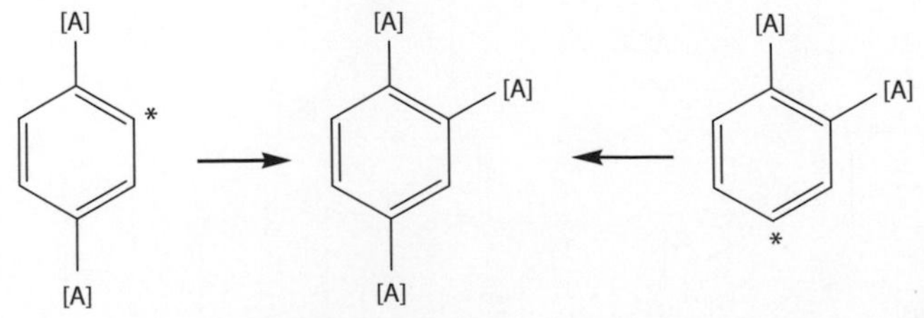

(II) 1-2-3 Combinations

(III) 1-3-5 Combinations

A.10 GLOSSARY OF TERMS USED IN ORGANIC SYNTHESIS

Accelerant: A substance added to a reaction mixture that increases the rate of production of product; it may or may not be a catalyst depending on whether its chemical structure is altered over the course of the reaction.

Active catalyst: The catalyst structure that is actually the one involved in lowering the energy barrier for a reaction step.

Additive: A substance added to a reaction mixture that improves reaction performance toward a desired product but whose role is not well defined.

Borsm—Based on recovered starting material: A calculation of reaction yield that includes both the intended target product and unreacted starting material in a chemical reaction as the desired products; this is usually reported in papers when the true reaction yield to the intended target compound is lower than 50%.

By-product of a reaction: A product formed in a reaction between reagents as a direct mechanistic consequence of producing the target product assuming a balanced chemical equation that accounts for the production of that target product.

Campaign: A term taken from military vocabulary to describe the often difficult labor involved in achieving a total synthesis of a target molecule having a complex structure.

Cascade: A sequence of reaction steps taking place consecutively typically in a "one-pot" reaction in such a way that each reaction occurs as a consequence of the preceding one akin to a cascading waterfall.

Catalyst loading: A term describing the mass amount of catalyst relative to the mass of substrate in a given reaction.

Chemoselectivity: A general term used to describe desired selectivity in carrying out a reaction according to some specific chemical group over all others in a given molecule.

Chiral pool: Source compounds that have at least one stereogenic center that are typically used as starting materials for the total stereoselective synthesis of chiral target molecules.

Click: A term used to describe a one-step [2 + 3] cycloaddition reaction between an azide and an acetylenic group.

Cocatalyst: A secondary chemical substance that imparts catalytic activity in addition to the main reaction catalyst; the cocatalyst often boosts the catalytic power of the main catalyst.

Complexity: A poorly defined term that is used to loosely describe chemical structures according to the presence of certain motifs such as complicated ring systems and stereochemical groups.

Convergent: It refers to a plan consisting of at least two branches each composed of a consecutive sequence of reaction steps that converge at some point along the synthesis route.

Conversion: It refers to the fractional amount of starting material that gets converted to products in a chemical reaction given by $(m_{final} - m_{initial})/m_{initial}$ where the m terms refer to the masses of starting material at the beginning and end of a reaction.

Detour: A change of direction along a planned synthetic route often done when there are problems encountered in the original plan.

Diastereomeric ratio: The mass ratio of one diastereomeric product to another, for example, m_{RR}/m_{RS}.

Divergent: It refers to a plan consisting of a multiple branches each composed of a consecutive sequence of reaction steps that diverge from a common intermediate and lead to different target products.

Domino: A "one-pot" process involving two or more consecutive reactions as a consequence of the functionality formed by bond forming or fragmentation in the previous step; (see *cascade*).

DOS—Diversity-oriented synthesis: Synthesis planning from a common set of starting materials to multiple target products; linked to construction of compound libraries in a combinatorial sense.

Enantiomeric excess: The ratio given by $ee = (m_R - m_S)/(m_R + m_S)$, where the terms refer to the masses of R and S enantiomeric products of a reaction; the equation as written refers to the enantiomeric excess with respect to the R enantiomer.

Enantiomeric ratio: The mass ratio of one enantiomeric product to another, for example, m_R/m_S.

End game: A term borrowed from chess that refers to the final steps in a total synthesis toward a given target molecule.

Formal synthesis: A synthesis plan to a key reaction intermediate on the way to a target molecule that is different from a prior published route.

Hill climbing: A term borrowed from mountaineering that is used to describe the arduous and treacherous journey when one embarks on a total synthesis of a complex target molecule (see *campaign*).

Hub intermediate: A key intermediate product in a convergent synthesis plan that appears more than once in the synthesis route; the structure of this intermediate is the same as a recurring structural motif found in the final target product.

"In our hands": A generic polite phrase used by authors in research papers when they are unable to reproduce the results of prior published work.

Linear: It refers to a plan consisting of a single branch composed of a consecutive sequence of reaction steps.

Moiety: A general term used to designate a group of atoms, usually a functional group, in a given chemical structure.

Molar yield: Ratio of moles of product collected to moles of limiting reagent used.

Mole percent: Usually used to describe quantitatively catalyst loading given by $(n_{catalyst}/n_{limiting\ reagent}) \times 100\%$, where n refers to the number of moles of each species

Motif: A substructure that usually is common to a set of complex structures.

Multicomponent: A chemical reaction involving at least three starting molecules that react in a single step; the starting materials may or may not be different in structure and the order of addition of the reactants may or may not matter in obtaining the intended final product.

"Occurred uneventfully": A phrase that refers to the occurrence of a chemical reaction taking place as intended without complications from competing side reactions.

"One pot": A term that refers to a chemical process involving more than one reaction step taking place in a single reaction vessel without having to isolate the intervening intermediates (see *cascade, domino, multicomponent, tandem*).

Plan: It refers generally to a synthesis route to a given target molecule.

Precatalyst: The catalyst structure that precedes the active catalyst that is actually the one involved in lowering the energy barrier for a reaction step.

Prefixes: *Allo-* (a close relative of analog structure), *amphi-* (refers to 2,6-positions in naphthalene analogs; also used to describe certain stereoisomers of dioximes), *apo-* (related to or formed from analog structure), ar- (aromatic analogue), *cis-* (groups oriented on the same side with respect to each other), *ent-* (enantiomer of), *epi-* (opposite configuration at a given stereocenter in analog structure; indicating a bridge or intramolecular connection; refers to 1,6 positions in naphthalene analogs), *eso-* (refers to substitution groups attached to a ring), *exo-* (substitution groups attached to a side chain), *homo-* (additional methylene group in analog structure), *iso-* (isomer of analogue structure), *nor-* (removal of a methyl group in analog structure), *omega-* (substitution group on last carbon atom of a side chain), *pseudo-* (indicates an isomerism or some other relation to given structure), *rac-* (racemate of), *sec-* (secondary—indicating a replacement to the second degree), *seco-* (a ring is cut in analog structure), *sym-* (symmetrical structural analog), *tert-* (tertiary—indicating a replacement to the third degree), *trans-* (groups oriented on opposite sides with respect to each other).

Privileged structure: A substructural feature that confers desirable (often drug-like) properties on compounds containing that feature.

Proclivity: A strong inclination or predisposition toward reaction.

Promoter: Some chemical entity that kick-starts a reaction to occur.

Protecting group: A functional group that is used to momentarily cap another so that it is protected in downstream steps from possible competing side reactions; this is often done when there is more than one reactive group of one kind in a given molecule and selectivity is required.

Quadrants—SW, NE, NW, SE, N, S, E, W: A term usually applied to very complex structures where they are subdivided roughly according to the cardinal directions in an effort to plan syntheses of the respective substructures; this is a classic way of parsing a difficult problem into smaller hopefully solvable problems; the hope is that these subunits can be linked together to produce the final complex target product in convergent steps.

Reaction step: It refers to interval between a given isolated intermediate and the next consecutive isolated intermediate in a synthesis plan.

Reaction yield: Ratio of moles of target product collected to moles of target product expected with respect to limiting reagent as per the stoichiometry of the given balanced chemical equation (see *molar yield*).

Reflexive synthesis: A convergent synthesis plan that uses a common intermediate in multiple branches (see *hub intermediate*).

Regioselectivity: Desired selectivity in carrying out a reaction according to a specific region or group over all others in a given molecule.

Retrosynthetic analysis: A reverse strategy analysis used in synthesis planning that begins with the end target molecular structure and through a series of bond disconnects leads to simpler structures that could be used as starting materials in the forward sense.

Route: see *plan*.

Sacrificial reactions: Nonproductive reactions in a synthesis plan that do not form target bonds appearing in the target product structure.

Scaffold: A substructure of a complex structure that is constant and common to a variety of similar complex structures; a term usually used in conjunction with the construction of compound libraries.

Scope table: A table of experimental results showing various reaction conditions in an effort to find the optimum set of variables that will produce the desired product effectively, such as reaction time, reaction temperature, reaction solvent, catalyst, and ligand.

Screening: A combinatorial trial-and-error method of finding a lead compound structure or a set of optimum reaction variables; this is usually done when there is little information available that can help to predict the desired outcome.

Selectivity: A term used to describe reactivity according to a specific region or group or structural motif in a given molecular structure (see *chemoselectivity, regioselectivity, stereoselectivity*).

Semi-synthesis: A short synthesis plan that begins from a natural product starting material whose structure is similar to that of the final target molecule.

Sequential: A term describing the consecutive progression from one reaction step to the next in a synthesis plan; this usually applies to a linear stretch or branch along the route.

Side product of a reaction: A product formed in a reaction between reagents, usually undesired, that arises from a competing reaction pathway other than the one that produces the intended target product and its associated by-products.

Solvent switching: A technique used in process chemistry to change a reaction solvent for another in a subsequent reaction after the first is completed; this is done without having to formally isolate and purify the first reaction intermediate.

Stereoselectivity: Desired selectivity in carrying out a reaction according to a specific stereochemical group over all others in a given molecule.

Strategy: A general term used to describe the planning of a synthesis route to a target molecule involving the kinds of reactions used and their sequence order.

Synthon: A chemical structural motif that is useful from a synthetic point of view in terms of having reactive groups that can be used to build more complex structures in total synthesis; usually having key electrophilic and nucleophilic centers in the same structure.

Tactic: An often clever method of dealing with a difficult challenge faced by chemists as they work out a total synthesis of a given target molecule.

Tandem: A type of "one-pot" reaction that involves a series of reactions occurring one after another in the same reaction vessel (see *cascade, domino*).

Target bond forming reactions: Productive reactions in a synthesis plan that result in the formation of bonds that appear in the structure of the target molecule.

Telescoping: A synthesis tactic often used in process chemistry that involves concatenation of reaction steps so that reaction intermediates are not isolated; this is an effective optimization strategy that reduces solvent demand.

TOS—Target oriented synthesis: Synthesis planning from multiple sets of starting materials to a given target molecule.

Total synthesis: A term that describes a man-made artificial route to a complex chemical structure starting from a set of simple building block molecules.

Two-directional synthesis: A synthesis strategy that involves reactions taking place at opposite ends of the same molecule simultaneously; this strategy is employed for target molecules that have high symmetry.

Warhead: A term borrowed from military vocabulary that is used to describe the substructure of a drug molecule that is responsible for its efficacy as a drug against a given biological target, usually the parts of the molecule that bind to key amino acid residues in the active site of an enzyme.

A.11 PRESIDENTIAL GREEN CHEMISTRY CHALLENGE AWARDS ADMINISTERED BY THE U.S. ENVIRONMENTAL PROTECTION AGENCY (EPA)

A.11.1 GREENER SYNTHETIC PATHWAYS AWARD CATEGORY

2009
Eastman Kodak Co.
A Solvent-Free Biocatalytic Process for Cosmetic and Personal Care Ingredients

2008
Battelle
Development and Commercialization of Biobased Toners

2007
Professor Kaichang Li, Oregon State University; Columbia Forest Products; Hercules Incorporated (now Ashland Inc.)
Development and Commercial Application of Environmentally Friendly Adhesives for Wood Composites

2006
Merck & Co., Inc.
Novel Green Synthesis for β-Amino Acids Produces the Active Ingredient in Januvia™

2005

Archer Daniels Midland Company
Novozymes
NovaLipid™: Low Trans Fats and Oils Produced by Enzymatic Interesterification of Vegetable Oils Using Lipozyme®

Merck & Co., Inc.
A Redesigned, Efficient Synthesis of Aprepitant, the Active Ingredient in Emend®: A New Therapy for Chemotherapy-Induced Emesis

2004

Bristol-Myers Squibb Co.
Development of a Green Synthesis for Taxol® Manufacture via Plant Cell Fermentation and Extraction

2003

Süd-Chemie Inc.
A Wastewater-Free Process for Synthesis of Solid Oxide Catalysts

2002

Pfizer, Inc.
Green Chemistry in the Redesign of the Sertraline Process

2001

Bayer CorporationBayer AG (technology acquired by LANXESS)
Baypure™ CX (Sodium Iminodisuccinate): An Environmentally Friendly and Readily Biodegradable Chelating Agent

2000

Roche Colorado Corporation
An Efficient Process for the Production of Cytovene® (ganciclovir), a Potent Antiviral Agent

1999

Lilly Research Laboratories
Practical Application of a Biocatalyst in Pharmaceutical Manufacturing (LY 300164)

1998

Flexsys America L.P.
Elimination of Chlorine in the Synthesis of 4-Aminodiphenylamine: A New Process That Utilizes Nucleophilic Aromatic Substitution for Hydrogen

1997

BHC Company (now BASF Corporation)
BHC Company Ibuprofen Process

1996

Monsanto Company
Catalytic Dehydrogenation of Diethanolamine (disodium iminodiacetate)

A.11.2 ACADEMIC AWARDS CATEGORY

2009

Professor Krzysztof Matyjaszewski, Carnegie Mellon University

Atom Transfer Radical Polymerization: Low-impact Polymerization Using a Copper Catalyst and
Environmentally Friendly Reducing Agents

2008

Professors Robert E. Maleczka, Jr. and Milton R. Smith, III, Michigan State University

Green Chemistry for Preparing Boronic Esters

2007

Professor Michael J. Krische, University of Texas at Austin

Hydrogen-Mediated Carbon–Carbon Bond Formation

2006

Professor Galen J. Suppes, University of Missouri-Columbia

Biobased Propylene Glycol and Monomers from Natural Glycerin

2005

Professor Robin D. Rogers, The University of Alabama

A Platform Strategy Using Ionic Liquids to Dissolve and Process Cellulose for Advanced New Materials

2004

Professors Charles A. Eckert and Charles L. Liotta, Georgia Institute of Technology

Benign Tunable Solvents Coupling Reaction and Separation Processes

2003

Professor Richard A. Gross, Polytechnic University

New Options for Mild and Selective Polymerizations Using Lipases

2002

Professor Eric J. Beckman, University of Pittsburgh

Design of Non-Fluorous, Highly CO_2-Soluble Materials

2001

Professor Chao-Jun Li, Tulane University

Quasi-Nature Catalysis: Developing Transition Metal Catalysis in Air and Water

2000

Professor Chi-Huey Wong, The Scripps Research Institute

Enzymes in Large-Scale Organic Synthesis

1999

Professor Terry Collins, Carnegie Mellon University

TAML™ Oxidant Activators: General Activation of Hydrogen Peroxide for Green Oxidation
Technologies

1998

Professor Barry M. Trost, Stanford University

The Development of the Concept of Atom Economy

1997
Dr. Karen M. Draths and Professor John W. Frost, Michigan State University
Use of Microbes as Environmentally Benign Synthetic Catalysts

1996
Professor Joseph M. DeSimone, University of North Carolina at Chapel Hill (UNC) and North Carolina State University (NCSU)
Design and Application of Surfactants for Carbon Dioxide

Professor Mark HoltzappleTexas A&M University
Conversion of Waste Biomass to Animal Feed, Chemicals, and Fuels

A.11.3   Greener Reaction Awards Category

2009
CEM Corporation
Innovative Analyzer Tags Proteins for Fast, Accurate Results without Hazardous Chemicals or High Temperatures

2008
Nalco Company
3D TRASAR® Technology

2007
Headwaters Technology Innovation
Direct Synthesis of Hydrogen Peroxide by Selective Nanocatalyst Technology

2006
Codexis, Inc.
Directed Evolution of Three Biocatalysts to Produce the Key Chiral Building Block for Atorvastatin, the Active Ingredient in Lipitor®

2005
BASF Corporation
A UV-Curable, One-Component, Low-VOC Refinish Primer: Driving Eco-Efficiency Improvements

2004
Buckman Laboratories International, Inc.
Optimyze®: A New Enzyme Technology to Improve Paper Recycling

2003
DuPont
Microbial Production of 1,3-Propanediol

2002
Cargill Dow LLC (now NatureWorks LLC)
NatureWorks™ PLA Process

2001
Novozymes North America, Inc.
BioPreparation™ of Cotton Textiles: A Cost-Effective, Environmentally Compatible Preparation Process

2000

Bayer CorporationBayer AG

Two-Component Waterborne Polyurethane Coatings

1999

Nalco Chemical Company

The Development and Commercialization of ULTIMER®: The First of a New Family of Water-Soluble
   Polymer Dispersions

1998

Argonne National Laboratory

Novel Membrane-Based Process for Producing Lactate Esters—Nontoxic Replacements for
   Halogenated and Toxic Solvents

1997

Imation (technology acquired by Eastman Kodak Company)

DryView™ Imaging Systems

1996

The Dow Chemical Company

100% Carbon Dioxide as a Blowing Agent for the Polystyrene Foam Sheet Packaging Market

A.11.4   DESIGNING GREENER CHEMICALS AWARD CATEGORY

2009

The Procter & Gamble Company

Cook Composites and Polymers Company

Chempol® MPS Resins and Sefose® Sucrose Esters Enable High-Performance Low-VOC Alkyd
   Paints and Coatings

2008

Dow AgroSciences

Spinetoram: Enhancing a Natural Product for Insect Control

2007

Cargill, Incorporated

BiOH™ Polyols

2006

S.C. Johnson & Son, Inc.

Greenlist™ Process to Reformulate Consumer Products

2005

Archer Daniels Midland Company

Archer RC™: A Nonvolatile, Reactive Coalescent for the Reduction of VOCs in Latex Paints

2004

Engelhard Corporation (now BASF Corporation)

Engelhard Rightfit™ Organic Pigments: Environmental Impact, Performance, and Value

2003

Shaw Industries, Inc.

EcoWorx™ Carpet Tile: A Cradle-to-Cradle Product

2002
Chemical Specialties, Inc. (CSI) (now Viance)
ACQ Preserve®: The Environmentally Advanced Wood Preservative

2001
PPG Industries
Yttrium as a Lead Substitute in Cationic Electrodeposition Coatings

2000
Dow AgroSciences LLC
Sentricon™ Termite Colony Elimination System, A New Paradigm for Termite Control

1999
Dow AgroSciences LLC
Spinosad: A New Natural Product for Insect Control

1998
Rohm and Haas Company (now The Dow Chemical Company)
Invention and Commercialization of a New Chemical Family of Insecticides Exemplified by
   CONFIRM™ Selective Caterpillar Control Agent and the Related Selective Insect Control
   Agents MACH 2™ and INTREPID™

1997
Albright & Wilson Americas (now Rhodia)
THPS Biocides: A New Class of Antimicrobial Chemistry (tetrakis(hydroxy-methyl)phosphonium sulfate)

1996
Rohm and Haas Company (now The Dow Chemical Company)
Designing an Environmentally Safe Marine Antifoulant (Sea-Nine™)

A.11.5 SMALL BUSINESS AWARD CATEGORY

2009
Virent Energy Systems, Inc.
BioForming® Process: Catalytic Conversion of Plant Sugars into Liquid Hydrocarbon Fuels

2008
SiGNa Chemistry, Inc.
New Stabilized Alkali Metals for Safer, Sustainable Syntheses

2007
NovaSterilis Inc.
Environmentally Benign Medical Sterilization Using Supercritical Carbon Dioxide

2006
Arkon ConsultantsNuPro Technologies, Inc. (now Eastman Kodak Company)
Environmentally Safe Solvents and Reclamation in the Flexographic Printing Industry

2005
Metabolix, Inc.
Producing Nature's Plastics Using Biotechnology

2004
Jeneil Biosurfactant Company
Rhamnolipid Biosurfactant: A Natural, Low-Toxicity Alternative to Synthetic Surfactants

2003
AgraQuest, Inc.
Serenade®: An Effective, Environmentally Friendly Biofungicide

2002
SC Fluids, Inc.
SCORR—Supercritical CO_2 Resist Remover

2001
EDEN Bioscience Corporation
Messenger®: A Green Chemistry Revolution in Plant Production and Food Safety

2000
RevTech, Inc.
Envirogluv™: A Technology for Decorating Glass and Ceramicware with Radiation-Curable, Environmentally Compliant Inks

1999
Biofine, Inc. (now BioMetics, Inc.)
Conversion of Low-Cost Biomass Wastes to Levulinic Acid and Derivatives

1998
PyrocoolTechnologies, Inc.
Technology for the Third Millennium: The Development and Commercial Introduction of an Environmentally Responsible Fire Extinguishment and Cooling Agent

1997
Legacy Systems, Inc.
Coldstrip™, A Revolutionary Organic Removal and Wet Cleaning Technology

1996
Donlar Corporation (now NanoChem Solutions, Inc.)
Production and Use of Thermal Polyaspartic Acid

A.12 WORKED OUT ATOM ECONOMIES FOR REACTIONS CITED IN TROST'S 1991 PAPER ON ATOM ECONOMY

Atom economies for reactions cited in Trost, B. *Science* **1991**, *254*, 1471.

A.12.1 CYCLOADDITIONS

Reaction 1 (AE = 1)

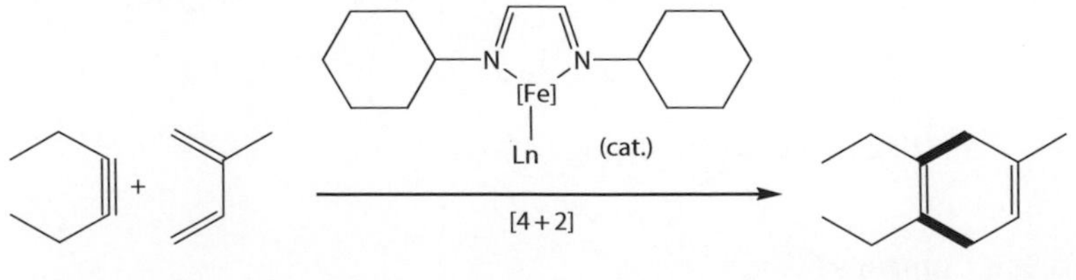

Reaction 2 (AE = 1)

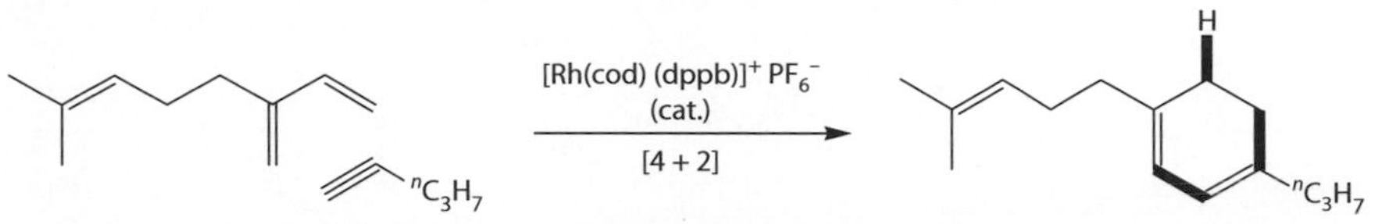

$$[Rh(cod)(dppb)]^+ PF_6^-$$
(cat.)

[4 + 2]

Reaction 3 (AE = 1)

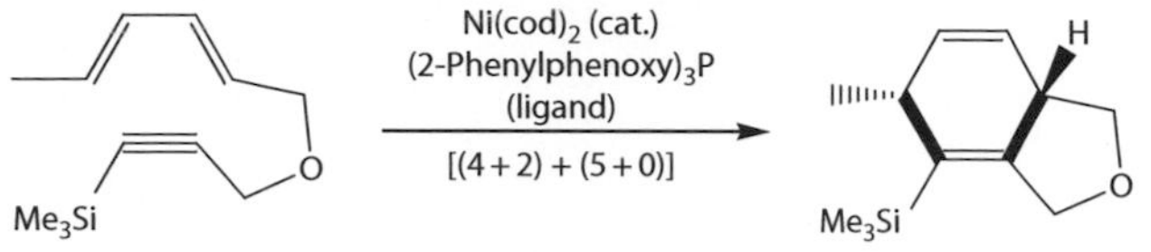

$Ni(cod)_2$ (cat.)
(2-Phenylphenoxy)$_3$P
(ligand)

$[(4+2) + (5+0)]$

Reaction 4 (AE = 1)

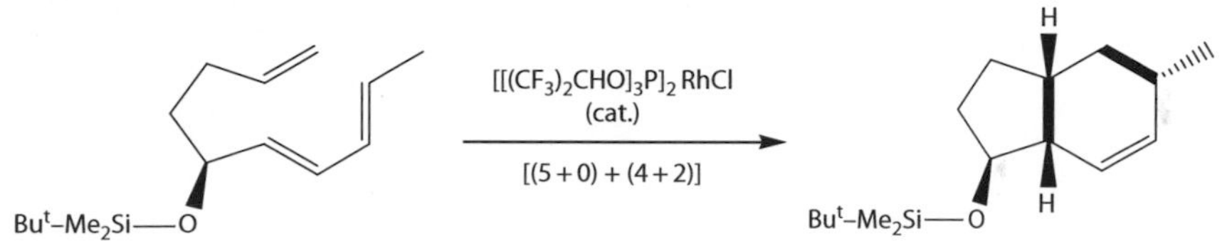

$[[(CF_3)_2CHO]_3P]_2 RhCl$
(cat.)

$[(5+0) + (4+2)]$

Reaction 5 (AE = 1)

$Co(acac)_3$ (cat.)
Et_2AlCl (cat.)
S,S-Chiraphos
(ligand)

[3 + 2]
[5 + 0]
[3 + 0]

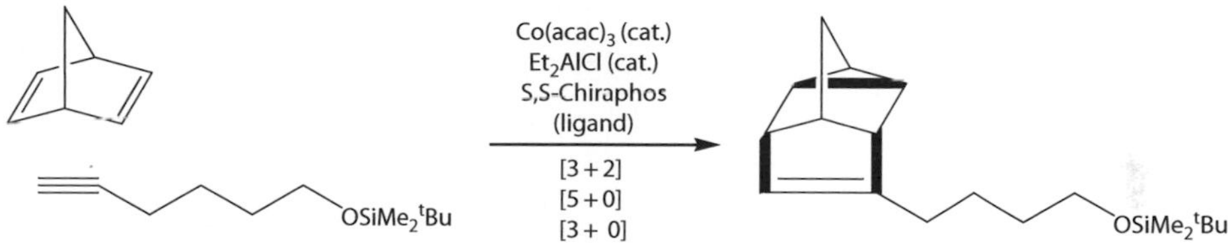

Reaction 6 (AE = 1)

$CpCo(CO)_2$
(cat.)

$[(2+2+2) + (4+0)]$

180°C

$[(4+2) + (6+0)]$

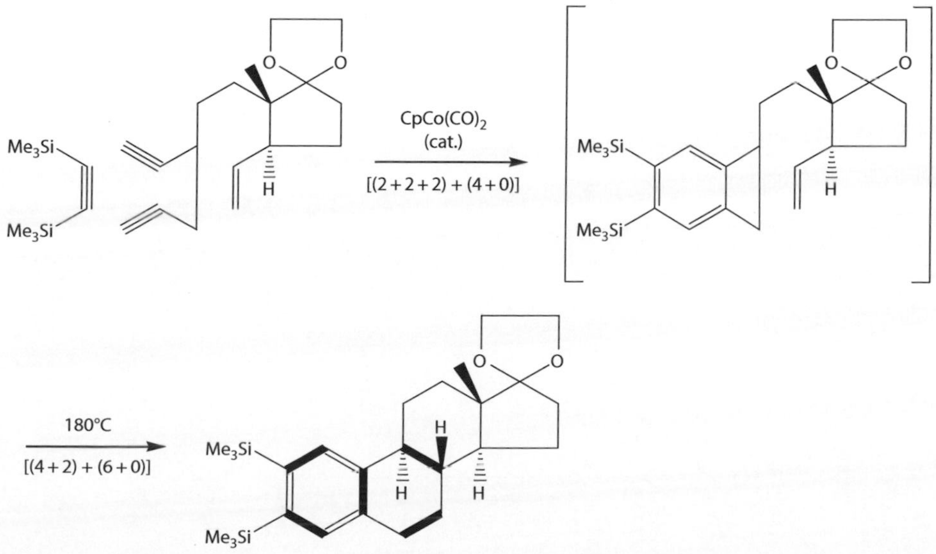

Reaction 7 (AE = 1)

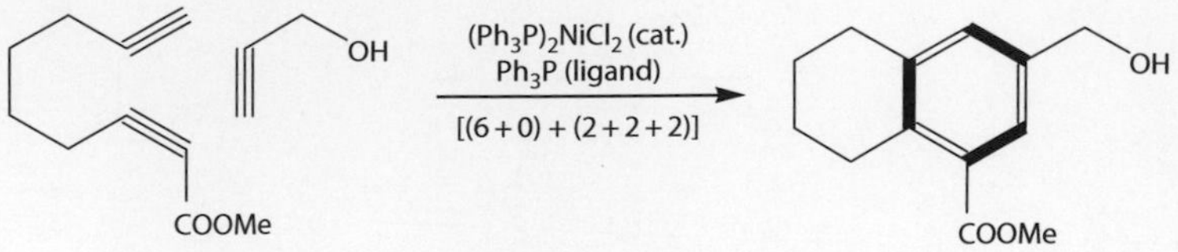

Reaction 8 (AE = 1)

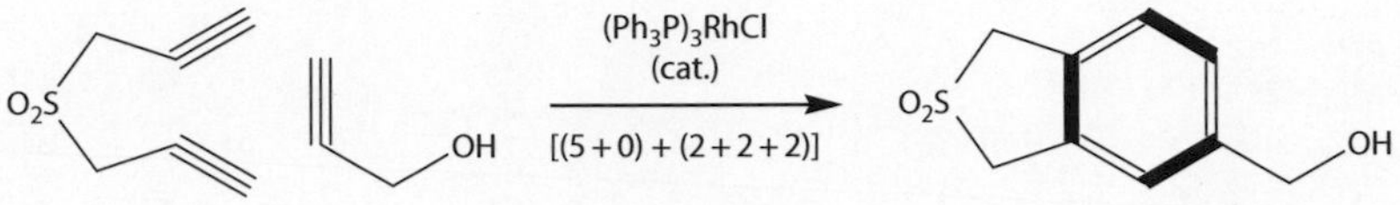

Reaction 9 (AE = 1)

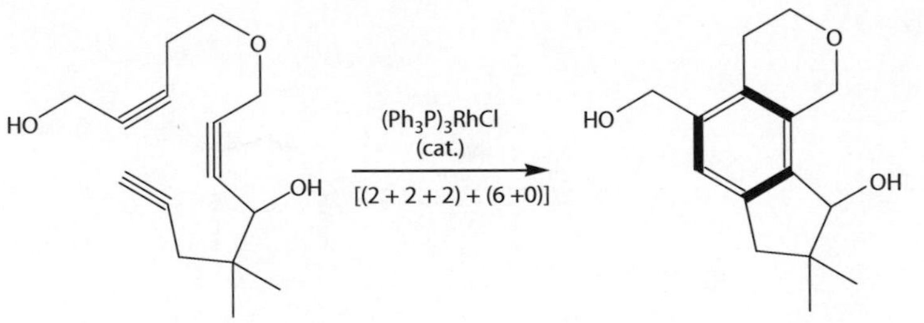

Reaction 10 (AE = 1)

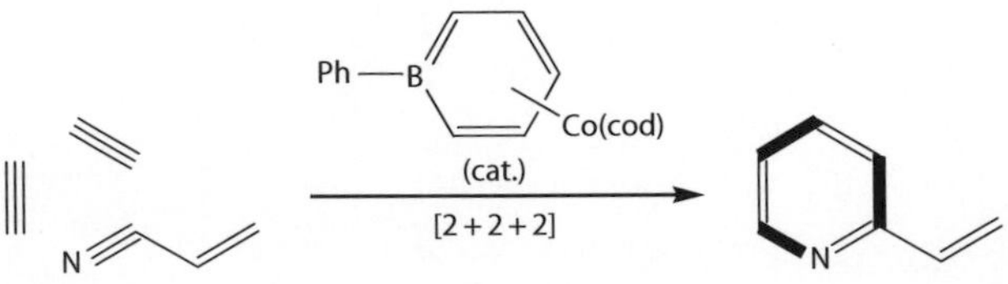

Reaction 11 (AE = 1)

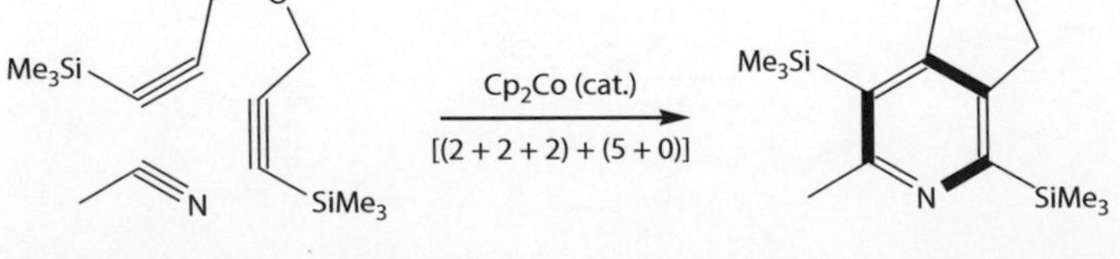

Reaction 12 (AE = 1)

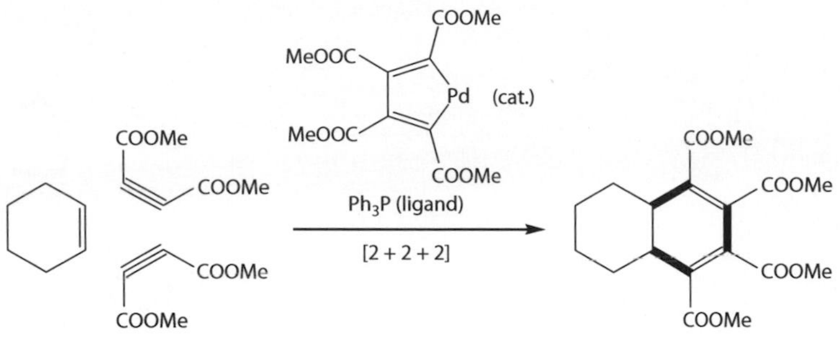

Reaction 13 (AE = 1)

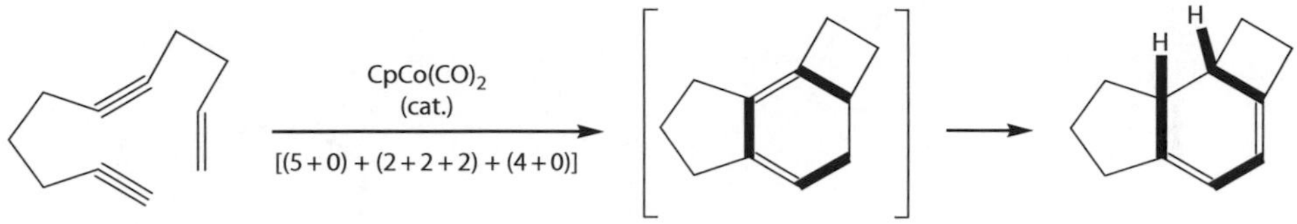

Reaction 14 (AE = 1)

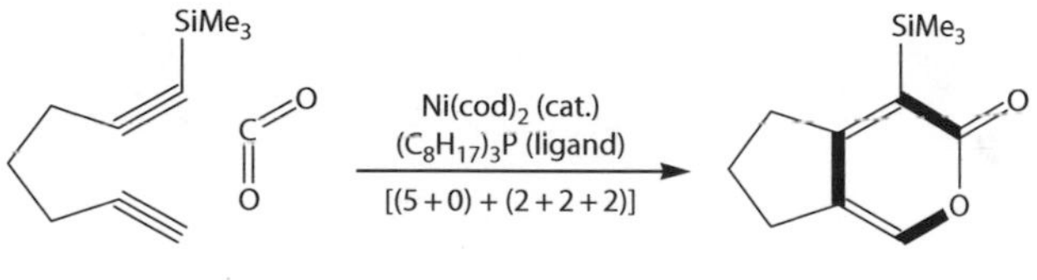

Reaction 15 (AE = 1)

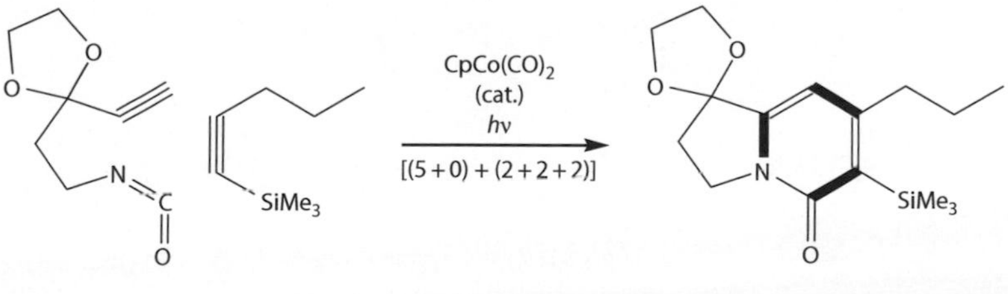

Reaction 16 (AE = 1)

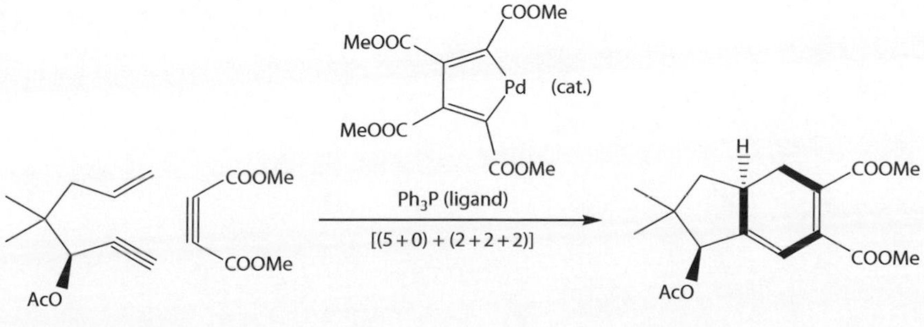

Reaction 17 (AE = 1)

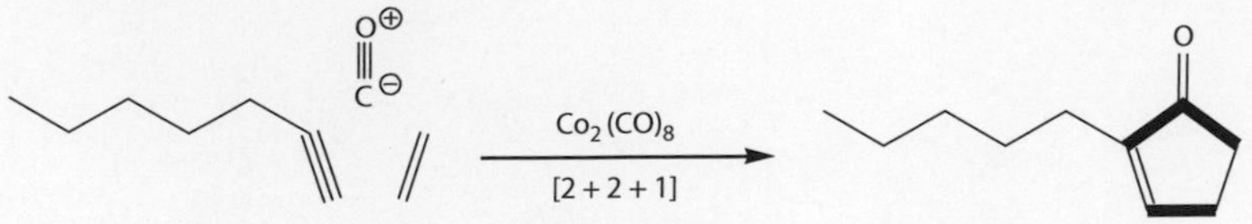

$$\xrightarrow[{[2+2+1]}]{Co_2(CO)_8}$$

Reaction 18 (AE = 1)

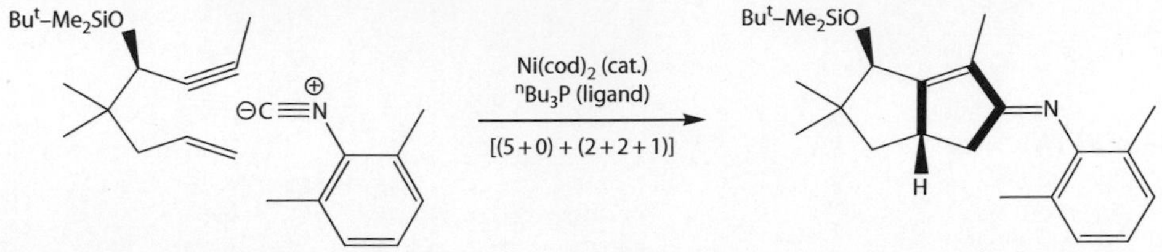

$$\xrightarrow[{[(5+0)+(2+2+1)]}]{\substack{Ni(cod)_2 \text{ (cat.)} \\ ^nBu_3P \text{ (ligand)}}}$$

Reaction 19 (AE = 1)

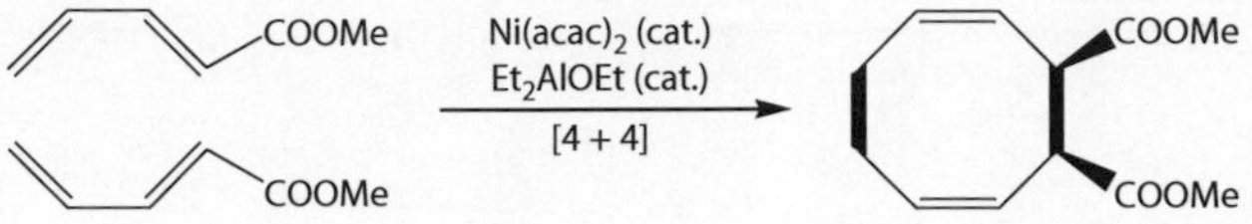

$$\xrightarrow[{[4+4]}]{\substack{Ni(acac)_2 \text{ (cat.)} \\ Et_2AlOEt \text{ (cat.)}}}$$

Reaction 20 (AE = 1)

$$\xrightarrow[{[(5+0)+(4+4)]}]{\substack{Ni(cod)_2 \text{ (cat.)} \\ Ph_3P \text{ (ligand)}}}$$

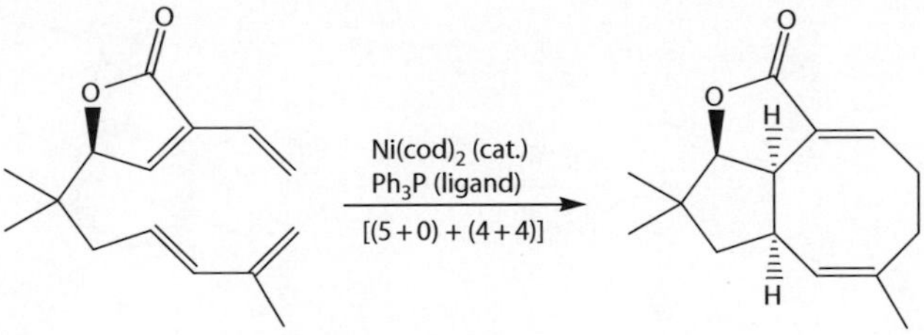

Reaction 21 (AE = 1)

$$\xrightarrow[{[(5+0)+(2+2)]}]{(2\text{-Methylphenoxy})_3P \text{ (ligand)}}$$

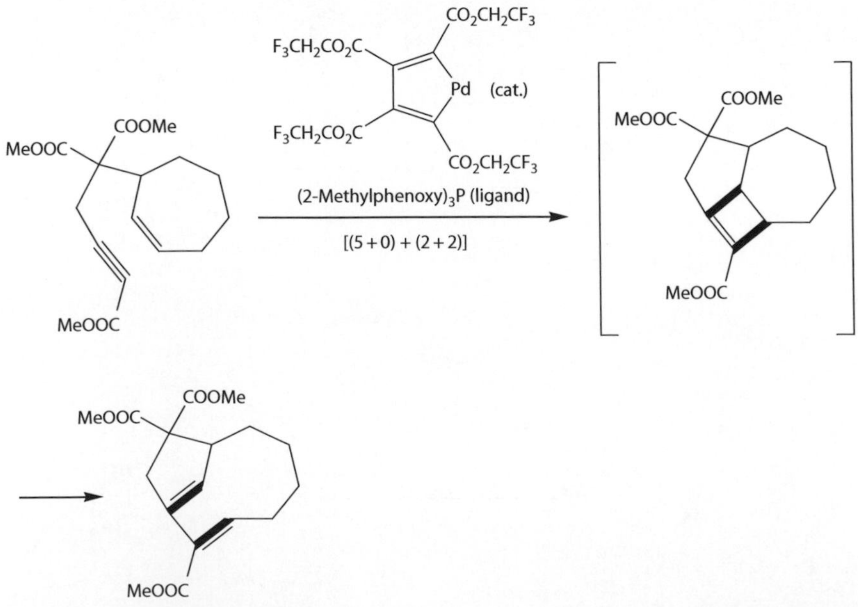

Reaction 22 (AE = 1)

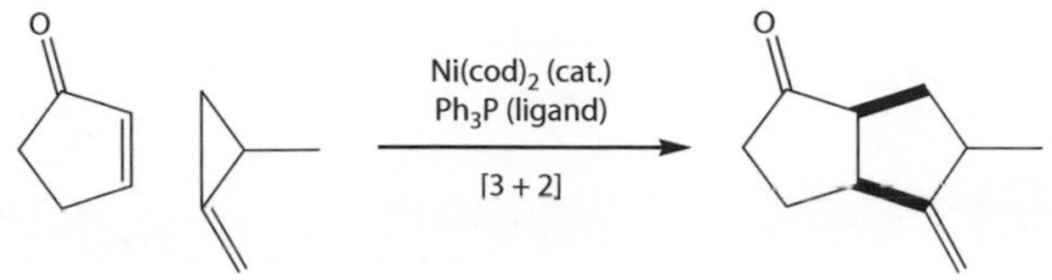

Reaction 23 (AE = 1)

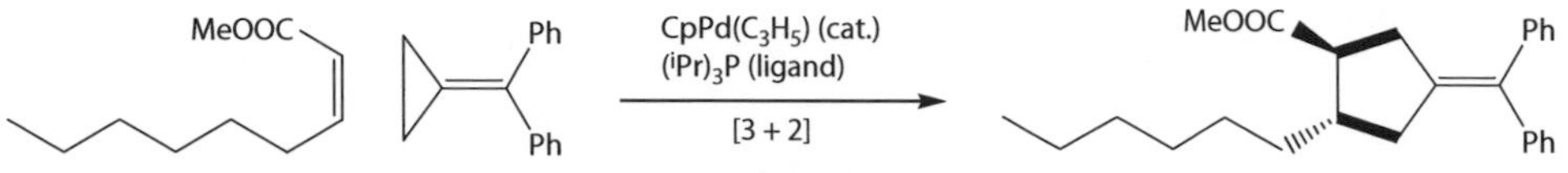

Reaction 24 (AE = 1)

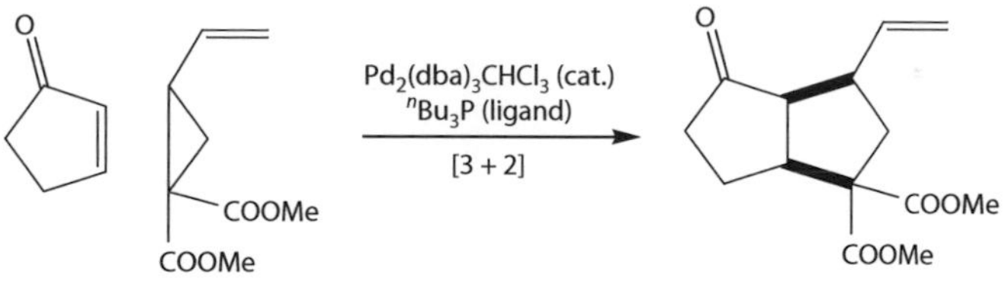

Reaction 25 (AE = 1)

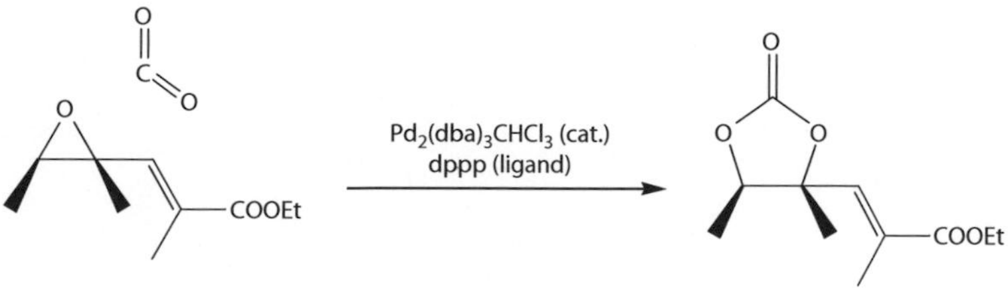

Prototropic Cycloisomerizations

Reaction 1 (AE = 1)

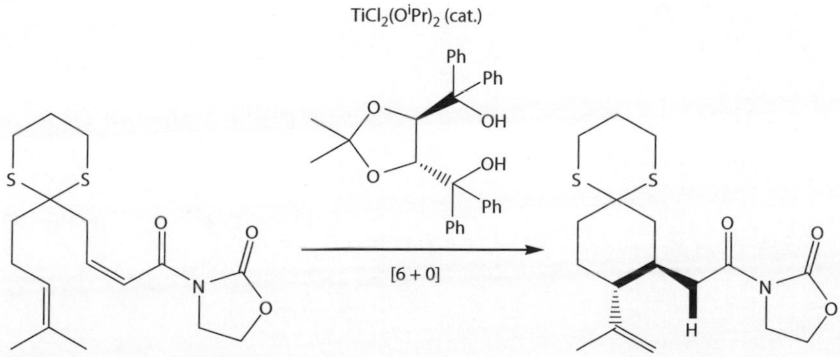

Reaction 2 (AE = 1)

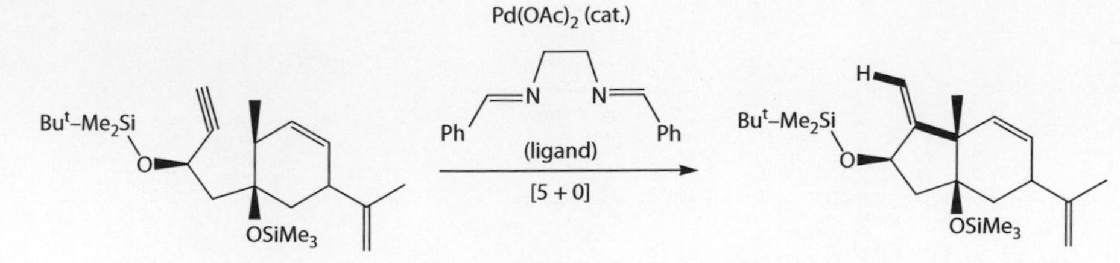

Reaction 3 (AE = 1)

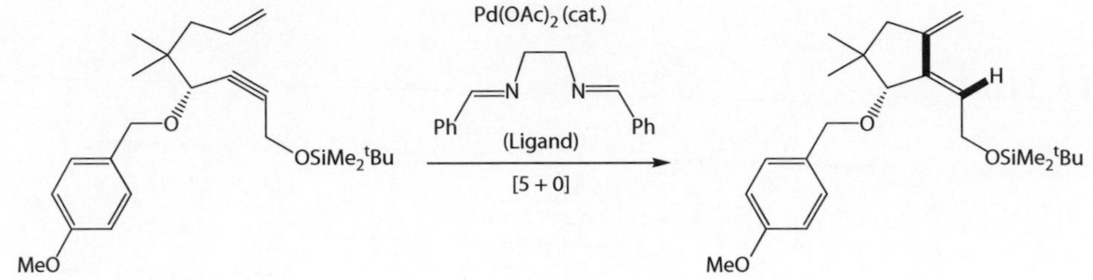

Reaction 4 (AE = 1)

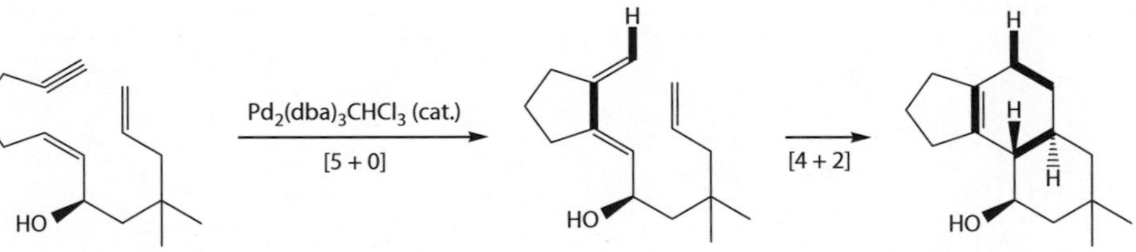

Reaction 5 (AE = 1)

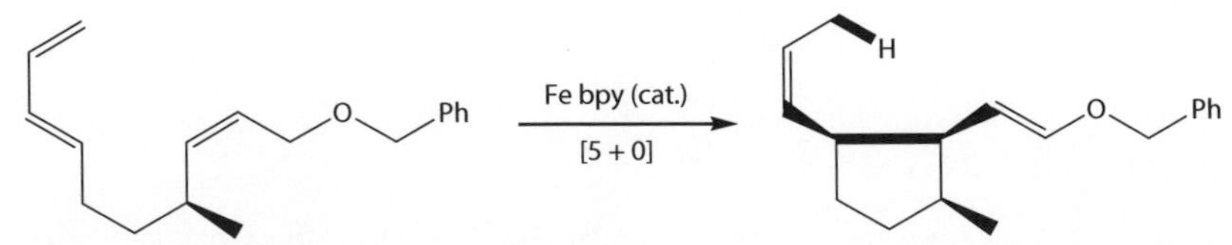

Reaction 6 (AE = 1)

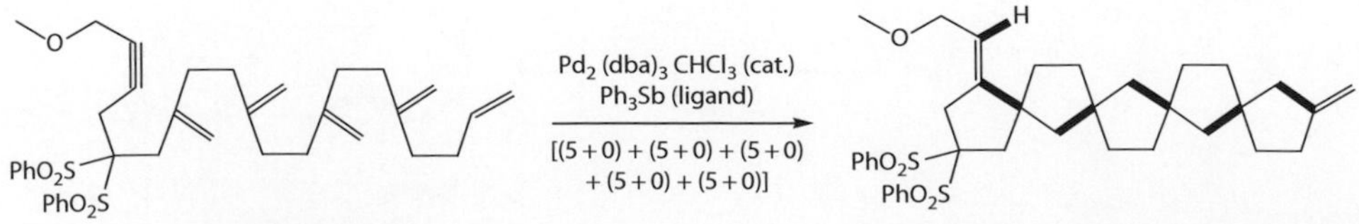

Reaction 7 (AE = 1)

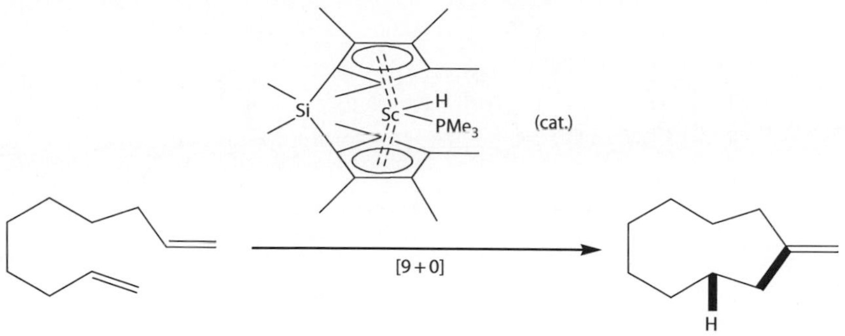
Si
Sc
H
PMe3
(cat.)
[9+0]
H

Reaction 8 (AE = 1)

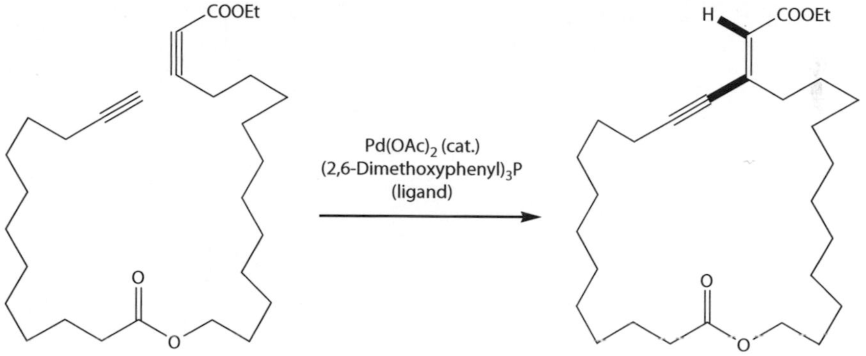
COOEt
O
O
Pd(OAc)2 (cat.)
(2,6-Dimethoxyphenyl)3P
(ligand)
H
COOEt
O
O

Reaction 9 (AE = 1)

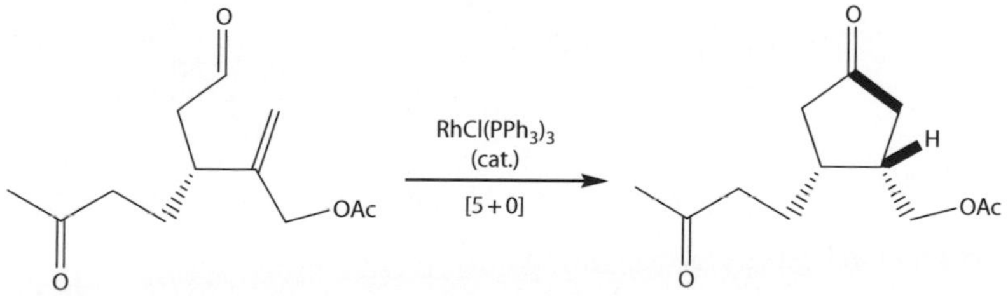
O
OAc
O
RhCl(PPh3)3
(cat.)
[5+0]
O
H
OAc
O

Reaction 10 (AE = 1)

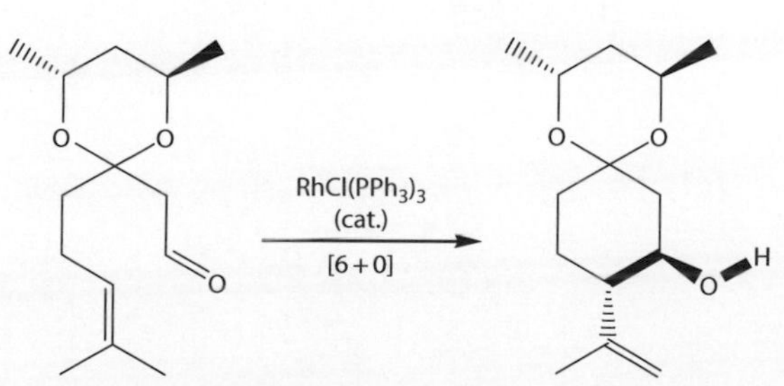
O
O
O
RhCl(PPh3)3
(cat.)
[6+0]
O
O
O
H

Reaction 11 (AE = 1)

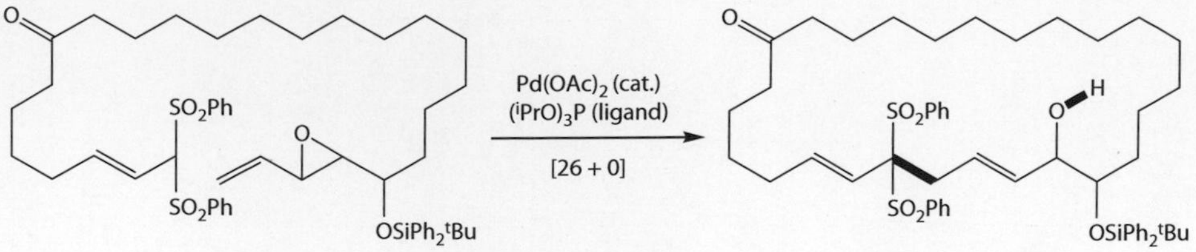

Pd(OAc)$_2$ (cat.)
(^{i}PrO)$_3$P (ligand)

[26 + 0]

Reaction 12 (AE = 1)

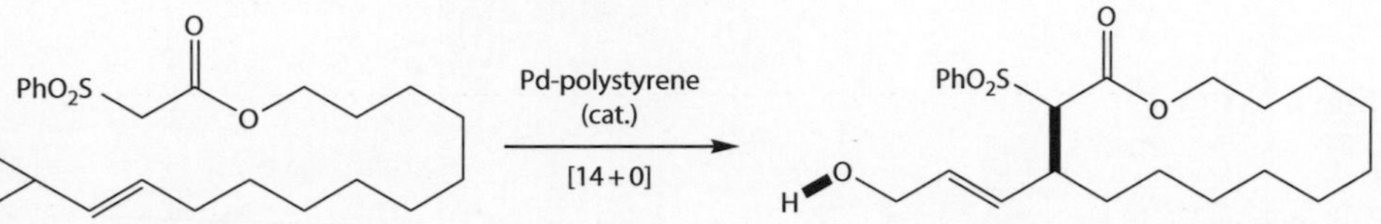

Pd-polystyrene
(cat.)

[14 + 0]

Reaction 13 (AE = 1)

RuCl$_2$(PPh$_3$)$_3$
(cat.)

[6 + 0]

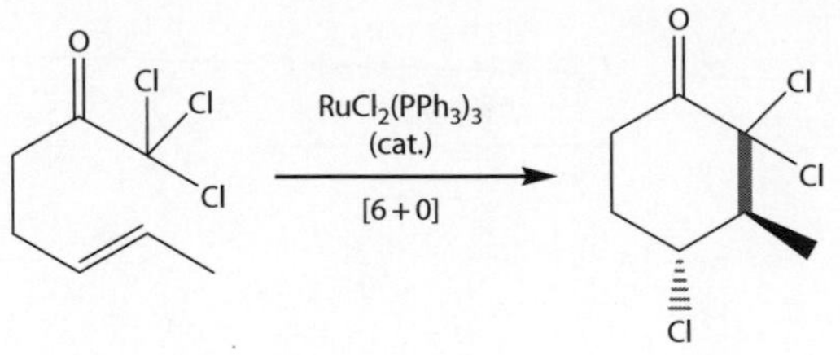

Reaction 14 (AE = 1)

Pd$_2$(dba)$_3$CHCl$_3$
(cat.)

[5 + 0]

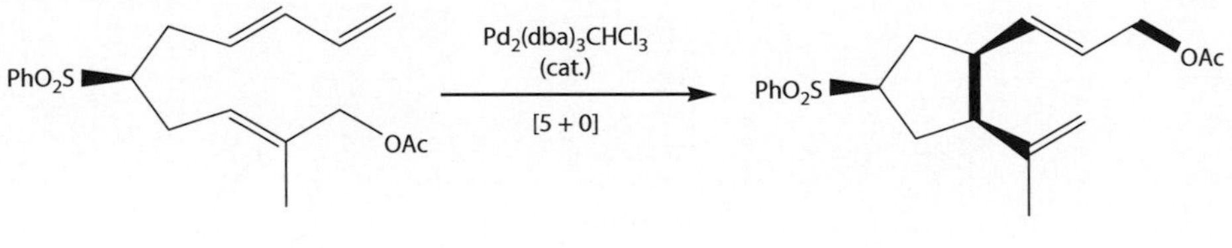

Appendix B: Challenging Redox Reactions from Synthesis Database

Abscisic acid: Mori

Mori, K. *Tetrahedron* 1974, *30*, 1065.

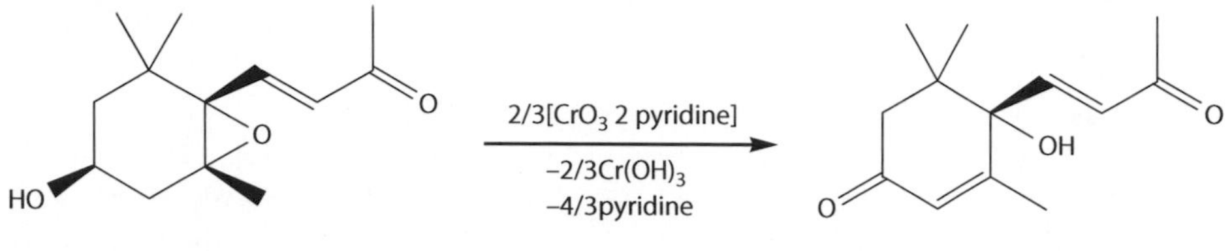

Abscisic acid: Shibasaki

Shibasaki, M.; Terashima, S.; Yamada, S. *Chem. Pharm. Bull.* 1976, *24*, 315.

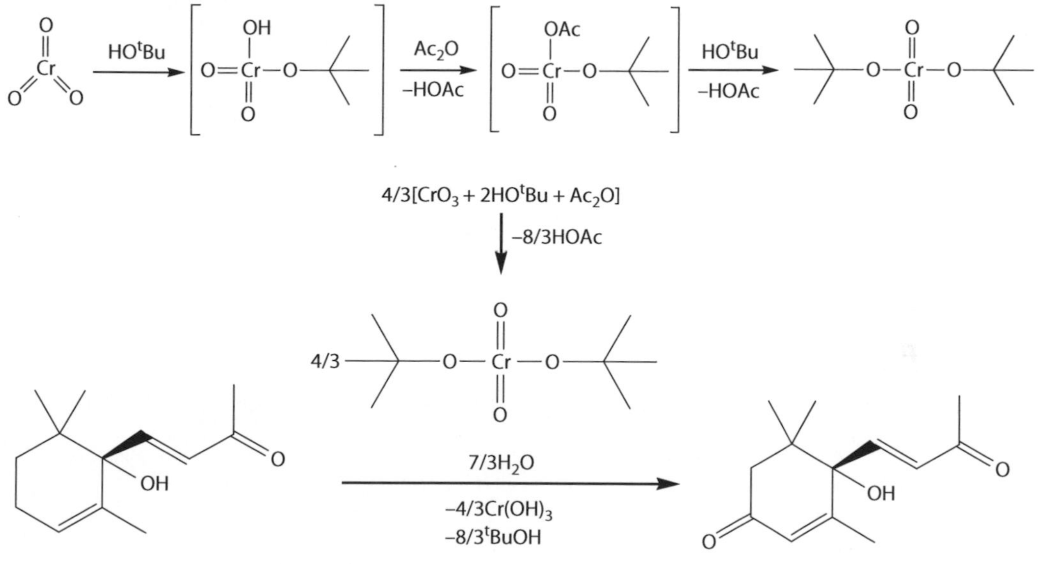

Abscisic acid: R.J. Reynolds Tobacco

Roberts, D.L.; Heckman, R.A.; Hege, B.P.; Bellin, S.A. *J. Org. Chem.* 1968, *33*, 3566.

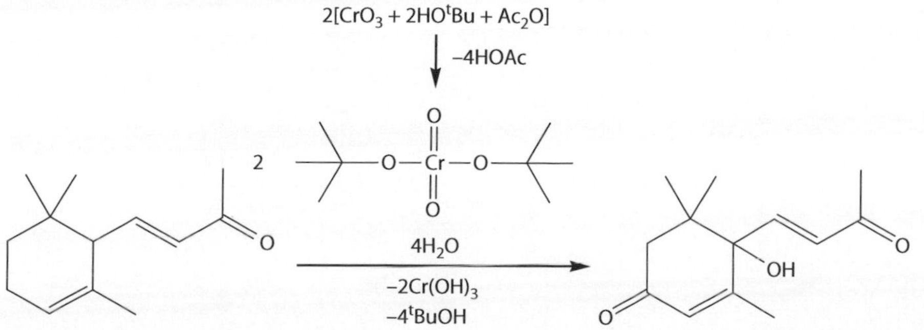

Adipic acid: Ellis

Ellis, B.A. *Org. Synth. Coll.* 1941, *1*, 18.

Faith, W.L.; Keyes, D.B.; Clark, R.L. *Industrial Chemicals*, 3rd edn., Wiley: New York, 1966, p. 44.

Lowenheim, F.A.; Moran, M.K. *Faith, Keyes, and Clark's Industrial Chemicals*, 4th edn., Wiley: New York, 1975, pp. 298, 304.

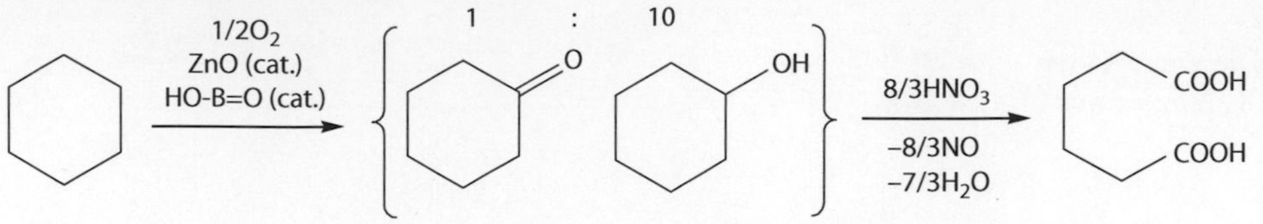

Adipic acid: Faith G2

Faith, W.L.; Keyes, D.B.; Clark, R.L. *Industrial Chemicals*, 3rd edn., Wiley: New York, 1966, p. 44.

Lowenheim, F.A.; Moran, M.K. *Faith, Keyes, and Clark's Industrial Chemicals*, 4th edn., Wiley: New York, 1975, p. 298, 304.

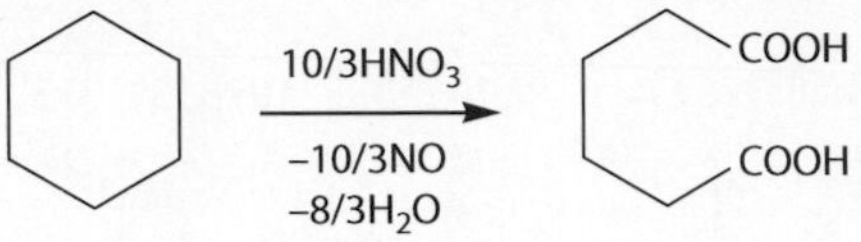

Adipic acid: Vogel

Furniss, B.S.; Hannaford, A.J.; Rogers, V.; Smith, P.W.G.; Tatchell, A.R. *Vogel's Textbook of Practical Organic Chemistry*, 4th edn., Longman: London, U.K., 1978, p. 476.

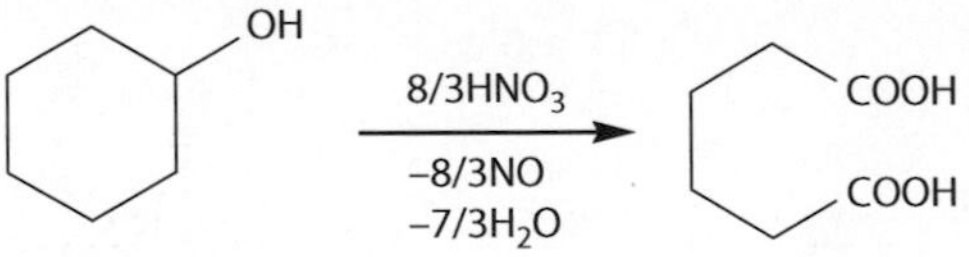

Alizarin: Graebe–Liebermann

Graebe, C.; Liebermann, C. *Chem. Ber.* 1869, *2*, 332.

Fierz-David, H.E. *The Fundamental Processes of Dye Chemistry*, J. & A. Churchill: London, U.K., 1921, p. 167.

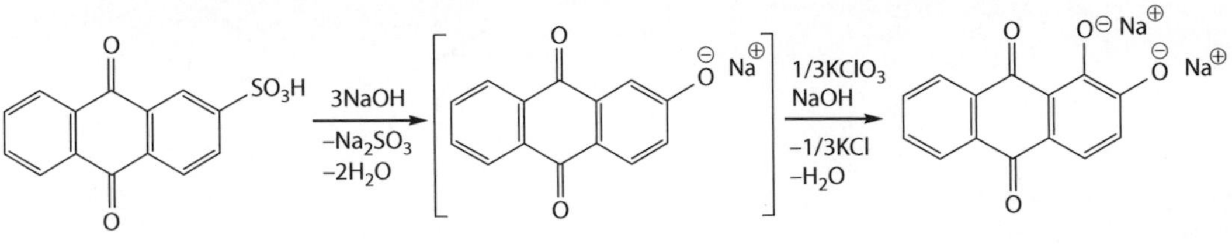

Artemisinin: Nowak

Nowak, D.M.; Lansbury, P.T. *Tetrahedron* 1998, *54*, 319.

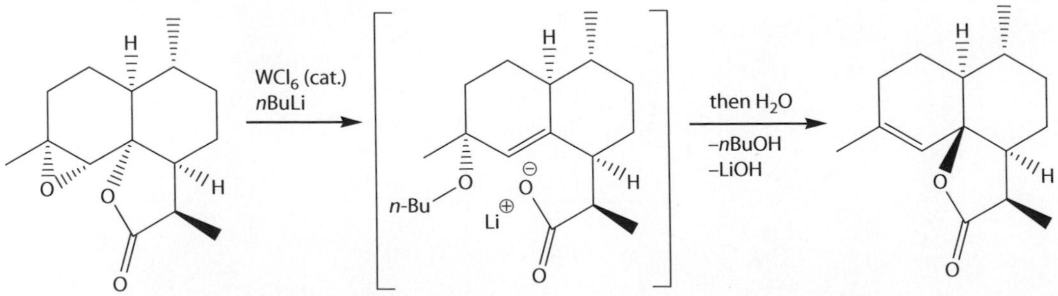

Cocaine: Pearson

Mans, D.M.; Pearson, W.H. *Org. Lett.* 2004, *6*, 3305.

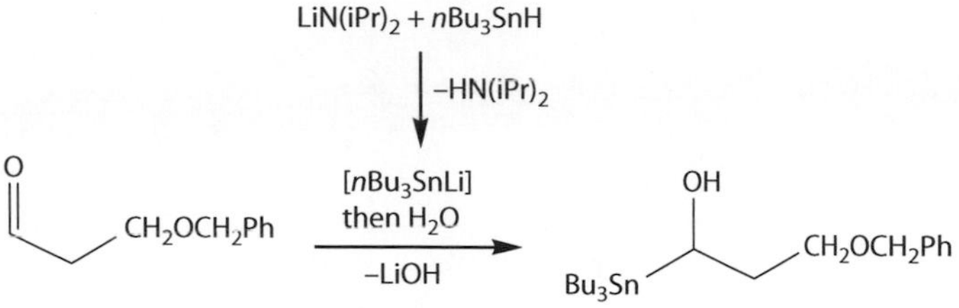
LiN(iPr)2 + nBu3SnH
–HN(iPr)2
[nBu3SnLi]
then H2O
–LiOH
O
CH2OCH2Ph
OH
Bu3Sn
CH2OCH2Ph

Cocaine: Rapoport

Lin, R.; Castells, J.; Rapoport, H. *J. Org. Chem.* 1998, *63*, 4069.

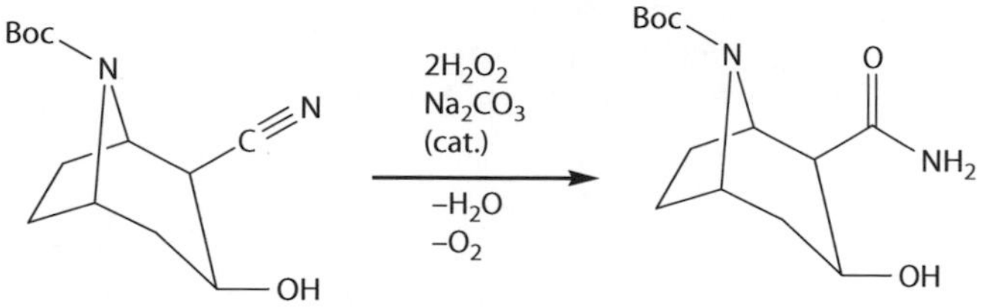
Boc
N
C
N
OH
2H2O2
Na2CO3
(cat.)
–H2O
–O2
Boc
N
O
NH2
OH

Corydaldine: Wiesner

Wiesner, K.; Valenta, Z.; Manson, A.J.; Stonner, F.W. *J. Am. Chem. Soc.* 1955, *77*, 675.

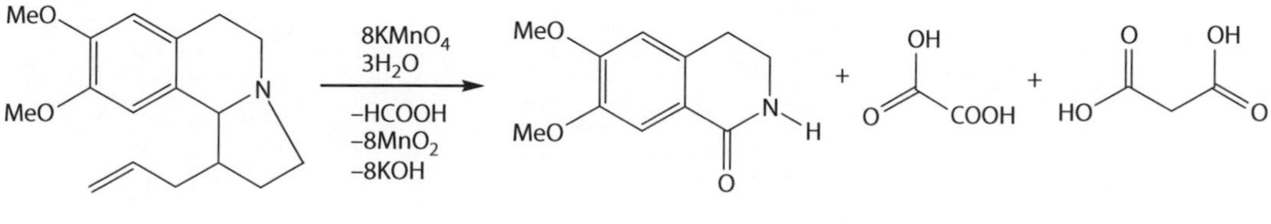
MeO
MeO
N
8KMnO4
3H2O
–HCOOH
–8MnO2
–8KOH
MeO
MeO
N
H
O
+
OH
O
COOH
+
O
OH
HO
O

Corydaline: Spath

Späth, E.; Kruta, A. *Chem. Ber.* 1929, *62*, 1024.

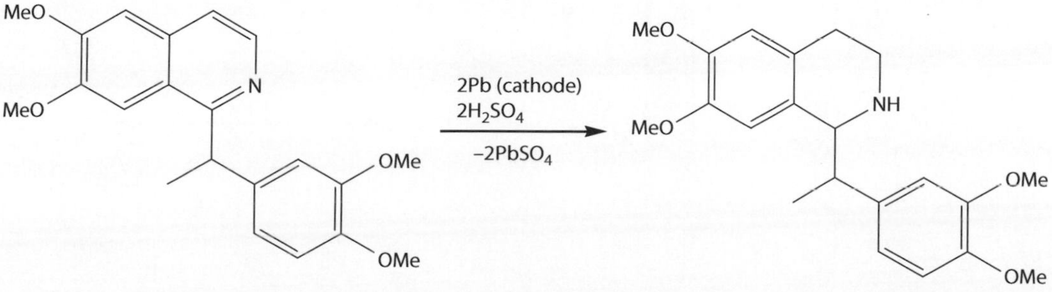
MeO
MeO
N
OMe
OMe
2Pb (cathode)
2H2SO4
–2PbSO4
MeO
MeO
NH
OMe
OMe

Cubane: Pettit

Barborak, J.C.; Watts, L.; Pettit, R. *J. Am. Chem. Soc.* 1966, *88*, 1328.

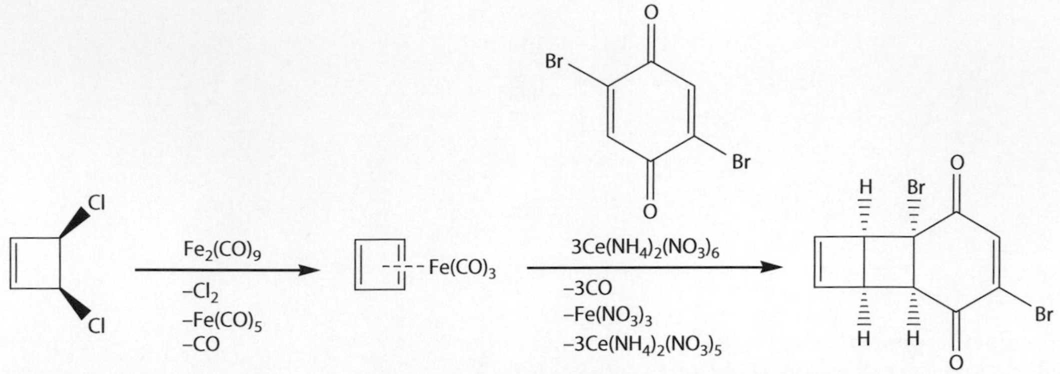

DEET: Dominion Rubber

Van Stryk, F.G. CA 716609 (Dominion Rubber, 1962).
Van Stryk, F.G. US 3198831 (US Rubber Company, 1965).
Emerson, W.S.; Lucas, V.E.; Heimsch, R.A. *J. Am. Chem. Soc.* 1949, *71*, 1742.

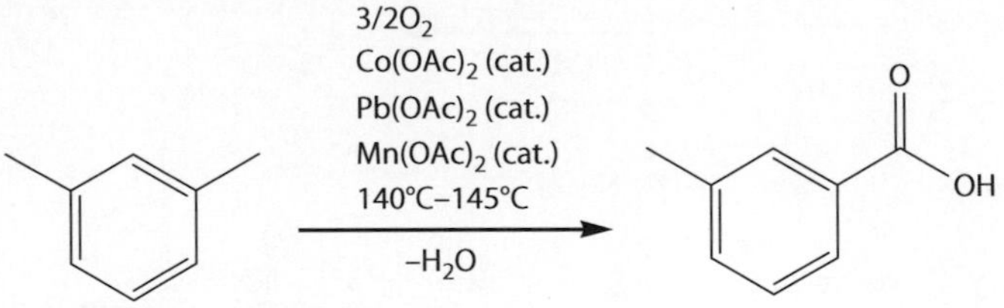

Drometrizole: Koutsimpelis

Koutsimpelis, A.G.; Screttas, C.G.; Igglessi-Markopoulou, O. *Heterocycles* 2005, *65*, 1393.

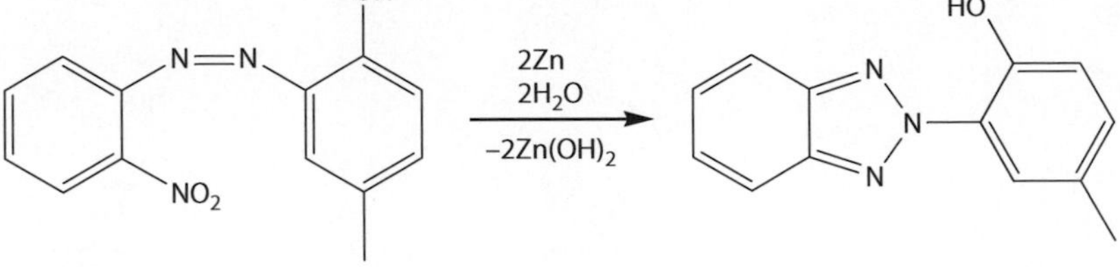

Estrone: Kocovsky

Kocovsky, P.; Baines, R.S. *Tetrahedron Lett.* 1993, *34*, 6139.
Kocovsky, P.; Baines, R.S. *J. Org. Chem.* 1994, *59*, 5439.

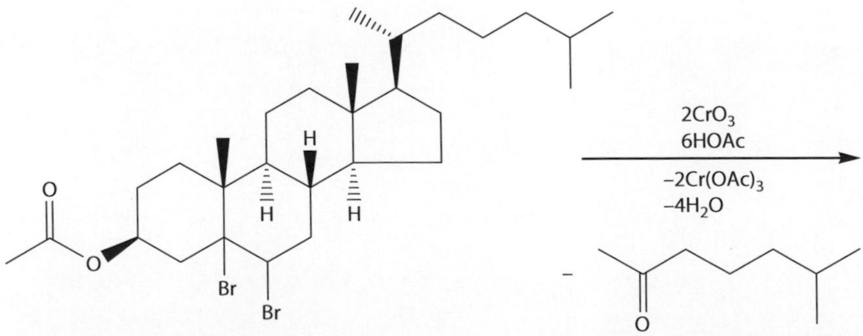

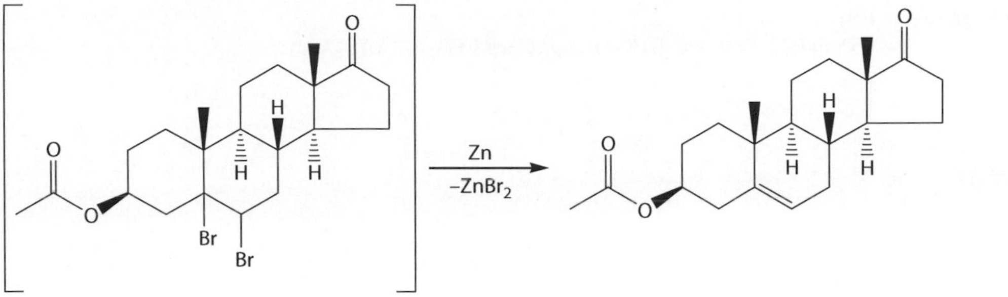

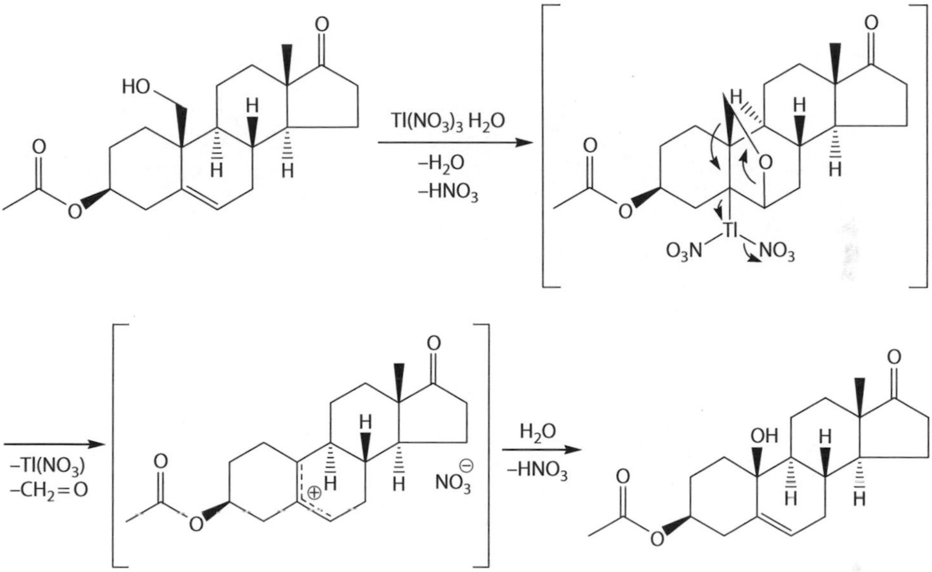

Estrone: Sih

Sih, C.J.; Lee, S.S.; Tsong, Y.Y.; Wang, K.C.; Chang, F.N. *J. Am. Chem. Soc.* 1965, *87*, 2765.

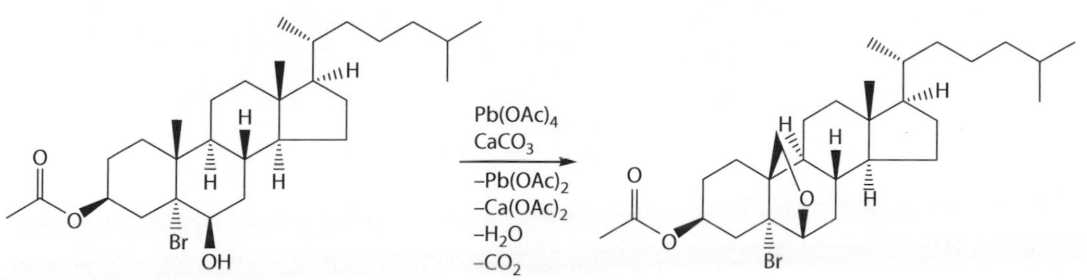

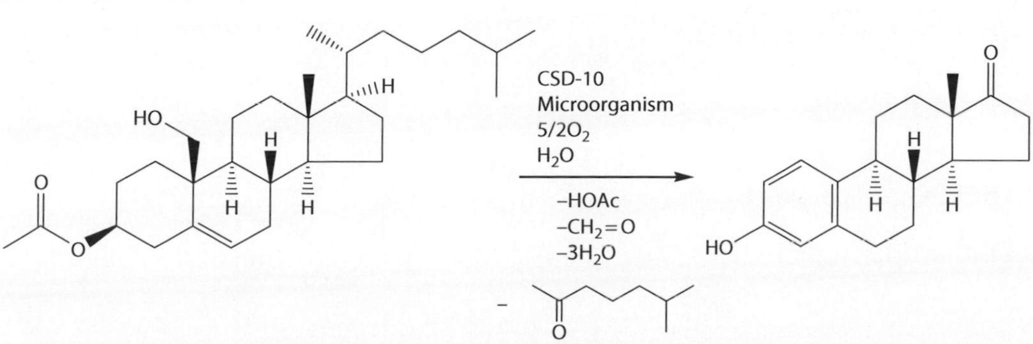

Ibuprofen: Pinhey

Kopinski, R.P.; Pinhey, J.T.; Rowe, B.A. *Aust. J. Chem.* 1984, *37*, 1245.

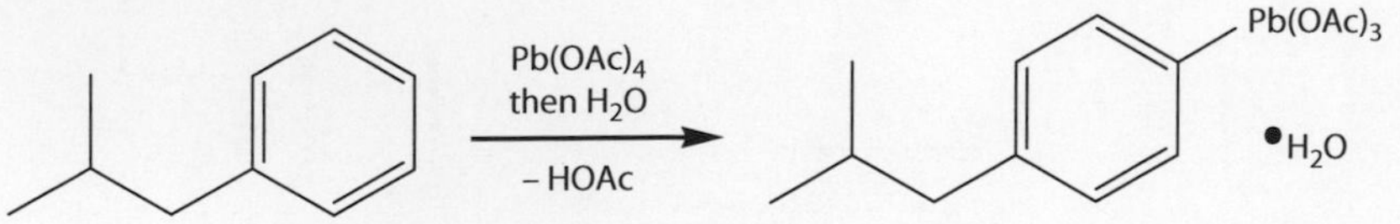

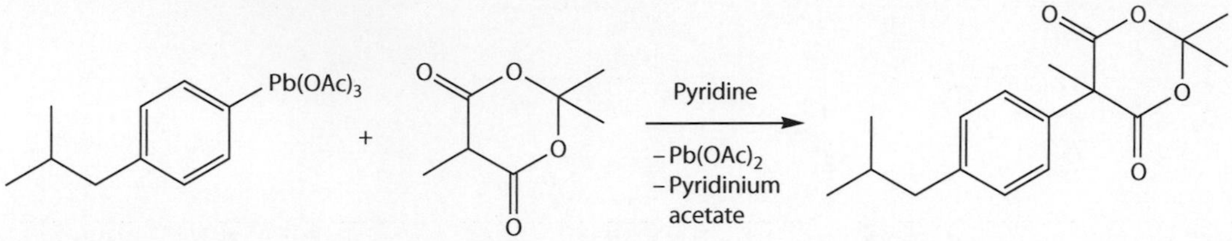

Indigo: Heumann–Fierz–David

DE 56273 (BASF, 1891).
DE 127178 (BASF, 1901).
Fierz-David, H.E.; Blangey, L. *The Fundamental Processes of Dye Chemistry*, Interscience Publishing:
    New York, 1949, pp. 174, 323.

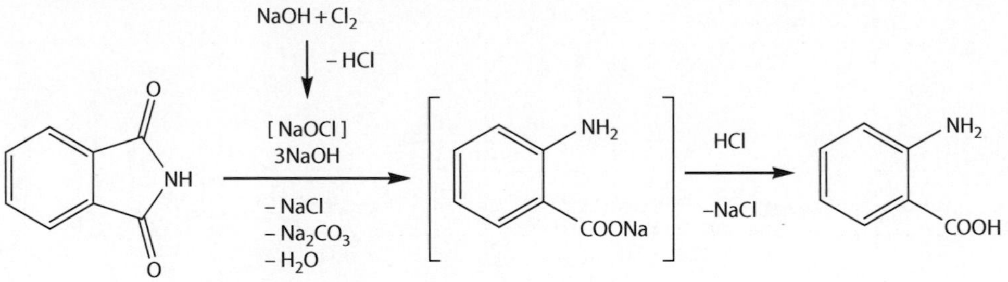

Heumann, K. *Chem. Ber.* 1890, *23*, 3437.
Cain, J.C. *The Synthetic Dyestuffs and the Intermediate Products from Which They Are Derived*, C. Griffin:
    London, U.K., 1920, pp. 294–296.

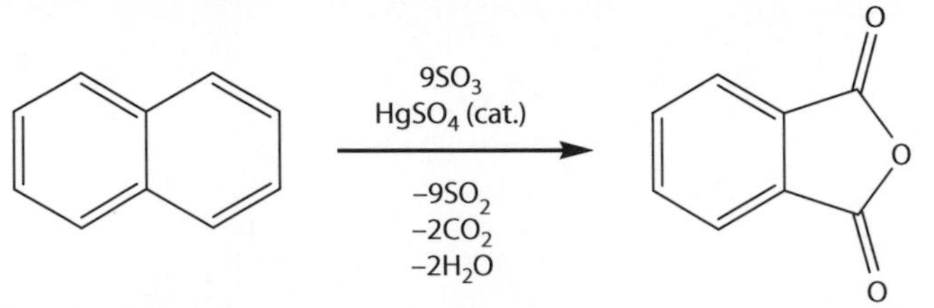

(+)-(2R,6S)-*cis*-α-Irone: Ohtsuka

Ohtsuka, Y.; Itoh, F.; Oishi, T. *Chem. Pharm. Bull.* 1991, *39*, 2540.

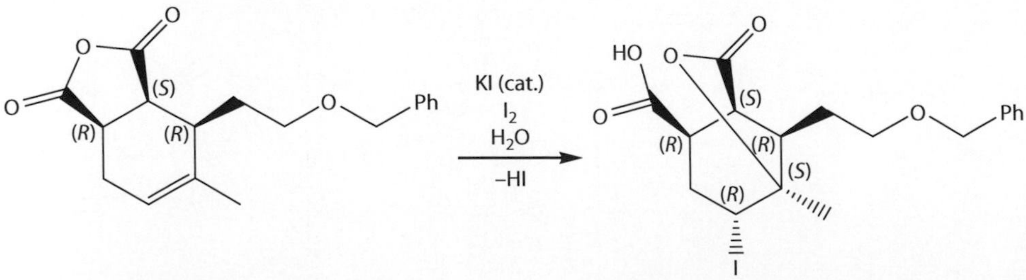

Juvabione: Miles

Miles, W.H.; Brinkman, H.R. *Tetrahedron Lett.* 1992, *33*, 589.

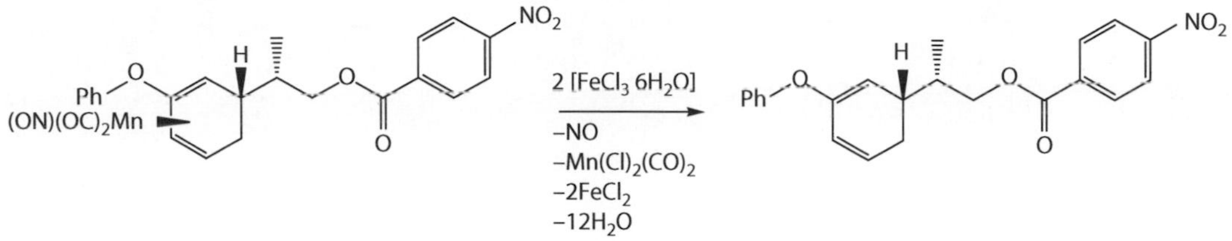

2 [FeCl$_3$ 6H$_2$O]

−NO
−Mn(Cl)$_2$(CO)$_2$
−2FeCl$_2$
−12H$_2$O

Juvabione: Ogasawara G1/Ogasawara G2

Kawamura, M.; Ogasawara, K. *Chem. Commun.* 1995, 2403.
Nagata, H.; Taniguchi, T.; Kawamura, M.; Ogasawara, K. *Tetrahedron Lett.* 1999, *40*, 4207.

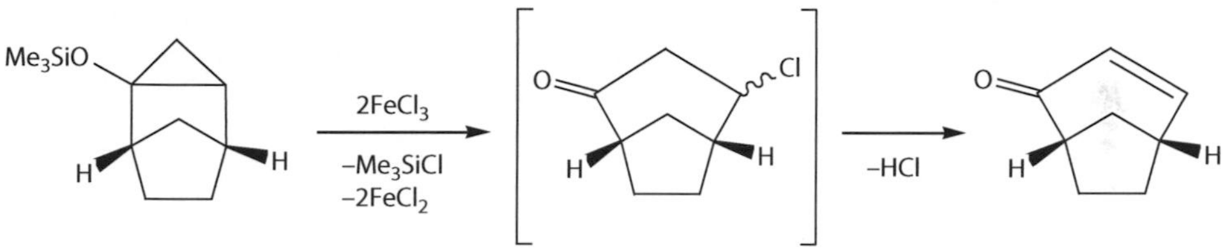

2FeCl$_3$

−Me$_3$SiCl
−2FeCl$_2$

−HCl

Kainic acid: Ogasawara G1

Nakada, Y.; Sugahara, T.; Ogasawara, K. *Tetrahedron Lett.* 1997, *38*, 857.
Haubenstock, H.; Mennitt, P.G.; Butler, P.E. *J. Org. Chem.* 1970, *35*, 3208.
Takano, S.; Iwabuchi, Y.; Takahashi, M.; Ogasawara, K. *Chem. Pharm. Bull.* 1986, *34*, 3445.

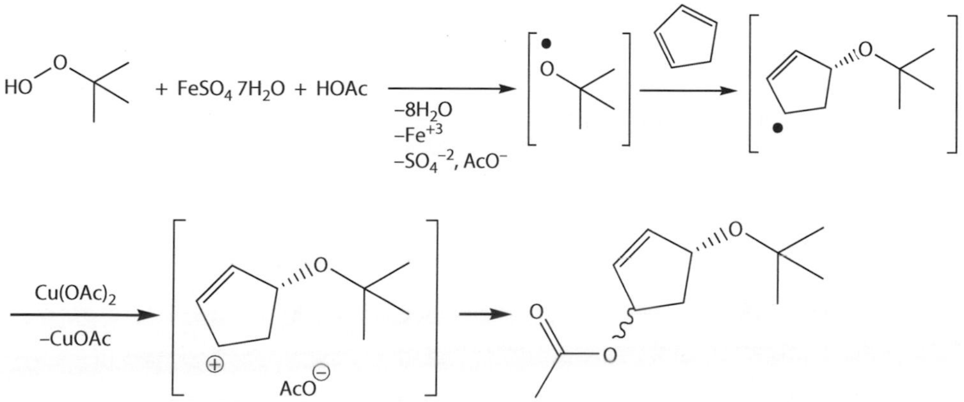

+ FeSO$_4$ 7H$_2$O + HOAc

−8H$_2$O
−Fe^{+3}
−SO$_4^{-2}$, AcO$^-$

Cu(OAc)$_2$

−CuOAc

Kainic acid: Ogasawara G3

Hirasawa, H.; Taniguchi, T.; Ogasawara, K. *Tetrahedron Lett.* 2001, *42*, 7587.

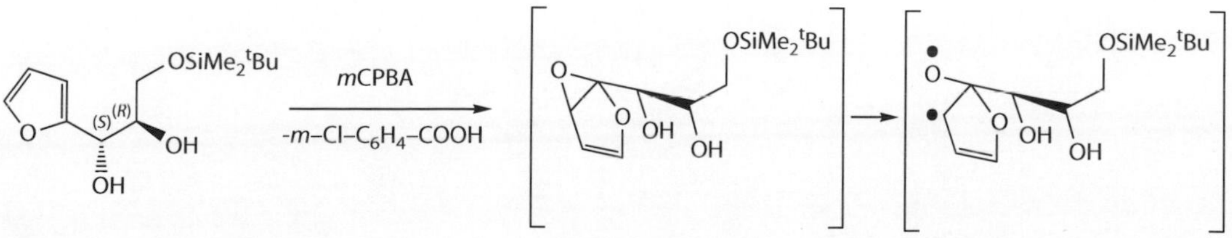

*m*CPBA

−*m*−Cl−C$_6$H$_4$−COOH

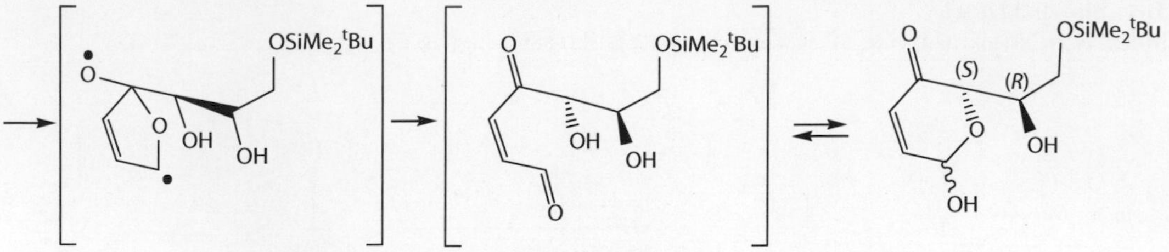

Kainic acid: Takano G3

Takano, S.; Inomata, K.; Ogasawara, K. *Chem. Commun.* 1992, 169.

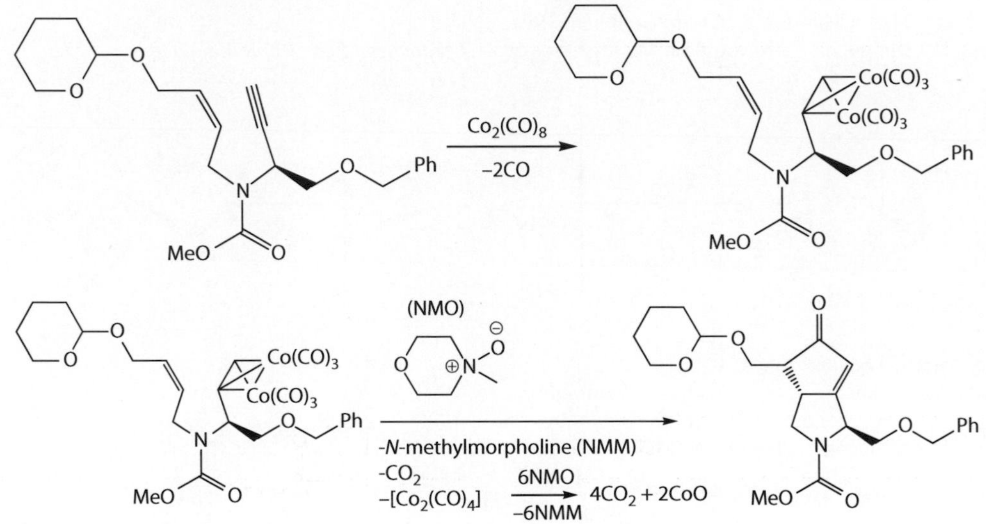

Lycopodine: Carter

Yang, H.; Carter, R.G. *J. Org. Chem.* 2010, *75*, 4929.

Yang, H.; Carter, R.G.; Zakharov, L.N. *J. Am. Chem. Soc.* 2008, *130*, 9238.

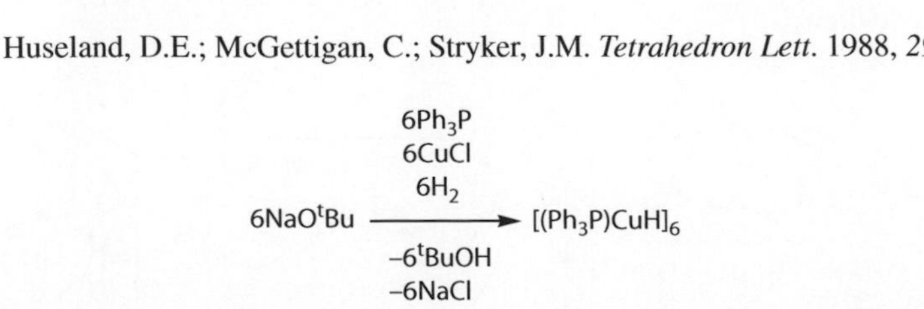

Brestensky, D.M.; Huseland, D.E.; McGettigan, C.; Stryker, J.M. *Tetrahedron Lett.* 1988, *29*, 3749.

$$6NaO^tBu \xrightarrow[-6^tBuOH,\ -6NaCl]{6Ph_3P,\ 6CuCl,\ 6H_2} [(Ph_3P)CuH]_6$$

Lycopodine: Mori

Mori, M.; Hori, K.; Akashi, M.; Hori, M.; Sato, Y.; Nishida, M. *Angew. Chem. Int. Ed.* 1998, *37*, 636.

$$2Ti(O^iPr)_4 + 6Me_3SiCl + N_2 + 8e^{\ominus} \longrightarrow 2TiCl_3 + 2(Me_3Si)_3N + 8O^iPr^{\ominus}$$

$$Li \longrightarrow Li^{\oplus} + e^{\ominus} \quad X\,8$$

$$2Ti(O^iPr)_4 + 6Me_3SiCl + N_2 + 8Li \longrightarrow 2TiCl_3 + 2(Me_3Si)_3N + 8LiO^iPr$$

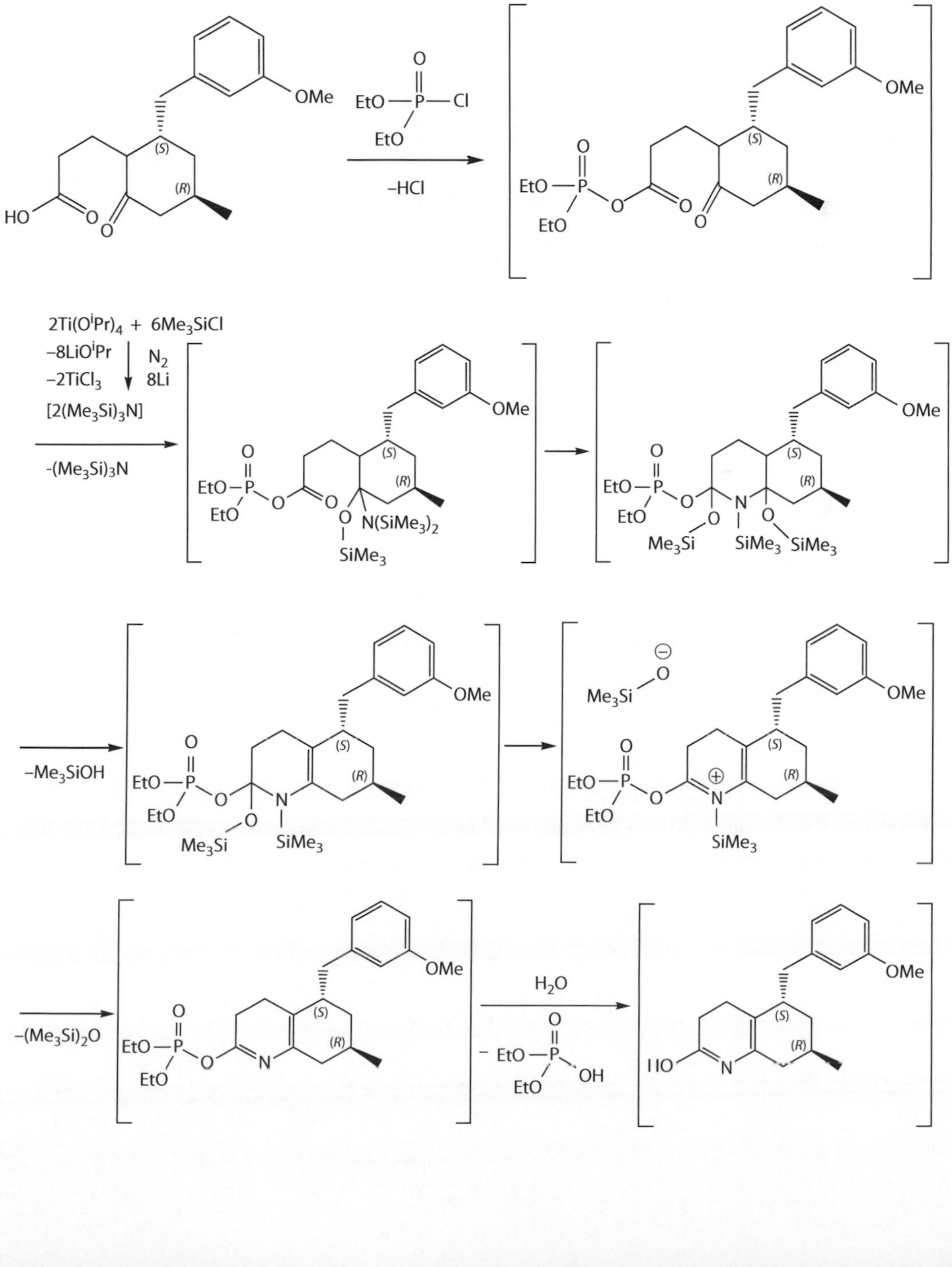

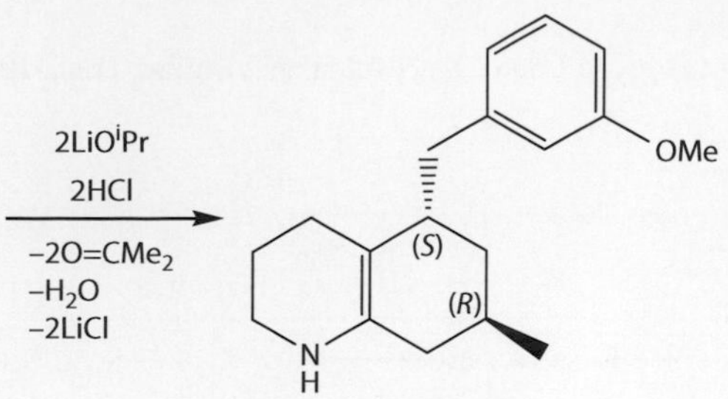
2LiO^{i}Pr
2HCl
−2O=CMe$_2$
−H$_2$O
−2LiCl
(S)
(R)
N
H
OMe

Mescaline: Rose

Rose-Munch, F.; Chavignon, R.; Tranchier, J.P.; Gagliardini, V.; Rose, E. *Inorg. Chim. Acta* 2000, *300–302*, 693.

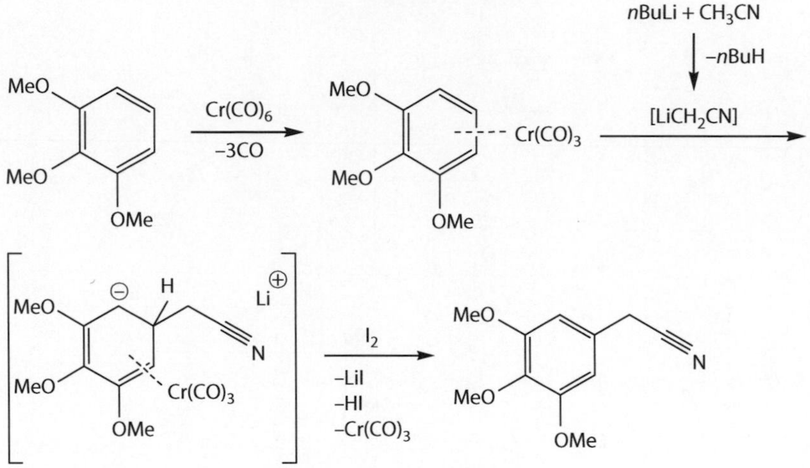
nBuLi + CH$_3$CN
−nBuH
MeO
MeO
OMe
Cr(CO)$_6$
−3CO
Cr(CO)$_3$
[LiCH$_2$CN]
H
Li
N
I$_2$
−LiI
−HI
−Cr(CO)$_3$

Wulff, W.D.; Yang, D.C. *J. Am. Chem. Soc.* 1984, *106*, 7565.

Lluch, A.M.; Sanchez-Baeza, F.; Camps, F.; Messeguer, A. *Tetrahedron Lett.* 1991, *32*, 5629.

Gilbert, M.; Ferrer, M.; Lluch, A.M.; Sanchez-Baeza, F.; Messeguer, A. *J. Org. Chem.* 1999, *64*, 1591.

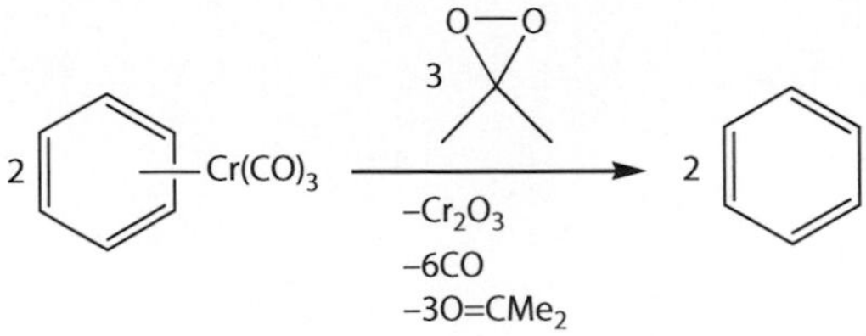
2
Cr(CO)$_3$
3
O
O
−Cr$_2$O$_3$
−6CO
−3O=CMe$_2$
2

Monastral fast blue B: Fierz-David

Fierz-David, H.E.; Blangey, L. *The Fundamental Processes of Dye Chemistry*, Interscience Publishing: New York, 1949, p. 338.

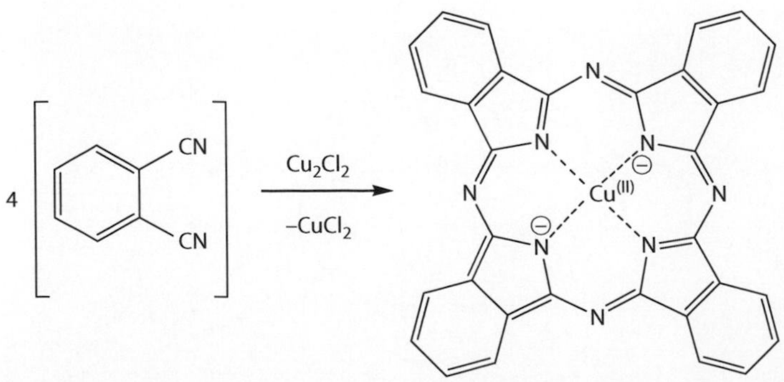
4
CN
CN
Cu$_2$Cl$_2$
−CuCl$_2$
N
Cu$^{(II)}$

Monastral fast blue B: Vogel

Vogel, A. *Textbook of Practical Organic Chemistry*, Longman: London, U.K., 1948, p. 852.

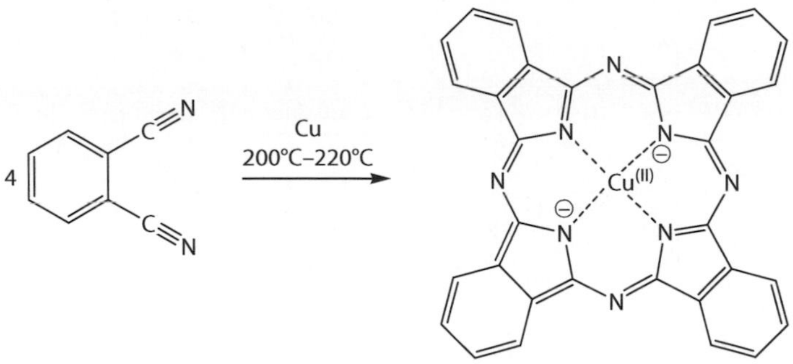

Nicotine: Pictet

Pictet, A.; Rotschy, A. *Chem. Ber.* 1904, *37*, 1225.
Pictet, A. *Chem. Ber.* 1900, *33*, 2355.
Pictet, A.; Crépieux, P. *Chem. Ber.* 1895, *28*, 1904.

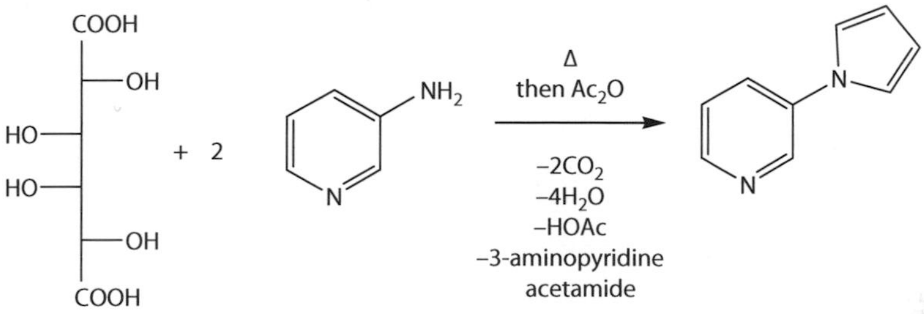

Nicotinic acid: Raja

Raja, R.; Thomas, J.M.; Greenhill-Hooper, M.; Ley, S.V.; Paz, F.A.A. *Chem. Eur. J.* 2008, *14*, 2340.

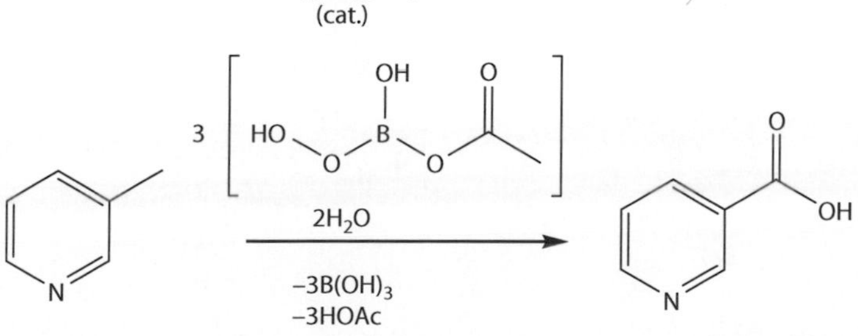

Roesler, R.; Schelle, S.; Gnann, M.; Zeiss, W. US 5462692 (Peroxid Chemie, 1995).

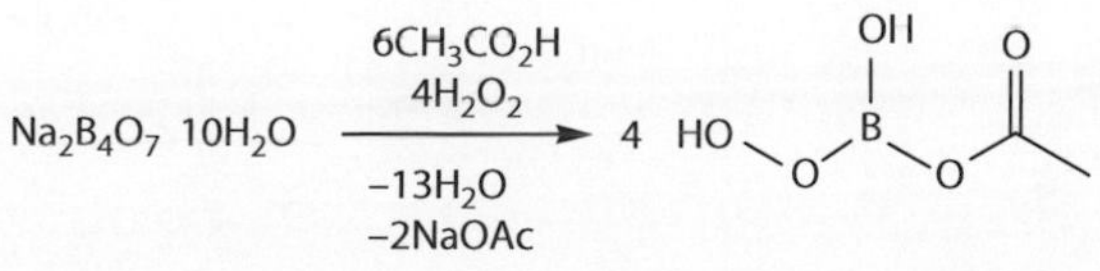

Oseltamivir: **Kongkathip**

Wichienukul, P.; Akkarasamiyo, S.; Kongkathip, N.; Kongkathip, B. *Tetrahedron Lett*. 2010, *51*, 3208.

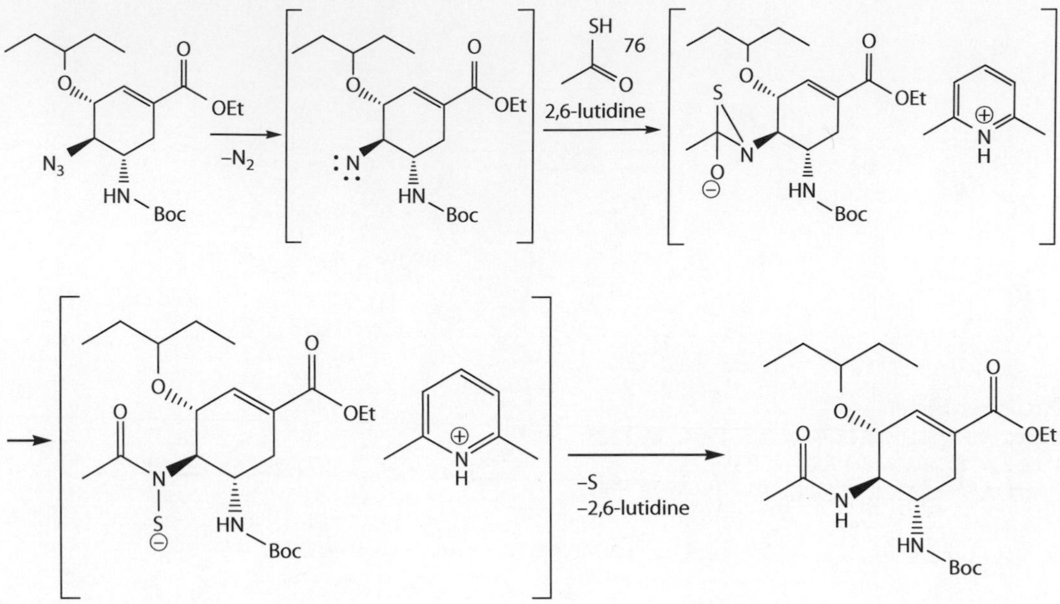

Oseltamivir: **Hudlicky G1**

Sullivan, B.; Carrera, I.; Drouin, M.; Hudlicky, T. *Angew. Chem. Int. Ed*. 2009, *48*, 4229.

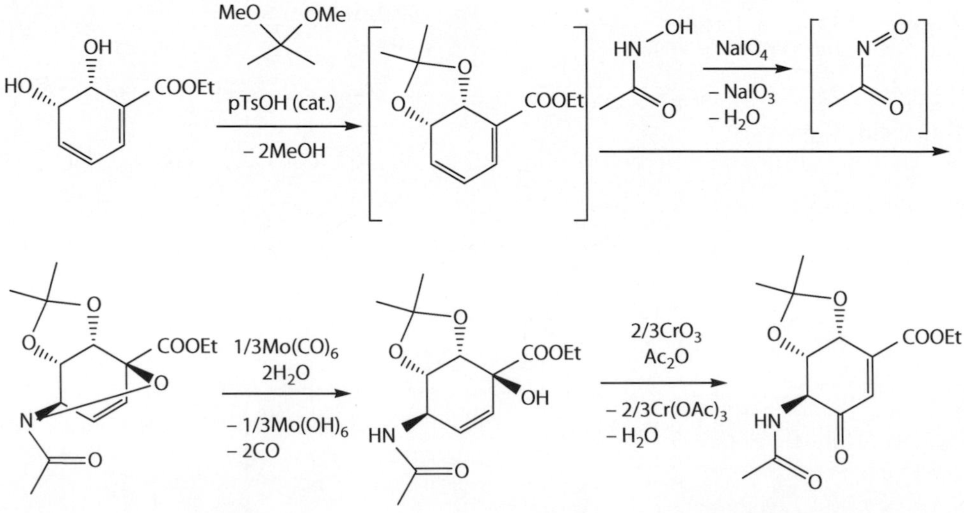

Oseltamivir: **Kann**

Bromfield, K.M.; Gradén, H.; Hagberg, D.P.; Olsson, T.; Kann, N. *Chem. Commun*. 2007, 3183.

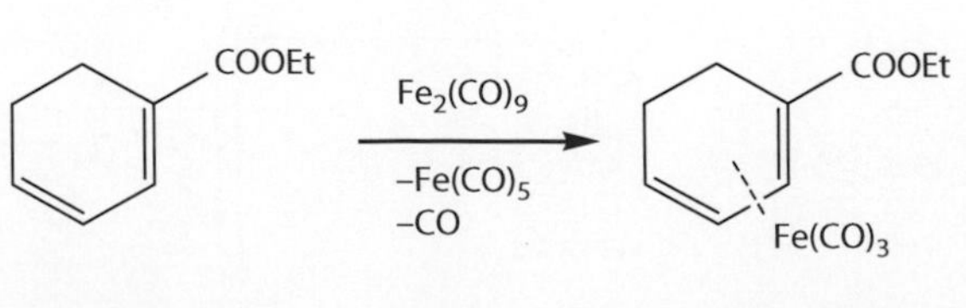

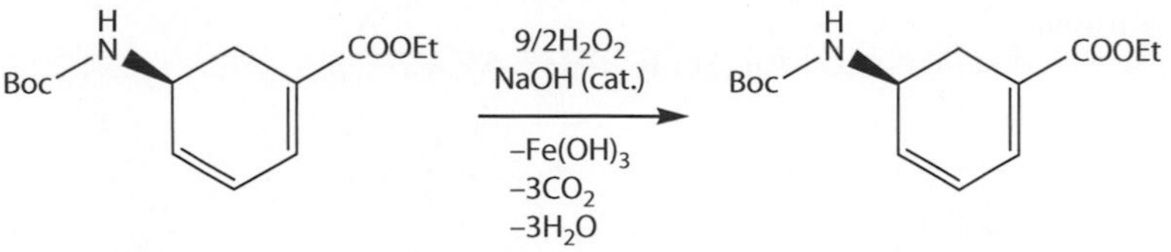

Oseltamivir: Hudlicky G1

Sullivan, B.; Carrera, I.; Drouin, M.; Hudlicky, T. *Angew. Chem. Int. Ed.* 2009, *48*, 4229.

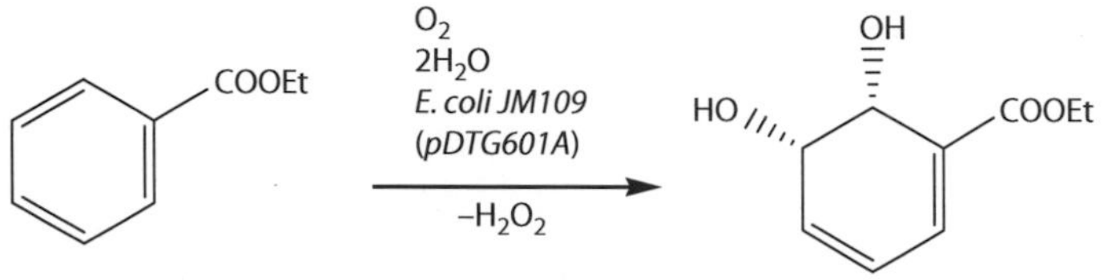

Paclitaxel: Danishefsky

Masters, J.J.; Link, J.T.; Snyder, L.B.; Young, W.B.; Danishefsky, S.J. *Angew. Chem. Int. Ed.* 1995, *34*, 1723.
Danishefsky, S.J.; Masters, J.J.; Young, W.B.; Link, J.T.; Snyder, L.B.; Magee, T.V.; Jung, D.K.; Isaacs, R.C.A.;
   Bornmann, W.G.; Alaimo, C.A.; Coburn, C.A.; Di Grandi, M.J. *J. Am. Chem. Soc.* 1996, *118*, 2843.

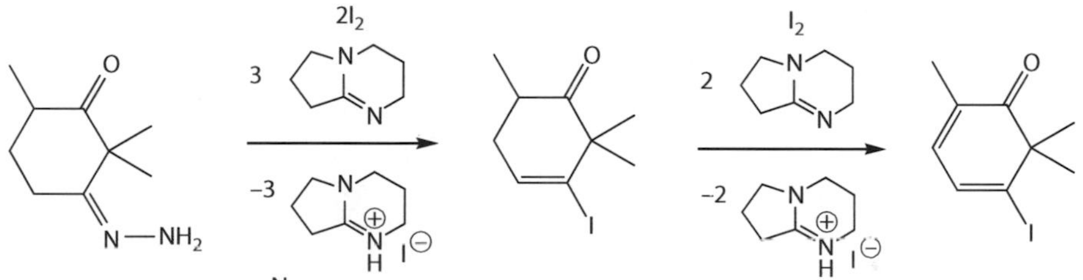

Papaverine: Dean

Dean, F.H. Ontario Research Foundation, unpublished results, 1974.

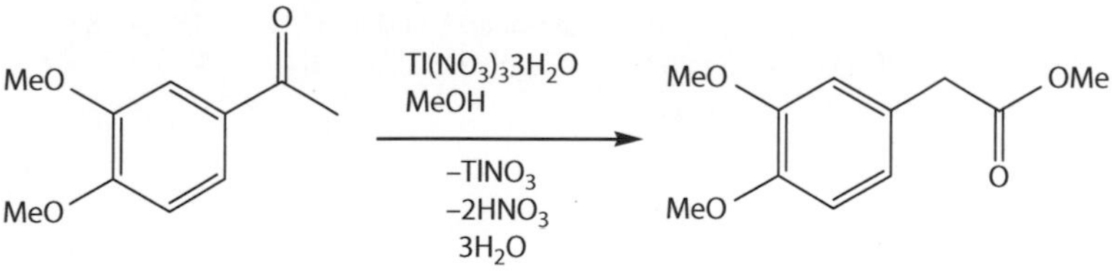

Saccharin: Fahlberg G1

Fahlberg, C.; List, A. DE 35211 (1884).
Fahlberg, C. US 319082 (1885).
Fahlberg, C. DE 64624 (1891).
Fahlberg, C. GB 189510955 (1895).
Fahlberg, C. GB 189617401 (1896).
GB 19098421 (Fahlberg, List & Co., 1909).

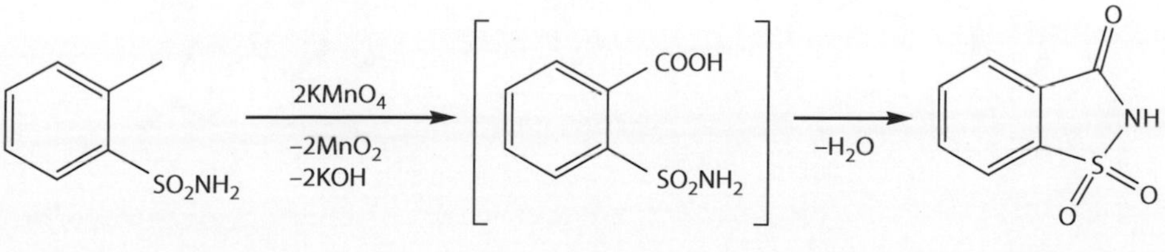

Saccharin: Maumee

Senn, O.F. US 2667503 (Maumee Development Co., 1954).

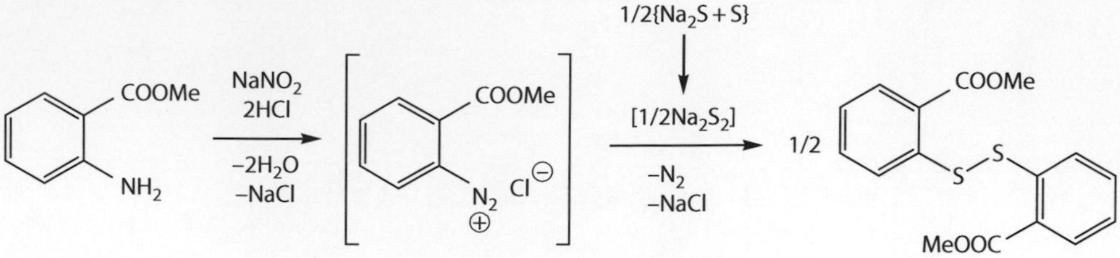

Saccharin: Sherwin-Williams

Riggin, R.M.; Kinzer, G.W. *Food Chem. Tox.* 1983, *21*, 1.

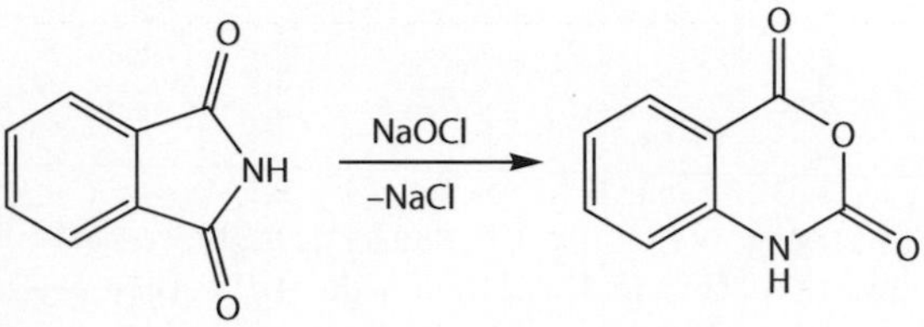

Saccharin: Rhone-Poulenc

Mounier, P. US 3759936 (Rhône-Poulenc, 1973).

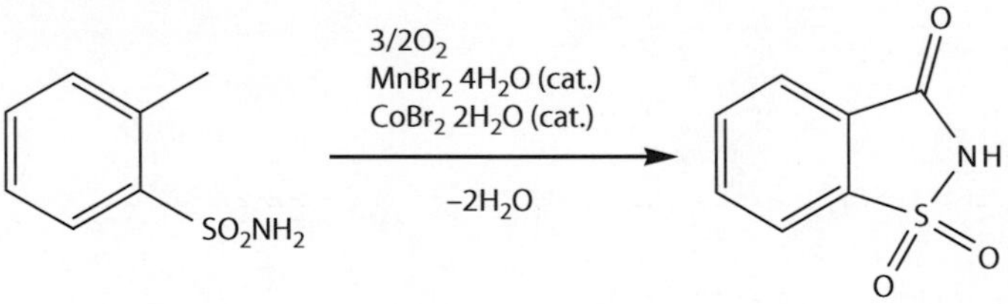

Saccharin: Vogel

Vogel, A. *Textbook of Practical Organic Chemistry*, Longman: London, U.K., 1948, p. 782.
Furniss, B.S.; Hannaford, A.J.; Rogers, V.; Smith, P.W.G.; Tatchell, A.R. *Vogel's Textbook of Practical Organic Chemistry*, Longman: London, U.K., 1978, p. 647.

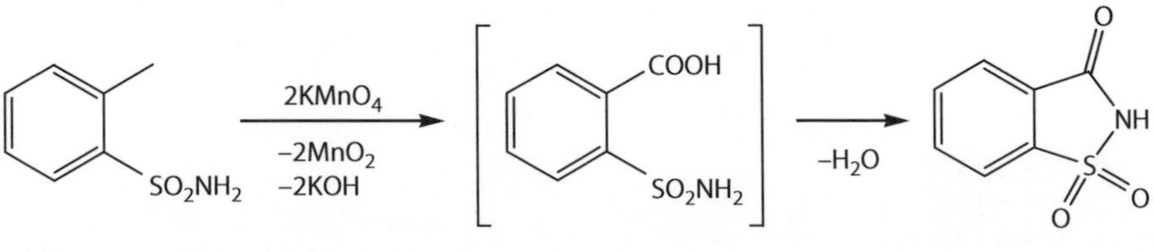

Strychnine: Bonjoch

Sole, D.; Bonjoch, J.; Garcia-Rubio, S.; Peidro, E.; Bosch, *J. Chem. Eur. J.* 2000, *6*, 655.
Sole, D.; Bonjoch, J.; Garcia-Rubio, S.; Peidro, E.; Bosch, *J. Angew. Chem. Int. Ed.* 1999, *38*, 395.

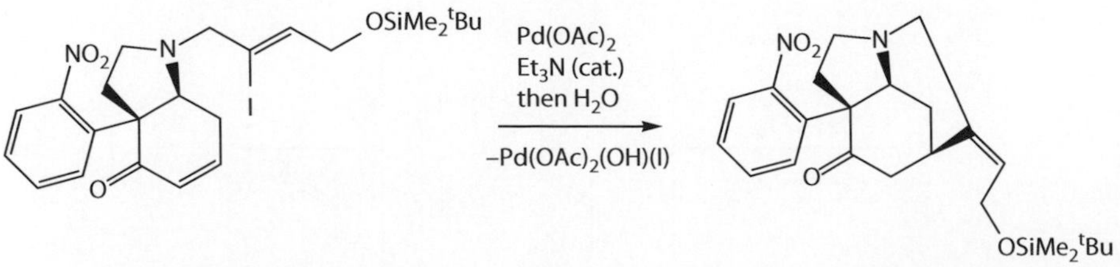

Strychnine: Reissig

Beemalmanns, C.; Reissig, H.U. *Angew. Chem. Int. Ed.* 2010, *49*, 8021.

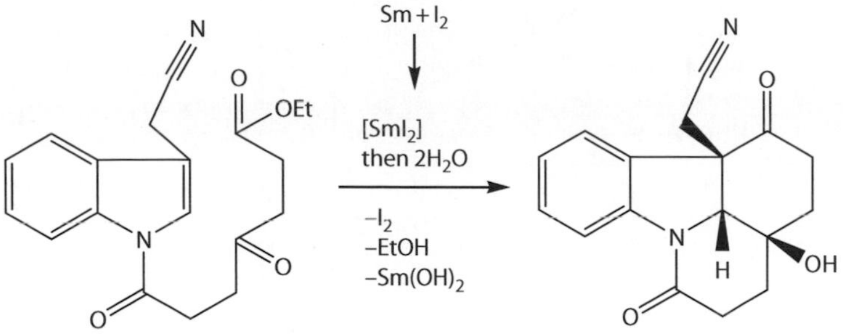

Strychine: Vollhardt

Eichberg, M.J.; Dorta, R.L.; Lamottke, K.; Vollhardt, K.P.C. *Org. Lett.* 2000, *2*, 2479.

Eichberg, M.J.; Dorta, R.L.; Grotjahn, D.B.; Lamottke, K.; Schmidt, M.; Vollhardt, K.P.C. *J. Am. Chem. Soc.* 2001, *123*, 9324.

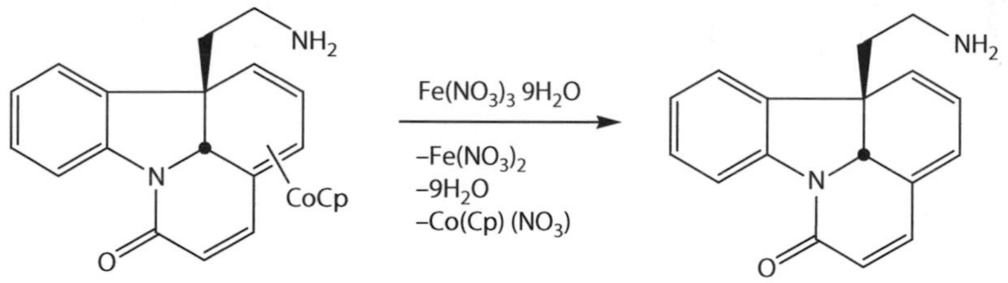

Terreic acid: Sheehan

Sheehan, J.C.; Lawson, W.B.; Gaul, R.J. *J. Am. Chem. Soc.* 1958, *80*, 5536.

Sheehan, J.C.; Lo, Y.S. *J. Med. Chem.* 1974, *17*, 371.

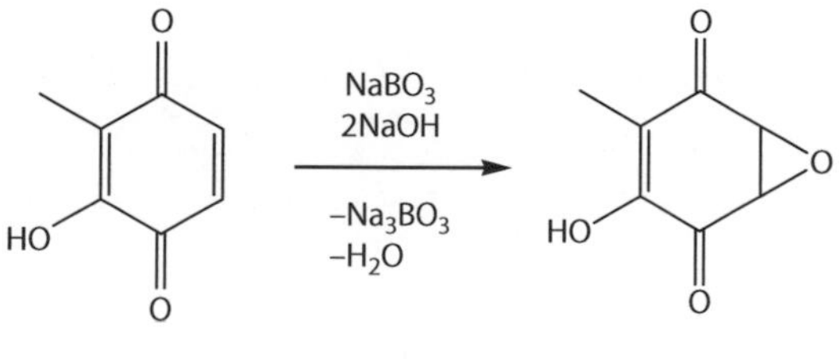

Tetrahydropalmatine: Pandey

Pandey, G.D.; Tiwari, K.P. *J. Ind. Chem. Soc.* 1979, *18B*, 545.

Levine, J.; Eble, T.E.; Fischbach, H. *J. Am. Chem. Soc.* 1948, *70*, 1930.

Rapoport, H.; Williams, A.R.; Cisney, M.E. *J. Am. Chem. Soc.* 1951, *73*, 1414.

Shaw, K.N.F.; Armstrong, M.D.; McMillan, A. *J. Org. Chem.* 1956, *21*, 1149.

Soicke, H.; Al-Hassan, G.; Frenzel, U.; Goerler, K. *Arch. Pharm.* 1988, *321*, 149.

Ghosh, S.; Datta, I.; Chakraborty, R.; Das, T.K.; Sengupta, J.; Sarkar, D.C. *Tetrahedron* 1989, *45*, 1441.

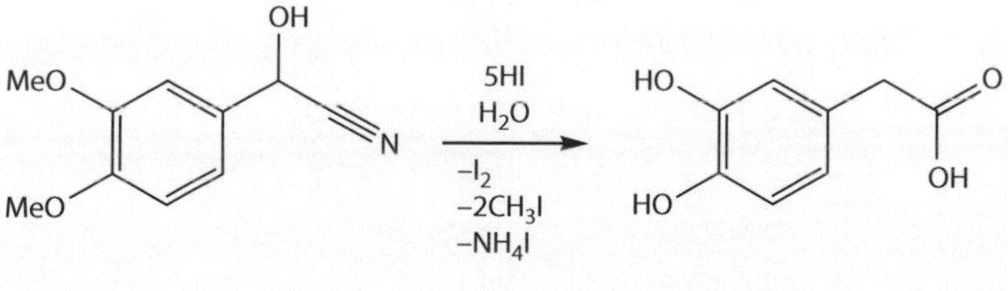

Alpha-Thujone: Oppolzer

Oppolzer, W.; Pimm, A.; Stammen, B.; Hume, W.E. *Helv. Chim. Acta* 1997, *80*, 623.

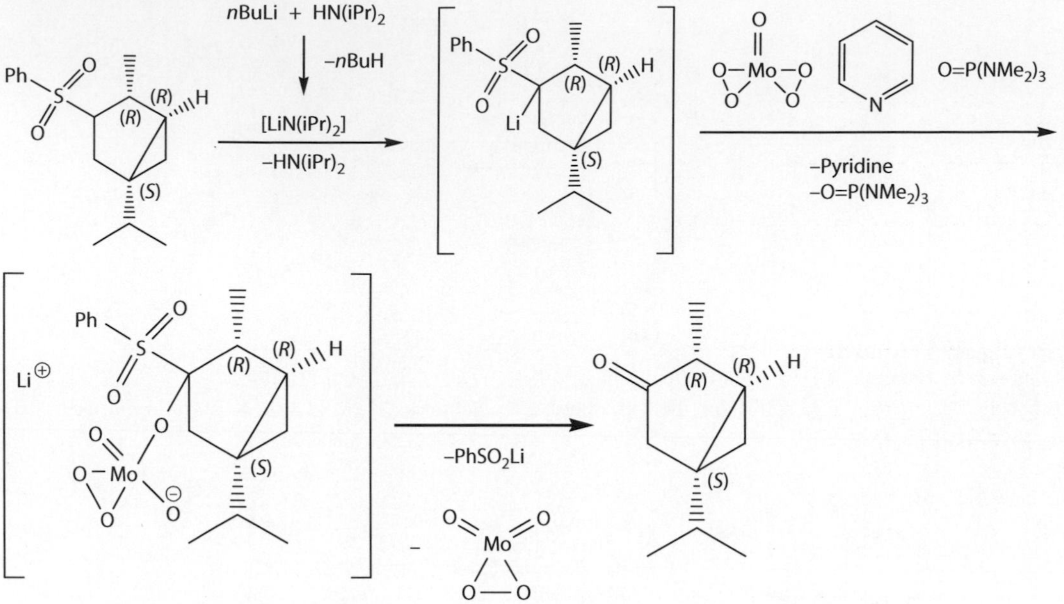

Vedejs, E.; Engler, D.A.; Telschow, T.E. *J. Org. Chem.* 1978, *43*, 188.

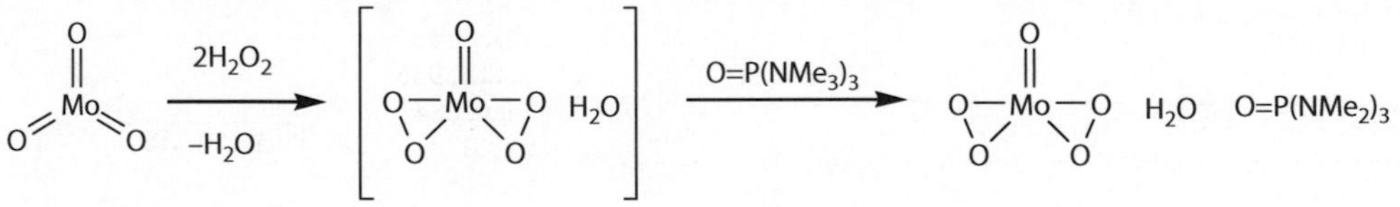

Vitamin D3: Lythgoe

Dawson, T.M.; Dixon, J.; Littlewood, P.S.; Lythgoe, B.; Saksena, A.K. *J. Chem. Soc. (C)* 1971, 2960.
Dawson, T.M.; Dixon, J.; Littlewood, P.S.; Lythgoe, B. *J. Chem. Soc. (C)* 1971, 2352.
Dixon, J.; Lythgoe, B.; Siddiqui, I.A.; Tideswell, J. *J. Chem. Soc. (C)* 1971, 1301.
Littlewood, P.S.; Lythgoe, B.; Saksena, A.K. *J. Chem. Soc. (C)* 1971, 2955.
Bolton, I.J.; Harrison, R.G.; Lythgoe, B.; Manwaring, R.S. *J. Chem. Soc. (C)* 1971, 2944.

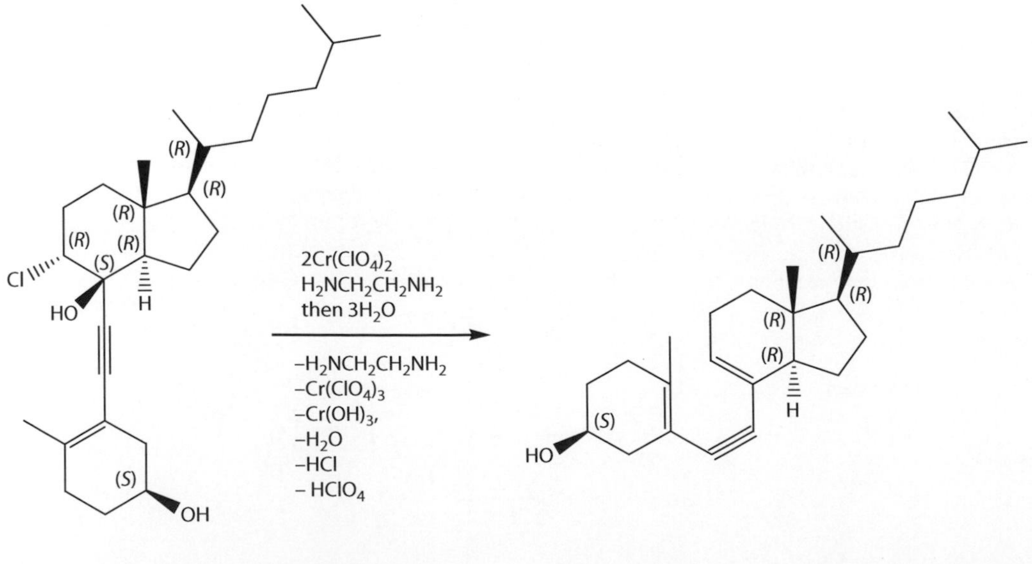

Index

Printed and bound by CPI Group (UK) Ltd, Croydon, CR0 4YY

23/10/2024

01778268-0006